The new edition of Biology continues to set

Figure 2-6 This katydid obtains the energy it needs to live from the food it eats. How does the green plant the katydid is munching on obtain its energy?

S*till the program by which all others are judged, Biology by Miller and Levine provides superior concept development through interactive text and visuals.*

Its engaging format provides the excitement students want and the solid content teachers need. This beautiful new edition with all its many components can provide students and teachers alike with a positive, interactive experience they will never forget.

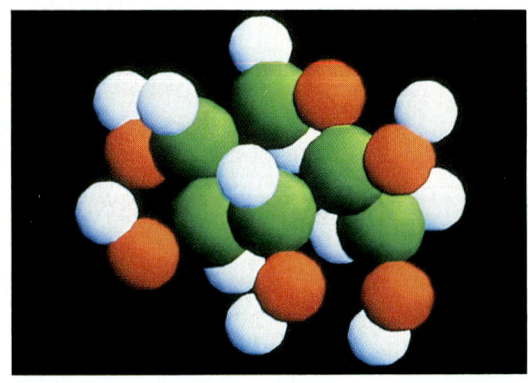

by Miller and Levine the standard.

Breakthrough interactive text and visuals make the difference.

- **Colorful, oversized photography and original art** engage even the most "science-phobic" students.

- **Unsurpassed writing and concept development** bring biology to life with student-friendly analogies.

- **More hands-on activities** give students discovery learning experiences that leave a lasting impression.

On the invention of the light microscope:

" Imagine living all your life as the only family on your street. Then, on a morning like any other, you open the front door and there are houses all around you, cars and bicycles on the street, neighbors tending their gardens, children walking to school. Where did they come from? What if the answer turned out to be that they were always there—you just couldn't see them?

How would your view of the world change? "

This new edition brings with it a complete selection With more integrated media and

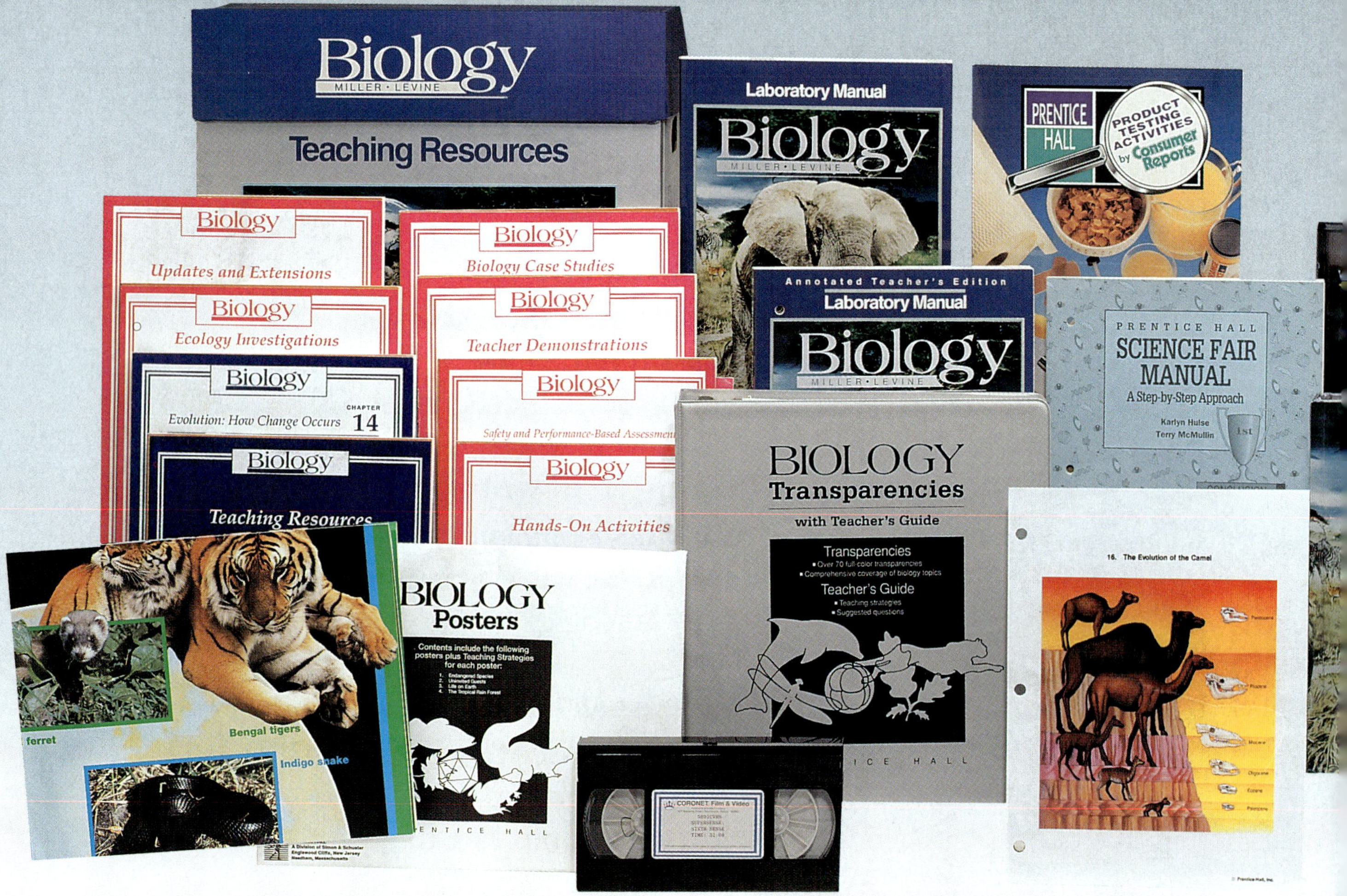

Student Edition—Even more student-friendly writing style, outstanding visuals, discovery activities, and real-life connections to other sciences and subject areas that made the original *Biology* the leading biology program.

Teacher's Edition—Only Prentice Hall offers wraparound teaching notes on the same page as the reduced student text, with cooperative learning activities, biology study skills, teaching strategies, and much more.

Teaching Resources—Over 1,500 pages of teaching options keyed to ability levels and referenced in the Teacher's Edition, including a variety of skill worksheets and activities, teacher demonstrations, case studies, ecology investigations, tests, and more.

Classroom Manager—Every stage of a well-designed lesson is organized for you, along with references to those program materials that will aid in accomplishing lesson skills and objectives.

of updated, easy-to-implement components.
technology than ever before.

Laboratory Manual, Student and Annotated Teacher's Editions—Written by high school biology teachers, the Laboratory Manual supplies an additional 99 labs to choose from and an additional eight introductory labs focused on skill development.

Study Guide, Student and Annotated Teacher's Editions—This self-contained workbook provides review materials for each section of a chapter.

- Prentice Hall Product-Testing Activities by *Consumer Reports*
- Transparencies
- Biology Posters
- Biotechnology Workbook
- Biotechnology Solutions Manual
- Science Fair Manual
- Computer Test Bank Disks (Apple, MS-DOS, Mac)

- Videos/Videodiscs
- Level III Interactive Videodiscs and Level III Interactive Videodiscs/CD-ROM
- Media Guide

Some words from the authors about the most engaging biology program in America today.

As biologists deeply involved with teaching, we realized several years ago that science education was heading for trouble. Despite the fact that science was more important than ever before in daily life, few students knew enough about science to appreciate its value. As you know, that situation has deteriorated to the point that it is now recognized as a national emergency.

We decided to do what we could to reverse that trend by writing a new textbook. Why another biology text? Because most texts today, rather than enhancing your teaching and aiding students, make matters worse. They are so crammed full of facts, lists and diagrams that they manage to hide some of the most important *real* facts about science. What are those facts? That biology is *exciting*. That science is a process of discovery that students can *understand*, rather than a body of trivia they must memorize. And that scientfic literacy is a vital survival tool for every citizen of the modern world.

Teacher's Edition

Biology
MILLER • LEVINE

Kenneth R. Miller, Ph.D.

Professor of Biology
Brown University
Providence, Rhode Island

Joseph Levine, Ph.D.

Science Writer and Consultant
Adjunct Assistant Professor of Biology
Boston College
Boston, Massachusetts

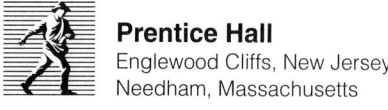

Prentice Hall
Englewood Cliffs, New Jersey
Needham, Massachusetts

THIRD EDITION
©1995, 1993, 1991 by Prentice-Hall, Inc.,
Englewood Cliffs, New Jersey 07632. A Paramount
Communications Company. All rights reserved.
No part of this book may be reproduced or transmitted
in any form or by any means, electronic or mechanical, including photocopying, recording, or by any
information storage and retrieval system, without
permission in writing from the publisher. Printed in
the United States of America.

0-13-803081-2

2 3 4 5 6 7 8 9 10   98 97 96 95 94

Teacher's Guide

Contents of Teacher's Edition

Other Components of *Biology* by Miller and Levine	T4
Philosophy	T6
To the Teacher	T6
Alternative Approaches Grid	T7
Overview of Science Education	T8
Features of the Student Text	T11
Themes and Biology Teaching	T13
Features of the Teacher's Edition	T14
Cooperative Learning	T17
Science Safety Guidelines	T19
NABT Guidelines for the Use of Live Animals	T21
List of Suppliers	T23
Comprehensive List of Student Edition Laboratory Materials	T24
Laboratory Skills Assessment Record Chart	T32

Other Components of *Biology* by Miller and Levine

Biology by Miller and Levine has been designed as a complete, comprehensive program. The student text presents the topics most widely covered in biology courses in an academically solid, student-friendly manner. A wide range of auxiliary materials, described below, is available to augment the program and enhance the learning experience.

TEACHER'S EDITION

This Teacher's Edition contains an enormous diversity of helpful materials, including the answers to all text and end-of-chapter questions. It has been designed to be useful to you, the teacher, whether you have many years of experience or are just about to teach your first class. One particularly convenient feature of this Teacher's Edition is its wraparound format, which allows you to see at a glance a two-page spread in the Student Edition and all the materials that pertain to that particular spread. The features of this Teacher's Edition are detailed on page T14.

TEACHING RESOURCES

The Teaching Resources provides different kinds of worksheets and handouts for each chapter, including Laboratory Investigation worksheets, which duplicate the Laboratory Investigations in the Student Edition and provide additional critical thinking questions. The Teaching Resources also includes hands-on activities, case studies, ecology investigations, demonstrations, traditional tests and performance-based assessments, and Updates and Extensions articles on selected topics written by Ken Miller and Joe Levine especially for the Teaching Resources.

The Teaching Resources is conveniently packaged in booklets in a sturdy file box. The booklet pages are perforated for easy removal, so that they can be photocopied. To simplify storing loose pages and organizing your own materials for the chapters, the booklet covers unfold to become tabbed, letter-sized file folders that can be stored in the original box or, if you prefer, in your own file cabinet.

CLASSROOM MANAGER

The Classroom Manager is a practical management tool to help you meet the educational challenges you face today. It contains reproducible lesson plans for each section in the Student Edition, program organizers that show the resources in all program components for each lesson, a pacing guide, and teaching tips.

LABORATORY MANUAL AND ANNOTATED TEACHER'S EDITION

The Laboratory Manual begins with material on safety in the biology laboratory and eight investigations designed to teach and reinforce basic laboratory skills. This introductory material is followed by 99 laboratory investigations, each of which is keyed to a particular chapter in the student text. There is at least one laboratory investigation for each chapter.

The Annotated Teacher's Edition provides answers to all questions posed in the laboratory investigations. It also gives set-up instructions, suggestions for shortening specific investigations if your lab time is limited, and names and addresses of companies that supply special equipment, if any is required.

STUDY GUIDE AND ANNOTATED TEACHER'S EDITION

The Study Guide is a self-contained workbook that helps students to review and reinforce the biology concepts they have learned in class. For each section, the Study Guide provides a brief synopsis and an assortment of activities, including concept mapping.

The Annotated Teacher's Edition gives the answers to all Study Guide exercises. It also provides a detailed sample concept map for each chapter.

PRODUCT TESTING ACTIVITIES

These fun, hands-on, highly motivational, and inexpensive activities were developed in a joint venture by Consumers Union, publisher of *Consumer Reports*, and Prentice Hall. In the Product Testing Activities, students test products using the same methods as the scientists who work for *Consumer Reports*, and then are challenged to devise their own tests. In the process, students experience science concepts firsthand and develop the skills and skepticism of an informed consumer.

The Teaching Guide describes the curriculum skills and concepts developed in each Product Testing Activity and provides answers when appropriate. It also suggests questions and topics for class discussions based on the activities.

BIOLOGY TRANSPARENCIES WITH TEACHER'S GUIDE

The 73 four-color Biology Transparencies appeal to students, especially visual learners—and save you much aggravation if you lack the time, materials, or artistic ability to draw your own overhead transparencies! The convenient three-ring binder also holds the Teacher's Guide, which suggests strategies for using the transparencies and questions to ask about each transparency image.

BIOLOGY POSTERS

The colorful and motivating Biology Posters are both valuable teaching aids and attractive decorations for your classroom. Teaching suggestions are provided for each poster in the set.

Teacher's Guide

BIOTECHNOLOGY WORKBOOK AND SOLUTIONS MANUAL

The Biotechnology Workbook consists of 15 articles, each of which focuses on a particular aspect of biotechnology, and accompanying worksheets. The Solutions Manual provides the answers to the exercises in the worksheets.

SCIENCE FAIR MANUAL

Science fairs excite student interest, encourage independent study, foster creativity, and promote parent and community involvement in science education. The Science Fair Manual is a practical month-by-month guide to organizing a science fair in your school or district.

COMPUTER TEST BANK WITH DIAL-A-TEST™ SERVICE

The Computer Test Bank provides you with unparalleled flexibility in creating tests. Each printed Chapter Test in the Computer Test Bank has five categories of questions: Multiple Choice, True or False, Completion, Using Science Skills (interpreting graphs and other visuals), and Critical Thinking and Application (essays). Select the questions you wish to use from an array of more than 80 per chapter. Software allows you to create customized tests, complete with illustrations photocopied from Illustration Masters. You can prepare several versions of the same test, add your own questions to the data bank, and modify the questions provided. Dial-A-Test™ gives you the option of phoning in question numbers and receiving a master copy of your test through the mail.

VIDEOS/VIDEODISCS

If a picture is worth a thousand words, just imagine how much teaching and learning can be accomplished through the use of moving images! Many media products are available in both VHS videotape and videodisc formats. So no matter what your technology capabilities are, Prentice Hall provides choices to meet your needs. Both videotapes and videodiscs give you the flexibility to show short excerpts to illustrate a concept in a lesson or to show a movie in its entirety. Bar-code correlations in the Biology Media Guide and in the Teacher's Edition make it a snap to access images by using a bar-code reader attached to your videodisc player. If you do not have a bar-code reader, or prefer not to use it, frame numbers are also provided so that you can call up specific frames and sequences using your videodisc remote control.

In this Teacher's Edition, the bar codes for certain videodiscs are given in stereo, which causes the English and Spanish audio tracks to play simultaneously. Use the appropriate bar code below to select the desired language.

Audio 1/L (ENGLISH)

Audio 2/R (SPANISH)

LEVEL I VIDEODISCS

Videodiscs are an exciting addition to any classroom. All our titles have been especially selected for Miller and Levine *Biology* because of their dynamic visuals and high interest level to students.

LEVEL III INTERACTIVE VIDEODISCS

Students (and teachers, too!) may use Level III videodiscs at two levels of interactivity. First, they may use a remote control to access desired frames or sequences. Second, they may use the interactive software provided with the videodisc, which allows them to start or stop the videodisc at any point to look up words, prepare multimedia reports, answer quizzes, or run experiments (with the results of the experiment played out on the videodisc, computer, or both). As well as providing a highly motivating independent activity for students, interactive videodiscs can also be used within the more structured context of a lesson.

The software for the interactive videodiscs is available in Apple IIGS, MAC, and IBM versions. The specific system requirements are detailed in the Prentice Hall School Division 6–12 Health and Science catalog.

LEVEL III INTERACTIVE VIDEODISCS/CD ROM

Level III interactive videodiscs/CD ROM provide you with the latest advances in educational technology. Each level III interactive videodisc/CD ROM package comes with a videodisc, a CD-ROM disc, and software. These media products provide unparalleled integration of animation, graphics, audio, and video. Working independently or in more structured settings, students may explore the world of a tropical rain forest in *Amazonia*, discover the secrets of different animals in *The Virtual BioPark*, or put together the pieces of an ecological puzzle in *Paul ParkRanger and the Mystery of the Disappearing Ducks*. To run these products, you need to have a MAC LC or LCII, a 386-based or higher IBM, Tandy, Zenith, or Dell; a minimum of 4 megabytes of RAM; a CD-ROM drive; and a videodisc player.

BIOLOGY MEDIA GUIDE

The Biology Media Guide consists of printed bar-code correlations for selected videodiscs and interactive videodiscs that are available from Prentice Hall and/or Coronet. Each correlation includes the relevant page number in the textbook, the broad concept illustrated by the short movie sequence or still image, a brief description, the title of the videodisc, and the side of the videodisc on which the movie sequence or still image is found. Each correlation also gives the frame numbers for the beginning and end of a movie sequence, the still image, or the first visual in a series of still images. In addition, a bar code is provided for each correlation.

Selected bar-code correlations from the Biology Media Guide appear throughout this Teacher's Edition. Although the bar codes are small, they can still be scanned by any standard bar-code reader. Frame numbers are also provided, in case you do not have a bar-code reader or prefer not to use it.

Philosophy

When the authors of your book, Kenneth Miller and Joseph Levine, approached Prentice Hall about writing a secondary biology textbook, they had two major concerns in mind: First, they wanted to write a book that stressed conceptual development and provided insight into the thinking processes behind scientific discovery. They did not want to write an encyclopedic biology tome that stressed memorization of terminology. Second, they wanted to write a book that teaches the evolutionary relationships among organisms.

When the science editors at Prentice Hall were approached by the authors, they had two important concerns as well. Market research had shown the need for a biology text that provided a conceptual outlook toward science, rather than a text that stressed terminology. Furthermore, biology teachers wanted us to make sure that the book would take an evolutionary approach to biology.

As you read through the book, you will quickly discover that the marriage worked. While all important terminology students are expected to know is here, the book emphasizes the need to understand the basic concepts that provide the framework of the biological sciences. In addition, whenever possible, important experiments and discoveries are not just announced as fact. Instead, students are provided with insight into the thinking processes of scientists and the ways science and technology often merge to provide answers to scientific problems.

You will also discover that the topic of evolution is not confined to a single unit. Rather, evolution is presented as a unifying concept that interrelates all other areas of biology. In this way, evolution is interwoven throughout the textbook and provides a conceptual framework that often ties together seemingly unrelated areas of science.

To the Teacher

RATIONALE OF THE PROGRAM

Science education is a vital force in helping students recognize the critical importance of scientific developments in today's world—and tomorrow's. The authors and editors of Miller and Levine *Biology* have designed this textbook and its auxiliary materials to meet this educational goal. This program provides students with the basic knowledge of biology as it relates to them and to their own range of experiences. However, this program goes even further. It makes it possible for young people to use their abilities to develop an appreciation of the basic concepts in biology. Historical achievements in the field of biology, career paths, and some thoughts on the future will all contribute to students' growth and development.

BALANCED BIOLOGY PROGRAM

The Miller and Levine *Biology* program has been developed in accordance with basic principles of science education. The program offers a balance between textual and investigative material, with enough flexibility to suit individual teaching styles and classroom needs.

The textbook presents relevant and recent facts that are used to build science concepts. Illustrations, teaching captions, and in-text questions encourage students to participate, to draw on previously learned information, to make judgments, and to inquire, thereby forming a basis for conceptual learning.

SKILLS DEVELOPMENT IN BIOLOGY

The Miller and Levine *Biology* program provides for the full development of science skills that are inherent in biology, as well as in other branches of science. The skills package within the textbook includes Laboratory Investigations, Problem Solving in Biology special features, and Critical and Creative Thinking questions at the end of every chapter. Through these investigations and features, students gain firsthand experience with such learning skills and processes as observing, classifying, identifying, measuring, inferring, hypothesizing, interpreting, and predicting.

By utilizing the skills package provided in the textbook, as well as the wealth of skills activities in the ancillary materials, teachers can be assured that their students will receive a comprehensive program that reinforces and extends all the skills that are applicable to biology. Furthermore, since most skills activities require report writing and the recording of data, teachers can easily assess whether their students have developed proficiency in skills development.

Teacher's Guide

READABILITY AND STUDENT COMPREHENSION AND INTEREST

The authors and editors of Miller and Levine *Biology* have taken many steps to improve the readability of this textbook as a means of facilitating student comprehension and interest.

Miller and Levine *Biology* is designed with an open, clean format that uses exciting and relevant visuals to enhance student interest. The text is written in a style that appeals to and accommodates the wide range of comprehension levels found in the typical biology classroom. The readability of the text has been carefully controlled so that most students will be able to understand the science content and concepts presented.

Studies have shown that students often feel hindered and frustrated by scientific terms because the words are alien to them, are usually difficult to pronounce, and are hard to remember. To remedy this situation, Miller and Levine *Biology* introduces scientific terms in boldfaced type. Definitions accompany the new terms, and phonetic pronunciation guides are given where necessary.

In addition, photographs and illustrations, most of which are in full color, visually reinforce the text material, thereby aiding readability and comprehension. All photographs and illustrations include an explanatory caption and frequently an inductive question.

The answers to these caption questions can be discerned by the student in three ways:

1. Referring back to text material
2. Basing the response directly on the photograph or illustration
3. Using previously acquired knowledge

Other aids to readability and comprehension include single-concept paragraphs, frequent in-text questions, section checkup questions, chapter summaries, and end-of-chapter questions. These features help to make the text more readable by providing immediate learning evaluations and by continually eliciting student participation and response.

Since a clear guide to what students will be expected to learn in each chapter is a proven readability aid, Miller and Levine *Biology* includes an outline of chapter sections on the chapter-opener pages. Moreover, each numbered section within the text provides skill-oriented section objectives to further guide students. In each section, a key idea statement tying together the most important concepts in the section is included in boldfaced type. These bold-faced key ideas are extremely helpful for students who need some guidance as to the most important information in each section.

Alternative Approaches Grid

Miller and Levine *Biology* presents an evolutionary approach to the study of biology. However, some teachers want to stress a particular approach or aspect of biology in their course. The following alternative approaches grid provides a suggested chapter sequence for several different approaches to the study of biology.

APPROACH	CHAPTER SEQUENCE
Ecology	1–2, 16–36, 47–49
Botany	1–2, 5–6, 15, 20–25
Zoology	1–2, 5, 26–36
Microbiology	1–8, 16–18
Anatomy/Physiology	1–4, 5–12, 37–46
Systemic	1–4, 26–30, 31–36, 37–43

Overview of Science Education

READING AND LANGUAGE DEVELOPMENT IN THE SCIENCE CLASSROOM

The type of reading required of secondary students is different from that required of elementary students. The vocabulary is more difficult, the sentence structure is more complex, the writing is denser, and the concepts presented are more challenging. In addition, reading in the science area presents difficulties of its own. Students must be able to isolate important facts and organize them, form hypotheses based on these facts, test alternatives, and draw conclusions.

In order for students to utilize any text, appropriate readability is essential. The *Dale-Chall* formula indicates that this text is written on the reading level of most biology students. The introduction of new terms in boldfaced type, the pronunciation guides and definitions, and the review questions at the end of each section and chapter improve the readability of this textbook. However, you will still need to help your students develop the specialized skills they need to read in the science area.

Vocabulary in the Science Area

Reading in the science area presents students with a largely unfamiliar technical vocabulary. Terms such as *exothermic* and *homeostasis* do not come up in everyday conversation. In addition, terms with which students may be familiar in an everyday context, such as *mammal*, have a specialized meaning in a scientific context. Students may require some assistance in dealing with these terms. Try doing the following before giving a reading assignment from this textbook:

1. Identify the key concepts and vocabulary words in each chapter.
2. Pronounce all new words. Each new scientific term in each unit is introduced in boldfaced type and is followed by a phonetic pronunciation guide where necessary. It is important for you to pronounce these terms and any others with which your students may have trouble. The pronunciation key is located at the beginning of the Glossary.
3. Define all new and potentially difficult terms. New terms are defined in the text and, for most terms, again in the Glossary.
4. Draw students' attention to the charts, drawings, and photographs that will help them understand new words and concepts. All photographs and artwork have been carefully selected to help students visualize concepts presented throughout the textbook.

Reading in the Science Area

Besides helping students develop a science vocabulary, a science course should also help them develop skills in the following:

1. Reading for exact meaning
2. Identifying the main ideas expressed in chapter sections
3. Classifying information and organizing ideas obtained from reading the text
4. Noting cause-and-effect relationships
5. Gaining accurate information from the visual representations throughout the text
6. Understanding the scientific formulas and symbols in the text
7. Reading directions accurately, especially for carrying out Laboratory Investigations and activities suggested in the text
8. Locating and using different sources of information

Science teachers sometimes overlook the development of these specialized reading skills. These skills are necessary in order for students to develop into mature and independent readers both inside and outside the science classroom.

Writing in the Science Area

Developing good writing skills goes hand in hand with developing good reading skills. Both are important for effective communication.

Because our society relies heavily on written expression and printed material, it is important for students to possess effective writing skills. Your students will have frequent opportunities to display and develop their writing skills. When using this textbook, you should expect students to

1. Keep records of all investigations and activities. Many investigations and activities require students to write a short report detailing observations and conclusions made during an investigation or activity.
2. Record notes from text material and your lectures and discussions. For students to fully comprehend the material presented by such sources, they need to learn to organize and summarize information. Writing the information will help students retain and understand the material better.
3. Write the answers to Section Review questions, as well as the questions at the end of each chapter. Concept Mastery and Critical and Creative Thinking questions at the end of each chapter require students to write brief essays on topics related to the chapter. In addition, at least one Using the Writing Process

question is included in every chapter. All such exercises should be used not only for reinforcement and as an evaluative and diagnostic aid, but also as an opportunity to develop writing skills.

Speaking in the Science Area

Miller and Levine *Biology* offers many opportunities to develop communication skills. When students have a chance to express their own thoughts or to interpret the thoughts of others, oral and written communications improve and become a means for learning. Thus, speaking is another important language skill that can and should be further developed in the science classroom. Oral communication is the most effective and common means of communicating and is the basis for a sound program in reading and writing.

1. **Presentations**. Students of this age are curious about their surroundings and the natural events that occur in their environment. Have students make oral presentations to their classmates about any biology topics that interest them or the results obtained from investigations and activities. Encourage other students to participate by asking relevant questions.
2. **Discussion**. Encourage students to participate in class discussions. Techniques for encouraging student involvement in discussions are presented in a later section, Questioning in the Science Area.
3. **Dramatizations**. Have students act out concepts presented in the textbook, such as important discoveries of scientists or the impact pollution may have on a town or an individual.

Listening in the Science Area

Even though speaking is the most common form of communication, it is ineffective without a listener. Listening skills should be developed concurrently with speaking skills. Students need to learn to respect each other's viewpoints.

Encourage good listening habits in your students by being a good listener yourself. Pause after asking a question. If no student offers an answer, rephrase the question. Teaching effective listening is important in helping all your students become better science students.

Listening constitutes a major portion of the communication process. Listening skills can be developed by encouraging students to

1. Listen to questions posed by the teacher and other students. For example, before proceeding with a laboratory investigation, you may ask a question such as "What conditions do you think bring about a chemical change?" A question of this nature would involve the entire class in a discussion and would require students to listen to responses from their classmates.
2. Listen to directions for carrying out investigations. Students should be able to follow directions, especially safety precautions, in a science class.

    Have students practice the listening skills by giving them oral instructions on how to make something—a wet mount, for instance. Repeat the directions frequently and ask students to repeat the instructions as they follow them. This activity will prepare students for laboratory investigations by helping them follow a thought sequence and by emphasizing the need for accuracy in communication and interpretation.
3. Listen to explanations or descriptions of natural phenomena provided by you, other students, or guest speakers. You can help students listen properly by telling them what to listen for, how to listen to the material presented, and how to mentally organize or write down what they hear.

Questioning in the Science Area

Developing communication in your science classroom involves your participation. Much of a teacher's class time is spent asking questions. In fact, research has shown that teachers use questioning more frequently than any other single teaching technique. You ask questions to develop creative learning situations, evaluate students' progress, give directions, correct behavior, and initiate instructions.

It is important to understand and use good questioning techniques and strategies during the instructional process. Thought-provoking questions and improved questioning techniques can help you develop and sustain student interest, provide new ways to deal with subject matter, and give purpose to your student evaluations.

Besides the type of questions you ask, the number of questions you ask affects student response. It is easy to ask too many questions during an instruction period. Studies have found that some teachers ask questions at a rate of 180 questions per science lesson! This rapid-fire method of questioning and calling on a student to respond immediately after the question is asked leads to brief student responses.

Thus, wait time serves a twofold purpose in the classroom: (1) It provides an atmosphere more conducive to discussion and learning, and (2) students learn to use wait time to organize a more complete answer.

Along with waiting 3 to 5 seconds after asking a question, pausing after a student response is also helpful. This second pause, or silent time, increases the chances that the student will add to his or her response or that other students will add to the initial response. If you follow these simple techniques of waiting before and after a student's response, more students may become involved, you may not need to ask as many questions, and the questions you do ask will probably be at a higher cognitive level.

TEACHING HETEROGENEOUS CLASSES

Miller and Levine *Biology* has been designed to meet the needs of students of all ability levels. Through careful analysis of readability, the text has been monitored to ensure that biology students can read and comprehend the material presented in the text. Moreover, large photographs and illustrations, which reinforce and extend the material in the text, are important tools in helping students comprehend facts and concepts in science.

TEACHING "SPECIAL" STUDENTS

Certain state and federal laws have mandated that all students are to have access to the least restrictive learning environment possible. Thus, many "special" students—those with physical and mental disabilities—are being mainstreamed into nonspecialized classes. This action challenges the teacher to accommodate a much wider range of student abilities, needs, and interests.

Students with Learning Problems

Learning processes that include inferences and abstract reasoning are often more difficult for students who have learning problems. Such students include those who have some degree of mental retardation. In order to better help such students grasp facts and concepts in biology, it is important to provide daily learning goals at a pace that will allow the goals to be achieved. These students will benefit greatly from the use of concrete examples in the classroom that relate back to daily life. The need to reinforce lessons is also important to such students. Furthermore, since many of these students will have experienced failure in their studies, it is vital to provide as much positive reinforcement as possible. Emphasize success and minimize failure whenever you can.

Students with Visual Problems

Students who are blind, as well as those with limited sight, are more dependent on senses such as hearing than other students. As a result, such students should always be seated where they will be able to hear the teacher and their classmates most easily. Tape recording lessons will help these students study and go over material at their own pace. Also, classmates can be a great aid by providing descriptions of photographs and illustrations in the textbook.

Students with Hearing Problems

Students with hearing problems are far more dependent on the written word than other students. Usually, these students should be seated near the front of the room so that they can read the teacher's lips. The teacher should enunciate every word and avoid talking too quickly. All instructions and assignments should be written down for these students. Allow students who cannot hear well to copy the notes taken in class by classmates.

Students with Other Physical Problems

Students who have physical problems that require crutches or wheelchairs will need extra room to get around in the classroom. Take care to make sure such students do not try to stretch their limits beyond their physical capabilities, but do not treat them any more differently than necessary so that they will feel an integral part of the class.

Students who have physical problems due to disorders such as muscular dystrophy or other disorders that deter motor coordination will often have trouble in the laboratory setting. Holding flasks, pouring liquids, and using other equipment may be beyond their capabilities. If these students can write, it is often best to assign them the task of recording during investigations while their lab partners carry out the more physical aspects of the investigation.

Some students may have illnesses such as diabetes or epilepsy. In general, such students will not need any special care. However, the teacher should be aware of any special problems or symptoms these illnesses might present in order to obtain prompt medical attention when necessary.

Teacher's Guide

Features of the Student Text

Miller and Levine *Biology* is organized to let you choose which chapters to teach and in which order to teach them. Though the authors suggest you teach the chapters in order, you will discover that the textbook's 10 units are sufficiently self-contained to be taught in almost any order. On page T7 of this section, the Alternative Approaches Grid provides chapter sequences that stress a particular approach to biology, such as a systemic approach, or an approach emphasizing ecology or zoology. You are encouraged to tailor the textbook to suit your own approach to biology and to the specific needs of your students.

UNIT OPENERS

Each unit begins with a two-page spread that includes a large dramatic photograph or illustration designed to capture students' attention and interest. The unit-opening text explains the significance of the visual and how it relates to the topics discussed in the unit. On the facing page is a list of the unit's chapters.

The spread also includes the unit's Discovery Learning activity. This simple, hands-on activity requires little preparation or setup and uses materials that are commonly found in the home or classroom. Use the Discovery Learning activity to introduce the unit and pique student interest.

CHAPTER OPENERS

Each chapter opener includes a large photograph or illustration. The visual's caption and the chapter-opening text are written in a conversational, anecdotal style, one that will entice students to read further. You may wish to begin teaching each chapter with a class discussion that is based on the visuals and text on these pages. Suggestions for such discussions are included on the corresponding pages of the Teacher's Edition.

The chapter-opener spread also includes a Guide for Reading and a Journal Activity. The Guide for Reading lists the chapter's sections and what students are expected to learn in each section. Students will find their studying easier if they keep the Guide for Reading in mind. With the Journal Activity, students explore the chapter's topics and relate these topics to personal experiences, thoughts, and feelings. You may wish to have students share their journal entries with the class, or use the Journal Activity as a basis for class discussion.

CHAPTER SECTIONS

The chapter sections are numbered consecutively, and there are from two to seven sections in each chapter. The sections are further divided into subtopics, the titles of which are printed in blue type.

Just as at the beginning of the chapter, the beginning of each section includes a Guide for Reading. The section Guide for Reading consists of questions that cover the major points in the section and that students should be able to answer when they complete the section.

Whenever the section introduces an important scientific word, that word is printed in boldfaced type. These vocabulary words are reprinted in the Chapter Preview pages of the Teacher's Edition and in the Chapter Review pages of the Student Edition. Encourage students to recognize these boldfaced words and to learn their definitions. Each section also includes at least one sentence in boldfaced type; these sentences serve to highlight the key idea or ideas in the section.

As they read the section, encourage students to look at the textbook's numerous photographs, illustrations, and graphs. Each is designed to augment or reinforce the section's material. Many sections also include in-text questions. You will find these questions annotated and answered on the corresponding pages of the Teacher's Edition.

Concluding each section are the Section Review questions. These short-answer questions test students' knowledge of the section's important topics. The last question of each Section Review is a critical thinking question, which requires students to use analytical and science reasoning skills to solve a problem that is related to the section material.

SCIENCE, TECHNOLOGY, AND SOCIETY FEATURES

Each chapter includes at least one feature under the heading of Science, Technology, and Society. These features discuss timely, important, and sometimes controversial topics that are related to their chapter's material. These features are further categorized as either a Breakthrough, Issue, or Connection. The Issues features in particular lend themselves to critical thinking and can often be used as a basis for class discussion or debate.

PROBLEM SOLVING

Problem Solving features are included in many chapters. As the name implies, this feature provides data to the student and then calls upon the student to use problem-solving skills to arrive at an answer, hypothesis, or theory. The Problem Solving features have been carefully written so that students can answer all questions based on material they have read in the chapter, material presented in the feature, or simply by using critical thinking skills. These features do not rely on further library research or call upon a knowledge of science beyond the student's ability level.

LABORATORY INVESTIGATIONS

Each chapter contains one full-page Laboratory Investigation just before the chapter summary page. These investigations

T11

provide students with the opportunity to work in the laboratory and actively participate in investigating science problems. Most of the investigations are designed to reinforce concepts presented in the text, but a few are designed to supplement the material in the text. Easy-to-obtain materials are used in these Laboratory Investigations.

Each Laboratory Investigation clearly outlines the Problem to be investigated, the Materials needed, and the Procedure to follow. A section called Observations alerts students to any observations or data they are to collect and tells students how to organize their data. In general, data is organized in the form of charts or graphs. An Analysis and Conclusions section ties up the laboratory investigation, calling upon students to analyze their data and to draw various conclusions. Often the Analysis and Conclusions section asks students to use their data to reinforce or further establish a scientific concept or theory. In addition, special safety symbols alert students when important safety precautions must be observed.

CHAPTER REVIEW

Chapter Summary and Vocabulary

At the end of every chapter is the Chapter Summary. The summary is divided into groupings based on the main numbered sections in the chapter. Under each grouping is a list of key sentences that describe the most important concepts presented in the chapter. The Chapter Summary might be considered a detailed outline of the chapter content.

Following the summary is a list of the vocabulary words used in the chapter. The vocabulary words include all boldfaced terms from the chapter.

End-of-Chapter Questions

A wide variety of questions end each chapter. The first set of questions is called Content Review. Content Review questions test factual recall and include multiple choice, word relationships, and true or false sections. The true or false questions ask students not only to identify an incorrect statement, but to make it correct by substituting the correct word or phrase into the statement.

The next set of questions is called Concept Mastery. These short-answer questions test students' comprehension of the basic concepts presented in the chapter, often calling upon them to tie together two or more concepts in order to synthesize an answer.

The last set of questions is called Critical and Creative Thinking. The particular thinking skill the student will employ is printed in boldface after the question number. These questions call upon students to use higher order thinking skills. The last question in this section is always a Using the Writing Process question that generally employs cross-curriculum writing assignments with a strong scientific basis.

END-OF-UNIT MATERIAL

At the end of every unit is a two-page spread that includes a Careers in Biology feature and a From the Author feature. Each Careers in Biology presents three careers related to the topics covered in the unit. Careers that require a high school diploma through a Ph.D. are included. In addition to a brief description of the career and a photograph of a person in that career, an address for further information about each career is included. At the end of the Careers in Biology is a tip to students that will help them find employment in the future. For example, students are told how to apply for a job or how to write a résumé.

On the page following the Careers in Biology is a feature called From the Authors. This page provides a personal glimpse into how the unit's author views the material in that unit. These features are highly personal, informative, and motivational. Your students will enjoy reading these features.

END-OF-TEXTBOOK REFERENCE SECTION

At the end of the textbook are several features designed to aid students in their understanding of biology.

For Further Reading

A bibliography for each chapter is provided in a section called For Further Reading. The books listed in the bibliography will help students do further research on a topic they find interesting and reinforce the materials they have learned.

Appendices, Glossary, and Index

A variety of appendices are located in the reference section at the back of the textbook. Students will find a wealth of useful information in these appendices.

Following the appendices is the Glossary. Scientific terms introduced in the text are listed in the Glossary in alphabetical order. Each term is clearly defined.

The Index is located after the Glossary. The Index provides students with an easy-to-use reference listing of subjects covered in the text.

Teacher's Guide

Themes and Biology Teaching

Science themes and the thematic teaching of science have become the focus of intense interest and energetic discussion in educational circles. But don't panic over the proliferation of papers, presentations, panel discussions, and implementation plans! Although the amount of information available may appear overwhelming, the basic concept is easily mastered. In fact, you may already be teaching thematically and not even realize it.

Themes are simply major concepts that link facts and ideas together. Themes have often been described as the "support beams" of a cognitive framework, on which the facts, theories, and other elements of science knowledge are assembled and organized. But you might find it useful to think about themes in terms of art rather than architecture. Imagine, if you will, a pointillist painting. By itself, each tiny dot of color that makes up the painting is not particularly informative or memorable. Studying many individual dots, or even small clumps of dots, is often equally uninformative. The only way you can discover the picture is by stepping back and seeing the relationships among the dots, the patterns the dots form. Science, like a pointillist painting, is made up of bits of information. Out of context, these bits are virtually meaningless and easily forgotten. Themes show how the bits fit together so that students can "see the whole picture."

Seven underlying themes appear time and again throughout Miller and Levine *Biology*. These themes are Energy, Evolution, Patterns of Change, Scale and Structure, Systems and Interactions, Unity and Diversity, and Stability.

ENERGY

In informal use, energy can be described as the ability to perform an action. Putting a wheel into motion, forming a chemical bond, or moving a muscle are examples that fall under this theme. The ways in which living things obtain and use energy to grow, develop, and reproduce are important recurring topics in biology.

EVOLUTION

Evolution is change over time. As a theme, Evolution is not restricted to the history of life on Earth. It pervades topics as diverse as the life cycle of stars, the building up and wearing down of mountains, and the development of scientific theories. Topics in biology in which the theme of Evolution is apparent include succession in ecosystems, life cycles of organisms, speciation, and adaptation.

PATTERNS OF CHANGE

This theme involves trends, linear patterns of change, and cycles, repeating patterns of change. Patterns of Change also involves random events—changes that do not follow a pattern. In biology, trends include the decrease in biodiversity in increasingly smaller islands and the tendency for organ systems to become more complex as you move through the vertebrates from fishes to mammals. Cycles include seasonal migrations and the movement of materials such as carbon and nitrogen through the environment.

SCALE AND STRUCTURE

Just about everything, from atoms to rocks to organisms, can be described in terms of structure. How something's structure is studied and described depends on the scale on which it is being studied. On a molecular scale, a scientist might study a human in terms of proteins in the body. On a global scale, a scientist might study humans in terms of their impact on the ozone layer.

SYSTEMS AND INTERACTIONS

The theme of Scale and Structure focuses on the parts that make up an ecosystem, organism, star system, engine, or any other kind of system. The theme of Systems and Interactions focuses on how the parts of a system work together and how separate systems interact with one another. Predation, competition, and symbiosis within an ecological community are examples of interactions that occur within a biological system. The regulation of blood pressure in the human body can be studied as interactions among the nervous, endocrine, excretory, and circulatory systems.

UNITY AND DIVERSITY

There is a great diversity of objects and phenomena around us—certainly more than can be examined individually. But because things share certain characteristics, we can put them into groups or categories, thereby giving order and meaning to our world. In science, we can use similarities to talk about inorganic compounds, simple machines, silicate minerals, flowering plants, and other groups. Each group, united by significant similarities, contains things that differ from one another in important ways. By knowing the commonalities, we can better understand the nature of the different things that make up a group.

STABILITY

The theme of Stability includes the notions of balance and predictability as well as lack of change. One of the underlying assumptions of all science is stability; phenomena can be predicted and understood only if the rules that govern them do not change. There are many examples of stability in nonbiological sciences. For example, momentum is conserved, chemical systems reach dynamic equilibrium, the velocity of one type of earthquake wave is a constant fraction of the velocity of another type. Biology topics in which the theme of Stability is prominent include homeostasis, feedback mechanisms, and climax communities.

Features of the Teacher's Edition

The Teacher's Edition of the Miller and Levine *Biology* is the most complete, comprehensive, and pedagogically sound teacher's edition available for you, the biology teacher. The basic structure of this teacher's edition provides reduced student pages. This reduction allows a wide variety of teaching aids to be wrapped around the student page, but the reduction is not so great as to make reading difficult.

You will immediately note that on most pages the right and left margin columns have a blue background. All material found in these margin columns pertains to Background Information, Historical Notes, Ecology Notes, Facts and Figures, cross-curriculum Tie-Ins, and discussions of special features in the student text such as Science, Technology, and Society and Problem Solving. In addition relevant materials from the Laboratory Manual, Study Guide, Teaching Resources, and other ancillaries are referenced in the blue side columns.

You will also immediately notice that there is a boxed-in area at the bottom of most pages. This boxed-in area has been designed to provide teaching strategies specifically geared to the information presented in the text.

One important aspect of this wrap-around Teacher's Edition is that all relevant material is right there along with the student pages for easy reference. Ease of use was an important criterion in the development of this teacher's edition. Of equal importance is the fact that the teaching material and marginal column material have been carefully controlled so that all material relates directly to the two-page spread in the student edition. That is, all material on the teacher's edition page refers directly to those same student pages. You need not try to decipher which teaching instruction or background information is applicable to which student pages. Applicable material is included directly along with the student pages so you don't have to flip through pages to look for answers or teaching strategies. This makes the Miller and Levine *Biology* Teacher's Edition the most comprehensive and functional teacher's edition available.

UNIT OPENERS

Wrapped around the two-page unit openers are a wide variety of instructional materials and teaching information. In the blue-tinted side columns you will find a Unit Overview, which provides a short overview of the facts and important concepts covered in a particular unit. Following the overview are Unit Objectives, which are the broad objectives students should meet during their study of a particular unit. Also included in the side columns is a list of chapters in the unit and a brief description of the topics covered in each chapter.

At the bottom of each unit opener, located in the bordered strategy box, is a teaching strategy for introducing the unit. In general, the strategy calls upon the teacher to have students observe the unit-opener photograph, then read the accompanying text. Questions based on the photograph are often provided. Many are open-ended and require some degree of critical thinking. A basic design feature of all questions posed in this Teacher's Edition is that a small bullet is placed before each question. The question is set in boldfaced type so that all suggested questions are immediately obvious at a glance. The answers to such questions are placed in parentheses immediately after the questions.

The suggestions for the introductory discussion are followed by notes on the Discovery Learning activity. Because the Discovery Learning activity is designed to spark student interest in the unit material, you may wish to perform the activity even before you start your discussion of the text.

INTERLEAFED PAGES

Four pages of teacher material are interleaved before each chapter in the Teacher's Edition. The first two pages are a Chapter Planning Guide, which is in the form of a chart. The first column gives the name of the section, the pages on which it is found, and one or two underlying science themes that are particularly apparent in the section. (Of course, other themes will be present in the section, and you may wish to emphasize these, rather than the ones listed in the Planning Guide.)

The second column lists the hands-on activities found in the Student Edition, Laboratory Manual, and Teaching Resources. It also lists the teacher demonstrations found in the Teacher's Edition and Teaching Resources. Where appropriate, Product Testing Activities are suggested. Because you probably do not have enough time to do every activity that you would like to do, most of the Product Testing Activities are correlated to more than one section in the textbook—so you *will* have another chance!

The third column lists the other activities that correlate to the section, including pages in the Study Guide, chapters in the Biotechnology Workbook, and Biology Case Studies in the Teaching Resources. The last row in this column, which describes materials that relate to the Chapter Review, lists performance-based assessments, which are intended to be given at the completion of a unit, and chapter tests.

The fourth column lists the media and technology products that are relevant to the section. These include four-color transparencies, movies available on both videotape and videodisc, videodiscs, interactive videodisc programs, and cutting-edge interactive videodisc/CD ROM products. The last row in this column tells you the page on which the Computer Test Bank chapter test begins.

Following the two-page planning guide are the interleafed Chapter Preview pages. The Chapter Preview begins with a Chapter

Teacher's Guide

Overview. Next, each numbered section is introduced through the Section Focus material. Performance Objectives for each section are included, as well as a list of the boldfaced Science Terms in each section and the page numbers on which they are found. Finally, two suggested Teacher Demonstrations are included for every chapter in the Chapter Preview spread.

CHAPTER OPENERS

The blue-tinted columns that wrap around the chapter-opener pages begin with a Guided Enquiry suggestion. The Guided Enquiry provides you with a set of questions that you may want to ask students prior to reading the chapter. As such, it will key students into the main points they are to garner from the chapter and can be used as a pre-reading feature. When students have completed the chapter, have them answer the questions again and compare their new answers to their initial answers. In this way, the Guided Enquiry can be used for pre- and post-assessment.

Another special feature on the chapter-opener pages is called Cooperative Learning. Each chapter includes a cooperative-learning suggestion that generally links some aspect of the chapter content to science, technology, or society. In most cases, the cooperative-learning strategy involves a cross-curriculum exercise. For guidelines on using cooperative learning in your classroom, see Cooperative Learning on pages T17 to T18.

The third special feature is the Journal Activity notes. These suggest ways in which you might incorporate student responses to specific Journal Activities into your lesson, provide extra questions for discussion, or give other useful tips.

In the strategy box at the bottom of the chapter opener you will find a Chapter Introduction. Like the unit-opener teaching suggestions, the chapter-opener suggestions generally call upon students to observe the chapter-opener visual and to answer questions based on the visual and the introductory copy following the visual.

CHAPTER SECTION MATERIALS

Each major section in Miller and Levine *Biology* is numbered for easy reference. Below the student pages you will find the boxed-in teaching strategy suggestions. Each numbered section in the text begins with a teaching strategy called Focus/Motivation. The strategy employed in Focus/Motivation is to present students with an activity, a thought question, a demonstration, or some type of teaching tool to help interest and motivate them. Focus/Motivation strategies are also employed throughout the section. Following the motivational ideas that begin each section is a teaching strategy called Content Development. Content Development information is designed to help the teacher teach the basic facts and concepts presented in the chapter. Like Focus/Motivation strategies, Content Development ideas are interspersed throughout each section. Another feature that is located in the teaching strategy box is called Skills Development. Under this title is a list of the skills that will be developed, followed by an activity or a set of questions to help test basic science skills. Yet another strategy employed in the teaching strategy box is called Reinforcement/Reteaching. These features help the teacher reteach basic concepts and facts to students who may not always grasp the facts and concepts without some extra help. For students who are highly academic, a feature called Enrichment is also interspersed throughout each chapter. These Enrichment ideas provide the teacher with a way to help academic students go beyond the textbook material. At the end of each numbered section you will find the answers to all Section Review questions printed in red. A special Reinforcement/Reteaching suggestion follows each set of Section Review answers. This strategy suggests that teachers review and reteach any material pertaining to questions students have trouble answering. Finally, each major section ends with a Closure strategy. The Closure strategy provides a suggestion for closing up the lesson and tying together the important facts and concepts presented in a particular section of the textbook.

The blue-tinted side columns in each major section contain a wealth of extra material for the teacher. Features found in these columns include Background Information regarding topics presented in the textbook, Historical Notes that discuss the material from a historical perspective, and interesting Facts and Figures. In addition, teaching strategies, answers when applicable, and other teaching suggestions are found in the margins beside the special textbook features called Science, Technology, and Society and Problem Solving in Biology.

Since Biology cannot be taught in a vacuum, Miller and Levine *Biology* also provides numerous Tie-Ins to other areas of science and to other curriculum areas, such as history, government, and literature. These Tie-Ins help relate biology to other areas of science as well as to the students' everyday life. Ecology Notes tie in other branches of biology to the branch that is familiar to students from newspapers, magazines, and newscasts—ecology. Each Ecology Note shows how the concepts in the pages students are currently studying relate to current issues and topics in environmental science.

Two side-column features are designed to help you deal with diversity in the classroom and in society. ESL Strategies include tips for teaching students with limited English proficiency and vocabulary-building exercises for such students. Multicultural Strategies include anecdotes, biographical sketches, and student activities that celebrate the diversity of cultures in the United States and the world.

The feature called Teaching Support, which appears on a green background, lists the materials in the various print and media and technology components of the *Biology* program that correspond to the two-page spread. When possible, the Teaching Support box contains a bar-code correlation to a media product. If you have a bar-code reader attachment on your videodisc player, you can simply scan the bar code. The videodisc will then immediately go to the sequence that has been coded. Frame numbers are printed beneath the bar code so that you can manually access the movie sequence or still image.

Whenever questions appear in the text material or in the captions to visuals, an Annotation Key will be found in the margin of the Teacher's Edition. A numbered bullet beside the question is printed on the reduced student page. The answer to the question is keyed in by number in the Annotation Key. In addition, the science skill needed to answer the question is in parentheses.

LABORATORY INVESTIGATIONS

Laboratory Investigations in Miller and Levine *Biology* are always located at the back of the chapter, just before the Student Study Guide summary section. In the Teacher's Edition, a great deal of extra material is provided for the teacher for each investigation. The first feature for all investigations is called Before the Lab. This tells the teacher when to prepare any materials for the investigation and gives any special instructions that may be necessary to successfully complete the investigation. The next feature is called Pre-Lab Discussion and tells the teacher which concepts presented in the chapter should be reviewed prior to the investigation, as well as other general information to be discussed with students. Variables and hypotheses are brought out in the Pre-Lab Discussion whenever applicable.

A section called Teaching Strategy is next. It provides the teacher, when necessary, with a strategy to be employed while students complete the investigation. Next comes a section called Observations, in which answers to questions in the Observation section of the investigation are provided. The same is true for the section called Analysis and Conclusions, in which all answers are provided.

Finally, for each investigation there is a section called Going Further: Enrichment. This section may include additional activities that can be undertaken in the lab, critical thinking questions based on the lab, application questions related to the lab, as well as other enrichment ideas.

END-OF-CHAPTER QUESTIONS

The last two pages in each chapter contain the answers to all end-of-chapter questions, including Multiple Choice, True or False, Word Relationships, Concept Mastery, and Critical and Creative Thinking.

Cooperative Learning

The structuring of classroom activities that involve students in meaningful tasks is an increasingly complex problem for teachers. Traditionally, lessons have been organized that have students learning and being evaluated either on an *individual* basis, where their attainment of goals is not affected by the work of others, or on a *competitive* basis, in which individual students are evaluated in comparison to each other. Only rarely has a third organizational scheme for the learning environment been used: having students work cooperatively in small groups with the individual student accountable for the material studied. The group is rewarded based upon mastery by all group members or on a group product. This third organizational scheme is known as *cooperative learning*, and it has become one of the most carefully researched areas of study in education in recent years.

The wide range of skills in the average classroom creates a problem of challenging the academically able student while attempting to address particular problems of others in the same class who are disadvantaged by limited language proficiency, learning disabilities, or lack of motivation. Hundreds of research studies indicate that frequent use of cooperative-learning activities has been found to be one way to address these challenges.

Cooperative learning involves students working together in learning groups or teams. The objectives for the group are carefully defined, and positive interdependence among group members is necessary for the group to be successful. Individual students within the group are accountable for mastering the required material and may be responsible for teaching other group members a specific portion of the assignment if the lesson calls for a "jigsaw" approach.

A "jigsawed" activity involves breaking apart a large body of information into smaller components. Each member of a learning team is then responsible for mastering a specific part of the assignment well. It often works best to have the student from each group who is responsible for a certain section work together with the other students who are responsible for that section. They then can return to their team as the "expert" on that section and teach their group members the most important information from the section. Other experts on the team also teach their part of the assignment. All students are responsible for demonstrating mastery of the entire assignment.

Not all cooperative-learning activities involve the jigsaw technique. However, to be successful, all should involve a task in which students discuss the material assigned, help and encourage each other as they learn that material, and work interdependently to complete the task. It might also require groups to complete a product that demonstrates mastery through an application, analysis, synthesis, or evaluation activity. The social skills needed to work in a collaborative effort must be recognized as essential to the successful completion of group tasks. Interpersonal skills needed to work cooperatively are taught and roles are usually assigned to facilitate the group process. These roles would vary according to the assignment but might include the marker, recorder, summarizer, materials manager, checker, and/or facilitator. The teacher's role becomes one of monitoring, as students become ever more proficient at developing two essential modes of behavior—working together and sharing leadership roles.

The laboratory-based science curriculum offers a wonderful environment for students to work in cooperative-learning groups. The lab-team approach can be altered only slightly to become the cooperative-learning "home-base team" to which students are assigned by the teacher. It is suggested that these teams represent a balance of ability levels, ethnicity, and sex as students learn to work cooperatively. The home-base teams may be changed every six or nine weeks. Groups of four work best in this arrangement. Activities should be planned for both the laboratory and the classroom setting using these group assignments.

In addition to having students work in their preassigned teams, it is recommended that they occasionally be given opportunities to work in randomly assigned or informal groupings. Activities that may be completed in one class period may be effectively done by students in pairs, trios, or groups of four assigned by the teacher or occasionally by self-selected two-person partnerships.

A variety of methods for random assignments should be used. Grouping with playing cards involves randomness and creates groups of four. The teacher might want to direct students to sit in the area of the room indicated by the number or "face" of the card they are dealt. By counting out exactly the number of cards needed for an incoming class ahead of time, each student can be dealt a card and can move quickly to be seated and begin the task. If the class enrollment is not evenly divisible by four, jokers may be used for the odd number remaining; students who are dealt the joker cards fill in for absentees. The playing-card assignment also facilitates checking for individual accountability at the end of the assigned task time. The teacher draws one card from the deck and announces that the suit (clubs, etc.) of the drawn card will determine which student's paper from each group will be used to determine the group's grade.

Determining random group assignments may also be accomplished by having each student draw for placement from some kind of "fish bowl." Other quick assignment methods might involve writing each student's group on his or her assignment paper to be returned or on each copy of a class set of papers. Using teacher-made "color number" cards is an easy way to assign students to jigsaw teams. The color of the card indicates the team assignment; the number of the card represents that student's responsibility during the assignment. Variations of the "count-off" method may also be used for assignments to groups, even though this method does not really allow for randomness.

Remember that one of the essential components of cooperative learning is the concept of group reward. If every member of the learning team has demonstrated mastery according to the criteria established by the teacher, the group's successful work should be rewarded. Rewards might take the form of bonus points, choices for using free time in the classroom and laboratory, or tangible nonacademic rewards such as stickers and stamps on papers. Praise and recognition for the group by the teacher is the least expensive and perhaps the most effective reward. For those assignments in which learning teams are creating products such as those suggested for each chapter in this text, groups are competing against each other. Each group's feelings of interdependence are intensified as they strive to be fluent, efficient, and creative in their approach to the problem assigned. Rewards for such group efforts might involve showcasing particularly well-done products on the chalkboard or bulletin board. Bonus points might also be added to the group grades of particularly well-done products.

For each chapter of the textbook, a cooperative-learning group product assignment has been suggested. The procedures vary but always involve each group's use of collaborative efforts to complete a task that extends the chapter material. Many of the assignments involve creative-writing, brainstorming, use of the five senses, and opportunities to use visual imaging. All will require task analysis by the group and will encourage students to see the advantages of sharing and working efficiently and creatively together.

Even the most enthusiastic supporters of collaborative student efforts feel that cooperative learning should be balanced by some competitive and some individual learning activities, particularly for older students. The activities suggested in this text reflect such a balance. Teachers who have extensive training and experience with cooperative learning will see many opportunities for learning teams to work cooperatively to master new vocabulary, apply concepts, and share the wonder of discovery in both the classroom and the laboratory. For those teachers who are unfamiliar with the philosophies and strategies essential to planning and monitoring cooperative-learning activities, a list of references has been provided.

REFERENCES

DeVries, D. L., and R. E. Slavin. "Teams-Games-Tournament (TGT): Review of Ten Classroom Experiments," *Journal of Research and Development in Education.* 1978, Vol. 12, No. 1, pp. 28–38.

Glasser, William. *Control Theory in the Classroom.* New York: Harper & Row. 1986.

Johnson, David W., Roger T. Johnson, and Edythe J. Holubec. *Circles of Learning: Cooperation in the Classroom.* Edina, MN: Interaction Book Company. 1986.

Johnson, David W., Roger T. Johnson. *Learning Together and Alone: Cooperation, Competition, and Individualization.* Englewood Cliffs, NJ: Prentice Hall, 1975.

Slavin, Robert E., "Cooperative Learning," *Review of Educational Research.* 1980, Vol. 50, pp. 315–42.

Slavin, Robert E., "Cooperative Learning: Can Students Help Students Learn?" *Instructor.* 1987, Vol. 96, pp. 74–78.

Slavin, Robert E., Shlomo Sharan, Spencer Kagan, Rachel Hertz-Lazarowitz, Clark Web, Richard Schmuck. *Learning to Cooperate, Cooperating to Learn.* New York: Plenum Press. 1985.

Teacher's Guide

Science Safety Guidelines

SCIENCE SAFETY CLASSROOM DO'S AND DON'TS

It is essential that students follow safety guidelines whenever performing an investigation or activity. Make sure your students read the safety section in Chapter 1, as well as the list of safety rules in the Reference Section of their textbook. You may also want to read the following do's and don'ts to your class.

Do

1. Wear protective goggles when working with chemicals, burners, or any substance that might get into your eyes. Many materials in a lab can cause injury to the eyes and even blindness.
2. Learn what to do in case of specific accidents such as getting acid in your eyes or on your skin. (Rinse acids on your body with lots of water.)
3. Before starting any Laboratory Investigation or other experiment, make a list of the things that could go wrong that might hurt you. Then make a list of what you should do if the accident occurs.
4. Make sure you have a fire extinguisher nearby to put out a fire.
5. Work with a friend, when you can, under the supervision of a science teacher or adult who understands lab safety rules.
6. Work in a well-ventilated area.
7. Learn how to use first aid to quickly treat burns, cuts, and bruises. Seek help if you are injured.
8. Read directions for an experiment carefully two or more times. Follow the directions exactly as they are written.

Don't

1. Mix chemicals "for the fun of it." You might produce an explosive reaction that could seriously injure you.
2. Taste, touch, or smell any chemical that you don't know for a fact is harmless. Many chemicals are poisonous.
3. Heat any chemical that you are not instructed to heat. A chemical that is harmless when cool can be dangerous when heated.
4. Heat a liquid in a closed container. The expanding gases produced may blow the container apart, injuring you.
5. Perform an experiment in which you must connect wires to house current. You could electrocute yourself. (Use dry cells instead.)
6. Tilt a test tube toward yourself or anyone else (or hold it upright) when you are heating its contents. (Always tilt the tube away from yourself and others.)
7. Perform any experiment for which you do not have written instructions (in a text or from your teacher.)

FIELD TRIP SAFETY

Field trips are an exciting part of the year in biology and provide students with the opportunity to have firsthand evidence and experiences outside the classroom. The following tips will help you plan and execute a successful field trip.

1. Site Selection: It is up to you to select an appropriate field trip site for your students, depending upon your locale. To aid you in making the necessary arrangements, we suggest that you make a school file of available sites in your area. The file should contain specific instructions concerning whom to contact at the site, directions or a map to the site, fees (if any), the hours the site is open to the public, and availability of meal facilities and restrooms.

It is suggested that you make a visit to the site prior to the field trip to inspect the facilities. While you are there, locate and inventory work-study areas. Make a list of specific equipment your students will need and note the site restrictions and danger spots. Also note facilities for the handicapped if any of your students have physical limitations.

2. Planning the Trip: Be sure that the field trip is justified in view of the school's educational program and your individual lesson objectives. Request written permission for the trip from school personnel and keep this written permission on file. After being granted permission, send a written statement of your destination, departure and arrival times, mode of transportation, and necessary expenditures to each student's parent or guardian.

Meanwhile, provide time in class for advance research on the site. Correlate the projected field trip with your lessons and text material. Tell your students the why, where, and when of the field trip. Be sure to inform them of any special equipment or clothing they will need—special shoes, shorts, hand lenses, notebooks, and so on.

3. The Actual Field Trip: Make a head count of your students at each boarding and departure and periodically during the trip. Each adult should be provided with a list of the students he or she is to supervise and should remain with that group throughout the entire trip. While you are on the way to the site, discuss the investigation with your students. When you arrive at the site, keep the group together unless you have planned otherwise.

Make certain the students understand the purpose of the field trip. Have them make sketches, drawings, plans, or maps or take notes. Do not allow students to remove anything from its natural setting unless carefully selected items are taken for observation and returned to their natural habitats.

Most importantly, be enthusiastic but don't rush. Don't try to crowd too much into one field trip. Keep in contact with the individuals in the group and be alert and prepared for the "teachable moment."

4. Follow-up Activities: A good field trip provides a base experience for other class activities. While you are returning to the school, have students exchange ideas and discuss their experiences and observations at the site. Encourage students to ask questions and propose future activities related to the field trip. Schedule individual or group reports and have the students evaluate the trip. The vehicle for evaluation can be developed as a class activity.

Later, you may want to have your students prepare exhibits or displays using their sketches, maps, photographs, or other materials from the trip. Have them use the library to investigate questions arising from the trip. A number of library investigations can usually be proposed after a successful field trip. Remember, the learning value of a field trip depends largely on you and the type of follow-up activities you provide or encourage.

Teacher's Guide

NABT Guidelines for the Use of Live Animals

> **The National Association of Biology Teachers (NABT) has developed the following set of guidelines to be used when working with live animals.**

Living things are the subject of biology, and their direct study is an appropriate and necessary part of biology teaching. Textbook instruction alone cannot provide students with a basic understanding of life and life processes. We further recognize the importance of research to understanding life processes and providing information on health, disease, medical care and agriculture.

The abuse of any living organism for experimentation or any other purpose is intolerable in any segment of society. Because biology deals specifically with living things, professional biological educators must be especially cognizant of their responsibility to prevent inhumane treatment to living organisms in the name of science and research. This responsibility should extend beyond the confines of the teacher's classroom to the rest of the school and community.

The National Association of Biology Teachers, in speaking to the dilemma of providing a sound biological education, while addressing the problem of humane experimentation, presents the following guidelines on the use of live animals.

A. Biological experimentation should lead to and be consistent with a respect for life and all living things. Humane treatment and care of animals should be an integral part of any lesson that includes living animals.

B. All aspects of exercises and/or experiments dealing with living things must be within the comprehension and capabilities of the students involved. It is recognized that these parameters are necessarily vague, but it is expected that competent teachers of biology can recognize these limitations.

C. Lower orders of life such as bacteria, fungi, protozoans and invertebrates can reveal much basic biological information and are preferable as subjects for invasive studies wherever and whenever possible.

D. Vertebrate animals may be used as experimental organisms in the following situations:
   1. Observations of normal living patterns of wild animals in the free living state or in zoological parks, gardens or aquaria.
   2. Observations of normal living patterns of pets, fish or domestic animals.
   3. Observations of biological phenomena, i.e. including ovulation in frogs through hormone injections that do not cause discomfort or adverse effects to the animals.

E. Animals should be properly cared for as described in the following guidelines:
   1. Appropriate quarters for the animals being used should be provided in a place free from undue stresses. If housed in the classroom itself, animals should not be constantly subjected to disturbances that might be caused by students in the classroom or other upsetting activities.
   2. All animals used in teaching or research programs must receive proper care. Quarters should provide for sanitation, protection from the elements, and have sufficient space for normal behavioral and postural requirements of the species. Quarters shall be easily cleaned, ventilated and lighted. Proper temperature regulation shall be provided.
   3. Proper food and clean drinking water for those animals requiring water shall be available at all times in suitable containers.
   4. Animals' care shall be supervised by a science teacher experienced in proper animal care.
   5. If euthanasia is necessary, animals shall be sacrificed in an approved, humane manner by an adult experienced in the use of such procedures. Laboratory animals should not be released in the environment if they were not originally a part of the native fauna. The introduction of nonnative species which may become feral must be avoided.
   6. The procurement and use of wild or domestic animals must comply with existing local, state, or federal rules regarding same.

F. Animal studies should be carried out under the provisions of the following guidelines:
   1. All animal studies should be carried out under the direct supervision of a competent science teacher. It is the responsibility of that teacher to ensure that the student has the necessary comprehension for the study being done.
   2. Students should not be allowed to take animals home to carry out experimental studies. These studies should be done in a suitable area in the school.
   3. Students doing projects with vertebrate animals should adhere to the following:
      a. No experimental procedures should be attempted that would subject vertebrate animals to pain or distinct discomfort, or interfere with their health in any way. Pithing of live frogs should be carried out by a teacher experi-

enced in such procedures and should not be a part of the general class activity.

    b. Students should not perform surgery on living vertebrate animals except under the direct supervision of a qualified biomedical scientist.

4. Experimental procedures should not involve the use of microorganisms pathogenic to humans or other animals, ionizing radiation, carcinogens, drugs or chemicals at toxic levels, drugs known to produce adverse or teratogenic effects, pain causing drugs, alcohol in any form, electric shock, exercise until exhaustion or other distressing stimuli.
5. Behavioral studies should use only positive reinforcement in training studies.
6. Egg embryos subjected to experimental manipulation must be destroyed humanely at least two days prior to hatching. Normal egg embryos allowed to hatch must be treated humanely within these guidelines.
7. The administration of anesthetics should be carried out by a qualified science teacher competent in such procedures. (The legal ramifications of student use of anesthetics are complex and such use should be avoided.)

G. The use of living animals for science fair projects and displays shall be in accordance with these guidelines. In addition, no living vertebrate animals shall be used in displays for science fair exhibitions.

H. It is recognized that an exceptionally talented student may wish to conduct original research in the biological or medical sciences. In those cases where the research value of a specific project is obvious by its potential contribution to science, but its execution would be otherwise prohibited by the guidelines governing the selection of an appropriate experimental animal or procedure, exceptions can be obtained if

1. the project is approved by and carried out under the direct supervision of a qualified biomedical scientist or a designated adult supervisor in the field of the investigation; and
2. the project is carried out in an appropriate research facility and
3. the project is carried out with the utmost regard for the humane care and treatment of the animals involved in the project; and
4. a research plan is developed and approved by the qualified biomedical scientists prior to the start of any research.

List of Suppliers

LABORATORY MATERIALS AND ADDRESSES

Carolina Biological Supply Company
Burlington, NC 27215

DAMON/Educational Division
115 Fourth Avenue
Needham, MA 02194

Edmund Scientific Company
103 Gloucester Pike
Barrington, NJ 08007

Fisher Scientific Company
Stansi Educational Materials Division
1259 Wood Street
Chicago, IL 60622

Hubbard Scientific Company
2855 Shermer Road
Northbrook, IL 60062

La Pine Scientific Company
6001 Knox Avenue
Chicago, IL 60018

Nasco
Fort Atkinson, WI 53538

Prentice-Hall Equipment Division
10 Oriskany Drive
Tonawanda, NY 14150

Sargeant-Welch Scientific Company
7300 North Linder Avenue
Skokie, IL 60076

Science Kit, Inc.
777 East Park Drive
Tonawanda, NY 14150

Scientific Glass Apparatus Company
737 Broad Street
Bloomfield, NJ 07003

Turtox/Cambosco
Macmillan Science Company, Inc.
8200 South Hoyne Avenue
Chicago, IL 60620

Nebraska Scientific
3823 Leavenworth Street
Omaha, NE 68105

A-V SUPPLIERS AND ADDRESSES

Coronet/MTI Film and Video
108 Wilmot Road
Deerfield, IL 60015

Encyclopaedia Britannica
Educational Corporation
425 N. Michigan Avenue
Chicago, IL 60611

Eye Gate Media
146-01 Archer Avenue
Jamaica, NY 11435

Guidance Associates/Center
   for the Humanities
90 South Bedford Street
Mt. Kisco, NY 10546

National Geographic Society
Educational Services
Dept. 79
Washington, DC 20036

Phoenix/BFA Films
468 Park Avenue South
New York, NY 10016

Prentice Hall
School Division of Paramount Publishing
4350 Equity Drive
P.O. Box 2649
Columbus, OH 43272-4480

Walt Disney Educational Media Company
500 South Buena Vista Street
Burbank, CA 91521

COMPUTER SOFTWARE SUPPLIERS

Carolina Biological Supply Company
2700 York Road
Burlington, NC 27215

Datatech Software Systems
19312 East Eldorado Drive
Aurora, CA 80013

Educational Dimensions Group
P.O. Box 126
Stamford, CT 06904

Comprehensive List of Student Edition Laboratory Materials

Item	Quantity per Group	Chapter
Agar, nutrient		15
Agar, nutrient plates	2	44
Albumin solution		39
Aluminum, metal		3
Aluminum foil		24, 35, 48
Bags, small plastic food		15
Balance	1	3, 22
Balloon, round and large	1	40
Balloon, round and small	1	40
Banana		
ripe	1	22
unripe	1	22
Beaker		
50 mL	3	8
50 mL	2	1
100 mL	2	5, 49
100 mL	1	17, 48
150 mL	2	24, 26, 35, 41
400 mL	1	14, 22, 32, 39
600 mL	1	16, 28, 31, 47
1000 mL	1	47
Beaker tongs	1	16
Beef, raw	small piece	38
Benedict's solution		22, 39
Biuret reagent		39
Boxes, small	2	42
Bread		15, 19
Bromthymol blue solution		6
Bunsen burner	1	16, 49
Calculator	1	46
Cardboard	1 piece	27

T24

Teacher's Guide

Item	Quantity per Group	Chapter
Cellophane tape		35
Cereal, dry		35
Cheesecloth		26, 28
Chicken egg, raw	1	32
Chicken wing	1	13
Clock with second hand	1	8, 14, 31, 38, 46
Coffee (solution)		46
Comb (or brush)	1	33
Compass	1	35
Cornflakes		28
Corn seeds	15	24
Cotton absorbent		18, 24, 41
Coverslip	1	2, 8, 18, 26, 33, 38, 46, 48
	2	5, 15, 17, 19, 21, 23, 25
	4	47
Daisy blossom	1	25
Daphnia		46
Deflagrating spoon	1	49
Depression slide	1	26, 46
Dextrose solution		39
Dissecting needle	1	19, 32, 38
Dissecting pan	1	13
Dissecting tray	1	28, 35, 38
Earthworms, live, in a storage container	2	27
Echinoderms, assorted		29
Elodea	2	6
Elodea leaf	1	23

Item	Quantity per Group	Chapter
Ethanol, 70%		15
Ethanol-acetic acid fixative		8
Fern plant	1	21
Filter paper		24, 49
Fishnet	1	31, 41
Flasks		
125 mL	5	4
125 mL	2	6
125 mL	3	16
1000 mL	1	49
Food items, assorted		
(milk, nuts, hard candy, etc.)		39
Forceps	1	5, 8, 13, 15, 19, 23, 25
Frog, preserved	1	13
Funnel	1	26
Game markers, labeled	2	9
Gloves, heat resistant		16
Goldfish (in an aquarium)	1	31, 41
Graduated cylinder		
10 mL	1	1, 3, 4, 39, 48
100 mL	1	6, 48, 49
100 mL	2	22
Grass		47
dried		16
Hand lens	1	22, 25, 28, 33
Hot plate	1	16, 22, 32, 39
Hydrochloric acid		8
0.1 M		4
Hydrogen peroxide, 3%		4
Incubator		44

Teacher's Guide

Item	Quantity per Group	Chapter
Index cards	2	24
Iodine solution		19
Jars, baby food with lids		15
Kidney beans, red	100	14
Labels		15
Lamp, desk		27
or sunlight		35
Leaves		47
Light source		6
Light, electric or bright sunlight		33
Lima beans		
large	100	14
	10	17
	2	48
Liver, raw beef	small piece	4
Liver puree (catalase)		4
Lugol iodine solution		5, 39
Mealworms	10	25
	25	28
Medicine dropper	3	4
	2	8, 17, 19, 26, 38, 48
	1	2, 5, 15, 18, 21, 24, 25, 27, 33, 39, 41, 46, 47
Metal sample, Zn or Al		3
Meterstick	1	37
Methylene blue		17, 33, 48
vital		26
Microscope	1	2, 5, 8, 15, 17, 18, 19, 20, 21, 23, 25, 26, 30, 33, 38, 41, 43, 46, 47, 48
Milk carton, large	1	24
Mirror	1	11
Moss plant	1	21

T27

Item	Quantity per Group	Chapter
Mosses and other plants		15
Mushroom	1	19
Mustard seeds	200	1
	100	49
Onion, small	1	5
Onion root tips		8
Paintbrush, small	1	35
Paper		
white	1 sheet	34, 47
unlined	1 sheet	10, 11, 35
unlined	3 sheets	41
graph	1 sheet	12, 31, 42
typing	1 sheet	45
pH		4, 49
construction paper (white, red, dark blue, light blue, dark green, light green)		7
Paper bag, small brown	1	14
Paper cup	1	9
Paper towels		8, 19, 26, 27, 38, 44, 46
Pen	1	10
Pencil		
glass-marking	1	1, 4, 5, 8, 22, 39, 44, 49
with eraser	1	8, 11
Penny	1	42
Petri dish	1	4, 19, 24, 32, 35, 41
	2	49
Petri dishes and covers		15
	3	1
Pinch clamp	1	26

Item	Quantity per Group	Chapter
Pinto beans	250 g	14
Plate, glass	1	41
Polyethylene bottle 200 mL	1	40
Pond water		15, 18
Potato, small	1	5
Probe	1	8, 28
Protist identification guide	1	18
Protractor	1	34, 45
Reference books		47
Refrigerator		44
Ring clamp	1	26
Ring stand	1	26
Rubber band	1	26, 28
Rubber stoppers		
#5	2	6
#5	1	16
#8½	1	49
Rubber stopper, #2 one hole	1	40
Rubber stopper, #5 one hole with S-tube	1	16
Rubber tubing, 10 cm		26
Ruler	1	22, 35, 45
metric	1	7, 8, 10, 21, 34
transparent, plastic metric	1	43
Safety goggles		4, 8, 16, 49
Scalpel	1	5, 8, 13, 22, 23, 25
Scissors	1 pair	2, 7, 12, 13, 21, 23, 24, 25, 26, 33, 34, 35, 40, 42, 45

Item	Quantity per Group	Chapter
Skeleton, human (model or chart)	1	13
Slides		
glass microscope	2	5, 8, 15, 17, 19, 21, 41
	1	2, 33, 38, 48
	3	23, 25
	4	18, 47
prepared, of *Spirogyra*	1	20
prepared, showing the early development of a starfish and a clam	1	30
prepared, of a mammalian ovary and of mammalian sperm cells	1	43
Soap, hand		44
Sodium hydroxide, 0.1 M		4
Sodium nitrate, saturated	50 mL	5
Soil		26, 47
potting		24
Sour milk (or yogurt or sauerkraut)		15
Starch suspension		39
Steel wool		49
Stick, wooden, from a frozen dessert	1	10
Stirring rod	1	4, 49
Straw, drinking	1	6
Sugar solution (dextrose)	5 mL	22
Sulfur		49
Tape, transparent		7, 15, 24, 44
Teasing needle	1	5
Test tubes	4	22, 48
	2	39, 49
Test tube holder	1	22, 39
Test tube rack	1	22, 39, 48, 49

Teacher's Guide

Item	Quantity per Group	Chapter
Thermometer, Celsius	1	31
Thread		
dark		2
light		2
Toluidine blue		8, 38
Tongs	1 pair	32, 49
Toothpick, flat	1	17, 33
Tray	1	27
Tulip blossom	1	25
Vertebrates, assorted		36
Worms and other insects		15
Yew leaf	1	23
Zinc, metal		3

Laboratory Skills Assessment Record Chart

The Laboratory Skills Assessment Record Chart that follows is designed to help you assess students' progress in the laboratory setting. By filling out one chart per student for each laboratory investigation, you can monitor individual student performance in a particular laboratory investigation as well as each student's continuing progress as the year evolves.

Use a proficiency rating scale from 1 to 5, with 1 showing absence of laboratory skills proficiency and 5 showing complete proficiency. Carefully monitor students who achieve low proficiency ratings in certain categories at the beginning of the year for improvement. Further instruction may be required for students who do not improve as the year progresses.

Teacher's Guide

Student's Name _____ Class _____ Date _____

Laboratory Investigation Title _____

SKILL	PROFICIENCY RATING
Instructional Skills	
• Reads lab before class	
• Understands purpose of investigation	
• Follows all directions as written or spoken	
• Does not improvise without permission	
• Contributes to the lab group	
• Records all observations and data accurately	
• Achieves expected results	
• Answers questions accurately based on data	
• Relates investigation to textbook material	
Manipulative Skills	
• Sets up all equipment as instructed	
• Uses measurement instruments accurately	
Safety Skills	
• Understands and follows all safety regulations	
• Uses safety goggles and heat-resistant gloves when necessary	
• Treats glassware carefully	
• Handles and disposes of chemicals properly	
• Does not fool around in class	
• Cleans up work area	

Biology

This book is dedicated to the African elephant, the largest living land mammal, which was placed on the endangered species list in October 1989.

Biology

Kenneth R. Miller, Ph.D.
Professor of Biology
Brown University
Providence, Rhode Island

Joseph Levine, Ph.D.
Science Writer and Consultant
Adjunct Assistant Professor of Biology
Boston College
Boston, Massachusetts

 **Prentice Hall**
Englewood Cliffs, New Jersey
Needham, Massachusetts

COMPONENTS OF THE BIOLOGY PROGRAM

Student Text and Teacher's Edition
Teaching Resources
Classroom Manager
Laboratory Manual and Annotated Teacher's Edition
Study Guide and Annotated Teacher's Edition
Product Testing Activities
Biology Transparencies with Teacher's Guide
Biology Posters
Biotechnology Workbook and Solutions Manual
Science Fair Manual
Computer Test Bank with DIAL-A-TEST™ Service
Videos/Videodiscs
Level I Videodiscs
Level III Interactive Videodiscs
Level III Interactive Videodiscs/CD Rom
Biology Media Guide

The illustration on the cover depicts endangered African elephants and other wildlife of the African savanna.

Photo credits begin on page Reference 64.

THIRD EDITION

© 1995, 1993, 1991 by Prentice-Hall, Inc., Englewood Cliffs, New Jersey 07632. A Paramount Communications Company. All rights reserved. No part of this book may be reproduced or transmitted in any form or by any means, electronic or mechanical, including photocopying, recording, or by any information storage and retrieval system, without permission in writing from the publisher. Printed in the United States of America.

ISBN 0-13-8030995

2 3 4 5 6 7 8 9 10   98 97 96 95 94

Prentice Hall
A Paramount Communications Company
Englewood Cliffs, New Jersey 07632

STAFF CREDITS

Editorial	Harry Bakalian, Pamela E. Hirschfeld, Julia Fellows, Lorraine Smith-Phelan, Lois Arnold, Maureen Grassi, Ann Collins, Robert P. Letendre, Natania Mlawer, Elisa Mui Eiger, Christine A. Caputo, Joseph Berman, Rekha S. Sheorey
Art Direction	Arthur F. Soares
Production	Suse F. Bell, Betsy Bostwick, Gertrude Szyferblatt
Photo Research	Libby Forsyth, Emily Rose, Martha Conway
Marketing	Andrew Socha, Arthur C. Germano, Victoria Willows, Joel Gendler
Prepress Production	Laura Sanderson, Paula Massenaro, Denise Herckenrath
Manufacturing	Rhett Conklin, Loretta Moe
Consultants	Janelle Conarton *National Science Consultant* Jeannie Dennard *National Science Consultant* Kathleen French *National Science Consultant* Brenda Underwood *National Science Consultant*

Contributing Writers

Stephanie Baron
*Biology Teacher
San Diego Unified School District
San Diego, California*

Pamela Cunningham
*Former Biology Teacher
Hazlet, New Jersey*

Linda Densman
*Biology Teacher
and Secondary Science Consultant
Hurst-Euless-Bedford
Independent School District
Hurst, Texas*

Kathy French
*Biology Teacher
Hurst-Euless-Bedford
Independent School District
Bedford, Texas*

Janis W. Lariviere
*Biology Teacher
Anderson High School
Austin, Texas*

Marcia Mungenast
*Science Writer
Upper Montclair, New Jersey*

Theresa Flynn Nason
*Science Writer
Voorhees, New Jersey*

Sylvia Neivert
*Science Writer
San Diego, California*

Renate Otterbach
*Gifted and Talented
Educational Specialist
Wichita Falls, Texas*

Denise DiRienzo-Skalecky
*Biology Teacher
Bishop Ford Central Catholic
High School
Brooklyn, New York*

Evan Silberstein
*Biology Teacher
Spring Valley High School
Spring Valley, New York*

Myrna Silver
*Environmental Science
and Physics Teacher
J. J. Pearce High School
Richardson, Texas*

College Reviewers

Michael J. Balick, Ph.D.
*Director, Institute of Economic Botany
New York Botanical Garden
Bronx, New York*

Marvin Druger, Ph.D.
*Professor of Biology
Syracuse University
Syracuse, New York*

Mary-Jane Gething, Ph.D.
*Professor of Biochemistry
University of Texas
Southwestern Medical Center
Dallas, Texas*

Paul E. Hertz, Ph.D.
*Chairman, Department of Biological
Sciences
Barnard College
New York, New York*

Cheryl L. Mason, Ph.D.
*Assistant Professor of Biology
San Diego State University
San Diego, California*

Laurence D. Mueller, Ph.D.
*Associate Professor of Ecology
and Evolutionary Biology
University of California
Irvine, California*

Ada Olins, Ph.D.
*Biological Researcher
Department of Biology
Oak Ridge National Laboratory
Oak Ridge, Tennessee*

John Penick, Ph.D.
*Science Education Center
University of Iowa
Iowa City, Iowa*

Kenneth Sebens, Ph.D.
*Marine Science Center
Northeastern University,
Nahant, Massachusetts*

Jerry W. Shay, Ph.D.
*Associate Professor
Department of Cell Biology
and Neuroscience
University of Texas
Southwestern Medical Center
Dallas, Texas*

Judy Snyder, Ph.D.
*Department of Biology
Denver University
Boulder, Colorado*

Susan Speece, Ed.D.
*Biology Department Chairperson
Anderson University
Anderson, Indiana*

Secondary Reviewers

Tod Anderson
*Science Teacher
L. V. Stockard Middle School
Dallas, Texas*

Leslie Ferry Bettencourt
*Biology Teacher
Lincoln Senior High School
Lincoln, Rhode Island*

William A. Feddeler
*Science Educator
Warren Consolidated Schools
Warren, Michigan*

Steve Ferguson
*Biology Teacher
and Department Chairman
Lee's Summit High School
Lee's Summit, Missouri*

Larry Flammer
*Biology Teacher and Science Supervisor
Campbell Union High School District
San Jose, California*

Emile Hamberlin, Ph.D.
*Science Instructor
DuSable High School
Chicago, Illinois*

Frank R. Johns
*Biology Teacher
Lake Placid High School
Lake Placid, New York*

Dwight Kertzman
*Biology Teacher
and Department Chairman
Booker T. Washington High School
Tulsa, Oklahoma*

Priscilla J. Lee
*Science Department Chairperson
Venice High School
Los Angeles Unified School District
Los Angeles, California*

Mary Grace Lopez
*Biology Teacher
W. B. Ray High School
Corpus Christi, Texas*

Warren D. Maggard
*Biology Teacher
and Department Chairman
South Oldham High School
Crestwood, Kentucky*

Della M. McCaughan
*Science Department Chairperson
Biloxi High School
Biloxi, Mississippi*

Jarvis VNC Pahl
*Biology Teacher
Etiwanda High School
Etiwanda, California*

Sr. John Ann Proach, C.S.B.
*Biology Teacher
and Science Chairperson
Bishop Conwell High School
Levittown, Pennsylvania*

Therese A. Scott
*Biology Teacher
Sam Houston High School
Arlington ISD
Arlington, Texas*

Norma B. Trevino
*Biology Teacher
Foy H. Moody Health
and Science Center
Corpus Christi, Texas*

Gary J. Vitta
*Assistant to the Superintendent
West Essex Regional Schools
North Caldwell, New Jersey*

Glenna Wilkoff
*Science Chairperson
John Hay High School
Cleveland, Ohio*

Reading Consultant

Larry Swinburne, Director
Swinburne Readability Laboratory

Contents

UNIT 1 — The World of Life 2–83

Chapter 1 *The Nature of Science* 4–25
1-1 What Is Science? 5–6
1-2 The Scientific Method 7–13
1-3 Science: "Facts" and "Truth" 15–18
1-4 Safety in the Laboratory 18–19
1-5 The Spaceship Called Earth 20–21

Chapter 2 *Biology as a Science* 26–43
2-1 Characteristics of Living Things 27–31
2-2 Biology: The Study of Life 31–38

Chapter 3 *Introduction to Chemistry* 44–61
3-1 Nature of Matter 45–47
3-2 Composition of Matter 47–51
3-3 Interactions of Matter 52–54
3-4 Chemical Reactions 55–56

Chapter 4 *The Chemical Basis of Life* 62–81
4-1 Water 63–67
4-2 Chemical Compounds in Living Things 68–69
4-3 Compounds of Life 70–77
Careers in Biology 82
From the Authors 83

UNIT 2 Cells: The Basic Unit of Life
84–177

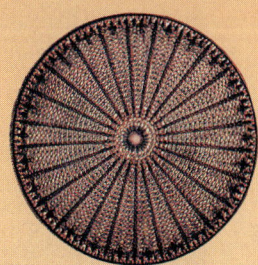

Chapter 5 *Cell Structure and Function* 86–111

- 5–1 The Cell Theory 87–88
- 5–2 Cell Structure 89–93
- 5–3 Cytoplasmic Organelles 94–99
- 5–4 Movement of Materials Through the Cell Membrane 99–104
- 5–5 Cell Specialization 104–106
- 5–6 Levels of Organization 106–107

Chapter 6 *Cell Energy: Photosynthesis and Respiration* 112–135

- 6–1 Photosynthesis: Capturing and Converting Energy 113–117
- 6–2 Photosynthesis: The Light and Dark Reactions 118–123
- 6–3 Glycolysis and Respiration 123–129
- 6–4 Fermentation 130–131

Chapter 7 *Nucleic Acids and Protein Synthesis* 136–157

- 7–1 DNA 137–145
- 7–2 RNA 146–148
- 7–3 Protein Synthesis 148–153

Chapter 8 *Cell Growth and Division* 158–175

- 8–1 Cell Growth 159–164
- 8–2 Cell Division: Mitosis and Cytokinesis 164–171

Careers in Biology 176
From the Authors 177

UNIT 3 — Continuity of Life 178–265

Chapter 9 Introduction to Genetics 180–203
9–1 The Work of Gregor Mendel 181–189
9–2 Applying Mendel's Principles 190–192
9–3 Meiosis 193–196

Chapter 10 Genes and Chromosomes 204–225
10–1 The Chromosome Theory of Heredity 205–211
10–2 Mutations 212–214
10–3 Regulation of Gene Expression 215–221

Chapter 11 Human Heredity 226–245
11–1 "It Runs in the Family" 227–229
11–2 The Inheritance of Human Traits 230–234
11–3 Sex-Linked Inheritance 235–239
11–4 Diagnosis of Genetic Disorders 239–241

Chapter 12 Genetic Engineering 246–263
12–1 Modifying the Living World 247–249
12–2 Genetic Engineering: Technology and Heredity 250–256
12–3 The New Human Genetics 256–259
Careers in Biology 264
From the Authors 265

UNIT 4 — Diversity of Life 266–335

Chapter 13 Evolution: Evidence of Change 268–289
13–1 Evolution and Life's Diversity 269–271
13–2 The Age of the Earth 272–277
13–3 The Fossil Record 278–282
13–4 Evidence from Living Organisms 283–285

Chapter 14 *Evolution: How Change Occurs* 290–317

14–1 Developing a Theory of Evolution 291–295
14–2 Evolution by Natural Selection 296–298
14–3 Genetics and Evolutionary Theory 299–303
14–4 The Development of New Species 304–310
14–5 Evolutionary Theory Evolves 310–313

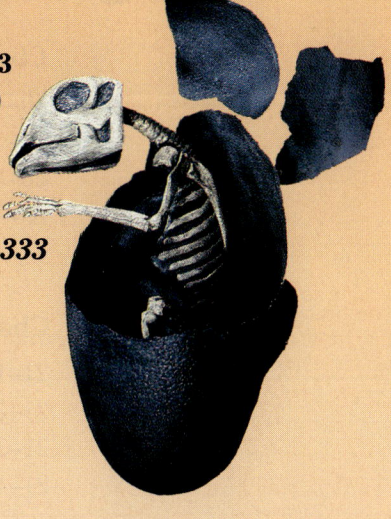

Chapter 15 *Classification Systems* 318–333

15–1 Why Classify? 319–320
15–2 Biological Classification 320–323
15–3 Taxonomy Today 323–325
15–4 The Five-Kingdom System 325–329
Careers in Biology 334
From the Authors 335

UNIT 5 *Life on Earth: Monerans, Protists, and Fungi* 336–429

Chapter 16 *The Origin of Life* 338–353

16–1 Spontaneous Generation 339–342
16–2 The First Signs of Life 342–346
16–3 The Road to Modern Organisms 347–349

Chapter 17 *Viruses and Monerans* 354–379

17–1 Viruses 355–360
17–2 Monerans—Prokaryotic Cells 360–372
17–3 Diseases Caused by Viruses and Monerans 372–375

Chapter 18 *Protists* 380–405

18–1 The Kingdom Protista 381–383
18–2 Animallike Protists 384–394
18–3 Plantlike Protists 394–401

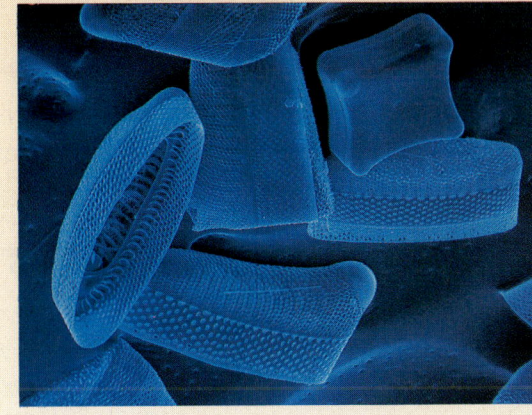

Chapter 19 *Fungi* 406–427

19–1 The Fungi 407–416
19–2 Fungi in Nature 417–423
Careers in Biology 428
From the Authors 429

ix

UNIT 6 Life on Earth: Plants

Chapter 20 *Multicellular Algae* 432–447
20–1 Characteristics of Algae 433–435
20–2 Groups of Algae 435–439
20–3 Reproduction in Algae 440–441
20–4 Where Algae Fit into the World 442–443

Chapter 21 *Mosses and Ferns* 448–465
21–1 Plants Invade the Land 449–451
21–2 The Mosses, Liverworts, and Hornworts 451–454
21–3 The Ferns and the First Vascular Plants 455–459
21–4 Where Mosses and Ferns Fit into the World 460–461

Chapter 22 *Plants with Seeds* 466–485
22–1 Seed Plants—The Spermopsida 467–470
22–2 Evolution of Seed Plants 471–475
22–3 Coevolution of Flowering Plants and Animals 476–481

Chapter 23 *Roots, Stems, and Leaves* 486–515
23–1 Soil: A Storehouse for Water and Nutrients 487–490
23–2 Specialized Tissues in Plants 491–493
23–3 Roots 494–498
23–4 Stems 499–502
23–5 Leaves 502–505
23–6 Transport in Plants 505–508
23–7 Adaptations of Plants to Different Environments 508–511

Chapter 24 *Plant Growth and Development* 516–531
24–1 Patterns of Growth 517–521
24–2 Control of Growth and Development 521–527

Chapter 25 *Reproduction in Seed Plants* 532–549
25–1 Cones and Flowers as Reproductive Organs 533–540
25–2 Seed Development 540–542
25–3 Vegetative Reproduction 542–545
Careers in Biology 550
From the Authors 551

UNIT 7   *Life on Earth: Invertebrate Animals*   552–675

Chapter 26   *Sponges, Cnidarians, and Unsegmented Worms*   554–583

26–1 Introduction to the Animal Kingdom   555–560
26–2 Sponges   560–563
26–3 Cnidarians   564–569
26–4 Unsegmented Worms   570–579

Chapter 27   *Mollusks and Annelids*   584–605

27–1 Mollusks   585–593
27–2 Annelids   594–601

Chapter 28   *Arthropods*   606–635

28–1 Introduction to Arthropods   607–616
28–2 Spiders and Their Relatives   617–620
28–3 Crustaceans   620–621
28–4 Insects and Their Relatives   622–628
28–5 How Arthropods Fit into the World   629–631

Chapter 29   *Echinoderms and Invertebrate Chordates*   636–651

29–1 Echinoderms   637–644
29–2 Invertebrate Chordates   645–647

Chapter 30   *Comparing Invertebrates*   652–673

30–1 Evolution of the Invertebrates   653–657
30–2 Form and Function in Invertebrates   658–669
Careers in Biology   674
From the Authors   675

UNIT 8 Life on Earth: Vertebrate Animals 676–805

Chapter 31 Fishes and Amphibians 678–705
31–1 Fishes 679–692
31–2 Amphibians 692–701

Chapter 32 Reptiles and Birds 706–735
32–1 Reptiles 707–719
32–2 The Evolution of Temperature Control 720–723
32–3 Birds 723–731

Chapter 33 Mammals 736–755
33–1 Mammals 737–745
33–2 Important Orders of Living Mammals 746–751

Chapter 34 Humans 756–769
34–1 Primates and Human Origins 757–759
34–2 Hominid Evolution: Human Ancestors and Relatives 759–765

Chapter 35 Animal Behavior 770–783
35–1 Elements of Behavior 771–775
35–2 Communication: Signals for Survival 775–777
35–3 The Evolution of Behavior 778–779

Chapter 36 Comparing Vertebrates 784–803
36–1 Evolution of the Vertebrates 785–789
36–2 Form and Function in Vertebrates 789–799
Careers in Biology 804
From the Authors 805

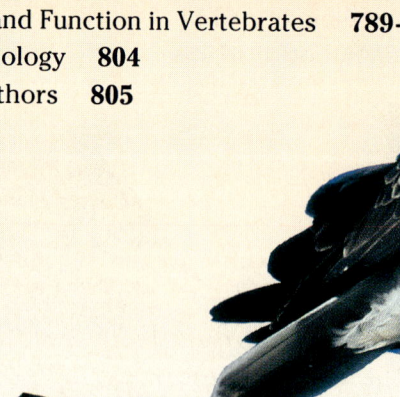

UNIT 9   Human Biology   806–1003

Chapter 37   Nervous System   808–835
37–1 The Nervous System   809–815
37–2 Divisions of the Nervous System   816
37–3 The Central Nervous System   817–824
37–4 The Peripheral Nervous System   825–826
37–5 The Senses   827–831

Chapter 38   Skeletal, Muscular, and Integumentary Systems   836–853
38–1 The Skeletal System   837–842
38–2 The Muscular System   842–846
38–3 The Integumentary System   847–849

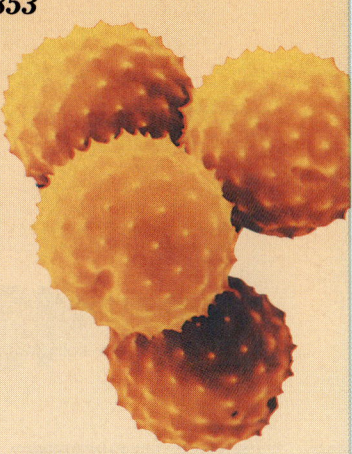

Chapter 39   Nutrition and Digestion   854–879
39–1 Food and Nutrition   855–866
39–2 The Process of Digestion   867–875

Chapter 40   Respiratory System   880–895
40–1 The Importance of Respiration   881–882
40–2 The Human Respiratory System   882–889
40–3 Control of the Respiratory System   890–891

Chapter 41   Circulatory and Excretory Systems   896–915
41–1 The Circulatory System   897–904
41–2 Blood   905–907
41–3 The Excretory System   908–911

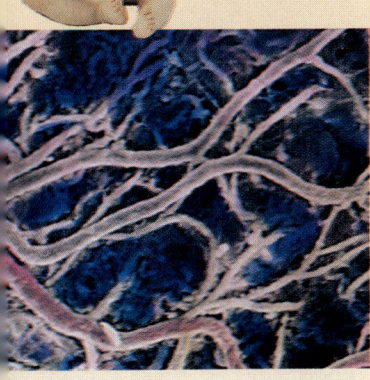

Chapter 42   Endocrine System   916–933
42–1 Endocrine Glands   917–926
42–2 Control of the Endocrine System   927–929

Chapter 43   Reproduction and Development   934–951
43–1 The Reproductive System   935–943
43–2 Human Development   943–947

Chapter 44 *Human Diseases* 952–967

44–1 The Nature of Disease   953–955
44–2 Agents of Disease   956–959
44–3 Cancer   960–963

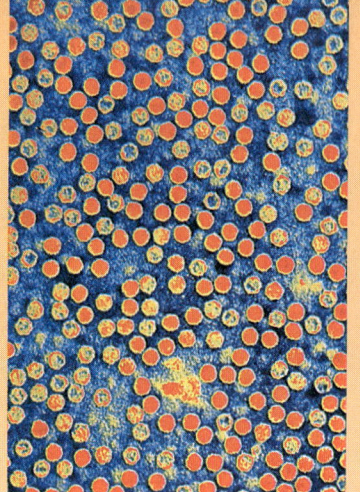

Chapter 45 *Immune System*   968–985

45–1 Nonspecific Defenses   969–971
45–2 Specific Defenses   972–976
45–3 Immune Disorders   977–981

Chapter 46 *Drugs, Alcohol, and Tobacco*   986–1001

46–1 Drugs   987–993
46–2 Alcohol   993–995
46–3 Tobacco   995–997
Careers in Biology   1002
From the Authors   1003

UNIT *Ecological Interactions*   1004–1077

Chapter 47 *The Biosphere*   1006–1031

47–1 Earth: A Living Planet   1007–1010
47–2 Land Biomes   1010–1016
47–3 Aquatic Biomes   1016–1020
47–4 Energy and Nutrients: Building the Web of Life   1021–1027

Chapter 48 *Populations and Communities*   1032–1047

48–1 Population Growth   1033–1035
48–2 Factors That Control Population Growth   1035–1040
48–3 Interactions Within and Between Communities   1040–1042

Chapter 49 *People and the Biosphere*   1048–1075

49–1 Human Population   1049–1052
49–2 Pollution   1052–1063
49–3 The Fate of the Earth   1063–1068
49–4 The Future of the Biosphere   1069–1071
Careers in Biology   1076
From the Authors   1077

Reference Section

For Further Reading Reference 1–6
Appendix A The Metric System Reference 7
Appendix B Science Safety Rules Reference 8–9
Appendix C Care and Use of the Microscope Reference 10–11
Appendix D The Laboratory Balance Reference 12–13
Appendix E The Periodic Table of the Elements Reference 14–15
Appendix F The Five-Kingdom Classification System Reference 16–21
Glossary Reference 22–42
Index Reference 43–63

Features

Problem Solving in Biology

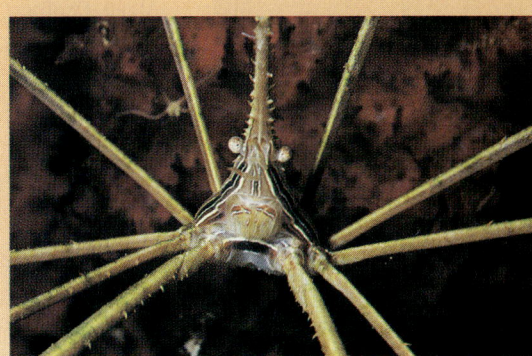

Dimensional Analysis 14–15
Structural Formulas 57
The Colors of Autumn 117
Scientific Notation 163
Solving Genetics Problems 197–199
Radioactive Dating 282
Classifying Common Objects 329
Food Poisoning 366
Interpreting Graphic Data About a Foraminifer 391
Identifying Arthropods 631
What Is It?—Analyzing Function Through Form 669
Analyzing *Homo erectus* Behavior 763
Identifying Vertebrates 799
The "Eyes" Have It! 829
Reading a Food Label 860–861
Analyzing Predator-Prey Population Models 1038

Science, Technology, and Society: Breakthroughs

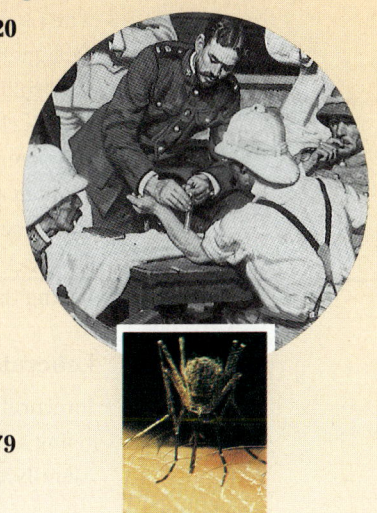

The Odd Couple: A Messy Lab and a Keen Mind 20
A Do-It-Yourself Poison 55
The Structure of Proteins 77
Gateways in the Nucleus 105
Finding a Better Way 122
The Ribozyme Revolution 151
A Useful Poison 171
Jumping Genes 221
Exploring the Unknown 229
The Race to Find a Killer 236
The Symbiotic Theory of Eukaryotic Origins 349
Cyclosporine 416
River Blindness: A Lifelong Battle Almost Won 579
The Secret Life of Salps 647
The Benefits of Living with Relatives 779

Nitric Oxide and the Brain   824
Finding the Gene for Duchenne Muscular Dystrophy   849
Fat Rats   866
Activation of Immune Cells Against Cancer   976

Science, Technology, and Society: Connections

Creating a New Breed of Cat   192
Manipulating Chromosome Numbers   214
DNA, Amber, Insects, and Dinosaurs   250
Dating with Carbon-14   275
The Littlest Miners   371
Algae, Ecology, and Public Health   443
Ferns, Bacteria, and Agriculture   460
Small Friends   490
Flowering by the Clock   526
Cloning Plants   545
Leeches: Modern Applications of Ancient Medicine   601
The Twisting Tentacle: No Bones About It   660
The Smallest Paleontologists   745
The Third Eye   797
An Invisible Poison   889
Replacing a Kidney   911
The Intimate Spread of Disease   947
The Magic Bullets   955
Lyme Disease: An Ounce of Prevention   959
Gypsy Moths: Nature to the Rescue   1043
Ecosystems and Local Economies: Keys to Conservation   1068

Science, Technology, and Society: Issues

Yellow Fever: The Scientific Method in Action   39
Are We Creating Superpests?   303
So Little Time, So Many Names   326
Are Prokaryotes Always Smaller?   363
The Line Between Plant and Animal   397
Designer Genes—Problem or Promise?   480
Controlling Agricultural Pests   628
Where Have All the Fishes Gone?   691
Dinosaurs: New Light on Old Bones   722
Women and the Development of Human Society   765
Anabolic Steroids   926
Crack: A Cheap Way to Die   991
Seeing the Forest   1020

Laboratory Investigations

Interpreting a Controlled Experiment   22
Using a Compound Microscope   40
Identifying Elements by Density   58

Testing Enzyme Activity   78
Observing Osmosis   108
Observing the Relationship Between Photosynthesis and Respiration   132
Constructing a Model of DNA Replication   154
Observing Onion Root Tips   172
Simulating a One-Factor Cross   200
Mapping Chromosomes   222
Observing Human Traits   242
Simulating DNA Fingerprints   260
Examining Homologous Structures   286
Simulating Natural Selection   314
Preparing a Collection of Organisms   330
Investigating Spontaneous Generation   350
Examining Bacteria   376
Examining Protists   402
Comparing a Mold and a Mushroom   424
Studying Filamentous Green Algae   444
Comparing Mosses and Ferns   462
Comparing Ripe and Unripe Fruits   482
Exploring the Structure of Leaves   512
Understanding Plant Tropisms   528
Comparing Different Types of Flowers   546
Collecting and Studying Roundworms   580
Observing Earthworm Responses   602
Observing Insect Metamorphosis   632
Identifying Echinoderms   648
Comparing Development in Echinoderms and Mollusks   670
Observing Respiration in Fish   702
Observing Terrestrial Adaptations of a Chicken Egg   732
Examining the Typical Mammalian Body Covering   752
Comparing Primates, From Gorillas to Humans   766
Exploring Mealworm Behavior   780
Comparing Vertebrates   800
Measuring Reaction Time   832
Examining Skeletal Muscle   850
Testing the Nutrient Content of Foods   876
Constructing a Model of the Respiratory System   892
Observing Circulation in a Fish Tail   912
Simulating the Negative-Feedback Process   930
Examining Mammalian Gametes   948
Observing Bacterial Growth   964
Simulating Clonal Selection of Antibodies   982
Observing the Effect of a Stimulant on the Heart Rate   998
Observing Succession in Aged Tap Water   1028
Examining Patterns of Population Growth in Bacteria   1044
Simulating the Effects of Acid Rain   1072

xvii

Welcome!

You are about to enter a wonderfully exciting, and sometimes unbelievable, world—the world of living things. In the pages of this textbook you will come to understand the nature of life as we know it today. You will also gain an appreciation of the scientific process that has resulted in the body of knowledge we know as biology.

Your journey through biology will be guided in large part by your teacher and this textbook.

There is much to learn between these covers! Great care has been taken to write and present the concepts of biology in a way that makes your learning easier.

But you have an important role to play in the process—to learn to use your textbook effectively. The following pages illustrate some of the features that were designed with you, the student, in mind.

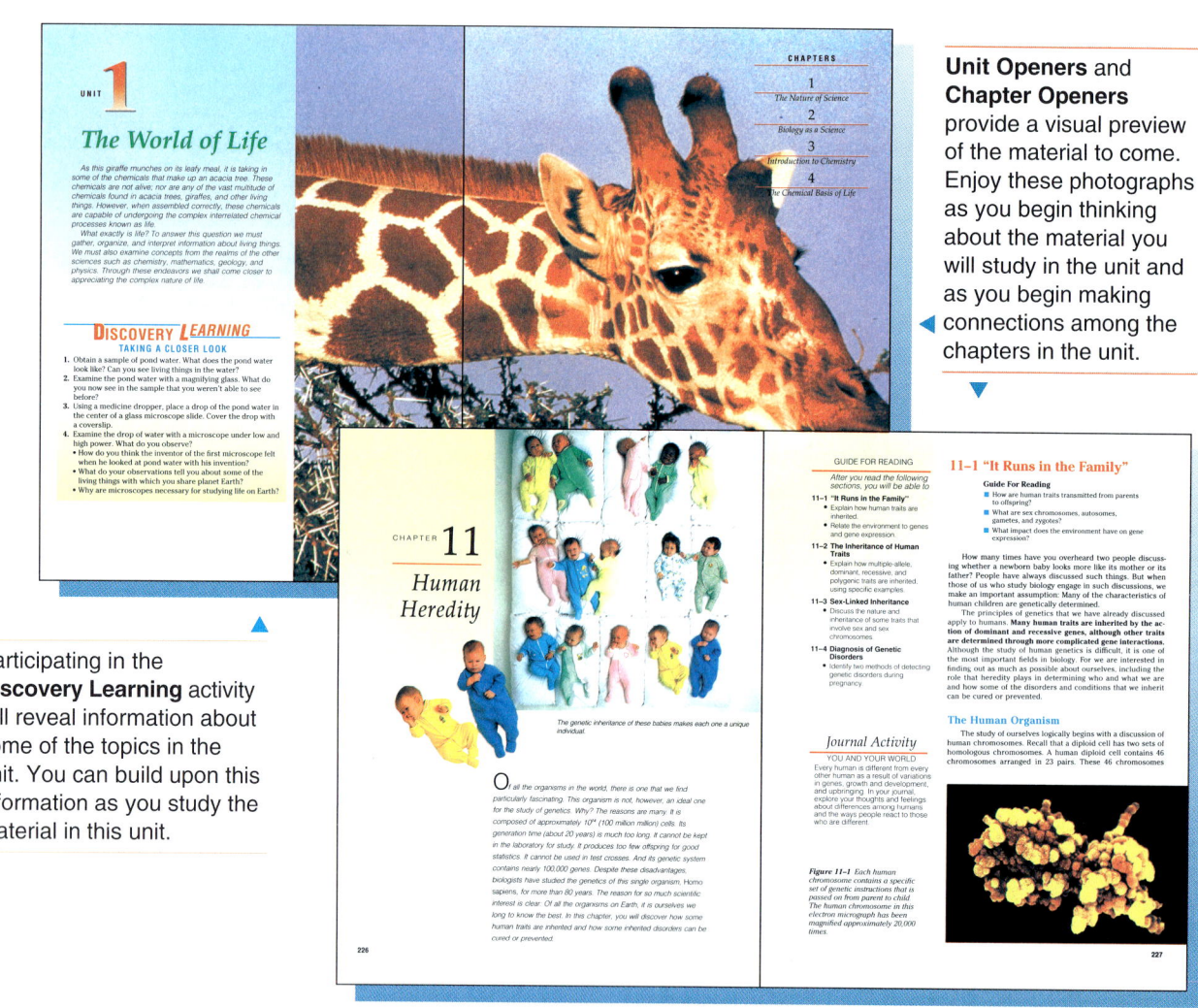

Unit Openers and **Chapter Openers** provide a visual preview of the material to come. Enjoy these photographs as you begin thinking about the material you will study in the unit and as you begin making connections among the chapters in the unit.

Participating in the **Discovery Learning** activity will reveal information about some of the topics in the unit. You can build upon this information as you study the material in this unit.

The **Guide for Reading** for the chapter represents the major divisions found in the chapter. Read the Guide for Reading in order to familiarize yourself with the organization and content of the chapter.

Use the **Journal Activity** to discover how much you already know about the topics about to be studied.

Skim the section using the subsection titles and the visuals with their captions. Then as you read more carefully, focus on the **Guide for Reading** for the section and the boldfaced **Key Idea**. These features will enhance your ability to organize the material you read.

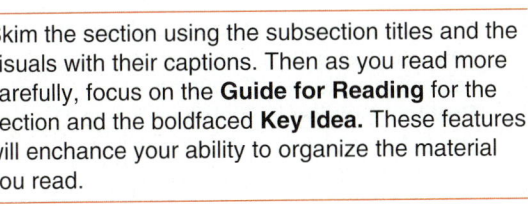

As you read each section, make a chapter outline based on the colored and boldfaced subsection titles. The handwritten outline you see below is an example of how Section 41–2 might be outlined.

CHAPTER OUTLINE

41-1 The Circulatory System
- The Heart
- How the Heart Works
- The Heartbeat
- Blood Vessels
- Pathways of Circulation
- Blood Pressure
- The Lymphatic System

41-2 Blood
- Blood Plasma
- Blood Cells

41-3 The Excretory System
- The Kidneys
- Control of Kidney Function

Check your understanding of concepts by completing the **Section Review** questions. The last question will either help you make the connection between the content and everyday life or another subject area or challenge you to use your thinking skills.

xix

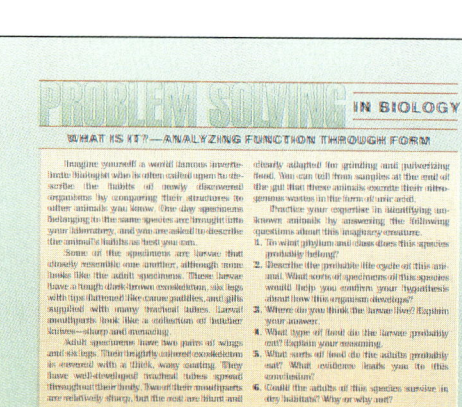

Although you will be studying biology for only one year, you will need to be scientifically literate all your life. **Problem Solving in Biology** will help you to further develop thinking skills you will use in all areas of learning.

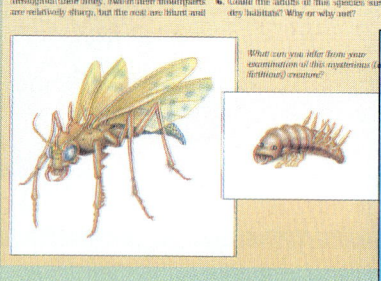

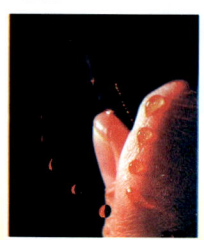

Use **Science, Technology, and Society** to further apply the biology you are learning and the thinking skills you are developing to evaluate current topics in your world. Topics examined may represent **Connections** to other subject areas, **Breakthroughs**, or **Issues**.

Experience biology as a process as well as a body of knowledge. Think about the scientific approach not only as you conduct **Laboratory Investigations** but as you read about the scientific discoveries of others.

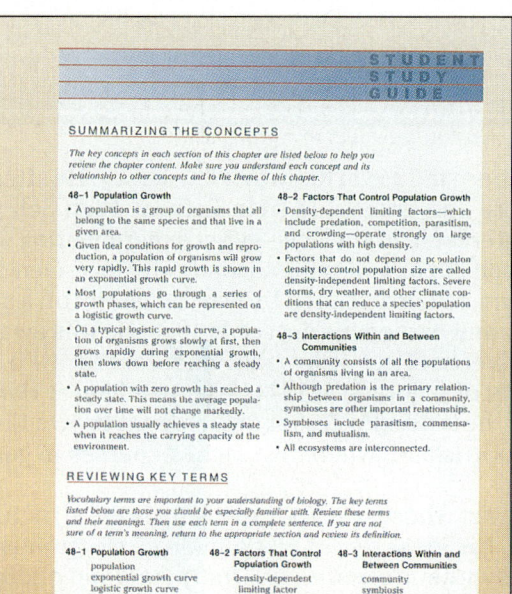

◀

The key concepts in each section of a chapter are listed in **Summarizing the Concepts** to help you review the chapter content. Make sure you understand each concept and its relationship to other concepts and to the theme of the chapter.

Vocabulary terms are important to your understanding of biology. The key terms listed in **Reviewing Key Terms** are those you should be especially familiar with. Review the terms and their meanings. Then use each term in a complete sentence. If you are not sure of a term's meaning, return to the appropriate section and review its definition.

▶

Test your knowledge of the facts presented in the chapter by answering the **Content Review** questions. These facts form the basis of fundamental concepts.

Evaluate your understanding of the concepts by answering the **Concept Mastery** questions.

Apply your factual knowledge and your concept understanding by answering the **Critical and Creative Thinking** questions.

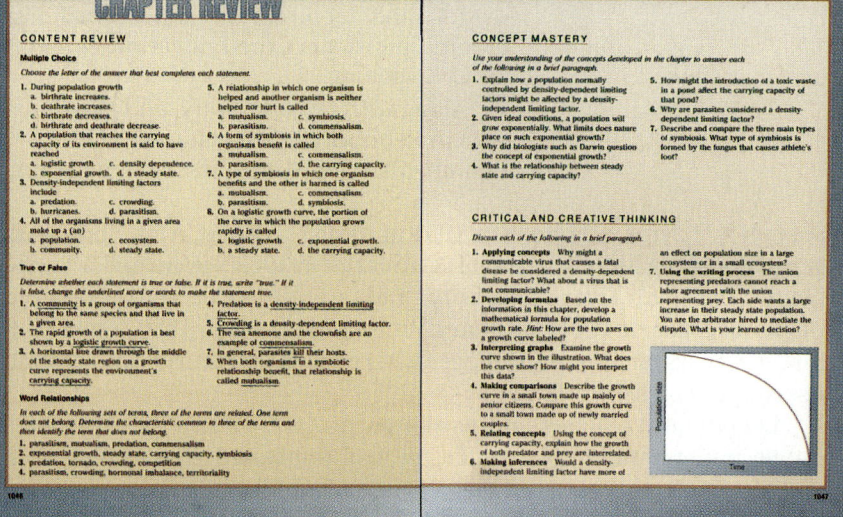

Your enjoyment and understanding of *Biology* by Kenneth Miller and Joseph Levine will be enhanced if you follow the suggestions presented here. We hope you will keep our three R's of biology in mind as you journey through the world of living things:

    Relevance—to connect biology with your everyday life.
    Respect—to develop a high regard for the environment.
    Responsibility—to help you make informed decisions.

ABOUT THE AUTHORS

Kenneth R. Miller was born in 1948 in Rahway, New Jersey. He attended the public schools there and graduated from Rahway High School in 1966. He was already interested in science—often experimenting with a chemistry set in his basement and winning second prize in a science fair with a project on *Euglena*. Miller attended Brown University on a scholarship to study biology and earned his degree in 1970. He was awarded a National Defense Education Act fellowship and earned a Ph.D. in biology at the University of Colorado in 1974. After teaching at Harvard University for six years, he is now Professor of Biology at Brown University in Providence, Rhode Island, where he teaches courses in general biology and cell biology.

Miller is a cell biologist whose research specialty is the structure of biological membranes. His most recent work involves electron microscope studies of photosynthetic membranes from plants and photosynthetic microorganisms. He has published more than 43 scientific papers in journals such as *Nature, The Journal of Cell Biology,* and *Scientific American*. Miller is also an editor for two scientific journals and has written and lectured on educational issues in biology. He is the former Chairman of the Education Committee of the American Society for Cell Biology.

Along with his wife, Jody, and two daughters, Miller lives on a small farm in Rehoboth, Massachusetts. He coaches Little League softball, is a competitor in the masters' swimming program, and enjoys bicycling.

Joseph S. Levine was born in 1951 in Mount Vernon, New York, and graduated from Mount Vernon High School in 1969. He earned a Bachelor's degree at Tufts University, a Master's degree at the Boston University Marine Program, and a Ph.D. at Harvard in 1980. He taught for five years at Boston College, where he is currently an Adjunct Assistant Professor of Biology.

Levine now teaches and works on projects aimed at improving public understanding of science. He has published research papers and articles in journals and magazines ranging from *Science* to *Smithsonian* and has written four books on aquatic subjects. He has designed exhibit programs for public aquaria in Texas, New Jersey, Florida, and Tennessee, and advised the California Museum of Science and Industry on an AIDS awareness exhibit. He has co-produced features for National Public Radio's *Morning Edition* and *All Things Considered*. During a fellowship in science journalism at public television station WGBH-TV in Boston, he co-produced the television program *Science Gazette*. Recently, he served as science editor and companion book author for *The Secret of Life*, an eight-hour prime-time WGBH television series on molecular biology.

Levine delights in the urban neighborhood of Dorchester in Boston, where a century-old home, gardens, and tropical fishes occupy much of his spare time. Levine also enjoys bicycling, windsurfing, skiing, SCUBA diving, and hiking.

TO THE STUDENT

Dear Reader:

We wrote this book so that you would have the opportunity to read about the science of biology. We hope that you will come away well-informed and enlightened about the living world around us. But we had a second goal as well: to convey some of the wonder and excitement felt by those who have been lucky enough to be involved in scientific work.

Science is not a dehumanized process carried out in white-walled laboratories. It is a human activity, subject to all the faults and frailties of our species. But most of all, there is a secret that scientific institutions and textbooks have often managed to conceal: **Science is fun!** Despite the constant frustrations of delicate equipment, difficult observations, and complicated experiments, scientists *enjoy* what they do.

The openness of science means that everything is subject to proof, and many of today's widely held ideas may have to be abandoned in the future. Scientific textbooks seem to be full of information—almost as if everything interesting has already been observed, measured, and cataloged. Wrong! Many people in the nineteenth century believed that science had progressed about as far as it could and that everything from then on would just involve filling in the details. Yet nearly everything in this book is based on discoveries in the twentieth century, and much of it from just the last ten years. Has biology solved all of its important scientific problems? Of course not!

We believe that our present knowledge represents only a fraction of what the world holds for us to discover. Quite frankly, the interesting work is just beginning—and nothing would reward us more than knowing that a few of you who have used this book might decide to come along for the ride.

KENNETH R. MILLER
JOSEPH S. LEVINE

UNIT 1
The World of Life

UNIT OVERVIEW

Unit 1 presents fundamental facts and concepts that are essential to understanding biological phenomena and principles and to perceiving the close connections between biology and other branches of science. The unit starts with an introduction to scientific method and measurement and a discussion of the nature of science. The next part of the unit challenges students to apply what they have learned to explore the characteristics of living things, the nature of biology, and the chemical principles that underlie the processes of life.

UNIT OBJECTIVES

1. Describe the steps in the scientific method.
2. Make measurements of length, mass, volume, and temperature using the metric system.
3. Describe the characteristics of living things.
4. Explain how microscopes work and how they are used.
5. Discuss the properties, phases, composition, and interactions of matter.
6. Explain how the chemistry of water and organic compounds affects living things.

UNIT 1
The World of Life

As this giraffe munches on its leafy meal, it is taking in some of the chemicals that make up an acacia tree. These chemicals are not alive; nor are any of the vast multitude of chemicals found in acacia trees, giraffes, and other living things. However, when assembled correctly, these chemicals are capable of undergoing the complex interrelated chemical processes known as life.

What exactly is life? To answer this question we must gather, organize, and interpret information about living things. We must also examine concepts from the realms of the other sciences such as chemistry, mathematics, geology, and physics. Through these endeavors we shall come closer to appreciating the complex nature of life.

DISCOVERY LEARNING
TAKING A CLOSER LOOK

1. Obtain a sample of pond water. What does the pond water look like? Can you see living things in the water?
2. Examine the pond water with a magnifying glass. What do you now see in the sample that you weren't able to see before?
3. Using a medicine dropper, place a drop of the pond water in the center of a glass microscope slide. Cover the drop with a coverslip.
4. Examine the drop of water with a microscope under low and high power. What do you observe?
   - How do you think the inventor of the first microscope felt when he looked at pond water with his invention?
   - What do your observations tell you about some of the living things with which you share planet Earth?
   - Why are microscopes necessary for studying life on Earth?

INTRODUCING UNIT 1

Using the Visuals

Have students read the copy on page 2 and examine the photograph on pages 2 and 3. You may wish to ask students the following questions about the photograph.

- **What does the photograph show?** (A giraffe eating the leaves of an acacia tree.)
- **How are the organisms in the photograph different?** (Accept all logical answers. Students will probably point out that the giraffe is an animal and the acacia is a plant.)
- **How are the giraffe and the acacia similar?** (Accept all logical answers. Students will probably point out that both are living things and that both are made up of chemicals.)
- **Why is it important to know some chemistry when studying biology?** (Accept all logical answers. Students may respond that chemicals are important to life, that life can be considered to be a set of complex chemical processes, and that all living things are made up of chemicals.)

CHAPTERS

1
The Nature of Science

2
Biology as a Science

3
Introduction to Chemistry

4
The Chemical Basis of Life

CHAPTER DESCRIPTIONS

1 The Nature of Science
Chapter 1 begins with a description of the processes involved in the scientific method and a discussion of the nature of science and scientific inquiry. Students are given practical introductions to the metric system and to laboratory safety. The chapter concludes by emphasizing the interdependence of all living things and the role of science in understanding and protecting life on Earth.

2 Biology as a Science
The characteristics of living things are described in the first part of Chapter 2. The second part of the chapter focuses on the science of biology. Branches of biology are described, and the kinds of questions asked by different kinds of biologists are examined. The microscope is discussed in some detail, and several major laboratory techniques are briefly described.

3 Introduction to Chemistry
Chapter 3 introduces some basic chemical concepts. The properties and phases of matter are described. The structure and composition of atoms, elements, and compounds are discussed. A comparison of ionic and covalent bonding is used to introduce the concept of chemical reactions.

4 The Chemical Basis of Life
Chemical principles are related to biology in Chapter 4. Mixtures, solutions, suspensions, acids, and bases are discussed in the context of the properties of water. Inorganic and organic compounds are compared in terms of their basic composition, roles in living things, and chemical properties, such as the ability to form polymers. The final section in this chapter is an introduction to the structure, function, and reactions of carbohydrates, lipids, proteins, and nucleic acids.

DISCOVERY LEARNING

TAKING A CLOSER LOOK

Begin your introduction to the unit by having students perform the Discovery Learning activity. Make sure that each student records his or her own observations and answers the questions in the activity. After the activity, encourage students to share their thoughts, observations, and other discoveries with their classmates.

CHAPTER 1 *The Nature of Science*

Section	Laboratory Investigations and Activities
1–1 What Is Science? pages 5–6 THEMES: Evolution, Unity and Diversity	**Teacher Edition** DEMONSTRATION: Observations and Questions, p. 4D

1–2 The Scientific Method
pages 7–15
THEMES: Scale and Structure, Systems and Interactions
	Student Edition
	    LABORATORY INVESTIGATION: Interpreting a Controlled Experiment, p. 22
	Laboratory Manual
	    A Metric Scavenger Hunt, p. 59
	Teaching Resources
	    HANDS-ON ACTIVITY: Observing a Fish, p. 5
	    HANDS-ON ACTIVITY: What Do Seeds Need to Grow?, p. 9
	    TEACHER DEMONSTRATION: Making and Using Observations, p. 1
	Product Testing Activities
	    Testing Jeans Testing Pens
	    Testing Paper Towels

1–3 Science: "Facts" and "Truth" pages 15–18 THEMES: Evolution, Stability	

1–4 Safety in the Laboratory
pages 18–20
THEME: Systems and Interactions
	Teacher Edition
	    DEMONSTRATION: Laboratory Safety, p. 4D
	Laboratory Manual
	    Observing the Uncertainty of Measurements, p. 63
	Teaching Resources
	    ACTIVITY: Incredible Measurements, p. 3

1–5 The Spaceship Called Earth pages 20–21 THEMES: Systems and Interactions, Stability	
Chapter Review pages 23–25	

Teacher Resources
Books

Grinnell, Frederick. *The Scientific Attitude.* Westview Press, 1987.

Judson, Horace Freeland. *The Search for Solutions.* Johns Hopkins, 1987.

Morris, Richard. *Dismantling the Universe: The Nature of Scientific Discovery.* Simon and Schuster, 1984.

CHAPTER PLANNING GUIDE

Other Activities	Media and Technology
Study Guide Section 1–1, p. 1	
Teaching Resources ACTIVITY: The Scientific Method, p. 5 BIOLOGY CASE STUDY: Decomposition of Fallen Leaves, p. 1 **Study Guide** Section 1–2, p. 3	**Biology Transparencies** The Metric System
Study Guide Section 1–3, p. 6	
Study Guide Section 1–4, p. 8	
Teaching Resources VOCABULARY REVIEW: Crossword, p. 1 **Study Guide** Section 1–5, p. 10	
Teaching Resources Chapter Test A, p. 11 Chapter Test B, p. 15	**Computer Test Bank** Chapter Test, p. 7

Audiovisuals
Scientific Methods & Values. 2 filmstrips with cassettes. Hawkhill Associates.
Problem Solving in Science. Film. Educational Activities.
Introduction to Biological Measurements: A General Introduction to the Metric System. Sound filmstrip. Ward.
Using the Scientific Method. Film. Coronet.

4B

CHAPTER 1: The Nature of Science

CHAPTER OVERVIEW

In this chapter students are introduced to science and the study of science. They will learn that the main goal of science is to understand the world around us. They will discover how science differs from other fields of study such as philosophy and the arts, and they will consider the meanings of the words *fact* and *truth* as they apply to science.

Students will learn that scientists use a process called the scientific method. By studying a sample experiment, students will be able to see how various steps of the scientific method are carried out. Students will also be introduced to the metric system, which is the universal system of measurement used by scientists.

In the last part of the chapter, the topics of study habits and laboratory safety will be discussed. Students will also be made aware of environmental concerns as the planet Earth is likened to a spaceship with a living cargo, carrying limited amounts of vital supplies.

1-1 WHAT IS SCIENCE?

Section Focus 1-1

The purpose of this section is to introduce science as a field of study. Students will learn that people have always asked questions about the world around them. They will discover that many cultures in the world have attempted to answer these questions through myth and legends. Students will then learn that there is another way to explain the world, and that is to assume that all events in nature have natural causes. This is the approach of science. Science endeavors to arrange observations or tests to learn what the causes of natural events are.

Performance Objectives 1-1

1. Describe how various cultures have attempted to explain natural events.
2. Discuss science as a way of explaining natural events.
3. State the main goal of science.

Science Terms 1-1

science   p. 6

1-2 THE SCIENTIFIC METHOD

Section Focus 1-2

The purpose of this section is to introduce students to the scientific method. Students will learn in detail about the six basic steps in the scientific method. They will also learn that an important aspect of the scientific process is the refusal to accept an explanation without evidence or proof.

Students will learn how a scientific experiment is designed. They will discover the importance of identifying the variable and of having a control setup and an experimental setup. They will also be introduced to the meaning of the word *theory* as it is used in science.

In this section the metric system of measurement will be introduced as the universal language of scientists. Metric units of length, volume, mass, and temperature will be discussed in detail. A Problem Solving section on the conversion of units is also included.

Performance Objectives 1-2

1. State and describe the steps in the scientific method.
2. Define the term *variable* and explain its importance in an experiment.
3. Compare an experimental setup with a control setup.
4. Discuss the importance of a universal system of measurement.
5. Identify and use the basic units of length, volume, mass, and temperature in the metric system.

Science Terms 1-2

scientific method   p. 7
hypothesis   p. 8
variable   p. 8
control setup   p. 8
experimental setup   p. 8
data   p. 9
theory   p. 10
metric system   p. 11
meter   p. 12
liter   p. 12
cubic centimeter   p. 12
mass   p. 12
weight   p. 12
kilogram   p. 13
Celsius scale   p. 13

1-3 SCIENCE: "FACTS" AND "TRUTH"

Section Focus 1-3

The purpose of this section is to emphasize that science is a constantly changing body of knowledge. Students will learn that many scientific "facts" of the past have been proven false. They will also come to realize that some of what they learn today may have to be changed in the future.

Students will be introduced to several ways of studying science successfully. These include understanding and relating information, rather than just memorizing facts, and having an appreciation for the scientific process that led to the discovery of the information they are learning.

The section concludes with a discussion of science and human values. Students will be made aware of the importance of a scientific education as they attempt to deal with the many complex issues in today's world.

Performance Objectives 1-3

1. Discuss ways in which scientific knowledge is constantly changing.
2. State several ways of studying science successfully.
3. Discuss the importance of understanding science in light of human values.

CHAPTER PREVIEW

1–4 SAFETY IN THE LABORATORY

Section Focus 1–4

The purpose of this section is to discuss the importance of laboratory work and to make students aware of the rules of laboratory safety. Students will learn that the most important rule in the laboratory is to follow the teacher's instructions and textbook directions exactly. Students will also be introduced to a chart of laboratory safety rules and symbols. These symbols appear in the laboratory investigations in this book and in its companion laboratory manual.

Performance Objectives 1–4

1. Discuss the importance of laboratory investigations in the study of science.
2. State the most important safety rule to follow in the laboratory.
3. Identify and describe laboratory safety symbols.

1–5 THE SPACESHIP CALLED EARTH

Section Focus 1–5

The purpose of this section is to heighten students' awareness of the importance of life and the need to protect the Earth's environment. Students will learn that humans have the power to change the Earth for better or for worse. They will also learn that the Earth has limited resources. This section leads into the rest of the text by telling students that they will learn about the basis of life and how closely all forms of life are connected. Students will also be challenged to work to preserve life and to protect the environment against destruction.

Performance Objectives 1–5

1. Discuss ways in which humans can change the Earth for better or for worse.
2. Compare Earth to a spaceship traveling through the universe.
3. Discuss the importance of preserving life and protecting the environment.

DISCOVERY LEARNING

TEACHER DEMONSTRATIONS
Modeling

Observations and Questions

The following demonstration can be used as an introduction to Chapter 1.

Several days or weeks before the demonstration, prepare a slice of bread or cheese so that by the day of the demonstration it has become moldy. Show the moldy bread or cheese to the class and ask:

- **What do you see?** (Possible answers: a slice of cheese with green spots; a slice of spoiled cheese; a slice of moldy bread.)
- **Can you describe in detail what you see on the cheese or bread?** (Answers will vary. Observations should include the color and texture of the mold, the extent to which it covers the cheese, whether the mold is in solid patches or small spots.)
- **What questions do you have after seeing the mold?** (Possible answers: What caused the mold? How long has the mold been there? Is the mold alive? Will the mold cover more of the bread or cheese in time? Is the mold good or bad to eat? Does all bread or cheese get moldy?)

Laboratory Safety

This demonstration can be performed as students study Section 1–4.

Take this opportunity to acquaint students with the laboratory in which they will be working. Point out the location of water, the nearest fire extinguisher, and first aid equipment. Explain to the class the procedures they are to follow in case of fire, accident, or injury. You may wish to make (or have a student make) a chart of these instructions to display in the lab. Display and show how to use safety goggles and heat–resistant gloves.

Show some of the equipment that students will be using in the lab. Glassware, microscopes, heating devices, chemicals, knives, or scalpels can be shown. For each piece of equipment you show, ask:

- **What is the safety symbol associated with this item?** (Answers will vary.)
- **What are some important safety rules to remember when working with this item?** (Answers will vary.)

CHAPTER 1

The Nature of Science

GUIDED ENQUIRY
Pose the following questions to students and have them record their responses. Point out that they will gain a better understanding of the key concepts if they read the chapter with these basic questions in mind. Upon completion of the chapter, pose the questions again. Ask students to compare their initial responses with those they have developed after reading the chapter.

- What is science?
- How is science different from other fields of study?
- What is the frame of mind of a scientist?
- How do scientists go about solving problems?
- What are "scientific facts"?
- How are science and human values related?
- How can one work safely in the laboratory?
- In what ways is the Earth like a giant spaceship?

INTRODUCING CHAPTER 1

Using the Visuals
Have students look at the picture on this page.
• **What do you see?** (Zebras.) Ask students to share one of their observations about the zebras. Elicit the responses that the zebras are in a large group and that they are all moving in the same direction.
Have students read the picture caption and the chapter-opener text.
• **What are the zebras doing?** (They are migrating from a place where the food supply is scanty or of poor quality to one where the grass is growing abundantly in the wake of the seasonal rains.)
Explain that the rains and the migrations of grazing animals such as zebras and wildebeest are events that mark the changing seasons in the plains of Africa.
• **What sorts of seasonal changes do you observe in your environment?** (Accept all logical answers. Encourage students to focus on seasonal changes that they themselves have observed.)
• **What questions do you have about seasonal changes in your environment?** (Accept all appropriate answers. You may wish to jot down student questions on the chalkboard.)
• **Can you think of some possible answers to each of these questions?** (Accept all logical answers.)
• **How might you go about finding out if your answers to these questions are correct?** (Possible answers: Use reference materials such as textbooks and ency-

CHAPTER 1

The Nature of Science

Mass migrations of some African animals, such as these zebras, follow the patterns of seasonal rains that bring life-giving water to parched lands.

Toward the end of August, you may have taken a walk through your neighborhood. That neighborhood may be one of buildings and pavement, lawns and trees, fields and pasture, or even sand and cactus. Such walks often inspire questions. Why do days get shorter at this time of year? Why does grass stop growing in the fall? Why do some plants flower in spring and others in autumn? How do geese know when to fly south for the winter? How do they even know which way is south?

Like yourself, scientists ask themselves similar questions about their world. Sometimes they find answers—as you do. Sometimes the answer remains to be discovered. And sometimes there is no answer. The important point is not finding the answer but asking the question. For making observations and asking questions is at the very heart of the process known as science.

GUIDE FOR READING

After you read the following sections, you will be able to

1–1 What Is Science?
- Identify the main goal of science.

1–2 The Scientific Method
- Describe the steps in the scientific method.
- Compare a control setup and an experimental setup.
- Identify the basic units of length, mass, volume, and temperature in the metric system.

1–3 Science "Facts" and "Truth"
- Relate the process of science and change.
- Identify successful study habits.

1–4 Safety in the Laboratory
- Identify and apply proper safety rules when working in the laboratory.

1–5 The Spaceship Called Earth
- Compare planet Earth to a spaceship traveling through the universe.

Journal Activity

YOU AND YOUR WORLD
Many people think that creativity is confined to the fine arts—painting or literature, for example. Is creativity important in science? What do you think? Express your feelings in your journal.

1–1 What Is Science?

Guide For Reading
■ What is the goal of science?

Asking questions about the world around us is part of human nature. How did life begin? Where did plants and animals come from? Why do animals behave as they do? Every culture in the world has tried to answer these questions—often through myths and legends.

The Pitjendara tribe of central Australia, for example, finds answers to questions about nature on an enormous mound of stone called Ayer's Rock. The rock, which towers more than 300 meters above the flat desert around it, has been carved by centuries of exposure to wind and rain. Erosion has etched gullies and caverns into the rock, and has produced striking formations on its surface. See Figure 1–1.

According to the Pitjendara, the images etched on Ayer's Rock tell stories that depict the adventures, the loves, and the battles of ten enormous creatures. The rock contains the likenesses of two snakes, Liru and Kunia. The Pitjendara believe that the two snakes fought an epic battle that created many features on the southern face of Ayer's Rock. Elsewhere on the rock, a sand-lizard man left his mark digging for water. Through these and other stories, the Pitjendara explain the formation of the world and the processes of birth, life, and death.

The stories of the rock also detail the connections between the tribe and animals important in their everyday lives: snakes,

Figure 1–1 To the Pitjendara, these etchings on Ayer's Rock represent the human brain.

COOPERATIVE LEARNING

Using preassigned lab groups or randomly selected teams, have groups produce the front page of the local newspaper for (date), 2025. On this date, a special edition, "Spaceship Earth," will be published. This edition will be a status report on the condition of "Spaceship Earth." Issues to be addressed by articles, cartoons, and illustrations might include
- Population (human, plant, and other animals)
- Environmental quality (air, water, land)
- Energy sources
- Waste disposal
- Medical technology

This activity can be done individually by each cooperative learning group or as a class, with each cooperative learning group contributing articles to the class newspaper.

Journal Activity

YOU AND YOUR WORLD
You may want to use the Journal Activity as a springboard for class discussion. Students should be instructed to keep their Journal Activity in their portfolio.

• **Why not?** (There is no way to gather evidence; there is no way to observe, for example, a god who is "invisible" or the humanness of a cloud that is "crying.")

Content Development
Point out to students that since the beginning of human history, people have tried to explain natural events. Also point out that ideas such as those students thought of to explain rain often became myths or legends.

• **Why do you think myths and legends appeal to people?** (Possible answers: they often explain things in human terms, such as attributing natural events to activities among supernatural beings; they appeal to the imagination; they explain something without requiring technical understanding or education.)

clopedias to find out if answers to these questions have been found; conduct experiments to test your answers.)

TEACHING STRATEGY 1–1

Focus/Motivation
Begin by asking students,
• **Suppose you knew nothing about science. How would you explain why it rains?** (Accept all answers that do not include scientific information; for example: the clouds are crying; there is a hidden river in the sky; an invisible rain god pours water on the Earth when angry.)
• **Suppose someone did not believe your explanation. Could you prove that it is right?** (No.)

ANNOTATION KEY

① Such people have little knowledge of other cultures or beliefs and tend to explain the natural world based solely on observations of nature. (Applying concepts)

② Accept all reasonable hypotheses, including changes in temperature, changes in length of day, internal clocks, and so on. (Hypothesizing)

TEACHING SUPPORT

Study Guide
- Section 1–1, p. 1

ESL STRATEGY

Cooperative-learning techniques and peer tutoring are extremely beneficial to ESL students. To encourage the learning and use of English, avoid assigning ESL students who share a native language to the same cooperative-learning group.

Select interested native or fluent speakers of English who are doing well in science to act as "buddies" to ESL students. The buddy's job is to provide assistance and encouragement to the ESL student. If possible, match buddies and ESL students so that the ESL students can help their buddies in another academic area, such as math. (**Note**: Instruct any multilingual buddies to use the native language only when necessary, such as defining difficult terms or concepts. Students learn English, as any language, by using it.)

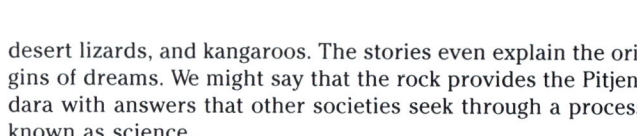

Figure 1–2 Ayer's Rock in central Australia is the largest individual rock in the world. Why might people who live their entire lives in this area develop myths and legends about Ayer's Rock? ①

Figure 1–3 These stone statues were created by people living on Easter Island in the Pacific Ocean. The statues may well represent the Easter Islanders' belief that the first humans hatched from eggs brought to Earth by a bird-man.

desert lizards, and kangaroos. The stories even explain the origins of dreams. We might say that the rock provides the Pitjendara with answers that other societies seek through a process known as science.

People like the Pitjendara live their entire lives in a single culture. Thus, they often find it difficult to imagine that their particular stories about the world might be in error. Today, however, we can visit and read about many cultures. And when we assemble stories from around the globe, it becomes obvious that all these stories cannot be true.

Is there some other way to explain the world around us? One way is to assume that all events in nature have natural causes. We can then try to arrange a series of observations or tests to learn what those causes are. **Science** is the word that we apply to this process. **The goal of science is to understand the world around us.** There are, however, many important fields of human endeavor that study the world around us but are not considered sciences. Such fields include language, history, art, music, and philosophy. The character that distinguishes science from nonscience is an approach known as the scientific method.

1–1 SECTION REVIEW

1. What is the goal of science?
2. **Connection—Literature** Compare the study of science with the study of literature. How do they differ? How are they similar?

1–1 (continued)

Enrichment

Have students work in small groups to create their own myths that explain various natural phenomena. For example, they might explain snow, thunder, earthquakes, or the change of seasons. Encourage students to be creative in the presentation of their myths, using such vehicles as pictures and dramatizations.

SECTION REVIEW 1–1

1. To understand the world around us.

2. They differ in approach. Science uses experimentation to help explain observations and questions regarding the natural world. The scientific method, however, is not an appropriate way to study or approach literature. They are similar in that both are fields of study that require expertise and training to fully comprehend.

1–2 The Scientific Method

The simplest definition of the **scientific method** was offered by biologist Claude Villee. He called it "Organized common sense." That is exactly what science should be. **In practice, the scientific method consists of several steps:**

- **Observing and stating a problem**
- **Forming a hypothesis**
- **Testing the hypothesis**
- **Recording and analyzing data**
- **Forming a conclusion**
- **Replicating the work**

To the true scientist, however, the scientific method is more a frame of mind—a frame of mind that involves curiosity. For without curiosity about nature, there would be no interest in why the sun appears to rise and set, why the seasons change, or why a snake sheds its skin. Another important characteristic of the scientific spirit is the refusal to accept an explanation without evidence or proof. This "prove it!" attitude encourages scientists to investigate phenomena and to develop new explanations and ideas.

Observing and Stating a Problem

The process of science starts with an observation. For example, we might notice that the leaves of maple trees turn bright red and yellow in autumn. As curious scientists, we would then be interested in discovering why this color change takes place.

Guide For Reading
- What are the steps in the scientific method?
- What is the difference between a control setup and an experimental setup?
- What are the basic units of length, mass, volume, and temperature in the metric system?

Figure 1–4 The changing colors of the leaves in Vermont is an event to which people travel from all over the country to enjoy. What hypothesis could you suggest to explain why the leaves of many trees change color in autumn?

HISTORICAL NOTE
ALCHEMISTS

An interesting pseudoscience that flourished during the Middle Ages was alchemy. Alchemy was a strange mixture of knowledge, magic, and philosophy. The main goal of the alchemist was to do what today's scientists know is impossible—to change ordinary metals into silver or gold. Alchemists were very popular, however, and most kings and princes had resident alchemists in their castles. Actually, alchemists did make a contribution to modern science. Some of their methods and laboratory equipment proved to be useful, and they possessed the curiosity that is the basic frame of mind of the scientist.

MULTICULTURAL STRATEGY

Scientists from the community can provide students with first-hand knowledge about careers in science. Identify some local people with careers related to science. As much as possible, mention individuals of different ethnic backgrounds, especially women, with whom students can relate. These neighbors will help students see that they too can enter careers in science.

Keep in mind that some findings of modern science as well as some types of scientific experiments may be incompatible with the beliefs of certain ethnic or religious groups and may violate taboos of students' cultures. The support of respected members of the community of different cultural backgrounds may help you alleviate concerns and promote understanding.

Reinforcement/Reteaching
For students who have trouble with the Section Review questions, go back to the discussion in the section and review the materials students have difficulty understanding.

Closure
Have students create posters or other visual materials that state and illustrate the main goal of science.

TEACHING STRATEGY 1–2
Focus/Motivation
Begin by asking students:
- **Who is a scientist?** (Accept all reasonable answers.)
- **Have you ever thought of yourself as a scientist? If so, when?** (Accept all answers.)
- **Do you know anyone who is a scientist by profession?** (Most will probably say yes.)
- **Can you explain some of the things this person does?** (Answers will vary.)

HISTORICAL NOTE
THE SCIENTIFIC METHOD

The first scientist credited with using the scientific method was the Italian, Galileo. In his investigation of falling objects, Galileo recorded a series of steps that he repeated as often as necessary in order to solve a problem. The steps were basically these: (1) general observation, (2) hypothesis, (3) mathematical analysis or deduction from hypothesis, (4) experimental test of deduction, and (5) revision of hypothesis in light of experiments. A simplified version of these steps formed the basis of the modern scientific method: observation, hypothesis, deduction, experimentation, and revision.

TIE-IN   LANGUAGE ARTS

The word science comes from the Latin word *scientia,* which means "to know." Have students discuss the meaning of this word as it relates to the main goal of science stated in this section.

Forming a Hypothesis

We proceed to gather information that helps us generate a **hypothesis**. A hypothesis is a possible explanation, a preliminary conclusion, or even a guess about some event in nature. For example, we can observe that maple leaves change color when the air gets colder. We might then hypothesize that color changes in maples are related to changes in temperature.

Testing the Hypothesis

Next we must test our hypothesis. Some hypotheses may be tested simply through further observation. As you will discover in Chapter 13, Charles Darwin spent most of his life in the field observing nature to test his hypotheses about evolution. Most hypotheses, however, must be tested by experiments. The "autumn leaves" hypothesis falls into this category.

Our observations indicate that leaves change color in the autumn. And our hypothesis is that the color change is related to changes in temperature. We must now perform a test to find out if this is a correct explanation. One experiment we might do is to take a small maple tree growing in a pot and place it in a growth chamber during the month of July. We would lower the temperature to typical autumn levels, about 14°C during the daytime and 3°C at night. If our hypothesis is correct, we might see the leaves begin to change color almost immediately.

Suppose we ran this experiment and the leaves did change color. Would we know for certain that changes in temperature alone were responsible for the color change? After all, there are many differences between the tree in our experiment and a tree in the woods. Did the leaves on our tree turn red because the tree was planted in a small pot? Or because the light inside the growth chamber was different from natural light?

These uncertainties point out the need for a more complicated experiment using at least two trees. We should choose two similar trees and pots of the same size, water the trees at the same time, and place them in identical chambers. We should then conduct an experiment in which one chamber is kept at normal summer temperatures while the other chamber is cooled to autumn temperatures.

Why must we go to the trouble of observing a second tree? This two-part test represents what is called a controlled experiment. Controlled experiments allow researchers to isolate and test the effects of a single factor, or experimental **variable**. In this case, the treatment of the first tree serves as our **control setup** by testing the effects of the pot and growth chamber. The treatment of the second tree, our **experimental setup**, is identical to the control setup in every respect except one—the variable of temperature. See Figure 1–5. Now if we observe a difference between the experimental setup and the control setup, we can be more certain that it is due to the lower temperatures in the experimental chamber.

Figure 1–5 Whenever possible, an experiment should have both a control setup and an experimental setup. Which part of the experiment contains the variable? What is the variable in this experiment? ❶

1–2 (continued)
Content Development

Point out to students that in the broadest sense, a scientist is anyone who makes observations and asks questions about nature and the world around us. In this sense, we are all scientists. Professional scientists have special training to seek answers to questions about events they observe in the world. A professional scientist also has a logical method of solving problems and of finding answers to questions.

Content Development

Emphasize to students that the function of a control setup is to eliminate the possibility of multiple variables. Stress that the control setup seeks to duplicate the experimental setup in every way except the variable that is being tested.

• **In the experiment described in the text, what are some of the factors in the experimental setup that are duplicated in the control setup?** (Size and type of tree; size of pot; type of chamber; watering schedule.)

• **What factor is not duplicated in the control setup?** (Temperature.)

Recording and Analyzing Data

If we were actually to perform this experiment, we would keep careful records of observations and information, or **data**. In this case, such data might include the time it took for each leaf to change color, the total number of leaves on each tree that changed color or fell off, and so on. We might arrange our data in the form of tables and graphs, such as those shown in Figure 1–6. You can see why researchers use these visual representations of experimental data to analyze their results.

EXPERIMENTAL SETUP: Autumn Temperatures											
Day	1	3	5	7	9	11	13	15	17	19	21
Number of Green Leaves	50	50	50	49	49	48	48	48	48	47	47
CONTROL SETUP: Spring Temperatures											
Day	1	3	5	7	9	11	13	15	17	19	21
Number of Green Leaves	50	50	49	49	49	49	49	47	47	47	47

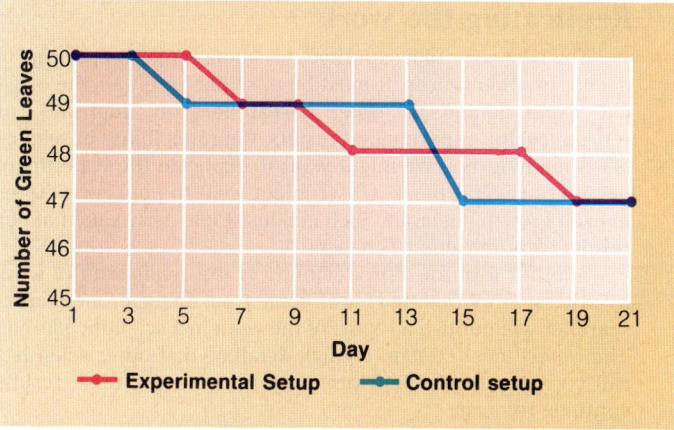

Figure 1–6 Information gathered from an experiment is usually organized in data tables (top). Then the data may be graphed in order to provide a visual representation that is easy to interpret (bottom). Using the data tables and the graph, what conclusions can you reach about the autumn leaves hypothesis?

Forming a Conclusion

If we find that the leaves on our experimental tree change color and those on the control tree do not, we may decide that we have confirmed our original hypothesis: Cold temperatures cause leaves to change color. If the leaves of neither tree change color or if the leaves of both trees change color, we might reject the hypothesis as wrong and start all over with a new idea. This is often the case in science because many guesses about the natural world prove incorrect. But this is also the way science advances, for it points the way to further experiments. If the leaves on neither tree change color, for example, it is possible that temperature alone is not responsible. Perhaps trees also respond to changes in the length of day, decreases in rainfall, or the action of strong winds. These new hypotheses suggest additional experiments that expand our understanding.

Although such experiments are probably impossible for you to perform yourself, scientists have been able to do them and have arrived at an answer to our original hypothesis: Do the colder temperatures of autumn cause a color change in the leaves of maple trees? If you have examined the data in Figure 1-6, you know part of their conclusion: Colder temperatures alone do not cause the color change. Scientists now know that colder temperatures combined with a change in the length of day both play a role in the color change in maple leaves. In addition, many scientists believe that a biological clock within the maple tree also plays a role. But how does this biological clock work? And what chemical triggers the clock? These are questions that as yet remain unanswered. If you should choose to pursue biology as a career, perhaps you will consider these questions once again.

Replicating the Work

The best scientific experiments can be replicated, or reproduced. In other words, it must be possible for either the original experimenter or other researchers to duplicate, or reproduce, the experimental results. Reproducibility in science takes many forms. We might want to replicate the experiment ourselves several times. Or we might want to put several trees in each growth chamber. Such actions would assure us that our results were not due to chance.

Another form of reproducibility is the replication of work by others. If interesting results come from an experiment, a researcher will publish a report of the work in a scientific journal. The report must contain enough detail so that other scientists can copy the experiment precisely to see if the same results continue to occur.

Hypotheses and Theories

When a hypothesis is tested and confirmed often enough that it is unlikely to be disproved by future tests, it may become worthy of being called a **theory**. In scientific usage, the word theory means a great deal more than it does in common speech. Scientific theories are not just hunches or hypotheses. They are powerful, time-tested concepts that make useful and dependable predictions about the natural world.

TIE-IN LANGUAGE ARTS

Discuss the word *theory* as it is used in everyday speech. One dictionary definition of the word lists *conjecture* and *speculation* as synonyms. Point out that people often use the word theory when they are really referring to a hypothesis— for example, "I have a theory about why the washing machine doesn't work."

MULTICULTURAL STRATEGY

Students may want to research the various tools used in different countries to make measurements. Discussing the everyday tools used for measurement may lead to a discussion on how basic systems of measurements evolved in different cultures. The ancient Chinese system of weights and measures, for example, included an acoustical dimension to measurement— the quantity of content in a vessel was defined not only by weight, but by the pitch that sounded when the object was struck.

TEACHING SUPPORT

Laboratory Manual
- A Metric Scavenger Hunt, p. 59

Teaching Resources
- Biology Case Study: Decomposition of Fallen Leaves, p. 1

MEDIA AND TECHNOLOGY
- Biology Transparencies
- 1: The Metric System

1–2 (continued)

Skills Development
Guided Practice
Skill: Forming a conclusion
Using the Visuals Before students read the answer to the original hypothesis in the text, have them use the data table and graph to draw their own conclusions about whether or not color change is related to change in temperature.

- **According to the graph on page 9, what happened to the number of green leaves on both trees during the 21-day period?** (Both decreased from 50 to 47.)
- **What does this lead you to conclude about the role of temperature change in leaves changing color?** (Temperature alone does not seem to be responsible for the change.)
- **On what do you base this conclusion?** (The results of the experimental setup were not much different from the results of the control setup.)
- **What would the data have been like in order for you to conclude that temperature is responsible for leaves changing color?** (By the end of 21 days, the number of leaves on the experimental tree should have been significantly fewer

The Scientific Method—An Everyday Experience

Because many experiments discussed in this book were performed by people called scientists, you might think that science is a special process used only by certain people and useful only under special circumstances. That is not true at all. We all use the scientific method every day!

Let's suppose, for example, that a car will not start. A mechanic will form and test hypotheses about the problem. Perhaps the battery is dead. One experiment would be to crank the starter to see what happens. If the starter motor works but the car still doesn't turn over, it could be that the car is out of gas. A glance at the fuel gauge would test that idea. Again and again, the mechanic would apply the scientific method until the problem is solved. The same technique can find a fault in an electrical circuit or balance a load of laundry in a washing machine. Remember, science is just organized common sense.

A Universal Language—The Metric System

Science works best when scientists everywhere read each other's papers, check each other's experiments, and argue about what those experiments mean. Because most experiments involve measurements, researchers need a universal system of measurement in which to present their findings. **Scientists use the metric system of length, volume, mass, and temperature when describing experiments and data.** The **metric system** is a decimal system based on certain standards and scaled on multiples of 10. The metric system is also known as the International System of Units, or SI.

Figure 1-7 The metric system is easy to use because it is based on units of ten. How many centimeters are there in 100 meters?

COMMON METRIC UNITS

Length	Mass
1 meter (m) = 100 centimeters (cm)	1 kilogram (kg) = 1000 grams (g)
1 meter = 1000 millimeters (mm)	1 gram = 1000 milligrams (mg)
1 meter = 1,000,000 micrometers (μm)	1000 kilograms = 1 metric ton(t)
1 meter = 1,000,000,000 nanometers (nm)	
1 meter = 10,000,000,000 angstroms (Å)	
1000 meters = 1 kilometer (km)	

Volume	Temperature
1 liter (L) = 1000 milliliters (mL) or 1000 cubic centimeters (cm³)	0°C = freezing point of water
	100°C = boiling point of water
kilo- = one thousand	micro- = one millionth
centi- = one hundredth	nano- = one billionth
milli- = one thousandth	

BACKGROUND INFORMATION
THE METRIC SYSTEM

The Standard International unit of length, the meter, was first introduced in France in 1791. The length of a meter was intended to be equal to 1/10,000,000 of the quadrant of the Earth's meridian passing through Paris. Since this length proved impossible to survey, French scientists decided to make instead a platinum bar to serve as the standard meter. This platinum bar, or *metre des archives*, defined the length of a meter until 1960. At this time, the meter was redefined as being equal to 1,650,763.73 wavelengths of radiation emitted in the transition from energy level $2p^{10}$ to $5d^5$ in an atom of krypton-86. In 1983, the meter was redefined yet again as being equal to the length of the path traveled by light during 1/299,792,458 of a second.

ANNOTATION KEY

❶ 10,000 centimeters in 100 meters. (Making calculations)

• Suppose many scientists perform another scientist's experiment and get widely differing results. What might you conclude about the first scientist's experiment? (That the original experiment was not well designed; that the procedure and setup were not clearly stated; that the experiment contained hidden variables.)

Reinforcement/Reteaching

Use the following activity to review the steps of the scientific method. Write the steps of the scientific method on a set of index cards, one step to a card. Place the cards face down on a desk or table. Have each student pick a card at random. Ask the students to line themselves up so that the steps they have drawn are in the correct order. Then have students take turns describing each step.

---

than the number of green leaves on the control tree.)

Content Development

After students have read the section in the text on Replicating the Work, ask:

• Suppose a scientist performs the same experiment as another scientist and the results do not agree. What might be the reason for this? (Possible answers: Perhaps there was some error in the first experiment; perhaps the second scientist did not correctly follow the procedure or duplicate the experimental setup; perhaps the first scientist's results were due to chance or some unusual factor; perhaps the second scientist unknowingly introduced a second variable.)

ANNOTATION KEY

① Meter; nanometers or micrometers. (Applying facts)
② One thousand. (Relating facts)
③ One thousand; one million. (Making conversions)
④ Tons or kilograms; grams or milligrams. (Applying facts)

TEACHING SUPPORT

Study Guide
• Section 1-2, p. 3

FACTS AND FIGURES

The Celsius temperature scale was originally devised by Anders Celsius (1701-44). Celsius designated 0° as the melting point of ice and 100° as the boiling point of water. Celsius temperature is also called centigrade temperature, which means based on a scale of 100.

1-2 (continued)

Skills Development
Guided Practice
Skill: Applying concepts
Write on the chalkboard a list of objects and distances varying in size from microscopic to enormous. Have students discuss which unit of metric measure would be most useful in measuring each object or distance. Possible examples include the size of a bacterium; the length of a bond between two atoms in a molecule; the distance from the Earth to the nearest star; the height of the school building; the height of the teacher; the width of a fingernail; the length of a belt worn by someone in the class; the width of a postage stamp; the distance from New York to Florida; the length of a football field.

Reinforcement/Reteaching
Prepare a set of flashcards that students can use to learn equivalent units of metric measure. For example:

Figure 1–8 Which unit of length would you use to measure this tall African giraffe (left)? This microscopic bacterium (right)? ①

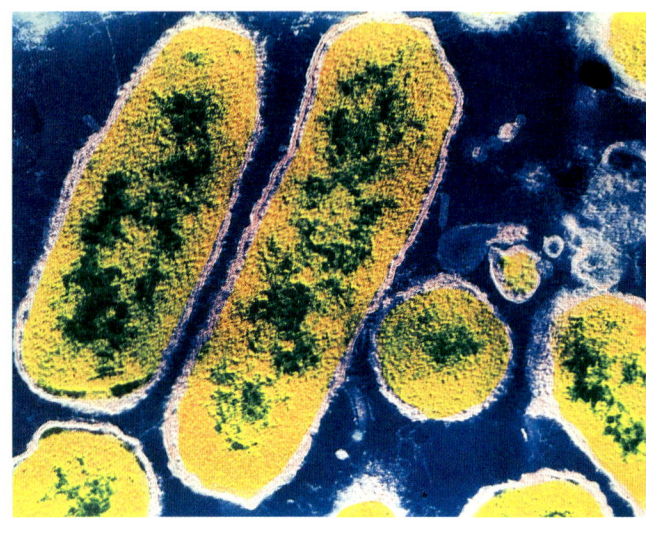

LENGTH The basic unit of length in the metric system is the **meter** (m). One meter is roughly equivalent to 39.4 inches, a little longer than a yard. To measure objects and distances much larger or smaller than a meter, the metric system uses units larger or smaller than a meter by multiples of 10. A centimeter (cm) is 1/100 of a meter—about the width of the nail on your pinky. As you may have guessed, the prefix *centi-* means one hundredth. A millimeter (mm) is 1/1000 of a meter. The prefix *milli-* means one thousandth. Even smaller units of length—such as those used to describe living cells, pieces of cells, and molecules—are shown in Figure 1–7 on page 11. To measure distances much greater than a meter, scientists use a unit called the kilometer (km). One kilometer contains 1000 meters. What does the prefix *kilo-* mean? ②

Figure 1–9 Solids are usually measured in cubic centimeters. One cubic centimeter is the volume of a cube measuring one centimeter on each side. Keep in mind the illustration is not drawn to scale.

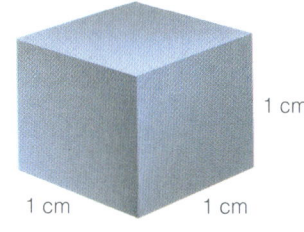

Cubic Centimeter (cc or cm³)

VOLUME Volume is the amount of space an object occupies. The basic metric units of volume are the **liter** (L) for liquids and the **cubic centimeter** (cc, or cm³) for solids. A liter contains slightly more liquid than a quart. To measure small volumes of liquids, scientists use fractions of a liter called milliliters (mL). There are 1000 milliliters in a single liter. A cubic centimeter is the volume of a solid that measures 1 cm by 1 cm by 1 cm. Keep in mind that 1 milliliter is equal in volume to 1 cubic centimeter, or 1 mL = 1 cc.

MASS AND WEIGHT **Mass** is a measure of the amount of matter in an object. **Weight** is a measure of the pull of gravity on that mass. In outer space, the weight of an object may vary with its position, but its mass always remains the same. On Earth's surface, however, an object's mass and weight can

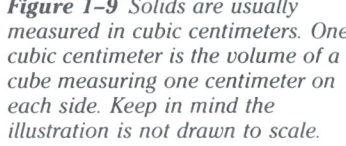

1 kilometer = _____ meters
1 liter = 1000 _____
1 gram = _____ milligrams

SECTION REVIEW 1-2

1. Observing and stating a problem; Forming a hypothesis; Testing the hypothesis; Recording and analyzing data; Forming a conclusion; Replicating the work.
2. A hypothesis is a possible explanation, a preliminary conclusion, or even a guess about an event in nature. A theory is a widely held, time-tested concept from which predictions about the natural world can be made.
3. The experimental setup contains the variable being tested.
4. No. Scientists may well follow the steps in a different order. For

usually be considered constant and are often used interchangeably. The basic metric unit scientists use to describe mass is the **kilogram** (kg). One kilogram is equal to approximately 2.2 pounds. The mass of small objects is measured in grams (g). One gram is 1/1000 of a kilogram. How many milligrams (mg) are there in a gram? In a kilogram?

Figure 1–10 *The African elephant is among the largest land creatures ever to walk the Earth (left). The harvest field mouse is one of the smallest mammals on Earth (right). What unit would you use to measure the mass of the elephant? The harvest mouse?*

TEMPERATURE The metric system measures temperature using the **Celsius** (SEHL-see-uhs) **scale** (°C). On this scale, water freezes at 0°C and boils at 100°C. Each Celsius degree, therefore, represents exactly 1/100 of the temperature range between the freezing and boiling points of water. Normal human body temperature is about 37°C, and comfortable room temperature is about 21°C.

Figure 1–11 *Temperature affects the lives of almost all living things. This caribou is well adapted to life in the cold. It trots along the frozen ground on large feet that act as snowshoes.*

1–2 SECTION REVIEW

1. List each step in the scientific method.
2. Define hypothesis. What is the difference between a hypothesis and a theory?
3. What is the only difference between a control setup and an experimental setup?
4. Must the steps in the scientific method always be followed in the same order? Explain your answer.
5. **Critical Thinking—Applying Concepts** If an experiment does not confirm the hypothesis it was designed to test, has that experiment "failed"?

PROBLEM SOLVING IN BIOLOGY

DIMENSIONAL ANALYSIS

Many times in science you will need to convert one unit of measurement to another. For example, you might be given a problem in kilometers but need the answer in centimeters. The process of converting one unit to another is called dimensional analysis.

Dimensional analysis involves determining three important facts: in what unit your measurement is given, in what unit it should be expressed, and what relationship between the units will allow you to make the conversion. The relationship between the units is called a conversion factor. A conversion factor expresses an exact relationship between the original unit and the desired unit. It is a fraction whose value is 1. For example,

$$\frac{60 \text{ minutes}}{1 \text{ hour}} \quad \text{or} \quad \frac{1 \text{ liter}}{1000 \text{ milliliters}}$$

Although the numerator (top) of the fraction and the denominator (bottom) of the fraction are different, their actual values are the same. You see that 60 minutes equals 1 hour and 1 liter equals 1000 milliliters.

Any number can be multiplied by 1 without changing its value. Multiplying a measurement by a conversion factor, then, does not change the value of the measurement. This means that you can convert the given unit to the desired unit by multiplying by an appropriate conversion factor. You must, however, choose the correct form of the conversion factor. The unit to be converted must cancel out and the desired unit remain. Thus, the denominator of the conversion factor must contain the same unit as the one given to enable them to be canceled. Similarly, the numerator must contain the unit to be in the answer.

The following example shows how a typical dimensional analysis problem is solved.

Sample Problem Your friend runs 2500 meters while you run 3 kilometers. Using kilometers, which one of you has run a longer distance?

Step 1: Determine the given unit and the desired unit.
given: meters
desired: kilometers

Step 2: Find the relationship between the units and consider the possible conversion factors.
1000 meters = 1 kilometer
therefore:
$$\frac{1 \text{ kilometer}}{1000 \text{ meters}} = 1 \text{ or } \frac{1000 \text{ meters}}{1 \text{ kilometer}} = 1$$

Step 3: Choose the conversion factor whose denominator has the same units as your given value.
$$\frac{1 \text{ kilometer}}{1000 \text{ meters}} = 1$$

Step 4: Write the original value next to the conversion factor with a multiplication sign between them. Cancel like terms (shown in red).
$$2500 \text{ meters} \times \frac{1 \text{ kilometer}}{1000 \text{ meters}}$$

Step 5: Multiply the resulting equation.
$$2500 \times \frac{1 \text{ kilometer}}{1000} =$$
$$\frac{2500 \text{ kilometers}}{1000} = 2.5 \text{ kilometers}$$

Your friend has run 2.5 kilometers and you have run 3 kilometers. You have run the longer distance!

PROBLEM SOLVING

Dimensional Analysis

Dimensional analysis is important because the data in science problems and in experiments are not always in the same units. Before beginning this feature you may wish to review with students the concept of a unit fraction (fractions equal to one) by writing some examples on the chalkboard: 2/2, 5/5, 3/3, 100/100. You may also wish to review the multiplication of fractions, emphasizing the cancellation of numerators and denominators.

Before students begin the Practice Problems, discuss the first problem and have students offer ideas as to which unit is the best one in which to express the data (the unit that is neither the greatest nor the smallest—in this case, grams). Then, as students work independently, circulate around the classroom to help those who are having trouble finding the correct conversion factor or who are having difficulty multiplying and canceling the fractions.

Answers to Practice Problems
1. 16,000 g; 2 g; 888 g; 0.155 g
2. too little; 0.25 L = 250 mL
3. 1800 cm
4. yes; 0.16 L = 160 mL
5. 31,536,000 sec; 315,360,000 sec

TEACHING STRATEGY 1–3

Focus/Motivation
Begin by asking students to look around the classroom and list as many facts as they can think of in 3 minutes. (Possible facts include: the number of windows; the color of the walls; the number of desks and chairs; the number of books on the teacher's desk; the number of people or boys or girls in the room; the time of day; the title of an article on the bulletin board; the material, such as wood, that a table is made of; the temperature in the room.)

Content Development
Use the Focus/Motivation activity to introduce the question,
• **What is a fact?** (Answers may vary but should include the idea that a fact is something that can be proved or verified by observation or measurement.)

Point out that the facts students listed about the classroom are easy to prove because everyone can see them.
• **What are some other types of facts besides the kind you just listed?** (Historical facts; scientific facts; facts about other areas of study, such as music or a foreign language; facts such as those one reads in the newspaper.)
• **Is it possible for a fact to change?** (It depends on the type of fact. A historical fact such as a date cannot change because the event has already happened; sci-

Practice Problems

Now try your hand at some dimensional analysis problems. You may choose to use a calculator or computer to do the arithmetic calculations.

1. In order to compare the results of several experiments, you need to have all your data in the same unit. Convert the following measurements to the same unit.
   16 kilograms   0.002 kilograms
   888 grams   155 milligrams
2. A recipe calls for 300 milliliters of water. You add 0.25 liters. Have you put in too much, too little, or the right amount?
3. Determine which of the following measurements is largest: 1800 centimeters, 2.1 meters, 0.0017 kilometers.
4. You are told that you need a jar with a volume of at least 150 cm^3. The label on the jar you find says 0.16 liters. Can you use it?
5. Calculate the number of seconds in 1 year. Then calculate the number of seconds in a decade.

1–3 Science: "Facts" and "Truth"

Scientific knowledge is a constantly changing body of observations. Many "scientific facts" of the past are now known to be false. For much of human history, for example, people "knew" that the Earth was flat, that the sun revolved around the Earth, and that rain fell through holes in heaven from a huge water tank in the sky. Those were the "facts" of an earlier time.

Now, of course, we "know" that the Earth is round (more or less), that the Earth and other planets revolve around the sun, and that rain falls from clouds made of water vapor. These are the scientific "facts" of today—the best explanations of the world around us that scientists have developed so far.

But new discoveries will be made. New theories will be born. Without a doubt, some of what you learn this year will have to be changed one day. Yet this does not mean that science has failed. On the contrary, it means that science continues to succeed in advancing our understanding of the world. Science is not a collection of eternal truths. Rather, it is a process, a way of looking at and understanding the world. And science will continue to change as long as humans wonder about the universe.

Guide For Reading

- How is the process of science related to change?
- What are some successful study habits?

Figure 1–12 As knowledge grows, facts can change. This print depicts a Hindu legend that tells that the Earth is supported by three elephants resting on the back of a giant tortoise.

TIE-IN LANGUAGE ARTS

Students may enjoy reading the play *Galileo* by Bertolt Brecht. This play depicts Italian scientist Galileo's struggle with the Church over scientific versus religious truth. Galileo's evidence that the sun, not the Earth, is at the center of the solar system disagreed with the Church's teaching of the time that the Earth is at the center of the universe.

1–3 (continued)
Content Development
In light of the idea that scientific facts can change, ask students:
• **If scientific facts may not be true after all, why learn them?** (Accept all reasonable answers.)

After students have expressed their opinions, emphasize the idea that science is a process and that scientific facts do not change randomly or capriciously. The knowledge that scientists have today becomes the basis for the knowledge that scientists will have tomorrow. Thus changes in scientific facts will always "make sense" to a person who understands science. Point out that the study of science is somewhat like a detective solving mystery. Some clues may prove to be false or lead to a wrong conclusion, but the detective does not give up. He or she continues to find new clues until the mystery is solved.

Enrichment
Have students think of some controversial issues, either local or national, that involve science. These issues may include the environment, nuclear power, "test-tube" babies, and the rights of terminally ill patients. Have students gather information about an issue that interests them, then form teams and have a debate.

Content Development
Using the Visuals Have students observe Figure 1–14. Ask students to think of some technological inventions or advancements that have changed people's lives in the last 25 years. Some things they might include are microwave ovens, compact-disk players, VCRs, satellite television, PCs, FAX machines, and laser scanners at supermarket checkout counters.

Figure 1–13 The giant panda (top, left), the white rhinoceros (top, right), and the mountain gorilla (bottom) all look very different. So it might seem that the best way to study these animals is to memorize individual facts about each one. However, if you organize them as a group of mammals, then you will find it easier to learn about them. For all mammals share certain characteristics, no matter how different they may appear. These three particular mammals have another important thing in common —they are endangered species.

How to Study Science

We, as scientists, sincerely hope that you will find your study of science to be both interesting and rewarding. For an understanding of science is essential to all of us in today's high-tech world. To make your learning easier and more enjoyable, we suggest you study science in a special way. First, do not try to memorize the contents of this book as a list of separate facts. Learning science this way will only make the process more difficult than it needs to be. Isolated facts will show little connection with each other in your mind. It will be as though you wrote each fact on a separate index card and then shuffled dozens of index cards together.

It makes much more sense to arrange facts you need to know in groups according to subject. As you read this book, work at understanding, rather than just memorizing, the topics we talk about. When you study the parts of the cell, for example, don't just memorize their names. Think about what each

Enrichment
An area of science that students may find fascinating is forensic science. Forensic science is the use of scientific analysis and evidence to aid in solving criminal cases. Scientists who work in this field provide detectives with

part does. Remember how each part works with other parts within the cell as a whole. Learning this way is like organizing the index cards of facts into groups that have meaning for you. Then when you need a particular fact, you can look for it under its proper subject heading. That will make it much easier to remember what you need.

Second, remember that science is a process. In addition to learning the scientific facts of today, try to appreciate the process of discovery. Try to see the thinking behind the experiments we describe. Try to understand the kinds of questions scientists ask as they struggle to make sense out of the world. Follow along as we describe the discovery of genes and the development of evolutionary theory. If you can learn to think the way scientists think, you will get a lot more than facts from this course. You will develop the ability to understand the process of science and to be at home in a world that will change constantly throughout your life.

Science and Human Values

An important goal in science is to be objective. But scientists are no different from the rest of us when it comes to emotions or personal opinions. Scientists, after all, are people too—people with likes, dislikes, and occasional biases.

In today's world, scientists have important things to say about questions that are raised regarding health, society, and the environment. Shall we build a nuclear power plant? Dam a river? Develop forest lands into homesteads? Should chemical wastes be buried, burned, or dumped at sea? Are there ways to use chemical wastes for our benefit? Should certain experiments involving humans be forbidden? All of these questions rely on science for at least part of their answers.

But scientific data can be misinterpreted or misapplied by scientists who want to prove a particular point. And decisions made by scientists with personal prejudices may or may not be in the public interest. What this means to you is that understanding science is even more important today than ever before. If enough people understand the nature of science, the dangers posed by misinterpreted or misleading information will be reduced.

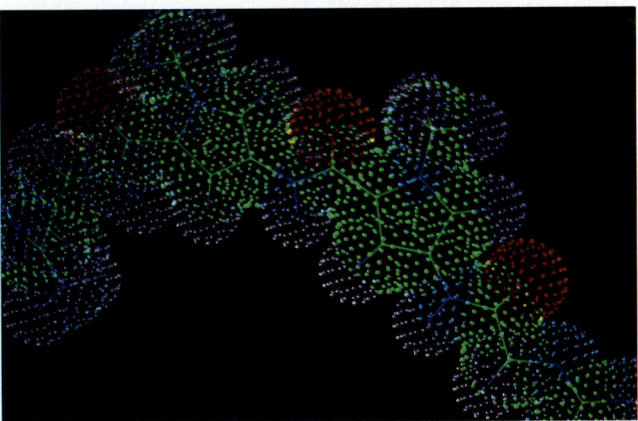

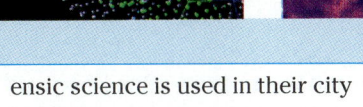

Figure 1–14 No doubt about it — we live in a high-tech world. Computers are now used to show the chemical structure of molecules such as this portion of a nucleic acid (left). Advances in physics, such as superconductors (right), may revolutionize the ways in which we travel and transmit electricity. As a citizen, you will have to make important decisions about the role of science in society. In order to do this, you must be able to read, understand, and evaluate scientific information.

BACKGROUND INFORMATION
SCIENCE AND PREJUDICE

There have been times in human history when scientific evidence or apparent evidence has been misused to serve the ends of racial prejudice and sexual bias. For example, the great biologist Louis Agassiz held the racist belief that non-European peoples were inferior to Europeans. Other scientists at the turn of the century shared his view and accepted data that was unsubstantiated or inaccurate to try to support their ideas.

In the late nineteenth century, a group of scientists called craniologists made measurements of brain and skull size to prove that women were intellectually inferior to men. These "scientific studies" were cited in attempts to deny women equal rights. Today scientists know that among humans, brain size has nothing to do with intelligence.

TEACHING SUPPORT
Study Guide
- Section 1–3, p. 6

clues by analyzing such evidence as blood or saliva found at the scene of a crime. Students who are interested in learning more about this subject can do research in the library. They can also contact their local police department to find out how forensic science is used in their city or town.

Skills Development
Guided Practice
Skill: Relating concepts
Using the Visuals Point out Figure 1–13.

• **What is an endangered species?** (A species of organism that is in danger of becoming extinct.)
• **What factors might cause an organism to become extinct?** (Possible answers include: change in the organism's environment; disruption of natural balance of food chains and food webs; human activities such as pollution; overhunting or overfishing by humans.)
• **What type of scientist might be especially concerned about endangered species?** (An environmental scientist or ecologist. Other answers are possible.)
• **What might these scientists try to do?** (Find ways to prevent the organism from becoming extinct; influence people such as hunters or industrialists who may be contributing to the problem of possible extinction.)

ANNOTATION KEY

① Safety goggles, etc. (Applying safety rules)

ESL STRATEGY

Some students may be familiar with international safety signs; however, as you explain the Lab Safety Rules and Symbols, using gestures to clarify significant features of each will ensure better understanding.

The words *flask* and *goggles* may be new to ESL students. Provide the spelling of each and instruct students to label both symbols on their copy of the Safety Chart. Explain that the flask symbol represents anything made of glass that is used in the lab. Show examples of laboratory glassware, such as microscope slides, beakers, and pipettes.

1–3 (continued)

SECTION REVIEW 1–3

1. Since biology is an ever-changing field of study, it is not possible to assume that what you have learned in the past will be accurate at a later date.
2. The best way to study science is to organize facts and concepts into an overall framework that ties individual facts and concepts together, rather than trying to memorize individual facts.
3. As citizens, we are all called upon to make decisions, some of which are based on scientific data and its interpretations. To make the best decision, we must be able to understand and interpret data.

Reinforcement/Reteaching

For those students who have difficulty with Section Review questions, go back to the discussion in the section and review the materials students have had difficulty understanding.

Closure

Have students simulate a round-table discussion in which they talk about the topics covered in this section—science facts and truth and science and human values. Have students play the roles of various "experts"—scientists, political leaders, philosophers, leaders of special interest groups, and so on. Students can use the information in the text, their own ideas, and information gained from additional research, if they wish.

TEACHING STRATEGY 1–4

Focus/Motivation

Begin by asking students:
• What do you think of when you hear the word *laboratory*? (Accept all logical answers.)
• What kinds of activities do you expect to do in the laboratory as you take this course? (Accept all reasonable answers at this time.)
• Do you think that some of

Luckily, men and women of science are taking action. Biologists, chemists, and physicists are speaking out about safety problems involved with nuclear power. Biologists and chemists are warning of the dangers posed by too many chemicals in food and water. And biologists are describing the serious threat to our environment posed by acid rain. We will touch on these and other matters in this book, and we hope that you will keep track of such issues, both in your local community and in the world at large.

1–3 SECTION REVIEW

1. Science will not sit still and allow itself to be memorized. You cannot "learn" biology, stick it away in some mental closet, and pull it out years later. Why is this true?
2. What is the best way to study science? Why?
3. **Critical Thinking—Appraising Conclusions** Why is it important that everyone understand the process of science and the nature of scientific facts?

1–4 Safety in the Laboratory

Biology does not happen in books or in scientists' minds. Biology is most alive and exciting where the experiments that test hypotheses actually take place—in the laboratory and in the field. That is why your course in biology will probably include several laboratory exercises. Only in the laboratory can you perform experiments to learn about the process of science firsthand. Only in the laboratory can you see, touch, and manipulate living systems to understand them more thoroughly.

In the laboratory—as in any place where there are chemicals, flames or heating elements, hot liquids, and electricity—it is important to follow certain basic safety precautions. These precautions are not directed only at you as students. Scientists with many years' experience know that laboratory safety is of prime importance. The experiments you will carry out this year have been tried and tested many times. When properly performed, they are both interesting and educational. However, if laboratory materials are handled carelessly, accidents can happen.

The single most important rule for you in the laboratory is simple: Always follow your teacher's instructions and the textbook directions exactly. If you are in doubt about what to do, always ask first. Use Figure 1–16 to familiarize yourself with safety symbols and the rules of laboratory safety. Read the rules and make certain that you understand them. If you are not sure that you understand a rule completely, ask your teacher to explain it.

Guide For Reading
■ What are some safety rules to follow when working in the laboratory?

Figure 1–15 *What important safety precautions are these students taking while working in the laboratory?* ①

LABORATORY SAFETY: RULES AND SYMBOLS

Glassware Safety

1. Whenever you see this symbol, you will know that you will be working with glassware that can easily be broken. Take particular care to handle such glassware safely. And never use broken glassware.
2. Never heat glassware that is not thoroughly dry. Never pick up any glassware unless you are sure it is not hot. If it is hot, use heat-resistant gloves.
3. Always clean glassware thoroughly before putting it away.

Fire Safety

1. Whenever you see this symbol, you will know that you will be working with fire. Never use any source of fire without wearing safety goggles.
2. Never heat anything—particularly chemicals—unless instructed to do so.
3. Never heat anything in a closed container.
4. Never reach across a flame.
5. Always use a clamp, tongs, or heat-resistant gloves to handle hot objects.
6. Always maintain a clean work area, particularly when using a flame.

Heat Safety

Whenever you see this symbol, you will know that you should put on heat-resistant gloves to avoid burning your hands.

Chemical Safety

1. Whenever you see this symbol, you will know that you will be working with chemicals that could be hazardous.
2. Never smell any chemical directly from its container. Always use your hand to waft some of the odors from the top of the container toward your nose—and only when instructed to do so.
3. Never mix chemicals unless instructed to do so.
4. Never touch or taste any chemical unless instructed to do so.
5. Keep all lids closed when chemicals are not in use. Dispose of all chemicals as instructed by your teacher.
6. Immediately rinse with water any chemicals, particularly acids, off your skin and clothes. Then notify your teacher.

Eye and Face Safety

1. Whenever you see this symbol, you will know that you will be performing an experiment in which you must take precautions to protect your eyes and face by wearing safety goggles.
2. Always point a test tube or bottle that is being heated away from you and others. Chemicals can splash or boil out of the heated test tube.

Sharp Instrument Safety

1. Whenever you see this symbol, you will know that you will be working with a sharp instrument.
2. Always use single-edged razors; double-edged razors are too dangerous.
3. Handle any sharp instrument with extreme care. Never cut any material toward you; always cut away from you.
4. Notify your teacher immediately if you are cut in the lab.

Electrical Safety

1. Whenever you see this symbol, you will know that you will be using electricity in the laboratory.
2. Never use long extension cords to plug in an electrical device. Do not plug too many different appliances into one socket or you may overload the socket and cause a fire.
3. Never touch an electrical appliance or outlet with wet hands.

Animal Safety

1. Whenever you see this symbol, you will know that you will be working with live animals.
2. Do not cause pain, discomfort, or injury to any animal.
3. Follow your teacher's directions when handling animals. Wash your hands thoroughly after handling animals or their cages.

Figure 1–16 You will see these safety symbols in the Laboratory Investigations in this textbook and in your Laboratory Manual. Make sure you understand the meaning of each symbol. If you are not sure what a symbol means, ask your teacher.

1–4 SECTION REVIEW

1. What is the most important general rule to follow when working in the laboratory?
2. Where is the nearest fire extinguisher located in your laboratory or classroom?
3. **Critical Thinking—Applying Concepts** What sort of instruments might you be more likely to find in a biology laboratory than in a chemistry laboratory?

TEACHING SUPPORT

Laboratory Manual

- The Uncertainty of Measurements, p. 63

Teaching Resources

- Activity: Incredible Measurements, p. 3

Study Guide

- Section 1–4, p. 8

these activities might be dangerous? In what ways? (Accept all answers. Students should think of some of the hazards of fire, sharp instruments, chemicals, animals, and electricity.)

Content Development

Explain to students that a laboratory is a place where scientists carry out experiments to test their ideas. Point out that laboratory activities may be quite different for various branches of science. A biologist, for example, may be making slides of living tissue to study under a microscope. A chemist may be using a sophisticated tool such as a spectroscope to analyze the structures of molecules. An astronomer may have a laboratory filled with telescopes and other instruments used in studying the stars and planets.

Emphasize to students the importance of laboratory work in a high school science course.

- **Why do you think laboratory work is important in the study of biology?** (Possible answers include: it gives students a chance to really do things, rather than just read about them; it provides firsthand experience of discovering a scientific fact; it gives students a chance to duplicate experiments carried out by other scientists.)

SECTION REVIEW 1–4

1. Always follow written or oral instructions.

2. Check student responses for accuracy.

3. Answers will vary but might include examples such as compound light microscope or dissecting tools.

Reinforcement/Reteaching

For those students who have difficulty with Section Review questions, go back to the discussion in the section and review it.

Closure

Describe to students various activities that they might carry out in the lab during a typical experiment. Then have students summarize the Lab Safety Rules and Symbols that would apply to each activity.

SCIENCE, TECHNOLOGY, AND SOCIETY
BREAKTHROUGH

The Odd Couple: A Messy Lab and a Keen Mind

Sir Alexander Fleming was a Scottish bacteriologist who lived from 1881 until 1955. He discovered lysozyme in 1922 and penicillin in 1928. In 1945, Fleming shared the Nobel prize in medicine for his work on penicillin. Fleming, who was a professor of bacteriology at the University of London, was knighted in 1944.

After students have read the feature, pose the following questions:

- **What was Fleming studying?** (Bacteria.)
- **What two "mistakes" happened as Fleming worked?** (A few drops from Fleming's runny nose fell on the bacteria; some bread mold had contaminated one of his plates.)
- **What did Fleming learn from these two accidents?** (That both substances—mucus and mold—contained something that killed bacteria. This led to the discovery of lysozyme and penicillin.)
- **What important quality of a scientist did Fleming display?** (The ability to observe that something unusual was happening and to be curious enough about it to find out what it was.)
- **Can you think of any other discoveries in science that were made by accident?** (Answers will vary. One example is Becquerel's discovery of radioactivity.)

SCIENCE, TECHNOLOGY, AND SOCIETY
BREAKTHROUGH

The Odd Couple: A Messy Lab and a Keen Mind

Some of us professional scientists, despite our efforts, aren't very neat workers in the lab. Oh, sure, we can keep things sterile when they need to be. But as a rule, some of us are a little messy and disorganized. Does that fit in science? Does it match our description of the scientific method? It may not match, but the messy scientist is in good company. One messy scientist was a man named Alexander Fleming.

History records at least two of Fleming's sloppy mistakes. The first was in 1922. He was suffering from a bad cold and accidentally allowed a few drops from his runny nose to drip on a plate of bacteria he was growing. Instead of discarding the dish, Fleming watched it for days. What happened? Lo and behold—the bacteria died! Fleming had discovered lysozyme, a chemical in tears and mucus that helps protect the body against bacteria.

More important work followed. Six years later, Fleming was growing more bacteria. He saw that some of his plates had been contaminated with a greenish mold (*Penicillium*) that grows on bread. Once again Fleming's sloppiness was to blame. But his careful eye and keen mind made something of the event. He saw a clear spot on the culture dish, indicating that the bacteria near the mold had died. The mold was making something that was killing the bacteria. That something turned out to be a chemical that has saved millions of lives: penicillin. Science works that way. It is a very human enterprise—one that even has room for us messy scientists!

Guide For Reading

■ Why is planet Earth often described as a spaceship?

1–5 The Spaceship Called Earth

Seen by astronauts from space, Earth is incredibly beautiful. It shines blue, white, and purple against the black, airless void. Since the beginning of human history, poets have written about the beauty of Earth. Artists have created paintings of Earth's magnificent wilderness lands and wild animals. Today, scientists from several fields are warning us that we must learn to protect our planet and preserve its living treasures. Our species has rapidly developed the ability to change the Earth much for the better and much for the worse. We have saved many species from extinction. At the same time, we have destroyed hundreds of other species. We have made gardens bloom in deserts and can now feed more people than ever before. But we have also turned fertile land into dust bowls. We have tremendously increased human life spans and can treat or cure many diseases that were once fatal. Yet we have also manufactured chemicals that once seemed useful but now threaten our health and the health of other species.

TEACHING STRATEGY 1–5

Focus/Motivation

Display photographs of the Earth as viewed from space. Then show some contrasting pictures of the Earth's environment. Let some of the pictures show the damage that has been done to the environment by such pollutants as litter and ocean dumping. Let the rest of the pictures show the beauty of the Earth—such as lakes, woods, mountains, and clean beaches.

- **How does the Earth look in the photographs taken from space?** (Answers may vary. Most students will probably comment on the beauty or awesomeness of the Earth.)
- **Consider the other two sets of pictures. Do both display true aspects of the Earth? How are these aspects different?** (Yes. One set of pictures shows how some of the environment has been ruined; the other set shows the unspoiled environment that still remains.)

Content Development

Point out to students that while the Earth seems huge and never-

Earlier in human history, the Earth seemed to be without end. There were always new wildernesses to settle, new resources to use, and plenty of places to dump our garbage. Now we know that there is limited land—and limited amounts of clean air, water, and other resources. There are also a limited number of places we can use as trash heaps. **The Earth is no longer a planet without end. It is more like a spaceship with a living cargo, carrying limited amounts of supplies.**

As you read this book, you will learn about the basis of life and about the process by which life has evolved. You will take a tour through the kingdoms of the different kinds of organisms with whom we share our planet. You will learn just how closely all forms of life on Earth are connected. We hope you will also learn how important it is to preserve that life. More than ever before in history, we must work to protect our environment against destruction. Properly applied, science can provide us with the knowledge and tools we need to meet this challenge. We think you will agree that the joys of a ride on spaceship Earth are more than worth the responsibilities we have as its most powerful passengers.

1–5 SECTION REVIEW

1. How does our modern view of Earth's resources differ from earlier impressions?
2. **Critical Thinking—Assessing Events** What are some of the great successes of human endeavors on Earth? What are some of the failures?

Figure 1–17 Humans have changed the Earth both for the better and for the worse. Vaccines help to prevent many diseases that were once fatal. Modern farming methods can now feed more people than ever before. At the same time, however, we have seen numerous species of living things destroyed by the actions of humans. And we have allowed toxic wastes to seep into our soil and waterways.

ending to us, it is really rather small compared with other objects in space. It is also rather isolated—the nearest neighboring planet is about 42 million kilometers away. Thus the image of the Earth as a spaceship is a good one. Like a spaceship, the Earth is limited to what it carries on board—we cannot just hop over to another planet to borrow supplies. And even if we could, other planets are so different from Earth that they probably would not have what we need.

Skill Development
Guided Practice
Skill: Relating concepts
• **How can studying biology make a person better able and more likely to protect the Earth's environment?** (Possible answers: By understanding living things and how they depend on each other, a person becomes aware of the consequences of certain actions. For example, overfishing or overhunting an area will upset the food chains and food webs in that area and may lead to the decrease or extinction of certain species. Also, studying living things gives a person a greater reverence for plants and animals. The person will be less likely to destroy the natural beauty of the environment with such thoughtless actions as littering.)

SECTION REVIEW 1–5

1. We no longer consider Earth's resources as unlimited.
2. Answers will vary. Accept all reasonable responses.

Reinforcement/Reteaching
For those students who have trouble with Section Review questions, go back to the discussion in the section and review the materials students have difficulty understanding.

Closure
Have students work in small groups to create posters showing Spaceship Earth. The posters should raise consciousness about the importance of preserving the environment. The posters might also relate this message to the study of biology. Have groups display their posters on a classroom bulletin board.

LABORATORY INVESTIGATION

INTERPRETING A CONTROLLED EXPERIMENT

Before the Lab
1. Divide the class into groups of three to six. (You may wish to maintain these same laboratory groupings throughout the year.)
2. Gather all materials at least one day prior to the investigation.
3. Use only 50-mL or 100-mL beakers. Other sizes will affect the water level.

Pre-Lab Discussion
Have a student read aloud the Problem.
- **Do you have any ideas about how the amount of water will affect the germination of the seeds?** (Accept all answers.)
- **Do you think there is a "right" or "ideal" amount of water for seed germination?** (Accept all answers.)

Have students write their answers to these questions in the form of a hypothesis. Ask students to put their hypothesis aside until the end of the investigation, then evaluate their ideas in light of their results.

- **Do you think there are other factors besides water that affect the rate of seed germination? If so, what?** (Possible answers include light, health of the seeds, type of container, whether the seeds are placed in water, soil, or other material.)

Have students read through the entire procedure. Then ask:
- **What is the obvious variable in this experiment?** (Amount of water.)
- **Is there any other possible variable?** (The two different containers.)
- **How are beakers and petri dishes different?** (The dishes are shallow, whereas beakers are much deeper.)
- **Do you think this factor could affect the experiment?** (Accept all answers.)

LABORATORY INVESTIGATION: INTERPRETING A CONTROLLED EXPERIMENT

PROBLEM
What effect does the amount of water given to seeds have on the rate of seed germination?

MATERIALS (per group)

    200 mustard seeds
    3 petri dishes with covers
    2 50-mL beakers
    graduated cylinder
    glass-marking pencil

PROCEDURE

Part A
1. In this investigation you will determine whether the amount of water given to mustard seeds has an effect on the number of seeds that germinate, or sprout. Before you begin, propose a hypothesis predicting the effect that varying amounts of water might have on mustard seed germination. The steps that follow will help you test your hypothesis.
2. Place 50 mustard seeds in each of two petri dishes.
3. Using the graduated cylinder, pour 5 mL of water into one petri dish. Then pour 30 mL of water into the other petri dish. Cover each petri dish. Use the glass-marking pencil to indicate the volume of water in each.
4. Set both petri dishes aside for 48 hours. After 48 hours, count the number of seeds in each petri dish that have begun to germinate. Record your observations in a data table similar to the one shown here.

Part B
1. Place 50 mustard seeds in each of the two beakers.
2. Using the graduated cylinder, pour 5 mL of water into one beaker and 30 mL of water into the other. Cover each beaker with the top or bottom of the remaining petri dish. Then, using the glass-marking pencil, indicate on the cover the volume of water in each beaker.

3. Set the beakers aside for 48 hours. After 48 hours, count the number of seeds in each beaker that have begun to germinate and record your results in your data table.

OBSERVATIONS

Type of Container	Volume of Water	
	5 mL	30 mL
Petri dish		
Beaker		

1. Did the mustard seeds float in any of the containers? If so, were these seeds more likely to germinate?
2. In which container did the germinated seeds have the longest roots?

ANALYSIS AND CONCLUSIONS
1. How did the number of germinated seeds in the petri dishes compare to the number of germinated seeds in the beakers?
2. Did the amount of water in the petri dishes appear to affect the number of germinated seeds? Did these results confirm your hypothesis?
3. Did the amount of water in the beakers appear to affect the number of germinated seeds? Did these results confirm your original hypothesis?
4. Sometimes the results of different parts of an experiment provide different interpretations. When this occurs, scientists often look to see if a hidden variable, which might affect the overall results of an investigation, was introduced. What hidden variable might account for the results you observed in the petri dishes and in the beakers? (*Hint:* Other than water, what substance may have played a role in your experiment?)

Skills Development
Students will use the following skills while completing this investigation.
1. Observing
2. Hypothesizing
3. Predicting
4. Manipulative
5. Comparing
6. Applying
7. Relating
8. Inferring
9. Measuring

Safety Tips
Remind students to handle glassware carefully. Have students report any broken glassware to the teacher immediately.

Teaching Strategy
1. Circulate throughout the room to make sure that students have correctly set up the four arrangements of seeds.

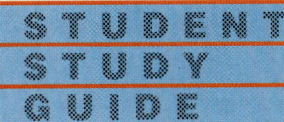

SUMMARIZING THE CONCEPTS

The key concepts in each section of this chapter are listed below to help you review the chapter content. Make sure you understand each concept and its relationship to other concepts and to the theme of this chapter.

1–1 What Is Science?
- Science is a process that uses observations and tests to identify the causes of events in nature.

1–2 The Scientific Method
- The main steps in the scientific method are observing and stating a problem, forming a hypothesis, testing the hypothesis through observation or experimentation, recording and analyzing data, forming a conclusion, and replicating the work.
- A hypothesis is a possible explanation about a particular event in nature.
- In order to ensure that the results of an experiment were due to the variable being tested, scientists must run both an experimental setup and a control setup.
- When a hypothesis has been tested and confirmed many times, it may then be considered a theory that can make dependable predictions about nature.
- The metric system is the system of measurement used by scientists all over the world.
- Although mass and weight are considered interchangeable on the surface of the Earth, mass is actually a measure of the amount of matter in an object and weight is a measure of the pull of gravity on that object.

1–3 Science: "Facts" and "Truth"
- Successful science study habits involve relating facts to each other and to the underlying concepts being discussed rather than simply memorizing a list of facts.

1–4 Safety in the Laboratory
- Always follow all written instructions or instructions from your teacher whenever working in the science laboratory.

1–5 The Spaceship Called Earth
- Planet Earth can be compared to a giant spaceship carrying a living cargo and limited amounts of food, water, and other resources.

REVIEWING KEY TERMS

Vocabulary terms are important to your understanding of biology. The key terms listed below are those you should be especially familiar with. Review these terms and their meaning. Then use each term in a complete sentence. If you are not sure of a term's meaning, return to the appropriate section and review its definition.

1–1 What Is Science?
science

1–2 The Scientific Method
scientific method
hypothesis
variable
control setup
experimental setup
data
theory
metric system
meter
liter
cubic centimeter
mass
weight
kilogram
Celsius scale

Analysis and Conclusions
1. More seeds germinated in the petri dishes than in the beakers.
2. A larger number of seeds germinated in the petri dish with the smaller volume of water. Answers will vary depending on the original hypothesis.
3. A small but similar amount of seeds germinated in each beaker. Answers will vary depending on the original hypothesis.
4. The hidden variable was the amount of air or oxygen the seeds get.

At this point you may want to tell students that this investigation was set up so that they might come to an incorrect conclusion regarding the role of water in seed germination. Point out that enough water was present in all containers for seed germination and that it was actually the hidden variable of oxygen that determined the different results. It is hoped that by doing this investigation in the first chapter, the concept of hidden variables affecting results will be impressed upon all students.

Going Further: Enrichment
Have students repeat the experiment with another variable, such as light or temperature. Have students formulate a hypothesis for the new investigation. You may wish to assign each group a different variable to test, then bring the results of all investigations together for class discussion. A chart could then be made showing the various factors that affect mustard seed germination.

---

2. Make sure that groups do not introduce a hidden variable when they set aside their dishes and beakers of seeds for 48 hours, for example, placing the seeds on a sunny windowsill. Decide on a similar place for all groups to store their seeds, such as on a shelf or in a corner of each lab station.

Observations
Possible sample data in the data table might be 34/12 for the petri dish and 8/8 for the beaker.
1. Some seeds were floating in the beaker with 30 mL of water. A larger proportion of these seeds germinated than those at the bottom of the beaker.
2. The longest roots were in the petri dish with 5mL of water.

CHAPTER REVIEW

CONTENT REVIEW

Multiple Choice
1. b   2. c   3. a   4. a
5. b   6. d   7. d   8. b

True or False
1. T
2. T
3. F   meters
4. F   liter
5. F   one-hundredth
6. F   often
7. T
8. F   milligram

Word Relationships
1. Replication
2. hypothesis
3. volume
4. centimeters
5. Mass

CONCEPT MASTERY

1. To be sure that the results of the experiment are due to the variable and not to some other factor.
2. Without a standard system of measurement, scientists would not be able to communicate, distribute their data, or reproduce experiments.
3. Science information is constantly changing. To remain up to date, science books must be revised every few years.
4. Like a spaceship, Earth travels through space and contains limited amounts of food, water, shelter, living things, and so on.
5. The approach that distinguishes science is a problem-solving approach called the scientific method. The scientific method contains the following steps: Observing and stating a problem; Forming a hypothesis; Testing the hypothesis; Recording and analyzing data; Forming a conclusion; Replicating the work.
6. a. milliliters b. centimeters c. milligrams or grams d. degrees Celsius e. tons or kilograms f. cubic centimeters

CHAPTER REVIEW

CONTENT REVIEW

Multiple Choice

Choose the letter of the answer that best completes each statement.

1. A proposed solution or explanation of a scientific event in nature is called a
   a. conclusion.   c. control setup.
   b. hypothesis.   d. theory.
2. The factor being tested in an experiment is the
   a. data.   c. variable.
   b. hypothesis.   d. theory.
3. The metric system is based or scaled on powers of
   a. 10.   c. 1000.
   b. 100.   d. none of these.
4. The basic unit of length in the metric system is the
   a. meter.   c. millimeter.
   b. kilometer.   d. centimeter.
5. The basic unit of mass in the metric system is the
   a. gram.   c. liter.
   b. kilogram.   d. pound.
6. A cubic centimeter is equivalent to a
   a. liter.   c. milligram.
   b. gram.   d. milliliter.
7. Which of the following relationships between an object on Earth and that same object on the moon is true?
   a. mass changes; weight remains the same
   b. neither mass nor weight changes
   c. both mass and weight change
   d. weight changes; mass remains the same
8. If you see the symbol of a flask in a laboratory experiment, you are being cautioned to
   a. put on goggles.
   b. take extra care with glassware.
   c. make sure your measurements are very precise.
   d. use heat-resistant gloves.

True or False

Determine whether each statement is true or false. If it is true, write "true." If it is false, change the underlined word or words to make the statement true.

1. Recorded observations are called <u>data</u>.
2. The <u>control experiment</u> is that part of the experiment without the variable.
3. A kilometer contains 1000 <u>centimeters</u>.
4. A <u>milliliter</u> contains 1000 cubic centimeters.
5. The prefix *centi-* means <u>one hundred</u>.
6. In science, facts <u>never</u> change.
7. The goal of <u>science</u> is to understand the world around us.
8. The <u>kilogram</u> is the unit of mass you would use to measure objects smaller than a gram.

Word Relationships

Replace the underlined definition with the correct vocabulary word.

1. <u>The ability to reproduce results</u> is an important part of any experiment.
2. After observing an event in nature, a scientist may develop a <u>possible explanation or a preliminary conclusion</u> regarding the cause of that event.
3. The liter is a measure of <u>the amount of space an object occupies</u>.
4. Scientists often measure length in <u>hundredths of a meter</u>.
5. <u>The amount of matter in an object</u> can be measured in milligrams.

CRITICAL AND CREATIVE THINKING

1. Accept all reasonable experimental designs, as long as they contain only one variable and have a control setup and an experimental setup.
2. Accept all reasonable predictions for scientific advances in the next century.
3. Accept all logical answers. The intended variable is probably the angle of the light shining on the plant. One possible source of error is a difference in light intensity between the experimental setup and the control setup. Light intensity falls off exponentially with the distance from the source and also with the distance from the center of the beam from that source. Thus

CONCEPT MASTERY

Use your understanding of the concepts developed in the chapter to answer each of the following in a brief paragraph.

1. Why is it important to have only one variable in an experiment?
2. Scientists throughout the world use a standard system of measurement. Why is a standard measurement system necessary?
3. Why must science textbooks be revised every few years?
4. Compare planet Earth to a giant spaceship.
5. Describe the approach that separates the process of science from other fields of study.
6. Which metric unit would you use to measure each of the following:
   a. volume of a glass of milk
   b. length of your textbook
   c. mass of a mouse
   d. temperature of the Gulf of Mexico
   e. mass of a killer whale
   f. volume of a marble

CRITICAL AND CREATIVE THINKING

Discuss each of the following in a brief paragraph.

1. **Designing an experiment** How would you use the scientific method to find the best place in your class to grow African violets?
2. **Making predictions** Develop a time line in which you predict some of the important advances that may occur in science during the next century.
3. **Interpreting diagrams** What is the intended variable in the experiment illustrated below? What are some possible sources of error?

4. **Recognizing fact and opinion** Suppose your friends believe that astrology is a science. How would you convince them that they are wrong?
5. **Relating concepts** The French biologist Louis Pasteur once said, "Chance favors only the mind that is prepared." Explain this statement by relating it to the process of science.
6. **Using the writing process** You have been elected to the Senate. The most important issue that has to be resolved is whether the United States will convert to the metric system. Write a 3-minute speech detailing your reasons for adopting or rejecting the metric system.
7. **Using the writing process** The year is 2050. Write a help-wanted ad that will encourage high school graduates to volunteer for a 2-year mission to planet Mars. The main task of the graduates will be to function as technicians in various scientific experiments. All graduates will be trained before the mission begins.

that it shines directly on the plant.

4. Accept all reasonable responses. Students should point out that astrology cannot be tested or verified by the scientific method and, as such, cannot be considered a science. Everything in science must be verifiable.
5. In many cases, an important discovery such as Fleming's discovery of penicillin is not necessarily the goal of an experiment. Rather, the curious mind may well take a chance happening and find that an important concept lies behind the event. An unprepared mind, on the other hand, will likely disregard the discrepant data and assume something was wrong with the experiment.
6. Accept all reasonable well-written responses.
7. Accept all reasonable well-written responses.

the plant in the experimental setup is receiving less light because its leaves are farther away from the lamp and it is located at the fringes of the beam of light from the lamp. Another possible source of error is a difference in temperature between the two setups. The lamp appears to be an incandescent floodlight, and such lights usually give off heat as well as light. Thus the control plant may be warmer that the experimental plant.

You may wish to discuss the answer to this question in class and encourage students to think of ways to eliminate these and other sources of error. For example, one might be able to eliminate differences in light intensity by tilting the experimental lamp so

CHAPTER 2 Biology as a Science

Section	Laboratory Investigations and Activities
2–1 Characteristics of Living Things pages 27–31 THEMES: Unity and Diversity, Scale and Structure	**Teacher Edition** DEMONSTRATION: Is It Alive? p. 26D **Laboratory Manual** Characteristics of Life, p. 67 **Teaching Resources** HANDS-ON ACTIVITY: Hydra Doing? p. 17
2–2 Biology: The Study of Life pages 31–39 THEME: Scale and Structure	**Student Edition** LABORATORY INVESTIGATION: Using a Compound Microscope, p. 40 **Teacher Edition** DEMONSTRATION: Effects of Gravity on Plants, p. 26D **Laboratory Manual** Measuring with a Microscope, p. 73 **Teaching Resources** HANDS-ON ACTIVITY: Life in a Drop of Water, p. 13
Chapter Review pages 41–43	

Teacher Resources

Books

Ambrose, E. J. *The Nature and Origin of the Biological World.* Halstead Press, 1982.

Burgess, Jeremy, Michael Marten, and Rosemary Taylor. *Microcosmos.* Cambridge University Press, 1987.

Mayr, Ernst. *The Growth of Biological Thought.* Harvard University Press, 1985.

Shih, Gene, and Richard Kessel. *Living Images: Biological Microstructures Revealed by Scanning Electron Microscopy.* Jones and Bartlett, 1982.

Wallace, Bruce, and George M. Simmons, Jr. *Biology for Living.* Johns Hopkins University Press, 1987.

CHAPTER PLANNING GUIDE

Other Activities	Media and Technology
Teaching Resources ACTIVITY: Name That Animal Group! p. 3 **Study Guide** Section 2–1, p. 13	**Biology Media Guide** page 1
Teaching Resources VOCABULARY REVIEW: Word Game, p. 1 ACTIVITY: Analyzing Science Terms, p. 5 ACTIVITY: Microscopic Proportions, p. 7 **Biotechnology Workbook** New Views on Old Bones, p. 97 **Study Guide** Section 2–2, p. 16	**Interactive Videodisc** In the Company of Whales **Biology Transparencies** 2: The Compound Light Microscope **Biology Media Guide** pages 1–2
Teaching Resources Chapter Test A, p. 13 Chapter Test B, p. 17	**Computer Test Bank** Chapter Test, p. 17

Audiovisuals

Science Laboratory Techniques. Video series. Media Design Associates, Inc.
Imaging a Hidden World: The Light Microscope. Film or video. Coronet.

Journey Into Microspace: A Photographic Odyssey. Sound filmstrip with cassette. Ward.

What Is Biology? Filmstrip with sound cassette. Ward.

CHAPTER 2 Biology as a Science

CHAPTER OVERVIEW

A study of living organisms might logically begin with a definition of life itself. For centuries scientists have been unable to agree on a definition or even on a list of characteristics that all living things share. There are, however, some generally accepted attributes. These are discussed in this chapter.

Biology is such a broad science that no individual is an expert in all of its aspects. There are many branches of biology. More valuable to student understanding than a mere naming of the fields is a look at biology from the point of view of questions or problems to be considered. The scope of the problem may involve biologists with many different specialties working together.

Finally, after considering the basis of biology, the characteristics of life, and the scope of biology, Chapter 2 discusses biologists and their tools. There is a brief discussion of different types of microscopes and their applications, followed by a sampling of modern experimental techniques.

2-1 CHARACTERISTICS OF LIVING THINGS

Section Focus 2-1

The purpose of this section is to give students a basic knowledge of characteristics that most scientists agree are common to all living things. First, living organisms are composed of units called cells. An organism may be composed of one or many cells, cells that may be very diverse in structure and function.

Second, living things can reproduce organisms of the same type. Reproduction is not necessary for the survival of the individual but must occur if the species is to continue. There are basically two types of reproduction. Asexual reproduction involves a single individual and is generally a simple process. Sexual reproduction requires that two cells from different individuals unite to form the first cell of a new individual.

Third, all organisms are capable of growth and development at some time during their life span. Unlike nonliving growth, the growth that occurs in living things transforms substances into living tissue. A part of the growth process for living organisms is development, aging, and eventually death.

Fourth, living organisms obtain energy from the environment and use it in unique ways. Plants perform a process called photosynthesis by utilizing the energy from sunlight. Animals take in energy in the form of food. There are two processes involved in the utilization of food. These two processes together summarize all the chemical reactions that occur in the body and make up the organism's metabolism.

Last, living things respond to their environment in many different ways. The ability to respond is called irritability and must be triggered by something in the environment called a stimulus. There are many different stimuli, including temperature, light, gravity, and so on. Basically, organisms respond in ways that improve their chances of survival in a process called homeostasis. Homeostasis refers to an organism's ability to maintain an internal balance that is stable enough to sustain life.

Performance Objectives 2-1

1. List the characteristics of living things.
2. Distinguish between sexual and asexual reproduction.
3. Define homeostasis.

Science Terms 2-1

unicellular   p. 28
multicellular   p. 28
sexual reproduction   p. 28
asexual reproduction   p. 28
anabolism   p. 29
catabolism   p. 30
metabolism   p. 30
homeostasis   p. 31

2-2 BIOLOGY: THE STUDY OF LIFE

Section Focus 2-2

The purpose of this section is to define "biologist" for students and to introduce the tools used by biologists. There are many fields of study within the science of biology. Some, like zoology or botany, are broad, while others are narrow. All biologists, however, ask questions.

Some biologists ask questions at the molecular level. Their questions may focus on a cell's DNA or proteins, or how drugs or chemicals alter a cell's molecular structure.

Other biologists ask questions at the cellular level. They may investigate how normal cells become cancerous, how cells grow and divide, or how cells communicate.

Questions at the multicellular level might be those involved with animal habits such as sleeping, eating, or mating. What mechanisms within the body cause these things to happen?

Biologists who are interested in studying groups of organisms are generally asking questions at the population level. They might be particularly concerned about the impact of human activities—building, producing waste, or using pesticides—on the various types of living things.

Questions that involve organisms and their environment on a worldwide scale may be posed by global ecologists studying the effects of burning coal and oil on the Earth's climate.

Finally, it should be noted that although there are many professional biologists studying the Earth and the myriad of organisms that call it home, many biologists

CHAPTER PREVIEW

are amateurs. The only qualifications are hard work, energy, and curiosity. There is no license required to be a good amateur biologist.

The tools of the biologist are many and varied. The microscopes alone come in a variety of types—from the familiar compound light microscope used by all high school biology students to the electron microscopes. Each type of microscope has unique problems and limitations associated with its use. Each one also performs functions for which it is ideally suited.

Some common laboratory techniques used by the biologist include centrifugation, microdissection, and the culturing of cells. Others will be discussed at various points throughout the text.

Performance Objectives 2–2

1. Discuss the levels of questions asked by biologists and give examples of each one.
2. Name the common types of microscopes and give a correct application for each one.
3. Explain the limitations of each type of microscope.
4. Briefly describe some common techniques used by biologists.

Science Terms 2–2

compound light microscope p. 34
limit of resolution p. 34
transmission electron microscope p. 36
scanning electron microscope p. 36
scanning probe microscope p. 37

DISCOVERY LEARNING

TEACHER DEMONSTRATIONS
Modeling
Is It Alive?

The following demonstration can be used to introduce Chapter 2.

Light a candle and show it to the class, pointing out the flame. The question to be answered is whether or not the flame is alive. Encourage students to list characteristics that they believe are those associated with living organisms.

As the discussion proceeds, test the actions of the flame against those that might be expected from living things. (For example, you might wave your hand by the flame and watch it waver. Is that a response to the environment?) The point of the demonstration is that it is difficult to define exactly what life is. Further points that tie into the chapter include the characteristics of the biologist and also the importance of the proper tools and techniques. Students know that the candle flame is not alive, but they may not be able to precisely explain that answer.

Effects of Gravity on Plants

This would be a good quick demonstration to do with the section on levels of questions (Section 2–2). You will already have discussed organisms responding to their environment. **You will need to set this demonstration up about a week before you plan to use it.**

Place one flowerpot containing a small plant in front of a window in bright light. Place a second plant on its side, facing away from the window. Put a third plant upside down on a ring stand. (You will need to secure the soil with cardboard and make sure that the pot will not fall.)

Observe all three plants and the growth of their stems. You might want to ask some preliminary questions about what stimulus the plant is responding to (gravity, but some students might say light) and what they hypothesize will happen if the plants are left in their positions.

Ask students: At which level of questioning is our study of this plant? (Students will probably come up with a variety of answers, but the most obvious ones are either the cellular or the multicellular level.)

26D

CHAPTER 2
Biology as a Science

GUIDED ENQUIRY

Pose the following questions to students and have them record their responses. Point out that they will gain a better understanding of the key concepts if they read the chapter with these basic questions in mind. Upon completion of the chapter, pose the questions again. Ask students to compare their initial responses with those they have developed after reading the chapter.

- What are the characteristics that distinguish life from nonlife?
- How long would it take for a group of organisms to become extinct if the whole group suddenly lost the ability to reproduce?
- How do different organisms get the energy they need to survive?
- What are some things for which organisms use energy?
- How is a study of cellular biology related to pollution studies?
- Why is curiosity an important characteristic of a biologist?

INTRODUCING CHAPTER 2

Using the Visuals
Have students look at the picture on this page and describe what they see. Ask them to classify the trees, ground, and snow as living or nonliving. Have them explain their answers. (Most will probably rely on past experiences or knowledge.) After probing enough to point out some of the difficulties, have them read the introduction below the picture.

CHAPTER 2
Biology as a Science

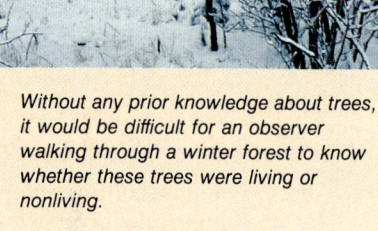

Without any prior knowledge about trees, it would be difficult for an observer walking through a winter forest to know whether these trees were living or nonliving.

How do you tell whether something is alive? With some living things, it's easy. You could check for breathing, or a pulse, or a response to a pinch or a poke. But such tests would not work for all living things. Is moss growing on a rock alive? How about rust spreading over a piece of metal? If you came upon a tree in the middle of winter when its leaves were gone, how would you tell whether the tree was living or dead?

What we are really asking in such questions is simple. We would like to know what distinguishes life from nonlife. The answer to such questions, however, is not always that clear. So we shall do the next best thing to providing the answer. In this chapter we shall list the characteristics that living things share.

- **If you conclude that most of the trees in the picture are alive, what clues would tell you when one has died?** (It may be uprooted and fallen; it may have lost all its branches and be just a trunk; it may be split or have lost its bark.)
- **What do you predict will happen to the trees in the spring? To the snow?** (Students may say that the trees will turn green, grow new leaves, bud, and so on, and that the snow will melt.)
- **Are growing, developing new structures, turning green, and melting characteristics you would associate with living things? All living things?** (Growth and development, as well as many other characteristics, are processes shared by living organisms. Help students begin to distinguish among those

GUIDE FOR READING

After you read the following sections, you will be able to

2-1 Characteristics of Living Things
- List and describe the characteristics of living things.
- Define homeostasis and explain why it is important to living things.

2-2 Biology: The Study of Life
- Discuss the different branches of biology and the levels of phenomena studied by biologists.
- Describe and compare different kinds of microscopes.
- Relate specific laboratory techniques to biology.

Journal Activity

YOU AND YOUR WORLD
What do you think living things would be like on another planet? Do you think human space explorers would immediately recognize these alien life forms as being alive? In your journal, explore your ideas in the form of a short story or essay.

2-1 Characteristics of Living Things

Guide For Reading
- What are the characteristics of living things?
- What is homeostasis?

Making up a list of the characteristics of living things is not as easy as it might sound. In fact, scientists have argued for centuries over the basic characteristics that separate life and nonlife. Some of these arguments are still unresolved. For example, in Chapter 17 you will discover that the line between life and nonlife becomes blurred when we consider whether or not viruses are living things.

Despite these arguments, there do seem to be some generally accepted characteristics common to all living things. **We can state with some confidence that all living things**

- **Are made up of one or more units called cells**
- **Reproduce**
- **Grow and develop**
- **Obtain and use energy**
- **Respond to their environment**

It will help in our understanding of living things to consider each of these characteristics in detail.

Living Things Are Made Up of Cells

Living things are made up of small self-contained units called cells. Each cell is a collection of living matter enclosed by a barrier that separates the cell from its surroundings. Most cells can perform all the functions we associate with life.

Cells are remarkably diverse. A single cell by itself can form an entire living organism. Organisms consisting of only a

Figure 2–1 This Lithops plant, commonly called the living stone, certainly does not appear to be alive (left). Yet you would have no trouble determining that it is a living organism if you saw it flowering (right).

COOPERATIVE LEARNING

Using preassigned lab groups or randomly selected teams, have groups plan a 10-minute videotape explaining how to use the compound light microscope to a class of fifth-graders. The videotape plan could follow a split-page column format with visuals described on the left and an audio script outlined on the right. The videotape could take the form of an actual demonstration, as song or rap, a demonstration by a famous scientist, or a familiar cartoon character's explanation. The tape plan should contain the following elements:

- A specific organism to be observed in the demonstration.
- Identification of the parts of the microscope.
- Step-by-step instructions for observing the organism.

Remind students that their audience is made up of 10- and 11-year-olds and that the videotape should be interesting and informative for that age group.

Journal Activity

YOU AND YOUR WORLD
Have students keep their journal entries in their portfolios. Ask for volunteers to read their short stories or essays to the class.

- **Must every organism reproduce to be alive?** (No.)
- **Why, then, is it considered an important characteristic of life?** (Essential to the survival of the species.)
- **What are some ways that a human (you) may respond to the environment?** (Answers will vary widely but might include heavier clothes in winter, studying for a test, blushing, becoming angry, or running away from a barking dog.)
- **Can a rock respond in the same or similar ways?** (No.)

characteristics exhibited only by some organisms, such as turning green.)

TEACHING STRATEGY 2-1

Focus/Motivation
As a class or in small groups, have students suggest answers to this question: What do all living things have in common?

Discuss their answers, and encourage students to argue for or against each answer with specific examples and counter-examples. From your discussion, develop the textbook's list of the characteristics of living things, which is given in the opening of this section.

ANNOTATION KEY

❶ 100 years. (Interpreting charts)

TEACHING SUPPORT

Laboratory Manual
• Characteristics of Life, p. 67

Teaching Resources
• Activity: Name That Animal Group! p. 3

MEDIA AND TECHNOLOGY
See the Biology Media Guide page 1 for bar-code correlations for this section.

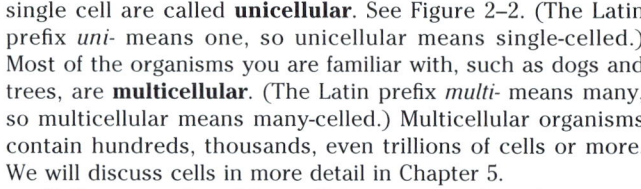

Figure 2–2 Biologists classify organisms as unicellular or multicellular. This unicellular protozoan (left) is a single-celled organism, whereas the multicellular tree sloth (right) is made up of trillions of cells.

Figure 2–3 Orangutans, like many other animals, reproduce sexually. In sexual reproduction, one cell from each parent unites to form the first cell of the new organism.

single cell are called **unicellular**. See Figure 2–2. (The Latin prefix *uni-* means one, so unicellular means single-celled.) Most of the organisms you are familiar with, such as dogs and trees, are **multicellular**. (The Latin prefix *multi-* means many, so multicellular means many-celled.) Multicellular organisms contain hundreds, thousands, even trillions of cells or more. We will discuss cells in more detail in Chapter 5.

Cells are not found in nonliving matter unless that matter was once alive. Wood, for example, is made up largely of the walls that once separated the individual cells in the living tree.

Living Things Reproduce

Living things can reproduce, or produce new organisms of the same type. Because all individual organisms eventually die, reproduction is necessary if a group of similar organisms (what we will later call a species) is to survive.

There are two basic kinds of reproduction: sexual and asexual. **Sexual reproduction** requires that two cells from different individuals unite to produce the first cell of a new organism. See Figure 2–3. You are reading this textbook because a cell from your mother united with a cell from your father to form that first cell that would grow and develop into you. Most familiar organisms—from maple trees to birds and bees—reproduce sexually. In **asexual reproduction**, a single organism can reproduce without the aid of another. (The prefix *a-* means without, so asexual means without sex.) Asexual reproduction can be very simple: Some single-celled organisms merely divide in two to form two organisms.

Living Things Grow and Develop

All living things, at one stage or another in their lives, are capable of growth. An acorn, when it sprouts, produces roots, stems, a trunk, and leaves that continue to grow for years. As it grows, the plant takes in substances from the air and soil and

---

2–1 (continued)

Content Development
Using the Visuals Have students examine the two photos in Figure 2–2.
• **Do you think the tree sloth's cells are similar to the protozoan's single cell?** (All cells are self-contained units, separated from the outside by a wall or membrane. But aside from this basic similarity, cells can differ tremendously both among organisms and within the same organism. The tree sloth's cells have widely different shapes, sizes, and inner structures, with each cell specially organized to perform its specific function. In contrast, the protozoan's single cell is organized to exist independently; this cell performs all the duties necessary for keeping itself alive.)

Content Development
Emphasize to students the difference in growth in living and nonliving things. A good comparison to make might be the growth of a child compared to that of a garbage heap. As a child eats food—meat, potatoes, fruits, and vegetables—he or she grows. If you were to throw the same foods into a pile, the garbage heap would also grow.
• **Based on the example given, compare the growth of living and nonliving things.** (Answers may include the concepts of

28

Figure 2-4 All living things grow and develop. Usually growth simply means getting larger, not changing form. But that is not always the case. This caterpillar (left) will grow and develop into an adult Cecropia *moth (right).*

transforms these substances into living tissue. And long after the tree stops getting larger, it continues to add new material to replace existing parts that wear out.

A snowball, on the other hand, may seem to "grow" if you roll it over fresh snow. But a snowball grows bigger only if someone adds new snow onto its surface. A snowball won't grow bigger by just sitting there. And it certainly cannot change liquid water or solid ice into new snow from which it can grow larger.

During growth, most living things go through a cycle of change called development. The single cell that starts an organism's life divides and changes again and again to form the many and varied cells of an adult organism. You are probably well aware of growth and development since you are now in the midst of one of the most intense spurts of growth and development that you will ever encounter in your life.

As development continues, organisms experience a process called aging. During aging, an organism becomes less efficient at the process of life. The ability to reproduce comes to an end. For virtually all organisms, death is the inevitable end of the life span of every individual. Death, too, is a process of change that separates living and nonliving things.

Living Things Obtain and Use Energy

Living things obtain energy from their environment, or their surroundings, and use that energy to grow, develop, and reproduce. All organisms require energy to build the substances that make up their cells. Any process in a living thing that involves putting together, or synthesizing, complex substances from simpler substances is called **anabolism** (uh-NAB-uh-lihz-uhm).

Plants obtain their energy from sunlight in a process called photosynthesis, which you will study in Chapter 6. (The prefix *photo-* refers to light, and the suffix *-synthesis* means put together. Thus photosynthesis means put together with light.)

Figure 2-5 According to this chart, what is the maximum life span of a blue whale? ❶

MAXIMUM LIFE SPANS

Organism	Life Span
Adult mayfly	1 day
Marigold	8 months
Mouse	1–2 years
Dog	17 years
Blue whale	100 years
Tortoise	152 years
Bristlecone pine	5500 years

MULTICULTURAL STRATEGY

As students read about energy's importance to living organisms, have them research how the sun was worshipped in many civilizations. Examples of such civilizations include the ancient Egyptians, Greeks, Sumerians, Mayas, Incas, and Aztecs.

More recently, creative people all over the world have featured the sun prominently in their work. Examples include American poet Emily Dickinson in her poem "The Sun," Russian composer Nikolai Rimsky-Korsakov in his "Hymn to the Sun," and the landscapes of Dutch painter Vincent van Gogh. Have students research these or other works, and have them share their discoveries with the class.

---

assimilation and organization of materials, development of specific structures, and/or overall growth rather than a "pile.")
• **Do organisms always grow and develop at the same rate?** (Obviously not.)
• **When do organisms stop grow-** ing/developing? (The process goes on at different rates but does not completely stop until death.)

Content Development
As stated at the beginning of this section, all living organisms obtain and use energy. But strictly interpreted, this characteristic applies to many nonliving things as well. Windmills, solar calculators, television sets, thunder clouds—all these objects absorb energy from their environments, concentrate it, and transmit it in new forms. However, what distinguishes living organisms is that they use energy for very specialized purposes. Emphasize that unlike objects, living organisms use energy for growth, development, and reproduction.

ANNOTATION KEY

❶ The green plant uses energy from the sun in a process called photosynthesis. (Making comparisons)

❷ Answers will vary but may include light, sound, touch, and so on. (Applying concepts)

❸ The bat is responding to the sight and possibly the sounds made by the frog. The frog's natural response is to retreat, probably by jumping away. (Making inferences)

❹ To help keep its body temperature from falling during the cold morning climate and to achieve homeostasis, the reptile basks in the sunlight, helping to warm its body. If it overheats, it will likely seek shade to help its body cool off. (Relating concepts)

Figure 2-6 *This katydid obtains the energy it needs to live from the food it eats. How does the green plant the katydid is munching on obtain its energy?* ❶

2–1 (continued)

Content Development

Using the Visuals Have students look at Figure 2–8. This picture shows an animal warming itself with sunlight, thus maintaining its temperature. Emphasize that to stay alive, all organisms must maintain certain essential conditions—temperature among them. The maintenance of internal stability in an organism is called homeostasis.

• **Suppose it's a hot day. What does your body do to cool off?** (It sweats. As sweat evaporates, it draws heat away from the body, thus helping to maintain body temperature.)

• **Suppose you keep sweating for an hour or so. How do you feel?** (Accept all logical answers. Elicit the idea that one would feel thirsty.)

• **What do you do in response to feeling thirsty?** (Drink.)

• **Why is this series of events— sweating on a hot day, feeling thirsty, and drinking—an example of homeostasis?** (It shows the body maintaining its supply of water. The body loses water through sweat, a loss that the brain interprets as feeling thirsty. This feeling is satisfied by drinking, the activity that replenishes the body's water supply.)

Animals cannot perform photosynthesis. Animals take in energy in the form of food. Food is broken down during a process called digestion, which you will study in Chapter 39. The final breakdown of complex substances into simpler ones, usually resulting in the release of energy, is called **catabolism** (kuh-TAB-uh-lihz-uhm).

Living things must practice both anabolism and catabolism at the same time, just as a business or a household must take some money in as income and pay some money out as expenses. The total sum of all chemical reactions in the body—the balance of anabolism and catabolism—is called **metabolism**.

Living Things Respond to Their Environment

Living things respond to their environment. Such responses can be rapid, usually through changes in behavior, or slow, usually through changes in metabolic processes or through growth. Anything in the environment that causes an organism to react is called a stimulus. Organisms react to many stimuli, including light, temperature, odor, sound, gravity, heat, water, and pressure. What stimuli are you responding to at this very moment? ❷

The ability of living things to react to stimuli is known as irritability. (No, that does not mean that living things are grouchy. At least not all the time!) Both plants and animals exhibit irritability and can react to a variety of stimuli. Plants, however, usually respond to stimuli more slowly than animals. Plant leaves and stems, for example, grow toward light and away from the pull of gravity. Plant roots, on the other hand, respond to gravity by growing down into the soil.

In general, living things respond to stimuli in ways that improve their chances for survival. **The process by which organisms respond to stimuli in ways that keep conditions in their body suitable for life is called homeostasis.** (The prefix

Figure 2-7 *Living things respond to stimuli from their environment. What stimuli is the bat responding to? What will be the logical response of the frog?* ❸

Emphasize that in humans, as in other organisms, homeostasis is controlled by involuntary, automatic responses.

SECTION REVIEW 2–1

1. Living things are made up of cells, reproduce, respond to their

30

Figure 2–8 This Australian reptile, called the frilled dragon, is basking on a rock in the sun. How does this behavior of the reptile help it achieve homeostasis? What might the reptile do if its body overheats?

homeo- means similar or same. The suffix *-stasis* means standing or stopping.) **Homeostasis** (hoh-mee-oh-STAY-sihs) refers to an organism's ability to maintain constant or stable conditions that are necessary for life. Just as a thermostat in your home turns on the heat when it gets down to a certain temperature, your body has a thermostat that maintains a constant internal temperature. If you get too hot, you sweat and cool off. And if you sweat for a long time, the resulting thirst persuades you to replace the water your body has lost.

You might point out that nonliving things also respond to the environment. However, the responses of nonliving things are purely mechanical (like a spring that jumps when compressed and released) and are not related to survival.

2–1 SECTION REVIEW

1. Describe five characteristics of living things.
2. Compare sexual reproduction and asexual reproduction.
3. Define metabolism, using the terms catabolism and anabolism in your definition.
4. **Critical Thinking—Making Generalizations** Try to think of a nonliving thing that satisfies each characteristic of living things. Does any nonliving thing have all the characteristics of life?

2–2 Biology: The Study of Life

Quite literally, biology means the study of life. (The prefix *bio-* means life, and the suffix *-logy* means study of.) Biology, then, is the science that seeks to understand, explain, and even control the living world. Biology, like any other science, advances by observing the world, asking questions, and forming hypotheses that can be tested by experiment. **A biologist is anyone who uses the scientific method to study living things**.

Guide For Reading
- What are the types of questions asked by different kinds of biologists?
- What kinds of microscopes are used by biologists? What are the advantages and disadvantages of using the different kinds of microscopes?
- What are some laboratory techniques used in biology?

Branches of Biology

The broad field of biology contains many branches, or divisions. Some divisions are quite general: zoologists study animals (the prefix *zoo-* means animal); botanists work with plants (the Greek prefix *botanikos-* means green plants); and microbiologists work with microscopic organisms (the prefix *micro-* means small). Other subdivisions of biology are more focused. Paleontologists, for example, work with extinct organisms (the prefix *paleo-* means ancient), and ethologists study animal behavior (the prefix *ethos-* means custom).

In the course of this textbook we will examine many fields of biology. Here, we can begin our investigation by considering examples of the types of questions that are asked by different kinds of biologists.

QUESTIONS AT THE MOLECULAR LEVEL Some biologists study life at the molecular level. Molecular biologists, for example, may study the basic chemical units of life. Molecular geneticists investigate the workings of DNA, the molecule that controls heredity and directs all the activities of the cell. Other researchers might study the effects of drugs on molecules in cells in order to understand why entire organisms react to those drugs as they do.

QUESTIONS AT THE CELLULAR LEVEL Some biologists study questions that deal with organisms at the cellular level. Cell biologists, for example, might study the way normal cells become cancer cells when exposed to radiation or to the chemicals found in cigarette smoke. Or they might try to explain how a single cell divides and changes to form all the cell types in an

Figure 2–9 Biology is filled with terms that may seem unfamiliar to you but are actually quite simple. Many scientific terms are derived from Latin or Greek words that may be added in front of another word as a prefix or after another word as a suffix. Using this chart, determine the meaning of the word cytology. ❶

Prefix	Meaning	Prefix	Meaning	Suffix	Meaning
anti–	against	herb–	pertaining to plants	–cyst	pouch
arth–	joint, jointed	hetero–	different	–derm	skin, layer
auto–	self	homeo–	same	–gen	producing
bio–	related to life	macro–	large	–itis	inflammation
chloro–	green	micro–	small	–logy	study
cyto–	cell	multi–	consisting of many units	–meter	measurement
di–	double	osteo–	bone	–osis	condition, disease
epi–	above	photo–	pertaining to light	–phase	stage
exo–	outer, external	plasm–	forming substance	–phage	eater
gastro–	stomach	proto–	first	–pod	foot
hemo–	blood	syn–	together	–stasis	stationary condition

ANNOTATION KEY

❶ Cytology is the study of cells. (Interpreting charts)
❷ Ethologists would study courtship behavior. (Relating facts)

TEACHING SUPPORT

Teaching Resources
- Activity: Analyzing Science Terms, p. 5

MEDIA AND TECHNOLOGY
📀 **Interactive Videodisc**
- *In the Company of Whales*
  Use this bar code to introduce the topic of identifying whales.

Play frames 23903 to 25557

See the Biology Media Guide pages 1–2 for additional bar-code correlations.

Before using the bar-code correlations printed in this Teacher's Edition, read the notes on page T5.

TEACHING STRATEGY 2–2

Focus/Motivation
To introduce this section, play a game of twenty questions with the class. Think of a familiar plant or animal, such as a dandelion, an ant or other insect, or a common bird or mammal. Tell students that you are thinking of a certain living organism and that they are allowed twenty yes-or-no questions to determine what this organism is. As the game progresses, you may suggest questions to the class; do not let them stray too far from the correct answer.

Tell students that whether they realized it or not, they were conducting a scientific investigation. They were presented with a problem, and they needed to ask the right questions to reach a solution. Emphasize that in science, the answers often are very easy—it's figuring out the right questions that is difficult.

Play another round of twenty questions. Try to steer the class toward an efficient approach to the game; they should begin with general questions and progress gradually to more specific ones.

To challenge students, select a more obscure organism, such as an eel, yeast, or crabgrass.

Content Development
As discussed in this section, scientists divide biology into many branches and subdivisions. But

adult organism. Other cell biologists might study how cells communicate with nearby cells.

QUESTIONS AT THE MULTICELLULAR LEVEL Going beyond individual cells, some biologists study multicellular organisms. Zoologists, for example, might be interested in the changes within animals that tell them when to sleep or eat or even when to mate. Paleontologists might try to explain how certain animals changed over time, or evolved. Ethologists might ask why the males of a particular kind of organism are more brightly colored than the females.

QUESTIONS AT THE POPULATION LEVEL Some biologists even go beyond individual organisms in their studies. These biologists are interested in groups of organisms that make up populations and how such populations interact with their environment. Some ecologists, for example, might want to know how the construction of a new road or dam, or the cutting down of forests, will affect nearby plant and animal life. Other ecologists might be concerned with the effects of pesticides or industrial wastes on organisms that live in our waterways.

QUESTIONS AT THE GLOBAL LEVEL Many biologists take a more worldwide view of biology and are concerned with organisms and their environment on a global scale. Global ecologists, for example, might try to estimate the effects on the Earth's climate of burning coal and oil. Or they might try to explain why the fishing off New England is excellent one year and poor the next.

Whether studying questions at the molecular level, the global level, or any level in between, biologists are making important contributions. They are both studying and trying to preserve the wonderful things that are alive on planet Earth—not just for their own use, but for the use of those who will live on this planet after us.

By now it may appear as if biology is a field strictly for scientists with a long list of college degrees. It is true that most biologists do have a bachelor's degree and often go on to graduate school. But it is just as true that anyone can be a biologist. The only real qualifications are hard work, energy, and curiosity. A biologist must be curious about life and have the energy to ask questions in a scientific way—and then try to answer those questions. As you will read in this textbook, many of the greatest names in the history of biology were amateurs, meaning that they did not practice science as a "job." Such people include Charles Darwin, who established the theory of evolution, and Gregor Mendel, who discovered the basic units of heredity. Today other amateurs, including high school and college students, continue to make important contributions to scientific research and the science of biology. No license is required!

Figure 2–10 Many bird species have elaborate courtship (mating) behaviors. These albatross are exhibiting a form of courtship behavior called sky pointing. What type of biologist might study courtship behavior? ❷

Figure 2–11 These Japanese citizens are taking part in an antipollution march in Tokyo. The effects of air pollution on health are among the global questions biologists often try to assess.

for students and experts alike, making progress in one branch of biology invariably requires knowledge from another branch. For this reason, scientists the world over share their findings and often work together to solve multifaceted problems.

• **Why would a scientist working to formulate contact lens solutions need to confer with a molecular biologist?** (He or she would need information about the way various ingredients in the solutions will react with the tissues of the eye.)

• **Why would an ornithologist (biologist who studies birds) at the University of Texas be interested in knowing about forest lands being cleared in Colorado?** (The disappearance of the forest might affect migration patterns and breeding habits.)

ESL STRATEGY

Have students use both Figure 2–9 and a dictionary to find the meanings of the following words: cytoplasm, epidermal, gastropod, hemacytometer, homeostasis, macrophage, microbiology, photosynthesis, neogenesis, osteoarthritis, prototype.

MULTICULTURAL STRATEGY

JoAnn Tall has worked tirelessly to preserve her home, the Pine Ridge Indian Reservation in South Dakota. She has defended this land from nuclear weapons testing, the dumping of hazardous waste, and water pollution from a uranium mine. For these and other efforts, she recently was awarded the Goldman Environmental Prize.

Have students research the special problems that threaten numerous reservations and what Native Americans and others are doing to protect these lands.

Reinforcement/Reteaching
Review the five levels of questions biologists might ask. Have students suggest another question that might be asked for each level.

Tools of a Biologist

To accomplish their diverse goals, biologists may choose to use a wide variety of tools. In the laboratory, biologists may use pipettes and graduated cylinders to measure and transfer liquids. Solids, on the other hand, are usually measured on mechanical or electronic balances. Many experiments are performed inside enclosures, called hoods, that protect researchers from dangerous fumes or help to control contamination.

In almost all areas of biology, the computer has become an invaluable tool that can be used to perform complex tasks and analyze quantities of data. Many tools, however, are more specific to the type of biological work being undertaken. Global ecologists, for example, may use orbiting satellites to provide detailed maps of the temperature, moisture content, or vegetation of large areas. We shall discuss many more tools and the ways in which they are used throughout this textbook.

When you think of biology, there is probably one tool that comes to mind above all others. For almost all biologists need to examine organisms or parts of organisms that are too small to be seen with the unaided eye. **To study small organisms, researchers have developed several kinds of microscopes. Microscopes are instruments that produce larger-than-life images, pictures, or even videotapes.**

THE COMPOUND LIGHT MICROSCOPE The most commonly used microscope is the light microscope. Figure 2–12 shows a light microscope similar to those in your high school laboratory. Light microscopes are useful to biologists because these instruments make it possible to observe many kinds of cells and small organisms while they are still alive.

To view most objects with the light microscope, you sandwich the object between a transparent microscope slide and a thin, transparent coverslip. This "sandwich" is then placed on the stage of the microscope so that light passes through it into the lenses of the microscope. The lens at the bottom of the microscope tube, the one closest to the object being studied, is called the objective lens. The viewing lens at the top of the tube is called the ocular lens. Because the instrument uses both lenses to form an image, it is properly known as a **compound light microscope.** If the objective lens produces a magnification of 100 times, and the ocular lens a magnification of 10 times, the resulting image that you see will be magnified 1000 times ($100 \times 10 = 1000$).

LIMITS OF RESOLUTION There are limits to what we can see with the compound light microscope. As we increase the magnifying power of a light microscope, we see more and more detail—up to a certain point. Beyond that point, called the **limit of resolution**, objects get blurry and detail is lost. For standard light microscopes, the limit of resolution is about 0.2 micrometers. (A typical cell is about 10 micrometers across.)

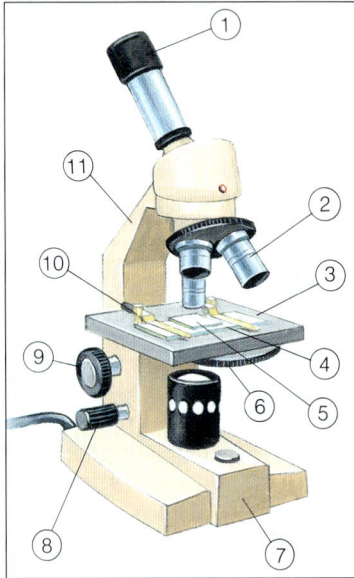

Figure 2–12 This diagram is of a typical compound light microscope. What is another word for the eyepiece? ❶

Compound Light Microscope
1. Ocular lens (eyepiece)
2. Objective lens 3. Stage
4. Glass slide 5. Coverslip
6. Diaphragm (regulates light intensity)
7. Base
8. Fine adjustment knob
9. Coarse adjustment knob
10. Stage clips
11. Arm

ANNOTATION KEY

❶ Ocular lens. (Interpreting illustrations)

❷ Accept all logical answers. Many students will point to the types of medical equipment used in popular science fiction shows such as *Star Trek*. (Making predictions)

TEACHING SUPPORT

Teaching Resources
- Activity: Microscopic Proportions, p. 7

MEDIA AND TECHNOLOGY
- Biology Transparencies
- 2: The Compound Light Microscope

FACTS AND FIGURES

Magnification is normally expressed in linear terms. In other words, an image of a cell magnified 1000 times has a diameter 1000 times greater than the cell itself.

2–2 (continued)

Content Development
As students study this section, point out the limits of what microscopes can do. All microscopes have limits of resolution, and the microscopes in your classroom likely will be flawed in some minor ways, perhaps by imperfect lenses or misalignment. Even so, your students should appreciate and value their compound microscopes. Stress that these microscopes are valuable tools for studying biology.

You also should discuss any other types of microscopes that you have in your classroom, as well as the new, technologically advanced microscopes described in the textbook.

• **If you are looking at feathers or insect legs under a microscope, how important is excellent resolution?** (Not very important because you are looking at overall structure.)

• **What if you are looking at slides of plant or animal cells?** (Resolution is now very important because detail is important.)

When the limit of resolution was first discovered, many people thought that if microscopes were made better, this problem would disappear. However, even a "perfect" microscope will have a limit of resolution. The reason for this has to do with the way light behaves. When light passes through a tiny opening or a lens, it is diffracted, or scattered in a way that makes it hard to form a clear image. When we look at something in the compound light microscope at 1000 times magnification, we have enlarged it just enough to see the limit of resolution of the best light microscopes we can make.

USING A COMPOUND LIGHT MICROSCOPE There are objects—such as dust, feathers, and pollen grains—that can be seen in the light microscope without any special preparation. But many cells and cell parts are so similar in appearance to their surroundings that they cannot be easily seen through the microscope. Researchers have developed several techniques to make such objects visible.

Many specimens are stained before they are observed under a microscope. Stains are used to color cells or parts of cells to make them clearly visible. Some stains color everything in a cell, whereas others color only a part of the cell. One such special stain, known as Feulgen, turns DNA a beautiful pinkish color. Some stains, which "stick" only to certain compounds, can be made to glow in the dark to highlight specific cell parts.

Because many stains kill living cells, special types of light microscopes that do not require staining are used to observe living specimens. Examples of these are the phase contrast microscope, the dark field microscope, and the Nomarski microscope. Each uses a different property of light rays to improve the contrast (clarity) of the image. See Figure 2–14.

ELECTRON MICROSCOPES Although light microscopes are very useful, their limit of resolution restricts their usefulness for studying very small objects such as viruses and individual molecules. Is there a way to see things smaller than light

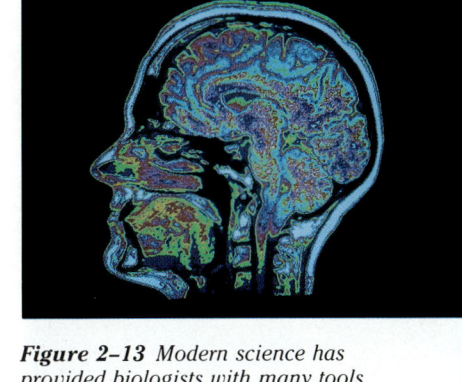

Figure 2–13 Modern science has provided biologists with many tools unknown in centuries past. This image of a patient's brain was made by a medical scanner called a nuclear magnetic resonance scanner (NMR). What sort of tools do you think biologists will be using a century from now? ❷

Figure 2–14 Notice the variations in this image of an alga as seen through a Nomarski microscope (left), a dark field microscope (center), and a phase contrast microscope (right).

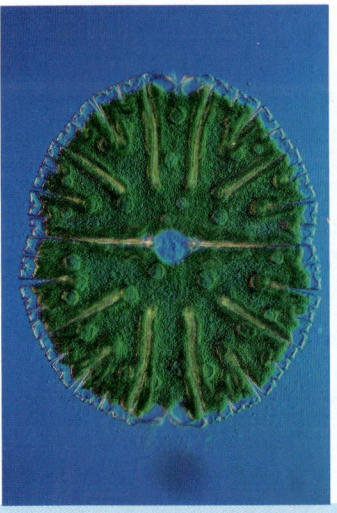 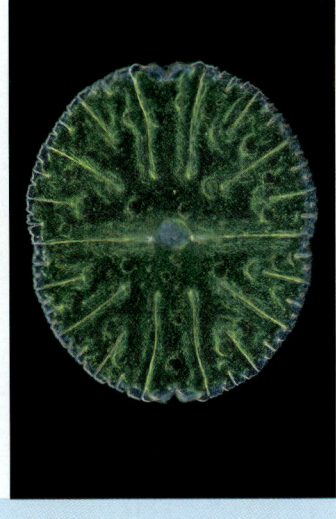 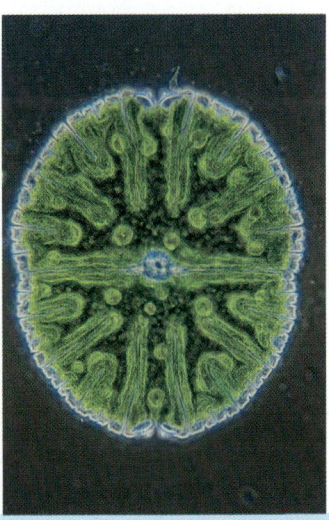

BACKGROUND INFORMATION
MICROSCOPES AND NEW TECHNOLOGY

Due to resolution problems, standard light microscopes are seldom used at powers higher than about 1000×. There are, however, several new technologies that have expanded the utility of the light microscope. Following are two of the most promising.

Electronic enhancement of video images, pioneered by Shinye Inoue at the Marine Biological Laboratory, uses high-resolution, low-light video cameras to produce images of living cells. Sophisticated electronic processing of those images dramatically improves their resolution and contrast. Inoue's microscopes have produced remarkable films of dividing cells (in which spindle fibers and chromosomes are clearly visible) and of fertilization (in which the moving acrosome is clearly visible).

The *scanning optical microscope,* a new and still experimental device developed by Aaron Lewis and Michael Isaacson at Cornell University, breaks the "resolution barrier" of light microscopes. This microscope places specimens within a few nanometers of an ultrathin "light pipe." The image is created by scanning back and forth over the specimen with a fine-tipped light-sensitive needle. The theoretical limit of resolution of this device is between 30 and 50 nanometers. Theoretically, it will allow researchers to see individual molecules within living cells.

- **Name some obvious advantages to looking at living rather than dead specimens under the microscope.** (Living color, movement, and so on.)

Reinforcement/Reteaching
This is the time to get out the microscopes and have a hands-on review of the parts and their functions. Most students will remember some things from previous science courses, but there is likely to be a wide range of proficiencies in the class. Have students work with partners to practice naming parts, giving functions, and demonstrating proper handling and usage.

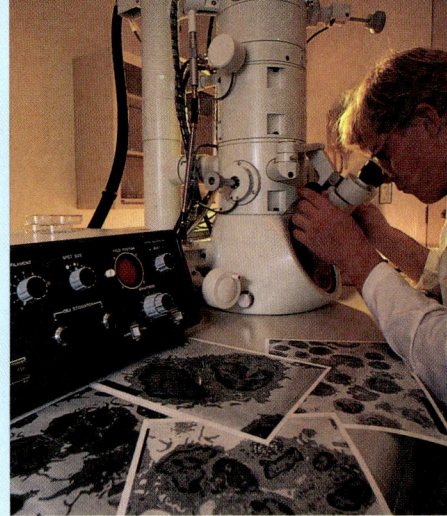

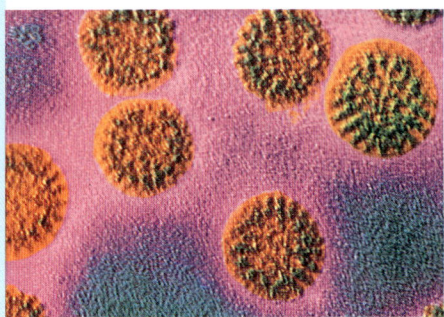

Figure 2–15 A transmission electron microscope (TEM) sends a beam of electrons through an object to produce an image (top). This image of viruses, which has had color added by a computer, was taken through a TEM (bottom).

Figure 2–16 A scanning electron microscope (SEM) bounces electrons off the surface of an object to form an image. This image of bacteria, which has been colored green through the use of a computer, was taken by using an SEM.

can reveal? Indeed, there is. In the 1920s, physicists in Germany realized that electromagnets could bend streams of electrons in much the same way that glass lenses bend beams of light. They then learned to use electromagnets to build devices called electron microscopes. These same physicists used electromagnets to bend electrons to produce another tool you are probably familiar with: television.

The limit of resolution of electron microscopes is about 1000 times finer than the light microscope. There are several different types of electron microscopes.

Transmission electron microscopes (TEMs) shine a beam of electrons at a sample and then magnify the image of that sample onto a fluorescent screen at the bottom of the microscope. The electron beam can also be used to expose photographic film to produce a permanent image of the specimen. See Figure 2–15.

Scanning electron microscopes (SEMs) get their name from a pencillike beam of electrons that scans back and forth across the surface of a specimen. Electrons that bounce off the specimen are picked up by detectors that provide the information to form an image on a television screen. Rather than showing details inside living things, SEMs show realistic (and often dramatic) three-dimensional pictures of their surfaces. See Figure 2–16.

LIMITATIONS OF ELECTRON MICROSCOPES Electron microscopes are extremely useful, but they do have serious drawbacks. Because electrons are charged particles, they do not penetrate air, let alone thick specimens! Therefore specimens must be placed in a vacuum, or a chamber from which all the air has been removed, and samples for TEM work must be cut into very thin slices. TEM samples are usually stained to increase their visibility. Samples for both TEM and SEM work must be completely dried out before they are placed in the vacuum inside the microscope. As you might infer, living cells cannot be observed in the electron microscope—they are killed by the sample-preparation processes.

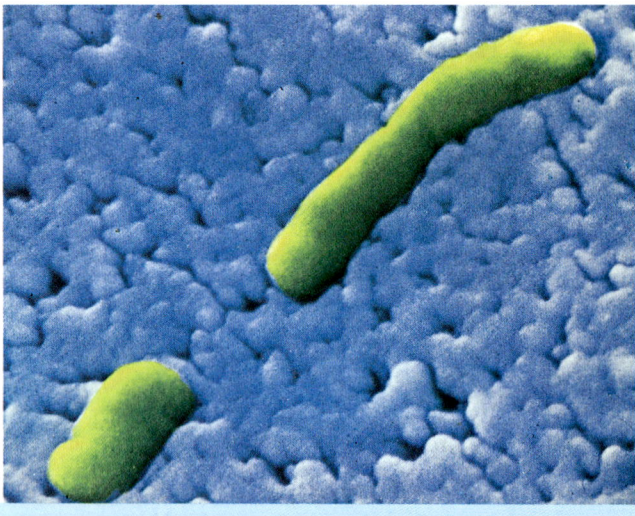

TEACHING SUPPORT

Laboratory Manual
- Measuring with a Microscope, p. 73

Teaching Resources
- Hands-On Activity: Life in a Drop of Water, p. 13

Biotechnology Workbook
- New Views on Old Bones, p. 97

2–2 (continued)

Content Development
Using the Visuals Remind students that biologists use different tools for different purposes. Emphasize that the latest or most expensive microscopes or other apparatus are not necessarily what best serve every scientific investigation.

Have students examine the figures on pages 36 and 37, and have them read the captions. Make sure students realize that these micrographs do not show their subjects' true colors. Rather, computers were used to artificially color the pictures to make structures easier to see. Students may already be familiar with this concept; many black-and-white movies have been "colorized" to make them more appealing.

Skills Development
Guided Practice
Skill: Relating concepts
Give students a list of the following topics, and ask them which microscope, if any, would best serve the topic's investigation.
- **The feeding habits of unicel-

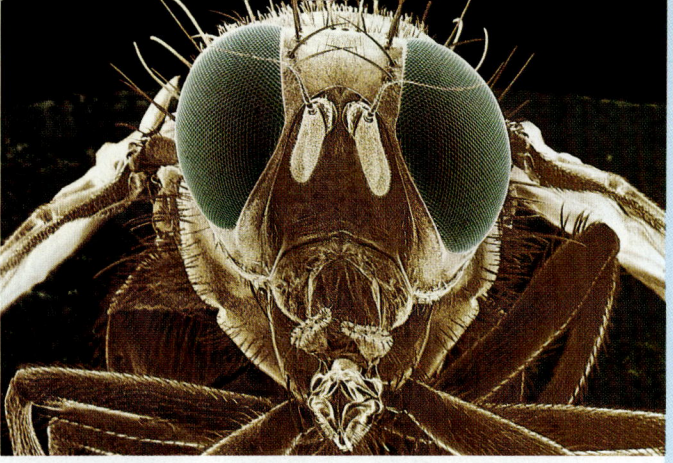

Figure 2–17 On this fruit fly (left), notice the two large red structures, which are the eyes. Then look at the head of the fruit fly as seen through a scanning electron microscope, magnified 60 times (right). The two rounded structures on each side of the head are the eyes. A high power view in the scanning electron microscope produced this detailed three-dimensional image of the eye of the fruit fly (bottom).

PROBE MICROSCOPES In the 1980s, researchers developed a new class of microscopes that do not use lenses to produce images. Because these instruments trace the surfaces of a sample with a fine tip known as a probe, they are called scanning probe microscopes. Scanning probe microscopes have revolutionized the study of surfaces and have even made it possible to observe single atoms. Unlike electron microscopes, scanning probe microscopes do not require that specimens be placed in a vacuum. Now researchers are eagerly searching for ways to use these new instruments for studies in biology.

Laboratory Techniques of a Biologist

You have already read about the laboratory technique called staining, in which parts of a cell are stained so that they appear more visible under a microscope. Throughout this textbook, we will discuss many other laboratory techniques used by biologists. But it will be helpful now for you to learn about some common laboratory techniques.

Figure 2–18 Scanning probe microscopes produce their images by moving a probe across a sample's surface and recording information about the surface's shape on a computer. This scanning probe image shows a thin film of iodine on a platinum plate. Each purple sphere is a single iodine atom.

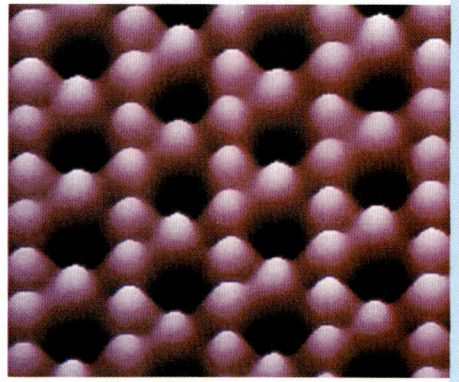

HISTORICAL NOTE
SLOW PROGRESS

It was Leonardo da Vinci (1452–1519) who first suggested that lenses could be used to magnify a tiny object. However, close to a century passed before the first compound microscope was actually made. Even more curious by today's standards, it took 50 more years before reports of the invention were published!

---

lular protozoa (A light microscope, because the organisms are small and would need to be studied alive.)
• **The surface of a red blood cell** (Scanning electron or probe microscopes are best for looking at surfaces.)
• **The atoms of a glucose molecule** (A scanning probe microscope, because it has the power to show atoms.)
• **The feeding habits of a house cat** (No microscope is necessary to observe the behavior of an animal as large as a cat.)

Enrichment
Have interested students research the way specimens are prepared for an electron microscope. Students may present their findings to the class, perhaps with a mock demonstration.

Figure 2–19 Blood samples are placed into a centrifuge in order to separate the various components of the blood.

CENTRIFUGATION Suppose a scientist wants to study one particular part of a cell or the same cell part from many similar cells. How can parts of a cell be obtained? One method is called centrifugation (sehn-trihf-yoo-GAY-shuhn). See Figure 2–19. One common centrifugation technique is to place the cells under study in a blender to break them apart. Breaking apart cells in a blender is called cell fractionation. The broken bits of cells are then placed in a liquid in a tube. The tube is inserted into a centrifuge, which is a device that can spin the tube up to 20,000 times per minute. While spinning, the cell parts begin to separate—with the heaviest parts settling near the bottom of the tube and the lightest parts rising toward the top. A scientist can then remove the specific part of the cell to be studied by selecting the appropriate layer.

MICROMANIPULATION Another technique to remove parts of a cell is called microdissection, which is a form of micromanipulation. Micromanipulation can also be used to insert material into a living cell. In micromanipulation, special tools that are so small they can be used only by looking through a microscope are used to dissect, remove, insert, or otherwise manipulate specific parts of a cell.

CELL CULTURES Sometimes scientists want to study a particular kind of cell but to do so they need large numbers of that exact cell. To obtain many identical copies of that cell, the scientist might prepare a cell culture. In this technique, a single cell is placed in a dish that contains the nutrients the cell needs. The cell is allowed to reproduce so that in time an entire population is grown from that single original cell.

Figure 2–20 Micromanipulation techniques allow researchers to operate on single cells. Here, a human egg cell is held in place by a pipette while a fine glass probe is used to scratch its surface. Scratching the egg's surface allows sperm to enter more easily and makes fertilization more certain.

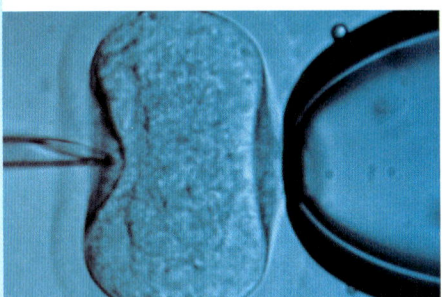

2–2 SECTION REVIEW

1. List four branches of biology. What sorts of questions might researchers in each branch study?
2. What type of research could be performed only by using a compound light microscope?
3. Name an advantage and a disadvantage of the light microscope and each of the electron microscopes.
4. Compare a TEM and a SEM. What type of information does each provide?
5. How does the resolution of a microscope affect the observations made by a microbiologist?
6. **Critical Thinking—Expressing an Opinion** It has been said that many great discoveries lie in wait for the tools needed to make them. What does this statement mean to you? If possible, include an example in your answer.

SCIENCE, TECHNOLOGY, AND SOCIETY ISSUE

Yellow Fever: The Scientific Method in Action

Near the turn of the century, American soldiers stationed in the vicinity of Havana, Cuba, began to die in great numbers. They died not from enemy bullets but from a disease known as yellow fever. At one point, more soldiers died from the disease than from fighting in the Spanish-American War that had brought them to Cuba in the first place! The situation became so serious that the United States sent a commission headed by Dr. Walter Reed.

At first, people thought the disease was spread by contact with infected individuals or through infected materials such as bedsheets and clothing. So they boiled the clothes and sheets of sick people before using them again and prevented other soldiers from coming into contact with the infected soldiers. But none of these measures was of any help.

One Cuban doctor, Carlos Finlay, believed that the disease was spread by mosquitoes, which were plentiful in Cuba. Reed did not immediately agree, but he devised what turned out to be a terrifying experiment to test Finlay's hypothesis. Ideally, Reed would have used animals for his experiment. But because no animals that could contract yellow fever were available, the lives of several extraordinarily brave human beings were put at risk.

The commission assembled two groups of volunteers. One group spent twenty difficult, anxious days wearing the filthy clothing of yellow fever patients, sleeping in their soiled sheets, and eating from plates they had used. During this time, however, they were completely screened in and thereby protected from mosquitoes. Not one of these soldiers contracted yellow fever.

The second group of volunteers used only fresh clothing and linens and remained totally isolated from yellow fever patients. These volunteers, however, allowed themselves to be bitten by mosquitoes that had previously bitten yellow fever patients. Because members of the commission thought it unfair to ask others to take a risk they would not take themselves, three of them participated in this group. (Reed himself wanted to take part, but his associates would not let him.) Many of the volunteers in the second group, including all three doctors from the commission, developed yellow fever. One of them, Dr. Jesse Lazear, died.

Reed's experiment had risked human lives—and had cost one. But it had tested an idea and had proved that mosquitoes carried the infection. As a result, war was declared against the mosquito. In ninety days, Havana was completely free of yellow fever. This courageous example of the scientific method in action saved the lives of thousands of people who might have died from the disease.

Walter Reed's confirmation of the way in which yellow fever was spread saved countless lives. But most scientists would not necessarily agree that humans should be used in such experiments. Are there diseases so dangerous that finding a cure or treatment is worth placing humans at risk? Or is no discovery worth the cost in human lives? What about only using volunteers, as Reed did? What about only using people with fatal diseases? These are difficult questions to consider. There may not be a right or wrong answer. What is your opinion on this issue?

SCIENCE, TECHNOLOGY, AND SOCIETY ISSUE

Yellow Fever: The Scientific Method in Action

After students read this article on yellow fever, have them discuss the questions posed in the article's last paragraph. Emphasize that these questions address ethical issues that remain significant today. For example, AIDS researchers and their patients continually wrestle with how aggressively they should test new drugs. On a different front, some animal-rights activists believe that scientists should not perform experiments on animals, regardless of the potential benefits.

---

microscope provides a 3-D image of its surface features.

5. As resolution increases, the available data in an image becomes greater and more detail is provided. However, at a certain point called the limit of resolution, the image becomes more blurry as it is magnified.

6. Answers will vary, but most students will point out that microorganisms could not be discovered until the invention of the compound microscope and that structures within bacteria and viruses could not be observed clearly until the invention of the electron microscope.

Reinforcement/Reteaching

If students have trouble answering the Section Review questions, go over the parts of the section that they did not understand.

Closure

Ask students if any branch of biology especially interests them, and why. You may want to discuss your own special interests in biology or your experiences studying biology in high school, college, or beyond.

LABORATORY INVESTIGATION

USING A COMPOUND MICROSCOPE

Before the Lab
1. This lab requires little advance setup, but you will need to assemble the same number of scissors, droppers, slides, and coverslips as you have microscopes.
2. Divide the class into groups according to the number of microscopes available.

Pre-Lab Discussion
The students in your class will probably have had some experience using the microscope, but there are several things you will need to review to ensure that the lab goes smoothly.
1. Parts and functions—if not reviewed in an earlier lesson, be sure that students are completely comfortable with names so that they will understand your directions.
2. Preparation of a wet-mount slide—a quick demonstration and discussion of how to position the coverslip to minimize air bubbles will eliminate problems in this lab and others throughout the year.
3. Care and handling of the microscope—demonstrate to students the proper use of lens paper to clean the lenses and caution the students about spilling fluids on the stage. Students should also be reminded to carry the microscope with both hands.
4. Focusing and light adjustment—discuss and demonstrate proper focusing of the microscope by always moving the body tube upward. Emphasize not focusing down while looking into the eyepiece. Only use the fine-adjustment knob when viewing on high power. Also point out to students that changing the amount of light can affect clarity.

Students should read through the lab, noting the steps in the procedure. You can check their understanding with the following questions:
- **What is the purpose of this lab?** (To learn to properly use and focus the microscope. To also learn about depth of focus.)
- **What is the total magnification of an object seen on low power under your microscope?** (Usually 100—multiply the magnification of the eyepiece by that of the objective lens.)
- **Which objective do you always use first in locating a specimen?** (Low power.)
- **Why do you use a coverslip?** (To protect the lens from contact with the specimen.)
- **Why do you move the body tube upward when focusing?** (To avoid breaking the coverslip or slide and to prevent damage to the objective.)
- **When the slide is moved to**

LABORATORY INVESTIGATION: USING A COMPOUND MICROSCOPE

PROBLEM
What is the proper procedure for using a compound microscope?

MATERIALS (per group)

compound microscope	light thread
glass slide	scissors
coverslip	medicine dropper
dark thread	

PROCEDURE

1. Study Figure 2-12 on page 34 and the Appendix on the use of a microscope at the back of this book until you know the names and functions of each part of the microscope.
2. The magnification of the eyepiece, or ocular lens, is written on the eyepiece. In general, the eyepiece magnifies 10 times.
3. Now find the low-power objective lens and the high-power objective lens. The magnification of each objective lens is written on the lens. In most microscopes, the low-power lens magnifies 10 times and the high-power lens magnifies 43 times. The total magnifying power of the microscope is found by multiplying the magnification of the eyepiece by the objective lens being used.
4. Cut a piece of dark thread about 1 cm long. Do the same with light thread. Place the threads on a clean glass slide so that the threads cross each other at right angles.
5. Use the medicine dropper to place a drop of water on the threads. Then touch one edge of a coverslip to the drop of water and lower the coverslip over the threads.
6. Place the slide on the stage, using the stage clips to hold the slide in position.
7. Position the low-power objective lens over the slide so that the lens is just above the top of the slide. **CAUTION:** *Never allow the objective lens to touch the slide.*
8. Look through the eyepiece. Adjust the mirror and the diaphragm so that the field is bright.
9. If the threads are out of focus, use the coarse adjustment knob to raise the objective lens until the threads are clearly seen. Then use the fine adjustment knob to sharpen the focus.
10. Adjust the slide so that the place where the threads cross is in the center of your field of vision.
11. After you have viewed the slide under low power, raise the body tube of the microscope so that the objective lens is well clear of the slide. Rotate the high-power objective lens into place above the slide, making sure the objective lens does not touch the slide. Lower the high-power objective lens until it is just above the slide.
12. Again, use the coarse adjustment knob to raise the objective lens until the threads are in focus. Then use the fine adjustment knob to sharpen the focus.
13. Move the slide to the left, to the right, forward, and back. Note the direction the threads appear to move each time you move the slide.

OBSERVATIONS
1. What happens to the brightness when the magnification is changed from low power to high power?
2. When you use the fine adjustment knob, how many threads are in focus at a time?
3. When you move the slide in any direction, how do the threads appear to move?

ANALYSIS AND CONCLUSIONS
1. Why is it important to raise the body tube before rotating the high-power lens into place?
2. Would you be able to see the threads if they were larger and thicker? Explain your answer.

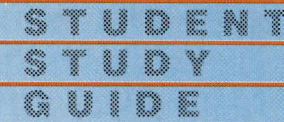

SUMMARIZING THE CONCEPTS

The key concepts in each section of this chapter are listed below to help you review the chapter content. Make sure you understand each concept and its relationship to other concepts and to the theme of this chapter.

2-1 Characteristics of Living Things

- All living things are made up of one or more cells.
- All living things reproduce, either sexually or asexually.
- Living things grow and develop, using energy and raw materials they obtain from the environment.
- Anabolism is any chemical process in a living thing in which complex substances are built up, or synthesized, from less complex substances.
- Catabolism refers to chemical reactions in living things in which complex substances are broken down into simpler ones.
- The sum total of anabolism and catabolism is called metabolism.
- Living things respond to stimuli from their environment. Stimuli include heat, light, water, sound, odor, and pressure.
- To survive, living things must keep conditions in their bodies relatively stable. Homeostasis refers to an organism's ability to maintain constant conditions within its body.

2-2 Biology: The Study of Life

- Different branches of biology deal with questions at the molecular level, cellular level, multicellular level, population level, and global level.
- Biologists use a wide variety of tools; in particular, various types of microscopes.
- The compound light microscope is used to examine living things. It uses light passing through an object to create a magnified image of that object.
- At some point, called the limit of resolution, the object being magnified loses clarity and becomes blurred.
- Electron microscopes use electrons, rather than light, to produce magnified images. To prepare a specimen for viewing under the electron microscope, the specimen must be killed.

REVIEWING KEY TERMS

Vocabulary terms are important to your understanding of biology. The key terms listed below are those you should be especially familiar with. Review these terms and their meanings. Then use each term in a complete sentence. If you are not sure of a term's meaning, return to the appropriate section and review its definition.

2-1 Characteristics of Living Things
unicellular
multicellular
sexual reproduction
asexual reproduction
anabolism
catabolism
metabolism
homeostasis

2-2 Biology: The Study of Life
compound light microscope
limit of resolution
transmission electron microscope
scanning electron microscope

41

the right, what happens to the image? (Moves to the left.)

Skills Development
Students will use the following skills in this investigation.
1. Safety
2. Observing
3. Manipulative
4. Applying
5. Hypothesizing

Safety Tips
Be sure that students clearly understand the proper use of the microscope. Students should be reminded that slides and coverslips are glass and break easily. Any accidents should be reported immediately to you.

Teaching Strategy
As students practice using the microscope, you will want to be sure that they are using correct focusing techniques. You will also want to talk with them as you circulate to see that they are understanding the concept of "depth of focus." Students frequently think of specimens as flat. Grasping the idea of depth now may help them with later labs. Make sure students also understand how specimens are oriented on the microscope. Lastly, encourage students to vary the amount of light on both low and high power so that they can see the difference it makes. If you have time at the end of the lab, have students demonstrate their skills for you.

Observations
1. The brightness decreases.
2. Only one thread should be in focus at a time.
3. The threads appear to move in the opposite direction of the slide's movement.

Analysis and Conclusions
1. To prevent the high-power lens from striking the coverslip, breaking the coverslip, and destroying the specimen. Such an action could also scratch or otherwise damage the lens.
2. No, the threads would likely block the light if they were much thicker.

Going Further: Enrichment
Pond water or a live protozoan culture offer wonderful opportunities for students to practice their microscope skills. Prepared slides of protozoans or algae are also good to use when learning about depth of focus.
- **What adjustments will you have to make if viewing living materials?** (Change the light or power, move the slide around—remember what you know about orientation—and use the fine adjustment.)
- **When looking for organisms in pond water or a culture, what power objective do you use?** (Low power.)

CHAPTER REVIEW

CONTENT REVIEW

Multiple Choice

1. c 2. d 3. b
4. b 5. d 6. b
7. a 8. d 9. b

True or False

1. F All
2. F photosynthesis
3. T
4. F liquids
5. F ocular
6. T
7. T
8. F scanning electron microscope

Word Relationships

1. Cell wall does not belong. The other words refer to an organism's number of cells.
2. Irritability does not belong. The other words refer to a type of stimulus.
3. Chemistry does not belong. The other words refer to a branch or division of biology.
4. Electron beam does not belong. The other words refer to a compound microscope.
5. Limit of resolution does not belong. The other words refer to a type of microscope.

CONCEPT MASTERY

1. Students' lists should include the fact that living things are made up of cells, grow and develop, reproduce, obtain and use energy, and respond to their environment.
2. In asexual reproduction a single organism can reproduce without the aid of another. In sexual reproduction two cells from different organisms unite to produce the cell that will grow into a new organism.
3. Irritability is the ability of living things to react to a stimulus from their environment. Examples will vary but could include blinking when something is thrown at the eyes, sweating when the temperature rises, shivering when temperatures go below freezing.
4. Many parts of the cell are not evident when observed under the microscope. Special stains are extremely helpful in making certain cell parts stand out from the rest of the cell.
5. Transmission electron microscopes shine a beam of electrons at a specimen and then magnify the image produced onto a fluorescent screen.
6. Accept all reasonable responses. Having read the chapter, most students will agree with this statement. They should understand how different types of microscopes play different roles in biological research.

CHAPTER REVIEW

CONTENT REVIEW

Multiple Choice

Choose the letter of the answer that best completes each statement.

1. Single-celled organisms are also called
   a. multicellular.   c. unicellular.
   b. bacteria.   d. viruses.
2. Reproduction that involves the union of cells from two parents is called
   a. asexual reproduction.
   b. irritability.
   c. cell division.
   d. sexual reproduction.
3. Chemical reactions in an organism that build, or synthesize, complex molecules from less complex molecules are called
   a. metabolism.   c. catabolism.
   b. anabolism.   d. digestion.
4. The ability of an organism to maintain constant conditions within the body is called
   a. irritability.   c. metabolism.
   b. homeostasis.   d. stimulus.
5. The branch of biology concerned with animal behavior is
   a. microbiology.   c. botany.
   b. zoology.   d. ethology.
6. To observe the surface features of a bacterium, you would use a
   a. compound light microscope.
   b. scanning electron microscope.
   c. magnifying lens.
   d. transmission electron microscope.
7. To observe living organisms in a drop of water, you would use a
   a. compound light microscope.
   b. scanning electron microscope.
   c. scanning probe microscope.
   d. transmission electron microscope.
8. To observe the internal features of a bacterium, you would use a
   a. compound light microscope.
   b. scanning electron microscope.
   c. magnifying lens.
   d. transmission electron microscope.
9. To see atoms on a surface, use a (an)
   a. compound light microscope.
   b. scanning probe microscope.
   c. electron microscope.
   d. magnifying lens.

True or False

Determine whether each statement is true or false. If it is true, write "true." If it is false, change the underlined word or words to make the statement true.

1. <u>Few</u> cells can perform all the functions we associate with life.
2. Plants obtain their energy from sunlight in a process called <u>metabolism</u>.
3. <u>Stimuli</u> include heat, light, pressure, and temperature.
4. Pipettes are used to transfer <u>solids</u> from one container to another.
5. The lens at the top of a compound light microscope is called the <u>objective</u> lens.
6. Compound light microscopes are used to observe <u>living</u> organisms.
7. In a <u>transmission electron microscope</u>, the image is magnified onto a fluorescent screen at the bottom of the microscope.
8. Surface features of very small organisms are best viewed using a <u>dark field microscope</u>.

Word Relationships

In each of the following sets of terms, three of the terms are related. One term does not belong. Determine the characteristic common to three of the terms and then identify the term that does not belong.

1. unicellular, multicellular, cell wall, single-celled
2. temperature, pressure, irritability, gravity
3. zoology, botany, chemistry, ecology
4. ocular lens, electron beam, objective lens, glass slide
5. SEM, compound light microscope, limit of resolution, TEM

CONCEPT MASTERY

Use your understanding of the concepts developed in the chapter to answer each of the following in a brief paragraph.

1. List and describe the basic characteristics of living things.
2. Compare sexual reproduction with asexual reproduction.
3. What is irritability? Give three examples of a stimulus and a possible response in humans.
4. Discuss the importance of cell-staining techniques in the study of living things.
5. Describe the way a transmission electron microscope produces an image.
6. No one microscope is better than another. Each type has an important role to play in biological research. Do you agree or disagree with these statements? Explain your answer.

CRITICAL AND CREATIVE THINKING

Discuss each of the following in a brief paragraph.

1. **Relating concepts** Why is biology considered a science?
2. **Making generalizations** List the attributes you think scientists should possess in order to be successful in their work.
3. **Applying concepts** How would you decide if an object were living or not?
4. **Making inferences** Why might biologists consider metabolism to be the single most important characteristic of life?
5. **Relating cause and effect** An organism's response to a stimulus can be a method of protection. How do these 3 stimuli-response situations protect the organism?
   a. Pulling a hand away from a hot iron
   b. Squinting in a bright light
   c. Producing tears if dirt gets in the eyes
6. **Making comparisons** Compare the growth of an icicle with the growth of a living thing.
7. **Identifying relationships** Which characteristic of living things is important to the survival of the species rather than to the organism itself? Explain your answer.
8. **Using the writing process** You are a biologist and have the opportunity to travel back in time to study an extinct species of your choice. You may take just one modern tool for your study. Which tool would you choose and why?
9. **Using the writing process** If microscopes could talk, what stories would they tell? Write a microscope story. (*Hint:* First decide if you are a compound light microscope or an electron microscope.)

43

CRITICAL AND CREATIVE THINKING

1. Biology, the study of living things, is a science because it employs the scientific method; it relies on data provided by verifiable experimentation rather than beliefs that cannot be tested.
2. Accept all reasonable answers. Most students will point out the need to enjoy science, the spirit of curiosity, the desire to know, and so on.
3. Accept all logical answers. Most students will suggest seeing if the object meets the basic characteristics of living things.
4. Since metabolism is the sum total of all chemical reactions in the body, it is logical to say that it is the single most important characteristic of life. It could be considered to include growth, reproduction, development, use of energy, and so on.
5. **a.** keeps you from being burned; **b.** keeps bright sunlight out of your eyes, protecting your retina; **c.** helps wash dirt out of the eye, as well as keeping the membranes of the eye moist
6. Accept all reasonable responses. Most students will determine that an icicle does in fact have some characteristics of living things, such as growth, but they should realize there is a difference between the growth of an icicle and the growth of a living thing.
7. Reproduction is important to the species. Obviously, a living thing can survive without reproducing, but the entire species would die off if none of its members reproduced.
8. Accept all logical responses. Students should explain why they want the tool, not just tell which tool they select.
9. Accept all well-written, creative responses.

CHAPTER 3 Introduction to Chemistry

Section	Laboratory Investigations and Activities
3–1 Nature of Matter pages 45–47 THEMES: Scale and Structure, Unity and Diversity	**Student Edition** LABORATORY INVESTIGATION: Identifying Elements by Density, p. 58 **Teacher Edition** DEMONSTRATION: Observing Changes in Matter, p. 44D **Teaching Resources** HANDS-ON ACTIVITY: Hunting for Treasure in Trash, p. 29 HANDS-ON ACTIVITY: Dazzling Displays of Densities, p. 21 **Product Testing Activities** Testing Popcorn
3–2 Composition of Matter pages 47–51 THEMES: Scale and Structure, Unity and Diversity	**Teacher Edition** DEMONSTRATION: The Periodic Table, p. 44D **Laboratory Manual** Making Predictions Using Indirect Evidence, p. 77 **Product Testing Activities** Testing Bottled Water Testing Yogurts Testing Sports Drinks
3–3 Interactions of Matter pages 52–55 THEME: Systems and Interactions	**Teaching Resources** HANDS-ON ACTIVITY: Up in Smoke, p. 27
3–4 Chemical Reactions pages 55–57 THEME: Systems and Interactions	**Laboratory Manual** Physical and Chemical Changes, p. 81 **Teaching Resources** HANDS-ON ACTIVITY: Popcorn Hop, p. 25 **Product Testing Activities** Testing Antacids Testing Orange Juice
Chapter Review pages 59–61	

Teacher Resources

Books
Corwin, Charles H. *Chemistry: Concepts and Connections.* Prentice Hall, 1994.
Vemulapalli, G.K. *Physical Chemistry.* Prentice Hall, 1993.

Audiovisuals
Learning About Chemicals. 2 filmstrips with cassettes. National Geographic Society Educational Services.
Introduction to Chemical Bonding. Sound filmstrip. Ward.
Polar Covalence. Sound filmstrip. Ward.
Introduction to Atomic Structure. Sound filmstrip. Ward.

CHAPTER PLANNING GUIDE

Other Activities	Media and Technology
Teaching Resources ACTIVITY: Observing Phase Changes, p. 3 **Study Guide** Section 3–1, p. 21	**Video/Videodisc** Periodic Table and Periodicity **Biology Media Guide** page 3
Teaching Resources ACTIVITY: Counting Atoms, p. 5 **Study Guide** Section 3–2, p. 24	**Video/Videodisc** Periodic Table and Periodicity Elements, Compounds, and Mixtures **Biology Media Guide** pages 3–4
Teaching Resources ACTIVITY: Bonding and Chemical Formulas, p. 7 **Study Guide** Section 3–3, p. 26	**Video/Videodisc** Chemical Bonding and Atomic Structure **Biology Transparencies** 3: Chemical Bonding **Biology Media Guide** page 4
Teaching Resources VOCABULARY REVIEW: Cross-a-Clue, p. 1 **Study Guide** Section 3–4, p. 29	
Teaching Resources Chapter Test A, p. 13 Chapter Test B, p. 17	**Computer Test Bank** Chapter Test, p. 27

Basic Chemistry for the Biologist. 2 filmstrips with cassettes. Prentice Hall Media.
The Chemistry of Life. Video or filmstrip. Biology Media.

CHAPTER 3 Introduction to Chemistry

CHAPTER OVERVIEW

This chapter provides students with the basic concepts of chemistry that they will need for the successful study of biology. Students will be introduced to the nature of matter, the atomic structure of matter, chemical bonding, and chemical reactions.

Students will begin by learning about the basic properties of matter, including the distinction between physical and chemical properties. Students will then learn that the basic unit of matter is the atom. Atomic theory and structure will be discussed, including the concepts of atomic number and mass number.

Students will learn that matter composed of one kind of atom is called an element. Radioactive isotopes of elements will be described, with special emphasis placed on the way these isotopes are used in life science.

Students will discover that atoms of different elements combine to form chemical compounds. The process by which this occurs is known as chemical bonding. Students will learn how atoms gain, lose, or share electrons to form ionic or covalent bonds.

The final section of the chapter is devoted to chemical reactions. Students will be introduced to the use of chemical equations to describe chemical reactions. They will also learn how energy plays a role in determining whether or not a particular reaction will occur.

3-1 NATURE OF MATTER

Section Focus 3-1

The purpose of this section is to introduce students to the properties of matter. Students will learn that scientists define matter as anything that has mass and volume.

Students will learn that matter has both physical and chemical properties. Physical properties are those properties that can be observed and measured without changing the identity of the substance. Chemical properties are those properties that describe a substance's ability to change into another substance as a result of a chemical change.

Students will learn that matter can exist in three different states, or phases. These phases are solid, liquid, and gas. Students will learn that phase change is a physical change because the basic identity of the substance is not altered.

Performance Objectives 3-1

1. Define matter.
2. Identify and describe several important properties of matter.
3. Compare physical and chemical properties of matter.
4. Identify the three phases of matter and describe phase changes.

Science Terms 3-1

physical property   p. 45
chemical property   p. 46
phase   p. 46

3-2 COMPOSITION OF MATTER

Section Focus 3-2

The purpose of this section is to introduce students to the atom as the basic unit of matter. A brief introduction describes how Greek philosopher Democritus first proposed the idea of atoms 3000 years ago. Then students will discover what scientists know today about the structure of atoms.

Students will learn that an atom consists of protons, neutrons, and electrons. Protons and neutrons are contained in the nucleus of an atom. Electrons travel around the nucleus in a series of distinct energy levels, or orbits. Students will learn that the number of protons in an atom is called the atomic number, and that the total number of protons and neutrons in the nucleus is called the mass number.

Students will discover that some substances consist of only one type of atom. These substances are called elements. Students will learn that some elements have isotopes. Isotopes are atoms of the same element that have the same number of protons but different numbers of neutrons in the nucleus. Some of the scientific and medical uses of radioactive isotopes will be discussed.

In the last part of the section, students will learn that atoms of different elements combine to produce chemical compounds. Students will be introduced to the use of chemical symbols to represent elements and the use of chemical formulas to represent compounds.

Performance Objectives 3-2

1. Define an atom as the basic unit of matter.
2. Describe the structure of an atom.
3. Define atomic number and mass number.
4. Describe and compare chemical elements and compounds.
5. Define isotope and discuss uses of radioactive isotopes.

Science Terms 3-2

atom   p. 47
nucleus   p. 48
proton   p. 48
electron   p. 48
energy level   p. 48
atomic number   p. 48
mass number   p. 49
element   p. 49
isotope   p. 50
compound   p. 51

CHAPTER PREVIEW

3-3 INTERACTIONS OF MATTER

Section Focus 3-3

The purpose of this section is to explain how atoms of elements combine to form compounds. Students will learn that the process by which atoms combine is called chemical bonding.

Students will discover that the way an atom bonds is determined by the number of electrons found in its outermost energy level. Atoms seek to fill their outermost energy levels by gaining, losing, or sharing electrons. When atoms are transferred, that is, gained or lost from one atom to another, an ionic bond is formed. When atoms share electrons, a covalent bond is formed.

Students will learn that if an atom has a full set of electrons in its outermost orbit, it is unreactive—it will not combine with other atoms to form compounds.

Performance Objective 3-3

1. Define chemical bonding.
2. Explain how electron arrangement determines the way an atom bonds.
3. Explain why some atoms are unreactive.
4. Describe and compare ionic and covalent bonds.

Science Terms 3-3

chemical bonding   p. 52
ionic bond   p. 53
ion   p. 53
covalent bond   p. 53
molecule   p. 54

3-4 CHEMICAL REACTIONS

Section Focus 3-4

The purpose of this section is to discuss the processes that cause chemical change. Any process in which a chemical change takes place is called a chemical reaction.

Students will learn that chemical reactions occur all around them every day. They will discover that two kinds of substances are always present in chemical reactions—the substances that existed before the change and the new substances formed by the change. Students will learn that the first set of substances is called reactants and the second set is called products.

Students will learn that chemical reactions can be represented visually by chemical equations. They will also learn about the role that energy plays in chemical reactions. They will discover that reactions that release energy will occur spontaneously, whereas those that require energy will not occur without a source of energy.

Performance Objectives 3-4

1. Identify the reactants and products in a chemical reaction.
2. Demonstrate how chemical reactions can be represented visually by chemical equations.
3. Explain how the flow of energy determines whether a chemical reaction will occur.

Science Terms 3-4

chemical reaction   p. 54

DISCOVERY LEARNING

TEACHER DEMONSTRATIONS
Modeling

Observing Changes in Matter

The following demonstration can be used as an introduction to Chapter 3.

In this demonstration students will observe the burning of a match. For the demonstration you will need several long kitchen matches.

Display a kitchen match to the class and ask students,

- **What is this match made of?** (Wood, plus some kind of burning material at the tip.)
- **Would you say that this match is solid, liquid, or gas?** (Solid.)

Now strike the match and have students observe it as it burns.

- **What do you see?** (Fire, flame, smoke.) Extinguish the match.
- **Does the match look the same as it did before? How has it changed?** (No. It is shorter, the wood is charred, and the tip has either blackened or fallen off.)
- **What do you think happened to the wood of the match?** (Answers may vary. The correct answer is that it changed into carbon dioxide and water, with some of the charred wood remaining as carbon.)
- **Was energy involved in the change that happened to the match? If so, what kind?** (Yes. The initial energy was the striking of the match, then heat and light were produced.)
- **Can we get the match back to the way it was before?** (No.)

Ask students to keep this demonstration in mind as they study this chapter. Then, after they have finished the chapter, have them use what they have learned to describe more accurately what happened to the match.

The Periodic Table

This demonstration can be performed as students study Section 3–2.

Take this opportunity to familiarize students with the periodic table of the elements. Display a wall-sized copy of the periodic table. If possible, have desk-sized photocopies made to distribute to each student.

Emphasize to students that the periodic table shows all of the elements known to scientists. Point out that each element is represented by a chemical symbol. Explain that some of these symbols are abbreviations for the names of elements as we know them, such as C for carbon, O for oxygen, and H for hydrogen. Many other symbols, however, derive from the Latin names for the elements. For example, gold is abbreviated Au, the symbol for lead is Pb, and the symbol for silver is Ag.

Show students how to use the periodic table to find the atomic number and mass number of each element. Then say to students,

- **Note the way the elements are arranged in columns. Why are the elements placed in columns?** (Elements in the same column have similar properties. Usually, the elements in a column also have the same number of electrons in the outermost energy level.)

CHAPTER 3

Introduction to Chemistry

GUIDED ENQUIRY

Pose the following questions to students and have them record their responses. Point out that students will gain a better understanding of the key concepts if they read the chapter with these basic questions in mind. Upon completion of the chapter, pose the questions again. Ask students to compare their initial responses with those they have developed after the chapter.

- What is the nature of matter?
- How does matter change from one substance into another?
- Why are atoms important in the study of matter?
- How do atoms combine with other atoms?
- What causes chemical changes to take place?
- Why do some chemical reactions occur spontaneously?

INTRODUCING CHAPTER 3

Using the Visuals

Begin the chapter by having students observe the chapter-opener photographs.
- **What do you see?** (Most students will observe the butterfly on the flower. Some students will realize that the inset photograph is a computer-generated image of a molecule.)

Point out that the "bug" is a nymphalid butterfly and the flower is a zinnia. The computer image is of a molecule of a body fluid—in this case, pus.
- **Why do you think an image of an animal and a plant is placed beside a molecule?** (To reinforce the fact that all living things are made up of chemical compounds.)

Now have students read the chapter-opener introductory matter. Stress the concept that on some levels the study of chemistry and the study of biology interact, and that to truly understand living things and their complex systems we must have a basic understanding of chemical properties and the nature of the atom. This will likely curtail the typical student response to an introductory chapter in a biology text, "Why do we have to study chemistry?"

CHAPTER 3

Introduction to Chemistry

An understanding of life—in all its glorious forms—depends upon an understanding of the atomic nature of matter.

For many years, biology was a science that respected a fundamental distinction between the living and the nonliving worlds. Biologists studied organisms; physicists and chemists studied materials. In the twentieth century, however, this distinction has begun to fade. Biologists have gradually realized that a complete description of life must begin at the atomic level and then work up through the complex chemistry of the living cell. Indeed, one of the great scientific stories of this century is the manner in which the boundaries between biology and the physical sciences have become blurred. Biologists now appreciate that the starting point for an understanding of the nature of life is an understanding of the atom. In this chapter, you will do exactly as biologists have done— you will learn about the atomic structure of living and nonliving matter.

GUIDE FOR READING

After you read the following sections, you will be able to

3–1 Nature of Matter
- Identify several important properties of matter.
- Compare physical and chemical properties.

3–2 Composition of Matter
- Describe the structure of an atom.
- Compare elements and compounds.
- Discuss the uses of radioactive isotopes.

3–3 Interactions of Matter
- Explain how electron arrangement determines an atom's reactivity.
- Describe the two types of chemical bonds.

3–4 Chemical Reactions
- Identify the substances involved in a chemical reaction.
- Describe the role of energy in chemical reactions.

Journal Activity

YOU AND YOUR WORLD

Although you may not realize it, there are chemical reactions going on all around you. They are even occurring inside your body right now. In your journal, write a letter to a friend who wants to know why he or she has to study chemistry in a biology course.

3–1 Nature of Matter

Guide For Reading

■ What are some important properties of matter?
■ What are the phases of matter?

What are living things made of? Are there special substances that are found in living things but not in nonliving material? Is there a special "spirit" or "essence" that living things possess? Does life have a physical and chemical basis that we can hope to understand and describe in the same way we do something that is not alive, like an automobile engine or a calculator? To answer these questions, we must first examine the world around us . . . a world made up of matter and energy.

Properties of Matter

Matter is all around us. Almost everything we see, touch, taste, or smell is matter. And all forms of matter have properties, or characteristics, by which they are identified.

Certain properties of matter are **physical properties**. Two very important physical properties are used by scientists to define matter: Matter is anything that has mass and volume. Mass is the quantity of matter in an object. Volume is the amount of space matter takes up.

Mass is related to another important property of matter: weight. An object has weight because it has mass, but mass and weight are not the same thing. Weight is a measure of the force of gravity on an object. The greater the mass of an object, the greater the force of gravity on it, and the greater its weight. Thus we can say that the weight of an object is directly proportional to its mass.

Figure 3–1 The physical properties of matter are often all you need to observe in order to tell substances apart. Because of their physical properties, the rose and the cactus are easily distinguishable.

COOPERATIVE LEARNING

Using preassigned lab groups or randomly selected teams, have groups prepare clues that describe the physical properties of assigned objects. Each group's clues should appeal to the five senses, identify the object's phase of matter, and include references that compare the mass, weight, and volume of the mystery item to known objects. Clues prepared by each group should be used by other groups to guess the identity of the "mystery matter." Items you might want to use include a cube of ice, watermelon, anchovies, avocado, kiwi fruit, toothbrush, refrigerator, and garbage truck. Encourage groups to provide accurate clues that reference all five senses; describe physical properties, not functions; and are not so obvious that they allow for easy identification. You might want to entitle this activity, "What's the Matter?"

Journal Activity

YOU AND YOUR WORLD

Students might mention the chemical reactions involved in burning fossil fuels to heat a school, or the reactions that are occurring as food is digested in the body. You may want to use the Journal Activity as a springboard for class discussion. Students should be instructed to keep their Journal Activity in their portfolio.

Content Development

Continue the Focus/Motivation discussion by explaining to students that both objects represent different types of matter. Like all matter, both objects have volume and take up space. Both objects also have certain other characteristics such as shape, color, taste, texture, and hardness. These two objects are also similar in that both are made from material that is or once was living.

TEACHING STRATEGY 3–1

Focus/Motivation

Display to the class two dissimilar objects such as an apple and a wooden block.

• **What do these two objects have in common?** (Answers may vary. Possible answers: Both are solids; both are small enough to hold in your hand; both have weight; both take up space; both are made from living or once-living material.)

• **How are these two objects different?** (Possible answers: One is very hard and the other is soft enough to bite into; one is red in color and the other is brown; one is round and the other is rectangular; one is a food and the other is not.)

45

ANNOTATION KEY

❶ Color, texture, hardness. Yes, for example, they could have similar shapes. (Inferring)

❷ Nut and bolt: rusting or slow oxidation; fireworks: combustion; grilling: combustion or fast oxidation plus decomposition. (Relating)

TEACHING SUPPORT

Teaching Resources
- Activity: Observing Phase Changes, p. 3
- Hands-On Activity: Hunting for Treasure in Trash, p. 29
- Hands-On Activity: Dazzling Displays of Densities, p. 21

Study Guide
- Section 3–1, p. 21

Product Testing Activities
- Testing Popcorn

MEDIA AND TECHNOLOGY

Video/Videodisc
- *Periodic Table and Periodicity* Use this bar code to introduce chemistry.

Play frames 1853 to 3100

See the Biology Media Guide, page 3 for additional bar-code correlations for this section.

3–1 (continued)

Skills Development

Guided Practice

Skill: Applying concepts

Write on the chalkboard a list of examples of changes in matter. Then ask students to state whether each change is physical or chemical and why. Some possible examples include

1. painting an object bright red (physical)
2. silver tarnishing (chemical)
3. food digesting in the human body (chemical)
4. gasoline burning to power an automobile (chemical)
5. chopping up vegetables to make a salad (physical)

SECTION REVIEW 3–1

1. Mass is the amount of matter in an object; weight is a measure of the pull of gravity on an object; volume is the amount of space an object takes up.
2. A physical property is one that can be observed or measured without changing the substance, whereas a chemical property is a property that describes a substance's ability to change. A physical change is a change that does not alter a substance's basic composition, whereas a chemical change results in new substances.
3. Physical properties include size, color, shape, texture,

Figure 3–2 What physical properties distinguish coal from sugar? Do these two substances have any physical properties in common? ❶

Figure 3–3 Chemical properties describe how a substance changes into other new substances as a result of a chemical change. What chemical changes can you identify in these photographs of a rusty nut and bolt (left), a fireworks display (center), and a delicious meal grilling on the barbecue (right)? ❷

Matter has other physical properties. These include color, odor, shape, texture, taste, and hardness. You know from experience that a lump of coal has a certain color, texture, and hardness. A lump of sugar has a different color, texture, and hardness. It even has a special taste. Two other important physical properties are melting point and boiling point. **Physical properties of matter can be observed and measured without permanently changing the identity of the matter.**

Matter has **chemical properties** also. **Chemical properties describe a substance's ability to change into another new substance as a result of a chemical change.** A chemical change is a process in which a substance is permanently altered. When a chemical change is completed, it is difficult, if not impossible, to reverse the process and get the starting material back. When coal burns, it undergoes a chemical change with oxygen to form other gases. Once a lump of coal has burned, it is no longer coal. And the gases produced cannot be converted back to coal. The chemical properties of a substance are determined by learning what sorts of chemical reactions the substance can undergo.

Phases of Matter

Ice, liquid water, and water vapor may seem very different to you. Certainly they have different appearances and uses. But actually they are all made of exactly the same substance in different states. These states are called **phases**. Phase is an important physical property of matter. Ice is the solid phase of water. Liquid water is the liquid phase. And water vapor is the gas phase.

The change from one phase of matter to another is a physical change because the substance is not altered. Water, regardless of its phase, is still water. Physical and chemical changes are common in the world around us. And because living things, including ourselves, are made of matter, an understanding of these physical and chemical changes is important to biology.

3-1 SECTION REVIEW

1. Define the following terms: mass, weight, volume.
2. Compare a physical property and a chemical property. A physical change and a chemical change.
3. **Critical Thinking—Applying Concepts** Using a piece of paper as an example, distinguish between the physical and chemical properties of matter.

Figure 3–4 Phase is an important physical property of matter. Here you see water in its three phases: solid, liquid, and gas.

3-2 Composition of Matter

Nearly 3000 years ago, the Greek philosopher Democritus challenged his students with a series of questions on the nature of matter. Democritus took a piece of salt, divided it into two fragments, and asked his students if each half was still salt. They all agreed that both halves were salt. He divided the halves again. "Still salt," his students said. Democritus then posed some difficult questions. Could the salt continue to be divided into smaller and smaller fragments forever? Was there a point at which the salt fragments could no longer be divided and still be called salt? The students were uncertain. Democritus, however, argued that all forms of matter were made up of basic, indivisible particles, which he called **atoms**. The word atom comes from the Greek word *atomos*, meaning unable to be cut.

The approach of Democritus was philosophical rather than experimental. In part, this means that he was content to think about the problem but go no further. Scientists, however, must go further. They will speculate much as Democritus did, but then they will think of a way to test their ideas in an experiment. In the last 200 years, scientists have carefully studied Democritus' concept of the atomic nature of matter and have concluded that he was basically right. We now know that matter is indeed made up of small particles—not because it makes philosophical sense but because the evidence proves it. But we also now know that the atom itself is divisible and that particles smaller than the atom do exist!

The Atom

The basic unit of matter is the atom. Atoms are very, very small. Indeed, 100 million atoms placed side by side would form a row only 1 centimeter long—about the width of your pinky! Despite its extremely small size, the atom contains many

Guide For Reading

- What is the structure of an atom?
- How do elements and compounds compare?
- How are radioactive isotopes used?

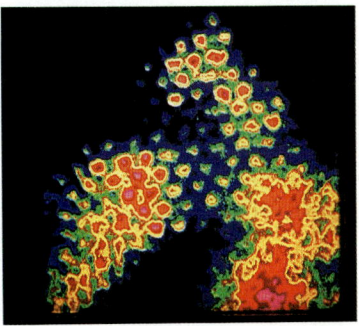

Figure 3–5 This false-color photograph of uranium atoms was taken with a scanning transmission electron microscope (STEM). Each colored dot represents a single uranium atom.

BACKGROUND INFORMATION
DENSITY

Density is a useful property when you want to distinguish one substance from another. For example, the metal tin is relatively "light" (low density), whereas the metal gold is relatively "heavy" (high density). If a piece of tin jewelry is coated to look like gold, the fraud will be uncovered if the density of the jewelry is measured.

The following formula can be used to find the density of an object or sample of matter:

$$\text{density} = \frac{\text{mass}}{\text{volume}}$$

Mass is usually expressed in grams; volume, in milliliters or cubic centimeters. So density is expressed in grams per milliliter (g/mL) or grams per cubic centimeter (g/cm^3).

to predict the chemical properties of each substance.

TEACHING STRATEGY 3-2

Focus/Motivation
Break a piece of chalk in half.
- **Is each piece still chalk?** (Yes.)
  Break one of the pieces in half again and yet again if possible.
- **Is each piece still chalk?** (Yes.)
- **Suppose I could continue breaking the chalk in half hundreds, even thousands, of times. Would each piece still be chalk?** (Yes.)
- **Do you think there would come a point where the chalk couldn't be divided any more and still be chalk?** (Accept all answers.)

---

whether the paper is lined, and so on. Chemical properties include its chemical composition, its flammability, and so on.

Reinforcement/Reteaching
For students who have trouble answering the Section Review questions, go back to the part of the section that contains the relevant material.

Closure
Give students about 5 minutes to look around the classroom. Ask them to observe as many different kinds of matter as they can. Have them write down their observations. Then conduct a class discussion in which students share their observations and state the physical properties of the matter they have observed. Also challenge the class

smaller particles, known as subatomic particles. Scientists know about the existence of at least 200 different kinds of subatomic particles. The three principal subatomic particles are the proton, neutron, and electron.

ATOMIC STRUCTURE The center of the atom is called the **nucleus**. Although the nucleus is about a hundred thousand times smaller than the entire atom, it makes up 99.9 percent of the mass of the atom. The nucleus contains two different kinds of subatomic particles held together by special atomic forces. One of these particles is the **proton**. The proton is a positively charged particle. The other particle found in the nucleus is the **neutron**. The neutron is an electrically neutral particle. It has no charge at all. The proton and neutron are nearly equal in mass (1 atomic mass unit, or amu).

In addition to protons and neutrons, the atom contains another kind of subatomic particle called the **electron**. The electron is a negatively charged particle with a mass about 2000 times less than that of either the proton or neutron (1/1836 amu). Under normal circumstances, the number of negatively charged electrons in an atom is equal to the number of positively charged protons. Therefore, the atom as a whole is neutral; it is neither negatively nor positively charged.

Electrons are not found in the nucleus. They travel at high speeds throughout the atom in a series of distinct **energy levels** that surround the nucleus. The existence of electrons in distinct energy levels is extremely important in determining the chemical properties of an atom, as you will learn shortly.

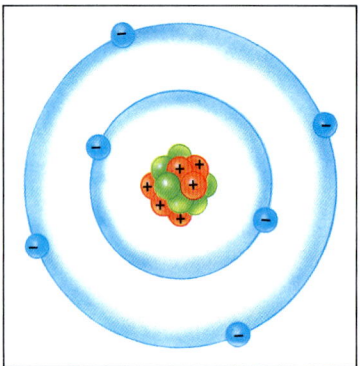

Figure 3–6 This artist's conception of an atom shows the nucleus, with its protons and neutrons, and the series of energy levels, with their electrons, that surround it (top). The illustration shows an atom of the element carbon (bottom).

ATOMIC NUMBER AND MASS NUMBER The number of protons in the nucleus of an atom is called the **atomic number**. The atomic number identifies an atom. Atoms of the same substance have the same atomic number. But atoms of different substances have different atomic numbers. Thus the atomic number of an atom of a substance is a unique quantity. An atom of hydrogen, which has only 1 proton, has an atomic number of 1. All hydrogen atoms—but only hydrogen atoms—have an atomic number of 1. Helium, with 2 protons in the nucleus of

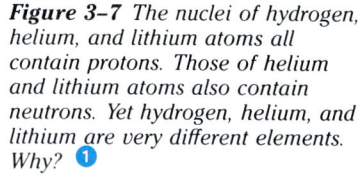

Figure 3–7 The nuclei of hydrogen, helium, and lithium atoms all contain protons. Those of helium and lithium atoms also contain neutrons. Yet hydrogen, helium, and lithium are very different elements. Why?

atomic number—is unique for each type of atom. Point out that no two elements can ever have the same atomic number.

Explain how the mass number of an atom differs from the atomic number. Point out that the mass number is equal to the number of protons plus the number of neutrons.

• **If a carbon atom has 6 protons and 6 neutrons in the nucleus, what is the mass number of carbon?** (12.)

• **What is the atomic number?** (6.)

every atom, has an atomic number of 2. Uranium has an atomic number of 92. How many protons are in a uranium atom? ❷

Although all subatomic particles contribute to the mass of an atom, protons and neutrons are much more massive than electrons. This explains why most of the mass of an atom is concentrated in the nucleus. Scientists often refer to the total number of protons and neutrons in the nucleus as the **mass number**. Hydrogen has a mass number of 1 because the only particle in the nucleus of a hydrogen atom is a proton. A helium atom has 2 protons and 2 neutrons in the nucleus. So the mass number of helium is 4.

Chemical Elements

As you read before, every substance in the world is made up of atoms. **Some substances, known as elements, consist entirely of one type of atom.** At present, scientists have identified 109 different **elements**. Of these 109 elements, 90 types are found in nature. The remaining 19 have been artificially produced by physicists using very special laboratory equipment. Each element can be represented by a chemical symbol. A chemical symbol is a shorthand way of representing an element. Each symbol consists of one or two letters, usually taken from the element's name. For example, the symbol for the element iodine is I. The symbol for hydrogen is H; for carbon, C. The Latin name of an element was often used to create its symbol. For example, the symbol for lead is Pb, from the Latin word *plumbus*. You may recognize that word as the source for the English word plumbing, reflecting the fact that water pipes were once made of lead. Figure 3–8 lists some common elements and their symbols.

Most of the chemical elements are solids. Common examples include carbon, sulfur, sodium, calcium, and potassium. Some elements—such as oxygen, nitrogen, and chlorine—are gases. Only a few elements are liquids at room temperature. Mercury and bromine are the most common examples.

ISOTOPES As you have read, each element has an atomic number that is determined by the number of protons in the nucleus of its atoms. The atomic number of an element never changes. This means that the number of protons in the nucleus of every atom of the element is always the same. However, the number of neutrons can vary from one atom of the element to the next. For example, the most common form of hydrogen has 1 proton and no neutrons in its nucleus. Another form of hydrogen, sometimes known as deuterium, has 1 proton and 1 neutron in its nucleus. And a third form, tritium, has 1 proton and 2 neutrons in its nucleus. Although they have different mass numbers, each form of hydrogen has an atomic number of 1, a single proton in the nucleus, and a single electron. Each form is hydrogen. Atoms of the same element that have the

ELEMENTS IN THE HUMAN BODY

Element	Symbol	Mass (%)
Oxygen	O	65.0
Carbon	C	18.5
Hydrogen	H	9.5
Nitrogen	N	3.3
Phosphorus	P	1.0
Sulfur	S	0.3
Sodium	Na	0.2
Magnesium	Mg	0.1
Silicon	Si	trace
Fluorine	F	trace

Figure 3–8 This chart shows some of the elements in the human body, their symbol, and their percentage by mass. Which element makes up the greatest percentage? Which elements are found in only trace quantities? ❸

HISTORICAL NOTE
THOMSON, RUTHERFORD, AND BOHR

The first inkling that scientists had of the existence of subatomic particles came in 1897. English scientist J.J. Thomson proposed that if negatively charged particles exist within the atom, then positively charged particles must also exist to balance the negative charge. Thomson pictured the atom as similar to a "pudding" in which negatively charged particles were scattered evenly throughout positively charged material.

In 1908, another British scientist, Ernest Rutherford, decided to test Thomson's idea. He fired a stream of positively charged particles at a thin sheet of gold foil. The pattern of deflection that he observed showed him that positively charged material is not spread evenly throughout an atom but is concentrated in one very small area. Rutherford proposed correctly that an atom's positive charge is concentrated in a small, dense center, which he called the nucleus.

Thanks to Rutherford, scientists knew that positively charged particles were in the nucleus of an atom and negatively charged particles were outside the nucleus. Their next task was to discover just where the electrons were located. In 1913, Danish scientist Niels Bohr proposed a model of the atom that placed each electron in a specific energy level, or orbit, around the nucleus. Bohr pictured the interior of an atom as resembling the orbits of the planets around the sun. Bohr's model was close to being correct, but modern scientists now know that electrons do not move in paths that are as definite as planetary orbits. Scientists can only predict where an electron is most likely to be found. The probable location of an electron is based on how much energy the electron has.

---

following:
- **If an atom has 8 protons in its nucleus, how many electrons must it have?** (8.)
- **If an atom has 8 protons in its nucleus, how many neutrons must it have?** (Cannot tell.)
- **Suppose I tell you that the mass number of the atom is 16. Now can you tell the number of neutrons?** (Yes, 8.)
- **An atom has 11 electrons and a mass number of 23. How many neutrons does it have? What is its atomic number?** (12; 11.)

Content Development
Emphasize that in isotopes, only the number of neutrons is different, not the number of protons or electrons.

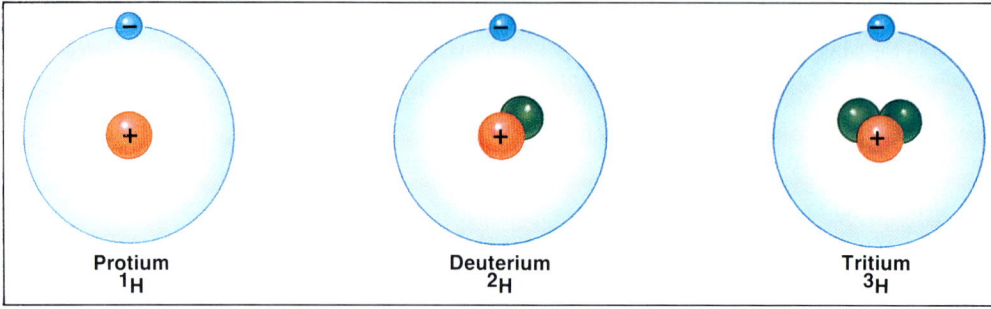

Figure 3–9 *The three isotopes of hydrogen are protium, deuterium, and tritium. Each isotope is still hydrogen, because it has a single proton and a single electron. Which isotope contains 2 neutrons? What is the atomic number of each isotope?*

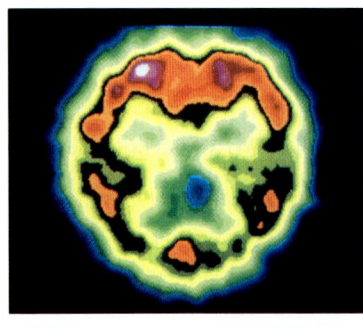

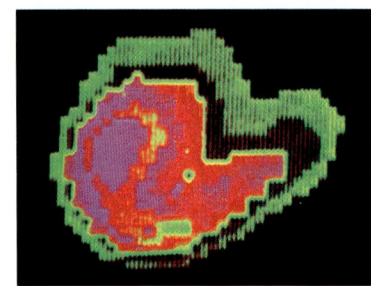

Figure 3–10 *Radioactive isotopes can be used as tracer elements to produce images such as this axial section through a normal brain (top) and this image of the heart (bottom).*

same number of protons but different numbers of neutrons are known as **isotopes** of that element.

Isotopes of an element are represented by adding the number that indicates the mass number of that isotope to the atomic symbol. Thus, ordinary hydrogen is ^1H, deuterium is ^2H, and tritium is ^3H.

RADIOACTIVE ISOTOPES The nuclei of some atoms are unstable and will from time to time break down, releasing matter and/or energy that we call radiation. Atoms that emit radiation are said to be radioactive. Many elements have at least one radioactive isotope. All the isotopes of elements with atomic numbers greater than 83 are radioactive.

Radioactive isotopes have many practical uses. They are used to study living organisms, to diagnose and treat diseases, to sterilize foods, and to measure the ages of certain rocks.

Radioactive isotopes are frequently used as tracers. A tracer is a radioactive element whose pathway through the steps of a chemical reaction can be followed. An example of a tracer is phosphorus-32.

The nonradioactive element phosphorus is used in small amounts by both plants and animals. If phosphorus-32 is given to an organism, the organism will use the radioactive phosphorus just as it does the nonradioactive phosphorus. However, the path of the radioactive element can be traced, allowing scientists to learn how plants and animals use phosphorus.

Tracers are extremely valuable in diagnosing diseases. Radioactive iodine, iodine-131, can be used to study the function of the thyroid gland, which absorbs iodine. Sodium-24 can be used to detect diseases of the circulatory system. Iron-59 can be used to study blood circulation.

Radioactive isotopes are also used to treat certain diseases. When administered carefully and in the proper amounts, radiation can kill cancer cells with minimal damage to healthy tissue. Cobalt-60 is used extensively in cancer radiation treatments. Carbon-14 has been used to treat brain tumors.

Radioactive isotopes can also be used to kill bacteria that cause food to spoil. Radiation is used to preserve the food that astronauts eat while on the moon and in orbit.

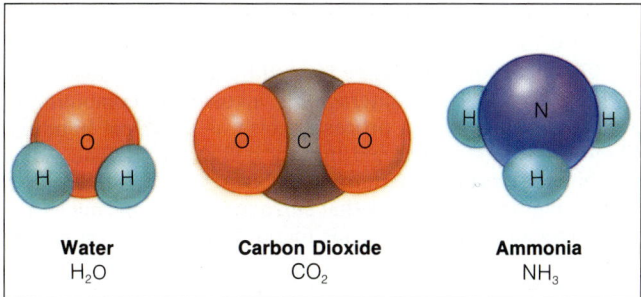

Figure 3–11 As you can see from this diagram, compounds are made of two or more different atoms combined in definite proportions. The chemical formula for each compound indicates the elements that make up the compound and the number of atoms of each element.

Another use of radioactive isotopes is in measuring the ages of certain rocks and the fossils they may contain. This use has increased our understanding of the evolution of life on Earth.

One of the difficulties in using radioactive materials is that these substances must be handled with great care. Radiation can damage or kill living things. So it is important that proper safety precautions be taken whenever radioactive isotopes are used. And it is also important that you be aware of and concerned about radioactive materials in the environment.

Chemical Compounds

When elements combine to form substances consisting of two or more different atoms, chemical compounds are produced. A chemical **compound** involves the combination of two or more different atoms in definite proportions. Most materials in the living world are compounds.

Just as elements are represented by chemical symbols, so too are compounds represented by a kind of shorthand. A chemical compound is represented by a chemical formula. A chemical formula consists of the chemical symbols for the elements that make up the compound. Water, which contains 2 atoms of hydrogen and 1 atom of oxygen, has the chemical formula H_2O. Table salt, made of sodium and chlorine, has the formula NaCl. What elements are present in the compound sulfuric acid, whose chemical formula is H_2SO_4? ❷

3–2 SECTION REVIEW

1. Describe the structure of an atom. What is meant by atomic number? Mass number?
2. What is an element? How are elements represented?
3. What is an isotope? What are some uses of radioactive isotopes?
4. **Critical Thinking—Relating Concepts** How is a compound different from an element?

TIE-IN MATH

A chemical formula describes the ratio of different types of atoms that make up the molecule. For example, the formula for carbon dioxide is CO_2. This means that in every carbon dioxide molecule, the ratio of carbon atoms to oxygen atoms is 1:2. In a molecule of sugar, $C_6H_{12}O_6$, the ratio of carbon atoms to hydrogen atoms to oxygen atoms is 6:12:6 or 1:2:1.

BACKGROUND INFORMATION
INERT GASES

Elements that have atoms with filled outermost energy levels are helium, neon, argon, krypton, xenon, and radon. All are found in Group 8A (IUPAC 18) of the periodic table. These elements, which are all gases, are often called the inert gases or noble gases because they do not normally combine with other elements.

MULTICULTURAL STRATEGY

During the seventeenth and eighteenth centuries, Central and South America were leading sites for metallurgy and mining research. Have students investigate the work of Juan D'Elhuyar, Fausto D'Elhuyar, and Andres Manuel del Rio.

TEACHING STRATEGY: 3-3

Focus/Motivation

At the front of the classroom, arrange 2 circles of 8 chairs each. Have the circles at least 1 meter apart so that they are distinctly separate from each other. Ask for 7 volunteers to take seats in the first circle of chairs. Then ask for 9 volunteers to take seats in the second circle of chairs. As 1 student is left without a chair, ask this person to stand outside the circle and walk around the edge, as if looking for a place to sit.

- **How would you describe the situation that exists in these 2 circles?** (Possible answer: The first circle has 1 empty chair; the second circle has all seats filled plus 1 extra person looking for a place to sit.)
- **How might the extra person's problem be solved?** (He or she could go and sit in the extra chair in the first circle.)

Content Development

Explain to students that the outermost energy level in an atom is somewhat like a circle of 8 chairs, and the electrons are like the people sitting in the chairs. An atom has room in its outermost energy level for 8 electrons, and it is most stable when all 8 places are filled with no electrons left over. In most atoms, however, 1 or more "chairs" is empty, or else

Guide For Reading
- How is an atom's reactivity determined?
- What are ionic bonds?
- What are covalent bonds?

3–3 Interactions of Matter

Chemical compounds are formed by the interactions of individual atoms. These interactions involve the combining of atoms of elements in a process known as **chemical bonding**. Chemical bonds are formed in very definite ways. The atoms combine according to certain rules. Such rules are determined by the number of electrons that surround the atomic nucleus—more specifically, the electrons found in the outermost energy level.

Each energy level in an atom can hold only a certain number of electrons. The first, or innermost, energy level can hold only 2 electrons. The second energy level can hold 8 electrons, as can the third when it is an outermost energy level. Figure 3–12 illustrates the arrangement of electrons in the first three energy levels.

When the outermost energy level of an atom contains the maximum number of electrons, the level is full, or complete. Atoms that have filled outermost energy levels are very stable, or unreactive. Such atoms usually do not combine with other atoms to form compounds. They do not form chemical bonds.

In order to achieve stability, an atom will either gain, lose, or share electrons. In other words, an atom will bond with another atom if the bonding gives both atoms complete outermost energy levels.

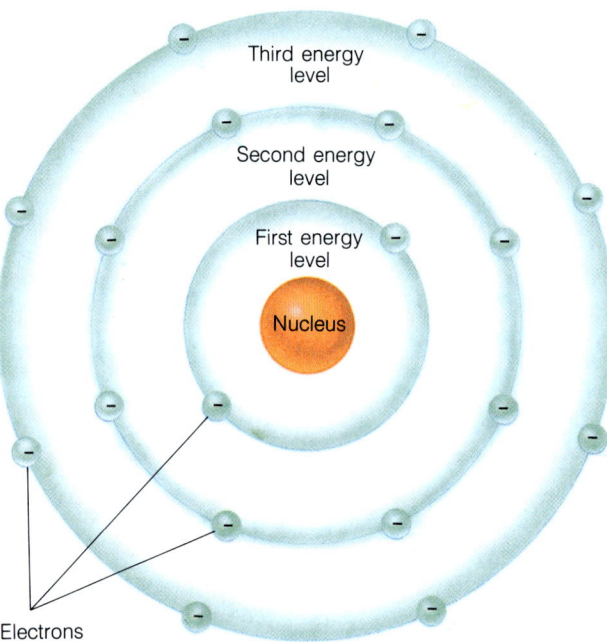

Figure 3–12 Each energy level in an atom can hold only a certain number of electrons. When the outermost energy level is holding its maximum number of electrons, it is complete, and the atom is stable.

Ionic Bonds

One way a complete outermost energy level can be achieved is by the transfer of electrons from one atom to another. A bond that involves a transfer of electrons is called an **ionic bond**. An ionic bond, or electron-transfer bond, gets its name from the word **ion**, which means charged particle. Ions are formed when an ionic bond occurs.

A sodium atom has only 1 electron in its outermost energy level, so it can lose that electron very easily. The loss of an electron from a sodium atom produces a sodium ion (Na^+), which has a positive charge. Similarly, chlorine needs only 1 electron to fill its outermost energy level, so it grabs an extra electron easily—possibly from a nearby sodium atom. The addition of an electron to a chlorine atom produces a negatively charged chloride ion (Cl^-). These two ions, now oppositely charged, have a strong attraction for each other—much like opposite poles of a magnet. The strong attraction between oppositely charged ions that have been formed by the transfer of electrons holds the ions together in an ionic bond.

Covalent Bonds

A chemical bond formed by the sharing of electrons is known as a **covalent bond**. By sharing electrons, each atom

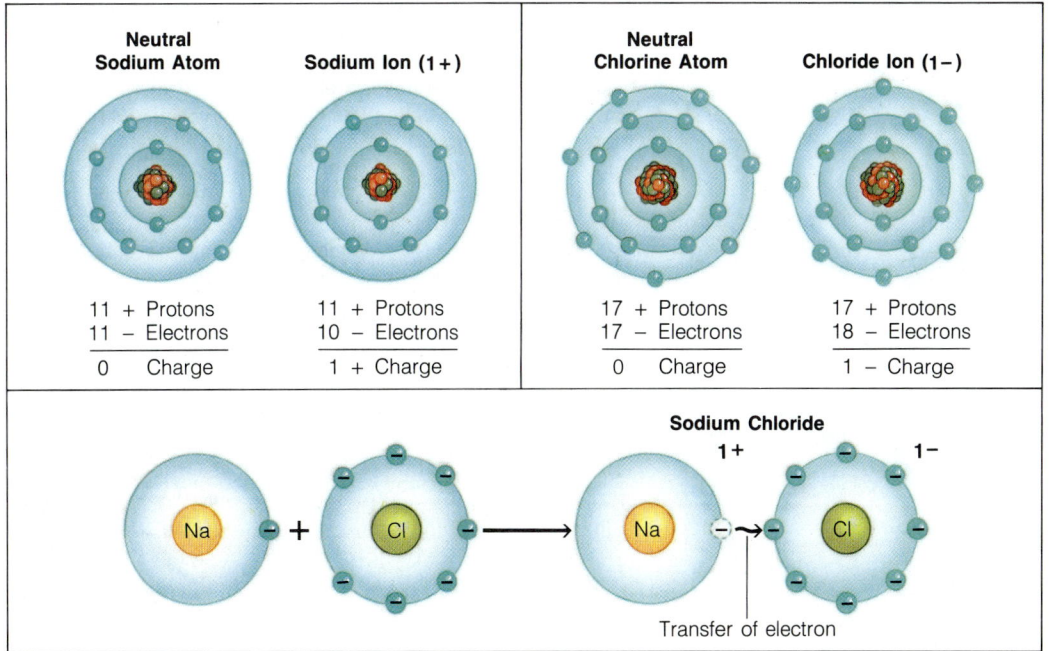

Figure 3–13 The formation of a positive sodium ion involves the loss of an electron by a sodium atom (top left). The formation of a negative chloride ion involves the gain of an electron by a chlorine atom (top right). The transfer of an electron produces oppositely charged ions that are strongly attracted and held together in an ionic bond (bottom). Because only the electrons in the outermost energy level are involved in bonding, only this level has been shown in the bottom diagram.

BACKGROUND INFORMATION
VALENCE NUMBERS

Atoms do not combine randomly in just any proportion. Every bond is determined by the number of electrons in the outermost energy level, and that number is constant for each element. As a result, each element has its characteristic bonding capacity, or "valence." The valence number indicates how many bonds an atom can make. For example, hydrogen has 1 outer electron, so its valence is 1. Oxygen lacks 2 outer electrons, so its valence is 2.

Elements with the same valence can be found in the same vertical column, or group, of the periodic table. For example, Group 1A (IUPAC 1) elements are all metals that have 1 electron in their outermost energy level. Each of these elements, then, has a valence of 1. Elements in Group 2A (IUPAC 2) all have 2 electrons in their outermost energy level, so the valence of these elements is 2.

HISTORICAL NOTE
ANCIENT GREEKS

Some ancient Greeks believed that all matter was composed of four basic elements: earth, fire, air, and water. Various substances were combinations of these four elements. For example, they believed that steam could be created by combining fire and water.

---

there are 1 or more extra "people" (electrons) outside the circle of 8. What happens is that an atom with 1 extra electron looks for an atom with 1 empty place in its outermost energy level. An atom with 2 extra electrons looks for an atom with 2 empty places—or for 2 atoms with 1 empty place each.

Content Development

It is important for students to understand that ionic bonds can involve the transfer of more than 1 electron, and they can also involve more than 2 atoms. For example, an atom of calcium has 2 electrons in its outermost energy level. It can transfer both electrons to 1 atom of oxygen, which has space for 2 electrons in its outermost energy level. Or, it can transfer 1 electron each to 2 atoms of chlorine, each of which has space for 1 electron in its outermost energy level. Point out that the formula for the first compound would be CaO, whereas the formula for the second compound would be $CaCl_2$. The first formula represents an ionic bond between 2 atoms, whereas the second represents 2 ionic bonds among 3 atoms.

ANNOTATION KEY

❶ Two dashes. (Inferring)
❷ 2. (Interpreting formulas)

ECOLOGY NOTE
CARBON DIOXIDE

As is well known, carbon dioxide contributes to the greenhouse effect. What is less known, however, is that it may also produce changes in wildlife ecosystems by directly affecting plant development. According to recent research at the University of Michigan, an abundance of carbon dioxide increases the rate of photosynthesis, root development, and plant growth. Experiments demonstrate that seedlings photosynthesize at twice the normal rate and root growth increases by 50% if levels of carbon dioxide are doubled.

Even more significant is the fact that the increased plant growth causes an increase in the levels of carbon and nitrogen in the soil. This could directly influence an ecosystem.

Although such changes in an ecosystem may not seem harmful, they could eventually produce detrimental effects. For example, unwanted plants such as weeds might one day become difficult to control.

TEACHING SUPPORT

Teaching Resources
- Activity: Bonding and Chemical Formulas, p. 7
- Hands-On Activity: Up in Smoke, p. 27

Study Guide
- Section 3–3, p. 26

MEDIA AND TECHNOLOGY
📺 **Biology Transparencies**
- 3: Chemical Bonding

🎞 **Video/Videodisc**
- *Chemical Bonding and Atomic Structure*

See the Biology Media Guide page 4 for additional bar-code correlations.

Figure 3–14 The bonding in molecules can be represented by electron-dot formulas and by structural formulas. An electron-dot formula shows the electron pairs involved in the bonding. A single bond involves one pair of electrons; a double bond involves two. In a structural formula, a dash (–) is used to represent a pair of shared electrons, or a single bond. How is a double bond represented? ❶

Figure 3–15 A covalent bond involves the sharing of electrons. How many double bonds are there in a molecule of carbon dioxide? ❷

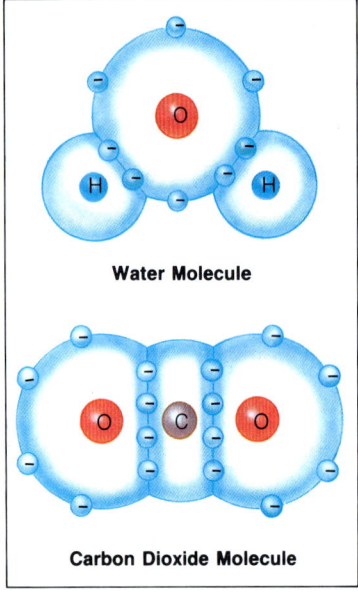

Water Molecule

Carbon Dioxide Molecule

---

fills up its outermost energy level. So the shared electrons are in the outermost energy level of both atoms at the same time. This produces a strong bond that is characteristic of most of the chemicals found in living organisms.

Covalent bonds can be single, double, or triple, depending on the number of electrons that are shared in the bond. The covalent bond between hydrogen and oxygen in water (H_2O) is a single bond. A single pair of electrons is shared between the oxygen atom and each of the 2 hydrogen atoms. In the compound carbon dioxide (CO_2), the carbon atom shares two pairs of electrons, or a total of 4 electrons, with each of the oxygen atoms in the compound. The carbon-oxygen bond in carbon dioxide is a double bond.

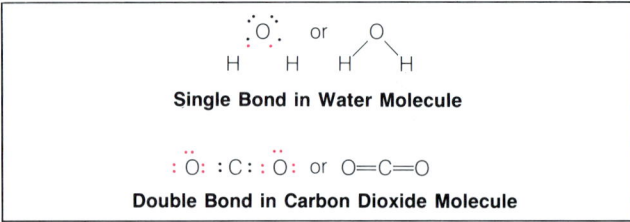

In a covalent bond, a relatively small number of atoms are involved in the sharing of electrons. The combination of atoms that results forms a separate unit called a **molecule**. A molecule is the smallest particle of a covalently bonded compound. In addition to water and carbon dioxide, some other common covalent compounds are sugar ($C_6H_{12}O_6$) and ammonia (NH_3).

Regardless of the type of bond formed, atoms change their physical and chemical properties when they form a compound. For example, sodium is a silvery metal that reacts explosively with water. Chlorine is a poisonous greenish gas, which was responsible for the deaths of many soldiers in World War I. Neither element in its pure form can be used by your body. Yet they combine to form sodium chloride (NaCl), or table salt, a solid that dissolves easily in water, is not poisonous, and is essential to most living things. It also tastes great on a pretzel!

3–3 SECTION REVIEW

1. Describe the arrangement of electrons in the first three energy levels.
2. Compare an ionic bond to a covalent bond.
3. **Critical Thinking—Applying Concepts** An oxygen atom has 6 electrons in its outermost energy level. A hydrogen atom has 1. Describe how the covalent bonds in the water molecule enable each atom to achieve stability.

3–3 (continued)

Reinforcement/Reteaching
The following analogy may help students remember the difference between covalent and ionic bonds. Ask students to imagine that two people are caught in the rain and only one has an umbrella. Suppose the person with the umbrella offers to share it. Now the umbrella covers both people. That is what happens in a covalent bond—2 atoms share electrons, and the electrons occupy the outermost energy levels of both atoms at the same time. Now ask students to suppose that the person with the umbrella goes inside and gives the umbrella to the other person. This is what happens in an ionic bond. Electrons are transferred from 1 atom to another.

SCIENCE, TECHNOLOGY, AND SOCIETY

BREAKTHROUGH

A Do-It-Yourself Poison

Each year they cause more than $5 billion in damage to crops and livestock. They are a threat to human health because they can transmit certain diseases. And in order to eliminate them, poisonous substances that may endanger our health as well as the health of our environment must be used. What are these pests who have created economic, environmental, and safety problems? Insects!

Farmers and gardeners use insecticides to protect their plants and animals against insects. Insecticides are compounds that kill insects in a variety of ways. Some kill insects that eat the chemical compounds. Others kill insects upon contact. Certain insecticides produce lethal vapors and gases. Still others kill by infecting insects with deadly diseases.

Although many insecticides are effective in the battles against insect infestation, they often have profound effects on the environment and other living things. Until recently, people had to balance the benefits of insecticide use against possible damage to planet Earth. Now a new class of insecticides that affect only their target and pose little danger to us has been developed.

Scientists working at the State University of New York at Stony Brook combined polysterols—special compounds manufactured by plants—with the element fluorine to produce a new group of compounds called 29-fluorophytosterols. These compounds are harmless to most organisms—ourselves included! But inside an insect's body, special chemicals called enzymes can act on the fluorophytosterols and break them down to produce fluoroacetate, a deadly poison. For insects unfortunate enough to have the special enzymes, this new class of insecticide spells doom!

In and of themselves, 29-fluorophytosterols are harmless, presenting little danger to most living things. But the chemical composition and bodily processes of insects render these compounds deadly! By applying a knowledge of basic chemistry to the workings of living systems, scientists were able to develop exactly the compound they sought: a safe and effective insecticide.

3–4 Chemical Reactions

You have learned that two types of bonds can exist between atoms and that compounds with different chemical and physical properties can be formed by the transfer or sharing of electrons. Whenever a chemical bond is formed, a chemical change takes place. Any process in which a chemical change occurs is known as a **chemical reaction**.

Chemical reactions occur all the time and are very much a part of your daily life. Some chemical reactions occur slowly, such as the combination of iron and oxygen to form rust. Others occur quickly. When hydrogen gas is ignited in the presence of oxygen, the reaction is rapid and explosive.

In any chemical reaction there are always two kinds of substances: the substances that are present before the change and the substances that are formed by the change. The elements or

Guide For Reading

- What are the names of the substances involved in a chemical reaction?
- Why is energy an important factor in chemical reactions?

55

TEACHING SUPPORT

Laboratory Manual
- Physical and Chemical Changes, p. 67

Teaching Resources
- Vocabulary Review: Cross-a-Clue, p. 1
- Hands-On Activity: Popcorn Hop, p. 25

Study Guide
- Section 3–4, p. 29

Product Testing Activities
- Testing Antacids
- Testing Orange Juice

3–4 (continued)

Content Development

Demonstrate to students how coefficients and subscripts indicate the balance of atoms in a chemical equation. Write on the chalkboard the equation:

$$2H_2O \rightarrow O_2 + 2H_2$$

Explain that the large 2 before water on the left side of the equation indicates 2 molecules of water. Since 1 water molecule contains 2 hydrogen atoms, 2 molecules must contain 2×2, or 4 hydrogen atoms. Since 1 molecule of water contains 1 oxygen atom, 2 molecules of water must contain 2×1, or 2 oxygen atoms. Thus, in this equation, there are 4 hydrogen atoms and 2 oxygen atoms present in the reactants. Say to students,
- Now look at the right side of the equation. How many hydrogen atoms do you see? (4.)
- How can you tell? (Each molecule of hydrogen contains 2 hydrogen atoms, and there are 2 hydrogen molecules present.)
- How many oxygen atoms do you see? (2.)
- How do you know? (One molecule of oxygen contains 2 atoms of oxygen, and there is 1 oxygen molecule present.)
- How do the number of hydrogen atoms on the right side of the equation compare with the number of hydrogen atoms on the left side? Of oxygen atoms? (They are equal.)

SECTION REVIEW 3–4

1. Any process in which a chemical change occurs. An equation uses symbols and numbers to represent the reactants and products in a chemical reaction.
2. Reactants are the substances that enter a chemical reaction whereas products are the substances produced by a chemical reaction.

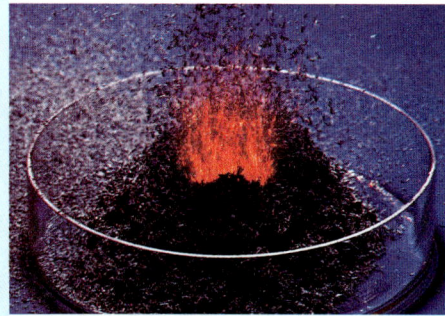

Figure 3–16 Chemical reactions are many and varied. Here you see two examples: the heating of ammonium dichromate (top) and the decaying of apples (bottom). A chemical reaction always involves a change in the properties and in the energy of the substances involved.

compounds that enter into a chemical reaction are known as reactants. The elements or compounds produced by a chemical reaction are known as products.

A chemical reaction can be described by using a shorthand method known as a chemical equation. In a chemical equation, symbols and formulas are used to represent reactants and products. And an arrow, which is read "yields," shows the direction of the reaction. Let us use the chemical reaction involving the combination of hydrogen and oxygen to form water to illustrate a chemical equation.

$$\underset{\text{oxygen}}{O_2} + \underset{\text{hydrogen}}{2H_2} \rightarrow \underset{\text{water}}{2H_2O}$$

Chemical reactions are reversible. In principle, any chemical reaction can run in either direction. For example, it is possible for the reaction we have just described to run in reverse.

$$\underset{\text{water}}{2H_2O} \rightarrow \underset{\text{oxygen}}{O_2} + \underset{\text{hydrogen}}{2H_2}$$

This generally does not occur however. Why? Because there is an important difference between the two chemical reactions. When oxygen gas and hydrogen gas burn to form water vapor, an enormous amount of energy is released in the form of heat, light, and sound. But water, on the other hand, does not spontaneously decompose to form hydrogen gas and oxygen gas. In fact, the only convenient way to drive the reaction in this direction is to run an electric current through the water to decompose it. Thus, in one direction the reaction produces energy; in the other direction, it requires energy.

Chemists have learned that the most important factor in determining whether a reaction will occur is the flow of energy. Chemical reactions that release energy will occur spontaneously. Chemical reactions that require energy will not occur without a source of energy.

What significance does this have for you and the other living things that share your world? All living things carry on both kinds of reactions in order to stay alive. The energy needed to grow tall, to read, to think, and even to dream comes from chemical reactions that occur in your body when you digest the food you eat. In the following chapters, you will explore these reactions and the flow of energy necessary for life.

3–4 SECTION REVIEW

1. What is a chemical reaction? A chemical equation?
2. What is a reactant? A product?
3. Describe the role of energy in chemical reactions.
4. **Critical Thinking—Making Inferences** Why is it important that energy-absorbing and energy-releasing reactions take place in living things?

PROBLEM SOLVING IN BIOLOGY

STRUCTURAL FORMULAS

Molecules are too small to be seen, yet we draw their pictures anyway. The reason for doing this is simple: In living things, a great many important things happen at the molecular level, and we need a way of representing these events. How shall we draw these pictures of molecules? And once drawn, what do they mean? Let's begin by considering glucose, a simple sugar.

One way to represent glucose is to write its molecular formula:

$$C_6H_{12}O_6$$

This formula tells us how many atoms of each element are in a molecule of glucose. But it does not indicate how the atoms are arranged. To do that, we need to draw a structural formula for the glucose molecule:

Chemists often try to simplify structural formulas. One way of doing this is to leave nearly all of the carbons out of the picture. However, everybody knows they're still there! Now the molecule looks like this:

Sometimes chemists draw the same molecule in an even simpler fashion, like this:

Each drawing is useful because it provides important information. For example, we can see that the glucose molecule is folded to form a ring and that the -OH groups (hydroxyl groups) project from one side or the other of that ring into space. None of the structural formulas, however, actually looks like a real molecule of glucose:

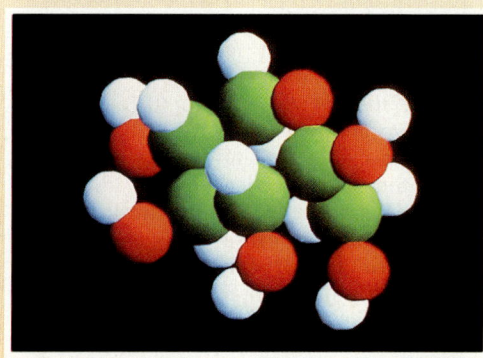

The key idea to remember about chemical formulas and structural formulas is that they are representations. Representations are useful and informative. They help us to visualize things that are hard to explain.

Now try your hand at drawing structural formulas for the following compounds: CH_4 methane; CCl_2F_2 Freon; $C_2H_4(OH)_2$ ethylene glycol or antifreeze; C_2H_6 ethane.

PROBLEM SOLVING

Structural Formulas

Introduce the feature by asking students,
- **What are some situations in which you might use a diagram to see how something is put together?** (Possible answers; building a model car or airplane; putting together do-it-yourself furniture; fixing a broken TV set or appliance.)
- **Why is a diagram more helpful than just words?** (It shows at a glance the overall structure of the object and how all the parts are related.) Point out that in a similar way, structural formulas of molecules give an overview of how the molecule is put together. These drawings make it easier for scientists to understand and predict how a molecule will behave in a chemical reaction.

Answers
1. methane

   H
   |
   H—C—H
   |
   H

2. freon

   Cl
   |
   F—C—Cl
   |
   F

3. ethylene glycol

   H H
   | |
   H—C—C—H
   | |
   HO OH

4. ethane

   H H
   | |
   H—C—C—H
   | |
   H H

---

3. If a chemical reaction releases energy, it will occur spontaneously. If it requires energy, it will not occur until that energy (activation energy) is added.
4. Living things must break down materials such as foods in reactions that generally release energy. At the same time, living things must build up or synthesize materials in reactions that usually require the addition of energy. Both types of reactions are necessary for survival.

Reinforcement/Reteaching
If students have trouble answering any of the Section Review questions, go back to the part of the section that contains the material with which they are having difficulty.

Closure
Write several chemical reactions on the chalkboard and have students identify the products and the reactants. Then have them predict whether they think that each reaction will require energy or release energy and why.

LABORATORY INVESTIGATION

IDENTIFYING ELEMENTS BY DENSITY

Before the Lab

1. At least one day before the investigation, gather enough materials for your class, assuming 6 students per group.
2. Make a record of the identities and densities of the unknown element samples that you plan to give to each group. If the appearance of the samples is too great of a clue to their identities, you may wish to paint them with waterproof paint.

Pre-Lab Discussion

Introduce students to the concept of density by displaying two objects of similar size and shape but of very different densities, such as a golf ball and a Ping-Pong ball.

- **How are these two objects similar?** (Both are round; both are white; both are balls used in a sport.)
- **How are they different?** (One is much heavier than the other; one is hollow whereas the other is solid inside.)

Place both balls in a beaker of water.

- **What do you see?** (The golf ball sinks whereas the Ping-Pong ball floats.)
- **Why do you think this happens?** (Accept all reasonable answers.)

Point out to students that the golf ball sinks because it is more dense than the Ping-Pong ball. The golf ball is no larger than the Ping-Pong ball, but it contains much more matter.

Write the formula for density on the chalkboard. Emphasize that the volume of both balls is about the same. The mass of the golf ball, however, is much greater.

Explain to students that in this investigation, they will measure the densities of two known elements and the density of an unknown element. After you have given each group their samples, ask students to predict which element they expect will have the greater density. Also ask them if they have any clues as to the identity of their unknown element.

Skills Development

Students will use the following skills while completing this laboratory investigation.

1. Measuring
2. Observing
3. Comparing
4. Recording
5. Calculating
6. Identifying
7. Inferring
8. Relating

LABORATORY INVESTIGATION: IDENTIFYING ELEMENTS BY DENSITY

PROBLEM

How do the densities of different elements compare?

MATERIALS (per group)

aluminum metal
zinc metal
unknown metal sample
graduated cylinder
balance

PROCEDURE

1. Density is mass per unit volume. In this laboratory investigation, you will measure the density of two elements, zinc and aluminum. Copy the data table shown below onto a separate sheet of paper to record your observations.
2. Using a balance, measure the mass of a small sample of aluminum metal (about 40 g). Record the mass in the data table.
3. Measure the volume of the aluminum metal using the following procedure: Fill a 100-mL graduated cylinder halfway with water and record the initial volume in your data table. Put the aluminum metal into the graduated cylinder. This will cause the water level to rise. Record the final volume of water. Find the volume of the aluminum by subtracting the initial volume of the water from the final volume of the water. Record the result.
4. Calculate the density of the aluminum by dividing the aluminum's mass by its volume: D = M/V. Record the result.
5. Find the density of a small sample of zinc metal by measuring its mass and volume according to the same procedure you used for aluminum.
6. Measure the density of an unknown sample of metal. Use your results to determine whether the unknown metal is aluminum or zinc.

OBSERVATIONS

1. How do the densities of aluminum and zinc compare?
2. What are some observable properties of aluminum and zinc?

ANALYSIS AND CONCLUSIONS

1. Is the unknown metal aluminum or zinc? How do you know?
2. Is density a physical or a chemical property? How can you tell?
3. Can density be used to help identify elements? Explain your answer.

Element	Mass (g)	Volume (mL)			Density (g/mL)
		Initial	Final	Metal	
Aluminum					
Zinc					
Unknown					

STUDENT STUDY GUIDE

SUMMARIZING THE CONCEPTS

The key concepts in each section of this chapter are listed below to help you review the chapter content. Make sure you understand each concept and its relationship to other concepts and to the theme of this chapter.

3–1 Nature of Matter

- Two important properties—mass and volume—are used to define matter.
- Physical properties of matter can be observed and measured without permanently changing the identity of the matter. Chemical properties describe a substance's ability to change into another new substance as a result of a chemical change.

3–2 Composition of Matter

- The basic unit of matter is the atom. The nucleus of the atom contains positively charged protons and electrically neutral neutrons. Energy levels outside the nucleus contain negatively charged electrons.
- The number of protons in the nucleus of an atom is called the atomic number.
- Substances that consist entirely of one type of atom are called elements. Atoms of the same element that have the same number of protons but different numbers of neutrons are called isotopes.
- A chemical compound consists of two or more different atoms chemically combined in definite proportions.

3–3 Interactions of Matter

- The combining of atoms of elements is known as chemical bonding. An ionic bond involves a transfer of electrons between atoms. A covalent bond involves the sharing of electrons.

3–4 Chemical Reactions

- In any chemical reaction, reactants combine to yield products.
- Although chemical reactions are reversible in principle, they occur spontaneously only in the direction that produces energy.

REVIEWING KEY TERMS

Vocabulary terms are important to your understanding of biology. The key terms listed below are those you should be especially familiar with. Review these terms and their meanings. Then use each term in a complete sentence. If you are not sure of a term's meaning, return to the appropriate section and review the definition.

3–1 Nature of Matter
physical property
chemical property
phase

3–2 Composition of Matter
atom
nucleus
proton
neutron
electron
energy level
atomic number
mass number

element
isotope
compound

3-3 Interactions of Matter
chemical bonding
ionic bond

ion
covalent bond
molecule

3-4 Chemical Reactions
chemical reaction

Observations

1. Aluminum is less dense than zinc.
2. They are both shiny, silvery metals. They are solid and do not dissolve in water.

Analysis and Conclusions

1. Answers will vary. You can tell the identity of the unknown because it has the same density as either zinc or aluminum.
2. It is a physical property since it is not necessary to form new substances to determine the density.
3. Density can be used to identify unknown elements because each element has a characteristic density.

Going Further: Enrichment

Have students create an experiment in which a piece of real gold or silver is discovered amid a collection of fakes. Groups can use jewelry or metal samples. Discuss with students how other metals can be made to look like gold or silver—for example, jewelry can be gold plated or metal samples can be coated with silver paint. Also discuss how density can be used to determine that a piece of metal is not gold or silver. Students may enjoy creating detective stories to go with their experiments. For example, a stolen gold ring may be hidden in a dime store display of tin rings.

Safety Tips

Remind students to be careful when handling the graduated cylinders. Emphasize that any broken glassware should be reported immediately.

Teaching Strategy

1. Explain that an object placed in water displaces a volume of water equal to its own volume. Make sure that students know to subtract the initial volume of water from the volume of the water plus the metal.
2. Make sure that students divide mass by volume, not the other way around. Stress that the units of density are grams/milliliter.

CHAPTER REVIEW

CONTENT REVIEW

Multiple Choice

Choose the letter of the answer that best completes each statement.

1. Substances that enter into a chemical reaction are known as
   a. reactants.
   c. products.
   b. catalysts.
   d. isotopes.
2. Two important physical properties used to define matter are
   a. taste and smell.
   b. mass and volume.
   c. weight and boiling point.
   d. mass and freezing point.
3. The nucleus of an atom contains
   a. electrons and protons.
   b. electrons and neutrons.
   c. protons, electrons, and neutrons.
   d. protons and neutrons.
4. A substance consisting of two or more different elements chemically combined is a(an)
   a. molecule.
   b. crystal.
   c. compound.
   d. isotope.
5. An atom with an atomic number of 8 has
   a. 8 protons.
   b. 4 protons plus 4 neutrons.
   c. 4 protons plus 4 electrons.
   d. 4 neutrons plus 4 electrons.
6. Which is true of carbon-12 and carbon-14?
   a. They have the same number of neutrons.
   b. They have the same number of protons.
   c. They have the same mass number.
   d. They are isotopes of different elements.
7. A shorthand way of representing an element is a chemical
   a. formula.
   b. reaction.
   c. property.
   d. symbol.
8. Radioactive isotopes are used
   a. as tracers.
   b. to diagnose disease.
   c. to treat disease.
   d. to do all of these.

True or False

Determine whether each statement is true or false. If it is true, write "true." If it is false, change the underlined word or words to make the statement true.

1. <u>Chemical properties</u> describe a substance's ability to undergo a chemical change.
2. The sharing of electrons is characteristic of <u>ionic</u> bonds.
3. Any process in which a chemical change occurs is known as a chemical <u>formula</u>.
4. The number of protons in the nucleus of an atom is called the <u>mass number</u>.
5. Chemical reactions that <u>absorb</u> energy will occur spontaneously.
6. The amount of space an object takes up is known as <u>weight</u>.
7. Atoms that have <u>complete</u> outermost energy levels are said to be very stable.
8. The positively charged subatomic particle is the <u>electron</u>.

Answer Key

CONTENT REVIEW

Multiple Choice
1. a 2. b 3. d 4. c
5. a 6. b 7. d 8. d

True or False
1. T
2. F covalent
3. F reaction
4. F atomic number
5. F release
6. F volume
7. T
8. F proton

Word Relationships
1. terms associated with the center (nucleus) of an atom; energy levels are located outside the nucleus.
2. elements; carbon dioxide is a compound.
3. phases of matter; mass is the quantity of matter in an object.
4. terms associated with the second and third energy levels; 2 is the number of electrons that can be held in the first energy level.
5. terms associated with the sharing of electrons; electron transfer.
6. physical properties; ability to burn is a chemical property.
7. terms associated with chemical reactions; physical change, unlike chemical change, does not alter the identity of matter.
8. terms associated with the development of the concept of the atom; divisible particle.

CONCEPT MASTERY

1. Physical properties, unlike chemical properties, can be observed and measured without permanently changing the identity of the matter.
2. Physical: a, b, f. Chemical: c, d, e, g.
3. Proton and neutron in nucleus. Proton has a positive charge and a mass of 1 amu. Neutron has no charge and a mass of 1 amu. Electron found in energy levels outside the nucleus. Has a negative charge and a mass of 1/1836 amu.
4. Atomic number is the number of protons in the nucleus, which distinguishes the type of atom.
5. Nitrogen-15 has 1 more neutron (8) than nitrogen-14 (7).
6. Atoms whose nuclei are unstable and will from time to time break down, releasing matter and or energy called radiation. Uses include treating disease, sterilizing foods, studying living organisms, and measuring ages of rocks.

CRITICAL AND CREATIVE THINKING

1. Advantages: Students should review the numerous uses of such isotopes. Disadvantages: Radioactive materials can be dangerous if not handled

Word Relationships

In each of the following sets of terms, three of the terms are related. One term does not belong. Determine the characteristic common to three of the terms and then identify the term that does not belong.

1. energy level, proton, nucleus, neutron
2. oxygen, sodium, carbon dioxide, mercury
3. solid, mass, liquid, gas
4. 2, second energy level, 8, third energy level
5. molecule, electron transfer, covalent bond, water
6. color, phase, ability to burn, texture
7. products, physical change, reactants, yields
8. Democritus, atom, divisible particle, Greek philosopher

CONCEPT MASTERY

Use your understanding of the concepts developed in the chapter to answer each of the following in a brief paragraph.

1. Describe two important differences between physical and chemical properties.
2. Identify each of the following changes as either a physical change or a chemical change:
   a. melting butter
   b. boiling water
   c. digesting food
   d. baking brownies
   e. exploding TNT
   f. dissolving sugar
   g. burning fuel oil
3. Describe the subatomic particles found in the atom. Be sure to include their charge, mass, and location in your description.
4. What is the atomic number of an element? What is its significance?
5. Nitrogen-14 and nitrogen-15 are isotopes of the element nitrogen. Describe how atoms of these isotopes differ from each other.
6. What is a radioactive isotope? Describe three uses of radioactive isotopes.

CRITICAL AND CREATIVE THINKING

Discuss each of the following in a brief paragraph.

1. **Relating facts** Explain why the use of radioactive isotopes has both advantages and disadvantages.
2. **Developing a model** Draw a diagram of the electron arrangement for the following elements based on their atomic number: lithium (3), carbon (6), fluorine (9), magnesium (12), phosphorus (15), argon (18).
3. **Identifying patterns** Explain why the element neon, with an atomic number of 10, is unreactive.
4. **Synthesizing concepts** Explain the role of energy in chemical reactions.
5. **Applying concepts** A basic principle of electricity is "like charges repel and opposite charges attract." How does the organization of particles in an atom illustrate this principle? How does it refute it?
6. **Using the writing process** The time is 3000 years ago. The Greek philosopher Democritus has proposed that all matter is made up of atoms. Write a brief speech supporting or refuting the concept of indivisible atoms.

properly.

2. Lithium: 2, 1; carbon: 2, 4; fluorine: 2, 7; magnesium: 2, 8, 2; phosphorus: 2, 8, 5; argon: 2, 8, 8.
3. Its outermost energy level is filled, making it very stable.
4. Energy is the driving force for all chemical reactions. If a reaction gives off energy, it will occur spontaneously. If it requires energy, it will not occur without a source of energy.
5. Illustrates it in that negatively charged electrons are held in the atom by a strong attraction to positively charged protons. It refutes it by the fact that positively charged protons in the nucleus do not repel each other with enough force to cause the nucleus to break apart. (They are held together by the strong force.)
6. Students' essays should be checked for creativity and writing skills rather than scientific accuracy. However, they should be logical and well thought out.

CHAPTER 4 The Chemical Basis of Life

Section	Laboratory Investigations and Activities
4–1 Water pages 63–67 THEME: Scale and Structure	**Teacher Edition** DEMONSTRATION: Acids and Bases, p. 62D **Laboratory Manual** Using Acid-Base Indicators to Test Unknown Substances, p. 85 **Product Testing Activities** Testing Bottled Water Testing Glues Testing Shampoos
4–2 Chemical Compounds in Living Things pages 68–69 THEME: Scale and Structure	**Teacher Edition** DEMONSTRATION: Preparation of Nylon, p. 62D **Product Testing Activities** Testing Bubble Gum Testing Food Wraps Testing Nail Enamel
4–3 Compounds of Life pages 70–77 THEMES: Scale and Structure, Unity and Diversity	**Student Edition** Laboratory Investigation: Testing Enzyme Activity, p. 78 **Laboratory Manual** Identifying Organic Compounds, p. 89 **Teaching Resources** HANDS-ON ACTIVITY: These "Fuelish" Things, p. 31 TEACHER DEMONSTRATION: The Effect of Temperature on Enzyme Action, p. 3 **Product Testing Activities** Testing Cereals Testing Sports Drinks
Chapter Review pages 79–81	

Teacher Resources

Books

Binford, Jesse S., Jr. *Foundations in Chemistry*. Book Publishers, Inc., 1985.
Brescia, Frank, et al. *General Chemistry*. Harcourt Brace Jovanovich, 1988.
Brown, T.L., H.E. LeMay, Jr., and B.E. Bursten. *Chemistry: The Central Science*. Prentice Hall, 1991.

Chang, Raymond. *Chemistry*, 3rd ed. Random House, 1987.
Kennedy, John F., ed. *Carbohydrate Chemistry*. Oxford University Press, 1988.
Stryer, L. *Biochemistry*, 2nd ed. Freeman, 1981.

Williams, David J. *Biochemistry: An Illustrated Outline*. Lippincott, 1988.

CHAPTER PLANNING GUIDE

Other Activities	Media and Technology

Teaching Resources
    ACTIVITY: Solutions and Temperature, p. 3
Study Guide
    Section 4–1, p. 33
	Video/Videodisc
	    Elements, Compounds, and Mixtures
	    Acids, Bases, and Salts
	Biology Media Guide
	    pages 5–6

Study Guide Section 4–2, p. 36	

Teaching Resources
    VOCABULARY REVIEW: Word Game, p. 1
    ACTIVITY: Identifying Chemical Compounds, p. 5
Study Guide
    Section 4–3, p. 38
	Biology Transparencies
	    4: Dehydration Synthesis
	Biology Media Guide
	    page 6

Teaching Resources
    Chapter Test A, p. 11
    Chapter Test B, p. 15
    Performance-Based Assessment, Unit 1
	Computer Test Bank
	    Chapter Test, p. 37

Audiovisuals

Acids, Bases, and Salts, 2nd ed. Film or video. Coronet Films.
Solutions: Ionic and Molecular, 2nd ed. Film or video. Coronet.
Formulas and Equations. Film or video. Coronet.
Elements, Compounds and Mixtures. 2nd ed. Film or video. Coronet Films.
All About Acids and Bases. Video. Focus Media.
The Chemistry of Carbohydrates and Lipids. Video or filmstrip. Biology Media.

CHAPTER 4 The Chemical Basis of Life

CHAPTER OVERVIEW

In the last chapter students were given a general overview of chemistry. In this chapter students will be introduced to the specific aspects of chemistry that are most important to living things. Students will learn that there is no one substance that makes something "alive." There are, however, certain substances that are found in many or all living things.

Students will begin by learning about water, which is the most abundant compound in nearly all living organisms. They will learn that the most important property of the water molecule is that it is slightly charged on each end. Mixtures, solutions, suspensions, acids, and bases will be discussed.

In the second section of the chapter students will be introduced to the terms inorganic and organic. They will learn that organic compounds are carbon-containing compounds. Various properties of organic compounds, including polymerization, will be discussed.

The last section of the chapter introduces students to four important groups of compounds found in living things: carbohydrates, lipids, proteins, and nucleic acids. Students will learn how the first three compounds are formed and how they are used by living things. They will discover how enzymes, which in most cases are proteins, speed chemical reactions in organisms. The role of nucleic acids in storing and transmitting genetic information will also be briefly discussed.

4-1 WATER

Section Focus 4-1

The purpose of this section is to introduce students to the properties of water and to emphasize the important role water plays in living organisms. Students will discover that water is unique in many ways, with physical and chemical properties found in no other material.

Students will learn that the water molecule's most important property is due to an uneven distribution of electrons. This causes the molecule to be slightly charged at each end. Because of this property, water is especially good at attracting other charged molecules as well as ions to form mixtures.

Students will learn that mixtures formed with water can be of two important types: solutions and suspensions. Students will learn the terms solvent and solute and will discover the process by which solute materials enter a solution. Students will also be introduced to two special types of solutions: acids and bases. Neutralization reactions and the pH scale will also be discussed.

Performance Objectives 4-1

1. Discuss the unique properties of water.
2. Explain how water molecules attract other molecules or ions in solution.
3. Describe and compare mixtures, solutions, and suspensions.
4. Define and describe acids, bases, and neutralization.
5. Explain how the pH scale measures the strength of acids and bases.

Science Terms 4-1

mixture   p. 64
solution   p. 65
solvent   p. 65
solute   p. 65
acid   p. 65
base   p. 66
neutralization reaction   p. 66
pH scale   p. 66
suspension   p. 66

4-2 CHEMICAL COMPOUNDS IN LIVING THINGS

Section Focus 4-2

The purpose of this section is to distinguish between inorganic and organic compounds found in living things. Students will learn that only 11 of the 92 known elements are commonly found in living organisms. Students will discover that scientists divide the compounds formed from these elements into two groups.

Students will learn that organic compounds contain the element carbon, whereas inorganic compounds do not. They will discover that carbon has many important properties, including its ability to bond easily with common elements. They will also discover that carbon has the ability to form chains of almost limitless length by bonding to other carbon atoms.

Students will learn that many organic compounds are formed by a process called polymerization. In polymerization, complex molecules are formed by joining many smaller molecules together.

Performance Objectives 4-2

1. List the most abundant elements found in living things.
2. Define and compare inorganic and organic compounds.
3. Discuss the unique properties of carbon.
4. Explain the process of polymerization.

Science Terms 4-2

organic compound   p. 68
inorganic compound   p. 68
polymerization   p. 69
monomer   p. 69
polymer   p. 69
macromolecules   p. 69

4-3 COMPOUNDS OF LIFE

Section Focus 4-3

The purpose of this section is to discuss four important groups of organic compounds found in living things. These compounds are carbohydrates, lipids, proteins, and nucleic acids.

Students will learn that carbohydrates, or sugars, are important to living things because they contain a great deal of energy. Students will learn how complex carbohydrates are formed by a polymerization process called dehydration synthesis. They will also learn how polysaccharides are formed by joining together many simple sugars.

Students will discover that lipids have three major roles in living organisms: They can store energy, they form biological

CHAPTER PREVIEW

membranes, and they can serve as chemical messengers. Students will learn that many lipids are formed from fatty acids and glycerol. Saturated and unsaturated lipids will be discussed, as will sterols and phospholipids.

Students will learn that proteins are at the very center of life, performing numerous functions in a living organism. Students will learn that proteins contain nitrogen, carbon, hydrogen, and oxygen. They will discover that proteins are polymers of amino acids joined by peptide bonds.

Students will be introduced to the role of enzymes. They will learn that enzymes act as catalysts for chemical reactions in living things. Students will discover that enzymes speed chemical reactions by binding to the reactants, which are known as substrates, at a region known as the active site.

In the last part of the section students will be briefly introduced to nucleic acids. They will learn that nucleic acids are polymers of smaller monomers called nucleotides. They will also learn that the two basic types of nucleic acids are DNA and RNA.

Performance Objectives 4–3

1. List the four important groups of organic compounds found in living things.
2. Describe the structure and function of carbohydrates, lipids, and proteins.
3. Describe the function of enzymes and explain how they work.
4. Describe the structure and function of nucleic acids.

Science Terms 4–3

carbohydrate   p. 70
monosaccharide   p. 70
dehydration synthesis   p. 71
disaccharide   p. 71
polysaccharide   p. 71
hydrolysis   p. 72
lipid   p. 72
cholesterol   p. 74

protein   p. 74
amino acid   p. 74
peptide bond   p. 74
catalyst   p. 75
enzyme   p. 76
substrate   p. 76
active site   p. 76
nucleic acid   p. 76
nucleotide   p. 76
ribonucleic acid (RNA)   p. 76
deoxyribonucleic acid (DNA)   p. 76

DISCOVERY LEARNING

TEACHER DEMONSTRATIONS
Modeling

Acids and Bases
The following demonstration can be used as an introduction to Chapter 4. For the demonstration you will need some phenolphthalein indicator, a medicine dropper, and samples of a strong acid and a strong base.

Display the acid and base but do not indicate to students what the substances are. (For convenience, you may wish to label the liquids Sample 1 and Sample 2.) Place several drops of phenolphthalein in the acid sample.

• **What do you see happening?** (Nothing; the drops of liquid added seem to have no color.)

Now place a few drops of phenolphthalein in the base.

• **What do you see?** (The liquid turns bright pink.)

• **Based on what you have seen, do you know what the liquids are in Sample 1 and Sample 2?** (Accept all answers. Some students may recognize the reaction of the acid and base with phenolphthalein, although other students may be unfamiliar with this indicator.)

• **Do you think that substances like Sample 1 and Sample 2 are found in living things?** (Accept all answers.)

Preparation of Nylon
This demonstration can be performed after students have studied polymerization in Section 4–2. For the demonstration you will need: adipyl chloride solution (0.25 M in hexane), hexamethylene diamine solution (0.50 M in 0.50 M NaOH), acetone in alcohol, paper clip, small beakers, graduated cylinder.

Explain to students that you are going to demonstrate the preparation of the polymer nylon. Point out that nylon is one of many synthetic fibers produced by polymer chemistry.

• **Can you think of other synthetic fibers that are popular today?** (Possible answers: dacron, polyester, rayon. Explain that these synthetic fibers are also polymers.)

Begin the demonstration by bending apart the paper clip, leaving a hook at one end. Pour 5 mL hexamethylene diamine solution into one beaker. Slowly add 5 mL adipyl chloride solution. A white film should form as the two solutions meet.

Put the paper clip into the beaker, hooking around the white film. Then slowly lift the paper clip out of the beaker.

• **What do you see coming out of the beaker?** (A white substance.)
• **What is the substance?** (Nylon.)

Continue to remove all of the nylon from the beaker. Place the nylon in a clean beaker, wash it, then rinse it with acetone and let it dry.

CHAPTER 4

The Chemical Basis of Life

GUIDED ENQUIRY
Pose the following questions to students and have them record their responses. Point out that they will gain a better understanding of the key concepts if they read the chapter with these basic questions in mind. Upon completion of the chapter, pose the questions again. Ask students to compare their initial responses with those they have developed after reading the chapter.

- Why is water such a unique substance?
- How are acids different from bases?
- How does carbon form compounds?
- How do living things use carbohydrates and lipids?
- Why are proteins important to living things?
- How do enzymes catalyze reactions in living organisms?

INTRODUCING CHAPTER 4

Using the Visuals
Begin Chapter 4 by having students examine the chapter-opener photographs on page 62. These photographs are computer-generated models of protein molecules.

- **What is shown in these photographs?** (If students have read the caption, they will point out that the photographs show the molecular structure of a protein.)

Point out that proteins are among the most important compounds in living things.

- **What other compounds are found in living things?** (Answers may vary but should include water, sugars and starches [carbohydrates], and fats. Some students might mention genetic material such as DNA or RNA as well.)
- **If we simply mix together the basic compounds found in living things, will we create life?** (No, the process of living, or life, is far more complex than a simple mixture of chemical compounds.)

Point out that in this chapter students will examine the compounds that make up living things because, by understanding the compounds of life, they will gain a better understanding of life itself.

TEACHING STRATEGY 4–1

Focus/Motivation
Prepare a display of five different liquid substances, including a glass of water. The other substances might include: a glass of fruit juice or colored carbonated beverage;

CHAPTER 4

The Chemical Basis of Life

A variety of important chemical substances are combined in specific ways to produce protein molecules, which are among the basic compounds of life.

There are no special elements found in living things. In fact, the molecules that make up a living cell do not, in and of themselves, have the property of life. Yet there can be no doubt that certain elements organized into a variety of molecules do indeed account for life. How, you might ask, is this possible? In a way, we will spend the rest of this book searching for an answer, an answer that will remain incomplete in the end. For life itself is not completely understood. But we can begin. And one way to start is by understanding life not as the property of one particular molecule but as something made possible by a large group of molecules. In this chapter you will learn about the molecules that give life to you and to all other organisms on Earth.

GUIDE FOR READING

After you read the following sections, you will be able to

4–1 Water
- List some important properties of water.
- Describe the nature of mixtures, solutions, and suspensions.
- Define acids, bases, neutralization, and pH scale.

4–2 Chemical Compounds in Living Things
- Identify the four most abundant elements in living things.
- Compare inorganic compounds and organic compounds.
- Describe some important properties of carbon.
- Explain the importance of polymerization.

4–3 Compounds of Life
- Identify the four groups of organic compounds found in living things.
- Describe the structure and function of each group of compounds of life.
- Explain how enzymes work and why they are important to living things.

Journal Activity

YOU AND YOUR WORLD "Better Living Through Chemistry" was a slogan used in the past by a major chemical corporation. What do you think this slogan means? How has chemistry changed the ways we live? Write your thoughts in your journal.

4–1 Water

Guide For Reading
- What is one important property of water?
- How do mixtures, solutions, and suspensions differ?
- What is pH?

Someone once said that if there is magic on this planet, then it is to be found in water. Water? To most of us, water is so ordinary that it is of little interest and certainly does not seem magical. Yet how wrong we are!

Water is one of the few naturally occurring compounds that is liquid at the temperatures found on much of the Earth's surface. Unlike most other materials, water expands slightly as it makes the phase change from liquid to solid. This explains why ice floats at the surfaces of lakes and rivers rather than sinking to the bottom—a situation that might be disastrous for fish and plant life, to say nothing of the sport of skating!

Water is a most unusual molecule, with physical and chemical properties found in no other material. Water covers more than 75 percent of the Earth's surface and is the most abundant compound in nearly all living organisms.

The most important property of the water molecule is that due to an uneven distribution of electrons, it is slightly

Figure 4–1 The importance of water to all living things cannot be denied—whether it's as a vital body fluid or as home to penguins in Antarctica. The Earth and its inhabitants cannot survive without water. Every attempt must be made to preserve this resource necessary to life.

COOPERATIVE LEARNING

Using preassigned lab groups or randomly selected teams, have groups prepare and perform a 2-minute scene from their favorite soap opera. The script must include specific references to the characteristics and functions of the following compounds of life: water, carbohydrates, lipids, proteins, enzymes, and nucleic acids. Encourage students to be creative. The script might cast the listed substances as the main characters in someone's body or follow a more traditional storyline. Students should use exaggerated emotions and body movements. A list of the required compounds on the chalkboard or an overhead could serve as a visual reinforcement as well as a checklist for the evaluation of each group's performance.

Journal Activity

YOU AND YOUR WORLD

You may want to use the journal activity as the basis of a class discussion. Explore both the positive and the negative ways the chemical industry affects our lives. Point out the chemical products found in your school or classroom, which could include ink, fabrics, cleaning solutions, and food preservatives and additives.

---

a jar of paint or ink; a container of cleaning fluid; a can of motor oil. After students have observed the display, ask,
• Suppose I tell you that one of these liquids is a "magic substance" with very unusual properties. Which substance would you pick? (Accept all answers. You may find it effective to tally students' responses to see which liquid "wins.")

Content Development
Tell students that they may be surprised to learn that the most magical of the five substances is "ordinary" water. Point out that water has physical and chemical properties that no other substance has. First, water is one of the few naturally occurring compounds that is liquid at the temperatures found on most of the Earth's surface. Second, most substances contract when they freeze, but water expands—which is why ice cubes float in a glass of water rather than sink to the bottom. A third factor that makes water unique is that it is the greatest solvent in the world. Water has this important characteristic because of the ability of its molecules to attract other molecules and ions. Finally, water is essential to living things and is the most abundant compound found in most living organisms.

63

MULTICULTURAL STRATEGY

A native of Durham, North Carolina, Dr. James W. Mitchell is head of the Analytical Chemistry Department at Bell Laboratories in Murray Hill, New Jersey. Dr. Mitchell is best known for his work in developing techniques for identifying trace elements in high-purity materials. He has also developed processes for producing ultra-high-purity chemicals.

Dr. Mitchell earned his bachelor's degree in Chemistry from North Carolina Agriculture and Technical State University in Greensboro and his Ph.D. in Analytical Chemistry from Iowa State University. In 1981, he was awarded the Percy L. Julian Outstanding Research Award. You may wish to have students research the names of other outstanding African American scientists who have received this award.

ANNOTATION KEY

1. Positive: hydrogen end; Negative: oxygen end. (Interpreting illustrations)
2. Accept all correct answers. (Relating facts)

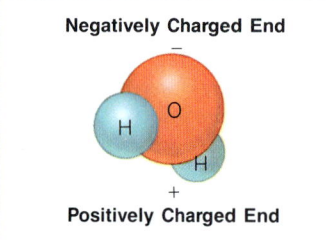

Figure 4–2 The polarity of a water molecule explains why water is able to dissolve thousands of substances —whether they be ionically or covalently bonded. Which end of a water molecule is positive? Negative? ❶

4–1 (continued)

Content Development
Using the Visuals Emphasize that a water molecule is made up of 1 oxygen atom and 2 hydrogen atoms.
• **Which atoms are bonded to each other?** (Each hydrogen atom is bonded to the oxygen atom.)
• **What type of bonds are these?** (Covalent.)
• **What do you recall about the nature of covalent bonds?** (Electrons are shared between two atoms.)

Direct students' attention to Figure 4–2.
• **Which part of a water molecule has a negative charge?** (The part nearest the oxygen atom.)
• **Which part has a positive charge?** (The part nearest the hydrogen atoms.)
• **Why do you think the charge**

charged on each end. See Figure 4–2. In this way, each water molecule is like a little magnet, with one pole near the nucleus of the oxygen atom and the other pole near the nuclei of the two hydrogen atoms. This polarity produces a strong attraction between individual water molecules and between water molecules and other charged molecules.

Mixtures

The slight charges of water molecules make them especially good at forming **mixtures**. A mixture is a substance composed of two or more elements or compounds that are mixed together but not chemically combined. Salt and pepper stirred together constitute a mixture. So do sugar and sand. Unlike chemical compounds, the substances that make up mixtures can be added in varying amounts and are not linked by chemical bonds.

Earth's atmosphere is a mixture of gases; soil is a mixture of individual solid particles; and living things are in part com-

Figure 4–3 By combining powdered iron (top) with powdered sulfur (center), an iron-sulfur mixture is formed. But because the substances making up this mixture are not linked by chemical bonds, they can be separated easily by a magnet (bottom). The gold miner can separate heavy pieces of gold from rock, sand, and dirt by shaking the mixture in a pan of water and letting the gold settle to the bottom (right).

arranges itself in this way? (Answers may vary. Students may recognize that an oxygen atom has more electrons than a hydrogen atom, so it is more likely to attract, rather than give up, electrons. The most correct answer is that a difference in electronegativity exists between oxygen and hydrogen.)

Skills Development
Guided Practice
Skills: Applying concepts, inferring
Using the Visuals Ask students:

posed of mixtures involving water. The mixtures that are made with water can be of two important types: solutions and suspensions.

Solutions

If a cube of sugar is placed in a glass of warm water, the movement of the water molecules gradually breaks off single molecules from the sugar cube. These sugar molecules then become dispersed in the water and the sugar dissolves. See Figure 4-4. Before long, the sugar cube has completely disappeared and the sugar molecules are uniformly spread throughout the water, forming a **solution**. The sweet taste of the solution tells us that the sugar is still there. The sugar has undergone a physical but not a chemical change in forming a solution with water.

You have probably made solutions of water and many other materials, such as sugar, salt, tea, and cocoa. In these solutions, water is the **solvent**, or substance that does the dissolving. The substance that is dissolved—salt or tea, for example—is the **solute**. All solutions consist of a solvent and a solute. Without exaggeration, water is the greatest solvent in the world. This distinction is due largely to the charges at either end of the water molecule, which attract water molecules to other molecules whether they are positively or negatively charged.

ACIDS When ionically bonded compounds dissolve in water, they often dissociate, or break apart, into individual ions. Common table salt (sodium chloride) is a good example.

$$NaCl \rightarrow Na^+ + Cl^-$$
sodium chloride   sodium ion   chloride ion

The dissociation of sodium chloride produces the positive sodium ion (Na^+) and the negative chloride ion (Cl^-). Many other compounds follow this general pattern when they dissolve in water. One group of compounds, however, deserves special attention. Consider the following dissociation reaction:

$$HCl \rightarrow H^+ + Cl^-$$
hydrochloric acid   hydrogen ion   chloride ion

At first glance this compound dissociates in the same way that salt does. Just like salt, it produces a chloride ion (Cl^-). But the hydrogen ion (H^+) that is produced is very different from the sodium ion (Na^+). The H^+ is the most chemically reactive ion known. It is a single proton, lacking any electrons. This absence of electrons enables the hydrogen ion to attack the chemical bonds in a wide variety of molecules.

Compounds that release hydrogen ions into solution are known as **acids**, and we know HCl as hydrochloric acid. Can you name some other examples of acids? ❷

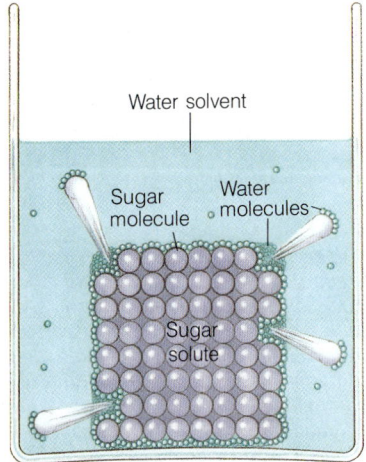

Figure 4–4 The solution process first involves the separation of solute particles from the surface of the solid solute. Then the solute molecules enter the solution. Finally, the solute molecules are attracted to the solvent molecules.

Figure 4–5 Acids are found in many of the foods we eat. Fruits such as apples and pears contain malic acid. Oranges, grapefruits, and lemons taste sour because they contain citric acid. These fruits also contain vitamin C, which is ascorbic acid.

BASES Bases are compounds that release hydroxide ions (OH^-) into solution. Sodium hydroxide is an example of a base.

$$NaOH \rightarrow Na^+ + OH^-$$
sodium hydroxide　　sodium ion　　hydroxide ion

Strong acids and bases are highly reactive chemical compounds. They can attack and break a variety of chemical bonds, thus making them potentially dangerous to living tissue.

NEUTRALIZATION AND pH Mixing a strong acid and a strong base results in a reaction in which hydrogen ions and hydroxide ions react to form water. This type of reaction is called a **neutralization reaction**. Can you guess the reason for this name?

$$H^+ + OH^- \rightarrow H_2O$$
hydrogen ion　　hydroxide ion　　water

If the quantities of acid and base being mixed are in perfect balance, the neutralized solution that results is neither acid nor base. The concentrations of hydrogen ions and hydroxide ions are equal. The relative concentrations of hydrogen ions and hydroxide ions in a solution is an important indicator of the properties of the solution. A measurement system known as the **pH scale** indicates the relative concentrations of these two ions. See Figure 4–6. Why do we pay so much attention to the hydrogen ion? It is because the hydrogen ion contains no electrons surrounding its nucleus, and is therefore one of the most reactive ions in nature.

The pH scale ranges from 0 to 14. Pure water, in which the concentrations of H^+ and OH^- ions are equal, has a pH value of 7.0. Acids have pH values of less than 7. Strong acids, such as those that help digest food in the stomach, have pH values of 1 to 3. Bases have pH values greater than 7. Strong bases, like ammonia or lye, have high pH values, ranging from 11 to 14. The pH values within most cells range from 6.5 to 7.5. These near-neutral values are maintained by dissolved compounds that help prevent sharp, sudden swings of pH, which might cause damaging chemical changes within living tissues.

Suspensions

Some materials do not break into individual molecules when placed in water but still form pieces so small that they will not settle to the bottom of a container. The movement of water molecules keeps these small particles suspended. Such mixtures of water and nondissolved material are known as **suspensions**.

In living things, both solutions and suspensions are very important. A perfect example is the blood that circulates through your body. Blood, as you might have guessed, is mostly water. In that water are many dissolved compounds, including salt. So blood is a solution. But blood is also a suspen-

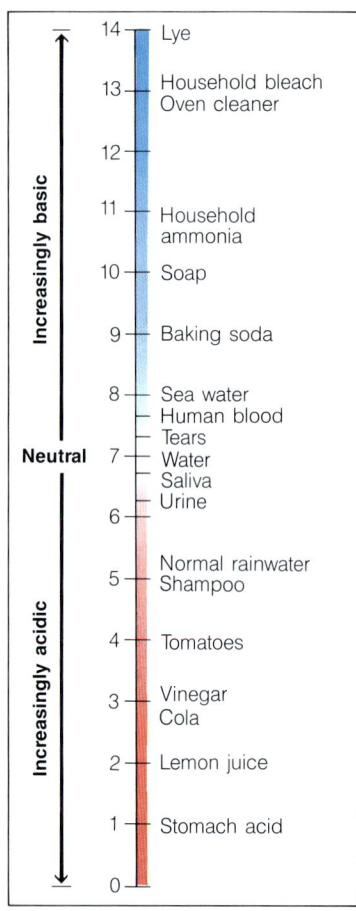

Figure 4–6 On the pH scale, 7 is neutral, acids are between 0 and 7, and bases are between 7 and 14. A difference of 1 between consecutive numbers on the scale is actually a difference of 10 times the relative strengths of the acids or bases. What cleaner is more basic? What body fluid is most acidic?

positive and negative ions. Emphasize that acids and bases are ionic compounds and that they are identified by the kind of ions they release in solution.

• **What kind of ions do acids release in solution?** (Hydrogen, or H^+, ions.)

• **What kind of ions do bases release in solution?** (Hydroxide, or OH^-, ions.)

SECTION REVIEW 4–1

1. Polar molecule, liquid at temperatures found on most of the Earth's surface, expands as it goes from liquid to solid, best solvent. Its polarity.
2. Substance composed of two or more elements or compounds that are mixed together but not chemically combined. Solutions and suspensions. In

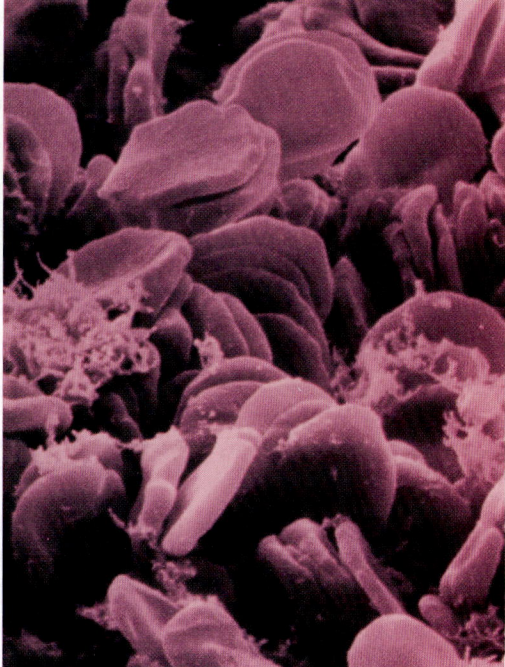

sion. It contains tiny structures that aid in clotting, as well as living cells and other suspended particles. See Figure 4–7. These components do not dissolve in blood. Instead, they form a suspension that circulates through the body—an extremely vital suspension made possible by the most important molecule found in living things: water.

Figure 4–7 *Mixtures of water and nondissolved material are called suspensions. Oil and water (left) is a common suspension. Perhaps less familiar to you is the suspension circulating through your body—blood! Blood contains clotting structures, living cells, and other particles, all suspended in a solution of water and other dissolved compounds (right).*

4–1 SECTION REVIEW

1. What are some important properties of water? What property accounts for its being the best solvent?
2. What is a mixture? What are two important types of mixtures? How do these two types differ?
3. What is a solution? A solvent? A solute?
4. Describe two important differences between acids and bases.
5. What is a neutralization reaction? Give an example.
6. **Critical Thinking—Making Predictions** Hydrogen fluoride (HF) is dissolved in pure water. Predict whether the pH of the solution will be greater than or less than 7.0.

67

ANNOTATION KEY

❶ Two electrons in the first energy level and 4 electrons in the second energy level. (Applying facts)

ECOLOGY NOTE
OIL TANKER SPILLS

Oil and water do not mix, and oil tanker spills have caused several ecological disasters over the years. The flow of oil into the sea causes both long-term and short-term damage. Fish die in large numbers, and seals, otters, and sea birds also are affected. The oil poisons animals. It also clings to fur and feathers, which weighs down and may drown animals. European otters sometimes die from the cold as their fur becomes matted with oil.

An oil slick affects the whole ecosystem. At best, we can only partially clean up oil spills. The only way to ensure the safety of marine life is to prevent spills from happening.

TEACHING STRATEGY 4–2

Focus/Motivation

Begin by asking students,
• **Suppose you walked into an ice cream store and the owner offered you a free sample of ethyl butyrate ice cream. Would you take it? What do you think ethyl butyrate is?** (Accept all answers.)

Content Development

Explain to students that ethyl butyrate is pineapple—or to be more precise, the compound that gives pineapple its distinctive flavor. Point out that ethyl butyrate is an organic compound.
• **What are organic compounds?** (Compounds that contain carbon.)

Explain that ethyl butyrate is one of a large group of organic compounds called esters. Esters give taste and odor to many natural and artificial substances.

Guide For Reading

■ What are the four most abundant elements in living things?
■ How does an organic compound differ from an inorganic compound?
■ How do the properties of the element carbon make it important to life?
■ Why is polymerization important to living things?

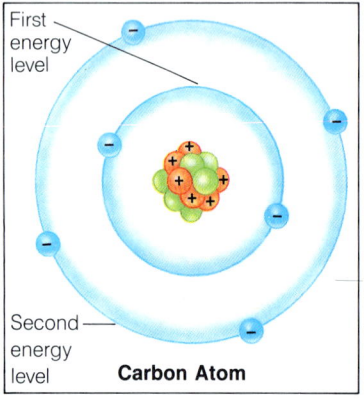

Figure 4–8 A carbon atom contains 6 protons, 6 neutrons, and 6 electrons. What is the arrangement of these 6 electrons? ❶

Content Development

List on the chalkboard the major elements that are found in living things—hydrogen, oxygen, nitrogen, and carbon. Have students discuss the things that they know about these elements. For example, the first three elements are gases; and nitrogen and oxygen make up most of the earth's atmosphere. Carbon is a solid that exists in many familiar forms—among them, graphite, coal, charcoal, and diamonds.

4–2 Chemical Compounds in Living Things

Although the Earth's crust contains 90 naturally occurring chemical elements, only 11 of these elements are common in living organisms. Another 20 are found in trace amounts. Just four elements—carbon, nitrogen, oxygen, and hydrogen—make up 96.3 percent of the total weight of the human body. **In varying combinations, the elements carbon, hydrogen, oxygen, and nitrogen make up practically all the chemical compounds in living things.** To make the study of these and all other chemical compounds easier, scientists have divided them into two groups: **organic compounds**, which contain carbon, and **inorganic compounds**, which do not.

Inorganic Compounds

Inorganic compounds are primarily those compounds that do not contain carbon. One exception to this definition is carbon dioxide, which although it does contain carbon is an inorganic compound. The natural world is dominated by such compounds. Water is inorganic, as are the minerals that make up most of the sand, soil, and stone of the Earth's landmasses.

Living things contain a great many inorganic compounds, ranging from water to carbon dioxide to calcium phosphate, a mineral from which bones are formed. The group of compounds known as salts that help to balance the pH of the blood are largely inorganic.

Organic Compounds

Organic compounds are carbon-containing compounds. A special branch of chemistry called organic chemistry deals with the chemistry of carbon and its more than 2 million compounds. Why is carbon so special?

Carbon is a unique element because of its remarkable ability to form covalent bonds that are strong and stable. You will recall that covalent bonds involve the sharing of electrons. Carbon has 6 electrons, 2 in the first energy level and 4 in the second. So only 4 of the 8 positions in its outermost energy level are filled. This means that carbon can form four single covalent bonds. The simplest compound that can be formed from carbon is methane, CH_4. Carbon can also form covalent bonds with oxygen, nitrogen, phosphorus, and sulfur atoms. The ability to bond easily and form compounds with these common elements would be enough to make carbon an interesting element. But there's more!

Carbon can form chains of almost unlimited length by bonding to other carbon atoms. The bonds between carbon atoms in these straight chains can be single, double, or triple

Reinforcement/Reteaching

Review with students the properties of covalent bonds. Also review the idea that atoms seek to share electrons in order to obtain a stable outermost energy level of 8 electrons.

covalent bonds—or combinations of these bonds. No other element can equal carbon in this respect. These chains can be closed on themselves to form rings. The ring structures may include single or double bonds, or a mixture of both. This gives even more variety to the kinds of molecules that carbon can form. See Figure 4-9.

Figure 4-9 Because of its remarkable ability to form a variety of covalent bonds, carbon is an unparalleled element. Here you see some of the different types and arrangements of carbon bonds.

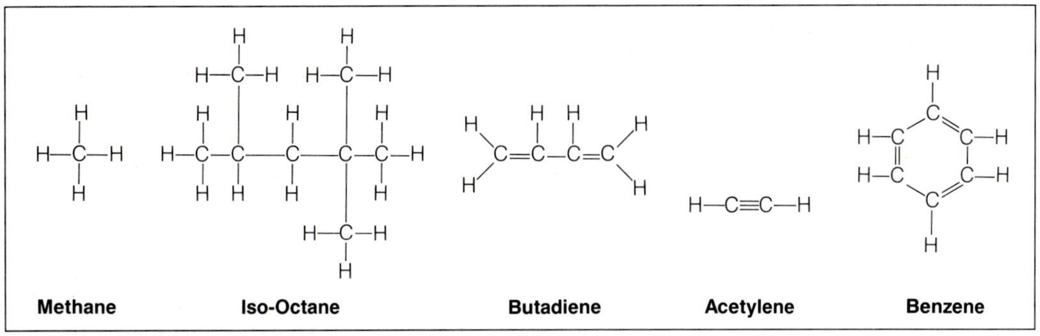

Methane   Iso-Octane   Butadiene   Acetylene   Benzene

Polymerization

Many carbon-based compounds are formed by a chemical process known as **polymerization**, in which large compounds are constructed by joining together smaller compounds. The smaller compounds, or **monomers**, are joined together by chemical bonds to form **polymers**. Many polymers are so large that they are called **macromolecules**. As used here, the prefix *macro-* means giant.

Polymerization provides a way to form complex molecules by joining monomers together. The chemical diversity that polymerization allows living things is similar to the diversity that our alphabet allows us. Although there are only 26 letters in the alphabet, our ability to join them together (polymerize them) to form words gives us an almost infinite variety of possible words (molecules).

Figure 4-10 A polymer is made up of a series of monomers.

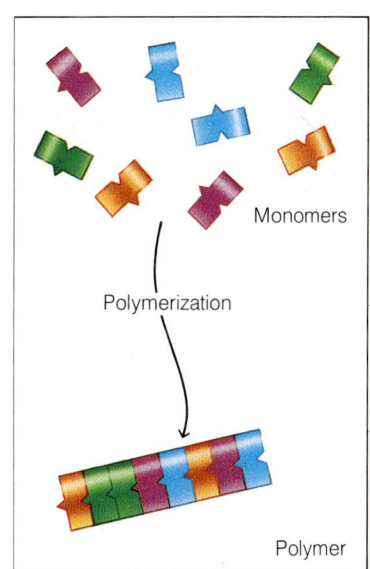

4-2 SECTION REVIEW

1. What are the four most abundant elements in living things?
2. What is an inorganic compound? An organic compound?
3. Is the chemical composition of the human body similar to the composition of the Earth's crust? Explain.
4. What special properties of carbon make it such an important element in living things?
5. **Critical Thinking—Describing Processes** Describe the process of polymerization.

BACKGROUND INFORMATION
DIGESTING CELLULOSE

Why can't we human beings eat paper and wood? At first glance, it seems that we should be able to. These substances are made up primarily of cellulose, and cellulose is composed of glucose—a simple sugar that human beings digest all the time when they eat starches. The key factor is the way the glucose molecules are joined together. Simple sugars are joined in starches by what chemists call alpha bonds. In cellulose, however, the glucose molecules are joined by beta bonds. The human digestive system has the enzyme necessary to break apart alpha bonds but not beta bonds. Thus munching on substances made of cellulose provides no food value for humans.

ANNOTATION KEY

❶ The arrangement of the individual atoms. (Interpreting diagrams)
❷ Disaccharide. (Interpreting diagrams)

TEACHING STRATEGY 4–3

Focus/Motivation

Begin by asking students,
• **What have you eaten so far today?** (Accept all answers.)
• **What foods did you eat yesterday after you left school?** (Accept all answers.)
• **How do you think your body used each of the foods you ate?** (Answers will vary. Some students may give general answers such as "the food was used for energy"; others may realize that certain foods contain proteins, carbohydrates, or fats, and that each of these has specific uses.)

Content Development

Point out that three of the most important types of organic compounds found in living things are present in the foods we eat. These three types of compounds are carbohydrates, lipids, and proteins. Further point out that carbohydrates are the substances that we often refer to as sugars and starches, while lipids are familiar in the form of fats and oils.

Guide For Reading

■ What four groups of organic compounds are found in living things?
■ What are the structure and function of each group of compounds of life?
■ How do enzymes work, and why are they important to living things?

Figure 4–11 Glucose, fructose, and galactose are single sugars, or monosaccharides. Sugars form a group of organic compounds known as carbohydrates. The formula for all three of these single sugars is $C_6H_{12}O_6$. What makes them different from one another? ❶

Skills Development
Guided Practice
Skill: Interpreting diagrams
Using the Visuals Have students observe closely Figure 4–11.
• **What differences do you see between the molecule of glucose and the molecule of fructose?** (In glucose, the C=O is on the top carbon, whereas in fructose, it is on the second carbon. In all other respects, the molecule of glucose and the molecule of fructose are identical.)
• **How is the molecule of galactose different from the molecule**

4–3 Compounds of Life

The number of possible organic compounds is almost limitless. Fortunately, however, it is possible to classify many important organic compounds found in living things into four groups. **The four groups of organic compounds found in living things are carbohydrates, lipids, proteins, and nucleic acids.** By knowing the characteristics of just these groups, you will know a great deal about the chemistry of living things.

Carbohydrates

You are probably quite familiar with the group of organic compounds known as **carbohydrates**. Carbohydrates, you see, are the molecules that we often call sugars and starches. Carbohydrates are made up of carbon, hydrogen, and oxygen atoms. Note that carbohydrates contain two hydrogen atoms for each oxygen atom. This is the same ratio of hydrogen and oxygen atoms that you find in a molecule of water. The simplest carbohydrates are called **monosaccharides**, meaning single sugars. Examples of monosaccharides include glucose, galactose, and fructose. Glucose is the sugar green plants produce. Galactose is found in milk. And fructose, the sweetest of these simple sugars, is found in fruits. The formula for these three simple sugars is $C_6H_{12}O_6$. What makes them different from one another is the arrangement of the individual atoms. See Figure 4–11.

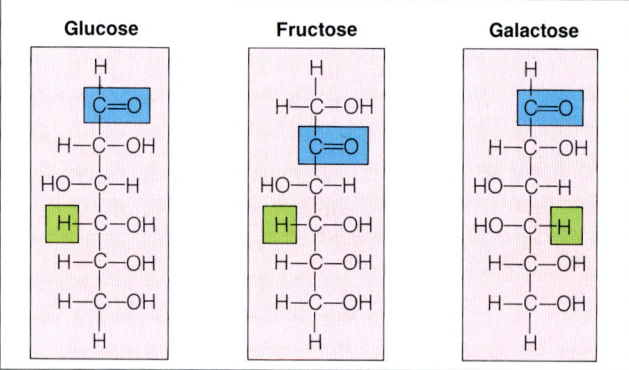

Sugars are important because they contain a great deal of energy. This energy is stored in the bonds that make up the carbohydrate molecules. When the bonds are broken, the energy is released. Nearly all organisms use glucose as one of their basic energy sources. In Chapter 6, you will learn how the energy from sugar molecules is used by living things.

DEHYDRATION SYNTHESIS Complex carbohydrates are made by a process of polymerization in which two or more

70

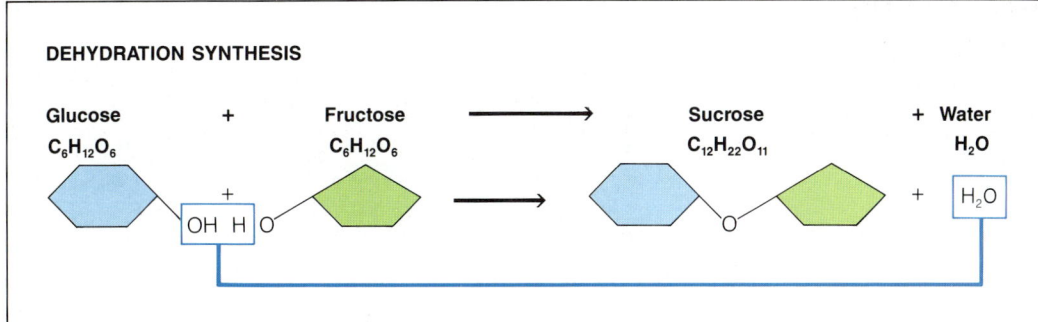

Figure 4–12 *The dehydration synthesis of a molecule of glucose and a molecule of fructose produces a molecule of sucrose and a molecule of water. What type of sugar is sucrose?* ❷

monosaccharides are combined to form larger molecules. The chemical bond that links two simple sugars is formed between the –OH groups present in each molecule. As you can see from Figure 4–12, one OH from one molecule combines with the H of the OH from the other molecule. When the bond is complete, a molecule of water is removed from the two monosaccharides. Because of the loss of water, the joining of two sugars is known as **dehydration synthesis**. Dehydration means loss of water, and synthesis means putting together.

The compound formed from the joining of two single sugars in dehydration synthesis is called a **disaccharide**, or double sugar. Ordinary table sugar, or sucrose ($C_{12}H_{22}O_{11}$), is a disaccharide. Other disaccharides include maltose (malt sugar) and lactose (milk sugar).

POLYSACCHARIDES Very large molecules can be formed by joining together many monosaccharide units. Such compounds are known as **polysaccharides**. Polysaccharides are the form in which living things store excess sugar. One important polysaccharide is starch. Plants store excess sugar in the form of starch, which is present in potatoes and grains. Starch is a very large molecule formed by joining together hundreds of glucose molecules. Animals store their excess sugar in the form of glycogen in the liver and muscles. Glycogen is an even larger molecule consisting of hundreds or even thousands of glucose molecules. Glycogen is sometimes called animal starch. Do you see why? Because they are polymers of single sugars, both starch and glycogen help store energy in living things. Another polysaccharide is cellulose, which is found only in plants. Cellulose helps to support a plant by giving it strength and rigidity. Cellulose is the major component of wood. As such, it is often used as a building material or a printing material. The page upon which these words are printed is made principally of cellulose.

Figure 4–13 *Cellulose fibers (top) are extremely strong and thus able to support the enormous mass of these giant sequoia trees (bottom).*

FACTS AND FIGURES

Have you ever wondered why cold margarine spreads more easily than cold butter? The reason is a difference between saturated and unsaturated fats. The double bonds in unsaturated fatty acids, which are found in margarine, produce little "kinks" in the hydrocarbon chains. These kinks make it difficult for margarine to solidify at low temperatures. Saturated fats, on the other hand, have no such kinks. Thus butter forms a stiff, solid mass when refrigerated.

TIE-IN HEALTH

Based on research, many doctors and scientists feel that people should limit their intake of saturated fats. It is believed that eating too much saturated fat—that is, animal fat—contributes to heart disease. That is why people with heart or other circulatory diseases are often told to replace butter with margarine and to eat less red meat.

The average American diet contains about 40% fats, close to twice the necessary amount. Over half the Calories in eggs, peanut butter, and many meats and cheeses come from fats.

4–3 (continued)

Content Development
Explain to students how carbohydrates are oxidized for energy in the human body. Point out that before carbohydrates can be oxidized, the complex carbohydrates must be broken down by the digestive system into simple sugars.
• **What is this process called?** (Hydrolysis.)
Explain that once complex carbohydrates have been broken down, the simple sugars combine with oxygen according to the following equation:

$$C_6H_{12}O_6 + 6O_2 \rightarrow CO_2 + 6H_2O + energy$$

• **Where does the oxygen for this reaction come from?** (Air.)
• **What is the most important product in this reaction?** (Energy.)

• **What is it used for?** (To carry out all body functions.)
• **What happens to the carbon dioxide that is a product of this reaction?** (It is exhaled in the breathing process.)
• **Can the human body use carbon dioxide?** (No.)

Stress that the equation sums up the life-sustaining process that utilizes oxygen and releases carbon dioxide.

Content Development
Using the Visuals Direct students' attention to Figure 4–15.

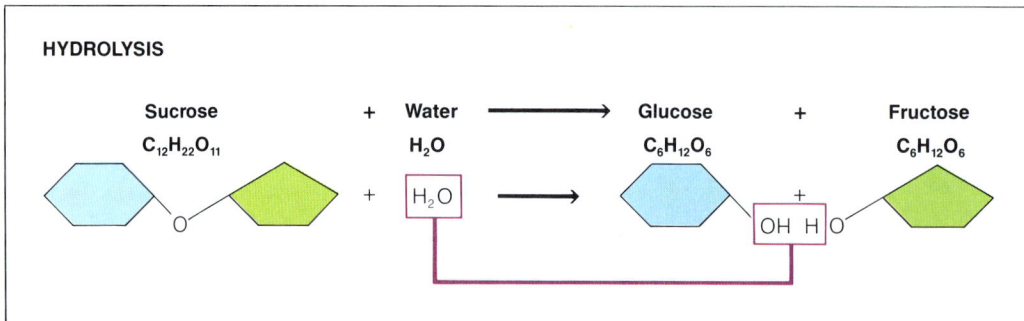

Figure 4–14 The hydrolysis of a molecule of sucrose produces a molecule of glucose and a molecule of fructose.

HYDROLYSIS When polysaccharides are split apart to again form monosaccharides, the dehydration synthesis reaction is reversed. This reverse reaction is known as **hydrolysis**. Hydrolysis, which means water splitting, is an appropriate name because a molecule of water is consumed by the chemical reaction that splits the bond between adjacent monosaccharides. Figure 4–14 illustrates the hydrolysis reaction.

Lipids

Lipids are organic compounds that are waxy or oily. Lipids have three major roles in living organisms. Like carbohydrates, lipids can be used to store energy. Lipids are used to form biological membranes. And certain lipids are used as chemical messengers. The common names by which we know lipids are fats, oils, and waxes. Generally, fats and waxes are solid at room temperature, whereas oils are liquid.

Many important lipids are formed from combinations of fatty acids and glycerol. Fatty acids are long chains of hydrogen and carbon atoms that have a carboxyl group attached at one

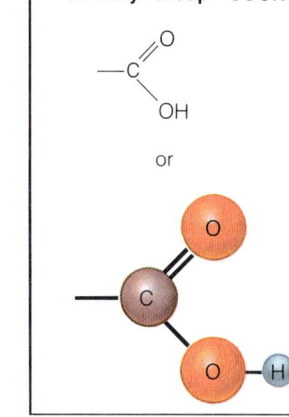

Figure 4–15 All fatty acids contain long chains of carbon and hydrogen atoms to which a carboxyl group is attached. A carboxyl group consists of 1 carbon atom, 1 hydrogen atom, and 2 oxygen atoms (left). Fatty acids are important in the formation of lipids. Lipids are organic compounds that are most familiar as fats, oils, and waxes (right).

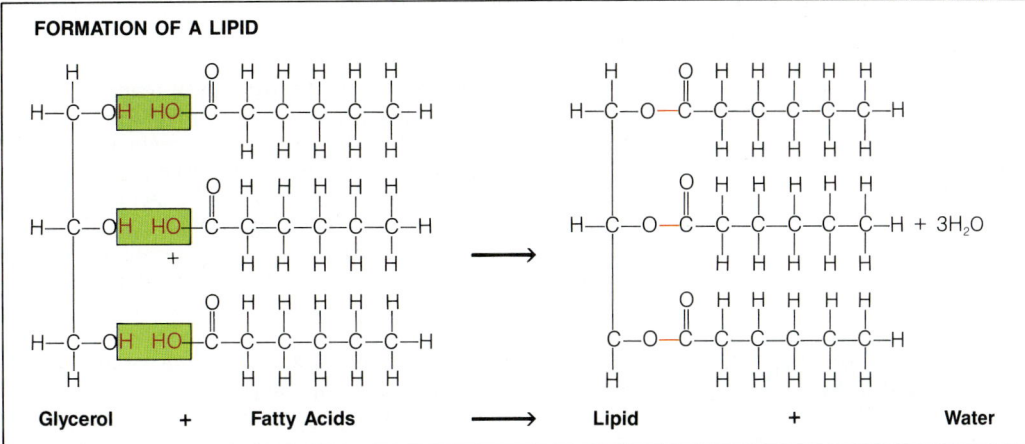

FORMATION OF A LIPID

Glycerol + Fatty Acids → Lipid + Water

end. A carboxyl group is a chemical group consisting of one carbon atom, one hydrogen atom, and two oxygen atoms (–COOH). See Figure 4-15. Glycerol, which is an organic alcohol, contains three carbon atoms, each of which is attached to a hydroxyl (–OH) group. Many lipids are formed by the attachment of two or three fatty acids to glycerol. Figure 4-16 shows how fatty acids combine with glycerol to form a lipid. Can you recognize the chemical reaction that is responsible for the formation of the lipid? ❷

SATURATED AND UNSATURATED LIPIDS If every carbon atom in a fatty acid chain is joined to another carbon atom by a single bond, the fatty acid is said to be saturated because it contains the maximum number of hydrogen atoms. If a pair of carbon atoms is joined by a double bond, the fatty acid is said to be unsaturated. Because of the double bond, it does not contain the maximum number of hydrogen atoms. If a fatty acid contains several double bonds, it is said to be polyunsaturated. Figure 4-18 on page 74 shows an example of each of these types of fatty acids.

Lipids made from saturated fatty acids are called saturated fats. Such fats are commonly found in meats and most dairy products. Lipids made from polyunsaturated fatty acids are called polyunsaturated fats. If that term seems familiar, it should. Polyunsaturated fats tend to be liquid at room temperature and are used in many cooking oils, such as sesame, peanut, and corn oil. Replacing saturated fats in the diet with polyunsaturated fats may help to prevent heart disease, a connection we will explore more completely in Chapter 41.

Both plants and animals use lipids as a means of storing energy. Because lipids contain far fewer oxygen atoms than carbohydrates do, they have less mass per unit of chemical energy than carbohydrates. In other words, when lipids are

Figure 4-16 The formation of a lipid involves the combination of three fatty-acid molecules with one glycerol molecule. What type of chemical reaction is taking place here? ❶

Figure 4-17 Animals use lipids as a means of storing energy. When black bears retreat to their dens in late autumn, they lie down and often cease to move, eat, drink, and eliminate wastes for about 5 months. Burning up nearly 4000 calories per day, a bear's metabolism during hibernation operates at 50 to 80 percent its normal rate. The energy the bear needs comes primarily from its stored lipids.

BACKGROUND INFORMATION
SOAP

An interesting product made from fats is soap. Soaps are made by boiling oils or solid fats with solutions of sodium hydroxide or potassium hydroxide. This process is called saponification. During saponification, the fat molecules split apart, resulting in an acid portion and an alcohol portion. The acid portion of the molecule combines chemically with the sodium or potassium ions in the solution, forming a salt of the organic acid. It is this acid salt that is used as soap.

During the Middle Ages, soaps were made by boiling animal fat or olive oil in water containing ashes from wood or seaweed. This method worked because these ashes contain sodium and potassium bases.

ANNOTATION KEY

❶ Dehydration synthesis. (Interpreting diagrams)
❷ Dehydration synthesis. (Relating facts)

• **What is similar in these two reactions?** (Atoms of hydrogen and oxygen are removed, forming molecules of water.)
• **What is the general name of this type of reaction?** (Dehydration synthesis.)

Content Development
Emphasize to students that saturated fats generally come from animals, while unsaturated fats are generally found in plants. Also point out that fats are among the large group of organic compounds called esters.

• **How are the bonds formed in a lipid molecule?** (By the splitting off of water when the –OH group on the glycerol and –COOH group on the fatty acid react with each other.)

Write –COOH on the chalkboard and draw its structure, emphasizing the C=O and the –OH. Explain that the functional group –COOH identifies organic acids. Point out that various types of organic compounds can be identified by functional groups. One such functional group is the hydroxyl group, –OH, which identifies alcohols.
• **Where on these pages do you see one or more –OH group?** (On the glycerol molecule in Figure 4-16.)
• **What does this tell you about glycerol?** (It is an alcohol.)

Skills Development
Guided Practice
Skill: Comparing
Have students compare the formation of a lipid with the formation of a disaccharide.

ESL STRATEGY

Have students use the dictionary to locate the original meaning of the terms listed below. Instruct students to make flashcards out of index cards, putting the term on one side and the original meaning on the other. Then instruct students to combine the flashcards in various orders to form longer terms in the chapter, such as

poly + saccharide, de + oxy + ribo + nucleic, hydro + lysis

Tell students that they may need to add or subtract a letter or two to form certain terms in the chapter.

Terms

hydro-	mono-
di-	tri-
sub-	nucle-
micro-	oxy-
carbo-	nitro-
sacchar-	-lysis
macro-	solv-, solu-
-gen	de-
-some, soma-	poly-
-pep-	ribo-
-mer	-ide

ANNOTATION KEY

❶ Oleic: has a double bond between carbon atoms; linoleic: has two double bonds between carbon atoms. (Interpreting diagrams)

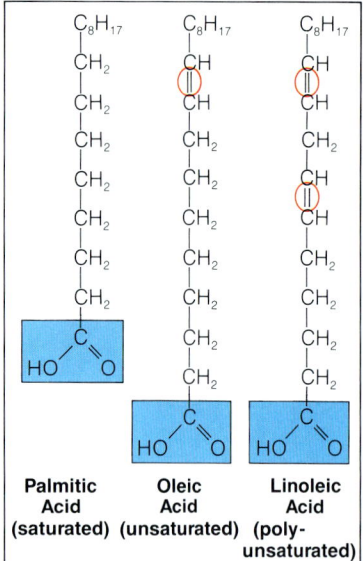

Figure 4-18 Palmitic acid is a saturated fatty acid. Why do we say that oleic acid is unsaturated? Why is linoleic acid polyunsaturated? ❶

Figure 4-19 All amino acids have an amino group ($-NH_2$) on one end and a carboxyl group ($-COOH$) on the other. They differ in a region of the molecule known as an R group.

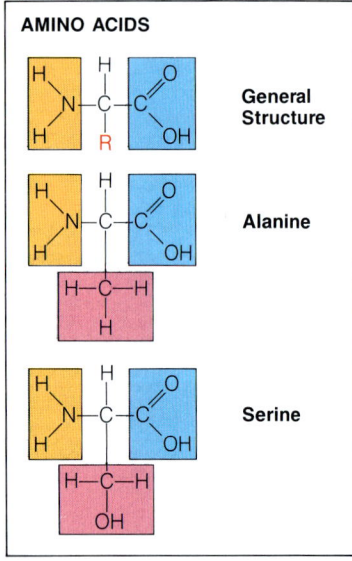

broken down, they produce more energy gram for gram than carbohydrates do.

STEROLS AND PHOSPHOLIPIDS Two other kinds of lipids are particularly important in living organisms. They are sterols and phospholipids. One of the most common sterols is a molecule called **cholesterol**. Cholesterol is an important part of many cells, but excessive cholesterol in the diet is a risk factor in heart disease. Sterol lipids play a number of important roles in building cells and carrying messages from one part of the body to another. We will learn more about sterols later in this textbook.

Phospholipids are molecules consisting of parts that dissolve well in water and parts that do not dissolve well in water. The portions of a phospholipid molecule that do not dissolve in water are oily. What happens to a molecule made of two very different parts? When phospholipid molecules are mixed with water, they may form small balloonlike structures known as liposomes. Each liposome is formed from a double layer, or bilayer, of lipid molecules. A liposome forms spontaneously, or without any outside help. It forms merely by the attraction of the oily parts of the lipid molecules for each other and by the attraction of the other parts of the lipid molecules for the surrounding water. The ability to form bilayers spontaneously is an important property of lipids that enables them to play a key role in forming cell membranes.

Proteins

Proteins are organic compounds that contain nitrogen in addition to carbon, hydrogen, and oxygen. Proteins are polymers of **amino acids**. An amino acid has an amino group ($-NH_2$) on one end and a carboxyl group ($-COOH$) on the other. These groups can form covalent bonds with each other. As a result of these bonds, very long chains of amino acids can be put together. All amino acids have a similar chemical structure, but they differ in a region of the molecule known as an R group. See Figure 4-19. There are more than 20 different amino acids, each of which contains a different R group.

PEPTIDES The covalent bond that joins two amino acids is known as a **peptide bond**. A molecule of water is lost when a peptide bond is formed between two amino acids. This reaction is another example of dehydration synthesis. A dipeptide is two amino acids joined by a peptide bond. A tripeptide contains three amino acids. And, as you might expect, a polypeptide is a long chain of amino acids.

PROTEIN STRUCTURE A complete protein contains one or more polypeptide chains and may contain a few other chemical groups that are important to its proper function. Proteins

---

4-3 (continued)

Reinforcement/Reteaching
Have students make a list of the fats and oils that they can find around their homes. Included should be such things as cooking oils, butter, wax, margarine, and lard, as well as products that contain fats, such as cheese or meat. Have students note on their lists whether each fat or oil is saturated or unsaturated.

Content Development
Emphasize to students that lipids have several important functions in the body, only one of which is to provide energy. Stress that sterols and phospholipids are important in the building of cells and cell membranes. Also point out that although some Americans do eat a diet excessive in fats, others who are weight or health conscious may try to eliminate fats completely.

• **Do you think it's a good idea to eat a diet with little or no fat?** (Accept all answers.)
• **What do you think might happen in the body if there were no fats available?** (Possible an-

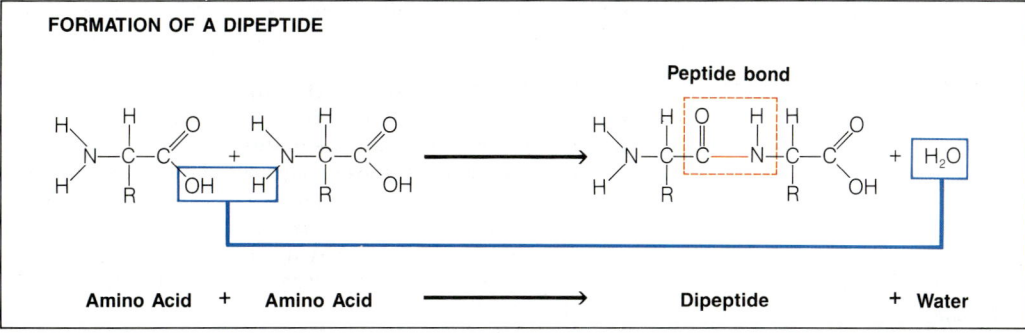

Figure 4-20 Peptide bonds—which result in the formation of dipeptides, tripeptides, and polypeptides—form during the dehydration synthesis of amino acids.

have numerous roles: They help to carry out chemical reactions; they pump small molecules in and out of cells; and they are even responsible for the ability of cells to move. The functions of proteins are at the very center of life itself.

Enzymes

Chemical reactions make life possible. Hundreds of chemical reactions are involved in a process as simple as digesting a chocolate bar. If these chemical reactions proceeded too slowly, not only would the chocolate bar remain in the stomach for a long time, but the ordinary activities of life would come to a halt as well. Since this is not the case, some substances in the body must be responsible for speeding up the process.

A substance that speeds up the rate of a chemical reaction is called a **catalyst**. Catalysts are not changed by the reactions

Figure 4–21 Proteins are found in a variety of substances. These photographs show some of the more common proteins: collagen used in tennis racket strings (top, left), keratin in a peacock feather (top, right), silk from a spider's web (bottom, left), and human hair (bottom, right).

BACKGROUND INFORMATION
TYPES OF PROTEINS

Proteins are essential in the growth and maintenance of cells in both structure and function. The basic types of proteins are as follows:
1. Proteins that are structural, such as the keratins of fingernails, skin, and hair; the collagens of connective tissue; and the lipoproteins of cell membrane systems.
2. Proteins that are hormones (regulators of metabolic processes), such as insulin.
3. Proteins that transport oxygen, for example, hemoglobin.
4. Proteins that make up chromosomes, such as histones.
5. Proteins that are organic catalysts, or enzymes.

BACKGROUND INFORMATION
POLYPEPTIDES

The distinction between a protein and a polypeptide is a subtle one. Most biochemists like to define a protein as a functional unit. A simple protein may consist of just one polypeptide, and in such cases the terms *protein* and *polypeptide* are synonymous. However, more complex proteins often consist of several polypeptides, as well as other molecules. The other molecules may include metal ions or even organic compounds, like the porphyrin rings that bind iron in hemoglobin.

swer: the building of cells and cell membranes might suffer; cells in the body might become dysfunctional or diseased in some way; certain body functions might be impaired.)

Reinforcement/Reteaching
Using the Visuals Have students observe Figure 4–20.
• **What types of molecules form a dipeptide?** (Amino acids.)
• **What parts of the molecules come together?** (The amino group of one molecule joins with the acid group of another molecule.)
• **What atoms are split off as the molecules join?** (2 H's and 1 O.)
• **What do these atoms form?** (A molecule of water.)
• **Have you seen this type of reaction before? When?** (Yes. In the formation of a disaccharide and in the formation of a lipid.)
• **In general, what is this type of reaction called?** (Dehydration synthesis.)
• **What is the bond called that joins the two amino acids?** (Dipeptide bond.)
• **What atoms are involved in this bond?** (C, H, O, and N.)
• **Are all the bonds single covalent bonds?** (No.)
• **Which bond is double?** (The C-O bond.)

75

ANNOTATION KEY

① To catalyze, or speed up, a chemical reaction. (Relating facts)

TEACHING SUPPORT

Laboratory Manual
- Identifying Organic Compounds, p. 89

Teaching Resources
- Vocabulary Review: Word Game, p. 1

Study Guide
- Section 4–3, p. 38

4–3 (continued)

Content Development

Discuss with students the role of a catalyst in a chemical reaction. The "start-up" energy of a reaction is called the *activation energy*. Until the activation energy is reached, a chemical reaction will not occur. A catalyst speeds up a reaction by lowering the activation energy. Emphasize that a catalyst is not consumed or changed during a chemical reaction.

Content Development

Point out to students that the names of enzymes often provide clues to their function. Usually, an enzyme is named for the reaction that it causes or for the substance upon which it acts. For example, lipase is the enzyme that breaks down lipids. Sucrase acts on the substrate sucrose. DNA polymerase is named for the reaction that it causes—the synthesis of DNA molecules.

SECTION REVIEW 4–3

1. Carbohydrates (sugar), lipids (cholesterol), proteins (enzymes), nucleic acids (DNA).
2. Monosaccharide: simple sugar; disaccharide: two simple sugars linked together; polysaccharide: many sugars linked together.

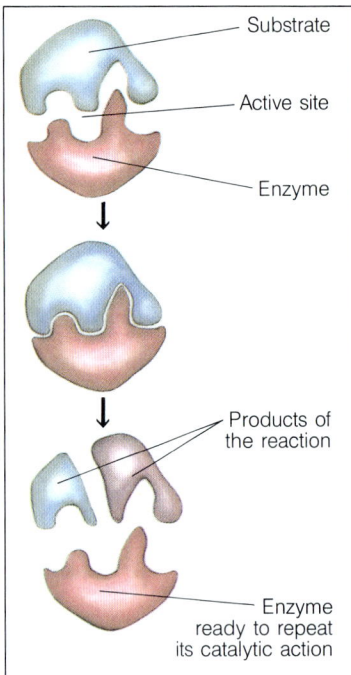

Figure 4–22 In one hypothesis of enzyme action, an enzyme and its substrate bind at a region known as the active site in a manner similar to the way in which two pieces of a jigsaw puzzle fit together. What is the role of an enzyme? ①

Figure 4-23 Nucleic acids are polymers of nucleotides, which are molecules built up from three basic parts: a special 5-carbon sugar, a phosphate group, and a nitrogenous base.

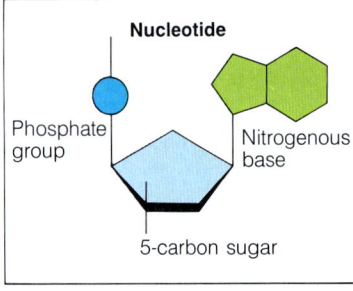

they promote, and therefore they are not used up during the reaction. Catalysts work by lowering the "start-up" energy of a reaction. Chemists often seek catalysts that will speed up reactions important to industry. Living organisms have gone the industrial chemist one better—they contain their own special catalysts, which are known as **enzymes.**

With a few important exceptions, enzymes are proteins. Understanding their function is an important part of the study of proteins. Simple cells may have as many as 2000 different enzymes, each one catalyzing a different reaction. An enzyme may accelerate a reaction by a factor of 10^{10}—that is, making it happen 10,000,000,000 times faster! Thus a reaction that might take as long as 1500 years without an enzyme can be completed in just 5 seconds!

Enzymes speed up a reaction by binding to the reactants, which, as you may recall, are the substances that enter into a chemical reaction. The reactants that are affected by an enzyme are known as **substrates**. Substrates bind to enzymes at a region known as the **active site**.

The way in which a chemical reaction is catalyzed varies from one enzyme to another. An enzyme can catalyze a reaction by holding two substrates in positions in which they can react with each other. Or an enzyme can catalyze a reaction by twisting a substrate molecule slightly so that a chemical bond is weakened and broken.

Enzymes are very specific. A particular enzyme can catalyze only one particular chemical reaction involving specific substrates. Scientists theorize that this has something to do with the shape of the enzyme's active site. In fact, the fit between an enzyme's active site and its substrate is often compared to that of a lock and key.

Enzymes are important in regulating chemical pathways, synthesizing materials needed by cells, releasing energy, and transferring information. You will discover that enzymes are involved in digestion, respiration, reproduction, vision, movement, thought, and even in the making of other enzymes.

Nucleic Acids

Nucleic acids are large complex organic molecules composed of carbon, oxygen, hydrogen, nitrogen, and phosphorus atoms. Nucleic acids are polymers of individual monomers known as **nucleotides**. Nucleotides are molecules built up from three basic parts: a special 5-carbon sugar, a phosphate group, and a molecule generally known as a nitrogenous base. See Figure 4–23. Individual nucleotides can be linked together by covalent bonds to form a polynucleotide. There are two basic kinds of nucleic acids: **ribonucleic acid (RNA)**, which contains the sugar ribose, and **deoxyribonucleic acid (DNA)**, which contains the sugar deoxyribose. Despite their name, nucleic acids are not strongly acidic.

3. Lipids are waxy or oily organic compounds. Lipids are used to store energy, in the formation of biological membranes, and as chemical messengers.
4. Proteins are polymers of amino acids, each of which has a –NH₂ at one end and a –COOH at the other end.
5. An enzyme is a catalyst that speeds up the rate of chemical reactions and enables the important metabolic reactions in living things to occur quickly.
6. Nucleic acids are polymers of monomers called nucleotides. Each nucleotide contains a 5-carbon sugar, a phosphate group, and a nitrogenous base. DNA and RNA are two nucleic acids.

Nucleic acids store and transmit the genetic information that is responsible for life itself. How they do this, and how that information is decoded and transferred, is a fascinating story that we will leave for Chapter 7.

4-3 SECTION REVIEW

1. What are the four groups of organic compounds found in living things? Give an example of each.
2. Distinguish between monosaccharides, disaccharides, and polysaccharides.
3. What are lipids? How are they important to living things?
4. Describe the structure of a protein.
5. What is an enzyme? What is its function in living things?
6. **Critical Thinking—Relating Facts** Describe the structure and function of nucleic acids. What are two important nucleic acids?

SCIENCE, TECHNOLOGY, AND SOCIETY — BREAKTHROUGH

The Structure of Proteins

Although proteins are among the largest of all macromolecules, it is sometimes possible to determine their detailed structure, including the position of every atom. This is done with the aid of a technique called X-ray crystallography.

The first, and often the trickiest, step in the procedure is to grow small crystals of a protein—often no larger than a grain of salt (0.1 mm). These protein crystals are then placed in a finely tuned beam of X-rays. As the X-rays pass through the crystal, they are scattered in a pattern that is recorded on film and then analyzed by computers. These X-ray scattering patterns contain enough information to allow scientists to build a complete model of the protein. The accompanying photograph, showing the arrangement of proteins on the surface of a common cold virus, was obtained by this technique. X-ray crystallography is a powerful scientific tool that enables us to visualize protein structure and understand the biological activities of complex molecules.

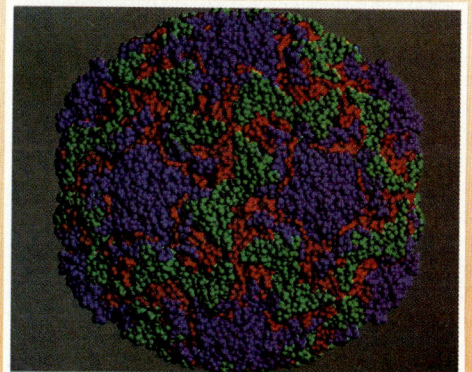

SCIENCE, TECHNOLOGY, AND SOCIETY — BREAKTHROUGH

The Structure of Proteins

The example of X-ray crystallography described and pictured in this feature is from the work of Michael Rossmann at Purdue University. In 1985, Rossmann developed a procedure for viewing rhinovirus 14, a common cold virus. Rossmann began his work by making a crystal of pure virus. Then he used X-ray crystallography to generate images of slices of the virus. One problem that Rossmann and his research team faced was that a virus is three-dimensional, while the film used in X-ray crystallography is two-dimensional. In order to overcome this limitation, the team took pictures of the virus from different angles, then put them together in a three-dimensional picture. This technique produced 6 million pieces of information about the structure of rhinovirus 14. The result was a computer-generated model of rhinovirus 14.

Rossmann's research has been used to find out how the virus attacks living cells. From the computer model, scientists learned that the surface of the virus consists of ridges and valleys. It is believed that the valleys enable the virus to attach itself to a cell. Antibodies cannot fit into the valleys, so the virus is left undisturbed. Scientists hope to develop drugs that will fight viral attacks, but two factors make this difficult. One factor is that every cold virus differs in structure, and there are about 100 different viruses that cause the common cold alone. The second factor is that viruses frequently change by mutation—which means that the surface structure studied today may be different by tomorrow.

Reinforcement/Reteaching
If some students are having difficulty in answering the Section Review questions, review the material with which they are having difficulty.

Closure
Divide the class into four groups. Assign each group one of the following topics: carbohydrates; lipids; enzymes; proteins and nucleic acids. Ask each group to write 6 to 8 questions on index cards about the topic they were assigned. When the groups are finished, put all of the questions face down on a desk or table. Have members of the class draw questions at random to answer, then discuss.

LABORATORY INVESTIGATION

TESTING ENZYME ACTIVITY

Before the Lab

1. At least one day prior to the investigation, gather enough materials for your class, assigning 6 students per group.
2. Aqueous solutions of 3% hydrogen peroxide can be obtained at a drugstore. The 3% concentration is by mass, which is equal to a 0.90 M solution.
3. Prepare the liver puree by placing fresh raw beef liver in a blender with a small amount of water. Blend at high speed for several minutes until the puree is smooth. Keep the puree refrigerated. For optimum catalase activity, do not use frozen liver.

Pre-Lab Discussion

Write the formula for hydrogen peroxide, H_2O_2, on the chalkboard. Then write the following equation for its decomposition:

$$2H_2O_2 \rightarrow 2H_2O + O_2$$

- **If hydrogen peroxide is a liquid, what might you expect to see when it decomposes?** (Accept all reasonable answers. The correct answer is that bubbles of gas will be visible as oxygen is released and water remains.)

Read aloud or have a student read aloud the Problem at the top of the investigation. Explain to students that the function of the enzyme catalase is to speed up the decomposition of hydrogen peroxide.

Also review at this time the concept of pH. Remind students that pH is a measure of acidity or basicity. If students are unfamiliar with using pH-sensitive papers to measure pH, take time now to demonstrate their use.

- **Do you have any ideas about how pH might affect enzyme activity? Do you think that there is an ideal pH for enzymes?** (Accept all reasonable answers.)

LABORATORY INVESTIGATION

TESTING ENZYME ACTIVITY

PROBLEM
How does pH influence the activity of the enzyme catalase?

MATERIALS (per group)

5 125-mL flasks	0.1 M hydrochloric
graduated cylinder	acid
petri dish	3% hydrogen peroxide
3 medicine droppers	liver puree (catalase)
stirring rod	small piece of raw
hydrion (pH) papers	beef liver
0.1 M sodium	glass-marking pencil
hydroxide	safety goggles

PROCEDURE

1. Prepare an appropriate data table on a separate sheet of paper.
2. Put on safety goggles. Put a small piece of raw liver in an open petri dish. Using a medicine dropper, put a drop of hydrogen peroxide on the liver. **CAUTION:** *Hydrogen peroxide can be irritating to skin and eyes. If you spill any on yourself or your clothes, wash it off immediately.* Observe what happens. Liver contains the enzyme catalase, which breaks down hydrogen peroxide formed in cells. When hydrogen peroxide is broken down by catalase, bubbles of oxygen gas are released.
3. Using the glass-marking pencil, number the flasks from 1 through 5. With a clean medicine dropper, put 10 drops of water in flask 1. Using a graduated cylinder, measure out 5 mL of hydrochloric acid and add it to the water in the flask. **CAUTION:** *When diluting an acid, always pour the acid into the water.* Rinse the dropper and the cylinder.
4. Prepare the following solutions for the remaining flasks. Use procedures similar to those used in step 3. Flask 2: 10 drops of hydrochloric acid added to 5 mL of water. Flask 3: 5.5 mL of water only. Flask 4: 10 drops of sodium hydroxide added to 5 mL of water. Flask 5: 5 mL of sodium hydroxide added to 10 drops of water.
5. Using the stirring rod, put a drop of the solution from flask 1 on a piece of hydrion paper to measure the solution's pH. Record the results in the data table. Repeat the procedure for the solutions in flasks 2, 3, 4, and 5.
6. Using the clean graduated cylinder, add 5 mL of hydrogen peroxide to each flask. Look for bubbles. Identify the amount of activity in each flask by assigning it a number from 0 to 3, when 0 means no bubbles, 1 means few bubbles, and 3 means many bubbles.
7. Rinse out the five flasks thoroughly. Repeat steps 3 through 5.
8. Using a clean medicine dropper, add 5 drops of liver puree to each solution and mix well by swirling the flask.
9. Using the clean graduated cylinder, add 5 mL of hydrogen peroxide to each flask. Identify the amount of activity in each flask from 0 to 3, as you did in step 6.

OBSERVATIONS

1. What happened when hydrogen peroxide was added to each flask without liver puree?
2. What happened when hydrogen peroxide was added to each flask with liver puree?
3. In which flask(s) was the bubble activity greatest? What was the pH of the solution(s)?

ANALYSIS AND CONCLUSIONS

1. What is the purpose of the flasks without liver puree?
2. At what pH does catalase function best? How do you know?
3. If a strong acid has a pH of 1, a strong base, 14, and water, 7, does catalase work best in an acid, base, or neutral environment? Explain.
4. What conclusions can you draw about the pH in body cells?
5. Predict ways in which pH affects the chemical reactions in living cells.

Have students present their answers in the form of a hypothesis. Tell students to write down their hypotheses, then compare them with their conclusions at the end of the investigation.

Have students read through the entire investigation procedure.

- **What is the control setup in this experiment?** (The flasks without the liver puree.)
- **What is the experimental setup?** (The flasks with the liver puree added.)
- **Why is the control setup important?** (To see if adding acid or base has an effect on hydrogen peroxide without introducing the enzyme.)

SUMMARIZING THE CONCEPTS

The key concepts in each section of this chapter are listed below to help you review the chapter content. Make sure you understand each concept and its relationship to other concepts and to the theme of this chapter.

4–1 Water
- Water is the most abundant compound in the majority of living organisms.
- A mixture is a substance composed of two or more compounds mixed together but not chemically combined.
- A solution consists of a solvent and solute.
- Compounds that release hydrogen ions (H^+) in solution are acids. Those that release hydroxide ions (OH^-) are bases.

4–2 Chemical Compounds in Living Things
- The four most abundant elements in living things are carbon, hydrogen, oxygen, and nitrogen.
- Inorganic compounds are primarily those that do not contain carbon. Organic compounds contain carbon.
- Polymerization is the process in which polymers are made by joining monomers.

4–3 Compounds of Life
- The four groups of organic compounds found in living things are carbohydrates, lipids, proteins, and nucleic acids.
- The polymerization of two monosaccharides to form a disaccharide occurs as a result of dehydration synthesis. Hydrolysis is the reverse reaction.
- Lipids are important sources of energy and compounds of biological membranes.
- Proteins are polymers of amino acids.
- Enzymes are, with a few exceptions, proteins. Enzymes are catalysts.
- Nucleic acids are polymers of nucleotides.

REVIEWING KEY TERMS

Vocabulary terms are important to your understanding of biology. The key terms listed below are those you should be especially familiar with. Review these terms and their meanings. Then use each term in a complete sentence. If you are not sure of a term's meaning, return to the appropriate section and review its definition.

4–1 Water
- mixture
- solution
- solvent
- solute
- acid
- base
- neutralization reaction
- pH scale
- suspension

4–2 Chemical Compounds in Living Things
- organic compound
- inorganic compound
- polymerization
- monomer
- polymer
- macromolecule

4–3 Compounds of Life
- carbohydrate
- monosaccharide
- dehydration synthesis
- disaccharide
- polysaccharide
- hydrolysis
- lipid
- cholesterol
- protein
- amino acid
- peptide bond
- catalyst
- enzyme
- substrate
- active site
- nucleic acid
- nucleotide
- ribonucleic acid (RNA)
- deoxyribonucleic acid (DNA)

Skills Development
Students will use the following skills while completing this laboratory investigation:
1. Observing
2. Measuring
3. Comparing
4. Hypothesizing
5. Inferring
6. Recording Data
7. Manipulative
8. Relating

Safety Tips
Have a student read aloud the CAUTION about diluting an acid or a base. Also review with students the rules for chemical safety. Remind students that they must wear safety goggles when working with acids or bases.

Teaching Strategy
1. Make sure that students understand the scale of 0 to 3 for observing bubbles. Since the scale is somewhat subjective, observe students as they work, to make sure that their assessments of the amount of bubbles are reasonable.
2. Circulate throughout the room and help any group that is having trouble with any aspect of the procedure.
3. Separate drops of liver puree may be hard to distinguish. Stress that students should put the same amount of puree in each flask.

Observations
1. There was no activity; no bubbles formed.
2. Bubbles formed.
3. Answers may vary slightly. The bubble activity should be greatest in flask 3 (pH 7) and flask 4 (pH 9). Students will probably observe more activity in flask 3 than in flask 4.

Analysis and Conclusions
1. They serve as controls.
2. Around pH 7, where most bubbling occurred.
3. Neutral environment, since bubble activity was greatest at pH 7.
4. The pH of body cells is probably close to 7.
5. The rate of chemical reactions is affected by pH because pH influences the activity of enzymes that control reaction. As such, a shift in the pH of the body will likely cause many metabolic reactions to slow down or even stop.

Going Further: Enrichment
An interesting demonstration that you can perform for the class is to place a glowing wooden splint in a beaker of hydrogen peroxide that has been catalyzed. The splint should burst into flame. Students should relate their observations to the fact that the presence of released oxygen gas ignites the splint. If a glowing splint is placed in uncatalyzed hydrogen peroxide, nothing happens.

CHAPTER REVIEW

CONTENT REVIEW

Multiple Choice
1. b 2. d 3. a 4. b
5. c 6. c 7. a 8. d

True or False
1. F suspension
2. F acids
3. T
4. F monosaccharides
5. T
6. T
7. F polyunsaturated
8. T

Word Relationships
1. Terms used to describe mixtures in which one substance is dissolved in another solution; suspension is a mixture in which undissolved particles of a substance are suspended in a liquid.
2. Polysaccharides; monosaccharides are simple sugars.
3. Lipids; proteins are polymers of amino acids.
4. Parts of the enzyme reaction; nucleic acid is a polymer of nucleotides.

CONCEPT MASTERY

1. Since a water molecule has a positively charged end and a negatively charged end, molecules of many substances are attracted to one end or the other and thus form a solute when placed in water.
2. In a solution, the molecules are spread uniformly throughout (dissolved). In a suspension, the substance placed in the solvent does not break apart into individual molecules (does not dissolve), but does stay suspended in the solvent and does not settle to the bottom.
3. Carbon forms covalent bonds that are strong and stable. A carbon atom can form four single covalent bonds, or it can form double bonds. Carbon atoms can form long chains with each carbon atom bonded to another carbon atom to form the structure of the chain. Other atoms can be bonded to the carbon atoms in the chain, forming a wide variety of compounds that are used in living things.
4. Check students' diagrams to make sure they show that a molecule of water is removed during dehydration synthesis.
5. Saturated: All carbon atoms have single bonds; unsaturated: A pair of carbon atoms has a double bond; polyunsaturated: At least two different pairs of carbon atoms have double bonds.
6. Enzymes are catalysts that speed up the rate of chemical reactions. Reactants, known as substrates, bind to enzymes at active sites. In doing so, they lower the "start-up" energy required for the reaction to occur. Enzymes are very specific and

---

CHAPTER REVIEW

CONTENT REVIEW

Multiple Choice

Choose the letter of the answer that best completes each statement.

1. The most abundant compound in most living things is
   a. carbon dioxide.
   b. water.
   c. sodium chloride.
   d. hydrochloric acid.
2. A strong acid has a pH of
   a. 14.
   b. 7.
   c. 5.
   d. 2.
3. Which group of elements combine to form practically all the chemical compounds in living things?
   a. carbon, hydrogen, oxygen, nitrogen
   b. carbon, hydrogen, phosphorus, nitrogen
   c. carbon, sodium, chlorine, oxygen
   d. sulfur, phosphorus, carbon, oxygen
4. Proteins are polymers of
   a. fatty acids.
   b. amino acids.
   c. sterols.
   d. nucleic acids.
5. The region on an enzyme to which a substrate binds is called the
   a. catalyst.
   b. activated complex.
   c. active site.
   d. carboxyl group.
6. The function of nucleic acids is related to
   a. energy release.
   b. enzyme formation.
   c. transmission of genetic information.
   d. catalyzing chemical reactions.
7. When phospholipid molecules are mixed with water, they form small balloonlike structures called
   a. liposomes.
   b. vacuoles.
   c. lysosomes.
   d. dipeptides.
8. The polysaccharide found only in plants is
   a. glucose.
   b. glycogen.
   c. cholesterol.
   d. cellulose.

True or False

Determine whether each statement is true or false. If it is true, write "true." If it is false, change the underlined word or words to make the statement true.

1. A mixture of oil and vinegar is an example of a <u>solution</u>.
2. Compounds that release hydrogen ions into solution are known as <u>bases</u>.
3. Compounds that contain carbon are called <u>organic</u> compounds.
4. The simplest carbohydrates are called <u>disaccharides</u>.
5. A common sterol whose excessive intake is related to heart disease is <u>cholesterol</u>.
6. Individual monomers of nucleic acids are known as <u>nucleotides</u>.
7. Including <u>saturated</u> fats in the diet may help to prevent heart disease.
8. Polysaccharides are split apart to form monosaccharides in a reaction called <u>hydrolysis</u>.

Word Relationships

In each of the following sets of terms, three of the terms are related. One term does not belong. Determine the characteristic common to three of the terms and then identify the term that does not belong.

1. solution, suspension, solute, solvent
2. monosaccharide, starch, glycogen, cellulose
3. fats, oils, waxes, proteins
4. substrate, enzyme, active site, nucleic acid

CONCEPT MASTERY

Use your understanding of the concepts developed in the chapter to answer each of the following in a brief paragraph.

1. Explain how the polarity of water molecules makes water the best solvent.
2. Compare solutions and suspensions.
3. Describe the properties of carbon that make carbon compounds so numerous.
4. Explain why the formation of a lipid molecule is a dehydration synthesis reaction. Use a diagram to help explain your answer.
5. Distinguish between a saturated, unsaturated, and polyunsaturated fatty acid.
6. Explain the importance of enzymes to living things. Be sure to include the following terms in your answer: catalyst, substrate, active site, "start-up energy," lock and key.

CRITICAL AND CREATIVE THINKING

Discuss each of the following in a brief paragraph.

1. **Applying concepts** One of the digestive juices in the human stomach is hydrochloric acid (HCl). Sometimes an excess of the acid causes stomach discomfort. In such a case, a person might take an antacid such as magnesium hydroxide (Mg(OH)$_2$). Explain why this substance works to relieve the discomfort. What reaction is taking place?
2. **Identifying patterns** Explain why the name carbohydrate is an indication of the chemical composition of any sugar.
3. **Relating facts** Give two reasons why dehydration synthesis and hydrolysis are opposite chemical reactions.
4. **Applying concepts** Relate the structure of a phospholipid to the property that makes it so important in forming cell membranes.
5. **Developing a model** Materials generally dissolve more quickly in a solvent if its temperature is increased. Why do you think this is so?
6. **Identifying relationships** Describe the role that polymerization plays in the formation of carbohydrates, proteins, and nucleic acids.
7. **Applying concepts** The chemical process by which a saturated fat is converted to an unsaturated fat is called dehydrogenation. Explain why this is an appropriate name.
8. **Relating concepts** Despite their name, many amino acids actually raise the pH of a solution in which they are dissolved. How could this be true?
9. **Using the writing process** At the yearly convention of chemical compounds, a great debate is raging. The organic compounds, represented by the carbohydrates, claim they are the most important compounds in living things. The inorganic compounds, championed by water molecules, claim that life could not exist without them. Write a 5-minute speech in which you take the side of the organic or the inorganic compounds. Or, as an alternative, you may file a minority report in which you argue both sides are incorrect.
10. **Using the writing process** Write a short story entitled, "The Day Ice Stopped Floating."

---

bind to a substrate at a specific site, much as a specific key fits into a specific lock.

CRITICAL AND CREATIVE THINKING

1. A neutralization reaction takes place, forming water in the process, and relieving the acidity and discomfort in the stomach.
2. The general chemical formula for sugars, $(C)_n(H_2O)_n$, makes it appear as if they might consist of a carbon (C, or *carbo*-) portion and a water (H_2O, or *-hydrate*) portion.
3. In dehydration synthesis, a molecule of water is released as two substances combine. In hydrolysis, a molecule of water is used when two substances are combined. Dehydration synthesis results in a product that is more complex than either of the reactants. In hydrolysis, the reactants are generally more complex than the product.
4. A phospholipid is a molecule consisting of parts that dissolve in water and parts that do not. When mixed with water, they form balloonlike molecules called liposomes. Each liposome is formed from a double layer of lipid molecules, which is similar in structure to cell membranes. The formation of liposomes from phospholipids occurs spontaneously, enabling them to play a key role in the formation of membranes.
5. The dissolving process involves collisions between solute and solvent molecules. As temperature increases, the molecular motion of various molecules increases, making the likelihood of collisions greater.
6. Polymerization is a process in which small units (monomers) are linked together to form long molecules made up of bonded monomers (polymers). A carbohydrate such as starch is a polymer of individual monomers called sugars. A protein is a polymer made up of monomers called amino acids. A nucleic acid is a polymer made up of monomers called nucleotides.
7. As a saturated fat is converted into an unsaturated fat, a double bond must form between 2 carbon atoms. At that time, 2 hydrogen atoms are removed, and thus the name dehydrogenation.
8. If an amino acid's amino group is strongly basic, the amino acid will release hydroxide (OH$^-$) ions when it dissolves, thereby raising the pH of the solution.
9. Check students' writing and science skills as you read their reports. Accept any logical responses.
10. Check each short story for writing skills.

CAREERS IN BIOLOGY

Textbook Illustrator, Biology Teacher, Biochemist

Students interested in a biological career should be instructed to write for further information to the address listed beneath each career description. However, in consideration of the organizations that provide career information, please have only one student write to an organization. Do not instruct the entire class to write to every organization for the same career information. You may want to use the information provided to start a biology career file for the use of all your students.

UNIT 1   CAREERS IN BIOLOGY

Textbook Illustrator

Textbook illustrators create the artwork that accompanies the information in a textbook. Their illustrations serve to express ideas and clarify explanations contained in the text.

To become a textbook illustrator, a person needs to take general art classes and have artistic ability. Many artists obtain preparation beyond high school at specialized art schools. Furthermore, some illustrators choose areas such as medical or scientific art that require training in biology or the physical sciences.

To receive additional information write to the Association of Medical Illustrators, 2692 Huguenot Springs Rd., Midlothian, VA 23113.

Biology Teacher

Biology teachers help students explore the complex world of living things. By presenting topics through lectures, demonstrations, laboratory work, and field trips, they help students learn the importance of the scientific method and the value of research. Biology teachers also monitor the progress of individual students by evaluating assignments and exams.

To become a biology teacher, a Bachelor's degree with courses in biology and education is required. In addition, experience as a student teacher is necessary to obtain a state teaching license.

For more information write to the National Association of Biology Teachers, 11250 Roger Bacon Dr., Reston, VA 22090.

Biochemist

Biochemists study chemical processes within living things. They use highly complex equipment to better understand life processes such as growth, reproduction, and metabolism. Some biochemists study the effects of different chemicals on the human body, whereas others develop methods to help doctors diagnose and treat disease.

Biochemists must have a Bachelor's degree in chemistry or biochemistry and an advanced degree in a particular area of biochemistry.

For information about this career write to the American Society of Biological Chemists, 9650 Rockville Pike, Bethesda, MD 20014.

HOW TO FIND OUT ABOUT CAREERS

School and community libraries are a good source of information about different types of jobs and careers. There you will find a number of sources that not only define specific job titles but also describe the nature of the work, educational and training requirements, the employment outlook for that field, working conditions, and other necessary information. Two such publications are the Occupational Outlook Handbook, *published by the United States government, and* The Encyclopedia of Careers and Vocational Guidance.

In addition, many corporations provide career information to interested students. Your school counseling office may have information about some of these companies.

FROM THE AUTHORS

Covalent bonds, ionic bonds, peptide bonds. Solutions, suspensions, pH. At this point you might be tempted to ask, "What does any of this have to do with biology?" Good question. You might even try it on your teacher. We can guarantee your teacher has heard it before. In fact, neither one of us has *ever* taught a biology course without being asked that same question. So by this time you might suspect we've got a good answer. Well, we have!

Biology is based on chemistry, and to really understand biology you have to understand chemistry. Chemistry has something to do with the way cells communicate, how plants respond to light, how bees find flowers, and even how you read this page. Every day, biologists discover new ways in which chemistry is connected to biology.

Why are we telling you this? Well, both of us have a confession to make: Chemistry has never been our favorite subject. But the truth is that with each passing year, the chemistry we learned in school has become *more* important, not less—even though we've spent our careers as biologists. Although you might not believe it now, you will use the chemistry you've learned in this unit just about every day in biology class for the rest of the year. Chemistry. You'll be glad you learned it!

FROM THE AUTHORS

**Kenneth R. Miller
and Joseph Levine**

This first message is from both of the authors. Point out to students that this is a personal message from the authors to them directly. You may want to hold a class discussion on the topic addressed in this feature. Remind students that the different branches of science are closely linked.

UNIT 2
Cells: The Basic Unit of Life

UNIT OVERVIEW

In Unit 2 students are introduced to the specific attributes of individual cells and to the complex interactions among groups of cells. First they learn about the functions of the structures that make up cells. Next they are taught how cells work together to create different levels of organization: tissues, organs, and organ systems. They then go on to learn how cells capture, store, and transform energy through the processes of photosynthesis and respiration. They also read about how the genetic code operates and is stored in individual cells and how it is replicated and transferred to new cells. Finally, they learn how new cells are produced in order to allow organisms to grow and reproduce.

UNIT OBJECTIVES

1. Explain the cell theory and describe the structures and activities of cells.
2. Compare eukaryotes with prokaryotes, plant cells with animal cells, and unicellular organisms with multicellular organisms.
3. Describe the processes by which materials enter or leave individual cells.
4. Explain how cells capture, store, and transform energy.
5. Describe the structures and functions of DNA and RNA and relate them to protein synthesis.
6. Relate cell growth to cell division and describe mitosis.

UNIT 2
Cells: The Basic Unit of Life

Scanning its surroundings for its next meal, a grasshopper spots an appetizing plant with red-and-green fringed flowers. The grasshopper hops onto the leaves of the plant, then upon a flower—and suddenly a trap snaps shut around the unlucky insect. The "flower" is not a flower at all, but the specialized insect-trapping leaves of a Venus' flytrap! The grasshopper struggles, but in vain. Tiny structures within and beneath the red lining of the trap begin to produce chemicals that break down the body of the grasshopper to produce raw materials. Similar tiny structures busily take in the raw materials, distribute them, process them, transport the products to where they are needed, and get rid of waste materials. What are these wondrous tiny structures? They are the fundamental units of all living organisms—cells.

Every cell is a highly organized structure that is responsible for the form and function of an organism. Some organisms consist of only one cell. Others—including grasshoppers and Venus' flytraps—consist of trillions of cells organized into interconnected organ systems. By understanding the nature and complexity of life at the cellular level, we can begin to understand and appreciate the marvels of living things.

DISCOVERY LEARNING

A DIFFERENT BREATHALYZER TEST

When most living things break down food to produce energy, they produce carbon dioxide and another substance. What is this mystery substance? Find out by doing the following.

1. Breathe on a mirror. What happens? What substance forms on the mirror? Where do you think this substance comes from?
2. Cover the top of a small plant with a plastic bag. Secure the bag around the stem at the base of the plant with a twist tie. Place the plant in sunlight. What substance forms on the inside of the bag? Where do you think this substance comes from?
- What substance, along with carbon dioxide, do most living things produce when they break down food?

INTRODUCING UNIT 2

Using the Visuals

Have students examine the photograph on pages 84 and 85.
• **How many different types of organisms appear in the photograph? What are they?** (Two: a grasshopper—an animal; and a Venus' flytrap—a plant.)
• **Compare the types of movements of both organisms shown in the photograph.** (Students may describe the inferred movements of the grasshopper as it struggles for freedom. The plant moves when the trap closes.)
• **What does the plant need to live?** (The plant requires sun, water, and nutrients. In most places, plants get nutrients from the soil. Venus' flytraps and other carnivorous plants get nutrients from digesting the prey

CHAPTERS

5
Cell Structure and Function

6
Cell Energy:
Photosynthesis and Respiration

7
Nucleic Acids and
Protein Synthesis

8
Cell Growth and Division

CHAPTER DESCRIPTIONS

5 Cell Structure and Function
Chapter 5 introduces the cell theory, then describes major cell structures. Next, it presents the processes by which materials move through the cell membrane. Finally, it discusses cells that are specialized to perform a particular function and the four levels of organization in multicellular organisms.

6 Cell Energy: Photosynthesis and Respiration
Chapter 6 begins by describing how organisms capture and convert energy through the process of photosynthesis. The production of usable energy through glycolysis and respiration is then explained. The two different types of fermentation are also described and related to glycolysis.

7 Nucleic Acids and Protein Synthesis
The genetic code is introduced in Chapter 7. The structure of DNA and the process of DNA replication is covered, as is the structure of RNA and the process of transcription. Finally, the relationship between the two nucleic acids is described in the process of protein synthesis.

8 Cell Growth and Division
Chapter 8 begins with a discussion of the effects of, rates of, and controls on cell growth. The stages of cell division, mitosis, and cytokinesis are then explored in detail.

they trap.)
• **Do you think there are any other activities occurring in the organisms shown in this photograph? Why are these activities not visible?** (Yes, there are many activities occurring in the organisms shown in this photograph. Many of these activities occur at the cellular level and cannot be seen in the photograph.)

DISCOVERY LEARNING

A DIFFERENT BREATHALIZER TEST

Begin your introduction to the unit by having students perform the Discovery Activity. This activity can also be used as a cooperative learning experience.
1. Water vapor forms on the mirror. It results when the water vapor in warm breath condenses on the relatively cold mirror.
2. Water vapor forms on the inside of the bag. It is given off by the leaves of green plants.
• Most living things give off water vapor as a waste product when they break down food.

CHAPTER 5 Cell Structure and Function

Section	Laboratory Investigations and Activities
5–1 The Cell Theory pages 87–88 THEMES: Scale and Structure, Unity and Diversity	**Teacher Edition** DEMONSTRATION: Cell Structure, p. 86D
5–2 Cell Structure pages 89–93 THEMES: Scale and Structure, Unity and Diversity	**Laboratory Manual** Comparing Plant and Animal Cells, p. 99 **Teaching Resources** HANDS-ON ACTIVITY: Now You See It—Now You Don't, p. 45
5–3 Cytoplasmic Organelles pages 94–99 THEME: Scale and Structure	**Laboratory Manual** Characteristics of Prokaryotic and Eukaryotic Cells, p. 95
5–4 Movement of Materials Through the Cell Membrane pages 99–104 THEMES: Patterns of Change, Energy	**Student Edition** LABORATORY INVESTIGATION: Observing Osmosis, p. 108 **Teacher Edition** DEMONSTRATION: The Process of Osmosis, p. 86D **Teaching Resources** ACTIVITY: Analyzing a Laboratory Investigation, p. 5 HANDS-ON ACTIVITY: Across a Crowded Cell, p. 39 HANDS-ON ACTIVITY: Coming and Going, p. 43 **Product Testing Activities** Testing Bandages Testing Lip Balms Testing Food Wraps
5–5 Cell Specialization pages 104–106 THEME: Scale and Structure	
5–6 Levels of Organization pages 106–107 THEME: Scale and Structure	**Teaching Resources** HANDS-ON ACTIVITY: A Human Cell vs. an Ameba, p. 35
Chapter Review pages 109–111	

CHAPTER PLANNING GUIDE

Other Activities	Media and Technology
Study Guide Section 5–1, p. 43	**Biology Media Guide** page 7
Study Guide Section 5–2, p. 45	**Biology Media Guide** page 7
Teaching Resources ACTIVITY: Comparing Cells, p. 3 **Study Guide** Section 5–3, p. 47	
Teaching Resources ACTIVITY: Passive and Active Transport, p. 7 **Study Guide** Section 5–4, p. 49	
Study Guide Section 5–5, p. 51	
Teaching Resources VOCABULARY REVIEW: Crossword, p. 1 **Study Guide** Section 5–6, p. 52	
Teaching Resources Chapter Test A, p. 13 Chapter Test B, p. 17	**Computer Test Bank** Chapter Test, p. 49

CHAPTER 5: Cell Structure and Function

CHAPTER OVERVIEW

Any examination of the phenomenon that we call life must include a study of cells—their structure and function. Because cells have many structures and functions in common, a definition and description would seem to be a simple matter. However, the variations in specialization and organization can make an in-depth study complex. The more information scientists gather about cells, the more questions arise.

In this chapter students will be given some historical background that will help them to understand and appreciate the findings that led to the development of the cell theory. The structures that are common to most cells are discussed individually in some detail, highlighting both structure and function. Next, there is information about cytoplasmic organelles; again structure and function are emphasized. After discussing cell structure, the important concept of material movement into and out of a cell is presented. The discussion includes both passive and active means of transport.

Next is a section on cell specialization that uses three types of human cells as examples. This section emphasizes the relationship between structure and function. Finally, the chapter concludes with a discussion of the levels of cell organization that exist in a multicellular organism.

5-1 THE CELL THEORY

Section Focus 5-1

The purpose of this section is to give students a historical perspective on the events leading up to the discovery of the cell and the development of the cell theory.

Performance Objectives 5-1

1. Discuss in detail the discovery of the "cell" by Robert Hooke.
2. Discuss the contributions of other scientists leading to the development of the cell theory.
3. State the cell theory.

Science Terms 5-1

cell   p. 87
cell theory   p. 88

5-2 CELL STRUCTURE

Section Focus 5-2

The purpose of this section is to give students information about the similarities and differences that exist among living cells.

One important point is that the cells of animals, plants, and related organisms have three basic structures. The cell membrane, the nucleus, and the cytoplasm are discussed in this section. There is also information about the structure and function of the cell wall. The text is supported by drawings of typical cells.

The distinction between eukaryotic and prokaryotic cells is made when the nucleus is discussed. Scientists consider this division more important than the distinction between plant and animal cells.

Performance Objectives 5-2

1. Name the three structures common to most cells.
2. Discuss the structure and function of the nucleus.
3. Distinguish between the cell membrane and the cell wall in both structure and function.
4. State the basic difference between eukaryotic and prokaryotic cells.

Science Terms 5-2

cell membrane   p. 90
cell wall   p. 91
nucleus   p. 92
eukaryote   p. 92
prokaryote   p. 92
nuclear envelope   p. 92
nucleolus   p. 92
chromosome   p. 93
cytoplasm   p. 93

5-3 CYTOPLASMIC ORGANELLES

Section Focus 5-3

The purpose of this section is to discuss the various organelles found in the cytoplasm. The mitochondria and chloroplasts are grouped together because of their similar function—to change energy from one form to another—and because of their similar structure. Both are systems of folded membranes. Students are reminded that chloroplasts are found only in plants.

Ribosomes are composed of RNA and protein and are the structures in which proteins are made. The endoplasmic reticulum and Golgi apparatus are also discussed together because of their similar structure and functions, which include the synthesis and modification of cell products.

Lysosomes, formed by the Golgi apparatus, contain enzymes that digest certain materials in the cell. The process of endocytosis is introduced here. Vacuoles and plastids are both used as storage tanks in cells holding a variety of materials. Finally, there is a discussion of the microtubules and microfilaments that make up the cytoskeleton. The idea of "protoplasm" is refuted.

Performance Objectives 5-3

1. Define organelle.
2. Compare the functions of the mitochondria and chloroplasts.
3. Discuss the function of the two types of endoplasmic reticulum.
4. Name the two major components of the cytoskeleton and discuss their importance.

Science Terms 5-3

organelle   p. 94
mitochondrion   p. 94
chloroplast   p. 94
ribosome   p. 95
endoplasmic reticulum   p. 95
Golgi apparatus   p. 96
lysosome   p. 97
vacuole   p. 97
plastid   p. 97
cytoskeleton   p. 98

CHAPTER PREVIEW

5-4 MOVEMENT OF MATERIALS THROUGH THE CELL MEMBRANE

Section Focus 5-4

The purpose of this section is to give students an understanding of the processes by which materials enter and leave a cell through the membrane. Many substances move across the cell membrane by diffusion from an area of lesser concentration of molecules. Factors that affect the movement of substances by diffusion include concentration of molecules and permeability of the membrane.

Water passes very quickly through most selectively permeable membranes. This process is called osmosis. More specialized types of movement include facilitated diffusion, involving carrier proteins, and active transport. Diffusion, osmosis, and even facilitated diffusion operate with the concentration difference. Active transport, on the other hand, is an energy-requiring process that involves the movement of substances against the gradient. One type of active transport involves transport macromolecules that move molecules such as calcium, potassium, or sodium ions across the membrane through membrane-associated pumps.

A second type of active transport involves the movement of large amounts of materials. Examples of these processes include endocytosis, phagocytosis, pinocytosis, and exocytosis.

Performance Objectives 5-4

1. Discuss the processes of diffusion and osmosis.
2. Distinguish between passive and active transport.

Science Terms 5-4

diffusion   p. 90
selectively permeable   p. 100
osmosis   p. 100
facilitated diffusion   p. 102
active transport   p. 102

5-5 CELL SPECIALIZATION

Section Focus 5-5

The purpose of this short section is to give students a general understanding of the ways that cells may be uniquely suited to perform particular functions within multicellular organisms. Three human body cells are used as examples of cell specialization. The cells of the pancreas are likened to small protein factories. Their enormous number of protein-synthesizing organelles is related to function. The second example is the light-sensitive cells of the eye. The third is the specialized cells of the respiratory tract. Again, in all cases structure is related to function.

Performance Objectives 5-5

1. Define cell specialization and discuss its importance to all multicellular organisms.

Science Terms 5-5

cell specialization   p. 104

5-6 LEVELS OF ORGANIZATION

Section Focus 5-6

The purpose of this section is to explain the levels of organization of cells found in multicellular organisms. The first level—cells—is organized into specialized groups called tissues. Groups of tissues organized to work together to perform one or more specific functions are called organs. An organ system is made up of a group of organs working together to perform certain functions. This organization, and the division of labor that exists, is what makes multicellular life possible.

Performance Objectives 5-6

1. Explain the four levels of organization that make up multicellular organisms.
2. Give two examples of each level of organization.

Science Terms 5-6

tissue   p. 106
organ   p. 107
organ system   p. 107

DISCOVERY LEARNING

TEACHER DEMONSTRATIONS
Modeling

Cell Structure

The following demonstration can be used as an introduction to Chapter 5.

Show students several photographs or projected images of various types of cells. You might use photographic slides or a microprojector or videoprojector for this activity. Have students take note of similarities and differences among the cells.

If you have access to electron micrographs (*Scientific American* magazine, university science departments, or biological supply houses), students might be challenged to find and identify structures. This is difficult at first, but with a little practice, it can be done more easily.

After students have discussed key structures and have made their comparisons, you might want to lead them into a discussion of the importance of the microscope and the historical background leading to the discovery of the cell.

The Process of Osmosis

This demonstration on the movement of materials through the cell membrane can be done with Section 5-4.

Place a small amount of sugar solution in a test tube with a semipermeable membrane covering one end. Suitable membranes include dialysis tubing, sausage casing, the thin membrane from inside an eggshell, or even the skin of frogs that may have been used for dissection. Turn the tube upside down in a container of distilled water. Leave overnight. Osmosis will result in a movement of water molecules across the membrane into the sugar solution. As a result, the volume of sugar solution will increase until a column of liquid in the tube rises several centimeters above the level of the water.

This is a good time to talk about concentration differences. It is also important that students understand that it is water molecules that move by osmosis rather than sugar, salt, or other substances that they tend to associate with concentration.

86D

CHAPTER 5
Cell Structure and Function

GUIDED ENQUIRY
Pose the following questions to students and have them record their responses. Point out that they will gain a better understanding of the key concepts if they read the chapter with these basic questions in mind. Upon completion of the chapter, pose the questions again. Ask students to compare their initial responses with those they have developed after reading the chapter.

- How is the invention of the microscope related to the development of the cell theory?
- Do plant and animal cells have more basic similarities or differences?
- How are internal structures related to cell functions?
- How do materials enter and leave living cells through the cell membrane?
- Why is cell specialization important to multicellular organisms?

INTRODUCING CHAPTER 5
Using the Visuals
Point out the photographs on page 86. Be sure that students understand that the main photo shows human cells magnified many times by a microscope much more powerful than the first microscopes. Have students read the introduction. Stress the fact that though we have many cells now, we started as single cells. Both a single-celled organism and a human zygote are functional units of life in the same way that atoms are functional units of matter. In Chapter 5 the emphasis will be on relating structure and function.

Check student knowledge and understanding with questions such as these.

- **Can you tell what basic life function is occurring in the photographs?** (Accept all reasonable answers. You might want to talk about other functions carried on by cells.)
- **Do you recall the "parts" of an atom?** (The nucleus, protons, neutrons, and electrons.)
- **Can you name any parts of a cell?** (Students will probably know about the nucleus and the cell membrane; some may be familiar with organelles.)
- **What scientific instrument has made the study of cell structure**

CHAPTER 5
Cell Structure and Function

Some organisms are made up of many cells. These human cells must work with one another and with the billions of other cells in the body to carry out life functions. Each Micrasterias alga cell (inset) is an individual organism capable of carrying out all its life functions.

How many cells are there in the human body? More than 100 trillion (10^{14}, or 100,000,000,000,000)! We have 30 billion cells in our brain alone and about 20 trillion red blood cells. There are roughly 155,000 cells in every square centimeter of our skin. Many microscopic organisms, including bacteria, consist of just a single cell. Despite our enormous complexity, we all begin our lives as single cells.

Cells are the "atoms" of biology, the fundamental units of living organisms, the smallest units that can reasonably be thought of as being alive. Every question in biology—from ecology to inheritance, from behavior to evolution—must be answered partly at the level of the cell. Understanding the cell is the key to understanding biology. In this chapter, we shall begin our study of the cell.

GUIDE FOR READING

After you read the following sections, you will be able to

5–1 The Cell Theory
- Discuss the cell theory.

5–2 Cell Structure
- Identify and give the function of the three basic structures of most cells.
- Distinguish between prokaryotes and eukaryotes.

5–3 Cytoplasmic Organelles
- List the major cytoplasmic organelles and describe their functions.
- Compare the structure of plant and animal cells.

5–4 Movement of Materials Through the Cell Membrane
- Describe the processes by which materials move through the cell membrane.

5–5 Cell Specialization
- Relate cell specialization to cell structure.

5–6 Levels of Organization
- Describe the four levels of organization in a complex multicellular organism.

Journal Activity

YOU AND YOUR WORLD
When did you first learn about cells? How did you feel when you found out that you were made of cells? More than 100 trillion of them, in fact! Describe your thoughts and feelings in your journal.

Figure 5–1 The invention of the telescope enabled astronomers to study distant objects such as this spiral galaxy. The microscope, on the other hand, opened up the world of the very small to biologists. Before the invention of the microscope, a microscopic alga, such as this diatom, could not have been observed.

5–1 The Cell Theory

Guide For Reading

■ How did van Leeuwenhoek, Hooke, Schleiden, Schwann, and Virchow contribute to the development of the cell theory?
■ What are the parts of the cell theory?

Each of us, at one time or another, has tried to look closely at something. You may have picked up a coin and tried to read the initials of its designer, cut closely into the coin's surface. You may have tried to read the details on the face of a stamp, or stared at a blade of grass until the tiniest detail was clear.

Such curiosity led early investigators to examine living things under lenses and microscopes in the hope of getting a better glimpse of their structure. Little by little, their findings led to the most fundamental of all discoveries about the nature of living things: All living things are made of **cells**. Cells are the basic units of structure and function in living things.

The first lenses were used in Europe hundreds of years ago by merchants who needed to determine the quality of cloth. They used their magnifying lenses to examine the quality of the thread and the precision of the weave in a bolt of cloth. From these simple glass lenses, combinations of lenses were put together. In Holland in the early 1600s, two useful instruments were constructed. One was the telescope, which enabled people to see objects at a distance. The telescope made the distant stars in the sky visible. The other instrument, the microscope, made the very small objects in nature visible.

The person who is given credit for developing the first microscope was Anton van Leeuwenhoek (LAY-vuhn-hook), a

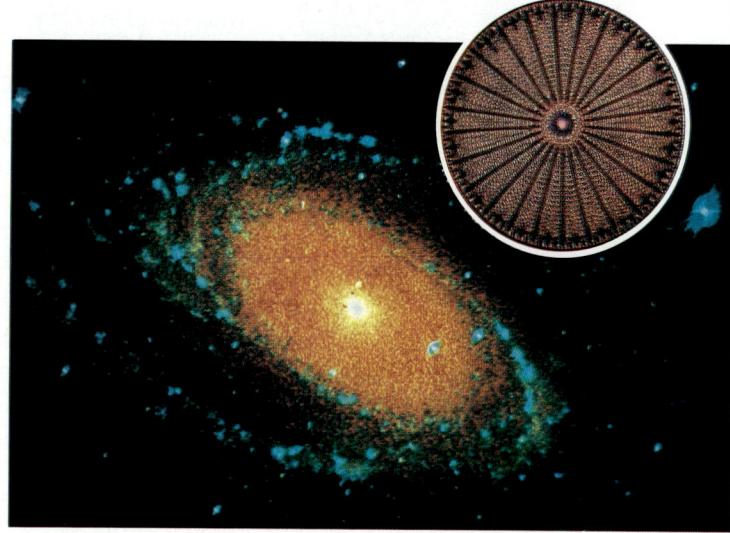

COOPERATIVE LEARNING

Using preassigned lab groups or randomly selected teams, have groups assume that, like characters in some popular motion pictures, they have been shrunk to microscopic size and find themselves living in a human cell. Each group should describe their new home. The descriptions should include:
- Type of cell in which they live and where their cell is located
- Organelles found in the cell
- Explanation of how each organelle helps the cell maintain its life
- Explanation of how their cell is specialized to carry out its particular function

Final products could be a story, a written report, an audiotape, or a correctly labeled drawing.

Journal Activity

YOU AND YOUR WORLD
You may also want to use the Journal Activity as a springboard for class discussion. Most students will be amazed at the number of cells in a human body. Students should be instructed to keep their Journal Activity in their portfolio.

---

possible? (Microscope.)
- **What type of microscope do you use in this class?** (Light microscope.)
- **In this chapter you will see several electron micrographs of cells. What do you know about these cells?** (They are dead.)

TEACHING STRATEGY 5–1

Focus/Motivation
Using the Visuals Have students look at Figure 5–1 and read the caption. Both the telescope and the microscope came into use after the first lenses were developed hundreds of years ago.
- **What human characteristic led to the development of both the telescope and the microscope?** (Curiosity.)
- **Think of other inventions or discoveries that were the result of human curiosity.** (Answers will vary, but one example is the discovery of electricity by Benjamin Franklin.)

Tell students that the history leading to the development of the cell theory is interesting in terms of biological advances, but it should also be a reminder of the nature of scientists as people who are curious and interested in learning.

HISTORICAL NOTE
EARLY COMPOUND MICROSCOPES

Two brothers named Janssen actually put lenses together at opposite ends of a tube around 1590. Technically, their apparatus was a compound microscope. It was undoubtedly the use to which von Leeuwenhoek put his "microscope" that has earned him credit for inventing the device. Anton von Leeuwenhoek was not born until 1632!

TEACHING SUPPORT

Study Guide
• Section 5–1, p. 43

MEDIA AND TECHNOLOGY
See the Biology Media Guide page 7 for bar-code correlations for this section.

5–1 (continued)

Content Development

Discuss the contribution of each of the scientists leading up to the formulation of the cell theory. Be sure that students understand the importance of the individual accomplishments. It is also important that they understand the time span involved.
• **How did each observation influence the next scientist's work?** (Answers will vary, but generally they are pretty obvious. The microscope was essential to the discovery of the cell. Brown's discovery of the nucleus was made after he knew that cells existed. Schwann elaborated Schleiden's findings, and so on.)

SECTION REVIEW 5–1

1. *Leeuwenhoek:* developed the microscope through which cells could be seen; *Hooke:* observed dead cells in cork, and coined the term *cell*; *Schleiden:* stated that all plants have cells; *Schwann:* stated that all animals have cells; *Virchow:* stated that all cells come from preexisting cells.
2. Although Hooke observed dead cells, he did apply his observations to living cells.
3. Without the microscope, cells could not be observed. As such, the cell theory (if it had ever been discovered) would have to be inferred from indirect observation (much like the parts of an atom).

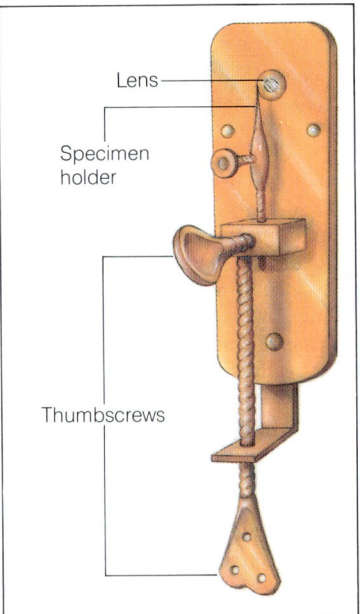

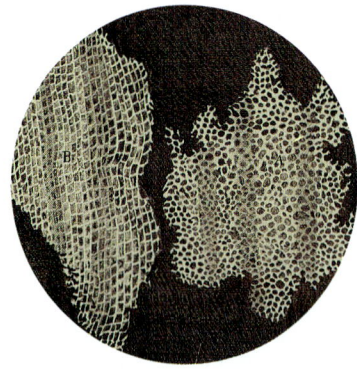

Figure 5–2 Van Leeuwenhoek's simple microscope (top) could magnify objects a few hundred times. Robert Hooke made this drawing of cork cells (bottom) using a microscope that he built. Hooke, however, was not looking at living cells; what he saw were the cell walls that surround living cork cells.

Dutch biologist. His invention enabled him to see things that no one had ever seen before. He could see tiny living organisms whose world consisted of a drop of water. Van Leeuwenhoek carefully observed the tiny living things in pond water and made detailed drawings of each kind of organism.

Van Leeuwenhoek's work interested other people in building microscopes. Before long, pioneers in several countries were experimenting with these new instruments. One such person was the Englishman Robert Hooke, who used one of his microscopes to look at thin slices of plant stems, wood, and pieces of cork. Looking at the cork, Hooke saw that it was composed of thousands of tiny chambers. He called these chambers cells because they reminded him of the small rooms called cells in a monastery.

Unfortunately, Hooke was not looking at living cells. He was looking at the nonliving outer walls of what had once been living plant cells. Nonetheless, Hooke's discovery was significant because it opened up the study of cells.

Gradually over the next 200 years, other scientists began to discover that cells were not only found in plants but in other living things too. In 1833, Robert Brown, a Scottish scientist, observed that many cells seemed to have a dark structure near the center of the cell. We now call this structure the nucleus. Five years later, German botanist Matthias Schleiden stated that all plants are made of cells. The next year, Theodor Schwann discovered that all animals are made of cells too. In 1855, Rudolf Virchow, a German physician, added one more element to the developing theory of cells. Based on research, he stated that all cells arise from the division of preexisting cells.

Today, the observations and conclusions of these scientists are summarized into the **cell theory**. The cell theory forms the basic framework in which biologists have tried to understand living things ever since. **The cell theory states:**

- All living things are composed of cells.
- Cells are the basic units of structure and function in living things.
- All cells come from preexisting cells.

5–1 SECTION REVIEW

1. What contributions did van Leeuwenhoek, Hooke, Schleiden, Schwann, and Virchow make to the development of the cell theory?
2. When Hooke first used the term cell, did he intend it to apply to living material? Explain your answer.
3. **Critical Thinking—Identifying Relationships** What role did the invention of the microscope play in the development of the cell theory?

Reinforcement/Reteaching

At this point, for those students who have difficulty with the Section Review questions, go back to the section material and review problem areas.

5–2 Cell Structure

There is enormous variety in the size and shape of different cells. The smallest cells, belonging to a group of organisms known as *Mycoplasma* (migh-koh-PLAZ-mah), are only about 0.2 micrometers in diameter. A micrometer is equal to one millionth of a meter. *Mycoplasma* are so small that they often are beyond the resolving power of light microscopes. Larger cells include the giant ameba *Chaos chaos*, which is about 1000 micrometers in diameter. Larger still are the yolks of bird eggs, which are actually single cells containing stored food for the developing bird. For the most part, cells are between 5 and 50 micrometers in diameter. Physical limits on the flow of information through the cell and on the flow of materials into and out of the cell prevent most cells from being much larger than this.

Despite differences in size and shape, there are certain structures that are common to most cells. **The cells of animals, plants, and related organisms have three basic structures: the cell membrane, or outer boundary of the cell; the nucleus, or control center; and the cytoplasm, or material between the cell membrane and the nucleus.** Let's examine a typical plant cell and an animal cell to learn about some of these basic structures.

Guide For Reading

■ What are the functions of the three basic structures of most cells?
■ How do prokaryotes and eukaryotes differ?

Figure 5–3 *This diagram shows a typical animal cell. Note that the structures shown are not to scale. (This is true of almost all cell diagrams.) Most structures have been enlarged so that they can be clearly seen. In addition, some structures are more extensive or more numerous in an actual cell than they are in this diagram.*

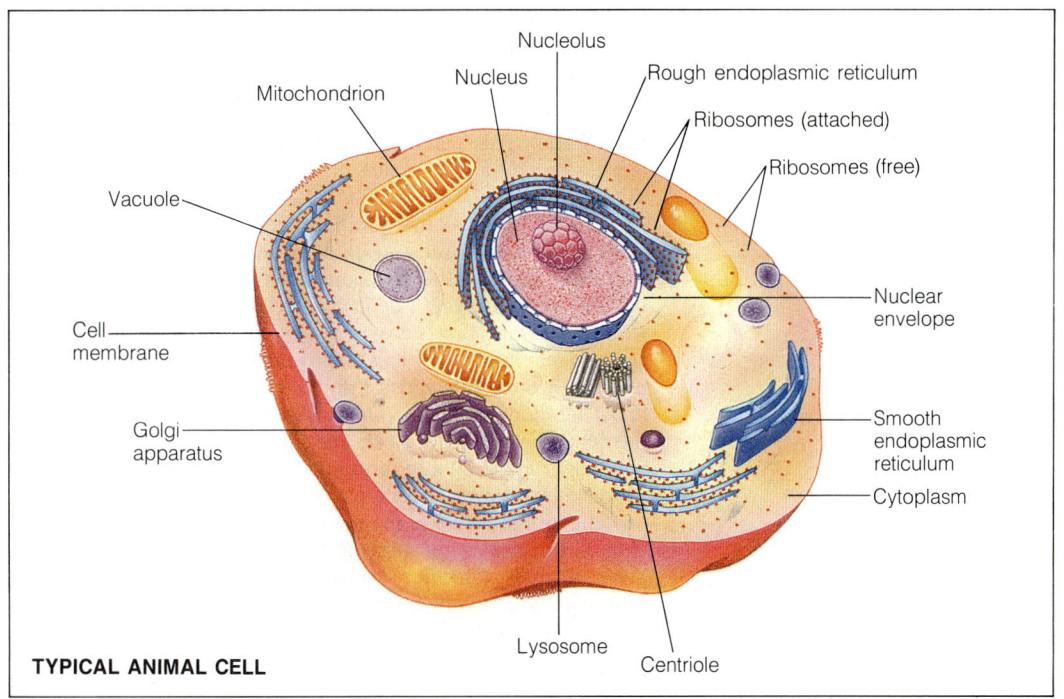

TYPICAL ANIMAL CELL

BACKGROUND INFORMATION
EGG YOLKS

While it's true that egg yolks are single cells, students should not be encouraged to think of the yellow material in the egg yolk as cytoplasm. In fact, the cellular portion of a chicken oocyte is restricted to a small, whitish disk of material that sits atop a carefully separated yolk. If the egg is fertilized, the nucleus in this small disk starts to divide and gradually produces a group of cells that invade the yolk, forming blood vessels that carry nourishment from the yolk into the developing embryo. It's more correct, therefore, to think of the yolk as an enormous pile of storage material sitting beneath a small clump of nucleus and true cytoplasm.

model. The last question that students are asked should be similar to this one.
• **Based on your observation of the model, is it a plant or an animal cell?** (Answers will vary depending on the model.)

Point out that the structures in the diagram, as in almost all cell diagrams, are not to scale. You may wish to remind students of this fact as they study cell structure. You may also wish to point out that cell diagrams do not show the actual colors of cell structures.

Content Development
Using the Visuals It is interesting to students to talk about the diversity of size among cells. It is more important that they understand the relationship between the size and the ability to get necessary materials across the surface area of the membrane.

The cell membrane is one of the structures common to all cells, as is the cytoplasm. Point out the cell membrane, cytoplasm, and nucleus on the drawing of the animal cell (Figure 5–3).

Closure
Prepare a set of large-print flashcards (they need not be elaborate) with the names of the scientists mentioned in Section 5–1 written on them, one per card. Similarly, list the contributions on the cards. For review, ask students to match the scientists with their accomplishment.

TEACHING STRATEGY 5–2
Focus/Motivation
Show students a three-dimensional model of a cell. Explain that cells are sometimes studied using models or diagrams of "typical" cells. While no cell is typical, most cells do have some commonalities. Check students' prior knowledge by quizzing them on the names and functions of the identifiable structures in the

FACTS AND FIGURES

In animal cells, cholesterol is an important part of the cell membrane. It reduces permeability and strengthens the membrane.

ANNOTATION KEY

❶ Cell wall.
(Interpreting illustrations)

TEACHING SUPPORT

Laboratory Manual
- Comparing Plant and Animal Cells, p. 99

Teaching Resources
- Hands-On Activity: Now You See It—Now You Don't, p. 45

MEDIA AND TECHNOLOGY

See the Biology Media Guide page 7 for bar-code correlations for this section.

ESL STRATEGY

Provide pairs of students with an unlabeled outline drawing of a plant or animal cell. You can also have students draw, but not label, the cells on pages 89 and 90.
Have students print identifying labels on small stick-on notes and place them appropriately on the drawing. The notes can be folded to conceal the words. Notes can be removed as students master the cell structures and can be replaced on the drawings for review. You can use notes of a different color for cell structures you wish to emphasize. This activity can be converted into a game, with the winner being the first student to remove the labels on all correctly identified structures.

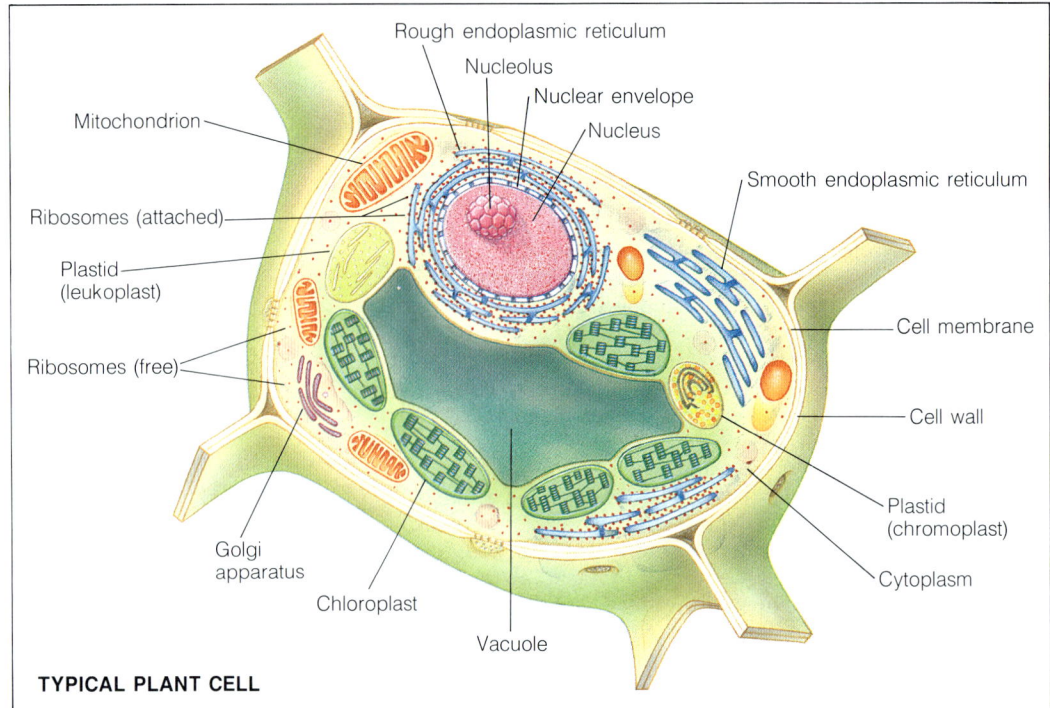

TYPICAL PLANT CELL

Figure 5–4 A typical plant cell is shown in this diagram. What is the outer boundary of a plant cell called? ❶

Cell Membrane

All cells are separated from their surroundings by a **cell membrane**. The cell membrane regulates what enters and leaves the cell and also aids in the protection and support of the cell.

In a way, the cell membrane is similar to the walls that surround a house. As these walls help to protect the house from what is outside, so the cell membrane seals off the cell from its outside environment. But if you lived inside the house, you would still want to receive messages, fuel, and power from outside. So telephone, gas, and electric lines would have to be able to pass through the walls of the house. You would also want to bring in food and take out the trash. Thus doors would be needed. The needs of a cell are similar. It must communicate with other cells, take in food and water, and eliminate wastes. All of these processes take place through the cell membrane.

The cell membrane is composed of several kinds of molecules. The most important of these are the lipids. A double layer of lipid molecules, known as a bilayer, forms the basic unit from which cell membranes are constructed. You can see the structure of the cell membrane in Figure 5–5.

Proteins and carbohydrates are also associated with the cell membrane. Some proteins stick to the surface of the lipid

90

5–2 (continued)

Content Development

There are two very important points to make about the cell membrane. First, students should understand the importance of a functioning cell membrane, especially in the context of protection for the cell. The analogy to the walls of a house should help with that understanding. You might want to carry the analogy a bit further. Like walls, the cell membranes are not perfect barriers—some poisons, like cyanide, do enter the cell. Second, students should realize that the structure of the membrane is directly related to function. One example is the molecular pump. Another is the idea of "identification card" carbohydrates, which has many important implications.

• **How does the structure of cell membranes play a role in organ transplants?** (Antigens on the cell membrane cause the rejection syndrome to occur when tissues are not closely matched.)

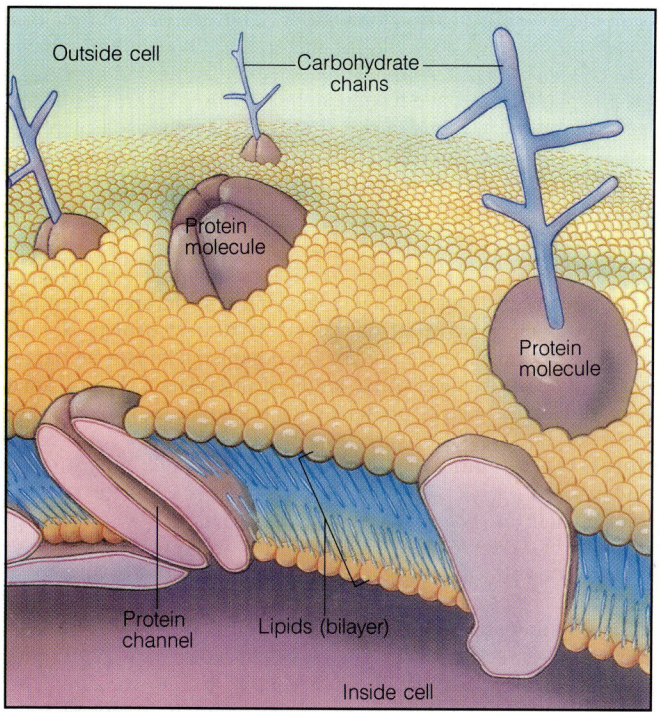

Figure 5–5 *According to a widely accepted model of membrane structure called the fluid-mosaic model, cell membranes are formed from double layers of lipids (bilayers) in which proteins are embedded. Some of the proteins form channels that allow certain molecules to pass into and out of the cell. Others resemble small pumps that push molecules from one side of the membrane to the other.*

bilayer, whereas others are free to move around within the layer. Some of the free-moving proteins act as channels through which molecules can pass. Others act like small pumps, actively pushing molecules from one side of the membrane to the other. The carbohydrates are attached to proteins or lipids at the membrane surface. Many of these carbohydrates act like chemical identification cards, allowing cells to recognize and interact with each other.

Cell Wall

In organisms such as plants, algae, and some bacteria, the cell membrane is surrounded by a **cell wall**. In other words, the cell wall lies outside the cell membrane. The cell wall helps to protect and support the cell. Because the cell wall is very porous, water, oxygen, carbon dioxide, and other substances can pass through easily.

If we looked at an electron micrograph of a plant cell, we would discover that the cell wall is made up of two or more layers. These layers form in a series of steps. The first layer to form develops where two plant cells meet. See Figure 5–6. This layer contains a gluey substance, called pectin, that helps hold the cells together. Each of the cells then forms a primary cell

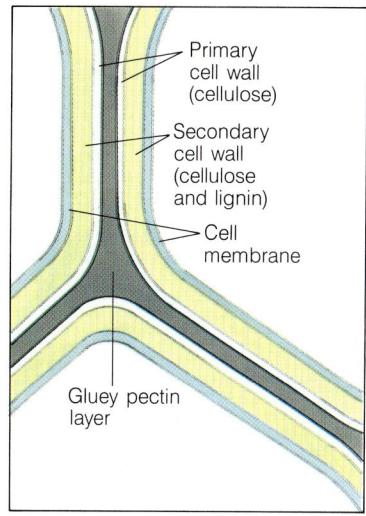

Figure 5–6 *The primary cell wall forms the outer boundary of plants, algae, and some bacteria. The secondary cell wall, however, generally forms only in woody stems.*

BACKGROUND INFORMATION
CELL MARKERS

The carbohydrate molecules associated with the surface of cell membranes are particularly important in cell recognition. It seems that nearly all cells have special carbohydrate molecules on their surfaces that are unique to their own cell type, their own individual, and their own species. Other cells carry special molecules that seem to be able to attach to these "markers" and allow a host of special events to occur: The cells of one individual will recognize the cells of another as "outsiders"; cells from kidney tissue will sort through cells of a dozen other tissue types to find other kidney cells and form a small clump of kidney tissue; viruses that infect the cells of a dog will find canine cells among those of hundreds of other species by means of the carbohydrate molecules on the surfaces of the canine cell membranes. The existence of these membrane-bound "markers" is an important feature of the cell membrane that makes possible the development and function of large multicellular organisms—like ourselves!

for the difference? (The living yeast cell membranes do not allow the dye to enter the cell. Boiling kills the cells and breaks down the membranes. The dye can now enter the cells.)

Reinforcement/Reteaching

Students sometimes have difficulty distinguishing the cell membrane and the cell wall. Emphasize that all cells have a membrane but not all have a cell wall. Here is an easy way to compare cells that have only a primary cell wall and those that also have the secondary wall. Show the students a stalk of celery (primary only) and the branch of a tree (secondary also). This and a little demonstration of relative flexibility should aid their understanding.

Skills Development
Guided Practice
Skill: Making observations
For students who have difficulty with the concept of the cell membrane acting as a protective barrier, try this demonstration. Prepare a yeast culture (dry or cake yeast, a small amount of sugar, and water). Put a small amount in a test tube and boil it for about one minute. Add a few drops of Congo Red dye to the unboiled and the boiled solutions. Place a drop of each solution on a microscope slide, cover with a coverslip, and view. Direct students to look closely at the yeast cells on each slide.
• **What color are the cells of unboiled yeast?** (Light blue.)
• **What about the boiled yeast cells?** (Red or pink.)
• **What explanation can you give**

BACKGROUND INFORMATION
PROKARYOTES AND EUKARYOTES

Prokaryotes differ from eukaryotes in several ways besides lack of a nucleus. They have no mitochondria, lysosomes, vacuoles, endoplasmic reticulum, Golgi apparatus, or chloroplasts. In other words, prokaryotes have no membrane-bound organelles.

Prokaryotic cells do have DNA, though it is less complex than that found in eukaryotic cells. Photosynthetic prokaryotes have photosynthetic membranes that are not enclosed in chloroplasts. Prokaryotes do have ribosomes and cell walls. However, the cell walls are usually composed of carbohydrates and polypeptides rather than cellulose. Even the flagella of prokaryotic cells have a different structure.

5-2 (continued)
Content Development

The discussion of eukaryotes and prokaryotes is an important one. When you are making the point that prokaryotes have no nucleus, be sure that students know that they do have nuclear material (DNA). You might also emphasize again that the difference between prokaryotic and eukaryotic cells is considered more significant than the differences between plant and animal cells.

The importance of the nucleus is pretty obvious. A brief discussion here can set the stage for the later study of genetics. If students can relate the coding for the production of molecules to life functions, genetics will not be such a foreign concept.

Students who understand the structure and function of the cell membrane should have little difficulty realizing that the nuclear envelope performs the same function for the nucleus.

The nucleoli and the chromosomes are specialized structures found within the nucleus. Both are related to the process of protein synthesis that will be discussed in Chapter 7.

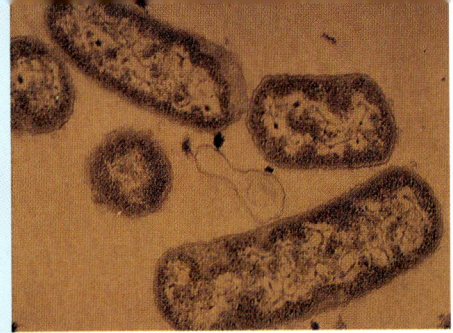
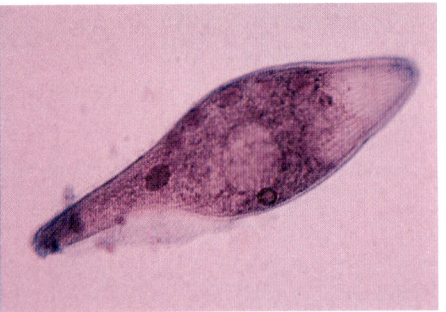

Figure 5–7 *One way of classifying living things is by the presence or absence of a nucleus. Prokaryotes, such as bacteria, do not have a nucleus (top). Eukaryotes, such as a blepharisma, contain a nucleus surrounded by a nuclear envelope (bottom).*

wall on its side of this gluey layer. The primary cell wall is made up of cellulose, a fibrous material. Cellulose fibers make the cell wall elastic, so that it can stretch as the cell grows.

In plants that have woody stems, another layer, called the secondary cell wall, develops. This wall is composed of cellulose and lignin (LIHG-nihn). Lignin makes cellulose more rigid. Wood consists mainly of secondary cell walls.

Nucleus

In many cells we can see a large dark structure, called the **nucleus,** which was first described by Robert Brown. Not all cells have nuclei. Small unicellular organisms known as bacteria, as well as several other kinds of organisms, do not have nuclei. The absence or presence of a nucleus can be used to divide organisms into two general categories. **Prokaryotes** (proh-KAIR-ee-ohts) are organisms whose cells lack nuclei. **Eukaryotes** (yoo-KAIR-ee-ohts) are organisms whose cells contain nuclei. *Karyon* means nucleus, *pro-* means before, and *eu-* means true. Can you see why the terms are appropriate?

Prokaryotic organisms, which include bacteria and their relatives, are usually small and unicellular. Eukaryotic organisms include both unicellular and multicellular forms. The distinction between prokaryotes and eukaryotes is a basic one, and we will return to it many times as we consider the diversity of living things. In fact, scientists consider this distinction far more important than the distinction between plant and animal cells!

Many of the scientists who first examined cells under a microscope suspected that the nucleus was doing something important. What could it be? The nucleus has been found to be the information center of the cell and contains DNA (deoxyribonucleic acid). The instructions for making thousands of different molecules are found in the nucleus. These instructions are decoded and executed by a process that we will discuss in Chapter 7. The nucleus also directs all the activities that occur in a living cell.

NUCLEAR ENVELOPE The nucleus of a eukaryotic cell is generally 2 to 5 micrometers in diameter. Surrounding the nucleus are two membranes that form the **nuclear envelope**. These two membranes form the boundary around the nucleus. In the nuclear envelope are dozens of nuclear pores, or small openings. The molecules that move in and out of the nucleus pass through the nuclear pores.

NUCLEOLUS Most nuclei contain a small region called the **nucleolus** (noo-KLEE-uh-luhs) that is made up of RNA (ribonucleic acid) and proteins. The nucleolus is the structure in which ribosomes are made. Ribosomes, as you will soon learn, aid in the production of proteins within the cell.

The cytoplasm is a semifluid mixture that occupies the space in the cell not taken up by the nucleus. There are many small structures in the cytoplasm that perform various cell functions.

- **Can a eukaryotic cell function normally without a nucleus?**

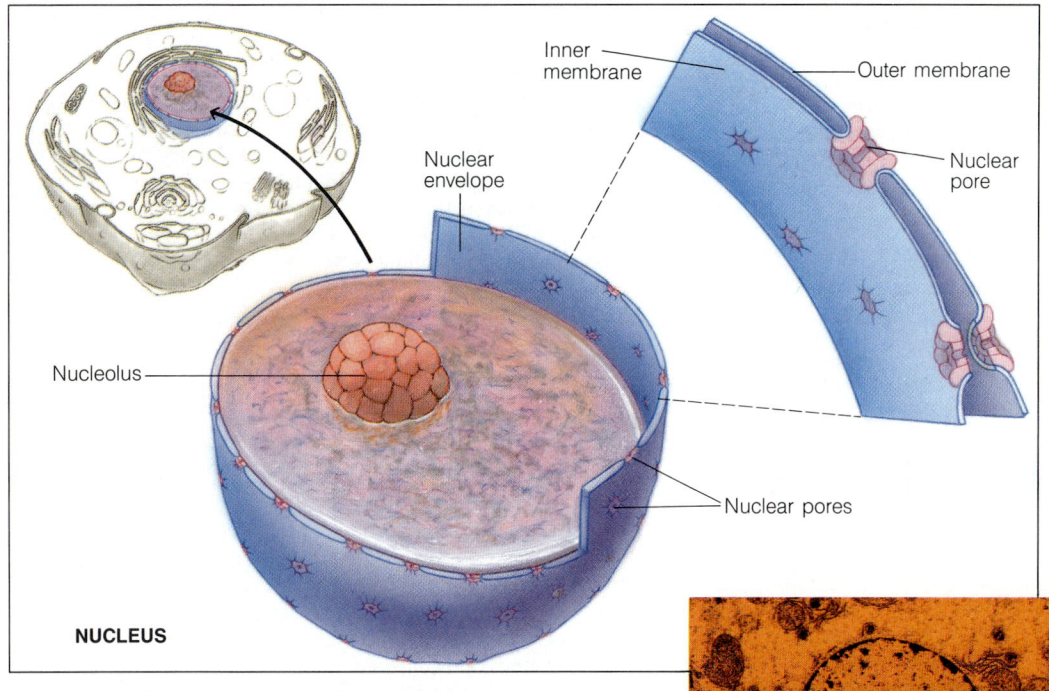

Figure 5–8 The nucleus directs all the activities of a cell. Notice the various structures that make up the nucleus and the appearance of the nucleus in the micrograph.

CHROMOSOMES The DNA in the nucleus of eukaryotic cells is attached to special proteins and forms large structures called **chromosomes**. Chromosomes contain the genetic information that must be passed to each new generation of cells.

Cytoplasm

Because the nucleus sits in the center of many eukaryotic cells, we can divide the space within a cell into two compartments: the nucleus and the **cytoplasm**. The cytoplasm is the area between the nucleus and the cell membrane. The cytoplasm contains many important structures that we shall discuss in the next section.

5–2 SECTION REVIEW

1. What are the three basic structures found in most cells?
2. What is the function of the cell membrane?
3. **Critical Thinking—Assessing Concepts** Distinguish between prokaryotes and eukaryotes. Why is this distinction important?

FACTS AND FIGURES

Mitochondria have their own DNA, RNA, and ribosomes. They are self-replicating, arising only by the division of existing mitochondria. The cell cannot make them from raw materials.

TEACHING SUPPORT

Laboratory Manual
- Characteristics of Prokaryotic and Eukaryotic Cells, p. 95

Guide For Reading
- What is an organelle?
- What are the functions of the major cytoplasmic organelles?
- How does the structure of a plant cell differ from that of an animal cell?

5–3 Cytoplasmic Organelles

Even in the low-power images that we can produce with the light microscope, it is clear that there are structures inside the cytoplasm of the cell. The structures in the cytoplasm are generally called **organelles** (or-guh-NEHLZ). **An organelle is a tiny structure that performs a specialized function in the cell.** Each organelle (little organ) has a special job that helps maintain the cell's life. Let's look at some of the organelles that are found in plant and animal cells.

Mitochondria and Chloroplasts: Power Stations

All living things require a reliable source of energy. On Earth that source is usually the sun or food substances. The **mitochondrion** (might-oh-KAHN-dree-uhn; plural: mitochondria) and the **chloroplast** are key organelles that change energy from one form to another. Mitochondria change the chemical energy stored in food into compounds that are more convenient for the cell to use. Chloroplasts trap the energy of sunlight and convert it into chemical energy. The reactions that take place in both of these organelles are closely related, and we will examine them in more detail in Chapter 6.

The mitochondrion contains two special membranes. The outer membrane surrounds the organelle, and the inner membrane has many folds that increase the surface area of the mitochondrion.

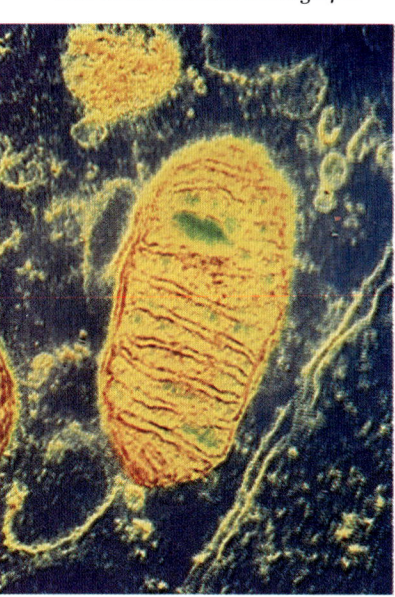

Figure 5–9 Mitochondria are the powerhouses of the cell. They provide the cell with the energy it needs to survive. Note the structure of a mitochondrion in the diagram and in the electron micrograph.

94

TEACHING STRATEGY 5–3

Focus/Motivation
Have students organize a sheet of paper into a blank features matrix. The title of the matrix will be "The Cell." The headings across the top of the chart might include structure, description, function, location, and type of cell. Help students get started by putting cell membrane, cell wall, nucleus, and so on under the heading Structure. As you proceed through the section, students can complete the matrix.

Content Development
The organelles in this section are grouped on the basis of similarity of function. To the knowledgeable biologist, mitochondria and chloroplasts have unique functions within cells. To the beginning biologist, they have the common function of changing energy from one form to another.

Students should be reminded that only cells that use sunlight to make food contain chloroplasts. Besides plant cells, these include some protists and algae

The structure of the chloroplast is similar. It is surrounded by two envelopelike membranes and contains a third kind of membrane, where the radiant energy of the sun is actually changed into chemical energy. Chloroplasts are found only in plant cells and algae. As we will see later, both mitochondria and chloroplasts have a degree of independence from the rest of the cell. This has led to some discussion that they may be descended from independent organisms.

Ribosomes: Protein Factories

Ribosomes are the structures in which proteins are made. Cells that are active in protein synthesis are often crowded with ribosomes. Ribosomes are composed of RNA and protein. Some ribosomes are attached to membranes; some are found free in the cytoplasm. Ribosomes are among the smallest of organelles. They are no larger than 25 nanometers in diameter. A nanometer is equal to one billionth of a meter.

Endoplasmic Reticulum and Golgi Apparatus: Manufacturers and Shippers

Many cells are filled with a complex network of sacs known as the **endoplasmic reticulum** (ehn-doh-PLAZ-mihk rih-TIHK-yuh-luhm), or ER. The endoplasmic reticulum transports materials through the inside of the cell. There are two types of endoplasmic reticulum. In the smooth endoplasmic reticulum (smooth ER), the walls of the sacs look smooth

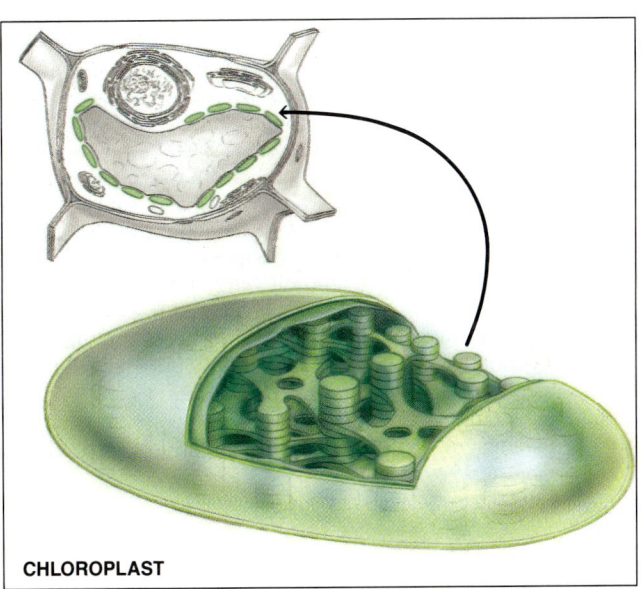

CHLOROPLAST

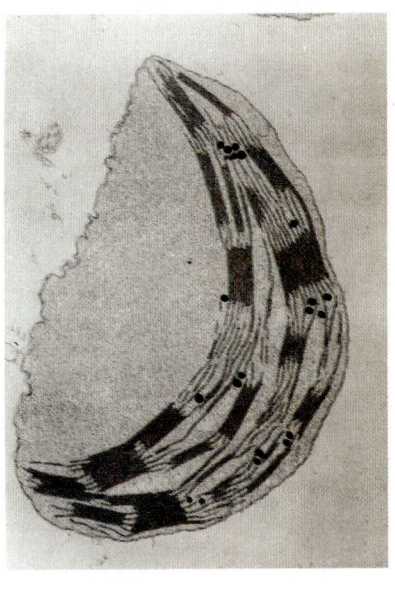

Figure 5–10 Chloroplasts are organelles that use the radiant energy of sunlight to produce the chemical energy organisms need to survive. Notice the envelopelike membranes within the chloroplasts.

BACKGROUND INFORMATION
ORGANELLES

Some biologists enjoy arguing about just what should and should not be called an organelle. One school of thought holds that a true organelle is membrane-bound; therefore, structures such as microtubules and ribosomes do not count as organelles. Another holds a broader view and suggests that any functional structure within the cell should be considered an organelle. Although this view seems sensible, its logical extension would give organelle status to a variety of structures within other organelles, such as the chromosomes within the nucleus or the photosynthetic membranes within the chloroplast. In this book we've tried to stay clear of such semantic controversies by applying the term *organelle* to those structures that most cell biologists consider organelles. We hope that this avoids needless controversies regarding definitions of a word that most scientists use with little difficulty.

---

but definitely exclude animal-like organisms.
   Students may be interested to learn that there is a relationship between the amount of "work" done by a cell and the number of mitochondria.
• Would you expect to find more mitochondria in a skin cell or a muscle cell? (Muscle.)
• What is the source of the energy produced by the mitochondria? (Food.)
   The liver performs many vital functions—much work—for the body. Liver cells contain huge numbers of mitochondria—as many as 1000 per cell.

Content Development
When discussing ribosomes, you can once again emphasize function. Tell students that the number of ribosomes in a cell is related to how much protein a cell makes since protein synthesis occurs at the ribosomes.

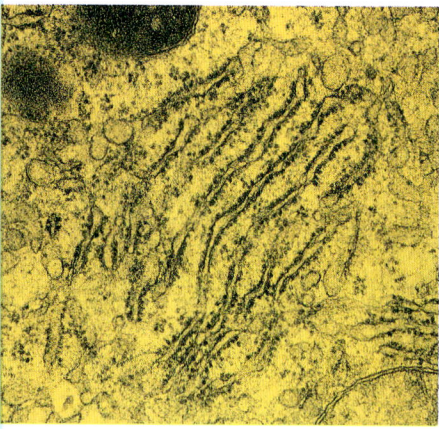

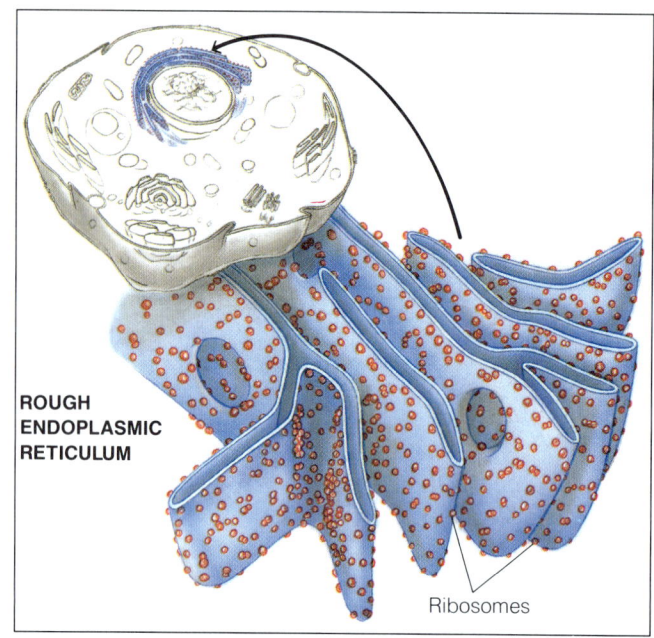

Figure 5–11 *The endoplasmic reticulum (ER) is a series of sacs into which certain proteins and other substances are inserted to separate them from the cytoplasm. The ER also transports materials throughout the cell. Which type of ER has ribosomes attached to its surface, as shown here?*

Figure 5–12 *The Golgi apparatus modifies proteins by attaching carbohydrates or lipids to the proteins. The modified proteins can then be transported from one part of the cell to another.*

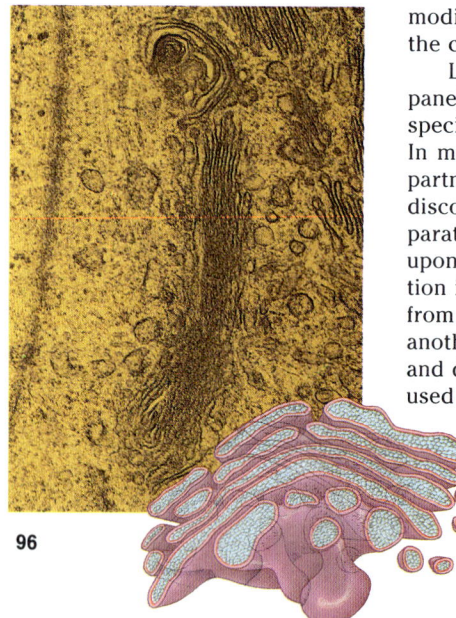

and are not studded with particles or granules. In some cells, special enzymes and chemicals are stored in the smooth endoplasmic reticulum.

The other form of endoplasmic reticulum is involved in the synthesis of proteins. This form is called the rough endoplasmic reticulum (rough ER) because the ribosomes that are stuck to its surface give it a rough appearance. Newly made proteins are inserted into the rough ER, where they may be chemically modified. Many proteins that are released, or exported, from the cell are synthesized on the rough ER.

Like automobiles to which enthusiasts attach chrome panels and hood ornaments, proteins often are modified by special enzymes that attach carbohydrates and lipids to them. In most cases, the proteins are first moved into special compartments known as the **Golgi apparatus**, because they were discovered by the Italian scientist Camillo Golgi. The Golgi apparatus looks like a flattened stack of membranes piled one upon the other, not unlike a stack of pancakes. After modification in the Golgi apparatus, the proteins may then be released from the cell or take up positions in other parts of the cell. Put another way, the Golgi apparatus modifies, collects, packages, and distributes molecules made at one location of the cell and used at another.

Lysosomes: Cleanup Crews

Some white blood cells in your blood are capable of swallowing whole bacteria. Other cells, such as the single-celled ameba, are capable of engulfing clumps of yeast, sugar, and even tiny pieces of bread. Many cells will seem to engulf microscopic particles of food, foreign material, or even other cells. This process is called **endocytosis** (ehn-doh-sigh-TOH-sihs). Endocytosis is a way in which materials that are too large to pass through the cell membrane get into the cell.

When a cell encircles a particle, the cell membrane forms a pocket around the foreign material. The foreign material must now be digested, or broken down. That is the job of the **lysosomes** (LIGH-suh-sohmz). Lysosomes are small membrane-bordered structures that contain chemicals and enzymes necessary for digesting certain materials in the cell. Lysosomes are formed by the Golgi apparatus. Plant cells do not have lysosomes.

Lysosomes are also involved in breaking down organelles that have outlived their usefulness. In a way, lysosomes can be thought of as the cell's cleanup crews. As such, lysosomes perform the vital function of removing "junk" that might otherwise accumulate and clutter up the cell. Why do you think human diseases that cause the lysosomes to work improperly, such as Tay-Sachs disease, can be serious and even fatal? ❷

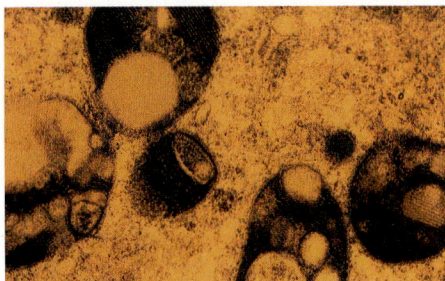

Figure 5–13 The round, dark-colored organelles are lysosomes that have been magnified approximately 95,000 times. Lysosomes contain enzymes that can digest substances in the cell, such as foreign materials and organelles that have outlived their usefulness. For this reason, lysosomes can be thought of as the cell's cleanup crews.

Vacuoles and Plastids: Storage Tanks

Many cells contain saclike structures, called **vacuoles** (VAK-yoo-ohlz), that store materials such as water, salts, proteins, and carbohydrates. In many plant cells there is a single large central vacuole filled with liquid. The pressure of the liquid-filled vacuole in these cells makes it possible for plants to grow quickly and to support heavy structures such as leaves and flowers. Why do you think a plant that has lost a large amount of water will begin to wilt?

Plastids are plant organelles that may take many forms, one of which is the chloroplast, an organelle we have already mentioned. Many plastids are involved in the storage of food and pigments. Some examples of plastids are leukoplasts (LOO-koh-plasts), which store starch granules, and chromoplasts (KROH-muh-plasts), which store pigment molecules. The red color in the skin of a ripe tomato comes from pigments produced in chromoplasts.

Cytoskeleton: Framework

Cells are found in a bewildering variety of shapes: some round, some cubical, some long and slender. In addition, most cells are capable of some type of movement. Even stationary cells can move the organelles around inside of them. How do they accomplish such motion? And what allows cells to keep their unusual shapes?

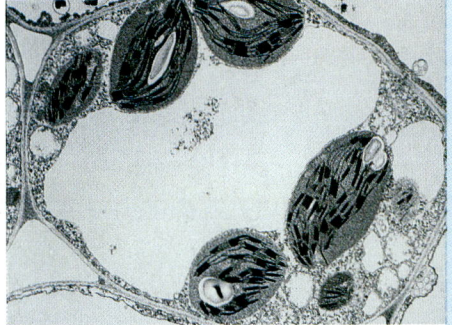

Figure 5–14 The large saclike structure in this leaf cell is a vacuole, which has been magnified approximately 16,000 times. Vacuoles act as cellular storage tanks and may hold water, salts, proteins, and carbohydrates.

HISTORICAL NOTE
CAMILLO GOLGI

Camillo Golgi's discovery of a structure near the cell nucleus was not widely credited by his contemporaries. In fact, the structure he observed was called Golgi's apparatus in derision. For years, cytologists were divided as to the existence of this organelle. The application of the electron microscope to the study of the cell settled the issue: The Golgi apparatus has a distinct and characteristic structure that can easily be recognized in the EM. Plant biologists quickly realized that the dictyosome, a structure that had also been observed only with special stains, was in fact the Golgi apparatus. The name dictyosome is seldom used today, in recognition of the fact that both plant and animal cells contain the Golgi apparatus.

Skills Development
Guided Practice
Skill: Hypothesizing

The process of endocytosis can best be explained by drawing simple diagrams on the chalkboard or overhead transparencies. Stress that the enzymes produced by the lysosomes are "targeted" at the foreign material rather than generally released throughout the cell. Since the cytoplasm is digestible, as are the organelles, general distribution could destroy the cell. Just for fun, have students hypothesize the relationships between lysosomes and (a) the development of a frog from a tadpole and (b) the process of aging.

Content Development

Vacuoles do occur in both plant and animal cells, but animals usually have several small vacuoles rather than the central structure that occupies a large part of the plant cell. Plastids are found only in plant cells and store pigments as well as starch. All plastids, like chloroplasts, have a system of internal membranes especially adapted for storage.

TEACHING SUPPORT

Study Guide
• Section 5–3, p. 47

5–3 (continued)

Content Development
Using the Visuals Point out Figure 5–15, which shows the cytoskeleton of a cell. The diagram shows that even without the cytoplasm, the basic shape of the cell remains. Review the many functions of the microbules and the microfilaments listed in the text. Students might be interested to know that most of what biologists know of the cytoskeleton they have learned within the last 25 years.

Reinforcement/Reteaching
Have students complete their features matrix. Encourage them to use the features matrix to review the material on cell structure and function.

Enrichment
Assign to students the task of constructing their own three-dimensional cell models using materials of their own choosing. Encourage them to stay close to shapes, relative sizes, and so on. This activity works well as an individual or group project.

SECTION REVIEW 5–3

1. An organelle is a tiny structure (usually membrane-bound)

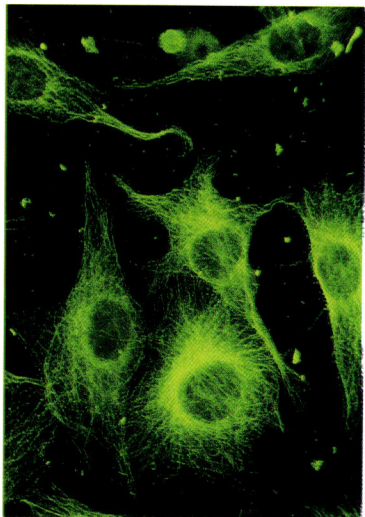 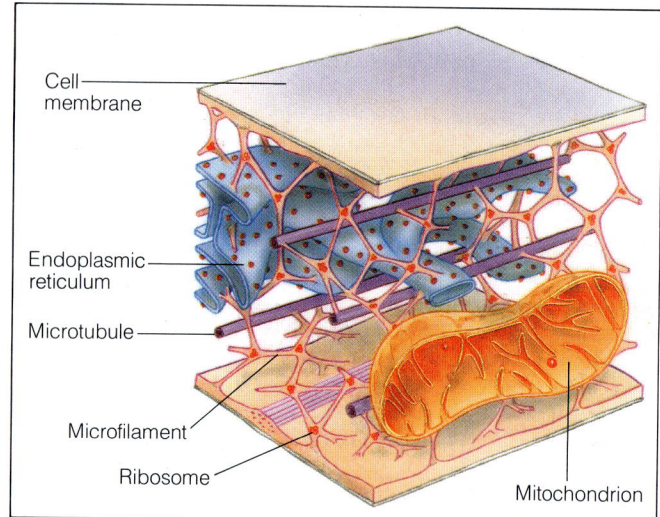

Figure 5–15 This diagram shows the cytoskeleton of the cell. Notice that the endoplasmic reticulum, mitochondria, and ribosomes are all held in place by the microfilaments and microtubules that form the cytoskeleton. The photograph shows the actual microtubules that make up the cytoskeleton.

Not too many years ago, scientists were content to suggest that the cytoplasm was filled with a material, called protoplasm, that gave the cell many unique properties. As we have learned more about the composition of the cell, the idea of "protoplasm" has lost acceptance. In its place, biologists have uncovered a framework known as the **cytoskeleton**. The cytoskeleton is composed of a variety of filaments and fibers that support cell structure and drive cell movement.

One of the main components of the cytoskeleton are microtubules, hollow tubules made out of proteins. Microtubules provide support for cell shape, help move organelles through the cell, and play a special role in cell division by forming centrioles (SEHN-tree-ohlz). Centrioles are found in animal cells and many other eukaryotic cells but not in plant cells.

In some cells, the microtubules support hairlike projections from the cell surface known as cilia (SIHL-ee-uh; singular: cilium) and flagella (fluh-JEHL-uh; singular: flagellum). Cilia are short threadlike structures that help unicellular organisms move. They also aid in the movement of substances along the cell's surface. Flagella are longer whiplike structures that help unicellular organisms move about. Cilia and flagella contain nine pairs of microtubules arranged around a pair in the center. These microtubules are linked to each other, and the bridges that connect them generate the force to produce motion.

Another principal component of the cytoskeleton are microfilaments—long, thin fibers that function in the movement and support of the cell. They also permit movement of the cytoplasm within the cell. This is called cytoplasmic streaming. The proteins that make up microfilaments are also found in muscle cells, which are specialized for contraction.

that performs a specialized task for the cell. Examples include mitochondria and chloroplasts.
2. Enzymes contained within the lysosome are used to break down particles within the cell.
3. Mitochondria and chloroplasts are both membrane-bound organelles involved in energy production and storage. Chloroplasts use energy from sunlight in a process called photosynthesis to produce food, which contains energy stored in its chemical bonds. Mitochondria use cellular respiration to produce energy found in the chemical bonds in ATP.
4. The cytoskeleton is made up primarily of microtubules and microfilaments.

Reinforcement/Reteaching
At this time you might want to

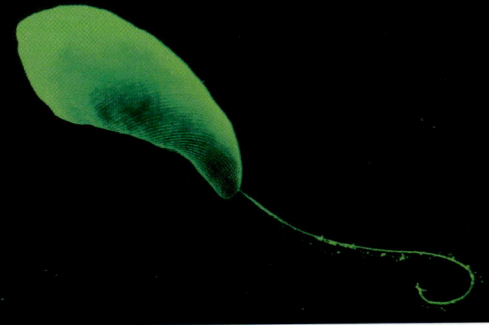

Figure 5-16 The microtubules of the cytoskeleton support hairlike projections called cilia and flagella. Cilia, such as those on this single-celled organism (left), are short threadlike structures that help the cell move about by beating back and forth. Flagella, such as the one on the single-celled *Euglena* (right), help the cell move by beating in a whiplike fashion.

5-3 Section Review

1. What are organelles? Give two examples of organelles.
2. How do lysosomes function to digest particles?
3. Compare mitochondria and chloroplasts.
4. **Critical Thinking—Summarizing Information** What components make up the cytoskeleton?

5-4 Movement of Materials Through the Cell Membrane

Each individual cell exists in a liquid environment. Even the cells of multicellular organisms such as a maple tree or a human are bathed in liquid. The cells of our bodies are bathed in a liquid that was once part of blood. The presence of a liquid environment makes it easier for materials such as food, oxygen, and water to move into and out of the cell. There are several ways in which materials enter and leave the cell.

Guide For Reading
- How do diffusion and osmosis move materials into and out of cells?
- How do active and passive transport differ from each other?
- What are endocytosis, phagocytosis, pinocytosis, and exocytosis?

Diffusion

Although the cell membrane is a barrier, it must not be too effective. Materials must pass into and out of the cell. How is this movement accomplished?

We know that molecules move constantly, colliding with one another and tending to spread out randomly through space. This random motion results in the tendency of molecules to move from a region where they are more numerous to a region where they are less numerous. In other words, molecules move from an area of higher concentration to an area of lower concentration.

The driving force behind the movement of many substances across the cell membrane is called **diffusion** (dih-FYOO-zhuhn). **Diffusion is the process by which molecules of a substance move from areas of higher concentration of that substance to areas of lower concentration.**

99

BACKGROUND INFORMATION
PENICILLIN

Penicillin, one of the most important antibiotic drugs in the history of medicine, depends on osmosis for its killing action. Penicillin inhibits an enzyme with which many bacteria produce chemical cross-links in their cell walls. This leads to the formation of a weakened cell wall that cannot stand the stress of osmotic pressure. Gradually, the cell wall becomes weaker and weaker until it breaks, and the bacterium bursts under the inrush of water caused by osmosis.

5–4 (continued)

Content Development

Using the Visuals While the term *equilibrium* may not be too familiar to students, they can usually grasp it quickly when they think about "equal amounts"—a concept with which they are familiar. It is necessary that they understand equilibrium now, in order to understand osmosis. Figure 5–17 will help students visualize the concept.

In discussing selective permeability, you might want to refer to Section 5–2 on the cell membrane. Tell students that some molecules, like proteins, are too large and complex to move through the membrane. They do not dissolve in the lipid layer. Here is a question to challenge your students.

• **In what way is having a cell membrane impermeable to large molecules necessary to the survival of a cell?** (Permeability is a two-way process. If the membrane allowed passage of proteins into a cell, it would also allow some of the cell proteins to leave.)

Reinforcement/Reteaching

Some students might understand the concept of selective permeability better with a visual demonstration. Pour various materials through a kitchen strainer or collander. (Sand, salt, sugar, water, a cup of marbles, and so on—almost anything you have on hand.) Be sure to select some substances that will go through and some that will not. Students will get the point and remember it better.

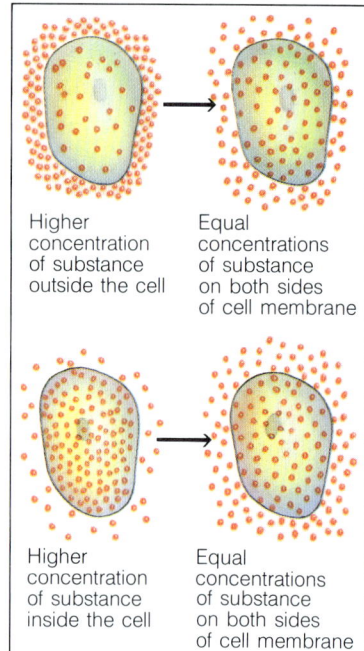

Figure 5–17 *Diffusion is the movement of the molecules of a substance from areas where they are more concentrated to areas where they are less concentrated. What point has been reached when the concentration of a substance is equal on both sides of a cell membrane?* ❶

What factors determine whether diffusion occurs across a membrane? If two substances are present in unequal amounts on either side of a membrane, each substance will tend to move toward the area of lower concentration until equilibrium is reached. See Figure 5–17. Equilibrium occurs when the concentrations of the substances on both sides of the membrane are the same. At this point, we may say that an equilibrium point for each substance has been reached. Individual molecules continue to move rapidly between the two sides at equilibrium, but because roughly equal numbers of molecules move in each direction, there is no further change in concentration.

Another important factor that determines whether diffusion occurs across a membrane is permeability (per-mee-uh-BIHL-uh-tee). If a particular substance is able to diffuse across a membrane, then we say that the membrane is permeable to that substance. A membrane is said to be impermeable to those things that cannot pass across it. As you might expect, biological membranes are permeable to some things and impermeable to others. Therefore, we can describe most biological membranes as **selectively permeable**.

Osmosis

Although compounds that are able to dissolve in the lipid bilayer of the cell membrane pass through the membrane easily, most other molecules do not. There is, however, one important exception to that rule: water. Water molecules pass through most cell membranes very rapidly. The fact that water molecules can pass through membranes so easily has important consequences for the cell.

The diffusion of water molecules through a selectively permeable membrane is called **osmosis** (ahs-MOH-sihs). If a membrane separates two solutions, in which direction will the water molecules move? To answer this question, let us suppose that we place a concentrated sugar solution on one side of a selectively permeable membrane and a dilute sugar solution on the other. As you have just learned, water will pass through the membrane by diffusion from the area of high water concentration to the area of low water concentration. In which direction is that?

Look at Figure 5–18. On the side of the membrane with the concentrated sugar solution, fewer water molecules will strike the membrane because they will be crowded out by the sugar molecules. Thus, this side is the area of low water concentration. On the side of the membrane with the dilute sugar solution, more water molecules will strike the membrane. This side is the area of high water concentration. As a result, more water moves through the membrane from the side that has the low sugar concentration to the side that has the high sugar concentration than moves through in the opposite direction. This results in a net movement of water from the side with the dilute

Content Development

Using the Visuals Students first need to understand that the presence of water in a cell is vital to the survival of the cell, but too much water moving into the cell can destroy the cell.

Refer to Figure 5–18. Stress re-

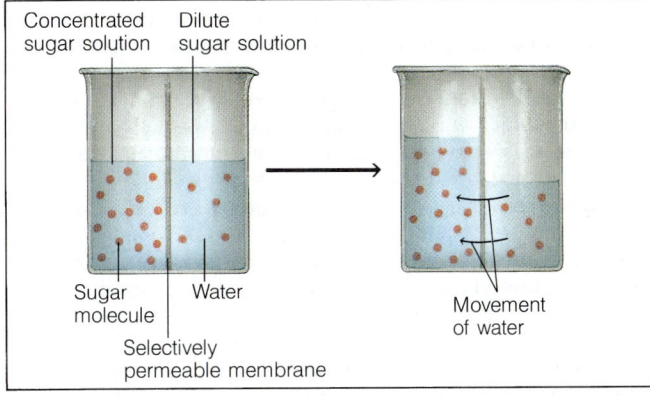

Figure 5–18 Osmosis is the diffusion of water molecules through a selectively permeable membrane from an area of greater concentration of water to an area of lesser concentration of water. In this diagram, there are more water molecules on the side containing the dilute sugar solution. As a result, the water molecules move to the side containing the concentrated sugar solution. Osmosis continues until equilibrium is reached.

sugar solution to the side with the concentrated sugar solution.

The force exerted by osmosis, or osmotic pressure, tends to move water across membranes from a more dilute solution into a more concentrated solution. If, however, two solutions contain exactly the same amount of dissolved material, there is no osmotic pressure across a membrane separating them because the concentrations of dissolved materials are in equilibrium.

Osmotic pressure can cause serious problems for a cell. Because the cytoplasm is filled with salts, sugars, proteins, and other molecules, it will almost always have a much lower concentration of water than is found in fresh water. Therefore, there should be a net movement of water into a typical cell if it is surrounded by fresh water. If water moves in freely, the volume of a cell will increase until the cell becomes swollen and bursts like an overinflated balloon.

Figure 5–19 Normal red blood cells (left) will shrink if too much water leaves the cells due to osmosis (center). If too much water enters the cells during osmosis, the red blood cells will swell (right).

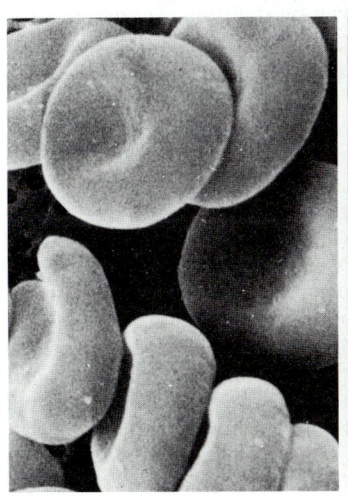 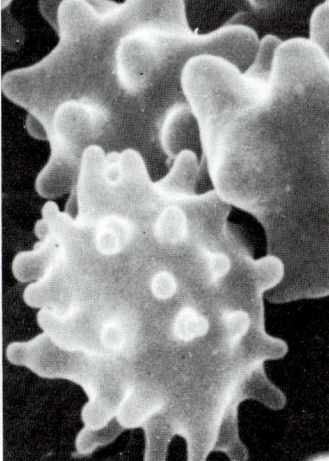 

101

BACKGROUND INFORMATION
FACILITATED DIFFUSION

Facilitated diffusion is important in speeding up the rate of diffusion of materials into a cell. It may be just as necessary to increase the rate at which substances such as glucose leave a cell. In facilitated diffusion, glucose moves back and forth from the bloodstream to the cells of the liver as concentrations vary.

The rate at which facilitated diffusion occurs is sometimes influenced by hormones. Insulin, for example, increases the uptake of glucose in some human tissues. In muscle cells, the stimulus seems to be activity.

5-4 (continued)
Content Development
It is the structure of the cell membrane itself that makes it permeable to water, so it cannot act as a barrier in any way. If excess water is surrounding the cell, how then can the cell protect itself? Discuss with students the various ways in which cells deal with this problem.

Skills Development
Guided Practice
Skill: Making inferences
Place a small number of paramecia in a petri dish on a microprojector. Have students observe the paramecia as you discuss contractile vacuoles.

Flood the environment of the paramecia with distilled water. As students continue to observe, point out the action of the contractile vacuoles.
• **What was added to the dish?** (Pure water.)
• **How do you know?** (Action of the contractile vacuoles increased.)
• **What will eventually happen to the paramecia?** (Explode.)
• **Why?** (The vacuoles cannot keep up with the inward movement of water.)
• **What would happen if a small amount of salt water were added?** (Probably vacuole action would return to normal.)

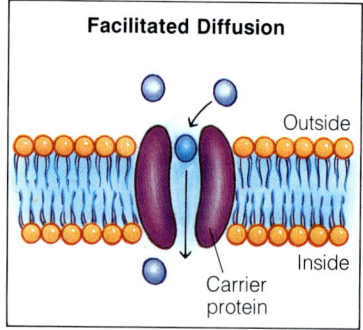

Figure 5-20 Most molecules that do not dissolve in the lipid bilayer of the cell membrane cannot pass through by diffusion. One way of getting such molecules into the cell is by facilitated diffusion. This process, which does not require any energy, uses a carrier molecule to transport a substance into the cell through the cell membrane.

The structure of the lipid bilayer makes cell membranes freely permeable to water. Cells deal with the problem of osmotic pressure in a variety of ways.

In some instances, the cells of many organisms do not come in contact with fresh water. Instead, the cells are bathed in fluids, such as blood, that have concentrations of dissolved materials roughly equal to the cells themselves.

Other cells, such as plant cells and bacteria, deal with osmotic pressure in another way. They are surrounded by cell walls that prevent the cells from expanding, even under tremendous osmotic pressure. However, the increased osmotic pressure makes the cells extremely vulnerable to injuries to the cell wall.

Still other cells employ a mechanism to pump out the water that is forced in by osmosis. For example, some unicellular organisms have a structure called a contractile (kuhn-TRAK-tihl) vacuole. By contracting rhythmically, the contractile vacuole pumps water out of the cell.

Facilitated Diffusion

Osmosis and diffusion are forms of passive transport across the cell membrane because energy is not needed for these processes. Some materials, including alcohols and small lipids, can pass directly through the membrane because they can dissolve in the lipid bilayer. But many molecules are transported across a membrane in the direction of lowest concentration by a carrier protein. This process is called **facilitated diffusion**. In red blood cells, for example, a carrier protein in the cell membrane transports glucose from one side of the membrane to the other. The glucose-transporter protein facilitates, or helps in, the diffusion of glucose.

Although facilitated diffusion is fast, specific, and does not require energy, it is still driven by diffusion. Therefore, a net movement of material across a cell membrane by facilitated diffusion can occur only if a concentration difference exists across that membrane.

Active Transport

Active transport is an energy-requiring process that enables material to move across a cell membrane against a concentration difference. There are two types of active transport.

In one type, individual molecules are carried through membrane-associated pumps. Special transport macromolecules that exist in the cell membrane move molecules across the membrane. Among the molecules that are transported are calcium, potassium, and sodium ions. Normally they are not able to diffuse across the membrane. The molecular pumps that carry out this transport require chemical energy, thus making this process a form of active transport. All cells that have been

Content Development
Students should understand that facilitated diffusion is still the movement of molecules *with* the concentration gradient. The difference is that it occurs more rapidly than diffusion and osmosis. The carrier protein, to which

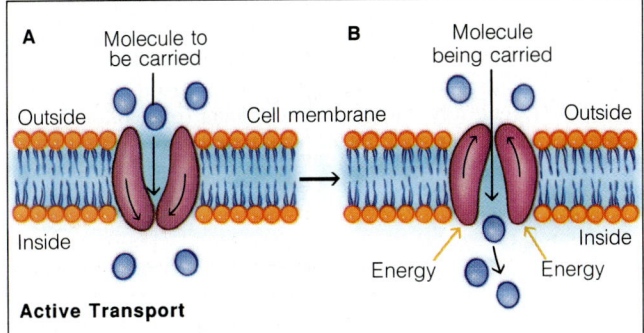

Figure 5–21 Very often a substance must pass through the cell membrane even if the concentration of that substance is greater inside the cell than it is outside the cell. In such cases, the cell must expend energy in a process called active transport. In one form of active transport, energy is used to change the shape of the cell membrane and to actively pull the substance through the membrane into the cell.

studied seem to transport at least a few molecules in this way. The use of energy in these systems enables cells to concentrate molecules, even when the normal forces of diffusion might tend to move those molecules in the opposite direction.

In the second type of active transport, large amounts of material are transported through movements of the cell membrane. One of these movements is called endocytosis. As you may remember, endocytosis is the process of taking material into the cell by means of infoldings, or pockets, of the cell membrane. The pocket that results breaks loose from the outer portion of the cell membrane and forms a vacuole within the cytoplasm. Large molecules, clumps of food, and even whole cells can be taken up in this way.

When large particles are taken into the cell by endocytosis, the process is called phagocytosis (fayg-oh-sigh-TOH-sihs). In phagocytosis, extensions of cytoplasm surround and engulf large particles. Amebas use this method of taking in food.

In a process similar to endocytosis, many cells take up liquid from the surrounding environment in a similar way. Tiny pockets form along the cell membrane, fill with liquid, and pinch off to form vacuoles within the cell. This process is known as pinocytosis (pighn-oh-sigh-TOH-sihs).

Figure 5–22 Another form of active transport is called phagocytosis. During phagocytosis, a large particle is surrounded by pockets of the cell membrane (right). Once the particle is surrounded, the pocket breaks away from the cell membrane and forms a vacuole within the cell. Amebas are one type of organism that takes in food and other materials through phagocytosis (left).

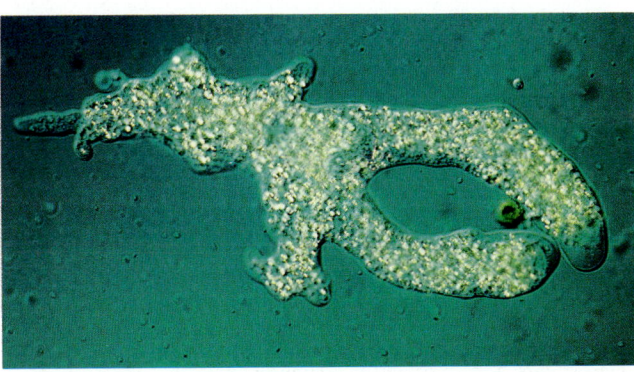

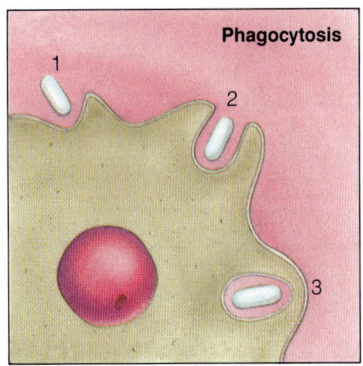

TEACHING SUPPORT

Study Guide
- Section 5–4, p. 49

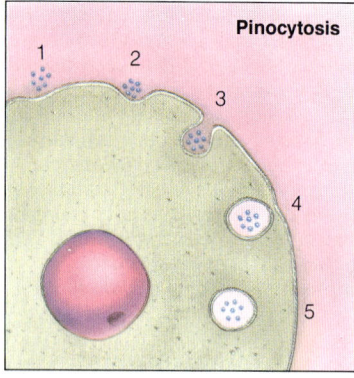

Figure 5–23 Like phagocytosis, pinocytosis is a form of active transport. In pinocytosis, however, cells take in liquids rather than solid particles.

As you might expect, cells are also capable of sending material out of the cell. The removal of water by means of a contractile vacuole is one example of this kind of active transport. When other large molecules are removed from the cell, the process is generally known as exocytosis (ehk-soh-sigh-TOH-sihs). As exocytosis occurs, the membrane surrounding the material fuses with the cell membrane, forcing the contents out of the cell.

5–4 SECTION REVIEW

1. What is diffusion?
2. Why is osmosis a form of diffusion?
3. Molecules that are soluble in lipids tend to move across cell membranes more quickly than those that are not. Based on this information, what could you conclude about the structure of the cell membrane?
4. **Critical Thinking—Making Inferences** The red blood cell contains a higher concentration of K^+ ions than does the liquid that surrounds it. Are K^+ ions most likely to be moved across the membrane by facilitated diffusion or active transport? Explain.

Guide For Reading

■ How is cell specialization expressed by cell structure?

Figure 5–24 This pancreas cell is adapted for enzyme production.

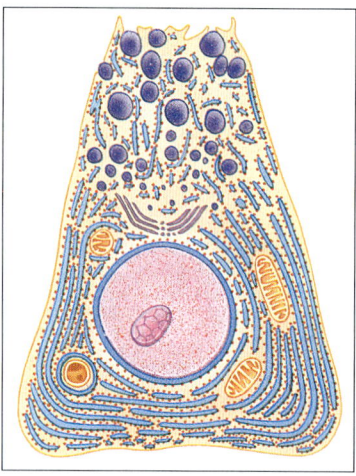

5–5 Cell Specialization

Cell specialization is one of the key characteristics of cells in a multicellular organism. **By cell specialization, we mean that cells are often uniquely suited to perform a particular function within the organism.** For example, some cells are specialized to move. Some cells are specialized to react to their environment—the world around them. Some cells are specialized to make certain products. We will now look briefly at three cells, each specialized for a different function.

Factory in Miniature

Located just below the stomach is a small structure called the pancreas. Portions of the pancreas contain cells that are specialized to produce digestive enzymes. Although these cells have most of the organelles found in other cells, they contain enormous amounts of the organelles involved in protein synthesis. In fact, one cell may contain 100 times the amount of rough endoplasmic reticulum found in other cell types. Each cell is filled with ribosomes as well as rough ER. These cells also contain large Golgi apparatus filled with vacuoles that are loaded with soon-to-be-released protein. Cells of the pancreas are dominated by organelles needed for protein synthesis.

5–4 (continued)

Content Development
Remind students that cells do not only take in substances. Cells frequently need to remove materials such as excess water or wastes. Remind them of the contractile vacuoles. This process is exocytosis. The prefixes *endo-* and *exo-* should help students remember the terms and their meanings.

SECTION REVIEW 5–4

1. The process by which molecules move from areas of higher concentration to areas of lower concentration.
2. Osmosis is the diffusion of water across a membrane.
3. Cell membranes contain lipids.
4. Active transport, since the ions move from lower concentration to higher concentration.

Reinforcement/Reteaching
This is a difficult section for students. For those having trouble answering the Section Review questions, go back to the discussion in the section and review the material students have had difficulty understanding.

Closure
Make up a fictitious list of substances. Have students tell the method of transport the cell would use and why. It is not necessary that they be completely accurate. What you are looking for is correct reasoning. (Basically, any gases would diffuse; water would move by osmosis; for other substances, use your judgment about the correctness of students' reasoning.)

TEACHING STRATEGY 5–5

Focus/Motivation
Have students think of something that they do really well. Think

SCIENCE, TECHNOLOGY, AND SOCIETY

BREAKTHROUGH

Gateways in the Nucleus

Molecules must enter and leave the nucleus on a regular basis. They do so by passing through the nuclear pores that connect the nucleus to the cytoplasm. Obviously, these pores are positioned to act as border checkpoints, regulating what goes in and what comes out.

Until very recently, too little was known about the structure of the pores to understand their function. However, a recent breakthrough has provided the critical first step toward understanding the function of nuclear pores.

Jenny Hinshaw, Bridget Carragher, and Ronald Milligan at the Scripps Research Foundation in California have produced a three-dimensional image that reveals the structure of the nuclear pores. To obtain this image, Hinshaw and her associates took hundreds of detailed electron micrographs of pores at various angles. They then used computer techniques

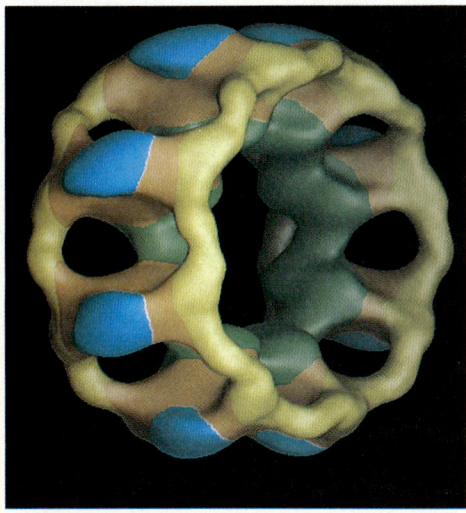

A nuclear pore is an intricate and elegant structure.

to merge the images in a way that allowed them to form a complete three-dimensional image of the pore. This image reveals some surprising details. The proteins that line the pore form a shape that resembles a lacy basket or an elegant bracelet. Alongside the main channel in the center of the pore are a number of smaller openings—eight in all. Other researchers are now trying to test the idea that ions and very small molecules pass through the side openings when large molecules are moving through the central channel.

A complete understanding of how nuclear pores function is still in the future. However, the new image of the pore constructed by Hinshaw and her associates will speed along this important work.

Jenny Hinshaw and Ronald Milligan were part of the team that determined the structure of a nuclear pore.

SCIENCE, TECHNOLOGY, AND SOCIETY
BREAKTHROUGH
Gateways in the Nucleus

Eukaryotes and prokaryotes function differently in many ways. The fact that DNA in prokaryotes is not enclosed within a nuclear structure means that there is less restriction on the process of protein synthesis. In eukaryotes, molecules enter and leave the nucleus through nuclear pores that connect the nucleus to the cytoplasm. Scientists have long wondered how the pores regulate what molecules pass into and out of the nucleus.

Little was known about the actual structure of the pores until recently. Jenny Hinshaw, Bridget Carragher, and Ron Milligan, working at Scripps Research Institute, produced the first three-dimensional image of the structure of the nuclear pores. It is the structure of the pore that provides the first clues to how the pores function.

- **How do materials enter and leave the nucleus of a cell?** (Through the pores in the nuclear membrane.)
- **Why is this important?** (Accept all logical answers. Students should infer that it is important to separate the nucleoplasm from the cytoplasm, but that materials must be moved between these areas.)
- **How does the work of Jenny Hinshaw and the team of scientists provide a clue to how the nuclear pores work?** (The work of the Scripps Institute team that produced images of the structure of nuclear pores also provided clues to the pores' functioning.)

105

---

about what special characteristic(s) they have that makes it possible. You can make this activity as serious or as funny as you like, and they will still get the point. The students would probably enjoy sharing their "specializations." You might want to set the tone by giving the first example or two.

Content Development
Using the Visuals You might want to point out that digestive enzymes are proteins. Review the structures that are involved in protein synthesis. Refer to the diagrams of the animal cell and the plant cell (Figures 5–3 and 5–4), noting the rough endoplasmic reticulum, the Golgi apparatus, and the free ribosomes.

- **Can you think of other cells that might also contain large numbers of organelles needed for protein synthesis?** (Any cells involved in producing enzymes or other proteins—stomach-lining cells or liver cells, for example.)

BACKGROUND INFORMATION
ORGANS AND SYSTEMS

Many organs can be classified as belonging to more than one system. The pancreas is part of the digestive system and the endocrine system. The heart is now known to release hormones, making it an endocrine gland as well as the key organ of the circulatory system. Bones contain tissue important to the immune system as well as to the skeletal system.

TEACHING SUPPORT
Study Guide
• Section 5–5, p. 51

5–5 (continued)
Content Development
Explain to students that they will learn more about vision later in the textbook. The important point here is that the cells of the eye are uniquely adapted for receiving light.
• **What other light-receiving structure has "stacked" membranes?** (Chloroplast.)

SECTION REVIEW 5–5
1. Cell specialization refers to the fact that many cells are specialized to perform a specific function in an organism.
2. Ribosomes; mitochondria.

Figure 5–25 The stacked membranes in the tip of this eye cell are specialized for sensing light.

Guide For Reading
■ What are the four levels of organization in a multicellular organism?

Reinforcement/Reteaching
For those students who have difficulty with the Section Review questions, go back to the discussion in the section and review the material that is not completely understood.

A Light-Sensitive Cell
You are able to see because your eyes are sensitive to light. But only a few cell types in the eye are actually sensitive to light. These light-sensitive cells are composed of two very different parts. The lower part of the cell is packed with mitochondria, perhaps four or five times as many as in a typical cell. This is an indication that the cell uses a lot of energy. The upper part contains small flattened membranes piled upon each other like a large stack of pancakes.

The membranes, which are known as disks, contain a pigment called rhodopsin (roh-DAHP-sihn). Rhodopsin absorbs light and signals the rest of the cell that light has struck the disk. This action causes messages to be sent to other cells. These changes result in the sensation we call vision. The presence of many membranes ensures that even the smallest unit of light can be detected by this light-sensitive cell.

The Street Sweepers
The air we breathe is filled with dust, smoke, and even small bacteria. Why doesn't all this material collect in the lungs and clog its passageways? Lining these passageways are special cells that release a mixture of water, carbohydrates, and salts, called mucus. The particles that are inhaled are trapped in this sticky mucus. Underneath this layer of mucus is another group of specialized cells that have cilia. As the cilia move, they create a sweeping action. This action keeps the most vital passageways in the body clean and open for business.

5–5 SECTION REVIEW
1. What is cell specialization?
2. **Critical Thinking—Developing a Hypothesis** What cell structures might the enzyme-producing cells in the pancreas contain? The cells in a muscle?

5–6 Levels of Organization

In order to describe a multicellular organism such as the human, biologists have developed levels of organization. These levels make it easier to classify and describe the cells within an organism. **The levels of organization in a multicellular organism include cells, tissues, organs, and organ systems that make up the organism.**

Tissues
In multicellular organisms, cells—the first level of organization—are organized in specialized groups called **tissues**. A

Closure
Show students photographs, diagrams, or slides (you might even list them) of other specialized cells. Discuss their specialization and what makes them uniquely suited to their function. (Examples might include palisade parenchyma—photosynthesis—uniform columnar shape; human red blood cells—carrying oxygen—disklike shape; epithelial tissue of any kind—protection—closely packed, regular appearance.)

tissue is a group of similar cells that perform similar functions. Because they are organizations of cells, tissues are the second level of organization. The cells that produce digestive enzymes in the pancreas are one kind of tissue. So are the cells in the eye that respond to light and the cells that line your air passages. Most animals, which are multicellular organisms, have four main types of tissue. They include muscle, epithelial, nerve, and connective tissue.

Organs

Although hundreds or even thousands of cells may be involved in forming a tissue, many tasks within the body are too complicated to be carried out by just one type of tissue. In these cases, an **organ**, or a group of tissues that work together to perform a specific function, is needed. Organs make up the third level of organization. Many types of tissue may be used to form a particular organ. For example, each muscle in your body is an individual organ. However, within a muscle there is a lot more than muscle tissue. There is nerve tissue and connective tissue, a special tissue that connects different parts of the body. Each tissue performs an essential task to help the organ function successfully.

Organ Systems

In many cases, even a complex organ is not sufficient to complete a series of specialized tasks. As a result, an **organ system**, or a group of organs, works together to perform a certain function. Organ systems are the fourth level of organization. There are ten organ systems in the body, including the muscular system, the skeletal system, the circulatory system, and the nervous system. You will discover more about the organ systems of the human body in Unit 9.

The organization of the cells of the body into tissues, organs, and organ systems makes possible a division of labor among those cells that makes multicellular life possible. Specialized cells such as nerve and muscle cells are able to exist precisely because other cells are specialized to obtain the food and oxygen that those cells need. This overall specialization and interdependence is one of the remarkable attributes of living things. Appreciating this is an important step in understanding the nature of living things.

5–6 SECTION REVIEW

1. Describe the four levels of organization.
2. How do an organ and a tissue differ?
3. **Critical Thinking—Assessing Concepts** Why is human blood classified as connective tissue?

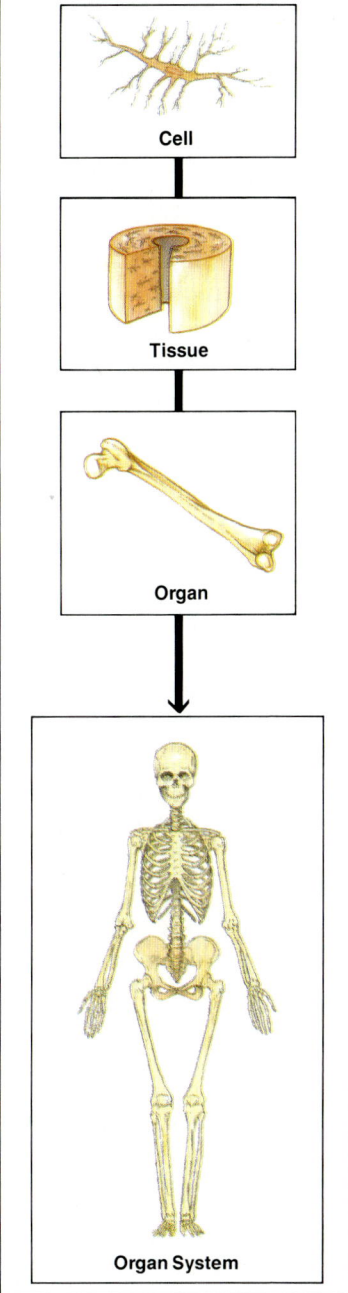

Figure 5–26 This diagram shows four levels of organization in the human body.

107

LABORATORY INVESTIGATION

OBSERVING OSMOSIS

Before the Lab
1. Assemble all the materials that will be needed by each group.
2. You might want to design an activity for the students to do while they are waiting the 30 minutes for osmosis to occur.
3. Depending on the experience level of the class, you may need to review microscope use.
4. Saturated sodium chloride can be used instead of saturated sodium nitrate. However, sodium nitrate provides 67% more dissolved sodium. To prepare a solution, add solute until the excess settles to the bottom and does not dissolve.
5. Discernable results can be obtained in 15 minutes if time is limited. However, results are more striking after a half-hour wait.

Pre-Lab Discussion
Review the definition of osmosis. You might want to introduce the terms *hypertonic* and *hypotonic* at this time if you have not previously done so. Students generally have a difficult time understanding that *hypo-* and *hyper-* refer to the concentration of water molecules, so you might want to omit the terms and just explain the concept.

Students also need some explanation of flexibility as it applies to vegetables. In other words, you might need to explain or demonstrate "normal." They sometimes find a number scale helpful.

Skills Development
Students will use the following skills while completing this investigation.
1. Observing
2. Comparing
3. Relating
4. Hypothesizing
5. Manipulative
6. Applying
7. Safety
8. Diagramming

LABORATORY INVESTIGATION

OBSERVING OSMOSIS

PROBLEM
How can the process of osmosis be observed?

MATERIALS (per group)

glass-marking pencil	forceps
2 100-mL beakers	medicine dropper
50 mL water	2 glass slides
50 mL saturated sodium nitrate solution	teasing needle
scalpel	Lugol solution
small onion	2 coverslips
small potato	microscope

PROCEDURE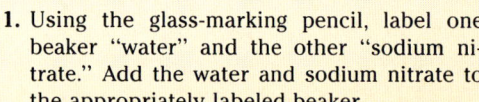

1. Using the glass-marking pencil, label one beaker "water" and the other "sodium nitrate." Add the water and sodium nitrate to the appropriately labeled beaker.
2. Using a scalpel, carefully cut the onion into quarters. **CAUTION:** *Be very careful when using a scalpel.* Each quarter will separate neatly into layers. Cut one layer into two pieces of equal size.
3. Again using the scalpel, carefully cut two slices from the potato. The slices should be about 2 to 3 mm in thickness. To determine the flexibility of the pieces of onion and potato, bend them back and forth a few times. Note their flexibility.
4. Place one piece of onion and one slice of potato in each beaker. Allow them to remain undisturbed in the beakers for 30 minutes.
5. After 30 minutes, remove the onion pieces and potato slices from the beakers with forceps. Note the degree of flexibility of each.
6. Return each onion piece and potato slice to its respective beaker.
7. Using a medicine dropper, place a drop of water in the center of a clean glass slide.
8. With the forceps, remove the onion piece from the beaker marked "sodium nitrate."

Remove the thin membrane from the inner surface of the onion by bending the onion inward and lifting the membrane with the forceps. Then place the thin onion membrane in the drop of water on the glass slide.
9. Using the teasing needle, try to flatten the onion membrane as much as you can. Add a drop of Lugol solution to the onion membrane and water. Cover with a coverslip.
10. Observe the onion membrane under the low power of the microscope. Then observe the membrane under high power. Note the appearance of cytoplasm. Draw a labeled diagram of a few onion cells.
11. Repeat steps 7 through 10 using the piece of onion from the beaker marked "water."

OBSERVATIONS
1. What was the degree of flexibility of the onion and potato before they were placed in water and in the sodium nitrate solution?
2. What was the degree of flexibility of the onion and potato after 30 minutes in water? After 30 minutes in the sodium nitrate solution?
3. Describe the differences between the cells of the onion membrane that were placed in water and the cells that were placed in the sodium nitrate solution.

ANALYSIS AND CONCLUSIONS
1. What is the purpose of the Lugol solution?
2. Which beaker contained the higher concentration of water?
3. Relate the degrees of flexibility of the potato and the onion dry, in water, and in the sodium nitrate solution to the process of osmosis.
4. Relate the differences in degrees of flexibility of the onion to the appearance of the cells observed under the microscope.

Safety Tips
Caution students about scalpels. It might be wise to demonstrate placing the vegetables on the table to cut them, rather than holding them in their hands. Testing needles can also be dangerous as students try to perform the painstaking job of removing the membrane.

If they still have little experience with the microscope, remind them about proper focusing techniques.

SUMMARIZING THE CONCEPTS

The key concepts in each section of this chapter are listed below to help you review the chapter content. Make sure you understand each concept and its relationship to other concepts and to the theme of this chapter.

5-1 The Cell Theory
- The cell theory states: All living things are composed of cells; cells are the basic units of structure and function in living things; all cells come from preexisting cells.

5-2 Cell Structure
- Most cells have a cell membrane, nucleus, and cytoplasm.
- Scientists divide cells into eukaryotic cells and prokaryotic cells.

5-3 Cytoplasmic Organelles
- Within the cytoplasm are organelles, or tiny structures that perform specialized functions. They include mitochondria, chloroplasts, ribosomes, endoplasmic reticulum, Golgi apparatus, lysosomes, vacuoles, plastids, and cytoskeleton.

5-4 Movement of Materials Through the Cell Membrane
- Diffusion is the process by which molecules move from areas of higher concentration to areas of lower concentration.
- Active transport is an energy-requiring process that enables material to move across a cell membrane against a concentration difference.

5-5 Cell Specialization
- Cell specialization is the process by which cells are uniquely suited to perform a particular function within the organism.

5-6 Levels of Organization
- In multicellular organisms, such as most animals, cells are organized into tissues, organs, and organ systems.

REVIEWING KEY TERMS

Vocabulary terms are important to your understanding of biology. The key terms listed below are those you should be especially familiar with. Review these terms and their meanings. Then use each term in a complete sentence. If you are not sure of a term's meaning, return to the appropriate section and review its definition.

5-1 The Cell Theory
cell
cell theory

5-2 Cell Structure
cell membrane
cell wall
nucleus
eukaryote
prokaryote
nuclear envelope
nucleolus
chromosome
cytoplasm

5-3 Cytoplasmic Organelles
organelle
mitochondrion
chloroplast
ribosome
endoplasmic reticulum
Golgi apparatus
lysosome
vacuole
plastid
cytoskeleton

5-4 Movement of Materials Through the Cell Membrane
diffusion
selectively permeable
osmosis
facilitated diffusion
active transport

5-5 Cell Specialization
cell specialization

5-6 Levels of Organization
tissue
organ
organ system

Teaching Strategy
There are several things that you will want to check:
1. Students in the same group should be cutting their vegetables as close to the same size and thickness as possible. Too much variation will affect the results.
2. Be sure that the vegetables are completely submerged in the liquid.
3. Removing the membrane from an onion is difficult for a beginner. Some groups are going to need help unless they have an unlimited amount of time.
4. After they get the onion membrane under the microscope, students may need help in seeing the difference between the two specimens; they do not know what they should be looking for unless you have previously instructed them. (They will probably still need help.)
5. Encourage students to make their diagrams large enough to easily see detail. Therefore, they must *look* at the detail.

Observations
1. Neither were particularly flexible.
2. Stiffer after being placed in water; softer after placed in sodium nitrate.
3. The onion cells from water look normal. In those from the sodium nitrate solution, the cytoplasm has shrunk and pulled away from the cell wall.

Analysis and Conclusions
1. To stain the cells and make the structures more visible.
2. The beaker of tap water.
3. The potato and onion in the sodium nitrate became soft and the cytoplasm in their cells shrank because water left the cells by osmosis. The potato and onion in the tap water became stiff because water entered the cells by osmosis.
4. Onion cells become more stiff and less flexible when they fill with water. Cells appear full under the microscope. Onion cells appear to have shrunk under the microscope, again due to osmosis, when placed in the sodium nitrate solution.

Going Further: Enrichment
To continue the study of osmosis, students can place a tiny leaf of *Elodea* on a slide. Next, they are to add a drop of the sodium nitrate solution and observe carefully. (As water leaves the cell, the chloroplasts all fall to the center of the cell.) Next, they are to flood the slide with water and again observe as the water reenters the cell.

CHAPTER REVIEW

CONTENT REVIEW

Multiple Choice
1. a   2. c   3. d   4. a
5. d   6. c   7. c   8. a

True or False
1. T
2. F   cell membrane
3. F   prokaryotes
4. T
5. T
6. F   passive transport
7. T
8. F   organ

Word Relationships
1. All are parts of a typical animal cell. Cell walls are not found in animals.
2. All are forms of passive transport. Endocytosis is a form of active transport.
3. All are levels of organization. A ribosome is an organelle involved in protein synthesis.
4. All are organelles. Cytoplasm is not an organelle but rather the material between the cell membrane and the nucleus in which many organelles are located.

CONCEPT MASTERY

1. *Leeuwenhoek:* invented microscope, which allowed scientists to see cells; *Hooke:* coined the term cell after observing dead cells under a microscope; *Schleiden:* stated that all plants have cells; *Schwann:* stated that all animals have cells; *Virchow:* stated that all cells come from preexisting cells.
2. For the cell to function, materials must be able to pass into and out of the cell. However, if the membrane were not selectively permeable, materials needed within the cell would pass through the membrane and leave the cell. Furthermore, materials not needed by the cell would enter the cell through the cell membrane.
3. The nucleus, found in eukaryotic cells, is surrounded by a membrane called the nuclear envelope (nuclear membrane); contains a small region made up of RNA and proteins called the nucleolus, in which ribosomes are produced; contains the chromosomes that pass on genetic information to each new generation of cells.
4. Through nuclear pores.
5. Mitochondria are organelles that change the chemical energy stored in food into compounds that are more convenient for the cell to use (ATP). Mitochondria contain two membranes. The outer membrane surrounds the mitochondria. The inner membrane contains many folds that increase the surface area

CHAPTER REVIEW

CONTENT REVIEW

Multiple Choice

Choose the letter of the answer that best completes each statement.

1. The basic unit of structure and function in living things is the
   a. cell.
   b. tissue.
   c. organ.
   d. organ system.
2. What is the outer boundary of an animal cell called?
   a. cell wall
   b. nuclear envelope
   c. cell membrane
   d. cytoplasm
3. The control center of the cell is the
   a. lysosome.
   b. ribosome.
   c. mitochondrion.
   d. nucleus.
4. Organisms whose cells do not have a nucleus are called
   a. prokaryotes.
   b. eukaryotes.
   c. organelles.
   d. plants.
5. What structure is the site of protein synthesis?
   a. nucleus
   b. lysosome
   c. mitochondrion
   d. ribosome
6. The process by which molecules of a substance move from areas of higher concentration of that substance to areas of lower concentration is
   a. active transport.
   b. endocytosis.
   c. diffusion.
   d. cell specialization.
7. Which is an example of active transport?
   a. diffusion
   b. osmosis
   c. endocytosis
   d. facilitated diffusion
8. A group of similar cells that perform a similar function is called a (an)
   a. tissue.
   b. organ.
   c. organ system.
   d. organism.

True or False

Determine whether each statement is true or false. If it is true, write "true." If it is false, change the underlined word or words to make the statement true.

1. Van Leeuwenhoek is given credit for developing the first microscope.
2. All cells have a cell wall.
3. Eukaryotes are organisms whose cells do not have a nucleus.
4. Microtubules are the main components of the cytoskeleton.
5. Osmosis is the diffusion of water.
6. Facilitated diffusion is a form of active transport.
7. Cell specialization is the process by which cells are suited to perform a particular function.
8. A group of tissues that work together is called an organ system.

Word Relationships

In each of the following sets of terms, three of the terms are related. One term does not belong. Determine the characteristic common to three of the terms and then identify the term that does not belong.

1. cytoplasm, cell membrane, cell wall, nuclear envelope
2. diffusion, osmosis, endocytosis, facilitated diffusion
3. tissue, ribosome, organ, organ system
4. mitochondrion, cytoplasm, ribosome, vacuole

CONCEPT MASTERY

Use your understanding of the concepts developed in the chapter to answer each of the following in a brief paragraph.

1. What were the contributions of van Leeuwenhoek, Hooke, Schleiden, Schwann, and Virchow to the development of the cell theory?
2. Why is it important for the cell membrane to be selectively permeable?
3. Describe the makeup of the nucleus.
4. How do materials enter and leave the nucleus?
5. Describe the structure of a mitochondrion. What is its function?
6. What are the functions of vacuoles? How do they differ in plant and animal cells?
7. How do plant and animal cells differ?
8. What is the difference between diffusion and osmosis?
9. What is osmotic pressure? Why is it important?
10. Explain the relationship between cells, tissues, organs, and organ systems.

CRITICAL AND CREATIVE THINKING

Discuss each of the following in a brief paragraph.

1. **Applying concepts** The beaker in the diagram has a selectively permeable membrane separating two solutions. Assume that the salt molecules are small enough to pass through the membrane but the starch molecules are too large to pass through. Will the water level on either side of the membrane change? Explain your answer.

   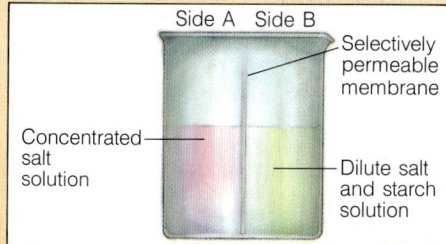

2. **Designing an experiment** You are given vegetable coloring and three beakers. The first beaker is filled with water at room temperature, the second beaker is filled with ice water, and the third with hot water. Design an experiment that will enable you to determine the effects of temperature on the rate of diffusion. Be sure to state your hypothesis and to include a control.
3. **Applying concepts** An animal cell contains about 10 to 20 Golgi apparatus, whereas a plant cell contains several hundred. Why do you think there is such a difference in the number of these organelles in each cell type?
4. **Using the writing process** Write a science fiction story entitled: "The Day Diffusion Stood Still."
5. **Using the writing process** One day, unicellular organisms got tired of being called simple organisms by the multicellular organisms. The unicellular organisms felt that they were rather complex individuals and should be recognized as such. In order to gain this recognition, they challenged the multicellular organisms to a debate. Pretend that you are a unicellular organism. What arguments would you use to defend your position?

CHAPTER 6 Cell Energy: Photosynthesis and Respiration

Section	Laboratory Investigations and Activities

6–1 Photosynthesis: Capturing and Converting Energy
pages 113–117
THEMES: Energy, Systems and Interactions
	Teacher Edition
	  DEMONSTRATION: Fluorescence of Chlorophyll, p. 112D
	  DEMONSTRATION: Priestley's Candle-Burning Experiment, p. 112D
	Laboratory Manual
	  Plant Pigments, p. 105
	Teaching Resources
	  HANDS-ON ACTIVITY: Pocketful of Posies, p. 49

6–2 Photosynthesis: The Light and Dark Reactions
pages 118–123
THEMES: Energy, Systems and Interactions
	Laboratory Manual
	  What Plants Do With Sunlight, p. 109
	Teaching Resources
	  HANDS-ON ACTIVITY: Lighten Up, p. 53

6–3 Glycolysis and Respiration
pages 123–129
THEMES: Energy, Systems and Interactions
	Student Edition
	  LABORATORY INVESTIGATION: Observing the Relationship Between Photosynthesis and Respiration, p. 132
	Teaching Resources
	  HANDS-ON ACTIVITY: Friends or Foes? p. 51

6–4 Fermentation
pages 130–131
THEMES: Energy, Systems and Interactions
	Teaching Resources
	  HANDS-ON ACTIVITY: Yeasts in Action, p. 57
	  ECOLOGY INVESTIGATION: The Role of Decomposers in the Environment, p. 1
	Product Testing Activities
	  Testing Yogurts

Chapter Review pages 133–135	

Teacher Resources

Books
Barber, J. *Light Reactions.* Elsevier, 1987.

Clayton, R.D. *Photosynthesis: Physical Mechanisms and Chemical Patterns.* Cambridge University Press, 1980.

Douce, R., and D.A. Day, eds. *Higher Plant Cell Respiration.* Springer-Verlag, 1985.

Gregory, R.P. *Biochemistry of Photosynthesis,* 3rd ed. Wiley, 1987.

Kendrick, R.E., and G.H. Kronenberg, eds. *Photomorphogenesis in Plants.* Kluwer Academic, 1986.

Palmer, John M., ed. *The Physiology and Biochemistry of Plant Respiration.* Cambridge University Press, 1984.

Sheeler, Phillip, and Donald Bianchi. *Cell and Molecular Biology,* 3rd ed. Wiley, 1987.

CHAPTER PLANNING GUIDE

Other Activities	Media and Technology

Teaching Resources
    ACTIVITY: The Absorption of Chlorophyll, p. 3
Study Guide
    Section 6–1, p. 55
	Biology Media Guide
	    page 8

Study Guide
    Section 6–2, p. 57
	Biology Transparencies
	    7: Photosynthesis
	    8: The Light and Dark Reactions of Photosynthesis
	Biology Media Guide
	    page 8

Teaching Resources
    ACTIVITY: Analyzing Photosynthesis and Respiration, p. 5
Study Guide
    Section 6–3, p. 59
	Biology Transparencies
	    9: Respiration
	Biology Media Guide
	    pages 8–9

Teaching Resources
    VOCABULARY REVIEW: Cross-a-Clue, p. 1
Study Guide
    Section 6–4, p. 63	

Teaching Resources
    Chapter Test A, p. 11
    Chapter Test B, p. 15
	Computer Test Bank
	    Chapter Test, p. 61

Audiovisuals

The Wonders of the Cell: A Living Factory. 3 filmstrips with cassettes. Sunburst Communications, Inc.
Introduction to Photosynthesis. Sound filmstrip. Ward.
Photosynthesis Details I.
Photosynthesis Details II.
Introduction to Carbohydrates. Sound filmstrip. Ward.
Photosynthesis: The Biochemical Process. Film. Coronet.
Energy for Life: Photosynthesis and Respiration. Sound. The Center for Humanities, Inc.
How Green Plants Make and Use Food, rev. Filmstrip or video. Coronet.
Photosynthesis: The Flow of Energy from Sun to Man. Filmstrip with cassette. Carolina Biological Supply Company.
Photosynthesis: Energy from Light. 4 filmstrips with cassettes. Science and Mankind.

CHAPTER 6 Cell Energy: Photosynthesis and Respiration

CHAPTER OVERVIEW

In this chapter students will discover how living things capture, store, and transform energy. They will learn that these most essential activities depend on two biochemical pathways: photosynthesis and respiration.

Students will learn that photosynthesis is the process by which green plants convert the sun's radiant energy into chemical energy. They will discover that by a set of reactions called the light reactions, plants produce the energy-storing compound ATP, and that by a set of reactions called the dark reactions, plants produce glucose.

The process by which cells break down glucose molecules to produce energy is called respiration. Students will learn that respiration consists of two phases: anaerobic, or without oxygen, and aerobic, or with oxygen. The major reactions in aerobic respiration—the Krebs cycle and the electron transport chain—will be discussed in detail.

In the last section of the chapter, students will learn about fermentation. Fermentation is the energy-producing process that is carried out when oxygen is not available.

6–1 PHOTOSYNTHESIS: CAPTURING AND CONVERTING ENERGY

Section Focus 6–1

The purpose of this section is to discuss the process of photosynthesis. Students will begin by learning about some of the historical experiments that led to an understanding of photosynthesis. They will read how the experiments of Van Helmont, Priestley, and Ingen-Housz contributed to scientists' understanding of how plants, in the presence of light, transform carbon dioxide and water into carbohydrates and oxygen.

Students will study in detail how chlorophyll, the green pigment in plants, absorbs sunlight. They will also learn about the production of the most important energy-storing compound in cells, ATP.

Performance Objectives 6–1

1. Discuss photosynthesis.
2. Describe experiments that led to an understanding of photosynthesis.
3. Distinguish between autotrophs and heterotrophs.
4. Explain how pigments absorb and reflect light.
5. Describe and discuss the relationship among ATP, ADP, and AMP.

Science Terms 6–1

photosynthesis   p. 113
glucose   p. 115
autotroph   p. 115
heterotroph   p. 115
pigment   p. 115
chlorophyll   p. 115
adenosine triphosphate (ATP)   p. 116

6–2 PHOTOSYNTHESIS: THE LIGHT AND DARK REACTIONS

Section Focus 6–2

The purpose of this section is to discuss in detail the two main parts of the photosynthetic process: light reactions and dark reactions. Students will learn that the light reactions require light, while the dark reactions take place independent of light.

Students will discover that the light reactions consist of four basic processes: light absorption, electron transport, oxygen production, and ATP formation. Students will learn that the light reactions use water, ADP, and $NADP^+$ to produce oxygen, ATP, and NADPH.

Students will next discover that the ATP and NADPH produced by the light reactions provide the necessary energy for the dark reactions. They will learn that the main purpose of the dark reactions, also called the Calvin cycle, is to produce the chemical "building block" PGAL.

Performance Objectives 6–2

1. Identify the products of the light reactions of photosynthesis.
2. Discuss the dark reactions of photosynthesis.
3. Identify the products of the dark reactions of photosynthesis.
4. Relate the dark reactions of photosynthesis to the light reactions.

Science Terms 6–2

light reactions   p. 118
dark reactions   p. 118
photosynthetic membrane   p. 118
photosystem   p. 118
electron transport   p. 119
Calvin cycle   p. 121

6–3 GLYCOLYSIS AND RESPIRATION

Section Focus 6–3

The purpose of this section is to examine two processes by which organisms release energy from glucose. These two processes are glycolysis and respiration.

Students will learn that glycolysis takes place in the cytoplasm of the cell. In glycolysis, enzymes catalyze chemical reactions that change glucose, step by step, into different molecules.

Students will next learn that when oxygen is available, the process of respiration can take place in the cell's mitochondria. Students will study in detail the two sets of reactions that make up respiration: the Krebs cycle and the electron transport chain. They will learn that it is in the chemical reactions of respiration that much of the remaining energy in glucose is released.

Performance Objective 6–3

1. Describe the process of glycolysis.
2. Describe the process of respiration.
3. Discuss the reactions that take place in the Krebs cycle.
4. Discuss the importance of the electron transport chain.
5. Relate the processes of breathing and cellular respiration.

Science Terms 6–3

calorie   p. 123
glycolysis   p. 124
respiration   p.124
aerobic   p. 124

CHAPTER PREVIEW

6-4 FERMENTATION

Section Focus 6-4

The purpose of this section is to discuss the process of fermentation. Fermentation is the means by which cells are able to carry on energy production in the absence of oxygen.

Students will learn that one type of fermentation is lactic acid fermentation. In lactic acid fermentation, the pyruvic acid produced during glycolysis is converted into lactic acid and NAD^+. It is lactic acid fermentation that takes place in muscle cells during vigorous exercise.

Students will learn that another type of fermentation is alcoholic fermentation. In alcoholic fermentation, pyruvic acid is broken down to produce alcohol, carbon dioxide, and NAD^+. Alcoholic fermentation occurs in yeasts and certain other microorganisms.

Performance Objectives 6-4

1. **Define fermentation and discuss its importance.**
2. **Relate fermentation to glycolysis.**
3. **Describe and compare the processes of lactic acid fermentation and alcoholic fermentation.**

Science Terms 6-4

fermentation p. 130
anaerobic p. 130
lactic acid fermentation p. 130
alcoholic fermentation p. 131

DISCOVERY LEARNING

TEACHER DEMONSTRATIONS
Modeling

Fluorescence of Chlorophyll

The following demonstration can be used as an introduction to Chapter 6.

One of the most important properties of chlorophyll is its ability to absorb light. Without chlorophyll, plants would have no way of obtaining the sunlight that they need to carry on photosynthesis. In this dramatic demonstration, students will see a fascinating aspect of chlorophyll's absorption of light—its ability to undergo fluorescence.

For the demonstration, you will need about 100 grams of fresh spinach leaves; water; a blender or a mortar and pestle; about 400 milliliters of acetone or ethyl alcohol; a filtering cloth; several beakers; and a source of bright light such as a strong flashlight or a slide projector.

Using a blender or mortar, grind up the spinach leaves with a little bit of water until you have a thick paste. Put the paste into a beaker, then pour in the acetone or ethyl alcohol to extract the chlorophyll from the broken cells. Next, pour the slurry through the cloth and into another beaker in order to filter out the clear green liquid. Now darken the room and shine an intense beam of light through the liquid. The chlorophyll will fluoresce a brilliant red.

Priestley's Candle-Burning Experiment

This demonstration, which should be done over a two- or three-day period, can be performed as students study Section 6-1.

One of the historic experiments that led to scientists' understanding of photosynthesis was Priestley's candle-burning experiment. This simple experiment, which is described in the text, can be duplicated for students to observe. For the demonstration, you will need a large beaker, a plate, a small candle, and a small potted plant, preferably mint.

For the first part of the demonstration, light the candle, place it on the plate, and place the beaker upside down on the plate so that it forms a dome over the candle. Have students observe what happens. (After burning a short time, the candle goes out.) Now add the small potted plant to the plate under the beaker and place the entire setup near a sunny window. Allow the beaker and plant to remain near the window for two or three days.

After this time has passed, return to the beaker and light the candle, being careful not to let too much air get under the beaker as the candle is lit. Have students observe what happens. (The candle should light and burn for a short time.)

- **What substance was consumed as the candle burned the first time?** (Oxygen.)
- **What must have been added to the beaker to enable the candle to burn the second time?** (Oxygen.)
- **Where did this oxygen come from?** (The plant that was placed under the beaker.)
- **What do you think would happen if we put the plant and beaker back on the windowsill for a few days, then once again tried to light the candle?** (It would burn again.)
- **Suppose we put the plant and beaker into a dark closet for a few days. Do you think the candle would light again? Why or why not?** (No, because no oxygen would be produced by the plant.)
- **What does this tell you about plants and oxygen?** (Plants produce oxygen on a continuous basis in the presence of sunlight.)

CHAPTER 6

Cell Energy: Photosynthesis and Respiration

GUIDED ENQUIRY

Pose the following questions to students and have them record their responses. Point out that they will gain a better understanding of the key concepts if they read the chapter with these basic questions in mind. Upon completion of the chapter, pose the questions again. Ask students to compare their initial responses with those they have developed after reading the chapter.

- How do green plants obtain sunlight?
- How do green plants convert sunlight into usable energy?
- How do organisms store energy?
- How is glucose broken down to release energy?
- What is the role of oxygen in respiration?
- How can respiration take place without oxygen?

CHAPTER 6

Cell Energy: Photosynthesis and Respiration

It may seem that little is happening in this peaceful sunny field. However, the trees, flowers, and grasses are actively involved in capturing energy from sunlight and transforming it into energy that can be used by all types of living things.

In contrast to the roar and fury of hurricanes and floods, a gentle sunrise over forest and field seems uneventful. But the movement of sunlight across a landscape is only the first event in the daily struggle of living things to obtain energy. The starting point for that struggle is the energy of sunlight and its conversion to chemical energy.

What kinds of organisms can trap the energy in sunlight? How is that energy changed into a form other organisms can utilize? In this chapter we shall examine these questions and, in so doing, discover one of the most important secrets of life: how organisms capture, store, and transform energy!

INTRODUCING CHAPTER 6

Using the Visuals

Have students observe the chapter-opener photograph.
• **What do you see happening in this picture?** (Answers may vary; some may say that trees and grass are growing, or that the sun is shining; some may say that puffy white clouds are floating by; still others may say that "nothing" is happening.)

Explain to students that although this scene looks very serene and commonplace, there is a tremendous amount of unseen activity going on inside every leaf, stem, and blade of grass. It is this activity that produces the energy on which living things depend. Tell students that they will study two important biochemical pathways that make it possible for organisms to obtain, store, and transform the energy that they need. These pathways are photosynthesis and respiration.

GUIDE FOR READING

After you read the following sections, you will be able to

6–1 Photosynthesis: Capturing and Converting Energy
- Describe the experiments that contributed to the understanding of photosynthesis.
- Discuss the requirements for photosynthesis.

6–2 Photosynthesis: The Light and Dark Reactions
- Discuss the light reactions of photosynthesis.
- Relate the dark reactions of photosynthesis to the light reactions.

6–3 Glycolysis and Respiration
- Discuss the process of glycolysis.
- Describe respiration.
- Explain how breathing is related to respiration.

6–4 Fermentation
- Relate fermentation to glycolysis.
- Compare lactic acid fermentation and alcoholic fermentation.

Journal Activity

YOU AND YOUR WORLD
What do you think is the relationship between sunlight and green plants? In your journal, answer this question and describe the observations that led you to your answer.

Figure 6–1 The very culture of a human society may revolve around the way energy is used to heat homes, cook food, light buildings and streets, and provide for other human needs. In nature, as in human societies, some of the most important activities concern the processes in which energy is produced and distributed.

6–1 Photosynthesis: Capturing and Converting Energy

Guide For Reading
■ What is photosynthesis?
■ What are the requirements for photosynthesis?

Energy is the ability to do work, and there is always work to be done, even at the level of the cell. In this chapter we will explore some of the ways in which living things capture, convert, store, and use energy. We will concentrate our attention on the way in which energy is used within the individual cell.

Photosynthesis

The study of energy and living things begins with photosynthesis. **In the process of photosynthesis, plants convert the energy of sunlight into the energy in the chemical bonds of carbohydrates—sugars and starches.** Put another way, plants use the energy of sunlight to produce carbohydrates in a process called **photosynthesis**.

An understanding of photosynthesis was first developed from studies of plant growth. For years gardeners had asked a perplexing question about plant growth: When a seedling with a mass of only a few grams grows into a tall tree with a mass of several tons, where does the tree's increase in mass come from? From the soil? From the water? From the air? In the seventeenth century, the Dutch physician Jan Van Helmont devised an experiment to find out.

COOPERATIVE LEARNING

Using preassigned lab groups or randomly selected teams, have groups assume that they are members of a hypothetical United Nations commission. They are studying the problem of rain forest destruction and related effects on the balance of the carbon cycle. Students might need to be reminded of the following points:
- Reasons for rain forest destruction — room for growth and revenue from timber
- Methods of destruction — cutting and burning
- Effects on carbon cycle — burning increases amount of CO_2 in the atmosphere; cutting reduces O_2-producing plants

Each group is to propose at least three alternatives to cutting and burning the rain forests. In a chart or grid format, groups should identify each alternative and predict the economic, social, political, and ecological effects of each one.

Journal Activity

YOU AND YOUR WORLD
Students should recognize that sunlight is important to plants. They may observe that plants grow toward a source of light, that few plants grow in shady areas, and that grass covered by lawn furniture or other items turns yellow and dies. You may want to use volunteers' responses to the Journal Activity as a starting point for your first lesson on photosynthesis. Students should be instructed to keep their Journal Activity in their portfolio.

TEACHING STRATEGY 6–1

Focus/Motivation
Display a light bulb, a flashlight battery, and a small green plant.
• **What do these three objects have in common?** (Accept all answers.)

• **How are these objects different?** (Accept all answers.)

Content Development
Continue the Motivation discussion by telling students that all three of these objects are energy sources—and that all are able to transform energy from one form into another. The light bulb takes electrical energy and transforms it into light and heat energy; the battery uses potential chemical energy to produce electrical energy; and the plant captures sunlight and changes it into stored chemical energy. Point out that the obvious difference between the objects is that the plant is living while the light bulb and battery are not. Further point out that unlike a light bulb or a battery, the green plant will not burn out. It will continue to produce energy as long as it is alive.

HISTORICAL NOTE
JOSEPH PRIESTLEY

Joseph Priestley was an English theologian and scientist who lived from 1733 until 1804. Priestley's techniques for studying gases led to his discovery of sulfur dioxide and ammonia, as well as something he called "dephlogisticated air"—the gas that was later named oxygen by Antoine Lavoisier. Priestley is also credited with the discovery of "laughing gas" and the fact that rubber could be used to erase pencil marks. Priestley immigrated to the United States in 1794 after his house, library, and laboratory were destroyed due to his support of the French Revolution.

FACTS AND FIGURES

A sequence of 80 chemical reactions makes up the complete process of photosynthesis.

6-1 (continued)

Content Development
Write the basic equation for photosynthesis on the chalkboard. Explain to students that this equation incorporates two main stages of photosynthesis. In the first stage, water molecules are split into hydrogen and oxygen, and the oxygen is released into the atmosphere. In the second stage, the hydrogen combines with carbon dioxide to form the carbohydrate glucose. Tell students to keep these two stages of photosynthesis in mind, since they will be discussed in detail in the next section.

Reinforcement/Reteaching
Review with students how the work of three scientists led to an understanding of photosynthesis. Van Helmont accounted for the role of water in producing carbohydrate. Priestley proved that plants produce oxygen. Ingen-Housz demonstrated that light is essential in order for plants to produce oxygen. Refer to the equation for photosynthesis and ask students:
- **What one important ingredient in photosynthesis did none of these three scientists recognize?** (Carbon dioxide.)
- **Why is carbon dioxide essential in the process of photosynthesis?** (It provides the carbon for carbohydrate.)

Figure 6–2 *As a tiny seedling grows into a large tree, it uses water and air to make the chemicals needed for growth.*

Figure 6–3 *A burning candle soon uses up the oxygen in a glass jar and goes out. In the presence of light, a plant placed inside the jar releases oxygen. This allows the candle to be relighted. The candle continues to burn as long as enough oxygen is available.*

Van Helmont carefully found the mass of a pot of dry soil and a small seedling. Then he planted the seedling in the pot of soil. He took care of it and watered it regularly for five years. At the end of five years, the seedling, which by then was a small tree, had gained about 75 kilograms. However, the mass of the soil was almost unchanged. Van Helmont concluded that most of the mass must have come from water because that was the only thing that he had added to the pot.

Van Helmont's experiment accounts for the *hydrate*, or water, portion of the carbohydrate produced by photosynthesis. But where does the carbon of the *carbo* portion come from? Although Van Helmont did not realize it, carbon dioxide in the air made a major contribution to the mass of his tree. And it is the carbon in carbon dioxide that is used to make carbohydrates in photosynthesis. Even though Van Helmont had only part of the story, he had made a major contribution to science.

Almost a hundred years after Van Helmont's experiment, Joseph Priestley performed an experiment that would give another insight into the process of photosynthesis. Priestley took a candle, placed a glass jar over it, and watched as the flame gradually died out. Something in the air, Priestley reasoned, was necessary to keep a candle flame burning. When that substance was used up, the candle went out. Today we call this substance oxygen.

Priestley then found that if he placed a sprig of mint under the jar and allowed a few days to pass, the candle could be relighted and would remain lighted for a while. The mint plant had produced the substance required for burning. In other words, it released oxygen. Later, the Dutch scientist Jan Ingenhousz showed that the effect observed by Priestley occurred only when the plant was exposed to light. This means that light is necessary for plants to produce oxygen.

Requirements for Photosynthesis

The experiments performed by Van Helmont, Priestley, Ingenhousz, and other scientists reveal that in the presence of light, plants transform carbon dioxide and water into carbohydrates and release oxygen. This gives us the basic outline of the process of photosynthesis:

$$CO_2 + H_2O \xrightarrow{light} (CH_2O)_n + O_2$$

carbon dioxide + water $\xrightarrow{light}$ carbohydrate + oxygen

Because photosynthesis usually produces a particular carbohydrate—the sugar **glucose** ($C_6H_{12}O_6$)—we can rewrite and balance the equation for photosynthesis as follows:

$$6\ CO_2 + 6\ H_2O \xrightarrow{light} C_6H_{12}O_6 + 6\ O_2$$

carbon dioxide + water $\xrightarrow{light}$ glucose + oxygen

Do not be fooled by how simple this equation looks. In photosynthesis, a great deal happens between the beginning and the end of the equation. You do not get glucose and oxygen just by putting water and carbon dioxide together. Let us, then, examine some of the other requirements of photosynthesis.

SUNLIGHT Nearly all organisms on Earth depend on the sun for energy. Some organisms, such as green plants, can use the sun's energy directly. As a result, green plants are called **autotrophs** (AW-toh-trohfs). Autotrophs are organisms that are able to use a source of energy, such as sunlight, to produce food directly from simple inorganic molecules in the environment. Other organisms, such as animals, cannot use the sun's energy directly. These organisms, known as **heterotrophs**, (HEHT-er-oh-trohfs) obtain energy from the foods they eat. Heterotrophs may eat autotrophs, other heterotrophs, or both.

The sun bathes the Earth in a steady stream of light. What our eyes perceive as colorless "white" light is actually a mixture of different wavelengths of light. Many of these wavelengths are visible to our eyes and make up what is known as the visible spectrum. Our eyes perceive the different wavelengths of the visible spectrum as different colors.

PIGMENTS The process of photosynthesis begins when light is absorbed by **pigments** in the plant cell. Pigments are colored substances that absorb or reflect light. The principal pigment of green plants is known as **chlorophyll**. As you can see from the graph in Figure 6–4, chlorophyll absorbs red and blue light but does not absorb light in the middle region of the spectrum very well. Instead, it reflects these wavelengths.

ENERGY-STORING COMPOUNDS What happens when sunlight is absorbed by matter? In some cases, the energy of sunlight is transferred to the electrons in matter, raising them to a higher energy level. In a modern solar cell, the high-energy

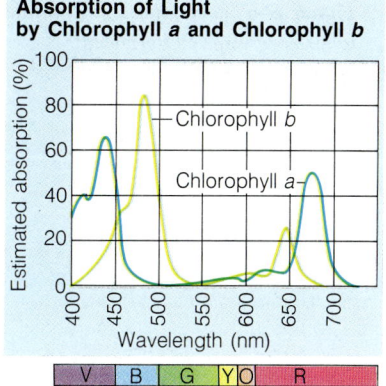

Figure 6–4 *Chlorophyll absorbs red and blue light quite well. Different forms of chlorophyll have their peak absorption at different wavelengths.*

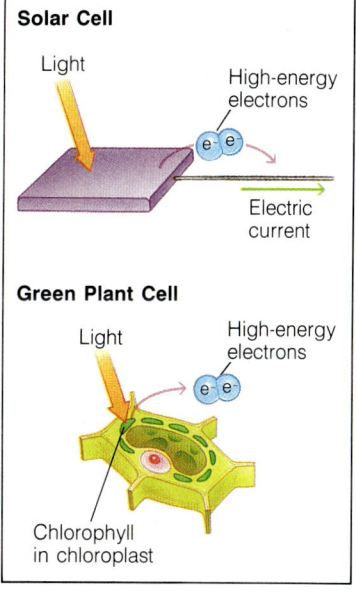

Figure 6–5 *In a solar cell, materials such as silicon produce high-energy electrons that are channeled through a wire, producing an electric current. In photosynthesis, chlorophyll produces high-energy electrons whose energy is trapped in chemical bonds.*

BACKGROUND INFORMATION
AUTOTROPHIC PLANTS

It is interesting to note that plants are autotrophs only when it comes to carbon. Plants are not autotrophic for nitrogen, which, in the form of ammonia or nitrates, is an essential plant nutrient. Plants depend on microorganisms in soil to provide nitrogen. That is why plants require fertile soil in which to grow, or require fertilizer in less rich soil.

HISTORICAL NOTE
THE SOURCE OF OXYGEN

For many years, scientists were uncertain of the source of the oxygen released in photosynthesis. Some suspected that water was the source of the oxygen, whereas others suggested CO_2. The first real clue to the mystery was offered by C. B. Van Neil in the 1930s. Van Neil pointed out that some photosynthetic bacteria produce pure sulfur from hydrogen sulfide, H_2S. Since the molecular structure of H_2S is similar to that of water, H_2O, Van Neil reasoned that the release of sulfur from hydrogen sulfide might be analogous to the release of oxygen from water.

Nearly 20 years later, experiments with radioactive tracer molecules showed that Van Neil was right. When water containing radioactive oxygen was used to support photosynthesis, radioactive oxygen gas was released. In contrast, scientists found that radioactive oxygen in CO_2 was incorporated into carbohydrates.

TEACHING SUPPORT

Teaching Resources
- Activity: The Absorption of Chlorophyll, p. 3

MEDIA AND TECHNOLOGY
See the Biology Media Guide page 8 for bar-code correlations for this section.

Skills Development
Guided Practice
Skills: Making inferences, relating concepts
After students have read about Van Helmont's experiment, ask:
- **What general type of organic compound are plants and trees made of?** (Carbohydrate.)
- **Do you remember from an earlier chapter which carbohydrate makes up most of the mass of a tree?** (Cellulose.)
- **What part of cellulose did Van Helmont's experiment account for?** (The *hydrate* part; that is, the part that is provided by water.)
- **What might Van Helmont have concluded about the material that makes up a tree's mass if much of the soil in the pot had disappeared?** (That a tree's mass is made up of soil or something in soil.)

ESL STRATEGY

Have students write these formulas in sentence form.
1. ATP → ADP + P + energy
2. AMP + P + energy → ADP

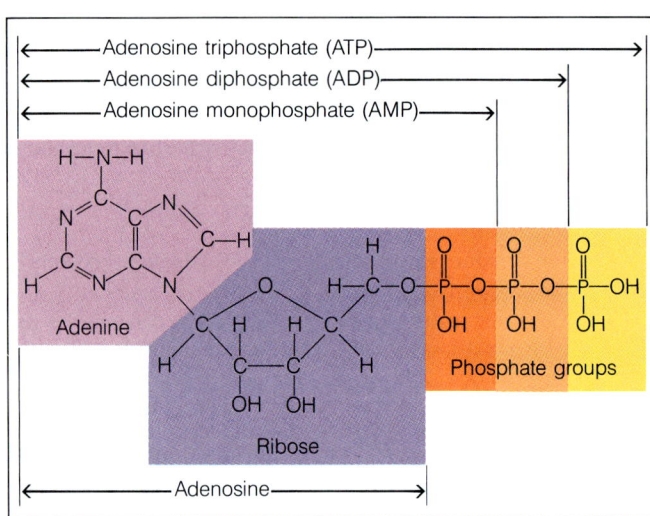

Figure 6–6 Adenosine triphosphate, or ATP, is the most important energy-storing compound in living things.

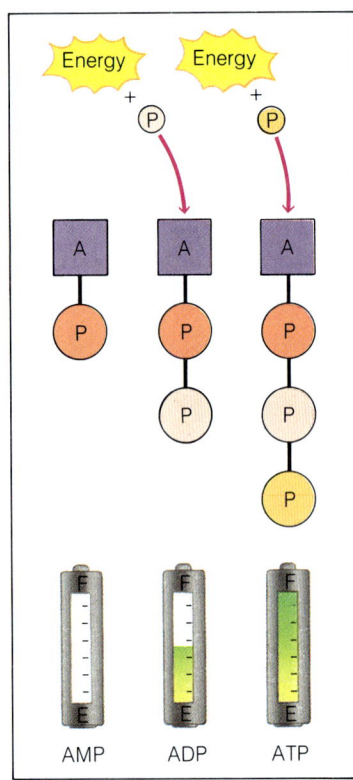

Figure 6–7 Energy is required to attach the second and third phosphate groups to AMP. Energy is released when the phosphate groups are removed.

electrons are then used to produce an electric current that can travel through a wire and power a calculator, charge a battery, or provide power for an electric utility. See Figure 6–5.

What happens when a green plant absorbs sunlight? As in a solar cell, electrons are raised to a higher energy level. In this case, the electrons belong to the pigment chlorophyll. The high-energy electrons do not travel along wires in green plants. Instead, the high-energy electrons are trapped in chemical bonds. There are two ways in which some of the energy of sunlight is trapped in chemical bonds.

The first way is the simpler of the two. In this process, a pair of high-energy electrons is passed directly to an electron carrier. An electron carrier is a molecule that can accept a pair of high-energy electrons and later transfer them along with most of their energy to another compound. Plants use an electron carrier called $NADP^+$. When $NADP^+$ accepts a pair of high-energy electrons, it is converted to NADPH. The conversion of $NADP^+$ to NADPH is one way in which some of the energy of sunlight can be trapped in chemical form.

The second way in which light energy is trapped in chemical form involves **adenosine triphosphate,** or **ATP**. As shown in Figure 6–6, an ATP molecule consists of adenine, a 5-carbon sugar called ribose, and three phosphate groups. During photosynthesis, green plants produce ATP, which is an energy-storing compound used by every living cell.

The first step in the production of ATP involves its close relative adenosine monophosphate (AMP). When a second phosphate group (abbreviated as P) is bonded to AMP, energy is stored in the bonds between the phosphate groups, and a compound called ADP (adenosine diphosphate) is formed. When a third phosphate group is bonded to ADP, still more energy is stored, and ATP is formed. Thus you can think of ATP

6–1 (continued)

Reinforcement/Reteaching
Review with students the requirements for photosynthesis: water, carbon dioxide, and sunlight. Also point out that temperature must be within a favorable range.

Skills Development
Guided Practice
Skill: Applying concepts
Explain to students that the Indian pipe plant contains no chlorophyll. (Try to obtain a picture of this plant — it is rather interesting to look at, with milky white stems.)
• **What can you conclude about the ability of the Indian pipe plant to make its own food?** (It cannot make its own food.)
• **Why?** (It cannot absorb light to carry out photosynthesis.)
Point out that this plant is a heterotroph and obtains nourishment from a particular fungus that lives off green plants. The fungus absorbs glucose from the green plants, and the Indian pipe absorbs the glucose from the fungus.

SECTION REVIEW 6–1
1. The process in which plants use energy from the sun, carbon dioxide, and water to produce food.
2. Proved that the source of material was water and not soil. It did not prove that much of the mass of a growing plant comes from carbon dioxide in the air.
3. $NADP^+$ and ATP are important energy-storing compounds.
4. Plants use the energy from sunlight to make food. As the efficiency of the chlorophyll absorption increases, more of the energy of sunlight is available and more food can be produced, assuming that the other raw materials are available.

Reinforcement/Reteaching
If some students are having trouble answering the Section Review questions, go back to the part of the section that contains

as a fully charged battery, ADP as a half-charged battery, and AMP as an uncharged battery. See Figure 6–7.

How is the energy that is stored in these molecules released? Simply by breaking the chemical bonds between the phosphate groups. The energy thus provided by ATP can be used for just about every type of cellular activity, including active transport across cell membranes, protein synthesis, and muscle contraction. The breakdown of ATP as well as its synthesis is controlled by enzymes, which are located in the cell.

6–1 SECTION REVIEW

1. Define photosynthesis.
2. Describe Van Helmont's experiment.
3. What is $NADP^+$? ATP?
4. **Critical Thinking—Making Inferences** Why is chlorophyll's efficiency in absorbing light energy important for photosynthesis?

PROBLEM SOLVING IN BIOLOGY

THE COLORS OF AUTUMN

In many parts of the country, the beautiful colors of autumn leaves provide inspiration to artists and poets as the countryside turns into a lovely blaze of red, orange, and yellow. What is the source of these colors? Do plants produce them in response to the changing of the seasons?

The pigments in plant cells, including chlorophyll, can be separated from one another by a technique known as paper chromatography. Study the results of the paper chromatography experiment shown on this page. The green chlorophyll spots, red-orange carotene spots, and yellow xanthophyll spots are labeled on samples taken from the leaves of a single maple tree in August, September, and October.

Using these three chromatographs as evidence, can you explain what seems to happen as the leaves change color as summer ends and autumn begins?

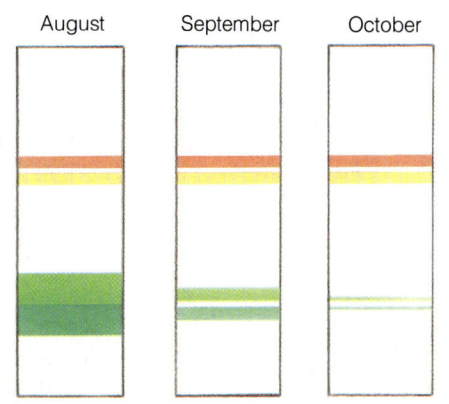

Paper chromatography is based on the fact that different substances (such as plant pigments) in a solvent will move through a piece of paper at different rates.

PROBLEM SOLVING
The Colors of Autumn
Chromatography is a technique of chemical analysis in which the mixture to be analyzed is contained in a mobile phase and is passed over an absorbent, stationary phase. In paper chromatography, the mobile phase is a liquid and the stationary medium is porous paper (filter paper). The absorbent in the paper is cellulose. When a drop of the mixture is placed on the paper and dissolved in a solvent, different substances contained in the sample move across the paper by capillary action at different rates. The result is that the different materials in the mixture are separated out.

You may wish to point out to students that two of the pigments that bring about colors in autumn leaves are the orange carotenes and the yellow xanthophylls. These pigments absorb wavelengths of light that are not absorbed by chlorophyll.

Students should infer from the illustration that the green pigment decreases as autumn approaches. As the green pigment decreases, the other pigments become visible.

TEACHING SUPPORT
Laboratory Manual
• Plant Pigments, p. 105

Teaching Resources
• Hands-On Activity: Pocketful of Posies, p. 49

Study Guide
• Section 6–1, p. 55

---

the material with which they are having difficulty.

Closure
Divide the class into three groups. Challenge each group to create and present to the class a "mixed-media review" of one of the following topics: Historical Experiments in Photosynthesis; Photosynthesis and Its Requirements; ATP—Structure and Function. Explain that a "mixed-media review" should include as many different methods of presentation as possible—pictures, diagrams, speeches, or skits (acting out a historic experiment would be effective), music, movement.

BACKGROUND INFORMATION
CHLOROPLASTS AND CHLOROPHYLL

Chloroplasts are a type of plastid. Plastids are found only in the cytoplasm of plant cells or in the cytoplasm of colored plantlike organisms. Plastids vary in color according to the pigment they contain. It is the pigment chlorophyll that gives a green color to chloroplasts.

Chloroplasts can be cup shaped, disk shaped, or spiral shaped. Some plant cells have many chloroplasts, while others have one large chloroplast that almost fills the cell.

Like all plastids, a chloroplast is enclosed in a membrane. This membrane consists of a layer of lipids between two layers of proteins.

A chloroplast contains numerous bodies called grana. Grana are composed of several sheets of protein between which are stacked molecules of chlorophyll, carotene, and lipids. The grouping of chlorophyll molecules in grana makes them more efficient in absorbing light energy.

Chlorophyll is a complex molecule in the shape of a ring, with an atom of magnesium at the center. It is the arrangement of bonds holding the atoms together in the chlorophyll molecule that allows electrons to move to higher energy levels when excited by sunlight.

The structure of the chlorophyll molecule is remarkably similar to that of hemoglobin. Hemoglobin has essentially the same ring structure, except for an atom of iron, rather than magnesium, at its center. This marked similarity has caused some scientists to conjecture that plants and animals must have arisen from a common ancestor.

Guide For Reading
- What events occur during the light reactions of photosynthesis?
- What is the relationship between the dark reactions of photosynthesis and the light reactions?

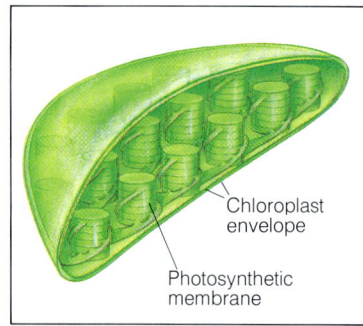

Figure 6–8 The chloroplast is the site of photosynthesis. The light reactions take place in the saclike photosynthetic membranes. The dark reactions take place in the areas surrounding the photosynthetic membranes.

6–2 Photosynthesis: The Light and Dark Reactions

As you have just read, the production of NADPH and ATP requires sunlight. The series of reactions that produce these energy-storing compounds are known as the **light reactions**—one of the two stages of photosynthesis. **In the light reactions, the energy of sunlight is captured and used to make energy-storing compounds.** If NADPH and ATP could store large amounts of energy over long periods of time, the process of photosynthesis might stop with their production. This is not the case, however.

Another set of reactions called the **dark reactions**—the second stage of photosynthesis—uses the energy stored in NADPH and ATP to produce glucose. Glucose is more stable than either NADPH or ATP and is able to store more energy than either of these compounds. In fact, a single molecule of glucose stores approximately 100 times more energy than a phosphate group in ATP does. The dark reactions are so named because they do not require light. However, the dark reactions can—and do—occur in the light.

The Light Reactions

Photosynthesis takes place in an organelle known as the chloroplast. Within the chloroplast are saclike **photosynthetic membranes** that contain chlorophyll. The light reactions take place within the photosynthetic membranes, whereas the dark reactions occur outside the photosynthetic membranes.

The light reactions can be divided into four basic processes: light absorption, electron transport, oxygen production, and ATP formation. We will see how each of these closely linked processes combine to make the products of the light reactions.

LIGHT ABSORPTION The photosynthetic membrane contains clusters of pigment molecules, or **photosystems**, that are able to capture the energy of sunlight. There are two photosystems in green plants: photosystem I and photosystem II.

Each photosystem contains several hundred chlorophyll molecules as well as a number of accessory pigments. These accessory pigments absorb light in the regions of the spectrum where chlorophyll does not, thus allowing more of the available light energy to be used. After light energy is absorbed by one of the pigment molecules in a photosystem, the energy is passed from one pigment molecule to the next until it reaches a special pair of chlorophyll molecules in the reaction center of the photosystem. In the reaction center, high-energy electrons are released and are passed to the first of many electron carriers. Although all the photosystem pigments can absorb light, only the special pair of chlorophyll molecules can process it.

TEACHING STRATEGY: 6–2

Focus/Motivation
Simulate a version of the game of hot potato. Ask 8 to 10 student volunteers to form a line. Take a potato or other designated "hot" object and toss it to the first student in line. Have students pass the hot potato as fast as possible down the line until it lands (and stays) with the last student.

Content Development
Use the Focus/Motivation activity to introduce the idea of excited, high-energy electrons being passed rapidly from one electron-carrier molecule to another. Point out that it is the energy contained in these electrons that provides the "start-up" energy needed to power the many reactions of photosynthesis that are to follow. Also point out that the last student in line in the game of hot potato is like the enzyme that finally passes the electrons to the electron carrier $NADP^+$.

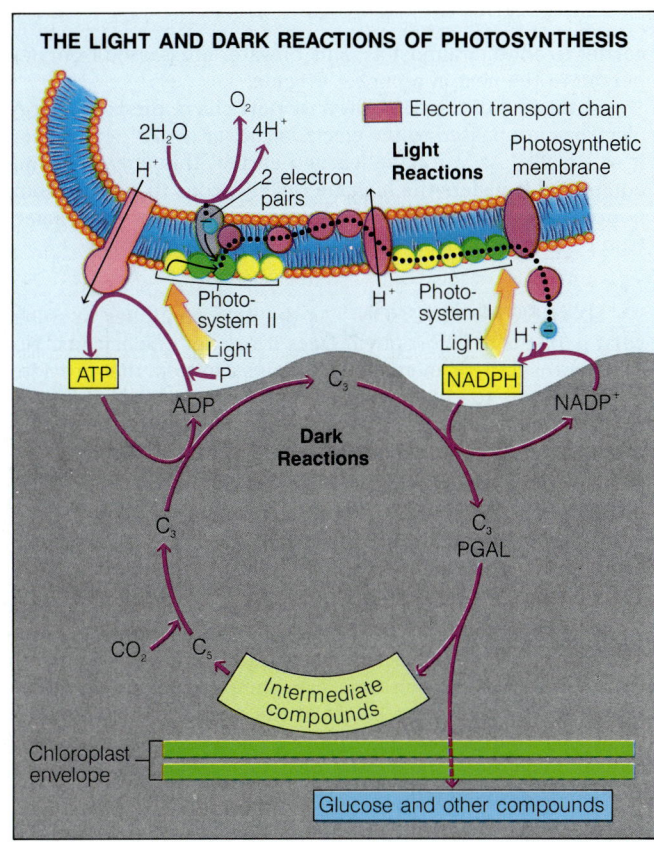

Figure 6–9 *Clusters of pigment molecules known as photosystems are embedded in the photosynthetic membrane. Photosystems capture the energy of sunlight, which is used to make energy-storing compounds. The energy in these compounds, in turn, is used to make biologically important compounds such as glucose.*

ELECTRON TRANSPORT Upon their release from the reaction center, the high-energy electrons are transferred along a series of electron carriers in the photosynthetic membrane. The process is called **electron transport** and the electron carriers themselves are known as the electron transport chain. The high-energy electrons are passed from one electron

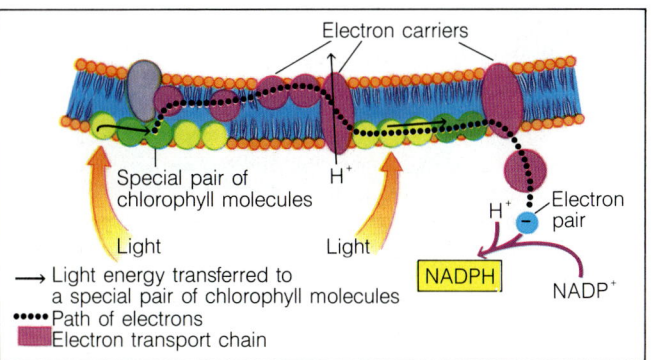

Figure 6–10 *The high-energy electrons produced by light are transferred from chlorophyll to the electron transport chain. The electrons are passed from one carrier in the chain to the next. At the end of the chain, they are used to make the energy-storing compound NADPH.*

HISTORICAL NOTE
THEODORE ENGLEMANN

In the late nineteenth century, German scientist Theodore W. Engelmann investigated the effect of different-colored light on the rate of photosynthesis. Engelmann measured changes in the rate of photosynthesis by measuring changes in the amount of oxygen produced. To determine the amount of oxygen produced, he placed a strand of green alga and some aerobic bacteria in water. Engelmann reasoned that the bacteria, which need oxygen, would move toward the area of greatest oxygen concentration. Then Engelmann shone on the alga a beam of light that had been passed through a triangular prism. This meant that all the colors of the spectrum were shining on the strand of alga. Engelmann observed that the greatest number of bacteria moved toward the regions of red light, and that many also moved to the region of blue-violet light. Engelmann therefore concluded that red light is most effective in promoting photosynthesis, and that blue-violet light is second most effective.

6-2 (continued)

Content Development
Using the Visuals Emphasize to students that the splitting of water molecules is an extremely important reaction in the light-dependent phase of photosynthesis. The process by which water molecules are split is called photolysis, meaning split by light energy. (Some students may be familiar with the decomposition of water by electrolysis, a process during which electrical energy splits water molecules.)

Direct students' attention to Figure 6–11.

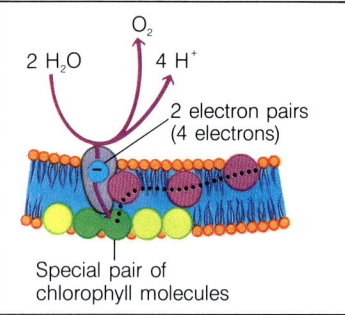

Figure 6–11 The breakdown of water produces electrons that replace the electrons lost by the special pair of chlorophyll molecules. This process also results in the formation of hydrogen ions and oxygen gas.

• **Where does the energy come from to split water molecules?** (From excited electrons in the electron transport chain.)
• **How many molecules of water are split at one time?** (2.)
• **What products are produced as the 2 molecules are split?** (Oxygen, hydrogen ions, and electrons.)

Point out that oxygen is first released as 2 atoms of oxygen; these quickly combine to form a molecule of oxygen gas.

• **How many hydrogen ions are produced?** (4.)

carrier to another almost as pails of water are passed from one person to the next in a bucket brigade.

At the end of the electron transport chain, the high-energy electrons are passed to the electron carrier NADP$^+$, converting it to NADPH. As you have learned earlier, this energy-storing compound transfers the high-energy electrons to another compound. In just a moment, you will see how the energy-rich electron carrier NADPH will be used in the dark reactions.

OXYGEN PRODUCTION As the light continues to shine, what happens to chlorophyll? Does it run out of electrons? No. The photosynthetic membrane contains a system that provides new electrons to chlorophyll to replace the ones that wound up in NADPH. These electrons are taken from water. Four electrons are removed from 2 water molecules (H_2O), leaving 4 hydrogen ions (H^+) and 2 oxygen atoms (O). The 2 oxygen atoms form a single molecule of oxygen gas (O_2) that leaves the chloroplast and is eventually released into the air.

ATP FORMATION The hydrogen ions left behind when water is "split" are released inside the photosynthetic membrane. In addition, as electrons are passed from chlorophyll to NADP$^+$, more hydrogen ions are pumped across the membrane. After a while, the inside of the membrane fills up with positively charged hydrogen ions. This makes the outside of the photosynthetic membrane negatively charged and the inside positively charged.

The difference in charges across the membrane is a source of energy. An enzyme in the photosynthetic membrane makes use of this energy to attach a phosphate molecule to ADP, forming ATP. This is the second way in which the energy of sunlight is trapped in chemical form.

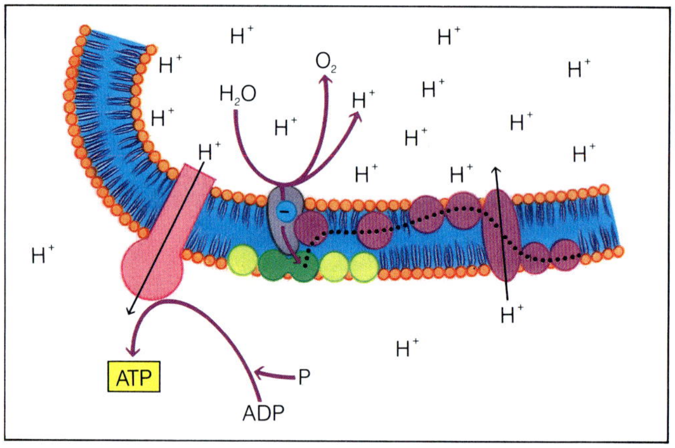

Figure 6–12 The buildup of hydrogen ions (produced in the breakdown of water) inside the photosynthetic membrane produces a difference in charges across the membrane. This powers the formation of the energy-storing compound ATP.

• **How many electrons are released?** (4.)

Point out that the products of the photolysis of water are vital to photosynthesis. First, water molecules provide new electrons to replace those that have already completed the electron

A SUMMARY OF THE LIGHT REACTIONS As we have seen, the light reactions use water, ADP, and $NADP^+$; they produce O_2 (which is of no help in photosynthesis but is of great use to us!) and the energy-storing molecules ATP and NADPH. The next stage of photosynthesis, the dark reactions, will convert these energy-storing molecules to a more convenient form.

The Dark Reactions

Do not be confused by the term dark reactions. The dark reactions generally take place in sunlight. However, light does not play a role in the dark reactions. If ATP and NADPH are supplied, the dark reactions can be carried out in a test tube, even in total darkness and without ever being exposed to light!

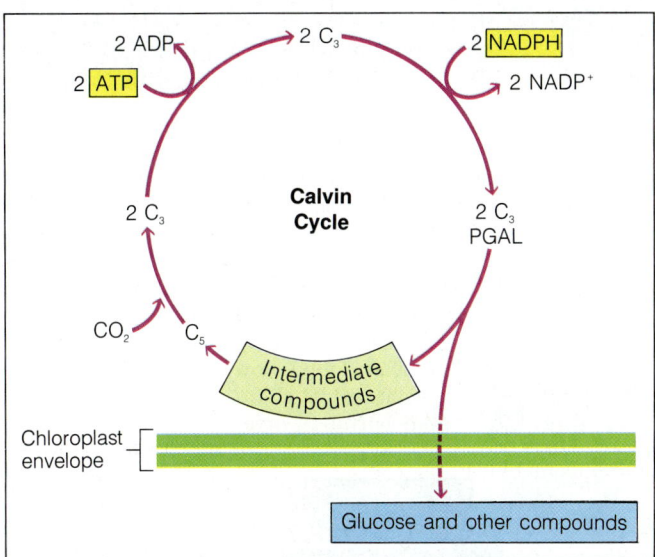

Figure 6–13 *The dark reactions use carbon dioxide and the energy-storing compounds produced in the light reactions to form PGAL. PGAL is a chemical "building block" that can be used to make other organic compounds, such as the sugar glucose.*

The series of chemical changes that make up the dark reactions is particularly critical to living things. In this part of photosynthesis, the simple inorganic molecule carbon dioxide is used to make a complex organic molecule. This molecule can be thought of as a building block that can be used to make other biologically important molecules, including glucose. The dark reactions form a cycle, or circular series of reactions. Because the chemistry of this remarkable cycle was worked out by the American scientist Melvin Calvin, the dark reactions are also known as the **Calvin cycle**.

Throughout this description of the Calvin cycle, you will find it helpful to refer to Figure 6–13. In the first reaction of the Calvin cycle, a 5-carbon sugar (C_5) combines with CO_2 to form two 3-carbon compounds (C_3). The first reaction of the Calvin

HISTORICAL NOTE
LIGHT AND DARK REACTIONS

Experiments by several scientists led to the discovery and understanding of the light and dark phases of photosynthesis. In the early 1900s, a British plant physiologist named F. F. Blackman discovered that the amount of oxygen given off by a plant exposed to a flashing light was the same as the amount of oxygen given off by a similar plant exposed to continuous light. This meant that the rate of photosynthesis was unaffected by short periods of darkness. Blackman reasoned that there must be a phase of photosynthesis that does not require light. He hypothesized that photosynthesis occurs in two phases, a light-dependent phase and a light-independent phase.

In 1937, a British biochemist named Robin Hill studied the light-dependent phase of photosynthesis. Hill removed the chloroplasts from some plant cells. He found that when the chloroplasts were placed in a solution containing iron salts and exposed to light, the mixture produced large amounts of oxygen. This showed that the release of oxygen occurs in the absence of carbon dioxide. Hill's experiment provided evidence that oxygen is produced during the light phase of photosynthesis, and that carbon dioxide is not used during this phase.

---

transport chain and landed in NADPH. Second, water molecules yield hydrogen ions. These hydrogen ions will eventually be carried into the dark reactions to provide the hydrogen needed to make carbohydrate. In the meantime, these hydrogen ions play an important role in ATP formation by creating a charge gradient across the photosynthetic membrane. Third, water molecules release atoms of oxygen that quickly combine to form molecules of oxygen gas. This oxygen, which is released into the atmosphere, is vital to animal life on Earth.

Reinforcement/Reteaching

Have students make diagrams to show 2 water molecules being split into 2 atoms of oxygen (which then combine to form O_2), 4 H^+ ions, and 4 electrons. The models should clearly show the hydrogen and oxygen atoms in each water molecule. An effective way to diagram this reaction is to use electron dot formulas to represent the water molecules. This way, students can clearly show not only the removal of the hydrogen and oxygen atoms, but also the removal of the two pairs of electrons.

SCIENCE, TECHNOLOGY, AND SOCIETY
BREAKTHROUGH

Finding a Better Way

Water, carbon dioxide, chlorophyll, and light are the simple "ingredients" necessary for the single most important chemical reaction on Earth. The ability of many plant species to use light energy to make sugars has piqued the curiosity of scientists for many years. However, green plants are not the only photosynthetic organisms on Earth. Some bacterial species are also able to use light as a source of energy. One such species was used by Hartmut Michel to unlock some of the secrets of a photosynthetic reaction center.

Michel developed a process to crystallize a reaction center protein, a task thought impossible. Michel was able to use X-ray crystallography to produce an image of the crystal. Michel's work forms the basis of our understanding of photosynthetic reaction centers.

Ask students to read this Breakthrough feature. Use questions such as the following to assess their understanding of the material.

- **What organism did Michel use in his studies?** (Michel used a photosynthetic bacterium.)
- **What substance did Michel use to crystallize the membrane protein?** (He used a detergent molecule.)
- **What was Hartmut Michel's reward for his work?** (In addition to an understanding of the important photosynthetic reaction centers, Harmut Michel was awarded a Nobel prize in chemistry at age 41.)

---

cycle tends to be relatively slow. In order to compensate for this, the chloroplast is loaded with an enzyme that catalyzes this reaction. This enzyme is known as rubisco. There is far more rubisco in an active plant cell than any other protein molecule. Because there are many more plant cells than animal cells on Earth, we can make a remarkable statement about rubisco: It is the most abundant protein in the world!

In the next two reactions of the Calvin cycle, using the chemical energy available in ATP and NADPH, these 3-carbon compounds are converted to PGAL (**p**hospho**g**lycer**al**dehyde). Most PGAL molecules are recycled for the dark reactions, but 1

SCIENCE, TECHNOLOGY, AND SOCIETY
BREAKTHROUGH

Finding a Better Way

For years, scientists were puzzled about what events took place in the core of a photosystem. Because there was no way to look directly inside the photosynthetic membrane, scientists had no way of knowing if the structure of a reaction center held clues as to how electrons were transferred from chlorophyll to acceptor molecules.

Hartmut Michel, a young German researcher, was determined to change all that by crystallizing a photosynthetic reaction center. If crystals of the reaction center protein could be produced, then its structure could be determined by passing X-rays through the crystal and analyzing how the X-rays scattered, a process known as X-ray crystallography. The only problem with Michel's plan was that the reaction center was a membrane protein. No one had ever crystallized a membrane protein before—it was thought to be impossible.

Nonetheless, Michel had an idea. An accomplished chemist, he synthesized a small detergent molecule that he hoped would bind to the oily membrane protein and make it easier to crystallize. His idea worked, and within a few months he had produced nearly perfect crystals of a bacterial photosynthetic reaction center.

Because the crystals were so complex, it took nearly three years to decode the X-ray scattering pattern they produced. However, the final pictures were well worth the wait. They showed in detail how a special pair of chlorophyll molecules is held within the reaction center and made it clear how excited electrons are passed directly from the pigment to nearby electron carriers.

Michel's work now forms the basis of our understanding of photosynthetic reaction centers. Not only did this discovery impress his colleagues, it impressed others as well. In 1988, at the age of 41, Michel became one of the youngest people ever to win the Nobel Prize in Chemistry.

6–2 (continued)

SECTION REVIEW 6–2

1. Oxygen, ATP, and NADPH.
2. Light reactions: photosynthetic membranes of the chloroplast; dark reactions: outside the photosynthetic membranes, but in the chloroplast.
3. Calvin cycle is a series of reactions that produce PGAL. The light reactions are important because they provide the raw materials for the Calvin cycle.

Reteaching/Reinforcement

If students are having trouble answering the section review questions, go back to the part of the section that contains the material with which they are having trouble. Review the material, then help students in answering the questions, if necessary.

Closure

Write the basic equation for photosynthesis on the chalkboard:

$$6CO_2 + 6H_2O \rightarrow C_6H_{12}O_6 + 6O_2$$

Have students relate the reactants and products of the light and dark reactions to this basic

out of every 6 is used to make glucose or other end products. How many times must the Calvin cycle go around to produce 1 molecule of glucose? (*Hint:* The formula for glucose is $C_6H_{12}O_6$.)

Although we have emphasized the production of glucose in the Calvin cycle, the intermediate compounds of the cycle are also important because they can be used by the chloroplast in many ways. Some of the intermediates may be used to form sugars other than glucose; some of them may be used to make amino acids; and some can be converted to lipids. The dark reactions of photosynthesis provide the raw materials to produce almost everything the cell needs.

6-2 SECTION REVIEW

1. What are the products of the light reactions?
2. Where in the cell do the light reactions take place? The dark reactions?
3. **Critical Thinking—Applying Concepts** Describe the Calvin cycle. Why are the light reactions important to the Calvin cycle?

6-3 Glycolysis and Respiration

The ability of autotrophs to produce glucose and other food molecules reflects the fact that photosynthesis is able to trap some of the energy of sunlight in chemical bonds. In a way, the energy stored by photosynthesis is like money deposited in a savings account: It is available for future needs. But organisms—autotrophs and heterotrophs alike—must be capable of making "withdrawals" from that savings account; that is, they must be able to release energy by breaking down food molecules. In this section we will examine two processes used to release energy from glucose.

Glycolysis—Breaking Down Glucose

Glucose ($C_6H_{12}O_6$) is a simple 6-carbon sugar. If glucose is broken down completely in the presence of oxygen, carbon dioxide and water are produced:

$$C_6H_{12}O_6 + 6O_2 \rightarrow 6CO_2 + 6H_2O$$
glucose + oxygen → carbon dioxide + water

This reaction gives off 3811 **calories** per gram of glucose. A calorie is the amount of heat energy required to raise the temperature of 1 gram of water 1 degree Celsius. As you can see, glucose obviously contains a lot of energy. (Those of you who watch your calories should not be alarmed. The Calorie associated with food and diet books is actually a kilocalorie, which is equal to 1000 calories.)

Guide For Reading
- What is glycolysis?
- What is respiration?
- How is breathing related to respiration?

123

BACKGROUND INFORMATION
GLYCOLYSIS AND RESPIRATION

To say that glycolysis and respiration are separate processes is somewhat misleading. The entire process of breaking down glucose to release energy and form molecules of ATP is called cellular respiration. It is glycolysis that is the first phase of cellular respiration. Glycolysis is often referred to as the anaerobic phase of cellular respiration, since it takes place independent of oxygen.

The second phase of cellular respiration consists of the Krebs cycle and the electron transport chain. These processes are often referred to as the aerobic phase of cellular respiration, since they require oxygen.

Glycolysis, or the anaerobic phase of cellular respiration, takes place in the cytoplasm of the cell. The Krebs cycle and electron transport, or the aerobic phase of cellular respiration (which the text simply calls "respiration"), takes place in the mitochondria of the cell.

TEACHING SUPPORT

MEDIA AND TECHNOLOGY
- Biology Transparencies
- 9: Respiration.
See the Biology Media Guide pages 8–9 for bar-code correlations for this section.

6–3 (continued)
Content Development
Continue the Motivation discussion by explaining that just as the energy to power a car is stored in gasoline, the energy needed to power a living thing is stored in carbohydrate. Carbohydrate—particularly glucose—is the fuel that organisms need in order to live.

Refer once again to the gasoline can or pump and ask:
- **Is it enough to get a car going just by having the gasoline here in the can or in the pump?** (No.)
- **Is it enough just to have the gasoline sitting in the tank of the car?** (No.)
- **What must happen to the gasoline in order for the car to run?** (It must be burned—broken down to release energy.)

Explain that the same is true of carbohydrate—just storing it in cells does not give an organism energy. The carbohydrate must be broken down in order for energy to be released. This happens in a series of processes called cellular respiration.

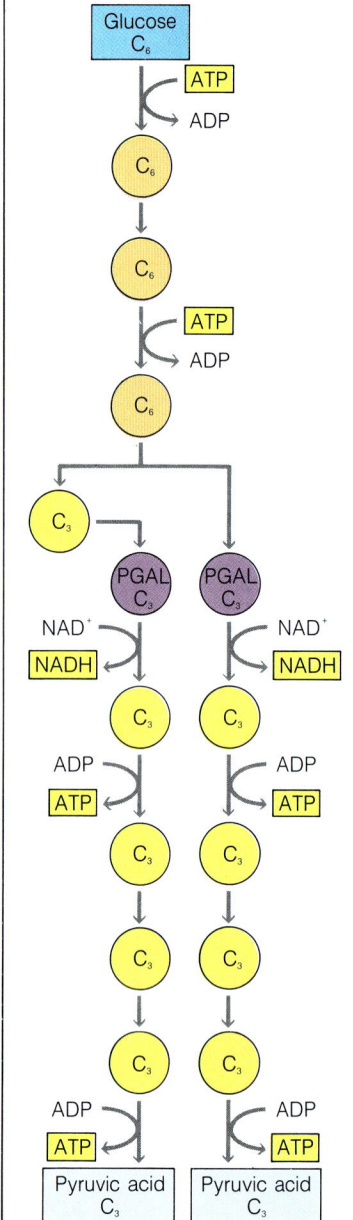

Figure 6–14 In the process of glycolysis, glucose is broken down and transformed into other molecules. In addition, energy is released in the form of 2 molecules of ATP.

A cell takes apart glucose slowly and captures the energy a little at a time. Remember, the energy stored in chemical bonds can be used when the bonds are broken. The first stage in this process is known as **glycolysis** (gligh-KAHL-ih-sihs), which means glucose-breaking. Glycolysis takes place in the cytoplasm of a cell. **In glycolysis, a series of enzymes catalyzes chemical reactions that change glucose, one step at a time, into different molecules.**

In the first step, a glucose molecule undergoes several chemical reactions that ultimately split the 6-carbon sugar into two 3-carbon PGAL molecules. See Figure 6–14. Two molecules of ATP are used up in the process. This means that the cell must use up a bit of its stored chemical energy to begin the breakdown of glucose.

Several more chemical steps then occur that transform the 2 PGAL molecules into 2 molecules of pyruvic acid, a 3-carbon compound. The energy from the 2 PGAL molecules is used to make 4 molecules of ADP and 2 molecules of NADH. NADH is an energy-storing compound similar to the NADPH that is formed in the light reactions of photosynthesis.

In the process of glycolysis, 4 molecules of ATP are synthesized from 4 molecules of ADP. Because 2 ATP molecules are used to start the process, we can do a little arithmetic and conclude that there is a net gain of 2 molecules of ATP during glycolysis.

The 2 ATP molecules produced in glycolysis represent no more than 2 percent of the total chemical energy in glucose. What happens to the rest? The answer depends on the presence of oxygen.

Respiration

If oxygen is available, a process known as **respiration** can take place. Because it requires oxygen, respiration is called an **aerobic** process. The word aerobic means with air—specifically, the part of the air called oxygen. **Respiration is the process that involves oxygen and breaks down food molecules to release energy.** Respiration uses the pyruvic acid formed in glycolysis. In breaking down pyruvic acid, respiration captures much of the remaining energy from glucose in the form of 34 additional molecules of ATP.

The term respiration may have a familiar ring. It is often used as a synonym for breathing. Because of this, some biologists use the term cellular respiration instead of respiration to refer to energy-releasing pathways within the cell. The double meaning of respiration points out a crucial connection between cell and organism: The energy-releasing respiratory pathways within the cells require oxygen, which animals take in by breathing—another type of respiration. Although plants do not breathe in the usual sense, they also need to take in oxygen for cellular respiration.

Reinforcement/Reteaching
Write on the chalkboard:

$$C_6H_{12}O_6 + 6O_2 \rightarrow 6CO_2 + 6H_2O$$

Emphasize that the complete breakdown of glucose is the reverse of glucose formation.
- **What were the main reactants**

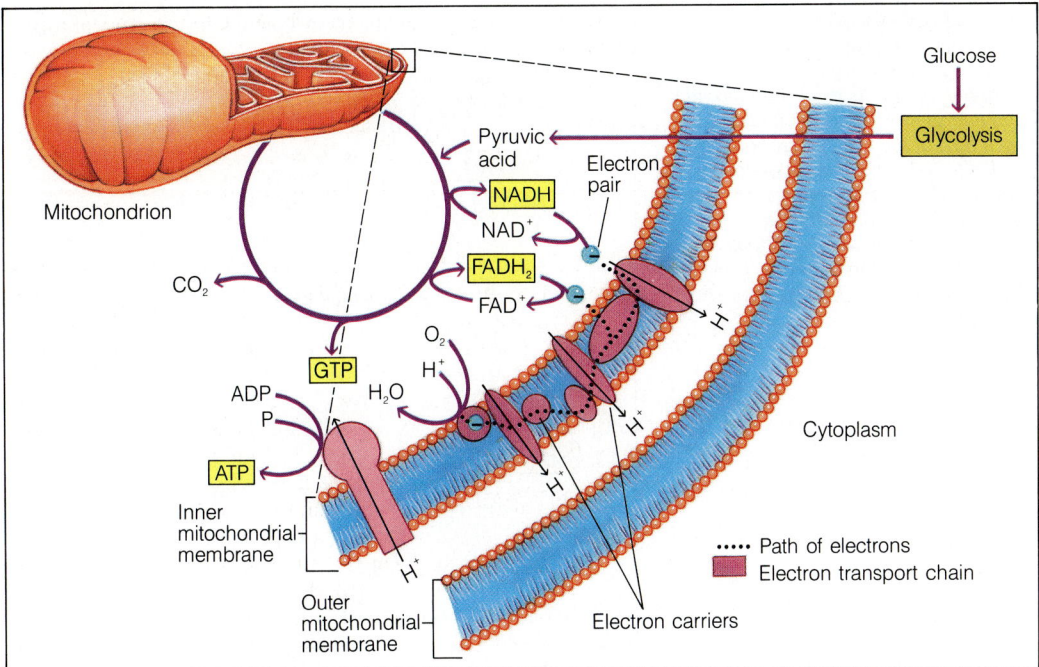

Figure 6–15 Respiration, an aerobic process, takes place in the mitochondrion. The Krebs cycle takes place in the area enclosed by the inner mitochondrial membrane. Electron transport and ATP formation involve complex molecules embedded in the inner membrane.

Respiration, like photosynthesis, is associated with a cell organelle. The chemical reactions of respiration take place in a cell's mitochondria. Recall that a mitochondrion consists of two membranes, an outer membrane and a folded inner membrane. The first set of reactions in respiration takes place in the area enclosed by the inner membrane. The second set takes place within the inner membrane itself. See Figure 6–15.

THE KREBS CYCLE The first set of reactions in respiration is the **Krebs cycle**, which is named for its discoverer, Hans Krebs. Unlike glycolysis, the Krebs cycle does not produce a final end product. Instead, it is a continuing series of reactions.

The pyruvic acid produced during glycolysis travels from the cytoplasm to the mitochondrion. In the first reaction of respiration, pyruvic acid is broken down into carbon dioxide and a 2-carbon acetyl group, which is bound briefly to a large complex known as coenzyme A. The acetyl-coenzyme A complex then passes the 2 carbons of the acetyl group into the Krebs cycle, where they join with a 4-carbon compound to produce citric acid, a 6-carbon compound. Because citric acid is the first compound formed in this series of reactions, the Krebs cycle is also known as the citric acid cycle.

There are nine reactions and nine intermediates in the Krebs cycle. At two places in the cycle, CO_2 is released. At four

HISTORICAL NOTE
SIR HANS ADOLF KREBS

Sir Hans Adolf Krebs was an English biochemist who was born in Germany in 1900. He immigrated to England in 1933. In 1953, Krebs shared with Fritz Lipmann the Nobel prize in physiology or medicine for his studies of intermediary metabolism. Among these studies was the cycle of chemical reactions now referred to as the Krebs cycle, which is the major source of energy in living organisms. Krebs died in 1981.

FACTS AND FIGURES

Scientists estimate that the total production of glucose by the Earth's plants is about 91 billion metric tons per year.

into two molecules of PGAL, then further changes these molecules into two molecules of pyruvic acid. Thus the end product of the anaerobic phase of cellular respiration is pyruvic acid. This anaerobic phase produces a net gain of two molecules of ATP.

In order for the aerobic phase of cellular respiration to take place, the molecules of pyruvic acid must be further broken down. The reactions that occur in this phase cannot take place without oxygen. The aerobic phase of cellular respiration is crucial to energy production since much of the energy that was stored in the original glucose molecules is still contained in the molecules of pyruvic acid.

• in photosynthesis? (Water and carbon dioxide.)
• What were the main products in photosynthesis? (Carbohydrate, or glucose, and oxygen.)
• What are the reactants in this equation? (Glucose and oxygen.)
• What are the products in this equation? (Water and carbon dioxide.)
• What did light give to the process of photosynthesis? (Energy.)
• What is released when glucose is broken down completely? (Energy.)

Content Development
Explain to students that the breakdown of glucose consists of an anaerobic (without oxygen) phase and an aerobic (with oxygen) phase. The anaerobic phase of cellular respiration breaks a molecule of glucose

BACKGROUND INFORMATION
MITOCHONDRIA

Mitochondria are complex structures that occur as either spherical or rod-shaped bodies. They have a tendency to change their size and shape. Mitochondria measure about 0.4 microns in diameter and are among the largest organelles in the cell. In a healthy cell, the mitochondria are in constant motion and tend to congregate near the nucleus.

Each mitochondrion is made up of an outer membrane and an inner membrane. The inner membrane has many folds, which are called cristae. A semifluid matrix fills the space between the cristae. This matrix contains DNA and ribosomes, which are involved with making proteins for growth and for the reproduction of the mitochondria. Biochemical studies have shown that most of the oxidation reactions in a cell take place along the cristae inside the mitochondria.

Mitochondria are often called the "powerhouse" of the cell because their main function is to continue the breakdown of sugars that begins in the cytoplasm. The mitochondria supply most of the energy that a cell needs. For this reason, the number of mitochondria in a cell varies according to the cell's need for energy. Cells that are active and need a great deal of energy have more mitochondria than do cells that are inactive and need less energy. For example, muscle cells have more mitochondria than bone cells, and a mature red blood cell often has no mitochondria. When a cell becomes old, the mitochondria seem to disappear from the cytoplasm. Some cells, such as a human liver cell, contain more than 1000 mitochondria.

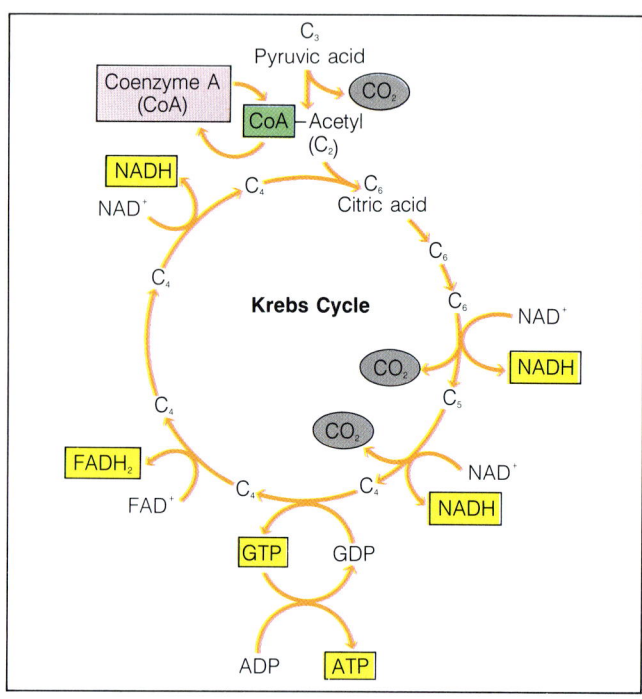

Figure 6–16 *The Krebs cycle, a continuing series of nine reactions involving nine intermediate compounds, produces carbon dioxide and energy-storing compounds. With each turn of the cycle, 2 carbons are added to a 4-carbon compound to produce the 6-carbon molecule citric acid. The 2 carbons are bound temporarily to a complex known as coenzyme A (abbreviated as CoA), forming a complex known as acetyl-coenzyme A.*

places in the cycle, a pair of high-energy electrons are accepted by electron carriers, changing NAD^+ to NADH and FAD to $FADH_2$. See Figure 6–16. At one place in the cycle, a molecule of GDP (similar to ADP) is converted to GTP (similar to ATP). The GTP may then be used to make ATP from ADP. We can summarize the events in the Krebs cycle as follows:

- 2 carbon atoms added (from the breakdown of pyruvic acid)
- 2 carbon atoms removed (in 2 molecules of CO_2)
- 3 molecules of NAD^+ converted to NADH
- 1 molecule of FAD converted to $FADH_2$
- 1 molecule of GFP converted to GTP

Step by step, the carbon atoms in glucose wind up as carbon dioxide. The carbon dioxide produced by the Krebs cycle is a waste product that is released from the cell. But what does the cell do with the other products of the Krebs cycle: NADH and $FADH_2$? And where does oxygen fit into the process?

ELECTRON TRANSPORT IN THE MITOCHONDRION
The high-energy electrons from NADH and $FADH_2$ are passed to a series of electron transport enzymes in the inner membrane of the mitochondrion. These enzymes form an electron transport chain along which electrons are passed. At the end of this chain is an enzyme that combines electrons from the electron

---

6–3 (continued)

Reinforcement/Reteaching
Students should recognize that gycolysis, or the anaerobic phase of cellular respiration, is a series of steps in which glucose is changed first into PGAL, then later into pyruvic acid. They should also recognize that without oxygen, the process of cellular respiration will stop at this point. It is the breakdown of pyruvic acid that begins the aerobic phase of cellular respiration.

Content Development
Emphasize that glycolysis, or the anaerobic phase of cellular respiration, takes place in the cytoplasm of the cell. The end product of glycolysis is pyruvic acid. Explain to students that a transition must now occur in order for the aerobic phase of cellular respiration to begin. The transition consists of pyruvic acid being transported from the cytoplasm to the fluid of the mitochondria. This is an active transport, requiring 1 molecule of ATP per molecule of pyruvic acid.

transport chain, hydrogen ions (H^+) from the fluid inside the cell, and oxygen (O_2) to form water (H_2O). This is where oxygen comes in. Oxygen is the final electron acceptor in the process of respiration. Thus, oxygen is essential for obtaining energy from both NADH and $FADH_2$.

ATP FORMATION Recall that the photosynthetic electron transport chain was associated with the movement of hydrogen ions (H^+) across the photosynthetic membrane in the chloroplast. Electron transport in the mitochondrion causes a similar movement of hydrogen ions. As they accept electrons, some of the enzymes in the electron transport chain pump a hydrogen ion from the inside of the inner membrane to the outside. As in photosynthesis, the movement of hydrogen ions powers the formation of ATP. On average, the movement of a pair of electrons down the electron transport chain produces enough energy to form 3 ATP molecules from ADP.

Electron transport produces a difference in electric charge across the membrane. There are more hydrogen ions outside the inner membrane of the mitochondrion than inside it. Thus, the outside of the membrane is more positively charged than the inside. The difference in charge created by the imbalance of hydrogen ions supplies the energy to make ATP from ADP.

Like the photosynthetic membrane, the inner mitochondrial membrane has two special properties that make this process work: The electron transport chains in the membrane are arranged so that hydrogen ions are pumped in one direction across the membrane; and the membrane does not allow ions to "leak" back across (otherwise there would not be enough of a charge difference to provide energy for ATP synthesis).

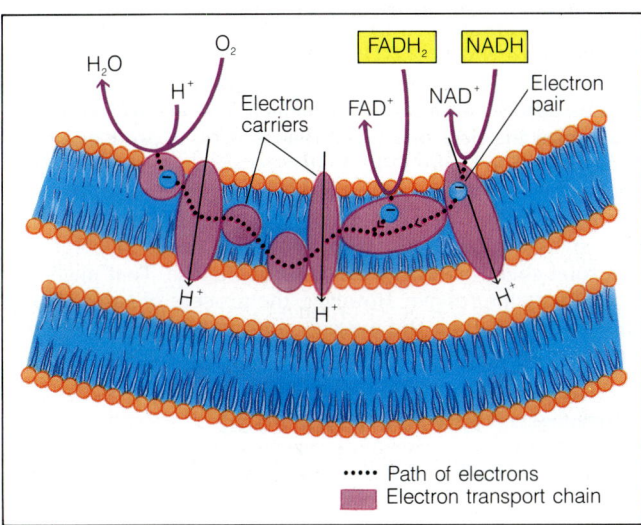

Figure 6–17 *The high-energy electrons in the energy-storing compounds NADH and $FADH_2$ are transferred to the molecules of the electron transport chain, which is located in the inner mitochondrial membrane. At the end of the chain, the electrons react with hydrogen ions and oxygen to form water.*

TIE-IN PHYSICAL SCIENCE

The difference in electric charge that occurs across the mitochondrial or photosynthetic membrane is called a transmembrane electrochemical potential. Students who have studied physics or electronics will no doubt be able to relate this term to the potential difference or voltage that exists in an electrical circuit. Like transmembrane electrochemical potential, potential difference or voltage is also a source of energy.

Point out that once the pyruvic acid reaches the mitochondria, it is changed into acetic acid within the inner mitochondrial membrane. Stress that it is the acetic acid that enters the Krebs cycle.

Skills Development
Guided Practice
Skill: Interpreting diagrams
Using the Visuals Direct students' attention to Figure 6–16.
• **What is meant by the word cycle?** (A circular process; something that happens over and over again; a process that has no end.)

Point out that in a cycle of chemical reactions, the series of reactions begins and ends with the same substance.

• **What is the substance that begins the series of reactions called the Krebs cycle?** (Acetic acid.)
• **Where does this substance come from?** (From the pyruvic acid that has entered the mitochondria from the cytoplasm.)
• **Follow the sequence of reactions around the cycle. What compound is reformed as the sequence comes back to the starting point?** (Acetic acid.)
• **What substances are products of the Krebs cycle—that is, which substances are shown in the diagram as "spinning off" and not continuing in the cycle?** (Carbon dioxide, NADH, $FADH_2$, and GTP, which is then used to make ATP.)

Point out that carbon dioxide is released as a waste product, just as oxygen was released as a waste product in photosynthesis. Emphasize, however, that NADH and $FADH_2$ are the important products that will be carried into the next phase of aerobic respiration—the electron transport chain.

TEACHING SUPPORT

Teaching Resources
- Activity: Analyzing Photosynthesis and Respiration, p. 5
- Hands-On Activity: Friends or Foes? p. 51

Study Guide
- Section 6–3, p. 59

BACKGROUND INFORMATION
"LOST ENERGY"

The missing 63% of the available energy from glucose is "lost" as heat. This is similar to what happens in a gasoline engine, where the production of heat is so great that a cooling system must be installed to prevent the engine from overheating. In human beings, however, the heat produced by glucose oxidation is not wasted—in fact, it is necessary, for it is the major source of body heat. That is why a person feels warmer after vigorous exercise—the rapid utilization of large amounts of glucose causes the body temperature to rise.

6–3 (continued)

Content Development
Emphasize to students that the process of cellular respiration is the same in plants and in animals. The difference is in the way each type of organism obtains the glucose that it needs for the respiration process.

Skills Development
Guided Practice
Skill: Comparing
Have students compare the electron transport system in cellular respiration with that in photosynthesis.
- **In what ways are these two electron transport systems similar?** (In both systems, high-energy electrons are passed from one carrier molecule to another along the chain; in both systems, movement of hydrogen ions across the membrane provides energy for ATP formation; in both systems, the transport takes place within a membrane in an organelle.)
- **Can you see any differences in the system?** (The major difference is in what happens at the end of the chain. In photosynthesis, electrons that reach the end of the chain are used to reduce $NADP^+$ to form NADPH. In cellular respiration, the final electron acceptor is oxygen, and the end product is water.)

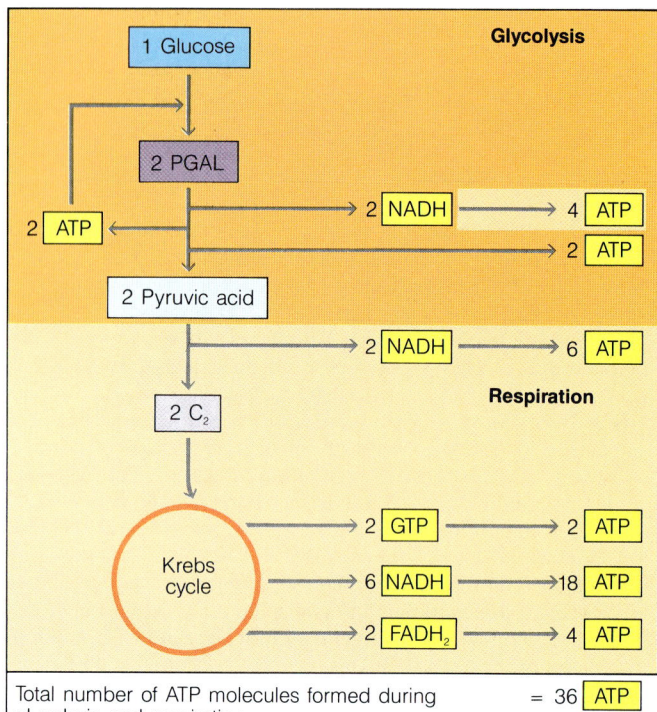

Figure 6–18 The complete breakdown of glucose through glycolysis and respiration results in the production of chemical energy in the form of 36 molecules of ATP.

THE TOTALS How much chemical energy is produced by the complete breakdown of glucose? Remember, glycolysis produces a net gain of 2 ATP molecules and passes high-energy electrons to 4 NADH molecules. The energy in these NADH molecules can be used to form 10 ATP molecules. In respiration, 6 more NADH, 2 GTP, and 2 $FADH_2$ molecules are produced in the Krebs cycle. The energy in the 2 GTP molecules can be used to form 2 ATP molecules. The electron transport chain uses the electrons in NADH and $FADH_2$ to produce 22 ATP molecules. See Figure 6–18. When the results of each of these biochemical pathways are considered, we find that a total of 36 molecules of ATP are produced from each glucose molecule. The energy in these 36 ATP molecules is about 37 percent of the total chemical energy available in glucose. That might not sound very impressive. However, the process of respiration is more efficient than the average automobile engine when it comes to converting fuel to usable energy.

Obtaining Energy from Food

Obviously very few of us eat a diet composed exclusively of glucose. But how is energy produced from foods other than glucose?

SECTION REVIEW 6–3

1. Process in which glucose is broken down into pyruvic acid.
2. Carbon dioxide; released as waste; NADH and $FADH_2$; used in electron transport; GTP: like ATP, provides energy for chemi-

Complex carbohydrates are broken down into simple sugars that are then converted into glucose. At that point, the pathways we have discussed in this chapter can be used to produce energy. Most lipids and many proteins can be broken down into molecules that can enter glycolysis or the Krebs cycle at one of several places. Like a furnace that can burn wood, coal, or oil, the cell can generate chemical energy in the form of ATP from just about any source.

Breathing and Respiration

After all of this chemical detail, it is reasonable to point out why we have used the term respiration to refer to energy-releasing pathways. Remember that the final acceptor for all electrons produced in respiration is oxygen. Without oxygen, electron transport cannot operate, the Krebs cycle stops, and the synthesis of ATP in the mitochondrion stops. If we are starved for oxygen, our cells will attempt to make the ATP they require by glycolysis alone. But for most cells this is not sufficient.

With each breath we take, air flows into our lungs. Since ancient times, humans have realized that a constant supply of air into the lungs is necessary. But only in this century have we been able to understand why that air is necessary—oxygen has a critical role to play in the mitochondria of every cell!

Energy in Balance

Photosynthesis and respiration can be thought of as opposite processes. Earlier, we compared the chemical energy in glucose to money in a savings account. Photosynthesis is the process that "deposits" energy. Respiration "withdraws" energy. The equations for photosynthesis and for the complete breakdown of glucose (through glycolysis and respiration) are the reverse of each other. The products of photosynthesis are the reactants of glucose breakdown, and the products of glucose breakdown are the reactants of photosynthesis.

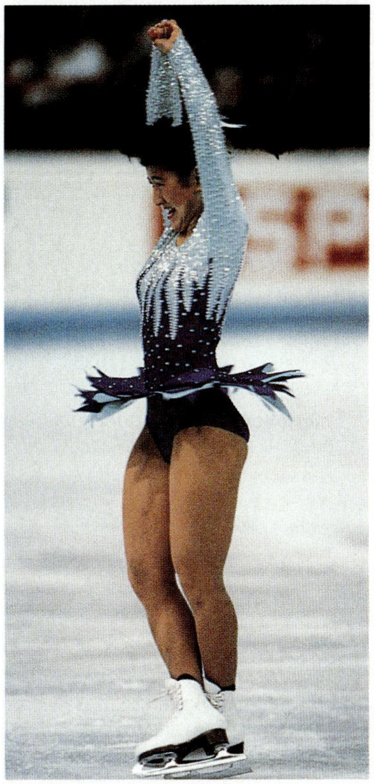

Figure 6–19 *As figure skater Midori Ito breathes, she takes in oxygen. The oxygen is used in the process of respiration, which provides energy in the form of ATP—energy that enables the skater to perform her feats on ice!*

6–3 SECTION REVIEW

1. What is glycolysis?
2. List the products of the Krebs cycle. What happens to each of these products?
3. Why is breathing necessary for cellular respiration in animals?
4. **Critical Thinking—Relating Concepts** None of the steps of the Krebs cycle involves oxygen. However, the Krebs cycle is considered to be an aerobic process. Explain why.

MULTICULTURAL STRATEGY

Ask students to think of something that they ate today, and have them make a diagram of the energy flow that was required to make the food available to them. Remind them that all energy flows should begin with the sun.

This may be a good opportunity to speak about meals as they traditionally occur in other cultures. You may consider asking students to find out about different eating traditions in other cultures, or have students speak about their own traditions. Have students include foods customarily served at breakfast, lunch, and dinner. For example, typical choices of a Japanese breakfast might include soybean paste soup, rice, crisp seaweed, grilled salted fish, and green tea. Typical choices of a French breakfast might include coffee with hot milk and a croissant. In Mexico, hot chocolate with *churros* is popular. In Germany, typical choices might include a variety of breads, sausages, cheeses, and jams, in addition to cereal and eggs.

cal reactions.
3. Breathing takes in oxygen, which is the final electron carrier in the electron transport chain.
4. For the end products of the Krebs cycle to be utilized, oxygen is required. For example, oxygen is required for NADH to be converted to NAD^+.

Reinforcement/Reteaching

If students have difficulty answering the Section Review questions, go back to the part of the text that contains the material with which they are having trouble. Review the material; then have the students try the questions again.

Closure

Have students make charts to summarize the two main phases of cellular respiration; the anaerobic (glycolysis) and the aerobic. In the anaerobic phase, students should emphasize the transformation of glucose into pyruvic acid and the net gain of 2 ATP molecules. They should also recognize that glycolysis takes place in the cytoplasm of the cell. In the aerobic phase, students should emphasize the changing of pryuvic acid into acetic acid, then the two major reactions that follow—the Krebs cycle and the electron transport chain. They should note the breakdown of glucose into carbon dioxide in the Krebs cycle and the production of 2 ATP molecules, and the production of 34 ATP molecules in the electron transport chain. Students should also indicate that the aerobic phase of respiration takes place within the mitochondria.

Guide For Reading

- How are fermentation and glycolysis related?
- How do lactic acid fermentation and alcoholic fermentation compare?

6–4 Fermentation

If oxygen is present, a cell can use the process of respiration to extract a great amount of energy from glucose. But what happens when no oxygen is available? Does this mean that we can no longer get energy from glucose?

As you may remember, glycolysis produces 2 ATP molecules per molecule of glucose and does not require oxygen.

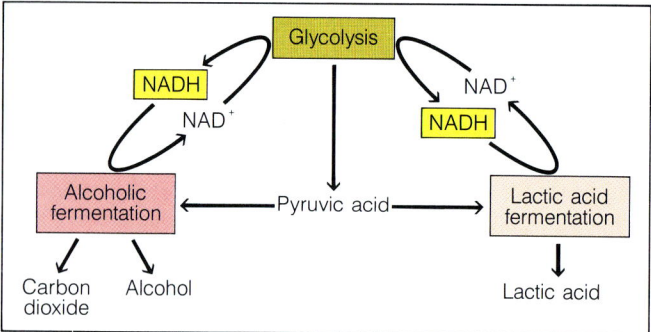

Figure 6–20 The process of glycolysis produces energy and converts NAD^+ to NADH. Fermentation is an anaerobic process that converts NADH back to NAD^+, thus enabling cells to produce energy under anaerobic conditions.

This enables cells to produce a limited amount of chemical energy in the form of ATP. Glycolysis also produces high-energy electrons that convert NAD^+ to NADH. In order for glycolysis to continue, NADH must be converted back to NAD^+.

This is where the process of **fermentation** comes in. Fermentation is **anaerobic**, which means that it does not require oxygen. In fermentation, NADH is converted to NAD^+ by adding the extra electrons in NADH to an organic molecule that acts as an electron acceptor. **Thus, fermentation enables cells to carry out energy production in the absence of oxygen.** The combination of glycolysis and fermentation produces 2 ATP molecules from a molecule of glucose.

In bacteria, many different organic molecules can serve as the final electron acceptor in fermentation. However, most eukaryotic cells use one of two fermentation pathways to change NADH to NAD^+. The two pathways are lactic acid fermentation and alcoholic fermentation.

Lactic Acid Fermentation

In many cells, the pyruvic acid that accumulates as a result of glycolysis can be converted to lactic acid. Because this type of fermentation produces lactic acid, it is known as **lactic acid fermentation**. This process regenerates NAD^+ so that glycolysis can continue:

pyruvic acid + NADH → lactic acid + NAD^+

Lactic acid is produced in muscles during rapid exercise when the body cannot supply enough oxygen to the tissues to

Figure 6–21 When the body cannot supply enough oxygen to muscles during vigorous exercise, muscle cells begin to produce energy by lactic acid fermentation. Even Florence Griffith-Joyner is subject to the aches and pains that result from the buildup of lactic acid in her muscles.

carry out cellular respiration. The cell must then depend upon fermentation to obtain the energy that it needs. Organisms that rely solely on fermentation to obtain energy are called anaerobes.

The second condition that produces fermentation is the inability of the cell to obtain sufficient oxygen, which prevents the cell from carrying out aerobic respiration. This is what happens during vigorous exercise when the body cannot supply oxygen to muscle cells fast enough. The muscle cells go into what is called oxygen debt. During the period of oxygen debt, the muscle cells rely on lactic acid fermentation to obtain energy from glucose. Once sufficient oxygen becomes available and the condition of oxygen debt recedes, the process of fermen-

produce all of the ATP that is required. The buildup of lactic acid causes a burning, painful sensation that every athlete is familiar with. When you exercise vigorously by running, swimming, or riding a bicycle as fast as you can, the large muscles of your arms and legs quickly run out of oxygen. Your muscle cells rapidly begin to produce ATP by fermentation. During this process, lactic acid is produced. This is why muscles may feel sore after only a few seconds of very rapid activity.

Alcoholic Fermentation

Another type of fermentation occurs in yeasts and a few other microorganisms. In this process, pyruvic acid (which is a 3-carbon compound) is broken down to produce a 2-carbon alcohol and carbon dioxide (CO_2). Because this type of fermentation produces alcohol, it is known as **alcoholic fermentation**. As in lactic acid fermentation, the process that alters pyruvic acid also changes NADH back into NAD^+:

pyruvic acid + NADH → alcohol + CO_2 + NAD^+

Alcoholic fermentation is of particular importance to bakers and brewers. The carbon dioxide produced by yeast during fermentation causes dough to rise and forms the air spaces you see in a slice of bread. The CO_2 released by fermentation is the source of bubbles in beer and sparkling wines. To brewers, alcohol is a welcome byproduct of fermentation. However, it is not desirable from a yeast cell's point of view. Alcohol is toxic. When the level of alcohol reaches about 12 percent, yeast cells die. Thus alcoholic beverages must be processed if higher concentrations of alcohol are desired.

Figure 6-22 When yeast cells (left) undergo alcoholic fermentation, they produce alcohol and carbon dioxide. The carbon dioxide produces the air spaces in bread and the bubbles in champagne.

6-4 SECTION REVIEW

1. What is fermentation? What are two types of fermentation?
2. How are respiration and fermentation similar? How are they different?
3. **Connection—Food Science** When oxygen is present, yeast cells undergo respiration. Explain why brewers must take special care to prevent air from entering fermentation tanks.

HISTORICAL NOTE
LOUIS PASTEUR

Before he became famous for his studies of disease, Louis Pasteur studied fermentation on behalf of the French wine industry. He discovered something that became known as the "Pasteur effect." The Pasteur effect can be described as follows: When yeast and grape juice are sealed into a fermentation vessel, the yeast begins to grow until all of the oxygen in the vessel is used up. At that point, the yeast cells switch over to fermentation, and the production of alcohol begins. Pasteur discovered that the rate at which sugar is used up by the yeast increases rapidly once the oxygen is exhausted.

Pasteur himself could not explain the Pasteur effect. The key can be found, however, by comparing the number of ATP molecules produced in fermentation to the number of ATP molecules produced in aerobic respiration. For every molecule of glucose, aerobic respiration yields 36 molecules of ATP. In contrast, anaerobic respiration yields only 2 molecules of ATP for every molecule of glucose. Thus it is not surprising that fermentation uses glucose at a much more rapid rate than does aerobic respiration.

LABORATORY INVESTIGATION

OBSERVING THE RELATIONSHIP BETWEEN PHOTOSYNTHESIS AND RESPIRATION

Before the Lab

1. At least one day prior to the investigation, gather enough materials for your class, assuming 6 students per group.
2. To prepare bromthymol blue (BTB) solution, first make a 0.1% stock solution by dissolving 0.5 g of BTB in 50 mL of water. Then add enough of this stock solution to a liter of water to give it a blue-green color. Finally, add a drop or two of ammonium hydroxide to turn the solution a definite blue.

Pre-Lab Discussion

Write on the chalkboard the two basic equations for photosynthesis and cellular respiration. Have a student read aloud the Problem that is stated at the beginning of the investigation.

• **Based on these equations, what do you think is the relationship between photosynthesis and respiration?** (Possible answer: Respiration produces carbon dioxide, which is used in photosynthesis; photosynthesis produces oxygen, which is used in respiration.)

Have students read through the entire procedure for the laboratory investigation.

• **What is being tested in this investigation?** (The effect of photosynthesis on CO_2, which is a product of respiration.)
• **What is the experimental setup?** (The flask that will be put into the light.)
• **What is the control?** (The flask that will be put into the dark.)
• **Why is the control needed?** (To make sure that something other than photosynthesis is not changing the composition of the gas in the flask.)
• **What do you think is going to happen in this investigation?**

(Accept all reasonable answers. Have students write out their ideas in the form of a hypothesis.)

Skills Development

Students will use the following skills while completing this laboratory investigation.

LABORATORY INVESTIGATION

OBSERVING THE RELATIONSHIP BETWEEN PHOTOSYNTHESIS AND RESPIRATION

PROBLEM

What is the relationship between the processes of photosynthesis and respiration?

MATERIALS (per group)

2 125-mL flasks
2 #5 rubber stoppers
100-mL graduated cylinder
bromthymol blue solution
2 *Elodea*
light source
drinking straw

PROCEDURE

1. Using a graduated cylinder, measure out 100 mL of bromthymol blue solution for each of the two flasks.
   CAUTION: *Bromthymol blue is a dye and can stain your hands and clothing.*
2. Insert one end of a drinking straw into the bromthymol blue in one of the flasks. Gently blow through the straw. Keep blowing until there is a change in the appearance of the bromthymol blue solution. Repeat this procedure with the other flask.
3. Place a sprig of *Elodea* into each flask. Stopper the flasks.
4. Place one flask in the dark for 24 hours. Place the other flask on a sunny windowsill for the same amount of time. Artificial light may be used to supplement the sunlight.
5. After 24 hours, examine each flask. Note any change in the appearance of the bromthymol blue solution.

OBSERVATIONS

1. What was the color of the bromthymol blue solution before you exhaled into it? After you exhaled into it?
2. What was the color of the bromthymol blue solution in the flask that was placed in the dark for 24 hours? In the flask that was placed in the light for 24 hours?

ANALYSIS AND CONCLUSIONS

1. What substance was released into the bromthymol blue solution when you exhaled into it? How is this substance produced?
2. Explain why the color of the bromthymol blue solution changed after you exhaled into it.
3. Why was *Elodea* placed in both flasks?
4. Which flask is the control? Describe additional controls that you might use for this experiment.
5. Why are the results for the two flasks different?
6. How are photosynthesis and respiration related?

1. Predicting
2. Comparing
3. Hypothesizing
4. Relating
5. Inferring
6. Manipulative

Safety Tips

1. Remind students of the rules for glassware safety. Make sure students do not work with broken or cracked glassware.
2. Have a student read aloud the CAUTION for working with bromthymol blue.

STUDENT STUDY GUIDE

SUMMARIZING THE CONCEPTS

The key concepts in each section of this chapter are listed below to help you review the chapter content. Make sure you understand each concept and its relationship to other concepts and to the theme of this chapter.

6-1 Photosynthesis: Capturing and Converting Energy

- In the process of photosynthesis, green plants capture the energy in sunlight and convert it into chemical energy. The balanced overall equation for photosynthesis is $6\ CO_2 + 6\ H_2O \xrightarrow{light} C_6H_{12}O_6 + 6\ O_2$.

6-2 Photosynthesis: The Light and Dark Reactions

- The light reactions can be broken down into four parts: light absorption, electron transport, oxygen production, and ATP synthesis.
- The dark reactions occur outside the photosynthetic membrane in the chloroplast. In the Calvin cycle, CO_2 is used to make PGAL, which can then be used to form glucose and other important molecules.

6-3 Glycolysis and Respiration

- Glycolysis is a series of chemical reactions that ultimately produces 2 molecules of ATP and 2 molecules of pyruvic acid.
- Respiration is an aerobic process that breaks down food molecules to release energy. This energy is stored in the form of ATP.

6-4 Fermentation

- Fermentation is an anaerobic process that regenerates NAD^+ for use in glycolysis. In fermentation, organic molecules take electrons from NADH, changing it back to NAD^+
- Photosynthesis stores energy; respiration releases energy. Products of photosynthesis are used in respiration. Products of respiration are raw materials for photosynthesis.

REVIEWING KEY TERMS

Vocabulary terms are important to your understanding of biology. The key terms listed below are those you should be especially familiar with. Review these terms and their meanings. Then use each term in a complete sentence. If you are not sure of a term's meaning, return to the appropriate section and review its definition.

6-1 Photosynthesis: Capturing and Converting Energy
photosynthesis
glucose
autotroph
heterotroph
pigment
chlorophyll
adenosine triphosphate (ATP)

6-2 Photosynthesis: The Light and Dark Reactions
light reactions
dark reactions
photosynthetic membrane
photosystem
electron transport
Calvin cycle

6-3 Glycolysis and Respiration
calorie
glycolysis
respiration
aerobic
Krebs cycle

6-4 Fermentation
fermentation
anaerobic
lactic acid fermentation
alcoholic fermentation

---

are combined with oxygen to produce water, carbon dioxide, and energy.

2. Exhaled carbon dioxide reacts with bromthymol blue, causing it to turn yellow.

3. To ensure that conditions in both flasks were the same, except for the experimental variable (light/photosynthesis).

4. The control was the flask placed in the dark. Student answers will vary but may include controls for temperature and the presence of *Elodea*.

5. Photosynthesis used up the carbon dioxide, which occurred in the flask that was placed in the light. This caused the bromthymol blue solution to return to its original color.

6. Photosynthesis uses carbon dioxide; respiration produces carbon dioxide.

Going Further: Enrichment
Have students design an experiment in which they investigate the effect of temperature on photosynthesis. (Until a maximum is reached, photosynthesis increases as temperature increases. If the temperature becomes high enough to denature enzymes, photosynthesis stops.)

---

3. Tell students to be careful to exhale only, not inhale, when using the straw.

Teaching Strategy
1. Circulate throughout the classroom and help any groups that are having trouble, either with the setup or with the interpretation of results.
2. Make sure students remember to stopper the flasks as soon as the *Elodea* is added.
3. If intense sunlight is not available, light supplementation will be necessary.

Observations
1. Blue; yellow
2. Yellow; blue

Analysis and Conclusions
1. Carbon dioxide is produced during the process of respiration, in which food molecules

CHAPTER REVIEW

CONTENT REVIEW

Multiple Choice
1. c 2. b 3. a 4. d
5. c 6. a 7. d 8. b

True or False
1. F light reactions of photosynthesis
2. F heterotrophs
3. T
4. F dark reactions
5. F lesser
6. F respiration
7. F alcohol
8. T

Word Relationships
A.
1. Parts of the light reactions; Calvin cycle is part of the dark reactions
2. Energy-releasing processes; photosynthesis captures and stores energy
3. Compounds involved in photosynthesis; energy is not a compound

B.
4. water, carbon dioxide, and energy
5. heterotroph
6. glycolysis
7. mitochondrion

CONCEPT MASTERY

1. Water provides electrons in the light reactions; carbon dioxide is involved in the formation of PGAL in the dark reactions; light energy excites electrons in the photosystems in the light reactions.
2. As hydrogen ions are pumped from one side of the membrane to another, ADP gains a phosphate and becomes ATP.
3. Answers will vary.
4. Breathing provides the oxygen needed in respiration.
5. The reactants of photosynthesis are the products of respiration and vice versa. As such, they are reverse processes.
6. Lactic acid fermentation: reactants are pyruvic acid and NADH; products are lactic acid and NAD^+. Alcoholic fermentation: reactants are pyruvic acid and NADH; products are alcohol, carbon dioxide, and NAD^+.

CONTENT REVIEW

Multiple Choice

Choose the letter of the answer that best completes each statement.

1. Which process is anaerobic?
   a. Krebs cycle
   b. mitochondrial electron transport
   c. glycolysis
   d. respiration
2. An experiment that showed that part of a growing plant's mass comes from water was performed by
   a. Priestley.   c. Ingenhousz.
   b. Van Helmont.   d. Calvin.
3. Autotrophs include
   a. ferns.   c. yeasts.
   b. camels.   d. birds.
4. Photosynthesis produces
   a. carbon dioxide.   c. water.
   b. alcohol.   d. glucose.
5. In respiration, the final electron acceptor is
   a. lactic acid.   c. oxygen.
   b. alcohol.   d. NAD^+.
6. Which of these is produced by the dark reactions?
   a. PGAL   c. water
   b. oxygen   d. ATP
7. A type of pigment found in green plants is
   a. NAD^+.   c. ATP.
   b. NADPH.   d. chlorophyll.
8. Which of these is not produced in the light reactions?
   a. NADPH   c. ATP
   b. glucose   d. oxygen

True or False

Determine whether each statement is true or false. If it is true, write "true." If it is false, change the underlined word or words to make the statement true.

1. In the <u>dark reactions of photosynthesis</u>, the energy in sunlight is captured and used to make ATP and NADPH.
2. Because they cannot make their own food, <u>autotrophs</u> rely on the food-making ability of green plants.
3. <u>Photosystems</u> are clusters of pigment molecules that are able to capture the energy of sunlight.
4. The <u>light reactions</u> occur outside the photosynthetic membrane.
5. ATP is formed when enzymes move hydrogen ions across a membrane to an area of <u>greater</u> concentration.
6. <u>Glycolysis</u> is an aerobic process that breaks down food molecules to produce energy.
7. When there is no oxygen present, yeast cells break down glucose to form <u>lactic acid</u> and carbon dioxide.
8. <u>Approximately 36</u> molecules of ATP are formed during glycolysis.

Word Relationships

A. *In each of the following sets of terms, three of the terms are related. One term does not belong. Determine the characteristic common to three of the terms and then identify the term that does not belong.*

1. Calvin cycle, light absorption, electron transport, oxygen production
2. fermentation, respiration, glycolysis, photosynthesis
3. glucose, oxygen, water, energy

CRITICAL AND CREATIVE THINKING

1. Answers will vary. Students should note the similarities between the electron transport chains in the light reactions of photosynthesis and in respiration. They should also note that the Krebs cycle and glycolysis reverse the reactions of the Calvin cycle.
2. Students should present this material in outline form. Light reactions: light absorp-

B. An analogy is a relationship between two pairs of words or phrases generally written in the following manner: a:b::c:d. The symbol : is read "is to," and the symbol :: is read "as." For example, cat:animal::rose:plant is read "cat is to animal as rose is to plant."

In the analogies that follow, a word or phrase is missing. Complete each analogy by providing the missing word or phrase.

4. food and oxygen:respiration::_____:photosynthesis
5. oak tree:autotroph::human:_____
6. PGAL:Calvin cycle::pyruvic acid:_____
7. photosynthesis:chloroplast::respiration:_____

CONCEPT MASTERY

Use your understanding of the concepts developed in the chapter to answer each of the following in a brief paragraph.

1. List the requirements for photosynthesis. Briefly describe how each is involved in the light reactions or the dark reactions.
2. Describe the process of ATP formation.
3. Name ten living things that you see every day. Identify each of these as an autotroph or a heterotroph.
4. What is the relationship between breathing and respiration?
5. How are photosynthesis and respiration related to each other?
6. Compare the reactants and products of two types of fermentation.

CRITICAL AND CREATIVE THINKING

Discuss each of the following in a brief paragraph.

1. **Making comparisons** Compare the chemical steps of the light and dark reactions of photosynthesis with those of glycolysis, the Krebs cycle, and the respiratory electron transport chain.
2. **Making an outline** Outline the process of photosynthesis.
3. **Applying concepts** Some desert animals such as the kangaroo rat never have to drink water. Explain how kangaroo rats can obtain the water they need to survive from the dry seeds they eat.
4. **Making inferences** Some scientists think that the dinosaurs may have become extinct because an asteroid struck the Earth and sent large amounts of dust into the upper atmosphere. The dust remained in the atmosphere a long time. Explain how this would have resulted in the dinosaurs dying off.
5. **Assessing concepts** Which is better: respiration or fermentation? Explain your answer.
6. **Using the writing process** You are a carbon atom. You have just returned to the air as carbon dioxide after being involved in the processes of photosynthesis and respiration. Describe what happened to you on your chemical journey. Where did you go? What other atoms (carbon and otherwise) did you find yourself bonded to? What molecules did you meet along the way, and what were they like?

---

tion, electron transport, oxygen production, and ATP formation. Dark reactions: Calvin cycle and formation of glucose.

3. Water is produced during the process of respiration, in which food (dry seeds) is broken down. The kangaroo rat is extremely efficient in conserving the water produced by metabolism.

4. Without adequate sunlight, plants could not perform photosynthesis, and herbivores, such as plant-eating dinosaurs, could not obtain enough food to survive. In turn, carnivores that ate herbivores and other carnivores died off from a lack of food.

5. Answers will vary, but students should construct a logical argument for their position. There is no single correct answer, and imaginative students who argue that neither is better should not be penalized.

6. Student descriptions should be well-written and convey the facts and concepts presented in the chapter in a logical manner.

CHAPTER 7 Nucleic Acids and Protein Synthesis

Section	Laboratory Investigations and Activities
7–1 DNA pages 137–145 THEMES: Scale and Structure, Stability	**Student Edition** LABORATORY INVESTIGATION: Constructing a Model of DNA Replication, p. 154 **Teacher Edition** DEMONSTRATION: Investigation, p. 136D
7–2 RNA pages 146–148 THEMES: Scale and Structure, Stability	**Teaching Resources** ACTIVITY: The Role of RNA, p.3
7–3 Protein Synthesis pages 148–153 THEMES: Systems and Interactions, Stability	**Teacher Edition** DEMONSTRATION: Patterns, p. 136D **Laboratory Manual** Assembling a Protein Molecule, p. 113 Simulating Protein Synthesis, p. 117
Chapter Review pages 155–157	

Teacher Resources

Books

Beljanski, M. *The Regulation of DNA Replication and Transcription.* S. Karger, 1983.

Bryan, J. A., and V. L. Dunham. *DNA Replication in Plants.* CRC Press, 1988.

Dickerson, R., and I. Geis. *The Structure and Action of Proteins,* 2nd ed. W. A. Benjamin, 1985.

Dulbecco, Renato. *The Design of Life.* Yale University Press, 1987.

Friefelder, D. *Molecular Biology.* Van Nostrand Reinhold, 1983.

Gribbon, John. *In Search of the Double Helix.* Bantam, 1987.

Kornberg, Warren, ed. National Science Foundation Staff. *DNA: The Master Molecule.* Avery Publishers, 1982.

CHAPTER PLANNING GUIDE

Other Activities	Media and Technology
Study Guide Section 7–1, p. 67	**Biology Transparencies** 10: DNA Replication
Study Guide Section 7–2, p. 70	
Teaching Resources VOCABULARY REVIEW: Crossword, p. 1 ACTIVITY: Protein Synthesis, p. 7 **Study Guide** Section 7–3, p. 71	**Biology Transparencies** 11: Protein Synthesis
Teaching Resources Chapter Test A, p. 13 Chapter Test B, p. 17	**Computer Test Bank** Chapter Test, p. 71

Audiovisuals

DNA and the Protein Express. Film or video. Lucerne Media.
Introduction to the Nucleotides. Sound filmstrip. Ward.
Introduction to the Amino Acids. Sound filmstrip. Ward.
Amino Acid Structure and the Peptide Link. Proteins and Deamination.
An Introduction to DNA and Protein Synthesis. Video or filmstrip. Carolina Biological Supply Company.
Protein Synthesis. Video or filmstrip. Biology Media.

136B

CHAPTER 7 Nucleic Acids and Protein Synthesis

CHAPTER OVERVIEW

The genetic code is the way in which the cell stores information that is passed from generation to generation. The search for answers about the code and the way it works began in 1928 with the work of Frederick Griffith. After much research performed by many scientists, DNA was identified as the molecule that carried the genetic code. Then began the quest for information about the structure and functioning of DNA itself.

The well-known double helix model developed by James Watson and Francis Crick was based on the work of Rosalind Franklin and her colleague Maurice Wilkins. The importance of this work that answered so many questions about DNA was recognized in 1962 when the Nobel prize for medicine or physiology was awarded to these researchers.

Before a cell divides, the cell DNA must be duplicated in a process called replication that is carried out by a series of enzymes. The Watson-Crick model explained the process of producing DNA molecules identical to the original.

The model, however, did not explain how the information contained in the molecule is used by the cell. This process involves a second nucleic acid called RNA. Transcription, the copying of DNA to form a complementary RNA molecule, called messenger RNA, is the first step in the synthesis of cell proteins.

The information passed from DNA to RNA is in the form of a code involving the nitrogenous bases found in both molecules. Specific combinations of bases signal the placement of particular amino acids into polypeptide chains. This complicated process involves two other types of RNA known as transfer RNA and ribosomal RNA. DNA directs the formation of all three kinds; thus the program for protein synthesis is passed from generation to generation through replication.

7-1 DNA

Section Focus 7-1

The purpose of this section is to present the historical background and research studies that have led to our current understanding of DNA structure and the process of replication.

The genetic code is the program that living cells use to perform many vital functions. It is the way that information is stored and passed from one generation to another.

In 1928, Frederick Griffith, while studying the way in which types of bacteria cause pneumonia, recognized a phenomenon that is now known as transformation. He experimentally showed that one strain of bacteria could be transformed into another strain. Griffith hypothesized that there was a transforming factor involved.

In 1944, a team of scientists at the Rockefeller Institute repeated Griffith's experiments and identified DNA as the transforming factor. DNA did, in fact, store and transmit genetic information from one generation to the next. Eight years later the Hershey-Chase experiments with bacteriophages confirmed once again that DNA was the molecule that contained the genetic code. The remaining question was, How could DNA store information and duplicate itself? A look at the structure offers an explanation. Students are referred to Chapter 4 for a review of the structure.

The study of the DNA molecule began in the early 1950s. Rosalind Franklin, using a purified sample of DNA, made X-ray diffraction pictures of the molecule. At about the same time, James Watson and Francis Crick were trying to build a three-dimensional model of DNA. Shortly after seeing Franklin's pictures, the double helix model was conceived. The model explained much about the structure of DNA, including the placement of the nitrogenous bases and the formation of hydrogen bonds. Working with information gained from the experiments of Edwin Chargaff, an American biochemist, the model was perfected.

This section concludes with a brief explanation of the process of replication.

Performance Objectives 7-1

1. Discuss the experiments leading to the identification of DNA as the molecule that carries the genetic code.
2. Describe the steps leading to the development of the double helix model of DNA.
3. Describe the structure of DNA.
4. Explain the process of DNA replication.

Science Terms 7-1

genetic code   p. 137
transformation   p. 139
DNA   p. 139
nucleotide   p. 141
adenine   p. 141
guanine   p. 141
cytosine   p. 141
thymine   p. 141
base pairing   p. 143
replication   p. 144

7-2 RNA

Section Focus 7-2

The purpose of this section is to explain the structure and synthesis of RNA. RNA is the nucleic acid that acts as a messenger between the DNA in the nucleus and the ribosomes in the process of making proteins from amino acids in the cell. RNA is similar to DNA in structure. It too is a long chain made up of nucleotides. Each nucleotide consists of a sugar, a phosphate group, and a nitrogen-containing base.

There are, however, three major differences between RNA and DNA. The deoxyribose sugar in DNA is replaced by ribose in RNA. The RNA molecule usually occurs as a single strand rather than the double strand structure typical of DNA. Finally, one of the nitrogenous bases that are found in DNA, thymine, is replaced by a similar base, uracil, in RNA. There are several forms of RNA found in the cell, but three main types are involved in the business of making proteins.

Transcription is the term given to the way that an RNA strand is made from a DNA pattern or template. The enzyme RNA polymerase plays an important role in the

CHAPTER PREVIEW

process. The base-pairing mechanism ensures an accurate copy in transcription just as it does in replication.

Performance Objectives 7-2

1. List the structural similarities and differences between DNA and RNA.
2. Explain the process of transcription.
3. Discuss the role of messenger RNA.

Science Terms 7-2

RNA   p. 146
uracil   p. 146
transcription   p. 146
messenger RNA   p. 147

7-3 PROTEIN SYNTHESIS

Section Focus 7-3

The purpose of this section is to explain the process of protein synthesis. Several important functions of proteins are listed. Proteins are made up of smaller molecules of amino acids strung together into long chains called polypeptides. The way that DNA and RNA translate their information into the process of protein synthesis is explained. The four nitrogenous bases found in RNA are arranged in triplets to produce different code words. These code words, called codons, correspond to specific amino acids. In addition, some triplets act as start or stop signals.

Translation involves the decoding of the message carried by the messenger RNA. This complex mechanism involves two other types of RNA—transfer RNA and ribosomal RNA. The transfer RNA, the complement to messenger RNA, is the anticodon. It is to the anticodon that the individual molecules of amino acid are attached. The ribosomes, consisting of ribosomal RNA and protein, release the new proteins. The ribosomes are essential to making the genetic code work.

Performance Objectives 7-3

1. Explain the terms codon and anticodon.
2. Describe the process of translation.
3. Discuss the importance of ribosomes.
4. List some of the functions of proteins in the cell.

Science Terms 7-3

codon   p. 149
translation   p. 150
transfer RNA   p. 150
ribosomal RNA   p. 150
anticodon   p. 150

DISCOVERY LEARNING

TEACHER DEMONSTRATIONS
Modeling

Investigation

The following demonstration can be used as an introduction to Chapter 7.

The first section of this chapter concentrates on the "discovery" of DNA, and its structure and function. This is a good time to once again affirm that science is a process.

Show students a "black box." Even if you have previously used this technique, students will enjoy it again. The black box should contain 3 to 5 small familiar objects and must be sealed. Allow students to pass the box around and make observations. List their observations on the board or overhead so that everyone can see them and add something new. After every student has had a chance to make observations, you might want to allow them to guess what the objects are. Be sure that they tell you what clue led to their conclusion. You may choose to confirm their right answers. If they do not guess all of the objects correctly, you might want to leave these as "questions" analogous to unanswered scientific questions.

You are now ready to discuss the research studies from the textbook. Be sure to tell the students that each group of scientists related their observations and conclusions to earlier experimental evidence just as the students have done, and that there was much "educated guessing" involved in both the students' and the scientists' work.

Patterns

This demonstration can be done with Section 7-3 to help students more clearly understand the process of protein synthesis.

You may make this exercise as simple or as elaborate as you wish, and students should still grasp the basic concept. You will need a pattern (a dress pattern will work, but one for a stuffed animal or other craft item will probably be more useful as an analogy). You will also want a piece of white paper and all the materials to make your garment or craft. Set out the pattern and white paper on the table and your fabric, threads, and trims on another.

Explain to the students that you are going to make something and use the pattern. You might want to make several of the same item and do not want to ruin your original pattern, so you are making a working pattern on white paper. After you have cut out some pieces of the white paper pattern, move to the table and begin assembling parts of the item. Do not at this point try to draw any direct analogies. Keep all of the demonstration items as they are and begin your discussion of the process of protein synthesis—making no further reference to the "pattern" at this time.

136D

CHAPTER 7
Nucleic Acids and Protein Synthesis

GUIDED ENQUIRY
Pose the following questions to students and have them record their responses. Point out to students that they will gain a better understanding of the key concepts if they read the chapter with these basic questions in mind. Upon completion of the chapter, pose the questions again. Ask students to compare their initial responses with those they have developed after reading the chapter.

- How do organisms pass on unique traits to their offspring?
- Would every cell in an organism contain the same amount of DNA?
- What about the amount of DNA in a sperm or egg cell?
- Why is it important that replication of DNA be accurate?

INTRODUCING CHAPTER 7

Using the Visuals
Direct students' attention to the visual of the DNA molecule on this page. Read the caption below the picture with them. Help them to see the "pieces" that are alike or different.

Ask students to think about working a puzzle. Have them try to recall the process of looking for a "piece" that fits in a particular place in the puzzle. Ask them to try to remember the excitement and pleasure they felt when a "discovery" was made.

Have students read the introduction to the chapter. Discuss the excitement that the scientists must have felt when they made their discoveries.
- **Can you see a DNA molecule with your unaided eye?** (No.)
- **What evidence supported the presence of some type of "genetic code"?** (Organisms have offspring like themselves.)
- **When performing an experiment, what are some of the important steps that a scientist must take?** (Record the procedure, use a control, have only one variable at a time, make careful observations, record accurate data.)
- **When a scientist makes a discovery or draws a conclusion based on experimentation, what must other scientists be able to do?** (Repeat the experiment with the same results.)

CHAPTER 7
Nucleic Acids and Protein Synthesis

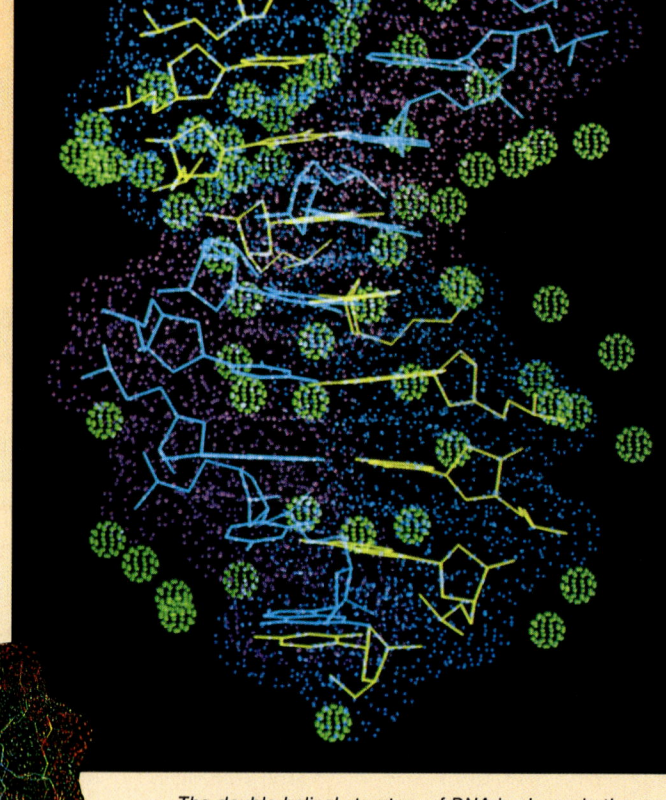

The double-helical structure of DNA is shown in these computer-generated images. The larger image shows a dark-blue structure and a purple structure coiling around each other. These structures represent the two linked strands of nucleotides in DNA. The smaller image shows DNA's double helix as it appears from above.

In the middle of the twentieth century, a great revolution began to take place in biology. Like many revolutions, its beginnings were modest. Its pioneers were a handful of investigators who sought to understand how information was transferred from one generation of life to the next. The discovery that nucleic acids carry and transfer that information has altered our view of living things. In this chapter we will begin to study the nucleic acids DNA and RNA, the genetic code they contain, and the way in which that code is put into action during protein synthesis.

GUIDE FOR READING

After you read the following sections, you will be able to

7–1 DNA
- List the contributions of various scientists to the idea that DNA carries the genetic code.
- Describe the structure and function of DNA.
- Summarize the process of DNA replication.

7–2 RNA
- State the function of RNA.
- Compare the structures of RNA and DNA.
- Describe the process of transcription.

7–3 Protein Synthesis
- Explain the term codon.
- Describe the process of translation.
- Identify the three main types of RNA.

Journal Activity

YOU AND YOUR WORLD The deciphering of the genetic code is one of the most remarkable scientific achievements of the twentieth century. How do you think you would feel if you were a member of the team that "cracked" this code? Write about your feelings in your journal.

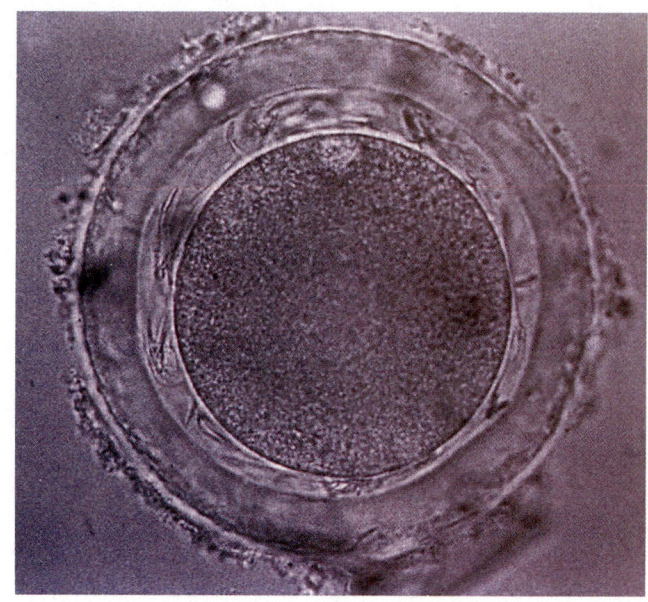

Figure 7–1 The entire genetic program needed to produce a human is contained in a single fertilized egg cell such as this.

7–1 DNA

Guide For Reading
- What contributions did various scientists make to the idea that DNA carries the genetic code?
- What is the structure and function of DNA?
- How does DNA replicate?

By now it should be clear that living cells are able to do some remarkable things. We might go so far as to say that cells "know" a great deal about the business of life: how to produce ATP, how to build cilia and centrioles, how to produce enzymes and membranes. It is as though cells are preinstructed by a code, or programmed, about what to do and how to do it.

A program, or code, in living cells must be able to duplicate itself quickly and accurately and must also have a means of being decoded and put into effect. In this chapter we will begin to learn about the nature of the cellular program—what biologists know and don't know about the ways in which the program is constructed, duplicated, and carried out.

The Genetic Code

Biologists call the program of the cell the **genetic code**. The word genetic refers to anything that relates to heredity. The genetic code, therefore, is the way in which cells store the program that they seem to pass from one generation of an organism to the next generation.

COOPERATIVE LEARNING

Using preassigned lab groups or randomly selected teams, have groups illustrate the process of protein synthesis as described in the analogy on page 148 (second paragraph of Section 7–3). Encourage students to visualize the process of protein synthesis in a concrete form to which they can relate. Their illustration might take the form of an organizational flow chart, a concept map, a factory floor plan, or a computer program.

Journal Activity

YOU AND YOUR WORLD Many scientists consider the joy of discovering something new or being the first to solve a puzzle to be the greatest reward of a career in science. Encourage students to draw on their own experiences of discovery as they work on this activity. Students should be instructed to keep their Journal Activity in their portfolio.

TEACHING STRATEGY 7–1

Focus/Motivation

Divide students into pairs. Have one student of the pair write a short message to the partner. One or two sentences is enough. Have each one devise a simple code. Have students give their partners the message both in code and in plain language. The partners' task is to work backward to reconstruct the code.

Remind students that this is essentially what the scientists in Section 7–1 were doing. They vaguely knew that some thing or process allowed organisms to pass on traits to offspring and that they had to "break" the code.

You might help them with their task by posing questions similar to these.

- **What is a code?** (A set of symbols or signals used to pass on information.)
- **What are some of the characteristics of a good code?** (Answers will vary but might include consistency, ability to be used and interpreted by more than one person.)
- **What steps will you follow to break your code?** (Look for patterns, make other observations.)
- **Did you see some pattern before you broke the entire code?** (Yes or no.)
- **If so, will this new information help you to reconstruct the entire code?** (Yes, it should.)

HISTORICAL NOTE
GRIFFITH'S EXPERIMENT

Griffith's first and foremost concern in these experiments was to ensure that pneumonia bacteria did not produce a poison, or toxin. Many other diseases, including diphtheria, are caused by bacterial toxins. His experiments with heat-killed bacteria were really nothing more than offshoots of his attempts to search for a toxin. In retrospect, he agreed that his discovery of transformation may have happened precisely because he was so eager to rule out the toxin hypothesis of pneumonia.

Is there a molecule that carries the genetic code? As often happens in science, an opportunity to answer this question presented itself to someone who was actually interested in something else.

In 1928, the British scientist Frederick Griffith was studying the way in which certain types of bacteria cause the disease pneumonia. Griffith had in his laboratory two slightly different strains, or types, of pneumonia bacteria. Both strains grew very well in culture plates in his lab, but only one strain actually caused the disease. The disease-causing strain of bacteria grew into smooth colonies on culture plates, whereas the harmless strain produced colonies with rough edges. The differences in appearance made the two strains easy to distinguish.

When Griffith injected mice with the disease-causing strain of bacteria, the mice got pneumonia and died. When mice were injected with the harmless strain, they did not get pneumonia and they did not die. And when mice were injected with the disease-causing strain that had been killed by heat, these mice too survived. By performing this third experiment, Griffith proved to himself that the cause of pneumonia was not a chemical poison released by the disease-causing bacteria.

TRANSFORMATION Next Griffith did an experiment that produced an astonishing result. He injected mice with a mixture of live cells from the harmless strain and heat-killed cells from the disease-causing strain. To Griffith's surprise, the mice developed pneumonia!

Figure 7–2 The results of Griffith's experiments showed that "something" had been transferred from the heat-killed disease-causing bacteria to the live harmless bacteria, transforming them into live disease-causing bacteria. This something was later isolated and identified as DNA.

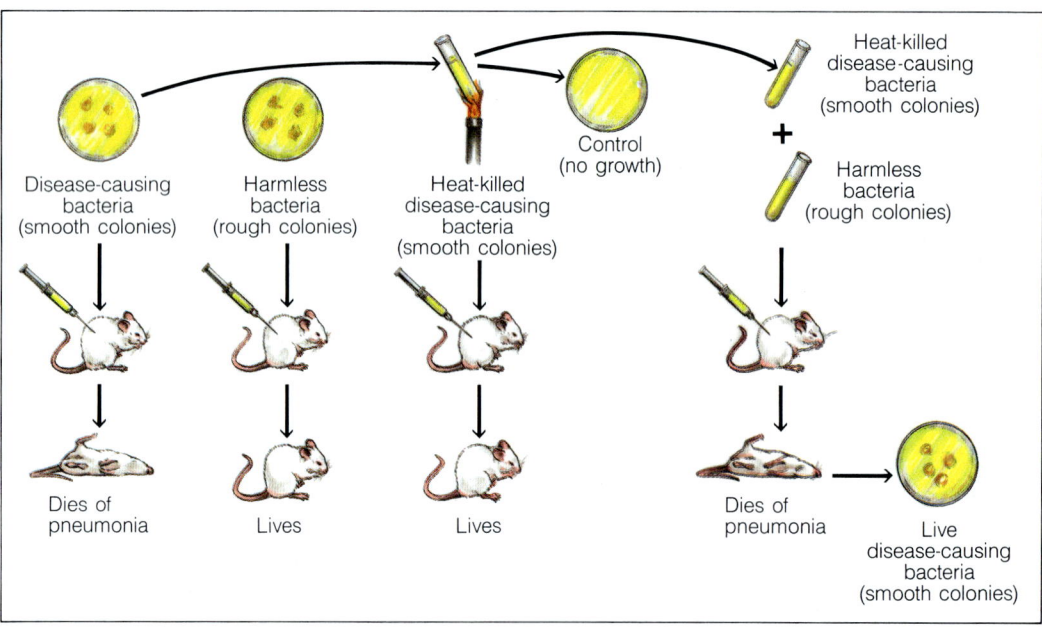

7–1 (continued)

Content Development

Tell students that following his discoveries, Griffith published his findings in several scientific journals. The group of scientists at the Rockefeller Institute read some of his articles and decided to test his results. Using this information as a starting point, discuss how important accurate observation and communication of data are to any scientific experiment.

Emphasize to students how important the work of Avery, McCarty, and MacLeod really was. It was the very first evidence that DNA had any influence on heredity.

Reinforcement/Reteaching

Using the Visuals As you talk about Griffith's experiments and transformations, be sure to make use of Figure 7–2 for your visual learners. This step-by-step presentation will be easier for them to follow.

As a point of interest, tell students that Griffith's original hypothesis was that the "poison" that caused pneumonia was in the smooth coating of the deadly bacteria. That was a logical supposition since the obvious difference in the two bacteria was in the appearance of the colonies in the culture dish. He proved his own hypothesis incorrect.

Based on what we now know about DNA, transformation seems reasonable. Remind students that Griffith and other scientists in 1928 did not have that knowledge. You might want to spend a few minutes discussing how surprising it would be to put

Somehow Griffith's heat–killed strain had passed on its disease-causing ability to the live harmless strain. It was almost as if a farmer had scattered ground beef in a henhouse and then returned the next morning to discover that some of the eggs had begun to hatch into calves instead of baby chicks!

To confuse matters even more, Griffith recovered bacteria from the animals that had developed pneumonia. When these bacteria were grown on culture plates, they formed smooth colonies characteristic of the disease-causing strain. For all practical purposes, one strain of bacteria had been transformed into another. Griffith called this process **transformation.**

THE TRANSFORMING FACTOR Griffith hypothesized that when the live harmless bacteria and the heat-killed bacteria were mixed together, a factor was transferred from the heat-killed cells into the live cells. In 1944, a group of scientists at the Rockefeller Institute in New York City led by Oswald Avery, Maclyn McCarty, and Colin MacLeod decided to repeat Griffith's work and see if they could discover which molecules were Griffith's transforming factor, that is, which molecules were responsible for transformation.

Avery and his colleagues made an extract, or juice, from the heat-killed bacteria. When they treated the extract with enzymes that destroy lipids, proteins, and carbohydrates, they discovered that transformation still occurred. Obviously these molecules were not responsible for the transformation. If they were, transformation would not have occurred because the molecules would have been destroyed by the enzymes.

Avery and the other scientists repeated the experiment, this time using enzymes that would break down RNA (ribonucleic acid). The scientists found that again transformation took place. But when they performed the experiment again, using enzymes that would break down **DNA** (deoxyribonucleic acid), transformation did not occur. DNA was the transforming factor! **DNA is the nucleic acid that stores and transmits the genetic information from one generation of an organism to the next.** In other words, DNA carries the genetic code.

BACTERIOPHAGES The work of Avery and his colleagues clearly demonstrated the role of DNA in the transfer of genetic information. Scientists, however, are notorious for being a skeptical group, and it sometimes takes many successful experiments to convince everyone.

One of the most important of these experiments was performed in 1952 by two American scientists, Alfred Hershey and Martha Chase. Hershey and Chase were interested in the kinds of viruses that infect bacteria. Such viruses are known as bacteriophages (bak-TEER-ee-uh-fayj-uhz), which means bacteria eaters. Bacteriophages are composed of a DNA core and a protein coat. They attach themselves to the surface of a bacterium and then inject a material into the bacterium.

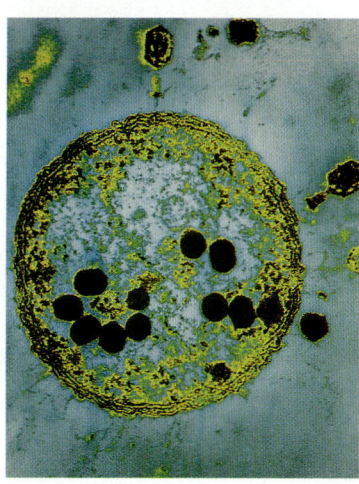

Figure 7–3 This electron micrograph shows some T-2 bacteriophages attached to the bacterium E. coli (the large yellow-green circle). Once inside, the bacteriophages take over the bacterium's DNA and instruct it to make many copies of themselves (black ovals in the bacterium).

ECOLOGY NOTE
TRANSFORMATION AND GENETIC POLLUTION

In 1993, researchers at the University of Florida reported an interesting discovery—marine bacteria can take up DNA plasmids that are floating in sea water.

This news implies that when genetically modified bacteria are released into the environment, their DNA can be taken up by wild bacteria, possibly with dire consequences. However, the University of Florida researchers also observed something that might mean that a worst-case "genetic pollution" scenario might never come to pass. After absorbing plasmids from their environment, recipient bacteria reshuffle the DNA. This sort of processing may deactivate genes that might be harmful if they got into the wrong organism.

ANNOTATION KEY

❶ Deoxyribose. (Relating facts)

MULTICULTURAL STRATEGY

Spanish-American biochemist Severo Ochoa made significant contributions to our understanding of how DNA works. Based on Ochoa's work, Kornberg created the first synthetic DNA. In recognition of their accomplishments, the two men shared the 1959 Nobel prize in medicine or physiology. Interested students may want to find out more about the life and work of Severo Ochoa.

7–1 (continued)

Content Development

Using the Visuals The real question that was asked in the Hershey-Chase experiments was, What part of the bacteriophage was being injected into the bacteria? The method was radioactive tagging. Be sure that your students have a basic understanding of how this process works. Figure 7–4 might help them understand. As you discuss the experiment with the students, pose the following questions to check their understanding.

• **How does a bacteriophage work?** (It injects material into a bacterium and makes many copies of the bacteriophage.)
• **What are the two parts of the bacteriophage?** (DNA core and protein coat.)

Once inside the bacterium, the injected material begins to reproduce, making many copies of the bacteriophage. Soon the bacterium bursts, and several hundred bacteriophages are released to infect other cells. Because the material injected into the bacterium produces new bacteriophages, it must contain the genetic code.

In their experiments, Hershey and Chase set out to learn whether the protein coat, the DNA, or both of these parts of the infecting virus was the material that entered the bacterium. First they prepared two batches of the virus. They added radioactive sulfur-35 to one batch and radioactive phosphorus-32 to the other batch. Sulfur-35 and phosphorus-32 are radioactive isotopes. You will recall from Chapter 3 that radioactive isotopes can be used as tracers to follow the pathway of certain materials.

By adding the radioactive isotopes to the viruses, Hershey and Chase were "labeling" the viruses' protein and DNA. The protein was labeled with sulfur-35, and the DNA was labeled with phosphorus-32. This was an excellent strategy because proteins contain little or no phosphorus, whereas DNA does not contain sulfur. If sulfur-35 was found in the bacteria, it would mean that the viruses' protein was injected into the bacteria. If phosphorus-32 was found in the bacteria, then it was the DNA that had been injected.

Figure 7–4 In a set of simple experiments, Hershey and Chase prepared two separate samples of bacteriophages—one in which the DNA was labeled with radioactive phosphorus-32 and the other in which the protein coat was labeled with radioactive sulfur-35. As a result, Hershey and Chase discovered that the DNA of a bacteriophage is injected into a bacterium, whereas the bacteriophage's protein coat remains outside the bacterium.

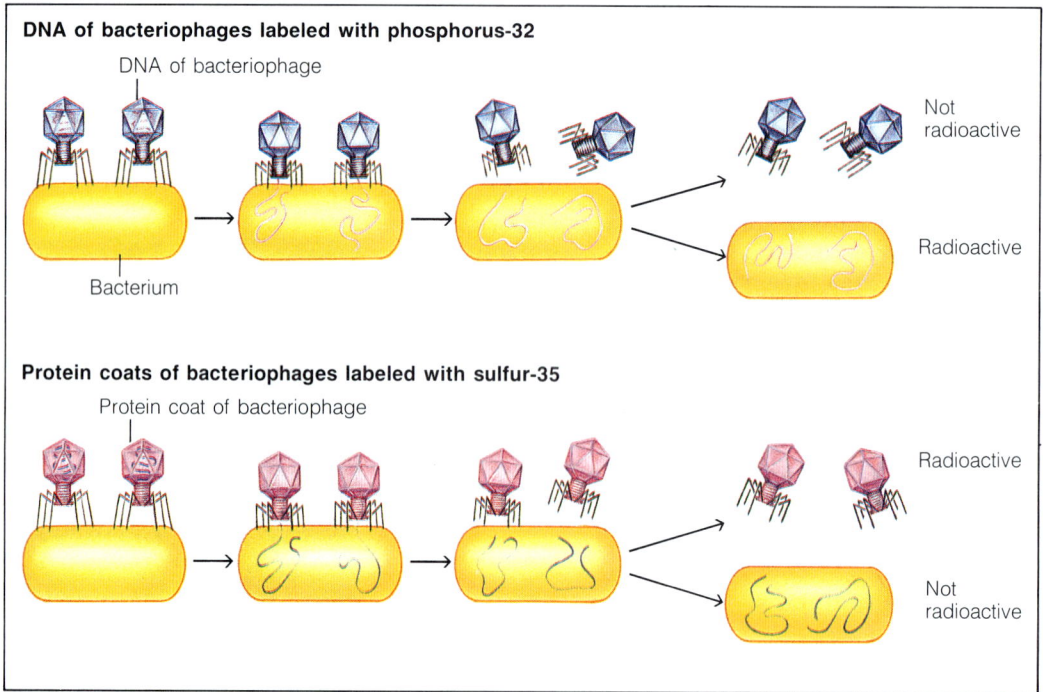

• **What were Hershey and Chase trying to learn?** (What part of the virus was being injected into the bacteria?)
• **Why were sulfur-35 and phosphorus-32 good choices of isotopes to use in the experiments?** (DNA contains no sulfur and proteins contain little or no phosphorus.)
• **What happened in the experiment?** (Very little sulfur entered the cell and most of the phosphorus went inside the bacterium.)
• **What did this prove?** (That it was DNA that was entering the bacterium.)
• **How is this experiment related to the experiment performed by Avery and his colleagues?** (Once again, it was confirmed that DNA contained the genetic code.)

The two scientists mixed the radioactively labeled viruses with the bacteria. They waited a few minutes for the viruses to attach and inject the genetic material. Then they separated the viruses from the bacteria by placing the mixture in a mechanical blender and then centrifuging the mixture.

Hershey and Chase discovered that nearly all the radioactive sulfur remained within the viruses. By contrast, nearly all the radioactive phosphorous had entered the bacteria!

From these results, it was clear that the viruses' DNA enters the bacteria, and the protein coats remain outside the bacteria. This was convincing evidence that DNA contains the genetic information. When considered along with the earlier experiments on the transformation of pneumonia bacteria, the Hershey-Chase experiments showed conclusively that DNA was the molecule that carried the genetic code.

The Structure of DNA

Even if DNA was shown to be the crucial molecule for the passing on of genetic information, a question of overwhelming importance remained. How could a molecule such as DNA store information and duplicate itself easily—two significant tasks? Let's take a look at the structure of DNA to see if that explains how it accomplishes these tasks.

As you will recall from Chapter 4, DNA is a polymer formed from units called **nucleotides**. Each nucleotide is a molecule made up of three basic parts: a 5-carbon sugar called deoxyribose (dee-ahks-ee-RIGH-bohz), a phosphate group, and a nitrogenous, or nitrogen-containing, base.

DNA contains four nitrogenous bases. Two of the nitrogenous bases, **adenine** (AD-uh-neen) and **guanine** (GWAH-neen), belong to a group of compounds known as purines (PYOOR-eenz). The remaining two, **cytosine** (SIGHT-oh-seen) and **thymine** (THIGH-meen), are known as pyrimidines (pih-RIHM-uh-deenz).

Individual nucleotides are joined together to form a long chain. Notice in Figure 7–10 on page 143 that the sugars and phosphate groups form the backbone of the chain, and the nitrogenous bases stick out from the chain.

X-RAY EVIDENCE In the early 1950s, the British scientist Rosalind Franklin turned her attention to the DNA molecule. She purified a large amount of DNA and then stretched the DNA fibers in a thin glass tube so that most of the strands were parallel. Then she aimed a narrow X-ray beam on them and recorded the pattern on film. When X-rays pass through matter, they are scattered, or diffracted. This X-ray scattering produces a pattern that provides important clues to the structure of many molecules.

Franklin worked hard to prepare better and better samples until the X-ray patterns became clear. The result of her work is

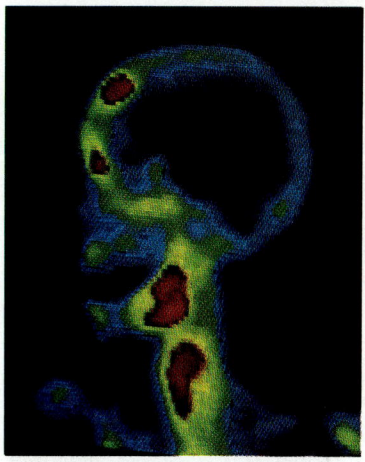

Figure 7–5 Radioactive isotopes are used not only to follow the pathway of certain materials, but also to detect the presence of some types of cancer. This photograph shows how a radioactive isotope becomes more concentrated in a cancerous part of the body, in this case in bone, than in a healthy part of the body. The cancerous part appears in red.

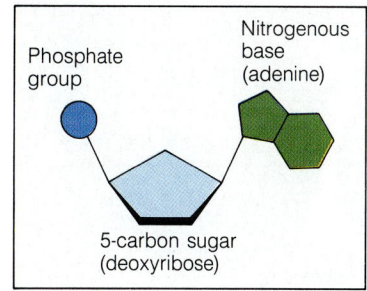

Figure 7–6 DNA is a polymer formed from units called nucleotides. A nucleotide is composed of three basic parts: a phosphate group, a 5-carbon sugar, and a nitrogenous base. What is DNA's 5-carbon sugar called? ❶

BACKGROUND INFORMATION
CHARGAFF'S RULE

The base-pairing mechanism also explained a piece of data that had been gathered by Erwin Chargaff. He had noticed that in any sample of DNA, the amounts of A and T bases were always the same, and the amounts of G and C were identical to each other. This relationship, which came to be known as Chargaff's rule, was true for DNA samples that varied widely in their overall base composition. Watson and Crick's base-pairing rules explained Chargaff's rule, and this was yet another reason why the model was so readily accepted.

MULTICULTURAL STRATEGY

The story of the discovery of the structure of DNA is one of competition and cooperation. Many researchers were competing to be the first to propose a viable structure for DNA. Although the British scientists James Watson, Sir Francis Crick, and Maurice Wilkins received the Nobel prize for their work, their achievements were dependent on the accumulated work of many scientists. Most scientific discoveries are the result of collaborative efforts. Scientists from different countries often work together and share their findings.

7–1 (continued)
Enrichment
Using the Visuals As you discuss the X-ray diffraction pattern, point out the various bits of evidence to be seen in Figure 7–8. The picture will probably be difficult for beginning students to decipher. Developing the analogy of crime scene clues might be helpful.

Content Development
Students might be interested to know that the discovery of DNA structure was a hot topic in biological research at the time that Franklin, Wilkins, Watson, and Crick were doing their work. Scientists all over the world were looking for answers. One model, though not a very good one, had already been proposed by Linus Pauling.

Throughout biological history there is evidence of this hot topic phenomenon.
• **Can you think of other instances when a large part of the scientific community worked or is working on solving a problem or making a discovery?** (Answers will vary but may include finding a cure for polio or AIDS or mapping the human genome.)

Content Development
Using the Visuals Until the time that Watson saw Franklin's X-ray pattern, he and Crick had

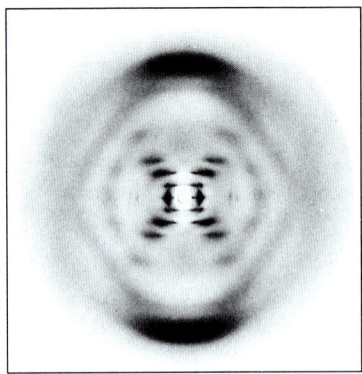

Figure 7–7 This X-ray diffraction photograph of DNA was taken by Rosalind Franklin in the early 1950s. The X-shaped pattern in the center indicates that the structure of DNA is helical.

Figure 7–8 This photograph of Watson and Crick with one of their first models of DNA was taken in 1953.

the X-ray pattern shown in Figure 7–7. This pattern does not prove anything in itself, but it does provide some very important clues about the structure of DNA. One important clue is that the fibers that make up DNA are twisted, like the strands of a rope. This is shown by the small X near the center of the pattern. The other important clue is that large groups of molecules in the fiber are spaced out at regular intervals along the length of the fiber.

Taken alone, neither of these facts was enough to determine the structure of DNA. The X-ray pattern was like a fingerprint or a scrap of cloth at the scene of a crime. Crucial evidence, perhaps, but it would only make sense when put into a larger picture by a clever detective—or an inventive scientist!

BUILDING A MODEL OF DNA At the same time that Franklin and her colleague, Maurice Wilkins, were doing their research, two young scientists working in Cambridge, England, were also trying to determine the structure of DNA. One of the scientists was Francis Crick, a British physicist. The other was James Watson, a 25-year-old American biochemist.

Watson and Crick had been trying to solve the mystery of DNA structure by building three-dimensional models of the atomic groups in DNA. They twisted and stretched the models in various ways to see if any of the structures that formed made any sense. Watson and Crick had some interesting ideas, but nothing to test them against.

Then, during a visit to London, Watson was able to observe Franklin's remarkable X-ray pattern of DNA. At once Watson and Crick realized that there was something important in that pattern. They immediately set out to use the clues that Franklin had provided. Within weeks, Watson and Crick had figured out the structure of DNA.

THE DOUBLE HELIX Working with these clues, Watson and Crick began, quite literally, to play with their models of the DNA fiber. What they needed to do was to twist their model into a shape that would account for Franklin's X-ray pattern. Before long, they developed a shape that seemed to make sense. They called this shape a helix because it was similar to a spiral, or the way in which the threads are arranged in a screw. Using Franklin's idea that there were probably two strands of DNA, Watson and Crick imagined that the strands were twisted around each other, forming a double helix.

Watson and Crick's model explained one more characteristic about DNA's structure. The nitrogenous bases on each of the strands of DNA are positioned exactly opposite each other. This positioning allows weak hydrogen bonds to form between the nitrogenous bases adenine (A) and thymine (T), and between cytosine (C) and guanine (G).

Another interesting piece of information that helped Watson and Crick to work out their model of DNA's structure was provided by Erwin Chargaff, an American biochemist.

142

Chargaff observed that in any sample of DNA, the number of adenine molecules was equal to the number of thymine molecules. The same was true for the number of cytosine and guanine molecules. Chargaff's observation enabled Watson and Crick to determine that adenine bonds only to thymine and cytosine bonds only to guanine. The two scientists further reasoned that the attraction of these bases for each other is very specific. The attraction between such bases is known as **base pairing**. Base pairing is the force that holds the two strands of the DNA double helix together. As you can see, Watson and Crick set out to solve one puzzle and found the answer to another puzzle!

In 1953, after making careful drawings of their model of DNA, Watson and Crick submitted their findings to a scientific journal. Their model, although speculative in some areas, was almost immediately accepted by scientists the world over. Why? Because, as we shall see, it contained a feature that explained a great mystery: how DNA could copy itself!

The importance of this work on DNA was acknowledged in 1962 by the awarding of the Nobel prize for medicine or physiology—the highest prize the international community can give for a scientific discovery—to its discoverers. Because Rosalind Franklin had died in 1958 and Nobel prizes are given only to living scientists, the prize was shared by Watson, Crick, and Franklin's associate, Maurice Wilkins.

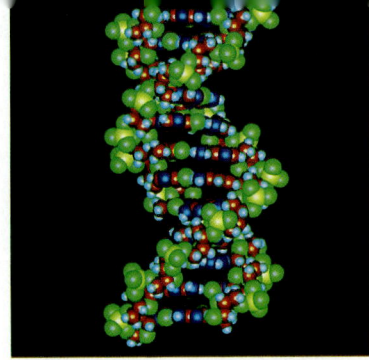

Figure 7–9 In this computer-generated model of DNA, the paired nitrogenous bases form the rungs, and the connecting sugar-phosphate molecules form the side rails.

Figure 7–10 The two strands of DNA are held together by hydrogen bonds. Because of the total number of hydrogen bonds that each can form with the other, adenine can pair only with thymine and cytosine can pair only with guanine.

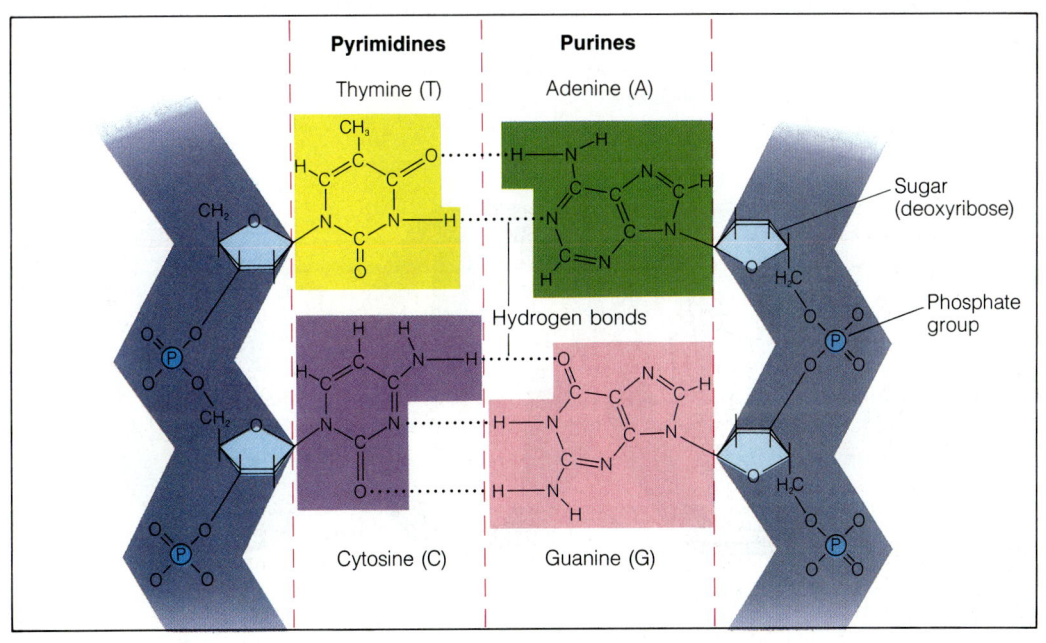

FACTS AND FIGURES

The models used by Watson and Crick were made from wooden balls and rubber balls connected by sticks and wires.

HISTORICAL NOTE
WATSON AND CRICK

When Watson and Crick were ready to announce their double helix model in 1953, they made drawings of DNA and sent their paper to *Nature* magazine. They ended their first paper by writing, "It has not escaped our notice that the specific pairing we have postulated immediately suggests a possible copying mechanism for the genetic material." Gunter Stent, a well-known molecular biologist, has written that this statement "can surely lay claim to being one of the most coy statements in the literature of science." Within a few weeks, Watson and Crick had written another paper describing the copying mechanism.

FACTS AND FIGURES

A toss of the coin determined that Watson would get top billing in the naming of the Watson-Crick model.

---

been primarily involved in making "ball and stick" models, accounting for the known biochemical structure—long chains of smaller molecules. It was only after they had seen the pattern that they twisted their model into the now well-known double helix. It was because of this important contribution that Franklin's colleague Maurice Wilkins shared in the Nobel prize.

Remind students that good models have to support all the known data. Point out that the Watson-Crick model of DNA structure was completely consistent with Erwin Chargaff's observation about base-pairing.

• **What if Chargaff's rule and the Watson-Crick model had been inconsistent?** (The model would have been invalidated because Chargaff's experimental evidence was irrefutable.)

Not only did the model answer most questions about the structure of DNA, it also provided important clues about the process of replication. Point out Figures 7–8 and 7–9.

ANNOTATION KEY

① Weak hydrogen bonds. (Relating concepts)

TEACHING SUPPORT

Study Guide
- Section 7–1, p. 67

MEDIA AND TECHNOLOGY
📽 **Biology Transparencies**
- 10: DNA Replication

7–1 (continued)

Content Development
Before you begin to discuss the process of replication, remind students where they would find DNA and why this duplication must occur. It will help them relate these events if you talk about chromosomes and mitosis before discussing the even more abstract concept of replication of DNA molecules.

The more visuals you can use, the more easily students will understand this concept. Emphasize that the entire process is really a chemical reaction controlled by enzymes.

Reinforcement/Reteaching
Give students several examples of "unzipping" and the formation of complementary strands. Allow them ample opportunity to practice base-pairing for themselves. One good way is to pair students, letting one student write the first strand of DNA and the other supply the answers forming the complementary strand. Monitor to be sure they understand the process.

Enrichment
Interested students might enjoy reading *The Double Helix* by James Watson. It is a lively account of Watson's frame of mind during the building of the double helix model. It is excellent, easy reading. Students might notice that Franklin's important contribution to the development of the model is minimized.

The Replication of DNA

Because each of the two strands of the DNA double helix has all the information, by the mechanism of base pairing, to reconstruct the other half, the strands are said to be complementary. To illustrate this concept, we can use the analogy of a dollar bill that has been torn in half. Each half of the dollar bill tells us just what the other half must look like, even though the two halves are not identical.

Although there are four nitrogenous bases in DNA, the situation is really quite similar to that of the dollar bill. As you can see from Figure 7–11, even in a long and complicated DNA molecule, each half can specifically direct the sequence of the other half by complementary base pairing. This allows us to imagine a very simple scheme for copying the double helix, as shown in Figure 7–12. In effect, each strand of the double helix of DNA serves as a template, or pattern, against which a new strand is made.

Before a cell divides, it must duplicate its DNA. This ensures that each resulting cell will have a complete set of DNA molecules. This copying process is known as **replication**. DNA replication, or DNA synthesis, is carried out by a series of enzymes. These enzymes separate, or "unzip," the two strands of the double helix, insert the appropriate bases, and produce covalent sugar-phosphate links to extend the growing DNA chains. The enzymes even "proofread" the bases that have been inserted to ensure that they are paired correctly.

As proposed by Watson and Crick, DNA replication begins

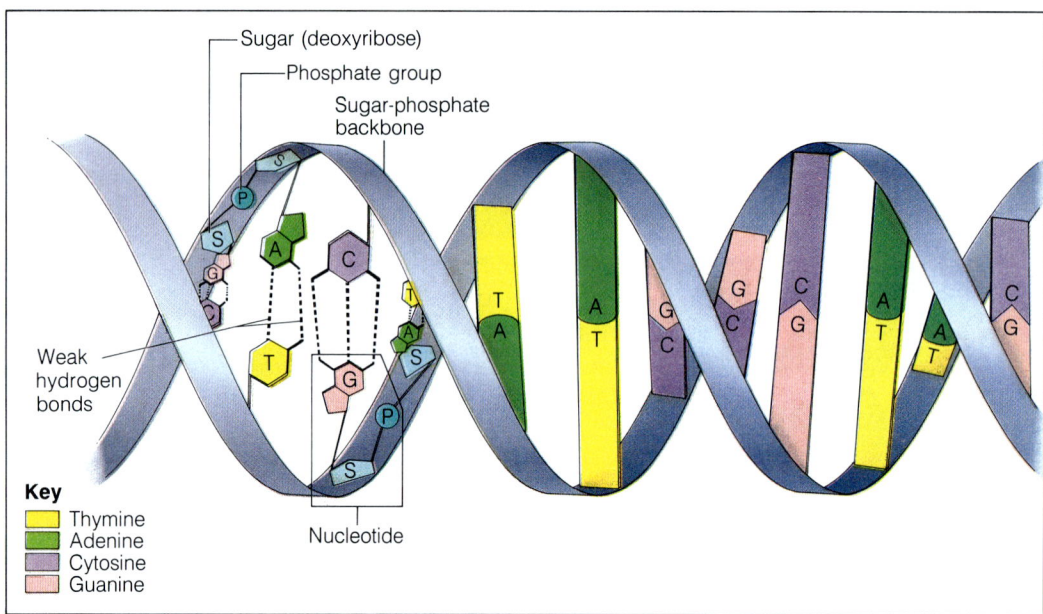

Figure 7–11 The structure of the DNA double helix resembles a twisted ladder. The sugar-phosphate backbones form the sides of the ladder and the nitrogenous bases form the rungs of the ladder. What holds the nitrogenous bases together? ①

144

SECTION REVIEW 7–1

1. Griffith's experiments indicated that the material from the heat-killed disease-causing bacteria had transformed the live harmless bacteria into the disease-causing type. Based on this information, Avery, MacLeod, and McCarty concluded that DNA is the carrier of genetic information.

2. Hershey and Chase tagged the DNA of some bacteriophages with radioactive phosphorus, and the protein coats of other bacteriophages with radioactive sulfur. The bacteriophages then were exposed to bacteria. Hershey and Chase

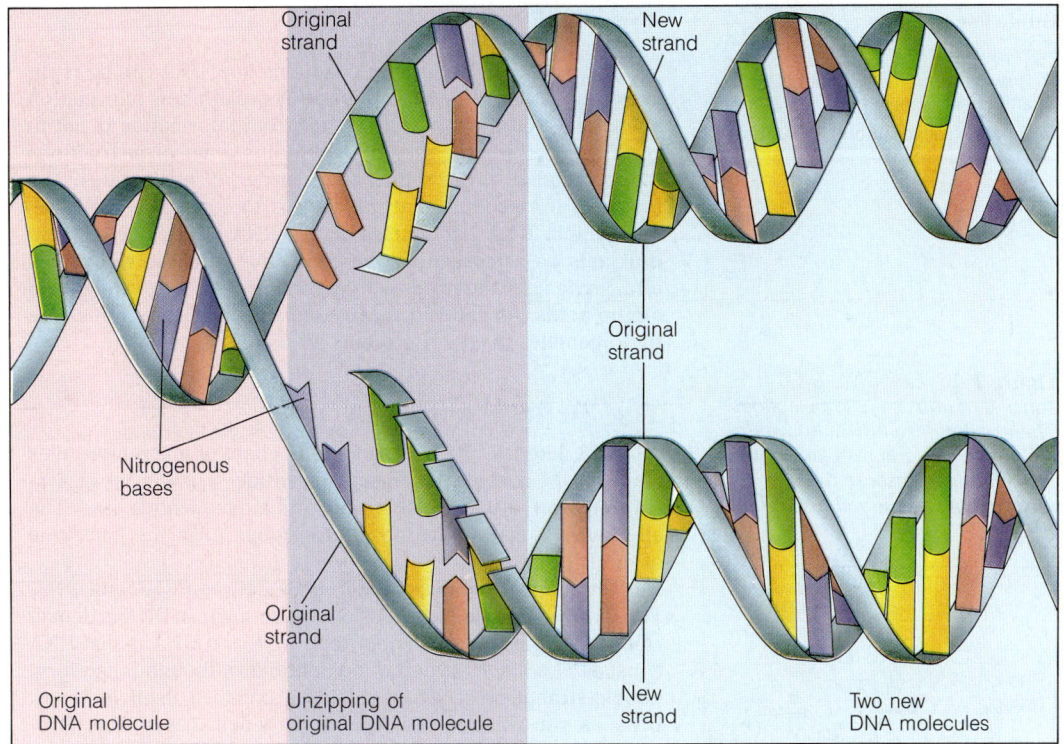

Figure 7–12 When DNA undergoes replication, each of the original strands unzips down the middle as the base pairs separate at the hydrogen bonds. Each of the original strands serves as the template along which a new complementary strand of DNA forms.

when a molecule of DNA "unzips." The unzipping occurs when the hydrogen bonds between the base pairs are broken and the two strands of the molecule unwind. Each of the separated strands serves as a template for the attachment of complementary bases. For example, a strand that has the bases T–A–C–G–T–T produces a strand with the complementary bases A–T–G–C–A–A. In this way, two DNA molecules identical to each other and to the original molecule are made.

7–1 SECTION REVIEW

1. Why was Griffith's work on transformation important in identifying DNA as the genetic material?
2. How did Hershey and Chase show that DNA was the molecule that carried the genetic code?
3. What is the function of DNA?
4. What are the parts of a DNA nucleotide?
5. Describe the shape of a DNA molecule.
6. **Critical Thinking—Making Inferences** Why is DNA replication necessary to life?

145

found that the bacteria cells contained large amounts of radioactive phosphorus but little sulfur. From the results of this experiment, the scientists concluded that the DNA of the bacteriophage enters the bacterial cell, and that the DNA carries the genetic material for making more bacteriophages.

3. DNA is the nucleic acid that stores and transmits the genetic information from one generation of an organism to the next.

4. A DNA nucleotide consists of a 5-carbon sugar called deoxyribose, a phosphate group, and a nitrogenous base.

5. The DNA molecule consists of two parallel strands of sugar-phosphate groups. Pairs of nitrogenous bases link the two strands together, forming a double helix.

BACKGROUND INFORMATION
GENETIC INTEGRITY

Many biologists have speculated as to why DNA and RNA differ the way they do. One might ask, for example, why a messenger DNA copy is not made to carry a nucleotide sequence to the ribosome. One reason may be that the chemical differences between DNA and RNA make it easier for enzymes, including DNA polymerase and RNA polymerase, to distinguish between the two molecules. The ability to distinguish DNA and RNA makes it far easier, biochemically, to protect DNA against environmental and physical damage. In short, the existence of two types of nucleic acid may be a device to ensure the integrity of genetic information.

FACTS AND FIGURES

Different types of body cells of the same organism contain the same amount of DNA, whereas sperm and egg cells, as you would suspect, contain about one half as much.

Guide For Reading
- What is the function of RNA?
- How do the structures of DNA and RNA compare?
- What is transcription?

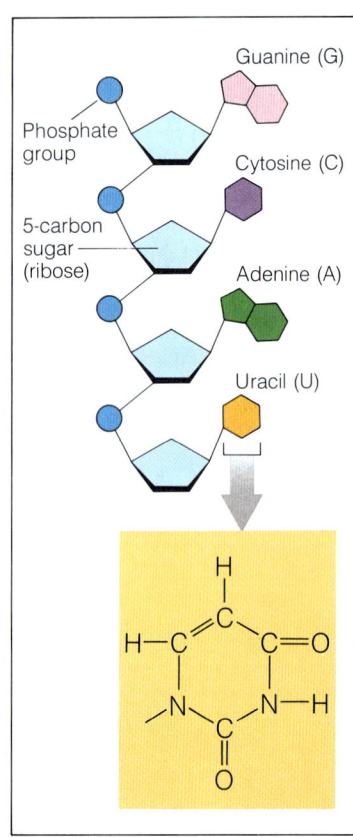

Figure 7–13 Like DNA, RNA consists of units called nucleotides. The nucleotides in RNA, however, contain the 5-carbon sugar ribose instead of deoxyribose and the nitrogenous base uracil instead of thymine.

7–2 RNA

The double helix structure explains how DNA can be replicated, or copied. However, it does not explain how information is contained in the molecule or how that information is put to good use. As we will see, DNA contains a set of instructions that are coded in the sequence, or order, of nucleotides. The first step in decoding that message is to copy part of the sequence into **RNA** (ribonucleic acid). **RNA is the nucleic acid that acts as a messenger between DNA and the ribosomes and carries out the process by which proteins are made from amino acids.** As you will recall from Chapter 5, ribosomes are the organelles in which proteins are made.

The Structure of RNA

RNA, like DNA, consists of a long chain of macromolecules made up of nucleotides. Each nucleotide is made up of a 5-carbon sugar, a phosphate group, and a nitrogenous base. The alternating sugars and phosphate groups form the backbone of the RNA chain.

There are three major differences between RNA and DNA. The sugar in RNA is ribose, whereas the sugar in DNA is deoxyribose. Another difference between RNA and DNA is that RNA consists of a single strand of nucleotides, although it can form double-stranded sections by folding back on itself in loops. DNA, as you will recall, is double-stranded. Lastly, the nitrogenous bases found in DNA are adenine, thymine, cytosine, and guanine. RNA also contains adenine, cytosine, and guanine, but **uracil** (YOOR-uh-sihl) is present instead of thymine. Like DNA, RNA follows the base-pairing rules. Adenine bonds to uracil, and cytosine bonds to guanine.

Although a cell contains many different forms of RNA, there are three main types that are involved in expressing the genetic code. Each of the three main types of RNA will be discussed later in this section.

In its own way, an RNA molecule is a disposable copy of a segment of DNA. The ability to copy a DNA base sequence into RNA makes it possible for a specific place on the DNA molecule to produce hundreds or even thousands of RNA molecules with the same information as DNA.

Transcription: RNA Synthesis

As you will recall, DNA replication is also known as DNA synthesis because the molecule being synthesized turns out to be the same as the molecule being copied. In RNA synthesis, the molecule being copied is just one of the two strands of a DNA molecule. Thus the molecule being synthesized is different from the molecule being copied. The term **transcription** is used to describe this process. **Transcription is the process by**

TEACHING STRATEGY 7–2

Focus/Motivation

Remind students of the activity on breaking codes used to introduce Section 7–1. Prompt them to think about the relationship between DNA and RNA by asking questions like these.
- If you were given a simple coded message, do you think you could eventually break the code? (Probably so.)
- Would everyone in the class break the code or interpret it in the same way? (Probably not; but if everyone succeeded in breaking the code, they would probably interpret the message in the same way.)
- Would you agree that the code "works" only with the proper interpreter? (Some discussion, but the answer is yes.)

Explain to students that RNA is the interpreter of the DNA code. That brings up a good question. Why does the cell need a messenger? Why not just use DNA itself? There are two probable reasons. One is probably the cell's way of safeguarding its original copy of DNA. The other probably has to do with the number of copies needed to produce the many molecules of protein needed by the cell.

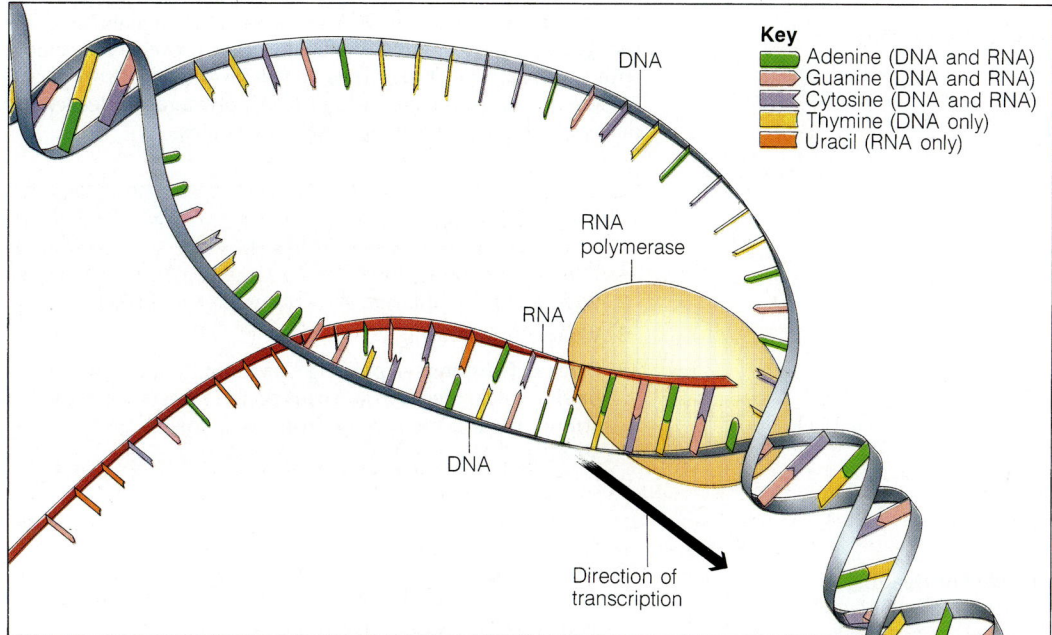

Figure 7–14 During transcription, the enzyme RNA polymerase attaches to an area on the DNA molecule, causing it to open up. As the RNA polymerase moves along the DNA molecule, the two strands of DNA separate. Then RNA nucleotides link to the complementary bases on DNA. Which of RNA's nitrogenous bases pairs with adenine in DNA? ❶

which a molecule of DNA is copied into a complementary strand of RNA. In other words, transcription is the process of transferring information from DNA to RNA.

Why is it necessary for DNA to transfer its genetic information to RNA? Recall from Chapter 5 that DNA is found in the nucleus and ribosomes are located in the cytoplasm. Because DNA does not leave the nucleus, a messenger, or carrier, must bring the genetic information from the DNA in the nucleus out to the ribosomes in the cytoplasm. The molecule that performs this function is **messenger RNA** (mRNA), one of the three main types of RNA.

In order to more fully understand how transcription takes place, we must discuss the role that an enzyme known as RNA polymerase (PAHL-ih-mer-ayz) plays in this process. In Chapter 4 you learned that an enzyme is specific, that is, it works on only one substance. For this reason, part of its name is usually derived from the substance on which it works. Then the suffix -*ase* is added. Thus RNA polymerase works on the polymers RNA and DNA.

During transcription, RNA polymerase attaches to special places on the DNA molecule, separates the two strands of the double helix, and synthesizes a messenger RNA strand. The messenger RNA strand is complementary to one of the DNA strands. The base-pairing mechanism ensures that the messenger RNA will be a complementary copy of the DNA strand that serves as its template.

ANNOTATION KEY

❶ Uracil. (Relating facts)

TEACHING SUPPORT

Teaching Resources
- Activity: The Role of RNA, p. 3

Study Guide
- Section 7–2, p. 70

Content Development
You will want to emphasize the structural similarities and differences between RNA and DNA. The only difference that seems to bother students is the substitution of uracil for thymine. You might want to tell them that uracil and thymine are very similar purine bases, so either one can pair with adenine.

Reinforcement/Reteaching
At this point have the students do some practice with base pairing DNA to RNA, emphasizing the uracil–adenine pair. Remind students that adenine pairs with thymine in DNA and with uracil in RNA.

Content Development
Using the Visuals Refer students to Figure 7–14 (and you may want to turn back to Figure 7–12) as you discuss the process of transcription. Be sure to emphasize the important role of RNA polymerase and its specificity.

One question that students frequently ask about RNA transcription from the double strand of DNA is whether or not both sides of the DNA are copied. The answer is that only one strand appears to be transcribed, with the other being nonsense, serving only as the complement to the template.

ANNOTATION KEY

① UAA, UAG, UGA
(Interpreting charts)

BACKGROUND INFORMATION
AMINO ACIDS

The genetic code in beginning biology books is usually presented as the 20 amino acids. Actually, in the chemical sense there is an infinite number of possible amino acids, and more than 40 amino acids are known in nature. Furthermore, one of the 20 amino acids, proline, is not an amino acid at all. Actually, it's an imino acid. Some of the most common amino acids in the body, such as the hydroxypyroline found in collagen, are not included in the 20 amino acids listed in the code table. Therefore, it's important to emphasize that the 20 in the code do not exhaust the chemical possibilities of amino acids!

7–2 (continued)

SECTION REVIEW 7–2

1. RNA is the nucleic acid that acts as a messenger between DNA and the ribosomes and carries out the process by which proteins are made from amino acids.
2. RNA is single-stranded, and DNA is double-stranded. In RNA, uracil replaces thymine, which is found in DNA. RNA contains the sugar, ribose, and DNA contains deoxyribose.
3. Transcription is the process by which a molecule of DNA is copied into a complementary strand of RNA.
4. U-G-G-C-A-G-U-G; A-G-C-G-U-G-C-A.

Reinforcement/Reteaching

For those students who have difficulty with the Section Review questions, go back to the discussion in the chapter and review the material students have had difficulty understanding.

Closure

Prepare a few questions on index cards to cover the important points in the section. Distribute the cards to members of the class. Have them read the question and give the answer. One way to randomly distribute the cards would be to make a set composed of question cards and blanks, with one for each member of the class. You may have a method you prefer. The purpose is to review the main points of the chapter.

Special sequences in DNA serve as "start signals" and are recognized by RNA polymerase and other proteins associated with transcription. Other areas on the DNA molecule are recognized as termination sites where RNA polymerase releases the newly synthesized messenger RNA molecules.

7–2 SECTION REVIEW

1. What is the function of RNA?
2. What are three differences between RNA and DNA?
3. What is transcription?
4. **Critical Thinking—Applying Concepts** Using the base-pairing rules, identify the bases on the messenger RNA strand that are transcribed from the following DNA strands: A–C–C–G–T–C–A–C; T–C–G–C–A–C–G–T.

Guide For Reading

- What is a codon?
- What is translation?
- What are the three main types of RNA?

Figure 7–15 If a single nucleotide coded for one amino acid, only four amino acids could be specified (top). If two nucleotides specified for one amino acid, there could be 16 amino acids (bottom)—still not enough to code for all 20 amino acids contained in proteins.

Four Code Letters	
A	G
C	U

Sixteen Doublets from the Four Code Letters			
AA	AC	AG	AU
CA	CC	CG	CU
GA	GC	GG	GU
UA	UC	UG	UU

7–3 Protein Synthesis

The information that DNA transfers to messenger RNA is in the form of a code. This code is determined by the way in which the four nitrogenous bases are arranged in DNA. To understand how the code works, we must answer two questions: What kind of information is contained in DNA, and how is that information decoded?

The nitrogenous bases in DNA contain information that directs protein synthesis. Why proteins and not other molecules? you might ask. The answer can be found in the diversity of things that proteins are capable of doing. Because most enzymes are proteins, proteins control biochemical pathways within the cell. Not only do proteins direct the synthesis of lipids, carbohydrates, and nucleotides, but they are also responsible for cell structure and cell movement. Like the manager of a factory, DNA does not work on the assembly line but can control what the cell factory makes by issuing orders to the organelles (workers). Together, DNA and its assistant, RNA, are directly responsible for making proteins. As you can see, DNA and RNA are like nucleic-acid executives who run the entire cell factory.

The Nature of the Genetic Code

As you will recall from Chapter 4, proteins are made by stringing amino acids together to form long chains called polypeptides. Each polypeptide contains a combination of any or all of the 20 different amino acids. How, then, can a particular order of nitrogenous bases in DNA and RNA be translated into a particular order of amino acids in a polypeptide?

TEACHING STRATEGY 7–3

Focus/Motivation

Begin your discussion of this section by talking about the following scenario.

The principal of the school, Mrs. Smith, would like the

As you know, DNA and RNA each contain different nitrogenous bases (DNA contains A, T, C, G; RNA contains A, U, C, G); hence, different nucleotides. For this reason, the genetic code must have a four-letter "alphabet." In order to code for the 20 different amino acids, more than one nucleotide must make up the code word for each amino acid. If code words were two nucleotides long, there would be 4^2, or 16, different code words. This is not enough for 20 amino acids. The four nucleotides arranged in triplets, or threes, however, produce 4^3, or 64, different code words. This is more than enough to produce a different code word for each amino acid. Therefore, the smallest size for a code word in DNA is three nucleotides.

The code words of the DNA nucleotides are copied onto a strand of messenger RNA. Each combination of three nucleotides on the messenger RNA is called a **codon** (KOH-dahn), or three-letter code word. **Each codon specifies a particular amino acid that is to be placed in the polypeptide chain.** See Figure 7–17. It is interesting to note that there is more than one codon for each amino acid. For example, the amino acid leucine (LOO-seen) has six different codons. There is also one codon, AUG, that can either specify the amino acid methionine (muh-THIGH-uh-neen) or serve as a starter for the synthesis of a protein. For this reason, AUG is called an "initiator" codon. Notice also that there are three "stop" codons. They do not code for an amino acid. Instead, these codons act like the period at the end of a sentence: They signify the end of a polypeptide.

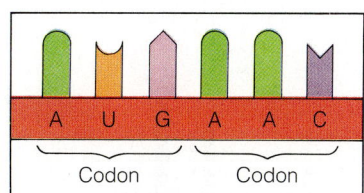

Figure 7–16 Each combination of three nucleotides on messenger RNA is called a codon, or three-letter code word.

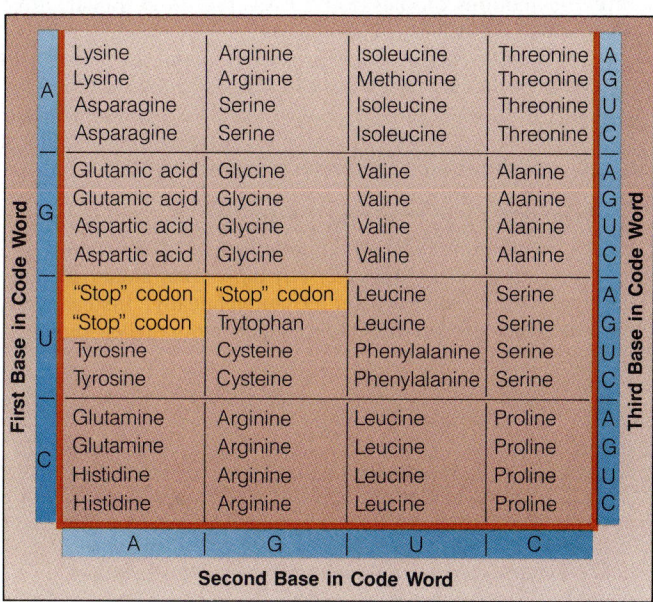

Figure 7–17 The genetic code consists of 64 codons along with their corresponding amino acids. These codons are found on messenger RNA. Of the 64 codons, 61 specify a particular amino acid. The other three are stop codons, which signify the end of a polypeptide chain. What are the three-letter code words for each of the stop codons?

BACKGROUND INFORMATION
tRNA

There is an important little fact that is almost always concealed from beginning students at the high school and even the college level because instructors feel that it complicates matters. You might be interested in knowing this fact and challenging your best students with it. From the genetic code table, we should expect that each cell has 61 different types of tRNAs, one for each of the 61 working codons—right? Wrong! Most cells produce somewhere between 22 and 30 kinds of tRNA. How can they get by with less than 61? The answer is found in the fact that the pairing between codon and anticodon is not quite as perfect as we generally suggest in an introductory book like this one. In fact, the codon and anticodon do not pair very well in the third base, producing an unevenness that molecular biologists call "wobble." This allows a single tRNA molecule to read several codons that differ in the third base.

The genetic code table shows the results of this mismatching. Challenge your students to find out which of the three bases in a codon matter the least in determining which tRNA binds. A careful check of the code table will show that the third base is the least important.

section, they should be able to easily complete the analogy.

Content Development

"Why proteins?" is an interesting discussion topic for you and your students. You can help them to better understand by talking about the many functions of proteins in the cell and by relating protein synthesis to heredity.

teacher Mr. Jones to write a letter to the parents of his students announcing the upcoming Open House. Mrs. Smith asks her secretary, Miss Brown, to take the message to Mr. Jones. Mr. Jones agrees to write the letter, but he lacks certain pieces of information. He sends John to find out the date of the Open House. Susan is to learn about the program for the evening. Eric is asked to get a list of the parents' names and addresses. Once he has all of the necessary information, Mr. Jones writes the letter, and it is sent to all the parents.

Now that they are somewhat familiar with the DNA and RNA codes and their functions, students can begin to label the people in the scenario as being representative of DNA, RNA, and so on. As you proceed through this

TEACHING SUPPORT

Laboratory Manual
- Assembling a Protein Molecule, p.113

Teaching Resources
- Activity: Protein Synthesis, p. 7

MEDIA AND TECHNOLOGY
📺 Biology Transparencies
- 11: Protein Synthesis

BACKGROUND INFORMATION
tRNA-SYNTHETASES

Some students will quickly notice that once the amino acid bound to a tRNA is used in protein synthesis, something must then attach a new amino acid to the tRNA. That something is one of a series of enzymes called tRNA-synthetases. These enzymes attach the correct amino acid to each form of tRNA, and they are absolutely essential to protein synthesis. In effect, it is the tRNA-synthetases that "know" the genetic code, since the amino acid that they attach to each form of tRNA ultimately determines which amino acid will be inserted for a particular codon.

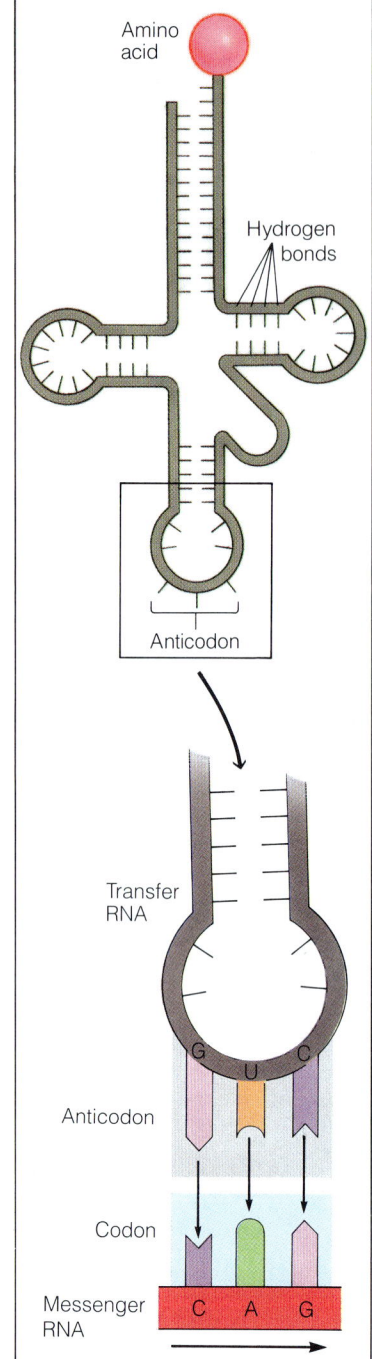

Translation

How does a messenger RNA molecule actually produce a polypeptide? The decoding of a messenger RNA message into a polypeptide chain (protein) is known as **translation**. Translation is an appropriate word for this process because it emphasizes that the message is being translated from the language of nucleic acids into a polypeptide. The messenger RNA molecule does not produce a polypeptide by itself. Instead, there is an elaborate mechanism that involves the two other main types of RNA—**transfer RNA** (tRNA) and **ribosomal RNA** (rRNA)—and the cytoplasmic organelle known as the ribosome.

Transfer RNA carries amino acids to the ribosomes, where the amino acids are joined together to form polypeptides. Transfer RNA is a single strand of RNA that loops back on itself. See Figure 7–18. There are different transfer RNA molecules for each of the 20 amino acids. Ribosomal RNA makes up the major part of the ribosomes.

THE ROLE OF TRANSFER RNA In order to translate the information from a single codon of messenger RNA, such as AUG, we would have to find out which amino acid is coded for by AUG. Look at Figure 7–17 on page 149. As you just read, the codon AUG codes for the amino acid methionine. Methionine is then brought to the polypeptide chain by transfer RNA.

Look at Figure 7–18. You will notice that there are three exposed bases on each transfer RNA molecule. These nucleotides will base pair with a codon on messenger RNA. Because the three nucleotides on transfer RNA are complementary to the three nucleotides on messenger RNA, the three transfer RNA nucleotides are called the **anticodon**.

Attached to each transfer RNA molecule is the amino acid specified by the codon to which it base pairs. By matching the transfer RNA anticodon to the messenger RNA codon, the correct amino acid is put into place. Each transfer RNA acts like a tiny beacon for its specific amino acid.

THE ROLE OF THE RIBOSOME Messenger RNA molecules do not automatically line up transfer RNA molecules and link their amino acids together any more than model airplane parts glue themselves together automatically. Instead, this process of protein synthesis takes place in organelles known as

Figure 7–18 Transfer RNA is a single strand of RNA that loops back on itself. Each transfer RNA molecule has two important sites of attachment. One site, called the anticodon, binds to the codon on the messenger RNA molecule. The other site attaches to a particular amino acid. During protein synthesis, the anticodon of a transfer RNA molecule base pairs with the appropriate messenger RNA codon.

7–3 (continued)

Content Development
DNA replication and RNA transcription are both processes that students seem to have little trouble understanding. With translation, however, students may become confused. This is a good time for you (or the students) to start illustrating the step-by-step process on the board, overhead, or on paper. The more visual and/or manipulative you can make this section, the better students will understand the concept.

Enrichment
Interested students will enjoy making a more elaborate display of process of protein synthesis from DNA to polypeptide. Encourage them to use their imagination—try unique materials or create a game.

Reinforcement/Reteaching
This is a wonderful time for peer tutoring. Pair the students (you might want to give a little thought to this) and have them explain protein synthesis to one another. Encourage them to critique their partners carefully. Monitor each

SCIENCE, TECHNOLOGY, AND SOCIETY
Breakthrough

The Ribozyme Revolution

All enzymes, the biological catalysts that speed up chemical reactions, are proteins. At least that's what biologists had thought until a little more than 10 years ago.

In 1981, Arthur Zaug and Paula Grabowski were graduate students in chemistry at the University of Colorado, working in the laboratory of Thomas Cech. The Cech laboratory was interested in working out the mechanism by which RNA molecules were processed (cut and spliced) into their final form. Naturally, they suspected that the processing activities were carried out by enzymes. Thus the young students found themselves working away at the lab bench, purifying an extract that was enriched in the RNA-processing reactions. They expected to find a protein in the extracts.

To their surprise, the most active RNA-processing extracts contained no protein at all! This finding seemed to make no sense. If RNA processing was going on at such a high rate, where were the processing enzymes? Gradually, the young scientists reached an unavoidable conclusion: The processing activity was being carried out by RNA itself. RNA could catalyze a chemical reaction. The enzyme was RNA!

At first, the self-processing RNA appeared to be a chemical curiosity—a strange exception to the rule that all enzymes are proteins. But

 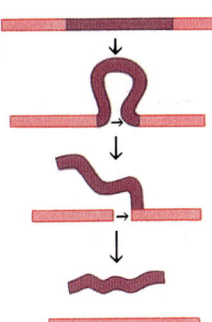

Paula Grabowski is a member of the team that discovered ribozymes. Self-splicing is one reaction catalyzed by RNA.

soon one biologist after another discovered RNA molecules that could catalyze other chemical reactions. As this book is being written, so many examples of RNA catalysis are known that the term "ribozyme" has been proposed to describe an RNA molecule that acts as an enzyme.

As a result of their careful work, Zaug, Grabowski, and Cech have shaken up one of the basic assumptions of biochemistry. In doing so, they have provided a whole new world for biologists to explore—the world of RNA.

---

ribosomes. Ribosomes are made up of two subunits, a large one and a smaller one. Each subunit consists of ribosomal RNA and proteins (about 70 different types).

The first part of protein synthesis occurs when the two subunits of the ribosome bind to a molecule of messenger RNA. Then the initiator codon AUG binds to the first anticodon of transfer RNA, signaling the beginning of a polypeptide chain. Soon the anticodon of another transfer RNA binds to the next messenger RNA codon. This second transfer RNA carries the second amino acid that will be placed into the chain of the polypeptide.

As each anticodon and codon bind together, a peptide bond forms between the two amino acids. You will recall from

---

group to ensure that accurate information is being shared.

Reinforcement/Reteaching

This chapter has so many terms that sound alike and actually have similar meanings that you will probably need to do a vocabulary lesson at this point for some of your students. Flash cards always work well for this.

Content Development

As you begin to emphasize the role of the ribosomes, be sure that students understand that protein synthesis could not occur at all without them. Refer back to drawings or models of the cell so that students can recall previously learned information about ribosomes.

SCIENCE, TECHNOLOGY, AND SOCIETY
BREAKTHROUGH

The Ribozyme Revolution
Perhaps the greatest strength of science is its ability to change when evidence shows that change is needed. For many years, scientists assumed that all enzymes—chemicals that alter the speed of chemical reactions—were proteins. The work of Arthur Zaug and Paula Grabowski working in the laboratory of Thomas Altman at the University of Colorado proved otherwise.

In 1981 these scientists found that RNA itself could catalyze a chemical reaction. Soon other scientists found still more reactions that are catalyzed by RNA. Today many RNA catalysts are known.

Although not covered in this feature, students might like to learn of an additional line of study that has been pursued since the discovery of RNA catalysts. Some scientists hypothesize that RNA is the molecule that was responsible for the origin of life on Earth. This important sequence of evolutionary events might have started with the development of RNA molecules that were able to catalyze their own synthesis. Next, RNA molecules that formed the first lipidlike molecules and then RNA molecules that were able to synthesize protein molecules evolved. Finally, DNA appeared. Because it is more stable than RNA, DNA could serve as a storehouse of genetic information. Although the job of storing information was taken up by DNA, the work of carrying out the directions contained in DNA continued to be performed by RNA.

- **What group of compounds did scientists believe acted as catalysts?** (Proteins)
- **What happened to change this assumption?** (The work of Zaug and Grabowski showed that RNA molecules could also catalyze reactions.)
- **What are this new group of catalysts called?** (Ribozymes)

ANNOTATION KEY

① Ribosome. (Relating facts)

ESL STRATEGY

Have students underline their choices:

1. Adenine, guanine, cytosine, and (uracil, thymine) are the four nitrogenous bases of DNA.
2. Before a cell divides, (reparation, replication) occurs in the (RNA, DNA). This important process is known as (RNA, DNA) synthesis.
3. (DNA, RNA) is found only in the nucleus of a cell.
4. (Translation, transformation, transcription) occurs when information is passed from DNA to an RNA messenger.
5. (Translation, transformation, transcription) leads to (polymer, protein) synthesis as the RNA messenger carries the coded message from inside the nucleus outside into the cytoplasm.

BACKGROUND INFORMATION
MITOCHONDRIAL DNA

Until a few years ago the genetic code was thought to be universal; the same code that applied to human cells also applied to bacteria and viruses. In the early 1980s, however, an important exception was discovered. Mitochondria, the energy-transducing organelles, which have their own DNA and synthesize few of their own proteins, have some minor differences in their genetic code. Four codons in mitochondria differ from most other organisms. Because of this interesting exception, it is best said that the code is nearly universal.

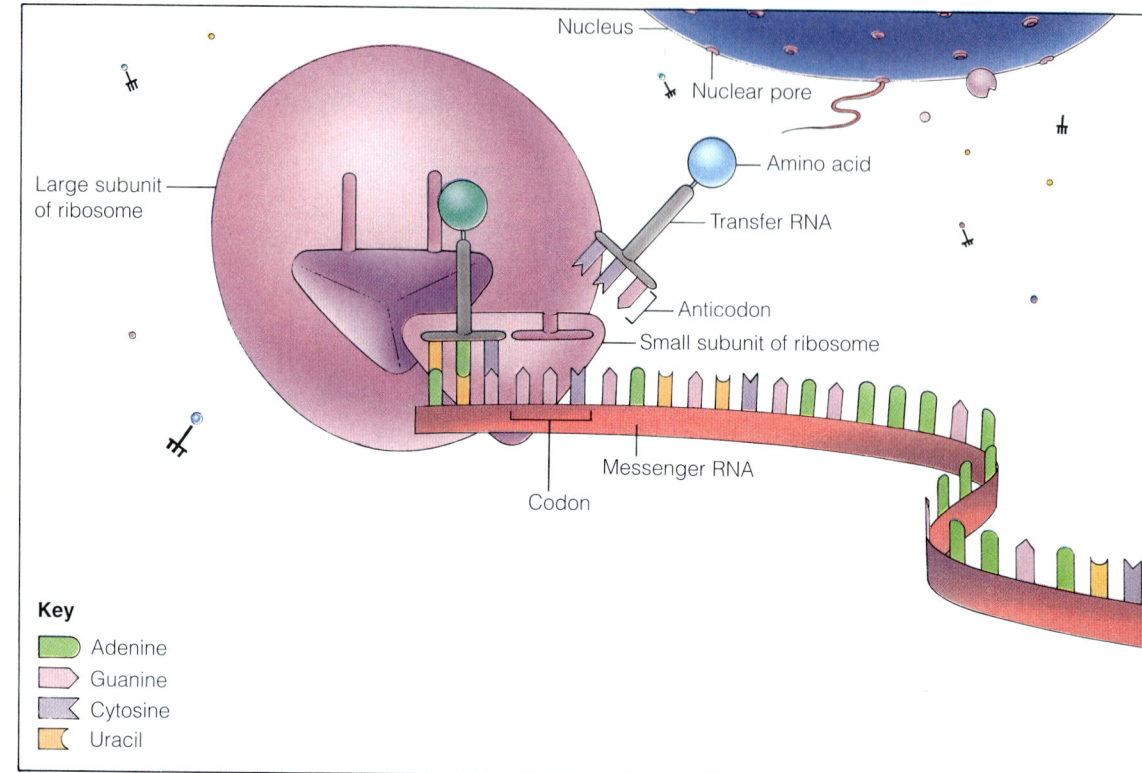

Figure 7–19 During translation, the language of the nucleic acids (DNA and RNA) is transferred to the language of the amino acids. In which cellular organelle does this process take place? ①

Chapter 4 that a peptide bond is the covalent bond that joins two amino acids together. The polypeptide chain continues to grow until the ribosome reaches a stop codon on the messenger RNA. A stop codon is a codon for which no transfer RNA molecules exist. When the stop codon reaches the ribosome, the ribosome releases the newly formed polypeptide and messenger RNA, completing the process of translation.

As you have seen, DNA directs the synthesis of three different kinds of RNA: transfer RNA, ribosomal RNA, and messenger RNA. Each kind of RNA helps to make the genetic code work. To better understand the different roles of DNA and RNA in directing protein synthesis, you might find it helpful to consider the two types of plans used by builders. A master plan has all the information needed to construct a building. But builders never bring the valuable master plan to the building site, where it might be damaged or lost. Instead, they prepare inexpensive, disposable copies of the master plan called blueprints. The master plan is safely stored in an office, and the blueprints are taken to the job site. Similarly, the cell prepares RNA "blueprints" of the vital DNA "master plan." DNA remains in the safety of the nucleus, while RNA goes to the protein-building sites in the cytoplasm (the ribosomes).

7–3 (continued)

Content Development
Using the Visuals As you talk about the ribosome as the site of the RNA "reunion," review the functions of DNA and the various types of RNA one more time using Figure 7–19. You will also want to review peptide bonding. Peptide bonding will be an easier concept for students now that they can see an application. If you have access to atomic models, put together three of four models of different amino acids. Demonstrate the removal of water molecules during the bonding process.

SECTION REVIEW 7–3

1. A codon is a sequence of three nucleotides on mRNA. An anticodon is a sequence of three nucleotides on tRNA.
2. The polypeptide chain would continue to grow until the ribosome reached a stop codon on the mRNA.
3. Translation is the decoding of a mRNA message into a polypeptide chain (protein).

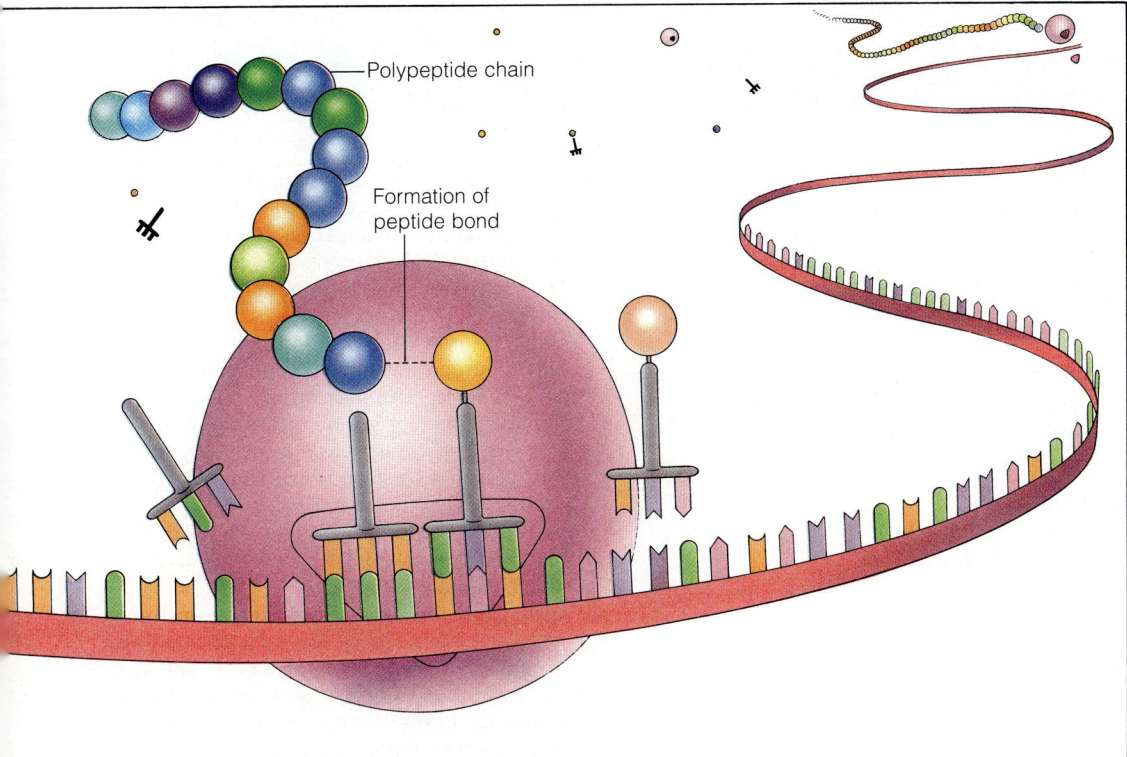

7-3 SECTION REVIEW

1. What is a codon? An anticodon?
2. UAG is a stop codon. What might happen if the uracil in this codon was changed to cytosine?
3. What is translation? Transcription?
4. In your own words, summarize the process of protein synthesis. (Be sure to describe the functions of all the nucleic acids and cell structures involved.)
5. What is the role of transfer RNA in translation?
6. If a code on a DNA molecule is CTA for a specific amino acid, what is the messenger RNA codon? The transfer RNA anticodon?
7. **Critical Thinking—Making Calculations** A certain protein is made up of 100 amino acids. What is the smallest number of bases in the messenger RNA molecule that is needed to carry the code for the synthesis of this protein?

Transcription is the process by which the DNA message is copied into RNA.
4. During translation, messenger RNA binds to the ribosomes on which ribosomal RNA is found. Amino acids in the cytoplasm are picked up by transfer RNA and are carried to messenger RNA. The anticodons in transfer RNA attach to the proper codons in messenger RNA. Thus the messenger RNA acts as the pattern for protein synthesis. In this way, amino acids are brought together in the correct sequence to form a protein molecule.
5. Transfer RNA picks up individual amino acids in the cytoplasm and brings them to the ribosomes.
6. GAU; CUA.
7. One start codon containing 3 bases equals 3. Three bases per amino acid equals 3 times 100, or 300. One stop codon containing 3 bases equals 3. The total equals 306.

LABORATORY INVESTIGATION

CONSTRUCTING A MODEL OF DNA REPLICATION

Before the Lab
There are several things you can do ahead of time to make the lab go more smoothly.
1. Be sure that you have adequate supplies.
2. Individual tape dispensers for each group will help prevent confusion and bottlenecks.
3. If you are using large sheets or roll paper, cut it into pieces of an appropriate size for each group.
4. If this lab is going to carry over to another class period, provide folders or envelopes for pieces.

Keep in mind that complementary bases are different values of the same hue to make them easier for students to remember. The only exception is uracil, which is a different color. If these colors are not available, select alternatives following the same logic.

Pre-Lab Discussion
Review the structure of DNA and the rules for base pairing. Emphasize the regularity of the structure and the importance of making the model carefully to reflect that feature. Review the process of DNA replication. Refer to the chart and go over colors, dimensions, and numbers. Finally, emphasize the importance of following the directions step by step.

Skills Development
Students will use the following skills while completing this investigation.
1. Observing
2. Manipulative
3. Comparing
4. Relating
5. Applying
6. Safety

LABORATORY INVESTIGATION

CONSTRUCTING A MODEL OF DNA REPLICATION

PROBLEM
How does DNA replicate?

MATERIALS (per group)

colored construction paper:
- white
- red
- dark blue
- light blue
- dark green
- light green

metric ruler
scissors
transparent tape

PROCEDURE

1. With the scissors, cut out the pieces of colored construction paper as indicated in the chart. **CAUTION:** *Be careful when using scissors.* Write the name of each part of your DNA model on the appropriate piece of construction paper.
2. To construct a model of a DNA nucleotide, tape a phosphate group, a sugar, and a guanine (G) together in the same manner as shown in Figure 7–6 on page 141. Then assemble eight additional nucleotide models with the following nitrogenous bases: T, T, A, C, A, A, T, C.

Part of DNA Model	Color	Size (cm x cm)	Number
Sugar	White	2 x 2	54
Phosphate group	Red	1 x 2	54
Adenine (A)	Dark blue	1 x 2	18
Thymine (T)	Light blue	1 x 2	18
Guanine (G)	Dark green	1 x 2	9
Cytosine (C)	Light green	1 x 2	9

154

3. Attach the nine nucleotide models together in the sequence given in step 2 (G, T, T, A, C, A, A, T, C) by taping the sugar on one nucleotide to the phosphate group on the next nucleotide. This will form the single strand of your DNA model.
4. Using the rules for base pairing, construct a second strand of the DNA model that is complementary to the first strand. Tape the nucleotides together as you did in step 3. Do not tape the two strands of DNA together.
5. Separate the two strands and use each to construct a new complementary strand.
6. Reassemble the original double strand of DNA. Tape this model of DNA together at its nitrogenous bases.
7. Again using the base-pairing rules, construct a complementary strand for each of the single strands of DNA you made in step 5. Tape these two new DNA models together at their nitrogenous bases. Compare the new double-stranded DNA models with your original DNA model.

OBSERVATIONS
1. What is the sequence of nitrogenous bases in the complementary strand of your original DNA model?
2. How many models of DNA did you construct from your original model of DNA? What is the base sequence for each of these new models?

ANALYSIS AND CONCLUSIONS
1. Why do you think complementary base pairing is necessary for replication?
2. Compare the base sequences of the original DNA model with the replicated models. Compare the base sequences of the replicated models.
3. If each triplet of nitrogenous bases on DNA codes for a specific amino acid, how many amino acids will your original DNA model code for?

Safety Tips
Be careful with scissors.

Teaching Strategy
You might need to help some students with their organizational skills. As you circulate around the room, spot-check to be sure that students are correctly reading the chart and following directions.

Observations
1. Original sequence: CAATGTTAG.
2. Constructed two models. The first sequence was G-C, T-A, T-A, A-T, C-G, A-T, A-T, T-A, C-G. The second sequence was C-G, A-T, A-T, T-A, G-C, T-A, T-A, A-T, G-C.

Analysis and Conclusions
1. Complementary base pairing ensures that each new strand of

STUDENT STUDY GUIDE

SUMMARIZING THE CONCEPTS

The key concepts in each section of this chapter are listed below to help you review the chapter content. Make sure you understand each concept and its relationship to other concepts and to the theme of this chapter.

7–1 DNA

- DNA is the nucleic acid that stores and transmits the genetic information from one generation of an organism to the next.
- DNA is a polymer that is made up of units called nucleotides. Each nucleotide is a molecule made up of three parts: a 5-carbon sugar called deoxyribose, a phosphate group, and a nitrogenous base.
- DNA has four nitrogenous bases: adenine, thymine, cytosine, and guanine. Adenine and guanine belong to a group of compounds called purines. Thymine and cytosine belong to a group of compounds called pyrimidines.
- During replication, the DNA molecule unzips, or separates, into two strands. Each of the separated strands serves as a template, or pattern, for the attachment of complementary nucleotides.

7–2 RNA

- There are three major differences between DNA and RNA: RNA contains the sugar ribose instead of deoxyribose; RNA is usually single-stranded instead of double-stranded; and RNA contains the nitrogenous base uracil instead of thymine.
- During transcription, the DNA code is transferred to messenger RNA, which carries the code out of the nucleus into the cytoplasm.

7–3 Protein Synthesis

- During translation, messenger RNA binds to the ribosomes on which ribosomal RNA is found. Amino acids in the cytoplasm are picked up by transfer RNA and are carried to messenger RNA. The anticodons in transfer RNA attach to the proper codons in messenger RNA. Thus the messenger RNA acts as the pattern for protein synthesis. In this way, amino acids are brought together in the correct sequence to form a protein molecule.

REVIEWING KEY TERMS

Vocabulary terms are important to your understanding of biology. The key terms listed below are those you should be especially familiar with. Review these terms and their meaning. Then use each term in a complete sentence. If you are not sure of a term's meaning, return to the appropriate section and review its definition.

7–1 DNA
- genetic code
- transformation
- DNA
- nucleotide
- adenine
- guanine
- cytosine
- thymine
- base pairing
- replication

7–2 RNA
- RNA
- uracil
- transcription
- messenger RNA

7–3 Protein Synthesis
- codon
- translation
- transfer RNA
- ribosomal RNA
- anticodon

DNA is an exact complement of one of the previously existing strands. This means that each new strand is also an exact copy, or replica, of one of the previously existing strands.

2. They are complementary to each other.

3. Three amino acids.

Going Further: Enrichment
There are many other materials that could be used to make a DNA model that more closely approximates the actual model. Have students design and construct models using their own materials. Ask them to explain their improvements over the paper model.

CHAPTER REVIEW

CONTENT REVIEW

Multiple Choice
1. a 2. c 3. b 4. d
5. c 6. b 7. d 8. c

True or False
1. F nucleotide
2. F thymine
3. F guanine
4. F nitrogenous bases
5. T
6. T
7. F ribosomes
8. T

Word Relationships

A.
1. Deoxyribose and phosphate are two of the basic parts of a DNA nucleotide. Uracil.
2. Amino acid, polypeptide, and protein are molecules. Translation.
3. Ribose and adenine are two of the basic parts of an RNA molecule. Thymine.
4. A-T, C-G, A-U follow the base-pairing rules. C-T.

B.
5. Base pairing
6. Codon
7. Transfer RNA
8. Anticodon

CONCEPT MASTERY

1. Griffith injected mice with two strains of bacteria. One strain causes pneumonia and the other is harmless. The mice injected with the live disease-causing bacteria died from pneumonia, whereas those injected with the live harmless bacteria survived. Mice also lived if injected with the heat-killed disease-causing bacteria. However, when Griffith injected mice with a mixture of the heat-killed disease-causing bacteria and the live harmless bacteria, the mice died of pneumonia. The dead mice were found to contain the live disease-causing bacteria.
2. Bacteriophages are viruses that infect bacteria and are composed of a DNA core and a protein coat. By tagging the DNA core with radioactive phosphorus, and the protein coat with radioactive sulfur, scientists were able to discover that the infected bacterial cells contained large amounts of phosphorus. This led them to conclude that the DNA of the bacteriophage entered the bacteria.
3. Hershey and Chase were able to conclude that the DNA of a bacteriophage enters a bacterial cell and carries the genetic information for the making of more bacteriophages.
4. Franklin's work provided Watson and Crick with X-ray patterns showing that the fibers that made up DNA were twisted, and that large groups in

---

CHAPTER REVIEW

CONTENT REVIEW

Multiple Choice

Choose the letter of the answer that best completes each statement.

1. The process of changing one strain of bacteria into another is called
   a. transformation.   c. replication.
   b. translation.   d. transcription.
2. DNA is a
   a. carbohydrate.   c. nucleic acid.
   b. lipid.   d. sterol.
3. A nucleotide of DNA would contain
   a. uracil, deoxyribose, and phosphate.
   b. phosphate, deoxyribose, and thymine.
   c. nitrogenous base, phosphate, and glucose.
   d. adenine, ribose, and phosphate.
4. Watson and Crick described the DNA molecule as a
   a. branching chain.   c. single strand.
   b. straight chain.   d. double helix.
5. Between which types of compounds in a double-stranded DNA molecule must the bonds break before replication takes place?
   a. phosphate–base   c. adenine–thymine
   b. sugar–phosphate   d. sugar–base
6. To which organelles does messenger RNA attach?
   a. chloroplasts   c. mitochondria
   b. ribosomes   d. lysosomes
7. The coded information in a DNA molecule directly determines the formation of
   a. polysaccharides.   c. monosaccharides.
   b. lipids.   d. polypeptides.
8. If the code for an amino acid is ATG on the DNA molecule, this code on the transfer RNA molecule may be written as
   a. ATG.   c. AUG.
   b. CTG.   d. CTA.

True or False

Determine whether each statement is true or false. If it is true, write "true." If it is false, change the underlined word or words to make the statement true.

1. A molecular group consisting of a sugar molecule, a phosphate group, and a nitrogenous base is a <u>nucleic acid</u>.
2. In a DNA molecule, a base pair would be composed of adenine and <u>guanine</u>.
3. <u>Cytosine</u> and adenine are purines.
4. During the replication of DNA, bonds are broken between the <u>phosphate groups</u>.
5. RNA contains the nitrogenous base <u>uracil</u>.
6. There are <u>20</u> different amino acids in a cell.
7. The coded message carried by messenger RNA is translated into polypeptides at organelles called <u>chloroplasts</u>.
8. The decoding of a messenger RNA message into a polypeptide chain is known as <u>translation</u>.

Word Relationships

A. *In each of the following sets of terms, three of the terms are related. One term does not belong. Determine the characteristic common to three of the terms and then identify the term that does not belong.*

1. deoxyribose, phosphate, DNA, uracil
2. amino acid, polypeptide, protein, translation
3. ribose, RNA, thymine, adenine
4. A—T, C—G, A—U, C—T

B. *Replace the underlined definition with the correct vocabulary word.*

5. The attraction between nitrogenous bases is the force that holds the two strands of DNA together.
6. Each combination of three nucleotides on messenger RNA specifies a specific amino acid that is to be placed in the polypeptide chain.
7. The RNA that carries amino acids to the ribosomes is a single strand of RNA that loops back on itself.
8. As a ribosome moves along a strand of messenger RNA, each codon is paired with its three nucleotides on transfer RNA that are complementary to the three nucleotides on messenger RNA.

CONCEPT MASTERY

Use your understanding of the concepts developed in the chapter to answer each of the following in a brief paragraph.

1. Describe Griffith's experiments involving bacterial transformation.
2. How did the study of bacteriophages lead scientists to support the idea that DNA carries hereditary material?
3. What was the contribution made by Hershey and Chase to understanding the genetic code?
4. How did Franklin's work contribute to the discovery of the structure of DNA?
5. Describe the contribution of Watson and Crick.
6. Describe DNA replication. What is its importance?
7. List three differences between DNA and RNA.
8. How is messenger RNA formed during transcription?
9. What is the function of the ribosomes?
10. What is the relationship between an amino acid and a codon?

CRITICAL AND CREATIVE THINKING

Discuss each of the following in a brief paragraph.

1. **Identifying patterns** Oxytocin is a hormone that helps to regulate blood pressure and stimulates the uterus to contract during childbirth. Following is a DNA sequence that could code for a part of a molecule of oxytocin: ACA ATA TAG CTT TTG ACG GGG AAC CCC ATT. Write the sequence of messenger RNA codons that would result from the transcription of this portion of DNA.
2. **Relating facts** How many guanines are in a DNA molecule 1000 base pairs long if 20% of the molecule consists of thymines?
3. **Applying concepts** Write out the messenger RNA sequence that would be transcribed from the following strand of DNA: T-A-C-A-A-G-T-A-C-T-T-G-T-T-T-C-T-T. Then using Figure 7–17 on page 149, write the amino acid sequence that would be translated when the messenger RNA combines with a ribosome.
4. **Using the writing process** Write a short story that follows a protein molecule from its point of origin in a corn kernel, through its use as cattle food, to a position of importance as a protein in your body.

deoxyribose found in DNA. In RNA, thymine is replaced by uracil. Unlike DNA, RNA molecules are single-stranded.
8. During transcription, RNA polymerase attaches to certain places on the DNA molecule, separates the two strands of the double helix, and synthesizes mRNA. The mRNA is complementary to the bases on one of the unzipped DNA strands.
9. The ribosomes are the sites of protein synthesis.
10. Each codon specifies a particular amino acid that is to be placed on the polypeptide chain.

CRITICAL AND CREATIVE THINKING

1. UGU UAU AUC GAA AAC UGC CCC UUG GGG UAA.
2. 600 guanines. Note that there are 2000 base pairs in a DNA molecule 1000 base pairs long. 20 percent, or 400, are thymine, so 20 percent (400) must be adenine. Of the remaining 60 percent of the molecule, half—that is, 30 percent—must be guanine and half cytosine. 30% of 2000 is 600.
3. A-U-G-U-U-C-A-U-G-A-A-C-A-A-A-G-A-A. Tyrosine-aspargine-tyrosine-leucine-phenylalanine-leucine.
4. The short story should be well-written and include information on how proteins are made. Students should show a clear understanding of the processes of replication, translation, and transcription.

---

the fiber are spaced out at regular intervals along the length of the fiber.
5. Watson and Crick developed a model for the structure of DNA. Their model of DNA consisted of two parallel strands of sugar–phosphate groups that were linked together by pairs of nitrogenous bases. The whole structure was then coiled into a double helix.
6. The double DNA strand separates to form two single strands, like the opening of a zipper. Complementary nucleotides become bonded to the exposed nitrogenous bases, thereby forming two complete double strands identical to the original.
7. RNA nucleotides contain the sugar, ribose, instead of the

CHAPTER 8 Cell Growth and Division

Section	Laboratory Investigations and Activities
8–1 Cell Growth pages 159–164 THEMES: Patterns of Change, Systems and Interactions	**Teacher Edition** DEMONSTRATION: Surface-to-Volume Ratio, p. 158D **Laboratory Manual** Investigating the Limits of Cell Growth, p. 121
8–2 Cell Division: Mitosis and Cytokinesis pages 164–171 THEMES: Patterns of Change, Scale and Structure	**Student Edition** LABORATORY INVESTIGATION: Observing Onion Root Tips, p. 172 **Teacher Edition** DEMONSTRATION: Mitosis, p. 158D **Laboratory Manual** Determining the Time Needed for Mitosis, p. 125
Chapter Review pages 173–175	

Teacher Resources

Books

Adolph, Kenneth W., ed. *Chromosomes and Chromatin*. 3 vols. CRC Press, 1987.

Bradbury, E. Morton, et al. *DNA, Chromatin, and Chromosomes*. Arco, 1985.

Johanson, John E., Jr., ed. *Aging and Cell Function*. Plenum Publishers, 1984.

John, P.C.L. *The Cell Cycle*. Cambridge University Press, 1981.

Murray, Andrew, and Tim Hunt. *The Cell Cycle*. Oxford University Press, 1993.

Sharma, A.K., and Archana Sharma, eds. *Chromosome and Cell Genetics*. Gordon and Breach, 1985.

Van Holde, K.E. *Chromatin*. Springer-Verlag, 1988.

Zimmerman, A.M., and A. Forer, eds. *Mitosis/Cytokinesis*. Academic Press, 1981.

CHAPTER PLANNING GUIDE

Other Activities	Media and Technology

Teaching Resources
    ACTIVITY: Cell Reproduction, p. 5
Study Guide
    Section 8–1, p. 75	

Teaching Resources
    VOCABULARY REVIEW: Cross-a-Clue, p. 1
    ACTIVITY: The Cell Cycle, p. 3
Study Guide
    Section 8–2, p. 77
	Biology Transparencies
	    12: Interphase, Mitosis, and Cytokinesis in an Animal Cell

Teaching Resources
    Chapter Test A, p. 11
    Chapter Test B, p. 15
    Performance-Based Assessment, Unit 2
	Computer Test Bank
	    Chapter Test, p. 81

Audiovisuals

Cell Division and the Life Cycle. 2 filmstrips with cassettes. Human Relations Media.
Introduction to Mitosis. Sound filmstrip. Ward.
Measurement and Scientific Notation. Sound filmstrip. Ward.
Mitosis, 2nd ed. Film. Encyclopaedia Britannica Educational Corporation.
Cell Division—Mitosis and Cytokinesis. Carolina Biological Supply Company. Video or filmstrip.
The Cell Cycle, Mitosis, and Cell Division. Video. Biology Media.

CHAPTER 8 Cell Growth and Division

CHAPTER OVERVIEW

This chapter explores the reasons for the occurrence of cell division as well as the process itself. Organisms grow in size as a result of an increase in the number of their cells rather than the size of their cells. The increase comes about through the precise and controlled process of cell division, resulting in the production of two daughter cells.

The rate of cell division in multicellular organisms depends on the nature of the particular cell as well as the circumstances. For example, cells at the site of an injury to the skin will undergo rapid division to promote the process of healing, while heart and nerve cells divide little, if at all.

The cell cycle is the period from the beginning of one mitosis to the beginning of the next. It is comprised of mitosis, the period during which the cell is actively dividing, and interphase, during which other cell processes occur.

Mitosis itself is divided into four stages: prophase, metaphase, anaphase, and telophase, during which the replicated chromosomes, known as chromatids, are precisely divided between two daughter cells.

Interphase is divided into three stages known as G_1, S, and G_2, during which, respectively, cell growth and development occurs, DNA is synthesized, and the cell produces the materials needed for the next cell division.

During most of its cell cycle, a typical cell is in interphase, since mitosis requires only a small percentage of the total time.

8-1 CELL GROWTH

Section Focus 8-1

The purpose of this section is to discuss the factors that influence cell growth. Cells cannot grow by simply increasing in size because of a limiting factor, the surface-to-volume ratio. As a cell enlarges, its volume increases several hundred times faster than its surface area, so the ratio quickly becomes very small. This results in inefficiencies in the cell's internal processes, slowing its growth. Cell division solves this problem.

In single-celled organisms such as bacteria, cell division may occur as frequently as every 20 minutes. In multicelled organisms, different tissue types differ in their frequency of growth and division.

Cells seem to possess internal controls over the process of cell division. For example, when isolated cells are cultured, they continue to grow until they make contact with other cells. Similarly, cells on the edges of an injured area of the body divide rapidly, adding to the healing process.

Uncontrolled cell growth in multicellular organisms may lead to disorders such as cancer. Much of the current focus in cancer research is on the mechanism of cell growth and division.

Performance Objectives 8-1

1. Explain why cells cannot simply grow by increasing in size.
2. Explain the term *surface-to-volume ratio*.
3. Describe the different rates of growth of different cell types.
4. Describe the factors that affect cell growth.
5. Describe the possible results of uncontrolled cell growth.

Science Terms 8-1

cell division   p. 161

8-2 CELL DIVISION: MITOSIS AND CYTOKINESIS

Section Focus 8-2

The purpose of this section is to describe and discuss the process of cell division in eukaryotic cells. In addition, the nuclear structures that are involved in cell division are described.

In eukaryotic cells, genetic information is carried in the form of DNA. The DNA, together with protein, forms chromatin, a dark-staining material found throughout the nucleus. Since the DNA double helix is actually many times longer than chromatin, it is folded or coiled so that it can "fit" within the nucleus. Evidence from research seems to show that when cell division begins, the DNA coils tightly around proteins called histones, forming nucleosomes. Interaction among the nucleosomes during cell division causes even more coiling and condensation of the chromatin, resulting in readily visible chromosomes.

The cells of each species contain their own specific number and kinds of chromosomes. At the beginning of mitosis, each chromosome actually consists of two identical sister chromatids attached at the centromere.

The term *cell cycle* is used to describe the continuous process in which cells grow, prepare for division, and divide into two daughter cells. It consists of two stages. One is mitosis, or the M phase of the cell cycle, in which the nucleus divides to form two nuclei. Mitosis is divided into four phases: prophase, metaphase, anaphase, and telophase.

Prophase, the longest phase, begins with the appearance of chromosomes, microscopically visible as two chromatids. It is also during this time that the two centrioles (present only in animal cells) take up positions on opposite sides of the nucleus and the spindle forms from the microtubule protein in the centrioles. At the end of prophase, the nucleolus and the nuclear envelope disappear.

During metaphase, the second and shortest phase of nuclear division, chromosomes line up across the equator of the cell and the centromeres are attached to the spindles.

Anaphase begins when the centromeres split, causing the sister chromatids to separate into chromosomes. The chromosomes from each pair of chromatids move in opposite directions toward the poles of the spindle. The mechanism of this motion is not yet thoroughly understood. When the movement of the chromosomes stops, anaphase is complete.

In telophase, the final phase of mitosis, the chromosomes uncoil to become indistinct chromatin once again, the spindle disappears, and the nucleolus becomes visible again. At the same time, the nuclear envelope re-forms.

CHAPTER PREVIEW

When the formation of two nuclei is complete, cytokinesis, or cytoplasmic division, begins. In animal cells, the cell membrane moves inward, pinching the cell into two halves. In plant cells, a cell plate forms between the nuclei. In this area a cell membrane and cell wall will form.

The complex yet precise nature of mitosis results in each new cell receiving exactly one copy of each chromosome, ensuring its viability.

When a cell is not actively dividing, it is said to be in interphase. During interphase, proteins, RNA, and organelles are synthesized and DNA replicates. Although the processes occurring during interphase are complex, they occur in specific sequences of events known as the G_1, S, and G_2 phases. In the G_1 phase, cell growth and development occur; in the S phase, DNA replicates; and in the G_2 phase, organelles and materials required for the next mitosis are synthesized. It is during interphase that most cell activity actually occurs.

Performance Objectives 8–2

1. Contrast mitosis and cytokinesis.
2. List the four stages of the cell cycle.
3. Describe the four phases of mitosis.
4. Describe the events that occur during cytokinesis.
5. Describe the events that occur during interphase.
6. Explain the importance of each new cell receiving exactly one copy of each chromosome.

Science Terms 8–2

mitosis p. 164
cytokinesis p. 164
chromosome p. 164
chromatin p. 165
chromatid p. 166
centromere p. 166
cell cycle p. 166
interphase p. 166
prophase p. 167
centriole p. 168
spindle p. 168
metaphase p. 168
anaphase p. 169
telophase p. 170

Discovery Learning

TEACHER DEMONSTRATIONS
Modeling

Surface-to-Volume Ratio
The following demonstration may be used as an introduction to Chapter 8.

Have several (at least 3) spherical balloons available. Blow up the first balloon and explain to students that they will compare the surface area and volume of the balloon as it grows in size.

Have a student assist you in finding the circumference of the balloon by looping a string around it, then measuring the string against a ruler. The radius may then be determined from the formula $C = 2\pi r$. Once the radius is known, the surface area may be found from the formula Surface area = $4\pi r^2$, and the volume may be found from the formula

$$\text{Volume} = \frac{4\pi r^3}{3}.$$

Repeat at least two more times with the remaining balloons. It will be obvious that the volume of the balloon is increasing at a much faster rate than the surface area.

- **If the balloons represent a growing cell, how will the large increase in volume affect it?** (Intercellular and intracellular transport will become difficult and the cell will not be able to function properly.)
- **What does this tell us about how large an individual cell may grow?** (It cannot continue to enlarge indefinitely. It must eventually undergo cell division.)

The authors have purposely included this demonstration, even though it is somewhat contradictory to the information presented in the chapter. Students learn in the chapter that a 10-fold increase in diameter causes a 1000-fold increase in volume. Using the formulas in the demonstration, however, a 10-fold increase would cause a volume increase of 333 times, not 1000. Challenge students to determine the cause of this difference. (The demonstration assumes a spherical cell whereas the text assumes a cubic cell.) Point out that this shows some of the difficulties in scientific measurement, but that in either case, the important point is that an increase in diameter (or length, width, and height) causes a much larger increase in volume than surface area.

Mitosis

You may want to perform this demonstration while you are discussing Section 8–2 with students.

Arrange 8 or 10 pipe cleaners, preferably of different colors, to form 4 or 5 identical pairs, with the pairs held together by "centromeres" made of modeling clay. Draw a large "cell membrane" on the chalkboard and inside it, a smaller "nuclear membrane."

Ask a student to come forward to arrange the pairs of pipe cleaners (chromatids), by sticking them on the board as they would appear at the beginning of prophase. Then ask other students to come forward to rearrange the "chromatids" to represent the various stages of mitosis. Allow students at the board to seek help from others, as necessary. The result should be two daughter cells drawn on the board, each containing the same number and kinds of chromosomes.

CHAPTER 8

Cell Growth and Division

GUIDED ENQUIRY
Pose the following questions to students and have them record their responses. Point out that they will gain a better understanding of the key concepts if they read the chapter with these basic questions in mind. Upon completion of the chapter, pose the questions again. Ask students to compare their initial responses with those they have developed after reading the chapter.
- How does a multicellular organism increase in size?
- What are the limits, if any, on how large an individual cell may become?
- How do individual cells increase in number?
- How does a cell "know" when to divide?
- When a cell divides, what ensures that the new cells will be exactly like the original one?

INTRODUCING CHAPTER 8
Using the Visuals
Ask students to read the paragraph that introduces this chapter and to observe the photograph of the two separating cells. Point out that the two cells illustrated are the result of a just-completed cell division.
- **How does a multicellular organism grow in size?** (There is an increase in the kinds and numbers of cells within it.)
- **Do individual cells continue to grow and multiply even** though an organism has reached its full size? (Yes.)
- **Why must individual cells continue to grow and multiply even though the organism will not?** (New cells are needed to replace cells that wear out or are destroyed by an injury.)

CHAPTER 8

Cell Growth and Division

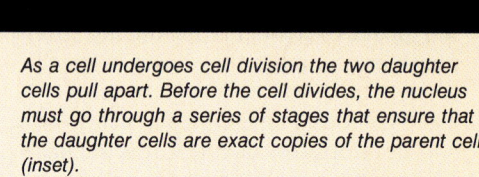

As a cell undergoes cell division the two daughter cells pull apart. Before the cell divides, the nucleus must go through a series of stages that ensure that the daughter cells are exact copies of the parent cell (inset).

One of the major characteristics of a living thing is the ability to grow. Growth is obvious when it involves an increase in size. We all notice the growth of grass and flowers in the spring and summer or the growth of a newborn animal into an adult. The growth of an organism results from an increase in the number and the size of the cells within it. But growth occurs even in organisms that are no longer increasing in size—continued cell growth is essential for life.

How do cells grow? Is there a limit to the growth of a cell? In this chapter we will examine how cells grow in size and increase in number, and why growth is so essential to living things.

- **What does the photograph show?** (It shows a cell that has just completed the process of cell division, or mitosis.)
- **How do you think the original "parent" cell compared in appearance to the two "daughter" cells shown in the photograph?** (It was larger in size, but genetically identical to the daughter cells.)
- **Why are the daughter cells genetically identical to the parent cell?** (Students may have difficulty answering at this point. Accept any reasonable answer,

GUIDE FOR READING

After you read the following sections, you will be able to

8-1 Cell Growth
- Describe cell growth.
- Define cell division.
- Relate cell growth to cell division.

8-2 Cell Division: Mitosis and Cytokinesis
- Define mitosis and cytokinesis.
- Describe the cell cycle and the changes that take place during interphase.
- Discuss the events and significance of mitosis.

Journal Activity

YOU AND YOUR WORLD In your journal, describe how you have changed as you've grown. Illustrate your description with drawings and/or photographs.

8-1 Cell Growth

Guide For Reading
- How does growth occur?
- What factors limit and control cell growth?
- What are the consequences of uncontrolled cell growth?

As you already know, living things are made up of cells. And living things grow, or increase in size. Does a living thing grow because its cells get larger and larger or does it grow because it produces more and more cells? **In most cases, a living thing grows because it produces more and more cells.** The cells of a human adult are no larger than the cells of a human baby, but there are certainly more of them.

Does growth have to occur in this way? Or could an organism grow simply by allowing its cells to get larger and larger? Let's consider what would happen if this could occur.

Limits of Cell Growth

You will recall from Chapter 5 that it is through the cell membrane that food, oxygen, and water enter the cell and waste products leave the cell. How quickly this exchange takes place depends on the surface area of the cell, or the total area of the cell membrane. How quickly food and oxygen are used up and waste products are produced depends on the cell volume, or the amount of space within the cell.

If we were to take a typical cell and double its diameter, what would happen to the amount of membrane (surface area) compared to the amount of material inside the cell (cell volume)? Look at Figure 8-2 on page 160. Notice that as the cell

Figure 8-1 As living things grow they produce more cells. Although this adult white Bengal tiger is larger than its cub, the sizes of its cells are the same as those of the cub.

159

TEACHING SUPPORT

Teaching Resources
- Activity: Cell Reproduction, p. 5

TIE-IN MATH

You may want to review the concept of ratio with your students in discussing surface-to-volume ratio. A ratio is simply a comparison between two variables, expressed mathematically as division. The surface-to-volume ratio is obtained by dividing the surface area by the volume. The fact that the ratio becomes smaller as a cell increases in size simply means that the volume, or internal contents of the cell, is increasing at a faster rate than the area of the surface.

MULTICULTURAL STRATEGY

Have students research the life and work of mathematician and scientist Norman T. Grier, who is the recipient of a NASA recognition award for contribution to the Plasma Interaction Experiment Satellite Team. Students may also wish to learn about other outstanding African Americans who have combined careers in mathematics and science, including Mary W. Jackson, James L. Harris, and James L. Jennings.

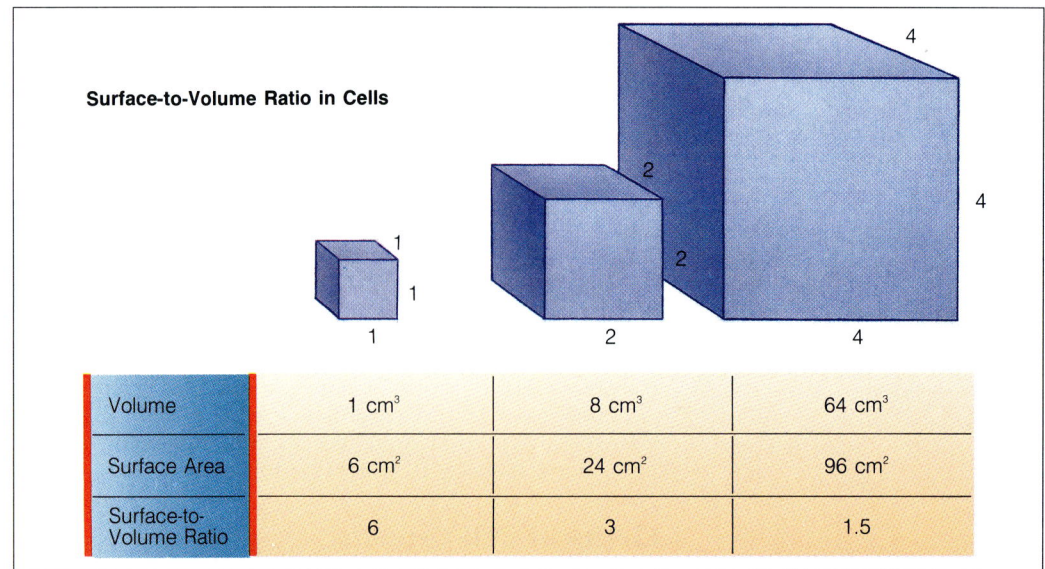

Figure 8–2 If the length of each side of a cube-shaped cell doubles from 1 to 2 to 4, the volume increases more rapidly than does the surface area. As a result, the ratio of surface area to volume decreases and the cell with the largest volume has a more difficult time getting materials in and waste products out.

increases in size, its volume increases at a faster rate than its surface area. For example, if the diameter of the cell increases 10 times, its surface area increases 100 times. Its volume, however, increases 1000 times!

The fact that surface area and volume do not increase at the same rate creates problems for the cell. The larger cell will have a more difficult time getting oxygen and nutrients in and waste products out.

To help you understand why the larger cell has a more difficult time than the smaller cell, let's compare the cells to a small and a large office building. Suppose the small office building that is serviced by a two-lane highway is replaced by a large office building. Now more people work in the office building. But the highway leading to the large building has not increased in size. It is the same two-lane highway that serviced the small office building. As a result, the people in the large building will have difficulty with traffic getting to the office in the morning and leaving the office in the evening.

A larger cell will experience similar problems—materials will have trouble entering and leaving the cell. This is one reason why cells do not grow much larger even if the organism of which they are a part does. There is another reason too.

As you will recall from Chapter 7, information for the cell's function and survival is stored in the sequence of nitrogenous bases in DNA. In eukaryotic cells, the DNA is stored in the nucleus of the cell. When the cell is small, copies of DNA that are stored in the nucleus are able to produce enough messenger RNA to make all the proteins the cell needs. But even though the cell increases in size, it does not make extra copies of DNA.

8–1 (continued)

Reinforcement/Reteaching Using the Visuals The relative increases in the size of the cell membrane compared to the cell contents may also be expressed as the surface-to-volume ratio. As the cell increases in diameter, this ratio decreases. Call students' attention to Figure 8–2. It can be seen that the surface-to-volume ratio decreases as the cube becomes larger. Refer to the balloon demonstration done earlier. Have students calculate the surface-to-volume ratio for the different-sized balloons used. (If you feel the class is capable, have them do the calculation on their own. Otherwise, go through it with them.)

Content Development
- How does a growing cell solve the problem of its increasing inefficiency as it enlarges? (A cell eventually undergoes cell division, dividing into two daughter cells.)
- What determines the rate at

(There are a few exceptions that we will learn about later.) If a cell were to grow without limit, an "information crisis" would occur. Like a town that has tripled in size but has not added a single book to its library, the cell must now make greater demands on its available genetic library.

In many cases, the amount of messenger RNA that is produced in the cell increases—but within limits. After a time, the cell's DNA may no longer be able to make enough RNA to supply the increasing needs of the growing cell. The cell must slow down its growth, thus becoming less efficient. The cell undergoes a process called **cell division** to solve these problems. **Cell division is the process whereby the cell divides into two daughter cells.** We will discuss cell division in more detail later in the chapter.

Rates of Cell Growth

Cells can grow at astonishing rates. For example, the bacterium *Escherichia coli* (ehsh-uh-RIHK-ee-uh KOH-ligh), or *E. coli*, is a single-celled organism that can easily double its volume in about 30 minutes. It can then divide to form two new cells. If conditions are ideal, each of these cells can grow to form two new cells in the next 30 minutes.

Ideal conditions for this kind of growth can never be maintained for very long, however. A quick look at the consequences of rapid growth will explain why. In just one day, a single cell would grow into a 14-kilogram mass of bacteria. In three days, the mass of the cells would equal the mass of the Earth! Real conditions, or the circumstances that cells normally face, are very different.

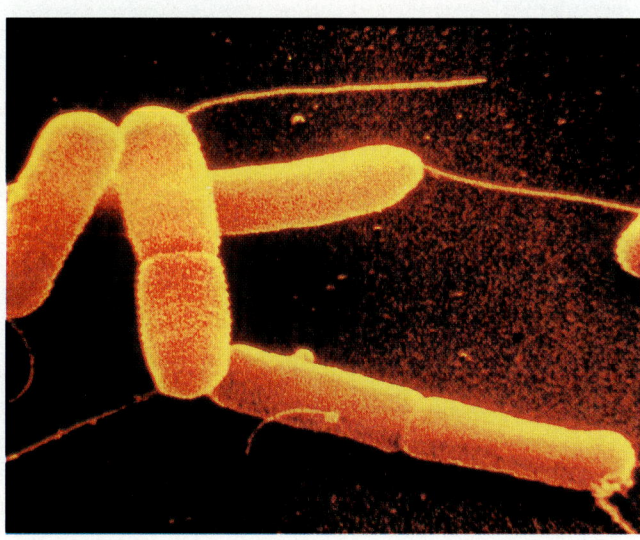

Figure 8-3 Single-celled prokaryotic organisms reproduce by cell division. In this scanning electron micrograph of bacteria (E. coli), you can see two bacteria undergoing cell division.

BACKGROUND INFORMATION
ASEXUAL REPRODUCTION

Chapter 8 deals only with cell growth and cell division, including mitosis and cytokinesis. Textbooks often deal with the various types of asexual reproduction, including binary fission, budding, spore formation, regeneration, and vegetative propagation in the same chapter. However, the authors of this textbook feel that including these topics at this point would introduce a great number of peripheral concepts, distracting students from focusing on the central idea of cell reproduction. In this text, asexual reproduction will be considered in the chapters dealing with the various organisms that reproduce asexually.

MULTICULTURAL STRATEGY

Many educational and career opportunities exist in science for women who are persons of color. Students can learn about these opportunities by contacting the National Network of Minority Women in Science (MWIS), Office of Opportunities in Science, American Association for the Advancement of Science, 1333 H Street NW, Washington, DC; (202) 326-6674.

---

which cells divide? (This is determined by the nature of the cell itself. It may vary from as little as once every 30 minutes to virtually never.)
- **If bacteria such as *E. coli* can divide every 30 minutes, why haven't they taken over the world?** (The main reason is that they quickly use up the available nutrient supply, and the rate of growth slows. There are other possible answers.)

Skills Development
Guided Practice
Skill: Making calculations
Ask students to verify that starting with a single *E. coli* bacterium with a mass of 1×10^{-9} gram dividing every 30 minutes, after 24 hours, one would have bacteria with a total mass of about 14 kilograms.
- **How many individual bacterial cells would this be?** (It would be 2.8×10^{14} individual bacteria.)

TEACHING SUPPORT

Laboratory Manual
- Investigating the Limits of Cell Growth, p. 121

BACKGROUND INFORMATION
CANCER

Cancer is caused by the uncontrolled multiplication of groups of cells. Because the rate of multiplication of cells, along with their other functions, is controlled by the genetic code contained in the cells, DNA, the cancerous growth must be the result of genetic change in these cells. Much of the focus in cancer research has been on the cause of this genetic change, and there is evidence that in many cases, viruses are the culprits. Although many viruses contain only RNA, it has been found that they are capable, within a host cell, of producing DNA, using the viral RNA as a template. This DNA becomes part of the host cell's chromosomes, altering the normal function of the host cell.

Figure 8–4 The scientist in this photograph is growing cells from a pine tree on culture plates. Many plants—African violets, daylilies, orchids, ferns, and pine trees, to name just a few—can be produced from single cells grown in culture. You have probably seen some of these high-tech plants for sale in a flower shop or growing in a garden.

Controls on Cell Growth

One of the most striking aspects of cell behavior in a multicellular organism is how carefully cell growth and cell division are controlled. Cells in certain places in the body, such as the heart and the nervous system, rarely divide—if they divide at all. In contrast, the cells of the skin and digestive tract grow and divide rapidly throughout life, providing new cells to replace those that are worn out or broken down due to daily wear and tear.

We can observe the effects of controlled cell growth in the laboratory by placing some cells in a petri dish containing nutrient broth. The nutrient broth provides food for the cells. Most cells will grow until they form a thin layer covering the bottom of the dish. Then the cells will stop growing. Why do they stop growing? When cells come into contact with other cells, they respond by not growing. At present, scientists are trying to understand how this process works.

Controls on cell growth and cell division can be turned on and off. When an injury—such as a cut in the skin or a break in a bone—occurs, cells at the edges of the injury are stimulated to divide rapidly. This action produces new cells, starting the process of healing. When the healing process nears completion, the rate of cell division slows down, controls on growth seem to be reimposed, and everything returns to normal.

Uncontrolled Cell Growth

The consequences of uncontrolled cell growth in a multicellular organism are severe. Cancer, a disorder in which some

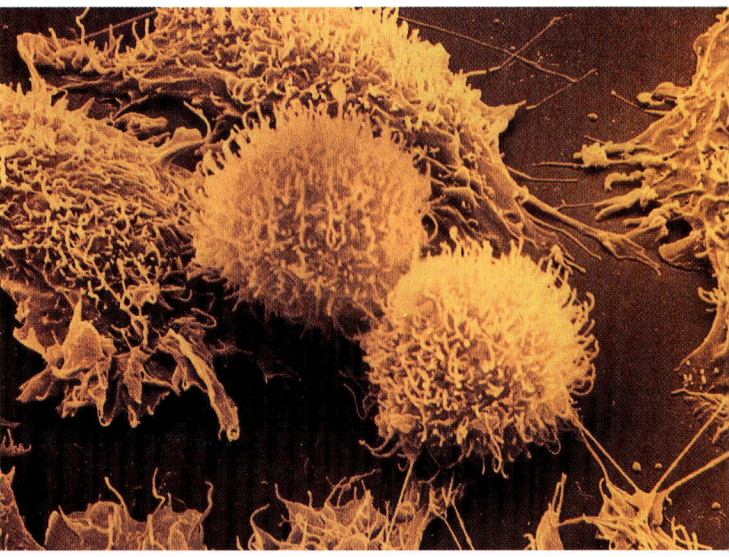

Figure 8–5 Unlike normal cells, cancer cells do not stop growing and dividing even if they come in contact with other cells. Actually, cancer cells, such as those shown in this scanning electron micrograph, have lost their ability to control their own rate of growth.

8–1 (continued)

Enrichment
Some students may want to explore the factors that control the rate of cell growth and division as a research project. Materials for a project of this sort are readily available from commercial supply companies. A less elaborate study of rates of cell growth may also be accomplished by "cloning" pieces of seed potatoes (containing an "eye") or carrot tops in shallow dishes of water. Different nutrient solutions containing nitrates or phosphates might be used, as well as plant growth hormones such as gibberellins or indoleacetic acid.

Content Development
Discuss with students the importance of controlled cell growth for the maintenance of good health in an organism.
- **If cells are not replaced when they are worn out, or not replaced at the site of an in-**

PROBLEM SOLVING IN BIOLOGY

SCIENTIFIC NOTATION

How many cells would you suspect there are in the human body? There are approximately 100,000,000,000,000 cells. And all these cells arise from a human egg cell that is only 0.0001 meter in diameter!

As you can see, there are numbers associated with biology that are so large that they are also difficult to write and understand. For this reason, scientific notation is used. Scientific notation is also known as exponential notation because it uses the exponential form of a number.

The exponential form of a number is made up of two parts: $M \times 10^n$
The first part, M, is a number between 1 and 10. The second part, n, is an exponent, or power, to which the base 10 is raised. An exponent indicates how many times the base is to be multiplied by itself.

Because the base is 10, what the exponent actually does is locate the decimal point in the number being represented. It does this by indicating how many places the decimal point in M must be moved. The decimal point is moved to the right if the exponent is positive and to the left if the exponent is negative. A negative exponent is a negative power of 10. It is shorthand for 1 divided by a power of 10.

To convert a large number to scientific notation, follow these steps.

1. Determine M by moving the decimal point in the number (100,000,000,000,000) to the left or the right so that only one digit is to the left of it.

   1 0 0,0 0 0, 0 0 0, 0 0 0, 0 0 0.
   1.0 0, 0 0 0, 0 0 0, 0 0 0, 0 0 0

2. Determine n by counting the number of places the decimal point has been moved. If moved to the left, n is positive; if moved to the right, n is negative.

   1.0 0, 0 0 0, 0 0 0, 0 0 0, 0 0 0   14 places to the left

3. Then write the number 1 multiplied by the base 10 with its exponent.

   $1 \times 10^{14} = 100,000,000,000,000$

To convert a small number to scientific notation, follow these steps.

1. Determine M by moving the decimal point in the number (0.0001) to the left or the right so that only one digit is to the left of it.

   0.0 0 0 1
   0 0 0 0 1.

2. Determine n by counting the number of places the decimal point has been moved. If moved to the left, n is positive; if moved to the right, n is negative.

   0.0 0 0 1   4 places to the right

3. Then write the number 1 multiplied by the base 10 with its exponent.

   $1 \times 10^{-4} = 0.0001$

Complete the following exercises. Check your answers by using a calculator or computer, if one is available.

A. Expressing Numbers in Scientific Notation

1. The average length of a mitochondrion is 0.00000001 cm.
2. A rat contains approximately 10,000,000,000 cells.

B. Converting Scientific Notation into Numbers

1. The average diameter of a bacterium is 2×10^{-6} cm.
2. The resolving power of an electron microscope is 1×10^6 times greater than that of the human eye.

PROBLEM SOLVING

Scientific Notation
An understanding of the meaning of numbers written in scientific, or exponential, notation is a useful skill for the biology student. Many of the quantities discussed in biology are in the realm of either the very large or the very small, requiring the use of great numbers of zeros. Naming numbers with many zeros is also awkward. Are we talking about zillions, or perhaps trillionths? The use of scientific notation helps us to avoid all of these difficulties.

Answers
A. Expressing numbers in scientific notation
   1. 1.0×10^{-8} cm
   2. 1.0×10^{10} cells
B. Converting scientific notation into numbers
   1. 0.000002 cm
   2. 1,000,000 times greater

---

jury, what might be some of the consequences for the organism? (Wounds would not heal, internal bleeding, nonfunctioning organs, and, ultimately, death would occur.)

You might also point out that damaged or destroyed nerve cells are rarely replaced. For this reason, nerve damage is usually permanent. On the other hand, if cells grow and divide too rapidly, tumors may develop, some of which might be cancerous.

cells have lost the ability to control their own rate of growth, is one such example. When cancer cells are placed in a culture of living tissue, they do not stop growing even though they come into contact with other cells. Cancer cells will continue to grow and divide until the supply of nutrients is exhausted.

Cancer is a serious disorder that claims many lives and affects all of us, directly or indirectly. To cell biologists, cancer provides valuable information concerning the importance of controls on cellular growth.

8–1 SECTION REVIEW

1. What is growth? What controls cell growth?
2. What is cell division?
3. **Critical Thinking—Making Calculations** A newly discovered strain of bacteria divides once every 60 minutes under ideal conditions. After 24 hours, a single bacterium produces cells that have a mass of 1 kilogram. From this information, calculate the average mass of a single bacterium.

Guide For Reading
- What are mitosis and cytokinesis?
- What is the cell cycle? What changes take place during interphase?
- What are the events and the significance of mitosis?

8–2 Cell Division: Mitosis and Cytokinesis

In this section we will discuss how eukaryotic cells divide and form two cells. The division of eukaryotic cells occurs in two main stages. The first stage of cell division is called **mitosis** (migh-TOH-sihs). **Mitosis is the process by which the nucleus of the cell is divided into two nuclei, each with the same number and kinds of chromosomes as the parent cell.**

The second stage of cell division is known as **cytokinesis** (sight-oh-kih-NEE-sihs). **Cytokinesis is the process by which the cytoplasm divides, thus forming two distinct cells.**

Because the structure of a eukaryotic cell is complex, the process of cell division in this type of cell is complex too. The reason for this complexity may be found in the need to separate large amounts of DNA accurately and efficiently. With as many as 100 chromosomes to take care of, a mechanism must exist in the eukaryotic cell to make certain that each new cell gets one (and only one) copy of each chromosome. A mistake in the process could make it impossible for one or both daughter cells to remain alive.

Chromosomes

You will recall from Chapter 5 that **chromosomes** are structures in the cell that contain the genetic information that is passed on from one generation of cells to the next. The word

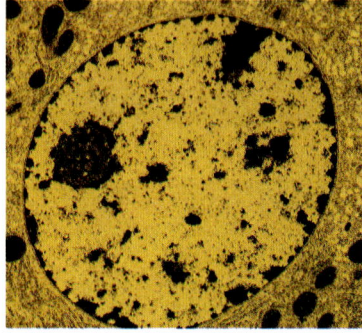

Figure 8–6 When a cell is not dividing, the chromatin forms condensed and dispersed regions.

chromosome, meaning colored body, was derived from the fact that when a dye was added to the cell, the chromosomes picked up the color of the dye and were easily seen through the light microscope during cell division. Unfortunately, chromosomes are not visible in most cells except during mitosis.

Chromosomes contain the genetic information in the form of DNA. In prokaryotic cells, the chromosomes are made up of long circular molecules of DNA. In eukaryotic cells, the chromosomes are made up of distinct lengths of DNA. The cells of every organism contain a specific number of chromosomes. Human cells, for example, contain 46 chromosomes; goldfish cells contain 94 chromosomes. See Figure 8–7.

COMPOSITION OF CHROMATIN Chromosomes are made up of a material called **chromatin** (KROH-muh-tihn). During the period of time between cell divisions, chromatin forms condensed and dispersed regions. See Figure 8–6. During the early stages of mitosis, the chromatin condenses and the chromosomes become visible.

Chromatin is composed of DNA and protein. Much of this protein is involved in the folding of DNA so that it can fit within the nucleus. This is an important job because the DNA double helix that makes up a chromosome is much longer than the chromosome itself. In fact, the total length of DNA in a typical human chromosome is about 10,000 times the length of the chromosome! How does all that DNA fit into the chromosome?

In 1973, the American scientists Don Olins, Ada Olins, and Christopher Woodcock discovered that the chromosomes' DNA was coiled around special proteins called histones. Together, the DNA and histone molecules formed beadlike structures

Organism	Chromosome Number
Ameba	50
Carrot	18
Cat	32
Chimpanzee	48
Dog	78
Earthworm	36
Goldfish	94
Human	46
Lettuce	18

Figure 8–7 This chart gives the chromosome numbers of some familiar organisms. Which organism's chromosome number is closest to that of an ameba?

Figure 8–8 Chromosomes contain highly coiled and supercoiled strands of DNA. Notice that a small part of the supercoil is made up of nucleosomes (inset).

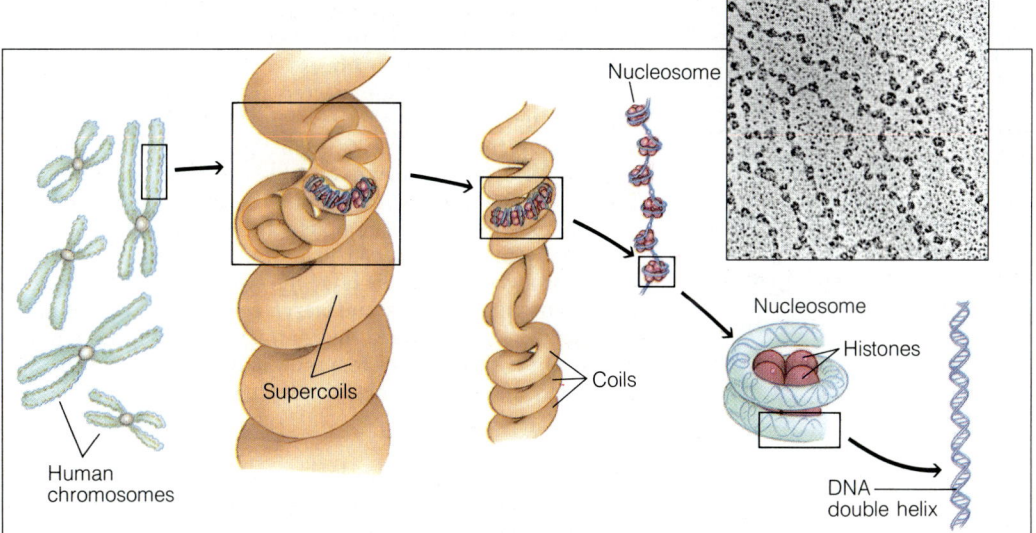

FACTS AND FIGURES

There is no actual difference between a chromosome and a chromatid. The different terms are used because biologists do not like to say that the number of chromosomes doubles during the S phase, since as a result of mitosis the chromosome number does not change. To avoid the difficulty, biologists will say that although the number of chromosomes always remains the same, after the S phase each chromosome is composed of two chromatids. When does a chromatid become a chromosome? It is generally accepted that once the sister chromatids separate at the beginning of anaphase, each chromatid is a chromosome.

165

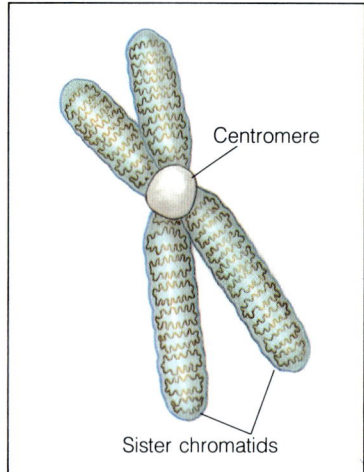

called nucleosomes (NOO-klee-oh-sohmz). The nucleosomes interact with one another to form a thick fiber, which is shortened by a system of loops and coils. The result is the tightly packed chromosomes that are seen through a light microscope in dividing cells. The tight packing of nucleosomes may be necessary in order to form a compact structure that can be separated during mitosis.

CHROMOSOME STRUCTURE After DNA replication, the chromosomes become visible by condensing. This is the beginning of mitosis. The chromosome contains two **chromatids** (KROH-muh-tihdz), or identical parts, which are often called sister chromatids. Each pair of chromatids is attached at an area called the **centromere** (SEHN-troh-meer). Centromeres are usually located near the middle of the chromatids, although some centromeres lie near the ends of the chromatids. A human cell entering mitosis contains 46 chromosomes, each of which consists of two chromatids.

The Cell Cycle

At one time biologists described the life of a cell as one mitosis after another separated by an "in-between" period of growth called interphase. We now know more and can represent some of the events in the life of a cell by using a concept known as the **cell cycle**. The cell cycle is the period from the beginning of one mitosis to the beginning of the next. **During a cell cycle, a cell grows, prepares for division, and divides to form two daughter cells, each of which begins the cycle anew.**

The cell cycle includes mitosis, a period of active division, and interphase, a period of nondivision during which other processes take place. It also includes a process in which cytoplasm and its contents divide, forming the two daughter cells. You will learn about this process, called cytokinesis, later in this section. Mitosis is represented as the M phase. During mitosis, the nucleus divides into two nuclei. Interphase is usually divided into three phases: G_1 (gap 1), S (DNA synthesis), and G_2 (gap 2). G_1 and G_2 are not really "gaps" in which nothing takes place. They are periods of intense growth and activity.

The time required to complete a single turn of the cycle is the time required for a cell to reproduce itself. Not all cells move through the cell cycle at the same rate. In the human body, most muscle cells and nerve cells do not divide at all once they have developed. In contrast, the cells that line the organs of the digestive system grow and divide rapidly. In fact, these cells may pass through a complete cycle every six hours.

Interphase

Within the normal cell cycle, **interphase**, or the period between cell divisions, can be quite long, whereas the actual divi-

Figure 8–9 *The diagram of a chromosome shows that it consists of two chromatids attached by means of a centromere. A human chromosome is shown as it appears through an electron microscope.*

BACKGROUND INFORMATION
THE CELL CYCLE

A precisely controlled sequence of events allows the cell to pass from one phase of the cell cycle to the next. For example, in the G_1 phase, DNA is replicated, followed by DNA synthesis in the S phase. If for some reason the cell is not going to divide, then DNA is not replicated in the G_1 phase and the cell is "locked" into this part of the cycle. Another example is the manufacture of proteins. Most cellular protein is made in all phases of interphase; the exceptions are certain proteins that are important in setting the "timing" of the cell cycle, and these are made only in specific phases, either G_1 or G_2.

8–2 (continued)

Enrichment
More capable students may be asked to do further research into the role of histones in the coiling of the DNA helix as discovered by Olins and Woodcock.

Reinforcement/Reteaching
Two pipe cleaners wound around each other once, with a push pin through the area where they join, will help students visualize the idea that at the beginning of mitosis, each chromosome consists of two chromatids joined at the centromere.

Content Development
Emphasize that the cell cycle includes both mitosis and interphase, and that although the cell is actively dividing only during mitosis, interphase is also an important stage. Although the events that occur during interphase are not easily visible, it is during this time that the cell actually carries out its main functions and is synthesizing the materials that are vital for growth, development, and cell division. To stress the importance of the processes occurring

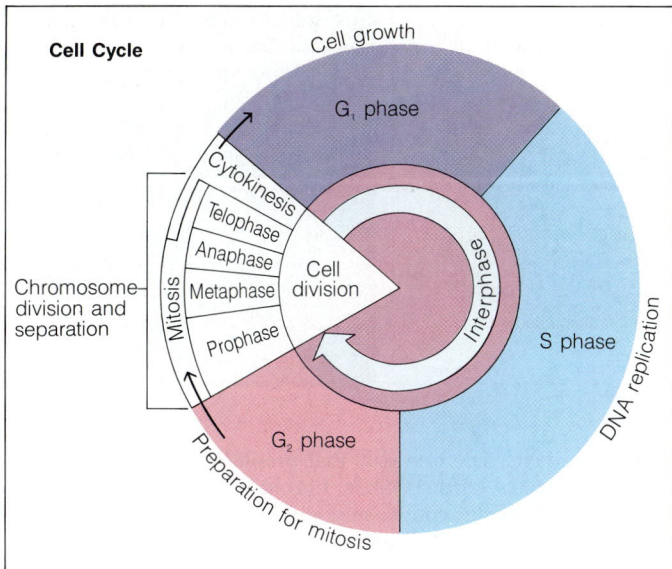

Figure 8–10 The cell cycle includes mitosis, cytokinesis, and the three phases of interphase. What are the names of the three phases of interphase? ❶

Figure 8–11 Interphase is the part of the cell cycle that occurs before mitosis can take place. Notice that the chromatin in the nucleus appears as an indistinct mass of threadlike structures. Just prior to cell division, the centrioles duplicate and begin to move apart, as shown in the diagram.

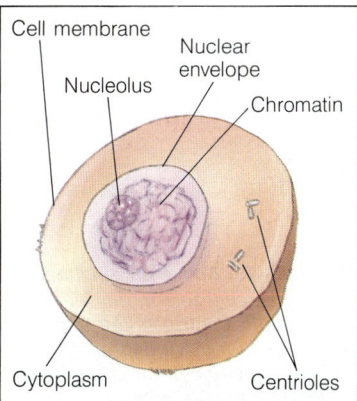

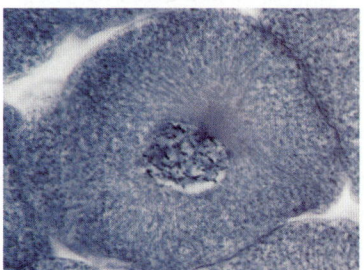

sion of the cell takes place quickly. As you read, interphase is divided into three phases: G_1, S, and G_2. Each of these phases is characterized by specific events.

The G_1, or gap 1, phase is a period of activity in which cellular growth and development take place. G_1 is followed by the S, or DNA synthesis, phase. During this phase, DNA replication takes place. Several proteins, including those associated with the chromosomes, are synthesized during the S phase.

Finally, when the S phase is completed, the cell enters the G_2, or gap 2, phase. This phase is usually the shortest of the three phases of interphase. The G_2 phase involves the synthesis of organelles and materials required for cell division.

During interphase, the nucleus is active in synthesizing messenger RNA in order to direct cellular activities. Although the cell seems to be "quiet" during interphase, it is actually a period of intense activity. Proteins are made; DNA is copied; ATP is made and utilized. In multicellular organisms, cells that are specialized for jobs such as secretion, movement, and signaling do most of their work during interphase.

Prophase

Recall that mitosis, or the M phase of the cell cycle, is the process by which the cell's nucleus divides into two nuclei. Mitosis may last anywhere from a few minutes to several days, depending on the type of cell. **Mitosis is divided into four phases: prophase, metaphase, anaphase, and telophase.**

The first phase of mitosis, **prophase**, is the longest phase, frequently taking 50 to 60 percent of the total time required to

TEACHING SUPPORT

Laboratory Manual
- Determining the Time Needed for Mitosis, p. 125

ESL STRATEGY

Stress the importance of learning about prefixes and suffixes to more easily understand the meaning of many scientific and nonscientific words. Tell students that many long words are much easier to understand if you break them up into their components. Reinforce the meaning of the term chromosome by showing its derivation. Both *chromo* and *chroma* are of Greek origin and mean "color." *Soma* is also from the Greek and means "body." Thus the word chromosome means "color body," although chromosomes are not visible until they have been stained.

Give students the following lists of prefixes and suffixes. Have them practice "decoding" the vocabulary words below.

Prefix
ana = up, chroma = color, cyto = cell, inter = between, meta = change or center, mitos = thread, pro = forward, telo = end

Suffix
id = compound, in = protein, kinesis = motion, osis = state, phase = stage

Vocabulary Word
anaphase, chromatid, chromatin, cytokinesis, interphase, metaphase, mitosis, prophase

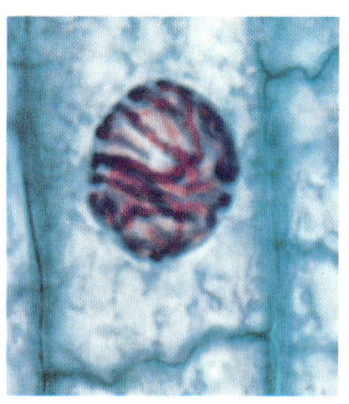

Figure 8–12 During prophase, the first stage of mitosis, the chromatin condenses into distinct chromosomes and the nucleolus disappears. The photomicrographs show prophase in a plant cell (left) and an animal cell (right).

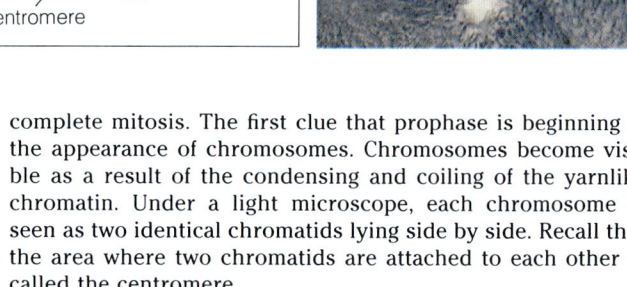

Figure 8–13 Notice the ringlike shape of a centriole in this electron micrograph. Centrioles are cellular structures that are found near the nuclear envelope of an animal cell and play a part in mitosis.

complete mitosis. The first clue that prophase is beginning is the appearance of chromosomes. Chromosomes become visible as a result of the condensing and coiling of the yarnlike chromatin. Under a light microscope, each chromosome is seen as two identical chromatids lying side by side. Recall that the area where two chromatids are attached to each other is called the centromere.

During prophase, the **centrioles**, two tiny structures located in the cytoplasm near the nuclear envelope, separate from each other and take up positions on the opposite sides of the nucleus. Centrioles are structures that contain tubulin, a microtubule protein.

During prophase, the condensed chromosomes become attached to fibers in the **spindle** at a point near the centromere of each chromatid. The spindle, a meshlike structure that helps move the chromosomes apart, develops from the centrioles. Recall from Chapter 5 that plant cells do not contain centrioles. However, plant cells form a spindle that is almost identical in structure to that of animal cells. Like the centrioles, the fibers of the spindle are composed of microtubules.

Near the end of prophase, the coiling of the chromosomes becomes tighter. In addition, the nucleolus disappears and the nuclear envelope breaks down.

Metaphase

As prophase ends, **metaphase**, or the second phase of mitosis, begins. Metaphase is the shortest phase of mitosis. It often lasts only a few minutes.

During metaphase, the chromosomes line up across the center, or equator, of the cell. Microtubules connect the centromere of each chromosome to the poles of the spindle. Because of their starlike arrangement around the poles of the spindle, these microtubules are called asters, which is the Greek word for star. See Figure 8–14.

8–2 (continued)

Reinforcement/Reteaching
Instead of the usual exercise of simply drawing and/or labeling diagrams of the stages of mitosis, the following exercise might be used.

Give each team of students 3 pairs of pipe cleaners (each pair of the same color) and a length of string sufficiently long to form a "cell membrane" around the pipe cleaners. Ask selected students to describe the appearance and activities of the pipe cleaner "chromosomes" during each of the stages of mitosis: prophase, metaphase, anaphase, and telophase. Have all the teams move their chromosomes to the correct positions under your observation. After some practice moving the chromosomes, have students close their books and draw labeled diagrams of the stages of mitosis using their pipe cleaner and string "cell" as a model.

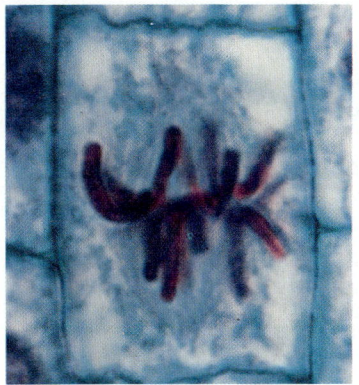 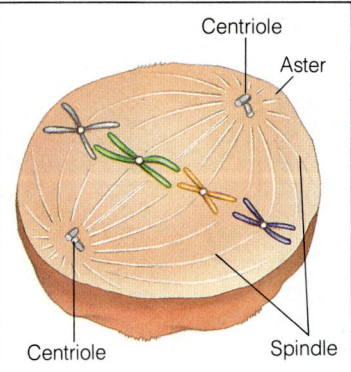 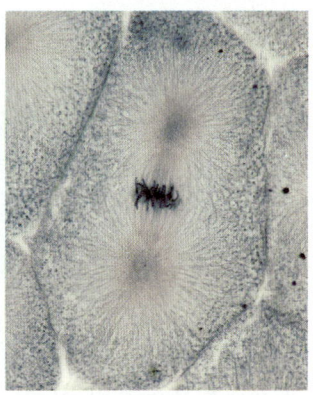

Anaphase

Anaphase, the third phase of mitosis, begins when the centromeres that join the sister chromatids split. This action causes the sister chromatids to separate, becoming individual chromosomes. As the two groups of chromosomes separate, the spindle itself grows longer. The chromosomes continue to move until they have separated into two groups near the poles of the spindle. See Figure 8–15. Anaphase ends when the movement of chromosomes stops.

For years, cell biologists have wondered what provides the force that separates the chromosomes during anaphase. Surprisingly, the answer is still not known. There is evidence that some proteins form bridges between the microtubules extending from opposite poles and use energy from ATP to push each microtubule toward its own pole. However, there is also a suggestion that the rapid assembly and breakdown of microtubules may cause chromosome movement. For the time being, this remains one of the cell's most intriguing riddles.

Figure 8–14 Notice how the chromatid pairs line up across the center of the plant cell (left) and the animal cell (right). This arrangement of chromatids marks metaphase, or the second phase of mitosis.

Figure 8–15 Anaphase, the third phase of mitosis, begins when the centromeres that link the sister chromatids split, allowing each chromatid to become an individual chromosome. Centrioles are present in the animal cell (right) but not in the plant cell (left).

BACKGROUND INFORMATION
ANAPHASE

Much controversy exists over the mechanism of chromosome movement during anaphase. In the past several years microtubule cross-bridging, microtubule assembly/disassembly, and actin-mediated force generation have all been proposed as mechanisms for anaphase movement. One of the problems in determining the mechanism is the fact that only an extremely small force is required to move a chromosome. A few investigators have calculated that the energy of no more than 30 or 40 ATP's would be sufficient. The small amount of energy used makes it extremely difficult to pin down where that energy is coming from, and it is possible that anaphase will remain an interesting problem for many years.

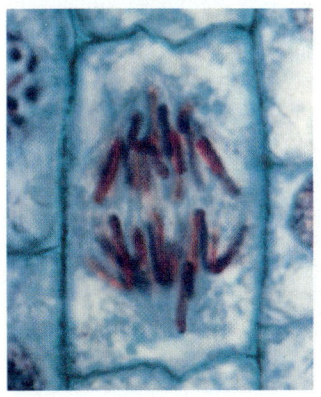 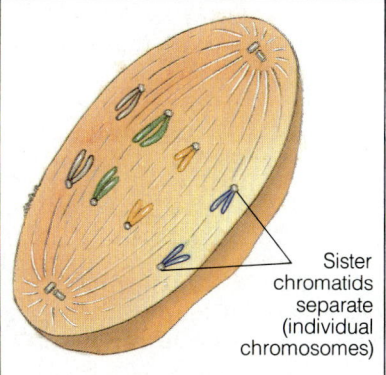 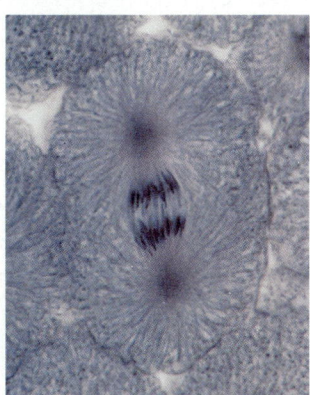

Focus/Motivation
A larger version of the pipe-cleaner mitosis model described previously may be made as a demonstration model visible to the entire class. Use a large piece of cardboard covered with fabric or felt. Several pieces of Velcro could be used to attach the string, leaving some slack to pinch in the string "membrane" during telophase. Removable pieces of electrical wire (attached with Velcro) could be used for the spindle, and pipe cleaners glued to small pieces of Velcro could be used to represent the chromosomes.

Content Development
Stress to students that the stages of mitosis do not all take the same amount of time to occur. You may have students "time" the four phases out loud and in unison— "Pro-o-o-o-phase" (very slowly), metaphase/anaphase (very quickly), telophase (normal speed).

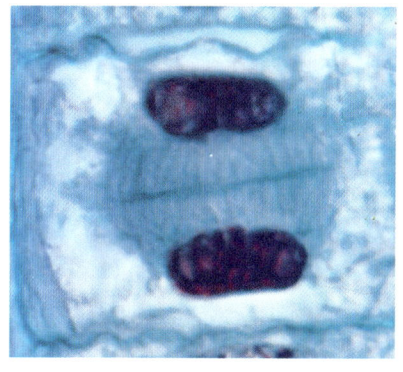
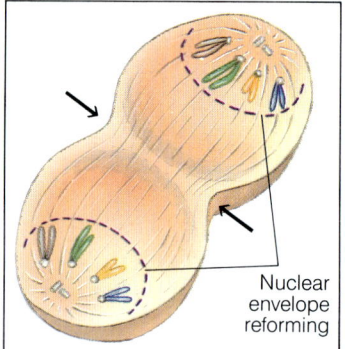
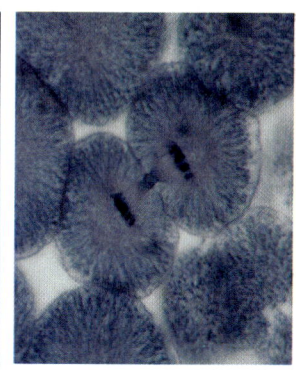

Figure 8–16 *In the final phase of mitosis, called telophase, two new daughter cells begin to form. Notice the two distinct groups of chromosomes in the plant cell (left) and in the animal cell (right).*

Telophase

Telophase is the final phase of mitosis. The chromosomes, which have been distinct and condensed all during mitosis, now begin to uncoil into a tangle of chromatin. This occurs in the two regions where the nuclei of the daughter cells will form. The nuclear envelope reforms around the chromatin, the spindle begins to break apart, and a nucleolus becomes visible in each daughter nucleus. Mitosis is now complete. However, the process of cell division is not.

Cytokinesis

As a result of mitosis, two nuclei—each with a duplicate set of chromosomes—are formed. Now it remains for the cytoplasm of the cell to divide. This is accomplished as cytokinesis quickly follows mitosis. Cytokinesis, you will recall, is the division of the cytoplasm into two individual cells.

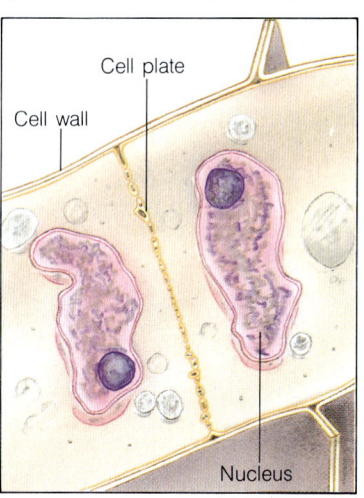

Figure 8–17 *Cytokinesis is the division of the cytoplasm and its contents into two individual daughter cells. In plant cells, the cytoplasm is divided by a cell plate.*

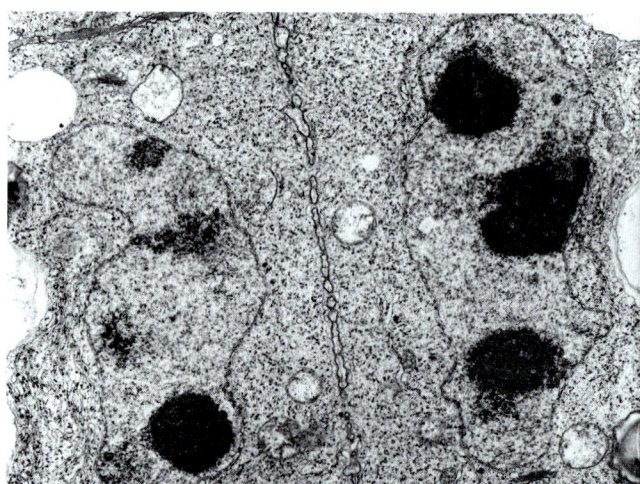

with the result of an orderly separation of chromosomes.

SECTION REVIEW 8–2

1. Mitosis is the process by which the nucleus of the cell is divided into two nuclei, each with the same number and kinds of chromosomes as the parent cell. Cytokinesis is the process by which the cytoplasm divides, thus forming two distinct cells.

2. A chromosome is made up of chromatin, which is composed of DNA and protein.

3. The cell cycle includes mitosis (a period of active cell division), interphase (a period of nondivision during which other processes take place), and cytokinesis (a process in which the

SCIENCE, TECHNOLOGY, AND SOCIETY

BREAKTHROUGH

A Useful Poison

The meadow saffron, *Colchicum autumnale*, is a flowering plant that grows wild in the meadowlands of England, Ireland, and middle and southern Europe. Its lovely flowers, which range in color from lavender to white, bloom in the fall. But the meadow saffron's flowers are deceiving. Though beautiful, the colorful meadow saffron contains a poison known as colchicine (KAHL-chih-seen).

In the laboratory, however, colchicine is a very useful drug. Colchicine stops cell division, making it important in the study of chromosomes. When colchicine is added to a culture of growing cells, it attaches to the microtubules in the spindle and causes them to disintegrate. As a result, the spindle is destroyed and mitosis is stopped at metaphase. When the cells are examined under a light microscope, each chromosome is visible as a pair of chromatids that are joined at the centromere.

Laboratory technicians routinely use colchicine because it enables them to count and measure chromosomes. Cell biologists use the drug as a method of studying the assembly and disassembly of microtubules. Scientists were able to find an important use for a very poisonous drug.

Cytokinesis can take place in a number of ways. In most animal cells, the cell membrane moves inward until the cytoplasm is pinched into two nearly equal parts, each part containing its own nucleus and cytoplasmic organelles. In plants, a structure known as the cell plate forms midway between the divided nuclei. The cell plate gradually develops into a separating membrane. A cell wall begins to appear in the cell plate.

8-2 SECTION REVIEW

1. What is mitosis? Cytokinesis?
2. Describe the structure of a chromosome.
3. Name the three main stages of the cell cycle.
4. Briefly describe the three phases of interphase.
5. **Critical Thinking—Summarizing Information** What are the four phases of mitosis?

171

LABORATORY INVESTIGATION

OBSERVING ONION ROOT TIPS

Before the Lab
1. One week to 10 days before the lab: Insert toothpicks into an onion bulb and suspend it in a beaker of water so that the bottom of the bulb is under water. Allow roots to grow to a length of at least 1 cm.
2. One day before the lab: Cut a 3-mm piece from the tip of each root and place it in a beaker of ethanol-acetic acid fixative. Try to do this at noontime because the occurrence of mitosis peaks at this time (as well as at midnight).
3. Prepare the fixative with 75 mL of ethanol to every 25 mL of glacial acetic acid. Use 5 M (18%) hydrochloric acid. Work only where there is sufficient ventilation, as the fixative is volatile.
4. Try to keep the roots in the fixative for at least 4 hours but not longer than 48 hours.

Pre-Lab Discussion
Show students the onion bulb and ask why the root tips are the only parts that will be used in this investigation. (Active mitosis is occurring in the rapidly growing roots.) Review the phases of mitosis and ask students to develop a hypothesis as to the relative number of cells they might expect to observe in each phase. Explain that the fixative will "stop the action" of the dividing cells and preserve them, while the purpose of the acid is to soften the cell walls to make their root-tip "squash" easier.

Skills Development
Students will use the following skills while completing this laboratory investigation.
1. Manipulative
2. Observing
3. Hypothesizing
4. Interpreting
5. Classifying
6. Inferring
7. Comparing
8. Diagramming
9. Generalizing

Safety Tips
Caution students to wear safety goggles until they are ready to observe their slide under the microscope. Review the safety procedures for the safe handling of acids and scalpels and caution students to use the forceps, not their fingers, to pick up the root tips. Use plastic coverslips, as glass may very likely break when "squashing" the root tip. Caution students that toluidine blue can stain clothing.

Teaching Strategy
1. Caution students to use *tiny* pieces of root tip. The success of this lab depends on their abil-

---

LABORATORY INVESTIGATION

OBSERVING ONION ROOT TIPS

PROBLEM
What are the stages of the cell cycle in a plant cell?

MATERIALS (per group)

onion root tips	timer or clock with second hand
safety goggles	2 glass slides
metric ruler	2 medicine droppers
scalpel	toluidine blue
ethanol-acetic acid fixative	probe
3 50-mL beakers	paper towels
glass-marking pencil	coverslip
hydrochloric acid	pencil with eraser
forceps	microscope

PROCEDURE

1. Label the three 50-mL beakers with the glass-marking pencil as follows: fixative, hydrochloric acid, water. Put on your safety goggles and add 10 mL of each liquid to the appropriately labeled beaker. **CAUTION:** *Because acids can burn the skin, handle them with care.* Place the beaker containing the water aside for now.
2. Using the forceps, place the onion root tips in the hydrochloric acid for 4 minutes.
3. After 4 minutes, use the forceps to remove the root tips from the hydrochloric acid. Place them in the fixative for 4 minutes.
4. After 4 minutes, use the forceps to remove the root tips from the fixative and place them on a clean glass slide. Holding the root tips with the forceps, use the scalpel to cut off about 1 mm from each root tip. **CAUTION:** *Be careful when using a scalpel.* Discard the rest of the root tips.
5. Using the forceps, place one root tip on a clean glass slide and then place the glass slide on a paper towel.
6. With a medicine dropper, place a few drops of toluidine blue on the root tip. Allow the toluidine blue to remain on the root tip for 2 minutes. After 2 minutes, use a paper towel to absorb the excess liquid.
7. Using a clean medicine dropper, place two drops of water on the root tip. Cover it with a coverslip. Use the probe to lower the coverslip over the slide.
8. Place a paper towel over the slide and with the eraser end of the pencil gently press down on the area covered by the coverslip. This will squash the root tip. **CAUTION:** *Do not press so hard as to break the coverslip or the glass slide.*
9. With the microscope, examine the onion root tip under low power to find cells in various stages of the cell cycle. (**Note:** If you cannot clearly distinguish the stages of the cell cycle, repeat steps 5 through 9 using another root tip.)
10. After you have located the various stages of the cell cycle under low power, switch to high power. Draw and label a cell from each stage of the cell cycle. Be sure to identify each stage.

OBSERVATIONS
1. Describe what is happening in each stage of the cell cycle.
2. What stage occurs most frequently?
3. How is interphase different from the other stages?
4. What is the color of the chromosomes? Of other cell structures?

ANALYSIS AND CONCLUSIONS
1. What is the purpose of the toluidine blue?
2. In which area of the onion root tip do most of the cells appear to be undergoing mitosis?
3. How can you tell that mitosis is a continuous process?
4. Do some stages of the cell cycle occur more frequently than others? Identify the stage(s) and explain why.

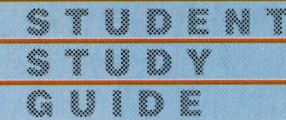

SUMMARIZING THE CONCEPTS

The key concepts in each section of this chapter are listed below to help you review the chapter content. Make sure you understand each concept and its relationship to other concepts and to the theme of this chapter.

8-1 Cell Growth

- Living things grow because they produce more and more cells.
- As a cell increases in size, its volume increases at a faster rate than its surface area.
- The amount of material entering and leaving a cell and the amount of DNA in the nucleus help to limit the size to which a cell can grow.
- Cell division is the process whereby the cell divides into two daughter cells.
- Cancer is a disorder in which some cells have lost the ability to control their rate of growth.

8-2 Cell Division: Mitosis and Cytokinesis

- Cell division in eukaryotic cells occurs in two main stages: mitosis and cytokinesis.
- Mitosis is the process by which the nucleus of the cell is divided into two nuclei, each with the same number and kinds of chromosomes as the parent cell.
- Cytokinesis is the process by which the cytoplasm divides, forming two distinct cells.
- Chromosomes are cellular structures that contain the genetic information that is passed on from one generation to the next. These structures are made up of chromatin, which is composed of DNA and protein.
- The two identical parts of a chromosome, called chromatids, are attached to each other at a centromere.
- During the cell cycle, a cell grows, prepares for division, and divides to form two daughter cells, which repeat the cycle.
- The cell cycle consists of three main stages: mitosis, interphase, and cytokinesis.
- Interphase, the period between cell divisions, is divided into three phases: G_1, S, and G_2.
- Mitosis has four phases: prophase, metaphase, anaphase, and telophase.
- Cytokinesis ends the process of cell division. In animal cells, the cell membrane moves inward until the cytoplasm is pinched in two, each half containing its own nucleus and cytoplasmic organelles. In plant cells, a cell plate forms midway between the divided nuclei, then a cell wall appears in the cell plate.

REVIEWING KEY TERMS

Vocabulary terms are important to your understanding of biology. The key terms listed below are those you should be especially familiar with. Review these terms and their meanings. Then use each term in a complete sentence. If you are not sure of a term's meaning, return to the appropriate section and review its definition.

8-1 Cell Growth
cell division

8-2 Cell Division: Mitosis and Cytokinesis
mitosis
cytokinesis
chromosome
chromatin
chromatid
centromere
cell cycle
interphase
prophase
centriole
spindle
metaphase
anaphase
telophase

Observations

1. During interphase, a cell with a complete nucleus is visible. During prophase, chromosomes appear and the nucleolus and nuclear envelope disappear. During metaphase, the chromosomes line up across the cell's center. During anaphase, the chromosomes separate. During telophase, the chromosomes are at opposite ends of the cell, the nuclear envelope and nucleolus reappear, and a cell plate forms.
2. Interphase.
3. During interphase, there seems to be no active division going on.
4. The chromosomes are purple. The other cell structures are colorless.

Analysis and Conclusions

1. Toluidine blue stains the chromosomes and makes them more visible.
2. At the very end of the onion root tip.
3. All of the cells are in a different phase of mitosis.
4. Interphase appears more frequently than any of the other phases because cells remain in interphase longer than in any other phase.

Going Further: Enrichment

1. Prepared slides of animal cell mitosis should be available for students to observe. *Ascaris* (a roundworm) or whitefish embryos are frequently used for this purpose. You may want students to draw the animal cells in the various stages of mitosis, emphasizing those structures that animal and plant cells do not have in common.
2. More capable students, as a special project, may be interested in doing a squash preparation of the salivary glands of a *Drosophila* larva in order to study its giant chromosomes. Instructions for the procedure may be found in genetics laboratory manuals or may be purchased from the supply companies from which you obtain the larvae.

---

ity to "squash" the tip so that it is one cell thick: if this is not done, it will be impossible to see the cells clearly.

2. When observing the root tip under the microscope, students will be more likely to see individual cells clearly at the edges rather than at the center.

3. If the cells seem poorly stained, try repeating step 6 before putting on the coverslip.

4. Have some commercially prepared slides of onion root tip available for students who cannot obtain a good "squash" or who cannot find all the mitotic stages on their slide (metaphase may be particularly difficult to find).

5. Students should avoid touching the root tip when they blot the excess toluidine blue.

CHAPTER REVIEW

CONTENT REVIEW

Multiple Choice
1. a 2. b 3. c 4. c
5. b 6. c 7. d 8. b

True or False
1. T
2. F 10
3. F interphase
4. T
5. T
6. F prophase
7. F anaphase
8. F plant

Word Relationships

A.
1. Anaphase, metaphase, and telophase are phases of mitosis. Interphase is not.
2. Chromatid and centromere are parts of a chromosome. A centriole is not.
3. Spindle and aster are microtubules; a chromosome is not.
4. Cell division, cytokinesis, and mitosis are all processes. An aster is a structure formed during cell division.

B.
5. Chromosomes
6. Prophase
7. Spindle
8. telophase

CONCEPT MASTERY

1. In most cases, living things grow because they produce more and more cells. Cell growth is limited by the amounts of food, oxygen, and water that enter the cell and waste products that leave.
2. As cells grow, their needs grow. If these needs are not met, then the cell slows down and becomes less efficient. To solve this problem, cells divide.
3. Chromatin, which is composed of DNA and protein, makes up the chromosomes. Chromosomes are structures that contain the genetic information. Chromatids are the individual strands that make up chromosomes. Centromeres are the structures that attach pairs of chromatids together.
4. The cell cycle includes mitosis (a period of active cell division), interphase (a period of nondivision during which other processes take place), and cytokinesis (a process in which the cytoplasm and its contents divide, forming two cells).
5. During interphase, cellular growth and development and DNA replication take place (G_1 phase); several proteins, including those associated with the chromosomes, are synthesized (S phase); and organelles and materials needed for cell division are synthesized (G_2 phase). Interphase is the part of the

CHAPTER REVIEW

CONTENT REVIEW

Multiple Choice

Choose the letter of the answer that best completes each statement.

1. Uncontrolled cell division is known as
   a. cancer.   c. cytokinesis.
   b. mitosis.   d. growth.
2. Chromatids are held together by a (an)
   a. centriole.   c. spindle.
   b. centromere. d. aster.
3. Each chromosome strand is called a (an)
   a. centriole.   c. chromatid.
   b. centromere. d. aster.
4. A cell has 12 chromosomes. How many chromosomes will each daughter cell have?
   a. 24   c. 12
   b. 6   d. 4
5. The phase of mitosis that is characterized by the lining up of chromosomes along the center of the cell is known as
   a. prophase.   c. anaphase.
   b. metaphase.   d. telophase.
6. During normal mitosis, which occurs first?
   a. spindle formation
   b. growth and development of daughter cells
   c. chromosome duplication
   d. cytoplasmic division of the cell
7. A cell that is undergoing mitosis is examined with a light microscope. The cell is most likely an animal cell if the
   a. chromosome pairs separate from each other.
   b. chromosomes twist about each other.
   c. nucleoli disappear.
   d. centrioles migrate.
8. A structure found during plant cell division that is not found during animal cell division is a
   a. centromere. c. cell membrane.
   b. cell plate. d. spindle.

True or False

Determine whether each statement is true or false. If it is true, write "true." If it is false, change the underlined word or words to make the statement true.

1. As a cell increases in size, its volume increases at a <u>faster</u> rate than its surface area.
2. As a result of mitosis, a cell having 10 chromosomes gives rise to two cells, each of which contains <u>20</u> chromosomes.
3. DNA replication occurs during <u>prophase</u>.
4. <u>Animal</u> cells contain centrioles.
5. <u>Interphase</u> is usually divided into three phases: G_1, S, and G_2.
6. <u>Anaphase</u> is the first phase of mitosis.
7. During <u>prophase</u>, the centrioles separate from each other and take up positions on the opposite sides of the nucleus.
8. In <u>animal</u> cells, a cell plate forms midway between the divided nuclei.

Word Relationships

A. *In each of the following sets of terms, three of the terms are related. One term does not belong. Determine the characteristic common to three of the terms and then identify the term that does not belong.*

1. anaphase, interphase, metaphase, telophase
2. chromatid, centromere, chromosome, centriole
3. spindle, aster, microtubule, chromosome
4. cell division, cytokinesis, mitosis, aster

B. *Replace the underlined definition with the correct vocabulary word.*
5. The structures in the cell that contain the genetic information are not visible in most cells except during mitosis.
6. The first phase of mitosis begins when the chromosomes become visible.
7. The meshlike structure that helps move the chromosomes apart develops from the centrioles.
8. In the final phase of mitosis, the chromosomes are found at opposite poles of the cell.

CONCEPT MASTERY

Use your understanding of the concepts developed in the chapter to answer each of the following in a brief paragraph.

1. How do living things grow? What are some factors that limit the growth of a cell?
2. Why do cells divide?
3. Distinguish between chromatin, chromosomes, chromatids, and centromeres.
4. List and describe the phases of the cell cycle.
5. What occurs during interphase? How are interphase and the cell cycle related?
6. Identify and describe the phases of mitosis.
7. Distinguish between cell division in an animal cell and cell division in a plant cell.
8. Why is cell division so complicated in eukaryotic organisms?
9. Compare spindles and asters.

CRITICAL AND CREATIVE THINKING

Discuss each of the following in a brief paragraph.

1. **Relating facts** The diagram below shows a phase of mitosis. Identify the phase of mitosis and indicate whether the cell most resembles that of a rose or a tiger. Give a reason for your answer.

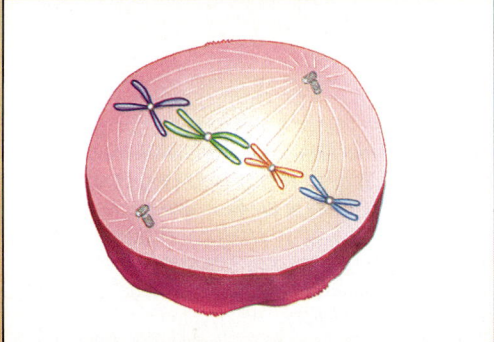

2. **Making calculations** A cell is 4μm long, 3μm wide, and 3μm high. Find the ratio of volume to surface area.
3. **Applying concepts** Is mitosis occurring in your body right now? If so, where? Explain your answer.
4. **Relating concepts** After muscle and nerve cells form in the body, they seldom undergo mitosis. Explain how this affects the human body.
5. **Relating cause and effect** What structure within the cell is most likely to play a major role in the development of generations of cancer cells? Explain your answer.
6. **Using the writing process** Suppose you were small enough to hitch a ride on a chromosome located in a cell that divides by mitosis. Describe what you would see happening during this process.

CAREERS IN BIOLOGY

Electron Microscopist, Forensic Laboratory Technician, Cell Biologist

Students interested in a biological career should be instructed to write for further information to the address listed beneath each career description. However, in consideration of the organizations that provide career information, please have only one student write to an organization. Do not instruct the entire class to write to every organization for the same career information. You may want to use the information provided to start a biology career file for your students.

UNIT 2

CAREERS IN BIOLOGY

Electron Microscopist

Electron microscopes use beams of electrons to magnify objects. People who operate electron microscopes to view tiny cell structures are called electron microscopists. They produce photographs called micrographs that allow researchers to learn more about the normal activities of cells.

To become an electron microscopist, the requirements are a high school education and extensive training in the use of an electron microscope.

For information write to the National Society for Histotechnology, P.O. Box 36, Larnham, MD 20706.

Forensic Laboratory Technician

Forensic laboratory technicians use scientific principles and instruments to analyze evidence found at the scene of a crime. They find clues by examining a variety of biological materials that includes hair, blood, fingernails, bones, and tissues. They also study fingerprints, firearms, and bullets. Forensic technicians might be called upon to help reconstruct the scene of a crime or to testify in court.

Most forensic laboratory technicians have a Bachelor's degree with a good background in biology and chemistry.

To receive career information write to the American Academy of Forensic Sciences, 225 S. Academy Blvd., Colorado Springs, CO 80910.

Cell Biologist

Cell biologists examine the structure of cells in plants and animals. They study individual parts of different cells to learn how each part functions and how it is affected by chemical and physical factors. Using microscopes, stains, and other instruments, cell biologists can observe the growth and division of various cells.

Cell biologists usually have medical or Ph.D. degrees. They work in hospitals, universities, and research laboratories.

For information contact the American Society for Cell Biology, 9650 Rockville Pike, Bethesda, MD 20814.

HOW TO CHOOSE A TECHNICAL SCHOOL

Technicians are workers who are trained beyond high school in a specialized field of technology. A technical education mixes theory with practice and usually takes one to three years, depending on the field. Technicians receive either a certificate or an associate degree.

When choosing a technical school, seek reliable evaluations from past graduates and job placement statistics. Although many schools advertise, only the better schools are accredited by the government.

For information write to the National Association of Trade and Technical Schools, 2251 Wisconsin Ave., NW, Washington, DC 20007 and ask for a copy of the Handbook of Trade and Technical Careers and Training.

FROM THE AUTHORS

The first time I heard the term DNA mentioned was when I was in junior high. Although our textbook barely mentioned DNA, my science teacher had made a point of keeping up to date. He put up newspaper clippings about the double helix model; told us about the Nobel prize that was shared by Watson, Crick, and Wilkins; and even built a crude model of the molecule. He did his best to let us know that biology didn't stop when our textbook was written.

Way back then (when John F. Kennedy was in his last year in the White House) it was already clear that DNA was the future. The question was not whether DNA was important—it was clear that it was—but how the information in DNA was coded, expressed, and controlled. Three decades later, we know enough to answer many of the questions that seemed so puzzling to me in junior high. I could do better in Mr. Zong's mid-term exam today, but I think it's too late to improve my grade. One thing I can't improve on, however, is the sense he gave me of the future—the feeling that science was changing tomorrow.

Even as they change, things stay the same. It's tempting to think that we've solved all of the big questions, but it would be just as wrong to hold that attitude as it was in 1963. So look around, and pay attention to those newspaper clippings and science magazines in your classroom. No book contains the last word about biology. Tomorrow is up there on the wall.

Ken Miller

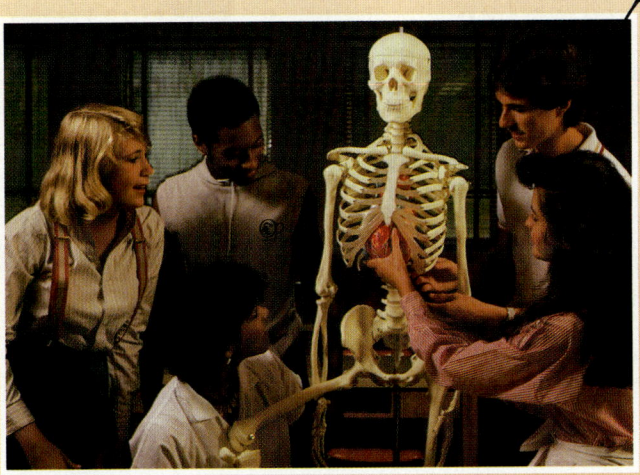
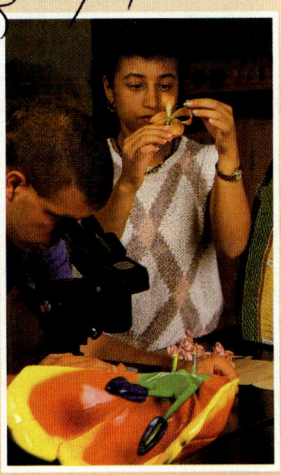

FROM THE AUTHORS

Kenneth R. Miller

In this feature Ken Miller describes some of his early experiences in biology. In addition, he tries to provide students with a sense of the excitement and ever-changing world of biology. You might want to have students predict some of the changes that will occur in the near future and how those changes will impact the study of biology for future students.

UNIT 3
Continuity of Life

UNIT OVERVIEW

Unit 3 examines the science of genetics from its beginnings in Mendel's experiments to current research in and applications of molecular genetics. First, students learn about basic patterns of inheritance and principles of heredity and discover how these relate to probability and to meiosis. Next, students focus on genetic phenomena at the level of genes and chromosomes. Students then apply what they have learned to understanding specific examples of human heredity. Finally, they learn about artificial methods of manipulating inheritance, such as breeding techniques and genetic engineering, and they read about how current and future applications of molecular genetics may impact human lives.

UNIT OBJECTIVES

1. Relate the principles of dominance, segregation, and independent assortment to patterns of inheritance and to meiosis.
2. Solve genetics problems using a Punnett square.
3. Explain how changes in chromosomes affect heredity.
4. Discuss some mechanisms for modifying and regulating gene expression.
5. Describe how certain human traits are transmitted from one generation to the next.
6. Describe how breeding techniques affect future populations of organisms.
7. Discuss major techniques used in molecular genetics.

UNIT 3
Continuity of Life

The spiraling molecule called DNA is a common thread that binds all organisms on Earth together. Bacteria and other tiny single-celled organisms contain DNA, as do mushrooms, apple trees, snails, birds, and humans. In fact, all living things require DNA to grow, develop, and carry out the functions of life.

DNA is also necessary for life to continue from one generation to the next. Encoded in the pattern of chemicals within the molecule are all the instructions needed to transform a single fertilized egg into a complex, many-celled organism. Some of the most fascinating aspects of biology involve the transmittal of DNA from parents to offspring and the ways in which DNA controls the development of living things.

DISCOVERY LEARNING
LIKES AND UNLIKES

1. Collect magazine photographs of different individuals of the same kind of organism. For example, you may collect photographs of cats, dogs, horses, or roses. (NOTE: *Obtain permission before cutting up magazines.*) Alternatively, you may make your own drawings based on photographs in books or magazines, or take your own photographs.
2. Select 5 or 6 easily observable traits of the organism you selected. How do the individuals differ for each trait? How are they similar?
3. Prepare a bulletin board display and a brief report to summarize your findings.

INTRODUCING UNIT 3

Using the Visuals

Have students read the unit introduction on page 178 and examine the illustration on pages 178 and 179. The object that spirals across the page is an artist's rendition of a DNA molecule. The photographs along the strand represent each of the five kingdoms of organisms. These photographs show a color-enhanced transmission electron micrograph of the bacterium *Clostridium perfringens* (top left); a color-enhanced scanning electron micrograph of a diatom (bottom left); a scarlet macaw (top center); an apple tree in bloom (bottom center); and small-gilled mushrooms on a moss-covered log (right).

CHAPTERS
9
Introduction to Genetics
10
Genes and Chromosomes
11
Human Heredity
12
Genetic Engineering

CHAPTER DESCRIPTIONS

9 Introduction to Genetics
Chapter 9 introduces the science of genetics. Students learn about Mendel's famous experiments and discover how to use Mendel's principles and the rules of probability to solve genetics problems. They also learn about meiosis and discover its role in reproduction and inheritance.

10 Genes and Chromosomes
In Chapter 10 students learn about the chromosome theory of heredity and discover how gene linkage and crossing-over relate to gene mapping. They examine gene linkage, crossing-over, sex linkage, mutations, and gene interactions. Finally, they explore mechanisms for gene regulation in prokaryotes and in eukaryotes.

11 Human Heredity
In Chapter 11 students apply the principles of genetics to human heredity. They learn about the inheritance of benign traits such as ABO blood group as well as of genetic disorders such as sickle cell anemia, Down syndrome, and Huntington disease. They learn about techniques of prenatal diagnosis and briefly explore the ethical issues involved with these techniques.

12 Genetic Engineering
In Chapter 12 students learn about artificial methods for influencing heredity, such as breeding strategies, inducing mutations, and genetic engineering. They examine the techniques used in genetic engineering and learn about several recent innovations in recombinant DNA technology.

- **What is the purpose of this illustration?** (To show that DNA is a part of all life on Earth.)
- **What is DNA?** (Deoxyribonucleic acid: a long, spiraling, complex molecule.)
- **Why is DNA important?** (DNA codes for all the proteins and enzymes that organisms need to grow, function, and develop. DNA is also the molecule of heredity; it is transmitted from generation to generation.)

DISCOVERY LEARNING
LIKES AND UNLIKES

Begin your introduction to the unit by having students perform the Discovery Activity. You may wish to have extra old magazines or catalogs available. Make sure each student participates in the activity and answers its questions. Afterward, encourage students to share their photographs and observations with the class.

CHAPTER 9 Introduction to Genetics

Section	Laboratory Investigations and Activities
9–1 The Work of Gregor Mendel pages 181–189 THEME: Patterns of Change	**Teacher Edition** DEMONSTRATION: Picture This, p. 180D **Laboratory Manual** The Principles of Genetics, p. 129
9–2 Applying Mendel's Principles pages 190–192 THEMES: Patterns of Change, Stability	**Student Edition** LABORATORY INVESTIGATION: Simulating a One-Factor Cross, p. 200 **Teacher Edition** DEMONSTRATION: Genetic Coin Toss: Heads or Tails? p. 180D **Laboratory Manual** Biochemical Genetics, p. 135 **Teaching Resources** HANDS-ON ACTIVITY: Flip Out! p. 63
9–3 Meiosis pages 193–199 THEMES: Patterns of Change, Scale and Structure	**Teaching Resources** HANDS-ON ACTIVITY: A Model of Meiosis, p. 61
Chapter Review pages 201–203	

Teacher Resources

Books

Ayala, F.J., and K.R. Lewis. *Modern Genetics*. Benjamin-Cummings, 1980.

Goodenough, U. *Genetics*. Saunders, 1984.

Ptshne, Mark. *A Genetic Switch*. Blackwell Scientific Publications and Cell Press, 1986.

Schlegel, Robert A., et al., eds. *Molecular Regulation of Nuclear Events in Mitosis and Meiosis*. Academic Press, 1987.

Strickenberger, M.W. *Genetics*, 3rd ed. Macmillan, 1986.

J. H. Miller, D. T. Suzuki, A. J. Griffiths, R. C. Lewontin and W. M. Gelbart. *An Introduction to Genetic Analysis*, 5th ed. Freeman, 1993.

Russell, Peter J. *Genetics*. Little, Brown, 1986.

CHAPTER PLANNING GUIDE

Other Activities	Media and Technology
Study Guide Section 9–1, p. 83	**Biology Media Guide** page 10
Teaching Resources ACTIVITY: Using Punnett Squares to Predict the Outcomes of Crosses, p. 3 ACTIVITY: Analyzing Genetic Data, p. 11 **Study Guide** Section 9–2, p. 87	**Biology Media Guide** page 10
Teaching Resources VOCABULARY REVIEW: Word Game, p. 1 ACTIVITY: Should This Dog Be Called Spot? p. 15 **Study Guide** Section 9–3, p. 90	**Biology Transparencies** 13: Meiosis
Teaching Resources Chapter Test A, p. 21 Chapter Test B, p. 25	**Computer Test Bank** Chapter Test, p. 91

Audiovisuals

Genetics and Heredity. 2 filmstrips with cassettes. Ward.
Mendel's Laws. Film or video. Coronet.
Chromosomes and Genes (Meiosis). Film or video. Coronet.
Introduction to Meiosis. Sound filmstrip. Ward.
Introduction to Mendelian Inheritance. Sound filmstrip. Ward.
Meiosis. Film. Encyclopaedia Britannica Educational Corporation.
An Introduction to Heredity. 2 sound filmstrips. National Geographic Society.

CHAPTER 9 Introduction to Genetics

CHAPTER OVERVIEW

How are genetic traits passed from one generation to the next? This is a question that has always intrigued and fascinated but has been answered only in recent times. There is still much to be learned.

The principles of heredity discovered by Gregor Mendel serve as a foundation for the study of genetics. Working quietly in a monastery garden, Mendel's meticulous observations on heredity in garden peas revealed how traits are inherited as distinct factors. He was also able to demonstrate how one factor could completely mask the effect of a second factor. Mendel also suggested that pairs of factors separated during the formation of reproductive cells, so that the offspring inherited only one factor from each parent. Finally, he demonstrated that factors were inherited independently of one another.

Today, Mendel's factors are known as genes. We can make certain predictions about the probability of inheriting one gene or another. Students will learn that they can make predictions also by applying Mendel's principles through a device known as a Punnett square.

The sex cells, or gametes, that carry the genetic information are formed by a special kind of cell division known as meiosis. The number of chromosomes, the structures that contain the genetic information, is reduced by half during meiotic division. By fertilization the full number of chromosomes is restored. These two processes—meiosis and fertilization—allow for infinite variety in the selection and recombination of genetic traits.

9–1 THE WORK OF GREGOR MENDEL

Section Focus 9–1

The purpose of this section is to introduce students to the pioneering studies in genetics by Gregor Mendel. By reviewing Mendel's breeding experiments with garden pea plants, this section shows how the results of these experiments led to Mendel's conclusions about heredity: that traits are controlled by factors (today known as genes); that some genes are dominant, others are recessive; that pairs of genes segregate during gamete formation and do so independently of other gene pairs.

Performance Objectives 9–1

1. Describe the experiments of Gregor Mendel.
2. Use a Punnett square to explain how pairs of genes segregate during an F_1 cross.
3. Use a two-factor cross to demonstrate how genes assort independently.

Science Terms 9–1

heredity   p. 181
genetics   p. 181
self-pollination   p. 182
cross-pollination   p. 182
purebred   p. 183
trait   p. 183
hybrid   p. 184
gene   p. 184
allele   p. 184
dominance   p. 184
dominant   p. 184
segregation   p. 186
Punnett square   p. 186
gamete   p. 186
phenotype   p. 186
genotype   p. 187
homozygous   p. 187
heterozygous   p. 187
independent assortment   p. 188

9–2 APPLYING MENDEL'S PRINCIPLES

Section Focus 9–2

The purpose of this section is to demonstrate how probability can be used to predict the expected outcome of genetic cross. The Punnett square is a device that can be used to calculate expected genetic ratios in one-factor and two-factor genetic crosses.

Performance Objectives 9–2

1. Discuss how the expected results of a genetic cross may differ from the observed results.
2. Identify the factors that may cause the observed results of a genetic cross to differ from the expected results.
3. Demonstrate how a Punnett square can be used to predict the results in a one-factor and two-factor genetic cross.

Science Terms 9–2

probability   p. 190

CHAPTER PREVIEW

9-3 MEIOSIS

Section Focus 9-3

The purpose of this section is to explain the need for meiosis, the reduction division of chromosomes, during gamete formation. An examination of the phases of meiosis illustrates how gametes are formed containing the haploid number of chromosomes selected at random from each homologous pair. The section concludes with a comparison of mitosis and meiosis, stressing how they differ.

Performance Objectives 9-3

1. Explain the need for reduction division of chromosomes during gamete formation.
2. Discuss how the phases of meiosis provide for the orderly reduction of chromosome number from diploid to haploid with random combinations of chromosomes.
3. Explain how the process of meiosis reveals the mechanism behind Mendel's conclusions about segregation and independent assortment on a cellular level.

Science Terms 9-3

homologous   p. 193
diploid   p. 193
haploid   p. 194
meiosis   p. 195
crossing-over   p. 195

DISCOVERY LEARNING

TEACHER DEMONSTRATIONS
Modeling

Picture This

The following demonstration can be used as an introduction to Chapter 9.

On the day before beginning this chapter, ask for several student volunteers in the class to bring in a photograph of the person they believe they most closely resemble.

The results should provide a variety of relations: parents, siblings, grandparents, aunts, uncles—and perhaps even a famous personality or two.

Permit each student volunteer to briefly explain the reasons for his or her choice. From these responses, the following ideas can be developed:

• There are traits that are passed from one generation to the next.
• Some traits do not appear in every generation. They may seem to skip a generation.
• Other traits seem to appear for the first time in one member (or generation) of the family.

(This should not be a mandatory activity for every student. Sensitivity should be shown toward students who may not want to participate. Not all members of every family are biologically related!)

Genetic Coin Toss: Heads or Tails?
You may want to perform this demonstration at the beginning of Section 9-2.

Exhibit a coin for the class to see and then select a student to assist you with this demonstration. Permit the student to examine the coin to confirm that it has both a heads side and a tails side.

• **What are the chances that the heads side will turn up when this coin is tossed?** (Students will answer quite readily 50:50.) Explain that these odds are our expected results and can also be expressed as 1:1 or $1/2$.

Permit the student to toss the coin and record the result on the chalkboard.

• **Now, if this coin is tossed again, what are the chances that the same side will turn up again?** (The chances still remain 1:1 or $1/2$.)

Every time the coin is tossed, the odds of turning up a particular side will revert to 1:1.

Permit the student to toss the coin for at least 9 more trials and record each result. Compare the actual results to the students' prediction of $1/2$ heads and $1/2$ tails.

• **As we increased the number of trials, how did our actual ratio of results compare to the expected ratio of results?** (The greater the number of trials, the closer the results to 1:1. If that was not the case, continue to toss the coin for several more trials and then compare this new result.)

CHAPTER 9

Introduction to Genetics

GUIDED ENQUIRY

Pose the following questions to students and have them record their responses. Point out to students that they will gain a better understanding of the key concepts if they read the chapter with these basic questions in mind. Upon completion of the chapter, pose the questions again. Ask students to compare their initial responses with those they have developed after reading the chapter.
- How are hereditary traits passed from parents to offspring?
- Why do some traits "disappear" in one generation and reappear in the next?
- Can you predict which traits will appear in the offspring?
- What must happen to the genetic material of a parent organism before its sex cells are formed?

CHAPTER 9

Introduction to Genetics

There are few things as graceful and magnificent as a running horse. A good combination of many inherited traits—fine-boned legs, swiftness, and stamina, for example—helps thoroughbred race horses compete at Churchill Downs in the Kentucky Derby.

Inheritance is something that we are all familiar with; each of us has a biological inheritance. As a child you may have been told that you "inherited" your eyes from your grandfather, your hair from your grandmother, your intelligence from your mother, and your artistic talents from your father. Everyday experience shows us that offspring resemble their parents.

But why do we take after our parents and grandparents? Why do people expect great things from a newborn foal if his father won the Kentucky Derby? Why do the kittens born of a yellow tabby and a black cat have fur colored somewhere between yellow and black? In this chapter you will discover the answers to these questions as you explore the nature of biological inheritance.

INTRODUCING CHAPTER 9

Using the Visuals
Have students observe the photograph on this page.
- **What is happening in this photograph?** (The winner is crossing the finish line in a horse race.)
- **These horses are known as "thoroughbreds." What does that name refer to?** (The parents of each horse are known and specially selected for their abilities as race horses.)
- **Why do horse breeders try to select the best horses to pro-**duce the next generation of foals? (They hope that the foals will inherit the desirable traits of the parents that made them winners.)

Point out to students that it is not always easy to predict which of the parents' traits will be inherited by offspring. Have students examine the guinea pigs in Figure 9–1.
- **Which of the mother's traits can you identify in her offspring?** (Coat color and texture.)
- **Do the babies have any characteristics that do not appear in**

GUIDE FOR READING

After you read the following sections, you will be able to

9–1 The Work of Gregor Mendel
- Discuss Mendel's experiments.
- Describe dominance, segregation, and independent assortment.

9–2 Applying Mendel's Principles
- Relate probability to genetics.
- Solve genetics problems using a Punnett square.

9–3 Meiosis
- Describe the process of meiosis.
- Compare meiosis and mitosis.

Journal Activity

YOU AND YOUR WORLD
For 14 years, Gregor Mendel conducted experiments to unlock the secrets of heredity. He used all his free time and performed many painstaking—even tedious—activities to achieve his goal. Is there some endeavor you feel committed to in a similar way? In your journal, write about the goal, hobby, or cause to which you are the most dedicated.

9–1 The Work of Gregor Mendel

Guide For Reading

- What are some of the experiments that Mendel performed?
- What do the terms dominance, segregation, and independent assortment mean?

Biological inheritance, or **heredity**, is the key to differences between species. Cats give birth to kittens, dogs produce puppies, and oak trees produce acorns from which, as the saying goes, mighty oaks may grow. Heredity, however, is much more than the way in which a few superficial characteristics are passed from one generation to another. Heredity is at the very center of what makes each species unique, as well as what makes us human. The branch of biology that studies heredity is called **genetics**.

Early Ideas About Heredity

Until the nineteenth century, the most common explanation for the resemblances between parent and offspring was the theory of blending inheritance. People reasoned that because both a male and a female were involved in producing offspring, each parent contributed factors that determined inheritance—factors that were blended in their offspring. The nature of these factors was unknown.

At first, the theory of blending inheritance seemed a reasonable explanation. It is common to see a little bit of both parents in a child. So it seemed fair to say that the characteristics of the mother and father have blended in making the new life. But in the last century biologists began to look at the details of heredity. When they did, they began to develop a very different

Figure 9–1 Why don't the two baby guinea pigs look like each other? Why are they different from their mother? The science of genetics helps us answer questions such as these.

COOPERATIVE LEARNING

Using preassigned lab groups or randomly selected teams, have groups prepare to participate in a review game on meiosis. Groups should generate a minimum of five questions about an assigned section of the text. You might want several groups to generate questions about mitosis to allow for review and comparison of these processes. The actual game might take the form of a teacher-made game board on the overhead projector or chalkboard. The game could contain the following elements: a starting point, an ultimate goal, positive and negative "chance" cards or steps, dice for movement, a shortcut to victory (made appropriately difficult), and an opportunity for competition between teams. Students should be aware of the rules of the game and the reward for winning. Each group should prepare a team symbol or game piece. This game activity might serve as a model for future cooperative learning assignments in which student groups produce games to be played by other teams.

Journal Activity

YOU AND YOUR WORLD
Have students keep their journal assignments in their portfolios. Ask for volunteers to share their entries with the class.

---

the mother? (Other coat colors and textures.)
- **Where might these traits have come from?** (The father or a grandparent.)

TEACHING STRATEGY 9–1

Focus/Motivation
Ask students to imagine what it would be like if human beings could grow to be only one of two heights: 1 meter tall or 2 meters tall.

- **If both parents were 1 meter tall, what height would you predict their offspring to be?** (1 meter tall.)
- **If both parents were 2 meters tall, what height would you predict their offspring to be?** (2 meters tall.)

- **However, if one parent were 1 meter tall and the other were 2 meters tall, what height would you predict their offspring to be?** (Remember, there are no individuals who are 1½ meters tall! We would have to observe the offspring to see how tall they would finally grow.)

181

MULTICULTURAL STRATEGY

Suggest that students look around the classroom and notice the ways in which each person is unique. Point out that many of a person's characteristics are inherited from previous generations. Remind students that although we may differ dramatically in appearance, as humans we are all more alike than we are different.

HISTORICAL NOTE
MENDEL AND THE THEORY OF EVOLUTION

This might be a good place to prepare students for one of the reasons why Mendelian genetics was so important. By the 1880s a number of scientists, critical of the theory of evolution, had pointed out a key flaw in Darwin's reasoning. Even if natural selection could "choose" favorable variations among individuals of a particular species, those variations might be "blended away" after just a few generations. This might have been a fatal flaw for the concept of evolution were it not for the existence of genes. As we now realize, genes do not blend. They are passed from one generation to the next without being altered, and this fact answered Darwin's critics.

9–1 (continued)

Content Development
Using the Visuals Have students examine Figure 9–2.
- **What is pollination?** (Pollination is the transfer of pollen from male to female flower parts, an important event in the sexual reproduction of flowering plants. The result of pollination is the formation of a new seed.)
- **How does cross-pollination differ from self-pollination?** (In cross-pollination, pollen is transferred between parts of two different flowers. In self-pollination, pollen is transferred between parts of the same flower.)
- **How do pea plants normally reproduce?** (Self-pollination.)
- **Why did Mendel prevent the peas from self-pollinating?** (Mendel wanted to study the offspring from crosses of different plants. By cross-pollinating flowers and preventing them from self-pollinating, Mendel could control and be certain of each seed's parentage.)

Skills Development
Guided Practice
Skill: Interpreting charts
Using the Visuals Have students examine Figure 9–3, which shows the seven pea traits that Mendel studied. Point out that each trait has two contrasting forms.

---

view. The work of the Austrian monk Gregor Mendel was particularly important in changing people's views about how characteristics are passed from one generation to the next.

Gregor Mendel

Gregor Mendel was born in 1822 to peasant parents in what is now the Czech Republic. He did very well in school and entered a monastery in the town of Brno at age 21. Four years later he was ordained a priest. The monastery was a center of scientific learning. In 1851, Mendel was sent to the University of Vienna to study science and mathematics. He returned two years later and spent the next fourteen years working in the monastery and teaching at the high school in Brno.

In addition to his teaching duties, Mendel had charge of the monastery garden. It was in this ordinary garden that he was to do the work that revolutionized biological science.

From his studies in biology, Mendel had gained an understanding of the sexual mechanisms of the pea plant. Pea flowers have both male and female parts. Normally, pollen from the male part of the pea flower fertilizes the female egg cells of the very same flower. Because the pollen produced by the plant fertilizes the egg cells of that very same plant, peas are said to produce seeds by **self-pollination**. Seeds produced by self-pollination inherit all of their characteristics from the single plant that bore them.

Mendel learned that self-pollination could be prevented. He carefully cut the male parts off all the flowers of one plant and the female parts off all the flowers of another plant. He then pollinated the two plants by dusting the pollen from one plant onto the flowers of the other plant. The fertilization of a plant's egg cells by the pollen of another plant is known as **cross-pollination**. Cross-pollination produces seeds that are the offspring of two different plants. With this technique, Mendel was able to cross plants with different characteristics.

Figure 9–2 Mendel used the garden pea in his experiments on heredity (top). Garden peas usually reproduce by self-pollination because the male and female reproductive parts are tightly enclosed within the flower's petals (bottom left). One method Mendel used to cross-pollinate his pea plants was to cut off the male parts of a flower (thus preventing self-pollination) and then dust the pollen from another plant onto that flower (bottom right).

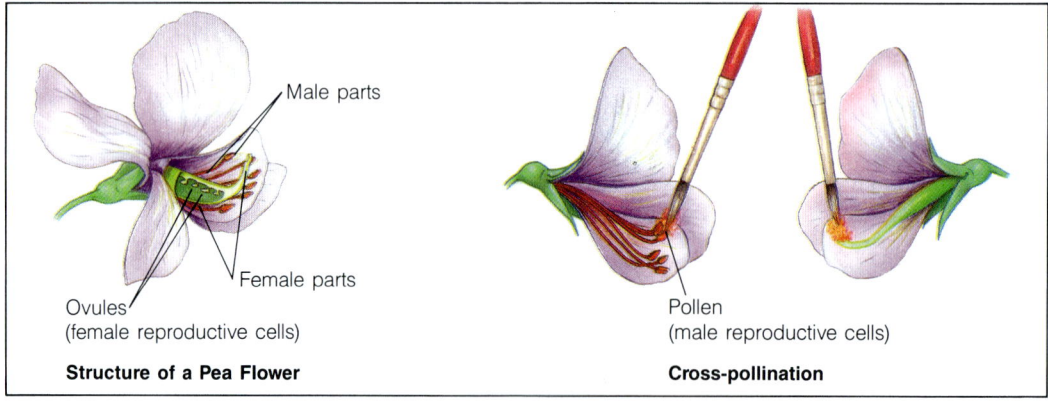

Structure of a Pea Flower — Male parts, Female parts, Ovules (female reproductive cells)

Cross-pollination — Pollen (male reproductive cells)

182

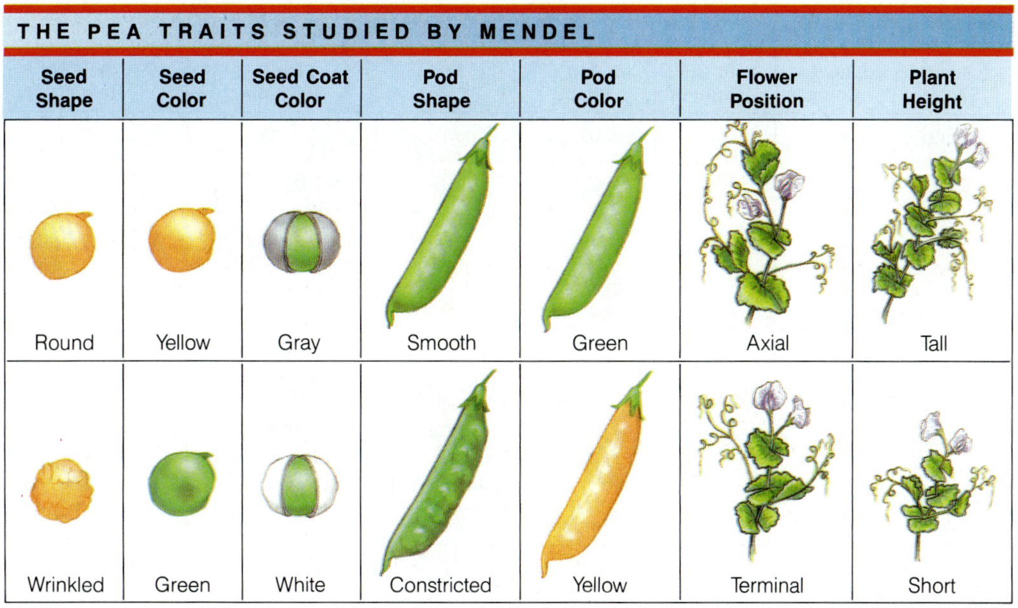

THE PEA TRAITS STUDIED BY MENDEL

Seed Shape	Seed Color	Seed Coat Color	Pod Shape	Pod Color	Flower Position	Plant Height
Round	Yellow	Gray	Smooth	Green	Axial	Tall
Wrinkled	Green	White	Constricted	Yellow	Terminal	Short

One of the gifts that Mendel received when he took charge of the monastery garden was a stock of peas developed by earlier gardeners. These peas were **purebred**. This means that if they were allowed to self-pollinate, the purebred peas would produce offspring that were identical to themselves. One line of plants would produce only tall plants, another only short plants. One line would produce only green seeds, another only yellow seeds. These purebred plants were the basis of Mendel's experiments.

In many respects, the most important decision Mendel made was to study just a few isolated **traits**, or characteristics, that could easily be observed. In the case of the peas, one trait was the size of the plant, another the shape of the pod, another the color of the seed. Figure 9–3 shows the seven traits that Mendel studied. By deciding to restrict his observations to just a few such traits, Mendel made his job of measuring the effects of heredity much easier.

Figure 9–3 The seven pea traits studied by Mendel are seed shape, seed color, seed coat color, pod shape, pod color, flower position, and plant height. Mendel chose these seven traits partly because each trait has only two contrasting characters, or forms.

Genes and Dominance

What would you do if you were interested in heredity, had several kinds of purebred peas, and also knew how to cause cross-pollination? No, pea soup is not a good answer.

Mendel decided to see what would happen if he crossed pea plants with different characters for the same trait. A character is a form of a trait. For example, the flower-position trait has two characters: axial and terminal. Mendel crossed the tall

HISTORICAL NOTE
WILLIAM BATESON

The study of heredity was named genetics in 1905 by the English scientist William Bateson, who was also responsible for adding the term *allele* to our genetics vocabulary.

BACKGROUND INFORMATION
HOW TALL IS TALL?

When studying Mendelian genetics, we routinely refer to "tall" pea plants and "short" pea plants. But actually how tall—and how short—were these plants?

A tall pea plant averages 2 to 2.5 meters in height, whereas a short pea plant grows to a height of only 0.3 meter.

MULTICULTURAL STRATEGY

Speaking about racial prejudice, Harold Marowitz, professor of biochemistry at Yale University, said: "When we hate an individual for his race, we are hating his genes over which he has no control.... When you get down to it, what we are hating are DNA molecules." Discuss this quote with students.

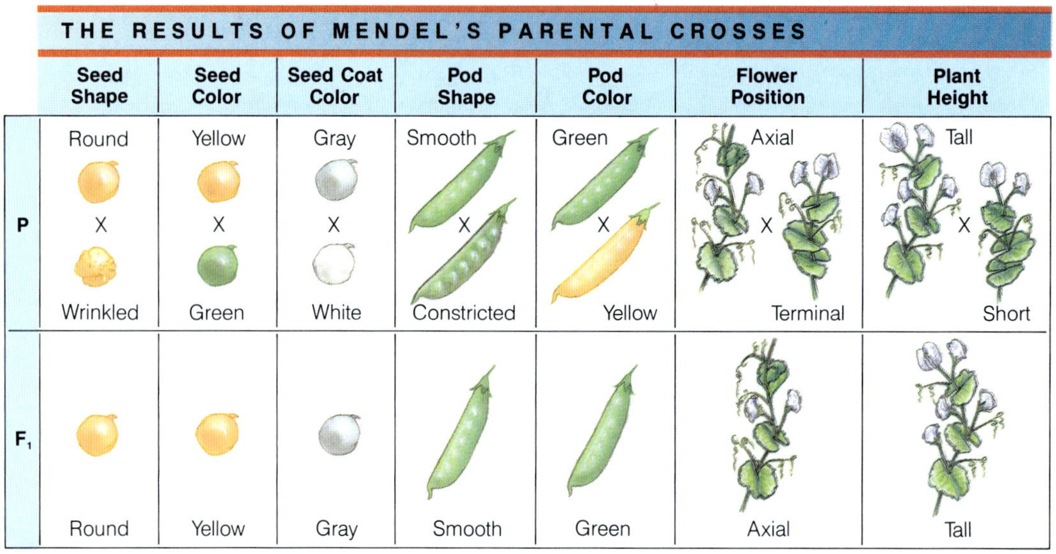

THE RESULTS OF MENDEL'S PARENTAL CROSSES

	Seed Shape	Seed Color	Seed Coat Color	Pod Shape	Pod Color	Flower Position	Plant Height
P	Round X Wrinkled	Yellow X Green	Gray X White	Smooth X Constricted	Green X Yellow	Axial X Terminal	Tall X Short
F₁	Round	Yellow	Gray	Smooth	Green	Axial	Tall

Figure 9–4 When Mendel crossed plants with contrasting characters for the same trait, the resulting offspring had only one of the characters. From the results of these experiments, Mendel concluded that factors that do not blend control the inheritance of traits and that some of these factors are dominant, whereas others are recessive.

plants with the short ones; the plants with yellow seeds with those with green seeds; and so on. From the crosses in his pea plants, Mendel obtained seeds that he then grew into plants. These plants were **hybrids**, or organisms produced by crossing parents with differing characters.

What were those hybrid plants like? Did the characters of the parent plants blend in the offspring? To Mendel's surprise, the plants were not half tall, nor were the seeds they produced half yellow. Instead, all of the offspring had the character of only one of the parents. The plants resulting from his crosses were all tall or produced only yellow seeds. The other character had apparently disappeared.

From this set of experiments, Mendel was able to draw two conclusions. The first is that individual factors, which do not blend with one another, control each trait of a living thing. Mendel used the word *Merkmal* to refer to these factors. *Merkmal* means character in German. Today the factors that control traits are called **genes**. Each of the traits Mendel studied was controlled by one gene that occurred in two contrasting forms. These contrasting forms produced the different characters of each trait. For example, the gene for plant height occurs in a tall form and a short form. The different forms of a gene are now called **alleles** (uh-LEELZ).

The second of Mendel's conclusions is often called the principle of **dominance**: Some factors (alleles) are **dominant**, whereas others are **recessive**. The effects of a dominant allele are seen even if it is present with a contrasting recessive allele. The effects of a recessive allele are not observed when the dominant allele is present. In Mendel's experiments, the tall

---

9–1 (continued)

Content Development
Using the Visuals Have students examine Figure 9–4.
• **What are some examples of dominant traits in pea plants?** (Round seed shape, yellow seed color, gray seed coat color, smooth pods, green pods, axial flowers.)

Introduce the concept of P generation and F₁ generation.
• **What kinds of offspring would be expected if the F₁ hybrid plants were permitted to reproduce among themselves?** (Accept all answers for now.)
• **What name would be given to the offspring of the F₁ generation?** (F₂ generation.)

Skills Development
Guided Practice
Skill: Interpreting Charts
Using the Visuals Have students examine Figure 9–5. Make sure they can interpret the results of the F₁ and F₂ generations.
• **Why were all of the plants in**

and yellow alleles were dominant, whereas the short and green alleles were recessive. Although dominance is seen in the inheritance of many traits, it does not apply to all genes.

Segregation

Mendel was not content to stop his experimentation at this point. He wanted the answer to another question: What happened to the recessive characters? To answer this question, he allowed all seven kinds of hybrid plants to reproduce by self-pollination.

To keep the different groups of seeds and plants clear in his mind, Mendel gave them different names. He referred to the purebred parental plants as the P generation (P for parental). To the first generation of plants produced by cross-pollination, he gave the name F_1, which stood for first filial generation (from the Latin word *filius*, which means son). If the F_1 plants were crossed among themselves, he called the offspring F_2, for second filial generation, and so forth.

THE F_1 CROSS The results of the F_1 cross were remarkable. The recessive characters had not disappeared! Some of the F_2 plants produced by each of the F_1 crosses showed the recessive trait. This proved to Mendel that the alleles responsible for the recessive characters had not disappeared. But why did the recessive alleles disappear in the F_1 generation and reappear in the F_2? To answer this question, let's take a closer look at one of Mendel's crosses. Keep in mind that the concepts that apply to the crosses involving the trait of plant height also apply to the crosses for the other six traits.

EXPLAINING THE F_1 CROSS To begin with, Mendel assumed that the presence of the dominant tall allele had masked

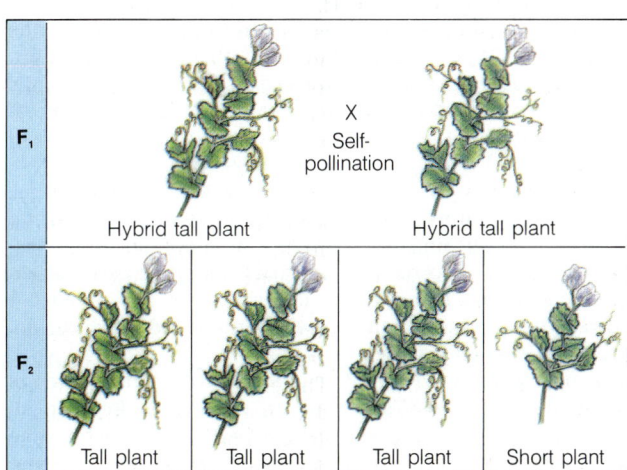

Figure 9–5 When the F_1 hybrid plants were allowed to reproduce by self-pollination, some of the resulting F_2 offspring had the recessive character.

185

TEACHING SUPPORT
Laboratory Manual
- The Principles of Genetics, p. 129

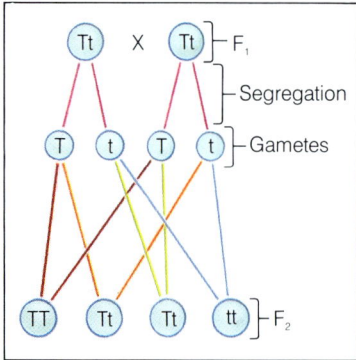

Figure 9–6 Segregation of alleles occurs during gamete formation. The alleles are paired up again when gametes fuse during the process of fertilization.

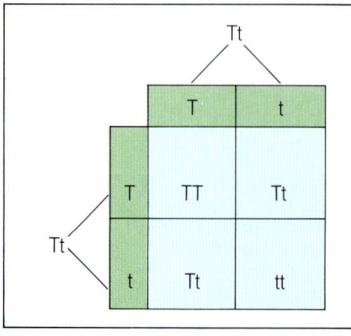

Figure 9–7 This Punnett square shows a cross between two hybrid tall (Tt) pea plants.

the recessive short allele in the F_1 generation. But the fact that the recessive allele was not masked in some of the F_2 plants indicated that the short allele had managed to get away from the tall allele. But how did this separation, or **segregation**, of alleles occur? Mendel suggested that during the formation of the reproductive cells (pollen and eggs), the tall and short alleles in the F_1 plants were segregated, or separated, from each other. Did that suggestion make sense?

Let's assume, as perhaps Mendel did, that the F_1 plants inherited an allele for tallness from one parent and an allele for shortness from the other parent. Because the tall allele is dominant, the F_1 plants appear tall. When the F_1 plant flowers, the two alleles must be separated from each other so that each reproductive cell carries only a single copy of each gene. Therefore, each F_1 plant produces two types of reproductive cells—those with the tall allele and those with the short allele.

If segregation occurs the way Mendel thought it did, then the possible gene combinations in the offspring that result from a cross can be determined by drawing a diagram known as a **Punnett square**. Biologists represent a particular allele by using a symbol. The dominant allele is represented by a capital letter. The recessive allele is represented by the corresponding lowercase letter. In this case, *T* represents the dominant tall allele and *t* represents the recessive short allele.

The Punnett square in Figure 9–7 shows the types of reproductive cells, or **gametes** (GAM-eets), produced by each F_1 parent along the top and left-hand side of the square. It also shows each possible gene combination for the F_2 offspring in the four boxes that make up the square.

You can see the probable results of the cross of two F_1 plants from the Punnett square: 1/4 of the F_2 plants have two tall alleles (*TT*); 2/4, or 1/2, of the F_2 plants have one tall allele and one short allele (*Tt*), and 1/4 of the F_2 plants have two short alleles (*tt*). Because tall is dominant over short, 3/4 of the F_2 plants should be tall and 1/4 of the F_2 plants should be short. These numbers are often expressed as ratios. For example, there are 3 tall plants for every 1 short plant in the F_2 generation. Thus the ratio of tall plants to short plants is 3:1. If, of course, Mendel's model of segregation is correct.

Did the data from Mendel's experiments fit his model? Yes. The predicted ratio—3 dominant to 1 recessive—showed up consistently, indicating that Mendel's assumptions about segregation had been correct. For each of his seven crosses, about 3/4 of the plants showed the dominant trait and about 1/4 showed the recessive trait. Segregation did indeed occur according to Mendel's model.

Look again at the Punnett square. You will see that three of the possible combinations result in tall plants. Because all these plants appear tall, we can say that they have the same **phenotype**, or physical characteristics. They do not, however,

9–1 (continued)

Focus/Motivation
Because garden pea plants reproduce so quickly, you can simulate some of Mendel's experiments in class. It is unlikely that you will know the genotype of pea plants you purchase, so there will be a bit of experimental "mystery" in your demonstration in that you will not necessarily know the results of various crosses. Point that fact out to students.

Begin by taking the pollen from one plant and placing it on the stigmas of another plant. Remove the anthers from the second plant so that self-pollination cannot occur. Have students suggest possible characteristics of the resulting F_1 generation. Students should write their suggested characteristics down and compare them to the actual results. At that time, have students try to determine the genotypes of the original plants.

If time permits, repeat the investigation to produce an F_2 generation.

Reinforcement/Reteaching
Obtaining the correct results from a completed Punnett square is as easy—or as difficult—as reading what you have written.

Encourage students to choose letters that are easy to read when selecting symbols to represent pairs of alleles. It is a good idea to use letters that look very different in uppercase and lowercase, for example, A and a. A choice such as C and c may prove confusing.

Skills Development
Guided Practice
Skill: Solving the Punnett square

Using the Visuals Reproduce Figure 9–7 on the chalkboard. Stress that the dominant allele is always represented by a capital

have the same **genotype**, or genetic makeup. The genotype of 1/3 of the tall plants is *TT*, whereas the genotype of 2/3 of the tall plants is *Tt*.

Organisms that have two identical alleles for a particular trait (*TT* or *tt* in our example) are said to be **homozygous** (hoh-moh-ZIGH-guhs) (*homo-* means same; *-zygous* refers to alleles). Organisms that have two different alleles for the same trait are **heterozygous** (heht-er-oh-ZIGH-guhs)(*hetero-* means different). In other words, homozygous organisms are purebred for a particular trait and heterozygous organisms are hybrid for a particular trait.

Independent Assortment

After establishing that alleles segregate during the formation of gametes (reproductive cells), Mendel began to explore the question of whether they do so independently. In other words, does the segregation of one pair of alleles affect the segregation of another pair of alleles? For example, does the gene that determines whether a seed is round or wrinkled in shape have anything to do with the gene for seed color? Must a round seed also be yellow? To answer these questions, Mendel first crossed purebred plants that produced round yellow seeds with purebred plants that produced wrinkled green seeds.

THE TWO-FACTOR CROSS: F$_1$ In this cross, the two kinds of plants would be symbolized like this:

    Round yellow seeds   *RRYY*

    Wrinkled green seeds   *rryy*

Because two traits are involved in this experiment, it is called a two-factor cross. As you examine this cross, keep in mind that you are looking at the kind of seeds the plant produces. (These seeds are not necessarily the same as the seeds from which the plants grew!)

The plant that bears round yellow seeds produces gametes that contain the alleles *R* and *Y*, or *RY* gametes. The plant that bears wrinkled green seeds produces *ry* gametes. An *RY* gamete and an *ry* gamete combine to form a fertilized egg with the genotype *RrYy*. Thus only one kind of plant will show up in the F$_1$ generation—plants that are heterozygous, or hybrid, for both traits. What is the phenotype of the F$_1$ plants? That is, what will the seeds produced by the F$_1$ plants look like? Because we know that round and yellow are dominant traits, we can conclude that the F$_1$ plants will produce seeds that are round and yellow. Remember that the concept of dominance tells us that the dominant traits will show up in a hybrid, whereas the recessive traits will seem to disappear.

This cross does not indicate whether genes assort, or segregate, independently. However, it provides the hybrid plants

Figure 9–8 Although these plants have different genotypes (TT and Tt), they have the same phenotype (tall).

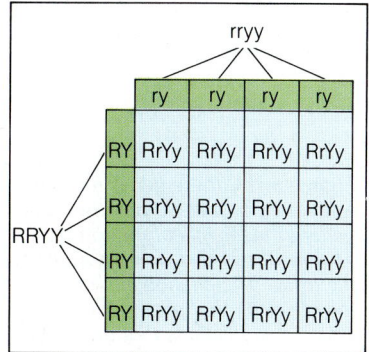

Figure 9–9 When an individual that is homozygous dominant for two traits is crossed with an individual that is recessive for the same two traits, all of the offspring are heterozygous dominant for those two traits.

HISTORICAL NOTE
MENDEL'S FIGURES

Although Mendel is considered the father of genetics and a scientist of great importance, many current scientists believe he fudged some of his results. Mendel's results are so perfect that they defy the predictions of theoretical probability, as well as most attempts at reproducing them. No proof can be found, but one guess is that Mendel slightly adjusted or exaggerated his figures to meet the results he expected.

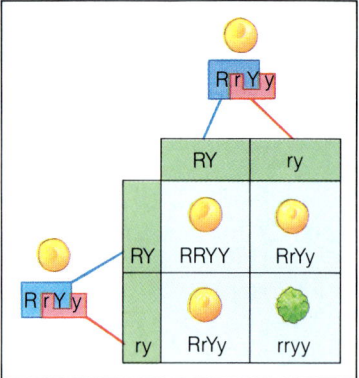

Figure 9–10 If the genes for these two traits were connected in some way, only two combinations of the traits would be possible in their gametes. This would produce just two possible genotypes in the F_2 generation.

needed for the next cross—the cross of F_1 plants to produce the F_2 generation. The seeds from the F_2 plants will show whether the genes for seed shape and seed color have anything to do with one another. Now for the real experiment.

THE TWO-FACTOR CROSS: F_2 What will happen when F_1 plants are crossed with each other? If the genes for seed shape and color are connected in some way, then the dominant *R* and *Y* alleles (which came from one parent) and the recessive *r* and *y* alleles (which came from the other parent) will be segregated as matched sets into the gametes. Thus, the gametes could only contain one of two possible allele combinations: *RY* or *ry*. As you can see in Figure 9–10, all round seeds will be yellow and all wrinkled seeds will be green.

If the genes are not connected, then they should segregate independently, or undergo **independent assortment**. This produces four possible types of gametes: *RY, Ry, rY,* and *ry*. In addition, if the genes assort independently, some of the seeds produced by the F_2 plants will have new combinations of traits —they may be wrinkled and yellow, or round and green.

This two-factor cross is examined in the Punnett square in Figure 9–11. Now that we have four possible gamete types (and sixteen possible offspring) the square is especially useful. If the genes for seed shape and seed color are inherited independently, then the seeds produced by the offspring should be exactly as predicted by the Punnett square:

9/16	yellow round seeds	(315)
3/16	yellow wrinkled seeds	(101)
3/16	green round seeds	(108)
1/16	green wrinkled seeds	(32)

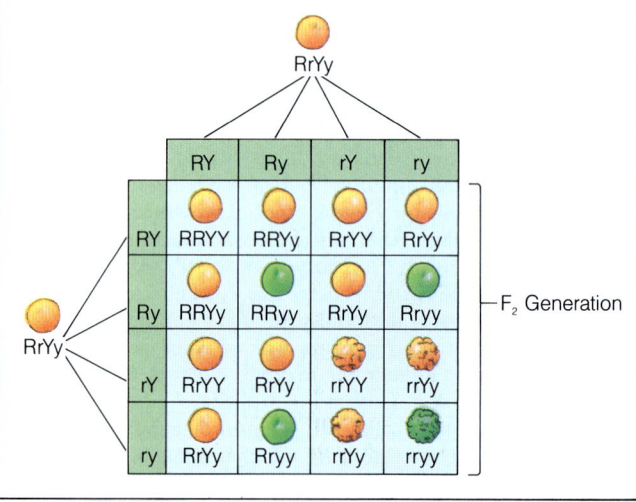

Figure 9–11 If the genes for two traits are not linked, then they will undergo independent assortment. This results in new combinations of the traits. The phenotypic ratio of the offspring that results from crossing individuals that are hybrid for two traits is 9:3:3:1.

188

9–1 (continued)
Content Development
Ask students these questions and keep track of their answers on the chalkboard.
• **What is a two-factor cross?** (A cross where the researcher keeps track of two traits.)
• **Suppose we crossed an *RRYY* plant with an *rryy* plant. What would be the results?** (An *RrYy* plant; a hybrid for both traits.)
• **What types of gametes could an *RrYy* plant produce?** (Four types: *RY, Ry, rY,* and *yy*.)
• **Suppose I told you that this plant produced only *RY* and *ry* gametes. What would this mean?** (That the genes for these two traits were linked.)

Skills Development
Guided Practice
Skill: Interpreting Charts
Using the Visuals Have students examine Figures 9–10 and 9–11. If time permits, recopy both figures on the chalkboard, emphasizing which gamete combinations produce each offspring.
• **What caused the different results of these two crosses?** (For Figure 9–10, the two genes are linked. For Figure 9–11, the genes sort independently.)
• **In which cross do we see the greater variety of offspring? Why?** (The cross shown in Figure 9–11. Only when the genes sort independently do we see all the combinations of traits.)

Reinforcement/Reteaching
Review the summary at the end of this section. Students should

Mendel actually carried out this exact experiment, and his results, shown in parentheses, were very close to the 9:3:3:1 ratio that the Punnett square predicts. From these results, Mendel concluded that genes could segregate independently during the formation of gametes. In other words, genes could undergo independent assortment. As we shall see later, there is an important exception to independent assortment. Genes located on the same chromosome are linked and may not undergo independent assortment.

Figure 9–12 *This young chameleon has obtained half of its genes—and a piggyback ride—from one of its parents.*

A Summary of Mendel's Work

Mendel's work on the genetics of peas can be summarized in four basic statements:

- **The factors that control heredity are individual units known as genes. In organisms that reproduce sexually, genes are inherited from each parent.**
- **In cases in which two or more forms of the gene for a single trait exist, some forms of the gene may be dominant and others may be recessive.**
- **The two forms of each gene are segregated during the formation of reproductive cells.**
- **The genes for different traits may assort independently of one another.**

9–1 SECTION REVIEW

1. Describe Mendel's experiments.
2. What is dominance? Segregation?
3. What is independent assortment?
4. Define the terms genotype and phenotype.
5. **Critical Thinking—Relating Concepts** Why were purebred peas important for Mendel's experiments?

SECTION REVIEW 9–1

1. Mendel hypothesized that certain pea plant traits were inherited through independent factors, now called genes. Mendel crossed purebred strains, producing hybrid offspring that exhibited only the dominant trait. He then let the hybrids reproduce, and the recessive trait reappeared in one-fourth of the offspring.
2. When an organism has two different alleles of the same gene, the it is dominant allele that is expressed. Segregation is the separation of corresponding alleles during gamete formation.
3. Independent assortment is the unlinked, random segregation of different genes.
4. An organism's genotype is its genetic makeup; its phenotype is its physical characteristics.
5. Mendel used purebred peas as reliable sources of trait factors, or genes.

9–2 Applying Mendel's Principles

Mendel's careful record keeping and his knowledge of mathematics enabled him to see the patterns of heredity. His paper describing his experiments and conclusions was published in 1866 in the *Proceedings of the Brno Society of Natural Science*. Unfortunately, Mendel's ideas about heredity and his applications of mathematics and statistics to biology were far ahead of their time. Mendel's fellow scientists failed to understand and recognize the importance of his work. In fact, a prominent botanist encouraged Mendel to start working with a different kind of plant and to give up on the hybrid peas. Mendel's pioneering work in genetics remained unappreciated during his lifetime.

In the early 1900s, several scientists working independently rediscovered Mendel's paper. They realized that it correctly described the basic principles of genetics. More than twenty years after his death, Mendel's experiments and conclusions were recognized as important breakthroughs in biology.

Guide For Reading
- How is probability related to genetics?
- How are Punnett squares used to solve genetics problems?

Genetics and Probability

One of the most innovative things Mendel did was to apply the mathematical concept of **probability** to biology. **Probability is the likelihood that a particular event will occur.** Probability is determined by the following formula.

Probability = $\dfrac{\text{The number of times a particular event occurs}}{\text{The number of opportunities for the event to occur (the number of trials)}}$

Consider, for example, the flipping of a coin. When a coin is flipped, one of two possible events can occur: The coin can land heads up or it can land tails up. The probability of the coin coming up heads is 1/2, or 1:1. In other words, you will probably get heads (the particular event) one time out of every two times you flip the coin (the trials). And you will probably get tails one time out of every two times you flip the coin. Therefore, if you were to flip the coin 10 times, the most likely outcome would be 5 heads and 5 tails. However, there is a chance that you would wind up with a different outcome—perhaps 4 heads and 6 tails, or 10 heads and 0 tails.

These different outcomes reveal an important rule of probability: You only get the expected ratio for large numbers of trials. The larger the number of trials, the closer you get to the expected ratios. If you were to flip the coin hundreds of times, your final results would be much closer to 1:1. If you were to flip the coin thousands of times, your final results would be even closer to 1:1. With this in mind, why do you think Mendel worked with thousands of pea plants?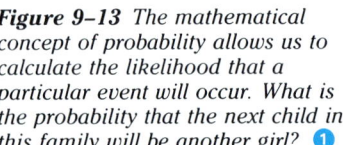

Figure 9–13 The mathematical concept of probability allows us to calculate the likelihood that a particular event will occur. What is the probability that the next child in this family will be another girl?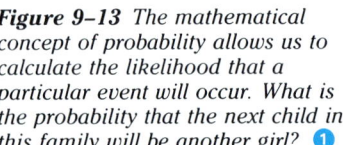

Suppose you flipped a coin and it came up heads ten times in a row. What are the chances of getting heads again on your next flip? Because of all the times the coin came up heads, you might think it is more likely that the coin will now come up tails rather than heads. But this is not the case. There is another important rule of probability: Previous events do not affect future outcomes. Each flip of the coin is a separate, independent event. For each flip of the coin, the probability of getting heads is always 1/2, or 1:1. It does not matter what happened on previous flips of the coin.

The rules of probability apply to genetics as well as to flipping a coin. Expected genetic ratios may not show up when only a few pea plants (or other organisms) are considered. The larger the number of organisms examined, the closer the numbers will get to the expected values. In addition, genetic ratios do not indicate what the outcome of a single event will be. Because previous events do not affect future outcomes, it cannot be assumed that a particular event will occur because it seems overdue. Despite such limitations, genetic ratios are still very useful because they make it possible to predict the most likely outcome for a large number of events.

Using the Punnett Square

The Punnett square is a handy device for analyzing the results of an experimental cross, and it's a good idea to become familiar with its use. Practice using the Punnett square in the one-factor and two-factor genetics problems that follow.

ONE-FACTOR CROSS In pea plants, tall (T) is dominant over short (t). You have a tall plant. Design a cross to see if this plant is homozygous (TT) or heterozygous (Tt).
Solution:
Cross your tall plant with a short plant. The cross of an organism of unknown genotype and a homozygous recessive individual is called a **test cross**. Figure 9–14 shows how a test cross works.

As you can see in the Punnett squares, if any of the offspring resulting from a test cross shows the recessive phenotype, then the unknown parent must be heterozygous.

TWO-FACTOR CROSS In pea plants, green pods (G) are dominant over yellow pods (g), and smooth pods (N) are dominant over constricted pods (n). A plant heterozygous for both traits ($GgNn$) is crossed with a plant that has yellow constricted pods ($ggnn$). What are the probable genotypic and phenotypic ratios for this cross?
Solution:
Genotypic ratio = $4GgNn:4Ggnn:4ggNn:4ggnn$ = 4:4:4:4 = 1:1:1:1
Phenotypic ratio = 4 green smooth:4 green constricted:4 yellow smooth:4 yellow constricted = 4:4:4:4 = 1:1:1:1

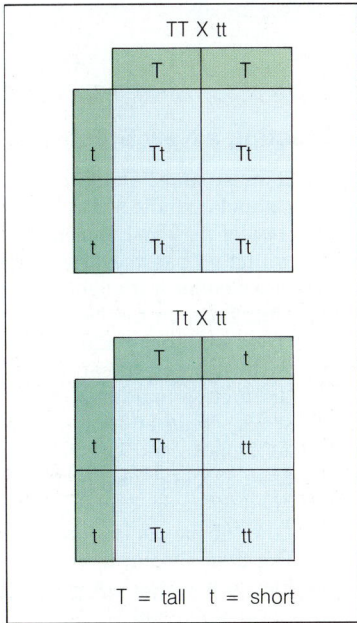

Figure 9–14 A test cross is used to determine the genotype of an individual showing a dominant trait. If any of the offspring resulting from a test cross has the recessive phenotype, then the unknown parent is heterozygous.

Figure 9–15 This Punnett square illustrates a cross between a pea plant that has the genotype GgNn and one that has the genotype ggnn.

GgNn X ggnn	GN	Gn	gN	gn
gn	GgNn	Ggnn	ggNn	ggnn
gn	GgNn	Ggnn	ggNn	ggnn
gn	GgNn	Ggnn	ggNn	ggnn
gn	GgNn	Ggnn	ggNn	ggnn

G = green   g = yellow
N = smooth   n = constricted

191

TEACHING SUPPORT

Teaching Resources
- Activity: Using Punnett Squares to Predict the Outcomes of Crosses, p. 3
- Activity: Analyzing Genetic Data, p. 11
- Hands-On Activity: Flip Out! p. 63

shape ($ggnn$), the genotype of the dominant parent can be inferred from the results. Only a plant heterozygous for both traits will yield the result $4GgNn$:$4Ggnn$:$4ggNn$:$4ggnn$ (or $1GgNn$:$1Ggnn$:$1ggNn$:$1ggnn$.)
- **What results would be expected if the dominant phenotype parent had the genotype $GGNn$?** ($8GgNn$:$8Ggnn$ or $1GgNn$:$1Ggnn$.)
- **What is the phenotypic ratio expected in this cross?** (1 green smooth:1 green constricted.)
- **How would those results differ if the genotype was $GgNN$?** ($8GgNn$:$8ggNn$ or $1GgNn$:$1ggNn$.)
- **What is the phenotypic ratio expected in this second example?** (1 green smooth:1 yellow smooth.)

---

in a cross between a pure tall pea plant with a pure short? (The phenotypes of the offspring will be 100 percent tall.)
- **What are the phenotypes expected among the offspring in a cross between a hybrid tall pea plant with a pure short?** (The offspring are expected to be 50 percent tall and 50 percent short.)

The appearance of any recessive offspring in a test cross would indicate that the tall parent was hybrid.

Enrichment

Using the Visuals A test cross is possible even when more than one trait is being studied. Review the results of the two-factor cross in Figure 9–15. Because one parent was homozygous recessive for both pod color and

191

SCIENCE, TECHNOLOGY, AND SOCIETY CONNECTION

Creating a New Breed of Cat

Siamese and Persian cats are two of the most popular kinds of purebred cats. They are also two of the most ancient breeds, with histories that go back several centuries. Siamese cats are slender, with light-colored bodies and dark markings on the face, legs, and tail. Persian cats are stocky and have long luxurious fur. A third breed of cat, the Himalayan, has the distinctive markings of the Siamese and the stocky build and long fur of the Persian. Unlike Siamese and Persian cats, Himalayans have been around for a relatively short time—less than 60 years. And unlike most breeds of cats, the Himalayan is an artificial breed.

The first Himalayan was born in 1935, the product of five years of genetics experiments at Harvard Medical School. The researchers were trying to determine how certain cat traits, such as long fur and Siamese markings, are inherited.

The scientists first crossed Siamese cats with gray, black, or striped Persian cats. The offspring of these crosses all had short hair and were gray, black, or striped in color. When two of the short-haired cats were mated, one of the female offspring had long black fur. This female, when mated to her male parent, produced a long-haired kitten with Siamese markings—the first Himalayan cat.

From their crosses, the scientists determined that short hair was dominant over long hair and that solid or striped colors were dominant over Siamese markings. Although the scientists had created a new type of cat, their role in the new Himalayan breed was over. After all, they had discovered what they had set out to find. Cat breeders in the United States and Britain took over where the scientists had left off. These breeders focused their attention, however, on the effects of other genes, such as those for general body type. By the early 1960s, more than 25 years after the birth of the first Himalayan, the breed was officially recognized by all of the major cat associations in the United States and Britain.

The Himalayan is an artificial breed of cat. It was created during the course of genetics experiments.

9–2 SECTION REVIEW

1. What is probability?
2. What is a test cross?
3. What kind of cross produces a 9:3:3:1 ratio of offspring?
4. **Critical Thinking—Relating Concepts** A cross of F_1 plants heterozygous for height should produce a ratio of tall to short plants of 3:1. Three seeds from such a cross produced three tall plants. What are the chances that a fourth seed will produce a short plant?

9-3 Meiosis

Guide For Reading
- What is meiosis?
- What are the phases of meiosis?
- How do meiosis and mitosis differ?

The principles of genetics described by Mendel require that organisms inherit a single copy of each gene from each of their parents. These two copies are then segregated from each other during the formation of gametes. An obvious way to test Mendel's ideas would be to see if something similar to segregation takes place during the formation of gametes.

But how are gametes formed? In Chapter 8 you learned about mitosis, a process that involves the separation of chromosomes and the formation of new cells. Could gametes be formed by mitosis?

The answer to this question is no. If gametes were formed by mitosis in the common fruit fly, *Drosophila melanogaster*, each gamete would contain 8 chromosomes—as do the other cells of the fly. When sperm and egg fused in the process of fertilization, the first cell of the offspring would contain 16 (or 8 + 8) chromosomes. This is twice the number of chromosomes in the cells of the parents. If the number of chromosomes doubled in each generation, before long the cells would contain an unwieldy number of chromosomes.

Chromosome Number

The 8 chromosomes in a *Drosophila* cell can be divided into two sets: One set contains 4 chromosomes from the male parent and the other set contains 4 chromosomes from the female parent. Each chromosome in the male set has a corresponding chromosome in the female set. These corresponding chromosomes are said to be **homologous**. (The chromosomes themselves are called homologs.) A cell that contains both sets of homologous chromosomes (one set from each parent) is said to be **diploid**.

The diploid number is sometimes represented by the symbol 2N. Thus for *Drosophila* 2N=8. A diploid cell contains two complete sets of chromosomes and two complete sets of genes.

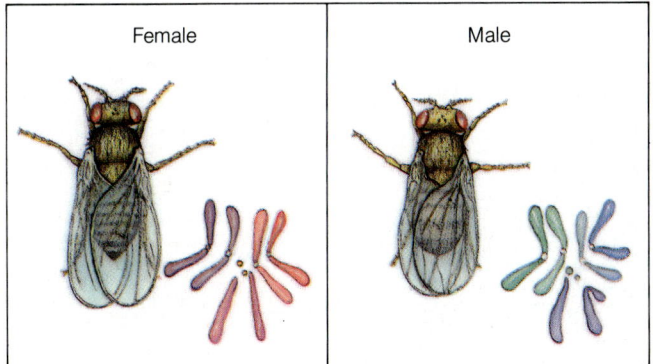

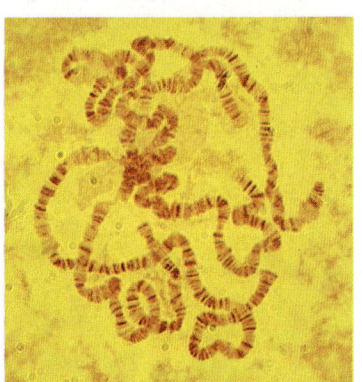

Figure 9–16 *Each body cell in a fruit fly contains 8 chromosomes— 4 from the male parent and 4 from the female parent (left). This photograph shows the 4 pairs of giant chromosomes of the fruit fly* Drosophila *(right). These giant chromosomes consist of many chromatids attached in parallel.*

1. In fruit flies, gray body trait is dominant over ebony body trait. Two gray fruit flies are mated and produce 188 gray-bodied and 56 ebony-bodied offspring. What were the parental genotypes? (*G*=gene for gray body; *g*=gene for ebony body. 188:56 is approximately 3:1. Parental genotypes were *Gg* × *Gg*.)
2. From a single ear of corn, a farmer planted 200 kernels that produced 140 tall plants and 40 short plants. What were the genotypes of these offspring? (*T*=gene for tallness; *t*=gene for shortness. 140:40 is approximately a 3:1 ratio. 20 kernels did not germinate. The tall plants had the genotypes *TT* and *Tt*; the short plants, *tt*.)

TEACHING SUPPORT

Teaching Resources
- Activity: Should This Dog Be Called Spot? p. 15

MEDIA AND TECHNOLOGY
- Biology Transparencies
- 13: Meiosis

9–3 (continued)

Content Development
Emphasize that the terms *diploid* and *haploid* can be misleading. Since we refer to 2N as the diploid number of chromosomes, we logically should call N the monoploid number. Similarly, the word haploid should describe $\frac{1}{2}$ N.

Unfortunately, the terminology developed differently. Make sure that students understand that diploid is 2N and haploid is N, and that in this instance logic does not apply.

Content Development
Using the Visuals Have students examine Figure 9–17.

- **What is the purpose of meiosis?** (The production of haploid gametes.)
- **What happens to the cell's chromosomes in prophase I?** (Homologous chromosomes pair off, forming tetrads.)
- **In metaphase I?** (Tetrads line up in the center of the cell.)
- **In anaphase I?** (The cell divides the tetrads. It moves half the chromosomes to one end, the other half to the opposite end.)
- **In telophase I?** (A wall or membrane forms between the cell's two ends. The result is 2 haploid cells.)
- **What happens in meiosis II?** (Meiosis II is identical to mitotic division, discussed in Chapter 8. The only difference is that in meiosis II, the parent cell is haploid.)
- **What is the end product of meiosis?** (4 haploid gametes.)

Skills Development
Guided Practice
Skill: Drawing Conclusions
Using the Visuals Refer students to Figure 9–17.

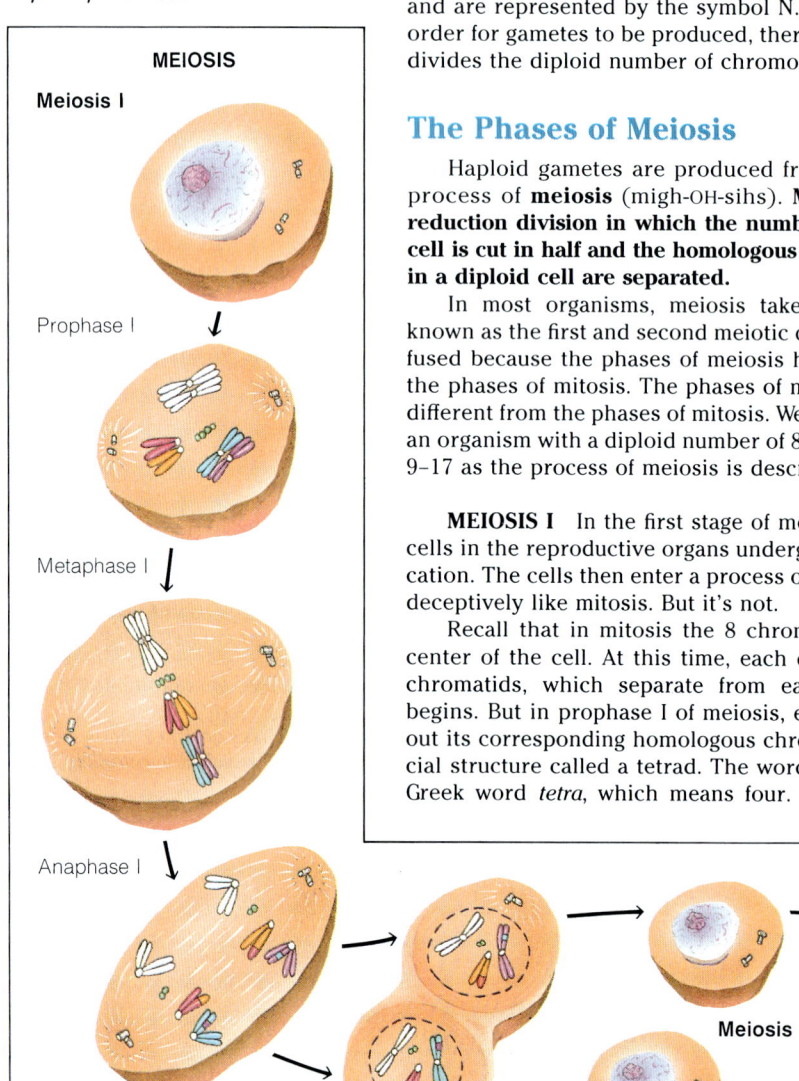

Figure 9–17 In the process of meiosis, the number of chromosomes per cell is cut in half and homologous chromosomes are separated. Meiosis produces four haploid gamete cells that are genetically different from one another and from the original diploid parent cell.

This agrees with Mendel's idea that all of an organism's cells (except the gametes) contain two alleles for a given trait. Mendel's model of segregation implies that gametes, unlike other cells, should contain only a single set of genes because alleles are separated during the process of gamete formation. This is exactly the case. The gametes of organisms that reproduce sexually contain a single set of chromosomes (and genes). Cells that contain a single set of chromosomes are said to be **haploid** and are represented by the symbol N. For *Drosophila*, N=4. In order for gametes to be produced, there must be a process that divides the diploid number of chromosomes in half.

The Phases of Meiosis

Haploid gametes are produced from diploid cells by the process of **meiosis** (migh-OH-sihs). **Meiosis is a process of reduction division in which the number of chromosomes per cell is cut in half and the homologous chromosomes that exist in a diploid cell are separated.**

In most organisms, meiosis takes place in two stages, known as the first and second meiotic divisions. Do not be confused because the phases of meiosis have the same names as the phases of mitosis. The phases of meiosis are actually very different from the phases of mitosis. We will examine meiosis in an organism with a diploid number of 8 (2N=8). Refer to Figure 9–17 as the process of meiosis is described.

MEIOSIS I In the first stage of meiosis, meiosis I, special cells in the reproductive organs undergo a round of DNA replication. The cells then enter a process of cell division that looks deceptively like mitosis. But it's not.

Recall that in mitosis the 8 chromosomes line up in the center of the cell. At this time, each chromosome contains 2 chromatids, which separate from each other as anaphase begins. But in prophase I of meiosis, each chromosome seeks out its corresponding homologous chromosome to form a special structure called a tetrad. The word tetrad comes from the Greek word *tetra*, which means four. And as you can see in

- The textbook defines meiosis as a process of reduction division. How does the phrase "reduction division" apply to meiosis? (The original cell *divides* into 4 cells, and the number of chromosomes is *reduced* to the haploid number.)

- Suppose that during anaphase I, one of the tetrads failed to separate and both of its chromosomes moved to the same end of the cell. Predict how this would affect the gametes. (Half of the gametes would have an extra chromosome and half would have one chromosome missing. This scenario does occasionally occur in meiosis and is called nondisjunction. Several genetic diseases that result from nondisjunction are discussed in Chapter 11.)

- **In Figure 9–17, are we looking at the formation of egg or sperm**

Figure 9–17, there are 4 chromatids in a tetrad. Tetrads, rather than individual chromosomes, line up in the center of the cell during metaphase I of meiosis.

As the homologous chromosomes pair up and form tetrads in meiosis I, they may exchange portions of their chromatids in a process called **crossing-over**. Crossing-over results in the exchange of genes between homologous chromosomes and produces new combinations of genes.

What happens next? The homologous chromosomes separate and two new cells are formed. Although each cell now has 8 chromatids (just as it would after normal mitosis) something is different. Look at Figure 9–17 closely. The two cells no longer have two complete sets of the 4 chromosomes. Instead, the maternal and paternal chromosomes have been shuffled like a deck of cards. The two cells produced by meiosis I have sets of chromosomes (and genes) that are different from each other. These sets are also different from the set in the cell that began the division.

MEIOSIS II The two cells produced by meiosis I now enter meiosis II. Unlike a cell undergoing a second mitotic division, neither cell goes through a round of DNA replication before entering the second meiotic division. Thus each of the cells' chromosomes contains 2 chromatids. During metaphase II of meiosis, 4 chromosomes line up in the center of each cell. In anaphase II, the paired chromatids separate. Each of the four daughter cells produced in meiosis II receives 4 chromatids. The four daughter cells contain the haploid number (N)—4 chromosomes each. The amount of genetic material has been reduced. In addition, the combinations of chromosomes in each gamete have been made at random.

Meiosis and Genetics

Chromosomes pair and separate during meiosis exactly as Mendel would have predicted for the structures that carry genes. Meiosis I results in segregation and independent assortment. During anaphase I of meiosis, the homologous chromosomes separate and are segregated to

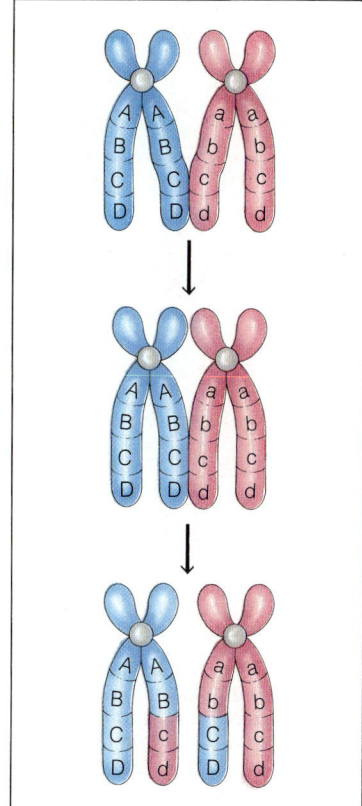

Figure 9–18 During prophase I of meiosis, homologous chromosomes may exchange portions of their chromatids in a process called crossing-over.

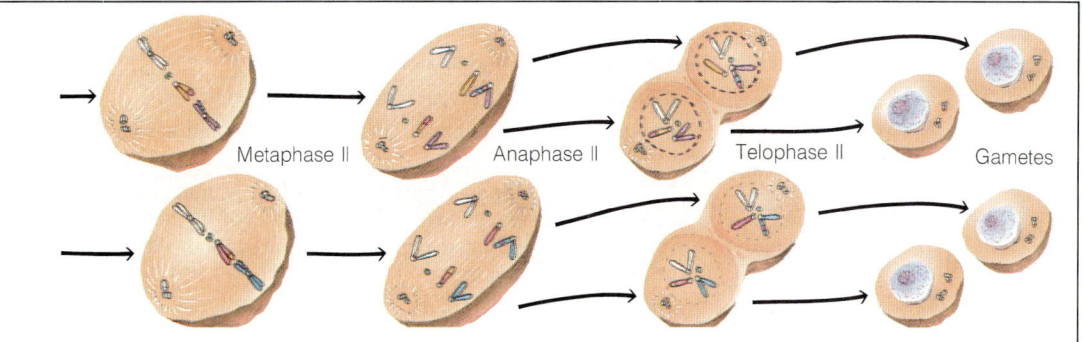

Metaphase II   Anaphase II   Telophase II   Gametes

TEACHING SUPPORT

Teaching Resources
- Vocabulary Review: Word Game, p. 1
- Hands-On Activity: A Model of Meiosis, p. 61

Study Guide
- Section 9–3, p. 90

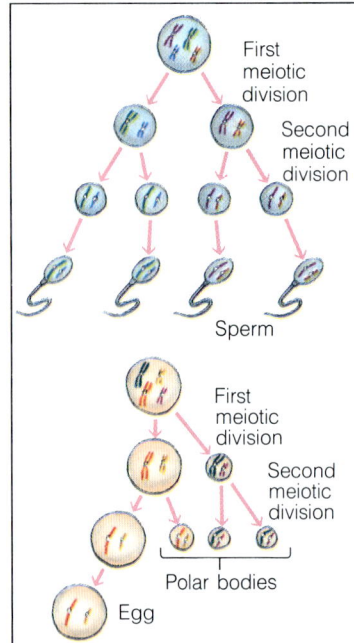

Figure 9–19 In males, meiosis results in four gametes that are the same size as one another. In females, the four cells produced in meiosis are different sizes—three are small and one is large. The large cell is usually the only one of the four female cells that participates in reproduction.

different cells. This also segregates the homologous forms of a gene, or alleles, that are located on these chromosomes. This agrees with Mendel's principles. The chromosomes themselves assort independently during meiosis I. The separation of chromosomes during the first meiotic division is completely random. Which cell receives the maternal copy of a chromosome and which cell receives the paternal copy is strictly a matter of chance.

Gamete Formation

In male animals, the haploid gametes produced by meiosis are called sperm. In higher plants, pollen grains contain haploid sperm cells. In female animals, generally only one of the cells produced by meiosis is used for reproduction. This female gamete is called an egg (animals) or ovule (higher plants). In female animals, the cell divisions at the end of meiosis I and meiosis II are uneven, so that the egg or ovule receives most of the cytoplasm. The other three cells produced in the female during meiosis are known as polar bodies and usually do not participate in reproduction.

Comparing Mitosis and Meiosis

In a way, it's too bad that the words mitosis and meiosis sound so similar and have the same names for their phases because the two processes are very different. Mitosis results in the production of two genetically identical cells. A diploid cell that divides by mitosis gives rise to two diploid daughter cells. The daughter cells have sets of chromosomes (and genes) identical to each other and to the original parent cell.

Meiosis, on the other hand, begins with a diploid cell but produces four haploid cells. These cells are genetically different from the diploid cell and from one another. This is because homologous chromosomes are separated during the first meiotic division and because crossing-over results in the production of new gene combinations on the chromosomes.

9–3 SECTION REVIEW

1. What is meiosis?
2. Define diploid and haploid.
3. In which meiotic division does segregation occur?
4. What are the principal differences between mitosis and meiosis?
5. **Critical Thinking—Applying Concepts** In human cells, 2N=46. How many chromosomes would you expect to find in a sperm cell? In an egg cell? In a white blood cell? Explain your answers.

9–3 (continued)

Content Development
Using the Visuals Have students examine Figure 9–19.
- **How is meiosis in females different from meiosis in males?** (In females, only one large egg cell is formed. The other three cells are very small and are known as polar bodies. Although the chromosomes are divided equally among the four cells, only one cell receives most of the cytoplasm. This is the cell that will become the female gamete, the egg.)

Enrichment
Here are some additional genetics problems for your students to try.
1. Black color in horses is dominant over chestnut color. If a pure black horse is mated to a chestnut horse, what is the probability that the offspring will be chestnut colored? (B = gene for black color; b = gene for chestnut color. $BB \times bb \rightarrow 100\%\ Bb$. 0% will be chestnut; 100% will be black.)
2. A common recessive trait in dogs is deafness. A pure line of normal-hearing dogs was crossed with a pure line of deaf dogs. F_1 and F_2 generations were produced. (H = gene for normal hearing; h = gene for deafness.)
   a. What percentage of the F_1 generation is expected to have normal hearing? ($HH \times hh \rightarrow 100\%\ Hh$. 100% have normal hearing.)
   b. What is the phenotype ratio of the F_2 generation? ($Hh \times Hh \rightarrow$ 25% HH, 50% Hh, 25% hh. 3 hearing [HH and Hh]:1 deaf [hh].)
3. Two mice with black fur were crossed and produced offspring with brown fur and offspring with black fur. What were the most probable genotypes of the parents? (B = gene for black fur; b = gene for brown fur. There are three possible combinations of parents: $BB \times BB$; $BB \times Bb$; and $Bb \times Bb$. Only the last combination [$Bb \times Bb$] will yield any offspring having brown fur [bb].)
4. Silky feathers in fowl is a trait caused by a gene that is recessive to that for normal feathers. (F = gene for normal feathers; f = gene for silky feathers.)
   a. A cross between two heterozygous birds yielded 68 offspring. How many would be expected to

PROBLEM SOLVING IN BIOLOGY

SOLVING GENETICS PROBLEMS

One-Factor Crosses: Sample Problem
In pea plants, red flowers are dominant over white flowers. A heterozygous red flower is allowed to self-pollinate. What are the probable genotypic and phenotypic ratios in the offspring of this plant?

SOLUTION

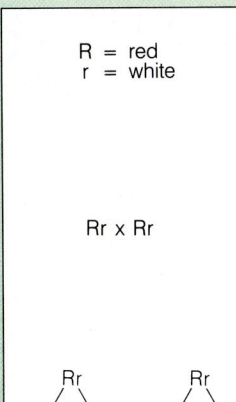

Step 1 **Choose a letter to represent the genes in the cross.**
Use a letter whose capital form does not look too similar to its lowercase form. This will make it easier for you to read your finished Punnett square. Except for that requirement, it is not important which letter you select. In this case, let's use *R* for the dominant red allele and *r* for the recessive white allele.

Step 2 **Write the genotypes of the parents.**
This step is often written as an abbreviation of the cross being studied. The X between the parents' genotypes is read "is crossed with." In this case, *Rr* X *Rr* is read "*Rr* is crossed with *Rr*." Although only one parent is involved in this problem, you must still write it as a cross in which you account for a male parent and a female parent.

Step 3 **Determine the possible gametes (reproductive cells) that the parents can produce.**
Remember that alleles are segregated during the formation of gametes. Each gamete has 1/2 the number of alleles in the parent.

Step 4 **Enter the possible gametes at the top and side of the Punnett square.**

Step 5 **Complete the Punnett square by writing the alleles from the gametes in the appropriate boxes.**
This step represents the process of fertilization. The allele from the gamete above the box and the allele from the gamete to the side of the box are combined inside each of the four boxes. If there is a combination of capital letter and lowercase letter in a box, write the capital letter first. The letters inside the boxes represent the probable genotypes of the offspring resulting from the cross. In this example, 1/4 of the offspring are genotype *RR*, 1/2 are *Rr*, and 1/4 are *rr*.

197

---

be silky and how many normal? (*Ff* × *Ff* → 75% normal and 25% silky, or 51 normal and 17 silky.)

SECTION REVIEW 9–3

1. The process in which haploid gametes are formed from the reduction division of a diploid mother cell.
2. Diploid: possessing two sets of homologous chromosomes; 2N. Haploid: possessing only one set of chromosomes; N.
3. Anaphase I.
4. Mitosis produces two diploid cells that are genetically identical to each other and to the mother cell. Meiosis produces four haploid cells that are different from one another and from the diploid mother cell.
5. Human sperm and egg cells are haploid (N); each contains 23 chromosomes. Human white blood cells, like other body cells, are diploid (2N); each contains 46 chromosomes.

BACKGROUND INFORMATION
MITOSIS AND GAMETES

In a few species of animals, such as bees and ants, males arise from unfertilized eggs and are haploid. These haploid males produce gametes through mitosis rather than meiosis. However, the eggs in females are produced by meiosis, so the chromosome number remains the same from one generation to the next.

TIE-IN LANGUAGE ARTS

Aristotle, a philosopher of Ancient Greece, taught that traits were inherited through the blood. We still use terms this idea inspired, such as "bloodline" and "blood relative."

Reinforcement/Reteaching
After students complete the Section Review questions, review any material on meiosis that gave them trouble.

Closure
Pose the following question:
• **Why is meiosis essential for the survival of a species?** (It ensures that the chromosome number will be held constant from one generation to the next and provides for one member of each pair of homologous chromosomes from each parent to be passed to the offspring.)

PROBLEM SOLVING

Solving Genetics Problems

Being able to solve genetics problems indicates that a student can put Mendel's genetics principles to work. By analyzing the information given in a word problem and assigning appropriate symbols, students demonstrate that they can distinguish the dominant from the recessive trait. In selecting the correct combinations of genes for the parental generation, students show that they comprehend segregation. And, finally, to tie all of these ideas together into a two-factor cross is to demonstrate an understanding of independent assortment.

It might be a good idea to work along with students in solving the two-factor cross problem in a step-by-step fashion. Some students will grasp it very quickly after seeing the solution only once. Direct them to additional practice problems at the end of this feature. Other students will need to see several problems solved before being able to work independently. Circulate through the room as students work on their practice problems in order to identify those who need assistance.

CONTINUED

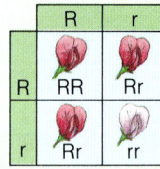

Genotypic ratio = 1:2:1
Phenotypic ratio = 3:1

Step 6 Determine the phenotypes of the offspring.
Remember that phenotype refers to the physical appearance of an organism. The principle of dominance makes it possible to determine the phenotype that corresponds to each genotype inside the Punnett square. In this example, 3/4 of the offspring have red flowers and 1/4 of the offspring have white flowers.

Step 7 Using the results of Steps 5 and 6, answer the problem.
Usually you will be asked to summarize the results of the cross by providing genotypic and phenotypic ratios. When writing such ratios, the numbers for the dominant genotype(s) or phenotype(s) come first. In this example, 1/4 of the offspring are genotype *RR*, 1/2 (or 2/4) are *Rr*, and 1/4 are *rr*. The genotypic ratio is therefore 1/4:2/4:1/4, or 1:2:1. Three fourths of the offspring have red flowers and 1/4 have white flowers. The phenotypic ratio is therefore 3/4:1/4, or 3:1.

Two-Factor Crosses: Sample Problem

In the fruit fly *Drosophila melanogaster*, wings (*A*) are dominant over a lack of wings (*a*) and red eyes (*E*) are dominant over sepia (brownish) eyes (*e*). A wingless fly that is heterozygous for eye color is crossed with a fly that is heterozygous for both eye color and presence of wings. What are the genotypic and phenotypic ratios for this cross? What fraction of the offspring from this cross will be wingless and have sepia eyes? What fraction will have the genotype *AaEe*?

SOLUTION

A = wings
a = wingless
E = red
e = sepia

aaEe x AaEe

aaEe   AaEe
aE ae aE ae   AE Ae aE ae

	aE	ae	aE	ae
AE				
Ae				
aE				
ae				

Step 1 Choose letters to represent the genes in the cross.
In this particular problem, the letters to use are given to you. In the event that they are not, follow the same suggestions for choosing letters stated in Step 1 of the One-Factor Cross.

Step 2 Write the genotypes of the parents.

Step 3 Determine the possible gametes that the parents can produce.

Step 4 Enter the possible gametes at the top and side of the Punnett square.

	aE	ae	aE	ae
AE	AaEE	AaEe	AaEE	AaEe
Ae	AaEe	Aaee	AaEe	Aaee
aE	aaEE	aaEe	aaEE	aaEe
ae	aaEe	aaee	aaEe	aaee

Step 5 **Complete the Punnett square by writing the alleles from the gametes in the appropriate boxes.**
The alleles from the gamete above the box and the alleles from the gamete to the side of the box are combined inside each of the boxes. Write the capital letter first for each pair of alleles. The letters inside each box represent the probable genotypes of the offspring resulting from the cross. In this example, 2/16 of the offspring are *AaEE*, 4/16 are *AaEe*, 2/16 are *Aaee*, 2/16 are *aaEE*, 4/16 are *aaEe*, 2/16 are *aaee*.

Step 6 **Determine the phenotypes of the offspring.**
In this example, 6/16 are winged and red-eyed, 2/16 are winged and sepia-eyed, 6/16 are wingless and red-eyed, 2/16 are wingless and sepia-eyed.

Step 7 **Using the results of Steps 5 and 6, answer the problem.**
Note that in this example, as in many of the genetics problems you will encounter, you are asked for more than just the ratios resulting from the cross. This is one reason why it is important to read genetics problems carefully. In this example, the genotypic ratio is 2/16:4/16:2/16:2/16:4/16:2/16 = 2:4:2:2:4:2 = 1:2:1:1:2:1. The phenotypic ratio is 6/16:2/16:6/16:2/16 = 6:2:6:2 = 3:1:3:1. And 4/16, or 1/4, of the offspring have the genotype *AaEe*.

One-Factor Crosses: Practice Problems

1. Pollen from a pea plant with white flowers is used to fertilize the ovules (female gametes) of a heterozygous plant. What are the possible phenotypes in the offspring from this cross?
2. You have a pea plant with red flowers. Design a cross to determine if this plant is homozygous or heterozygous. Use a Punnett square to show all possible crosses.

Two-Factor Crosses: Practice Problems

1. A purebred wingless red-eyed fruit fly is crossed with a purebred winged sepia-eyed fruit fly to produce F_1 flies. The F_1 flies are crossed to produce F_2 flies. What is the phenotypic ratio of the F_2 flies?
2. You have a winged red-eyed fruit fly. Design a cross to determine whether the fly is heterozygous for either or both traits. Use a Punnett square to show all possible crosses.

Answers to Practice Problems
Check students' Punnett squares for accuracy.

One-Factor Crosses
1. White flowers, red flowers
2. The unknown plant should be crossed with one that is homozygous recessive (has white flowers). Check students' Punnett squares to make sure they understand the concept of a test cross.

Two-Factor Crosses
1. 9 winged, red-eyed:3 winged, sepia-eyed:3 wingless, red-eyed:1 wingless, sepia-eyed.
2. The fly should be crossed with a wingless sepia-eyed fly. Check the students' Punnett squares to make sure they understand the concept of a test cross.

LABORATORY INVESTIGATION

SIMULATING A ONE-FACTOR CROSS

Before the Lab
1. You will need 1 paper cup and 2 game markers for each student in the class.
2. The game markers, such as those used in bingo or tiddlywinks, can be purchased from a toy or hobby store. Prepare them in advance of class by applying masking tape to both sides of each marker. Label one side *A* and the reserve side *a*. (If game markers are not available, suitable substitutes would be pennies or pieces of acetate, construction paper, or matteboard.)

Pre-Lab Discussion
- **If we used only one marker for this laboratory activity, what percentage of tosses would give us the result *A*?** (50%) **And *a*?** (50%)

  Point out that each marker represents the genetic contribution of a hybrid organism. The chances of landing on either side are equal 1:1 odds.

  By tossing two markers, you will be simulating the mating of two hybrids. There are three possible outcomes: Both markers land on the *A* side (genotype *AA*), or both land on the *a* side (genotype *aa*), or one shows *A* and the other *a* (genotype *Aa*).

- **Since we are using two markers for today's lab, what percentage of the time would you expect *AA*?** (Answers may vary.) **And *Aa*? And *aa*?** (Answers may vary.)

Skills Development
Students will use the following skills while completing this laboratory investigation.
1. Observing
2. Predicting
3. Recording
4. Comparing
5. Analyzing

LABORATORY INVESTIGATION: SIMULATING A ONE-FACTOR CROSS

PROBLEM
How is probability applied to genetics?

MATERIALS *(per student)*

> 2 labeled game markers
> paper cup

PROCEDURE
1. On a separate sheet of paper, prepare a completed data table similar to the one below.
2. In pea plants, yellow seeds (*A*) are dominant over green seeds (*a*). Using a Punnett square, determine the probable color of the seeds produced by pea plants whose parents are heterozygous (*Aa*) for the seed-color trait. Record the expected genotypic and phenotypic ratios in the appropriate places in your data table.
3. Each labeled marker represents the alleles in the heterozygous plant. Tossing the labeled markers together represents the crossing of heterozygous plants.
4. Put the 2 labeled markers into the cup. Holding one hand over the mouth of the cup, shake the cup to toss the markers. Empty the cup onto your desk or laboratory table. Record the results of each of 10 tosses by making a tally mark in the appropriate box in your data table.
5. Toss the markers 100 times and record the results in your data table.
6. One student in the class will compile the data for 100 tosses by 10 students and report the results. Record this information.
7. Count the tally marks for each genotype in each series of tosses (10, 100, 1000). Record these totals in the appropriate box.
8. Determine the total number of seeds with the yellow phenotype for each series of tosses.
9. Using the data, calculate the genotypic and phenotypic ratios for each series of tosses. This is done by dividing each number in the ratio by the ratio's smallest number and rounding off to the nearest tenth's place. For example, suppose you obtained 23 *AA*, 51 *Aa*, and 26 *aa* for a series of 100 tosses. This gives you the ratio 1:2.2:1.1 (23/23 = 1, 51/23 = 2.2, 26/23 = 1.1)

OBSERVATIONS
1. Which genotype was obtained most often?
2. What were your expected genotypic and phenotypic ratios?
3. What were your genotypic and phenotypic ratios for a series of 10 tosses? 100 tosses? 1000 tosses?

ANALYSIS AND CONCLUSIONS
1. How do the experimental ratios compare with the expected ratios?
2. Which series of tosses produced the experimental ratios that were closest to the expected ratios? Explain your results.
3. How does probability apply to the results of your experiment? How does probability apply to the study of genetics?

Offspring Phenotype	Yellow				Green		Total Number of Yellow Seeds (AA + Aa)	Expected Genotypic Ratio	Expected Phenotypic Ratio	Experimental Genotypic Ratio	Experimental Phenotypic Ratio
Offspring Genotype	AA		Aa		aa						
	Tally	Total	Tally	Total	Tally	Total					
10 Tosses											
100 Tosses											
1000 Tosses	✗		✗		✗						

Safety Tips
It is a good idea to keep several spare makers handy in the event that some may roll away during the student activity.

Teaching Strategy
Choose 10 students and assign the task of compiling the data after the class has finished working. The 10 students who are chosen to report their data should be directed to write their results on a slip of paper with their name.

Observations
1. *Aa*.
2. Genotypic ratio: 1:2:1; phenotypic ratio: 3:1.
3. Student answers will vary.

Analysis and Conclusions
1. Student answers may vary.

STUDENT STUDY GUIDE

SUMMARIZING THE CONCEPTS

The key concepts in each section of this chapter are listed below to help you review the chapter content. Make sure you understand each concept and its relationship to other concepts and to the theme of this chapter.

9–1 The Work of Gregor Mendel
- Mendel's genetic experiments involved the mathematical analysis of the offspring resulting from crosses between pea plants.
- Genes are individual factors that control heredity.
- The different forms of a gene are called alleles. Some alleles are dominant; others are recessive.
- Because of dominance, the effects of the recessive gene are not observed when the dominant gene is present.
- Alleles for the same trait are separated during the process of segregation.
- The genes for different traits may undergo independent assortment.
- An organism's physical characteristics, which reflect its genetic makeup, are its phenotype. An organism's genetic makeup is its genotype.
- Organisms that have two identical alleles for a particular trait are said to be homozygous, or purebred, for that trait. Organisms that have two different alleles for the same trait are said to be heterozygous, or hybrid, for that trait.

9–2 Applying Mendel's Principles
- Probability is the likelihood that a particular event will occur. The rules of probability apply to genetics.
- Genetics problems can be solved by using a Punnett square.

9–3 Meiosis
- A cell that contains both sets of homologous chromosomes is said to be diploid (2N). A cell that contains a single set of chromosomes is said to be haploid (N).
- Meiosis is a form of cell division that results in the formation of haploid gametes. The cells formed in meiosis have different sets of chromosomes from one another and from the diploid mother cell.

REVIEWING KEY TERMS

Vocabulary terms are important to your understanding of biology. The key terms listed below are those you should be especially familiar with. Review these terms and their meanings. Then use each term in a complete sentence. If you are not sure of a term's meaning, return to the appropriate section and review its definition.

9–1 The Work of Gregor Mendel
- heredity
- genetics
- self-pollination
- cross-pollination
- purebred
- trait
- hybrid
- gene
- allele
- dominance
- dominant
- recessive
- segregation
- Punnett square
- gamete
- phenotype
- genotype
- homozygous
- heterozygous
- independent assortment

9–2 Applying Mendel's Principles
- probability
- test cross

9–3 Meiosis
- homologous
- diploid
- haploid
- meiosis
- crossing-over

The series of 100 tosses should produce ratios that are fairly close to the expected values, and the series of 1000 tosses should produce ratios that are even closer to the expected values. The series of 10 tosses may or may not produce ratios that resemble the expected values.

2. Student answers may vary. In almost all cases, students should find that the series of 1000 tosses produced ratios that were closest to the expected values because, according to the rules of probability, the larger the number of trials, the closer you get to the expected ratios. If the series of 10 tosses or 100 tosses produced better ratios, students should attempt to explain the source of error.

3. The results of the experiment confirm two rules of probability: (1) The larger the number of trials, the closer you get to the expected ratios, and (2) previous events do not affect future outcomes. The results of genetics crosses follow these two rules of probability.

Going Further: Enrichment

Part 1
How close are your class distributions to the expected results? How close did your data come to Mendel's results? Chi square values could be calculated by students with some advanced mathematical ability.

Part 2
This lab can be extended to investigate two-factor crosses. In addition to using two markers each labeled *A* and *a*, prepare two more markers each labeled *B* and *b*. By tossing all four at one time, your students can simulate the results of a two-factor cross.

Part 3
Markers made of acetate can be used to show the results of crosses involving incomplete dominance. Students can see the heterozygous phenotype by putting one piece of acetate over another. For example, if you are working with yellow and red acetate, 1 yellow piece and 1 red piece produces a heterozygote with the intermediate phenotype of orange.

To perform this activity successfully, put 25 pieces of acetate of one color (red) and 25 pieces of a second color (yellow) into an opaque envelope. Students should be instructed to draw 2 pieces at random, record their results, and return the pieces to the envelope.

You will need one envelope and 50 pieces of acetate (25 of each color) for each student.

CHAPTER REVIEW

CONTENT REVIEW

Multiple Choice
1. d   2. b   3. a   4. d
5. b   6. d   7. b   8. b

True or False
1. F   cross-pollination
2. F   purebred
3. T
4. F   dominant
5. F F_2 generation
6. T
7. F   meiosis
8. F   large

Word Relationships
1. N
2. seed color
3. heterozygous
4. genotype
5. pollen
6. recessive

CONCEPT MASTERY

1. Because he founded the science of genetics.
2. DNA replicates; double-stranded chromosomes condense; homologous chromosomes pair; chromosome pairs line up in the center of the cell; crossing-over may occur; homologous chromosomes separate; two haploid cells formed; chromosomes line up in center of cell; centromeres replicate; single-stranded chromosomes separate; four haploid cells (gametes) formed.
3. Genes: Factors that control heredity are individual units that are inherited from each parent. Dominance: Some genes are dominant, others are recessive. Segregation: Alleles are separated during gamete formation. Independent Assortment: Genes that are not linked are segregated randomly during gametes formation.
4. Probability is the likelihood that a particular event will occur. The outcomes of genetic crosses can be predicted using the rules of probability.
5. The separation of homologous chromosomes during anaphase I of meiosis results in segregation and independent assortment.

CHAPTER REVIEW

CONTENT REVIEW

Multiple Choice

Choose the letter of the answer that best completes each statement.

1. Alleles for the same trait are separated from each other during the process of
    a. mitosis.   c. cross-pollination.
    b. meiosis II.   d. segregation.
2. Organisms that have two identical alleles for a particular trait are said to be
    a. haploid.   c. heterozygous.
    b. homozygous.   d. diploid.
3. Pea plants usually produce seeds through
    a. self-pollination. c. mitosis.
    b. cross-pollination. d. meiosis.
4. Because body cells contain two sets of chromosomes, they are
    a. homozygous.   c. haploid.
    b. heterozygous.   d. diploid.
5. An organism's physical appearance is its
    a. genotype.   c. heredity.
    b. phenotype.   d. homolog.
6. What is the exchange of genes between homologous chromosomes called?
    a. tetrad   c. independent assortment
    b. segregation   d. crossing-over
7. The phenotypic ratio in the offspring resulting from the cross $Tt \times Tt$ is
    a. 1:2:1.   c. 1:1.
    b. 3:1.   d. 9:3:3:1.
8. Some F_2 individuals show the recessive character because of
    a. dominance.   c. crossing-over.
    b. segregation.   d. independent assortment.

True or False

Determine whether each statement is true or false. If it is true, write "true." If it is false, change the underlined word or words to make the statement true.

1. The fertilization of a plant's ovules (egg cells) by the pollen of another plant is called <u>self-pollination</u>.
2. If allowed to self-pollinate, <u>hybrid</u> peas will produce offspring exactly like themselves.
3. Meiosis results in <u>haploid</u> cells.
4. The effects of a <u>recessive</u> allele are seen in a heterozygous organism.
5. The offspring produced by crossing F_1 plants are known as the <u>P generation</u>.
6. Reproductive cells are also called <u>gametes</u>.
7. Homologous chromosomes pair up during the first stage of <u>mitosis</u>.
8. Expected genetic ratios show up only when <u>small</u> numbers of offspring are considered.

Word Relationships

An analogy is a relationship between two pairs of words or phrases generally written in the following manner: a:b::c:d. The symbol : is read "is to," and the symbol :: is read "as." For example, cat:animal::rose:plant is read "cat is to animal as rose is to plant."

In the analogies that follow, a word or phrase is missing. Complete each analogy by providing the missing word or phrase.

1. mitosis:2N::meiosis: _____
2. allele:gene::green seeds: _____
3. purebred:hybrid::homozygous: _____
4. white fur:*ff*::phenotype: _____
5. egg:ovule::sperm: _____
6. *A*:dominant::*a*: _____

CRITICAL AND CREATIVE THINKING

1. Answers will vary, depending on which letters students decide to use to represent alleles; both parents are heterozygous (e.g., *Aa*).
2. Answers will vary, depending on which letters students decide to use to represent alleles; the unknown plant is homozygous recessive (e.g., *aa*) and has wrinkled seeds.
3. The offspring resulting from the cross may inherit the qual-

CONCEPT MASTERY

Use your understanding of the concepts developed in the chapter to answer each of the following in a brief paragraph.

1. Why is Gregor Mendel sometimes called the "father of genetics"?
2. Describe the process of meiosis.
3. List the four basic principles of genetics that Mendel discovered in his experiments. Briefly describe each of these principles.
4. What is probability? How does probability relate to genetics?
5. How do the cellular events in meiosis account for Mendel's observations?

CRITICAL AND CREATIVE THINKING

Discuss each of the following in a brief paragraph.

1. **Applying concepts** In dogs, short hair is dominant over long hair. Two short-haired dogs are the parents of a litter of eight puppies. Six puppies have short hair and two have long hair. What are the genotypes of the parents?
2. **Applying concepts** Purebred tall plants with round seeds are crossed with purebred short plants with wrinkled seeds. Some of the offspring are then crossed with an unknown plant. The offspring from this second cross are 1/2 tall round and 1/2 tall wrinkled. What are the genotype and phenotype of the unknown plant?
3. **Making inferences** Explain why horse breeders will pay a lot of money to breed one of their horses to a horse that has won the Kentucky Derby.
4. **Making calculations** The probability of two independent events occurring simultaneously is the product of each of their probabilities. For example, if you flip two coins, the chance that both will come up heads is 1/4 (= 1/2 x 1/2). Suppose that you have ten Ping-Pong balls in a jar, each marked with a numeral from 0 to 9. If you mix them thoroughly and take out one, the chances are 1 in 10 (1/10) that you will get a particular numeral. Now suppose you have six such jars. What is your chance of drawing the number 531963?
5. **Making generalizations** In guinea pigs, a rough coat (*R*) is dominant over a smooth coat (*r*). A heterozygous guinea pig (*Rr*) and a homozygous recessive guinea pig (*rr*) have a total of nine offspring. Each has a smooth coat. Explain these results.

	R	r
r	Rr	rr
r	Rr	rr

6. **Designing an experiment** In sheep, white wool (*A*) is dominant over black wool (*a*). How would you determine the genotype of a white ram, or male sheep?
7. **Using the writing process** In the 1950s, a television series presented an imaginary voyage back in time. Each week a famous person in history was visited. The show was designed to provide insight into the work of the person being interviewed and the time in which he or she lived. You are in charge of producing such a program for today's television viewers. The first person you will visit is Gregor Mendel. Write a script for this program.

---

ities that enabled their parent to win the Kentucky Derby.
4. One in a million; 1/10 × 1/10 × 1/10 × 1/10 × 1/10 × 1/10 = 1/1,000,000.
5. The expected ratios show up only when large numbers of offspring are considered.
6. Use a test cross.
7. Answers will vary. Check student responses for creativity as well as accuracy.

CHAPTER 10 Genes and Chromosomes

Section	Laboratory Investigations and Activities
10–1 The Chromosome Theory of Heredity pages 205–211 THEMES: Stability, Patterns of Change	**Student Edition** LABORATORY INVESTIGATION: Mapping Chromosomes, p. 222 **Teacher Edition** DEMONSTRATION: Isolating DNA, p. 204D **Laboratory Manual** Human Sex Chromosomes, p. 139
10–2 Mutations pages 212–214 THEMES: Patterns of Change, Unity and Diversity	**Laboratory Manual** Karyotypes, p. 143 **Teaching Resources** TEACHER DEMONSTRATION: Causing Lethal Mutations in Yeast, p. 7
10–3 Regulation of Gene Expression pages 215–221 THEME: Systems and Interactions	**Teacher Edition** DEMONSTRATION: Combinations and Genetic Variety, p. 204D
Chapter Review pages 223–225	

Teacher Resources

Books

Callan, Harold G. *Lampbrush Chromosomes*, Springer-Verlag, 1986.

Dillon, Lawrence S. *The Gene: Its Structure, Function, and Evolution*. Plenum Publishers, 1987.

Hennig, W., ed. *Structure and Function of Eukaryotic Chromosomes*. Springer-Verlag, 1987.

Lewin, Benjamin. *Genes*. Wiley, 1987.

Muench, K. H. *Medical Genetics*. Elsevier, 1987.

Seeberg, Erling, and Kjell Kleppe, eds. *Chromosome Damage and Repair*. Plenum Publishers, 1982.

Watson, James D., and Nancy H. Hopkins. *Molecular Biology of the Gene*, complete, 4th ed. Benjamin-Cummings, 1988.

Watson, James D., John Tooze, and David T. Kurtz. *Recombinant DNA: A Short Course*. Freeman, 1983.

CHAPTER PLANNING GUIDE

Other Activities	Media and Technology
Biotechnology Workbook Biotechnology Defined, p. 1 **Study Guide** Section 10–1, p. 95	
Study Guide Section 10–2, p. 98	
Teaching Resources VOCABULARY REVIEW: Word Scramble, p. 1 ACTIVITY: Pure Gold? p. 3 ACTIVITY: Genetic Patterns in Tribbles, p. 5 **Study Guide** Section 10–3, p. 100	**Biology Transparencies** 14: Gene Expression in Prokaryotes
Teaching Resources Chapter Test A, p. 13 Chapter Test B, p. 17	**Computer Test Bank** Chapter Test, p. 101

Audiovisuals

The Gene. 2 filmstrips with cassettes. Hawkhill Associates, Inc.
Genetic Biology. Film or video. Coronet.
Linkage, Crossing-Over and Chromosome Maps. Sound filmstrip. Ward.
Patterns of Development. Film or video. Coronet.
The Story of the Gene. Filmstrip with cassette, Hawkhill Associates, Inc.
Regulation of Protein Synthesis: The Operon Hypothesis. Film or video. Biology Media.

CHAPTER 10 Genes and Chromosomes

CHAPTER OVERVIEW

In this chapter students will continue the study of genetics that they began in Chapter 9. Students will learn how Mendel's identification of the factors that control heredity paved the way for an understanding of genetics at the cellular level.

In the first section of the chapter students will learn about the chromosome theory. This important theory states that genes are located on chromosomes and that each gene occupies a specific place on a chromosome. As the section unfolds, students will discover how gene linkage, crossing-over, and sex linkage contribute to the inherited characteristics of an organism.

In the second section of the chapter students will learn about gene mutation. They will discover that mutations can occur at two levels—the level of chromosomes and the level of genes. The process known as nondisjunction and the condition of polyploidy will be among topics discussed.

In the third and final section of the chapter students will learn how the expression and activity of genes are regulated and controlled. Among the topics discussed will be dominance and incomplete dominance, gene expressions, operons, repressors, and gene expression in eukaryotes.

10-1 THE CHROMOSOME THEORY OF HEREDITY

Section Focus 10-1

The purpose of this section is to answer the important question that remained unaddressed in Mendel's work: Where are genes located? Students will learn that the answer to this question was provided by a young scientist named Walter Sutton. Sutton discovered that genes are located on chromosomes in the nucleus of cells. This discovery led to the formulation of the chromosome theory of heredity, which states that genes are located on chromosomes and that each gene occupies a specific place on a chromosome.

In this section students will also learn how gene linkage, crossing-over, and sex-linked traits affect heredity. The process of gene mapping will also be discussed.

Performance Objectives 10-1

1. State the chromosome theory of heredity.
2. Explain how gene linkage affects inherited traits.
3. Describe the process of crossing-over and explain how it increases genetic variety.
4. Describe gene mapping.
5. Describe the process of sex determination and the patterns of inheritance for sex-linked traits.

Science Terms 10-1

chromosome theory of heredity   p. 206
linked gene   p. 206
linkage group   p. 208
recombinant   p. 208
sex chromosome   p. 209
autosome   p. 209
X chromosome   p. 209
Y chromosome   p. 210
sex-linked   p. 210

10-2 MUTATIONS

Section Focus 10-2

The purpose of this section is to discuss mutations and their effects. A mutation is a sudden change in the DNA pattern caused by a mistake in the duplication or transmission of genetic information.

Students will learn that mutations can occur at two levels: at the level of chromosomes and at the level of genes. Chromosomal mutations involve segments of chromosomes, whole chromosomes, or sets of chromosomes. Gene mutations involve individual genes.

Students will learn about the four types of mutations that can change the structure of chromosomes: deletions, duplications, inversions, and translocations. They will also learn about nondisjunction, which is the failure of homologous chromosomes to separate normally during meiosis.

Students will discover how gene mutations can seriously affect gene function. Point mutations and frameshift mutations will be discussed.

Performance Objectives 10-2

1. Define the word *mutation*.
2. Distinguish between germ and somatic mutations.
3. Compare chromosomal and gene mutations.
4. Describe the various types of chromosomal mutations.
5. Describe the various types of gene mutations.

Science Terms 10-2

mutation   p. 212
chromosomal mutation   p. 212
nondisjunction   p. 213
polyploidy   p. 213
gene mutation   p. 213
point mutation   p. 213
frameshift mutation   p. 213

10-3 REGULATION OF GENE EXPRESSION

Section Focus 10-3

The purpose of this section is to discuss the factors that regulate and control gene expression. These factors include interactions between different genes and between genes and their environment.

Students will learn that dominance is one example of how genes interact with each other. Dominant and recessive alleles will be discussed, as will the phenomena of incomplete dominance, codominance, and polygenic inheritance.

Students will learn in detail how gene expression is controlled in prokaryotes. They will discover that the mechanism of gene control depends on an operon (which consists of a cluster of genes), a promoter, and an operator. The activation and repression of the operon will be discussed.

Students will discover that the control of gene expression in eukaryotes is much more complex than it is in prokaryotes. Among the topics discussed will be the role of substances called inducers and the importance of exon and intron sequences.

CHAPTER PREVIEW

Performance Objectives 10-3

1. Discuss the various factors that influence gene expression.
2. Define and discuss dominance, incomplete dominance, and codominance.
3. Compare the regulation of gene expression in prokaryotes and eukaryotes.
4. Identify the components of an operon and explain how an operon is controlled.
5. Discuss the roles of exons and introns in the transcription and processing of mRNA.

Science Terms 10-3

incomplete dominance   p. 216
codominance   p. 216
polygenic   p. 216
operon   p. 217
operator   p. 217
promoter   p. 217
inducer   p. 217
repressor   p. 218
exon   p. 220
intron   p. 220

DISCOVERY LEARNING

TEACHER DEMONSTRATIONS
Modeling

Isolating DNA
The following demonstration can be used as an introduction to Chapter 10.

Although this demonstration requires a low-speed centrifuge, the use of chloroform (which should be done in a fume hood), and calf thymus (which must be special-ordered from a local butcher), it produces a very rich DNA extract that precipitates into long, visible fibers when mixed with alcohol.

Using a pair of scissors, mince about 5.0 gm calf thymus into small (0.5 cm or less) cubes. Drop the thymus cubes into a beaker and mix with 100 mL 6% NaCl solution. Homogenize the tissue by using a glass tissue homogenizer or by pouring the tissue mixture into a blender and mixing at medium speed for 10 to 20 seconds. Pour the homogenate through 4 to 6 layers of cotton gauze or cheesecloth to remove debris. Withdraw about 10 mL of the filtered extract into a test tube and place the container holding the remainder on ice.

Add 1 mL 10% SDS (sodium dodecyl sulfate, also known as sodium lauryl sulfate) to the tube. SDS, a common detergent, helps to separate the DNA from protein. Under a fume hood, add 10 mL chloroform to the tube. (The chloroform denatures protein, but not DNA.) Cover tightly. Mix by shaking the tube at 30-second intervals for 15 minutes. Pour the mixture into a couple of glass centrifuge tubes. Cover.

Centrifuge for 4 minutes at $2000 \times g$ (this is near the top speed of most desktop centrifuges). When you remove the tubes from the centrifuge, you should see three layers in each tube. The bottom is chloroform, the middle is precipitated protein and other contaminants, and the top is an aqueous DNA extract.

Use a Pasteur pipette to remove the top layer and place it in a clean test tube. Using a fresh pipette, carefully dribble ice-cold ethanol into the tube to form a layer above the extract. (Although ethanol is best, isopropanol works, too.) You should see a cloudy layer where the water and alcohol meet. This is the DNA. If the DNA content of your extract is high enough, you should be able to twirl a glass rod through the water-alcohol interface and pick up thick, stringy fibers of DNA.

The DNA can be stored on the rod immersed in alcohol for several days. For further analysis, it can be removed from the alcohol solution by centrifugation.

This protocol can be used to extract DNA from other tissues, such as liver. However, the resulting DNA precipitate will not produce the filaments that make this demonstration dramatic and interesting to students. Fortunately—since the starting material is important for this demonstration—butchers are generally quite willing to order calf thymuses on request. They usually call the slaughterhouses with which they do business and ask them to pack in a few thymuses with the next shipment of organ meat.

Combinations and Genetic Variety
This demonstration can be performed as students study Section 10-3.

The purpose of the demonstration is to provide a somewhat fanciful model of how genetic variety is achieved through random combinations. For the demonstrations you will need five different-colored (or style) hats, five different-colored scarves, and five different-colored pairs of gloves.

Set out the 15 articles of clothing on a desk or table at the front of the room. Ask a student volunteer to come and put on one hat, one scarf, and one pair of gloves. Have the rest of the class write down the combination. Then have the student return the hat, scarf, and gloves to the table and sit down. Repeat the process with several more volunteers, asking each one to choose a different combination. After about 10 students have participated, give the class about 5 minutes to write down as many other possible combinations as they can think of.

- **How many different combinations do you think are possible to make from five hats, five scarves, and five pairs of gloves?** (125; 625 if the gloves don't necessarily match but each combination contains a right glove and a left glove.)

Point out that certain genes in cells are formed by processes that are not very different from the random selection of a hat, scarf, and gloves—except that the number of possibilities for gene variations can be in the thousands. Genes that are near each other can move, cross over, and undergo mutations such as translocation. All these processes increase genetic variety.

CHAPTER 10
Genes and Chromosomes

GUIDED ENQUIRY
Pose the following questions to students and have them record their responses. Point out that they will gain a better understanding of the key concepts if they read the chapter with these basic questions in mind. Upon completion of the chapter, pose the questions again. Ask students to compare their initial responses with those they have developed after reading the chapter.

- How did Mendel's work fit in with later discoveries made by cell biologists?
- What evidence led to the discovery that genes are located on chromosomes?
- How are linked genes different from genes that assort independently?
- How does crossing-over increase genetic variety?
- How do sex chromosomes influence other genetic traits?
- What effects do gene mutations have on organisms?
- How are the expression and activity of genes controlled?
- How does gene expression in prokaryotes differ from gene expression in eukaryotes?

INTRODUCING CHAPTER 10
Using the Visuals
Direct students' attention to the chapter-opener photo.
- **What kind of animal do you see in this photograph?** (A tiger.)
- **How is this tiger different from most tigers?** (Instead of being orange, this tiger is white.)
- **What do you think caused this tiger to be white?** (Answers may vary. After reading the caption, students should realize that this tiger inherited an altered gene for white coat color.)
- **How and why does a gene become altered?** (Accept all answers. Point out that this question will be partly answered in this chapter.)
- **Do you think that this tiger's offspring will also have white coats?** (Accept all legitimate answers from students. Since the gene for coat color is inheritable, it is possible that at least some of the tiger's offspring will have white coats, although this gene may not actually be the dominant gene.)

Have students read the chapter-opener text.
- **What things are mentioned that you have already studied?** (Answers may include how the genetic code is written into DNA

CHAPTER 10

Genes and Chromosomes

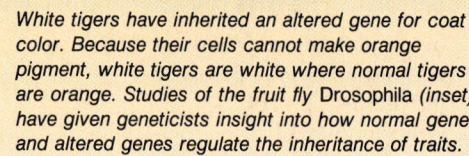

White tigers have inherited an altered gene for coat color. Because their cells cannot make orange pigment, white tigers are white where normal tigers are orange. Studies of the fruit fly *Drosophila* (inset) have given geneticists insight into how normal genes and altered genes regulate the inheritance of traits.

In the preceding chapters we set the stage for a detailed discussion of genes and their activities. We have seen that DNA is the genetic material. We know how the genetic code is written into DNA and how that code is translated into protein. We understand many of the things that proteins can do, and we appreciate their importance. We have recently learned how the science of genetics was developed, how the location of genes can be determined, how the chromosome is put together, and the rules that govern the passage of genetic information from one generation to the next. With all of these things behind us, we are like pilots with all of our ground training and study completed. Now we will begin to fly. We will see how the gene actually works!

GUIDE FOR READING

After you read the following sections, you will be able to

10-1 The Chromosome Theory of Heredity
- Relate genes to chromosomes.
- Explain how gene linkage and crossing-over affect heredity.
- Describe the patterns of inheritance for sex-linked traits.

10-2 Mutations
- Describe different kinds of mutations.
- Explain how mutations can affect heredity.

10-3 Regulation of Gene Expression
- Discuss gene interactions that influence gene expression.
- Explain how an operon is turned on and off.
- Describe introns and exons.

Journal Activity

YOUR AND YOUR WORLD Examine the photograph on the previous page. Do you think white tigers can survive for long in nature? In your journal, write a short story or a poem about a family of white tigers who live in a forest.

Figure 10-1 *These chromosomes are from a mouse cell. Chromosomes, which are located in the nucleus of a cell, are precisely separated during cell division—an indication that they contain something extremely critical to the cell.*

10-1 The Chromosome Theory of Heredity

Guide For Reading
- Where are genes located?
- How does gene linkage affect heredity?
- How does crossing-over affect heredity?
- What is a sex-linked trait?

History records Gregor Mendel as the founder of the science of genetics. Yet Mendel's work was incomplete because he never asked an important question that was the logical outcome of his work: Where in the cell are the factors that control heredity? In other words, where are the genes?

We can hardly blame Mendel for this shortcoming. Even if he had been inclined to ask, he would not have been able to search for genes among the bewildering variety of structures and components within cells. Shortly after completing his genetic experiments, Mendel was promoted to abbot, or head of the monastery. His administrative duties kept him too busy to continue his experiments. Fortunately, other scientists who had the time had begun to study cell structures and the processes of cell division. Their observations would prove to have enormous impact on the study of heredity.

Genes and Chromosomes

By the time Mendel's work was rediscovered in 1900, cell biologists had discovered most of the major structures within cells. They had also recorded the sequences of events that

205

BACKGROUND INFORMATION
GENES AND ALLELES

The distinction between a gene and an allele is one of the most important concepts in genetics. A gene is one of the controlling genetic elements that determines or helps to determine a trait. An allele is one of the possible forms of a gene. Although Mendel concentrated on a set of genes that have two alleles each, many genes do in fact have 3, 4, or more alleles. Genes with 20 possible alleles are not unheard of.

HISTORICAL NOTE
THE ORIGIN OF THE GENE

The first person to use the word *gene* was the Danish plant scientist Wilhelm Johannsen. In 1911, Johannsen used the word *gene* in place of the words *factor, trait,* and *characteristic*.

10–1 (continued)

Content Development
Emphasize to students that Sutton's development of the chromosome theory is an example of how the work of one scientist builds on the work of another scientist. Without Mendel's extensive experimentation and conclusions about the "factors" of heredity, Sutton might not have realized the significance of what he saw when he observed chromosomes. And, without Sutton's work, Mendel's ideas might never have been adequately proven or explained. In essence, Mendel paved the way for the chromosome theory, and the chromosome theory explained and lent substance to the work of Mendel.

Reinforcement/Reteaching
• What factors caused scientists to suspect that genes are located in the nucleus of cells? (Large size and central location of nucleus; fact that nucleus contains chromosomes, which participate in significant ways during cell division.)

• What evidence supported the idea that genes are located on chromosomes? (Careful and precise separation of chromosomes during mitosis; fact that during mitosis, diploid cells with two copies of each chromosome divide to form haploid cells with one copy of each chromosome; this fits in with Mendel's hypothesis that an organism has two factors for each inherited trait and that these factors separate during gamete formation.)

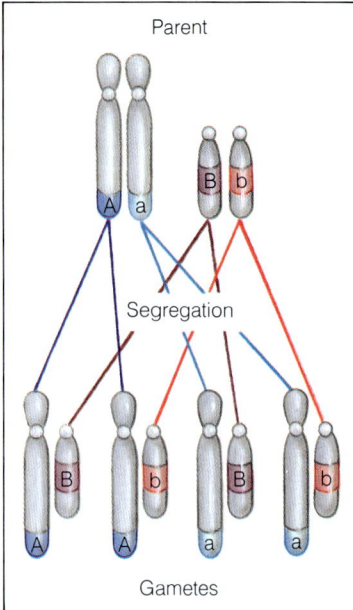

Figure 10–2 According to Mendel's hypothesis, the two factors for each trait are segregated during gamete formation. Thus gametes have only one factor for each trait. This hypothesis is supported by the observations that homologous chromosomes are separated during meiosis and that gametes contain only one of the chromosomes of each homologous pair.

occur during mitosis and meiosis. With this information, it seemed logical that the nucleus was the place for genes to be located. In addition to its large size and central location (which make it look important), the nucleus contains chromosomes—structures that behave in complex and interesting ways during cell divisions. The careful and precise mechanism by which chromosomes are separated during mitosis clearly suggests that they are structures vital to the cell. Could this be because they carry the cell's genes?

Further evidence that genes are located on chromosomes is the behavior of chromosomes during meiosis, the process by which reproductive cells are formed. In meiosis, diploid cells, which have 2 copies of each chromosome, divide to form haploid cells, which have 1 copy of each chromosome. This fits in with Mendel's hypothesis that an organism has 2 factors for each trait and that reproductive cells have only 1 factor for each trait. During the first part of meiosis, homologous chromosomes are separated from each other. This fits in with Mendel's hypothesis that the 2 factors for each trait are segregated, or separated, during the formation of gametes.

Walter Sutton, a young graduate student at Columbia University, arrived at the answer to the question of gene location in 1902. **The factors (genes) described by Mendel are located on chromosomes.** The tools with which Sutton made this discovery were a microscope—which he used to observe chromosomes during sperm formation in grasshoppers—and his own imagination, in which he compared the chromosomes to Mendel's factors. When the numbers and movements of chromosomes were analyzed, it was clear to Sutton that chromosomes behaved exactly as one would expect of the carriers of genetic information.

Sutton's **chromosome theory of heredity** states that genes are located on the chromosomes and each gene occupies a specific place on a chromosome. A gene may exist in several forms, or alleles. Each chromosome, however, contains just one of the alleles for each of its genes.

Gene Linkage

The fact that genes are located on chromosomes is important. Genes on a chromosome are linked together. This means that they are inherited together. In other words, **linked genes** do not undergo independent assortment. Recall Mendel's experiment in which the genes for seed color and seed shape were shown to assort independently. Unlike linked genes, genes that are located on separate chromosomes do, in fact, assort independently during meiosis.

One of the earliest examples of linked genes was discovered in the first part of this century by the American geneticist Thomas Hunt Morgan. Morgan studied the tiny fruit fly, *Drosophila melanogaster*, which can produce a new generation every

Enrichment
Interested students may wish to find out about the work of Theodor Boveri. Boveri was a German scientist who conducted genetics experiments similar to Sutton's and reached the same conclusions as Sutton.

four weeks. This short generation time makes *Drosophila* an ideal organism to study because traits in succeeding generations can be observed relatively quickly.

THE EFFECTS OF GENE LINKAGE Morgan crossed purebred flies that had gray bodies and normal wings with purebred flies that had black bodies and small wings. Because gray (*G*) is dominant over black (*g*), and normal wings (*W*) are dominant over small wings (*w*), all of the F_1 flies should have been gray with normal wings (genotype *GgWw*). That is exactly what Morgan observed.

However, when the F_1 flies (*GgWw*) were crossed with black small-winged flies (*ggww*), Morgan did not observe the expected results. If the principle of independent assortment were true for the *GgWw* X *ggww* cross, Morgan would have observed 25 percent (1/4) gray normal-winged, 25 percent (1/4) black small-winged, 25 percent (1/4) gray small-winged, and 25 percent (1/4) black normal-winged. Instead, Morgan obtained very different results for the cross.

As you can see in Figure 10-4, Morgan's actual results differed significantly from those predicted. Most gray-bodied flies had normal wings, and most black-bodied flies had small wings. These results indicated that the gene for body color and the gene for wing size were somehow connected, or linked. Morgan concluded that the two genes were linked by a physical bond in such a way that they could not assort independently.

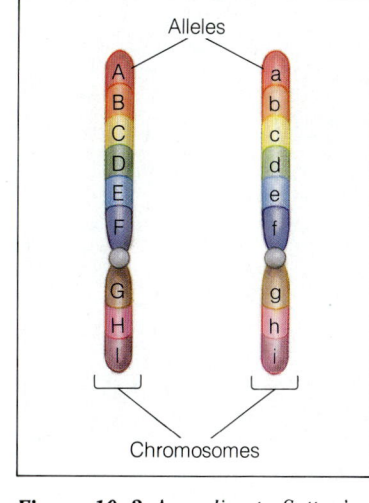

Figure 10-3 According to Sutton's chromosome theory of heredity, genes are located on the chromosomes. Homologous chromosomes have alleles for the same traits. These alleles may be the same on both homologous chromosomes or they may be different.

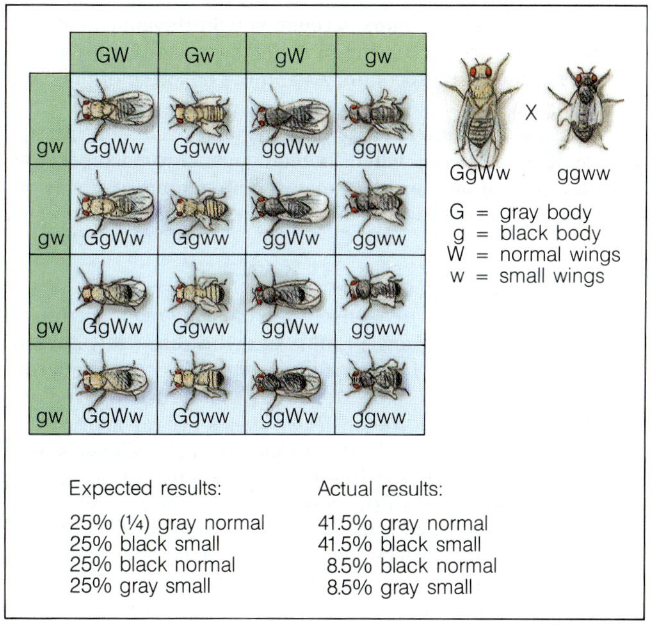

Figure 10-4 If the principle of independent assortment were true for the genes for body color and wing size in the fruit fly Drosophila, the offspring of the cross GgWw × ggww would have a 1:1:1:1 phenotypic ratio. Because the actual results did not resemble the expected results, Morgan concluded that the genes for body color and wing size were linked.

Expected results:
25% (¼) gray normal
25% black small
25% black normal
25% gray small

Actual results:
41.5% gray normal
41.5% black small
8.5% black normal
8.5% gray small

HISTORICAL NOTE
WALTER SUTTON

Walter Sutton determined the location of genes while studying meiosis in the cells of a grasshopper. Sutton noted that grasshoppers have 12 pairs of chromosomes. During meiosis, he saw that each gamete received one chromosome from each pair. He also noted that the chromosomes separated randomly and independently of one another. Finally, he noted that the union of the male and female gametes resulted in a zygote with the original number of chromosomes—that is, 24 chromosomes in 12 pairs.

Sutton concluded that Mendel's hypotheses could be explained in the following way:

1. A pair of factors determines each characteristic in an organism.
2. A gamete contains one factor from each pair.
3. Factors assort independently during meiosis.
4. Factors pair during fertilization to restore the original number of factors.
5. Individual factors remain unchanged from one generation to the next.

Not all scientists accepted Sutton's ideas. Some scientists argued that genes or hereditary factors did not exist at all. Others argued that Sutton must be wrong since organisms have many more traits than chromosomes. Sutton concluded, however, that each chromosome contains factors for more than one trait and that these factors are small particles located on chromosomes.

TEACHING SUPPORT
Biotechnology Workbook
- Biotechnology Defined, p. 1

Content Development
Make sure that students understand clearly the relationship between a gene and an allele. Point out that a gene determines a certain trait, such as hair color or eye color. An allele is one possible form of that gene or trait — for example, blonde hair or blue eyes. Thus one gene, such as a gene for hair color, may have several alleles, such as blonde, brown, red, auburn, and so on.

When a gene exists on homologous chromosomes, the alleles may be the same on both. For example, both chromosomes may contain an allele for brown hair. It is also possible that the alleles on homologous chromosomes will be different—for example, one chromosome may contain an allele for brown hair, and the other, one for blonde hair.

ANNOTATION KEY

❶ 19 (Applying concepts)

BACKGROUND INFORMATION
PREDICTING RECOMBINANTS

Because of the random nature of any single fertilization event, accurate measurements of recombination frequencies can be only determined in crosses that produce hundreds or thousands of offspring. Even in the fruit fly, which may produce 300 flies from a single mating pair, the statistical information often is not good enough to map closely linked genes. In an organism with far fewer offspring, the situation is much more difficult.

FACTS AND FIGURES

A human cell has about 100,000 different genes attached in a single line on each chromosome.

10–1 (continued)

Content Development

Point out to students that Morgan chose the fruit fly for his genetic studies for three basic reasons. First, fruit flies are easy to breed. Second, they produce a new generation every 4 weeks and may produce as many as 300 offspring from a single mating. Third, the chromosomes in the fruit fly are easy to study, since they are among the largest chromosomes found in animal cells.

Reinforcement/Reteaching

Use the following analogy to reinforce the concept of gene linkage groups or "packages."
• What are some products that often come in packages containing several different colors or flavors? (Possible answers: socks, colored pencils, M&Ms and certain other candies.)
• What happens if you want only one of the colors or flavors that happens to be in the package? (You have to buy the whole package to get the one you want.)
• What else do you get besides the color or flavor you want? (All the other colors or flavors that are packaged with it.)

Point out that linkage groups are like packages of different-colored or different-flavored products. They tend to be inherited together, just as some products are sold together. And, like the packaged products, if you inherit one trait, such as freckles, you are likely to inherit all the other traits that are "packaged" with it, whatever they might be.

Content Development
Using the Visuals Emphasize to students that crossing-over in-

Figure 10–5 A cobra has 38 chromosomes. How many linkage groups would this make? ❶

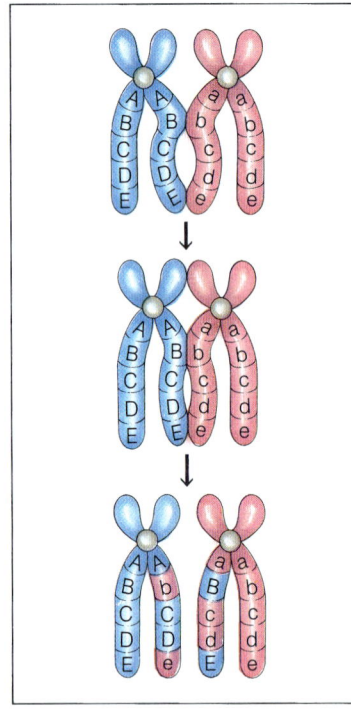

Figure 10–6 During prophase I of meiosis, homologous chromosomes may exchange sections of their chromatids in a process called crossing-over.

LINKAGE GROUPS As Morgan and his associates studied more and more genes, they found that the genes fell into distinct **linkage groups**, or "packages" of genes that always tended to be inherited together. The linkage groups themselves assorted independently, but all the genes on one group were inherited together. The linkage groups, of course, were chromosomes. Because homologous chromosomes contain the same genes, there is one linkage group for every homologous pair of chromosomes. *Drosophila* has four linkage groups. It also has four pairs of chromosomes. Corn has ten linkage groups and ten pairs of chromosomes.

Crossing-Over

Although the idea of linkage groups explains some of the results of the fruit fly crosses, it does not provide a complete explanation. Look at the results of the test cross between the *GgWw* and *ggww* flies again. Although 83 percent (41.5% + 41.5% = 83%) of the flies have gene combinations like their parents, 17 percent have new combinations. The 17 percent are, in the language of geneticists, **recombinants**—individuals with new combinations of genes.

If the genes for body color and wing size are linked, why aren't they linked all the time? Morgan and his associate, Alfred Sturtevant, proposed that the linkages could be broken some of the time. If two homologous chromosomes were positioned side by side, sections of the two chromosomes might cross, break, and reattach. As you can see in Figure 10–6, this process would rearrange the genes on the chromosome and produce new linkage groups. Morgan and Sturtevant called this hypothetical process crossing-over and suggested that it might be the reason for the recombinants in the offspring of the *Drosophila* test cross.

Crossing-over does indeed take place. It occurs during the first meiotic division, when homologous chromosomes are paired. Crossing-over produces new combinations of alleles on each chromosome and thus increases genetic variety.

Gene Mapping

Sturtevant further reasoned that crossing-over occurs at random along the linkage groups, and the distance between two genes determines how often crossing-over occurs between them. If two genes are close together, then crossing-over between them is rare. However, if two genes are far apart, then crossing-over between them is more common. Knowing the frequency with which crossing-over between two genes occurs makes it possible to map the positions of genes on a chromosome. This is precisely what Sturtevant tried to do with the genes in *Drosophila*'s linkage groups. And by 1915, just two years after he had begun, Sturtevant had mapped 85 genes in *Drosophila*.

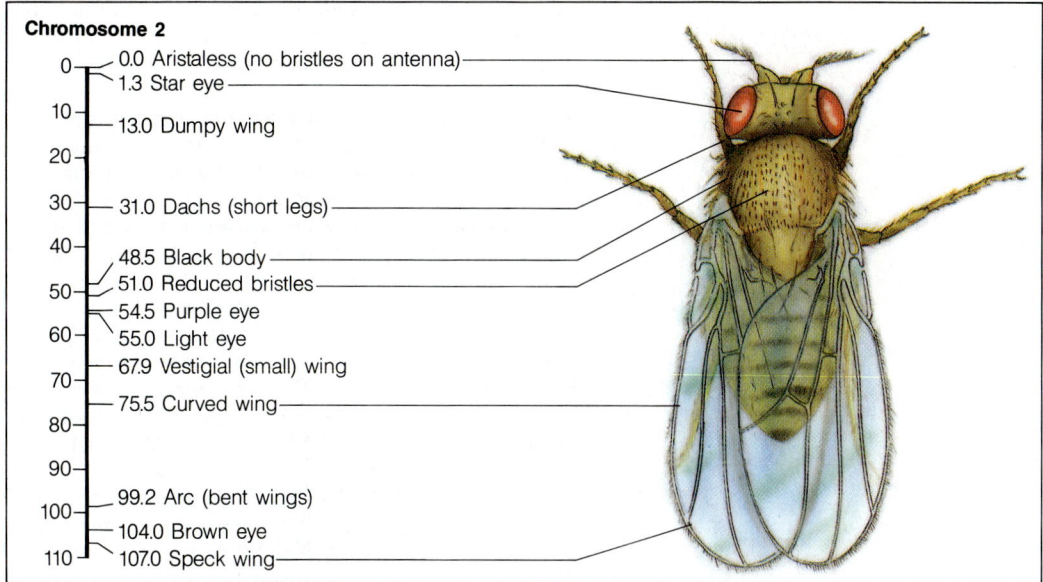

Figure 10–7 Knowing the frequency with which crossing-over occurs between genes makes it possible to map the positions of genes on chromosomes.

Morgan and Sturtevant were correct in their ideas about linkage groups, crossing-over, and the formation of gene maps. Today we have detailed maps of *Drosophila* that pinpoint the locations of more than 1500 different genes.

Sex Linkage

There is an exception to the rule that every chromosome has a corresponding homologous chromosome. This exception was discovered in 1905 by American biologist Nettie Stevens. She noticed that the cells of the female mealworm contain "20 large chromosomes, while those of the male contain 19 large ones and 1 small one." One of the pairs of chromosomes in the male mealworm is not, in the strictest sense of the term, homologous. The 2 chromosomes in the pair have very different shapes. Stevens found the same situation in *Drosophila*. The cells of the female *Drosophila* contain 8 chromosomes that can be arranged in 4 pairs of 2. The cells of the male, however, contain 3 sets of chromosomes that pair up nicely and 2 remaining chromosomes that do not match. One chromosome looks like a member of a pair of chromosomes found in the female, but the other looks completely different. It is small, shaped like a hook, and in no way similar to any of the other 7 chromosomes.

These seemingly mismatched chromosomes are the **sex chromosomes**. The other chromosomes, which are the same in both males and females, are called **autosomes**. The female *Drosophila* has two matching sex chromosomes—**X chromosomes**. The male has two dissimilar sex chromosomes—one X

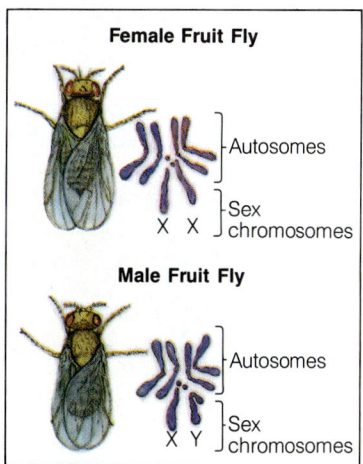

Figure 10–8 A female fruit fly has 2 matching sex chromosomes—X chromosomes. A male fruit fly has sex chromosomes that do not match—1 X chromosome and 1 Y chromosome.

HISTORICAL NOTE
SEX CHROMOSOMES

Sex chromosomes were not discovered by Morgan, although he did perform the first experiments to trace the inheritance of sex-linked genes. Sex chromosomes were first discovered in mealworms by Nettie M. Stevens in 1905. Stevens noticed that female cells contained "20 large chromosomes, while those of the male contained 19 large ones and 1 small one." The one small chromosome was, of course, the Y chromosome.

Nettie Stevens was a rather remarkable woman who attended a state teachers' college in Massachusetts, then taught for several years until she had saved enough money to attend Stanford University to study biology. She then studied for her Ph.D. at Bryn Mawr College in the laboratory of T. H. Morgan. Her promising scientific career was cut short by her death at age 41 from cancer.

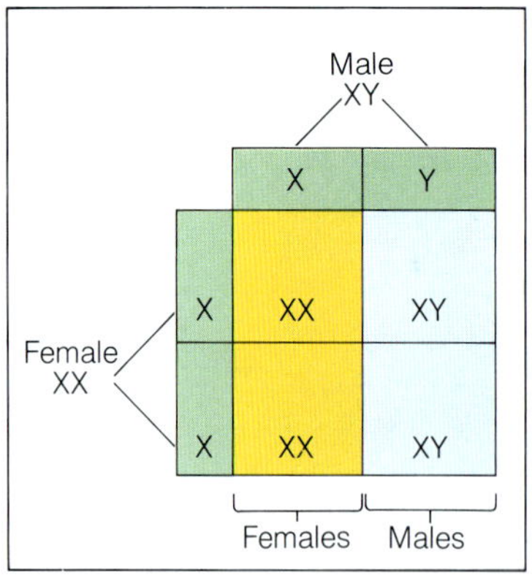

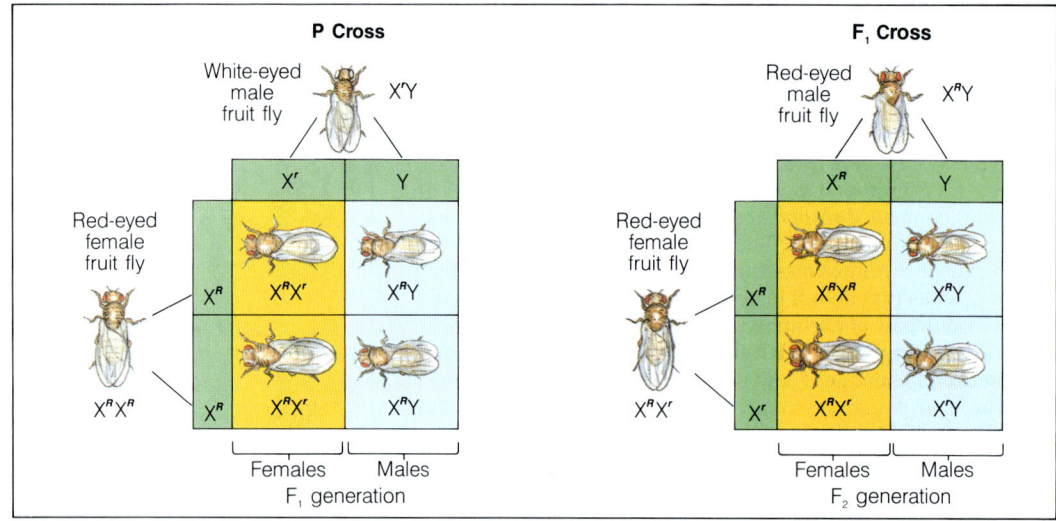

Figure 10–9 In animals such as fruit flies and humans, the sex chromosomes in the male's gametes usually determine the sex of the offspring.

Figure 10–10 Morgan discovered the first sex-linked gene in experiments with the fruit fly Drosophila. *The recessive allele, symbolized as X^r, causes white eyes in males—genotype X^rY.*

chromosome and one **Y chromosome**. It is the Y chromosome that is small and hook-shaped. If we consider just sex chromosomes, we can refer to the male as XY and the female as XX.

SEX DETERMINATION When female gametes are produced, meiosis separates one of the X chromosomes into each egg cell. In the male, meiosis separates the X and Y chromosomes so that 50 percent of the sperm cells carry a Y chromosome and 50 percent carry an X chromosome. When a Y sperm fertilizes an egg, a male fly (XY) is produced. When an X sperm fertilizes an egg, a female fly (XX) is produced. In a sense, the male is responsible for the sex of its offspring.

Many animals, including humans, follow the same system. We have 22 pairs of autosomes and 1 pair of sex chromosomes: XX in females and XY in males. Because about 50 percent of male sperm have the X chromosome and about 50 percent have the Y, the ratio of males to females is about 50:50, or 1:1.

GENES ON SEX CHROMOSOMES In addition to determining the sex of an individual, the sex chromosomes carry genes that affect other traits. A gene located on one of the sex chromosomes is said to be **sex-linked**. Morgan discovered the first sex-linked gene in his work with *Drosophila*.

Morgan crossed a group of white-eyed male flies with a group of red-eyed female flies. The allele for red eyes (R) is dominant over the allele for white eyes (r). Just as you might have expected, the hybrid flies of the F_1 generation all had red eyes. See Figure 10–10. Morgan allowed the F_1 flies to interbreed and produce a new generation of flies (the F_2 generation). Mendelian genetics predicts a 3 to 1 ratio of red-eyed flies to white-eyed flies, and that was exactly the ratio Morgan

10–1 (continued)

Content Development
Using the Visuals Reproduce on the chalkboard or on an overhead transparency the blank Punnett square for the P cross shown in Figure 10–10. Have students fill in the blanks that will result from crossing a red-eyed female, genotype X^RX^R, with a white-eyed male, genotype X^rY.
• **Why do none of the flies in the F_1 generation have white eyes?** (White eyes is the recessive trait, and in this cross, every fly that receives the *r* allele also receives an *R* allele, which will dominate.) Next, reproduce the Punnett square for the F_1 cross and have students fill in the blanks.
• **Why does one group in the F_2 generation have white eyes?** (The flies receiving the *r* allele do not receive any other allele for this trait—so the allele for white eyes is expressed.)
• **Why did Morgan suspect from this cross that the allele for white eyes must be on the X chromosome rather than the Y chromosome?** (If it were on the Y chromosome, the flies receiving it would also receive an X^R chromosome—and since *R* is dominant, none of the flies would have white eyes.

SECTION REVIEW 10–1

1. Genes are located on the chromosome; chromosomes can be thought of as strings of genes.

obtained. But Morgan noticed something strange about the F_2 flies. All of the white-eyed flies were male! There was not a single white-eyed female in the bunch.

Why should half of the males but none of the females have white eyes? Could it be that the gene for eye color was on one of the sex chromosomes?

Morgan realized that he could account for his results if the gene for eye color was on the X chromosome. Refer to the Punnett square in Figure 10-10 as you take a closer look at how the F_2 generation is formed.

The female flies produce two kinds of egg cells. One kind of egg cell has the R allele on its X chromosome (we can call those egg cells X^R); the other kind of egg cell has the r allele on its X chromosome (X^r). The male flies make two kinds of sperm cells: One kind of sperm cell has the X chromosome and the other kind of sperm cell has the Y chromosome. The X chromosome has the R allele (X^R) on it. The Y chromosome, however, has neither the r nor the R allele on it.

As you can see from the Punnett square, only one combination of chromosomes will produce a white-eyed fly—$X^r Y$. And every fly with that combination of chromosomes is male.

Morgan tested his hypothesis that the gene for eye color was sex-linked in a famous experiment. He crossed the F_2 white-eyed males ($X^r Y$) with F_1 ($X^R X^r$) females. The Punnett square in Figure 10-11 shows the probable results of the cross if, and only if, the gene for eye color is indeed on the X chromosome. According to the Punnett square, 1/2 of the offspring should have red eyes and 1/2 should have white eyes. In addition, the male offspring should be 1/2 red-eyed and 1/2 white-eyed. And the female offspring should also be 1/2 red-eyed and 1/2 white-eyed. Such results were exactly what Morgan obtained. He had confirmed his idea that the gene for eye color was sex-linked.

Sex-linked genes are not only found in *Drosophila*. They are important in humans as well. As you will see in Chapter 11, several important human genes are located on the X chromosome. Such genes include the genes for color vision and blood clotting. As with the eye color gene in *Drosophila*, the effects of recessive alleles tend to show up more frequently in males than in females.

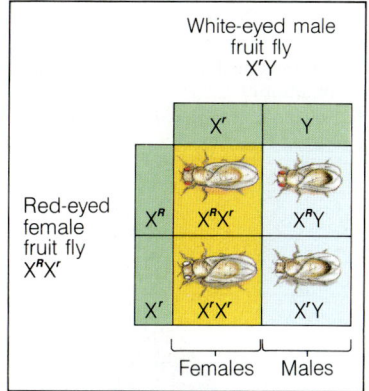

Figure 10-11 *The sex-linked X^r allele causes white eyes in females that have a "double dose" of the allele—genotype $X^r X^r$. White-eyed females inherit one X^r allele from their male parent (who must be $X^r Y$) and one from their female parent (who must be either $X^R X^r$ or $X^r X^r$).*

10-1 SECTION REVIEW

1. How are genes related to chromosomes?
2. How does crossing-over make genetic mapping possible?
3. What are sex chromosomes? Autosomes?
4. **Critical Thinking—Relating Concepts** Why are the effects of recessive sex-linked alleles seen more often in males than in females?

HISTORICAL NOTE
CROSSING-OVER

Mendel believed that he was observing an independent assortment of genes for seven different traits of the pea plant. Scientists now realize, however, that the genes for two of these traits are very far apart on the same chromosome—that is, they are linked! These genes are affected by crossing-over so frequently that they appear to assort independently.

TEACHING SUPPORT

Laboratory Manual
- Human Sex Chromosomes, p. 139

Teaching Resources
- Hands-On Activity: Stalking the Wild Fruit Fly, p. 65

Study Guide
- Section 10-1, p. 95

TEACHING SUPPORT

Laboratory Manual
- Karyotypes, p. 143

HISTORICAL NOTE
BEADLE AND TATUM

In the 1940s, two American scientists named George W. Beadle and Edward L. Tatum learned much about mutations and the actions of genes by studying *Neurospora*, a common pink mold that grows on bread and other substances.

Beadle and Tatum exposed some spores of this mold to X-rays, and found that some of the spores did not grow. The scientists concluded that mutations caused by the X-rays were responsible for the spores not growing.

At this time, scientists did not know what the substances were that had been affected by the mutations. Beadle and Tatum conducted more experiments and discovered that the substances that could not be synthesized by mutant spores were enzymes. This led to the famous "one gene, one enzyme" hypothesis, which states that one gene and one gene only controls the synthesis in each enzyme in an organism. An example of a mutation that suppresses the production of an enzyme is albinism. In an albino organism, the enzyme needed to make a pigment is not produced. In animals, this results in a lack of the pigment melanin, which gives color to skin, hair, and eyes. As a result, albino animals have white hair and pinkish skin and eyes.

Guide For Reading
- What is the difference between a chromosomal mutation and a gene mutation?
- How do mutations affect heredity?

Figure 10–12 Mutations in genes that regulate development resulted in extra hind legs on this frog.

10–2 Mutations

As you will recall from Chapter 7, the processes involved in duplicating genetic information and transmitting it to the next generation are complex and precise. Although mistakes are rare, they do occur. These mistakes are called **mutations**, from a Latin word that means change. Not all mutations are harmful. Many mutations either have no effect or cause slight, harmless changes. And once in a while a mutation may be beneficial to an organism.

Mutations may occur in any cell. Mutations that affect the reproductive cells, or germ cells, are called germ mutations. The variant characters studied in genetics (such as white eyes in *Drosophila* or wrinkled seeds in pea plants) are the result of germ mutations. Mutations that affect the other cells of the body are called somatic (from the Greek word *soma*, which means body) mutations. Because they do not affect the reproductive cells, somatic mutations are not inheritable. Many cancers are caused by somatic mutations. Both somatic and germ mutations can occur at two levels—the level of chromosomes and the level of genes. **Chromosomal mutations involve segments of chromosomes, whole chromosomes, and even entire sets of chromosomes. Gene mutations involve individual genes.**

Chromosomal Mutations

Whenever a **chromosomal mutation** occurs, there is a change in the number or structure of chromosomes. There are four types of chromosomal mutations that involve a change in the structure of a chromosome: deletions, duplications, inversions, and translocations. A deletion involves the loss of part of a chromosome.

The opposite of a deletion is a duplication, in which a segment of a chromosome is repeated.

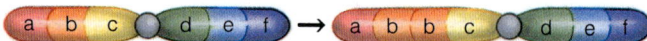

When part of a chromosome becomes oriented in the reverse of its usual direction, the result is an inversion.

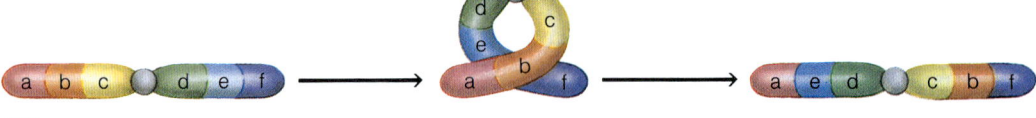

212

TEACHING STRATEGY 10–2

Focus/Motivation
Using the Visuals Direct students' attention to Figure 10–12.
- Do you notice anything unusual about this frog? If so, what? (Yes. Extra hind legs.)
- How do you think the frog got this way? (Accept all reasonable answers.)
- Do you think this characteristic is good or bad for the frog? (Accept all answers.)

Content Development
Explain to students that the frog's extra hind legs are the result of a mutation. Point out that a mutation is a mistake that occurs in the duplicating and/or transmitting of genetic information. Mutations cause genetic changes by changing the hereditary messages of DNA molecules. Further point out that mutations can be harmful, helpful, or neutral with respect to an organism's overall well-being. It is often the case, however, that a mutation is harmful, since mutant organ-

A translocation occurs when part of one chromosome breaks off and attaches to another, nonhomologous chromosome. In most cases, nonhomologous chromosomes exchange segments, so that two translocations occur at the same time.

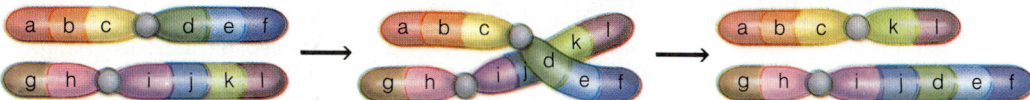

Chromosomal mutations that involve whole chromosomes or complete sets of chromosomes result from a process known as **nondisjunction**. Nondisjunction is the failure of homologous chromosomes to separate normally during meiosis. Nondisjunction literally means not coming apart.

When one chromosome is involved, nondisjunction results in an extra copy of a chromosome in one cell and a loss of that chromosome from the other. A number of human disorders that result from nondisjunction will be described in Chapter 11.

Nondisjunction can involve more than one chromosome. When all the homologous chromosomes fail to separate, the result may be a dramatic increase in chromosome number, producing triploid (3N) or tetraploid (4N) organisms. The condition in which an organism has extra sets of chromosomes is called **polyploidy**. Polyploidy is almost always fatal in animals. However, polyploid plants are often larger and hardier than normal plants.

Mutations in Genes

Mutations can occur in individual genes and can seriously affect gene function. Any chemical change that affects the DNA molecule has the potential to produce **gene mutations**. Some gene mutations result from a change involving many nucleotides within a gene; some involve only one nucleotide. The smallest changes, known as **point mutations**, affect no more than a single nucleotide.

Remembering how a DNA base sequence determines the amino acid sequence of a polypeptide, can you predict the possible effects of a point mutation in which one base is substituted for another? Because base substitutions usually affect only a single codon, only one amino acid is affected. In some cases, the sequence of amino acids that the gene codes for may not be changed at all, since different codons may code for the same amino acid. When the point mutation involves the insertion or deletion of a nucleotide, however, the situation may be much more serious. Remember that mRNA is read in groups of three bases. If a single base is inserted or deleted, the groupings are shifted for every codon following the point mutation. Such **frameshift mutations** can completely change the polypeptide product produced by a gene.

Figure 10–13 A frameshift mutation, such as that caused by the addition of a single nucleotide, can greatly change the polypeptide product of a gene (top). This can make it as useless and nonsensical as a sentence in which the deletion of a letter and a frameshift has occurred (bottom).

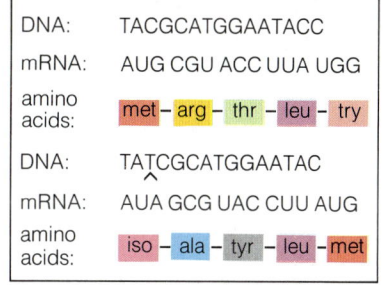

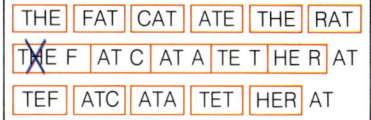

SCIENCE, TECHNOLOGY, AND SOCIETY CONNECTION

Manipulating Chromosome Numbers

When an organism has more than two sets of chromosomes, it is said to be polyploid. Polyploidy occurs during the formation of gametes. Once a cell becomes polyploid, the cells that it produces during mitosis will also be polyploid.

Polyploidy is usually fatal in animals, but it can be beneficial in plants, increasing the genetic variety. Many of the characteristics of polyploid plants are desirable to plant breeders. For example, cultivated bananas are triploid, meaning that they have three sets of chromosomes instead of two. These triploid bananas grow faster and are hardier than wild, diploid bananas.

Polyploid plants that have odd numbers of chromosome sets such as 3, 5, and 7 are usually sterile, meaning that they cannot reproduce. This characteristic is often considered desirable because it produces seedless fruits.

Although polyploidy often occurs naturally in plants, it can be induced by the use of chemicals such as colchicine. Colchicine is obtained from the seeds and corms of the autumn crocus, a member of the lily family.

SCIENCE, TECHNOLOGY, AND SOCIETY CONNECTION

Manipulating Chromosome Numbers

From Mendel's work we have come to expect that organisms will be diploid (2N). In other words, they will have two sets of chromosomes—one set inherited from each parent. However, plants with the "wrong" number of sets of chromosomes do exist. And, interestingly enough, these plants are extremely useful in agricultural research.

For example, a plant with only one set of chromosomes shows the effects of all its genes, including potentially desirable recessive genes. A plant with only one set of chromosomes is obtained by taking cells formed in meiosis and growing them under special conditions. However, plants with only one set of chromosomes cannot reproduce sexually because meiosis cannot occur normally in their cells. If these plants show a desirable characteristic, they can be made into diploid plants by treating them with the drug colchicine. Colchicine prevents spindle formation during mitosis and thus causes the number of chromosomes in a cell to double. The diploid plants produced by treatment with colchicine are purebred for the desired traits and can reproduce normally.

By manipulating chromosome numbers with colchicine, it was possible to produce high-protein Triticale.

Colchicine can also be used to change the number of sets of chromosomes in a naturally occurring plant. For example, colchicine has been used to create strawberry and blueberry plants that are polyploid, or have more than two sets of chromosomes. Polyploid plants produce larger fruit than plants with the normal number of sets of chromosomes. The ability to change the number of sets of chromosomes makes it possible to cross plants that have different diploid numbers. The high-protein grain *Triticale* was created by crossing wheat (2N=42) and rye (2N=14). The plants grown from the hybrid seeds were sterile; they could not produce offspring. However, when the hybrid plants were treated with colchicine, the chromosome number was doubled and meiosis could occur normally. Thus the second-generation plants were fertile.

Manipulating chromosome numbers is useful in studying plant genetics. It is also a technique that allows researchers to create new and potentially better crops.

10–2 SECTION REVIEW

1. Compare a chromosomal mutation and a gene mutation.
2. What is a somatic mutation? How does it differ from a germ mutation?
3. **Critical Thinking—Relating Facts** How does nondisjunction result in chromosomal mutations?

214

10–2 (continued)

SECTION REVIEW 10–2

1. A chromosomal mutation occurs on a much larger scale than a gene mutation—it may affect a segment of a chromosome, a whole chromosome, or entire sets of chromosomes. A gene mutation involves just one gene.
2. A somatic mutation is one that affects body cells and is not inheritable, whereas a germ mutation affects the reproductive cells and is inheritable.
3. Nondisjunction results in cells that have the wrong number of chromosomes.

Reinforcement/Reteaching

If students are having difficulty answering the Section Review questions, go back to the part of the section that includes the material with which they are having trouble. Review the material, then have them try again to answer the questions.

10–3 Regulation of Gene Expression

Like the individual cells of a large organism, individual genes do not function in isolation. The expression and activities of genes, like those of cells, are regulated and controlled, thereby enabling a complex genetic system to function smoothly. **As biologists have intensified their studies of gene activity, it has become clear that interactions between different genes and between genes and their environment are critically important.**

This section focuses on some of the ways in which genes are regulated by conditions within cells. However, genes are also influenced by environmental conditions such as temperature and light.

Gene Interactions

Dominance is the simplest example of how genes interact with each other. In Chapter 9, you learned that the effects of the dominant allele are seen even when the recessive allele is present. But what causes dominance?

Remember that a gene is a section of DNA, and DNA codes for a polypeptide, or string of amino acids. In many cases, the dominant allele codes for a polypeptide that works, whereas the recessive allele codes for a polypeptide that does not work. For example, suppose that the allele B codes for an enzyme that makes a black pigment in a mouse's fur and allele b codes for a defective enzyme that cannot make the pigment. A mouse that has the genotype bb will have white fur because it lacks the enzyme that makes the black pigment. But a mouse that has the genotype BB or Bb will have black fur because it possesses the enzyme that makes the black pigment. Although each cell in the Bb animal has just one copy of the functioning allele, that single copy can code for thousands of mRNA molecules. And each mRNA molecule can code for thousands of enzymes. This is the reason the B allele is dominant over the b allele.

INCOMPLETE DOMINANCE In 1760 the German scientist Josef Kölreuter reported on experiments in which he crossed white carnations (rr) with red carnations (RR). Kölreuter found that all of the offspring from his crosses had pink flowers (Rr). In other words, the hybrids had a phenotype that was intermediate between those of the parents. At first glance, it might appear as if the parents' genes had blended together. But when Kölreuter crossed his pink F_1 hybrids with each other to form an F_2 generation, the parents' phenotypes reappeared. In the F_2 generation, 1/4 of the plants had red flowers, 1/2 had pink flowers, and 1/4 had white flowers. (This 1:2:1 ratio should be familiar to you from Mendel's crosses.)

Guide For Reading
- How do gene interactions affect gene expression?
- How is an operon turned on and off?
- What is an intron? An exon?

Figure 10–14 For some organisms, such as these caterpillars, gene expression is regulated by chemicals in the food they eat. Each caterpillar of this species has the genetic ability to develop either the twiglike or the flowerlike form. However, caterpillars that eat gray-green oak twigs look like twigs (top). And caterpillars that eat fuzzy gold oak flowers look like flowers (bottom).

HISTORICAL NOTE
JACOB AND MONOD

French scientists François Jacob and Jacques Monod studied the bacterium *E. coli*, which lives in the alimentary canal of mammals. Jacob and Monod demonstrated that the bacterial genes coding for lactose-metabolizing enzymes are expressed only when lactose is present. When *E. coli* cells were grown on a lactose-free medium, the bacteria did not produce the enzymes. They did so within minutes, however, after being placed on a lactose-enriched medium. Thanks to the work of these scientists, we know today that a region of *E. coli* DNA includes a promoter, an operator, and three adjacent genes associated with lactose metabolism. These are known as the lactose operon.

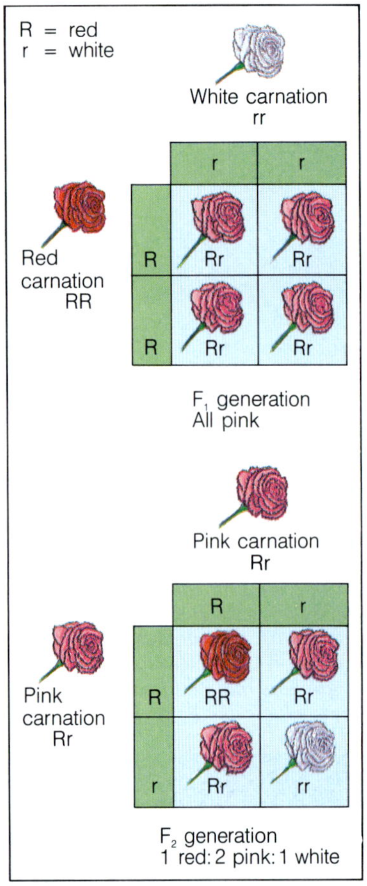

R = red
r = white

White carnation
rr

Red carnation
RR

F₁ generation
All pink

Pink carnation
Rr

F₂ generation
1 red: 2 pink: 1 white

Figure 10–15 *Flower color in carnations is an example of incomplete dominance. The heterozygous phenotype, pink, is somewhere in between the homozygous phenotypes, red and white.*

In carnations, the *R* allele, which codes for an enzyme that makes red pigment, is incompletely dominant over the *r* allele, which codes for a defective enzyme that cannot make pigment. In **incomplete dominance** the active allele does not compensate for the inactive allele, and the heterozygous phenotype is somewhere in between the homozygous phenotypes.

CODOMINANCE Many genes display **codominance**, a condition in which both alleles of a gene are expressed. In other words, both alleles are active. (Remember that in incomplete dominance only one of the alleles is active.) Codominant alleles are written as capital letters with subscripts (for example, B_1 and B_2) or superscripts (for example, R and R').

Codominance is seen in many organisms. For example, red hair (H^R) is codominant with white hair (H^W) in cattle. Cattle that have the genotype $H^R H^W$ are roan, or pinkish brown, in color because their coats are a mixture of red and white hairs. Much the same thing happens in certain varieties of chickens. Black feathers (F^B) are codominant with white feathers (F^W). Erminette chickens ($F^B F^W$) are speckled black and white.

POLYGENIC INHERITANCE It would be a great mistake to assume that all traits are produced by single genes. Many traits are produced by the interaction of many genes. Traits controlled by two or more genes are said to be **polygenic** (*poly-* means many). For example, at least three enzymes—each of which is produced by a different gene—are involved in making the reddish-brown pigment in the eyes of fruit flies. Different combinations of enzymes (which may be present in the normal form or altered or absent due to a mutation) produce different eye colors. More complicated traits, such as the shape of your nose or the color and markings on an animal's coat, are the result of interactions between large numbers of genes.

Gene Expression in Prokaryotes

In Chapter 7 you learned that individual genes on a chromosome serve as a template, or pattern, for the production of mRNA. In turn, mRNA serves as the instructions for the production of a protein (polypeptide). The genes of a single organism cannot all be activated at the same time. A cell that activated all of its genes at once would make a great many molecules that it did not need and would waste energy and raw materials in doing so. However, when the cell does need the product of a gene, it must be able to produce that product quickly and in adequate amounts.

When the product of a gene (a specific protein) is being actively produced by a cell, we say that the gene is being expressed. Within a single organism, some genes are rarely expressed, some are constantly expressed, and some are expressed for a time and then turned off. But how does a cell

---

10–2 (continued)

Content Development
Using the Visuals Continue the Motivation discussion by directing students' attention to the pictures of the caterpillars in Figure 10–14.
• **What determines the appearance of each type of caterpillar?** (The kind of food it eats.)
• **Why is the caterpillar's appearance altered by the kind of food it eats?** (The chemicals in the food regulate the expression of genes in the caterpillar.)
• **Does the food you eat have the same effect?** (For the most part, no.)
• **Can other factors influence the expression of genes in your cells?** (Accept all answers. The correct answer is yes.)

Point out to students that many factors, both internal and external, can affect gene expression in an organism. In humans, external factors such as viruses and radiation and internal factors such as hormones can affect gene expression. Factors that affect gene expression in organisms in general include temperature, light, conditions in cells, interaction with other genes, and chemicals from food.

Focus/Motivation
Draw a Punnett square on the chalkboard. Label along the top and side of the square the symbols for an *RR* × *rr* cross. Tell students that the letters *R* and *r* represent red and white flower color. Have a student volunteer fill in the squares with the correct genotypes (*Rr, Rr, Rr, Rr*).
• **What phenotypes would you expect to see in the flowers?** (Most will probably say all red.)
• **Why?** (Red is the dominant allele.)
• **Suppose I tell you that not one**

"know" when to make a protein and when not to make it? In other words, how does a cell "know" which genes to turn on and which to turn off?

THE OPERON The mechanisms by which genes are turned on and off are quite complex. In this section we will examine one way genes are regulated in prokaryotes (organisms without a nucleus). This particular mechanism was discovered by the French scientists François Jacob and Jacques Monod in 1961. They noticed that genes that work together are often clustered together on a small area of a prokaryote's chromosome. Jacob and Monod also determined that there are regions on a chromosome that lie near these gene clusters but that do not code for the production of proteins. These regions are, however, involved in the regulation and expression of nearby gene clusters. These regions and the gene cluster they regulate are called an **operon** because they operate together.

An operon consists of the following parts: a cluster of genes that work together; a region of the chromosome near the cluster of genes called the **operator**; and a region of the chromosome next to the operator called the **promoter**. As you can see in Figure 10–17, the operator and promoter regions overlap slightly.

Figure 10–16 In certain varieties of chickens, black and white feather colors are caused by codominant alleles. Thus the heterozygous phenotype, speckled black and white, is a result of the expression of both alleles.

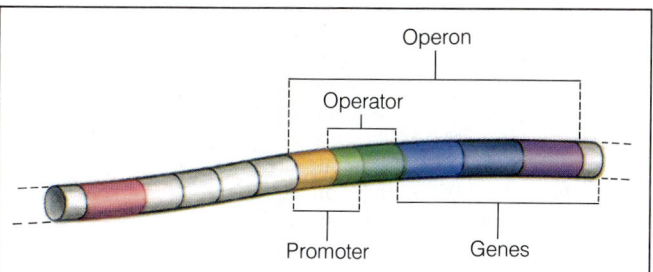

Figure 10–17 Some genes in prokaryotes are regulated by an operon. An operon consists of several sections of DNA that are operated together. The operon studied by Jacob and Monod, which breaks down the sugar lactose, consists of a promoter, an operator, and a cluster of genes that code for lactose-metabolizing enzymes.

The gene cluster in the operon that Jacob and Monod studied produces enzymes that break down lactose, a sugar that bacteria can use as a food source. Bacteria do not produce these lactose-breaking enzymes in large amounts unless lactose is present. In other words, the very presence of lactose induces the production of the enzymes necessary to break down lactose for use as food. In this operon system, lactose is called the **inducer** because it induces the production of enzymes.

In order to make the enzymes, RNA polymerase must move along the genes on the chromosome, producing mRNA in the process. Before the RNA polymerase can get to the desired genes, it must first attach to the promoter region near the genes. We might think of the promoter as a "Start Here!" instruction to RNA polymerase. Once the RNA polymerase attaches to the promoter, it can move along the chromosome, past

---

BACKGROUND INFORMATION
GENETIC CODE AND EVOLUTION

The basic genetic code is identical for all living things, from the simplest bacterium to the most complex organism, a human. Thus, from an evolutionary point of view, it seems likely that the genetic code dates back to the most ancient cells from which all existing species arose. This absence of variation among species suggests that the genetic code was formed before cells diverged into separate evolutionary lines.

In light of this evidence, it is interesting to note that a slightly altered genetic code exists in the mitochondria of some species. A mitochondrion, you will recall, is an organelle that specializes in ATP formation. It has its own DNA, and in some ways operates independently of the cells in which it resides. The mitochondrial codes are almost the same as the one used by cells, but a few codons have different meanings.

Many biologists feel that these slightly different codes indicate that the original ancestors of mitochondria were free-living cells, similar to modern bacteria. The scientists hypothesize that, in some way, these cells took up residence in the cytoplasm of other kinds of cells, possibly establishing a symbiotic relationship of guest and host. Eventually, the guest became dependent on the host and lost its independent function—thus the role of the mitochondrion as an organelle arose. If this is the case, then the mitochondrial code might be a remnant of a primitive code from a species that has long since disappeared.

---

flower resulting from this cross is red, but that all the flowers are pink. How would you explain that? (Accept all reasonable answers. The answer is that in this particular cross, the red allele is not completely dominant over the white allele.)

Content Development
Discuss with students Kolreuter's experiments with red and white carnations.
• How did Kolreuter know that the genes of the parent flowers had not blended together, but that the alleles for red color and white color still existed separately? (When he crossed the pink flowers, the red and the white flowers reappeared.)
• What genotypes caused the red and white phenotypes to reappear? (The *RR* and *rr* genotypes.)

• What would have been the result of the pink flower cross if the genes of the parents had blended together? (Probably all of the next generation of flowers would have been pink.)

the operator region, to the genes. When the RNA polymerase reaches the genes, it can produce mRNA, which "instructs" the ribosomes to make the enzymes. When this process is taking place, we say the genes are activated, or being expressed. The enzymes coded for by the genes can then perform their task in the cell—in this case, breaking down the inducer lactose.

Remember that Jacob and Monod observed that the enzymes coded for by the genes in the operon were not produced in the absence of the inducer (lactose). Does this mean that the cell has a way to turn the operon off? Indeed it does.

THE REPRESSOR The cell produces a special protein called a **repressor**. When the repressor nears the operator region of an operon, it attaches itself to the operator so that it sits between the promoter and the genes. The repressor's position blocks the access of RNA polymerase to the genes. Like the guard outside a locked factory, the repressor prevents the workers (RNA polymerase) from getting to their jobs (making mRNA). In other words, the repressor turns the genes of the operon off.

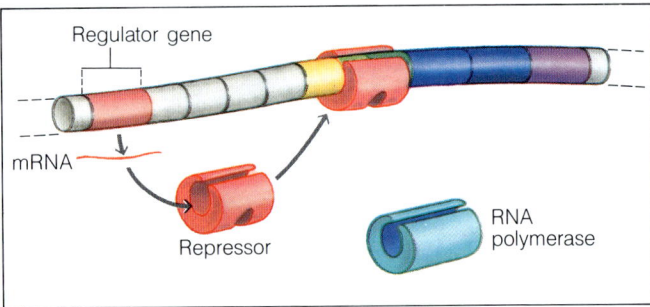

Figure 10-18 The genes in this operon are turned off when a protein called the repressor binds to the operator and thus prevents the transcription of mRNA.

You might wonder how the repressor "knows" which operon to attach to and thereby turn off. The actual mechanism is quite elegant and specific. Each repressor is shaped to fit a specific region of DNA on the chromosome. It can attach only to the specific operator on the operon it regulates. Thus each repressor turns off a specific operon.

GENE ACTIVATION You now know that a gene cluster is part of an operon that includes an operator and a promoter. And you also know that the genes are turned off when the repressor binds to the operator. But there is more to the story. How is the operon turned back on when it is needed?

When the inducer enters the cell, it binds to the repressor. And something quite remarkable happens: The repressor changes shape and can no longer bind to the operator. The repressor actually falls off the operator.

tional unit of gene expression control in prokaryotes. This functional unit is called an operon. An operon consists of a cluster of genes that work together, a promoter, and an operator.

Emphasize that controls over gene expression in prokaryotes are concerned mainly with altering the overall rate of transcription—that is, the number of mRNA molecules transcribed per unit of time from a given gene. The higher the rate of transcription, the greater the expression of the gene. Point out that one of the most important factors in determining the rate of transcription is the promoter. Some promoters bind RNA polymerase more strongly than do others. Genes with strong promoters are transcribed at a faster rate than are genes with weak promoters.

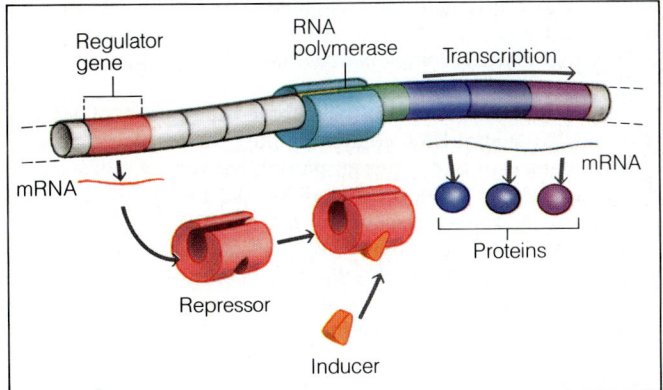

Figure 10–19 *The genes in this operon are turned on when lactose (the inducer) binds to the repressor and causes it to fall off the operator. This allows RNA polymerase to bind to the promoter, move across the operator, and transcribe mRNA from the cluster of genes. The operon system ensures that genes are activated only when their products are needed. In this case, the lactose-metabolizing enzymes are produced only when lactose is present.*

When the repressor falls off the operator, the RNA polymerase can bind to the promoter, move across to the genes, and produce messenger RNA. The mRNA codes for the enzymes that are used to break down the inducer. When the cell runs out of the inducer, the repressor can bind to the operator again, and the operon is turned off. The complete system is automatic and self-regulating. The presence of the inducer causes the cell to make the enzymes needed to use it. And when the inducer disappears, the enzymes are no longer made.

The repressor-operon system provides us with one model of how a group of genes can be regulated so that they are expressed only when they are needed. A number of other gene-regulation systems in prokaryotes have been identified. Some of these gene systems use repressors in a manner similar to that just described. Others use proteins that enhance the rate of transcription. In other systems, regulation occurs at the level of protein synthesis. Regardless of the actual system involved, the result is the same: Cells are able to turn their genes on and off, using the genes only when needed.

Gene Expression in Eukaryotes

The cells of eukaryotes contain a nucleus and membrane-enclosed organelles. Gene regulation in eukaryotes is more complex than in prokaryotes. Increased understanding of control systems in eukaryotes indicates that there are several systems of gene regulation. One system involves a substance called an inducer. Like the inducer in the operon system, inducers in eukaryotes induce, or cause, the activation of genes. In eukaryotes, inducers bind directly to DNA and either start or increase transcription of particular genes.

In 1976 Philip Sharp and Susan Berget discovered that mRNA produced during transcription may be altered before it is used to make protein (polypeptide) during translation. You may recall that during transcription, the base sequence in a

TEACHING SUPPORT

Teaching Resources
- Vocabulary Review: Word Scramble, p. 1

Study Guide
- Section 10–3, p. 100

BACKGROUND INFORMATION
INTRONS

What purpose do introns serve? Scientists do not really know the answer to that question, although many possible answers have been suggested. Some scientists believe that introns may serve as recognition signals for gene expression, but there is little evidence to support this hypothesis. Other scientists have suggested that introns may allow new kinds of proteins to be put together by causing minor variations in DNA sequences at the splicing sites. If this is the case, then introns contribute to genetic variety. This idea is supported by evidence that there is much more variation in introns than in exons of the same gene. It is also supported by observations of a kind of "multiple-choice" gene in some organisms. For example, a particular virus has a gene that contains several stop codons. Depending on how the mRNA from the gene is spliced together (stop codons being removed in the process), the virus can make a number of different proteins. A few cases similar to this have been observed in animal cells.

Yet another idea about introns is that they serve no purpose at all—that they are just DNA "junk" that the cell cannot eliminate. Connected to this hypothesis is the suggestion that introns may be remnants of a less efficient genetic system.

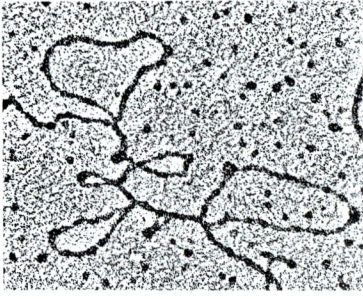

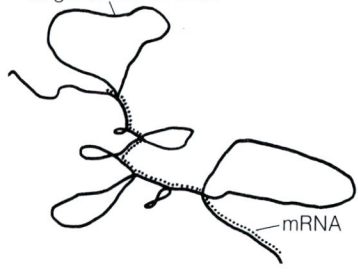

Figure 10–20 *The sequence of bases on the mRNA for this egg-white protein does not exactly match that of its corresponding gene. The DNA sequences that are not complementary to the mRNA appear as loops called introns.*

Figure 10–21 *RNA polymerase transcribes an entire gene, introns and exons alike. Before the transcribed RNA, or pre-mRNA, can leave the nucleus and code for protein, the introns must be removed and a chemical "cap" and "tail" must be attached.*

gene is transcribed into the complementary base sequence on mRNA. Most scientists assumed that a sequence of bases on mRNA was exactly complementary to a DNA sequence for a gene.

Sharp and Berget carefully compared the mRNA of a protein with a piece of DNA containing the gene for that protein. They expected to find an exact pairing between the bases on the gene and the bases on the mRNA that helps produce the protein. Instead, they discovered that the DNA fragment contained large sequences that were not complementary to the base sequence on the mRNA. What was going on?

Sharp and Berget quickly realized that the presence of DNA sequences that are not complementary to mRNA sequences implies that the gene is in "pieces." In other words, DNA sequences that code for protein (the gene pieces) are separated by DNA sequences that do not code for protein. Today we know that many eukaryotic genes contain sequences that are complementary to mRNA and sequences that are not. The sequences that are complementary code for protein. These "expressed" sequences are called **exons**. The segments that are not complementary do not code for protein. These "intervening" sequences are called **introns**.

When RNA polymerase moves along a gene, it transcribes the entire gene. This means that the RNA produced by transcription, or pre-mRNA, contains introns. Before the cell can produce protein, the pre-mRNA must be processed into functional mRNA. During this processing, the introns on the pre-mRNA are removed and the exons are spliced back together. In addition, a chemical "cap" and "tail" are attached to the RNA. At this point, the pre-mRNA can be called mRNA.

The conversion of pre-mRNA into mRNA has further significance because the processing takes place entirely within the nucleus. This means that mRNA must complete its processing before it leaves the nucleus. There may be some kind of a molecular watchman at the gate of the nucleus that allows only processed molecules to leave. Such a system would regulate gene expression by preventing the protein products of a gene from being produced.

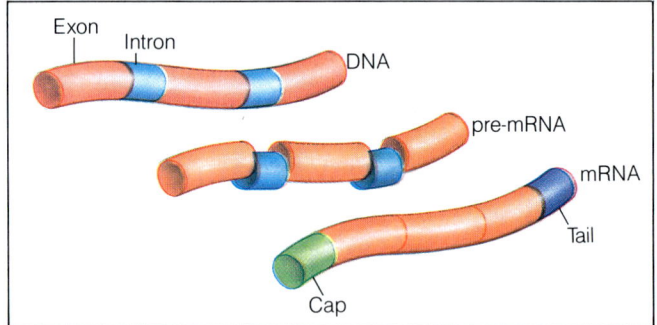

10–3 (continued)

SECTION REVIEW 10-3

1. Dominance, incomplete dominance, codominance, and polygenic inheritance are ways gene interaction occurs.

2. Both result in a phenotype that is a "cross" between the parental phenotypes. In incomplete dominance, an active allele does not entirely compensate for an inactive allele. In codominance, both alleles are expressed—they are both active.

3. A trait that is determined by the interaction of more than one gene.

4. Because introns are removed from mRNA before it leaves the nucleus, they do not affect protein synthesis. Thus mutations in introns are unlikely to have any effect.

SCIENCE, TECHNOLOGY, AND SOCIETY / Breakthrough

Jumping Genes

In the 1940s a young scientist named Barbara McClintock began her study of corn. She was particularly interested in why certain corn kernels developed a spotty, or speckled, appearance. She knew that the color of each kernel was determined genetically, and that every one of the cells in the kernel's coat had the same genetic makeup (genotype). Yet the speckled colors in some kernels indicated that genes were being unpredictably switched on and off in one cell or another. She could offer only one reasonable explanation for this observation: There are special controlling elements that can be transposed, or moved from one place to another, on a chromosome. These transposable elements, which are sometimes called jumping genes, can insert themselves into different parts of a chromosome and turn the genes near them on and off.

McClintock's theory of transposable elements was far ahead of its time and found little support in the scientific community. But McClintock patiently continued her studies. In time, she was able to provide experimental proof that she was correct about jumping genes. Finally, in 1975, techniques of molecular biology had advanced to the point where scientists could actually look for jumping genes. Jumping genes were found in organisms as diverse as bacteria, insects, mammals (including humans), and corn.

Jumping genes may play a role in gene regulation in some organisms, such as corn. Jumping genes are also involved in the formation of certain proteins that show a great number of variations, such as antibodies—proteins that defend the body against disease-causing organisms. New jumping genes are constantly being discovered, and it will be a long time before scientists are able to explain all their different functions.

Although she died in 1992, Barbara McClintock lived to see the importance of her discovery appreciated. She received many awards during her lifetime, including the Nobel prize, the highest award that can be presented to a scientist.

10–3 SECTION REVIEW

1. How do gene interactions affect gene expression?
2. Compare incomplete dominance and codominance.
3. What is a polygenic trait?
4. **Critical Thinking—Evaluating Statements** "Mutations in introns are less likely to affect phenotype than mutations in exons." Defend or refute this statement.

SCIENCE, TECHNOLOGY, AND SOCIETY / BREAKTHROUGH

Jumping Genes

The work of Barbara McClintock challenged the idea that only mutations such as deletions and inversions can change the number or order of genes on chromosomes. McClintock proposed the existence of what have come to be called transposable elements in the genes of corn plants. Transposable elements are parts of DNA that frequently "jump," or transpose themselves, to new locations in the same DNA molecule or a different molecule. Often these jumping genes inactivate the genes where they are inserted. In corn plants, the inactivated genes are those that affect pigment synthesis—so the color of the corn kernels is altered. The transposition of genes occurs early in plant growth, resulting in mutations that affect pigment synthesis in all daughter cells produced by subsequent cell divisions.

Support for Barbara McClintock's work did not come until nearly 20 years after she had reached her conclusions. The event that turned the tide was the discovery of transposable elements in the *E. coli* bacterium. At this time, these elements were given the name transposons. In the *E. coli*, each transposon consists of one or more genes that are flanked on both sides by a short series of nucleotides called an insertion sequence. A transposon codes for an enzyme, transposase, that catalyzes its own insertion into random locations in the bacterial chromosome. Such insertions cause a variety of changes in phenotype.

Reinforcement/Reteaching
If students are having trouble answering the Section Review questions, go back to the part of the section that contains the material with which they are having difficulty.

Closure
Divide the class into small groups. Ask each group to write 8 to 10 questions that deal with the content of this section. At least 3 of the questions should be concept or critical thinking questions, rather than factual recall. When all the groups are finished, have the groups exchange questions to answer, then return the questions to the original group for checking. Choose some of the more interesting questions for class discussion.

LABORATORY INVESTIGATION

MAPPING CHROMOSOMES

Before the Lab
1. At least one day prior to the investigation, gather enough materials for your class, assuming 6 students per group.
2. Prepare the wooden stick for each group ahead of time as follows: Using a felt-tipped marker, color one long edge of the stick.

Pre-Lab Discussion
Review with students the process of crossing-over. Have them recall that crossing-over changes the combination of alleles on chromosomes. Also have them recall that crossing-over causes genes that are normally linked to be inherited independently.

• **What factor is cited in the text as influencing how frequently crossing-over occurs?** (How far apart genes are on the chromosome.)

• **Would you expect crossing-over to occur frequently between genes that are close together on a chromosome?** (No.)

• **How about if the genes are far apart?** (Yes. Crossing-over would happen frequently, perhaps even to the extent that the genes appear to assort independently.)

Have a student read aloud the Problem that is stated at the beginning of the investigation. Ask students to write an answer to the question in their lab notebooks. At the end of the investigation, have them evaluate their ideas in light of what they have learned during the lab.

Have students read through the entire procedure. Discuss with them the idea that the toss of a stick and its landing on the line represents a random event.

• **Why is a random event used to simulate crossing-over?** (Because crossing-over is believed to occur randomly.)

LABORATORY INVESTIGATION: MAPPING CHROMOSOMES

PROBLEM
How can rates of crossing-over be used to map chromosomes?

MATERIALS (per group)

wooden stick from a frozen-dessert bar
metric ruler   pen
sheet of unlined paper

PROCEDURE
1. Using a pen and a ruler, draw a vertical line 15 cm long in the center of a clean sheet of unlined paper. Make a small horizontal mark at the bottom of the line. Measuring from the bottom of the line, add horizontal marks at 1 cm, 3 cm, 6 cm, 10 cm, and 15 cm.
2. Label each horizontal mark alphabetically (A through F), starting from the bottom.
3. The vertical line on the paper represents a chromosome that has six genes on it—A, B, C, D, E, and F. This chromosome's homologous chromosome is represented by the colored edge of the wooden stick.
4. Adjust the sheet of paper so that its bottom edge is about 15 cm from the edge of your lab table or desk. Move your chair back so that the front edge of your seat is about 30 cm from the edge of your lab table or desk.
5. Toss the wooden stick, underhand, toward the vertical line until the stick lands across the line. The landing of the stick across the line represents crossing-over.
6. When crossing-over occurs, look at the colored edge of the stick to determine which genes have been separated. Make a tally mark in the appropriate place in the data table for each gene that has become separated from gene A. For example, if the colored edge of the stick lands between D and E, make tally marks for genes E and F because they have been separated from gene A as a result of crossing-over.
7. Toss the stick and tally the results until crossing-over has occurred 100 times.
8. Count up the number of tally marks for each of the five genes. In the appropriate place in your data table, record the number of times each gene was separated from gene A.
9. Calculate the frequency of crossing-over by dividing the number of times each gene was separated from gene A by 100. Record the results of your calculations.
10. Calculate the location of each gene by multiplying the frequency of its crossing-over by 15 and rounding off to the nearest integer. Record the calculated gene locations.

OBSERVATIONS

Genes Separated from Gene A	Times Separated from Gene A		Frequency of Crossing-over	Gene Locations	
	Tally	Number		Calculated	Actual
B					1
C					3
D					6
E					10
F					15

1. Did every toss result in crossing-over?
2. Which genes became separated from gene A most frequently? Most infrequently?

ANALYSIS AND CONCLUSIONS
1. What is the relationship between the frequency of gene separation due to crossing-over and the distance between genes?
2. Are your calculated gene locations exactly the same as the actual gene locations? If so, discuss why the experiment went as expected. If not, discuss possible sources of error.
3. How can the frequencies of crossing-over be used to map chromosomes?

Skills Development
Students will use the following skills while completing this investigation.
1. Measuring
2. Observing
3. Predicting
4. Relating
5. Hypothesizing
6. Comparing
7. Applying
8. Calculating

Teaching Strategy
1. Circulate to see that each group has correctly drawn and measured the line that represents the chromosome and 6 genes.
2. Make sure that students understand clearly the tally procedure and how to determine which genes have separated from gene A.

STUDENT STUDY GUIDE

SUMMARIZING THE CONCEPTS

The key concepts in each section of this chapter are listed below to help you review the chapter content. Make sure you understand each concept and its relationship to other concepts and to the theme of this chapter.

10-1 The Chromosome Theory of Heredity
- Genes are located on the chromosomes.
- Genes that are located on the same chromosome tend to be inherited together.
- Gene linkages may be broken by crossing-over. The farther apart genes are located on a chromosome, the more likely that they will be separated by a cross-over event.
- Sex-linked genes are located on the sex chromosomes.

10-2 Mutations
- Mutations are changes in the genetic material. There are two kinds of mutations. Chromosomal mutations involve segments of chromosomes, whole chromosomes, or entire sets of chromosomes. Gene mutations involve individual genes.

10-3 Regulation of Gene Expression
- Interactions between genes affect gene expression.
- Genes can be turned on when they are needed and turned off when they are not.
- Many genes contain sequences that are not expressed. These sequences are removed when the RNA transcribed from DNA is made into mRNA.

REVIEWING KEY TERMS

Vocabulary terms are important to your understanding of biology. The key terms listed below are those you should be especially familiar with. Review these terms and their meanings. Then use each term in a complete sentence. If you are not sure of a term's meaning, return to the appropriate section and review its definition.

10-1 The Chromosome Theory of Heredity
chromosome theory of heredity
linked genes
linkage group
recombinant
sex chromosome
autosome
X chromosome
Y chromosome
sex-linked

10-2 Mutations
mutation
chromosomal mutation
nondisjunction
polyploidy
gene mutation
point mutation
frameshift mutation

10-3 Regulation of Gene Expression
incomplete dominance
codominance
polygenic
operon
operator
promoter
inducer
repressor
exon
intron

---

3. Help any groups that are having trouble calculating the frequency of crossing-over or the location of each gene on the chromosome.

4. This investigation may be done individually or in small groups. If they work in groups, have students take turns tossing the stick.

5. Tossing the stick underhand makes it difficult to control where the stick lands and thus allows crossing-over to occur almost at random. To further reduce bias, you may wish to have students turn their papers 180 degrees after crossing-over has occurred 50 times.

6. You may wish to pool class data for the number of times each gene is separated from gene A and show how a larger amount of data affects results.

7. Stress that perfect results are much less important than understanding why an experiment went wrong or why it went right.

Observations
1. No.
2. F was separated most frequently, followed by E. B was separated most infrequently, followed by C.

Analysis and Conclusions
1. Genes are separated more frequently when they are more distant from one another.
2. Answers will vary. In most cases, students will find that the calculated gene locations are very close to the actual gene locations. Errors may arise as a result of nonrandom tosses, for example, if the person tossing the stick manages to always have it land between E and F.
3. The frequencies of crossing-over can be used to estimate the distance between genes because the greater the distance between genes, the greater the frequency of crossing-over. Thus the relative location of genes can be determined and a map that shows distances between genes in units of frequency of crossing-over can be constructed.

Going Further: Enrichment
Students may enjoy designing a similar investigation that uses an actual gene map from an organism such as the fruit fly. Students can label the alleles A, B, C, and so on according to the traits that they produce. Then, as they throw the stick and tally the events of crossing-over, they can predict the actual characteristics that the organism may or may not have. For example, some flies might have the linked traits of a black body and speckled wings, while others may have only one of these traits due to the separation that results from crossing-over.

CHAPTER REVIEW

CONTENT REVIEW

Multiple Choice

Choose the letter of the answer that best completes each statement.

1. An organism has an extra copy of a chromosome in all its cells. This is an example of
   a. polyploidy.
   b. a chromosomal mutation.
   c. a gene mutation.
   d. a somatic mutation.
2. Both alleles are expressed in
   a. codominance.   c. polyploidy.
   b. nondisjunction. d. dominance.
3. The expressed allele does not make up for the inactive allele in
   a. polyploidy.
   b. translocation.
   c. codominance.
   d. incomplete dominance.
4. Which kind of mutation would result if an extra nucleotide were inserted into a gene?
   a. germ   c. chromosomal
   b. frameshift   d. deletion
5. Traits that are caused by the interaction of many genes are said to be
   a. polyploid.   c. polygenic.
   b. linked.   d. autosomal.
6. The gene product is not produced when the repressor binds to
   a. the operator.
   b. RNA polymerase.
   c. lactose.
   d. the enzyme gene.
7. A point mutation can best be classified as a
   a. chromosomal mutation.
   b. germ mutation.
   c. deletion mutation.
   d. gene mutation.
8. Gene linkages may be broken by
   a. nondisjunction.
   b. duplication.
   c. crossing-over.
   d. polyploidy.

True or False

Determine whether each statement is true or false. If it is true, write "true." If it is false, change the underlined word or words to make the statement true.

1. Crossing-over is more likely to occur between genes that are <u>close together</u>.
2. A <u>somatic</u> mutation affects sperm cells or egg cells and can be inherited.
3. The effects of recessive sex-linked alleles are seen more often in <u>males</u>.
4. Changes in DNA are called <u>mutations</u>.
5. Segments of nonhomologous chromosomes are exchanged in <u>crossing-over</u>.
6. Sex-linked genes are located on the <u>autosomes</u>.
7. <u>Polyploidy</u> is the condition of having extra sets of chromosomes.
8. A <u>point</u> mutation affects a single nucleotide.

Word Relationships

A. *In each of the following sets of terms, three of the terms are related. One term does not belong. Determine the characteristic common to three of the terms and then identify the term that does not belong.*

1. inversion, translocation, duplication, nondisjunction
2. operator, promoter, repressor, inducer
3. dominance, codominance, incomplete dominance, polygenic inheritance
4. gene mutation, point mutation, frameshift mutation, chromosomal mutation

---

CONTENT REVIEW

Multiple Choice
1. a   2. a   3. d   4. b
5. c   6. a   7. d   8. c

True or False
1. F far apart
2. F germ
3. T
4. T
5. F translocation
6. F sex chromosomes
7. T
8. T

Word Relationships

A.
1. types of chromosomal mutations that involve segments of chromosomes; nondisjunction involves whole chromosomes
2. parts of an operon; a repressor is a protein produced by a gene that is not part of the operon
3. interactions between the alleles for a given gene; polygenic inheritance involves the interaction of several genes
4. small-scale mutations involving a single gene; chromosomal mutation

B.
5. autosome
6. linkage
7. germ mutation
8. exon

CONCEPT MASTERY

1. Deletion: part of chromosome lost. Duplication: part of chromosome repeated. Inversion: part of chromosome becomes turned around. Translocation: part of chromosome breaks off and becomes attached to a nonhomologous chromosome.
2. The inducer binds to the repressor, preventing it from turning off the operon. When the inducer is absent, the repressor can bind to the operon and turn it off.
3. First, mRNA may require processing before it leaves the nucleus. Second, jumping genes may activate (or deactivate) nearby genes. Other answers are possible.
4. A linkage group is a set of genes that are inherited together. The behavior of linkage groups during gamete formation and fertilization indicates that they are the same as chromosomes.

CRITICAL AND CREATIVE THINKING

1. No. Introns, which are removed before proteins are made, have no effect on the gene's product and thus do not affect gene expression.

B. *Give the vocabulary word whose meaning is opposite that of the following words.*

5. sex chromosome
6. crossing-over
7. somatic mutation
8. intron

CONCEPT MASTERY

Use your understanding of the concepts developed in the chapter to answer each of the following in a brief paragraph.

1. List and describe the four types of chromosomal mutations that result in a change in the structure of a chromosome.
2. Explain why the operon is turned on when the inducer is present and turned off in the absence of the inducer.
3. Describe two ways genes are regulated in eukaryotes.
4. What is a linkage group? What kind of evidence suggests that linkage groups are the same as chromosomes?

CRITICAL AND CREATIVE THINKING

Discuss each of the following in a brief paragraph.

1. **Making predictions** Would a mutation in an intron affect gene expression? Explain.
2. **Relating concepts** How are crossing-over and linkage related to gene mapping?
3. **Problem solving** Examine the data for the crosses in cats shown below. How are the black and yellow coat colors inherited? (*Hint:* Male cats are XY and female cats are XX.) Using a Punnett square, predict the results of a cross involving a black male and a tortoise-shell female.
4. **Applying concepts** Two of the genes that Mendel studied are located on the same chromosome. Despite this, they show independent assortment in test crosses. What can you conclude about their relative positions on the chromosome?
5. **Using the writing process** Pretend you are the sex-linked gene for colorblindness. Describe the effects you have on each generation as you are passed down through two generations.

Parents	Offspring
black male X yellow female	½ yellow males ½ tortoise-shell (black and yellow) females
yellow male X black female	½ black males ½ tortoise-shell females
yellow male X tortoise-shell female	¼ black males ¼ yellow males ¼ yellow females ¼ tortoise-shell females

2. The frequency of crossing-over reflects the physical distance between genes. Genes that are located on the same chromosome are linked.
3. Black and yellow coat colors are sex-linked. A cross involving a black male (X^BY) and a tortoise-shell female (X^BX^Y) would produce ¼ black males (X^BY), ¼ yellow males (X^YY), ¼ black females (X^BX^B), and ¼ tortoise-shell females (X^BX^Y). Check students' Punnett squares.
4. Genes that are located very far apart on the chromosome are often separated by crossing-over and thus show independent assortment.
5. Student essays should be creative, well-written, and correct in terms of basic content.

CHAPTER 11 *Human Heredity*

Section	Laboratory Investigations and Activities
11–1 "It Runs in the Family" pages 227–229 THEMES: Systems and Interactions, Patterns of Change	**Student Edition** LABORATORY INVESTIGATION: Observing Human Traits, p. 242 **Teacher Edition** DEMONSTRATION: Observing Inherited Traits, p. 226D **Laboratory Manual** Constructing a Human Pedigree, p. 151
11–2 The Inheritance of Human Traits pages 230–234 THEMES: Patterns of Change	**Teacher Edition** DEMONSTRATION: Observing a Polygenic Trait, p. 226D **Laboratory Manual** Investigating Inherited Human Traits, p. 157
11–3 Sex-Linked Inheritance pages 235–239 THEME: Patterns of Change	
11–4 Diagnosis of Genetic Disorders pages 239–241 THEME: Patterns of Change	
Chapter Review pages 243–245	

Teacher Resources

Books

Ball, Susie, ed. *Strategies in Genetic Counseling: The Challenge of the Future*. Human Sciences Press, 1988.

Burns, George W., and Paul J. Bottino. *The Science of Genetics*, 6th ed. Macmillan, 1989.

Chen, Harold. *Handbook of Medical Genetics*. Green, 1988.

Delson, Eric. *Ancestors: The Hard Evidence*. A. R. Liss, 1985.

Gardner, Eldon J. *Human Heredity*. Wiley, 1983.

Nora, James J., and Clarke F. Fraser. *Medical Genetics: Principles and Practice*, 3rd ed. Lea and Febiger, 1988.

Singer, Sam. *Human Genetics: An Introduction to the Principles of Heredity*, 2nd ed. Freeman, 1985.

Thompson, Elizabeth A. *Pedigree Analysis in Human Genetics*. Johns Hopkins, 1986.

CHAPTER PLANNING GUIDE

Other Activities	Media and Technology

Teaching Resources
    ACTIVITY: Human Characteristics, p. 3
Study Guide
    Section 11–1, p. 105	

Study Guide Section 11–2, p. 106	

Teaching Resources
    ACTIVITY: Applied Genetics, p. 7
Study Guide
    Section 11–3, p. 110	

Teaching Resources
    VOCABULARY REVIEW: Crossword, p. 1
Study Guide
    Section 11–4, p. 113	

Teaching Resources
    Chapter Test A, p. 11
    Chapter Test B, p. 15
	Computer Test Bank
	    Chapter Test, p. 111

Audiovisuals

Human Heredity. Film or video. Coronet.
Heredity, Health and Genetic Disorders. 2 filmstrips with cassettes. Ward. (Caro)

Genetics—Observing Patterns of Inheritance. Video or 2 filmstrips with cassettes. Science and Mankind, Inc.
Heredity and Mutation. Video. Films for Humanities and Sciences, Inc.

CHAPTER 11 Human Heredity

CHAPTER OVERVIEW

In this chapter students will continue their study of genetics by learning about the genetic system of the most fascinating of organisms—the human being. Students will discover how some human traits are inherited and how some inherited disorders can be cured or prevented.

Students will learn that many human traits are inherited by the action of dominant and recessive traits, but that other traits are determined by more complicated gene interactions. They will also learn how the environment can influence gene expression.

Students will learn in detail about particular genes that affect human beings in significant and dramatic ways. Such genes include those for human blood groups, Huntington disease, and sickle cell anemia. Students will also learn about polygenic traits—such as height, weight, and skin color—which are controlled by a number of genes.

Students will continue from Chapter 10 their study of sex determination and sex-linked inheritance. Students will learn about the effects of nondisjunction in humans and about sex-linked genetic disorders such as colorblindness and hemophilia. They will also read about sex-influenced traits such as baldness.

In the final section of the chapter, students will learn how some genetic disorders are diagnosed. Included in this section is a discussion of Down syndrome and of some important techniques in the prenatal diagnosis of genetic disorders.

11–1 "IT RUNS IN THE FAMILY"

Section Focus 11–1

The purpose of this section is to introduce students to the ways in which human traits are inherited. Students will discover that the principles of genetics they have already studied apply to humans as well. They will learn that many human traits are inherited by the action of dominant and recessive genes, but that some traits are determined by more complicated gene interactions.

Students will learn how human chromosomes function in the inheritance of traits and that of the 46 chromosomes found in the human diploid cell, 2 are sex chromosomes and the other 44 are autosomes.

In the last part of the section, students will recall that the phenotype of an organism is only partly determined by genotype. They will discover that some human traits are strongly influenced by other factors, such as nutrition and exercise.

Performance Objectives 11–1

1. **Explain how human traits are inherited.**
2. **Discuss the role of chromosomes in human inheritance.**
3. **Distinguish between sex chromosomes and autosomes.**
4. **Discuss how nongenetic factors affect human traits.**

Science Terms 11–1

gamete   p. 228
zygote   p. 228

11–2 THE INHERITANCE OF HUMAN TRAITS

Section Focus 11–2

The purpose of this section is to discuss ways in which particular genes affect humans. These genes are those that determine blood type and those that cause Huntington disease and sickle cell anemia.

Students will learn that blood type is a trait that is determined by multiple alleles. This means that three or more alleles of the same gene code for a single trait. Students will learn how Austrian physician Karl Landsteiner discovered that human blood could be classified into four general groups. These groups are determined by a single gene with three possible alleles: I^A, I^B, and i.

Students will discover that abnormal genes cause the disorders of Huntington disease and sickle cell anemia. Huntington disease is characterized by a progressive degeneration of the nervous system that usually begins in early middle age. Sickle cell anemia is an inherited blood disease that causes red blood cells to become sickle shaped.

In the last part of the section, students will read about polygenic traits. Polygenic traits are human traits such as skin color and height that are controlled by a number of genes.

Performance Objectives 11–2

1. **Explain what is meant by multiple alleles.**
2. **List the human blood groups and explain their genetic basis.**
3. **Describe Huntington disease and explain how it is inherited.**
4. **Describe sickle cell anemia and discuss its genetic and molecular basis.**
5. **Define and give examples of polygenic traits.**

Science Terms 11–2

multiple allele   p. 230
polygenic trait   p. 234

11–3 SEX-LINKED INHERITANCE

Section Focus 11–3

The purpose of this section is to explain how sex is determined in humans and to describe some sex-linked genetic disorders. Students will learn that the presence or absence of a Y chromosome determines the sex of a human being. They will also learn that the failure of sex chromosomes to separate properly during meiosis results in disorders such as Turner syndrome and Klinefelter syndrome.

Students will learn how genes for certain disorders are carried on the X chromosome. Such disorders include colorblindness, hemophilia, and muscular dystrophy. Students will also discover that certain traits are sex-influenced. These traits, such as baldness, are controlled by genes that are expressed differently in males and females.

Performance Objectives 11–3

1. **Explain how sex is determined in humans.**

CHAPTER PREVIEW

2. Describe nondisjunction and name two nondisjunction disorders.
3. List and describe several sex-linked genetic disorders.
4. Discuss and give an example of sex-influenced traits.

Science Terms 11–3

sex-linked   p. 237
sex-influenced   p. 239

11–4 DIAGNOSIS OF GENETIC DISORDERS

Section Focus 11–4

The purpose of this section is to discuss some of the ways in which medical science is seeking to diagnose, prevent, and treat genetic disorders. This section also includes a discussion of some of the ethical considerations related to the diagnosis of genetic disorders.

Students will learn that some genetic disorders, such as Down syndrome, can be detected by examining a person's chromosomes. Persons with Down syndrome have an extra copy of chromosome 21. Down syndrome results in varying degrees of mental retardation and in an increased susceptibility to many diseases.

Students will discover that Down syndrome and other genetic disorders can be detected during pregnancy. Processes involved in prenatal diagnosis include amniocentesis and chorionic villus biopsy.

Performance Objectives 11–4

1. Explain how examination of chromosomes can indicate the presence of genetic disorders.
2. Describe Down syndrome and explain how it is inherited.
3. Identify and describe two methods used to detect genetic disorders during pregnancy.
4. Discuss some of the ethical considerations that arise when serious genetic disorders are detected during pregnancy.

Science Terms 11–4

amniocentesis   p. 240
chorionic villus biopsy   p. 241

DISCOVERY LEARNING

TEACHER DEMONSTRATIONS
Modeling

Observing Inherited Traits
The following demonstration can be used as an introduction to Chapter 11.

Ask for a group of about five volunteers to stand or sit facing the class. Ask the rest of the students to observe these students carefully and list as many characteristics as they can see that are inherited. (Possible characteristics include: eye color, hair color, skin color, height, body type, presence or absence of freckles, facial features.) Repeat the process with another group of volunteers until all or most students have been observed by the others. Ask students to share their lists of observations.

• **How many of the people you observed had one or more characteristic that was the same?** (Answers will vary. A trait such as freckles may appear in many students, although the number of freckles may differ in each; several students may have blonde hair, although the shade of blonde may differ in each.)
• **Which characteristics were unique in each person?** (Answers may vary. Facial features or exact shade of skin color may be unique in each person.)
• **From your observations, what can you conclude about genetic variety in humans?** (Accept all answers. Most will probably say that there is a great, almost infinite, genetic variety in humans.)

Observing a Polygenic Trait
This demonstration, which requires student participation, can be performed as students study the last part of Section 11–2. For the demonstration, students will need a large sheet of paper, a ruler, and some colored pencils.

Divide the class into groups of three or four. Ask one student in each group to trace his or her hand on a sheet of paper. Have students close their fingers and clearly mark the bottom of the hand. Ask students to draw a straight line from the tip of the middle finger to the base of the hand, then measure and record the length of the line. Next ask them to draw a line across the widest part of the hand, then measure and record its length.

Now ask another student in each group to place his or her hand on top of the drawing that the other student has made. Trace the outline of the hand with a different color pencil. Draw the two lines as before and measure and record their lengths. Repeat the procedure until each member in the group has traced his or her hand and recorded the measure of its length and width.

• **How did the lengths of the hands in your group compare?** (Answers will vary.)
• **How did the widths of the hands in your groups compare?** (Answers will vary.) Now have all the groups display their drawings and measurements. List on the chalkboard the measurements of length and width for each hand.
• **Would you say that hand size in humans is either small or large? Or is there more variation?** (There is more variation.)
• **Why do you think so?** (The hands of people in the class vary quite a bit in size.)
• **Do you think that human hands have just one or two different shapes?** (No.)
• **How did you reach that conclusion?** (Each drawing is somewhat different from the others; many different variations exist in whether a hand is long or short, narrow or wide.)

Point out that such variation indicates that hand size is a polygenic trait—that is, a trait that is determined by a number of genes.

• **If hand size were determined by just one gene with two alleles, what do you think would have been the result of this experiment?** (Only two different sizes and shapes of hands would have been found among members of the class.)

226D

CHAPTER 11

Human Heredity

GUIDED ENQUIRY

Pose the following questions to students and have them record their responses. Point out that they will gain a better understanding of the key concepts if they read the chapter with these basic questions in mind. Upon completion of the chapter, pose the questions again. Ask students to compare their initial responses with those they have developed after reading the chapter.

- How do the general principles of genetics apply to humans?
- Why are humans difficult subjects for genetic study?
- How are human blood types inherited?
- What type of person is most likely to inherit a gene for sickle cell anemia?
- What conditions can result from the nondisjunction of human chromosomes?
- How does a person inherit a gene for colorblindness or hemophilia?
- How can genetic disorders be diagnosed?
- How can doctors determine if an unborn child has a genetic defect?

CHAPTER 11

Human Heredity

The genetic inheritance of these babies makes each one a unique individual.

Of all the organisms in the world, there is one that we find particularly fascinating. This organism is not, however, an ideal one for the study of genetics. Why? The reasons are many. It is composed of approximately 10^{14} (100 million million) cells. Its generation time (about 20 years) is much too long. It cannot be kept in the laboratory for study. It produces too few offspring for good statistics. It cannot be used in test crosses. And its genetic system contains nearly 100,000 genes. Despite these disadvantages, biologists have studied the genetics of this single organism, *Homo sapiens*, for more than 80 years. The reason for so much scientific interest is clear: Of all the organisms on Earth, it is ourselves we long to know the best. In this chapter, you will discover how some human traits are inherited and how some inherited disorders can be cured or prevented.

INTRODUCING CHAPTER 11

Using the Visuals
Have students look at the chapter-opener picture.
- **What do you see in this picture?** (Many different human babies.)
- **In what ways are all these babies alike?** (Possible answer: They all probably have two legs and two arms, 10 fingers and 10 toes, heart and lungs, eyes, ears, nose, and mouth—in short, they are all typical humans.)
- **How are they different?** (Possible answers: They have different skin and hair colors; they are somewhat different in size; they have different facial features; they probably have different personalities or dispositions.)
- **How many of these differences do you think are inherited?** (Answers may vary. Probably all these factors are inherited.)
- **Choose any baby in the picture. What can you tell about the baby's parents?** (Answers will vary. An obvious response is to choose an Asian child and say that at least one of this child's parents is Asian.)

GUIDE FOR READING

After you read the following sections, you will be able to

11-1 "It Runs in the Family"
- Explain how human traits are inherited.
- Relate the environment to genes and gene expression.

11-2 The Inheritance of Human Traits
- Explain how multiple-allele, dominant, recessive, and polygenic traits are inherited, using specific examples.

11-3 Sex-Linked Inheritance
- Discuss the nature and inheritance of some traits that involve sex and sex chromosomes.

11-4 Diagnosis of Genetic Disorders
- Identify two methods of detecting genetic disorders during pregnancy.

Journal Activity

YOU AND YOUR WORLD

Every human is different from every other human as a result of variations in genes, growth and development, and upbringing. In your journal, explore your thoughts and feelings about differences among humans and the ways people react to those who are different.

Figure 11-1 Each human chromosome contains a specific set of genetic instructions that is passed on from parent to child. The human chromosome in this electron micrograph has been magnified approximately 20,000 times.

11-1 "It Runs in the Family"

Guide For Reading

- How are human traits transmitted from parents to offspring?
- What are sex chromosomes, autosomes, gametes, and zygotes?
- What impact does the environment have on gene expression?

How many times have you overheard two people discussing whether a newborn baby looks more like its mother or its father? People have always discussed such things. But when those of us who study biology engage in such discussions, we make an important assumption: Many of the characteristics of human children are genetically determined.

The principles of genetics that we have already discussed apply to humans. **Many human traits are inherited by the action of dominant and recessive genes, although other traits are determined through more complicated gene interactions.** Although the study of human genetics is difficult, it is one of the most important fields in biology. For we are interested in finding out as much as possible about ourselves, including the role that heredity plays in determining who and what we are and how some of the disorders and conditions that we inherit can be cured or prevented.

The Human Organism

The study of ourselves logically begins with a discussion of human chromosomes. Recall that a diploid cell has two sets of homologous chromosomes. A human diploid cell contains 46 chromosomes arranged in 23 pairs. These 46 chromosomes

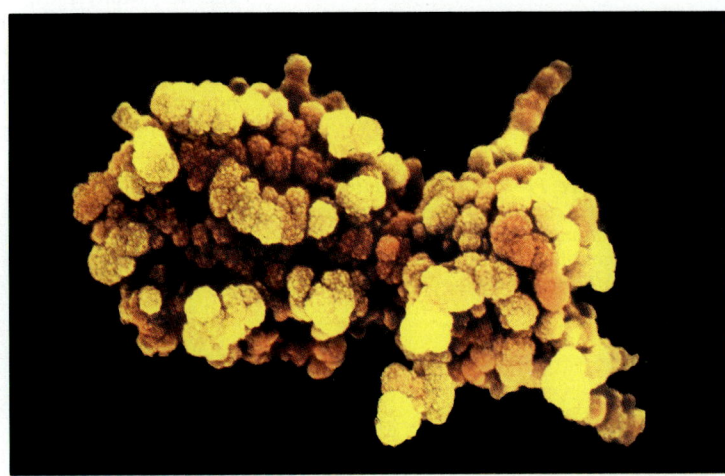

COOPERATIVE LEARNING

Using preassigned lab groups or randomly selected teams, have groups complete an alternatives grid on the issue of prenatal testing. (Refer to Section 11–4.) The alternatives for which groups will be predicting consequences are

A. Prenatal testing is illegal.
B. Prenatal testing is voluntary.
C. Prenatal testing is required.

Political, social, and economic consequences should be predicted for each alternative in a grid or chart format.

Journal Activity

YOU AND YOUR WORLD

Have students keep their journal assignments in their portfolios. Encourage students to honestly examine their own attitudes and the attitudes of those around them toward people of different races, religions, nationalities, or physical abilities.

this? (Humans take too long to mature; they do not produce many offspring in a single generation; they cannot be subjected to various laboratory experiments; the human genetic system is complex, containing nearly 100,000 genes.)
- **In spite of these difficulties, why do scientists persist in the study of human genetics?** (Humans want to understand themselves.)
- **Do you think that studying the genetic principles of other organisms, such as the fruit fly, helps scientists learn about human genetics?** (Answers may vary. The correct answer is yes.)

TEACHING STRATEGY 11–1

Focus/Motivation

Begin by asking students:
- **How many of you know someone who looks like his or her parents?** (Accept all answers.)
- **In what ways does this person look like his or her parents?** (Answers may include facial features, hair and skin color, height, general body type.)

Ask students to imagine these babies growing into children and eventually becoming adults.
- **Can you think of some genetic factors that are not obviously visible in these babies but that will show up later in their lives?** (Possible answers: intelligence; talents such as the ability to sing or draw; adult height and body type; various psychological factors; perhaps a genetic disease that does not manifest itself until adulthood; weaknesses in certain body systems such as digestive or respiratory problems.)

Have students read the chapter-opener text.
- **What can you conclude from the text about the suitability of humans for genetic study?** (They are not suitable for genetic study.)
- **What are some reasons for**

227

contain 6 billion nucleotide pairs of DNA—6 billion individual characters of the genetic code. To get an idea of how long a complete human DNA sequence actually is, consider the following: In this textbook, there are approximately 3000 letters on each page. If a complete human DNA sequence were to be written in the same-size type as this textbook, it would comprise a book more than 1 million pages long!

As you will recall from Chapter 9, the principles of genetics described by Mendel require that organisms inherit a single copy of each gene from each parent. In humans, the **gametes**, or reproductive cells, contain a single copy of each gene. Gametes (sperm and eggs) are formed in the reproductive organs by the process of meiosis. Each egg cell and each sperm cell contain 23 chromosomes, or the haploid number of chromosomes. During fertilization, sperm and egg unite and a **zygote** (ZIGH-goht), or fertilized egg, is produced. The zygote contains 46 chromosomes (23 pairs), or the diploid number of chromosomes characteristic of the organism.

Of the 46 chromosomes found in a human diploid cell, two are the sex chromosomes, X and Y. The remaining 44 chromosomes are the autosomes. The inheritance of dominant and recessive human genes carried on any of the 44 autosomes follows Mendel's principles. Human genes carried on the sex chromosomes are sex-linked.

The simplest patterns of inheritance occur in traits that are influenced by a single gene. Fortunately, many such genes have now been described. Human genetics is advancing so rapidly that scarcely a week goes by without a new human gene being reported and mapped.

Human Traits

You will recall from Chapter 10 that the phenotype of an organism is only partly determined by its genotype. There are some traits that are strongly influenced by environmental, or nongenetic, factors. Environmental factors include nutrition and exercise. For example, recent studies show that improvements in infant and childhood nutrition in the twentieth century in the United States and Europe have greatly increased

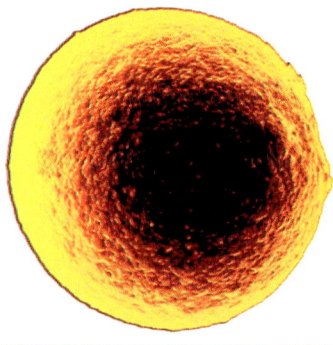
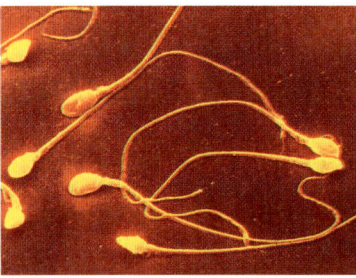

Figure 11–2 In humans, the egg (top) and the sperm (bottom) contain 23 chromosomes each. During fertilization, these gametes unite to produce a zygote that contains 46 chromosomes.

Figure 11–3 Notice the difference in the heights of these teenagers. Although the genes for height are inherited, the trait is strongly influenced by environmental factors, such as diet.

SCIENCE, TECHNOLOGY, AND SOCIETY

BREAKTHROUGH

Exploring the Unknown

We usually think of unexplored territory as being distant or remote. But what is perhaps the most interesting and mysterious of all this planet's unexplored territories lies right within us: The human genome. Despite years of study, we know very little about the DNA sequences in our own chromosomes.

The Human Genome Project is an effort designed to change that. Scientists throughout the word are cooperating in a systematic effort to map the genes and read the DNA sequences for each human chromosome. It is a monumental task. The first part of the job, pinpointing genetic "markers" on each chromosome to be used as reference points, was accomplished in 1993. But the main part of the job, reading and analyzing DNA sequences, will take as much as a decade more to complete.

What will the human genome project tell us? No one can say for certain. But the years ahead are sure to be full of surprises as little by little we discover more about the once unexplored territory of our own cells.

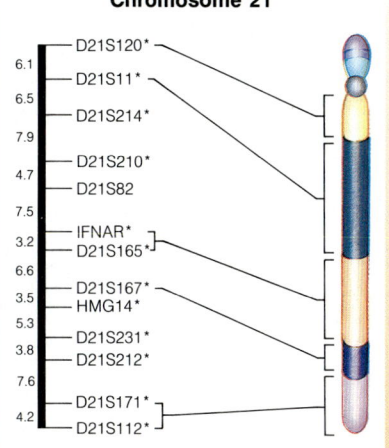

This preliminary map shows the locations of 13 markers on chromosome 21.

---

the average height of these populations over the nineteenth-century averages. Famine-stricken countries, however, were found to produce a generation with stunted growth.

Although it is important to consider the influence of the environment on the expression of some genes, it must be understood that environmental effects on gene expression are not inherited; genes are. Genes that are denied a proper environment in which to reach full expression in one generation can, in a proper environment, achieve full potential in a later generation.

11–1 SECTION REVIEW

1. How are human traits inherited?
2. What is a gamete? A zygote?
3. How many chromosomes are in a human diploid cell? A human haploid cell?
4. Distinguish between sex chromosomes and autosomes.
5. **Critical Thinking—Relating Cause and Effect** How does the environment influence gene expression?

ANNOTATION KEY

① Because their blood cells have neither type A nor type B antigens, their body may react badly to transfusions that contain these foreign antigens. (Making inferences)

② I^A and I^B are codominant, i is recessive. (Interpreting charts)

③ Because blood types A and B are each produced by two different genotypes. (Applying concepts)

BACKGROUND INFORMATION
ACHONDROPLASIA

Another genetic disorder that is caused by an autosomal dominant allele is achondroplasia. Achondroplasia is a type of dwarfism in which the affected person never reaches a height greater than 4 feet 4 inches tall. When long bones develop in an affected child, cartilage forms in such a way that the arms and legs end up being disproportionately short. About 1 in every 10,000 individuals is affected by achondroplasia.

TEACHING STRATEGY 11-2

Focus/Motivation

Ask students:
- **How many of you have ever had a blood test?** (Answers will vary.)
- **Do you recall why you had the blood test?** (Answers may vary. Some may have had a blood test when hospitalized; others may have had their blood tested during a physical examination or as part of a science or health project.)
- **Do you know what your blood type is?** (Accept all answers. For those who do know, have them share their blood type, perhaps writing it on the chalkboard.)
- **For those of you who know your blood type, do you know why you have that particular blood type?** (Answers may vary. Most students will guess that blood type is inherited.)

Guide For Reading

■ What are multiple alleles?
■ What are some examples of dominant, recessive, and polygenic traits in humans? How are they inherited?

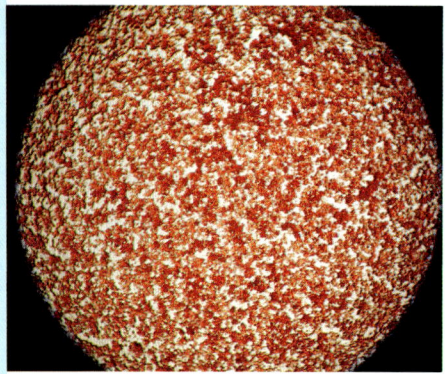

Figure 11-4 *This photomicrograph shows the agglutination, or clumping, of human red blood cells that can occur when blood of two different types is mixed.*

230

11-2 The Inheritance of Human Traits

At the writing of this textbook, more than 3000 human genes have been described and studied. We cannot list all of these genes, but we will examine a few that affect us in significant and even dramatic ways.

Human Blood Groups

In Chapters 9 and 10, you read about genes that have two contrasting forms, or alleles. It is possible, however, for genes to have **multiple alleles**, or more than two forms. **A gene that has three or more alleles is said to have multiple alleles.** Although many alleles may exist, it is important to remember that only two alleles are present in a diploid (2N) organism. **ABO and Rh blood groups are examples of human traits determined by multiple alleles.**

ABO BLOOD GROUPS In 1900, the Austrian physician Karl Landsteiner discovered that human blood could be classified into four general groups, or types, known today as the Landsteiner blood groups. These four blood types are determined by the presence or absence of specific chemical substances in the blood.

Landsteiner observed that when he mixed together blood from different people, the red blood cells sometimes agglutinated, or clumped together. After a great deal of work, Landsteiner discovered that the red blood cells could carry two different antigens, which he called A and B. Antigens are molecules that can be recognized by the immune system.

The presence or absence of the A and B antigens produces four possible blood types: A, B, AB, and O. A person with type A blood has red blood cells that carry only antigen A. A person with type B blood has red blood cells that carry only antigen B. A person with type AB blood has both antigens, and a person with type O blood has neither.

Landsteiner's ABO blood groups are especially important in blood transfusions. A transfusion of the wrong type can cause a violent, even fatal, reaction in the body as the immune system responds to an antigen not found on its own cells. For example, if type B blood is given to a person with type A or type O blood, a reaction will occur against the red blood cells carrying the B antigen. People with type AB blood can receive blood from any of the four types because they already have both possible antigens on their blood cells. Can you explain why people with type O blood can receive only type O transfusions? ①

The ABO blood groups are determined by a single gene with three alleles: I^A, I^B, and i. Figure 11-5 shows how these alleles determine the blood group. What are the relationships among the three alleles? (*Hint*: Review Section 10-3.) ②

Content Development

Review with students the definitions of genotype, phenotype, and dominant and recessive alleles. Review also the idea of codominance, that two different alleles of the same gene can be expressed concurrently.

Introduce students to the idea of multiple alleles, in which more than two alleles exist for a given gene. Emphasize the three alleles that determine human blood groups, and discuss each of their possible genotypes and corresponding phenotypes.

Rh BLOOD GROUPS In addition to the ABO antigens, there is another antigen on the red blood cells, called the Rh antigen (named after the rhesus monkey in which the antigen was first discovered by Landsteiner and Alexander Wiener). People who have the Rh antigen on their red blood cells are said to be Rh positive (Rh^+). People who do not have the Rh antigen on their red blood cells are said to be Rh negative (Rh^-). It is interesting to note that although the Rh blood groups are commonly referred to as either positive or negative, there are actually more than two alleles that determine these blood groups. In fact, there are eight common alleles and many that are rarer.

In blood banks, the ABO and Rh blood groups are often expressed together in symbols such as AB^- or O^+. The gene for the Rh antigen, however, displays simple dominance. Genotype $Rh^+ Rh^+$ and $Rh^- Rh^+$ produces the Rh positive phenotype. Only the double recessive genotype $Rh^- Rh^-$ produces Rh negative blood.

Genotypes	Phenotypes (blood types)
ii	O
$I^A I^A$ or $I^A i$	A
$I^B I^B$ or $I^B i$	B
$I^A I^B$	AB

Figure 11–5 *This chart illustrates the relationship between genotype and phenotype for each of the ABO blood groups. Because there are 3 possible alleles for the ABO gene, there are 6 possible genotypes. Can you explain why there are only 4 phenotypes?* ❸

Huntington Disease

Many of the tens of thousands of human genes are so essential to normal functioning of cells and tissues that we do not notice them unless something goes wrong and a genetic disease results. **Huntington disease, which is produced by a single dominant allele, is an example of genetic disease.** People who have this disease show no symptoms until they are in their thirties or forties, when the gradual damage to their nervous system begins. As you will recall from Chapter 10, the effects of a dominant allele are expressed even when the recessive allele is present. So people who have the dominant allele for Huntington disease have the disease and suffer painful progressive loss of muscle control and mental function until death occurs.

The gene for Huntington disease, located on chromosome 4, has recently been isolated using the techniques of molecular biology. You can learn more about the latest developments in our understanding of Huntington disease in the Science, Technology, and Society feature on page 236.

Sickle Cell Anemia

In 1904, Doctor James Herrick noticed an unusual ailment afflicting one of his young patients. His patient, a 20-year-old black college student, had been complaining of weakness and dizzy spells and had open sores on his legs. The Chicago physician examined the student's blood cells under a microscope and discovered that many of the red blood cells were bent and twisted into shapes that resembled sickles (farm tools used to cut grain). The normal shape of a red blood cell is that of a round, flattened disk. Doctor Herrick guessed correctly that these unusually shaped cells were the cause of his patient's

Figure 11–6 *This photograph shows a rhesus monkey, the type of monkey after which the Rh blood group was named.*

FACTS AND FIGURES

The frequency of Huntington disease has been higher in South Africa than in any other part of the world. Researchers have discovered that all affected persons in South Africa were directly or indirectly descended from a certain Dutchman who settled there in 1658. This phenomenon of one or a few individuals with a genetic abnormality causing the establishment of a new population is known as the founder effect. The founder effect is most likely to occur in remote areas where the total population is relatively small.

Skills Development
Guided Practice
Skill: Interpreting charts
Using the Visuals Direct students' attention to Figure 11–5.
- **What are the three alleles that determine human blood type?** (I^A, I^B, and i.)
- **What is the phenotype of a person who has the genotype $I^A i$?** (Type A blood.)
- **What other combination of alleles produces this phenotype?** ($I^A I^A$.)
- **In the $I^A I^B$ genotype, is one allele dominant and the other recessive?** (No.)
- **What is the relationship between these two alleles?** (They are codominant; both are expressed in the phenotype.)

Emphasize that while the I^A and I^B alleles are codominant, the i allele is recessive. Only persons with two i alleles will have type O blood.

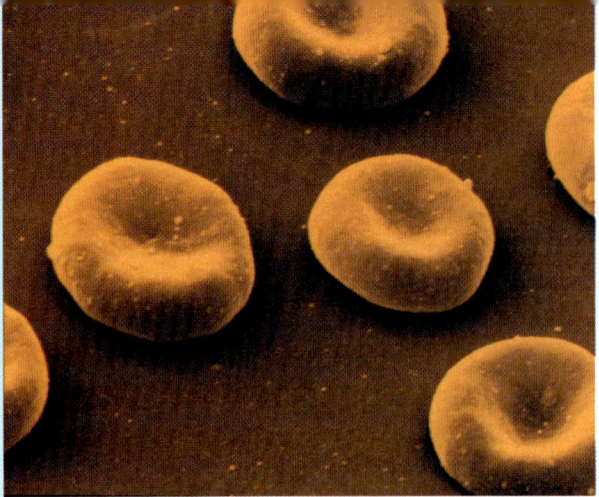

Figure 11–7 *These scanning electron micrographs show human red blood cells that contain normal hemoglobin (left) and a red blood cell that contains the abnormal hemoglobin characteristic of sickle cell anemia (right).*

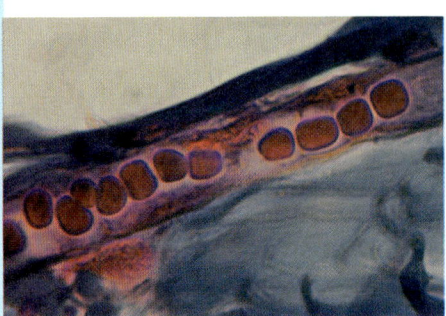

Figure 11–8 *Capillaries, which are the narrowest blood vessels, are so small that they permit only one red blood cell at a time to pass through. What happens to sickle-shaped red blood cells as they move through the capillaries?* ❶

problems, and he gave the disease the name by which we know it today—sickle cell anemia.

THE CAUSE OF SICKLE CELL ANEMIA Sickle cell anemia is caused by a change in one of the polypeptides found in hemoglobin. Hemoglobin is the protein that carries oxygen in red blood cells.

When a person who has sickle cell anemia is deprived of oxygen (from heavy exercise, holding one's breath, or even nervousness or anxiety), the hemoglobin molecules join together and form fibers. These fibers cause the red blood cells to undergo the dramatic changes in shape that Doctor Herrick observed. The sickle-shaped red blood cells are more rigid and tend to become stuck in the capillaries, the narrowest blood vessels in the body. See Figure 11–8. As a result, the movement of blood through these vessels is stopped and damage to cells and tissues occur. Serious injury or death may result.

THE GENETICS OF SICKLE CELL ANEMIA The allele for normal hemoglobin (H^A) is codominant with the sickle cell allele (H^S). People who are heterozygous ($H^A H^S$) are said to be sickle cell carriers. Because roughly half of the hemoglobin molecules in the blood of carriers is normal, these people suffer few ill effects of the disorder. People who are homozygous ($H^S H^S$) are said to be sickle cell sufferers. Because all of their hemoglobin molecules are affected by the sickle cell allele, these people may be severely afflicted by the disease.

THE MOLECULAR BASIS OF SICKLE CELL ANEMIA The allele for sickle cell hemoglobin differs from the allele for normal hemoglobin by a single nucleotide. Why does this seemingly small difference cause so much trouble?

The substitution of one nucleotide in the allele results in the substitution of a different amino acid in the sickle cell hemoglobin protein. This change makes hemoglobin less soluble in blood. Most of the time the condition does not present a problem. But when the body is under even minor stress, the hemoglobin in a large proportion of red blood cells will come out of solution as crystals. This crystallization of hemoglobin causes the sickle-shaped cells to appear, accompanied by their medical consequences.

THE DISTRIBUTION OF SICKLE CELL ANEMIA In the United States, people of African ancestry are the most common carriers of the sickle cell trait. In the rest of the world, sickle cell anemia is found in the tropical regions of Africa and Asia. Approximately 10 percent of Americans of African ancestry and as many as 40 percent of the population in some parts of Africa carry the trait. Why is sickle cell anemia so common in some regions and virtually unknown in others (such as Northern Europe and Asia)?

The answer to this question provides us with a surprising lesson in evolution. People who are heterozygous for sickle cell anemia ($H^A H^S$) are partially resistant to malaria, a serious disease that affects red blood cells. Sickle cell hemoglobin is thought to offer this resistance because sickled cells are frequently removed from the circulation and destroyed, killing any malaria parasites with them. People who are homozygous for normal hemoglobin ($H^A H^A$), on the other hand, have no resistance to malaria.

The incidence of sickle cell anemia parallels the incidence of malaria throughout the tropical areas of the world. The sickle cell trait probably developed in several populations throughout the world as a simple mutation, or change in genetic material. That mutation conferred an advantage wherever malaria was common, and thus it was favored by natural selection. Sickle cell anemia has persisted wherever it has helped its carriers survive malaria.

	H^A	H^S
H^A	$H^A H^A$	$H^A H^S$
H^S	$H^A H^S$	$H^S H^S$

H^A = allele for normal hemoglobin
H^S = allele for sickle cell hemoglobin

Figure 11–9 *This Punnett square shows the cross between two people who are heterozygous for sickle cell anemia. What are the possible genotypes of their offspring?* ❷

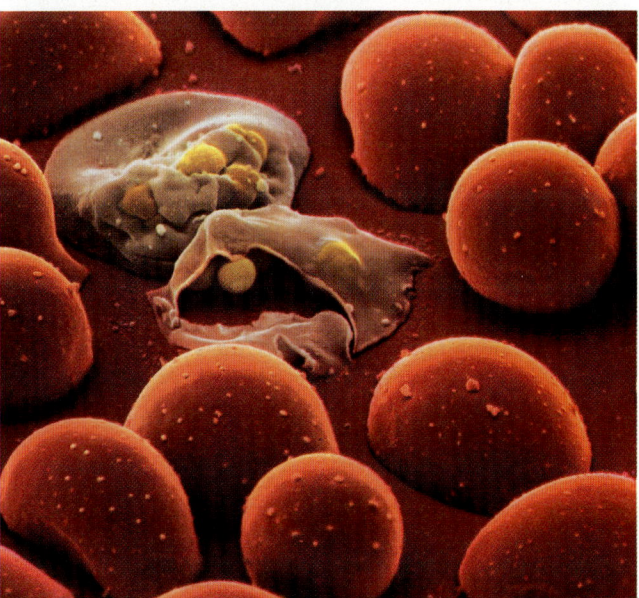

Figure 11–10 *Two of the red blood cells in this scanning electron micrograph have been invaded by the parasite that causes malaria. People who are heterozygous for sickle cell anemia are partially resistant to malaria.*

BACKGROUND INFORMATION
SICKLE CELL ANEMIA

Because every cell in the body needs oxygen, it is not surprising that sickle cell anemia is a very serious and debilitating disease—and one that can turn lethal at any time.

The disease's chronic symptoms include slow growth, poor wound healing, swollen feet and hands, and frequent colds and other infections. But the disease is at its worst during dramatic events called sickle cell crises. In these crises, sickled red blood cells clump together and block a blood vessel, often curtailing oxygen to vital organs. These crises are especially common—and painful—in young children, but they are potentially fatal at any age.

In the past, individuals with sickle cell anemia might have felt lucky to live into their teenage years. Today, better preventive care and improved management of the crises have helped many patients live healthy and productive lives into their forties, fifties, and beyond.

bin, H^A, is codominant with the sickle cell allele, H^S. $H^A H^S$ individuals, or sickle cell carriers, suffer few ill effects. $H^S H^S$ individuals have sickle cell anemia.
• **Knowing that the H^S gene produces abnormal hemoglobin, can you explain why Dr. Herrick's young patient felt weak and dizzy?** (Hemoglobin is the carrier of oxygen in the blood. Because much of the hemoglobin in sickle cells comes out of solution when the body is under even minor stress, there is much less hemoglobin available to carry oxygen at a time when the body needs it most. Dr. Herrick's patient was probably suffering from a deprivation of oxygen to his brain, muscle cells, and other vital organs and tissues.)

• **In what other ways does the sickle cell shape affect the sufferer of sickle cell anemia?** (Sickle-shaped cells can easily become stuck in the narrow blood vessels called capillaries. Blood flow is stopped, and serious injury or death may result.)

Figure 11–11 The color of human skin ranges from very light to very dark, depending upon the amount of melanin (dark pigment) present in the skin cells. Very dark-skinned people have alleles that code for production of melanin at all the gene positions for skin color. In very light-skinned people fewer of these positions are occupied by alleles that code for melanin production.

Polygenic Traits

Human traits that are controlled by a number of genes are called **polygenic** (pahl-uh-JEHN-ihk) **traits**. Examples of polygenic human traits include height, body weight, and skin color.

Unlike the simple skin pigmentation system of many animals, which allows a single gene to determine color, the human skin pigmentation system is rather complex. In humans, at least four different genes control skin color, and several of these genes may have multiple alleles. This means that the inheritance of skin color in our species can be somewhat unpredictable. And because skin color is a polygenic trait, children of the same mother and father may have quite different patterns of skin color—each pattern determined by which combination of genes they have inherited from their parents.

The color of human skin ranges from very dark to very light, depending upon the amount of melanin, a dark-colored pigment, present in skin cells. Very dark-skinned people have alleles that code for the production of melanin at all their gene positions for skin color. Light-skinned people, on the other hand, have alleles that code for the production of melanin at fewer gene positions for skin color. The range of human skin color, spread across the globe with no clear dividing line, is an illustration of what a wonderfully diverse species we are.

11–2 SECTION REVIEW

1. List several human traits determined by multiple alleles.
2. How is Huntington disease inherited?
3. What is sickle cell anemia?
4. What are polygenic traits? Give two examples of polygenic traits.
5. **Critical Thinking—Making Predictions** If a man with type O blood marries a woman with type AB blood, what are the possible genotypes of their offspring?

234

child will be a sickle cell sufferer and 50% that the child will be a carrier; for $H^AH^S \times H^AH^S$, P = 25% that the child will be a sickle cell sufferer and 50% that the child will be a carrier; for $H^AH^S \times H^AH^A$, P = 0% that the child will be a sickle cell sufferer and 50% that the child will be a carrier.)

SECTION REVIEW 11–2

1. ABO and Rh blood groups.
2. Huntington disease is inherited by the presence of a single dominant allele.
3. Sickle cell anemia is an inherited disease in which there is a change in the structure of hemoglobin. Under certain conditions, red blood cells that contain the variant hemoglobin will assume a sickle shape.
4. Polygenic traits are those

11–3 Sex-Linked Inheritance

As you will recall from Chapter 10, genes that are located on the sex chromosomes of an organism are inherited in a sex-linked pattern. As in many organisms, the sex in humans is determined by the X and Y chromosomes. In females, meiosis produces egg cells that contain one X chromosome and 22 autosomes. In males, meiosis produces sperm cells of which half contain one Y chromosome and 22 autosomes and the other half contain one X chromosome and 22 autosomes.

The sex of a person is determined by whether an egg cell is fertilized by an X-carrying sperm or a Y-carrying sperm. Males are normally 46XY, meaning that they have a total of 46 chromosomes, including an X chromosome and a Y chromosome. Females are normally 46XX.

Sex Determination

How do the X and Y chromosomes determine whether the sex of the zygote will be male or female? In *Drosophila* (fruit flies), which also have an XY system of sex determination, geneticists have found that sex is determined by the number of X chromosomes. A fruit fly that has a single X chromosome is male, regardless of whether it has a Y chromosome. A fruit fly that has two X chromosomes is female, regardless of whether a Y chromosome is present. This is not so in humans.

THE HUMAN XY SYSTEM Although meiosis is a precise mechanism that separates the two sex chromosomes of a diploid cell into single chromosomes of haploid gamete cells, errors sometimes do take place. The most common of these errors is nondisjunction. **Nondisjunction is the failure of chromosomes to separate properly during one of the stages of meiosis.** See Figure 11–12.

Nondisjunction can produce gametes that contain either two sex chromosomes or no sex chromosomes—a direct contrast to the normal condition of one sex chromosome. When one of these gametes joins with a normal gamete during fertilization, the result is a person with an abnormal number of sex chromosomes.

NONDISJUNCTION DISORDERS Roughly 1 birth in 1000 is affected by an abnormality involving nondisjunction of the sex chromosomes. The most common abnormalities are Turner syndrome and Klinefelter syndrome.

People who have Turner syndrome are female in appearance but their female sex organs do not develop at puberty and they are sterile, or unable to have children. Turner syndrome is abbreviated 45X or 45XO, where O denotes the absence of a second sex chromosome. People with Klinefelter syndrome are

Guide For Reading

- How is sex determined in humans?
- What are some disorders that result from nondisjunction of the sex chromosomes?
- What are some examples of sex-linked disorders in humans?
- How are sex-influenced traits inherited?

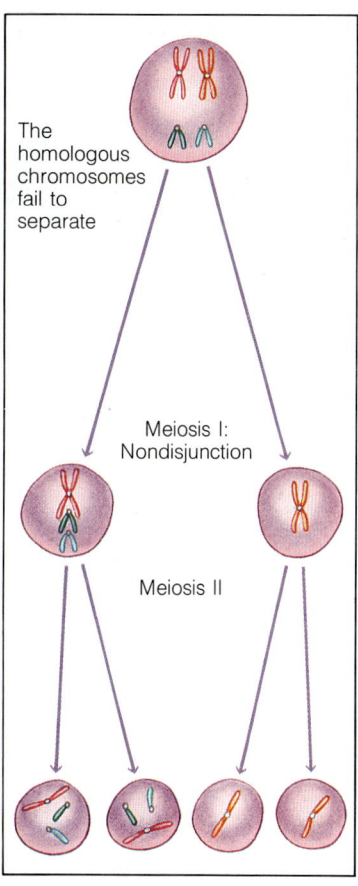

Figure 11–12 When homologous chromosomes fail to separate during meiosis, nondisjunction results.

SCIENCE, TECHNOLOGY, AND SOCIETY

BREAKTHROUGH

The Race to Find a Killer

Refer students to page 231 for more information on Huntington Disease (HD). Other recent developments in HD research are described in the Teaching Resources' Updates and Extensions article "Life's Newest Mystery."

HD involves a slow and progressive destruction of brain cells. Death usually occurs 10 to 20 years after the first onset of symptoms, which usually occurs between the ages of 30 and 50. Because the symptoms of HD appear so late, people may marry and have children before discovering that they have this disease—and worse yet, that they may have passed it on to their children.

When genetic tests for HD first became available, experts thought the tests would be in great demand. After all, the tests could enable the children of HD sufferers to be more certain about important decisions in their lives, such as having children. But many potential HD victims declined to have the tests. Some explained that they had adjusted to the uncertainty in their lives, and feared that a positive test would destroy them psychologically. Others worried about how their families and friends might react. Still others said that they simply did not wish to know.

Ask students how they would react if faced with such a choice. Would they want to know whether or not they carried the HD allele?

SCIENCE, TECHNOLOGY, AND SOCIETY

BREAKTHROUGH

The Race to Find a Killer

Huntington disease (HD), which you read about on page 231, is unusual because it is caused by a dominant allele and because its symptoms do not appear until middle age. In 1983, researchers found that the HD gene was located somewhere near the tip of chromosome 4. Everyone assumed that the exact location of the gene would be discovered within a couple of years. Research groups around the world joined forces in the race to pinpoint the gene. But as it turned out, the race was not a sprint but a marathon of almost 10 years!

In 1993, one of the research groups in the international team finally crossed the finish line. James Gusella and his associates at Massachusetts General Hospital announced that they had isolated the gene and determined its DNA sequence. The researchers also discovered an unusual feature of the gene that may hold the key to understanding how it works—and thus how to "fix" it. The feature is an expandable region near the middle of the DNA sequence. Normally this region is no more than 60 bases long. But in the HD allele, this region is greatly expanded—as many as 250 bases long.

The discovery of the HD allele has already produced a better test to determine if a person has inherited HD. More importantly, by finding the HD allele and a possible clue to the way it works, the researchers have moved closer to a genuine understanding of a puzzling and tragic human genetic disorder.

In 1993, this team of scientists at Massachusetts General Hospital triumphantly announced that they had isolated and sequenced the allele that causes Huntington disease.

---

male in appearance, and they, too, are sterile. The abbreviation for Klinefelter syndrome is 47XXY.

It is interesting to note that there have been no reported instances of babies being born without an X chromosome. Embryos without an X chromosome do not seem to develop properly because, as we will soon discover, the X chromosome contains a number of genes that are vital for normal development.

What can we learn from these abnormalities of the sex chromosomes? First, an X chromosome is absolutely essential for survival. Second, sex seems to be determined by the presence or absence of a Y chromosome and not by the number of X chromosomes. For example, there have been reported cases of people who have genotypes 48XXXY and 49XXXXY and are male in appearance.

11–3 (continued)

Content Development
Review with students the concept of sex-linked characteristics from Chapter 10. Point out that it was Morgan's experiments with fruit flies that gave the first evidence of sex-linked genes carried on the X chromosome. Explain that sex-linked genes are also present in humans and that, as in the fruit fly, they are carried on the X chromosome.

Reinforcement/Reteaching
Point out that although the fruit fly *Drosophila* has an X-Y system of sex determination, it is not the same as the human X-Y system. Sex in the fruit fly is determined by the number of X chromosomes, while sex in human beings is determined by the presence or absence of a Y chromosome. A person having one Y chromosome will be male, even if that person has additional X chromosomes due to nondisjunction. In a similar manner, a person having no Y chromosome

After years of uncertainty, the reason for the genetic importance of the Y chromosome is finally becoming clear: The Y chromosome contains a gene that switches on the male pattern of growth during embryological development. If this gene is absent, the embryo follows a female pattern of growth.

Sex-Linked Genetic Disorders

Genes that are carried on either the X or the Y chromosome are said to be **sex-linked**. In humans, the small Y chromosome carries very few genes. The much larger X chromosome contains a number of genes that are vital to proper growth and development. And as you just read, it seems to be impossible for humans to develop without the genes of the X chromosome.

It is particularly easy to spot recessive defects in alleles located on the X chromosome because the alleles are expressed more commonly in males than in females. What is the reason for this? Recall that males have one X chromosome. Thus all X-linked alleles are expressed in males, even if they are recessive. In order for a recessive allele to be expressed in females, there must be two copies of it, one on each of the two X chromosomes. If one of the X chromosomes contains a dominant allele, it will mask the expression of the recessive allele.

COLORBLINDNESS Colorblindness is a recessive disorder in which a person cannot distinguish between certain colors. **Most types of colorblindness are caused by sex-linked genes located on the X chromosome.** The alleles for colorblindness render people unable to make some of the pigments in the eye necessary for color vision.

The most common type of colorblindness is red-green colorblindness. People afflicted with this trait have difficulty distinguishing the lighter shades of red and green. The frequency of this type of colorblindness varies among different populations. In most Caucasian populations, about 8 percent of the males are affected but only about 1 percent of the females are.

In humans, color vision depends on the varying sensitivity of three groups of specialized nerve cells (cones) in the retina of the eye. One group is sensitive to blue light, one to red light, and one to green light. Colors of any given shade excite a specific level of activity from each of the three groups of nerve cells. We will learn more about color vision in Chapter 37.

Because the gene for color vision is carried on the X chromosome, the dominant allele for normal color vision is represented as X^C and the recessive allele for red-green colorblindness is represented as X^c. Homozygous (X^CX^C) and heterozygous (X^CX^c) females have normal color vision. A female who is heterozygous for colorblindness is said to be a carrier because she carries the recessive allele but does not express it. Although she is not colorblind, she is capable of

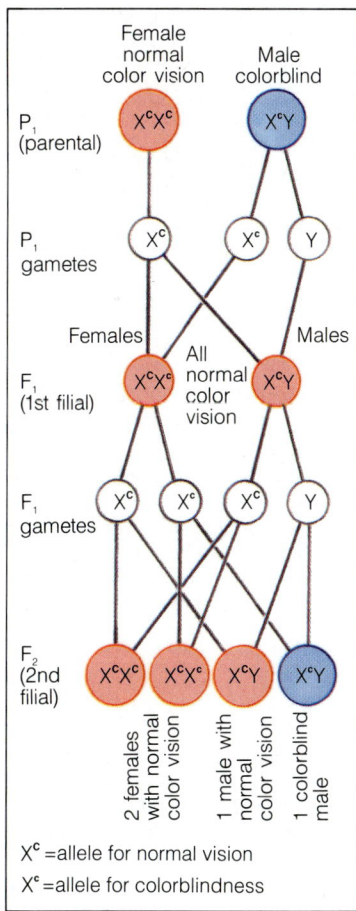

Figure 11-13 Colorblindness is an example of a sex-linked trait. In this cross, a female with normal color vision marries a colorblind male (P_1 generation). What are the phenotypes of the offspring in the F_2 generation?

will be female, no matter how many or how few X chromosomes are present.

Content Development
Write the abbreviations for Turner syndrome (45XO) and Klinefelter syndrome (47XXY) on the chalkboard.

• **What union of gametes would produce a person with Turner syndrome?** (The union of a normal gamete that contains one X chromosome and an abnormal gamete that contains no sex chromosomes.)

• **Why is the number in this abbreviation 45 rather than 46?** (The person with Turner syndrome has one less chromosome than normal humans, due to the missing sex chromosome in the one gamete.)

• **What union of gametes would produce a person with Klinefelter syndrome?** (Answers may vary. Some may say a normal gamete containing an X chromosome with an abnormal gamete containing X and Y; others may say an abnormal gamete containing XX with a normal gamete containing Y.)

Explain to students that although both possibilities are logical answers to the question, it is actually an abnormal male gamete, XY, that causes Klinefelter syndrome. Thus, Klinefelter syndrome is caused by the union of a normal X gamete with an abnormal XY gamete.

• **Why is the number in the abbreviation for Klinefelter syndrome 47 rather than 46?** (The person with this syndrome has one more chromosome than normal, due to the extra X chromosome.)

TEACHING SUPPORT
Teaching Resources
• Activity: Applied Genetics, p. 7

BACKGROUND INFORMATION
PSEUDOHERMAPHRODITES

In certain rare, recessive disorders, individuals have the gonads of one sex but the external genitalia of the other. Testicular feminization syndrome (TFS), also known as androgen insensitivity syndrome, results in an individual who is externally a normal female, despite the presence of a Y chromosome. Female androgenital syndrome (FAS) causes varying degrees of masculinization. The outward signs of FAS are usually detected and surgically corrected during infancy.

ANNOTATION KEY

❶ Normal female, normal male, colorblind male. (Interpreting diagrams)

ANNOTATION KEY

❶ $X^H X^h$ and $X^H Y$.
(Interpreting charts)

HISTORICAL NOTE
OFF WITH THEIR HEADS

In light of the fact that sperm cells determine the sex of human offspring, it is ironic to note that throughout history, kings and princes have often divorced (and even beheaded) their wives for not producing male heirs. A very famous example was Henry VIII of England, who divorced two of his six wives for not presenting him with a son. In more recent times, the Shah of Iran divorced his first wife after she bore him two daughters, then succeeded in producing boys with a later wife. One cannot help but wonder how the course of world history might have been changed had kings such as Henry VIII known a few basic facts about human biology!

BACKGROUND INFORMATION
BALDNESS

Female baldness is extremely rare, whereas male baldness is quite common. Geneticists have debated the question as to why there is not a higher incidence of female baldness. Some feel that male hormones are important in causing the expression of the baldness allele; others feel that female hormones serve as a protection against baldness.

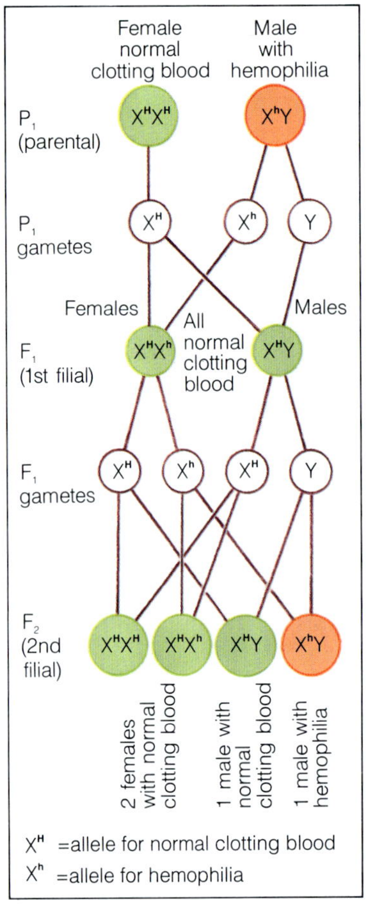

Figure 11–14 Hemophilia is an example of a sex-linked trait that results in the blood not clotting normally. In this cross, a female with normal-clotting blood marries a male with hemophilia. What are the genotypes of their offspring? ❶

passing on the allele for colorblindness to her offspring. Only homozygous recessive females ($X^c X^c$) are colorblind. Because males have only one X chromosome, they are either colorblind ($X^c Y$) or have normal color vision ($X^C Y$).

HEMOPHILIA Another recessive allele on the X chromosome produces a disorder called hemophilia (hee-moh-FIHL-ee-uh), or "bleeder's disease." In hemophilia, the protein antihemophilic factor (AHF) necessary for normal blood clotting is missing. Not only is hemophilia a rarer disorder than colorblindness, affecting only about 1 male in 10,000 and only about 1 female in 100,000,000, it is also more serious.

People with hemophilia can bleed to death from seemingly minor cuts and may suffer internal bleeding from bumps or bruises. Fortunately, hemophilia can be treated by injecting AHF isolated from donated blood. The donated blood must be carefully screened for infectious diseases, but the AHF injections relieve the most serious effects of hemophilia.

MUSCULAR DYSTROPHY Muscular dystrophy (DIHS-troh-fee) is an inherited disease that results in the progressive wasting away of skeletal muscle. Children with muscular dystrophy (MD) rarely live past early adulthood. The most common form of MD is caused by a defective version of the gene that codes for a muscle protein known as dystrophin. This gene is located on the X chromosome. Researchers are now using molecular techniques to insert healthy copies of the dystrophin gene into muscle cells. They hope that this new approach will soon make it possible to treat MD.

Sex-Influenced Traits

Many traits that may seem to be sex-linked, such as male pattern baldness, are actually caused by genes located on autosomes, not on sex chromosomes. Why then is baldness so much more common in men than it is in women? Male pattern

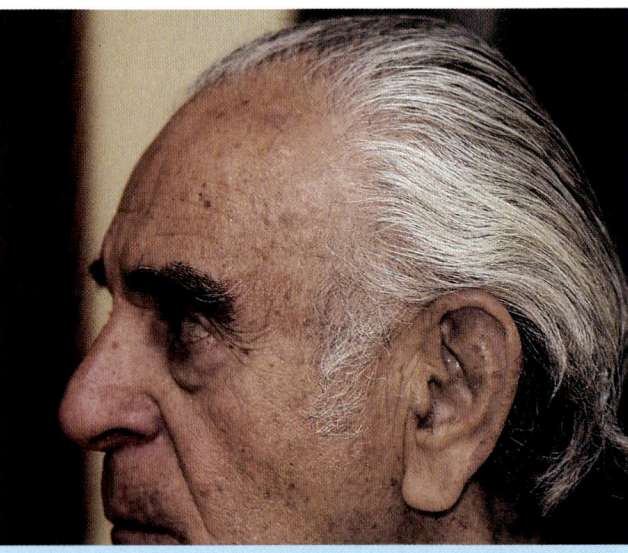

Figure 11–15 Male pattern baldness is a sex-influenced trait. The expression of this trait is thought to be enhanced by the presence of male sex hormones.

11–3 (continued)

SECTION REVIEW 11–3

1. The sex of a person is determined by whether an egg cell is fertilized by an X-carrying sperm or a Y-carrying sperm.
2. In Turner syndrome, people are female in appearance but their female sex organs do not develop at puberty and they are sterile. In Klinefelter syndrome, people are male in appearance, and they, too, are sterile.
3. Colorblindness, hemophilia, and a common form of muscular dystrophy are examples of sex-linked traits.
4. Sex-linked traits are those traits that are caused by genes carried on either the X or the Y chromosomes. Sex-influenced traits are caused by genes whose expressions differ in males and females.

Reinforcement/Reteaching
If students are having trouble answering the Section Review questions, review the material with which they are having difficulty.

baldness is a sex-influenced trait. **A sex-influenced trait is a trait that is caused by a gene whose expression differs in males and females.**

Many genetic studies indicate that baldness is controlled by a single gene with two alleles, one for normal hair (B) and one for baldness (b). Men and women who have the genotype BB have normal hair, whereas those relatively rare people who have the genotype bb (two baldness alleles) tend to be bald, whether they are male or female. The difference between the sexes occurs in the heterozygous condition, Bb. Males who are heterozygous tend to be bald, whereas females do not. We are not certain how the sex of a person influences the expression of these alleles, although it is possible that the male sex hormones may provide an answer.

11–3 SECTION REVIEW

1. How is sex determined in humans?
2. Describe two conditions that are caused by nondisjunction of the sex chromosomes.
3. What are some examples of sex-linked traits?
4. **Critical Thinking—Making Comparisons** Distinguish between sex-linked and sex-influenced traits.

11–4 Diagnosis of Genetic Disorders

Guide For Reading
- What is Down syndrome? How is it inherited?
- How can genetic disorders be diagnosed before birth?

Humans have been aware of genetic disorders throughout history. For centuries, religious and ethnic laws and customs have reflected a clear understanding on the part of people of the genetic nature of certain disorders. For years, physicians have longed for a way to detect and treat genetic disorders. Today, for some disorders detection is as simple as an examination of a person's chromosomes.

A Chromosomal Abnormality— Down Syndrome

You have just learned about several examples of chromosomal abnormalities. Failures of meiosis, including nondisjunction, cause abnormal combinations of chromosomes to be produced, including those responsible for Turner syndrome and Klinefelter syndrome. Nondisjunction affects autosomes, too.

An example of nondisjunction of autosomes is a condition known as Down syndrome (trisomy 21). In Down syndrome, there is an extra copy of chromosome 21. The presence of this

239

Figure 11–16 *Down syndrome results from the nondisjunction of chromosome 21. Although people with Down syndrome are physically challenged, they can live normal, active lives.*

extra chromosome can be detected in a careful examination of cells under the light microscope.

Down syndrome results in mental retardation that ranges from mild to severe. It is also characterized by an increased susceptibility to many diseases. In the United States, 1 baby in 800 is born with Down syndrome.

It is not clear why an extra copy of one chromosome should cause so much trouble, but scientists have recently discovered rare cases of Down syndrome in which only 46 chromosomes are present. In each of these cases, a small portion of chromosome 21 is attached to another chromosome. Scientists assume that this small portion must contain the genes that cause Down syndrome. A number of other genetic disorders are also produced by chromosomal abnormalities. Like Down syndrome, they too can be diagnosed by microscopic examination of chromosomes.

Prenatal Diagnosis

Down syndrome and other genetic disorders can now be diagnosed before birth by analyzing cells from the developing embryo. One technique, which is known as **amniocentesis**

nondisjunction must be so severe that the unborn baby cannot survive.)

Continue to explain that milder forms of nondisjunction in autosomes result in an unborn child that is brought to full term but is born with certain genetic disorders. Probably the most common of these is Down syndrome. Down syndrome is the result of a child receiving an extra copy of chromosome 21. Point out that this condition is also referred to as trisomy-21, meaning that there are three, rather than two, copies of chromosome 21.

SECTION REVIEW 11–4

1. Down syndrome is an inherited condition caused by an extra copy of chromosome 21.

(am-nee-oh-sehn-TEE-sihs), requires the removal of a small amount of fluid from the sac surrounding the embryo. Cells from the fluid are carefully grown in the laboratory for a few days, treated with a chemical that prevents cell division, and then carefully broken and examined. A karyotype (KAIR-ee-oh-tighp), which is a display of all the chromosomes in a cell nucleus, is then prepared to make certain that the chromosomes of the developing embryo are normal.

In **chorionic villus biopsy** (kor-ee-AHN-ihk VIHL-uhs BIGH-ahp-see), a sample of embryonic cells is removed directly from the membrane surrounding the embryo. Results are obtained more rapidly from this technique than from amniocentesis. However, recent studies have linked limb defects in babies to CVB tests done before the tenth week of pregnancy.

Both amniocentesis and chorionic villus biopsy make it possible to detect Down syndrome and other chromosomal abnormalities. Both techniques are considered safe for the mother and the developing baby when proper precautions are observed. Scientists are now developing procedures that can detect other genetic disorders. Some of these procedures test for biochemical abnormalities in the embryonic cells, whereas others test for the presence of certain DNA sequences that are characteristic of particular defective genes. The rapid advance of these techniques has made it possible to detect more than 100 genetic disorders from embryonic cells.

Ethical Considerations

The emerging ability to identify genetic disorders before birth has already begun to force parents and physicians to face ethical issues that past generations could never have imagined. How should parents react to the news that their child might be born with a serious or fatal genetic disorder? What factors—medical, economic, social, and ethical—should be considered in such cases, and who should make the decision?

Biology has given us great gifts of knowledge and understanding, but science alone cannot tell us what is proper in such difficult cases. Answers to ethical and moral questions do not come from the laboratory, but from the great resources of the human spirit.

11–4 SECTION REVIEW

1. What is Down syndrome?
2. What is amniocentesis? Chorionic villus biopsy?
3. **Critical Thinking—Applying Concepts** How is a karyotype useful in determining genetic disorders?

BACKGROUND INFORMATION
CYSTIC FIBROSIS (CF) AND TAY-SACHS DISEASE (TD)

Cystic fibrosis is an autosomal recessive disease. In the United States, CF is the most common fatal genetic disorder among people of European ancestry. In this disease the cells lining the respiratory and digestive systems produce abnormally large amounts of mucus. The long-term effects of this aberration are devastating. Children with CF constantly battle infections, and they struggle for breath against a respiratory system that seems determined to suffocate them. As a result, few CF patients live into their twenties. A test for the presence of the CF gene may be available in the near future.

Tay-Sachs disease (TD) is another autosomal recessive disease. It is most common in the United States among Jews of European ancestry. TD children suffer progressive destruction of the nervous system, which results in death at a very early age. A test is now available to determine whether an adult is a carrier of the TD gene. Since TD children can be born only to parents who are both carriers, genetic counselors urge people who suspect that they may have the gene to be tested before marriage.

---

2. In amniocentesis, a small amount of fluid from the sac surrounding an embryo is removed, grown in a laboratory for a few days, and then examined for abnormalities. In chorionic villus biopsy, a sample of embryonic cells is removed directly from the sac surrounding the embryo and examined.

3. A karyotype is useful in determining genetic disorders because it displays all of the chromosomes in a cell nucleus. In this way, the chromosomes can be examined for any defects.

Reinforcement/Reteaching
If students are having difficulty answering the Section Review questions, go back to the part of the section that contains the material with which they are having difficulty. Review the material, then provide further assistance.

Closure
Have students simulate a roundtable discussion in which they discuss the diagnosis of genetic disorders and the ethical questions that such diagnosis raises. Students may enjoy playing various roles in the discussion, such as that of a physician, a member of the clergy, a parent of a genetically impaired child, a genetic counselor, and so on. Encourage students to prepare for the discussion by going to the library and reading recent articles or books on the topic.

LABORATORY INVESTIGATION

OBSERVING HUMAN TRAITS

Before the Lab
1. At least one day prior to the investigation, make sure there are enough mirrors available for your class.
2. You may wish to put the data chart on the chalkboard a day ahead of time, in order to motivate student interest in the investigation.

Pre-Lab Discussion
Have students read through the laboratory procedure. Make sure that they know what to look for in each phenotype. You may wish to point out each phenotype (particularly the earlobes and the widow's peak) to the class and answer any questions students may have about their identification.

- **How does this laboratory investigation differ from some of the other investigations you have done?** (Possible answer: It is a survey, rather than an experiment.)
- **What are surveys used for?** (Answers may vary. Possible answers include: finding out certain characteristics of a population; finding out how people think about a particular issue; finding out what people's habits are, such as what programs they watch on TV.)
- **Can you think of a factor that makes a class of high school students a good group for a survey?** (Possible answer: Certain variables are eliminated since everyone is about the same age and lives in the same area.)
- **In what way might a class of high school students not be a good group for this particular survey?** (Possible answer: It is a rather small number of people; the data may not accurately reflect a larger human population.)

LABORATORY INVESTIGATION

OBSERVING HUMAN TRAITS

PROBLEM
What are some patterns of human heredity?

MATERIALS (per student)

sheet of paper   pencil   mirror

PROCEDURE

1. Copy the data table onto the sheet of paper.
2. In the column labeled Trait, circle your phenotype for each trait listed. A description of the dominant allele for each trait follows.
   Tongue rolling (R) is the ability to roll the tongue up at the edges.
   Widow's peak (W) is a hairline that forms a V in the center of the forehead.
   Free earlobes (F) are those that hang below the point of attachment to the head.
   Dimples (D) are indentations on the cheeks.
   Freckles (F) are brownish spots on the skin.
3. Transfer the data from your data table onto the chalkboard, where your teacher has constructed a chart to collect class data.
4. After your classmates have recorded their data on the chalkboard, record the information in the appropriate place in your data table.
5. To determine the percentage of students demonstrating each trait, divide the number of students who have the trait by the total number of students in the class and then multiply this number by 100. Record the data in the appropriate place in your data table.

OBSERVATIONS
1. For each trait, which occurs more frequently: the dominant or recessive allele?
2. Which trait is the most common in your class? The least common?
3. What is the ratio of the percentages for each of the traits?

ANALYSIS AND CONCLUSIONS
1. Do dominant traits occur more often than recessive traits? Explain your answer.
2. What would happen to your results if you were to perform this investigation with five other classes and recorded their data?

Trait		Number of Students Demonstrating Dominant Phenotype	Number of Students Demonstrating Recessive Phenotype	Percentage Demonstrating Dominant Phenotype	Percentage Demonstrating Recessive Phenotype
Dominant	Recessive				
Tongue roller (R)	Nonroller (r)				
Widow's peak (W)	Straight hairline (w)				
Free earlobes (E)	Attached earlobes (e)				
Dimples (D)	No dimples (d)				
Freckles (F)	No freckles (f)				

Skills Development
Students will use the following skills while completing this investigation.
1. Observing
2. Comparing
3. Relating
4. Applying
5. Predicting
6. Hypothesizing

Safety Tips
When handling mirrors, remind students to be careful of the glass.

Teaching Strategy
1. Emphasize to students that no phenotype listed is more or less desirable than another. Discourage any kind of comparisons or observations that might make a student feel self-conscious about a particular

STUDENT STUDY GUIDE

SUMMARIZING THE CONCEPTS

The key concepts in each section of this chapter are listed below to help you review the chapter content. Make sure you understand each concept and its relationship to other concepts and to the theme of this chapter.

11–1 "It Runs in the Family"
- Many human traits are inherited by the action of dominant and recessive genes, although other traits are determined through more complicated gene interactions.
- Each gamete (sperm cell or egg cell) contains 23 chromosomes.
- During fertilization, sperm and egg unite and a zygote is produced that contains 46 chromosomes.
- Of the 46 chromosomes found in the human diploid cell, 2 are the sex chromosomes (X and Y) and the remaining 44 are called autosomes.

11–2 The Inheritance of Human Traits
- Multiple alleles are three or more alleles of the same gene that code for a single trait. ABO and Rh blood groups are determined by multiple alleles.
- Sickle cell anemia is a disorder that causes the normally disk-shaped red blood cells to become sickle-shaped. The gene for normal hemoglobin is codominant with the sickle cell gene.
- Polygenic traits are traits that are controlled by a number of genes.

11–3 Sex-Linked Inheritance
- Nondisjunction is the failure of chromosomes to separate during one of the stages of meiosis.
- Genes that are carried on either the X or the Y chromosome are said to be sex-linked.
- A sex-influenced trait is a trait that is caused by a gene whose expression differs in males and females.

11–4 Diagnosis of Genetic Disorders
- Down syndrome is an example of a condition caused by the nondisjunction of autosomes.
- Many genetic disorders can now be diagnosed before birth because of such genetic screening techniques as amniocentesis and chorionic villus biopsy.

REVIEWING KEY TERMS

Vocabulary terms are important to your understanding of biology. The key terms listed below are those you should be especially familiar with. Review these terms and their meanings. Then use each term in a complete sentence. If you are not sure of a term's meaning, return to the appropriate section and review its definition.

11–1 "It Runs in the Family"	11–2 The Inheritance of Human Traits	11–3 Sex-Linked Inheritance	11–4 Diagnosis of Genetic Disorders
gamete	multiple allele	sex-linked	amniocentesis
zygote	polygenic trait	sex-influenced	chorionic villus biopsy

---

trait, such as the shape of his or her earlobes or the amount of freckles he or she has.
2. Circulate throughout the room to make sure that students are properly identifying each phenotype.
3. Assist any students who are having trouble calculating percentages.

Observations
1. The dominant allele.
2. Answers will vary.
3. Answers will vary.

Analysis and Conclusions
1. Answers will vary, but students may find that the small sample size affects the expected results.
2. The larger the size of the sample, the closer the results will be to what is expected.

Going Further: Enrichment
Students can expand their survey in one of two ways:
1. Have students combine their data with the data of other biology classes who may be doing the same investigation or enlist the cooperation of several teachers in your area of the building to allow your students to conduct the survey in their classes.
2. Have students expand the survey in their own class by adding the following phenotypes: cleft chin—present or not present; dark hair or light hair; eye color—brown or not brown; right-handedness or left-handedness.

CHAPTER REVIEW

CONTENT REVIEW

Multiple Choice

1. b 2. b 3. a 4. a
5. d 6. c 7. d 8. b

True or False

1. F zygote
2. F A
3. F positive
4. T
5. T
6. T
7. F 21
8. T

CONCEPT MASTERY

1. Their bodies contain too many cells; their generation time is too long; they cannot be kept in the laboratory for study; they produce too few offspring for good statistics; they cannot be used in test crosses; and their genetic systems contain nearly 100,000 genes.
2. Type O blood contains little antigen A or antigen B. As a result, type O blood can be given to people with any of the other blood types without causing any reaction in the recipient's immune system. Type O^+ blood, however, should be given only to people who have Rh^+ blood (A^+, B^+, AB^+, O^+).
3. Sickle cell anemia is caused by a change in one of the polypeptides found in hemoglobin, which is the protein that carries oxygen in red blood cells.
4. In the United States, people of African ancestry are the most common carriers of the sickle cell trait. In the rest of the world, sickle cell anemia is found in tropical regions of Africa and Asia. About 10 percent of Americans of African ancestry and as many as 40 percent of the population in some parts of Africa carry the trait.
5. The sickle cell trait probably developed in many people throughout the world as a single mutation. That mutation conferred an advantage wherever malaria was common, and thus it was favored by natural selection. Sickle cell anemia has persisted wherever it has helped carriers survive malaria.
6. Colorblindness is caused by a recessive allele that is linked to the X chromosome. In males, a single recessive allele (X^cY) can cause colorblindness. In females, two recessive genes must be inherited (X^cX^c).
7. There have been no reported incidences of babies being born without an X chromosome. Embryos without an X chromosome do not seem to develop properly because the

---

CHAPTER REVIEW

CONTENT REVIEW

Multiple Choice

Choose the letter of the answer that best completes each statement.

1. An example of a trait that is determined by multiple alleles is
   a. Huntington disease.
   b. ABO blood groups.
   c. Down syndrome.
   d. hemophilia.
2. Three brothers have blood types A, B, and O. What are the chances that the parents of these three will produce a fourth child whose blood type is AB?
   a. 0 percent
   b. 25 percent
   c. 50 percent
   d. 100 percent
3. A disorder that results from a change in an amino acid in the hemoglobin molecule is
   a. sickle cell anemia.
   b. hemophilia.
   c. Down syndrome.
   d. Turner syndrome.
4. Which is an example of a polygenic trait?
   a. skin color
   b. hemophilia
   c. ABO blood groups
   d. Huntington disease
5. Which parental pair could produce a colorblind female?
   a. homozygous normal vision mother and colorblind father
   b. colorblind mother and normal vision father
   c. heterozygous normal vision mother and normal vision father
   d. heterozygous normal vision mother and colorblind father
6. Which is an example of a sex-influenced trait?
   a. Down syndrome
   b. colorblindness
   c. male pattern baldness
   d. skin color
7. What is the total number of chromosomes in a diploid cell of a person with Down syndrome?
   a. 22
   b. 23
   c. 44
   d. 47
8. Which of the following techniques is directly associated with the diagnoses of certain genetic disorders by the examination of chromosomes?
   a. blood typing
   b. karyotyping
   c. chemical screening
   d. microsurgery

True or False

Determine whether each statement is true or false. If it is true, write "true." If it is false, change the underlined word or words to make the statement true.

1. During fertilization, sperm and egg unite and produce a <u>gamete</u>.
2. A person with type A blood has red blood cells that contain antigen <u>B</u>.
3. A genotype Rh^+Rh^- will produce the Rh <u>negative</u> phenotype.
4. People who are heterozygous for sickle cell anemia are partially resistant to malaria.
5. Based on the pattern of inheritance of sex-linked traits, if a male has hemophilia, he has <u>one gene</u> for this trait on the sex chromosomes in each of his diploid cells.
6. The dominant allele for normal color vision is represented as <u>X^c</u>.
7. In Down syndrome, there is an extra copy of chromosome <u>16</u>.
8. Chorionic villus <u>biopsy</u> is a genetic screening technique in which a small amount of fluid is removed from the sac surrounding the developing embryo.

CONCEPT MASTERY

Use your understanding of the concepts developed in the chapter to answer each of the following in a brief paragraph.

1. Why are humans not considered good organisms for genetic studies?
2. Explain why type O blood is sometimes referred to as the universal donor. Is this true in the case of O$^+$ blood?
3. What is the cause of sickle cell anemia?
4. Why is it easy to believe that the sickle cell trait may have arisen independently in several places throughout the world?
5. What is the evolutionary connection between the incidences of sickle cell anemia and the incidences of malaria?
6. Why are there more colorblind males than there are colorblind females?
7. What evidence suggests that an X chromosome is essential for normal development? Is the same true of the Y chromosome?
8. The genes for red-green colorblindness and hemophilia are located on the X chromosome. How might you determine whether these genes are closely linked?
9. How would you determine whether a human gene was sex-linked?

CRITICAL AND CREATIVE THINKING

Discuss each of the following in a brief paragraph.

1. **Applying concepts** Cystic fibrosis (CF) is caused by an autosomal recessive gene and is a fatal disorder that affects people of European ancestry. CF is characterized by a production of thick mucus that clogs the lungs and parts of the digestive system, making it difficult to breathe or eat. Suppose two people without a history of CF in their immediate families have a child with CF. They would like to have another child, but they are concerned about whether the second child will have CF. Can you help them predict the chances of having a second child with CF?
2. **Making diagrams** Polydactyly is a human characteristic in which a person has extra digits (fingers or toes) on his or her hands or feet. The trait for polydactyly is dominant over the trait for five digits on the hands and/or feet. Suppose a man who is heterozygous for this trait marries a woman with the normal number of digits. What are the possible genotypes and phenotypes of their offspring? Draw a Punnett square showing the possible results of this cross.
3. **Relating concepts** Nine of ten children in a family are right-handed. Right-handedness is a dominant trait. Assume that this trait is controlled by a pair of genes.
   a. What is the genotype of a left-handed child?
   b. What are the possible genotypes of the right-handed children?
   c. What are the possible genotypes of the parents?
   d. Could these parents have produced only right-handed children? Explain.
   e. If these parents had produced only one child, would this child have been left- or right-handed? Explain your answer.
4. **Making inferences** Discuss the impact of genetic screening techniques on society.
5. **Using the writing process** Your career ambition is to be a science reporter. You are sent by your school newspaper to interview a physician who works with human genetic disorders. Prepare a script of the questions you would like answered.

245

CRITICAL AND CREATIVE THINKING

1. Because cystic fibrosis (CF) is caused by an autosomal recessive gene, both recessive genes must be present in order for an individual to have CF. In this case, both parents are heterozygous for the disorder. The chances that they will have a second child with CF is 25%. This is the same for the first child or, for that matter, any child.
2. Check students' Punnett squares. Genotypes: *Pp* and *pp*; Phenotypes: polydactyly and normal.
3. a. homozygous recessive.
   b. homozygous dominant or heterozygous.
   c. Both are heterozygous.
   d. No. They have a 75% chance of producing right-handed children and a 25% chance of producing left-handed children.
   e. Probably right-handed because the chances of producing a child with this trait are three times greater (75 to 25) than producing a left-handed child.
4. Accept all logical, well thought-out responses.
5. Some of the questions that students may wish to pose to the physician include: What are the most commonly occurring genetic disorders? How are they inherited? What effects do they have on the body? Can they be detected by genetic screening methods? Can they be treated? If so, how?

X chromosome contains a number of genes that are vital for normal development. The Y chromosome contains a gene that switches on the male pattern of growth during embryological development. If this gene is absent, the embryo follows a female pattern of growth.

8. To determine whether the genes for red-green colorblindness and hemophilia, which are found on the X chromosome, are closely linked you would have to study family trees (pedigrees) in which one of the two traits occurred. Then you would have to find out if the other trait occurred also. If both traits occurred in the family trees, then the traits are closely linked.

9. You would have to find out if the genes are expressed more commonly in males than in females.

CHAPTER 12 Genetic Engineering

Section	Laboratory Investigations and Activities
12–1 Modifying the Living World pages 247–250 THEMES: Patterns of Change, Unity and Diversity	**Teacher Edition** DEMONSTRATION: Genetically Altered Examples, p. 246D
12–2 Genetic Engineering: Technology and Heredity pages 250–256 THEMES: Patterns of Change, Scale and Structure	**Laboratory Manual** Cloning, p. 163 **Teaching Resources** ACTIVITY: DNA Recombination, p. 3
12–3 The New Human Genetics pages 256–259 THEMES: Patterns of Change, Systems and Interactions	**Student Edition** LABORATORY INVESTIGATION: Simulating DNA Fingerprints, p. 260 **Teacher Edition** DEMONSTRATION: Call In an Expert, p. 246D **Teaching Resources** ACTIVITY: Chromosome Mapping, p. 5
Chapter Review pages 261–263	

Teacher Resources

Books

Antebi, Elizabeth, and David Fishlock. *Biotechnology Strategies for Life*. MIT Press, 1986.

Bains, Williams. *Genetic Engineering for Almost Everyone*. Penguin, 1988.

Hall, Stephen S. *Invisible Frontiers: The Race to Synthesize the First Human Gene*. Atlantic Monthly Press, 1987.

Hyde, Margaret O., and Lawrence E. Hyde. *Cloning and the New Genetics*. Enslow, 1984.

Kirby, Lorne T. *DNA Fingerprinting: An Introduction*. Oxford University Press, 1990.

McKinnell, Robert G. *Cloning: Of Frogs, Mice, and Other Animals*, 2nd ed. University of Minnesota Press, 1984.

Milunsky, Aubrey, ed. *Choices Not Chances: Controlling Your Genetic Heritage*. Little, Brown, 1989.

Nichols, Eve K. *Human Gene Therapy*. Harvard University Press, 1988.

Nossal, G.J.V. *Reshaping Life*. Cambridge University Press, 1985.

CHAPTER PLANNING GUIDE

Other Activities	Media and Technology
Study Guide Section 12–1, p. 117	
Biotechnology Workbook A Gene for All Seasons, p. 49 **Study Guide** Section 12–2, p. 120	**Biology Transparencies** 15: Cloning
Teaching Resources VOCABULARY REVIEW: Word Game, p. 1 **Study Guide** Section 12–3, p. 122	
Teaching Resources Chapter Test A, p. 11 Chapter Test B, p. 15 Performance-Based Assessment, Unit 3	**Computer Test Bank** Chapter Test, p. 123

Audiovisuals

DNA: Laboratory of Life. Film or video. National Geographic Society, Educational Services.

New Forms of Life: Gene Splicing and Genetic Engineering. Filmstrip with cassette. Knowledge Unlimited.

Of the Earth: Agriculture and the New Biology. Video. Industrial Biotechnology Association.

Improving Plants and Animals. Film or video. Coronet.

Genetic Engineering and Protein Synthesis. Film or video. Benchmark Films, Inc.

CHAPTER 12 Genetic Engineering

CHAPTER OVERVIEW

Humans seem to be constantly searching for ways to alter and improve the world in which we live. One way of accomplishing that goal would seem to be not only by changing the physical world around us but the development of "better" living organisms.

This is certainly not a new idea. For thousands of years humans have prized the biggest and best farm animals or the most disease-resistant varieties of plants and animals. Those special organisms have been used as breeding stock in the hope of perpetuating their outstanding qualities. This obvious way of improving species is called selective breeding.

Once a desirable trait or organism has been produced, it must be maintained and passed on in future generations. It cannot be denied that this process is useful, but it is also slow and chancy. Sometimes humans, in their impatience, wish to speed up the process or seek specific results. In the last two decades, a new way to alter the genetics of organisms has been born. This new science is called genetic engineering.

This new field of science is advancing rapidly and is directly related to the development of the technology that supports it. Scientists have been successful in manipulating the DNA in bacterial cells and many plants and animals.

But what of genetic engineering in humans? There is no doubt that there are many benefits to be derived from DNA manipulation in humans. There are diseases that might be cured, treated, or even prevented. Research has already given us an effective tool to be used in solving crimes. The logical conclusion is to proceed as quickly as possible with genetic engineering research—but where does it stop?

There are some serious ethical issues that must be considered. Experimentation on humans, for any purpose, is one of these issues. Beyond the prospect of curing disease, there is the potential for altering—even designing—the human body. Cloning of genetically identical copies is another issue. Is this the purpose of science? Who should be involved in making these and other decisions?

12-1 MODIFYING THE LIVING WORLD

Section Focus 12-1

The purpose of this section is to introduce and discuss the traditional methods of improving the productivity of domesticated species. The oldest and most obvious way to improve a species is to use selective breeding techniques. Basically, this involves identifying those organisms with particularly desirable traits and using those same organisms as parents of the next generation. Through selective breeding, better crops can be produced, particular flower colors developed, or higher milk yield attained. Perhaps the world's best-known selective breeder was Luther Burbank.

Once selective breeding has produced the desired result, there must be a way of maintaining that trait in the population. Inbreeding is the obvious answer. However, there are some problems with this process. The more inbred a line becomes, the greater the potential of reproducing harmful recessive traits.

Another type of selective breeding process that yields organisms that are often very hardy is that of hybridization. Although not as direct as inbreeding, hybridization has proven to be a valuable method of improving food crops, particularly corn.

Selective breeding is confined to traits that already appear in a population. If a favorable mutation occurs, breeders can take advantage of it, but the process is slow and haphazard. Mutagens, which include radiation and certain chemicals, cause mutations to occur. Many mutations are harmful or even lethal, but the greater the number of induced mutations, the greater the chances of desirable traits appearing. This technique has been especially useful with bacteria.

Performance Objectives 12-1

1. Define selective breeding.
2. Discuss the advantages and disadvantages of inbreeding.
3. Explain the process of hybridization and the phenomenon known as hybrid vigor.
4. Define the term *mutagen*.
5. Explain mutagenesis in bacteria.

Science Terms 12-1

selective breeding   p. 247
inbreeding   p. 248
hybridization   p. 248
mutagen   p. 249

12-2 GENETIC ENGINEERING: TECHNOLOGY AND HEREDITY

Section Focus 12-2

The purpose of this section is to explain some of the techniques and applications of genetic engineering. All of the selection processes discussed in Section 12-1 are actually indirect ways of altering the DNA of particular organisms. Within the last two decades, molecular biologists have developed new ways of directly changing an organism's DNA. This new set of manipulations is called genetic engineering.

The science of genetic engineering could not—and cannot—develop without the technological advances that support it. The discovery of restriction enzymes allows scientists to "cut" DNA at specific points along the strand. There are some 75 of these proteins known. Each recognizes and cuts DNA at a particular sequence with an amazing degree of accuracy.

The DNA fragments cannot function alone but must become a part of the genetic material of a living cell. One method used to combine DNA fragments with cell DNA involves the use of plasmids. Using restriction enzymes on both the fragment and the plasmid, the two pieces of DNA can be combined to form a new plasmid. These molecules of combined DNA are called recombinant DNA. They are also known as chimeras, after the ancient

CHAPTER PREVIEW

mythological beast with the characteristics of many animals.

The technical problem that remains is that of insertion back into a cell. This process is easiest with bacterial cells, but several techniques have been successful with other organisms. The specific process used in bacteria is sometimes called DNA cloning. The techniques used for sequencing DNA, including radioactive labeling and electrophoresis, are briefly explained in this section.

The remainder of this section is devoted to a discussion of the applications of genetic engineering. Genetically engineered bacteria can produce large quantities of human growth hormone, insulin, and interferon. All of these human proteins can be produced accurately and inexpensively by bacteria containing recombinant DNA.

Plants can now be grown from cells containing recombinant DNA. The goals for genetic engineering in plants include the production of plants that manufacture their own natural insecticides, produce their own nitrogen nutrients, or have a higher food value. Genetic engineering in animals has taken a number of interesting directions, including the use of transgenic farm animals to increase meat and milk yields.

Performance Objectives 12-2

1. Compare selective breeding and genetic engineering.
2. Name and discuss four techniques used in genetic engineering.
3. List three medical advances that are the result of genetic engineering in bacteria.
4. List several goals of genetic engineering in plants and animals.

Science Terms 12-2

genetic engineering   p. 250
restriction enzyme   p. 251
plasmid   p. 252
recombinant DNA   p. 252
clone   p. 253
transgenic   p. 253

12-3 THE NEW HUMAN GENETICS

Section Focus 12-3

The purpose of this section is to introduce students to the possibilities and problems associated with genetic engineering in humans. There are several fascinating possibilities when one begins to look at this topic. Early detection and prevention of genetic diseases is one that immediately comes to mind.

The decoding of the entire human genome—all the genes possessed by humans—is an amazing subject to contemplate. The amount of information and the value of that knowledge are astounding.

Another benefit of genetic engineering, surprisingly enough, adds a powerful tool to the arsenal of law enforcement agencies. DNA fingerprinting has been successfully used to convict a number of suspects and, conversely, to prove a suspect's innocence.

The remainder of this section is devoted to issues involving ethical questions of human genetic engineering. There are several questions posed that must be answered before scientists go much further with human experimentation.

Performance Objectives 12-3

1. Describe several applications of analyzing and sequencing human DNA.
2. Explain the problems involved in sequencing the entire human genome.
3. Discuss some of the ethical issues involved in human genetic engineering.

Science Terms 12-3

genome   p. 256
DNA fingerprinting   p. 257

DISCOVERY LEARNING

TEACHER DEMONSTRATIONS
Modeling

Genetically Altered Examples

The following demonstration may be used as an introduction to Chapter 12.

Show the students examples of genetically altered organisms. There are many possibilities, but seedless fruits might be the easiest to obtain. If you live in a rural area, you should have many examples available. Allow students some time to discuss what the genetic improvements are and hypothesize ways that the alteration might have occurred. You have already discussed mutations (Chapter 10) so that possibility is sure to arise. Continue along that line through an outline of the selective breeding process. Some students may bring in other ideas. The aim is to get students excited about this important topic.

Call In an Expert

This demonstration may be used with Section 12-3 on genetic engineering.

Have an expert speak to your class about some of the advances being made in genetic engineering. If you are fortunate to have a willing genetics expert, take advantage of this community resource. A nearby university is the best place to look for such an expert.

If you cannot obtain an expert, you must become the expert. Read everything you can find on the topic, put on your lab coat, give a short talk on the subject, and allow time for lots of questions and discussion.

Even if the expert cannot answer all the questions—who can?—it will be fun, interesting, and different for the students.

CHAPTER 12

Genetic Engineering

GUIDED ENQUIRY

Pose the following questions to students and have them record their responses. Point out that they will gain a better understanding of the key concepts if they read the chapter with these basic questions in mind. Upon completion of the chapter, pose the questions again. Ask students to compare their initial responses with those they have developed after reading the chapter.
- What human needs and problems might be affected by the ability to alter other organisms?
- How could artificially induced mutations increase the effectiveness of a selective breeding program?
- What problems could you imagine would be involved in genetic engineering?
- What moral or ethical questions would you pose about a genetic engineering project involving humans?

INTRODUCING CHAPTER 12

Using the Visuals
Have students examine the picture of the mythological chimera and read the introduction to the chapter. Have students explain in their own words the similarities between the ancient chimera and a DNA chimera.
- **Why would scientists produce a plant that glows in the dark?** (No real practical application, but it was a spectacular example of animal genes being placed in plant organisms.)
- **How could joining DNA from two different kinds of organisms enable scientists to design new organisms?** (DNA is the molecule of heredity; therefore, changing DNA would produce a new type of organism. By carefully selecting pieces of DNA to join together into a chimera, scientists may be able to create organisms with new and desirable combinations of characteristics—combinations impossible to obtain through breeding techniques.)
- **Can you think of any practical reason for placing human genes in mice?** (Answers will vary, but medical research on AIDS or cancer will probably be the most common answer.)

CHAPTER 12

Genetic Engineering

In mythology, the chimera is a fire-breathing monster that is part lion, part goat, and part serpent. In genetics, the word chimera refers to a piece of DNA (or even an entire organism) that is made up of DNA from different sources.

In Greek mythology, the chimera was a frightening beast with the head of a lion, the body of a goat, and the tail of a serpent. The idea of combining the characteristics of many animals into one, however, did not originate with the Greeks. The Sphinx of Egypt and the winged lions of Assyria are just two examples of composite monsters that predate the Greek myths. Although we do not take these stories seriously today, it is ironic to note that biology has progressed to the point where these old myths have been dusted off. Today the word chimera has a different meaning: It refers to fragments of DNA from two organisms that have been joined to make a single DNA molecule. Although these DNA chimeras have not yet merged the serpent, the goat, and the lion, they have carried human genes into mice, produced plants that glow in the dark, and made it possible to design new organisms to suit human needs.

246

TEACHING STRATEGY 12–1

Focus/Motivation
Start a bulletin board or notebook of articles on genetically altered organisms and news items on current advances. Take some time to talk about the articles

GUIDE FOR READING

After you read the following sections, you will be able to

12–1 Modifying the Living World
- Describe breeding strategies that have been used to modify living things.
- Explain how mutations in organisms can be useful to humans.

12–2 Genetic Engineering: Technology and Heredity
- Describe how DNA is isolated, cut, spliced, and handled.
- Explain how DNA is used to transform cells and organisms.

12–3 The New Human Genetics
- Describe some possible applications of analyzing and sequencing human DNA.
- Discuss ethical issues involving new genetic techniques.

Journal Activity

YOU AND YOUR WORLD

Imagine that you are a molecular biologist and have all the latest technological equipment at your disposal. How would you use your knowledge and skill to make the world a better place to live? Why did you choose to make this particular improvement? In your journal, answer these questions in the form of an illustrated essay or short story.

Figure 12–1 The small horse is a full-grown miniature horse. Miniature horses, which are 86 centimeters high at the shoulder (or less), are the product of selective breeding. First produced in the mid-1700s, they were originally intended to pull ore carts in the low, narrow passages of mines. However, they were soon adopted as exotic pets by wealthy families.

12–1 Modifying the Living World

Guide For Reading

■ What do the terms selective breeding, inbreeding, and hybridization mean?
■ How can mutations be useful to humans?

We humans are rarely, if ever, satisfied with the world around us. We are always trying to improve it. Humans are not content with merely altering the course of a river, building cities, or sailing the seas. For thousands of years we've been trying to do something equally remarkable: to make "better" living things.

Breeding Strategies

Farmers and ranchers throughout the world have long tried to improve the organisms with which they work. **By selecting the most productive plants or animals to produce the next generation, people have found that the productivity of a domesticated species can gradually be increased.** Such an increase in productivity results from using breeding strategies such as selective breeding and two special forms of selective breeding: inbreeding and hybridization.

SELECTIVE BREEDING The oldest and most obvious way of improving a species is by **selective breeding**, or selecting a few individuals to serve as parents for the next generation. By crossing only those individual plants or animals that have a

247

COOPERATIVE LEARNING

Divide the class into five groups. Each group is to assume that they are the editorial staff of a national daily newspaper. Posing the question, "Should alterations of the human genome be allowed and encouraged, and, if so, who should control such alterations?" have groups produce an editorial page for their newspaper. When completed, the page should contain the following components:

- Background information (facts behind the issue)
- Editorial cartoon (pro or con)
- Guest editorial in support of genetic engineering on humans
- Guest editorial opposed to genetic engineering on humans
- Hypothetical quotes from man-on-the-street surveys

Authors of guest editorials and man-on-the-street survey respondents should be identified. It might be helpful to show students samples of actual editorial pages of local and national newspapers as models. If time does not permit having each group complete the entire page, you might randomly assign each of the five components to one group to produce a class page.

Journal Activity
YOU AND YOUR WORLD

You may want to use the Journal Activity as a springboard for class discussion. When students have completed the chapter, have them reexamine their journal entry. Given what they now know about genetic engineering, would they choose a different improvement? Why or why not?

Students should be instructed to keep their Journal Activity in their portfolio.

you have collected. Encourage students to participate in this information search. (This could be a regular class assignment or one done for extra credit.) Throughout the study of Chapter 12, take time to talk about real-life events related to the chapter. Each time you have a discussion, emphasize the effects of the advances on everyday life.

Content Development
Using the Visuals Have students examine Figure 12–1 and read the caption.

• **Do you suppose that the miniature horse would ever have developed naturally?** (Probably not one that small because it would not have been able to compete in a natural environment.)
• **What benefits and/or disadvantages are there to human intervention in plant and animal breeding?** (Answers will vary but might include better milk, meat, and vegetable crops as benefits. Some of the disadvantages would include loss of diversity and the more frequent appearance of deleterious recessive traits.)

247

ESL STRATEGY

Verify that students understand the term "breeding" and its synonym "crossing." Explain that *selective* breeding involves *choosing*, or *selecting*, the individuals that will produce the next generation. *In*breeding involves crossing individuals with*in* a group in which all members have basically the same genes. *Hybrid*ization involves the production of a *hybrid*, an organism in which traits from genetically dissimilar parents are combined.

MULTICULTURAL STRATEGY

Students may be interested in finding out how marriage customs in different cultures encourage or discourage inbreeding. If you have students from parts of the world in which inbreeding is commonplace, encourage them to relate their culture's attitude on this topic. Their viewpoints may serve to enlighten those in the class whose cultures forbid or discourage inbreeding.

Figure 12-2 *Selective breeding has produced domestic animals such as pigs (top) that are very different from their wild ancestors (bottom).*

desired characteristic, the breeder can improve the yield of a crop plant, increase the milk production of cattle, or develop flowers of a particular color. Most present-day crop plants were first developed by selective breeding.

Luther Burbank of California (1849–1926) was perhaps the world's foremost selective breeder. Burbank produced more than 250 new varieties of fruit by selective breeding, but his work also extended to other plants, including the daisy and the famous Burbank Potato.

INBREEDING Once a breeder has successfully produced an organism with a useful set of characteristics, the next concern is to maintain a stock of similar organisms. The most direct method is **inbreeding**, or crossing individuals with similar characteristics so that those characteristics will appear in their offspring. Often the individuals crossed in inbreeding are closely related. The many varieties of purebred dogs—including German shepherds, toy poodles, and Great Danes—are maintained by inbreeding. Puppies are considered purebreds only if both their parents are registered members of the same breed.

Although inbreeding is useful in retaining a certain set of characteristics, it does have its risks. Because most of the members of an inbred line are genetically similar, the chances that recessive genetic defects will show up after repeated inbreeding become much greater. Problems in many dog breeds, including deformities in the joints and progressive blindness in German shepherds and golden retrievers, have resulted from repeated inbreeding.

HYBRIDIZATION One of the most useful of the breeder's techniques is **hybridization**, a cross between dissimilar individuals. Hybridization often involves crossing members of different (but related) species. Hybrids, the individuals produced by such crosses, are often hardier than either of the parents. This phenomenon is known as hybrid vigor. All breeds of corn that are grown commercially are hybrids, and every year corn breeders produce new hybrids that combine such basic characteristics as disease resistance, yield per acre, and nutritional value. Modern hybrid corn produces as much as ten times the crop per acre of older varieties of corn.

Mutations: Producing New Kinds of Organisms

As useful as selective breeding is, it is confined to characteristics that already exist in the population. But as you may recall from earlier chapters, mutations, which are inheritable changes in DNA, can sometimes produce organisms with new characteristics. If the new characteristics are desirable, breeders can use selective breeding to produce an entire population possessing these characteristics.

12–1 (continued)

Content Development

The topic of inbreeding is one that many students find interesting. As you begin to discuss it, many students will relate to the negative aspects, especially those who have pets that have been adversely affected. Emphasize that it is a natural occurrence to get both positive and negative results. This phenomenon is quite common in many areas of scientific experimentation. As "thinking" individuals, they will often find it necessary to weigh both the benefits and the disadvantages in life situations. That is precisely what breeders must do in developing traits by intervention.

Content Development

Mutations were discussed in Chapter 10. Remind students what a mutation is and how it relates to a breeding program. Emphasize that mutations must occur for different traits to appear in a population. Breeders simply take advantage of the desirable ones that do happen. They might, however, increase the chances of mutations occurring with mutagens. Note that

Figure 12–3 Inbreeding has resulted in an increased susceptibility to diseases and deformities in cheetahs (left), dogs, and other organisms. Fortunately for this golden retriever puppy (right), which was bred by Ken Miller, its ancestors have been certified to be free of certain inherited defects. The puppy can be expected to lead a happy—and healthy—life.

Of course, a breeder may not want to wait for a beneficial mutation to appear naturally. In such a case, a breeder may decide to artificially increase the chances of mutations occurring in a group of organisms. This can be done with agents or substances known as **mutagens**. Mutagens, which include radiation and chemicals, cause mutations. Many mutations are harmful; but with luck and perseverance, a few mutants (individuals with mutations) with desirable characteristics not found in the original population may be produced.

Mutagenesis, or using mutagens to increase the mutation rate, is particularly useful with bacteria. Their small size enables millions of organisms to undergo mutagenesis at the same time, thereby increasing the chances of producing a useful mutant. Using this technique, scientists have been able to develop hundreds of useful bacterial strains. It has even been possible to produce bacteria that can digest oil and are thus useful in cleaning up oil spills.

Figure 12–4 Seedless oranges and hairless mice are the result of mutations.

12–1 SECTION REVIEW

1. What is selective breeding? Inbreeding? Hybridization?
2. What are some advantages of breeding strategies?
3. Why might breeders want to cause mutations in plants?
4. **Connection—Social Studies** Many human societies have laws preventing marriage between close blood relatives. How are these restrictions related to inbreeding?

plant and animal mutations considered useful by humans may not be favorable from the organism's point of view. For example, seedless fruits generally require human assistance to reproduce, and temperature-sensitive fruit flies, immune-deficient mice, and nutrient-requiring (auxotrophic) bacteria cannot survive for long outside the laboratory.

SECTION REVIEW 12–1

1. Selective breeding: allowing only those organisms with desired characteristics to reproduce. Inbreeding: crossing similar individuals that are often closely related. Hybridization: crossing dissimilar individuals that may belong to different varieties or species.
2. They increase the frequency of a desired characteristic in a population or produce new combinations of characteristics.
3. Because some favorable mutations may be produced.
4. Restricting marriages between closely related individuals discourages inbreeding and helps prevent some of the genetic problems associated with inbreeding.

SCIENCE, TECHNOLOGY, AND SOCIETY CONNECTION

DNA, Amber, Insects, and Dinosaurs

If you've read Michael Crichton's novel *Jurassic Park* or seen the movie of the same name, you know that the story revolves around a scientific attempt to reproduce actual dinosaurs that roamed the Earth 100 million years ago. The fictional scientists accomplish this feat by recovering bits of dinosaur DNA from the gemstone amber.

Amber, which is produced from sticky tree sap that has hardened over time, occasionally contains the remains of insects that were trapped when the amber was still sap. In *Jurassic Park*, the scientists obtain DNA from dinosaur blood within the digestive tracts of mosquitoes. The mosquitoes and their last meal, so the story goes, have been perfectly preserved in amber for many millions of years.

Is this part of the story realistic? Maybe. David Grimaldi at the American Museum of Natural History recently recovered DNA samples from insects trapped in 40-million-year-old amber. There's only a tiny amount of fragmented DNA in ancient insects, but a technique called polymerase chain reaction (PCR) allows researchers to make enough duplicates of pieces of DNA to have enough material to study. With the help of PCR, Grimaldi and his associates were able to study a few genes and compare them to those in modern relatives of the ancient insects.

These researchers are a long way from being able to "clone" these insects, let alone dinosaurs. However, their work makes it clear that the study of ancient DNA is possible.

Guide For Reading

- What is genetic engineering? How does it affect DNA?
- How do the various techniques of genetic engineering work?
- What are some applications of genetic engineering?

12–2 Genetic Engineering: Technology and Heredity

As you have just seen, there is nothing new about using scientific techniques to produce new varieties of plants and animals. In the past, however, these efforts usually involved selecting organisms with the most favorable combinations of genes and using breeding techniques to produce new combinations. Today it is possible to go further—to directly change the genetic material of living organisms and, in effect, design organisms by manipulating their DNA.

In the last two decades molecular biologists have developed a powerful new set of techniques that affect DNA directly. For the first time biologists can engineer a set of genetic changes directly into an organism's DNA. Appropriately enough, this new form of manipulation is called **genetic engineering**.

The Techniques of Genetic Engineering

Genetic engineering could not have come about without the development of a technology to support the process. The

250

SCIENCE, TECHNOLOGY, AND SOCIETY CONNECTION

DNA, Amber, Insects, and Dinosaurs

You will probably find that most students have seen the movie *Jurassic Park*, which is based on the novel by Michael Crichton.

Have students ponder for a moment the difficulties involved in extracting DNA from the fossilized remains of a tiny insect.

- **How much DNA would you expect to obtain from a fossil insect?** (Very little.)

Point out that PCR enables scientists to make huge numbers of copies of DNA. Because the bacteria used to copy the DNA strand reproduce so quickly, in a few hours the amount of DNA present in a sample is increased about one million times. Students may find it helpful to think of PCR as an amplifier that transforms an inaudible whisper into a clear, unmistakable shout.

Students might be interested to learn that PCR has been used to identify persons guilty of committing a crime. PCR can be used to amplify the DNA in blood, skin, or semen—even if the evidence is limited to a tiny smear of biological material. This DNA can then be used to identify a criminal through DNA fingerprinting.

Remind students that a theme park with real dinosaurs is still nothing more than a flight of fancy. Only DNA from the fossilized insects has been removed and copied. No DNA from the insects' meals has been recovered, and no dinosaur DNA has been found.

TEACHING STRATEGY 12–2

Focus/Motivation

Have students think of a situation when a friend had a "crush" on another person.

- **Did you just watch and wait to see what would happen?** (Sometimes yes; sometimes no.)
- **Did you ever give your friend advice about what to do?** (Yes or no.)
- **Did you get involved by talking to the other person yourself or doing something else to help the romance along?** (Answers will vary.)

Make an analogy between the three options presented in the questions above and natural breeding, selective breeding, and genetic engineering. Make the point that genetic engineering is direct manipulation of DNA.

Content Development

You may wish to remind students of the characteristics of enzymes, particularly their specificity. Point out that restriction enzymes are also known as endonucleases.

- **Why is the term endonuclease**

Wright brothers could not have built an airplane without the existence of a small gasoline engine and lightweight metal tubing. Such equipment was actually developed for the automobile and the bicycle; but once available, it was adapted to the needs of a flying machine. The story of genetic engineering is much the same.

Imagine for a moment the tools you would need to take a gene from one organism and place it into another. First, you would need a way to carefully cut the DNA containing the gene away from the genes surrounding it. Second, you would have to find a way to combine that gene with a piece of DNA from the recipient organism; that is, the organism that will receive the DNA. Third, you would have to insert the combined DNA into the new organism. Finally, it would be useful to have a way to read the sequences of nucleotide bases in the gene in order to analyze the genes that you are manipulating. Each of these tools is now available, making it possible to insert a gene from one organism into another.

RESTRICTION ENZYMES Genes can be cut at specific DNA sequences by proteins known as **restriction enzymes.** More than 75 different kinds of restriction enzymes are known, and each one "recognizes" and cuts DNA at a particular sequence. The cutting sites of three typical restriction enzymes are shown in Figure 12–6. Each one of the restriction enzymes shown "recognizes" a site of four to six nucleotide pairs and then makes a cut across both strands of DNA.

The accuracy of these enzymes is breathtaking. They will not cut any sequence other than the one they recognize, even if five out of six bases are identical to their recognition site. Restriction enzymes make it possible to cut DNA into fragments that can be isolated, separated, and analyzed.

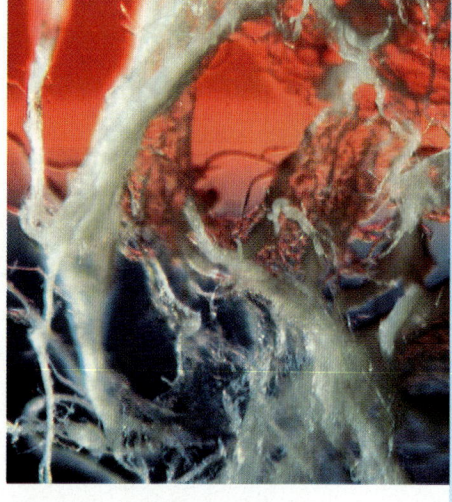

Figure 12–5 These strands of artificial DNA were synthesized from DNA extracted from animal cells. The techniques of genetic engineering allow scientists to manipulate and change DNA directly.

Figure 12–6 Different restriction enzymes "recognize" different sequences of bases on DNA molecules. Each restriction enzyme cuts DNA at the sites at which its recognition sequence occurs.

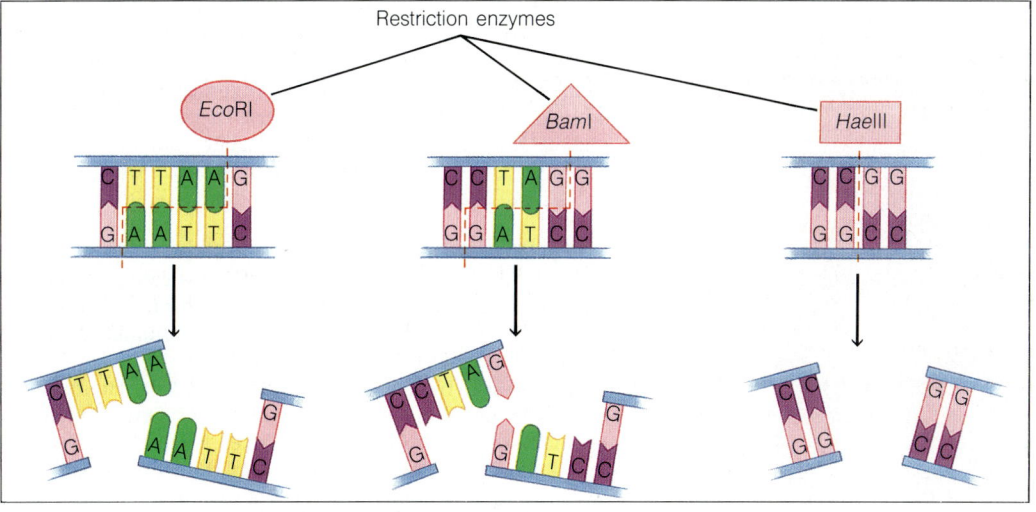

HISTORICAL NOTE
RESTRICTION ENZYMES

The discovery of restriction enzymes is an excellent lesson in the importance of basic research. In the early 1970s several scientists observed that some bacteria seemed to protect themselves against viral infection in a strange way. These bacteria seemed to have special enzymes that could recognize special sequences of bases in viral DNA. These enzymes would then cut the DNA at those places. The result was that the infecting viral DNA was cut to pieces and destroyed. The enzymes were a tremendously effective means of defense against viruses. The enzymes did not affect the bacterium's own DNA because the restriction sequences did not occur in the bacterial DNA.

As work on restriction enzymes continued, biologists realized that these enzymes might be useful in other ways. Today, it is clear that the entire revolution in molecular biology would not have been possible without restriction enzymes, a chance discovery of basic research on a problem with no apparent practical significance.

---

appropriate? (The prefix *endo-* means inside, the root word *-nucle-* refers to nucleic acids, and the suffix *-ase* is used to indicate an enzyme. Thus an endonuclease is an enzyme that recognizes specific nucleotide sequences inside a DNA molecule and cuts the DNA at that site.)

Explain that restriction enzymes got their name because they restrict the multiplication of bacteriophages. (You may wish to briefly review the material on bacteriophages in Chapter 7, pages 139–140.)

• **How do you think restriction enzymes prevent viral infections in bacteria?** (The enzymes "chop up" the DNA of bacteriophages, destroying it.)

**Reinforcement/Reteaching
Using the Visuals** Visual learners will find it helpful to examine Figure 12–6 as you go over the material on restriction enzymes. You may find that DNA models are also useful props when reviewing this material.

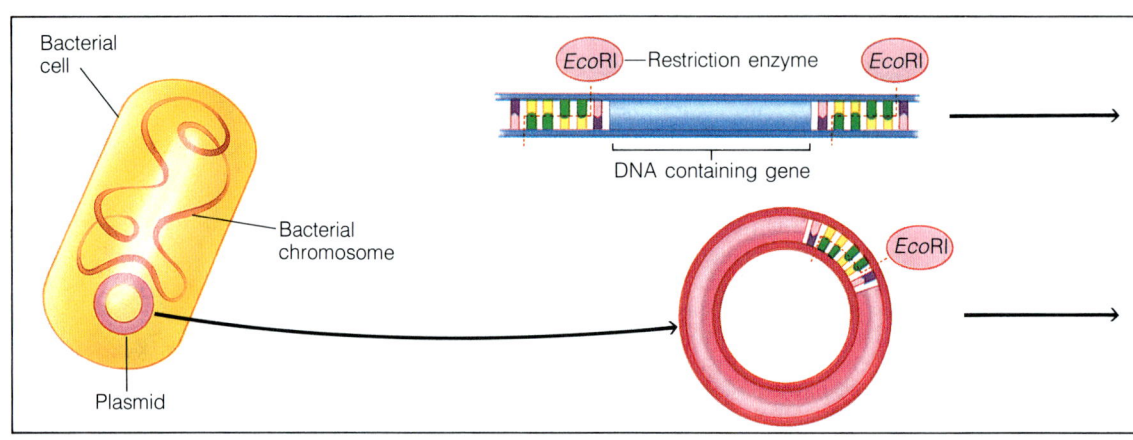

TEACHING SUPPORT

Laboratory Manual
- Cloning, p. 163

Teaching Resources
- Activity: DNA Recombination, p. 3

MEDIA AND TECHNOLOGY

Biology Transparencies
- 15: Cloning

BACKGROUND INFORMATION
DNA INSERTION

There is one aspect of the DNA insertion technique that is still not completely understood. When DNA is injected into animal and plant cells, sometimes it is inserted into a host cell chromosome and sometimes it is not. Although researchers are trying to develop ways of targeting DNA to a particular place on one of the chromosomes, as yet we do not understand exactly what happens inside the cell to cause the injected DNA to take up permanent residence. Therefore, in most experiments researchers insert DNA into hundreds or thousands of cells, hoping that the new DNA will be successfully maintained in a few of them.

DNA RECOMBINATION DNA fragments cannot function all by themselves. They must become part of the genetic material of living cells before the genes they contain can be activated. In the second step of genetic engineering, DNA fragments are incorporated into part of the recipient cell's genetic material.

For example, DNA fragments may be combined with bacterial DNA so that they can later be inserted into a bacterial cell. Bacteria often contain small circular DNA molecules known as **plasmids** in addition to their chromosomes. These plasmids can be removed from bacterial cells and cut with the same restriction enzyme used to produce the DNA fragments. The cuts made by the restriction enzyme produce matching "sticky ends" on the DNA fragments and the cut plasmids. These sticky ends are the sites at which a DNA fragment and a plasmid can be joined end to end, thereby forming a new plasmid that contains a piece of foreign DNA. See Figure 12–7.

Like the mythical chimera (kigh-MAIR-ah), which you read about in the chapter opener, the combined DNA formed by fusing a DNA fragment and a plasmid consists of parts from different kinds of organisms. In genetic engineering, molecules of combined DNA are known as chimeras because they are produced by combining DNA from different species. Combined DNA is also known as **recombinant DNA**, since DNA from two sources have been recombined to produce it.

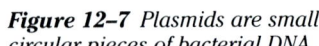

Figure 12–7 Plasmids are small circular pieces of bacterial DNA.

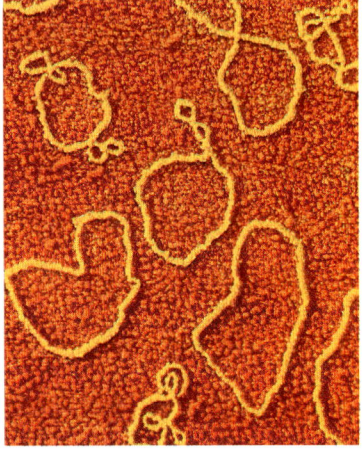

DNA INSERTION In the first two steps of genetic engineering, DNA fragments containing the desired gene are obtained and then inserted into DNA that has been removed from the recipient cell, thereby producing recombinant DNA. But how is this DNA put back into living cells?

It is easiest to transfer DNA into bacterial cells. The recombinant DNA is mixed in with millions of bacteria suspended in a dense salt solution. After a few minutes, several bacteria will take up the DNA. These bacteria can then be isolated and grown into large colonies that contain the recombinant DNA.

12–2 (continued)

Content Development
Using the Visuals As you begin talking about DNA recombination, you may get questions involving how scientists know what piece of DNA is what. Without getting into too much scientific detail, explain that the DNA pieces are thoroughly analyzed by the process of electrophoresis.

The concept of plasmids (extra bits of DNA) existing in cells may not be as difficult for students to conceptualize. Use Figure 12–7 to help them visualize these structures.

Tell students to think about the work *recombinant* and what it really means. Help them to realize how truly amazing this whole process is.

Enrichment

There are commercially prepared kits on DNA restriction analysis available from some biological supply houses. Although the process is fairly sophisticated, it is not impossible for a young scientist. For really serious stu-

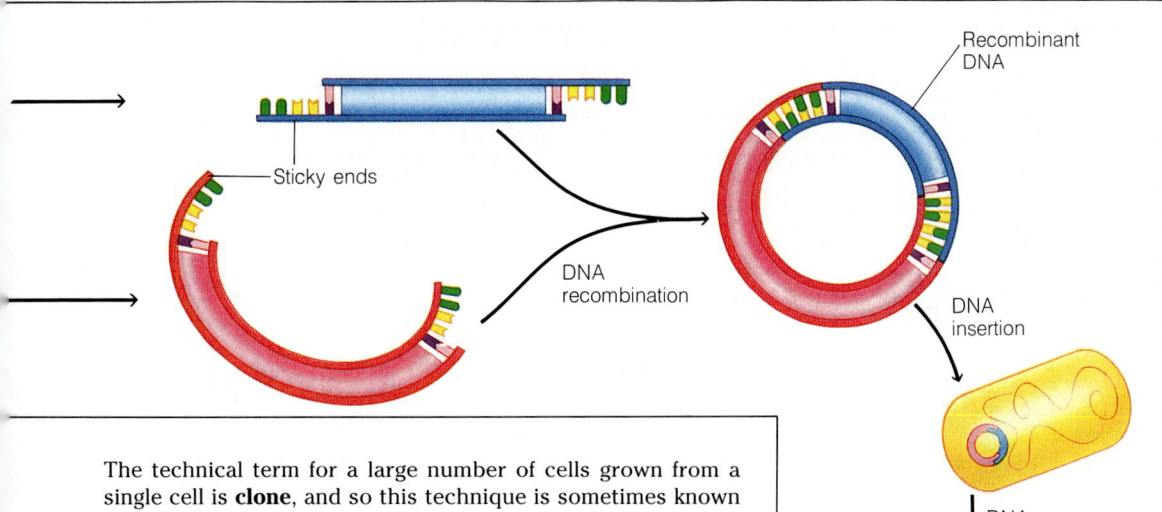

The technical term for a large number of cells grown from a single cell is **clone**, and so this technique is sometimes known as DNA cloning.

Using different techniques, recombinant DNA can be inserted into cells other than bacteria—for example, cells from yeasts, animals, and plants. These techniques include microinjection with a glass needle, fusion with plasmid-like DNA, and a new procedure in which DNA is attached to fine wirelike pellets that are then shot into cells with a microscopic gun.

DNA SEQUENCING Sequencing a piece of DNA, or reading the sequence of DNA bases along its length, can now be accomplished. Only one of the two strands of the DNA double helix is used in the process of DNA sequencing. However, many copies of this one strand are needed. These multiple copies can be produced through the process of DNA cloning.

In one form of DNA sequencing, a radioactive label (usually phosphorus) is added to single-stranded DNA. The DNA is divided into four groups that undergo different chemical treatments. The chemical treatments break the DNA into pieces that when separated reveal the positions of the bases on the original strand. The DNA pieces are separated by an electrical field in a process called electrophoresis (ee-lehk-troh-fuh-REE-sihs). Refer to Figure 12–9 for a more detailed explanation.

Engineering New Organisms

Recombinant DNA technology has advanced rapidly in the past few years. Techniques now exist for cutting and splicing DNA molecules, for inserting DNA into cells of a wide variety of organisms, and for controlling foreign genes moved from one species into another. Organisms that contain such foreign genes are said to be **transgenic**.

TRANSGENIC BACTERIA When a gene coding for a human protein is properly inserted into bacteria, the

Figure 12–8 *In genetic engineering, foreign genes are inserted into plasmids. The altered plasmids are then used to carry the foreign genes into a bacterial cell.*

HISTORICAL NOTE
EARLY DNA RECOMBINANTS

Genetic engineering in bacteria is more than 60 years old. Recall the experiments of Frederick Griffith on bacterial transformation, the same experiments that Avery used to implicate DNA as the genetic material. These clearly involved recombinant DNA, even though Griffith did not realize it at the time. Looking back on his extract of heat-killed virulent bacteria, we can realize that DNA fragments must have been taken up from this extract into harmless bacteria, and a few of these cells then expressed the genes that made them pathogenic.

---

dents, this might be an excellent project, involving principles of chemistry, physics, and biology.

Content Development
Using the Visuals The insertion of DNA into a cell is more of a technical problem than a scientific one. Several methods are used, including microinjection. Encourage students to think about how these methods were devised.

The term *clone* is perhaps familiar to students, but they may have a lot of misconceptions about the actual definition. Explain what is meant by DNA cloning. Figure 12–8 will help.

Reinforcement/Reteaching
This is a good time to relate several previously learned ideas. You have already discussed base sequences in DNA and radioactive tagging in Chapter 7. You might also want to make the point—perhaps again—how the different branches of science are interrelated.

ECOLOGY NOTE
INVASION OR COEXISTENCE

When creating new, hardier strains of plants by genetic engineering, the potential effects on natural ecosystems must be considered. Plants are usually engineered so that they have characteristics that make them unusually well equipped to grow and survive. These traits include high productivity, disease resistance, and increased growth rates.

Although such traits are positive for plant growers, some of them can have negative implications for natural habitats. Domesticated plants can and do "jump the fence" and colonize wild areas outside gardens and cultivated fields. If these fence-jumping plants have exceptional characteristics to grow and survive, they may invade and take over local ecosystems. For example, purple loosestrife, a flower that is not gene-altered, has escaped from gardens and severely disrupted the natural balance in wetlands habitats in the Midwest. If genetically engineered plants, which possess even better characteristics than naturally occurring plants, were to "escape," the results might be catastrophic.

This is just one of the dangers associated with producing genetically altered organisms. Because several potential problems have been identified, much research is now being carried out to ensure that danger to natural ecosystems is minimized.

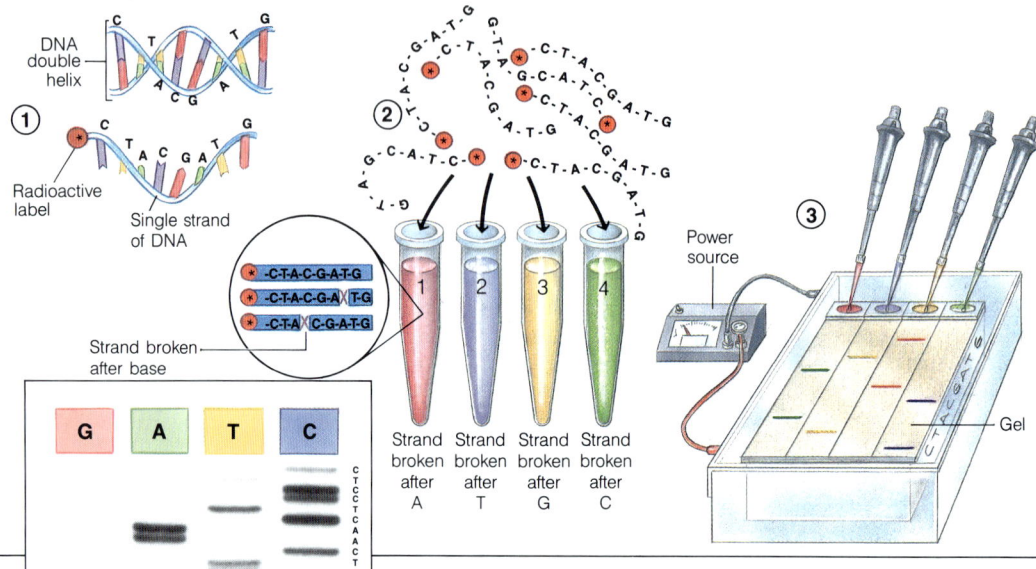

Figure 12–9 In DNA sequencing, DNA is treated to produce pieces that are then separated by electrophoresis. The bands on the gel that result from this process reveal the sequence of the bases in DNA (inset).

recombinant cells can be used to produce large amounts of the protein quickly and inexpensively. As recently as 15 years ago, human growth hormone, a protein needed for normal growth and development, was rare and extremely costly. (It could only be obtained in small amounts from cadavers, or dead bodies used for scientific purposes.) Thus people who do not produce normal amounts of this hormone themselves could not always be guaranteed a supply. Genetically engineered bacteria that contain the human growth hormone gene now produce large quantities of this protein, making it plentiful enough to treat everyone who needs it. Other genetically engineered bacteria produce human insulin, a hormone used to treat diabetes, and interferon, a protein that blocks the growth of viruses and may be useful in cancer treatment.

TRANSGENIC PLANTS DNA can be injected into plant cells directly or attached to plasmids of certain species of bacteria that infect plant cells. Plant cell biologists have developed techniques that enable a complete transgenic plant to be grown from the cells containing recombinant DNA. The goals of genetic engineering in plants include the production of plants that manufacture natural insecticides and the production of plants

---

12–2 (continued)

Skills Development
Guided Practice
Skill: Interpreting charts
Using the Visuals This section has several diagrams that condense a lot of material into a small space. Use these visuals to help students learn to interpret this information. Give them an opportunity to tell the story of DNA sequencing in their own words.

Content Development
The section on genetic engineering in bacteria should be very interesting to students. It is amazing that human genes can be placed inside bacterial cells and cause bacteria to produce a variety of useful human proteins.

• **What characteristics of bacteria make them suitable as a growth place for human DNA?** (Answers will vary but may include ease of culturing and rapid reproduction.)

• **Can you think of other ways that genetically engineered bacteria might be used to produce useful products?** (Answers may

that contain genes that enable them to produce their own nitrogen nutrients, thus eliminating the need for fertilizers. Australian researchers have successfully transferred into alfalfa the gene for an enzyme that aids in the synthesis of amino acids and thereby increases the protein content of the plants. These engineered plants will be a new source of protein-rich feed for Australian sheep.

Can genes from animals be made to work in plants? Steven Howell and his associates at the University of California at San Diego provided the answer to that question in 1986. They isolated the gene for luciferase, the enzyme that makes fireflies glow, and inserted it into tobacco cells. When whole plants were grown from the recombinant cells and the gene was activated, the plants glowed in the dark!

Research laboratories around the world are now eagerly experimenting with transgenic plants to meet a variety of human desires. One lab has developed a transgenic tomato that does not spoil as quickly as natural varieties. Another has produced a plant containing the enzymes needed to manufacture a kind of plastic that previously could only be produced from petroleum.

One of the newest techniques involves the introduction of DNA sequences that turn off existing genes in the plant. As you can see in Figure 12–11, the technique is especially useful in controlling flower color. Transgenic flowers and plants may soon be found in florist shops and garden centers everywhere.

TRANSGENIC ANIMALS DNA can be introduced into animal reproductive cells in a number of ways, including direct injection. As is the case in plants, there seem to be no fundamental barriers to the expression of genes from different species. Researchers have produced a variety of transgenic insects, roundworms, and vertebrates.

Some transgenic animals will be useful in farming and ranching. In one experiment, researchers introduced the growth hormone gene from rainbow trout into carp, thus producing a bigger, faster-growing fish. In other experiments, researchers are trying to produce farm animals that are more efficient in their use of feed and more resistant to disease.

Other transgenic animals already are being used in scientific research. The genes of the virus that causes AIDS have

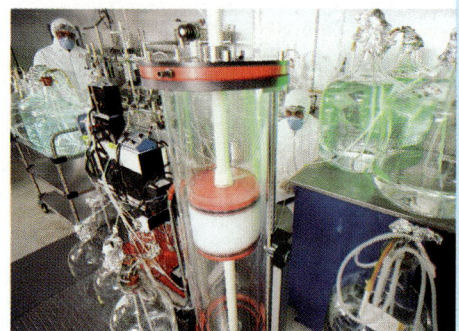

Figure 12–10 The hepatitis B vaccine being tested here for purity was produced by genetically engineered bacteria.

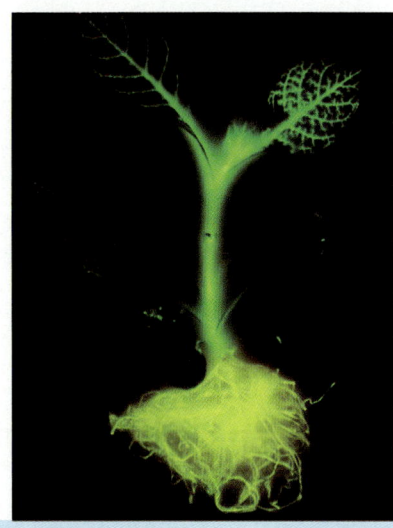

Figure 12–11 These transgenic petunias contain a "new" gene that inhibits pigment production (left). The petunias' color patterns vary because the new gene's effects depend on where it is inserted into the petunia DNA. This transgenic tobacco plant contains firefly genes (right). When the firefly genes are activated, the plant glows!

vary but may include drug research and development.)

Content Development
Discuss the goals and accomplishments of genetic engineering in plants with the students.
• **What seems to be the general goal of genetic engineering in plants?** (Increased food production for livestock as well as for humans.)
• **What is the usefulness of developing a glow-in-the-dark plant?** (No real practical use.)
• **Why do you suppose that biologists bothered to produce something with no practical use?** (Answers will vary but should include the idea that they wanted to prove that it could be done. This discovery added to the total pool of knowledge. These principles can be applied to other processes.)
• **What characteristic of DNA made the glow-in-the-dark plant possible?** (The universal nature of the molecule—very similar in all organisms.)

TEACHING SUPPORT

Study Guide
- Section 12–2, p. 120

Biotechnology Workbook
- A Gene for All Seasons, p. 49

12–2 (continued)

Content Development
Emphasize the variety of applications that have been tried with transgenic animals. Encourage students to try to think of other possible ways that genetic engineering could be used in the future.

SECTION REVIEW 12–2

1. Restriction enzymes cut DNA at specific sequences of nucleotides. In genetic engineering they are used to cut DNA into fragments with "sticky ends."
2. Recombinant DNA: DNA produced by fusing DNA from more than one source. Plasmid: a small circular piece of bacterial DNA. Clone: a large number of cells grown from a single cell. In genetic engineering, plasmids are often fused with a piece of foreign DNA, producing recombinant DNA. Many copies of the recombinant DNA are obtained when genetically engineered cells are allowed to replicate, forming a clone.
3. A transgenic plant contains foreign genes that were introduced by techniques of genetic engineering. In a hybrid, genes are inherited from the parents when the fertilized egg is formed.
4. They can be produced relatively inexpensively in large quantities, are the "real thing," and are relatively pure.

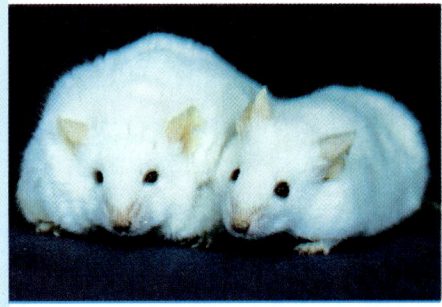

Figure 12–12 This giant mouse, shown next to a normal mouse, is transgenic. The giant mouse developed from a mouse egg that was injected with recombinant DNA containing the gene for rat growth hormone.

Guide For Reading
- What are some present and future applications of analyzing and sequencing human DNA?
- What are some ethical issues involved in human genetic engineering?

been introduced into mice to create a strain that can be used for research on AIDS itself. The genes of the human immune system have been given to other mice so that they can be used to investigate human immunity to disease.

12–2 SECTION REVIEW

1. What are restriction enzymes? How are they used in genetic engineering?
2. Define recombinant DNA, plasmid, and clone. How do these terms apply to genetic engineering?
3. How does a transgenic plant differ from a hybrid?
4. **Connection—Medicine** What are some of the advantages of genetically engineered proteins such as human growth hormone?

12–3 The New Human Genetics

The rapid development of molecular biology has produced a number of other developments. For the first time it has become possible to think about curing genetic diseases. We have also begun to wonder about the practicality of decoding the entire human **genome**, which is all of the genes possessed by humans. And we have started to apply molecular biology to personal identification and the diagnosis of disease.

Analyzing Human DNA

As we saw in Chapter 11, researchers have already developed tests for genetic disorders, which compare human DNA to DNA sequences known to produce the disorders—Huntington disease and sickle cell anemia, for example. In the near future, DNA obtained from prenatal sampling may be tested for the presence of a great many other disorders, enabling physicians to treat some disorders before birth and to be prepared to treat others shortly after birth.

Researchers have also begun to look for genes that might predispose individuals to other medical problems, such as heart disease, diabetes, and cancer. If tests that identify individuals at risk can be developed, early medical attention would be able to prolong many lives.

DNA Fingerprinting

The great complexity of the human genome ensures that no individual is exactly like any other. In the last three years, medical researchers have used this biological fact to add a powerful

Reinforcement/Reteaching
For those students who have difficulty with the Section Review questions, go back to the discussion in the section and review the materials students have had difficulty understanding.

Closure
Using the Visuals Turn back to the beginning page of Section 12–2. Use the visuals in this section to review the major points. Have a student explain the significance of each figure to the rest of the class.

TEACHING STRATEGY 12–3

Focus/Motivation
Begin this section with an exploration of the students' knowledge, beliefs, and feelings about human genetic research. Organize a debate with teams supporting po-

new tool to criminal investigations. The tool is known as **DNA fingerprinting**. DNA fingerprinting takes advantage of the fact that large portions of the human genome are made up of repeated sequences of varying lengths that do not code for proteins. It has turned out that individuals may have completely different numbers of these repeated units between actual working genes. You may have 15 repeats between two genes; the person sitting next to you in class might have 6 repeats between the same two genes; and someone else may have 33.

Here's how DNA fingerprinting works: A small sample of human DNA is cut with a restriction enzyme. The resulting fragments are separated by size through the process of electrophoresis. Fragments that contain the repeats are then labeled with DNA "probe," which binds to the DNA sequence of the repeats. This produces a series of bands that by their position show the lengths of the fragments containing the repeats. If enough combinations of restriction enzymes and DNA probes are used, a pattern of bands is produced that can be distinguished from the pattern of any other individual in the world.

DNA samples can be obtained from blood, sperm, and even hair strands with small pieces of tissue at the base. It is clear that DNA fingerprinting will be useful in solving crimes. Several rape suspects have already been convicted in the United States

Figure 12–13 DNA fingerprinting can be used to match up a suspect with blood, sperm, or other bits of DNA-containing material left at the scene of a crime.

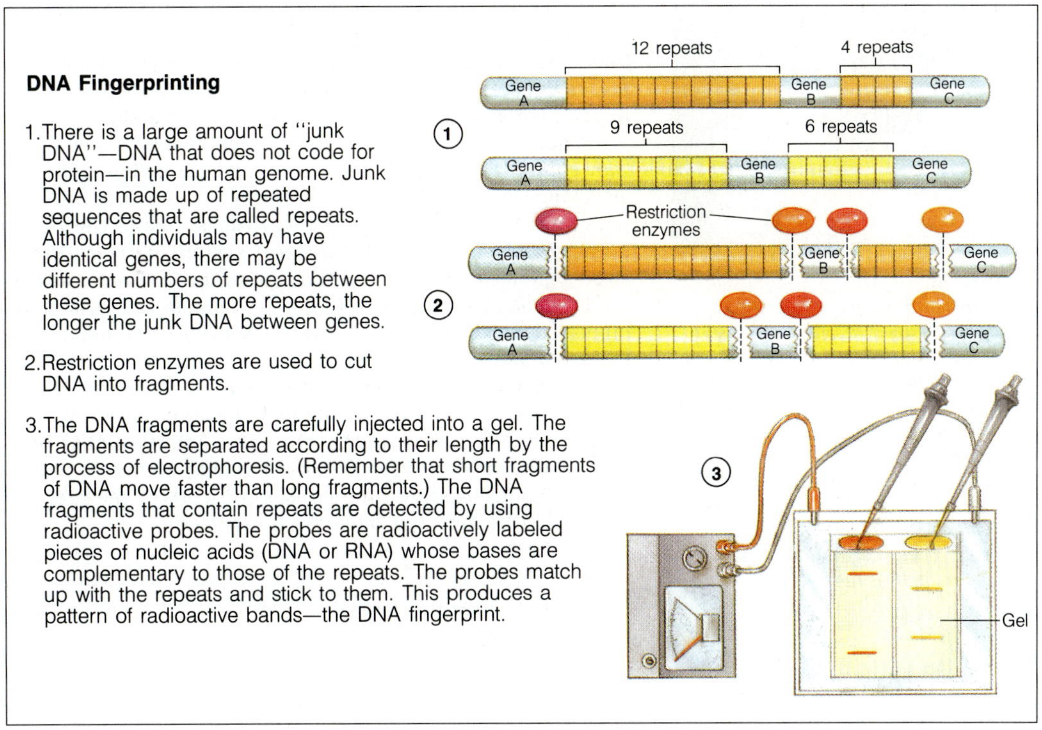

DNA Fingerprinting

1. There is a large amount of "junk DNA"—DNA that does not code for protein—in the human genome. Junk DNA is made up of repeated sequences that are called repeats. Although individuals may have identical genes, there may be different numbers of repeats between these genes. The more repeats, the longer the junk DNA between genes.

2. Restriction enzymes are used to cut DNA into fragments.

3. The DNA fragments are carefully injected into a gel. The fragments are separated according to their length by the process of electrophoresis. (Remember that short fragments of DNA move faster than long fragments.) The DNA fragments that contain repeats are detected by using radioactive probes. The probes are radioactively labeled pieces of nucleic acids (DNA or RNA) whose bases are complementary to those of the repeats. The probes match up with the repeats and stick to them. This produces a pattern of radioactive bands—the DNA fingerprint.

ECOLOGY NOTE
DNA FINGERPRINTING AND CONDORS

It is all too easy for an endangered species to become dangerously inbred. In fact, some endangered species, such as the cheetah, are so lacking in genetic diversity (as a result of inbreeding) that they may be doomed to extinction even though there are a relatively large number of individuals.

One goal of captive breeding programs for endangered species is to keep the gene pool as large as possible by avoiding inbreeding and making sure that individuals that have uncommon genes are bred more often than those that have common genes. But how do you know which genes each animal possesses?

This is where DNA fingerprinting comes in. Although this technique was developed to help solve crimes, it can also be used to help save endangered species.

With only 27 condors remaining in 1987, California Condor Recovery Program (CCRP) scientists needed all the help they could get. So they took a blood sample from each bird and used it to prepare a DNA fingerprint. By comparing the pattern of bands in the DNA fingerprints, they were able to determine which condors were close relatives. The scientists then paired off birds that had dissimilar "fingerprints"—and thus were not closely related—and allowed them to breed.

In 1993, six years after the last wild condor was captured for the CCRP, the condor population had more than doubled to 76. There were even enough condors that a few individuals with redundant genes could be released into the wild. Once more, the condor soars above the rugged mountains of California.

---

sitions for and against the research, which is continuing and expanding. Add an interesting twist to the discussion by having students defend the position contrary to their true belief.

Content Development
Begin your discussion of the new human genetics by talking about the term *genome*. Talk about some of the things that science will know and be able to do if the genome is fully decoded. Allow students to brainstorm for a while. Then talk specifically about disease research and DNA fingerprinting.

Focus/Motivation
Using the Visuals There have been numerous articles in newspapers and magazines about DNA fingerprinting. Read and discuss any of these that are part of your collection. Use the visuals on this page to help explain the techniques used in the process.

HISTORICAL NOTE
CANCER-FIGHTING CELLS

The first federally approved transfer of cells containing foreign genes into a human was accomplished in May 1989 at the Clinical Center of the National Institutes of Health in Bethesda, Maryland.

Cancer-fighting cells that had been altered by insertion of a foreign gene were infused into the bloodstream of a cancer patient who had volunteered for the experiment. The primary purpose of the genetic manipulation was to make the cells easily identifiable so that doctors could track them in the patient's body. The patient was not expected to benefit directly from the manipulation.

The cells, tumor-infiltrating lymphocytes, had been taken from the patient's cancerous tissue and treated in the laboratory to increase their numbers and thus their ability to attack the cancer tissue.

Because the procedure involved genetic alterations in the human cells that are used in the transfers, the proposal to do the medical experiments was studied and discussed for 7 months by several committees of the National Institutes. Federal guidelines require stringent studies of any attempted gene transfers.

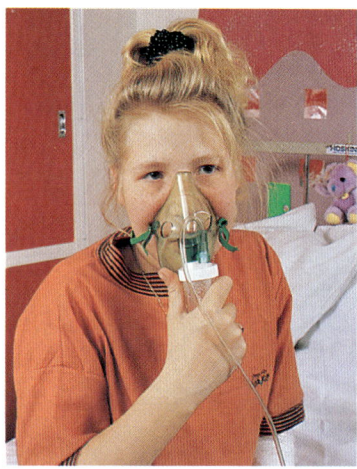

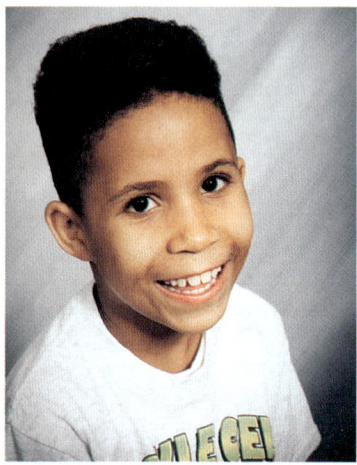

Figure 12–14 *This girl with cystic fibrosis is inhaling the fine spray from a nebuliser in order to loosen the mucus that clogs her lungs (top). This boy with sickle-cell anemia must be careful to avoid over-exertion, which can cause his red blood cells to change shape, block small blood vessels, and stop carrying enough oxygen (bottom). Researchers are currently testing forms of "gene therapy" that may someday cure diseases such as these.*

and Great Britain because DNA fingerprinting showed that their DNA patterns matched blood or sperm samples recovered from their victims. And in at least one case, a wrongly accused person has been set free.

Genetic Engineering of Humans

Because humans, too, are animals there is no *technical* barrier to the insertion of foreign genes into human cells. The production of transgenic animals—inserting DNA into fertilized eggs and then transplanting the eggs back into the female reproductive tract—serves as a model for how transgenic humans could be produced.

A review of human genetic disorders indicates that there are many cases in which molecular techniques could be used as effective treatments and as possible cures. Cystic fibrosis, Tay-Sachs disease, and sickle-cell anemia are just a few examples of genetic disorders that could, in principle, be treated by transplanting copies of a normal gene into human cells.

There seem to be at least two practical methods that could be used for human "gene therapy." One, as shown in Figure 12–15, involves removing cells from an individual and growing them in culture. These cells would then be injected with DNA containing a gene able to correct the disorder. The engineered cells would then be injected back into the individual where they would produce proteins to cure the disorder. This approach may be tried in a few years with human blood-forming bone-marrow cells to treat sickle-cell anemia. It is hoped that the engineered cells would produce enough normal hemoglobin to cancel out the ill effects of the sickle-cell allele.

A more radical approach involves modifying viruses so that they cannot cause disease and then further modifying them by attaching DNA containing a desired gene to the viral DNA. The patient is then infected with the viruses. Inside the human body, the viruses act as a transport system, carrying the desired gene into the cells and thus correcting a genetic defect.

How close are we to the day when human DNA will be manipulated by genetic engineering? Closer than you think. For, in fact, that day is already here. After the viral technique was shown to work in mice, the first human trials began in 1993. In these experiments, viruses that infect the air passageways were genetically engineered and then sprayed into the lungs of patients with cystic fibrosis (CF). Researchers hope that the normal alleles inserted into the viruses will correct the problems caused by the defective alleles that cause CF.

As this book goes to press, scientific results are not yet in. However, it is safe to predict that attempts to use genetic engineering to correct human genetic disorders will continue. Sooner or later, they will be successful.

12–3 (continued)

Reinforcement/Reteaching
If you have students who are interested in this topic, encourage them to contact local law enforcement agencies or attorneys who deal with criminal cases. Ask for interviews concerning DNA fingerprinting. The students should ask the persons they interview what they know about the process and if they know of cases where it has been used.

Content Development
If you began the chapter with a discussion of the ethics of human genetic engineering, you may want to recap your earlier discussions and add any others presented in this section.

SECTION REVIEW 12–3

1. It allows people to make informed decisions about health and family-planning issues.
2. DNA fingerprinting is a technique used to identify individuals. It produces a unique series of bands from samples of a person's DNA.
3. Some issues in human genetic engineering include: what traits people should be allowed to change; whether the "cloning" of people should be permitted; and who should regulate experimentation in and the practice of human genetic engineering. Other answers are possible.

Ethical Issues

Despite these possibilities, there are problems, risks, and doubts that have persuaded many scientists that the time is not yet right to carry out these procedures on human beings. Of course it would be marvelous to be able to cure hemophilia or other genetic diseases. But could we be sure that experimentation on humans would involve only diseases? If human cells can be manipulated in this way, should we try to engineer taller people or change their eye color, hair texture, sex, blood type, ear shape? What will happen to the human species and to our conception of ourselves if we gain the opportunity to design our bodies? What will be the consequences if we develop the ability to "clone" ourselves by making identical genetic copies of our own cells? These are questions that science will rapidly force us to come to grips with.

If we acquire the ability to alter ourselves, we shall also acquire awesome responsibilities. How shall we decide which genes should be transplanted, altered, or redesigned? Who shall determine whether experiments with genetic engineering should be done? Scientists? A government agency? The local community?

As we said in Chapter 1, the purpose of science is to gain a better understanding of the nature of life. The more we understand life, the more we shall be able to manipulate it. As our power over nature increases, our society shall have to learn to use wisely the tools that science has given us. And we shall have to develop a thoughtful and ethical consensus of what should and should not be done with the human genome. Scientists alone should not be expected to assume all of the responsibility for these decisions. Society in its entirety should have to deal with these questions. To do anything less would be to lose control of our most precious gifts: our intellect and our humanity.

12-3 SECTION REVIEW

1. What are some potential benefits of genetic analysis?
2. What is DNA fingerprinting? How does it work?
3. What are some of the principal ethical issues involved in human genetic engineering?
4. **Critical Thinking—Making Inferences** If human bone marrow cells were removed, altered genetically, and reimplanted, would the change be passed on to a patient's children? Explain.

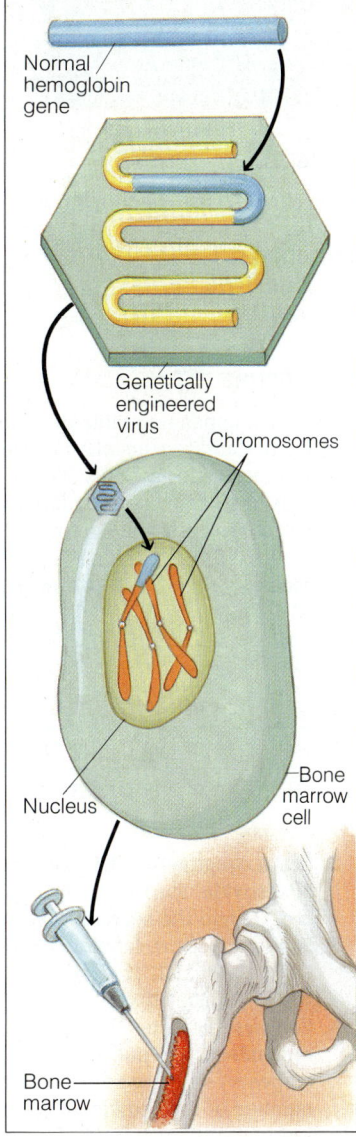

Figure 12–15 Gene therapy may someday make it possible to "correct" cells that contain defective genes and thus cure genetic diseases.

LABORATORY INVESTIGATION

SIMULATING DNA FINGERPRINTS

Before the Lab
Assemble the materials that you will need for each student or group. If you are not familiar with this lab, do it yourself before you try to explain it to the class. The strips and data table that you prepare can be used as visuals to help clarify your explanation.

Pre-Lab Discussion
Review the similarities that exist between blood typing, fingerprinting, and DNA fingerprinting. Review the ways that DNA fingerprinting can be used in criminal investigations. You might also want to review the section of the text on DNA fingerprinting.

After explaining the purpose of the lab, a step-by-step explanation using visuals will make the lab go more smoothly. After the explanation, you can check the students' understanding with the following questions.

- **What is the purpose of this lab?** (To simulate the way that DNA fingerprinting is used to identify criminals.)
- **What does each strip of graph represent?** (The DNA segment of the individual.)
- **What do the marks on the data tables represent?** (The DNA fingerprints or the marks that tagged DNA would leave on photographic film.)

Skills Development
Students will use the following skills while completing this laboratory investigation.
1. Observing
2. Manipulative
3. Applying
4. Analyzing
5. Comparing
6. Relating
7. Classifying

LABORATORY INVESTIGATION

SIMULATING DNA FINGERPRINTS

PROBLEM
How can DNA be used for the identification of criminals?

MATERIALS (per student)

scissors   graph paper

PROCEDURE

1. The accompanying table describes the pattern of genes and repeats of a particular segment of DNA for five individuals. This information will be used to construct models of the DNA segment for each of the five individuals.

Individual	DNA Sequence
1	G1 G2 **10R** G3 **4R** G4 G5 **7R** G6 **6R** G7
2	G1 G2 **1R** G3 **15R** G4 G5 **3R** G6 **2R** G7
3	G1 G2 **11R** G3 **8R** G4 G5 **5R** G6 **3R** G7
4	G1 G2 **6R** G3 **2R** G4 G5 **9R** G6 **7R** G7
5	G1 G2 **4R** G3 **5R** G4 G5 **12R** G6 **4R** G7

Key: G = gene   R = repeats

2. Cut a lengthwise strip of graph paper two boxes wide. Starting at the top, label the first box *G1* and the second box *G2*.
3. For each repeat, label a box *R*. For example, for individual 1, the ten boxes following *G2* would be labeled *R*.
4. Continue labeling the boxes through to *G7*, remembering to put the proper number of *R*'s (repeats) in the correct locations. Write the number of the individual on the back of the strip.
5. Fold a clean sheet of graph paper into sixths lengthwise. Unfold the paper and number the boxes up the left-hand edge, starting at the bottom, from 1 through 25. This sheet will be used to record the DNA fingerprints of individuals 1 through 5 and of an unknown criminal. Label the first of the six columns *1* for individual 1; the second, *2* for individual 2; and so on. Label the sixth column *CRIMINAL*.

6. Select one of the five individuals to be a criminal. *Note: Do not tell your classmates which individual you have selected.* Make a duplicate strip of DNA for that individual following the procedure outlined in steps 2 through 4.
7. Cut the strip representing the criminal's DNA between genes 4 and 5. This represents the cutting of DNA by restriction enzymes.
8. Arrange the two pieces of the DNA strip in order of size, with the larger first. This represents the separation of DNA pieces through electrophoresis.
9. Count the number of boxes in the longer piece of DNA. Color in a box in the column of your data table marked *CRIMINAL* directly across from this number. Repeat this procedure using the shorter fragment. You should have two marks in the column. These represent the marks that DNA tagged with a radioactive probe would leave on a photographic plate—a DNA fingerprint.
10. Exchange the DNA fingerprint of your criminal with one of your classmates.
11. Prepare DNA fingerprints for individuals 1 through 5.

OBSERVATIONS
1. Did any of the five individuals have the same DNA fingerprints?
2. Which individual was the criminal?

ANALYSIS AND CONCLUSIONS
1. If the possible number of repeats between two genes varies between 1 and 20, what is the probability of having exactly 3 repeats?
2. Why is it unlikely that two individuals would have the same DNA fingerprint?
3. If blood or hair samples are recovered at the scene of a crime, how could they be used to make a positive identification of a criminal?

Safety Tips
Students should be cautioned to always use scissors with care.

Teaching Strategy
As students begin this lab, be sure that they are following the instructions correctly. Impress on each group that accuracy is as much a part of police work as of science.

Observations
1. No. (In rare instances the answer may be yes.)
2. Answers will vary.

Analysis and Conclusions
1. 1/20.
2. A high number of possible repeats and a large number of locations at which repeats occur ensure that the probability that two individuals will have the same pattern of repeats is low.

STUDENT STUDY GUIDE

SUMMARIZING THE CONCEPTS

The key concepts in each section of this chapter are listed below to help you review the chapter content. Make sure you understand each concept and its relationship to other concepts and to the theme of this chapter.

12–1 Modifying the Living World

- By using carefully selected individuals as parents for the next generation, people can improve and change domesticated plants and animals.
- Inbreeding involves crosses between similar individuals.
- Hybridization involves crossing dissimilar individuals.
- Random mutations, which can be caused artificially, can produce useful variations in organisms.

12–2 Genetic Engineering: Technology and Heredity

- Genetic engineering changes DNA through direct manipulation.
- Restriction enzymes are used to cut DNA. The cut DNA can be combined with other DNA molecules and inserted into organisms in various ways.
- There are techniques for reading the sequence of bases in DNA.
- Genetic engineering uses DNA to transform cells and organisms.

12–3 The New Human Genetics

- Human DNA can be analyzed to detect genetic diseases.
- DNA fingerprinting can be used to solve crimes and identify people.
- The technology exists to determine the sequence of the entire human genome.
- There are many ethical considerations that must be taken into account before new genetic techniques are applied to humans.

REVIEWING KEY TERMS

Vocabulary terms are important to your understanding of biology. The key terms listed below are those you should be especially familiar with. Review these terms and their meanings. Then use each term in a complete sentence. If you are not sure of a term's meaning, return to the appropriate section and review its definition.

12–1 Modifying the Living World
selective breeding
inbreeding
hybridization
mutagen

12–2 Genetic Engineering: Technology and Heredity
genetic engineering
restriction enzyme
plasmid
recombinant DNA
clone
transgenic

12–3 The New Human Genetics
genome
DNA fingerprinting

---

3. DNA can be extracted from blood, and the tissues on the ends of hairs can be used to make a DNA fingerprint. This fingerprint can be compared to the DNA fingerprints of a suspect for positive identification.

Going Further: Enrichment
Do a creative writing assignment describing a scenario in which a crime was committed and evidence was left behind that will allow DNA fingerprinting. Play the part of the detective and explain how you would use this technique to help solve the crime.

CHAPTER REVIEW

CONTENT REVIEW

Multiple Choice

Choose the letter of the answer that best completes each statement.

1. Crossing individuals that have similar characteristics and are often closely related is called
   a. hybridization.
   b. inbreeding.
   c. mutagenesis.
   d. genetic engineering.
2. A piece of DNA that is produced by combining DNA fragments from different sources is called
   a. hybrid DNA.
   b. cloned DNA.
   c. plasmid DNA.
   d. recombinant DNA.
3. A large number of cells grown from a single cell is called a
   a. clone.
   b. hybrid.
   c. chimera.
   d. plasmid.
4. A cross between dissimilar individuals is
   a. cloning.
   b. mutagenesis.
   c. hybridization.
   d. genetic engineering.
5. Organisms that contain foreign genes are
   a. mutagenized.
   b. transgenic.
   c. hybridized.
   d. inbred.
6. What is the name of the process of engineering changes directly into an organism's DNA?
   a. mutagenesis
   b. hybridizing
   c. selective breeding
   d. genetic engineering
7. Which technique takes advantage of repeated DNA sequences that do not code for protein?
   a. DNA fingerprinting
   b. DNA sequencing
   c. genetic engineering
   d. cloning
8. The process that uses electricity to separate DNA fragments according to their length is
   a. electrophoresis.
   b. DNA sequencing.
   c. genetic engineering.
   d. mutagenesis.

True or False

Determine whether each statement is true or false. If it is true, write "true." If it is false, change the underlined word or words to make the statement true.

1. Offspring resulting from a cross between <u>similar</u> individuals often show hybrid vigor.
2. A <u>plasmid</u> is a cell or organism whose DNA comes from different species.
3. Crossing similar individuals is called <u>selective breeding</u>.
4. Increasing the mutation rate is called <u>genetic engineering</u>.
5. <u>Transgenic</u> organisms contain foreign genes.
6. A <u>chimera</u> is a large number of cells grown from a single cell.
7. A circular piece of bacterial DNA is called a <u>clone</u>.
8. <u>DNA fingerprinting</u> produces a pattern of bands that is unique for each person.

Word Relationships

In each of the following sets of terms, three of the terms are related. One term does not belong. Determine the characteristic common to three of the terms and then identify the term that does not belong.

1. hybridization, genetic engineering, selective breeding, inbreeding
2. restriction enzymes, recombinant DNA, DNA cloning, mutagenesis
3. clone, chimera, transgenic, recombinant DNA

CONTENT REVIEW

Multiple Choice
1. b 2. d 3. a 4. c
5. b 6. d 7. a 8. a

True or False
1. F dissimilar
2. F chimera
3. F inbreeding
4. F mutagenesis
5. T
6. F clone
7. F plasmid
8. T

Word Relationships
1. types of breeding techniques, which are ways of affecting DNA indirectly; genetic engineering involves the direct manipulation of DNA
2. terms associated with genetic engineering; mutagenesis is the process of inducing mutations, and is associated with selective breeding
3. terms referring to DNA from at least two different sources; clone is a group of cells derived from and genetically identical to a single original cell

CONCEPT MASTERY

1. Both involve selecting which organisms will produce the next generation.
2. A restriction enzyme is used to cut out the gene that codes for a desired protein from human DNA; the same restriction enzyme is used to cut a bacterial plasmid, producing "sticky ends"; the human gene is inserted into the plasmid; the plasmid is taken up by a bacterial cell and thus incorporated in the bacterial genome; the bacterial cell is allowed to replicate, forming a clone; the clone of altered bacterial cells produces the desired human protein.
3. It occasionally produces useful mutations.
4. Almost all adult organisms, including frogs, can be thought of as a clone—all the cells in an organism's body are descended from a single original cell—the zygote. Each of the frog embryos is derived from one of the adult frog cells. Thus the genetically identical frog embryos can be considered a clone.

CRITICAL AND CREATIVE THINKING

1. Answers will vary, but all correct answers will involve crossing the different varieties of roses and selecting some of the resulting off-

CONCEPT MASTERY

Use your understanding of the concepts developed in the chapter to answer each of the following in a brief paragraph.

1. Explain why inbreeding and hybridization are sometimes considered to be forms of selective breeding.
2. Describe the processes involved in creating bacteria that produce human proteins.
3. How is mutagenesis useful to humans?
4. The nuclei from certain cells in an adult frog can be placed into frog eggs from which the nuclei have been removed. This produces genetically identical frog embryos. Explain why these embryos are considered to be a clone.

CRITICAL AND CREATIVE THINKING

Discuss each of the following in a brief paragraph.

1. **Applying concepts** Suppose a plant breeder has thornless rose bushes that have pink flowers, thorny rose bushes that have sweet-smelling yellow flowers, and thorny rose bushes that have purple flowers. How might the plant breeder develop a purebred variety of thornless sweet-smelling purple roses?
2. **Problem solving** The following fragments are obtained when a gene that consists of 10 codons is cut by restriction enzymes. What is the sequence of bases in the gene? (*Hint:* Look for overlapping sequences on the fragments.)
3. **Making comparisons** Compare the advantages and disadvantages of breeding techniques and genetic engineering.
4. **Applying technology** Describe the processes involved in isolating, cloning, and sequencing a gene.
5. **Expressing an opinion** Should genetic engineering ever be done on humans? If so, under what circumstances? Explain your answer.
6. **Using the writing process** If you could create the "ideal organism" by combining the traits of two organisms, what two organisms would you select and why?

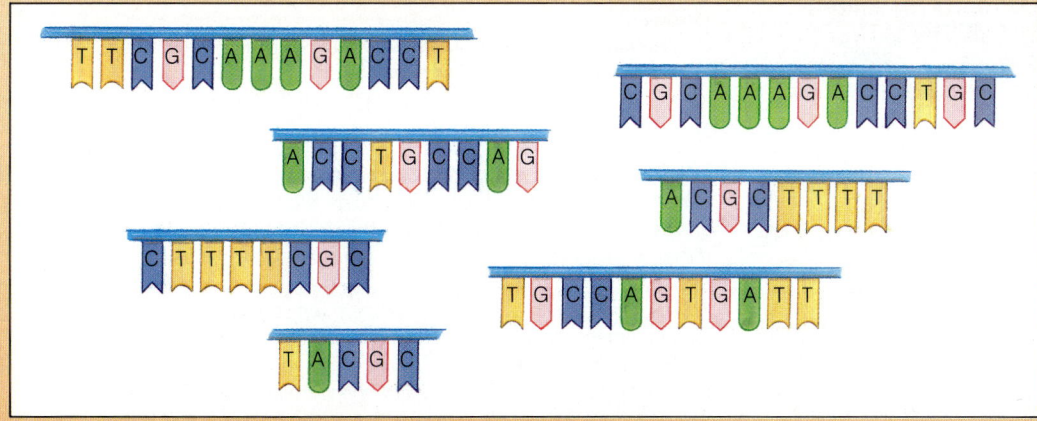

263

CAREERS IN BIOLOGY

Holographic Technician, Plant Breeder, Molecular Geneticist

Students interested in a biological career should be instructed to write for further information to the address listed beneath each career description. However, in consideration of the organizations that provide career information, please have only one student write to an organization. Do not instruct the entire class to write to every organization for the same career information. You may want to use the information provided to start a biology career file for student use.

UNIT 3

CAREERS IN BIOLOGY

Holographic Technician

Holographic technicians work with laser beams to create three-dimensional images called holograms. The technicians operate the equipment, care for the mirrors and lenses, and develop the resulting film.

Holograms are used today in a variety of fields. They are particularly helpful in science and medicine, where computer-generated images allow researchers and doctors to view the structures of parts of living organisms.

A holographic technician must complete a high school education and have some additional training at a technical or junior college.

For information on this career write to the Museum of Holography, 11 Mercer St., New York, NY 10013.

Plant Breeder

Plant breeders use a knowledge of genetics and biology to develop desirable traits in many kinds of plants. Traits that would be considered desirable include the ability to resist disease, to thrive in different soils and climates, and to produce large numbers of offspring. Plant breeders work in experimental nurseries, where they cross-breed various plants to observe and record the characteristics of the offspring.

A position as a plant breeder requires at least a Bachelor's degree in agronomy or horticulture.

For more information write to the American Society of Agronomy, 677 South Segoe Rd., Madison, WI 53711.

Molecular Geneticist

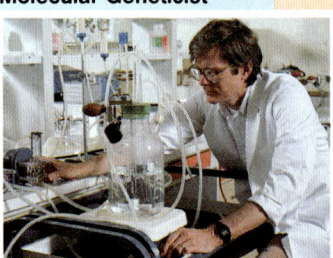

Molecular geneticists study patterns of heredity in various forms of life. They are concerned with how biological traits originate and are passed down from one generation to another. Some molecular geneticists specialize in studying diseases that are linked to genes and chromosomes.

To become a molecular geneticist, a person must earn a Bachelor's degree in chemistry or biology and a Ph.D. in genetics.

For more information write to the American Genetic Association, 818 18th St., NW, Washington, DC 20006.

HOW TO FIND THE RIGHT COLLEGE

When choosing a college, first consider what type of career or academic program interests you. Also decide what geographical location and student-body size would make you most comfortable. Other things to consider are financial constraints, availability of housing, and ease with which you can travel back and forth to school. Once you have a list of potential colleges, compare the courses they offer, their job placement records, and their extracurricular activities.

In addition to talking with counselors, you might want to read some college guides to learn about specific institutions. Two popular examples are Peterson's Guide to Four-Year Colleges *and* Barron's Profiles of American Colleges.

FROM THE AUTHORS

"You can choose your friends but not your relatives."

The first time that phrase made sense to me was when my baby cousin ruined a perfectly good trip to a restaurant by deciding that his meal would look better on the floor. Over the years my cousin turned out just fine, of course, and he even developed table manners. But for a few minutes I would rather have been related to a family of worms.

We don't choose our relatives, and we can't choose our genes. Each one of us has a biological inheritance that includes our blood type, our eye color, our appearance, and even our height. It's easy to feel trapped by that inheritance, limited by the biological traits that seem so important and so far beyond our abilities to change them. But don't make the mistake of being overwhelmed by your own genetics, especially in your studies. None of us reaches more than a fraction of the intellectual potential that we inherit. The limits on what we achieve in life stem less from our genes than from ourselves.

The biological capacities that we all inherit, regardless of our individual differences, are more than enough to give each of us the tools to achieve great things in whatever we set out to do. When you think about the men and women who have pioneered scientific research, dreamed up inventions, written great literature or music, what do you see? Great genes or great efforts? The answer, without a doubt, is the latter. What we inherit is the beginning of what we can be. It is not the end of it.

Ken Miller

At her first halter lesson, three-week old Allegra poses with owner Ken Miller. Whether Allegra becomes a good riding horse or even part of a championship equestrian team depends not so much on her genes as on her training. Inheritance provides a starting point but does not dictate a future for either horses or humans.

FROM THE AUTHORS

Kenneth R. Miller

In this article Ken Miller imparts an important idea to students. He points out that we can do very little about which genes we inherit, but that we are all born with a potential that few of us realize. Put another way, if you do badly in school—or in life—do not blame your genes; blame yourself.

UNIT 4
Diversity of Life

UNIT OVERVIEW

In Unit 4, evidence for evolution, theories about evolution, and a classification scheme based on evolution are presented. First the development of the modern theory of evolution is described in its historical context. Next, students are asked to examine the various types of evidence for evolution. Students then discover how the various types of evidence for evolution have been interpreted in theories about the nature and pace of evolution. Finally, students learn about taxonomy and binomial nomenclature. They are also introduced to the five-kingdom classification scheme and learn about the distinguishing characteristics of each kingdom.

UNIT OBJECTIVES

1. Explain how scientists determine the age of the Earth and of fossils.
2. Describe the fossil, embryological, anatomical, and biochemical evidence for evolution.
3. Discuss the development of Darwin's theory of evolution.
4. Relate evolutionary theory to genetics.
5. Explain how new species develop.
6. Describe some modern modifications of Darwin's theory of evolution.
7. Explain how organisms are classified.

UNIT 4
Diversity of Life

In the eerie stillness of the evening, the light of the setting sun forms blood-red puddles among the lifeless bones of a dinosaur. Something moves unexpectedly in the shadows of the giant ribs—small shapes appear and vanish like ghosts. With whiskers quivering, a tiny furry creature pokes its nose into the light and then scampers onto the dinosaur's skull. Reassured, other furry animals emerge from the shadows.

The small furry creatures chasing one another among the bones and squabbling over the last scraps of dinosaur meat may not seem all that impressive. However, they and their relatives were the ancestors of elephants, horses, whales, and other mammals. Some of their descendants became intelligent enough to contemplate the extraordinary diversity of life, theorize about the processes that formed many kinds of creatures from a single original kind, perform experiments to test theories, and imagine events that occurred millions of years ago.

DISCOVERY LEARNING
ALL MIXED UP

1. Using a cup of mixed nuts or trail mix, develop a classification system for the individual pieces that make up the snack.
2. Prepare a brief report that explains your system of classification. Include a diagram that represents your classification system.
   - Why do you think it is helpful to classify objects?
   - What are some objects that you classify in everyday life? How do you classify these objects, and why?

INTRODUCING UNIT 4

Using the Visuals
Have students read the unit introduction on page 266 and examine the illustration on pages 266 and 267, which shows early mammals scampering on and around the skeleton of a *Triceratops* dinosaur. You may wish to ask the following questions about the illustration.

• **What does this illustration show?** (Early mammals on and around a dinosaur skeleton.)
• **What do you think would be a good title for this illustration?** (Accept all reasonable answers. One possibility is "The Dawn of Mammals." Students should recognize that this illustration depicts the end of the reign of the dinosaurs and the beginning of the age of mammals.)

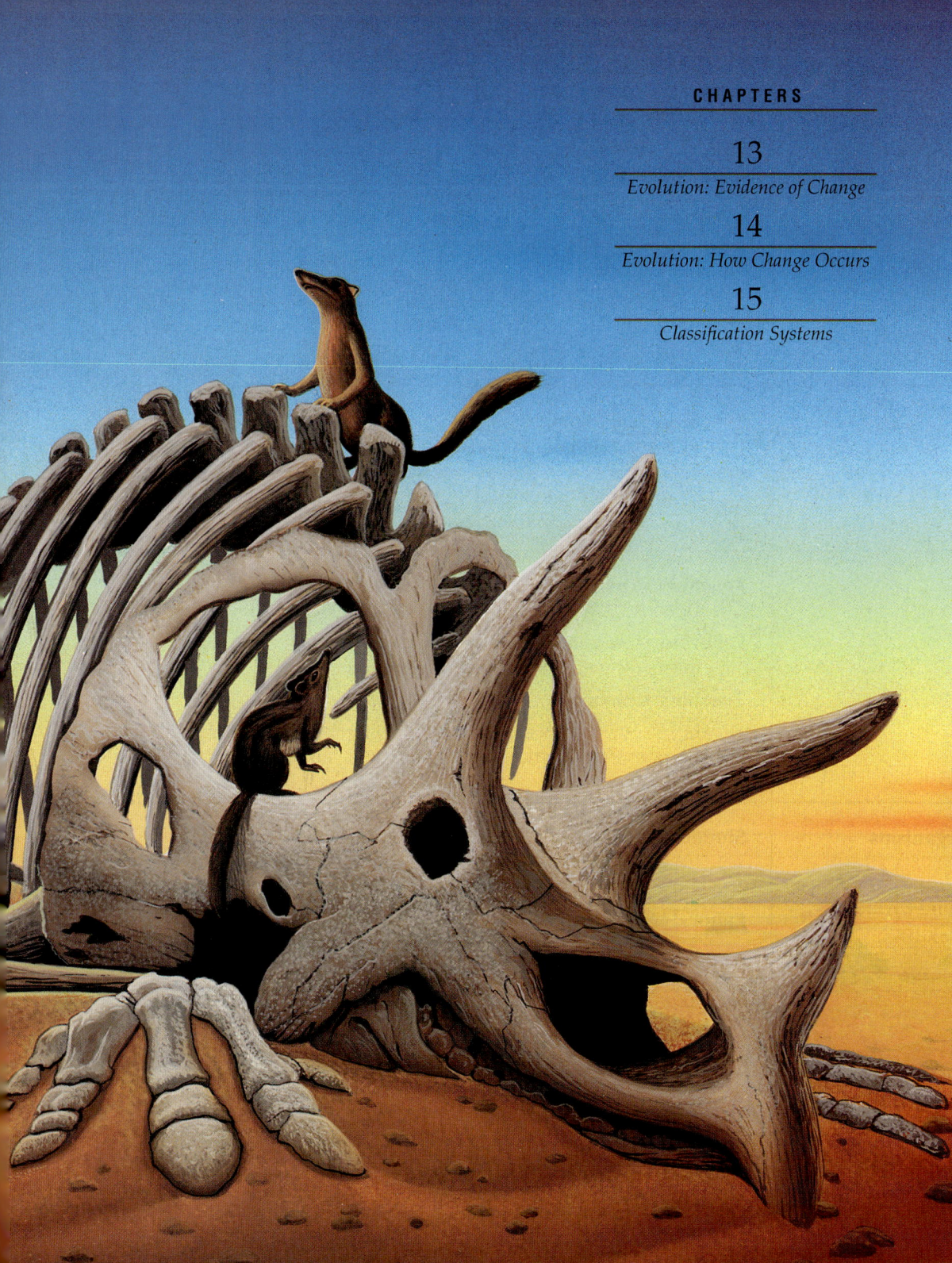

CHAPTERS

13
Evolution: Evidence of Change

14
Evolution: How Change Occurs

15
Classification Systems

CHAPTER DESCRIPTIONS

13 Evolution: Evidence of Change
Chapter 13 introduces students to evidence that indicates that living things have evolved. Students first learn how the types of evidence Darwin encountered on his voyage led to the formulation of his theory of evolution. Next they find out how rocks—and fossils within rocks—are dated. Finally, they examine the geological, embryological, anatomical, and biochemical evidence for evolution.

14 Evolution: How Change Occurs
Chapter 14 describes the development of theories that explain how the process of evolution occurs. Students focus on the theory of evolution by natural selection and examine one example of natural selection in action. Next, students discover how evolution relates to genetics. Finally, they learn about new theories that modify Darwin's original ideas.

15 Classification Systems
In Chapter 15 students read about the need for classification systems and about some early systems. They then learn about Linnaeus's system for naming and classifying organisms and about the different taxonomic categories. Then they explore the link between evolutionary theory and taxonomy. Finally, students study the five-kingdom classification scheme and learn about the major characteristics of each kingdom.

• **What do these early mammals look like to you?** (Accept all reasonable answers. Students may suggest that the mammals look like rats, mice, or squirrels.)
• **Why are these early mammals important?** (They are the ancestors of animals—including humans—that play a dominant role in modern ecosystems.)
• **What are some examples of modern animals that descended from these early mammals?** (Try to get students to mention a wide variety of modern-day mammals.)

Discovery Learning
ALL MIXED UP

Begin your introduction to the unit by having students perform the Discovery Learning activity.

Answers may include: to create order, to make generalizations about classes of objects. Students may classify books, records, clothing, and foods to make individual items easier to find or to save space. Other responses are possible.

CHAPTER 13 Evolution: Evidence of Change

Section	Laboratory Investigations and Activities
13–1 Evolution and Life's Diversity pages 269–271 THEMES: Evolution, Unity and Diversity	**Teacher Edition** DEMONSTRATION: Darwin: The Man, p. 268D
13–2 The Age of the Earth pages 272–277 THEMES: Evolution, Patterns of Change	**Teaching Resources** ACTIVITY: Radioactive Decay, p. 7
13–3 The Fossil Record pages 278–282 THEMES: Evolution, Patterns of Change	**Teacher Edition** DEMONSTRATION: Forming Sediments, p. 268D **Laboratory Manual** Interpreting Events from Fossil Evidence, p. 167 **Teaching Resources** HANDS-ON ACTIVITY: Turned to Stone, p. 67 ECOLOGY INVESTIGATION: Making Casts of Animal Tracks, p. 3
13–4 Evidence from Living Organisms pages 283–285 THEMES: Evolution, Unity and Diversity	**Student Edition** LABORATORY INVESTIGATION: Examining Homologous Structures, p. 286 **Laboratory Manual** Amino Acid Sequences and Evolutionary Relationships, p. 173
Chapter Review pages 287–289	

Teacher Resources

Books

Arduini, Paolo, and Giorgio Teruzzi. *Simon and Schuster's Guide to Fossils*. Simon and Schuster, 1986.

Darwin, C.R. *The Voyage on the* Beagle. Natural History Press, 1962.

Darwin, C. *On the Origin of Species: A Facsimile of the First Edition*. Harvard University Press, 1975.

Dobzhansky, T., F. Ayala, G. Stebbins, and J. Valentine. *Evolution*. Freeman, 1977.

Hancock, Judith. *Variety of Life*. Weston Walch, 1987.

Kornberg, W., ed. National Science Foundation Staff. *Evolution: New Perspectives*. Avery Pub., 1982.

Livingstone, David N. *Darwin's Forgotten Defenders*. William B. Eerdmans Publishing, 1987.

Simpson, George Gaylord. *Fossils and the History of Life*. (Scientific American Library) Freeman, 1983.

Strahler, Arthur N. *Science and Earth History—Evolution/Creation Controversy*. Prometheus, 1987.

CHAPTER PLANNING GUIDE

Other Activities	Media and Technology
Study Guide Section 13–1, p. 125	
Teaching Resources ACTIVITY: The Geologic Time Scale, p. 3 **Study Guide** Section 13–2, p. 127	**Interactive Videodisc** In the Company of Whales **Biology Media Guide** page 11
Teaching Resources ACTIVITY: Interpreting an Ancient Puzzle, p. 11 **Study Guide** Section 13–3, p. 129	
Teaching Resources VOCABULARY REVIEW: Crossword, p. 1 **Study Guide** Section 13–4, p. 130	**Interactive Videodisc** In the Company of Whales **Biology Media Guide** pages 11–12
Teaching Resources Chapter Test A, p. 19 Chapter Test B, p. 23	**Computer Test Bank** Chapter Test, p. 135

Audiovisuals

Introducing Evolution. 6 filmstrips. Ward.
Patterns of Inheritance. Film or video. Coronet.
Patterns of Diversity. Film or video. Coronet.
Patterns of Evolution. Film or video. Coronet.
Radioactive Dating. Film or video. Coronet.
Darwin's Bulldog. Film. Time-Life Video.
Story in the Rocks. Shell Oil Company, Film Division.

268B

CHAPTER 13 Evolution: Evidence of Change

CHAPTER OVERVIEW

Evolution is the process by which modern organisms have descended from ancient organisms. In 1831, Charles Darwin set sail on a worldwide voyage that would change scientific thought forever. Throughout this five-year journey, Darwin collected and studied specimens of plants and animals from all over the world. In addition to living specimens, Darwin found evidence of organisms that no longer existed.

Nearly 30 years later, Darwin published his explanation of all the wonders he had observed. He introduced the concepts of fitness, adaptation, and common descent.

Darwin and other scientists have accumulated substantial evidence that evolution has occurred. Geologists Hutton and Lyell greatly influenced Darwin's thinking. They believed that the Earth was very old and that the land was constantly moving and shifting. Other geologists added evidence and more questions with the discovery of fossils.

The geological time scale is used to date the Earth's history. Both relative dating, using the position of rock layers, and absolute dating, using radioactive dating, are used to tell the age of rocks, fossils, and the Earth itself.

Though fossils form in a variety of ways, most are found in sedimentary rock. At best, it is a chancy process that leaves many missing pieces in the puzzle of history. Even without all the parts, the fossil record shows that many changes have occurred over time.

In addition to fossil evidence, there are traces of history within the bodies of living organisms. The early embryos of many different animals look very similar. The presence of homologous structures and vestigial organs are also evidence of evolutionary change. Finally, the presence of many of the same compounds in all living organisms is an important piece of information supporting Darwin's conclusions.

13-1 EVOLUTION AND LIFE'S DIVERSITY

Section Focus 13-1

The purpose of this section is to introduce the topic of evolution. There is a brief account of Darwin's journey on the *Beagle*. Over five years' time, Darwin collected specimens from all over the world. He also found evidence that many more organisms had vanished from the Earth. Darwin called the combination of physical traits and behaviors that helps organisms survive and reproduce in their environment fitness. Fitness arises through a process called adaptation.

Nearly 30 years after Darwin's voyage, he published his studies. He maintained that modern species have been produced through evolution, a long, slow process of change over time. He further asserted that all species have descended from common ancestors.

Performance Objectives 13-1

1. Explain evolution in terms of Darwin's observations and studies.
2. Relate Darwin's principles of fitness and adaptation.

Science Terms 13-1

evolution   p. 269
fitness   p. 270
common descent   p. 271
adaptation   p. 271

13-2 THE AGE OF THE EARTH

Section Focus 13-2

The purpose of this section is to discuss the evidence of evolution that is to be found in the rock of the Earth itself. The geologists James Hutton and Charles Lyell proposed that the Earth was very old and had changed slowly and gradually over time. At the same time, other scientists began to discover the preserved remains of ancient organisms. Some of these fossils resembled living organisms; others did not.

Scientists date the Earth's past with the geological time scale. Each layer of rock represents a different period of time. Relative dating can be used to determine the age of fossils relative to other fossils in different layers of rock. A more exact way of dating is by radioactive dating, a procedure based on the rate of decay of radioactive elements in rocks. There is a brief discussion of half-life presented in this section. Because the actual age of rock can be determined by this method, it is called absolute dating.

Performance Objectives 13-2

1. Relate the contributions of James Hutton and Charles Lyell to Darwin's work.
2. Explain relative dating of fossils.
3. Discuss radioactive dating, explaining decay and half-life in your discussion.
4. Explain the time divisions used on the geological time scale.

Science Terms 13-2

fossil   p. 273
geologic time scale   p. 274
relative dating   p. 274
radioactive element   p. 274
half-life   p. 274
absolute dating   p. 276
era   p. 276
period   p. 276
epoch   p. 276

CHAPTER PREVIEW

13-3 THE FOSSIL RECORD

Section Focus 13-3

The purpose of this section is to introduce students to ways that fossils form and what can be learned from fossils.

Fossils are formed in a variety of ways, but most of them are found in sedimentary rock because of the way that the rock is formed. The fossil record is incomplete for several reasons. Many organisms leave imperfect fossils or none at all. Scientists must sometimes reassemble the remains of ancient organisms from small bits and pieces.

The fossils that do exist are sometimes difficult to date by radioactive dating processes. From the existing information contained in fossils, paleontologists know a great deal. They know that many changes in climate and geography have occurred. The many species of living organisms on Earth have seen change after change.

Performance Objectives 13-3

1. Discuss the way that fossils are preserved in sedimentary rock.
2. Describe the difficulties involved in obtaining evidence of evolution from fossils.
3. Discuss some changes on Earth that are supported by the fossil record.

Science Terms 13-3

sedimentary rock   p. 278
paleontologist   p. 280
fossil record   p. 281

13-4 EVIDENCE FROM LIVING ORGANISMS

Section Focus 13-4

The purpose of this section is to show that fossils are not the only evidence in support of evolution. Living organisms can also be used to prove that changes have occurred.

As early as the late nineteenth century, scientists noticed that the early embryos of many different organisms looked so much alike that they were difficult to tell apart. These similarities indicate that similar genes are active. As they grow and develop, the embryos became more dissimilar because of genes that have changed through evolution.

In embryos, the cells that develop into limbs look quite similar. These cells grow into limbs that are different because of evolutionary changes that have altered the structure and appearance of the bones. The presence of vestigial organs that no longer serve any useful function is also evidence that changes have happened over time. Finally, the many complex biochemical compounds shared by all living organisms are strong indications of common ancestry. In fact, all of the homologies discussed are best explained by Darwin's conclusion that living organisms have gradually evolved from a common ancestor.

Performance Objectives 13-4

1. Explain how similarities in embryos are evidence of shared ancestry.
2. Describe what homologous structures are and how they can be used to support evolution.
3. List several examples of vestigial organs that indicate that evolutionary change has occurred.
4. Relate the presence of biochemical compounds found in all organisms to a shared evolutionary relationship.

Science Terms 13-4

embryo   p. 283
homologous structure   p. 284
vestigial organ   p. 284

DISCOVERY LEARNING

TEACHER DEMONSTRATIONS
Modeling

Darwin: The Man

This activity can be used to introduce Chapter 13.

Students frequently have many misconceptions about Darwin's contributions to science. Evolution is a difficult concept for many students to handle. Help them to understand the breadth and depth of Darwin's work by bringing in biographical information of Charles Darwin. You might also want to quote excerpts from *The Voyage of the Beagle*. To really involve the students, try role-playing some of Darwin's discoveries. Two possibilities are the discovery of the South American beetles or the Galapagos Island finches. The purpose of the activity is to present Darwin as an intelligent, curious young man. After all, he was only 22 years old when the *Beagle* set sail. His job at the time was companion to the captain and self-appointed naturalist—he did not set out to revolutionize scientific thought.

Forming Sediments

This activity can be used to introduce Section 13-3.

You will need several beakers, half-filled with water. Place a leaf or small shell in the bottom of each beaker. Have students observe and record the settling action of each mixture as you add samples of sand, soil, dirt, or clay. Observe what happens to the beakers over time. Check them at the end of the period and for the next day or two. Discuss what has happened to the leaves and shells. They have obviously been covered with the sediment. Point out that this is what happens to organisms or parts of organisms that fall into ponds or rivers that have a lot of sediment. Over time and under pressure from the weight of large amounts of water, geologists believe that these sedimentary deposits become rock. Speculate about what conditions would allow things to become fossilized.

CHAPTER 13

Evolution: Evidence of Change

GUIDED ENQUIRY

Pose the following questions to students and have them record their responses. Point out that they will gain a better understanding of the key concepts if they read the chapter with these basic questions in mind. Upon completion of the chapter, pose the questions again. Ask students to compare their initial responses with those they have developed after reading the chapter.

- How do evolutionary changes occur?
- Why do species of living things become extinct?
- How is adaptation related to what Darwin called fitness?
- How did Darwin arrive at his conclusion of common descent?
- How does geological and biological evidence support the theory of evolution?

INTRODUCING CHAPTER 13

Using the Visuals

This exercise emphasizes the process skills of observation and communication of data. Relate this activity to Darwin's use of keen observation on his voyage. Science is based on observation, and all the evidence presented in Chapter 13 is based on this skill. Divide the class into small groups or pairs. Have them cover the caption and text under the figure that introduces the chapter. In their own words, have each group write a description of the organisms and the action in the picture. Since the organisms should be unfamiliar to students, encourage them to use as many descriptive terms as possible. After they have completed the assignment, have them share their work with the class. Now read the caption. Discuss how well they did as observers of the ancient world. You might want them to briefly explore how scientists know what some of these animals were like—or do students think it is only fiction?

CHAPTER 13

Evolution: Evidence of Change

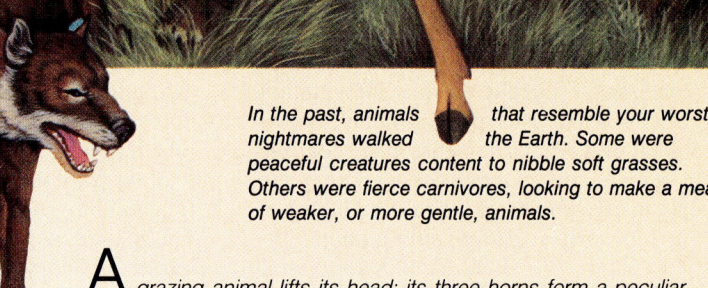

In the past, animals that resemble your worst nightmares walked the Earth. Some were peaceful creatures content to nibble soft grasses. Others were fierce carnivores, looking to make a meal of weaker, or more gentle, animals.

A grazing animal lifts its head; its three horns form a peculiar silhouette against the evening sky. Another type of grazer scratches itself with the forked horn on the end of its nose. Suddenly, a pack of sleek, bearlike predators bursts from the underbrush. The herds of grazers erupt into a flurry of moving hooves and horns. The predators pursue, single out their quarry, and close in. Their prey kicks and slashes with its triple horns, but to no avail. It is over in moments. The pack begins to feed.

Even though this scene is imaginary, the strange, remarkable animals were not. They lived some 7 million years ago in what is now Texas. How do we know of these ancient animals? Where did they come from? And why are they no longer alive? In this chapter you will learn about methods scientists use to reconstruct a picture of ancient life.

TEACHING STRATEGY 13-1

Focus/Motivation

Read the introductory paragraphs in this section. Review with students the basic premises of science. Contrast these with beliefs. The beliefs held by stu-

GUIDE FOR READING

After you read the following sections, you will be able to

13–1 Evolution and Life's Diversity
- Relate Darwin's observations to his explanation of evolution.
- Explain how fitness is related to adaptation.

13–2 The Age of the Earth
- Discuss how scientists determined that the Earth was much more than a few thousand years old.
- Distinguish between relative and absolute dating.
- Explain how radioactive elements in the Earth's rocks act as a natural clock.

13–3 The Fossil Record
- Describe how sedimentary rocks are formed.
- Explain how fossils provide evidence of evolution.
- Discuss why there may be problems in interpreting fossil evidence.

13–4 Evidence from Living Organisms
- Describe how similarities in embryos support the concept of common descent.
- Explain how homologous and vestigial structures indicate that evolution has occurred.
- Discuss how biochemical similarities are evidence of evolution.

Journal Activity

YOU AND YOUR WORLD

Darwin kept voluminous notes in the diary he recorded during his five-year-long voyage. His diary helped him remember details he later used to write his books. Keep a record of your observations for a period of time. Note the people you met, what they wore, what you ate, what you thought. Like Darwin, there might be a book in your future.

13–1 Evolution and Life's Diversity

Guide For Reading
- What is "fitness"?
- How do adaptations contribute to fitness?

The idea that life on Earth has changed over time, or evolved, is very old. But just believing that change occurs is not enough to make evolution a science. In science, you will recall, observation, questioning, and constant testing of hypotheses must replace belief. Scientists have accumulated considerable evidence to show that organisms alive today have been produced by a long process of change over time. The process by which modern organisms have descended from ancient organisms is called **evolution**.

One man, Charles Darwin, contributed more to our understanding of the process of evolution than anyone else. For this reason, we will begin this chapter by looking at the natural phenomena that convinced Darwin that evolution occurred.

Darwin's Dilemma

Two days after Christmas in 1831, a young Englishman named Charles Robert Darwin (1809–1882) set sail on HMS *Beagle* for a cruise around the world. Although no one knew it then, this voyage would revolutionize scientific thought.

Darwin was well educated and had a strong interest in natural history. He also had keen powers of observation and an analytical mind. Over five years' time, the *Beagle* took Darwin to several continents and many remote islands. Darwin went

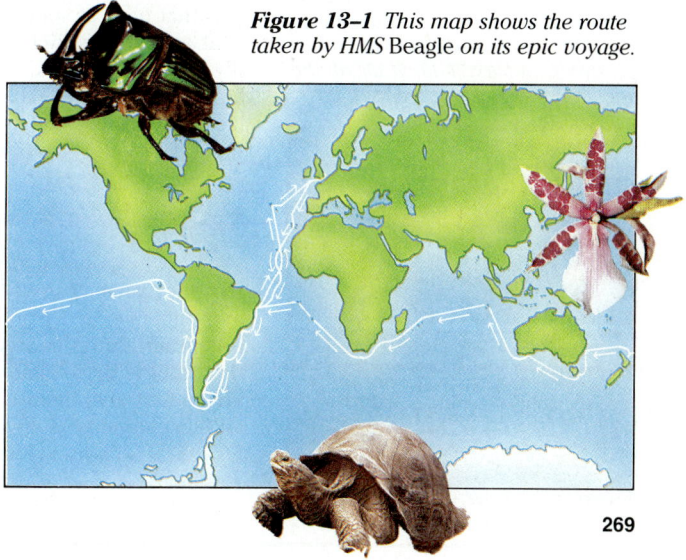

Figure 13–1 This map shows the route taken by HMS Beagle *on its epic voyage.*

269

BACKGROUND INFORMATION
DARWIN AND RELIGION

Despite efforts by antievolutionists to paint Darwin as an atheist, he was a very religious man. He had a degree in theology from Cambridge and even thought about spending his life as a country preacher. He remained a devout Christian all of his life. He saw no conflict—and no less awe or majesty—in a God that ruled through natural laws rather than through supernatural means.

George Gaylord Simpson wrote, "Scientists certainly know more about God than the theologians do, for scientists study the works of God and theologians only study what has been felt and written about God."

Figure 13–2 Roxie Laybourne is one of the world's experts on birds. She stands amidst file drawers that contain only a small part of the Smithsonian Institution's collection of specimens. As large as this collection is, however, it represents only a small fraction of the species of birds alive today. The specimens are used to compare birds collected in the wild with named species.

Figure 13–3 Reproductive behaviors can contribute to a species' survival. A female cuckoo laid her egg and abandoned it in this bird's nest. After the cuckoo hatched, it killed its foster brothers and sisters by pushing them out of the nest. Now alone and constantly hungry, it demands all the attention of its foster parents.

ashore whenever the ship anchored. At each new place, he collected animal and plant specimens that he added to an ever-growing collection. At sea, between bouts of seasickness, Darwin examined his specimens and filled notebooks with his thoughts and observations. He also spent many hours reading the most current scientific books.

Throughout the voyage, Darwin witnessed countless wonders of nature for which his bright young mind demanded an explanation. Those mysteries of life will spark your curiosity, too, if you stop to think about them.

The Diversity of Life

Our planet houses living organisms of every imaginable shape, size, and habit. This variety of living things is called the diversity of life. Only a tiny fraction of these organisms lived in Darwin's England—or in your hometown. But if you travel to or read about different countries, you will discover, as Darwin did, that the diversity of life is staggering.

When Darwin traveled in South America, he found more different forms of life than he had ever dreamed of. For example, in just one day spent in a Brazilian forest he collected 68 different species of beetles—even though he wasn't really looking for beetles! Darwin's observations helped him to realize that an enormous number of species inhabit the Earth. Even today, as scientists search land, water, and air, there is no precise count of the number of different kinds of organisms that exist. Estimates range from 3 million to more than 20 million different living species.

Darwin soon realized that the diversity of life he observed was only one part of a much larger puzzle. For as he traveled, he found evidence that even more organisms had vanished from the Earth. Today, researchers estimate that of all the species that have at some time lived on Earth, more than 99.9 percent are now extinct! If that estimate is correct, several hundred million species have come and gone during Earth's long history. Where have all the marchers in this endless parade of life come from? Why have so many of them disappeared over time? These are two of the questions Darwin tried to answer.

Fitness: To Survive and Reproduce

Darwin was also impressed by the many different ways in which organisms survive and produce offspring. He noted that most animals and plants have body parts and behaviors that do certain things very well. **The physical traits and behaviors that enable organisms to survive and reproduce in their environment give them what Darwin called fitness.** But how did all these organisms develop the structures that give them their fitness? And why are there so many different techniques for survival?

13–1 (continued)
Content Development
For five years Darwin collected plants and animals from every stop. He examined his specimens and made records. Have students think about some of the things that Darwin would need to do in order to make a useful and usable collection.

Skills Development
Guided Practice
Skill: Relating experiences
Talking with students about the diversity of life can be frustrating because, like Darwin, they frequently have limited experience. If you do have students who have traveled to different areas, encourage them to talk about plants and animals they observed that seemed unusual to them. Emphasize that 10 million is a very large number of estimated living species and that 99.9% of all organisms that have ever lived are now extinct! Point out the two questions at the end of the section to set the stage for the sections that follow.

Content Development
To even begin to understand Darwin, students must understand what he meant by the terms *fitness* and *adaptation*. After defining the terms, use lots of examples to clarify students' thinking. Use the following example to help them grasp the point. Emphasize that Darwin was referring to naturally occurring genetic adaptation and the example is about artificially applied behaviors, but the point of "fitness" is the same.

Imagine that you live in a very primitive society. Every member of the group has to hunt in the wilderness for food in order to live. Each person needs to be completely self-sufficient.

• **How would poor vision affect your life?** (You would be less able to hunt and compete suc-

These are very difficult questions, so it is not surprising that Darwin wasn't able to provide answers overnight. It was not until 1859, nearly 30 years after he began his voyage on the *Beagle,* that Darwin published his explanations in a book called *The Origin of Species by Means of Natural Selection.* This book changed the way people think about the living world.

In *The Origin of Species,* Darwin maintained that modern organisms have been produced through evolution. Evolution is a long, slow process of change in species over time. Darwin argued that just as each new organism comes from preexisting organisms, each species has descended from other species over time. If you look back far enough in time, you will see that all species have shared, or common, ancestors. Since species have descended from common ancestors, Darwin called this principle **common descent.**

Darwin also argued that fitness arises through a process called **adaptation.** Successful adaptations enable organisms to become better suited to their environment, better able to survive and reproduce. Darwin also used the word adaptation to describe any inherited characteristic that increases an animal's or plant's fitness for survival. Thus, the long neck and legs of a giraffe are adaptations that permit giraffes to feed on the leaves of trees. With these adaptations, giraffes can eat leaves too high for most grazing animals to eat and thus are better able to survive and reproduce, passing their genes on to their young.

Figure 13–4 *The giraffe dines on leaves of trees out of the reach of even the tallest zebra's head. The sphinx moth uses a long feeding tube to reach the nectar in a flower. The vampire bat punctures the skin of another animal with its razor-sharp teeth and then greedily laps up blood from the wound.*

13–1 SECTION REVIEW

1. What is fitness?
2. How did Darwin's voyage on the *Beagle* influence his thoughts about life on Earth?
3. How is the diversity of life related to evolution?
4. **Critical Thinking—Relating Concepts** How is adaptation related to fitness?

271

FACTS AND FIGURES

Darwin was not hired as the ship's naturalist but to serve as a companion for the ship's captain, Robert Fitzroy. Captain Fitzroy was forbidden by British custom to socialize with his crew because he was a member of the upper class. In order to remain sane on a five-year voyage, he needed a socially acceptable person to talk to. Charles Darwin was that person.

BACKGROUND INFORMATION
SEMANTIC CONVENIENCES

Throughout this chapter and other parts of the book, semantic conveniences to describe the evolutionary process have been used, for example, "this bird evolved a larger beak to feed on these large seeds." Do not infer by this that some Lamarckian phenomenon of species intent is at work. It is simply a language shortcut to the much longer "over many generations, because of the higher fitness of individual variants with large beaks, the genotype associated with large beaks came to dominate the population."

TEACHING SUPPORT
MEDIA AND TECHNOLOGY
 Interactive Videodisc
- In the Company of Whales
Use this bar code to show a picture of whale fossils.

Search to frame 25928

See the Biology Media Guide page 11 for additional barcode correlations.

Guide For Reading
- What is the age of the Earth?
- What is the difference between relative and absolute dating?
- What is a half-life?

Figure 13–5 *These stone arches were formed by the relentless forces of nature. Tiny particles of windblown sand constantly hit the rock and wear it away. Darwin learned from his reading that most geological changes do not happen overnight—or even during a single human lifetime. Instead, they take many, many years.*

13–2 The Age of the Earth

Darwin and other scientists have accumulated a vast amount of evidence that proves that evolution has occurred. Some of the evidence certifies that planet Earth is more than 4 billion years old. Other evidence makes it clear that both Earth and the life on it have changed dramatically over time. Still other evidence supports the principle of common descent and emphasizes the importance of adaptation to the environment. Much of the evidence is found in the rocks of the Earth itself. And it is this evidence that we will now examine.

Evidence in Stone

In the past, many people believed the Earth was relatively young—only a few thousand years old. They also believed that the Earth had remained unchanged over this time. Rocks and major geological features, they thought, had been produced suddenly by catastrophic events that humans rarely (if ever) witnessed and, even if they did, could not understand. But other people have offered a different explanation, an explanation based on evidence stored in the rocks of the Earth itself.

In the eighteenth and nineteenth centuries, scientists began to examine the Earth in great detail. And they offered the hypothesis that the Earth was indeed very old and had changed slowly over a long period of time by natural forces like weather. It was the work of these scientists that profoundly influenced Darwin's thoughts.

Evidence that supported the idea that the Earth was very old first came from geologist James Hutton in 1788. Hutton proposed that rocks, mountains, and valleys had been changed gradually by rain, heat, cold, the activity of volcanoes, and other natural forces. Because most of these processes operate slowly, Hutton argued, the Earth had to be much more than a few thousand years old.

In 1830, just before Darwin began his voyage, the geologist Charles Lyell carried these arguments further. Lyell agreed that the Earth had changed slowly and gradually over time. Lyell also argued that scientists must always explain past events in terms of events and processes they could observe themselves. That, Lyell insisted, was the only way the scientific method could work. Lyell's work, as you will discover in the next chapter, was an important influence on Darwin's thinking.

The evidence proved to Hutton and Lyell that the Earth was very old. Further evidence suggested to them that the land is constantly moving and shifting: Forces beneath the Earth's surface twist and bend some rock layers, bury others, and even push up some parts of the sea floor into mountain ranges. For these scientists, the Earth had indeed changed over the long period of its existence.

At the same time, other scientists found evidence that life on Earth had also changed over time. While examining the Earth's rocks, geologists—professional and amateur—began to

---

TEACHING STRATEGY 13–2
Focus/Motivation
Tell students that some of the evidence collected by scientists indicates that the Earth is about 4.5 billion years old. They also have a great deal of evidence to show that organisms have changed over that period of time. They must have reliable ways to date the Earth and its inhabitants. Section 13–2 deals with the methods that scientists use for dating the Earth's rocks and living things.

Set up a simple demonstration to illustrate the relative dating process. Start with an empty clear container (a large beaker or plastic shoe box will work). Add layers of dry materials that can be distinguished from one another. Use whatever materials are at hand—sand, salt, corn meal, soil, and so on. Keep this container and its contents intact for later use. Discuss with students the layers of rock that make up the Earth.
- **If your container represented the Earth, which layer would be**

make some startling discoveries. In the stones they examined they found **fossils**. Fossils are the preserved remains of ancient organisms. Some of these fossils resembled organisms still alive. Others did not. These fossils raised many questions that would remain unanswered for some time. Even though they could not explain the meaning of all the fossils they found, these early geologists made a great contribution. They created an interest in and a sense of wonder about the Earth and the life that lived upon it.

The Geologic Time Scale: A Clock in the Rocks

Earth's story is not complete without a "clock" to tell us when things happened. Both biologists and geologists date the

Figure 13–6 These ancient, fossilized relatives of the modern camel have been arranged in chronological order according to the layers of rock in which their bones were found. Provided that the order of the layers has not been disturbed, fossils found in lower layers are older than those found in the layers above.

BACKGROUND INFORMATION
HUTTON AND LYELL

The ideas of both Hutton and Lyell were critical to Darwin's thinking about the nature and rate of change over time. In Darwin's day, creationist thinking insisted that neither the Earth nor the organisms on it had changed since their creation. Therefore, Darwin had to overcome a deep-seated belief in the fixity of nature. Lyell's work convinced him that the Earth had changed over time, setting him up to realize that organisms had changed, too.

Later on, however, Darwin's exclusive belief in gradualism led him into certain problems. Whether or not the punctuated equilibrium theory gains wide-spread acceptance, it is clear that rates of evolutionary change vary over time and among species.

ly by the elements and natural forces. Therefore, the Earth must be much older than a few thousand years.)
• **What important argument did Lyell make?** (Scientists must explain past events in terms of events and processes that they could observe themselves.)
• **What are fossils?** (The preserved remains or traces of ancient organisms.)

Content Development
Using the Visuals Talk with your students about the uses of clocks. Geologists make similar use of the geological time scale. Refer to the demonstration you did earlier, again pointing out the layers in the container. Discuss the evolution of the camel shown in Figure 13–6 to demonstrate the importance of the geologic time scale. Emphasize that the layering of rock is only one factor used to develop the geological time scale. Other factors will be discussed later.

---

the oldest? (The one on the bottom.)
• **How do you know?** (It would have been deposited first.)

Content Development
Discuss and contrast the two arguments concerning the age of the Earth that are presented in the text. The following questions can be used to direct your class discussion.
• **What did those people who thought that the Earth was only a few thousand years old use as the basis of their argument?** (They thought that the major geological features had been produced suddenly by catastrophic events that humans rarely saw or understood.)
• **What was Hutton's proposal?** (He said that rocks and mountains had been changed gradual-

273

BACKGROUND INFORMATION
ERAS, PERIODS, AND EPOCHS

The almost immeasurably long geological history of the Earth wasn't divided arbitrarily. When the study of geology was in its formative years, the beginnings and ends of major time periods were set according to major biological events revealed in particular geological stratum. The boundary between the Precambrian and the Paleozoic eras, for example, was set at a geological stratum that recorded a great flowering of early multicellular life (the "Cambrian explosion"). And the transition between the Mesozoic and Cenozoic eras saw one of the most extensive mass extinctions in the history of life on Earth (the "Cretaceous-Tertiary—K-T—boundary event"). Thus, although we *measure* geologic time in fixed units (years), we often *describe* epochs, periods, and eras in terms of the organisms that characterize them.

TEACHING SUPPORT

Teaching Resources
- Activity: Radioactive Decay, p. 7

Earth's past with the help of a record in the rocks called the **geologic time scale.**

More than 100 years ago, researchers noticed that certain layers of rock often appeared in the same vertical order wherever they were found. It is the position of the layers relative to each other that determines their age. This knowledge helped geologists assemble a column of rocks in which each layer represented a different period of time. Geologists knew that the lower rock layers were deposited before the upper layers. Thus, lower layers are older than upper layers, provided that the layers have not been disturbed since they were formed. In addition, fossils found in lower layers are older than fossils found in the layers above them. **Relative dating** is a technique used by scientists to determine the age of fossils relative to other fossils in different layers of rock. However, because geologists did not know how long it took for the layers to form, they could not determine the actual age of the fossils.

Radioactive Dating

Near the middle of this century, our growing understanding of radioactivity provided scientists with a tool that could determine the actual age of rocks. Rocks are made up of many different elements. In certain rocks, some of these elements are radioactive. **Radioactive elements** decay, or break down, into nonradioactive elements at a very steady rate. Scientists measure this rate of radioactive decay in a unit called a **half-life.** A half-life is the length of time required for half the radioactive atoms in a sample to decay. This means that after one half-life, one half of the radioactive atoms in a sample have decayed. At the end of the next half-life, one half of the remaining radioactive atoms have decayed. In other words, one quarter of the original number of radioactive atoms remain after the second half-life reduces the remaining radioactive atoms by half.

Each radioactive element has a different half-life. Uranium-238, for example, has a half-life of 4.5 billion years. During that time, one half of the uranium-238 atoms in a rock sample decay into lead-206. Potassium-40 has a half-life of 1.3 billion years. During that time, one half of the potassium-40 atoms decay to argon-40, an inert gas that remains trapped inside the potassium crystals. Still another radioactive element, carbon-14, has a much shorter half-life of about 5770 years. During that period, half the carbon-14 decays to nitrogen-14.

Elements with different half-lives provide natural "clocks" that "tick" at different rates. When properly interpreted, these clocks help scientists date rocks and specimens of different ages. Here is how this process works.

Suppose geologists have uncovered what they think is a very old piece of rock—one that might date back to the birth of our planet. To determine the age of their sample, they measure and compare the amounts of uranium-238 and lead-206 it contains. Next, they determine how much lead has been

Figure 13–7 Scientists use the natural decay of radioactive elements to date certain fossils. (The bracketed numbers give a fractional representation of the radioactive element and its decay element in the sample.) It is the constancy of a radioactive element's decay that makes radioactive dating accurate.

Decay of a Radioactive Element with a Half-Life of 1 Million Years in a Fossil

Time (millions of years ago)	Amount of Radioactive Element (kg)	Amount of Decay Element (kg)
4	1 [1]	0 [0]
3	0.5 [1/2]	0.5 [1/2]
2	0.25 [1/4]	0.75 [3/4]
1	0.125 [1/8]	0.875 [7/8]
Present	0.0625 [1/16]	0.9375 [15/16]

▨ = Radioactive element
▨ = Decay element

13–2 (continued)

Content Development
Continue to use your container of materials to simulate the column of rocks that make up the surface of the Earth. Define relative dating and encourage students to think about the word *relative* and what it means. Discuss what relative dating can tell about the fossils found in various rock layers. The information that cannot be gained by relative dating, such as exact age, should also be emphasized.

Content Development
Students will need to understand radioactive decay and half-life in order to make sense of radioactive dating. They may have studied this topic in an earlier course so that a brief review might be sufficient.

SCIENCE, TECHNOLOGY, AND SOCIETY CONNECTION

Carbon–14: To Catch a Thief!

Many collectors purchase art and objects made by ancient peoples. But other people make remarkable forgeries designed to cheat museums and private art collectors. Scientists, however, can use the same sorts of techniques they use to date rocks and fossils to date artifacts—thus helping to protect collectors from making costly and embarrassing mistakes.

One technique for dating artifacts uses radioactive carbon-14, which is found in the remnants of once-living organisms. Carbon-14 dating works not only on bones, but also on objects carved from wood or made from clay and straw.

For example, pottery bowls and statues often contain bits of straw that were originally used to hold the clay together. Because carbon-14 decays at a steady rate over time to form nitrogen-14, investigators can determine how long ago a sample of straw died. They assume, of course, that the straw was alive until shortly before it was used to make the pottery object. In this way, radioactive dating can catch thieves and forgers of ancient art—and force a piece of pottery to confess its true age!

Archaeologists often date artifacts using techniques that measure radioactive decay.

produced by radioactive decay since the rock was formed. Because scientists know the half-life of uranium-238, they can calculate the rock's age.

Because uranium-238 and potassium-40 have such long half-lifes, they are quite useful in dating very old samples. But they are not much use in dating samples that are younger than about 100,000 years. This is why researchers use carbon-14.

Carbon-14 is particularly useful because it can be used to date material that was once alive, such as bones, or to date objects that contain once-living material. Because carbon-14 is present in the atmosphere, living things take it into their bodies while they are alive. After an organism dies, it no longer takes in carbon-14. By comparing the amounts of carbon-14 and nitrogen-14 in a sample, researchers can estimate the age of the sample. Because it has a relatively short half-life, carbon-14 isn't really useful in dating samples that are more than 60,000 years old, or roughly ten of the element's half-lives. After this period of time, there is really too little carbon-14 left to measure accurately.

Skills Development
Guided Practice
Skill: Graphing
Using the Visuals Use the information presented in Figure 13–7 to produce a two-color bar graph to show radioactive decay of an element with a half-life of 1 million years. You could also use the information presented in the textbook on page 275 about other radioactive materials to make a bar graph comparing the half-life of those materials.

SCIENCE, TECHNOLOGY, AND SOCIETY CONNECTION

Carbon–14: To Catch a Thief!
Carbon–14 is often used to date fossils that are less than 60,000 years old. This is approximately ten half-lives of this element. Other objects can be dated by the same process if they contain any substance that has once been alive. This is frequently the case with ancient pottery or statues that contain bits of straw.

This method of dating can protect collectors and museums from unscrupulous individuals who make forgeries that they wish to sell for large profits.

There are many ways that ancient artifacts and priceless art objects can be authenticated. Most of the experts in this field have some scientific background that they use in their work. This article could serve as a springboard for a research assignment for interested students. Have them research the science behind various authentication methods.

MULTICULTURAL STRATEGY

Dr. Chien-Shiung Wu is a native of Shanghai, China, who came to the United States to study at the University of California. Her field of research is beta decay. Beta decay takes place when a radioactive element breaks down and emits beta particles, or electrons.

Dr. Wu is best known for her work on a theory about the behavior of nuclear particles that was developed by two other Asian scientists, Dr. Tsung Dao Lee and Dr. Chen Ning Yang. Dr. Wu was able to provide proof for the theory.

Dr. Wu has taught at Smith College and Princeton University. She became a member of the Columbia University physics faculty in 1952.

TEACHING SUPPORT

Teaching Resources
- Activity: The Geologic Time Scale, p. 3

Study Guide
- Section 13–2, p. 127

Half-Lives of Radioactive Elements

Element	Half-life
Rubidium-87	50 billion years
Thorium-232	13.9 billion years
Uranium-238	4.5 billion years
Potassium-40	1.3 billion years
Uranium-235	713 million years
Carbon-14	5770 years

Figure 13-8 As you can see from this chart, the half-lives of many radioactive elements are much longer than a human's whole life.

Radioactive dating is important because it enables researchers to calculate the actual age of a sample. This is called **absolute dating**. The evidence provided by radioactive clocks, along with observations of long-term geological processes, has enabled geologists to compile a remarkably accurate history of our planet.

Using these data, scientists have determined that the Earth is about 4.5 billion years old. By combining absolute dating, relative dating, and observations of important events in the history of life on Earth, scientists have divided the 4.5 billion years into large units called **eras**. Eras are further divided into **periods,** which are in turn divided into **epochs**. Unlike the periods of time we use daily, the components of geological time do not have standard lengths. See Figure 13–9.

13–2 SECTION REVIEW

1. What is the age of the Earth?
2. **Critical Thinking—Applying Concepts** How might relative dating provide inaccurate data?

Figure 13-9 GEOLOGIC HISTORY OF THE EARTH

Era	Period	Epoch	End Date (millions of years ago)	Notes
Cenozoic	Quaternary	Recent		Humans are the dominant form of life; civilization begins and spreads
		Pleistocene	0.01	"The Ice Age"; modern humans present; mammoths and other animals become extinct
	Tertiary	Pliocene	2.5	Fossils of ancient humans near end of epoch; many birds, mammals, and sea life similar to modern types; climate cools
		Miocene	5	Many grazing animals; flowering plants and trees resemble modern types
		Oligocene	25	Fossils of primitive apes; elephants, camels, and horses develop; climate generally mild
		Eocene	38	Fossils of "dawn horse" (*Hyracotherium*); grasslands and forests present; many small mammals; larger mammals such as whales, rhinoceroses, and monkeys begin to develop

13–2 (continued)

Skills Development
Guided Practice
Skill: Making charts

Using the Visuals Use the information in Figure 13–9 as the basis for a class or group activity. Have students produce their own geological time scale. Each group will need a piece of long paper. (If rolled paper is not available, cut strips of graph or construction paper and tape it end to end.) Space the entries to represent the relative lengths of times. Encourage students to add illustrations of the types of organisms that lived during the various periods. You might want to have different groups responsible for different geological periods and then assemble all the pieces to make a class project. You would then end up with one very long mural.

SECTION REVIEW 13–2

1. The Earth is estimated to be about 4.5 billion years old.
2. Relative dating depends, to a certain extent, on the position of the fossil sample in the rocks. One way this method of dating may prove inaccurate is if movements in the Earth push rock layers that contain old fossils above layers that contain younger fossils. Thus, older fossils may be found above

Era	Period	Epoch	End Date (millions of years ago)	Notes
Cenozoic	Tertiary	Paleocene	55	Beginning of "Age of Mammals."; flowering plants and small mammals abundant; many different climates exist
Mesozoic	Cretaceous		65	First flowering plants; placental mammals develop; dinosaurs die out, as do many marine animals, at end of period
Mesozoic	Jurassic		135	The Rocky Mountains rise; first birds; palms and cone-bearing trees dominant; largest dinosaurs thrive; primitive mammals develop
Mesozoic	Triassic		195	"Age of Reptiles" begins; first dinosaurs; first mammals; corals, insects, and fishes resemble modern types
Paleozoic	Permian		245	First cone-bearing plants; ferns, fishes, amphibians, and reptiles flourish; many marine invertebrates, including trilobites, die out
Paleozoic	Carboniferous		285	Ice covers large areas of the Earth; swamps cover lowlands; first mosses; great coal-forming forests form; seed ferns grow; first reptiles and winged insects appear
Paleozoic	Devonian		345	First forests grow in swampy areas; fishes flourish; first amphibians, sharks, and insects develop
Paleozoic	Silurian		400	"Age of Fishes" begins; coral reefs form; jawed fishes develop; first land plants appear; first air-breathing animals, including land arthropods
Paleozoic	Ordovician		430	First fishes (jawless) appear; invertebrates flourish in the sea
Paleozoic	Cambrian		500	"Age of Invertebrates" begins; trilobites, brachiopods, sponges, and other marine invertebrates are present
Precambrian			580	Earth's history begins; first life forms in the sea; first prokaryotes (bacteria) appear; as time passes, first eukaryotes appear

ESL STRATEGY

Write the following life forms on the chalkboard. Ask students to order the life forms chronologically. Use the number 1 to indicate the oldest form, 2 for the next oldest, etc.
____ "Age of Fishes"
____ bacteria in the Precambrian Era
____ appearance of modern humans
____ dinosaurs die out
____ trilobites and other invertebrates are present
____ formation of coal-forming forests during the Carboniferous Period

younger ones. The relative position of these fossils may provide an inaccurate picture of the ages of the two fossils.

Reinforcement/Reteaching

If some students are having difficulty answering the Section Review questions, go back to the appropriate part of the section and review those topics that the students are having difficulty understanding.

Closure

Ask three student volunteers to come to the front of the class. Each one is to review the major points of one of the three parts of Section 13–2. If they miss an important concept, prompt them with questions.

TEACHING SUPPORT

Laboratory Manual
- Interpreting Events from Fossil Evidence, p. 167

Teaching Resources
- Hands-on Activity: Turned to Stone, p. 67
- Ecology Investigation: Making Casts of Animal Tracks, p. 3

Guide For Reading
- Why is the fossil record incomplete?
- How are sedimentary rocks formed?
- What kinds of changes are shown in the fossil record?

13-3 The Fossil Record

Since the early nineteenth century, biologists have learned about animals and plants that lived long ago by examining preserved traces of those organisms, or fossils. As you can see in Figure 13–10, there are many different kinds of fossils. A fossil can be as large and complete as an entire perfectly preserved animal or plant, or it might be as small and incomplete as a tiny fragment of a jawbone or leaf. There are fossil footprints, fossil eggs, and even fossilized animal droppings. Some of the fossils Darwin found in South America represent animals so strange that they resemble creatures from science fiction films more than they do any plant or animal alive today. How did these organisms leave their remains in stone? How do fossil remains help to explain the history of life on Earth?

How Fossils Form

Fossils have been formed in a variety of ways, all of which depend a great deal on chance. In cold places, animals sometimes fell into crevasses in ice or became trapped in snow fields. Insects and other small animals were occasionally trapped in the sticky tree sap that eventually hardened into amber. Still other fossils were formed when animals became mired in peat bogs, certain kinds of quicksand, or tar pits. In all these cases, the material that surrounded the dead animal helped to protect it from decay and acted to preserve it as evidence of past life.

Most fossils are found in **sedimentary rock**. Sedimentary rocks are formed when exposure to rain, heat, and cold breaks down existing rocks into small particles of sand, silt, and clay.

Figure 13–10 *This hatchling dinosaur did not live to become an adult, but its bones and even the delicate shell of its egg remain. The trilobite remains illustrate two kinds of fossils, a cast (center) and a mold (right).*

TEACHING STRATEGY 13-3

Focus/Motivation

Using the Visuals Display several examples of fossils. If possible, choose some that are nearly complete and some that are fragments. Point out the pictures in Figure 13–10.

- Do any of the fossils resemble organisms that still exist today? (Answers will vary.)
- How and why do you think that these particular remains were preserved and not destroyed by the forces of nature? (Answers will vary.)
- What can you guess about the lives of these organisms from their fossils? (Answers will vary. They might guess that certain organisms lived in water; some may have been predators based on their teeth; and so on.)

Content Development

Discuss the various ways that fossils are formed. If you have access to specimens or pictures of fossil digs, share them with your students. Discuss ways that scientists preserve the fragile fossils that are discovered. You may

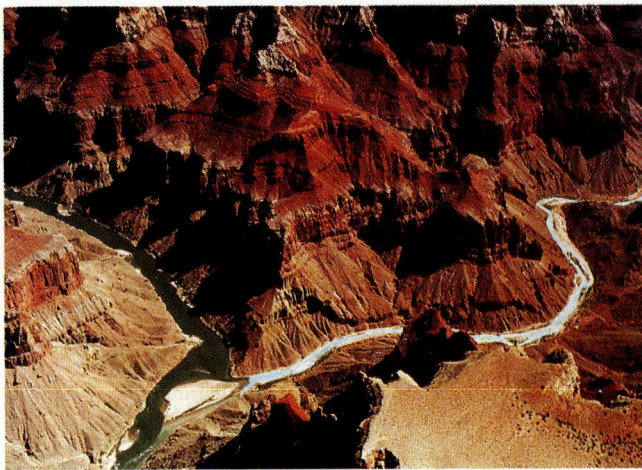

Figure 13–11 *The Colorado River, flowing along the bottom of the Grand Canyon, has cut through layer upon layer of rock over millions of years, exposing fossils long buried in sedimentary rock.*

Figure 13–12 *Fossils are usually found in sedimentary rocks. Because lower sedimentary rock layers are older than upper rock layers, scientists can determine the sequence of changes in life forms on the Earth.*

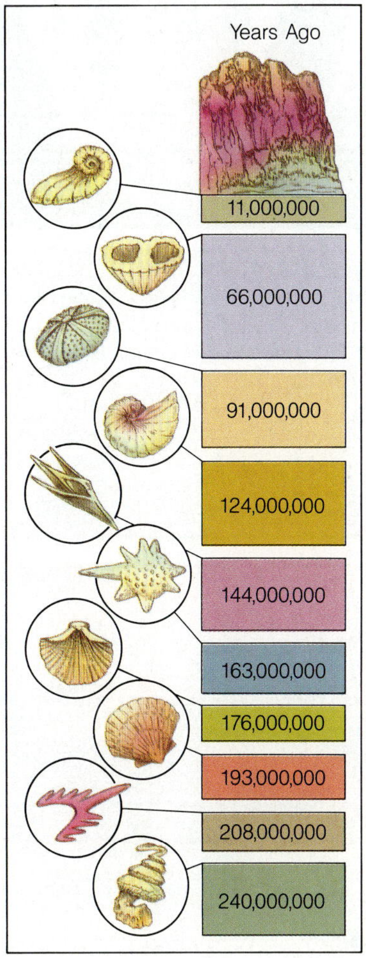

These particles are carried by streams and rivers into lakes or seas. Because the particles are heavier than water, they eventually settle to the bottom of the lake or sea. Here they build up in layer upon layer of sediments. Dead organisms, carried in the water, also eventually fall to the bottom. The organisms can become embedded in the sediment layers. As sediments pile up, pressure on the lower layers compresses the sediments and slowly turns them into rock, which thus preserves the remains of the dead organisms.

In some cases the small particles of rock that buried plant or animal remains preserved the organism's soft parts. In other cases, the hard parts of plants or animals were preserved when wood, shells, or bones were replaced with long-lasting mineral compounds. These fossils are petrified, or turned to rock.

Fossil Evidence: Problems in Assembling the Puzzle

You might think of fossils as pieces of a jigsaw puzzle. All the pieces, assembled correctly, would provide a complete picture of the history of life on Earth. But we do not have all the pieces, and we probably never will.

The chancy process by which organisms are fossilized means that the fossil record is not as complete as we would like it to be. For every organism that leaves a proper fossil, many die and vanish without leaving a trace. Because sedimentary rocks form only in certain bodies of water, organisms that live in mountains and deserts may never become part of the fossil record.

Finding fossils embedded in tons of rock is difficult, if not impossible. However, the natural forces that help make sedimentary rock may reveal hidden fossils. Rocks that contain

FACTS AND FIGURES

The oldest fossils are the remains of bacteria. These fossils are about 3.5 billion years old.

have students who are interested in this topic and have visited fossil sites.

Focus/Motivation

Students become more interested in fossils and the study of paleontology when they learn more about the procedures used. It is a relatively simple activity to make molds using plaster of Paris or modeling clay. Place an object such as a shell covered with petroleum jelly between two layers of clay or plaster. Remove the object, leaving the mold. Mix up the casts and molds and have students try to match them up correctly.

Content Development

Using the Visuals Emphasize that there are many pieces of the fossil record that are missing.

Spend some time talking about how unusual it is for organisms to fossilize. Next, discuss how difficult it is to find and preserve those fossils that do exist. With all these negative factors, it is amazing that the record is as complete as it is. Use Figure 13–12 to emphasize what can be learned from fossils about the evolution of life on Earth.

fossils may be exposed by weather. Bit by bit, the upper, younger layers are worn away by wind and rain, and the older layers beneath are exposed. The Grand Canyon is an example of a place where many layers of rock have been exposed—in this case, by the moving water of the Colorado River. When a fossil is exposed, a fortunate (and observant) scientist may happen along at the right time and remove it for study.

The quality of fossil preservation also varies. Some fossils are preserved so perfectly that we can see the microscopic structure of tiny bones and feathers. Other fossils are not preserved as well and so raise fascinating questions about their meaning and importance. Often scientists must reconstruct an extinct species from a few fossil bits and pieces of bone, leaves, or stems. Fossil reconstruction requires a thorough knowledge of the anatomy of living animals and plants as well as a great amount of skill.

Placing absolute dates on fossils isn't always easy, and scientists often use several methods to check their results. As we mentioned earlier, radioactive dating using potassium-40 works well with old fossils, whereas carbon-14 dating is quite accurate in dating more recent fossils.

What the Fossil Record Tells Us

Scientists who study fossils are called **paleontologists** (pay-lee-uhn-TAHL-uh-jihsts). Over the years, paleontologists

Figure 13–13 About 1 million years ago, this sabertooth cat became trapped in the La Brea tar pits, in what is now California (top). Over time, mud hardened into stone, preserving this impression of a delicate fern (center). Encased in amber, or fossilized tree sap, this lizard has been perfectly preserved (bottom left). This sea lily, a relative of starfish, lived in Earth's oceans about 350 million years ago (bottom center). Each of these microscopic yellow cylinders is a fossil bacterial cell (bottom right).

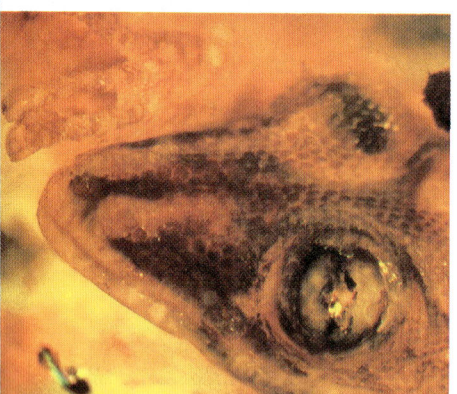  

Figure 13–14 Approximately 310 million years ago, places on Earth looked like this. At that time the dragonfly had a wingspan of about one meter. Compare this to a modern dragonfly that flits delicately above the surface of a pond.

Figure 13–15 The first animal commonly called a "horse" was much smaller than modern horses. The other animals in this series represent several stages in the evolution of body size and hoof structure.

have collected millions of fossils to make up the **fossil record**. The fossil record represents the preserved collective history of the Earth's organisms. Although incomplete, the fossil record has long inspired scientists. As the naturalist Loren Eisley once wrote, ". . . every bone that one holds in one's hands is a fallen kingdom, a veritable ruined world, a unique object that will never return through time."

Our picture of ancient life has many missing pieces. Still, paleontologists have assembled good evolutionary histories for many animal groups. Figure 13–15, for example, shows several ancient animals whose evolutionary line over 50 million years gave rise to the modern horse. The oldest of these, the "dawn horse," was much smaller than modern horses and had very different foot and leg bones. The increase in size and the development of the hooves that facilitate running in these animals have been very dramatic.

The fossil record also tells of major changes that occurred in Earth's climate and geography. Fossil shark teeth have been found in Arizona, indicating that the deserts of the American Southwest were once covered by ancient seas. Giant fossil ferns found in Canada show that North America once had a much warmer, tropical climate. Every period, every epoch in Earth's history had a different climate and contained different kinds of organisms that were adapted to it.

But species do not last forever. As Earth's environments changed over time, many species died out. In the very spot where you now sit and read this book, giant dragonflies with wings that measured almost a meter across may have flitted over swamps filled with giant ferns. Dinosaurs appeared, thrived for a time, and eventually disappeared into Earth's fossil record. The huge fossil skeletons they left behind, when reconstructed, still have the power to amuse and amaze us. **The fossil record shows that change followed change on Earth.**

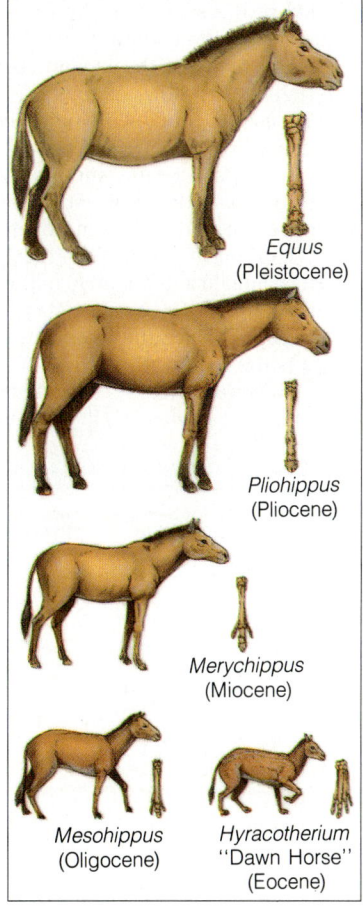

Equus (Pleistocene)
Pliohippus (Pliocene)
Merychippus (Miocene)
Mesohippus (Oligocene)
Hyracotherium "Dawn Horse" (Eocene)

281

PROBLEM SOLVING IN BIOLOGY

RADIOACTIVE DATING

You are a famous paleontologist and an expert in radioactive-dating techniques. One day, two visitors to your laboratory present you with two different fossils. One fossil is a dinosaur footprint, the other a human jawbone. Both were found at the bottom of a deep valley cut by a stream through cliffs of sedimentary rock. Your guests are very excited. Because they found these fossils next to each other near the stream bed, they feel they have found conclusive evidence that humans and dinosaurs lived at the same time. You are asked to date the samples to confirm their claims.

You first test the human jawbone. You determine that it now contains 1/16 the amount of carbon-14 it contained when alive. How old is the jawbone?

You next examine the fossil footprint. You discover that the fresh mud that the dinosaur stepped in had just been covered with a thin layer of volcanic ash. You study the amount of potassium-40 and argon-40 in the ash. The ratio shows that 1/10 of one potassium-40 half-life has passed since the footprint was made. How old is the footprint?

Were your visitors' conclusions about these fossils' ages correct? If they were not, how could you explain the fact that they were found together at the bottom of the valley? Refer to Figures 13–7 and 13–8 and pages 274 and 276 for helpful information.

13–3 SECTION REVIEW

1. Why is the fossil record incomplete?
2. Describe how sedimentary rocks are formed.
3. What is a fossil? How are fossils formed?
4. What kind of changes are reflected in the fossil record?
5. **Critical Thinking—Summarizing Information** What are some problems faced by paleontologists when reconstructing the history of life on Earth?

13–4 Evidence from Living Organisms

Fossils of extinct organisms are not the only evidence that shows the ongoing process of evolution. All living organisms carry within their bodies traces of the history that links them to their ancestors.

Similarities in Early Development

In the late nineteenth century, scientists noticed that the **embryos** of many different animals looked so similar that it was difficult to tell them apart. Embryos are organisms at early stages of development. Today, no scientist would say that a human embryo is identical to a fish or a bird embryo. However, as you can see in Figure 13–16, all of these embryos are similar in appearance during early stages of development. But why are they so similar? And what do these similarities tell us?

Similarities in early development indicate that similar genes are at work. All genes in an organism are not active at the same time. But those that are active during the early development of fish, birds, humans, and related animals are the shared heritage from a common ancestor. The common ancestor of these different animals had a particular sequence of genes that controlled its early development. And this sequence of genes has been passed on to the species that descended from it.

As they grow and develop, the embryos gradually become more and more dissimilar. These differences in form are caused by genes that have changed during the course of evolution. Changes in form are produced by mutations, or changes in the genetic blueprint contained within an organism's DNA. Remember that this DNA blueprint is very complicated. Mutations that affect early stages of development are likely to be lethal, or deadly. An organism carrying such a mutation dies while it is an embryo, and its genes are not passed on. As a consequence, the portions of DNA that control the early stages of development remain relatively unchanged. Thus the embryos of different kinds of animals resemble each other.

Mutations that cause less drastic, as well as potentially useful, changes in structure are likely to occur at later stages of growth and development. Thus the later stages in the development of related organisms begin to show marked differences. An organism with this kind of mutation may survive to reproduce and pass the changes in its DNA to its offspring.

Similarities in Body Structure

In the embryos of many animals—humans, birds, horses, and whales, for example—the clumps of cells that develop into limbs look quite similar. But as the embryos mature, the limbs grow into arms, wings, legs, and flippers that differ greatly in form and function. These different forelimbs evolved in a series

Guide For Reading

- What is the best explanation for the structural and biochemical similarities that exist in living organisms?
- How do similarities in embryo development support the concept of common descent?
- What is a homologous structure?
- How does biochemistry provide evidence for evolution?

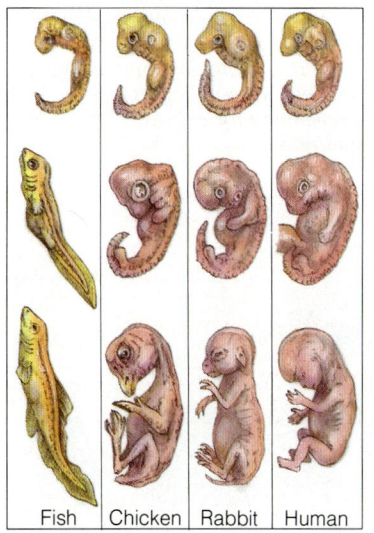

Figure 13–16 During certain embryological stages, vastly different organisms show similarities. During later stages of development, profound changes occur. Thus the adults bear little resemblance to one another.

Fish Chicken Rabbit Human

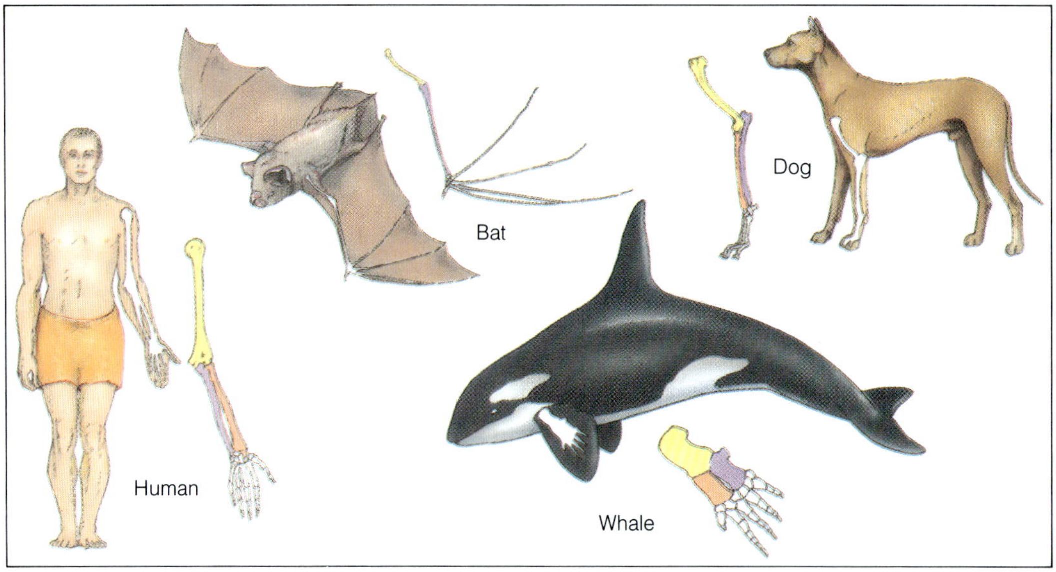

Figure 13–17 Superbly adapted to performing different tasks, the limbs of various organisms are remarkably similar in structure. Here you can see the homologous limbs of four mammals. How is the form of each limb adapted for different movements?

Figure 13–18 Albinism (lack of color) is the result of a gene mutation. In this photograph a normally colored penguin is "chatting" with an albino penguin. Perhaps it's scolding its pale companion for being out of uniform.

of evolutionary changes that altered the structure and appearance of the arm and leg bones of ancient animals. As you can see in Figure 13–17, the changes in the final structure of the limbs of different species are adaptations that enable each organism to survive in a different environment. Structures such as these that develop from the same body parts are called **homologous structures**.

Many animals have organs that seem to serve little or no purpose. These are called **vestigial organs**. Vestigial organs may resemble miniature arms, legs, tails, or other structures.

Why should organisms carry organs that have little or no function? As you will see in the next chapter, evolutionary change occurs as species develop new adaptations. The changes may involve the modification of primitive limbs into wings or flippers. But evolution may also involve the loss or reduction of structures whose main function is no longer valuable. Structures that no longer serve their original purpose often become smaller, but they may never disappear completely.

Snakes, for example, evolved from four-legged ancestors. Today, some pythons and boa constrictors have tiny bones that are all that remain of these legs. These vestigial legs have no function in walking, which was their original purpose.

Humans also have vestigial organs. We have a set of miniature tailbones at the base of our spine—which obviously no longer support a tail. In most people, the muscles that move the ears are vestigial. Humans also have a small sac called the appendix that leads off the large intestine. The appendix is a vestigial organ that does not seem to serve a function in digestion today. In fact, an appendix can become infected, and sometimes must be removed surgically.

Today, a large functioning appendix is found in some animals, such as the koala, that eat primarily plant materials. So it is probable that our appendix is left over from a time during which our ancestors needed this organ to digest their food.

Similarities in Chemical Compounds

All organisms—from bacteria to humans—share many biochemical details. All organisms use DNA and/or RNA to carry information from one generation to the next and to control growth and development. The DNA of all eukaryotic organisms always has the same basic structure and replicates in the same way. The RNAs of various species may act a little differently, but all RNAs are similar in structure from one species to the next. Remember, too, that ATP (adenosine triphosphate) is an energy carrier found in all living systems. A wide variety of complicated proteins, such as cytochrome c, are also shared by many organisms.

The more closely related two species are, the more closely their important chemical compounds resemble each other. In Chapter 15 we will discuss the importance of similar proteins in studying evolutionary relationships among species.

What Homologies Tell Us

Similarities in structure and biochemistry provide powerful evidence that all living things evolved from common ancestors. **The structural and biochemical similarities among living organisms are best explained by Darwin's conclusion: Living organisms evolved through gradual modification of earlier forms—descent from a common ancestor.** In a previous chapter you learned that life's chemical pathways are extremely complex. In a later chapter you will learn that tissues and organs are equally complex. If organisms had arisen independently of one another, there would be very little chance that they would have similar structures and biochemistries. The very complexity of life and its processes supports Darwin's conclusion.

Figure 13–19 *If you look closely, you can see the vestigial leg bones of this python, which serve no function in locomotion—the snake can move quite well without them! Another example of a vestigial organ is the human appendix. In a koala, however, the appendix is a functioning organ that helps the animal to digest leaves.*

13–4 SECTION REVIEW

1. What is the best explanation for the structural and biochemical similarities that exist in living organisms?
2. What is a vestigial organ?
3. How does evidence from embryology support the concept that all animals evolved from common ancestors?
4. **Critical Thinking—Applying Concepts** How does biochemistry support the idea that all living things evolved from common ancestors?

ECOLOGY NOTE
VESTIGIAL ORGANS

If a change in an organism's habits or environment causes a structure to become unnecessary, natural selection will no longer work to keep that structure in the population. If a mutation occurs that eliminates the structure without otherwise harming the organism, the structure can disappear. But remember that genetic variation is random. Just because a structure is not used in an animal does not mean that there will be mutations to get rid of that structure. It is likely, however, that there is genetic variation in the size of that structure. Genes that cause a useless structure to get smaller will certainly not hurt the organism. Those genes may even help the organism if they avoid wasting energy in growing structures that serve no purpose. For this reason, genetic changes that cause useless structures to get smaller can take over the population.

LABORATORY INVESTIGATION

EXAMINING HOMOLOGOUS STRUCTURES

Before the Lab
Assemble the number of dissecting pans and instruments needed for the number of groups that you have. If you are missing models or charts of the human skeleton, assemble those materials also. If you are using a full-size skeleton, it can be shared by several groups. Fresh chicken wings should be purchased at the market no more than two days prior to the activity. If frozen wings are to be used, be sure to allow ample time for them to thaw completely.

Dissection Alternative
If you or your students would prefer not to do a dissection in class, you can substitute accurate anatomical drawings or models of the frog and chicken skeletons. Students can complete most of the investigation using such drawings or models.

Pre-Lab Discussion
You will need to once again review what homologous structures are and what they mean in terms of support for evolution. If you intend to reuse the frogs for later internal dissection, remind students not to do any cutting except on the forelimb, as directed. Cutting muscle and bone can be a little difficult for novices, so you might want to demonstrate this operation before the students begin.

Skills Development
Students will use the following skills while completing this investigation.
1. Manipulative
2. Safety
3. Observing
4. Comparing
5. Relating
6. Diagramming

LABORATORY INVESTIGATION: EXAMINING HOMOLOGOUS STRUCTURES

PROBLEM
How do homologous structures support the theory of common descent?

MATERIALS (per group)

preserved frog	dissecting pan
chicken wing	scalpel
human skeleton	scissors
(model or chart)	forceps

PROCEDURE

1. Put a preserved frog in a dissecting pan. Using your scissors, cut the skin around the base of the frog's arm where it joins the body. See the accompanying figure. Then grasp the cut edge of the skin with forceps and pull it down the arm until the skin is removed. The skin should come off the arm easily.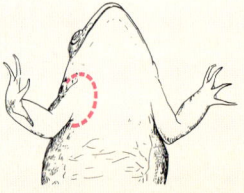
2. Pull the frog's arm up toward the head. Use your scissors to cut the muscles around the base of the arm. When you have completed your cut, the bones of the arm and shoulder should be visible. Pull back on the arm. The arm should come loose from the shoulder, but you may have to use your scissors to cut the connections that hold the arm in place.
3. With your scissors, cut the ends of the muscles away from the bone on the severed frog arm. Use your forceps to peel the muscles away from the bones. Remove as much of the muscles from the frog's arm as you can. Then, using a scalpel, gently and carefully scrape the bones clean. **CAUTION:** *Use care when using a scalpel. Always cut or scrape away from your body.*
4. Draw a diagram that shows the arrangement of the bones in a frog's arm.
5. Put a chicken wing in a dissecting pan. Use your scissors to cut the skin toward the wingtip. The dotted line in the accompanying figure indicates where you should cut. Using the forceps, remove all the skin from the chicken wing.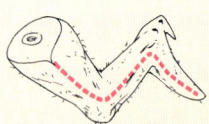
6. Carefully cut the muscles away from the bones using your scissors. With a scalpel, carefully scrape the bones clean. **CAUTION:** *Use care when using a scalpel. Always cut or scrape away from your body.*
7. Draw a diagram that shows the arrangement of the bones in a chicken wing.
8. Use the model or chart of the human skeleton to draw the arrangement of the bones in the human arm.

OBSERVATIONS
1. How many bones did you find in each of the limbs?
2. In what ways are the structures of the limbs similar?
3. In what ways are the shapes of the bones similar?

ANALYSIS AND CONCLUSIONS
1. List some functions of each of the three different limbs you have examined.
2. Even though the limbs of these organisms are very different in outward appearance and function, their internal structure is remarkably similar. How can you explain this observation?
3. From an evolutionary standpoint, what would it mean if these bones were very different in structure?

Safety Tips
Any time that students use dissection instruments, they should be cautioned to use them properly and with great care. This lab experiment is particularly tricky when they begin removing the muscle from the bone.

Teaching Strategy
1. Removing skin and muscle from the bones is a difficult job for anyone. Students should be carefully monitored during this experiment. As you circulate, you might have to assist some students with this procedure. Once the muscles are removed, show students how to position the bones so that they can be compared more easily.
2. Point out the groove in the frog's forearm, which almost separates the forearm into two bones.

STUDENT STUDY GUIDE

SUMMARIZING THE CONCEPTS

The key concepts in each section of this chapter are listed below to help you review the chapter content. Make sure you understand each concept and its relationship to other concepts and to the theme of this chapter.

13–1 Evolution and Life's Diversity
- The observation of the diversity of living things on Earth contributed to the formation of Darwin's theory of evolution.
- Fitness is the combination of physical traits and behaviors, or adaptations, that help organisms survive.

13–2 The Age of the Earth
- Scientists use the decay of radioactive elements to date the Earth's rocks.
- The position of rock layers as well as the decay of radioactive elements have enabled scientists to set up a geological time scale.
- Dating techniques have shown that the Earth is about 4 1/2 billion years old.

13–3 The Fossil Record
- Fossils, the remains of ancient organisms, are preserved in sedimentary rock layers.
- Although the fossil record is not complete, fossil evidence shows that life on Earth has changed over time.

13–4 Evidence from Living Organisms
- Similarities in the structure and development of embryos provide evidence of descent from a common ancestor.
- Homologous structures and vestigial organs are additional evidence of descent from a common ancestor.
- The biochemistries of living organisms support the hypothesis that these organisms evolved from a common ancestor.

REVIEWING KEY TERMS

Vocabulary terms are important to your understanding of biology. The key terms listed below are those you should be especially familiar with. Review these terms and their meanings. Then use each term in a complete sentence. If you are not sure of a term's meaning, return to the appropriate section and review its definition.

13–1 Evolution and Life's Diversity
evolution
fitness
common descent
adaptation

13–2 The Age of the Earth
fossil
geologic time scale
relative dating
radioactive element
half-life
absolute dating
era
period
epoch

13–3 The Fossil Record
sedimentary rock
paleontologist
fossil record

13–4 Evidence from Living Organisms
embryo
homologous structure
vestigial organ

287

---

3. Help students to find the separation between the radius and ulna of their own forearm.

Observations
1. Answers will vary.
2. All consisted of three main sections. The upper arm or wing was one large bone. The forearm or wing was two parallel bones or one bone with a deep groove. The hand or wing tip was many small bones.
3. They were all narrow at the center, elongated and broader at the ends.

Analysis and Conclusions
1. Frog's arm aids standing and swimming. Bird's wing is for flying. Human arm and hand are for grasping and manipulating.
2. The different limbs use homologous structures, or structures that all evolved from a common ancestral structure. As such, they should appear similar internally.
3. One would then assume that the organisms were not closely related in terms of sharing a common ancestor.

Going Further: Enrichment
Extend this lab by comparing the frog, chicken, and human forelimb with the forelimbs of several other skeletons. A bat-wing skeleton, if available, makes an interesting comparison. Make diagrams of these limbs also and answer the same questions. Conclude the lab investigation by writing a paragraph that summarizes your findings. Include both similarities and differences.

CHAPTER REVIEW

CONTENT REVIEW

Multiple Choice
1. b 2. d 3. a 4. d
5. d 6. a 7. b 8. b

True or False
1. T
2. F adaptation
3. F epochs
4. T
5. F absolute dating
6. F 4 ½ billion
7. T
8. F homologous organs

Word Relationships

A.
1. human vestigial organs; heart is a functional organ
2. relate to ways the age of fossils are determined; double dating is done by modern humans!
3. lengths of geologic time; half-life is the unit of radioactive decay
4. terms used to describe an organism's ability to survive in an evolutionary sense; fossil is the preserved remains of an ancient organism

B.
5. fossil
6. layers
7. relative dating
8. homologous organs

CONCEPT MASTERY

1. Since radioactive elements decay at a steady rate, scientists can calculate the age of rock samples by comparing the amounts of the various elements found in them.
2. A geologic time scale summarizes the history of life on Earth. This time scale is based on the fossil record in the Earth.
3. A vestigial organ is present in an organism, yet has no known function. An example of a vestigial organ in humans is the appendix. Since this organ is found in animals that eat plants, scientists speculate that the appendix may have functioned in early humans who ate more plant material.
4. Common descent means that a group of dissimilar organisms descended from a common ancestor.
5. A fossil is any remains of past life. A fossil may be the tracks of an animal. A fossil may be a cast or mold of an organism or the remains of organisms preserved in amber, tar, or frozen in ice.
6. Sedimentary rocks form when existing rocks are eroded into small particles. These small particles are carried in water and eventually fall to the bottom of a body of water, where they form layers. These layers are subject to pressures that slowly change them into rocks.

CHAPTER REVIEW

CONTENT REVIEW

Multiple Choice

Choose the letter of the answer that best completes each statement.

1. The length of time it takes for one half of a radioactive element to decay is a (an)
   a. era.
   b. half-life.
   c. year.
   d. epoch.
2. Another term for radioactive dating is
   a. relative dating.
   b. geologic time scale.
   c. calendar dating.
   d. absolute dating.
3. The ability of an organism to survive and reproduce is known as
   a. fitness.
   b. evolution.
   c. diversity.
   d. adaptation.
4. Fitness arises through a process called
   a. common descent.
   b. half-life.
   c. homologous change.
   d. adaptation.
5. Evidence for the age of the Earth is provided by all of the following except
   a. fossils.
   b. radioactive dating.
   c. the rate of sediment formation.
   d. vestigial organs.
6. Most fossils are found in
   a. sedimentary rock.
   b. amber.
   c. ice.
   d. tar pits.
7. A vestigial organ is
   a. always useful.
   b. useless.
   c. not evidence for evolution.
   d. sometimes useful.
8. Change in species over time is called
   a. fitness.
   b. evolution.
   c. diversity.
   d. relative dating.

True or False

Determine whether each statement is true or false. If it is true, write "true." If it is false, change the underlined word or words to make the statement true.

1. <u>Paleontologists</u> are scientists who study fossils.
2. The ability of an organism to survive and reproduce is known as <u>diversity</u>.
3. In the Earth's history, periods are divided into <u>eras</u>.
4. <u>Embryos</u> are early stages in an organism's development.
5. <u>Relative dating</u> is the same as radioactive dating.
6. Today, scientists have evidence that shows that the Earth is <u>1 million</u> years old.
7. The theory that organisms share a common ancestor is known as <u>common descent</u>.
8. A human's arm and a dog's leg are examples of <u>vestigial organs</u>.

Word Relationships

A. *In each of the following sets of terms, three of the terms are related. One term does not belong. Determine the characteristic common to three of the terms and then identify the term that does not belong.*

1. appendix, tailbones, ear muscles, heart
2. absolute dating, double dating, radioactive dating, relative dating
3. half-life, era, epoch, period
4. diversity, fitness, fossil, adaptation

B. An analogy is a relationship between two pairs of words or phrases generally written in the following manner: a:b::c:d. The symbol : is read "is to," and the symbol :: is read "as." For example, cat:animal::rose:plant is read "cat is to animal as rose is to plant."

In the analogies that follow, a word or phrase is missing. Complete each analogy by providing the missing word or phrase.

5. biologist:organism::paleontologist:_____
6. era:periods::sedimentary rock:_____
7. half-life:absolute dating::rock position:_____
8. appendix:vestigial organ::bat wing and whale flipper:_____

CONCEPT MASTERY

Use your understanding of the concepts developed in the chapter to answer each of the following in a brief paragraph.

1. Explain how radioactivity is used to date rock samples.
2. What is the geologic time scale? How was it developed?
3. What is a vestigial organ? Give an example of a vestigial organ in humans.
4. What does the phrase common descent mean?
5. What is a fossil?
6. How is sedimentary rock formed?

CRITICAL AND CREATIVE THINKING

Discuss each of the following in a brief paragraph.

1. **Summarizing information** Evolutionary biologists say that there is good reason for gaps in the fossil record. Can you explain why some extinct animals and plants were never fossilized?
2. **Applying concepts** A giraffe's long neck enables it to eat the leaves of trees. How does this adaptation help the giraffe survive?
3. **Sequencing events** How did the work of geologists help Darwin formulate his theory of evolution?
4. **Applying technology** Discuss the limitations of radioactive dating.
5. **Making inferences** How did the diversity of life that Darwin observed during the five-year voyage of the *Beagle* contribute to the development of his theory?
6. **Using the writing process** You have been given the opportunity to become the first person to travel back in time to the age when dinosaurs roamed the Earth. Describe your feelings as you observe these ancient creatures walking through the area you now call home.
7. **Using the writing process** This time you will travel forward in time several million years. Use your imagination to describe the kinds of organisms now living on Earth. (*Hint:* Remember that any organism you describe must be well adapted to the environment of Earth in the distant future.)

CRITICAL AND CREATIVE THINKING

1. Some organisms never became fossils because they lived in areas away from water and thus their remains did not become part of sedimentary rock formations. Also, organisms that consisted mostly of soft tissues did not usually become fossils. Other answers are possible.
2. The long neck of the giraffe allows this animal to eat leaves far above the ground. Other animals that eat plants cannot reach tall tree branches—thus the giraffe is assured a good food supply with little competition from other animals.
3. In the seventeenth and eighteenth centuries, geologists began to notice evidence in the Earth that indicated that the Earth was much older than many people thought. Evidence also showed that over long periods of time, great changes occurred over much of the Earth. This evidence of great age inspired Darwin to view the evolution of organisms on Earth as part of a long series of changes.
4. Some rock samples do not contain radioactive materials, and thus they cannot be dated by this method. Slight differences in the rate of decay may mean a little less than total accuracy. Also, some fossils are too young to be dated with accuracy using radioactive techniques.
5. Darwin was amazed by the diversity of living things he observed. But he was even more astounded by the number of fossil organisms he found. Darwin began to think about ways to explain the great diversity of life on Earth. His theories were an attempt to explain his observations and the evidence he gathered.
6. Check student writings. Make sure that the information they present about past life forms is accurate. They should refer to other sources as well as this textbook for information.
7. Check student writings. Make sure the organisms they describe in their story are appropriate to the fictional future environment of the Earth the students write about.

CHAPTER 14 Evolution: How Change Occurs

Section	Laboratory Investigations and Activities
14–1 Developing a Theory of Evolution pages 291–295 THEMES: Evolution, Systems and Interactions	**Teacher Edition** DEMONSTRATION: Mutation as an Agent of Change, p. 290D
14–2 Evolution by Natural Selection pages 296–298 THEMES: Evolution, Patterns of Change	**Teacher Edition** DEMONSTRATION: Survival of the Fittest, p. 290D **Laboratory Manual** A Human Adaptation, p. 181
14–3 Genetics and Evolutionary Theory pages 299–303 THEMES: Evolution, Scale and Structure	**Student Edition** LABORATORY INVESTIGATION: Simulating Natural Selection, p. 314 **Teaching Resources** HANDS-ON ACTIVITY: Where Are They? p. 75 ECOLOGY INVESTIGATION: Interspecific Competition and Growth Rates, p. 5
14–4 The Development of New Species pages 304–310 THEMES: Evolution, Patterns of Change	**Laboratory Manual** Variation Within a Population, p. 187
14–5 Evolutionary Theory Evolves pages 310–313 THEMES: Evolution, Stability	**Teaching Resources** HANDS-ON ACTIVITY: Variety Is the Spice of Life, p. 71
Chapter Review pages 315–317	

Teacher Resources

Books

Eldredge, Niles. *Macroevolutionary Dynamics: Species, Niches, and Adaptive Peaks.* McGraw-Hill, 1989.

Gould, Stephen Jay. *Eight Little Piggies: Reflections in Natural History.* Norton, 1993.

Hanson, Robert W., ed. *Science and Creation.* McGraw-Hill, 1985.

Lewin, Roger. *The Thread of Life: The Smithsonian Looks at Evolution.* Smithsonian Books, 1982.

CHAPTER PLANNING GUIDE

Other Activities	Media and Technology
Study Guide Section 14-1, p. 135	
Teaching Resources BIOLOGY CASE STUDY: Finding a Niche in a New Environment, p. 3 **Study Guide** Section 14–2, p. 138	**Interactive Videodisc** On Dry Land: The Desert Biome **Biology Media Guide** page 13
Study Guide Section 14–3, p. 139	**Interactive Videodisc** In the Company of Whales **Biology Media Guide** page 13
Teaching Resources ACTIVITY: Finches in the Galápagos, p. 3 ACTIVITY: Making a Model Molecular Clock, p. 5 **Study Guide** Section 14–4, p. 140	**Interactive Videodisc** On Dry Land: The Desert Biome **Interactive Videodisc** Patterns of Evolution **Biology Transparencies** 17: Adaptive Radiation **Biology Media Guide** pages 13–14
Teaching Resources VOCABULARY REVIEW: Cross-a-Clue, p. 1 ACTIVITY: Investigating Models of Evolution, p. 7 **Study Guide** Section 14–5, p. 143	**Interactive Videodisc** In the Company of Whales **Biology Media Guide** page 14
Teaching Resources Chapter Test A, p. 13 Chapter Test B, p. 17	**Computer Test Bank** Chapter Test, p. 147

Audiovisuals

The Story of Evolution. Filmstrip with cassette. Hawkhill Associates.

Species and Evolution. Film or video. The Media Guild.

Evolutionary Biology. Film or video. Coronet.

Darwin's Finches. Filmstrip. Time-Life Video.

CHAPTER 14 Evolution: How Change Occurs

CHAPTER OVERVIEW

Evolutionary theory is the foundation on which the rest of biological science is built. Scientists have realized for over a century that plants and animals have changed over time. Thoughts on how evolution happens have been revised as more evidence has accumulated. Lamarck proposed a theory of evolution long before Darwin. He based his explanation on three assumptions that we now know to be incorrect. Lamarck believed that organisms changed because they had an inborn urge to better themselves; that organisms could change the shape of their bodies through use; and that once characteristics were acquired, they could be passed on to their offspring.

Genetics and evolutionary theory are inseparable parts of biology. Fitness, adaptation, species, and the evolutionary process can all be defined in terms of genetics. There are some basic ways in which organisms interact within their environment. *Niche* is the biological term for the place that an organism occupies in that environment. No two species can occupy the same niche in the same location at the same time for a long period of time. This phenomenon can lead to the formation of new species. The evolution of new species from old ones is called speciation. This occurs only when populations are isolated. Darwin's finches provide an example of speciation.

When one species gives rise to others, it is called adaptive radiation or divergent evolution. Adaptive radiations among different organisms often produce species that are similar in appearance and behavior. This is known as convergent evolution, which has produced many of the analogous structures found in organisms today.

Evolutionary theory continues to be modified with time. Several aspects of evolutionary theory are discussed in the final section of this chapter. The concepts of genetic drift, a stable gene pool, gradualism, and punctuated equilibria are introduced.

14–1 DEVELOPING A THEORY OF EVOLUTION

Section Focus 14–1

The purpose of this section is to present the historical background underlying the development of evolutionary theory. Evolutionary theory underlies the rest of biological science.

The theory of evolution originally proposed by Darwin has been revised over the years. Current evolutionary theory is a collection of carefully reasoned and tested hypotheses.

As Darwin began to formulate his ideas about evolution, he was influenced by the geologist Charles Lyell, who demonstrated that the Earth was very old. This was essential to Darwin because it confirmed his theory that organisms had evolved over long periods of time. Darwin was also influenced by farmers who were engaged in breeding programs involving artificial selection. He was sure that some similar process must occur in nature. He called the process *natural selection*. The third important influence on Darwin's thinking was the work of Thomas Malthus. According to Malthus, people were born at a faster rate than they were dying, so there had to be some outside control on the growth of the human population. Darwin also observed that a high birthrate and a shortage of life's necessities force organisms into competition for survival. Some individuals survive because they have characteristics well-suited for their environment. Darwin called this principle *survival of the fittest*.

Performance Objectives 14–1

1. Define the term *evolutionary theory*.
2. Discuss three major influences on the development of Darwin's theory of evolution.

Science Terms 14–1

artificial selection   p. 294

14–2 EVOLUTION BY NATURAL SELECTION

Section Focus 14–2

The purpose of this section is to introduce the idea of natural selection. Darwin observed that there were variations among wild plants and animals. He also knew that high birthrates and shortages of necessities for survival forced organisms to compete with one another in a struggle for survival. The organisms with characteristics well-suited to the environment are able to survive and produce offspring; those less well-suited do not.

Performance Objectives 14–2

1. Explain what Darwin meant by natural selection.
2. Discuss the principle of survival of the fittest.

Science Terms 14–2

natural selection   p. 296
survival of the fittest   p. 297

14–3 GENETICS AND EVOLUTIONARY THEORY

Section Focus 14–3

The purpose of this section is to show the relationship between genetics and evolutionary theory. Darwin did not know anything about genetics as he was developing his theory of evolution. Today, fitness, adaptation, species, and the process of evolutionary change are all defined in genetic terms. Genes are the source of random variation, upon which natural selection is based. Mutations cause some variation and additional variation occurs during meiosis. Natural selection operates only on the physical and behavioral variation among individuals. Biologists study groups of plants and animals called populations. The members of the population can interbreed; therefore, they and their offspring share a common group of genes called a gene pool. Evolution is any change in the frequencies of alleles that occurs in the gene pool of a population. Fitness, in evolutionary terms, is the successful adaptations of an organism to its environment. An adaptation can be defined as a genetically controlled characteristic of an organism that increases its fitness.

CHAPTER PREVIEW

Performance Objectives 14–3
1. Explain the relationship between genes and variations.
2. Discuss natural selection in terms of variation.
3. Discuss evolution as it occurs in a population of organisms.
4. Define evolutionary fitness.
5. State a genetic definition for species.

Science Terms 14–3
population   p. 300
gene pool   p. 300
relative frequency   p. 300

14–4 THE DEVELOPMENT OF NEW SPECIES

Section Focus 14–4
The purpose of this section is to explain the process of speciation. The place that an organism occupies in a biological system is called a niche. No two species can occupy the same niche in the same location at the same time and survive for a long period of time. Any species or population that occupies an empty niche will be better able to survive.

Scientists have found that new species are usually formed only when populations become isolated and are unable to interbreed. Reproductive isolation is the agent for formation of new species. As natural selection and adaptation to the new environment occur, the gene pools become more dissimilar, finally producing totally separate species.

This process in which one species gives rise to many new species is called adaptive radiation or divergent evolution. Convergent evolution produces species that are similar in appearance and behavior.

Performance Objectives 14–4
1. Define niche in biological terms.
2. State the biological "rule" that applies to species and their niches.
3. Explain the process of speciation.
4. Using Darwin's finches as an example, explain how speciation might occur.

Science Terms 14–4
speciation   p. 304
niche   p. 304
reproductive isolation   p. 305
adaptive radiation   p. 308
divergent evolution   p. 308
convergent evolution   p. 308
analogous structure   p. 308

14–5 EVOLUTIONARY THEORY EVOLVES

Section Focus 14–5
The purpose of this section is to discuss the various modifications to evolutionary theory that have occurred since Darwin. Genetic drift is a random change in gene frequency—evolution in the absence of natural selection. Recently, new ideas about the pace of evolution have sparked much discussion and controversy.

Performance Objectives 14–5
1. Explain genetic drift.
2. Define gradualism.
3. List three reasons for the occurrence of rapid evolution.
4. Discuss punctuated equilibrium.

Science Terms 14–5
genetic drift   p. 311
gradualism   p. 312
equilibrium   p. 312
mass extinction   p. 313
punctuated equilibrium   p. 313

DISCOVERY LEARNING

TEACHER DEMONSTRATIONS
Modeling

Mutation as an Agent of Change
This demonstration can be used to introduce Chapter 14.

The theme of this chapter is the process of change and how it occurs. An integral part of variation that leads to significant change is mutation among organisms. To illustrate this point, prepare several culture dishes or nutrient agar. Streak them liberally with bacteria such as *Escherichia coli*. Place 2 or 3 common antibiotic disks on each plate. Incubate the plates at 37°C for 24 to 48 hours. Have students check the plates for mutant bacteria growing in the inhibition zone around the disks where most of the bacteria have been killed. The more plates you have prepared, the greater the probability of finding mutants. Ask students to explain why a bacterium's resistance to antibiotics would be a beneficial adaptation. Discuss how the bacterium would become resistant to antibiotics.

Survival of the Fittest
This demonstration can be used at the conclusion of Section 14–1 or near the beginning of Section 14–2.

Use a paper punch to make tiny dots of paper in different colors. Use red, pale yellow or blue, and white. Glue several of the dots to a large piece of white poster board. Be sure that each individual dot can be seen.

Tell students to count as many red dots as they can in 10 seconds. Repeat the process with the pale color and then with the white. The students should have been able to locate more of the darker dots than the lighter ones. Discuss color as a characteristic that made the dots more "fit to survive" in their environment. Relate this to Malthusian Doctrine.

CHAPTER 14

Evolution: How Change Occurs

GUIDED ENQUIRY
Pose the following questions to students and have them record their responses. Point out that they will gain a better understanding of the key concepts if they read the chapter with these basic questions in mind. Upon completion of the chapter, pose the questions again. Ask students to compare their initial responses with those they have developed after reading the chapter.

- How was the theory of evolution, as we know it, developed?
- How is genetics related to the evolutionary process?
- How do new species arise?
- Do all organisms evolve in the same way?
- Why is evolutionary theory important to the study of biology?

INTRODUCING CHAPTER 14

Using the Visuals
Tell students to think of a time when they have been forced to stay in one place for an extended time—maybe in the hospital or confined at home by bad weather. Assuming that they had no TV or telephone, what did they or would they do with their idle time? They will give a variety of answers—silly to serious. The ones for which you are looking include reading, studying, writing, and thinking. After you have accepted all of their answers, write them on the board or the overhead. Have them look at the chapter-opener picture and read the caption. Ask students what they think Darwin did with his spare time. Remind them that he was the captain's companion (and the captain was busy most of the time) and that the voyage lasted five years. Through the discussion, it is hoped that students will realize that Darwin probably had similar activities during his confinement on ship.

Read the introduction to the chapter.

CHAPTER 14

Evolution: How Change Occurs

It was in this ship that Charles Darwin sailed around South America. By today's standards, it was a small ship, and time on board was spent in cramped quarters.

There is grandeur in this view of life, with its several powers, having been originally breathed into a few forms or one; and that, whilst this planet has gone cycling according to the fixed laws of gravity, from so simple a beginning endless forms most beautiful and most wonderful have been, and are being, evolved.

So wrote Charles Darwin in his conclusion to On the Origin of Species, the book in which he attempted to explain what he among many had observed: that life on Earth has changed over time. But how does change occur? And how can one species evolve into another? It is this explanation in part that you will learn about in the pages that follow.

290

- **What is your reaction to the excerpt from *On the Origin of Species*?** (Answers will vary. Some students may have heard religious overtones. Remind them that Darwin had a degree in theology.)

- **To what observations is Darwin referring?** (Answers will vary. They should include that Darwin had observed many species of organisms and much variation among them. He also saw fossils.)

GUIDE FOR READING

After you read the following sections, you will be able to

14–1 Developing a Theory of Evolution
- Discuss the development and importance of evolutionary theory.

14–2 Evolution by Natural Selection
- Describe natural selection.

14–3 Genetics and Evolutionary Theory
- Relate evolution, natural selection, fitness, and adaptation to genetics.

14–4 The Development of New Species
- Describe speciation and relate speciation to niche availability.

14–5 Evolutionary Theory Evolves
- Explain new ideas about the pace of evolutionary change and the role of chance in evolution.

Journal Activity

YOU AND YOUR WORLD

Few people, it seems, actually understand what evolutionary theory is. Nevertheless, many people have strong opinions about it. In your journal, write down as much as you know about evolutionary theory and the controversy surrounding it. When you have finished this chapter, reread your journal entry. What aspects of evolutionary theory did you know pretty well? What misconceptions did you have?

Figure 14–1 Although the mouse and the pig do not look at all alike, an analysis of their DNA shows that these two animals are more closely related than appearances would indicate.

14–1 Developing a Theory of Evolution

Guide For Reading
■ What is evolutionary theory, and why is it important?
■ What were the problems with Lamarck's theory of evolution?
■ How was Darwin's thinking about evolution influenced by the ideas of others?

Evolutionary theory is the foundation on which the rest of biological science is built. In fact, the biologist Theodor Dobzhansky once wrote that nothing in biology makes sense except in the light of evolution. Much research in genetics, ecology, and medicine is based on evolutionary theory.

The fact that plants and animals have changed over time was obvious to Darwin and has been clear to scientists throughout the last century. Observing that evolution has occurred is relatively simple. Explaining how and why evolution occurs is more difficult.

Certain aspects of Darwin's original theory of evolution have been revised by biologists in the years since the publication of *On the Origin of Species*. But the revisions do not mean that evolutionary change itself is debatable or that evolutionary theory is merely a collection of vague hunches that are not supported by evidence. Evolutionary change is undeniable. **Evolutionary theory is a collection of carefully reasoned and tested hypotheses about how evolutionary change occurs.**

By way of comparison, consider that even today physicists do not completely understand gravity, although there is no doubt in anyone's mind that gravity exists. There are at least two competing modern theories that explain how gravity works. Both theories make important useful predictions of natural events. For example, there is no question that if you jump into the air, you will end up on the ground below. It makes no difference whether you understand—or even believe in—gravity. What goes up must come down. Just as definitely, life on Earth evolves, or changes over time. Explaining the fine points of evolutionary change will continue to be one of the great challenges of biology.

- The last chapter concentrated on presenting evidence. What is the focus of this chapter? (Chapter 13 pointed out evidence of evolution. Chapter 14 will explain how evolution occurs.)

TEACHING STRATEGY 14–1

Focus/Motivation

The concept of evolutionary theory as the foundation of biological science is an important one, and it can tie the entire course together. You will want to refer to it many times during the year. At this point, take time to develop the notion of foundations.

- **What is a foundation?** (According to *Webster's New World Dictionary*, it is the part on which other parts rest for support.)
- **Why is the foundation impor-**

COOPERATIVE LEARNING

Using preassigned lab groups or randomly selected teams, have groups design an organism that might evolve to fill the following niche. In AD 3000, the Earth's atmosphere has changed dramatically. It contains compounds harmful to humans—sulfur dioxide, carbon monoxide, lead, and so on. At the same time, O_2-producing plants have become virtually extinct. A new source of natural O_2 is needed! Groups will design an organism that has evolved to fill this niche. Each group should draw an organism and explain how it is successfully adapted to life in this niche. They should also construct an evolution tree for each organism.

Journal Activity

YOU AND YOUR WORLD

Tell students that you do not expect them to have any expertise on evolution at this point. They simply should write down what they know now about evolutionary theory and later compare their entries with what they learn in this chapter.

tant? (Answers will vary. It is important to have support.)
- **What kinds of things have foundations?** (Answers will vary. Buildings, roads, schools, hospitals, and nonprofit organizations are sometimes supported by foundations.)
- **In the foundation for a major highway, is there any built-in support?** (There are steel rods to give it additional strength.)
- **What supports the theory of evolution—the foundation for all other biological sciences?** (Answers will vary. Any or all of the evidence presented in Chapter 13 would be correct.)

291

HISTORICAL NOTE
ALFRED WALLACE

Just as Darwin was ready to publish his work, another English naturalist, Alfred Wallace, came up with an idea that was virtually identical to Darwin's. Darwin and Wallace presented their papers at a scientific meeting. Both Wallace and the world at large agreed that Darwin deserved the lion's share of the credit for proposing the theory of evolution by natural selection because his work was more thorough.

TIE-IN LANGUAGE ARTS

Personification is frequently studied in literature. Is not what Lamarck believed as his first assumption really the same thing? He was endowing plants and animals with the human desire to become better. This is an example imposing human reaction onto nonhuman subjects.

14–1 (continued)

Content Development
Review with students what a theory is. Discuss the importance of the evidence underlying the theory. Review some of the evidence from Chapter 13 to support evolution.

Also explain that scientists do not dispute the fact that evolution has occurred. Fossil and biochemical data provide ample evidence to prove that evolution is a fact. The term *theory*, in the case of evolution, refers to the mechanism by which evolution occurs. Several varying but not contradictory theories are presented in this chapter.

Reinforcement/ Reteaching
Lamarck died shortly before Darwin set sail on the *Beagle*. Because Lamarck's theory of evolution was the prevalent one of the day, it is safe to assume that Darwin was familiar with it. He may even have accepted it at one time. The assumptions of Lamarck—that we now know to be incorrect—were logical ones. Even disproved, many people today, including some of your students perhaps, may find them somewhat acceptable. Students, especially those with limited experience, may find it quite easy to impose the human desire for betterment onto other organisms. You might want to remind them that the desire to change and be better is a high-level thought process that requires the ability to reason. Ask them if they think that bacteria, plants, or even starfish are capable of reasoning. The textbook's example involving the Wright Brothers is also a good one to help you persuade students who are reluctant to agree on the fallacy of this assumption. Remember, you may be fighting the Americanism that

An Early Explanation for Evolutionary Change

Jean Baptiste de Lamarck (1744–1829) was among the first scientists to recognize that living things changed over time. And long before Darwin, Lamarck also realized that organisms were somehow adapted to their environments. In explaining how adaptation occurred, however, Lamarck relied on three assumptions we now know to be incorrect.

A DESIRE TO CHANGE Lamarck thought that organisms change because they have an inborn urge to better themselves and become more fit for their environments. In Lamarck's view, for instance, the ancestors of birds acquired an urge to fly. Over many generations, birds' constant efforts to become airborne led to the development of wings. What a pity for the Wright Brothers that this element of Lamarck's theory proved not to be true!

USE AND DISUSE Lamarck also believed that change occurred because organisms could alter their shape by using their bodies in new ways. Organs could increase in size or change in shape depending on the needs of the organism. For example, by trying to use their front limbs for flying, birds could eventually transform those limbs into wings. In the opposite way, Lamarck believed that if an animal did not use a particular part of its body, that body part would decrease in size and might finally disappear.

Figure 14–2 When this adult Doberman pinscher was still a young puppy, her ears were clipped so that they would stand up on her head. The operation occurred long before she bore the puppy resting beside her. Yet the puppy's ears look just like his mother's did when she was born. This is one of the many forms of evidence that traits acquired during an organism's lifetime are not passed on to the next generation.

PASSING ON ACQUIRED TRAITS Included in Lamarck's reasoning was the belief, shared by many biologists of that time, that acquired characteristics were inherited. He thought that if an animal acquired a body structure (such as long arms or feathers) during its lifetime, it could pass that change on to its offspring. By the same reasoning, structures that became smaller from disuse would eventually disappear.

Although later discoveries showed that Lamarck's explanation of evolution was incorrect, he is still credited with being one of the first people to devise a theory of evolution and adaptation. He is also credited with bringing the concept of evolution to the attention of scientists. Thus Lamarck paved the way for Darwin's theory of evolution.

Lamarck's ideas may seem strange to you now, yet his theory was consistent with knowledge of that time. It was not until a century after Lamarck proposed his theory that an improved understanding of genetics and the principles of heredity showed that the mechanisms he proposed would not work.

Lamarck, you see, knew nothing about genes. As you know now, only genes and changes in genes—not alterations in body structure—are passed from parents to offspring. There is no evidence that experience during its life can cause specific changes in an organism's genes. Years of proper exercise and diet, for example, can turn a weakling into a champion weight lifter. But that weight lifter's children cannot benefit genetically from the parent's pumping iron. If the children do not exercise and eat a proper diet, they will not develop large muscles, even if their parents were world champions!

Ideas That Shaped Darwin's Theory of Evolution

Personal experience on the *Beagle*'s voyage awakened Darwin's interest in explaining the diversity and fitness of life on Earth. But both during his trip and after his return, Darwin's thinking was also influenced by the books he read and by discussions with geologists, farmers, and others.

THE INFLUENCE OF GEOLOGY: LYELL'S IDEAS As you will remember from Chapter 13, the geologist Charles Lyell demonstrated that the Earth was very old and that it had changed over time. After reading Lyell's book *Principles of Geology*, Darwin became convinced that the Earth was much older than most people of his time believed. This was an important idea for Darwin. For in order to explain evolution—to even recognize that evolution had occurred—it was essential for Darwin to realize that the Earth was very old. The long periods of time it would have taken for millions of species to have evolved from a common ancestor could be accounted for only if the Earth was very old.

Figure 14–3 Volcanoes can alter the Earth's face. This volcano, emerging from beneath the ocean, resulted in the formation of a new island. Within a short period of time, living organisms will discover this newly formed island and begin to exploit the opportunities that exist there. A volcanic eruption can produce a completely opposite effect, destroying an area of land and all its inhabitants.

BACKGROUND INFORMATION
DARWIN AND ACQUIRED CHARACTERISTICS

Darwin believed that there was some inheritance of acquired characteristics. However, he felt that natural selection was a much more powerful force in the evolution of organisms.

Figure 14–4 The cabbage (top), Brussels sprouts (center), and cauliflower (bottom) are all varieties of the same plant species that have been "selected" over time to produce familiar food crops.

Figure 14–5 Variation in a species is quite common in nature. These ladybug beetles show different markings.

294

Lyell's writing also caused Darwin to appreciate the geological phenomena he observed on his journey. In Chile, Darwin saw a spectacular volcanic eruption. Shortly thereafter, he observed that an earthquake had lifted a stretch of shoreline three meters higher than it had been before. With Lyell's writings fresh in his mind, Darwin came to realize that geological phenomena such as the ones he had observed could transform the face of the Earth over time. And if the Earth itself could change over time, so too could life on the Earth.

THE INFLUENCE OF FARMERS: ARTIFICIAL SELECTION
While assembling his thoughts back in England, Darwin spoke extensively with plant and animal breeders. He learned that farmers altered and improved their crops and livestock through breeding programs. But how, Darwin wondered, did such programs work?

Farmers told Darwin that domesticated animals and plants vary a great deal. For example, in every corn field, some plants are larger than average; others are smaller than average. Certain cows produce a large amount of milk; other cows produce a small amount of milk. Here and there among a flock of white chickens, a gray or black chicken appears. The farmers convinced Darwin that many of these variations were often passed on to the animals' offspring. In other words, these were inheritable variations.

Darwin realized that farmers could not cause variation to occur. Variation either happened naturally or it did not. But once farmers encountered variation, they could use it to their advantage. They noted the variations they found and decided which organisms to use as breeding stock. Individuals with undesirable variations—scrawny bulls or cows that produced little milk, for example—were not allowed to mate. Superior animals—husky bulls or cows that produced much milk—would be mated as often as possible.

This process, which Darwin called **artificial selection**, allowed only the individuals who suited the farmers' needs to produce offspring. Over the years, breeders have used artificial selection to produce plants and animals that are much more

14–1 (continued)

Content Development
Take a few minutes to talk about how Darwin's observations of geological phenomena influenced his thinking.

Content Development
You may want to go back to Chapter 12 to briefly review selective breeding as practiced by farmers.
• **How does the process of selective breeding work?** (Farmers are aware of the variations that exist among domesticated plants and animals. They choose those individuals with desirable characteristics as breeding stock.)
• **Why did Darwin call this process artificial selection?** (Those organisms that were superior were selected as breeding stock, rather than letting the stock interbreed naturally.)
• **What was Darwin's conclusion about selection?** (He concluded that a similar process must take place in a natural environment as well.)
• **What puzzled Darwin about the selection process?** (How did it operate?)

Content Development
Using the Visuals Discuss the Malthusian Doctrine as it applies to humans. Discuss the examples in the textbook and in Figure 14–6 that indicate that the doctrine also applies to organisms other than humans in their natural setting. Emphasize the question that concludes the section. Ask students to think about it until the next lesson.

SECTION REVIEW 14–1

1. Evolutionary theory is the foundation of biological

suited to human needs than—and often dramatically different in appearance from—their original parent stock. **In artificial selection, the intervention of humans ensures that only individuals with the more desirable traits produce offspring.**

Darwin became convinced that a process similar to artificial selection must be at work in nature. This process would allow only those organisms best suited to their environment to survive and reproduce. But in nature there is no human intervention; so how, Darwin wondered, could such a process operate?

THE INFLUENCE OF MALTHUS: POPULATION CONTROLS
An important influence on Darwin was the work of the economist Thomas Malthus (1766–1834). Malthus observed that babies were being born at a faster rate than people were dying. If the human population continued to increase in that way, Malthus reasoned, sooner or later there wouldn't be enough living space and food. The only conditions that would prevent the endless growth of human populations, Malthus observed, were famine, disease, and war. In time, these unpleasant observations were called the Malthusian Doctrine.

Darwin realized that the Malthusian Doctrine applied even more to animals and plants than to humans, for most other species produce far more offspring than we do. For example, every summer each mature maple tree produces thousands of seeds. Marine animals, such as the common mussel, produce millions of eggs each time they spawn. If all the offspring of just one of these maple trees or mussels survived, they would overcrowd the area in which they lived. If each offspring then produced as many offspring as its parents, and if all those offspring reproduced, there would soon be so many maple trees or mussels that they would cover the Earth or fill the oceans!

Obviously, the oceans are not filled with mussels and the continents are not covered with maple trees. Most baby mussels die during their first year of life. Most maple seeds never grow into mature trees. Thousands upon thousands of individuals of each species die, and only a few survive. Even fewer successfully raise offspring. That much is clear. But what determines which individuals survive and reproduce?

Figure 14–6 Some animals and plants produce enormous numbers of offspring. Eggs in this praying mantis egg case have begun to hatch (top). If all the young survived to reproduce, you can imagine how the number of mantises in the world would be affected. Each sunflower in this field is capable of producing hundreds of seeds (bottom). If each seed survived and reproduced, there would be uncountable numbers of sunflowers.

14–1 SECTION REVIEW

1. What is the importance of evolutionary theory?
2. How did Lamarck explain evolution? What are the major problems with his explanation?
3. What is artificial selection? How did this concept influence Darwin's thinking?
4. **Critical Thinking—Identifying Relationships** Would Darwin have developed his theory of evolution if he had not read the works of Lyell and Malthus? Explain.

295

science. It helps biologists make sense of the natural world.
2. Lamarck was among the first scientists to realize that change occurred in living things over time. He thought that change occurred because organisms had a desire to change and become more fit to live in a particular environment. He also believed that animals could acquire traits during their lifetime that could be passed on to their offspring.
3. Artificial selection occurs when farmers and plants breeders select animals and plants to crossbreed. By doing this, a breeder hopes to combine the best traits of the parents so that the offspring will be more fit than their parents. Darwin was aware of artificial selection

14–2 Evolution by Natural Selection

Guide For Reading

- What is the driving force of evolution according to Darwin's theory?
- How does natural selection work?

Ultimately, Darwin recognized in nature a process that operates in a manner similar to the way artificial selection worked on farms and in fields. Darwin called this process **natural selection** and explained its action in terms of several important observations.

Darwin observed that wild animals and plants showed variations just as domesticated animals and plants did. His field notebooks were filled with records of height, weight, color, claw size, tail length, and other characteristics among members of the same species. Darwin did not understand the reasons for these variations, but he realized that many of them were inherited.

Darwin observed that high birthrates and a shortage of life's necessities forced organisms into a constant "struggle for existence," both against the environment and against each other. Plant stems grow tall in search of sunlight; plant roots grow deep into the soil in search of water and nutrients. Animals compete for food and space in which to build nests and raise young. But who among all the contenders wins the struggle for existence?

Figure 14–7 *For all organisms, life is a constant struggle to survive. This smaller lizard has just come to the end of that struggle in the jaws of a gecko.*

296

Darwin knew that each individual differs from all the other members of its species. Sometimes the differences are easy to observe; sometimes the differences are subtle. **Individuals whose characteristics are well-suited to their environment survive. Individuals whose characteristics are not well-suited to their environment either die or leave fewer offspring.** This principle Darwin called **survival of the fittest.**

Natural selection thus operates in a manner similar to artificial selection, but over much longer periods of time and without any goal or purpose. On farms, breeders intentionally select cows that give more milk or corn with bigger ears to be parents to the next generation. In nature, the struggle for existence permits only those individuals well-suited to their environment to survive and reproduce. The fact that less-fit individuals of a species do not survive helps keep the species from covering the Earth.

Peppered Moths: Natural Selection in Action

England's peppered moth provides an example of natural selection in action. It also offers us a chance to study the sorts of experiments that can be used to test evolutionary theory. The story is as follows. The peppered moth spends much of the daytime resting on the bark of oak trees. In the beginning of the nineteenth century, the trunks of most oak trees in England were light brown speckled with green. Most of the peppered moths of that time were mottled light brown too. There were always a few dark-colored moths around, but light-colored moths were always the most common.

Then the Industrial Revolution began in England. Pollution (mostly soot from burning coal) stained London's tree trunks dark brown. At about the same time, biologists noticed that more and more moths with dark coloration were appearing. Why was the population changing color in this way?

The evolutionary hypothesis suggested by observation was straightforward. Birds are the major predators of peppered moths. It is much harder for birds to see, catch, and eat moths that blend in with the color of the tree bark than it is for them to spot moths whose color makes a strong contrast with the tree trunks. The moths that blend in with their background are said to be camouflaged.

As the tree trunks darkened, the rarer, dark-colored moths were better camouflaged and harder for birds to spot. Being harder to spot, the darker individuals were now better able to survive. The darker forms had greater fitness than the lighter forms. More of the darker moths survived and reproduced, passing on the genes for dark color to their offspring, and the moth population evolved darker coloration.

But a hypothesis that looks good is not enough. Scientific hypotheses must be tested by experiment whenever possible.

Figure 14–8 Before the Industrial Revolution, soot was rare in the English countryside. A light-colored moth was difficult to see against the clean tree bark (top). After several years, during which the bark was darkened by the soot of burning coal, a light-colored moth stood out against the darkened tree bark (bottom). In each photograph, which moth would most likely be noticed by a hungry bird?

ANNOTATION KEY

① The moth that most stands out against the tree's background. (Making inferences)

TIE-IN LITERATURE

Darwin's influence can be seen in works ranging from John Fowles's *The French Lieutenant's Woman* to episodes of television science-fiction series, including *Star Trek* and *Doctor Who*. Mature students may enjoy reading Kurt Vonnegut's novel *Galápagos*. In this book, tourists intending to recreate Darwin's travels are instead stranded on the Galápagos Islands, where their descendants eventually evolve into a new species. Vonnegut's "new humans" are amphibious, have smaller brains, and have seallike fur for warmth and protection.

Have students compare the processes of artificial and natural selection.

Content Development
Using the Visuals Refer students to Figure 14–8 and discuss how the story of the peppered moths illustrates natural selection. Emphasize that the evolutionary hypothesis the textbook states is obvious: The darkening tree trunks made darker moths more fit then lighter moths, so the moth population evolved darker coloration.

You also may wish to emphasize how Lamarck's principles do not apply to this example. The moths by themselves did not choose to become darker. Rather, the environment allowed more darker than lighter moths to survive and reproduce, and so these moths' characteristics passed to the next generation.

> **BACKGROUND INFORMATION**
> **DARWIN'S THEORY**
>
> This is a quick summary of Darwin's theory of evolution by natural selection.
>
> 1. All living species are descended from species that lived before them. This is the concept of common descent.
> 2. No two members of a species are exactly alike. The physical characteristics of plants and animals vary, and much of this variation can be inherited. This is the concept of inheritable variation.
> 3. Far more organisms are born than ever grow to adulthood.
> 4. Organisms compete with each other and with nature in a constant struggle to survive and reproduce.
> 5. Those organisms best fitted, or adapted, to their environment will have the best chance to survive and successfully produce another generation. The favorable inherited variations that allowed these individuals to survive will be passed on to their offspring. This is the concept of survival of the fittest.
> 6. Because of statements 2 through 5, organisms will evolve or change over time. Generation after generation of this kind of selection causes organisms to become better and better adapted to their environment. Because this process is controlled by condition in nature, Darwin called it evolution by natural selection.

Figure 14–9 *Looking like something it's not can be helpful to an organism. It is difficult to spot the insect disguised as a thorn (top, left), the toad that resembles a fallen leaf (top, right), and the moth that looks like a plant leaf complete with diseased areas (bottom).*

British ecologist H.B.D. Kettlewell devised just such a test for this hypothesis. Kettlewell learned how to capture both light- and dark-colored forms of the peppered moth and then managed to raise them in captivity. He also learned to mark living moths in such a way that birds could not see the marks.

Kettlewell then released equal numbers of light- and dark-colored moths in two types of areas. In one area, trees were normally colored. In the other area, they were blackened by soot. Later on, he recaptured, sorted, and counted all the marked moths he could. What type of results do you think Kettlewell needed to either prove or disprove the hypothesis?

Kettlewell found that in unpolluted areas, more of his light-colored moths had survived. In soot-blackened areas, more of his dark-colored moths had survived. Thus Kettlewell showed that in each environment the moths that were better camouflaged had the higher survival rate. It was logical to conclude that when soot darkened the tree trunks in an area, natural selection caused the dark-colored moths to become more common. Today Kettlewell's work is considered to be a classic demonstration of natural selection in action.

14–2 SECTION REVIEW

1. What is natural selection? What observations led Darwin to develop this concept?
2. Define survival of the fittest. How are the concepts of natural selection and survival of the fittest related?
3. **Critical Thinking—Relating Cause and Effect** Explain how natural selection might produce a modern giraffe from short-necked ancestors.

> **14–2 (continued)**
>
> **Content Development**
> Have students list the steps of the scientific method. Explain each step as it applies to the work of Kettlewell. Emphasize his conclusion.
>
> **SECTION REVIEW 14–2**
>
> 1. Natural selection is the process by which the most fit organisms reproduce. Variations that make an organism more fit could be passed on to offspring in a process that does not require human intervention. Darwin observed that all organisms are in a constant struggle to survive against both the environment and each other. Those organisms that were best able to survive and reproduce were selected "naturally."
> 2. Survival of the fittest refers to reproductive success. All organisms will die; reproductive success means that a species will survive. Those organisms most fit will reproduce, passing on their traits to their offspring. Survival of the fittest is the mechanism through which natural selection operates.
> 3. In an area where giraffes compete with other herbivores for food, a short-necked giraffe

14–3 Genetics and Evolutionary Theory

Guide For Reading
- What causes genetic variations? What do these variations have to do with evolution by natural selection?
- How are evolution, fitness, and adaptation described in genetic terms?

In developing his theory of evolution, Darwin worked under a serious handicap. He had no idea how the inheritable traits so important to his theory were passed from one generation to the next. For although Mendel had formulated his genetic principles during Darwin's lifetime, his work remained unknown to most scientists until the early part of this century. The rediscovery of Mendel's work and the growth of our knowledge of genetics enable us to explain the mechanism of evolution more completely than Darwin could. Genetics and evolutionary theory are inseparable. Today we define fitness, adaptation, species, and the process of evolutionary change in genetic terms.

Genes: Units of Variation

Genes, the carriers of inheritable characteristics, are also the source of the random variation upon which natural selection operates. Mutations cause some variation. Much additional variation arises during meiosis as the parents' chromosomes are copied, shuffled like a deck of playing cards, and dealt out to the gametes.

It is important to remember that genetic variation does not occur because an animal needs or wants to evolve—an idea central to Lamarck's theory. Sometimes genetic variation occurs; sometimes it doesn't. There is no way for an organism to cause a particular change in its DNA. There is also no way for an organism to prevent variations that do occur.

Raw Material for Natural Selection

In the evolutionary struggle for existence, entire organisms, not individual genes, either survive and reproduce or do not. How then does natural selection operate? Natural selection can operate only on the phenotypic variation among individuals. As you learned in Chapter 9, an organism's phenotype includes all the physical and behavioral characteristics produced by the interaction of genotype and environment.

You can sample phenotypic variation by measuring the height of all the students in your class. Using mathematics, you can calculate an average height for this group. Many students will be just a little taller or shorter than average. However, a few very tall or very short individuals may be in your class. If you graph the number of individuals of each height, you will get a curve similar to the one shown in Figure 14–11 on page 300. This phenotypic variation is produced by a combination of genetic instructions and environmental influences, such as nutrition and exercise. If your classmates are not malnourished, most (though not all) of the variations in height you observe

Figure 14–10 It is a zebra's genes that determine the exact pattern of the animal's stripes. Notice that there are slight variations in the stripes of each animal.

299

would be at a disadvantage. If a giraffe with a slightly longer neck was born, it could reach plant materials in tall shrubs and low trees. A giraffe with a longer neck would be assured of a food supply that other herbivores couldn't exploit. Thus, this giraffe would have a better chance of survival and passing on its genes for a longer neck to its offspring.

Reinforcement/Reteaching
If students are having difficulty with the Section Review questions, turn back to the appropriate material and review it.

Closure
Students will need their notes on the opening musical chairs' activity. Discuss, at this time, what happened in the game and relate

---

TEACHING SUPPORT

Study Guide
- Section 14–2, p. 138

it to Darwin's principle of natural selection. Draw parallels to the peppered moth experiment where possible and appropriate.

TEACHING STRATEGY 14–3

Focus/Motivation
Put a small, colorful object into a paper bag. Use a ball, stuffed animal, piece of fruit, or any familiar object. Have a student describe the object to the class based only on what he or she can sense by touch. Have a second student look at the object and describe it, using all of his or her senses. Make an analogy between not being able to see the object and trying to explain Darwin's theory without knowing anything about genetics.

Content Development
Discuss genes as the units of variation. Students may never have looked at genetics from an evolutionary perspective.

Skills Development
Guided Practice
Skill: Graphing
Have students prepare a graph of the individual height in the class. A bar graph works well for this type of information.

299

ANNOTATION KEY

❶ The frequency depended on the background color of the trees, which darkened after the Industrial Revolution. This allowed the dark-colored moths better camouflage than the light-colored moths. (Applying concepts)

TEACHING SUPPORT

Teaching Resources
- Hands-On Activity: Where Are They? p. 75
- Ecology Investigation: Interspecific Competition and Growth Rates, p. 5

MEDIA AND TECHNOLOGY

 Interactive Videodisc
- *In the Company of Whales*
Use this bar code to show how scientists study the DNA in a representative gene pool.

Play frames 25559 to 26936

See the Biology Media Guide page 13 for additional bar-code correlations.

Figure 14–11 This photograph shows height distribution in a population of U.S. Army recruits. As you can see, most of the recruits fall in the range of average heights in the center of the curve. There are relatively few very tall or very short recruits.

can be said to result from differences in genotype. Of course, you can also observe many other kinds of phenotypic variation among your classmates. For example, variations in skin, hair, and eye color, and variations in the shapes of noses, the curves of lips, and the amount of body hair can be observed.

In nature, organisms show as many variations as humans, although most humans are not aware of this. For example, to the casual observer, one zebra looks much like any other zebra. But when researchers study the characteristics of many individuals of a species, they find the same sort of distribution for each characteristic that you saw in human height. It is this sort of variation in organisms that provides the raw material for natural selection.

Evolution as Genetic Change

In order to describe the evolution of plants and animals, modern evolutionary biologists study groups of organisms called **populations**. A population is a collection of individuals of the same species in a given area whose members can breed with one another. For example, all the fishes of a certain species in a single pond could be considered one population. Individuals in another, separate pond would belong to a different population, even if that pond was close by.

Because all members of a population can interbreed, they and their offspring share a common group of genes, called a **gene pool**. Each gene pool contains a number of alleles—or forms of a certain gene at a given point on a chromosome—for each inheritable trait, including alleles for recessive traits. The number of times an allele occurs in a gene pool compared with the number of times other alleles for the same gene occur is called the **relative frequency** of the allele.

Sexual reproduction alone does not change the relative frequency of alleles in a population. To understand why, you can compare the combinations of alleles produced by sexual reproduction to the different hands you get when you shuffle and deal a deck of playing cards. Shuffling and reshuffling produce

300

14–3 (continued)

Content Development
Emphasize that variation is just as common among plants and animals as it is among people. You can clearly illustrate your point with a bag of peanuts in their shells. Form small groups and have each group examine five to six peanuts and record their observations. Ask students if they found any two peanut shells that were exactly alike.

Content Development
Students may have difficulty understanding evolution in terms of genetic change. To illustrate this concept, perform with them the following demonstration with M&M candies.

Give each student or group of students a bag of M&Ms and have them count the pieces of each color. Tell students that the M&Ms represent a gene pool for a certain furry, seven-legged animal, and that each candy represents the alleles that determine the color of this animal. In other

an enormous variety of different hands. But shuffling alone will not change the relative numbers of aces, kings, fours, or jokers in the deck.

With this in mind, we can define evolution in another way. **Evolution is any change in the relative frequencies of alleles in the gene pool of a population.** And, as you can see in Figure 14–13, when the relative frequencies of alleles in a population change, the curves that describe the distribution of traits controlled by those alleles also change. In the case of the peppered moths, as the alleles for dark color increased, more dark-colored moths appeared in the population. This is the visible result of evolutionary change.

Genes, Fitness, and Adaptation

Each time an organism reproduces, it passes copies of its genes to its offspring. Thus we can define evolutionary fitness as the success an organism has in passing on its genes to the next generation. And we can define an adaptation as any genetically controlled characteristic of an organism that increases its fitness.

Let's return to our discussion of human weight lifters for a moment. Muscles acquired as a result of exercise are not passed on to offspring. Thus they cannot be considered an evolutionary adaptation and cannot contribute to evolutionary fitness. A gene that somehow allowed an individual to develop stronger muscles by doing less work or by eating less food, on the other hand, might be a useful adaptation under certain circumstances. This gene could be passed on to offspring.

A Genetic Definition for Species

In the past, biologists defined a species as a group of organisms that looked alike. Species were defined according to precise physical descriptions, and differences among individuals were seen as imperfections or mistakes. This definition, however, did not recognize that variation in a population is the rule rather than the exception.

Figure 14–12 These flamingoes represent a tiny part of a much larger group of interbreeding individuals of the same species, or population. Because all members of a population can interbreed, they share a common group of genes, or gene pool.

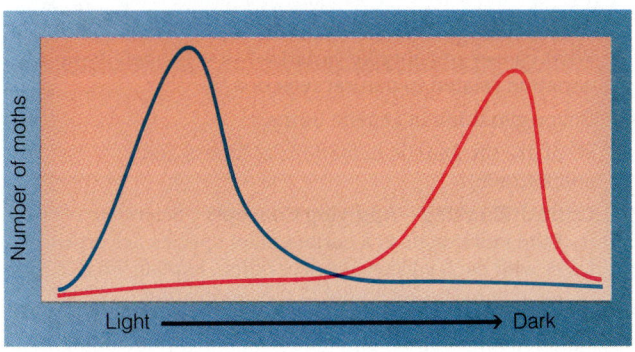

Figure 14–13 Sometimes the frequency of an allele changes in a population. The light-colored moths occurred with greater frequency before the Industrial Revolution. After the Industrial Revolution, the frequency of dark-colored moths became greater. How do you explain these changes? ❶

301

Figure 14–14 A species is a population of organisms that breed with one another and share a common gene pool. If this baby hippopotamus survives, it can pass on its genes to another generation.

We now define a species as a group of similar-looking (though not identical) organisms that breed with one another and produce fertile offspring in the natural environment. This definition is important because it allows us to determine what it means to belong to the same or different species.

Because members of a species can breed with one another, they share a common gene pool. Because of that shared gene pool, a genetic change that occurs in one individual can spread through the population as that individual and its offspring mate with other individuals. If the genetic change increases fitness, that gene will eventually be found in many individuals in the population. Members of a species can thus evolve together and interact with their environment in similar ways.

14–3 SECTION REVIEW

1. What causes phenotypic variation? How is phenotypic variation related to natural selection?
2. Define evolution in genetic terms.
3. What are the genetic definitions of fitness and adaptation?
4. **Critical Thinking—Applying Concepts** Scientists notice that the individuals in a certain plant species are growing taller with each successive generation. Explain what is happening to the gene pool of this population.

SECTION REVIEW 14–3

1. Gene mutations cause phenotypic variations that may make organisms more fit. These organisms will reproduce and pass on these traits to their offspring.
2. Evolution can be defined as a change in the relative frequency of alleles in the gene pool of a population.
3. In genetic terms, fitness is the success of an organism in passing on its genes to the next generation. An adaptation is any genetically controlled characteristic of an organism that increases its fitness.
4. The frequency of the gene for tallness is increasing in this

SCIENCE, TECHNOLOGY, AND SOCIETY ISSUE

Are We Creating Superpests?

Evolution did not just happen millions of years ago. Evolutionary change occurs around us constantly. Scientists today have observed many examples of evolutionary change that have occurred in living organisms.

Many species of insects and microorganisms damage crops, spoil food, or make us ill. Farmers spray their crops with poisons that kill harmful insects. Doctors and veterinarians use antibiotics and other medicines to kill disease-causing bacteria and other microorganisms.

However, scientists have observed that many of the insect species are not killed by insecticides that killed others of that species several years ago. And some microorganisms are no longer eliminated by antibiotics. The resistance of certain pest species to chemicals that once controlled them are examples of evolutionary change.

How can this resistance be explained? The work of Darwin provides an answer. Remember, Darwin suggested that there is great variation among organisms. Suppose there were a few insects that were not killed by an insecticide that killed other members of their species. Suppose there existed a bacterium that was unaffected by an antibiotic proven to control other members of its species. The surviving organisms would be more fit. They could pass on their genes to their offspring.

Normally these resistant organisms are uncommon in a population. However, the use of insecticides and antibiotics changes the environment. These chemicals kill organisms that are not resistant. But at the same time, these chemicals increase the fitness to survive of resistant organisms. With less competition, these fit individuals survive and reproduce. And because they reproduce quickly, the genes that made them more fit are rapidly spread in a population. Soon many more resistant individuals are found in the population. And the chemicals that once controlled a pest population are no longer effective.

This example of evolution in action poses problems for human society. New chemical controls must constantly be developed. However, some physicians feel that we will run out of new and effective antibiotics before microorganisms run out of variations. To control crop pests, farmers now use chemicals that are potentially more dangerous—chemicals that may also threaten humans and the environment. So there is much work for future biologists.

303

SCIENCE, TECHNOLOGY, AND SOCIETY ISSUE

Are We Creating Superpests?

Scientists of today can observe evolutionary change in progress. Many species of insects and microorganisms that were once limited by insecticides or antibiotic have become resistant and are no longer affected by these substances.

The work of Darwin answers the question of how this resistance has come about. Variation among the species account for this phenomenon. If an individual in a population is resistant to an insecticide or antibiotic, it will survive and pass on that resistance to the next generation. In other words, those individuals are more fit.

Because of this recurring resistance, scientists must continue to develop new and more powerful chemicals. This brings up two areas of concern. First, some scientists believe that we will run out of chemicals that effectively control these organisms before the organisms run out of variations. The other concern is for the environment. The chemicals are stronger and more dangerous.

In small groups or as a class, have students discuss the issues and questions raised in this article. What regulations, if any, should be imposed on new drugs and chemicals? Should farmers always spray their crops with the most powerful insecticides? Should doctors always use the most powerful antibiotic to fight a disease? What would the future be like if we succeeded in creating "superpests"?

---

plant species. Thus, over time, the plant species is becoming taller.

Reinforcement/Reteaching

If some students in the class are having difficulty answering one or more of the Section Review questions, review those parts of the section that are the source of the difficulty.

Closure

Using the Visuals Have students reexamine either Figure 14–11 or the graphs they made for the height distribution of the class. Discuss how these variations in height are reflected in the alleles for height in the gene pool. Have students imagine how human height might evolve in the future.

14–4 The Development of New Species

Guide For Reading
- What is a niche?
- How do new species develop from existing species?
- What are the results of divergent evolution? Of convergent evolution?

We are now nearly ready to explain how new species evolve from old ones, a process biologists call **speciation**. But before we can explain how speciation occurs and how it can lead to diversity, we must first understand some basic concepts about the way species interact in their environment.

The Niche: How to Make a Living

Organisms, like members of a human community, need to survive and acquire the necessities of life. But like people crowded into a city, organisms would have difficulty surviving if they all tried to do the same kinds of work, eat the same kinds of food, and live in the same place. Isn't it hard to imagine, for example, an entire city of butchers or tailors? And certainly you wouldn't want the population of an entire city living in your house!

In human cities, thousands of people survive near one another. They have different jobs, they shop in different stores, and they live in different places. Animals and plants do much the same thing. The combination of an organism's "profession" and the place in which it lives is called its **niche** (NIHCH). If two species occupy the same niche in the same location at the same time, they will compete with each other for food and space. One of the species will not survive. **No two species can occupy the same niche in the same location for a long period of time**. Chances are, one of the species will be more efficient than the other. The more efficient species will survive, reproduce, and drive the less efficient species to extinction.

Figure 14–15 *Reproductive isolation can lead to the development of a new species. The fish on the left is a member of the same species as the fish on the right, although it is quite different in appearance. In time, if it is reproductively isolated, it may evolve into another species.*

If two species occupy different niches, however, they will not compete with each other as much. With less competition, there is less chance that one species will cause the other to become extinct. So in the evolutionary struggle for existence, any species (or a population within a species) that occupies an unoccupied niche will be better able to survive. We will soon see how this phenomenon can lead to the formation of entirely new species.

The Process of Speciation

Remember that biologists define a species as a group of organisms that can breed with one another and produce fertile offspring in a natural environment. This definition means that individuals in the same species share a common gene pool. Individuals in different species have different gene pools.

Scientists have learned that new species usually form only when populations are isolated, or separated. This separation of populations so that they do not interbreed to produce fertile offspring is called **reproductive isolation**. If two populations are not reproductively isolated, their gene pools will blend with each other. No new species will be formed. Reproductive isolation is the agent for the formation of new species.

Reproductive isolation may occur in a variety of ways. Geographic barriers such as rivers, mountains, and even roads may separate populations and prevent them from interbreeding. Differences in courtship behavior or fertile periods may result in organisms that breed only with individuals that are most similar to themselves.

Once reproductive isolation occurs, natural selection usually increases the differences between the separated populations. As the populations become better adapted to different

Figure 14–16 Hundreds of years ago, the dodo was quite common on the island of Mauritius. This large bird, unable to fly, made its nest on the ground. In time, settlers arrived with dogs and other domestic animals. The helpless dodos were killed and their nests destroyed. The expression "dead as a dodo" is used today to refer to any organism that is now extinct.

Figure 14–17 The male bower bird builds a nest, or bower, on the ground to entice females to mate with him. Males compete for females by decorating their bowers with shiny, colorful objects such as shells, flowers, and bits of paper. The bower that attracts the most females earns the greatest reproductive success for its maker. Because different species of bower birds build different types of bowers, this mating behavior helps to ensure reproductive isolation between species.

BACKGROUND INFORMATION
THE HARDY-WEINBERG PRINCIPLE

Discuss the Hardy-Weinberg Principle with your most able students. A summary of the Hardy-Weinberg Principle is as follows:

The relative frequencies of the alleles in a population's gene pool will remain the same if

1. The population is large enough so that chance alone (genetic drift) will not cause changes in the gene pool.

2. Mutations do not occur.

3. All genotypes have equal fitness (no natural selection operating on the population).

4. There are no organisms entering or leaving the population in a way that would bring in or eliminate genes.

5. All mating in the population is completely at random.

When we examine natural populations of animals, we find that all of these conditions are rarely present at the same time. For this reason, it is unusual for gene pools of any but the largest populations to remain the same for long.

---

ductive isolation, you need to impress upon students—many of whom are romantic about love and courtship at this age—that mating among organisms in a natural environment is an instinctive behavior necessary to the survival of the species. Mating will occur within the isolated population just as it would in a larger population, but it will involve a far smaller gene pool. Over a period of time, populations that are isolated from one another will become more and more dissimilar because the two gene pools are no longer mixing with each other. Students will probably be able to think of instances where this has resulted from human intervention, such as among breeds of dogs.

ANNOTATION KEY

❶ Each beak is well-adapted to the type of food found on the island where the particular species lives.
(Interpreting diagrams)

❷ The two shells permit different ranges of head movement and thus let the two tortoises reach different food supplies.
(Applying concepts)

BACKGROUND INFORMATION
MUSICAL SPECIATION

Diverging physical traits do not cause all speciation. In Hawaii, a variety of courtship songs has separated the native *Drosophila* fruit fly into more than 500 species! Some Hawaiian *Drosophila* sound more like cicadas than flies. Others make a cricketlike noise. Still others make a sound like that of a North American fruit fly, but they create the sound by vibrating their abdomen instead of their wings.

---

environments, their separate gene pools gradually become more dissimilar. Now the populations are separated not only by the physical or behavioral barriers that once separated them, but by vastly different genes. If the populations remain separated for a long time, their gene pools eventually become so different that their reproductive isolation becomes permanent. When this occurs, the groups of organisms are no longer separate populations. They have become separate species.

Darwin's Finches: An Example of Speciation

We can now use our understanding of evolution to explain the fascinating case of Darwin's finches, a group of 14 bird species on the Galapagos Islands. All these finch species evolved from a single ancestral species. Yet each of the 14 species exhibits body structures and behaviors that enable it to live in a different niche. For example, each species shows adaptations that allow it to feed differently. Some of the finch species eat small seeds, whereas others crack open much larger seeds or seeds with thicker shells. Some species pick ticks—small insectlike animals—off the islands' tortoises and iguanas. One finch species uses twigs or cactus spines to remove insects from inside dead wood. And some finches, often called vampire finches, drink the blood of large sea birds after pecking them at the base of their tail! How did so many strange and unusual finch species evolve on these islands? The evolution of the various species of finches on the Galapagos Islands shows how geographic and behavioral barriers and reproductive isolation eventually lead to the formation of new species.

Figure 14–18 The many kinds of finches that Darwin observed on the Galapagos Islands evolved from a single species that emigrated from the South American mainland some kilometers away. How have the shapes of these birds' beaks contributed to their survival? ❶

Name	Vegetarian tree finch	Large insectivorous tree finch	Woodpecker finch	Cactus ground finch	Sharp-beaked ground finch	Large ground finch
Shape of Bill	Parrotlike bill	Grasping bill	Uses cactus spines	Large crushing bill	Pointed crushing bill	Large crushing bill
Main Food	Fruit	Insects	Insects	Cactus	Seeds	Seeds
Habitat	Trees	Trees	Trees	Ground	Ground	Ground

14–4 (continued)

Content Development
Emphasize that separated species become more and more dissimilar.

• If two populations become dissimilar in body structure and behavior, particularly mating behavior, what do you think will happen to them? (They will no longer mate with one another.)

• If they no longer mate in the natural environment, what could happen? (They could become two separate, distinct species.)

Content Development
Using the Visuals Talk briefly about finches in general. Some of your students may be totally unaware of the existence of these small birds. If you have pictures of all 14 species, it might help you to clarify the topic of speciation. A good way to begin would be to point out some similarities and differences that are most obvious among the species. Encourage students to speculate about why some of the differences might have been beneficial. Use Figure 14–18 to point

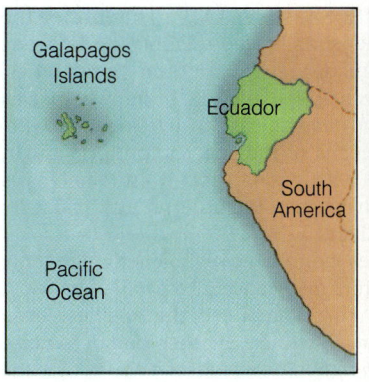

STEP 1: FOUNDING FATHERS AND MOTHERS Darwin's finches are descendants of a few ancestral finches that found their way to the Galapagos Islands from the South American mainland. Finches are small birds that do not usually fly far over open water. They may have gotten lost, or they may have been blown off course by a storm. In any case, once they arrived at one of the Galapagos Islands, which we can call Island A, they managed to survive and reproduce.

STEP 2: SEPARATION OF POPULATIONS Then, some birds from Island A crossed to another island in the Galapagos group. We will call this Island B. Remember, these birds do not like to fly over open water. So the populations on Islands A and B were essentially isolated from each other. Even though they were still members of the same species, the ocean between them prevented the blending of their gene pools.

STEP 3: CHANGES IN THE GENE POOL Over time, the populations on each island became adapted to the needs of their environments. For example, suppose the plants growing on Island A produced mainly small thin-shelled seeds, whereas the plants on Island B had larger thick-shelled seeds. Individual finches in the Island B population with larger, heavier beaks would be able to crack open and eat the seeds more easily. So birds with large beaks would be better able to survive on Island B. Over time, natural selection could have caused that population to evolve larger beaks.

STEP 4: REPRODUCTIVE ISOLATION Now imagine that a few birds from Island A cross to Island B. Will the birds from Island A be able to breed with the birds on Island B? Probably not. It happens that the finches prefer to mate with birds that have the same size beak they do. Thus differences in beak size, combined with mating behavior, act as a mechanism for reproductive isolation. The gene pools of the two bird groups do not mix. The two populations have become separate species.

Figure 14–19 This map shows the location of the Galapagos Islands off the coast of Ecuador (left). Variation exists even among the giant tortoises that live on the Galapagos Islands. The shell of one type of tortoise is raised in the front (right), enabling it to lift its head farther off the ground than the tortoise with a shell rounded in the front (center). How does the shell shape contribute to the survival of each tortoise? ❷

BACKGROUND INFORMATION
SPECIATION IN ACTION

Speciation had done its work on Darwin's finches long before Darwin or anyone else theorized about it. But scientists may have the chance to document from start to finish a speciation event that could be taking place today.

Before the 1800s the fruit fly *Rhagoletis pomonella* lived on the fruit of the North American hawthorn tree. When apple trees were introduced to the continent, *R. pomonella* began infesting them as well. Today, this fruit fly species exists as two host-specific populations, one for each type of fruit tree. The two populations do not interbreed and already they have been shown to have some genetic differences. Very possibly, the populations could diverge far enough to qualify as different species.

ECOLOGY NOTE
HAWAII

Before humans settled Hawaii 1500 years ago, a new species arrived on the islands once in some 100,000 years! Yet both deliberately and by accident, humans have introduced a wide variety of new organisms to Hawaii, including some that have done great harm to the islands' ecosystems.

One ecologist attributes a tremendous amount of damage to *Schizachyrium condensatum*, a grass plant ranchers had brought to Hawaii in the mid-twentieth century. This grass is not only very flammable, but it recovers from fires far more quickly than do native Hawaiian plants. These characteristics explain why from 1967 to 1989, the islands have suffered twice their normal number of large fires, and why these fires have been irreversibly destroying the Hawaiian dry forest, the home of birds and plants found nowhere else on Earth.

In October 1992, the U.S. Government passed three laws to further protect the Hawaiian ecosystem. Although much damage has already been done, the rate of destruction should now at least be slowed.

Figure 14–20 *Five million years of adaptive radiation in Hawaiian honeycreepers resulted in a wide array of beak shapes and as many as 43 species. Human actions have resulted in the extinction of most of these species.*

STEP 5: SHARING THE SAME ISLAND The fate of the two species on Island B depends on the relationship between the birds and their environment. There are three possibilities: coexistence, extinction, or further evolution. If the two species occupy different niches, they can coexist, or live together without changing. But if the niches of the species are too similar, the species will compete with each other. If one species is much better at making a living than the other, it may cause its competitor to become extinct.

If, however, one species exhibits enough genetic variation, the competition it encounters may cause it to evolve further. A new species may result. For example, if the species' beak changes size again, it will be able to eat another kind of food for which it does not have to compete. Scientists believe that the 14 species of finches on the Galapagos Islands evolved from a single ancestor when speciation happened in this way again and again on different islands over time.

Speciation and Adaptive Radiation

The process we have just described on the Galapagos Islands, in which one species gives rise to many species, is called **adaptive radiation.** The process of adaptive radiation is also known as **divergent evolution.** In adaptive radiation, a number of different species diverge, or move away, from a common ancestral form, much as the spokes of a bicycle wheel radiate from the hub. During a period of adaptive radiation, organisms evolve a variety of characteristics that enable them to survive in different niches. Throughout the history of life on Earth, adaptive radiations have occurred many times and in many places. Adaptive radiation occurred on the Hawaiian Islands among a group of birds called Hawaiian honeycreepers. The dinosaurs experienced an adaptive radiation in their day, only to eventually become extinct. The mammals alive today were produced by another wave of adaptive radiation.

Evidence of past adaptive radiations can be observed in many organisms. The homologous structures discussed in the last chapter are evidence of past adaptive radiations in which the similar body parts of related organisms evolved to perform different tasks.

Adaptive radiations among different organisms often produce species that are similar in appearance and behavior. This phenomenon is known as **convergent evolution.** Convergent evolution has produced many of the **analogous structures** in organisms today. Analogous structures are similar in appearance and function, but they have different origins. Because they have different origins, analogous structures usually have very different internal structures. For example, the wings of butterflies, birds, and bats are analogous structures that allow the organisms to fly. However, a closer examination of these wings

---

14–4 (continued)

Content Development
Discuss step 5, completing the theory of how speciation occurred. Develop the three possibilities of coexistence, extinction, or further evolution.
• **Under what circumstances can two species coexist successfully?** (If they occupy different niches and thus do not compete for existence.)
• **What will happen if the niches of two species are too similar?** (They will have to compete for survival, and one species may become extinct.)
• **How might further evolution occur?** (A species may evolve as a result of selection forces other than competition or in response to a new competitor.)

Enrichment
Bat and bird wings can be considered both homologous and analogous: homologous because they both evolved from the ancestral tetrapod limb; analogous because their internal structure is quite different. The hand bones make up most of a bat's wing, but only the tip of a bird's wing.

Content Development
Adaptive radiation, or divergent evolution, is difficult for some students to understand. Refer to Chapter 13 and use the examples

shows that a butterfly's wing is made of a thin nonliving membrane with an intricate network of supports. A bird's wing is made of skin, muscles, and arm bones. And a bat's wing is made of skin stretched between elongated finger bones.

Figure 14–21 *Adaptive radiation is the process by which many different species develop from a common ancestor. As you can see from this illustration, some of the descendants of the cotylosaur do not resemble it in the least.*

of homologous structures to refresh their memories and to serve as past evidences of adaptive radiation. Be sure that students have a clear understanding of adaptive radiation before introducing the concept of convergent evolution and analogous structures. The wings of birds, bats, and butterflies are good examples to use for this concept. It is especially effective if you have skeletons of a bat and a bird and a dissecting microscope to examine the butterfly wings.

Enrichment
Using the Visuals Use Figure 14–21 as the basis for a features matrix illustrating divergent and convergent evolution. Assign a number to each descendant. Put the numbers on the left-hand side of the matrix. Place the features to be noted across the top. You might use features such as wings, four legs, flippers, no legs, obvious teeth, beaks, claws, and so on. Have students indicate the presence or absence of features with plus or minus signs. When the matrix is complete, students will have enough information to describe, compare, and make inferences about the various organisms.

TEACHING SUPPORT

Study Guide
• Section 14–4, p. 140

TIE-IN LANGUAGE ARTS

"Living fossil" is the popular term for an organism from a very ancient species. But while some species may undergo fewer changes than others, no species survives millions of years without changes in its gene pool. The term "living fossil" has no real scientific meaning; no living organism is truly identical to its fossilized forebears.

14–4 (continued)

SECTION REVIEW 14–4

1. A niche consists of the place an organism occupies and the way in which it lives. Since no two species can live long in the exact same niche, organisms have a better chance of surviving if they occupy different niches. Competition for available niches thus leads to the formation of new species, or speciation.
2. Speciation occurs only when populations are reproductively isolated. The isolated populations develop different genetic makeups and eventually form new species.
3. Adaptive radiation occurs when one species gives rise to many different species. The finches Darwin observed in the Galápagos Islands are an example of adaptive radiation. It is believed that the several different species of finches found on the islands today resulted from a single finch species that originally settled on one of the islands.
4. It is the first species that usually undergoes adaptive radiation because many different niches are available. Adaptive radiation occurs as natural selection favors genetic diversity. And the filling of different niches lessens competition among species.

Figure 14–22 The wings of a butterfly and the wings of a bat are analogous structures—similar in appearance and function. Although the wings of both the butterfly and the bat show adaptations for flight, they are made up of different tissues.

14–4 SECTION REVIEW

1. What is a niche? How do niches contribute to speciation?
2. How are speciation and reproductive isolation related?
3. What is adaptive radiation? Explain how Darwin's finches illustrate this process.
4. **Critical Thinking—Identifying Patterns** The first species to reach a newly formed volcanic island often undergoes an adaptive radiation. Explain this observation.

Guide For Reading
■ How can evolution occur in the absence of natural selection?
■ What are some phenomena that affect the pace of evolution?
■ What is the theory of punctuated equilibria?

14–5 Evolutionary Theory Evolves

It should now be apparent to you that evolutionary theory has been modified over the years. With the contributions of scientists such as Lamarck and Darwin and a better understanding of heredity, scientists can now explain how variation occurs and define evolutionary concepts in terms of genetics. But even today, evolutionary theory continues to evolve as scientists formulate theories about the details of evolutionary change.

Genetic Drift

Natural selection is not always necessary for genetic change to occur. **With the aid of theories and genetic experiments, biologists have realized that gene pools can change —in other words, evolution can occur—in the absence of natural selection.** This does not mean that natural selection is not important. However, biologists now realize that chance plays an even larger part in evolutionary change than Darwin thought.

Reinforcement/Reteaching
For those students who are having difficulty answering one or more of the Section Review questions, review the appropriate section material.

Geneticists have shown that an allele can become common in a population by chance. This kind of random change in the frequency of a gene is called **genetic drift**.

How does genetic drift work? One possibility is that an individual with a particular allele may produce more offspring than other members of its species—not because it is better adapted but just by chance. It is also possible for environmental events to wipe out many individuals who do not carry a particular allele. For example, the distribution of some alleles in the population of mountain goats in Washington State may have changed when Mount St. Helens erupted in 1980, killing many mountain goats in the area of the volcano. Thus, in very special circumstances, a new or previously rare allele may become common in a population after only a few generations. Genetic drift occurs most efficiently in small populations because chance events, such as a volcanic eruption, are less likely to affect all members of a very large population. Genetic drift could also have played a role in the evolution of Darwin's finches, since each new population was founded by relatively few birds.

Genetic drift implies that all characteristics of an organism do not have to contribute to its fitness. For example, consider the differences between the Indian rhinoceros (which has one horn) and the African rhinoceros (which has two). Both rhinoceros species use their horns to fight predators and to joust among themselves, so having a horn or two is useful. But it is not clear whether having two horns is better for survival than having one horn. If the two types of rhinoceros lived in the same area, the rhinoceros with one horn would probably have the same fitness as the rhinoceros with two horns. Thus the extra horn does not necessarily contribute to fitness. Most likely, the ancestral populations that gave rise to the two modern rhinoceros species developed slightly different horn systems just by chance. Natural selection provided a distinct advantage to individuals with horns; but the two populations developed different numbers of horns because of random genetic drift. Genetic drift probably led also to the evolution of one hump on African camels and two humps on Asian camels.

Figure 14–23 Natural disasters such as the explosive eruption of Mount St. Helens may drastically shrink a gene pool by killing many individuals and restricting gene flow among the survivors. Such chance events may set the stage for genetic drift.

Figure 14–24 The Indian rhinoceros (left) has a single horn; the African rhinoceros (right) has two. The number of horns may not contribute to survival, since one horn is as good as two in defending animals as grand as these.

Unchanging Gene Pools

Modern evolutionary biologists recognize that although natural selection and genetic drift are both powerful forces of change, they do not cause genetic alterations in all species all the time under all conditions. And because sexual reproduction by itself does not change the frequency of alleles in a population, it is possible for the gene pool of a species to remain the same for a long time.

Every now and then there arises a species, particularly well adapted to an environment, that does not change over time. If no new species enter into competition with that species and if certain other conditions are met, that species may remain nearly unchanged for long periods of time. One example of such a species is the horseshoe crab, *Limulus*, whose living members are nearly identical to ancestors that lived hundreds of millions of years ago. Such organisms are often called living fossils. Though relatively rare among both plants and animals, living fossils are fascinating indications that under some conditions evolution can slow down.

Gradual and Rapid Evolutionary Change

Darwin, convinced by the work of Lyell of the slow and steady nature of geologic change, felt that biological change was also slow and steady. The theory that evolutionary change occurs slowly and gradually is known as **gradualism**. In many cases the fossil record shows that a particular group of organisms has indeed changed gradually over time.

But there is also evidence that many other species did not change very much from the time they appeared in the fossil record to the time they disappeared. In other words, much of the time these groups of animals and plants are in a state of **equilibrium** (ee-kwih-LIHB-ree-uhm), which means they do not change very much.

But every now and then, something happens to upset the equilibrium. At several points in the fossil record, changes in animals and plants occurred over relatively short periods of time. Some biologists argue that these rapid changes—rather than long, slow changes—are what create new species. Remember that when we say "short" and "rapid" we are talking about the geological time scale. Short periods of time for geologists can be hundreds of thousands, even millions, of years!

Rapid evolution after long periods of equilibrium can occur in several ways. It may occur when a small population of a species becomes isolated from the main part of the population. This small population can evolve more rapidly than the larger one because genetic changes can spread more quickly among fewer individuals. Or rapid evolution may occur when a small group of organisms migrates to a new environment, as happened with the Galapagos finches. The organisms then evolve rapidly to fill available niches.

Figure 14–25 This plant-eating dinosaur, which once munched plants on ancient Earth, is today extinct. The environment on Earth changed, but, alas, the dinosaur did not.

SECTION REVIEW 14-5

1. Genetic drift is a random change in the frequency of an allele in a population. Genetic drift alters the frequency of alleles in a gene pool.
2. First, a small population of a species could become isolated. In small populations, genetic changes can spread more rapidly. Second, a group of organ-

Rapid evolution may also result from dramatic changes on the Earth. Every now and then, many species have vanished in a phenomenon known as a **mass extinction**. Some mass extinctions were caused by changes in global climates that altered many environments. The causes of other mass extinctions remain uncertain. But whatever their causes, the effects of mass extinctions are clear. When many species die, many niches are left unoccupied. The species that remain suddenly find lots of empty niches. Groups of animals with enough genetic variability can undergo adaptive radiations. These adaptive radiations can produce a large number of new species to fill those empty niches.

Scientists use the term **punctuated equilibria** to describe this pattern of long stable periods interrupted by brief periods of change. Punctuated equilibria theory, which has generated much debate, is still controversial among biologists today. It is clear, however, that evolution has often proceeded at different rates for different organisms at different times during the long history of life. But whatever the pace of change might have been, it is certain that organisms have evolved over time.

The Significance of Evolutionary Theory

Evolutionary theory is, in the minds of many biologists, the foundation on which all biological science is built. Only because all living organisms are related through common descent can we talk about universal characteristics of life. Only because the physiological properties of all multicellular organisms are so similar can we study other animals to learn how our own bodies operate. And only through application of evolutionary theory can we truly understand the way that organisms interact with each other and with their environments.

But the influence of evolutionary thought extends far beyond biology. Philosopher J. Collins has written that "there are no living sciences, human attitudes, or institutional powers that remain unaffected by the ideas . . . released by Darwin's work." We cannot even touch on these remote disciplines in this book, although we hope you will be inspired to read about them later in life. We will, however, use the remainder of this book to discuss the products of evolution: Earth's "most beautiful and wonderful" living organisms.

14-5 SECTION REVIEW

1. What is genetic drift? How does genetic drift affect a gene pool?
2. What three factors can cause an increase in the rate of evolution of a species?
3. **Critical Thinking—Describing Concepts** Describe the theory of punctuated equilibria.

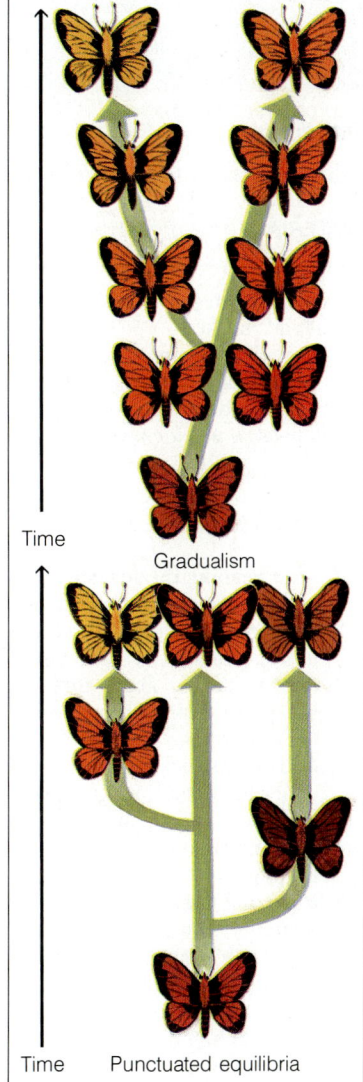

Figure 14–26 This illustration shows two possible ways in which a population of butterflies changes over time. The insects may change color gradually, with each succeeding generation showing only small color changes. Or the color of the population may remain fairly stable for long periods of time and then make great changes suddenly.

BACKGROUND INFORMATION
EVOLUTION TODAY

Organisms alive today are well-adapted to today's environments. Natural selection has played an important role in making them as well-adapted as they are. In fact, it would be easy to say that natural selection has made organisms perfectly suited to their environment. If organisms have been getting more and more fit for millions of years, evolution should be nearly at a standstill by now. However, by looking carefully at the geological record, biologists can see that evolution is not even slowing down.

What keeps evolution going? First, organisms always cause change in their environments. Second, the fossil record and geology indicate that the world's climate is constantly changing. These major changes mean that animals never have a single, stable environment to adapt to. Even while natural selection is shaping the organism to live in its environment, the environment is constantly changing. Animals and plants must continually adapt to new conditions just to keep up.

The raw material for evolutionary change is the amount of genetic variation in a species. If a species has enough genetic variation for it to evolve quickly enough to keep up with the environment, it will survive. If it does not, it will become extinct.

---

isms might migrate to a new environment, where they could rapidly evolve to fill available niches. Third, the environment could change dramatically and cause a mass extinction. Such an event would open up niches for the surviving organisms.

3. The theory states that evolution occurs when long periods of relatively little change are interrupted, or punctuated, by brief periods of rapid change.

Reinforcement/Reteaching
If students have difficulty answering the Section Review questions, turn back to the appropriate material.

Closure
Spend a few minutes reviewing the various facets of current evolutionary thought or have several students present and review the various theories for the class.

LABORATORY INVESTIGATION

SIMULATING NATURAL SELECTION

Before the Lab
Assemble all the materials needed for each group. Count the lima beans and kidney beans to avoid wasting lab time. Put them in one envelope per group. Make a transparency of the data sheet to use during the pre-lab discussion.

Pre-Lab Discussion
Review natural selection. Discuss the laws of chance, which account for variations. Give each student one of each type of bean for examination prior to the lab investigation.

Be sure that students understand what each bean represents. Go over the data table and explain how it is to be completed. Tell students to have the same member of the group take all the beans for each generation out of the sack because using different students could alter the results. Give students directions for handling the materials after the lab is over.

Skills Development
Students will use the following skills while completing this laboratory investigation.
1. Manipulative
2. Comparing
3. Relating
4. Inferring
5. Analyzing

Safety Tips
There are no dangerous materials used in this lab, but you might want to remind students that you will not tolerate horseplay.

Teaching Strategy
Monitor the class to check procedure. Be sure students are accurately following directions and that they are completing the data table correctly. Encourage them to compare the simulation to what would actually be happening in nature.

LABORATORY INVESTIGATION

SIMULATING NATURAL SELECTION

PROBLEM
What happens to harmful genes over time?

MATERIALS (per group)

small brown paper bag
400-mL beaker
250 g pinto beans
100 large lima beans
100 red kidney beans
stopwatch or clock with second hand

PROCEDURE

1. After reading the investigation carefully, prepare a data table on a separate sheet of paper.
2. Examine and note the differences and similarities among the three types of beans.
3. Fill the beaker about three-fourths full with the pinto beans. Then pour the beans into the paper bag.
4. Add 50 lima beans and 50 kidney beans to the paper bag. The lima and kidney beans represent organisms. The pinto beans represent the environment in which the organisms are hiding.
5. Have one member of your group time you for 3 minutes. During the 3 minutes, remove one bean at a time from the paper bag. Without looking, try to remove as many lima and kidney beans as you can. Use the shapes of the beans to identify them.
6. Record the number of lima beans and kidney beans that you removed in the appropriate place in the data table.
7. To determine the number of lima beans and kidney beans that remain in the paper bag, subtract the number of lima beans removed from the starting number. Do the same for the kidney beans.
8. Add the numbers of remaining lima and kidney beans together. Record this information in the appropriate place in the data table.
9. To find the frequencies of lima beans and kidney beans that remain in the paper bag, divide the numbers of each of the remaining beans by the total number of remaining lima and kidney beans. Round these numbers off to the nearest hundredths place. Record this information in the appropriate place in the data table.
10. The frequencies of each of the beans that remain in the paper bag represent the distribution of genes in the population. To determine the starting number of lima beans and kidney beans that are present in the next generation, multiply the frequency of each bean by 100. Record this information. Place the calculated numbers of lima beans and kidney beans in the bag. The number of lima beans plus the number of kidney beans should equal 100.
11. Repeat steps 5 through 9 until the information in the data table is complete for five generations.

OBSERVATIONS

1. What was the ratio of lima beans to kidney beans?
2. Based on your data, was one type of bean removed from the bag more frequently than the other? If so, which type?
3. What happened to the frequencies of the lima beans in the population over five generations? The frequencies of the kidney beans?
4. What happened to the total number of beans remaining in the bag after each generation?

ANALYSIS AND CONCLUSIONS

1. Which beans represent the beneficial genes in a population? The harmful genes? How do you know?
2. Explain why the frequency of remaining genes changes in each generation.
3. What do you think happens to harmful genes over time?

Observations
1. The ratio was 50:50 or 1:1.
2. Lima beans were removed more often than kidney beans.
3. The percentage of lima beans decreased while the percentage of kidney beans increased. The frequency of lima beans in the environment decreased over time, while the frequency of kidney beans in the environment increased.
4. The total number of remaining beans increased with each generation.

Analysis and Conclusions
1. The lima beans represent organisms with harmful genes, while the kidney beans represent organisms with helpful genes. We know this because the frequency of harmful lima beans decreased in the environment.

STUDENT STUDY GUIDE

SUMMARIZING THE CONCEPTS

The key concepts in each section of this chapter are listed below to help you review the chapter content. Make sure you understand each concept and its relationship to other concepts and to the theme of this chapter.

14–1 Developing a Theory of Evolution
- Lamarck was one of the first people to propose a theory of evolution and adaptation. He thought that organisms respond to the needs of their environment and pass on acquired characteristics to their offspring.

14–2 Evolution by Natural Selection
- In natural selection, only the organisms whose characteristics are well-suited for their environment survive and reproduce.

14–3 Genetics and Evolutionary Theory
- Gene mutations and gene recombinations provide the variations upon which natural selection acts.
- Evolution is a change in the relative frequencies of alleles in the gene pool of a population.

14–4 The Development of New Species
- A niche is the combination of an organism's "profession" and the place in which it lives. If two species occupy the same niche in the same location at the same time, they will compete with each other for food and space.
- Reproductive isolation is necessary for speciation.
- In adaptive radiation, one species gives rise to many new species.

14–5 Evolutionary Theory Evolves
- Genetic drift—a random change in the frequency of a gene—occurs most efficiently in small populations.
- The theory of punctuated equilibria states that there are long periods of stability punctuated by brief periods of rapid change.

REVIEWING KEY TERMS

Vocabulary terms are important to your understanding of biology. The key terms listed below are those you should be especially familiar with. Review these terms and their meanings. Then use each term in a complete sentence. If you are not sure of a term's meaning, return to the appropriate section and review its definition.

14–1 Developing a Theory of Evolution
- artificial selection

14–2 Evolution by Natural Selection
- natural selection
- survival of the fittest

14–3 Genetics and Evolutionary Theory
- population
- gene pool
- relative frequency

14–4 The Development of New Species
- speciation
- niche
- reproductive isolation
- adaptive radiation
- divergent evolution
- convergent evolution
- analogous structure

14–5 Evolutionary Theory Evolves
- genetic drift
- gradualism
- equilibrium
- mass extinction
- punctuated equilibria

(Answers will vary.)
- **If your results differ from your predictions, explain why.** (Answers will vary.)
- **In a natural setting, what factors would affect the results?** (Answers may include availability of food, competition, and so on.)

2. The total number of remaining beans increases after each generation because the beans get harder to find as the number of lima beans decreases.

3. Organisms with harmful genes tend to die out over time; therefore, the frequency of harmful genes in the gene pool decreases.

Going Further: Enrichment
Have students predict how many generations it would take, under the circumstances presented in the lab, for one group to become extinct. Test the prediction by continuing the experiment until one set of beans is completely removed from the bag.
- **How many generations did it take?** (Answers will vary.)
- **How did your prediction compare with your results?**

CHAPTER REVIEW

CONTENT REVIEW

Multiple Choice
1. a 2. d 3. c 4. b
5. b 6. a 7. d 8. c

True or False
1. F punctuated equilibria
2. F Darwin
3. T
4. F Evolution
5. T
6. F Convergent evolution
7. F genetic drift
8. F natural selection

Word Relationships
1. punctuated equilibria
2. divergent evolution
3. artificial selection
4. reproductive isolation

CONCEPT MASTERY

1. Lyell proposed that the Earth was much older than previously thought and that changes occurred in the Earth over time. Lamarck was one of the first persons to propose that organisms changed over time, and that traits in organisms could be passed on to succeeding generations. Malthus provided the idea that many organisms are produced but that the size of a population is limited by disasters such as famine and war.
2. Lamarck thought organisms could change if they willed themselves to change and that traits acquired during an organism's life could be passed on to offspring. Both of these assumptions later proved to be incorrect.
3. Punctuated equilibria is the theory that describes the process of evolution as a long period of few changes punctuated by brief periods of great change.
4. Farmers can change a population of chickens by artificial selection. They can select those chickens that produce large quantities of eggs and cross them with roosters that come from a population of hens that were good producers. This artificial selection should result in a population of chickens that produce large quantities of eggs.
5. Reproductive isolation prevents the genes of the isolated population from being mixed with the genes of other populations, thereby producing a separate gene pool.
6. Changes in genes often produce changes in phenotype, sometimes resulting in organisms that are more fit to survive. Natural selection favors such individuals, and they pass on their genes to new generations.
7. Survival of the fittest is the principle that states that those individuals with characteristics well-suited to their environment will survive and reproduce, thus passing on their traits to their offspring.
8. Two species occupying the same niche will compete with each other and the death of one or both of the species may result.

CHAPTER REVIEW

CONTENT REVIEW

Multiple Choice

Choose the letter of the answer that best completes each statement.

1. Darwin was familiar with the works of all of the following except
   a. Mendel.   c. Lamarck.
   b. Lyell.   d. Malthus.
2. Which of the following is needed for new species to form?
   a. a niche
   b. homologous structures
   c. analogous structures
   d. reproductive isolation
3. Farmers change the gene pool of a population by
   a. adaptive radiation.
   b. natural selection.
   c. artificial selection.
   d. convergent evolution.
4. The source of random variation on which natural selection operates are changes in
   a. a niche.
   b. genes.
   c. relative frequency.
   d. the survival of the fittest.
5. An example of analogous structures are a
   a. whale's flipper and a bat's wing.
   b. bird's wing and a butterfly's wing.
   c. hawk's wing and a robin's wing.
   d. dog's leg and a horse's leg.
6. Which of the following ideas proposed by Lamarck was later found to be incorrect?
   a. Acquired characteristics can be inherited.
   b. Analogous structures can be inherited.
   c. Living things change over time.
   d. The Earth is very young.
7. Malthus thought that all of the following would prevent the endless growth of the human population except
   a. famine.   c. disease.
   b. war.   d. evolution.
8. Natural selection is also known as
   a. adaptive radiation.
   b. convergent evolution.
   c. survival of the fittest.
   d. divergent evolution.

True or False

Determine whether each statement is true or false. If it is true, write "true." If it is false, change the underlined word or words to make the statement true.

1. The theory of <u>gradualism</u> states that the fossil record shows long periods of stability and short periods of rapid evolution.
2. <u>Lamarck</u> and Wallace both developed a theory of evolution by natural selection.
3. Members of a <u>population</u> share a common group of genes, called a gene pool.
4. <u>Speciation</u> is any change in the relative frequencies of alleles in the gene pool of a population.
5. The combination of an organism's "profession" and the place in which it lives is called its <u>niche</u>.
6. <u>Divergent evolution</u> has produced many of the analogous structures in organisms today.
7. Random change in the frequency of a gene is called <u>relative frequency</u>.
8. More dark-colored peppered moths were found when soot darkened the color of tree trunks. This is an example of <u>artificial selection</u>.

CRITICAL AND CREATIVE THINKING

1. Punctuated equilibria theory states that long periods of relatively little change are punctuated by brief periods of rapid change. Gaps in the fossil record can be explained if rapid changes occurred when conditions were not appropriate for fossil formation.
2. In the wild, turkeys that are unable to fly would be at a disadvantage since predators and hunters would be able to

Word Relationships

Give the vocabulary word whose meaning is opposite that of the following words.

1. gradualism
2. convergent evolution
3. natural selection
4. interbreeding

CONCEPT MASTERY

Use your understanding of the concepts developed in the chapter to answer each of the following in a brief paragraph.

1. Explain how the work of Lamarck, Lyell, and Malthus influenced Darwin's thinking.
2. What are two errors in Lamarck's theory of evolution?
3. What is punctuated equilibria?
4. How can farmers change a population of chickens?
5. Why is reproductive isolation needed for a new species to form?
6. How do genes provide the raw material for natural selection?
7. What is survival of the fittest?
8. Two organisms cannot occupy the same niche at the same time. Explain why this is so.

CRITICAL AND CREATIVE THINKING

Discuss each of the following in a brief paragraph.

1. **Applying concepts** How does punctuated equilibria try to explain gaps in the fossil record?
2. **Making predictions** Domesticated turkeys cannot fly. This is an advantage to a farmer who raises thousands of birds. What would happen to these birds if they escaped from the farm into the woods? Explain your answer.
3. **Evaluating theories** Is protecting endangered species defying natural selection? Explain.
4. **Relating cause and effect** The giant panda occupies a very small niche by eating only one kind of food: bamboo. How can being adapted to such a small niche actually endanger this species' survival?
5. **Applying concepts** How might having a small gene pool cause a species to become extinct?
6. **Using the writing process** A friend of yours invents a time machine through which you both embark on a trip far into the future. When you arrive at your destination, you discover that Earth is much warmer and the sunlight more intense. You see many plants and animals that are unlike any you have ever seen before. Write a short story that describes the new life forms that are able to survive under these different conditions. You may wish to accompany your story with one or more drawings depicting these unusual plants and animals.

---

kill them quite easily. If they did survive, however, the flightless turkeys might pass on this trait to their offspring. These birds would also be unlikely to survive for long. All in all, not a happy prospect for the turkeys.

3. In some ways it might be considered so. However, it is hard for species to compete in a world whose fate is largely determined by one population, the human one. Endangered species are protected and, at the same time, jeopardized by the actions of humans. Protecting endangered species, even if it disrupts natural processes, gives some species their chance of survival.

4. The panda's survival is endangered when its food supply is limited. Actually, this occurs with some regularity. Bamboo, a favorite food of the panda, is a giant member of the grass family. All members of the same bamboo species flower and make seeds at the same time and then die. When these species die, some pandas starve. This further threatens this already endangered species.

5. A small gene pool consists of relatively few individuals with similar genes. If the niche occupied by these organisms is altered, the species may not have the ability to change and survive. The species may therefore become extinct. An example of such alteration may be the introduction of a new disease. If a species has a large gene pool, some members of the population may be immune to the disease. These members could survive and reproduce, thus ensuring the survival of the species. If all members of the population are immune to the new disease, the population may die. The species might then become extinct.

6. Check students' work to make sure that their writings conform to scientific fact. Students should base their fictional descriptions on animals and plants that exist today and that are found in warm areas on Earth. However, commend students who imagine new and wildly different organisms as long as the organism's traits contribute to survival in a warm environment.

CHAPTER 15 Classification Systems

Section	Laboratory Investigations and Activities
15–1 Why Classify? pages 319–320 THEME: Unity and Diversity	**Teacher Edition** DEMONSTRATION: Diversity of Life, p. 318D **Teaching Resources** HANDS-ON ACTIVITY: Sorting It Out, p. 81
15–2 Biological Classification pages 320–323 THEMES: Unity and Diversity, Scale and Structure	**Laboratory Manual** Classifying Organisms, p. 193
15–3 Taxonomy Today pages 323–325 THEMES: Evolution, Unity and Diversity	**Student Edition** LABORATORY INVESTIGATION: Preparing a Collection of Organisms, p. 330 **Teacher Edition** DEMONSTRATION: Two-Kingdom System of Classification, p. 318D
15–4 The Five-Kingdom System pages 325–329 THEMES: Evolution, Unity and Diversity	**Laboratory Manual** Using and Constructing a Classification Key, p. 203
Chapter Review pages 331–333	

Teacher Resources

Books

Abbott, Lois A., et al. *Taxonomic Analysis in Biology: Computers, Models, and Databases.* Columbia University Press, 1985.

Hickman, C. P., L. S. Roberts, and F. M. Hickman. *Integrated Principles of Zoology,* 8th ed. Times Mirror/Mosby College Publishing, 1988.

Humphries, C. J., ed. *Ontogeny and Systematics.* Columbia University Press, 1988.

Jeffrey, C. *An Introduction to Plant Taxonomy,* 2nd ed. Cambridge University Press, 1982.

Linnaeus, Carolus. *Systema Naturae.* 1735. Coronet Books, 1964.

Parker, Sybil P., ed. *Synopsis and Classification of Living Organisms,* 2 vols. McGraw-Hill, 1982.

Ross, H. H. *Biological Systematics.* Addison-Wesley, 1974.

CHAPTER PLANNING GUIDE

Other Activities	Media and Technology

Teaching Resources
    ACTIVITY: Fun with Fictitious Animals, p. 3
Study Guide
    Section 15–1, p. 147	

Teaching Resources
    ACTIVITY: Analyzing Relationships Within a Classification System, p. 5
Study Guide
    Section 15–2, p. 148
	Interactive Videodisc
	    In the Company of Whales
	Biology Media Guide
	    page 15

Study Guide Section 15–3, p.150	

Teaching Resources
    VOCABULARY REVIEW: Word Game, p. 1
Study Guide
    Section 15–4, p. 151	

Teaching Resources
    Chapter Test A, p. 13
    Chapter Test B, p. 17
    Performance-Based Assessment, Unit 4
	Computer Test Bank
	    Chapter Test, p. 159

Audiovisuals

Classifying in Science. Video. Agency for Instructional Television.
Classifying Plants and Animals. Film or video. Coronet.
The Five-Kingdom Classification. Film or video. Benchmark Films.

Plant Classification: How Botanists Classify Plants. Sound filmstrip. Ward.
Taxonomy: How Living Organisms Differ. Slides or filmstrip and cassette. Science and Mankind, Inc.

How Living Things Are Classified. Sound filmstrip. Encyclopedia Britannica Educational Corporation.

CHAPTER 15 Classification Systems

CHAPTER OVERVIEW

To date, scientists have identified over 2½ million species of organisms on Earth. Such great diversity among living things called for a system of naming and ordering the organisms in a logical manner. In the eighteenth century such a system was devised by a Swedish botanist named Carolus Linnaeus.

Linnaeus's system of naming organisms is called binomial nomenclature. In this system each organism is assigned a two-part scientific name that identifies its genius and species.

Linnaeus grouped organisms according to similar body structures. The groups to which organisms are assigned are called taxa. The taxa that make up Linnaeus's classification system are kingdom, phylum, class, order, family, genus, and species.

Taxonomy, the science of naming and grouping living things, illustrates evolutionary relationships among organisms. Modern biochemists consider similarities and differences in the chemical compounds of organisms to create a biochemical taxonomy of living things.

Linnaeus's two-kingdom classification system has evolved to a five-kingdom classification system widely used today. The five-kingdom system more accurately groups organisms according to evolutionary trends.

15–1 WHY CLASSIFY?

Section Focus 15–1

The purpose of this section is to identify the need for a system of naming and organizing Earth's organisms. To date, the number of identified species is over 2½ million, and the total is still growing. In order to understand this great diversity of life, a system was needed to name and order the organisms. To be useful to scientists all over the Earth, the system had to assign a universally accepted name to each organism and a set of universally accepted rules for grouping the organisms.

Performance Objectives 15–1

1. Identify the need for classification systems.
2. Describe characteristics of a good biological system of classification.

15–2 BIOLOGICAL CLASSIFICATION

Section Focus 15–2

The purpose of this section is to identify the need for a universally accepted method of naming and grouping organisms. In the eighteenth century Carolus Linnaeus developed the system of binomial nomenclature. This system assigns a two-part scientific name to every organism. The first part is the genus name and the second part is the species name.

Linnaeus also devised a method of grouping organisms based on structural characteristics. The groups to which Linnaeus assigned organisms are called taxa. The taxa in Linnaeus's classification system are kingdom, phylum, class, order, family, genus, and species. Members of the same taxon share certain similar traits.

Performance Objectives 15–2

1. Explain the system of binomial nomenclature.
2. Identify the taxa in the classification system devised by Linnaeus.
3. Compare the traits of two organisms, each from a different taxon.

Science Terms 15–2

binomial nomenclature   p. 321
taxonomy   p. 321
taxon   p. 322
species   p. 322
genus   p. 322
family   p. 322
order   p. 322
class   p. 322
phylum   p. 322
kingdom   p. 322

15–3 TAXONOMY TODAY

Section Focus 15–3

The purpose of this section is to explain how taxonomy provides information regarding evolutionary relationships of organisms. Taxonomy is the science of naming organisms and assigning them to groups. Taxonomists identify and study homologous structures in organisms. Species shown to be closely related are classified together. Organisms are grouped according to their evolutionary relationships.

Biochemists classify organisms according to the similarities and differences of their chemical compounds. The study of molecules of DNA and RNA from various organisms yields information about their evolutionary relationships.

Performance Objectives 15–3

1. Identify a taxon with a clear biological identity.
2. Explain how taxa show evolutionary relationships between organisms.
3. Discuss the importance of biochemistry to taxonomists.

CHAPTER PREVIEW

15-4 THE FIVE-KINGDOM SYSTEM

Section Focus 15-4

The purpose of this section is to identify the structure of the five-kingdom system of classification. The discovery of new life forms since the time of Linnaeus led to the development of this system of classification. The five kingdoms in this generally accepted classification system are Monera, Protista, Fungi, Plantae, and Animalia. All prokaryotes are placed in the kingdom Monera. Most single-celled eukaryotic organisms are placed in the kingdom Protista. Members of the kingdom Fungi are heterotrophic and build cell walls that do not contain cellulose. Most members of the kingdom Plantae are multicellular, have cell walls that contain cellulose, and are autotrophic. Members of the kingdom Animalia are multicellular, heterotrophic, and have cell membranes without cell walls. In the five-kingdom system, organisms are grouped more accurately, based on evolutionary trends.

Performance Objectives 15-4

1. Explain why the most generally accepted classification system contains five kingdoms.
2. Identify the characteristics of the organisms belonging to each of the five kingdoms.

Science Terms 15-4

Monera   p. 327
Protista   p. 327
Fungi   p. 327
Plantae   p. 328
Animalia   p. 328

Discovery Learning

TEACHER DEMONSTRATIONS
Modeling

Diversity of Life

The following demonstration can be used as an introduction to Chapter 15.

Have all class members stand. Read each of the following physical characteristics aloud. Only those students possessing each trait should remain standing.

- Over 5 feet tall
- Brown eyes
- Female
- Left-handed

The result will be an extremely small group of students. Discuss whether those students standing or those sitting are more alike. Guide students' understanding that great differences exist among living things. Grouping organisms according to similar traits makes it easier to understand the diversity of life.

Two-Kingdom System of Classification

You may wish to perform this demonstration after students complete Section 15-3.

Have students identify the two kingdoms in Linnaeus's system of classification. Then show pictures of various plants and animals (such as a lion, rose, oak tree, bird, turtle, grass, or dog). Ask students to identify the kingdom to which each organism belongs in the two-kingdom system.

Next, show students a picture of a paramecium. Ask students to classify this organism according to the two-kingdom system. Guide students' understanding that the development of microscopes helped scientists discover types of organisms that were unknown when Linnaeus created his classification system. As a result, the two-kingdom system was adapted to include additional organisms.

CHAPTER 15
Classification Systems

GUIDED ENQUIRY

Pose the following questions to students and have them record their responses. Point out that they will gain a better understanding of the key concepts if they read the chapter with these basic questions in mind. Upon completion of the chapter, pose the questions again. Ask students to compare their initial responses with those they have developed after reading the chapter.

- Why is it necessary for scientists worldwide to use the same name for a particular organism?
- How does grouping organisms aid the investigation of Earth's diverse life forms?
- How has scientific knowledge of Earth's life forms changed in the past 300 years?
- How are evolutionary relationships among organisms determined?

INTRODUCING CHAPTER 15

Direct students' attention to photographs of various animals. Determine which of the animals pictured have been observed firsthand by members of the class. Explain that the total number of organisms directly viewed by every class member represents only a small portion of the life forms found on Earth.
- **How are the organisms related?** (They are all animals.)
- **How are the organisms different?** (Answers may include different species; different habitats; herbivores, carnivores, omnivores; different methods of reproduction; vertebrates and invertebrates.)
- **How could the animals shown in the photographs be divided into groups?** (According to similar traits.)
- **What information about the evolutionary relationships among organisms is revealed by studying the traits of organisms?** (Homologous structures in organisms indicate similar evolutionary ancestors.)

Have students read the introduction to the chapter.

CHAPTER 15
Classification Systems

The ruffed lemur, like other lemurs, lives only on Madagascar, an island off the east coast of Africa. Orchids, on the other hand, are found throughout the world. Both of these organisms are part of the great diversity of life on planet Earth.

Staring out at you from the pages of this textbook, you will find a bewildering number and variety of organisms. There are furry organisms, slimy organisms, and scaly organisms. Some are so tiny that they drift on the wind and can barely be seen with the aid of an electron microscope. Others are much larger than elephants and have a mass of many tons. Yet all the plants, animals, and other life forms that we'll introduce to you in the next twenty chapters represent only a fraction of the millions of species with whom we share our planet. How can biologists keep track of all these organisms? How do they find names for them? These tasks are indeed difficult, but the vital work does get done—as you shall now begin to learn.

TEACHING STRATEGY 15–1

Focus/Motivation
Using the Visuals Have students observe Figure 15–1 and read the caption.
- **What type of climate is found in a tropical rain forest?** (Warm,

GUIDE FOR READING

After you read the following sections, you will be able to

15–1 Why Classify?
- Discuss the usefulness of classification systems.
- List the characteristics of a good biological classification system.

15–2 Biological Classification
- Describe the importance of the classification system developed by Carolus Linnaeus.
- Identify the different taxa that make up the classification system developed by Linnaeus.

15–3 Taxonomy Today
- Discuss how taxa show evolutionary relationships among different organisms.
- Explain how modern scientific techniques contribute to the classification of organisms.

15–4 The Five-Kingdom System
- Explain why the five-kingdom system more accurately represents evolutionary trends.
- Discuss characteristics of organisms placed in each of the five kingdoms.

Journal Activity

YOU AND YOUR WORLD
Look around your home, school, or neighborhood. Are there any examples of a classification system in use? If so, write the examples in your journal. If there are no examples, can you suggest some?

Figure 15–1 There are so many different species of organisms in the tropical rain forests that these areas are called the "nurseries of the globe." Because many rain forest organisms have never been collected, studied, and classified, scientists view with alarm the destruction of these areas.

15–1 Why Classify?

Guide For Reading
■ Why are classification systems useful?
■ What are some characteristics of a good classification system?

Scientists have identified more than 2.5 million kinds of organisms. Given that feat, you may be surprised to learn that this catalog of life isn't even close to being complete! Some biologists estimate that there may be another 20 million or so unknown species still out there. These species include insects living in tropical rain forests, odd creatures living in the unexplored depths of the sea, and microorganisms living all around us.

No one can think about that many organisms at once, and certainly no one can keep track of them by their names alone. The only way to study this great diversity of organisms is to divide living things into manageable groups. But what kinds of groups? **To work with the diversity of life we need a system of biological classification that names and orders organisms in a logical manner.**

Biological classification systems have two important features. First, they assign a universally accepted name to each organism. In this way, an American scientist, for example, can talk to colleagues in Germany or India and be sure that everyone is discussing the same organism. Second, biological classification systems place organisms into groups that have real biological meaning. When we mentioned insects a moment ago, we gave you a rough idea about the type of animals we were discussing. When biologists place organisms into useful groups such as insects (or fishes, etc.), they can expect members of each group to

wet climate.)
- **How are the organisms shown in the figure alike?** (Students should realize that organisms possess adaptations suitable for survival in a rain forest.)
- **What effect could the destruction of rain forests have on the evolutionary record?** (Destruction of rain forests could reduce populations of organisms adapted to the environment. Unknown species could become extinct, causing gaps in the evolutionary record.)

Content Development
Develop the idea that dividing organisms into small groups aids the investigation of Earth's life forms. However, it is important for scientists worldwide to use similar classification schemes.
- **How could the organisms**

COOPERATIVE LEARNING

Complete this activity after reading the Science, Technology, and Society feature *So Little Time, So Many Names* on page 326.

Using preassigned lab groups or randomly selected teams, have groups prepare illustrated obituaries for assigned species such as bald eagle, African elephant, panda, California condor, grizzly bear, black rhinoceros, snow leopard, or humans. As an introduction to this assignment, students should be reminded that all major news agencies maintain a file of biographical information on prominent people in order to quickly prepare an obituary upon their death. Groups should include the following information in the obituary:
- Scientific name
- Habitat
- Niche
- Survivors ("family" members)
- Projected date of extinction
- Projected cause of extinction
- Notable contributions
- Illustration

Journal Activity

YOU AND YOUR WORLD
You may also want to use the Journal Activity as a springboard for class discussion. Students should be able to suggest that libraries, grocery stores, record and book stores, and even clothing stores classify things. Classification systems make it easier to find books on a particular subject or merchandise in a particular style or color. Students should be instructed to keep their Journal Activity in their portfolio.

shown in the picture be grouped? (Accept any reasonable answer.)
- **Why is it necessary for scientists around the world to use a similar grouping system?** (A universally accepted classification system eliminates confusion and permits scientists to understand each other's data.)

TEACHING SUPPORT

Teaching Resources
- Activity: Fun with Fictitious Animals, p. 3
- Hands-On Activity: Sorting It Out, p. 81

Study Guide
- Section 15–1, p. 147

15–1 (continued)

SECTION REVIEW 15–1

1. Classification systems name and order living organisms in a logical manner so that relationships between organisms are evident. A good classification system helps in understanding the natural world. A good classification system assigns a single universally accepted name to each organism and places organisms in groups that have real biological meaning.
2. If both scientists use a single name for an organism, they can be certain that they are discussing the same organism.
3. The Dewey Decimal System is used to classify books in a library. This system groups books in subject areas, making it easier for people to locate books on a particular subject.

Reinforcement/Reteaching

For those students who have trouble with the Section Review questions, go back to the discussion in the section and review the material.

Closure

Have students explain the function of a good classification system.

TEACHING STRATEGY 15–2

Focus/Motivation

Using the Visuals Have students observe Figure 15–2 and read the caption. Emphasize that while all three organisms are worms, they are not exactly alike.

- **How are these organisms alike?** (Possible answers include animals or invertebrates.)
- **How are these organisms different?** (Possible answers include different body shapes and different habitats.)
- **What criteria would you use to divide these organisms into groups?** (Accept any reasonable answers.)

Content Development

Using the Visuals Use the Motivation discussion to lead into the idea that each of the worms

Figure 15–2 These three organisms are all commonly referred to as worms because they share certain basic features. Yet one of the three animals differs from the others in several important ways. For this reason, the worms in the photographs at the left and center are classified together in one very large group whereas the worm in the photograph at the right is placed in another group.

share important traits. Of course to be really useful, scientific groupings must be more precise than just "insects" (or "fishes"). The trick is to create a useful classification system based on a universally accepted set of rules for organizing the groups.

15–1 SECTION REVIEW

1. Why are classification systems useful? What are two characteristics of a good classification system?
2. What advantage is there for two scientists on opposite sides of the world to use a single name for a particular organism?
3. **Connection—You and Your World** Give an example of a system you use to classify objects or people around you.

Guide For Reading
- Why was the classification system developed by Carolus Linnaeus important?
- What taxa made up the classification system developed by Linnaeus?

15–2 Biological Classification

By the eighteenth century, European scientists, responding to the need for a universally recognized naming system, no longer used common names in local languages to describe organisms. Instead, they used names based on Latin or ancient Greek words because these languages were understood by scientists everywhere.

These early scientific names described the physical characteristics of a species in great detail and were often twenty words long! For example, the English translation of the name of a particular tree might have been "Oak with deeply divided leaves that have no hairs on their undersides and no teeth around their edges."

This cumbersome system of naming organisms had another major drawback. It was difficult to standardize names of organisms because different scientists sometimes chose to describe different characteristics of the same species. Thus the same organism might have had different names.

shown in Figure 15–2 has a unique scientific name. The first part of the name is the genus name and the second part is the species name. As a result, scientists worldwide know that a *Lumbricus terrestris* is an earthworm and a *Taenia pisiformis* is a tapeworm.

The Naming System of Carolus Linnaeus

Some order was made out of all this confusion by the work of Carolus Linnaeus. Linnaeus, a Swedish botanist, developed a system for naming plants and animals that is still in use today. This system is known as **binomial nomenclature** (bigh-NOH-mee-uhl NOH-muhn-klay-cher).

In his system of binomial nomenclature, Linnaeus gave each organism a two-part scientific name. For example, the tree we call the red maple is called *Acer rubrum*. The first part of the name, *Acer*, is the genus name (JEE-nuhs; plural: genera, JEHN-er-uh). A genus name refers to the relatively small group of organisms to which a particular type of organism belongs. All maple trees carry the genus name *Acer*, the Latin word for maple. The second part of the name, *rubrum*, is the species name. The species name is usually a Latin description of some important characteristic of the organism. Red maples are called *Acer rubrum* because *rubrum* is the Latin word for red. Another kind of maple, which has no English common name, has a leaf that resembles a human hand. That maple is called *Acer palmatum*. *Palmatum* comes from the Latin word for hand.

Notice that when we use the Latin name for an organism, we always capitalize the genus name but not the species name. We also print the entire name in italics. This rule helps us recognize scientific names as we read.

Today this system of two Latin names is used by scientists everywhere. Even in a scientific book written in Chinese, you will see the names of any organisms described in the text given in this form. An international committee makes certain that once a scientific name has been chosen, it is used consistently. That committee, as well as other scientists, also makes sure that there is a carefully selected specimen of each species on file for reference. By maintaining a library of organisms in zoology and botany museums, scientists can compare the specimen they are examining with a named preserved specimen. If the specimens match, the scientist knows that the species being examined has already been described and named. If the specimen does not match anything on file, the scientist may have found a new species—in which case he or she has the privilege of naming it.

The Classification System of Linnaeus

After naming organisms, Linnaeus grouped them together according to the body structures they shared. He did this by examining the structural characteristics of organisms and deciding which structures were the most important for classifying the organisms. Organisms that shared important characteristics were classified in the same group. The groups to which Linnaeus assigned organisms are called taxa (singular: taxon), and the science of naming organisms and assigning them to these groups is called **taxonomy**.

Figure 15–3 The scientific name of an organism consists of two parts—the genus name and the species name. As is the case of the spotted skunk in the photograph, these names are often quite descriptive. Spilogale putorius, its scientific name, means smelly, spotted weasel—a fairly accurate description, indeed!

Figure 15–4 This herbarium sheet will become part of a major botanical garden's collection. The collection is used by scientists to compare unfamiliar plants with known plants in order to identify them.

HISTORICAL NOTE
CAROLUS LINNAEUS

Linnaeus did not intentionally establish the system of binomial nomenclature. He actually preferred to use the longer descriptive name for plants in his works. However, he would write a shorter two-word name for each organism in the margins of his notes. Other researchers found this shorthand so useful that they adopted it.

ESL STRATEGY

Have students use complete sentences to explain the order Linnaeus used in naming organisms. Next, mention the term "surname" and reinforce by explaining that Linnaeus's order is similar to using a person's surname, or family name, first. Ask if this is the custom in any of their native countries. (Many East Asian countries share this custom.) If this is not customary in their cultures, point out that it is a common practice used when filling out many types of business or government forms in this country.

BACKGROUND INFORMATION
THE NAMES OF SPECIES

Strictly speaking, the second part of a scientific name is the "specific epithet" or "trivial name," and a species' name consists of both the genus name and the specific epithet. However, we have decided to use the term "species name" because it is easier for students to understand and remember.

Enrichment
Ask students to determine the scientific name of five members of the animal kingdom. Compare the findings of class members to determine whether animals belonging to the same genus were identified. For example, a *Panthera leo* (lion), a *Panthera pardus* (leopard), and a *Panthera onca* (jaguar) all belong to the same genus.

Content Development
Develop the idea that a particular organism may have different English names but only one scientific name. The scientific name for a panther and a leopard is *Panthera pardus*.

• **What does this indicate about panthers and leopards?** (They are the same species.)

Figure 15–5 This illustration shows the classification groups that contain a grizzly bear.

The smallest **taxon** is the **species**, which we have previously defined as a population of organisms that share similar characteristics and that can breed with one another. If two species share many features but are clearly separate biological units, they are classified as different species within the same **genus**. Genus is the next largest taxon within the Linnaean system of classification. All the various species included in the same genus have many common characteristics. The common house cat, for example, is named *Felis domesticus*. The genus *Felis* to which the house cat belongs contains other species, such as the familiar mountain lion, *Felis concolor*. All members of the genus *Felis* share many characteristics. For example, they all have similar teeth, feet, and claws.

There are other catlike animals, however, that differ enough from those in genus *Felis* that they are placed in different genera. Lions (*Panthera leo*) and tigers (*Panthera tigris*) belong to the genus *Panthera*. And cheetahs (*Acinonyx jubatus*), although similar to lions and tigers, belong to a different genus —the genus *Acinonyx*. Groups of genera such as these, which share many common characteristics, are gathered into larger units called **families**. A family is a larger taxon than a genus. All genera of catlike animals belong in the family Felidae.

Several families of similar organisms make up the next largest taxon—an **order**. Cats (family Felidae) are placed in the same order as dogs (family Canidae). The order to which these two families, as well as several others, belong is called Carnivora. All members of the order Carnivora are carnivores, or meat-eaters.

Orders are grouped into **classes**. All members of the order Carnivora are warmblooded, have body hair, and produce milk for their young. For these reasons, they are placed together with humans (order Primates) and other similar animals in the class Mammalia.

In turn, several classes are placed in a **phylum** (FIGH-luhm), which includes a large number of very different organisms. These organisms, nevertheless, share some important basic characteristics. For example, mammals are placed in the phylum Chordata along with birds, fishes, and reptiles because all of them share certain similar characteristics.

In the classification system that was designed by Linnaeus, all phyla belonged in one of two giant taxa called **kingdoms**. Animals formed the kingdom Animalia, and plants formed the kingdom Plantae. The system developed by Linnaeus looks like this:

  Kingdom
   Phylum
    Class
     Order
      Family
       Genus
        Species

15–2 SECTION REVIEW

1. Why is the system of binomial nomenclature a good way to name organisms?
2. What is the smallest taxon? What is the largest taxon? Which of these taxa is the most specific?
3. **Critical Thinking—Applying Concepts** Two groups of organisms are in different genera but they are included in the same family. What does this information tell you about the two groups?

Figure 15–6 *Although these animals are all members of the cat family (Felidae), scientists group them into different genera. The tiger (left) is placed in the genus* Panthera, *whereas the ocelot (top center), the puma (bottom center), and the house cat (bottom right) are placed in the genus* Felis. *The cheetah (top right) belongs to a third genus,* Acinonyx.

15–3 Taxonomy Today

Guide For Reading
- How do taxa show evolutionary relationships among different organisms?
- How do modern scientific techniques contribute to the classification of organisms?

Taxonomy, particularly the grouping of organisms into higher taxa, is not as simple as it might seem. Ideas about the arrangement of organisms into families, orders, phyla, and kingdoms have changed dramatically since the time of Linnaeus. How and why has taxonomy changed so much?

Remember that despite the importance of taxonomy to biologists, the only taxon that has a clear biological identity is the species. Members of a species share a common gene pool because they breed with one another. So members of a species form a very real biological unit. We might even say that organisms themselves determine which individuals belong to their species and which do not.

The taxa above the level of species, however, do not have a clear biological identity. This is because taxonomists, or scientists who classify organisms, draw the lines between one genus and another and between one family and another. Of course, taxonomists try to create taxa that group organisms according to biologically important characteristics. But different scientists have different ideas about which characteristics are biologically most important. As a result, organisms have sometimes been "moved" from one taxon to another.

Figure 15–7 *The pygmy chimpanzee (left) and the common chimpanzee (right) belong to different species. They do not breed with each other and do not share a gene pool.*

Figure 15–8 *This photograph is one of the few taken of a living coelacanth. Coelacanths are believed to be relatives of the early fish that developed into four-legged land mammals.*

Taxonomy and Evolutionary Relationships

Today, evolutionary theory teaches that living species have evolved from earlier species. This unifying biological principle thus provides both a purpose and a guiding philosophy to modern classification systems. For this reason, taxonomists attempt to group organisms in ways that show their evolutionary relationships. Taxonomists do this by identifying and studying homologous structures in adult organisms, in developing embryos, and in well-preserved fossils. **Species shown to be closely related are classified together. Other species that may look alike but possess analogous structures only are classified in different groups.**

Deciding which structures are most important is not always easy though, and researchers often disagree on how to classify certain organisms. In writing this textbook, we have adopted one of the classification systems most widely accepted among biologists.

Biochemical Taxonomy

All forms of life share organic molecules that are almost—but not exactly—identical from species to species. Taxonomists use these molecular similarities and differences to classify organisms in much the same way as anatomists use comparisons among visible body structures.

What sorts of biochemical similarities exist? As you probably already know, all forms of life (except some viruses) carry genetic information in the form of DNA. Because all DNA is descended from the DNA carried by the earliest life forms, the DNA of all organisms shares a common genetic code. And because the genes of living organisms descended from the genes of common ancestors, many genes in many different organisms strongly resemble one another. These similar genes direct the synthesis of similar proteins.

One such protein is cytochrome *c*. Virtually every organism uses cytochrome *c* in its electron transport chain. However, each species' cytochrome *c* differs slightly from the lutionary paths would likely have homologous structures and similar DNA and RNA.)

Enrichment

Cytochromes are iron-containing proteins found in the electron transport chain of a cell's mitochondria. Through this chain, electrons are transferred from foodstuff molecules to oxygen. While there are three types of cytochromes, *a*, *b*, and *c*, the amino acid sequence of cytochrome *c* has been found to be quite similar in related species. Encourage students to research the function of cytochromes in the electron transport chain.

SECTION REVIEW 15-3

1. The species is the only taxon with a clear biological identity

cytochrome *c* of other species. The differences among cytochrome *c* were produced by mutations that occurred after the ancestors of living species diverged. If two species diverged hundreds of millions of years ago, there has been lots of time for mutations to alter the structure of their cytochrome *c* genes. But if two species shared common ancestors until fairly recently, their genes and proteins are likely to be more similar.

To help classify organisms into groups, therefore, we can compare either the nucleotide sequences of their DNA and RNA or the amino acid sequences of their proteins. Figure 15–9 shows a wide range of organisms grouped according to the similarities and differences among their cytochrome *c* molecules.

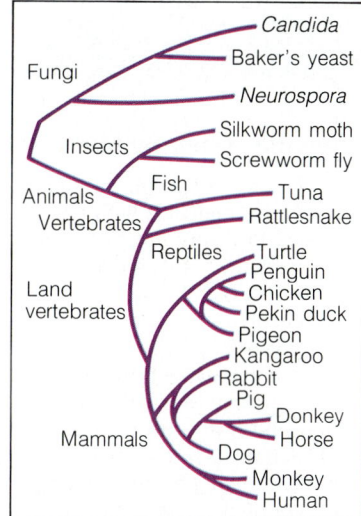

Figure 15–9 This diagram groups organisms according to the similarities and differences in the versions of cytochrome c that they possess.

15–3 SECTION REVIEW

1. Which taxon has a clear biological identity? Explain your answer.
2. Why did evolutionary theory prove important in taxonomy?
3. **Critical Thinking—Relating Concepts** How can a study of biochemistry help taxonomists?

15–4 The Five-Kingdom System

Linnaeus created his taxa in the eighteenth century, basing his system on the knowledge available at that time. As biologists gathered more and more information over the years, it became clear that Linnaeus's two kingdoms were not sufficient to logically include all organisms.

For example, microorganisms, which were discovered only after the development of microscopes, look and act significantly different from plants and animals. Some of these microorganisms lack nuclei, mitochondria, and chloroplasts. As you may remember, organisms that lack a nucleus are known as prokaryotes. Scientists now consider these prokaryotes to be fundamentally different from other living things, so they have placed them in a separate kingdom: Monera. Most scientists also feel that many single-celled eukaryotes are different enough from multicelled eukaryotes that they belong in yet another kingdom: Protista. Finally, molds and yeasts, once included in the plant kingdom, are now considered different enough from green plants to be in their own kingdom: Fungi.

As a result of discoveries of new life forms and changing ideas about those characteristics of greatest importance in classifying organisms, the most generally accepted classification system now contains five kingdoms. **The five kingdoms are Monera, Protista, Fungi, Plantae, and Animalia.**

Guide For Reading

■ Why does the five-kingdom system more accurately represent evolutionary trends?
■ What are the characteristics of organisms that make up each of the five kingdoms?

Figure 15–10 These five kingdoms have been used as a basis for classification in this textbook.

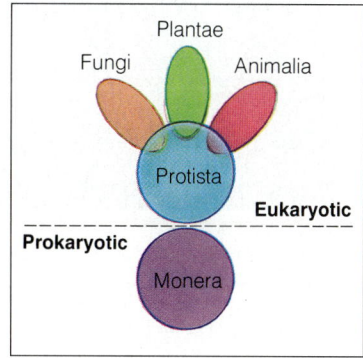

because members of a species breed with one another and therefore have a common gene pool. All other taxa contain members that are grouped by the taxonomists' human perception of the importance of similarities and differences between organisms.
2. Evolutionary theory proved important because scientists could group organisms in ways that show their evolutionary relationships. Closely related species are grouped together.
3. Biochemistry, analyzing the chemicals that are found in organisms, demonstrates concrete ways in which organisms are related. A chemical analysis of genetic materials can also show relationships between different organisms.

325

SCIENCE, TECHNOLOGY, AND SOCIETY ISSUE

So Little Time, So Many Names

The destruction of the tropical rain forests has biologists worried that much of their subject matter will disappear before being catalogued. According to Thomas E. Lovejoy of the Smithsonian Institution, between 50 and 100 acres of tropical forest are being destroyed every minute. This widespread deforestation and development is occurring in a region that is home to a majority of Earth's species.

Certain steps have been taken by the scientific community to reduce this problem. The Smithsonian Institution is considering establishing a think tank to study the environmental impact of deforestation and plan strategies for biological sampling in threatened areas. In October 1988, Congress passed a joint resolution urging the President to vigorously pursue discussions on an international agreement to protect biological diversity which led to the "Earth Summit" of June 1992. In addition, the National Science Board has instituted a task force on this subject.

Meanwhile, biologists are turning to computers as a method of speeding up the cataloguing process. Computerized databases are also viewed as a way of compensating for the lack of taxonomists available for the task. The Association of Systematic Collections hopes to make major portions of U.S. collection data accessible by computer within the next ten years.

SCIENCE, TECHNOLOGY, AND SOCIETY ISSUE

So Little Time, So Many Names

Will we have time to identify and classify all of the organisms with whom we share this planet? Probably not. For even with the modern methods used to classify organisms, time is running out. Throughout the world, species are becoming extinct. It has been estimated that 15 to 20 percent of the species on Earth will become extinct in the next 30 to 40 years. In one part of the world—the tropical rain forests—the rate of extinction has become critical. Scientists are concerned that organisms there are being destroyed faster than they can be classified. Tropical rain forests are home to more species than all other parts of the world combined. But only about 15 percent of these species have been identified.

Increasing economic pressures are threatening huge areas of tropical rain forests, and with these areas the future of life on Earth. In many parts of the world tropical rain forests are being destroyed so that farm crops can be grown and farm animals can be raised. Areas of tropical rain forests are usually cleared by burning the vegetation to the ground. Many other species—such as insects, birds, and small mammals—are destroyed as the flames level the plant life. If the rate of destruction of tropical rain forests continues, many species of organisms will become extinct before their value can be calculated and appreciated.

There is one especially sad footnote to the destruction of tropical rain forests. The soil there is very poor and can support farming for only a few years. When the soil, depleted of nutrients, can no longer support the growth of farm crops, it is abandoned. New areas must then be burned. Thus more and more land, and the life on it, is destroyed for very limited benefits.

The destruction of rain forests is a serious issue. Biologists argue that we must do everything possible to save the organisms that live in these rain forests, many of which exist nowhere else on Earth. But people who live in these areas argue that they cannot preserve the rain forest without sacrificing their own survival. As in most issues, both sides have a valid point.

In many areas, a shortage of space to grow crops forces people to cut down and burn rain forests. The soil in these areas contains few nutrients, and so more areas must be cut down and burned to continue the production of food crops.

15–4 (continued)

Content Development

Explain that all members of the kingdom Monera are prokaryotes. All members of the other four kingdoms are eukaryotes.

- **Which traits are shared by all members of the kingdom Monera?** (They are prokaryotes. As such, they lack a nucleus and membrane-bound organelles.)
- **How do monerans reproduce?** (By binary fission.)
- **How does a protist differ from a moneran?** (Protists are eukaryotes. They possess membrane-enclosed organelles, a nucleus, and mitochondria.)
- **How are protists grouped?** (Animallike, plantlike, and funguslike protists.)
- **Which do scientists believe evolved first—monerans or**

You should keep in mind that not all scientists agree on this grouping. More research may someday show that other classification systems make more sense. But right now most scientists view this five-kingdom classification system as a useful tool for studying organisms, which is what taxonomy is all about. In order to understand why the five kingdoms are arranged as they are, it helps to think of them as representing a simple evolutionary tree. Following are brief descriptions of the five classification kingdoms.

Monera

All prokaryotes are placed in the kingdom **Monera**. You will learn in the next chapter about the strong biochemical and fossil evidence that indicates that prokaryotes were the earliest life forms on Earth. Like prokaryotes alive today, the ancient prokaryotes lacked nuclei, mitochondria, and chloroplasts, and reproduced by splitting apart in a process called binary fission.

Monerans are therefore placed at the base of our evolutionary tree. This does not mean that other organisms evolved from living prokaryotes. Rather, it means that other organisms probably evolved from extinct organisms that were very similar to modern prokaryotes.

Protista

In the five-kingdom classification system, the kingdom **Protista** contains all the single-celled eukaryotic organisms. Eukaryotes, remember, differ from prokaryotes because they possess a nucleus and membrane-bound organelles. Some eukaryotes (including some protists) also have chloroplasts.

The kingdom Protista is further divided into three groups: animallike protists, plantlike protists, and funguslike protists. Ancestors of animallike protists may have evolved into animals. Ancestors of plantlike protists may have evolved into land plants. And ancestors of funguslike protists may have evolved into the living fungi.

Recall, however, that taxonomic groups are determined by biologists, not by the organisms that make up the groups. In the next several chapters you will see that the division between the protists and the three multicellular kingdoms is not at all clear-cut. Many species straddle the line between single-celled and multicellular plants, animals, and fungi. These organisms are important examples of in-between steps in evolution that link the modern groups of organisms together.

Fungi

Members of the kingdom **Fungi** build cell walls that do not contain cellulose. Fungi are heterotrophic. Heterotrophs do not carry on photosynthesis. And although fungi have many nuclei,

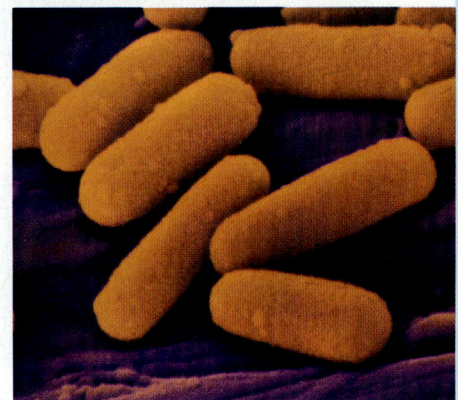

Figure 15–11 *These rod-shaped bacteria, photographed on the head of a pin, are placed in the kingdom Monera. The bacteria have been magnified more than 4000 times to make them visible.*

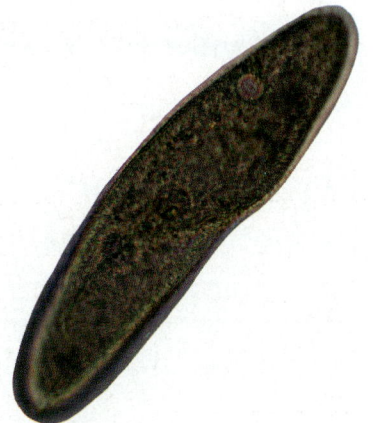

Figure 15–12 *The fragile appearance of this paramecium belies its strong constitution. Parameciums, which are placed in the kingdom Protista, are widely distributed in many bodies of water throughout the world. Can you see why the paramecium is called the "slipper organism"? Its shape looks much like the impression made by a slipper on wet sand.*

TEACHING SUPPORT

Laboratory Manual
• Using and Constructing a Classification Key, p. 203

BACKGROUND INFORMATION
KINGDOMS: A GREAT DEBATE

The five-kingdom system of classification was suggested by R. H. Whittaker in 1969. A major difference in this classification scheme lies in grouping fungi as a separate kingdom. Since fungi lack photosynthetic pigments and obtain their nutrients by absorbing them from the environment, Whittaker believed they should be placed in a separate kingdom. The five kingdoms in Whittaker's scheme are divided between prokaryotes (kingdom Monera), unicellular eukaryotes (kingdom Protista), and three kingdoms of multicellular eukaryotes (kingdoms Plantae, Animalia, and Fungi).

Our remark "not all scientists agree on this grouping" is a monumental understatement. Kingdom-level systematics is in turmoil. Recent investigations in molecular phylogeny have led some experts to divide kingdom Monera into two, three, or more kingdoms—and consolidate all eukaryotes into a single kingdom. And although everyone seems to agree that kingdom Protista is a completely artificial grouping, there is no widely accepted consensus on how protists *should* be classified.

So for the time being, we think that the five-kingdom system is still the best way to go at the high school level.

---

protists? (Monerans.)

Content Development
Guide students in understanding that members of each kingdom have unique characteristics, which are used as a basis for classification.

• **What are some traits of members of the kingdom Fungi?** (Heterotrophs have cell walls at some stage in their life cycle.)
• **How are protists and fungi similar?** (Both are eukaryotes.)
• **How are protists and fungi different?** (Some protists can carry on photosynthesis. All protists are unicellular. All fungi are multicellular heterotrophs.)

Figure 15–13 The fly agaric (left), a beautiful and colorful member of the kingdom Fungi, is poisonous. The small maple tree in the foreground and the moss plants growing on the tree (right) are both members of the kingdom Plantae.

they do not always have separate cells divided by complete cell walls. For these reasons, the fungi are not included with the plants and are placed in their own kingdom.

Plantae

Members of the kingdom **Plantae** are multicellular, have cell walls that contain cellulose, and are autotrophic. Autotrophic plants are able to carry on photosynthesis using chlorophyll. The plant kingdom includes all the plants you have come to know by now, such as the flowering plants, mosses, and ferns. In our classification system, the plant kingdom also includes multicellular algae.

Animalia

Members of the kingdom **Animalia** are multicellular, heterotrophic, and have cell membranes without cell walls. As you will see in later chapters, there is incredible diversity within the animal kingdom.

Figure 15–14 Both of these organisms are placed in the kingdom Animalia. The orange sponge (left) grows attached to the ocean bottom, unable to move from place to place. Unlike the sponge, the huge elephant (right) is able to move about.

328

contain photosynthetic pigments and animals do not.)

Skills Development
Guided Practice
Skill: Classification
Have students make a list of 15 organisms that they have seen firsthand. (Before making their lists, students should recall organisms they have seen while visiting a zoo, park, beach, the mountains, and so on.) After students have completed their lists, have them classify each entry according to its kingdom.

Students can check the accuracy of their classifications with a taxonomic guide.

SECTION REVIEW 15–4

1. The five-kingdom classification system became more

PROBLEM SOLVING IN BIOLOGY

CLASSIFYING COMMON OBJECTS

Imagine that you are setting up a hardware store and have just received an assortment of items used to fasten things together. To help your customers find the items they need, you must devise a sensible way to group and arrange these objects on your shelves. In grouping the fasteners, you will proceed in much the same way that scientists do in classifying organisms.

In order to make sensible groupings, you should first examine the similarities among these objects as well as the differences. You must determine how these similarities and differences relate to the objects' functions. Finally, you must decide which functions are most important to your customers.

Even a quick glance reveals several possible ways to group your hardware. You may feel that the most important feature of these objects is the tool used to drive them. Some of the fasteners can be driven with a screwdriver; others with a hammer. The shape of the fasteners' head will indicate how they are driven. Thus you could group the fasteners by the shape of their head.

You may feel, however, that a more important feature of the fasteners is whether they are meant to be used to fasten wood or metal. Or perhaps what the fasteners are made of is the most important feature to you. Or whether they have threads.

Notice that there is no single correct way to classify these fasteners. Each classification scheme serves a different purpose for you and your customers.

With this in mind, choose one approach to begin this exercise and create a classification system for these objects. Your smallest taxon should contain only items that are very similar to one another. Your largest taxon might include all the objects used as fasteners. If partway through this exercise you feel your system is not the best one, save your work and begin developing another strategy. Compare your results with those of your classmates, noting similarities and differences in the various schemes.

What information about the objects and about the classification process have you learned from this exercise? Do you now understand why taxonomists often disagree about the classification of organisms? Do you now have a greater appreciation for why classification is useful?

15–4 SECTION REVIEW

1. Why did the five-kingdom system of classification become widely used?
2. Why are the fungi placed in a separate kingdom?
3. **Critical Thinking—Applying Concepts** A single-celled organism could be placed in the Monera or the Protista kingdom. What factor would be the most significant for determining into which kingdom this organism should be placed?

329

LABORATORY INVESTIGATION

PREPARING A COLLECTION OF ORGANISMS

Before the Lab

1. Have students collect specimens for the lab test one day before the investigation. You may want to divide the class into small groups and assign to each group a particular type of specimen to collect. Remind students to collect enough specimens for all class members.
2. A moist piece of bread must be placed in an open petri dish about 1 week before the investigation. This will allow time for organisms to grow on the bread. Keep the bread moist.
3. Prepare agar by mixing 1 gram of lactose, 1 gram of peptone, 20 grams of agar, and 1000 milliliters of water. Heat in a water bath until dissolved. Then cool to 45°C, pour into petri dishes, and cover. After 2 hours, sterilize the dishes.
4. If pond water is not available, prepare leaf cultures by adding leaves and a small amount of garden soil to water in a loosely covered container. Leave it in a cool spot for 3 weeks.

Pre-Lab Discussion

Review some basic traits of organisms that belong to each of the five kingdoms. Emphasize that when classifying an organism taxonomists compare its structures with those of previously classified organisms. While an organism may show traits of many kingdoms, it is grouped in the kingdom with which it shares the greatest number of homologous structures or biochemical similarities.

LABORATORY INVESTIGATION

PREPARING A COLLECTION OF ORGANISMS

PROBLEM
How do you prepare a collection of organisms that represents each of the five kingdoms?

MATERIALS (per group)

microscope	bread
medicine dropper	sour milk, sauerkraut,
glass slides	or yogurt
coverslips	pond water
forceps	small plastic food
baby food jars	bags
with lids	transparent tape
70% ethanol	labels
petri dishes and	mosses and other
covers	plants
nutrient agar	worms and insects

PROCEDURE

1. List characteristics of organisms in each of the five kingdoms of the classification system used in this textbook. You may refer to material already presented in this chapter. After you examine each organism in this investigation, decide which kingdom it belongs to. Be prepared to defend your decisions.
2. Use a medicine dropper to put a drop of liquid from sauerkraut, yogurt, or sour milk on a glass slide. Cover the drop with a coverslip. Examine your specimen under the high-power objective of the microscope. Draw what you observe. Label the drawing with the kingdom of the organism.
3. Use a medicine dropper to place a drop of pond water on a glass slide. Place a coverslip over the drop. Use the microscope to examine the pond water. Draw the organisms you observe. Label the drawing with the kingdom of the organism.
4. Moisten a piece of bread and place it in an open petri dish. Leave the dish on a windowsill for several days. After that time, use a forceps to remove some of the "fuzzy" material that has grown on the bread. Place this material on a glass slide. Add a drop of tap water and cover the drop with a coverslip. Examine this slide with the low-power objective. Draw what you observe. Remove another sample from the bread and place this sample in a jar. Fill the jar with ethanol. Cover the jar tightly and label it with the name of the kingdom to which the organism belongs.
5. Examine several small plants, leaves, or flowers. Put each sample in a separate baby food jar. Fill the jar with ethanol. Cover tightly. Label each jar and display.
6. Look for worms, grubs, and pill bugs in the soil. Find some insects on plants in areas near your home or school. Place each specimen in a baby food jar. Fill the jar with ethanol and cover tightly. Label each jar and display.

OBSERVATIONS

1. For organisms in which kingdoms did you require a microscope for observation?
2. What similarities and differences do you observe among organisms from different kingdoms?
3. For which kingdom was it easiest to find a variety of organisms? Most difficult?

ANALYSIS AND CONCLUSIONS

1. Why are photographs or drawings necessary for making a display of protists and monerans?
2. What difficulties did you encounter in classifying organisms?

Skills Development
Students will use the following skills while completing this investigation.
1. Observing
2. Comparing
3. Relating
4. Classifying

Safety Tips
Caution students to be careful not to release animal specimens into the classroom.

Teaching Strategy
Be sure students understand how to use the taxonomic key to classify the specimens.

STUDENT STUDY GUIDE

SUMMARIZING THE CONCEPTS

The key concepts in each section of this chapter are listed below to help you review the chapter content. Make sure you understand each concept and its relationship to other concepts and to the theme of this chapter.

15-1 Why Classify?
- A classification system allows us to name and order living organisms in a logical manner.
- A good classification system provides a universally accepted name for each organism.

15-2 Biological Classification
- Because they assigned a long, complicated series of names to an organism, early classification systems were very difficult to learn and cumbersome to use.
- Carolus Linnaeus gave each organism he classified a simple two-part scientific name.

15-3 Taxonomy Today
- Taxonomy is the science that names organisms and assigns the organisms to groups known as taxa.
- Taxa show evolutionary relationships among organisms.
- Today, biochemists use similarities and differences among chemical compounds to classify organisms. Molecules of DNA and RNA are often used to demonstrate evolutionary relationships among different kinds of organisms.

15-4 The Five-Kingdom System
- Because living things are so diverse and complex, Linnaeus's two-kingdom classification system is considered inadequate today in accurately representing relationships between all living things.
- A five-kingdom classification system has been developed and is widely used. This classification system shows relationships between different groups of organisms. This five-kingdom system more accurately groups organisms based on evolutionary trends.
- The Monera, Protista, Fungi, Plantae, and Animalia are five kingdoms that are used to group organisms.

REVIEWING KEY TERMS

Vocabulary terms are important to your understanding of biology. The key terms listed below are those you should be especially familiar with. Review these terms and their meanings. Then use each term in a complete sentence. If you are not sure of a term's meaning, return to the appropriate section and review its definition.

15-2 Biological Classification
- binomial nomenclature
- taxonomy
- taxon
- species
- genus
- family
- order
- class
- phylum
- kingdom

15-4 The Five-Kingdom System
- Monera
- Protista
- Fungi
- Plantae
- Animalia

Observations
1. Protists and monerans
2. All organisms are made up of cells. Plants, animals, and fungi are multicellular, but plants contain chlorophyll. Monerans and protists are single-celled, but monerans lack a nucleus.
3. Plants were easiest to find. Monerans and protists were hardest to find.

Analysis and Conclusions
1. The actual organisms are too small to be seen in a display without a microscope.
2. Not all characteristics are readily observable.

Going Further: Enrichment
To continue investigating the classification procedure, students may wish to obtain specimens from a different ecosystem. For example, if a beach or lake is accessible to the students, have them obtain specimens from these areas. Repeat the investigation using the new specimens. After repeating the investigation, ask students the following questions:
- **Do organisms vary between ecosystems?** (Yes.)
- **Were members of each kingdom represented in your second investigation?** (Students will likely have collected specimens from all five kingdoms from any type of ecosystem.)
- **Which kingdom was represented by the greatest number of specimens?** (Answers will vary.)

CHAPTER REVIEW

CONTENT REVIEW

Multiple Choice
1. a 2. c 3. d 4. a
5. b 6. a 7. d 8. a

True or False
1. F eukaryotes
2. T
3. F not photosynthetic
4. F multicellular
5. T
6. F genus
7. F kingdom
8. T

Word Relationships

A.
1. taxa; group is not a taxon
2. eukaryotic kingdoms; prokaryotes lack a nucleus
3. biochemical characteristics used in systematics; height is not a biochemical characteristic
4. names of taxonomic kingdoms; Carnivora is the name of an order in class Mammalia

B.
5. genus
6. taxonomy
7. kingdom
8. arm bones (other homologous organs may be listed)

CONCEPT MASTERY

1. Early classification systems were difficult to use for several reasons. Many early systems used many names to identify a particular organism, and the names were difficult to pronounce and even harder to learn. Early classification systems that had two major kingdoms did not accurately reflect important differences in organisms.
2. Easy to use, few names, and show evolutionary and other relationships between taxa.
3. Species, genus, family, order, class, phyla, and kingdom.
4. It allows people all over the world to apply a universally accepted name to a single organism, and shows evolutionary relationships among different organisms.
5. An organism with three common names would be difficult to identify clearly.

CHAPTER REVIEW

CONTENT REVIEW

Multiple Choice

Choose the letter of the answer that best completes each statement.

1. The largest taxon is a
    a. kingdom.
    b. species.
    c. family.
    d. phylum.
2. A good classification system does all of the following except
    a. show relationships.
    b. show evolutionary trends.
    c. create confusion.
    d. use one name for an organism.
3. The two-name system for classifying organisms was developed by
    a. Charles Darwin.
    b. Thomas Edison.
    c. a Swedish king.
    d. Carolus Linnaeus.
4. *Acer rubrum* and *Acer palmatum* are both names of different kinds of maple trees. What is the genus name for all maple trees?
    a. *Acer*
    b. *rubrum*
    c. *palmatum*
    d. Cannot be determined from the information given.
5. The science of naming organisms and placing them in groups is called
    a. biology.
    b. taxonomy.
    c. ornithology.
    d. ecology.
6. Each of the following is the name of a taxon except
    a. group.
    b. genus.
    c. family.
    d. kingdom.
7. The only taxon that has a clear biological identity is the
    a. kingdom.
    b. phylum.
    c. genus.
    d. species.
8. The kingdom that includes all prokaryotes is the
    a. Monera.
    b. Fungi.
    c. Protista.
    d. Plantae.

True or False

Determine whether each statement is true or false. If it is true, write "true." If it is false, change the underlined word or words to make the statement true.

1. All members of the kingdom Protista are <u>prokaryotes</u>.
2. <u>Linnaeus</u> developed the two-name system for naming organisms.
3. Fungi are <u>photosynthetic</u>, heterotrophic organisms.
4. All members of the plant kingdom are <u>unicellular</u>.
5. A good classification system uses <u>standardized</u> names.
6. *Acer rubrum* is the scientific name for the red maple. *Acer* is the <u>species</u> name for this organism.
7. The largest taxa is the <u>species</u>.
8. Several families of similar organisms make up <u>an order</u>.

Word Relationships

A. *In each of the following sets of terms, three of the terms are related. One term does not belong. Determine the characteristic common to three of the terms and then identify the term that does not belong.*

1. family, phylum, group, species
2. plants, animals, fungi, prokaryotes
3. biochemistry, cytochrome *c*, DNA and RNA, height
4. Monera, Protista, Fungi, Carnivora

CRITICAL AND CREATIVE THINKING

1. The classification system developed by Linnaeus was much easier to use. It consisted of only two names whereas other systems used many. It used Latin forms for names that were understood by many people throughout the world.
2. They permit scientists to use new criteria to classify organisms. For example, the microscope has enabled tax-

B. An analogy is a relationship between two pairs of words or phrases generally written in the following manner: a:b::c:d. The symbol : is read "is to," and the symbol :: is read "as." For example, cat:animal::rose:plant is read "cat is to animal as rose is to plant."

In the analogies that follow, a word or phrase is missing. Complete each analogy by providing the missing word or phrase.

5. genus:species::family:_____
6. Darwin:evolution::Linnaeus:_____
7. orders:classes::phyla:_____
8. biochemistry:cytochrome *c*::homologous organ:_____

CONCEPT MASTERY

Use your understanding of the concepts developed in the chapter to answer each of the following in a brief paragraph.

1. Why were some of the early classification systems difficult to use?
2. What are three characteristics of a good classification system?
3. List the taxa in the classification system in current use from the smallest to the largest.
4. How does a workable classification system promote scientific understanding?
5. Suppose an organism had three common names in different parts of the United States. How might these names lead to confusion?

CRITICAL AND CREATIVE THINKING

Discuss each of the following in a brief paragraph.

1. **Making comparisons** In what ways was the classification system developed by Linnaeus an advantage over previous classification systems?
2. **Applying technology** How has the development of new technologies changed the ways we classify organisms?
3. **Applying concepts** Suppose you discovered a new single-celled organism. This organism had a nucleus, mitochondria, and a giant chloroplast. In what kingdom would you place this organism? What are your reasons?
4. **Making inferences** It has been said that organisms decide which individuals belong to their species and which do not. What does this statement mean?
5. **Assessing concepts** Libraries use the Dewey Decimal System to group books by similarities. How do the major groupings in this system help you locate research materials more quickly?
6. **Using the writing process** It has been estimated that there are more unknown organisms in the tropical rain forests than there are known organisms in the world. Scientists are concerned that these rain forests might be destroyed before the organisms in them can be classified. Write an editorial for a television news program protesting the destruction of rain forests. Offer reasons why rain forests should be protected.

CAREERS IN BIOLOGY

Museum Curator's Assistant, Taxonomist, Archaeologist

Students interested in a biological career should be instructed to write for further information to the address listed beneath each career description. However, in consideration of the organizations that provide career information, please have only one student write to an organization. Do not instruct the entire class to write to every organization for the same career information. You may want to use the information provided to start a biology career file for the use of all your students.

UNIT 4 — CAREERS IN BIOLOGY

Museum Curator's Assistant

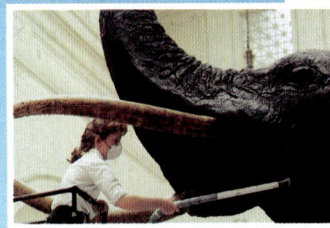

Museum curator's assistants help prepare museum collections by cleaning, sorting, labeling, and numbering animal and plant specimens under the supervision of the museum curator. They also help to design exhibits by arranging specimens in glass cases or by constructing entire fossil skeletons.

Museum curator's assistants must be organized, pay attention to detail, and develop a solid knowledge of the collection with which they wish to work.

To receive information write to the American Association of Museums, 1055 Thomas Jefferson St., NW, Washington, DC 20007.

Taxonomist

Taxonomists identify and name organisms so that they can be organized into groups. Taxonomists must understand the relationships among various organisms. The work of taxonomists helps biologists identify the characteristics of an organism by knowing the group to which it belongs.

Taxonomists need a Bachelor's degree with a strong background in the biological sciences.

For additional information contact the Society of Systematic Zoology, Smithsonian Institution (NHB), Washington, DC 20560.

Archaeologist

Archaeologists are involved in gathering and analyzing remains of earlier cultures. They use special techniques to find materials such as pottery, clothing, tools, and weapons from houses and cities that existed long ago. With this information, they are often able to determine the customs, languages, traditions, and religious beliefs of earlier societies.

Archaeologists need a Ph.D. degree to qualify for positions in colleges, museums, and government agencies.

For information write to the Society for American Archaeology, 1511 K St., NW, Ste. 716, Washington, DC 20005.

HOW TO USE THE WANT ADS

Help-wanted ads are a good place to begin a job search. They list positions that are available and provide some idea of which fields are hiring people, what salaries are being offered, and what qualifications are necessary.

Want ads can be found in a variety of publications. Newspapers usually have a section listing job openings under the title of the position. Trade magazines, which are periodicals related to a specific field, also list job openings.

As you look through the want ads, be sure to check all the categories that apply to you. Answer ads that give specific information about the position and the company that placed the ad. Respond promptly with a neatly typed letter and make sure you include all the information requested in the ad.

FROM THE AUTHORS

Every now and then, a non-scientist friend will ask me "Why do biologists make such a big deal about evolution?" That's a very important question that deserves a thorough answer.

The shelves in my office are crammed with books about every living thing you can imagine. There are books about plants, animals, and viruses. There are books about how the human brain works, how termites build nests, how winds and ocean currents influence the lives of fishes, and how genes direct the synthesis of proteins. At first glance, this material has as many different forms as the organisms you'll meet in the chapters to come. What common element ties all of this together? What makes zoology, botany, paleontology, genetics, and medicine all part of biology?

This is where evolution comes in. Charles Darwin, by finding order in the chaos of the living world, helped to create modern biology. Evolutionary theory makes sense out of genetics, ecology, physiology, and even biochemistry. It also enables those fields to relate to one another in vital ways. When molecular biologists uncovered the workings of DNA, for example, they discovered both the living record of evolutionary change and the shared evolutionary heritage that unites all of life.

The more we learn, the more we realize that all organisms—from the tallest redwoods to the tiniest bacteria to each and every human being on Earth—are members of one enormous family. Whether scientists work to cure cancer or save rain forests, they are guided by the knowledge that all living things—all of us—share a common past and a common destiny on this planet. If you remember nothing else from this course ten years from now, remember this—and your year will have been well spent.

Joe Levine

The tarsier (a type of primate), trees, and ferns of this southeast Asian tropical rain forest are members of the family of life on Earth—and so are you.

FROM THE AUTHORS

Joseph Levine

In this feature Joe Levine reminds students of a common theme that is woven throughout the textbook: Evolutionary theory shows us that all living things share a common history. The author also points out that this concept is perhaps the most important biological concept students will carry with them as they study science in future years. You may want to initiate a class discussion about the way evolutionary theory has or will change scientific studies in areas other than biology.

UNIT 5

Life on Earth: Monerans, Protists, and Fungi

UNIT OVERVIEW

In Unit 5 students learn about the origin of life on Earth and are introduced to viruses and three of the five kingdoms of organisms. As they study Unit 5, students will discover that "simple" organisms are not all that simple and that many of the living things traditionally studied in microbiology are far from microscopic. Students first examine explanations for the origin of life and explore the earliest stages of the evolution of living things. They then study the major characteristics of viruses, monerans, protists, and fungi and learn how these living things fit into the world.

UNIT OBJECTIVES

1. Discuss the experiments that disproved the hypothesis of spontaneous generation.
2. Describe the earliest stages and major evolutionary innovations in the development of life on Earth.
3. Explain how viruses affect the cells of living organisms.
4. Compare the structure, ways of obtaining energy, forms of reproduction, and impact on other living things in various types of bacteria.
5. List and describe nine phyla of protists.
6. Compare structure and reproduction in the five phyla of fungi.

UNIT 5

Life on Earth: Monerans, Protists, and Fungi

I discovered in a tiny drop of water, incredibly many very little animalcules and these of diverse sorts and sizes.

So reported Anton Van Leeuwenhoek in 1675 after examining a sample of water from a well. Van Leeuwenhoek was certainly among the first to examine the secret world of microscopic organisms that share our planet.

Today, the world of the very small is huge—visible to prying eyes through wonderful microscopes that reveal the tiniest organisms and the smallest details of their cells.

DISCOVERY LEARNING

LOOKING AT YOGURT

Could there be living organisms inside your food? You might be surprised at the answer, which you can determine by taking a close look at yogurt.

1. Add water to some plain yogurt to make a thin mixture.
2. With a medicine dropper, place a drop of the yogurt mixture on a glass slide.
3. Use another dropper to add one drop of methylene blue to the slide.
4. Carefully place a coverslip over the drop on the slide.
5. Observe the slide under the low-and high-power objectives of a microscope.
   • Record what you see in words and drawings.
   • Why do you think you had to use methylene blue?

336

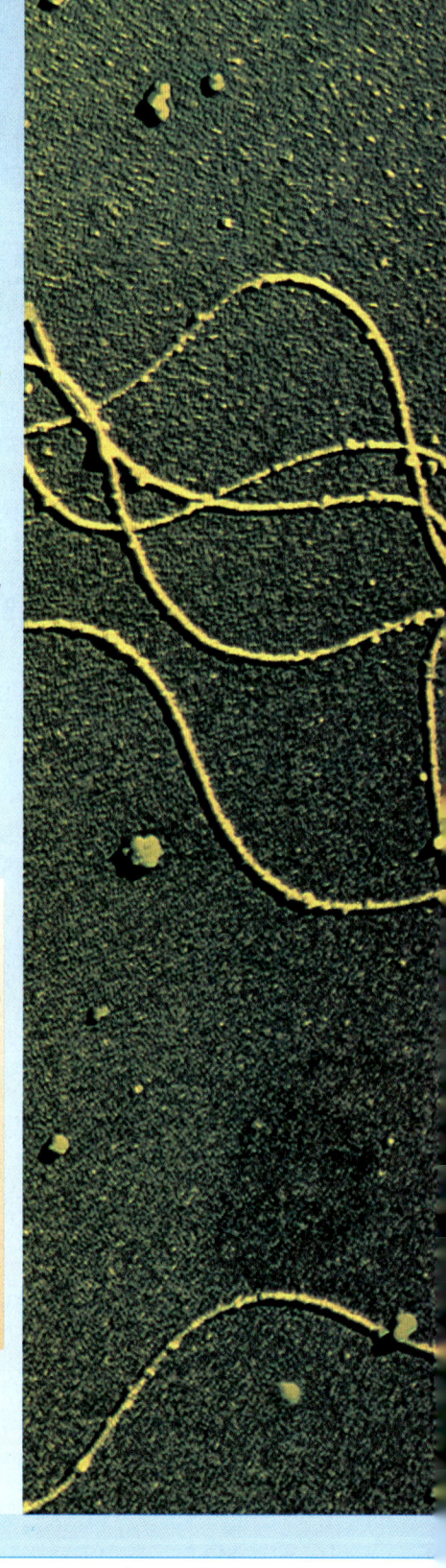

INTRODUCING UNIT 5

Using the Visuals
Have students read the unit introduction and examine the photograph that illustrates it.
• **What do you think this photograph shows?** (Students should be able to infer that it shows a microscopic organism.)

Explain that the visual is an electron micrograph of the bacterium *Listeria monocytogenes,* which causes a form of meningitis. You may want to point out that the photograph is color-enhanced to make structures clearer; electron micrographs are actually black and white.
• **Do you think Leeuwenhoek was surprised to see tiny organisms in water?** (Yes. Because these organisms are not visible to the unaided eye, their existence was not suspected.)

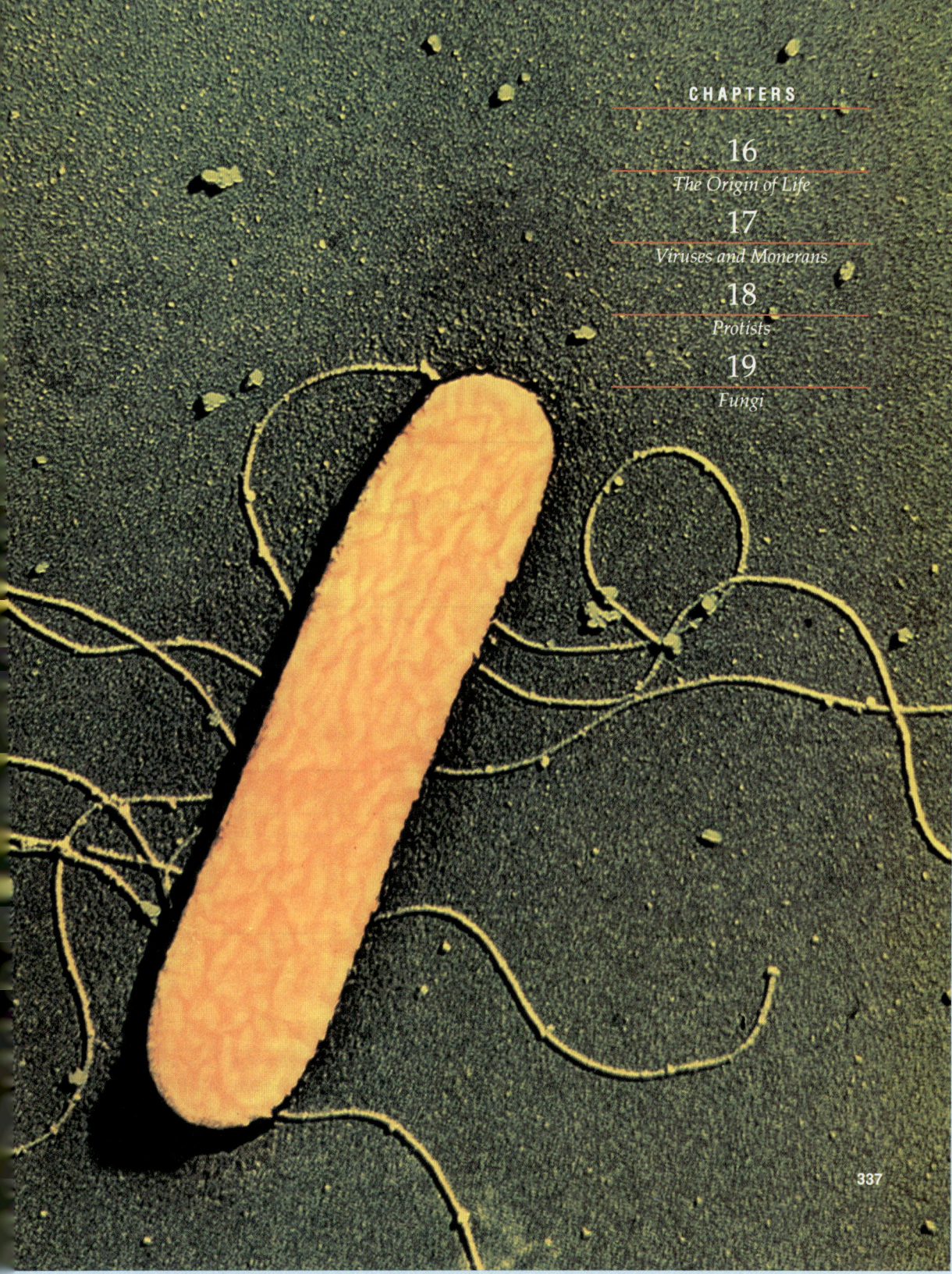

CHAPTERS

16 The Origin of Life
17 Viruses and Monerans
18 Protists
19 Fungi

CHAPTER DESCRIPTIONS

16 The Origin of Life
In Chapter 16 students learn about ancient and modern explanations for the origin of living things. First they read about the hypothesis of spontaneous generation. Next they learn about the chemical evolution of life on Earth—from relatively simple molecules in organic "soup" to proto-life in the form of complex organic molecules to living cells. They also examine some of the major evolutionary innovations, such as aerobic respiration, sexual reproduction, multicellularity, and photosynthesis.

17 Viruses and Monerans
In Chapter 17 students learn about the structure and life cycles of viruses. They also examine the relationship of viruses to living cells, discover why viruses are considered to straddle the line between life and nonlife, and read about the origin of viruses. Next, students learn about the different phyla of bacteria, study methods of identifying bacteria, and examine how bacteria obtain energy, reproduce, and affect other organisms and the environment. Finally, students are introduced to the disease causing properties of viruses and monerans.

18 Protists
Chapter 18 introduces students to unicellular eukaryotic organisms: protists. Students first read about the difficulty of classifying protists and examine a hypothesis about the origin of protists. They then learn about the structure and physiology of the various animallike and plantlike phyla of protists.

19 Fungi
In Chapter 19 students are first introduced to the general characteristics of fungi. They then explore the structure and life cycles of the various fungi phyla.

DISCOVERY LEARNING

LOOKING AT YOGURT

Students will probably be surprised to observe rod-shaped *Lactobacillus* cells in the yogurt. The methylene blue is a stain that makes the bacteria easier to see.

Lactobacillus obtains the energy it needs by digesting lactose (milk sugar) to produce lactic acid, carbon dioxide, and water. When enough lactic acid accumulates to lower the pH to about 4.5, globules of the milk protein casein clump together, thickening milk into yogurt.

Direct students' attention to the list of ingredients on the side of a yogurt container. Point out that "active cultures" are cultures of yogurt-making bacteria.

CHAPTER 16 The Origin of Life

Section	Laboratory Investigations and Activities
16–1 Spontaneous Generation pages 339–342 THEMES: Evolution, Stability	**Student Edition** LABORATORY INVESTIGATION: Investigating Spontaneous Generation, p. 350 **Teacher Edition** DEMONSTRATION: Life on Earth, p. 338D
16–2 The First Signs of Life pages 342–346 THEMES: Evolution, Patterns of Change	**Teacher Edition** DEMONSTRATION: Spontaneous Generation, p. 338D **Laboratory Manual** Making Coacervates, p. 209
16–3 The Road to Modern Organisms pages 347–349 THEMES: Evolution, Patterns of Change	
Chapter Review pages 351–353	

Teacher Resources

Books

Bendall, D. S. *Evolution from Molecules to Men.* Cambridge University Press, 1986.

Margulis, Lynn, and Dorion Sagan. *Microcosmos: Four Billion Years of Evolution from Our Microbial Ancestors.* Simon & Schuster, 1991.

Pellegrino, Charles R., and Jesse A. Stoff. *Darwin's Universe: Origins and Crises in the History of Life.* Van Nostrand Reinhold, 1983.

Smith, Peter J., ed. *The Earth.* Macmillan, 1986.

Tributsch, Helmut. *How Life Learned to Live: Adaptation in Nature.* MIT Press, 1983.

Weiner, Jonathan. *Planet Earth.* Bantam Books, 1986.

CHAPTER PLANNING GUIDE

Other Activities	Media and Technology

Teaching Resources
  Activity: Pasteur's Experiments, p. 3
  Activity: Testing the Hypothesis of Spontaneous Generation, p. 7
Study Guide
  Section 16–1, p. 155	

Teaching Resources
  Activity: Evaluating a Theory About Cells, p. 9
Study Guide
  Section 16–2, p. 157
	Biology Transparencies
	  18: Stanley Miller's Experiment

Teaching Resources
  Vocabulary Review: Word Game, p. 1
Study Guide
  Section 16–3, p. 158	

Teaching Resources
  Chapter Test A, p. 17
  Chapter Test B, p. 21
	Computer Test Bank
	  Chapter Test, p. 171

Audiovisuals

The Earth's Atmosphere. Film or video. Coronet.

Three Billion Years of Life: The Drama of Evolution. 3 slide sets and cassettes. Science and Mankind.

Origin of Cellular Life. Video. Biology Media.

Water, Birth, Planet Earth: The Sea. Film or video. Coronet.

Water, Birth, Planet Earth: The Land. Film or video. Coronet.

CHAPTER 16 The Origin of Life

CHAPTER OVERVIEW

For many centuries, people believed that living things were able to develop from nonliving things. The hypothesis that life arises from nonlife is called *spontaneous generation.* Through the work of Italian scientists Francesco Redi and Lazzaro Spallanzani and the French scientist Louis Pasteur, the hypothesis of spontaneous generation was finally disproved.

By studying the environmental conditions on ancient Earth, scientists have attempted to discover the origin of life. Laboratory experiments have re-created the formation of certain organic compounds on ancient Earth. These compounds contained many of the basic building blocks of life. Microfossils indicate that the first life forms on Earth were anaerobic prokaryotes, similar to modern bacteria. In time, some cells developed the ability to produce food through a primitive form of photosynthesis.

As organisms that used water and produced oxygen as a waste product during photosynthesis developed, the Earth's atmosphere slowly accumulated oxygen gas. As oxygen became plentiful, organisms utilizing the more efficient process of aerobic metabolism replaced anaerobic organisms. The first eukaryotic cells containing membrane-bound organelles evolved around 1.4 billion years ago. The development of sexual reproduction in eukaryotes greatly speeded up the process of evolution.

16–1 SPONTANEOUS GENERATION

Section Focus 16–1

The purpose of this section is to describe the hypothesis of spontaneous generation and the work of three scientists who sought to disprove the hypothesis. The hypothesis of spontaneous generation states that life can arise from nonliving matter.

In 1668, the Italian scientist Francesco Redi stated that maggots did not develop from decaying meat, as was previously thought. In a series of experiments, Redi proved that the maggots hatched from eggs laid by flies. In the 1700s, another Italian scientist, Lazzaro Spallanzani, demonstrated that microorganisms did not appear in sealed jars of gravy, which suggested that such organisms entered unsealed jars from the air.

It was not until 1864 that the hypothesis of spontaneous generation was finally disproved by the French scientist Louis Pasteur. Pasteur used flasks with long curved necks to prove that microorganisms travelled with dust and other particles through the air and did not develop spontaneously—life comes only from life.

Performance Objectives 16–1

1. State the hypothesis of spontaneous generation.
2. Describe the experiments of Redi, Needham, and Spallanzani.
3. Describe Pasteur's experiment that finally disproved the hypothesis of spontaneous generation.

Science Terms 16–1
spontaneous generation   p. 339

16–2 THE FIRST SIGNS OF LIFE

Section Focus 16–2

The purpose of this section is to describe various hypotheses of the evolution of the first life forms on Earth. The atmosphere on ancient Earth was very different from our modern atmosphere. Microfossils indicate that the first life forms were prokaryotes, similar to modern bacteria, which appeared about 3.5 billion years ago.

Laboratory experiments have simulated the formation of many organic compounds on ancient Earth. These compounds form the basic building blocks of life. Organic compounds may also have been brought to Earth by meteorites and comets.

The first true cells were prokaryotic heterotrophic anaerobes. As food sources dwindled, some cells developed the ability to produce their own food through the process of photosynthesis.

Performance Objectives 16–2

1. Describe the environmental conditions on ancient Earth.
2. Discuss various hypotheses of the evolution of the first life forms.
3. Identify the characteristics of the first true cells.
4. Describe the evolution of the process of photosynthesis.

Science Terms 16–2
microfossil   p. 343
anaerobe   p. 345

16–3 THE ROAD TO MODERN ORGANISMS

Section Focus 16–3

The purpose of this section is to describe the evolution of eukaryotic cells that are capable of aerobic metabolism and sexual reproduction. During the development of organisms that used water and produced oxygen during photosynthesis, the Earth's atmosphere slowly accumulated oxygen gas. Once oxygen filled the atmosphere, anaerobes could no longer live on the Earth's surface. As oxygen became plentiful, organisms displaying an energy-rich aerobic metabolism evolved. Some of the oxygen produced through photosynthesis formed the protective ozone layer found in the Earth's upper atmosphere.

Around 1.4 billion years ago, the first eukaryotic cells containing membrane-bound organelles evolved. The development of sexual reproduction greatly speeded up the course of evolution by providing more opportunities for genetic variation in organisms.

Performance Objectives 16–3

1. Relate the process of photosynthesis to the development of the Earth's atmosphere.
2. Describe the evolution of energy-rich aerobic metabolism.
3. Discuss the development of the ozone layer and its importance to living things.
4. Define eukaryote.
5. Discuss the importance of the evolution of sexual reproduction.

CHAPTER PREVIEW

DISCOVERY LEARNING

TEACHER DEMONSTRATIONS

Modeling

Life on Earth

The following demonstration can be used as an introduction to Chapter 16.

Draw a time line or clock indicating a 24-hour cycle beginning at midnight. Compare the 24-hour cycle to the history of the Earth. Point out that the oldest known fossils appeared at about 6 A.M., the oldest known nucleated cells between 4 and 5 P.M., the oldest known complex organisms between 8 and 9 P.M., the oldest plants between 9 and 10 P.M., and humans in the last 30 seconds. This activity will help demonstrate that the variety of life on Earth arose relatively recently.

Spontaneous Generation

You may want to perform this demonstration after students complete Section 16–1.

Set up a demonstration of Francesco Redi's experiments. If peeled bananas are placed in the jars instead of meat, then fruit flies will be attracted to them. Also consider setting up a demonstration of Pasteur's swan-necked flasks. An S-shaped piece of glass tubing fitted into a one-hole rubber stopper that has been securely placed into a Florence flask can be used to simulate a swan-necked flask.

CHAPTER 16
The Origin of Life

GUIDED ENQUIRY

Pose the following questions to students and have them record their responses. Point out that they will gain a better understanding of the key concepts if they read the chapter with these basic questions in mind. Upon completion of the chapter, pose the questions again. Ask students to compare their initial responses with those they have developed after reading the chapter.

- Is it possible for living things to arise from nonliving matter?
- What was the atmosphere of ancient Earth like?
- How might life on Earth have begun?
- Why is the process of photosynthesis necessary for the survival of all living things?
- What is the importance of the Earth's ozone layer?
- How is sexual reproduction important to the development of life on Earth?

INTRODUCING CHAPTER 16

Using the Visuals
Begin your introduction of Chapter 16 by having students examine the photographs on page 338.
- **Do you think there is any life in the places shown in these photographs? Why or why not?** (Accept all answers.)
- **What do living things need to survive?** (Accept all reasonable answers.)

Then have students read the chapter introduction on page 338.

- **In what other areas of study would the words of T. S. Eliot be appropriate?** (Accept all logical answers.)
- **Why is it impossible to know exactly how life on Earth first began?** (Accept all logical answers.)

TEACHING STRATEGY 16–1

Focus/Motivation
Before beginning this section, take a walk around the school area with the class. Have students make a list of the living and nonliving objects they see, labeling each item in the list as either living or nonliving. Also tell students to jot down the properties of each object. Have students save their lists until you have the opportunity to review the characteristics of living and nonliving objects.

CHAPTER 16
The Origin of Life

Tests performed by the Viking spacecraft on Martian soil indicated that there is no life on Mars today. Astronauts who walked on the moon confirmed what scientists already knew: There is no life on our lunar neighbor.

> We shall not cease from exploration.
> And the end of all our exploring
> Will be to arrive where we started
> And know the place for the first time.
> — T.S. Eliot, *Four Quartets*

In this work by T.S. Eliot, the author suggests that to truly understand ourselves and the world about us, we must return to our roots and discover where we came from and how we came to be. Although the author is not speaking about biology, his words have meaning to scientists, as well as students of literature.

This chapter deals with the origin of life on planet Earth. Here you will discover that there are many current hypotheses regarding the formation of life. But you will also learn that, as in much of science, there are many unanswered questions as well.

GUIDE FOR READING

After you read the following sections, you will be able to

16–1 Spontaneous Generation
- Define spontaneous generation.
- Describe the experiments of Redi, Spallanzani, and Pasteur.

16–2 The First Signs of Life
- Describe conditions on ancient Earth.
- Define microfossil.
- Discuss various hypotheses on the evolution of cells.

16–3 The Road to Modern Organisms
- Relate oxygen production in photosynthesis to the evolution of aerobic metabolism.
- Describe the importance of membrane-bound organelles to the evolution of life on Earth.

Journal Activity

YOU AND YOUR WORLD
Traveling back through time is only a dream. Write a short story or poem about what it would be like to travel back to a period of time that is of interest to you.

16–1 Spontaneous Generation

Guide For Reading
- What is spontaneous generation?
- How did the experiments of Redi, Spallanzani, and Pasteur disprove the theory of spontaneous generation?

Scientists have always wondered how life on Earth began. For many centuries, they believed that life simply "arose" from nonliving matter. They believed, in fact, that life arose from nonlife all the time. Mice, for example, arose from piles of grain. Bees were "produced" in the carcasses of cattle. Those who disputed this idea were ridiculed. The English naturalist Ross expressed a common feeling when he wrote, "To question that beetles and wasps were generated in cow dung is to question reason, sense, and experience." **The hypothesis that life arises regularly from nonlife is called spontaneous generation.**

Spontaneous Generation: True or False?

Lazzaro Spallanzani was born in Italy in 1729—three years before the birth of George Washington. As he grew up, he developed an interest in the natural world and began to consider one of the burning questions that people argued about in his school: **spontaneous generation.** Those who believed in spontaneous generation often supported their argument with the example of rotting meat. As meat spoiled, it was said to give rise to maggots, which eventually changed into flies. The rotting of

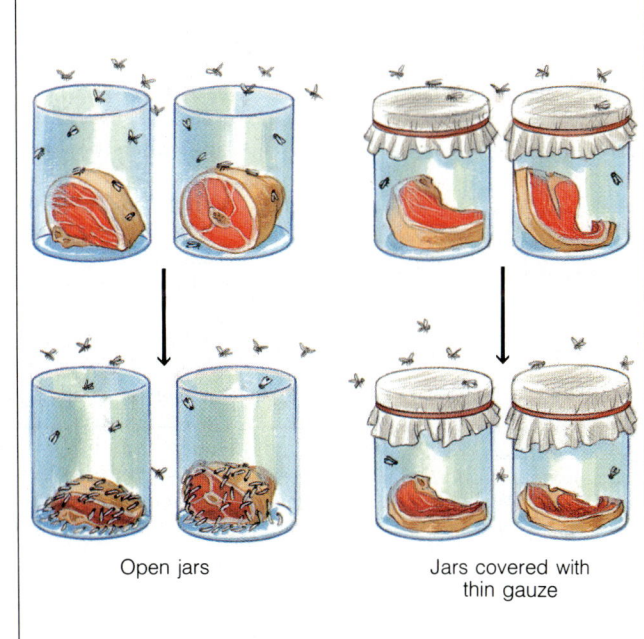

Figure 16–1 Redi's experiments helped disprove spontaneous generation. Maggots appeared only when adult flies were able to enter the jars and deposit their eggs.

Open jars   Jars covered with thin gauze

COOPERATIVE LEARNING

Students have just completed the reading of 4 billion years of Earth's history from a scientific viewpoint. Using preassigned lab groups or randomly selected teams, have groups predict the next major development in the scientific history of the Earth. Each group should assume the role of encyclopedia writers and complete the entry on History of the Earth for the (year) edition of (name of encyclopedia). They should be encouraged to make geological, ecological, meteorological, and biological predictions of future developments.

Journal Activity

YOU AND YOUR WORLD
Have students discuss what time period they selected and why. If several students are very interested in a particular historical time period, you might consider making arrangements with the students' social studies teachers for the students to cooperate on a group project.

MULTICULTURAL STRATEGY

Throughout history, various cultures have developed alternative explanations concerning the origin of life. Interested students may want to research these alternative theories, focusing particularly on the ideas arising in ancient Egypt and India as well as various concepts from Native American cultures.

Content Development
Point out that many years ago people believed that life could come from nonliving matter. This idea is called the hypothesis of spontaneous generation.

Tell students that in 1668, an Italian physician named Francesco Redi (1626–97) disproved this hypothesis. Explain that Redi proved that living things can arise only from living things.

Reinforcement/Reteaching
Have students review the characteristics of living things presented in Chapter 2 of their textbook. As these characteristics are reviewed, have students refer to the list of living and nonliving things they prepared in the Focus/Motivation activity of this section and correct their initial observations if necessary.

339

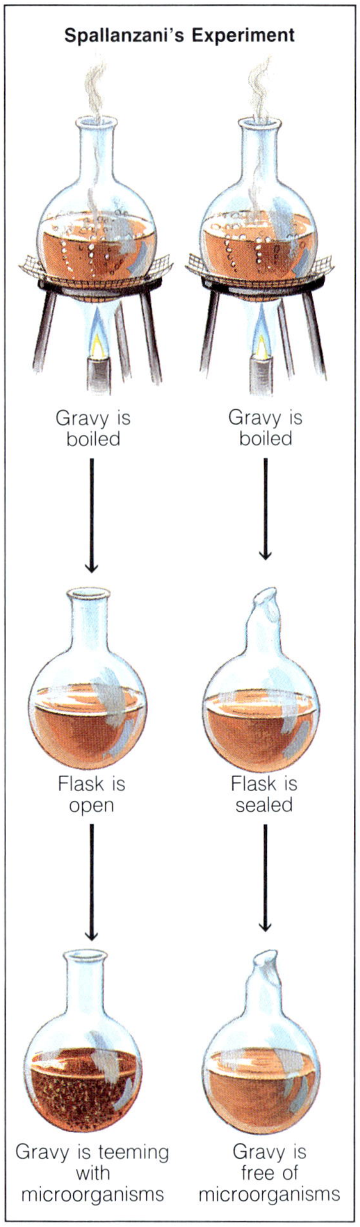

meat in the warm Italian climate was a common experience before refrigeration. Everyone seemed to agree that the maggots were produced from the meat.

Spallanzani was already skeptical of the idea that life could arise from nonlife when he came upon a small book written by another Italian scientist, Francesco Redi. In his book, Redi hypothesized that the maggots actually arose not from the meat itself but from eggs. He believed that flies laid their eggs, which were too small to be seen with the unaided eye, directly on the meat. The eggs then developed into maggots, which later became adult flies. Redi described the simple experiment he had performed to test his spontaneous generation hypothesis. See Figure 16–1 on page 339.

Redi placed pieces of meat in several jars. Half of the jars he left open to the air so that flies could land on the meat. The other jars were covered with a thin gauze. Although these jars were open to the air, the gauze prevented flies from landing on the meat. After a few days, the meat in all the jars had spoiled. But maggots were found only on the meat in the uncovered jars. Redi concluded that the maggots did not arise spontaneously; rather, the maggots developed from the eggs laid by the flies.

Redi's conclusions were attacked by another eighteenth-century scientist, the Englishman John Needham. Needham claimed that spontaneous generation could occur under the right conditions and cited his own experiment as proof. He sealed some gravy in a bottle. Then he heated the bottle, killing (he claimed) all living things in the gravy. After several days, he examined the contents of the bottle under a microscope and found it swarming with microscopic organisms. "These little animals," he argued, "can only have come from the juice of the gravy."

The work of both Redi and Needham was well known to Spallanzani, and it is to his role in spontaneous generation that we will now return. Spallanzani believed that Needham was wrong and that all the organisms had not been completely killed when the gravy was heated. He could have written Needham with his objections. Instead, he chose to perform an experiment to prove his hypothesis. See Figure 16–2.

Spallanzani prepared some gravy identical to that used by Needham. He placed half of the gravy into one jar and half into another jar. He then boiled the gravy in both jars thoroughly. Spallanzani sealed one of the jars tightly and left the other jar open to the air.

After a few days, Spallanzani noticed that the gravy in the open jar was teeming with microscopic organisms. The gravy in the sealed jar contained no living things. Spallanzani concluded that the microorganisms did not develop from the gravy—from nonlife—but entered the jar from the air. If this was not the case, he argued, then both jars should contain microorganisms.

Figure 16–2 Spallanzani's experiment showed that microorganisms do not grow in nutrient broth that has been boiled and placed in sealed flasks.

16–1 (continued)

Skills Development
Guided Practice
Skill: Interpreting illustrations
Using the Visuals Have students observe Figure 16–1 on page 339 showing Redi's experiment.

• **How does the meat appear in all the first jars?** (Fresh and new.)
• **What is happening in the first pair of jars?** (Flies are entering the jars.)
• **What is happening in the second pair of jars?** (Flies are unable to enter the jars.)
• **What will eventually happen to the meat in the uncovered jars?** (Maggots will hatch from eggs laid by flies.)
• **What will eventually happen to the meat in the covered jars?** (Maggots will not appear on the meat.)
• **What can you predict about maggots developing on rotten meat from these illustrations?** (The flies must come in contact with the meat to lay their eggs. If the flies do not come in contact with the meat, maggots will not form on the meat. Living things come only from the same kind of living things.)

Content Development
Using the Visuals Have students observe Figure 16–2 as you discuss the experiment conducted by Lazzaro Spallanzani (1729–99) that further disproved spontaneous generation. Point out that despite the work of Francesco Redi (1626–97) and Spallanzani, many people still believed in spontaneous generation.

Pasteur and Spontaneous Generation

Despite the work of Redi and Spallanzani, many people still believed in spontaneous generation. They argued that air was necessary for spontaneous generation and because Spallanzani had kept air out of the sealed jar, spontaneous generation could not have occurred in that part of his experiment. It was not until 1864, and the elegant experiment of French scientist Louis Pasteur, that the hypothesis of spontaneous generation was finally disproved.

Pasteur placed nutrient broth, a substance similar to Needham's gravy, in a flask that had a long, curved neck. Although the end of the neck was open to the air, the curve in the neck served to trap dust and other airborne particles. Pasteur boiled the flask thoroughly to kill any microorganisms it might contain. But he did not seal the open end of the flask. Pasteur waited an entire year. In all that time, no microorganisms could be found in the flask. See Figure 16-4.

Pasteur took his experiment even further. He broke off the neck of the flask, allowing air, dust, and other particles to enter the broth. In just one day the flask was clouded from the growth of microorganisms. Pasteur had clearly shown that the microorganisms had entered the flask along with dust particles from the air. Pasteur, like Redi and Spallanzani before him, had shown that life comes only from life.

Figure 16-3 According to the spontaneous generation hypothesis, living things, such as a wasp, could arise from inorganic matter.

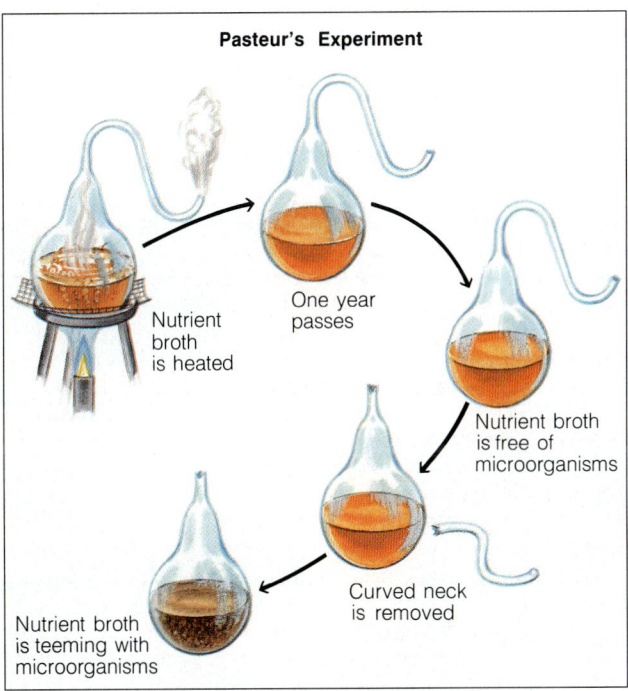

Figure 16-4 Louis Pasteur put an end to the spontaneous generation hypothesis. He showed that microorganisms did not arise in a boiled nutrient broth, even if the broth was exposed to the air, as long as dust particles and other airborne matter were not allowed to enter the broth.

MULTICULTURAL STRATEGY

There is a Native American legend from the Lakota (Sioux) tribe called "Tunkashila," or "Grandfather Rock," that attempts to explain the origins of land. According to the legend, at one time all things existed as spirits. The spirits tried to live on the sun but it was too hot for them, so they came to the Earth. They couldn't find any land on the Earth because it was totally covered by water. Then out of the water arose a huge, hot rock. The rock eventually formed land where life began.

Have students investigate various legends about the creation of the Earth and compare these legends with scientific theory. How are they similar? How are they different?

16–1 (continued)

SECTION REVIEW 16–1

1. Spontaneous generation is a hypothesis that life arises regularly from nonlife.
2. Spallanzani showed that organisms will not arise spontaneously in a nutrient broth if the broth is heated sufficiently and then sealed from the atmosphere.
3. Scientists argued that the lack of air kept spontaneous generation from occurring.
4. Pasteur showed that microorganisms would not appear in a heated nutrient broth even in the presence of air as long as the curved neck of the flask prevented dust and microorganisms from entering the broth.

Reinforcement/Reteaching

For those students who have difficulty with the Section Review questions, go back to the discussion in the section and review the material describing the experiments of Redi, Spallanzani, Needham, and Pasteur.

Closure

After reviewing the experiments of Redi, Spallanzani, Needham, and Pasteur, encourage students to examine the findings of the four scientists in terms of the logic of their scientific thought and methods. Reinforce the concept that hypotheses such as spontaneous generation change over time.

TEACHING STRATEGY 16–2

Focus/Motivation

Pour egg white at room temperature into a clear glass beaker.
- **What color is the liquid?** (Accept all logical answers, such as clear, or light yellow.)

Move the beaker around to swirl the egg white.
- **What is the consistency of the liquid?** (Accept all answers.)

16–1 SECTION REVIEW

1. Define spontaneous generation.
2. What evidence did Spallanzani give to refute Needham's experiment?
3. In Redi's first experiment he used only open jars and sealed jars. What arguments might scientists have come up with that caused Redi to redo his experiment with open jars and jars covered with gauze?
4. **Critical Thinking—Applying Concepts** How did Pasteur's use of a flask with a long, curved neck finally disprove spontaneous generation?

Guide For Reading

- What gases made up Earth's early atmosphere?
- What are two sources for Earth's first organic molecules?
- Why is the evolution of photosynthesis important?

16–2 The First Signs of Life

If life can come only from life, how did life on Earth first arise? For that matter, does life exist (or has it ever existed) on Mars? Or Venus? By studying the beginnings of life, we examine our own origins.

Our planet was born approximately 4.6 billion years ago as a great cloud of gas and dust condensed into a sphere. As gravity pulled this matter tightly together, heat from great pressure and radioactivity melted first the planet's interior and then most of its mass. As far as we can tell, Earth cooled enough to allow the first solid rocks to form on its surface about 4 billion years ago. For millions of years afterward, violent planet-wide volcanic activity shook the crust. At the same time, an intense meteor shower bombarded Earth with missiles from space.

Figure 16–5 *You can see from this illustration depicting primitive Earth that most living things would not easily survive, even if the atmosphere were similar to the atmosphere you breathe today.*

What was Earth like 4 billion years ago? Had you been able to visit, you wouldn't have recognized the place. In fact, you wouldn't have been able to survive, even for an instant! For the first atmosphere that covered our planet was lost to space before the Earth cooled. And there was no liquid water at all.

Where did our atmosphere come from? We know from studying volcanoes that eruptions pour out carbon dioxide, nitrogen, and other gases. We also know that meteorites carry water (in the form of ice) and many carbon-containing compounds. So it is reasonable to propose that between 4 billion and 3.8 billion years ago, a combination of volcanic activity and a constant stream of meteorites released the gases that created Earth's atmosphere.

What was that early atmosphere like? **The Earth's first atmosphere most likely contained carbon monoxide (CO), carbon dioxide (CO_2), and hydrogen (H_2), mixed with some nitrogen (N_2) and possibly other gases, such as ammonia (NH_3) and methane (CH_4).** Notice that there is no free oxygen in this list—precisely the reason this atmosphere could not have supported life as we know it. Geological evidence supports this idea: Rocks from this time contain almost no compounds that would have formed if oxygen had been present.

Where did the oceans of ancient Earth come from? Oceans could not exist at first because Earth's surface was extremely hot. Any rain that fell upon it would immediately boil away. But, about 3.8 billion years ago, Earth's surface cooled enough for water to remain a liquid on the ground. Thunderstorms drenched the planet for many thousands of years, and oceans began to fill. We know this because the earliest sedimentary rocks, which are laid down in water, have been dated to this time period.

No one can say with certainty exactly when life first formed on ancient Earth. But paleontologists working near Lake Superior have found microscopic fossils, called **microfossils**, that have been dated as far back as 3.5 billion years. **Microfossils provide outlines of ancient cells that have been preserved in enough detail to identify them as prokaryotes, similar to bacteria alive today.** See Figure 16–7.

Somehow these earliest life forms appeared within half a billion years after the formation of Earth's first rocks. How might that have happened?

Starting from Scratch: The Molecules of Life

Experiments first performed in 1953 by American scientists Stanley Miller and Harold Urey provide a fascinating glimpse of events leading to the appearance of Earth's first life forms. Miller re-created what he thought might have been Earth's earliest atmosphere by mixing methane, ammonia, water, and hydrogen in a flask. He then exposed the contents of the flask to

Figure 16–6 Astronauts are able to work in space because they carry their "atmosphere" with them. The atmosphere that supports them today is far different from the atmosphere that existed on early Earth.

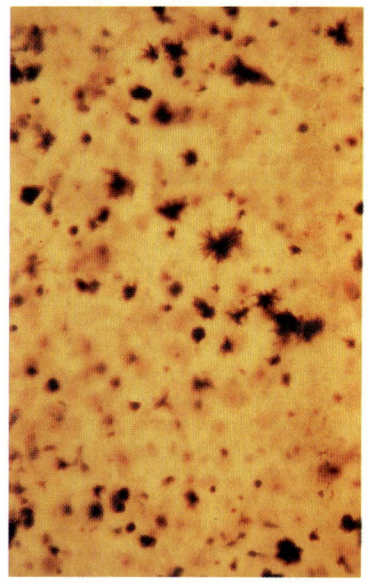

Figure 16–7 You can see 2-billion-year old microfossil bacteria in this thin slice of rock.

BACKGROUND INFORMATION
RNA AND THE BEGINNINGS OF LIFE ON EARTH

Was RNA the original molecule of heredity? According to some hypotheses, it was.

According to one hypothesis, RNA evolution was rapid because replication in primitive RNA was much more prone to errors—mutations—than replication in modern DNA. Due to natural selection, RNA variants that were most efficient at replicating and editing themselves predominated.

Eventually, RNA reached a level of complexity and efficiency at which mutations were much more likely to cause problems than improvements. At this point, natural selection favored stability rather than change. Using DNA as the molecule of information storage provided necessary stability.

But where did DNA come from? One intriguing hypothesis suggests that the first DNA might have been synthesized using the enzyme reverse transcriptase. Although this enzyme was discovered fairly recently, it appears to have been in existence for a very long time. Today, reverse transcriptase is used by retroviruses such as HIV to transcribe RNA into DNA.

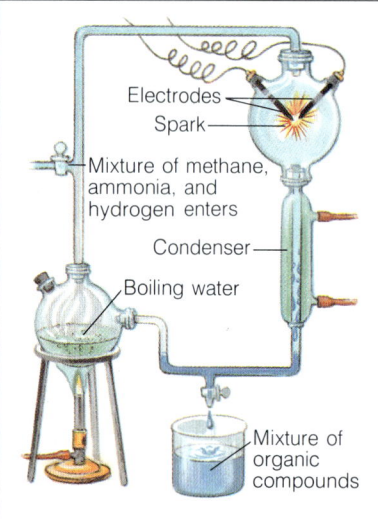

Figure 16–8 An experiment performed by Stanley Miller (top) and Harold Urey first demonstrated how organic matter may have formed in Earth's primitive atmosphere. By re-creating the early atmosphere (ammonia, water, hydrogen, and methane) and passing an electric spark (lightning) through the mixture, Miller and Urey proved that organic matter such as amino acids could have formed spontaneously (bottom).

electric sparks to simulate sunlight and lightning on primitive Earth. See Figure 16–8.

In just a few days, a "soup" of organic molecules formed. In it, Miller found urea, acetic acid, lactic acid, and several amino acids. Other researchers have repeated Miller's experiment using gas mixtures that reflect current views of the composition of the early atmosphere. All of these experiments produced organic molecules, including ATP and adenine.

What's more, astronomers have determined that meteors, comets, and cosmic dust are filled with organic molecules, ice, and other elements that are essential to life. For example, scientists discovered lipids and the five nitrogenous bases found in DNA and RNA in a large meteorite that crashed to Earth in 1969. No one knows how these compounds formed. However, meteors could have carried organic molecules to Earth from outer space. These molecules could have become mixed in with those already forming on the young Earth. Thus, over the course of millions of years, some of the building blocks of life could have accumulated in great quantities on Earth.

The Formation of Complex Molecules

A collection of bases, amino acids, and other organic molecules, however, is certainly not life. What might have happened next? Russian scientist Alexander Oparin and American scientist Sidney Fox have shown that the organic soup on the early Earth would not necessarily have remained a mix of simple molecules. In the absence of oxygen, for example, amino acids tend to link together on their own to form short protein chains. Other compounds can link together to form simple carbohydrates, alcohols, and lipids.

But there's more. Collections of these molecules tend to gather into tiny round droplets. Some of these droplets grow by themselves, and others even reproduce. Still others can break down glucose. We wouldn't say these droplets are "alive," but we might call them "proto-life" because they have begun to perform some tasks necessary for life.

From Proto-life to Cells

We are still left with the difficult task of explaining how the complex system of protein synthesis evolved from this soup of organic molecules. Today, DNA can make proteins only with the help of several enzymes and several kinds of RNA. And DNA can replicate itself only with the help of another batch of enzymes. But these enzymes and RNA are assembled by DNA! Can you see the problem? No part of this system can exist without the others. So how could the whole thing have gotten started in the first place? No one knows for certain, but scientists have offered some interesting hypotheses.

G. Cairns-Smith and J. Bernal note that amino acids and nucleic acids (DNA and RNA) stick to the repeating structures

---

16–2 (continued)

Reinforcement/Reteaching
Have students review the characteristics and importance of organic compounds as described in Chapter 4.

Reinforcement/Reteaching
Review the hypothesis of spontaneous generation described in Section 16–1. Point out that Redi, Spallanzani, and Pasteur showed only that life cannot come from nonlife under the conditions they

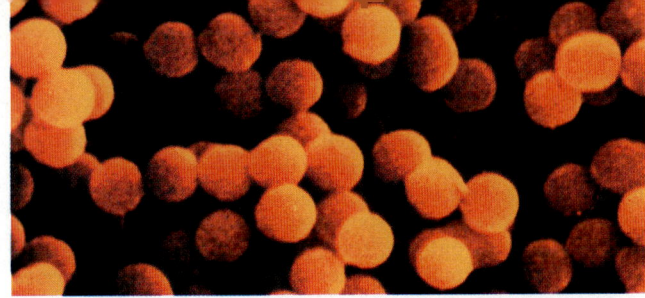

of clay crystals. Held together in a regular pattern on clay crystals, these molecules combine to form proteins and polynucleotides. Other researchers note that some kinds of RNA can join amino acids into protein chains without help from protein enzymes. What's more, some forms of RNA can copy themselves and can "edit" other RNAs, adding and deleting nucleotides.

These experiments support a hypothesis first suggested in 1968 by Francis Crick and Leslie Orgel. Crick and Orgel suggested that RNA, rather than DNA, functioned as life's first information storage system. According to this hypothesis, life based on RNA could have started when RNA fragments began to copy and edit themselves and assemble proteins. Over time, these RNAs could have evolved to the point where they produced protein enzymes that took over the work of bringing about chemical reactions. Later, the job of storing genetic information could have similarly been passed on to DNA. In this way, over millions of years, RNA, DNA, and proteins could have evolved into the complex system that characterizes life today.

Researchers note that the chemical reactions thought to have produced the first life on Earth still occur in nature in very special places—wherever volcanic activity combines with water. For example, near volcanic vents on the bottom of the sea, molten rock heats sea water to very high temperatures. When this water gushes out of the vents, it is filled with energy-rich sulfur compounds and rushes past deposits of clay.

When these conditions are duplicated in the laboratory, both amino acids and stretches of RNA are spontaneously synthesized. As you will see in the next chapter, the oldest living types of prokaryotes—bacteria that survive by obtaining energy from sulfur compounds—still live near both surface and undersea hot vents. Some biologists suspect that these bacteria are living links to the very first forms of life on Earth.

The First True Cells

Although the origin of the first true cells is uncertain, we can identify several of their characteristics. They were prokaryotes that resembled types of bacteria alive today. They were heterotrophs that obtained their food and energy from the organic molecules in the soup that surrounded them. And they must have been **anaerobes**. Anaerobes are organisms that can live without oxygen. Why can we be certain these first cells were anaerobes? ❶

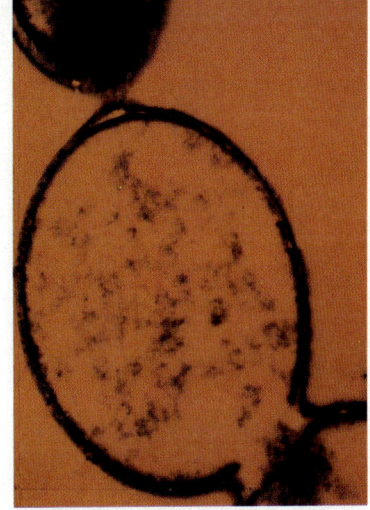

Figure 16–9 These droplets (left), magnified 3000 times, were created in the laboratory of Sidney Fox. Although the droplets are not alive, some can actually reproduce by dividing into two separate droplets (right).

Figure 16–10 One hypothesis about the origin of life suggests that living things evolved around hot sea vents. Today, bacteria that can use the sulfur compounds as a source of energy live in areas near deep-sea vents.

ANNOTATION KEY

❶ There was no oxygen in the atmosphere so they must have been anaerobes. (Making inferences)

BACKGROUND INFORMATION
ALEXANDER OPARIN

In 1923, the Russian scientist Alexander Oparin proposed that proteinlike substances in the prehistoric oceans may have formed aggregates of large molecules, which he called coacervates. These complex structures were surrounded by a membranelike "shell" of water molecules. Coacervates have been formed in the laboratory from proteins and other organic molecules.

In 1960, the American biochemist Sidney Fox showed that amino acids could be heated at temperatures above 100°C to produce proteins. In water, these proteinlike particles formed clusters of membrane-bound, complex units that Fox called microspheres. Although coacervates and microspheres are not living organisms, they reinforce the possible nonbiological origin of the first primitive life forms.

TEACHING SUPPORT

Laboratory Manual
- Making Coacervates, p. 209

---

used. They did not show that life could never arise spontaneously.

Content Development
Tell students that astronomers have shown that organic compounds also form in space and theorize that meteorites and other objects in space may have carried some organic compounds to Earth. These compounds then mixed with the organic soup that had already formed in the Earth's oceans. Point out that the first true cells were prokaryotic heterotrophic anaerobes.

345

ESL STRATEGY

Have students imagine what life must have been like on Earth before and after the time of photosynthesis. Then ask questions.
- How long ago did photosynthesis begin?
- What kind of life existed before photosynthesis?
- What was life like after photosynthesis?

You may wish to have students provide written responses and include illustrations with their answers. Suggest that they share their work with the class.

TEACHING SUPPORT

Teaching Resources
- Activity: Evaluating a Theory About Cells, p. 9

Study Guide
- Section 16–2, p. 157

16–2 (continued)

Reinforcement/Reteaching
Have students review the process of photosynthesis as discussed in Chapter 6.

Content Development
Point out that some cells developed the ability to harness energy from the sun in a primitive form of photosynthesis. Hydrogen sulfide (H_2S) was probably substituted for water (H_2O) in ancient photosynthesis.

Content Development
Stress to students that the presence of bacteria and oxygen in today's environment makes it unlikely that life could again rise from nonlife.

SECTION REVIEW 16–2

1. Water vapor, carbon monoxide, carbon dioxide, nitrogen, hydrogen sulfide, and hydrogen cyanide.
2. Microfossils are microscopic fossils of organisms that lived on ancient Earth, usually no more than an outline of an ancient cell.
3. The gases in Earth's early atmosphere, along with energy from lightning, could have formed organic molecules that accumulated in an organic soup. Another possible source is from outer space, primarily carried by meteorites.
4. If all organisms were heterotropic, eventually all the organic matter in the organic soup would be used up and there would be no source of food for organisms.

Reinforcement/Reteaching
Some students may have difficulty with the Section Review questions. Have these students review the material they had difficulty understanding.

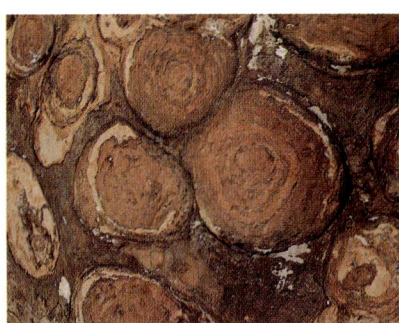

Figure 16–11 *The first autotrophs grew in layered mats called* stromatolites. *Fossils of such stromatolites can be found in rock layers throughout the world (top). However, Shark Bay, Australia, is one of the few places on Earth where living stromatolites still exist (bottom).*

346

The Evolution of Photosynthesis

The first heterotrophic cells could have survived without difficulty for a long time because there were plenty of organic molecules for them to "eat." But as time went on, the complex molecules in the organic soup would have begun to run out. In order for life to continue, some organisms would have had to develop a way to make complex molecules from simpler ones. In addition, the intense pressure of natural selection would have favored organisms that could harness an outside source of energy for their own purposes. The stage was set for the appearance of the first autotrophs.

At some point an ancient form of photosynthesis evolved. Photosynthesis in early cells, however, was very different from the photosynthesis that occurs in modern plants, which you read about in Chapter 6. The first true cells probably used hydrogen sulfide (H_2S) the way modern photosynthetic organisms use water (H_2O).

These first autotrophs were enormously successful, spread rapidly, and were commonplace on Earth about 3.4 billion years ago. They grew in layered, matlike formations called stromatolites (the prefix *stroma-* means layer). Today, living stromatolites can be found only in special habitats such as Shark Bay, Australia. See Figure 16–11. However, fossils of stromatolites have been found in many parts of the world.

Life from Nonlife

In Section 16–1 you read about some experiments that disproved the hypothesis of spontaneous generation. "Hey, what's going on?" you might exclaim. If we just said that life did arise from nonlife billions of years ago, why couldn't it happen again? The answer is simple: Today's Earth is a very different planet from the one that existed billions of years ago. On primitive Earth, there were no bacteria to break down organic compounds. Nor was there any oxygen to react with the organic compounds. As a result, organic compounds could accumulate over millions of years, forming that original organic soup. Today, however, such compounds cannot remain intact in the natural world for a long enough period of time to give life another start.

16–2 SECTION REVIEW

1. List five gases in Earth's first atmosphere.
2. What is a microfossil?
3. Name two sources for Earth's first organic molecules.
4. **Connection—Botany** Why did photosynthesis (or something like it) have to evolve if life was to continue past its earliest stages?

16–3 The Road to Modern Organisms

Guide For Reading
- Why is the development of sexual reproduction important to the history and development of life on Earth?
- In what ways is the ozone layer important in the development of life on Earth?

Once life evolved on Earth, things would never be the same. For over millions of years, life has changed the Earth in ways that have affected our planet dramatically. The first great change occurred roughly 2.2 billion years ago when a more modern form of photosynthesis evolved. By substituting H_2O for H_2S in their metabolic pathways, photosynthetic organisms released a deadly new gas into the atmosphere. That gas was oxygen—a waste product of photosynthesis!

Because you rely on oxygen to survive, you might be surprised to learn that it can be deadly. However, oxygen is a very reactive gas that destroys organic compounds. So imagine the catastrophe that struck Earth's earliest life forms. Over a period of 500 million years, a waste product (oxygen) produced by some organisms transformed Earth from a totally anaerobic planet into a planet whose atmosphere is nearly 1/5 oxygen.

Because oxygen was deadly to anaerobes, such organisms were forever banished from the planet's surface. Today, organisms that cannot tolerate oxygen survive only deep in mud or in other places where the atmosphere does not reach. The very first case of living organisms producing a kind of pollution that made the entire Earth uninhabitable for many forms of life had occurred. Let us hope that we as a species do not make decisions that have similar results!

One effect of oxygen in the atmosphere, however, was beneficial to those organisms that survived. The first atmosphere had allowed ultraviolet radiation from the sun to strike the Earth's surface. Ultraviolet radiation is damaging, even toxic, to many life forms. But as oxygen gas (O_2) from photosynthesis reached the upper atmosphere, some of it was broken apart by ultraviolet radiation into individual oxygen atoms (O). These atoms quickly recombined with oxygen molecules to form the gas ozone (O_3). In time, an ozone layer formed in the Earth's atmosphere. This ozone layer absorbs much of the ultraviolet radiation from the sun, shielding living things from the dangerous rays. In Unit 10 of this textbook, you will read how the burning of fossil fuels and the release of certain compounds into our atmosphere is slowly but surely destroying the ozone layer so very important to life on our planet.

The Evolution of Aerobic Metabolism

The addition of oxygen to the atmosphere began a new chapter in the history of life on Earth. That chapter started with the evolution of organisms that not only survive in oxygen but utilize it in their metabolic pathways. Metabolism is the sum total of all the chemical reactions that occur in a living thing. These new aerobic pathways allowed organisms to obtain 18 times more energy from every sugar molecule than

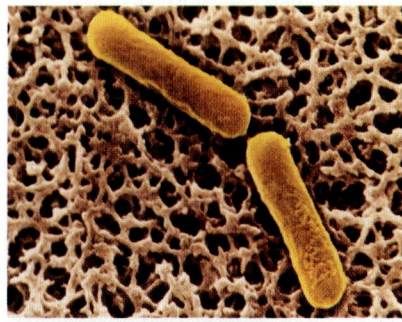

Figure 16–12 Once Earth's atmosphere contained oxygen, anaerobic bacteria such as these were banished to places where the atmosphere could not reach.

BACKGROUND INFORMATION
MODERN ANAEROBES

Note that in addition to anaerobic muds and other microenvironments devoid of oxygen, anaerobic bacteria can thrive in certain parts of the bodies of multicellular organisms. The bacteria that cause tetanus, for example, thrive in the lowered oxygen environment of infected deep puncture wounds. Other anaerobes function in parts of animal guts where there are low levels of oxygen.

ers break down in waste dumps, they release synthetic chemicals that attack and destroy ozone molecules. Our fast-paced, fast-food society is upsetting an atmospheric balance that has protected us from ultraviolet radiation for millions of years.
• **What other occurrences, both human and natural, may contribute to the destruction of the Earth's atmosphere and climate?** (Answers may include pollution, burning of fossil fuels, deforestation, and volcanic activity.)

Content Development
In this era of new public awareness of global ecological crises, the discussion of photosynthesis and the Earth's environment offers a golden opportunity. Emphasize that here was a case in which millions of microorganisms, working slowly and steadily over time, transformed the environment of the entire planet into one that poisoned them and most of their relatives. Only those organisms that first learned to tolerate and later to utilize oxygen could survive. Use this episode to make the point that life is a powerful phenomenon. It has changed the global environment before and can do it again—unless we are all very careful.

Closure
Review the environmental conditions on ancient Earth. Have students discuss the various hypotheses of the evolution of the first primitive cells.

TEACHING STRATEGY 16–3
Focus/Motivation
Hold up a plastic foam food container obtained from a local fast-food restaurant. Point out that each time we buy food packed in one of these containers, we increase our chances of getting a sunburn and possibly skin cancer.

Explain that the ozone layer in our atmosphere absorbs powerful ultraviolet rays from the sun, shielding living things from these dangerous rays. Point out that as the plastic foam food contain-

TEACHING SUPPORT

Teaching Resources
- Vocabulary Review: Word Game, p. 1

Study Guide
- Section 16–3, p. 158

ECOLOGY NOTE
CHANGES OVER TIME

Life, the effects of living things on Earth's physical environment, and geological changes in the Earth over time interact constantly. One may even compare the biosphere and the geosphere to two partners in a dance, in which each move by one elicits a corresponding move from the other. Some examples:

1. The anaerobic environment on early Earth allowed the evolution of life from nonliving organic molecules.
2. The evolution of photosynthesis changed Earth's surface and oceans from anaerobic environments to aerobic ones.
3. Free oxygen in the atmosphere created the ozone layer that shields us from dangerous ultraviolet radiation.
4. Movements of continents over time carried evolving organisms around, alternately isolating them and mixing them together.

This system of mutual effects is important to emphasize because humans are a large part of it today. As both world leaders and the mass media are realizing, we are greatly affecting the biosphere by our activities.

---

anaerobic pathways did. As you may recall from Chapter 6, these new aerobic pathways are part of the process of obtaining energy called cellular respiration.

The Evolution of Eukaryotic Cells

Between 1.4 and 1.6 billion years ago, the first eukaryotic cells evolved, fully adapted to an aerobic world. Eukaryotes have a nucleus that contains DNA. The outer membrane of the nucleus is called the nuclear envelope. Eukaryotic cells also carry other membrane-bound organelles such as mitochondria and chloroplasts.

The Evolution of Sexual Reproduction

Among the most important steps in the evolution of eukaryotic life was the emergence of sexual reproduction. The advent of sexual reproduction catapulted the process of evolution forward at far greater speeds than ever before. But why did sexual reproduction speed up the evolutionary process? Isn't it just another form of reproduction?

Most prokaryotes reproduce asexually. Often they simply duplicate their genetic material and divide into two new cells. (This process, called binary fission, will be discussed in detail in the next chapter.) Although this is an efficient and effective form of reproduction, it yields daughter cells that are exact duplicates of the original parent cell. As such, this type of reproduction restricts genetic variation to mistakes or mutations in DNA. As you read in Chapter 14, genetic variation is crucial to the process of adaptive radiation and the evolution of new species.

Sexual reproduction, on the other hand, shuffles and reshuffles genes in each generation, much like a person shuffling a deck of cards. The offspring of sexually reproducing organisms, therefore, never resemble their parents (or each other) exactly. This increase in genetic variation greatly increases the chances of evolutionary change in a species due to natural selection.

The evolution of sexual reproduction, along with the development of the membrane-bound organelles mitochondria and chloroplasts, were of enormous importance to the history and development of life on Earth. If not for these developments, multicellular organisms may not have evolved.

The Evolution of Multicellular Life

A few hundred million years after the evolution of sexual reproduction, evolving life forms crossed another great threshold: the development of multicellular organisms from single-celled organisms. In the blink of an evolutionary eye, these first multicellular organisms experienced a great adaptive radiation. Earth's parade of life was well on its way.

Figure 16–13 In asexual reproduction, such as the division of a bacterium into two new bacteria, each new cell is an exact copy of the original cell (top). In sexual reproduction, however, offspring contain genes from each parent, and genetic variation is increased. How boring it would be if we all contained the exact same genes and looked exactly alike (bottom).

16–3 (continued)

Content Development
Tell students that once oxygen was plentiful on Earth, the more efficient form of aerobic metabolism utilizing cellular respiration evolved and replaced anaerobic metabolism. Around 1.4 billion years ago, the first eukaryotic cells evolved.

Content Development
Point out that the development of sexual reproduction greatly speeded up the course of evolution by providing more chances for genetic variation in organisms. Single-celled organisms eventually evolved into multicellular organisms.

SECTION REVIEW 16–3

1. Sexual reproduction greatly increases genetic variability. The greater the genetic variability, the more likely it is that evolution will proceed at a quicker pace than if most organisms in a species had the exact same genetic makeup, which occurs as a result of

SCIENCE, TECHNOLOGY, AND SOCIETY — BREAKTHROUGH

The Symbiotic Theory of Eukaryotic Origins

For many years biologists have wondered how eukaryotic cells evolved from prokaryotic cells. Eukaryotic cells contain membrane-bound organelles and a nucleus surrounded by a nuclear envelope, or membrane.

Of particular interest to scientists are the organelles called mitochondria and chloroplasts. Why? Although these organelles usually act like ordinary parts of a cell, they contain their own DNA. That DNA is different from the DNA found within the nucleus of the cell. These organelles also reproduce on their own when the cell divides.

Some years ago biologists noted that mitochondria and chloroplasts strongly resemble living prokaryotes. Mitochondria resemble certain aerobic bacteria, whereas chloroplasts resemble certain photosynthetic bacteria. One American scientist, Lynn Margulis, has championed an intriguing hypothesis about the evolution of eukaryotic cells.

Margulis feels that eukaryotic cells evolved when ancient aerobic prokaryotes similar to modern chloroplasts and mitochondria took up residence within other prokaryotic cells. Over time, a long-lasting symbiosis developed. Symbiosis refers to any relationship in which two organisms live closely together. This ancient symbiosis was particularly helpful to both organisms. The organisms, which evolved into mitochondria and chloroplasts, now lived within the nutrient-rich cytoplasm of their host cell. The host cell containing mitochondria-type prokaryotes could now produce energy faster and more efficiently because it could utilize oxygen in its metabolic pathways. If the host cell contained chloroplast-type prokaryotes, it could now use the energy from the sun to produce food. In time, of course, mitochondria and chloroplasts came to function more and more as part of the cell structure in eukaryotes.

It was not easy, at first, for scientists to accept the idea that eukaryotic cells developed as communities. But both structural and chemical evidence strongly supports the theory proposed by Margulis. The symbiotic theory of eukaryotic origins is now accepted by most biologists. However, there are still many unanswered questions in the search for eukaryotic origins. We still do not know, for example, how the earliest eukaryotes developed the nuclear envelope that surrounds their DNA.

16–3 SECTION REVIEW

1. How did the development of sexual reproduction speed up the process of evolution?
2. What compound replaced H_2S in the photosynthetic process?
3. **Connection—Ecology** Why are people concerned with protecting the ozone layer?

---

asexual reproduction.
2. H_2O
3. The ozone layer protects living organisms from the harmful ultraviolet radiation given off by the sun.

Reinforcement/Reteaching
If some students have difficulty with the Section Review questions, have them return to the discussion in the section.

Closure
Have students relate the production of oxygen in photosynthesis to the evolution of organisms with aerobic metabolism. Stress the importance of eukaryotic organisms with membrane-bound organelles and sexual reproduction to the evolution of modern life forms on Earth.

SCIENCE, TECHNOLOGY, AND SOCIETY — BREAKTHROUGH

The Symbiotic Theory of Eukaryotic Origins

Have students look at photographs of a chloroplast and a blue-green bacterium.

- **How are these two structures similar?** (Accept all logical answers.)

Now have students look at photographs of a mitochondrion and a bacterium.

- **How are these two structures similar?** (Accept all logical answers.)

Explain that in the late 1800s, biologists noticed the similarities between mitochondria, chloroplasts, and living prokaryotes. The symbiotic hypothesis of eukaryote evolution later fell out of favor when no supporting evidence could be found.

In the late 1950s, Lynn Margulis started accumulating evidence to resurrect the symbiotic hypothesis. Margulis continues to compile evidence for this hypothesis and one of its extensions: that centrioles have evolved from swimming bacteria.

The following question can be used as a topic for class discussion or assigned as written homework: Certain prokaryotes today live in close association with eukaryotes. Could this information be used as evidence to support the symbiotic hypothesis of eukaryotic origins? Support your answer.

LABORATORY INVESTIGATION

INVESTIGATING SPONTANEOUS GENERATION

Before the Lab

1. Divide the class into groups of 3 to 6 students.
2. Gather all materials at least one day before the investigation. You should have enough supplies to meet your class's needs, assuming 3 to 6 students per group.
3. Sterilize all glassware by placing it in an autoclave or pressure cooker for 15 minutes at 15 pounds of pressure. If an autoclave or pressure cooker is not available, boil the glassware to sterilize it.
4. Use a flame to heat and bend a 15-centimeter piece of 6-millimeter diameter straight glass tubing into an S-shape. Each student group will need one S-shaped tube.

Pre-Lab Discussion

Have students read the complete laboratory procedure. Discuss the investigation by asking questions similar to these:
- **What is the purpose of this laboratory investigation?** (To determine if spontaneous generation can occur.)
- **What hypothesis is being tested in this investigation?** (Accept all logical answers. Lead students to suggest a hypothesis similar to "Spontaneous generation cannot occur on Earth today.")
- **What is the experimental variable in this investigation?** (The degree to which the water/grass solution is exposed to the environment.)

Skills Development

Students will use the following skills while completing this laboratory investigation.
1. Manipulative
2. Observing
3. Inferring
4. Recording
5. Applying
6. Safety
7. Hypothesizing
8. Relating
9. Comparing

LABORATORY INVESTIGATION

INVESTIGATING SPONTANEOUS GENERATION

PROBLEM

Does spontaneous generation occur on Earth today?

MATERIALS (per group)

600-mL beaker	hot plate
3 125-mL flasks	safety goggles
#5 rubber stopper	beaker tongs
#5 one-hole rubber stopper with S-tube	heat-resistant gloves
	dried grass
	Bunsen burner

PROCEDURE

1. Place two handfuls of dried grass in the 600-mL beaker. Pour water into the beaker until the water level is about 1 cm below the top of the beaker.
2. Put on the safety goggles. Carefully place the beaker on the hot plate. Set the hot plate to its highest level. **CAUTION:** *Do not touch the top of the hot plate.*
3. Allow the water to boil for about 30 minutes. **CAUTION:** *If the water level drops below the midpoint of the beaker, have your teacher replenish the water with some more boiling water.*
4. After the water has boiled for 30 minutes, turn off the hot plate. Light the Bunsen burner. Pass the front of the tongs through the flame of the burner several times. Turn off the burner. Now use the tongs to remove all of the grass from the water in the beaker. Discard the grass.
5. At this point, ask your teacher for three sterile flasks, a rubber stopper without a hole, and a rubber stopper with an S-tube in the hole.
6. Put on the heat-resistant gloves. Pick up the hot beaker and pour 100 mL of the hot grass solution into each flask. Do not allow any grass to fall into the flasks.
7. As quickly as possible, place the rubber stoppers into two of the flasks. Do not stopper the third flask. Set the flasks aside and observe each flask twice a week for three weeks.
8. On a separate sheet of paper, prepare a data table similar to the one shown. Record your observations in the data table. If you observe thin threadlike structures, you are probably seeing the growth of mold. If the solution in the flask becomes clouded, you are probably seeing evidence of bacterial growth.

OBSERVATIONS

1. Describe what you saw in each flask after three weeks.
2. How long did it take before you saw living things in any of the flasks?

ANALYSIS AND CONCLUSIONS

1. What purpose did the grass serve in this investigation?
2. Why was it necessary to boil the water containing the grass before adding the grass/water solution to the flasks?
3. Why was it necessary to use sterile flasks and stoppers?
4. If living things appeared in any of the flasks, where did the living things come from?
5. Do the results of this investigation provide evidence for or against spontaneous generation?

Week	Observation	Appearance of Liquid in Each Flask		
		Stopper	No Stopper	Stopper with S-tube
1	1			
	2			

350

Safety Tips

To avoid contamination of materials, students should wash their hands thoroughly with soap and water before beginning the investigation. Have students wear safety goggles while working with the hot plate and Bunsen burner. Caution students to be very careful when working with the hot plate, Bunsen burner, and glassware. Some students may be allergic to dried grass. Have these students wear dust masks while observing or working with dried grass.

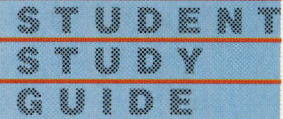

SUMMARIZING THE CONCEPTS

The key concepts in each section of this chapter are listed below to help you review the chapter content. Make sure you understand each concept and its relationship to other concepts and to the theme of this chapter.

16–1 Spontaneous Generation

- The hypothesis that life arises from nonlife is called spontaneous generation.
- Louis Pasteur, a French scientist, put an end to the spontaneous generation controversy when he showed that a nutrient broth that had been thoroughly heated did not have any signs of microorganisms even when left open to the air. Pasteur had allowed air but not dust or other particles to reach the broth. When he did allow dust and other particles to enter the broth, microorganisms soon appeared. Pasteur had proved that the microorganisms in the broth did not develop spontaneously.

16–2 The First Signs of Life

- The atmosphere on ancient Earth was very different from our modern atmosphere. It contained water vapor, carbon monoxide, carbon dioxide, nitrogen, hydrogen sulfide, and hydrogen cyanide. The atmosphere did not contain free oxygen gas.
- Microfossils indicate that the first life forms were prokaryotes, similar to modern bacteria.
- Many organic compounds, including amino acids and ATP, could have formed when ultraviolet rays and lightning reacted with the gases in the early atmosphere. Laboratory experiments have recreated the formation of these compounds on early Earth. The organic compounds formed an organic "soup" containing the basic building blocks of life.
- The first true cells were prokaryotic heterotrophic anaerobes.
- In time, some cells developed the ability to harness energy from the sun in a primitive form of photosynthesis.

16–3 The Road to Modern Organisms

- Once organisms that used water and produced oxygen as a waste product during photosynthesis developed, the atmosphere slowly accumulated oxygen gas.
- Once oxygen was plentiful, aerobic metabolism utilizing cellular respiration evolved. Aerobic metabolism provided more energy than earlier forms of anaerobic metabolism.
- Around 1.4 billion years ago, eukaryotic cells containing membrane-bound organelles evolved.

REVIEWING KEY TERMS

Vocabulary terms are important to your understanding of biology. The key terms listed below are those you should be especially familiar with. Review these terms and their meanings. Then use each term in a complete sentence. If you are not sure of a term's meaning, return to the appropriate section and review its definition.

16–1 Spontaneous Generation
spontaneous generation

16–2 The First Signs of Life
microfossil
anaerobe

3. You may wish to point out the difficulty of conducting experiments such as this one, primarily because of sterilization problems.

Observations
1. The closed flask and the S-tube were clear. The open flask was cloudy (or moldy).
2. In general, it will take about 3 weeks to see evidence of living things.

Analysis and Conclusions
1. The grass provided nutrients so that the mixture simulated a nutrient broth.
2. Boiling the water destroyed any microorganisms in the water.
3. If the flasks and stoppers were not sterile, they could introduce microorganisms into the solution.
4. From organisms carried by the air.
5. The experiment provides evidence against spontaneous generation.

Going Further: Enrichment
Obtain three half-pint (237-mL) bottles containing banana agar. Cover one bottle with cheesecloth and another with plastic wrap. Use a rubber band to hold the cheesecloth and plastic wrap in place. Allow the third bottle to remain open. Place these bottles under a large open-top bell jar. Obtain a stock bottle of fruit flies and invert the mouth of the bottle into the opening in the bell jar. After the fruit flies have entered the bell jar, cover the opening of the bell jar with cotton. Observe the contents of the jars over a two-week period. Explain how this experiment echoes Redi's experiment on spontaneous generation.

Teaching Strategy
1. Circulate throughout the laboratory to make certain students are on task and carefully following the directions for safe laboratory procedures.
2. The hidden purpose of this investigation is to reinforce laboratory techniques. All flasks can be easily contaminated at several points in this investigation, regardless of the precautions taken by both students and teacher.

- **Where in this investigation might contamination of the flasks have occurred?** (Accept all logical answers.)
- **Why must all of the instruments used in a hospital operating room first be sterilized?** (To prevent bacterial contamination and infection of the patient.)

CHAPTER REVIEW

CONTENT REVIEW

Multiple Choice
1. c 2. b 3. b 4. a
5. c 6. c 7. d 8. b

True or False
1. F life/nonlife
2. T
3. T
4. F absence
5. T
6. F autotrophs
7. T
8. F sexually

Word Relationships
A.
1. related to early atmosphere; oxygen did not exist in the early atmosphere
2. compounds in living things; organelle is a cell component
3. related to nucleic acids; amino acid is related to protein
4. related to first cells; aerobic cells came later
5. related to prokaryotes; eukaryote is sexual, multicelled

B.
6. spontaneous generation
7. Microfossils
8. anaerobic

CONCEPT MASTERY

1. They were anaerobic since there was no free oxygen in the atmosphere. They were heterotrophs because they used the organic soup for nutrients. Fossil evidence indicates they were prokaryotes.
2. Redi showed that maggots do not arise spontaneously on decaying meat by exposing the meat to the atmosphere and preventing flies from reaching the meat and laying their eggs on it. Needham tried to prove spontaneous generation by boiling a nutrient broth and showing that organisms arose from the broth. Spallanzani disproved Needham's hypothesis by showing that Needham had not boiled the broth thoroughly enough to kill all the microorganisms in it. Pasteur showed that spontaneous generation would not occur in a nutrient broth even if the broth was exposed to the atmosphere as long as organisms in the air could not enter the broth.
3. Sexual reproduction results in more variability since the offspring have the opportunity to inherit genes from two parents.
4. By tipping the flasks, Pasteur had allowed airborne organisms to enter the broth.
5. The atmosphere was likely formed because of volcanic eruptions and compounds carried by meteorites. Oceans formed when Earth's surface cooled enough to allow liquid water to form on the surface.

CHAPTER REVIEW

CONTENT REVIEW

Multiple Choice

Choose the letter of the answer that best completes each statement.

1. The hypothesis that mice can arise from spoiled grain is called
   a. evolution.
   b. microfossil.
   c. spontaneous generation.
   d. metabolism.
2. One scientist who believed in spontaneous generation was
   a. Redi.   c. Pasteur.
   b. Needham.   d. Spallanzani.
3. Earth's early atmosphere did not contain free
   a. nitrogen.   c. carbon dioxide.
   b. oxygen.   d. hydrogen cyanide.
4. Microfossils indicate that the first living cells were not
   a. prokaryotes.   c. eukaryotes.
   b. heterotrophic. d. anaerobes.
5. Modern photosynthetic organisms have replaced H_2S with
   a. HCN.   c. H_2O.
   b. CO_2.   d. O_2.
6. In the atmosphere, oxygen forms a layer of O_3, or ozone, that protects organisms from
   a. sunlight.
   b. infrared radiation.
   c. ultraviolet radiation.
   d. hydrogen cyanide.
7. Cells with membrane-bound organelles are called
   a. prokaryotes.   c. chloroplasts.
   b. mitochondria.   d. eukaryotes.
8. Sexual reproduction can speed up evolution because it provides more
   a. chromosomes.   c. identical cells.
   b. genetic variation. d. organelles.

True or False

Determine whether each statement is true or false. If it is true, write "true." If it is false, change the underlined word or words to make the statement true.

1. The hypothesis that <u>nonlife</u> arises from <u>life</u> is called spontaneous generation.
2. Redi showed that the <u>flies</u> that developed on raw meat did not arise spontaneously.
3. Earth formed around <u>4.6 billion</u> years ago.
4. In the <u>presence</u> of oxygen, amino acids spontaneously link to form short chains.
5. The first true cells were <u>prokaryotes</u>.
6. The first <u>heterotrophs</u> were similar to modern-day stromatolites.
7. The ozone layer protects living things from <u>ultraviolet radiation</u> from the sun.
8. Genetic variation increases when organisms reproduce <u>asexually</u>.

Word Relationships

A. *In each of the following sets of terms, three of the terms are related. One term does not belong. Determine the characteristic common to the three terms and then identify the term that does not belong.*

1. early atmosphere, hydrogen sulfide, oxygen, nitrogen
2. organelle, amino acid, lipid, carbohydrate
3. RNA, DNA, amino acid, nucleic acid
4. aerobic, first true cells, heterotrophic, anaerobic
5. eukaryote, asexual, prokaryote, single-celled

B. *Replace the underlined definition with the correct vocabulary word.*

6. Pasteur helped disprove the life arises from nonlife hypothesis.
7. Microscopic fossils provide outlines of ancient cells in rocks.
8. The first true cells were organisms that can live without oxygen.

CONCEPT MASTERY

Use your understanding of the concepts developed in the chapter to answer each of the following in a brief paragraph.

1. Explain why scientists believe the first true cells were anaerobic heterotrophic prokaryotes.
2. Discuss the experiments of Redi, Needham, Spallanzani, and Pasteur as they relate to spontaneous generation.
3. Which is more likely to result in increased variety among organisms, sexual reproduction or asexual reproduction? Why?
4. In one early experiment, Pasteur used flasks that had curved necks. He tipped some of the flasks so that the nutrient broth ran into the neck and then back into the body. Pasteur later observed microorganisms in these flasks. Explain this observation.
5. Discuss how scientists believe the Earth's early atmosphere and oceans formed.
6. Describe the symbiosis theory of eukaryotic development.

CRITICAL AND CREATIVE THINKING

Discuss each of the following in a brief paragraph.

1. **Sequencing events** Draw a time line that begins with the formation of the Earth and ends with the development of multicellular life. Make sure every significant event discussed in the chapter is included.
2. **Applying facts** Describe the ways in which the evolution of photosynthesis changed not only living things but the environment of Earth as well.
3. **Making predictions** Predict how modern life on Earth would have evolved if organisms did not begin using H_2O instead of H_2S in photosynthesis.
4. **Relating cause and effect** When people believed in spontaneous generation, a scientist developed this recipe for producing mice: Place a few wheat grains and a dirty shirt in an open pot; wait 3 weeks. Suggest a reason why this recipe may have worked. How could you prove that the mice were not due to spontaneous generation?
5. **Drawing conclusions** Although scientists have re-created some of the events that led to the formation of complex organic compounds, they do not believe that similar events could occur in the natural world today. Explain why not.
6. **Making inferences** Suppose autotrophic organisms had not evolved. What would life on Earth be like today?
7. **Using the writing process** You are asked to develop a television program for young children that explains the origin of life on Earth. Write a script for this show. You might like to videotape your presentation.
8. **Using the writing process** Did you ever wonder what it would have been like to be the first cell on Earth? Pretend you are that first cell. Keep a written diary of your first week on Earth.

---

3. Accept all logical predictions. Most predictions will be based on the fact that the atmosphere would not have oxygen and aerobic organisms would not have evolved.
4. Since mice would logically be attracted to the soiled shirt and wheat grains, it is quite likely that one would find mice in the open pot after a given period of time. By putting a mouse-proof screen over the pot, you could show that mice did not arise spontaneously but entered the pot in response to the food and nesting material inside it.
5. The events that led to the formation of life occurred in an oxygen-free atmosphere. Our atmosphere today contains free oxygen and, thus, the conditions for the formation of life no longer exist on Earth.
6. Without autotrophic organisms, oxygen would not have formed in the atmosphere. It is likely that the organic soup would have been used up by now and there might be no life on Earth. If there was, it would probably be similar to the first prokaryotic cells.
7. Students' scripts should reflect the content in the chapter and be both imaginative and appropriate for the suggested age level.
8. Accept all logical diary entries. Student responses should be reasonably detailed and reflect the material learned in Chapter 16.

---

6. This theory states that organelles such as mitochondria and chloroplasts were once free-living organisms that entered host cells and eventually became a part of the cellular structure of the host.

CRITICAL AND CREATIVE THINKING

1. Check students' time lines for accuracy.
2. The evolution of photosynthesis allowed an oxygen atmosphere to form since oxygen is a byproduct of photosynthesis. This not only changed the Earth's environment, it allowed the evolution of aerobic organisms and eventually led to the demise of anaerobic organisms in most places on Earth.

CHAPTER 17 Viruses and Monerans

Section	Laboratory Investigations and Activities
17–1 Viruses pages 355–360 THEMES: Scale and Structure, Patterns of Change	**Teacher Edition** DEMONSTRATION: "Curing" a Cold, p. 354D
17–2 Monerans—Prokaryotic Cells pages 360–372 THEMES: Scale and Structure, Unity and Diversity	**Student Edition** LABORATORY INVESTIGATION: Examining Bacteria, p. 376 **Teacher Edition** DEMONSTRATION: Observing Bacteria, p. 354D **Laboratory Manual** Identifying Bacteria, p. 213 **Teaching Resources** HANDS-ON ACTIVITY: Yuck! What Are Those Bacteria Doing in My Yogurt? p. 85 **Product Testing Activities** Testing Yogurts
17–3 Diseases Caused by Viruses and Monerans pages 372–375 THEME: Systems and Interactions	**Laboratory Manual** Controlling Bacterial Growth, p. 219
Chapter Review pages 377–379	

Teacher Resources

Books

Brock, T. D., D. W. Smith, and M. T. Madigan. *Biology of Microorganisms,* 6th ed. Prentice Hall, 1991.

Phillips, R. Stephen. *Malaria.* State Mutual Book, 1983.

Scott, Andrew. *Pirates of the Cell: The Story of Viruses from Molecule to Microbe.* Blackwell, 1985.

Singleton, Paul, and Diana Sainsbury. *Introduction to Bacteria.* Wiley, 1981.

Stanier, Roger Y., John L. Ingraham, Mark Wheelis, and Page R. Painter. *The Microbial World.* Prentice Hall, 1986.

CHAPTER PLANNING GUIDE

Other Activities	Media and Technology
Teaching Resources 　Activity: The Development of Virology, p. 3 **Study Guide** 　Section 17–1, p. 161	
Teaching Resources 　Activity: Identifying Unknown Bacteria, p. 7 **Biotechnology Workbook** 　From Laboratory to Supermarket, p. 13 **Study Guide** 　Section 17–2, p. 164	**Biology Transparencies** 　19: The Structure of a Prokaryote
Teaching Resources 　Vocabulary Review: Crossword, p. 1 　Activity: Examining Bacterial Contamination, p. 9 **Study Guide** 　Section 17–3, p. 166	
Teaching Resources 　Chapter Test A, p. 15 　Chapter Test B, p. 19	**Computer Test Bank** 　Chapter Test, p. 183

Audiovisuals

Viruses. Film or video. Lucerne Media.
Viruses, 2nd ed. Film or video. Coronet.
Simple Organisms: Bacteria, rev. Film or video. Coronet.
Introduction to Bacteria. Filmstrip with cassette. Ward.
A New Look at Bacteria. Film or video. Ward.
Bacteria: Invisible Friends and Foes. 3 filmstrips with cassettes.
Viruses: The Mysterious Enemy. 2 filmstrips with cassettes. Carolina Biological Supply Company.
Protists: Threshold of Life. Film or video. National Geographic.

CHAPTER 17 Viruses and Monerans

CHAPTER OVERVIEW

Viruses are parasitic noncellular particles that can invade living cells. When a virus attaches to the right kind of host cell, it injects its genetic material into the cell. Viruses may cause either a lytic infection or a lysogenic infection. Experts still disagree as to whether viruses are alive.

Monerans are bacteria, or unicellular prokaryotes (cells that lack a nucleus and membrane-enclosed organelles). Bacteria are divided into four phyla: Eubacteria, Cyanobacteria, Archaebacteria, and Prochlorobacteria. There are three basic shapes of bacteria: rod-shaped bacilli, spherical cocci, and spiral-shaped spirilla. Nutrition and respiration vary among bacteria. Bacteria interact with other living things in helpful or harmful ways. In the environment, bacteria are important because some are saprophytes and others fix nitrogen.

Both viruses and monerans cause a number of human diseases. Viruses are also known to cause cancer in animals, such as chickens. One possible approach to the treatment of viral diseases is the use of interferons. Antibiotics are used in the treatment of many bacterial diseases. Sterilization and food-processing methods help control the growth of bacteria.

17-1 VIRUSES

Section Focus 17-1

The purpose of this section is to introduce the structure and life cycles of viruses, explain why it is not clear whether viruses should be considered living things, and present a hypothesis about the origin of viruses.

Viruses consist of a protein coat, or capsid, that surrounds a core of nucleic acid (genetic material). Viruses may infect a host cell in one of two ways. In a lytic infection, the viral genetic material takes over the protein-making machinery of the cell, causing the cell to manufacture thousands of virus particles. Eventually, the infected cell lyses, releasing virus particles that can then infect other cells. In a lysogenic infection, viral DNA is incorporated in the host cell's DNA as a prophage. The prophage may remain part of the host's DNA for many generations. In some cases, the prophage may eventually become active. Viruses probably developed from living cells; they may have evolved from the genetic material of living cells.

Performance Objectives 17-1

1. Define virus.
2. Describe the structure of a virus.
3. Compare lysogenic and lytic viral infections.
4. Relate viruses to living cells.
5. Discuss the origin of viruses.

Science Terms 17-1

virus   p. 356
bacteriophage   p. 356
lytic infection   p. 358
lysogenic infection   p. 358
prophage   p. 358
parasite   p. 359

17-2 MONERANS— PROKARYOTIC CELLS

Section Focus 17-2

The purpose of this section is to introduce structure and function in bacteria, explain how bacteria are classified, describe how bacteria are identified, and identify the roles of bacteria in the environment.

Bacteria are unicellular prokaryotes that are enclosed within a cell wall. Bacteria, which belong to the kingdom Monera, are divided into four phyla. Phylum Eubacteria is the largest of the bacterial phyla. Phylum Cyanobacteria consists of photosynthetic organisms that were once called blue-green algae. Phylum Archaebacteria includes methanogens. Phylum Prochlorobacteria consists of photosynthetic organisms that greatly resemble chloroplasts. Bacteria are identified by examining their shape, movement, and cell wall.

Some bacteria are phototrophic or chemotrophic autotrophs; others are heterotrophs. Some are obligate aerobes; others are obligate anaerobes; still others are facultative anaerobes. Bacteria reproduce through binary fission. Under certain conditions, bacteria may exchange genetic material in the process of conjugation or they may form spores.

Bacteria are important for many reasons. They are used in food production and in other industries and are involved in a number of symbiotic relationships with other organisms. They help to recycle nutrients in the environment and fix nitrogen.

Performance Objectives 17-2

1. Define bacterium.
2. Describe how bacteria are classified.
3. Explain how bacteria are identified.
4. Discuss nutrition and respiration in bacteria.
5. Compare binary fission, conjugation, and spore formation.
6. Explain the place of bacteria in the world.

Science Terms 17-2

prokaryote   p. 360
bacteria   p. 361
methanogen   p. 362
bacillus   p. 363
coccus   p. 364
spirillum   p. 364
phototrophic autotroph   p. 365
chemotrophic autotroph   p. 365
chemotrophic heterotroph   p. 365
phototrophic heterotroph   p. 365
obligate aerobe   p. 367
obligate anaerobe   p. 367
toxin   p. 367
facultative anaerobe   p. 367
binary fission   p. 368
conjugation   p. 368
endospore   p. 368
symbiosis   p. 369
saprophyte   p. 370
nitrogen fixation   p. 372

CHAPTER PREVIEW

17-3 DISEASES CAUSED BY VIRUSES AND MONERANS

Section Focus 17-3

The purpose of this section is to list some of the diseases caused by viruses and monerans and to describe some of the ways in which such diseases are treated and prevented.

Viruses cause human diseases such as smallpox, polio, measles, AIDS, influenza, and the common cold. Viruses also cause cancer in animals. Certain viral diseases may be prevented with vaccines. The use of interferons, which make it difficult for viruses to infect new cells, is one possible approach in the treatment of diseases caused by viruses.

Human bacterial diseases include tuberculosis, tetanus, cholera, and bubonic plague. Some bacterial infections can be prevented with vaccines. Antibiotics can assist the immune system after the onset of a bacterial infection by destroying bacteria. Bacterial infections can be prevented by sterilization through heat or chemicals, refrigeration, thorough cooking of food, canning, and the use of chemicals. These treatments kill bacteria or slow their growth.

Performance Objectives 17-3

1. Define pathogen.
2. List some human viral diseases.
3. Relate cancer and interferon to viruses.
4. List some human bacterial diseases.
5. Describe how bacterial diseases are treated.
6. Explain how sterilization and food processing control the growth of harmful bacteria.

Science Terms 17-3

pathogen   p. 372
antibiotic   p. 374

DISCOVERY LEARNING

TEACHER DEMONSTRATIONS

Modeling

"Curing" a Cold

The following demonstration can be used as an introduction to Chapter 17.

On a piece of poster board, prepare a display showing a large number of advertisements and packages of cold remedies. Good sources are old magazines and your medicine cabinet. Call attention to the display.

- **What do these products have in common?** (They are all cold remedies.)
- **What is the purpose of these products?** (To "cure" a cold.)
- **Which one does the best job of "curing" your cold?** (Expect a wide variety of responses. Many responses will reflect customer loyalty and advertising.)
- **How does this product "cure" a cold?** (Lowers fever, relieves aches and pains, reduces congestion, stops cough, and so on.)
- **Is this a sign that you are cured?** (Most students will say yes. Some may suspect that this is a trick question—which it is!)

Point out that these medications merely provide relief from symptoms; they do not cure a cold. A cure would have to disable the virus that causes the cold. Have students note the various ingredients and relate them to their activity. Stress that these ingredients only treat symptoms. You may also want to point out that many of the ingredients are exactly the same in each remedy. Have students discuss advertising practices. Explain the difference between relief and cure. Wrap up the discussion by noting that cold remedies simply make us more comfortable as we wait for a cold to run its course. You may also want to point out that many of the symptoms of a cold are actually caused by the body's response to the infection.

Observing Bacteria

You may wish to perform this demonstration before students begin Section 17-2.

Tell students that you are preparing a microscope slide to examine fresh yogurt under the microscope. As you prepare the slide, involve students by asking questions about the process.

Add water to a tiny amount of plain yogurt to make a thin cloudy mixture.

- **Why do you think I'm adding water to the yogurt?** (Accept all reasonable answers. The sample has to be thinned in order for light to pass through and for individual particles to be more visible.)

Place a drop of the yogurt mixture on a microscope slide. Stain the sample with a drop of methylene blue.

- **Why do you think I'm adding methylene blue?** (To stain something.)

Put on a coverslip.

- **Why is it important to put a coverslip on the slide?** (Accept all reasonable answers.)

Have students look at the slide you have prepared under the microscope. If your class is large, you may wish to have just one or two students observe the slide, describing what they see to the rest of the class.

- **Describe what you see.** (Blue ovals or cylinders.)

Tell students that the cylinders are *Lactobacillus* bacteria.

- **Why might there be bacteria in the yogurt?** (Accept all reasonable answers. Students might suggest that the yogurt is spoiled or that bacteria is used to make yogurt.)

Explain that *Lactobacillus* is used to make milk into yogurt. Point out that students who suggested that the yogurt might be spoiled are not completely off the mark. Note that bacteria come in three different shapes. Draw these shapes on the chalkboard.

- **If there are only three basic shapes for bacteria, how can we know what kind of bacteria is in the yogurt?** (Accept all reasonable answers.)

Tell students that they will learn about different kinds of bacteria and how to identify different types in this section.

CHAPTER 17
Viruses and Monerans

GUIDED ENQUIRY

Pose the following questions to students and have them record their responses. Point out to students that they will gain a better understanding of the key concepts if they read the chapter with these basic questions in mind. Upon completion of the chapter, pose the questions again. Ask students to compare their initial responses with those they have developed after reading the chapter.

- What are viruses? Are viruses living organisms?
- How do viruses infect living cells?
- Why are bacteria classified as monerans?
- What is a prokaryote?
- What are the three basic shapes of bacteria?
- What is an autotroph? What is a heterotroph?
- How do bacteria reproduce?
- What are some common viral and bacterial diseases?

INTRODUCING CHAPTER 17

Using the Visuals
Begin your discussion of the chapter by having students study the chapter-opener photographs. Tell students not to read the caption for these photographs.

- **What do you see?** (Answers will vary but may include: a computer image of a cell or chemical compound; a mosaic; a model of a microscopic organism. Because of the chapter title, many students will infer that the photographs are of viruses or monerans.)

Now have students read the caption. Point out that although we think of viruses as causing disease only in humans, viruses can in fact cause disease in almost every organism on Earth, including plants and animals.

- **Why are these computer-generated images of viruses rather than actual color photographs?** (Viruses are extremely small and cannot be easily photographed through any microscope, even electron microscopes.)

- **Of what use are computer models?** (Scientists use computer models to study possible viral structures, the way viruses may bind to host cells, and for many other purposes.)

- **Are viruses alive?** (Accept all answers at this point. Make it clear to students that this topic is still hotly debated and open to interpretation.)

Point out that whether alive or not, viruses have an effect on living things. Tell students that the focus of this chapter will be on viruses and monerans. Have stu-

CHAPTER 17
Viruses and Monerans

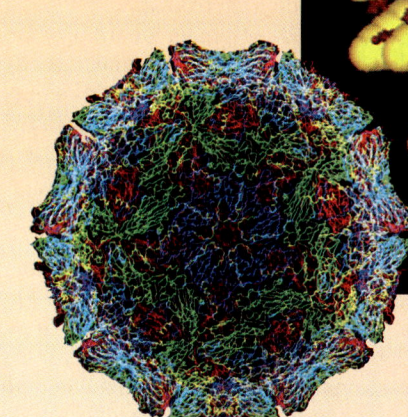

These photographs show the computer-generated images of the semiliki forest virus (right) and the foot-and-mouth virus (left).

We are accustomed to thinking of ourselves as complex organisms—and that view is correct. By comparison, there must be simpler organisms—and that view too is correct. According to this standard, the simplest organisms should be the single-celled bacteria. Even simpler than bacteria must be viruses, which cannot live outside a living cell.

As useful as this reasoning is, it has two flaws. First, no species should be thought of as simple. Even the tiniest bacterium has a complexity of organization that is almost beyond description. Second, every species has been shaped by millions of years of natural selection. The result is a world filled with spectacular examples of life's ability to adapt to and master challenges. What are viruses and bacteria? How are they classified? How do they adapt to their surroundings? Read on to find out.

354

GUIDE FOR READING

After you read the following sections, you will be able to

17–1 Viruses
- Describe the structure of viruses.
- Discuss two methods by which viruses infect living cells.

17–2 Monerans—Prokaryotic Cells
- Describe the four major phyla of monerans.
- Compare autotrophic and heterotrophic monerans.
- Describe growth and reproduction in monerans.
- Identify the role of monerans in the environment.

17–3 Diseases Caused by Viruses and Monerans
- List some diseases caused by viruses and bacteria.
- Describe two methods of controlling the growth of bacteria.

17–1 Viruses

Guide For Reading
- What is a virus?
- How do viral life cycles differ?
- What is the relationship between viruses and their hosts?

Imagine for a moment that you have been presented with a great challenge. A disease has begun to destroy certain crops. The leaves of diseased plants are covered with large bleached spots that form a pattern farmers call a mosaic. As the disease progresses, the leaves turn yellow, wither, and fall off, killing the plants.

To determine what is causing the disease, you take some leaves from a diseased plant and crush them until a juice is extracted. Then you place a few drops of the juice on the surfaces of leaves of healthy plants. A few days later, you discover that wherever you have placed the juice on the healthy leaves, a mosaic pattern has appeared. You reason that the cause of the disease must be in the juice of the infected plant.

You then search for a microorganism that might be responsible for the disease, but none can be found in the juice. In fact, even when the juice is passed through a filter with pores so fine that not even cells can pass through, the juice still causes the disease. When you look at a small amount of the filtered juice under the light microscope, you see no evidence of cells. The juice, which is capable of transferring the disease from one plant to another, must contain disease-causing particles so

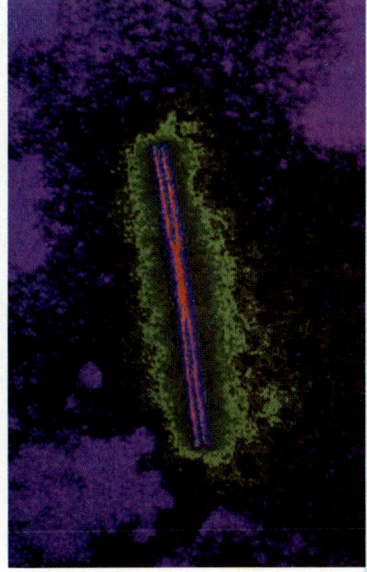

Figure 17–1 Tobacco mosaic virus (TMV) causes the leaves of tobacco plants to develop a pattern of spots called a mosaic (left). A TMV particle, magnified approximately 41,800 times, appears as a thin purple tube in the color-enhanced micrograph (right).

Journal Activity

YOU AND YOUR WORLD

Chances are, you have been sick at least once in your life thanks to the effects of viruses or bacteria! What was the most memorable time you were sick, and why? What were your symptoms? How was your illness treated? Describe your recollections in your journal.

BACKGROUND INFORMATION
VIROIDS AND PRIONS

Viruses are extremely small. But there are infectious particles that are even smaller than viruses.

Viroids are circular single-stranded pieces of RNA about 250 to 400 nucleotides in length. Unlike viruses, viroids lack a protein coat. Viroids cause a number of plant diseases, including potato spindle tuber disease. To date, all known viroids affect plants.

A number of animal diseases once thought to be caused by viruses or viroids are now attributed to prions. Prions, which were first described in the mid-1980s by Stanley Prusiner of the University of California at San Francisco, are infectious particles that consist only of protein. Prions are found in extracts from the brains of diseased animals, and they are thought to cause scrapie in sheep and goats, a scrapielike disorder in cattle, and Creutzfeldt-Jakob disease in humans. They may also be involved in other human diseases such as multiple sclerosis, Alzheimer's disease, Lou Gehrig's disease, and Parkinson's disease. Prions are thought to reproduce by binding to the host's DNA and activating a normally dormant gene that encodes for the prion.

17–1 (continued)

Content Development
Ask students to relate their most unpleasant bouts with viruses. Point out that the word *virus* comes from a Latin word meaning poison.
• **Why is the origin of the word *virus* appropriate?** (The effects of a viral infection may resemble those of poison.)

Skills Development
Guided Practice
Skill: Making diagrams
Using the Visuals Have students make a three-dimensional model of a bacteriophage based on Figure 17–2 in the textbook, using materials such as pipe cleaners, florist's wire, drinking straws, and construction paper.

Focus/Motivation
Give each student an unshelled sunflower seed. (*Note:* Other easily shelled seeds such as peanuts, pumpkin seeds, and pistachio nuts will also work well. Modify questions according to the model used.)

• **In what ways is the structure of a virus like the structure of a sunflower seed?** (Accept all logical answers. Students should recognize that both sunflower seeds and viruses consist of a protective outer shell that encases vital contents.)

Point out that a sunflower seed can serve as a model of a virus.

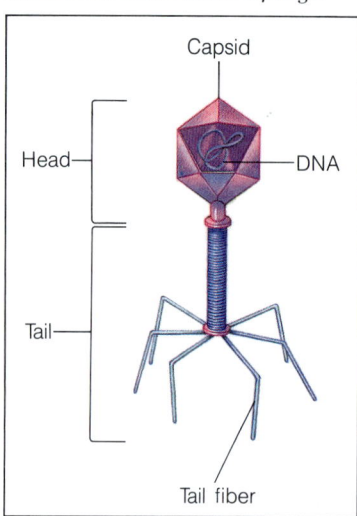

Figure 17–2 A bacteriophage is a virus that infects bacteria. Compare the structures shown in the diagram of the bacteriophage to those in an actual bacteriophage.

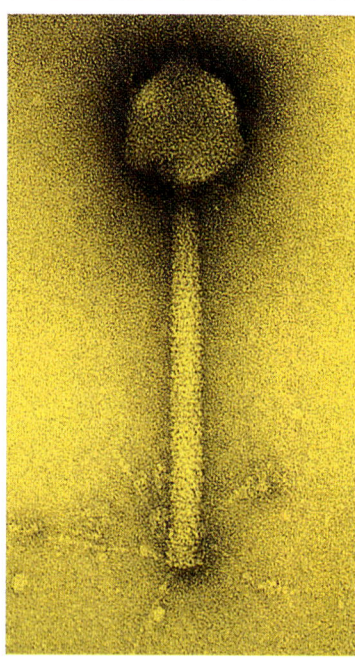

small that they are not visible under the light microscope. Although you cannot see the disease-causing particles, you decide to give them the name viruses, from the Latin word meaning poison.

With a few exceptions, most of these events actually took place. About 100 years ago in what is now Ukraine, an epidemic of tobacco mosaic disease occurred that seriously threatened the tobacco crop. The disease-causing nature of the juice from infected tobacco leaves was discovered by the Russian biologist Dimitri Iwanowski. A few years later, the Dutch scientist Martinus Beijerinck determined that tiny particles in the juice caused the disease. He named these particles **viruses.**

What Is a Virus?

You have just read how scientists hypothesized the existence of viruses, which they thought were cells even smaller than one-celled bacteria. This idea persisted until 1935 when the nature of a virus was discovered by the American biochemist Wendell Stanley. He had set out to chemically isolate the particle responsible for the tobacco mosaic disease. Stanley identified the particle as the tobacco mosaic virus (TMV).

Since Stanley's discovery, many viruses have been identified, largely through the use of the electron microscope, which was invented in the 1930s. We now appreciate the fact that viruses have distinct structures that are complex and fascinating. **A virus is a noncellular particle made up of genetic material and protein that can invade living cells.**

STRUCTURE OF A VIRUS A typical virus is composed of a core of nucleic acid surrounded by a protein coat called a capsid. The capsid protects the nucleic acid core. Depending on the virus, the nucleic acid core is either DNA or RNA but never both. The core may contain several genes to several hundred genes.

A more complex structure occurs in certain viruses known as **bacteriophages**. You may recall from Chapter 7 that bacteriophages are viruses that invade bacteria. A bacteriophage has a head region, composed of a capsid (protein coat), a nucleic acid core, and a tail. Bacteriophages are interesting and relatively easy to study because their hosts (bacteria) multiply quickly.

One well-studied bacteriophage, known as T4, has a core of DNA contained within a protein coat. A number of other proteins (about 30 in all) form the other parts of the virus, including the tail fibers. The tail fibers are the structures by which the virus attaches itself to a bacterium.

Viruses come in a variety of shapes. Some, such as the tobacco mosaic virus, are rod-shaped. Others, such as the bacteriophages, are tadpole-shaped. Still others are many-sided, helical, or cubelike. Figure 17–3 shows some of these shapes.

Viruses vary in size from approximately 20 to 400 nanometers. A nanometer is one billionth of a meter. The tobacco mosaic virus is about 300 nanometers long, whereas the virus that causes polio is about 20 nanometers in diameter.

SPECIFICITY OF A VIRUS Usually, specific viruses will infect specific organisms. For example, a plant virus cannot infect an animal. There are some viruses that will infect only humans. Others, such as the virus that causes rabies, infect all mammals and some birds. Still others infect only coldblooded animals (animals with body temperatures that change with the surrounding air). There are even some viruses that will infect species of animals that are closely related. For example, viruses that infect mice may infect rats. So you can see that viruses are capable of infecting virtually every kind of organism, including mammals, birds, insects, and plants.

Life Cycle of a Lytic Virus

In order to reproduce, viruses must invade, or infect, a living host cell. However, not all viruses invade living cells in exactly the same way. When T4 bacteriophages invade living cells, they cause the cells to lyse, or burst. Thus T4 viruses are known as lytic (LIHT-ihk) viruses.

INFECTION A virus is activated by chance contact with the right kind of host cell. In the case of T4, molecules on its tail fibers attach to the surface of a bacterium. The virus then injects its DNA into the cell. In most cases, the complete virus particle itself never enters the cell.

GROWTH Soon after entering the host cell, the DNA of the virus goes into action. In most cases, the host cell cannot tell the difference between its own DNA and the DNA of the virus. Consequently, the very same enzyme RNA polymerase that makes messenger RNA from the cell's own DNA begins to make messenger RNA from the genes of the virus. This viral messenger RNA now acts like a molecular wrecking crew, shutting down and taking over the infected host cell. Some of these viral genes turn off the synthesis of molecules that are important to the infected cell. One viral gene actually produces an enzyme that destroys the host cell's own DNA but does not harm the viral DNA!

REPLICATION As the virus takes over, it uses the materials of the host cell to make thousands of copies of its own protein coat and DNA. Soon the host cell becomes filled with hundreds of viral DNA molecules. When *Escherichia coli*, or *E. coli*, the bacterium found in the human intestine, is infected by a T4 bacteriophage, this sequence of infection, growth, and replication can happen in as brief a time as 25 minutes!

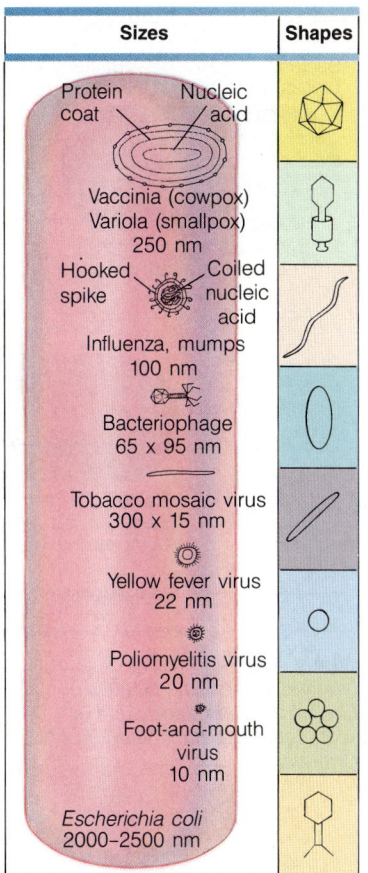

Figure 17–3 Viruses come in a variety of sizes and shapes. Notice the size of the bacterium E. coli as compared to the sizes of the viruses.

ANNOTATION KEY

❶ Molecules on the tail fibers attach to the surface of the bacterium.
(Relating facts)

BACKGROUND INFORMATION
HOW VIRUSES ENTER A CELL

Although bacteriophages typically inject their DNA into the host cell, not all viruses invade a host cell in this manner. Many animal viruses, such as the semiliki forest virus, enter the host through endocytosis—the binding of the virus to the cell membrane, inducing the cell to take in the virus particle. Once inside the host cell, the virus sheds its protein coat and either undergoes replication or becomes part of the host's DNA. Interestingly, a few animal viruses, such as those responsible for rabies, AIDS, and influenza, leave the host cell through budding, which can be thought of as the opposite of endocytosis.

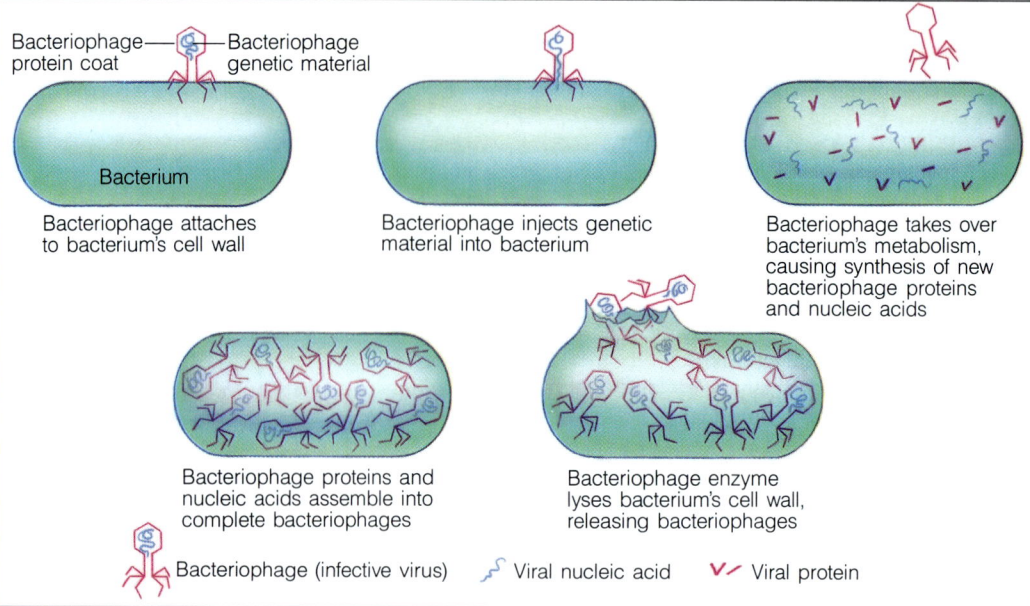

Figure 17–4 In the life cycle of a lytic virus, the virus invades a bacterium, reproduces, and is scattered when the bacterium lyses, or breaks.

Figure 17–5 This electron micrograph shows bacteriophages attacking the bacterium E. coli. How do viruses attach themselves to the bacterium? ❶

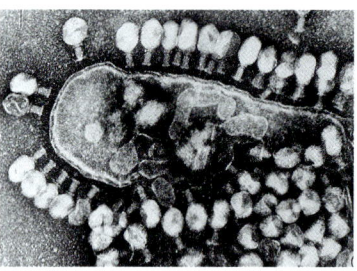

During the final stage of reproduction, the DNA molecules serve as the starting points around which new virus particles are assembled. Before long, the infected cell lyses (bursts) and releases hundreds of virus particles that may now infect other cells. Because the host cell is lysed and destroyed, this process is called a **lytic infection**. Lytic infections are one way in which viruses can infect host cells.

The life cycle of a lytic virus such as T4 consists of repeated acts of infection, growth, and cell lysis. We may imagine the virus as a desperado moving into a town in the Old West. First, the desperado eliminates the town's existing authority (host cell DNA). Then the desperado demands to be outfitted with new weapons, horses, and riding equipment by terrorizing the local merchants and businesspeople (using the machinery of the host cell to make proteins). Finally, the desperado recruits more outlaws and forms a gang that leaves the town and attacks new communities (the host cell bursts, releasing hundreds of virus particles).

Lysogenic Infection

Another way in which a virus infects a cell is known as a **lysogenic** (ligh-soh-JEHN-ihk) **infection**. In a lysogenic infection, the virus does not reproduce and lyse its host cell—at least not right away! Instead, the DNA of the virus enters the cell and is inserted into the DNA of the host cell. Once inserted into the host cell's DNA, the viral DNA is known as a **prophage**.

17–1 (continued)

Enrichment
Tell students that some people experience recurring viral infections, such as cold sores (fever blisters), whenever they are under stress or spend too much time in the sun.
• **How can you explain this phenomenon?** (Factors such as ultraviolet radiation and body chemicals can cause a provirus [prophage] to enter an active, lytic state.)
Note that some researchers suspect that certain forms of cancer in humans may be related to lysogenic viruses.
• **If a certain type of lung cancer is caused by a lysogenic virus, how might smoking cause this kind of cancer?** (Smoking introduces chemicals that might activate the virus.)
• **How do you know whether you have a lysogenic virus in your cells?** (You cannot tell unless the virus becomes active—in which case it is no longer a lysogenic virus.)

Content Development
Point out that retroviruses are just one kind of RNA virus.
• **How might other RNA viruses**

The prophage may remain part of the DNA of the host cell for many generations. An example of a lysogenic virus is the bacteriophage *lambda*, which infects *E. coli*.

PROPHAGE ACTIVITY The presence of the prophage can block the entry of other viruses into the cell and may even add useful DNA to the host cell's DNA. For example, a lambda virus can insert the DNA necessary for the synthesis of important amino acids into the DNA of *E. coli*. As long as the lambda virus remains in the prophage state, *E. coli* can use the viral genes to make these amino acids.

A virus may not stay in the prophage form indefinitely. Eventually, the DNA of the prophage will become active, remove itself from the DNA of the host cell, and direct the synthesis of new virus particles. A series of genes in the prophage itself maintains the lysogenic state. Factors such as sudden changes in temperature and availability of nutrients can turn on these genes and activate the virus.

RETROVIRUSES One important class of viruses are the **retroviruses**. Retroviruses contain RNA as their genetic information. When retroviruses infect a cell, they produce a DNA copy of their RNA genes. This DNA, much like a prophage, is inserted into the DNA of the host cell. Retroviruses received their name from the fact that their genetic information is copied backward—that is, from RNA to DNA instead of from DNA to RNA. The prefix *retro-* means backward. Retroviruses are responsible for some types of cancer in animals and humans. One type of retrovirus produces a disease called AIDS.

Viruses and Living Cells

As you have just learned, viruses must infect living cells in order to carry out their functions of growth and reproduction. They also depend upon their hosts for respiration, nutrition, and all of the other functions that occur in living things. Thus viruses are **parasites**. A parasite is an organism that depends entirely upon another living organism for its existence in such a way that it harms that organism.

Are viruses alive? If we require that living things be made up of cells and be able to live independently, then viruses are not alive. However, when they are able to infect living cells, viruses can grow, reproduce, regulate gene expression, and even evolve. Viruses have so many of the characteristics of living things that it seems only fair to consider them as part of the system of life on Earth.

Because it is possible to study the genes that viruses bring into cells when they infect them, viruses have been extremely valuable in genetic research. And, as we saw in Chapter 12, some viruses are now being used in gene therapy. It is possible that modified viruses may one day be routine medical tools.

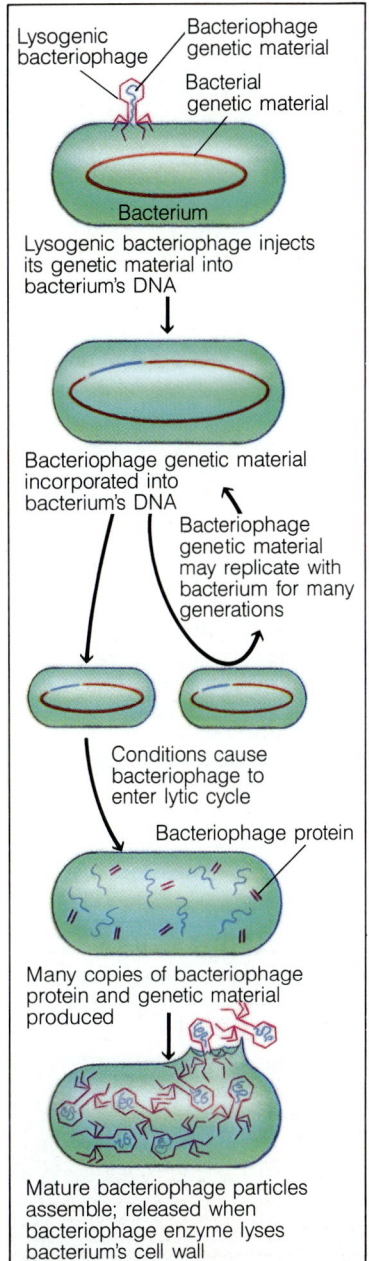

Figure 17–6 *In a lysogenic infection, the DNA of the bacteriophage enters the host cell and is inserted into its DNA.*

BACKGROUND INFORMATION
PROPHAGES AND PROVIRUSES

The term *prophage* is used only for bacteriophages. When other viruses, especially retroviruses, infect a cell, the portion of viral DNA that is inserted into the host DNA is known as a *provirus*. This term was not introduced in the text so that the student would not be burdened with too many terms. One of the surprising discoveries in recent studies on viruses is that proviral genes are sometimes nearly identical copies of genes that are normally found in cells. This lends credence to the idea that viruses are not "outside invaders" but pieces of cellular DNA that have been made part of an infectious particle.

FACTS AND FIGURES

The complete cycle of a virus infection from DNA injection to the lysis of the host cell takes about 20 minutes.

---

work? (Accept all logical answers. If students have trouble suggesting mechanisms, review the process of protein synthesis, emphasizing the roles of nucleic acids. Some RNA viruses, such as tobacco mosaic virus and rhinoviruses—which cause the common cold—act directly, as mRNA. Others, such as those that cause rabies and influenza, contain RNA that is complementary to the mRNA that directs the formation of the virus. The viral RNA, like DNA, acts as a template for mRNA.)

Content Development
Viruses are interesting to geneticists and microbiologists because they can be used in genetic engineering.

Review lysogenic infections with your students.
• **How might lysogenic viruses be used in genetic engineering?** (As students may recall from Chapter 12, viruses can be used to transfer genes into cells. Genetically altered viruses are currently being tested as vaccines against AIDS, tuberculosis, and certain forms of skin cancer.)

BACKGROUND INFORMATION
CLASSIFICATION OF MONERANS

There is widespread disagreement about how to classify monerans. Many of these arguments center on new evidence about nucleic acid sequences, which suggest that some of the archaebacteria are as different from bacteria as bacteria are different from eukaryotes. Many scientists now prefer to call all members of this kingdom "bacteria," and that is what we have done here. Many books have given one of two different names to the phylum Eubacteria: Schizomycetes ("fission fungi") or Schizophyta ("fission plants"); however, most scientists now reject these names because these organisms are neither plant nor fungus.

In order to make students aware of the existence of the archaebacteria and the prochlorobacteria, we have chosen a scheme that lists these groups as separate phyla alongside the traditional phyla of Eubacteria and Cyanobacteria (blue-green bacteria). We believe that this arrangement is the best of many possible schemes in which to present the diversity of this group of organisms to students, although it is important to emphasize that it is not the only possible scheme.

TEACHING SUPPORT

Study Guide
• Section 17–1, p. 161

Origin of Viruses

Although viruses are smaller and simpler than the smallest cells, they could not have been much like the first living things. Viruses are completely dependent upon living cells for growth and reproduction, and they cannot live outside their host cells. Thus it seems more likely that viruses developed after living cells. In fact, the first viruses may have evolved from the genetic material of living cells and have continued to evolve, along with the cells they infect, over billions of years.

17–1 SECTION REVIEW

1. What is a virus?
2. List and describe the parts of a bacteriophage.
3. Describe two methods of viral infection.
4. **Critical Thinking—Applying Concepts** How can a virus be helpful to its host?

Guide For Reading
■ How are monerans classified?
■ How do monerans obtain energy?
■ How do monerans grow and reproduce?
■ How do monerans affect other living things?

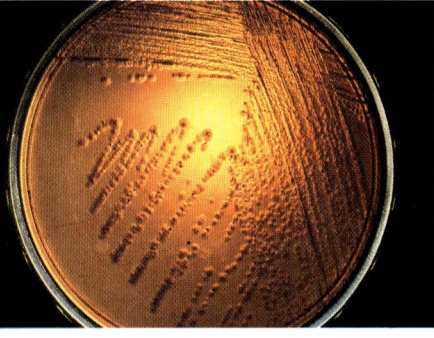

Figure 17–7 With a nutrient-rich culture medium on which to grow, these bacteria have produced thousands of colonies.

17–2 Monerans—Prokaryotic Cells

Imagine living all your life as the only family on your street. Then, on a morning like any other, you open the front door and there are houses all around you, cars and bicycles on the street, neighbors tending their gardens, children walking to school. Where did they come from? What if the answer turned out to be that they were always there—you just couldn't see them? In fact, they lived on your street for years and years before your house was even built. How would your view of the world change? What would it be like to go, almost overnight, from being the only family on the block to just one family in a crowded community? A bit of a shock?

Because of Robert Hooke and Anton van Leeuwenhoek, the human species had just such a shock. The invention of the light microscope opened our eyes to what the world around us is really like. And it opened our eyes almost overnight. Suddenly we saw that the block is very crowded!

Microscopic life covers nearly every square centimeter of planet Earth. What form does that microscopic life take? As you learned in Chapter 5, there are cells of every size and shape imaginable, even in a drop of pond water. The smallest and most common of these cells are the **prokaryotes**. Prokaryotes are cells that do not have a nucleus.

Where do we find prokaryotes? Everywhere! Prokaryotes exist in almost every place on Earth. They grow in numbers so great that they form colonies you can see with the unaided eye.

360

17–1 (continued)

SECTION REVIEW 17–1

1. A virus is a noncellular particle made up of genetic material and protein that can invade living cells.

2. A bacteriophage has a head region, composed of a capsid (protein coat), a nucleic acid core, and a tail.

3. In a lytic infection, the virus takes over the cell and its materials and uses them to make copies of the virus's protein coat and DNA. Before long, the infected cell lyses and releases hundreds of viruses. In a lysogenic infection, the DNA of a virus is inserted into the DNA of the host cell. The viral DNA is now known as a prophage, and it may remain part of the host cell's DNA for many generations. During this time, the viral DNA remains harmless. However, it may eventually replicate, producing new viruses that lyse and destroy the cell.

4. Viruses have the ability to transfer genetic information

Classification of Monerans

All prokaryotes are placed in the kingdom Monera. The monerans are the first large group of organisms that we shall consider as we examine each of the five kingdoms of living things. In this textbook we have divided the kingdom Monera into four phyla. These phyla are Eubacteria (yoo-bak-TEER-ee-uh), Cyanobacteria (sigh-uh-noh-bak-TEER-ee-uh), Archaebacteria (ahr-kee-bak-TEER-ee-uh), and Prochlorobacteria (proh-klor-oh-bak-TEER-ee-uh). Although there are important differences among these four groups of organisms, each group shares enough similarities with the others to allow them to be called **bacteria**, or one-celled prokaryotes.

Bacteria range in size from 1 to 10 micrometers (one micrometer is equal to one thousandth of a millimeter). Bacteria are much smaller than eukaryotic cells, or cells with a nucleus, which generally range from 10 to 100 micrometers in diameter. The reason for the difference in size is that bacteria and other monerans do not contain the complex range of membrane-enclosed organelles that are found in most eukaryotic cells.

EUBACTERIA The phylum Eubacteria ("true" bacteria) is the largest of all the moneran phyla. Members of this phylum have always been referred to as bacteria. Eubacteria are generally surrounded by a cell wall composed of complex carbohydrates. The cell wall protects the bacterium from injury. Within the cell wall is a cell membrane that surrounds the cytoplasm. Some eubacteria are surrounded by two cell membranes. In some organisms, long whiplike flagella protrude from the membrane through the cell wall. Flagella are used for movement.

Within the phylum Eubacteria is a wide variety of organisms that have many different lifestyles. Some species live in

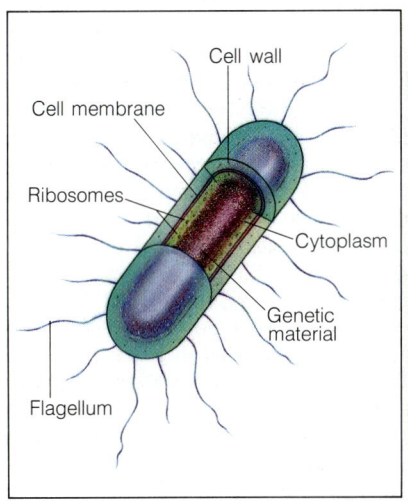 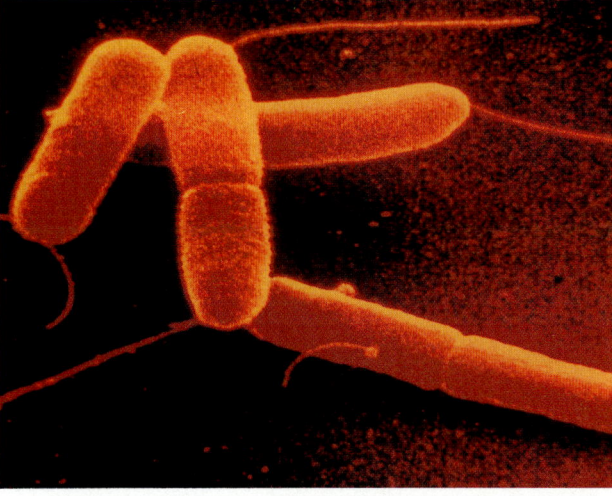

Figure 17–8 A bacterium such as E. coli (right) has the basic structure typical of most bacteria: cell wall, cell membrane, region of genetic material, and cytoplasm. Note the flagella projecting from the cell surface.

HISTORICAL NOTE
PHOTOSYNTHETIC BACTERIA

The photosynthetic members of the phylum Eubacteria are often ignored, possibly because monerans were for a long time divided into two groups: blue-green algae (photosynthetic) and bacteria (nonphotosynthetic). However, photosynthetic eubacteria have had an impact on science—a 1988 Nobel prize was awarded for the crystallization of a reaction center from a photosynthetic eubacterium.

ESL STRATEGY

Because the pronunciation guides in dictionaries and textbooks are almost impossible for ESL students to use, have students devise their own guides. Most guides will be useful only to the individuals who devised them. However, a guide written in Korean characters can generally be used by all speakers of Korean, as that language's alphabet is phonetic.

Both native speakers of English and ESL students may be amused by the observation that the word *fish* may be respelled *ghoti*—*gh* as in cough, *o* as in women, and *ti* as in action!

TEACHING STRATEGY 17–2

Focus/Motivation
Collect some common items that can be used to model the three basic shapes of bacteria. For example, cocci can be represented by marbles, beads, spheres of modeling clay, or malted milk balls. Bacilli may be represented by unsharpened pencils, pieces of chalk, or short dowels. Spirilla may be represented by springs or pipe cleaners that have been shaped into spirals by wrapping them around a pencil.

from one host cell to another. In this way, the virus can transfer helpful genes between cells. As the cells divide, they produce daughter cells that carry the altered genetic code.

Reinforcement/Reteaching
If students cannot answer the Section Review questions, review the material.

Closure
Suppose the resolution for a debate was "Resolved: that viruses should be banned." Assign half of the class to the affirmative side and the other half to the negative side. Have each student prepare a brief written or oral argument defending his or her position.

361

Figure 17–9 Bacteria can survive in many environments that support no other forms of life, such as in a near-boiling hot spring called Morning Glory Pool in Yellowstone National Park, Wyoming.

the soil. Others infect larger organisms and produce disease. Still others are photosynthetic. Photosynthetic bacteria are those bacteria capable of making their own food by using light energy. Later in this chapter we will examine some of the important roles that eubacteria play in the natural environment and some of the ways in which they affect us.

CYANOBACTERIA The bacteria that belong to the phylum Cyanobacteria are photosynthetic. Cyanobacteria are also known as blue-green bacteria. At one time, cyanobacteria were called blue-green algae, but today we use the word algae only for eukaryotes. In fact, only a few of the blue-green bacteria are a blue-green color. Those monerans that are blue-green in color contain a blue pigment called phycocyanin. They also contain chlorophyll *a*, which you will recall from Chapter 6 is green. The presence of these two pigments gives the name blue-green to the entire group of cyanobacteria. The presence of other pigments, however, may change the color of these monerans to yellow, brown, or even red.

Cyanobacteria contain membranes that carry out the light reactions of photosynthesis. These membranes contain the photosynthetic pigments and are quite different from and simpler than the chloroplasts (organelles that trap light energy and convert it to chemical energy) in plant cells.

Cyanobacteria are found throughout the world—in fresh and salt water and on land. A few species can survive in extremely hot water, such as that in hot springs. Others can survive in the Arctic, where they can even grow on snow. In fact, cyanobacteria are often the very first species to recolonize the site of a natural disaster, such as a volcanic eruption.

ARCHAEBACTERIA AND PROCHLOROBACTERIA Recent studies of monerans have led to the establishment of two new phyla that include organisms that differ from eubacteria and cyanobacteria. One phylum, Archaebacteria, includes organisms that live in extremely harsh environments. For example, one group of archaebacteria lives in oxygen-free environments such as thick mud and the digestive tracts of animals. These archaebacteria are called **methanogens** because they produce methane gas. Other archaebacteria live in extremely salty environments, such as the Great Salt Lake in Utah, or in extremely hot environments, such as hot springs where temperatures approach the boiling point of water.

The prochlorobacteria are a newly discovered group of photosynthetic organisms that contain chlorophyll *a* and *b* as their principal pigments. The presence of these pigments makes prochlorobacteria more similar to chloroplasts of green plants than to cyanobacteria. For this reason, prochlorobacteria are sometimes called Prochlorophyta (*-phyta* means plants) to emphasize this similarity. To date, only two species of prochlorobacteria have been discovered.

Identifying Monerans

Identifying living organisms can be a simple task. If we were given an unknown plant or animal, we would search through the photographs and diagrams in a reference book until we found one that resembled our unknown specimen. Such a method works for organisms that we can identify by appearance. But what about bacteria? How can they be identified?

CELL SHAPE One way in which bacteria can be identified is by their shape. **Bacteria have three basic shapes: rod, sphere, and spiral.** Rod-shaped bacteria are called **bacilli**

SCIENCE, TECHNOLOGY, AND SOCIETY — ISSUE

Are Prokaryotes Always Smaller?

Bacteria are smaller than other cells, right? After all, the study of bacteria is part of a field called microbiology, so bacteria must be very, very small. Indeed, most of them are. But there are exceptions. A few large bacteria are almost as big as a small eukaryotic cell—still pretty small. In 1993, however, Esther Angert, a student in Norman Pace's lab in Indiana, went fishing and found the whopper of all exceptions.

Angert and Pace found the bacterium in the digestive system of surgeonfish that live off the coast of Australia. Other researchers had thought that these large cells must be eukaryotic, since they are about the same length as the period at the end of this sentence. However, Angert's work showed conclusively that these enormous cells are, in fact, bacteria. They contain a number of cell structures and DNA sequences that show that they are closely related to other, much smaller, bacteria.

What do these gigantic cells do? And why are they so big? These are questions that Angert and Pace are trying to answer. For the time being, they have the satisfaction of knowing that their bacterium goes into the record books as being by far the largest prokaryote ever discovered. And that's no fish story!

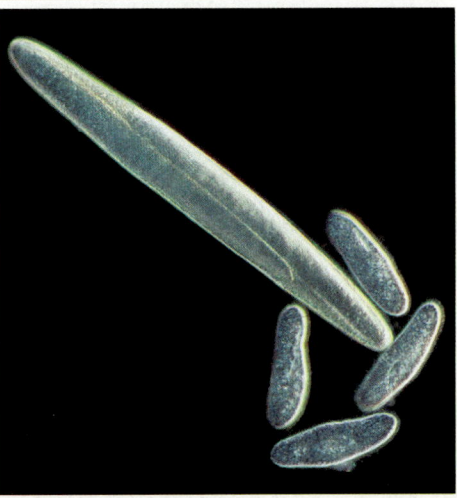

Paramecia are large eukaryotic cells, but the four paramecia in this micrograph are dwarfed by the giant bacterium.

SCIENCE, TECHNOLOGY, AND SOCIETY — ISSUE

Are Prokaryotes Always Smaller?

This chapter deals with the world of the very small—a world that became "real" with the invention of microscopes capable of making the invisible visible. Several years ago, you could have said that all monerans are microscopic, too small to be viewed with unaided eyes. Today this statement would be incorrect, made false by the work of Esther Angert and Norman Pace. A constant search to understand the natural world is one of the great beauties of science. This search may prove, as it did in this case, that today's truths are tomorrow's misconceptions.

- **How large was the organism Angert and Pace found?** (It is about as large as the period at the end of this sentence.)
- **Where did they find this organism?** (They found it in the digestive system of a species of surgeon fish that lives off the coast of Australia.)
- **How did the researchers decide that the organism they found was a moneran?** (They analyzed its cell structures and its DNA sequences.)

- **How does the prefix describe the bacteria in this phylum?** (Accept all logical answers.)

You may wish to help students develop mnemonics for these terms. For example, "prochlorobacteria resemble the endosymbionts that came before chloroplasts," "cyanide makes you turn as blue as cyanobacteria," and so on.

Skills Development
Guided Practice
Skill: Summarizing information
Have students prepare a table summarizing the information on the classification of bacteria.

Content Development
Show each of the models you prepared for the Focus/Motivation on page 361, one at a time. Have students identify the type of bacteria each model represents. On the chalkboard, write

diplo- = pair
strepto- = chain
staphylo- = cluster

Have student volunteers assemble different types of bacterial colonies—streptococci, diplobacilli, and so on—using the models. Alternatively, have students draw different types of colonies and identify each one.

> **BACKGROUND INFORMATION**
> **HOW THE GRAM STAIN WORKS**
>
> The cell walls of Gram-negative cells are covered with a layer of lipopolysaccharide, a combination of carbohydrate and lipid. This layer enables the bacterium to exclude the stain. Gram-positive bacteria have no lipopolysaccharide layer and take up the stain quite well.
>
> In general, Gram-positive bacteria will also take in drugs prescribed by physicians to fight bacterial infections. For this reason, Gram-positive infections are easier for doctors to treat than Gram-negative infections—the ability of Gram-negative bacteria to exclude stain also makes it easier for them to exclude drugs.

> **FACTS AND FIGURES**
>
> Bacteria that obtain energy by oxidizing inorganic materials—such as sulfur, iron, or hydrogen—are sometimes called lithotrophs, which literally means rock-eaters.

> **TEACHING SUPPORT**
> **Laboratory Manual**
> • Identifying Bacteria, p. 213
>
> **Teaching Resources**
> • Activity: Identifying Unknown Bacteria, p. 7
> • Hands-On Activity: Yuck! What Are Those Bacteria Doing in My Yogurt? p. 85
>
> **MEDIA AND TECHNOLOGY**
> ▫ **Biology Transparencies**
> 19: The Structure of a Prokaryote

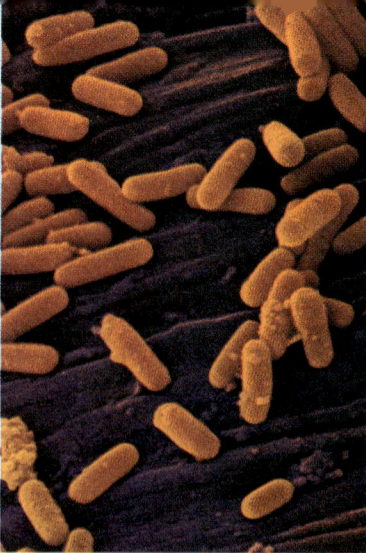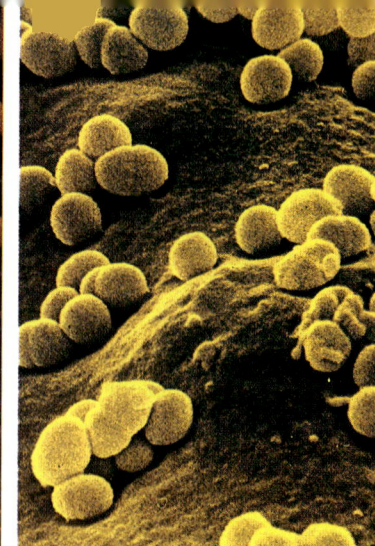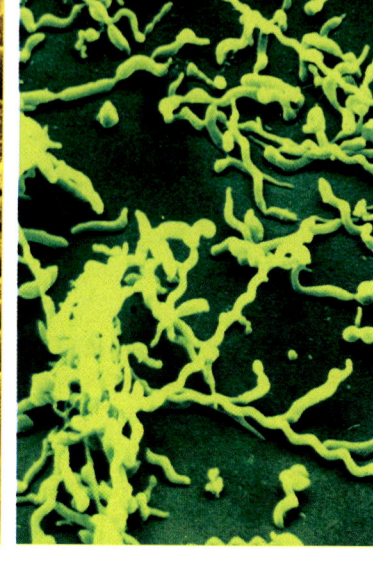

Figure 17–10 Bacteria have three basic shapes. Rod-shaped bacteria are called bacilli (left), spherical bacteria are called cocci (center), and spiral-shaped bacteria are called spirilla (right).

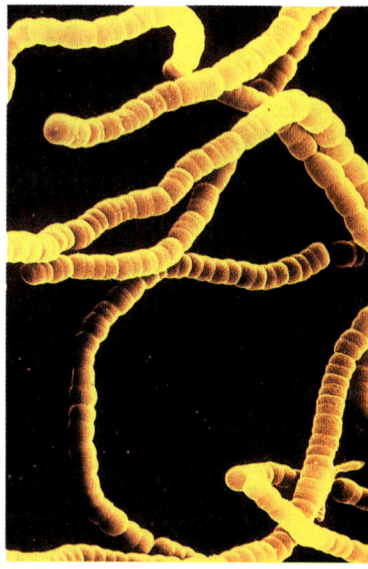

Figure 17–11 Some spherical bacteria like these streptococci form long chains.

(buh-SIHL-igh; singular: bacillus). Spherical bacteria are called **cocci** (KAHK-sigh; singular: coccus). And spiral-shaped bacteria are called **spirilla** (spigh-RIHL-uh; singular: spirillum). See Figure 17–10.

Individual bacterial cells can also arrange themselves in a number of different ways. For example, cocci sometimes grow in colonies containing two cells. Many cocci, including the disease-causing bacteria *Streptococcus* and *Pneumococcus*, may form long chains. A few others, such as *Staphylococcus*, form large clumps or clusters. These differences are very helpful in distinguishing one kind of bacteria from another.

Unfortunately, many bacteria look the same under the microscope. So we need to find another characteristic by which to distinguish one type from another. Fortunately, there are three other characteristics of bacteria that improve our ability to identify them: their cell walls, the kind of movement they are capable of, and how they obtain energy.

CELL WALL The chemical nature of bacterial cell walls can be studied by means of a method called Gram staining—which is named after its inventor, the Danish physician Hans Christian Gram. Gram's stain consists of two dyes—crystal violet (purple) and safranine (red).

When Gram added his stain to bacteria, he noticed that the bacteria took up either the purple dye or the red dye. The bacterial cells with only one thick layer of carbohydrate and protein molecules outside the cell membrane took up the crystal violet. They appeared purple under the light microscope. These bacteria are called Gram-positive bacteria. The bacterial cells with a second, outer layer of lipid and carbohydrate molecules took up the safranine. They appeared red under the microscope. These bacteria are called Gram-negative bacteria.

364

> **17–2 (continued)**
> **Enrichment**
> You may wish to explain the process of Gram staining. Basically, a Gram stain involves staining, decolorizing, and counterstaining cells. First, the cells are treated with crystal violet, which stains all the cells purple. The cells are then treated with an iodine solution, which reacts with the crystal violet to fix it in the cells. (At this point, all of the cells are still purple.) Next, the cells are briefly treated with alcohol or acetone. This decolorizes the Gram-negative cells but not the Gram-positive cells. The cells are then counterstained with safranin, which dyes the Gram-negative cells red.
> • **What color are the cells after they are treated with alcohol?** (Gram-positive: purple. Gram-negative: colorless.)
> • **Why are the cells counterstained?** (Colorless objects are hard to see.)
> • **Why is it important that crystal violet is a darker stain than safranin?** (Since all the cells pick

BACTERIAL MOVEMENT We can also identify bacteria by studying how they move. Some bacteria are propelled by means of one or more flagella. Others lash, snake, or spiral forward. Still others glide slowly along a layer of slimelike material that they secrete themselves. And there are some bacteria that do not move at all.

How Monerans Obtain Energy

Although the structure of monerans is rather simple, their lifestyles are remarkably complex. No characteristic of monerans illustrates this point better than the ways in which they obtain energy.

AUTOTROPHS Monerans that trap the energy of sunlight in a manner similar to green plants are called **phototrophic autotrophs.** Examples of phototrophic autotrophs include the cyanobacteria and some photosynthetic eubacteria.

Monerans that live in harsh environments and obtain energy from inorganic molecules are called **chemotrophic autotrophs.** The inorganic molecules that are used by chemotrophic autotrophs include hydrogen sulfide, nitrites, sulfur, and iron. *Nitrosomonas* is an example of a chemotrophic autotroph that uses ammonia and oxygen to produce energy.

HETEROTROPHS Many bacteria obtain energy by taking in organic molecules and then breaking them down and absorbing them. These bacteria are called **chemotrophic heterotrophs.** Most bacteria, as well as most animals, are chemotrophic heterotrophs.

Because we are chemotrophic heterotrophs ourselves, many bacteria compete with us for food sources. For example, *Salmonella* is a bacteria that grows in foods such as raw meat, poultry, and eggs. If these foods are not properly cooked, *Salmonella* will get to your dinner table before you do! Once there, these bacteria will not only "eat" some of the food ahead of time, but they will release poisons into the food. Food poisoning can result. The symptoms of food poisoning range from an upset stomach to serious illness.

There is another group of heterotrophic bacteria that has a most unusual means of obtaining energy. These bacteria are photosynthetic—they are able to use sunlight for energy. But they also need organic compounds for nutrition. These bacteria are called **phototrophic heterotrophs.** There is nothing quite like these organisms in the rest of the living world.

Bacterial Respiration

Like all organisms, bacteria need a constant supply of energy to perform all their life activities. This energy is supplied by the processes of respiration and fermentation. Respiration

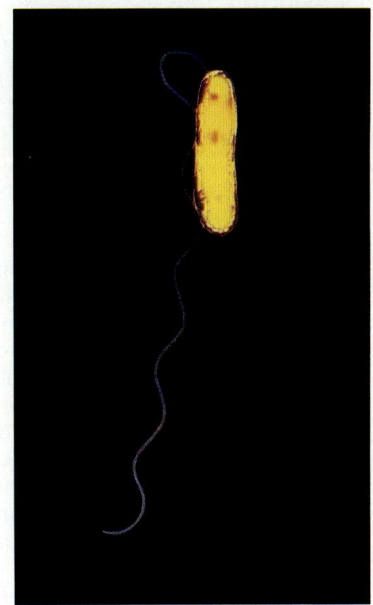

Figure 17–12 Many types of bacteria, such as the bacterium that causes Legionnaires' disease, move by means of a whiplike flagellum.

BACKGROUND INFORMATION
HOW ANAEROBES OBTAIN ENERGY

There are four main ways that anaerobes obtain energy: fermentation, nitrate reduction, carbonate reduction, and sulfate reduction.

In fermentation, which is performed by many bacteria as well as by fungi such as yeasts, an energy-rich molecule such as glucose is split, releasing energy. This process was described in detail in Chapter 6.

In nitrate reduction, which occurs in a number of bacteria that are facultative anaerobes, the oxygen in the nitrate ion is used to oxidize an organic compound and so obtain energy.

In carbonate reduction, the oxygen in carbon dioxide or carbonate is used to oxidize hydrogen produced by other microorganisms and so obtain energy. This method is used by methanogens.

In sulfate reduction, the oxygen in the sulfate ion is used to oxidize organic matter or hydrogen and so obtain energy. One of the products of this reaction under acidic conditions is hydrogen sulfide (H_2S), a foul-smelling gas that is poisonous to most living things.

PROBLEM SOLVING

Food Poisoning

The purpose of this activity is to familiarize students with the phenomenon of food poisoning and to give students an opportunity to apply what they have learned from reading to real-life situations.

The patient in Case Study 1 has symptoms that indicate a salmonella infection. The patient in Case Study 2 has symptoms that indicate staphylococci infection.

17–2 (continued)

Content Development

Show students slides or photographs of Gram-positive and Gram-negative bacteria.

- **What color are the bacteria?** (Gram-positive: pink or red. Gram-negative: blue or purple.)
- **Which bacteria are Gram-positive? Gram-negative?** (Gram-positive: pink or red. Gram-negative: blue or purple.)
- **Why do you think Gram staining is known as a differential stain?** (Because it stains different bacteria differently.)

Note that stains such as methylene blue, which color the object to make it visible, are known as positive stains.

- **When is a differential stain more useful than a positive stain?** (When you need to see differences between similarly shaped objects.)
- **When is it sufficient to use a positive stain?** (When you just need to make some structures visible.)

Skills Development

Guided Practice
Skill: Making calculations
Have students calculate the geometric progression that results from a bacterial cell doubling every 20 minutes for 5 hours. (1; 2; 4; 8; 16; 32; 64; 128; 256; 512; 1024; 2048; 4096; 8192; 16,384; 32,768) Ask students to make a graph of the growth of bacteria for the first 2 hours.

Enrichment

After students complete the Problem Solving feature called Food Poisoning, you may wish to share the following information with them.

About 33 million people in the United States are affected by microbial food poisoning each year. Much of the time, the effects of food poisoning are attributed to "stomach flu" or other causes. However, complications from food poisoning can be fatal, especially to infants and the elderly.

Several types of foods in particular are associated with food poisoning: seafood, poultry, and

---

PROBLEM SOLVING IN BIOLOGY
FOOD POISONING

Every year, thousands of cases of bacterial food poisoning are reported. In each case, a medical detective is assigned to find out how the person got food poisoning. Once the cause of the food poisoning has been determined, the medical detective can move to correct the conditions that led to the food poisoning.

Two types of bacteria that cause food poisoning are salmonella and staphylococci. A medical detective knows that these bacteria produce very different symptoms. So in order to determine which bacteria is the culprit, the detective will ask a series of important questions. The first thing the detective will want to know is exactly what the patient ate in the 24 hours prior to becoming ill and where the food was eaten. Other important information that the detective will gather includes how long after eating the patient became ill and whether the patient developed a fever. The detective will also want to know if the patient developed chills. Armed with answers to these questions, the detective can determine what caused the food poisoning and which meal contained the tainted food. But how?

The medical detective knows many details about these two types of bacterial food poisoning. For example, staphylococci produce a toxin, or poison, that is secreted into the food source as the bacteria multiply. Once a person eats the tainted food, the toxin will be carried throughout his or her body by the bloodstream. Within a few hours after the food has been ingested, the toxin will usually cause symptoms that include diarrhea, vomiting, nausea, and abdominal cramps. Fortunately, recovery is usually complete 24 to 48 hours after the onset of the symptoms.

Like staphylococci, salmonella produce a toxin. This toxin, however, is contained in the bacteria's cell walls and is released only when the bacteria lyse, or burst. Because of this difference, the symptoms produced by salmonella are somewhat unlike those produced by staphylococci. For example, it takes longer for a person to feel the effects of salmonella, often 12 hours or more. Salmonella infections almost always cause diarrhea. And they also generally result in a fever, chills, frequent vomiting, and abdominal pain. It may also take a patient quite a bit longer to recover from a case of salmonella food poisoning.

Now it's time for you to play medical detective.

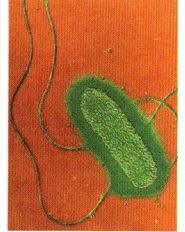

Salmonella bacterium magnified 12,600 times

Case Study 1   A patient with food poisoning reports that he ate his last meal at about 6 P.M. Although he felt fine the next morning, the patient became very sick at work. Due to severe abdominal pain and vomiting, the patient returned home. The patient also had a fever, chills, and severe diarrhea. He still felt sick the next day and did not fully recover for several days.

Case Study 2   You interview a patient who is suffering from food poisoning. However, this patient shows signs of recovery and feels well enough to go back to work. You discover that the last time the patient ate was about 6 P.M. the night before. While watching television later that evening, the patient became ill and had extremely severe abdominal cramps. The patient had a mild case of diarrhea that has subsided. There was no fever.

Analysis   Analyze each case study to determine whether the food poisoning was caused by salmonella or staphylococci. Support your diagnoses based on the data provided.

is the process that involves oxygen and breaks down food molecules to release energy. Fermentation, on the other hand, enables cells to carry out energy production without oxygen.

Organisms that require a constant supply of oxygen in order to live are called **obligate aerobes**. We, and many species of bacteria, are obligate aerobes. Some bacteria, however, do not require oxygen, and in fact may be poisoned by it! These bacteria are called **obligate anaerobes**. Obligate anaerobes must live in the absence of oxygen.

An example of an obligate anaerobe is the bacterium *Clostridium botulinum*, which is often found in soil. Because *Clostridium* is unable to grow in the presence of oxygen, it normally causes very few problems. However, if these bacteria find their way into a place that is free of air (air contains oxygen) and filled with food material, they will grow very quickly. As they grow, the bacteria produce **toxins**, or poisons, that cause botulism. Botulism is a rare form of food poisoning that interferes with nerve activity and can cause paralysis and, if the breathing muscles are paralyzed, death. A perfect place for these bacteria to grow is in the space inside a can of food. Most commercially prepared canned foods are safe because the bacteria and their toxins have been destroyed by heating the foods for a long time before the cans are sealed. However, botulism is always a danger when food is canned at home. Thus experienced canners thoroughly heat food before sealing it in jars.

A third group of bacteria are those that can survive with or without oxygen. They are known as **facultative anaerobes**. Facultative anaerobes do not require oxygen, but neither are they poisoned by its presence. What does such diversity imply? It means that bacteria can live in virtually every place on the surface of planet Earth. And indeed they do! Bacteria are found in freshwater lakes and ponds, at the bottom of the ocean, at the tops of the highest mountains, in the most sterile hospital rooms, and even in our own digestive systems!

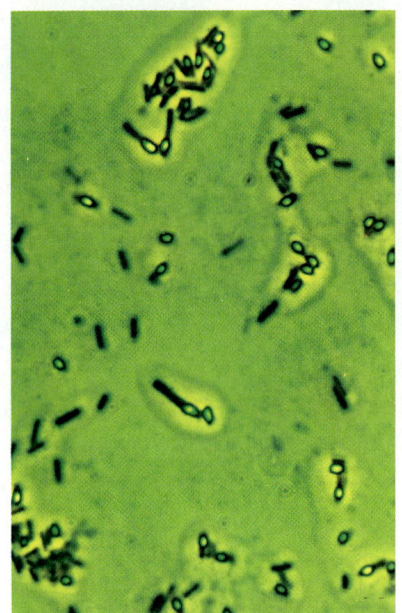

Figure 17–13 Botulism, a kind of food poisoning, is caused by the bacterium Clostridium botulinum. The small round structures on some of the bacteria are endospores.

Bacterial Growth and Reproduction

When conditions are favorable, bacteria can grow and reproduce at astonishing rates. Some types of bacteria can reproduce as often as every 20 minutes! If unlimited space and food were available to a single bacterium and if all of its offspring divided every 20 minutes, then in just 48 hours (2 days) they would reach a mass approximately 4000 times the mass of the Earth! Fortunately for us, this does not happen. In nature, the growth of bacteria is held in check by the availability of food and the production of waste products. However, bacteria do reproduce, and they do so in a number of ways.

BINARY FISSION When a bacterium has grown so that it has nearly doubled in size, it replicates its DNA and divides in half, producing two identical daughter cells. This type of

BACKGROUND INFORMATION
CONJUGATION

Strictly speaking, conjugation is not a form of sexual reproduction since it produces no additional organisms. In fact, conjugation may initially reduce the bacterial population because the donor cell typically dies after it transfers its DNA.

FACTS AND FIGURES

Bacteria have an incredibly high rate of metabolism. Lactose-digesting bacteria can break down 1,000 to 10,000 times their weight in lactose each hour, whereas it would take a human 30 to 40 years to metabolize his or her own weight in lactose.

FACTS AND FIGURES

Up to 200 million tons of nitrogen are fixed each year, almost entirely by bacteria.

TEACHING SUPPORT

Product Testing Activities
- Testing Yogurts

Biotechnology Workbook
- From Laboratory to Supermarket, p. 13

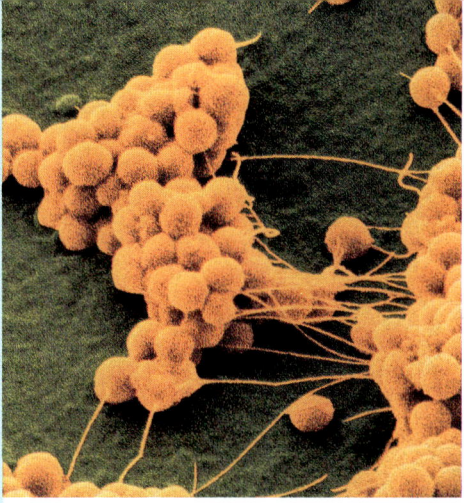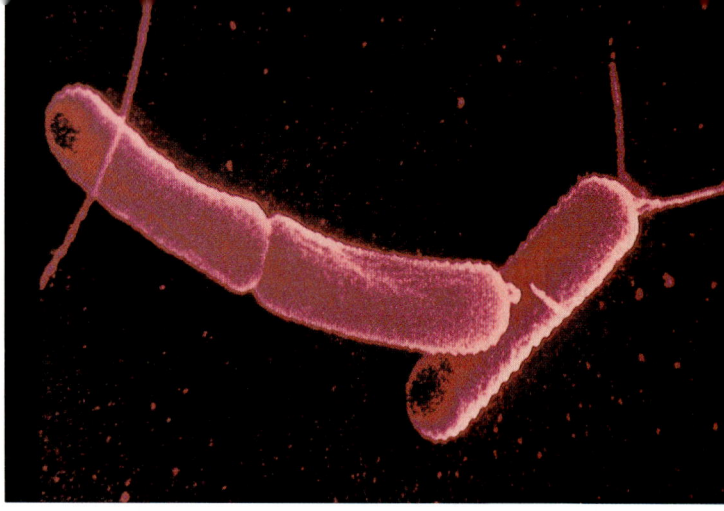

Figure 17–14 Most bacteria reproduce by binary fission, producing two identical cells (right). However, some bacteria reproduce by conjugation, or the transfer of parts of their genetic information from one cell to another through a protein bridge (left).

reproduction is known as **binary fission**. Because binary fission does not involve the exchange or recombination of genetic information, it is an asexual form of reproduction. The bacterium *E. coli* undergoes binary fission. See Figure 17–14.

CONJUGATION Although many bacteria reproduce only through asexual binary fission, others take part in some form of sexual reproduction. Sexual reproduction involves the exchange of genetic information. One form of sexual reproduction that occurs in some bacteria is known as **conjugation**. See Figure 17–14.

During conjugation, a long bridge of protein forms between and connects two bacterial cells. Part of the genetic information from one cell, called the donor, is transferred to the other cell, called the recipient, through this bridge. When the process of conjugation is complete, the recipient cell has a different set of genes from those it had before conjugation occurred. The new combinations of genes that result from conjugation increase the genetic diversity in that population of bacteria. Genetic diversity helps to ensure that even if the environment changes, a few bacteria may have the right combinations of genes to survive.

SPORE FORMATION When growth conditions become unfavorable, many bacteria form structures called spores. One type of spore, called an **endospore**, is formed when a bacterium produces a thick internal wall that encloses its DNA and a portion of its cytoplasm.

The endospore can remain dormant for months or even years, waiting for more favorable growth conditions. When conditions improve, the endospore will open and the bacterium will begin to grow again. Strictly speaking, spore formation in bacteria is not a form of reproduction because it does not

17–2 (continued)

Content Development
Using the Visuals Direct students' attention to Figure 17–14. Tell students that the proper name for the conjugation bridges is *pili* (singular: *pilus*), a term that comes from the Latin word for hair.
- **Why is the term *pilus* appropriate for a conjugation bridge?** (It looks like a hair.)

Tell students that many colonial bacteria have pili even when they are not undergoing conjugation.
- **What do you think may be the purpose of these pili?** (Accept all logical answers. Pili are thought to function in communication between the cells in a colony.)

Focus/Motivation
Using the Visuals Direct students' attention to Figure 17–15. Inform students that bacterial spores are extremely tough. Living spores have been found in the deepest ocean trenches and at altitudes as high as 27,000 meters above the Earth's surface. Some spores can survive being boiled for an hour; some

result in the formation of new bacterial cells. However, the ability to form spores makes it possible for some bacteria to survive harsh conditions that would otherwise kill them.

Importance of Monerans

Many of the remarkable properties of monerans provide us with products upon which we depend every day. For example, bacteria are used in the production of a wide variety of foods and beverages, such as cheese, yogurt, buttermilk, and sour cream. Some bacteria are used to make pickles and sauerkraut, and some make vinegar from wine.

Bacteria are also used in industry. One type of bacteria can digest petroleum, which makes them helpful in cleaning up small oil spills. Some bacteria remove waste products and poisons from water. Others can even help to mine minerals from the ground. Still others have been useful in synthesizing drugs and chemicals through techniques of genetic engineering.

Many kinds of bacteria develop a close relationship with other organisms in which the bacteria or the other organism or both benefit. Such a relationship is called **symbiosis** (sihm-bigh-OH-sihs). The symbiotic relationships that bacteria develop with other organisms are particularly important. Monerans form symbiotic relationships with organisms from all of the other four kingdoms.

Our intestines are inhabited by large numbers of bacteria, including *E. coli*. Indeed, the species name *coli* was derived from the fact that these bacteria were discovered in the human colon, or large intestine. In the intestines, the bacteria are provided with a warm safe home, plenty of food, and free transportation. We, in turn, get help in digesting our food. These bacteria also make a number of vitamins that we cannot produce on our own. So both we and the bacteria benefit from this symbiotic relationship.

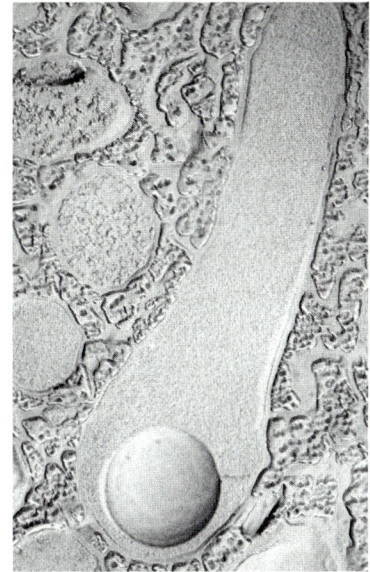

Figure 17–15 *The round circle at the bottom of this electron micrograph of a bacterium is an endospore. Endospores enable bacteria to survive unfavorable conditions, such as high temperature.*

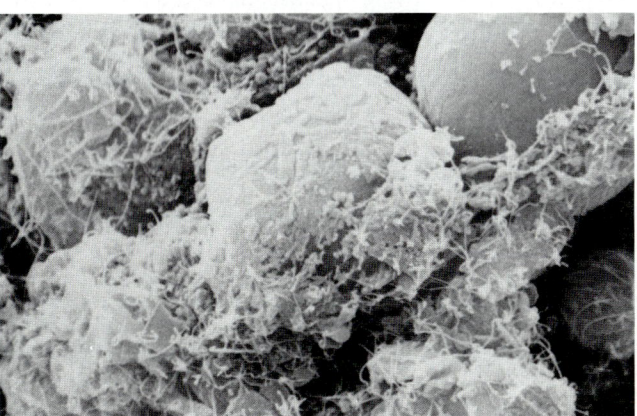

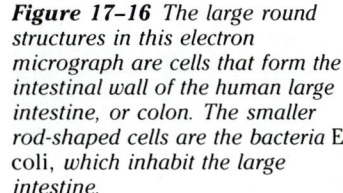

Figure 17–16 *The large round structures in this electron micrograph are cells that form the intestinal wall of the human large intestine, or colon. The smaller rod-shaped cells are the bacteria* E. coli, *which inhabit the large intestine.*

ECOLOGY NOTE
BACTERIA AND THE NITROGEN CYCLE

The role of bacteria in the nitrogen cycle is not limited to nitrogen fixation. Photosynthetic bacteria use the nitrates formed in nitrogen fixation to form nitrogenous organic compounds. Bacteria such as *Clostridium* and *Pseudomonas* produce enzymes that break down complex organic compounds, such as proteins, into simpler compounds, such as amino acids. Other bacteria are involved in ammonification— breaking down amino acids to form ammonia. Some bacteria are involved in the process of nitrification. For example, the genus *Nitrosomonas* changes ammonia into nitrite, which *Nitrobacter* can convert into nitrate. Under anaerobic conditions, certain bacteria perform the reverse of this process, changing nitrate to ammonia. Other bacteria, such as *Thiobacillus denitrificans,* are involved in the process of denitrification, in which nitrates are changed to nitrogen gas or nitrous oxide.

FACTS AND FIGURES

Although antibiotics are a recent discovery, beneficial microbes were used to fight harmful ones as early as 2500 years ago. A Chinese manuscript from this time period recommends that boils be treated with a poultice of moldy soybean curds.

Animals such as cattle are also dependent upon the symbiotic relationship they have with the bacteria in their intestines. You see, no vertebrate (animal with a backbone) can produce the enzymes necessary to break down cellulose, the principal carbohydrate in grass and hay. Bacteria living in the digestive systems of such animals can make these enzymes, thus allowing the animals to digest their food properly.

Bacteria in the Environment

Sometimes we are bold enough to consider ourselves the principal actors on the stage of life. We tend to place other organisms in supporting roles, like the minor actors in a play. But no drama can begin without the dozens of workers who are never seen on stage. The bacteria are like these unseen stagehands. We seldom think about them, but they are absolutely vital to maintaining the kind of living world we see about us.

NUTRIENT FLOW Every living thing depends on a supply of raw materials for growth. If these materials were lost forever when an organism died, then life could not continue. Before long, plants would drain the soil of the minerals they need, plant growth would stop, and the animals that depend on plants for food would starve.

Bacteria recycle and decompose, or break down, dead material. When a tree dies and falls to the forest floor, it begins to undergo many changes. Over the course of a few summers, the bark peels off and the wood begins to weaken because it becomes infested with insects. Then the tree crumbles into the soil. Over time, the whole substance of the tree disappears. What happens to all of the material that made up the tree?

From the moment the tree dies, armies of bacteria attack and digest the dead wood, breaking it down into simpler substances. These bacteria are called **saprophytes** (SAP-ruh-fights). Saprophytes are organisms that use the complex molecules of a once-living organism as their source of energy and nutrition. Gradually, the material of the tree is recycled, enriching the soil in which it grew. Although bacteria play a major role in this process, some eukaryotic organisms, such as insects and fungi, have a supporting role.

Figure 17–17 Most bacteria are heterotrophs, or organisms that obtain food from the organic compounds of other organisms. Many of these heterotrophs live as saprophytes, decomposing dead organisms such as this tree.

SEWAGE DECOMPOSITION Humans take advantage of the ability of bacteria to decompose material in the treatment of sewage. One of the critical steps in sewage treatment is carried out by a diverse mixture of bacteria that is added directly to the waste water. Waste water contains human waste, discarded food, organic garbage, and even chemical waste. Bacteria grow rapidly in this mixture. As they grow, they break down the complex compounds in the sewage into simpler compounds. This process produces purified water, nitrogen gas and carbon dioxide gas, and leftover products that can be used as crop fertilizers.

17–2 (continued)
Content Development
Scientists can gather a lot of information by monitoring the numbers and types of bacteria in the environment. For example, water supplies are often checked for the presence of *E. coli*.
- **Where is *E. coli* normally found?** (In the human intestine.)
- **What does it mean when large amounts of *E. coli* are found in water?** (The water is being contaminated by sewage.)

Point out that *E. coli* is mostly harmless.
- **Why are health officials concerned when *E. coli* is found in water supplies?** (Because it indicates that other, more harmful, bacteria in sewage are likely to be present.)

Focus/Motivation
Obtain specimens of native copper (crystalline elemental copper) and a copper-containing rock. Possibilities for the latter include malachite or azurite, which are quite colorful and are composed of copper carbonate,

SCIENCE, TECHNOLOGY, AND SOCIETY CONNECTION

The Littlest Miners

Copper is one of the most important minerals in the Earth—our modern civilization needs large amounts of it. Most, however, is low-grade copper ore, which means that the ore has a low percentage of copper in it. This is especially true of the copper ore deposits located in the United States. A way must be found to get the small amount of copper out of the ore in a pure enough form to make it worth mining. Does such a job require big, strong, burly miners? Not at all! The best miners for the job are bacteria, and the process is called bioleaching.

The process of bioleaching begins by stacking large piles of low-grade copper ore out in the open. Then water that has been made acidic is sprayed on the surfaces of the piles. The acidified water helps a bacterium called *Thiobacillus ferrooxidans* to grow. *Thiobacillus*, which occurs naturally in the copper ore, helps to break down sulfate-containing minerals, releasing sulfuric acid and ferric iron. The sulfuric acid and ferric iron react with copper-containing minerals in the ore to produce copper sulfate, which is water soluble. The copper sulfate dissolves in the water, forming copper ions. The dissolved copper sulfate is washed out of the pile and leaks through the bottom of the heap, where it is collected in basins. The copper ions are then separated from the copper sulfate solution to produce metallic copper.

At a copper leaching dump in Bingham Canyon, Utah, low-grade ore is extracted using the bacterium Thiobacillus ferrooxidans.

Thiobacillus does what human miners cannot—it extracts copper from low-grade copper ore. Presently, about 10 percent of the copper produced in the United States is mined by using bacteria. A few mining companies have begun to apply the same biotechnological process to extract gold from sulfur-containing gold ore, giving the term gold bug a whole new meaning!

NITROGEN FIXATION All organisms on our planet are totally dependent on monerans for nitrogen. All green plants need nitrogen to make amino acids, which are the building blocks of proteins. And because animals eat plants, plant proteins are ultimately the source of proteins for animals.

Although our atmosphere is made up of approximately 80 percent nitrogen gas (N_2), plants are not able to use the nitrogen gas. Neither can most other organisms. Living organisms generally require that nitrogen be "fixed" chemically in the form of ammonia (NH_3) and related nitrogen compounds. Chemists can make synthetic nitrogen-containing fertilizers by mixing nitrogen gas and hydrogen gas, heating the mixture to 500°C, and then squeezing it to 300 times the normal atmospheric pressure. The process is expensive, time-consuming, and sometimes dangerous.

SCIENCE, TECHNOLOGY, AND SOCIETY CONNECTION

The Littlest Miners

Bioleaching has been used for about a decade in copper mining. Some of the reactions involved in this process include the following:

- Monovalent copper is oxidized by *Thiobacillus ferrooxidans* under acidic conditions.
- Ores such as covellite are oxidized with ferric iron, and the ferric iron is regenerated through bacterial oxidation.
- Copper ions react with scrap iron to produce elemental copper.
- *T. ferrooxidans* is also useful in uranium mining. In one of the first steps of uranium leaching, the insoluble U^{4+} ion is converted into the soluble U^{6+} ion when the uranium oxide in uranium ore is oxidized with ferric iron (usually in the form of associated pyrite in the ores) to form soluble uranium compounds. *T. ferrooxidans* is then used to change the ferrous ions back into ferric ions.

In 1987, some of the first trials of bioleaching in gold mining were carried out in Canada. In this process, *T. ferrooxidans* is used to break down the sulfurous rock in which gold particles are embedded, making it easier to retrieve the gold. (Gold is removed from ore by dissolving the gold in cyanide or another leaching agent, adding charcoal to bind with the gold ions, and then chemically washing the gold from the charcoal.) The tests showed that bioleaching improved the yield of gold from ore. About 96% of the gold in the ore was recovered, whereas only about 65 to 70% of the gold is obtained when ores are treated by conventional methods.

---

and chalcocite or covellite, which are black crystals of copper sulfide. A geology or earth science teacher might be a good source.

- **How are these rocks different from one another?** (Answers will vary, depending on the specimens you obtain. Native copper forms branching crystals that are copper colored; malachite often looks like a cluster of bubbles and has many green bands; azurite forms vivid blue crystals.)
- **How is pure copper retrieved from copper ores such as these?** (By crushing the ore and melting it down.)

Point out that the melting point of pure copper is about 1083°C and that of copper ores can be higher. (Chalcocite—Cu_2S—has a melting point of about 1100°C.)

TEACHING SUPPORT

Study Guide
- Section 17–2, p. 164

17–2 (continued)

SECTION REVIEW 17–2

1. *Eubacteria:* have cell walls composed of complex carbohydrates; some live in soil; others produce disease; and still others are photosynthetic. *Cyanobacteria:* photosynthetic and contain phycocyanin and chlorophyll *a*; found in fresh and salt water and on land. *Archaebacteria:* live in extremely harsh environments such as in the digestive tracts of animals and in hot springs; some are methanogens. *Prochlorobacteria:* members of this phylum are photosynthetic and contain chlorophyll *a* and *b*.
2. Rod-shaped bacteria are called bacilli; spherical bacteria are called cocci; and spiral bacteria are called spirilla.
3. Autotrophic monerans are bacteria that obtain their energy by trapping the energy of sunlight in a manner similar to green plants. Heterotrophic monerans are bacteria that obtain their energy by taking in organic molecules and then breaking them down so that they can be absorbed. Obligate aerobes are organisms that require a constant supply of oxygen. Obligate anaerobes are organisms that must live in the absence of oxygen. Facultative anaerobes are organisms that do not need oxygen but are not poisoned by its presence.
4. In binary fission, a bacterium replicates its DNA and divides in half, producing two identical daughter cells. In conjugation, a long bridge of protein forms between and converts two bacterial cells. Part of the genetic information from one cell, called the donor, is transferred to the other cell, called the recipient, through this bridge.
5. Bacteria are used in the production of food and beverages; some can digest petroleum; others remove waste products; still others help mine minerals from the ground and help synthesize drugs and chemicals.
6. Because all organisms are totally dependent on monerans for converting nitrogen into nitrogen compounds, these organisms would die if the monerans lost their ability to fix nitrogen.

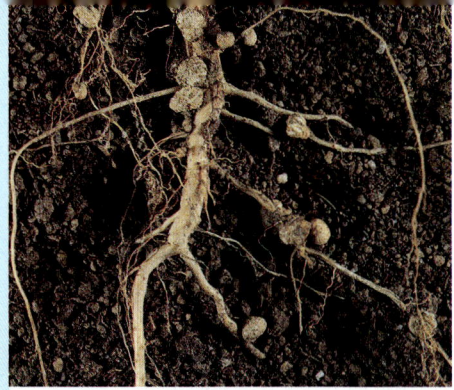

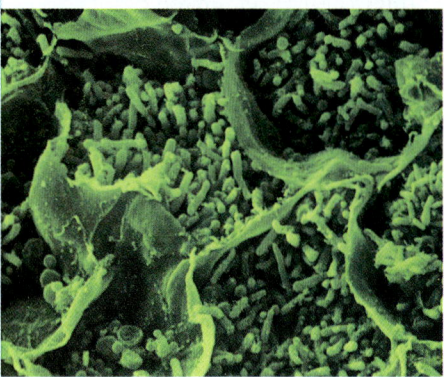

Figure 17–18 The knoblike structures growing on the roots of this soybean plant are called nodules (top). Within these nodules are the nitrogen-fixing bacteria Rhizobium, *which have a characteristic rod-shaped appearance (bottom).*

Guide For Reading
- What are some diseases caused by viruses and bacteria? How are these diseases prevented and cured?
- How is the growth of bacteria controlled?

In contrast to this difficult process, many cyanobacteria and other bacteria can take nitrogen from the air and convert it to a form that plants can use. This conversion process is known as **nitrogen fixation**. Monerans are the only organisms capable of performing nitrogen fixation.

Many plants have symbiotic relationships with nitrogen-fixing bacteria. The soybean, which hosts the bacterium *Rhizobium*, is among the best known. *Rhizobium* grows in nodules, or knobs, that form on the roots of the soybean plant. The soybean plant provides a home and a source of nutrients for the nitrogen-fixing *Rhizobium*; the bacterium fixes nitrogen directly from the air into ammonia for the plant. All plants benefit from the nitrogen-fixing ability of monerans, but soybeans are a step ahead. With a little help from their "friends," soybeans have their own fertilizer factories built right into their roots.

As you have learned, eukaryotes are dependent upon monerans to fix nitrogen and release it into the environment. And it is because of the nitrogen-fixing ability of these organisms that more than 170 million metric tons of nitrogen are released into the environment every year.

17–2 SECTION REVIEW

1. Describe the four major phyla of the kingdom Monera.
2. Compare the three basic shapes of bacteria.
3. Distinguish between autotrophic and heterotrophic monerans. Between obligate aerobes, obligate anaerobes, and facultative anaerobes.
4. Describe binary fission and conjugation.
5. List some ways in which bacteria are important.
6. **Critical Thinking—Making Predictions** Suppose monerans lost the ability to fix nitrogen. How would this affect other organisms?

17–3 Diseases Caused by Viruses and Monerans

Contrary to popular belief, only a small number of viruses and monerans are capable of producing disease in humans. Despite their small numbers, these **pathogens**, or disease-producing agents, are responsible for much human suffering.

From the point of view of a microorganism, however, disease is nothing more than a conflict of lifestyles. All viruses infect living cells, and disease results when the infection causes harm to the host. All bacteria require nutrients and energy, and disease results only when bacteria interfere with the host in obtaining them.

Reinforcement/Reteaching
If students cannot answer the Section Review questions, review the material in the section that is giving them trouble.

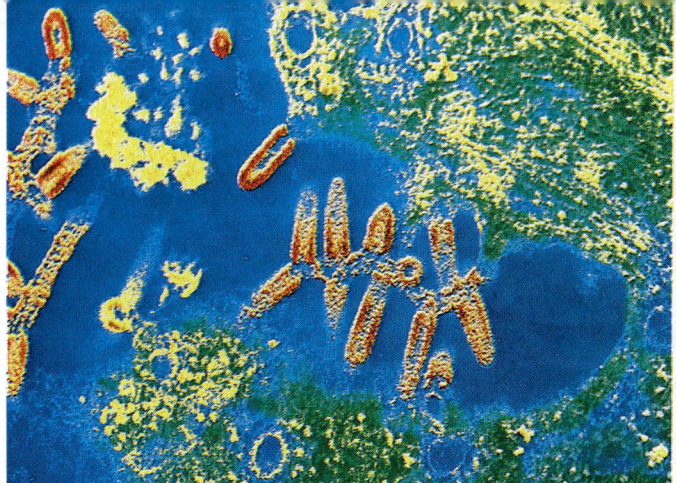
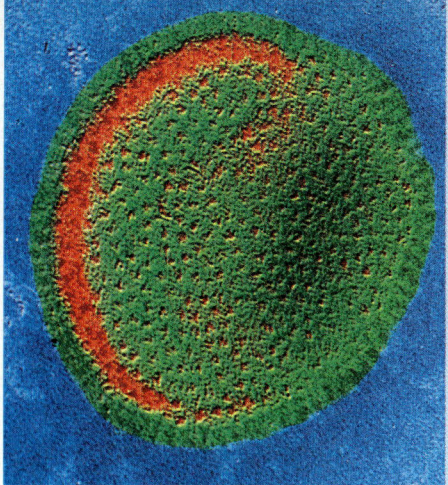

Viruses and Disease

Viruses are the cause of such human diseases as smallpox, polio, measles, AIDS, mumps, influenza, yellow fever, rabies, and the common cold. In most viral infections, viruses attack cells of the body in the same way that the T4 bacteriophage attacks *E. coli*. As the virus reproduces, it destroys the cells that it infects, causing the symptoms of the disease.

Although we tend to think that some viral diseases are curable, the only successful protection against most of them lies in preventing their infection. In order to do this, the body's own immune system must be stimulated to prevent the infection. A vaccine is a substance that contains the weakened or killed disease-causing virus. When injected into the body, the vaccine provides an immunity to the disease. Diseases such as smallpox and polio have all but been eliminated because vaccines were developed to stop their spread. As powerful as they are, however, vaccines can only provide protection if they are used before an infection begins. Once a viral infection starts, there is often little that medical science can do to stop the progress of the disease. However, sometimes the symptoms of the infection can be treated.

INTERFERONS One possible approach in the treatment of viral diseases is the use of substances called interferons. Interferons are small proteins that are produced by the body's cells when the cells are infected by viruses. When interferons are released from virus-infected cells, they seem to make it more difficult for the viruses to infect other cells. The word interferon is derived from the fact that these proteins interfere with the growth of the virus. The specific way in which these proteins work is not yet entirely understood. Until recently, interferons cost millions of dollars a milliliter to isolate and purify. But new techniques using genetically engineered bacteria have made the production of interferons less expensive and more plentiful.

Figure 17–19 Viruses are the cause of many human diseases. The bullet-shaped rabies virus particles are about 180 nm in length (left). The spherical influenza virus is about 100 nm in diameter (right).

Figure 17–20 These flasks contain interferons, which were produced by genetically engineered bacteria. Interferons, proteins made by virus-infected cells, inhibit the growth of viruses.

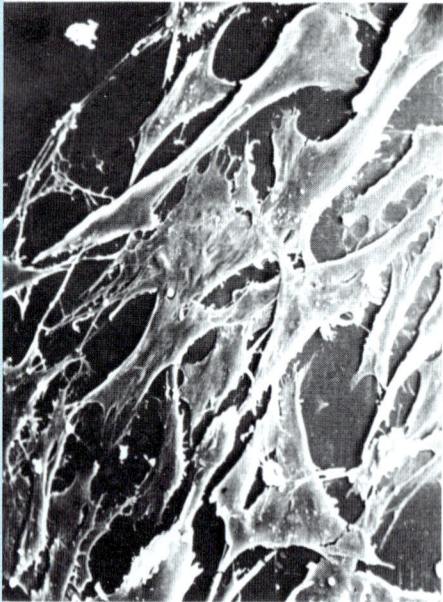

Figure 17–21 These cells of chicken connective tissue clearly show the effects of invasion by a virus. Notice that the normal cells are flat (bottom), whereas those infected with Rous sarcoma virus are round (top).

CANCER Certain viruses cause cancer in animals. These cancer-causing viruses are known as oncogenic (*onco-genic* means tumor-making) viruses. One example is the Rous sarcoma virus (RSV), discovered by the American physician Peyton Rous. RSV causes cancer in chickens and other domestic fowl. As we shall see in Chapter 44, this virus adds certain genes to the infected cell that seem to turn it into a cancer cell. Although it is clear that not all cancers are caused by viruses (cancer is generally not spread by a person-to-person infective process), a few cancers are. This has caused scientists to study cancer-causing viruses closely.

Bacteria and Disease

Of all the bacteria in the world, there are only a few that produce disease. The lifestyles of these few disease-causing organisms have, unfortunately, made us think of disease whenever we think of the word bacteria.

The French chemist Louis Pasteur was the first person to show convincingly that bacteria cause disease. Pasteur established what has become known as the germ theory of disease when he showed that bacteria were responsible for a number of human and animal diseases.

Some of the diseases caused by pathogenic bacteria include diphtheria, tuberculosis, typhoid fever, tetanus, Hansen disease, syphilis, cholera, and bubonic plague. Bacteria cause these diseases in one of two general ways. They may damage the cells and tissues of the infected organism directly by breaking down its living cells to use for food. Or they may release toxins (poisons) that travel throughout the body, interfering with the normal activity of the host.

Most bacteria can be grown in a culture dish, outside host tissues and cells. One class of bacteria known as rickettsias (rih-KEHT-see-ahz) are a curious exception to this rule. In order for most rickettsias to grow, they must be inside a living cell. In this respect, rickettsias are similar to viruses. Rickettsias seem to possess "leaky" cell walls and membranes that allow them to live inside a living cell and absorb nutrients from the cell's cytoplasm. Rickettsias include the organisms that cause Rocky Mountain spotted fever, typhus, and Q fever.

Although we shall discuss the medical measures that are used to fight disease in Chapter 45, it is worth pointing out that many of these diseases can be prevented by stimulating the body's immune system through the use of vaccines. If an infection does occur, however, there are many more effective measures to fight the infection if it is bacterial than if it is viral. These measures include a number of drugs and natural compounds, known as **antibiotics**, that can attack and destroy bacteria. One of the major reasons for the dramatic increase in life expectancy in the last two centuries is an increased understanding of how to prevent and cure bacterial infections.

Controlling Bacteria

Although most bacteria are harmless and many are beneficial, the risks of bacterial infection are great enough to warrant efforts to control bacterial growth.

STERILIZATION The growth of potentially dangerous bacteria can be controlled by sterilization. This process destroys living bacteria by subjecting them either to great heat or to chemical action. Heating is the simplest way to control bacterial growth. Bacteria cannot survive high temperatures for a long time, so most can be killed in boiling water.

An entire hospital, of course, cannot be dropped into boiling water. But it can be sterilized, one room at a time, by using disinfectants. A disinfectant is a chemical solution that kills bacteria. Disinfectants are also used in the home to clean bathrooms, kitchens, and other rooms where bacteria may grow and spoil food or cause disease.

FOOD PROCESSING As you have learned, bacteria are everywhere, including in our food. If we are not careful, the bacteria will begin to "eat" our food before we do. In doing so, the bacteria will cause the food to spoil. One method to stop food from spoiling is to refrigerate those foods in which bacteria might grow. Bacteria, like most organisms, grow slowly at low temperatures. For this reason, food that is stored at a lower temperature will keep longer because the bacteria will take much longer to grow and cause damage. In addition, many kinds of food are sterilized by boiling, frying, or steaming them. Each of these cooking techniques raises the temperature of the food to a point where all the bacteria are killed.

If the food is to be preserved for a long time by canning, the sterilized food must be immediately placed into sterile glass jars or metal cans and sealed. Food that has been properly canned will last almost indefinitely. Finally, a number of chemical treatments will inhibit the growth of bacteria in food. These include treating food with everyday chemicals such as salt, vinegar, or sugar. Salted meat, pickled vegetables, and jam are examples of chemically preserved foods.

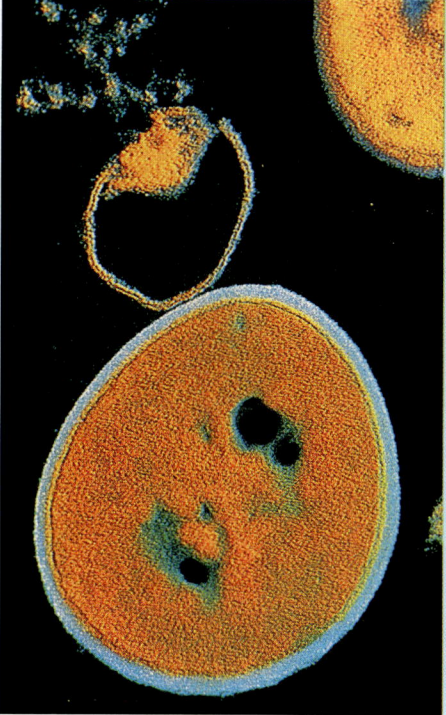

Figure 17–22 The addition of an antibiotic, or a substance that destroys bacteria, has caused the bacterium Staphylococcus aureus to burst.

MULTICULTURAL STRATEGY

William A. Hinton, MD, invented the Hinton test, a method to detect syphilis. Dr. Hinton also collaborated with Dr. J.A.V. Davies on what is now called the Davies-Hinton test for detecting the same disease.

Born in Chicago in 1883, Dr. Hinton received his undergraduate degree from Harvard University in 1905, then went on to complete Harvard Medical School in 1912. In 1949, Dr. Hinton became the first African American to hold the position of a professor at Harvard Medical School.

In addition to being recognized as one of the world's leading authorities on venereal disease, Dr. Hinton was also known for his work as director of the Boston Dispensary Laboratory, a position that he held from 1916 until shortly before his death in 1959. It was in this position that Dr. Hinton began a training program for women. From this program developed one of the leading institutions for the training of medical technicians in the United States.

17–3 SECTION REVIEW

1. What are some diseases caused by viruses? By bacteria?
2. How can viral and bacterial diseases be prevented? How can they be treated?
3. Describe several methods of controlling bacterial growth.
4. **Connection—You and Your World** Why is it important to cook food thoroughly before canning?

Closure

Students may be surprised to find out that tooth decay is caused by a contagious bacterial infection.
- **What is plaque?** (A film on the teeth that contains bacteria.)

Tell students that plaque contains a number of different kinds of bacteria. When they break down sugars, some of these bacteria produce lactic acid, a substance that dissolves the calcium in teeth.
- **Why does brushing help prevent cavities?** (Reduces the amount of cavity-causing bacteria clinging to the teeth.)
- **Many mouthwashes contain a lot of alcohol. Why do you think these mouthwashes help prevent cavities?** (Accept all logical answers. Alcohol may kill bacteria or inhibit their growth.)
- **What other methods of controlling bacterial infections might help prevent cavities?** (Encourage students to use the concepts they have learned in this section to answer the question.)

LABORATORY INVESTIGATION

EXAMINING BACTERIA

Before the Lab
1. Prepare methylene blue stain by adding 10 milliliters of stock solution to 90 milliliters of water for every 100 milliliters of stain. Stock solutions of methylene blue may be prepared by adding 1.48 grams of methylene blue powder to 100 milliliters of 95% ethyl alcohol and letting it stand for two days. During this time, stir frequently. Filter the newly prepared stock solution at the end of the two-day period.
2. Note that step 1 of the Procedure must be completed 2 days before the lab is to be performed. Step 2 of the Procedure must be done the day before the lab.
3. Divide the class into groups of two to four students, depending on the number of microscopes available.

Pre-Lab Discussion
Have students read the complete laboratory Procedure. Review the proper procedure for using a microscope. Discuss the investigation using questions similar to the following:
- **What is the purpose of this investigation?** (To examine some of the characteristics of bacteria.)
- **What are bacteria?** (Unicellular prokaryotic organisms.)
- **What are the sources of the bacteria that you are to examine?** (Water in which lima beans have been soaking; the inside of the mouth.)

Skills Development
Students will use the following skills while completing this laboratory investigation.
1. Observing
2. Comparing
3. Manipulative
4. Relating
5. Safety

LABORATORY INVESTIGATION

EXAMINING BACTERIA

PROBLEM
What are some of the characteristics of bacteria?

MATERIALS (per group)

100-mL beaker
10 lima beans
2 medicine droppers
2 glass slides
2 coverslips
microscope
methylene blue
flat toothpick

PROCEDURE

1. Fill the beaker halfway with water. Put the lima beans in the beaker and then put the beaker and its contents in a warm place where they will remain undisturbed overnight.
2. After 24 hours, examine the beaker to make sure that the beans are covered with water. If necessary, add more water to cover the beans. Again put the beaker of beans in a warm place overnight.
3. After 48 hours, the water in the beaker should appear cloudy because of the presence of a large number of bacteria. Using a medicine dropper, remove a drop of water from the beaker and place it in the center of a clean glass slide.
4. Cover the drop of water with a coverslip. Examine the drop of water under the low-power objective of the microscope.
5. Examine the drop of water under the high-power objective. **CAUTION:** *When switching to the high-power objective, you should always look at the objective from the side of your microscope so that the objective does not hit or damage the slide.* Notice the shapes and methods of movement of the bacteria.
6. With the second medicine dropper, place a drop of methylene blue in the center of a clean glass slide.
7. Using the broad end of a flat toothpick, gently scrape the inside of your cheek. The scraping should contain some of your cheek cells. Using the same end of the toothpick, mix the scraping from your cheek with the drop of methylene blue.
8. Add a drop of water from the beaker containing the beans to the mixture of cheek scraping and methylene blue. Cover this mixture with a clean coverslip.
9. Locate a cheek cell under the low-power objective of the microscope. Then switch to the high-power objective and use the fine adjustment to locate some bacteria near a cheek cell.
10. Observe the sizes of the bacteria, the cheek cell, and the nucleus of the cheek cell.

OBSERVATIONS
1. Describe the colors and shapes of the bacteria.
2. Are the bacteria arranged singly, in pairs, in chains, or in clusters?
3. Are the bacteria capable of movement? If so, describe the ways in which they move.
4. Compare the size of the bacteria to the sizes of the cheek cell and the nucleus of the cheek cell.

ANALYSIS AND CONCLUSIONS
1. What was the source of the bacteria? How were they able to grow in the beaker containing water and lima beans?
2. Explain why methylene blue was added to the scraping from the cheek.
3. Bacteria are prokaryotic cells, whereas cheek cells are eukaryotic cells. Explain this statement.

376

Safety Tips
Caution students that methylene blue will stain clothing, skin, and anything else it touches. Tell students to be careful when obtaining the cheek scraping. Remind students to handle all equipment with care.

Teaching Strategy
1. Demonstrate the technique students should use to obtain cheek epithelial cells.
2. Help students who are having difficulty locating bacteria or focusing the microscope.

Observations
1. Most of the bacteria are colorless and rod-shaped.
2. For the most part, the bacteria are arranged singly, although some are in pairs. Rarely are they seen in chains or clusters.
3. If any movement was seen, it

STUDENT STUDY GUIDE

SUMMARIZING THE CONCEPTS

The key concepts in each section of this chapter are listed below to help you review the chapter content. Make sure you understand each concept and its relationship to other concepts and to the theme of this chapter.

17-1 Viruses

- A virus is a noncellular particle made up of genetic material and protein.
- Viruses cannot carry out any life processes unless they are within a living host cell.
- Infection and destruction of the host cell by a virus is called a lytic infection.
- In a lysogenic infection, the virus does not reproduce at once and lyse the host cell.

17-2 Monerans—Prokaryotic Cells

- Bacteria are prokaryotes, or cells that do not have a nucleus.
- Rod-shaped bacteria are called bacilli; spherical bacteria are called cocci; and spiral-shaped bacteria are called spirilla.
- Obligate aerobes need oxygen to live. Obligate anaerobes must live in the absence of oxygen. Facultative anaerobes can survive with or without oxygen.
- Some bacteria reproduce by binary fission or conjugation.
- Many bacteria form a symbiotic relationship with other organisms. Some are saprophytes.
- Many bacteria can take nitrogen from the air and convert it into a form that plants can use to make proteins.

17-3 Diseases Caused by Viruses and Monerans

- Although vaccines are the only successful protection against some viral diseases, interferons are being considered as another possible treatment. Antibiotics are substances that can attack and destroy bacteria.

REVIEWING KEY TERMS

Vocabulary terms are important to your understanding of biology. The key terms listed below are those you should be especially familiar with. Review these terms and their meanings. Then use each term in a complete sentence. If you are not sure of a term's meaning, return to the appropriate section and review its definition.

17-1 Viruses
- virus
- bacteriophage
- lytic infection
- lysogenic infection
- prophage
- retrovirus
- parasite

17-2 Monerans—Prokaryotic Cells
- prokaryote
- bacterium
- methanogen
- bacillus
- coccus
- spirillum
- phototrophic autotroph
- chemotrophic autotroph
- chemotrophic heterotroph
- phototrophic heterotroph
- obligate aerobe
- obligate anaerobe
- toxin
- facultative anaerobe
- binary fission
- conjugation
- endospore
- symbiosis
- saprophyte
- nitrogen fixation

17-3 Diseases Caused by Viruses and Monerans
- pathogen
- antibiotic

---

was due to flagella lashing back and forth.
4. The bacteria are smaller than both the cheek cell and nucleus.

Analysis and Conclusions
1. The source of the bacteria is the air, which contains spores. These spores fell into the water and used the food (beans) and the moisture to grow and reproduce.

2. The methylene blue was added to stain the nucleus of the cheek cell.
3. Because bacteria lack a nucleus, they are classified as prokaryotic cells. Because cheek cells have a nucleus, they are classified as eukaryotic cells.

Going Further: Enrichment
1. Have students examine prepared slides of various types of bacteria.
2. Students might leave out small containers of juice, milk, or other substances to compare the types and amounts of bacteria that grow in each.

CHAPTER REVIEW

CONTENT REVIEW

Multiple Choice

Choose the letter of the answer that best completes each statement.

1. Viruses contain
   a. cell walls.
   b. cell membranes.
   c. nuclei.
   d. protein coats.
2. Viruses that invade cells and cause them to burst are said to be
   a. parasitic.
   b. lysogenic.
   c. saprophytic.
   d. lytic.
3. Methanogens are members of the phylum
   a. Cyanobacteria.
   b. Archaebacteria.
   c. Prochlorobacteria.
   d. Eubacteria.
4. A rod-shaped bacterium is known as a
   a. spirillum.
   b. bacillus.
   c. coccus.
   d. virus.
5. Organisms that need a constant supply of oxygen in order to live are called
   a. obligate anaerobes.
   b. facultative anaerobes.
   c. chemotrophic autotrophs.
   d. obligate aerobes.
6. A structure that forms when a bacterium produces a thick internal wall that encloses its DNA and part of its cytoplasm is called a (an)
   a. endospore.
   b. capsid.
   c. prophage.
   d. spirillum.
7. Organisms that use the complex molecules of once-living organisms for energy and nutrition are called
   a. parasites.
   b. viruses.
   c. saprophytes.
   d. eukaryotes.
8. An example of a disease caused by a bacterium is
   a. influenza.
   b. measles.
   c. AIDS.
   d. syphilis.

True or False

Determine whether each statement is true or false. If it is true, write "true." If it is false, change the underlined word or words to make the statement true.

1. In a <u>lysogenic</u> infection, the virus does not reproduce and lyse its host cell immediately.
2. A virus is composed of a <u>nucleus</u> surrounded by a protein coat.
3. Bacteria are <u>eukaryotes</u>.
4. <u>Spirilla</u> are spherical bacteria.
5. Bacteria that can live with or without oxygen are known as <u>obligate</u> anaerobes.
6. Monerans that trap the energy of sunlight in a manner similar to green plants are called <u>chemotrophic</u> autotrophs.
7. In bacteria, <u>spore formation</u> involves the transferring of genetic material from one cell to another cell.
8. <u>Methanogens</u> are disease-causing agents.

Word Relationships

A. *In each of the following sets of terms, three of the terms are related. One term does not belong. Determine the characteristic common to three of the terms and then identify the term that does not belong.*

1. obligate anaerobe, facultative anaerobe, phototrophic autotroph, obligate aerobe
2. *E. coli*, T4, *Rhizobium*, *Salmonella*
3. prophage, bacillus, spirillum, coccus
4. measles, polio, rabies, tetanus

B. An analogy is a relationship between two pairs of words or phrases generally written in the following manner: a:b::c:d. The symbol : is read "is to," and the symbol :: is read "as." For example, cat:animal::rose:plant is read "cat is to animal as rose is to plant."

In the analogies that follow, a word or phrase is missing. Complete each analogy by providing the missing word or phrase.

5. eukaryote:human::prokaryote:_____
6. rod-shaped:bacillus::spherical:_____
7. oxygen:obligate aerobe::no oxygen:_____
8. eubacteria:true bacteria::cyanobacteria:_____

CONCEPT MASTERY

Use your understanding of the concepts developed in the chapter to answer each of the following in a brief paragraph.

1. Why are viruses considered parasites?
2. What is the relationship between cell wall structure and whether a bacterium is Gram positive or Gram negative?
3. What is the difference between bacterial autotrophs and heterotrophs?
4. Describe a symbiotic relationship between bacteria and another organism.
5. What are some ways in which bacteria are important to the environment?
6. Compare the different methods of bacterial reproduction.

CRITICAL AND CREATIVE THINKING

Discuss each of the following in a brief paragraph.

1. **Interpreting graphs** Describe the growth of the bacteria shown in the graph. Explain why growth levels off in stage 3.

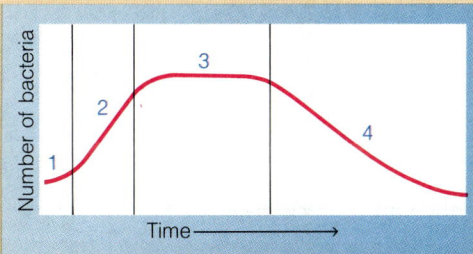

2. **Relating concepts** Bacteria can be grown in the laboratory on synthetic media. Can viruses be grown in this way? Can viruses be grown on cultures of bacteria?

3. **Using the writing process** As you know, many diseases are caused by microorganisms. Imagine that you have the ability to develop a chemical capable of wiping out all viruses and monerans on Earth. Write an advertising campaign for your new chemical in which you describe its benefits and dangers.

4. **Using the writing process** In *The War of the Worlds*, a wonderful book written by H. G. Wells, Earth is invaded by aliens. No weapons can kill the invaders, and the Earth seems doomed. The Earth is saved, however, when the invaders die from diseases they contract here. Using a similar premise, write a story about people from Earth voyaging to another planet some time in the future.

379

CRITICAL AND CREATIVE THINKING

1. In stages 1 and 2, the growth level is increasing because the bacteria have sufficient food and moisture. In stage 3, the amount of food and moisture is becoming limited, so the growth of the bacteria levels off. In stage 4, there is a decrease in bacterial growth because conditions for growth are no longer suitable, so the bacteria begin to die off.

2. Viruses can reproduce only within living things. As a result, viruses can grow on cultures of bacteria but not on synthetic media.

3. Student advertising campaigns should show knowledge of the characteristics of viruses and bacteria. They should include the idea that although some viruses and bacteria cause disease, the eradication of all viruses and bacteria can also cause the eradication of all living things.

4. Students' stories should be creative and well-written and should demonstrate an understanding of the characteristics of disease-causing organisms.

CHAPTER 18 Protists

Section	Laboratory Investigations and Activities
18–1 The Kingdom Protista pages 381–383 THEMES: Unity and Diversity, Evolution	**Teacher Edition** DEMONSTRATION: An Old-Fashioned Hay Infusion, p. 380D
18–2 Animallike Protists pages 384–394 THEMES: Unity and Diversity, Scale and Structure	**Teacher Edition** DEMONSTRATION: Symbiotic Flagellates, p. 380D **Laboratory Manual** Adaptations of the Paramecium, p. 225 **Teaching Resources** HANDS-ON ACTIVITY: Putting the Squeeze On, p. 89 ECOLOGY INVESTIGATION: Separating and Culturing Soil Protozoa, p. 7 TEACHER DEMONSTRATION: The Effect of Various Stimuli on Paramecia, p.11
18–3 Plantlike Protists pages 394–401 THEMES: Unity and Diversity, Scale and Structure	**Student Edition** LABORATORY INVESTIGATION: Examining Protists, p. 402 **Laboratory Manual** Comparing Protists, p. 231 **Teaching Resources** HANDS-ON ACTIVITY: Shedding a Little Light on Euglena, p. 93
Chapter Review pages 403–405	

Teacher Resources

Books

Farmer, J. N. *The Protozoa.* Mosby, 1980.

Fenchel, Tom. *Ecology of Protozoa: The Biology of Freeliving Phagotrophic Protists.* Science Tech Publishers and Springer-Verlag, 1987.

Gall, Joseph G. *The Molecular Biology of Ciliated Protozoa.* Academic Press, 1986.

Gortz, H. D., ed. *Paramecium.* Springer-Verlag, 1988.

Margulis, Lynn, and Karen Schwartz. *Five Kingdoms*, 2nd ed. Freeman, 1987.

Starr, M. P., et al., eds. *The Prokaryotes.* 2 vols. Springer-Verlag, 1981.

Witcherman, Ralph. *The Biology of Paramecium*, 2nd ed. Plenum, 1986.

CHAPTER PLANNING GUIDE

Other Activities	Media and Technology

Teaching Resources
    ACTIVITY: Arranging Protists, p.3
Study Guide
    Section 18–1, p. 169	

Teaching Resources
    ACTIVITY: Experimental Models of Protists, p. 5
Study Guide
    Section 18–2, p. 170
	Biology Transparencies
	    20: Protists

Teaching Resources
    VOCABULARY REVIEW: Word Game, p. 1
    ACTIVITY: Investigating Cellular Slime Molds, p. 7
Study Guide
    Section 18–3, p. 174
	Interactive Videodisc
	    Planet Earth: The Blue Planet
	Videodisc
	    Aquatic Ecosystems: Estuaries and Marine
	    Aquatic Ecosystems: Freshwater Wetlands and Freshwater
	Biology Media Guide
	    page 16

Teaching Resources
    Chapter Test A, p. 15
    Chapter Test B, p. 19
	Computer Test Bank
	    Chapter Test, p. 193

Audiovisuals

Protozoa: Structures and Functions. Film or video. Coronet.
From Protistans to First Multicellular Animals. Film. Benchmark Films.

Modern Biology Series: The Paramecium. Film. Benchmark Films.
The Protozoa. Sound filmstrip. Carolina Biological Supply Company.

Protist Physiology. Film or video. Ward.
Protist Behavior. Film or video. Ward.
Protist Reproduction. Film or video. Ward.

CHAPTER 18 Protists

CHAPTER OVERVIEW

The first kingdom of eukaryotic organisms to be considered is the kingdom Protista. This group, whose name comes from the Greek, meaning "first," consists of those organisms that can be described as single-celled and eukaryotic. As a group, they are not nearly as old as the monerans. The oldest fossils of protists are only about 1.5 billion years old. Protists are extremely diverse and difficult to classify because they have characteristics in common with animals, plants, and fungi. The more than 115,000 species might be said to be classified by exclusion. There is still much disagreement among scientists about this group.

At one time, animallike protists were called protozoa and classified separately from the plantlike protists. However, the similarities are so great that it no longer makes sense to separate these two groups into different kingdoms. We now recognize four phyla of animallike organisms within the kingdom Protista. *Paramecium* is the most familiar member of the phylum Ciliophora. Ciliophora are so named because they all have cilia that they use for movement. They also generally have two different kinds of nuclei. A second phylum of animallike protists is Zoomastigina. These organisms move through the water by means of flagella and so are commonly called flagellates. Many flagellates are free-living and get their food by absorption through the cell membrane. Others live within the bodies of other organisms as parasites. The members of phylum Sporozoa make up the third group of animallike protists. They are all parasitic and nonmotile. They reproduce by means of spores. A typical sporozoan is *Plasmodium*, the organism that causes the human disease malaria. The last group of protists with animal characteristics is phylum Sarcodina. This phylum gets its name from *sarcode*, a word coined in the nineteenth century to refer to the homogeneous jelly from which these cells were believed to be composed.

There are also plantlike protists that make up an additional five phyla of unicellular organisms. They are called plantlike because most of them contain chlorophyll and carry on photosynthesis. Three of the phyla are considered to be simple types of algae found in water or very damp areas. They are sometimes classified as plants for this reason. The other two phyla are not photosynthetic and have many funguslike characteristics. Thus, the slime molds are sometimes classified as fungi.

Phylum Euglenophyta contains members that are closely related to the zoomastiginans. They are sometimes placed in the same phylum. The reason for separating them is the presence of chloroplasts in euglenophytes. The best-known members of this group are euglenas. The second phylum of plantlike protists is Pyrrophyta, or dinoflagellates. Most, but not all, of the dinoflagellates are photosynthetic. The group has some interesting and unique characteristics. There are three types of organisms found in the third phylum, Chrysophyta. They include the yellow-green algae, the golden-brown algae, and diatoms. This group also has some characteristics that make them unusual. However, the most unusual of the protists make up the last two phyla. Phylum Acrasiomycota consists of the cellular slime molds. The last group, phylum Myxomycota, is made up of the acellular slime molds. Both of these last two groups are very atypical in life cycle and difficult to classify.

18–1 THE KINGDOM PROTISTA

Section Focus 18–1

The purpose of this section is to explain what organisms are placed in this kingdom. The section also gives a brief overview of the classification of protists and some suppositions about the origin of the organisms in this group. Protists are defined as being unicellular eukaryotic organisms. This is a much younger group than the monerans. The oldest fossils of protists are only 1.5 billion years old, indicating that the evolution of the first eukaryote may have taken nearly 2 billion years. The classification of this kingdom, consisting of 115,000 species of extremely diverse organisms, is a source of contention among biologists, who still cannot agree on how it should be done or even what groups of living things should be included.

Performance Objectives 18–1

1. List the characteristics of members of kingdom Protista.
2. Explain the problems with the classification of protists.
3. Explain the Endosymbiont Hypothesis.

Science Terms 18–1

Protista p. 381
protist p. 381
Endosymbiont Hypothesis p. 382

18–2 ANIMALLIKE PROTISTS

Section Focus 18–2

The purpose of this section is to introduce the animallike protists and their importance to other organisms. The first phylum of animallike protists presented is Ciliophora, or ciliates. They all have short hairlike projections, called cilia, that they use for locomotion. There are more than 7000 species of ciliates. One of the best-known groups are the paramecia. This section gives quite a bit of information about the structure and behavior of paramecia. It includes a discussion of the pellicle, trichocysts, the two nuclei, and the contractile vacuoles. There is also an explanation of conjugation. The second phylum, Zoomastigina, of animallike protists includes those single-celled organisms that move through the water by means of flagella. Zoomastiginans generally get their food by absorbing it through the cell membrane. The phylum includes some free-living forms and some that are parasitic on other organisms. Phylum Sporozoa is the third group of animallike protists. The members of this phylum are nonmotile and parasitic on a wide variety of host organisms. A representative member of this phylum is *Plasmodium*, the organism that causes the disease malaria. The life cycle of the *Plasmodium* and some information about malaria are presented in this section. The last group of animallike protists introduced is the phylum Sarcodina. The name of this group comes from the word *sarcode*, used to describe the homogeneous "jelly" from which these cells were once thought to be composed. They move by means of pseudopodia. Ameba is often presented as representative of this phylum. Foraminifers are also discussed.

Performance Objectives 18–2

1. Discuss the four phyla of animallike protists, including characteristics and examples of each group.
2. Discuss the role of animallike protists in the world.

Science Terms 18–2

Ciliophora p. 384
ciliate p. 384
cilium p. 384
paramecium p. 384
pellicle p. 384
trichocyst p. 384
macronucleus p. 384
micronucleus p. 384
gullet p. 384
food vacuole p. 384
anal pore p. 385
contractile vacuole p. 385
Zoomastigina p. 386
flagellum p. 386
flagellate p. 386
Sporozoa p. 387
Sarcodina p. 389
pseudopod p. 389
ameba p. 389

18–3 PLANTLIKE PROTISTS

Section Focus 18–3

The purpose of this section is to introduce the five phyla of plantlike protists. Three phyla, Euglenophyta, Pyrrophyta, and Chrysophyta, are considered to be types of algae; therefore, they are sometimes considered to be plants. The other two groups, Acrasiomycota and Myxomycota, are the slime molds and are sometimes classified as fungi. The euglenophytes are very closely related to the zoomastiginans. They are flagellates with chloroplasts. The best-known members of this phylum are euglenas. This section includes some discussion of the structure and behavior of this organism. Pyrrophytes are commonly called the fire protists. The phylum is made up of a group of organisms known as dinoflagellates. They are mostly photosynthetic and move by means of two flagella. Many of the dinoflagellates are surrounded by thick plates that give them an armored appearance. This group has two unique characteristics that set them apart. Many of them are luminescent. When agitated, they give off light, creating an interesting effect in dark water. This property is the source of the phylum name. The second interesting characteristic of dinoflagellates is that, unlike all other eukaryotic cells, they have no histones around the DNA in their cells. The reason for this is still a mystery. The third group of plantlike protists is Chrysophyta. This phylum includes the yellow-green algae, the golden-brown algae, and the diatoms. This group also has some unique characteristics. The slime molds are divided into two phyla: Acrasiomycota, cellular slime molds, and Myxomycota, acellular slime molds. There is some discussion of the life cycle of both groups in Section 18–3.

Performance Objectives 18–3

1. Discuss the five phyla of plantlike protists, including characteristics and examples in your discussion.
2. Discuss the importance of plantlike protists in the world.

Science Terms 18–3

Euglenophyta p. 394
euglena p. 395
Pyrrophyta p. 396
dinoflagellate p. 396
Chrysophyta p. 398
diatom p. 398
slime mold p. 398
Acrasiomycota p. 398
Myxomycota p. 399
plasmodium p. 399
bloom p. 400
phytoplankton p. 401

CHAPTER PREVIEW

DISCOVERY LEARNING

TEACHING DEMONSTRATIONS
Modeling

An Old-fashioned Hay Infusion
This demonstration can be used to introduce Chapter 18.

About a week before beginning this chapter, put a handful of dried grass or hay in a beaker of distilled water. Put it aside in a dark, warm place for several days. By the time you are ready to begin to talk about protists, there should be many in your culture. (Remember to check it before class, of course.) Put a sample on a slide and use a microprojector to view the specimens as a group. This is a good time to begin noting characteristics. Save the hay infusion culture to compare with pond water samples later.

Symbiotic Flagellates
This demonstration can be used with the material on animallike flagellates in Section 18–2.

The symbiotic organisms that live in the gut of the termite are mentioned in the text. It is relatively easy to make a squash slide of a termite. Pull the head off the termite, bringing the intestinal tract with it. (You can simply squash the whole organism, but locating the flagellates will be more difficult.) Squash the digestive tract in a drop of distilled water and look for the movement of the flagellates. You might have to make two or three slides before you get one that is easy to view, so do this before class. Once you have located the flagellates, either project them on a screen or allow the students to look into the microscope. This makes your discussion of symbiotic relationships between protists and other organisms much more dramatic.

CHAPTER 18
Protists

GUIDED ENQUIRY

Pose the following questions to students and have them record their responses. Point out that they will gain a better understanding of the key concepts if they read the chapter with these basic questions in mind. Upon completion of the chapter, pose the questions again. Ask students to compare their initial responses with those they have developed after reading the chapter.

- Why was phylum Protista created?
- Why is the classification of protists a point of contention for biologists?
- How are protists important to other organisms?

INTRODUCING CHAPTER 18

Using the Visuals

Use a microprojector to show slides of *Paramecium, Euglena,* and diatoms. Show photographs or cultures of slime molds. Look at the chapter-opener photograph and read the caption and introductory information. Have students brainstorm the characteristics of a kingdom that would include all of the organisms highlighted. Remember, you are interested in the kingdom as a whole, not the individuals. Students might notice that it would be a large and diverse group. They might also say that there are a number of different lifestyles. Finally, they probably will notice that the organisms seem to be simple and single-celled. From this discussion you should be able to develop the idea that this is a kingdom created by exclusion.

CHAPTER 18
Protists

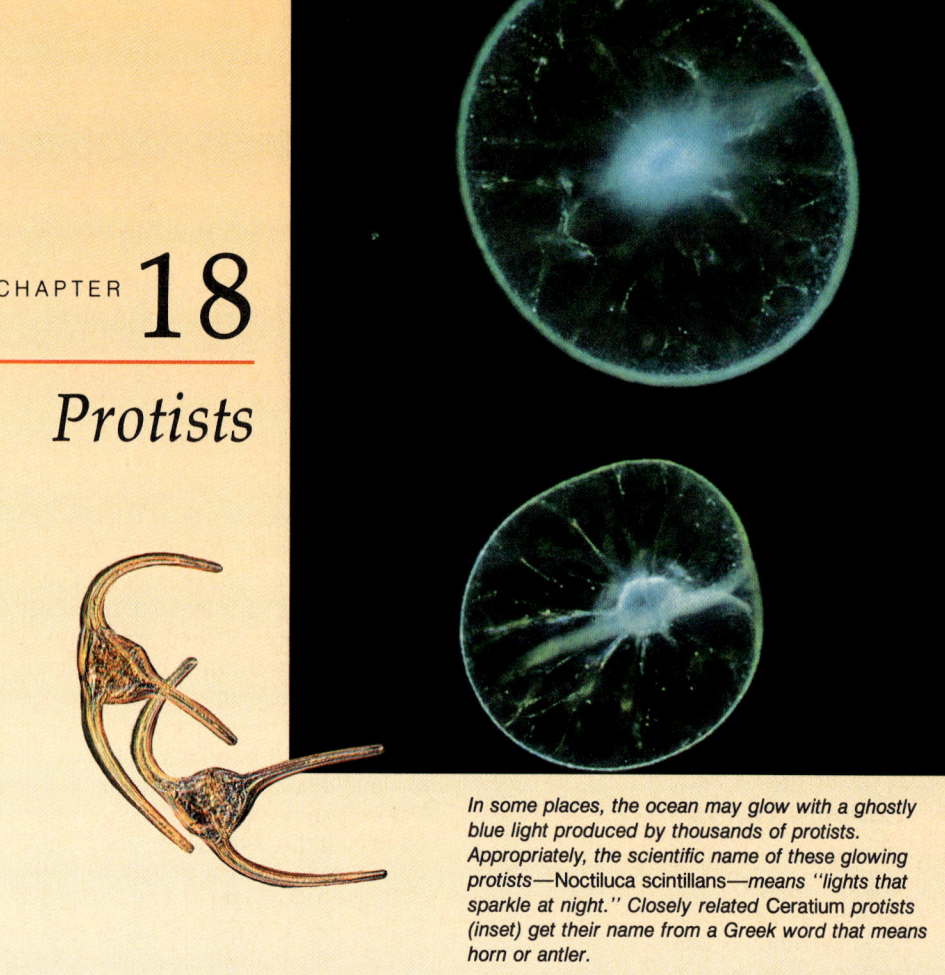

In some places, the ocean may glow with a ghostly blue light produced by thousands of protists. Appropriately, the scientific name of these glowing protists—*Noctiluca scintillans*—means "lights that sparkle at night." Closely related *Ceratium* protists (inset) get their name from a Greek word that means horn or antler.

O n a moonless night, a fishing boat motors toward a New England harbor. The dark water sparkles in its wake, glittering in a thousand places and leaving a ghostly glowing tail as it makes its way to port. Half a world away, the sun is high over a village in equatorial Africa. Although it is harvest time and many members of his family are in the fields working, a young man shudders in his bed, delirious from the burning fever of malaria. These two scenes, as different as they are, have one fundamental biological fact in common: The glowing ocean and the ravages of malaria are both produced by protists—organisms that belong to the kingdom Protista, the first kingdom of eukaryotic organisms.

TEACHING STRATEGY 18–1

Focus/Motivation

Ask students to think about a single cell such as the cheek cells that they looked at under the microscope earlier in the year. Have them recall that they were unable to see them at all without the magnification of the microscope. Tell them that there are 115,000 species of protists, single-celled organisms. Try to imagine with the students some of the complications involved in classifying protists.

GUIDE FOR READING

After you read the following sections, you will be able to

18–1 The Kingdom Protista
- Describe the major characteristics of protists.
- Discuss the Endosymbiont Hypothesis.

18–2 Animallike Protists
- Name and describe the four phyla of animallike protists.
- Discuss how animallike protists fit into the world.

18–3 Plantlike Protists
- Name and describe the five phyla of plantlike protists.
- Discuss how plantlike protists fit into the world.

Journal Activity

YOU AND YOUR WORLD
When did you first become aware of the existence of organisms so tiny that they could not be seen with the unaided eye? In your journal, describe your thoughts and feelings after making this discovery.

18–1 The Kingdom Protista

Guide For Reading
- What are the major characteristics of protists?
- What is the Endosymbiont Hypothesis?

The first kingdom of eukaryotic organisms (organisms whose cells contain nuclei and membrane-enclosed organelles) that we will consider is the kingdom **Protista**. The name is appropriate, for it is derived from a Greek word that means first. The term **protist** refers to any member of the kingdom Protista. **Protists are defined as being unicellular, or single-celled, eukaryotic organisms.** However, as you will discover later in this chapter, a few types of protists stretch the concept of unicellular. Many protists are solitary, which means that they live as individual cells. However, other protists are colonial, which means that they live in groups of individuals of the same species that are attached to one another.

As you learned in Chapter 16, the oldest fossils of monerans (single-celled organisms that lack a nucleus and membrane-enclosed organelles) are more than 3.5 billion years old. Compared with the kingdom Monera, the kingdom Protista is relatively young—the oldest fossils of protists are only about 1.5 billion years old. This indicates that the evolution of the first eukaryote may have taken nearly 2 billion years.

Classification of Protists

The kingdom Protista is an extremely diverse group that includes more than 115,000 species. In the past, many of these species were extremely difficult to classify because they had characteristics in common with more than one of the three kingdoms of multicellular organisms: Animalia (animals), Plantae (plants), and Fungi. The kingdom Protista was created

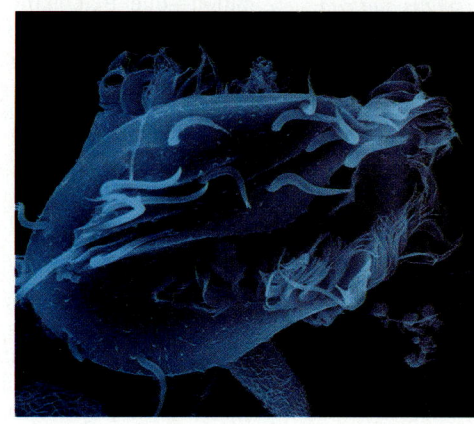 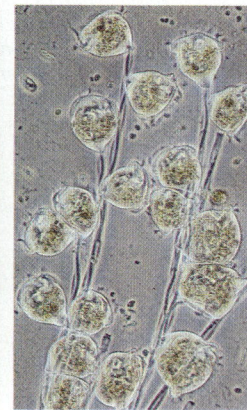

Figure 18–1 This solitary protist zips through the water by means of swimming bristles formed of fused cilia (left). These colonial protists also have cilia (right). However, they lead a less active existence because they are attached to one another and to a base such as a stone or a water plant.

COOPERATIVE LEARNING

Divide the class into eight groups. Groups are to assume the role of educational consultants to the World Health Organization, a branch of the United Nations. Randomly assign one of the following protist-caused diseases to each group: malaria, African sleeping sickness, amebic dysentery, and leishmaniasis. Using the school library or reference materials in your classroom, each group should produce an informative pamphlet about their assigned disease for volunteers working in developing nations. The pamphlet should include the following information: disease-causing organism, method of infection, symptoms, treatment, and preventive measures.

Journal Activity

YOU AND YOUR WORLD
You may also want to use the Journal Activity as a springboard for class discussion. Some students might mention that the first time they became aware of microscopic organisms was in an earlier life science class. Others may have been introduced when they viewed a television special. Other answers are possible. Students should be instructed to keep their Journal Activity in their portfolio.

Content Development

If scientists are confused by the classification of this kingdom, it is safe to assume that students are not going to be too sure of what organisms belong in kingdom Protista either. Stress the general characteristics of the kingdom. You will probably have to review the concept of eukaryotic cells.

- **What is the origin of the name Protista?** (It comes from the Greek word for first.)
- **List at least two characteristics of protists.** (They are unicellular and eukaryotic.)
- **How are protists different from monerans?** (They are eukaryotic instead of prokaryotic.)
- **What does colonial mean?** (It means that they live as individual cells attached to one another.)
- **Why are they not simply called multicellular organisms?** (They do not function as a unit but only as individuals living together.)
- **What does it mean that the oldest fossil protists are 1.5 billion years old and the oldest fossil monerans are 3.5 billion years old?** (It apparently took about 2 billion years for the protists to evolve.)
- **Protists have characteristics in common with what three kingdoms of multicellular organisms?** (Plants, animals, and fungi.)

TEACHING SUPPORT

Teaching Resources
- Activity: Arranging Protists, p. 3

Study Guide
- Section 18–1, p. 169

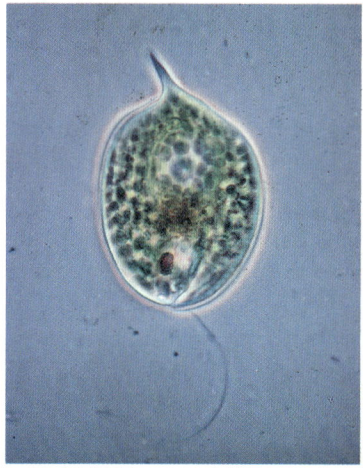

Figure 18–2 Protists often share characteristics with more than one multicellular kingdom. Some are photosynthetic like plants and can move like animals (bottom). Others are similar to fungi in life cycle and appearance but move like animals and have cells that are more like those of animals than those of fungi (top).

primarily to solve the problem of classifying these difficult organisms and only partly because the protists may share an evolutionary ancestry. In fact, the American biologist Lynn Margulis wrote that the kingdom Protista "is defined by exclusion: its members are neither animals . . . , plants . . . , fungi . . . , nor prokaryotes."

Scientists do not agree on how organisms in the kingdom Protista should be classified. They do not even agree on which organisms should be considered protists! However, most would agree that there is much we can learn from protists. In this chapter we will examine the major kinds of protists and see what roles they play in the living world.

Evolution of Protists

Where did the first protists come from? Did prokaryotic cells gradually evolve nuclei? For many years, most biologists considered this a reasonable explanation. In recent years, however, biologist Lynn Margulis has revived an alternative explanation—one that many biologists now find persuasive. She suggests that the first protist cell was formed by a symbiosis among several prokaryotes. (Symbiosis literally means living together and is defined as the living together in close association of dissimilar organisms.)

Margulis noted that a number of organelles in eukaryotic cells are very similar in structure to prokaryotes. For example, mitochondria and chloroplasts closely resemble bacteria and blue-green bacteria, respectively. The flagella and cilia of many eukaryotic cells are similar to a group of bacteria known as the spirochetes. The similarities between organelles and prokaryotes, Margulis reasoned, are not merely coincidental—organelles are descended from symbiotic prokaryotes.

According to Margulis's **Endosymbiont Hypothesis**, these prokaryotes lived within another moneran as endosymbionts (symbiotic organisms that live within another organism, which is called the host organism). The endosymbionts and their

18–1 (continued)

Content Development
If you have old biology textbooks that have fewer than five kingdoms, show these to the students. Explain that there is still much disagreement among biologists about the classification of several groups, the protists among them. Remind students that classification systems are created by humans for convenience and can be changed at any time. Go on to emphasize the statement by Margulis that says that the kingdom Protista was created by exclusion. Ask students to think about what this means and what problems it might bring up. Consider why protists do not really fit into the plant, animal, fungi, or moneran kingdoms.

Content Development
Using the Visuals Talk with students about the traditional idea that prokaryotic cells, over long periods of time, evolved nuclei to become protists. Now spend some time explaining Margulis's Endosymbiont Hypothesis. You will have to explain symbiosis as a beneficial relationship for both organisms. Use Figure 18–3 to help you explain Margulis's hypothesis. Discuss the protist *Cyanophora paradoxa* and its similarity to the hypothesis. Stress that, although similar, it is not the same. The hypothesis can never be proven, but the existence of *Cyanophora* and its sym-

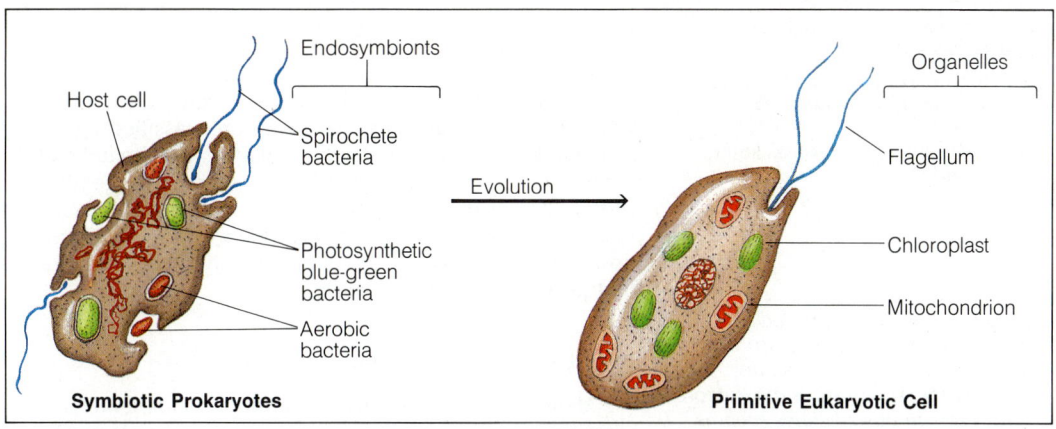

Figure 18–3 According to the Endosymbiont Hypothesis, the organelles in eukaryotic cells evolved from symbiotic prokaryotes that lived inside a host cell.

host cell formed an effective team—each member benefited from the relationship. But eventually the endosymbionts lost their independence, were unable to live without one another or outside the host cell, and gave rise to the organelles that we observe today in eukaryotic cells.

We will never be sure that Margulis's Endosymbiont Hypothesis is correct. But it does provide a model for how the first eukaryotic cell—the first protist—may have developed. Is there something in nature that supports this model? Are there cells alive today that suggest this is a reasonable idea?

The answer to each of these questions is yes. The protist *Cyanophora paradoxa* was first thought to be a kind of alga. However, it was soon learned that the "chloroplasts" within it were not chloroplasts at all. They were blue-green bacteria that could be removed from the cell in which they lived as endosymbionts. They could be grown outside the rest of the cell! The relationship between *Cyanophora paradoxa* and its blue-green bacteria may be similar to the relationships that produced the first protists. Thus when we examine a present-day eukaryotic cell, we may actually be dealing with the descendants of more than one organism!

18–1 SECTION REVIEW

1. What are protists? When did protists first appear on Earth?
2. What evidence has led scientists to believe that symbiosis played an important role in the evolution of protists?
3. **Critical Thinking—Applying Concepts** How might the Endosymbiont Hypothesis explain the double membranes that surround organelles such as the mitochondrion, chloroplast, and nucleus?

BACKGROUND INFORMATION
CONTRACTILE VACUOLES

Water passes through the cell membrane continually due to osmotic pressure. Because of this, the contractile vacuole must keep filling with water and then discharging it to the outside much like collecting water from a leaky boat in a bucket.

Most marine protists do not have contractile vacuoles. Because of the amount of salt dissolved in the oceans, the osmotic pressure on marine cells is much less. They do not need a specialized organelle to remove water.

Other materials, including oxygen, carbon dioxide, salts, and other small molecules can diffuse both into and out of the cell via the cell membrane.

TEACHING STRATEGY 18–2

Focus/Motivation
This section is about animallike protists. Students have not yet done much with the characteristics of animals, except perhaps when taxonomy was studied. This would be a good time to think with the students about all the animal characteristics they can recall. As they think and list some properties, start now to establish those characteristics common to all (or at least most) animals and those possessed only by some. This will help you and the students with material later in the course. The characteristics that you will want to stress at this point are motility, heterotrophic nutrition, and, if it is listed, the lack of a cell wall.

Content Development
Protozoa may be a familiar term to students. When the name *protozoa* was coined, these organisms were considered "first animals," as the name indicates. Discuss why they are now placed in kingdom Protista, but have students take note that they are still referred to as animallike.

Reinforcement/Reteaching
Using the Visuals As you begin to explore the phylogeny of protists, you may want to emphasize biological concepts rather than facts and terms. For example, you can look at modes of locomotion, protective devices, reproduction, food-getting, and economic importance of the various phyla. Putting this information on a features matrix will help your students organize a lot of material in a usable form.

As you discuss cilia, refer to earlier information on the microtubule. Review it briefly to help your students see this example of a pattern that repeats itself in biological systems.

Guide For Reading
- What are the major characteristics of the four phyla of animallike protists?
- How do animallike protists fit into the world?

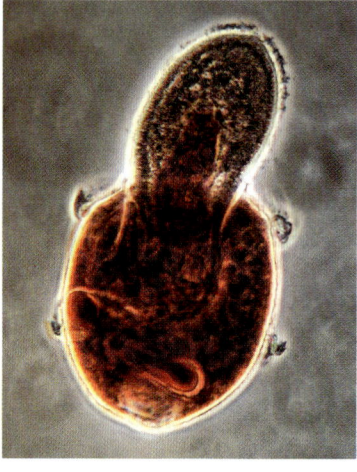

Figure 18–4 Ciliates include trumpet-shaped Stentor *(top)* and egg-shaped Didinium *(bottom). As you can see,* Didinium *is a carnivore that feeds on other ciliates such as* Paramecium.

18–2 Animallike Protists

At one time, many of the protists were called protozoa, which means first animals. Protozoa were classified separately from more plantlike protists. But biologists have found that the similarities between some animallike protists and the plantlike protists are so great that it no longer makes sense to place each in a separate kingdom. Four phyla within the kingdom Protista, however, are known as the animallike protists. These are the first protists that we shall examine.

Ciliophora: Cilia-bearing Protists

Members of the phylum **Ciliophora** (sihl-ee-AHF-uh-rah) are either solitary or colonial organisms. These organisms are often known as **ciliates** (SIHL-ee-ihts) because they have **cilia** (singular: cilium). (Ciliophora literally means cilia-bearing.) Cilia are short hairlike projections that produce movement. The internal structure of cilia consists of microtubulelike structures. The beating of cilia, like the pull of hundreds of oars in an ancient ship, propels the cell rapidly through water. Ciliates are found in both fresh and salt water—many may live in a lake or stream near your home.

More than 7000 species of ciliates are known. Most ciliates are free-living, which means that they do not exist as parasites or symbionts. A well-studied example of the ciliates is found in the genus *Paramecium*.

A **paramecium** is a large organism (as unicellular organisms go)—as much as 350 micrometers in length. In the electron microscope, the cell membrane and cilia can be observed closely. A paramecium's cell membrane and associated underlying structures make up a complex living outer layer called the **pellicle** (PEHL-ih-kuhl). The pellicle is folded in a repeating pattern that gives the surface of the cell a quiltlike appearance. Embedded in the pellicle are a series of tiny flask-shaped structures known as **trichocysts** (TRIHK-oh-sihsts). Trichocysts are used for defense. When a paramecium is confronted by serious danger, the trichocysts discharge. The spiny projections produced in this way can injure a nearby cell as well as cover a paramecium with protective bristles.

Like almost all ciliates, a paramecium possesses two different kinds of nuclei. Each cell normally has a **macronucleus** and a smaller **micronucleus**. We will examine the roles that these two kinds of nuclei play when we consider the process of reproduction in ciliates.

A paramecium obtains food by using its cilia to force water into the **gullet**, an indentation in one side of the cell. Particles that include bits of food such as bacteria are trapped in the gullet and then forced into cavities called **food vacuoles** that form at the base of the gullet. The food vacuoles break off into the cytoplasm and eventually fuse with lysosomes, which are

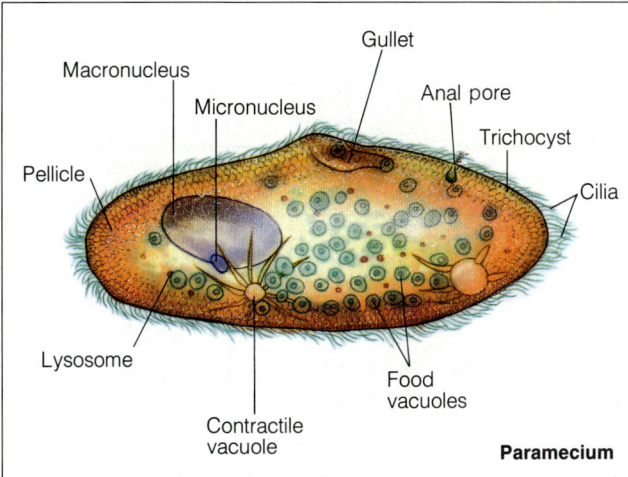

Figure 18–5 The essential life functions of a paramecium are divided among many organelles (left), several of which can be seen in this live paramecium (right). The small green circles are food vacuoles that contain bits of algae.

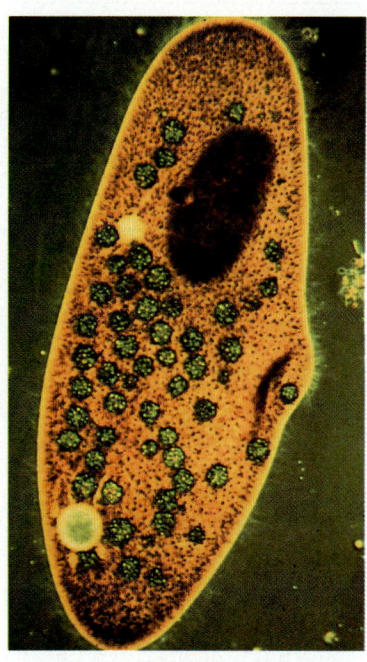

organelles that contain digestive enzymes. Thus the material in the food vacuoles is digested and the organism obtains the nourishment it requires. Waste materials are emptied into the environment when the food vacuole fuses with a region of the cell membrane called the **anal pore**.

Excess water (which moves into a cell in fresh water because of osmosis) is collected in other vacuoles. These vacuoles empty into canals that are arranged in a star-shaped pattern around structures known as **contractile vacuoles**. When a contractile vacuole is filled with water, it "contracts" quickly and pumps water out of the cell.

Under most conditions, a paramecium reproduces asexually by means of a form of mitotic cell division called binary fission. During this process, a single paramecium elongates, its gullet splits into two, and the cell divides in half crosswise. Binary fission results in two cells that are genetically identical.

Under certain circumstances, including starvation and temperature stress, paramecia will engage in a form of sexual reproduction known as conjugation. Refer to Figure 18–7 on page 386 as you read about the process of conjugation. In the first stages of conjugation, two paramecia attach themselves to each other. Their macronuclei disintegrate, and their diploid (2N) micronuclei undergo meiosis. When meiosis is complete, each paramecium contains 4 haploid (N) micronuclei. In the next stages of conjugation, 3 of the 4 micronuclei disintegrate. The 1 remaining micronucleus then divides to form 2 genetically identical haploid micronuclei. The paramecia exchange one set

Figure 18–6 A paramecium can discharge its trichocysts to produce a shield of protective bristles.

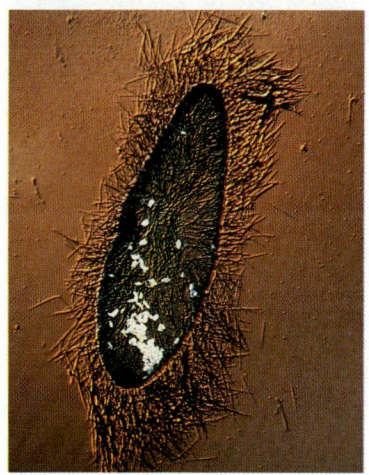

FACTS AND FIGURES

In some ciliates, some of the cilia have fused together to form specialized feeding structures called undulating membranes or membranelles. In other ciliates, the cilia have fused to form bristlelike structures that are used for "walking." A few ciliates have cilia only for brief periods during their life cycle.

FACTS AND FIGURES

Some ciliates, such as *Didinium*, have toxic trichocysts. These toxic trichocysts are used for capturing prey and for defense. Discharged trichocysts may also be used to anchor a ciliate while it is feeding.

gests its food, have your students think about their own bodies and the processes as they occur in humans.
- **In this process in the *Paramecium*, what is the function of cilia?** (They sweep the food into the gullet.)
- **What structure do you use for the same purpose?** (The hand or arm.)
- **Where does digestion occur in the *Paramecium*?** (Inside the food vacuoles inside the cell.)
- **Where does digestion occur in the human body?** (In the digestive tract.)
- **What is the role of the lysosome in digestion in the *Paramecium*?** (It contains digestive enzymes used to digest the food.)
- **Do humans have digestive enzymes?** (Yes.)
- **Summarize the similarities and differences between digestion in the *Paramecium* and human digestion.** (Answers will vary but should reflect the answers already given.)

Stress that cilia can be used for different functions by different organisms, including movement and food-getting.

Be sure that students realize that there is great diversity within the ciliates. Use Figure 18–4 to help you make this point clear.

Paramecium is used frequently as a representative type, but it is not the only ciliate. As you talk about *Paramecium*, use all of the visual aids you have available. The ideal, of course, is to have live, moving specimens projected on a screen for students to observe as you discuss paramecia with them. Use the figures in the textbook to illustrate the internal structures. Food-getting and digestion are important activities for all organisms and can easily be studied. As you discuss the way the *Paramecium* gets and di-

BACKGROUND INFORMATION
CONJUGATION

Conjugation is an interesting aspect of ciliate reproduction, but many students find it confusing. During the process, the two cells exchange part of their micronuclear "libraries" in order to form new combinations of genetic information. Once these new combinations are formed, the cell destroys its old macronucleus and makes a new one from the new set of information. Each cell leaves the conjugation event with a genetic makeup that is different from the one with which it entered. The macronuclei seem to contain the genes that must function on a daily basis to keep the cell alive. For the sake of efficiency, those genes have been copied hundreds of times. The existence of two kinds of nuclei seems to help the cell to express the genes required all the time and also allows it to keep a repository of genetic information that may be passed on to future generations.

18–2 (continued)
Enrichment
Review asexual and sexual reproduction. Discuss binary fission. Stress that the result is two cells that are genetically alike. Think with the students about the advantages of sexual reproduction. Discuss why paramecia would ever need a form of sexual reproduction. (Remember to stress Darwin's principles of natural selection and survival of the fittest.)

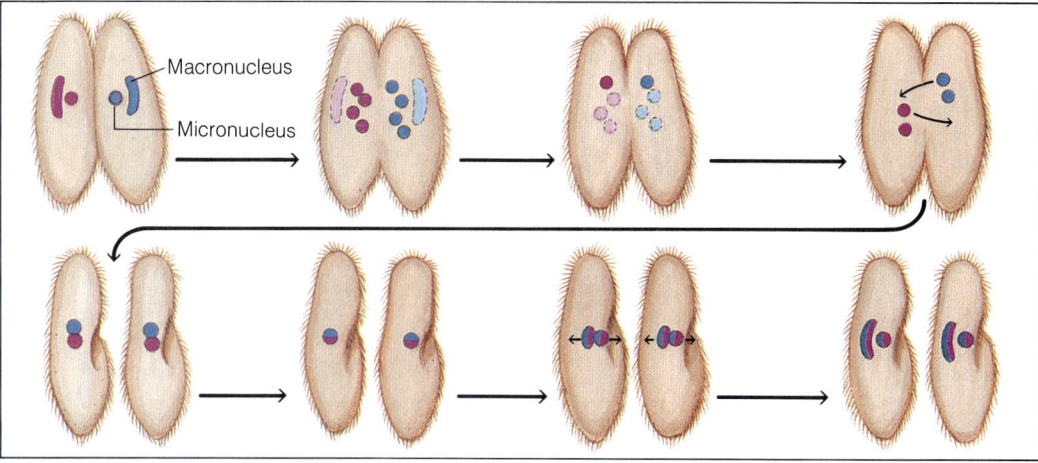

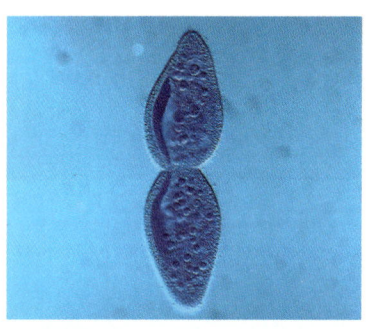

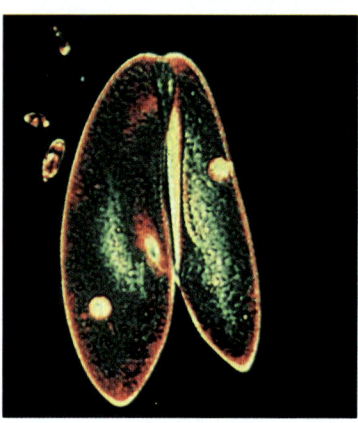

Figure 18–7 In the process of conjugation, two paramecia join together and share genetic information. Conjugation produces new combinations of genes—combinations that may give paramecia a better chance of survival.

of micronuclei, so that each cell has 1 micronucleus obtained from the other cell and 1 micronucleus of its own. In the final stage of conjugation, the paramecia separate from each other. The 2 haploid micronuclei in each paramecium fuse to form a new diploid micronucleus. From the new micronucleus, a new macronucleus is formed. The two paramecia that participated in conjugation are now genetically identical.

Strictly speaking, conjugation is not reproduction. No new cells are produced—2 cells enter conjugation and 2 leave it. Nonetheless, it is still a sexual process. New combinations of genetic information are produced. Within a large population, the process of conjugation helps to create genetic diversity and ensures the ultimate survival of the species.

Zoomastigina: Animallike Protists with Flagella

The phylum **Zoomastigina** (zoh-oh-mas-tuh-GIGH-nuh) consists of animallike protists that move through the water by means of **flagella**. Flagella are long, whiplike projections that have an internal structure identical to that of cilia. The number of flagella varies from one zoomastiginan to the next, ranging from one to four or more. Because they have flagella, zoomastiginans are called **flagellates**. The term flagellate is also used for plantlike protists and unicellular algae that have flagella. Because of this, zoomastiginans are sometimes called zooflagellates, which means animal flagellates.

Zoomastiginans are generally able to absorb food through their cell membranes, which are not enclosed in shells or cell

Figure 18–8 Paramecia usually reproduce asexually through binary fission (top). Under certain circumstances, paramecia will undergo a form of sexual reproduction called conjugation (bottom).

The pellicle in *Paramecium* and other free-swimming ciliates consists of the plasma membrane, flattened membranous vesicles (alveoli) just underneath the plasma membrane, and a variety of other structures, such as trichocysts and the bases of cilia. The placement of the alveoli beneath the plasma membrane gives the pellicle its characteristic ridges and bumps. Because of the alveoli, the pellicle can be thought of as having three layers of membrane: plasma membrane, outer alveolar membrane, and inner alveolar membrane.

In other protists, the structure of the pellicle may be quite different. For example, the pellicle in *Euglena* consists of interlocking protein strips just beneath the plasma membrane.

Content Development
Using the Visuals Compare the processes of binary fission and conjugation using Figures 18–7 and 18–8. Stress that conjugation does not result in additional cells but that there are new cells produced. Emphasize again that this

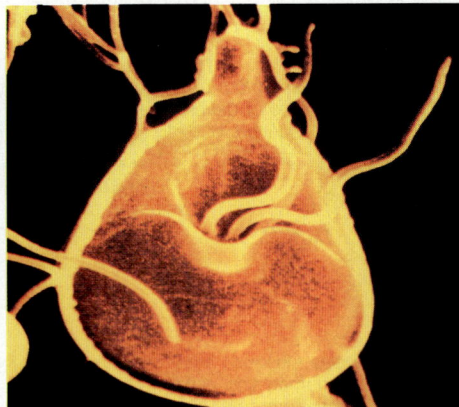

walls. Some zoomastiginans have found special environments in which they are able to find enough food to absorb. Others live within the bodies of other organisms, taking advantage of the food that the larger organism provides.

Zoomastiginans can reproduce asexually by binary fission, although most also have a sexual life cycle as well. During sexual reproduction, gamete cells are produced by meiosis. Sometimes meiosis is triggered by a change in the food supply or in the amount of oxygen in the water. In some species, meiosis occurs only at certain times of the year. The gametes formed by meiosis fuse together, forming an organism with a new combination of genetic information.

Some zoomastiginans are found in lakes and ponds. Others exist as parasites or symbionts of other organisms. You will learn about the relationships of these zoomastiginans to other organisms shortly, when we consider how the animallike protists fit into the world.

Sporozoa: Spore-producing Parasitic Protists

The members of the phylum **Sporozoa** (spohr-oh-ZOH-uh) are nonmotile, which means that they do not move. All sporozoans are parasitic; that is, they live in a host organism and cause it harm. Sporozoans are parasites on a wide variety of other organisms, including worms, insects, fish, birds, and humans. Many sporozoans have complex life cycles that involve more than one host. Sporozoans reproduce by means of spores, which are cells or groups of cells enclosed in a protective membrane. Under the right conditions, spores are able to attach themselves to a host cell, penetrate it, and then live within it as parasites. A typical sporozoan is *Plasmodium*, which causes the human disease malaria.

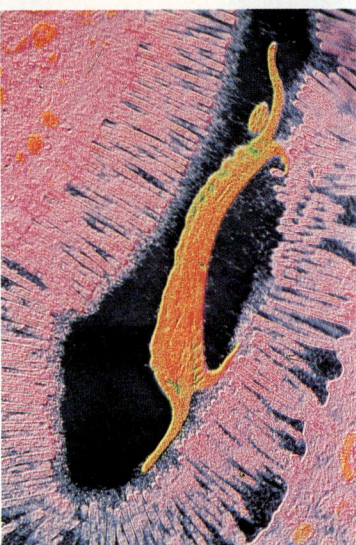

Figure 18–9 Many zoomastiginans are parasites. Giardia (top, right and bottom) attaches to the lining of the small intestine in humans, causing much irritation and digestive disturbance. Trichomonas (top, left) causes intestinal and venereal diseases in humans. It is also responsible for a number of diseases in livestock and poultry.

HISTORICAL NOTE
CILIA VERSUS FLAGELLA

Many students believe that there is an unexplained, important difference between a cilium and a flagellum. Actually, the difference is completely arbitrary. The two words, coined during the nineteenth century, refer to hair (cilia) and whips (flagella) and were originally applied because of the difference in length between the two structures. Smaller, more numerous structures were called cilia and longer structures were known as flagella. Today we recognize that cilia and flagella have the same internal structure and produce motion by the same mechanism. The two terms persist, but the structures are identical organelles.

---

process, which is much more complex and energy-consuming than binary fission, is very important to the survival of the species.

Content Development
Compare flagella and cilia as modes of locomotion. Again point out that the flagellates as a group are very diverse. Even the number of flagella varies. Alert students to the close resemblance between plantlike flagellates and animallike flagellates. This will be discussed in more detail later in the chapter.

The main difference to be emphasized here is the food-getting behavior. Discuss the reproductive behaviors of the flagellates with students. The sexual reproductive cycle is an example of true sexual reproduction. Contrast this with conjugation. Remind students about symbiotic relationships, including beneficial ones and harmful ones like parasitism. Stress that zoomastiginans can be free-living, symbiotic, or parasitic. If you are doing a features matrix, you may have to supplement the information given in the textbook or wait until a later part of the chapter to complete it.

TEACHING SUPPORT

Laboratory Manual
- Adaptations of the Paramecium, p. 225

Teaching Resources
- Hands-On Activity: Putting the Squeeze On, p. 89

MEDIA AND TECHNOLOGY
- Biology Transparencies
- 20: Protists

HISTORICAL NOTE
MALARIA

Malaria is an important disease from a historical perspective. In terms of total human deaths that have been attributed to infectious diseases, malaria ranks among the greatest killers. Even here in the United States, occasional cases are detected. This was particularly true among the military personnel returning from Vietnam.

MULTICULTURAL STRATEGY

Have students locate areas of the world where malaria is a serious health problem. Have students present their findings in the form of a map. Students may also wish to investigate types of prevention and treatment of malaria used in various locations.

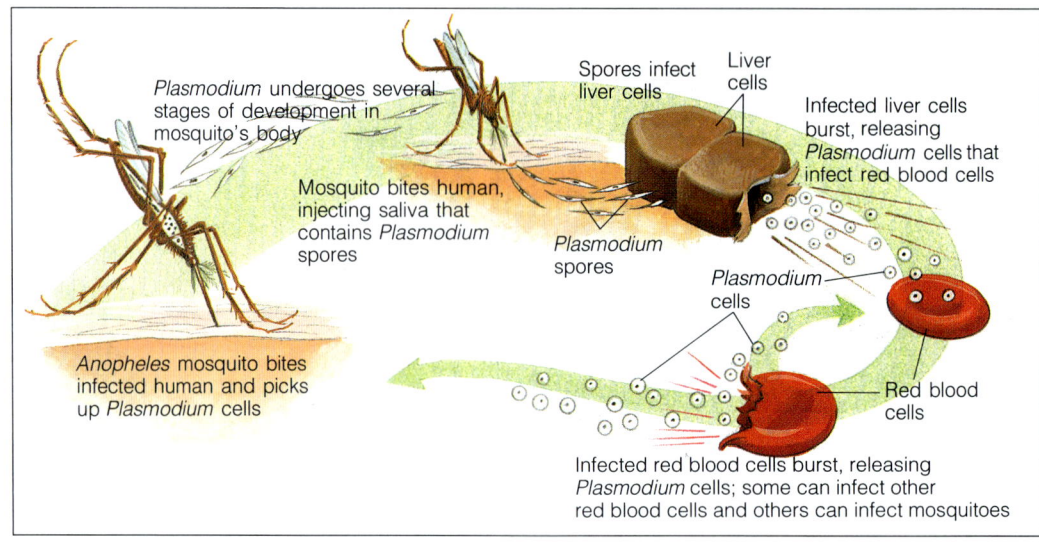

Figure 18–10 Plasmodium *causes the human disease malaria.*

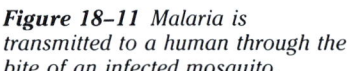

Figure 18–11 Malaria is transmitted to a human through the bite of an infected mosquito.

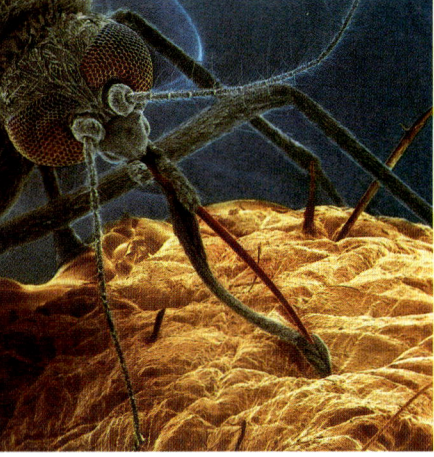

Malaria is carried by the *Anopheles* (uh-NAHF-uh-leez) mosquito. When an infected mosquito bites a human, some of its saliva, which contains spores of the parasite, is injected into the bloodstream. Once inside the body, *Plasmodium* infects liver cells and then red blood cells. *Plasmodium* grows rapidly within the infected cells and eventually causes these cells to burst at intervals of 48 or 72 hours. When millions of parasite-filled red blood cells burst, they dump large amounts of toxins into the bloodstream. The toxins produce chills and fever—the symptoms of malaria.

The disease is transmitted back to the mosquito if a mosquito bites a human infected with malaria. The blood that the mosquito swallows contains *Plasmodium*. In the insect's digestive system, *Plasmodium* grows rapidly and penetrates the insect's entire body, including the salivary glands. After a time, the infected insect contains active *Plasmodium* spores in its saliva and is able to pass the infection on to another human. Malaria is a serious disease that weakens an infected individual and may lead to death. In areas of the world where *Anopheles* mosquitoes flourish, malaria is a serious problem. Every year, more than 250 million people suffer from malaria, and more than 2 million people die from it. Although drugs such as chloroquinine are effective against some forms of the disease, many strains of malaria-causing *Plasmodium* sporozoans are resistant to these drugs. To date, the most effective way to control malaria is to destroy breeding areas for *Anopheles* mosquitoes. This interrupts the life cycle of *Plasmodium* and thus prevents the spread of malaria.

18–2 (continued)

Focus/Motivation
Emphasize that sporozoans are quite different from the two phyla already studied. Students are sometimes interested to learn that "lower" animals such as worms and insects can be affected by parasites. This is a good opportunity to help students look at a bigger picture of the world. Encourage them to be aware of things that do not affect them personally. Some things in nature appear to have no impact on humans at all, whereas others are directly or indirectly important.

Content Development
Using the Visuals Emphasize again that the members of phylum Sporozoa are very diverse. *Plasmodium* is used as one ex-

Sarcodina: Protists with False Feet

The phylum **Sarcodina** (sahr-kuh-DIGH-nuh) contains protists that use temporary projections of cytoplasm to move and feed. Such a projection is called a **pseudopod** (SOO-doh-pahd), which literally means false foot. Pseudopods are usually thought of as being rounded and broad. However, some sarcodines have thin, strandlike pseudopods and others have weblike pseudopods. The name Sarcodina comes from the word sarcode, which was coined in the nineteenth century to describe the homogeneous "jelly" from which these cells were thought to be composed.

One major family of Sarcodina is the **amebas**. Amebas are flexible, active cells without cell walls, flagella, cilia, and even a definite shape. Amebas move by means of thick pseudopods, which they extend out of the central mass of the cell. The cytoplasm of the cell streams into the pseudopod, and the rest of the cell follows. This motion is known as ameboid movement.

An ameba is capable of capturing and eating particles of food and even other cells. It does so by first surrounding its meal with streaming cytoplasm and then taking it inside the cell to form a food vacuole. Once inside the cell, the material is digested rapidly and the nutrients are passed along to the rest of the cell. Amebas reproduce by means of binary fission—one large ameba divides by mitosis to produce two smaller, but genetically identical, amebas.

Amebas are not the only members of the phylum Sarcodina. The phylum also includes three groups known as heliozoans, radiolarians, and foraminifers (for-uh-MIHN-ih-ferz). Most of these protists are beautiful organisms that produce external shells to help support their unusual shapes. Although some heliozoans and radiolarians do not have shells, many produce delicate shells of silica (SiO_2), a glasslike substance.

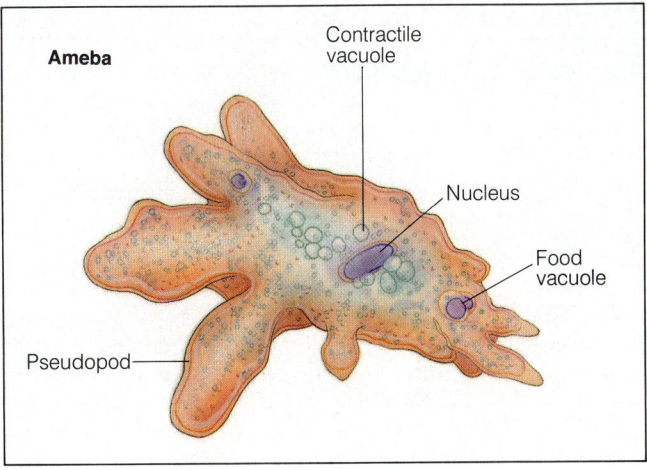
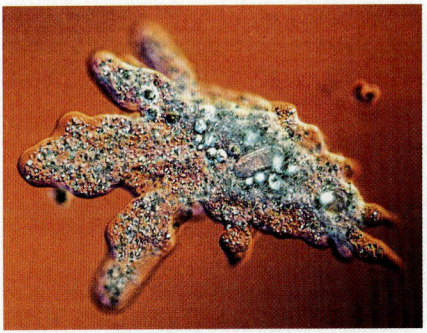

Figure 18–12 *Although a live ameba may at first appear to be a featureless blob (right), careful study reveals its internal structure is well-organized.*

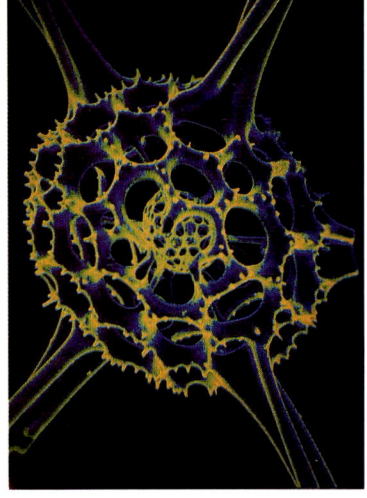

Figure 18–13 Radiolarians have delicate shells of silica.

Foraminifers secrete shells of calcium carbonate ($CaCO_3$). These protists are abundant in the warmer regions of the oceans. As foraminifers die, the calcium carbonate from their shells accumulates on the ocean bottom. In some regions, thick deposits of foraminifer skeletons have formed on the ocean floor. The white chalk cliffs of Dover, England, are huge deposits of foraminifer skeletons that were raised above sea level by geological processes.

Most species of foraminifers are known not from living specimens but from fossils of their skeletons. Because they have continued to change and evolve over millions of years, the species of foraminifers found in sedimentary rocks are useful measures of the ages of such rocks. By examining foraminifer fossils from rock samples, geologists can determine the age of the samples. This makes it possible to date certain other fossils and also to predict where oil may be found. Because the richest oil deposits were formed at certain times in the Earth's history, foraminifer fossils are valuable clues to the presence of oil in rocks.

Summary of the Animallike Protists

- **Members of the phylum Ciliophora, such as *Paramecium*, are known as ciliates. Almost all ciliates use cilia for movement.**
- **Members of the phylum Zoomastigina are known as flagellates because they use flagella for movement.**
- **Members of the phylum Sporozoa, such as *Plasmodium*, reproduce by means of spores. Sporozoans are nonmotile and parasitic.**
- **The phylum Sarcodina includes amebas and foraminifers. Sarcodines use pseudopods for feeding and movement.**

Figure 18–14 A live foraminifer, which resembles a clump of tinsel, captures food with its threadlike pseudopods (left). Foraminifers produce shells of calcium carbonate (right).

PROBLEM SOLVING IN BIOLOGY

INTERPRETING GRAPHIC DATA ABOUT A FORAMINIFER

Globigerinoides sacculifer is a foraminifer that lives near the surface of the ocean. The shell of *G. sacculifer* consists of several bubble-like chambers that are covered with long iridescent spines. Numerous perforations in the chambers of its shell permit *G. sacculifer* to extend many threadlike pseudopods among its spines. These pseudopods are used to capture microscopic animals and protists. In the laboratory, *G. sacculifer* is fed day-old brine shrimp.

Study these graphs, which show the results of experiments involving the growth of *G. sacculifer*'s shell under different conditions. Answer the questions that follow.

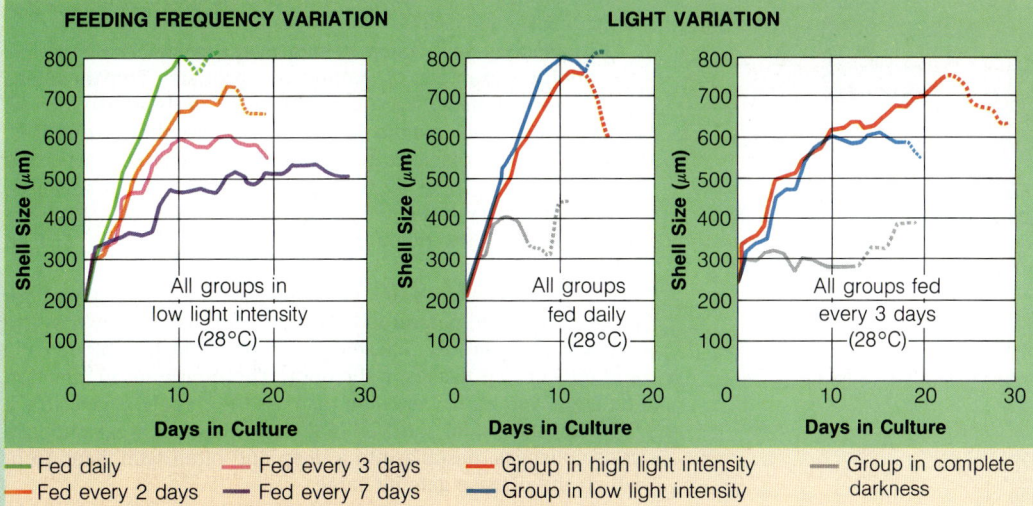

The shell of *Globigernoides sacculifer* grows at different rates under different conditions. It stops growing when the protist dies or undergoes gamete formation. (In gamete formation, the body of the foraminifer divides into many tiny cells that leave through the pores of the shell.)

1. What variables were tested in these experiments?
2. According to these graphs, what is the maximum shell size in *G. sacculifer*?
3. Explain why the data represented by the dotted lines are less significant than the data shown by the rest of the curve.
4. *G. sacculifer* has symbiotic dinoflagellates that live in its cytoplasm. How might this fact explain why the groups of foraminifers raised in complete darkness died or underwent premature gamete formation?
5. How does light intensity affect final shell size?
6. How does feeding frequency affect final shell size?
7. What is the dominant factor in shell growth in *G. sacculifer*? Explain your answer.

PROBLEM SOLVING

Interpreting Graphic Data About a Foraminifer

This activity involves the presentation of data in graph form. The students are asked to answer questions based on their ability to interpret the graphs. This is an important but difficult skill for many students. After discussing foraminifers and looking at pictures, if available, go over the graphs with students. If some students are having particular difficulty with this exercise, assign them partners so that they have the opportunity to discuss and explain the graphs to one another.

After the students have completed the exercise, brainstorm other variables that might be tested, the expected results, and how they could be represented in graph form.

Answers to questions

1. Light intensity; feeding frequency.
2. About 800 μm.
3. These represent data for only two or three individuals.
4. *G. sacculifer* probably depends on its photosynthetic symbiotes for part of its nutritional needs.
5. The foraminifers raised in high light intensity grew slightly larger than those raised in low light intensity. However, as long as there is enough light for some photosynthesis, light intensity affects shell growth very little.
6. The more often the foraminifers were fed, the larger was the final shell size.
7. Feeding frequency is the dominant factor in shell growth. Variations in feeding frequency produce greater difference in shell size than variation in light intensity.

How Animallike Protists Fit into the World

The animallike protists are found throughout the world. They are some of the most common organisms in the oceans, and they are also abundant in fresh water, on land, and in the bodies of larger organisms.

HARMFUL RELATIONSHIPS Unfortunately for us and for other organisms, there are a great many parasitic protists. Parasitic protists affect plants, all types of animals (including humans), and even other protists. You have already read about one important protist parasite and pathogen (something that causes disease)—the genus of sporozoans called *Plasmodium*, which causes malaria.

Zoomastiginans, which belong to the genus *Trypanosoma*, are another example of pathogenic protists. Trypanosomes (trih-PAN-oh-sohmz) live in the blood of vertebrates such as humans and other mammals and cause a number of diseases. Although these diseases have different common names, any disease caused by trypanosomes is called trypanosomiasis, which means trypanosome infection. One form of trypanosomiasis that affects humans is commonly known as African sleeping sickness.

The trypanosomes that cause African sleeping sickness are passed from one person to another by an insect known as the tsetse (TSEET-see) fly. These trypanosomes destroy blood cells and infect other tissues in the body. The symptoms of infection include fever, chills, and skin rash. As the trypanosomes attack the nervous system, infected individuals become weak and lose consciousness, passing into a deep and often fatal sleep from which the disease gets its name.

Some trypanosome species infect domestic livestock. In areas infested with the tsetse fly, which include vast regions of central Africa, the raising of cattle is virtually impossible. The control of the tsetse fly and the protist pathogens that it spreads is a major goal of scientists in Africa and around the world. The only trypanosome native to the Western Hemisphere is *T. cruzi*, which causes Chagas disease. This disease can result in heart failure because the trypanosomes invade and weaken muscle, especially heart muscle. Some historians believe that Charles Darwin contracted this disease during his visit to South America and that it was this parasite that made him ill much of his life.

A third kind of pathogenic protist is similar in appearance to the harmless amebas that you may find in a nearby pond or examine under a microscope. In certain regions of the world, many people are infected with species of *Entamoeba*, which cause a disease known as amebic dysentery. The parasitic amebas that cause this disease live in the intestines, where they absorb food from the host. They also attack the wall of the

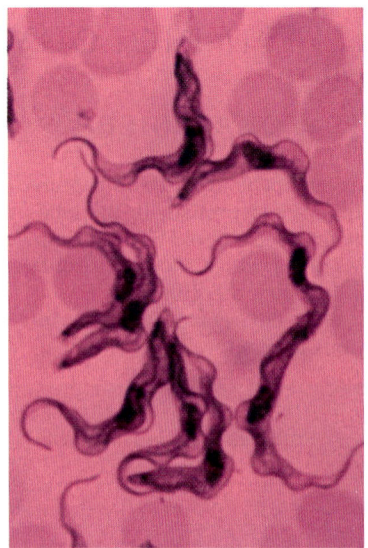

Figure 18–16 *Trypanosomes are parasites that live in the blood of vertebrates and cause a number of diseases, such as African sleeping sickness and Chagas disease.*

intestine itself, destroying parts of it in the process and causing severe bleeding. These amebas are passed out of the body in feces. In places where sanitation is poor, the amebas may then find their way into supplies of food and water. In some areas of the world, amebic dysentery is a major health problem, weakening the human population and contributing to the spread of other diseases.

HELPFUL RELATIONSHIPS Although many animallike protists are responsible for disease, there are many that are beneficial to other organisms. An interesting example of a beneficial protist is *Trichonympha*, a zoomastiginan that lives within the digestive systems of the termite and the wood roach. Termites eat wood, the major component of which is the carbohydrate cellulose. But termites do not have enzymes to break apart the chemical bonds that hold the simple sugars in cellulose together. Neither do we, incidentally, so it does us little good to munch on a woodchip. How, then, does a termite digest cellulose? It doesn't. *Trichonympha* does.

Young termites ingest some of the feces of an adult termite, thereby swallowing several thousand *Trichonympha*, which then take up residence in the termites' digestive system. Experiments have shown that insects deprived of these protists cannot live more than a few days. *Trichonympha* and other organisms living in the termites' gut manufacture cellulase, the enzyme that the termites need to break the chemical bonds in cellulose. Without these symbionts, termites and wood roaches would be no more able to digest wood than we are.

Figure 18–17 Particles of wood appear as irregular grains inside the body of Trichonympha, *a wood-digesting protist (left).* Trichonympha *lives in the digestive system of insects such as termites (right), making it possible for them to obtain nutrients from the wood they eat.*

TEACHING SUPPORT

Study Guide
- Section 18–2, p. 170

18–2 (continued)

Content Development

Be sure that students take note of the short paragraph at the top of page 394. Not only does it talk about another reason that protists are important, but in a larger sense, it brings to mind the interrelationships among all of Earth's organisms.

SECTION REVIEW 18–2

1. Student examples will vary. Ciliophora: *Paramecium*. Zoomastigina: *Trypanosoma*. Sporozoa: *Plasmodium*. Sarcodina: *Ameba*.
2. Almost all ciliates use cilia to move. Zoomastiginans use flagella. Sporozoans are nonmotile. Sarcodines typically use pseudopods to move.
3. Conjugation: ciliates join together; macronucleus breaks down; micronucleus divides; micronuclei are exchanged; micronuclei fuse; new macronucleus forms. Strictly speaking, conjugation is not reproduction—no additional cells are formed.
4. Some help wood-eating animals to digest their food; others are a source of food for tiny organisms.

Guide For Reading
- What are the major characteristics of the five phyla of plantlike protists?
- How do plantlike protists fit into the world?

Reinforcement/Reteaching

For those students who have difficulty with the Section Review questions, go back to the discussion of the material students have difficulty understanding.

Closure

Review the information in the features matrix. Divide the class into small groups. Have each group develop a paragraph on one phyla of protists. Share the paragraphs with the class as a review of Section 18–2.

The animallike protists play another major role in the living world. Enormous numbers of protists living in the seas are food for tiny multicellular animals that in turn serve as food for larger animals. A similar role is played by protists in freshwater lakes, streams, and ponds. Without such tiny organisms, the larger fish would have no food supply. The animallike protists thus perform an essential function for all other living things.

18–2 SECTION REVIEW

1. List the four phyla of animallike protists. Give an example of each.
2. Compare the forms of locomotion used by the four phyla of animallike protists.
3. Describe the process of conjugation. Is conjugation a form of reproduction? Explain your answer.
4. **Connection—Ecology** In what ways are animallike protists helpful to other living things?

18–3 Plantlike Protists

In addition to the four phyla of animallike protists, we recognize five phyla of plantlike protists. Like other protists, these organisms are unicellular and most of them are motile. We call them plantlike because most contain the pigment chlorophyll and carry out photosynthesis. Like zoomastiginans, many plantlike protists are called flagellates because they have flagella. These organisms are known as phytoflagellates, which means plant flagellates, to distinguish them from zoomastiginans (zooflagellates).

Three of the phyla of plantlike protists—Euglenophyta, Pyrrophyta, and Chrysophyta—are considered to be types of algae. These simple plantlike organisms, found in water or damp places, lack true roots, stems, and leaves. As a result, these three phyla are sometimes classified as plants. The unusual slime molds, which are placed in phyla Acrasiomycota and Myxomycota, are not photosynthetic and have many funguslike characteristics. They are sometimes known as funguslike protists, or are even classified as fungi.

Euglenophyta: Flagellates with Chloroplasts

The members of the phylum **Euglenophyta** (yoo-glee-nuh-FIGHT-uh) are closely related to zoomastiginans. In some classification schemes, euglenophytes and zoomastiginans are actually considered to be in the same phylum. The main reason

TEACHING STRATEGY 18–3

Focus/Motivation

Bring a houseplant to class and set it on your desk or some other visible location. Talk about the characteristics of the plant. The most obvious one, of course, is

for grouping them together is that except for the fact that euglenophytes possess chloroplasts, the two phyla of protists closely resemble each other.

The most famous members of the phylum Euglenophyta belong to the genus from which the entire phylum takes its name: *Euglena*. As you can see in Figure 18–18, a **euglena** is a long cell that has a pouch that contains two flagella at its front end. The longer of these two flagella extends far out of the euglena's pouch and is used to propel the cell forward through the water. A euglena is an excellent swimmer and can move very quickly in this manner. However, when a euglena is forced against a surface—for example, when it is squeezed down on a glass laboratory slide—it is able to move in a different manner. The euglena changes shape rapidly and crawls along the surface by a process called euglenoid movement. Thus a euglena is able to move along in a distinctly animallike fashion.

A red eyespot at the front end of the cell (the end with the flagella) helps a euglena find the brightest areas of its immediate environment. Finding sunlight is important to a euglena because it is filled with between 10 and 20 oval chloroplasts. A euglena is a full-fledged phototrophic autotroph, or an organism that makes its own food from light and simple raw materials, and thus is able to carry out the light and dark reactions of photosynthesis.

When sunlight is not available, a euglena can also live as a heterotroph, or an organism that eats food made by other organisms. If dissolved nutrients are available in the water, a euglena can absorb them and get along in darkness with no ill effects. In nature this gives euglenas the ability to live as saprophytes, or organisms that absorb the nutrients available in decayed organic material.

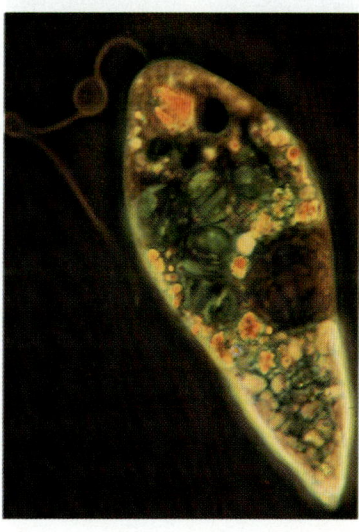

Figure 18–18 A euglena's organelles include a large prominent nucleus, whiplike flagella, green chloroplasts, and grainy yellow food-storage structures.

that it is green, so you will want to mention the chloroplasts and chlorophyll. You may also want to mention lack of motility and the presence of cell walls. Point out that this particular plant is multicellular, which is one characteristic not typical of the plantlike protists. The other characteristics are typical of the group you are about to study. However, not all properties will appear in all organisms. Again, we are talking about a group that was created by exclusion, and students should keep that in mind.

Content Development
Present an overview of the five groups of plantlike protists. When you talk about the slime molds, tell students that these funguslike organisms were once classified as plants because of the presence of the plantlike cell walls. In some current classification systems, they are still considered fungi.

Reinforcement/Reteaching
Using the Visuals Continue with the features matrix, adding the five groups of plantlike protists. You may want to spend some additional time on euglenophytes. The dilemma of flagellate classification is a good example of the dynamic nature of science. The more biologists learn, the more complex some situations become. Not all information leads to answers: Sometimes new questions arise. Be sure to emphasize the dual nutritional capabilities of *Euglena*. Have students think about what a great adaptation that would be for humans. Use Figure 18–18 to point out the structures found in *Euglena*. You might want to point out the similarities and differences between *Euglena* and *Paramecium*.

395

TEACHING SUPPORT

Laboratory Manual
- Comparing Protists, p. 231

Teaching Resources
- Hands-On Activity: Shedding a Little Light on Euglena, p. 93

MEDIA AND TECHNOLOGY

 Videodisc
- *Aquatic Ecosystems: Freshwater Wetlands and Freshwater*
Use this bar code to introduce the class to phytoplankton that form the basis of a food chain in a lake.

Play frames 21504 to 22242

See the Biology Media Guide page 16 for additional bar-code correlations for this section.

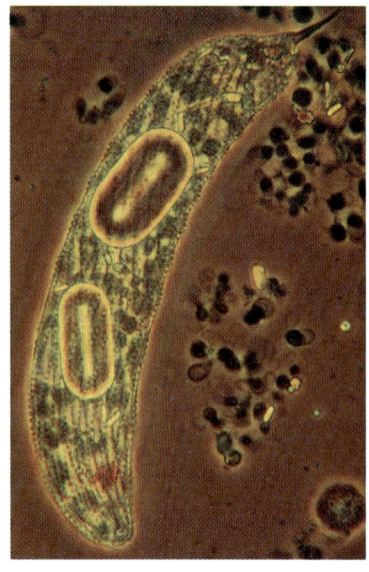

Figure 18–19 A euglena is a photosynthetic protist that swims by means of flagella. The ridges that spiral around the body of the euglena are part of its pellicle.

The pellicle (the cell membrane and associated structures) of *Euglena* is of special interest to many biologists because of its unusual structure. The pellicle consists of a series of ribbonlike ridges that spiral around the surface of the organism. See Figure 18–19. Underneath each ridge is a small sac and a set of microtubules. The microtubules and the sacs probably are important in maintaining the shape of the cell when it is swimming and in generating the force for euglenoid movement.

A euglena reproduces asexually by binary fission. In this process, a single euglena divides in two, beginning at the end with the flagella and finally separating at the base. Just prior to cell division, the cell doubles the number of ridges in its pellicle by producing tiny new ridges between the old ones. Because of their adaptability—there are few organisms able to exist as both autotrophs and heterotrophs—euglenas are able to live in many different places and are quite common.

Pyrrophyta: Fire Protists

The members of phylum **Pyrrophyta** (pigh-roh-FIGHT-uh) are a group of organisms known as the **dinoflagellates**. Most dinoflagellates are photosynthetic, although a few have lost their chloroplasts and exist as heterotrophs. The cells usually swim by means of two flagella. In most dinoflagellates, one flagellum wraps around the organism like a belt, whereas the other trails behind like a tail. Many dinoflagellates are surrounded by thick plates that give them a strange armored appearance. Dinoflagellates reproduce asexually by binary fission.

The dinoflagellates have two interesting properties that set them apart from most other protists and, indeed, most other organisms. First, a great many dinoflagellate species are luminescent. When agitated by sudden movement in the water, they give off light. Many areas of the ocean are so filled with dinoflagellates that the movement of an oar or the wake of a boat will cause the dark water to shimmer with a ghostly blue light. This is a remarkable sight, and one that gives the phylum its name (Pyrrophyta means fire plants).

The second interesting property of dinoflagellates has to do with their genetic material. Like other organisms, dinoflagellates store genetic information in the form of DNA. In the case of all other eukaryotic cells, however, that DNA is tightly bound with special proteins known as histones. Dinoflagellates do not have histones. In fact, they are the only eukaryotes that do not. The reason for this difference, as well as an explanation of how the roles of the histone proteins are carried out without them in dinoflagellates, remains a mystery.

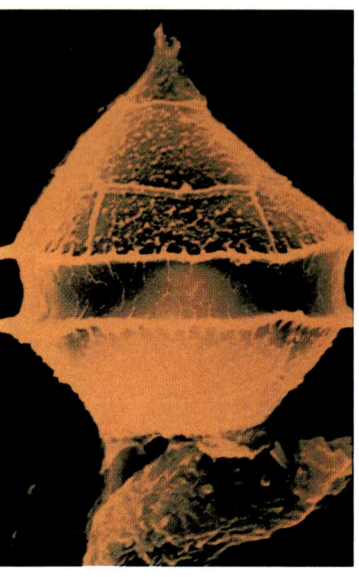

Figure 18–20 Dinoflagellates are plantlike protists. They often have thick protective shells that give them bizarre shapes.

18–3 (continued)

Content Development
Once again, there really is no substitute for being able to directly observe these organisms. Especially as you talk about the pellicle and the force for movement, the students will understand much more clearly if they see it. By this time, students should be able to explain binary fission to you, so it can be mentioned briefly. As a point of curiosity, mention the ridges in the pellicle. You might ask students to infer a meaning for this cell activity.

Content Development
The dinoflagellates are an interesting group of protists. As far as most students are concerned, they are unknown. However, many of them have seen or heard literary references to the "shimmering sea," a phenomenon caused by the luminescent dinoflagellates. You can also point out that this quality is the one that gives the phylum its rather dramatic name. The second

SCIENCE, TECHNOLOGY, AND SOCIETY ISSUE

The Line Between Plant and Animal

Because of its mix of autotroph and heterotroph, *Euglena* is often said to straddle the division between plant and animal. Today, scientists who classify organisms generally solve the plant or animal problem by placing *Euglena* in a separate phylum among the Protista. However, it is worth noting just how indefinite the line between plant and animal can be at the level of a single cell.

An important member of the phylum Zoomastigina is *Astasia*. The resemblance between *Astasia* and *Euglena* is striking. The shapes of the cells are similar, the manner in which they move is similar, even the organization of their cell membranes is similar. In many older systems of classification, one was classified as an animal (*Astasia*) and one as a plant (*Euglena*). The only clear-cut difference between them is that *Euglena* possesses chloroplasts. Should the presence of chloroplasts be enough to place an organism in a separate kingdom?

Most scientists have concluded that it is not enough. A simple experiment gives some of the reasons why. Usually *Euglena* is grown at a temperature of about 25°C—the temperature of a slightly warm room. If *Euglena* is grown in the laboratory at a higher than normal temperature (for example, at about 37°C), something remarkable happens. *Euglena* loses its chloroplasts! After several generations of growth at the elevated temperature, the green *Euglena* cells will have become colorless—their chloroplasts will have vanished. The colorless *Euglena* cells, which can only be distinguished from *Astasia* cells by an expert, never regain their chloroplasts.

We now understand the reason for this phenomenon. The enzymes that replicate chloroplast DNA do not function at the elevated temperature, whereas those that replicate nuclear DNA do. Hence the cell can keep on dividing, but the chloroplasts do not. As the cells grow, the average of 16 chloroplasts per cell is quickly reduced to 8, then 4, then 2, then finally none!

Regardless of the cause, scientists quickly realized that the presence or absence of chloroplasts could not serve as a basis for placing organisms in two different kingdoms—not when chloroplasts can be so easily lost! Instead, there is now an interesting debate among scientists regarding the relationship between *Euglena* and the colorless flagellates of the phylum Zoomastigina. Are zoomastiginans descended from *Euglena*-like organisms that lost their chloroplasts? Or are the euglenophytes actually zoomastiginans that acquired chloroplasts? An interesting question.

Euglena (top) and *Astasia* (bottom) are virtually identical—except that one has chloroplasts.

TEACHING SUPPORT

Teaching Resources
- Activity: Investigating Cellular Slime Molds, p. 7

MEDIA AND TECHNOLOGY
- Videodisc
- *Aquatic Ecosystems: Estuaries and Marine*

Use this bar code to introduce the class to phytoplankton that form the basis of food chains in the oceans.

Play frames 40438 to 40905

See the Biology Media Guide page 16 for additional bar-code correlations for this section.

ECOLOGY NOTE
DIATOMS

The hard shell of diatoms remain long after individuals have died and form dense deposits in certain regions of the ocean bottom. These deposits are mined and are called diatomite (or diatomaceous earth). Diatomite is used in polishes and cleaning agents, as well as rubber and plastic products. It is also used as a filtering agent and is useful for insulation and soundproofing.

18-3 (continued)

Content Development
Using the Visuals If you have several houseplants with variations of green in the leaves, you can use them to explain the name golden plants. Most students never consider that plants are different colors of green. They might misinterpret the name to mean that there is no chlorophyll. Be sure that they know that the chlorophyll is just a different color. Once again, you will seem to be repeating yourself when you tell students that this is a very large and diverse group. List the three types of organisms that make up the phylum. Use Figure 18–21 and any other available visuals in your explanation. Emphasize the unique characteristics of the group. Students might not fully appreciate the significance of the differences, but it is important that they realize that the differences do exist and make these organisms truly unique. Spend a few extra minutes talking about diatoms because they are so beautiful but unknown to the students. Some students may be familiar with diatomaceous earth used in metal polishes and insulations. In some places there are large deposits of this substance made from millions of shells of diatoms. The shells, made primarily of silicon, are deposited after the diatoms die.

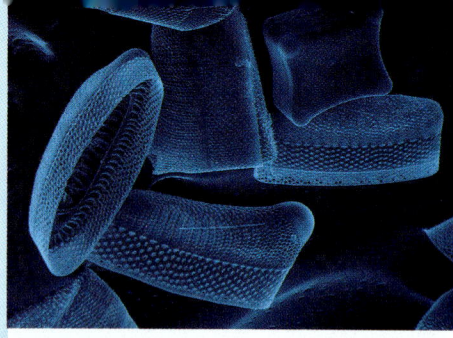
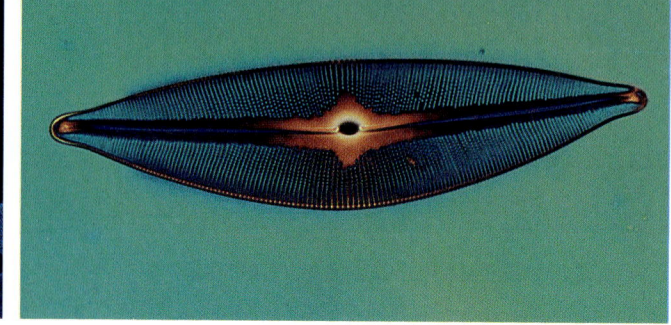

Figure 18–21 *Diatom cells are enclosed within delicate and ornate two-part shells of silica. These glassy shells are quite beautiful.*

Chrysophyta: Golden Protists

There are three general kinds of organisms found within the phylum **Chrysophyta** (krihs-uh-FIGHT-uh): yellow-green algae, golden-brown algae, and diatoms. The name of the phylum, which means golden plants, refers to the one or two gold-green chloroplasts found in the cells of most species of chrysophytes.

The cell walls of chrysophytes contain the carbohydrate pectin rather than cellulose, and they generally store food in the form of oil rather than starch. Chrysophytes are extremely diverse. They reproduce asexually and sexually. Most are solitary, but some form threadlike colonies. Some have flagella and others do not. And some live within cell walls of glass! All but about 2000 of the species in this phylum belong to the **diatoms**.

The diatoms are cells that produce intricate cell walls rich in silicon (Si)—the main ingredient in glass. These walls are shaped like the two sides of a petri dish or flat pillbox, with one side fitting snugly into the other. They are etched with fine lines and designs that seem to be carved into the glasslike brilliance of both sides. Even a modest sampling of the range of diatom species will serve to establish that they are among the most beautiful organisms on Earth. Diatoms are photosynthetic and are among the most abundant species in the oceans.

The Slime Molds: Unusual Protists

Slime molds are found near rich sources of food such as rotting wood, piles of compost, and even thick wet lawns. The slime molds are organisms that are extremely difficult to classify. At one stage of their life cycle they appear as amebalike cells. At other stages they produce moldlike masses that give rise to spores. As a result, slime molds have been classified as amebas and as fungi. They are now placed in two phyla within the kingdom Protista.

ACRASIOMYCOTA: CELLULAR SLIME MOLDS Cellular slime molds belong to the phylum **Acrasiomycota** (uh-kras-ee-oh-migh-KOH-tuh). Cellular slime molds begin their life cycle as individual cells that look very much like amebas. In fact, cellular slime molds spend most of their lives as free-living cells not easily distinguishable from soil amebas. These ameboid, or amebalike, cells reproduce very rapidly. When the food supply is exhausted, groups of ameboid cells gather together to produce a large mass of cells that begins to function as a single organism. This unusual behavior forces scientists to stretch the definition of protists. Protists are defined as being unicellular —but here is a group of protists acting like a primitive multicellular organism!

These solid masses of cells may migrate for several centimeters. They then form a reproductive structure called a fruiting body that produces spores by mitosis. These spores give rise to ameboid cells that repeat the cycle.

Cellular slime molds are an interesting system for biologists who study how cells communicate. The formation of an intricate structure such as the fruiting body from what was previously a mass of independent cells is a most intriguing process. It has kept biologists busy for decades, and its secrets are still not fully understood.

MYXOMYCOTA: ACELLULAR SLIME MOLDS Acellular slime molds belong to the phylum **Myxomycota** (mihks-uh-migh-KOH-tuh). Like a cellular slime mold, an acellular slime mold begins its life cycle as an amebalike cell. However, acellular slime molds produce structures known as **plasmodia** (singular: plasmodium) that contain thousands of nuclei enclosed in a single cell membrane. In contrast to the cellular slime molds, the large plasmodium of an acellular slime mold is a single multinucleate cell. A plasmodium may grow as large as several centimeters in diameter.

Eventually small structures known as fruiting bodies spring up from the plasmodium. The fruiting bodies produce haploid spores by meiosis. These spores scatter to the ground where they germinate into flagellated cells. These flagellated cells fuse to produce diploid ameboid cells that repeat the cycle.

Figure 18–22 When the food supply runs out, cellular slime mold cells come together. The collection of cells that results from this process looks and acts much like a simple multicellular organism.

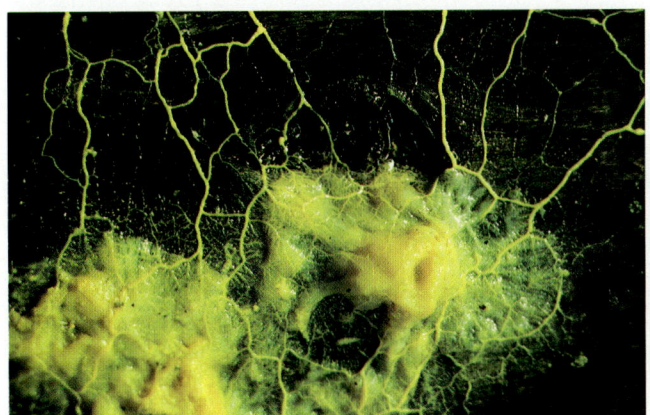

Figure 18–23 At one point in its life cycle, an acellular slime mold forms a structure called a plasmodium, which contains many nuclei and creeps about like a giant ameba (left). Later, reproductive structures called fruiting bodies are produced by the plasmodium (right).

FACTS AND FIGURES

Deposits of diatomaceous earth are often found on land that was covered by the oceans. One deposit of this kind, found in California, is over 450 meters thick.

ESL STRATEGY

To help students learn the spelling of the three plantlike protists mentioned in this section and to help them assimilate some characteristics of organisms in the three groups, have them unscramble the following clusters of letters:

    ryschoyatph
    ropytraphy
    phaytuglenae

Then ask students to use the unscrambled names as headings for a three-column chart. Have students place the following characteristics under the correct heading in their chart.

    reddish eyespot
    red tide
    two-part glassy shell
    pellicle
    pouch with two flagella
    toxin
    producer
    contain silicon
    chloroplasts

Again, students may question how this is done. The best explanation is that it is like a giant multicellular ameba. Fruiting body is another term that may require additional explanation.

Content Development

The acellular slime molds are most interesting organisms from a biological point of view. However, students may not find them quite so fascinating. You will want to spend some time talking about the plasmodium. This is not a frequently occurring phenomenon in biology, but it does appear occasionally. They might recall the term from the genus name used earlier in the chapter.

Enrichment

If you have students who are interested in drawing, ask them to draw large, detailed pictures of diatoms that they view under the microscope. If they have artistic talent and a good eye for detail, they can get some extra credit and you will have some beautiful biological illustrations.

Content Development

The cellular slime molds are frequently studied and seldom understood. As you go step by step through the life cycle of these organisms, there are many biological terms and concepts to be covered. Students may well question how the large mass of cells forms from single independent cells. Biologists are still questioning that also. You will have to explain migration as the term is used.

Summary of the Plantlike Protists

- Members of the phylum Euglenophyta are very similar to zoomastiginans. Euglenophytes, such as *Euglena,* are photosynthetic flagellates.
- Members of the phylum Pyrrophyta are called dinoflagellates. Most dinoflagellates are photosynthetic, and some are luminescent.
- The phylum Chrysophyta is a diverse group of protists that have gold-colored chloroplasts. Most of the species of chrysophytes are diatoms, photosynthetic protists that lack flagella and live in beautiful glasslike "boxes."
- Slime molds are unusual organisms that are difficult to classify. Cellular slime molds belong to the phylum Acrasiomycota. Acellular slime molds belong to the phylum Myxomycota.

How the Plantlike Protists Fit into the World

Like the animallike protists, plantlike protists are found throughout the world in bodies of fresh water, in the ocean, and on land. Unlike animallike protists, most plantlike protists are autotrophs rather than heterotrophs. Although plantlike protists can be harmful to other organisms, few plantlike protists are truly parasitic.

HARMFUL RELATIONSHIPS In lakes and ponds, euglenophytes are among the most common organisms. In areas into which large amounts of sewage are discharged, euglenophytes may thrive. Because they are able to absorb organic material directly and use it for food, they grow rapidly in such regions and their presence may actually turn the water of a lake or slow-moving stream a murky green. They play a vital role in helping to recycle sewage and other waste materials. But when the amount of waste dumped into a body of water is excessive, the euglenophytes and other green organisms may grow into enormous masses of cells known as **blooms**. While not harmful in themselves, these blooms quickly run out of nutrients and the cells begin to die in great numbers, compounding the problem of disposing of waste matter.

Great blooms of the dinoflagellate *Gonyaulax polyhedron* have occurred in recent years on the east coast of the United States, although scientists are not sure of the reasons. This species contains a toxin that can cause paralysis and even death if ingested in large amounts. Fortunately for all but the most allergic people, it simply is not possible to swallow enough of the "red tide" of *Gonyaulax* while swimming to be affected by the toxin. Thus the blooms of dinoflagellates are not harmful to most swimmers. However, shellfish such as clams

Figure 18–24 *Red tide (top) is produced by blooms of the dinoflagellate* Gonyaulax *(bottom).* Gonyaulax *contains a toxin that can become concentrated in the tissues of shellfish such as clams and oysters, making them unfit to eat. This toxin can also kill fish and other marine animals.*

and oysters filter enormous amounts of sea water in order to trap organisms like *Gonyaulax* for food and thus become filled with the toxin. Eating shellfish from areas infected with red tide can cause serious illness. In addition, the toxin can kill fish and weaken or even kill dolphins.

HELPFUL RELATIONSHIPS The plantlike protists form some spectacular symbiotic relationships with other organisms. Many types of coral contain intercellular dinoflagellates. These dinoflagellates allow the tiny animals that form the coral to use the products of photosynthesis, allowing coral to grow in areas where nutrients are few. In turn, the dinoflagellates can use many of the waste products of the coral organisms before they are diluted by diffusion through sea water.

Other dinoflagellates make their homes with other organisms. In the giant clam *Tridacna gigas*, a special tissue called the mantle contains large numbers of symbiotic photosynthetic protists. These dinoflagellates are held in a position from which they are able to gather as much sunlight as possible and increase the nutrient benefit to the organism.

Plantlike protists play a major ecological role: They make up a considerable part of the **phytoplankton**. The term phytoplankton is applied to any small photosynthetic organism found in great numbers near the surface of the ocean. Thus it can apply to any photosynthetic organism, regardless of kingdom or phylum.

The importance of the phytoplankton for other forms of life cannot be underestimated. More than 70 percent of the photosynthesis that occurs on Earth goes on near the surface of the oceans. The result of this photosynthesis is that the rest of the organisms on our planet are provided with enormous amounts of oxygen and food. The phytoplankton provide a direct source of nourishment for organisms as diverse as shrimp and whales. And even land animals such as humans obtain nourishment indirectly from the phytoplankton. When you eat a tuna fish sandwich, you are eating fish that fed on smaller fish that fed on still smaller animals that fed on phytoplankton!

18–3 SECTION REVIEW

1. List the five phyla of plantlike protists. Give an example of an organism in each phylum.
2. What are slime molds? Why is it appropriate for the slime molds to be considered protists?
3. Why is red tide particularly damaging to the shellfish industry?
4. **Critical Thinking—Expressing an Opinion** "Life on Earth depends on plantlike protists." Defend or refute this statement.

Figure 18–25 Symbiotic photosynthetic protists live inside the tissues of animals such as sea anemones (top) and giant clams (bottom).

Figure 18–26 Phytoplankton provide enormous amounts of food and oxygen for other organisms on Earth.

ECOLOGY NOTE
PHYTOPLANKTON

Phytoplankton includes the photosynthetic organisms that float near the surface of water bodies and serve as the first stage in the aquatic food chains. These organisms can survive only near the surface of the water where there is light, so most have some adaptations to help them stay there. When they die, they sink to the bottom and form oil-rich sediments. The numbers and kinds of organisms that make up the phytoplankton vary seasonally with the availability of nutrients and light.

LABORATORY INVESTIGATION

EXAMINING PROTISTS

Before the Lab
1. Assemble the microscopes and other materials that each group will need for the lab. Be sure that you have an extra supply of slides and coverslips. Check your pond water samples so that you will know what protists the students are likely to find. Pure cultures of common protists are excellent comparative references if they are available.
2. If pond water is not available, prepare a leaf infusion with a small amount of soil, dried leaves and grass, and water. Leave the infusion in a cool place to limit bacterial growth. Succession leading to large paramecia may take up to three weeks.

Pre-Lab Discussion
There are several tips that will help students be more successful with this lab.
1. Review the characteristics of protists. In any pond water sample, students are likely to find a lot of nonprotists as well as protists. You want them to be able to distinguish between the two.
2. Review the proper way to make wet mount slides.
3. Discuss water bubbles and their appearance. This lab is exciting for the students but it can be frustrating because they do not know exactly what they will find. Students sometimes assume that water bubbles are special.
4. Instruct students to get a sample of pond water in their droppers. Hold the dropper still with the tip down for a few seconds before putting a sample on the slide. A small drop is easier to handle than a flooded slide.
5. After the slide is prepared, place it on the microscope stage and move it around very slowly under the objective until protists are located.
6. The drawings of protists are more useful if they are large enough to be seen clearly. Encourage students to make large, clear drawings based on their observations. That imposes the necessity of making good observations.
7. Show students how they are to use the identification guides.

Safety Tips
There is nothing particularly dangerous about this lab, but students should be reminded that they are working with glass. Precautions should be taken and procedures for handling broken glass reviewed.

LABORATORY INVESTIGATION: EXAMINING PROTISTS

PROBLEM
What characteristics of protists can be observed with a microscope?

MATERIALS (per group)

pond water	cotton
coverslips	microscope
microscope slides	protist identification
medicine dropper	guide

PROCEDURE

1. Using the medicine dropper, put a drop of pond water in the center of a clean glass microscope slide.
2. Pull apart a small piece of cotton and put a few threads in the drop of pond water. **Note:** *Only a few threads are necessary.*
3. Cover the drop of pond water with a coverslip. With your microscope set on low power, look for signs of life in the water.
4. When you find microorganisms, switch to medium power and focus with the fine adjustment. Observe what happens when one of the microorganisms bumps into the thread.
5. Try to get a microorganism in the center of the field of view. Switch to high power and focus with the fine adjustment. **Note:** *Use only the fine adjustment to focus when using the high-power objective. When using the high-power objective, do not focus downward (do not bring the objective closer to the slide) while looking through the microscope.*
6. Draw the protist on a separate sheet of paper. Next to your drawing, write down your observations of the protist's appearance and behavior. Note its shape and color and the organelles that are visible. Try to determine how it moves—a "fluttery" edge on the protist, for example, is a sure sign of cilia.
7. Repeat steps 5 and 6 with as many different protists as you can find.
8. Using a protist identification guide, try to identify the protists in your drawings.

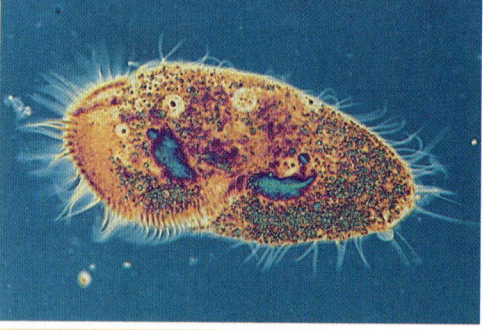

OBSERVATIONS
1. What types of movement did you observe in the protists?
2. What organelles did you observe in the protists?
3. Describe what protists do when they bump into an obstacle such as a thread.
4. What characteristics did the protists have in common?
5. How were the protists different from one another?

ANALYSIS AND CONCLUSIONS
1. What important characteristics of protists are difficult or impossible to observe with a light microscope?
2. Why was the cotton put into the drop of pond water? Predict what you might have observed if there were no cotton in the pond water.
3. Was it easy to identify the protists you observed? Explain.
4. Suggest what might be done to make it easier to observe and identify protists.

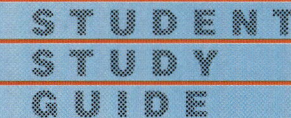

STUDENT STUDY GUIDE

SUMMARIZING THE CONCEPTS

The key concepts in each section of this chapter are listed below to help you review the chapter content. Make sure you understand each concept and its relationship to other concepts and to the theme of this chapter.

18–1 The Kingdom Protista

- Protists, or members of the kingdom Protista, are unicellular eukaryotic organisms.
- The Endosymbiont Hypothesis states that the first eukaryotic cell was formed by a symbiosis among several prokaryotes.

18–2 Animallike Protists

- Ciliates possess cilia.
- Zoomastiginans possess flagella.
- Sporozoans are nonmotile parasitic protists.
- Sarcodines use pseudopods for locomotion and feeding.
- Some animallike protists are symbionts of other organisms. A few species are pathogens. Animallike protists are a source of food for small organisms that in turn are food for larger organisms.

18–3 Plantlike Protists

- Euglenophytes are photosynthetic flagellates that are similar in form to zoomastiginans.
- Pyrrophytes are also known as dinoflagellates. Some dinoflagellates are luminescent.
- Most of the species in the phylum Chrysophyta are diatoms.
- Slime molds are also known as funguslike protists. Unlike most plantlike protists, slime molds are not photosynthetic.
- Cellular slime molds form a mass of cells that function like a single organism at one point in their life cycle.
- Acellular slime molds form plasmodia at one point in their life cycle.
- Photosynthetic plantlike protists form an important part of the phytoplankton, which supplies much of the Earth's oxygen.

REVIEWING KEY TERMS

Vocabulary terms are important to your understanding of biology. The key terms listed below are those you should be especially familiar with. Review these terms and their meanings. Then use each term in a complete sentence. If you are not sure of a term's meaning, return to the appropriate section and review its definition.

18–1 The Kingdom Protista
Protista
protist
Endosymbiont Hypothesis

18–2 Animallike Protists
Ciliophora
ciliate
cilium
paramecium
pellicle
trichocyst
macronucleus
micronucleus
gullet
food vacuole
anal pore
contractile vacuole
Zoomastigina
flagellum
flagellate
Sporozoa
Sarcodina
pseudopod
ameba

18–3 Plantlike Protists
Euglenophyta
euglena
Pyrrophyta
dinoflagellate
Chrysophyta
diatom
slime mold
Acrasiomycota
Myxomycota
plasmodium
bloom
phytoplankton

Skills Development

Students will use the following skills in the completion of this laboratory investigation.
1. Manipulative
2. Observation
3. Diagramming
4. Comparing
5. Applying
6. Inferring

Observations

1. Answers will vary. Protists move with the aid of cilia, flagella, or pseudopods.
2. Answers will vary. The nucleus and vacuoles should be visible. Some students may also observe chloroplasts.
3. Most protists back up and try again from a different angle.
4. Answers will vary. Students should note that the protists were unicellular, motile, and had organelles such as nuclei and vacuoles.
5. Answers will vary, depending on the protists observed. Students will likely point out modes of locomotion, size and shape, and so on.

Analysis and Conclusions

1. Most of the small internal structures and organelles such as mitochondria and ribosomes will not be seen with a light microscope. Accept all other reasonable answers, including the fact that we cannot see processes such as photosynthesis under the microscope.
2. The cotton was used to slow down the protists. Without the cotton, fast-moving protists would zoom out of the field of view before they could be adequately observed.
3. Answers will vary. Common protists such as paramecium may have been easy to identify. Less common protists may have been difficult, or even impossible, to identify in the short amount of time devoted to the investigation.
4. Answers will vary. Most students will suggest slowing down the protists even more, perhaps using chemicals or physical barriers.

Going Further: Enrichment

Have students design their own population studies with protists. One example would be to take the pond water or mixed cultures and view them over a period of time. Count the number of each of the species that crosses the field in a given period of time, depending on the richness of the culture. Over a period of days, the populations will change. Have students graph the data. There are a number of other options. Encourage students to use their imaginations and "scientific training" to come up with an original experiment.

CHAPTER REVIEW

CONTENT REVIEW

Multiple Choice
1. d   2. c   3. a   4. b
5. d   6. d   7. a   8. c

True or False
1. T
2. F   Acellular slime molds
3. F   Plantlike protists
4. F   micronuclei
5. F   Sarcodines
6. F   pellicle
7. F   trichocysts
8. T

Word Relationships

A.
1. plantlike protists; ameba is an animallike protist
2. parts of a ciliate; flagella are found on flagellates
3. diseases caused by animallike protists; red tide is a condition caused by plantlike protists
4. sarcodines; sporozoan is a member of a different phylum, Sporozoa

B.
5. Amebas
6. saprobes (saprophytes and detritus feeders are also acceptable)
7. protists
8. phytoplankton

CONCEPT MASTERY

1. It explains how the first eukaryotic cells may have come into existence.
2. Student answers should be logical and well thought out. Most students will agree that these categories are useful.
3. Because they are made up of a glasslike substance, diatom shells are abrasive and reflect light well.
4. Many protists do not fit well in existing categories. Some classification schemes do not recognize relationships between different species of protists. Other answers are possible.

CHAPTER REVIEW

CONTENT REVIEW

Multiple Choice

Choose the letter of the answer that best completes each statement.

1. All protists are
   a. solitary.   c. motile.
   b. colonial.   d. eukaryotic.
2. Which organism causes malaria?
   a. *Paramecium*   c. *Plasmodium*
   b. *Trypanosoma*   d. *Euglena*
3. Short hairlike projections that produce movement in certain protists are
   a. cilia.   c. flagella.
   b. pseudopods.   d. microtubules.
4. Diatoms belong to the phylum
   a. Ciliophora.   c. Pyrrophyta.
   b. Chrysophyta.   d. Myxomycota.
5. Which of the following organisms are not placed in the phylum Sarcodina?
   a. amebas   c. heliozoans
   b. radiolarians   d. flagellates
6. A euglena moves by means of
   a. pseudopods.   c. spores.
   b. cilia.   d. flagella.
7. Which organism is not associated with a disease in humans?
   a. *Trichonympha*   c. *Trypanosoma*
   b. *Entamoeba*   d. *Plasmodium*
8. A paramecium excretes excess water through the
   a. gullet.   c. contractile vacuole.
   b. trichocysts.   d. micronucleus.

True or False

Determine whether each statement is true or false. If it is true, write "true." If it is false, change the underlined word or words to make the statement true.

1. Red tides are caused by <u>dinoflagellates</u>.
2. <u>Cellular slime molds</u> produce plasmodia.
3. <u>Animallike protists</u> make up a considerable part of the phytoplankton.
4. During conjugation, paramecia exchange <u>macronuclei</u>.
5. <u>Sporozoans</u> use pseudopods to move.
6. In some protists, the cell membrane and associated structures make up the <u>flagellum</u>.
7. A paramecium uses flask-shaped structures called <u>contractile vacuoles</u> for defense.
8. Some species in the phylum <u>Pyrrophyta</u> are luminescent.

Word Relationships

A. *In each of the following sets of terms, three of the terms are related. One term does not belong. Determine the characteristic common to three of the terms and then identify the term that does not belong.*

1. dinoflagellate, diatom, ameba, euglena
2. macronucleus, micronucleus, flagella, cilia
3. red tide, malaria, African sleeping sickness, amebic dysentery
4. sporozoan, foraminifer, radiolarian, heliozoan

404

---

5. Answers will vary. Check student tables. Ciliophora; animallike; *Paramecium, Stentor*; cilia; food for other organisms, some eat other protists. Zoomastigina; animallike; *Trichonympha*, trypanosomes; flagella; some help digest wood, some cause diseases. Sporozoa; animallike; *Plasmodium*; nonmotile; cause diseases such as malaria. Sarcodina; animallike; amebas, foraminifers, heliozoans, radiolarians; pseudopods; some cause disease, fossil foraminifers used as chalk. Euglenophyta; plantlike; *Euglena*; flagella; part of phytoplankton. Pyrrophyta; plantlike; *Gonyaulax*; flagella; some cause red tides, some symbiotes. Chrysophyta; plantlike; diatoms; (some are

B. *Replace the underlined definition with the correct vocabulary word.*
5. Flexible, active cells without cell walls, flagella, cilia, and even a definite shape engulf prey by using pseudopods.
6. Slime molds are heterotrophic organisms that feed on dead or decaying organic material.
7. The oldest fossils of members of the kingdom that consists of unicellular eukaryotic organisms are about 1.5 billion years old.
8. Diatoms, chrysophytes, and euglenophytes are an important part of the small photosynthetic organisms that are found in great numbers near the surface of the ocean.

CONCEPT MASTERY

Use your understanding of the concepts developed in the chapter to answer each of the following in a brief paragraph.

1. What is the significance of the Endosymbiont Hypothesis?
2. Are the categories animallike, plantlike, or funguslike useful in classifying protists? Explain your answer.
3. Why are diatom shells often used in toothpaste and reflective paint?
4. Explain why protists are difficult to classify.
5. Make a table that contains the following information about each protist phylum: Name of Phylum; Animallike or Plantlike; Representative Members; Means of Locomotion; Relationships with Other Organisms.

CRITICAL AND CREATIVE THINKING

Discuss each of the following in a brief paragraph.

1. **Developing a hypothesis** A scientist observes that termites that are fed a certain antibiotic die of starvation after a few days. The scientist also notices that certain protists that live inside the termite's gut are affected by the antibiotic in a peculiar way: Although the protists continue to thrive, they lose a certain kind of structure in their cytoplasm. Develop a hypothesis to explain these observations.

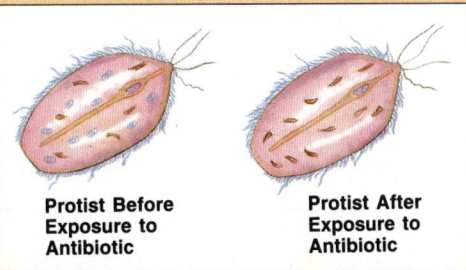

Protist Before Exposure to Antibiotic   Protist After Exposure to Antibiotic

2. **Relating cause and effect** How might water pollution result in a red tide?
3. **Making predictions** Growing "holes" in the Earth's ozone layer may increase the amount of radiation that reaches the surface of the ocean. If this radiation were to affect the growth of phytoplankton, what long-term consequences might this have on the Earth's atmosphere?
4. **Using the writing process** Imagine that you could shrink down to microscopic size and fit inside a paramecium. Describe the adventures you and the paramecium have in a small pond one summer day.

CHAPTER 19 Fungi

Section	Laboratory Investigations and Activities
19–1 The Fungi pages 407–416 THEME: Unity and Diversity	**Student Edition** LABORATORY INVESTIGATION: Comparing a Mold and a Mushroom, p. 424 **Teacher Edition** DEMONSTRATION: Making Spore Prints, p. 406D **Laboratory Manual** The Structures of Fungi, p. 237 **Teaching Resources** HANDS-ON ACTIVITY: Spreading Spores, p. 95 HANDS-ON ACTIVITY: Yeast Meets Best, p. 97 ECOLOGY INVESTIGATION: Relationships Between Bacteria and Fungi in Soil, p. 9
19–2 Fungi in Nature pages 417–423 THEME: Systems and Interactions	**Teacher Edition** DEMONSTRATION: Activity of Yeast, p. 406D **Laboratory Manual** Mold Growth, p. 243 **Teaching Resources** ECOLOGY INVESTIGATION: Decomposers in Soil, p. 11 **Product Testing Activities** Testing Food Wraps
Chapter Review pages 425–427	

Teacher Resources

Books

Alexopoulos, C.J., and C.W. Mims. *Introductory Mycology,* 3rd ed. Wiley, 1979.

Bold, H.C., C.J. Alexopoulos, and T. Delevoryas. *Morphology of Plants and Fungi,* 4th ed. Harper and Row, 1980.

Miller, Orson K., Jr. *Mushrooms of North America.* Dutton, 1977.

McKnight, Kent H., and Vera McKnight. *A Field Guide to Mushrooms of North America.* Houghton Mifflin, 1987.

Pirozynski, K.A., and David Hawksworth, eds. *The Coevolution of Fungi with Plants and Animals.* Academic Press, 1988.

Ray, P.M., T. Steeves, and S.A. Fultz. *Botany.* Saunders, 1983.

Rayner, A., and L. Boddy. *Fungal Decomposition of Wood. Its Biology and Ecology.* Wiley, 1988.

CHAPTER PLANNING GUIDE

Other Activities	Media and Technology
Teaching Resources ACTIVITY: Structures of Fungi, p. 3 **Study Guide** Section 19–1, p. 179	**Transparency Binder** 21: The Life Cycle of a Mushroom
Teaching Resources VOCABULARY REVIEW: Word Scramble, p. 1 ACTIVITY: Relationships with Fungi, p. 5 **Study Guide** Section 19–2, p. 182	
Teaching Resources Chapter Test A, p. 11 Chapter Test B, p. 15 Performance-Based Assessment, Unit 5	**Computer Test Bank** Chapter Test, p. 203

Audiovisuals

Fungi. Film. Benchmark Films, Inc.
Fungi: The One Hundred Thousand. Film or video. Coronet.
Molds and How They Grow. Film or video. Coronet.
Fungi. Film or video. Ward.

Fungi (Diversity and Ecology). Filmstrip. Ward.
Microscopic Fungi. Film. McGraw-Hill.
The Fungi. Filmstrip with cassette. Biology Media.

The Impact of Fungi on Man and His Environment. Filmstrip with cassette. Carolina Biological Supply Co.
Fungi and Man. Film or video. Benchmark Films, Inc.

CHAPTER 19 Fungi

CHAPTER OVERVIEW

In this chapter students will be introduced to the organisms that make up the kingdom Fungi. The chapter is divided into two main sections. The first section deals with the characteristics of fungi in general and the characteristics of the five phyla of kingdom Fungi in particular. The second section deals with the role of fungi in nature and their impact on humans.

Students will learn that fungi are eukaryotic heterotrophs. They will discover that many fungi are saprophytes, or organisms that obtain nourishment from decaying organic matter. Other fungi live as parasites, and still others live as symbionts.

Students will learn that fungi are classified according to their methods of reproduction and their basic structures. Discussed in detail will be the characteristics of the five phyla: Oomycota, Zygomycota, Ascomycota, Basidiomycota, and Deuteromycota. Among the more familiar fungi included in this discussion are molds, mushrooms, and yeasts.

Students will discover that fungi are important in nature as agents of organic decay. They will learn that by speeding up the breakdown of dead organisms, fungi help to recycle essential nutrients and chemicals.

The importance of fungi as disease-producing organisms will be discussed, particularly as they affect plants and humans. Included in this discussion will be plant diseases such as potato blight and wheat rust, as well as human afflictions such as ringworm and athlete's foot.

19–1 THE FUNGI
Section Focus 19–1

The purpose of this section is to describe the basic characteristics of fungi and to discuss the characteristics of each phylum in particular. Topics such as basic structure, reproduction, and method of obtaining food will be discussed.

Students will discover that except for yeasts, which are unicellular, fungi are made up of a tangled mass of filaments called hyphae. Together, the hyphae form a thick mass called a mycelium. Students will also discover that most fungi reproduce both asexually and sexually. Asexual reproduction is accomplished either by the production of spores or by the fragmentation of the hyphae. In many fungi, sexual reproduction involves two different mating types, one of which is referred to as + (plus) and the other as – (minus).

Students will learn that the phylum Oomycota includes protistlike fungi such as the water mold. Common molds, such as bread mold, belong to the phylum Zygomycota. Students will discover that sac fungi are members of the phylum Ascomycota. This phylum includes yeasts and morels. Perhaps the most familiar phylum to students will be Basidiomycota, or the club fungi. This phylum includes the many varieties of mushrooms. The last phylum to be discussed will be Deuteromycota, or the imperfect fungi. Included in this phylum are many of the fungi that cause diseases.

Performance Objectives 19–1
1. List and discuss the general characteristics of fungi.
2. Describe the structure of a typical fungus.
3. Identify the various ways in which fungi obtain food.
4. Discuss the various ways in which fungi reproduce.
5. Identify and compare the five phyla of fungi.

Science Terms 19–1
fungus   p. 407
decomposer   p. 408
mycelium   p. 408
hypha   p. 408
sporangium   p. 408
sporangiophore   p. 408
gametangium   p. 408
zygospore   p. 410
rhizoid   p. 411
stolon   p. 411
conidiophore   p. 411
conidium   p. 411
ascus   p. 412
ascospore   p. 412
basidium   p. 413
basidiospore   p. 413

19–2 FUNGI IN NATURE
Section Focus 19–2

The purpose of this section is to discuss the ecological significance of fungi and to view the effects that fungi have on human life. The role of fungi as recyclers of organic material will be discussed. This recycling is important because it prevents the loss of chemical energy and it returns nutrients to the soil.

Fungi are found almost everywhere on Earth and in almost every kind of environment. An interesting aspect of fungi that will be discussed is the many ways in which fungi disperse spores in order to accomplish asexual reproduction. Another topic of discussion will be the participation of fungi in symbiotic relationships with other organisms.

Students will learn that fungi affect humans in many ways, some of which are beneficial. For example, yeasts are important to people because of their roles in baking and brewing. Mushrooms—the kind that are not poisonous—are important as a desirable food item.

Effects of fungi that are not beneficial to humans include plant diseases that damage crops. Among the plant diseases that will be discussed are potato blight, wheat rust, and corn smut. Students will learn that although most fungal diseases are associated with plants, some fungi cause diseases in humans. Among the more familiar conditions caused by fungi are athlete's foot, ringworm, and thrush.

CHAPTER PREVIEW

Performance Objectives 19-2

1. Discuss the ecological significance of fungi.
2. Describe various methods of spore dispersal.
3. Describe some symbiotic relationships that involve fungi.
4. Discuss some of the ways in which fungi are beneficial to humans.
5. Discuss some plant and human diseases caused by fungi.

Science Terms 19-2

lichen   p. 418
mycorrhiza   p. 413

DISCOVERY LEARNING

TEACHER DEMONSTRATIONS
Modeling

Making Spore Prints

The following demonstration can be used as an introduction to Chapter 19.

For the demonstration, you will need some fresh mushrooms from a grocery store, sheets of white paper, plastic cups, and hand lenses or stereomicroscopes.

Explain to students that in this chapter one of the organisms they will study is the mushroom. Point out that mushrooms produce hundreds of thousands of spores every few days. By placing a mushroom cap on a sheet of white paper, it is possible to obtain a copy of the mushroom's gill structure. These designs, which are called spore prints, are often used to identify various types of mushrooms.

To make a spore print, carefully remove the stalks from the caps of the mushrooms. Then lay each cap, gills down, on a sheet of white paper. Place a cover, such as a plastic cup, over the mushrooms and allow them to stand at least 24 hours. Then carefully remove the covers and the mushroom caps. Have students use a hand lens or stereomicroscope to examine the prints. You may wish to divide the class into small groups and give each group a different print to examine—then groups can compare the prints for similarities and differences.

Activity of Yeast

This demonstration can be performed as students study the part of Section 19-2 that deals with fungi and human life.

For the demonstration, you will need a package of dry yeast, warm tap water, a beaker, molasses, aluminum foil, microscopes, coverslips, and methylene blue.

Begin by showing students the package of dry yeast. Point out that the dry yeast inside this package is a collection of dormant yeast spores.

• **Why do you think that we say these spores are dormant?** (Answers may vary. Correct answer is that conditions are not favorable for growth or reproduction.)

• **What would be required for the yeast to become active?** (Answers may vary. Many students may know that yeast will become active if mixed with bread or cake dough—and these environments contain the essential ingredients of water and carbohydrates.)

Prepare a culture of yeast cells by providing the essential ingredients needed to activate the yeast. In a beaker, mix 5 milliliters of molasses and 500 milliliters of warm tap water. Add half of the package of dry yeast and stir the mixture. Then cover the top loosely with aluminum foil and allow it to stand for about 30 minutes in a warm place.

After the half hour is up, have students use microscopes to view the yeast cells under low and high power. A drop of methylene blue added to the culture under the coverslip may enable the cells to be seen better. Students should see buds when they view the yeast cells.

• **What do these buds represent?** (Yeast cells reproduce by budding; these buds will develop into new yeast cells.)

CHAPTER 19
Fungi

GUIDED ENQUIRY
Pose the following questions to students and have them record their responses. Point out that they will gain a better understanding of the key concepts if they read the chapter with these basic questions in mind. Upon completion of the chapter, pose the questions again. Ask students to compare their initial responses with those they have developed after reading the chapter.
- How do fungi differ from other organisms?
- How does the lifestyle of a fungus enable it to obtain food?
- What factors make each phylum of fungi distinctive?
- What is the ecological significance of fungi?
- How do fungi affect humans?

INTRODUCING CHAPTER 19

Using the Visuals
Have students look at the photographs on this page.
- **What do you see in the main picture on this page?** (Students probably will say that they see a flowering plant.)
- **Read the caption. What is really shown in the picture?** (There are no real flowers in this picture! Instead, a fungus has infected a rock cress plant, causing the plant to produce "fungus-flowers.")

Have students read the chapter-opener text.
- **Are "fungus-flowers" made of fungus?** (No, they are not. Rather, the fungus is inside the plant, where it causes the plant to produce these flower imitations. The "flowers" are actually deformed rock cress leaves.
- **Why would a fungus go to all the trouble of making a plant imitate a fancy flower?** (As shown in the inset, the fungus-flowers' color and scent attract passing butterflies or other insects. Fungal spores stick to these insects, and the insects distribute them to new locations.)
- **Do you think this fungus hurts or helps the rock cress plant?** (The fungus hurts the rock cress plant. The fungus draws off the plant's nutrients, and it suppresses the real rock cress flowers.)
- **Would you guess that fungi could flower on their own?** (Fungi do not flower, nor do they resemble flowers. This picture shows only a very unusual fungal infection, one that caused a plant's leaves to imitate an attractive flower.)

As students will learn in this chapter, fungi are organized very differently from plants, which is why fungi are classified in their own kingdom.

CHAPTER 19
Fungi

These "fungus-flowers" are so convincing that they attract flies and butterflies away from the genuine flowers on nearby plants.

Springtime in the Rocky Mountains presents many beautiful vistas as great patches of wildflowers bloom beneath deep blue skies. Barbara Roy, a California biologist, was the first person to notice that some stands of a plant known as rock cress were producing large, fragrant, spectacular flowers that differed quite dramatically from the small flowers normally borne by the plants.

The large "flowers," Roy soon discovered, weren't flowers at all. Rather, they were deformities caused by the rust fungus *Puccinia*. The "petals" of the fake flowers are coated with fungus spores and a sweet fluid that attracts insects. The insects, while feasting, pick up the spores and later carry them to other plants.

The rust fungus is only one type of fungus. In this chapter, you will examine some of the characteristics of fungi and discover some of the roles they play in our environment.

19–1 The Fungi

Guide For Reading

- What are fungi?
- What are some characteristics of each of the five phyla of fungi?
- How do fungi obtain food?
- How do fungi reproduce?

For many of us, the most common encounters with **fungi** (FUHN-jigh; singular: fungus) are unwanted ones: molds spoil our fruits and breads, mildew weakens our fabrics, and athlete's foot attacks our skin. In tropical areas, more than 50 percent of the food that is produced is spoiled by fungi before it can be eaten. And in temperate regions, trees such as the elm die of a fungal disease known as Dutch elm disease.

For reasons such as these, when we think of fungi, we think of death and decay. Fungi, however, are among the most interesting organisms in the living world. Not only do they help shape the natural environment, they also provide us with food—and in so doing, they display some remarkable and exotic lifestyles.

Characteristics of Fungi

Fungi are eukaryotic heterotrophs. You will recall from Chapter 17 that heterotrophs depend on other organisms for food. Many fungi are saprophytes, or organisms that obtain food from decaying organic matter. Others are parasites, which are organisms that live directly on the body of a plant or animal host and in so doing harm that organism. Still other fungi are symbionts, or organisms that live in close association with an organism of another species.

Figure 19–1 All fungi are heterotrophs, or organisms that obtain food from organic compounds produced by other organisms. Many of these heterotrophs, such as the bracket fungi shown here, live as saprophytes, decomposing dead matter.

TEACHING SUPPORT

Teaching Resources
- Activity: Structures of Fungi, p. 3
- Hands-On Activity: Spreading Spores, p. 95
- Ecology Investigation: Relationships Between Bacteria and Fungi in Soil, p. 9

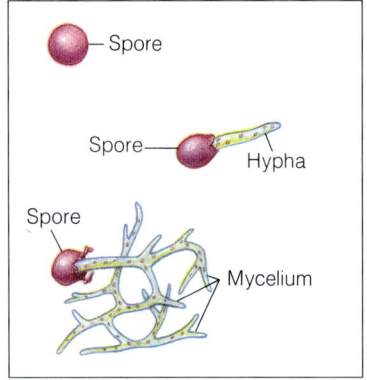

Figure 19–2 A fungus develops from a spore that grows into a threadlike hypha. The hypha grows rapidly and branches until it resembles a tangled mass of threads called a mycelium.

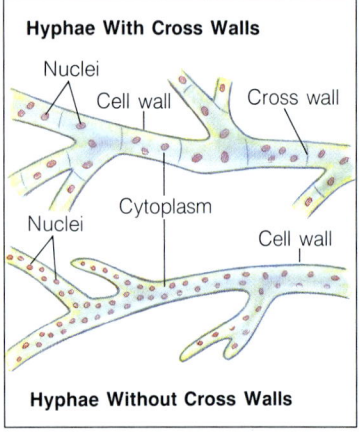

Figure 19–3 Some hyphae are divided by cross walls that contain one or more nuclei. Other hyphae are coenocytic, meaning they lack cross walls.

Fungi do not ingest their food. Instead, they absorb it through their cell walls and cell membranes. Fungi release digestive enzymes into their environment. The enzymes break down leaves, fruit, or other organic material into simple molecules, which then diffuse across the cell walls and cell membranes. This method of obtaining food makes fungi very important in nature: They produce powerful digestive enzymes that speed the breakdown of dead organisms, helping to recycle nutrients and essential chemicals. Together with the bacteria, fungi are the major **decomposers,** or organisms of decay.

Except for yeasts, which are unicellular, the body of a typical fungus is made up of many tiny filaments tangled together into a thick mass called a **mycelium** (migh-SEE-lee-uhm). The individual filaments are called **hyphae** (HIGH-fee; singular: hypha). In many fungi, the hyphae are divided by cross walls into cells containing one or more nuclei. See Figure 19–3. The cell walls of most hyphae are made up of chitin, a complex carbohydrate that is also found in the external skeleton of insects. The cell walls of other hyphae contain cellulose, the complex carbohydrate that makes up the cell walls in plants. The mycelium, or tangled mass of hyphae, is well suited to absorbing food because it permits a larger surface area to come in contact with the food source.

Most fungi reproduce both asexually and sexually. Asexual reproduction occurs either by the production of spores or by the fragmentation of the hyphae (each fragment becomes a new fungus). In some fungi, spores are produced in structures called **sporangia** (spoh-RAN-jee-uh; singular: sporangium). Sporangia are found at the tops of specialized hyphae called **sporangiophores.**

In many fungi, sexual reproduction involves two different mating types. One mating type is referred to as + (plus) and the other is referred to as − (minus). When the hyphae of opposite mating types meet, each hypha forms a **gametangium** (gam-uh-TAN-jee-uhm), or gamete-forming structure. Then the two gametangia fuse, and some of the nuclei pair and join to form zygote nuclei.

During the greater part of their life cycle, the nuclei of most fungi are haploid (N). Diploid (2N) nuclei form during sexual reproduction. Shortly after the nuclei fuse, meiosis (reduction division) occurs and produces haploid nuclei that dominate the remainder of the life cycle of fungi.

Fungi are classified according to their methods of reproduction and their basic structure. At one time fungi were classified either in the kingdom Plantae or in the kingdom Protista. Today, fungi are placed in their own kingdom, the Fungi.

We have divided the kingdom Fungi into five phyla: Oomycota, Zygomycota, Ascomycota, Basidiomycota, and Deuteromycota. Notice that the name of each phylum ends in -*mycota*. This suffix is derived from the Greek word for mushroom, which is *mykes*. *Mykes* is also the root for mycelium.

19–1 (continued)

Content Development

Using the Visuals Have students examine Figure 19–1.

- **In what ways do the fungi shown in this photograph appear similar to plants?** (The fungi grow from fixed positions, and they have what look like stems.)
- **In what ways do these fungi appear different from plants?** (They are orange. Fungi lack chlorophyll, the pigment that gives plants their green color. Also, fungi lack the roots, stems, and leaves that characterize most plants.)
- **Out of what are these fungi growing?** (A dead tree.)
- **What does this tell you about the fungi's source of energy?** (These fungi are saprophytes; they live by decomposing dead organic matter.)
- **Why do you think that fungi are classified in a separate kingdom?** (Possible answer: They are quite different from organisms in other kingdoms.)

Explain to students that at one time, some biologists classified fungi in the kingdom Plantae. They discarded this classification because of fungi's many differences from plants, some of which the class just discussed.

Oomycota—Protistlike Fungi

Because the fungi in the phylum Oomycota (oh-oh-migh-KOHT-ah) are so closely related to the plantlike protists, many scientists include them as one of the phyla within the kingdom Protista. Members of this phylum, called oomycetes, commonly form a white fuzz on aquarium fish or on organic matter sitting in water. Although oomycetes are commonly known as "water molds," a few are able to grow on land under damp, humid conditions. Even though these fungi are not common on land, they do cause a number of serious diseases among crop plants, including potato blight. We will consider these diseases when we examine how fungi fit into the environment.

The cell walls of oomycetes are made of cellulose. It is through these thin cell walls that the water molds absorb food. Oomycetes are the only fungi that produce motile spores. These spores swim through water and raindrops to new sources of food. The hyphae of oomycetes lack cross walls. As a result, the hyphae are multinucleate (have many nuclei).

The life cycle of a water mold is shown in Figure 19–4. Notice the two types of reproduction that can occur: asexual and sexual. In asexual reproduction, portions of the hyphae develop into sporangia (spore cases). Each sporangium produces flagellated spores that swim away from the sporangium in search of food. When food is found, the spores develop into hyphae, which grow into new organisms.

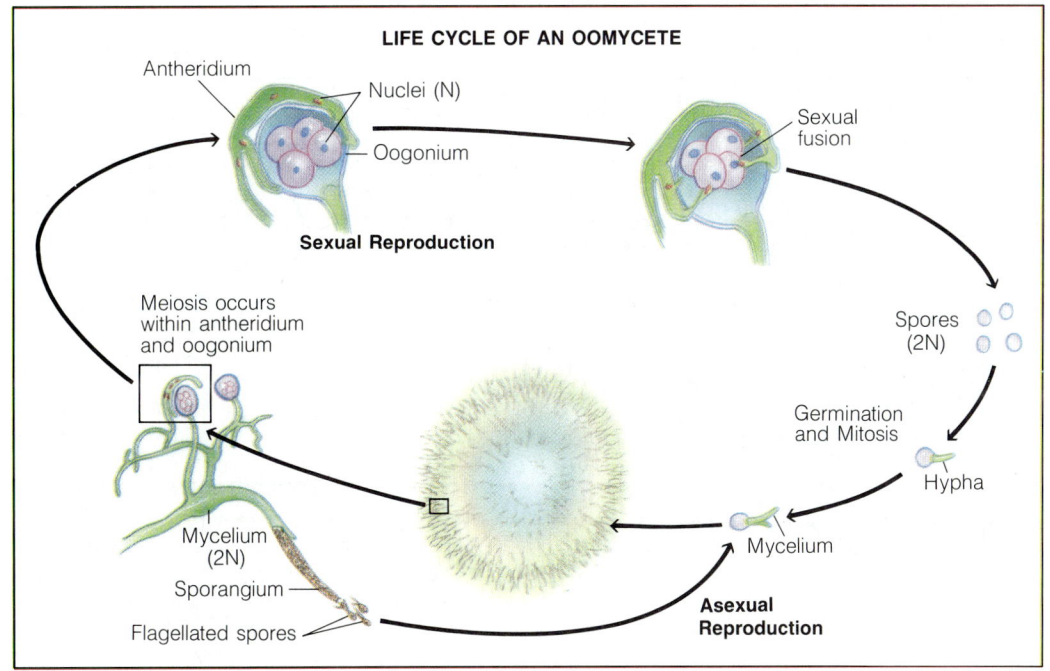

Figure 19–4 The water mold, an oomycete, reproduces both asexually and sexually. During asexual reproduction, flagellated spores are produced by the diploid (2N) mycelium. These spores grow into new mycelia. During sexual reproduction, the male gamete fuses with the female gamete.

BACKGROUND INFORMATION
HYPHAE

Hyphae consist of tubelike structures that contain cytoplasm and hundreds of nuclei. In some fungi, hyphae are divided into sections by cross walls called septa. In the septa are perforations that allow cytoplasm and nuclei to move freely throughout the hyphae. Other fungi have no septa dividing the hyphae.

HISTORICAL NOTE
FUNGI ANCESTORS

Fungi are descended from the most ancient organisms on Earth. It is believed that the first fungi arose during the Proterozoic Era, which preceded the Paleozoic Era. By the beginning of the Cambrian Period, fungi, bacteria, and protistans made up life on Earth. At this time, fungi were essentially aquatic. Through millions of years of evolution and adaptation, fungi made the transition from water to land, and in so doing managed to cover the Earth on every continent.

Other biologists classified fungi with certain unicellular organisms, ones now classified in the kingdom Protista. But because of fungi's many differences from these and other organisms, biologists today place fungi in their own kingdom.

Skills Development
Guided Practice
Skill: Comparing
Have students compare the unusual way in which fungi obtain nourishment from food with the way that humans obtain nourishment from food.

• In what way is the breakdown of food by fungi similar to the breakdown of food in humans? (Both have strong enzymes that help to break down food.)

Reinforcement/Reteaching
The authors of this text do not feel that the life cycles of the various fungi should be used as an exercise in memorization of terms. Rather, the emphasis should be on the fact that successful organisms may use very different methods to ensure that reproduction occurs.

BACKGROUND INFORMATION
CLAVICEPS PURPUREA

An interesting sac fungi is *Claviceps purpurea*. *C. purpurea* is a parasite that lives on rye and other grains. This fungus produces alkaloids that can stimulate smooth muscle and block sympathetic nerve pathways in animals that eat the infected grain. Depending on the amount ingested, these alkaloids can be either helpful or harmful to the animals (including humans) that consume them. For example, in small doses, the alkaloids have several uses in human medicine. They can be used to treat migraine headaches and they are widely used after childbirth to return the uterus to normal size and to prevent hemorrhaging. When the alkaloids are eaten in large amounts, however—as might occur when rye flour is contaminated—a disease called ergotism develops. The symptoms of ergotism include hysteria, hallucination, convulsions, vomiting, diarrhea, and dehydration. Lesions may develop on the body's limbs and gangrene may set in. Severe cases of ergotism are fatal.

19–1 (continued)

Reinforcement/Reteaching
Using the Visuals Review with students the process of sexual reproduction in oomycetes. Point out that one structure, called the antheridium, produces sperm cells. These sperm cells fertilize egg cells that are produced by another structure, called the oogonium. Emphasize that the fertilization of egg cells takes place in the oogonium.
• **What are formed when sperm cells and egg cells unite?** (Zygotes.)

Figure 19–5 The black bread mold *Rhizopus stolonifer*, a zygomycete, is commonly found growing on bread. The round black-colored structures at the top of the threadlike hyphae are the sporangia, or spore cases.

Sexual reproduction takes place in specialized structures that are formed by the hyphae. One of these structures, called the antheridium (an-thuh-RIHD-ee-uhm), produces sperm cells (male gametes). The other structure, called the oogonium, produces egg cells (female gametes). Fertilization occurs within the oogonium and, like spores, the zygotes that form develop into new organisms.

Zygomycota—Common Molds

Fungi that belong to the phylum Zygomycota (zigh-goh-migh-KOHT-uh) are called zygomycetes and are terrestrial organisms. During sexual reproduction, they form a thick-walled zygote known as a **zygospore**. The hyphae of zygomycetes lack cross walls although there are cross walls present that isolate the reproductive structures from the rest of the hypha. We have all had some experiences—most often unpleasant ones—with members of this phylum. These common molds are the molds that grow on meat, cheese, and bread.

An example of a zygomycete is the black bread mold *Rhizopus stolonifer*. You can grow this mold yourself by exposing a slice of freshly baked bread (not the processed kind) to some airborne dust. Then keep the bread from drying out by putting it in a covered container and placing it in a warm spot.

Figure 19–6 Sexual reproduction in black bread mold occurs when two hyphae from different mating types fuse, forming gametangia. The gametangia develop into a zygospore, which grows into new hyphae that form sporangia. Asexual reproduction occurs when the spores are discharged from the sporangia.

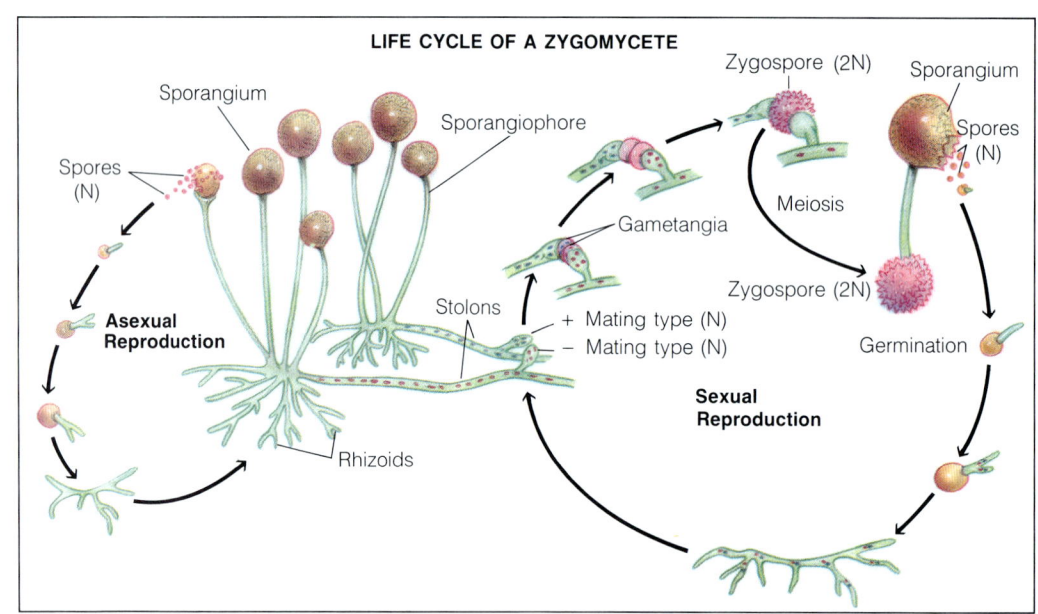

Use Figure 19–4 to point out that both sexual and asexual reproduction lead to the same result: Both zygotes and spores germinate into tubelike structures that become hyphae, which grow and differentiate into the mycelium of a new organism.

Skills Development
Guided Practice
Skill: Inferring
Tell students that about 50,000 different species of fungi have been identified, although scientists believe that there may actually be four times that number.

Also point out that ancient fungi were among the first organisms on Earth at a time when all life on Earth was aquatic.
• **Why do you think that so many different species of fungi have developed over time?** (Answers may vary. Possible an-

In a few days, if you use a magnifying glass to examine the fuzz that grows on the warm bread you will see tangles of delicate hyphae, or mycelia. Actually, you would be seeing more than one kind of hypha. The rootlike hyphae that penetrate the surface of the slice of bread are called **rhizoids.** Rhizoids anchor the fungus to the bread (much as roots anchor a plant), release digestive enzymes, and absorb digested organic material. The stemlike hyphae that run along the surface of the bread are called **stolons.** And the hyphae that push up into the air are the sporangiophores, which form sporangia at their tips.

During asexual reproduction, sporangia produce spores. A single sporangium may contain as many as 40,000 spores. When fully developed, the sporangium opens, scattering the spores to the wind. Under proper conditions of warmth and moisture, the spores germinate, producing new masses of hyphae.

Sexual reproduction occurs in bread molds and other zygomycetes when two hyphae from different mating types come together, forming gametangia (gamete-producing structures). Haploid gametes are produced in the gametangia. Gametes of one mating type fuse with gametes of the opposite mating type, forming diploid (2N) nuclei. A thick wall develops around the nuclei, producing a zygospore. The tough, resistant zygospore may remain dormant for months. Eventually, when conditions become favorable, the zygospore germinates, undergoes meiosis, and develops into a hypha. The hypha then forms a sporangium and releases spores. Each spore can develop into a new mycelium.

In zygomycetes, as in other organisms, the main function of a sexual reproductive process is to produce new combinations of genetic information. The sexual reproductive process is an effective way to maintain genetic diversity in a species.

Ascomycota—Sac Fungi

The phylum Ascomycota (as-kuh-migh-KOHT-uh) is the largest phylum of the kingdom Fungi. There are more than 30,000 species of ascomycetes, as members of this phylum are called. The nuclei in the hyphae of ascomycetes are separated by cross walls so that individual cells do exist within the organism. In the cross walls, however, there are tiny openings through which the cytoplasm and the nuclei can move. Some ascomycetes, such as the morels, are large enough to be visible when they grow above the ground. Others, such as yeasts, are microscopic.

The life cycle of an ascomycete usually includes both asexual and sexual reproduction. Asexual spores are formed at the tip of specialized hyphae called **conidiophores** (koh-NIHD-ee-uh-forz). Because these spores are very fine, they are called **conidia** (koh-NIHD-ee-uh; singular: conidium) from the Greek word *konis*, which means dust.

Figure 19–7 Ascomycetes are the largest phylum of fungi, containing 30,000 species. Among the ascomycetes are the common morel (bottom) and a type of cup fungus (top).

FACTS AND FIGURES

Some species of fungi can be both saprophytic and parasitic. For example, the fungus that causes Dutch elm disease lives on the tree as a parasite while the tree is alive, then continues to live on the wood as a saprophyte after the tree is dead.

HISTORICAL NOTE
ERGOTISM

Epidemics of ergotism caused by *C. purpurea* were common in Europe during the Middle Ages when rye was a principal crop. One such epidemic had political consequences for Peter the Great, the Russian czar who was bent on conquering ports along the Black Sea to add to his vast empire. While his soldiers were laying siege to the ports, they ate mostly rye bread and fed rye to their horses. Because the rye was contaminated with *C. purpurea,* the soldiers went into convulsions and the horses developed "blind staggers."

Later in history, ergotism was known to occur in the New World. Some historians have speculated that outbreaks of ergotism may have been used as an excuse to initiate the Salem witchhunts in colonial Massachusetts.

---

swers might include the idea that fungi have been remarkable in their ability to evolve and adapt to the changing Earth, and this process of evolution has brought about the development of many different varieties of fungi. In addition, fungi have managed to grow and survive in almost every kind of climate and environment. It would seem logical that the need for different adaptations would give rise to many different species of fungi.)

Content Development
While students are studying the phylum Zygomycota, have them grow samples of mold for observation. Tell students to dampen a piece of bread and leave it in a dark place for several days. (They should use a bread that does not contain preservatives.) Then have students use hand lenses to examine the molds and draw and/or describe what they see.

Mold can also be grown on fruit. Have students line a covered container with moist blotting paper, then place an orange in the container and put the cover on. After storing the container in a dark place for several days, a green, black, or white mold will appear.

MULTICULTURAL STRATEGY

The edible fungi include more than the familiar varieties in the supermarket. Have students write short reports on some of the world's different types of edible mushrooms. Interesting choices include the following:
- Snow mushrooms, which resemble a flower in bloom and can be bought in Chinese markets.
- Shiitake mushrooms, which grow in Asia and are now sold in the United States.
- Chanterelles, golden-colored mushrooms found throughout the world.
- Truffles, the world's most valued and expensive mushroom, especially important in French cuisine.

Sexual reproduction in ascomycetes involves the formation of an **ascus,** or tiny sac. The ascomycetes are named for this reproductive structure. In most ascomycetes, sexual reproduction occurs between two different mating types (+ and −), which produce gametangia. The gametangia grow together to allow the haploid (N) nuclei to fuse. The cell that results from this fusion begins to develop into a structure that forms the ascus. See Figure 19–8. At first the cell has two nuclei, indicating that the nuclei of the two mating types do not fuse right away. When fusion does eventually occur, a diploid (2N) zygote is formed. The fusion is quickly followed by meiosis, producing 4 haploid cells. In most ascomycetes, meiosis is followed by a round (or two) of mitosis, so that 8 or 16 cells are found within the ascus. The cells produced within the ascus are known as **ascospores.** Like conidia, ascospores are capable of growing into new organisms.

The fruiting bodies of ascomycetes can be spectacular. A fruiting body is the part of the fungus that you see above the ground. It contains the spore-producing structures. The morel is an edible ascomycete in which the fruiting body bearing the asci has become the largest visible part of the organism.

The yeasts, which are unicellular, are one of the most interesting groups of ascomycetes. Most of their reproduction is asexual and takes place by mitosis and by budding. Budding is the formation of a smaller cell from a larger one.

Under the right circumstances, yeasts also reproduce sexually. They form asci that contain ascospores. Most scientists

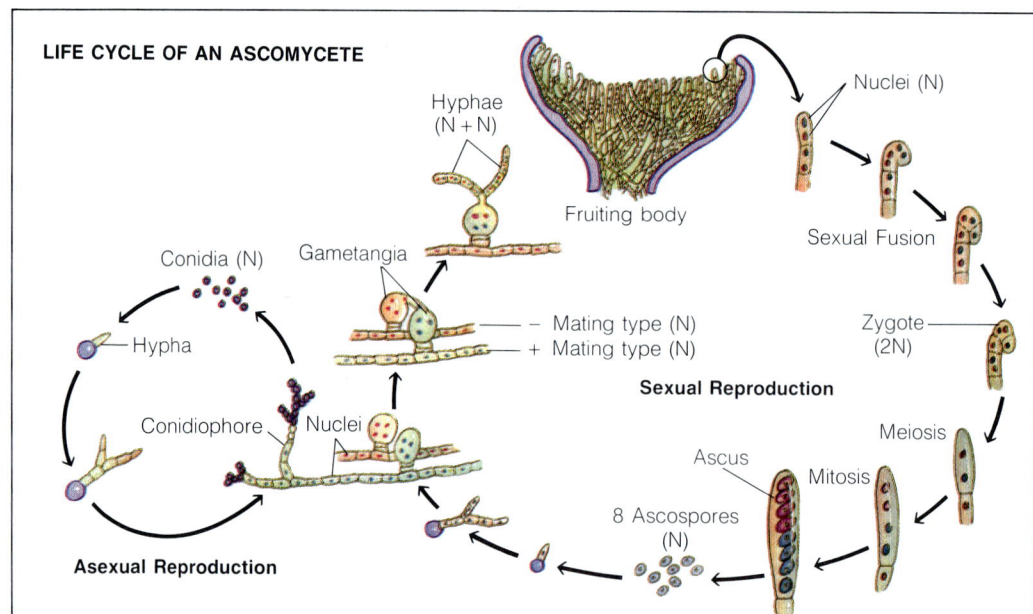

Figure 19–8 As in most fungi, the life cycle of ascomycetes includes both asexual and sexual reproduction. During asexual reproduction, spores called conidia are formed at the tip of conidiophores, or special hyphae. During sexual reproduction, an ascus, or tiny sac that contains ascospores, forms.

19–1 (continued)

Content Development
Point out to students that in zygomycetes, it is a chemical attraction between + and − hyphae that causes them to fuse. Also emphasize that the result of this fusion is a zygospore, which is characteristic of the members of this phylum.
- **What makes a zygospore distinctive?** (It has a thick wall.)
- **What does this thick wall do?** (It protects the zygote during a period of dormancy.)
- **Why does the zygote of a zygomycete often remain dormant for a period of time?** (It is waiting for favorable growth conditions.)
- **What might these favorable growth conditions be?** (Possible answers: proper temperature, necessary moisture, and a suitable medium in which to grow.)

Skills Development
Guided Practice
Skill: Comparing
Have students compare reproduction in zygomycetes with reproduction in oomycetes.
- **In what ways is reproduction alike in these two groups?** (Members of both phyla can reproduce either sexually or asexually; in both groups asexual reproduction is accomplished by the dispersal of spores.)
- **How is reproduction different in the two groups?** (In oomycetes, sperm cells enter the

believe that yeast evolved from more complicated (and more typical) ascomycetes that lost the ability to form hyphae and became unicellular.

You might think of yeast as a lifeless dry powder used to make bread and rolls. But the dry granules actually contain ascospores, which become active in a moist environment. To see this for yourself, take a teaspoon of dry yeast and add it to about 50 milliliters of warm water that contains two teaspoons of sugar. When you examine a drop of this mixture under a microscope in about twenty minutes, you will be able to see cell division in the rapidly growing yeast cells.

Basidiomycota—Club Fungi

Most of the organisms that we call mushrooms belong to the phylum Basidiomycota (buh-sihd-ee-uh-migh-KOHT-uh) and are known as basidiomycetes. The phylum gets its name from a specialized reproductive structure that resembles a club. This spore-producing structure is called a **basidium** (buh-SIHD-ee-uhm; plural: basidia). In mushrooms, basidia are found in the cap.

Basidiomycetes undergo what is probably the most elaborate life cycle of all the fungi. A **basidiospore** germinates to produce haploid primary mycelia. The haploid primary mycelia of different mating types fuse. A secondary mycelia containing two nuclei—one nucleus from each mating type—is formed. (The nuclei themselves do not fuse at this stage.)

Figure 19–9 All basidiomycetes, or club fungi, are composed of masses of hyphae. The coral fungus (top, left), shaggy mane fungus (right), and jelly fungus (bottom, left) illustrate the many different shapes that masses of hyphae can form.

oogonium and fertilize egg cells to form zygotes. In zygomycetes, + and – hyphae conjugate to give rise to a zygote. Another important difference is that the zygote of a zygomycete usually undergoes a period of dormancy before it germinates.)

Reinforcement/Reteaching
Emphasize that ascomycetes, like zygomycetes, are named for their distinctive reproductive structure.
• **What is the distinctive reproductive structure in ascomycetes?** (A tiny sac called an ascus.)

Point out that the ascus usually contains eight ascospores, and that these are formed by sexual reproduction.

Emphasize to students that in ascomycetes, the ascus is associated with sexual reproduction. Asexual reproduction is accomplished by asexual spores (not to be confused with ascospores) produced at the tips of specialized hyphae. Also point out that asexual reproduction in yeasts is accomplished by budding; sexual reproduction in yeasts is accomplished by means of ascospores, as in other ascomycetes.

ESL STRATEGY

Have students complete the paragraph below with the terms listed. Students should choose the best term for each blank in the paragraph.

hyphae   chemicals   heterotrophs
yeast   organisms   mushrooms
molds   symbiotic   multicellular
spores   digested   unicellular

All fungi are _____. They obtain food from _____ relationships, the remains of dead _____, and by absorbing the food that _____ (released by the fungi) have _____. Many fungi reproduce through _____. (Answers: heterotrophs, symbiotic, organisms, chemicals, digested, spores.)

BACKGROUND INFORMATION
MUSHROOMS AS MEAT-EATERS

Strange as it may sound, about ten species of mushrooms prey upon animals for food. These mushrooms attack nematodes, which are roundworms commonly found in soil and rotting wood.

Some of the mushroom species have special cells in the hyphae that produce a sticky adhesive in which the roundworms get stuck. The cells of the mushroom then quickly invade and kill the animal. Its body is broken down and used to supply nutrients, especially nitrogen, that the mushrooms need to survive.

Other mushrooms paralyze roundworms with a special poison. Then they grow through openings in the worm's body wall and destroy it before it can crawl away.

Figure 19–10 This photograph shows the underside of a parasol mushroom. Notice the gills and the stalk. The ringlike structure on the stalk is called the annulus.

The secondary mycelia can grow in the soil for years, reaching an enormous size. (A few mycelia have been found to be hundreds of meters across, making them perhaps the largest organisms in the world!) When the right combination of moisture and nutrients occurs, a spore-producing fruiting body pushes above the ground. We recognize these fruiting bodies as mushrooms.

The mushroom (fruiting body) begins as a mass of growing hyphae that forms a button, or thick bulge, at the soil's surface. The bulge expands with astounding speed and force, producing fully developed mushrooms overnight. This rapid growth occurs because the cytoplasm from thousands of hyphae in the soil quickly streams into the growing mushroom, enlarging it and producing a great amount of force.

When the mushroom cap opens, it exposes hundreds of tiny gills on its underside. Each gill is lined with basidia. Within a few days, the two nuclei in each basidium fuse to form a true diploid (2N) zygote cell. The diploid cells quickly undergo meiosis, forming clusters of haploid basidiospores. The basidiospores form at the edge of each basidium and, within a few hours, are ready to be scattered. Mushrooms are truly amazing reproductive structures—a single mushroom can produce as many as one billion spores!

In addition to common mushrooms, this phylum includes shelf (bracket) fungi, which grow near the surfaces of dead or decaying trees. The visible bracketlike structure that forms is actually a reproductive structure, and it too is an amazing producer of spores. Puffballs, toadstools, jelly fungi, and plant parasites known as rusts are other examples of basidiomycetes.

Figure 19–11 The most familiar basidiomycetes are the mushrooms. The mushroom cap, which contains basidia, is made up of masses of tightly packed hyphae. Basidia are the club-shaped structures that produce the basidiospores.

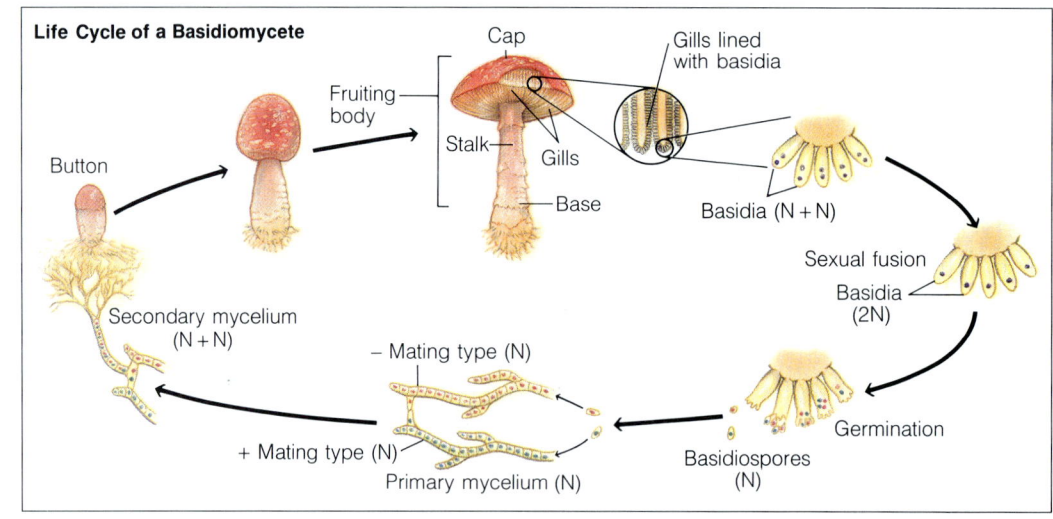

Reinforcement/Reteaching

Review with students the special reproductive structures that characterize zygomycetes, ascomycetes, and basidiomycetes. Remind students that each of these groups of fungi is named for its reproductive structure: zygomycetes for the zygospore; ascomycetes for the ascus; and basidiomycetes for the basidium.

• **Describe a zygospore. How is it formed?** (A zygospore is formed when gametes of opposite mating types fuse during sexual reproduction. A thick wall develops around the diploid nuclei, enabling the zygote to remain dormant for months.)

• **In what ways are an ascus and a basidium similar?** (Within both structures, spores are produced that are capable of growing into new organisms.)

Deuteromycota—Imperfect Fungi

The phylum Deuteromycota (doo-ter-uh-migh-KOHT-uh) includes fungi that cannot be placed in any of the other phyla because their sexual reproduction has never been observed. The word imperfect is a botanical term referring to a lack of sexual reproduction; hence, the name imperfect fungi.

A great majority of the deuteromycetes (as they are also known) closely resemble ascomycetes. Others are similar to basidiomycetes. And a few are much like zygomycetes. An example of a deuteromycete that is similar to ascomycetes is *Penicillium*. *Penicillium* is a mold that frequently grows on fruit and is the source of the antibiotic penicillin. *Penicillium* forms large mycelia on the surfaces of its food source. And like ascomycetes, *Penicillium* reproduces asexually by means of conidia. Biologists believe that *Penicillium* may have developed from a type of ascomycete that lost the ability to carry out the sexual phase of its life cycle.

The deuteromycetes include some of the most infamous members of the kingdom Fungi: those that are responsible for ringworm, athlete's foot, and other skin infections that affect humans. Other deuteromycetes cause several plant diseases, including black spot of roses and early tomato blight.

Figure 19–12 Aspergillus niger, *a deuteromycete, contains spore-bearing structures called conidiophores (inset), which form at the tips of hyphae.*

Figure 19–13 Five phyla of fungi

Phylum	Examples	Characteristics	Reproduction	
			Asexual	Sexual
Oomycota (protistlike fungi)	Water molds, downy mildew, potato blight fungus	Cell walls with cellulose; coenocytic diploid hyphae	Flagellated oospores in sporangia	Fusion of gametes in gametangia resulting in oospores
Zygomycota (common molds)	*Rhizopus* (black bread mold), *Pilobolus* (a dung fungus)	Cell walls with chitin; coenocytic hyphae	Unflagellated spores in sporangiophores	Fusion of gametes in gametangia resulting in zygospores
Ascomycota (sac fungi)	Yeasts, morels, truffles, *Neurospora* (red bread mold)	Hyphae divided by perforated cross walls; short stage in which cells have two nuclei	Conidia on conidiophores	Fusion of hyphae resulting in ascospores in ascus
Basidiomycota (club fungi)	Mushrooms, puffballs, bracket fungi, rusts, jelly fungi, toadstools	Hyphae divided by perforated cross walls; long stage in which cells have two nuclei	None or conidia on conidiophores	Fusion of cells on tips of hyphae resulting in basidiospores on a basidium
Deuteromycota (imperfect fungi)	*Penicillium*, *Aspergillus*, ringworm and athlete's foot fungus, black spot on roses fungus, tomato blight fungus, cucumber scab fungus	Some resemble ascomycetes; others similar to basidiomycetes; few like zygomycetes	Conidia on conidiophores	None known

HISTORICAL NOTE
PENICILLIN

Among the most famous Deuteromycota is *Penicillium notatum*, from which the antibiotic penicillin is made. The medical value of this fungus was discovered by accident in 1928 by British microbiologist Sir Alexander Fleming.

At the time, Fleming was studying cultures of the bacterium *Staphylococcus*. One day, Fleming noticed that a bluish mold had begun to form on some of his culture plates. Some scientists might have discarded the cultures and perhaps felt frustrated at a ruined experiment. Fleming, however, took time to look at the mold and observe what was happening. Much to his surprise, Fleming noticed that the small patches of mold appeared to be killing the bacteria! Fleming had a hunch that the bacteria were dying from a substance produced by the mold—and his hunch turned out to be right.

Fleming and his co-workers set out to purify some of this substance, which they named penicillin. Tests on sick humans proved that penicillin could cure serious bacterial diseases. Some diseases cured by penicillin include strep throat, staph infection, and pneumonia.

The discovery of penicillin has proven to be one of the most significant medical events in human history. The availability of penicillin—and of the other antibiotics that followed—has increased life expectancy and all but eliminated the fatal effects of numerous diseases and infections.

In the ascus, these spores are called ascospores; in the basidium they are called basidiospores.)

Content Development
Explain to students that the mushroom has rootlike hyphae that extend into the ground. The purpose of these hyphae is to anchor the mushroom and to act as a means of obtaining nourishment. The hyphae release enzymes that break down the food source, then the digested food is absorbed.

• **What other familiar fungus that you have studied has rootlike hyphae that penetrate into the surface on which the fungus is growing?** (Bread mold.)
• **What are the rootlike hyphae of bread mold called?** (Rhizoids.)

SCIENCE, TECHNOLOGY, AND SOCIETY
BREAKTHROUGH

Cyclosporine

The main obstacle to successful transplantation is the rejection of foreign tissue by the immune system of the recipient. Rejection occurs because transplanted tissue from another person contains antigens that stimulate an immune response on the part of the recipient's lymphocytes.

To minimize the possibility of rejection, an antigenic typing system, called the HLA system, is used. This system is similar to blood typing and is used to determine the degree of tissue compatibility between a potential donor and a recipient. Doctors also use immunosuppressive drugs to prevent the rejection of a transplanted organ. Immunosuppressive drugs interfere with the production of antibodies by the recipient's immune system. Cyclosporine is an example of an immunosuppressive drug.

Before students read this feature, remind them of the discussion of human blood groups in Chapter 11. Have students recall that it is essential that a person receiving a blood transfusion be given the correct blood type; otherwise, his or her immune system will reject the antigens of the foreign blood.

SCIENCE, TECHNOLOGY, AND SOCIETY
BREAKTHROUGH

Cyclosporine

When disease or injury destroys one of a patient's vital organs, physicians often try to transplant the organ from a donor into the patient. This procedure, however, poses a serious and often life-threatening problem. In most transplants, the recipient's immune system recognizes the transplanted organ tissue as foreign and attacks it, causing the rejection of the organ. It was not until 1972 that a substance was discovered that would suppress the immune system's response to a foreign organ.

While working in Switzerland, Jean Borel, a Swiss immunologist, isolated a new strain of deuteromycetes, called *Tolypocladium inflatum,* from soil samples that were obtained in Wisconsin and Norway. This strain produced a substance, later named cyclosporine, that was capable of suppressing the immune system's response to transplanted organs. In other words, the substance was remarkably effective in preventing transplant rejection.

Cyclosporine has revolutionized the field of medical transplantation. Before it became available in 1979, fewer than half of all kidney transplants were successful. With the advent of cyclosporine, the percentage of survival has risen to 90 percent.

Cyclosporine is indeed a wonder drug—but it is not perfect. It causes severe side effects and is not successful in all cases. However, the discovery of cyclosporine has opened up a new frontier in medicine, and we cannot help but wonder how many other pleasant surprises the fungi have in store for us in the future.

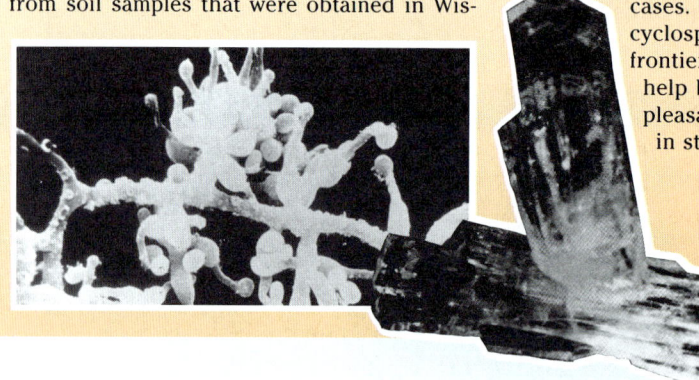

Synthetic crystals of cyclosporine (right) and fungus Tolypocladium inflatum *that produces cyclosporine (left)*

19–1 SECTION REVIEW

1. What is a fungus? Describe the structure of a typical fungus.
2. How do fungi obtain food?
3. Describe the five phyla of fungi.
4. What are some methods by which fungi reproduce?
5. **Critical Thinking—Applying Concepts** Sensitive tests show that tissue from several mushrooms gathered near the base of a tree are genetically identical. How might you explain this finding?

19–1 (continued)

SECTION REVIEW 19–1

1. Fungi are eukaryotic heterotrophs. They either decompose organic matter or live as parasites or symbiotes with other organisms. The body of a typical multicellular fungus is made up of many tiny filaments tangled together in a thick mass called a mycelium. The individual filaments are called hyphae. In many fungi, the hyphae are divided by cross walls into cells containing one or more nuclei.

2. Fungi release digestive enzymes into the environment. These enzymes break down food into simple molecules, which then diffuse across the cell walls and cell membranes.

3. The five phyla of fungi are Oomycota, Zygomycota, Ascomycota, Basidiomycota, and Deuteromycota. See Figure 19–13 on page 415 for their characteristics.

4. Most fungi reproduce both asexually and sexually. Asexual reproduction occurs either by the production of spores or by the fragmentation of the hyphae. In many fungi, sexual reproduction involves two different mating types: + and −. When the hyphae of opposite mating types meet, each hypha forms a gametangium. Then the two ga-

19–2 Fungi in Nature

As you have learned, fungi live by feeding on living organisms or on the remains of dead ones. Although this may paint a grim picture of fungi—linking them with death and decay—they are actually some of the most beautiful organisms on Earth. In this section we will take a very human-centered view of the kingdom Fungi as we examine what effects these organisms have on us and the rest of the living world.

Guide For Reading

- What is the principal role fungi play in the environment?
- How do fungi form symbiotic relationships with other organisms?
- In what ways are humans affected by fungi?

Ecological Significance

The principal role fungi play in the environment is to decompose and recycle living material. Imagine a world in which fungi do not exist: The ground would be littered with leaves, fallen wood, and the bodies of dead animals. What impact would this have on organisms living in this world?

First, you may recall that the material of which a living organism is composed is rich in chemical energy. Because this energy exists, we can make a crackling fire out of wood or a good snack out of an apple. If such material does not undergo decay, the energy it contains will be lost. Second, many organisms, particularly green plants, require small amounts of trace elements and nutrients in order to survive. During their development, green plants remove these materials from the soil. If the materials are not eventually returned, the soil will soon be depleted and the destruction of plants, as well as animals whose lives depend on the plants, will result.

WHERE ARE FUNGI FOUND? There are remarkably few places on Earth where one species of fungus or another does not make its home. Even more amazing is the fact that fungal spores are found in almost every environment. Indeed, this is why molds seem to spring up in any location that has the right combination of moisture and food.

In many places, large mycelia occupy a nearly permanent place in the environment and last for many years. A mushroom develops from a mycelium located just below the ground. As the mycelium grows, new mushrooms pop up from the mycelium wherever nutrients are available. This is why strands of mushrooms are often part of the same organism.

As time goes by, the available nutrients near the center of the mycelium become depleted, causing new mushrooms to sprout only at the edges of the mycelium. This produces a ring of mushrooms called a fairy ring. People once thought fairies dancing in circles during warm nights produced these rings, so they called them fairy rings. Over many years, fairy rings can become enormous, forming rings 10 to 20 meters in diameter.

SPORE DISPERSAL Many fungi, including most common mushrooms, produce dry, almost weightless, spores that are

Figure 19–14 *Together with bacteria, fungi—such as this mushroom—are the major decomposers and recyclers of living material on Earth.*

417

BACKGROUND INFORMATION
PILOBOLUS

Pilobolus is a zygomycete that grows in the dung of grazing animals. The hyphae of the *Pilobolus* penetrate the surface of the dung, absorbing the large amounts of undigested nutrients that are present. In a short time, special sporangiophores (spore-carrying hyphae) grow out of the dung and begin to point themselves toward the sunlight.

Pilobolus has a kind of built-in mechanism that enables its hyphae to "see" where the sun is. At the top of the fruiting body of the *Pilobolus* is a sporangium that contains the spores it is about to release. Just below the sporangium is a small swelling with very thin walls that is filled with clear liquid. This swelling acts as a lens, concentrating the sun's rays at one point in the bottom of the swelling. The hyphae that make up the sporangiophore are able to detect these rays, and they grow a little faster on one side than the other. This bends the sporangiophore toward the light.

Once the sporangiophore is aimed toward the light, *Pilobolus* has a unique method of releasing its spores. The spore-bearing hyphae explode, a few at a time, and fire their spores in the direction of the sunlight. The spores are fired with an initial speed of about 50 kilometers per hour, and they may travel as far away from the fungus as 1 meter.

Figure 19–15 *The giant puffball contains as many as 7 trillion spores. In the common puffball, the dispersal of spores can be triggered by the slightest touch, even by a raindrop.*

Figure 19–16 *Fungi have remarkable ways of dispersing their spores. The lacy stinkhorn attracts flies by producing a spore-containing fluid that has the odor of rotting flesh. The spores pass unharmed through the flies' digestive system and are deposited over great distances. Pilobolus (inset) fires its sporangia at an initial speed of 50 kilometers per hour—as far as 1 meter.*

easily scattered by the wind. On a clear day, a few liters of fresh air may contain hundreds of spores from many species of fungi. Some of these species have remarkable ways of getting their spores into places where they are likely to grow.

For fungi, spore placement is crucial. A single spore has a slim chance of finding the proper combination of temperature, moisture, and food so that it can germinate. Even under the best of circumstances, the odds of a spore producing a mature organism can be more than one in a billion. So you can see why anything that might help reduce those odds is considered a selective advantage to that species.

The puffballs (basidiomycetes) go to extremes to produce their spores. A mature puffball is virtually a warehouse of spores. The simple action of a raindrop falling on a puffball can release thousands of spores in a small cloud of dust.

Other species of fungi trick animals into dispersing their spores for them. The stinkhorns (*Phallus*) go so far as to mimic rotting meat. The surface of a stinkhorn is covered with a fluid that has the odor of rotting flesh. Flies are drawn to the stinkhorn. Then they land on the stinkhorn to taste the sticky fluid. Once ingested, the spore-containing fluid will pass unharmed out of the flies' digestive systems, depositing spores over great distances.

Symbiotic Relationships

Many fungi associate with members of other species in symbiotic relationships. In some of these relationships, such as early tomato blight, fungi are harmful. But in other cases, fungi form relationships in which both partners benefit. Such is the situation with the **lichens** (LIGH-kuhnz) and **mycorrhizae** (migh-koh-RIGH-zee).

Lichens are symbiotic partnerships between a fungus and a photosynthetic organism. The fungus in the relationship is usually an ascomycete, although it can be a basidiomycete. The photosynthetic organism is either a cyanobacterium (blue-green bacterium) or a green alga.

19–2 (continued)

Content Development
• Why are fungi particularly well-suited to the task of decomposing and recycling living material? (They release into the environment digestive enzymes that break down organic matter; many fungi are saprophytes, meaning that they feed specifically on dead organic material.)

Point out that the role of fungi is made even more valuable by the fact that these organisms live almost everywhere on Earth and in almost every environment. This means that the entire Earth continually benefits from their recycling activities. Also point out that one of the reasons fungi have survived so long on Earth may be because they play such a vital role in the recycling of energy and nutrients in nature.

Content Development
Emphasize to students that dispersing spores is an important aspect of fungal activity. Various

Because they are extremely resistant to drought and cold, lichens grow in places where few other organisms can survive—on dry, bare rock in deserts and on the tops of mountains. Lichens are able to survive in these harsh environments because of the relationship between the two partner organisms. The alga carries out photosynthesis, providing the fungus with a source of organic nutrients. The fungus, in turn, provides the alga with water and minerals that it has collected from the surfaces on which it grows.

Lichens are often the first organisms to enter barren environments, gradually breaking down the rocks upon which they grow. In this way, lichens help in the early stages of soil formation and eventually form an environment that is hospitable to other organisms.

Another symbiotic relationship, called mycorrhizae, forms between fungi and green plants (mycorrhiza means fungus root in Greek). The tiny hyphae of the fungi aid plants in absorbing water, minerals, and nutrients. They do this by producing a network that covers the roots of the plants and increases the effective surface area of the root system. The plants, in turn, provide the fungi with the products of photosynthesis.

An example of mycorrhizae involves orchids, which are considered by many as the most beautiful of the flowering plants. The seeds of orchids germinate in nature only in the presence of a certain species of fungi. These fungi penetrate the seed, providing it with moisture and food during the early stages of the orchid's growth.

The symbiotic relationships between green plants and fungi have existed for millions of years. Some of the earliest fossils of land plants contain evidence of fungi. This suggests that fungi may have played a crucial role in the colonization of the land by green plants.

Fungi and Human Life

Two of the oldest discoveries of civilization are the techniques for making bread and alcohol. Interestingly enough, both techniques rely on a cooperative effort between humans and fungi and provide an important example of the many ways in which humans have made use of this living kingdom.

Because of the role yeasts play in baking and brewing, one might argue that they are the most important fungi to humans. The common yeasts used for baking and brewing are members of the genus *Saccharomyces* (sak-uh-roh-MIGH-seez). To grow these yeasts, a rich nutrient mixture containing very little oxygen is prepared. In brewing, it is a vat of grape juice or barley malt. In baking, it is a mound of thick dough. The yeasts within the mixture quickly begin the process of alcoholic fermentation in order to obtain enough energy to survive. The byproducts of alcoholic fermentation are carbon dioxide and alcohol. The carbon dioxide gas makes bread rise (by producing bubbles

Figure 19–17 Lichens generally grow in three forms. Crustose lichens are flat (top), foliose lichens resemble leaves (center), and fruticose lichens grow upright (bottom).

TEACHING SUPPORT

Laboratory Manual
- Mold Growth, p. 243

Teaching Resources
- Ecology Investigation: Decomposers in Soil, p. 11

TIE-IN   BIOCHEMISTRY

The deathcap mushroom (*Amanita phalloides*) produces two poisons—amanitin and phalloiden. Both of these poisons concentrate in the liver and attack liver cells. Amanitin binds to the enzyme RNA polymerase and prevents the transcription of RNA. Phalloiden affects molecules in the cytoskeletons of liver cells. The cells begin to lose their shapes, and some rupture and die. The combination of these two poisons produces death in a few days. So far, there is no effective medical treatment that can stop the effects of the deathcap's poisons.

19-2 (continued)

Enrichment
Arrange for students to visit a garden center, nursery, or the agricultural department of a local college or university to learn more about the role of fungi in plant growth. Of particular interest would be a visit to an orchid grower, if there is one in your area.

Content Development
Point out to students that symbiotic relationships between fungi and green plants provide the fungus with a food source and the other organism with a way to take in water and nutrients.

• **Why is it important to the fungus to have the other organism provide it with food?** (Fungi cannot make their own food; they must absorb it from outside sources.)

• **What is the source of the food that is provided by this other organism?** (The food is produced by photosynthesis.)

• **How does the fungus help the green plant absorb water and nutrients?** (By attaching itself to the green plant, particularly at the roots, the fungus increases the surface area of the plant so that it can absorb more water and minerals.)

Figure 19–18 In France, pigs are used in the search for truffles. Truffles (left) are considered by many people to be a rare and delicious treat.

Figure 19–19 Although the death cap mushroom looks harmless, eating only one cap of it can prove fatal.

within the dough) and beverages bubble. The alcohol is used in alcoholic beverages or as a fuel.

As you may recall from Chapter 12, yeasts are now used for genetic engineering. Because they are eukaryotes, yeasts often process the protein products of genes cloned from other eukaryotes more efficiently than bacteria (prokaryotes) do. It is not impossible to imagine that sometime in the near future, genetically engineered yeasts may be used to produce a wide variety of biologically important compounds.

Some types of fungi have long been considered a delicacy by humans. One example are the mushrooms (basidiomycetes). There are approximately 10,000 different species of mushrooms found throughout the world. Many of these mushrooms are cultivated and prepared by people and then sold in supermarkets and specialty food shops. These species of mushrooms are easy to grow, taste good when properly cooked, and do not pose a danger to anyone who eats them.

Wild mushrooms are a different story: Although some are edible, many are poisonous. You may have heard someone say that only toadstools are poisonous, whereas mushrooms are safe to eat. Unfortunately, this is not true. For toadstools are mushrooms. Furthermore, poisonous mushrooms do not belong to just one order or family of basidiomycetes.

Because many species of poisonous mushrooms look almost identical to edible mushrooms, it is best not to pick or eat any mushrooms found in the wild. Instead, mushroom gathering should be left to experts who can positively identify each mushroom they collect. The result of eating a poisonous mushroom is severe illness and sometimes death.

History records that the Roman Emperor Claudius was given a plate of mushrooms, known as death caps (*Amanita phalloides*), by his wife and stepson Nero in a plot to remove him from the throne. He ate heartily. Although the death caps were delicious, the meal served its purpose and the throne was empty the next day.

Diseases Caused by Fungi

Not all fungi are suited to human needs. Some species of fungi cause tremendous losses of food and crops every year, and some cause disease in humans.

POTATO BLIGHT In their own way, the plant diseases caused by fungi have influenced history. In 1845, the potato crops of Ireland and Europe were devastated by a fungus that destroyed the foliage of the plant and infected the potatoes themselves. The culprit was the oomycete *Phytophthora infestans*, which causes the disease known as late potato blight.

Potatoes that are infected with the blight may appear normal at harvest time. But within a few weeks, the fungus makes its way into the potato, reducing it to a spongy sac of spores and dust. During the years that followed the potato blight infection, more than one million people in Ireland died of starvation as the result of the destruction of their main food source. Many others left Ireland and emigrated to the United States rather than face a similar fate.

WHEAT RUST Another fungal disease, called rust, affects wheat, one of the most important crops grown in North America. During the early part of this century, farmers in the Midwest watched helplessly as their plants developed tiny rustlike spots on their leaves. These spots gradually expanded and killed the plants before they could be used to produce grain. A similar incident occurred in the 1930s, when great plagues of rust disease added to the economic misery of the Great Depression. During this time, farmers not only lost their crops but their farms as well.

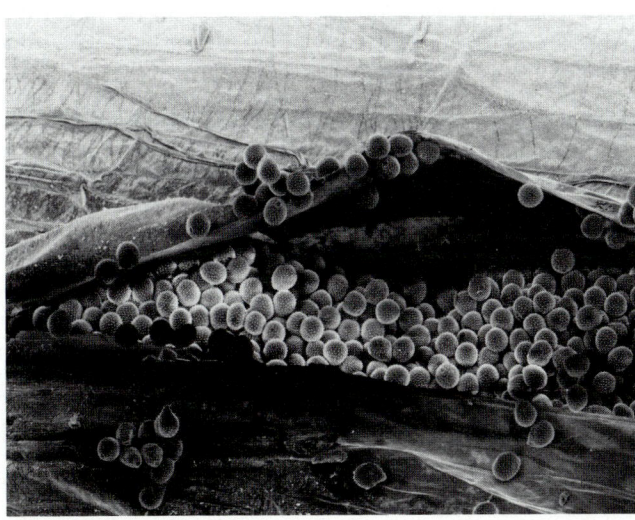

Figure 19–20 This electron micrograph shows a cluster of spores of the wheat rust fungus bursting through the plant's epidermis, or outer covering.

BACKGROUND INFORMATION
MYCORRHIZAE

The symbiotic relationships that exist between mycorrhizae and higher plants were discovered when orchid growers found that their plants required soil containing certain fungi in order to grow. Plant experts have since found that nearly all plants require associations with fungi. It is estimated that in North America, about 90% of all plant families maintain mycorrhizae. Nearly all conifers, such as the pine, fir, and spruce, require mycorrhizae for proper growth. Foresters often inoculate young seedlings with mycorrhizae before planting them.

The symbiotic relationship that exists between plants and mycorrhizae is a classic one. The fungus assists the plant in gathering water and nutrients, and the plant provides the fungus with the products of photosynthesis as food.

Symbiosis between plants and fungi is believed to be among the oldest relationships between living things on Earth. Recognizable fossils of mycorrhizae have been found in some of the earliest fossils of land plants. This has led many evolutionary biologists to suggest that the evolution of plants has taken place for millions of years with such symbioses. In fact, the formation of mycorrhizae may have been a critical step in the movement of plant life from the seas onto land.

Focus/Motivation
If you have not already done so, perform the demonstration called Activity of Yeast, which is found on page 406D. After students have observed the activated yeast, emphasize that the processes of baking and brewing both provide a rich nutrient solution in which yeast can grow.
- **What is the nutrient solution in the brewing process?** (A vat of grape juice or barley malt.)
- **What is the nutrient solution in the baking process?** (The dough.)
- **What essential ingredients for yeast growth are present in both solutions?** (Sugar and water.)
- **Sometimes a baker will mix yeast with water or milk before making dough. Why might he or she do this?** (To activate the yeast and get it started germinating before the baking begins; this way the dough will rise faster.)
- **Suppose that air is allowed to leak into a bottle in which grape juice is fermenting to make wine. What do you think will happen to the wine?** (The wine will be spoiled.)
- **Why?** (The admission of oxygen into the mixture will stop the yeast from carrying out anaerobic respiration—and it is anaerobic respiration that produces the alcohol to make the wine.)

HISTORICAL NOTE
BARBERRY

The problems caused by the barberry plant as a participant in the life cycle of wheat rust have been recognized for a long time. It is recorded that the colonial legislature of the Massachusetts Bay Colony passed the following act in 1760: "Whoever... hath any barberry bushes growing in his or their land... shall cause the same to be extirpated or destroyed on or before the 13th day of June, AD 1760."

TEACHING SUPPORT

Teaching Resources
- Vocabulary Review: Word Scramble, p. 1

Study Guide
- Section Review 19–2, p. 182

19–2 (continued)

Content Development

Emphasize to students that diseases caused by fungi attack people, plants, and animals. In order to prevent or cure fungal diseases, chemical preparations known as fungicides are used. Some fungal infections in humans can be treated with drugs that are taken orally, such as sulfa drugs or other antibiotics.

Stress that "an ounce of prevention is worth a pound of cure," because fungi are extremely persistent and difficult to get rid of. The minimum time needed to cure a fungal infection is 10 to 14 days, and some infections take 6 months to a year to cure.

Point out that fungal diseases in plants are generally the most difficult to treat. In fact, experts state that most agricultural fungicides are effective only if applied before a fungal disease strikes. For this reason, it is important to treat crops, trees, and garden plants with fungicides as a preventive measure.

SECTION REVIEW 19–2

1. The major role of fungi in the environment is to decompose and recycle living material.
2. Fungi disperse spores carried by wind and by animals.
3. Lichens and mycorrhizae are two symbiotic relationships in which fungi play a role. A lichen consists of a fungus (ascomycete or basidiomycete) and a

---

Rusts are caused by a type of basidiomycete that needs two different plants in order to complete its life cycle. Spores produced by the rust in the barberry plant are carried by the wind into wheat fields. There the spores germinate and infect wheat plants. The patches of rust produce a second type of spore that infects other plants, allowing the disease to spread through a field of wheat at an alarming rate.

Later in the year, often after the wheat crop has been ruined, new kinds of spores are produced by the rust. These black-colored spores are tough enough to survive through the winter. In the spring, they go through a sexual phase and produce spores that infect the barberry plant. Once on the barberry leaves, the rust produces the spores that infect the wheat plant, and the cycle continues. Fortunately, the life cycle of the rust can be broken and the disease brought under control by destroying the barberry plant.

OTHER PLANT DISEASES Fungi that attack crop plants cause other diseases such as corn smut, which destroys the corn kernels, and mildews, which infect a wide variety of fruits. It is estimated that fungal diseases are responsible for the loss of approximately 15 percent of the crops grown in temperate regions of the world. In tropical areas, where high humidity favors fungal growth, the loss of crops is sometimes as high as 50 percent. As you can see, these organisms are in direct competition with us for our own food supply.

HUMAN DISEASES Although most fungal diseases are associated with plants rather than animals, there are several fungi that cause disease in humans. These pathogenic (disease-causing) organisms are deuteromycetes (imperfect fungi). One type can infect the areas between the toes, causing athlete's foot. The fungus forms a mycelium directly within the outer layers of skin. This produces a red, inflamed sore from which the spores can easily spread from person to person.

When the same fungus infects the skin of the scalp, it produces a red scaling sore known as ringworm. Contrary to popular belief, ringworm is not caused by a worm; it is caused by a fungus. Ringworm can be passed from person to person by the exchange of hats, combs, and athletic headgear. The fungi that cause athlete's foot and ringworm can be destroyed by the application of fungicides, or chemicals that kill fungi.

Another type of fungal disease that infects humans is caused by the yeast *Candida albicans*. This fungus grows in moist regions of the body, such as the mouth and the urinary tract. Usually its growth is kept in check by competition from bacteria and by the body's immune system. This normal balance can be upset by many factors, including the use of antibiotics, which kill bacteria, or by damage to the immune system. When this happens, *Candida* may produce thrush, a serious and painful mouth infection, or infections of the urinary tract.

Figure 19–21 *This ant has been killed by a fungus called* Cordyceps lloydii. *Once the fungus's tiny spore enters the insect's body, it begins to multiply. Within days, the fungus digests all but the ant's outer covering. The umbrellalike structures growing out of the ant's body are the fungus's fruiting bodies.*

ANIMAL DISEASES As serious as human fungal diseases can be, few approach the deadliness of *Cordyceps lloydii*. This fungus infects ants in forests near the basin of the Amazon River in Venezuela. Microscopic spores become lodged in the ant, where they germinate and produce enzymes that slowly penetrate the insect's tough exoskeleton (external skeleton). Once the spores have gained entry, they multiply in the insect's blood, digesting all its cells and tissues until the insect dies. To complete the process of digestion, hyphae develop, cloaking the decaying exoskeleton in a web of fungal material. Reproductive structures, which will produce more spores that will spread the infection, then emerge from the ant's remains.

19–2 SECTION REVIEW

1. What is the major role of fungi in the environment?
2. What are some methods by which fungi disperse their spores?
3. Describe the role of fungi in two important symbiotic relationships.
4. What are some beneficial effects of fungi? Some harmful effects?
5. **Critical Thinking—Relating Concepts** Why are fungi a more serious problem in tropical regions of the world than they are in temperate regions?

BACKGROUND INFORMATION
SECRET GARDENS

Humans are not the only organisms who prize certain fungi as food. Amazingly enough, there are species of ants who actually cultivate tiny gardens of fungi in order to provide themselves with gourmet food! Perhaps the most famous of these ant species are the parasol ants (genus: *Atta*). These ants cut off pieces of leaves and blades of grass, then carry them back to their nests by the score. A single ant carrying a large leaf looks a lot like a person with a parasol—thus, the derivation of their common name.

One might suspect that the ants eat the leaves that they bring back, but this is not the case. Instead, the leaves are carefully inoculated with spores and then tended by worker ants until a rich mycelium forms upon them. The hyphae from these mycelia become food for the ants. As the hyphae are eaten, spores are harvested to be used on the next crop of leaves. Normally, the mycelia produced by these spores would give rise to mushrooms, but the ants prevent that from happening—they manage to keep the crop growing at just the right pace to supply their food needs. Obviously, the ants have achieved a rather remarkable feat of agriculture!

---

photosynthetic organism (cyanobacterium or a green alga). Mycorrhizae consist of a fungus and a green plant.
4. Fungi are used for making bread and alcohol, are used in genetic engineering, and are a source of food. Some fungi, on the other hand, are poisonous, and others cause disease.
5. In tropical areas, the high humidity favors the growth of fungi. As a result, fungi that cause crop diseases grow faster, killing the crop.

Reinforcement/Reteaching
If students are having trouble answering the Section Review questions, go back to the part of the section that contains the material with which they are having difficulty.

Closure
Conduct a class discussion in which students consider the many ways in which fungi affect humans.

LABORATORY INVESTIGATION

COMPARING A MOLD AND A MUSHROOM

Before the Lab
1. At least one day prior to the investigation, gather enough materials for your class, assuming six students per group.
2. To prepare the iodine solution, add 1 gram of iodine crystals and 3.5 grams of potassium iodine to 300 milliliters of distilled water. Stir until the solids are dissolved.
3. The growth of the bread mold can be hastened by using bread that does not contain preservatives.

Pre-Lab Discussion
Review with students the material in the textbook about bread molds and mushrooms.
- **Do you remember a characteristic that is common to both organisms that was emphasized in the textbook?** (Both have rootlike hyphae that penetrate the surface on which the mold or mushroom is growing.)
- **What is the purpose of these rootlike hyphae?** (To anchor the organism; to release digestive enzymes; to absorb the digested food.)

Have students read through the entire investigation procedure. Then ask a student to read aloud the Problem at the top of the page. Ask students what types of similarities and differences between a mold and a mushroom they expect to find. Have students state these predictions in the form of a hypothesis. Have them write down their hypotheses, then evaluate the accuracy of their predictions at the close of the investigation.

LABORATORY INVESTIGATION: COMPARING A MOLD AND A MUSHROOM

PROBLEM
What are some similarities and some differences between a mold and a mushroom?

MATERIALS (per group)

bread
petri dish
2 medicine droppers
2 glass slides
dissecting needle
2 coverslips
microscope
iodine solution
paper towel
mushroom
forceps

PROCEDURE

1. Moisten a piece of bread with tap water. Place the moistened bread in the bottom of the petri dish.
2. Allow the petri dish to remain uncovered for 30 minutes. After 30 minutes, put the cover on the petri dish.
3. Place the petri dish in a warm, dark place where it will remain undisturbed for 1 to 2 weeks. Examine the bread daily for mold.
4. After the bread becomes moldy, use a medicine dropper to place a drop of water in the center of a glass slide.
5. With the dissecting needle, separate a small piece of the mold from the bread. **CAUTION:** *Be careful when using a dissecting needle.* Add the mold to the water on the glass slide and cover with a coverslip.
6. Locate some hyphae (threadlike filaments) with the low-power objective of the microscope. Then switch to the high-power objective and use the fine adjustment to locate some hyphae. Notice their color.
7. With the other medicine dropper, place a drop of iodine solution at one edge of the coverslip. Hold a piece of paper towel at the opposite edge of the coverslip to draw the iodine solution across the coverslip.
8. Examine the hyphae again under the low and high powers of the microscope. Observe the shape and arrangement of the hyphae. Notice the sporangia, or bulb-shaped structures, at the ends of some of the hyphae. Make a labeled diagram of the structures of the mold.
9. Place a drop of water in the center of the second glass slide.
10. Break the stalk off the mushroom slightly below the place where the stalk meets the cap. Insert the dissecting needle just under the surface of the stalk and carefully remove a small flap of the stalk.
11. With the forceps, peel off a thin layer of mushroom that runs parallel to the stalk. This layer contains the secondary mycelia (mass of hyphae).
12. Place the thin layer of mycelia in the water on the glass slide. Flatten the layer before covering it with a coverslip.
13. Repeat steps 6 through 8 using the mass of hyphae of the mushroom.

OBSERVATIONS
1. How many different kinds of mold do you see growing on the bread?
2. What is the color of the hyphae in the bread mold? In the mushroom stalk?
3. Describe the shape and arrangement of the structures of the mold and the mushroom.

ANALYSIS AND CONCLUSIONS
1. Explain why the bread was exposed to the air.
2. Why was the bread allowed to remain undisturbed in a warm, dark place for several weeks?
3. What was the purpose of examining the unstained mold and mushroom structures under the microscope?
4. How are a mold and a mushroom similar? How are they different?

Skills Development
Students will use the following skills while completing this laboratory investigation.
1. Observing
2. Comparing
3. Predicting
4. Hypothesizing
5. Manipulative
6. Relating
7. Applying

Safety Tips
1. Have a student read aloud the CAUTION about using a dissecting needle in step 5.
2. Remind students to be careful when handling glass slides. Broken slides should be reported and discarded.

Teaching Strategy
1. Make sure students understand how to break the stalk of

STUDENT STUDY GUIDE

SUMMARIZING THE CONCEPTS

The key concepts in each section of this chapter are listed below to help you review the chapter content. Make sure you understand each concept and its relationship to other concepts and to the theme of this chapter.

19-1 The Fungi

- Fungi are eukaryotic heterotrophs that are placed in the kingdom Fungi. They may be saprophytes, parasites, or symbionts.
- Together with bacteria, fungi are the major decomposers, or organisms of decay.
- Hyphae secrete digestive enzymes that break down food into simpler molecules and absorb them into their cells.
- The body of a fungus consists of tiny filaments tangled together into a thick mass called a mycelium. The individual filaments are called hyphae.
- Most fungi reproduce asexually and sexually.
- The kingdom Fungi is divided into five phyla: Oomycota, Zygomycota, Ascomycota, Basidiomycota, and Deuteromycota.
- Oomycetes reproduce asexually by producing flagellated spores in structures called sporangia, or spore cases. Sexual reproduction results in the formation of male and female gametes. The nuclei of these gametes fuse, forming a diploid cell.
- Zygomycetes, ascomycetes, basidiomycetes, and deuteromycetes reproduce asexually by spores, which develop in sporangia. With the exception of deuteromycetes, these fungi reproduce sexually when two mating types come into contact, producing cells in the gametangia. The sexual part of the life cycle of deuteromycetes has never been observed.
- Ascomycetes produce spores sexually in an ascus, or tiny sac. Basidiomycetes produce spores in club-shaped basidia.

19-2 Fungi in Nature

- The principal role that fungi have in the environment is to decompose and recycle living material.
- Lichens are symbiotic partnerships between a fungus and a photosynthetic organism. The fungus is usually an ascomycete, and the photosyntheic organism is either a cyanobacterium or a green alga.
- Some fungi cause tremendous losses of food and crops, and some cause disease in humans and other animals.

REVIEWING KEY TERMS

Vocabulary terms are important to your understanding of biology. The key terms listed below are those you should be especially familiar with. Review these terms and their meanings. Then use each term in a complete sentence. If you are not sure of a term's meaning, return to the appropriate section and review its definition.

19-1 The Fungi

fungus
decomposer
mycelium
hypha
sporangium
sporangiophore
gametangium
zygospore
rhizoid
stolon
conidiophore
conidium
ascus
ascospore
basidium
basidiospore

19-2 Fungi in Nature

lichen
mycorrhiza

---

elongated and arranged end to end, forming filaments. In the mushroom, the cells, too, are elongated and arranged end to end, forming filaments that lie parallel to each other.

Analysis and Conclusions

1. The air contains millions of mold spores, which develop hyphae when conditions are favorable (moisture and proper temperature).
2. Fungi grow best in warm, dark places.
3. To determine the color of the mold and mushroom structures before adding stain.
4. Molds and mushrooms are heterotrophs; form threadlike hyphae, which are colorless; and produce spores. Molds have small sporangia (spore-producing structures) at the tip of a single hypha. Mushrooms have large umbrella-shaped spore-producing structures at the top of their parallel hyphae.

Going Further: Enrichment

Have students grow mold on an orange or other piece of fruit in addition to growing bread mold. Then have students observe both types of mold under a microscope to determine how the two molds are similar and how they are different. Students can also compare the fruit mold to the mushroom.

CHAPTER REVIEW

CONTENT REVIEW

Multiple Choice
1. b   2. b   3. a   4. a
5. d   6. c   7. b   8. c

True or False
1. F   heterotrophs
2. F   spore formation
3. F   zygospores
4. T
5. F   Basidiomycota
6. T
7. F   lichen
8. T

---

the mushroom and peel off a thin layer to place on a slide. It may be helpful to demonstrate this procedure to the class before groups begin working on their own.
2. Review, if necessary, the techniques of slide preparation.
3. Circulate throughout the room and help any groups that are having difficulty.

Observations

1. Students should be able to recognize the presence of different molds by their color. The most common bread mold is the black bread mold *Rhizopus stolonifer*, which looks like gray fuzz.
2. The hyphae in the bread mold and in the mushroom stalk are colorless.
3. In the mold, the cells are

Word Relationships

A.

1. Ascomycetes, basidiomycetes, and zygomycetes produce nonmotile spores. Oomycetes produce motile spores.
2. Zygomycetes, ascomycetes, and basidiomycetes reproduce sexually. Sexual reproduction has never been observed in deuteromycetes.
3. Athlete's foot, thrush, and ringworm are caused by deuteromycetes. Rust is caused by a basidiomycete.
4. Stolon, rhizoid, and sporangiophore are kinds of hyphae. A zygospore is a type of spore.
5. Yeast, morel, and truffle are ascomycetes. Mushroom is a basidiomycete.

B.

6. club fungus
7. basidiomycete
8. mycorrhiza

CONCEPT MASTERY

1. Fungi are eukaryotic heterotrophs that externally digest and then absorb food through their cell walls and cell membranes. Most fungi are multicellular and contain thick masses of hyphae called mycelia. Most fungi reproduce both asexually and sexually.
2. Oomycetes are the only phylum of fungi that contain motile spores. Deuteromycetes have no observable means of sexual reproduction.
3. During asexual reproduction, *Rhizopus stolonifer* sporangia produce spores. During sexual reproduction, two hyphae from different mating types meet and form gametangia, which produce haploid gametes. These gametes join and form diploid nuclei, which develop into a zygospore. The zygospore germinates, undergoes meiosis, and develops into a hypha, which forms a sporangium.
4. Fungi are classified according to their methods of reproduction and their basic structure.
5. Because they are eukaryotes, yeasts often process the protein products of genes from other eukaryotes more efficiently than bacteria do. As a result, they are useful in genetic research.
6. Lichens are often the first organisms to enter barren environments, gradually breaking down the rocks upon which they grow. In this way, lichens help in the early stages of soil formation, forming an environment that is hospitable to other organisms. Mycorrhizae help green plants in absorbing water, minerals, and nutrients. They do this by producing a network of hyphae that covers the plants' roots, increasing their surface area.
7. When oxygen is not present, yeasts get their energy by undergoing fermentation. During fermentation, not only energy

CHAPTER REVIEW

CONTENT REVIEW

Multiple Choice

Choose the letter of the answer that best completes each statement.

1. Fungi consist of tiny filaments called
   a. asci.   c. basidia.
   b. hyphae.   d. sporangia.
2. Fungi obtain nutrients by
   a. photosynthesis.
   b. external digestion of food.
   c. ingestion of small organisms.
   d. absorption through cilia.
3. What are the small rootlike hyphae in bread mold called?
   a. rhizoids   c. basidia
   b. mycelia   d. caps
4. Yeasts are
   a. ascomycetes.   c. zygomycetes.
   b. basidiomycetes.   d. deuteromycetes.
5. Basidiomycetes are also known as
   a. sac fungi.   c. water molds.
   b. imperfect fungi. d. club fungi.
6. Lichens are symbiotic partnerships between a
   a. fungus and a plant.
   b. green alga and a cyanobacterium.
   c. fungus and a cyanobacterium.
   d. green alga and a plant.
7. Which fungal disease destroyed Ireland's main source of food in 1845?
   a. wheat rust
   b. potato blight
   c. corn smut
   d. cucumber scab
8. Athlete's foot is caused by a (an)
   a. oomycete.   c. deuteromycete.
   b. basidiomycete. d. ascomycete.

True or False

Determine whether each statement is true or false. If it is true, write "true." If it is false, change the underlined word or words to make the statement true.

1. Fungi are <u>autotrophs</u>.
2. Fungi reproduce by <u>binary fission</u>.
3. During asexual reproduction, zygomycetes form thick-walled zygotes called <u>basidiospores</u>.
4. In the bread mold, the hyphae that run along the surface of the bread are called <u>stolons</u>.
5. Mushrooms belong to the phylum <u>Oomycota</u>.
6. Sexual reproduction has never been observed in <u>deuteromycetes</u>.
7. A <u>morel</u> is a symbiotic partnership between a fungus and a photosynthetic organism.
8. <u>Yeasts</u> are used in the baking and brewing industries.

Word Relationships

A. *In each of the following sets of terms, three of the terms are related. One term does not belong. Determine the characteristic common to three of the terms and then identify the term that does not belong.*

1. oomycetes, ascomycetes, basidiomycetes, zygomycetes
2. zygomycetes, ascomycetes, basidiomycetes, deuteromycetes
3. athlete's foot, rust, thrush, ringworm
4. stolon, rhizoid, sporangiophore, zygospore
5. mushroom, morel, truffle, yeast

B. *An analogy is a relationship between two pairs of words or phrases generally written in the following manner: a:b::c:d. The symbol : is read "is to," and the symbol :: is read "as." For example, cat:animal::rose:plant is read "cat is to animal as rose is to plant."*

In the analogies that follow, a word or phrase is missing. Complete each analogy by providing the missing word or phrase.

6. ascus:sac fungus::basidium: _____
7. water mold:oomycete::club fungus: _____
8. fungus and green alga:lichen::fungus and green plant: _____

CONCEPT MASTERY

Use your understanding of the concepts developed in the chapter to answer each of the following in a brief paragraph.

1. Discuss the general characteristics of fungi.
2. How do oomycetes differ from the other members of the kingdom Fungi? How do deuteromycetes differ?
3. Discuss reproduction in the bread mold *Rhizopus stolonifer*.
4. Explain the basis for the classification of fungi.
5. Why are yeasts useful in genetic research?
6. What is the ecological importance of lichens and mycorrhizae?
7. Why is the absence of oxygen important for fermentation by yeast?
8. Describe the life cycle of wheat rust.

CRITICAL AND CREATIVE THINKING

Discuss each of the following in a brief paragraph.

1. **Identifying relationships** Heavily polluted fresh water contains few fungi. How might this affect life in a lake?
2. **Applying concepts** Explain why there are more fungi in a forest than in a field.
3. **Making predictions** What would be the effect on humans of a fungicide capable of killing all types of fungi?
4. **Classifying fungi** Develop your own system of classification for the fungi. Draw a diagram to represent your system.
5. **Relating concepts** While on a walk in the forest, you come upon some fungi, similar to those shown in the photograph, growing from the trunk of a fallen tree. You examine the underside of one of the fungi with a magnifying glass and notice some club-shaped structures. To which phylum do these fungi belong? Explain your answer.

6. **Using the writing process** A debate is raging in your classroom. Some people argue that because fungi cause human disease and damage crops, they should be eradicated from Earth. Their case seems compelling. But you are responsible for defending the opposing viewpoint. Let the fungi be, you maintain. Write the script for the argument you would present in the debate.

CRITICAL AND CREATIVE THINKING

1. The principal role of fungi in the environment is to decompose and recycle living material. If a body of water contains few fungi, little decomposition and recycling of living material will occur. As a result, the lake will become even more polluted.
2. Fungi are heterotrophic parasites and saprophytes. They flourish in areas that are moist, receive less light, and have organic debris, such as leaf litter and rotting trees. Because forests have all these characteristics and fields do not, more fungi will grow in forests.
3. Some possible answers include lower food prices because of the absence of fungal plant infections, increase in bacterial infections because of the absence of antibiotic produced by *Penicillium* mold, increase in unemployment because people lose jobs associated with fungus, and increase in the amount of decaying matter because fungi will not be available to decompose and recycle it.
4. Students should devise a system that reflects a good understanding of fungal characteristics.
5. The presence of club-shaped structures on the fungus indicates that it belongs to the phylum Basidiomycota. The fungus shown here is a bracket, or shelf, fungus.
6. Students should include points discussed in the chapter to strengthen their case.

is produced but ethanol and carbon dioxide as well.

8. Wheat rust must infect two different plants in order to complete its life cycle. Spores produced by the rust in the barberry plant are carried by the wind into wheat fields. There the spores germinate and infect wheat plants. The patches of rust produce a second type of spore that infects other plants, allowing the disease to spread through a field of wheat. Later in the year, after the wheat crop has been destroyed, new types of spores are produced by the rust. In the spring, they go through a sexual phase and produce the spores that infect the barberry plant, and the cycle continues.

CAREERS IN BIOLOGY

Biological Technician, Microbiologist, Virologist

Students interested in a biological career should be instructed to write for further information to the address listed beneath each career description. However, in consideration of the organizations that provide career information, please have only one student write to an organization. Do not instruct the entire class to write to every organization for the same career information. You may want to use the information provided to start a biology career file for the use of all your students.

UNIT 5

CAREERS IN BIOLOGY

Biological Technician

Biological technicians assist biologists by performing laboratory experiments on biological substances and living organisms. Biological technicians use a variety of laboratory techniques and equipment that includes microscopes, chemical scales, and centrifuges. The data they gather enable biologists to develop new substances—such as medications, food additives, and insecticides—and to reach conclusions about how organisms function.

Biological technicians require at least a high school education, but further training in biology is recommended.

For more information write to the American Chemical Society, 1155 16th St., NW, Washington, DC 20036.

Microbiologist

Microbiologists study the growth, structure, and development of microscopic organisms such as bacteria, viruses, molds, and algae. They work in laboratories with electron microscopes, computers, and other complex equipment.

Many microbiologists specialize in studying specific types of organisms. Some study only those microbes that cause disease, others study those that cause pollution, and still others study those that harm animals.

Microbiologists must have at least a Bachelor's degree in biology or microbiology. Many continue their education to earn advanced degrees.

For information contact the American Society for Microbiology, 1913 I St., NW, Washington, DC 20006.

Virologist

Virologists isolate and study viruses in order to learn how they cause disease. With this information, virologists develop and test ways to fight or control different viruses.

Virologists also determine which viruses can be useful to us. For example, certain viruses have been identified that kill insects harmful to crops without hurting the plant.

Virologists perform research for the government, hospitals, pharmaceutical companies, and universities. They are required to have an advanced degree in virology, usually a Ph.D.

For more information write to the American Public Health Association, 1015 15th St., NW, Washington, DC 20005.

HOW TO FIND A SUMMER JOB

Various summer jobs are available to high school students. Your local newspaper probably lists summer job openings specifically meant for students in a separate section of the classified ads.

Local and state governments often provide summer work programs for students. Contact your local government to find out if such programs are available in your area. Many companies and local businesses also hire students to work part time or full time during the summer. Your school or community library may have a list of such employers. It is important to begin looking for a summer job before you are actually ready to begin working.

FROM THE AUTHORS

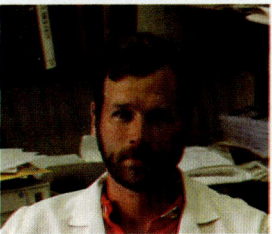

I remember thinking that it was the easiest question on the exam. I had to write a short essay comparing the animallike and plantlike protists. Our teacher then asked us to comment on whether the distinctions between "plantlike" and "animallike" organisms were valid. I wrote a terrific answer. Honest, I did. A **B** for sure . . . maybe an **A**, I thought. But next week the exam came back with a big **0** at the top of the question. No comments, nothing marked wrong. Just a zero and two innocent words underlined in a simple sentence:

> In many of these <u>simple organisms</u>, the distinction between plant and animal is not valid.

After class I approached my teacher. I got all my courage together and asked about the grade. He didn't answer. Instead he motioned me back to the small prep room behind the classroom where he kept a few microscopes and specimens. He put a drop of water under the microscope, telling me that he had gotten it from the pond, which was a short walk from the high school. He made me look. I could recognize a couple of the organisms from our labs in class, but most of them were a complete mystery. There were clusters of green cells in groups of two and three. There were little ciliated cells that swam around like bullets. There were a couple of massive slow-moving amebas. There were marvelous flower-shaped cells with cups in the center and long stalks busily filtering everything in sight in search of food.

"Can you explain to me how those little cells know that they've bumped into something and then reverse direction," he asked. "No," I said. "How about those cone-shaped cells? Are those green clusters yellow or green algae? Is the red spot in the little cell an eyespot or a vacuole? How does an eyespot work?" I admitted that I knew the answers to none of his questions. "Fine," he said. "Then you will never call these organisms 'simple' again, Mr. Miller?" "No, sir," I said, as I rushed out. And since that day, I never have.

Ken Miller

FROM THE AUTHORS

Kenneth R. Miller

In this article Ken Miller describes an experience in high school that has influenced his thinking ever since. Ken relates how he first came to understand that microorganisms were small, but not simple. His learning this point through a mistake on a test will be appreciated by students, and you can point out that making a mistake on a test is actually an opportunity to go back and relearn the information.

UNIT 6

Life on Earth: Plants

UNIT OVERVIEW

In Unit 6 students are introduced to the characteristics and life cycles of plants. They first study aquatic plants such as multicellular algae. Next they learn about mosses and ferns and consider the adaptations necessary for plants to live on land. Finally, they discover the terrestrial seed plants—their structures, growth patterns, and reproductive processes.

UNIT OBJECTIVES

1. Compare the characteristics and reproductive cycles of different groups of multicellular algae.
2. Analyze adaptations necessary for life on land.
3. Describe mosses, liverworts, hornworts, and ferns.
4. Identify the structures and functions of seed plants.
5. Discuss patterns of growth in seed plants and factors that affect growth.
6. Compare the life cycles of gymnosperms and angiosperms.
7. Describe reproduction in seed plants, including pollination, fertilization, seed formation, and germination.

UNIT 6

Life on Earth: Plants

A hiker on a summer's walk can't help but notice the wonderful green color of planet Earth. This color comes in large part from plants. Plants provide beauty for the eye—and much more.

Plants constantly replenish our supply of oxygen, which they give off during photosynthesis. During the same process, they make nutrients and, in turn, become food sources for other living things. Plants also give off water, which cools the temperature of the air.

Plants cover much of the surface of the Earth, existing in a multitude of forms and surviving in a variety of environments. It is plants that make it possible for other forms of life to share this planet.

DISCOVERY LEARNING

PLANTS ARE POPPING UP ALL OVER

1. Obtain the following materials: a clean, small milk carton; a large plastic can lid (or a small, clean foam meat tray); a nail; soil; 6 to 8 fresh, unpopped popcorn kernels.
2. Carefully poke 4 or 5 holes in the bottom of the milk carton with the nail.
3. Add water to the soil so that it is moist but not soggy.
4. Open up the flaps at the top of the milk carton, then fill the carton about three-fourths full with moist soil.
5. Plant the popcorn kernels about 1 centimeter deep. Try to space them evenly.
6. Put the carton on the can lid (or meat tray), and place it in a sunny window. Water the soil every other day so that it stays moist, but not wet.
   - What happened to the popcorn kernels you planted? Make a drawing of your observations.

INTRODUCING UNIT 6

Using the Visuals

Have students examine the photographs on pages 430 and 431. Ask students the following questions:
- **What do you see in this photograph?** (A cactus plant and its flower.)
- **Do you think this cactus would survive long in a different environment, such as a hardwood forest?** (No. Cacti are specially adapted to living in a dry, desert environment.)
- **Are you surprised that cacti have such attractive flowers?** (Accept all answers.)
- **What other plants have flowers?** (Accept all true answers. Students should realize that a wide variety of plants—but not all of them—produce flowers.)
- **What is one thing this cactus has in common with other plants?** (Plants have chlorophyll, which they use in photosynthesis.)

CHAPTERS

20 *Multicellular Algae*

21 *Mosses and Ferns*

22 *Plants with Seeds*

23 *Roots, Stems, and Leaves*

24 *Plant Growth and Development*

25 *Reproduction in Seed Plants*

CHAPTER DESCRIPTIONS

20 Multicellular Algae
Chapter 20 describes how multicellular algae are adapted to a water environment. The three groups of multicellular algae and the reproductive cycles of various species are also described.

21 Mosses and Ferns
The adaptations required by plants on land are first discussed in Chapter 21. The physical characteristics and life cycles of mosses, liverworts, hornworts, and ferns are then compared.

22 Plants with Seeds
Chapter 22 begins with an explanation of how seed plants are designed for life on land. Next, the evolution of seed plants is explored. Finally, the various processes of pollination and seed dispersal are compared.

23 Roots, Stems, and Leaves
Chapter 23 opens with a description of soil structure in relation to seed plants. The five main types of specialized plant tissues are then compared. The structures and functions of roots, stems, and leaves are described. The chapter closes with descriptions of adaptations of plants to different environments.

24 Plant Growth and Development
Patterns of growth in different types of plants and plant structures are described in Chapter 24. The second part of the chapter treats the role of factors that affect growth in seed plants, such as hormones, auxin, cytokinins, gibberellin, and tropisms.

25 Reproduction in Seed Plants
Chapter 25 compares the life cycle of gymnosperms with that of angiosperms and female gametophytes with male gametophytes. Methods of pollination and fertilization are also explored.

DISCOVERY LEARNING

PLANTS ARE POPPING UP ALL OVER

Introduce this unit with the Discovery Activity. Have students observe the soil every day as you proceed with this unit. Students should record their own observations and illustrations and keep them in their portfolios.

Encourage students to be as accurate and precise as possible as they study the sprouting seeds.

Discuss this activity again when you cover Chapter 25, Reproduction in Seed Plants.

CHAPTER 20 Multicellular Algae

Section	Laboratory Investigations and Activities
20–1 Characteristics of Algae pages 433–435 THEMES: Evolution, Unity and Diversity	**Teacher Edition** DEMONSTRATION: Introducing the Algae, p. 432D **Teaching Resources** HANDS-ON ACTIVITY: If We Lived Near the Ocean, We'd See Seaweed, p. 101
20–2 Groups of Algae pages 435–439 THEMES: Unity and Diversity, Scale and Structure	**Student Edition** LABORATORY INVESTIGATION: Studying Filamentous Green Algae, p. 444 **Teacher Edition** DEMONSTRATION: Pigments in Algae, p. 432D **Laboratory Manual** Characteristics of Green Algae, p. 247
20–3 Reproduction in Algae pages 440–441 THEMES: Patterns of Change, Unity and Diversity	
20–4 Where Algae Fit into the World pages 442–443 THEME: Systems and Interactions	**Laboratory Manual** Examining an Algal Bloom, p. 251 **Teaching Resources** HANDS-ON ACTIVITY: Seaweed Sweets, p. 103
Chapter Review pages 445–447	

Teacher Resources

Books

Becker, E. W., ed. *Production and Use of Microalgae.* Lubrecht and Cramer, 1985.

Bold, H. C., and M. J. Wynne. *Introduction to the Algae,* 2nd ed. Prentice Hall, 1985.

Fogg, G. E., W. D. Stewart, P. Fay, and A. E. Wasby. *The Blue-Green Algae.* Academic Press, 1973.

Fryxell, Greta A., ed. *Survival Strategies of the Algae.* Cambridge University Press, 1983.

Irvine, David E., and David M. John, eds. *Systematics of the Green Algae.* Academic Press, 1985.

Kiermeyer, O., ed. *Cytomorphogenesis in Plants.* Springer-Verlag, 1981.

Lee, R. E. *Phycology.* Cambridge University Press, 1980.

Lewin, Ralph A. *The Biology of Algae and Diverse Other Verses.* Boxwood, 1987.

Weissner, W., and D. J. Robinson, eds. *Algal Development.* Springer-Verlag, 1987.

CHAPTER PLANNING GUIDE

Other Activities	Media and Technology
Study Guide Section 20–1, p. 187	**Videodisc** Aquatic Ecosystems: Estuaries and Marine **Biology Media Guide** page 17
Teaching Resources ACTIVITY: Determining the Role of the Cell Nucleus, p. 3 **Study Guide** Section 20–2, p. 190	
Teaching Resources ACTIVITY: Reproduction in Algae and Fungi, p. 5 **Study Guide** Section 20–3, p. 193	**Biology Transparencies** 22: The Life Cycle of an Alga
Teaching Resources VOCABULARY REVIEW: Word Game, p. 1 **Study Guide** Section 20–4, p. 196	
Teaching Resources Chapter Test A, p. 11 Chapter Test B, p. 15	**Computer Test Bank** Chapter Test, p. 213

Audiovisuals

The Color of Life. Video. Landmark Films, Inc.
Simple Organisms: Algae and Fungi, rev. Film or video. Coronet.
A New Look at Algae. Film or video. Ward.

Algae (Diversity and Ecology). Filmstrip with cassette. Ward.
Introduction to Algae. Filmstrip with cassette. Carolina Biological Supply Company.

The Algae. Filmstrip with cassette. Biology Media.
Making Seaweeds Worth Eating. Filmstrip or video. Biology Media.

CHAPTER 20 Multicellular Algae

CHAPTER OVERVIEW

Most algae—photosynthetic organisms found in water and damp environments throughout the world—are multicellular. Although taxonomists disagree as to the exact classification of algae, multicellular algae, along with plantlike unicellular algae, are treated here as members of the kingdom Plantae.

As water-dwelling organisms, the major limiting factor to the growth of algae is the availability of light for photosynthesis. In response to this, algae have evolved different forms of chlorophyll, as well as accessory pigments, in order to make the most efficient use of the available wavelengths of light penetrating to different depths in the water. The colors resulting from the chlorophyll types and the accessory pigments form the basis for the classification of algae into phyla.

As do all members of the plant kingdom, algae reproduce by means of a complex reproductive cycle involving alternation of generations. From this cycle, algae gain the advantages of both sexual and asexual reproduction.

Algae are of vital importance to all life on Earth. Since between 50 and 75 percent of all photosynthesis is performed by algae, they are not only the primary producers in global food chains but also the major source of the Earth's oxygen.

20-1 CHARACTERISTICS OF ALGAE

Section Focus 20-1

The purpose of this section is to introduce students to the general characteristics of algae, the rationale for their classification, and their adaptations to aquatic habitats.

Algae are aquatic, photosynthetic, and mostly multicellular organisms lacking internal structures for conduction of materials, as well as roots, stems, or leaves. All algae contain chlorophyll *a*, and many have a complex reproductive cycle involving alternation of generations.

Multicellular algae are classified in this text as members of the kingdom Plantae along with unicellular algae with plantlike characteristics.

The aquatic habitat of algae does away with the need for structures for support and for protection against dehydration. However, this habitat requires adaptations to the limited light supply of the underwater environment.

Performance Objectives 20-1

1. Describe the characteristics of algae.
2. Explain why algae lack certain structures found in terrestrial plants.
3. Explain how algae have adapted to the limited availability of light under water.

Science Terms 20-1

accessory pigment   p. 435

20-2 GROUPS OF ALGAE

Section Focus 20-2

The purpose of this section is to introduce students to the scheme of classification of algae and to discuss the characteristics of three groups of multicellular algae.

Chlorophyta, or green algae, are found in both freshwater and marine environments. Many unicellular chlorophyta form colonies of organisms with little specialization, whereas others are multicellular with a high degree of specialization. All multicellular green algae have cellulose cell walls, contain chlorophylls *a* and *b*, and store food as starch.

Phaeophyta, or brown algae, are marine plants found in cool coastal waters. They include most of the seaweeds. Kelp, a giant alga, may grow more than 60 meters long.

Rhodophyta, the red algae, are found in marine waters all over the world and have adapted to living on the surface as well as at depths as great as 170 meters.

Performance Objectives 20-2

1. Describe the characteristics of each of the three groups of algae.
2. Describe the types of green algae and their evolutionary importance.
3. Contrast the degree of specialization in colonial algae with true multicellular algae.

Science Terms 20-2

colony   p. 436
filament   p. 438
holdfast cell   p. 438

20-3 REPRODUCTION IN ALGAE

Section Focus 20-3

The purpose of this section is to introduce the general concept of alternation of generations and to discuss the specific ways in which various representative types of algae reproduce.

In alternation of generations, the production of diploid structures with the full complement of chromosomes for a particular species alternates with the production of haploid structures that have half the full chromosome number. In sexual reproduction the haploid structures are called gametes, and in asexual reproduction they are called zoospores.

The advantage of reproduction by alternation of generations is that it allows organisms to have the benefits of both asexual and sexual reproduction. Organisms can reproduce quickly asexually, and they gain genetic variation from the production of zygotes, which remain dormant until conditions are favorable.

In sexual reproduction, algae may produce gametes that appear to be alike, a condition called isogamy. Production of two different types of gametes is called heterogamy.

Performance Objectives 20-3

1. Describe alternation of generations.
2. Discuss the advantages of reproduction by alternation of generations.
3. Describe reproduction in algae such as *Chlamydomonas, Ulva,* and *Fucus.*

CHAPTER PREVIEW

Science Terms 20-3
alternation of generations p. 440
zoospore p. 440
isogamy p. 440
syngamy p. 440
sporophyte p. 441
gametophyte p. 441
heterogamy p. 441
egg p. 441
sperm p. 441

20-4 WHERE ALGAE FIT INTO THE WORLD

Section Focus 20-4
The purpose of this section is to show the variety of ways in which algae play a part in our lives.

As photosynthesizers, algae are vital to all life on Earth. Also called the "grasses of the seas," they capture the sun's energy in chemical bonds and make it available to other living things. Because of their abundance and as a direct result of their photosynthesis, algae provide between 50 and 75 percent of the Earth's free oxygen.

Substances extracted from algae are used in a wide variety of ways, from medicines and food additives to cosmetics and many different industrial products.

Performance Objectives 20-4
1. Explain the importance of algae as a source of food for other organisms.
2. Explain the importance of algae in the production of atmospheric oxygen.
3. Describe some of the uses of algae in medicines, food additives, and industrial products.

DISCOVERY LEARNING

TEACHER DEMONSTRATIONS
Modeling

Introducing the Algae
The following demonstration can be used as an introduction to Chapter 20.

Both freshwater and marine algae should be readily available from a store that sells aquariums and related supplies. If possible, obtain samples of such common algae as *Spirogyra, Oedogonium, Ulva, Fucus,* and *Porphyra*. If actual specimens are not available, obtain pictures of algae from magazines such as *National Geographic* or *Omni*, or refer students to the illustrations of algae that appear throughout this chapter.

- **What do all the algae appear to have in common?** (They are found in or near water.)
- **How do the algae obtain food?** (By photosynthesis.)
- **In order to carry on photosynthesis, where in the water must the algae live?** (They need light for photosynthesis and so must live on or close to the surface of the water.)

Pigments in Algae
You may want to perform this paper chromatography demonstration after students complete Section 20-2.

Before class, under a vent hood, use a water bath to boil some samples of green, brown, and red algae (in separate beakers) in denatured ethyl alcohol or acetone to dissolve the pigments in each. Take a disk of filter paper large enough to cover the top of each beaker and cut two parallel slits about one-half centimeter apart. Bend the paper strip downward so that it extends into the liquid and acts as a wick. Within a class period, the pigments will be differentially absorbed by the paper so that streaks of the different pigments can be seen.

- **What pigment(s) seem to be present in every type of algae?** (Chlorophylls—specifically, *a* and *b*.)
- **Why aren't all algae green?** (The other pigments mask the chlorophyll.)
- **Why do algae have pigments other than chlorophyll?** (The other pigments are capable of absorbing wavelengths of light that chlorophyll cannot, making photosynthesis more efficient.)

CHAPTER 20
Multicellular Algae

GUIDED ENQUIRY

Pose the following questions to students and have them record their responses. Point out that they will gain a better understanding of the key concepts if they read the chapter with these basic questions in mind. Upon completion of the chapter, pose the questions again. Ask students to compare their initial responses with those they have developed after reading the chapter.

- What are algae?
- Where are algae found?
- What are some specific kinds of algae?
- How do algae obtain food?
- How do algae reproduce?
- Why are algae important to other living things?
- How are algae used by humans?

INTRODUCTING CHAPTER 20

Using the Visuals

Ask students to look at the chapter-opener photograph.
- **Why are the sea otters wrapped in seaweed?** (The seaweed, a giant kelp, keeps the otters from floating away on the waves.)

Point out to students that the plants that we commonly call seaweed are multicellular algae. Ask students to think about the last time they visited a lake or the seashore and have them describe any algae they might have found there.

- **Where did you see the algae?** (On the shore close to the water or in the water, either floating or anchored to the bottom.)
- **What color(s) were the algae?** (Most students will have seen green or brown algae; a few may have seen red algae.)
- **How were the algae different from other plants?** (They have no roots, stems, or leaves. Accept other reasonable answers.)
- **What do all algae seem to have in common?** (They all live in or very close to water; they all are photosynthetic.)
- **Did any of the algae you saw have any unusual structures?** (Students may mention the swim bladders of *Fucus* or the holdfasts that many algae have that enable them to anchor themselves to objects.)

CHAPTER 20
Multicellular Algae

Sea otters, among the most appealing species of animals, make their home in the kelp forests off the coast of California. They wrap themselves in strands of algae to keep from floating away on the ocean waves.

Along the coasts of Washington, Oregon, and northern California, pine and redwood forests crown rocky cliffs that dive steeply into the pounding surf. Beneath the ocean waves, invisible to the casual observer, another type of forest thrives—a forest of giant greenish-brown kelp. Kelp are huge seaweeds that grow as long or longer than the tallest forest trees. Growing with amazing speed, kelp offer both home and food to a wide variety of fish, shellfish, and sea otters.

The kelp forest is but one scene from the little-known world of life beneath the sea, a world fed and sheltered by algae. What are algae? Where do they live? In this chapter we shall examine this extraordinary group of plants and explore the many ways in which algae are adapted to life on Earth.

GUIDE FOR READING

After you read the following sections, you will be able to

20–1 Characteristics of Algae
- Discuss the characteristics of algae.
- Describe the adaptations of algae to life in a water environment.

20–2 Groups of Algae
- Identify the characteristics of three algae groups.
- Recognize differences between colonial and multicellular algae.

20–3 Reproduction in Algae
- Define alternation of generations.
- Discuss various ways groups of algae reproduce.

20–4 Where Algae Fit into the World
- Discuss the importance of algae to life on Earth.
- List several commercial applications of algae.

Journal Activity

YOU AND YOUR WORLD
Have you ever eaten foods that contained algae? Would you eat foods that contain algae? Record your thoughts and feelings in your journal. When you complete your reading of this chapter, review your journal entry. Have your opinions changed? Why or why not?

20–1 Characteristics of Algae

Guide For Reading
- What are the major characteristics of algae?
- How are algae adapted to their physical environment?

Algae are photosynthetic organisms that live in streams, ponds, lakes, swamps, and all the oceans of the world. Algae also live on damp tree trunks and rocks and in moist soil. The stringy green filaments you may have seen in a local pond are algae. So are the giant brown seaweeds that wash up on a beach after a storm. And if you keep tropical fish, you may have noticed algae covering the walls of the tank.

From these examples, you may already have determined one important characteristic of algae: Algae must live in or near a source of water. **Unlike land plants, most algae lack an internal system of tubes to move water and materials from one part of the plant to another.** The water in which algae live bathes their cells with carbon dioxide, oxygen, and nutrients, and carries away wastes produced by the cells.

Algae can be unicellular or multicellular and, therefore, vary considerably in size. Most algae are multicellular. Some species of multicellular algae, such as the giant kelp, can grow to more than 60 meters in length. Most unicellular algae are microscopic, resembling plantlike protists.

Figure 20–1 Giant kelp are often washed up on shore when storms tear the plants loose from the holdfasts that anchor them to the bottom. These giant algae can be as long as a football field.

COOPERATIVE LEARNING

Using preassigned lab groups or randomly selected teams, have groups assume the role of a foods research and development committee. Each group should create an algal food product. Encourage groups to keep the following in mind: type of food (snack, health, diet, etc.), sensory appeal of food, market for their product, ingredients (at least one type of algae must be used and identified for the consumer as the primary ingredient), packaging, and cost. The groups' final product will be a full-page newspaper or magazine ad or a 30-second TV commercial to introduce the public to their algal food product.

Journal Activity

YOU AND YOUR WORLD
Students will be surprised to learn of the widespread presence of algae in their food, as well as algae's many other uses and important roles. Have students keep this journal entry in their portfolio.

TEACHING STRATEGY 20–1

Focus/Motivation
Use the specimens or the photographs of algae from the demonstration that introduces the chapter.
- **What organisms with which you are already familiar do algae resemble most closely?** (They resemble plants.)
- **How are these algae different from other plants?** (Algae live in or near water; they have neither roots, stems, nor leaves.)
- **How are algae similar to other plants?** (Algae contain chlorophyll and obtain their food by photosynthesis.)

Content Development
Emphasize that algae lack an internal system of conducting tubes. You may want to review the process of diffusion with students.
- **What substances do the cells of algae need?** (Oxygen, carbon dioxide, and minerals.)
- **How do the cells of algae obtain the substances they need?** (These substances diffuse directly into the cells from the surrounding water.)

BACKGROUND INFORMATION
CLASSIFICATION OF ALGAE

Taxonomists often debate which organisms belong in the kingdom Protista. As defined in Chapter 15, the protists include all unicellular, eukaryotic organisms. But as it does with algae, this strict definition serves to divide a group of organisms that otherwise appear very much related.

We chose to include algae here—at the beginning of the unit on plants—because biologists typically regard algae as very elementary plants, and most species of algae are multicellular. But as you may emphasize to students, the five-kingdom system is something biologists created, not the organisms themselves. In many cases, the boundaries between kingdoms are ill-defined.

Figure 20–2 Different kinds of algae, along with many kinds of small animals, live in tide pools. With each crashing wave, water brings food for the animals and plants. Wastes produced by the organisms are carried away as the ocean water seeps from the tide pool.

The cells of all algae have a cell wall. Algae never develop the specialized root, stem, and leaf structures found in land plants. All algae contain chlorophyll a, one of several forms of chlorophyll. Some species of algae contain another form of chlorophyll in addition to chlorophyll a. Many species of algae have a complicated reproductive cycle in which stages of sexual reproduction alternate with stages of asexual reproduction.

There is some disagreement among scientists about the classification of algae. Because some species of algae are unicellular, some scientists place all algae in the kingdom Protista. But because many species of algae are multicellular, other scientists place all algae in the kingdom Plantae.

In this textbook we have chosen to use the following classification: The single-celled Chrysophytes (diatoms), Euglenophytes (euglenas), and Pyrrophytes (dinoflagellates) are placed in the kingdom Protista. (You read about these types of algae in Chapter 18.) But because of their chemistry and reproductive cycles, multicellular algae are placed in the kingdom Plantae. This chapter deals primarily with these multicellular species of algae. However, as you will notice in reading the chapter, certain single-celled algae have been included. These algae are more plantlike than protistlike and thus are grouped together with the multicellular algae they most resemble. You should also notice in reading this chapter that although some botanists call the major plant groups divisions, we call them phyla—as we do the major groups of all other organisms.

Adaptations of Algae to Life Under Water

Most algae live under water and show different adaptations from those of land plants. For example, underwater plants do not need protection from drying out. Thus many kinds of algae have very thin (often only two cells thick) leaflike structures that lack a waterproof covering. These thin structures can exchange oxygen, carbon dioxide, and nutrients directly with the water around them. As you already learned, algae have no specialized tissues to carry such materials throughout their body.

Because algae are supported by water, they do not need stemlike structures to keep them from falling over, such as land plants do. And, as you will soon learn, sexual reproduction can be more easily accomplished in water because reproductive cells can swim through water and the fragile young plants do not dry out.

Chlorophyll and Accessory Pigments

Life under water poses one major problem for plants: a lack of light for photosynthesis. As you will remember, light is necessary for the food-making process, and it is chlorophyll that traps the energy of sunlight. However, water absorbs much of

---

20–1 (continued)

Reinforcement/Reteaching
Remind students that diffusion is a fairly slow process. As a result, most algae are never more than a few cells thick. If the algae were thicker than this, diffusion would not occur quickly enough to meet the algae's needs.

Enrichment
Have interested students research the different types of accessory pigments found in algae. Students may investigate how these pigments function, their evolution, or their differences and similarities with chlorophyll.

Skills Development
Guided Practice
Skills: Comparing, contrasting
Have students prepare a two-column chart. Label one column Aquatic Plants, and the other Land (or Terrestrial) Plants. Along the left side of the chart have students list the processes necessary for life. Students should indicate the structure in each type of plant that performs or accomplishes the process, noting that in some instances, no special structure is present.

SECTION REVIEW 20–1

1. Algae are photosynthetic organisms that live in water or in

this energy as sunlight passes through it. In particular, seawater absorbs large amounts of energy corresponding to the red and violet wavelengths of sunlight. And it is exactly these two wavelengths that chlorophyll *a* uses best. Because seawater absorbs most of the red and violet wavelengths, light becomes much dimmer and bluer in color as the depth of water increases. The dim blue light that penetrates into deep water contains very little light energy that chlorophyll *a* can use.

In adapting to the challenge of life with little light, various groups of algae have evolved different forms of chlorophyll. Each form of chlorophyll—chlorophyll *a*, chlorophyll *b*, chlorophyll *c*, and chlorophyll *d*— absorbs different wavelengths of light. As you have read, all algae contain chlorophyll *a*. Some species contain chlorophyll *a* in combination with *b*, *c*, or *d*. The result of this evolution of different forms of chlorophyll is that more of the energy of sunlight available to algae can be used.

Algae have also evolved compounds that absorb different wavelengths of light than chlorophyll absorbs. These light-absorbing compounds are called **accessory pigments.** Accessory pigments pass the energy they absorb on to the algae's photosynthetic machinery. For example, some accessory pigments make blue light more useful for photosynthesis. So, algae that contain these accessory pigments can live in deeper water than plants that contain only chlorophyll. Because accessory pigments also reflect different wavelengths of light than chlorophyll, they give algae a wide range of colors.

Figure 20–3 The color of this beautiful plumelike red alga results from combinations of accessory pigments and chlorophyll. If you look closely, you can see tiny hooks on the alga. The alga uses the hooks to attach itself onto stationary algae, thus ensuring that it will remain in one place.

20–1 SECTION REVIEW

1. What are two characteristics of algae?
2. In which kingdom are multicellular algae placed?
3. Why are some unicellular algae classified as plants?
4. **Critical Thinking—Relating Facts** How are algae adapted to life in water?

20–2 Groups of Algae

The colors provided by chlorophyll and accessory pigments, as well as the form in which food is stored, are characteristics used to classify algae into different groups. **There are three groups of multicellular algae: green algae, brown algae, and red algae.**

Green algae are members of phylum Chlorophyta (KLOR-oh-fight-ah). All green algae contain chlorophylls *a* and *b*. It is these chlorophylls that give the algae their green color. All green algae store food in the form of starch.

Guide For Reading

■ What are the distinguishing characteristics of the three groups of multicellular algae?
■ How might multicellular plants have evolved from protists?

435

Figure 20–4 As you can see, the colors of algae are quite varied. The red algae (left), brown algae (right), and green algae (center) illustrate how different combinations of pigments can result in dramatically different colors.

Figure 20–5 The green alga Ulva is called sea lettuce because the blades of this plant are flat and resemble salad greens.

Brown algae are members of phylum Phaeophyta (fee-AH-fuh-tuh). Brown algae contain chlorophylls *a* and *c,* as well as a brown accessory pigment called fucoxanthin (fyoo-koh-ZAN-thihn). The combination of fucoxanthin and chlorophyll *c* gives these plants their yellow-brown color. Brown algae store food in the form of special starches and oils.

Red algae are members of phylum Rhodophyta (roh-DAH-fuh-tuh). All red algae contain chlorophyll *a.* Some species of red algae also contain chlorophyll *d.* All red algae contain reddish accessory pigments called phycobilins (figh-koh-BIHL-ihnz). Phycobilins are very efficient at absorbing the energy of blue light and making it available for photosynthesis. Thus red algae can live deeper in the ocean than other kinds of algae. Depending on the amount of phycobilins they carry, red algae can be pink, red, purple, or even black. Red algae store food in the form of a special kind of starch.

Chlorophyta—The Green Algae

Green algae are found primarily in moist areas on land and in fresh water. Some species of green algae live in the oceans. Green algae have evolved many forms in adapting to these widely different environments. Green algae may live as single cells. Some species of single-celled green algae form a **colony,** which is a group of cells that are joined together and show few specialized structures, or structures that perform a particular function. Other green algae are multicellular and have well-developed specialized structures. All species of green algae have reproductive cycles that include both sexual and asexual reproduction.

The multicellular green algae have cellulose in their cell walls, contain chlorophylls *a* and *b,* and store food in the form of starch, just like land plants. One stage in the life cycle of mosses—small land plants you will learn about in the next chapter—looks remarkably like a tangled mass of green algae strands. All these characteristics lead scientists to believe that the ancestors of modern land plants looked a lot like certain species of living green algae. Unfortunately, algae rarely form fossils, so there is no single specific fossil that scientists can call an ancestor of both living algae and mosses. However, scientists believe that mosses and green algae shared such a common algaelike ancestor millions of years ago.

CHLAMYDOMONAS—A SINGLE-CELLED GREEN ALGAE *Chlamydomonas* (kla-mee-doh-MOH-nuhs), which is a single-celled green alga, grows in ponds, ditches, and wet soil. *Chlamydomonas* is a small egg-shaped cell with two flagella. A light-sensitive area called the eyespot cannot actually see but can sense whether the organism is in bright light or darkness.

Chlamydomonas has a large cup-shaped chloroplast. At the base of the chloroplast is a small pyrenoid (PIGH-reh-noid), an organelle that synthesizes and stores starch. *Chlamydomonas* lacks the large vacuoles found in the cells of land plants. Instead it has two small contractile vacuoles. Unlike land plants, *Chlamydomonas* has a cell wall that does not contain cellulose.

As you can see, *Chlamydomonas* has characteristics of both the algae grouped in the Protist kingdom and land plants. This combination of characteristics has led botanists to believe that *Chlamydomonas* is a good example of one step in the evolution of multicellular plants from unicellular protists.

COLONIAL GREEN ALGAE Several species of green algae provide an idea of how multicellular plants may have evolved. From single-celled species such as *Chlamydomonas*, species such as *Gonium* (GOH-nee-uhm) may have evolved. *Gonium* is a colonial alga composed of between 4 and 32 cells. In a colony, many identical cells live together although each cell still functions independently. The cells do not form specialized tissues. If the colony is broken apart, each cell can live and grow into a new colony.

Other species of green algae form larger colonies. The beautiful genus *Volvox* (VAHL-vahks) is one example. *Volvox* can form colonies consisting of as few as 500 or as many as 50,000 cells. Observation of *Volvox* provides two interesting details of the way this organism functions.

First, the cells in a *Volvox* colony are connected to one another by strands of cytoplasm. Thus the cells that make up this colony can communicate. Communication is necessary for the *Volvox* colony to swim: When the cells on one side of the colony "pull" with their flagella, the cells on the other side of the colony have to "push."

Second, although most cells in a *Volvox* colony are identical, a few cells are specialized for reproduction. These cells, which produce gametes, are the first step in the development

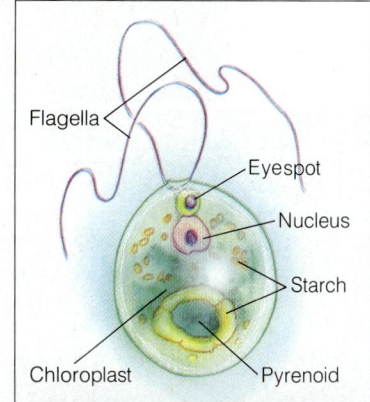
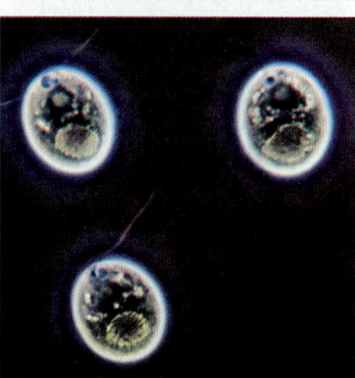

Figure 20–6 The single-celled alga Chlamydomonas *is a favorite subject of laboratory study. This alga is able to move about when it beats its two flagella back and forth in the water. It also has an eyespot that is sensitive to light.*

BACKGROUND INFORMATION
PLANT EVOLUTION

It is generally believed that the *Chlorophyta*, or green algae, are the group from which land plants evolved. Among the algae, only the *Chlorophyta* have cellulose in their cell walls, contain chlorophylls *a* and *b*, and store their food in the form of starch, all of which are also characteristics of the land plants. Because algae do not form fossils, we do not have direct evidence for an evolutionary relationship. However, one stage in the life cycle of mosses looks remarkably like a tangle of green algal filaments. Perhaps both mosses and the modern multicellular green algae descended from a common algaelike ancestor.

FACTS AND FIGURES

Not all single-celled green algae are microscopic. *Acetabularia* is a giant single-celled green alga that resembles a toadstool. Some giant species may grow as much as 10 centimeters long. *Acetabularia* is usually found in tropical seas but may range as far north as the Mediterranean.

Reinforcement/Reteaching
• **What kinds of specialization are shown in the cells of *Volvox*?** (Certain cells perform different tasks when the organism swims, and *Volvox* also forms specialized reproductive cells that produce gametes. But in most respects, *Volvox* cells are not specialized.)

Content Development
Using the Visuals As students view living *Chlamydomonas* or a photograph, have them compare what they see to the diagram in Figure 20–6.

If a microprojector is being used, the flagella will more likely be seen if the light is cut down. Also, if the pyrenoid is not visible, add a dilute solution of potassium iodide to the slide containing the algae. The pyrenoid should be stained a blue-black color, making it readily visible. *Volvox* may be viewed with a microprojector or a stereo microscope. If possible, have students observe the "swimming" motion of this colonial alga and note the interaction of the cells.

BACKGROUND INFORMATION
TYPES OF ALGAE

The characteristics of *Ulva* demonstrate an evolutionary link between simpler green algae and more complex land plants. Although only two cells thick, *Ulva* is truly multicellular, forming such specialized structures as holdfasts. While *Ulva* exhibits alternation of generations like the algae and simpler land plants, it exhibits heterogamy, with one nonmotile gamete (the "egg") slightly larger than the motile gamete (the "sperm").

Brown algae are classified together because they are all multicellular, contain chlorophylls *a* and *c* and the brown pigment fucoxanthin, and store their food in the form of starches and oils. The combination of fucoxanthin and chlorophyll *c* gives these plants their yellowish-brown color.

Red algae are classified together because they all contain chlorophylls *a* and *d* and the reddish pigments called phycobilins, and they store their food in the same form of starch.

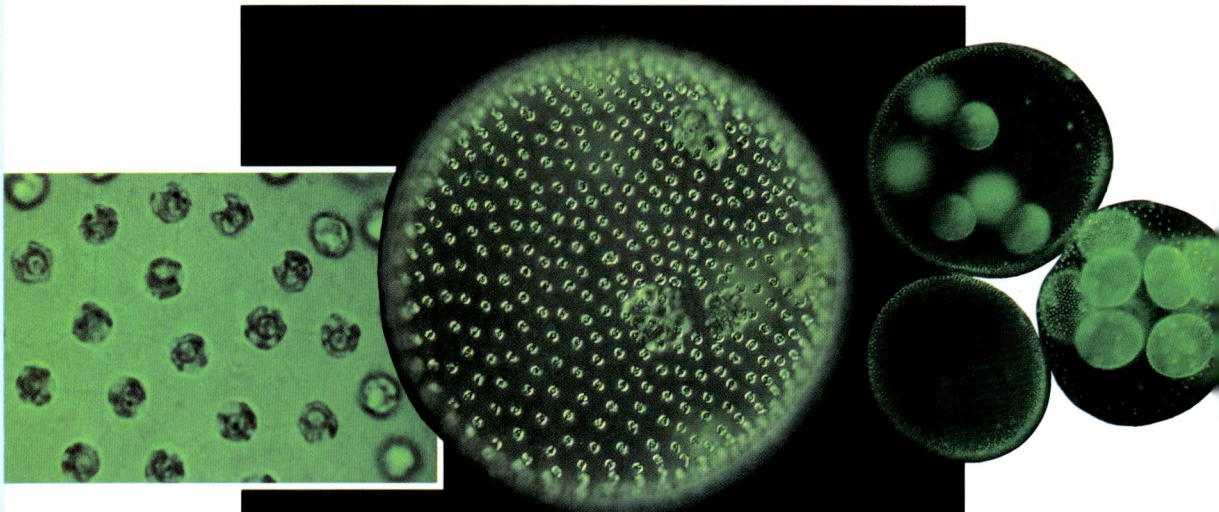

Figure 20–7 Beautiful globes of the green alga Volvox can often be seen in bodies of fresh water (center). Each cell of this colonial alga is connected to other cells by strands of cytoplasm (left). New colonies often develop within a colony. Eventually the old colony ruptures, freeing the smaller colonies developing within (right).

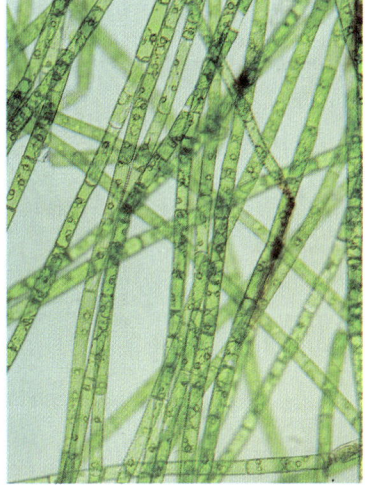

Figure 20–8 Long strands of the green alga Oedogonium often grow in ponds. This alga can reproduce sexually by producing gametes. New algae can also form when pieces of the strands break off and develop into new plants.

of specialized tissues that become more common in plants adapted to life on land. Because it shows some cell specialization, *Volvox* straddles the fence between colonial and multicellular life.

SPIROGYRA AND OEDOGONIUM—THREADLIKE GREEN ALGAE Many green algae form long threadlike colonies called **filaments,** the cells of which are shaped like soda cans stacked end on end. *Spirogyra* (spigh-roh-JIGH-ruh) and *Oedogonium* (ee-duh-GOH-nee-uhm) are two common examples of freshwater filamentous green algae. Filamentous algae can grow and reproduce asexually. For example, if the algae filaments are broken, the cells of each piece can continue to divide and grow. Many filamentous green algae can also reproduce sexually. *Oedogonium*, like *Volvox*, forms two different specialized reproductive cells, or gametes. Each *Oedogonium* filament is attached to the bottom of a lake or pond by another kind of specialized cell called a **holdfast cell.**

ULVA—A MULTICELLULAR GREEN ALGA *Ulva* (UHL-vuh), or "sea lettuce," is a bright-green multicellular marine alga that is commonly found along rocky seacoasts. Although *Ulva* plants are only two cells thick, they are tough enough to survive the pounding of waves on the shores where they live. A group of specialized cells at the base of the plant form holdfasts that attach *Ulva* to the rocks.

20–2 (continued)

Focus/Motivation

Spirogyra and *Oedogonium* are two of the most common freshwater algae and are very likely to be found in local ponds. Although *Oedogonium* is usually found attached to a submerged stone or water plant, *Spirogyra* tends to be free-floating. If possible, obtain some specimens of these two algae and have students observe them under the low-power objective of the light microscope or under a stereo microscope. The spiral chloroplast of *Spirogyra* makes it easily distinguishable from the linear chloroplast of *Oedogonium*.

Reinforcement/Reteaching

Review with your students how cells become increasingly specialized in function as we progress from single-celled algae to colonial types to true multicellular algae. The main characteristic that distinguishes colonial algae from those that are truly multicellular is that the latter have many different kinds

Phaeophyta—The Brown Algae

Brown algae are important marine plants that are found in cool shallow coastal waters of temperate or arctic areas. The brown algae have very complicated structures, although they are not as highly developed as land plants. Most of what are commonly called sea weeds are species of brown algae. The largest alga in the world is a form of giant kelp that can grow more than 60 meters long. Another brown alga called *Sargassum* (sahr-GAS-suhm) forms huge floating mats many kilometers long in an area of the Atlantic Ocean near Bermuda known as the Sargasso Sea. Bunches of *Sargassum* often drift on currents to beaches in the Caribbean and southern United States.

One common brown alga is *Fucus* (FYOO-kuhs), or rockweed, which lives along the rocky coast of the eastern United States. Each *Fucus* plant has a holdfast that glues the plant to the bottom. The body of the plant consists of flattened stemlike structures called stipes, leaflike structures called blades, and gas-filled swellings called bladders. Many species of brown algae have bladders, which keep the plants floating upright in the water.

Rhodophyta—The Red Algae

The *Rhodophyta* are another important group of marine algae that can be found in waters from the far north to the tropics. These algae can grow anywhere from the ocean's surface to depths of up to 170 meters. *Rhodophyta* can exist at such extreme depths because they have special pigments that enable them to trap whatever energy is contained in the small amount of light that penetrates there. Most species of red algae are multicellular, and all species have complicated life cycles.

One common red alga is *Chondrus crispus,* or Irish moss. It grows in tide pools and on rocky coastlines. Some red algae, known as the coralline algae, play an important role in the formation of coral reefs. In Japan, the red alga *Porphyra* (por-FIHR-uh) is grown on special marine farms. Dried *Porphyra*—called *nori* in Japanese—is used to wrap portions of rice to make sushi rolls.

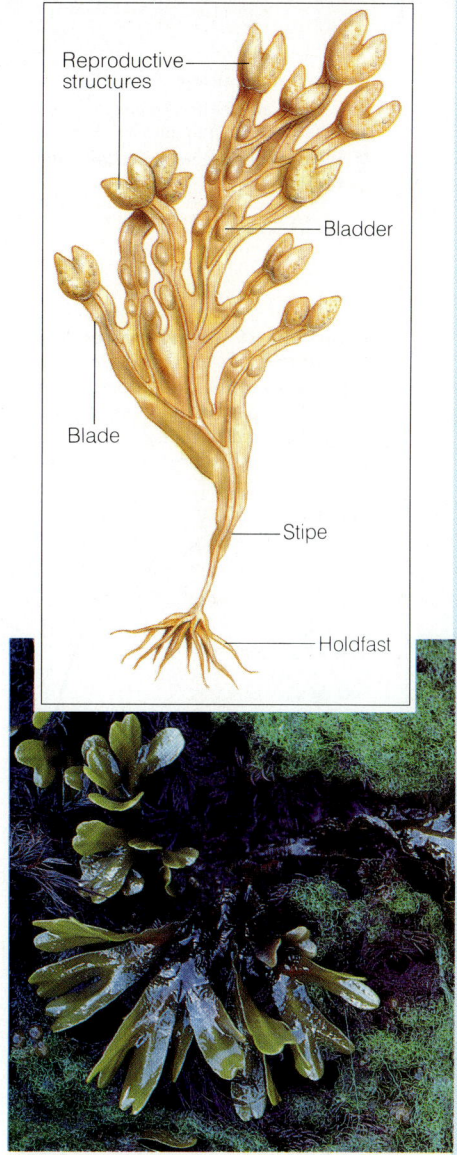

Figure 20–9 Because the brown alga Fucus commonly grows attached to rocks along the shoreline, it is called rockweed.

20–2 SECTION REVIEW

1. What important factors are used to group algae?
2. What is one important difference between *Chlamydomonas* and *Volvox*?
3. **Critical Thinking—Relating Cause and Effect** Red algae often live in deep water. What important adaptation do red algae show that enables them to do this?

Guide For Reading

- What is alternation of generations? What is its significance?
- What stages are seen in algae life cycles?

20–3 Reproduction in Algae

The life cycles of most algae include both a diploid and a haploid generation. Diploid cells have the normal number of chromosomes for a particular species, whereas haploid cells have half the normal number of chromosomes. The switching back and forth between the production of diploid and haploid cells is called **alternation of generations.** In addition to alternating generations, most species of algae also shift back and forth between sexual reproduction that involves the production of gametes and asexual reproduction that involves haploid cells called **zoospores** (ZOH-oh-sporz).

Complex life cycles that involve alternation of generations are characteristic of all members of the plant kingdom. These life cycles are much more complicated than the simple kinds of sexual reproduction that occur in familiar animals such as birds and mammals.

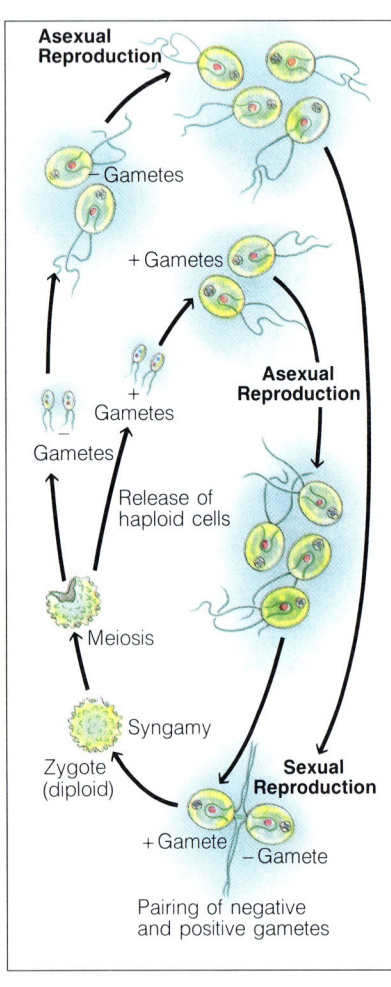

Figure 20–10 As you can see from this diagram of the life cycle of Chlamydomonas, *this green alga reproduces asexually by producing zoospores and sexually by producing zygospores.*

Reproduction in *Chlamydomonas*

The single-celled *Chlamydomonas* spends most of its life in the haploid stage. As long as its living conditions are suitable, this haploid cell reproduces asexually by mitosis. Each time the cell divides, it produces identical haploid zoospores. Smaller than the parent cell, the zoospores soon mature and are able to reproduce asexually. This sequence may be repeated over and over again.

If conditions become unfavorable, *Chlamydomonas* can switch to a stage that reproduces sexually. See Figure 20–10. The haploid cells continue to undergo mitosis, but instead of releasing zoospores, the cells release gametes. Different parent cells produce gametes of two different types, which we call (+) and (−). To our eyes, these gametes appear to be identical. The condition in which the gametes of an organism appear identical is called **isogamy** (igh-SAHG-ah-mee). *Iso* means equal; *gamy* refers to cells that are involved in sexual reproduction. So isogamy means identical reproductive cells.

During sexual reproduction, the gametes gather in large groups. Then (+) and (−) gametes form pairs that soon move away from the group. The paired gametes join flagella and spin around in the water. Both members of the pair then shed their cell walls and fuse, forming a diploid zygote. The fusing of gametes is called **syngamy** (SIHN-gah-mee).

The zygote sinks to the bottom of the pond or ditch and grows a thick protective wall. Within this protective wall *Chlamydomonas* can survive freezing or drying conditions that would ordinarily kill it. When conditions once again become favorable, the zygote begins to grow. It divides by meiosis to produce four flagellated haploid cells. These haploid cells can swim away, mature, and reproduce asexually.

of reproducing by alternation of generations. To help students contrast alternation of generations in *Chlamydomonas, Ulva,* and *Fucus,* ask them to refer to Figures 20–10, 20–11, and 20–12, which illustrate the life cycles of these organisms. If possible, obtain microscope slides or samples of the actual organisms for students to observe.

SECTION REVIEW 20–3

1. Alternation of generations is the switching back and forth between the production of diploid and the production of haploid cells.
2. Asexual reproduction results in the production of offspring with the same genetic makeup as their parents. In many cases, asexual reproduction results

Reproduction in *Ulva*

The life cycle of *Ulva* involves an alternation of generations in which both the diploid and the haploid stages are multicellular plants. See Figure 20–11. The diploid plant is called the **sporophyte**, or "spore producer," because it produces spores. The haploid plant is called the **gametophyte** (gah-MEET-oh-fight) because it produces gametes.

Ulva actually has two different types of gametophytes. Each type produces a different kind of gamete, one of which is larger than the other. The production of two different kinds of gametes is called **heterogamy.**

When the two different gametes fuse, the resulting diploid zygote does not enter a resting stage. Instead, it begins to grow into a multicellular diploid sporophyte. Specialized cells within the sporophyte reproduce asexually by undergoing meiosis and releasing haploid zoospores. These zoospores then divide by mitosis to grow into the two different types of multicellular gametophytes. The two different types of gametophytes produce their gametes and the cycle continues. The only tricky part of the *Ulva* life cycle, at least for humans, is that all three multicellular plants—the two gametophytes and the sporophyte—look exactly the same to the unaided eye!

Reproduction in *Fucus*

The brown alga *Fucus* demonstrates both alternation of generations and heterogamy. Here, the two gametes are radically different from each another. The female gamete, or **egg,** is large and cannot swim. The male gamete, or **sperm**, is small, has a flagella, and can swim. Differences between male and female gametes develop further in plants adapted to land.

Fucus resembles land plants in that the multicellular haploid gametophyte is missing. The diploid sporophyte plant is the only multicellular part of the life cycle. There are two different types of specialized reproductive areas on the tips of the *Fucus* blades. One area produces eggs; the other area produces sperm. Both eggs and sperm are released into the water. If some of the sperm manage to swim to the drifting eggs, fertilization occurs and a zygote is formed. The zygote sinks, and with some luck, lands on a rock to which it will attach itself and grow by mitotic division into a new diploid sporophyte.

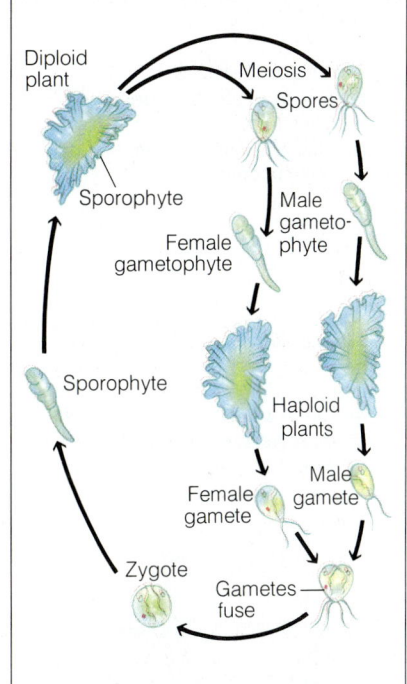

Figure 20–11 *Both generations of the green alga* Ulva *are multicellular plants.*

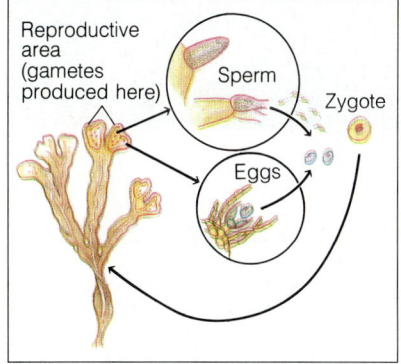

Figure 20–12 *Reproductive structures form on the tips of the blades of the brown alga* Fucus. *Most species of brown algae reproduce sexually.*

20–3 SECTION REVIEW

1. What is alternation of generations?
2. What is an advantage of asexual reproduction?
3. **Critical Thinking—Developing a Theory** How is alternation of generations an effective way of ensuring that fit individuals survive?

BACKGROUND INFORMATION
ASEXUAL VERSUS SEXUAL REPRODUCTION

You may wonder why the reproductive cycles of plants have evolved in such complex ways. In order for a species to survive, its method of reproduction must be well-adapted to its environment. If it is not, the species will become extinct. For this reason, different methods of reproduction are important targets of natural selection. However, each method of reproduction has advantages and disadvantages.

For example, asexual reproduction permits an organism to make genetically identical copies of itself very quickly. In addition, asexual reproduction requires only a single individual. If no other members of the species are present, asexual reproduction can still ensure the production of new individuals. Asexual reproduction, however, limits the genetic variety of a population. The only variations that can appear during asexual reproduction are those that arise from mutations or when mistakes occur during the duplication of DNA.

Sexual reproduction, on the other hand, increases genetic variety within a population. Offspring produced through sexual reproduction are never exact genetic duplicates of either parent. This increased genetic variation allows populations to evolve more rapidly in response to changing environments. Sexual reproduction in plants has an additional advantage. Often, though not always, it produces dormant cells capable of surviving harsh conditions that would kill a growing plant. These dormant cells begin to grow when conditions once again become favorable.

---

in the production of many identical copies.
3. During the sexual stage of reproduction, organisms with genetic diversity are produced. Some of these organisms may be more fit to survive. When these organisms reproduce asexually, they are able to produce many copies of themselves that are genetically identical. Thus, a population of individuals carrying the genes that make them more fit can be produced quickly.

Reinforcement/Reteaching
If students have difficulty answering the Section Review questions, refer them to the appropriate material in the section and review it with them.

Closure
A crossword puzzle might offer students an interesting way to summarize and interrelate the material in this section. Puzzles may easily be made from graph paper with large squares.

TEACHING SUPPORT

Laboratory Manual
- Examining an Algal Bloom, p. 251

Teaching Resources
- Vocabulary Review: Word Game, p. 1
- Hands-On Activity: Seaweed Sweets, p. 103

Study Guide
- Section 20–4, p. 196

MULTICULTURAL STRATEGY

Seaweed is edible and is part of the cuisine of Japan and other countries. Have students look for seaweed foods at a specialty market, or try the Hands-On Activity, Seaweed Sweets, listed in the Teaching Support box on this page.

BACKGROUND INFORMATION
ANTIBIOTIC PRODUCTION IN ALGAE

Some algae are capable of producing antibiotics, which are substances capable of inhibiting the growth of microorganisms. For example, *Chlorella vulgaris*, a freshwater green alga, produces an antibiotic called chlorellin. This may be used to treat infections caused by the bacterium *Staphylococcus aureus*, as well as contaminated water supplies.

TEACHING STRATEGY 20–4
Focus/Motivation

- If all the algae in the world were to die, how would it affect other living things on Earth? (Most of the other species on Earth would also die, since algae are the primary sources of both food and oxygen for all other life forms.)

Guide For Reading
- How do algae affect other living things?
- What are some ways in which people use algae?

Figure 20–13 Although polar bears are carnivores, they will eat algae. People eat algae, too. Algae are often used to thicken frozen dairy desserts—much to the delight of these two youngsters.

Figure 20–14 One important product of algae is agar, a substance used to thicken growth media. Colonies of bacteria are growing on agar in the petri dishes shown here.

442

Content Development
Discuss the fact that algae carry out at least as much photosynthesis, if not more, than the terrestrial plants. Since about 72 percent of the Earth is covered with water, there are simply greater quantities of algae compared to other photosynthesizers.

20–4 Where Algae Fit into the World

Single-celled algae, together with blue-green bacteria and some green flagellates, are food for most of the life in the oceans. Because they are such an important source of food, algae have been called the grasses of the seas. Algae also provide homes for many species of animals. The huge brown kelp forests along the coasts of the United States are home not only to sea otters but to many animal species.

Life as we know it would never have evolved without algae, for they produce much of Earth's free oxygen through photosynthesis. Scientists calculate that between 50 and 75 percent of all the photosynthesis that occurs on Earth is performed by algae. This fact alone makes algae one of the most important groups of organisms on the entire planet.

Over the years, people have learned to use algae—and the chemicals produced by algae—in many different ways. Many species of algae are rich in vitamin C and iron. Other chemicals in algae are used to treat stomach ulcers, lung ailments, high blood pressure, arthritis, and other health problems.

Algae are also used in the manufacture of many food products. Algae are used to make ice cream smooth and candy bars last longer. Algae are used in pickle relishes, salad dressings, chip dips, pancake syrups, egg nogs, and canned chow mein. In Japan and other parts of Asia, algae farms produce crops of red and brown algae that are eaten by millions of people. Algae are eaten as a vegetable and used as flavorings in soups, meat dishes, and candy. Toothpastes, adhesives, hand lotions, and finger paints all contain algae.

Modern industry has even more uses for algae. Chemicals from algae are used to make plastics, waxes, transistors, deodorants, paints, lubricants, and even artificial wood. Algae products are found in poultry feed, cake batters, pie fillers, bakery jellies, and doughnut glazes. Algae even have an important use in scientific laboratories. The compound agar-agar, derived from certain seaweeds, thickens the nutrient mixtures scientists use to grow bacteria and other microorganisms. As you can see, our lives would be very different without algae.

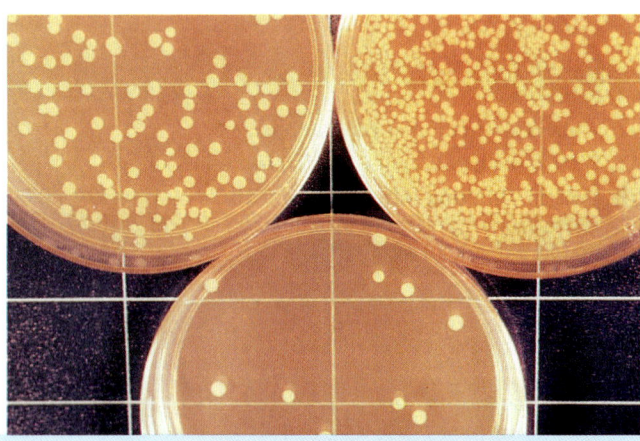

SECTION REVIEW 20–4

1. Algae change the atmosphere of the Earth because as they undergo photosynthesis, they produce oxygen. Since there are so many algae in the Earth's water, scientists have calculated that algae produce between 50 and 75 percent of the Earth's free oxygen.
2. If all the algae on Earth suddenly became extinct, the available oxygen supply would

SCIENCE, TECHNOLOGY, AND SOCIETY CONNECTION

Algae, Ecology, and Public Health

Researchers are studying ways to use algae to clean sewage produced by cities and towns. Sewage, or waste water, contains the liquid and solid wastes each of us produces every day. Since there is no way to avoid producing sewage, we must learn to dispose of it without harming the environment. Untreated sewage released into streams and ponds has several unpleasant consequences. One is that algae in the streams and ponds begin to grow rapidly. Algae may grow so rapidly, in fact, that they cover the surface of the water. This increased growth is called an algal bloom. An algal bloom may harm other forms of life that live in the water.

Why does sewage cause algal blooms? Human wastes—like the wastes of all animals—contain substances that plants use as nutrients for growth. As algae grow, they remove these nutrients from sewage. In other words, sewage acts as fertilizer for algae.

Some researchers think they have found a way to put this characteristic of algae to use in treating sewage. Their plan works as follows: First the sewage is diluted. Then the sewage is trickled through large tanks of algae. As the algae grow, they remove nutrients contained in the sewage. Eventually, the treated water can be released carefully into the environment. Because it contains a smaller concentration of harmful wastes, this water does not threaten the environment as much as untreated sewage does. If researchers can grow the right kinds of algae, not only will they be able to treat sewage, but they may also be able to produce plants for use in industry.

As they grow, algae take in chemicals and other substances from water. Various species of algae are being used in pilot programs to design new methods of treating sewage.

SCIENCE, TECHNOLOGY, AND SOCIETY CONNECTION

Algae, Ecology, and Public Health

Algal bloom is the direct result of cultural eutrophication, the release of inorganic nutrients from human sources such as farms and sewage into the water. It is the most widespread problem today in our lakes, with roughly seven out of every eight lakes experiencing eutrophication. It is also occurring in our coastal waters, where many fish and shellfish spend at least part of their life cycle. Besides depleting the lakes of oxygen as they decompose, many types of algae release toxic substances into the water, which are then taken up by other organisms. When humans eat these organisms, food poisoning may be the unpleasant result.

Have students research the causes of algal bloom and the measures that might be taken to prevent it from occurring.

20–4 SECTION REVIEW

1. How have algae changed the atmosphere on Earth?
2. How would life on Earth be different if all algae suddenly became extinct?
3. What two foods contain algae products?
4. **Critical Thinking—Relating Concepts** How have algae contributed to our understanding of human disease?

---

greatly diminish. Since algae form the base of many food pyramids, the network of feeding relationships within many ecosystems would collapse.
3. Algae are used in the production of many foods, including ice cream and salad dressing.
4. Certain algae produce agar-agar. This material is used to thicken certain nutrient mixtures that are used to grow bacteria and other microorganisms. Without algae, it would be extremely difficult to study many disease-causing microorganisms.

Reinforcement/Reteaching
Use the Section Review questions to check for mastery of the concepts in this section. If students have any difficulties, refer them to the related material.

Closure
Have students list the ways in which algae are involved, either directly or indirectly, in our everyday lives. Compare responses by listing them on an overhead transparency. Students should appreciate not only the length of the list but also the overwhelming number of algae's positive contributions.

LABORATORY INVESTIGATION

STUDYING FILAMENTOUS GREEN ALGAE

PROBLEM

What are the characteristics of *Spirogyra*?

MATERIALS (per group)

microscope
prepared slide of *Spirogyra*

PROCEDURE

1. Using a microscope, examine a prepared slide of *Spirogyra* under the low-power objective. Notice the general shape of *Spirogyra*. On a separate sheet of paper, draw a diagram of a filament of *Spirogyra*. Indicate the appearance of individual cells. Note the arrangement of the chloroplasts.
2. Move your slide until you locate two parallel filaments that are attached to each other by small "bridges," or tubes. This attachment is a form of sexual reproduction called conjugation. Switch to the medium-power objective and focus. Draw the portion of the filaments that shows the conjugation tubes. If possible, locate a portion of the filaments where material is passing from one cell to another through the tubes. Draw these cells.
3. Switch to the high-power objective. Use the fine adjustment to focus on a nucleus in one of the cells. Draw a diagram of a single cell and its nucleus. Label the nucleus.
4. Leaving the high-power objective in place, move the slide to focus on a portion of a chloroplast. Draw a diagram of a chloroplast.

OBSERVATIONS

1. Are there different kinds of cells in *Spirogyra*?
2. What is the shape of *Spirogyra*?
3. Based on your observations, describe what happens during conjugation in *Spirogyra*.
4. How is the nucleus held in place?
5. What is the shape of the chloroplasts?

ANALYSIS AND CONCLUSIONS

1. Are *Spirogyra* multicellular or colonial algae? Explain your answer.
2. What is the function of the conjugation tubes?
3. How are *Spirogyra* adapted to life in water? Would these plants survive on land? Explain your answer.

STUDENT STUDY GUIDE

SUMMARIZING THE CONCEPTS

The key concepts in each section of this chapter are listed below to help you review the chapter content. Make sure you understand each concept and its relationship to other concepts and to the theme of this chapter.

20-1 Characteristics of Algae

- Algae must live in or near a source of water.
- Algae vary in size, shape, and color. Some species of algae are unicellular, other species form colonies, and still other species are multicellular.
- Algae show many different adaptations to life in water. Unlike land plants, most algae lack a system of internal tubes to move water and nutrients from one part of the plant to another. Since algae live in water, they do not need protection from drying out or stemlike structures to keep them upright.
- Algae contain accessory pigments in addition to different forms of chlorophyll. These pigments give algae their color. Scientists group algae on the basis of the chlorophylls and pigments they contain, as well as the form in which they store food.

20-2 Groups of Algae

- There are three groups of multicellular algae: green algae, red algae, and brown algae. Each group receives its color from combinations of special pigments and chlorophyll.
- The three different groups of algae store the food they make as special starches and/or oils.

20-3 Reproduction in Algae

- Algae have complex life cycles that involve alternation of generations.
- Many algae produce special cells during reproduction that enable the algae to survive harsh environmental conditions. These cells can resume growth when environmental conditions improve.

20-4 Where Algae Fit into the World

- Algae produce much of the Earth's free oxygen during photosynthesis. Without this oxygen, life on Earth as we know it would never have evolved.
- Algae provide food for animals and people. Algae are also used in the manufacture of many foods and industrial products.

REVIEWING KEY TERMS

Vocabulary terms are important to your understanding of biology. The key terms listed below are those you should be especially familiar with. Review these terms and their meanings. Then use each term in a complete sentence. If you are not sure of a term's meaning, return to the appropriate section and review its definition.

20-1 Characteristics of Algae
accessory pigment

20-2 Groups of Algae
colony
filament
holdfast cell

20-3 Reproduction in Algae
alternation of generations
zoospore
isogamy
syngamy
sporophyte
gametophyte
heterogamy
egg
sperm

---

the cells into the water. They would not survive on land since they would quickly dry out and would be unable to obtain the nutrients they require to perform the essential functions of living things.

Going Further: Enrichment

1. Have students observe fresh specimens of *Spirogyra* on a wet-mount slide or under a stereo microscope. Conjugation will be less likely to occur in these samples, but they may be more interesting for students.

2. Have students observe other types of algae, such as *Oedogonium*, to compare the structure and reproduction of other algae with those of *Spirogyra*.

CHAPTER REVIEW

CONTENT REVIEW

Multiple Choice

Choose the letter of the answer that best completes each statement.

1. Unlike most plants, algae lack
   a. chloroplasts.
   b. accessory pigments.
   c. a nucleus.
   d. an internal system of tubes.
2. Which of the following algae are able to live in deep ocean water?
   a. red algae
   b. green algae
   c. *Volvox*
   d. *Spirogyra*
3. Algae take in nutrients through their
   a. holdfast.
   b. stipe.
   c. blade.
   d. cell membrane.
4. Some single-celled algae are included in the kingdom
   a. Monera.
   b. Protista.
   c. Animalia.
   d. Fungi.
5. Red algae have a special pigment that enables them to absorb energy from
   a. red light.
   b. yellow light.
   c. violet light.
   d. blue light.
6. Green algae live in each of the following environments except
   a. dry land.
   b. a pond.
   c. moist areas on land.
   d. a lake.
7. Filaments of some forms of green algae are attached to the bottom of a lake or pond by a (an)
   a. stipe.
   b. holdfast cell.
   c. air bladder.
   d. egg.
8. The largest algae in the world are the
   a. red algae.
   b. filamentous algae.
   c. green algae.
   d. brown algae.

True or False

Determine whether each statement is true or false. If it is true, write "true." If it is false, change the underlined word or words to make the statement true.

1. <u>Alternation of generations</u> is a characteristic of all plants.
2. <u>Chlorophyll</u> can make blue light more useful for photosynthesis.
3. <u>Fucus</u> are single-celled green algae.
4. <u>Kelp</u> are the largest algae in the world and can grow longer than 60 meters.
5. Filamentous algae can reproduce <u>sexually</u> when they break apart.
6. A <u>colony</u> is a group of cells that are joined together and that show few specialized structures.
7. Special light-absorbing pigments found in some kinds of algae are called <u>gametes</u>.
8. Most algae get necessary gases and nutrients from the <u>air</u>.

Word Relationships

In each of the following sets of terms, three of the terms are related. One term does not belong. Determine the characteristic common to three of the terms and then identify the term that does not belong.

1. eggs, sperm, spores, stipes
2. red, brown, green, yellow
3. *Chlamydomonas, Volvox, Spirogyra, Fucus*
4. ponds, lakes, rivers, oceans

CHAPTER REVIEW

CONTENT REVIEW

Multiple Choice
1. d 2. a 3. c 4. b
5. d 6. a 7. b 8. d

True or False
1. T
2. F Accessory pigments, or phycobilins
3. F *Chlamydomonas*
4. T
5. F asexually
6. T
7. F phycobilins
8. F water

Word Relationships
1. types of reproductive cells; stipes are flattened stemlike structures that make up the body of certain types of algae
2. the colors of major algae groups; yellow
3. kinds of green algae; *Fucus* is a brown alga
4. bodies of fresh water; the ocean contains salt water

CONCEPT MASTERY

1. Algae are an important food source for many organisms. Algae are also responsible for the production of much of the Earth's supply of oxygen. Other answers are possible.
2. Scientists classify algae based on the pigments they contain and whether they are unicellular or multicellular.
3. Some kinds of red algae contain accessory pigments that are able to trap the energy in the small amount of light that is able to penetrate great depths of water. These red algae are able to perform photosynthesis using this energy.
4. Cells in a colony function independently; they do not form specialized tissues. Cells in a multicellular organism form specialized tissues and cannot function independently.
5. Alternation of generations is a life cycle that switches back and forth between a haploid and a diploid generation. Alternation of generations is common to all plants.
6. Some scientists place all algae in the kingdom Protista because they feel that all algae are unlike land plants and more closely resemble protists. The protist kingdom has become a catchall category that includes many diverse life forms.
7. Algae differ from land plants because they do not have a system of internal tubes that conduct water and other materials to their cells. Algae are also much more dependent upon water for life. All algae live in water or in places that remain very damp.
8. Accessory pigments also enable algae to trap the energy in different wavelengths of

CONCEPT MASTERY

Use your understanding of the concepts developed in the chapter to answer each of the following in a brief paragraph.

1. Explain the importance of algae to life.
2. Describe two characteristics that help scientists classify algae.
3. How are some kinds of red algae adapted to life in deep ocean water?
4. What is the main difference between a colony and a multicellular organism?
5. What is alternation of generations?
6. Why do some scientists place all algae in the kingdom Protista?
7. How are algae different from land plants?
8. Accessory pigments do more than provide color for algae. What important function do accessory pigments perform?

CRITICAL AND CREATIVE THINKING

Discuss each of the following in a brief paragraph.

1. **Applying concepts** How does having a method of both asexual reproduction and sexual reproduction contribute to an organism's survival?
2. **Making predictions** Suppose all species of algae on Earth died out. What effects would this have on other organisms?
3. **Relating cause and effect** Suppose you were able to design a new species of algae that could live in deep water. What important characteristics would you include? Why?
4. **Applying technology** You have been chosen to work on a new space probe that will carry humans deep into space. A friend suggests that it would be important to have a population of algae on this trip. How could these algae help human space travelers on a long voyage in space?
5. **Analyzing concepts** You know that all life comes from life. Your friend observes that a pond in the area has dried up and appears to lack life. Later, this pond fills with water again. Algae begin to grow. Your friend is puzzled. How could you explain the growth of algae in this pond?
6. **Relating concepts** Some scientists think that a form of algae was an ancestor of certain land plants. What are some changes that would have to occur in order to make algae able to survive on land?
7. **Using the writing process** The huge kelp forests along the west coast of the United States are home to the sea otter. There are relatively few of these animals alive in the wild. The otters are fond of eating abalone and other shellfish. Because of their eating preferences, they often come into conflict with local people who earn a living by capturing and selling the shellfish. These local people want to eradicate the kelp. They feel that the sea otters will move away if the kelp is destroyed. You are the lawyer hired by the government to protect the kelp and the otters. The local newspaper wants you to write an article detailing your position. Here is your chance to influence people. Write an article for the paper. Save the otters!

---

light. Thus, the accessory pigments enable algae to live in many diverse environments, including the deep water of the ocean.

CRITICAL AND CREATIVE THINKING

1. During sexual reproduction, genetic material is exchanged between organisms. This results in a recombination of genes. Sexual reproduction contributes to diversity in members of a species. Asexual reproduction produces offspring that are genetically identical to the parent.
2. Many organisms that depend upon algae for food would die. The amount of oxygen in the atmosphere would most probably decrease, since algae are responsible for the formation of much of the Earth's oxygen as a byproduct of photosynthesis.
3. An alga that could live in deep water would have to contain pigments that were able to trap the energy in the wavelengths of light that were able to penetrate there. This algae species would probably resemble a species of red algae that is alive today.
4. With a source of light energy, the algae on a space flight could become a food source. The algae would produce oxygen used by astronauts and use the carbon dioxide they produce. Some of the other wastes produced by astronauts could be used by algae as fertilizer.
5. Algae are able to make special spores that are able to withstand conditions of drying. These spores may have remained in a dormant state in the dry pond. They may have been carried by wind or deposited by animals in the pond.
6. Algae would have to develop a way to keep them from losing water, such as a waterproof coating on their blades. They would also have to evolve a way to take in water from the ground. They would have to evolve something like true roots, stems, or leaves.
7. Check student writing for clarity and correctness.

CHAPTER 21 Mosses and Ferns

Section	Laboratory Investigations and Activities
21–1 Plants Invade the Land pages 449–451 THEMES: Evolution, Scale and Structure	**Teacher Edition** DEMONSTRATION: Land Plant Discovery Tour, p. 448D
21–2 The Mosses, Liverworts, and Hornworts pages 451–454 THEMES: Scale and Structure, Unity and Diversity	**Laboratory Manual** Comparing Marine and Land Nonvascular Plants, p. 255 **Teaching Resources** HANDS-ON ACTIVITY: Growing Moss, p. 105
21–3 The Ferns and the First Vascular Plants pages 455–459 THEMES: Evolution, Scale and Structure	**Teacher Edition** DEMONSTRATION: Alternation of Generations in Ferns, p. 448D **Laboratory Manual** Special Characteristics of Ferns, p. 261
21–4 Where Mosses and Ferns Fit into The World pages 460–461 THEME: Systems and Interactions	**Student Edition** LABORATORY INVESTIGATION: Comparing Mosses and Ferns, p. 462
Chapter Review pages 463–465	

Teacher Resources

Books

Abbe, Elfriede. *The Fern Herbal: Including the Ferns, the Horsetails, and the Club Mosses.* Cornell University Press, 1985.

Conrad, Henry S., and Paul L. Redfearn, Jr. *How to Know the Mosses and Liverworts,* 2nd ed. Wm. C. Brown, 1979.

Dyer, A. F., ed. *The Experimental Biology of Ferns.* Academic Press, 1979.

Hallowell, Anne E., and Barbara Hallowell. *Fern Finder.* Nature Study, 1981.

Richardson, P.M., T. A. Steeves, and S. A. Fultz. *The Biology of Mosses.* Wiley, 1981.

Streams, John, ed. *Treasures of Nature: Ferns.* Crossing Press, 1987.

Taylor, Ronald J., and Alan E. Leviton, eds. *Mosses of North America.* American Association for the Advancement of Science, 1980.

Taylor, Thomas N., and Edith L. Taylor. *The Biology and Evolution of Fossil Plants.* Prentice Hall, 1993.

Wills, J. C. *A Dictionary of the Flowering Plants and Ferns,* 8th ed. Cambridge University Press, 1985.

CHAPTER PLANNING GUIDE

Other Activities	Media and Technology
Study Guide Section 21–1, p. 199	
Study Guide Section 21–2, p. 200	**Biology Transparencies** 23: The Life Cycle of a Moss
Teaching Resources ACTIVITY: Distinguishing Between Mosses and Ferns, p. 3 ACTIVITY: Alternation of Generations in Tracheophytes, p. 5 **Study Guide** Section 21–3, p. 203	**Biology Transparencies** 24: The Life Cycle of a Fern
Teaching Resources VOCABULARY REVIEW: Crossword, p. 1 **Study Guide** Section 21–4, p. 205	
Teaching Resources Chapter Test A, p. 13 Chapter Test B, p. 17	**Computer Test Bank** Chapter Test, p. 223

Audiovisuals

Mosses, Liverworts and Ferns. Film or video. Coronet.
Mosses and Liverworts (Life Cycles and Ecology). Filmstrip with cassette. Ward.
Ferns and Horsetails (Life Cycles and Ecology). Filmstrip with cassette. Ward.
Fern Life Cycle. Filmstrip with cassette. Carolina Biological Supply Company.
Moss Life Cycle. Filmstrip with cassette. Carolina Biological Supply Company.
The Bryophytes and Ferns. Filmstrip with cassette. Biology Media.
The Liverworts. Filmstrip with cassette. Carolina Biological Supply Company.

CHAPTER 21 Mosses and Ferns

CHAPTER OVERVIEW

How were aquatic plants able to colonize the land? Through evolution, many adaptations in structure and function—such as transport and reproduction—were necessary before plants could thrive in a dry terrestrial environment.

The bryophytes, such as mosses, liverworts, and hornworts, were successful because of their small size and their location in a moist habitat. Lacking roots, leaves, and vascular tissue, they make their home in wetlands, rain forests, and wherever the soil remains damp. They still require a moist environment in which to reproduce.

The primitive tracheophytes, such as club mosses, horsetails, and ferns, evolved methods to stand upright and support their leaves for photosynthesis. These plants developed methods for absorption, transport, and retention of water, enabling them to live in drier terrestrial environments than the mosses. Still, they depend upon water for reproduction.

21-1 PLANTS INVADE THE LAND

Section Focus 21-1

The purpose of this section is to familiarize students with the adaptations that plants evolved in order to survive on land. Students are also introduced to the two major phyla of the plant kingdom, Bryophyta and Tracheophyta.

The section closes with a summary of the adaptations aquatic plants made in order to evolve into land plants.

Performance Objectives 21-1

1. Describe the adaptations that land plants evolved in order to acquire, transport, and conserve water.
2. Explain the necessity for adaptations by the ancestors of land plants to reproduce in a dry environment.
3. Introduce the phyla Bryophyta and Tracheophyta.

21-2 THE MOSSES, LIVERWORTS, AND HORNWORTS

Section Focus 21-2

The purpose of this section is to acquaint students with members of the phylum Bryophyta: mosses, liverworts, and hornworts.

The bryophytes are adapted to living only in wet habitats. They do not have vascular tissue, true roots, or any adaptation to prevent water loss from their surfaces. The bryophytes also require water for reproduction.

The bryophytes have a life cycle that involves alternation of generations. The typical moss plant is the haploid gametophyte plant that produces sperm and eggs. The diploid zygote grows into a sporophyte that remains attached to the gametophyte. The sporophyte plant produces haploid spores by meiosis.

Performance Objectives 21-2

1. Identify the members of the phylum Bryophyta: mosses, liverworts, and hornworts.
2. Describe bryophyte morphology.
3. Explain the adaptations of the bryophytes for life in a wet terrestrial environment.
4. Review alternation of generations in the life cycle of a plant.
5. Describe alternation of generations in the life cycle of the mosses, liverworts, and hornworts.
6. Explain why bryophytes require a moist environment for reproduction.

Science Terms 21-2

rhizoid   p. 452
antheridium   p. 453
archegonium   p. 453
protonema   p. 454

21-3 THE FERNS AND THE FIRST VASCULAR PLANTS

Section Focus 21-3

The purpose of this section is to acquaint students with the tracheophytes, the first true land plants. Primitive members of this phylum that can still be found today are the club mosses, horsetails, and ferns.

With the development of a true vascular system and the evolution of true roots and true leaves, the tracheophytes successfully evolved methods for survival in a dry terrestrial environment.

The life cycle of the tracheophytes also involves alternation of generations. In plants such as ferns, the large leafy sporophyte plant produces haploid spores. Under proper conditions, the spores will germinate into haploid gametophyte plants. Each small green gametophyte grows independently of the sporophyte and will produce gametes. Fertilization can occur only when there is sufficient moisture on the surface of the ground. The diploid zygote will grow into the sporophyte plant that we typically identify as a fern.

Performance Objectives 21-3

1. Identify the primitive tracheophytes: club mosses, horsetails, and ferns.
2. Describe the morphology of the ferns.
3. Explain the adaptations evolved by the tracheophytes for life in a dry terrestrial environment.
4. Describe alternation of generations in the life cycle of the ferns.
5. Explain why ferns require a moist environment for reproduction.

Science Terms 21-3

vascular tissue   p. 455
xylem   p. 455
phloem   p. 455
tracheid   p. 455
vascular cylinder   p. 455
vein   p. 456
cuticle   p. 456
rhizome   p. 457
frond   p. 457
sporangium   p. 458
sorus   p. 458
prothallium   p. 458

CHAPTER PREVIEW

21-4 WHERE MOSSES AND FERNS FIT INTO THE WORLD

Section Focus 21-4

The purpose of this section is to identify the types of natural habitats in which mosses and ferns can be found. Mosses of several species are of economic value to gardeners and can also be used as a source of fuel. Ferns are a prized horticultural plant, and some species are suitable for human consumption.

Performance Objectives 21-4

1. Identify the natural habitats of mosses and ferns.
2. Describe the economic importance and many uses of mosses, especially sphagnum moss.
3. Identify the uses humans make of ferns.

DISCOVERY LEARNING

TEACHER DEMONSTRATIONS
Modeling

Land Plant Discovery Tour
The following activity can be used to introduce Chapter 21.

Take students on a guided tour of the exterior of your school building. You will be looking for the many places that terrestrial plants can grow. You can begin by looking in some of the obvious locations first—the lawn or garden or in the pits surrounding planted trees. Then look in some of the less obvious places—in pavement cracks, on the shady sides of the building or walls, on rocks, on trees, or near a source of standing water.

Examine each plant and have the students note the following:
1. Does it have leaves?
2. Does it have veins?
3. Where was the plant growing?
4. What is the approximate size of the plant?
5. Is the plant mosslike or does it have a green stem or a woody stem?

Use your discretion in removing some of the plant samples for additional closer observations in your classroom. (You may want to familiarize yourself with local regulations on plant collection.) Here are some activities your students may enjoy:

1. **Microscopic**—Make wet mounts from algae or moss scrapings from tree bark or walls.
2. **Dissecting microscope**—Examine the mosslike plants and small plants with green stems in order to better see their characteristics.
3. **Tracings**—Select leaves from plants with veins for the students to trace. Once the outline is completed, the students should draw in the pattern of the veins.

- **Which kinds of plants are more numerous?** (Plants with stems.)
- **Where are mosslike plants found?** (In shady locations, cracks in pavement, places where water collects.)
- **Did all the plants have leaves?** (Only those plants with stems had leaves.)

Alternation of Generations in Ferns
You may want to use this activity in Section 21-3.

Students may try to grow their own fern gametophyte plants in a homemade terrarium. Have students obtain an empty 2-liter soda bottle that is made of colorless plastic and has an opaque plastic base. Remove the base from the bottom of the bottle. Fill the base with peat moss to within 2 centimeters of the top and moisten thoroughly with water.

Cut the transparent portion of the bottle in half along its horizontal axis. Discard the top portion containing the cap. The bottom half will become the dome of the terrarium.

Collect fern spores from a frond of an actively growing fern plant. Sprinkle the spores over the moistened peat moss. Cover with the dome of the terrarium. In about three weeks, prothallia, the fern gametophytes, will be visible.

The terrarium should not be kept in direct sunlight. It should be monitored periodically to correct for improper water balance. There should always be some moisture clinging to the dome of the terrarium. If the terrarium is completely fogged and the contents are not visible, the top should be removed for a few minutes. If the dome appears to be dry, water should be added to the peat moss using a spray bottle and the dome replaced immediately.

In time, the fronds of the sporophyte plant can be seen growing from the gametophyte plant.

- **Describe the appearance of the prothallium, the fern gametophyte.** (Small, green, heart-shaped.)
- **Does the gametophyte have a diploid or haploid number of chromosomes?** (Haploid.)
- **What structures are found on the prothallium?** (Antheridia and archegonia—male and female organs, respectively—can be found on the underside.)

448D

CHAPTER **21**

Mosses and Ferns

GUIDED ENQUIRY

Pose the following questions to students and have them record their responses. Point out that they will gain a better understanding of the key concepts if they read the chapter with these basic questions in mind. Upon completion of the chapter, pose the questions again. Ask students to compare their initial responses with those they have developed after reading the chapter.
- What adaptations are necessary for plants to survive in a terrestrial environment?
- How can land plants prevent excessive water loss?
- What do land plants need in order to transport water from the ground to their topmost leaves?
- How are plants able to reproduce on land?

CHAPTER **21**

Mosses and Ferns

Rain forests do not occur in tropical areas only. Olympic National Park in Washington State is a temperate rain forest. Bathed by damp Pacific air currents, ferns and tiny mosses (inset) grow well.

Life began in the sea, and for millions of years living things remained in the sea. Yet, slowly, life emerged onto land. On empty ancient continents, new opportunities existed. But new challenges to survive and reproduce awaited as well.

The first land plants were the first multicellular organisms to meet these challenges. To these plants we owe a great debt: Had they not colonized the land, animals would not have been able to follow. In this chapter you will glimpse the struggle of early plants to survive on land—a struggle whose evidence is visible in the form of living plants that remain suspended in a life halfway between water and land.

INTRODUCING CHAPTER 21

Using the Visuals
Ask one or two student volunteers to describe what images come to mind when asked to describe a rain forest. Compare their responses with the photograph on this page. Have students read the caption and identify the location of this photograph.
- **What features identify this photograph as a rain forest?** (Answers will vary but may include dense plant growth that carpets the forest floor and plants growing from trees.)
- **Is it surprising to see that a rain forest is found in Washington State? How is it possible for a rain forest to grow in Washington State?** (Moist air is supplied from the Pacific Ocean.)
- **What kinds of plants are well suited for living in such a moist environment?** (Answers will vary but may include mosses and ferns.)

GUIDE FOR READING

After you read the following sections, you will be able to

21–1 Plants Invade the Land
- Describe some of the adaptations plants need to survive on land.

21–2 The Mosses, Liverworts, and Hornworts
- Identify the characteristics of the three main groups of bryophytes.
- Describe some adaptations shown by bryophytes that enable them to survive on land.
- Identify patterns of reproduction in bryophytes.

21–3 The Ferns and the First Vascular Plants
- Recognize the importance of vascular tissue to land plants.
- Identify characteristics of club mosses and horsetails.
- Discuss ways in which ferns resemble other land plants.
- Describe alternation of generations in ferns.

21–4 Where Mosses and Ferns Fit into the World
- Describe ways in which certain characteristics of mosses make these plants useful to people.
- List ways in which ferns are used by people.

Journal Activity

YOU AND YOUR WORLD
Consider the expression: A rolling stone gathers no moss. Is there any truth to this saying? Explain why or why not in your journal.

21–1 Plants Invade the Land

Guide For Reading
- What adaptations do plants need to survive on land?
- What are the main phyla of land plants?

Because the first multicellular organisms evolved in water, their entire lives were designed around an aquatic environment. All the processes that ensured survival—from photosynthesis to sexual reproduction—took place in water.

But over time, some organisms adapted to life in drier environments. In the following pages you will learn about some living plants and some extinct plants that represent stages in this process of adaptation. These plants illustrate steps in the evolution of structures to acquire, transport, and conserve water. They also demonstrate how land plants evolved reproductive cycles that enable them to survive in terrestrial environments, or land environments.

The First Land Plants

The fossil record does not provide much information about the very earliest stages of the evolution of land plants. Remember that it is the hard parts of organisms, such as shells and bones, that form the best fossils. Because the first land plants were soft-bodied, they have left few fossils. But we do have enough evidence about early plant life on land to say several things with certainty.

Figure 21–1 Dating from the Carboniferous Period, this fossil fern looks very much like its still-growing relatives. As you can see, these plants have changed little over time.

COOPERATIVE LEARNING

Using preassigned lab groups or randomly selected teams, have groups imagine that they live in the late eighteenth century in an area in which the only fuel available is peat. They should make specific references to the five senses in describing how peat permeates their life. Groups may choose to describe their life in a story, a skit, or a poem. If available, it would be helpful to display and burn a piece of peat.

Journal Activity

YOU AND YOUR WORLD
Encourage students to take a creative approach to this assignment. Ask them what moss could mean to a "rolling stone." Have volunteers read their responses to the class. Have students keep their journal entries in their portfolios.

TEACHING STRATEGY 21–1

Focus/Motivation
Using the Visuals Display fossil specimens (or pictures or slides) of ferns. Review how fossils, such as those you are displaying, were formed. Review how scientists determine the approximate age of fossils. The age of the fern specimen in Figure 21–1 is probably about 340 million years old.

Content Development
The first multicellular plants were aquatic and completely adapted to their watery environment. In order to make the transition to land, the ancestors of today's terrestrial plants needed to adapt to a dry environment. Their evolution included many solutions to problems of acquiring, transporting, and conserving water.

449

BACKGROUND INFORMATION
PLANTS WERE HERE FIRST!

When students think of life first emerging from the sea, they invariably envision an amphibian-like creature skulking around the shore. In truth, much smaller animals—such as insects—were the first to colonize dry land. But plants arrived on land before any animal species. Without plants, land animals would have had nothing to eat.

Some biologists believe that plants did not make the transition to land by themselves. Their theory is that plants coevolved with symbiotic fungi, developing the very first mycorrhizae. This plant/fungi partnership allowed necessary minerals to be extracted from the sterile, inorganic soils of the young Earth.

Figure 21–2 Ferns can grow in water as well as on land. This fern, Marsilia, *resembles a floating four-leaf clover.*

The adaptation of plants to life on land was a long, slow process. Algae that could live out of water at least part of the time evolved 500 to 600 million years ago. From these plant pioneers, at least two separate groups of algaelike land plants evolved between 450 and 500 million years ago.

One group developed into the phylum Bryophyta (brigh-oh-FIGHT-uh), which includes mosses, liverworts, and hornworts. The other group evolved into the phylum Tracheophyta (tray-kee-oh-FIGHT-uh), which includes the ferns and the rest of the higher plants. Both the bryophytes and the tracheophytes faced the same set of problems in adapting to terrestrial environments, but each group evolved its own set of solutions.

Demands of Life on Land

The adaptations that enabled aquatic organisms to survive in dry environments were not simple. To understand how important these adaptations were, let us examine some of the requirements of life on land.

- All cells need a constant supply of water. For this reason, land plants must obtain water and deliver it to all of their cells, even those cells that grow above ground in dry air. Once plants provide water to their tissues, they must protect that water against loss by evaporation to the atmosphere.
- The parts of the plant that make food for the plant must be exposed to as much sunlight as possible. Aquatic plants that float on the surface of the water have no problem obtaining

Figure 21–3 Mosses grow well in the shade of trees, as this carpet of mosses on the floor of a pine forest illustrates. You can be certain that the ground beneath the pines remains relatively damp, for mosses grow best under damp conditions.

21–1 (continued)

Focus/Motivation

Display moss plants and a fern for the students to examine. These plants represent the plant kingdom's two major phyla. The moss belongs to the phylum Bryophyta, and the fern to the phylum Tracheophyta.

• **How are these plants different from one another in the way they are constructed?** (Mosses are small; ferns are much larger. Ferns have leaves and veins; mosses do not.)

• **Are they similar in any ways?** (Both are green plants and require the same materials to survive and photosynthesize—oxygen, carbon dioxide, sunlight, and water.)

Content Development

Emphasize the many differences between a plant's needs on land and in water, and review the textbook's list of problems that aquatic plants faced as they adapted to land. In this chapter and in later ones, students will learn exactly how today's land plants meet each of these challenges.

SECTION REVIEW 21–1

1. Three problems faced by land plants are the need for a supply of water, the need to support their leaves so that they may be exposed to light, and their need to move materials to and from cells.

2. The bryophytes and the tracheophytes are the two main groups of land plants. Mosses,

sunlight because there is little water above them to interfere with the absorption of the sun's energy. Land plants, however, need rigid supports to hold their leaves up to the sun in ways that expose the leaves to sunlight.
- Land plants take up water and nutrients in roots but make food in leaves. To supply all cells with the necessities of life, land plants must transport water and nutrients upward and the products of photosynthesis downward.
- Land plants must exchange water and carbon dioxide with the environment without losing too much water in the process.
- Fully terrestrial plants must be able to reproduce in environments that lack standing water in which the sperm can swim. In many terrestrial situations, the zygotes and young embryos of land plants are in danger of drying out.

The bryophytes have partially solved these problems. Bryophytes no longer need to be constantly submerged in water, but they do need to remain wet most of the time. The simplest tracheophytes—the ferns—have evolved further toward complete independence from water. But as you will see, ferns still have not solved all the problems posed by a terrestrial life.

21-1 SECTION REVIEW

1. What are two of the problems faced by plants that live on land?
2. What are the names of the two main phyla of land plants? Give examples of each.
3. **Critical Thinking—Applying Concepts** Your friend finds a small plant growing in the desert. He identifies the plant as a moss. Explain why this plant is probably not a moss.

21-2 The Mosses, Liverworts, and Hornworts

The phylum Bryophyta includes the mosses, liverworts, and hornworts. **Bryophytes, like the algae from which they evolved, have life cycles that involve an alternation of generations between a haploid gametophyte and a diploid sporophyte. Also like the algae, bryophytes need water for reproduction to occur.** Thus bryophytes can thrive only in wet areas, or in areas where rainfall is plentiful at least part of the year. Bryophytes grow most abundantly in swamps, marshes, near streams, and in rain forests in tropical areas and along the western coast of the United States.

Guide For Reading
- What are the characteristics of the three main groups of bryophytes?
- What adaptations of bryophytes enable them to survive on land?
- How do bryophytes reproduce?

451

TEACHING SUPPORT

Laboratory Manual
- Comparing Marine and Land Nonvascular Plants, p. 255

BACKGROUND INFORMATION
MOSSES IN SURPRISING PLACES

As the textbook emphasizes, mosses need abundant water to grow and reproduce. Yet many mosses are found in areas with seasonal droughts. Even more surprising, some mosses live in tundras and other frigid climates, where liquid water may be unavailable for months at a time.

Mosses survive such habitats by entering a state similar to suspended animation. Their tissues virtually dehydrate when water is not available, and they neither grow nor reproduce. But although they may not prosper without water, mosses have adapted to periods without it, and that has let them live in some surprising places.

21–2 (continued)
Content Development
- **What conclusions can you draw about the size of bryophytes from the examples you have examined?** (They are small plants, less than a few centimeters tall.)
- **Do they carry on photosynthesis?** (Yes, they are green and therefore they must contain chlorophyll.)
- **Describe the appearance of a single moss plant.** (It looks like a tiny stem with miniature leaves. At the bottom of the stem there are tiny rootlike structures.)
- **Describe the appearance of a single liverwort.** (It looks like a tiny flat leaf growing along the surface of the ground.)

Skills Development
Guided Practice
Skill: Making a wet mount
Have students take a closer look at a moss plant by making a wet mount. Use forceps to remove a leaflike structure from a moss plant, or examine the entire plant if it is small. Place the specimen on a clean slide. (A depression slide may be necessary if a whole plant is mounted.) Cover the specimen with one or two drops of water. Lower the coverslip into place.

By examining their slides under a microscope, students should be able to see and identify chloroplasts in the cells of a moss plant.

Figure 21–4 The tiny brown structures on the tip of these moss plants (left) are the sporophyte plants. When the spores are ripe, they are shed from the brown capsules like pepper from a shaker. Looking much like fallen leaves, these liverworts (right) have raised what appear like little green umbrellas. These structures produce gametes.

Bryophytes vary in appearance. Some look like miniature evergreen trees; others, like the softest green carpet; still others, like leaves of a higher plant lying on the ground. Regardless of variations in appearance, almost all bryophytes are less than a few centimeters tall.

The moss plants you might observe on a walk through the woods are actually clumps of haploid moss gametophytes growing close together. Each moss plant has a thin upright shoot that looks like a stem with tiny leaves. Because the plant does not have tubes that conduct water and other substances, however, these are not true leaves and stems. From the base of the shoot grow a number of thin branches called **rhizoids** that penetrate into the ground and act like roots to securely anchor the plant.

The odd little plants called liverworts are bryophytes too. These plants are scarcer than mosses and need to live in places that remain wet constantly. Liverwort gametophytes look like flat green leaves growing along the ground. When these plants mature, the gametophytes produce structures that look like tiny green umbrellas. These "umbrellas" carry the structures that produce eggs and sperm.

The gametophytes of hornworts look very much like the gametophytes of liverworts. The hornwort sporophyte, however, differs from the liverwort sporophyte. Instead of looking like a tiny umbrella, the hornwort sporophyte looks like a tiny horn, which is why this plant received its common name.

Physical Characteristics of Bryophytes

Bryophytes are well adapted to life in wet habitats, where they often grow much better than do the higher plants that you will learn about in the next chapter. But outside of wet habitats, bryophytes do not usually grow well because they lack several critical adaptations to life in dry places.

Bryophytes lack the water-conducting tubes that are found in higher plants. In bryophytes, water passes from cell to cell by osmosis and by means of surface tension around the stems. These methods of transporting water work well over short distances only. This is one reason bryophytes never grow tall.

Bryophytes lack a protective surface covering to keep water from evaporating from their cells. Because their "leaves" are only one cell thick, the plants lose the water they contain very quickly if the surrounding air is dry.

Bryophytes lack true roots. True roots contain water-conducting tubes that enable a plant to absorb and transport water efficiently. Instead of roots, bryophytes have rhizoids that anchor them in the ground. Rhizoids, however, do not play a major role in the absorption and transport of water and minerals.

Bryophytes have sperm cells that must swim through water to fertilize the eggs. The sperm cells use their flagella to propel themselves. For this reason, bryophytes must live in areas that are wet for at least part of the year. Some bryophytes can survive dry periods, but to do so they must stop growing.

Alternation of Generations in Mosses

The life cycle of the moss *Mnium*, a typical bryophyte, is shown in Figure 21–6 on page 454. At the tips of the gametophytes are reproductive structures similar to those of several species of algae. One structure, the **antheridium** (an-ther-IHD-ee-uhm), produces tiny flagellated sperm cells. Another structure, the **archegonium** (ahr-kuh-GOH-nee-uhm), produces eggs. Unlike the reproductive structures of algae, the reproductive structures of mosses are designed to protect the gametes from drying out. Thus the eggs of mosses have a better chance of surviving during dry conditions.

Some species of mosses have both male and female reproductive organs on one gametophyte; other species have male and female reproductive structures on separate gametophytes. Mosses can reproduce sexually only when standing water is present. Sperm can swim to the archegonium only when the gametophytes are covered with rainwater or dew. When a sperm swims to an egg, syngamy (the fusing of gametes) occurs and a diploid zygote is produced.

When the zygote germinates, or begins to grow, it produces a diploid sporophyte. As it grows, the sporophyte is supplied with water and nutrients by the gametophyte. Moss sporophytes cannot live independent of the gametophyte from which they grow. This is one way in which bryophytes differ from all other land plants. The mature sporophyte is composed of a "foot" that remains stuck in the gametophyte—a long stalk—and a capsule that looks like a salt shaker. Inside the capsule,

Figure 21–5 This illustration (top) shows the parts of a typical moss plant. How many of these parts can you locate in the photograph? ❶

ANNOTATION KEY

❶ Students should note the capsule and stalk, or the sporophyte portion. (Interpreting photographs)

BACKGROUND INFORMATION
SPHAGNUM MOSS

The moss genus *Sphagnum*, also called peat moss, grows especially well in cold freshwater ponds, where it may form floating green mats that cover the surface. *Sphagnum* moss species make the water around them very acidic, creating special environments called sphagnum bogs. These bogs are inhabited by many rare plants, including orchids and carnivorous plants such as pitcher plants and sundews.

Together with the roots of other plants, the *Sphagnum* mat is sometimes strong enough to support the weight of a full-grown human. If you jump up and down on one of these mats, you can see the surface tremble, as fronts spread out from your location like waves. Such places are often called quaking bogs for this reason.

Content Development
Bryophytes are well adapted for life in a wet terrestrial habitat. Since they do not have water-conducting tubes, they do not grow very tall. They lack a protective surface and lose water to the air easily. Instead of roots, bryophytes have rhizoids. These structures do not absorb water but only anchor the plant in the ground. Their sperm cells require water in order to be able to swim to the egg cells.

Focus/Motivation
Have students examine the living or preserved moss plants once again. The green leaflike structures make up the moss gametophyte plant, which is haploid (N). The gametophyte will produce either egg cells or sperm cells at its top. Now direct their attention to the brownish stalked structure growing on top of some of the moss plants. This is the diploid (2N) sporophyte plant, which lives directly on the haploid gametophyte plant. After fertilization occurs, the zygote will grow to become the sporophyte plant. Meiosis occurs in the capsule on the top. When the capsule ripens, it will release haploid spores.

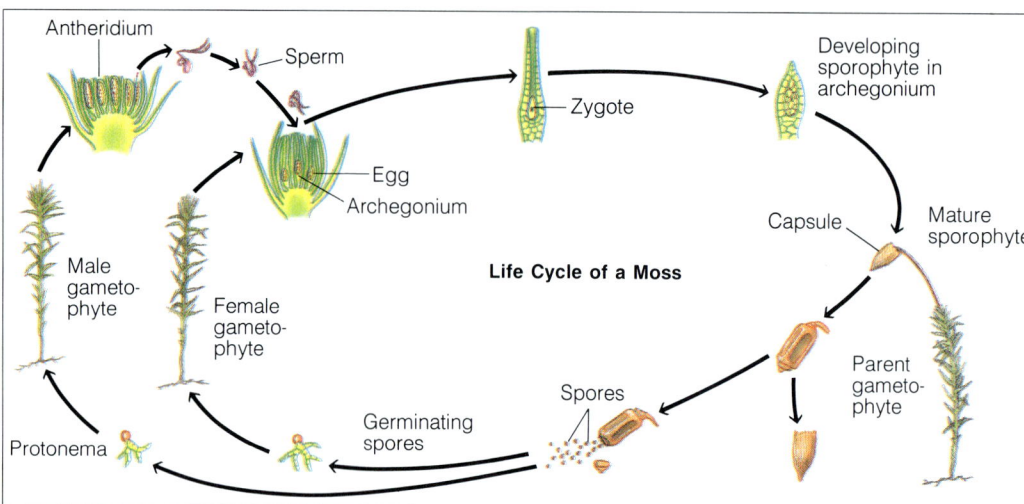

Figure 21–6 Moss plants are usually short and grow close to the ground. These tiny plants need a supply of standing water in order for sperm to swim to and fertilize an egg.

haploid spores are produced by meiosis. When the capsule ripens, special pores—and in some cases the whole top of the capsule—open. The spores are shaken out, to be carried off by wind and water.

If a spore lands in a moist place, it germinates and grows into a mass of tangled green filaments called a **protonema**. Moss protonemas look remarkably like filamentous green algae. (This resemblance is evidence that mosses evolved from either ancient green algae or from an ancestor common to both mosses and algae.) As the protonema grows, it forms rhizoids that grow into the ground and shoots that grow into the air. These shoots develop into the familiar moss gametophytes, and the cycle continues.

We can summarize the life cycle of mosses as follows:

1. The haploid gametophyte is the dominant, obvious stage. It is in fact the stage commonly thought of as a moss plant.
2. Standing water is needed for sperm to swim to and fertilize eggs.
3. The diploid sporophyte is small and can grow only with nourishment provided by the gametophyte.

21–2 SECTION REVIEW

1. List the characteristics of each bryophyte group.
2. What are two adaptations that enable bryophytes to survive on land?
3. **Critical Thinking—Relating Facts** What is an archegonium? An antheridium? Why are these structures important in the life cycle of a moss?

21-3 The Ferns and the First Vascular Plants

Guide For Reading

- Why is vascular tissue important to land plants?
- What are the characteristics of club mosses, horsetails, and ferns?
- What are the stages in the life cycle of ferns?

Remember that although bryophytes live on land, they depend upon an abundant supply of water to survive. **The members of the phylum Tracheophyta are "true" land plants because they have evolved ways of freeing themselves from dependence on wet environments.**

Among the most important adaptations of tracheophytes are specialized tissues called **vascular tissues**. Vascular tissues transport water and the products of photosynthesis throughout the plant. There are two types of vascular tissue: **xylem** and **phloem**. Xylem tissue is associated with the movement of water from the roots to all parts of the plant. Phloem tissue is responsible for the transport of nutrients and the products of photosynthesis.

One important type of cell present in vascular tissue is the **tracheid**. Tracheid cells carry water from roots in the soil to leaves in the air. Thus they are the most important type of cells in xylem tissue. Tracheid cells have thick, strong cell walls that strengthen stems and help plants stand up against the pull of gravity. All plants in the phylum Tracheophyta have tracheids; in fact, the phylum is named after this type of cell.

The other kind of vascular tissue, phloem tissue, carries important nutrients and the products of photosynthesis from place to place within a plant. Both xylem and phloem tissue will be discussed in more detail in Chapters 22 and 23.

With the development of vascular tissue, tracheophytes have evolved true roots and true leaves. True roots have vascular tissues gathered in a central area of the root that is called the **vascular cylinder**. True leaves are photosynthetic organs

Figure 21-7 Ferns are able to grow much taller than mosses because they have an internal system of water-conducting tubes. This tree fern is growing in a rain forest in New Caledonia. Although tree ferns were once widespread, today they are limited to the tropical areas on Earth.

ECOLOGY NOTE

FERNS AS FRESHWATER PESTS

The freshwater fern *Ceratopteris* is found in Brazil and other countries of South and Central America. *Ceratopteris* floats on the water's surface, and its prolific growth has literally blanketed many lakes, reservoirs, and canals. Though intriguing to look at, *Ceratopteris* clogs water pipes, disrupts navigation, and chokes off other aquatic life.

Have interested students research the troubles caused by this fern or other aquatic plants.

• **What structures are present in the ferns that were not found in the bryophytes?** (Ferns have leaves; each leaf is called a frond.)
• **Look even more closely at one frond and each of the smaller leaflets that make up the frond. What can be found running up the middle of each frond and into each leaflet?** (There is a large vein in the middle of each frond that branches out into each leaflet.)

Content Development

Ferns are members of the phylum Tracheophyta, the true land plants. The members of this phylum have leaves, roots, and veins—or vascular tissue—running throughout the plant. Specialized tissue is present for transporting water and nutrients. Xylem tissue moves water from the roots to all parts of the plant through specialized cells known as tracheids. Phloem tissue carries nutrients throughout the plant.

The leaves are the photosynthetic organs of the fern plant. Each leaf has veins made up of vascular tissue. To prevent water loss to the air, each leaf has a waxy covering called a cuticle.

Closure

If liverwort plants are available, have the students examine them. The leafy green structures are the gametophyte plants. Any umbrellalike structures growing from them are the sporophyte plants.

If hornwort plants are available, ask the students to compare them to the mosses and liverworts. The leafy structures are the gametophytes, and the sporophyte resembles a tiny horn growing up from the gametophyte.

TEACHING STRATEGY 21-3

Focus/Motivation

Display a fern plant (or several of different species, if available) for students to observe. Ask the students to examine the fronds of the fern closely.

455

TEACHING SUPPORT

Laboratory Manual
- Special Characteristics of Ferns, p. 261

Teaching Resources
- Activity: Distinguishing Between Mosses and Ferns, p. 3

ESL STRATEGY

Have students fill in the blanks in the following paragraph. Plants of the phylum _____ are the first true land plants. These plants have _____ _____ for internal transportation. The cells that carry water from roots to leaves are called _____, and they are found in the _____. The tissue called _____ transports nutrients and other compounds. These plants' leaves have a _____, a waxy covering that prevents loss of _____.

(Answers: Tracheophyta, vascular tissues, tracheids, xylem, phloem, cuticle, water)

Figure 21-8 *The* Psilotum *plant is commonly called the whisk fern. Scientists believe that the first true land plants resembled this organism.*

that contain one or more bundles of vascular tissue gathered into **veins**. The leaves of tracheophytes usually have a waxy covering called a **cuticle** that helps prevent water loss by evaporation. These structures will also be discussed in more detail in Chapters 22 and 23. But it is important that you have an understanding of these terms as we begin a discussion of early vascular plants.

The First Vascular Plants

Fossils of psilophytes, the first vascular plants, were first found early in this century. These small creeping plants had primitive xylem and phloem tissues, but they lacked true roots and true leaves. Although most botanists think the psilophytes are extinct, some believe that two species of living plants (classified as ferns) are actually living psilophytes. At present, neither group of botanists can prove conclusively that they are correct. But both groups wonder why no psilophyte fossils more recent than the Devonian Period (400 million years ago) have been found.

Club Mosses and Horsetails

The club mosses (lycophytes) and horsetails (sphenophytes) alive today are the only living descendants of large and ancient groups of land plants. The first of these primitive tracheophytes appeared more than 400 million years ago. Over the next 100 to 200 million years, many more species evolved. Some ancient lycophytes and sphenophytes grew into huge trees—up to 40 meters tall. The Earth's very first forests were made up of vast numbers of these plants. At one point in the Earth's history, the entire area of what is now Pennsylvania was covered with a dense tropical jungle of these plants. It is the fossilized remains of these primitive tracheophytes that were transformed into Pennsylvania's huge beds of coal.

Over time, however, the climate of the Earth changed. (Pennsylvania does not have a tropical climate today!) For some reason, these primitive plants could not compete with new types of plants that evolved with the changing climate. Forests of lycophytes and sphenophytes were replaced by forests of entirely new plants. The few species of lycophytes and sphenophytes alive today are relatively small plants that live in moist woodlands and near stream beds and marshes.

Lycopodium (ligh-koh-POH-dee-uhm), the common club moss, looks like a miniature pine tree about 9 centimeters tall. Another common name for *Lycopodium* is "ground pine." *Lycopodium* has small scalelike leaves that cling to the stems.

The only living genus of sphenophytes is *Equisetum* (ehk-wih-SEET-uhm), a plant that grows about 1 meter tall. *Equisetum* is commonly called horsetail or scouring rush because its stems contain crystals of silica, which are quite abrasive. During Colonial times, horsetails were commonly used to scrape,

21-3 (continued)

Focus/Motivation
Display samples (living or preserved) of the club moss *Lycopodium* and the horsetail *Equisetum* for the students to examine.

- Why was *Lycopodium* given the common name ground pine? (It looks like a miniature evergreen tree and has structures that look like elongated pine cones. In some species, the bristly branches resemble those of a Norfolk pine.)

- *Lycopodium* belongs to a group of plants known as the club mosses. What may have been the origin of that name? (Each plant appears to have cones at the tips of each branch, making each branch resemble a club.)
- Where are leaves located on

Figure 21–9 This Lycopodium *plant (left), which resembles a small evergreen tree, is better known by its common name ground pine. It is often used as a holiday decoration. Horsetails (right) incorporate silica, the main component of sand, in their stems. Silica gives the stems a rough texture, which is why these plants were once used as a scouring material for cleaning pots and pans.*

or scour, pots and pans. If you should some day set up camp near some *Equisetum*, you will know you can use this plant to clean your pots and pans. Like the lycophytes, horsetails have true leaves, stems, and roots. The leaves of horsetails are arranged in whorls at joints along the stems.

Physical Characteristics of Ferns

Ferns probably evolved about 400 million years ago, at about the same time as the lycophytes and sphenophytes. Ferns were an important part of the lycophyte forests that covered the ancient Earth. Ferns, however, have been more successful at competing with other plants that have appeared during the Earth's long history. Today more than 11,000 species of ferns are still alive!

In many respects, ferns are well-developed tracheophytes. They have true vascular tissues, strong roots, creeping or underground stems called **rhizomes**, and large leaves called **fronds**. Ferns that commonly grow in the United States range in height from a few centimeters to about one meter.

Ferns are most abundant in wet, or at least seasonally wet, habitats around the world. Ferns grow throughout the United States, but they grow best in the rain forests of the Pacific Northwest, and in wet tropical areas. In tropical forests, some species of ferns grow as large as small trees.

Figure 21–10 The leaves of ferns are covered by a waxy coating that prevents water loss. The waxy coating also causes drops of water to bead up on the leaf surface, much like wax on a car causes water to form beads.

ESL STRATEGY

Using Figure 21–12 as a model, have students make up flashcards for the stages of the fern's life cycle. In pairs or small groups, have students name and illustrate each stage of the cycle on ten large index cards. When all have finished, have them stack their flashcards in random order, exchange cards with other groups, and arrange the exchanged cards in the proper order for the fern's life cycle.

21–3 (continued)

Focus/Motivation
Students may try to grow their own fern gametophyte plants. The instructions for this activity can be found in the Chapter Preview on page 448D.

Content Development
Like mosses, ferns depend on a watery environment for reproduction, and they need standing water for sperm to swim to eggs.
• **In what ways are ferns better adapted to life on land than mosses?** (Ferns can grow in drier places because they have vascular tissue, roots, and cuticle.)

Focus/Motivation
Have students examine the undersides of fronds from a fern plant that produces sori. Have students touch the sori and examine the spores that are released.

Content Development
The plant that we usually associate with the name *fern* is the diploid (2N) sporophyte plant. It is the function of the sporophyte plant to produce spores by meiosis.

When the spores are released, they will develop into a monoploid (N) gametophyte plant. The small green heart-shaped fern gametophyte plant is known as a prothallium. It produces antheridia and archegonia on its lower side. When the ground is covered with a film of water, the sperm that are produced in the antheridia are able to swim to the eggs produced in the archegonia, and fertilization occurs. The diploid zygote will begin to develop into the sporophyte plant, completing the cycle.

Alternation of Generations in Ferns

Like the life cycles of all other plants, those of ferns involve alternation of generations. The plants that are recognizable as ferns are the diploid sporophytes. Because of their well-developed vascular tissues, these sporophytes can grow in drier places than can bryophyte sporophytes. But sexual reproduction in ferns still depends upon the presence of standing water for sperm to swim to eggs.

Fern sporophytes produce haploid spores on the underside of their fronds. Spores are produced in tiny containers called **sporangia**. Sporangia do not occur individually but are grouped into large clusters called **sori** (singular: sorus).

When spores are ripe, they are released from the sporangia and may be carried by wind and water over long distances. If environmental conditions are right for the spores to germinate, they develop into haploid gametophytes. The gametophyte first grows a set of rootlike rhizoids. Then it flattens out into a thin heart-shaped green structure called a **prothallium** (proh-THAL-ee-uhm). Antheridia and archegonia, which produce gametes, are found on the underside of the prothallium. If there is a moist woods near your home where you have seen ferns growing, take a close look at the ground near the base of the plants. See if you can spot the tiny prothallia among the mature plants.

When the antheridia are mature, sperm are released. Fertilization can take place when the ground and the prothallia are covered with a thin film of water. As in bryophytes, fern sperm have to swim to the archegonia to fertilize the eggs.

The diploid zygote produced by fertilization immediately begins to grow into a new sporophyte plant. The developing sporophyte quickly puts out its first fronds and then its creeping stems, or rhizomes. As the sporophyte grows, the gametophyte withers away. Fern sporophytes often live for many years. In some species, the fronds produced in the spring die in

Figure 21–11 Clusters of sporangia form on the underside of fern leaves. As you can see, sporangia can form many varied, beautiful patterns.

SECTION REVIEW 21–3

1. Vascular tissue is xylem and phloem tissue. In the vascular tissue of plants, water and other

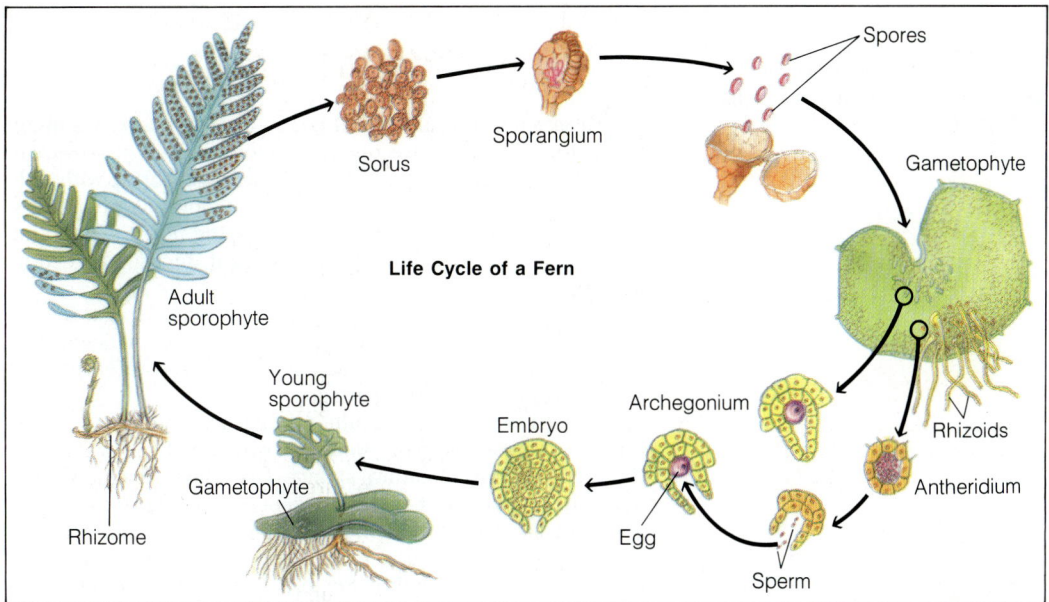

the fall, but the rhizomes live through the winter and sprout again the following spring.

We can summarize the life cycle of ferns as follows:
1. Ferns employ alternation of generations, but in ferns the diploid sporophyte is the dominant, obvious stage. The gametophyte is tiny and lives for only a short time.
2. The sporophyte is a well-developed land plant with true vascular tissues. The gametophyte lacks vascular tissues, is very tiny and delicate, and can grow only in moist areas.
3. Sexual reproduction in ferns still requires water because sperm from the antheridia must swim to the archegonia to fertilize eggs.

It should be obvious to you that ferns still need abundant water to reproduce sexually. In the next chapter you will see how the evolution of the seed has freed the higher tracheophytes from this dependence on water.

21–3 SECTION REVIEW

1. What is vascular tissue?
2. How are ferns adapted to life on land?
3. What generation in ferns is most obvious? What substance is needed by ferns to reproduce sexually?
4. **Critical Thinking—Applying Concepts** Even though ferns survive under many of the same environmental conditions as mosses, ferns are able to grow much larger than mosses. Why is this so?

Figure 21–12 In the life cycle of a typical fern, the heart-shaped gametophyte plant (bottom) is small and requires dampness for the sperm it produces to fertilize an egg. The young sporophyte grows from the gametophyte plant. In ferns, the sporophyte plant is large and obvious.

SCIENCE, TECHNOLOGY, AND SOCIETY
CONNECTION

Ferns, Bacteria, and Fertilizer
• **Why do green plants need nitrogen?** (To make amino acids, the building blocks of proteins.)

The relationship between the aquatic fern *Azolla* and the blue-green bacteria *Anabaena* illustrates two interesting ecological concepts. First, it demonstrates the importance of nitrogen fixation, the initial step in the nitrogen cycle. Although nitrogen is by far the most abundant substance in the atmosphere, it is in a gaseous form that plants cannot use. What certain bacteria do is oxidize gaseous nitrogen into nitrates, which plants can readily absorb and use. This oxidation of nitrogen gas is called nitrogen fixation, and blue-green bacteria are among the few organisms that perform this vital function.

Second, the *Azolla* and *Anabaena* illustrate mutualism, an important symbiotic relationship. The bacteria fixes nitrogen for the fern, and the fern provides a good growing environment for the bacteria. In mutualistic relationships such as this one, both organisms benefit from their partnership.

TEACHING STRATEGY 21-4

Focus/Motivation
Display dried sphagnum moss for students to observe. Have a student weigh the sample on a triple-beam balance.

Sphagnum moss is a natural sponge and can absorb many times its own weight in water. Soak the sample in a beaker of water for several minutes. Allow the sample to drain. Have the student weigh the sample once again.

Guide For Reading
■ To which types of environments are mosses and ferns well adapted?
■ How are mosses and ferns useful to people?

21-4 Where Mosses and Ferns Fit into the World

Mosses and ferns are well adapted to certain types of environments. Mosses are quite common in areas that remain damp for much of the year. Ferns, which can thrive with only little light, are often found living in the shadows of forest trees, where direct sunlight hardly penetrates the forest's leafy umbrella. But wherever conditions are right, mosses and ferns grow abundantly.

Both mosses and ferns are important plants to gardeners. Several kinds of mosses are grown in gardens for decorative purposes. For example, mosses are often used to carpet the ground in Japanese-style gardens.

Mosses are frequently added to garden soil. Dried sphagnum moss absorbs many times its own weight in water and thus acts as a sort of natural sponge. Over time sphagnum moss decomposes into peat moss. Gardeners add peat moss to the soil because it improves the soil's ability to retain water. In addition, peat moss has a low pH, so when added to the soil it increases the soil's acidity. Some plants, such as azaleas, will grow well only if planted in acid soil. Sphagnum peat moss is also used to add organic material to sandy soil.

SCIENCE, TECHNOLOGY, AND SOCIETY CONNECTION

Ferns, Bacteria, and Agriculture

Rice farmers in Southeast Asia have learned to make good use of the floating water fern *Azolla*. In nature, *Azolla* does not grow alone but rather in association with *Anabaena*, a blue-green bacteria. Colonies of *Anabaena* live within tiny cavities in the flat *Azolla* fronds, where they grow much more rapidly than they do when they live in water.

Like several other monerans, *Anabaena* can take nitrogen out of the air and "fix" it into a form that other plants can use. Rice farmers plant the fern and its accompanying bacteria along with rice in their paddies. The nitrogen that the blue-green bacteria fix makes the use of expensive chemical fertilizers unnecessary. In fact, the nitrogen fixed by *Azolla* can increase the rice paddy yield by 150 percent.

Azolla is so important to the people of Southeast Asia that a temple in Vietnam has been dedicated to this useful plant.

The water fern Azolla *is often grown in rice paddies. This plant is able to convert nitrogen in the air into fertilizer that can be used by rice plants.*

• **Compare the original dry weight of the sphagnum moss to the weight of the sample after it was soaked in water. How much water was it able to absorb in comparison to its original weight?** (Answers will vary according to the situation.)

Content Development
Describe the many uses for sphagnum moss: natural sponge, flavoring, fertilizer, and fuel.

Focus/Motivation
Locate some fiddlehead fronds on an actively growing fern plant. Permit students to have a closer look and examine the plant.

SECTION REVIEW 21-4

1. You would expect to find mosses growing in areas that remain damp. In most cases

Figure 21-13 Maidenhair ferns are one of the more beautiful ferns (left). The leaves are produced on thin stems that quiver in even the most gentle breeze. Maidenhair ferns are frequently grown in gardens or as a houseplant. Mosses often carpet the floor of Japanese gardens (right). Although moss plants appear quite similar to one another at first glance, different species vary in color and shape.

Many different varieties of ferns are planted and cultivated by gardeners for their ornamental value. Although they do not produce flowers, fern fronds can be quite beautiful.

At one time, mosses were ground up and used by Native Americans to treat burns and bruises. Aside from its many uses in gardening, sphagnum moss is also used to add flavor to Scotch whisky. The moss is burned by brewers. The smoke produced by the burning moss gives Scotch whisky its characteristic "smoky" flavor.

Certain species of moss form peat. Peat forms after mosses die and are subjected to enormous pressure for long periods of time. Peat is actually a kind of coal that is cut from the ground and burned as a fuel.

A few types of ferns are eaten by humans. In the early spring, fern fronds emerge from the ground. When they are just beginning to grow, the fronds look very much like the top part of a violin. For this reason, the fern fronds are called fiddleheads. If picked when they are young and cooked when they are fresh, fiddlehead greens are considered a delicacy. Unless you are certain which ferns are edible, it is best to purchase fiddleheads in a supermarket or at a vegetable stand.

21-4 SECTION REVIEW

1. In what kinds of areas would you expect to find mosses growing in nature?
2. What characteristics of mosses make them useful?
3. **Connection—You and Your World** What are two ways in which ferns are used by people?

Figure 21-14 These unfurling fern fronds resemble the top of a violin and are thus called fiddleheads. At this stage they are quite tender and can be eaten. However, you should not eat wild plants.

LABORATORY INVESTIGATION

COMPARING MOSSES AND FERNS

Before the Lab
1. Decide whether students are to work individually or in groups of two.
2. Gather all materials at least one day prior to the investigation. Obtain at least one living fern and a large cluster of moss plants for students' use.

Pre-Lab Discussion
Have students read the complete laboratory Procedure. Discuss the purpose of the investigation.
- **What structures do ferns possess that mosses do not have?** (Leaves, roots, stem, vascular tissue, cuticle.)
- **What is the purpose of determining whether a plant has a shiny surface?** (A shiny surface indicates that the plant has a waxy cuticle.)
- **What is the purpose of bending the plant back and forth?** (To see if it has any supporting tissue to hold its leaves upright.)

Skills Development
Students will use the following skills while completing this investigation.
1. Manipulative
2. Observing
3. Comparing
4. Diagramming
5. Measuring
6. Recording
7. Inferring

Safety Tips
Glass slides should be checked before use to determine that they have no rough or jagged edges. Caution students to use care in handling the glass slides. Use plastic coverslips.

Remind students to use only the low-power objective in examining their slides.

LABORATORY INVESTIGATION: COMPARING MOSSES AND FERNS

PROBLEM
Are ferns better adapted to live on land than mosses?

MATERIALS (per group)

2 microscope slides	metric ruler
2 coverslips	scissors
medicine dropper	fern plant
microscope	moss plant

PROCEDURE

1. Examine a fern frond carefully. Notice whether the top surface of the frond is shiny or dull. Notice whether the bottom surface of the frond is shiny or dull. Draw a diagram of the frond on a separate piece of paper.
2. Bend the fern frond gently back and forth. Notice whether the frond bends easily.
3. Use a ruler to measure the length and width of the frond. Record your observations.
4. Remove a few moss plants from the clump of moss provided by your teacher. Bend one plant gently back and forth. Notice whether the moss bends easily.
5. Use a ruler to measure the length and width of a moss plant.
6. Cut a small piece (about 5 mm long) from the tip of one of the leaflets of the fern frond. Place it face down on a clean microscope slide. Use the medicine dropper to place a drop of water on top of the piece of fern. Cover the fern with a coverslip.
7. Examine the fern leaflet under the low-power objective of your microscope. Focus on the midline of the leaflet. Notice whether there are veins in the leaflet. Draw a diagram of what you observe.
8. Remove a single moss plant from the clump of moss. Place the plant in the center of another clean microscope slide. Use a medicine dropper to place a drop of water on top of the moss plant. Cover the moss with a coverslip.
9. Examine the moss "leaflet" under the low-power objective of your microscope. Draw a diagram of what you observe.

OBSERVATIONS
1. What are the dimensions of the fern frond? Of the moss plant? Which plant grows larger?
2. Which surface of the fern frond is shinier?
3. Which is firmer, the fern frond or the moss plant?
4. Which plant has veins in its leaves?

ANALYSIS AND CONCLUSIONS
1. Why is the fern able to grow larger than the moss?
2. How can you explain the firmness of the fern frond?
3. What do you think makes the surface of the fern frond shiny? How is this an adaptation to life on land?
4. Which of the plants shows adaptations that make it better able to survive on land? Explain.

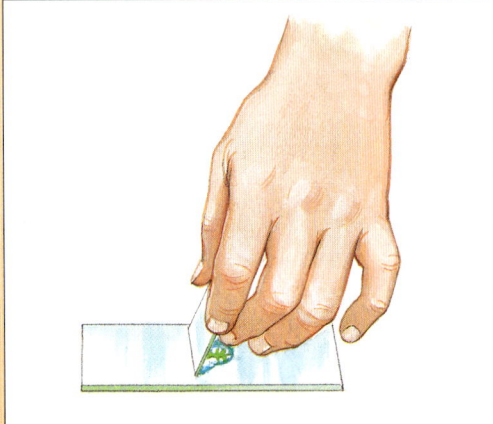

Teaching Strategy
1. Station yourself by the plant material from time to time. Students usually take too large a sample for microscopic observation and may attempt to fit a clump of moss under a coverslip, instead of a single moss plant.
2. When circulating through the room, make sure students use only the low-power objectives for their observations.

Observations
1. Answers will vary. A fern frond might have dimensions of 15 centimeters by 10 centimeters. A sprig of moss might have dimensions of 1 centimeter by 0.1 centimeter.
2. Upper surface is shinier.
3. The fern frond is firmer than a sprig of moss.
4. The fern has veins.

STUDENT STUDY GUIDE

SUMMARIZING THE CONCEPTS

The key concepts in each section of this chapter are listed below to help you review the chapter content. Make sure you understand each concept and its relationship to other concepts and to the theme of this chapter.

21-1 Plants Invade the Land

- At least two separate groups of algaelike land plants evolved between 450 and 500 million years ago. One group developed into the Bryophyta; the other group developed into the Tracheophyta.
- Land plants have certain adaptations that enable them to live in a dry environment. These adaptations prevent water loss from the plant; expose the plant to the sunlight; take up water and nutrients from the soil; and move water and nutrients, along with the products of photosynthesis, throughout the plant. Special adaptations have evolved to permit plants to reproduce in land environments.

21-2 The Mosses, Liverworts, and Hornworts

- Like the algaelike organisms from which they evolved, mosses, liverworts, and hornworts have a complex life cycle that involves an alternation of generations between a haploid gametophyte and a diploid sporophyte.

- Bryophytes lack the water-conducting tubes that are found in higher plants. Without these tubes, bryophytes can never grow tall.

21-3 The Ferns and the First Vascular Plants

- The ferns were among the first land plants to develop vascular tissue: xylem and phloem. Vascular tissue is a system of tubes that move water and other materials throughout the plant. A well-developed vascular system enables ferns to grow tall. Some ferns may even grow as tall as a small tree.
- Ferns have true roots, stems, and leaves. They also have a thick waxy covering called a cuticle. The cuticle helps prevent water loss from the cells.

21-4 Where Mosses and Ferns Fit into the World

- Because mosses and ferns are well adapted to life in certain environments, they are grown in many gardens.

REVIEWING KEY TERMS

Vocabulary terms are important to your understanding of biology. The key terms listed below are those you should be especially familiar with. Review these terms and their meanings. Then use each term in a complete sentence. If you are not sure of a term's meaning, return to the appropriate section and review its definition.

21-2 The Mosses, Liverworts, and Hornworts
- rhizoid
- antheridium
- archegonium
- protonema

21-3 The Ferns and the First Vascular Plants
- vascular tissue
- xylem
- phloem
- tracheid
- vascular cylinder
- vein
- cuticle
- rhizome
- frond
- sporangium
- sorus
- prothallium

Analysis and Conclusions

1. Ferns have vascular tissue to help deliver water over greater distances and to support larger stems.
2. Vascular tissue (xylem and phloem) provide strength and firmness to the fern.
3. A waxy coating makes the fern frond shiny and helps to prevent evaporation of water.
4. Ferns are better adapted to the dry conditions on land. They have vascular tissue to help transport water and a cuticle to prevent water loss.

Going Further: Enrichment

1. This investigation can be extended to include the sporangia of mosses and ferns. Under the low-power objectives of their microscopes, have students examine the sporophyte moss plant and spores from a ripe moss sporangium, as well as sporangia from a fern's sori.
2. Several different species of mosses can be examined. In some mosses, such as *Funaria*, the leaflike structures are only one cell thick except along the midrib. In other species, such as *Polytrichum*, the leaflike structures are several cells thick. Compare the thickness of these plants with the thickness of a fern plant.

CHAPTER REVIEW

CONTENT REVIEW

Multiple Choice
1. a 2. b 3. d 4. a
5. c 6. c 7. b 8. a

True or False
1. F bryophytes
2. F gametophytes
3. T
4. F fronds
5. T
6. T
7. F few
8. F eggs

Word Relationships
A.
1. structures involved with spore production; sperm is a gamete
2. parts of the vascular system in plants; cuticle is a waterproof coating
3. haploid cells; zygote is a diploid cell. Alternatively, structures involved with sexual reproduction; spore is a cell that develops into a gametophyte
4. haploid structures or terms associated with gametophyte generation; sporophyte is the diploid, spore-producing stage in a plant life cycle
5. parts of a fern; rhizoid is a rootlike structure in a bryophyte

B.
6. antheridium
7. sporophyte
8. spores

CONCEPT MASTERY

1. Standing water is necessary for sperm to swim in, to fertilize the eggs.
2. The moss sporophyte is not very obvious. It resembles a small pepper shaker at the end of a thin stalk. It grows attached to the moss gametophyte plant.
3. After the spore begins to grow, it produces a small heart-shaped plant called the prothallium, which represents the fern gametophyte generation. Archegonia and antheridia are produced on the prothallium. Sperm produced in the antheridia swim to an archegonium, where they fertilize the eggs. When the fertilized egg begins to grow, it develops into a sporophyte plant.
4. Mosses and ferns are both grown as houseplants and garden plants. Sphagnum moss is added to soil to improve the soil's quality. The fuel peat was originally moss. Some ferns are edible.
5. Ferns have vascular tissue to conduct water and other materials. Ferns have a waxy layer, the cuticle, that covers their leaves and prevents water loss.
6. Eggs and sperm are produced

CHAPTER REVIEW

CONTENT REVIEW

Multiple Choice

Choose the letter of the answer that best completes each statement.

1. All of the following plants are bryophytes except
   a. ferns.   c. liverworts.
   b. mosses. d. hornworts.
2. Fern leaves are called
   a. sori.   c. rhizomes.
   b. fronds. d. spores.
3. The most obvious stage of a moss is the
   a. sporophyte. c. protonema.
   b. parent.   d. gametophyte.
4. Mosses are used for all of the following except
   a. food.   c. soil additive.
   b. garden plants. d. fuel.
5. Mosses do not grow in
   a. swamps.   c. deserts.
   b. marshes. d. rain forests.
6. Each of the following can be found on a fern sporophyte except a
   a. sorus.   c. prothallium.
   b. frond.   d. rhizome.
7. The moss sporophyte lives
   a. a solitary life.
   b. attached to the gametophyte.
   c. attached to a spore.
   d. attached to a leaf.
8. The waxy covering on the leaves of a tracheophyte is called the
   a. cuticle. c. xylem.
   b. sori.   d. phloem.

True or False

Determine whether each statement is true or false. If it is true, write "true." If it is false, change the underlined word or words to make the statement true.

1. Mosses are <u>tracheophytes</u>.
2. Moss <u>sporophytes</u> are the most obvious stage of the moss life cycle.
3. Fern <u>gametophytes</u> are small heart-shaped structures.
4. Fern leaves are called <u>sori</u>.
5. <u>Sexual reproduction</u> in ferns depends on the presence of water.
6. <u>Xylem</u> tissue conducts water in a plant stem.
7. There are <u>many</u> fossils of early land plants.
8. In mosses, the archegonium produces <u>sperm</u>.

Word Relationships

A. *In each of the following sets of terms, three of the terms are related. One term does not belong. Determine the characteristic common to three of the terms and then identify the term that does not belong.*

1. sori, sporangium, spore, sperm
2. xylem, phloem, tracheids, cuticle
3. sperm, egg, zygote, spore
4. antheridium, archegonium, gametophyte, sporophyte
5. rhizoid, frond, rhizome, vascular cylinder

B. *Give the vocabulary word whose meaning is opposite that of the following words.*

6. archegonium   7. gametophyte   8. gametes

CONCEPT MASTERY

Use your understanding of the concepts developed in the chapter to answer each of the following in a brief paragraph.

1. Why is water needed for reproduction to occur in mosses?
2. Describe the appearance of the moss sporophyte.
3. Briefly describe sexual reproduction in a fern.
4. What are two uses of mosses and ferns?
5. Mosses must live in areas that remain damp for much of the time. Ferns can live in drier environments. What adaptations do ferns show that enable them to survive in areas that would not support moss plants?
6. Briefly summarize the life cycle of a typical moss plant.

CRITICAL AND CREATIVE THINKING

Discuss each of the following in a brief paragraph.

1. **Applying concepts** Moss plants are small. Ferns can grow as tall as a small tree. Explain why this is so.
2. **Relating concepts** Suppose you wanted to grow a garden of mosses in your backyard. What kinds of conditions would you have to provide to make these plants grow well?
3. **Applying concepts** What stage in a fern's life cycle would require more water to survive? Why?
4. **Identifying patterns** This photograph shows the structure of a tracheophyte. What structure is it? To what kind of plant does this structure belong? Is this a part of a sporophyte or a gametophyte plant?
5. **Making predictions** A friend of yours lives in a desert area of New Mexico. She wants to grow a garden of mosses. Is this a good idea? What will probably happen to her garden?
6. **Using the writing process** Imagine that you are a moss plant living in the Olympic Forest in Washington State. Every day, moist fogs roll in from the Pacific Ocean. One day the prevailing winds that blow from the west abruptly change direction. Now the winds blow from the east. Write a brief autobiography that describes your life before the winds changed direction, explaining the transformation that would occur in you and your forest home as a result.

465

---

in structures at the tips of a gametophyte moss plant. A sperm swims to and fertilizes an egg. The fertilized egg begins to grow, producing a sporophyte plant, which grows attached to the gametophyte plant. Spores are produced by the sporophyte plant in a capsule. When the spores ripen, they are shed. If they fall to a suitable place on moist ground, the spores begin to grow, producing the generation of gametophyte plants.

CRITICAL AND CREATIVE THINKING

1. Ferns can grow tall because they have a vascular system that moves water and other materials around the plant. The vascular tissue also helps to support the stem. Ferns also have true roots that are able to take in water and anchor the plant efficiently.
2. You would have to keep the yard moist all of the time. Mosses grow best in a soil that is acidic. They also grow best in shade.
3. The gametophyte generation in a fern is small and delicate. It would dry out quickly without water. The sporophyte fern plant is larger and better able to withstand dry growing conditions.
4. This is an unfurling fern frond. It is the stem and new leaves of a fern. It is part of the sporophyte plant, the larger and more obvious plant in a fern.
5. This is probably not a good idea. The hot desert sun in New Mexico would dry out moss plants rapidly, and they would probably die. Also, without standing water, the moss plants could not reproduce sexually. She might be successful in a greenhouse underneath trees on a shady riverbank. But otherwise, she should plant cactuses—plants that are adapted to life in a desert.
6. Check students' autobiography for clarity of expression. This story should reflect the dramatic changes that would occur as this area dried out. Winds from the east would not carry the moisture-laden air of winds from the west. Students should also point out that they would eventually not be able to reproduce and that the flora of the area would gradually change to plants that are better able to survive in drier conditions. A sad tale, indeed.

CHAPTER 22 Plants with Seeds

Section	Laboratory Investigations and Activities
22–1 Seed Plants—The Spermopsida pages 467–470 THEME: Unity and Diversity	**Teacher Edition** DEMONSTRATION: Grow Your Own, p. 466D **Laboratory Manual** Seed-Bearing Plant Tissues, p. 267
22–2 Evolution of Seed Plants pages 471–475 THEMES: Evolution, Unity and Diversity	**Teacher Edition** DEMONSTRATION: Plant Organs, p. 466D **Laboratory Manual** Comparing Monocots and Dicots, p. 273 **Teaching Resources** HANDS-ON ACTIVITY: Flat Flowers, p. 107
22–3 Coevolution of Flowering Plants and Animals pages 476–481 THEMES: Evolution, Systems and Interactions	**Student Edition** LABORATORY INVESTIGATION: Comparing Ripe and Unripe Fruits, p. 482
Chapter Review pages 483–485	

Teacher Resources

Books

Bewley, J. Derek, and Michael Black, eds. *Seeds: Physiology of Development and Germination.* Plenum Publishing Corp., 1985.

Britton, Nathaniel L., and Addison Brown. *An Illustrated Flora of the Northern United States and Canada*, 3 vols. Dover, 1970.

Essau, Katherine. *Anatomy of Seed Plants*, 2nd ed. John Wiley and Sons, 1977.

Heywood, V. H., ed. *Flowering Plants of the World.* Prentice Hall, 1985.

Murray, David, ed. *Seed Dispersal.* Academic Press, 1987.

Peterson, Roger Tory. *Peterson's First Guide to Wildflowers of Northeastern and Northcentral North America.* Houghton Mifflin, 1986.

CHAPTER PLANNING GUIDE

Other Activities	Media and Technology

Teaching Resources
    ACTIVITY: Plants Designed for Life on Land, p. 3
Study Guide
    Section 22–1, p. 209
	Video/Videodisc
	    Coniferous Forest and Tropical Rain Forest
	Biology Transparencies
	    25: Vascular Tissue
	Biology Media Guide
	    page 18

Study Guide
    Section 22–2, p. 211
	Video/Videodisc
	    Coniferous Forest and Tropical Rain Forest
	Biology Media Guide
	    page 18

Teaching Resources
    VOCABULARY REVIEW: Cross-a-Clue, p. 1
    ACTIVITY: Flower Pollination and Seed Dispersal, p. 5
Study Guide
    Section 22–3, p. 214
	Interactive Videodisc
	    On Dry Land: The Desert Biome
	Video/Videodisc
	    Coniferous Forest and Tropical Rain Forest
	    Seeing Sense
	Biology Media Guide
	    pages 18–19

Teaching Resources
    Chapter Test A, p. 13
    Chapter Test B, p. 17
	Computer Test Bank
	    Chapter Test, p. 235

Audiovisuals

Pollination: The Insect Connection. Video. Carolina Biological Supply.
Seeds. Filmstrip with cassette. Carolina Biological Supply Company.

Pollination Mechanisms. Film or video. Coronet.
Seed Dispersal Mechanisms. Filmstrip with cassette. Ward.

Seed Dispersal. Film or video. Coronet.
Flowers. Film or video. Coronet.
Evergreens. Film or video. Coronet.

CHAPTER 22 Plants with Seeds

CHAPTER OVERVIEW

This chapter contains a great amount of information about seed plants, from their structure to their adaptations for survival. The specific plant structures discussed are those directly related to adaptations made by plants that allow them to live successfully on land. These include roots, stems, leaves, and seeds. Vascular tissue is also discussed as an adaptation to life on land. The unique problems of reproduction on land are presented. Although seed plants, like other plant groups, reproduce by alternation of generations, they must be able to carry out their life cycles without a source of standing water. The gametophyte stage is greatly reduced, while the recognizable plant is always the sporophyte generation. Flowers and cones are briefly discussed, as is the process of pollination.

The history of plant evolution is marked by several examples of adaptive radiation. Each time a beneficial adaptation appeared in a group of plants, that group gave rise to many new species of plants with that adaptation. These organisms then filled previously empty niches and thus survived. As the Earth's environment changed over time, some plant species thrived and others became extinct. The first seed-bearing plants, sometimes called seed ferns, appeared during the Devonian Period. Except for the fact that they reproduced by seeds instead of spores, they closely resembled common ferns of the period. No living seed ferns exist today.

The most ancient surviving seed plants are the gymnosperms. They belong to three classes: Cycadae, Ginkgoae, and Coniferae. The distinguishing characteristic of these plants is the presence of specialized male and female reproductive structures called scales. There is a brief discussion of the cycads and ginkgoes, with more detail given about the conifers.

Angiosperms, the flowering plants, are presented in some detail. This is the most widespread group of land plants, with more than a quarter of a million species living in virtually every type of environment. The two subclasses of angiosperms are introduced. There is a summary of the differences between the monocots and dicots.

The chapter concludes with a discussion of the coevolution of flowering plants and animals. There are a number of examples presented, including interactions between various types of insects, birds, and mammals with flowering plants. Finally, the adaptations of fruits for seed dispersal are explained. The relationship between fruit color and taste with the maturing of the seed is discussed.

22-1 SEED PLANTS— THE SPERMOPSIDA

Section Focus 22-1

The purpose of this section is to introduce students to plants that live and reproduce on land. Both the advantages and disadvantages of land living are discussed. The seed plants, members of the subphylum Spermopsida, have many adaptations that allow them to succeed in land environments. These adaptations are the result of genetic variations that have been perpetuated over time because they increased the fitness of the plants for survival. Some of these adaptations include roots, stems, leaves, and seeds, as well as highly developed vascular tissues. This section includes brief introductions to roots, stems, leaves, and their functions. The two types of vascular tissue—xylem and phloem—are discussed. They are briefly compared in terms of structure and function.

Seed plants, like other plants, show alternation of generations. The sporophyte generation is the plant that is most visible; the gametophyte is very tiny. This tiny gametophyte grows and matures within the structure of the sporophyte called the flower or cone. These specialized reproductive structures eliminate the need for standing water in the reproduction process. The carrying of pollen grains to the female gametophyte is called pollination. The entire male gametophyte, along with the sperm that it produces, is contained within a pollen grain. Seeds are the structures that protect the zygotes of seed plants. After fertilization, the zygote becomes an embryo that grows and matures into a tiny plant inside the seed.

Performance Objectives 22-1

1. Describe the advantages and disadvantages of life on land for plants.
2. Name several adaptations that plants have made to life on land.
3. Explain how the evolution of adaptations in land plants has occurred.
4. Discuss the functions of roots, stems, and leaves.
5. Briefly compare the structure and function of xylem and phloem.
6. Describe the adaptations of plants that allow them to reproduce in an environment free of standing water.

Science Terms 22-1

pollen grain p. 470
pollination p. 470
embryo p. 470
seed coat p. 470

22-2 EVOLUTION OF SEED PLANTS

Section Focus 22-2

The purpose of this section is to describe the evolutionary history of land plants. This history shows several cases of adaptive radiation. Each time a group of plants evolved a useful adaptation, that group gave rise to many new species. This allowed some new species to fill previously empty niches. Over time and as the Earth's environment changed, those species that were better adapted survived, while others became extinct. As areas of the Earth became drier, the spore-bearing plants, like mosses and ferns, found it more difficult to survive, and many species became extinct. The first seed-bearing plants appeared during the Devonian Period. They resembled ferns but showed one significant differ-

CHAPTER PREVIEW

ence: They reproduce by using seeds instead of spores. The seed ferns were rapidly replaced by other plant species, and none exist today.

The most ancient surviving seed plants belong to three classes. In these plants, leaves have evolved into specialized reproductive structures called scales. Scales are grouped in larger structures called cones. Once the seeds are produced, they sit exposed on the scales; thus, the name of the group is gymnosperm, which means naked seed. There were once large numbers of the first class, the cycads—beautiful palmlike plants—but now only nine genera exist in tropical and subtropical parts of the world. The second class, the ginkgoes, were common during the era of the dinosaurs, but today, only a single species remains. This may be the oldest seed plant alive. The third and largest class of gymnosperms are the conifers.

Conifers have made several adaptations that have allowed them to survive. The leaves of these plants are long and thin. They are usually called needles. The group is commonly called evergreens, but in some species this is not accurate because they do shed their needles. Like other gymnosperms, the conifers produce two kinds of cones. Sometimes the male and female cones are on the same plants; other species have separate male and female plants.

Angiosperms are the flowering plants. They all reproduce sexually with flowers in a process involving pollination. The angiosperm seeds are protected inside a structure called a fruit. This is a scientific term and refers to many structures commonly called vegetables as well as those known as fruits. The angiosperms are the most widespread of all land plants, numbering more than a quarter of a million species. They have adapted to life in many different land environments. Angiosperms are divided into two subclasses. The differences between monocots and dicots are explained in the text.

Performance Objectives 22–2

1. Briefly explain the evolutionary history of seed plants.
2. Identify three groups of existing gymnosperms.
3. Discuss the adaptations and reproductive processes of conifers.
4. Explain why angiosperms are the most successful of the land plants.
5. List the differences between monocots and dicots.

Science Terms 22–2

scale   p. 471
gymnosperm   p. 472
pollen cone   p. 473
angiosperm   p. 473
flower   p. 473
fruit   p. 473
monocot   p. 474
dicot   p. 474
cotyledon   p. 474
vascular bundle   p. 474

22–3 COEVOLUTION OF FLOWERING PLANTS AND ANIMALS

Section Focus 22–3

The purpose of this section is to show the interrelationship that exists between the success of land plants and various groups of animals. This is possible only because of a phenomenon called coevolution, the process by which two organisms evolve structures and behaviors in response to changes in each other over time. A brief evolutionary history of plants is used to illustrate how this could have occurred.

Pollination is essential to the reproduction of flowering plants. Wind pollination is discussed. There is also an explanation of vector pollination—pollination by the action of animals. Some of the flower adaptations of various plants to animal pollination are given as examples. These generally involve shape, color, and fragrance. Seed dispersal methods are also good examples of plant adaptations to life on land. In many cases, coevolution with animals is important in this process also. Attractive color, good taste, and nutritious compounds frequently play a part in seed dispersal.

Performance Objectives 22–3

1. Discuss the process and importance of plant-animal coevolution.
2. Describe several methods of pollination in seed plants, including the adaptations of plants that facilitate the process.

Science Terms 22–3

coevolution   p. 476
vector pollination   p. 478
seed dispersal   p. 479

DISCOVERY LEARNING

TEACHER DEMONSTRATIONS
Modeling

Grow Your Own

This activity can be used to introduce Chapter 22.

Perhaps the best way to get students interested in plants is to allow them to grow their own seedlings. This is a simple process—an activity that every child has probably done in school at one time or another. You may be surprised at the excitement that it will generate among older students. Provide potting soil, small containers, and seeds. Any seeds will do, but those that germinate quickly are more fun—small flowers, beans, corn, and so on. Put these in a warm, light area and allow students brief opportunities to observe and care for them every day. You can make this as basic or elaborate as you wish. The seedlings can be used in later chapters to study plant structure, although some students may want to take them home.

Plant Organs

This demonstration can be used to introduce the topic of plant organs.

Bring several samples of leaves, stems, and exposed roots to class. Spend some time talking about each of the specimens, allowing students to examine them. If you want to make this activity more formal, set up a number of examination areas around the room. Put some information on a card accompanying each specimen as well as some specific things for which to look. Have students circulate around the room in pairs or small groups to make their observations. Be sure to follow the activity with a discussion of what the students learned or remembered from previous experiences.

CHAPTER **22**

Plants with Seeds

GUIDED ENQUIRY

Pose the following questions to students and have them record their responses. Point out that they will gain a better understanding of the key concepts if they read the chapter with these basic questions in mind. Upon completion of the chapter, pose the questions again. Ask students to compare their initial responses with those they have developed after reading the chapter.

- For a plant, how is life on land different from life in water?
- How are the cells of plants organized into tissues and organs?
- How have plants changed over time?
- Why are flowering plants so widespread?
- How are plant evolution and animal evolution interrelated?

CHAPTER **22**

Plants with Seeds

This bee is busy gathering nectar from flowers. Pollen produced by the flowers sticks to the bee's body. An oak tree produces more than enough acorns to satisfy hungry squirrels and more than enough to produce new oaks.

Try to imagine what life would be like without plants. It's a rather difficult image to conjure up, especially because without plants there would be no animals. Almost every animal on the face of the Earth ultimately depends on food produced by plants. And just as importantly, plants shape environments in which animals live.

Humans and other land animals are able to benefit from plants only because members of one certain plant group have evolved in ways that allow them to live in a variety of different places. Most mosses and ferns cannot survive in many habitats because they need an almost constant supply of water. But seed plants—which include nearly all the plants you encounter—have, as a result of many evolutionary changes, been freed from dependence on water. It is this evolutionary story that you will uncover in this chapter.

INTRODUCING CHAPTER 22

Using the Visuals
Bring in several specimens of seeds, fruits, flowers, and cones. Explain that these are all examples of adaptations of plants to life on land. Have students examine the chapter-opener photograph and read the caption. The interactions between plants and animals are also examples of adaptations to life on land. Explain that both types of involved organisms are dependent on each other. Read the chapter-opener information with the students. After you have discussed this material, you will be ready to introduce more information about the advantages and disadvantages of life on land.

GUIDE FOR READING

After you read the following sections, you will be able to

22–1 Seed Plants— The Spermopsida
- Describe several adaptations of seed plants to life on land.
- Identify the functions of roots, stems, and leaves.
- Explain why reproduction in seed plants is not dependent upon water.

22–2 Evolution of Seed Plants
- Describe the evolution of seed plants.
- List several characteristics of gymnosperms and angiosperms.
- Compare monocots and dicots.

22–3 Coevolution of Flowering Plants and Animals
- Describe the process of pollination in seed plants.
- Explain plant-animal coevolution.
- Discuss the importance of seed dispersal to the success of the seed plants.

Journal Activity

YOU AND YOUR WORLD

Poets have long written about the beauty of plants. Why don't you try, too? Write a short poem about a flower, tree, or other plant you see on the way to school each day.

Figure 22–1 Fields of sunflowers follow the daily movement of the sun. Here thousands of plants grow in conditions that are quite favorable. But plants often grow in less hospitable places, such as a tiny crack in the surface of a road.

22–1 Seed Plants— The Spermopsida

Guide For Reading

■ In what ways are seed plants able to survive on land?
■ What are the functions of roots, stems, and leaves?
■ In what ways are plants adapted to reproduce on land?

Compared to life in water, life on land offers several benefits to plants. Life on land provides abundant sunlight for photosynthesis. On land there is continuous free movement of gaseous carbon dioxide and oxygen, which plants use during photosynthesis and respiration.

But life on land also presents significant problems to plants. Water and nutrients are available to most land plants only from the soil. On land, dry air draws water from exposed plant tissues by the process of evaporation. On land, photosynthetic tissues must be held upright to capture sunlight. And unlike the reproductive cycles of mosses and ferns, the reproductive cycles of most land plants must work without standing water.

Seed Plants—Designed for Life on Land

Seed plants, members of the subphylum Spermopsida, exhibit numerous adaptations that allow them to survive the difficulties of life on land. Note that seed plants did not evolve

COOPERATIVE LEARNING

Using preassigned lab groups or randomly selected teams, have groups create an illustrated children's story for 7-year-olds. Their storyline should center around one of the following ideas: plants for food, plants for shelter and clothing, or plants for medicines. A children's story should contain the following elements:
- Characters to which the audience can relate
- A plot that introduces some type of problem and then solves the problem
- An interesting setting

Remind groups that 7-year-olds have a short attention span, respond well to action and illustration, and have limited reading abilities.

Journal Activity

YOU AND YOUR WORLD

You may want to use famous poems such as Kilmer's "Trees," Sandburg's "Grass," or Wordsworth's "Daffodils" as well as Journal Activity poems to start a discussion on the relationships between humans and plants. Students should be instructed to keep their Journal Activity in their portfolio.

TEACHING STRATEGY 22–1

Focus/Motivation

Set two beakers on your desk or on a table so that all students can see them. Fill one with water and leave one with air. Ask students to imagine themselves in each environment. Discuss such things as the availability of oxygen and other gases; the possible loss of moisture in dry air; the aspects of self-support vs. floating. In each case talk about how humans are affected and then how plants are affected. Help students realize that plants living on land have problems different from those living in water, just as humans or other animals would have.

Content Development

You may want to spend just a few minutes reviewing plant classification so that students will have all of the relationships among the taxa right. They might be confused by the designation of subphylum Spermopsida.

TEACHING SUPPORT

Laboratory Manual
- Seed-Bearing Plant Tissues, p. 267

MEDIA AND TECHNOLOGY

📽 **Biology Transparencies**
- 25: Vascular Tissue

📹 **Video/Videodisc**
- *Coniferous Forest and Tropical Rain Forest*

Use this bar code to introduce the topic of roots.

Play frames 27876 to 28350

See the Biology Media Guide page 18 for additional bar-code correlations for this video/videodisc.

ESL STRATEGY

Provide students with an assortment of drawings, photographs, models, or actual specimens of different seed plants. Have them do the following:
1. Select a plant.
2. Identify the plant.
3. Make a labeled drawing of the plant in which the pathways taken by the xylem and phloem are color-coded and identified.

Figure 22–2 *Roots, such as these of a corn plant, anchor the plant in the soil (top). The stem of the white pine is strong enough to support the plant for many meters above the ground (bottom, left). The leaves of most plants are green, the color of chlorophyll. However, leaves such as those of the brilliantly colored croton often show other colors besides green (bottom, right).*

these adaptations because they "wanted" to or because the processes of evolution somehow "knew" that such adaptations would be useful on dry land. Rather, in every generation of plants, the types of genetic variations we discussed in earlier chapters produced individuals with different characteristics. Over time, those individuals with characteristics best suited to their environments survived and produced offspring.

In this way, over hundreds of millions of years, the ancestors of seed plants evolved a variety of new adaptations that enabled them to survive in many places in which mosses and ferns could not. These ancient plants evolved well-developed vascular tissues that conduct water and nutrients between roots and leaves. They evolved roots, stems, leaves, and structures that enable them to live everywhere—from frigid mountains to scorching deserts. And, seed plants, as their name implies, evolved seeds—the key adaptation in a new form of sexual reproduction that does not require standing water. Let us briefly examine these adaptations one at a time.

Roots, Stems, and Leaves

Just like the cells in your body, the cells in a plant are organized into different tissues and organs. The three main organs in a plant are roots, stems, and leaves. Each organ shows adaptations that make the plant better able to survive.

ROOTS Roots perform several important functions. They absorb water and dissolved nutrients from moist soil. They anchor plants in the ground. Roots also hold plants upright and prevent them from being knocked over by wind and rain. Roots are able to do all these jobs because as they grow, they develop complex branching networks that penetrate the soil and grow between soil particles.

STEMS Stems hold a plant's leaves up to the sun. Although plenty of sunlight reaches the Earth, plants compete

---

22–1 (continued)

Content Development

As you discuss this material, incorporate some previously learned concepts to make this topic more meaningful for students. Follow your discussion with questions similar to the following, which should help students tie everything together.
- **What is meant by adaptation?** (Genetically controlled characteristics or behaviors that increase fitness.)
- **What is the importance of a genetic variation?** (Variations among individuals in species are caused by genetic differences and appear in a random fashion. Without them, no adaptations would be possible and no differences in fitness would exist.)
- **How did Darwin use the term *fitness*?** (He defined fitness as the capacity to survive in a particular environment.)
- **Name some of the adaptations that plants have made to life on land.** (Roots, stems, leaves, seeds, fruits, vascular tissue.)
- **What is considered the key adaptation?** (The seed.)
- **Why?** (Because it allowed plants to reproduce without a source of standing water.)

with one another for this solar energy. Many plants have tall stems and branches that reach above other plants around them. To support such tall plants, stems must be very sturdy.

LEAVES Leaves are the organs in which plants capture the sun's energy—a process vital to photosynthesis. Leaves evolved because plants that had broad, flat surfaces over which to spread their chlorophyll were able to capture more solar energy than plants that did not have such surfaces. So over time, in most habitats, plants with leaves had higher fitness—and produced more offspring—than plants without leaves. But those broad, flat leaves also exposed a great deal of tissue to the dryness of the air. These tissues must be protected against water loss to dry air. That's why most leaves are covered with a waxy coating called the cuticle. Because water cannot pass through the cuticle, this coating slows down the rate of evaporation of water from leaf tissues. Adjustable openings in the cuticle help conserve water while allowing oxygen and carbon dioxide to enter and leave the leaf as needed.

Vascular Tissue

As plants evolved longer and longer stems, the distance between their leaves and roots increased. The leaves of a tall tree might be 100 meters above the ground. Thus tall plants face an important challenge: Water must be lifted from roots to leaves, and compounds produced in leaves must be sent down to roots. Over time, the evolutionary forces of variation, chance, and natural selection produced a well-developed vascular system. This remarkable two-way plumbing system consists of two kinds of specialized tissue: xylem and phloem.

XYLEM Xylem is the vascular tissue primarily responsible for carrying water and dissolved nutrients from the roots to stems and leaves. Because xylem cells often have thick cell walls, they also provide strength to the woody parts of large plants such as trees. Oddly enough, most xylem cells grow to maturity and die before they function as water carriers.

PHLOEM Phloem tissue carries the products of photosynthesis and certain other substances from one part of the plant to another. Whereas xylem cells conduct water in only one direction (upward), phloem cells can carry their contents either upward or downward. Unlike xylem cells, functioning phloem cells are alive and filled with cytoplasm.

Reproduction Free from Water

Like other plants, seed plants have alternation of generations. However, the life cycles of seed plants are well adapted to the rigors of life on land. All of the seed plants you see

Figure 22–3 Growing tall can be an advantage to a plant's survival. Tall plants receive more of the sun's light and are less likely to be shaded by other plants. Vascular tissues transport water from the roots to leaves at the tallest part of a plant.

TEACHING SUPPORT

Teaching Resources
- Activity: Plants Designed for Life on Land, p. 3

Study Guide
- Section 22-1, p. 209

22–1 (continued)

Content Development

Review the principles of alternation of generations. Students sometimes learn about this reproductive cycle when mosses and ferns are studied but fail to realize that seed plants also exhibit this type of life cycle. Stress that the plant that they see and recognize is the sporophyte generation. Be sure students understand that flowers and cones are both specialized structures that allow plants to reproduce without standing water. You may have to spend a few minutes explaining the significance of that capability. Pollination is a familiar term, but students rarely have any concept of what the process really involves. Spend a few minutes talking about the structures involved in the process. For example, even if students understand that fertilization is accomplished by the sperm uniting with the egg, they seldom relate that in any way to plant reproduction. They usually have no idea that the sperm is found in a pollen grain. Help them to understand the basics now, so that a more detailed discussion later will mean more to them.

Again, students may have some idea of what a seed is but still have many misconceptions. As you discuss the seed, be sure to emphasize that the seed is really a wonderful adaptation to life on land. So many potential hazards that could affect a tiny growing embryo have been eliminated. You might want to talk about the built-in food supply, the protective coating, and the capacity for lengthy dormancy until conditions are favorable. It really works!

SECTION REVIEW 22–1

1. Three adaptations of seed plants that enable them to live on land are waterproof coverings on their leaves, a system of internal tubes that conduct materials throughout the plant's body, and reproductive structures that make standing water unnecessary.

2. Roots anchor a plant and take in water and minerals from the soil. Stems hold leaves up and expose them to the sun. Leaves gather energy from sunlight for use in photosynthesis.

3. Xylem and phloem tissue are both conducting tissues in a plant. Both tissues reach from a

Figure 22–4 *Texas bluebonnets are a wildflower that grows in huge numbers. Flowers are a plant's reproductive structures.*

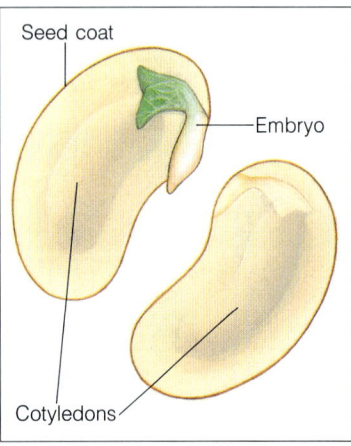

Figure 22–5 *Seeds are a promise and a plant's insurance. A seed contains the promise of a plant to come and the insurance that a species will have a chance to survive.*

around you are members of the sporophyte generation. By comparison, the gametophytes of seed plants are tiny, consisting of only a few cells. This size difference can be seen as the final result of an evolutionary trend in plants in which the gametophyte becomes smaller as the sporophyte becomes larger.

FLOWERS AND CONES The tiny gametophytes of seed plants grow and mature within the parts of the sporophyte we call flowers and cones. Flowers and cones are special reproductive structures of seed plants, which we shall discuss later. Because they develop within the sporophyte plant, neither the gametophytes nor the gametes need standing water to function. Thus the special reproductive structures of seed plants (flowers and cones) can be considered important adaptations that have contributed to the success of these plants.

POLLINATION The entire male gametophyte of seed plants is contained in a tiny structure called a **pollen grain**. Sperm produced by this gametophyte do not swim through water to fertilize the eggs. Instead, the entire pollen grain is carried to the female gametophyte by wind, insects, birds, small animals, and sometimes even by bats. The carrying of pollen to the female gametophyte is called **pollination**. Pollination is an important process that we shall discuss shortly.

SEEDS Seeds are structures that protect the zygotes of seed plants. After fertilization, the zygote grows into a tiny plant called an **embryo**. The embryo, still within the seed, stops growing while it is still quite small. When the embryo begins to grow again later, it uses a supply of stored food inside the seed. A **seed coat** surrounds the embryo and protects it and the food supply from drying out. Inside the seed coat, the embryo can remain dormant for weeks, months, or even years. Seeds can survive long periods of bitter cold, extreme heat, or drought—beginning to grow only when conditions are once again right. Thus the formation of seeds allows seed plants to survive and increase their number in habitats where mosses and ferns cannot.

22–1 SECTION REVIEW

1. What are three adaptations of seed plants that enable them to live on land?
2. What are the functions of roots, stems, and leaves?
3. How are xylem and phloem tissues similar? How are they different?
4. **Connection—You and Your World** What is a seed? What are two ways seeds provide food for people?

22–2 Evolution of Seed Plants

Guide For Reading

- How do useful adaptations give rise to new plant species?
- What are some characteristics of gymnosperms?
- What are some characteristics of angiosperms?
- How do monocots differ from dicots?

The history of plant evolution is marked by several great adaptive radiations. **Each time a group of plants evolved a useful new adaptation (such as vascular tissue or seeds), that group of plants gave rise to many new species.** Because of the new adaptation, some new species were able to survive in previously empty niches. For other new species, the new adaptation made them better suited to their environments than existing species that did not possess the new adaptation. Over time, the better adapted species survived and the older species became extinct.

It is important to remember that Earth's environments did not remain constant through time. Over a period of millions of years, landmasses moved and mountain ranges rose. In some cases, plant species produced by an adaptive radiation continued to evolve in ways that enabled them to survive as their environment changed. Such species survived for long periods. In other cases, plant species could not survive changing environments. These species became extinct.

Mosses and ferns, for example, underwent major adaptive radiations during the Devonian and Carboniferous periods, 300 to 400 million years ago. During these periods, land environments were much wetter than they are today. Tree ferns, tree lycopods, and other spore-bearers grew into lush forests that covered much of the Earth.

But over a period of millions of years, continents became much drier, making it harder for spore-bearing plants to survive and reproduce. For that reason, many moss and fern species became extinct. They were replaced by seed plants whose adaptations equipped them to deal with drier conditions. To help you understand how seed plants became successful, we shall now trace the evolution of these fascinating organisms.

Seed Ferns

The first seed-bearing plants, which appeared during the Devonian Period, resembled ferns. But these plants were different from ordinary ferns in one very important respect: They reproduced by using seeds instead of spores. Fossils of these so-called seed ferns document several evolutionary stages in the development of seed plants. Although seed ferns were quite successful for a time, they were rapidly replaced by other plant species. Today, no seed ferns survive.

Gymnosperms

The most ancient surviving seed plants belong to three classes: the Cycadae, Ginkgoae, and Coniferae. In plants of these classes, a number of leaves have evolved into specialized male and female reproductive structures called **scales**. Scales

Figure 22–6 Seed ferns are part of the fossil record. They represent a link between ferns that do not form seeds and seed plants that do. This ancient plant had leaves that resemble the leaves of modern ferns.

plant's roots to the tip of its branches. Xylem tissue, which is no longer alive, carries water and dissolved nutrients from the roots to the leaves. Phloem conducts the products of photosynthesis throughout the plant. The materials carried in phloem can be moved upward or downward in a plant. Phloem cells are alive.

4. A seed is a structure that protects the zygote of a seed plant. Seeds also contain a supply of food for the developing plant. People eat the seeds of certain plants—peas and beans, for example—directly. Seeds can sprout, and people eat the sprouted seeds. Or seeds can produce plants that people eat.

BACKGROUND INFORMATION
LIVING GINKGOES

Ginkgo trees, with their unusual fan-shaped leaves, existed long before humans. For some reason, they are highly resistant to damage from polluted air. Because of this trait, many have been planted in cities in recent years. The ginkgoes produce male and female cones on separate plants. The female ginkgo seeds are surrounded by a fleshy seed coat that produces butyric acid, which has the strong odor of rancid butter. For this reason, when ginkgo trees are planted as ornamentals, it is best to plant only male trees to avoid a very smelly fruit fall.

Figure 22–7 Confusingly named the sago palm, this cycad is not a palm at all (left). Cycads grow primarily in warm and temperate areas. Cycads produce reproductive structures that look like giant pine cones (right).

Figure 22–8 The ginkgo is often planted on city streets because it can tolerate the air pollution produced by city traffic.

are grouped into larger structures called male and female cones. Male cones produce male gametophytes called pollen. Female cones produce female gametophytes called eggs. Later, the female cones hold seeds that develop on their scales. Each seed is protected by a seed coat, but the seed is not covered by the cone. Because their seeds sit "naked" on the scales, cycads, ginkgoes, and conifers are called naked seed plants, or **gymnosperms** (*gymno-* means naked; *-sperm* means seed).

CYCADS Cycads are beautiful palmlike plants that first appear in the fossil record during the Triassic Period, 225 million years ago. Huge forests of cycads thrived when dinosaurs roamed the Earth. Many biologists think that some species of dinosaurs ate the young leaves and seeds of cycads. Today, only nine genera of cycads, including the confusingly named sago palm, remain. Cycads can be found growing naturally in tropical and subtropical places such as Mexico, the West Indies, Florida, and parts of Asia, Africa, and Australia.

GINKGOES Ginkgoes were common when dinosaurs were alive, but today only a single species, *Ginkgo biloba*, remains. The living ginkgo species looks almost exactly like its fossil ancestors, so it is truly a living fossil. In fact, *Ginkgo biloba* may be the oldest seed plant species alive today. This single species may have survived only because the Chinese have grown it in their gardens for thousands of years.

Conifers: Cone Bearers

Conifers, commonly called evergreens, are the most abundant gymnosperms today. They are also the most familiar and important. Pines, spruce, fir, cedars, sequoias, redwoods, and yews are all conifers. Some conifers, such as the dawn redwood, date back 400 million years to the Devonian Period—well before the time of the cycads. But although other classes of gymnosperms are largely extinct, conifers still cover vast

---

22–2 (continued)

Content Development
Name the three classes of gymnosperms. Emphasize that in some cases there are few existing members today, but the groups were once very large and highly successful. Be sure to point out to students that the scales that make up the cones are modified leaves that have evolved into reproductive structures. This is really a foreign concept to the students, but it is one that they need to begin to recognize.

Focus/Motivation
Show the students samples of male and female cones. Explain why the group is known as gymnosperms. Spend a few minutes talking about cycads and ginkgoes. If you have a sample of a ginkgo leaf, show it to the class. Some of the students may have seen these trees and not realized what they were.

Enrichment
The history of the *Ginkgo biloba* is long and interesting. If you have students who are so in-

areas of North America, China, Europe, and Australia. Conifers grow on mountains, in sandy soil, and in cool moist areas along the northeast and northwest coasts of North America. Some conifers live more than 4000 years and can grow more than 100 meters tall.

ADAPTATIONS The leaves of conifers are long and thin, and are often called needles. Although the name evergreen is commonly used for these plants, it is not really accurate because needles do not remain on conifers forever. A few species of conifers, like larches and bald cypresses, lose their needles every fall. The needles of other conifer species remain on the plant for between 2 and 14 years. These conifers seem as if they are "evergreen" because older needles drop off gradually all year long and the trees are never completely bare.

REPRODUCTION Like other gymnosperms, most conifers produce two kinds of cones. The scales that form these cones carry structures called sporangia that produce male and female gametophytes. Both male and female gametophytes are very small. Male cones, called **pollen cones**, produce male gametophytes in the form of pollen grains. Female cones, called seed cones, house the female gametophytes that produce ovules. Some species of conifers produce male and female cones on the same plant, whereas other species have separate male and female plants.

Each spring, pollen cones release millions of dustlike pollen grains that are carried by the wind. Many of these pollen grains fall to the ground or land in water and are wasted. But some pollen grains drift onto seed cones (female cones), where they may be caught by a sticky secretion. When a pollen grain lands near a female gametophyte, it produces sperm cells by mitosis. These sperm cells burst out of the pollen grain and fertilize ovules. After fertilization, zygotes grow into seeds on the surfaces of the scales that make up the seed cones. It may take months or even years for seeds on the female cones to mature. In time, and if they land on good soil, the mature seeds may develop into new conifers.

Angiosperms: Flowering Plants

Angiosperms are the flowering plants. All angiosperms reproduce sexually through their **flowers** in a process that involves pollination. Unlike the seeds of gymnosperms, the seeds of angiosperms are not carried naked on the flower parts. Instead, angiosperm seeds are contained within a protective wall that develops into a structure called a **fruit**. The scientific term fruit refers not only to the plant structures normally called fruits but also to many structures often called vegetables. Thus, by definition, apples, oranges, beans, pea pods, pumpkins, tomatoes, and eggplants are all fruits.

Figure 22–9 Pine cones may be either male or female. Male cones (top) produce windborne pollen that is carried to female cones (bottom). Female cones nurture and protect the developing seeds, which often take two years to mature.

ANNOTATION KEY

① Monocotyledonae (Interpreting charts)

TEACHING SUPPORT

Laboratory Manual
- Comparing Monocots and Dicots, p. 273

Teaching Resources
- Hands-On Activity: Flat Flowers, p. 107

Study Guide
- Section 22–2, p. 211

FACTS AND FIGURES

The majority of the world's human population obtains most of its food from a single angiosperm family: the grasses. All cereal grains—oats, wheat, barley, and rye, for example—are grasses. Corn, another grass, was at one time the most important crop of Native Americans. Today, it still feeds both humans and cattle. Rice, a grass that grows in wet places, is the single most important food item in China, Japan, and Southeast Asia.

22–2 (continued)

Content Development
Using the Visuals Show students several samples of plants that are monocots and several that are dicots. You will also need flowers of both types. After you have discussed the reason for the subclass names and the early development of these plants, use the samples to illustrate the noticeable differences in the two groups. You might want to use pictures or projected slides of stem and root cross sections to illustrate the not-so-noticeable differences.

The features matrix shown in Figure 22–12 will help you summarize this information. As you point out specific differences, be sure to use lots of examples. Students may not realize that grasses are monocots, that most of the trees with which they are familiar are dicots, and so forth. In other words, relate the examples to their scientific terms.

Figure 22–10 These pear flowers are a form of floral advertising that attracts bees and other insects. The insects pollinate the flowers. Six weeks after pollination has occurred, the developing pears are still quite small. In time they will ripen.

Figure 22–11 Flowers can vary in appearance. This orchid flower is colorful and has petals and sepals of different shapes.

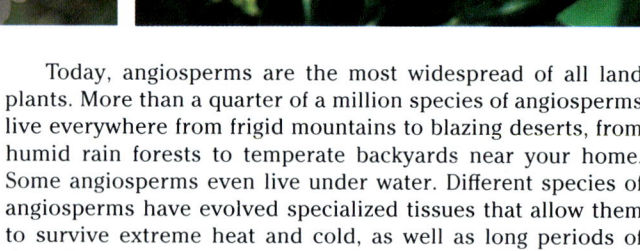

Today, angiosperms are the most widespread of all land plants. More than a quarter of a million species of angiosperms live everywhere from frigid mountains to blazing deserts, from humid rain forests to temperate backyards near your home. Some angiosperms even live under water. Different species of angiosperms have evolved specialized tissues that allow them to survive extreme heat and cold, as well as long periods of drought.

Angiosperms can be separated into two subclasses: the Monocotyledonae (mahn-oh-kaht-'l-EED-'n-ee), called monocots for short, and the Dicotyledonae (digh-kaht-'l-EED-'n-ee), called dicots for short. The **monocots** include corn, wheat, lilies, daffodils, orchids, and palms. The **dicots** include plants such as roses, clover, tomatoes, oaks, and daisies.

There are several differences between monocots and dicots. The simplest difference has to do with the number of leaves the embryo plant has when it first begins to grow, or germinate. The leaves of the embryo are called **cotyledons**, or seed leaves. Monocotyledons have one seed leaf (*mono-* means one). Dicotyledons start off with two seed leaves (*di-* means two). In some species, cotyledons are filled with food for the germinating plant. In other species, the cotyledons are the first leaves to carry on photosynthesis for the germinating plant.

Figure 22–12 shows several characteristics of monocots and dicots. These differences are summarized below:

1. Veins in monocot leaves usually lie parallel to one another. Veins in dicot leaves form a branching network.
2. In monocot flowers, petals and other flower parts are usually found in threes or multiples of three (3, 6, 9, and so on). In dicot flowers, petals and other flower parts occur in fours or fives or in multiples of four (4, 8, 12) or five (5, 10, 15).
3. In monocot stems, xylem and phloem tissues are gathered into **vascular bundles** that are scattered throughout the stem. In dicot stems, these vascular bundles are arranged in a ring near the outside of the stem.

SECTION REVIEW 22–2

1. Useful adaptations make plants better suited to their environment than existing plants that do not possess the adaptation. If a plant population's useful adaptations and other features cause it to become reproductively isolated from other populations, then it is a new species. New species may supplant existing, less well-adapted species.

2. A gymnosperm is a plant that reproduces by forming seeds that are not covered but sit

	Monocots	Dicots
Leaves	Veins in leaves of most monocots are parallel to each other.	Veins in leaves form a branching network.
Flower	Flower parts in threes or multiples of three.	Flower parts in fours or fives or multiples of four or five.
Vascular bundles in stem	Vascular bundles are scattered in a cross section of a stem.	Vascular bundles are arranged in a ring in a cross section of a stem.
Vascular bundles in root	Bundles of xylem and phloem alternate with one another in a circle.	A single mass of xylem forms an "X" in the center of the root; phloem bundles are located between the arms of the "X."
Stem thickness	Stems of most monocots do not grow thicker from year to year.	Stems can grow thicker from year to year.

Figure 22–12 Flowering plants are placed into two main subclasses, Monocotyledonae and Dicotyledonae. This chart identifies the differences between these two classes. Which class contains plants whose leaves have veins that are parallel to one another?

4. In monocot roots, bundles of xylem and phloem alternate with each other in a circular arrangement, like the spokes of a bicycle wheel. In dicot roots, a single mass of xylem tissue forms an X in the center of the root, and bundles of phloem tissue are positioned between the arms of the "X."

5. Most monocots have stems and roots that do not grow thicker from year to year. For this reason there are very few treelike monocots. Palms are one of the few treelike monocots. Some dicot stems and roots can grow thicker from year to year. Most of the flowering trees you see are dicots.

22–2 SECTION REVIEW

1. How do useful adaptations give rise to new plant species?
2. Compare gymnosperms and angiosperms.
3. Which generation is more obvious in seed plants? How do the relative sizes of these generations follow a trend in the evolution of plant reproduction?
4. **Critical Thinking—Applying Concepts** Suppose you found a plant whose leaves have parallel veins and whose flowers have six petals. Is this plant a monocot or a dicot? What is your reasoning?

Figure 22–13 This tiny bean seed has pushed its stem above the soil surface and into the light. Just below the leaves at the top of the plant, the two bean-shaped cotyledons remain attached to the stem. Later, when the plant is large enough to make its own food, the cotyledons will shrivel and fall off.

BACKGROUND INFORMATION
MEDICINAL PLANTS

Seed plants have been an important source of medicine since the dawn of civilization. For example, the ancient Greeks discovered that chewing the leaves of a certain willow tree (genus *Salix*) cured aches and pains. Later, doctors found that those willows contain a compound called salicylic acid. Many centuries later, we still use this compound, commonly called aspirin.

Other plants have provided pain killers, muscle relaxants, drugs that raise or lower blood pressure, drugs that fight malaria and other diseases, and drugs that are used to treat heart problems. Recently, researchers have found chemicals in plants that show great promise in treating leukemia as well as other forms of cancer. A small tropical flower called the rosy periwinkle (genus *Vinca*) synthesizes two chemicals—vincristine and vinblastine—that seem to prevent cancer cells from dividing. These chemicals are produced by the rosy periwinkle in a series of chemical reactions that may require 20 or more steps. As yet, chemists have not learned to make these medicines synthetically, so all research must be done on chemicals extracted from the plants themselves.

Reinforcement/Reteaching
For those students who have difficulty with the Section Review questions, go back to the discussion in the section and review the material students have had difficulty understanding.

Closure
Make a set of flashcards using the important terms and group names from Section 22–2. Ask students (volunteers or "volunteered") to talk about each of the terms or groups, thereby reviewing the important points of the section.

naked on a scale. Angiosperms are flowering plants. Unlike gymnosperms, the seeds of angiosperms are enclosed within a fruit.

3. During the evolution of land plants, the sporophyte generation has become most obvious, whereas the gametophyte generation has become increasingly smaller. In mosses, the gametophyte generation is much more obvious than the sporophyte generation. In ferns, the gametophyte is much reduced in size, whereas the sporophyte is much larger. In seed plants, the gametophyte generation is very small and inconspicuous.

4. This plant is a monocot. Parallel veins and a flower that contains six parts are characteristics of monocots.

ECOLOGY NOTE
NATURE'S MEDICINE

Humans are not the only creatures that use plants for therapeutic purposes. Some wild animals also use certain plants to cure or prevent disease. For example, chimpanzees in Tanzania swallow the leaves of plants of the spiderwort family and *Aspilia*. According to a scientist at the University of California, these plants have chemicals that destroy bacteria, fungi, and parasites. The chimpanzees swallow the leaves whole, just as humans might swallow tablets. This means that most of the medication present in the leaves is released in the intestines, where it is needed. Similarly, baboons in Ethiopia eat a fruit that is effective in destroying parasites.

In addition to attacking parasites and other pathogens, chemicals in the medicinal plants may also affect reproduction in a desirable way. Researchers have noticed that the plants eaten by Brazilian muriqui monkeys at certain times of the year seem to influence the monkeys' fertility, and that Costa Rican howler monkeys eat different plants that may determine the sex of their offspring!

Guide For Reading

- What is the importance of pollination?
- How do plants and animals affect each other's evolution?
- How does seed dispersal contribute to the success of seed plants?

22-3 Coevolution of Flowering Plants and Animals

Watching bees travel from flower to flower is such a common experience that most of us probably do not think about it. We take for granted the fact that flowers are brightly colored and beautifully perfumed. Rarely do we wonder why fruits are tasty and nutritious as well as colorful. But how and why did insects begin exhibiting flower-visiting behavior? When did animals begin to eat fruits and seeds? And why have plant flowers and fruits evolved into their present forms?

The process by which two organisms evolve structures and behaviors in response to changes in each other over time is called **coevolution**. Some of the most fascinating examples of coevolution involve relationships between angiosperm flowers and fruits and a wide variety of animal species.

To understand plant-animal coevolution, we must look once again at the evolutionary history of plants. The first flowering plants probably evolved during the early Cretaceous Period, about 125 million years ago. At that time, gymnosperms and other plants formed huge forests. Dinosaurs were the dominant land animals. During the Cretaceous Period, the first birds and mammals began to appear in the fossil record. Flying insects, particularly beetles of several types, became common. Thus the first flowering plants evolved at about the same time as the earliest mammals, a short time after the earliest birds, and a good while after the earliest insects.

Then, toward the end of the Cretaceous Period, the Earth's climate changed dramatically. Dinosaurs and many gymno-

Figure 22–14 Many different animals pollinate plants. Bees, such as this honeybee covered with pollen, are perhaps the most common (right). Bees are responsible for the pollination of many of the plant varieties that produce the fruits we eat. Bananas, like this one growing in Southeast Asia, are often pollinated by bats, not by bees (left).

TEACHING STRATEGY 22-3
Focus/Motivation
Begin by showing students an interesting example of coevolution. Get a bird-of-paradise flower from the florist. Examine this flower with your students. Look at the flower cluster as a whole, then zero in on a single flower. Either demonstrate or let the students discover the landing platform that spreads apart to coat birds' feet with sticky pollen. Point out the stigma and its location, which encourages pollen deposition. Orchids also have some interesting plant-pollinator relationships. The essential point is that the species involved rely heavily on one another.

Content Development
This is a topic that you can spend days discussing if you are so inclined. There are so many interesting examples. Students may have difficulty understanding the idea of coevolution. You may want to spend time talking about this. It might be easier if you stress that the adaptations, both plant and animal, did not necessarily help them merely to survive but rather to be more successful.

Reinforcement/Reteaching
Most students have some basic idea of what pollination means. They may not understand the

sperms became extinct. This mass extinction opened up many niches for other organisms. New adaptive radiations of both animals and plants occurred. New species of birds and mammals evolved and filled niches vacated by the dinosaurs. New species of angiosperms replaced disappearing gymnosperms. And many new species of insects—including moths, bees, and butterflies—evolved.

The coincidence of angiosperm evolution with the evolution of modern insects, birds, and mammals is very important. Flowers and fruits are specialized reproductive structures that could evolve only in the presence of insects, birds, and mammals. Let us now see how and why this is so.

Flower Pollination

Pollination is essential to the reproduction of flowering plants. Over millions of years, a variety of ways to ensure that pollination will occur has evolved. For example, some plants are pollinated by the wind. Wind-pollinated plants include willow trees, ragweed, and grasses such as corn and wheat. The tiny pollen grains of these plants fall off their flowers without difficulty, making it easy for them to be carried by the wind to other flowers. Wind-pollinated plants usually have small, plain simple flowers with little or no fragrance.

But most angiosperms are not pollinated by the wind. Most flowering plants are pollinated by insects, birds, or mammals that carry pollen from one flower to another. In return, the plants provide the pollinators with food. The food may take the form of pollen or a liquid called nectar, which may contain up to 25 percent glucose, or a combination of pollen and nectar.

Figure 22–15 Hummingbirds are able to flap their wings so fast that they hover in place. This hummingbird is drinking nectar from a flower. Because hummingbirds are able to see red and orange quite well, they are attracted to these flower colors.

BACKGROUND INFORMATION
COEVOLUTION AND ORCHIDS

One of the most unusual examples of coevolution of plants and animals is the orchid pollinated by any one of several insect species in which the male insects emerge in the spring before the females. These flowers have evolved both shapes and odors that mimic the stimuli presented by the female insects. The males, eager to mate, fly around searching for females. They happen upon these insect-mimicking flowers and try to mate with one after another. In the process, they cross-pollinate the orchids.

TEACHING SUPPORT

MEDIA & TECHNOLOGY
Interactive Videodisc
- *On Dry Land: The Desert Biome*
Use this bar code to show pollination by bats.

Play frames 35050 to 36852

Video/Videodisc
- *Seeing Sense*
Use this bar code to introduce the topic of bee vision.

Play frames 25557 to 27887

- *Coniferous Forest and Tropical Rain Forest*
Use this bar code to show pollination by birds.

Play frames 33237 to 33687

See the Biology Media Guide pages 18–19 for additional bar-code correlations for these media.

---

process, but they know the end results. Discuss the idea of efficiency here. Point out that while wind pollination seems wasteful, it requires less energy from the plant because of the small colorless flowers produced. Remind students that it is often wind-pollinated species that cause problems to humans during the allergy seasons.

Enrichment
Students may be interested in this topic of plant-pollinator relationships. Encourage them to research the topic either at the library or by visiting with a nursery person or horticulturist. If they can locate a person involved in growing orchids, they will obtain some interesting information.

ECOLOGY NOTE
THE DOMINO EFFECT

Plants and their animal "partners" may be so highly adapted to each other that the extinction of one may cause the extinction of the other.

This tragic process is not restricted to distant nations—it is under way in the United States. In Hawaii, the decline of drepanidid birds (shown in Figure 14–20 on page 308) has lead to the decline of the plants on which they feed. And the only remaining pollinators of *Brighamia* plants are a few botanists with the courage to rappel down sheer, crumbly cliffs to reach the rare plants.

Interested students may wish to read the articles on Hawaiian plants and birds in the August/September and October/November 1993 issues of *National Wildlife*.

Figure 22–16 This flower looks different under natural sunlight (top) than it does under ultraviolet light (bottom). Insects can perceive ultraviolet light whereas humans cannot. The pattern that shows up under ultraviolet light may attract insects to the center of the flower, where the flower's reproductive structures are found. This makes it more likely that the insect will pollinate the plant.

It is easy to imagine how pollinators such as bees first learned to visit certain flowers. When a bee finds food on a particular flower, it remembers clearly the color, shape, and odor of that flower. So if a bee finds edible pollen on a flower of a particular type, it will search for more flowers of that same type. While feeding on different flowers, a bee may accidentally pick up extra pollen that it then carries to the next flower it visits. Because the bees remember the color and odor of flowers so well, it is probable that pollen picked up from one flower will be deposited on another flower of the same species.

This kind of interaction between animals and plants increases the evolutionary fitness of both organisms. Insects benefit by learning to identify dependable sources of food. Plants benefit because this kind of **vector pollination**, or pollination by the actions of animals, is a very efficient way of getting the male gametophyte to the female gametophyte. Vector pollination is much more efficient than wind pollination, which wastes enormous amounts of pollen.

Of course, flowers that depend upon specific animals to pollinate them could only have evolved after those animals evolved. When angiosperms first appeared, this sort of relationship began accidentally. But over time the coevolutionary relationship strengthened because it proved beneficial to the survival of both plants and animals. Coevolutionary relationships can be very specific. The following examples of flower-pollinator pairs illustrate this fact.

One common pollinator is the honeybee. To attract bees to their flowers, many plants have brightly colored flower petals that bees can see well. Because bees can see ultraviolet, blue, and yellow light the best, these are the colors of most bee-pollinated flowers. We cannot see ultraviolet light under ordinary circumstances. But special film can make this color visible to our eyes. In Figure 22–16 you can see a picture of a flower taken in ultraviolet light. The petals of some flowers even have markings that point to the center of the flower. These markings are like a secret sign for bees alone to see! The markings direct the bee to the center of the flower—the source of nectar. On its way to the food, the bee might pollinate the flower, thus ensuring the survival of the plant species. Flowers that are pollinated by bees usually have some kind of landing platform because bees gather nectar only when they are standing, not when they are flying.

Flowers that have coevolved with animals other than bees show different methods of attracting pollinators. For example, some flowers are pollinated by night-flying moths that cannot see color but have an excellent sense of smell. The petals of these flowers are often plain and white, but the flowers themselves are very fragrant—especially at night. (We use many of these floral fragrances—jasmine, for example—in perfumes.) Moth-pollinated flowers usually do not have landing platforms because unlike bees, moths feed while hovering in midair. The

---

22–3 (continued)

Content Development
Using the Visuals Direct students' attention to Figure 22–16.
- **What do these photographs show?** (A flower under natural light and ultraviolet light.)
- **What is the significance of the patterns on the flower?** (The markings that appear under ultraviolet light may attract insects to the center of the flower.)

You may wish to explain to more advanced students that the ultraviolet photograph does not show how an insect actually experiences the pattern, for the special techniques that reveal ultraviolet to humans distort other colors. An insect might perceive the center of the flower as a blend of yellow and ultraviolet—a color we humans cannot even imagine!

Focus/Motivation
Using the Visuals The relationship between bees and flowers is, of course, the classic example of plant-pollinator interaction. However, some students are practically unaware of it. Go through the passage in the text that gives the various examples and use Figure 22–16 to further clarify the way bees locate and

nectar of moth-pollinated flowers is usually contained deep within the flower, where only the long tongue of a moth can reach it.

Several species of flowers are pollinated by flies that lay their eggs in the bodies of dead and decaying animals. You certainly would not want to grow these flowers in your house because they smell like rotting meat! The smell produced by the flowers attracts the flies that are looking for a place to lay their eggs. The flowers of these plants even heat up when they are ready to be pollinated, thus intensifying the smell they produce to lure additional flies that may act as vector pollinators.

Some flowers are pollinated by birds. Birds have a very poor sense of smell but a good sense of sight. Birds can easily see the colors orange and red. Not surprisingly, bird-pollinated flowers, such as the beautiful bird-of-paradise flower, are a reddish-orange color. These flowers usually have no fragrance.

Seed Dispersal

Just as flowers have different methods that ensure pollination, angiosperm fruits have adaptations that help scatter seeds away from the parent plant. The process of distributing seeds away from parent plants is **seed dispersal.** Seed dispersal is very important to plants. Why? If the seeds of a plant are not dispersed but instead fall to the ground beneath the parent plant, the seedlings will compete with one another and with the parent plant for sunlight, water, and nutrients. This competition will reduce the chances of survival for the growing seeds. Seed dispersal also enables plants to colonize new environments. Although adult plants cannot move around, their seeds can be carried to new environments.

Figure 22–17 The stapelia flower, also called the carrion flower, smells like a piece of rotting meat. Although not attractive to us, the smell proves alluring to a fly looking for a place to lay her eggs.

Figure 22–18 The seeds of the milkweed (left) and the dandelion (right) are carried by the wind.

SCIENCE, TECHNOLOGY, AND SOCIETY ISSUE

Designer Genes—Problem or Promise?

Genetic engineering holds enormous promise for agriculture. Current lines of research may lead to the commercial production of plants that have more nutritious seeds or fruits, contain a more complete array of amino acids, manufacture their own herbicides or insecticides, or manufacture drugs for human use. Every day, more genetic-manipulation techniques evolve from experimental procedures to assembly-line routines. Certainly, the potential for a new agricultural revolution exists.

Yet plant genetic engineering, like any revolutionary technology, raises important issues that affect all of us. Encourage your class to do some serious thinking about the potential costs and benefits of "designer-gene" plants. Students need to be taught that they have a responsibility to find out the facts behind issues, and to use scientific knowledge and good judgment when they become voters and decision makers.

You may wish to link this feature with a role-playing exercise based on current local or national developments in "designer-gene" plants. Role playing provides students with an opportunity to practice decision-making skills and to explore the many facets of an issue. Possible roles in an exploration of, say, gene-spliced tomatoes might include: a parent, a consumer advocate, a manufacturer's representative, an environmentalist, a farmer, a supermarket owner, a chemist, a restaurant owner, a trucker, and representatives from the FDA, EPA, and other relevant governmental agencies.

SCIENCE, TECHNOLOGY, AND SOCIETY

Designer Genes—Problem or Promise?

At one agricultural laboratory, a single tomato plant in a cage full of hungry caterpillars remains untouched while its neighbor is stripped bare of leaves. The cells of the untouched plant are able to manufacture an insecticide because it has genes transplanted from a bacterium.

At another greenhouse, two rows of cotton plants grow side by side. The benches they grow in have been treated with an herbicide, a chemical used to kill weeds. In one bench the cotton plants are stunted and dying—much like the weeds the powerful herbicide kills. In the other bench, the cotton plants thrive. The thriving plants carry a gene that confers resistance to that particular herbicide, a gene that was grafted onto the plants' genome by genetic engineers.

These are just two new and improved plants produced by the application of genetic engineering, which makes it possible to design and produce plants that have traits that people could once only dream about. People who support this new field assure us that a new agricultural revolution has begun. However, other researchers warn that we must be careful about the ways in which genetic engineering is used. What sorts of problems could occur? Some researchers worry that accidental cross-pollination could produce "super weeds" immune to insects or herbicides.

Some ecologists worry that if herbicide-resistant varieties of plants (such as cotton) are made available, farmers will be encouraged to spray more or stronger poisons on their fields.

So far, genetic engineers point out that no problems with genetically altered organisms have occurred. Should genetic engineering be restricted in organisms that are moved outside of the laboratory? What do you think?

Several different methods of seed dispersal have been observed in angiosperms. The seeds and fruits of some angiosperms, like those of dandelions, are carried by the wind. In other angiosperms, pressure builds inside the fruit and finally forces seeds out of the ripe fruit like bullets from a gun. The common garden plant *Impatiens* has fruits that spring open when touched, scattering the seeds over substantial distances.

SECTION REVIEW 22-3

1. Pollination is essential to the reproduction of flowering plants. Pollination is the process by which sperm cells come into proximity with egg cells (in the ovules).

2. The plant's flowers are pollinated and the animal pollinator gets food from the plants in the form of nectar or pollen. Thus, each organism benefits from its relationship with the other.

3. Seed dispersal is the movement of seeds away from the parent plant. It is important for several reasons. Seeds that are dispersed may grow in a new area, thus increasing the range of the plant. Also, plants that grow from dispersed seeds may have a better chance of surviv-

Many fruits have coevolved with animal species that help disperse the fruits' seeds. For example, some fruits have sharp barbs that catch in fur or feathers, allowing the seeds inside to hitch rides on mammals or birds. As they move from place to place, such animals may enter a new environment. If the seeds fall off the animals and land on a spot that provides good growing conditions, they will develop into new plants. In this way, plants are carried to new environments.

Some fruits have attractive colors, pleasant tastes, and contain a variety of nutritious compounds. These fruits and the seeds inside them are eaten by mammals and birds. The fleshy, nourishing, and tasty pulp of the fruit is digested by the animal, but the seeds, which are protected by tough seed coats, are not. These seeds pass through the digestive tract of the animal without being damaged. While inside the animal, seeds may be carried over great distances. Eventually the seeds are deposited, along with a convenient dose of natural fertilizer, in a new location where they can grow.

Have you ever wondered why so many unripe fruits are green and have a bitter taste? Think about the function of fruits in relation to the evolutionary fitness of plants. Inside the unripened fruits the seeds are still maturing. If the fruits are eaten too soon, the immature seeds will not be able to grow. The plant's fitness for survival would decrease. But plants manufacture bitter-tasting compounds that they pump into fruits as the fruits develop. These bitter-tasting compounds discourage animals from eating fruits that are not ripe. The green color of unripe fruits makes it more likely that animals will not notice the fruits hidden among the green leaves of plants. When the seeds are mature, plants either remove the bitter-tasting compounds from the fruits or chemically break down the compounds completely. Plants then pump sugars into the fruits. At the same time, the fruits change color. The brightly colored fruits are more easily noticed by birds and other animals. The distribution of seeds in fruits is yet another example of plant-animal coevolution.

Figure 22–19 The tiny seeds of the cocklebur have many hooks (top). The hooks catch onto the fur of animals and are carried to new environments. When the seeds are ripe, raspberries turn a bright red and can easily be seen by birds and other animals (bottom). After the fruits are eaten, the indigestible seeds pass through the animal and are deposited, along with other solid wastes, in a new environment.

22–3 SECTION REVIEW

1. Why is pollination important?
2. Explain how plant-animal coevolution has led to the development of relationships between vector pollinators and flowers.
3. What is seed dispersal? Why is it important?
4. **Critical Thinking—Relating Concepts** Explain how fruits are dispersed by animals. How does fruit dispersal contribute to seed dispersal?

LABORATORY INVESTIGATION

COMPARING RIPE AND UNRIPE FRUITS

Before the Lab
1. This lab requires quite a bit of equipment and setting up. Be sure to assemble enough equipment for the number of groups that you will have. Mix enough sugar solution for the experiment. Clean all glassware to avoid false tests.
2. Ripe bananas should be yellow with some brown specks. Unripe bananas should be as green as possible.
3. Prepare the sugar solution by adding 5 g of dextrose to 95 mL of water. This will give you a 5% dextrose solution.
4. If possible, water baths should be set up at the beginning of the period so that water is boiling when the Benedict's test is performed.

Pre-Lab Discussion
Review the last part of Section 22–3, the discussion of the ripening of fruit. You should establish the relationship between the ripening of fruit and the presence of sugar. You may want to briefly explain that Benedict's solution tests only for the presence of reducing sugars and not the more complex sugars. For example, it would not yield a positive result on a sucrose-sugar solution. Demonstrate the color changes that should be expected. This indicator is sensitive to dilutions, and students will need help in recognizing and understanding the variations in color that they will get. Rubber-banding the four test tubes together makes them easier to handle when they are heated. Finally, emphasize to students that they are to follow instructions closely, make careful observations, and accurately record their data. Variations will affect their results.

LABORATORY INVESTIGATION
COMPARING RIPE AND UNRIPE FRUITS

PROBLEM
Why do fruits get ripe?

MATERIALS (per group)

unripe banana	hot plate
ripe banana	ruler
balance	scalpel
400-mL beaker	4 test tubes
Benedict's solution	test tube holder
sugar (dextrose) solution	test tube rack
	hand lens
2 100-mL graduated cylinders	glass-marking pencil

PROCEDURE

1. Fill a 400-mL beaker halfway with water. Place the beaker on a hot plate. Turn the hot plate on high.
2. Use a glass-marking pencil to label four test tubes. Label the first test tube C, for control; the second S, for sugar; the third, R, for ripe banana; and the fourth, U, for unripe banana.
3. Use a graduated cylinder to put 5 mL of water into the test tubes labeled C, R, and U. Place 5 mL of a sugar (dextrose) solution into the test tube labeled S.
4. With a clean graduated cylinder, add 5 mL of Benedict's solution to each of the test tubes.
5. Observe the color and appearance of the unripe banana. Peel it. Use a scalpel to cut a slice, or cross section, 5 mm thick. CAUTION: *Always cut away from yourself and others*.
6. Cut the slice of banana in half along its diameter. Then make a cut parallel to the diameter, about 5 mm from the cut edge, as shown in the accompanying illustration.
7. Measure the mass of the cut piece. It should have a mass of about 1 g. Put this piece of banana into the test tube marked U.
8. Repeat steps 5 to 7 with the ripe banana. Make sure the mass of the piece of ripe banana is the same as the mass of the unripe banana. Place this piece in the test tube marked R.

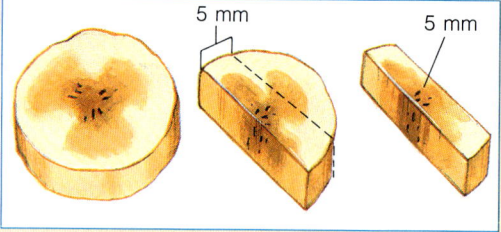

9. Place the test tubes in the beaker of boiling water on the hot plate. CAUTION: *Use the test tube holder. Place the tubes carefully.*
10. Observe the four test tubes. When the test tube that contains the sugar solution changes color, observe the color of the other test tubes.
11. Use the test tube holder to remove the test tubes from the beaker. Place the test tubes in the test tube rack. Turn off the hot plate and allow the beaker to cool.
12. Make several more slices of the ripe banana. Use a hand lens to examine the region near the center of each slice.

OBSERVATIONS
1. What did the peel of the unripe banana look like? The ripe banana?
2. In which test tubes did the greatest change occur?
3. Describe the structures you observed in the center of the banana slices.

ANALYSIS AND CONCLUSIONS
1. What do the results of the tests with Benedict's solution show?
2. What are the structures in the center of a banana?
3. How do animals help disperse banana seeds?
4. What changes occur when a banana ripens?
5. Why would an animal be more likely to find and eat ripe bananas than unripe bananas?

Skills Development
Students will use the following skills in the completion of this laboratory investigation.
1. Manipulative
2. Measuring
3. Observing
4. Comparing
5. Relating
6. Inferring
7. Analyzing
8. Safety

Safety Tips
Students should wear goggles when doing any lab involving fire or the heating of liquids. They should also remember that glass remains hot for a long time but does not look hot. Finally, glass and scalpels are always potentially hazardous, and students should be alerted to proper use and care.

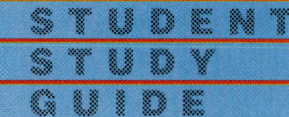

SUMMARIZING THE CONCEPTS

The key concepts in each section of this chapter are listed below to help you review the chapter content. Make sure you understand each concept and its relationship to other concepts and to the theme of this chapter.

22–1 Seed Plants—The Spermopsida

- Seed plants have roots, stems, and leaves that show adaptations that enable them to perform different functions.
- Seed plants are able to reproduce without the need for standing water. Seed plants produce seeds that are able to survive periods of time that are unfavorable for growth.

22–2 Evolution of Seed Plants

- The gymnosperms are the most ancient group of surviving seed plants. The name gymnosperm means naked seed.
- The most common gymnosperms are the conifers. Conifer means cone-bearing. Most conifers produce cones, which are special reproductive organs.
- All flowering plants belong to the angiosperms. Flowers are the angiosperms' reproductive organs.
- There are two main subclasses of angiosperms: monocots and dicots. Monocots have one seed leaf; dicots have two. The veins in monocot leaves are parallel to one another. The veins of dicots form a branching network in the leaves. The flower parts of monocots occur in threes or multiples of three. The flower parts of dicots occur in fours or fives or multiples of four or five. The vascular bundles in dicots form a ring around the stem. The vascular bundles of monocots are scattered around the stem.

22–3 Coevolution of Flowering Plants and Animals

- Some flowering plants are pollinated by the wind. These plants shed vast amounts of pollen into the air.
- The process by which two organisms evolve structures and behaviors in relation to or complementary to one another is called coevolution.
- Many animals are pollinators of flowers, or agents that transfer pollen from one flower to another.

REVIEWING KEY TERMS

Vocabulary terms are important to your understanding of biology. The key terms listed below are those you should be especially familiar with. Review these terms and their meanings. Then use each term in a complete sentence. If you are not sure of a term's meaning, return to the appropriate section and review its definition.

22–1 Seed Plants— The Spermopsida
pollen grain
pollination
embryo
seed coat

22–2 Evolution of Seed Plants
scale
gymnosperm
pollen cone
angiosperm

flower
fruit
monocot
dicot
cotyledon
vascular bundle

22–3 Coevolution of Flowering Plants and Animals
coevolution
vector pollination
seed dispersal

Observations
1. Ripe: yellow with brown specks. Unripe: greenish-yellow.
2. In the test tubes with sugar and ripe banana.
3. The structures were small brown ovals.

Analysis and Conclusions
1. Ripe bananas contain more sugar then unripe bananas.
2. Seeds.
3. By picking bananas and discarding them from place to place as they eat.
4. When bananas ripen, they get sweeter and turn yellow.
5. Ripe bananas are easier to spot, and they smell and taste better.

Going Further: Enrichment
Repeat the same experiment with other fruits and vegetables to determine their ripeness. Vegetables that contain large amounts of sugar, like some peas, can be a real challenge if they retain their green color. After testing several ripe and unripe fruits, students should be prepared to make some general comments and conclusions about sugar content and variation.

Teaching Strategy
1. As you move around the room, monitor to be sure that students are following directions. Also, caution them about safety rules. This lab requires careful observation. Students may need help with determinations of relative amounts of sugar. Be sure test tubes are clearly labeled to avoid mix-ups.
2. If the water bath is boiling, color changes appear within one minute after the test tubes are placed in the bath. Do not let students leave the test tubes in the heat longer than necessary, as starchy material can break down, giving a positive result.

CHAPTER REVIEW

CONTENT REVIEW

Multiple Choice
1. c 2. d 3. a 4. c
5. d 6. a 7. a 8. c

True or False
1. F Coevolution
2. T
3. F Roots
4. F Xylem
5. F insects or flies
6. T
7. T
8. F stationary

Word Relationships

A.
1. flowers
2. seeds
3. products of photosynthesis
4. seeds

B.
5. monocot features; net veins are a dicot feature
6. animal pollinators; wind is not an animal
7. fruits; potato is a tuber

CONCEPT MASTERY

1. Seed dispersal is the movement of seeds away from the plant that produced them. It lessens competition between plants and also widens the growing area of the plant.
2. A cotyledon is a seed leaf. It contains stored food that is used by the plant after it germinates.
3. Answers will vary but may include food, fibers used for clothing, wood for houses, and seed plants that produce oxygen.
4. Because both the squash and the tomato produce seed-bearing fruit.
5. Roots anchor these great trees in the ground. The redwood's vascular tissue carries water from the roots to the top of the plant and food materials from the leaves to the rest of the plant. The vascular tissue also helps support the huge stem.
6. The gametes produced by seed plants do not have to swim through standing water for fertilization to occur. Seed plants have waterproof layers on their leaves that prevent the evaporation of water.
7. A conifer is a gymnosperm that produces cones made of scales. The scales of the female cone hold the seeds. Angiosperms are flowering plants whose seeds are contained within fruits.
8. Wind can act as a pollinating agent. Some plants produce pollen grains that are carried by the wind. These pollen grains may eventually fall upon a pistil. Wind pollination is wasteful. Much of the pollen produced does not land on a pistil. Animal pollination is less

---

CHAPTER REVIEW

CONTENT REVIEW

Multiple Choice

Choose the letter of the answer that best completes each statement.

1. Flowering plants are in the class
   a. cotyledonae.   c. angiospermae.
   b. gymnospermae.   d. coniferae.
2. A red flower is most probably pollinated by a (an)
   a. bat.   c. insect.
   b. gust of wind.   d. bird.
3. Each of the following is an adaptation of plants to a life on land except
   a. tall stems.   c. xylem and phloem.
   b. a waxy cuticle.   d. seeds.
4. The entire male gametophyte of a seed plant is contained within the
   a. embryo.   c. pollen grain.
   b. fruit.   d. xylem.
5. The first seed-bearing plants were the
   a. ferns.   c. conifers.
   b. mosses.   d. seed ferns.
6. You examine a flower and find six petals. This flower is most likely from a
   a. monocot.   c. dicot.
   b. conifer.   d. fern.
7. Each of the following is a fruit except a
   a. potato.   c. squash.
   b. tomato.   d. strawberry.
8. Inside a seed coat, an embryo
   a. continues to grow.
   b. is kept warm.
   c. is protected from drying out.
   d. awaits fertilization.

True or False

Determine whether each statement is true or false. If it is true, write "true." If it is false, change the underlined word or words to make the statement true.

1. <u>Codevelopment</u> is the process by which two organisms evolve structures and behaviors complementary to each other.
2. <u>Leaves</u> are the organs in which most plants make food.
3. <u>Stems</u> absorb water and dissolved nutrients from the soil.
4. <u>Phloem</u> carries water up a plant stem.
5. A flower that smells like rotten meat most likely attracts <u>birds</u> for pollination.
6. The process of distributing seeds away from the parent plant is called <u>seed dispersal</u>.
7. A plant whose leaf veins form a branching network is most probably a <u>dicot</u>.
8. Bees gather nectar while <u>flying</u>.

Word Relationships

A. *An analogy is a relationship between two pairs of words or phrases generally written in the following manner: a:b::c:d. The symbol : is read "is to," and the symbol :: is read "as." For example, cat:animal::rose:plant is read "cat is to animal as rose is to plant."*

In the analogies that follow, a word or phrase is missing. Complete each analogy by providing the missing word or phrase.

1. gymnosperm:cones::angiosperm:_____
2. fruit:seeds::cones:_____
3. xylem:water::phloem:_____
4. ferns:spores::conifers:_____

B. *In each of the following sets of terms, three of the terms are related. One term does not belong. Determine the characteristic common to three of the terms and then identify the term that does not belong.*

5. net veins, parallel veins, one cotyledon, nine petals
6. bee, bird, bat, wind
7. strawberry, blueberry, apple, potato

CONCEPT MASTERY

Use your understanding of the concepts developed in the chapter to answer each of the following in a brief paragraph.

1. What is seed dispersal? How does it contribute to the survival of a plant species?
2. What is a cotyledon?
3. How do seed plants help humans survive?
4. Why do botanists consider a tomato and a squash fruits?
5. How do roots and vascular tissues contribute to a redwood tree's great size?
6. How are seed plants better able to survive drier conditions than mosses and ferns?
7. What is a conifer? How does a conifer differ from an angiosperm?
8. What is wind pollination? How does wind pollination differ from vector pollination?
9. Why is it important that seeds provide food for the embryo plant?

CRITICAL AND CREATIVE THINKING

Discuss each of the following in a brief paragraph.

1. **Applying concepts** In nature, flowers have a limited range of colors. In a garden, however, flowers can have many more colors. Apply your knowledge of pollination and artificial selection to explain why.
2. **Making predictions** In the future, a terrible, fatal disease is found to affect all monocots. Predict the effect of this disease on the human population.
3. **Relating cause and effect** Scientists invent a new insecticide that can kill all the insects in the world. What important harmful effect would this have on plants?
4. **Interpreting diagrams** Examine the plant in this photograph. How many cotyledons would the seeds of this plant have? Explain your reasoning.
5. **Applying concepts** A farmer decides not to plant her fields one year. Later in the year heavier than normal rains fall on the field. Now the farmer wishes she had planted her crops. Why do you think she changed her mind?
6. **Applying concepts** Making a cut through the bark of a tree in a complete circle around the trunk often results in the death of the tree. Using your knowledge of vascular tissue, explain why this might happen.
7. **Relating facts** The seeds of a gymnosperm are probably not likely to be dispersed by animals, whereas the seeds of angiosperms are likely to be dispersed by animals. Explain why this is so.
8. **Using the writing process** Suppose all gymnosperms died out tomorrow. Write a story that details ways in which your life would be changed.

485

wasteful because the pollen is carried to and deposited directly on the pistil.

9. Many seeds grow at or beneath the soil level. They need the food that is stored in the seed to enable them to grow until they can break through the soil, grow leaves, and produce their own food.

CRITICAL AND CREATIVE THINKING

1. In nature, flowers tend to display only the colors and patterns that best attract pollinators. Garden flowers have a greater variety of colors, patterns, and forms as a result of artificial selection.
2. Most of the food plants—for example, corn and wheat—that humans depend on for food are monocots. There would probably be widespread famine if all monocots became extinct. Other answers are possible.
3. It would be a disaster for many plants that are dependent upon insect vectors for pollination. Since it can be assumed that bees would also be killed, many plants would not be able to produce fruit.
4. It would have one cotyledon because it is a monocot. This can be determined by counting the flower parts that occur in multiples of three.
5. The roots of crop plants hold the soil in place. The heavy rains might have washed away valuable topsoil. She may also have wished she planted her fields because the better-than-average rains might have guaranteed a better-than-average harvest.
6. A cut completely around a tree will cut the delicate vascular tissues. This cut would interfere with the movement of materials within the plant, and it would most likely die.
7. The seeds of gymnosperms are contained within a cone. The seeds of angiosperms are contained within a fruit. The fruit would be more likely to appeal to an animal and thus would more likely be eaten. Animals are attracted to the bright colors, smell, and sweet taste of fruit. Thus, the seeds of an angiosperm might have a better chance of being dispersed by animals.
8. Check students' compositions for clarity and correctness.

CHAPTER 23 Roots, Stems, and Leaves

Section	Laboratory Investigations and Activities
23–1 Soil: A Storehouse for Water and Nutrients pages 487–490 THEMES: Systems and Interactions, Scale and Structure	**Teacher Edition** DEMONSTRATION: Examination of Seedlings, p. 486D **Teaching Resources** HANDS-ON ACTIVITY: Making Soil, p. 115 ECOLOGY INVESTIGATION: Measuring Soil Temperature, p. 15 ECOLOGY INVESTIGATION: Calculating the Water-Holding Capacity of Soil, p. 19
23–2 Specialized Tissues in Plants pages 491–493 THEMES: Scale and Structure, Patterns of Change	
23–3 Roots pages 494–498 THEMES: Scale and Structure, Energy	**Teacher Edition** DEMONSTRATION: Roots, p. 486D **Teaching Resources** HANDS-ON ACTIVITY: A Hairy Situation, p. 109 ECOLOGY INVESTIGATION: Respiration in Roots, p. 23 TEACHER DEMONSTRATION: Observing Root Pressure, p. 15
23–4 Stems pages 499–502 THEMES: Scale and Structure, Unity and Diversity	**Laboratory Manual** Root and Stem Structures, p. 277 **Teaching Resources** ECOLOGY INVESTIGATION: Intraspecific Competition, p. 13
23–5 Leaves pages 502–505 THEMES: Scale and Structure, Stability	**Student Edition** LABORATORY INVESTIGATION: Exploring the Structure of Leaves, p. 512 **Laboratory Manual** Leaf Structures, p. 285 **Teaching Resources** HANDS-ON ACTIVITY: Bubbling Leaves, p. 113
23–6 Transport in Plants pages 505–508 THEMES: Scale and Structure, Systems and Interactions	**Laboratory Manual** Estimating the Number of Stomata in a Leaf, p. 293 **Teaching Resources** HANDS-ON ACTIVITY: Whence the Water? p. 117 ECOLOGY INVESTIGATION: Determining the Permanent Wilting Percentage, p. 25 ECOLOGY INVESTIGATION: Observing the Capillarity of Soil, p. 27 **Product Testing Activities** Testing Paper Towels
23–7 Adaptations of Plants to Different Environments pages 508–511 THEMES: Evolution, Unity and Diversity	**Teaching Resources** ECOLOGY INVESTIGATION: Measuring the Water Content of Soil, p. 17 ECOLOGY INVESTIGATION: Measuring and Comparing the Pore Space in Soils, p. 21
Chapter Review pages 513–515	

CHAPTER PLANNING GUIDE

Other Activities	Media and Technology
Teaching Resources BIOLOGY CASE STUDY: Decomposition of Oak Leaves, p. 5 BIOLOGY CASE STUDY: Daily Temperature Changes Above and In Soil, p. 7 **Study Guide** Section 23–1, p. 219	
Study Guide Section 23–2, p. 221	
Teaching Resources ACTIVITY: Illustrating Transport in Plants, p. 3 **Study Guide** Section 23–3, p. 222	**Interactive Videodisc** On Dry Land: The Desert Biome **Biology Media Guide** page 20
Study Guide Section 23–4, p. 224	**Interactive Videodisc** On Dry Land: The Desert Biome **Biology Media Guide** page 20
Study Guide Section 23–5, p. 225	**Interactive Videodisc** On Dry Land: The Desert Biome **Video/Videodisc** Biomes: Coniferous Forest and Tropical Rain Forest **Biology Media Guide** pages 20–21
Teaching Resources ACTIVITY: Observing Transpiration, p. 7 **Study Guide** Section 23–6, p. 227	
Teaching Resources VOCABULARY REVIEW: Crossword, p. 1 **Study Guide** Section 23–7, p. 229	**Interactive Videodisc** On Dry Land: The Desert Biome **Video/Videodisc** Biomes: Coniferous Forest and Tropical Rain Forest Biomes: Desert and Tundra **Videodisc** Aquatic Ecosystems: Freshwater Wetlands and Freshwater Aquatic Ecosystems: Estuaries and Marine **Biology Media Guide** pages 21–25
Teaching Resources Chapter Test A, p. 15 Chapter Test B, p. 19	**Computer Test Bank** Chapter Test, p. 245

CHAPTER 23 Roots, Stems, and Leaves

CHAPTER OVERVIEW

Seed plants have several different kinds of tissues specialized to perform different tasks efficiently. Meristematic tissue, which occurs in several places in a plant, is the only plant tissue that produces new cells by mitosis. The meristem is responsible for all the growth in a plant. Epidermal tissues form the surface layers of leaves, stems, and roots. Parenchyma tissue is composed of thin-walled cells that perform a variety of functions. Sclerenchyma and the two types of vascular tissue, xylem and phloem, are also discussed.

Some plants have root systems with large primary roots called taproots. Others have fibrous root systems with many small secondary roots. The tissues of roots include the epidermis, the cortex, and the vascular cylinder. Each of these tissues is discussed with respect to its function in water and nutrient uptake.

Stems have the two major functions of holding up leaves to the light and conducting materials between roots and leaves. Stems have four basic types of tissues that make these processes more efficient. Stems show many modifications for a variety of purposes.

The leaves of plants carry on photosynthesis, producing a variety of sugars, starches, and fats. Leaves contain several specialized tissues, including epidermal cells, parenchyma, and vascular tissues. The processes of gas exchange and water conservation must be balanced within the leaves of plants if photosynthesis is to occur. Stomata, found in the epidermal layer of leaves, regulate these processes. Most of the photosynthesis in leaves occurs in the mesophyll tissue, which is described in this chapter.

A major problem in any plant is the movement of water from the soil to the leaves. This is the function of the xylem. Obviously, the larger the plant, the greater the problem. Adhesion, cohesion, and capillarity are discussed, as is transpiration pull. Transport of materials in the phloem is more complex because both organic compounds and inorganic ions are transported. Further, transport in the phloem involves the movement of materials in more than one direction. The exact mechanism by which phloem transport works is not completely understood, but there are some hypotheses explaining how this energy-requiring process occurs.

23-1 SOIL: A STOREHOUSE FOR WATER AND NUTRIENTS

Section Focus 23-1

The purpose of this section is to give students some understanding of soil types. Soil is called the "bank" of nutrients that plants use for growth. Soil is defined as a complex mixture of materials. Different soil types are briefly discussed and soil profiles are explained and illustrated.

Performance Objectives 23-1

1. Discuss the importance of soil for plant growth.
2. Compare and contrast types of soil.

23-2 SPECIALIZED TISSUES IN PLANTS

Section Focus 23-2

The purpose of this section is to introduce the different kinds of specialized tissues found in seed plants. Meristematic tissue is found in the tips of growing stems and roots, allowing them to grow in length. The cork cambium and vascular cambium, other types of meristematic tissue, allow roots and stems to branch and grow thicker. Epidermal tissue forms the outer layer of leaves, stems, and roots. Epidermal tissue may protect plants from water loss or help in the absorption of water. Parenchyma tissue is composed of thin-walled cells found in leaves, stems, and roots. Parenchyma cells may be photosynthetic or may serve primarily as storage cells. Sclerenchyma cells have thick cell walls that strengthen and support plant tissues.

Performance Objectives 23-2

1. Discuss the structure and function of plant tissues.

Science Terms 23-2

meristematic tissue   p. 491
apical meristem   p. 491
cork cambium   p. 491
vascular cambium   p. 491
pericycle   p. 491
parenchyma   p. 492
sclerenchyma   p.492
tracheid   p. 492
vessel element   p. 493
sieve tube element   p. 493
companion cell   p. 493

23-3 ROOTS

Section Focus 23-3

The purpose of this section is to explain root structure and function. The tissues of mature roots can be divided into three groups. The first, the epidermis, is a thin layer of cells that takes in water and nutrients. The second group of plant tissues presented are those that make up the cortex. The parenchyma cells of the cortex are involved in the movement of water in plants. The third group of tissues make up the vascular cylinder. Root pressure in the vascular cylinder forces water into the xylem. As the pressure increases, water is forced higher and higher in the xylem.

Performance Objectives 23-3

1. Explain the process of water movement from the soil to the xylem.
2. Discuss the importance of root hairs to water absorption.

Science Terms 23-3

taproot   p. 494
fibrous root   p. 494
epidermis   p. 494
cortex   p. 494
vascular cylinder   p. 495
Casparian strip   p. 497

23-4 STEMS

Section Focus 23-4

The purpose of this section is to discuss the structure and function of stems and some of their unusual adaptations. Stems have two basic functions: They hold leaves up in the sunlight and they conduct various substances between roots and leaves. Stems have four basic types of tissue: parenchyma, vascular, cambium, and cork tissues. These tissues are arranged differently in monocots and dicots.

Performance Objectives 23–4

1. List the functions of plant stems.
2. Describe the arrangement of the vascular tissue in the stem of a young dicot.
3. Discuss several types of stems that are modified for food storage.

Science Terms 23–4

dormancy p. 500
rhizome p. 501
tuber p. 501
bulb p. 501

23–5 LEAVES

Section Focus 23–5

The purpose of this section is to acquaint students with leaf structure and function. In order to collect sunlight, most leaves have large thin, flattened sections called blades that occur in a variety of shapes and sizes, depending on the environment to which they are adapted. Like roots and stems, leaves have an outer covering of epidermal cells, inner layers of parenchyma cells, and vascular tissues. The epidermal layer of many leaves is covered by a waxy coating called a cuticle. However, leaves are still able to exchange water and gases between the air and their interior tissues. There are specialized cells in the epidermis called guard cells that surround small openings. These openings, called stomata, facilitate gas exchange. Most of the photosynthesis in a leaf takes place in the mesophyll tissues. The mesophyll has two distinct types of cells that make up the palisade layer and the spongy mesophyll. Both are important to the photosynthesis process.

Performance Objectives 23–5

1. Describe the epidermis of the leaf.
2. Discuss the exchange of water and gases that occurs between the plant leaves and the atmosphere.
3. Discuss the importance of the mesophyll tissue.

Science Terms 23–5

petiole p. 502
stoma p. 504
guard cell p. 504
mesophyll p. 504
palisade layer p. 504
spongy mesophyll p. 504

23–6 TRANSPORT IN PLANTS

Section Focus 23–6

The purpose of this section is to explain the transport of materials inside a plant. Xylem tissue forms a continuous set of tubes that move water from the roots to the leaves. Root pressure plays an important role in this process but cannot alone account for the movement of water and dissolved materials over great distances in plants. Obviously, other forces are at work. Adhesion, cohesion, and capillarity and their relationship to one another are explained. Capillarity accounts for some water movement in plants that are not very tall. In very tall plants, another force, transpiration pull, is also at work. Water evaporates continually from the surface of leaves. This evaporation, called transpiration, causes water to move upward in the plant to replace that which is lost. The phloem transports organic molecules in plants. During the day, products of photosynthesis are moved from the leaves into the stems and roots, where they are used or stored. The phloem also transports sugars and food for storage within the plant on a seasonal basis as they are needed. The exact mechanism by which phloem transport works is not completely understood, but it is known that it is a process that requires a source of energy.

Performance Objectives 23–6

1. Discuss the various theories concerning the movement of water through a plant.
2. List several functions of the phloem.
3. Describe the way that phloem transport works.

Science Terms 23–6

transpiration p. 506
transpiration pull p. 506

23–7 ADAPTATIONS OF PLANTS TO DIFFERENT ENVIRONMENTS

Section Focus 23–7

The purpose of this section is to describe some of the special and unusual adaptations found in angiosperms. These adaptations show how roots, stems, and leaves have been modified through evolutionary change to serve different functions.

Performance Objectives 23–7

1. Discuss some of the adaptations that make plants successful in different environments.

DISCOVERY LEARNING

TEACHER DEMONSTRATION
Modeling

Examination of Seedlings

This activity can be used to introduce Chapter 23.

If the students grew seedlings as suggested in Chapter 22, you could use them for this demonstration. Otherwise, any seedlings will do. Remove all of the soil or sawdust from the roots and gently rinse. Lay the seedlings on paper towels for the students to examine. Indicate the roots, stems, and leaves of the plants. Point out to students that the root and stem are continuous. You will want them to understand that inside the plant is a continuous system of tubes from the leaves to the roots. This is similar to the transport system in animals. The transport system and the three plant organs they observe are those they will be studying in this chapter.

Roots

This demonstration can be used with Section 23–3.

Show students samples of fibrous roots and taproots that you can get from the school grounds or your yard. Discuss the advantages and disadvantages of each type of root system. Now show the students radishes and carrots. Discuss the special functions of these taproots.

CHAPTER 23

Roots, Stems, and Leaves

GUIDED ENQUIRY
Pose the following questions to students and have them record their responses. Point out that they will gain a better understanding of the key concepts if they read the chapter with these basic questions in mind. Upon completion of the chapter, pose the questions again. Ask students to compare their initial responses with those they have developed after reading the chapter.
- How is the type of soil related to the kinds of plants that grow in an area?
- Do plants have specialized tissues and organs similar to those found in animals?
- How does a tall plant get water from the roots to the top?
- How do the roots of a plant get food if they do not carry out photosynthesis?
- Are the veins in plant leaves similar in function to the veins in animals?
- How do plants grow?

CHAPTER 23

Roots, Stems, and Leaves

Growing through a crack in the top of a mountain, this pine tree demonstrates the ability of plants to grow under adverse conditions. The sunflower, too, growing in the ever-moving desert sands, shows the same tenacity.

On a mountain ridge, the twisted roots of an ancient pine tree cling to solid rock in subzero temperatures. In a desert, a sagebrush plant survives in the blazing sun for months without rain. On a dune by the seashore, a clump of beach grass flowers in poor, salty sand. And in the Everglades, mangrove trees live with their roots submerged in rich, organic mud.

These plants have adaptations that enable them to live under difficult conditions. How do these plants obtain the water and nutrients they need to survive in their harsh surroundings? For that matter, how does any plant extract what it needs from soil and air? The answers to these questions will become clear in this chapter.

INTRODUCING CHAPTER 23
Using the Visuals
Perform a simple experiment that students have done many times before. The day before this lesson, place one or more cut white carnations in water containing food coloring. Set these aside where the students cannot see them. In class, repeat the procedure. Discuss what will happen to the flowers. After the discussion, show them the flowers that you prepared ahead of time. Examine them under a dissecting scope.

• **What happened to the colored water?** (It moved into the petals of the flower.)
• **How did this occur?** (There are vessels in the stems into which water traveled to the leaves.)
• **How could the plant take in water without roots?** (The vessels are continuous from the roots to the stems to the leaves.) Look at the chapter-opener photographs and read the caption. Read the introductory information. Discuss how the photographs and the information relate to the experiment.

GUIDE FOR READING

After you read the following sections, you will be able to

23–1 Soil: A Storehouse for Water and Nutrients
- Discuss soil's importance.
- Describe soil's composition.
- List some nutrients in soil that are essential for plants.

23–2 Specialized Tissues in Plants
- Discuss meristematic tissue.
- Describe the functions of different plant tissues.

23–3 Roots
- Discuss root structure and function.
- Explain how water and nutrients move into and through roots.

23–4 Stems
- Describe stem functions.
- Discuss stem tissues.

23–5 Leaves
- Describe leaf structure.
- Explain how gas exchange and water conservation are balanced in a leaf.

23–6 Transport in Plants
- Explain how water moves through a plant.
- Describe phloem transport.

23–7 Adaptations of Plants to Different Environments
- Discuss plant adaptations.

Journal Activity

YOU AND YOUR WORLD
In your journal, make a drawing of your favorite plant. Explain why it is your favorite.

23–1 Soil: A Storehouse for Water and Nutrients

Guide For Reading

■ What is soil? How does it affect plants?
■ How does soil vary in composition and structure?
■ What nutrients are essential for plant growth?

Seed plants are autotrophs; they make the organic molecules they need from raw materials such as water, carbon dioxide, and several inorganic nutrients. Where do they get these raw materials? Carbon dioxide is removed from the air by the leaves. Water and inorganic nutrients are taken in from the soil by plant roots. Soil is a "bank" of nutrients plants use to grow.

Types of Soil

An understanding of soil will help explain how plants function. What exactly is soil? **Soil is a complex mixture of sand, silt, clay, and bits of decaying animal and plant tissue.** Soil in different places and at different depths contains varying amounts of these ingredients. The ingredients define the soil and determine, to a large extent, the kinds of plants that can grow in it.

Sandy soil is composed mostly of sand grains. Because sand grains are large and irregularly shaped, there are large empty spaces between them. That's why water poured on sand drains through it so quickly.

Clay soil is composed mostly of very fine clay particles that have extremely small spaces between them. Horticulturists—professional growers of plants—use the word heavy to describe soil that contains large amounts of clay particles. Water poured on clay soil sinks in slowly.

Figure 23–1 *Wildflowers grow well in the dappled forest shade. The soil beneath the trees contains decaying leaves and other organic materials that are used by plants as natural fertilizer.*

COOPERATIVE LEARNING

Read the following scenario to the students:

There has been much publicity about the contamination of soil and water by toxic wastes leaking from existing dump sites. You live in a community that is economically dependent on agriculture. A large international chemical company has applied for a license and hopes to begin construction of a toxic waste dump immediately following a round of public hearings.

Using preassigned lab groups of randomly selected teams, have groups prepare an alternatives grid (see the alternatives grid in the Cooperative-Learning assignment for Chapter 6) using the following alternatives:

- Build according to the company's plan
- Build with modifications (groups must specify modifications)
- Do not build

Political, social, economic, and ecological consequences should be predicted for each alternative. As time permits, randomly assign each group to represent one of the three alternatives at the public hearing. Remind students to use the concepts presented in this chapter.

Journal Activity

YOU AND YOUR WORLD
You may want to use student drawings to make a classroom bulletin-board display. Students should be instructed to keep their Journal Activity in their portfolio.

TEACHING STRATEGY 23–1

Focus/Motivation
Set up several beakers with samples of different types of dry soils—clay, sand, and loam. Use the purest samples available so that differences are obvious. Ask students to examine the soil to determine particle size and overall appearance. Pour a small amount of water into each soil sample and observe the movement of water into the soil.

• **Which soil type has the largest particles?** (Particles of organic matter in the loam are the largest. Depending on your samples, the particles of sand may appear to be the largest.)

• **What happens when water is poured into the sample?** (Answers will vary but should indicate that water goes straight through the sand but stays on top of the clay. The loam may act in different ways, depending on its composition.)

• **Which soil do you think would be easiest for root growth?** (Sand or loam.)

• **Most difficult?** (Clay.)

• **Are these pure samples typical of soil that you would find in a garden or your yard?** (No, that is usually a mixture of soil types.)

487

ANNOTATION KEY

① The leaves of the iron-deficient plant are pale and appear sickly, while the leaves of the healthy plant are dark green. (Making observations)

ECOLOGY NOTE
SOIL, CLIMATE, AND LIFE

Soil, climate, and plants interact in ways that dramatically affect the response of ecosystems to disturbance. Across the American Midwest, for example, deep, sandy soil was covered for many years by native grasses whose leaves and roots decayed to produce thick, humus-laden soil. That soil is superbly suited to production of corn, wheat, and other grains that can be sustained for decades with proper management. In most tropical rain forests, on the other hand, heavy rainfall, high temperatures, and high humidity cause organic matter to break down so quickly that humus is restricted to a thin surface layer. When vegetation is removed for farming, humus quickly vanishes, the remaining soil is leached and exhausted of nutrients, and dense sediment grains compact into clay so hard that it can break a plow. Such land turns to wasteland in about five years from the time it is cleared.

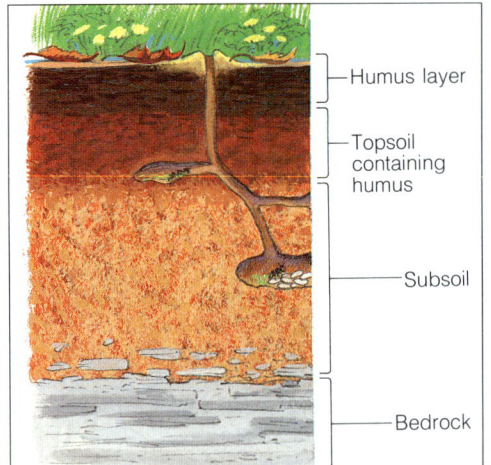 

Figure 23–2 Soil is composed of layers of materials such as humus, sand, clay, minerals, and rocks. The topmost layer is usually richest in organic material. Together the layers make up the soil profile.

Loamy soil contains decaying organic matter formed from plant and animal tissue, in addition to particles of sand and clay. Particles of organic matter are usually larger than sand or clay particles. Organic matter helps increase the soil's ability to hold air and water. Nitrogen, phosphorus, and other nutrients that can be used by plants are released from the soil as the organic matter is broken down by fungi and bacteria.

Soil Profiles

If you cut a trench in the ground, you will see that the soil is arranged in layers, with each layer corresponding to a different kind of soil. The particular order of layers is called a soil profile. See Figure 23–2. Most organic matter is on top, where it forms the humus layer. The humus layer in forest soil can be several centimeters thick. In grasslands, however, the humus layer may be much thinner.

Under the humus layer is the topsoil, which is a mixture of humus, sand, clay, and minerals. The thickness of topsoil varies with the location. Topsoil in the fertile plains of the Midwest is often a meter thick, whereas in northeastern forests it may be only a few centimeters deep. Plant roots grow best in the humus and topsoil layers.

Beneath the topsoil is the subsoil, a mixture of rocks and inorganic soil particles often tightly packed together. Subsoil offers little room for air, water, and plant roots. Beneath the subsoil is bedrock, a layer of rock that usually cannot be penetrated by plant roots.

Almost half the total volume of good garden soil actually consists of open spaces between soil particles. These open spaces are important for several reasons. They are home to beneficial bacteria, fungi, and protozoa that help plants grow.

23–1 (continued)

Content Development
Define the term *soil* and emphasize that typical soils are mixtures of clay, sand, and loam. Relate this to the demonstration you did to introduce this section. Discuss the types of soils and the characteristics of plants (particularly the root systems) that would be successful in these soils. Be sure to discuss the importance of organic matter in the soil. Point out the role of fungi and bacteria as an interesting sidelight.

Content Development
Using the Visuals Study each layer of the soil profile with the students. Use Figure 23–2 to help students understand a soil profile.

SECTION REVIEW 23–1

1. Soil is a complex mixture of sand, silt, clay, and bits of decaying animal and plant tissue. Soil is important for plant growth because it contains minerals and water necessary to the plants.
2. Roots grow in the spaces between soil particles. It is in the spaces that water and air, both essential to the growth of roots,

The tiny pockets of water and air are essential to the good health of plant roots and soil microorganisms. Air pockets allow roots to breathe. Water pockets are vital because nutrients are available to plants only when they are dissolved in water. Finally, roots do not actually grow "in" soil. Roots really grow into and through the spaces between soil particles. If there are not enough spaces, roots cannot grow properly.

Essential Nutrients

To grow, flower, and produce seeds, plants require a variety of inorganic nutrients in addition to carbon dioxide and water. These inorganic nutrients are located in the soil. And it is from the soil's bank of nutrients that plants withdraw their needs. The most important of these nutrients are nitrogen, phosphorus, potassium, calcium, magnesium, and various trace elements.

Nitrogen is essential for proper leaf growth and color. A lack of nitrogen will stunt plant growth and cause the foliage to turn a sickly yellow color.

Phosphorus is used to make DNA in all cells and is important in the development of roots, stems, flowers, and seeds. Too little phosphorus causes poor flowering and stunted growth.

Potassium (sometimes called potash) plays a critical role in the making of proteins and carbohydrates by plants. Potassium also plays a role in the development of roots, stems, and flowers, and helps plants resist cold and disease. The first sign of a potassium deficiency is usually stunted roots, which soon result in stunted leaves.

Calcium is used in cell metabolism and is necessary for cell growth and division and cell wall strength.

Magnesium is a critical part of the chlorophyll molecule. A lack of magnesium will prevent the manufacture of chlorophyll and may eventually kill the plant.

Trace elements, which plants require in small quantities, include sulfur, iron, zinc, molybdenum, boron, copper, manganese, and chlorine. To maintain proper plant growth, trace elements are needed in very small amounts. In fact, large amounts of trace elements in the soil can be poisonous.

Figure 23–3 Plants need minerals from the soil in order to grow well. The rhododendron in the top photograph shows the effects of iron deficiency. The rhododendron in the bottom photograph received a good supply of needed minerals. What differences can you observe between the two plants? ❶

23–1 SECTION REVIEW

1. What is soil? Why is soil important for plant growth?
2. Why are the spaces between soil particles important?
3. What are some of the essential nutrients found in soil? Why is each important for plants?
4. **Critical Thinking—Making Inferences** Which type of soil is best for plant growth? Explain your answer.

SCIENCE, TECHNOLOGY, AND SOCIETY CONNECTION

Small Friends

This feature talks about the microorganisms that are beneficial to plants. Some plants have "fertilizer factories" on their roots that are actually bacteria. Many of these bacteria are from the genus *Rhizobium* and can take nitrogen from the air and convert it into compounds that can be used by plants. These bacteria commonly grow on legumes such as peas and beans. Other plants grow well only when their roots are infected by a fungus that improves the plant's ability to absorb important minerals from the soil. This association is called mycorrhiza. These are only two examples of the benefits plants get from microorganisms. In the future, scientists may find other such benefits.

Read and discuss the feature with your students. Read the background information about mycorrhizae to the students. Point out that plants could probably survive without microorganisms but would not be as successful. To illustrate the importance of these symbiotic relationships, discuss the nitrogen cycle and explain where nitrogen-fixing bacteria and fungi fit into the cycle. The nitrogen cycle is always easier to understand with illustrations, especially those made by the students.

SCIENCE, TECHNOLOGY, AND SOCIETY CONNECTION

Small Friends

In our daily struggle to live a healthy life, we tend to view all microorganisms as harmful. Microorganisms are the cause of many human and plant diseases. That is a fact. However, many of the microorganisms we rightly view with alarm have relatives that make important contributions to human and plant life. The benefits of microorganisms to humans are discussed elsewhere in this textbook. Here you will read about the benefits to plants.

You know that plants need to take in vital minerals and other inorganic compounds from the soil. These compounds are so important that farmers often spend millions of dollars to add them, in the form of fertilizers, to their soils. However, some plants carry their own "fertilizer factories" right on their roots. The factories are actually hard-working bacteria. The bacteria, many from the genus *Rhizobium*, are able to take nitrogen from the air and convert it into a form that plants can use. (You may remember that nitrogen makes up most of the air we breathe.) In this way the bacteria "fertilize" the plants on which they live. Farmers have found that these bacteria commonly grow on legumes such as peas and beans. Unlike most crop plants, which deplete, or use up, the minerals in the soil, legumes actually improve the soil in which they grow.

Other plants grow well only when they have a fungus infecting their roots. This may seem strange, but the fungus is able to improve the plant's ability to absorb important minerals from the soil. The association of plant roots and fungi is called mychorrhizae. This word comes from two Greek words that mean fungus and root. Pines, orchids, blueberries, and rhododendrons are but a few of the plants that form mychorrhizae. In fact, these plants do not grow well without the fungi!

These are only two of the benefits accrued to plants from microorganisms. Scientists are currently working to improve plant growth by exploiting these unseen relationships. In the future they may be able to harness these tiny microscopic farmers that live in soil.

Bacteria that live in these nodules on the root of a pea plant are able to change atmospheric nitrogen into a form that can be used by plants.

490

TEACHING STRATEGY 23–2

Focus/Motivation

Using the Visuals Ask students to list some of the life functions of animals. Students will come up with many varied answers, but the ones that you will want to highlight include food-getting, transport of materials, and gas exchange. Tell the students that plants also have specialized tissues and organs to perform these tasks. Point out Figure 23–4 and discuss the structures that are labeled. There may be some terms on the drawing that you will have to define because they are unfamiliar.

Content Development

As you begin to develop the content in Section 23–2, you might find that a features matrix works as well as notes for your students. There is a considerable amount of unfamiliar material that needs to be organized. You might have features that you wish to include, but use a minimum of one column for description and one for examples.

23–2 Specialized Tissues in Plants

Guide For Reading
- What is meristematic tissue, and why is it important?
- What are the functions of different types of plant tissues?

Seed plants have several different kinds of tissue, each of which is specialized to perform different tasks. This specialization allows each tissue to perform efficiently. Later in this chapter when we examine the structure of roots, stems, and leaves, you will see how the tissues in plants are exquisitely adapted to perform different functions.

Meristematic Tissue

Mature plant cells can grow in length, and they can also expand to store water or food. But most mature plant cells do not divide to form new cells. Yet plants do grow—which means new cells must be formed.

Meristematic (mair-ih-steh-MAT-ihk) **tissue** is the only plant tissue that produces new cells by mitosis. Meristematic cells divide rapidly and have thin cell walls. At first, all cells produced in meristematic tissue look alike. As meristematic cells mature, however, they differentiate into tissues of various kinds.

Meristematic tissue is found in several places. At the end, or tip, of each growing stem and root is an **apical meristem**. Apical meristems enable stems and roots to grow in length. In fact, plants have what is known as an open type of growth. They continue to grow from their tips as long as they live. For example, trees increase their height above ground and the length of their roots below ground for the duration of their lives. This is in contrast to humans and animals, who reach a certain size and then cease to grow (in height, not necessarily in width).

Two other kinds of meristematic tissue—**cork cambium** and **vascular cambium**—allow stems and roots to branch as well as to grow thicker. Cork cambium produces the outer covering of stems. Vascular cambium produces vascular tissues and increases the thickness of stems over time. Another kind of cambium, the **pericycle**, is found in roots. The pericycle enables roots to grow thicker and makes it possible for roots to branch.

Surface Tissue

Epidermal (the prefix *epi-* means on or on the outside of; *dermal* means of the skin) tissues form the outer, or surface, layers of leaves, stems, and roots. Some epidermal tissues, like the cork tissue in tree bark, protect plants from water loss. The epidermal tissue of roots, on the other hand, helps in the absorption of water.

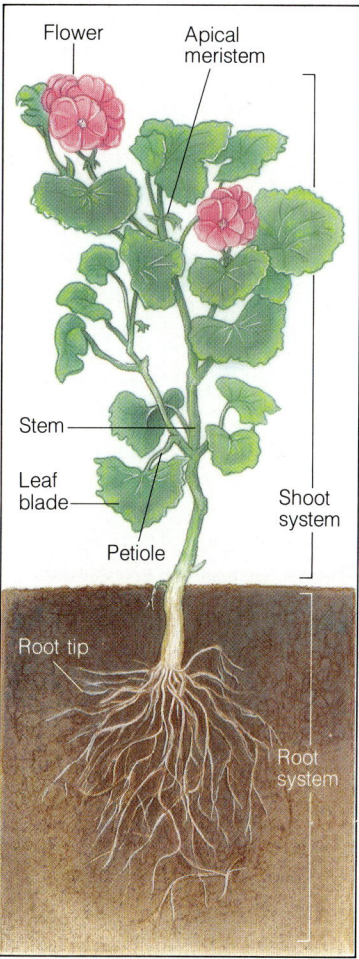

Figure 23–4 Plants are made up of relatively few kinds of tissue. New cells are formed in meristematic tissue found at the tips of both the stem and the roots.

ECOLOGY NOTE
SYMBIOTIC ASSOCIATIONS

An adaptation that enables some plants to live in poor soil is the development of mycorrhizae, symbiotic associations between the roots of plants and some fungi found in the soil. The fungus grows in or around the root and sometimes penetrates into the cells of the root cortex. Trees are the best examples of plants that exhibit mycorrhiza.

The classic study of this phenomenon was done by A. B. Hatch. He compared pine seedlings that had formed mycorrhizae with those that had not. He found that the seedlings with mycorrhizae absorbed nitrogen, phosphorus, and potassium more rapidly. They also grew more rapidly. Some of the fungi in mycorrhizae are believed to increase the solubility of soil nutrients by secreting substances into the soil or to increase the rate of conversion of nutrients into compounds that the plant can use. These special activities of the fungi are believed to be the important advantages of mycorrhizae to plants.

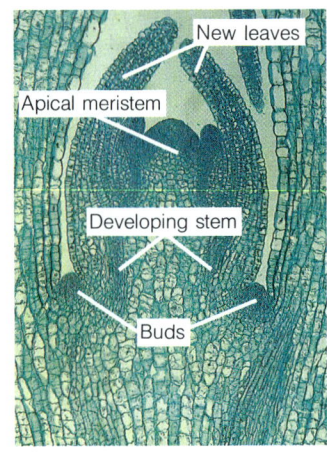
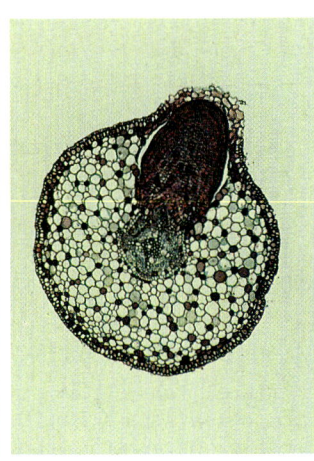

Figure 23-5 The growing tip of a flax plant in this photomicrograph shows newly dividing cells in the dome-shaped cluster of cells near the top of the stem (left). Over the cluster of cells, forming an arch that resembles two arms held overhead, are two young leaves. Actively dividing cells are also found in roots, as this photomicrograph of a branch root being formed in the root of a willow tree shows (right).

Parenchyma

Parenchyma (puh-REHNG-kih-muh) tissue is composed of thin-walled cells found in leaves, stems, and roots. These cells often have large central vacuoles surrounded by a thin layer of cytoplasm. Most of the plant roots we eat—such as potatoes and radishes—are composed primarily of parenchyma cells. Certain parenchyma cells in leaves and stems contain chlorophyll. It is in these cells that most photosynthesis takes place. Some parenchyma cells use vacuoles to store the products of photosynthesis.

Sclerenchyma

Sclerenchyma (sklih-REHNG-kih-muh) cells have tough, thick cell walls that strengthen and support plant tissues. Some sclerenchyma cells are star-shaped and are scattered throughout parenchyma tissue. Other sclerenchyma cells grow into long, narrow fibers. Linen, a fine cloth, is made from the sclerenchyma fibers of the flax plant.

Vascular Tissue

Xylem tissue conducts water. All seed plants have a type of xylem cell called a tracheid. Tracheids are long, narrow cells with walls that are impermeable to water. These walls, however, are pierced by openings that connect neighboring cells to one another. When tracheids mature, they die, and their cytoplasm disintegrates. This leaves a network of hollow connected cells through which water can pass. See Figure 23-6.

Angiosperms have yet another kind of xylem cell called a **vessel element**. Vessel elements are much wider than tracheids, but like tracheids, they mature and die before they

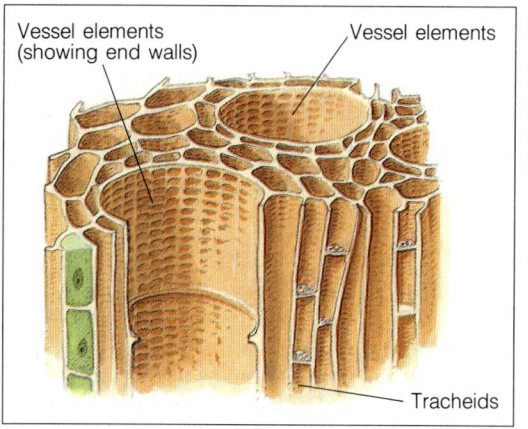
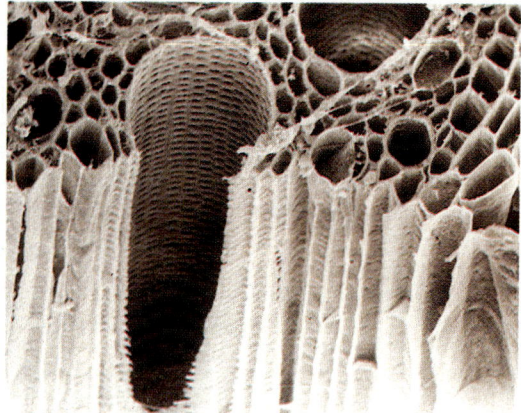

Figure 23–6 *In vascular plants, xylem tissue conducts water from roots to the rest of the plant. The photomicrograph shows the hollow dead cells that make up xylem tissue.*

conduct water. Vessel elements are arranged end to end on top of one another like a stack of tin cans. The cell walls at both ends are lost when the cells die. At that time, the stacked vessel elements become continuous tubes through which water can move freely.

Phloem tissue conducts a variety of plant products. The main phloem cells are the **sieve tube elements**. These cells are arranged end to end, like the vessel elements, to form sieve tubes. The end walls of sieve tube elements have many small holes in them. Materials can move from one adjacent cell to another through these holes. As sieve tube elements mature, they lose their nuclei and most of the other organelles in their cytoplasm. The remaining organelles hug the inside of the cell wall. The rest of the cytoplasm can be moved from cell to cell to carry substances around in the plant.

Companion cells are phloem cells that surround sieve tube elements. Companion cells keep their nuclei and other organelles throughout their lifetime. Companion cells are believed to control the activity of the sieve tube elements.

Figure 23–7 *Phloem tissue conducts a variety of materials, mostly carbohydrates, throughout a plant.*

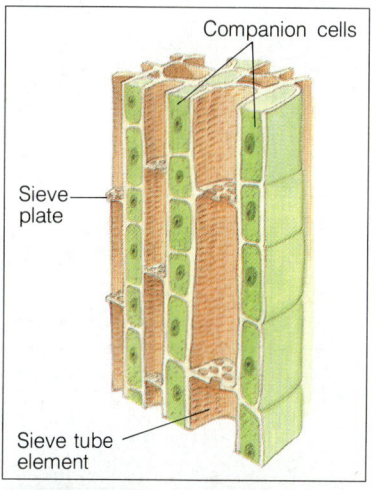

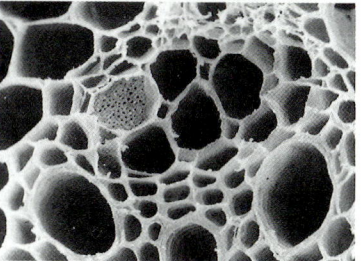

23–2 SECTION REVIEW

1. What is meristematic tissue? Why is it important?
2. What is the function of parenchyma tissue? Sclerenchyma tissue?
3. What are the functions of the two types of vascular tissue?
4. **Critical Thinking—Assessing Relationships** What is the most important type of phloem cell? Why?

493

MULTICULTURAL STRATEGY

Plant images abound in the cultural heritages of most peoples. In Norse mythology, the branches of the ash tree Yggdrasil support the universe even as the tree's roots are gnawed by the monster Nithogg. Judeo-Christian tradition has it that the Bible is a "tree of life to those who hold fast to it." According to a Polynesian legend, the coconut tree—which provides many useful things—was given to humans as a reward for gifting a stranger with food, water, shelter, and other necessities. A Native American myth from the Southwest tells how the first people climbed up into the world on the branches of a tree. Indian philosophers have noted that people are like leaves on a tree—although we may perceive ourselves as separate from one another, we all are connected and belong to something greater than ourselves.

Encourage students to explore the myths and legends of their own cultural traditions and share what they have learned with the class.

Guide For Reading

- What are the main tissues in a mature root?
- How are taproots and fibrous roots different? How are they similar?
- How do the parts of a root work together to absorb and transport water and nutrients?

23–3 Roots

As soon as a seedling begins to grow, it puts out its first root: the primary root. Other roots, called secondary roots, branch out from this primary root. See Figure 23–8. When the root first begins to grow, it resembles all other plant roots. Later in its development, the pattern of growth changes.

Taproots and Fibrous Roots

In some plants the primary root grows longer and thicker, whereas secondary roots remain small. This type of primary root is called a **taproot**. Taproots of oak and hickory trees grow so long that they can reach water far below the earth's surface. Carrots, dandelions, beets, and radishes have short, thick taproots that store sugars or starches. Taproots make it extremely difficult to remove certain plants from a lawn. For example, if you pull on the green leaves of a dandelion, it breaks off above the root. The taproot that remains in the ground can grow new leaves. Sometimes even a small piece of a taproot can produce a new plant.

In plants in which taproots do not form, secondary roots grow and branch. In these plants no single root grows larger than the rest. Such plants, of which grasses are an example, are said to have **fibrous roots**. The fibrous root system of a single rye plant—a plant in the grass family—can have as many as 14 million secondary roots. If these roots were laid end to end in a single line, they would have a total length of over 600 kilometers—much longer than the distance from New York City to Washington, D.C.! The extensive root systems produced by some plants are extremely important in holding topsoil so that it does not get washed away by rains.

Root Tissues

The tissues of mature roots can be divided into three groups: epidermis, cortex, and vascular cylinder. The arrangement of these tissues differs between monocots and dicots, as you can see in Figure 23–9.

Each of these tissues performs important functions that are directly related to its structure. The **epidermis** is a thin layer of cells that take in water and nutrients. The **cortex** transports water and nutrients inward through the root and may store

Figure 23–8 Roots of plants vary greatly in appearance. The long taproot of a dandelion (top) makes it extremely difficult to eradicate this plant from a lawn. If even the smallest piece of the brittle taproot is left in the soil, the dandelion will grow anew. The fibrous root system of an onion (bottom) is different from the taproot of a dandelion. Onions can be pulled from the ground with ease.

TEACHING STRATEGY 23–3

Focus/Motivation

Note: Place several radish seeds between damp paper towels in culture dishes 2 to 3 days before you do this lesson.

Begin the lesson by arranging the students into small groups. Remove several radish seedlings from the culture dishes and place them individually on paper towels so that students can examine the roots. Give each group a hand lens to aid their observations of roots and root hairs.

- **Is this an example of a taproot or fibrous root?** (Taproot.)
- **Describe the appearance of the root hairs.** (White, cottony fibers.)
- **What do you suppose is the function of the root hairs?** (Answers will vary but might include mention of the increased surface area.)

Content Development

Compare the structural characteristics of each type of root. Stress the importance of each type of root in the biological sense (the big picture). Talk about the advantages and disadvantages of each type to the plant.

Reinforcement/Reteaching

As you talk about the groups of tissues that make up the root,

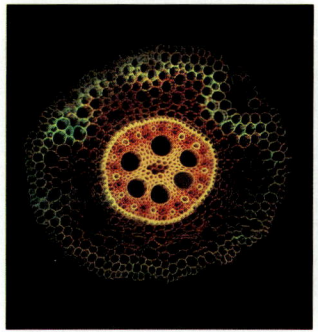
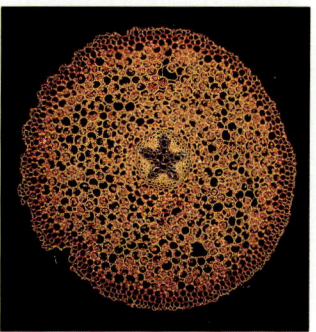

Figure 23–9 Cross sections of the root of a corn plant (left) and the root of a buttercup (right) are shown in these photomicrographs. Most dicot roots have a central column of xylem cells with radiating arms. Which plant is the dicot? ❶

sugars or starches. The **vascular cylinder** contains xylem and phloem. We will discuss these tissues and their functions, in order, from the outside of the root to its center.

The Epidermis: Uptake of Water and Nutrients

The outermost layer of cells, the epidermis, absorbs water and nutrients from the soil. Epidermal cells grow slender projections called root hairs. Although individual root hairs are very small, there are an enormous number of them on a typical plant. The rye plant you just read about can have more than 14 billion root hairs!

Root hairs penetrate the spaces between soil particles. It is through the delicate tissues of the root hairs that water and nutrients enter plant roots. Although each individual root hair has a small surface area, the combined surface area of the huge numbers of root hairs is tremendous. That single rye plant's 14 billion root hairs provide the plant with a root surface area of more than 400 square meters! Root hairs thus enable roots to make close contact with a large area of soil.

NUTRIENT ABSORPTION Root hairs absorb dissolved nutrients from soil spaces through the process of active transport. Recall from Chapter 5 that active transport is a process in which a cell expends energy to move something from one side of a membrane to the other. In roots, active transport is necessary because nutrient ions are present in soil water in lower concentrations than they are in epidermal cells. In fact, these ions would tend to move out of root hairs by diffusion if active transport did not pull them inside.

As you may remember, active transport requires ATP (as a source of energy) and oxygen. Thus roots need a constant supply of oxygen to survive. Normally, roots obtain oxygen from the air in soil spaces. But if soil spaces are filled with water, the roots of most land plants cannot obtain the oxygen they need. This is one reason why floods destroy farm crops so quickly. It

---

ANNOTATION KEY
❶ Buttercup. (Making inferences)

BACKGROUND INFORMATION
ROOTS AND THE UNDERGROUND

Many plants produce roots above ground as well as below. Corn plants, for example, form roots that emerge from the stems and grow toward the ground. Orchids frequently grow perched high on tree trunks and branches. They attach themselves to the tree with roots that secrete a kind of cement. Orchids actually grow on other plants. They gain no nourishment from their unintended hosts. They get nutrients from leaves that fall and decompose near their roots.

tion of the process. The last paragraph on page 495 talks about the need for oxygen and the flooding of crops. Students will probably be interested in this idea because it may not have occurred to many of them. Draw the analogy of "drowning" a plant, an experience many of them have had.

Enrichment
Some of your lab-oriented students will enjoy making simple wet-mount slides of root hairs. With a little stain, they can probably see that the root hairs are actually extensions of the epidermal cells. Students usually find this interesting because they generally assume that the root hairs are composed of several cells.

---

you are once again going to be in unfamiliar territory for most students. This section of the chapter has a lot of detail that can be clarified with a visual presentation. Develop concept maps as you work your way through the various tissues and their function of taking in water and nutrients. The epidermis will be a familiar topic, but the concept of water and nutrient uptake is rather difficult.

Emphasize again the importance of root hairs and their relationship to increased surface area. As you begin to talk about nutrient uptake, go back to Chapter 5 and thoroughly review active transport. It has been several weeks, and students probably did not thoroughly understand the concept before, so use this opportunity to show an applica-

TEACHING SUPPORT

Teaching Resources
- Activity: Illustrating Transport in Plants, p. 3
- Hands-On Activity: A Hairy Situation, p. 109
- Ecology Investigation: Respiration in Roots, p. 23

MEDIA AND TECHNOLOGY

 Interactive Videodisc
- *On Dry Land: The Desert Biome*

Use this bar code to show a picture of different types of root systems in desert plants.

Search to frame 31124

See the Biology Media Guide page 20 for additional barcode correlations for this interactive videodisc.

---

is also the reason why you can kill your houseplants by overwatering them. Later in this chapter we will see how aquatic and swamp plants have evolved solutions to the problem of having their roots constantly in water.

WATER ABSORPTION Despite the importance of water to living cells, there is no active transport mechanism that can grab water molecules on one side of a cell membrane and drag them to the other side. Yet a single corn plant takes in more than 200 liters of water in a growing season that lasts little more than three months! How can plants move so much water into their roots?

To move water across membranes, plants use the power of osmosis. Recall that water will move across a membrane by osmosis from an area of high concentration of water molecules to an area of low concentration of water molecules. The cytoplasm of root epidermal cells is filled with amino acids, sugars, and other dissolved compounds. There are relatively few water molecules in root epidermal cells. The water in soil spaces usually contains only a small amount of dissolved minerals and many molecules of water. Thus the concentration of water molecules is higher in the soil spaces and lower in the root epidermal cells. Because of this difference in water concentration and because the thin cell walls of root hairs are permeable to water, water moves by osmosis from the soil spaces into the cytoplasm of root epidermal cells. See Figure 23–10.

Note that this process will not work if soil water carries a high concentration of dissolved minerals. In fact, if the concentration of water in the soil spaces is too low, water may even move out of root hairs and back into the soil. This situation is called root burn. Root burn can occur in nature if soil is saturated with salty water. Root burn can affect house, garden, and farm plants if too much fertilizer is added to the soil.

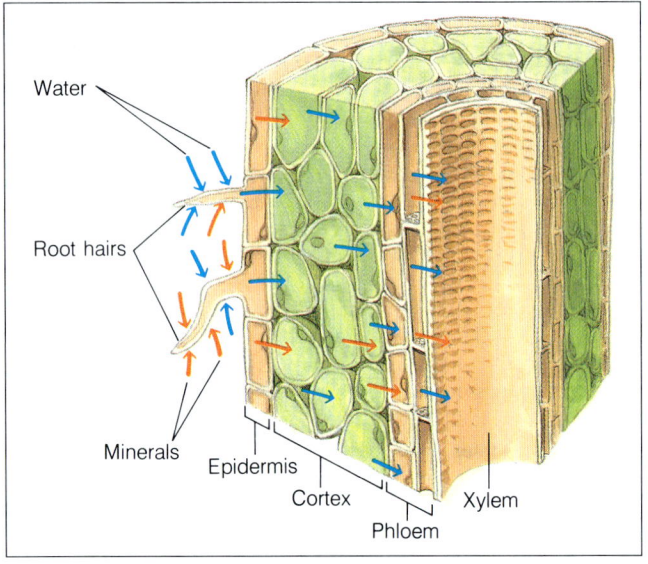

Figure 23–10 It is through the tiny hairs on plant roots that most molecules of water and minerals enter a plant. Water moves from the outside of the roots into the xylem, where it is then transported throughout the plant.

23–3 (continued)

Reinforcement/Reteaching
This is a good example of "applied osmosis." However, it has probably been several weeks since you have talked about osmosis. Go back to Section 5–4 and review hypotonic and hypertonic conditions. Use the information given in the textbook as your final example and let students see for themselves in which direction the water will move. Be sure you talk about the practical application of this information, specifically about the overuse of fertilizers.

Content Development
Using the Visuals Be sure at this point that students understand that you are moving from the outside of the root—remind them that the epidermis is the surface tissue—toward the innermost part of the root. Use Figure 23–11 to help them understand the relative positions of the various tissues in the root. Remind students about the characteristics of the parenchyma cells and their

The Cortex: Transport of Water and Nutrients

The parenchyma cells of the root cortex perform the next step in the movement of water in plants. The cell walls and cell membranes of root parenchyma cells are permeable to water. Thus water moves easily through and around these cells. Root cortex cells also have active transport mechanisms in their membranes. Active transport in these root cortex cells keeps water and nutrients moving deeper toward the center of the root.

How does this process work? First, active transport moves dissolved nutrients through the cortex toward the center of the root. As the concentration of nutrient ions in the center of the root increases, the relative concentration of water molecules in the cells in that region decreases.

Next, water in the outer cortex moves by osmosis to "follow" the nutrient ions and to "equalize" the water concentration in all cortex cells. Thus, both water and nutrients move toward the center of the root.

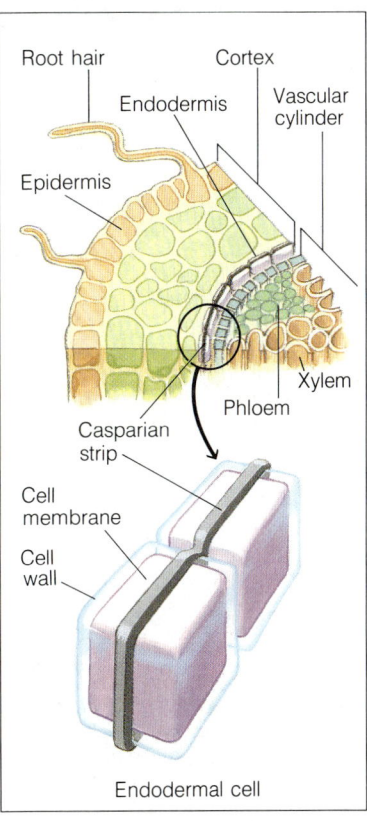

Figure 23–11 Cells in the endodermis are surrounded by Casparian strips. These strips are waterproof and thus prevent water molecules from seeping back between the cells.

The Endodermis: One-Way Passage

At the inner boundary of the cortex is a single layer of cells called the endodermis (*endo-* means inside; *dermis* means skin). The endodermis, which looks like a circle in cross section, encloses the vascular cylinder and stretches up and down the entire length of the root. The endodermis is composed of many individual cells, each shaped like a brick. Each of these cells is surrounded on four sides by a waterproof strip called the **Casparian strip**. To imagine how the Casparian strip looks, imagine a brick with a thick, sticky rubber band stretched around four sides. See Figure 23–11. Now imagine many such bricks placed edge to edge to build a cylinder. The rubber bands around the individual cells touch each other and stick together like mortar between the bricks in a brick wall.

The Casparian strips of the endodermal cells are not permeable to water. Thus water and nutrients cannot move around endodermal cells—only through them. This is the key to understanding how the root works. Endodermal cells use active transport to pump dissolved nutrients into the vascular cylinder. As dissolved nutrients build up inside the vascular cylinder, water moves through the endodermal cells by osmosis to equalize the relative concentration of water molecules in the tissues.

But neither water nor dissolved nutrients can cross the Casparian strips that seal the spaces between endodermal cells. So both substances are trapped in the vascular cylinder. Dissolved nutrients cannot go back through the endodermal cells because those cells use active transport to force the

TEACHING SUPPORT

Study Guide
• Section 23–3, p. 222

23–3 (continued)

Enrichment

Using the Visuals Some students in the class might be interested in duplicating the experiment depicted in Figure 23-12. If they are successful, ask them to explain to the class what they did and what is happening.

SECTION REVIEW 23–3

1. A taproot is a primary root that is long and thick. Taproots can grow far beneath the surface of the soil to reach water. Plants with fibrous roots do not have one particular root growing longer and thicker than its other roots. Instead, plants with fibrous roots have secondary roots that grow and branch.
2. Dandelions have short taproots that break apart easily. When a dandelion is pulled from a lawn, the taproot breaks. Pieces of the taproot remaining in the ground are able to grow a new top. That is why it is hard to eliminate dandelions from a lawn.
3. Roots need oxygen for the energy it provides, especially when ATP is broken down.
4. Root hairs greatly increase the surface area of roots, which allows more materials, such as water and minerals, to pass into the plant.
5. A one-way passage of materials is important because if there were a two-way passage, water and minerals could pass out of the roots as well as into the roots. A one-way passage prevents materials from leaving the plant.

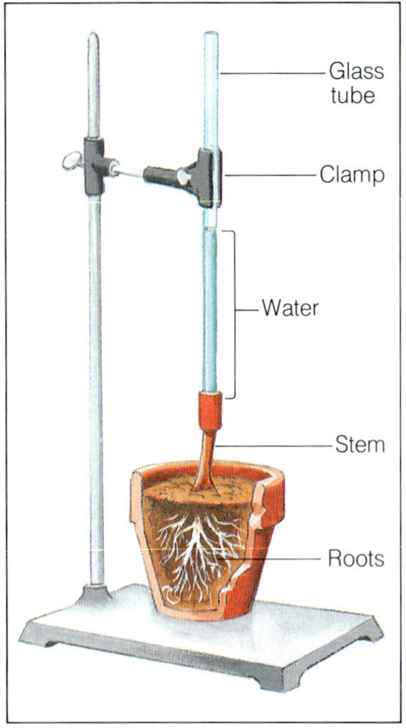

Reinforcement/Reteaching
For those students who have difficulty with a Section Review question, go back to the discussion in the section and review the material students have had difficulty understanding.

Closure
Summarize the information in this section by putting the headings Water Absorption and Nutrient Uptake on the chalkboard. Describe the process under each heading. Then elicit a list of the root tissues that are involved in the process, in order. Be sure students remember to list root hairs and describe their function.

Figure 23–12 Pressure exerted by roots can move water a distance up the tube. However, root pressure alone cannot force water to the top of a tall plant.

nutrients inside. Water cannot move back through the endodermal cells because to do so would mean moving from an area of low concentration of water molecules to an area of high concentration of water molecules. Thus there is a one-way passage of materials into the vascular cylinder in plant roots.

Root Pressure: Pushing Water and Nutrients Upward

As more nutrients are pumped into the vascular cylinder, more molecules of water follow. Because neither nutrients nor water can move back into the cortex, pressure known as root pressure builds up inside the vascular cylinder. This increased root pressure forces water into the xylem. As you will soon see, root pressure is part of the driving force behind the movement of water from roots to leaves.

As root pressure in the vascular cylinder forces water into the xylem, and as more and more water moves from the cortex into the vascular cylinder, the water in the xylem is forced upward through the root into the stem. In short plants such as grasses and strawberries, root pressure alone can push water through stems from roots to leaves. In tall plants such as trees, however, root pressure alone cannot accomplish this process.

You can see root pressure in action if you cut the stem of certain plants close to the ground and place a glass tube over the stump. A liquid called sap will continue to come out of the stump and rise up the glass tube for some distance.

How do we know that active transport by root cells creates root pressure? If we poison the root or cut off its supply of oxygen, the movement of sap stops. Because once the endodermal cells in the root stop pumping dissolved nutrients into the vascular cylinder, water stops moving in by osmosis.

23–3 SECTION REVIEW

1. Compare a taproot and a fibrous root.
2. Why is it difficult to remove dandelions from a lawn by pulling them out by the leaves?
3. Why do roots need a constant supply of oxygen?
4. What is the importance of root hairs in the absorption of water and nutrients?
5. **Critical Thinking—Making Inferences** Why is it important that the root epidermis permits only a one-way passage of materials?

TEACHING STRATEGY 23–4

Focus/Motivation
Bring in as many diverse samples of stems as you can find—all sizes and textures. Be sure to include some modified stems, such as bulbs. Read the first para-

23-4 Stems

Stems vary greatly in size and shape from one plant species to another. Some stems are tiny; others are a hundred meters tall. Some grow entirely underground; others reach far into the air. **Regardless of size and shape, all stems have two important functions: They hold leaves up in the sunlight and they conduct various substances between roots and leaves.** Some stems may also store water and nutrients.

Stems have four basic types of tissue: parenchyma tissue (pith), vascular tissue (xylem and phloem), cambium tissue (vascular cambium and cork cambium), and cork tissue (outer bark). Looking at cross sections of monocot and dicot stems, such as those in Figure 23–13, you can see that some of these tissues are arranged differently in stems than they are in roots.

Parenchyma

In monocot stems, parenchyma, or pith, is distributed throughout the stem and is often used for storage. In dicot stems, a core of pith is laid down in the center of the stem, where it is surrounded by layers of xylem.

Vascular Tissue in Woody Stems

Vascular tissue in stems conducts water, nutrients, and various plant products up and down the plant. To accomplish this, xylem and phloem tissues form continuous tubes from the roots through the stems to the leaves. Stem xylem is connected to root xylem, and stem phloem is connected to root phloem.

In young dicot stems, bundles of xylem, vascular cambium, and phloem are arranged in a ring. The xylem faces the interior of the stem, whereas the phloem faces the exterior. In woody dicots, such as trees, new xylem and phloem cells are produced as the plant grows. Other tissue joins the xylem and phloem bundles into two complete rings. The xylem ring is closest to the center of the stem, and the phloem ring surrounds the xylem.

It is actually xylem tissue that makes up the rings of trees. By counting the number of rings in a cross section of a tree, you can estimate the age of the tree. The size of the rings can provide information about weather conditions in the area over time. Thick rings indicate that weather conditions were favorable for tree growth; thin rings indicate that weather conditions were not. For example, in a year with abundant rainfall, a tree might make thick rings. In a year when rainfall was scarce, the rings might be thin. As woody stems grow thicker from year to year, the older xylem near the center of the stem becomes what is known as heartwood. Heartwood no longer conducts

Guide For Reading
- What are some general and specialized functions of stems?
- What are the functions of the different tissues in a stem?

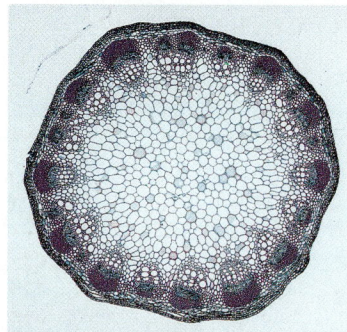
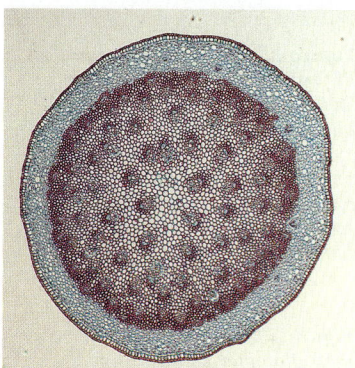

Figure 23–13 The arrangement of vascular tissue in the stem of a dicot differs from that in the stem of a monocot. In a dicot, vascular tissue is arranged in a ring. In a monocot, it is scattered in bundles throughout the stem. Is the sunflower (top) a monocot or a dicot?

ANNOTATION KEY

① Cambium layer. (Interpreting diagrams)

BACKGROUND INFORMATION
ANNUAL RINGS

The annual rings of a tree are sometimes used to provide information about the weather conditions of an area. In some trees the inner part of an annual ring, the springwood, is lighter in color. There is usually more moisture available to the tree in the spring. In response, the tree produces a high concentration of the growth hormone auxin, and new cells in the xylem grow rapidly, becoming quite large before they die. In summer, the tracheids and vessels grow more slowly, due to less water and auxin, and stay smaller. Cell walls occupy a greater proportion of the area, so the summer wood appears darker.

A wide band of growth in an annual ring reflects a good growing season (more moisture), while a narrow band forms in a dry year. By comparing the pattern of tree rings in building timber with the patterns found in old trees in the same area, archaeologists and anthropologists have been able to date structures made by humans.

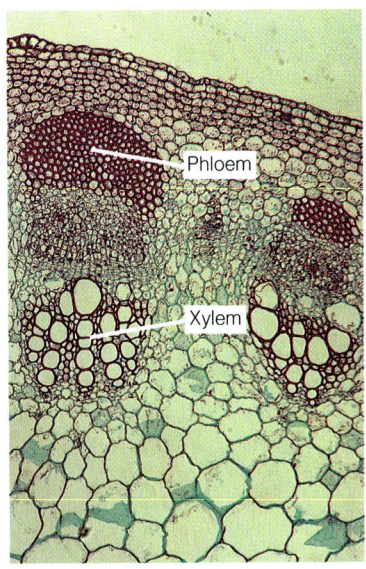

Figure 23–14 A small portion of the vascular bundles in a dicot stem can be seen in this photomicrograph. The much larger xylem cells are found at the edge of the vascular bundle and point toward the center of the stem. The smaller phloem cells face the outside of the stem.

Figure 23–15 Trees can increase their diameter as they grow. In which layer does the division of cells account for this increase? ①

water. The outer rings of living cells, known as sapwood, transport liquids.

Phloem tissue forms the inner part of what is called bark. Phloem carries sugars and other products of photosynthesis from the leaves to other plant parts that do not make their own food. For this reason, if the phloem in tree bark is removed in a ring around the trunk, the roots of the tree will starve and the tree will die.

Rings of xylem and phloem are separated by a thin layer of tissue called the vascular cambium. The vascular cambium makes more xylem and phloem cells, enabling the stem to grow thicker. You will learn more about this process in Chapter 24.

Cork Cambium and Cork

Outside the phloem tissue in woody dicots is the cork cambium, which produces cork tissue—the outer bark of trees. Cork cells have thick walls and usually contain fats, oils, or waxes. These waterproof substances help prevent the loss of water from the stem. The outermost cork cells are usually dead. As the stem increases in size, this dead bark often cracks and flakes off in strips or patches, depending on the tree species.

Stems Modified for Food Storage and Dormancy

In addition to support and the transport of materials from roots to leaves, stems also perform other functions. Plants often store food in their stems during their growth period. Plants need to store food to enable them to survive **dormancy**. During dormancy, a plant's growth slows or stops. Dormancy occurs during cold winters or long dry periods that may last for months or even years. When conditions once again support

500

23–4 (continued)

Content Development

Discuss the sapwood and its importance to the plant. Be sure that students understand the location of the phloem in a stem. Scrape some of the bark off a twig so that they can see the green underneath. Discuss the girdling process and the effect it has on trees. Remind students that the vascular cambium is a meristematic tissue.
• **What special ability does the vascular cambium have if it is a meristematic tissue?** (It can produce new cells by mitosis.)

Enrichment

You may have students who are interested in doing some library research for a class demonstration. Two good topics for research/demonstration are maple tree tapping and girdling.

Skills Development

Guided Practice
Skill: Collecting specimens
Cork and cork cambium are interesting topics for students. Most of them know what cork is, of course, but few realize how many varieties of trees produce

plant growth, the dormant plants begin to grow, using the food they previously stored. The plant uses the food stored in the stem until its new growth can make enough food. Several kinds of modified stems are used by plants for this purpose. Refer to Figure 23–16 as you read about these modified stems.

Rhizomes are thick, fleshy, creeping stems that grow either along or just beneath the surface of the ground. Along the length of rhizomes are buds from which leaves and stems can grow. When the frosts of winter kill the above-ground parts of plants, the rhizomes survive to grow again the following spring. Garden plants such as irises, canna lilies, and many species of grasses survive cold winters and grow year after year because of rhizomes.

Tubers are modified underground stems that are swollen with stored food, usually in the form of starch. Many tubers, such as potatoes, hardly look like stems at all. But tubers have one or more prominent "eyes," which are actually lateral buds. From these buds grow new above-ground stems and leaves.

Bulbs, such as those of tulips and daffodils, have underground stems too. The stem at the center of the bulb is small. Most of the food stored in the bulb is located in the layers of short, thick leaves that wrap around and protect the stem.

Figure 23–16 The rhizome of the iris is a thick creeping stem (top, left). Tubers—underground stems that store food—are characteristic of the potato (bottom, left). Bulbs, like the daffodil, are also underground stems, although the stem is small, cannot be seen, and is found in the center of the bulb (top, right). Corms, such as this gladiolus (bottom, right), are similar to a bulb in structure but they have thinner leaves.

501

some kind of cork on their trunks. You might ask students to bring in small samples of bark from trees in their yard and spend a few minutes looking for cork in the samples.

Content Development
Using the Visuals Explain the term *dormancy* for those students who are unfamiliar with it. Discuss each type of modified stem, using all the photographs in Figure 23–16 in order to distinguish among them.

TEACHING SUPPORT

Laboratory Manual
- Root and Stem Structures, p. 277

Teaching Resources
- Ecology Investigation: Intraspecific Competition, p. 13

Study Guide
- Section 23–4, p. 224

MEDIA AND TECHNOLOGY
Interactive Videodisc
- *On Dry Land: The Desert Biome*

Use this bar code to introduce the topic of stem adaptations.

Play frames 4244 to 4566

See the Biology Media Guide page 20 for additional bar-code correlations for this interactive videodisc.

MULTICULTURAL STRATEGY

Ask students what they know about spices. Spices are seeds, buds, dried flowers, bark, or roots of different plants. They originate in different parts of the world but are mainly found in tropical regions. Each ethnic cuisine uses its own assortment of spices to create many of the unique flavors of that cuisine. Cinnamon, for instance, is the bark of a tree brought by Arab traders from China to the Middle East centuries ago. From there, it spread to other parts of the world. Turmeric, the ground rhizome of a plant, comes from India. Nutmeg is from the East Indies, and saffron, a very expensive spice, comes from Spain.

Ask students to name some spices used for cooking in their own homes. Can they name (and perhaps describe) some ethnic dishes in which cardamom, turmeric, ginger, or chili peppers are used? Can they name some typical spices used in these cuisines: Mexican, Chinese, Italian, Scandinavian, Creole, Greek?

501

Corms are similar to bulbs, but they have much thinner leaves. Most of the stored food in corms is located in the stem itself. Thin, scalelike leaves surround the stem and serve mainly to protect it. Crocuses and gladioli are common garden plants that produce corms.

23–4 SECTION REVIEW

1. What are the functions of plant stems?
2. How does vascular tissue contribute to the strength of plant stems?
3. What are tree rings? What important information do tree rings provide?
4. **Critical Thinking—Relating Concepts** How do rhizomes and corms contribute to a plant's survival?

23–5 Leaves

Guide For Reading
■ What are the parts of a leaf? What are their functions?

The leaves of green plants are the world's oldest solar energy collectors. Leaves are also the world's most important manufacturers of food. Sugars, starches, and oils manufactured by plants are sources of virtually all food for all land animals. Even animals that do not eat plants must eat other animals that do.

Leaf Structure

To collect sunlight, most leaves have large thin, flattened sections called blades. The blade is attached to the stem by a thin structure called the **petiole**.

Leaf blades occur in an incredible variety of shapes and sizes, depending on the environment to which they are adapted. Simple leaves have only one blade and one petiole. Compound leaves have several blades, or leaflets, that are joined together and to the stem by several petioles. These leaflets can be joined together in several ways. The leaflets of some compound leaves spread out like the fingers of a hand. The leaflets of others grow in pairs along a long central petiole. In still another pattern, leaflets are arranged on petioles that, in turn, are arranged along a long central petiole.

Figure 23–17 Leaves are arranged differently in different plants. The oak leaf is solitary and held to a stem by a single petiole (top). Both the strawberry and the poison ivy have three leaflets attached to a single petiole (bottom).

Leaves contain several specialized tissues. Like roots and stems, leaves have an outer covering of epidermal cells, inner layers of parenchyma cells, and vascular tissues.

Epidermis: Controlling Water Loss

All leaves are covered on the top and bottom by a layer of tough cube-shaped epidermal cells that do not contain chloroplasts. The epidermal layer of many leaves is also covered by a waterproof waxy coating called the cuticle. The cuticle and epidermal cells together form a waterproof barrier that protects delicate tissues inside the leaf by slowing down the loss of water through evaporation.

But no leaf could survive completely sealed off from the atmosphere that surrounds it. Plants need to "breathe" just as we do. Leaves must take in carbon dioxide and give off oxygen during photosynthesis. And in order to use the food they make, leaves must take in oxygen and give off carbon dioxide (just as animals do). Plants need to exchange enormous amounts of gases with the surrounding atmosphere. But in order for a gas to enter the cells of a leaf, it must be dissolved in a thin film of water. So all leaf cells that need to exchange gases with the air must be kept wet.

But you know what happens to a wet surface exposed to the air—the water evaporates and the surface dries out! To keep the surface wet, more water must constantly be added to it. The more exposed the tissue is to the air, the faster the water will evaporate and the more quickly it will need to be replaced.

If all the moist tissues of leaves were exposed to the air, leaves would lose water much faster than roots could replace it. The plant would wilt and die. This is exactly the problem that mosses have. Without cuticles and vascular tissues, mosses must grow in wet soil and have their green parts in moist air all the time or they will dry out.

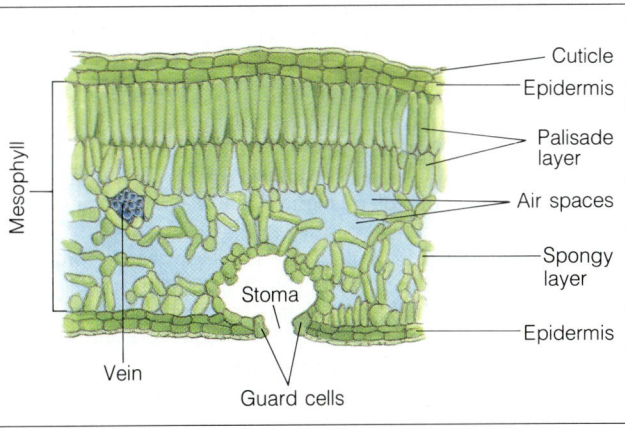
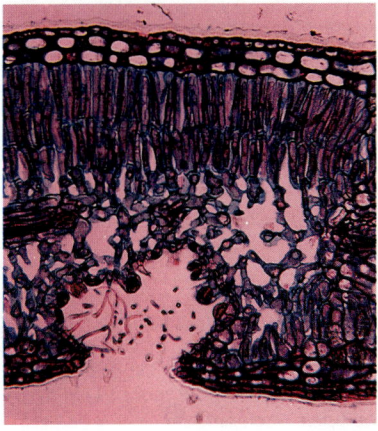

Figure 23-18 Some of the most important manufacturing sites on Earth are found in the leaves of plants. The cells in plant leaves are able to use light energy to make carbohydrates. What is the opening at the bottom of this plant leaf called?

ANNOTATION KEY

❶ Closed. (Applying concepts)

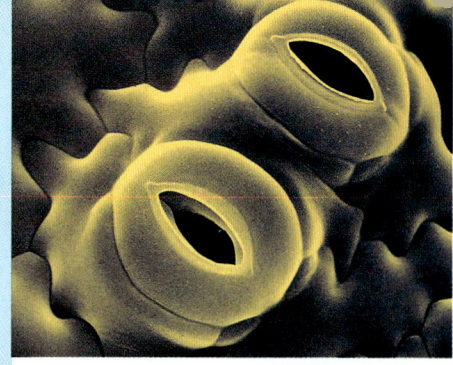
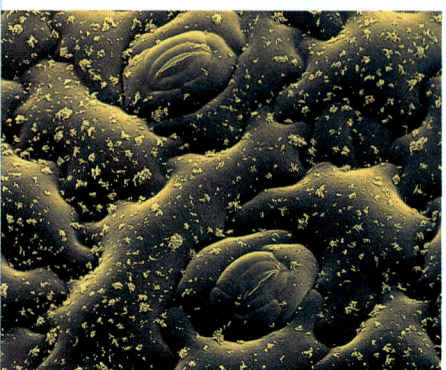

Seed plants solve this problem by striking a balance between their need for gas exchange and their need to conserve water. Plants control the loss of water from their leaves by allowing air in and out of their waterproof covering only through small openings called **stomata** (singular: stoma). Stomata open to allow the leaf to "breathe" and close when the leaf does not need much air exchange. Stomata also close when the plant loses too much water and begins to wilt. Most of the stomata of terrestrial plants are located on the undersides of their leaves, but there are some stomata along green stems too.

Stomata are formed by two specialized epidermal cells called **guard cells**. The stomata open and close in response to changes in water pressure within the guard cells. When water pressure within the guard cells is high, the thin outer walls of the cells are forced into a curved shape. This pulls the thick inner walls of the guard cells away from each other, opening the stoma. When water pressure within the guard cells decreases, the springiness of the inner walls pulls them together. The stoma closes.

Because water is lost through the same stomata that let in carbon dioxide, guard cells have a difficult job. On hot, dry days, open stomata will allow large amounts of water to leave the leaf. But if the stomata are closed all the time, the inside of the leaf will run out of carbon dioxide, and photosynthesis will slow down or stop. Each type of plant has guard cells that balance water loss against the need for carbon dioxide.

Vascular Tissue: The Veins of a Leaf

The vascular system of leaves is directly connected to the vascular tissues of stems. In leaves, xylem and phloem tissues are gathered together into bundles that run out of the stem and into the petiole. As the bundles enter the leaf blade, they are surrounded by parenchyma and sclerenchyma cells. All these tissues together form the veins of a leaf. In most monocot leaves, veins run parallel to one another. In dicot leaves, they may run in different patterns.

Mesophyll Tissue: The Food Factory of the Leaf

Most leaf tissue is composed of specialized cells called leaf **mesophyll**. Mesophyll cells contain many chloroplasts and perform most of the plant's photosynthesis. Just under the upper epidermal covering of the leaf is a layer of tall, column-shaped mesophyll cells called the **palisade layer**. Under the palisade layer is a thick layer called the **spongy mesophyll**. The spongy mesophyll gets its name from the fact that it really does look like a sponge. Cells of the spongy mesophyll are arranged in a network with a great many air spaces between them. These air spaces connect with the stomata and allow carbon dioxide and oxygen to diffuse into and out of the leaf.

Figure 23–19 Stomata open (top) and close (center) to let gases in and out of the leaves. On a hot day, are the stomata more likely to be open or closed? ❶

504

23–5 (continued)

Focus/Motivation
Using the Visuals The stomata and guard cells are interesting phenomena for teachers and students alike. Spend a little time talking about how "smart" these structures are. The protective devices and adaptations of animals are frequently discussed—this is a good plant example. Be sure to use the visuals on this page to illustrate the operation of the guard cell–stoma structure. Filling a balloon with water and emptying it can help explain the concept of turgor.

Content Development
When you have talked about the vascular tissue in leaves, you will have discussed the whole circulatory system of the plant from roots to stems to leaves. Emphasize that the veins are continuous, which is necessary to transport needed materials in both directions—to and from roots and leaves.

Content Development
• **How do the location and structure of the palisade layer enhance its function?** (Because the palisade layer is located just below the upper surface of the leaf, the intensity of the sunlight is not diminished by passing through many layers of structures. The cells of this layer are columnar and densely packed, which ensures that as many cells as possible are exposed to light. You may want to point out that cytoplasmic movement within the cells ensures that each chloroplast within a cell has its "moment in the sun.")
• **How does the structure of the spongy mesophyll enhance its function?** (The air spaces and relatively large amounts of exposed cell membrane facilitate gas exchange.)

Remind students that gas exchange, like other forms of diffusion in living things, requires a thin, moist membrane.

SECTION REVIEW 23–5

1. A simple leaf consists of one blade and one petiole. A com-

The surfaces of spongy mesophyll cells are kept moist so that gases can enter and leave the cells easily. The water necessary to do this is supplied by the branching network of leaf veins that runs through the mesophyll. The opening and closing of stomata regulate the amount of contact the air in the spongy mesophyll spaces has with the drier air outside. Although the stomata cut down on the loss of moisture through evaporation, a substantial amount of water is still lost to the outside through evaporation.

23–5 SECTION REVIEW

1. Compare simple leaves and compound leaves. How are these shapes related to solar energy collecting?
2. What is the function of the epidermis and cuticle layers? What is the function of the openings in these layers?
3. Describe the structure and function of mesophyll tissue.
4. **Critical Thinking—Relating Facts** Describe how vascular plants control gas exchange and water loss.

23–6 Transport in Plants

You now know how the tissues of a typical plant are put together—from the root tips all the way up to the leaves. Remember that water and dissolved nutrients from the soil are absorbed through root hairs by active transport. The pressure created by water entering the tissues of a root can push water upward in a plant stem, but only for a short distance. How then does water reach the topmost needles of a redwood tree 90 meters above the ground?

Water Movement in Xylem

Xylem tissue forms a continuous set of tubes that stretch from roots through stems and out into the spongy mesophyll of leaves. **Root pressure forces water into the xylem, but root pressure alone cannot account for the movement of water and dissolved materials in plants.** Obviously, other powerful forces must also be at work.

ADHESION, COHESION, AND CAPILLARITY If water is not pushed up, could it perhaps be pulled up? It is indeed. Two characteristics of water make this pulling action possible.

Water molecules are attracted to one another by a force called cohesion. Water molecules are also attracted to other molecules by a force called adhesion. The combination of cohesion and adhesion explains the phenomenon known as capillarity, which is the movement of water upward in a small solid tube. Capillarity can be observed when water climbs up a

Guide For Reading
- How does water move through a plant?
- How are materials transported by phloem?

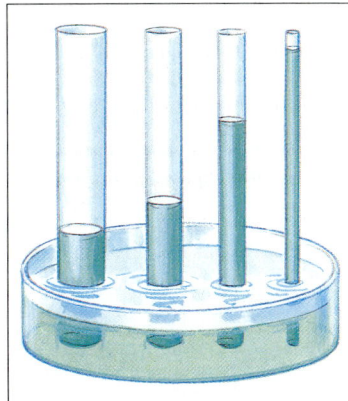

Figure 23–20 Capillarity—the ability of water molecules to stick to one another and to the walls of a tube—accounts in part for the movement of water up xylem tissue.

pound leaf has several blades joined to each other and to the stem by several petioles. In each type, the thin, wide structure allows for the maximum exposure of the leaf to sunlight.
2. The epidermis is the outermost layer of cells in a leaf. The epidermis is covered by a waterproof, waxy coating called the cuticle that prevents water from evaporating from the leaf. Special openings in the epidermis permit a controlled flow of air into the leaf and water and oxygen out of the leaf.
3. Mesophyll cells contain many choloroplasts. It is in mesophyll tissue that most of the plant's photosynthesis takes place. Air spaces in the spongy mesophyll connect to stomata and allow the diffusion of materials into and out of the leaf.

HISTORICAL NOTE
EARLY EXPERIMENTS

Some of the earliest attempts to determine the functions of xylem and phloem were performed by Marcello Malpighi in 1679. After girdling a tree, Malpighi discovered a swelling that oozed a sweet fluid in the bark just above the stripped area. For many days the leaves seemed to be unaffected, but they eventually wilted and died. Malpighi concluded that the phloem, which had been severed, transported food to the roots. As the food reserves in the lower part of the tree diminished, the roots died. Because the leaves remained healthy for a long period of time, Malpighi surmised that the xylem, which had remained intact, transported water up from the roots.

In 1727, Stephen Hales, a clergyman, performed a series of experiments that convinced him that transpiration was an activity of the leaves, rather than merely the activity of the xylem, as Malpighi had thought. Hales showed that more water was pulled up from the roots on a bright, sunny day than on a cloudy one. In the 1890s, Dixon and Joly did similar experiments showing the role of water cohesion.

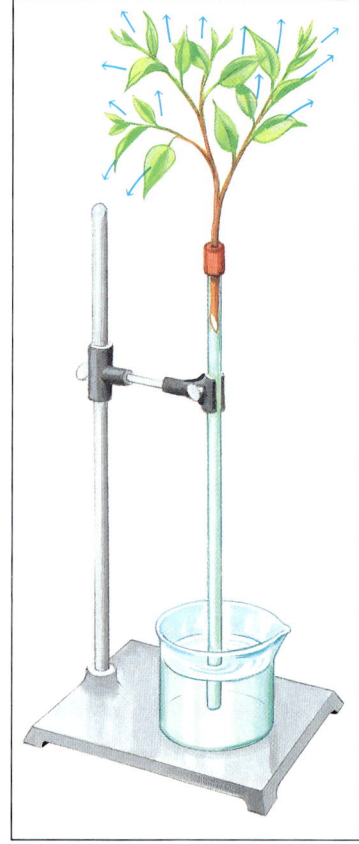

Figure 23–21 This apparatus shows that transpiration—the movement of water molecules out of leaves—can pull water up a thin tube. The faster water evaporates, the faster it is pulled up in the tube.

small glass tube placed in a beaker. Although gravity pulls the water in the tube down, the force of adhesion between the water molecules and the walls of the glass tube pulls the water up. At the same time, the water molecules cling to one another by cohesion, pulling molecule after molecule up inside the tube. The smaller the diameter of the tube, the higher the water will climb. Although capillarity accounts for the movement of water in tiny xylem vessels in short plants, it is much too weak a force to move water up the xylem tubes in a redwood. Another force must be at work.

TRANSPIRATION PULL The final pull necessary to get water up into leaves is provided by the power of evaporation itself. This idea may sound strange, but evaporation is a powerful process. Except in rare cases when the atmosphere is totally saturated with water vapor, water will always evaporate from a wet surface. Therefore, water will almost always evaporate from leaves into the air. The evaporation of water from plant leaves is called **transpiration**.

Remember all those wet cells in the spongy mesophyll layer of leaves? As water evaporates from those cells, their water content drops. When this happens, water from nearby mesophyll cells moves by osmosis into the spongy mesophyll. This in turn causes the water content of these other mesophyll cells to drop below the water content of the cells that surround the veins. Once again water moves by osmosis to replace water lost by evaporation—this time from the cells surrounding the veins into the mesophyll.

As the cells surrounding the veins lose their water, water moves by osmosis out of the xylem into these cells. Finally, as water is removed from xylem vessels in the leaves, adhesion and cohesion pull other water molecules up the tubes from the roots through the stems. Because this entire process is driven by transpiration, it is called **transpiration pull**.

Transpiration pull is a very powerful force. A good-sized tree can move more than 1800 liters of water from the ground into the atmosphere in a single day. The hotter and drier the air, and the windier the day, the greater the amount of water lost by transpiration and the greater the amount of water drawn up from the roots.

Transport of Materials in Phloem

Phloem transports most organic molecules in plants. During the day, phloem transports products of photosynthesis out of the leaves and down into the stems and roots, where they are either stored or used.

But this is not the only job phloem does. For example, many plants pump sugars into their fruits. Such action often requires moving sugars out of leaves or roots into stems and then through stems to the fruits. All this movement of sugars occurs in the phloem. In addition, many plants living in cold climates

---

23–6 (continued)

Focus/Motivation
One way to illustrate transpiration and thus begin a discussion of transpiration pull is to take a houseplant, water it normally, put a clear plastic bag over the plant, and set it in a well-lighted area to observe it for a few days. As the moisture accumulates on the inside of the bag, you can discuss transpiration and the fact that the water is coming up through the plant from the roots. This should help students understand that as each drop evaporates from the leaves, another moves up from the roots in a continuous column.

Enrichment
Using the Visuals Figure 23–21 illustrates an interesting experiment that some of your more motivated students might enjoy performing. Ask them to write up a lab report that includes observations over several days.

Content Development
The exact mechanism of phloem transport is not fully understood, but it is known that this is an example of active transport. Remind students that you have already talked about this. Discuss the hypothesis that is pre-

pump food down into their roots for winter storage. This stored food must be moved back up into the trunk and branches before the plant begins to grow again in the spring—another job done in the phloem. It is this rich, sugary sap moving through the phloem that is tapped from maple trees in late winter and early spring to make maple syrup.

Botanists at one time thought that the function of phloem tissue was limited to the transport of organic compounds only. But phloem also transports inorganic ions. And this transport is not always away from the leaves. For example, phosphorus is needed in large amounts by young, actively growing leaf tissues. Xylem moves phosphorus from the roots up through the stems into the growing tips of the plant. But when growth in those tissues slows down, phloem helps recycle extra phosphorus. Phloem transport moves phosphorus out of the growing leaves and back down into the main stem of the plant. From there it can be carried up into the growing tip when it is needed.

How Phloem Transport Works

The exact mechanism by which phloem transport works is not completely understood. We know that the process requires energy. We also know that material moves through phloem cells much faster than it possibly could if it was powered by diffusion alone. For these reasons, it is clear that some kind of active transport is involved.

In order to explain one hypothesis, it is necessary to use the following terminology: When plants move sugars from their leaves to their roots, the leaf mesophyll tissue is called the source of the sugars and the parenchyma tissue of the root is called the sink. Active transport moves sugars into the phloem at the source. Similar active transport moves sugars out of the phloem into storage tissue at the sink.

When sugars are pumped into the phloem at the source, phloem tissue contains more dissolved minerals and less water than the surrounding tissue. Thus water moves into the phloem by osmosis. Pressure increases in the phloem at the source. When sugars are pumped out of the phloem at the sink, water also moves out. The movement of water out of the phloem at the sink causes a drop in pressure at the sink. Because pressure builds up at the source and drops at the sink, the contents of the sieve tube elements are forced down from source to sink.

If the plant needs to move sugars from the roots to the leaves, it simply reverses the pumping action at these two locations. The root tissue thus becomes the source, and the leaf

Figure 23–23 This illustration shows a current hypothesis that explains the movement of sugar molecules in phloem tissue. As you can see, materials move from the source, where they are available in great supply, to the sink, where they are scarce.

Figure 23–22 Tree sap often has industrial uses. After being collected, the sap of a pine tree is made into turpentine, a chemical often used as a solvent for paint.

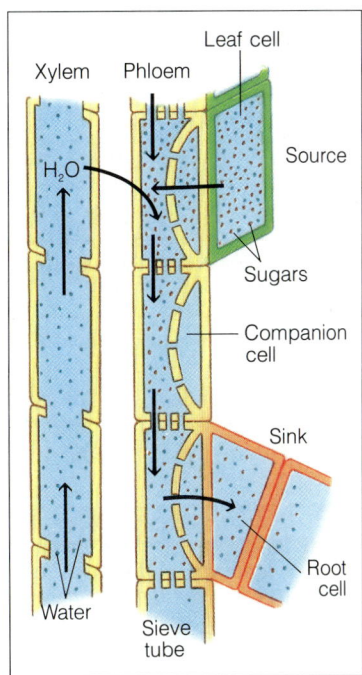

sented in the textbook, but emphasize that it is only one of several hypotheses concerning phloem transport. This is a scientific question that is still being investigated and studied.

Enrichment
Ask students to do some research on the various methods being used to investigate phloem transport. Some of this research is interesting, and you may want them to share the information with the class.

BACKGROUND INFORMATION
PHLOEM TRANSPORT

The exact mechanism by which phloem transport occurs is not fully known. It is difficult to investigate because phloem is extremely delicate. One technique for the study of phloem is the use of aphids. These are tiny insects that feed on plants. An aphid's mouthpart forms a long tube that it can insert into a plant in such a way that the end of the tube enters a single sieve element. The contents of the element are under pressure. The fluid in the phloem goes into the mouthpart tube and on through the aphid's gut with such force that the feeding aphids often have a drop of "honeydew" on their posterior ends. The fluid that comes from the phloem can be collected and analyzed. By using several aphids on different parts of the plant, a scientist can introduce substances into the phloem at certain points and study their speed and direction of flow.

TEACHING SUPPORT
Laboratory Manual
- Estimating the Number of Stomata in a Leaf, p. 293

Teaching Resources
- Activity: Observing Transpiration, p. 7
- Hands-On Activity: Whence the Water? p. 117
- Ecology Investigation: Determining the Permanent Wilting Percentage, p. 25
- Ecology Investigation: Observing the Capillarity of Soil, p. 27

Study Guide
- Section 23–6, p. 227

Product Testing Activities
- Testing Paper Towels

ANNOTATION KEY

❶ The water lily exchanges gases with the air at its upper surface. Stomata on the bottom of the leaf are useless for exchanging gases since they are under the water level.
(Making inferences)

becomes the sink. The phloem then carries sugars upward. Any dissolved compounds can be moved through the phloem in either direction by changing the location of the source and the sink. This hypothesis about phloem function is called the pressure flow hypothesis.

23-6 SECTION REVIEW

1. What role do cohesion and adhesion of water molecules play in the movement of water in xylem tissue?
2. Describe the process of transpiration pull.
3. **Critical Thinking—Summarizing Information** How does the pressure flow hypothesis explain how phloem functions?

Guide For Reading

■ What are some ways in which plants are adapted to live in different environments?

23-7 Adaptations of Plants to Different Environments

Flowering plants grow in a variety of places: in deserts, in ponds, on mountaintops, in salt water, in arctic regions, and in the tropics. **Angiosperms can survive in these areas because through natural selection the basic designs of their roots, stems, and leaves have evolved to fulfill the particular needs of each location.** In this section we will see how roots, stems, and leaves have been modified through evolutionary change to serve different functions. In many cases, different families of plants show similar adaptations to similar environmental conditions.

Desert Plants

Desert plants must survive where strong sun and daytime heat combine with sandy soil and infrequent rainfall. Instead of staying near the surface, rainwater sinks rapidly through desert soils. The hot, dry air quickly removes moisture from any wet surface, making life difficult for plants.

Figure 23-24 Unlike the stomata of most plants, those of water lilies are clustered on the top of their leaves. How is this an adaptation for life in water? ❶

508

23-6 (continued)

SECTION REVIEW 23-6

1. Water molecules are attracted to each other by the force of cohesion and to other molecules by the force of adhesion. The combination of these two forces explains how water rises in a narrow tube. The water molecules that stick to the sides of a xylem vessel pull other water molecules along with them. Adhesion and cohesion explain the movement of water up short plants only, however, not tall ones.

2. As water evaporates from a leaf, the number of water molecules in the spongy mesophyll decreases. Water moves by diffusion from the veins in the leaf to the spongy mesophyll. Water moves from xylem vessels to fill the empty veins. In this way, water is pulled up the xylem vessels by the power of evaporation, which is the driving force behind transpiration pull.

3. When sugars are pumped into the phloem at the source, the phloemn contains more dissolved materials and less water than the surrounding tissue. Water moves into the phloem by osmosis. Pressure increases in the phloem at the source.

When sugars are pumped out of the phloem at the sink, water also moves out. Because pressure builds up at the source and is decreased at the sink, materials in the sieve tube elements of the phloem are forced from the source to the sink.

Reinforcement/Reteaching

If students have difficulty answering the Section Review questions, turn back to the appropriate material and review it.

Two families of desert plants are cactuses (family Cactaceae), which evolved in the Americas, and the crown-of-thorns family (family Euphorbiaceae), which evolved in Africa. Figure 23-25 shows examples of these families of desert plants. Both families have either a root system that spreads out for long distances just under the soil surface or a root system that reaches deep down into the soil. In each of these root systems, the roots have many root hairs that quickly pick up water after a rainstorm, before the water sinks too far down in the soil.

Most water loss in plants occurs through transpiration from the leaves. Both families of desert plants either have leaves that are very small or have no leaves at all. In cactuses, leaves have been reduced to thin, sharp spines. In euphorbias, leaves either are tiny or can be dropped if the weather gets too dry. Both plant families have thick green stems, where photosynthesis occurs. These stems are also adapted to store water. The stems of cactuses, in particular, swell during rainy periods, as the plants greedily store water, and shrivel during dry spells, when the plants are forced to use up their water reserves.

Seeds of many desert plants can remain dormant for years, germinating only when sufficient moisture guarantees them a chance for survival. Other desert plants have bulbs, tubers, or rhizomes that can remain dormant for several years if necessary. When rain does come, the plants grow with amazing speed. They mature, flower, and set seed in a matter of weeks or even days—before the water disappears.

Plants Adapted to Life in Water

Some plants, such as waterlilies, have the opposite problem of desert plants. Their roots and stems grow in water or in mud that is saturated with water and nearly devoid of oxygen. To supply oxygen to their roots, these plants have large open spaces in the long petioles that reach from their leaves down to the roots at the bottom. These open spaces are filled with air through which oxygen can diffuse to the roots.

Many other plants show similar adaptations. Several species of mangrove trees grow in shallow water along tropical seacoasts. Stately baldcypress trees thrive in freshwater

Figure 23-25 Plants that live in dry environments show adaptations to conserve water. The cactus leaves have been reduced to spines (top). The euphorbia has leaves that drop off in dry seasons (bottom).

Figure 23-26 Mangrove roots, which look like a series of stilts, support the plant. In time, a mangrove plant grows to cover a huge area.

TEACHING SUPPORT

Teaching Resources
- Vocabulary Review: Crossword, p. 1

Study Guide
- Section 23-7, p. 229

MEDIA AND TECHNOLOGY

🌿 **Video/Videodisc**
- *Biomes: Coniferous Forest and Tropical Rain Forest*

Use this bar code to show tropical vines and epiphytes.

Play frames 30477 to 31512

🌿 **Videodisc**
- *Aquatic Ecosystems: Estuaries and Marine*

Use this bar code to introduce the topic of salt-tolerant plants.

Play frames 15934 to 16827

See the Biology Media Guide pages 21–25 for additional bar-code correlations.

23-7 (continued)

Content Development

Discuss salt tolerance in terms of osmosis and upsetting the balance. Most students at some time or other have seen salt used as a weedkiller. It is truly an example of a unique adaptation for a plant to have the ability to pump out excess salt! You might tell students that some sea birds that survive by drinking salt water have adaptations that remove excess salt from their bodies.

Content Development

Students find carnivorous plants fascinating, although they tend to view this topic as a gory or gross subject. You will need to emphasize the characteristic as a necessity for survival in nutrient-deficient soil.

Enrichment

Students will enjoy researching the different types of carnivorous plants and sharing the information. You might even want to grow some samples in your classroom. (Follow the directions carefully.)

SECTION REVIEW 23-7

1. Desert plants may lack leaves, which limits water loss.

---

swamps in the southern United States. The roots of these trees, like those of waterlilies, must survive in mud that has little or no oxygen. Mangroves survive by growing special roots called air roots. Air roots have large air spaces in them, just like waterlily stems. These spaces conduct air down to the buried roots so the root tissues can respire normally. Baldcypress trees grow structures called knees, which protrude above the water. The knees bring oxygen-rich air down to the roots.

Salt-Tolerant Plants

In addition to getting air to their roots, mangroves have another problem. As you have seen, when plant roots take in dissolved minerals, a difference in the concentration of water molecules is created between the root cells and the surrounding soil. This concentration difference causes water to enter the root cells by osmosis. For plants growing in salt water, this means taking in much more salt than the plant can use. The roots of plants such as mangroves can tolerate salt concentrations that would quickly burn the root hairs off most plants. And the leaves of these plants have special cells that actively pump salt out of the plant tissues and onto the leaf surfaces, where it is washed off by rain.

Climbing Plants

In environments such as tropical rain forests, the struggle for sunlight is fierce. Here any plant that can quickly climb up the trunk of another plant to escape the darkness of the forest floor has a considerable advantage over its neighbors. Vines are plants that have evolved specialized root, stem, and leaf structures that enable them to climb on rocks or other plants.

The stems of many climbing plants do not grow straight up. Rather, the growing tips point sideways. See Figure 23-27. As the tip of a shoot grows, it spins around in a circle. In this way, the plant "looks" for something to climb on. When the tip encounters an object, it quickly wraps around it.

Some climbing plants have long, twisting leaf tips or petioles that wrap tightly around small objects. Other plants have extra growths called tendrils that emerge near the base of the leaf petioles and wrap tightly around any object they encounter.

Still other vines, such as the strangler fig, grow aerial roots along the entire length of their stem. These roots attach to rocks or trees and help anchor the climber in place. Sometimes these aerial roots can be so numerous and so large that they end up strangling the host plant.

Plants That Eat Animals

Some plants live in very wet places called bogs, where there is little or no nitrogen present in the watery soil. Because conditions are too wet and too acid in bogs, bacteria of decay

Figure 23-27 *The stem of the morning glory twines around and around its support. Twining stems enable plants to grow "taller" by climbing up the stems of other naturally tall plants so that they can capture more of the sun's energy.*

cannot survive. Without these bacteria, neither plant nor animal material is broken down into the nutrients plants can use.

A number of plants have overcome this difficulty by evolving specialized leaves that trap and digest insects. The best known of these carnivorous plants is the Venus' flytrap. This plant has bright-red leaf blades that are hinged at the middle. Insects are attracted to the leaf by the bright color. If an insect touches the trigger hairs on the leaf, the leaf folds up suddenly —trapping the animal inside. During a period of several days, the leaf secretes enzymes that digest the insect and release nitrogen for the plant to use.

There are several other kinds of carnivorous plants. Pitcher plants drown their prey in brightly colored pitchers that hold rainwater and digestive enzymes. Sundews trap insects on leaf hairs tipped with sticky secretions. There are even underwater carnivorous plants, such as bladderworts, that prey upon tiny water insects. See Figure 23–28.

Figure 23–28 Some plants show special adaptations to catch and digest insects. The pitcher plant (top, left) and the sundew (bottom) are land plants that trap insects. The bladderwort (top, right) is a water plant able to catch its own food.

23–7 SECTION REVIEW

1. What are two adaptations shown by desert plants?
2. Why do climbing plants survive better in tropical rain forests than do some nonclimbing plants?
3. Describe the adaptations of a mangrove tree to a saltwater environment.
4. **Critical Thinking—Applying Concepts** How is insect eating an adaptation that contributes to the survival of the Venus' flytrap?

LABORATORY INVESTIGATION

EXPLORING THE STRUCTURE OF LEAVES

Before the Lab

1. Assemble all the equipment and materials needed for the number of groups you will have. Collect *Elodea* from a pond or purchase it from an aquarium supply shop. Yew leaves can be obtained from nurseries or florists. If you have not prepared and observed leaf cross sections recently, perform this lab ahead of time to preview any technical problems and identification difficulties.
2. Prepare the iodine solution by dissolving 10 grams of potassium iodine in 100 milliliters of water. Then add 5 grams of iodine.
3. As an alternative to yew leaves, other small leaves such as forsythia can be used. Geranium leaves also work well because they are thick. In general, broad leaves such as maple leaves do not work well.

Pre-Lab Discussion

Review basic leaf structure and function and demonstrate the preparation of leaf cross sections.

Remind students to use just a small drop of water on the slide.

Give specific instructions detailing the quality of drawings you expect.

Alert students to any technical problems you may have identified when you performed the lab tasks for yourself. Discuss laboratory cleanup and disposal procedures.

LABORATORY INVESTIGATION   EXPLORING THE STRUCTURE OF LEAVES

PROBLEM

How are leaves suited to their function?

MATERIALS (per group)

3 clean microscope slides
2 coverslips
microscope
forceps
1 clean medicine dropper
scalpel
leaves: *Elodea* and yew *(Taxus)*

PROCEDURE

1. Use a medicine dropper to put a drop of water in the center of a slide.
2. Pull a small leaf from the tip of a sprig of *Elodea* in the drop of water on the slide. Cover the leaf with a coverslip.
3. Place the slide on your microscope. Focus the low-power objective on the *Elodea* leaf. Observe the leaf. Switch to the high-power objective. Focus with the fine adjustment. Look for pigmented structures in the cells. Note their shape and color. Remove the slide.
4. Place another slide on a flat surface. Put a yew leaf on the slide. Cut the tip off the leaf by pressing straight down with a scalpel about 1 mm from the end of the leaf. **CAUTION:** *The scalpel is very sharp. Always cut away from yourself and others.* Discard the tip of the leaf.

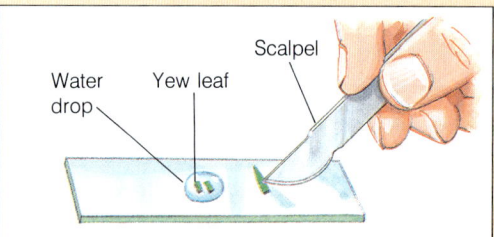

Shave off 6 thin slices, or cross sections, from the cut end of the leaf. See the accompanying illustration. Make the slices as thin as possible by pressing the scalpel straight down as close to the cut end of the leaf as allowable.

5. Use a medicine dropper to put a drop of water in the center of another slide. Slide the leaf cross sections over to the drop of water. Position the cross sections so that they lie next to each other. Cover the leaf cross sections with a coverslip.
6. Place the slide on a microscope. Focus the low-power objective on the thinnest cross section on the slide. Switch to the medium-power objective.
7. Draw a diagram of the leaf cross section. Locate the following structures and label them on your drawing: epidermis; palisade layer; spongy layer; vein. Remove the slide.

OBSERVATIONS

1. What is the shape and color of the pigmented structures in the *Elodea* leaf?
2. Describe the appearance of the yew leaf cross sections. What is the shape and color of the cells?

ANALYSIS AND CONCLUSIONS

1. What are the pigmented structures in the *Elodea* leaf? What is the function of these structures?
2. What is the function of each labeled structure in the yew leaf?
3. Which parts of the leaf help in food production? In delivering water?
4. How do the shape, structures, and positions of the structures you observed contribute to the functioning of a leaf?

Skills Development

Students will use the following skills in the completion of this laboratory experiment.
1. Manipulative
2. Observing
3. Diagramming
4. Labeling
5. Safety

Safety Tips

Students should be cautioned to work carefully with the scalpels. They should be reminded that slides and coverslips are glass, so care should be taken.

Teaching Strategy

Monitor students as they prepare the slides. Some groups will need more direct instruction and help. Assist students in choosing a good cross section to draw.

Observations

1. Round and green.
2. Thin transparent layer of cells at the top. Below them are long parallel green cells. Below these are rounded loosely packed green cells. In the center is a circle of tightly packed cells.

STUDENT STUDY GUIDE

SUMMARIZING THE CONCEPTS

The key concepts in each section of this chapter are listed below to help you review the chapter content. Make sure you understand each concept and its relationship to other concepts and to the theme of this chapter.

23–1 Soil: A Storehouse for Water and Nutrients
- Plants take in the water and inorganic materials they need from the soil.

23–2 Specialized Tissues in Plants
- Vascular tissue, xylem and phloem, is the plant's circulatory system.

23–3 Roots
- Roots anchor a plant in the ground and enable the plant to take in water and nutrients.

23–4 Stems
- Stems hold leaves up to sunlight and conduct water and other substances from roots to leaves.

23–5 Leaves
- Leaves are solar energy collectors as well as the food-making center in most plants.

23–6 Transport in Plants
- Root pressure and transpiration pull are the forces that move water in xylem tubes.
- Scientists offer the pressure flow hypothesis to explain phloem transport. In this hypothesis, materials move from a source, an area of high concentration of materials, to a sink, an area of low concentration of materials.

23–7 Adaptations of Plants to Different Environments
- Flowering plants show a great many adaptations to different environments.

REVIEWING KEY TERMS

Vocabulary terms are important to your understanding of biology. The key terms listed below are those you should be especially familiar with. Review these terms and their meanings. Then use each term in a complete sentence. If you are not sure of a term's meaning, return to the appropriate section and review its definition.

23–2 Specialized Tissues in Plants
- meristematic tissue
- apical meristem
- cork cambium
- vascular cambium
- pericycle
- parenchyma
- sclerenchyma
- vessel element
- sieve tube element
- companion cell

23–3 Roots
- taproot
- fibrous root
- epidermis
- cortex
- vascular cylinder
- Casparian strip

23–4 Stems
- dormancy
- rhizome
- tuber
- bulb
- corm

23–5 Leaves
- petiole
- stoma
- guard cell
- mesophyll
- palisade layer
- spongy mesophyll

23–6 Transport in Plants
- transpiration
- transpiration pull

513

maximum amount of light that passes through the thin outer layer. The vascular tissues are centrally located, allowing distribution of materials to all the cells in the leaf.

Going Further: Enrichment
Two enrichment activities might be performed.
1. Look at cross sections of other types of leaves. Diagram and compare them with the yew leaves.
2. Look at the epidermal layer of lettuce leaves. Observe, diagram, and label the stomata, guard cells, and epidermal cells.

Analysis and Conclusions
1. Chloroplasts, which contain chlorophyll to absorb light, and photosynthetic membranes to produce food.
2. and 3. The green elongated cells of the palisade layer and the loosely packed cells of the spongy layer are used in photosynthesis. The tightly packed circle of cells is a vein, which helps to transport water. The loosely packed, spongy layer is used in gaseous exchange.
4. The outer epidermal layer is thin and transparent, allowing for the easy passage of materials from the stomata and the absorption of light by cells beneath the epidermal level. The elongated cells packed tightly beneath the epidermal layer contain chloroplasts and are well positioned to obtain the

513

CHAPTER REVIEW

CONTENT REVIEW

Multiple Choice

1. b 2. a 3. d 4. a
5. c 6. a 7. d 8. b

True or False

1. F leaves
2. F evaporation/transpiration
3. T
4. F cuticle
5. F Phloem
6. T
7. F xylem
8. F stems

Word Relationships

1. products of photosynthesis
2. support
3. leaves
4. active transport

CONCEPT MASTERY

1. Plants continue to grow from their tips as long as they are alive. They continue to increase in size. This is an open growth pattern.
2. You would be more successful removing weeds that have a fibrous root system. Plants with a taproot are much harder to pull from the ground, and if a small piece of a taproot is left behind, the weed can regrow. It is easy to pull plants with fibrous roots from the soil.
3. Many desert plants lack leaves, have extensive root systems, and short life cycles that can be completed quickly after a rainfall.
4. Leaves are thin and flat. Most leaves have a large surface area. Thus, sunlight strikes a large area of photosynthetic tissue.
5. Root hairs increase the surface area of roots and thus increase the amount of water and minerals that can enter the root.
6. Stomata are openings in the surface of a leaf. In most plants, stomata are found on the lower leaf surface. In some plants that live in water, stomata are on the upper surface of the leaf. Guard cells that are found on both sides of a stoma open and close the opening. The guard cells regulate the amount of material that passes into and out of the openings.
7. In order for gases to move into or out of a cell, the membrane surrounding the cell must remain moist. Thus, carbon dioxide moves through the moist membranes of cells into the spongy mesophyll and oxygen moves out of these cells.

CHAPTER REVIEW

CONTENT REVIEW

Multiple Choice

Choose the letter of the answer that best completes each statement.

1. In order to grow well, roots need
   a. minerals.   c. carbon dioxide.
   b. oxygen.   d. water.
2. Openings that permit gases to enter and leave the leaf are called
   a. stomata.   c. palisade cells.
   b. cuticles.   d. chloroplasts.
3. A nutrient that plants use for proper leaf growth and color is
   a. zinc.   c. calcium.
   b. potassium.   d. nitrogen.
4. The type of plant tissue that divides by mitosis is
   a. meristematic.   c. phloem.
   b. xylem.   d. surface.
5. The inner part of bark is made of
   a. xylem.   c. phloem.
   b. parenchyma.   d. cuticle.
6. A meristematic tissue that increases the thickness of plant stems over time is
   a. vascular cambium. c. pericycle.
   b. apical meristem.   d. epidermal meristem.
7. Thin-walled cells that store the products of photosynthesis are
   a. tracheids.   c. sclerenchyma cells.
   b. companion cells. d. parenchyma cells.
8. A primary root that grows longer and thicker is
   a. a fibrous root.   c. easy to remove.
   b. a taproot.   d. not found in plants.

True or False

Determine whether each statement is true or false. If it is true, write "true." If it is false, change the underlined word or words to make the statement true.

1. Carbon dioxide enters a plant through the <u>roots</u>.
2. Plants lose water through the leaves by <u>perspiration</u>.
3. Trees grow taller because cells in the <u>apical meristem</u> are able to divide.
4. The waterproof covering on the outside of the leaf is the <u>cambium</u>.
5. <u>Xylem</u> tissue conducts the products of photosynthesis up through a plant's stem.
6. There is a <u>one-way</u> passage of materials into the vascular cylinder in plant roots.
7. The rings of a tree are made up of <u>phloem</u> tissue.
8. Tubers are underground <u>roots</u>.

Word Relationships

An analogy is a relationship between two pairs of words or phrases generally written in the following manner: a:b::c:d. The symbol : is read "is to," and the symbol :: is read "as." For example, cat:animal::rose:plant is read "cat is to animal as rose is to plant."

In the analogies that follow, a word or phrase is missing. Complete each analogy by providing the missing word or phrase.

1. xylem:water::phloem: _____
2. parenchyma:storage::sclerenchyma: _____
3. root hairs:roots::mesophyll: _____
4. water:osmosis::nutrients: _____

CONCEPT MASTERY

Use your understanding of the concepts developed in the chapter to answer each of the following in a brief paragraph.

1. Plants are said to have an open growth pattern. What does this mean?
2. If you were weeding a lawn, would you be more successful at removing weeds with a taproot or weeds with fibrous roots? Explain your answer.
3. What are three adaptations shown by plants that are able to survive in a desert?
4. Leaves are able to trap solar energy. How does their shape enable them to perform this job efficiently?
5. How are root hairs important to plants?
6. What are stomata? How do they work?
7. Why is it important for tissues in the spongy mesophyll of the leaf to remain constantly moist?

CRITICAL AND CREATIVE THINKING

Discuss each of the following in a brief paragraph.

1. **Applying concepts** Certain Native Americans used bark cut from birch trees to make canoes. Removing the bark did not kill the trees. What precautions did they take when they removed the bark?
2. **Relating concepts** The leaves of cactuses have been modified into thorns. What two functions does this modification have?
3. **Drawing conclusions** Suppose you were going away on vacation and you couldn't find anyone to water your houseplants. You knew that plants lose water through the stomata in their leaves, so you decided to cover the leaves with petroleum jelly to prevent water loss. When you returned home, you found your plants had died. What is the most logical explanation for this?
4. **Designing an experiment** Your friend says it's necessary to fertilize plants every time they are watered. His plants grow well. You decide to test his claim. Design an experiment to test your friend's hypothesis.
5. **Relating cause and effect** A corn farmer decides to save money by not fertilizing the fields. The crop grows well the first year but diminishes each succeeding year. How would you explain this situation?
6. **Relating cause and effect** In Japan, the art of growing miniature trees is highly valued. By cutting the roots and tips of the branches, the tree remains small. The trunk of the tree, however, continues to increase in diameter. How do you explain the ever-increasing diameter of the trunk?
7. **Applying concepts** During the nineteenth century, people often raised ferns and other delicate plants that normally require a great deal of water in enclosed glass containers called Wardian cases. Plants in Wardian cases did not have to be watered for years. What is the most logical explanation for this phenomenon?
8. **Applying concepts** In cold northern climates many trees lose their leaves in autumn. How is this an adaptation that helps the trees survive the cold of winter?
9. **Using the writing process** Many people find insectivorous plants interesting. Most of these plants are collected from the wild. Some scientists are concerned that too many of these plants are being collected. Write a letter to your congressional representative proposing a law to protect these plants.

CRITICAL AND CREATIVE THINKING

1. They were careful not to cut the delicate layer of phloem tissue beneath the bark. They also may not have cut the strip of bark in a complete circle around the trunk.
2. Unlike leaves, thorns do not lose water to the environment. Thorns also protect the cactus from being eaten by animals.
3. The petroleum jelly also blocked the stomata in the leaves. The plants could not take in carbon dioxide or give off oxygen and water vapor. Without carbon dioxide, the plants could not make food and died.
4. Accept all answers that include an unfertilized control plant and a fertilized plant grown under identical conditions. Also, students should include references to maintaining accurate records of plant growth and to the amount and kinds of fertilizers used. Plants grow well under diverse conditions; however, by comparing the growth of the two plants, it should be possible to determine the effect of fertilizers.
5. The corn plants eventually use up the minerals present in the soil. Year after year, the corn yield diminishes as minerals in the soil are used up. A clever farmer who wanted to save money spent on expensive fertilizers would rotate or alternate the corn crop with legumes that are able to enrich the soil by converting atmospheric nitrogen into compounds that can be used by other plants.
6. The person training the miniature tree trims off the apical meristems at the tips of shoots and roots. This keeps the tree short. However, the person does not touch the vascular cambium in the stem, so the stem continues to increase in thickness.
7. The water that evaporated from plant leaves was contained in the Wardian case, fell back to the soil, and could be used again by the plants.
8. Plants lose water through their leaves. In cold climates, the water in the ground is frozen. If a tree continued to lose water from its leaves, it could not take in the frozen water present in the soil. The lack of leaves prevents water loss at a time of year when it would be difficult for a plant to replace it.
9. Check letters for correctness and clarity of expression.

CHAPTER 24 Plant Growth and Development

Section	Laboratory Investigations and Activities
24–1 Patterns of Growth pages 517–521 THEMES: Scale and Structure, Patterns of Change	**Teacher Edition** DEMONSTRATION: Growth Patterns, p. 516D **Teacher Edition** DEMONSTRATION: Examining Stems, p. 516D **Teaching Resources** ECOLOGY INVESTIGATION: Interspecific Competition and Chemical Warfare, p. 31
24–2 Control of Growth and Development pages 521–527 THEMES: Systems and Interactions, Scale and Structure	**Student Edition** LABORATORY INVESTIGATION: Understanding Plant Tropisms, p. 528 **Laboratory Manual** Plant Hormones, p. 299 Germination Inhibitors, p. 307 Observing Plant Responses, p. 311 **Teaching Resources** HANDS-ON ACTIVITY: Lean to the Light, p. 119 HANDS-ON ACTIVITY: You *Can* Force a Flower to Bloom, p. 121 ECOLOGY INVESTIGATION: Effects of Minerals on Plant Growth, p. 29
Chapter Review pages 529–531	

Teacher Resources

Books

Art, Henry. *A Garden of Wildflowers: 101 Species and How to Grow Them.* Storey Communication, 1986.

Baker, N. R., et al. *Control of Leaf Growth.* Cambridge University Press, 1985.

Burgess, J. *An Introduction to Plant Cell Development.* Cambridge University Press, 1985.

Dennis, David T. *The Biochemistry of Energy Utilization in Plants.* Chapman and Hall, 1987.

Nickell, L. G. *Plant Growth Regulators—Agricultural Uses.* Springer-Verlag, 1982.

Wareing, P. F., and I.D.J. Phillips. *Growth and Differentiation in Plants,* 3rd ed. Pergamon, 1981.

Wilkins, Malcolm. *Plantwatching: How Plants Remember, Tell Time, Form Partnerships, and More.* Facts on File, 1988.

CHAPTER PLANNING GUIDE

Other Activities	Media and Technology
Study Guide Section 24–1, p. 233	
Teaching Resources VOCABULARY REVIEW: Cross-a-Clue, p. 1 ACTIVITY: External Factors That Affect Plant Growth, p. 3 ACTIVITY: The Influence of Latitude and Altitude on Plants, p. 5 **Biotechnology Workbook** The Plant and Caterpillar Wars, p. 63 **Study Guide** Section 24–2, p. 235	
Teaching Resources Chapter Test A, p. 11 Chapter Test B, p. 15	**Computer Test Bank** Chapter Test, p. 257

Audiovisuals

Get Ready, Get Set, Grow! Film or video. Bullfrog Films.

Plants That Grow from Leaves, Stems, and Roots, rev. Film or video. Coronet.

Plant Tropisms and Other Movements. Film or video. Coronet.

Growth of Plants. Film. Encyclopaedia Britannica Educational Corp.

Plant Growth Regulators. Video filmstrip. Carolina Biological Supply Company.

CHAPTER 24 Plant Growth and Development

CHAPTER OVERVIEW

Plants are classified as annuals, biennials, or perennials on the basis of how long they live and the time they take to grow from seed to maturity. Growth patterns in stems and roots of perennials are discussed in detail.

Stems of dicots grow in length as cells in the apical meristem divide and enlarge. The increase in thickness as well as new tissues are produced by their vascular and cork cambiums. The stems of most monocots do not grow in width. Like stems, roots grow in length as the cells in their apical meristems produce new cells.

Plants grow in response to a number of environmental factors, and their growth responses are under the influence of various hormones. Depending on such variable factors as their concentrations and interactions, hormones bring about a variety of responses in different plant tissues. Auxins are an important class of hormones that cause cells to elongate. Their activity stimulates the stems to grow toward light and the roots to grow downward. Auxins also limit the growth of lateral buds from which new branches develop. Other hormones discussed in the chapter include the cytokinins and gibberellins. Cytokinins stimulate cell division and gibberellins promote rapid stem elongation.

Hormones also influence the responses to environmental stimuli known as tropisms. Plants respond to light, gravity, touch, and several other stimuli. During winter months, many plants enter a period of dormancy. The changes that occur at this time are controlled by the interaction of several hormones. Another process that occurs on a seasonal basis is the production of flowers. Flowering appears to be based on changes in the periods of daylight and darkness. These changes induce flowering by bringing about changes in compounds called phytochromes.

24–1 PATTERNS OF GROWTH

Section Focus 24–1

The purpose of Section 24–1 is to compare growth patterns in monocot and dicot stems and to explain primary growth in roots. Students are first introduced to three variations in plant life cycles. Those that grow from seed to maturity and then die in a single growing season are classified as annuals. Biennials have a life cycle spanning two years, and perennials may live for many years.

Perennial stems grow in both length and thickness from year to year. They lengthen as the cells divide in the apical meristem at the tip. All tissues derived from the apical meristem are called primary tissues. Dividing cells in other meristematic regions, called vascular cambium and cork cambium, produce an increase in thickness.

In woody dicot stems, a layer of vascular cambium lies between primary xylem and phloem tissues. Dividing cells here give rise to new layers of vascular tissue called secondary xylem and secondary phloem. Secondary xylem that forms on the inner side of the vascular cambium becomes the wood of the tree. New wood is laid down in concentric rings during each growing season. Secondary phloem cells are laid down on the outside of the vascular cambium. This new phloem forms the inner part of the tree bark. Since most monocots lack a vascular cambium, monocot stems generally do not grow thicker after they have reached their final height.

Roots, like stems, lengthen as their apical meristem produces new cells at the tip. However, most of the actual growth in length is due to cells growing larger in the zone of elongation just above the apical meristem.

Performance Objectives 24–1

1. Compare and list examples of annuals, biennials, and perennials.
2. Distinguish between woody and herbaceous plants.
3. Describe the formation of primary tissues by the apical meristem.
4. Compare growth patterns in dicot and monocot stems.
5. Describe primary growth in roots.

Science Terms 24–1

annual   p. 517
biennial   p. 517
perennial   p. 517
herbaceous   p. 518
zone of elongation   p. 518
zone of maturation   p. 518
primary tissue   p. 518
secondary xylem   p. 519
secondary phloem   p. 519
sapwood   p. 519
heartwood   p. 519
annual ring   p. 519
root cap   p. 520

24–2 CONTROL OF GROWTH AND DEVELOPMENT

Section Focus 24–2

In this section students will learn how hormones control the growth and development of plants. As in animals, plant hormones that are manufactured in one part of the organism affect activities in other parts. The part of the organism that is affected is known as the target organ. Depending on such factors as the concentration and the presence of other hormones, a given hormone can have varying effects on a target.

Auxin is a hormone that regulates growth, but it produces different effects on different plant tissues. One effect is the bending of stems toward light and away from the pull of gravity. High concentrations of auxin stimulate elongation of plant cells. If light strikes a stem mainly from one side, auxins accumulating on the shaded side stimulate a rapid elongation of cells on that side. Such uneven cell growth bends the stem toward the light. However, in roots, a high auxin concentration has the opposite effect. Here, auxin actually inhibits cell elongation and causes the root to bend downward. Auxin can also act as a growth inhibitor. High concentrations in the apical meristem inhibit the growth of lateral buds. The effect, called apical dominance, prevents lateral buds from producing side branches.

CHAPTER PREVIEW

Other plant hormones include the cytokinins and gibberellins. Cytokinins stimulate cell division, and they often produce the opposite effect of auxin. For example, they cause dormant seed to sprout, and they stimulate lateral buds to grow. Gibberellins can cause a very quick elongation of stems that brings about a dramatic increase in size.

Hormones also control several plant responses to environmental stimuli. These responses, known as tropisms, cause the plant to move toward or away from such stimuli as light, gravity, and touch. Auxin and other hormones also interact to control winter dormancy in deciduous plants. Such changes in hormone levels cause leaves to change color and eventually drop from trees. Changes in hormone production also cause protective bud scales to form about the meristems during dormancy.

The response of plants to changes in periods of light and darkness is called photoperiodicity. Evidence suggests that changes in the length of day or night cause different plants to produce flowers in different seasons. Some plants flower when days are short, others when days are long. Still others do not appear to be affected at all by changes in day or night length.

Performance Objectives 24-2

1. Describe the effects of hormones on plant growth and development.
2. Compare the effect of auxins, cytokinins, and gibberellins on different target organs.
3. Discuss the role of hormones in winter dormancy.
4. Identify three tropisms exhibited by plants.
5. Describe the effect of changes in day and night on flowering.

Science Terms 24-2

hormone p. 521
target organ p. 522
auxin p. 522
lateral bud p. 523
cytokinin p. 524
gibberellin p. 524
tropism p. 525
phototropism p. 525
gravitropism p. 525
thigmotropism p. 525
abscission layer p. 525
bud scale p. 525
photoperiodicity p. 527

DISCOVERY LEARNING

TEACHER DEMONSTRATIONS
Modeling

Growth Patterns

The following demonstration can be used to introduce Chapter 24.

To review some previously taught concepts and to stimulate thinking on some forthcoming topics, display a common houseplant, such as a geranium, from which the soil has been removed from the roots. Also be sure the stem can be plainly seen.

• **How does growth occur in this plant?** (Answers will vary, but lead the class to suggest that growth is the result of cell division and cell enlargement.)

• **Where do you think these kinds of cell activities take place in this plant?** (Lead students to suggest that growth occurs in the meristematic regions at the tips of the roots and stems.)

Next, show a small woody twig and a tree branch or fireplace log that has a larger diameter.

• **In what other way does this kind of plant stem apparently grow?** (Students should answer that some plants grow in thickness as well as in length.)

Do not attempt to discuss details of primary and secondary growth at this time. Tell the class that questions they may have about growth patterns will be answered as they read the chapter. Finally, refer again to the exposed roots and the stem of the plant.

• **As you know, this plant's stem and roots grow in opposite directions. Why do you think the roots grow down, while the stem grows up?** (Answers will vary.)

At this time do not go into details of the hormonal control of plant development. Rather, let the class speculate as to why different plant parts exhibit different responses to environmental factors.

Examining Stems

You may want to perform the following demonstration after students read about the growth in stems in Section 24-1.

Gather a few twigs with terminal buds and lateral buds that can easily be seen. Distribute them for examination by small groups. Direct their observations by asking questions similar to those that follow.

• **Where will this stem grow in length?** (It will lengthen at the tip.)

• **What name is given to the region of rapidly dividing cells at the tip?** (It is called the apical meristem.)

Point out that the meristematic cells are located in the bud. The bud itself is made up of newly formed, unopened leaves. If the bud is dormant, it will be covered by a ring of bud scales. Direct students to look for bud scales covering the bud. Explain that woody stems produce bud scales to protect the bud at the end of a growing season. The bud scales fall away in spring when new growth begins.

• **When bud scales fall off, a set of rings called bud scale scars are left on the stem. Can bud scale scars be seen on your stem?** (Answers will vary.)

• **How can you determine the amount of growth that occurred during a year?** (Since bud scale scars mark the location of former terminal buds, the distance between them indicate one year's growth.)

Next call attention to the lateral buds on the side of the stem.

• **What do you think develops from the lateral buds?** (New branches, leaves, and sometimes flowers develop from lateral buds.)

If different types of twigs are examined by the group, have them compare similarities and differences in the buds. If time permits, students can cut through a terminal bud longitudinally and examine the tissues with a hand lens or stereo microscope. The shoot apex, or meristematic region, may be seen in the lower central portion of the bud.

CHAPTER 24

Plant Growth and Development

GUIDED ENQUIRY
Pose the following questions to students and have them record their responses. Point out that they will gain a better understanding of the key concepts if they read the chapter with these basic questions in mind. Upon completion of the chapter, pose the questions again. Ask students to compare their initial responses with those they have developed after reading the chapter.
- Why do annual rings develop in a growing tree?
- Why does the woody part of a stem grow thicker but not the bark?
- What causes a plant's stem to bend and grow toward the light?
- Why does cutting off the tip of a stem cause a plant to branch outward?
- Why do leaves change color during autumn?
- Why do some flowers bloom in the spring, while others bloom in the fall?

INTRODUCING CHAPTER 24

Using the Visuals
Direct the class to look at the pictures of the oak tree and the acorns on page 516.
- **What is the relationship between the tree and the acorns in the smaller picture?** (Acorns contain a seed from which an oak tree can develop.)
- **How are oak seeds different from other seeds?** (Answers will vary. Allow a few minutes for students to compare acorns with other familiar seeds.)

You may want to point out that an acorn is actually a type of dry fruit commonly referred to as a nut. Botanists define nuts as one-seeded fruits with a hard pericarp (shell). An acorn contains a single seed.
- **Squirrels and other animals often gather and store acorns. In what way might this benefit the oaks?** (Squirrels disperse the acorns to new locations that may favor their germination and development.)

Next, ask the class to read the paragraph below the photographs.
- **How will an oak tree change as it develops from a seed to maturity?** (Answers will vary. Lead students in a discussion of some observable changes that occur during plant growth and development. But do not attempt to discuss the control of these processes at this time.)
- **What changes occur in an oak**

CHAPTER 24

Plant Growth and Development

An oak forest is home for many different animals. Some eat the acorns produced by oaks, others make their home in tree trunks, and still others find safety in the lush oak foliage.

You have probably heard the old saying, "Mighty oaks from tiny acorns grow." But have you ever thought about the amazing changes that take place during the long life of such a tree? A tiny embryo oak plant sits within an acorn for months, maybe even years. Then one spring, it sprouts. Its roots grow deep into the soil in search of water and nutrients. Its leaves reach toward the sun—toward the light energy that powers the life of the plant. In autumn, growth slows. The oak's leaves turn color and fall to the ground. The entire plant undergoes the changes necessary to survive the approaching cold winter weather.

Many years later, the tree matures. Suddenly it produces hundreds, perhaps even thousands, of flowers. Some of the flowers produce seeds that mature into acorns. When the acorns ripen, they fall to the ground, where they may be buried or eaten by squirrels. With luck, the process of growth begins again.

516

GUIDE FOR READING

After you read the following sections, you will be able to

24–1 Patterns of Growth
- Compare annual, biennial, and perennial plants.
- Compare the growth of monocot and dicot stems.
- Explain how roots increase in length.

24–2 Control of Growth and Development
- Describe the effects and importance of hormones on plant growth.
- Identify several plant tropisms.
- Compare long-day, short-day, and day-neutral plants.

Journal Activity

YOU AND YOUR WORLD
To see how a "maze"ing plants are, construct a maze inside a shoebox. Use any design that you want for your maze, but make sure it has some twists and turns in it. Make a 6 cm x 2 cm hole in the shoebox at one end of the maze and place a small plant at the other end. Put the lid on the shoebox and turn the opening toward bright light. Observe your plant for six weeks. In your journal, record your weekly observations along with drawings showing the movement of the plant.

Figure 24–1 The zinnia (left) is one of the many plants that live for only one growing season. Plants such as the foxglove (right) take two growing seasons to complete their life cycle.

24–1 Patterns of Growth

Guide For Reading
- What are the characteristics of annual, biennial, and perennial plants?
- How do stems and roots grow?

Plants grow in different ways. If you have ever grown plants from seeds, you may be aware that some plants mature and produce flowers in the same year they are sown. Other plants take longer. **Plants are classified into three main groups—annuals, biennials, and perennials—depending on how long it takes them to produce flowers and how long they live.**

Annuals, Biennials, and Perennials

Some plants grow from seed to maturity, flower, produce seeds, and die all in the course of one growing season. These plants are called **annuals** (from the Latin word *annus*, which means year). Many common plants—such as marigolds, corn, and peas—are annuals.

Biennials (*bi* is the Latin word for two) are plants that usually live for two years. During the first growing season, biennials grow roots, stems, and leaves. The leaves and stems die back to the ground in winter, but the below-ground roots remain alive. The following season, new leaves and stems grow from the roots. In this second season, however, the plant produces flowers. Once the flowers produce seeds, the plant dies. Common biennial plants include some species of foxglove, sugar beets, carrots, and turnips.

Perennials are plants that live for more than two years. Peonies, popular perennial garden plants, often outlive the person who planted them. Other perennials, such as dawn redwoods, have trunks and branches that live and grow larger for hundreds, even thousands, of years. Trees and shrubs are

---

tree on a seasonal basis? (Lead the class in a discussion of such seasonal changes as the development of leaves in spring, their change of color in autumn, and winter dormancy. Again, do not go into the factors that control these processes at this time.)

TEACHING STRATEGY 24–1

Focus/Motivation
Show the class a packet of common annual flower or garden seeds. Then call attention to any trees or shrubs that can be viewed from classroom windows. If actual trees or shrubs cannot be seen, then show some pictures of trees.

• **Nurseries and garden stores feature seeds for gardeners to plant every spring and summer. Why do they usually not sell seeds for trees?** (Many common

COOPERATIVE LEARNING

Using preassigned lab groups or randomly selected teams, have groups propose a plant for beautifying the campus through a landscaping project. Have each group identify one area of the campus in which they would place a flower bed. Groups should make a scale drawing of the flower bed on graph paper and determine the types of plants they will include. Groups should consider the following:
- Size and shape of flower bed
- Local climate
- Amount and type of sunlight at this location
- Characteristics of annuals, biennials, and perennials
- Cost of plants
- Maintenance and care

If possible, have available for group use catalogs from local nurseries, ads from local newspapers, and books on gardening. If time permits, divide available chalkboard space and allow groups to diagram their flower beds using colored chalk.

Journal Activity

YOU AND YOUR WORLD
Have students make plant mazes on their own or in small groups, or design one yourself for the entire class. Ask students to predict the outcome of this six-week-long project. Have students keep their observations and drawings in their portfolio.

garden plants are annuals that must be replaced every year. Trees do not need to be replaced each year, and are usually too slow-growing for most gardeners to start them from seed.)

After students have read page 517, write the terms *annual*, *biennial*, and *perennial* on the chalkboard. Then call on someone to define each term. Invite students to obtain seed or nursery catalogs and list examples of annuals, biennials, and perennials. To conserve resources, have students request only one catalog from any company; do not allow every student to request the same catalog!

TEACHING SUPPORT

Teaching Resources
- Ecology Investigation: Interspecific Competition and Chemical Warfare, p. 31

BACKGROUND INFORMATION
PRIMARY TISSUES

The primary tissues of a stem are formed from three meristematic tissues produced within the apical meristem. The outermost layer that is formed at the apex of the meristem is called the protoderm. It gives rise to the epidermis. The largest portion to develop from the apical meristem is the ground meristem. The ground meristem, in turn, develops into the pith at the center of the stem and cortex just beneath the epidermis. The pith and cortex are largely storage tissues. A third strand of cells produced by the apical meristem is the procambium. It differentiates into phloem and xylem.

24-1 (continued)

Content Development
Write the terms *herbaceous* and *woody* on the chalkboard.
- **How do herbaceous plants differ from woody ones?** (Herbaceous plants have little or no wood in their stems.)
- **What happens to herbaceous plants during cold winters?** (Answers may vary. Many herbaceous plants are annuals that complete their life cycle before winter. However, some die back to their roots and produce new stems and leaves in the spring.)

Ask for examples of herbaceous plants. All grasses including the cereal grains, familiar weeds, garden flowers, and vegetables are herbaceous.

Reinforcement/Reteaching
Help students recall how growth occurs in plants by referring to Chapter 23. Ask questions similar to those that follow.
- **Why do stems grow at their tips?** (A region of rapidly dividing cells called the apical meristem is located at the tip.)
- **After new cells have been formed, where in the plant do they grow larger?** (They grow larger in the zone of elongation behind the meristem.)
- **What takes place in the zone of maturation?** (Cells differentiate into other types of cells.)
- **What name is given to tissues produced by the apical meristem?** (It is called primary tissue.)

Content Development
Be sure that students understand the difference between primary

---

Figure 24-2 The chrysanthemum (left) and the peony (right) are perennials, which means they are able to grow for many years. Cold winter temperatures cause the top parts of the plants to die down to the ground. The roots, locked in the frozen soil, remain alive and will begin to grow again with the warmth of spring.

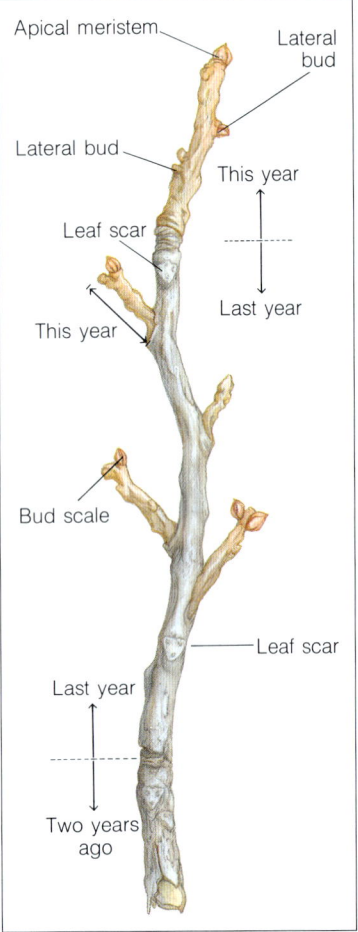

perennials because they live for indefinite periods of time. Trees and shrubs whose stems contain woody tissue are called **woody plants**.

Plants whose stems have little or no woody tissue are called **herbaceous** (her-BAY-shuhs) plants. In cold climates, herbaceous plants die back to their roots every year. For example, perennials such as chrysanthemums and tulips grow new leaves and stems every spring. In tropical climates, the above-ground parts of both woody and herbaceous plants can grow throughout the year.

Growth of Stems

As you may recall from Chapter 23, stems grow longer at their tip as cells in the apical meristem divide. These newly formed cells then grow larger in an area behind the meristem called the **zone of elongation**. Farther back from the zone of elongation is the **zone of maturation**. Here cells differentiate into the various cells that make up a plant stem. In herbaceous plants, almost the entire stem is composed of tissue produced by the apical meristem. Tissue produced by the apical meristem is called **primary tissue**.

It should be clear to you that if a perennial plant is to grow larger year after year, its stems must increase in thickness as well as in length. Yet, as you may remember, only meristematic tissue can produce new cells for growth. In perennial plant stems, new tissue is added by special meristematic tissues called the vascular cambium and cork cambium. The addition

Figure 24-3 A twig taken from a tree in winter shows how the apical meristem—the place where plant cells divide—is protected from the cold by thick bud scales. Twigs from different plant species have different markings, which can be used by botanists to identify the species—even without the benefit of leaves.

of new tissue in these cambium layers increases the thickness of the stem, which is needed to support a larger plant.

Growth in Dicot Stems

In woody dicot stems, a layer of vascular cambium located between the xylem and phloem tissues remains alive through the plant's first winter. In the following spring, this vascular cambium becomes active as its cells begin to divide. Xylem cells are formed on the surface of the vascular cambium that faces the center of the stem. Phloem cells are formed on the surface that faces the outside of the stem. These new layers of vascular tissue are called **secondary xylem** and **secondary phloem**. Both secondary layers spread sideways from each vascular bundle in the cambium layer, soon uniting the circle of vascular bundles into a solid ring. The xylem tissue is on the inside of the ring, and the phloem tissue is on the outside. The vascular cambium remains between the xylem and phloem.

XYLEM GROWTH The new ring of xylem tissue formed during every growing season plus the older xylem become the wood of trees. Although the cells of the older xylem layers are dead, they still conduct water for several years. Xylem is called **sapwood** as long as it conducts water. After some time, however, the oldest xylem cells become clogged with tars and resins and are no longer able to conduct water. This xylem tissue, which no longer conducts water but gives strength and support to the tree, is called **heartwood**.

You have probably seen tree rings, or **annual rings**, in the wood of trees. Tree rings are formed because the vascular cambium makes large xylem cells during the spring and smaller xylem cells during late summer and fall. The large cells of spring wood look lighter than the smaller cells of summer wood. Thus the tree accumulates set after set of alternating light and dark rings, one set for every year the tree is alive.

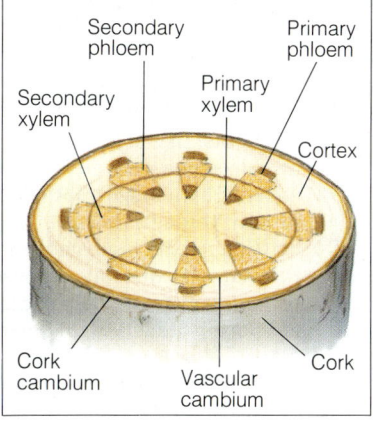

Figure 24–4 Each tree ring represents a season's growth. During a single growing season, secondary xylem and phloem are produced by the vascular cambium. Note the variations in the sizes of the rings.

TEACHING SUPPORT

Study Guide
• Section 24–1, p. 233

24–1 (continued)

Content Development

Emphasize the fact that the vascular cambium gives rise to secondary phloem cells as well as secondary xylem. However, biologists are not sure why the cells on one side of the cambium become xylem and those on the other side become phloem. Point out that in a mature woody stem, the vascular cambium is a thin ring at the boundary between the wood and the bark.

Also be sure that students understand the role played by the cork cambium in secondary growth. As a stem increases in diameter, layers of old tissues, such as epidermis and primary phloem, are crushed and split. Cork cambium forms at the splits in the stem where it produces cork cells. These cells form a layer that protects the stem against excessive water loss.

SECTION REVIEW 24–1

1. An annual plant completes its life cycle in one growing season. At the end of a single growing season, an annual will have produced seeds that will grow next year. A perennial often takes several years to mature. Then it continues to grow, year after year, flowering and producing seeds.
2. Most monocots have an apical meristem at the tips of the stems. They do not have a cambium layer in the stem. Without a cambium layer, the stems of monocots do not grow thicker.
3. Apical meristems at the tips of the stems and the roots contain cells that divide actively. Later, these cells elongate and mature. It is in the zone of elongation that stems and roots increase in length.
4. Trees continue to grow taller, year after year, because the tip of each shoot has an apical meristem that contains cells that are actively dividing. Thus, tree branches continue to increase in length as long as the tree is alive. Humans fortunately do not have an apical meristem. Their growth stops after a finite period of time. If humans continued to grow as long as they lived, basketball would be a much easier game to play than it is!

Figure 24–5 Beavers are known for their sharp teeth, which have produced the results you see here. The bark and the delicate underlying phloem tissue of the tree have been gnawed away. The tree has been girdled and will eventually die.

PHLOEM GROWTH The vascular cambium makes new phloem cells at the same time it makes new xylem cells. These secondary phloem cells are added to the inner surface of the previous year's phloem. The phloem layer, however, never becomes as thick as the xylem layer. One reason for this is that the cambium makes only one new phloem cell for every six or eight xylem cells. Another reason is that phloem cells have thinner walls than xylem cells and are crushed as the stem grows thicker.

Phloem tissue forms the inner half of tree bark. Here it carries sugars and other products of photosynthesis from leaves to those plant parts that do not make their own food. For this reason, a tree will die if the phloem in tree bark is damaged or removed in a ring around the trunk.

Growth in Monocot Stems

Recall that in monocots (corn and lilies, for example) xylem and phloem tissues are arranged in vascular bundles scattered throughout the stem. In most monocots, these bundles lack a vascular cambium, so no new xylem and phloem cells can be produced. This means that once the apical meristem of a monocot produces a stem, that stem cannot grow thicker. Few monocots grow more than several meters tall.

But there are a few tall monocots. Palms are one example. The few monocot trees remain short for several years. During this time, they produce leaf after leaf, without growing very tall. While they are producing leaves, however, the apical meristem as well as the stem below it grow wider. Finally, the meristem produces a stem strong enough to serve as a trunk. At that point, the monocot tree starts to grow taller. Unlike dicot stems, the stem of a palm does not increase greatly in width once the tree begins to grow tall.

Growth of Roots

Roots, like stems, grow in length as their apical meristem produces new cells near the root tip. The fragile new cells are covered by a tough **root cap** that protects the root as it forces its way through the soil. As the root grows, the root cap secretes a slippery substance that lubricates the progress of the root through the soil. Because cells at the very tip of the root cap are constantly being scraped away, new root cap cells are continually added by the meristem.

Figure 24–6 Palms are one of the few monocot trees. Palms remain short for a time, as their apical meristem increases in width (inset). However, the tiny palm will eventually grow tall.

Figure 24–7 *Only the cells in the root tip divide. In the area just behind the root tip, the newly divided cells increase in length, pushing the root tip farther into the soil. The root cap, located just ahead of the root tip, protects the dividing cells as they are pushed forward.*

Most of the actual increase in root length occurs in the zone of elongation immediately behind the meristem. There, the small cells produced by the meristem grow longer. In the zone of maturation, these cells take on the structures and functions of mature root cells.

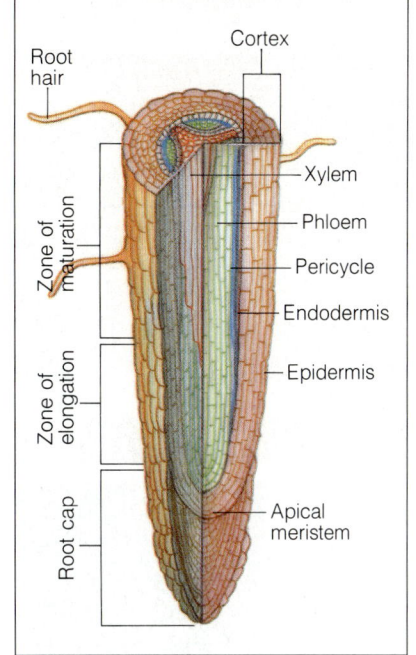

24–1 SECTION REVIEW

1. How does an annual plant differ from a perennial plant?
2. Why don't most monocot stems grow thicker?
3. How do stems and roots grow in length?
4. **Critical Thinking—Applying Concepts** The tallest humans stop growing eventually, yet a tall tree increases its height year after year. Why are plants able to continue to grow taller as long as they live?

24–2 Control of Growth and Development

Plants grow in response to important environmental factors such as light, moisture, temperature, and gravity. But just how do roots "know" to grow down and stems "know" to grow up toward light? The answers to these questions involve the actions of special chemicals that direct, or control, the growth of a plant.

Guide For Reading
- What role do hormones have in the growth and development of plants?
- What are phototropism, gravitropism, and thigmotropism?
- How do short-day, long-day, and day-neutral plants differ?

Hormones and the Control of Plant Growth

In plants, as in animals, the division, growth, maturation, and differentiation of cells is controlled by a diverse group of chemical substances called hormones. A **hormone** is a substance produced in one part of an organism that affects another part of the organism. Plant hormones are manufactured primarily in apical meristems, in young leaves, and in growing seeds and developing fruits. Plant hormones control a plant's branching pattern, the rate at which its stems elongate, and its responses to environmental conditions.

How does a stem respond if a plant is knocked over? How do roots grow when they come up against a rock in the soil? How does a plant adjust its growth rate when winter or a dry season approaches? How do plants flower at the proper season? All these phenomena are controlled by hormones.

The Nature of Hormone Action

The part of the organism affected by a hormone is known as the target tissue or **target organ**. The effects of a given hormone on a target tissue or target organ can vary a great deal for several reasons.

One reason is that the same concentration of a particular hormone can have two different effects on two different target tissues. For example, the identical concentration of a hormone can stimulate growth in stem tissues but inhibit growth in root tissues. Another reason is that different concentrations of a particular hormone can produce different effects on the same target organ. For example, low concentrations of a hormone can cause meristematic cells to divide, but high concentrations of that same hormone can prevent division. This phenomenon will help explain branching in plants. And a third reason is that two or more hormones can interact in a variety of ways. For example, the effect of one particular hormone may depend on whether or not another particular hormone is present. Systems of interacting hormones are important in regulating reproduction in plants.

The Role of Auxin

One well-known plant hormone is indoleacetic acid, commonly called **auxin**. Auxin can produce different effects on different plant tissues. High concentrations of auxin stimulate elongation of stem cells and inhibit elongation of root cells. Let us see how auxin directs the growth of stems and roots.

AUXIN AND STEM GROWTH Auxin produced in the apical meristem of a stem causes the stem to grow toward light and away from the pull of gravity. This effect was first noticed by Charles Darwin and his son Francis in experiments they performed in the 1880s. The Darwins observed that light caused "some influence" (they did not know what the influence was) to be transmitted from the tip of a growing stem down to the zone of cell elongation. This mysterious influence, the Darwins felt, caused the stem to bend and grow toward the source of light.

Other experiments performed between 1910 and 1920 by the Danish scientist P. Boysen-Jensen and the Hungarian researcher A. Paal showed that the "influence" the Darwins had observed was a chemical manufactured in the apical meristem. This chemical, the hormone we call auxin, is produced in the apical meristem and moves down the stem.

Today we understand that high concentrations of auxin stimulate young stem cells to elongate. If the source of light is directly overhead, and if the stem is growing vertically, the concentration of auxin will be the same on all sides of the stem. All the cells in the zone of elongation will grow at the same rate and the stem will grow straight up.

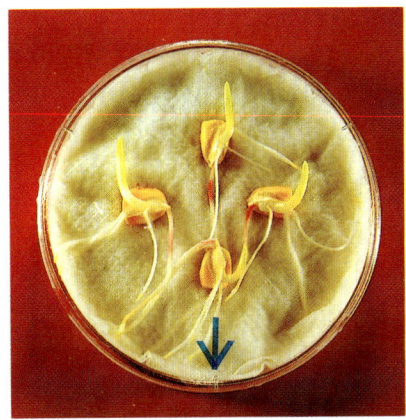

Figure 24–8 The roots of the corn seeds in this petri dish are growing in the same direction even though the seeds have been planted upside down (top). Stems grow toward the light (bottom). Why is it important that roots and stems grow the way they do? ❶

Figure 24-9 Plants will bend toward a source of light. This fact is explained by the stimulating effect of plant hormones on cell growth. How do plant hormones cause a plant stem to bend?

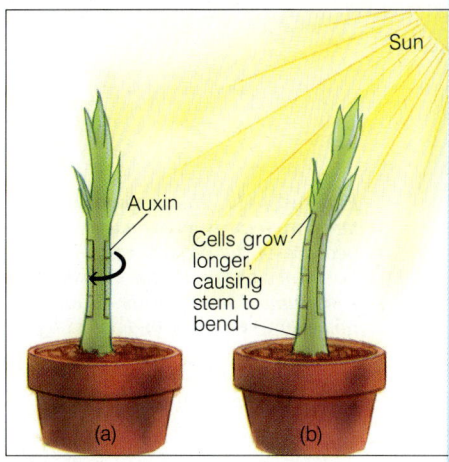

If the source of light is to one side of the stem, however, auxin will move away from the light to the shaded part of the stem. The higher concentration of auxin in the shaded side of the stem will stimulate cells in that side to elongate more than cells in the lighted side. This will cause the stem to bend toward the light.

If a plant is knocked on its side, auxin will accumulate in the lower side of the stem, or the side along the ground, because this side is darker than the upper side. It is the shaded side of the stem. The auxin-stimulated cells in the lower side will elongate more than the cells in the upper side. The stem will curve up, even in the absence of light.

AUXIN AND ROOT GROWTH Auxin produced in the apical meristem of roots causes the roots to grow away from light and toward the pull of gravity. If a growing root is exposed to light, auxin will concentrate in the shaded side of the root, just as it did in the shaded side of the stem. But high concentrations of auxin inhibit the elongation of root cells. Thus cells in the side of the root where auxin concentration is high will not elongate. Cells in the lighted side of the root will elongate normally, causing the root to bend away from the light.

If a growing root is directed sideways by an obstacle in the soil such as a rock, auxin will accumulate in the lower side of the root. Once again, high concentrations of auxin inhibit root cell elongation. The uninhibited root cells will elongate more than the auxin-inhibited root cells, and the root will bend downward.

AUXIN AND BRANCHING In addition to controlling cell elongation, auxin can also cause cell division in meristematic regions to stop or start. As a stem grows in length, cells in its apical meristem periodically produce other meristematic areas. These meristematic areas, called **lateral buds**, are on the sides of the stem. See Figure 24-11 on page 524. If these lateral buds begin to grow, they give rise to side branches that grow from the main stem.

If you grow plants at home or in the classroom, you know that most lateral buds do not start growing right away. Instead, they begin to grow only after the main growing tip of the plant has grown a distance above them. However, you can force these lateral buds into growth at any time by "pinching," or removing, the apical meristem. Why should this be the case?

Again, the action of auxin provides an explanation for the growth of lateral buds. Scientists have learned that high concentrations of auxin inhibit the growth of lateral buds, whereas low concentrations of auxin stimulate growth. Thus the high

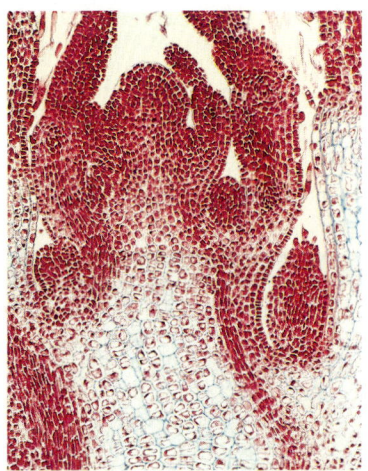

Figure 24-10 A longitudinal section through the tip of a plant stem can be seen in this photomicrograph. Lateral buds can be seen just behind the apical meristem. These lateral buds will remain dormant as long as the tip of the plant remains intact nearby.

BACKGROUND INFORMATION
ROOT GROWTH

Biologists do not know exactly what causes the downward growth of roots. It is known that small amounts of auxin inhibit root cell elongation. Thus, downward growth is thought by some to be due to the accumulation of auxin on the lower side of a horizontally growing root. As cells on the upper surface elongate more rapidly than those on the lower surface, the root bends downward. Yet how does a root sense its orientation with respect to gravity, and what causes the downward effect of auxin?

Some plant biologists believe the answer lies in colorless plastids, known as amyloplasts, that always seem to be adjacent to the cell wall on the cell's downward side. If a vertically growing plant is laid on its side, the amyloplasts slide down to the bottom. It is believed that the amyloplasts may carry high levels of auxin to the lower cell wall, thus inhibiting its growth.

Other biologists believe that auxin is not involved at all in normal root curvature. Rather, they think that another hormone, abscissic acid, moves up from the root tip to control curvature. The precise regulatory chemicals and mechanisms influencing root curvature are an unsolved puzzle.

523

(It will grow straight up, since auxin is evenly distributed on all sides of the stem.)
• **What will happen to the seedling grown with light on one side?** (Auxin moves to the side away from the light, causing the stem to grow unevenly.)

• **What about the bean seedling grown in the dark?** (Accept all logical responses. A seedling grown in dark will be very tall and spindly. Encourage students to verify this unexpected result through experimentation.)

Skills Development
Guided Practice
Skill: Making comparisons
• **How does light affect auxin concentration in both stems and roots?** (In both, auxin moves away from the lighted side and becomes concentrated on the shady side.)

• **Why does the higher auxin concentration on the shady side cause different responses in stems and roots?** (High auxin concentration stimulates elongation of stem cells, but it inhibits elongation of root cells.)

523

TEACHING SUPPORT

Laboratory Manual
- Germination Inhibitors, p. 307

Teaching Resources
- Activity: External Factors That Affect Plant Growth, p. 3
- Activity: The Influence of Latitude and Altitude upon Plants, p. 5
- Ecology Investigation: Effects of Minerals on Plant Growth, p. 29

MULTICULTURAL STRATEGY

Japanese botanist Ewiti Kurosawa and his colleagues discovered gibberellins in the 1920s, but it was not until well after World War II that their work was published in America and Europe. Have interested students report on the different constraints, both political and practical, that have separated scientists in the past and present.

24–2 (continued)

Content Development
Using the Visuals Write the term *apical dominance* on the chalkboard and call on someone to define it. Then call attention to Figure 24–11.
- **Why is the plant on the right more bushy?** (Removal of the terminal bud from that plant keeps it from producing auxin, which would limit the growth of lateral buds.)

Point out that in many deciduous trees and other flowering plants there is no single dominant apical tip. Each major branch has a center of apical dominance at the tip. Therefore, many familiar trees have a rounded shape. On the other hand, pruning or topping a tree can stimulate the production of additional lateral stems, giving it a more bushy appearance.

Reinforcement/Reteaching
Have interested students do a simple experiment to observe the effect of gibberellins on plant growth. First, plant wheat seeds in soil in two flats or other small containers, such as cut-off plastic milk jugs. After the seedlings grow to about 3 centimeters high, students should start watering one container with a gibberellin solution and the other with plain water.

Gibberellic acid is available from biological supply companies. To use it, dissolve about 25 milligrams in a few milliliters of 70% ethyl alcohol, then mix with 1 liter of distilled water. Spray solutions of gibberellic acid may also be available at nurseries or

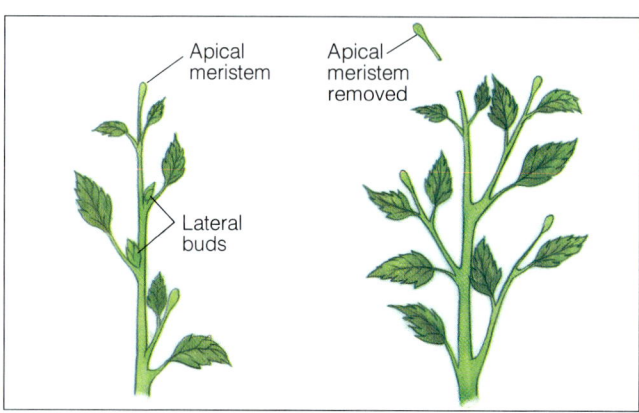

Figure 24–11 The dormant lateral buds are kept from growing because of the production of hormones by the apical meristem. If the apical meristem is removed, however, the lateral buds will begin to grow. Pinching the apical meristem is one way a plant can be made to grow more bushy.

Figure 24–12 The effects produced by the growth hormone gibberellin are dramatic. The short seedlings have not been treated with the hormone; the tall seedlings have.

concentration of auxin produced by the apical meristem inhibits the growth of lateral buds near the tip of the shoot. This inhibition is called apical dominance. When the apical meristem grows away from the lateral buds, or when it is removed, the auxin concentration in the lateral buds drops. Apical dominance ends, and the lateral buds begin to grow.

Cytokinins

Growing roots manufacture hormones called **cytokinins**. These hormones move upward in plants. Some cytokinins are also produced in developing fruits and seeds. Cytokinins stimulate cell division and cause dormant seeds to sprout. Several of the effects produced by cytokinins are the opposite of the effects produced by auxins. For example, whereas auxins stimulate cell elongation, cytokinins inhibit elongation and cause cells to grow thicker. Whereas auxins inhibit the growth of lateral buds, cytokinins stimulate lateral bud growth. Cytokinins have different effects on different tissues.

In growing plants, therefore, the relative concentrations of auxin, cytokinin, and other hormones determine how the plant will grow. You will learn in later chapters that the interactions of hormones are common in animals as well as in plants.

Gibberellin

In the 1920s, Japanese biologists studying certain rice plants made an important discovery. The rice plants they were observing seemed to be afflicted by a "disease" that made them grow unusually tall and spindly. Eventually the rice plants grew so tall that they could no longer be supported by their stems, and so they fell over. The scientists found that the plants' extraordinary growth was caused by a substance produced by the fungus *Gibberella fujikuroi*. They named the growth-producing substance **gibberellin** (jihb-er-EHL-ihn) after the fungus that produced it. Later, researchers learned that there are several different gibberellins. They also learned that gibberellins are hormones manufactured by all higher plants.

Tropisms

The responses of plants to environmental stimuli are called **tropisms** (TROH-pihz-uhmz) from the Greek word that means turning. If a plant grows toward a stimulus, it is said to have a positive tropism. If it grows away from a stimulus, it is said to have a negative tropism. There are several different kinds of tropisms. A plant's response to light is called **phototropism**. A response to gravity is called **gravitropism**. A response to touch is called **thigmotropism**.

For example, as soon as the primary root emerges from a germinating seed, it grows down and away from light. The root thus shows negative phototropism (away from the stimulus, light) and positive gravitropism (toward the stimulus, gravity). The stem, on the other hand, immediately starts growing toward the surface of the soil and toward light. The stem has a positive phototropism (toward the stimulus, light) and a negative gravitropism (away from the stimulus, gravity). From the very beginning of a plant's life, these positive and negative tropisms direct the plant's growth. And as you learned, these relatively simple responses of plants are controlled by hormones.

Winter Dormancy

Auxin and other hormones work together to control the growth, dormancy, and death of leaves in deciduous plants. During the growing season, auxin is produced in the leaves. At summer's end, days become shorter and nights become longer. The change in the length of light and dark periods causes a change in the chemistry of a protein called phytochrome. This change in phytochrome causes several things to happen to the plant's production of hormones. Auxin production drops, but the production of the hormones abscisic (ab-SIHS-ihk) acid and ethylene gas increases.

These changes in hormone production have several effects. One is that the cells that join leaf petioles to their stems become weak, forming a band called the **abscission layer**. Another effect is that chlorophyll synthesis in leaves slows down and stops. The chlorophyll that remains in the leaves is slowly broken down by the plant or destroyed by sunlight. When this happens, brightly colored accessory pigments (such as those you learned about in the chapter on algae) become visible. During the summer, the green color of chlorophyll hides these pigments. But when the chlorophyll is no longer present in the leaves, these once-hidden pigments appear, providing the beautiful colors we associate with autumn foliage.

A third effect of changes in hormone production is that cell division in apical meristems changes. Instead of producing leaves, the meristems produce thick, waxy **bud scales**. These bud scales wrap around the apical meristems, protecting the terminal buds. Enclosed in its coat of scales, a terminal bud can

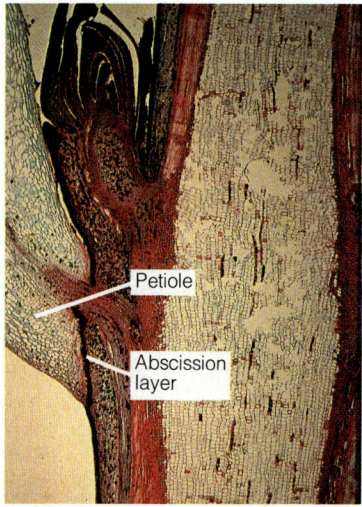

Figure 24–13 In the fall, when winter's chill approaches, a tree prepares to shed its leaves. Here you can see the formation of the abscission layer at the base of a leaf petiole. When the abscission layer is complete, even the most gentle breeze will cause the cells to tear and the leaf to fall to the ground.

SCIENCE, TECHNOLOGY, AND SOCIETY CONNECTION

Flowering by the Clock

Suggest that student teams conduct research on photoperiodicity and flowering. Some teams might visit greenhouses and flower shops to talk to someone about flowering plants that are sold around certain holidays. Have them prepare a list or an illustrated poster of plants whose flowering has to be timed for special sales. You might also invite a horticulturist to speak to the class on techniques used in modifying the flowering process.

As a project, students might investigate whether certain common potted plants, such as coleus or geraniums, need long or short periods of dark to flower normally. A more advanced project might involve growing short-day plants, such as chrysanthemums, and then attempting to delay their flowering by interrupting their dark period.

---

survive the coldest winter days. At the onset of winter, xylem and phloem tissues pump themselves full of ions and organic compounds. These molecules act like antifreeze in a car, preventing the tree's sap from freezing.

During this time, the abscission layers at the bases of the leaves continue to develop. Eventually only a network of fragile vascular bundles holds the leaves on the stems. Sooner or later, a gust of wind or a rainstorm breaks the delicate abscission layers apart and the leaves fall. Where the leaves were once attached to the stems, only leaf scars remain as evidence of a season of growth.

SCIENCE, TECHNOLOGY, AND SOCIETY CONNECTION

Flowering by the Clock

For many people, flowers have special meanings. Flowers are often given as gifts, as tokens of affection, and to wish friends well. But the life of a grower of gift plants and flowers is not an easy one. In addition to the problem of insects and diseases that may harm the crop, there is the question of timing. A plant produced for Valentine's Day would be of little use if it flowered even a day late. Poinsettia plants grown for winter holiday decoration would be worthless if they flowered in June. Fortunately, you can buy roses on February 14 and poinsettias on December 25. Like clockwork, these flowers and plants are ready when they can be sold most profitably. How does this happen?

Growers of plants take advantage of day length to produce their plants at the proper times. For example, chrysanthemums flower in nature during the autumn. They are, as you might suspect, short-day plants—they need long nights in order to flower. To coax chrysanthemums into flowering for, say, Mother's Day in June, growers cover the plants with lightproof cloths during late-spring afternoons. This prevents light from reaching the plants and thereby lengthens the amount of time they are in the dark. In effect, the plants are tricked into behaving as if the days are short.

By covering some plants and adding additional light to others, growers can almost guarantee that they will have plants and flowers available at suitable times. By taking advantage of the natural cycles of plant growth and flowering, consumers can be assured that they will have plants and flowers when they want them.

Long after the sun has ceased to shine, the plants in this greenhouse receive supplemental lighting to encourage, or in some cases delay, flowering.

24–2 (continued)

SECTION REVIEW 24–2

1. A hormone is a substance produced in one part of an organism that affects another part of an organism. In plants, hormones control the division, growth, maturation, and differentiation of their cells. Plants manufacture hormones primarily in the apical meristems, in young leaves, and in growing seeds and developing fruits.

2. Auxins cause cells to elongate in stems. Auxin moves to the side of a stem away from light. Increased levels of auxin on the shady side of a stem cause these cells to increase their length. The longer cells on one side of the stem cause the stem to bend toward the light. In roots, auxin inhibits the growth of cells. It still concentrates on the side of the roots away from light. These cells are inhibited from growing longer. The cells on the opposite side of the root elongate, causing the root to grow away from light.

3. Auxin produced in the apical meristem prevents lateral buds close to apical meristem from growing. If the apical meristem

Day Length and Flowering

"To everything there is a season," a popular song tells us, and nowhere is this more evident than in the regular cycles of growth in plants. Year after year, some plants flower in the spring, others in summer, and still others in the fall. How do plants time their flowering so precisely?

You know that there are fewer hours of daylight in winter than there are in summer. In the Northern Hemisphere, the shortest day occurs on or about December 21 and the longest day occurs on or about June 21. This means that from midwinter through spring, days increase in length.

In the early 1920s, W. W. Garner and H. A. Allard, of the United States Department of Agriculture, learned that tobacco plants time their flowering according to the number of hours of light and darkness they receive. It soon became clear that many other plants also respond to periods of light and darkness. This response is called **photoperiodicity**.

At first, researchers assumed that plants respond primarily to the number of hours of light they receive. Plants such as chrysanthemums and poinsettias, which flower when days are short, were called short-day plants. Petunias, clover, and hollyhocks, which flower when days are long, were called long-day plants. The flowering of some plants, such as corn and tomatoes, did not seem to be affected by day length to any great extent. These plants were called day-neutral plants.

Today we know that the important factor in photoperiodicity is the length of the night or darkness period, not the length of the day or light period. Short-day plants actually flower because the nights are long. Long-day plants flower because the nights are short.

How do changes in the length of night cause plants to flower? As you learned earlier, changes in the length of light and darkness periods cause changes in compounds called phytochromes. In a manner that is still not understood, these changes in phytochrome chemistry stimulate plants to flower.

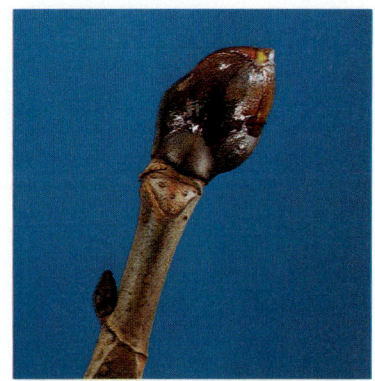

Figure 24–14 During the winter, tight, thick bud scales encase the apical meristem on a twig (top). Most deciduous plants that grow in cold winter areas need a period of below-zero temperatures before their buds will begin to grow. However, when warm spring days encourage new growth, the bud scales open to reveal the new year's leaves and flowers (bottom).

24–2 SECTION REVIEW

1. What are hormones? What is their role in plant growth and development? Where are most plant hormones produced?
2. How do different auxin concentrations affect cell elongation in stems? In roots? How do these effects explain phototropism and gravitropism in roots and stems?
3. What is the role of auxin in branching?
4. **Connection — Earth Science** What environmental cue do plants use to time their flowering? How is this cue related to short-day plants and long-day plants?

LABORATORY INVESTIGATION

UNDERSTANDING PLANT TROPISMS

Before the Lab
1. As described in step 1 of the Procedure, soak the corn seeds for 2 days before the investigation. Assemble all other necessary materials 1 day before the investigation.
2. If 150-milliliter beakers are in short supply, half-pint milk or juice cartons could be substituted. Be sure that absorbent cotton is used, or white paper toweling if cotton is not available.
3. Students may work in teams of 2 to 4.

Pre-Lab Discussion
Have students read the entire Procedure prior to coming to lab. Review the definitions of geotropism and phototropism if necessary; then discuss the Procedure by asking these questions.
• **What is the purpose of this lab?** (The purpose is to determine the effects of gravity and light on root growth and stem or leaf growth.)
• **In the first part of the experiment, the soaked seeds will be placed in a horizontal position in a petri dish. Formulate a hypothesis for this part of the experiment.** (Answers will vary. Some may suggest that the roots will grow down and the leaves will grow up.)
• **In the second part of the experiment, the seeds will be planted in soil inside a carton with a window cut in the side. What is your hypothesis for this part of the experiment?** (Students may predict that the seedlings will grow toward the window.)
• **What do you think is the purpose of placing foil caps over some of the seedlings?** (Students may suggest that the cap prevents light from reaching the tip of the seedling.)

LABORATORY INVESTIGATION

UNDERSTANDING PLANT TROPISMS

PROBLEM
How do roots find water, and how do leaves find light?

MATERIALS (per group)

aluminum foil	2 index cards
2 150-mL beakers	medicine dropper
transparent tape	milk carton
15 corn seeds	petri dish
cotton	scissors
filter paper	potting soil

PROCEDURE

1. Place some water in a beaker. Add 15 corn seeds. Soak the seeds for 2 days.
2. Select 4 swollen corn seeds and place them in the bottom of a petri dish. Arrange the seeds in a straight line across the center of the dish, with each seed at right angles to the adjacent seed. See the accompanying illustration.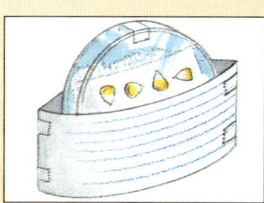
3. Put a piece of wet filter paper over the seeds. Cover the filter paper with enough wet cotton so that when the petri dish is closed, the seeds are held tightly in place. Cover the petri dish and tape it closed.
4. Make a stand for the petri dish from two index cards, as shown in the accompanying illustration. Fold the index cards in half. Then bend the cards so that the folds form right angles. Tape the cards together back to back. Slip the petri dish between the cards.
5. Fill the other 150-mL beaker with potting soil. Select 6 swollen corn seeds. Plant the seeds just below the surface of the soil, as far away from each other as possible. Moisten the soil.
6. Place the petri dish and the beaker in a dark place until the seeds have sprouted. This will take 2 or 3 days. Keep the soil evenly moist.
7. Examine the seeds in the petri dish after they have sprouted. Observe the directions in which the roots and leaves grow.
8. Make 3 small caps by wrapping small pieces of aluminum foil around the tip of a pen or pencil. Gently place these caps on half of the corn seedlings in the beaker.
9. Use scissors to cut the top off the large milk carton. Turn the milk carton upside down. Cut a small window on one side of the carton. Cut the window so that its base will be at the same level as the top of the beaker.
10. Place the beaker in a brightly lighted spot. Cover it with the milk carton. Position the carton so that light can reach the seedlings.
11. Leave the corn seedlings in bright light for about 4 days. During this time, water the plants with a medicine dropper if necessary.

OBSERVATIONS
1. In which direction are the leaves of the corn seeds in the petri dish growing? The roots?
2. In which direction are the leaves of the covered corn plants in the beaker growing?
3. In which direction are the leaves of the uncovered plants in the beaker growing?

ANALYSIS AND CONCLUSIONS
1. Why were the seeds in the petri dish grown in the dark?
2. What was the main influence on the direction of growth of the leaves and the roots of the corn plants in the petri dish?
3. How did the aluminum caps influence the direction of leaf growth in the corn plants?
4. What influenced the direction of growth of the leaves of the corn plants in the beaker?

Skills Development
Students will use the following skills while completing this laboratory investigation.
1. Applying
2. Comparing
3. Concluding
4. Experimenting
5. Hypothesizing
6. Observing
7. Predicting
8. Relating
9. Interpreting

Safety Tips
Remind students to be careful when working with scissors.

Teaching Strategy
1. Check to see that students have placed the four seeds in the petri dish at right angles.

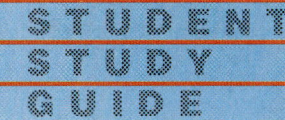

SUMMARIZING THE CONCEPTS

The key concepts in each section of this chapter are listed below to help you review the chapter content. Make sure you understand each concept and its relationship to other concepts and to the theme of this chapter.

24–1 Patterns of Growth

- There are three main groups of plants: annuals, biennials, and perennials.
- Stems increase in length as cells in their tip divide. These cells elongate in the zone of elongation behind the stem tip. The same cells differentiate in the zone of maturation, which occurs farther back from the tip.
- Roots increase in length as cells in the apical meristem in the root tips divide. These newly formed root cells elongate in a zone of elongation immediately behind the root tip. And they differentiate in a zone of maturation that is found farther from the tip.

24–2 Control of Growth and Development

- Hormones—special chemicals made by plants—direct the growth of plants.
- Auxin is an important plant hormone. High concentrations of auxin stimulate stem cells to grow longer. The same high concentrations of auxin can inhibit the elongation of root cells. Auxin also controls plant branching, since auxin inhibits the growth of lateral buds.
- Cytokinins are hormones produced in plant roots and developing fruits and seeds. These hormones stimulate cell division and cause seeds to sprout.
- Gibberellins are hormones produced by all higher plants.
- A tropism is the response of a plant to an environmental stimulus. A positive tropism is a response toward the stimulus. A negative tropism is a response away from the stimulus.
- During cold winters, many plants are dormant. Dormancy in plants is coordinated by the actions of several hormones. Dormancy is a reaction to environmental conditions that do not favor active plant growth.

REVIEWING KEY TERMS

Vocabulary terms are important to your understanding of biology. The key terms listed below are those you should be especially familiar with. Review these terms and their meanings. Then use each term in a complete sentence. If you are not sure of a term's meaning, return to the appropriate section and review its definition.

24–1 Patterns of Growth

annual
biennial
perennial
herbaceous
zone of elongation
zone of maturation
primary tissue
secondary xylem
secondary phloem
sapwood
heartwood
annual ring
root cap

24–2 Control of Growth and Development

hormone
target organ
auxin
lateral bud
cytokinin
gibberellin
tropism
phototropism
gravitropism
thigmotropism
abscission layer
bud scale
photoperiodicity

---

2. Caution students not to plant their corn seeds too deeply in the soil. The seeds should be planted about as deep as the length of the seed.
3. If only translucent plastic milk jugs are available, have students substitute some other kind of opaque container. A shoebox set up on one end with its lid taped in place could be used instead.

Observations
1. The leaves grew upward and the roots downward.
2. Straight up.
3. Toward the light.

Analysis and Conclusions
1. So that the light would not be a variable affecting the direction of root and leaf growth.
2. Gravity, since all the seeds received water from behind and were all grown in darkness. The only difference among the seeds was their orientation with respect to gravity. Despite the orientation of the seeds, all the leaves grew upward and all the roots grew downward.
3. The cap prevented the growing part of the leaf from getting any light, so the plant grew straight up instead of bending toward the light.
4. Light and gravity.

Going Further: Enrichment
To further investigate the effects of geotropism on root and leaf development, ask the teams to turn their petri dishes upside down after making their initial observations. After several more days, ask them to note any changes in the direction of the growth of the root and leaves.

- **What changes did you notice after the petri dish was inverted?** (Answers may vary, but in all likelihood both the roots and leaves will reverse direction.)
- **Suppose a gardener planted all corn seeds upside down. Will this affect the way the roots and leaves will grow?** (No, the roots should still grow downward and the leaves grow upward.)

CHAPTER REVIEW

CONTENT REVIEW

Multiple Choice
1. b 2. c 3. d 4. c
5. d 6. b 7. c 8. c

True of False
1. F auxin or indoleacetic acid
2. T
3. F zone of elongation
4. T
5. F photoperiodicity
6. T
7. T
8. F cambium layer

Word Relationships
1. plant hormones; xylem is a type of vascular tissue
2. annual plants; dawn redwood is a perennial plant
3. responses associated with the onset of winter dormancy; increase in the number of water molecules in xylem and phloem occurs with the breaking of dormancy
4. phenomena associated with flowering in long-day plants; long nights

CONCEPT MASTERY

1. The responses of plants to environmental stimuli are called tropisms. Stems show a positive tropism when they grow toward light and a negative one when they grow away from gravity. Roots show a positive tropism when they grow toward gravity and a negative tropism when they grow away from light.
2. A herbaceous plant is a plant whose stems have little or no woody tissue. Chrysanthemums and tulips are examples of herbaceous plants.
3. Auxin concentrates on the side of a stem away from light. It causes cells in the shady part of the stem to increase in length. These cells, longer than cells on the opposite side of the stem, cause the stem to bend toward the light.
4. Photoperiodicity is the effect that day length—or, more accurately, the length of time a plant is in darkness—has on flowering. Short-day plants make their flowers when nights are long. Long-day plants make their flowers when nights are short.
5. A biennial is a plant that requires two growing seasons to mature and flower. A biennial plant dies after it flowers, in its second year. A perennial plant, on the other hand, will continue to grow and flower for many years.
6. Cells in the zone of maturation begin to differentiate and form the different tissues that make up a plant's stem.
7. Primary plant tissues are the tissues produced by the apical meristem.

CHAPTER REVIEW

CONTENT REVIEW

Multiple Choice

Choose the letter of the answer that best completes each statement.

1. The response of plants to periods of light and dark is called
   a. gravitropism. c. thigmotropism.
   b. photoperiodicity. d. gibberella.
2. A plant that grows, matures, makes seeds, and dies within one year is a (an)
   a. biennial. c. annual.
   b. tree. d. perennial.
3. A part of a plant that shows positive phototropism and negative gravitropism is a
   a. leaf. b. root. c. flower. d. stem.
4. One way to coax a plant to branch is to
   a. fertilize it.
   b. trim the roots.
   c. remove the growing tip.
   d. keep the plant in the dark.
5. A plant that is laid on its side will
   a. die.
   b. grow very slowly.
   c. flower.
   d. bend its stem upward.
6. A chemical that directs the growth of a plant is called a (an)
   a. apical meristem. c. sapwood.
   b. hormone. d. heartwood.
7. An example of a monocot tree is a
   a. pine. c. palm.
   b. lily. d. maple.
8. A band of weak cells that forms where a leaf petiole joins a stem is called a (an)
   a. chlorophyll layer. c. abscission layer.
   b. meristem. d. cambium layer.

True or False

Determine whether each statement is true or false. If it is true, write "true." If it is false, change the underlined word or words to make the statement true.

1. The hormone that makes cells in the apical meristem divide is called <u>acetic acid</u>.
2. Brightly colored accessory pigments appear in the autumn when <u>chlorophyll</u> in the leaves breaks down.
3. Cells increase in length in the <u>zone of maturation</u>.
4. <u>Bud scales</u> protect the apical meristem from cold temperatures.
5. The response of plants to the length of day and night is called <u>phototropism</u>.
6. In autumn, the layer of weak cells that forms between the leaf petiole and the stem is called the <u>abscission layer</u>.
7. A <u>hormone</u> is a substance made in one part of an organism that affects another part.
8. The trunks of most trees increase in width as cells in the <u>apical meristem</u> divide.

Word Relationships

In each of the following sets of terms, three of the terms are related. One term does not belong. Determine the characteristic common to three of the terms and then identify the term that does not belong.

1. xylem, auxin, gibberellin, cytokinin
2. marigold, corn, dawn redwood, peas
3. breakdown of chlorophyll, formation of abscission layer, increase in the number of water molecules in xylem and phloem, formation of bud scales
4. summer, long days, short nights, long nights

CONCEPT MASTERY

Use your understanding of the concepts developed in the chapter to answer each of the following in a brief paragraph.

1. What is a tropism? Give an example of both a positive tropism and a negative tropism for a stem. For a root.
2. What is a herbaceous plant? Give an example of one.
3. How does auxin make a plant grow toward a light source?
4. What is photoperiodicity? How does it affect flowering plants?
5. What is a biennial? How does a biennial plant differ from a perennial plant?
6. What happens in the zone of maturation in a plant stem?
7. What is primary plant tissue?

CRITICAL AND CREATIVE THINKING

Discuss each of the following in a brief paragraph.

1. **Designing an experiment** One of your classmates suggests that all root cells have the ability to increase their length. Another classmate suggests that the cells in a root increase in length only in the zone of elongation just behind the root tip. Design an experiment to show which hypothesis is correct.
2. **Applying concepts** The sweet bay in the photograph was grown as a "standard." As you can see, a standard plant is grown to resemble a small tree. Next to it is a "normal" sweet bay plant. Use your knowledge of plant growth to determine the techniques used in growing a standard sweet bay.

3. **Applying concepts** People grow houseplants on windowsills. Many books advise giving houseplants a quarter turn every week. Why is this good advice to follow if you want to grow attractive plants?
4. **Relating cause and effect** Some dormant plants need an extended period of cold before their buds begin to grow. Some dormant plants even have to undergo two periods of freezing temperatures before their buds start growing again. Why is an extended period of cold or two freezing periods necessary for growth to resume?
5. **Making inferences** The black walnut tree grows in the northeastern part of the United States. Scientists have observed that few plants grow near the trunk of a black walnut. Offer an explanation for this observation.
6. **Using the writing process** Plants use the energy of light to make food. The flowering of some plants is affected by the length of day and night. Suppose light affected people in similar ways. Write a story about a person whose growth was affected by light.

CRITICAL AND CREATIVE THINKING

1. You can see the zone of elongation in a plant root by performing the following simple investigation. Take a newly sprouted bean seedling. Mark off equidistant lines in ink along the length of the root. In a few days, an examination of the root will show that the lines are now spaced farther apart in the zone of elongation. The lines in other parts of the root will be the same distance apart. Other solutions are possible.
2. To produce a standard plant you must first select a plant with a tall straight stem. Let the stem grow to the length you want. Then pinch off the tip of the stem. The plant will now begin to branch. Continue to pinch the new shoots and they will continue to branch. Eventually a ball-shaped tuft of leaves will be produced. Any shoots that appear on the stem below the top ball of leaves should be removed.
3. Window light comes from one direction only. A plant growing in a window will grow toward the light. Turning the plant will keep exposing a new part of the stem to the light. The stem will continue to grow straight.
4. Plants often set seeds in the fall. These seeds fall to the ground. If they sprouted right away, the new young plants would be killed by the cold temperatures of winter. Alternating periods of cold and warm temperatures encourage a seed to remain dormant until spring weather favors plant growth.
5. The roots of the black walnut give off a chemical that prevents other plants from growing nearby. This is why few plants grow near a black walnut. Other plants also give off chemicals that prevent the growth of other plants. For example, sunflower seeds inhibit the growth of other plants. This is why people who feed birds often notice that few plants grow beneath a bird feeder where the ground is littered with the hulls of sunflower seeds.
6. Accept all reasonable stories. Check students' writing for clarity of expression.

CHAPTER 25 Reproduction in Seed Plants

Section	Laboratory Investigations and Activities
25–1 Cones and Flowers as Reproductive Organs pages 533–540 THEMES: Scale and Structure, Evolution	**Student Edition** LABORATORY INVESTIGATION: Comparing Different Types of Flowers, p. 546 **Teacher Edition** DEMONSTRATION: Reproductive Structures, p. 532D **Laboratory Manual** Reproductive Structures in a Flower, p. 317
25–2 Seed Development pages 540–542 THEMES: Evolution, Systems and Interactions	**Teacher Edition** DEMONSTRATION: Fruits, p. 532D **Laboratory Manual** Fruits and Seeds, p. 323 **Teaching Resources** HANDS-ON ACTIVITY: Pet Plants—Or Are They Plant Pets? p. 133
25–3 Vegetative Reproduction pages 542–545 THEME: Stability	**Laboratory Manual** Germination and Seedling Development, p. 329 **Teaching Resources** HANDS-ON ACTIVITY: Clone a Plant, p. 123
Chapter Review pages 547–549	

Teacher Resources

Books

Beck, Charles B., ed. *Origin and Evolution of Gymnosperms*. Columbia University Press, 1988.

Buckles, Mary P. *The Flowers Around Us: A Photographic Essay on their Reproductive Structures*. University of Missouri Press, 1985.

Doust, Jan L., ed. *Plant Reproductive Ecology: Patterns and Strategies*. Oxford University Press, 1988.

Huxley, Anthony. *Plant and Planet*, rev. ed. Penguin, 1987.

———. *The Green Inheritance*. Anchor Press, 1985.

Meeuse, Bastiaan, and Sean Morris. *The Sex Life of Flowers*. Facts on File, 1984.

Salisbury, F. B., and I. Ross. *Plant Physiology*, 3rd ed. Wadsworth, 1985.

Scott, Jane. *Botany in the Field*. Prentice-Hall, 1984.

CHAPTER PLANNING GUIDE

Other Activities	Media and Technology

Teaching Resources
  ACTIVITY: The Life Cycle of Angiosperms, p. 3
Study Guide
  Section 25–1, p. 241
	Biology Transparencies
	  27: The Life Cycle of a Gymnosperm
	  28: The Life Cycle of an Angiosperm

Study Guide
  Section 25–2, p. 245
	Interactive Videodisc
	  On Dry Land: The Desert Biome
	Video/Videodisc
	  Biomes: Desert and Tundra
	Biology Media Guide
	  page 26

Teaching Resources
  VOCABULARY REVIEW: Crossword, p. 1
  ACTIVITY: Artificial Vegetative Reproduction, p. 5
Biotechnology Workbook
  Variety—The Spice of Life, Tomatoes, Peppers, etc. p. 73
Study Guide
  Section 25–3, p. 246	

Teaching Resources
  Chapter Test A, p. 11
  Chapter Test B, p. 15
  Performance-Based Assessment, Unit 6
	Computer Test Bank
	  Chapter Test, p. 267

Audiovisuals

Growth of Flowers, 2nd ed. Film or video. Coronet.
Pollinating Mechanisms. Filmstrip with cassette. Ward.
Gymnosperms (Life Cycles and Ecology). Filmstrip with cassette. Ward

Flowering Plants (Reproduction). Filmstrip with cassette. Ward.
Seeds: How They Germinate. Film or video. Coronet.
How Flowers Make Seeds. Film or video. Coronet.

Flowers: Structure and Function. Film or video. Coronet.
The Gymnosperms. Video filmstrip. Biology Media.

CHAPTER 25 Reproduction in Seed Plants

CHAPTER OVERVIEW

Seed plants are a predominant life form found in various habitats throughout the Earth. The success of these plants is chiefly due to specialized structures by which seed plants reproduce sexually. Such structures are located in the cones of gymnosperms and in the flowers of angiosperms.

The evolution of these structures combined with the production of seeds enable these plants to reproduce on dry land. In addition, the development of seeds has contributed greatly to the survival of gymnosperms and angiosperms in a variety of environments.

Sexual reproduction in seed plants increases the variation in their populations. Such variations increase the probability of survival of the species. Seed dormancy also contributes to the survival of a species by preventing the seed from germinating in unfavorable environmental conditions.

Asexual reproduction in seed plants produces offspring identical to the parent plant. Although vegetative reproduction yields few variations among a species, it does produce plants that are well adapted to a particular environment. Horticulturists employ methods of artificial vegetative reproduction to produce plants with specific traits.

25-1 CONES AND FLOWERS AS REPRODUCTIVE ORGANS

Section Focus 25-1

The purpose of this section is to introduce the cones of gymnosperms and the flowers of angiosperms as the plant structures specialized for sexual reproduction. In gymnosperms, male cones contain microsporangia, which produce pollen grains. Female cones contain megasporangia, from which ovules are produced. The reproduction process begins when pollen grains released from the male cones fall near an ovule of a female cone. Fertilization occurs when sperm travels through a pollen tube to an egg located inside an ovule. The resulting zygote grows into a seed-covered embryo.

The parts of a flower have unique structures and purposes. The sepals enclose and protect a developing flower. The brightly colored flowers attract pollinators. The pistil contains one or more megasporangia where ovules are produced. Female gametophytes, called embryo sacs, are produced in the ovules. Male gametophytes, called pollen, are produced in the anther.

Sexual reproduction in angiosperms begins when a pollen grain falls on the stigma of a flower. A pollen tube forms and grows down the style to the ovule. Double fertilization occurs when sperm nuclei enter the embryonic sac. One sperm nucleus fuses with the egg nucleus to create a zygote. The other sperm nucleus fuses with the polar nuclei to form a food supply for the developing organism. A seed coat then develops around the embryo and endosperm. The success of angiosperms on land is due in part to the development of seeds that provide an embryo with nourishment and protection.

Performance Objectives 25-1

1. Describe reproduction in gymnosperms.
2. Identify the function of various parts of a flower.
3. Explain the process of double fertilization.
4. Discuss the importance of seeds to the survival of a plant species.

Science Terms 25-1

sepal   p. 535
calyz   p. 535
petal   p. 535
corolla   p. 535
stamen   p. 535
filament   p. 535
anther   p. 535
carpel   p. 536
pistil   p. 536
ovary   p. 536
style   p. 536
stigma   p. 536
ovule   p. 536
embryo sac   p. 536
polar nuclei   p. 536
egg nucleus   p. 536
pollen chamber   p. 536
tube nucleus   p. 537
generative nucleus   p. 537
pollination   p. 537
self-pollination   p. 537
cross-pollination   p. 538
pollen tube   p. 538
sperm nuclei   p. 538
double fertilization   p. 538
endosperm   p. 538
fruit   p. 539
cotyledon   p. 539
epicotyl   p. 539
hypocotyl   p. 539
radicle   p. 539

25-2 SEED DEVELOPMENT

Section Focus 25-2

The purpose of this section is to further develop the topic of seeds. When a seed begins to sprout, it is said to germinate. Germination occurs when a seed has absorbed enough water to cause its endosperm and cotyledons to swell. The swelling causes the seed coat to crack. The radicle breaks through and becomes the primary root of the plant.

Not all seeds germinate when they first mature. Some seeds enter a period of dormancy in which the embryo is alive but not growing. Dormancy enables some seeds to postpone germination until environmental conditions are most favorable for survival.

Performance Objectives 25-2

1. Describe germination.
2. Identify the importance of dormancy to the survival of a plant species.

Science Terms 25-2

dormancy   p. 541

CHAPTER PREVIEW

25-3 VEGETATIVE REPRODUCTION

Section Focus 25-3

The purpose of this section is to describe how some angiosperms reproduce asexually through vegetative reproduction. This process produces offspring that are genetically identical to the parent. Some angiosperms reproduce asexually by producing tiny plants in their leaves or stems. The stems of other angiosperms produce roots when they touch the ground. Horticulturists can reproduce plants through asexual methods of reproduction. Taking cuttings, layering, grafting, and budding are methods of artificial vegetative reproduction.

Performance Objectives 25-3

1. Describe natural methods of asexual reproduction in angiosperms.
2. Explain methods of artificial vegetative reproduction.

Science Terms 25-3

scion   p. 544
stock   p. 544

DISCOVERY LEARNING

TEACHER DEMONSTRATIONS
Modeling

Reproductive Structures
The following demonstration may be used as an introduction to Chapter 25.

Ask students to bring flowers and/or cones to class. Have students use a magnifying glass to observe the specimens.

Discuss the physical traits of each specimen. Have students record their observations in words and drawings.

Challenge students to predict which traits of each specimen play a role in its reproduction process.

Upon completion of the chapter, have students check the accuracy of their predictions.

Fruits
You may want to perform this demonstration after students complete Section 25–1.

Display an apple, peach, or tomato and a bean pod, ear of corn, or acorn to the class. Ask students how the specimens are similar. (They are all fruits.)

Develop the idea that each item displayed is actually an enlarged and ripened ovary.

Ask students which fruits have soft, fleshy outer walls. (Apple, peach, tomato.) These fruits are classified as fleshy fruits, whereas the bean pod, acorn, and ear of corn are called dry fruits.

Cut the fruits in half to display their seeds. Ask students to describe what they see. (Matured, fertilized seeds.) Develop the idea that if any of these seeds were placed in ideal environmental conditions, germination would occur.

CHAPTER 25

Reproduction in Seed Plants

GUIDED ENQUIRY
Pose the following questions to students and have them record their responses. Point out that they will gain a better understanding of key concepts if they read the chapter with these basic questions in mind. Upon completion of the chapter, pose the questions again. Ask students to compare their initial responses with those they have developed after reading the chapter.
- How does the production of seeds contribute to the survival of a plant species?
- Can fertilization occur before pollination in seed plants?
- How does seed dormancy protect an embryo?
- How are plants reproduced in artificial vegetative reproduction?

CHAPTER 25

Reproduction in Seed Plants

Annual desert flowers complete their life cycle following a brief desert rain. Here, California poppies light the desert floor with golden flowers. Flowers produce pollen grains, such as the dimpled ones belonging to a rose (inset).

You have probably seen many different kinds of pine cones during your lifetime. Cones smaller than a golf ball or larger than your hand are produced by various species of gymnosperms. You have also probably noticed many different kinds of flowers. Flowers vary from the familiar beauty of a fragrant rose to the bizarre shapes and exotic colors of tropical gingers and bananas. But have you ever stopped to think about why plants produce these structures? Do you know what benefits cones and flowers offer plants? In this chapter you will gain an understanding of why plants invest the time and energy to produce cones and flowers.

532

INTRODUCING CHAPTER 25

Using the Visuals
Have students read the introduction to the chapter and study the accompanying photograph.
- **How do cones and flowers differ in structure?** (Possible answers include: Cones are hard and brittle and flowers are soft and pliable.)
- **What are examples of plants that have cones?** (Possible answers include: evergreens, such as the pine, spruce, fir, and redwood trees.)
- **In what type of habitat have you viewed cone-bearing plants?** (Answers may range from very humid rainforests to cold subarctic regions.)
- **What are examples of flowering plants?** (Possible answers include: rose, cactus, orchid, lily, and carnation.)
- **In what type of habitat have you viewed flowering plants?** (Answers may include almost all types of habitats.)
- **How are cone-bearing plants and flowering plants similar?** (Photosynthesis can occur in both types of plants. Tell students that in this chapter they will discover other similarities in the structures of a cone and a flower.)

532

GUIDE FOR READING

After you read the following sections, you will be able to

25-1 Cones and Flowers as Reproductive Organs
- Explain reproduction in gymnosperms.
- Describe the structure of a flower.
- Discuss reproduction in flowering plants.

25-2 Seed Development
- Explain the germination of seeds.
- Discuss dormancy in seeds.

25-3 Vegetative Reproduction
- Define vegetative reproduction.
- Describe how plants are reproduced by cuttings, layering, and grafting.

Journal Activity

YOU AND YOUR WORLD
Have you ever watched a dandelion seed float along in the air to land at wind's whim in a place far removed from the flower that produced it? Imagine that you are small enough to ride on a dandelion seed. In a short paragraph, describe your feelings about the places you pass over on your voyage.

25-1 Cones and Flowers as Reproductive Organs

Guide For Reading
■ What parts of seed plants are adapted for sexual reproduction?
■ What is the function of each part of a flower?

The cones of gymnosperms and the flowers of angiosperms are plant structures that are specialized for the purpose of sexual reproduction. **It is in the cones and flowers that the vital process that ensures the continuation of the species takes place.** Thus, cones and flowers are as important to the survival of plant species as roots, stems, and leaves are to the survival of individual plants.

As you may recall, all plants have life cycles in which a diploid sporophyte generation alternates with a haploid gametophyte generation. Gametophyte plants produce male and female gametes. When male gametes and female gametes join, they form a zygote. The zygote develops into the next sporophyte generation. In seed plants, the sporophyte generation is large and obvious. The gametophyte generation is small and often hidden within the cones or flowers.

As you read about the process of reproduction in seed plants, keep this important concept in mind: The development of cones and flowers and the production of seeds have enabled seed plants to reproduce without being dependent upon standing water. In this way, seed plants differ dramatically from mosses and ferns. The methods of reproduction evolved by seed plants help them survive the dry conditions of life on land better than mosses and ferns.

Figure 25-1 Female pine cones bear the seeds produced by pine plants. As you can see, pine cones vary greatly from the long, thin cones produced by the Eastern white pine to the short, squat cones produced by the Virginia pine.

533

ANNOTATION KEY

① Female: stigma, style, ovary; Male: anther, filament. (Interpreting illustrations)

BACKGROUND INFORMATION
GYMNOSPERMS

Gymnosperms are often referred to as "naked seed" plants. This does not mean that gymnosperms lack a seed coat. Rather, the term comes from the fact that the seeds of a gymnosperm develop on the surface of the plant structure. In some gymnosperms, seed-bearing structures are located within cones. However, the cones do not actually enclose the seeds but rather hide them from view. In addition, the seeds of all gymnosperms are "naked" in that they are not enclosed by a fruit, as in angiosperms.

FACTS AND FIGURES

In many gymnosperms, a year or more may pass between pollination and fertilization. It can take that long for a grain of pollen to penetrate the ovule via the pollen tube. In addition, for many gymnosperms, several years pass between fertilization and the shedding of seeds.

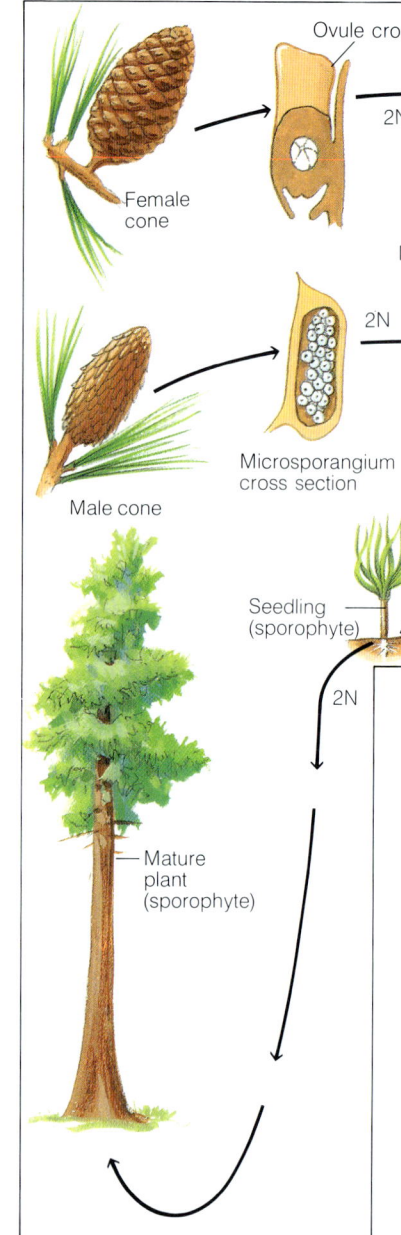

Figure 25–2 A pine tree produces both male and female cones. Male cones produce pollen; female cones produce ovules. If an ovule is fertilized, it becomes a zygote that is nourished by the female cone. In time the zygote develops into a seed.

Life Cycle of Gymnosperms

Familiar gymnosperms such as pine trees are diploid sporophytes, each of which has grown from a zygote contained within a seed. Years after the seed germinates, a pine tree matures and produces male and female cones. Male cones carry structures called microsporangia. The microsporangia produce male gametophytes called pollen grains. Female cones carry structures called megasporangia, which produce female gametophytes. Female gametophytes, in turn, produce ovules, or the structures in which egg cells form.

Pollen grains released from male cones are carried by the wind to female cones, where they may be caught by a sticky secretion. If a pollen grain lands near an ovule, the grain splits open and begins to grow a structure called a pollen tube. Two haploid sperm are located within the pollen tube. Inside an ovule is an unfertilized egg. The pollen tube grows into the ovule, and eventually the two sperm break out of the tube. One sperm fertilizes the egg; the other sperm disintegrates. The zygote that is formed grows into an embryo encased within what later develops into a seed. The seed is rather a neat package, for contained within it is an embryo plant as well as a supply of food for the embryo when it begins to grow.

Life Cycle of Angiosperms

Angiosperms, or flowering plants, are the dominant form of plant life on Earth today. During the long process of evolution, many kinds of angiosperms have developed. Today,

---

25–1 (continued)

Content Development
Using the Visuals Have students study Figure 25–2. Explain that this diagram illustrates the life cycle of a common conifer, the pine. In most conifers, male and female cones are located on the same tree. However, in cycads and ginkgoes, male and female cones are located on separate plants. Have students use the figure as a guide in describing sexual reproduction in conifers. Encourage class discussion on the benefits and disadvantages of the presence of male and female cones on the same tree as opposed to separate trees.

Enrichment
Have students investigate evolutionary patterns of gymnosperms to determine the relationship between gymnosperms that produce male and female cones on separate trees versus those that produce them on the same tree.

angiosperms grow in most land environments and in many watery ones. The complicated life cycle of angiosperms must receive most of the credit for the tremendous evolutionary success of these plants. Angiosperms have evolved a life cycle that liberates the reproductive stages of these plants from standing water. Thus, angiosperms were able to spread over the face of the land, colonizing most of planet Earth.

You may not think of a flower as anything but a decorative object, but flowers are visible evidence of the angiosperms' success. Each flower represents proof of a plant's survival and offers assurance that a plant species will produce more of its own kind.

Structure of a Flower

The following description is of a typical flower. This flower produces both male and female gametes. Many plants, however, produce flowers that differ somewhat from this general plan. In some plants, the male and female gametes are produced in separate flowers on the same plant. Corn, for example, produces male gametophytes in flowers at the top of the plant and female gametophytes in flowers located along the stem below the male flowers. Other plants, such as willows, produce male and female gametes on separate plants. Many other variations occur in the structure of flowers. As you read about the parts of a typical flower, use Figure 25–4 to locate them. You will thus be familiar with the names and functions of the various flower parts, even if a flower you examine differs somewhat from the generalized description.

Flowers are actually miniature stems that produce four kinds of specialized leaves: sepals, petals, stamens, and carpels. These specialized leaves are arranged in circles and have been modified during the course of evolution to serve different purposes related to reproduction. The outermost circle of flower parts consists of several **sepals**. In many flowers, sepals are green and actually resemble leaves. Sepals enclose the flower bud before it opens and protect the flower while it is developing. All the sepals in a flower together form the **calyx**.

Petals make up the second circle of flower parts. Petals, which are often brightly colored, are found just inside the sepals. All of the petals in a flower form the **corolla**. In some plant species, both petals and sepals are similarly colored. Brightly colored flower parts act as a kind of "flower advertisement," attracting insects and other pollinators to the flower. Because they produce no gametophytes, the sepals and petals of a flower are called sterile leaves.

Fertile leaves are located inside the petals. The fertile leaves contain the structures that produce male and female gametophytes. The first circle of these fertile leaves consists of **stamens.** Each stamen has a long, thin **filament** that supports an **anther**. Inside the anther are microsporangia in which the

Figure 25–3 *The flower and the structures associated with the flower contain the reproductive organs of a plant.*

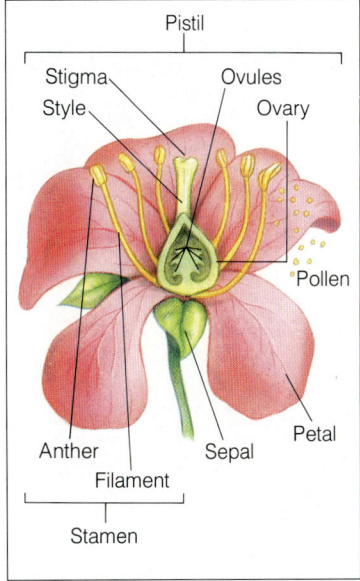

Figure 25–4 *This illustration shows the parts of a typical flower. Keep in mind that individual flowers may not have all the parts shown here. What parts make up the female reproductive organ? The male reproductive organ?* ❶

BACKGROUND INFORMATION
SEPALS AND PETALS

In many flowers, both sepals and petals are brightly colored. In others, sepals are smaller and thicker than petals and are plain green in color. In these cases, sepals protect the more fragile flower bud from damage.

In some textbooks, the flower's sterile leaves are called "nonessential" flower parts. However, in the majority of vector-pollinated flowers, the petals and sepals are just as essential to the flower's total reproductive effort as are the spore-bearing "essential" flower parts.

BACKGROUND INFORMATION
PERFECT AND IMPERFECT FLOWERS

Most angiosperm flowers contain both pistils and stamens; these flowers are called perfect flowers. But in certain species, individual plants produce either flowers with stamens only (staminate flowers) or flowers with pistils only (pistillate flowers). These unisexual flowers are called imperfect flowers and are found in such plants as date palms, willows, and poplars.

Content Development
Develop the idea that angiosperm flowers consist of miniature stems that carry four different kinds of specialized leaves: sepals, petals, stamens, and carpels. During the course of evolution, the structures adapted to suit a particular function in the process of reproduction. All of these specialized leaves may have adapted in different ways during the coevolution of particular plants and their pollinators.

• **What is the role of sepals and petals in the reproduction of angiosperms?** (Sepals protect a developing bud. Brightly colored sepals and petals attract pollinators to the flower.)

TEACHING SUPPORT

Laboratory Manual
- Reproductive Structures in a Flower, p. 317

Teaching Resources
- Activity: The Life Cycle of Angiosperms, p. 3

BACKGROUND INFORMATION
MEGASPORES AND MICROSPORES

Although their prefixes indicate otherwise, the size of megaspores and microspores in seed plants is not very different. In fact, a functional megaspore is about the size of a microspore. Although the terms are not indicative of gross size differences, they have come to indicate differences in the type of gametophyte produced by spores.

Figure 25-5 The stamens in this lily flower are not quite ready to release their pollen (top). At this stage, the stamens are smooth. Later, when pollen is ripe and ready to be released, it covers the outside of each stamen (bottom).

male gametophytes, the microspores, are produced. In most species of angiosperm, each flower has several stamens.

Carpels comprise the centermost circle of flower parts. Carpels are produced from fertile leaves that have rolled up. This rolling places megasporangia—the structures in which female gametophytes are produced—inside the female leaves, rather than on the outer surface of cones, as they are in gymnosperms. A single flower may contain one or more carpels. When multiple carpels are present in a flower, they may be either separate or fused. One or more carpels form the **pistil**. The pistil consists of a base called the **ovary**, a stalk called the **style**, and the **stigma**, located at the top of the style. In some plants the style is short; in other plants it can be quite long. For example, the style of a corn plant (actually the long thread often called a corn silk) can be more than 30 centimeters long. The stigma is the surface upon which pollen is deposited by wind or animal pollinators. In many plants the stigma is sticky or contains many small projections that help catch pollen.

The Female Gametophyte

Located inside each ovary is one or more megasporangia called **ovules**. A single diploid (2N) cell called the megaspore mother cell grows inside each ovule. Each megaspore mother cell produces a female gametophyte in a series of steps. First, the megaspore mother cell undergoes meiosis, producing four haploid (N) cells, three of which die. Next, the remaining haploid cell divides mitotically to produce eight nuclei. These eight nuclei and the membrane that surrounds them are called the **embryo sac**. The embryo sac is the entire female gametophyte. Note that this gametophyte is much smaller and simpler than the female gametophytes of mosses and ferns.

Inside the embryo sac, the eight nuclei move around. Eventually two nuclei locate themselves in the center of the sac, and three nuclei clump together at each end. The nuclei in the center are called the **polar nuclei**. Finally, one of the three nuclei in the group closest to the opening in the ovule enlarges to become the **egg nucleus**. The two other nuclei flank the egg nucleus. The three nuclei cells at the opposite end of the embryo sac die. The female gametophyte now contains a female gamete (egg nucleus) ready to be fertilized.

The Male Gametophyte

The male gametophyte is even smaller than the female gametophyte. Inside the anthers, microsporangia called **pollen chambers** produce many diploid (2N) microspore mother cells. Each microspore mother cell divides by meiosis to produce four haploid (N) microspores. Each microspore ultimately becomes a single pollen grain. The wall of each pollen grain thickens to protect the pollen grain's contents from dryness and physical damage when it is released from the anther.

25-1 (continued)

Content Development

Continue the discussion of the function of specialized leaves in the reproduction of flowers.

• **Which types of leaves are called fertile leaves? Why?** (Stamens and carpels are called fertile leaves. Stamens contain structures that produce male gametophytes and carpels contain structures that produce female gametophytes.)

• **Could reproduction occur in an angiosperm devoid of sterile leaves such as sepals and petals? Explain.** (Yes—if pollen is transferred from stamen to pistil. Wind-pollinated flowers in a benign environment, for example, probably have little need for sterile leaves. Sterile leaves increase the chances that a flower will produce seeds by attracting pollinators and protecting the developing flower bud before it opens.)

Enrichment

Students (particularly those who suffer from allergies) may be interested in investigating the relationship between pollen and

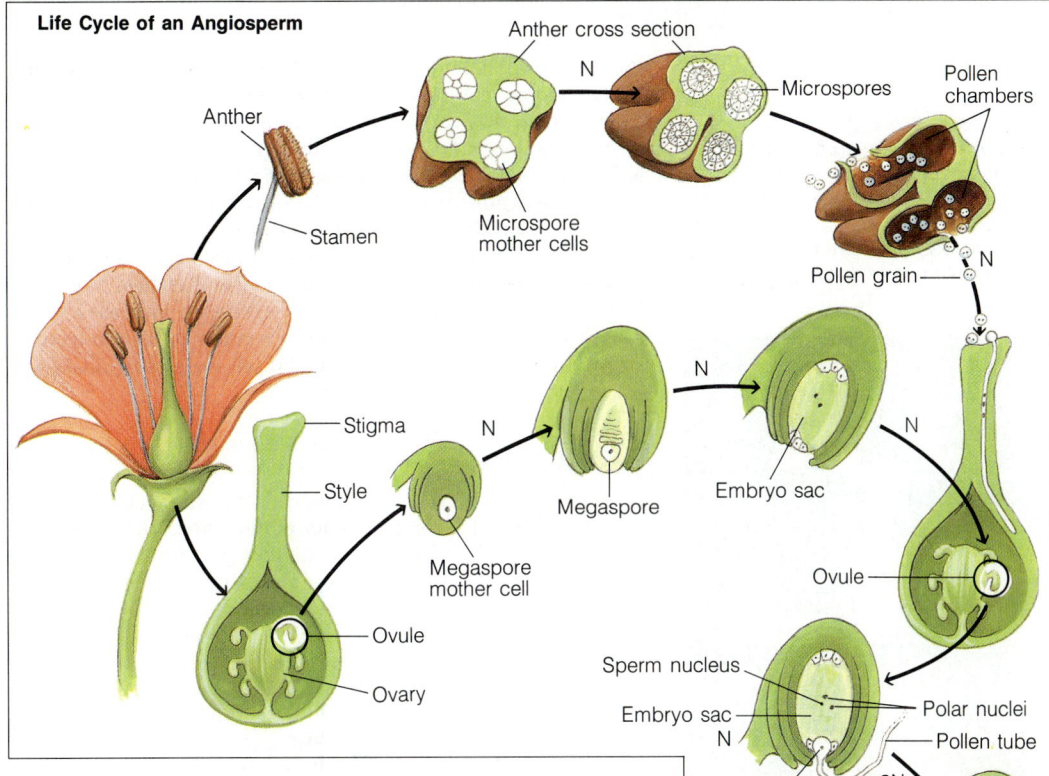

Life Cycle of an Angiosperm

The nucleus of the pollen grain undergoes one mitotic division, producing two haploid nuclei. One of these nuclei is called the **tube nucleus**; the other is called the **generative nucleus**. The tube nucleus disintegrates. The generative nucleus divides to form two sperm cells. The pollen grain, which is the entire male gametophyte, usually stops growing until it is deposited on a stigma.

Eventually the anther dries out, its pollen chambers split open, and mature pollen grains are released. This process can easily be observed in a lily flower. When a lily flower first opens, its anthers are large and smooth. After a few days (depending on environmental conditions such as temperature) the pollen in the chambers ripens. The chambers split, shrink, and turn inside out, exposing the dustlike mature pollen.

Pollination

At this point in the reproduction process pollen must be transferred from an anther to a stigma. The transfer of pollen from anther to stigma is called **pollination**. There are two types of pollination that occur in different plant species.

Some plants, such as the peas Mendel worked with, allow pollen to fall from the anther to the stigma of the same flower. This process is called **self-pollination**. Most plants, however,

Figure 25–6 Species of flowering plants have evolved complicated lifestyles to ensure their survival. The developing seeds of a flowering plant are protected and nourished inside the ovary located at the base of a flower.

allergic reactions. Ask students to predict the time of the year when a person who is allergic to pollen would most likely experience discomfort.

Content Development
Using the Visuals Direct students' attention to Figure 25–6. Have them use the diagram as a guide in explaining the formation of male and female gametophytes in angiosperms.
• How does the process a megaspore mother cell undergoes compare with that of a microspore mother cell? (Both types of mother cells divide by meiosis to produce four haploid cells.)
• What happens to the haploid cells produced by a megaspore mother? (Three of the cells die. The remaining cell divides by mitosis to produce eight nuclei.)
• What happens to the haploid cells produced by a microspore mother? (Each microspore becomes a pollen grain.)

Content Development
Guide students' understanding that a male gametophyte must join a female gametophyte if fertilization is to occur. In the process of pollination, pollen is transferred from an anther to a stigma.
• Could fertilization occur without pollination? Explain. (No, because without pollination, male and female gametes could not fuse.)

ESL STRATEGY

Have students write a "biography" of a mature seed plant of their choice. Their stories should include all of the important events of this plant's life, including its formation as a seed, transportation away from its parents, germination, and growth to maturity. Alternatively, students may use drawings and captions, presenting them either as a booklet or a poster. Encourage students to be imaginative and creative, perhaps by giving their plants names and personalities, or by incorporating current or historical events into the stories.

ANNOTATION KEY

❶ Corn has one cotyledon and the bean has two. Both contain an embryo plant and stored food. (Applying concepts)

BACKGROUND INFORMATION
AROMATIC EVOLUTION

In their quest to attract pollinators, plants have evolved an incredible variety of structures and strategies. But one African plant, *Anchomanes difformis,* may have topped them all: It does the botanical equivalent of burning incense! In the middle of its flowers, *A. difformis* has a foot-tall structure called a spadix. The plant's tissues are able to generate heat, and they warm the spadix to about 40°C (or 104°F)! The result is a sweet aroma emanating from the spadix, an aroma that attracts the small beetles that pollinate the flower.

BACKGROUND INFORMATION
FRUITS

Apples, watermelons, and tomatoes are all examples of fruits. These structures are actually matured ovaries that contain seeds. The number of seeds in a fruit is identical to the number of ovules present in the ovary from which it developed.

Sometimes a fruit develops from the sepals, petals, or receptacle of a flower, in conjunction with the ovary. Such a fruit is called an accessory fruit. An apple is an example of an accessory fruit. Only the core of the apple formed from the ovary. The bulk of the apple is actually an enlarged fleshy receptacle.

Figure 25–7 Some fruits, such as the apple and banana, are familiar to you. Some, such as the spikey durian, are not. Bright colors and sweet tastes attract animals that eat the fruits and help distribute the ripe seeds.

do not self-pollinate. The majority of flowering plants have evolved complicated methods of reproduction that ensure that seeds will form only when pollen from one flower is transferred to the stigma of a flower on another plant. The transfer of pollen from one flower to a flower on another plant is called **cross-pollination**. In Chapter 22 you read about a few of the many relationships that exist between flowers and pollinators.

Why is self-pollination uncommon in many plant species? Recall that sexual reproduction allows the exchange of genetic material between individuals. This exchange increases variation in offspring. Usually, the more variation there is in a population, the more likely it is that at least some individuals will survive to reproduce.

Fertilization

Once a pollen grain has landed on the stigma of an appropriate flower, it begins to grow a **pollen tube**. The generative nucleus within the pollen grain divides and forms two **sperm nuclei**. The pollen tube now contains a tube nucleus and two sperm nuclei. Following a chemical trail, the pollen tube grows down the style. Eventually the pollen tube reaches the ovary and enters the ovule through a small hole.

When the pollen tube reaches the female gametophyte (embryo sac), the sperm nuclei enter. Both nuclei participate in a process called **double fertilization**. Double fertilization occurs only in angiosperms. During this process, one sperm nucleus fuses with the egg nucleus to form the zygote. The other sperm nucleus fuses with the two polar nuclei. These three nuclei form the triploid (3N) **endosperm**. It is the endosperm that provides food for the embryo, which is produced when the zygote begins to grow. It is interesting to note that many animals eat the endosperm because it is so rich in important nutrients. Indeed, most of the food supply of humans is the endosperm of grasses. Three important examples are corn, wheat, and rice.

25–1 (continued)

Content Development

Guide students' understanding of fertilization by examining each step of this process.

• **What happens after a pollen grain has fallen on the stigma of an appropriate flower?** (The pollen grain grows a pollen tube.)
• **Through which organs of the flower does a pollen tube pass?** (It begins in the stigma and grows down the style, through the ovary, and into an ovule.)
• **What happens in double fertilization?** (One sperm nucleus fuses with the egg nucleus to form the zygote. The other nucleus fuses with the two polar nuclei to form the endosperm.)
• **Could double fertilization produce multiple zygotes?** (Multiple zygotes would be possible only if the ovule contained multiple egg nuclei.)

Content Development

Develop the idea that fertilization triggers rapid changes in various parts of the flower.

• **From which part of the flower**

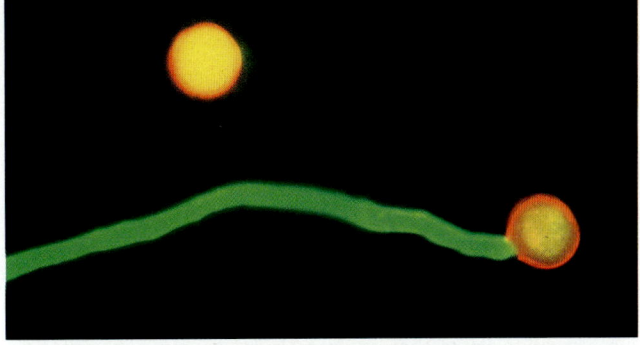

Figure 25–8 A pollen grain begins to grow when it lands on a suitable pistil. Here you can see a growing pollen tube produced by a pollen grain from a tomato plant.

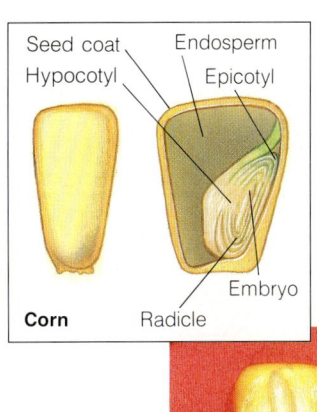

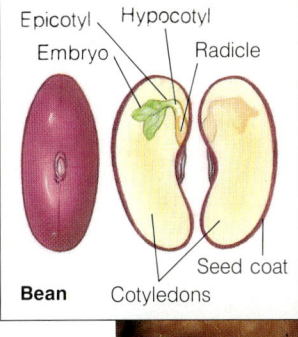

Figure 25–9 Corn is a monocot; a bean is a dicot. In what ways are corn seeds different from bean seeds? In what ways are they similar? ❶

Fertilization causes rapid changes to occur in the ovule, ovary, and other structures of the flower. Parts of the ovule toughen to form a seed coat. The seed coat protects the delicate embryo and its tiny food supply. The ovary wall thickens and joins with other parts of the flower stem to become the **fruit** that holds the seeds. A fertilized flower produces hormones that induce the plant to pour energy into the developing fruits and seeds. If a flower is not fertilized, these hormones are not produced. The flower withers and falls away.

Formation of Seeds

The development of seeds was a major factor in the success of angiosperms on land. Seeds provide nourishment and protection for delicate embryos. Although different plants produce seeds with different structures, most of the essential parts of all seeds are alike.

Angiosperm seeds have either one or two seed leaves called **cotyledons**. Cotyledons contain stored food that is used when a seed germinates, or begins to grow. Monocots, such as corn, have one cotyledon. Dicots, such as beans, have two.

The various parts of the embryo are named according to their point of attachment to the cotyledon or cotyledons. The length of the stem above the cotyledon(s) is called the **epicotyl** (*epi-* means above or on top of). The epicotyl develops into the plant's stem. At the tip of the epicotyl is the tissue that will become the apical meristem. The length of the stem below the cotyledon(s) is called the **hypocotyl** (*hypo-* means below). At the very base of the hypocotyl is a region called the **radicle**, which contains the apical meristem of the root. The radicle will become the primary root of the plant.

In many plants the food stored in the endosperm is almost completely used up by the time the seed is mature. In these seeds the food used by the embryo during germination is stored in large cotyledons. The two halves of a bean, for example, are actually two cotyledons. The small embryo plant can be observed if these two bean-shaped cotyledons are split

539

---

does the seed come? (Ovule.)
• **What is a fruit?** (A fruit is a matured ovary that holds seeds. Sometimes other parts of the flower's stem join the ovary in forming the fruit.)

Reinforcement/Reteaching
Most students are surprised to learn that nuts are actually fruits. Such fruits, composed of hard, dry tissue, are called dry fruits. In a dry fruit, the wall of the ovary hardens into a shell that encloses the seed.

Content Development
Using the Visuals Have students examine Figure 25–9. Discuss the similarities and differences between a corn seed and a bean seed. Although the number of cotyledons vary among plants, the basic structure of all seeds is similar.

TEACHING SUPPORT
Study Guide
• Section 25–1, p. 241

MEDIA AND TECHNOLOGY
📺 Biology Transparencies
• 27: The Life Cycle of a Gymnosperm
• 28: The Life Cycle of an Angiosperm

MULTICULTURAL STRATEGY

On October 14, 1834, the U.S. Patent Office granted Henry Blair a patent for a corn-planting machine. In their registry, the patent office identified Blair as "a colored man"—the only reference to race found in any patent records. For this reason, Blair is believed to be the first African American to receive a U.S. patent. Two years later, he received a patent for a similar device, a cotton-planting machine.

open. In other plants, such as corn and coconuts, much endosperm remains in the mature seed. In fact, the bulk of the space in these seeds is taken up by endosperm. In a coconut, the "milk" is liquid endosperm and the "meat" is solid endosperm. In seeds that retain a great deal of endosperm, the cotyledons look more like typical leaves produced by the plant.

Seed coats can be either thin and fragile or thick and woody. Thick seed coats protect seeds from dryness, salt water, and other adverse environmental conditions. As you learned in Chapter 22, many seeds are eaten by animals when the attractive fruits that contain them are eaten. Tough seed coats protect these seeds from the animal's teeth as well as from the strong chemicals present in its digestive system. These seeds often germinate after they are eliminated by the animal along with digestive wastes. In fact, the digestive wastes provide a bit of natural fertilizer that the plant can use as it begins to grow. Passing through an animal's digestive system provides an additional benefit to the seeds. The seeds usually pass out of the animal some distance away from where the fruit was eaten. This lessens competition for available food and water between the adult plant and its seeds. The animal also distributes seeds in other areas that may provide a suitable environment for the seeds' survival.

25–1 SECTION REVIEW

1. What parts of seed plants are adapted for sexual reproduction?
2. Give the location and function of the following flower parts: sepals, petals, stigma, anther, ovary.
3. What is pollination? What are two types?
4. **Critical Thinking—Applying Concepts** How does the formation of seeds contribute to the survival of a plant species?

25–2 Seed Development

Guide For Reading
- What happens when a seed germinates?
- How is seed dormancy an adaptation that contributes to plant survival?

One of the rites of spring for many people is planting a garden. During cold winter days, often when the soil is frozen beneath one's feet, avid gardeners begin to thumb through seed catalogues. The written descriptions and the colorful photographs that accompany them are a gardener's dream. Based on the promises offered by prose and pictures, gardeners eagerly order seeds. Rarely are they disappointed. Within a period of time—a period that always seems much too long—tiny plants begin to appear above the surface of the ground.

25–1 (continued)

SECTION REVIEW 25–1

1. The cones of gymnosperms and the flowers of angiosperms are seed plants' reproductive structures.
2. Sepals are the outer layer of the flower. They enclose and protect the developing buds. Petals are the next layer of the flower, located just within the sepals. The petals of most flowers are colorful and attractive to pollinators. The stigma is the sticky topmost structure on top of the style, which connects to the ovary. Pollen falls on the stigma and is held in place by the stigma's sticky surface. Located in the center of the flower, the ovary produces eggs and provides food for and protects the developing seeds. The anther is located at the top of the filament and produces microspores.
3. Pollination is the transfer of pollen from anther to stigma. Self-pollination occurs when the pollen from the anther falls on the stigma of the same flower. Cross-pollination occurs when pollen is transferred from one flower to another.
4. The formation of seeds ensures that the species will continue. Seeds contain the genetic material to produce another generation of plants. Many seeds can also withstand severe temperature conditions such as cold and dryness. They do not begin to grow until the conditions for good plant growth resume.

Reinforcement/Reteaching
Those students who have difficulty with the Section Review questions should review the pertinent material.

Closure
Have students compare the reproduction process in gymnosperms and in angiosperms. Then ask students to identify traits of angiosperms that contribute to their success.

TEACHING STRATEGY 25–2

Focus/Motivation
Using the Visuals Direct students' attention to Figure 25–10

Germination

When seeds germinate, they absorb water. The absorbed water causes the endosperm and cotyledons to swell, cracking open the seed coat. Through the cracked seed coat, the radicle emerges and grows into the primary root.

In most monocots the single cotyledon remains within the seed. The growing shoot emerges protected by a sheath. In some dicots the hypocotyl starts growing soon after the primary root starts growing. The hypocotyl forms an arch that pushes up through the soil. The cells in the hypocotyl are much tougher than the cells in the fragile apical meristem, and thus they can scrape against rocks and sand grains without being badly damaged. Once the arch breaks out of the soil and into the sunlight, the hypocotyl straightens and the cotyledons and epicotyl are pulled into the sunlight.

In other dicots the cotyledons remain in the soil. In these plants the epicotyl grows out of the ground, carrying the apical meristem with it. The epicotyl grows in an arch that protects the delicate shoot tip until it breaks through the surface of the soil. See Figure 25–10.

Seed Dormancy

Some seeds germinate so rapidly that they are practically instant plants. Bean seeds are a good example. With proper amounts of water and warmth, a newly planted mature bean seed rapidly develops into a bean plant. But many seeds will not grow when they first mature. Instead, these seeds enter a period of **dormancy**, a period during which the embryo is alive but not growing. The length of dormancy varies in different plant species. A number of environmental factors can cause a seed to end dormancy and germinate.

Seed dormancy serves several purposes for plants. The seeds of a coconut palm have a long dormant period. During

Figure 25–10 Beneath the ground, this bean seed absorbed a great deal of water from the soil—an important reason to water seeds well when you plant them. Eventually the seed pushed its way above the ground. Now the tiny leaves of the bean plant, using the energy of sunlight, can make their own food.

Figure 25–11 This coconut was carried on ocean currents far from its parent tree. During its voyage, the seed remained dormant. Now, after being swept by waves onto a warm beach, it begins to grow.

TEACHING SUPPORT

Teaching Resources
- Hands-On Activity: Pet Plants—Or Are They Plant Pets? p. 133

Study Guide
- Section 25–2, p. 245

25–2 (continued)

Content Development

• **How does seed dormancy contribute to the success of a plant species?** (Through dormancy, a seed delays germinating until environmental conditions are most suitable to the survival of the young plant. As a result, the species has a better chance of flourishing.)

• **What adaptation in certain types of pine seeds increases the likelihood that the species will survive forest fires?** (Although the fire might kill the mature trees, the heat from the fire causes their seeds to mature. The seeds germinate in the area where the mature trees once stood.)

Enrichment

Ask students to plan a garden for their home or school. Have them use seed catalogs and planting manuals to determine the types of seeds they would plant as well as when each type should be planted. Students should prepare a diagram of their garden, illustrating the planting and anticipated germination schedule.

SECTION REVIEW 25–2

1. When a seed germinates, it takes in water. The absorbed water causes the endosperm and the cotyledon to swell. This swelling breaks open the seed coat. The radicle emerges and develops into the root.
2. Seed dormancy serves several purposes. Dormant seeds can be spread from one area to another. In the new area, they may begin to grow. Dormancy allows seeds to survive difficult environmental conditions. It also ensures that some seeds will have a better chance of survival. For example, if all seeds of a particular plant sprouted at the same time and a change in weather made it too dry for the seeds to survive, all seeds would perish. But if some seeds did not germinate, they could begin to grow when wetter conditions resumed.
3. The growing season for most plants occurs during the warm months. If seeds that ripened in the fall began to germinate then, they would be killed by winter's first frost. Seeds that need a period of cold temperature before they germinate

---

dormancy coconuts can float across the sea for weeks or even months until they wash ashore on a beach far from the parent plant. Thus coconuts can be dispersed by water and waves to new areas where, if conditions are suitable, the coconuts may survive and grow. Another important purpose of dormancy is to allow seeds to wait to germinate until environmental conditions support plant growth. For example, it is best if seeds of many temperate plants do not germinate during the hot summer or the cold winter. During these periods, extremes of temperature would make it impossible for fragile seedlings to survive. Thus these seeds remain dormant during the winter and germinate in the spring, when conditions are best for growth. The long period of cold temperatures during which the seeds are dormant is required before growth can begin.

Other environmental conditions can end seed dormancy. Some plants, particularly several species of pine, would not survive if they began growing on the dark forest floor shaded by mature pines. But when a forest fire kills mature trees, some of these seeds not only survive but are stimulated by the heat to grow! These seeds then germinate in a light-filled environment cleared of competition by the fire.

25–2 SECTION REVIEW

1. What happens when a seed germinates?
2. Why do some seeds remain dormant before they germinate?
3. Why is it advantageous for some seeds to undergo a period of cold temperatures before they germinate?
4. **Connection—You and Your World** How does seed dormancy allow you to order packets of seeds through the mail?

Guide For Reading
■ How is vegetative reproduction an adaptation that contributes to plant survival?
■ How are plants reproduced by cuttings, layering, and grafting?

25–3 Vegetative Reproduction

Flowering plants, both wild and cultivated, reproduce sexually when they produce flowers and seeds. Sexual reproduction contributes to the genetic diversity of the species. Many kinds of flowering plants also reproduce asexually through vegetative reproduction. **Vegetative reproduction enables a plant that is well adapted to a particular environment to produce many offspring genetically identical to itself.**

Vegetative Reproduction in Nature

It is not uncommon for many angiosperms to produce new plants by reproducing asexually. Strawberries, for example, send out long trailing stems called stolons that produce roots

542

when they touch the ground. Bamboo plants grow long underground stems that can send up new shoots in several places. In fact, bamboo forests that cover huge areas are often the descendants of one bamboo plant that has, over a long time period, reproduced in this way.

Several species of angiosperms produce tiny plants on their leaves or along their stems. If the parent plant is knocked over, these plantlets can grow into new plants. They can also grow into new plants if they drop off the parent plant. Plants that reproduce in this way include *Tolmeia*, the "piggyback plant" often grown as a houseplant, and the kalanchoe, shown in Figure 25-12. New plants can also grow from the leaves of a parent plant if the leaves fall to the ground under conditions that allow them to root. The African violet and certain species of begonia often grow from leaves.

Artificial Vegetative Reproduction

Sometimes the characteristics of a particular plant are so attractive or beneficial that horticulturists want to make many exact copies of the plant. But the growers also want to avoid the variation that would result if the plant was reproduced sexually by seeds. In addition, new varieties of some plants, such as grapefruits and navel oranges, do not produce seeds. In either of these cases, horticulturists must reproduce these plants asexually by vegetative reproduction.

CUTTINGS One of the simplest ways to reproduce plants vegetatively is by cuttings. A grower "cuts" a length of stem that includes a number of lateral buds from the plant. That stem is then partially buried in soil or in a special rooting mixture. Some common plants, such as coleus, root so easily that no other treatment is necessary. The cuttings of many woody plants, however, do not develop roots easily. To help cuttings of these plants form roots, growers use mixtures of plant hormones called rooting powders. These powders contain auxins and other compounds that stimulate root growth.

LAYERING Layering is used with plants that take a long time to root as cuttings. The stem is cut partway through and the cut area is dusted with rooting powder. Then the stem is either wrapped in moistened moss or bent to the ground and buried. In this way, the treated stem receives water and nutrients from the parent plant while it develops its own roots. When rooting is completed, the rooted stem is separated from the parent plant and allowed to grow on its own. Several common houseplants, such as rubber plants, are often reproduced by layering.

GRAFTING AND BUDDING Grafting and budding are used to reproduce seedless plants and varieties of woody

Figure 25-12 Although it is a flowering plant and can produce seeds, this kalanchoe also produces small plants along the edges of its leaves.

Figure 25-13 Some plants can be reproduced when a piece of stem, called a cutting, is placed in water. This begonia cutting is already growing tiny roots at the base of the stem.

543

TEACHING SUPPORT

Laboratory Manual
- Germination and Seedling Development, p. 329

Teaching Resources
- Vocabulary Review: Crossword, p. 1
- Activity: Artificial Vegetative Reproduction, p. 5
- Hands-On Activity: Clone a Plant, p. 123

Study Guide
- Section 25–3, p. 246

Biotechnology Workbook
- Variety—The Spice of Life, Tomatoes, Peppers, etc., p. 73

25–3 (continued)

Content Development
Explore various methods of artificial vegetative reproduction through a class discussion.
- **What is a cutting?** (A length of plant stem that includes a number of lateral buds.)
- **Why are rooting powders added to some cuttings?** (Rooting powders contain auxins and other compounds that stimulate root growth in a cutting.)
- **What is layering?** (In layering, a stem is cut partway through and then dusted with rooting powder. The stem is then moistened or buried.)
- **What types of plants are reproduced by layering?** (Plants that would take a long time to root as cuttings, such as rubber plants.)
- **Describe the processes of budding and grafting.** (A piece of stem or a lateral bud is cut from a parent plant and attached to another plant. Thus the new plant has a strong root system from the start.)
- **What is a scion? A stock?** (A scion is a cut piece of stem or lateral bud from a parent plant. A stock is the plant to which the scion is attached.)
- **What is the difference between budding and grafting?** (In budding, buds are used as scions. In grafting, stems are used as scions.)
- **What must be done if grafting is to be successful?** (The vascular cambiums of the scion and the stock must be firmly connected.)

Figure 25–14 Plants can be grafted onto other plants. First, a small branch is inserted into the stem of the other plant (top). In making a graft, it is important that the cambium layers of the two pieces align properly (center). Then the two pieces are wrapped to hold them in position until they begin to grow together (bottom).

plants that do not produce strong root systems. In both grafting and budding, new plants are grown on plants that have strong root systems. To do this, a piece of stem or a lateral bud is cut from the parent plant and attached to another plant. The cut piece is called the **scion**, and the plant to which it is attached is called the **stock**. When stems are used as scions, the process is called grafting. When buds are used as scions, the process is called bud grafting or budding.

Grafting usually works best when plants are dormant because the wounds created can heal before new growth starts. In all cases, grafts are successful only if the vascular cambiums of scion and stock are firmly connected to each other.

Enrichment

Have students interview a local florist or farmer to determine the role of artificial vegetative reproduction in his or her business.

SCIENCE, TECHNOLOGY, AND SOCIETY CONNECTION

Cloning Plants

It is possible to produce large numbers of identical plants rapidly through a process called cloning. In the first step of the cloning process, a branch containing meristematic tissue is sterilized. Then, under sterile conditions, a portion of the meristem is removed and placed into a culture medium that contains essential plant nutrients. The growth of the meristematic tissue can be controlled by the careful addition of selected plant hormones to the culture medium.

In the next step, the tissue is encouraged to produce a mass of meristematic cells. These meristematic cells can be split apart by shaking or cutting. Each mass of cells will continue to grow as long as conditions are suitable.

Eventually auxins and cytokinins are added to induce the mass of meristematic cells to begin to develop other plant tissues. Soon, tiny plantlets begin to form. These plantlets are grown in a nutrient medium for a time and then carefully planted in soil. In this manner, thousands upon thousands of genetically identical plants can be produced. Plants cloned in this way are used both in agriculture and in research laboratories.

Plants are cloned in sterile test tubes. Here, small clumps of cells have been removed from (left to right) a sundew, Venus' flytrap, fern, African violet, and kalanchoe. The clumps of cells have been placed into individual test tubes. Each clump of cells will produce plants genetically identical to the plants from which they were removed.

25–3 SECTION REVIEW

1. How does vegetative reproduction contribute to plant survival?
2. What is a cutting? How do gardeners use cuttings to increase the number of their plants?
3. Compare grafting and budding.
4. **Critical Thinking—Applying Concepts** You want to reproduce one of your houseplants to give to your friend. This plant, however, does not make roots easily. How could you reproduce this plant?

545

SECTION REVIEW 25–3

1. A plant must expend a great deal of energy to flower, and its seeds usually stand just a small chance of germinating and maturing. Vegetative reproduction offers a more efficient way for plants to reproduce, allowing a single plant to establish copies of itself in its environment.
2. A cutting is a piece of plant stem that contains several buds. Gardeners cut off such a stem and place it in soil or water, where it can sprout roots. Eventually, the cutting becomes a functional copy of the original plant.
3. Grafting joins a piece of stem from one plant to another plant. Budding joins a lateral bud from one plant to another. In both cases, if the graft or bud begins to grow, a new plant is produced.
4. This plant could be reproduced in many ways, including sowing its seeds, cloning it from part of its meristem, grafting a piece of it onto another plant, and growing a new plant from a cutting. The latter strategy would work if the cutting was treated with rooting powders.

SCIENCE, TECHNOLOGY, AND SOCIETY CONNECTION

Cloning Plants

Through cloning, humans can artificially reproduce large numbers of identical plants. Generally, this is done to benefit people in some way. However, some argue that processes such as cloning put humans in the omnipotent position of disrupting the balance of nature to suit their own purposes. Have students research cloning, focusing on the benefits and disadvantages it poses to all the Earth's organisms. Then have them debate whether or not cloning is a desirable technology.

Reinforcement/Reteaching

Have students who are experiencing difficulty with the Section Review questions reread pages 542–544.

Closure

Ask students to identify the method of artificial vegetative reproduction used to reproduce plants bearing seedless fruits. (Grafting or budding.)

LABORATORY INVESTIGATION

COMPARING DIFFERENT TYPES OF FLOWERS

Before the Lab
1. Gather all the materials at least one day in advance. Be sure you have enough flower blossoms for each group of students.
2. Check with the school nurse to see if any class members are allergic to pollen. You may wish to excuse those students from the laboratory investigation.

Pre-Lab Discussion
In order to understand this investigation, students should have a knowledge of the structure of a flower. They should also understand the processes of pollination and fertilization.

Skills Development
Students will use the following skills when completing this investigation.
1. Observing
2. Comparing
3. Relating
4. Applying
5. Diagramming

Safety Tips
Caution students to be very careful when working with a scalpel and scissors.

Teaching Strategy
1. Be sure students understand that a daisy is a composite flower made up of many separate flowers. As such, it possesses multiple structures.
2. Be sure students remember to cut away from themselves when working with a scalpel.

Observations
1. Six petals and six stamens.
2. Yellow pollen grains fall onto the slide.
3. Small ovals (ovules).
4. A daisy has a yellow disk in the center with flat petallike rays around the outside. The center is composed of small cup-shaped flowers, and the perimeter is composed of flowers with long, flat blades.

Analysis and Conclusions
1. Tulips are monocots since they have three petallike sepals, three petals, six anthers, and parallel veins on the leaves.
2. A flower contains reproductive organs. Petals help to attract pollinators. Stamens produce pollen. The pistil contains the ovules, which develop into seeds after fertilization.
3. Both flowers have pistils and stamens, and both have petals and sepals to attract pollinators.

LABORATORY INVESTIGATION

COMPARING DIFFERENT TYPES OF FLOWERS

PROBLEM
What kinds of adaptations for reproduction do different kinds of flowers show?

MATERIALS (per group)

tulip blossom	3 glass slides
daisy blossom	2 coverslips
scalpel	medicine dropper
scissors	hand lens
forceps	microscope

PROCEDURE

1. Examine a tulip blossom. Count the number of petals. Look inside the blossom. The structure in the center of the flower is the pistil. Find the stamens and count them.
2. On a separate sheet of paper, draw the tulip flower. Label the parts.
3. Peel the petals off the tulip blossom. Carefully remove the stamens. Tap one of the stamens on a clean glass slide. Notice what happens.
4. Put the remainder of the flower down on a clean glass slide. Using a scalpel, cut across the ovary at the base of the pistil about one fourth of the way up from the bottom. **CAUTION:** *The scalpel is very sharp, so use care. Always cut away from you.* Discard the top part of the pistil.
5. With the cut portion of the pistil still on the glass slide, shave off several thin cross sections by pressing straight down with the scalpel as close as you can to the cut edge of the pistil. Lay the cross sections flat on the slide. With the medicine dropper, place a drop of water over each cross section. Cover the cross sections with a coverslip.
6. Examine the cross sections of the pistil under the low-power lens of the microscope.
7. Examine a daisy blossom. Try to locate the same structures you found in the tulip. The yellow central portion of a daisy is called a disk. Petallike structures surround the disk.
8. A daisy is a composite flower, which means it is made up of many tiny separate flowers. Each of the petallike structures is a single-ray flower. Each ray flower has a cup-shaped bottom with a tiny pistil poking up from it.
9. Look closely at the disk of the daisy. It, too, is composed of many separate flowers. Rub your thumb over the center of the disk to remove a few disk flowers. Examine a single disk flower with a hand lens. Use a forceps to remove the pistil. Use the scissors to cut straight down through the disk flower to open up the flower.
10. Put the open disk flower on a clean glass slide. Place a drop of water over the flower. Cover the flower with a coverslip.
11. Examine the slide under the low-power lens of the microscope. Draw your observations.

OBSERVATIONS
1. How many petals and stamens does a tulip have?
2. What happens when you tap a tulip stamen on a clean glass slide?
3. What did you observe inside the pistil of a tulip flower?
4. Describe the appearance of a daisy flower and the individual flowers that compose it.

ANALYSIS AND CONCLUSIONS
1. Based on your observations, is the tulip a monocot or a dicot? Explain your answer.
2. What is the function of a flower? What structures in a flower carry out this function?
3. Despite their differences in form, how are a tulip and a daisy suited to their function?

STUDENT STUDY GUIDE

SUMMARIZING THE CONCEPTS

The key concepts in each section of this chapter are listed below to help you review the chapter content. Make sure you understand each concept and its relationship to other concepts and to the theme of this chapter.

25–1 Cones and Flowers as Reproductive Organs

- The sporophyte generation in seed plants is most obvious; the gametophyte generation is small and often hidden.
- Gymnosperms make seeds in special structures called cones. Male cones produce pollen; female cones produce ovules. It is within the female cones that seeds develop.
- The parts of a flower have different forms and functions. The anthers produce pollen. The pistil consists of an ovary and the style.
- Seeds are an important reason for the survival of gymnosperms and angiosperms in land environments. They enable plants to be dispersed into different environments and to survive environmental conditions that would be difficult for adult plants.

25–2 Seed Development

- In flowers, sexual reproduction exchanges genetic material and increases variability in offspring. Asexual reproduction increases the number of plants, but the newly formed plants are genetically identical.

25–3 Vegetative Reproduction

- Many plants reproduce asexually. Horticulturists often produce new plants using asexual methods of reproduction.

REVIEWING KEY TERMS

Vocabulary terms are important to your understanding of biology. The key terms listed below are those you should be especially familiar with. Review these terms and their meanings. Then use each term in a complete sentence. If you are not sure of a term's meaning, return to the appropriate section and review its definition.

25–1 Cones and Flowers as Reproductive Organs

sepal
calyx
petal
corolla
stamen
filament
anther
carpel
pistil
ovary
style
stigma
ovule
embryo sac
polar nuclei
egg nucleus
pollen chamber
tube nucleus
generative nucleus
pollination
self-pollination
cross-pollination
pollen tube
sperm nuclei
double fertilization
endosperm
fruit
cotyledon
epicotyl
hypocotyl
radicle

25–2 Seed Development

dormancy

25–3 Vegetative Reproduction

scion
stock

Going Further: Enrichment
To continue the study of angiosperms, have students research the average germination period of the tulip and daisy. Students should also determine what environmental conditions are favorable for each type of plant. The laboratory investigation could be further developed to include a discussion of how these plants could be reproduced by artificial vegetation. Students may wish to try to produce new plants from cuttings of live tulip or daisy plants.

CHAPTER REVIEW

CONTENT REVIEW

Multiple Choice
1. a 2. b 3. b 4. d
5. a 6. b 7. d 8. a

True or False
1. F Angiosperms
2. F asexually
3. T
4. F anther
5. F sepals
6. T
7. T
8. F stock

Word Relationships
1. eggs or ovules
2. seeds
3. corolla
4. stem

CONCEPT MASTERY

1. You can take a cutting of a plant and root it in water or soil. This will produce a new plant asexually.
2. Petals attract pollinators. The pistil contains female reproductive cells. The stigma is the sticky top of the style that holds pollen grains. Sepals are tough green modified leaves that protect the developing flower buds. The anther produces pollen.
3. They use the food stored in the endosperm and the cotyledons to grow until the stem can be pushed above the ground into the light.
4. Cross-pollination produces variety in the genetic material of the fertilized seeds. The seeds are not genetically identical to either parent. This diversity may allow a plant species to survive in various conditions.
5. A pollen tube is produced when a pollen grain falls on a stigma and begins to grow. The pollen tube grows down into the ovary. Pollen move down the pollen tube and fertilize an egg. The fertilized egg can develop into a seed.
6. Seeds help a plant live through unfavorable climate conditions. Seeds also help in the distribution of a species of plants to new areas.
7. Double fertilization occurs when one sperm nucleus fuses with the egg nucleus to form the zygote. The other sperm nucleus fuses with the two polar nuclei to form the triploid endosperm.
8. A fruit consists of a thickened ovary wall and part of a stem that surrounds and protects the seeds. An apple, a tomato, and a string bean are actually all fruits.

CHAPTER REVIEW

CONTENT REVIEW

Multiple Choice

Choose the letter of the answer that best completes each statement.

1. The part of the flower that receives pollen is the
   a. stigma.   c. style.
   b. ovary.   d. scion.
2. The seed leaves in angiosperms are also called
   a. roots.   c. stigmas.
   b. cotyledons.   d. epicotyls.
3. The thickened part of the ovary wall that holds the seeds is called the
   a. petal.   c. anther.
   b. fruit.   d. cone.
4. The parts of a flower that are most involved with attracting pollinators are the
   a. pistils.   c. anthers.
   b. seeds.   d. petals.
5. The part of the embryo that develops into a root is the
   a. radicle.   c. epicotyl.
   b. hypocotyl.   d. endosperm.
6. One way to reproduce plants asexually is by
   a. planting seeds.   c. cross-pollination.
   b. making cuttings.   d. self-pollination.
7. In gymnosperms, seeds are produced in
   a. flowers.   c. pollen.
   b. male cones.   d. female cones.
8. The part of a seed that protects the seed from dryness, salt water, and other adverse environmental conditions is called the
   a. seed coat.   c. endosperm.
   b. radicle.   d. sepal.

True or False

Determine whether each statement is true or false. If it is true, write "true." If it is false, change the underlined word or words to make the statement true.

1. <u>Gymnosperms</u> produce seeds in flowers.
2. Cuttings are a way to reproduce plants <u>sexually</u>.
3. The <u>sporophyte</u> generation in seed plants is large and obvious.
4. Pollen is produced in the <u>stigma</u>.
5. The <u>petals</u> protect the developing flower.
6. All of a flower's parts are made up of modified <u>leaves</u>.
7. A corn silk is a long <u>style</u>.
8. The plant to which a graft is attached is called a <u>stigma</u>.

Word Relationships

An analogy is a relationship between two pairs of words or phrases generally written in the following manner: a:b::c:d. The symbol : is read "is to," and the symbol :: is read "as." For example, cat:animal::rose:plant is read "cat is to animal as rose is to plant."

In the analogies that follow, a word or phrase is missing. Complete each analogy by providing the missing word or phrase.

1. anther:pollen::pistil: _____
2. asexual reproduction:graft::sexual reproduction: _____
3. sepals:calyx::petals: _____
4. radicle:root::epicotyl: _____

CONCEPT MASTERY

Use your understanding of the concepts developed in the chapter to answer each of the following in a brief paragraph.

1. Describe one way a plant can be reproduced asexually.
2. Give the functions of the following plant parts: petals, pistil, stigma, sepals, anther.
3. Plants need light to make food. How do seeds planted beneath the ground grow?
4. How does cross-pollination help plants survive in many different environments?
5. What is a pollen tube? How is a pollen tube important in the formation of seeds?
6. In what two ways do seeds help a plant species survive?
7. What is double fertilization?
8. What is a fruit? Which of the following are fruits: apple, tomato, potato, string bean?

CRITICAL AND CREATIVE THINKING

Discuss each of the following in a brief paragraph.

1. **Applying concepts** Suppose you find a plant growing on a mountain. You find that this plant can only reproduce vegetatively. What disadvantage does this method of reproduction have to the plant?
2. **Relating concepts** You observe that all the flowers on a garden plant lack anthers. On another plant of the same species, you observe that all the flowers have anthers. What conclusion can you make about the plants' method of sexual reproduction? Does this method offer any advantages?
3. **Designing an experiment** A friend suggests that seeds do not need cotyledons to grow. You argue that cotyledons are important to seeds. Design an experiment that shows the effect on seed growth of removing cotyledons.
4. **Interpreting photographs** Identify the parts of the flower that are shown in this photograph. Give the function of each part.

5. **Applying concepts** Gardeners must freeze and thaw the seeds of some plants before the seeds will germinate. Why is this done? What does freezing and thawing do for a plant?
6. **Applying concepts** Many plants have small flowers, often green in color, that are hard to see. Ragweed, a common wild plant, is an example. These plants are often pollinated by wind, not by animals. How does the color and shape of the plants' flowers explain its method of pollination?
7. **Making comparisons** Compare sexual reproduction in gymnosperms with sexual reproduction in angiosperms.
8. **Using the writing process** Some cities give unused land to community groups to use as a garden. People plant seeds and produce food crops and flowers in the small plots of ground allocated to them. Write a proposal to your city or town government supporting this use of unused city land. Detail the benefits to the community.

549

answers with the flower diagram in Figure 25–4 on page 535.

5. These seeds need a period of cold before they can germinate. This prevents seeds from growing during the winter. Freezing and thawing the seeds causes the seeds to germinate.
6. Plants that are pollinated by the wind have many small flowers to catch the pollen floating by. These flowers do not rely on animals for pollination. They do not have large colorful petals to attract pollinators.
7. In gymnosperms, pollen is borne on the wind from male cones to female cones. The eggs are held on the scales of the female cones. They develop uncovered, hence the name gymnosperm (naked seed). In angiosperms, the seeds are contained in flowers. Pollen is also produced in the flowers, and the seeds develop within the ovary.
8. Check students' papers for clarity of expression. The community may benefit because people find gardening relaxing. They may also eat the foods produced or enjoy the flowers that are grown. Gardens can also beautify an area.

CRITICAL AND CREATIVE THINKING

1. Vegetative reproduction does not produce genetic diversity. If environmental conditions changed and the plant was no longer well-adapted to life under the changed conditions, it would die. Vegetative reproduction would not increase the range of this plant.
2. This plant reproduces by cross-pollination. This method of pollination results in genetic diversity and could mean that the plant could increase its distribution.
3. Choose seeds with large cotyledons and remove the cotyledons before planting. Leave the cotyledons on some seeds as a control.
4. Have students compare their

CAREERS IN BIOLOGY

Farmer, Landscape Architect, Plant Pathologist

Students interested in a biological career should be instructed to write for further information to the address listed beneath each career description. However, in consideration of the organizations that provide career information, please have only one student write to an organization. Do not instruct the entire class to write to every organization for the same career information. You may want to use the information provided to start a biology career file for the use of all your students.

UNIT 6 — CAREERS IN BIOLOGY

Farmer

Farmers produce most of the food we eat every day. Farmers often use complex equipment and breeding techniques to develop better crops and livestock and improve the efficiency of agricultural production.

There are no specific educational requirements to become a farmer, but a successful farmer should have a knowledge of soil preparation and cultivation, disease control, and machinery maintenance.

For information on farming write to the United States Department of Agriculture, Washington, DC 20250.

Landscape Architect

Landscape architects plan and design outdoor areas such as those around houses, schools, and offices. They analyze the physical features of the land and attempt to make the best use of those features without endangering the natural environment.

Landscape architects combine the skills of scientists, engineers, and architects as they study the land and draw detailed plans for its use.

Landscape architects must have a Bachelor's degree in landscape architecture. A license is also required in some states.

For more information on this career write to the American Society of Landscape Architects, 4401 Connecticut Ave., NW, Washington, DC 20008.

Plant Pathologist

Plant pathologists research diseases in plants. They attempt to identify the symptoms of different diseases, determine their causes, and develop ways to control these diseases. Plant pathologists also try to predict disease outbreaks by studying how different soils, climates, and geographical locations affect the spread of plant diseases.

Plant pathologists must complete degree programs beyond the college level.

For more information write to the American Society of Plant Physiologists, 15501 Monona Dr., Rockville, MD 20855.

HOW TO SELECT COMPANIES TO CONTACT

When searching for a job, you will probably want to send your résumé to a large number of companies. Unless you have specific employers in mind, you may find it rather confusing to choose those companies. However, there are publications in your school or community library that will be of help to you. These publications list major employers—either alphabetically, by industry, or by geographical location. They also give a brief description of the company and list the major positions or careers that are available at that company. Two such publications are *The National Job Bank* and *Dun's Employment Opportunity Directory.*

FROM THE AUTHORS

Although I am sometimes embarrassed to admit it, I have a special relationship with plants. There are plants in my bedroom, plants in my office, a whole porch full of plants at the front of my house, and vegetables and flowers grow in my backyard. Each morning, I check my plants. On hot days, I take breaks from work to water them. And I sit with them a while every evening.

Why do I enjoy plants so much? I relate to them in several ways. First, flowers brighten up even the dreariest of winter days. Second, as a biologist I admire their adaptations to various environments. In nature, some of my plants grow in a desert, while others grow in a rainforest. I have to remember the growing conditions each is adapted to when I water and feed them, and when I decide how much light to provide them with. Most importantly, I know that all life on Earth, including humans, depends on plants for survival. And that is a lesson vital for all to learn.

Joe Levine

FROM THE AUTHORS

Joseph Levine

In this essay, the author discusses his personal collection of plants, and he explores the reasons why he enjoys plants so much. For a class discussion, ask students what kinds of plants they have in their homes or gardens, if any of them take care of these plants, and what benefits and enjoyment they get from their plants.

UNIT 7

Life on Earth: Invertebrate Animals

UNIT OVERVIEW

Unit 7 introduces students to the major characteristics of animals and presents the major invertebrate phyla. Students study the animal phyla in roughly evolutionary order, starting with sponges, cnidarians, flatworms, and roundworms, continuing with the coelomate protostomes—mollusks, annelids, and arthropods—and concluding with echinoderms and chordates, the deuterostomes. The final chapter in this unit summarizes and ties together what students have learned in the previous chapters by comparing systems in different phyla.

UNIT OBJECTIVES

1. Discuss the seven essential life functions in animals.
2. Describe three basic trends in animal evolution.
3. Describe and give examples of each of the major invertebrate phyla.
4. Compare the ways different invertebrates carry out their essential life functions.

UNIT 7

Life on Earth: Invertebrate Animals

Hidden under a cabbage leaf, the beautiful symmetry of these butterfly eggs foretells trouble for a farmer. In time, tiny caterpillars will hatch. The leaf that hides them also provides their dinner. Their presence is given away only by the holes that they chew into the leaves—rendering the cabbages unattractive and unsaleable.

Caterpillars are just one example of invertebrates, or animals without backbones. Invertebrates range in size from creatures even smaller than the caterpillars shown here to squids many meters long. They vary in form, color, and habits as much as they do in size. In their incredible diversity, invertebrates provide an endless source of amusement, horror, beauty, and wonder.

DISCOVERY LEARNING

LIFE UNDER A ROCK

1. Look under a rock in moist soil. **CAUTION:** *Be careful. Use a field guide to familiarize yourself with the dangerous plants and animals in your area.*
2. Make drawings of the creatures you find. Next to each drawing, briefly describe the creature's size, color, and other interesting or noteworthy characteristics.
3. Carefully replace the rock in the same position when you are finished making your observations.
4. Wash your hands thoroughly.
   - What kinds of organisms did you find?
   - Why do you think these organisms live under a rock?

552

INTRODUCING UNIT 7

Using the Visuals

Have students read the unit introduction on page 552 and examine the photograph on page 553. The photograph shows some of the organisms that exist on the lower side of a plant leaf—in this case, butterfly eggs and some hatching caterpillars. These insects will probably be safe as long as they remain under the leaf where predators will not be likely to find them.

- **What does this photograph show?** (Students should be able to observe caterpillars and eggs.)
- **In what ways are caterpillars like a cat? What do all animals need to survive?** (Answers will vary. Students should suggest that all animals need food, water, air. Accept all answers.)
- **How do caterpillars differ from a cat?** (Accept all answers. Students should be led to respond that caterpillars do not have a backbone, but if they do not come up with this response, tell students that they will discover an important difference between cats and caterpillars in this unit.)

CHAPTERS
26
Sponges, Cnidarians, and Unsegmented Worms
27
Mollusks and Annelids
28
Arthropods
29
Echinoderms and Invertebrate Chordates
30
Comparing Invertebrates

CHAPTER DESCRIPTIONS

26 Sponges, Cnidarians, and Unsegmented Worms
In Chapter 26 students examine form and function in sponges, cnidarians, flatworms, and roundworms and discover how the members of these phyla fit into the world.

27 Mollusks and Annelids
In Chapter 27 students first learn about the general structure and function in mollusks and then examine the three major classes of mollusks in detail. Next, students study the life functions and general anatomy of annelids.

28 Arthropods
Phylum Arthropoda, the largest animal phylum, is the subject of Chapter 28. Students first learn about the evolution and general characteristics of arthropods. They then study the three living subphyla of arthropods. They explore insect societies and their communication in some detail. Finally, students examine the impact of arthropods on the environment and on human life.

29 Echinoderms and Invertebrate Chordates
In Chapter 29 students first study form and function in echinoderms. Next, students learn about characteristics of chordates and read about two invertebrate subphyla of chordates.

30 Comparing Invertebrates
In the first part of Chapter 30 students are introduced to the concept of a phylogenetic tree and examine some of the evidence that has been used to derive a phylogenetic tree of the animals. In the second part of this chapter students compare and contrast the ways in which invertebrates carry out their essential life functions, and they trace trends in invertebrate evolution.

DISCOVERY LEARNING
LIFE UNDER A ROCK

Begin your introduction to the unit by having students perform the Discovery Activity. Answers may include earthworms, bugs, salamanders, and grubs. Organisms that live under a rock might seek protection there, they might need to live in damp places, they may be sensitive to light. Other answers are possible.

553

CHAPTER 26 Sponges, Cnidarians, and Unsegmented Worms

Section	Laboratory Investigations and Activities
26–1 Introduction to the Animal Kingdom pages 555–560 THEMES: Unity and Diversity, Evolution	**Teacher Edition** DEMONSTRATION: Observing Animals and Plants, p. 554D
26–2 Sponges pages 560–563 THEMES: Scale and Structure, Unity and Diversity	**Teacher Edition** DEMONSTRATION: Observing a Sponge, p. 554D **Teaching Resources** TEACHER DEMONSTRATION: Reaggregation and Differentiation of Sponge Cells, p.17
26–3 Cnidarians pages 564–569 THEMES: Scale and Structure, Unity and Diversity	**Laboratory Manual** Sponges and Hydras, p. 335
26–4 Unsegmented Worms pages 570–579 THEMES: Scale and Structure, Unity and Diversity	**Student Edition** LABORATORY INVESTIGATION: Collecting and Studying Roundworms, p. 580 **Laboratory Manual** Flatworms and Roundworms, p. 341 **Teaching Resources** HANDS-ON ACTIVITY: Flat as a Worm, p. 135 ECOLOGY INVESTIGATION: Collecting Nematodes from Soil Samples, p. 33
Chapter Review pages 581–583	

Teacher Resources

Books

Evslin, Bernard. *The Hydra (Monsters of Mythology)*. Chelsea House, 1989.

Schmidt, Gerald D., and Larry S. Roberts. *Foundations of Parasitology*, 4th ed. Times Mirror/Mosby College Publishing, 1989.

Simpson, T. L. *The Cell Biology of Sponges*. Springer-Verlag, 1984.

Thompson, D'Arcy W. *Bibliography of Protozoa, Sponges, Coelenterata and Worms*. Longwood Publishing Group, 1977.

CHAPTER PLANNING GUIDE

Other Activities	Media and Technology
Study Guide Section 26–1, p. 249	

Teaching Resources
   ACTIVITY: Investigating Sponges, p. 3
Study Guide
   Section 26–2, p. 251
	Interactive Videodisc
	   Planet Earth: The Blue Planet
	Biology Transparencies
	   29: The Anatomy of a Sponge
	Biology Media Guide
	   page 27

Study Guide
Section 26–3, p. 253
	Interactive Videodisc
	   Planet Earth: The Blue Planet
	Biology Transparencies
	   30: The Life Cycle of a Jellyfish
	Biology Media Guide
	   page 27

Teaching Resources
   VOCABULARY REVIEW: Word Scramble, p. 1
   ACTIVITY: Investigating Bilharzia—A Major Health Problem, p. 5
Study Guide
   Section 26–4, p. 255	

Teaching Resources
   Chapter Test A, p. 11
   Chapter Test B, p. 15
	Computer Test Bank
	   Chapter Test, p. 277

Audiovisuals

Sponges and Coelenterates: Porous and Sac-like Animals. Film or video. Coronet.
Flatworms. Film or video. Ward.

Modern Biology Series: Simple Multicellular Animals: Sponges, Coelenterates, and Flatworms. Film. Benchmark Films.

The Biology of Flatworms. Film or video. Lucerne Media.
The Biology of Cnidaria (Coelenterates). Film or video. Lucerne Media.

CHAPTER 26 Sponges, Cnidarians, and Unsegmented Worms

CHAPTER OVERVIEW

With this chapter students begin their study of the animal kingdom. Although the focus of the chapter is on the simpler invertebrates, the chapter begins with a discussion of animals in general.

Students will learn that an animal is a multicellular eukaryotic heterotroph whose cells lack cell walls. They will learn that, in order to survive, an animal must feed, respire, circulate oxygen and nutrients, eliminate waste products, respond to environmental conditions, reproduce, and, in most cases, move from place to place.

Students will discover that among the simplest animals are the sponges. Sponges are invertebrates that have no specialized tissues or organ systems. Students will study the body plan of a typical sponge and the means by which these organisms perform essential functions. They will also learn about the ecological significance of sponges and their usefulness to humans.

In the third section of the chapter students will be introduced to the phylum Cnidaria. Students will learn that cnidarians include many fascinating animals such as corals, sea anemones, and jellyfish. The form and function of cnidarians will be discussed, as well as their ecological and economic importance.

An important characteristic of most animals is body symmetry. In the fourth section of the chapter, which focuses on unsegmented worms, students will learn that flatworms are the simplest animals that exhibit bilateral symmetry. In the same section, students will learn that roundworms are the simplest animals that have a digestive system with two openings. General characteristics of both flatworms and roundworms will be discussed, as well as the characteristics of some specific species.

26–1 INTRODUCTION TO THE ANIMAL KINGDOM

Section Focus 26–1

The purpose of this section is to introduce students to the study of animals. The section begins by making the distinction between vertebrate and invertebrate animals. Then a general definition of animals is given.

Students will learn that an important characteristic of animals is cell specialization and division of labor. It is this division of labor that enables an animal to perform functions that are essential to survival. These functions include feeding, respiration, internal transport, elimination of waste products, response to environmental conditions, movement, and reproduction. Also discussed in this section will be trends in animal evolution.

Performance Objectives 26–1

1. Give a general definition of animals.
2. Distinguish between vertebrate and invertebrate animals.
3. List and discuss the essential functions of animals.
4. Explain what is meant by cell specialization and division of labor.
5. Discuss some trends in animal evolution.

Science Terms 26–1

vertebrate   p. 555
invertebrate   p. 555
division of labor   p. 556
herbivore   p. 556
carnivore   p. 556
parasite   p. 557
filter feeder   p. 557
detritus feeder   p. 557
larva   p. 559
metamorphosis   p. 559
radial symmetry   p. 559
bilateral symmetry   p. 559
anterior   p. 559
posterior   p. 559
dorsal   p. 559
ventral   p. 559
cephalization   p. 560
ganglion   p. 560

26–2 SPONGES

Section Focus 26–2

The purpose of this section is to discuss the characteristics of sponges. Sponges belong to the phylum Porifera.

Students will learn that sponges inhabit almost all areas of the sea. These animals, which were once thought to be plants, have no specialized tissues or organ systems and nothing that resembles a mouth or gut.

Students will learn that sponges are filter feeders that sift microscopic particles of food from water. They will also learn that water flowing through a sponge serves as a respiratory, excretory, and internal transport system.

Students will learn that sponges have been important to humans for bathing purposes. Currently, scientists are discovering that some of the chemicals manufactured by sponges may be useful in fighting bacteria and in treating certain diseases.

Performance Objectives 26–2

1. Explain how sponges are different from other multicellular animals.
2. Describe the body plan of a typical sponge.
3. Explain how sponges perform essential functions.
4. Discuss the ecological importance of sponges.
5. Discuss some ways in which sponges are useful to humans.

Science Terms 26–2

Porifera   p. 560
collar cell   p. 561
osculum   p. 561
spicule   p. 561
amebocyte   p. 561
spongin   p. 561
gemmule   p. 562
budding   p. 562

26–3 CNIDARIANS

Section Focus 26–3

The purpose of this section is to introduce students to the phylum Cnidaria. Cnidarians are soft-bodied animals with stinging

554C

CHAPTER PREVIEW

tentacles arranged in circles around the mouth. Some familiar cnidarians include jellyfish, corals, and hydras.

Students will learn that all cnidarians exhibit radial symmetry and have specialized cells and tissues. They will also learn that a typical cnidarian has an internal space called a gastrovascular cavity in which digestion takes place.

Students will discover that almost all cnidarians capture and eat small animals by using stinging structures called nematocysts, which are located on their tentacles. They will also learn that cnidarians lack a central nervous system and muscle cells. There are, however, specialized epidermal cells that serve the same function as muscle cells.

Performance Objectives 26–3

1. Describe the general characteristics of cnidarians.
2. Name and describe organisms in the three classes of the phylum Cnidaria.
3. Explain how cnidarians perform essential functions.
4. Discuss the ecological and economic importance of cnidarians.

Science Terms 26–3

Cnidaria p. 564
polyp p. 564
medusa p. 564
gastrovascular cavity p. 564
nematocyst p. 565
hermaphrodite p. 566

26–4 UNSEGMENTED WORMS

Section Focus 26–4

The purpose of this section is to introduce students to two phyla of animals known as unsegmented worms. Unsegmented worms include flatworms (phylum Platyhelminthes) and roundworms (phylum Nematoda).

Students will learn that flatworms are the simplest animals with bilateral symmetry. In addition, most members of this phylum exhibit enough cephalization to have what can be called a head. Planarians, flukes, and tapeworms are among the flatworms that will be discussed.

Students will learn that roundworms are among the simplest animals that have a digestive system with two openings, a mouth and an anus. Several parasitic roundworms that cause diseases in humans will be discussed, including *Ascaris, Trichinella,* and hookworms.

Performance Objectives 26–4

1. Discuss the general characteristics of flatworms and roundworms.
2. Name and give examples of three classes of flatworms.
3. Explain how unsegmented worms perform essential functions.
4. Describe some human diseases caused by parasitic roundworms.

Science Terms 26–4

unsegmented worms p. 570
Platyhelminthes p. 570
flatworm p. 570
Nematoda p. 570
roundworm p. 570
pharynx p. 571

DISCOVERY LEARNING

TEACHER DEMONSTRATIONS
Modeling

Observing Animals and Plants
The following demonstration can be used as an introduction to Chapter 26.

Set up a display in the classroom of some common examples of plants and animals. Try to obtain as many living specimens as possible; use pictures when live specimens are not available. Some possible examples include

    algae floating in water
    clump of moss
    elodea
    potted geranium
    potted ivy
    potted fern
    tree branch
    earthworm
    snail
    cricket
    rayfish
    goldfish
    mouse
    hamster or gerbil
    bird

Have students take time to move about the room and observe the specimens and/or pictures. Encourage them to take notes on outstanding characteristics. Then have students return to their seats.

- **In what ways are all the things you have observed similar?** (Answers may vary. Some students may observe the obvious, such as all are alive or all are found on Earth; others may think of the less obvious, such as all are made up of more than one cell and all have a life cycle that will eventually end.)
- **In what way are the things you observed different?** (Answers may vary. Some students may immediately say that some are plants and others are animals; others may list more subtle characteristics, such as some can move and others cannot; some can make their own food and others cannot; some live in water and others live on land.)
- **If you had to classify these things into two groups, how would you do so?** (Answers may vary. Many students will probably say plants and animals.)

Tell students to keep in mind some of the differences between plants and animals that they have observed as they begin to study Section 26–1, Introduction to the Animal Kingdom.

Observing a Sponge
This demonstration can be performed as students study Section 26–2.

For the demonstration, arrange students in small groups. Each group should be provided with a natural sponge, a hand lens, a glass slide, and a microscope.

Ask students to first observe the surface and pores of the sponge with a hand lens. Have them make drawings of what they see. Then have students tear off a small piece of the sponge and place it on a glass microscope slide. Have students draw what they observe under the microscope.

Students will probably notice hard pointed structures called spicules that support the body of some sponges. Explain to students that these spicules make up the skeleton of the sponge. The spicules are made of either calcium carbonate or silica.

If students do not see any spicules, they will, however, see fibers. These fibers are made of a tan-colored protein called spongin. Many sponges have a skeleton made up of both mineral spicules and spongin fibers.

CHAPTER 26

Sponges, Cnidarians, and Unsegmented Worms

GUIDED ENQUIRY

Pose the following questions to students and have them record their responses. Point out that they will gain a better understanding of the key concepts if they read the chapter with these basic questions in mind. Upon completion of the chapter, pose the questions again. Ask students to compare their initial responses with those they have developed after reading the chapter.

- How do animals differ from other organisms?
- How do animals carry out essential life functions?
- How have animals evolved over time?
- How do sponges differ from other animals?
- How does the form and structure of a cnidarian enable it to carry out essential life functions?
- Why are cnidarians important to humans?
- What characteristics do unsegmented worms exhibit that simpler animals do not?
- How do roundworms cause diseases in humans?

CHAPTER 26

Sponges, Cnidarians, and Unsegmented Worms

Sponges, such as the yellow tube sponge and red bath sponge shown here, are the simplest type of animals. Although flatworms (inset) are the simplest animals that have bilateral symmetry, they are much more complex than sponges.

The world around us swarms with an incredible variety of animals, as you probably realize. What you may not be aware of, however, is that most animal species are not the birds and mammals that are most familiar to us. The vast majority are much smaller and far stranger in appearance. Some are as strange as anything you've ever seen in a science fiction movie. Many of them are also much more important than birds or mammals in the grand scheme of life on Earth. What are these animals? What do they look like and where are they found? How do they perform the essential functions common to all living things? How do they fit into the world? In this chapter we shall begin our exploration of the world of animals by first considering those animals without backbones—the invertebrates.

INTRODUCING CHAPTER 26

Using the Visuals
Begin by having students look at the chapter-opener photographs and read the caption.
- **What is pictured in the large photograph?** (Yellow and red sponges.)
- **According to the caption, a sponge is what type of organism?** (An animal.)
- **Do you think that a sponge looks like an animal? Why or why not?** (Accept all answers. Many students will probably say that it does not look like an animal because it appears to be growing in one place and does not have any of the usual characteristics of an animal.
- **What type of organism is pictured in the inset photograph?** (A flatworm.)
- **A flatworm is also an animal. Does it seem to have much in common with a sponge?** (Most students will probably say that it has little in common.)

Point out that although all animals share certain characteristics, members of the animal kingdom are extremely diverse—there really is no typical animal.

GUIDE FOR READING

After you read the following sections, you will be able to

26–1 Introduction to the Animal Kingdom
- List the essential functions of animal life.
- Describe some trends in animal evolution.

26–2 Sponges
- Describe the structure of a sponge.
- Discuss how sponges perform essential functions.

26–3 Cnidarians
- Describe the structure of a cnidarian.
- Discuss how cnidarians perform essential functions.
- Name and give examples of the three classes of cnidarians.

26–4 Unsegmented Worms
- Discuss how unsegmented worms perform essential functions.
- Name and give examples of the three classes of flatworms.
- Describe some diseases caused by parasitic roundworms.

Journal Activity

YOU AND YOUR WORLD
If you could be any kind of animal in the world, what would you want to be? Why? What do you imagine a day would be like as the animal of your choice? Explore your ideas in words and drawings in your journal.

Figure 26–1 A yak is a vertebrate (left). Its thick, shaggy coat helps it survive the cold winters in central Asia and Tibet, where it makes its home. A hickory horned devil is an invertebrate (right). Despite its frightening appearance, this caterpillar is quite harmless.

26–1 Introduction to the Animal Kingdom

Guide For Reading
- What is an animal?
- What are some trends in animal evolution?

Of all the kingdoms of organisms, the animal kingdom is the most diverse in form. Some animals have forms that are comfortingly familiar. Others resemble creatures from a nightmare or a horror movie. Some animals are so small that they can live inside our bodies. Others are many meters long and live in the depths of the sea. Animals can be black, white, beautifully colored, or nearly transparent. Animals walk, swim, crawl, burrow, and fly all around us. In every case, each animal performs the essential functions of life in its own special way.

You will soon become acquainted with several major divisions in the animal kingdom. One division that we refer to often is that between **vertebrates** and **invertebrates**. Vertebrates, such as humans, have a backbone, or vertebral column. Invertebrates, the subjects of this unit, have no backbone.

What Is an Animal?

As different as they are, all animals share certain basic characteristics. Animals are heterotrophs, which means that they do not make their own food. Instead, they obtain the nutrients and energy they need by feeding on organic compounds that have been made by other organisms. Animals are multicellular, which means that their bodies are composed of more than one cell. And animal cells are eukaryotic—they contain a nucleus and membrane-enclosed organelles. Unlike plant cells or fungus cells, animal cells do not have cell walls. **We can thus define an animal as a multicellular eukaryotic heterotroph whose cells lack cell walls.**

TEACHING STRATEGY 26–1

Focus/Motivation
Give each student a piece of drawing paper and a pencil or crayon. Tell students that they have 3 minutes to draw the first thing that comes to mind when they hear the word *animal*. When students are finished, ask:
- How many of you drew a familiar animal such as a dog, cat, or horse? (Accept all answers.)
- How many of you drew a wild animal such as a giraffe, lion, tiger, elephant, or zebra? (Accept all answers.)
- How many of you drew more than one type of animal? (Accept all answers.)
- Did anyone draw something that shows the general characteristics of an animal, rather than a specific animal? (Answers will vary. Accept all reasonable answers.)

Content Development
Continue the Motivation activity by having volunteers display their drawings to the class.
- **If you knew nothing about animals, what would you conclude about animals based on these drawings?** (Accept all answers.)

Point out to students that some of the characteristics that come to mind when we hear the word *animal* may actually be true of only a small group of animals, such as mammals or birds. These animals are what are known as vertebrates—animals with backbones. The vast majority of animals, however, are invertebrates—that is, animals without backbones.

COOPERATIVE LEARNING

Have groups of students "create" a multicellular eukaryotic heterotroph that lacks cell walls—an animal. Groups should be encouraged to be creative as they illustrate how their animal carries out the seven essential functions of life: feeding, respiration, internal transport, elimination of waste products, response to environmental conditions, movement, and reproduction. Each group should produce a labeled drawing that identifies the animal's symmetry and classifies it as a vertebrate or invertebrate. Groups should assign a scientific name to their "animal" and describe its size relative to a house cat. This assignment is designed to reinforce the concept that all animals must perform essential functions of life.

Journal Activity

YOU AND YOUR WORLD
Accept all answers. Although responses should be based on reality, creativity should also be rewarded. You may also want to use the Journal Activity as a springboard for class discussion.

555

FACTS AND FIGURES

Over 97% of all animal species are invertebrates.

FACTS AND FIGURES

The most complex invertebrate is the giant squid. This animal is over 18 meters long, and most of its organ systems are as sophisticated as those of humans.

HISTORICAL NOTE
EARLY INVERTEBRATES

The first invertebrate animals appeared on Earth during the Cambrian Period of the Paleozoic Era. Some of the simpler species have changed little since that time.

FACTS AND FIGURES

The simplest of all animals is *Trichoplax adhaerens*, an organless pancake-shaped marine animal less than 0.5 millimeters in diameter. It is the sole member of the phylum Placozoa.

26–1 (continued)

Reinforcement/Reteaching

Write on the chalkboard the definition of an animal: a multicellular eukaryotic heterotroph whose cells lack cell walls. Underline, or have a student underline, the following key terms: multicellular, eukaryotic, heterotroph, cell walls. Discuss the meaning of each term.
- **What do we mean when we say that an organism is multicellular?** (It is made up of more than one cell.)
- **What would be the opposite of a multicellular organism?** (A unicellular organism.)
- **Can you think of some unicellular organisms you have studied?** (Possible answers: yeasts, bacteria, protozoans.)
- **Are these organisms animals?** (No.) Why not? (They have only one cell.)
- **What is a eukaryote?** (An organism whose cells contain organelles and a true nucleus.)
- **What do we mean when we say that an animal is eukaryotic?** (Its cells are eukaryotic.)
- **Can you think of some organisms that are not eukaryotes?** (Possible answer: bacteria.)
- **What are these organisms called?** (Prokaryotes.)
- **Are prokaryotes animals?** (No.)
- **What is a heterotroph?** (An organism that is unable to make its own food.)
- **What is an organism that can make its own food?** (An autotroph.)

Figure 26–2 Animals get the nutrients and energy they need by eating organic compounds that have been made by other organisms. The squirrel is munching on a hazelnut, and the crayfish is nibbling on a worm.

Figure 26–3 Unicellular organisms do not have division of labor. They perform all life functions with only their single cell. This false-color micrograph shows a cross-section of the intricate shell that once housed the solitary cell of a foraminifer.

Cell Specialization and Division of Labor

The bodies of animals contain many types of specialized cells. Each specialized cell has a shape, physical structure, and chemical composition that make it uniquely suited to perform a particular function within a multicellular organism. For this reason, groups of specialized cells carry out different tasks for the organism—a phenomenon known as **division of labor**.

You may wonder what advantage there is in dividing up different tasks among specialized cells. After all, monerans and protists do just fine as single cells! But large numbers of cells growing together simply cannot function the way single cells do. Recall from Chapter 8 that cells require a certain amount of surface area to take in food and oxygen and remove wastes. Cells that grow together have little, if any, of their surface exposed to the environment. They would soon be starved for food and oxygen and smothered in carbon dioxide and other wastes if there were no efficient systems to carry out essential functions such as feeding, respiration, and elimination of wastes. In multicellular organisms, efficient systems require specialization. Specialized cells can perform their tasks more efficiently than unspecialized cells.

What Animals Must Do to Survive

In order to survive, animals must be able to perform a number of essential functions. For each animal group we study in the next several chapters, we shall examine these functions and describe the cells, tissues, organs, and organ systems that perform them. To help you make a checklist of those functions, we shall briefly describe them here.

FEEDING Animals have evolved a variety of ways to feed. **Herbivores**, or animals that eat plants, may feed on roots, stems, leaves, flowers, or fruits. Some herbivores even feed on the nutrient-rich fluids in plant vascular tissues. **Carnivores**, or

organisms that eat animals, may also feed on any part of their prey—fat, muscle, bone marrow, or even blood. **Parasites** live and feed either inside or attached to outer surfaces of other organisms, thereby doing harm to their hosts. Many aquatic animals, called **filter feeders**, strain tiny floating plants and animals from the water around them. And many animals feed not on living organisms but on tiny bits of decaying plants and animals called detritus (dee-TRIGHT-uhs). **Detritus feeders** are easy to overlook, but they are vitally important members of the living world.

RESPIRATION As you learned in Chapter 6, living cells consume oxygen and give off carbon dioxide in the process of cellular respiration. Thus entire animals must respire, or breathe, in order to take in and give off these gases. Small animals that live in water or in moist soil may respire through their skin. For large active animals, however, respiration through the skin is not efficient enough. The respiratory systems these animals have evolved take many different forms in adaptations suited to different habitats.

INTERNAL TRANSPORT Some aquatic animals, such as small worms, can function without an internal transport system. But once an animal reaches a certain size, it must somehow carry oxygen, nutrients, and waste products to and from cells deep within its body. Thus many multicellular animals have evolved a circulatory system in which a pumping organ called a heart forces a fluid called blood through a series of blood vessels. You will see in the next several chapters that circulatory systems can be simple or quite complex.

EXCRETION Cellular metabolism produces chemical wastes such as ammonia that are harmful and must be eliminated. Small aquatic animals depend on diffusion to carry wastes from their tissues into the surrounding water, which then carries the wastes away. But larger animals, both in water

Figure 26–4 Animals have many different modes of feeding. The puffin (left), which is holding a meal of sand eels in its beak, is a carnivore. The white structures on the back of the caterpillar (right) are cocoons of parasites that have devoured the insides of their host. Sea cucumbers (bottom, right) are detritus feeders.

BACKGROUND INFORMATION
THE DIFFERENCE BETWEEN PRIMITIVE AND SHODDY

It is easy for students, as they read material that relates to evolution, to think that anything described as ancient or primitive is also shoddy. When we talk about evolutionary trends from simple to complex, we do not imply that newer organisms are better, or that evolution had a purpose in producing them. Many ancient and "simple" body plans have persisted through eons of environmental change, while numerous more complicated plants and animals have evolved and then become extinct. If an ancient organism is sufficiently well-adapted to its environment, there is no reason that it "must" evolve. Favored by chance and natural selection, it can persist for long periods of time and even be more competitive than more recent arrivals.

• What organisms have you studied recently that are autotrophs? (Plants.)
• Does only the fact that an organism can't make its own food mean it is an animal? (No.)
• Can you think of some organisms that cannot make their own food but are not animals? (Possible answer: fungi.)
• What are cell walls? (Outermost boundary of a cell that is made of cellulose.)
• What types of organisms have cells with cell walls? (Plants; bacteria.)
• Because they have cell walls, these organisms are not animals. What other characteristics have we discussed that indicate that these organisms are not animals? (Plants are autotrophs; bacteria are prokaryotic and single-celled.)

Content Development
Point out to students that animals do share certain characteristics with plants. Perhaps the most significant of these in both vascular plants and in animals is the organization of cells into specialized tissues. These specialized tissues represent a division of labor among cells that contributes to the survival of the organism.

ANNOTATION KEY

① Indirect development. (Applying concepts)

BACKGROUND INFORMATION
BRANCHES OF ZOOLOGY

In biology, the study of animals is called zoology. Early zoologists classified animals according to physical resemblance, habitat, or economic use. The invention of the microscope and the use of modern experimental techniques expanded zoology as a field and gave rise to such branches as cytology, histology, embryology, physiology, and genetics. Modern zoologists study cell structure and function, as well as psychological, anthropological, and ecological aspects of animals.

Figure 26–5 Sense organs, such as eyes, help animals gather information about the environment. The ghost crab uses its stalked eyes to peek from its hiding place under the sand and see if the coast is clear (top). Six of the wolf spider's eight eyes can be seen from the front (bottom). The other two are on the side of its head.

and on land, must work to remove poisonous metabolic wastes. As we study animals from worms to mammals, we shall follow the development of the excretory systems that store and dispose of these wastes.

RESPONSE Animals must keep watch on their surroundings to find food, spot predators, and identify others of their own kind. To do this, animals use specialized cells called nerve cells, which hook up together to form a nervous system. Sense organs, such as eyes and ears, gather information from the environment by responding to light, sound, temperature, and other stimuli. The brain, which is the nervous system's control center, processes the information and regulates how the animal responds. The complexity of the nervous system varies greatly in animals.

MOVEMENT Some animals are sessile, which means that they live their entire adult lives attached to one spot. But many animals are motile, which means that they move around. To move, most animals use tissues called muscles that generate force by contracting. In the most successful groups of animals, muscles work together with a skeleton, or the system of solid support in the body. Insects and their relatives wear their skeletons on the outside of their bodies. These are called exoskeletons (*exo-* means outside). Reptiles, birds, and mammals have their skeletons inside their bodies. These are called endoskeletons (*endo-* means inside). We call the combination of an animal's muscles and skeleton its musculo-skeletal system.

REPRODUCTION Animals must reproduce or their species will not survive. Because reproduction is so important, and because animals use many different methods to reproduce, we

Figure 26–6 The sea urchin larva (inset) looks and acts nothing like the adult (right). What kind of development do sea urchins undergo? ①

26–1 (continued)

Content Development
Emphasize that animals must feed on the organic materials of other organisms. This can mean eating an autotroph directly, or it can mean eating another heterotroph that has previously eaten an autotroph. Human eating patterns illustrate both possibilities.
• **In what ways do humans eat autotrophs directly?** (By eating fruits, vegetables, seeds, nuts.)
• **In what ways do humans eat heterotrophs that have fed on autotrophs?** (By eating chicken or beef that has been grain fed; by eating fishes that have fed on plant matter.)
 Although the topic of food chains and food webs will be discussed in a later unit, it is worth pointing out to students that animals are ultimately dependent on the products of photosynthesis in plants as a source of food. Students should recall from an earlier chapter that the sun is the source of energy on Earth—and an important aspect of this is that many food chains begin with plants carrying out photosynthesis.

Enrichment
Students may enjoy reading and writing about some mythical animals that appear in legend or literature. Some examples include unicorns and dragons.

Content Development
Have students think of some of the different ways in which animals move. For example, various animals walk, fly, run, swim, creep, or crawl. Point out that so many different ways of locomotion indicate the great diversity

shall spend a lot of time studying reproduction. Some animals, such as jellyfish, switch back and forth between sexual and asexual reproduction. (Note that this is not the same as alternation of generations in plants, during which diploid (2N) and haploid (N) generations alternate. The sexual and asexual generations in animals are both diploid.) Many animals that reproduce sexually bear their young alive. Others lay eggs. The eggs of some species hatch into baby animals that look just like miniature adults. As they grow, these baby animals increase in size but do not change in overall form. This type of development is called direct development. In other species, eggs hatch into **larvae** (singular: larva), which are immature stages that look and act nothing like the adults. As larvae grow, they undergo a process called **metamorphosis** in which they change shape dramatically. This type of development is called indirect development.

Trends in Animal Evolution

As we explore the invertebrate phyla, keep in mind that these phyla share an evolutionary heritage. In Chapter 30, the relationships between the different phyla of invertebrates will be represented in an evolutionary tree of the animal kingdom. This evolutionary tree will show our best understanding of the way in which animal phyla are related to one another. For now, focus on tracing a few important evolutionary trends and patterns as you move from one animal phylum to the next.

The levels of organization become higher as animals become more complex in form. The essential functions of less complex animals are carried out on the cell or tissue level of organization. As you move on to more complex animals, you will observe a steady increase in the number of specialized tissues. You will also see those tissues joining together to form more and more specialized organs and organ systems.

Some of the simplest animals have radial symmetry; most complex animals have bilateral symmetry. Some of the simplest animals, such as sea anemones, have body parts that repeat around an imaginary line drawn through the center of their body. These animals exhibit **radial symmetry**. See Figure 26–7. Animals with radial symmetry never have any kind of real "head." Many of them are sessile, although some drift or move about in a more or less random pattern. Most complex invertebrates and all vertebrates have body parts (at least outside body parts such as arms and legs) that repeat on either side of an imaginary line drawn down the middle of their body. One side of the body is the mirror image of the other. These animals are said to have **bilateral symmetry**. Animals with bilateral symmetry have specialized front and back ends as well as upper and lower sides. The **anterior** is the front end and the **posterior** is the back end. The **dorsal** is the upper side and the **ventral** is the lower side.

Figure 26–7 Starfish have radial symmetry, which means that their body parts repeat around an imaginary line drawn through the center of the body.

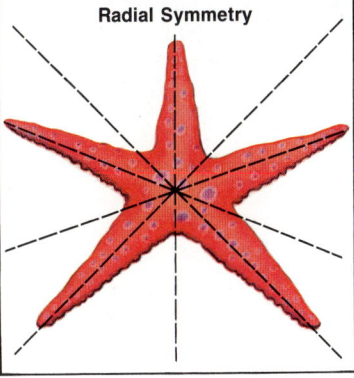

TEACHING SUPPORT

Study Guide
- Section 26–1, p. 249

BACKGROUND INFORMATION
EVOLUTION AND BODY FEATURES

In looking at evolutionary trends among animals, the presence or absence and degree of complexity of five basic body features can be observed. These features are body symmetry; type of gut; type of body cavities located between gut and body wall; segmentation, the division of the body into a series of units; and cephalization.

26–1 (continued)

SECTION REVIEW 26–1

1. A multicellular ingestive heterotroph whose cells lack cell walls. Accept all logical answers.
2. Feeding: eating food. Respiration: taking in oxygen and giving off carbon dioxide through cellular metabolism. Internal transport: carrying substances from one part of the body to another. Excretion: eliminating metabolic wastes. Response: reacting to environmental stimuli. Movement: changing the position or orientation of body parts or of the entire body. Reproduction: producing offspring. Make sure students define these terms in their own words.
3. Radial symmetry: body parts repeat around an imaginary line drawn through the center of the body. Bilateral symmetry: body parts repeat on either side of an imaginary line drawn down the center of the body.
4. The level of organization becomes higher as animals become more complex; functions are carried out on the level of organs and organ systems rather than that of cells and tissues. Most complex animals exhibit bilateral symmetry, whereas some of the most simple animals exhibit radial symmetry. Cephal-

Figure 26–8 Most of the more complex animals have bilateral symmetry, which means that the body parts repeat on either side of an imaginary line drawn down the center of the body.

Guide For Reading
- What is a sponge?
- How do sponges perform essential functions?
- How do sponges affect other organisms?

More complex animals tend to have a concentration of sense organs and nerve cells in their anterior (head) end. Because animals with bilateral symmetry usually move with their anterior end forward, this end encounters new parts of the environment first. As you might imagine, natural selection favors animals that can sense the nature of the environment into which they are moving before their entire body is exposed to the new environment. It is not wise to back up into a potentially dangerous situation! Thus sense organs tend to gather at the anterior end. As the sense organs collect up front, so do the nerve cells that process information and "decide" what the animal should do. Eventually, the anterior end is different enough from the posterior end that we call it a head. This gathering of sense organs and nerve cells into the head region is called **cephalization** (*cephalo-* means head).

Cephalization becomes more pronounced as animals become more complex. Nerve cells in the head gather into clusters that process the information gathered by the nervous system and control responses to stimuli. Small clusters of nerve cells are called **ganglia** (singular: ganglion). In the most complex animals, large numbers of nerve cells gather together to form larger structures called brains.

26–1 SECTION REVIEW

1. What is an animal? Why is it important to study animals?
2. List seven essential functions in animals. Define these functions in your own words.
3. Compare two different kinds of symmetry found in the animal kingdom.
4. Describe three basic trends in animal evolution.
5. **Critical Thinking—Applying Concepts** Why are specialized cells necessary in multicellular animals?

26–2 Sponges

Sponges are among the most ancient of all animals that are alive today. The first sponges date back to the beginning of the Cambrian Period (about 580 million years ago), when the first traces of multicellular animals appeared in the fossil record. Most sponges live in the sea, although a few live in freshwater lakes and streams. Sponges inhabit almost all areas of the sea —from the polar regions to the tropics and from the low-tide line down into water several hundred meters deep. Sponges belong to the phylum **Porifera** (por-IHF-er-ah). This name, which literally means pore-bearers, is appropriate because sponges have tiny openings all over their body.

Sponges were once thought to be plants, which is easy to understand in light of the fact that adult sponges are sessile

560

ization tends to improve as animals become more complex; the brain and anterior sense organs are generally most highly developed in more complex animals.
5. Specialized cells can carry out specific functions more efficiently than nonspecialized cells.

Reinforcement/Reteaching
If some students have trouble answering the Section Review questions, go back to the material with which they are having difficulty. Review the material and encourage students to ask questions. Then have students attempt the Section Review questions again.

Closure
Have students take out the drawings they made at the beginning of this section when you asked them to respond to the word

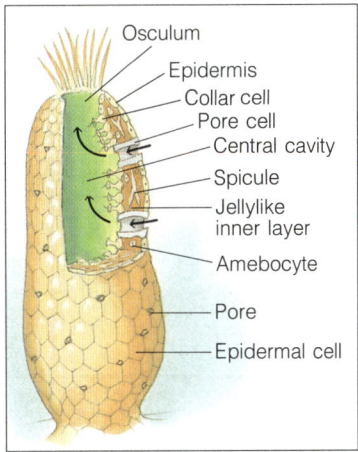

and show little detectable movement. As far as modern biologists are concerned, sponges are clearly multicellular animals—sponges are heterotrophic, have no cell walls, and contain several specialized cell types that live together. But sponges are very different from all other animals. **Sponges have nothing that even vaguely resembles a mouth or gut, and they have no specialized tissues or organ systems.** For these reasons, most biologists believe that sponges evolved from single-celled ancestors separately from other multicellular animals. The evolutionary line that gave rise to sponges was a dead end that produced no other groups of animals.

Form and Function in Sponges

The body plan of a typical sponge is simple. Refer to Figure 26–10 as you read about the structure of a sponge. The body of a sponge forms a wall around a central cavity. In this wall are thousands of openings, or pores. A steady current of water moves through these pores into the central cavity. This current is powered by the flagella of cells called **collar cells**. The water that gathers in the central cavity exits through a large hole called the **osculum** (AHS-kyoo-luhm). The current of water that flows through the body of a sponge delivers food and oxygen to the cells and carries away cellular waste products. The water also transports gametes or larvae out of the sponge's body.

Many sponges manufacture thin, spiny **spicules** that form the skeleton of the sponge. A special kind of cell called an **amebocyte** (ah-MEE-boh-sight) builds the spicules from either chalklike calcium carbonate ($CaCO_3$) or glasslike silica (SiO_2). These spicules interlock to form beautiful and delicate skeletons, such as the Venus' flower basket shown in Figure 26–11 on page 562. The softer but stronger sponge skeletons that we know as natural bath sponges consist of fibers of a protein called **spongin**. Some sponges have skeletons that are made up of both spongin and spicules.

Figure 26–9 Sponges come in a wide variety of shapes, colors, and sizes. Some, such as this basket sponge (center), are larger than humans!

Figure 26–10 The essential life functions of sponges are performed at the level of cells or tissues. There are no true organs in sponges. Each different type of cell in a sponge—epidermal cells, pore cells, collar cells, and amebocytes—performs specific functions.

561

Figure 26–11 The lacy skeleton of a glass sponge consists of thousands of spicules of silica.

Figure 26–12 In some sponges, the eggs are fertilized inside the body wall of the parent sponge (bottom). In others, the eggs are squirted into the surrounding water, where they may be fertilized (top).

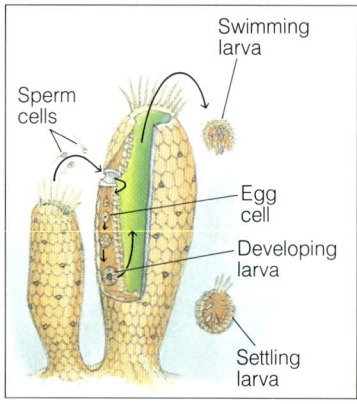

Sponges are filter feeders that sift microscopic particles of food from the water that passes through them. As the water moves through the sponge, tiny food particles stick to the collar cells. The trapped particles are then engulfed by the collar cells (endocytosis), where they may be digested. If the collar cells do not digest the food, they pass it on to the amebocytes. When the amebocytes are finished digesting the food particles, they wander around, delivering digested food to other parts of the sponge. Note that all digestion in sponges is intracellular; that is, it takes place inside cells.

The water flowing through a sponge simultaneously serves as its respiratory, excretory, and internal transport system. As water passes through the body wall, sponge cells remove oxygen from it and give off carbon dioxide into it. Metabolic wastes produced by cellular respiration (such as ammonia) are also released into the water, which carries them away. The amount of water that is pumped through a sponge is amazing. A sponge 10 centimeters in height and 1 centimeter in diameter was found to pump 22.5 liters of water per day through its body.

The water that flows through the body of the sponge also plays a role in sexual reproduction. Although eggs are kept inside the body wall of the sponge, sperm are released into the water flowing through the sponge and are thus carried out into the open water. If those sperm are taken in by another sponge, they are picked up by amebocytes and carried to that sponge's eggs, where fertilization occurs. The zygote (fertilized egg) that results develops into a larva that swims and can be carried by currents for a long distance before it settles down and grows into a new sponge.

Sponges reproduce asexually as well as sexually. Faced with cold winters, some freshwater sponges produce structures called **gemmules** (JEHM-yoolz). Gemmules are sphere-shaped collections of amebocytes surrounded by a tough layer of spicules. Gemmules can survive long periods of freezing temperatures and drought, which would kill adult sponges. When conditions again become favorable, gemmules grow into new sponges. Sponges can also reproduce asexually by **budding**. In this process, part of a sponge simply falls off the parent and grows into a new sponge.

Budding is one indication of the sponges' remarkable powers of regeneration (the ability to regrow a lost or damaged part). In fact, if you were to grind up a sponge, separate its cells by passing them through a filter, and place the cells in a container of water, the cells would clump together and grow into several new sponges! It is not surprising, therefore, that sponges can easily repair torn body parts.

How Sponges Fit into the World

Sponges are often the most common forms of life in dark places such as the walls of underwater caves and on dock pilings. Many other marine animals—certain kinds of worms, shrimp, snails, and starfish, for example—live on, in, and under sponges. Sponges are also involved in symbiotic relationships with organisms that are not animals. Certain sponges contain symbiotic bacteria, blue-green bacteria, or plantlike protists. The photosynthetic symbionts provide food and oxygen to the sponge and remove wastes. Although sponges produce spicules and protective chemicals that discourage most animals from feeding on them, sponges are important parts of the diets of certain snails, starfish, and fishes.

The family of sponges known as the boring sponges are particularly important in "cleaning up" the ocean floor. Special amebocytes in these sponges release chemicals that allow the sponges to bore, or drill, tunnels through old shells and pieces of coral. These tunnels weaken the shells and coral and thus help break them down.

Since the time of the Greeks and Romans, humans have used the dried and cleaned bodies of some sponges in bathing. Most sponges you see in supermarkets today are artificial, but natural bath sponges are still available. Recently, scientists have found uses for parts of the sponge other than its skeleton. In a series of exciting new developments, scientists are learning to use several chemicals manufactured by sponges.

Because sponges cannot move, they must protect themselves from their enemies in other ways. Bacteria, algal spores, and many tiny organisms are constantly looking for surfaces on which to settle. To protect themselves from being overgrown by these organisms, sponges manufacture numerous compounds that are toxic to such organisms. These chemicals also discourage many animals from chewing on sponges. Researchers have found that many of these chemicals are powerful antibiotics that can be used to fight bacteria and fungi that cause disease. Other sponge chemicals act against viruses almost as well as antibiotics fight bacteria. One compound taken from a Caribbean sponge may be useful against leukemia and herpes viruses. Another may help fight certain forms of arthritis. Still other sponge chemicals may be effective against the bacteria that cause strep throat and those that become resistant to penicillin. Although most of these drugs are still in the experimental stage, scientists hope that they will soon be ready for human use.

Figure 26–13 *Since ancient times, the soft skeletons of certain types of sponges have been used by humans for bathing.*

26–2 SECTION REVIEW

1. How do sponges differ from other animals? How do they feed, respire, and eliminate wastes?
2. How are sponges proving useful to medical science?
3. **Critical Thinking—Assessing Concepts** Why are sponges thought to be an evolutionary dead end?

ESL STRATEGY

Have students make a two-column chart with the heads Sponges and Cnidarians. Then have them write the following terms in the appropriate columns: sessile, motile, specialized tissues, reproduce sexually, reproduce asexually, mouth, carnivore, filter feeder. Some terms may be used more than once.

BACKGROUND INFORMATION
NEMATOCYSTS

A unique characteristic of cnidarians is the ability to produce nematocysts. A nematocyst is secreted within a type of cell called a cnidoblast. It consists of a capsule with an inverted tubular thread inside. Attached to the cell containing the nematocyst is a modified flagellum that serves as a trigger. When potential prey or a predator touches the trigger, or when the cell receives a particular chemical stimulus, the permeability of the capsule increases and water diffuses into it. This increases the pressure within the capsule, forcing the thread to turn inside out.

Once a nematocyst has been fired, its life is over—it cannot be used again. For this reason, cnidarians are constantly producing new nematocysts in order to ensure an ample supply.

Guide For Reading
- What is a cnidarian?
- How do cnidarians perform essential functions?
- How are cnidarians classified?
- How do cnidarians affect other living things?

Figure 26–14 Some cnidarians, such as sea nettles (top, left) and sea anemones (left), are solitary. Others, such as gorgonian coral polyps (right), are colonial.

26–3 Cnidarians

The phylum **Cnidaria** (nigh-DAIR-ee-ah) includes many animals with brilliant colors and unusual shapes. Delicate jellyfish float in ocean currents. Brightly colored sea anemones cling to rocks, looking more like underwater flowers than animals. These beautiful and fascinating animals are found all over the world, but most species live only in the sea.

What Is a Cnidarian?

Cnidarians are soft-bodied animals with stinging tentacles arranged in circles around their mouth. Some cnidarians live as single individuals. Others live as groups of dozens or even thousands of individuals connected into a colony. All cnidarians exhibit radial symmetry and have specialized cells and tissues. Many cnidarians have life cycles that include two different-looking stages, the sessile flowerlike **polyp** (PAH-lihp) and the motile bell-shaped **medusa** (meh-DOO-sah).

The body plans of a typical cnidarian polyp and a medusa are shown in Figure 26–15. As you can see, both polyps and medusae have a body wall that surrounds an internal space called the **gastrovascular cavity**. It is in the gastrovascular cavity that digestion takes place. The body wall consists of three layers: epidermis, mesoglea, and gastroderm. The epidermis is a layer of cells that covers the outer surface of the cnidarian's body. The gastroderm is a layer of cells that covers the inner surface, lining the gastrovascular cavity. Between these two cell layers is the mesoglea (mehz-oh-GLEE-ah). The mesoglea ranges from a thin noncellular membrane to a thick jellylike material that may contain wandering amebocytes. In general, the mesoglea is a thin layer in polyps and a thick layer in medusae.

TEACHING STRATEGY 26–3

Focus/Motivation

Using the Visuals Have students observe the diagram of the body plan of a cnidarian in Figure 26–15.
- **Do you see anything in this body plan that is distinctly different from the body plan of a sponge?** (The most important difference that students should notice are the two structures that the sponge does not have: tentacles and a gastrovascular cavity with a mouth opening.)

Content Development

Use students' observations of the cnidarian body plan to emphasize two important characteristics of this phylum: first, the presence of tentacles that contain stinging structures called nematocysts and, second, the presence of a hollow gut area called the gastrovascular cavity.

The first characteristic is important because the presence of nematocysts is unique to members of this phylum—thus, the name Cnidaria. Other groups of animals may have tentacles that

Form and Function in Cnidarians

Almost all cnidarians capture and eat small animals by using stinging structures called **nematocysts** (neh-MAT-oh-sihsts), which are located on their tentacles. Each nematocyst is a poison-filled sac containing a tightly coiled "spring-loaded" dart. When another animal touches a nematocyst, the dart uncoils as if it had exploded and buries itself in the skin of the animal. The dart carries with it enough poison to paralyze or kill the prey. Once the prey is rendered helpless, the cnidarian's tentacles push the food through the mouth and into the gastrovascular cavity. There the food is gradually broken up into tiny pieces. These food fragments are taken up by special cells in the gastroderm that digest them further. The nutrients are then transported throughout the body by diffusion. Any materials that cannot be digested are passed back out through the mouth, which is the only opening in the gastrovascular cavity, several hours later.

Although most cnidarians are considered carnivores, many do not actually "eat" much, thanks to an extraordinary symbiosis, which we talked about in Chapter 18. In many cnidarians, tiny photosynthetic protists grow right inside the living cells of the gastroderm. This relationship between autotrophic protist and heterotrophic animal works very efficiently. The photosynthetic protists use the carbon dioxide and other metabolic wastes produced by the cnidarian's cells to manufacture oxygen and organic compounds such as carbohydrates and proteins. The protists use some of the oxygen and organic compounds themselves and release the rest into the tissues of their cnidarian hosts. Many cnidarians depend on this symbiosis to such an extent that they can live only in bright sunlight! These cnidarians will slowly starve if kept in a darkened laboratory tank, even if they are fed pieces of shrimp and fish.

Because most cnidarians are only a few cell layers thick, they have not had to evolve many complicated body systems in order to survive. Some colonial cnidarians and some jellyfish have long, tube-shaped, branching gastrovascular cavities that help carry partially digested food through their bodies. Because these animals live in clean constantly flowing water, they can respire and eliminate waste products by diffusion directly through their body walls. There is no organized internal transport network or excretory system in cnidarians.

Cnidarians also lack a centralized nervous system and anything that could be called a brain. They have simple nervous systems called nerve nets. The nerve net is concentrated around the mouth, but it does spread throughout the body.

Information about the environment is transmitted to the rest of a cnidarian's nervous system by specialized sensory cells. Both polyps and medusae have sensory cells in the epidermis that detect chemicals from food and the touch of foreign objects. In medusae, some groups of sensory cells are organized into simple organs. These organs, which are called

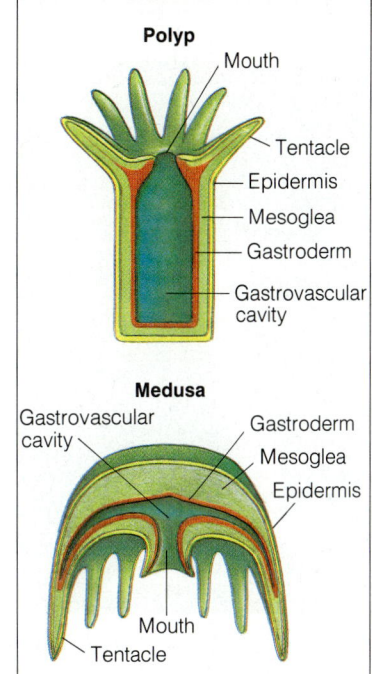

Figure 26–15 Two basic body forms are seen in cnidarians: the flowerlike polyp and the bell-shaped medusa.

Figure 26–16 The body wall of a cnidarian consists of three layers: epidermis, mesoglea, and gastroderm.

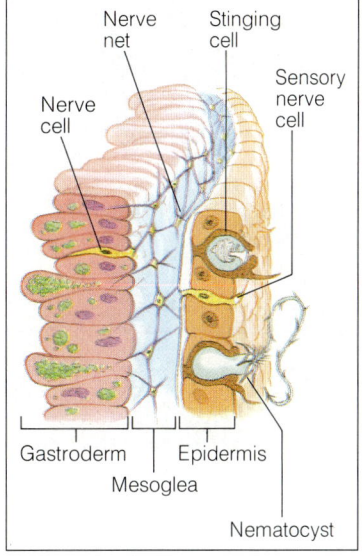

ECOLOGY NOTE
REEF-BUILDING CORALS AND SYMBIOSIS

Corals that build limestone skeletons have a close relationship with single-celled algae that live within their tissues as symbionts— a relationship that is a bit more multifaceted than we describe in the textbook. Many cnidarians (and some mollusks and other invertebrates) depend upon algal symbionts for a supply of substances that seem to act like vitamins; they aren't needed in enormous quantities, but they are necessary for good health.

In reef-building corals, one or more of these substances have a direct effect on the animals' ability to build their limestone skeletons. Without the contribution of the algae, coral animals cannot grow. That is why the phenomenon known as coral bleaching— in which the algae leave of their own accord or are tossed out by their hosts—is of great concern to reef biologists. Without that vital symbiosis, the entire reef will stop growing and eventually die.

TIE-IN   LANGUAGE ARTS

The phylum name Cnidaria is derived from the Greek word for nettle. The phylum was so named because of the painful stings that many of these organisms can give with their tentacles. Phylum Cnidaria has in the past been known as phylum Coelenterata. The word coelenterata means hollow intestine.

---

capture prey, but they are not equipped with nematocysts.

The second characteristic of this phylum, the presence of a gastovascular cavity, is important because cnidarians are the simplest animals that have a distinctive gut in which digestion takes place. Students will recall that in the phylum Porifera, there was no specific gut area—digestion took place in the individual cells. Tell students that in Section 26–4, they will encounter animals in which the gastrovascular cavity has two openings, one for ingestion and the other for excretion.

It should be pointed out that one of the measures of the increasing complexity among animal organisms is the type of gut. The simplest animals, the sponges, have no distinctive gut. The next simplest animals have a gut with only one opening, the mouth. On the next level of complexity, animals have what is often referred to as a complete gut, with both a mouth and an anus. In more complex animals, the gut contains organs and is part of an organ system.

TEACHING SUPPORT

Laboratory Manual
- Sponges and Hydras, p. 335

MEDIA AND TECHNOLOGY
- Biology Transparencies
- 30: The Life Cycle of a Jellyfish

MULTICULTURAL STRATEGY

In Greek mythology, the hydra was a monster with many heads. When one head was severed, another immediately grew in its place. Have students use reference materials to learn more about the myths of the hydra, including how it was finally vanquished. Also discuss with students how the characteristics of a hydra described in this chapter are similar to those of the hydra of myth.

FACTS AND FIGURES

The giant jellyfish *Cyanea artica* is one of the largest known invertebrates. Its body is 2 meters in diameter and its tentacles can be 36 meters long. A single specimen of this species can weigh up to 900 kilograms.

26–3 (continued)

Skills Development
Guided Practice
Skill: Interpreting diagrams
Using the Visuals Have students look back at the diagram of the cnidarian body plan in Figure 26–15.
- According to the diagram, what is the main difference between the polyp stage and the medusa stage? (A difference in shape.)
- How would you describe this difference? (The polyp is vertical and tubelike, and the medusa is wider and umbrella shaped.)

Figure 26–17 The buds at the base of this hydra's body will develop into new individuals that are genetically identical to their parent.

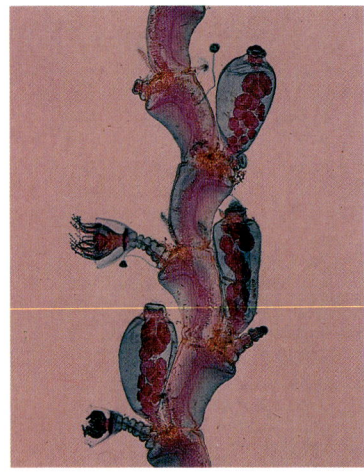

Figure 26–18 In this colonial hydrozoan, the polyps with tentacles are used in feeding and defense. The round buds found inside the reproductive polyps will eventually develop into medusae.

statocysts and ocelli, are arranged around the rim of a medusa's bell. Statocysts are involved with balance—they help an organism determine which way is up. Ocelli (oh-SEHL-igh; singular: ocellus), or eyespots, detect the presence of light.

Cnidarians lack the muscle cells that most other animals use to move about. But many of the epidermal cells in cnidarians can change shape when stimulated by the nervous system. Thus these cells serve the same function as muscles. Cnidarian polyps can expand, shrink, and move their tentacles by relaxing or contracting these epidermal cells. In medusae, contractions of the special epidermal cells change the bell-shaped body, causing it to "close" like a folding umbrella. The "closing" of the body pushes water out of the bell. This moves a medusa forward by jet propulsion.

Most cnidarians can reproduce both sexually and asexually. As you can see in Figure 26–17, polyps can produce new polyps asexually by budding. Budding begins with a swelling on the side of an existing individual. This swelling eventually grows into a complete polyp. Many polyps also reproduce asexually by budding off tiny medusae. When the medusae mature, they reproduce sexually by releasing gametes into the water. Depending on the species, fertilization occurs either in open water or inside an egg-carrying medusa. The zygote (fertilized egg) grows into a ciliated larva that swims around for some time. Later, the larva settles down, attaches to a hard surface, and changes into a polyp that begins the cycle again.

Hydras and Their Relatives

The class Hydrozoa (high-droh-ZOH-ah) is made up of cnidarians that spend most of their lives as polyps, although they usually have a short medusa stage. As you can see in Figure 26–18, most hydrozoan polyps grow in branching sessile colonies. Hydrozoan colonies range in length from a few centimeters to more than a meter. In each of these colonies, specialized polyps perform particular functions, such as feeding, reproduction, or defense. Reproductive polyps produce free-swimming medusae by budding. These medusae are usually less than 2 centimeters in diameter. Soon after they form, the medusae produce both eggs and sperm and then die.

The most common freshwater hydrozoans are the hydras. Hydras are not typical hydrozoans because they live as solitary polyps and lack the medusa stage in their life cycle. Unlike most other polyps, hydras can move around with a curious somersaulting movement. Hydras can reproduce either asexually by budding or sexually by producing eggs and sperm in their body walls. In most species of hydras, the sexes are separate. In other words, individuals are either male or female. However, a few species are hermaphroditic. A **hermaphrodite** is an individual that has both male and female reproductive organs and thus produces both sperm and eggs.

- Do both body forms have the same structures? What are they? (Yes. Mouth, tentacles, epidermis, mesoglea, gastroderm, gastrovascular cavity.)
- How are the tentacles arranged in both forms? (They encircle the mouth.)
- Why is this arrangement a practical one for the cnidarian? (The tentacles paralyze and capture prey and then take it to the mouth.)
- Which layer of cells lines the gastrovascular cavity? (The gastroderm.)
- Which layer of cells is on the exterior of the cnidarian? (The epidermis.)
- Where is the mesoglea found? (Between the gastroderm and the epidermis.)
- This layer is very thick in jellyfish. Can you explain why?

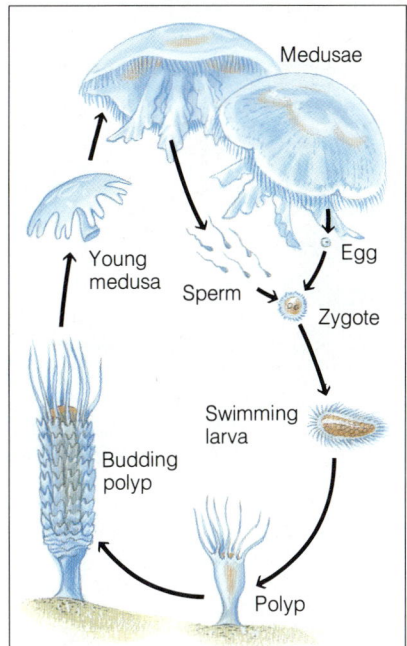

Figure 26–19 Many cnidarians, such as the jellyfish Aurelia, *have life cycles that include both medusa and polyp stages.*

One unusual hydrozoan is the Portuguese man-of-war. These animals form floating colonies that contain several specialized kinds of polyps. In each Portuguese man-of-war, one polyp forms a balloonlike float that keeps the colony on the surface. This float may be up to 30 centimeters long. Some of the polyps in the colony produce long stinging tentacles that hang several meters below the float and paralyze and capture prey. Some polyps digest the food held by the tentacles, and still others do nothing but make eggs and sperm. Portuguese man-of-war nematocysts are strong enough to sting humans very badly, so swimmers and beach-goers must take care when these animals are spotted near shore.

Jellyfish

The class Scyphozoa (sigh-foh-ZOH-ah) contains the true jellyfish. Jellyfish go through the same life-cycle stages as hydrozoans. However, in scyphozoans the medusa is large and long-lived, and the polyp is restricted to a tiny larval stage.

Some jellyfish, such as the lion's mane, which is found in the north Atlantic, often grow up to 2 meters in diameter. The largest jellyfish ever found was more than 3.6 meters in diameter and had tentacles more than 30 meters long. The nematocysts of most jellyfish are harmless to humans, but a few can cause painful stings. One tiny Australian jellyfish has a toxin powerful enough to cause death in 3 to 20 minutes!

Sea Anemones and Corals

The class Anthozoa (an-thoh-ZOH-ah) contains sea anemones and corals, which are among the most beautiful and ecologically important invertebrates. Anthozoans have only the polyp stage in their life cycles. Adult polyps reproduce sexually by producing eggs and sperm that are released into the water. The zygote grows into a ciliated larva that settles to the ocean bottom and becomes a new polyp. Many anthozoans also reproduce asexually by budding.

Sea anemones are solitary polyps that live in the sea from the low-tide line to great depths. Although sea anemones can catch food with the nematocysts on their tentacles, many shallow-water species depend heavily on their photosynthetic symbionts. Some sea anemones can grow up to a meter in diameter.

Figure 26–20 Sea fans (top) and sea pens (bottom) are two types of exotic colonial anthozoans. The purple-and-white feather stars clinging to the sea fan are relatives of starfish.

BACKGROUND INFORMATION
HYDRA

Hydras usually attach themselves to leaves or rocks with a sticky substance produced in their basal disks, which appear at the bottom of the animal opposite the mouth. Hydras can move from one place of attachment to another by means of a tumbling motion or by gliding on their basal disks. They can also form an air bubble at their bases and float upside down.

The hydra feeds on small organisms such as freshwater shrimp and water fleas. To lure its prey, the hydra waves its tentacles. When a prey organism touches one of these tentacles, nematocysts are released, which sting and paralyze the victim. The wounded prey then releases a chemical substance called glutathione. Glutathione provides the chemical stimulus that causes the tentacles of the hydra to contract and its mouth to open, with the result that the tentacles pull the prey into the mouth. The release of this chemical is very important, for if it does not happen, the hydra does not complete its feeding action.

Hydras have a simple nervous system that consists of nerve cells arranged in an irregular network called a nerve net. The nerve net, which is located in the mesoglea, is able to conduct impulses in all directions throughout the organism. It is the nerve net that coordinates the contraction of the tentacles and the feeding process.

Hydras also have sensory receptor cells. These special cells, which are located in the ectoderm and gastrodermis, are sensitive to chemicals, touch, and light.

(The mesoglea can be a thick jellylike material. In the jellyfish, it is the mesoglea that gives this animal its distinctive body texture.)
- **Is the mesoglea made up of cells?** (No, it is a noncellular material.)
- **What about the epidermis and the gastroderm?** (These layers are made up of cells.)

Enrichment

Students who enjoy art or creative writing can work together to create a "Cnidarian Underwater Fantasy." Have students include such animals as jellyfish, sea anemones, and the Portuguese man-of-war. For a story line, students can personify the organisms, having them talk, think, and act like humans, or they can describe the animals from the human perspective of scuba divers or people in a submarine. Students' fantasies should reflect the characteristics of cnidarians that they have learned about in this section.

ECOLOGY NOTE
WHY CARE ABOUT REEFS?

Coral reefs are among the world's most endangered ecosystems. Why? In part because their inhabitants are accustomed to constant, benign conditions and so are susceptible to nearly any change in water temperature, salinity, nutrient level, or sediment load.

For most of their existence, reefs bordered islands and continents with very small human populations. Under these circumstances, local inhabitants could harvest reef fishes for food and use coral rock for building materials with little adverse impact on the reef. But local population growth and an increase in oceanfront tourism over the last three decades have resulted in explosive development along once-quiet coral shores. As a result, reefs are in trouble everywhere, from the once pristine South Pacific to our very own Florida Keys—where the coral reefs are dying from sewage, pollution, siltation, over-harvesting, and construction.

Why should anyone care? From a biological perspective, reefs are the aquatic equivalent of tropical rain forests; they house an incredible diversity of organisms. Every major group of organisms discussed in this unit has at least one representative species living on reefs, and many groups have dozens or even hundreds of species there. And from a more human-centered viewpoint, as mentioned in the textbook, reefs act as living, self-repairing breakwaters that protect long stretches of tropical coastline against erosion.

Figure 26–21 *Sea anemones (bottom) are solitary polyps. The polyps of stony corals (top left and right) are similar in structure to sea anemones. Unlike sea anemones, stony corals produce hard skeletons of calcium carbonate. Most stony corals are colonial.*

Corals grow in shallow tropical waters around the world. Coral polyps are very similar in form to sea anemones. However, corals produce skeletons of calcium carbonate ($CaCO_3$), or limestone. Although a few corals are solitary, most are colonial. As a coral colony grows, new polyps are produced by budding, and more and more limestone is laid down. Coral colonies grow very slowly, but they may live for hundreds, or even thousands, of years. Together, countless coral colonies produce huge structures called coral reefs. Some of these reefs are enormous and contain more rock and living tissue than even the largest human cities. The Great Barrier Reef off the coast of Australia is more than 2000 kilometers long and some 80 kilometers wide.

How Cnidarians Fit into the World

Cnidarians form a number of interesting symbiotic relationships with other animals. Certain fish, shrimp, and other small animals live among the tentacles of large sea anemones. The sea anemone protects and provides scraps of food for these symbionts, which are unaffected by the sea anemone's nematocysts. In turn, the symbionts are thought to help clean the sea anemone and protect it from certain predators.

Corals and the reefs they form are extremely important in the ecology of tropical oceans. Because coral reefs are built from many separate coral colonies attached together, they contain tunnels, caves, and deep channels. In these recesses live some of the most beautiful and fascinating animals in the world.

Corals are important to humans in many ways. Coral reefs provide a home for food fishes and other edible animals, as well as for organisms that produce valuable shells, pearls, and other products. Reefs also protect the land from much of the action of waves. When coral reefs are destroyed or severely

26–3 (continued)

Skills Development
Guided Practice
Skills: Observing, making diagrams

Have students use a medicine dropper to place a live hydra in a small dish along with some of the water in which it lives.

Ask students to observe the hydra and locate its mouth, body cavity, and tentacles. Have students make diagrams of what they see.

Next, have students gently touch the hydra with a toothpick in three different places.
- **What happens when you touch the hydra? Why?** (Its tentacles contract as a reaction against possible danger or in response to food.)

Provide students with some *Daphnia* to place in the dish.
- **How does the hydra respond to the *Daphnia*?** (The hydra eats the *Daphnia* by using its tentacles to grab the prey and pull it into its mouth.)

Have students add a drop of vinegar to the dish.
- **How does the hydra react to the vinegar?** (The hydra contracts its tentacles.)
- **How do you think it does this?** (The vinegar provides a chemi-

Figure 26–22 Although large sea anemones often eat fish, this clownfish is perfectly safe because it is "immune" to sea anemone stings. In addition, the clownfish and sea anemone are engaged in a symbiotic relationship that is thought to benefit both organisms. The clownfish is protected from some of its enemies by the anemone's stinging tentacles. The anemone, in turn, is protected by the clownfish from several kinds of fishes that would otherwise snack on its tentacles.

damaged, large amounts of shoreline may be washed away. Fossil reefs offer important clues to geologists about the locations of oil deposits. Large blocks of coral have been used to build houses and to filter drinking water. Humans have long used certain corals to make jewelry and decorations.

Some cnidarians are used in medical research. Corals, like sponges, produce chemicals to protect themselves from being infected, overgrown, or settled upon by other organisms. Some of these chemicals may provide us with anti-cancer drugs, and others may help us learn more about cancer itself. The nerve toxins produced in cnidarian nematocysts are another powerful research tool. Whenever a compound poisons a biological system, studies of how the poison operates reveal a lot about how the system works. Cnidarians such as the sea wasp jellyfish produce several strong nerve poisons that have already helped scientists better understand nerve-cell function.

26–3 SECTION REVIEW

1. What is a cnidarian? What kind of symmetry do cnidarians have?
2. Give an example of each class of cnidarians.
3. Describe the life cycle of a typical cnidarian.
4. Discuss symbiotic relationships and other interactions between cnidarians and other living things.
5. **Critical Thinking—Making Inferences** A medusa usually has specialized sense organs. It may also have nerves that are organized into rings that encircle its body and structures that control body contractions. Explain why a medusa needs a more complex nervous system than a polyp. (*Hint:* How does the lifestyle of a medusa differ from that of a polyp?)

26–4 Unsegmented Worms

When most people think of worms, they think of earthworms—long, squiggly creatures that spend their time making tunnels in the ground. But there are many animals called worms that look nothing like earthworms. Many live in fresh water, a large number live in the ocean, and lots of them are important to humans. The two phyla of wormlike animals that we shall examine in this section are much simpler in structure than earthworms. They are known as **unsegmented worms** because their bodies are not divided into special segments. The phylum **Platyhelminthes** (pla-tee-hehl-MIHN-theez) consists of simple animals called **flatworms**. The phylum **Nematoda** (nee-mah-TOHD-ah) consists of long, thin worms called **roundworms**.

Flatworms

The members of the phylum Platyhelminthes are the simplest animals with bilateral symmetry. Most members of this phylum exhibit enough cephalization, or development of the anterior end, to have what we call a head. Because flatworms really are flat, the name of the phylum is quite appropriate (*platy-* means flat and *helminth* means worm). Many flatworms are no more than a few millimeters thick, although they may be up to 20 meters long. Flatworms have more developed organ systems than either sponges or cnidarians.

Figure 26–23 Members of the phylum Platyhelminthes, such as this spotted marine flatworm, are the simplest animals with bilateral symmetry.

types of organisms that are classified as worms. Point out that the familiar earthworm is actually a much more advanced animal than the worms students will be studying in this section. Explain that these worms are what are known as unsegmented worms, meaning that their bodies are not divided into parts, or segments. Of the two phyla of unsegmented worms that will be studied, the simplest are the flatworms, which are members of the phylum Platyhelminthes.

Continue by explaining that, despite their simplicity, the flatworms are considerably more advanced than the cnidarians. First, the flatworms exhibit bilateral symmetry. This means that an imaginary line drawn lengthwise from the head separates right and left halves that are identical.

Form and Function in Flatworms

Flatworms feed in either of two very different ways. Some are aquatic and free-living, which means that they wander around in streams, lakes, and oceans. These worms may be carnivores that feed on tiny aquatic animals, or they may be scavengers that feed on recently dead animals. (You can probably catch flatworms in a local stream by leaving a piece of liver in the water overnight.) Free-living flatworms have a gastrovascular cavity with one opening at the end of a muscular tube called a **pharynx** (FAIR-ihnks). See Figure 26–24. They use the pharynx to suck food into the gastrovascular cavity. The gastrovascular cavity forms an intestine with many branches along the entire length of the worm. In the intestines, enzymes help break down the food into small particles. These particles are taken inside the cells of the intestinal wall, where digestion is completed. Because the intestine branches into nearly all parts of the body, completely digested food can diffuse to other body tissues. Like cnidarians, flatworms expel undigested materials through the mouth.

Many other flatworms are parasites that feed on blood, tissue fluids, or pieces of cells inside the body of their host. Some of these animals have a pharynx that pumps food into a pair of dead-end intestinal sacs where the food is digested. But in many parasitic flatworms, the digestive tract is simpler than in free-living forms. Tapeworms, which live within the intestines of their host, do not have any digestive tract at all. They have hooks and/or suckers with which they latch onto the intestinal wall of the host. From this position, they can simply absorb the food that passes by—food that has already been broken down by the host's digestive enzymes.

Flatworms lack any kind of specialized circulatory or respiratory system. Because they are so flat, they can depend on diffusion to transport oxygen and nutrients to their tissues. And they can get rid of carbon dioxide and most other metabolic wastes by allowing them to diffuse out through their body walls. Freshwater flatworms such as planarians have structures called flame cells that help them get rid of extra water. Many flame cells join together to form a network that empties through tiny pores in the animal's skin.

Free-living flatworms have nervous systems that are much more developed than those of cnidarians and sponges. They have a definite head in which a simple brain is located. This brain is the control center of a simple nervous system that stretches throughout the body. One or more long nerve cords run from the brain down the length of the body on either side. Shorter nerve cords run across the body. Many flatworms have one or more pairs of light-sensitive organs called ocelli, or eyespots. These eyespots do not see objects as our eyes do; they

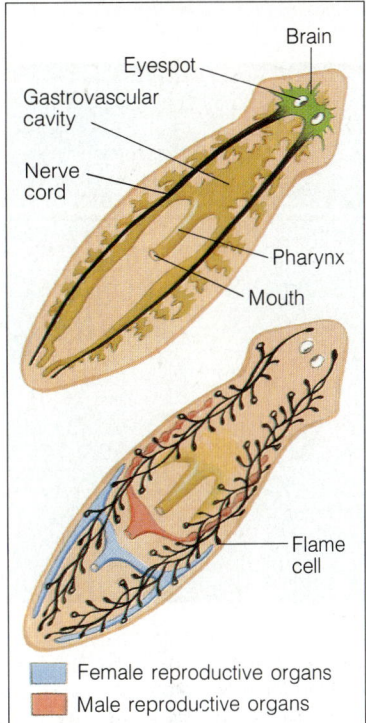

Figure 26–24 Flatworms, such as planarians, perform their essential life functions at the level of organ systems.

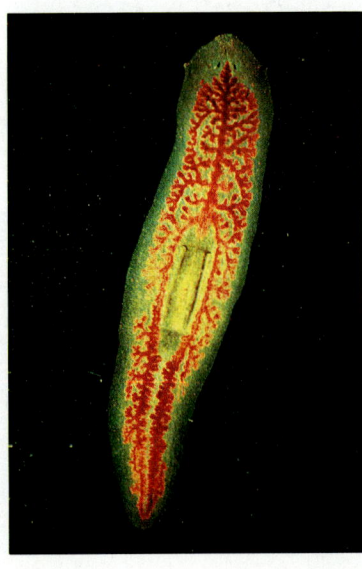

Figure 26–25 The branching gastrovascular cavity and the pharynx can be clearly seen in this planarian.

Bilateral symmetry is considered a more advanced trait in animals than radial symmetry, which was exhibited by the cnidarians.

A second trait that puts the flatworms ahead of the cnidarians is cephalization, or the development of a head area. The head in a flatworm is equipped with sensory structures and a rudimentary brain—thus making for a more highly developed nervous system.

Yet another level of advancement that can be seen in the flatworm is the fact that life functions are performed at the level of organs and organ systems. In the cnidarians, the level of cellular organization was for the most part no higher than that of tissues, with the exception of a few simple sensory organs.

BACKGROUND INFORMATION
FLATWORMS

Flatworms are acoelomate, meaning that they do not have a body cavity between the gut and the body wall.

BACKGROUND INFORMATION
FLAME CELLS

Freshwater turbellaria have an organ system that regulates the volume and salt concentration of their body fluid. This system depends on one or more units called protonephridia.

Each unit of protonephridia consists of branched tubules that extend from a pore at the body surface to many cup-shaped flame cells in the body tissues. Within the flame cells, tufts of cilia flicker—thus the name flame cells. When excess tissue fluid moves into the flame cells, the flickering of the cilia drives the fluid down the tubule system to the outside of the organism.

Reinforcement/Reteaching

Using the Visuals Have students list some of the important organs that are present in the flatworm. (You may wish to have students refer to Figure 26–24 to locate some of them.) These organs include the pharynx, the intestine, the brain, and the nerve cord. Point out that although the flatworm does not have a respiratory system or a circulatory system, it does have a digestive system and a nervous system.

Figure 26–26 An injury divided the head of this planarian in half, and the two halves regenerated their lost parts. Eventually, the two-headed planarian will split lengthwise to form two new planarians.

simply detect whether the animal is in light or in darkness. Most flatworms have cells that are sensitive to chemicals found in food, and other cells that tell the worm which way the water around them is flowing. These cells are usually scattered all over the body. The nervous system of free-living flatworms allows them to gather information from their environment—information that they use to locate food and to find dark hiding places beneath stones and logs during the day.

Parasitic flatworms often do not have much of a nervous system. As you can imagine, there is not much need for a nervous system in an organism that mainly hangs onto an intestinal wall and absorbs food! In fact, in tapeworms the nervous system has completely disappeared as the worms have adapted to their parasitic lifestyle.

Free-living flatworms usually use two means of locomotion at once. Cilia on their epidermal cells help them glide through the water and over the bottom. Muscle cells controlled by the nervous system allow them to twist and turn so that they are able to react to environmental conditions rapidly.

Reproduction in free-living flatworms can be either sexual or asexual. Most free-living flatworms are hermaphrodites, which means that they have both male and female organs. During sexual reproduction, the worms join in pairs. One worm delivers sperm to the other worm while receiving sperm from its partner at the same time. The eggs, which are laid in small clusters, hatch within a few weeks. Asexual reproduction by fission is also common among free-living flatworms. Most of these worms have incredible abilities of regeneration. In one form of asexual reproduction, a worm will simply "fall to pieces" and each piece will grow into a new worm! Parasitic flatworms do not reproduce asexually. They often have complicated life cycles, as you will see shortly.

Figure 26–27 Like planarians, marine flatworms belong to the class Turbellaria.

PLANARIANS The free-living flatworms belong to the class Turbellaria. The most familiar members of this class are planarians, the "cross-eyed" freshwater worms. Turbellarians vary greatly in color, form, and size. See Figure 26–27. Although most turbellarians are less than 1 centimeter in length, some giant land planarians, which are found in moist tropical areas, can attain lengths of more than 60 centimeters!

FLUKES The members of the class Trematoda are parasitic flatworms known as flukes. Some flukes are external parasites that live on the skin, mouth, gills, or other outside parts of a host. Most flukes, including the ones that affect humans, are internal parasites that infect the blood and organs. These flukes have complicated life cycles that involve at least two different host animals. Although many flukes are less than a centimeter long, the damage they cause to their host during their life cycle sounds like the script for a horror movie! Refer to Figure 26–28 as you read about the life cycle of a blood fluke. Keep in mind that the pattern of multiple hosts is typical of most parasitic flukes and, indeed, of many parasites in general.

Blood flukes are found primarily in Southeast Asia, North Africa, and other tropical areas. As you might expect, blood flukes live in the blood—specifically, the blood within the tiny blood vessels of the intestines. Humans are the primary hosts of blood flukes that belong to the genus *Schistosoma*. (The primary host of a parasite is the host organism in which adult parasites are found and in which sexual reproduction of the parasite occurs.)

Most flukes are hermaphrodites and undergo sexual reproduction in a manner similar to that of free-living flatworms. (However, the sexes are separate in *Schistosoma*.) Flukes produce many more eggs than free-living flatworms—about 10,000 to 100,000 times as many! Blood flukes lay so many eggs that the tiny blood vessels of the host's intestine break open. The broken blood vessels leak both blood and eggs into the intestine. The eggs are not digested by the host and thus become

Figure 26–28 The blood fluke Schistosoma mansoni *causes a serious human disease. The life cycle of the schistosome involves two hosts—humans and snails.*

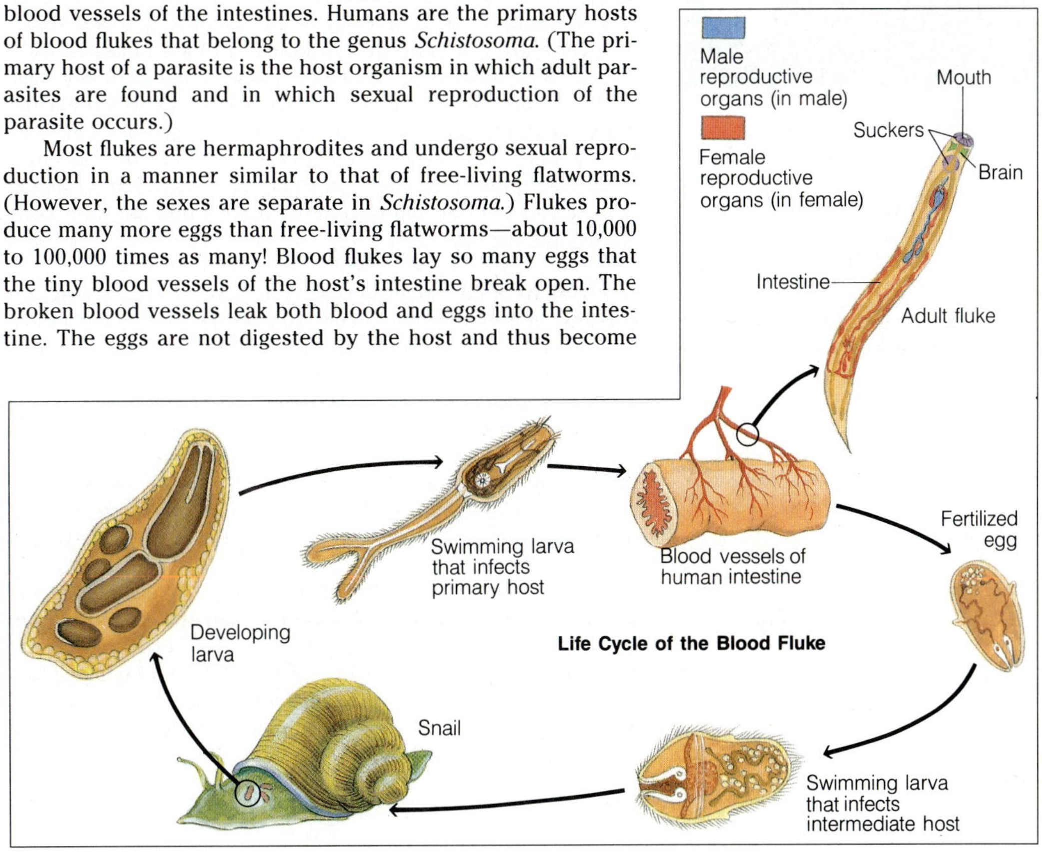

BACKGROUND INFORMATION
EVIDENCE OF INCREASED COMPLEXITY

The bodies of flatworms have clearly defined upper and lower surfaces, as well as clearly defined front and rear ends. The upper surface is called the dorsal surface, while the lower or underside surface is called the ventral surface. The front end (head area) of the animal is referred to as the anterior, and the rear is referred to as the posterior. These four areas of the body are most evident in more complex animals.

Another important characteristic of flatworms is that they have three cell layers—ectoderm, endoderm, and mesoderm. All three germ tissue layers form in the embryo. This indicates that flatworms are more complex than cnidarians, which have only two cell layers.

The ectoderm is the outer cell layer in the flatworm, and the endoderm is the inner layer. Between these two layers is the mesoderm. The mesoderm enables cells to develop independently of the ectoderm and the endoderm and gives rise in more complex animals to muscles, reproductive structures, bones, kidneys, and other internal organs and tissues.

the cells on other parts of the worm's body.)

Skills Development
Guided Practice
Skills: Predicting, observing
On the chalkboard, draw a diagram of a planarian cut into three sections—a head region, a middle region, and a tail region.
• **Suppose that we cut a worm into three pieces just as in this drawing. Assuming that we put each piece in a suitable environment, what do you think will happen?** (Answers may vary. The correct answer is that each part will regenerate.)

If time permits, take several specimens of live planarians and cut them in three pieces. This can be done by placing each planarian on a flat microscope slide, holding the slide over an ice cube to chill the worm, then using a scalpel or razor blade to cut the worm. Put each section in a separate petri dish labeled head, middle, tail. Fill each petri dish half full with aquarium water; cover the dishes and place them in a dark place. Have students observe the sections three times a week for three weeks and record what they see. Students can then compare their observations with their earlier predictions.

TEACHING SUPPORT

Teaching Resources
- Activity: Investigating Bilharzia—A Major Health Problem, p.5

BACKGROUND INFORMATION
FLATWORMS, CNIDARIANS, AND EVOLUTION

Scientists have discovered that the simplest turbellaria and the larval stages of flukes and tapeworms resemble the planula of the cnidarian life cycle. Planulae are formed from zygotes as part of the cnidarian reproductive process; eventually they grow into polyps.

This similarity between flatworms and planulae has led some scientists to hypothesize that ancient bilateral animals evolved from ancestors that were much like planulae. This may have occurred through increased cephalization and the emergence of tissues derived from the mesoderm.

If this idea is accurate, then planulalike organisms may have given rise to most groups of complex animals. So far, this idea has not been proven, but there is much evidence to support it.

26–4 (continued)

Content Development

Emphasize that tapeworms spread from one host to another by means of the proglottids. The proglottids, which contain mature fertilized eggs, break off the tapeworm and pass out of the host through the feces.

Skills Development

Guided Practice
Skills: Sequencing events, relating concepts

The following events in the life cycle of the beef tapeworm are written out of order. Have students arrange the events in the proper sequence by placing a number in front of each statement. The first statement, which is already numbered, is the beginning of the cycle.

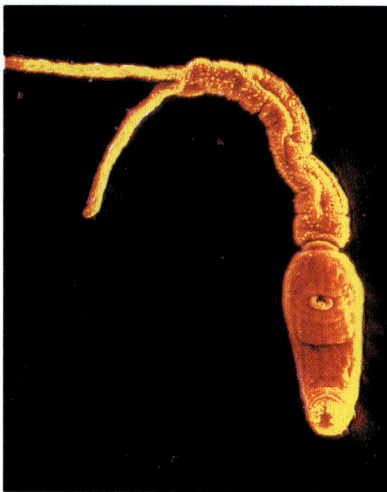

Figure 26–29 *In* Schistosoma mansoni, *the adult male is about 6 to 10 millimeters long and has a groove running the length of its body. The female, which is longer and thinner than the male, lives within this groove (top). If the schistosome larva shown here encounters a human, it will burrow through the skin, enter the bloodstream, and develop into an adult (bottom).*

part of the feces. In developed countries, where there are toilets and proper sewage systems, these eggs are usually destroyed in the sewage treatment process. But in many undeveloped parts of the world, human wastes are simply tossed into streams or even used as fertilizer.

Once the fluke eggs get into the water, they hatch into swimming larvae. When these larvae find a snail of the correct species, they burrow inside it and digest its tissues. The snail is an intermediate host for the fluke. Although sexual reproduction does not occur in an intermediate host, this host is still an essential part of the parasite's life cycle. In the intermediate host (in this case, a snail), the flukes reproduce asexually. The resulting new worms break out of the snail and swim around in the water. If they find a human, the worms bore through the skin and eat their way to the blood vessels. In the blood, they get carried around through the heart and lungs to the intestine, where they live as adults.

People infected with blood flukes get terribly sick. They become weak and often die—either as a direct result of the fluke infection or because they cannot recover from other diseases in their weakened condition. Blood flukes cause some of the most serious health problems in the world today. But because the species dangerous to humans live only in the tropics, most people in the United States know nothing about them—even though hundreds of millions of people suffer from blood flukes.

There are only one or two kinds of blood flukes in lakes and streams of the United States. These flukes normally have fishes or water birds as their primary hosts. If these worms find human swimmers, they try to burrow through the skin. This causes what is known as "swimmers itch." But because they are not adapted as human parasites, the worms cannot live in human bodies. The itch goes away after a time and the body repairs the damage.

TAPEWORMS Members of the class Cestoda are long, flat parasitic worms that live a very simple life. They have a head called a scolex (SKOH-leks) on which there are several suckers and a ring of hooks. These structures attach to the intestinal walls of humans and other animals. Inside the intestine, these worms are surrounded with food that their primary host has already digested for them. The worms absorb this food through their body walls. Adult human tapeworms can be up to 18 meters long! Tapeworms almost never kill their hosts, but they do use up a lot of food. For this reason, hosts may lose weight and become weak.

Behind the scolex of the tapeworm is a narrow neck region that is constantly dividing to form the many proglottids (proh-GLAH-tihds), or sections, that make up most of the body of the tapeworm. As you can see in Figure 26–30, the youngest and smallest proglottids are at the anterior (head) end of the tapeworm, and the largest and most mature proglottids are at the posterior (tail) end. Proglottids contain little more than male

_1_ Grass contaminated with tapeworm eggs is eaten by a cow.

___ Human eats poorly cooked meat.

___ Mature tapeworms produce eggs.

___ Eggs pass to cow's intestine and develop into immature worms.

___ Tapeworm eggs cling to grass and soil.

___ Human eliminates solid wastes containing tapeworm eggs.

___ Blood carries immature worms to cow's muscles (meat).

___ Immature worms develop

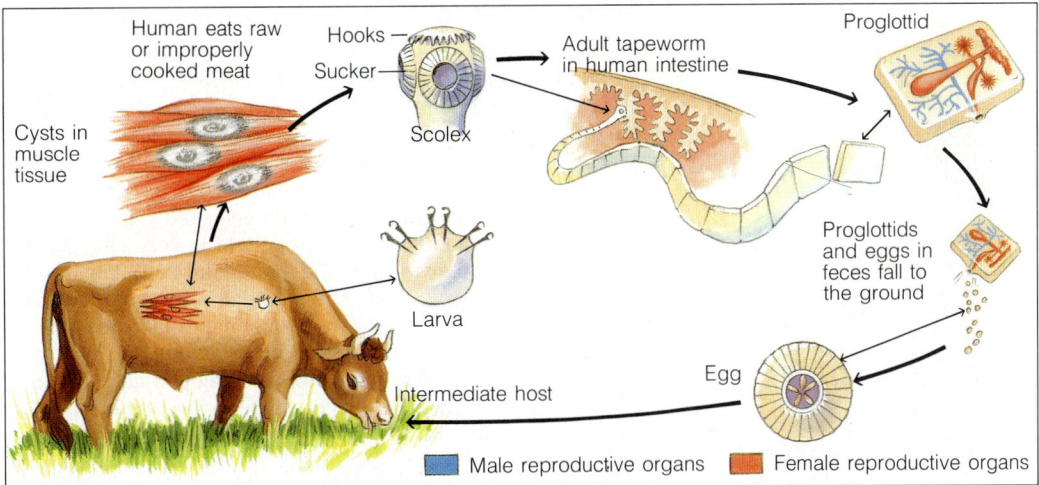

and female reproductive organs. Sperm produced by the testes, or male reproductive organs, can fertilize eggs in the proglottids of other tapeworms or of the same individual. Fertilized tapeworm eggs are released when mature proglottids break off the posterior end of the tapeworm and burst open. A mature proglottid may rupture either in the host's intestine or after it has been passed out of the host's body with the feces. A single proglottid may contain over 100,000 eggs, and a single worm can produce more than half a billion eggs each year!

If food or water contaminated with tapeworm eggs is consumed by cows, pigs, fishes, or other intermediate hosts, the eggs enter the intermediate host and there hatch into larvae. These larvae grow for a time and then burrow into the muscle tissue of the intermediate host and form a dormant protective stage called a cyst. If a human eats raw or incompletely cooked meat containing these cysts, the larvae become active within the human host. Once inside the intestine of the new host, they latch onto the intestinal wall and grow into adult worms.

Roundworms

Members of the phylum Nematoda, which are known as roundworms, are among the simplest animals to have a digestive system with two openings—a mouth and an anus. Food enters through the mouth, and undigested food leaves through the anus. Roundworms, which range in size from microscopic to a meter in length, may be the most numerous of all multicellular animals. It is difficult to imagine just how many roundworms there are around us all the time. A single rotting apple can contain as many as 90,000 roundworms! And a small bucketful of garden soil or pond water may house more than a million roundworms.

Figure 26–30 Cattle are secondary hosts to beef tapeworms; humans and other beef-eating animals are primary hosts.

Figure 26–31 The scolex, or head, of a tapeworm has suckers and other structures that enable it to attach to the inside of its host's intestine.

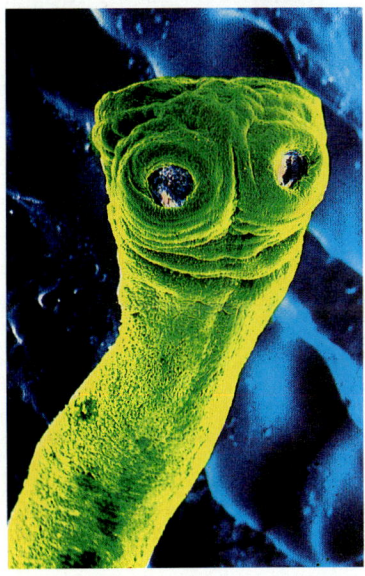

FACTS AND FIGURES

The largest tapeworm that affects humans is the broad fish tapeworm. This parasite can grow to about 18 meters long, with a body composed of up to 4000 sections. In addition to infecting humans, the broad fish tapeworm can also infect bears, dogs, foxes, and cats. Its presence in humans is most common in certain parts of the Baltic region, where almost the entire population is infected.

BACKGROUND INFORMATION
FLUKES AND PLANARIANS

As parasites, flukes exhibit many differences and some similarities when compared with the free-living planaria. Unique to the fluke is a thick outer layer of cells called a tegument. The tegument protects flukes from being digested by their hosts. Unlike free-living worms, flukes do not have muscles or cilia for movement. Instead, they have suckers that they use to attach themselves to their hosts. Flukes also lack specialized sense organs.

Flukes are similar to planarians in that they have similar excretory systems; they are also like planarians in that they are hermaphrodites. Unlike free-swimming worms, flukes cannot reproduce asexually. They must rely on a complicated life cycle that usually involves several hosts.

---

into mature tapeworms in human intestines.
___ Immature worms dig out of cow's intestine and enter cow's bloodstream.
(Answers: 5, 7, 2, 9, 8, 4, 6, 3)
After students have finished numbering the statements, ask:

• **How does the life cycle of the beef tapeworm point up the need for effective sewage treatment?** (Since the tapeworm eggs are eliminated in the feces of an infected human, proper treatment of sewage could destroy the eggs before they reach the grass or soil. This would break the cycle.)
• **What other precaution will break the cycle?** (Cooking meat thoroughly before it is eaten.)
• **At what stage in the cycle will this destroy the tapeworm?** (In the immature worm stage.)

• **Is there any way that the cycle can be broken by preventing cows from eating contaminated grass?** (There is no obvious way; however, students should feel free to offer any ideas they might have.)

BACKGROUND INFORMATION
RIBBON WORMS

Classified between phylum Platyhelminthes and phylum Nematoda are a group of animals known as ribbon worms (phylum Nemertea). At first glance, ribbon worms seem quite a bit like turbellarians. Their tissue organization is similar to flatworms in general, and like the free-living flatworms, they have external cilia. Yet ribbon worms are different from flatworms in that they have a complete gut and circulatory system and the sexes are nearly always separate. Also, nemerteans have a unique body structure—a proboscis that is used for capturing prey. The proboscis is equipped with glands that secrete a paralytic venom and with a penetrating device that is plunged into the prey.

BACKGROUND INFORMATION
THE PSEUDOCOEL

Instead of a true coelom lined with peritoneum, roundworms have a body cavity called a pseudocoelom, or false coelom. Most of the space in the pseudocoelom is taken up by reproductive organs. The fluid in the body cavity acts as a kind of circulatory system, distributing nutrients throughout the organism. The pseudocoelom also serves as a hydrostatic skeleton.

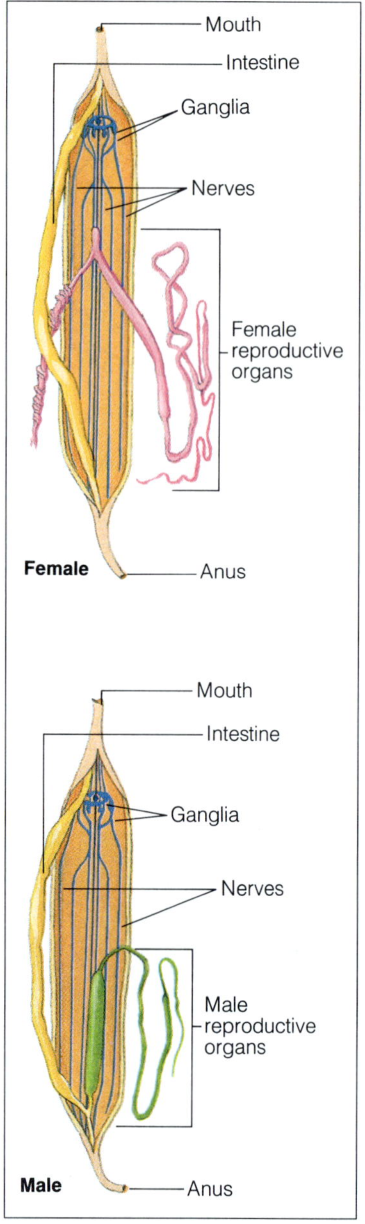

26–4 (continued)
Content Development

Several characteristics of roundworms are significant because they indicate a higher level of development than any of the animals studied thus far. One of these characteristics is a complete gut—that is, a digestive system with two openings, a mouth and an anus. The complete gut is a one-way gut in which food is taken in at the mouth and wastes are excreted at the anus.
• **How does this differ from the gut in flatworms and cnidarians?** (In these organisms, there is only one gut opening, the mouth. The mouth is used both to take in foods and to eliminate wastes. Thus the mouth serves as a two-way passage.)

Explain to students that the two-opening gut provides a more efficient digestive tract than the one-opening gut and offers greater possibilities for specializing in such tasks as grinding, digesting, and absorption. The two-opening gut is a characteristic of more highly developed animals.

Point out that another characteristic of roundworms that indicates a higher level of development is exclusively sexual reproduction. Explain that lower animals exhibit a combination of sexual and asexual reproduction, including regeneration.

Figure 26–32 *The internal organs of male and female ascarids are shown here. Ascarids, like other roundworms, have a digestive tract with two openings—a mouth and an anus.*

Form and Function in Roundworms

Most roundworms are free-living. Free-living roundworms are found in virtually all parts of the Earth—in soil, salt flats, and aquatic sediments; in polar regions and in the tropics; in fresh water, oceans, and hot springs. There are, however, many species of parasitic roundworms. Parasitic roundworms affect almost every kind of plant and animal.

All roundworms have a long tube-shaped digestive tract with openings at both ends. This system is very efficient because food can enter through the mouth and continue straight through the digestive tract. Any material in the food that cannot be digested leaves through an opening called the anus.

Free-living roundworms are often carnivores that catch and eat other small animals. Some soil-dwelling and aquatic forms eat small algae, fungi, or pieces of decaying organic matter. Some actually live on the organic matter itself. Others digest the bacteria and fungi that break down dead animals and plants. Many roundworms that live in the soil attach to the root hairs of green plants and suck out the plant juices. These parasitic worms cause tremendous damage to many crops all over the world. Roundworms are particularly fond of tomato plants. For this reason, many tomato plants have been specially bred to be resistant to roundworms. Other roundworms live inside plant tissues, where they cause considerable damage.

Like flatworms, roundworms breathe and excrete their metabolic wastes through their body walls. They have no internal transport system and thus depend on diffusion to carry nutrients and wastes through their body.

Roundworms have simple nervous systems. They have several ganglia, or groups of nerve cells, in the head region, but they lack anything that can really be called a brain. Although roundworms have several types of sense organs, these are simple structures that detect chemicals given off by prey or hosts. Several nerves extend from the ganglia in the head and run the length of the body. These nerves transmit sensory information and control movement. The muscles of roundworms run in strips down the length of their body walls. Aquatic roundworms contract these muscles to move like snakes through the water. Soil-dwelling roundworms simply push their way through the soil by thrashing around.

Roundworms reproduce sexually. Most species of roundworms have separate males and females, but a few species are hermaphroditic. Fertilization takes place inside the body of the female. Roundworms that are parasites on animals often have complex life cycles. Two or three hosts may be involved in the life cycle of some roundworms. In other roundworms, such as *Ascaris*, the stages of the life cycle take place in different organs of one host.

Ascaris is a parasitic roundworm that lives in humans. Species that are closely related to *Ascaris* affect horses, cattle, pigs, chickens, dogs, cats, and many other animals. *Ascaris* and its relatives, which are collectively known as ascarids, have life cycles that are similar to one another. One of the reasons puppies are wormed while they are young is to rid them of the ascarid that affects dogs.

Adult ascarid worms live in the intestines, where they produce many eggs that leave the host's body in the feces. If food or water contaminated with these feces is eaten by another host, the eggs hatch in the small intestine of the new host. The young worms burrow into the walls of the intestines and enter surrounding blood vessels. Carried around in the blood, the tiny worms end up in the lungs. Here they break out into the air passages and climb up into the throat, where they are swallowed. Carried back into the intestines, they mature and the cycle repeats itself.

How Unsegmented Worms Fit into the World

Unsegmented worms do not exert much positive influence on the daily lives of humans, and thus they are easy to ignore. Most unsegmented worms lead inoffensive lives. They eat small organisms and are eaten by larger organisms; some help aerate the soil with their burrows. However, unsegmented worms are generally known by the parasitic rather than the free-living members of their phylum. We have already talked about parasitic flatworms. In this section we shall focus our attention on parasitic roundworms, which are responsible for some of the most painful and horrific diseases known. Parasitic roundworms include hookworms, trichinosis-causing worms, filarial worms, eye worms, and a host of others too numerous to be mentioned here.

Hookworms are serious human intestinal parasites that are often found in the southern United States and are common in tropical countries. As many as one fourth of the people in the world today are infected with hookworm! Hookworm eggs hatch outside the body of the host and develop in the soil. If they find an unprotected foot, they use sharp teeth and hooks to burrow into the skin and enter the bloodstream. Like *Ascaris*, these worms travel through the blood to the lungs and then

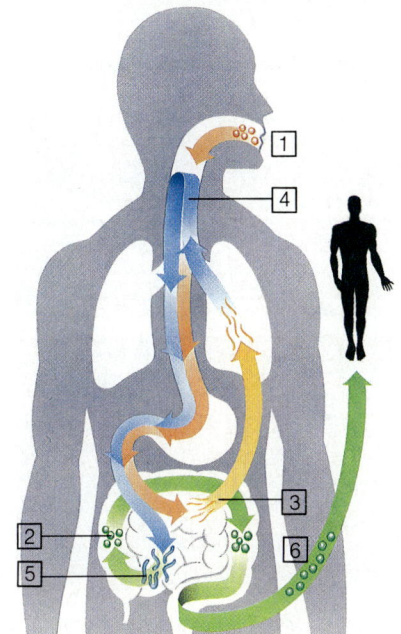

1	Eggs in food or water are ingested by host
2	Eggs hatch in small intestine
3	Larvae enter blood vessels and are carried to the lungs
4	Larvae travel to throat and are swallowed
5	Adult ascarid worms live in the small intestine
6	Eggs leave host in feces

Figure 26–33 The stages of the life cycle of the human ascarid, *Ascaris lumbricoides*, take place in several different host organs.

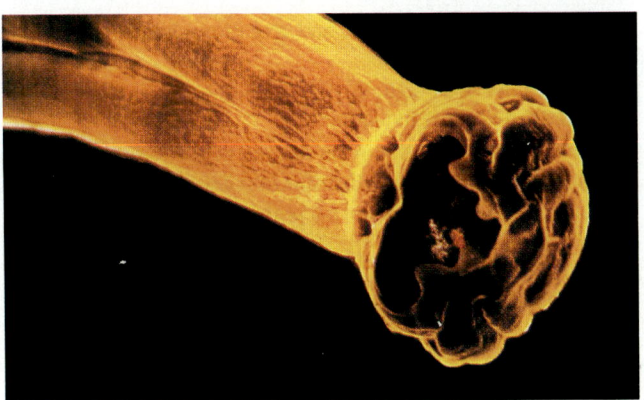

Figure 26–34 Hookworms use the sharp teeth and hooks on their anterior end to burrow through a host's skin.

BACKGROUND INFORMATION
SEXUAL REPRODUCTION IN ROUNDWORMS

The male and female sex organs in roundworms usually occur in different individuals. Male and female roundworms differ in size, shape, and color. The living arrangements of these worms can be very unusual. For example, in the case of one of the parasitic roundworms, the tiny male worm lives inside the reproductive structure of the female worm.

The female reproductive structures of roundworms include two ovaries, or egg-producing organs. The ovaries widen into oviducts, which are passages that lead into the uteri, or egg-storage organs. Each uterus opens into a tube, called a vagina, that leads to the outside of the body.

The male roundworm has a single sperm-producing organ called a testis. Sperm passes from the testis into a sperm duct. This duct leads into the cloaca. The cloaca is a chamber into which the digestive, excretory, and reproductive organs empty. During mating, sperm leaves the cloaca and enters the female through the vagina. Fertilization then occurs in the uterus.

Skills Development
Guided Practice
Skills: Observing, comparing
Set up a classroom display of live flatworms and roundworms so that students can observe and compare the two groups of animals. An appropriate flatworm would be a planarian; vinegar eels are roundworms that make excellent subjects for observation.

Place several planarians in a small amount of pond water or aquarium water in a petri dish. Have students use a hand lens or a stereo microscope to watch these animals move.

In order for students to observe the roundworms, place a few drops of vinegar eel culture in a depression slide or on a plain microscope slide. Do not use a coverslip. Have students observe the slide under low power.

- **How would you describe the movement of the planarians?** (They swim around the water.)
- **How would you describe the movement of the roundworms?** (They move with a rapid, jerky motion.)

TEACHING SUPPORT

Teaching Resources
- Vocabulary Review: Word Scramble, p. 1

Study Guide
- Section 26-4, p. 255

26-4 (continued)

Enrichment

Have interested students find out how some of the diseases caused by parasitic roundworms are treated medically. Students may also wish to contact veterinarians to find out which diseases caused by roundworms are likely to infect animals in your area.

SECTION REVIEW 26-4

1. The simplest kind of animal that exhibits bilateral symmetry, a member of phylum Platyhelminthes. Turbellaria: planarian. Trematoda: *Schistosoma*. Cestoda: tapeworm. Students' examples will vary.
2. Parasitic flatworms have special structures for attachment to a host and for getting from one host to another. In addition, the reproductive system tends to be more highly developed in a parasitic flatworm, whereas the nervous and digestive systems are more highly developed in free-living flatworms.
3. An animal with a cylindrical, tapered body and a two-opening digestive system that belongs to phylum Nematoda. Roundworms have a digestive system with two openings—a mouth and an anus—whereas flatworms have a digestive system with only one opening.
4. Feeding: may be carnivorous, herbivorous, detritus feeders, or parasites. Respiration: diffusion through the body surface. Internal transport: diffusion through the body surface; material may be moved about in the body cavity in roundworms. Excretion: diffusion through the body surface. Response: possesses a nerve cord and cerebral ganglia. Movement: many flatworms use cilia; roundworms use muscles that run the length of the body. Reproduction: some flatworms reproduce asexually by fission; sexual reproduction occurs in both flatworms and roundworms.
5. Cooking kills parasitic worms that may be in meat.

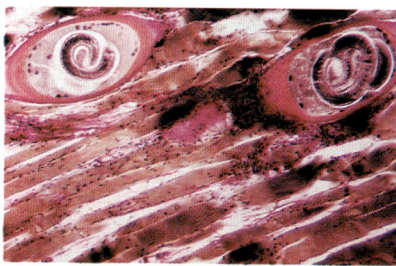
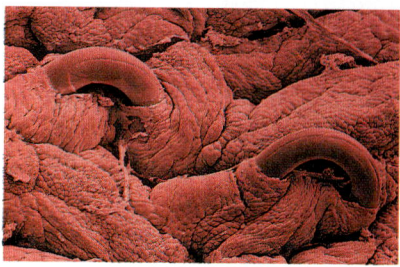

Figure 26-35 Trichinella *worms, which cause the disease trichinosis, form cysts in the muscle tissue of their host (top). These threadworms, tunneling through the tissues of a sheep's intestine, are parasitic roundworms (bottom).*

down the throat to the intestines. There, the adult worms dig into the intestinal wall and suck the blood of the host. These worms can devour enough blood to cause weakness and poor growth.

Trichinosis (trihk-ih-NOH-sihs) is a terrible disease caused by the roundworm *Trichinella*. Adult worms, which are hard to see without a microscope, live and mate in the intestines of the host. Females carrying fertilized eggs burrow into the intestinal wall, where each releases up to 1500 larvae. These larvae travel through the bloodstream, from which they eventually exit through small blood vessels, and then burrow into organs and tissues. This causes terrible pain for the host. The larvae then form cysts in the host's muscle tissue and become inactive.

The only way these encysted worms can complete their life cycle is if infected muscle tissue is eaten. This means that hosts for *Trichinella* must be carnivorous—animals that do not eat infected meat do not get trichinosis. Two very common hosts for *Trichinella* are rats and pigs. (Rats eat any meat they can find, and may even eat each other. Pigs regularly catch and eat rats and other small animals.) Humans get trichinosis almost exclusively by eating raw or incompletely cooked pork.

Filarial worms, which are found primarily in tropical regions of Asia, are threadlike worms that live in the blood and lymph vessels of birds and mammals such as humans. They are transmitted from one primary host to another through biting insects, especially mosquitoes. In severe infections, large numbers of filarial worms may block the passage of fluids within the lymph vessels. This causes elephantiasis, a condition in which an affected part of the body swells enormously. Fortunately, extreme cases of elephantiasis are now rare.

Eye worms are closely related to the filarial worms that cause elephantiasis. They are found in Africa and affect both humans and baboons. Eye worms live in and burrow through the tissues just below the skin of their host. In their travels, the worms occasionally move across the surface of the eye—hence the name eye worm.

26-4 SECTION REVIEW

1. What is a flatworm? Name and give examples of the three classes of flatworms.
2. How do the body structures of parasitic flatworms differ from those of free-living forms?
3. What is a roundworm? What are the major differences in structure between roundworms and flatworms?
4. How do unsegmented worms perform essential functions?
5. **Connection—Health** Explain why you should cook meat and fish thoroughly in areas that have parasitic worms.

Reinforcement/Reteaching

If students are having trouble answering the Section Review questions, go back to the material with which they are having difficulty.

SCIENCE, TECHNOLOGY, AND SOCIETY
BREAKTHROUGH

River Blindness: A Lifelong Battle Almost Won

The sight is a familiar one in many parts of West Africa: A child leads an adult along the banks of a river. The adult, like many others in the village, is blind—a victim of the disease onchocerciasis, or river blindness. It has been called river blindness because the tiny black flies that spread the disease breed in fast-moving water. River blindness affects an estimated 18 million people living in Africa and the Middle East, more than 300,000 of whom have been blinded.

River blindness is caused by a parasitic roundworm that enters the body when a black fly, which has picked up the roundworm by biting an infected human, bites another victim. The roundworm larvae deposited by the black fly quickly grow into threadlike adult worms, which can live under the skin for as long as 12 years. It is not the adult worms that cause this dreadful disease but their offspring—millions of microworms that swarm through the skin and eyes.

Blindness is not the only effect of this disease. As the microworms migrate under the skin, intolerable itching results. Over time, the skin begins to decay and often loses its pigment.

The scourge of river blindness has economic implications as well. When the rate of blindness in a village becomes significant, fearful young people abandon their homes. Farm production in fertile river valleys is curtailed because there are limited laborers to grow and harvest the crops.

Since 1974, when an ambitious effort to reduce the numbers of black flies was undertakin, the World Health Organization (WHO) has been battling this disease with limited success. Spraying with an ecologically safe insecticide has halted the transmission of river blindness in certain areas to some extent. But complications have developed. Some insects have become resistant to the available insecticides. And several areas once cleared of the black flies have been reinvaded as the insects prove to be more mobile than expected.

What is giving WHO and victims of river blindness cause to rejoice is the arrival of ivermectin. Developed in the 1970s as a weapon against worm parasites in livestock, ivermectin has been shown in a series of human trials to be an effective weapon against river blindness. Although ivermectin does not kill the parasitic roundworm, it does destroy the microworm offspring. And it also appears to inhibit, for a time, the production of more offspring.

Though not a total cure, ivermectin's advantages are obvious. Taken in pill form as infrequently as once a year, it protects those already infected from the worst symptoms. By temporarily ridding a victim's skin of microworms, ivermectin slows the transmission of the disease by preventing the flies that bite the victim from picking up the parasitic roundworm. And ivermectin is so safe that it can be dispensed in mass campaigns in isolated villages rarely visited by doctors.

With ivermectin now easily available, those affected by river blindness in one way or another can look to the future with hope. Although the drug cannot restore the sight of victims of the disease, it can spare hundreds of thousands of children from this scourge.

SCIENCE, TECHNOLOGY, AND SOCIETY
BREAKTHROUGH

River Blindness: A Lifelong Battle Almost Won

After students have read this feature, pose the following questions:

• **What intermediate host is used to transfer the parasitic roundworm that causes river blindness?** (A black fly.)

• **How does the black fly transfer the parasite?** (By biting an infected human, ingesting the roundworm larvae, then biting another human.)

• **Would you expect a person bitten by this fly to become blind right away? Why or why not?** (No, because it takes time for the adult worms to produce the offspring that cause the disease.)

• **Before the development of ivermectin, how did people attempt to prevent and control river blindness?** (By trying to eliminate the black flies.)

• **What does ivermectin do?** (It destroys the microworm offspring that live under the skin of victims.)

• **How does this prevent transmission of the disease?** (Black flies cannot pick up worms from the skin of an infected person who has taken ivermectin. Thus transmission of the disease is slowed or stopped.)

Closure

Write on the chalkboard the following list of topics:
- General characteristics of flatworms
- General characteristics of roundworms
- Evidence in flatworms of higher development as compared with cnidarians
- Evidence in roundworms of higher development as compared with flatworms
- Feeding and digestion in flatworms and roundworms
- Nervous system in flatworms and roundworms
- Reproduction in flatworms and roundworms
- Parasitic flatworms
- Parasitic roundworms
- Diseases caused by flatworms in humans
- Diseases caused by roundworms in humans
- Body plan of typical flatworm
- Body plan of typical roundworm

Have students work in pairs. Assign each pair one of the topics listed above, then ask them to prepare an oral, written, or visual summary of the topic. When everyone is finished, have a class discussion in which each pair shares their summary and answers questions posed by other members of the class.

LABORATORY INVESTIGATION

COLLECTING AND STUDYING ROUNDWORMS

Before the Lab
1. At least one day prior to the investigation, gather enough materials for your class, assigning 6 students to each group.
2. To prepare methylene blue stain, dissolve 0.3 g of methylene blue in 30 mL of absolute (95%) alcohol. Add 100 mL of distilled water. Set the solution aside for 24 hours.
3. Note that the investigation requires a 24-hour waiting period between steps 5 and 6. Be sure to plan your class time accordingly.

Pre-Lab Discussion
Read or have a student read the Problem at the beginning of the investigation. Encourage students to offer possible answers to the question. Then have students write their ideas in the form of a hypothesis. When the investigation is over, have students compare their hypothesis with the conclusions they have reached as a result of the investigation.

Have students read through the first five steps of the laboratory procedure. You may wish to demonstrate at this time the assembly of the apparatus for collecting roundworms.

Have students read through the remaining steps of the procedure. Answer any questions they might have. If necessary, review the proper technique for staining a slide with methylene blue.

Skills Development
Students will use the following skills while completing this laboratory investigation.
1. Observing
2. Relating
3. Hypothesizing
4. Diagramming
5. Manipulative
6. Inferring
7. Applying

Safety Tips
1. If glass funnels are being used, remind students to use caution when inserting the funnel into the rubber tubing in order to avoid cracking or breakage.
2. Remind students of the safety rules for handling glassware. Make sure that students know they must report any broken or cracked glassware immediately.
3. Remind students how to properly use the high-power objective of the microscope.

Teaching Strategy
1. As students complete the first five steps of the procedure, be sure that each group has set up the apparatus correctly.

LABORATORY INVESTIGATION
COLLECTING AND STUDYING ROUNDWORMS

PROBLEM
How do roundworms move?

MATERIALS (per group)

2 150-mL beakers	paper towel
cheesecloth	ring stand
coverslip	10-cm rubber tubing
depression slide	rubber band or
funnel	twist tie
pinch clamp	scissors
ring clamp	soil
2 medicine droppers	vital methylene blue
	microscope

PROCEDURE

1. Assemble the apparatus for collecting roundworms as shown in the accompanying diagram.
2. Using scissors, cut a piece of cheesecloth with dimensions of approximately 30 cm by 15 cm. Fold the cheesecloth over to make a square.
3. Put a handful of soil in the center of the cheesecloth and pull the corners together to make a small bag. Tie the bag closed with a rubber band or twist tie.
4. Using a beaker, pour some water into the funnel to make sure the pinch clamp does not leak. Once you are certain the pinch clamp works properly, place the bag of soil in the funnel. Fill the funnel the rest of the way with water, making sure that the bag is submerged.
5. Leave the apparatus undisturbed for about 24 hours.

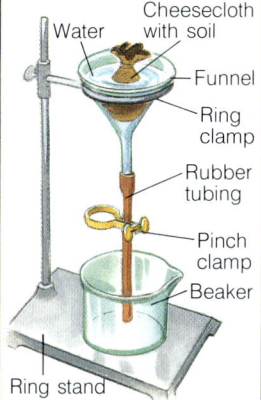

6. Open the pinch clamp briefly, allowing only a small amount of water to empty into the beaker below.
7. Using a dropper, put a few drops of water from the beaker in the center of a clean depression slide. Cover the water with a coverslip.
8. With the microscope set on low power, locate some roundworms on the slide. Observe how the roundworms move.
9. Using a clean dropper, put a drop of vital methylene blue at one edge of the coverslip. Hold a piece of paper towel at the opposite edge of the coverslip to draw the vital methylene blue underneath.
10. Locate a stained roundworm. Switch to high power and focus on the stained roundworm using the fine adjustment.
11. On a separate sheet of paper, draw a diagram of the roundworm you observed under high magnification.

OBSERVATIONS
1. Describe the appearance of a roundworm.
2. Describe how roundworms move.

ANALYSIS AND CONCLUSIONS
1. Based on the way the roundworms move, what can you infer about the arrangement of muscles in roundworms?
2. Are roundworm movements more effective in soil than they are in the water on the slide? Explain.
3. Explain how the apparatus used in this investigation helps in the collection of roundworms. (*Hint:* Do soil roundworms seem capable of swimming against gravity?)
4. Based on your answer to question 3, would you expect to find more or fewer roundworms in subsequent samples of water from the funnel? Explain.

STUDENT STUDY GUIDE

SUMMARIZING THE CONCEPTS

The key concepts in each section of this chapter are listed below to help you review the chapter content. Make sure you understand each concept and its relationship to other concepts and to the theme of this chapter.

26-1 Introduction to the Animal Kingdom

- Animals are multicellular eukaryotic heterotrophs whose cells lack cell walls. Invertebrates are animals that lack a backbone.
- Essential functions for life include feeding, respiration, internal transport, elimination of waste products, response to environmental conditions, movement, and reproduction.
- Evolutionary trends in animals include performing essential functions at higher levels of organization, moving from radial to bilateral symmetry, and increasing cephalization.

26-2 Sponges

- Sponges belong to the phylum Porifera. Sponges are simple organisms that lack tissues and organs.

26-3 Cnidarians

- Cnidarians are aquatic animals that exhibit radial symmetry and stinging structures called nematocysts on their tentacles. Many cnidarians have two body forms in their life cycles—a flowerlike polyp and a bell-shaped medusa.

26-4 Unsegmented Worms

- Unsegmented worms include phylum Platyhelminthes and phylum Nematoda.
- Flatworms are the simplest animals with bilateral symmetry.
- Roundworms have a digestive tract with two openings. Parasitic roundworms cause a variety of diseases in humans and other animals.

REVIEWING KEY TERMS

Vocabulary terms are important to your understanding of biology. The key terms listed below are those you should be especially familiar with. Review these terms and their meanings. Then use each term in a complete sentence. If you are not sure of a term's meaning, return to the appropriate section and review its definition.

26-1 Introduction to the Animal Kingdom
- vertebrate
- invertebrate
- division of labor
- herbivore
- carnivore
- parasite
- filter feeder
- detritus feeder
- larva
- metamorphosis
- radial symmetry
- bilateral symmetry
- anterior
- posterior
- dorsal
- ventral
- cephalization
- ganglion

26-2 Sponges
- Porifera
- collar cell
- osculum
- spicule
- amebocyte
- spongin
- gemmule
- budding

26-3 Cnidarians
- Cnidaria
- polyp
- medusa
- gastrovascular cavity
- nematocyst
- hermaphrodite

26-4 Unsegmented Worms
- unsegmented worm
- Platyhelminthes
- flatworm
- Nematoda
- roundworm
- pharynx

---

2. As students isolate and observe the roundworms as in steps 6 to 11, circulate throughout the room to help any group that is having trouble.

3. At the conclusion of the investigation, have students display their diagrams. Discuss any differences or variations that may be evident.

Observations

1. Roundworms are small, transparent, long, and threadlike. One end is slightly more rounded than the other.

2. Roundworms move by thrashing back and forth, bending from one side to the other.

Analysis and Conclusions

1. Roundworms have only longitudinal (lengthwise) muscles.

2. Yes. In the water, roundworms have nothing to press against, which makes their type of motion ineffective.

3. Since roundworms cannot swim well, when they enter the water in the funnel from the soil, they tend to sink to the bottom and become concentrated just above the hose clamp.

4. Fewer. Since most of the roundworms were concentrated near the base of the funnel, each subsequent sample would contain fewer than the previous one.

Going Further: Enrichment

Have students collect flatworms by tying a piece of string around a piece of liver and lowering it into a pond or lake. After a few hours, have them raise the string; flatworms should be attached to the liver. Have students design an investigation in which they observe and diagram the flatworms, then explore how flatworms respond to various stimuli.

CHAPTER REVIEW

CONTENT REVIEW

Multiple Choice

Choose the letter of the answer that best completes each statement.

1. All animals are
    a. unicellular.
    b. sessile.
    c. radially symmetric.
    d. heterotrophic.
2. A hydra is best described as a
    a. herbivore.
    b. carnivore.
    c. parasite.
    d. filter feeder.
3. In which animal would you expect to observe cephalization?
    a. jellyfish
    b. sponge
    c. roundworm
    d. sea anemone
4. Which animal is free-living?
    a. *Hydra*
    b. *Trichinella*
    c. *Schistosoma*
    d. *Ascaris*
5. Animals in the phylum Cnidaria include
    a. flukes.
    b. roundworms.
    c. medusae.
    d. sponges.
6. Which animal is most likely to possess ocelli, statocysts, and a nerve net?
    a. sponge
    b. jellyfish
    c. coral
    d. flatworm
7. Which animal lacks a digestive system and digestive organs?
    a. jellyfish
    b. hookworm
    c. planarian
    d. tapeworm
8. An immature animal that looks and acts nothing like the adult of that species is called a
    a. gemmule.
    b. larva.
    c. bud.
    d. proglottid.

True or False

Determine whether each statement is true or false. If it is true, write "true." If it is false, change the underlined word or words to make the statement true.

1. <u>Invertebrates</u> have a backbone.
2. Organisms that eat animals are called <u>herbivores</u>.
3. Flukes and tapeworms are best described as <u>detritus feeders</u>.
4. Trichinosis is usually caused by <u>eating flukes in raw fish</u>.
5. Planarians have <u>bilateral</u> symmetry.
6. <u>Sea anemones</u> are polyps that have skeletons of calcium carbonate (limestone).
7. Adult parasites undergo sexual reproduction in their <u>intermediate</u> host.
8. Jellyfish are placed in the class <u>Anthozoa</u>.

Word Relationships

In each of the following sets of terms, three of the terms are related. One term does not belong. Determine the characteristic common to three of the terms and then identify the term that does not belong.

1. spicule, ganglia, osculum, collar cell
2. nematocyst, epidermis, gastroderm, mesoglea
3. tapeworm, hookworm, ascarid, planarian
4. Porifera, Cestoda, Cnidaria, Nematoda
5. dorsal, ventral, anterior, sessile
6. Anthozoa, Protozoa, Scyphozoa, Hydrozoa
7. multicellular, heterotroph, eukaryotic, cell walls

CHAPTER REVIEW

CONTENT REVIEW

Multiple Choice
1. d 2. b 3. c 4. a
5. c 6. b 7. d 8. b

True or False
1. F Vertebrates
2. F carnivores
3. F parasites
4. F eating encysted roundworms in meat
5. T
6. F Corals
7. F primary
8. F Scyphozoa

Word Relationships
1. parts of a sponge; ganglia are clusters of nerve cells
2. body layers of a cnidarian; nematocyst is a stinging structure found in cnidarians
3. parasitic unsegmented worms; planarian is a free-living flatworm
4. phyla of invertebrates; Cestoda is a class in phylum Platyhelminthes
5. regions of the body; sessile refers to an organism that spends its adult life attached to one spot
6. classes of cnidarians; protozoa are animallike protists
7. characteristics of animals; cell walls

CONCEPT MASTERY

1. The drawings of the human and of the dog should have the following labels: bilateral symmetry, anterior, posterior, dorsal, ventral, motile. The drawing of the sea anemone should have the following labels: radial symmetry, sessile. Check students' drawings to make sure the placement of labels is correct.
2. The blue dye was drawn into the pores of the sponge by the water current created by the movement of the collar cells' flagella. The red dye was also taken into the sponge in this manner. However, the red dye was absorbed by the collar cells. If the sponge were cut open, it would probably have particles of red dye in the collar cells and amebocytes.
3. (a) Flukes have suckers, and tapeworms have suckers and/or hooks. (b) Unlike free-living flatworms, flukes and tapeworms typically lack eyes. (c) Flukes have modified mouthparts, and tapeworms absorb food through the body walls. (d) Both flukes and tapeworms produce many more offspring than free-living flatworms. Almost all of a tapeworm's body consists of reproductive organs. (e) Fluke larvae are adapted for swimming and burrowing into the body of a host. Tapeworms form cysts in the muscle tissue of their secondary hosts, which are then eaten by the primary host.

CONCEPT MASTERY

Use your understanding of the concepts developed in the chapter to answer each of the following in a brief paragraph.

1. Draw a human, a sea anemone, and a dog. Label each drawing using as many of the following terms as are appropriate: radial symmetry, bilateral symmetry, anterior, posterior, dorsal, ventral, sessile, motile.
2. Suppose you placed a harmless purple-colored mixture of red dye and blue dye in the water beside a vase-shaped sponge. After a while, you noticed blue dye coming out of the top of the sponge. Describe how the blue dye got from the outside environment into the sponge. Propose an explanation for what happened to the red dye. How might you determine if your explanation about the red dye is correct?
3. Explain how flukes and tapeworms display the following parasitic adaptations: (a) organs for attachment to the host, (b) reduced sense organs, (c) modifications in food-getting, (d) increased reproductive capabilities and well-developed reproductive organs, (e) larvae that allow the transfer from one host to another.
4. At one time, diet pills containing tapeworm eggs were sold. Why would such pills work? Why are such pills dangerous?
5. State three basic trends in animal evolution in your own words.

CRITICAL AND CREATIVE THINKING

Discuss each of the following in a brief paragraph.

1. **Interpreting diagrams** Refer to the diagram of the life cycle of a typical liver fluke to explain the following: To help prevent liver fluke infections, experts often recommend that ponds, irrigation ditches, and other bodies of water be treated with snail-killing pesticides. Why does killing snails prevent liver fluke infections in humans?
2. **Relating concepts** Flukes that are internal parasites are often facultative anaerobes. This means that although they can use cellular respiration to obtain energy from food, they usually use anaerobic processes (glycolysis and fermentation) instead. Explain how this metabolic switch hitting might be an adaptation of flukes to a parasitic lifestyle.
3. **Developing a hypothesis** You observe that a hydra that lives in fresh water often squirts water out of its mouth. Because this water does not contain particles, you assume that the hydra's behavior is not involved with the removal of solid wastes. How can you explain this behavior?
4. **Using the writing process** Write a humorous dialogue in which a person tries to explain to a tapeworm that there is no such thing as a free lunch.

583

4. These pills worked because tapeworms "steal" food from the host. However, these pills cause the complications associated with parasite infestations. In addition, larval tapeworms form cysts that can be harmful and even fatal if located in vital organs such as the brain or heart.
5. Essential life functions tend to be carried out at higher levels of organization in the more complex animals. Bilateral symmetry is associated with more complex animals (although there are exceptions). More complex animals tend to have a concentration of nerve tissues and sense organs in the head. Make sure students rephrase these statements in their own words.

CRITICAL AND CREATIVE THINKING

1. Killing snails deprives liver flukes of their secondary host and thus prevents the flukes from completing their life cycle.
2. There is not much oxygen available in the body of the host in which the flukes reside.
3. The hydra "spits out" excess water that tends to collect in its tissues through osmosis. The gastrovascular cavity serves the same function as a contractile vacuole in a protist.
4. Students' essays should be creative and well-written.

583

CHAPTER 27 Mollusks and Annelids

Section	Laboratory Investigations and Activities
27–1 Mollusks pages 585–593 Themes: Scale and Structure, Unity and Diversity	**Teacher Edition** DEMONSTRATION: Filter Feeding, p. 584D **Laboratory Manual** Examining a Clam, p. 345 **Teaching Resources** HANDS-ON ACTIVITY: Moving at a Snail's Pace, p. 139
27–2 Annelids pages 594–601 THEMES: Scale and Structure, Unity and Diversity	**Student Edition** LABORATORY INVESTIGATION: Observing Earthworm Responses, p. 602 **Teacher Edition** DEMONSTRATION: Observing Earthworms, p. 584D **Laboratory Manual** The Earthworm, p. 351 **Teaching Resources** ECOLOGY INVESTIGATION: Effects of Earthworms on Soil, p. 35
Chapter Review pages 603–605	

Teacher Resources

Books

Bratcher, Twila, and Walter Cernohorsky. *Living Terebras of the World.* Abbott, R. T., ed. American Malacologists, 1987.

Edwards, C. A., and J. F. Loffy. *Biology of Earthworms,* 2nd ed. Routledge Chapman and Hall, 1977.

Honer, Jay V., and E. Evan Brown. *Crustacean and Mollusk Aquaculture in the United States.* AVI, 1985.

Lee, Kenneth F. *Earthworms: Their Ecology Relationships with Soils and Land Use.* Academic Press, 1985.

Russell-Hunter, W. D., ed. *The Mollusca: Ecology,* vol. 6. Academic Press, 1983.

Wilbur, K. M., and C. M. Yonge. *The Mollusca: Reproduction,* vol. 7. Academic Press, 1984.

CHAPTER PLANNING GUIDE

Other Activities	Media and Technology
Teaching Resources ACTIVITY: Mollusks in Our World, p. 3 ACTIVITY: Investigating the Behavior of an Octopus, p. 5 **Study Guide** Section 27–1, p. 259	**Videodisc** Aquatic Ecosystems: Freshwater Wetlands and Freshwater Aquatic Ecosystems: Estuaries and Marine **Biology Transparencies** 31: The Anatomy of a Clam **Biology Media Guide** pages 28–29
Teaching Resources VOCABULARY REVIEW: Word Game, p. 1 **Study Guide** Section 27–2, p. 261	**Videodisc** Aquatic Ecosystems: Estuaries and Marine Aquatic Ecosystems: Freshwater Wetlands and Freshwater **Biology Transparencies** 32: The Anatomy of a Marine Sandworm **Biology Media Guide** page 29
Teaching Resources Chapter Test A, p. 13 Chapter Test B, p. 17	**Computer Test Bank** Chapter Test, p. 287

Audiovisuals

The Earthworm: Darwin's Plow. Film or video. Coronet.
Mollusks. Film or video. Ward.
Modern Biology Series: Mollusks: The Mussel Respiration and Digestion. Film. Benchmark Films.

Annelids. Film or video. Ward.
The Biology of Mollusks. Film or video. Lucerne Media.

The Biology of Annelids. Film or video. Lucerne Media.

CHAPTER 27 Mollusks and Annelids

CHAPTER OVERVIEW

In this chapter students will continue their study of invertebrate animals as they learn about mollusks and annelids. Biologists believe that members of these two groups are closely related, for both show marked similarities in their early developmental stages.

Students will learn that mollusks, which belong to phylum Mollusca, include such familiar animals as snails, slugs, clams, oysters, scallops, octopi, and squid. They will discover that among the various classes of mollusks are animals with one shell, such as snails; animals with two shells, such as clams; and animals with a small internal shell or no shell, such as octopi. Students will learn how different types of mollusks perform essential life functions, and they will also find out about the many ways in which mollusks are important to humans.

In the section devoted to annelids, students will learn about the members of phylum Annelida. These animals, which are also known as segmented worms, include the familiar earthworm as well as about 9000 other species, such as sandworms, bloodworms, and leeches. Students will learn about form and function in segmented worms, and they will learn in detail about the various body systems of the earthworm. They will also discover how annelids are important to humans and to living systems in general.

27–1 MOLLUSKS

Section Focus 27–1

The purpose of this section is to pose and answer the question, "What is a mollusk?" Students will discover that a wide diversity of animals make up the phylum Mollusca. They will learn that these animals are classified together in one phylum because all show similar features during early development and all exhibit different forms of the same basic body plan.

Students will learn how mollusks feed, carry out respiration, reproduce, carry out internal transport, eliminate wastes, and respond to environmental conditions. Included will be a discussion of the feeding organ called the radula that is found in some mollusks, as well as a discussion of the open or closed circulatory systems that are found in mollusks.

Students will learn that within the phylum Mollusca, animals are classified on the basis of the number and type of shells they have. One-shelled mollusks include snails, slugs, and their relatives; two-shelled mollusks include clams, oysters, and scallops; and tentacled mollusks that have little or no shell include octopi, squid, and the chambered nautilus.

Students will learn that mollusks affect humans in a variety of ways. Many mollusks are popular as food, and the oyster is important economically not only as a food source but as a producer of pearls. Students will discover that some mollusks have a negative impact on humans by serving as intermediate hosts for parasites and by causing damage to gardens and crops.

Performance Objectives 27–1

1. List the general characteristics of mollusks.
2. Identify some familiar examples of mollusks.
3. Explain how mollusks carry out essential life functions.
4. Compare and contrast gastropods, bivalves, and cephalopods.
5. Discuss ways in which mollusks affect humans.

Science Terms 27–1

mollusk   p. 585
foot   p. 586
mantle   p. 586
shell   p. 586
visceral mass   p. 586
radula   p. 587
gill   p. 587
open circulatory system   p. 588
closed circulatory system   p. 588
nephridium   p. 588
gastropod   p. 590
bivalve   p. 590
cephalopod   p. 592

27–2 ANNELIDS

Section Focus 27–2

The purpose of this section is to introduce students to members of the phylum Annelida. These animals are known as annelids, or segmented worms.

Students will discover that the familiar earthworm is but one of about 9000 species that are classified in this phylum. They will learn that annelids are characterized by a long segmented body and that they live both in water and on land.

Students will learn how annelids carry out essential life functions. They will study the earthworm's body systems in detail as an example of the typical form and function among this group of animals.

In addition to earthworms, other species, such as sandworms and leeches, will be discussed. The effect of leeches on humans will be covered in detail, and a fascinating feature at the end of the section will describe how leeches are now being used as valuable aids in modern medicine.

Performance Objectives 27–2

1. List and discuss the basic characteristics of annelids.
2. Explain how annelids carry out essential life functions.
3. Describe in detail the body systems of the earthworm.
4. Discuss some important characteristics of polychaetes and leeches.
5. Discuss how annelids affect living systems in general and humans in particular.

CHAPTER PREVIEW

Science Terms 27-2
annelid p. 594
polychaete p. 598
oligochaete p. 599
leech p. 599

DISCOVERY LEARNING

TEACHER DEMONSTRATIONS
Modeling

Filter Feeding
This demonstration can be performed as an introduction to Chapter 27 or as students learn about bivalve mollusks in Section 27-1. For the demonstration you will need a sample of water from a pond, lake, stream, or ocean; several beakers; a sieve; a coffee filter; a piece of screen or wire mesh; a piece of cheesecloth or similar fabric; and several microscopes.

Divide the pond water sample into several parts. Filter each part using one of the filtering devices just mentioned. Have students observe the materials that remain after the water has been filtered, first with the unaided eye, then with a microscope.

• **Do you think that any of these materials might be useful to a sea organism as food?** (Answers may vary. In most cases, the correct answer will be yes.)
• **How does the use of different filtering materials affect the type of particles that remain after filtering?** (Depending on the size of the openings in the filtering device, the remaining particles vary according to size.)

Have students relate what they have observed of the filter-feeding mechanism of the clam. Point out that the gills of the clam are able to trap particles of exactly the right size so that the clam can obtain the type of food it needs.

Observing Earthworms
The following demonstration can be used as an introduction to Section 27-2.

Earthworms are among the most familiar of organisms, yet probably few students have taken the time to observe earthworms closely. In this simple demonstration, students can observe, as a class or in groups, the basic physical characteristics of earthworms.

For the demonstration you will need a clear plastic box; some sand; some topsoil; pond water or, if pond water is not available, tap water that has been standing for a day; 6 to 12 earthworms; and clear plastic wrap.

Fill the plastic box with about 2 centimeters of sand. Place about 7 centimeters of loosely packed topsoil over the sand. Use the pond or tap water to slightly moisten the soil. (Add more water whenever the soil appears dry.) Place the earthworms on top of the soil. Cover the box with clear plastic wrap and put in a few air holes.

Have students observe the earthworms for several days. Ask them to make diagrams of earthworms and to note especially how earthworms move through the soil. After students have had ample time to observe and diagram the worms, ask these questions:

• **Could you tell the difference between the head and the tail of the earthworm? If so, how?** (Students should be able to see that the head is more rounded and the tail is more pointed, and that there is a lip at the head end extending over the mouth. Students should notice the light-colored, somewhat swollen band around the earthworm that is closer to the anterior end than to the posterior end. This band is called the clitellum.)
• **What other features did you notice about the earthworm's body?** (Students should notice the worm's segments and its *setae*, or bristles.)
• **What did you notice about the way earthworms move through soil?** (They can move forward or backward by using their setae.)

CHAPTER 27
Mollusks and Annelids

GUIDED ENQUIRY

Pose the following questions to students and have them record their responses. Point out that they will gain a better understanding of the key concepts if they read the chapter with these basic questions in mind. Upon completion of the chapter, pose the questions again. Ask students to compare their initial responses with those they have developed after reading the chapter.

- What distinguishes mollusks from other animals?
- How are mollusks and annelids related?
- How do mollusks perform essential life functions?
- How are mollusks classified?
- How do mollusks benefit or harm humans?
- How do annelids perform essential life functions?
- How are organ systems important in annelids?
- How do annelids affect humans and other living things?

INTRODUCING CHAPTER 27

Using the Visuals
Begin by having students observe the chapter-opener photograph.
- **What do the objects in the large photograph remind you of?** (Answers may vary. Answers may include feathers; feather duster; mop; flowers.)
- **Suppose I tell you that these objects are worms. Would you believe me? Why or why not?** (Accept all answers.)
- **Do they look like typical worms?** (Most students will probably say no.)
- **What makes them look different from the average everyday worm?** (They do not have a long, thin body; they are not crawling on the ground; they are not brown; they are full of feathers or something like feathers.)

Now have students read the caption.
- **What are the feathery structures that you see on these worms?** (Gills.)
- **Did you know that some worms have gills?** (Accept all answers.)

Direct students' attention to the inset photograph. Point out that in some ways, the slug resembles a worm more than do the feathery organisms in the large photograph. Explain that the sea slug, however, is a mollusk, not an annelid worm.

Have students read the chapter-opener text.
- **Which mollusks are mentioned in the first sentence?** (Clams,

CHAPTER 27
Mollusks and Annelids

The organisms that seem to be exotic flowers swaying in the breeze are actually annelid worms. The worms use their feathery gills for feeding and respiration. The spotted nudibranch, or sea slug, (inset) is a mollusk from the Great Barrier Reef in Australia.

Have you ever eaten fried clams, broiled scallops, or calamari in tomato sauce? Have you ever gone fishing with live worms as bait? If you have, you are already familiar with some of the more common members of the two phyla that we shall study in this chapter: mollusks and annelids (segmented worms). Both of these phyla are ancient, very large, and remarkably diverse. Both provide many examples of how evolution can mold a single basic animal body plan into many different shapes. And both remind us that animals with ancient and simple body plans can be very well adapted to their environments.

What are mollusks and annelids? How are they related to one another? How are they adapted to their environments? What relationships do they have with other living things? You will find the answers to these questions in the pages that follow.

GUIDE FOR READING

After you read the following sections, you will be able to

27-1 Mollusks
- Explain how mollusks perform their essential life functions.
- Describe and give examples of the three major classes of mollusks.
- Discuss how mollusks affect humans and other living things.

27-2 Annelids
- Describe how annelids perform their essential life functions.
- List and give examples of three classes of annelids.

Journal Activity

YOU AND YOUR WORLD

Have you ever observed a snail or a slug? In your journal, describe the animal and how it moves. What were you doing when you noticed the animal? What characteristic intrigues you most about the animal?

27-1 Mollusks

Guide For Reading
- What are mollusks, and how do they perform essential life functions?
- What are the three major classes of mollusks?
- How do mollusks fit into the world?

Members of the phylum Mollusca are known as **mollusks**. Mollusks evolved in the sea more than 600 million years ago and have experienced a long and successful adaptive radiation. Today there are more than 100,000 mollusk species, which are divided into seven classes. Mollusks live everywhere—from deep ocean trenches to mountain brooks to the tops of trees. They range in size from snails as small as a grain of sand to giant squids that may grow more than 20 meters long. And as you can see in Figure 27-1, mollusks come in a wide range of forms and colors.

What Is a Mollusk?

Why are animals that look and act so differently grouped in the same phylum? One reason mollusks are classified together is that they share similar developmental patterns. (As you may recall from Chapter 15, many animals are classified on the basis of shared features during early development.) Most mollusks have a special kind of larva called a trochophore (TROH-koh-for). See Figure 27-2 on page 586. Trochophore larvae swim in open water and feed on tiny floating plants.

Figure 27-1 Chitons are relatively primitive marine mollusks that have a shell made up of a number of plates (inset). Snails are more specialized mollusks that have a one-part shell. The tree snail is creeping over a red Heliconia flower.

COOPERATIVE LEARNING

Complete this activity after reading the Science, Technology, and Society feature entitled "Leeches: Modern Applications of Ancient Medicine" (page 601). Using randomly selected pairs or trios, have groups write an "update letter" to the Greek physician who first recorded the use of leeches as a medical treatment. Their letter should explain the history of leeching since his death, how leeching is used today, and why we now know that leeching is a medically sound procedure. Students should end their letter with predictions of how leeching might be used in the future.

Journal Activity

YOU AND YOUR WORLD

Ask volunteers to read their responses to the class. Have students keep their journal assignments in their portfolios.

---

scallops, and calamari, which is squid.)
- **Can you think of any other mollusk that is used as food?** (Oysters, octopi, snails.)
- **Do you think that all of these animals are alike?** (Accept all answers.)

- **Do you think that they all have a lot in common with the sea slug in the inset photograph?** (Accept all answers.)
Point out to students that as they study this chapter they will learn how such seemingly different animals are related and why they are classified in their respective phyla.

TEACHING STRATEGY 27-1

Focus/Motivation

Write the words *clam, oyster, snail, octopus,* and *squid* on the chalkboard. Have students write down as many facts as they can about these five animals. Then ask students to write a sentence at the bottom of the page telling what they think all these animals have in common.

Have students share their responses. Do not dispute any misconceptions now, but ask students to keep their answers in mind as they study the material in this section. Tell students that they will use their papers again for the Closure activity at the end of the section.

585

BACKGROUND INFORMATION
CLASSIFICATION OF MOLLUSKS

Among the characteristics used to classify mollusks are the presence or absence of a shell, the type of shell, and the type of foot. Based on these characteristics, phylum Mollusca is divided into seven living classes:

1. Monoplacophora: tiny deep water marine mollusks with a single symmetrical shell and a flat creeping foot
2. Polyplacophora: chitons
3. Aplacophora: solenogasters, small wormlike marine mollusks that lack shell, mantle, and foot in the usual sense, but possess homologous structures
4. Gastropoda: snails and slugs, including limpets, abalones, cowries, nudibranchs, and sea butterflies
5. Bivalvia: clams, oysters, and other mollusks that possess a shell with two valves that is hinged dorsally
6. Scaphopoda: burrowing marine mollusks commonly known as tusk shells or tooth shells
7. Cephalopoda: tentacled mollusks such as octopi, squids, and nautiluses

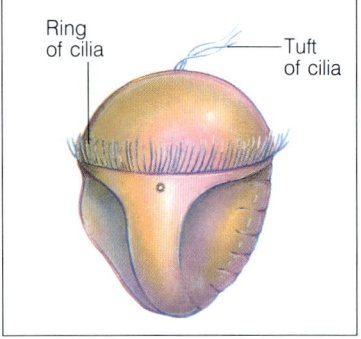

Figure 27–2 The trochophore larva of a chiton, like other trochophore larvae, has a tuft of cilia making up the "handle" on its top-shaped body and a band of cilia encircling its body.

Figure 27–3 The basic body parts of mollusks are the foot, mantle, shell, and visceral mass. Note that the form and function of the foot and shell vary greatly among mollusks.

Trochophore larvae are also seen in segmented worms, which belong to the phylum Annelida. Biologists believe that this indicates that mollusks and annelids evolved from a common ancestor that existed during the Precambrian Period (more than 580 million years ago) and had a trochophore larva. Because the phyla Mollusca and Annelida are closely related to each other, we shall discuss them both in this chapter.

Another reason mollusks are placed in a single phylum is that their different forms are the results of variations on the same basic body plan. **Mollusks are defined as soft-bodied animals that have an internal or external shell.** Their name is derived from the Latin word *molluscus*, meaning soft. Although a few present-day mollusks lack shells, they are thought to have evolved from shelled ancestors.

Form and Function in Mollusks

As you can see in Figure 27–3, the body plan of almost all mollusks consists of four basic parts: **foot, mantle, shell,** and **visceral mass**. The soft muscular foot usually contains the mouth and other structures associated with feeding. The foot takes many different shapes in mollusks: Flat surfaces adapted to crawling, spade-shaped structures for burrowing, and tentacles for capturing prey are a few examples. The mantle is a thin, delicate tissue layer that covers most of a mollusk's body, much like a cloak. The shell, which is found in almost all mollusks, is made by glands in the mantle that secrete calcium carbonate ($CaCO_3$). Just beneath the mantle in most mollusks is the visceral mass, which contains the internal organs.

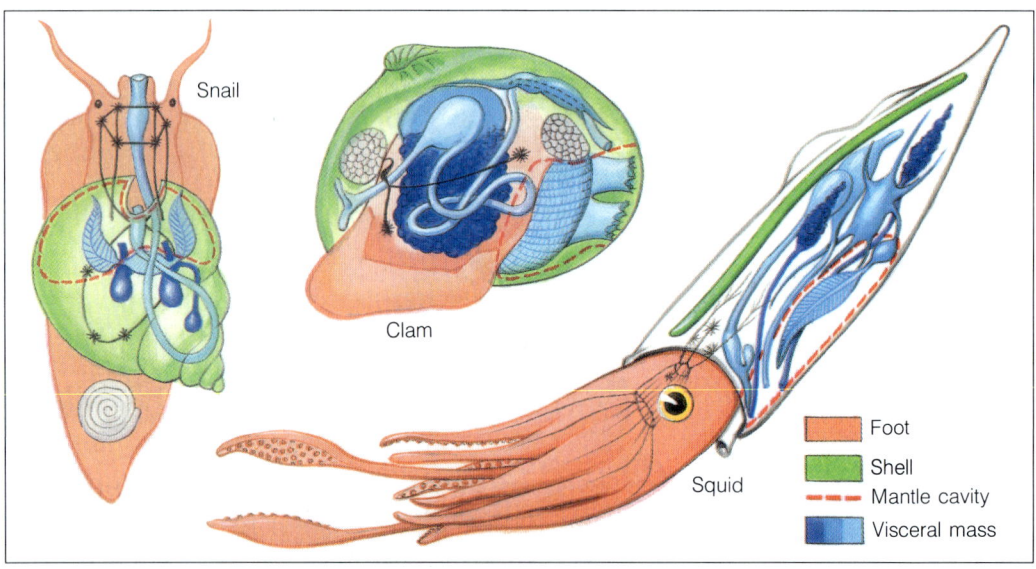

27–1 (continued)

Content Development

Using the Visuals Emphasize that the animals in the phylum Mollusca are united by two major factors: an early stage of development called a trochophore and a basic body plan that includes, or is believed to have once included, a foot, mantle, shell, and visceral mass.

Have students identify the basic body parts of a mollusk in Figure 27–3. Explain that the visceral mass contains the digestive system, the excretory system, the reproductive organs, and the heart. Point out that covering the visceral mass is the mantle. Stress that glands in the mantle secrete calcium carbonate.

• **Why is the secretion of calcium carbonate important to the mollusk?** (Calcium carbonate is the material that forms the mollusk shell.)

Direct students' attention to the dotted line in the diagram.

• **What does this dotted line represent?** (The mantle cavity.)

• **What is the mantle cavity?** (A space that exists between the

These basic body parts have taken on different forms as mollusks evolved adaptations to different habitats. The type of foot and the kind of shell that mollusks have are used to group them into classes. Later in this chapter we shall examine the three major classes of mollusks.

FEEDING Mollusks have evolved many types of feeding mechanisms and feed on many kinds of food. In fact it would be simpler to list the few things these animals do not eat than it would be to describe everything they can feed on! Every mode of feeding is seen in the phylum Mollusca. Most mollusks are herbivores, carnivores, or filter feeders, but a few species are detritus feeders and others are parasites.

Many mollusks—snails and slugs, for example—feed with a tongue-shaped structure called a **radula** (RAJ-oo-lah). The radula is a layer of flexible skin that carries hundreds of tiny teeth, which make it look and feel like sandpaper. Inside the radula is a stiff supporting rod of cartilage. When the mollusk feeds, it places the tip of the radula on the food and pulls the sandpapery skin back and forth over the cartilage. Mollusks that are herbivores use their radula to scrape algae off rocks and twigs in the water or to eat the buds, roots, and flowers of land plants. Mollusks that are carnivores use their radula to drill through the shells of other animals. Once they have made a hole through the shell, these carnivores extend their mouth and radula into the shell and tear up and swallow the prey's soft tissue. In the carnivorous snails called cone shells, the tiny rasping teeth of the radula have evolved into long hollow darts that are attached to poison glands. A cone shell uses these darts to stab and poison prey such as small fish.

Although they may have a radula, carnivorous mollusks such as octopi and certain sea slugs typically use sharp jaws to eat their prey. Like cone shells, some octopi produce poison to subdue their prey. Although cone shells and octopi generally feed on fish and other small animals, the poisons produced by some species are strong enough to hurt or even kill humans.

Mollusks such as clams, oysters, and scallops are filter feeders. They use their feathery **gills** to sift food from the water. As these animals pass water over their gills, phytoplankton (tiny photosynthetic organisms) in the water become trapped in a layer of sticky mucus. Cilia on the gills move the mixture of mucus and food into the mouth.

RESPIRATION Gills serve as organs of respiration as well as filters for food. In fact, in most species gills are used only for breathing. Aquatic mollusks such as snails, clams, and octopi breathe by using gills located inside their mantle cavities. But land snails and slugs breathe by using a specially adapted mantle cavity that is lined with many blood vessels. The mantle is wrinkled or folded to fit a larger surface within the limited

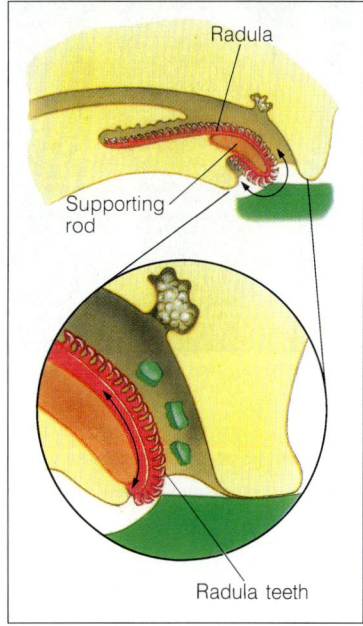

Figure 27–4 Many mollusks scrape bits of food into their mouth by pulling the tooth-covered skin of the radula back and forth over a supporting rod of cartilage. The scanning electron micrograph shows the teeth on the radula of a land snail.

BACKGROUND INFORMATION
CHITONS

Chitons are marine mollusks that have elongated bodies, a large broad foot, and a radula. Chitons eat algae, hydrozoans, and other low-growing organisms.

The dorsal shells of chitons display beautiful variations of pattern and color. The dorsal shells, which are divided into a series of eight plates, are also very practical. These plates make the shell flexible enough so that the chiton can roll up into a little ball when dislodged from its attachment. Thus it can protect itself until it can safely unroll and reattach elsewhere.

Another defense mechanism of the chiton is its ability to anchor itself to its substrate when it is disturbed or when it is exposed by a receding tide. The muscles in its foot pull the animal down tightly so that the edge of the mantle, which partly or completely covers the shell plates, can function like the rim of a suction cup. In this way, it becomes extremely difficult to dislodge the chiton.

HISTORICAL NOTE
FOSSILIZED MOLLUSKS

One of the ways in which scientists find relationships among organisms is by studying the fossil record. Mollusks, which are known to have existed since the Paleozoic Era, have left what is perhaps the most complete fossil record of any group of animals.

TEACHING SUPPORT

Teaching Resources
- Activity: Mollusks in Our World, p. 3
- Hands-On Activity: Moving at a Snail's Pace, p. 139

BACKGROUND INFORMATION
EVOLUTIONARY PATHWAYS

When the bilateral coelomate animals of the Cambrian Period began to diversify, they gave rise to two distinct lineages: protostomes and deuterostomes. This fork in the evolutionary path is still in evidence today, as nearly all complex coelomate animals belong to one or the other of these groups.

Mollusks and annelids, along with arthropods, fall into the protostome category. Echinoderms and chordates are deuterostomes. In general, the animals in both categories have a complete gut and coelom. The difference between the two lineages lies in the pattern of embryonic development. In protostomes, the first opening on the embryonic surface becomes a mouth and the anus forms later. In deuterostomes, the first embryonic opening becomes an anus and the mouth forms later.

The timing of cell specialization also differs in the two groups. In most protostomes, the destiny of a cell is fixed at the first embryonic cleavage; at this time it is determined what developmental path the cell will take and what its specialized function will be. In deuterostomes, cell specialization is determined somewhat later on.

Figure 27–5 The nudibranch (right) breathes through its skin and tuft of gills. Janthina (left), which uses a raft of air bubbles to float at the ocean surface, breathes with gills inside its shell. Many land snails (bottom) use their mantle cavity as a lung.

space of the cavity. This surface is constantly kept moist so that oxygen can enter the cells. Because the mantle loses water in dry air, most land snails and slugs must live in moist places. They prefer to move around at night, during rainstorms, and at other times when the air is humid.

INTERNAL TRANSPORT Oxygen that is taken in by the respiratory system and nutrients that are the products of digestion are carried by the blood to all parts of a mollusk's body. The blood is pumped by a simple heart through what is called an **open circulatory system**. "Open" does not mean that blood can spill to the outside of the animal! It means that blood does not always travel inside blood vessels. Instead, blood works its way through body tissues in open spaces called sinuses. These sinuses lead to vessels that pass first through the gills, where oxygen and carbon dioxide are exchanged, and then back to the heart. Open circulatory systems work well for slow-moving or sessile (attached to one spot) mollusks like snails and clams. But the flow of blood through sinuses is not efficient enough for fast-moving octopi and squids. Those animals have **closed circulatory systems**, in which blood always moves inside blood vessels.

EXCRETION Like other animals, mollusks must eliminate waste products. Undigested food becomes solid waste that leaves through the anus in the form of feces. Cellular metabolism produces nitrogen-containing waste in the form of ammonia. Because ammonia is poisonous, it must be removed from body fluids. Mollusks get rid of ammonia by using simple tube-shaped organs called **nephridia** (neh-FRIHD-ee-ah; singular: nephridium). Nephridia remove ammonia from the blood and release it to the outside.

27–1 (continued)

Skills Development
Guided Practice
Skill: Relating concepts
Stress the characteristics of a mollusk that indicate an increased level of complexity compared with previously studied phyla. The most important of these characteristics is the presence of a true coelom, or body cavity. Explain that a true coelom is formed from the mesoderm layer of cells.
• Did any of the animals you studied in the last chapter—the sponge, cnidarian, flatworm, or roundworm—have a true coelom? (No. However, the roundworm has a pseudocoelom.)
• Do mollusks have any other characteristics that indicate a higher level of development? (More highly developed organ systems and the fact that organ systems are used to carry out all essential life functions.)

Content Development
Discuss with students the difference between an open and a closed circulatory system.

RESPONSE Mollusks vary greatly in the complexities of their nervous systems and their abilities to respond to environmental conditions. Clams and other two-shelled mollusks, many of which lead basically inactive lives burrowing in mud or sand, have simple nervous systems. They have several small ganglia near the mouth, a few nerve cords, and simple sense organs such as chemical and touch receptors, statocysts (simple organs for balance), and ocelli (eyespots). Octopi and other tentacled mollusks, on the other hand, are active and intelligent predators that have the most highly developed nervous systems of all members of their phylum. Because of their well-developed brain, these animals can remember things for long periods of time, and they may even be more intelligent than some vertebrates. The numerous complex sense organs these mollusks possess help them distinguish shapes by sight and texture by touch. Octopi can be trained to perform different tasks in order to obtain a reward or avoid punishment. Because of these abilities, octopi are often studied by psychologists interested in the way animals learn.

REPRODUCTION As with almost all other essential functions, mollusks accomplish the function of reproduction in different ways. In most mollusks, the sexes are separate and fertilization is external. These mollusks—which include many snails, almost all two-shelled mollusks, and most of the species in the four minor classes of mollusks—release eggs and sperm into open water in enormous numbers. Eggs and sperm find each other by chance, and free-swimming larvae develop from the resulting fertilized eggs. In tentacled mollusks and certain snails, fertilization takes place inside the body of the female. Fertilization is also internal in some hermaphrodites (organisms that have both male and female reproductive organs). For example, many hermaphroditic snails get together in pairs and fertilize each other's eggs at the same time. Some other hermaphroditic mollusks, such as certain oysters, switch from one sex to the other. Sometimes they are male (and thus produce sperm) and sometimes female (and thus produce eggs)!

Figure 27–6 Like most mollusks, coquina clams (inset) have an open circulatory system. Cephalopod mollusks, such as cuttlefish, have a closed circulatory system. Clams move slowly, whereas cuttlefish can jet backwards through the water at high speeds.

Figure 27–7 A scallop gathers information about its environment with tiny round eyespots and sensory tentacles. Using those eyespots and other sense organs, scallops stay alert for enemies, such as starfish. Unlike most bivalves, scallops can sometimes escape from enemies by clapping their shells together rapidly, using jet propulsion to help them jump off the ocean bottom and scuttle away. See Figure 27–10.

ESL STRATEGY

Many colloquial expressions in the English language refer to mollusks, for example, "clammed up," "happy as a clam," "the world is my oyster," and "at a snail's pace." Have students discuss the meanings of these expressions and use them in sentences. Ask them if they can think of any other expressions that contain the names of mollusks.

- **Why do you think that an open circulatory system works for sessile mollusks but that a closed system is needed for more active mollusks?** (A fast-moving mollusk will need to have oxygen transported rapidly and in large amounts to vital organs during vigorous activity. The closed system accomplishes this best, since blood stays in the direct pathways of the blood vessels. In an open system, the flow of oxygen is slowed down as the blood works its way through numerous sinuses in the tissues.)

Also discuss with students the different methods of reproduction in mollusks. Emphasize that in most mollusks, fertilization is external.

- **What is meant by external fertilization?** (Fertilization of an egg does not occur inside the mollusk's body; instead, an egg is fertilized outside the mollusk's body in the surrounding water.)
- **What are some disadvantages of external fertilization?** (Possible answers: The chances of a particular egg becoming fertilized are not as great, since eggs can easily be lost to the surrounding water before fertilization can take place; the safety of the fertilized egg is at risk since it does not have the same protection that it would have if it were inside the animal's body.)
- **Can you think of any advantages of external fertilization?** (Possible answers: Many eggs can become fertilized at one time; genetic variety is accomplished by having numerous sperm interact with numerous eggs; the body of the animal producing the egg remains free and does not have to incubate an embryo.)

TEACHING SUPPORT

Laboratory Manual
- Examining a Clam, p. 345

MEDIA AND TECHNOLOGY

Biology Transparencies
- 31: The Anatomy of a Clam

Videodisc
- Aquatic Ecosystems: Estuaries and Marine
- Aquatic Ecosystems: Freshwater Wetlands and Freshwater

Use this bar code to show a snail laying its eggs.

Play frames 34350 to 35152

See the Biology Media Guide pages 28–29 for additional bar-code correlations.

MULTICULTURAL STRATEGY

In Japan, a mollusk called the *awabi* (also known as the western abalone) has been from ancient times regarded as a valuable gift. Its dried and sliced meat is wrapped in red and white papers and given to friends as a symbol of joy and happiness.

Ask students to share information about the gift-giving customs of their own cultural heritage. What traditional presents—natural or made by humans—might they give on festive occasions?

Snails, Slugs, and Their Relatives

Members of the class Gastropoda are called **gastropods** (GAS-troh-pahdz). The name gastropod literally means stomach foot. This name is quite appropriate because most gastropods move by means of a broad, muscular foot located on their ventral (stomach) side. Gastropods include the familiar pond snails and land slugs as well as more exotic mollusks such as abalones, sea butterflies, sea hares, and nudibranchs.

Many gastropods have a one-piece shell that protects their soft bodies. This shell may be simple and shieldlike, as in limpets, or coiled, as in snails. When threatened, many snails can pull up completely into their coiled shells. Some snails are additionally protected by a hard disk on their foot that forms a solid "door" at the mouth of their shell when they withdraw.

Some gastropods have small shells or, as is the case with slugs, lack shells completely. This would seem to make them easy prey for hungry predators. However, these gastropods are not entirely helpless. Most land slugs are protected by their behavior—they spend the daylight hours hiding under rocks and logs, hidden from birds and other animals that might eat them. Some sea hares have a special ink-producing gland that they use when threatened to squirt ink into the surrounding water. This confuses predators and allows the sea hare to escape under its "smoke screen." Some gastropods, such as sea butterflies, escape predators by swimming rapidly. Many nudibranchs, or sea slugs, have chemicals in their bodies that taste bad or are poisonous. When a predator nibbles on one of these bad-tasting morsels, it gets sick. In addition, some nudibranchs use nematocysts from the cnidarians they eat to sting predators. The bad-tasting, poisonous, stinging, or otherwise booby-trapped nudibranchs are usually brightly colored. The bright colors warn predators to stay away. If a predator ignores the warning colors and eats a nudibranch, the consequences usually guarantee that the predator will remember the bright

Figure 27-8 The ringed top snail (bottom, left) is found in the oceans of the Pacific Northwest. Despite their lack of a protective shell, the marine nudibranch (bottom, right) and terrestrial banana slug (top) are not likely to be eaten by predators—their bright colors and patterns indicate that these gastropods are poisonous.

27-1 (continued)

Reinforcement/Reteaching
Review with students the ways in which mollusks perform each of the essential life functions. List on the chalkboard the following categories: Feeding, Respiration, Internal Transport, Elimination, Response, and Reproduction. For each category, pose a question that will motivate students to summarize the most important information in one or two sentences. For example:

- **How do mollusks feed?** (Most mollusks are herbivores, carnivores, or filter feeders, but a few mollusks are parasites or detritus feeders. Many mollusks that are herbivores and carnivores, such as snails, are equipped with a special tongue-shaped feeding organ called a radula.)
- **How do mollusks breathe?** (Aquatic mollusks breathe by using gills; land mollusks have a specially adapted mantle cavity that serves as a kind of lung.)
- **What kinds of circulatory systems do mollusks have?** (Sessile mollusks such as the

colors of the nudibranch and avoid it in the future! (While this does not help the first nudibranch, it does protect others of its kind.) Thus shell-less gastropods do have means of protection.

Two-Shelled Mollusks

Members of the class Bivalvia (*bi-* means two; *valve* means shell) have two shells that are hinged together at the back and held together by one or two powerful muscles. Common **bivalves** include clams, oysters, and scallops. Bivalves may be tiny or as large as the giant clam *Tridacna*, which has been known to grow as large as 1.9 meters in length.

Although bivalve larvae are free-swimming, they soon settle down to a relatively quiet life on the bottom of a body of water. Some bivalves, such as clams, burrow in mud or sand. Others, such as mussels, secrete sticky threads to attach themselves to rocks. Although most adult bivalves are sessile, some, such as scallops, can move around rapidly by flapping their shells when threatened.

The mantles of bivalves, like those of most other mollusks, contain glands that manufacture the shells. These mantle glands also keep the shell's inside surfaces smooth and comfortable by secreting layers of mother-of-pearl. If a foreign object—a sand grain or small pebble, for example—gets caught between mantle and shell, the mantle glands cover it with this secretion. After many years these objects become completely coated and are called pearls.

Figure 27–9 When threatened, a sea hare releases purple ink into the water. This confuses predators and allows the sea hare to make its escape.

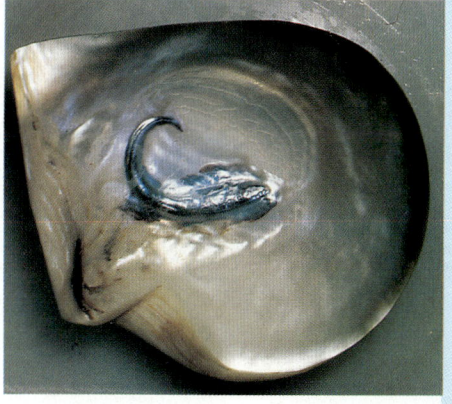

Figure 27–10 The internal structures of a clam, a typical bivalve, are shown in the diagram on the left. Another bivalve, the scallop, can swim by rapidly opening and closing its shell (top). Pearls—objects coated by smooth, shiny secretions of a bivalve's mantle—may be beautiful and valuable gems or they may be fascinating curios, like the pearl fish (bottom).

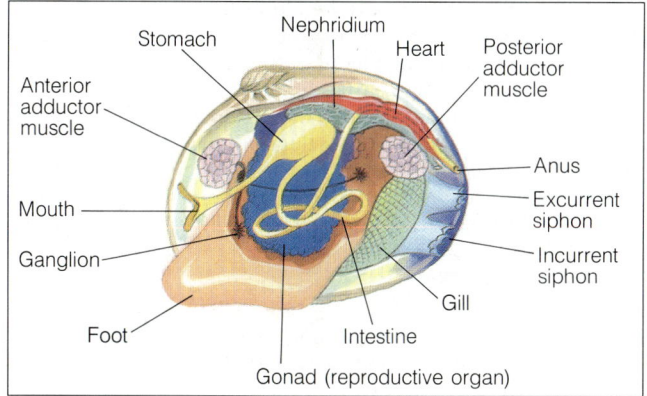

BACKGROUND INFORMATION
TORSION

The coiled shell of the snail and other gastropods is the result of an internal realignment process called torsion. During the trochophore stage of the gastropod's development, the animal's visceral mass (often referred to in gastropods as the visceral hump) begins to grow upward. This growth is uneven on the right and left sides. The uneven growth, coupled with the contraction of certain muscles, causes the posterior mantle cavity to twist around to the right side. At a critical moment, the body rotates a full 180° so that the back end of the body comes to rest just behind the head. The result is that the gastropod balances its internal organs above the rest of the body much as a human would carry a backpack.

The coiled shell is an advantage to the gastropod, for it provides a retreat for the animal's head in times of danger. The twisted body arrangement, however, has its drawbacks. As a result of torsion, the gastropod has its anus and kidney openings above the head. This creates something of a sanitation problem, because waste products tend to fall in the vicinity of the mouth and the gills. Most gastropods use cilia to create currents to sweep these areas clean.

591

---

clam and snail have open circulatory systems, while more active mollusks such as the octopus have closed circulatory systems.)
• **How do mollusks eliminate waste products?** (Solid wastes leave the body through the anus; liquid wastes are eliminated through tube-shaped organs called nephridia.)
• **What types of nervous systems do mollusks have?** (Clams and other mollusks that are basically inactive have simple nervous systems consisting of small ganglia near the mouth and a few simple sense organs. Octopi and other tentacled mollusks have highly developed nervous systems, with a well-developed brain and complex sense organs.)
• **How do mollusks reproduce?** (Most mollusks reproduce externally by ejecting eggs and sperm into open water. In some tentacled mollusks, fertilization of eggs takes place within the body of the female.)

TEACHING SUPPORT

Teaching Resources
- Activity: Investigating the Behavior of an Octopus, p. 5

Study Guide
- Section 27–1, p. 259

BACKGROUND INFORMATION
CEPHALOPODS AND VERTEBRATES

The eye of an octopus, with its lens and focusing muscle, is remarkably similar to the eyes of humans and other vertebrates. Also like vertebrates, cephalopods have highly developed sense organs for balance and touch. These similarities between cephalopods and vertebrates are examples of convergent evolution.

27–1 (continued)
Content Development
In discussing the negative impact of mollusks on humans, be sure to remind students of the life cycle of the fluke, which they studied in Chapter 26.

- **Which mollusk often plays an important role in the life cycle of a fluke?** (The snail.)
- **What is the role of the snail?** (The snail serves as an intermediate host for the fluke, before it enters the fish or animal that will eventually be eaten by a human.)

SECTION REVIEW 27–1

1. Mollusks are soft-bodied invertebrates with an internal or external shell. The class Gastropoda includes snails and slugs; the class Bivalvia includes clams, oysters, and scallops; and the class Cephalopoda includes squids and octopi.
2. Mollusks are used as food and in environmental and medical research. Snails and slugs may eat crops. Shipworms damage wooden boats and docks.
3. Mollusks use radula to scrape and tear food, to drill through the shells of prey, and, in the case of cone shells, to stab and poison their prey.
4. They run away, hide inside their shells, create "smoke screens" of ink, secrete poisonous or distasteful chemicals, and/or sting.
5. Oysters are filter feeders, and their loss would increase the particle concentration of the water. This increase might foul the water for other marine life, or it could spur the growth of algae or other plants. Oysters' loss might also stress the populations of their natural predators, such as starfish. As for humans, they would lose oysters as food, subjects of medical research, monitors of water quality, and a source of pearls.

Reinforcement/Reteaching
If students have trouble answering the Section Review questions, go back to the material with which they are having difficulty.

Tentacled Mollusks

Cephalopods (SEHF-uh-loh-pahdz)—members of the class Cephalopoda—are among the most active and interesting mollusks. This class includes octopi, squids, cuttlefish, and nautiluses. Cephalopoda means head-foot (*cephalo-* means head; *-pod* means foot). This name refers to the fact that a cephalopod's head is attached to its foot, which is divided into tentacles, or arms. Cephalopods range in size from tiny cuttlefish less than 2 centimeters long to giant squids, which are thought to grow to more than 20 meters long.

Most cephalopods have eight flexible tentacles equipped with a number of round sucking disks that are used to grab and hold fish and other prey. In addition to these tentacles, cuttlefish and squids also have two long, slender arms with suckers on the end. Nautiluses have many more tentacles (38 to 90) than other cephalopods. Their tentacles lack suckers but are made sticky by a mucuslike covering.

Although fossil evidence indicates that their ancestors had large external cone-shaped or coiled shells, most modern cephalopods have small internal shells or no shells at all. The only present-day cephalopods with shells are the few species of nautiluses. These cephalopods look much like fossil cephalopods from the beginning of the Cambrian Period, more than 500 million years ago. Cuttlefish have small shells that are found inside their bodies. The shells of some cuttlefish are thin and coiled, whereas others (which serve as the cuttlebone on which pet birds condition their beaks) are flat, platelike, and do not resemble shells at all. In both nautiluses and cuttlefish, gases in the shell help the cephalopod remain upright and allow it to float in the water. A squid's internal shell has evolved into a thin, flexible supporting rod known as a pen. Octopi have lost their shells completely.

Although most cephalopods lack a protective shell, they do have other means of protection. Most cephalopods can move quickly, either by swimming or crawling. They can also move by using a form of jet propulsion. The cephalopods draw water into their mantle cavities and then force that water out through the tubelike siphon. By pointing the siphon in different directions, they can shoot out a jet of water that propels them backward, away from danger. In addition, many cephalopods can release large amounts of dark-colored, foul-tasting ink when they are frightened. After squirting out a large cloud of ink, they make a hasty retreat. Perhaps most fascinating of all, octopi can quickly change color to match the colors of their surroundings. The match is often close enough that the octopi are nearly invisible.

Figure 27–11 The luminescent squid (top), chambered nautilus (center), and extremely venomous blue-ringed octopus (bottom) are examples of cephalopods.

How Mollusks Fit into the World

Mollusks play many different roles in living systems. For example, they feed on plants, prey on animals, and "clean up" their surroundings by eating detritus. Some of them are hosts to symbiotic algae or to parasites; others are themselves parasites. In addition, mollusks are an important source of food for many organisms, including humans.

Modern-day scientists have found some new uses for mollusks. Because filter-feeding bivalves concentrate dangerous pollutants and microorganisms in their tissues, careful checks of bivalves can warn biologists and public health officials of health problems long before scientists can detect these dangers in the open water. Besides acting as environmental monitors, mollusks also serve as subjects in biological research. Some current investigations are based on the observation that snails and other mollusks never seem to develop any form of cancer. If scientists can determine what protects the cells of these animals from cancer, they will gain valuable insights into how to fight cancer in humans.

Although mollusks are beneficial in many ways, they do have some negative relationships with humans. For example, land slugs and snails are plant eaters that can do much damage to gardens and crops. The bivalves called shipworms, which use their shells to drill their way slowly through pieces of wood in the water, are sometimes described as the termites of the sea. They settle on wood in large numbers and can reduce a good-sized log to a pile of wet sawdust over the course of a few years. Shipworms cause millions of dollars worth of damage to wooden boats and docks every year. Another problem with mollusks is associated with their use as food. Clams and oysters, which are among the few marine animals that are farmed in the sea, are filter feeders and thus gather and concentrate particles floating in the water—including bacteria, viruses, and the toxic protists that cause red tides. Eating bivalves that contain high concentrations of pathogens (things that cause disease), toxins, or pollutants can result in sickness or even death.

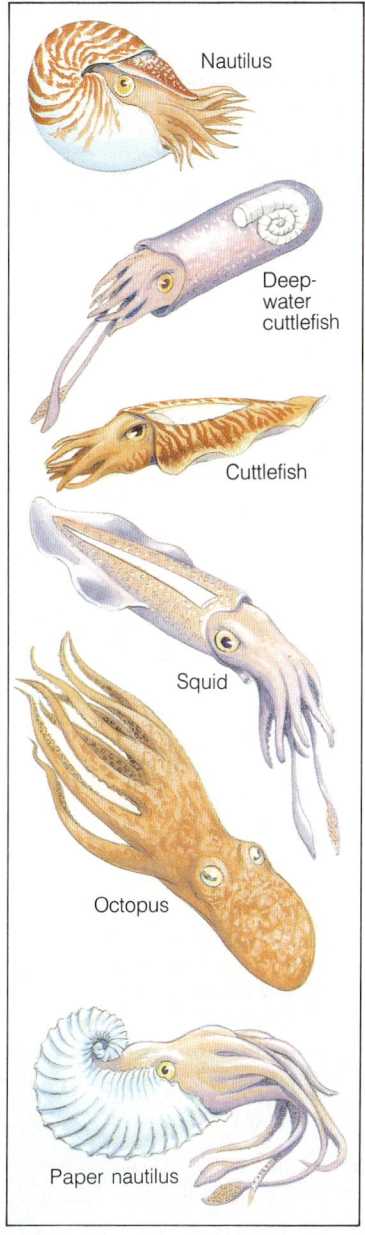

Figure 27–12 One trend in cephalopod evolution has been a reduction in the size of the shell. Most modern species have a small shell or no shell at all. The "shell" of the paper nautilus is actually an egg case.

27–1 SECTION REVIEW

1. What are mollusks? List the three major classes of mollusks and give an example of each.
2. Describe some of the ways mollusks affect humans.
3. What are some different ways mollusks use a radula?
4. How do mollusks protect themselves?
5. **Connection—Ecology** The number of oyster beds in Chesapeake Bay, which is an arm of the Atlantic Ocean, is dwindling rapidly. What effect would this have on the surrounding area? On the human community?

ECOLOGY NOTE
WORMS EAT MY GARBAGE!

Instead of seeing her garbage go to a landfill, Mary Appelhof feeds a good deal of it to a large colony of earthworms, which she keeps in a garbage can in her basement! As Appelhof discovered, composting with earthworms is simple, efficient, and when done properly, does not smell offensively. To begin your own earthworm compost, consult Appelhof's book, *Worms Eat My Garbage*.

ESL STRATEGY

Many slang and colloquial expressions in the English language refer to annelids, or worms. For example, "the worm in the apple" is often used to describe the one bad thing in an otherwise good situation or idea. To "worm your way out of something" means to wiggle out of unpleasant circumstances, usually by using devious or covert means. A person who is described as a "leech" is one who clings to another in an unappealing and usually noncontributory way.

Have students write sentences that use these and other "wormy" expressions.

Guide For Reading

■ What are annelids, and how do they perform essential life functions?
■ What are the three classes of annelids?
■ How do annelids fit into the world?

27–2 Annelids

Have you ever dug in a garden? If so, you have probably made the acquaintance of the long, thin, pink earthworm. The soft-bodied earthworm is the most common terrestrial, or land-dwelling, segmented worm. But this species is only one of approximately 9000 species of segmented worms that live in moist soil, in fresh water, and in the sea. Segmented worms, or annelids, live just about everywhere in the world. But because most segmented worms live in the sea, and many others spend their lives underground, only a few species are familiar to us.

What Is an Annelid?

Members of the phylum Annelida are known as **annelids**, or segmented worms. **An annelid is a round, wormlike animal that has a long, segmented body.** The name Annelida is derived from the Latin word *annellus*, which means little ring, and refers to the ringlike appearance of the body segments.

Annelids range in size from tiny aquatic worms less than half a millimeter long to giant earthworms more than 3 meters long. Although they also vary greatly in color, patterning, number of bristles, and other superficial features, most annelids are quite wormlike in appearance.

Form and Function in Annelids

The many segments of an annelid's body are separated by internal walls called septa (singular: septum). Most of the body segments are virtually identical to one another. However, some segments are modified to perform special functions. For example, the first few segments may carry one or more pairs of eyes, several pairs of antennae, and other sense organs.

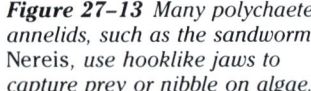

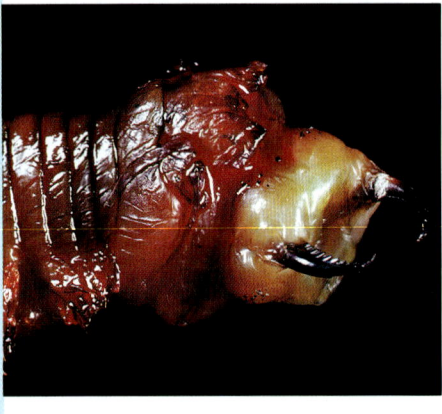

Figure 27–13 Many polychaete annelids, such as the sandworm Nereis, *use hooklike jaws to capture prey or nibble on algae.*

FEEDING The digestive tract, or gut, is a long tube within the body cavity of the worm that extends from the mouth to the anus (in the tip of the "tail"). Food enters through the mouth and travels through the gut, where it is digested. Like mollusks, annelids have evolved structures and behaviors that allow them to use a wide variety of foods.

One feeding organ that has evolved many different forms in different groups of annelids is the pharynx, or the muscular front end of the digestive tube. Many annelids can extend the pharynx through the mouth. In carnivorous annelids, this type of pharynx usually has two or more sharp jaws attached to it. When a suitable animal approaches, the worm lunges forward, rapidly extends the pharynx, and grabs the prey with its jaws. Jaws are also present in herbivores, which use them to tear off bits of algae. In some detritus feeders, the pharynx is covered with sticky mucus. When these worms extend the pharynx and press it against the sea-floor sediments, food particles stick to

TEACHING STRATEGY 27–2

Focus/Motivation
Display a map of the world and point to the continent of Antarctica.
• **What kind of animal lives everywhere in the world except for Antarctica?** (Answers may vary.)

Tell students that annelids, or unsegmented worms, live everywhere in the world except in Antarctica. Annelids live in fresh water, in ocean water, and on land. Point out that the most familiar representative of this phylum is the earthworm but that there are actually about 9000 other species of segmented worms.

Content Development
Emphasize that the body of an annelid is divided into many small segments, most of which appear to be almost identical. An important characteristic of annelids is that they have a body cavity, or coelom, filled with fluid. This gives segmented worms a body structure that is a "tube within a tube." Emphasize that compared with other worms, segmented worms have a more complete digestive system, a better developed circulatory system, and a more highly developed nervous system.

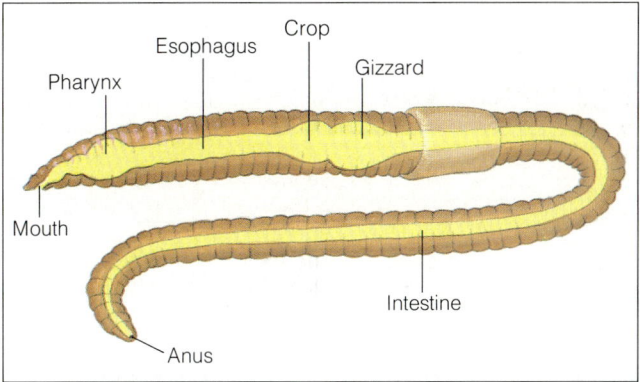

Figure 27-14 The digestive system of an earthworm is shown here. The pharynx pumps a mixture of food and soil into a tube called the esophagus. The food then moves through the crop, where it can be stored, and through the gizzard, where it is ground into smaller pieces. The food is digested in the intestine. Undigested materials pass through the intestine and are eliminated through the anus.

it. When the pharynx returns to its normal position, it carries these food particles back into the gut. In other detritus feeders, such as earthworms, the pharynx acts like a pump. It sucks a mixture of soil and detritus through the mouth and forces it down into the gut. In parasites, such as leeches, the pharynx is used to suck blood and tissue fluids from the host.

Annelids have a number of other structures that are used in feeding. For example, some annelids filter-feed by fanning water through their tubelike burrows and catching passing food particles in a mucus bag. In other filter-feeding annelids, such as the plume worm shown in Figure 27-15, the first segment forms featherlike structures that sift detritus and plankton from the surrounding water. These feeding structures are also used as gills for respiration.

RESPIRATION Aquatic annelids often breathe through gills. In some of these annelids, such as feather-duster worms, the large brightly colored feathery gills protrude from the opening of the worm's burrow or tube. In other annelids, small

Figure 27-15 The spaghetti worm (left) uses long tentacles to pluck bits of detritus from the ocean floor. In plume worms (right), a brush-shaped structure on the head is used in filter feeding and in respiration.

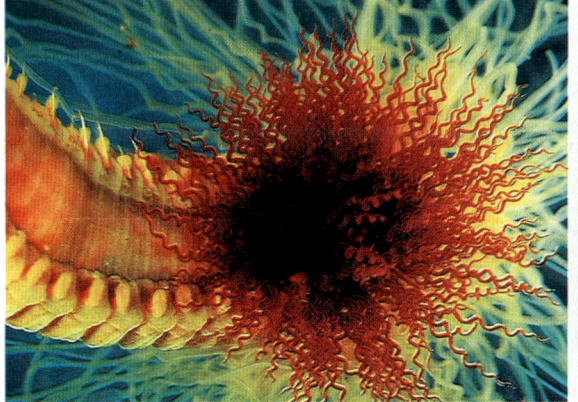

ANNOTATION KEY

① Five. (Interpreting illustrations)

BACKGROUND INFORMATION
MOVEMENT IN ANNELIDS

Annelids move through soil and sediment by using the power of their muscles and the liquid inside their body segments. Each body segment is sealed off from the segment next to it, which means that body fluids cannot move from one segment to another. When the longitudinal muscles contract and make the worm shorter, each segment, which is filled with fluid, has to become wider. In a similar manner, when the circular muscles contract and make the worm longer, each segment must become narrower. When the earthworm moves forward, its first few body segments elongate while the segments just behind them hold their position. Then, the first few segments shorten and widen. Alternating contractions and elongations continue along the length of the worm's body, enabling it to move through soil or sediment.

27-2 (continued)

Content Development

Emphasize that the segments of an annelid's body are separated from one another. This is important because it means that body fluid cannot move from one segment to another. Sometimes these fluid-filled segments are referred to as a hydrostatic skeleton. This means that the pressure of fluid in the segments provides a medium against which muscles can operate and helps to form and support the annelid's body.

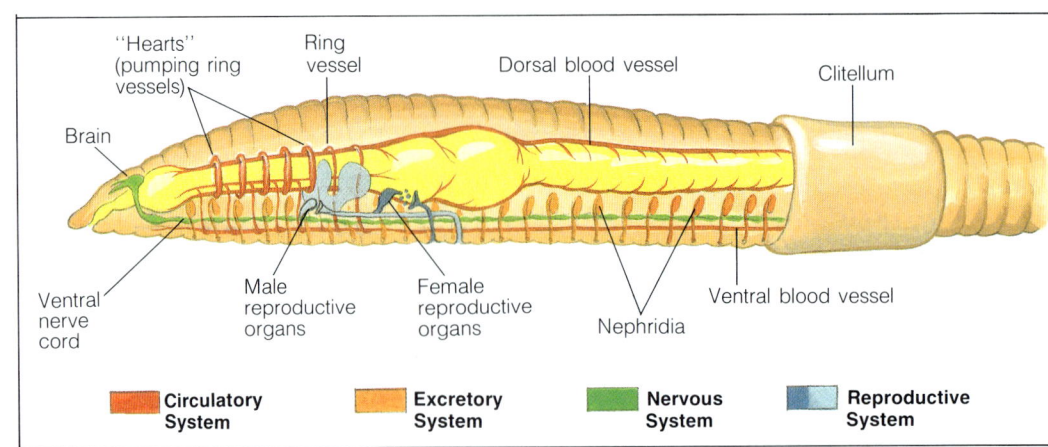

Figure 27–16 The circulatory, excretory, nervous, and reproductive systems of an earthworm are shown here. How many "hearts" does an earthworm have? ①

delicate gills are located on the sides of the body. The tube-dwelling annelids with this type of gill breathe by fanning water through their tubes. Many annelids take in oxygen and give off carbon dioxide through their skin. Because the skin must stay moist to make gas exchange possible, the worms die if the skin dries out. To help guard against this, terrestrial annelids, such as earthworms, secrete a thin protective coating called a cuticle to hold moisture around them.

INTERNAL TRANSPORT Annelids typically have closed circulatory systems organized around two blood vessels that run the length of their bodies. Blood moves toward the head of the worm in the dorsal (top side) vessel and toward the back of the worm in the ventral (bottom side) vessel. In each body segment is a pair of smaller vessels called ring vessels that connect the dorsal and ventral vessels and supply blood to the internal organs. In annelids such as earthworms, several of the ring vessels near the anterior (front) end of the worm are larger than the other ring vessels and have muscle tissue in their walls. These vessels are often called hearts because they contract rhythmically and help pump blood through the system. In other annelids, blood is moved through the body by muscle contractions when the worm moves.

EXCRETION Like other animals, annelids produce two kinds of wastes. Solid wastes pass out through the anus at the end of the gut. Wastes resulting from cellular metabolism are eliminated by nephridia (simple tube-shaped excretory organs). A pair of nephridia in each body segment removes waste products from the body fluids and carries them to the outside.

596

Reinforcement/Reteaching

Review with students the structure of the pharynx and the various ways in which it is adapted for feeding in annelids.

• How would you describe the **pharynx of an annelid?** (It is a muscular organ at the front end of the digestive tube.)

• **How is the pharynx adapted for feeding in carnivorous annelids?** (It has two or more sharp jaws. When a prey animal approaches, the pharynx can be extended through the mouth of the annelid to grab the prey.)

• **In what other type of feeders does the pharynx have jaws? What function do they serve?** (In herbivores, jaws on the pharynx are used to tear off bits of algae.)

• **What special adaptations does the pharynx have in annelids**

RESPONSE Many annelids are active animals with well-developed nervous systems. The brain sits on top of the gut at the front end of the body. Two large nerves pass around the gut and connect the brain with a pair of ganglia below. From these ganglia, a ventral nerve cord runs the entire length of the worm. Nerves from each segment of the worm enter and leave the nerve cord at a pair of small ganglia. These nerves help carry messages from sense organs and coordinate the movements of muscles.

Sense organs are best developed in the free-living marine species of annelids. Many of these annelids have sensory tentacles, statocysts, chemical receptors, and two or more pairs of eyes. Although the eyes are usually simple light detectors, in a few species the eyes can actually perceive objects. Most tube-dwelling species have light-sensitive cells either on their gills or near their mouths. These cells allow the animals to detect the shadows of predators passing overhead. When a shadow is detected, the worm pulls back into the shelter of its tube with amazing speed. In addition to specialized sense organs, these free-living marine worms also have various types of isolated sensory cells scattered along their epidermis. These cells respond to light, chemicals, and vibration.

Many other annelids have much simpler sensory systems. For example, earthworms have no specialized sense organs. They rely on simple sensory cells in the skin that are similar to those found in the skin of marine annelids.

Most free-living annelids do not have body structures that protect them from predators. Many depend on rapidly burrowing or swimming away from danger. Some, like earthworms, grab onto the walls of their burrows to make it harder to pull them out. Others, such as the marine fanworms, secrete protective tubes of calcium carbonate into which they withdraw if frightened. But some annelids do fight back. Several carnivorous annelids use their sharp jaws to attack animals that try to eat them. And the marine fireworms have tufts of poisonous bristles that easily break off and penetrate skin, causing painful sores and a burning sensation.

MOVEMENT Annelids have two major groups of muscles in their body walls. One group, called longitudinal muscles, runs from the front of the worm to the rear. When these muscles contract, they make the worm shorter. Another group of muscles runs in circles around the body of the worm. When these muscles contract, they make the worm skinnier. Marine annelids can swim by using these muscles to wriggle through the water. Burrowing annelids use their muscles to force their way through heavy sediment—not an easy thing for a soft-bodied animal to do!

REPRODUCTION Although a few annelids are able to reproduce asexually by budding, most annelids reproduce sexually. Some species have separate sexes and external

Figure 27–17 Sense organs are best developed in free-swimming annelids such as the paddleworm, which has a pair of beady eyes and a number of sensory tentacles on its head.

TEACHING SUPPORT

Laboratory Manual
- The Earthworm, p. 351

MEDIA AND TECHNOLOGY
Videodisc
- *Aquatic Ecosystems: Estuaries and Marine*

Use this bar code to introduce lug worms and life in estuaries.

Play frames 3635 to 3887

See the Biology Media Guide page 29 for additional bar-code correlations.

BACKGROUND INFORMATION
NEPHRIDIA

A system of nephridia regulates the volume and composition of body fluid in annelids. Sometimes the functional units of this system have flame cells, which suggests an evolutionary link between annelids and flatworms. More often, however, there is at the beginning of each nephridium a funnellike structure that collects fluid from a particular body segment. The funnel leads into a tubular portion of the nephridium, which then carries the fluid to a pore in the body wall. It is through this pore that the fluid is eliminated from the worm. It is interesting to note that the funnellike structure is located in one body segment but the pore into which it eventually empties is located in the body wall of the next segment in line.

---

that are detritus feeders? (In some, it is covered with sticky mucus so that it can withdraw particles from sea-floor sediments; in others, it acts like a pump, sucking a mixture of soil and detritus into the mouth and down into the gut.)

- **What is the function of the pharynx in parasitic annelids such as leeches?** (It is used to suck blood and tissue fluids from the host.)

Skills Development
Guided Practice
Skill: Applying concepts
Pose the following question to students:
- **After a heavy rain, worms are often found washed up on the sidewalk or other hard surface.** These worms are usually dead or dying. Can you explain why? (Accept all logical answers. Possible answer: The worms cannot extract enough oxygen when soil is waterlogged, and drown.)

> **BACKGROUND INFORMATION**
> **METAMERISM**
>
> In annelids, most body segments between the first and the last are nearly identical. This type of segmentation is called metamerism, and each segment is referred to as a metamere.
>
> Metamerism has evolved at least twice in the animal kingdom—among the ancestors of annelids and arthropods and among primitive chordates.
>
> In all groups exhibiting metameric segmentation, embryologically distinct segments may fuse to create complex anterior body parts. In polychaetes, for example, the "head" is composed of several segments. In arthropods, the fusion of segments creates the head and thorax. In chordates, the fusion of several segments to form the head has left traces of segmentation such as the pairs of cranial nerves.
>
> Some experts have suggested that metamerism may have allowed for a kind of evolutionary experimentation that eventually led to the specialization of segments. The reasoning behind this idea is that a mutation in only a few segments in a long line of identical segments would less likely be harmful to an organism. And, if it proved to be beneficial, the mutation could continue and eventually become a specialized segment or group of segments.

Figure 27–18 Although they look very different from each other, both the fanworm (left) and the fireworm (right) are polychaetes. The fanworm is a filter feeder that retreats into its tube when threatened. The fireworm defends itself with poisonous bristles that break off and penetrate skin at the slightest touch. The pain caused by these bristles gives the fireworm its name.

fertilization. This means that females and males release eggs and sperm, respectively, into the open water where fertilization takes place. Of course, the chances of fertilization taking place are enhanced if many worms in an area release their eggs and sperm at the same time. This is exactly what happens in some species. In the South Pacific, islanders eagerly await the autumn spawning season of the annelids called palolo worms. At a particular phase of the moon, hundreds of thousands of male and female palolo worms swarm at the surface of the water to release their eggs and sperm. Just before sunrise, the sea is literally covered with these worms. The islanders, who consider these worms a great delicacy, join sea birds and fishes that gather to feast on the spawning worms.

Some annelids, such as earthworms and leeches, are hermaphrodites that undergo internal fertilization. Although an individual worm produces both sperm and eggs, it rarely fertilizes its own eggs. Instead, worms pair up, attach themselves to each other, and exchange sperm. Each worm stores the sperm it has received in special sacs. When eggs are ready for fertilization, a band of thickened, specialized segments called the clitellum (cligh-TEHL-um) secretes a mucus ring into which eggs and sperm are released. The ring then slips off the worm's body and forms a cocoon that shelters the eggs.

Sandworms, Bloodworms, and Their Relatives

The class Polychaeta (*poly-* means many; *chaeta* refers to bristles) contains many common and important marine worms. **Polychaetes** (PAHL-ee-keets) are characterized by paired pad-

---

> **27–2 (continued)**
>
> **Content Development**
> Point out to students that there is much diversity among polychaetes. Different species of these worms crawl, burrow, swim freely, or live attached to one place. Some dig vertical or U-shaped burrows in soft mud or sand, and others build tubes for themselves out of calcium carbonate, organic secretions, or bits of sand and shells. Some polychaetes feed on other animals, some on algae, and some on organic matter in sediment.
>
> Explain that most of the tube-dwelling polychaetes have ciliated tentacles or feathery structures near the mouth. These organs secrete mucus that is able to trap bacteria, diatoms, and other small particles of food.
>
> **Reinforcement/Reteaching**
> Review with students the ways in which annelids reproduce.
> • **Do annelids reproduce asexually? If so, how?** (A few species can reproduce asexually by budding, but most annelids reproduce sexually.)

dlelike appendages on their body segments. These appendages are tipped with the bristles that give this class its name. In the sea mouse, shown in Figure 27–19, the bristles are so long that they extend over the back of the worm and look like hair or fur.

Polychaetes live in cracks and crevices in coral reefs, in sand, mud, and piles of rocks, and even out in the open water. Some burrow through or crawl over sediments. Others live almost entirely in tubes they build for themselves. Some polychaetes are dull in color and rather uninteresting; some are brightly colored, iridescent, or even luminescent.

Earthworms and Their Relatives

The class Oligochaeta contains earthworms and related species. Two **oligochaetes** (AHL-ih-goh-keets) that you might be familiar with are earthworms and tubifex worms. Earthworms are long pink worms that often show up on the surfaces of lawns and sidewalks after it rains, are dug up in gardens, or are sold as fishing bait. Tubifex worms are red threadlike aquatic worms that are sold as tropical-fish food in pet stores. Most oligochaetes live in soil or fresh water, although some species live in the ocean. As the name of the class indicates (*oligo-* means few), oligochates have fewer bristles than polychaetes. These bristles, which can be felt as a roughness on the ventral (bottom) side of an earthworm, help anchor it in its burrow.

Although earthworms spend most of their lives hidden under ground, an observant person may find evidence of their presence above ground in the form of squiggles of mud known as castings. Recall that an earthworm (which swallows just about anything it can get into its mouth) uses its pharynx to suck a mixture of detritus and soil particles into its mouth. As the mixture of food and soil passes through the intestine, part of it is digested. Sand grains, clay particles, and indigestible organic matter pass out through the anus in large quantities, producing castings. Some tropical earthworms produce enormous castings—as large as 18 centimeters long and 2 centimeters in diameter!

Leeches

The class Hirudinea contains the **leeches**, most of which live in moist tropical countries. Leeches are typically no more than 6 centimeters long, but there are some tropical species that are as long as 30 centimeters. Most leeches are freshwater organisms that exist as external parasites, drinking the blood and body fluids of their host. However, there are some marine and terrestrial leeches. And roughly one fourth of all leeches are carnivores rather than parasites. Carnivorous leeches, which feed on soft-bodied invertebrates such as snails, worms, and insect larva, either swallow their prey whole or suck all the soft parts from its body.

Figure 27–19 The long bristles of the sea mouse look like iridescent fur.

Figure 27–20 Earthworms are hermaphrodites that undergo internal fertilization. The sperm from one worm fertilizes the eggs of its partner, and vice versa.

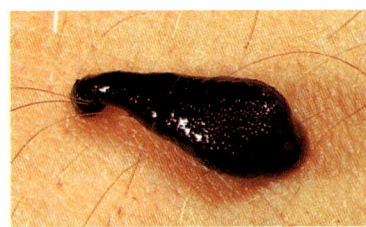

Figure 27–21 *A leech attaches to its host with suckers on its anterior and posterior ends (top). As it feeds, the body of the leech swells to accommodate as much as ten times its mass in blood (bottom).*

All leeches have powerful suckers at both ends of their bodies. These suckers—especially the anterior one, which usually surrounds the mouth—are used to attach a leech to its host. The posterior sucker is also used to anchor a leech to rocks, leaves, and other objects as it waits for a host to come by. Leeches penetrate the skin of their host in one of two ways. Some leeches use a muscular proboscis (proh-BAHS-ihs), or tubular organ, that they force into the tissue of their host. Others slice into the skin of their hosts with a razor-sharp pair of jaws. Once the wound has been made, the leech uses its muscular pharynx to suck blood from the area. Both types of leeches release a special secretion from their salivary glands to prevent the blood from clotting as they drink it. Some leeches also produce a substance that anesthetizes the wound—thus keeping the host from knowing it has been bitten!

During feeding, a leech can swallow as much as 10 times its weight in blood. Such a huge meal can take the leech up to 200 days to digest, with the help of symbiotic bacteria that live in its gut. A leech can live for a year before it must feed again.

How Annelids Fit into the World

Annelids are important in many habitats. Small polychaetes and their larvae are members of the ocean plankton, where they are food for many fishes, crabs, and lobsters. Bottom-dwelling polychaetes are important items in the diets of food fishes such as flounder.

Oligochaetes, particularly earthworms, perform an essential task in conditioning soil, as Charles Darwin noted in a lengthy and detailed study. By constantly burrowing through the ground, earthworms help aerate the soil. And by grinding and partially digesting the incredible amount of soil and detritus that passes through their guts, earthworms speed the return of nitrogen and other important nutrients from dead organisms to forms that can be used by plants. Without the continual efforts of these annelids, the structure and fertility of farm soils would degenerate quickly, lowering crop yields.

27–2 SECTION REVIEW

1. What is an annelid? List and give examples of three classes of annelids.
2. Discuss three adaptations for feeding in annelids.
3. Describe the structure of the digestive tract in an earthworm.
4. **Critical Thinking—Making Inferences** Explain why it is advantageous for an earthworm to have more light-sensitive cells in its anterior and posterior segments than in other parts of its body.

SCIENCE, TECHNOLOGY, AND SOCIETY CONNECTION

Leeches: Modern Applications of Ancient Medicine

There are few medical techniques as ancient as leeching, or applying leeches to a patient. The earliest known reference to leeching was written by a Greek physician more than 2200 years ago. And experts believe that leeching is much older than that!

Many people once believed that diseases could be cured by using leeches to remove blood from the patient. However, when people began to better understand the nature of disease, leeches ceased to be popular medical tools. After all, it seemed senseless to remove blood from a patient when it was clear that microorganisms—not "bad blood"—caused disease. But interestingly enough, leeches are once again in the medical spotlight.

One modern medical problem faced by surgeons is that blood tends to collect in body parts reattached by microsurgery. Here is where leeches come in handy. They are used to remove the excess blood until the blood vessels in the reattached part have healed.

The chemicals in leech saliva make it possible to use leeches for a variety of other medical purposes. These chemicals prevent blood from clotting, dissolve existing blood clots, expand blood vessels (to keep blood flowing), loosen the connections between cells (to help disperse the other chemicals), and anesthetize the area of the bite. Researchers are currently developing medicines based on chemicals extracted from leech saliva. These new medicines may soon be used to clear blocked blood vessels and to treat a variety of circulatory-system diseases.

Leeches also produce chemicals that harm bacteria—chemicals that they may inject into the host as they feed. Symbiotic bacteria inside the leech's gut produce an antibiotic that keeps stored blood fresh by killing bacteria. And the chemical in leech saliva that loosens or dissolves connections between host cells may also dissolve the protective coating on bacteria, thus making them vulnerable to an attack by the immune system.

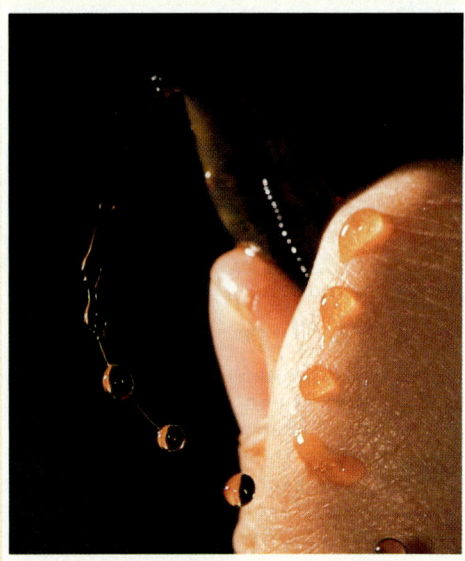

The leech shown here is being "milked" for its saliva. Researchers are currently developing medicines based on the chemicals found in leech saliva.

As medical researchers discover new uses for leeches, they are reminded that the ancient practice of leeching may not have been quite as senseless as it seemed. They are also reminded of the role evolution plays in shaping and refining the relationships between organisms (such as leeches and vertebrates). Of course, the leeches themselves didn't decide to perform a useful function for humans. But recall that parasites must evolve along with their hosts. Under pressure from natural selection, leeches have evolved adaptations that enable them to feed effectively on vertebrate hosts, including humans. Although leeches appeared on Earth long before humans, the chemicals they produce still affect us. Why? Due to common descent, our body chemistry is similar to that of other vertebrates—including the leech's original vertebrate host.

SCIENCE, TECHNOLOGY, AND SOCIETY CONNECTION

Leeches: Modern Applications of Ancient Medicine
After students have read the feature, pose the following questions for class discussion. Interested students may enjoy doing further research on the topic.

• **Leeches were not "designed" to feed on humans—they were active on Earth long before humans appeared. Why then do their "chemicals" work on us?** (Humans are related by evolution to the animals that leeches originally fed on. Similarities in body chemistry between humans and these early animals explain why the leech's chemicals are effective on humans.)

• **Why might leeches someday be useful in the treatment of heart attacks and strokes?** (Often these illnesses are associated with blood clots or blocked blood vessels. Chemicals in leech saliva have been shown to prevent blood from clotting, dissolve existing blood clots, and expand blood vessels to keep blood flowing.)

Reinforcement/Reteaching
If students are having trouble answering the Section Review questions, go back to the material with which they are having difficulty.

Closure
Have students work in small groups. Ask each group to make up 10 to 20 questions for a game called Annelid Worms—Fact or Fiction? The idea of the game is to make a statement about segmented worms that may or may not be true. For example: "Annelids can reproduce only sexually—fact or fiction?" In this case, the answer is fiction because a few species can reproduce asexually by budding.

When all groups have finished writing, pool the questions and eliminate any duplicates. Then place the remaining questions in a box or basket for members of the class to draw at random. To make the game more effective as a review activity, you might require students to give a reason to back up an answer of "fact," and a true statement to correct a question for which the answer is "fiction."

LABORATORY INVESTIGATION

OBSERVING EARTHWORM RESPONSES

Before the Lab

1. At least one day prior to the investigation, gather together enough materials for your class. Students may work in teams of up to 6.
2. To prepare storage containers for earthworms, use clear plastic boxes with several centimeters of sand on the bottom and several centimeters of topsoil over the sand. Boxes can be covered with plastic wrap.
3. To moisten the earthworms, use pond water or tap water that has been standing for a day.
4. Make sure that all the desk lamps are working. Have a few extra bulbs on hand.

Pre-Lab Discussion

Remind students of the demonstration on observing earthworms that was used to introduce this chapter. Point out that this investigation will pick up where the demonstration left off. In addition to observing earthworms, students will now test the earthworms' responses to several types of stimuli.

Read or have a student read the Problem at the beginning of the investigation.

• **How do you think moisture might be important to an earthworm?** (Answers may vary. Students should recall from the textbook that earthworms depend on moisture to carry out respiration through their skin.)

• **Do you think light is important to an earthworm?** (Answers may vary. Some students may think, incorrectly, that earthworms need light to "see" or to find food; others may realize, correctly, that earthworms prefer dark places where they are protected from predators.)

After students have had a chance to think about the importance of moisture and light, have them write predictions stating what they think the earthworms' responses will be. These predictions can be written in the form of a hypothesis. Students will then use the investigation to test their hypothesis.

LABORATORY INVESTIGATION

OBSERVING EARTHWORM RESPONSES

PROBLEM

How do live earthworms respond to moisture and light?

MATERIALS (per group)

2 live earthworms in a storage container
tray
paper towels
piece of cardboard
desk lamp
medicine dropper

PROCEDURE

1. Open the storage container and examine the earthworms. Record your observations of their physical characteristics. Fill the medicine dropper with water and use it to give your earthworms a "bath." **Note:** *Make sure you keep your earthworms moist by giving them frequent baths. If an earthworm's skin dries out, it dies.*
2. Fold a dry paper towel and place it on one side of your tray, as shown in the accompanying figure. Fold a dampened paper towel and place it on the other side of the tray.
3. Place the earthworms in the center of the tray, between the dry paper towel and the moist paper towel. Cover the tray with the piece of cardboard.
4. After 5 minutes, remove the cardboard and observe the location of the earthworms. Record your observations.
5. Return the earthworms to their storage container. Using the dropper, moisten the earthworms with water.
6. Cover the entire bottom of the tray with a damp paper towel.
7. Place the earthworms in the center of the tray.
8. Cover one half of the tray with the piece of cardboard. Position the lamp above the open side of the tray.
9. After 5 minutes, observe the location of the earthworms. Record your observations.
10. Return the earthworms to their storage container. Using the dropper, moisten the earthworms with water. Cover the container.

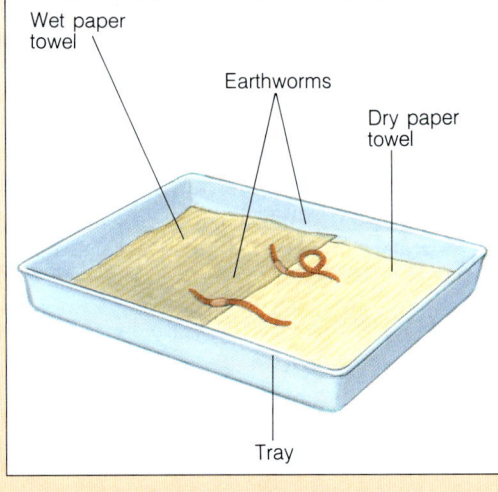

OBSERVATIONS

1. Which kind of surface did the earthworms prefer—moist or dry?
2. Do earthworms prefer light or darkness?
3. Describe the earthworms' color, texture, external features, and other physical characteristics.

ANALYSIS AND CONCLUSIONS

1. How does an earthworm's response to moisture help it survive?
2. Does an earthworm's response to light have any protective value? Explain.
3. How is an earthworm's body adapted for movement into and through soil?
4. Would you expect to find earthworms in hard soil? Explain.

Skills Development

Students will use the following skills while completing this laboratory investigation.
1. Observing
2. Predicting
3. Manipulative
4. Relating
5. Applying
6. Hypothesizing
7. Inferring

Safety Tips

Remind students of the rules for electrical safety when using the lamp. Emphasize that they

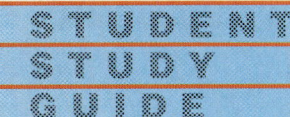

SUMMARIZING THE CONCEPTS

The key concepts in each section of this chapter are listed below to help you review the chapter content. Make sure you understand each concept and its relationship to other concepts and to the theme of this chapter.

27–1 Mollusks

- Mollusks are soft-bodied animals such as snails, clams, and squids that usually have an internal or external shell. The body plan of mollusks consists of four basic parts: foot, mantle, shell, and visceral mass.
- Most gastropods move by means of a broad, muscular ventral foot. Many gastropods have a one-piece shell.
- Bivalves have a hinged two-part shell. Although larvae are motile, most adult bivalves are sessile.
- Cephalopods have a well-developed nervous system, relatively advanced sense organs, and a closed circulatory system. A cephalopod's foot is divided into eight or more tentacles.
- Mollusks play many roles in the natural world. Many products that are important or valuable to humans are made by mollusks.

27–2 Annelids

- Annelids, which are also known as segmented worms, are round, wormlike animals with long, segmented bodies. An annelid's segments are very similar to one another and are separated by internal partitions.
- Polychaetes have a pair of paddlelike appendages on each segment. Most polychaetes are free-living marine worms.
- Oligochaetes have few bristles and lack appendages. Many are burrowing worms. Most live in fresh water or in soil.
- Leeches are typically blood-sucking external parasites that live in fresh water.
- Annelids interact in many different ways with other parts of the living world. Burrowing annelids such as earthworms are important in aerating soil.

REVIEWING KEY TERMS

Vocabulary terms are important to your understanding of biology. The key terms listed below are those you should be especially familiar with. Review these terms and their meanings. Then use each term in a complete sentence. If you are not sure of a term's meaning, return to the appropriate section and review its definition.

27–1 Mollusks
mollusk
foot
mantle
shell
visceral mass
radula
gill
open circulatory system
closed circulatory system
nephridium
gastropod
bivalve
cephalopod

27–2 Annelids
annelid
polychaete
oligochaete
leech

603

Observations
1. Moist.
2. Darkness.
3. Earthworm will be brown-orange color, moist and smooth to the touch, and segmented, and it will contain a distinct clitellum region. Other characteristics may be noted as well.

Analysis and Conclusions
1. Since an earthworm's skin must be kept moist, a positive response to moisture helps the earthworm to survive.
2. A negative response to light helps to keep the earthworm underground during the day, and this may help it to avoid drying out.
3. The long and slender body of the earthworm moves easily through the soil. The bristles on its outer surface provide more traction for increased mobility.
4. With its soft body, the earthworm would have difficulty penetrating hard soil.

Going Further: Enrichment
An excellent extension of this investigation would be for students to carry out a dissection of an earthworm. They could then make diagrams showing the location of digestive organs and other structures.

---

should NEVER touch the lamp or the cord with wet hands.

Teaching Strategy
1. As students begin the investigation, have them note the places in the procedure where they are told to moisten the earthworms (steps 1, 5, 10). Point out that this is essential to the survival of the worms and must be done at least as often as the procedure dictates. Stress that if the worms begin to look dry, students should moisten them immediately.

2. Circulate throughout the classroom and help any groups that are having trouble. Be sure that in steps 2 and 8, each group has correctly divided the tray to simulate moisture and dryness and light and darkness.

CHAPTER REVIEW

CONTENT REVIEW

Multiple Choice

Choose the letter of the answer that best completes each statement.

1. Which characteristic do many mollusks and annelids have in common?
   a. segmented body
   b. one- or two-part shell
   c. open circulatory system
   d. trochophore larvae
2. One major class of mollusks is
   a. Cephalopoda.   c. Oligochaeta.
   b. Hirudinea.   d. Polychaeta.
3. A mollusk that swims by flapping its broad, muscular foot is probably a
   a. bivalve.   c. cephalopod.
   b. gastropod.   d. polychaete.
4. Which organ is used for both respiration and filter feeding in some animals?
   a. nephridium   c. gill
   b. radula   d. ganglion
5. A bristly marine worm that has paired appendages on each segment belongs to Class
   a. Annelida.   c. Oligochaeta.
   b. Polychaeta.   d. Cephalopoda.
6. An oligochaete probably
   a. is a parasite.
   b. has paired appendages.
   c. has septa.
   d. has a mantle, foot, and visceral mass.
7. A scraping organ used for feeding is a
   a. nephridium.   c. mantle.
   b. pharynx.   d. radula.
8. In earthworms, the clitellum
   a. is involved in asexual reproduction.
   b. secretes a cocoon for the eggs.
   c. often has a pair of jaws.
   d. grinds food particles into smaller pieces.

True or False

Determine whether each statement is true or false. If it is true, write "true." If it is false, change the underlined word or words to make the statement true.

1. A pharynx is an organ used in excretion.
2. Softbodied animals that typically have a shell are known as oligochaetes.
3. Segmented worms belong to phylum Hirudinea.
4. Pearls and the shells of mollusks are formed by secretions from the radula.
5. Many leeches are blood-sucking parasites.
6. Cephalopods are characterized by a one-part shell and a broad, muscular foot.
7. Gastropods are usually sessile as adults.
8. Hermaphrodites usually undergo external fertilization.

Word Relationships

A. An analogy is a relationship between two pairs of words or phrases generally written in the following manner: a:b::c:d. The symbol : is read "is to," and the symbol :: is read "as." For example, cat:animal::rose:plant is read "cat is to animal as rose is to plant."
   In the analogies that follow, a word or phrase is missing. Complete each analogy by providing the missing word or phrase.

1. one-part shell:gastropod::two-part shell:_____
2. cocoon:clitellum::shell:_____
3. shell:snail::nematocyst and chemicals:_____
4. light detection:ocelli::balance:_____

CHAPTER REVIEW

CONTENT REVIEW

Multiple Choice
1. d 2. a 3. b 4. c
5. b 6. c 7. d 8. b

True or False
1. F nephridium
2. F mollusks
3. F Annelida
4. F mantle
5. T
6. F Gastropods
7. F Bivalves
8. F internal

Word Relationships
Part A
1. bivalve
2. mantle
3. nudibranch
4. statocyst

Part B
5. closed circulatory system
6. annelids
7. foot

CONCEPT MASTERY

1. Carnivores often have venom to kill or paralyze prey and mouthparts such as a beak and a radula for cutting up the body of their prey. Cephalopods have tentacles for capturing prey. The radula in herbivores is adapted for scraping up bits of plant material. Filter feeders such as clams have siphons for taking in and expelling water and sticky mucus on the gills that traps food particles.
2. Feeding: Polychaetes show a wide variety of feeding modes; earthworms, the only oligochaetes that students have learned about in detail, are nonselective detritus feeders. Respiration: Many polychaetes have gills; earthworms use their skin for gas exchange. Internal transport: Both polychaetes and oligochaetes have closed circulatory systems. Excretion: Both polychaetes and oligochaetes have nephridia. Response: Polychaetes have eyes, sensory tentacles, and other relatively well-developed sense organs; oligochaetes do not. Movement: Polychaetes have parapodia; oligochaetes use muscular contractions of the body wall. Reproduction: Some polychaetes reproduce asexually through fission or budding, but most species reproduce sexually and have separate sexes and external fertilization; oligochaetes such as earthworms are hermaphrodites that undergo sexual reproduction, with fertilization taking place internally.
3. Clams have flattened shells and spade-shaped feet, and many have siphons that can protrude above the surface

B. *Replace the underlined definition with the correct vocabulary word.*

5. Octopi have a circulatory system in which the blood is always contained in blood vessels.
6. Earthworms are members of the segmented worms phylum.
7. The part of a mollusk that contains the mouth and is often used in locomotion in cephalopods is divided into tentacles.

CONCEPT MASTERY

Use your understanding of the concepts developed in the chapter to answer each of the following in a brief paragraph.

1. How are mollusks adapted to different modes of feeding?
2. Compare the ways in which polychaetes and oligochaetes perform their essential functions.
3. How are clams adapted for burrowing in mud and sand?
4. How do mollusks fit into the world?
5. Explain why a person might purchase earthworms to put in a garden.

CRITICAL AND CREATIVE THINKING

Discuss each of the following in a brief paragraph.

1. **Assessing concepts** Although a number of animals are hermaphrodites, they rarely fertilize their own eggs. Explain why cross-fertilization is usually better than self-fertilization. Under what circumstances might self-fertilization be better than cross-fertilization?

Although many nudibranchs are simultaneously male and female, they do not fertilize their own eggs.

2. **Making inferences** Some oligochaetes can survive in areas that have little oxygen and can even tolerate a complete lack of oxygen for short periods of time. Some of these oligochaetes die when exposed to normal oxygen levels for a long period of time. What is probably the natural habitat of these oligochaetes? Explain.
3. **Developing a hypothesis** Female octopi die after brooding their eggs (tending and protecting eggs until they hatch). However, if certain glands near the brooding octopus's eyes are surgically removed, the octopus stops brooding, resumes feeding, and has a life span longer than the normal three to four years. Develop a hypothesis to explain this phenomenon. How might you go about testing your hypothesis?
4. **Using the writing process** Suppose that the topic of a debate is, "Resolved: It is better to be a free-swimming polychaete than a sessile one." Take either the affirmative or the negative stance and prepare a persuasive argument for your position.

605

CHAPTER 28 Arthropods

Section	Laboratory Investigations and Activities
28–1 Introduction to Arthropods pages 607–616 THEMES: Unity and Diversity, Scale and Structure	**Teacher Edition** DEMONSTRATION: The Diversity of Arthropods, p. 606D **Teaching Resources** ECOLOGY INVESTIGATION: Flotation Extraction of Arthropods from Soils, p. 37
28–2 Spiders and Their Relatives pages 617–620 THEMES: Unity and Diversity, Scale and Structure	**Teaching Resources** HANDS-ON ACTIVITY: Spinning Webs, p. 145
28–3 Crustaceans pages 620–621 THEMES: Unity and Diversity, Scale and Structure	**Laboratory Manual** Isopod Environments, p. 367 **Teaching Resources** HANDS-ON ACTIVITY: Off and Running, p. 143
28–4 Insects and Their Relatives pages 622–628 THEMES: Unity and Diversity, Scale and Structure	**Student Edition** LABORATORY INVESTIGATION: Observing Insect Metamorphosis, p. 632 **Teacher Edition** DEMONSTRATION: Insect Life Cycle, p. 606D **Laboratory Manual** The Grasshopper, p. 359
28–5 How Arthropods Fit into the World pages 629–631 THEMES: Systems and Interactions	**Teaching Resources** ECOLOGY INVESTIGATION: Observing Soil Arthropods, p. 39 ECOLOGY INVESTIGATION: Reactions of Arthropods to Different Textures, p. 41
Chapter Review pages 633–635	

CHAPTER PLANNING GUIDE

Other Activities	Media and Technology
Study Guide Section 28–1, p. 265	**Interactive Videodisc** Insects: Little Giants of the Earth On Dry Land: The Desert Biome **Video/Videodisc** Seeing Sense **Biology Media Guide** pages 30–33
Teaching Resources ACTIVITY: Studying Spider Webs, p. 3 **Study Guide** Section 28–2, p. 268	**Interactive Videodisc** Insects: Little Giants of the Earth **Biology Transparencies** 34: The Anatomy of a Spider **Biology Media Guide** page 33
Teaching Resources ACTIVITY: Investigating Crustaceans, p. 5 **Study Guide** Section 28–3, p. 270	**Interactive Videodisc** Insects: Little Giants of the Earth On Dry Land: The Desert Biome **Biology Transparencies** 35: The Anatomy of a Crayfish **Biology Media Guide** page 33
Study Guide Section 28–4, p. 272	**Interactive Videodisc** Insects: Little Giants of the Earth On Dry Land: The Desert Biome **Videodisc** Super Scents **Biology Transparencies** 33: The Anatomy of a Grasshopper **Biology Media Guide** pages 34–36
Teaching Resources VOCABULARY REVIEW: Cross-a-Clue, p. 1 BIOLOGY CASE STUDY: Census of Soil Arthropods, p. 9 **Study Guide** Section 28–5, p. 275	**Interactive Videodisc** Insects: Little Giants of the Earth On Dry Land: The Desert Biome **Biology Media Guide** pages 36–37
Teaching Resources Chapter Test A, p. 11 Chapter Test B, p. 15	**Computer Test Bank** Chapter Test, p. 297

CHAPTER 28 Arthropods

CHAPTER OVERVIEW

In terms of sheer numbers of organisms, varieties of species, and diversity of habitats, arthropods are the most successful animals on Earth. From their first appearance in the sea over 600 million years ago, arthropods have been adapting and evolving into the organisms we know today. Arthropods include animals such as insects and spiders, crabs and lobsters, and centipedes and millipedes.

The common features shared by all arthropods include a chitinous exoskeleton, jointed appendages, and a segmented body. Other characteristics of the arthropods, shared with other invertebrate groups, are a distinct head region with a dorsal brain, a ventral nerve cord, and an open circulatory system with a single-chambered heart. Because of the incredible variety of niches occupied by different groups of arthropods, many have highly specialized structures that are not found in other organisms. Some insects have developed complex patterns of growth and reproduction, communication, and social behavior.

In their interactions with humans and other organisms, arthropods play a wide variety of roles, including those of predator and prey, decomposer and symbiont. Not only are they an important food source by themselves, they also play a vital role in successful agricultural cultivation. However, many arthropods also produce deadly poisons, ruin crops, and transmit diseases. At some point in its life cycle, virtually every organism on Earth will encounter members of the phylum Arthropoda.

28-1 INTRODUCTION TO ARTHROPODS

Section Focus 28-1

The purpose of this section is to provide students with an appreciation for the diversity of arthropods and acquaint them with arthropod taxonomy, evolutionary development, and general structure and function.

The four subphyla of the Arthropoda are named and described, and their evolutionary development from various ancestral groups is discussed.

The characteristics shared by all arthropods include a chitinous exoskeleton, jointed appendages, and a segmented body. The tremendous variety of niches occupied by the arthropods has required specialization of function, and structures have become greatly modified among the different subphyla. For example, while aquatic arthropods have developed gills or book gills and terrestrial arthropods have tracheal tubes and book lungs, many arthropods have no specialized respiratory structures at all.

Performance Objectives 28-1

1. Describe the characteristics that all arthropods have in common.
2. Name the four arthropod subphyla and some of the representative organisms in each.
3. Discuss the specific adaptations of various arthropods for the performance of their life functions, including locomotion, feeding, respiration, internal transport, excretion, response to the environment, and reproduction, growth, and development.

Science Terms 28-1

arthropod   p. 607
trilobite   p. 607
chelicerate   p. 607
crustacean   p. 607
uniramian   p. 607
exoskeleton   p. 609
chitin   p. 609
molt   p. 615
metamorphosis   p. 615
pupa   p. 616

28-2 SPIDERS AND THEIR RELATIVES

Section Focus 28-2

The purpose of this section is to describe the members of the subphylum Chelicerata, characterized by a body divided into two parts, the cephalothorax and abdomen, and two pairs of appendages adapted as specialized mouthparts, the chelicerae and the pedipalps.

Among the members of this group are the horseshoe "crabs" (*Limulus*), which are "living fossils." The newly hatched young greatly resemble extinct trilobites.

The arachnids are the most familiar members of this subphylum and include the spiders, scorpions, ticks, and mites. All adults have four pairs of walking legs on their cephalothorax.

Spiders are predators whose chelicerae are hollow, fanglike, and capable of injecting paralyzing venom into their prey. They also have abdominal glands that produce silk.

Mites and ticks are small arachnids that are generally parasites of animals and plants. Scorpions are carnivores found in warm regions throughout the world. Their pedipalps are enlarged into claws, and they have a venomous barb at the end of their abdomen that is used to paralyze or kill prey.

Performance Objectives 28-2

1. Describe the characteristics of the subphylum Chelicerata.
2. Name the organisms that are members of the subphylum Chelicerata.
3. Describe some of the specialized structures for trapping prey and feeding that are found in chelicerates.

Science Terms 28-2

arachnid   p. 618

28-3 CRUSTACEANS

Section Focus 28-3

The purpose of this section is to describe the members of the class Crustacea, which are generally aquatic organisms that have an exoskeleton composed of calcium carbonate, two pairs of antennae, and mouth-parts called mandibles. Because crustaceans are found in marine, freshwater, and terrestrial habitats, their structure and function vary greatly. The appendages, including the antennae and mandibles, have been adapted for many different purposes. For example, the antennae may serve as sense organs for some crustaceans, whereas in others they may be adapted for filter feeding or locomotion.

The crayfish is used as the representative species, and the modifications of its appendages for specific functions are discussed in detail.

Performance Objectives 28–3
1. Describe the characteristics of the subphylum Crustacea.
2. Name some of the organisms that are members of the subphylum Crustacea.
3. Using the crayfish as the representative species, describe how the appendages of Crustacea have become modified.

Science Terms 28–3
mandible   p. 621

28–4 INSECTS AND THEIR RELATIVES
Section Focus 28–4
The purpose of this section is to describe the subphylum Uniramia, which includes the insects, centipedes, and millipedes. Although mostly terrestrial, these organisms are found in virtually every habitat and include more species than all other living animals on Earth. All uniramians have one pair of antennae and appendages that do not branch.

Centipedes and millipedes both resemble segmented worms, with many legs, but differ in that centipedes have one pair of legs per segment whereas millipedes have two. Centipedes are carnivores with a pair of poison claws in their head region, whereas millipedes are decomposers of plant material.

Although insects are extremely diverse in structure and function, they are all characterized by a body divided into a head, thorax, and abdomen, with three pairs of legs attached to the thorax. Most insects also have a single pair of antennae, a pair of compound eyes, two pairs of wings on the thorax, and tracheal tubes for breathing.

With almost a million species, insects exhibit a tremendous variety of adaptations for feeding, locomotion, and reproduction. In addition, many insects—most notably the termites, wasps, bees, and ants—live in societies with division of labor, a caste system, and complex methods of communication.

Performance Objectives 28–4
1. Identify the characteristics of the subphylum Uniramia.
2. Name the organisms that are members of the subphylum Uniramia.
3. Contrast the millipedes and the centipedes.
4. Describe different insect adaptations for feeding, movement, and reproduction.
5. Describe the complex social behavior within groups of insects such as termites, wasps, bees, and ants.

Science Terms 28–4
pheromone   p. 626

28–5 HOW ARTHROPODS FIT INTO THE WORLD
Section Focus 28–5
The purpose of this section is to describe the interrelationship among the arthropods and other organisms, including humans.

Arthropods interact with other organisms not only as both predator or prey, but also in a variety of symbiotic relationships.

Without insects to cross-pollinate crops, certain types of agriculture would be impossible. Moreover, crustaceans are an important human food source all over the world, and insects are eaten in Africa and Asia.

Chemicals extracted from arthropods, or synthetic versions of them, are being used in a wide variety of ways. Some examples include chitin, which is used in surgical dressings and sutures, and components of spider venom, which are being tested for use as pesticides.

Performance Objectives 28–5
1. Describe the roles played by arthropods in predator-prey and symbiotic relationships with other organisms.

DISCOVERY LEARNING

TEACHER DEMONSTRATIONS
Modeling
The Diversity of Arthropods
This demonstration might be used as an introduction to Chapter 28.

Obtain illustrations of various arthropods from magazines or books. It would be best to have them individually mounted. Or, as an alternative, use commercial 35-millimeter slides that show different arthropods. Include a picture of the extinct trilobite.

Ask students to observe the variety of arthropods.

• **Why should all of these different organisms be placed in the same phylum? What characteristics do they have in common?** (Students should notice the segmentation, the jointed appendages, and the exoskeleton.)

• **What kinds of variations exist in these common characteristics?** (Students should mention the degree of obvious segmentation, the number of appendages, the modifications of appendages, the degree of development of the head region, among others. Mention that these differences are the basis for placing arthropods into different subphyla.)

Insect Life Cycle
You may want to perform this demonstration while discussing Section 28–4.

To show that insects become inactive in colder weather, place a vial of *Drosophila* in crushed ice in a large beaker. Leave for 10 minutes, then observe. As the vial warms, flies should become more active. Discuss the differences between "warm-blooded" and "coldblooded" animals. Elicit the idea that flies and most other insects are coldblooded.

CHAPTER 28
Arthropods

GUIDED ENQUIRY

Pose the following questions to students and have them record their responses. Point out that they will gain a better understanding of the key concepts if they read the chapter with these basic questions in mind. Upon completion of the chapter, pose the questions again. Ask students to compare their initial responses with those they have developed after reading the chapter.

- What structural features do all arthropods have in common?
- What are the four subphyla into which arthropods are divided?
- What are some of the organisms in each of the four arthropod subphyla?
- How do different arthropods carry out their life functions in their particular environments?

CHAPTER 28
Arthropods

Arthropods display a wide range of forms and habits. The butterfly is a terrestrial herbivore. The crab (inset) is a marine carnivore.

Beneath the frigid sea off Alaska, a giant king crab scuttles along the ocean floor on legs nearly a meter long. In the lukewarm waters of a Louisiana bayou, two crayfish battle for control of a burrow beneath a cypress tree. And in the heart of a Florida swamp, 43 different species of mosquitoes swarm in such numbers that the buzzing of their wings fills the hot, humid air.

All these animals, and more than a million other species, belong to the phylum Arthropoda—the largest, most diverse, and arguably the most successful of all the animal phyla. Arthropods live in virtually every habitable environment on Earth—from ice fields in the polar regions to brine pools on the equator. Some arthropods are among the most destructive animals on Earth. Others are beneficial, and even essential, to the survival of other organisms.

What are arthropods? What common characteristics do they possess? How do they carry out their life functions? You will discover the answers to these questions in the pages that follow.

INTRODUCING CHAPTER 28

Ask students to read the introduction to the chapter, and then do the chapter-opener demonstration.

After this, ask students to spend about 5 minutes writing down the names of all arthropods that come to mind. Write the names on an overhead transparency or on the chalkboard.
- **You have read that there are more arthropods on Earth than any other phylum. What characteristics can we use to divide this long list of organisms into smaller groups?** (Students may mention habitat, food sources, or perhaps winged vs. nonwinged. Elicit that the structure and function of these organisms is the basis on which the organisms are separated into subphyla.)

Instruct students to mark the arthropods on their list with a *T* if they are terrestrial, with an *A* if they are aquatic.
- **What does the habitat of most arthropods seem to be?** (There are more terrestrial organisms.) Instruct students to put a check next to the arthropods that directly or indirectly have an effect on humans.
- **How many arthropods did you check?** (Virtually all will be checked.)

GUIDE FOR READING

After you read the following sections, you will be able to

28-1 Introduction to Arthropods
- Discuss the characteristics and classification of arthropods.
- Explain how arthropods perform essential life functions.

28-2 Spiders and Their Relatives
- Discuss the distinguishing characteristics of chelicerates.

28-3 Crustaceans
- Discuss the distinguishing characteristics of crustaceans.

28-4 Insects and Their Relatives
- Describe and give examples of three classes in the subphylum Uniramia.
- Discuss the anatomy of a typical insect.
- Explain ways insects communicate.

28-5 How Arthropods Fit into the World
- Describe how arthropods interact with other living things.

Journal Activity

YOU AND YOUR WORLD

Bugs! Creepy crawly insects. The mere thought of them makes some people cringe. But not all insects are harmful. Pretend that you are a press agent for insects. Write a short press release informing the public about the benefits of sharing our world with insects.

28-1 Introduction to Arthropods

Guide For Reading

■ What are the four subphyla of arthropods?
■ What are three important arthropod features?
■ How are arthropods adapted for performing life functions?
■ What is metamorphosis?

To describe even a fraction of the living **arthropods** would take several books. More than a million arthropod species have been described, and scientists are certain there are many more that have not as yet been found! As you will soon learn, members of this phylum vary enormously in size, shape, and habits.

Diversity and Evolution in Arthropods

You already know about many common arthropods. In fact, you have probably even eaten a number of them! Today most biologists divide arthropods into four subphyla:

- *Trilobita* This is thought to be the oldest subphylum of arthropods. **Trilobites** (TRIGH-loh-bights) were dwellers in ancient seas. They are now all extinct.
- *Chelicerata* **Chelicerates** (keh-LIHS-er-ayts) include spiders, ticks, mites, scorpions, and horseshoe crabs.
- *Crustacea* **Crustaceans** (kruhs-TAY-shuhnz) include such familiar (and edible) organisms as crabs and shrimp.
- *Uniramia* **Uniramians** (yoo-nih-RAY-mee-ahnz) include most arthropods: centipedes, millipedes, and all insects—including bees, moths, grasshoppers, flies, and beetles.

Why are there so many different kinds of arthropods? One reason is that they have been evolving on Earth for a long time. The first arthropods appeared in the sea more than 600 million years ago. Since that time, these animals have experienced several adaptive radiations. Some arthropods have remained in

Figure 28-1 The scorpion, a chelicerate, is feeding on a hawk moth, a uniramian (left). The white-booted shrimp is a crustacean (right).

BACKGROUND INFORMATION
ARTHROPOD CLASSIFICATION AND EVOLUTION

As you might expect, the classification and evolutionary history of a phylum as large and diverse as the arthropods is under discussion. We have followed the conventional system of dividing the group into four subphyla, but other possibilities are favored by different experts. Some researchers have insisted for years that the four main groups do not share a common ancestor and should therefore be considered as separate phyla. They even suggest that the name Arthropod be promoted to an umbrella "superphylum." Some molecular biologists argue for an opposite viewpoint. After studying similarities and differences among the mitichondrial genes that code for ribosomal RNA in these animals, at least one research team not only argues that all arthropods share a common ancestor, but insists that the onycophorans share the ancestor as well. These molecular data also suggest the crustaceans and insects are the two most closely related arthropod subphyla. These new molecular data are still being studied, and the conclusions formed from these data are still open to debate.

MULTICULTURAL STRATEGY

Have students research the life and work of Charles H. Turner, an African American entomologist who lived from 1861 to 1923. Also discuss with students the field of entomology in general and how entomological research has applications in agriculture, medicine, and environmental science.

Figure 28–2 The velvet worm is traditionally placed in its own phylum along with other organisms that have characteristics of both annelids and arthropods. Such animals support the hypothesis that modern annelids and uniramian arthropods descended from a common ancestor. However, some molecular biologists now feel that these animals should be placed in the phylum Arthropoda.

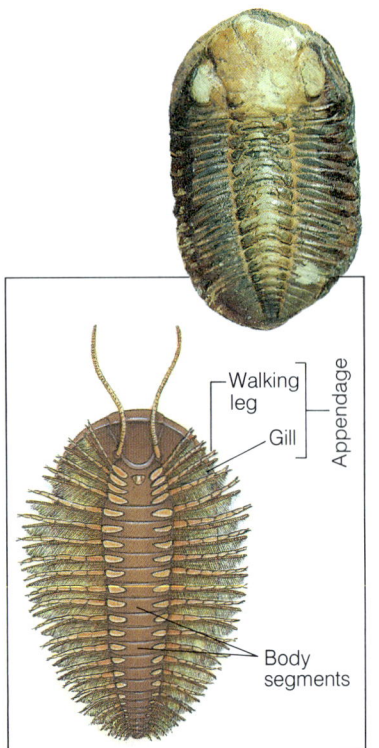

the water, where they have colonized all parts of the sea and most freshwater habitats. Others were among the very first members of the animal kingdom to colonize the land. The descendants of those pioneers were on hand when the first flowering plants appeared millions of years later.

The roots of the arthropod family tree are cloaked in mystery because the ancestors of the arthropods were soft-bodied animals that left few fossils. But by studying both living and fossil invertebrates, researchers have accumulated many clues to arthropod evolutionary history. Insects, centipedes, and millipedes seem to have evolved from ancestors that were closely related to the ancestors of modern annelid worms. Living evidence for this line of descent can be found in wormlike animals that live in the tropics. See Figure 28–2. Other arthropods, including crustaceans, spiders, and the extinct trilobites, evolved from more ancient and more distantly related ancestors.

The body form of the earliest arthropods is thought to be similar to that of the trilobites. A typical trilobite's body had a thick, tough outer covering and was composed of many segments, each of which bore a pair of appendages. Each appendage was divided to form two branches, one a walking leg and one a featherlike gill. See Figure 28–3.

Figure 28–3 The dorsal side of a fossil trilobite clearly shows the three lengthwise body lobes that give the animal its name (inset). An artist's rendering of the ventral side of a trilobite as it might have appeared when the animal was alive reveals numerous similarly shaped appendages.

28–1 (continued)

Reinforcement/Reteaching Using the Visuals It may be advisable to discuss with students the meaning of the term *appendage* at this point. Basically, an appendage is an "attachment" to a body segment of an arthropod. Refer students to Figure 28–2 and point out that each pair of legs of a velvet worm is a pair of appendages attached to a segment. You might mention at this point that in other arthropods that will be studied, segments have fused and appendages have become modified to perform many functions other than locomotion.

Most living arthropods exhibit two evolutionary trends away from the trilobite form. First, many have far fewer body segments. The many segments found in their embryos fuse into larger segments during development. Second, arthropod appendages have become increasingly specialized for feeding, locomotion, and other functions.

Form and Function in Arthropods

Although living arthropods are quite different from one another, all arthropods exhibit several key features. **The three most important arthropod features are a tough exoskeleton, a series of jointed appendages, and a segmented body.** Other characteristics of arthropods include a brain located in the dorsal part of the head, a ventral nerve cord, and an open circulatory system powered by a single heart.

THE ARTHROPOD BODY PLAN The **exoskeleton** (*exo-* means outside) is a system of external supporting structures that are made primarily of the carbohydrate **chitin** (KIGH-tihn). Some exoskeletons, such as those of most insects, are leathery and flexible. Others, such as those of ticks and lobsters, are extremely hard. These exoskeletons provide excellent protection from physical damage. The exosketetons of many terrestrial arthropods are waterproof. This adaptation restricts the loss of water from the body and makes it possible for arthropods to live in extremely dry environments such as deserts. The exoskeleton also helps arthropods move efficiently and adapt to their environment in many other ways.

Although the exoskeleton protects an arthropod's body like a suit of armor, it has the disadvantages you might expect from such a covering. Because an exoskeleton is a solid coating, not a living tissue, it cannot grow as the animal grows. (You will learn how arthropods deal with this problem shortly.) And movement can occur only at the joints of the "armor."

Figure 28–4 Some arthropods, such as the spiny lobster (center) and the tick (top), have extremely hard, tough exoskeletons. The mouthparts of a tick (inset) are adapted for biting and hanging onto a host. Other arthropods, such as the emperor gum moth caterpillar (bottom), have flexible, leathery exoskeletons.

TEACHING SUPPORT

Teaching Resources
- Ecology Investigation: Flotation Extraction of Arthropods from Soils, p. 33

MEDIA AND TECHNOLOGY
- Video/Videodisc
  - Seeing Sense
- Interactive Videodisc
  - On Dry Land: The Desert Biome
  - Insects: Little Giants of the Earth

Use this bar code to introduce types of arthropods.

Play frames 4402 to 5169

See the Biology Media Guide pages 30–33 for additional bar-code correlations for these media.

ECOLOGY NOTE
ARTHROPOD DECOMPOSERS

The importance of the role of certain arthropods as decomposers is often overlooked. Arthropods, along with other decay organisms or saprobes, help break down dead plants, animals, and animal wastes. Thus, the chemical compounds contained in these organisms are made available to nourish the growth of new organisms. For example, without termites and carpenter ants to break down wood into sawdust, forests would be choked with dead trees that would take many more years to decay than they do now.

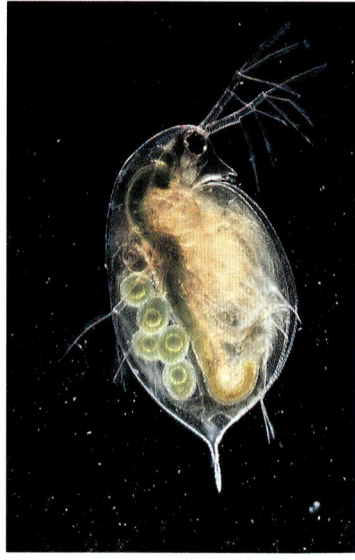

Figure 28–5 The waterflea is a tiny freshwater crustacean that lacks external segmentation and uses its antennae to propel it through the water. The round structures on the waterflea's back are its eggs.

All arthropods have jointed appendages (*arthro-* means joint; *-pod*, which literally means foot, refers to the appendages) that enable them to move. In primitive arthropods, such as trilobites, every body segment carries a single pair of appendages. But in species in which body segments are fused together, some appendages have been lost. Over millions of years, the remaining appendages have evolved into marvelously versatile adaptations to different environments. Arthropod appendages include antennae, claws, walking legs, wings, flippers, and other specialized structures.

All arthropods have segmented bodies. Some, such as millipedes and centipedes, have long, wormlike bodies with many visible segments. Others, such as insects, spiders, and crabs, have lost some segments in the course of evolution or had the segments fuse together to form a few large body parts.

FEEDING The appendages of arthropods have evolved in ways that enable these animals to eat almost any food you can imagine. Every mode of feeding is seen in arthropods—herbivores, carnivores, parasites, filter feeders, and detritus feeders. Although some herbivores, such as locusts, eat just about anything green, other herbivores are more selective. Some herbivores are specialized to eat specific parts of plants. Others feed exclusively on a particular kind of plant. Some carnivores, such as spiders, praying mantises, centipedes, and king crabs, catch and eat other animals. Other carnivores, such as many crabs and crayfish, feed primarily on animals that are already dead. External parasites—such as ticks, fleas, and lice—drink the blood and body fluids or nibble on the skin of other animals, including humans. Some internal parasites passively absorb nutrients through the body wall, whereas others eat away at the host from inside. Many marine arthropods are filter feeders that use comblike bristles on their mouthparts or legs to filter tiny plants and animals from the water.

Figure 28–6 The praying mantis (left) is a carnivore. The lubber grasshopper (right) is a herbivore.

28-1 (continued)

Focus/Motivation
Again referring students to the arthropod specimens, have them pay particular attention to the appendages.

Content Development
Point out or elicit from students that the appendages have become modified in different arthropods to perform mainly three functions: locomotion, feeding, and respiration. (In some arthropods, a small number of appendages are also sense organs.) The great variety of ways in which arthropods perform these functions is reflected in the structure of the appendages. Ask students to examine the arthropod specimens and describe how the appendages of the representatives of each subphylum are modified. (Since the subphyla have not yet been studied in detail, the descriptions may be fairly general.)

- **How are the appendages of the different arthropods modified for different types of locomotion?**

RESPIRATION Arthropods have evolved three basic types of respiratory structures—gills, book gills and book lungs, and tracheal tubes. Although most arthropods have only one of these types of respiratory structures, a few species have both book lungs and tracheal tubes. And some species completely lack specialized respiratory organs.

Many aquatic arthropods, such as crabs and shrimp, have gills that look like a row of feathers located just under cover of their exoskeleton. These gills are formed from part of the same appendages that form mouthparts and legs. Movement of the mouthparts and other appendages keeps a steady stream of water moving over the gills.

Book gills (which are found in horseshoe crabs) and book lungs (which are found in spiders and their relatives) are unique to these arthropods. In both these structures, several sheets of tissue are layered like pages in a book. The many tissue layers increase the surface area for gas exchange. A horseshoe crab's book gills are carried beneath its body, whereas a spider's book lungs are contained inside a sac within the body. An opening called the spiracle (SPIHR-ah-kuhl) connects the sac containing the book lungs with the fresh air outside.

Most terrestrial arthropods—insects, some spiders, and millipedes, for example—have another respiratory device found in no other animals. From spiracles, long branching tracheal tubes reach deep into the animals' tissues. The network of tracheal tubes supplies oxygen by diffusion to all body tissues. As the arthropods walk, fly, or crawl, the movements of their body muscles cause the tracheae to shrink and expand, pumping fresh air in and out of the spiracles. Tracheal tubes work well only in small animals; large animals require a more efficient way to deliver oxygen and remove carbon dioxide.

Figure 28–7 Although a fiddler crab spends much time on land, it uses gills for respiration. The male crab's large claw is used to attract females and to fight with other males.

Figure 28–8 The internal structures of a representative arthropod—a grasshopper—are shown here.

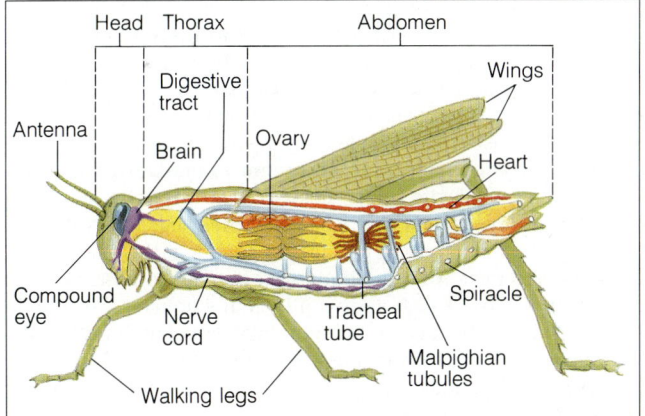

ESL STRATEGY

Make sure students remember that antonyms are words that are opposite in meaning and synonyms are words that have the same or similar meanings. When discussing the terms *exoskeleton* and *molting*, acquaint students with the concept of defining words by using antonyms and synonyms. Introduce *endoskeleton* as an antonym of *exoskeleton* and *shedding* as a synonym of *molting*.

Ask students to list the subphyla that make up the phylum Arthropoda. Have them choose one example from each subphylum and provide an illustration of it with labels indicating the three characteristics common to all arthropods.

BACKGROUND INFORMATION
INSECT RESPIRATION

Aside from their exoskeletons, the main limitation on the size of insects is their inefficient respiratory system. When an insect is at rest and its need to obtain oxygen and get rid of carbon dioxide is relatively low, it may keep all but one or two pairs of spiracles closed to conserve moisture. When active, with all spiracles open, small insects must rely on air pressure to force air into the tracheal tubes. Larger insects may be able to "pump" air in and out of the tracheal tubes and spiracles through movements of associated muscles.

(Although answers will vary, students should mention the various specialized structures.)

• **How do the anterior appendages, or mouthparts, of the arthropods show that they have different feeding habits?** (Although answers will vary, students should observe differences in the structure of the mouthparts of herbivores, carnivores, parasites, and filter feeders.)

• **In which of the arthropods have the appendages become modified for respiration?** (Crustaceans—students should be able to observe the gills by lifting the exoskeleton in the thoracic region of their specimens.)

Reinforcement/Reteaching

The system of tracheal tubes and spiracles is unique among invertebrates and deserves further discussion. Although extremely good at preventing water loss from the insect's body, the tracheal tubes are relatively inefficient in transporting gases. Because of their high level of activity, this limits the size of insects, since air could not reach the tissues of a large animal fast enough.

BACKGROUND INFORMATION
ARTHROPOD BEHAVIOR

One of the great differences between animals such as the arthropods and "higher" animals is the fact that arthropods' responses depend only on the various stimuli received by their nerves. Unlike higher animals, arthropods are not capable of "thought" or "decision making," and so their reactions in a particular situation are almost totally predictable.

Figure 28–9 The diamond beetle (left) uses Malpighian tubules to get rid of nitrogen-containing wastes, whereas the hermit crab (right) uses its green glands and gills.

INTERNAL TRANSPORT In arthropods, a well-developed heart pumps blood through an open circulatory system. In spiders and some insects, the heart is long and narrow and stretches along the abdomen. In lobsters and crayfish, the heart is smaller and lies about halfway down the body. When the heart contracts, it pumps blood through arteries that branch into smaller vessels and enter the tissues. There the blood leaves the vessels and moves through spaces in the tissue called sinuses. Eventually, the blood collects in a large cavity surrounding the heart, from which it re-enters the heart through small openings and is pumped around again.

EXCRETION In arthropods, as in many other animals, undigested food becomes solid waste that leaves through the anus. The nitrogen-containing wastes that result from cellular metabolism are removed in different ways in different arthropods. Most terrestrial arthropods, such as insects and spiders, dispose of nitrogen-containing wastes by using a set of Malpighian tubules (mal-PIHG-ee-an TOO-byools). Malpighian tubules, like other arthropod organs, are bathed in blood inside the body sinuses. The tubules remove wastes from this blood, concentrate them, and then add them to undigested food before it leaves via the anus. Terrestrial arthropods may have small excretory glands at the bases of their legs in addition to, or instead of, Malpighian tubules. In aquatic arthropods, cellular wastes diffuse from the body into the surrounding water at unarmored places such as the gills. Many aquatic arthropods, such as lobsters, also eliminate nitrogen-containing wastes through a pair of green glands located near the base of the antennae. These wastes are emptied to the outside through a pair of openings on the head.

28–1 (continued)

Content Development

Arthropods have developed a great variety of organs for the removal of nitrogenous wastes. Point out or elicit from students that the main differences occur because of the important need of terrestrial arthropods to conserve the water contained in their tissues. The nitrogenous wastes of arthropods such as the grasshopper are solid, dry crystals of uric acid that have become concentrated by the removal of water. The water then is able to remain within the body of the grasshopper.

Students should be reminded that since the open circulatory system in arthropods is inefficient, there is an extensive system of Malpighian tubules in insects that terminate in the blood sinuses. In aquatic arthropods, this inefficiency is not as much of a problem, since nitrogenous wastes may be excreted directly into the water.

Reinforcement/Reteaching

It may be helpful to have students set up a chart to sum-

RESPONSE Most arthropods have well-developed nervous systems. All have a brain that consists of a pair of ganglia in the head. These ganglia serve as central switchboards for incoming information and outgoing instructions to muscles. From the brain, a pair of nerves runs around the esophagus and connects the brain to a nerve cord that runs along the ventral part of the body. Along this nerve cord are several more ganglia, usually one for each original body segment. These ganglia serve as local command centers to coordinate the movement of legs and wings. (That's why many insects can still walk or flap their wings after their heads are cut off!) Where many body segments have fused together, as in insects, there are several ganglia for each major body part.

Arthropods have simple sense organs such as statocysts and chemical receptors. Most arthropods also have sophisticated sense organs such as compound eyes for gathering information from their environment. Compound eyes may have more than 2000 separate lenses and can detect color and motion very well. In fact, insects can see certain things better than we can. (That's one reason it is so hard to swat flies and mosquitoes!) For example, the blades of a quickly moving fan—just a blur to our eyes—are clearly visible to a fly. And many insects can see ultraviolet light, which is invisible to humans.

Both crustaceans and insects have a well-developed sense of taste, although their taste receptors are located in strange places. The chemical receptors associated with the senses of taste and smell are located on the mouthparts, as might be expected, but also on the antennae and legs! Flies, for example, know immediately whether a drop of water they step in contains salt or sugar. Crustaceans and insects have sensory hairs that detect movement in the air or water (another reason insects are hard to swat or catch). As an object moves toward them, they can feel the movement of the displaced air or water and respond appropriately. Many insects have well-developed ears that hear sounds above the human range. Insect ears are often in odd places. The eardrums in grasshoppers, for example, are behind their legs.

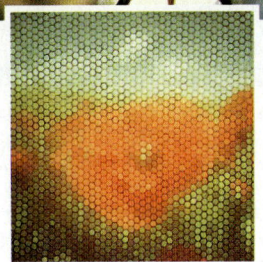

Figure 28–10 *The horsefly has huge compound eyes through which it sees poppy flowers much as they appear here (inset).*

Figure 28–11 *A harlequin beetle (left) uses its long antennae as "feelers." The red hourglass on the abdomen of the black widow spider (right) warns of the spider's venomous bite.*

HISTORICAL NOTE
WASPS AND ARISTOTLE

Aristotle, the great observational scientist, was the first to notice that wasps can remain alive and at almost normal activity levels even when their heads have been removed.

BACKGROUND INFORMATION
SENSITIVE SPIDERS

Many arthropods are sensitive to vibrations. Spiders pay close attention to vibrations in their webs, which they detect with their legs. This helps them to know when the web has snared a good dinner.

---

marize the various structural adaptations of the arthropods. On the left side, have them list either a representative organism or the name of the subphylum; at the top, have them indicate the function, such as respiration, excretion, and so on. In the chart, students should place the name of the appropriate structure.

Focus/Motivation
Point out that the beauty and complexity of a spider's web and the difficulty of swatting a fly are both the result of the fact that most arthropods have highly developed nervous systems.

Content Development
Discuss the fact that the nervous system of some of the arthropods, especially the insects, greatly resembles that of the annelid worms. Emphasize that the compound eye is unique to arthropods. Although a compound eye probably does not give as sharp an image as a simple eye, its main advantage is in the detection of the slightest movement of prey or an enemy.

BACKGROUND INFORMATION
FACT OR SCIENCE FICTION

Many arthropods have life cycles and reproductive habits that are so bizarre—and occasionally so repulsive—that they have inspired some of the most frightening science fiction movies of all time. An outstanding example is the creature that starred in the *Alien* films, clearly inspired by a combination of arthropod characteristics. For example, several species of wasps lay their eggs inside a caterpillar or other host insect that has been paralyzed but not killed. As the eggs hatch, the larvae devour their host from inside, much as the *Alien* creatures did. Moreover, like a queen bee, the giant Alien queen ruled the "nest" and, laying eggs in a steady stream, was both guarded and fed by her daughters.

Figure 28–12 Can you find the grasshopper in the photograph? The stick grasshopper's shape is a form of camouflage (left). The markings and behavior of this caterpillar trick insect-eating birds into thinking it is a bird-eating viper (center). The harmless hoverfly mimics a stinging honeybee (right).

An arthropod's well-developed sense organs help it detect and escape predators. The combination of these sense organs and a tough exoskeleton is enough to protect many arthropods. But some arthropods have additional means of protection. Scorpions, bees, and some ants have venomous stings, and many spiders and centipedes have venomous bites. Lobsters and crabs can attack potential enemies with powerful claws. And many insects and millipedes fight back using nasty chemicals. Some arthropods trick predators by creating a diversion. For example, some crabs can drop a claw or leg. This body part keeps on moving to distract predators while the rest of the animal scurries away. The crab then grows back the lost limb. Other arthropods use visual trickery to fool predators. Some hide through camouflage—matching the color and texture of their surroundings so closely that they seem to disappear. Others imitate the warning coloration of poisonous or dangerous species—a phenomenon called mimicry.

MOVEMENT Arthropods have well-developed muscle systems that are coordinated by the nervous system. Muscles generate force by contracting, then transfer that force to the exoskeleton. At each body joint, some muscles are positioned to flex the joint and other muscles to extend it. See Figure 28–13. The pull of muscles against the exoskeleton allows arthropods to beat their wings against the air to fly, push their legs against the ground to walk, or beat their flippers against the water to swim.

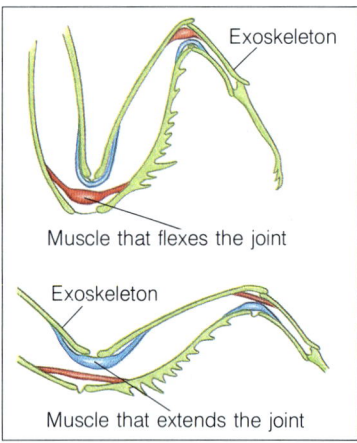

Figure 28–13 Muscles attached to the inside of an exoskeleton bend and straighten the joints.

REPRODUCTION Reproduction in most arthropods is simple. Males and females produce sperm and eggs, respectively, and fertilization usually takes place inside the body of the female. In spiders and some crustaceans, the male deposits a small packet of sperm that the female picks up. In most insects and crustaceans, however, the male uses a special reproductive organ to deposit sperm inside the female.

28–1 (continued)

Reinforcement/Reteaching
Because of the wide variety and complexity of the arthropod nervous system, you may want students to summarize the specialized sense and response organs in a chart. On the left side, have them list either a representative organism or the name of the subphylum; at the top, have them indicate the function, such as sight, sound, balance, protection, and so on. In the chart, students may place the name of the appropriate structure(s) or the response(s).

Focus/Motivation
Using the Visuals Have students observe Figures 28–14 and 28–15.
• **Why must arthropods go through the process of molting?** (They must shed their "outgrown" exoskeleton as they grow larger.)
• **What is meant by metamorphosis?** (A change in the body form or structure of an arthropod as it grows.)
• **What is the difference between incomplete and complete meta-**

Growth and Development in Arthropods

Exoskeletons, as useful as they are, present a problem in terms of growth. As a growing student, you can understand that problem. Imagine that you had a skin-tight suit of clothes tailored for you last year. Could you fit into it now? Probably not. You need larger clothes as you grow. Similarly, arthropods must replace their exoskeletons with larger ones in order to allow their bodies to increase in size as they mature. The problem is not a simple one, however, because the exoskeleton not only covers all appendages and sense organs, but also lines the gut as far down as the stomach. In order to grow, all arthropods must **molt**, or shed, their exoskeletons. This complicated process is controlled by several important hormones, the most important of which is called molting hormone.

When molting time is near, the epidermis (the layer of cells that covers the outside of the body) digests the inner part of the exoskeleton and absorbs much of the chitin in order to recycle the chemicals in it. After it secretes a new exoskeleton inside the old one, an arthropod pulls completely out of its old exoskeleton. Arthropods often eat what is left of the old exoskeleton. The animal then expands to its new, larger size, and the new exoskeleton (which is still soft) stretches to cover it. The animal must then wait for the new exoskeleton to harden, a process that can take from a few hours to a few days. During this time, the new shell stays soft and the animal is quite vulnerable. Thus arthropods hide from predators during molting.

Most arthropods molt several times between hatching and adulthood. In most cases the process of growth and development involves **metamorphosis**, or a dramatic change in form.

Some arthropods, such as grasshoppers, mites, and crustaceans, hatch from eggs into young animals that look much like the adults. However, these young animals lack functioning sexual organs and often lack other adult structures such as wings. As the young grow, they keep molting and getting larger until they reach adult size. Along the way, they gradually acquire the characteristics of adults. In insects, this kind of gradual change during development is called incomplete metamorphosis.

Figure 28–14 The adult cicada is emerging from the molted exoskeleton of an immature stage. Arthropods molt in order to increase in size and also to change from one body form to another in the process of metamorphosis.

Figure 28–15 The grasshopper undergoes incomplete metamorphosis, whereas the monarch butterfly undergoes complete metamorphosis.

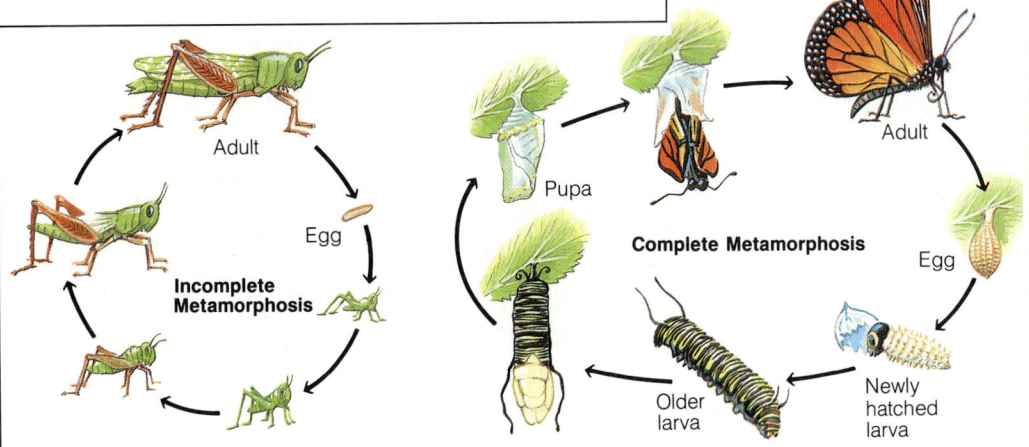

BACKGROUND INFORMATION
INCOMPLETE METAMORPHOSIS

Many arthropods—such as shrimp, lobsters, and most other crustaceans—exhibit a somewhat different form in incomplete metamorphosis. These animals hatch into larvae that look nothing at all like adults but do not pass through a pupal stage in which major structural rearrangements occur. Instead, these animals pass through a series of molts as they grow. Lobsters begin as free-swimming larvae that drift near the surface of the ocean. Undergoing three molts, the larva looks very different after each one. After the fourth molt, the baby lobsters look like miniature adults. It takes at least 7 years and many more molts for lobsters to grow to the one-and-a-quarter pound weight at which they are legal to catch and sell in most New England states.

MULTICULTURAL STRATEGY

Have students investigate the various arthropods that are used as food sources in different parts of the world. They might want to find out about the industry of crabbing, especially about the research that has been conducted to catch soft-shelled crabs during the proper time in their developmental cycle.

---

morphosis? (In incomplete metamorphosis, the young resemble a smaller version of the adult, although they may lack certain organs present in the adult. In complete metamorphosis, the organism's appearance changes radically as its development proceeds from one stage to the next.)
• **How does an organism "know" when to molt or undergo metamorphosis?** (The process is controlled by several hormones produced in the brain. The most important is called ecdysone, or molting hormone.)

Content Development
Discuss the stages of development in incomplete and complete metamorphosis, using specific organisms as examples. Although insect metamorphosis is usually the focus of discussion, it should be remembered that chelicerates and crustaceans also go through molting and incomplete metamorphosis.

Many insects, such as bees, moths, and beetles, undergo a four-stage process of development called complete metamorphosis. Refer to Figure 28–15 on page 615 as you read about the process of complete metamorphosis. The eggs of insects that undergo complete metamorphosis hatch into larvae that look nothing like their parents. As these larvae grow, they molt repeatedly, growing larger each time but changing little in appearance. When a larva reaches a certain age, it sheds its larval skin one last time and becomes a **pupa** (PYOO-pah; plural: pupae). During the pupal stage, the insect's body is totally rearranged—adult structures grow from tiny buds and larval structures are broken down to supply the raw materials for the adult structures. When metamorphosis is complete, the animal emerges as a fully grown adult with both internal and external body parts that are completely different from what it had before. Not only does this adult look like a totally different animal, it acts differently too.

Metamorphosis is controlled by a complicated interaction of several hormones, including molting hormone. In insects that undergo complete metamorphosis, the levels of juvenile hormone help regulate the stages of development. High levels of juvenile hormone keep an insect in its larval form each time it molts. As the insect matures, however, its production of juvenile hormone decreases. At some point, the level of juvenile hormone drops below a certain critical point. The next time the insect molts, it becomes a pupa. And when no juvenile hormone is produced, the insect undergoes a pupa-to-adult molt.

Because the balance of juvenile hormone, molting hormone, and other hormones is critical in arthropod development, it is possible to combat insects by tampering with their hormone levels. Certain plants defend themselves against herbivorous insects by producing chemicals that prevent molting, cause insects to develop at the wrong rate, or keep insects from becoming functional adults. In recent years, researchers have developed chemicals that act in a similar manner. These chemicals may eventually enable people to control crop-eating insects without using dangerous poisons.

Figure 28–16 Insect pupae are often surrounded by a protective covering. The bumblebee pupa in this photograph is surrounded by a wax case. Many caterpillars spin cocoons of silk.

28–1 SECTION REVIEW

1. What are three characteristics of arthropods? Name the four subphyla of arthropods.
2. Compare complete and incomplete metamorphosis.
3. Describe the different types of organs that are used in arthropod respiration.
4. **Critical Thinking—Making Inferences** Terrestrial arthropods often have valves that can open and close the spiracles. How are these valves an adaptation to life on land? *(Hint: What is the function of the stomata on leaves?)*

616

TEACHING SUPPORT

Study Guide
- Section 28–1, p. 265

TIE-IN FORENSIC SCIENCE

Insects and their larvae provide clues to criminal investigators about the circumstances of crimes—especially murders. Forensic entomologists examine the species of insects in a piece of evidence—a package of marijuana or a corpse, for example—to determine where the crime was committed. Since the larvae of many insects, such as blowflies, develop at an extremely regular rate, larvae removed from a corpse can be raised in carefully controlled conditions. The length of time it takes the larvae to develop into adults gives investigators a fairly accurate indication of when the parent fly laid her eggs on the corpse. And since the time it takes for a corpse to attract insects is also a known constant, the investigators can pinpoint the time of death.

28–1 (continued)

Enrichment
Environmental awareness has increased the pace at which biological controls of insect pests are being developed. Artificial insect hormones designed to prevent them from reaching sexual maturity has been one area showing great promise. Some students might want to do library or laboratory research into the development of "natural" pest controls.

SECTION REVIEW 28–1

1. A tough exoskeleton, a series of jointed appendages, and a segmented body.
2. Complete metamorphosis involves a major change in form from one stage of development to the next in insects and includes a pupal stage. Incomplete metamorphosis involves a gradual series of changes during development.
3. Gills: usually feathery structures located within the exoskeleton of aquatic arthropods. Tracheal tubes: a network of tubes that carry air throughout the body. Book gills: multiple pagelike sheets of respiratory tissue protected by tough covers. Book lungs: multiple pagelike sheets of respiratory tissue inside a body chamber that connects with the outside.
4. They help prevent water loss through the spiracles.

Reinforcement/Reteaching
If students have difficulty answering the Section Review ques-

28–2 Spiders and Their Relatives

Guide For Reading

- What are several distinguishing characteristics of chelicerates?
- In what ways do spiders use silk?

Spiders and their relatives—horseshoe crabs, ticks, and scorpions, for example—belong to the subphylum Chelicerata. **Chelicerates are arthropods that are characterized by a two-part body and mouthparts called chelicerae.** These arthropods also lack the sensory "feelers" that are found on the heads of most other arthropods.

All chelicerates have a body that is divided into two parts: the cephalothorax (sehf-ah-loh-THOR-aks) and the abdomen. The anterior end of the cephalothorax contains the brain, eyes, mouth and mouthparts, and esophagus. The posterior end of the cephalothorax carries the front part of the digestive system and several pairs of walking legs. The abdomen contains most of the internal organs. See Figure 28–17.

All chelicerates have two pairs of appendages attached near the mouth that are adapted as mouthparts. The first pair of mouthparts are called chelicerae (keh-LIHS-er-ee; singular: chelicera). The second pair of mouthparts, which are longer than the chelicerae, are called pedipalps (PEHD-ih-palps). Both sets of mouthparts are adapted to serve different purposes in feeding in different species.

Horseshoe Crabs

Among the oldest chelicerates are the horseshoe crabs. This name is somewhat misleading because these animals are not true crabs (which are crustaceans). Horseshoe crabs appeared in the Ordovician Period (more than 430 million years ago) and have not changed much since then—they are true "living fossils." Horseshoe crabs are heavily armor-plated,

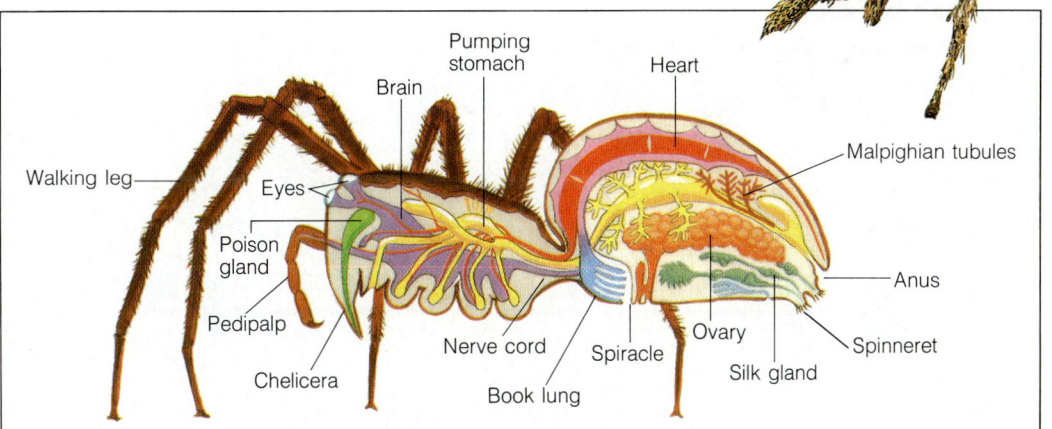

Figure 28–17 The internal structures of a typical spider are shown in this diagram. A jumping spider (inset) captures prey by pouncing on it, rather than by catching it in a web.

617

TEACHING SUPPORT

Teaching Resources
- Activity: Studying Spider Webs, p. 3
- Hands-On Activity: Spinning Webs, p. 145

MEDIA AND TECHNOLOGY
Biology Transparencies
- 34: The Anatomy of a Spider

Interactive Videodisc
- Insects: Little Giants of the Earth

See the Biology Media Guide page 33 for bar-code correlations for this interactive video.

BACKGROUND INFORMATION
SPIDER SILK: THE NEXT SECRET WEAPON?

No spider we know of has ever taken a course in biochemistry, yet many of them are able to accomplish a feat that baffles U.S. Army chemists. Somehow, spiders turn a mixture of water-soluble proteins into the remarkable (and insoluble) fiber called silk. Silk's abilities inspire envy as well as puzzlement among army researchers for several reasons. When a spider's web is struck by an insect, for example, its silk stretches a little, transforms some of the energy it absorbs into heat, and then returns to its former length gently enough so that it doesn't catapult the insect to freedom. Silk retains this elasticity down to −50 to −60°C, temperatures much lower than those tolerated by artificial fibers. In addition, silk is almost four times as stretchable as Kevlar—the artificial fiber that is used to make bulletproof vests. Researchers for the army now understand about 50% of the chemistry involved in making silk, but they still do not know how the original protein building blocks are assembled, crystallized, and cured by the spiders.

Figure 28–18 A horseshoe crab's tiny pincerlike chelicerae and five pairs of walking legs are visible when the animal is turned on its back. The platelike structures on the abdomen cover and protect the book gill's "pages." The long tail is not seen here because the horseshoe crab has pushed it into the sand to right itself.

have five pairs of walking legs, and long spikelike tails. They can grow up to 60 centimeters long—about the size (and shape) of a frying pan. When they first hatch, however, horseshoe crabs are only about 1 centimeter long. These newly hatched horseshoe crabs are called trilobite larvae because they look much like their extinct distant relatives.

Arachnids

The most familiar chelicerates are the **arachnids**, which include spiders, scorpions, ticks, and mites. All adult arachnids have four pairs of walking legs on their cephalothorax. Many arachnids are carnivores that have pedipalps adapted for capturing and holding prey and chelicerae adapted for biting and sucking out their soft parts.

SPIDERS Spiders are predators that usually feed on insects. However, a few large tropical spiders are capable of catching and eating small vertebrates, such as hummingbirds. Spiders capture their prey in a variety of ways. Some spiders ensnare their prey in webs. Others stalk and then pounce on the prey. And some lie in wait beneath the lid of a camouflaged underground burrow, leaping out to grab unlucky insects that venture too near.

Once a spider captures its prey, it uses its hollow fanglike chelicerae to inject paralyzing venom into it. When the prey is paralyzed, the spider's mouth introduces enzymes into the wounds made by the chelicerae. These enzymes break down the prey's tissues, enabling the spider to suck up the liquefied tissues with its esophagus and a specialized pumping stomach. The pumping stomach then forces the liquid food through the rest of the spider's digestive system.

Whether or not they spin webs, all spiders produce a strong, flexible protein called silk. Silk, which is produced in special glands located in the abdomen, is five times stronger

Figure 28–19 The wolf spider (left) ambushes prey from its silk-lined burrow. Large tarantulas (right) are capable of catching and devouring small vertebrates, such as lizards.

28–2 (continued)

Focus/Motivation
Using the Visuals Most chelicerate species are arachnids.
- **What kinds of organisms are arachnids?** (Arachnids include spiders, scorpions, ticks, and mites.)

Ask students to observe Figures 28–19 through 28–21.
- **What do arachnids seem to have in common?** (They have four pairs of walking legs.)

Content Development
- **How do arachnids obtain food?** (They are either predatory or parasitic.)

Point out that most of the special structures found in the arachnids are modifications of the basic arthropod body plan for a predatory or parasitic existence.

than steel. It is strong enough, in fact, to withstand the equivalent of the impact of a jet fighter every time a spider's web traps a fly. Spiders spin silk into webs, cocoons for eggs, wrappings for prey, and other structures by forcing liquid silk through organs called spinnerets. As the liquid silk is pulled out of the spinnerets, it hardens into a single strand. Interestingly, the complicated behavior of web-spinning seems to be "preprogrammed" into a spider's brain. The spiders of web-spinning species can build their intricate webs almost as soon as they hatch—without having to learn how.

MITES AND TICKS Mites and ticks are small arachnids, many of which are parasites on humans, on farm animals, and on important agricultural plants. Most species are smaller than 1 millimeter, but some ticks can be as large as 3 centimeters. In many mites and ticks, the chelicerae are needlelike structures that are used to pierce the skin of their hosts. These chelicerae may also have large teeth to help the parasite keep a firm hold on the host. The pedipalps are often equipped with claws for digging in and holding on. Some species, such as spider mites, damage houseplants and are major agricultural pests on crops such as cotton. Others—including chiggers, mange, and scabies mites—cause painful itching rashes in humans, dogs, and other mammals. A whole host of ticks parasitize humans and the animals we raise. Tick bites are not just annoying—they can be dangerous. In the United States, ticks can spread Rocky Mountain spotted fever and Lyme disease.

SCORPIONS Scorpions are widespread in warm areas around the world, including the southwestern United States. All scorpions are carnivores that prey on other invertebrates, usually insects. The pedipalps of scorpions are enormously enlarged into a pair of claws. The abdomen, which is long and segmented, terminates in a venomous barb used to sting prey. Usually, a scorpion grabs prey with its pedipalps, then whips

Figure 28–20 Some spiders build webs to capture prey.

Figure 28–21 Red velvet mites are similar in form to other members of their class (right). However, they are unusual in that they are not parasites and are relatively large (about 1 centimeter long). The loser of a fight between two scorpions will be stung and eaten by the winner (left). Biologists can locate scorpions at night by shining ultraviolet (UV) light on the desert floor. Under UV light, scorpions glow brightly in the dark.

TEACHING SUPPORT

Laboratory Manual
- Isopod Environments, p. 367

Teaching Resources
- Activity: Investigating Crustaceans, p. 5
- Hands-On Activity: Off and Running, p. 143

Study Guide
- Section 28–2, p. 268

MEDIA AND TECHNOLOGY

Biology Transparencies
- 35: The Anatomy of a Crayfish

Interactive Videodisc
- *Insects: Little Giants of the Earth*
- *On Dry Land: The Desert Biome*

See the Biology Media Guide page 33 for bar-code correlations for this section.

28–2 (continued)

SECTION REVIEW 28–2

1. Arthropods that possess mouthparts called chelicerae. Horseshoe crabs; horseshoe crab (*Limulus*). Arachnids: spider. Student examples may vary.
2. Silk is a protein produced by silk glands and extruded through spinnerets to form threads. Spiders use silk in many ways—making webs, immobilizing prey, and protecting eggs, for example.
3. Spiders: poison fangs that inject venom into prey. Ticks: needlelike piercing structures that have teeth so that the tick can keep a firm grip on the host.

Reinforcement/Reteaching
If students have difficulty completing the Section Review questions, go back to the appropriate material and review it with them.

Closure
Have students summarize the special characteristics of each group of organism in the subphylum Chelicerata.

---

its abdomen over its head to sting the prey, thus killing or paralyzing it. The scorpion then chews its meal with its chelicerae. Because scorpions like to crawl into moist, dark places, people in areas with scorpions should check inside their shoes before putting them on in the morning. Most North American scorpions have venom powerful enough to cause about as much pain as a wasp sting. However, the venom of one genus of scorpions that lives in Mexico, New Mexico, and Arizona has killed small children who were stung accidentally.

28–2 SECTION REVIEW

1. What are chelicerates? Name and give examples of the two main groups of chelicerates.
2. What is silk? How do spiders use silk?
3. **Critical Thinking—Summarizing Information** How are chelicerae modified for feeding in spiders? In ticks?

28–3 Crustaceans

The subphylum Crustacea contains over 35,000 species. Crustaceans are primarily aquatic, although there are some terrestrial species. Crustaceans range in size from microscopic water fleas less than 0.25 millimeter long to Japanese spider crabs that are thought to grow up to 6 meters across and lobsters that have a mass of more than 20 kilograms. And crustaceans vary in form as much as they vary in size!

Although crustaceans adapted to different conditions are quite dissimilar in form, all crustaceans share a number of structural similarities. **In general, crustaceans are characterized by a hard exoskeleton, two pairs of antennae, and mouthparts called mandibles.** As we examine a little of the enormous diversity of form and function in crustaceans, we will focus on a representative species, the crayfish. Refer to Figure 28–23 as you read about structure and function in crustaceans.

The main crustacean body parts are the head, thorax, and abdomen. In crayfish, as in many other crustaceans, the head and thorax have fused into a cephalothorax that is covered by a tough shell called the carapace. Unlike most other arthropods, many large crustaceans have calcium carbonate (limestone) in the exoskeleton. This is what makes the shells of crustaceans such as crabs and lobsters hard and stony.

In crustaceans, the first two pairs of appendages are "feelers" called antennae, which bear many sensory hairs. Antennae serve primarily as sense organs in crayfish, but in some other crustaceans they are used in filter feeding. Still other crustaceans, such as water fleas, use their antennae as oars to push them through the water.

Guide For Reading
- What are crustaceans?
- How are the body parts of crustaceans adaptations for survival?

Figure 28–22 The pill bug is a terrestrial crustacean. When threatened, a pill bug curls into a ball to protect its soft underside.

620

TEACHING STRATEGY 28–3

Focus/Motivation
Tell students that because of the large numbers of species in this subphylum, crustaceans are sometimes called the "insects of the sea." Show students photographs or slides of some of the organisms in the subphylum Crustacea and ask them to identify the organisms. Be sure to include some terrestrial crustaceans, such as the pillbug and the fiddler crab.

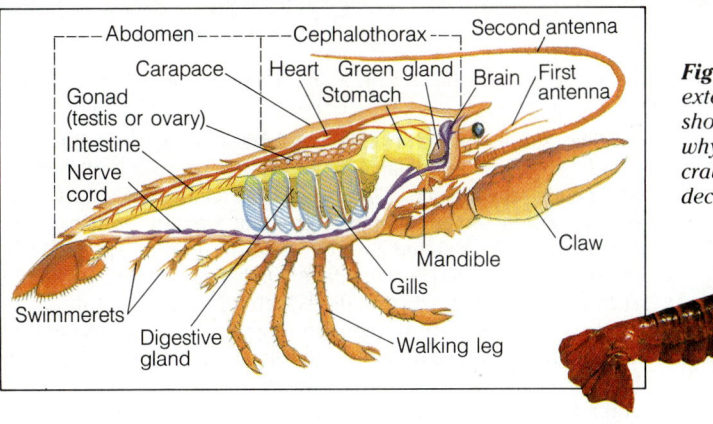

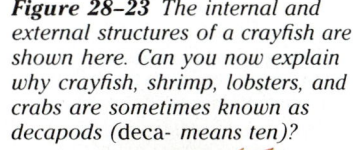

Figure 28–23 The internal and external structures of a crayfish are shown here. Can you now explain why crayfish, shrimp, lobsters, and crabs are sometimes known as decapods (deca- means ten)?

The third pair of appendages are mouthparts that are called **mandibles**. In many species of crustaceans, including crayfish, mandibles are short heavy structures designed for biting and grinding food. In other species, mandibles are bristly structures used in filter feeding, probelike structures used for finding and picking up detritus, or needlelike structures used to suck blood from a host.

The appendages on the thorax and abdomen vary greatly from one group of crustaceans to another. Some, such as barnacles, have delicate, feathery appendages for filter feeding; others have legs for walking or paddles for swimming. Appendages may be modified for internal fertilization, carrying eggs, spearing prey, burrowing, or many other functions.

As you can see in Figure 28–23, the appendages on a crayfish's thorax and abdomen are adapted for several different functions. A pair of large claws, which are used to catch prey and pick up, crush, and cut food, are located on the thorax. Four pairs of walking legs are also attached to the thorax. Flipperlike appendages called swimmerets, which are used for swimming, are located on the abdomen. A large pair of paddle-like appendages are found on the second-to-last abdominal segment. The paddlelike appendages and the final abdominal segment together form a large, flat tail. When the muscles of the abdomen contract, the crayfish's tail snaps forward. This provides a powerful swimming stroke that can rapidly pull the animal backward.

Figure 28–24 The abdomen of a crab is tucked beneath its cephalothorax. A female crab uses its abdomen and the swimmerets attached to it to carry its reddish-brown eggs. The blue and white semicircle below the eggs is made up of the last few segments of the tail.

28–3 SECTION REVIEW

1. What is a cephalothorax?
2. Describe the types of appendages on crayfish and give their functions.
3. **Critical Thinking—Applying Concepts** Suppose you want to catch a crayfish with a net. Should you try to scoop it up head first or tail first? Explain.

28–4 Insects and Their Relatives

The subphylum Uniramia contains more species than all other groups of animals alive today. It includes centipedes, millipedes, and insects. Uniramians are characterized by one pair of antennae and appendages that do not branch (*uni-* means one; *ramus* means branch). (Recall that the appendages in crustaceans and trilobites have two branches—usually a gill and a leg.) These arthropods, which display a multitude of forms and habits, are thought to have evolved on land about 400 million years ago. They inhabit almost every terrestrial habitat on Earth. In addition, some species live in fresh water and a few other species live in marine environments.

Centipedes and Millipedes

Centipedes and millipedes are many-legged animals. Compared to crustaceans and insects, these two classes of arthropods are quite small in number—there are approximately 3000 species of centipedes and 7500 species of millipedes. **Centipedes and millipedes are characterized by a long, wormlike body composed of many leg-bearing segments.** Because they lack closable spiracles and a waterproof coating on their exoskeleton, their bodies lose water easily. Thus they tend to live beneath rocks, in soil, or in other relatively moist areas.

Figure 28–25 A centipede (top) is a carnivore that has one pair of legs per body segment. A millipede (bottom) is a herbivore that has two pairs of legs per body segment.

CENTIPEDES Centipedes are carnivores that have, in addition to other mouthparts, a pair of poison claws in their head region. These poison claws are used to catch and stun or kill prey. Centipedes eat other arthropods, earthworms, toads, small snakes, and even mice. The North American centipedes that may be familiar to you are usually red-brown in color and about 3 to 6 centimeters long. Some tropical species are brightly colored and quite large—up to 26 centimeters long. Despite their name, which means 100 legs (*centi-* means hundred; *-pede* refers to legs), centipedes may have from 15 to 170 pairs of legs, depending on the species. Each segment that makes up the body of the centipede bears one pair of legs, except for the first segment (which bears the poison claws) and the last three segments (which are legless).

MILLIPEDES Although millipedes do not have a thousand legs (*milli-* means thousand), they do seem to have twice as many as centipedes. Each millipede body segment is formed from the fusion of two segments in the embryo and thus bears two pairs of legs. Millipedes are timid creatures that live in damp places under rocks and in decaying logs. They feed on dead and decaying plant material. When disturbed, many millipedes roll up into a ball to protect their softer undersides. Some can also defend themselves by secreting unpleasant or toxic chemicals.

Insects

We know of more than 900,000 insects, and new ones are discovered in the tropics all the time. Insects are extremely varied in body shape and habits. However, all members of this class share basic structural similarities. **Insects are characterized by a body that is divided into three parts—head, thorax, and abdomen—and that has three pairs of legs attached to the thorax.** In addition, a typical insect has one pair of antennae and one pair of compound eyes on the head, two pairs of wings on the thorax, and uses a system of tracheal tubes for respiration.

Insects get their name from the Latin word *insectum*, meaning notched, which refers to the division of their body into three main parts: head, thorax, and abdomen. In many insects, such as ants, the three body parts are clearly separated from each other by narrow connections. In other insects, such as grasshoppers, the divisions between the three body parts are not as sharply defined.

The essential life functions in insects are carried out in basically the same ways as they are in other arthropods. However, insects show a variety of interesting adaptations in feeding, movement, and behavior that deserve a closer look.

FEEDING Insects have three pairs of appendages that are used as mouthparts, including a pair of mandibles. Mouthparts can take on an enormous variety of shapes in species adapted to feed on different foods. For example, a grasshopper's mouthparts are designed to cut and chew plant tissues into a fine pulp. A female mosquito's mouthparts form a sharp tube that is used to pierce skin and suck blood. A butterfly's mouthparts

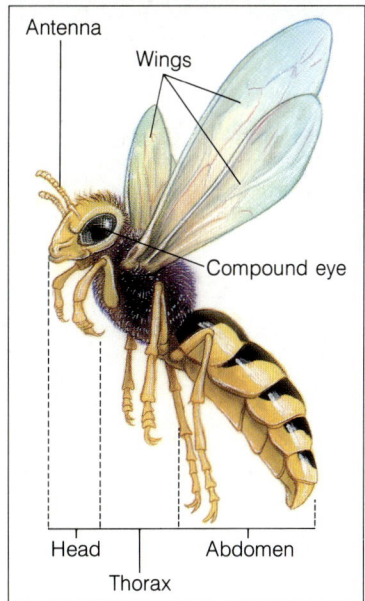

Figure 28–26 An insect is characterized by a three-part body, six legs, two pairs of wings, one pair of antennae, and one pair of compound eyes.

Figure 28–27 Although insect mouthparts are adapted for many different eating habits, they all evolved from the same basic structures.

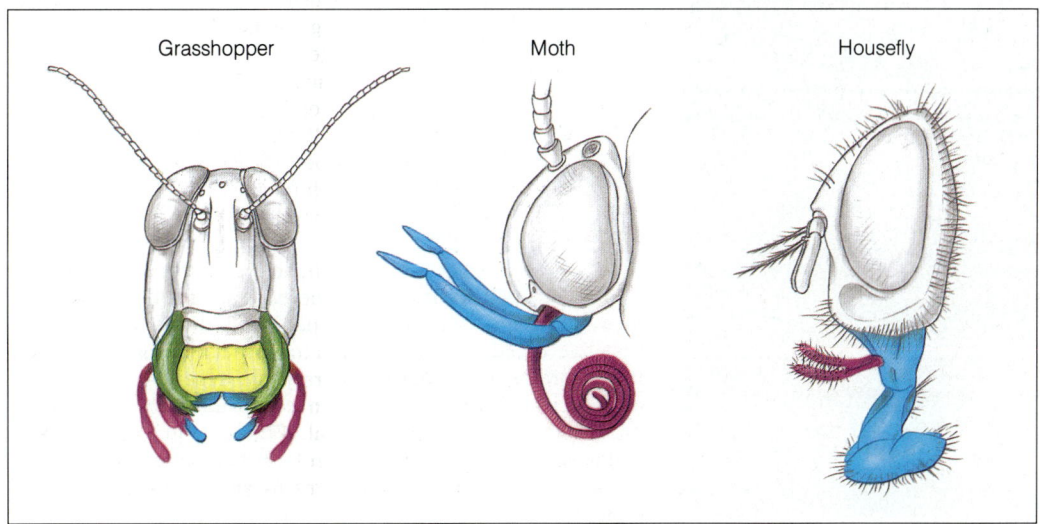

ECOLOGY NOTE
CONSERVING WATER

Many of the world's ant species live in hot climates for at least part of the year. They also carry food many times their weight and scurry around almost constantly. As a result, one might expect that they would dry out when traveling across scorching sands. However, they are protected by their gas-exchange system. Instead of using lungs for respiration, ants have a number of spiracles connected to the trachea, which forms a network of tubes within the ant's body.

Using infrared carbon dioxide analyzers, it has been found that ants first inhale for a few minutes and then exhale. During the long period of inhalation, oxygen enters the tissues, but carbon dioxide and water vapor are not released. This produces a vacuum, so more air enters while only a tiny amount of carbon dioxide and water vapor escape.

When the muscles around the spiracles become saturated with carbon dioxide, they relax. As a result, the spiracles open, allowing the carbon dioxide to leave the body. This occurs only after 5 to 20 minutes. If the spiracles opened more often much water vapor would also escape, causing rapid dehydration. This demonstrates an effective adaptation to survival in arid conditions.

28–4 (continued)

Content Development

Insects exhibit a high degree of specialization of their digestive systems.
- **Give some examples of the specialized digestive structures of some insects.** (Examples may include: the mosquito salivary enzyme, which prevents its prey's blood from clotting; digestive glands in the honeybee, which change plant nectar into honey; and wax-forming glands in the honeybee.)

Enrichment

Because of the tremendous variety of adaptations exhibited by insects in carrying out their life functions, you might want to have your students do library research to find out more about these adaptations. Some students might want to find out more about special organs for feeding or digesting food; others might be interested in the mechanism of insect flight; still others might explore the social behavior and communication that occur among insects.

---

are fused together to form a long tube that is used for sipping nectar. A bee has mouthparts that are used for chewing and gathering nectar. And a fly has a spongy mouthpart that is used to soak up food.

Insect adaptations for feeding are not restricted to the shapes of the mouthparts. Many insects produce saliva that contains digestive enzymes and helps break down food. The saliva of female mosquitoes, which is injected when the mosquito bites, contains chemicals that prevent blood from clotting. (Unfortunately for humans and other animals, mosquito saliva also contains chemicals that cause the body's familiar itching reaction. It may contain pathogens as well.) Honeybees have a number of adaptations for gathering, processing, and storing food. The legs and bodies of worker bees are covered with hairs that collect pollen. Chemicals in bee saliva help change nectar into a more digestible form—honey. And glands on the abdomen secrete wax, which is used to build storage chambers for food and other structures within a beehive.

MOVEMENT As you have just read, insects have three pairs of walking legs. These legs are often equipped with spines and hooks for holding onto things and for defense. In addition to being used for walking, the legs may be adapted for functions such as jumping (as in grasshoppers and fleas) or capturing and holding prey (as in praying mantises).

Along with birds and bats, insects are the only living organisms that are capable of unassisted flight. The flying ability of insects varies greatly. Butterflies fly quite slowly and have limited maneuverability. But certain flies, bees, and moths can hover, change direction rapidly, and dart off at speeds up to 53 kilometers an hour. In flying insects, most of the space in the thorax is taken up by the large muscles that operate the wings. The enormous amount of energy required by these muscles during flight is supplied by oversized mitochondria (which are about half the size of a human red blood cell). The wing muscles in many insects also have a special blood supply that helps retain heat produced by muscle activity. For example, bees can maintain a wing muscle temperature of up to 35 degrees Celsius. This means that bees can keep their flying muscles warmer than the outside temperature and operate efficiently even when it is cold outside.

INSECT SOCIETIES Many animals and protists form colonies, which are collections of individuals of the same species that live together. Several types of insects are unique among invertebrates in that they form a special type of colony known as a society. In a society, separate individuals are dependent on one another for survival. Insects that live in societies, such as many species of termites, wasps, bees, and ants, are called social insects. These insects have developed complicated societies that may be composed of from half a dozen to more than

Figure 28–28 Insects, along with birds and bats, are the only living organisms capable of unassisted flight. The hard brown wing covers on the may bug are modified forewings; in may bugs and other beetles, only the hindwings are used for flight.

Content Development

The legs and wings vary greatly in structure to perform different functions.
- **What are some of the ways in which insect's legs are modified to perform different functions?** (Insect's legs may be modified for

Figure 28-29 Leaf-cutter ant workers carry pieces of leaves and flower petals to their underground nest. Certain chambers in the nest are "farms" in which the ants grow edible fungi on the bits of vegetation.

7 million individuals. Within such societies there is division of labor: Different individuals perform the tasks necessary for the survival of the entire group. There are several castes, or types of individuals, within insect societies. Each caste has a body that is specialized for its functions and is therefore distinctly different from that of another caste. The basic castes are reproductive females, reproductive males, and workers.

The reproductive females, which are called queens, lay eggs that hatch into new individuals for the society. Most insect societies have only one queen, who is typically the largest individual in the colony. Termite queens, for example, may be 14 centimeters long (more than 10 times longer than a worker) and 3.5 centimeters wide. Most of a termite queen's body consists of a grotesquely swollen abdomen that contains enlarged reproductive organs. A termite queen can produce more than 30,000 eggs a day!

The reproductive males function only to fertilize the queen's eggs. In some insect societies, such as those of termites, a single reproductive male stays with the queen as a permanent member of the colony. In other societies, such as those of bees, the queen receives all the sperm she needs for her eggs after a single mating with one or more reproductive males. The successful males die after mating, and the unsuccessful males are ejected from the colony and soon perish.

The workers perform all the colony's tasks except for reproduction: They care for the queen, eggs, and young; they gather, store, and even grow food; they build, maintain, and defend the colony's home; and they perform all other necessary jobs. In societies of ants, bees, and wasps, the workers are all females; in those of termites, there are both male and female workers. Bee and wasp workers are capable of performing all of their societies' tasks for workers. Ant and termite workers are specialized and are able to carry out only their specific tasks, such as defending the colony or storing food.

Figure 28-30 Mature termite queens are approximately the size and shape of a hot dog. The large termite next to the queen is a reproductive male. The smaller brown termites with the large heads are called soldiers. The tiny white termites are workers.

BACKGROUND INFORMATION
HAPLODIPLOIDY IN HYMENOPTERANS

Hymenopterans, including ants, bees, and wasps, have an odd system of sexual reproduction called haplodiploidy. All males develop from unfertilized eggs and so are haploid. Fertilized eggs develop into diploid females. Queens can store sperm for long periods and can produce either fertilized or unfertilized eggs whenever they choose. In bee colonies, all important work is performed by female workers. For that reason, the queen produces only fertilized eggs most of the year. Males, or drones, do nothing other than mate with a queen. Since mating takes place mostly in the spring, queens produce only a limited number of unfertilized eggs, which develop into males at the end of winter.

Theoretically, female workers, as genetic females, could produce fertile eggs. However, as long as a queen is present in the hive, she suppresses the development of ovaries in the workers with the pheromone called queen substance.

TEACHING SUPPORT
Laboratory Manual
- The Grasshopper, p. 359

---

walking or jumping, for defense, or for capturing and holding prey.)

Have students give specific examples of insects that have these modifications.

Reinforcement/Reteaching
Review the role of the mitochondria in cell respiration and energy production and relate this to the fact that insects have greatly oversized mitochondria in the cells of their wing muscles.

Content Development
The so-called "social insects" exhibit a degree of organized group behavior that has no equal among the invertebrates. There is a caste system with true division of labor; most of the activity centers on maintaining the society and caring for the single reproductive female, the queen, whose sole function is to produce eggs. Besides the queen, insect societies have nonreproductive females, or workers, and males who fertilize the queen's egg, sometimes called drones.

TEACHING SUPPORT

Study Guide
- Section 28–4, p. 272

MEDIA AND TECHNOLOGY

▶ **Videodisc**
- *Super Scents*

Use this bar code to introduce the topic of pheromones.

Play frames 16797 to 18810

See the Biology Media Guide pages 36–37 for additional bar-code correlations for this section.

BACKGROUND INFORMATION
BEE STINGS

At the very end of the worker bees' abdomen are a pair of sharp, barbed darts and a venom gland that forms the stinger. Bees rarely use this stinger except in self-defense. When the stinger is embedded in the flesh of a human or other large animal, the barbs keep the stinger so firmly attached to the "stingee" that the entire stinger, with venom glands and part of the abdomen attached, breaks out of the bee's body. This causes so much damage to the abdomen that the bee dies.

28–4 (continued)

Content Development

• **What are some of the ways in which insects communicate with each other?** (Insects may communicate by making sounds; producing visual signals, such as a firefly's flash; or producing chemicals called pheromones.)

• **What is the main purpose of most insect communication?** (It is usually related to mating and reproduction.)

• **What differences are there in methods of communication in social insects compared with other insects?** (Social insects have much more complex forms of communication that involve not only reproduction but also highly specific information about food sources and the presence of danger.)

SECTION REVIEW 28–4

1. Millipedes: two pairs of legs per segment; herbivorous. Centipedes: one pair of legs per segment; carnivorous.

2. An insect's body is divided into three parts and possesses three pairs of walking legs on the thorax. A typical insect has one pair of antennae and one pair of compound eyes on its head, has two pairs of wings attached to the thorax, and uses tracheal tubes in respiration.

3. Crickets chirp to attract a mate. Ants leave a pheromone trail that enables them to travel back and forth between the ant nest and a source of food. Bees

Figure 28–31 A male luna moth's feathery antennae can detect pheromones released by a female several kilometers away.

INSECT COMMUNICATION All insects use sound, visual, chemical, and other types of signals for communication. Much of the communication done by nonsocial insects involves finding a mate. To attract females, male crickets chirp by rubbing their forewings together, and male cicadas buzz by vibrating special membranes on the abdomen. Male fireflies turn a light-producing organ in their abdomens on and off, producing a distinct series of flashes. When female fireflies (which are wingless and are known as glowworms) see the correct signal, they flash back a signal of their own and the males fly to them. This is not always a good thing from the male firefly's point of view—the carnivorous females of one genus of fireflies can mimic the signal of another genus and lure males to their death. Many female moths release chemicals that attract distant males to them. These chemicals are a type of **pheromone**, which is a specific chemical messenger that affects the behavior and/or development of other individuals of the same species.

Communication in social insects is generally more complex than in nonsocial insects. A sophisticated system of communication is necessary to organize a society. Each species of social insect has its own "language" of visual, touch, sound, and chemical signals that convey information among members of the colony.

Pheromones are particularly important in insect societies. Certain pheromones function as rapid short-term messages that signal alarm, the death of a member of the colony, or the presence of food. For example, when a worker ant finds food, she heads back to the nest, dragging her abdomen along the ground. As she does so, she leaves behind a trail of a special kind of pheromone. Her nestmates can detect her trail by using sensory hairs on their antennae and follow it back to the food. Other pheromones act as long-term controls over the colony. For example, a queen honeybee produces a pheromone, called queen substance, that prevents the development of rival queens. Queen substance makes worker bees unable to lay eggs. It also causes them to raise female larvae as workers, not as queens. However, when the amount of queen substance in the hive is low, worker bees feed a few female larvae a special diet. This causes the larvae to develop into queens.

Honeybees communicate with sound and movement as well as with pheromones. Worker bees are able to convey information about the type, quality, direction, and distance of a food source by "dancing." The language of the bee's dance was decoded by Austrian biologist Karl von Frisch. Von Frisch discovered that bees have two basic dances: a round dance and a waggle dance.

In the round dance, the bee that has found food circles first one way and then the other, over and over again. This dance tells the other bees that there is a source of food within 50 meters of the hive. The frequency with which the dancing bee

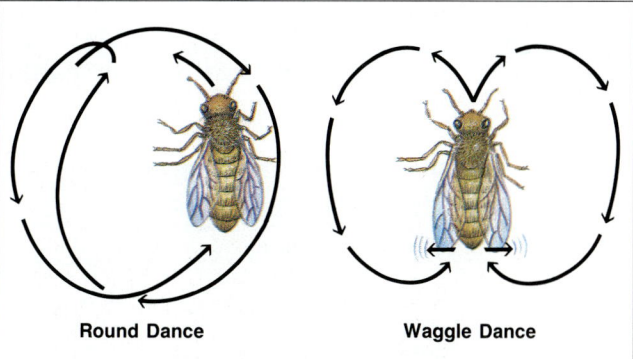

Figure 28–32 *Bees communicate information about food sources by using a language of movement. The round dance tells members of the hive that a source of food is nearby. The waggle dance gives information about a more distant food source.*

changes direction indicates the quality of the food source—the more frequent the changes in direction, the greater the energy value of the food. By smelling the dancer with the chemical receptors on their antennae, the other bees can determine what kind of flowers she has found.

In the waggle dance, the bee that has found food runs forward in a straight line while waggling her abdomen, circles around one way, runs in a straight line again, and circles around the other way. See Figure 28–32. The waggle dance tells the other bees that the food is more than 50 meters away. Most of the information about the food source is conveyed by the part of the waggle dance called the straight run. The longer the bee takes to perform the straight run and the more she waggles, the farther away the food. The direction of the straight run indicates in which direction the food is to be found. For example, if the dancer runs straight up the vertical honeycomb, the food source is in the same direction as the sun.

28–4 SECTION REVIEW

1. Compare the body plans and feeding habits of millipedes and centipedes.
2. Describe the basic body plan of an insect.
3. Give three specific examples of why and how insects communicate.
4. Explain how the mouthparts of bees, mosquitoes, and butterflies are adapted to different food sources.
5. How does the waggle dance of honeybees convey information about the location of a food source?
6. **Critical Thinking—Relating Cause and Effect** If all worker bees are females, why is the queen the only egg-layer in the colony?

BACKGROUND INFORMATION
ANNUAL CYCLES IN HONEYBEE COLONIES

Most of the time the queen bee does little more than eat honey and lay eggs that will develop into workers. In the spring, however, the queen's production of queen substance falls off. At this time, workers build two kinds of enlarged brood cells in the honeycomb. Into one set of the cells, the queen deposits unfertilized eggs that develop into male drones. Into the other set of enlarged cells, workers place special salivary gland secretions that turn their honey into "royal jelly." The fertilized eggs that the queen deposits in these cells develop rapidly into large pupae that emerge as new queens. In the meantime, the lack of queen substance causes the workers to lose interest in the old queen. Eventually, she leaves the hive with a few thousand of her daughters and a number of drones in a swarming flight to found a new colony. When the new queens emerge, they also go off on a swarming flight in which they mate with males from another hive. During mating, the queen receives several million sperm, which she stores for the entire three-to-five-year egg-laying period of her life. Eventually, only one new queen remains in the old nest.

Reinforcement/Reteaching
If students have difficulty answering any of the Section Review questions, refer them to the specific part of the section where the information appears and review the information with them.

Closure
Because of the large number of vocabulary words and concepts introduced in this section, a concept map or simple outline developed by the students would provide them with an organized summary of the key ideas.

"dance" to communicate the location of food to other members of the hive. Students' examples will vary.
4. Bees: chewing mouthparts for eating pollen and sipping structures for gathering nectar. Mosquitoes: females have sharp tube-shaped mouthparts for piercing skin, injecting chemicals, and drinking blood. Butterflies: straw-shaped mouthparts for drinking nectar.
5. The waggle dance indicates that the food source is more than 50 meters away; the direction of the straight run indicates the direction in which the food is located.
6. The queen releases a pheromone that makes workers sterile and that also prevents workers from raising new queens.

SCIENCE, TECHNOLOGY, AND SOCIETY ISSUE

Controlling Agricultural Pests

Many plant-eating insects are enemies of both crops and plants in nature. Yet although cornfields can be destroyed by insects, forests and grasslands are never wiped out. Why not? And how can the answer to that question help us protect food crops?

Pests rarely get out of control in nature for three main reasons. First, they are eaten by such natural enemies as ladybugs, spiders, parasitic wasps, and birds, and suffer from diseases caused by bacteria and fungi. Second, many plants produce compounds that taste bad to insects, poison them, or mimic the action of insect juvenile hormones. Larvae that eat these hormone mimics cannot pupate, never mature, and cannot reproduce. Third, natural environments contain several plant species mixed together, so insect pests have to work harder to find their favorite foods.

For many years now, farmers have taken their cues from poisonous plants, and have relied on toxic sprays such as DDT to kill pests. It is clear, however, that widespread, constant use of poison sprays has created more problems than it solved. Many pesticides are deadly to humans, livestock, wildlife, and the natural enemies of pests. Some chemicals stay in the environment for a long time, poisoning rivers and water supplies. Furthermore, many insecticides are no longer useful because pests have evolved resistance to them. In 1938, there were only 7 pests that were resistant to chemicals; by 1985 there were 447.

Clearly, different insect control methods are needed. Researchers have therefore focused their attention on pests' natural enemies and on ancient agricultural practices.

A great deal has been learned in recent years. Insect biologists are learning to breed and collect pest-eating insects, and to release them in fields at the right time to control pests. Biotechnology companies are learning to manufacture hormone mimics economi-

Adult ladybugs and their larvae feed on crop-eating pests. In the long run, the use of natural pest controls is safer and more efficient and effective than the use of artificial chemical pesticides.

cally. Farmers are learning to use these nontoxic sprays, which affect only the insects at which they are targeted and remain useful for a long time. (Pests cannot usually evolve resistance against hormone mimics, because that would involve extensive modifications to its complex internal control system.)

Finally, ecologists and anthropologists have discovered that certain traditional agricultural practices control plant pests and diseases remarkably well. Some practices encourage animals that prey on pests to live in farmers' fields. Others control harmful insects by starving them—through planting different crops in different years and leaving some fields unplanted every few seasons.

Today, agricultural experts favor combining all these methods into a strategy called *Integrated Pest Management* or IPM. IPM strives to control pests more naturally, while using dangerous poison sprays only in emergencies. This tactic has dramatically reduced the use of pesticides in many places around the world. Importantly, farmers are discovering that IPM is often cheaper in the long run. This new technology thus has the potential to improve our environment while saving farmers worldwide many millions of dollars each year.

28–5 How Arthropods Fit into the World

Guide For Reading

- How do arthropods interact with other organisms in nature?
- In what ways do arthropods affect humans?

As you might expect from such a large, diverse group of animals, arthropods play many roles in the natural world. They are a direct source of food for many carnivorous organisms—from protists such as radiolarians to plants such as Venus' flytraps to animals such as sea anemones, fishes, frogs, turtles, birds, whales, and humans. And they are also an indirect source of food for many other organisms.

The interactions of arthropods with other organisms are not limited to eating or being eaten. Two thirds of the world's flowering plants depend on insects to pollinate them. Some plants live even more intimately with arthropods. The bull's horn acacia tree has hollow swellings at the base of some of its thorns that house symbiotic ants and special structures that feed the ants. The ants protect the acacia by eating herbivorous insects and by driving away larger herbivores with their painful bites and stings. Animals are also involved in symbiotic relationships with arthropods. It is amazing to see a large fish allow a bite-sized cleaner shrimp to crawl on its body and even into its mouth. But by allowing the cleaner shrimp to go unharmed, the fish is cleaned of annoying parasites and bits of dead tissue, and the cleaner shrimp gets a meal.

Figure 28–33 The damselfly has been trapped by a carnivorous sundew plant. Although sundews and other carnivorous plants are photosynthetic, they need to "eat" insects and other small animals to obtain the nitrogen compounds they need to survive.

Figure 28–34 The cleaner shrimp and fish such as the queen angelfish, shown here, engage in a symbiotic relationship that benefits both organisms. The cleaner shrimp gets a meal by eating parasites on the fish and the fish gets rid of annoying pests. A number of fishes can sometimes be observed congregating around a shrimp's cleaning station, waiting for their turn to be cleaned!

BACKGROUND INFORMATION
CHEMICAL PESTICIDES

In some places the widespread use of chemical pesticides "won the battle but lost the war." For example, after years of spraying in the United States and Mexico, the numbers of cotton-eating pests such as the tobacco budworm and bollworm were greater than ever before because the pesticides had killed off their natural enemies. In addition, spraying transformed some minor pests into major ones—the corn rootworm, for example, developed resistance to most pesticides and is now one of the worst pests of corn in the United States.

ECOLOGY NOTE
PLAGUING LOCUSTS

A team of scientists in South Africa recently discovered a slow-acting pesticide against locusts that also helps the locusts' natural enemies. This pesticide, deltamethrin, acts by weakening the locust and killing it in 3 to 5 days. During this time, locust flies have the opportunity to lay their eggs on the locusts. Usually only 6 percent of the locust population is affected by locust flies, as the locusts brush off the flies before any eggs are laid. However, when the locusts are immobilized by the pesticide, 30 percent of their population is affected.

As this pesticide actually helps the locust flies, their numbers are likely to increase. This may mean that eventually the flies will control the locust population so that less pesticide is needed.

TEACHING SUPPORT

Teaching Resources
- Vocabulary Review: Cross-a-Clue, p. 1

Study Guide
- Section 28–5, p. 275

MEDIA AND TECHNOLOGY

 Interactive Videodisc
- *On Dry Land: The Desert Biome*
- *Insects: Little Giants of the Earth*

Use this bar code to introduce the topic of helpful insects.

Play frames 10204 to 13688

See the Biology Media Guide pages 36–37 for additional bar-code correlations.

BACKGROUND INFORMATION
ANTS AND WOUNDS

Biting ants were once used to suture wounds. The edges of the wound were held together and a number of ants were induced to bite the area. The ant's mandibles held the wound closed even after the ant's body was removed.

BACKGROUND INFORMATION
***LIMULUS* AND MEDICAL RESEARCH**

Recently, the blood of the horseshoe crab has become an important tool in medical research. One of the chemicals extracted from *Limulus* blood is called limulus amebocyte lysate, or LAL, for short. LAL is used in an important medical test to detect extremely small amounts of poisonous contaminants in drugs to be injected into patients. In this test, a small amount of LAL is mixed with a sample of a drug and the mixture is incubated for an hour. If the drug sample is contaminated, the mixture reacts by forming a solid precipitate. LAL, at $15,000 an ounce, is probably the most valuable chemical extracted to date from any marine animal.

Figure 28–35 *Millions of microscopic dust mites live in human homes. The mites feed on the tiny flakes of dead skin that are constantly being shed from the bodies of humans.*

Humans encounter arthropods almost everywhere. The pores of our skin are home to thousands of harmless microscopic mites that feed on dead skin and oil. And no matter how neat and clean we are, our homes, even our beds, contain millions of microscopic dust mites. For the most part, these tiny harmless mites are of little consequence. But many of the more visible arthropods are of significance because they are either useful to us or a great nuisance.

Arthropods contribute enormously to the richness of human life. Agriculture would be impossible without the bees, butterflies, wasps, moths, and flies that pollinate crops. Bees manufacture honey, and silkworms produce silk. In Southeast Asia and Japan, whole shrimp and shrimp paste are important sources of protein and major ingredients in cooking. In the United States, shrimp, crab, crayfish, and lobster are considered delicacies. In Africa and Asia, many people eat insects such as grasshoppers and termites. (These insects, which are quite nutritious, are said to taste rather good.) And many insects and spiders are predators or parasites that prey upon harmful species.

There are many useful chemicals that may be obtained from arthropods—far too many to list here. An extract of horseshoe-crab blood, for example, is used to test the purity of medications. Chitin extracted from crustacean shells is used to dress wounds and to make thread for surgical stitches. The chemical that makes fireflies glow is used in medical tests and as a marker in genetic engineering.

Many new applications of arthropod chemicals are currently being investigated. For example, chitin could be sprayed onto fruit and frozen food to prevent spoilage and to preserve flavor. The adhesive that barnacles use to attach themselves to rocks, which sets quickly and hardens under water into a permanent bond, could be useful in applications ranging from dentistry to underwater construction and repair. Chemicals in spider venom are being tested for potential applications as pesticides. And scientists are currently trying to produce genetically engineered spider silk that could be used in making products as diverse as aircraft, helmets, bulletproof vests, and surgical thread.

Not all arthropods are beneficial to humans, however. Insects (such as locusts and "medflies") and arachnids (such as mites and ticks) cause billions of dollars in damage each year to livestock and crops around the world. Mosquitoes inflict

Figure 28–36 *Some species of grasshoppers exist in two distinctly different forms: a dull-colored solitary grasshopper or a brightly colored gregarious locust (shown here). Locusts travel in immense swarms that may contain as many as 50 thousand million individuals. A swarm can devastate huge areas of crops. One swarm destroyed 167,000 tons of growing grain—enough to feed 1 million people for a year.*

28–5 (continued)

Content Development
Chemicals produced by arthropods are being used for many novel applications. Examples include the use of a chitin spray on food to prevent spoilage, an artificial adhesive similar to that produced by barnacles that would be used for underwater construction and repair, and chemical extracts from spider venom used as insecticides.

SECTION REVIEW 28–5

1. Some pollinate plants; others prey on certain agricultural pests.
2. Crustaceans such as krill are an important food source in marine ecosystems. Cleaner

PROBLEM SOLVING IN BIOLOGY

IDENTIFYING ARTHROPODS

Having heard that you are now an expert on arthropods, some of your friends bring you the creatures pictured below and ask you to identify them. Can you place each in the correct arthropod group? Be as specific as possible.

(*Hint:* Check the following important features: number of body segments, walking legs, and pairs of antennae; presence of wings; presence of claws. Compare the animals with the descriptions and photos in the text.)

annoying bites, and some species carry malaria and yellow fever. Biting flies carry diseases such as sleeping sickness and river blindness, and fleas can carry bubonic plague. Termites cause extensive damage to wooden structures. Locusts have destroyed crops from the time humans first began to farm. Boll weevils are notorious for the trouble they cause cotton farmers in the South. For years, farmers have spent billions of dollars on poisonous chemicals to save their crops from these pests.

28–5 SECTION REVIEW

1. Why are certain insects essential to agriculture?
2. How are arthropods beneficial to other living things? Give specific examples.
3. **Critical Thinking—Relating Concepts** Name three dangerous or destructive arthropods and explain how they cause problems for humans.

PROBLEM SOLVING

Identifying Arthropods

The purpose of this activity is to give students an opportunity to apply the concepts they have learned in this chapter to identify some unfamiliar arthropods.

The spiderlike arthropod walking on a red flower (left) is a harvestman, or daddy longlegs. Students should be able to identify this as an arachnid (and therefore a chelicerate) based on the observations that it has eight walking legs, has no antennae, and is terrestrial.

The marine arthropod shown against a background of red coral (center) is an arrow crab. Students should be able to identify this as a crustacean based on the observations that it has a stony exoskeleton and ten legs, and does not fit in any of the other subphyla. The antennae in this crab are fused together to form the spine in the center of the crab's head.

The blue-and-black insect (right, top) is a painted weevil. Students should be able to identify this as an insect (and therefore a uniramian) based on the observations that it has one pair of antennae, six legs, and a body divided into three main parts.

The black banjo-shaped insect (right, bottom) is a ghost walker beetle. Students should be able to identify this as an insect (and therefore a uniramian) based on the observations that it has one pair of antennae, six legs, and a body divided into three main parts.

---

shrimp remove parasites from fishes. Silkworms produce silk. Students' examples may vary.
3. Locusts: eat crops. Mosquitoes: have annoying bites and carry diseases. Termites: damage wooden structures. Students' examples may vary.

Reinforcement/Reteaching
If students have difficulty answering the Section Review questions, refer them to the specific part of the text where the concept is explained and review the ideas with them.

Closure
Have students list the positive, negative, mutually beneficial (symbiotic), and neutral relationships between arthropods and other organisms, including humans.

LABORATORY INVESTIGATION

OBSERVING INSECT METAMORPHOSIS

Before the Lab
1. Cut pieces of cheesecloth to fit a 600-mL beaker in advance.
2. Mealworms can be obtained from local pet stores at any time during the year.
3. The complete life cycle of the mealworm from egg to egg-laying adult is about 4 to 6 months. However, large, mature larvae will undergo metamorphosis in much less time than that.
4. When students complete their observations, you may wish to suggest that they release the insects and larvae in a wooded area, or you may wish to culture them for future use.

Pre-Lab Discussion
1. Review the stages of insect metamorphosis with your students so that they will know what to expect in their cultures.
2. Remind students that they may not see any adults for some time in their cultures. Why? (It takes 4 to 6 months for the life cycle to be complete. Not enough time may have passed.)
3. Does the mealworm (*Tenebrio*) undergo complete or incomplete metamorphosis? (Complete.)

Skills Development
Students will use the following skills while completing the laboratory investigation.
1. Observing
2. Comparing
3. Relating
4. Applying
5. Predicting
6. Inferring
7. Diagramming
8. Manipulative
9. Recording

LABORATORY INVESTIGATION

OBSERVING INSECT METAMORPHOSIS

PROBLEM
What changes occur as some insects grow and develop?

MATERIALS (per group)

600-mL beaker	hand lens
cheesecloth	25 mealworms
corn flakes	probe
dissecting tray	rubber band

PROCEDURE

1. Fill a 600-mL beaker halfway with corn flakes.
2. Put 25 mealworms into the beaker of corn flakes. Observe their behavior.
3. Using a probe, slide one of the mealworms into a dissecting tray. The mealworm is the larval stage of the *Tenebrio* beetle. Carefully examine the larva with a hand lens, noting the location and number of appendages. Draw a diagram of the larva.
4. Put the mealworm back into the beaker of corn flakes. Cover the beaker with a piece of cheesecloth and secure the cheesecloth in place with a rubber band.
5. Check the beaker every other day for a few months by moving the probe carefully through the corn flakes. Be careful not to injure the larvae. Look for any changes in the larvae with respect to size or shape. Do you find any lifeless shells that look like the exoskeletons of the larvae?
6. If you find a short, thick motionless football-shaped object among the corn flakes, carefully slide it into the dissecting tray with a probe. This is the pupal stage of the *Tenebrio* beetle. Carefully examine the pupa with a hand lens, noting the location and number of appendages. Draw a diagram of the pupa.
7. Put the pupa back into the beaker. Continue making observations until you find adult beetles.

632

8. Do not remove the beetles from the corn flakes, as they may fly. Instead, try to keep a beetle uncovered in the beaker and examine it with a hand lens. Note the location and number of appendages. Draw a diagram of the adult.

OBSERVATIONS
1. What did the mealworms do when you placed them in the corn flakes?
2. How many legs does each of the stages have? What other appendages does each of the stages have?
3. What happened to the sizes of the larvae over time?
4. Compare the larval, pupal, and adult stages in terms of appearance and behavior.

ANALYSIS AND CONCLUSIONS
1. What did the mealworms use for food during growth and development?
2. What evidence did you find of molting? Why is molting necessary?
3. What changes occur as *Tenebrio* beetles grow and develop?

Safety Tips
Students should be cautioned that the probes are sharp and potentially dangerous and should be used with care.

Observations
1. They burrowed into the corn flakes.
2. The larva and adult have six legs. Both the larva and adult have mouthparts and antennae. Only the adult has wings. The pupa has no visible appendages.
3. They grew larger.
4. Answers will vary but should reflect observation rather than simply reiterating the information in the chapter.

STUDENT STUDY GUIDE

SUMMARIZING THE CONCEPTS

The key concepts in each section of this chapter are listed below to help you review the chapter content. Make sure you understand each concept and its relationship to other concepts and to the theme of this chapter.

28-1 Introduction to Arthropods
- Arthropods are characterized by an exoskeleton of chitin, jointed appendages, and a segmented body.
- In order to grow, arthropods must periodically shed their exoskeletons in a process called molting.
- The process of growth and development in arthropods often involves metamorphosis.

28-2 Spiders and Their Relatives
- Chelicerates have a body that consists of two parts—cephalothorax and abdomen. Chelicerates have chelicerae and lack antennas.
- Arachnids, such as spiders, scorpions, and mites, are typically carnivores that have four pairs of walking legs.

28-3 Crustaceans
- Crustaceans, such as crabs and crayfish, are characterized by a stony exoskeleton, two pairs of antennae, and mandibles.

28-4 Insects and Their Relatives
- Uniramians include centipedes, millipedes, and insects.
- Centipedes are carnivores that have poison claws and possess one pair of legs per body segment. Millipedes are herbivores that have two pairs of legs per body segment.
- Insects have a body that is divided into three parts: head, thorax, and abdomen. They have three pairs of legs attached to the thorax.
- Members of insect societies are specialized for performing different functions.
- Insects communicate. Some forms of communication rely on pheromones.

28-5 How Arthropods Fit into the World
- Arthropods play many roles in the natural world.
- Some arthropods are of little significance to humans; others are important because they are useful or a great nuisance.

REVIEWING KEY TERMS

Vocabulary terms are important to your understanding of biology. The key terms listed below are those you should be especially familiar with. Review these terms and their meanings. Then use each term in a complete sentence. If you are not sure of a term's meaning, return to the appropriate section and review its definition.

28-1 Introduction to Arthropods
arthropod
trilobite
chelicerate
crustacean
uniramian
exoskeleton
chitin
molt
metamorphosis
pupa

28-2 Spiders and Their Relatives
arachnid

28-3 Crustaceans
mandible

28-4 Insects and Their Relatives
pheromone

Analysis and Conclusions
1. Corn flakes.
2. Remnants of the exoskeleton from some molts could be found among the corn flakes. Since the exoskeleton cannot grow, molting is necessary when the insect grows.
3. The insect changes from an egg to a wormlike larva to a motionless pupa to an adult.

Going Further: Enrichment
1. Students might try their own cultures of *Tribolium confusum*, the flour beetle, often found in packaged flour and cereals. These may be kept in jars with slightly moistened flour, oatmeal, or cornmeal. Metamorphosis takes 5 to 6 weeks.
2. To observe incomplete metamorphosis, purchase egg cases of praying mantis from a commercial supply house. Keep in a terrarium; when the temperature warms, hundreds of nymphs will emerge. Feed them with a diluted sugar solution or honey in petri dishes.

CHAPTER REVIEW

CONTENT REVIEW

Multiple Choice
1. a 2. c 3. c 4. a
5. b 6. d 7. d 8. b

True and False
1. F larval
2. F open
3. F Millipedes
4. F chelicerae
5. F Trilobita
6. F chelicerates
7. F chitin
8. T

Word Relationships
1. respiratory organs; pupa
2. organs used in excretion; chitin is a protein that makes up the exoskeleton
3. mouthparts; walking leg
4. terms referring to arthropod subphyla; insect refers to a class in subphylum Uniramia. Alternatively: arthropods with mandibles; chelicerate is an arthropod that has chelicerae rather than mandibles.

CONCEPT MASTERY

1. Beetle: egg hatches into a wormlike larva that looks nothing like an adult; larva grows and undergoes a series of molts in which it changes in size but changes little in form; eventually larva undergoes a molt into a dormant pupal stage; pupa undergoes one molt to become an adult beetle. Dragonfly: egg hatches into a tiny immature animal that greatly resembles an adult; through a series of molts the immature dragonfly grows and gradually acquires adult features such as wings and reproductive organs.
2. Mandibulata: crustaceans, centipedes, millipedes, and insects. This grouping is based on a feature that may have arisen separately in two different evolutionary lines; crustaceans are thought to be as different from uniramians as they are from chelicerates.
3. Answers will vary. Students should note that a crayfish has jointed appendages, a chitinous exoskeleton, and a segmented body. They should also point out that a crayfish has a ventral nerve cord, an anterior brain, and an open circulatory system.
4. The absence of juvenile hormone signals a pupa-to-adult molt. Such chemicals would cause a moth pupa to molt prematurely into an adult and would probably cause a larva to molt prematurely into a pupa. Both of these events would result in the death of a moth.

CHAPTER REVIEW

CONTENT REVIEW

Multiple Choice

Choose the letter of the answer that best completes each statement.

1. Which of these is an arachnid?
   a. scorpion   c. grasshopper
   b. horseshoe crab d. lobster
2. A free-living arthropod is certain to have
   a. antennae.   c. jointed appendages.
   b. chelicerae.   d. gills.
3. Insects are characterized by
   a. a stony exoskeleton containing calcium carbonate.
   b. chelicerae and pedipalps.
   c. three pairs of legs on the thorax.
   d. many body segments, each of which bears two pairs of legs.
4. In crustaceans, nitrogenous wastes are excreted with the help of
   a. green glands   c. Malpighian tubules.
   b. spiracles   d. pheromones.
5. A wormlike immature animal undergoes a resting stage during which it changes into an adult that has four wings and six legs. This animal is a (an)
   a. crustacean.   c. chelicerate.
   b. insect.   d. trilobite.
6. Most spiders breathe using
   a. mandibles.   c. Malpighian tubules.
   b. tracheal tubes. d. book lungs.
7. Which is most likely to be a herbivore?
   a. spider   c. tick
   b. centipede   d. millipede
8. Trilobites
   a. are primarily terrestrial.
   b. are extinct.
   c. have highly specialized appendages.
   d. communicate by "dancing."

True or False

Determine whether each statement is true or false. If it is true, write "true." If it is false, change the underlined word or words to make the statement true.

1. If the level of juvenile hormone in an insect's body is high, the insect is in the <u>pupal</u> stage.
2. Arthropods have a <u>closed</u> circulatory system.
3. <u>Centipedes</u> are herbivores that have two pairs of legs on each segment.
4. A spider uses fanglike <u>pedipalps</u> to inject venom into its prey.
5. All the members of the class <u>Uniramia</u> are now extinct.
6. Horseshoe crabs are classified as <u>crustaceans</u>.
7. Arthropods are characterized by an exoskeleton composed of <u>calcium carbonate</u>.
8. Arthropods must periodically <u>molt</u>, or shed, their exoskeletons.

Word Relationships

In each of the following sets of terms, three of the terms are related. One term does not belong. Determine the characteristic common to three of the terms and then identify the term that does not belong.

1. tracheal tube, book lung, spiracle, pupa
2. green gland, gill, Malpighian tubule, chitin
3. mandible, chelicera, pedipalp, walking leg
4. chelicerate, uniramian, crustacean, insect

CONCEPT MASTERY

Use your understanding of the concepts developed in the chapter to answer each of the following in a brief paragraph.

1. Beetles undergo complete metamorphosis and dragonflies undergo incomplete metamorphosis. Describe the major events of the life cycles of beetles and dragonflies. Be sure to include a comparison of their life cycles.
2. In some classification schemes, arthropods are divided into two subphyla—Chelicerata and Mandibulata—based on the type of mouthparts they possess (chelicera and mandibles, respectively). Which groups of arthropods belong to the subphylum Mandibulata according to this classification scheme? Explain why many experts do not favor this method of grouping arthropods.
3. Using a crayfish as your representative organism, discuss the distinguishing characteristics of arthropods.
4. Certain chemicals bind with juvenile hormone and make it inactive. Describe how exposure to such chemicals would affect the development of a moth.

CRITICAL AND CREATIVE THINKING

Discuss each of the following in a brief paragraph.

1. **Applying concepts** Explain why you will never see spiders three stories tall or ants big enough to eat New York (except in the movies).
2. **Relating concepts** Blue crabs usually have hard, stony shells. However, some blue crabs have thin, papery shells. These crabs are called soft-shell crabs and are a popular food for some people, who eat them whole—shell and all! Explain why some crabs are soft-shelled.
3. **Relating cause and effect** People who squash an annoying hornet are often unpleasantly surprised to find themselves suddenly under attack by dozens of hornets. Explain this phenomenon.
4. **Making inferences** Instead of spraying a field with chemicals, a plane disperses tens of thousands of tiny wasps over the growing plants. What is the most likely reason for such an action?
5. **Applying concepts** At the park one day, you observe a bee flying around an open can of soda. Soon after, you notice that there are a lot of bees buzzing around this can. However, there are no bees on other open cans of soda a few meters away.
   a. Explain how the bees probably found the first can of soda.
   b. Explain why the bees do not seem interested in the other cans of soda.
6. **Assessing concepts** Which do you think is a better arrangement for an insect society: having workers that can each perform all necessary tasks or having workers that are specialized for specific tasks? Explain your answer.
7. **Using the writing process** Certain crabs have a peculiar symbiotic relationship with coral: They cause branches of coral to grow around them to form a protective prison. The imprisoned crab obtains food and oxygen from the currents of water that flow through its coral cage. Write a short story or play in which one of these imprisoned crabs converses with a more typical crustacean.

4. The wasps are probably parasites of certain agricultural pests that are attacking the crops in that field.
5. a. A worker bee scouting for food found the can of soda, flew back to the hive, and reported its find through "dancing."
   b. One of two explanations might account for this. First, the scout bee only discovered the first can and the hive is more than 50 meters away. Since the waggle dance gives the precise direction of a food source, the other members of the hive literally made a beeline for the first can and have not yet found the other cans. Second, the other cans contain diet soda—which has virtually no calories and thus is worthless from a bee's point of view!
6. Students' essays should be logical, well thought out, and well written. Students should recognize that generalized workers are more versatile but specialized workers can perform their particular tasks more efficiently.
7. Students' stories or plays should be creative and well written.

CRITICAL AND CREATIVE THINKING

1. An arthropod of these proportions would probably be crushed to death by the mass of its own unsupported body when it molted. In addition, the respiratory systems and open circulatory system of arthropods would not be able to adequately transport materials around a monster arthropod's body.
2. Crabs have a soft shell soon after they molt because the new exoskeleton has not had time to become hardened with calcium carbonate deposits.
3. The body of a squashed hornet releases alarm pheromone, which makes nearby hornets aggressive.

CHAPTER 29 Echinoderms and Invertebrate Chordates

Section	Laboratory Investigations and Activities
29–1 Echinoderms pages 637–644 THEMES: Unity and Diversity, Scale and Structure	**Student Edition** LABORATORY INVESTIGATION: Identifying Echinoderms, p. 648 **Teacher Edition** DEMONSTRATION: Echinoderm Characteristics, p. 636D **Laboratory Manual** The Starfish, p. 373
29–2 Invertebrate Chordates pages 645–647 THEMES: Unity and Diversity, Scale and Structure	**Teacher Edition** DEMONSTRATION: Observing Invertebrate Chordates, p. 636D
Chapter Review pages 649–651	

Teacher Resources

Books

Chamberlain, John B., et al. *The Sea Urchin: Molecular Biology*, 3 vols. Irvington, 1973.

Jangoux, M., and J. M. Lawrence, eds. *Echinoderm Studies*, vol. 2. Brookfield Publications Co., 1987.

Lawrence, John. *A Functional Biology of Echinoderms*. Johns Hopkins, 1987.

Sumich, James L. *An Introduction to the Biology of Marine Life,* 3rd ed. Wm C Brown, 1984.

CHAPTER PLANNING GUIDE

Other Activities	Media and Technology
Teaching Resources ACTIVITY: Form and Function in Echinoderms, p. 3 ACTIVITY: Investigating Starfish, p. 5 **Study Guide** Section 29–1, p. 279	**Biology Transparencies** 36: The Anatomy of a Starfish
Teaching Resources VOCABULARY REVIEW: Word Scramble, p. 1 **Study Guide** Section 29–2, p. 282	
Teaching Resources Chapter Test A, p. 13 Chapter Test B, p. 17	**Computer Test Bank** Chapter Test, p. 307

Audiovisuals

Echinoderms and Mollusks. Film or video. Coronet.

Echinoderms. Film or video. Ward.

The Biology of Echinoderms. Film or video. Lucerne Media.

Octopus! Octopus! (with Jacques Cousteau) Film. Churchill Films.

CHAPTER 29 Echinoderms and Invertebrate Chordates

CHAPTER OVERVIEW

Echinoderms and invertebrate chordates seem quite unlike vertebrates such as fishes, frogs, snakes, birds, and humans. However, echinoderms and chordates—vertebrate as well as invertebrate—share a common pattern of embryonic development. This pattern, which is unlike those of all the other animal phyla examined in Chapters 26–28, indicates that echinoderms and chordates evolved from a common ancestor.

Echinoderms have many unusual features. Most have a crusty and spiny outer covering—actually an endoskeleton covered by a thin layer of epidermis. This endoskeleton develops from mesoderm, as does the endoskeleton of chordates. Echinoderms are also characterized by their unique water vascular system. No other animals have such a system of water-filled canals ending in rows of tube feet that project from the body. This unique system enables an echinoderm to build up and release pressure for movement and gripping things. Echinoderm larvae have bilateral symmetry, yet the adults exhibit a five-part radial, or pentaradial, symmetry that is unique to members of this phylum. Echinoderms are found in only one environment: the ocean.

Consisting of five classes, the phylum Echinodermata includes some of the most colorful and varied organisms known. As a whole, echinoderms are of minor economic importance. However, starfish feed on large numbers of commercially important shellfish; sea urchins have long been a source of eggs and sperm for embryological research.

Though not well known, the invertebrate chordates are of evolutionary interest. At some point in their development, all chordates are characterized by having a notochord, a hollow dorsal nerve cord, and pharyngeal slits. Most chordates do not exhibit these traits as adults, but they are seen in the larvae and/or adult of such invertebrate chordates as tunicates and lancelets. Biologists think that these animals may be quite similar to the ancestors of our own vertebrate subphylum.

29-1 ECHINODERMS

Section Focus 29-1

The purpose of this section is to introduce students to a phylum of invertebrates with some anatomical features found nowhere else in the animal kingdom. Called echinoderms, these animals have a long history dating back to the beginning of the Cambrian Period. As do animals in other phyla, echinoderms vary greatly in outward appearance. Yet certain traits are shared by most. Among these are a spiny skin, five-part radial symmetry, and a water vascular system used for locomotion and feeding. Echinoderms are found only in the oceans. Despite differences in their outward appearance, biologists believe that echinoderms are related to vertebrates. Both have an endoskeleton, and the larvae of some echinoderms and chordates are remarkably similar.

The means of carrying out basic life processes is compared among different echinoderms. Common to all is a unique system of internal tubes called a water vascular system. The tubes making up this system end in rows of tiny tube feet that project from the body. Pressure caused by water entering and leaving the tube feet enables echinoderms to pull themselves along the ocean floor. Some echinoderms, such as starfish, are carnivores that use the tube feet to open the shells of bivalve mollusks on which they feed. Other echinoderms are filter feeders or detritus feeders.

Respiration, internal transport, excretion, and response are similar in most echinoderms. However, echinoderms vary somewhat in their means of locomotion. Starfish use tube feet to creep about the ocean floor, but some, such as brittle stars, swim mainly by moving their arms. Sea cucumbers crawl by contracting muscles in their body wall.

In most echinoderms, the sexes are separate. Fertilization is external; eggs and sperm are shed into the water. The larva that develops from the zygote is bilaterally symmetrical but develops into an adult with radial symmetry. Many echinoderms have great powers of regeneration.

There are approximately 6000 species of echinoderms, which are grouped into five classes. The most familiar are the starfish, often called sea stars. Brittle stars and basket stars have long, snakelike or branching arms radiating from a central disk. Sea urchins and sand dollars have an endoskeleton made up of fused plates forming a shell called a test. Sea urchins have extremely long, sharp spines, but in sand dollars, the spines and tube feet are greatly reduced. Sea cucumbers have a leathery tubelike body with a fringe of tentacles at one end. Sea lilies are the most primitive echinoderms. They are attached to the ocean floor by a stalk.

Echinoderms are important carnivores that control the populations of other sea organisms. Some, such as starfish, are considered to be pests because of their voracious appetite for commercially important shellfish and for reef-building coral polyps. In some parts of the world, sea urchins and sea cucumbers are used as food for humans. Sea urchin eggs and embryos have been widely used in research in embryology. Presently, chemicals from starfish and sea cucumbers are being studied as possible sources of anticancer and antiviral drugs.

Performance Objectives 29-1

1. **Identify distinguishing characteristics of echinoderms.**
2. **Compare the ways by which the different classes of echinoderms carry out basic life functions.**
3. **List characteristics of the echinoderm classes.**
4. **Describe the ecological and economic importance of echinoderms.**

CHAPTER PREVIEW

Science Terms 29-1
echinoderm p. 637
water vascular system p. 638
tube feet p. 638

29-2 INVERTEBRATE CHORDATES

Section Focus 29-2
The purpose of this section is to introduce members of the phylum Chordata that are generally not well known. Most chordates are vertebrates. A few, such as tunicates and lancelets, are not. Because these invertebrate chordates link vertebrates and the rest of the animal kingdom, they are of great evolutionary interest.

Students are first introduced to the characteristics that all chordates share at some time in their life cycle. At some time all have a notochord, a hollow dorsal nerve cord, and pharyngeal slits or pouches. In most adult chordates the notochord is replaced by the backbone, and in those that are terrestrial, the pharyngeal (gill) slits are present only in early embryonic stages.

Tunicates are marine chordates that have a cellulose body covering called a tunic. Most adult tunicates are sessile filter feeders, but they develop from a tadpole-like larva that has the three chordate characteristics. Lancelets are small chordates that live in warm shallow oceans. In lancelets, the characteristics of chordates are seen in the adult.

Present-day vertebrates did not evolve directly from lancelets or tunicates. However, similarities in structure and embryological development indicate that both evolved from common ancestors.

Performance Objectives 29-2
1. Define chordates and describe their three major characteristics.
2. Compare and contrast tunicates and lancelets.
3. Discuss the evolutionary significance of invertebrate chordates.

Science Terms 29-2
chordate p. 645
notochord p. 645
hollow dorsal nerve cord p. 645
pharyngeal slit p. 645

DISCOVERY LEARNING

TEACHER DEMONSTRATIONS
Modeling

Echinoderm Characteristics
The following demonstration can be used as an introduction to Chapter 29.

Obtain a few preserved starfish from a biological supply company or souvenir shop in coastal areas. Rinse excess preservative from preserved specimens and place them in dissecting pans for teams of three or four students to examine. After the group has had sufficient time to examine the starfish, ask some questions similar to those that follow in order to focus their observations.

• **The starfish is a well-known representative of the phylum known as the echinoderms. In what kind of environment do you think echinoderms live?** (If students are not sure, point out that all echinoderms are marine animals.)
• **What are some typical animal traits that a starfish does not appear to have?** (Answers will vary. Students will probably notice the absence of a head and sense organs.)
• **Does the body appear to be segmented?** (No.)

You might want to mention that the absence of segmentation is one indication that echinoderms are not close relatives of annelids and arthropods.

• **Where is the starfish's mouth located?** (Students will probably notice the mouth opening at the center of the groove on the oral side.)
• **What type of symmetry does a starfish appear to have?** (Lead students to answer that starfish exhibit radial symmetry.)

Point out that although cnidarians also exhibit radial symmetry, echinoderms are unique in that they are the only animals with five-part symmetry. If students have forgotten the difference between bilateral and radial symmetry, explain how a starfish can be divided into mirror-image portions through several lengthwise planes.

• **How would you describe the body covering of this echinoderm?** (Students will probably mention the spiny covering.)

You may want to point out that even though it appears that a starfish has an exoskeleton, it actually has an endoskeleton. A very thin layer of epidermis covers the spiny endoskeleton in living starfish.

Observing Chordates
You may want to set up this demonstration after students have read Section 29-2.

Many students have never seen a lancelet or a tunicate such as a sea squirt. Preserved specimens of both are available from biological supply companies, and they are relatively inexpensive. Place some specimens of lancelets and tunicates at several stations around the classroom for students to examine. Because of their small size, lancelets should be displayed under a stereo microscope. You may want to dissect a few specimens and use pins with attached labels to call attention to important features.

If you have a video camera with a stereo microscope attachment, you can go over the structure of a lancelet and sea squirt with the entire class at once.

CHAPTER 29

Echinoderms and Invertebrate Chordates

GUIDED ENQUIRY

Pose the following questions to students and have them record their responses. Point out that they will gain a better understanding of the key concepts if they read the chapter with these basic questions in mind. Upon completion of the chapter, pose the questions again. Ask students to compare their initial responses to those they have developed after reading the chapter.

- Why do almost all adult echinoderms live on the ocean floor?
- What unique features set the echinoderms apart from all the other animal phyla?
- Why are echinoderms usually not a source of food for most animals?
- Why are invertebrate chordates of great evolutionary interest?

CHAPTER 29

Echinoderms and Invertebrate Chordates

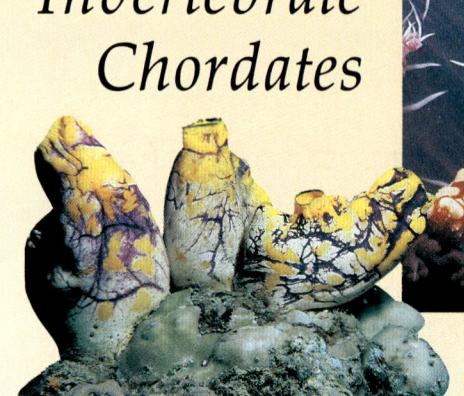

The delicate symmetrical appearance of this starfish obscures the fact that starfish are carnivores preying on other forms of sea life. The small tunicate shown in the inset is closely related to vertebrate animals, even though its appearance more closely resembles other simpler animals.

When most people see a starfish, they immediately think of the ocean—an appropriate reaction, considering that starfish and their relatives live only in the sea. Surprisingly, not all of the members of this phylum resemble stars. Some look like graceful long-stemmed flowers; others, like a peculiar cross between a polka-dotted pickle and a stalk of cauliflower. Some are as round and prickly as a pincushion. Others are flat bristly disks with holes and notches around their edges. Still others resemble armored feather dusters, pentagon-shaped cookies, or the curlicues and flourishes beneath an old-fashioned signature.

What other animals belong to this phylum? How are they all adapted to the ocean world? Why do scientists consider them to be closely related to the vertebrates? You will discover the answers to these questions in the pages that follow.

INTRODUCING CHAPTER 29

Using the Visuals
Direct students to look at the chapter-opener photographs on page 636 and read the accompanying caption and text. Encourage students to share their reactions by asking questions such as the following.
• **Were you surprised by anything you read?** (Accept all answers. Most students will be surprised that tunicates are close relatives of vertebrates.)
• **Have you ever seen these animals before? What can you tell us about them?** (Accept all responses. Few students will be familiar with tunicates. If you live in a coastal community, students may know tunicates by common names such as sea peach and sea squirt. All students should recognize the starfish; most will be able to describe how carnivorous species feed on commercially important shellfish, such as oysters and clams.)

TEACHING STRATEGY 29–1

Focus/Motivation
Ask students to look at Figure 29–1 and read the caption. Mention that several major characteristics of echinoderms are prominently displayed by the an-

GUIDE FOR READING

After you read the following sections, you will be able to

29–1 Echinoderms
- Relate the structure of echinoderms to essential life functions.
- Describe the characteristics of the classes of echinoderms.
- Explain how echinoderms fit into the world.

29–2 Invertebrate Chordates
- Name and discuss the three distinguishing characteristics of chordates.
- Describe the two subphyla of invertebrate chordates.

Journal Activity

YOU AND YOUR WORLD

Perhaps you have combed a beach for treasures of shells, starfish, or sea-urchin spines, scrambled over grassy dunes, swum in the ocean, or imagined what the seashore is like. In your journal, describe your experiences, whether they are real or imagined.

29–1 Echinoderms

Guide For Reading
- What are echinoderms?
- How do echinoderms carry out essential functions?
- How are echinoderms classified?
- How do echinoderms affect other living things?

Nearly everyone who has visited the seashore has seen starfish, sea urchins, sand dollars, or their remains washed up on the beach. These animals are members of the phylum Echinodermata (*echino-* means spiny; *dermis* means skin), a phylum that has a long and fascinating history stretching back to the beginning of the Cambrian Period, more than 580 million years ago.

What Is an Echinoderm?

As their name indicates, **echinoderms** (ee-KIGH-noh-dermz) are spiny-skinned animals. **In addition to having a spiny skin, echinoderms are characterized by five-part radial symmetry, an internal skeleton, a water vascular system, and suction-cuplike structures called tube feet.** The internal skeleton, or endoskeleton, is made up of hardened plates of calcium carbonate, which are often bumpy or spiny.

Figure 29–1 Echinoderm means spiny skin, which as you can see from this sea urchin (bottom) is an appropriate name. The sea cucumber (top) is also an echinoderm. Although its skin is smooth, it shows another characteristic of echinoderms—five-part symmetry.

imals show in Figure 29–1 and in the main photograph in the chapter opener.
• **What characteristic is exemplified by the sea urchin?** (Spiny skin. Some students may notice dark, threadlike tube feet and pentaradial symmetry.)

Point out that in some species of giant sea urchins, the spines may be 30 centimeters long. Explain that though a sea cucumber has a smooth rather than a spiny skin, microscopic examination reveals tiny calcareous plates in its leathery body wall.

• **What characteristic is exemplified by the starfish on page 636?** (Five-part radial, or pentaradial, symmetry.)
• **Look carefully at the orange lines on the sea cucumber. What do you observe?** (Accept all responses. Lead students to real-

ize that the orange "tentacles" are tube feet.)
• **How many rows of tube feet would you expect to find on the sea cucumber? How many feeding arms? Why?** (Five rows of tube feet; five many-branched feeding arms; echinoderms are characterized by pentaradial symmetry.)

Content Development

Since starfish and other echinoderms appear to be hard and rough, students often believe that they have an exoskeleton. Explain that their skeleton, like ours, is an endoskeleton of mesodermal origin. In echinoderms, it is covered by a thin layer of epidermis. The skeletal framework is made up of a number of calcareous plates joined by connective tissue. Like the bony skeletons of vertebrates, an echinoderm's endoskeleton contains spaces filled with fluids and living cells.
• **How is an echinoderm's skeleton different from that of an arthropod?** (An echinoderm has an endoskeleton composed of calcium carbonate plates. An arthropod has an exoskeleton.)

COOPERATIVE LEARNING

Using preassigned lab groups or randomly selected teams, have groups write an acrostic biopoem about one class of echinoderms. Randomly assign one of the classes described in Section 29–1 to each group. (See the Cooperative Learning annotation in Chapter 17 for a description of a biopoem.) If time permits, each biopoem should be illustrated.

Journal Activity

YOU AND YOUR WORLD

You may want to have volunteers share their essays with the class. Students should be instructed to keep their Journal Activity in their portfolio.

BACKGROUND INFORMATION
RADIAL SYMMETRY

The origin of echinoderms is uncertain, but it is believed that they evolved from bilateral free-swimming organisms. Most adult echinoderms are slow-moving, and some are permanently attached. Their radial symmetry may have evolved as an adaptation to a sessile existence since such a body plan enables an animal to interact with all sides of its environment equally.

TEACHING SUPPORT

Laboratory Manual
- The Starfish, p. 373

Teaching Resources
- Activity: Investigating Starfish, p. 5

MEDIA AND TECHNOLOGY
- Biology Transparencies
- 36: The Anatomy of a Starfish

29-1 (continued)

Content Development
Point out that although no echinoderms are found in fresh water, they are found in all oceans of the world and at a variety of depths. Starfish, which are also called sea stars, are commonly seen in great numbers clinging to rocks in tide pools along beaches. Other species have been dredged from the sea floor 5900 meters below the surface. Echinoderms are often the most numerous animals found in deep waters. Some sea cucumbers live at depths as great as 10,500 meters.

Reinforcement/Reteaching
Write the word *oral* on the chalkboard.
- **What does this word mean?** (Having to do with the mouth.)
- **Which surface of a starfish or sea urchin is the oral surface?** (The side with the mouth.)
Write *ab = away from* on the chalkboard.
- **Which is the aboral surface?** (The side opposite the mouth.)
You may want to have a volunteer point out the oral and aboral surfaces on an actual specimen. Alternatively, cut out a starfish-shaped piece of paper, write *mouth* in the center of one side, and have a volunteer label the oral and aboral surfaces.
- **On which surface are a starfish's tube feet?** (Oral.)
- **Its madreporite?** (Aboral.)

Figure 29-2 *Starfish, or sea star, species vary greatly. This bat star lives in the ocean off the coast of southern California. Lacking the thin arms of other starfish, this species resembles a pentagon.*

The **water vascular system**, which you will learn more about shortly, consists of an internal network of fluid-filled canals connected to external appendages called **tube feet**. The water vascular system is involved with many essential life functions in echinoderms, including feeding, respiration, internal transport, elimination of waste products, and movement.

Some echinoderms, such as starfish and sand dollars, live in shallow water and are thus familiar to beach-goers. Other echinoderms live only on coral reefs or on the floor of the deep ocean. Although echinoderms possess certain characteristics found in no other animals, living or extinct, they share several important features with members of our own phylum (Chordata). For example, certain stages in the development of echinoderm larvae are remarkably similar to stages seen in some members of the phylum Chordata. In addition, echinoderms have an internal skeleton (as do vertebrates) rather than an external skeleton (as do other invertebrates). For these reasons, biologists believe that among common invertebrates, echinoderms are most closely related to humans.

Echinoderm species vary greatly in appearance. Starfish exhibit a fragile beauty and perfection in shape that stirs wonder in most observers. Some sea cucumber species fascinate because their ugliness has a certain repulsive appeal. Regardless of appearance, however, echinoderms have adaptations that make them successful survivors in the world of the sea. As you read this section, keep in mind that echinoderms are survivors of history. Their success is confirmed by the fact that some echinoderms alive today look much like their ancient ancestors who lived in the seas millions and millions of years ago.

Form and Function in Echinoderms

Adult echinoderms have a body plan with five parts organized symmetrically around a center. As a result of this body plan, adult echinoderms typically have neither an anterior nor a posterior end and no brain. However, most echinoderms are two-sided. The side where the mouth is located is called the oral surface, and the opposite side is called the aboral surface.

All echinoderms have a unique system of internal tubes called a water vascular system. The water vascular system opens to the outside through a sievelike structure called the madreporite (ma-druh-POR-ight). In starfish, the madreporite connects to a tube called the ring canal that forms a circle around the animal's digestive system. From the ring canal, five radial canals extend into each body segment. Attached to each radial canal are hundreds of movable tube feet. The entire water vascular system operates like a series of living hydraulic pumps that can propel water in or out of the tube feet. When water is pushed into a tube foot, the tube foot expands. When water is pulled out, the cup on the end of the tube foot shrinks,

Skills Development
Guided Practice
Skill: Interpreting illustrations
Using the Visuals Direct students to look at Figure 29-3, and call on someone to read the caption.
- **What is the function of the wa-**

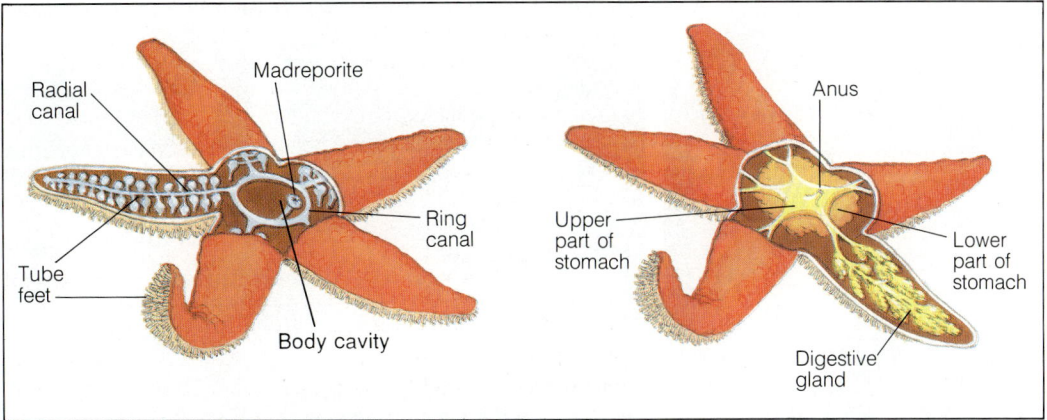

Figure 29–3 The pressure exerted by the water vascular system of the starfish (left) moves the animal along the ocean bottom. The digestive system of a starfish (right) breaks down food, which is then transported throughout the animal's body in the digestive glands and in the fluid within the body cavity.

creating a partial vacuum that holds onto whatever the foot is touching. In this way, the tube feet act like living suction cups. A single tube foot alone cannot accomplish much, but hundreds acting together create enormous force. All echinoderms "walk" with their tube feet, and some use their tube feet for feeding.

FEEDING Echinoderms have several methods of feeding. Carnivores, such as many species of starfish, use their tube feet to pry open the shells of bivalve mollusks such as clams and scallops. Once the bivalve's shell is opened, the starfish flips its stomach out of its mouth, pours out enzymes, and digests its prey in the prey's own shell. When the starfish has finished dining, it moves its stomach back into its mouth, leaving behind an empty shell as the only evidence of its deed. Starfish also eat snails, corals, and even other echinoderms. Herbivores, such as sea urchins, scrape algae from rocks by using their five-part jaw. Filter feeders, such as sea lilies, basket stars, and some brittle stars, use tube feet on flexible arms to capture plankton that float by on ocean currents. Detritus feeders, such as sea cucumbers, move much like a bulldozer across the ocean floor, taking in a mixture of sand and detritus. Then, like an earthworm, they digest the organic material and pass the sand grains out in their feces.

RESPIRATION Echinoderms, like other animals, need to exchange carbon dioxide for oxygen. In most species the thin-walled tissue of the tube feet forms the main respiratory surface. In some species small outgrowths called skin gills also function in gas exchange.

INTERNAL TRANSPORT The functions of transporting oxygen, food, and wastes—which are performed by the circulatory system in many animals—are shared by different systems

BACKGROUND INFORMATION
ECHINODERM CIRCULATION

A well-developed circulatory system appears to be absent in echinoderms. Rather, circulation of oxygen and nutrients within the coelomic cavity is accomplished primarily by the water vascular system and the beating of the cilia in the epidermis lining the cavity. A system of fluid-filled tissues called the hemal system runs parallel to the water vascular system, but it appears to have little to do with circulation. Biologists are unsure of its function. Some think that the hemal system may serve to produce and distribute special cells and hormones.

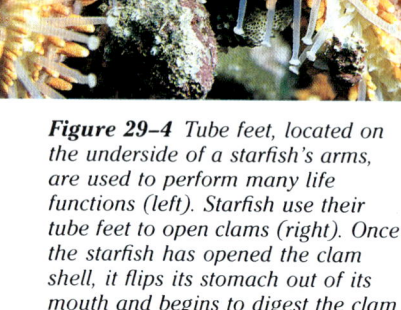

Figure 29–4 *Tube feet, located on the underside of a starfish's arms, are used to perform many life functions (left). Starfish use their tube feet to open clams (right). Once the starfish has opened the clam shell, it flips its stomach out of its mouth and begins to digest the clam right in the shell. No wonder starfish are not welcome in clam beds!*

in echinoderms. Because respiration (gas exchange) and the removal of metabolic wastes occur through skin gills and tube feet located all over the body, a system to deliver oxygen and carry away carbon dioxide and other wastes is not essential. The distribution of nutrients is performed primarily by the digestive glands and the fluid within the body cavity.

EXCRETION In almost all echinoderms, solid wastes are released through the anus in the form of feces. (The exceptions are brittle stars, which lack an anus and thus release undigested materials through the mouth.) Echinoderms, like many other marine invertebrates, excrete nitrogen-containing cellular wastes primarily in the form of ammonia. Wastes seem to be excreted in many of the same places around the body in which gas exchange takes place—the tube feet and the skin gills.

RESPONSE As you might expect in animals that have no head, echinoderms have primitive nervous systems. Most echinoderms have a nerve ring that surrounds the mouth and radial nerves that connect the ring with the body sections. Scattered sensory cells that are sensitive to chemicals released by potential food are also characteristic of most species. Starfish additionally have up to 200 light-sensitive cells clustered in eyespots at the tip of each arm. Although these structures have lenses, they do little more than tell the animal whether it is light or dark. Some echinoderms also possess statocysts (simple organs for balance that tell an organism whether it is right side up).

Although you might think that the tough, spiny skins of echinoderms protect them from predators, spines actually offer protection for only a few species—the crown-of-thorns starfish, for example. Many predators have learned ways around the

29–1 (continued)

Focus/Motivation

Using the Visuals Direct students' attention to Figure 29–4.

- **Have you ever tried to open a live clam or oyster with your bare hands?** (Answers will vary. If any students have attempted to open a bivalve mollusk barehanded, they probably found it to be a difficult task.)
- **How is a starfish able to open a clam's shell?** (The suction produced by the rows of tube feet gradually pulls the shell apart slightly.)

Explain that the tube feet of some species work in concert to produce a pull of as much as 53.9 newtons.

- **Once the shell is opened, how does the starfish feed?** (The stomach is everted into the space between the shells. Digestive enzymes of the starfish's stomach digest the clam within its own shell.)

Mention that an opening of only 0.1 millimeter is all that is necessary for the stomach to be thrust into the shell. Further, some starfish are capable of extending their stomach a distance of 10 centimeters.

Students may be interested to know that some starfish eat whole mollusks, shell and all. The soft parts are digested, and later the empty shell is ejected.

Content Development

Point out that for most echinoderms there are three levels of function within the nervous system. The best developed of these is the nerve ring around the mouth and the radial nerves extending to the body sections.

Lying above this system is another set of nerves that controls the body muscles. A third system located on the aboral side is connected to the sex organs. A

spiny defenses of echinoderms. For example, basket stars, feather stars, and spiny sea urchins are very slow moving. Clever fishes (and you will meet some clever fishes in Chapter 31) have learned to turn these animals over and attack them through their unprotected underside. For this reason, many echinoderms hide under rocks and in crevices by day, coming out to feed at night, when most predators are asleep.

MOVEMENT Most echinoderms use tube feet and thin layers of muscle fibers attached to the plates of the endoskeleton to move. An echinoderm's mobility is determined in part by the structure of its endoskeleton. In sand dollars and sea urchins, the plates are fused together to form a rigid box that encloses the animal's internal organs. These animals usually have movable spines attached to their endoskeleton, which they use along with their tube feet to creep from one place to another or to burrow in the sand. In starfish, brittle stars, and feather stars, the skeletal plates move around a series of flexible joints, enabling these echinoderms to use their arms for locomotion. Feather stars can swim for short distances by flapping their arms like wings, but starfish and brittle stars are only able to crawl. In sea cucumbers, the plates are reduced to tiny vestiges inside a soft, muscular body wall. The loss of the plates makes the body of sea cucumbers very flexible. Some species are able to crawl along the ocean floor like large, fat worms by contracting the muscles of the body wall.

REPRODUCTION Most echinoderms are either male or female, although some are hermaphrodites. In starfish, the sperm or eggs are produced in testes or ovaries, respectively, which fill the arms during the reproductive season. The animals shed their sperm and eggs into the water. Individual starfish detect gametes of their own species in the water, and they respond to that stimulus by releasing their own gametes. Fertilization takes place in open water, and larvae swim around for some time as members of the huge community of plankton that swarm in the ocean. Eventually the larvae, which have bilateral symmetry, swim to the ocean bottom, where they mature and metamorphose into adults that have radial symmetry.

Many starfish have incredible abilities to repair themselves when damaged. In fact, if a starfish is pulled into pieces, each piece can grow into a new animal as long as it contains a portion of the central part of the body. This ability of a starfish to regenerate itself has caused a great deal of trouble to people who earn their living fishing for bivalves (two-shelled mollusks). In the past, angry shellfishermen who were aware that starfish ate bivalves would tear the animals into two or three pieces and toss them overboard. Imagine their surprise when they noticed even more starfish in their bivalve beds. Today, shellfishermen know that starfish have the ability to regenerate and that every piece of torn starfish they throw back could develop into a completely new organism.

Figure 29–5 The basket star spreads its branching arms to filter particles of food from the water. When disturbed, the basket star curls up these arms and exposes the armored surface for protection.

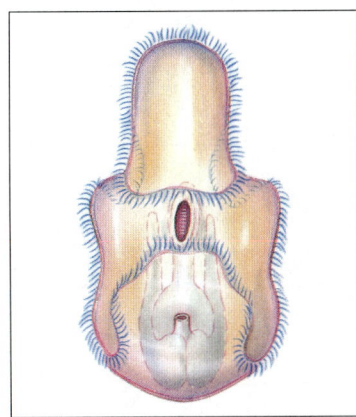

Figure 29–6 Unlike adults, which mostly crawl along the ocean bottom, echinoderm larvae are free-swimming. These larvae resemble closely the free-swimming larvae of invertebrate chordates.

nerve net connects these three systems with the body wall.

Emphasize the fact that in the absence of a head, it follows that echinoderms also lack a single central nerve center, such as a brain. Instead, they have several autonomous control centers at the junctions where radial nerves join the central nerve ring. Even though the nervous system and sense organs are poorly developed, research has shown that starfish can be conditioned to associate certain stimuli, such as light, with food.

Reinforcement/Reteaching
Review by having students answer *true* or *false* to the following statements:
1. The water vascular system is an internal network of fluid-filled canals connected to appendages called tube feet. (True.)
2. The mouth of an echinoderm is located on the aboral surface. (False. It is located on the oral surface.)
3. Echinoderms have lungs that function in gas exchange. (False. Gas exchange occurs through skin gills and tube feet.)
4. Both gas exchange and metabolic waste removal can occur through skin gills and tube feet. (True.)
5. Echinoderms have a well-developed central nervous system. (False. Echinoderms have a primitive nervous system.)
6. Most echinoderms are hermaphrodites. (False. In most echinoderms the sexes are separate.)
7. Echinoderm larvae exhibit radial symmetry. (False. They are bilaterally symmetrical.)

Content Development
Using the Visuals Call attention to Figure 29–6.
• **In what way are echinoderm larvae significantly different from adult forms?** (Lead students to answer that the larvae exhibit bilateral symmetry.)

Point out that all echinoderms except crinoids develop from a bilaterally symmetrical larva called a dipleurula. This free-living larva swims by means of a band of cilia about its body.

FACTS AND FIGURES

Brittle stars are the most abundant echinoderms, both in terms of numbers of species and of individuals. About 2000 species are found worldwide, from the seashore to depths as great as 6000 meters. In some places, millions of individuals live in clusters on the ocean floor.

29–1 (continued)

Skills Development
Guided Practice
Skill: Making comparisons
Using the Visuals Call students' attention to the photograph of the basket star shown in Figure 29–5 and the brittle star pictured in Figure 29–7. Explain that both of these echinoderms are members of class Ophiuroidea.
• **What are some major differences in the body plans of brittle stars and basket stars?** (Answers will vary, but students should notice that the arms of a brittle star are slender and unbranched. The arms of a basket star branch out.)
• **Despite their differences in appearance, can you think of any reason why they are grouped in the same class?** (Answers will vary. If necessary, point out that the arms are more set off from the central disk than they are in starfish. You may want to tell students that ophiuroids lack the groove found on the underside of asteroid arms and do not have a complete digestive system.)

Instead of using their tube feet to move along the ocean floor like other echinoderms, brittle stars use muscles in their arms to swim in a slithering type of motion. Usually one or two arms pull the animal along by moving like the arms of a swimmer doing the butterfly stroke.

Content Development
Sea urchins, sand dollars, and heart urchins are placed in the class Echinoidea. Students may wonder why sea urchins, with their extremely long spines and sand dollars, with very short spines, are grouped together. Explain that despite the difference in their spines, they share the same fundamental body structure. In both, the body is enclosed within a shell, or test, composed of overlapping calcareous plates. Also, neither has arms like other echinoderms. However, pentaradial symmetry is seen in other body parts. For example, urchins are composed of ten pairs of columns of plates. Five of these are homologous to the five arms of a starfish. In sand dollars, five furrowlike areas radiate from the central

The Echinoderm Classes

The almost 6000 species of living echinoderms are found in almost every ocean in the world. However, no echinoderms have ever entered fresh water, and they cannot survive for long on land. Although they share certain characteristics, echinoderm species are remarkably diverse in appearance. The following descriptions of echinoderm classes will provide a brief introduction to these animals.

STARFISH This class contains the common starfish, which are also known as sea stars. Starfish occur in many colors, and you may be surprised to learn that many species have more than five arms. Starfish creep slowly along the ocean bottom. Most are carnivorous, preying upon the bivalves they encounter as they move. Some species of starfish are important predators in rocky areas along the coast.

BRITTLE STARS These animals live in tropical seas, especially on coral reefs. They look much like a common starfish, but they have longer, more flexible arms and are thus able to move much more rapidly. In addition to using speed for protection, brittle stars protect themselves by shedding one or more of their arms when attacked. The detached parts keep wriggling violently, distracting predators, while the rest of the animal escapes. Brittle stars are filter and detritus feeders that hide by day and wander around in search of food only under the cover of night.

Figure 29–7 The brittle star gets its name from the fact that it can shed its arms when it is threatened (left). This distracts predators so that the brittle star can escape. In time, it will regrow the missing arm. Some starfish, such as the sun star, have more than five arms (right).

SEA URCHINS AND SAND DOLLARS This class includes disk-shaped sand dollars, oval heart urchins, and round sea urchins. Many of these animals, which are found in marine environments all over the world, are grazers that eat large quantities of algae. Others are detritus feeders. Heart urchins and sand dollars live hidden in burrows that they dig in sand or mud. Most sea urchins wedge themselves in crevices in rock during the day and come out only at night. However, many sea urchins have formidable defenses in the form of long, sharp spines. One type of sea urchin even has small blue poison sacs covering the tips of each spine, ensuring that wounds it inflicts will be painful!

SEA CUCUMBERS As their name implies, these echinoderms look like warty moving pickles with a mouth at one end and an anus at the other. Most sea cucumbers are detritus feeders. Although these animals are not numerous in shallow water, herds containing hundreds of thousands of them often cover areas of the sea floor at great depths. A few species of sea cucumbers expel sticky substances that attach to a predator. The predator, in all probability an attacking crustacean, is immobilized as it is glued into a helpless ball.

SEA LILIES AND FEATHER STARS These filter feeders, which have 50 or more long, feathery arms, comprise the most ancient class of echinoderms. Although sea lilies and feather stars are not common today, a rich fossil record indicates that

Figure 29–8 The slate urchin (top, right) has thick, strong spines that were once harvested for use as implements for writing on slateboards—thus, its name. Sea urchins have a lanternlike set of bony plates inside their body that power their jaws (inset). The sand dollar (top, left) gets its name from its flattened, coin-shaped appearance.

ECOLOGY NOTE
SEA CUCUMBERS IN THE GALÁPAGOS

On September 27, 1993, a group of angry Ecuadoran fishermen staged a brief invasion of the Charles Darwin Research Station on the Galápagos island of Santa Cruz. The cause of their protest? Sea cucumbers!

The director of Darwin station, American ecologist Dr. Chantal Blanton, and her team support a ban on sea cucumber fishing in the Galápagos. The scientists argue that such fishing would seriously disrupt the island's food chains, thus jeopardizing rare and endangered species. They point out that sea cucumber populations elsewhere in the Pacific were overfished beyond recovery when fishing was allowed. The fishermen argue that they need to make a living—which is much more difficult to do now that lobster fishing has been banned in the area. They resent the attempts of foreigners to prevent Galápagos residents from doing as they wish in their own islands.

fensive action distracts the predator while the sea cucumber slips away. A complete new set of innards is regenerated in a short time.

The peculiar behavior of expelling certain internal organs is termed *evisceration*. In addition to eviscerating body organs as a defensive mechanism, some sea cucumbers appear to eviscerate seasonally as a means of ridding the body cavity of accumulated wastes.

---

mouth region, and the short spines on the aboral surface form a starlike pattern.

Focus/Motivation
Students are usually fascinated by the defensive behaviors and remarkable regenerative powers of sea cucumbers.

• **In what unusual way do some sea cucumbers defend themselves against predators?** (They expel a sticky substance that entraps the predator.)

Explain that the sticky tubules that are ejected are branches of the sea cucumber's respiratory tract. These tubules can be quickly regenerated. A few species, when attacked by a predator such as a crab, can also expel portions of their digestive tract and gonads through a break in the body wall. This bizarre de-

TEACHING SUPPORT

Teaching Resources
- Activity: Form and Function in Echinoderms, p. 3

Study Guide
- Section 29-1, p. 279

29-1 (continued)

Enrichment
Now that students are intrigued by evisceration in sea cucumbers, encourage them to find out about other echinoderm defenses. Some of the more interesting include the prodigious amounts of slime secreted by the starfish *Pteraster tessalatus*, the envenomed spines of the sea urchin *Asthenosoma*, and the poison pedicellaria of the sea urchin *Toxopneustes*.

SECTION REVIEW 29-1

1. Echinoderms are spiny skinned animals that exhibit five-part radial symmetry, an internal skeleton, a water vascular system, and tube feet; starfish, or sea stars, brittle stars, sea urchins, sand dollars, sea cucumbers.
2. The suction-cuplike tube feet are used in locomotion in many echinoderms. Some starfish use their tube feet to pry open the shells of their bivalve prey. Oxygen diffuses into the body through the thin walls of the tube feet, and carbon dioxide and nitrogenous wastes diffuse out.
3. Tube feet enable starfish to move around. They are also used to open the shells of clams and oysters.

4. As long as a piece of a starfish contains a portion of the central part of the body, it can regenerate those parts. Tearing a starfish into pieces may mean that even more starfish will be found in an oyster bed.

Figure 29-9 Acting much like a living vacuum cleaner, this sea cucumber (top) moves along the ocean bottom swallowing organic material along with sand. Sea lilies (center) and feather stars (bottom) feed by filtering floating organic material from the water. The names reflect the delicate beauty of these animals.

they were once widely distributed. Sea lilies are sessile animals that are attached to the ocean bottom by a long, stemlike stalk. Modern sea lilies live at depths of 100 meters or more. Many feather stars live on coral reefs, where they perch on top of rocks at night and use their tube feet to catch floating plankton.

How Echinoderms Fit into the World

Echinoderms are numerous in most marine habitats. In many areas, starfish are important carnivores that control the populations of other animals. Sometimes their numbers rise or fall suddenly, causing major changes in the numbers of other forms of marine life. For example, several years ago the coral-eating crown-of-thorns starfish suddenly appeared in great numbers over wide areas of the Pacific Ocean. Within a short span of time, these starfish caused extensive damage to many coral reefs. The extent of their damage surprised and alarmed marine biologists, many of whom took drastic action to kill the starfish by injecting them with poisonous chemicals. We still do not know what caused this population explosion in the crown-of-thorns starfish or what will be its long-term effects on coral reefs.

In many coastal areas, sea urchins are important because they control the distribution of algae. However, if present in large numbers, they can threaten to literally "eat out of house and home" the other dwellers that share this habitat.

In various parts of the world, some echinoderms—for example, sea urchin eggs and sea cucumbers—are considered delicacies by some people. Many more echinoderms, however, are useful as research subjects and as possible sources of medicine. Several chemicals extracted from starfish and sea cucumbers are currently being studied as potential anti-cancer and anti-viral drugs. Sea urchins have been the subject of pioneering studies in embryology. These animals are easy to study because they produce large eggs that are fertilized externally and develop in plain sea water. Sea urchin embryos also make excellent subjects for testing the effects of drugs on cell division and development.

29-1 SECTION REVIEW

1. What is an echinoderm? Name five kinds of echinoderms.
2. How do tube feet help echinoderms to carry out their essential life functions?
3. How do starfish move? How do starfish open bivalves?
4. **Connection—Ecology** Why is tearing a starfish apart and throwing it back into the water not a good way to limit a starfish population?

Reinforcement/Reteaching
If students have trouble with any Section Review questions, go back to the pertinent material for review.

Closure
Working in teams of three or four, have students compose a list of traits that are unique to echinoderms. After completing their lists, compile a master list of distinguishing traits on the chalkboard.

29–2 Invertebrate Chordates

The phylum Chordata, to which fishes, frogs, birds, snakes, dogs, cows, and humans belong, will be the subject of many of the chapters to come. Most of the **chordates** (KOR-dayts) are vertebrates, which means that they have backbones, so they are placed in the subphylum Vertebrata. But there are also some invertebrate chordates. The invertebrate chordates are divided into two subphyla—tunicates and lancelets. Because they show possible links between vertebrates and the rest of the animal kingdom, the invertebrate chordates are of great evolutionary interest.

What Is a Chordate?

Members of the phylum Chordata are called chordates. **Chordates are animals that are characterized by a notochord, a hollow dorsal nerve cord, and pharyngeal (throat) slits.** Some chordates possess these distinguishing characteristics as adults; others, only as embryos. However, all chordates display these three characteristics at some stage of their life.

The first chordate characteristic, the **notochord**, is a long, flexible supporting rod that runs through at least part of the body, usually along the dorsal surface just beneath the nerve cord. Most chordates have a notochord only during the early part of embryonic life. In most vertebrates, the notochord is quickly replaced by the backbone.

The second chordate characteristic, the **hollow dorsal nerve cord**, runs along the dorsal surface just above the notochord. Remember that in most invertebrates, nerve cords run along ventral surfaces. In most chordates, the front end of this nerve cord develops into a large brain. Nerves leave this cord at regular intervals along the length of the animal and connect to internal organs, muscles, and sense organs.

The third chordate characteristic, **pharyngeal slits**, are paired structures in the pharyngeal (fuh-RIHN-jee-uhl), or throat, region of the body. (Remember that pharynx is another word for throat.) In aquatic chordates such as lancelets and fishes, the pharyngeal slits are gill slits that connect the pharyngeal cavity with the outside. The location of gills is very important. Many invertebrates have gills of some sort in various places, but only chordates have pharyngeal gills. In terrestrial chordates that use lungs for respiration, pharyngeal slits are present for only a brief time during the development of the embryo. These slits soon close up as the embryo develops. In chordates such as humans, pouches form in the pharyngeal region but never open up to form slits. For this reason, some scientists regard pharyngeal pouches, not slits, as the "true" chordate characteristic.

TUNICATES Tunicates are small marine chordates that eat plankton they filter from the water. They get their name

Guide For Reading

- What are chordates? What are invertebrate chordates?
- What are the distinguishing characteristics of chordates?
- Why are invertebrate chordates important to evolutionary biologists?

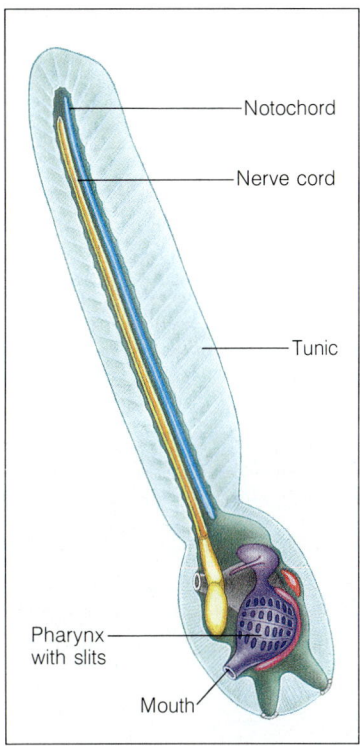

Figure 29–10 Although it seems like a simple animal, the tunicate is a chordate. It is, therefore, a relative of ours—although a very distant one.

ESL STRATEGY

Reinforce the concepts of synonyms and antonyms by having students fill in the blanks in the following exercise.

1. antonym of aboral:
   _____

2. synonym of pharynx:
   _____

3. antonym of vertebrate:
   _____

4. synonym of sea star:
   _____

5. antonym of motile:
   _____

chalkboard.

• **For what anatomical feature seen in the tunicate drawing do you think the phylum was named?** (Answers will vary. If students have not yet read the text, they may answer "nerve cord," since they are familiar with this structure, rather than the notochord.)

Call attention to the spelling of notochord and explain that this, not the nerve cord, is the structure for which the phylum is named.

Content Development

Be sure students realize that in the majority of chordates, the notochord is present only during the early stages of embryonic development. Though the notochord is replaced by cartilaginous or bony vertebrae in vertebrates, slight remnants of it still remain in the intervertebral disks.

• **How does the location of the nerve cord in chordates compare to that in arthropods and annelids?** (In chordates the nerve cord is dorsal; in arthropods and annelids it is ventral.)

You might want to mention that the nerve cord of an invertebrate is a solid rod, whereas it is a hollow tube in chordates.

TEACHING STRATEGY 29–2

Focus/Motivation

Using the Visuals Ask the class to look at Figure 29–10 and read the caption.

• **Would you have suspected that a tunicate is related to vertebrates such as humans?** (Answers will vary. Allow a few minutes for students to share their thoughts.)

• **Can you identify any characteristics that tunicates have in common with us?** (Students will probably mention such familiar features as the complete digestive tract and nerve cord.)

Mention that the tadpolelike larva remains in this form for less than 36 hours—only a few minutes in some species. In many species the larva does not even feed. Write Chordata on the

BACKGROUND INFORMATION
CHORDATE EVOLUTION

Although seemingly quite different, echinoderms and chordates are believed to have evolved from a common ancestor. Members of both phyla undergo similar embryonic development, and they are referred to as deuterostomes. Following fertilization, the embryos of deuterostomes undergo radial cleavage. That is, the cell divisions are at right angles to the embryo's polar axis. The blastopore, which is an opening formed during gastrulation in the embryo, becomes the anus, and the coelom is formed from outpocketings of the embryonic gut.

The exact pattern of echinoderm-chordate evolution is not known, and it may never be known. Yet many biologists infer that echinoderms, lower chordates, and vertebrates have evolved from a sessile filter-feeding deuterostome that produced free-swimming larvae.

TEACHING SUPPORT
Teaching Resources
- Vocabulary Review: Word Scramble, p. 1

Study Guide
- Section 29–2, p. 282

Figure 29–11 Sea squirts are tunicates. As adults, these organisms are sessile, living firmly attached to one place. However, the larvae of these animals, like the larvae of echinoderms, are free-swimming.

Figure 29–12 Lancelets are small fishlike creatures that often live with their body half buried in sand. They filter food particles from the water.

from a special body covering called the tunic. Only the tadpole-shaped larvae of tunicates have a notochord and a dorsal nerve chord. When most tunicate larvae mature, they undergo metamorphosis and become sessile adults that grow into colonies attached to a solid surface. Both larval tunicates and adults filter feed and breathe at the same time through a pharyngeal basket pierced by gill slits.

LANCELETS The small fishlike creatures called lancelets live in the sandy bottom of shallow tropical oceans. Unlike tunicates, adult lancelets have a definite head. They have a mouth that opens into a long pharyngeal region with up to a hundred pairs of gill slits. Lancelets feed by passing water through their pharynx, where food particles are caught in a sticky mucus. This mucus is swallowed into a digestive tract that starts at one end of the pharynx and continues straight through the animal to the anus, near the tail.

Lancelets have a simple, primitive heart that pumps blood through vessels in a closed circulatory system. Additionally, lancelets show evidence of segmentation in the arrangement of their nerves and muscles. A lancelet's muscles are organized into V-shaped units that are paired on either side of the body. Each muscle unit receives a branch from the main nerve cord. A similar segmented nerve and muscle organization is found in all living vertebrates. Unlike most vertebrates, lancelets have no jaw. Their mouth is composed entirely of soft tissues. Lancelets also lack appendages and can move only by bending their bodies back and forth.

How Invertebrate Chordates Fit into the World

In some ways, studying invertebrate chordates is like using a time machine to study the ancestors of our own subphylum. It is important to remember that living vertebrates did not evolve from living lancelets or tunicates. Both these subphyla have evolved over time. However, similarities in structure and embryological development indicate that vertebrates and invertebrate chordates evolved from common ancestors many millions of years ago.

29–2 SECTION REVIEW

1. What characteristics are found in a chordate?
2. What characteristics of tunicates and lancelets make them seem like close relatives of vertebrates?
3. **Critical Thinking—Making Comparisons** Which characteristics of tunicates and lancelets are unlike vertebrate characteristics?

646

29–2 (continued)
Content Development

In no other adult chordates are the phylum characteristics so well displayed as they are in a lancelet. Use a wall chart, model, pictures in reference books, or an overhead transparency to show the lancelet's internal anatomy.

- **What three characteristics of the phylum Chordata are present in adult lancelets?** (The notochord, dorsal nerve cord, and pharyngeal gill slits can be seen.)
- **Is a head present in these organisms?** (Lancelets have a definite head region.)

Point out that although a lancelet has a head, it lacks the brain and specialized sense organs that most chordates have at the anterior end.

- **How does a lancelet's circulatory system differ from that of an arthropod, bivalve, or gastropod?** (A lancelet has a closed circulatory system; an arthropod, bivalve, or gastropod has an open circulatory system.)

Mention that a closed circulatory system is a major feature of chordates. The circulatory pattern in lancelets is similar to that

SCIENCE, TECHNOLOGY, AND SOCIETY

Breakthrough

The Secret Life of Salps

Sometimes a remarkably simple change in the techniques biologists use to study the world causes us to alter our ideas about the way the living world works. For example, the invention of the microscope opened up the world of "unseen life." Recently, new methods have contributed to our understanding of the once mysterious open-water tunicates known as salps.

Salps are free-swimming animals that live in the open sea. Biologists have known about their existence for many years, but they knew little about their importance. This was due to the fact that research vessels had no way of collecting and identifying salps. These beautiful animals are so fragile that they literally fall to pieces if they are handled roughly. And that was exactly what happened in the collecting nets marine biologists used to gather plankton. Any salps that entered the net were squeezed into a clear, featureless mush.

Once scuba-diving scientists became sufficiently experienced (and sufficiently brave) to hop off their boats in the middle of the ocean, however, our knowledge of salps increased dramatically. It became clear that salps are everywhere. Giant herds of salps drift just beneath the surface. Certain species form huge snakelike colonies that stretch for many meters.

By collecting salps carefully, researchers have learned how they live. In many places, salps form important links in the ecology of the open sea. Salps eat certain plankton and are themselves food for other plankton, sea turtles, and certain fishes. This new knowledge has come to light because of a simple change in research techniques. Although these vital creatures have been nearly ignored for decades, marine biologists can now study them closely and discover how they fit into the web of life.

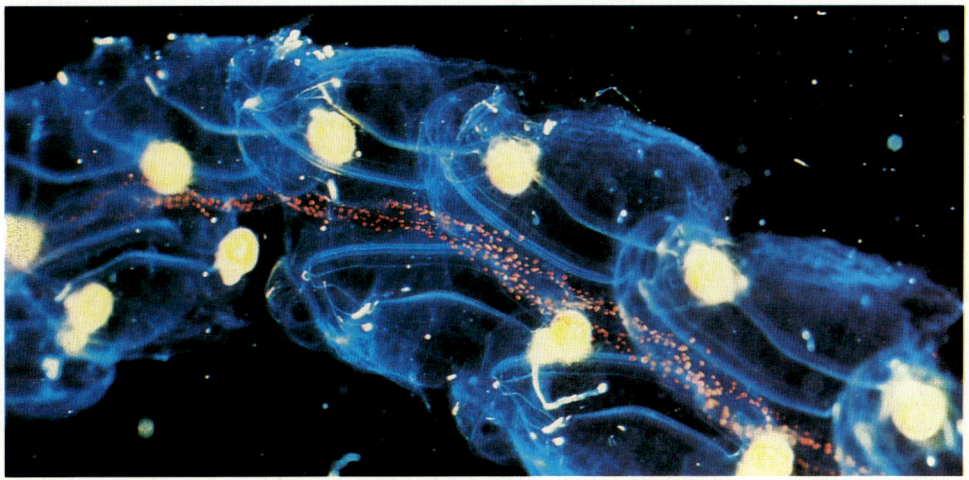

Salps are free-swimming invertebrate chordates found in the open ocean.

SCIENCE, TECHNOLOGY, AND SOCIETY

BREAKTHROUGH

The Secret Life of Salps

Salps are extremely numerous in warm ocean waters. As many as 25,000 per cubic meter are not uncommon near the surface. Their barrel-shaped bodies, ranging in length from 1.5 to 19 centimeters, are colorless and transparent. So, despite their numbers, salps often remain unseen.

Salps use bands of muscles encircling the body to carry in water from which they filter out food particles and oxygen. Thus, they are filter feeders, like other tunicates. The incoming current also serves to move them through the water by jet propulsion.

The life cycle of salps includes both sexual and asexual stages. The asexual stage called oozoids, produces numerous buds that develop into the sexually reproducing stage. The sexual generation consists of chains of hermaphroditic animals known as gonozooids. At times, these chains may exceed a length of 25 meters. Fertilization and embryonic development occur inside the body of the parent. On reaching maturity, a single asexual individual is released.

LABORATORY INVESTIGATION

IDENTIFYING ECHINODERMS

Before the Lab

1. If ample preserved specimens are not available, obtain pictures of representative echinoderms from nature or wildlife magazines, college biology texts, or library resources. Your librarian may be able to help you locate appropriate pictures. Another source of pictures might be from the students themselves. Offer extra credit for suitable echinoderm pictures brought in prior to the lab.
2. The numbered specimens or pictures can be placed at several stations around the classroom or lab. During the lab, students may circulate individually or in teams from station to station. If you have a video camera, color slides, filmstrips, or laser disk pictures of echinoderms, an alternative approach might be to have students remain seated and identify the specimens as they are projected.

Pre-Lab Discussion

Have students read the complete laboratory procedure and prepare their data table prior to coming to class. In some classes, an explanation or review of the procedures for using a classification key may be needed. If so, demonstrate the use of the key on the lab page by having the entire class key one of the echinoderms pictured in the chapter. Before using the key, be sure that students understand the difference between radial and bilateral symmetry. If necessary, review the definition of these terms and ask for examples of organisms that exhibit each type of symmetry.

LABORATORY INVESTIGATION: IDENTIFYING ECHINODERMS

PROBLEM

How can echinoderms be identified?

MATERIALS (per group)

assorted echinoderms
(pictures or preserved specimens)

PROCEDURE

1. On a separate piece of paper, draw a data table similar to the one shown here.
2. Your teacher will provide either pictures or preserved echinoderm specimens. Each specimen will be numbered.
3. Use the key to identify each numbered specimen. Start at step 1 and read descriptions A and B. Only one of the descriptions correctly applies to the specimen you are examining. At the end of a description is the identity of the specimen or directions to proceed to another step. Continue to follow the directions step by step until you identify the specimen.
4. After you identify the specimen, write its name next to its identification number in the data table. Then proceed to the next numbered specimen.

Specimen Identification Number	Identity of Specimen (from Identification Key)
1	
2	
3	
4	

OBSERVATIONS

1. What feature did all of the echinoderms you examined have in common?
2. How did the echinoderms you examined differ?
3. Did any echinoderms with visible differences have the same identity?

ANALYSIS AND CONCLUSIONS

1. How is the use of an identification key similar to the process of classification?
2. Why is it possible for two organisms that look different to have the same identity based on the key used in this investigation?

Identification Key

1	A. Has obvious radial symmetry	Go to 2
	B. Appears to have bilateral symmetry	Sea Cucumber
2	A. Has arms or branches	Go to 3
	B. Spherical, oval, or disk shaped	Go to 5
3	A. Arms in multiples of five	Go to 4
	B. Arms are branched and feathery	Go to 7
4	A. Arms are long, slender, and flexible	Brittle Star
	B. Arms are thick and less flexible	Starfish
5	A. Spherical; covered with spines	Sea Urchin
	B. Not spherical	Go to 6
6	A. Oval	Heart Urchin
	B. Flattened disk	Sand Dollar
7	A. Has a long stalk	Sea Lily
	B. Stalk short or absent	Feather Star

648

Skills Development

Students will use the following skills while completing this laboratory investigation.
1. Applying
2. Classifying
3. Comparing
4. Generalizing
5. Observing
6. Hypothesizing
7. Relating

Safety Tips

If preserved specimens are used, remind students that the preservative might be a skin irritant.

Teaching Strategy

1. Circulate about the room as students work in order to answer questions and ensure that students stay at the task.
2. Insist that students record the identity of each specimen as it is identified, rather than

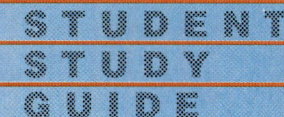

SUMMARIZING THE CONCEPTS

The key concepts in each section of this chapter are listed below to help you review the chapter content. Make sure you understand each concept and its relationship to other concepts and to the theme of this chapter.

29–1 Echinoderms

- Echinoderms are spiny-skinned animals with five-part radial symmetry, an internal skeleton, a water vascular system, and suction-cuplike structures called tube feet.
- Echinoderms are marine animals; no echinoderms live in fresh water or on land. Certain stages in the development of echinoderm larvae are similar to stages seen in members of the phylum Chordata.
- All echinoderms have a water vascular system that opens to the outside through a sievelike structure called the madreporite.
- In a starfish, the madreporite connects to a tube that leads to the ring canal, a part of the water vascular system, which forms a circle around the starfish's digestive system. Nutrients are moved around the animal in the digestive glands and the body cavity.
- Starfish reproduce externally by pouring eggs and sperm into the ocean water. The larvae that result from a fertilized egg float as part of the plankton.
- In spite of their fragile appearance, echinoderms are important predators in many environments.

29–2 Invertebrate Chordates

- Most chordates are vertebrates, which means they have backbones. However, a few chordate species are invertebrates. There are two subphyla of invertebrate chordates —tunicates and lancelets.
- At some stage of their life, all chordates possess these distinguishing characteristics: a notochord, a hollow dorsal nerve cord, and pharyngeal (throat) slits.
- Tunicates are small chordates that live in the ocean. Tunicate larvae resemble tadpoles and can move around in the water.
- Lancelets are small fishlike creatures that live in sandy ocean bottoms. Adult lancelets have a definite head.
- Invertebrate chordates are important because they indicate that vertebrate and invertebrate chordates evolved from common ancestors many millions of years ago.

REVIEWING KEY TERMS

Vocabulary terms are important to your understanding of biology. The key terms listed below are those you should be especially familiar with. Review these terms and their meanings. Then use each term in a complete sentence. If you are not sure of a term's meaning, return to the appropriate section and review its definition.

29–1 Echinoderms
echinoderm
water vascular system
tube feet

29–2 Invertebrate Chordates
chordate
notochord
hollow dorsal nerve cord
pharyngeal slit

Analysis and Conclusions

1. An identification key separates echinoderms into categories based on characteristics. With each step of the key, the categories become narrower and better defined until a single organism or category of organisms is identified.
2. Aside from the fact that there are individual differences among organisms of the same type, this key divides echinoderms into broad categories, not into species.

Going Further: Enrichment

1. Have students group the eight echinoderms that are listed in the right-hand column of the key into the class to which they belong. Then have them prepare a list of the major characteristics for each class.
2. As an alternative approach to this investigation, provide teams of students with echinoderm specimens or pictures that are already identified. Then have them develop a classification key based on the specimens' observable characteristics.

---

trying to remember for later.
3. Suggest that students log the steps they use as they work through the key. If they come up with the wrong identification, they can then retrace their steps to determine where they may have made the wrong choice.

Observations

1. They all had a spiny skin.
2. Some appeared plantlike with feathery arms, some were star-shaped, some were spherical, some were disk-shaped, some were oval, and some were pickle-shaped with few spines. Furthermore, the spines varied in shape and size, and there were variations in the size and color of the organisms.
3. Yes.

CHAPTER REVIEW

CONTENT REVIEW

Multiple Choice

1. a   2. a   3. b   4. c
5. b   6. d   7. c   8. d

True or False

1. F radial symmetry
2. T
3. T
4. F muscles
5. T
6. F water
7. F statocysts
8. T

Word Relationships

A.
1. chordate
2. filter feeder
3. balance
4. oral surface

B.
5. types of echinoderms; lancelet is an invertebrate chordate
6. parts of the water vascular system; skin
7. characteristics of all chordates; vertebrae are not found in invertebrate chordates
8. structures found in echinoderms; brain

CONCEPT MASTERY

1. Radial symmetry is the kind of symmetry exhibited by echinoderms. An animal having radial symmetry can be cut from one side of the body to the other—passing through the center of the animal—and result in two symmetrical halves. A starfish exhibits radial symmetry.
2. The starfish positions its mouth at the hinge of a clam and spreads its arms out over both of the clam's shells. It uses its tube feet to hold the shell open. Since the clam must contract its muscles to keep its shell tightly shut, it eventually tires. Its muscles relax, opening the shell a little at a time. The starfish flips its stomach out of its mouth, pours out digestive juices into the clam, and takes in nutrients from the partially digested clam.
3. A starfish can increase and decrease the suction in its tube feet by pumping water into and out of them. Thus, the water vascular system enables the starfish to move as the tube feet grasp and release the surface across which the starfish is moving.
4. Statocysts tell the starfish whether it is right side up. Light-sensitive organs in the arms of a starfish tell the animal whether it is in the light or the dark. Some species have cells that can sense chemicals in the water that are given off by potential prey.
5. Starfish shed eggs and sperm, produced by separate individuals, into the water. Starfish

---

CHAPTER REVIEW

CONTENT REVIEW

Multiple Choice

Choose the letter of the answer that best completes each statement.

1. A kind of echinoderm that is eaten by some people is a
   a. sea urchin.   c. starfish.
   b. sea lily.   d. lancelet.
2. To open a clam, a starfish uses its
   a. tube feet.   c. madreporite.
   b. brain.   d. stomach.
3. Echinoderms have
   a. a backbone.
   b. a long history on Earth.
   c. lungs.
   d. smooth skin.
4. Echinoderms show
   a. bilateral symmetry.
   b. top and bottom symmetry.
   c. radial symmetry.
   d. no symmetry.
5. Digested nutrients are moved around the body of a starfish in its
   a. skin gills.   c. bony plates.
   b. digestive glands. d. water vascular system.
6. Tunicates and lancelets are examples of
   a. vertebrates. c. echinoderms.
   b. fish.   d. chordates.
7. The side of an echinoderm where the mouth is located is called the
   a. aboral surface. c. oral surface.
   b. tunicate.   d. vascular surface.
8. Invertebrate chordates lack a
   a. larva.   c. nerve cord.
   b. notochord. d. backbone.

True or False

Determine whether each statement is true or false. If it is true, write "true." If it is false, change the underlined word or words to make the statement true.

1. All echinoderms have <u>bilateral symmetry</u>.
2. In echinoderms, <u>tube feet</u> and skin gills are used in respiration and excretion.
3. Lancelets have a primitive <u>heart</u>.
4. Echinoderms have <u>nerves</u> attached to plates in their endoskeleton.
5. A sea cucumber is a <u>herbivore</u>.
6. Tube feet are able to create suction when <u>air</u> is pumped out of them.
7. Some echinoderms have <u>madreporites</u> that tell them whether they are right side up.
8. If a piece of a starfish contains a portion of the central part of the body, the piece is able to <u>regenerate</u>.

Word Relationships

A. *An analogy is a relationship between two pairs of words or phrases generally written in the following manner: a:b::c:d. The symbol : is read "is to," and the symbol :: is read "as." For example, cat:animal::rose:plant is read "cat is to animal as rose is to plant."*

In the analogies that follow, a word or phrase is missing. Complete each analogy by providing the missing word or phrase.

1. starfish:echinoderm::tunicate:_____
2. sea cucumber:detritus feeder::feather star:_____
3. eyespots:light::statocysts:_____
4. madreporite:aboral surface::mouth:_____

650

B. *In each of the following sets of terms, three of the terms are related. One term does not belong. Determine the characteristic common to three of the terms and then identify the term that does not belong.*

5. starfish, sea lily, lancelet, sea urchin
6. ring canal, radial canal, tube feet, skin
7. notochord, hollow dorsal nerve cord, pharyngeal slits, vertebrae
8. tube feet, brain, water vascular system, madreporite

CONCEPT MASTERY

Use your understanding of the concepts developed in the chapter to answer each of the following in a brief paragraph.

1. What is radial symmetry? Name an animal that shows this kind of symmetry.
2. Briefly explain how a starfish eats a clam.
3. How does the water vascular system of a starfish help this animal to move?
4. What structures on a starfish tell this animal about its environment?
5. How do starfish reproduce?
6. How does a sea cucumber feed?
7. What characteristics does a lancelet share with vertebrate chordates?
8. Why is it not a good idea to break up a starfish and throw the pieces back into the water, especially if you fish for oysters?

CRITICAL AND CREATIVE THINKING

Discuss each of the following in a brief paragraph.

1. **Making predictions** Suppose that you are living alone on a small tropical island in the Pacific Ocean. This island is protected by a coral reef that surrounds it. One day while you are skin diving, you notice several crown-of-thorns starfish eating some of the coral animals that are part of your reef. Predict what might happen if the crown-of-thorns starfish increase in number.
2. **Making comparisons** Compare the form and function of a starfish and a sea cucumber. Describe the animals' adaptations for movement and feeding.
3. **Applying concepts** Explain why many fertilized starfish eggs never develop into adult starfish.
4. **Designing an experiment** Your friend tells you that starfish can regenerate themselves from even a small portion of an arm. You challenge this assumption. Design an experiment to prove who is correct.
5. **Using the writing process** Suppose that humans had the ability to regenerate themselves. For example, an arm might be able to grow a whole new body. Write a science fiction story that describes how this process might work for a person who was severely injured in an automobile accident.

CRITICAL AND CREATIVE THINKING

1. The crown-of-thorns starfish eats coral animals. Eventually the coral reef that surrounds the island may be destroyed. If this happens, ocean waves will sweep over the island. The island, lacking the protection of the coral reef, will probably disappear.
2. The starfish has arms that contain tube feet used for movement and feeding. It is star-shaped and is often called a sea star. The sea cucumber looks like a pickle that moves. Its five-part radial symmetry is most evident around its mouth. The starfish is a carnivore. The sea cucumber eats the detritus that it comes across as it crawls along the ocean floor.
3. Starfish release their eggs or sperm into the ocean water. In the water, the eggs may or may not be fertilized by sperm that have been released by other starfish. The larval forms of starfish float and swim in ocean water. Many of these larvae are eaten by predators. Therefore many fertilized starfish eggs never develop into adults.
4. Accept all reasonable experiments that include cutting up a starfish and observing it for a period of time to determine whether or not cut parts of the animal have the ability to regenerate.
5. Student responses should be well written, creative, and reflect an accurate understanding of regeneration.

can sense when gametes of their own species have been released by other individuals. This triggers the release of their own gametes.

6. A sea cucumber moves along the ocean bottom, sucking in detritus. It retains organic material that is digested and spits out inorganic undigestible material such as sand.
7. Lancelets have a nerve cord, a primitive heart, a closed circulatory system, and also show evidence of segmentation in the arrangement of nerves and muscles.
8. If only a portion of the central part of the starfish remains, the pieces will regenerate and there will be more starfish than before it was broken into pieces.

CHAPTER 30 Comparing Invertebrates

Section	Laboratory Investigations and Activities

30–1 Evolution of the Invertebrates
pages 653–657
 	
THEMES: Evolution, Unity and Diversity
	Student Edition
	  LABORATORY INVESTIGATION: Comparing Development in Echinoderms and Mollusks, p. 670
	Teacher Edition
	  DEMONSTRATION: An Invertebrate Family Tree, p. 652D
	Laboratory Manual
	  Comparing Invertebrate Body Plans, p. 379

30–2 Form and Function in Invertebrates
pages 658–669
 	
THEMES: Unity and Diversity, Scale and Structure
	Teacher Edition
	  DEMONSTRATION: Examining Invertebrates, p. 652D
	Laboratory Manual
	  Identifying Invertebrates, p. 385
	Teaching Resources
	  ECOLOGY INVESTIGATION: Collecting Soil Organisms, p. 43

Chapter Review pages 671–673	

Teacher Resources

Book

Ali, M. A. *Nervous Systems in Invertebrates.* Plenum, 1987.

Barnes, R. D. *Invertebrate Zoology*, 5th ed. Saunders, 1987.

Barth, Robert H., and Robert Broshears. *The Invertebrate World.* Saunders, 1982.

Buchsbaum, R., M. Buchsbaum, J. Pease, and V. Pease. *Animals Without Backbones*, 3rd ed. University of Chicago Press, 1987.

Gould, Stephen Jay. *Wonderful Life.* Norton, 1989.

Kozloff, Eugene. *Invertebrates.* Saunders, 1990.

Morris, S. Conway, et al., eds. *The Origins and Relationships of Lower Invertebrates.* Oxford University Press, 1985.

Pecknick, Jan A. *A Biology of the Invertebrates.* Wadsworth, 1985.

CHAPTER PLANNING GUIDE

Other Activities	Media and Technology

Teaching Resources
    ACTIVITY: Examining a Phylogenetic Tree, p. 3
Study Guide
    Section 30–1, p. 285
	Biology Transparencies
	    37: A Phylogenetic Tree

Teaching Resources
    VOCABULARY REVIEW: Word Game, p. 1
    ACTIVITY: Comparing Form and Function Among Invertebrates, p. 5
Study Guide
    Section 30–2, p. 288
	Interactive Videodisc/CD ROM
	    The Virtual BioPark

Teaching Resources
    Chapter Test A, p. 11
    Chapter Test B, p. 15
    Performance-Based Assessment, Unit 7
	Computer Test Bank
	    Chapter Test, p. 317

Audiovisuals

Invertebrates. Film or video. Coronet.
The Invertebrates. 4 filmstrips with cassettes. Ward.
Animals Without Backbones, 2nd ed. Film or video. Coronet.

Lower Invertebrates. Filmstrip with cassette. Ward.
Intermediate Invertebrates. Filmstrip with cassette. Ward.

Advanced Invertebrates. Filmstrip with cassette. Ward.

CHAPTER 30 Comparing Invertebrates

CHAPTER OVERVIEW

The evolutionary relationships among animals can be shown in the form of a phylogenetic tree. Animals are divided into several different groups according to shared developmental and structural patterns. Although the terms *deuterostome, protostome, acoelomate, pseudocoelomate,* and *coelomate* are not a part of true taxonomic nomenclature, they are useful for comparing and contrasting animal phyla.

Different groups of invertebrates carry out their essential life functions—movement, feeding, internal transport, respiration, excretion, response, and reproduction—in a variety of ways. The organs and other body structures that carry out these functions vary greatly in form from phylum to phylum and are often diverse within a single phylum. Invertebrate behavior, like invertebrate form, is quite varied. The variation in behavior among invertebrate phyla can be seen in the ways in which invertebrates care for their offspring.

30-1 EVOLUTION OF THE INVERTEBRATES

Section Focus 30-1

The purpose of this section is to familiarize students with the concept of the phylogenetic tree and to introduce them to basic divisions of the animal kingdom.

The base of the phylogenetic tree represents the common ancestor of all the groups shown on the tree. Branches that originate near the bottom of the tree represent groups that evolved earlier than those represented by branches higher up on the tree. Branches that do not reach the outside of the tree represent extinct groups.

Animals can be divided into groups based on the nature of their body cavity. Acoelomates lack a body cavity; pseudocoelomates have a body cavity that is only partially lined with mesoderm; coelomates have a body cavity that is completely lined with mesoderm. Animals can also be classified as protostomes or deuterostomes based on their developmental patterns. In protostomes, the mouth develops from the blastopore and cleavage is generally spiral. The mesoderm forms from cells that migrate from the endoderm, and the coelom is formed by the development of a hollow space in the mesoderm. In deuterostomes, the mouth develops from an opening that arises after the formation of the blastopore, which then becomes the anus. Cleavage is typically radial. In some deuterostomes, the mesoderm and coelom arise from outpocketings of the endoderm.

Performance Objectives 30-1

1. Explain how a phylogenetic tree depicts evolutionary relationships.
2. Describe the early development of an animal embryo from the start of cleavage to mesoderm formation.
3. Compare the development of a typical protostome and a typical deuterostome in cleavage, fate of the blastopore, mesoderm formation, and coelom formation.
4. Differentiate between acoelomates, pseudocoelomates, and coelomates.

Science Terms 30-1

phylogenetic tree   p. 653
protostome   p. 654
deuterostome   p. 654
acoelomate   p. 654
pseudocoelomate   p. 654
coelomate   p. 654
coelom   p. 657

30-2 FORM AND FUNCTION IN INVERTEBRATES

Section Focus 30-2

The purpose of this section is to tie together the concepts students have learned in the previous chapters of this unit and to present these concepts in a way that encourages them to compare systems and functions in the major invertebrate phyla.

Invertebrates may have an exoskeleton, an endoskeleton, or a hydrostatic skeleton. An invertebrate may be a carnivore, herbivore, detritus feeder, filter feeder, parasite, or display a combination of two or more of these feeding modes. Intracellular digestion is associated with simpler animals; extracellular digestion is usually found in more complex animals. Internal transport involves diffusion between cells, between cells and fluid in the body cavity, and/or between cells and blood. Circulatory systems may be open or closed. Some invertebrates exchange gases through the body surface; others have specialized respiratory organs such as gills, lungs, tracheal tubes, book gills, and book lungs. Nitrogenous wastes are excreted in the form of ammonia, urea, or uric acid. As animals become more complex, centralization of the nervous system and cephalization tend to increase and the sense organs tend to be more highly developed. Reproduction may be asexual or sexual. Most invertebrate species have separate sexes; a few are hermaphrodites. Fertilization may be internal or external. Some invertebrates care for their offspring; others do not.

CHAPTER PREVIEW

Performance Objectives 30–2

1. Describe how endoskeletons, exoskeletons, and hydrostatic skeletons are used in movement in different kinds of invertebrates.
2. Differentiate between intracellular and extracellular digestion.
3. Discuss ways specific organisms carry out the functions of internal transport.
4. Compare the respiratory and nervous systems of different kinds of invertebrates.
5. List the three major forms of nitrogenous wastes and explain how these wastes are adaptations to different conditions.
6. Discuss reproduction and parental care in invertebrates.

Science Terms 30–2

hydrostatic skeleton　p. 659
endoskeleton　p. 659
intracellular digestion　p. 659
extracellular digestion　p. 659
external fertilization　p. 667
internal fertilization　p. 667

DISCOVERY LEARNING

TEACHER DEMONSTRATIONS
Modeling

An Invertebrate Family Tree
The following demonstration can be used as an introduction to Chapter 30.
　Set up a classroom display of live invertebrates, preserved invertebrates, and pictures of invertebrates. Try to assemble as large and diverse a group of specimens as possible. Label the specimens.
• **Which invertebrates do you think are closely related? Why?** (Accept all logical answers. Insist that students rely on their own observations. If students answer that animals are insects, mollusks, arthropods, and so on, ask them to explain what those terms mean and what observations led them to those conclusions.)
• **What information do you think you need to positively determine the relationships among these invertebrates?** (Accept all logical answers. Students might suggest looking at fossils, finding out how the invertebrates carry out their life functions, and examining the internal structure of the invertebrates.)

Examining Invertebrates
You may wish to perform this demonstration before students begin Section 30–2.
　Obtain an earthworm and an arthropod such as a beetle, crayfish, or land hermit crab. You may also wish to obtain other invertebrates for students to examine. Ask more responsible students to help you collect invertebrates. Possible sources for invertebrates include pet stores, bait shops, backyards, and the school grounds.
　Have students examine the live invertebrates. Focus students' attention on comparing form and function in the invertebrates through the questions you ask. Encourage students to make informed guesses if they are not sure of the answer to a question, but require them to support their inferences from their observations.
• **How are these animals similar?** (Accept all logical answers.)
• **How are these animals different?** (Accept all logical answers.)
• **Where in nature are these animals found?** (Accept all logical answers.)
• **How do these animals move?** (Answers will vary, depending on the invertebrates you obtain.)
• **What do these animals eat?** (Accept all logical answers.)
• **How might these animals carry out the functions of internal transport? Gas exchange? Excretion?** (Accept all logical answers.)
• **How do these animals respond to their environment?** (Answers will vary, depending on the invertebrates you obtain.)

CHAPTER 30
Comparing Invertebrates

GUIDED ENQUIRY

Pose the following questions to the students and have them record their responses. Point out that they will gain a better understanding of the key concepts if they read the chapter with these basic questions in mind. Upon completion of the chapter, pose the questions again. Ask students to compare their initial responses with those they have developed after reading the chapter.

- How does early development differ among invertebrates?
- How do body cavities differ among invertebrates?
- How do similarities and differences among invertebrates help us determine the evolutionary relationships among phyla?
- How are the ways in which invertebrates carry out their essential life functions adaptations to different environments?

INTRODUCING CHAPTER 30

Using the Visuals

Have students read the text and examine the photographs on page 652.
- **What kind of animals are shown in the photographs?** (Students should be able to identify soft corals—among them, a sea fan—and sponges in the large photo. The inset photo shows a butterfly.)
- **How are they similar?** (They are invertebrates.)
- **How are they different?** (Accept all logical answers. Students should recognize that the animals in the large photo are marine and sessile, whereas the butterfly is terrestrial and motile. Students should also note that the animals belong to different phyla.)
- **Which invertebrate phyla are described in the text or represented in the photographs on page 652?** (Sponges: Porifera. Coral: Cnidaria. Flatworm: Platyhelminthes. Roundworms: Nematoda. Octopus: Mollusca. Polychaete: Annelida. Crustaceans: Arthropoda. Sea urchin: Echinodermata. Tunicates: Chordata. Butterfly: Insecta.)
- **Why do you think the writer described marine invertebrates rather than freshwater or terrestrial invertebrates?** (The writer probably wanted to give examples of every invertebrate phylum that was examined in the previous chapters, and echinoderms are found only in salt water.)

CHAPTER 30
Comparing Invertebrates

Soft corals and sponges, some of nature's loveliest invertebrates, are underwater havens for other creatures. The butterfly (inset), another type of invertebrate, is sipping nectar from brightly colored lantana flowers.

Beneath the shimmering surface of the ocean, a profusion of animals lives unseen by landbound eyes. Long, fingerlike sponges sway with the movement of waves far above. Tiny coral polyps shrink into themselves as a spotted flatworm swims by. Microscopic roundworms swarm in the miniature world between sand grains on the ocean floor. An octopus lurks in a crevice beneath the coral, its body slowly pulsing as it pumps water over its gills. A polychaete worm extends its feathery gills from the opening of its burrow and begins to strain tiny crustaceans and other bits of food from the water. Nearby, a spiny sea urchin slowly nibbles a path through the algae next to a colony of purple tunicates.

All of these wonderfully diverse animals are invertebrates. How did invertebrates evolve? How are different kinds of invertebrates related to one another? And what evidence do scientists examine to determine the nature of these relationships? You will discover the answers to these questions as you read the pages that follow.

GUIDE FOR READING

After you read the following sections, you will be able to

30–1 Evolution of the Invertebrates
- Explain how evolutionary relationships are shown on a phylogenetic tree.
- Compare the development of protostomes and deuterostomes.
- Compare acoelomate, pseudocoelomate, and coelomate animals.

30–2 Form and Function in Invertebrates
- Discuss the ways in which different invertebrate phyla carry out their essential life functions.
- Describe the evolution of various body systems in invertebrates.

Journal Activity

YOU AND YOUR WORLD In your journal, list a few common expressions that have to do with invertebrates. For example: You move at a snail's pace. Explain the meaning of each expression as you think it might relate to the invertebrate.

30–1 Evolution of the Invertebrates

Guide For Reading
- How does a phylogenetic tree show evolutionary relationships?
- How does the development of protostomes and deuterostomes differ?
- What are acoelomate, pseudocoelomate, and coelomate animals?

The evolutionary relationships between different groups of organisms can be shown in the form of a diagram called a **phylogenetic** (figh-loh-juh-NEHT-ihk) **tree**. *Phylo-* literally means tribe and refers to taxonomic groups; *-geny* means origin and development. Thus, as its name indicates, a phylogenetic tree shows our best understanding of which phyla originate from a common ancestor and approximately when evolutionary lines diverged. See Figure 30–3 on page 654. The base of the tree represents the common ancestor of all the groups shown on the tree. Branches that originate close to the bottom of the tree represent groups that evolved long ago; branches that originate near the top of the tree represent groups that evolved relatively recently. The tips of the branches represent living groups. Some phylogenetic trees show "dead" branches that do not reach the outside of the tree. Dead branches represent extinct evolutionary lines. There are no living groups from these lines.

How do scientists decide where animals belong on a phylogenetic tree? Recall from Chapter 13 that evidence for evolutionary relationships is found in the fossil record, in the body structure and chemical compounds of living organisms, and in the early development of organisms. Scientists examine many characteristics in each of these categories, looking for similarities and differences that indicate how closely organisms are related to one another.

Figure 30–1 The tropical katydid, here munching on a leaf, is one of the more endearing terrestrial invertebrates.

COOPERATIVE LEARNING

Have each student select a partner. Each pair of students should create an invertebrate "creature" to star in a science fiction movie. Students should keep in mind that their creature should have body structures that will enable it to perform the tasks necessary for survival (movement, feeding, internal transport, respiration, excretion, response, and reproduction). Remind them that their creature's body systems must be consistent with each other and its environment; for example, a desert-dwelling creature would not possess gills as its only organ of respiration. Students' finished product should be a drawing of their creature and a written explanation of how it carries out the essential functions of life. Their drawing could take the form of a poster that advertises their science fiction movie.

Journal Activity

YOU AND YOUR WORLD

Call students to read their favorite expression to the class and explain what it means and why they like it. Students should be instructed to keep their Journal Activity in their portfolio.

Content Development

Using the Visuals Explain that animals also have a family tree. Have students read the first three paragraphs of this section (on pages 653 and 654). Instruct them to examine Figure 30–3 on page 654 and answer the question in the caption. You may wish to ask questions based on the phylogenetic tree at this point.
- **Which animals were the first to evolve?** (Sponges)
- **Which organisms are the "grandparents" of animals?** (Protists.)
- **Which phyla of animals are protostomes?** (Mollusks, annelids, arthropods.)
- **Which phylum of animals is acoelomate?** (Flatworms.)

TEACHING STRATEGY 30–1

Focus/Motivation

Before students begin this section, have them make a diagram of their own family tree, starting with one pair of grandparents and including all the aunts, uncles, and cousins on that side of the family. You may want to draw a hypothetical tree to give them the general idea.
- **Why is this kind of diagram called a family tree?** (It looks like the branches of a tree. Depending on how they have drawn their diagrams, some students may have to turn their drawings upside down to see the similarity between what they have drawn and a tree.)
- **What sort of information is given in a family tree?** (Names of family members and how these people are related to one another.)

653

ANNOTATION KEY

❶ Lancelets, tunicates, and echinoderms. (Interpreting diagrams)

BACKGROUND INFORMATION
PROTOSTOMES AND DEUTEROSTOMES

Another major difference between protostomes and deuterostomes has to do with when genes are permanently "turned off" during development. Protostomes have determinate cleavage. By the time the embryo is in the two-cell stage, the fate of each cell has already been fixed. If separated, the cells cannot grow into complete embryos. If any of a developing protostome embryo's cells are destroyed or misplaced, the corresponding body region will be lost or displaced. Deuterostomes, on the other hand, have indeterminate cleavage. The developmental instructions arise gradually. As a result, if cells are lost, the embryo can still regulate its development and grow normally. If the cells of an early embryo are separated from one another, they can develop into undersized but otherwise normal embryos.

TEACHING SUPPORT

Teaching Resources
- Activity: Examining a Phylogenetic Tree, p. 3

MEDIA AND TECHNOLOGY
- Biology Transparencies
- 37: A Phylogenetic Tree

Figure 30–2 The bubblelike sea squirts, clustered like berries on a branch of red coral, are deuterostomes.

As you can see in Figure 30–3, there are several major branches on the phylogenetic tree—including **protostomes** and **deuterostomes**; **acoelomates**, **pseudocoelomates**, and **coelomates**. These branches represent basic evolutionary lines in animals with bilateral symmetry. **The division of animals into deuterostomes and protostomes is based on events in early development. The division of animals into acoelomates, pseudocoelomates, and coelomates is based on the structure of the body cavity.**

Early Development

Protostomes include flatworms, roundworms, annelids, mollusks, arthropods, and the members of most of the minor invertebrate phyla. Deuterostomes include echinoderms, several small phyla of strange-looking marine animals we have not discussed, and all members of our own phylum, Chordata. To understand the reasons for dividing animals into protostomes and deuterostomes, we must examine the earliest stages in the development of animals.

Figure 30–3 The phylogenetic tree shows our best understanding of the evolutionary relationships between different groups of organisms. Which three groups of invertebrates appear to be most closely related to vertebrates? ❶

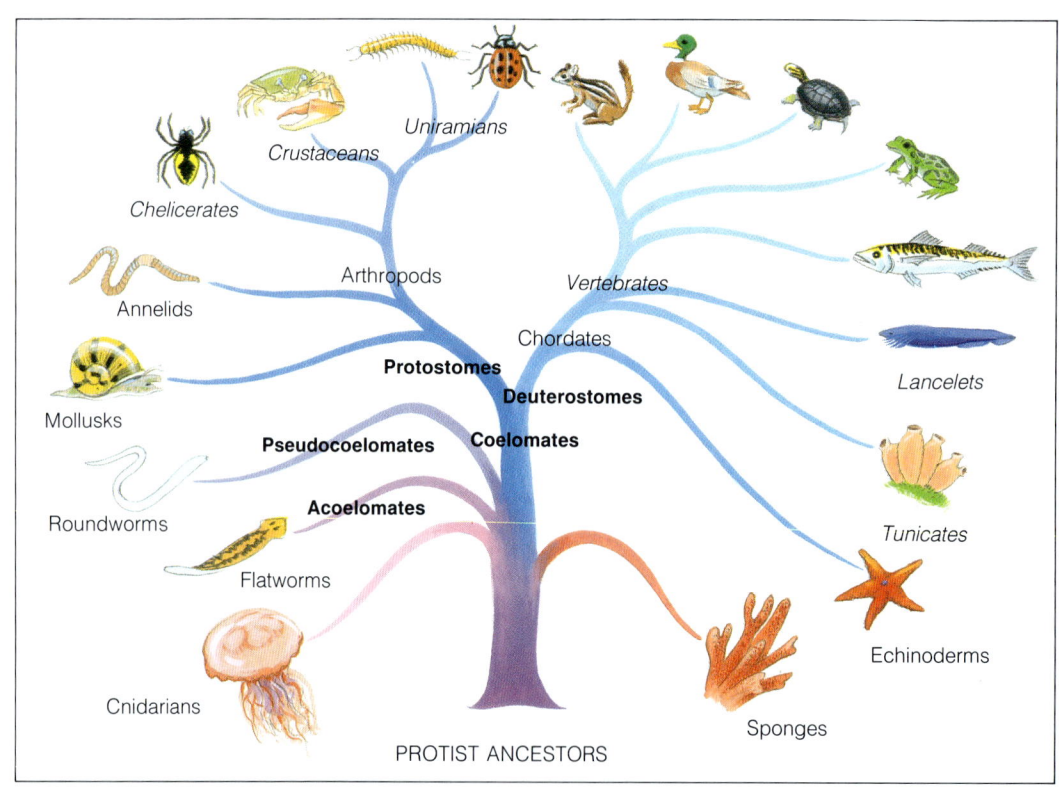

30–1 (continued)

Content Development
Stress that "higher" animals did not evolve directly from "lower" animals. Modern species descended from common ancestors that existed sometime in the past. For example, chordates are thought to have evolved from flatworm ancestors. This does not mean that parakeets are descended from planarians. Flatworms, like all other groups of animals, have changed over time. Planarians and parakeets may have evolved from a common ancestor that existed in the remote past, but parakeets do not have planarians in their family tree!

Reinforcement/Reteaching
Give each student a round balloon. Instruct students to partially blow up the balloon until it is roughly 6 or 7 centimeters in diameter and then tie a knot in the stem of the balloon.

- **What can you do with this balloon?** (Accept all reasonable answers.)

Tell students that the balloon

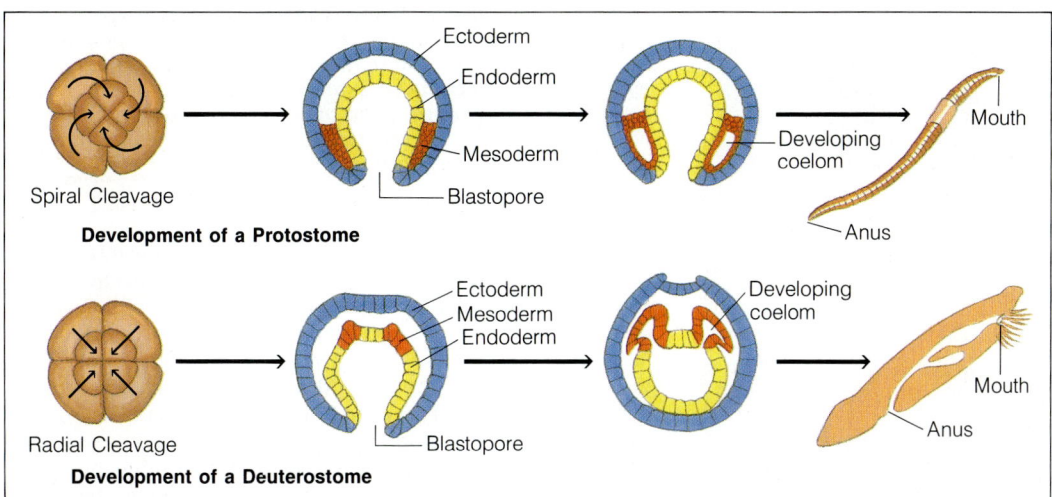

Figure 30–4 *In protostomes, the blastopore becomes the mouth. In deuterostomes, the blastopore becomes the anus.*

Soon after an egg has been fertilized, it begins a series of divisions. These divisions lead first to a two-cell stage and then to a four-cell stage. When the embryo grows from four cells to eight cells, the new cells can be arranged in different ways. In spiral cleavage, which occurs in almost all protostomes, the four new cells sit in between the four older cells. In radial cleavage, which occurs in almost all deuterostomes, the four new cells sit directly on top of the four older cells. It is important for you to note that neither of these patterns is "better" than the other. Radial and spiral cleavage are just two different patterns of growth that have evolved in the animal kingdom.

In both protostomes and deuterostomes, the cells of the embryo continue to divide until they form a hollow ball. Then the ball becomes flattened on one side and folds in on itself. This change in shape is similar to what you would get if you pushed your fist slowly into a partially inflated balloon. The layer of cells on the outside of the ball is called the ectoderm (*ecto-* means outside; *-derm* means skin). The layer of cells that has folded inside the ball is called the endoderm (*endo-* means inside). Both the ectoderm and the endoderm eventually develop into several different kinds of tissue.

The round central cavity enclosed by the endoderm will become the digestive tract of the developing embryo. The opening of this cavity to the outside is called the blastopore. And it is the fate of the blastopore that determines whether an animal is a protostome or a deuterostome. If the blastopore becomes the mouth, the animal is a protostome (*proto-* means first; *stoma* means mouth). If the blastopore becomes the anus and an opening that appears later becomes the mouth, the animal is a deuterostome (*deutero-* means second).

MULTICULTURAL STRATEGY

A haiku is a poem that consists of three lines. The first and third lines contain five syllables each, and the second line contains seven syllables. The haiku, which originated in Japan 300 years ago, tries to capture the essence of a single moment or image from nature. The last line often conveys a sense of revelation or insight.

The simplicity and elegance of the haiku appeals to poets from many eras and cultures. This haiku by African American writer Richard Wright (1908–1960), author of the novel *Native Son*, is an example.

> Make up your mind, snail!
> You are half inside your house
> And halfway out!*

Have students write a haiku or two on the invertebrate of their choice. To let them experience one kind of contest waged by the nobility of feudal Japan, give students a topic and 5 minutes in which to write their haiku.

---

can be used as a model of an embryo. Hold up a partially inflated balloon.

• **What stages of development does this represent?** (The hollow-ball stage.)

Have students carefully push the knot toward the center of the balloon and hold it there.

• **What stage of development are you simulating?** (The indented stage.)

• **What part of the balloon represents the ectoderm? The endoderm?** (The outside. The pushed-in part.)

Enrichment

You may wish to teach more advanced students the proper terms used in describing development. The cell divisions in the early embryo are known as cleavage. The hollow-ball stage is the blastula; the space enclosed by the blastula is the blastocoel. The embryo indents during the process of gastrulation to form the gastrula; the space enclosed by the gastrula and connected to the outside by the blastopore is the archenteron.

*"Make up your mind, snail!" by Richard Wright. Reprinted by permission of John Hawkins & Associates, Inc.

BACKGROUND INFORMATION
MESODERM FORMATION

There is a lot of variation in mesoderm formation in animals. For example, in certain sea urchins, part of the mesoderm originates from cells that migrate into the blastula even before gastrulation begins. In vertebrates the cells of the blastula are not arranged in a single layer one cell thick as they are in the lancelets and sea urchins. This makes the cell movements that result in the formation of the mesoderm rather complex. The mesoderm in vertebrate embryos usually results from the inward migration of a sheet of cells or a number of individual cells at a groove on the embryo's surface.

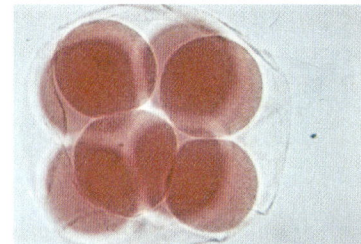

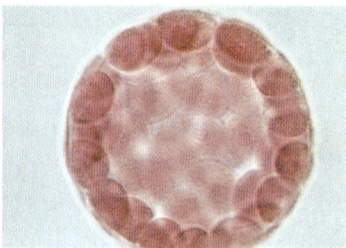

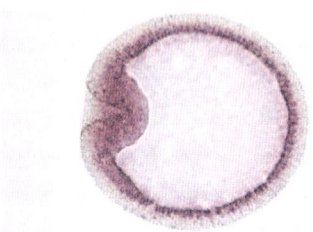

Figure 30–5 *As the embryos of animals develop, a single fertilized egg cell divides to form two cells, then four cells, then eight cells, and so on. The starfish embryo at the top is in the eight-cell stage. Later in development the cells form a hollow ball, as in the starfish embryo in the center. In the embryos of most animals, the hollow ball eventually begins to indent, as in the starfish embryo at the bottom.*

There is a third cell layer in embryos, called the mesoderm (*meso-* means middle), which is located between the endoderm and the ectoderm. In most protostomes, the bulk of the mesoderm is produced by a few cells in the area where the ectoderm meets the endoderm. In deuterostomes other than vertebrates, the mesoderm is typically produced from pouches of endoderm. See Figure 30–4 on page 655. Many important tissues, including muscles, develop from the mesoderm. And as you will learn shortly, the mesoderm is also important in defining the nature of an animal's body cavity—a characteristic that provides the basis for another division of the animal kingdom into major evolutionary lines.

Body Cavities

One of the most significant features of animal body plans is the presence or absence of a fluid-filled body cavity located between the digestive tract and the body wall. Body cavities are important for several reasons. They provide a space in which internal organs can be suspended so that they are not pressed on by muscles and twisted out of shape by body movements. Thus body systems can work in a more efficient and controlled manner. Body cavities also allow room for internal organs to develop and expand—for example, when ovaries fill with eggs. In addition, body cavities contain fluids that may be involved with internal transport, or the carrying of food, wastes, and other materials from one part of the body to another.

Some phyla, such as flatworms, have no body cavity at all. The body is basically a solid mass of mesoderm sandwiched between an inner layer of endoderm and an outer layer of ectoderm. Animals in these phyla are called acoelomates (ay-SEE-loh-mayts) (*a-* means without; *coelom* refers to the body cavity).

Figure 30–6 *The marine flatworm (left) has no body cavity—it is an acoelomate. The bubble-blowing ghost crab (right) has a body cavity that is completely lined with mesoderm—it is a coelomate.*

30–1 (continued)

Skills Development

Guided Practice

Skill: Making generalizations

Using the Visuals Direct students' attention to the top photograph in Figure 30–5.

• **What kind of cleavage is shown in this photograph?** (Radial.)

• **How do you know that this is radial cleavage?** (The cells are directly on top of or behind one another.)

• **Is this the embryo of a protostome or of a deuterostome? How can you tell?** (Deuterostome. It undergoes radial cleavage.)

SECTION REVIEW 30–1

1. A branching diagram showing the evolutionary relationships among organisms.

2. Deuterostomes and protostomes. Deuterostomes.

3. In protostomes, the blastopore become the mouth, cleavage is typically spiral, the mesoderm arises from a few cells near the blastopore, and the coelom forms from a split in a mass of mesodermal cells. In a deuterostome, the blastopore becomes the anus, cleavage is typically radial, and the mesoderm and coelom may result from pouches of endoderm.

4. Acoelomate: lack a body cavity. Pseudocoelomate: body

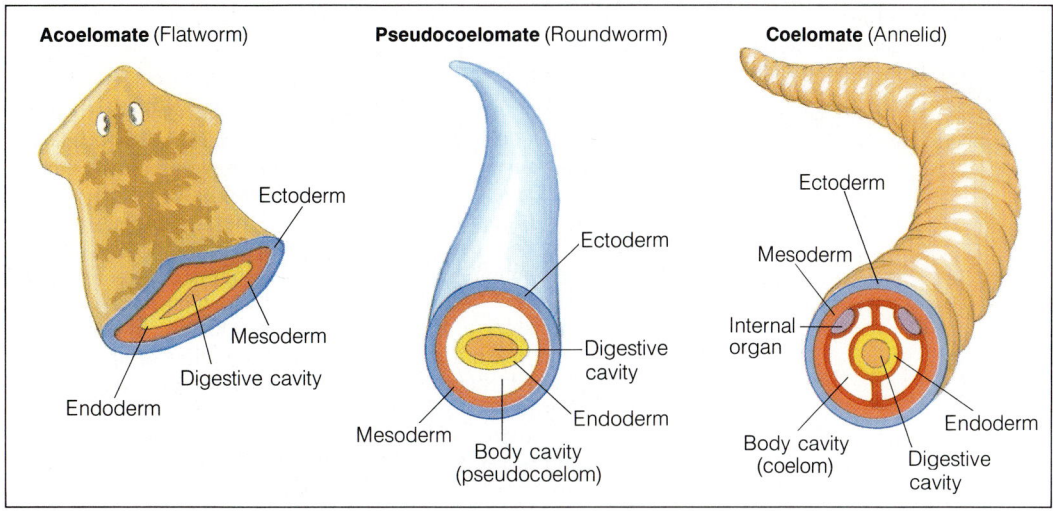

Other phyla, such as roundworms, have a body cavity that is partially (but not completely) lined with mesoderm. Animals with this kind of body cavity are called pseudocoelomates (soo-doh-SEE-loh-mayts) (*pseudo-* means false.)

Still other phyla have a true **coelom** (SEE-lohm), or body cavity that is completely lined with mesoderm. Animals with this kind of body cavity are called coelomates and are considered to be more advanced than acoelomates and pseudocoelomates. The complete mesoderm lining of the coelom makes it possible for the digestive tract to develop specialized regions and organs, allows for the formation of blood vessels (which are formed from mesoderm), and makes it easier for complex organ systems to develop.

Figure 30–7 The basic body cavities and three body layers of an acoelomate, pseudocoelomate, and coelomate are shown here. Note that a pseudocoelom is only partially lined with mesoderm, whereas a coelom is completely lined.

30–1 SECTION REVIEW

1. What is a phylogenetic tree?
2. What are the two main branches of the animal kingdom's evolutionary tree? On which branch do humans belong?
3. How do the names protostome and deuterostome relate to the differences in the development of animal embryos?
4. Compare acoelomate, pseudocoelomate, and coelomate animals.
5. **Critical Thinking—Relating Cause and Effect** Explain why the specialization of internal organs in coelomates is much greater than that in pseudocoelomates.

BACKGROUND INFORMATION
STARFISH ENDOSKELETON

The pincer-shaped structure in the center of Figure 30–8 is a fish-catching pedicellaria of the starfish *Stylasterias*. A number of these pedicellariae form a bumpy wreath around the base of the starfish's spines. When a small bottom-dwelling fish comes to rest on the surface of the starfish, the wreath rises around the spine and the "jaws" of the pedicellariae open to grab the fish.

In most starfish that possess pedicellariae, the pedicellariae are used for protection rather than feeding. They help keep the starfish clean by nipping small organisms that try to settle on its body surface.

TEACHING SUPPORT
Teaching Resources
- Ecology Investigation: Collecting Soil Organisms, p. 43

MEDIA AND TECHNOLOGY
- Interactive Videodisc/CD ROM
- *The Virtual BioPark*

TEACHING STRATEGY 30–2
Focus/Motivation
Using the Visuals Have students observe Figure 30–8 and read the caption.
- **What part of a crab makes up its exoskeleton?** (Its shell.)
- **What are some examples of invertebrates with exoskeletons?** (Arthropods and mollusks. Students will probably name specific animals.)
- **What are some examples of invertebrates with endoskeletons?** (Sponges and echinoderms. Students may name specific animals.)
- **Which invertebrates might have hydrostatic skeletons?** (Cnidarians, worms, sea cucumbers, and some mollusks. Students should recognize that motile animals that lack rigid supporting structures are likely to have hydrostatic skeletons.)

Content Development
Using the Visuals Direct students' attention to Figure 30–8.

Guide For Reading
- How do the different invertebrate phyla carry out their essential life functions?
- How did the various body systems in invertebrates evolve?

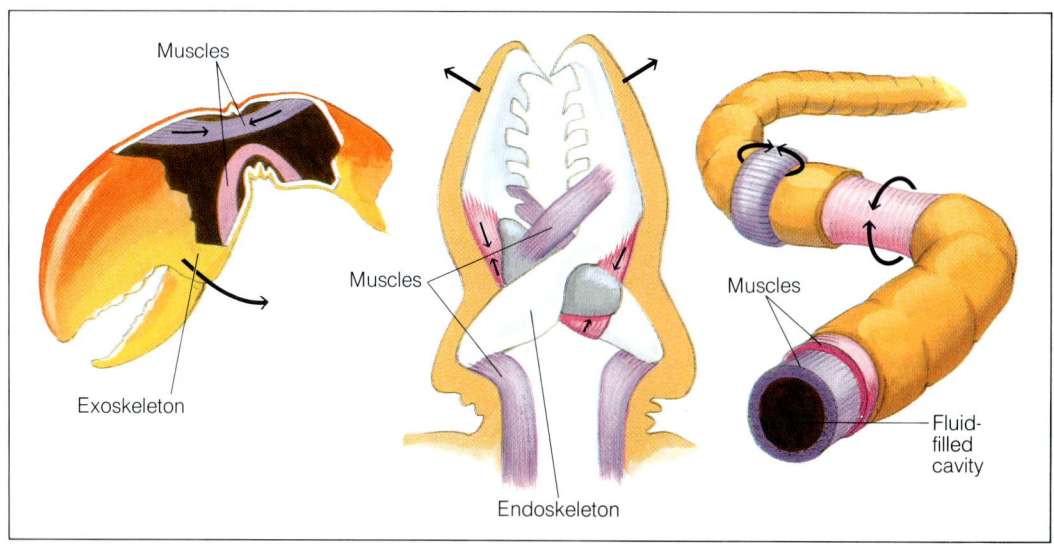

Figure 30–8 Three main kinds of skeletal systems are characteristic of most animals. In animals that have endoskeletons or exoskeletons, muscles pull against hard supporting structures to produce movement. In animals that have hydrostatic skeletons, muscles push against a fluid-filled body cavity when they contract.

- **What happens when the circular muscles on an earthworm's segment contract?** (The segment become longer and narrower.)
- **What happens when the longitudinal muscles on an earthworm's segment contract?** (The segment becomes shorter and fatter.)
- **What happens to the internal volume of an earthworm's segment when the muscles contract?** (Remains the same.)
- **What does an earthworm's segment contain?** (Accept all

30–2 Form and Function in Invertebrates

In many ways, each animal phylum represents an experiment in the design of body structures to perform the tasks necessary for survival. Of course, there has never been any kind of plan to these experiments because evolution works without either plan or purpose. Nevertheless, the appearance of each phylum in the fossil record represents the random evolutionary development of a basic body plan that is different in some way from other body plans. The rest of the history of each phylum is the story of further evolutionary changes in that plan.

We can learn a great deal about the nature of life by comparing body systems among invertebrate groups and by tracing the patterns of change as we move from one phylum to another. As we do so, it is important to keep this concept in mind: **Evolution is random and undirected.** A common misconception is that evolution has proceeded from one group of organisms toward a goal of perfection. This is definitely not true. Organisms are not better or worse than one another—they are simply different. And the ways in which organisms carry out their essential life functions are neither more nor less perfect than one another—they are merely different methods of accomplishing the tasks necessary for survival.

The body systems that perform the vital functions of life have taken many different forms in different phyla. Some systems are complex; others are simple. Some are efficient; others are not. It is important to remember, however, that more complicated and efficient systems are not necessarily "better" in any absolute sense than simpler systems. The fact that these systems are found in living animals is testimony to their success in performing the functions they have evolved to perform.

Movement

Almost all animals use specialized tissues called muscles to move. (There are a few exceptions: Sponges do not move at all; animals such as small flatworms use cilia to get from one place to another; and the contractile cells in cnidarians are not true muscles.) Without muscles, animals could not swim, fly, burrow, or run. In fact, most animals could not breathe, pump blood, or perform other life functions without muscles. Muscles work only by contracting. When muscles are stimulated, they generate force by getting shorter. When they are not stimulated, they relax.

In most animals, muscles work together with some sort of skeletal system that provides firm support. There are three main kinds of skeletal systems: **hydrostatic skeletons**, exoskeletons, and **endoskeletons**.

Hydrostatic skeletons, unlike endoskeletons and exoskeletons, do not contain hard structures, such as bones or chitin plates, for muscles to pull against. Instead, the muscles surround and are supported by a water-filled body cavity. When the muscles contract, they push against the water in the body cavity. Cnidarians, some flatworms, roundworms, some mollusks, and annelids are animals that have hydrostatic skeletons.

The term exoskeleton usually refers to the hard nonliving coating that encloses an arthropod's internal organs and muscles. However, the shells of mollusks can also be considered exoskeletons. Muscles attached to the inside of an arthropod's exoskeleton are used to bend and straighten the joints. Muscles attached to the shell in mollusks make it possible for snails to withdraw into their shell and for bivalves to close their two-part shell.

Endoskeletons are frameworks located inside the body of animals. Sponges, echinoderms, and vertebrates have endoskeletons. Animals with endoskeletons typically have muscles that attach to the outside surface of the endoskeleton. (The only exceptions to this rule are sponges, which lack muscles and cannot move.)

Feeding

There are many sources of food in nature. And as you have learned, many different modes of feeding and kinds of digestive systems have evolved in invertebrates that enable them to utilize these resources.

As you move through the invertebrate phyla from simpler animals such as sponges to more complex animals such as arthropods, you can observe three major evolutionary trends. First, simpler animals such as sponges, cnidarians, and flatworms break down their food primarily through **intracellular digestion**. More complex animals use **extracellular digestion**.

Figure 30–9 The thick, armorlike exoskeleton of this crustacean serves in movement and in protection.

BACKGROUND INFORMATION
MOVEMENT IN SPONGES

It can be argued that sponges do move, after a fashion. The epidermis of a sponge has a different structure from the epithelia of other animals—there is no basement membrane. This enables the cells of a sponge's epidermis to change shape more easily than epidermal cells in other animals. Because sponge epidermal cells are quite flexible, they can actually contract a bit, so that the entire body of the sponge gets slightly smaller.

ECOLOGY NOTE
AN UNUSUAL JAIL

Some glass sponges have an unusual symbiotic relationship with a certain species of shrimp. A young male and a young female shrimp enter the central cavity of the sponge. After a while, the shrimp grow too large to escape through the sieve plate covering the osculum of the sponge. The shrimp spend the rest of their lives imprisoned in the sponge, where they feed on plankton brought in by the sponge's water currents. In Japan, the sponge with its imprisoned shrimp is a traditional wedding present— a symbol of the harmony, faithfulness, and permanence of marriage.

Reinforcement/Reteaching
Remind students that earthworms have bristles (setae) on their ventral side that prevent them from slipping backward.
• **What happens when the circular muscles on the anterior end of an earthworm contract?** (The front end gets longer and thinner and is pushed forward.)
• **How do an earthworm's muscles help the posterior end "catch up" with the anterior end?** (The longitudinal muscles contract, pulling the rest of the body forward.)

---

correct answers. Elicit the words *fluid* or *water*.)
• **Which root comes from the Greek word for water?** (*Hydro-*.)
Write the word *hydrostatic* on the chalkboard.
• **What does the word *static* mean?** (Not moving, inactive, at rest. Students will probably also come up with the alternate meaning of radio or television interference.)
• **Why is the term *hydrostatic* appropriate for describing an earthworm's system for support and movement?** (It is based on the principle that the volume of water does not change.)
• **Would an earthworm's segments change shape if the volume of water changed when the muscles contracted?** (No.)

SCIENCE, TECHNOLOGY, AND SOCIETY CONNECTION

The Twisting Tentacle: No Bones About It

The biomechanical principle behind squid tentacles was worked out by a husband-and-wife research team in North Carolina. William Kier, a biologist at the University of North Carolina at Chapel Hill, and Kathleen Smith, an anatomist at Duke University, refer to the muscle structures that power movement and provide skeletal support as muscular hydrostats.

Muscular hydrostats operate in structures other than those that have a rigid skeleton or a hydrostatic cavity. A heart can keep on beating after it is disconnected from its blood flow because muscular hydrostats are involved in the expansion of the heart after contraction. It is also thought that muscular hydrostats are used in swimming by animals such as sharks, eels, and dolphins.

TIE-IN: MECHANICAL ENGINEERING

The work of Kier and Smith described in the Science, Technology, and Society feature is just one example of current research in the new field of comparative biomechanics, a discipline that fuses biology and engineering. In addition to revealing the mechanical principles underlying biological systems, biomechanics also seeks to explain why many natural forms are shaped the way they are.

SCIENCE, TECHNOLOGY AND SOCIETY CONNECTION

The Twisting Tentacle: No Bones About It

Until recently, squid tentacles, elephant trunks, and human tongues posed an unanswerable question to biologists. These animal body parts have neither exoskeletons nor endoskeletons. They also lack the fluid-filled cavities associated with hydrostatic skeletons. Without some sort of support to work against, the muscles in these body parts should not be able to work at all. And yet they work extremely well! Squid tentacles, for example, can shoot forward to catch prey in a mere 30 milliseconds. How do these body parts work?

Researchers have discovered that the muscles in squid tentacles (and similar structures) pull against one another to produce movement. Like all cells, muscle cells are made up primarily of water. Thus each muscle cell can act like a fluid-filled cavity in a hydrostatic skeleton, providing support and giving other muscle cells something to work against. The arrangement of muscles in squid tentacles gives the animal extremely precise control over movement. For example, a tentacle can bend or twist at any point along its length. In contrast, a human limb or an insect limb can bend only where there are joints.

These new lessons about muscle function have already found applications in industry. Engineers have designed robot arms based on the same principles as squid tentacles. These robot arms are more efficient than those modeled after human arms. They are also less prone to damage and more adaptable to working in narrow, awkward spaces.

Scientists have recently identified the mechanical principles that permit an octopus's tentacles to move.

As you may recall, the collar cells in sponges take in microscopic food particles by endocytosis and then pass them to wandering cells called amebocytes. The food is digested inside food vacuoles within the amebocytes, and the nutrients from the food are then passed on to the other cells of the sponge by diffusion from the amebocytes. This is an example of intracellular (*intra-* means inside) digestion because food is digested, or broken down, inside the cells. In extracellular (*extra-* means outside) digestion, food is broken down outside the cells—specifically, in a digestive tract. Extracellular digestion is an adaptation that enables animals to eat and digest pieces of food that are much larger than the animals' cells. In the majority of cnidarians and flatworms, most of the process of digestion is intracellular—it occurs inside the cells that line the gastrovascular cavity. However, food particles are partially broken

---

30–2 (continued)

Reinforcement/Reteaching

You may wish to review the different modes of feeding in invertebrates by asking questions similar to the following.

- **What do you call an animal that eats other animals?** (Carnivore.)
- **What do you call an animal that eats plants?** (Herbivore.)
- **What does a detritus feeder eat?** (Bits of decaying organic matter in soil or in sediments.)
- **What are some examples of filter feeders?** (Accept all correct answers. Filter feeders include feather duster worms, barnacles, sea lilies, and feather stars.)
- **What are some examples of parasites?** (Accept all correct answers. Parasites include flukes, tapeworms, ascarids, hookworms, leeches, ticks, and fleas.)

Content Development

Using the Visuals Stress that animals with a two-opening digestive system also have a tube-within-a-tube body plan.

down in the gastrovascular cavity. Mollusks, annelids, arthropods, echinoderms, and chordates typically rely on extracellular digestion.

Second, cnidarians and some flatworms have a simple digestive system that has a single opening through which food enters and through which solid wastes are expelled. More advanced digestive systems, such as those found in roundworms, mollusks, annelids, arthropods, echinoderms, and chordates, have two openings—a mouth at one end and an anus at the other. Animals with a two-opening digestive tract are said to have a tube-within-a-tube body plan. The inner tube is the digestive tract and the outer tube is the body wall. Between these two tubes is the body cavity.

Third, the digestive tract tends to acquire more and more specialized regions. Simpler animals, such as flatworms, have a gastrovascular cavity in which one part does not differ very much from another part. The tube-within-a-tube body plan in more complex animals, such as mollusks and arthropods, allows for specialization because food passes through in one direction. As the food travels from the mouth to the anus, it is processed by each specialized region in turn.

The digestive system is not the only system to become more specialized as you move from simpler animals to more complex animals. As you may recall from Chapter 26, this evolutionary trend is seen in most of the other systems responsible for performing essential life functions. Keep this concept in mind as you read about these other invertebrate systems.

Figure 30–10 The jellyfish (top) has a digestive system with only one opening. The plume worm (bottom) has a digestive system with two openings—a mouth and an anus.

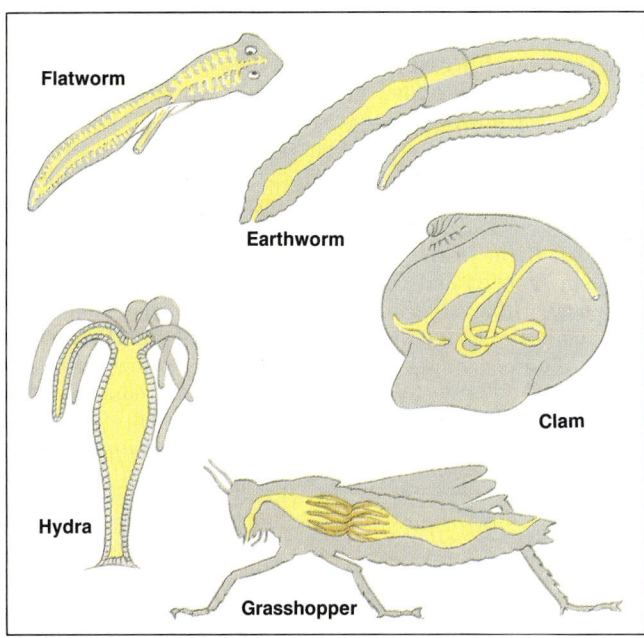

Figure 30–11 As animals become more complex, extracellular digestion and two-opening digestive systems become more common than intracellular digestion and one-opening systems. In addition, the regions of the digestive tract become more specialized.

BACKGROUND INFORMATION
RESPIRATORY PIGMENTS

Respiratory pigments bind to oxygen when the concentration of oxygen is relatively high and release bound oxygen when the concentration of oxygen is relatively low. In addition to helping to transport oxygen throughout the body, respiratory pigments also increase the blood and/or coelomic fluid's capacity for holding oxygen.

There are four basic respiratory pigments in animals: hemoglobin, chlorocruorin, hemerythrin, and hemocyanin. Each of these pigments exists in a large number of variant forms and may be dissolved in blood plasma or coelomic fluid or carried within cells. In some animals, more than one type of respiratory pigment may be present. In a few animals, such as insects, there are apparently no respiratory pigments.

Hemoglobin and chlorocruorin, which are very similar in structure, consist of multiple units that contain at their core an iron atom surrounded by a porphyrin ring. Small differences in the side chains attached to the porphyrin ring result in the major difference between hemoglobin and chlorocruorin: Hemoglobin is red; chlorocruorin is green. Hemoglobin is found in animals ranging from mollusks to vertebrates. Chlorocruorin is found in only a few animals, such as tube-dwelling polychaetes.

Like hemoglobin and chlorocruorin, hemerythrin is based on iron. However, it is a protein, not a porphyrin. In hemerythrin, which is found in animals such as brachiopods and polychaetes, an oxygen molecule is carried between two iron atoms.

Hemocyanin is a copper-containing protein similar to hemerythrin. It is found dissolved in the blood plasma or coelomic fluid of many invertebrates, including horseshoe crabs and most mollusks. In hemocyanin, a molecule of oxygen is carried between two copper atoms. Oxyhemocyanin is pale blue; deoxyhemocyanin is colorless.

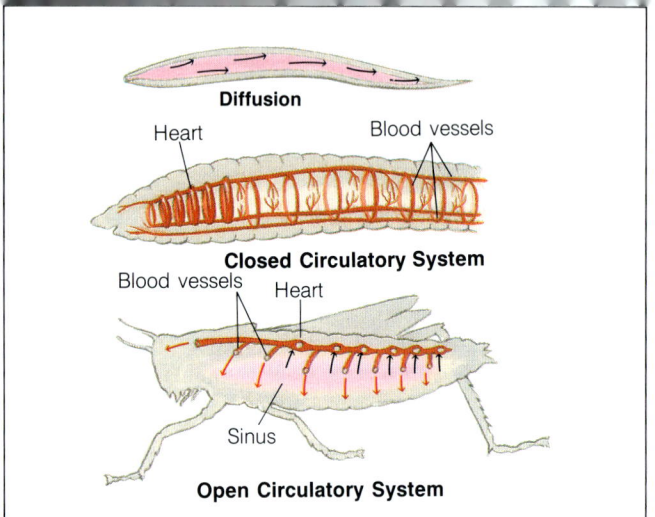

Figure 30–12 *The body cavity of a roundworm can be thought of as an extremely simple circulatory system. As the roundworm moves, fluid sloshes around the body cavity and dissolved materials are transported from one body region to another. A grasshopper has an open circulatory system, whereas an earthworm has a closed circulatory system.*

Internal Transport

All cells of multicellular animals must be supplied with oxygen and nutrients and must dispose of metabolic wastes. The smallest and thinnest multicellular animals manage to fulfill their internal transport needs through diffusion between their body surface and the environment. Although echinoderms are relatively large, they rely on this diffusion as well as diffusion between their body tissues and the fluid within their body cavity for their transport needs. Most complex multicellular animals, however, have a collection of pumps and tubes called a circulatory system. Typically, the pumps (hearts) force a fluid (blood) that carries food, oxygen, carbon dioxide, and other important substances through tubes (blood vessels) that extend throughout the body. There are two basic types of circulatory systems: open circulatory systems and closed circulatory systems. Open circulatory systems, which are found in arthropods and most mollusks, do not keep blood contained within blood vessels. At some point, the blood comes in direct contact with the tissues, collects in body sinuses, and then makes its way back to the heart. Closed circulatory systems, which are found in annelids and chordates, keep the blood completely contained within a network of blood vessels that stretches throughout the body. Materials diffuse from the blood to the tissues (and vice versa) through the walls of the blood vessels. Thus the blood normally does not come in direct contact with the tissues. In large active animals, a closed circulatory system offers greater control over blood flow and allows more efficient direction of blood to various parts of the body.

Respiration

In order to supply oxygen to and remove carbon dioxide from their tissues, animals must exchange these gases with the environment. Remember that diffusion of gases into and out of

---

30–2 (continued)
Reinforcement/Reteaching

Review the concepts of diffusion, osmosis, and active transport. You may wish to check students' mastery of these concepts by drawing simple diagrams on the chalkboard and asking questions about the diagrams.

For example, draw a circle on the chalkboard. Explain that the circle represents a cell. Write "50 CO_2" inside the circle and "5 CO_2" outside the circle.

• Where is the concentration of carbon dioxide greater—inside the cell or outside the cell? (Inside.)

• Which way will carbon dioxide diffuse? (Out of the cell.)

an animal's body requires a thin moist membrane. Thus two features are common to all respiratory systems. First, they almost always have structures that maximize the amount of surface area in contact with air or water. The more membrane exposed to the environment, the greater the amount of gas exchange that can occur. Second, they have some way of keeping the gas exchange surfaces moist so that diffusion can occur.

Aquatic animals have no problem in this regard—their watery environment keeps their respiratory surfaces moist. But terrestrial animals have a problem: Air contains little water and thus dries out the respiratory surfaces. To make matters worse, respiratory surfaces in terrestrial animals have a large surface area, so more membrane can dry out and more water can be lost from the body by diffusion. To prevent excessive water loss, the respiratory surfaces of terrestrial animals are kept moist with a coating of either water or mucus. In addition, the respiratory surfaces are often contained within the body (which is mostly water). Air is moistened as it travels through the body to the respiratory surface, reducing its drying effects.

Some animals that live in water or in very moist soil, such as cnidarians and flatworms, respire through their skin. Dry skin is more than just a cosmetic problem for these animals—it is death by suffocation! For most active animals larger than worms, however, respiration through the skin is not sufficient. Aquatic organisms—mollusks, crustaceans, some insects, and many annelids, for example—have gills that help them exchange gases with the water around them.

Terrestrial invertebrates have evolved several organs for breathing air. These include the highly modified mantle cavities of land snails, the book lungs of spiders, and the tracheal tubes of insects. See Figure 30–14.

Figure 30–13 The gills of a plume worm are made up of threadlike projections that help maximize the surface area available for respiration.

Figure 30–14 Invertebrates use a variety of structures for respiration. A crayfish has gills. A spider has book lungs. Many land snails have a mantle cavity that acts as a lung. A grasshopper has tracheal tubes.

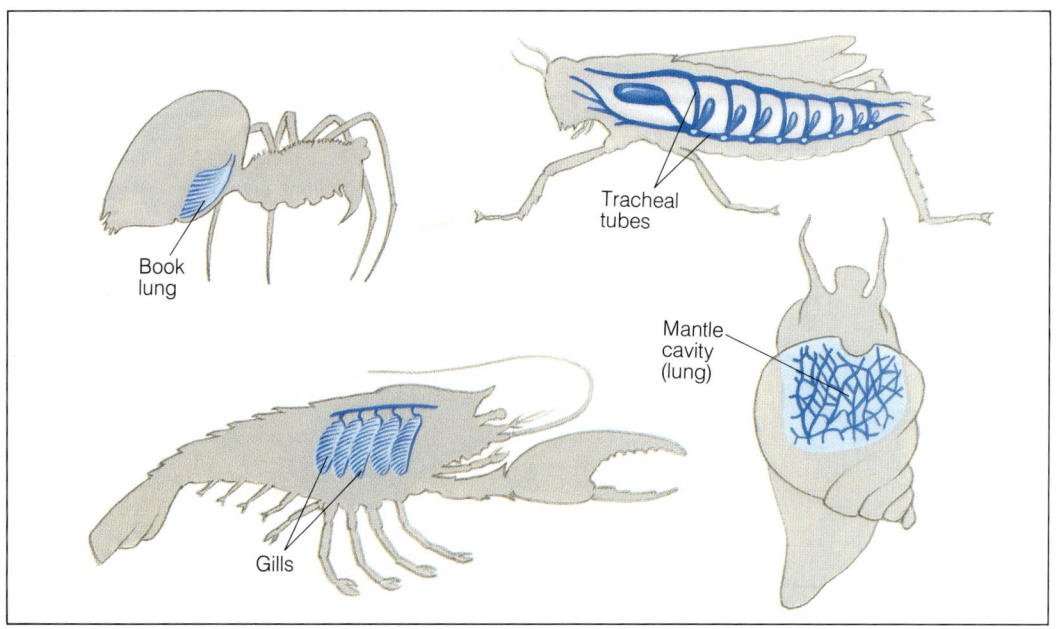

BACKGROUND INFORMATION
MYOGLOBIN

Respiratory pigments are not limited to the circulatory system. Myoglobin, an iron porphyrin that can be thought of as a simple version of hemoglobin, is found exclusively in skeletal muscle. Myoglobin, which has a higher affinity for oxygen than hemoglobin, serves as an oxygen reservoir in animals such as seals and whales. When the supply of oxygen from hemoglobin is exhausted, myoglobin releases its oxygen to supply skeletal muscles. This enables sea mammals to spend long periods of time under water.

oxygen higher in blood than in body cells? (Body cells use up oxygen in the process of cellular respiration.)

Skills Development
Guided Practice
Skill: Relating concepts
Help students visualize the interconnectedness of the systems that carry out essential life functions by asking questions such as the following.
• **What are the functions of an internal transport system?** (To deliver food and oxygen to cells and to carry away waste products such as carbon dioxide.)
• **What are the functions of a respiratory system?** (To take oxygen into the body and to remove carbon dioxide from the body.)
• **How are the functions of internal transport and respiration related to each other?** (The internal transport carries the oxygen taken in by the respiratory system to the body tissues and delivers carbon dioxide to a site where it can be removed from the body by the respiratory system.)

Content Development
Point out that functions such as internal transport and respiration are based primarily on the principle of diffusion.
• **What is diffusion?** (The principle that molecules tend to move from an area of higher concentration to an area of lower concentration.)
• **Which ways do carbon dioxide and oxygen diffuse in a respiratory organ?** (Carbon dioxide: out of the body/blood. Oxygen: into the body/blood.)
• **Which ways do carbon dioxide and oxygen diffuse in body tissues?** (Carbon dioxide: out of the cells/into the blood. Oxygen: into the cells/out of the blood.)
• **Why does oxygen diffuse out of blood into body cells?** (Concentration is higher in blood.)
• **Why is the concentration of**

> **BACKGROUND INFORMATION**
> **CHROMATOPHORES**
>
> A number of invertebrates—most notably many cephalopods and decapod crustaceans—change color as a response to various stimuli. These color changes are caused by special pigment-containing cells known as chromatophores.
>
> In cephalopods, chromatophores of different colors are arranged in several overlapping layers in the skin. When tiny muscles attached to the edges of a chromatophore contract, the chromatophore becomes a large, flat plate and the color of the pigments it contains becomes quite noticeable. When the muscles relax, the chromatophore shrinks and its pigments become concentrated into a barely visible speck. The expansion and contraction of the chromatophores—processes that appear to be under the control of hormones as well as the nervous system—allow cephalopods such as octopi and cuttlefish to change color to match their surroundings. They also cause such cephalopods to display distinctive color patterns when alarmed.
>
> In crustaceans, chromatophores are highly branched noncontractile cells that are located beneath a relatively transparent exoskeleton and contain white, black, blue, red, yellow, or brown pigments. The pigment colors are visible when pigment granules are dispersed throughout the chromatophore (stellate state); they are not visible when the granules are concentrated at the center of the cell (punctate state). Crustacean chromatophores may contain between one and four different colors of pigment, each of which can move independently of the others. The movement of pigment granules is controlled by paired hormones: One disperses pigment and the other concentrates pigment. Crustaceans use color changes to blend into their surroundings and to send visual signals to other members of their species.

Excretion

Multicellular animals, whether they live in water or on land, must control the amount of water in their tissues. Freshwater animals, for example, constantly take in water through the process of osmosis. In order not to blow up like water balloons, they must get rid of that extra water. Terrestrial and many marine animals, on the other hand, must protect their body tissues against water loss to the environment. At the same time, all animals—aquatic and terrestrial—must get rid of toxic nitrogenous (nitrogen-containing) wastes produced as a result of normal cellular metabolism. Excretory systems in invertebrates have evolved in ways that enable these animals to both regulate the amount of water in the body and get rid of nitrogenous wastes.

In all animals, the breakdown of amino acids during cellular metabolism produces ammonia (NH_3). Most aquatic invertebrates leave their nitrogenous wastes in this form, despite the fact that ammonia is highly toxic. They are able to do this because ammonia is also highly soluble; it dissolves readily in water. Many aquatic animals simply allow ammonia to diffuse through their body tissues and out into the surrounding water, which immediately dilutes it and carries it away.

Terrestrial animals, however, have a double problem: They must conserve body water and get rid of nitrogenous wastes at the same time. In order to do this, many invertebrates convert ammonia into a compound called urea. Urea is soluble in water and is much less toxic than ammonia. For this reason, urea can be concentrated by the excretory system. The waste product produced by the excretory system, which is called urine, is expelled from the body. Thus terrestrial animals can get rid of more wastes in less water than their aquatic counterparts.

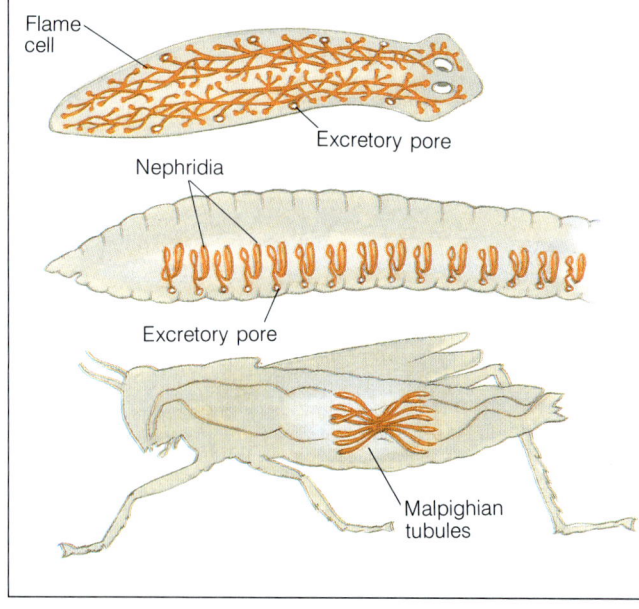

Figure 30–15 Flame cells help remove excess water from a planarian's body. In many arthropods, Malpighian tubules filter nitrogenous wastes from the blood, concentrate the wastes in the form of uric acid, and dump the uric acid into the intestine. In some animals—earthworms, for example—nephridia remove nitrogenous wastes and concentrate them in the form of urea.

> **30–2 (continued)**
>
> **Content Development**
> **Using the Visuals** Direct students' attention to Figure 30–15.
> • **How are flame cells different from nephridia and Malpighian tubules?** (Students should recognize that flame cells are used to get rid of water, whereas nephridia and Malpighian tubules are used to get rid of nitrogenous wastes.)
> • **How do flatworms get rid of nitrogenous wastes?** (Through diffusion.)
> • **What kinds of nitrogenous waste is excreted by a flatworm?** (Ammonia.)
> • **Why don't earthworms excrete ammonia?** (There is not enough water in their surroundings or in their body to dilute ammonia and wash it away.)

Some land animals, including insects, convert nitrogenous wastes to uric acid, a compound that is less toxic than ammonia but is also much less soluble in water. When a solution of uric acid is concentrated in the excretory system, uric acid forms solid crystals. The excretory system can then return most of the water to the body tissues and get rid of the uric acid in the form of a thick paste.

What sorts of excretory systems are found in invertebrates? Simpler animals—sponges, cnidarians, and roundworms, for example—tend to depend on diffusion through body surfaces to get rid of nitrogenous wastes in the form of ammonia. In freshwater flatworms, such as planarians, flame cells help get rid of excess water, but ammonia still leaves primarily through the body surfaces. Insects and some arachnids have Malpighian tubules, which absorb fluid from the blood in the body sinuses. Nitrogenous wastes are concentrated in the tubules and are added to the undigested wastes in the intestine as a paste of uric acid crystals. Annelids, mollusks, and chordates have tubelike excretory structures called nephridia. Nephridia take in body fluids, remove nitrogenous wastes, and return most of the water and important compounds from those fluids to the body. They then concentrate waste products in urine, which is discharged from the body.

As urine becomes more concentrated, more wastes can be eliminated in the same amount of water. Advanced kinds of excretory systems, such as Malpighian tubules and the nephridia in more complex animals, are better at producing concentrated urine than simpler kinds of excretory systems. Thus animals with more advanced excretory systems can live in drier habitats because such systems allow for water conservation.

Response

Nervous systems gather information from the environment, process information, and allow animals to respond to it. The simplest nervous systems, which are found in cnidarians and some of the more primitive flatworms, are called nerve nets. Nerve nets consist of individual nerve cells that form a netlike arrangement throughout the animal's body.

As you can see in Figure 30-17 on page 666, invertebrates show three obvious trends in the evolution of the nervous system: centralization, cephalization, and specialization. Refer to Figure 30-17 as you read about these evolutionary trends.

The cells that make up a hydra's nerve net are quite spread out, although they are concentrated in the tentacles and around the mouth. In animals such as jellyfish and flatworms, the nerve cells are more concentrated, or centralized. They form nerve cords or nerve rings in certain areas. These structures transmit signals more quickly and efficiently than nerve nets and help coordinate responses.

Most animals that have bilateral symmetry also have a concentration of nerve tissue in the anterior end of the body. As

Figure 30-16 Nervous systems enable invertebrates to respond to danger. The gooseneck barnacle, a sessile crustacean, jerks its feathery legs back to safety and snaps its shell shut (top). The sea butterfly, a gastropod mollusk, escapes a threat by "flying" rapidly through the water using its winglike foot (bottom).

FACTS AND FIGURES

Cells cannot survive high concentrations of ammonia. In order to be transported or stored without harming cells, 1 gram of ammonia must be diluted in 300 to 500 milliliters of water. In contrast, 1 gram of urea needs only 50 milliliters of water to be safely transported, and 1 gram of uric acid needs only 10 milliliters of water.

ESL STRATEGY

Give students sets of related terms that deal with invertebrate structure and function. Have students write short paragraphs in which they use the terms and compare and contrast the structures and functions. Possible sets of terms include the following.
1. coelomates, acoelomates, pseudocoelomates
2. endoskeleton, exoskeleton, hydrostatic skeleton
3. intracellular digestion, extracellular digestion
4. ammonia, urea, uric acid
5. nerve net, ganglia, brain

Focus/Motivation
Using the Visuals Have students observe Figure 30-16 and read the caption.
• **What concept is illustrated in this figure?** (Invertebrates respond to danger in different ways.)
• **How else do invertebrates respond in order to protect themselves from danger?** (Accept all correct answers. Students may recall that sea hares and cephalopods may use a "smoke screen" of ink to help them escape; that animals such as nudibranchs are distasteful or poisonous; and that many animals can drive away predators by biting, stinging, or pinching.)

Reinforcement/Reteaching
Stress that the excretory system is extremely important in homeostasis.
• **What is homeostasis?** (The maintenance of constant internal conditions.)
• **How do a flatworm's flame cells maintain the water balance in its body?** (They pump out excess water.)
• **How do nephridia and Malpighian tubules help maintain the water balance in animals?** (They concentrate nitrogenous wastes and thus help conserve water.)
• **How do excretory systems contribute to homeostasis?** (They maintain the water balance of the body. They also remove waste products.)

FACTS AND FIGURES

The eyes of squids, octopi, and cuttlefish are remarkably similar in structure to those of vertebrates. The cephalopods' eyes are surrounded and protected by a socket of cartilage plates and contain an iris, lens, and focusing muscles. The similarity between the vertebrate eye and the cephalopod eye is an example of convergent evolution.

BACKGROUND INFORMATION
APHID REPRODUCTION

Certain species of aphids undergo both ovoviviparous parthenogenesis and oviparous sexual reproduction. Female aphids that hatch in the spring give birth to female offspring that are identical to one another and to their parent. Throughout the spring and summer, succeeding generations are produced through parthenogenesis. This reproductive strategy enables aphids to quickly increase in number and rapidly colonize new areas. However, as the days grow shorter in autumn, the aphids adopt a new reproductive strategy and produce male as well as female offspring. The females in this generation mate with the males and then lay eggs. These eggs are able to survive over the winter, whereas adult aphids cannot. In addition, sexual reproduction produces new combinations of genes—combinations that may give some of the offspring a better chance of survival.

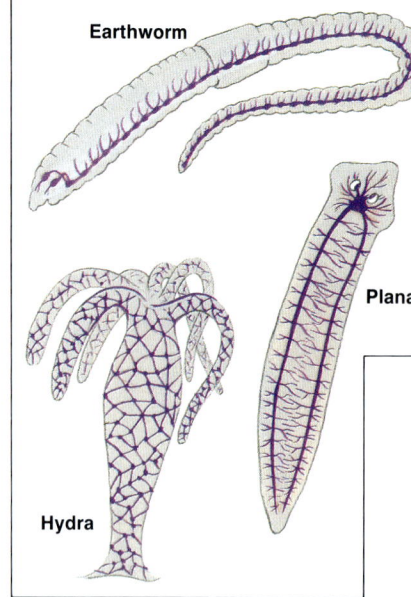

Figure 30–17 The nervous system in a hydra is relatively diffuse. In a planarian, the nervous system is concentrated in some areas, forming nerve cords and ganglia. More advanced animals tend to have a brain, specialized sense organs, and a highly developed nervous system.

you may recall from Chapter 26, the concentration of nerve cells and sense organs in the head region is known as cephalization. Cephalization tends to increase as animals become more complex. In simpler animals, such as flatworms, there are a few small clumps of nerve tissue, or ganglia, in the head. In more advanced animals—cephalopod mollusks and arthropods, for example—the ganglia are organized into a brain that controls and coordinates the nervous system.

In general, the more complex the animal, the more highly specialized its sense organs. For example, planarians have simple ocelli, or eyespots, that do little more than detect the presence of light. More complex animals, such as insects, have eyes that can detect motion and color and can form images. Complex animals may have a variety of specialized sense organs that detect light, sound, chemicals, movement, and sometimes even electricity to help them discover what is happening around them.

Reproduction and Development

Many simple invertebrates reproduce asexually through fragmentation or budding. In fragmentation, which occurs in several kinds of worms and cnidarians, an organism breaks into pieces. Each piece then regenerates into an entire individual. In budding, which occurs in sponges and cnidarians, new individuals are produced from outgrowths of the parent's body wall. A number of more complex animals, such as certain insects, also reproduce asexually. However, these animals do not undergo fragmentation or budding. Instead, their offspring develop from unfertilized eggs.

Asexual reproduction allows animals to produce offspring rapidly from a single individual. This enables species to quickly take advantage of new opportunities in the environment. The obvious disadvantage of asexual reproduction is that the offspring are genetically identical to the parent. And, as you may recall, a lack of genetic diversity makes populations less able to deal with changes in the environment.

Sexual reproduction maintains genetic diversity in a population. Although sexual reproduction does not create new

30–2 (continued)

Skills Development
Guided Practice
Skill: Interpreting diagrams
Using the Visuals Direct students' attention to Figure 30–17.
• **Which of these organisms have a nerve net?** (Hydra.)
• **Which of these organisms have obvious ventral nerve cords?** (Earthworm and grasshopper.)
• **Which of these organisms show cephalization?** (Planarian, earthworm, grasshopper, octopus.)

Focus/Motivation
Using the Visuals Direct students' attention to the photograph of the black widow spiders on page 667.
• **Why is this type of spider called the "black widow"?** (Some students should know that the female of this type of black-colored spider often eats her mate.)
• **Why would a male black widow spider undergo such a potentially fatal endeavor as a mating?** (Accept all logical answers.)

Point out that reproduction is necessary for the survival of a species, if not for individuals.

genes, it does result in new combinations of genes. Such combinations may improve an individual's chances of surviving and coping with change. Most of the more complex animals reproduce sexually. The majority of these animals have separate sexes—individuals are either males (sperm producers) or females (egg producers). But many animals, including certain mollusks and echinoderms, are hermaphrodites that produce both sperm and eggs at the same time. Hermaphrodites have some advantages over animals with separate sexes. One advantage is that anytime two sexually mature individuals meet each other, they can mate. (In species with separate sexes, of course, the individuals must be of different sexes or they cannot produce offspring together.) Another advantage is that one mating results in two batches of offspring (one per mated individual), rather than just one.

FERTILIZATION There are two basic ways in which sperm cells and egg cells are brought together in sexual reproduction: **external fertilization** and **internal fertilization**. External fertilization is generally associated with less complex animals, although there are examples of both external and internal fertilization in most animal phyla. In invertebrates that have external fertilization—jellyfish, clams, and sea urchins, for example—the eggs are fertilized outside the body. Adults release sperm and eggs into the surrounding water, and sperm swim to the eggs to fertilize them. Although this system is simple, it does have disadvantages. Clearly, as we saw earlier with plants, it can work only under water or in very wet places. It is also both risky and wasteful of sperm and eggs because many sperm cells never find an egg to fertilize and many eggs are never fertilized. To increase the chances of the eggs being fertilized, animals that use external fertilization release huge numbers of eggs and sperm when they spawn.

Internal fertilization is associated with more complex animals. In internal fertilization, the egg is fertilized inside the female's body. Usually, the males use specialized organs to deposit sperm inside the female's reproductive tract. Most land animals use internal fertilization.

Figure 30–18 Aphids reproduce sexually at certain times of the year and asexually at others. This aphid, which has undergone asexual reproduction, is giving birth to genetically identical offspring (left). Butterflies reproduce sexually. Here, a male deposits sperm inside the body of the female (right). Spiders also reproduce sexually. As you can see, the male black widow spider is much smaller and less strikingly colored than his mate (bottom).

TEACHING SUPPORT

Teaching Resources
- Vocabulary Review: Word Game, p. 1

Study Guide
- Section 30–2, p. 288

30–2 (continued)

SECTION REVIEW 30–2

1. Endoskeleton: system of rigid support structures embedded within the living tissues of the organism. Exoskeleton: system of rigid nonliving supporting structures that enclose the living tissues of the organism. Hydrostatic skeleton: fluid-filled chambers surrounded by muscle support the body.
2. As animals become more complex, the digestive system tends to have two openings rather than one. In addition, it also tends to have specialized regions and organs for processing food.
3. Echinoderms: material carried by the fluid contained in the water vascular system. Arthropods: open circulatory system; materials carried in blood that bathes the tissues. Chordates: closed circulatory system; materials diffuse into and out of the blood through the walls of blood vessels.
4. Asexual reproduction: parent produces offspring that are identical to one another and to the parent; allows for rapid colonization of a new area from only one original organism; lacks genetic diversity. Sexual reproduction: male and female gametes fuse to produce fertilized egg that develops into new

organism that is usually genetically different from its parents and siblings; produces new combinations of genes; usually requires two parents, a male and a female, which might not be easy to obtain in an area with a low population density.

Figure 30–19 *Some invertebrates care for their eggs and young. Its warning colors readily apparent, the harlequin bug stands guard over its eggs (top). The female wolf spider carries its numerous offspring on its back to keep them out of danger (bottom).*

PARENTAL CARE Many invertebrates do not take care of their fertilized eggs or young. The eggs are ignored as soon as they are laid. In many invertebrates, the fertilized eggs and resulting young live in the water and drift wherever currents may take them. Most of the young are eaten or are exposed to adverse environmental conditions and die. Perhaps this seems wasteful to you. But for animals in which the adults are cemented permanently in one spot, such as corals and barnacles, floating eggs and free-swimming larvae offer the only chance for the next generation to find a new spot in which to settle and spend the rest of their lives. Although each egg has only a tiny chance of surviving to adulthood, a few manage to make it. In other invertebrates, the females lay their eggs in a spot where they are hidden from predators or where food will be available for emerging young. These young have better chances of surviving than young that emerge in inhospitable places or in locations where no food is nearby.

Some invertebrates take care of their offspring. For example, queen bees lay their eggs in special compartments in a well-provisioned and protected hive. The small octopus called the paper nautilus builds a special shell to carry and protect her eggs. Some scorpions carry eggs and young around on their back. Some of the ways in which invertebrates care for their offspring may seem horrifying to humans. For example, the eggs of some species of mites hatch within the female's body. The larvae immediately begin to devour their mother from the inside! Within two days—while still inside their mother's nearly empty exoskeleton—the young mites mature, mate, and eat their way to the outside. The males die within a few hours. The females seek out prey in the form of insect eggs and begin to feed—even as their own offspring start chewing on their internal organs.

30–2 SECTION REVIEW

1. Name and describe the three main types of skeletal systems found in the animal kingdom.
2. Describe how the digestive system evolves as animals become more complex.
3. Compare internal transport in echinoderms, arthropods, and chordates.
4. Compare asexual and sexual reproduction. What are their advantages? Disadvantages?
5. Discuss three evolutionary trends in the development of the nervous system.
6. **Critical Thinking—Applying Concepts** What are the three forms of nitrogenous wastes excreted by animals? How is the waste related to the animal's environment?

5. Cephalization: concentration of nerves and sense organs at the anterior end of an organism. Centralization: concentration of nerves into a few major nerve cords and ganglia. Specialization: development of sensory organs for specific functions.

6. Ammonia, urea, and uric acid. Nitrogenous wastes can be released as ammonia only in aquatic animals, as a large amount of water is needed to safely dilute ammonia. Land animals produce urea and uric acid, which requires less water

PROBLEM SOLVING IN BIOLOGY

WHAT IS IT?—ANALYZING FUNCTION THROUGH FORM

Imagine yourself a world famous invertebrate biologist who is often called upon to describe the habits of newly discovered organisms by comparing their structures to other animals you know. One day specimens belonging to the same species are brought into your laboratory, and you are asked to describe the animal's habits as best you can.

Some of the specimens are larvae that closely resemble one another, although none looks like the adult specimens. These larvae have a tough dark-brown exoskeleton, six legs with tips flattened like canoe paddles, and gills supplied with many tracheal tubes. Larval mouthparts look like a collection of butcher knives—sharp and menacing.

Adult specimens have two pairs of wings and six legs. Their brightly colored exoskeleton is covered with a thick, waxy coating. They have well-developed tracheal tubes spread throughout their body. Two of their mouthparts are relatively sharp, but the rest are blunt and clearly adapted for grinding and pulverizing food. You can tell from samples at the end of the gut that these animals excrete their nitrogenous wastes in the form of uric acid.

Practice your expertise in identifying unknown animals by answering the following questions about this imaginary creature.

1. To what phylum and class does this species probably belong?
2. Describe the probable life cycle of this animal. What sorts of specimens of this species would help you confirm your hypothesis about how this organism develops?
3. Where do you think the larvae live? Explain your answer.
4. What type of food do the larvae probably eat? Explain your reasoning.
5. What sorts of food do the adults probably eat? What evidence leads you to this conclusion?
6. Could the adults of this species survive in dry habitats? Why or why not?

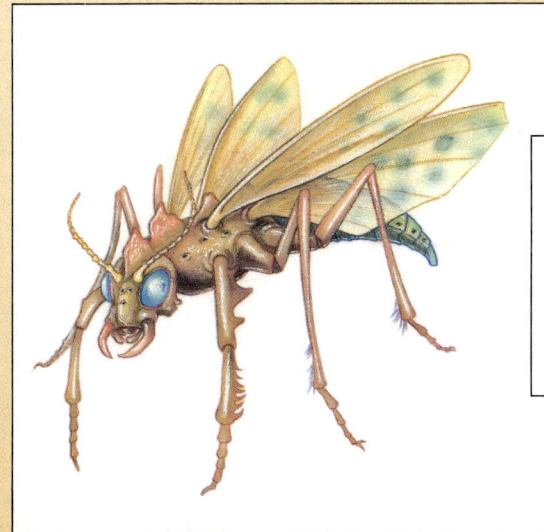

What can you infer from your examination of this mysterious (and fictitious) creature?

PROBLEM SOLVING

What is it?—Analyzing Function Through Form

The purpose of this activity is to give students an opportunity to make inferences about the habits of an unknown invertebrate based on a written description and a drawing based on that description. The organism described in this activity is purely a product of the authors' and the artist's imaginations, although it is based on a number of real insects.

1. Phylum Arthropoda, Class Insecta.
2. Life cycle probably involves complete metamorphosis: A larva develops into a pupa, which then develops into an adult. Specimens of the pupal stage would confirm this hypothesis.
3. Larvae live in water, as indicated by the gills and paddlelike swimming appendages.
4. Larvae are probably carnivores, as indicated by sharp slicing mouthparts. More advanced students may note that no definitive conclusions about larval feeding can be drawn without more information about the structure of the digestive tract. Sharp slicing teeth could belong to a herbivore if there were organs such as a gizzard and long intestine to process plant material.
5. Adults are probably herbivores—sharp mouthparts are used to cut plant material and blunt mouthparts are used to crush tough plant materials.
6. Yes, because the waxy coating on the exoskeleton and the excretion of uric acid both help prevent excessive water loss.

---

to be transported and/or stored. This helps conserve water in the body.

Reinforcement/Reteaching
If students cannot answer the Section Review questions, review the pertinent material.

Closure
Explain that the members of the class will all have the opportunity to participate in a popular television game show. Have each student prepare five "answers" on separate index cards. Then divide the class into three teams of approximately equal size.

Read an "answer," then ask the first person on team 1 to give the correct "question." If this person is not able to give the correct question, have the first person on team 2 try. If this person cannot provide the question, have the first person on team 3 try. If all three contestants fail to provide the question, allow another member of the class to provide the question and award no points.

LABORATORY INVESTIGATION

COMPARING DEVELOPMENT IN ECHINODERMS AND MOLLUSKS

PROBLEM
How do patterns of development compare in echinoderms and mollusks?

MATERIALS (per group)

> prepared slides showing the early development of a starfish and a clam
> microscope

PROCEDURE

1. Using the low-power objective of the microscope, examine a slide showing starfish development. Locate a single cell alone. This is an egg cell. Switch to the high-power objective and focus with the fine adjustment. Draw a diagram of the egg cell and label it.
2. After fertilization, the egg cell begins to divide. Switching back to low power, locate an embryo in the 2-cell stage of development. Examine it in detail under high power and draw a labeled diagram of it.
3. Repeat step 2 for each of the following stages of development: 4-cell, 8-cell, 16-cell, and 32-cell. In the 8-, 16-, and 32-celled stages, note whether the cells are directly above or behind one another or whether the cells appear to be between one another.
4. Notice the size of the individual cells and the size of the entire embryo as the number of cells in the embryo increases.

5. Eventually, the embryo develops into a solid ball-shaped cluster of cells in which individual cells can be seen. This cluster is called a morula. Locate a morula under low power. Then switch to high power to examine it. Draw a diagram of a morula and label it.
6. As cleavage (cell division) continues, the embryo becomes a hollow ball called a blastula. A blastula looks like a sphere with a dark circle of cells near the edge. Locate a blastula under low power. Then switch to high power. Draw a diagram of the blastula and label it.
7. As development continues, the sphere-shaped embryo indents on the bottom. This indented stage is called a gastrula. Locate a gastrula under low power. Then switch to high power. Draw a diagram of the gastrula and label it.
8. Repeat this procedure using the slide showing clam development. Note any similarities and differences between starfish development and clam development.

OBSERVATIONS
1. What similarities did you observe between echinoderm (starfish) development and mollusk (clam) development?
2. What differences did you observe between echinoderm cleavage and mollusk cleavage?
3. What happened to the size of the embryo as the number of cells increased?
4. What happened to the size of the individual cells as the number of cells in the embryo increased?

ANALYSIS AND CONCLUSIONS
1. Which organism undergoes spiral cleavage?
2. Which organism undergoes radial cleavage?
3. How do patterns of development compare in echinoderms and mollusks?

STUDENT STUDY GUIDE

SUMMARIZING THE CONCEPTS

The key concepts in each section of this chapter are listed below to help you review the chapter content. Make sure you understand each concept and its relationship to other concepts and to the theme of this chapter.

30-1 Evolution of the Invertebrates

- A phylogenetic tree shows the evolutionary relationships between different organisms.
- In protostomes, the blastopore becomes the mouth, cleavage typically is spiral, and the mesoderm usually arises from a few cells near the blastopore. In deuterostomes, the blastopore becomes the anus, cleavage typically is radial, and the mesoderm characteristically arises from pouches of endoderm.
- Acoelomates lack a body cavity. Pseudocoelomates have a body cavity that is partially lined with mesoderm. Coelomates have a body cavity that is completely lined with mesoderm, which is called a coelom.

30-2 Form and Function in Invertebrates

- There are three main types of skeletal systems: hydrostatic skeletons, exoskeletons, and endoskeletons.
- Simpler animals tend to use intracellular digestion. More complex animals tend to use extracellular digestion, in which food is broken down in a digestive tract.
- Simpler digestive systems have just one opening through which food is taken in and wastes are expelled. More complex digestive systems have two openings—a mouth and an anus.
- In an open circulatory system, blood comes in direct contact with body tissues. In a closed circulatory system, the blood is always contained within blood vessels.
- Respiratory systems require a thin moist membrane and a large surface area for effective gas exchange.
- Excretory systems regulate the amount of water in the body and dispose of nitrogenous wastes.
- Three trends in the evolution of the nervous system are centralization, cephalization, and specialization.
- Asexual reproduction makes it possible for one individual to rapidly produce many genetically identical offspring. Sexual reproduction creates new combinations of genes.
- Fertilization may be internal or external.

REVIEWING KEY TERMS

Vocabulary terms are important to your understanding of biology. The key terms listed below are those you should be especially familiar with. Review these terms and their meanings. Then use each term in a complete sentence. If you are not sure of a term's meaning, return to the appropriate section and review its definition.

30-1 Evolution of the Invertebrates
phylogenetic tree
protostome
deuterostome
acoelomate
pseudocoelomate
coelomate
coelom

30-2 Form and Function in Invertebrates
hydrostatic skeleton
endoskeleton
intracellular digestion
extracellular digestion
external fertilization
internal fertilization

Observations

1. Both echinoderms and mollusks go through the same basic stages of development.
2. In early echinoderm embryos, the cells were directly on top of or behind one another. In early mollusk embryos, the layers of cells were in between one another.
3. The size of the embryo remained constant.
4. The cell size decreased.

Analysis and Conclusions

1. Clam (mollusk).
2. Starfish (echinoderm).
3. They are basically the same from the egg stage through the gastrula stage except that echinoderms undergo radial cleavage and mollusks undergo spiral cleavage.

Going Further: Enrichment

Students might construct clay models of spiral and radial cleavage in early embryos.

Interested students might wish to do library research to find more information about the different patterns of early development. They may be startled to find out how many variations there are on basic patterns of development. They should be able to describe the following forms of cleavage after doing some research: equal, unequal, dextral (clockwise spiral), sinistral (counterclockwise spiral), holoblastic (complete), and meroblastic (incomplete). They may also find out about bilateral and rotational cleavage patterns.

CHAPTER REVIEW

CONTENT REVIEW

Multiple Choice
1. c 2. d 3. a 4. c
5. d 6. c 7. b 8. d

True or False
1. F coelom
2. F protostomes
3. F neither protostomes nor deuterostomes
4. F asexual
5. F uric acid
6. T
7. F mesoderm
8. F excretion

Word Relationships
1. phylogenetic tree
2. Urine
3. intracellular digestion

CONCEPT MASTERY

1. The most primitive animals lack a body cavity. More advanced animals have a body cavity that is partially lined with mesoderm (pseudocoelom). The most advanced animals have a body cavity that is completely lined with mesoderm (coelom). Acoelomate: jellyfish, planarian. Pseudocoelomate: hookworm. Coelomate: crayfish. Students' examples may vary.
2. Highly folded or feathery surfaces maximize the surface area for diffusion of respiratory gases.
3. It is an echinoderm, chordate, or a member of one of the minor deuterostome phyla. The blastopore in its embryo becomes the anus. It has a coelom. It probably has radial cleavage and a mesoderm that arises from pouches of endoderm.
4. Cephalization: A crayfish has a brain and specialized sense organs such as eyes and antennae in its head region. Centralization: It has one main nerve cord on its ventral side and a number of ganglia and minor nerve cords. Specialization: Among other sense organs, it has eyes for seeing and antennae for sensing objects.
5. They have intracellular digestion and thus cannot eat anything bigger than one of their cells.

CHAPTER REVIEW

CONTENT REVIEW

Multiple Choice

Choose the letter of the answer that best completes each statement.

1. Which of the following statements about a protostome is true?
   a. It is an acoelomate.
   b. It has radial cleavage.
   c. Its blastopore becomes its mouth.
   d. Its mesoderm arises from pouches of endoderm.
2. Roundworms, which have body cavities that are partially lined with mesoderm, are
   a. deuterostomes.   c. acoelomates.
   b. coelomates.   d. pseudocoelomates.
3. An animal that lives in an extremely dry climate probably
   a. has a very efficient excretory system.
   b. tends to gain water through osmosis.
   c. has a hydrostatic skeleton.
   d. excretes ammonia.
4. The nitrogenous waste that is least soluble in water is
   a. urine.   c. uric acid.
   b. ammonia.   d. urea.
5. Which animal has the greatest amount of cephalization?
   a. starfish   c. jellyfish
   b. flatworm   d. octopus
6. Sea urchins produce huge numbers of sperm cells or egg cells. These animals probably
   a. are hermaphrodites.
   b. have internal fertilization.
   c. have external fertilization.
   d. care for their young.
7. An animal with a tube-within-a-tube body plan always has a
   a. coelom.
   b. digestive tract with two openings.
   c. hydrostatic skeleton.
   d. highly specialized digestive tract.
8. Which animal has the most complex circulatory system?
   a. clam   c. flatworm
   b. sea anemone d. squid

True or False

Determine whether each statement is true or false. If it is true, write "true." If it is false, change the underlined word or words to make the statement true.

1. A body cavity that is completely lined with mesoderm is called a <u>blastopore</u>.
2. Most <u>deuterostomes</u> have spiral cleavage.
3. Acoelomate animals are <u>protostomes</u>.
4. Budding is a form of <u>sexual</u> reproduction.
5. Malpighian tubules remove <u>urea</u>.
6. Insects have <u>open</u> circulatory systems.
7. The middle layer of cells in an animal embryo is called the <u>ectoderm</u>.
8. Nephridia are used in <u>respiration</u>.

Word Relationships

Replace the underlined definition with the correct vocabulary word.

1. The animal kingdom's <u>diagram that shows the evolutionary relationships between groups of organisms</u> indicates that animals evolved from protists.
2. <u>The waste product produced by the excretory system</u> contains concentrated nitrogenous wastes.
3. Sponges have <u>the type of digestion in which food is broken down inside food vacuoles within cells rather than in a digestive cavity</u>.

6. Internal fertilization allows animals to reproduce on land. Although sperm cannot swim on dry land, they can swim to and fertilize eggs when placed in the moist environment inside an animal's body.
7. Typical protostomes: spiral cleavage, blastopore becomes the mouth, mesoderm arises from a few cells near the blastopore, coelom arises from a split in a solid mass of mesoderm. Typical deuterostomes: radial cleavage, blastopore becomes the anus,

CONCEPT MASTERY

Use your understanding of the concepts developed in the chapter to answer each of the following in a brief paragraph.

1. Discuss the evolution of the body cavity in animals. Give specific examples of animals with each kind of body plan.
2. Explain why lungs tend to have a highly folded inner surface and gills are often feathery in appearance. (*Hint:* What features are common to all respiratory surfaces?)
3. What sorts of inferences can you make if you are told that an animal is a deuterostome?
4. How do crayfish show the evolutionary trends of cephalization, centralization, and specialization? (*Hint:* Refer to Figure 28–23 on page 621.)
5. Explain why sponges are able to eat only microscopic particles of food.
6. Why can internal fertilization be considered an adaptation to living on land?
7. Compare early development in protostomes and deuterostomes.

CRITICAL AND CREATIVE THINKING

Discuss each of the following in a brief paragraph.

1. **Summarizing information** Construct a table that compares the nine invertebrate phyla you have studied in this unit with regard to the seven essential life functions. Be sure to include at least two examples of each phylum and any other information that will help you see the relationships between invertebrates.
2. **Relating concepts** At one time, cnidarians and echinoderms were placed in the same classification group because they both have radial symmetry. Explain why echinoderms are now thought to be more closely related to chordates.
3. **Making generalizations** Ctenophores are marine animals that have radial symmetry, a digestive tract with a single opening, and long, branching tentacles that have special sticky cells used to capture prey. The thin, transparent body wall consists of three layers: epidermis, mesoglea, and gastroderm. Although the mesoglea contains muscle cells, ctenophores typically move by using combs of fused cilia. Ctenophores range in size from about that of a pea to that of a golf ball.
   a. Where would you expect ctenophores to fit on the phylogenetic tree? Explain.
   b. How would you expect a ctenophore to carry out the functions of respiration, internal transport, and excretion? Explain.
4. **Developing a hypothesis** Formulate a hypothesis to explain why slugs are slimy. How might you test your hypothesis?
5. **Evaluating theories** Deuterostomes are thought to have evolved from protostomes. Give two reasons explaining why this is a reasonable theory. What information would you need to make a more informed evaluation of this theory?
6. **Using the writing process** Pretend that you are the invertebrate of your choice. Prepare a résumé that will inform a potential employer of your specialized skills.

---

in each Other Information box students should identify the phylum as deuterostome, protostome, acoelomate, pseudocoelomate, or coelomate. Check students' tables for accuracy. Students may need some help filling in a few of the boxes.

2. Echinoderms have a pattern of development that is more like that of chordates than that of cnidarians. In addition, the larvae have bilateral symmetry, indicating that the radial symmetry of adult echinoderms is of secondary origin.
3. **a.** Because ctenophores are quite similar to cnidarians, their branch would arise as a fork off the branch leading to cnidarians.
   b. Since a ctenophore is quite similar to a cnidarian in structure, it would carry out these functions as a cnidarian does—via diffusion through the body surface.
4. Accept all logical answers. One reason slugs are slimy is that the mucus coating helps prevent their body from drying out. Experiments to test this hypothesis might involve finding out what happens to slugs that lack slime.
5. Accept all logical answers. Students may note that deuterostomes are thought to be more complex—and thus later to evolve—than protostomes; reversal of the direction of the digestive tract so that an opening that used to become the mouth becomes the anus could have occurred. Students should recognize that they would need to know more about the origin of deuterostomes and patterns of development in many invertebrates before they could make an informed evaluation of this theory.
6. The content of student answers should be logical and correct. You may wish to explain the format of a résumé.

---

mesoderm and coelom arise from a pouch of endoderm.

CRITICAL AND CREATIVE THINKING

1. Tables should list invertebrate phyla (Porifera, Cnidaria, Platyhelminthes, Nematoda, Mollusca, Annelida, Arthropoda, Echinodermata, Chordata) along one axis. Along the other axis, the tables should list the seven life functions (Movement, Feeding, Internal Transport, Respiration, Excretion, Response, Reproduction), Examples, and Other Information. When applicable,

CAREERS IN BIOLOGY

Reptile Farmer, Chef, Entomologist

Students interested in these biological careers should be instructed to write for further information to the address listed beneath each career description. However, in consideration of the organizations that provide career information, please have only one student write to an organization. Do not instruct the entire class to write to every organization for the same career information. You may want to use the information provided to start a biology career file for student use.

UNIT 7 CAREERS IN BIOLOGY

Reptile Farmer

Reptile farmers breed and capture reptiles such as snakes and tortoises for preservation, exhibition, meat, and venom. They raise the reptiles in cages that are similar to the animals' natural habitat.

Many reptile farmers are skilled in extracting venom from live snakes. This venom can be used by scientists to develop chemical treatments for people who have been bitten by poisonous snakes.

Reptile farmers usually learn about their trade through experience.

For additional information contact the National Council of Farmer Cooperatives, 1800 Massachusetts Ave., NW, Washington, DC 20036.

Chef

Chefs are responsible for the preparation, cooking, and presentation of foods. They plan menus, order food supplies, and follow or create recipes.

Many chefs specialize in a particular food group. For example, certain chefs prepare only seafoods such as shrimp, crab, lobster, scallops, and mussels. Other chefs prepare only meat dishes, or pastries.

Although a high school diploma is not required to become a chef, it is an asset to job applicants. Many chefs also complete training programs at cooking schools.

For more information write to the American Culinary Federation, P.O. Box 3466, St. Augustine, FL 32084.

Entomologist

Entomologists study insects and the relationships between insects and plants and animals. Their work involves collecting, observing, and classifying different insects, as well as determining the effects insects can have on other organisms.

Some entomologists do research on insects that are harmful to other organisms, such as those that destroy crops. They develop and improve pesticides and are interested in methods to prevent the spread of such insect pests.

To become an entomologist, a Bachelor's degree in the biological sciences and a Ph.D. in entomology are necessary.

For information write to the Entomological Society of America, 4603 Calvert Rd., College Park, MD 20740.

HOW TO WRITE A COVER LETTER

When you contact an employer through the mail, you should write a cover letter to introduce yourself. You must realize that the person receiving your letter probably reads many letters like yours every day. Therefore, your cover letter must be impressive, organized, and brief. Your letter must be an original copy, neatly typed without errors.

It is best to begin your letter with a statement explaining the kind of position you are seeking. The second paragraph is usually used to sum up your qualifications for the position. The third paragraph is used to close the letter and ask for an interview at a convenient time.

FROM THE AUTHORS

I never really liked hydras. They were small, didn't do much, and were usually half dead by the time my high school biology class got a chance to see them. Insects didn't turn me on much either, to tell the truth. Bees and mosquitoes bit me, ants pestered me at picnics, and flies kept me awake at night when I went camping. Worse yet, I had to memorize their names and body parts for school tests. "What are these things good for?" I'd ask myself. "Who cares about them anyway?"

I didn't change my mind about invertebrates for several years. Then, in one 24-hour period, everything I once thought about them changed.

I was visiting a tropical island. Not a resort, mind you, but a biological station on a tiny piece of rock in the middle of the Caribbean. The very first night, during dinner, a huge centipede crawled out of the garden and bit the resident dog on the nose. As the poor mutt ran away yelping, I thought, "This is something worth paying attention to." (I later found out that the centipede's bite was venomous and quite painful.)

The next morning I saw a living coral reef for the first time. Sure, I'd seen pictures of them, but they hadn't prepared me for the fantasyland of colors and shapes that drifted in front of my eyes. Most of the animals I saw were invertebrates, and I struggled to remember their names. There were corals, sea anemones, starfish, sea cucumbers, and more. And all of these animals played important roles in the living system of the reef. I was amazed. I was also hooked. I would never think poorly of animals without backbones again.

Joe Levine

FROM THE AUTHORS

Joseph Levine

An expert in aquatic organisms and an aquarium designer, author Joe Levine describes in this feature his first encounter with the denizens of a coral reef. Joe describes how he came to understand the marvelous diversity of invertebrates and their importance to living things. As the recent photograph on this page shows, Joe has had many opportunities to closely observe invertebrates—and their vertebrate neighbors—in underwater habitats.

After students read the feature, you might want them to imagine that they are exploring a part of the Earth they have never seen and to describe the wonders they encounter.

UNIT 8

Life on Earth: Vertebrate Animals

UNIT OVERVIEW

In Unit 8 students are introduced to vertebrate animals. First they learn about the characteristics and evolutionary history of each of the vertebrate classes. Next they examine the characteristics of primates and read about human evolution. They then explore some of the basics of animal behavior. Finally, they look at an evolutionary tree of the vertebrates, compare the ways in which different vertebrates carry out their essential life functions, and examine trends in vertebrate evolution.

UNIT OBJECTIVES

1. Discuss form and function in each class of vertebrates.
2. Describe the evolutionary history of the vertebrates.
3. Compare methods of internal temperature regulation in vertebrates.
4. Discuss the characteristics and evolution of primates.
5. Explain how instincts and learning affect behavior.
6. Discuss communication in animals.
7. Compare the ways in which vertebrates carry out their essential life functions.

INTRODUCING UNIT 8

Using the Visuals
Begin your teaching of the unit by having students read the unit introduction on page 676 and examine the photographs on pages 676 and 677.

UNIT 8

Life on Earth: Vertebrate Animals

At first glance, fishes, amphibians, reptiles, birds, and mammals appear considerably different from one another. Some have feathers; others fins. Some fly; others swim. Some walk on land; others burrow beneath it. Some live on land; others in rivers, lakes, and oceans. Indeed, these variations and others are some of the characteristics biologists use to separate these animals into different classes.

All of the animals, however, have at least one important characteristic in common: They have special bones called vertebrae in their back that form a strong column that protects delicate nerve tissue and helps to support their body. This vertebral column, a unique characteristic shared by many different kinds of animals, forms the basis for classifying them in the same subphylum—the vertebrates, or animals with a backbone.

DISCOVERY LEARNING

PAUSE FOR PETS

Have you ever closely watched your pet or another animal for a period of time? How is an animal's movement related to its structure? Find out by doing this activity.

1. Over the course of an hour, closely observe a pet animal such as a dog, cat, bird, turtle, or fish.
2. Record your observations in the form of written notes, quick pencil sketches, or even photographs.
3. Assemble a portfolio of your observations to share with others. Compare your observations with students who have similar or different pets.
   - What do the animals look like? How are the animals similar in appearance? How are they different?
   - How do the animals behave? Why do you think the animals behave in this way? How is their behavior similar? How is it different?

- **What animals are pictured on these pages?** (Two orangutans.)
- **What do orangutans have in common with fishes, reptiles, and birds?** (All have vertebrae.)
- **What other animals have vertebrae?** (Answers will vary.)
- **What advantages do vertebrae give an animal?** (Vertebrae support the body and enclose and protect the nerve cord.)
- **From their picture, describe some other characteristics of orangutans.** (They have a hairy body, eyes with eyelids, a nose, and a mouth. Students may include other characteristics.)
- **Do you think that all vertebrates share these characteristics?** (Accept all answers for now. You may point out that body hair is an important characteristic of mammals.)

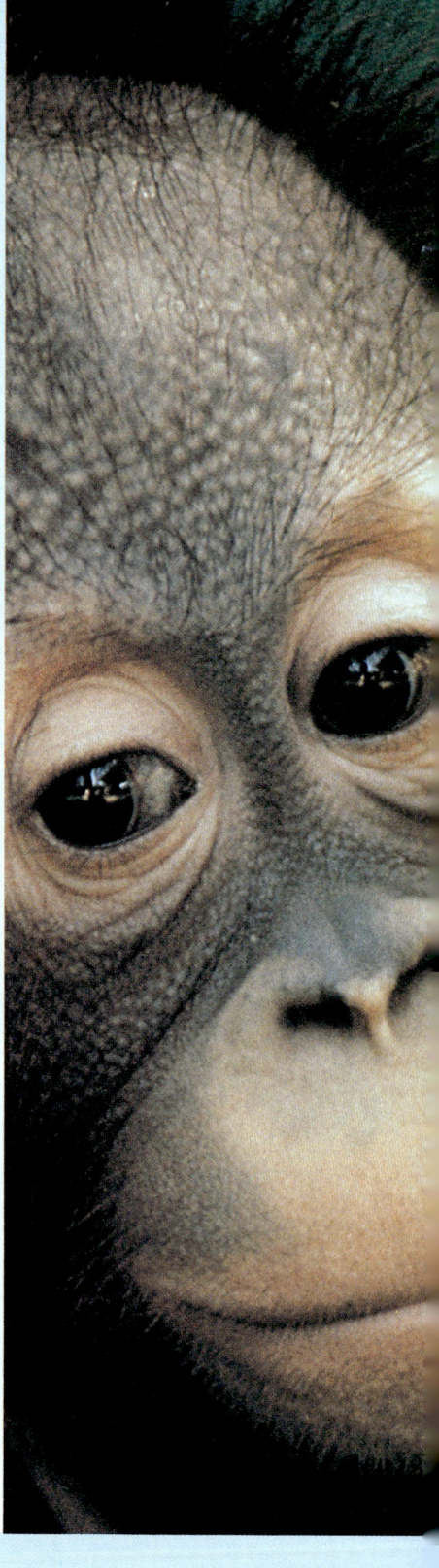

CHAPTERS

31 *Fishes and Amphibians*

32 *Reptiles and Birds*

33 *Mammals*

34 *Humans*

35 *Animal Behavior*

36 *Comparing Vertebrates*

CHAPTER DESCRIPTIONS

31 Fishes and Amphibians
Chapter 31 deals with the first classes of vertebrates to evolve—fishes and amphibians. First, students learn about the general characteristics, evolution, and major orders of fishes and examine the ways in which fishes carry out their essential life functions. Amphibians are discussed in the second part of the chapter. Students learn about the characteristics, evolution, and major orders of amphibians and study how amphibians carry out their essential life functions.

32 Reptiles and Birds
The characteristics, evolution, orders, and importance of reptiles are discussed in the first part of Chapter 32. The second part of this chapter deals with endothermy, an important evolutionary innovation that arose some time between the appearance of reptiles and the first birds. In the final part of the chapter students learn about the characteristics, evolution, and importance of birds.

33 Mammals
In Chapter 33 students study the characteristics of the major orders of mammals and examine the impact of mammals on the physical environment and on other living things.

34 Humans
Chapter 34 discusses human evolution. First, students learn about the characteristics and evolution of the primates. Next, students examine the evolution of the hominids, from *Australopithecus* to *Homo sapiens*.

35 Animal Behavior
Chapter 35 introduces students to some of the basics of animal behavior.

36 Comparing Vertebrates
Chapter 36 gives students the opportunity to use the information from previous chapters to trace evolutionary trends and compare the ways in which vertebrates carry out their essential life functions.

Discovery Learning

PAUSE FOR PETS

Students may study their own pets, the pets of friends or neighbors, or animals they see at a park or playground. Interested students may want to study the vertebrates at a zoo or aquarium.

Make sure that students record their own observations and answer the activity's questions. Have students share their observations and illustrations with the class.

677

CHAPTER 31 Fishes and Amphibians

Section	Laboratory Investigations and Activities
31–1 Fishes pages 679–692 THEMES: Evolution, Unity and Diversity	**Student Edition** LABORATORY INVESTIGATION: Observing Respiration in a Fish, p. 702 **Teacher Edition** DEMONSTRATION: Fishes, Amphibians, and Geologic Time, p. 678D DEMONSTRATION: Classroom Aquarium, p. 678D **Laboratory Manual** The Fish, p. 391 **Teaching Resources** HANDS-ON ACTIVITY: To Float or Not to Float? p. 151 TEACHER DEMONSTRATION: Investigating Schooling Behavior in Fish, p. 21
31–2 Amphibians pages 692–701 THEMES: Evolution, Unity and Diversity	**Laboratory Manual** The Frog, p. 397 **Teaching Resources** HANDS-ON ACTIVITY: A Frog of a Different Color, p. 149
Chapter Review pages 703–705	

Teacher Resources

Books

Bartlett, Richard D. *In Search of Reptiles and Amphibians.* Flora and Fauna, 1988.

Boschung, Herbert T., Jr., et al. *The Audubon Society Field Guide to North American Fishes, Whales, and Dolphins.* Knopf, 1983.

Duellman, William E., and Linda Trueb. *Biology of Amphibians.* McGraw-Hill Publications, 1985.

Halliday, Tim, and Kraig Adler, eds. *The Encyclopedia of Reptiles and Amphibians.* Facts on File, 1986.

Jahn, Johannes. *Turtles.* F. H. Publications, 1987.

Robins, C. Richard, and G. Carlton Ray. *A Field Guide to Atlantic Coast Fishes of North America.* Houghton Mifflin, 1986.

Steel, Rodney. *Sharks of the World.* Facts on File, 1985.

CHAPTER PLANNING GUIDE

Other Activities	Media and Technology

Teaching Resources
  ACTIVITY: Investigating Fish Respiration, p. 3
Study Guide
  Section 31–1, p. 295
	Interactive Videodisc
	  On Dry Land: The Desert Biome
	Videodisc
	  Aquatic Ecosystems: Freshwater Wetlands and Freshwater
	  Aquatic Ecosystems: Estuaries and Marine
	Video/Videodisc
	  Seeing Sense
	Biology Transparencies
	  38: The Internal Anatomy of a Fish
	Biology Media Guide
	  pages 38–40

Teaching Resources
  VOCABULARY REVIEW: Cross-a-Clue, p. 1
  ACTIVITY: Form and Function in Frogs, p. 7
Study Guide
  Section 31–2, p. 299
	Interactive Videodisc
	  On Dry Land: The Desert Biome
	Videodisc
	  Aquatic Ecosystems: Freshwater Wetlands and Freshwater
	Video/Videodisc
	  Biomes: Coniferous Forest and Tropical Rain Forest
	Biology Transparencies
	  39: The Digestive System of a Frog
	  40: The Respiratory, Circulatory, and Excretory Systems of a Frog
	  41: The Nervous and Muscular Systems of a Frog
	  42: The Reproductive Systems of a Female and a Male Frog
	Biology Media Guide
	  pages 40–41

Teaching Resources
  Chapter Test A, p. 13
  Chapter Test B, p. 17
	Computer Test Bank
	  Chapter Test, p. 327

Audiovisuals

Fascinating Fishes. Film or video. Marty Stouffer Productions.
Fish, rev. Film or video. Coronet.
Amphibians, rev. Film or video. Coronet.
Expansion of Life on Land: Amphibians and Reptiles. Film. Carolina Biological Supply Company.
Sharks. Film. Time-Life Video.

CHAPTER 31 Fishes and Amphibians

CHAPTER OVERVIEW

This chapter begins the study of vertebrate animals. The chapter opens with a discussion of vertebrates in general, then proceeds to describe the most ancient groups of vertebrates—fishes and amphibians.

The first fishes appeared on Earth over 500 million years ago. These fishes were jawless creatures that did not much resemble modern fishes. However, two descendants of these ancient organisms are still found on Earth today. These fishes, the lamprey and the hagfish, are members of a group known as jawless fishes.

The two other basic groups of fishes are the cartilaginous fishes and the bony fishes. Of these two groups, the bony fishes are by far the more numerous. In fact, the tens of thousands of species of bony fishes make up the largest class of vertebrate animals. Many familiar fishes, from guppies to perch to flounder, are found in this group. The cartilaginous fishes, which include skates, rays, and sharks, are characterized by a skeleton made of cartilage rather than bone.

The last section of the chapter introduces the interesting group of animals known as amphibians. Amphibians are both land and water dwellers, although the usual pattern is for this animal to be aquatic as a larva and terrestrial as an adult. The evolution, the form and function, and the ecological and economic roles of these animals are discussed at the end of the section.

31–1 FISHES

Section Focus 31–1

The purpose of this section is to introduce students to the characteristics of vertebrates in general and of fishes in particular. Students learn in detail about the three main groups of fishes, which are the jawless fishes, cartilaginous fishes, and bony fishes.

Fishes are the most ancient vertebrates, having first appeared on Earth over 500 million years ago. The first fishes were jawless fishes. Two descendants of this group, the lampreys and the hagfish, remain on Earth today. The first jawed fishes appeared about 400 million years ago. This group split into two evolutionary lines, the cartilaginous fishes and the bony fishes.

Fishes live in both fresh water and sea water in nearly every kind of climate. They exhibit many different methods of feeding and, depending on the species, their reproduction may be oviparous, ovoviviparous, or viviparous. Fishes breathe by means of gills, and they have a single-loop closed circulatory system with a heart that pumps blood throughout the body.

Fishes are important to humans in many ways. First, they are an important source of food in nearly every part of the world. Second, they play key roles in many ecological systems. And third, they are often prized as pets for their beautiful color and form.

Performance Objectives 31–1

1. List the main characteristics of vertebrates.
2. Discuss the basic characteristics of fishes.
3. Explain how fishes evolved.
4. Explain how fishes carry out essential life functions.

Science Terms 31–1

vertebral column   p. 679
jawless fish   p. 680
bony fish   p. 680
cartilaginous fish   p. 680
atrium   p. 684
ventricle   p. 684
aorta   p. 684
olfactory bulb   p. 685
cerebrum   p. 685
optic lobe   p. 685
cerebellum   p. 685
medulla   p. 685
oviparous   p. 687
ovoviviparous   p. 687
viviparous   p. 687

31–2 AMPHIBIANS

Section Focus 31–2

The purpose of this section is to discuss the characteristics of amphibians. Amphibians include frogs, toads, salamanders, and newts.

The term *amphibian* means "double life"; amphibians are adapted to both water and land. Most amphibians are aquatic as larvae and terrestrial as adults.

Most adult amphibians are carnivorous. Amphibians have a closed double-loop circulatory system with a three-chambered heart, and, depending on the species, amphibians may reproduce by external or internal fertilization.

Humans tend to have little interaction with amphibians, except for their use in medical research and of frogs' legs as a gourmet food item. In some parts of the world, hunters tip their arrows with toxins secreted by frogs. Amphibians do play an important role in ecological systems, however, particularly as predators of insects.

Performance Objectives 31–2

1. Discuss the general characteristics of amphibians.
2. Discuss the evolution of amphibians.
3. Describe and compare the larvae and adult stages of amphibians.
4. Explain how amphibians carry out essential life functions.

Science Terms 31–2

amphibian   p. 692

DISCOVERY LEARNING

TEACHER DEMONSTRATIONS
Modeling

Fishes, Amphibians, and Geologic Time
The following demonstration can be used as an introduction to Chapter 31.

Draw on the chalkboard or an overhead transparency a time line to show the geologic history of the Earth. Label clearly the four eras: Precambrian, Paleozoic, Mesozoic, and Cenozoic. Within the Paleozoic Era, label the Cambrian, Ordovician, Silurian, Devonian, Carboniferous, and Permian periods.

Point to the Cambrian Period and explain to students that it was near the end of this period—just over 500 million years ago—that the first jawless fishes appeared. These fishes continued to exist and change during the next two periods, the Ordovician and Silurian.

Point to the Silurian Period and explain that it was during this period—about 400 million years ago—that the first jawed fishes appeared. Continue by pointing to the Devonian Period and explaining that during this period, jawed fishes continued to develop; sharks appeared; and the first amphibians appeared. Also explain that by the end of the Devonian Period, which was about 350 million years ago, the ancient jawless fishes became extinct.

Point to the Mesozoic Era and explain that it was during the first period of this era—the Triassic—that modern fishes developed. This means that the development of modern fishes took place between 190 and 225 million years ago.

Classroom Aquarium
This demonstration can be set up for students to observe as they study Section 31–1.

For the demonstration you will need a rectangular 15- to 20-liter aquarium, an aquarium filter (optional), gravel, water plants, snails, guppies, guppy food, and an aquarium cover.

Rinse the aquarium with lukewarm water (no soap) and place it on a flat surface in indirect sunlight. Rinse the gravel and use it to cover the bottom of the aquarium to a height of about 3.5 centimeters. Fill the aquarium about two-thirds full with tap water. Place the water plants in the aquarium by pushing the roots into the gravel. If you have a filter, install it now and turn it on. Add more water, until the water level is about 5 centimeters from the top. Let the aquarium stand for two days.

After two days, add the snails and guppies to the aquarium. Add one guppy and one snail per 4 liters of water. Cover the aquarium and keep the temperature between 23°C and 27°C. Feed the fish a small amount of food every day. Add tap water that has stood for 24 hours to the aquarium as needed. Remove any plants or animals that die.

Have students observe the aquarium every day as they study this chapter. Have them record their observations. Ask them to make observations such as the following:

- **Do the guppies swim alone or in schools?**
- **What do you see when you observe the gill covers of the guppies?**
- **What is the reaction of the guppies to food?**

In order to create a more interesting aquarium, you may wish to purchase other types of fish from a pet store. Before combining fishes, however, ask a knowledgeable person at the store if the aquarium should be set up differently. You may also need to use a larger aquarium to avoid overcrowding.

CHAPTER 31
Fishes and Amphibians

The lionfish is one of the most exotic, beautiful, and venomous marine fishes. Amphibians such as salamanders also protect themselves with their toxins.

Just off the shores of a tropical island, small reef fish dart among coral formations, creating dazzling, ever-changing patterns of brilliant colors. A shark swims by, its open mouth revealing several rows of sharp triangular teeth. On the other side of the world, the decks of a fishing ship are covered with wriggling silver bodies as nets filled with herring are hauled in from a cold gray ocean. Using its fins like stubby legs, a catfish drags itself out of the mud of a drying pond in Africa and scuttles off to find more suitable surroundings. Thousands of meters beneath the ocean surface, nightmarish fish—which seem to consist of nothing more than needlelike teeth, oversized jaws, and enormous eyes—flicker with ghostly lights produced by their own body.

Fishes are the most ancient and diverse group of vertebrates, or animals with backbones. In the pages that follow, you will learn more about fishes and their relatives the amphibians—the first groups of vertebrates to appear on Earth.

GUIDE FOR READING	# 31–1 Fishes

After you read the following sections, you will be able to

31–1 Fishes
- Describe the distinguishing characteristics of vertebrates.
- Explain how fishes carry out their essential life functions.
- Describe the three basic groups of fishes and give an example of each.

31–2 Amphibians
- Describe how a typical amphibian carries out its essential life functions.
- Compare the two major living orders of amphibians.

Journal Activity

YOU AND YOUR WORLD
Suppose that your favorite pond is in danger of being drained and filled to build the parking lot for a new mall. Write a letter to your congressperson explaining why this construction should be stopped. (You might want to mention the loss of habitat for the fishes, frogs, and other animals that make the pond their home.)

31–1 Fishes

Guide For Reading
■ What are the distinguishing characteristics of vertebrates?
■ How are fishes adapted for a life in water?
■ What are the characteristics of the three basic groups of fishes?

The name Earth is not particularly appropriate for the planet on which we live, for more than two thirds of its surface is water, not earth. And just about anywhere there is water, there are fishes. At the edge of the ocean, blennies jump from rock to rock and occasionally dunk themselves in tide pools. Beneath the arctic ice live fishes whose bodies contain a biological antifreeze that prevents them from freezing solid in sea water colder than 0°C. And in shallow desert streams in the southwestern United States, pupfish tolerate temperatures that would cook almost any other animal.

What Is a Fish?

To clearly understand what fishes are and how they are related to other vertebrates, it is necessary to know something about the characteristics that unite fishes, amphibians, reptiles, birds, and mammals into the subphylum Vertebrata. You may recall from Chapter 29 that vertebrates belong to the phylum Chordata. This means that fishes and other vertebrates have at some time during their development a notochord, a hollow dorsal nerve cord, and pharyngeal slits. In most vertebrates, the notochord is replaced during development by a backbone, or **vertebral column**, which encloses and protects much of the nerve cord. In addition, most vertebrates have two sets of paired appendages, a closed circulatory system with a ventral heart, and either gills or lungs for breathing.

Fishes can be defined as aquatic vertebrates that are characterized by scales, fins, and pharyngeal gills. However, fishes are so varied that for almost every general statement

Figure 31–1 In ocean water deep below the reach of sunlight live fishes that sparkle with light produced by their own body. The lights on the viperfish may serve to attract prey or distract predators.

COOPERATIVE LEARNING

Working in pairs, have students review the chapter material on sharks (pages 688–689). They should compare this information with the commonly held view, popularized by recent films, that all sharks are "man-eaters." Students should pretend that they are concerned members of SOS (Save Our Sharks). SOS is dedicated to providing factual information to the general public and, in particular, to sport anglers who have been indiscriminately slaughtering sharks. Groups may choose one of the following activities:
- Compose a letter to the editor of the *American Angler Monthly* in which you correct common misconceptions about sharks as predators.
- Design an ad for a national weekly magazine of your choosing. This ad should be designed to correct the public's misconceptions about sharks.

Journal Activity

YOU AND YOUR WORLD
Ask volunteers to share their letters with the class. Have students keep their journal entries in their portfolios.

pendages protruding from it.)
- **Do you think you would want to encounter this fish? Why or why not?** (Students should say no because this fish is poisonous.)
  Direct students' attention to the inset photograph.
- **Is this animal a fish?** (No.)
- **How is it different from a fish?** (It has legs instead of fins; it has a tail; it is not shaped like a fish; and it does not seem to be swimming in water.)
- **Do you know what group of animals this animal belongs to?** (The animal is an amphibian.)

TEACHING STRATEGY 31–1

Focus/Motivation
Hold a globe in front of the class.
- **What fraction of the Earth is covered by water?** (About two thirds.)
- **If you were from another planet, would you conclude that this globe was made by land dwellers or water dwellers?** (Land dwellers. Whoever made the globe marked the land with cities and political boundaries. The water was left relatively unmarked.)
- **Are humans the masters of this planet?** (Perhaps not all of it. Emphasize that water remains inhabitable only to fishes and other marine organisms.)

BACKGROUND INFORMATION
EVOLUTION OF VERTEBRATES

The earliest known vertebrates were jawless fishlike animals called ostracoderms. They are so named because they had external body plates made of a material similar to bone. The brain was enclosed in a skull of bone or cartilage, and the notochord probably functioned as a supporting structure. Other early evolutionary lines of jawless vertebrates showed distinct units of cartilage or bone, which surrounded the notochord and supported the tissue bridges between gill slits. Eventually these led to the evolution of well-developed vertebrae in fishes.

Jaws evolved from the most forward gill supports in jawless fishes. After the development of jaws came the evolution of more sensitive eyes, olfactory receptors, and sensors that could detect vibrations. The brain and motor pathways also became more complex.

FACTS AND FIGURES

Scientists have no direct fossil evidence linking living vertebrates with lower chordates. The reason for this is that animals with no hard body parts often do not fossilize well.

31–1 (continued)
Content Development

Tell students that with more than two thirds of Earth's surface covered by water, it is not surprising that Earth looks like the "water planet." It is also not surprising that the most numerous group of vertebrate animals on Earth is aquatic.

Explain to students that the overall form and body plan of fishes tell quite a bit about the medium in which they live. For example, most fishes have

Figure 31–2 Representatives of the three main groups of living fishes are shown here. The lamprey is a parasitic jawless fish (bottom). The blue-spotted stingray is a venomous bottom-dwelling cartilaginous fish (right). The Potter's angelfish, which is found only in Hawaiian coral reefs, is a bony fish (left).

made about them there are exceptions. For example, some fishes do not have scales. One reason for the many differences among fishes is that four living classes of vertebrates make up this group of animals. Thus many fishes—sharks and lampreys, for example—are no more closely related to one another than humans are to frogs.

There are so many fishes, living and extinct, that their correct scientific classification is complicated. For our purposes, we can say that the living fishes fall into three main groups: **jawless fishes**, sharks and their relatives, and **bony fishes**. Sharks and their relatives are also known as **cartilaginous fishes** because their skeletons are made up of soft, flexible cartilage rather than bone.

Evolution of Fishes

Fishes are considered to be the most primitive living vertebrates. (This means that fishes were the first vertebrates to evolve and that they have many characteristics that are thought to have existed in their earliest ancestors. It does not mean that they are somehow inferior to other types of vertebrates.) Fishes did not evolve from such organisms as living lancelets or tunicates. But similarities in structure and embryological development show that fishes and modern invertebrate chordates probably did evolve from common invertebrate ancestors that lived many millions of years ago.

The first fishes—which are also the first vertebrates—were odd-looking jawless creatures whose bodies were covered with bony plates. They lived in the oceans of the late Cambrian Period, about 540 million years ago. For over 100 million years, fishes retained the basic armored jawless body plan. Then

streamlined bodies with a large proportion of skeletal muscle suited to forward motion. This is important because the density of water, which is about 800 times denser than air, makes movement through water difficult. Most fishes also have several pairs of fins, which are used to stabilize the body and propel and guide it through the water.

Point out that the muscles of fishes are not designed to work hard in counteracting the pull of gravity.

• **Can you explain why fishes do not have to work as hard to resist gravity as do animals that travel on land or in the air?** (The downward pull of gravity is not nearly as great in water as it is on Earth or in the air because water exerts an upward, buoyant force that helps to sup-

Early Jawless Fishes   Early Jawed Fishes

during the Ordovician and Silurian periods, fishes underwent a major adaptive radiation. Some of the groups that emerged from this adaptive radiation were jawless fishes that had very little armor—the ancestors of modern lampreys and hagfish. Others were armored jawless fishes in a variety of new forms. These fishes were ultimately evolutionary dead ends that became extinct around the end of the Devonian Period, about 350 million years ago. Still others were armored fishes that possessed a feeding adaptation that would revolutionize vertebrate evolution: These fishes had jaws.

Jaws are extremely important evolutionary innovations. Jawless fishes are limited to eating small particles of food that they can filter out of the water or suck up like a vacuum cleaner. Jaws made it possible for vertebrates to nibble on plants, munch on other animals, and defend themselves by biting.

Another evolutionary innovation seen in the early jawed fishes were paired pectoral (anterior) and pelvic (posterior) fins that were attached to girdles of cartilage or bone. These fins gave the fishes more control over their movement in the water. See Figure 31–4. In addition, the pectoral fins and girdle provided the raw material from which evolution shaped the forelimbs and shoulder bones of terrestrial vertebrates. Similarly, the pelvic fins and girdle were the origins of the hindlimbs and hip bones.

Although the early jawed fishes soon disappeared, they left behind two major classes that continued to evolve and still survive today. The first of these classes is the cartilaginous fishes, an old and successful group that includes sharks and rays. The second class is the bony fishes, a large and diverse assemblage that contains more than 97 percent of all living fish species.

Figure 31–3 Early jawless fishes, unlike modern jawless fishes, had bones and were often heavily armored. Most early jawed fishes also had bony armor.

Figure 31–4 The blackbar soldierfish has a typical assortment of fins for a bony fish. In this fish, the dorsal fin is divided into two distinct parts: a spiny anterior region and a relatively soft posterior region. Not all fins are present in all kinds of fishes.

TIE-IN LANGUAGE ARTS

Although there is general confusion about the proper use of the words *fish* and *fishes*, the rule followed by biologists is quite simple. When referring to a single individual of a single species, use *fish*. When referring to many individuals of a single species, the proper plural is also *fish*. However, when referring to several individuals of more than one species, the proper plural is *fishes*.

BACKGROUND INFORMATION
CLASSES OF VERTEBRATES

There are eight living classes of vertebrates. Four of these classes are fishes. The classes are as follows:
- Myxini—hagfishes
- Cephalaspidomorphi—lampreys
- Chondrichthyes—cartilaginous fishes
- Osteichthyes—bony fishes
- Amphibia—amphibians
- Reptilia—reptiles
- Aves—birds
- Mammalia—mammals

jaws open up a whole new world of feeding mechanisms. (It is significant that the two types of jawless fishes alive today, the lamprey and the hagfish, can feed only by sucking or scavenging.) Jaws enable an animal to compete for prey; they also increase the possibility of one animal eating another. It stands to reason that with the evolution of jaws, animals developed keener sense organs to recognize another animal as prey or predator.

port the weight of the body.) Point out that another adaptation for living in water is scales. Slow-moving fishes that live near the bottom of water usually have large scales that serve as armor. Fishes that swim closer to the surface often have small scales that serve as protection while being lightweight.

Content Development
Explain to students that there are several advantages to an animal in having vertebrae. First, the segmented vertebrae allow the animal's body to flex while still providing it with structure and support; second, they provide attachment sites for muscles; and third, they protect the nerve cord.

Explain that jaws also give an animal advantages. Obviously,

Form and Function in Fishes

Fishes have entered many environments and evolved adaptations that enable them to survive a tremendous variety of conditions. Here we can give only a brief survey of the many ways fishes accomplish the basic functions of life.

FEEDING Every mode of feeding is seen in fishes—there are herbivores, carnivores, parasites, filter feeders, and detritus feeders. In fact, a single fish may exhibit several modes of feeding, depending on what kind of food happens to be available. Certain carp, for example, eat just about anything—algae, water plants, worms, mollusks, arthropods, dead fish, and detritus. Some fishes—such as great white sharks, tunas, and barracuda—are carnivores. A few fishes are parasites. For example, pencil catfish live and lay their eggs in the gills of larger fishes. And the male in certain species of anglerfish attaches permanently to the much larger female and obtains nutrients from her blood. Still other fishes, such as lamprey larvae and manta rays, are filter feeders. Although their prey are tiny, many filter feeders are not—the filter-feeding whale shark, which grows as long as 18.5 meters, is the largest fish in the world.

The adaptations for feeding in fishes are often remarkable. The sawfish (a relative of sharks) kills and stuns prey by slashing into a school of small fish with a long snout edged with sharp teeth. The parrotfish has teeth fused into a short beak

Figure 31–5 Fishes are adapted to many modes of feeding. Some male anglerfishes are parasites that are nourished by the blood of their much larger mates. Note the two males attached to the belly of this female (top). The parrotfish (left) uses its "beak" to bite off chunks of algae covered coral. It digests the plant material and passes the crushed coral dust through its gut. The sawfish (right) slashes its way through schools of fishes, then doubles back to devour the dead or wounded prey.

that it uses to bite off chunks of living coral and additional teeth in its throat that grind the chunks of coral into sand. The archerfish shoots down insects by spitting drops of water at them. Anglerfish have wormlike or lighted lures that they use to entice prey. And some deep-sea fishes have enormous jaws that allow them to swallow prey larger than themselves!

Although some fishes have strong blunt teeth adapted for crushing clam and other mollusk shells, most fishes do not really chew their food. Instead, they tear their food into conveniently sized chunks or swallow their prey whole. The digestive system of a typical fish is shown in Figure 31-6.

From the mouth, the food passes through a short tube called the esophagus to the stomach, where it is partially broken down. In many fishes, the food is further processed in fingerlike pouches called pyloric ceca (pigh-LOR-ihk SEE-kah; singular: cecum), which are located at the point where the stomach and the intestine meet. The pyloric ceca secrete digestive enzymes and absorb nutrients from the digested food. The intestine receives partially digested food from the stomach and pyloric ceca and completes the process of digestion and nutrient absorption. In the intestine, the digestive enzymes from several other organs, such as the liver and pancreas, are added to the food. Any materials that remain undigested after passing through the intestine are eliminated through the anus.

The structure of a fish's intestine is adapted in ways that help fishes meet their nutritional needs. For example, herbivores typically have a much longer intestine than carnivores. (Incidentally, this is true of most types of animals.) This gives the animal more time and space to break down plant matter, which is difficult to digest. Lampreys, cartilaginous fishes, and a few bony fishes have a flap of tissue that spirals around the inside of part of the intestine, thus increasing the surface area for nutrient absorption.

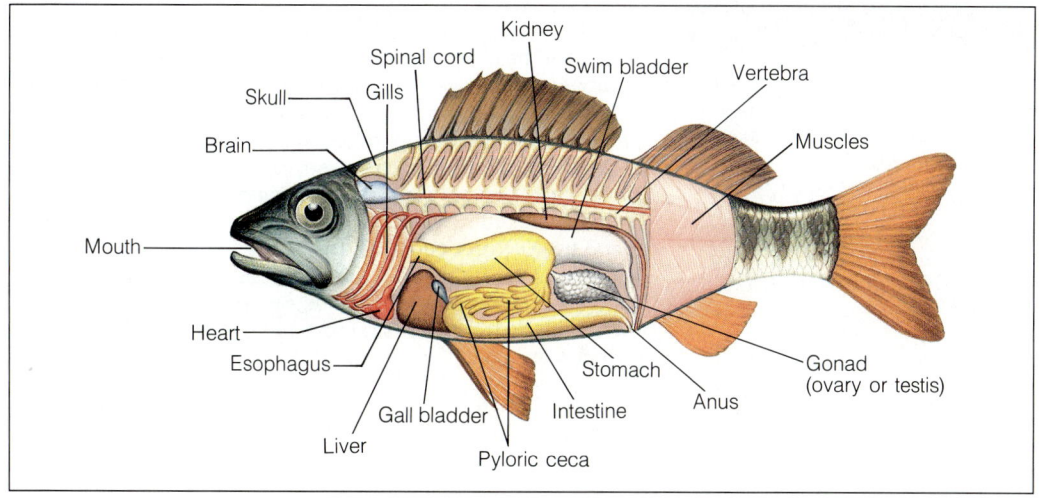

Figure 31-6 *The internal organs of a typical bony fish are shown here. What bones do the muscles in a fish pull against?* ❶

BACKGROUND INFORMATION
ANGLERFISH

The *Edriolynchnus schmidti* anglerfish shown in Figure 31-5 is one of several deep-sea anglerfishes that include parasitic males. The parasitic males are the only known examples of complete vertebrate parasitism, and the attached female and male together form what is essentially a self-fertilizing hermaphrodite. In addition, these fishes have somehow managed to overcome the problem of tissue rejection, which usually occurs when tissues from genetically different individuals are combined. This particular ability of anglerfishes makes them of great interest to medical researchers seeking to improve the rate of success of organ transplants in humans.

FACTS AND FIGURES

Like most deep-sea fishes, the anglerfish in Figure 31-5 is quite small—smaller, in fact, than the photograph in the student textbook! Although the female is only about 6 centimeters long, specimens of this size have been known to contain 9000 eggs.

In addition, some fishes exhibit a combination of two or more of these feeding modes.

Stress that various fishes are able to feed as they do because of particular adaptations.
• **What adaptation makes the sawfish successful in its quest for food?** (It has a long snout edged with sharp teeth that can stun and kill prey by slashing.)
• **What special adaptations for feeding does the parrotfish have?** (Teeth fused into a short beak that can be used to bite food, as well as teeth in its throat to grind the food.)
• **Describe the unusual feeding behavior shown in anglerfish.** (Some male anglerfish live as parasites of the much larger female. This means that anglerfish have adapted very different feeding mechanisms for each gender!)

• The textbook states that lamprey larvae, manta rays, and whale sharks are all filter feeders. **What is a filter feeder?** (A filter feeder lives off the tiny organisms that it filters from the water around it.)
• **What other filter feeders have we studied?** (Bivalve mollusks and certain echinoderms.)

ANNOTATION KEY

❶ Optic lobes would be smaller and olfactory lobes would be larger. (Making inferences)

FACTS AND FIGURES

In medieval times, it was widely believed that for every land animal there was a corresponding water animal. This is part of the reason for names such as catfish, sea horse, and dogfish. People were also thought to have corresponding water denizens. As a result, skates and rays, often trimmed to enhance their resemblance to humans wearing conical hats, were once sold to the unsuspecting as genuine "sea bishops."

BACKGROUND INFORMATION
FISHES THAT CONSERVE HEAT

Fish are commonly thought of as coldblooded, but tunas and mako sharks are exceptions to this rule. Thanks to a capillary network called the *rete mirabile*, these fishes are able to conserve the heat that they generate in their muscles. With this ability, tunas and makos can adapt to different temperatures at different ocean depths, and they can swim faster and longer than both their competition and their prey.

Please note that most biologists do not use the terms coldblooded and warmblooded. In Section 32–2, the textbook introduces two more precise classifications: *ectotherm* and *endotherm*.

31–1 (continued)

Content Development

Explain to students how a swim bladder functions in fishes. Point out that a submarine changes levels in a similar way, although the submarine changes its mass rather than its volume. Tanks inside the submarine either empty or fill with water in order to make the submarine's mass decrease or increase while its volume stays the same.

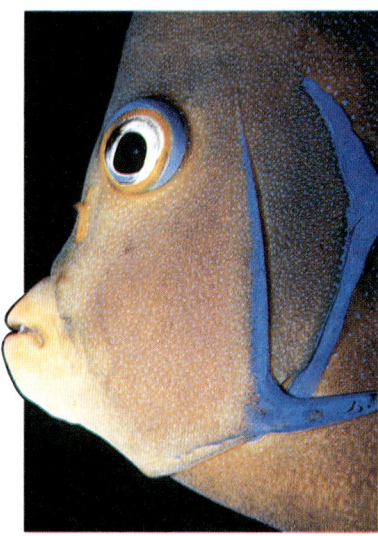

Figure 31–7 In some fishes, such as sharks, the gill chambers open to the outside through a number of slits (top). In other fishes, the gill chambers empty through a single opening that is covered by a protective flap. The curved blue stripe on this angelfish runs along the edge of the protective flap (bottom).

RESPIRATION Most fishes breathe with gills that are located on either side of the pharynx. The feathery gill filaments contain many capillaries and provide a large surface area for exchange of oxygen and carbon dioxide. Most fishes breathe by pumping water through the mouth, over the gill filaments, and out through slits in the sides of the pharynx. Some fishes, such as lampreys and sharks, have several gill slits on either side of the pharynx. Other fishes, such as almost all bony fishes, have a single opening on each side of their body through which water is "exhaled." This opening is usually hidden beneath a protective gill cover.

In many fishes the basic respiratory setup has been modified by evolutionary processes. For example, skates and rays are able to breathe while lying on the ocean floor even though their mouth is located on the underside of their body. Instead of taking in water (and gill-clogging mud and sand) through their mouth, they "inhale" water through special openings on the upper surface of their body.

A number of fishes—such as lungfish, gars, and Siamese fighting fish—have an adaptation that allows them to survive in oxygen-poor water or in areas where bodies of water often dry up. They have specialized organs that serve as lungs by obtaining oxygen from the air. In most air-breathing fishes, this organ is actually a modified swim bladder. A swim bladder, which is found in most bony fish, is a gas-filled sac that lies at the top of the body cavity just beneath the backbone. The majority of fishes use the swim bladder to regulate their buoyancy: Dissolved gases in the blood diffuse into and out of the swim bladder and permit the fish to swim at lesser or greater depths, respectively. Unlike typical fishes, fishes that use the swim bladder as a lung have a tube that connects the swim bladder to the mouth. Some air-breathing fishes are so dependent on getting part of their oxygen from the air that they will suffocate if prevented from reaching the surface of the water.

INTERNAL TRANSPORT Fishes typically have closed circulatory systems with a heart that pumps blood around the body. The heart consists of two muscular pumping chambers: an **atrium** (AY-tree-uhm; plural: atria) and a **ventricle** (VEHN-trihk-uhl). Blood from the body collects in the atrium, which pumps blood into the ventricle. The ventricle pumps blood out of the heart into a muscular vessel called the **aorta** (ay-OR-tah). Blood goes directly from the aorta into the fine capillary networks in the gills, where gas exchange occurs. From the gills, blood travels throughout the rest of the body tissues and internal organs. As blood leaves muscles and organs, it collects in veins that gather in a thin-walled sac called the sinus venosus (SIGH-nuhs veh-NOH-suhs). From this sinus, blood enters the atrium once again. See Figure 31–8.

EXCRETION Like many other aquatic animals, most fishes get rid of nitrogenous wastes in the form of ammonia.

• **What will happen to the submarine when the tanks fill with water?** (It will sink.) **Why?** (Its mass has increased, which means that its weight has increased; now the amount of water that it displaces will not be enough to counteract the downward pull of gravity on the submarine.)

Skills Development
Guided Practice
Skill: Making a model
Have students model the function of the swim bladder in fishes by performing the following experiment. Students can work in small groups. For the experiment, each group will need two identical balloons and a trough filled with water.

Have students blow up one of the balloons to its fullest and

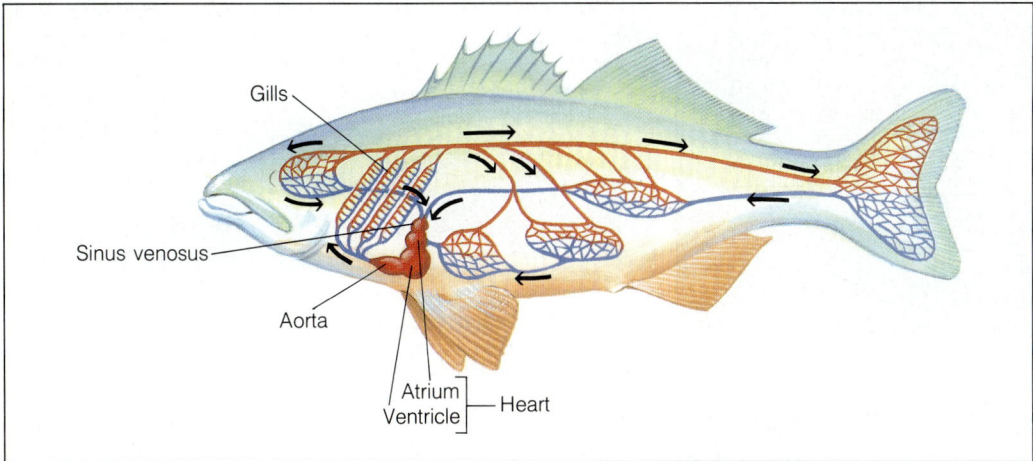

Some wastes diffuse through the gills into the surrounding water. Others are removed by the kidneys, which are excretory organs composed of many tubules that filter nitrogenous wastes from the blood and concentrate them.

Kidneys help fishes control the amount of water in their body. Because fishes in salt water tend to lose water by osmosis, the kidneys of marine fishes concentrate nitrogenous wastes and return as much water as possible to the body. The kidneys of freshwater fishes, on the other hand, pump out lots of dilute urine because in fresh water, a great deal of water continually enters by osmosis. One of the factors that determines which fishes are able to move from fresh to salt water (as salmon do) is their ability to control kidney function.

RESPONSE Fishes have a fairly well-developed nervous system organized around a brain. Fish brains, like those of other vertebrates, have several clearly visible parts. Refer to Figure 31-9 as you read about the parts of a fish's brain.

The most anterior parts of a fish's brain are the **olfactory bulbs**, which are connected by stalks to the two lobes of the cerebrum (ser-REE-bruhm). In fishes, the cerebrum is primarily involved with the sense of smell, although it also seems to be involved in such behaviors as taking care of young and exploring the environment. The **optic lobes** process information from the eyes. The **cerebellum** (sair-uh-BEHL-uhm) coordinates body movements. The **medulla** (mih-DUHL-ah) controls many internal organ functions and maintains balance.

Posterior to the brain is the spinal cord, which is in fact the hollow dorsal nerve cord that characterizes chordates. In cartilaginous and bony fishes, the spinal cord is enclosed and protected by the vertebral column. Between each set of vertebrae, a pair of spinal nerves exits the cord and connects with internal organs and muscles.

Figure 31-8 Almost all fishes have a closed circulatory system in which a two-chambered heart pumps oxygen-poor blood from the body to the gills. Oxygen-rich blood then travels from the gills to all parts of the body.

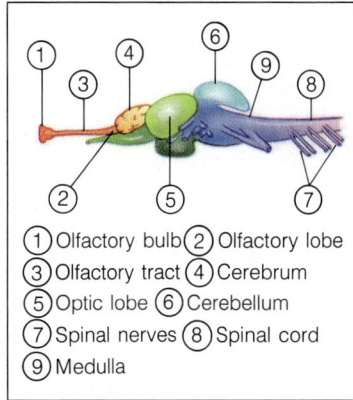

1. Olfactory bulb 2. Olfactory lobe
3. Olfactory tract 4. Cerebrum
5. Optic lobe 6. Cerebellum
7. Spinal nerves 8. Spinal cord
9. Medulla

Figure 31-9 The brain of a typical fish has several clearly visible parts. How might the sizes of the various parts of the brain differ in a blind cave fish that relies primarily on its sense of smell?

TEACHING SUPPORT

Laboratory Manual
- The Fish, p. 391

Teaching Resources
- Activity: Investigating Fish Respiration, p. 3
- Hands-On Activity: To Float or Not to Float? p. 151

MEDIA AND TECHNOLOGY
- Video/Videodisc
  - *Seeing Sense*
- Videodisc
  - *Aquatic Ecosystems: Freshwater Wetlands and Freshwater*

Use this bar code to introduce respiration in fishes.

Play frames 11935 to 12438

See the Biology Media Guide pages 38–40 for additional bar-code correlations for this section.

---

than it produced for the fully inflated balloon.)
- **Explain how the balloons illustrate the function of a swim bladder in fishes.** (When the bladder is full of gas, the fish floats near the surface of the water. When the bladder is empty or only partially filled with gas, the fish sinks deeper into the water.)

place it in the trough of water.
- **How much weight was added to the balloon when you filled it with air?** (Not very much.)
- **How much volume was added to the balloon?** (A lot—the balloon may be as much as ten times its original size.)

- **Does the balloon float or sink in the water?** (It floats.)
- **Why?** (It displaces a volume of water large enough to produce a buoyant force that is greater than its weight.)

Now have students fill the second balloon only partially full with air. Tell them to submerge it in the trough of water.
- **Why can this balloon be pushed deeper into the water than the first balloon?** (This balloon has a smaller volume, so it displaces less water. The water produces a lesser buoyant force

BACKGROUND INFORMATION
LOCOMOTION

Both fins and body shape help fishes to move from one place to another. Fishes with paired fins use them the way airplanes use wings and propellers to generate thrust—to direct the body up and down and to change direction. The few types of fishes that lack paired fins have trouble keeping themselves from spinning and rolling in the water. As a result, these fishes expend a lot of energy just moving from place to place.

BACKGROUND INFORMATION
SCALES

Most fishes are covered with scales. There are several different kinds of scales, but all share certain characteristics. Scales are firmly attached to the skin tissue from which they grow. The outer part of each scale overlaps the next scale, like shingles on the side of a house. As the fish grows, more material is added to the scales to enlarge them. In temperate climates, a layer of new material is added each year, forming a series of rings. Because one ring is added each growing season, scale rings can be counted to estimate a fish's age.

Figure 31–10 The sense organs in fishes are highly developed. A chimaera's huge silvery eyes enable it to see in the permanent dark of its deep-water home (left). The lateral line, which appears as a series of tiny dots in the pink stripe of the rainbow trout, detects water movements (right). Some fishes, such as the elephantfish (bottom), are able to detect electricity.

Most fishes have superbly designed sense organs that collect information about their environment. Almost all fishes active in daylight have well-developed eyes and color vision at least as good as our own. Fishes active only at night or in cloudy water have large eyes with big pupils that gather as much light as possible. These species do not see color well, but they see in the dark as well as cats do.

Many fishes possess extraordinary senses of taste and smell. Special cells called chemoreceptors (*chemo-* means chemical) are located all over the head and much of the rest of the body surface as well as in the nose and mouth. Many species, such as catfish, also carry chemoreceptors on their "whiskers." Salmon can distinguish between the odor of their own home stream and that of another stream while they are still far out at sea. And sharks can detect the presence of a drop of blood in 115 liters of sea water!

Most fishes have ears inside their head, but they cannot hear sounds well. They can, however, easily detect gentle currents and vibrations in the water. All around their head and down the sides of their body are a series of pores connected to canals beneath the skin that form a sensitive motion detector called the lateral line system. Fishes use their lateral lines to detect other fishes or prey swimming nearby.

In addition to having the senses that we are familiar with, some fishes—electric eels, catfish, and sharks, for example— are able to detect electricity. These electrical detectors are extremely sensitive: A shark can detect one millionth of a volt, which is less than the charge produced by the nerves in an animal's body. Electric eels and electric catfish produce a strong electrical field around their bodies that responds to the presence of nearby objects. Thus electric fishes can locate prey and avoid obstacles in murky water. In addition, some electric fishes such as electric eels can produce jolts of electricity (up to 650 volts!) that stun or kill prey and strongly discourage predators.

REPRODUCTION In most species of fishes, there are separate males and females. A number of fishes, however, are born as males but change to females as they grow older. Others start as females and later change into males. Unlike many invertebrates, few fishes function as both a male and a female at the same time.

31–1 (continued)

Skills Development
Guided Practice
Skills: Observing, relating concepts
Using the Visuals Direct students' attention to Figure 31–10.
• **In the picture on the left, what color are the eyes of the chimaera?** (Silvery-blue.)

Explain to students that the chimaera's eyes appear to be silver because light is reflecting off the tapetum, a shiny opaque layer behind the retina. By reflecting light back through the retina, the tapetum allows the light to strike one of the rod cells again and thus improves the animal's ability to see in the dark.
• **Do you think any land animals have tapetums?** (Answers may vary. Land animals that do have tapetums include cats, dogs, toads, and snakes.)
• **What makes you think that these animals have tapetums?** (At night, the eyes of these animals seem to glow as they reflect the light from car headlights or a flashlight. That is why animals such as cats see well in the dark.)

Many fishes are **oviparous** (oh-VIHP-ah-ruhs), which means they lay eggs. Most oviparous fishes have external fertilization. However, a few oviparous fishes—certain species of sharks and rays, for example—have internal fertilization and thus lay fertilized eggs.

Some oviparous fishes, such as cod, do not take care of their young. Such fish typically release hundreds or even millions of eggs, which increases the chances that a few offspring will survive to adulthood. Other oviparous fishes care for their offspring. For example, Siamese fighting fish build nests of bubbles, and sticklebacks build nests of twigs. Some cichlids and catfish hold their eggs and young in their mouths. And male seahorses hold fertilized eggs in a pouch until the eggs are ready to hatch. These species lay fewer eggs because the parental care means more of the young are likely to survive.

In some species of fishes that have internal fertilization, such as guppies, the eggs develop inside the female. The developing embryos, like those of oviparous fishes, are nourished by food stored in an attached yolk sac. Thus the young do not receive food directly from the mother's body. The young are typically "born" after they have absorbed the yolk sac and are ready to swim on their own. Species with this pattern of reproduction are **ovoviviparous** (oh-voh-vigh-VIHP-ah-ruhs). In other species, including several sharks, the young are actually nourished by the mother's body as they develop. These fishes are said to be **viviparous**, or truly live-bearing.

Many fishes, including several species you can keep in a home aquarium, exhibit fascinating mating behaviors. Guppy males dance up and down in front of females, trying to get the females interested in them. Cichlids often display beautiful colors to one another and build elaborate nests to attract a mate. A male stickleback will build a nest, drive all other males away, and perform an elaborate dance that shows passing females where his nest is.

Jawless Fishes

The jawless fishes alive today are divided into two classes: lampreys and hagfishes. Although modern jawless fishes are thought to have evolved from heavily armored, bony ancestors, both lampreys and hagfishes have no bones at all. In fact, they are the only vertebrates that do not have backbones as adults! Instead, their long, snakelike bodies are supported by a notochord.

LAMPREYS Lampreys are typically filter feeders as larvae and parasites as adults. An adult lamprey's head is taken up almost completely by a circular sucking disk with a round jawless mouth in the center. Adult lampreys live by attaching themselves to fishes (and occasionally whales and porpoises) and scraping away at the skin with their large teeth and a strong, rasping tongue. Lampreys then suck up the tissues and

Figure 31-11 Some newly hatched fishes, such as salmon, are nourished by a yolk sac on their belly.

Figure 31-12 Modern jawless fishes are divided into two classes: lampreys (top) and hagfishes (bottom).

BACKGROUND INFORMATION
DETECTION OF ELECTRICITY BY FISHES

The muscles of all animals create small electric currents as they operate. For that reason, every time an animal moves, even slightly, it generates electric currents. This means that even cleverly camouflaged animals give themselves away by the electric currents created as they breathe.

Sharks are known for their ability to detect electricity. This is useful to a shark when attacking prey because, just before it strikes, the shark must lift its head in such a way that for a moment it cannot see. At this time, the shark relies on its electric sense to aim its jaws. It is interesting to note that when a shark attempts to strike at dead meat, it usually misses—because the electric cue is not there.

Some freshwater species have an electricity detection mechanism that is even more elaborate than that of sharks. These animals not only detect electric fields—they also create them. They can detect the presence of nearby fishes and other objects by sensing the changes that occur in their own electric fields. These fishes also use this sensing apparatus to navigate through cloudy water when vision would be useless and to "talk" to other members of their species by modulating their individual fields.

Content Development
Discuss with students the different methods of reproduction in fishes. Emphasize that fishes can be oviparous, ovoviviparous, or viviparous.
- **How do oviparous fishes reproduce?** (They lay eggs.)
- **Are all oviparous fishes alike in the way they release and care for their eggs? If not, what are the differences?** (No. Some fishes release hundreds or millions of eggs but do not take care of them; others release fewer eggs but care for them by building nests or by carrying them until they are ready to hatch.)
- **What pattern of reproduction is found in fishes that are ovoviviparous?** (The eggs develop inside the female's body and fertilization is internal. The eggs are nourished by food stored in the yolk sac.)
- **What type of reproduction is found in fishes that are viviparous?** (The young are actually nourished by the mother's body as they develop within. They are born live.)

body fluids of their host. Lampreys rarely kill their host, but they leave it in a weakened condition with a large open wound that is easily infected.

HAGFISHES Hagfishes are probably the most primitive vertebrates alive today. They have pinkish-gray wormlike bodies and four or six short tentacles around the mouth. Hagfishes lack eyes, although they do have light-detecting regions scattered around their body. They feed on dead and dying fish by using a toothed tongue to scrape a hole into the fish's side. Hagfishes have some peculiar traits: They secrete incredible amounts of slime, have six hearts, possess an open circulatory system, and regularly tie themselves into half-knots!

Sharks and Their Relatives

The class Chondrichthyes (cahn-DRIHK-theez)—which contains sharks, rays, skates, and a few uncommon fishes such as sawfish and chimaeras—is an ancient and successful group. The name of this class (*chondros* means cartilage; *ichthys* means fish) refers to the fact that all members have an endoskeleton built entirely out of cartilage. Most of them also have toothlike scales covering the skin. These scales make sharkskin so rough that it is possible to use it as sandpaper.

Most of the 225 living shark species have large curved tails, torpedo-shaped bodies, and rounded snouts with a mouth underneath. One of the most noticeable characteristics of sharks is their enormous number of teeth. A typical shark has about 3000 teeth arranged in 6 to 20 rows. As teeth in the front rows are worn out or lost, new teeth are continually replacing them. A shark goes through about 20,000 teeth in its lifetime!

Figure 31–13 Cartilaginous fishes include sharks and rays. The wobbegong, or carpet shark, is a bottom dweller that feeds primarily on fishes (top). The leopard shark (right) is one of the most attractive sharks. Its teeth are adapted for crushing the shells of mollusks and crustaceans. The underside of some rays seems to have an almost human face (left).

You have probably heard a lot about shark feeding habits. But contrary to what you may have heard, not all sharks attack people. Some sharks are filter feeders; other have flat teeth adapted for crushing the shells of mollusks and crustaceans. And although there are a number of carnivorous sharks large enough to prey on humans, each year more people are killed by lightning than by sharks.

Unlike sharks, which are adapted for swimming rapidly through the water, rays and skates are adapted for living on the ocean floor. Rays and skates are flattened from top to bottom (you can think of them as squashed sharks), and they swim by flapping their large winglike pectoral fins. Most rays and skates reach a maximum length of about 1 meter, but some, such as manta rays, are up to 7 meters in length.

Bony Fishes

Bony fishes make up the class Osteichthyes (ahs-tee-IHK-theez) (*oste-* means bone). There are more species in this class than in any of the other vertebrate classes. About 40 percent of all vertebrates are bony fishes. Experts estimate that there are somewhere between 15,000 and 40,000 species alive today.

Almost all bony fishes belong to the enormous group called the ray-finned fishes. This group includes everything from guppies to groupers, salmon, and eels. The name ray-finned refers to the thin bony spines, or rays, that are connected by a thin layer of skin to form the fins. These fins are adapted to a wide variety of functions. Stonefishes, scorpionfishes, and lionfishes have fin rays that are modified into poison spines. Flying fishes, on leaping from the water, can glide with winglike pectoral fins. Mudskippers, which spend a lot of time out of the

Figure 31–14 Bony fishes come in a wide variety of forms and colors. The porcupine fish can inflate itself into a prickly ball when threatened (top). The moray eel has a narrow snakelike body (center). The bright colors of angelfish may be a means of communication within its species (right). The hawkfish's narrow snout enables it to pluck bits of food from crevices (left).

ECOLOGY NOTE
LAMPREYS

In the early nineteenth century, engineers dug a canal to connect Lake Ontario and Lake Erie and allow the passage of ships around Niagara Falls. Unfortunately, the canal also removed a natural barrier to sea lampreys, which swim into fresh water to mate and lay eggs. Soon the sea lampreys invaded Lakes Erie, Michigan, Huron, and Superior. With few natural enemies, the lampreys multiplied rapidly while preying on several important freshwater fishes such as lake trout. As a result, many species of fishes native to the Great Lakes were nearly eliminated.

The lamprey population in the Great Lakes was finally brought under control, but not without the expenditure of a great deal of time and money—a classic case of how human endeavors can have unexpected and potentially disastrous effects on the environment.

FACTS AND FIGURES

The largest living fish is the whale shark, which can grow up to 20 meters long and weigh up to 14 tons. The smallest fish is the pygmy goby, which measures less than 1.5 centimeters.

---

a shock absorber.

In humans, cartilage is present in the external ear and in parts of the nose; it is also present in the intervertebral disks, which cushion the vertebrae, and at the ends of bones at many joints. Most vertebrates have cartilage skeletons when they are embryos; these are replaced by bone during development.

Content Development
Point out to students that unlike most bony fishes, sharks do not have swim bladders. For this reason, it was once believed that sharks would sink if they did not keep swimming. Modern research, however, has revealed some interesting information about how a shark manages to keep afloat. It was discovered that sharks store fats and oils in very large quantities in their liver. Students may have noticed that oil floats on water because it is less dense than water. Thanks to this "storage tank" of lightweight fats and oils, a shark is able to offset some of its body weight and remain afloat at the same water level.

MULTICULTURAL STRATEGY

The Japanese admire the *koi*, or carp, for the way it fights upstream and its determination in overcoming obstacles. An old custom in Japan and Hawaii is for parents of sons to fly a colorful, carp-shaped kite on "Boy's Day," May 5. The kite symbolizes the hope that a son will grow to be steadfast and successful.

The Japanese also value the *kingyo*, or goldfish, for its graceful appearance and brilliant color. Goldfishes are bred both as a hobby and for commercial use and are kept in elaborate aquariums or artificial ponds.

Ask students if they find qualities to admire in certain fish. Point out that fish have lent their names to several sports teams, from the Hiroshima Toyo Carp to the Florida Marlins.

Figure 31–15 A few fishes manage quite nicely out of water for brief periods of time. African lungfishes (top) get most of their oxygen from the air, which they gulp into a simple sac that serves as a lung. Mudskippers (bottom) climb out of the water onto logs and rocks. As you can see, the mudskipper's bulging eyes are quite mobile, enabling it to appear as if it has eyes on the back of its head.

water, have fins that have evolved into a suction cup that the fish uses in climbing. Triggerfishes have dorsal fins that are usually folded but can be locked in an upright position to help wedge the fish into a hiding place.

Only seven living species of bony fishes are not classified as ray-finned fishes. These are the lungfishes and the coelacanth. These fishes are of interest because they give us an idea of what the lungs and limbs may have been like in the ancestors of terrestrial vertebrates.

The six species of lungfishes alive today are found in Australia, Africa, and South America. The African and South American species live in areas that are flooded during the rainy season but are practically baked during the dry season. When water is available, lungfish use their gills to eliminate carbon dioxide, but they get most of their oxygen by gulping air into a simple sac that functions as a lung. During the dry season, lungfish burrow in the mud and enter a dormant state.

The single species of coelacanth (SEE-lah-kanth) alive today, *Latimeria*, is the only surviving member of the lobe-finned fishes, which were quite common in Devonian times. Unlike ray-finned fishes, which have many bones in the bases of their fins, coelacanths have few bones in their fin bases. Several of these bones are clearly homologous to the limb bones of terrestrial vertebrates. Attached to those bones are a few large rays that form the fins. Ancient lobe-finned fishes seemed to have lived in swampy areas where shallow pools alternated with mud flats and sand bars. Like some of the "walking catfishes" alive today, those lobe-finned species probably used their pelvic and pectoral appendages to move from pool to pool. Unlike its predecessors, the modern coelacanth lives in water about 70 to 400 meters deep in a relatively small area of ocean off the western coast of Africa. However, scientists have observed captive coelacanths "walking" on the bottom of their tank, moving their stubby fins in the same way terrestrial vertebrates move their legs.

Coelacanths were thought to have disappeared with the dinosaurs about 70 million years ago. In 1938, however, fishermen sailing in the ocean off the coast of South Africa caught a strange blue fish that was 1.5 meters long and had stubby fins and a triple tail. The coelacanth was not extinct after all! Scientists were enormously excited to find living coelacanths because these animals represent a fascinating piece of evolutionary history: Coelacanths are the closest thing we know of to the ancestors of all land vertebrates.

31–1 (continued)

Content Development

Emphasize to students that bony fishes are divided into three subclasses: ray-finned fishes, lungfishes, and coelacanths. Point out that by far the largest of these subclasses are the ray-finned fishes, which include all but seven species of bony fishes.

Explain that lungfishes and coelacanths are important to biologists because they provide insight into the evolution of land-dwelling vertebrates.

• **What two structures in these fishes are especially important in studying the evolution of land-dwelling vertebrates?** (Fin bones that are homologous to limb bones and lungs.)

Reinforcement/Reteaching

Have students stretch their arms out horizontally and notice how the arms extend from the chest area. Explain that this chest area is called the pectoral region. Point out that paired fins called pectoral fins develop in this same

SCIENCE, TECHNOLOGY, AND SOCIETY ISSUE

Where Have All the Fishes Gone?

For most of history, humans have eaten fishes, a good source of protein that is both delicious and healthful. Even if you don't eat fish directly, you probably benefit from fish. Why? Because much of the chicken sold today is fed dried ground-up fish meal.

In the past, rivers and oceans seemed to be inexhaustible sources of food to satisfy the human appetite. But recent years have seen dramatic drops in the catches of fish and shellfish. A century ago, for example, salmon was so plentiful that it was dirt cheap. In fact, hired household servants often demanded a contract stating that they wouldn't be forced to eat this "trash fish" more than three times a week! Today, so few salmon survive that it has become an expensive food. Halibut, another delicious fish that was once common, is now so rare that it is almost never seen in restaurants or stores. And lobsters, so common in colonial times that Native Americans used them to fertilize their crops and bait their hooks, are so scarce today that they often sell for more than $7.00 a pound.

What has happened? Pollution and development have fouled water habitats. In many places, people catch fishes and shellfish faster than the remaining animals can reproduce. There is also an increasing demand for fish as scientists discover the healthful benefits of including fish in the diet. As a result, people

now work harder and harder to catch fewer and fewer fish. Both fish-eaters and the fishing industry are in trouble.

Part of this problem is due to a lack of scientific information about fishes. How many fishes live in the sea? What conditions do fish populations need to remain healthy? How quickly do fish grow? How quickly do they reach the age when they can reproduce? Answers to these and other questions are essential in judging how many fishes can be harvested without driving species of fish to extinction. People who manage the nation's fisheries are struggling to gather information before it is too late. They must also determine how to help commercially important fish species to survive while preserving the economic viability of the fishing industry. What do you think should be done?

How Fishes Fit into the World

With representatives in almost every body of water on our planet, fishes are vital parts of many biological systems. For many birds and mammals—including seagulls, raccoons, dolphins, and humans—fishes are important foods. As predators and herbivores, fishes help control the populations of the organisms they eat. Humans have learned to take advantage of this aspect of fish biology. Mosquito fish have been introduced into ponds and lakes in places far from their native home in North America because they consume large amounts of insect larvae. And grass carp and cichlids are used to keep waterways

---

SCIENCE, TECHNOLOGY, AND SOCIETY ISSUE

Where Have All the Fishes Gone?
Discuss this STS from a historical perspective. Read and translate into modern English this eighteenth-century account of lobster abundance in New England.

Lobsters in plenty in most places, very large ones, some being twenty pound in weight. These are taken in low water amongst the rocks. They are very good fish, the small ones being the best; their plenty makes them little esteemed and seldom eaten. The Indians get many of them every day for to bait their hooks withal and to eat when they can get no bass.

Have students discuss why this account is no longer true.
• **Why are lobsters no longer as plentiful as in colonial times?** (Because of coastal development, pollution of estuaries, and extreme overfishing.)
• **How can we replenish the lobster population?** (Students may suggest lobster farming, reduced lobster fishing, and antipollution efforts.)

Emphasize that many people earn their living from fishing and have invested thousands of dollars in fishing boats and other equipment. You may want to compare the fishing industry's troubles with those of the logging industry in the Pacific Northwest.

---

area of fish bodies.
• **What structures in land-dwelling vertebrates do pectoral fins eventually give rise to?** (Arms.)

Now ask students to place their hands just below their hips and note where the leg enters the pelvis. Explain that this area of the body is called the pelvic region. Point out that fish have paired fins called pelvic fins in the pelvic region of their body.
• **What structures in land-dwelling vertebrates did pelvic fins eventually give rise to?** (Legs.)

Enrichment

Have interested students work as a group to learn about the fishing industry in the United States. Encourage students to find out about the history of the fishing industry as well as the various types of fishes that are caught and sold in different parts of the United States today. Have students present their findings to the class in the form of an oral report. Also ask the group to prepare a map that shows where various types of fishes can be found.

TEACHING SUPPORT

Study Guide
- Section 31–1, p. 295

Figure 31–16 *For centuries, people have kept fishes for food and for pets. Selective breeding has resulted in strange-looking varieties of goldfish, such as bubble-eyes and lionheads.*

31–1 SECTION REVIEW

1. What are the main characteristics of vertebrates?
2. Discuss ways in which fishes are adapted for three different modes of feeding.
3. Name and describe three different reproductive strategies in fishes.
4. Describe the three basic groups of fishes. Give an example of a member of each group.
5. **Critical Thinking—Relating Concepts** Why are lobe-finned fishes and lungfishes important to evolutionary biologists?

Guide For Reading
- What is an amphibian?
- How does a typical amphibian carry out essential life functions?
- What are some physical characteristics of salamanders, frogs, and toads?

31–2 Amphibians

With about 4000 living species, **amphibians** are the smallest major group of vertebrates. Despite their small numbers, amphibians are a varied group. They range in size from tiny tropical tree frogs 1 centimeter long to enormous salamanders 170 centimeters long. Some amphibians, such as salamanders and newts, have long tails and scuttle about on four legs. Others, such as frogs and toads, have no tails and leap from one place to another with large hind legs. Still others have no legs at all and burrow in soil like giant worms.

Amphibians are to the animal kingdom what mosses and ferns are to the plant kingdom. They are descendants of ancestral organisms that evolved some—but not all—of the adaptations necessary for life on land. Although many of these animals spend a great deal of time on land, nearly all of them are restricted to moist areas, and most of them must return to water to breed.

What Is an Amphibian?

The name amphibian refers to the double life that most amphibians lead (*amphi-* means both; *bio-* means life). Amphibian larvae are fishlike aquatic animals that breathe through gills, whereas adult amphibians are terrestrial carnivores that breathe through lungs and skin. Of course, there are exceptions: Some amphibians are completely terrestrial; others are

---

31–1 (continued)

SECTION REVIEW 31–1

1. Like other chordates, vertebrates have a notochord, hollow dorsal nerve cord, and pharyngeal slits at some point during their development. In most vertebrates, the notochord is replaced by a vertebral column. Other characteristics of most vertebrates include paired appendages, a closed circulatory system with a ventral heart, and either gills or lungs.
2. Carnivorous sharks have sharp serrated teeth for cutting meat; herbivorous fishes usually have long intestines that help break down tough plant materials; and certain male anglerfishes are parasites whose jaws and teeth are adapted for attaching to a female. Students' answers may vary.
3. Oviparous: lay eggs. Ovoviviparous: eggs develop inside body of female but are not nourished directly by the female's body. Viviparous: embryos develop in and are nourished by the mother's body. Some students may discuss reproductive strategies in specific fishes.
4. Jawless fishes: lack jaws and paired fins, most primitive; lamprey. Cartilaginous fishes: skeleton of cartilage; shark. Bony fishes: skeleton of bone; salmon. Students' examples may vary.
5. They are similar to the ancestors of land vertebrates.

Reinforcement/Reteaching

If students have trouble answering the Section Review questions, go back to the material with which they are having difficulty.

Closure

Review with students the distinguishing characteristics of vertebrates. Do this by having students complete these sen-

Figure 31–17 *Representatives of the three orders of living amphibians—salamanders, frogs and toads, and legless amphibians—are shown in these photographs. The tree salamander, which lives in the rain forests of Central America, can cling to branches with its prehensile tail (left). The leaf frog, an inhabitant of the Amazon rain forest, is as good at climbing trees as it is at hopping (right, top). Legless amphibians (right, bottom), which are found in the tropics, prey on insects and other small animals they encounter as they tunnel through fallen leaves and under ground.*

Figure 31–18 *Most amphibians spend the first part of their life in water as gilled larvae (top). As adults, they usually live on land (bottom).*

completely aquatic. But the majority of amphibians live in water for the first part of their life and on land as adults.

The aquatic larva is one reason that most amphibians must live in moist areas. However, it is not the only reason. Another reason amphibians are strongly tied to water is that their eggs do not have a shell. Thus they tend to dry out unless they are laid in water or in very moist places. Yet another reason is that amphibian skin does not bear scales, fur, or any other protective covering that would help prevent drying out. (However, the skin does contain mucous glands whose secretions help keep the skin moist.) In addition, the skin of almost all adult amphibians is used in respiration and thus must remain moist. If the skin dries out, most amphibians will suffocate.

Amphibians can be defined as vertebrates that are aquatic as larvae and terrestrial as adults, breathe with lungs as adults, have a moist skin that contains many glands, and lack scales and claws. This functional definition is not perfect. As you will soon discover, there are exceptions to almost every "rule" about amphibians.

ESL STRATEGY

Have students photocopy pictures of different fishes and amphibians from an encyclopedia or magazine. Have students make flashcards for each fish and amphibian, writing the name of the organism on one side and taping its picture to the other.

---

followed two major evolutionary pathways—one became the cartilaginous fishes and the other became the bony fishes.

TEACHING STRATEGY 31–2

Focus/Motivation
Write the word *amphibian* on the chalkboard. Underline *amphi-* and *bi-*.

• **In this word, *bi-* represents *bio-*. What is the meaning of *bio-*?** (It means life.)

• **What is the meaning of the prefix *amphi-*?** (It means both.)

• **How would you describe an amphibian based on the meanings of these word parts?** (Answers may vary. The literal meaning is "both life"; students should infer that in some ways, amphibians lead a double life.)

---

tences: "A vertebrate is . . ." and "A vertebrate animal has . . ." Some possible answers include: A vertebrate is a chordate; a vertebrate is an animal in which the notochord has been replaced by a vertebral column; a vertebrate animal has two sets of paired appendages; a vertebrate animal has gills or lungs for breathing.

Next, have students make diagrams to illustrate the evolution of fishes. Students should show that the first fishes were jawless and that they developed into three basic evolutionary lines. One line consisted of jawless fishes that had little armor—these became the ancestors of the lamprey and hagfish. Another line was armored but was ultimately an evolutionary "dead end." The third line developed the adaptation of jaws. These jawed fishes

BACKGROUND INFORMATION
ANCIENT AMPHIBIANS

A number of early Permian animals are shown in Figure 31–19. From left to right: The fin-backed *Dimetrodon*, a reptile, stalks a *Cacops*, a relatively small (60 cm long) amphibian. The points on the boomerang-shaped skull of the meter-long *Diplocaulus*, here swimming in a stream, are used to thump rivals. A larval *Eryops* swims near the shore on the opposite side of the stream, where an adult *Platyhystrix* has captured a young *Diplocaulus*. *Platyhystrix's* multi-colored fin, like that of the *Dimetrodon*, was probably used to attract mates. In the right-hand corner, a rhipidistian fish comes up for a breath of air. This fish is quite similar to the ancestors of the amphibians.

BACKGROUND INFORMATION
AMPHIBIAN NOMENCLATURE

The terms *salamander*, *newt*, and *eft* can all be used to describe the same animal. *Salamander* is a more general term and refers to all members of the order Urodela (Caudata). *Newt* refers to certain genera (*Triturus*, *Taricha*, and *Notophthalmus*) of salamanders that return to the water each spring to breed. *Eft* refers to the terrestrial juvenile newts that belong to the genus *Notophthalmus*. Efts are often bright-red in color—an example of warning coloration.

Evolution of Amphibians

Amphibians first appeared around the end of the Devonian Period, about 360 million years ago. They probably evolved from lobe-finned fishes, similar to the modern coelacanth, that had bones in their fin bases and lungs.

Making the transition from water to land is no easy task. Gills are useless. Lungs that expand easily when the body is supported by water tend to collapse under the weight of other organs. Appendages that work fine under water are too weak to hold much weight on land. The result is inefficient movement and a tendency to scrape and damage the skin of the belly when traveling over rough surfaces. Because vibrations in air are weaker than those in water, it becomes difficult to detect sound and movement. And loss of water from the body is a constant danger.

Because natural selection favored individuals that were better able to live on land, early amphibians evolved in ways that surmounted these problems. For example, the bones of the limbs and limb girdles became stronger, permitting the first amphibians to move around on land more efficiently than lobe-finned fishes. The ribs formed a bony cage to support and protect the internal organs. Many early amphibians had scales that protected the skin on their undersides. Ears, which use a membrane to translate sound waves in the air into pressure waves in body fluid, were added to the lateral-line systems. And it is probable that mucous glands, eyelids, and other structures that protect sense organs and other parts of the body from drying out also developed in early amphibians.

Soon after they first appeared, amphibians underwent an adaptive radiation. Some of these ancient amphibians were huge—the largest is thought to have been about 4 meters long. Amphibians became so numerous that the Carboniferous Period (345 to 285 million years ago) is sometimes called the Age of Amphibians. Why were amphibians so successful?

When amphibians started crawling onto land, they entered an environment nearly empty of animal life. Land plants

---

31–2 (continued)

Content Development
Use the Focus/Motivation discussion to introduce the idea that amphibians are adapted to living both on land and in water. Point out that most amphibians are aquatic as larvae and terrestrial as adults, although there are exceptions. Also emphasize that even when living on land, amphibians live in moist areas.

• **What basic life functions of amphibians are dependent on moisture? Why?** (Reproduction because their eggs must be laid in water or damp places; respiration because they breathe partly through the skin, which must remain moist.)

Point out that the class of animals known as amphibians is divided into three orders. The first order is the salamanders; the second includes frogs and toads; and third is the apodans. Explain that apodans, which are commonly called caecilians, are the least familiar amphibians. They are legless animals with tiny eyes and look like large earthworms.

Figure 31–19 *On a spring day during the early Permian Period (about 270 million years ago), ancient amphibians may have swum in streams and basked in the sun much as modern amphibians do today.*

appeared early in the Silurian Period. By the time amphibians appeared during the Devonian, there were well-established forests of mosses and ferns. Arthropods too had gone ashore, and many species of insects had already evolved. This meant that any vertebrates whose legs and lungs allowed them to spend time on land had lots of food and no competitors. The first land dwellers had at their disposal an environment full of empty niches!

The heyday of the amphibians was short-lived, however. Climate changes ultimately caused many of the low, swampy amphibian habitats to disappear. Most of the amphibian groups became extinct by the end of the Permian Period (about 245 million years ago), leaving behind four groups of land vertebrates—reptiles, which evolved from amphibians early in the Carboniferous Period, and three orders of small amphibians.

Form and Function in Amphibians

Living amphibians have evolved many adaptations that help them overcome the problems of living both in water and on land. As we examine the essential life functions in amphibians, we will focus our attention on the structures found in frogs. Keep in mind that although the majority of amphibians perform their life functions in ways that are similar to that of a typical frog, there are a number of species that function in unusual ways.

FEEDING Tadpoles, the larvae of frogs and toads, are typically filter feeders, which devour tiny floating plants and bits of organic matter, or herbivores, which use teeth on their lower jaw to graze on attached algae. Some tadpoles eat so much so quickly that up to half of their body mass is in their guts! Tadpoles are mostly herbivorous, so their long, coiled intestines are extremely important in helping them break down their hard-to-digest plant food. Tadpoles, of course, have to grow quickly, for those that lag behind may starve or die if their puddle dries out.

BACKGROUND INFORMATION
AMPHIBIAN LOCOMOTION

Tadpoles swim in much the same manner as hagfishes and lampreys—by wriggling their bodies and tails. Since they lack paired fins, tadpoles are not very stable in the water. As tadpoles gradually change into adults, they acquire legs and webbed feet that work like paddles. This greatly improves their swimming ability. As their tails disappear, legs are used more and more until the animals climb out to land.

Salamanders move about on land in a way that combines a kind of swimming movement with walking. Because they do not have fine control of their legs, they move by throwing their bodies into S-shaped curves in order to move their legs around.

Frogs and toads are well-adapted for all kinds of terrestrial locomotion. Real hoppers, such as bullfrogs and leopard frogs, have long, muscular hind legs for jumping. Their front legs help to stabilize them and provide a good landing platform. Tropical tree frogs have miniature suction cups on the ends of their toes. These structures make them excellent climbers and allow them to cling to tiny branches in the rain forest where they live.

BACKGROUND INFORMATION
SALAMANDERS

Salamanders are unusual in that the adults maintain many larval features, including arrested tooth and bone development and the retention of external gills. Some salamanders make their home in caves, where they eat small invertebrates.

Most apodans are blind, and they have no ears. They do, however, have tentacles that help them sense their surroundings. Apodans tunnel through moist soil or live in aquatic habitats in the tropics.

Explain to students that the amphibians in the three orders differ mainly according to their method of locomotion. Salamanders have legs, but they walk with a swimming motion, bending their bodies from side to side. Frogs and toads have strong, well-developed legs that are used for jumping, and, in some species, climbing. Apodans are legless amphibians that must depend on swimming or tunneling for movement.

Another characteristic that distinguishes members of the different orders is the presence or lack of a tail. As wormlike animals, apodans lack a distinct tail. Salamanders have tails throughout their lifetime. Frogs and toads lack tails as adults, but in the larval stage as tadpoles, they do have tails.

MULTICULTURAL STRATEGY

Frogs and toads appear in the fairy tales and folklore of many cultures. In many stories, frogs are enchanted princes—creatures who may be ugly on the outside but who are inherently good underneath. Conversely, these tales typically portray toads in a purely negative way, as unflattering descriptives or as ingredients in witch's brews. One legend says that toads cause warts. (The legend is not true; warts are caused by viruses.)

Have interested students report on how frogs and toads are used in folk tales and other stories.

BACKGROUND INFORMATION
BREATHING IN FROGS

While its lungs are full, the frog keeps expanding and contracting the floor of its mouth cavity. This brings air in and out of the mouth through the nostrils. Some gas exchange takes place across the mouth tissues during this time. The continual movement of air in and out also clears any "stale" air remaining from the last breath. At some point, the glottis and mouth are opened and the lungs empty with a rush. The process then starts over again.

31–2 (continued)

Skills Development
Guided Practice
Skill: Making a chart

Have students make time lines to show the evolutionary development of amphibians. The time lines should show that amphibians appeared around the end of the Devonian Period, about 360 million years ago. The time lines should also show that land plants appeared prior to that time, early in the Silurian Period (about 430 million years ago). Students should show that amphibians were most numerous during the Carboniferous Period, about 285 to 345 million years ago; that most groups became extinct by the end of the Permian Period, about 245 million years ago; that three orders of small amphibians remained; and that reptiles,

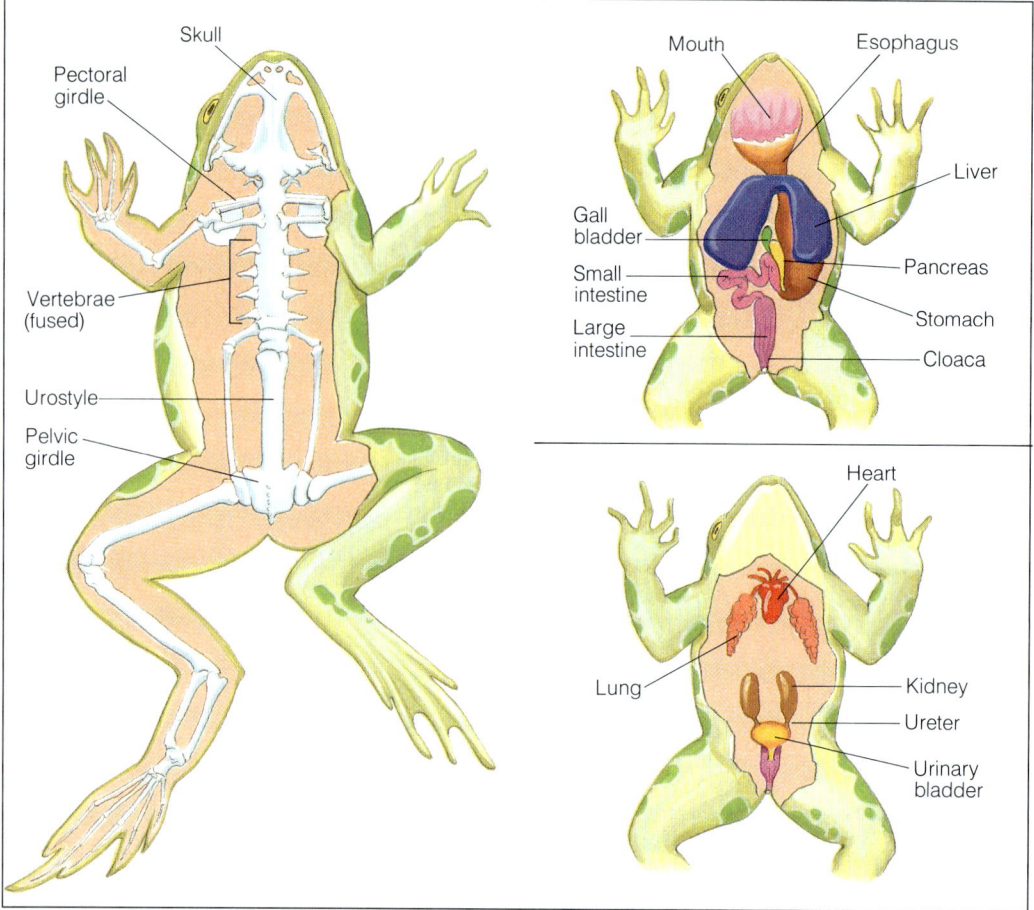

Figure 31–20 *The skeleton and major internal organs of a frog are shown in this illustration. Some organs have been removed or cut away so that as many internal structures as possible can be shown. For example, the fat bodies and part of the liver have been removed.*

696

Adult amphibians are almost entirely carnivorous. Salamanders and legless amphibians can only snap their jaws open and shut to catch prey, but many frogs have a long sticky tongue specialized to capture insects. Figure 31–20 shows the digestive system of a frog.

From the mouth, food slides down the esophagus into the stomach. The stomach connects with the small intestine, where digestive enzymes are manufactured and dissolved food is absorbed. Tubes connect the intestine with organs that also produce digestive enzymes, such as the liver, pancreas, and gall bladder. The small intestine leads to the large intestine, or colon. At the end of the large intestine is a muscular cavity called the cloaca (kloh-AY-kah), which stores wastes until they are expelled.

RESPIRATION Adult amphibians typically breathe using lungs, mouth cavities, and skin. In many amphibians, such as frogs and toads, the lungs are reasonably well developed: Their

which evolved from amphibians early in the Carboniferous Period, also remained.

Once students have completed their time lines, ask:
• **Why was it important to the development of amphibians that land plants appeared early in the Silurian Period?** (By the time amphibians were ready to live on land, plants were well established. These plants became important food sources for the amphibians.)
• **What was the other important food source for amphib-**

internal surfaces are richly supplied with capillaries, as well as with folds that increase their surface area. In other amphibians, the lungs are not as well developed. In fact, many terrestrial salamanders have no lungs at all!

The lining of the mouth cavity and the body skin of many adult amphibians are thin and richly supplied with blood vessels. Thus they can serve as a gas-exchange organ whenever these animals are under water or in extremely moist places. The skin is an important respiratory organ in amphibians—most carbon dioxide is removed through it. And in salamanders and frogs, a large percentage of oxygen is taken in through it. Tadpoles, salamander larvae, and a few types of adult salamanders breathe primarily through their gills. However, they can get rid of excess carbon dioxide through their skin.

Because they do not have the necessary chest and stomach muscles, frogs cannot inhale and exhale as we do. Instead, they fill their mouth cavity with air, close their mouth, and force air back through an opening called the glottis into the lungs. The glottis then closes to keep the air in the lungs for a short time.

Frogs can also direct some of the air they take in to a pair of expandable vocal sacs in the rear of their mouth. Frogs croak by forcing air from these sacs over vocal cords in their throat. By directing air back and forth through the throat between the vocal sacs and the lungs, frogs can even croak under water!

INTERNAL TRANSPORT The circulatory system in adult amphibians is closely linked to the development of lungs. In adult amphibians and other air-breathing vertebrates, the circulatory system forms what is known as a double loop. The first loop carries oxygen-poor blood from the heart to the lungs and takes oxygen-rich blood from the lungs back to the heart. The second loop transports oxygen-rich blood from the heart to the rest of the body and oxygen-poor blood from the body back to the heart.

The amphibian heart has three separate chambers: left atrium, right atrium, and ventricle. Blood returning from most of the body collects in a large vein called the vena cava. The vena cava and other veins draining the head and skin empty into the sinus venosus. The sinus venosus, in turn, empties into the right atrium. Blood returning from the lungs in the pulmonary vein enters into the left atrium.

When the atria contract, they empty their blood into a single ventricle. The ventricle then pumps blood into a single large vessel called the bulbus cordus. The bulbus cordus quickly divides into a series of aortic arches that lead to the major body arteries.

Tadpoles have two-chambered hearts and single-loop circulatory systems, much like bony fishes. In a single-loop system, blood travels from the heart to the gills to the rest of the body and back to the heart. When tadpoles mature into adults, the circulatory system changes into a double-loop system.

Figure 31–21 As air moves between the vocal sac and mouth in male toads it causes the vocal cords to vibrate. Although the resulting sounds may not be music to human ears, a female toad finds them quite attractive.

TEACHING SUPPORT

Teaching Resources
- Activity: Form and Function in Frogs, p. 7
- Hands-On Activity: A Frog of a Different Color, p. 149

BACKGROUND INFORMATION
REPRODUCTIVE ORGANS IN FROGS

Male frogs have a pair of testes that produce sperm. From the testes, sperm pass through a series of ducts into the cloaca. Some frogs have a storage sac, called the seminal vesicle, in which sperm can be stored for a time before mating.

Females have a pair of large ovaries that produce and release eggs. The eggs then pass down long tubes, called oviducts, to a storage area near the cloaca. Before the eggs are released, the oviduct walls surround them with a jellylike yolk.

Sex differences in frogs are almost entirely internal. Therefore, it is difficult to distinguish between male and female frogs in most species simply by looking at them.

31-2 (continued)

Content Development
Emphasize that amphibians undergo a metamorphosis from a larval stage to an adult stage.
- **What is meant by the term *metamorphosis*?** (A change in form.)
- **What type of metamorphosis do frogs undergo?** (They change from tadpoles into frogs.)

Point out that there are many important differences between frogs and tadpoles and that many changes occur as a tadpole grows into an adult frog. Explain that when frog eggs hatch into tadpoles, the tadpoles have short tails, no legs, lidless eyes, and a pair of suckers that enable them to hold onto objects. Three gills develop on the sides of the tadpole's head and a fold of skin develops over the gills. As would be expected from their form, tadpoles live in water.

Soon the suckers disappear and a pair of lips with horny teeth appear. Legs develop next, with the hind legs appearing first; then the front legs break through the gill covers. The mouth widens, and it develops a tongue and strong jaws. Lungs develop, as well as openings to the nasal cavity. The gills and tail disappear. The tadpole has become a frog that is adapted to living on land.

Explain that most tadpoles are herbivorous filter feeders that feed primarily on algae.
- **How is the feeding mode of tadpoles different from that of**

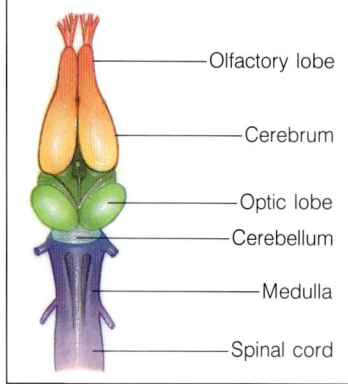

Figure 31–22 The brain of a frog has the same basic parts as that of other vertebrates, such as fishes.

EXCRETION Amphibians use kidneys to eliminate wastes from their bloodstream. The kidneys are dark-colored oval structures that lie against the dorsal part of the body wall. Structures in the kidneys filter nitrogenous wastes from the blood. The excretory product of the kidneys—urine—travels through tubes called ureters into the cloaca. From there it can be passed directly to the outside or it may be stored in a small urinary bladder.

RESPONSE Amphibians have well-developed nervous and sensory systems. Their eyes are large and often bulge outward from the sides of the head. Amphibians can move their eyes around in their sockets quite well. The surface of the eye is protected from damage under water and kept moist on land by a transparent nictitating (NIHK-tih-tayt-ihng) membrane. This membrane is located inside the regular eyelid, which can also be closed over the eye. Frogs have keen vision for spotting moving insects, but they probably do not see color as well as fishes do.

Although amphibian ears have no external sound collectors, they are often very sensitive. Many frogs and toads use croaks, peeps, and a variety of other calls to find a mate, so hearing is vital to their survival and reproduction. Many amphibian larvae and adults also have a lateral line system like that of fishes that detects water movements.

Amphibians respond to adverse conditions in their environment in a number of ways. Because amphibians do not have an internal mechanism for regulating their body temperature, they deal with seasons that are too hot or too cold by hiding

Figure 31–23 Frogs escape predators in many ways. The Amazon horned frog is almost invisible as it hides among dead leaves (top). Bullfrogs have long muscular hind legs that enable them to quickly leap away from enemies (right). Tree frogs live among the concealing leaves of plants (left). The miniature suction cups on the ends of their toes make them excellent climbers.

698

in a sheltered spot, such as an underground burrow, and entering a dormant state.

Predators might seem to be a major problem for clawless, soft-skinned amphibians. However, amphibians have many ways of protecting themselves. Some amphibians hide; others run away quickly; still others are well-camouflaged. Many produce distasteful or toxic chemicals that are secreted in their mucous coatings. The more toxic amphibians usually have warning coloration—bright patterns that tell potential predators that the brilliantly colored animal is poisonous or otherwise dangerous. In addition, some poisonous amphibians respond to a threat by waving their tail or freezing in a pose that shows off their warning colors.

REPRODUCTION When frogs reproduce, the male climbs onto the female's back and squeezes. In response to this stimulus, the female releases as many as 200 eggs, which the male then fertilizes. The embryos are surrounded with a sticky, transparent jelly that attaches the egg mass to underwater plants and nourishes the developing embryos. The eggs typically hatch into tadpoles after one to three weeks.

Not all amphibians have external fertilization and are oviparous like frogs. Many have internal fertilization and are either oviparous, ovoviviparous, or viviparous. Some salamanders have an unusual form of internal fertilization in which the male never needs to come into direct contact with the female! Instead, the male deposits a packet of sperm on the ground and through an elaborate courtship dance persuades the female to pick up the sperm packet with her cloaca.

Figure 31–24 Frogs typically go through an aquatic and a terrestrial phase during their life cycle. Frog eggs are fertilized externally (left) and generally develop in water (right). In most species the fertilized eggs hatch into aquatic larvae, or tadpoles (center). The tadpoles gradually grow limbs and lose their tails as they develop into terrestrial adults (top).

BACKGROUND INFORMATION
AMPHIBIANS AND TEMPERATIURE CHANGE

Many amphibians living in the temperate zone enter a dormant state, known as hibernation, during long cold winters. In preparation for this period of dormancy, the animals store energy in bodies of fat located just above the kidneys and within the livers. Then they bury themselves in the mud beneath ponds and streams, slowing their metabolism down until the warm weather comes in the spring.

Amphibians who live in places where summers are hot and dry avoid drying out by a process called estivating. When their water holes dry down to mud flats, these tropical species dig deep burrows and coat the inside of their new homes with a mixture of mucus and dead skin. Inside these watertight burrows, some species can live in a kind of suspended animation until rain comes again. A few species have been known to estivate for up to two years.

BACKGROUND INFORMATION
FROG EARS

Frogs and toads have large tympanic membranes on either side of the head. These membranes are connected to the inner ear by a long bone called the stapes, which conducts sound. The frog's inner ear can distinguish many different sounds.

adult frogs? (Tadpoles are herbivores. Adult frogs are basically carnivores, and they feed by using their tongues and jaws.)
• **The digestive system of the tadpole is different from that of the adult frog in that the intestine is much longer. Why is this so?** (A longer intestine is needed to digest plant material, which is harder to digest than meat.)

The circulatory system of a tadpole is different from that of the adult frog, too. Tadpoles have a single-loop circulatory system with a two-chambered heart. When the tadpole matures, the circulatory system becomes a double-loop system with a three-chambered heart.
• **This change in the circulatory system is related to the development of a particular organ in the adult frog. What is this organ, and how is it linked to a double-loop system?** (The lungs. The double-loop system develops so that one loop of the system can transport blood back and forth between the lungs and heart while the other loop transports blood back and forth between the heart and the rest of the body.)

Figure 31–25 *Salamanders are usually terrestrial as adults (left). However, certain species of salamanders, known as newts, are terrestrial only during an immature phase known as a red eft (right).*

Parental care in amphibians varies even more than their methods of fertilizing eggs. As in other animals, parental care is a way of ensuring that more young will survive. In many amphibians, it may also be an adaptation to a lack of suitable bodies of water in which aquatic larvae can develop. Some frogs incubate their young in their mouth, vocal sac, or stomach. Male midwife toads wrap sticky strings of fertilized eggs around their hind legs and carry them about until the eggs are ready to hatch. The Surinam toad and marsupial frog have special structures on their back in which the young develop. And in certain tree frogs, tadpoles cling to their parent's back with a suckerlike mouth and are carried between pools of rainwater that collect among the leaves of certain plants.

Salamanders

These amphibians keep their tails even as adults. Both adults and larvae are carnivores. Although many fossil salamanders were more than 3 meters long, most modern salamanders are about 15 centimeters long. Most hatch as fully aquatic larvae with gills. As adults they live in moist woods, where they tunnel under rocks and rotting logs.

Some salamanders, such as the mud puppy and the axolotl (AK-soh-laht'l), never lose their gills and live in water all their life. Some newts, like the crimson-spotted newt, switch back and forth between water and land. Starting as aquatic larvae, they emerge and live entirely on land in a form called the red eft. After a year or two, the red eft changes its colors to green with red spots and returns to the water to breed.

Frogs and Toads

Most of us are familiar with the frogs and toads that live all over the United States. Their mating calls fill the night air in many parts of the country. Of these animals, frogs are more closely tied to water. Even adult frogs spend much of their time in or near ponds and streams. Adult toads, on the other hand, often live in moist woods. Some toads have even managed to invade relatively dry places by using the water permeability of

---

SECTION REVIEW 31–2

1. Amphibians are vertebrates that are aquatic as larvae and terrestrial as adults, have a moist skin that contains many glands, and lack scales and claws.
2. As adults, most have lungs for breathing air and legs for walking or jumping on land. Most are restricted to places with ample supplies of water because their eggs and larvae need to develop in water. In addition, most amphibians have skin that must remain moist.
3. Feeding: adults are carnivorous, many with a sticky tongue for catching prey; herbivorous or filter-feeding tadpoles have a long intestine for digesting plant matter. Respiration: adults use the skin, the lungs, and the lining of the mouth in gas exchange; tadpoles use gills and skin. Internal transport: closed circulatory system, double-

their skin to good advantage. These animals burrow deep into moist soil and press their skin against the walls of their burrows. Their skin then functions just like the root hairs of plants. Because of osmotic pressure, water moves out of the soil into the toad.

Many toads and frogs produce potent toxins. For example, from glands behind its head, the marine toad can squirt toxins up to 1 meter, blinding or otherwise injuring predators. Some tropical tree frogs make a poison so powerful that it can kill humans and other large animals. Native tribes in the tropics often poison their arrow tips by rubbing them on these frogs. For this reason, these brightly colored amphibians are called poison arrow frogs. See Figure 31-26. One species of poison arrow frog produces a toxin so powerful that 0.00001 gram can kill an adult human.

How Amphibians Fit into the World

Adult amphibians prey on animals that have been most abundant for millions of years: insects. As tadpoles, amphibians devour large quantities of algae, powerfully affecting the energy balance of many bodies of water.

Humans do not often interact with amphibians, although frog legs are considered a delicacy by some. In tropical rain forests, native hunters tip their arrows with the toxic secretions of the poison arrow frogs. This enables them to kill large animals such as jaguars and deer with small weapons. In laboratories, researchers are studying the action of poison arrow frog toxins for clues to the way in which the nervous system works. Amphibians have also been the subject of studies on regeneration. It is still a mystery why salamanders can regenerate lost limbs, but closely related frogs cannot. By solving that mystery, researchers hope to develop new ways of treating humans who have lost limbs due to accidents or birth defects.

Figure 31-26 Toads are better adapted to life on land than frogs, primarily because their warty skin helps conserve water (top). Although water is plentiful in the tropical forests where poison arrow frogs live, puddles formed by rain last only briefly. Thus, the frogs occasionally give their tadpoles a piggyback ride to a new puddle (bottom).

31-2 SECTION REVIEW

1. What is an amphibian?
2. In what ways are amphibians adapted to life on land? What characteristics restrict most of them to water?
3. How does a frog carry out its essential life functions?
4. How are frogs and toads similar to salamanders? How are they different?
5. Major changes take place in a tadpole's digestive and circulatory systems during metamorphosis. Why do you think these changes are necessary?
6. **Connection—Ecology** In some areas, populations of toads and frogs are decreasing dramatically. What effects might this decrease have on other organisms?

LABORATORY INVESTIGATION

OBSERVING RESPIRATION IN A FISH

Before the Lab

1. At least one day before the investigation, gather enough materials for your class. Plan on up to 6 students per group.
2. Just before the investigation, prepare ice-water and warm-water baths for each group. Do this by filling one large beaker with ice water and another with warm water.
3. Make sure that all the timers work before distributing them to the class.

Pre-Lab Discussion

Explain to students that in fishes, the bones of the head are arranged so that when the mouth opens, the sides of the head move outward. At the same time, the operculae seal off the "back doors" to the gill chambers. These movements together pull in water through the mouth. When the fish closes the mouth, the gill slits are opened, and the sides of the head move back toward the center. These movements push out water over the gills. This push-pull pump is very efficient and allows the fish to keep water moving steadily over the gills.

Have students read the Problem at the beginning of this investigation. Ask them to state in the form of a hypothesis how they think temperature will affect the breathing rate of a goldfish. Write several student responses on the chalkboard, then compare them with the results of this investigation during the Analysis and Conclusions section.

Ask students to read through the Procedure, and then answer any questions they might have. Discuss as a class the best way to time the number of exhalations per minute. You may, for example, want to suggest that one student set the timer and say "go" while another student counts the exhalations. Students may also discuss whether they think it is best to have different members of the group perform each of the four trials for steps 3 and 8.

Review the graphing techniques students need for step 1 of the Analysis and Conclusions. Ask students:
- **What is the independent variable in this experiment?** (Temperature.)
- **On which axis is it placed?** (The horizontal.)
- **What is the independent variable?** (Breathing rate.)
- **On which axis is it placed?** (The vertical.)

Skills Development

Students will use the following skills while completing this investigation.

LABORATORY INVESTIGATION: OBSERVING RESPIRATION IN A FISH

PROBLEM
How does temperature affect the breathing rate of a goldfish?

MATERIALS (per group)

600-mL beaker	graph paper
fish net	timer
goldfish in an aquarium	Celsius thermometer

PROCEDURE

1. Fill the 600-mL beaker about halfway with water from the aquarium containing the goldfish. Remove a goldfish from the aquarium with a fish net and place it in the beaker of water. **CAUTION:** *Be careful when handling live animals.*
2. Place the thermometer in the beaker with the fish. Leave the fish undisturbed for 5 minutes to allow it to get used to its new surroundings. After this 5-minute period, record the water temperature.
3. With the aid of the timer, count the number of times the fish exhales per minute at room temperature. **Note:** *The gill covers at the side of the fish's head open each time the fish exhales.* Record your results.
4. Repeat step 3 for a total of four trials. Calculate the fish's average breathing rate. Record this number.
5. Gently place the beaker containing the goldfish in the ice-water bath your teacher has prepared. This will help simulate the cooler water that a goldfish encounters while swimming in its natural environment. Watch the temperature of the water in the beaker closely. When it is 3 degrees lower than room temperature, remove the beaker from the ice-water bath.
6. Measure the fish's breathing rate at the lower temperature for four trials, calculate the average breathing rate, and record.
7. Gently place the beaker containing the goldfish in the warm-water bath your teacher has prepared. This will help simulate the warm water that a goldfish encounters while swimming in its natural environment. Watch the temperature of the water closely. When it is 3 degrees higher than room temperature, remove the beaker from the warm water bath.
8. Measure the fish's breathing rate at the higher temperature for four trials, calculate the average breathing rate, and record.
9. When the water in the beaker has returned to room temperature, return the goldfish to the aquarium.

OBSERVATIONS

1. Describe how a goldfish breathes.
2. What is the average breathing rate of a goldfish at room temperature? At the lower temperature? At the higher?

ANALYSIS AND CONCLUSIONS

1. Prepare a graph of your results in which temperature is on the horizontal (x) axis and average breathing rate is on the vertical (y) axis.
2. How does temperature affect the breathing rate of a goldfish?
3. According to your graph, is a goldfish's breathing rate affected to the same extent by a 3-degree rise in temperature as by a 3-degree drop in temperature? Explain.
4. What are some possible sources of error in this investigation? Explain.

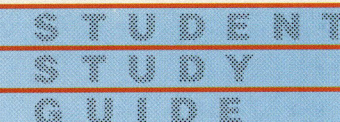

SUMMARIZING THE CONCEPTS

The key concepts in each section of this chapter are listed below to help you review the chapter content. Make sure you understand each concept and its relationship to other concepts and to the theme of this chapter.

31–1 Fishes

- Vertebrates have at some time during their development a notochord, a hollow dorsal nerve cord, and pharyngeal slits. In almost all vertebrates, the notochord is replaced by a vertebral column.
- Fishes are aquatic vertebrates that typically have scales, fins, and pharyngeal gills.
- The first fishes were armored and jawless. Later fishes evolved jaws and paired fins.
- Fishes may be oviparous (egg-laying), ovoviviparous (eggs are incubated inside the mother's body), or viviparous (embryos develop inside the mother and are nourished directly by the mother's body).
- Jawless fishes—lampreys and hagfishes—are eellike parasites and scavengers that lack paired fins, scales, and a vertebral column.
- Members of class Chondrichthyes have a skeleton of cartilage. Cartilaginous fishes include sharks, rays, sawfish, and chimaeras.
- Members of class Osteichthyes, or bony fishes, have skeletons of bone. Almost all living bony fishes are ray-finned fishes.

31–2 Amphibians

- Amphibians are vertebrates that have a moist skin with many glands, lack scales and claws, and are typically aquatic as larvae and terrestrial as adults. Most adult amphibians breathe with lungs.
- Amphibian adults and young are typically carnivores. However, tadpoles are herbivores and filter feeders.
- Adult amphibians usually use a combination of lungs, skin, and mouth cavities to breathe, whereas amphibian larvae usually use gills and their skin.
- Adult amphibians have a three-chambered heart and a double-loop circulatory system.
- Fertilization in amphibians may be internal or external. Amphibians may be oviparous, ovoviviparous, or viviparous.
- Salamanders have a tail even after they undergo metamorphosis into the adult form.
- Frogs and toads have large hind legs adapted for jumping. Adult frogs and toads usually lack tails.

REVIEWING KEY TERMS

Vocabulary terms are important to your understanding of biology. The key terms listed below are those you should be especially familiar with. Review these terms and their meanings. Then use each term in a complete sentence. If you are not sure of a term's meaning, return to the appropriate section and review its definition.

31–1 Fishes

vertebral column
jawless fish
bony fish
cartilaginous fish
atrium
ventricle
aorta
olfactory bulb
cerebrum
optic lobe
cerebellum
medulla
oviparous
ovoviviparous
viviparous

31–2 Amphibians

amphibian

703

CHAPTER REVIEW

CONTENT REVIEW

Multiple Choice
1. d   2. c   3. a   4. c
5. a   6. c   7. b   8. d

True or False
1. F   cartilaginous fishes
2. F   Fishes
3. F   medulla
4. F   herbivores
5. F   Oviparous
6. F   carnivores
7. F   Lateral lines
8. F   lobe-finned

Word Relationships
1. structures involved in movement; cloaca is a cavity into which the digestive, reproductive, and excretory tracts empty
2. parts of the circulatory system; glottis is the opening to the windpipe
3. fishes; salamander is an amphibian
4. parts of the brain; pyloric ceca are intestinal pouches
5. vertebrate animals; electricity. Alternatively, terms associated with a discussion of response in fishes; bullfrog. Catfish and sharks have special receptors that sense electricity, whereas bullfrogs do not.

CONCEPT MASTERY

1. Single-loop: Blood travels from the heart to the gills, then travels throughout the body before it returns to the heart. Double-loop: Blood travels from the heart to the lungs, returns to the heart, is pumped to the rest of the body, and finally returns to the heart.
2. Salmon: oviparous. Catfish: oviparous. Porbeagle shark: viviparous. Salmon produce the most offspring during their lifetime, although few survive to maturity. The large number of offspring offset the losses from lack of parental care.
3. Hagfishes: A hagfish is a jawless fish with a long narrow body that lacks paired fins and a backbone; it feeds on dead and dying fishes. Lampreys: A lamprey is a jawless fish with a long narrow body that lacks paired fins and a backbone; it is parasitic as an adult. Cartilaginous fishes: A stingray is a flat-bodied bottom-dwelling fish with a skeleton of cartilage that moves by flapping its winglike lateral fins. Bony fishes: A parrotfish is a brightly colored fish that has a skeleton of bone and is specialized for feeding on coral. Students' examples may vary.
4. Digestive: Intestine becomes shorter. Respiratory: Gills disappear; lungs develop. Circulatory: Becomes a double-loop system; heart becomes

---

CONTENT REVIEW

Multiple Choice

Choose the letter of the answer that best completes each statement.

1. Sharks can detect a drop of blood in a large pool of water using their
    a. lateral lines.   c. optic lobes.
    b. pyloric cecae.   d. chemoreceptors.
2. Animals in which the young are nourished by the mother's body as they develop are
    a. ovoviviparous.   c. viviparous.
    b. oviparous.   d. externally fertilized.
3. At the end of the large intestine of a frog is a muscular cavity called the
    a. cloaca.   c. gall bladder.
    b. pancreas.   d. colon.
4. In most aquatic vertebrates, the notochord is replaced during development by
    a. pharyngeal gills.   c. a vertebral column.
    b. fins.   d. cartilage.
5. Bony fishes make up the class
    a. Osteichthyes.   c. Chordata.
    b. Chondrichthyes.   d. Latimeria.
6. An adult amphibian's heart typically has
    a. one chamber.   c. three chambers.
    b. two chambers.   d. four chambers.
7. Information from a fish's eyes is processed by the part of the brain called the
    a. olfactory bulbs.   c. cerebrum.
    b. optic lobes.   d. cerebellum.
8. Which of the following statements about excretion in a freshwater fish is true?
    a. The kidneys remove salt.
    b. Water is lost from the gills by osmosis.
    c. Wastes are excreted as urea.
    d. The urine is dilute.

True or False

Determine whether each statement is true or false. If it is true, write "true." If it is false, change the underlined word or words to make the statement true.

1. Members of the class Chondrichthyes are also known as <u>jawless fishes</u>.
2. <u>Amphibians</u> are considered to be the most primitive living vertebrates.
3. The <u>cerebrum</u> in the brain of a fish controls many internal organs and maintains balance.
4. Tadpoles are typically filter feeders or <u>carnivores</u>.
5. <u>Viviparous</u> animals lay eggs.
6. Adult amphibians are typically <u>herbivores</u>.
7. <u>Olfactory bulbs</u> are used to detect motion in the water.
8. The coelacanth is a <u>ray-finned</u> fish.

Word Relationships

In each of the following sets of terms, three of the terms are related. One term does not belong. Determine the characteristic common to three of the terms and then identify the term that does not belong.

1. pelvic fin, girdle, cloaca, pectoral fin
2. atrium, glottis, bulbus cordus, aortic arch
3. lamprey, shark, salamander, salmon
4. pyloric ceca, olfactory bulbs, cerebrum, medulla
5. catfish, electricity, shark, bullfrog

CONCEPT MASTERY

Use your understanding of the concepts developed in the chapter to answer each of the following in a brief paragraph.

1. Compare double-loop and single-loop circulatory systems.
2. Pacific salmon die soon after spawning (releasing reproductive cells into the water). Certain catfish incubate their eggs and young in their mouth. Female porbeagle sharks produce infertile eggs that are eaten by the young developing inside their body. Identify each of these fishes as oviparous, viviparous, or ovoviviparous. Which of these fishes probably produces the most offspring in its lifetime? Explain.
3. Give the common name for each of the four living classes of fishes. Briefly describe a specific example of each class.
4. Explain how a tadpole's digestive, respiratory, and circulatory systems must be modified when it undergoes metamorphosis.
5. Some fishes, such as salmon, live in both fresh and salt water. Explain how the excretory systems of such fishes must be adapted to cope with this kind of lifestyle.

CRITICAL AND CREATIVE THINKING

Discuss each of the following in a brief paragraph.

1. **Making inferences** Members of the amphibian order Apoda are typically blind, wormlike, brightly colored, and viviparous. They have two rows of needlelike inward-curving teeth on the upper jaw and one or two rows on the lower. They do not have gills. The head is shaped like a spade; the bones of the skull are thick and sturdy.
   a. Where would you expect to find apodans?
   b. What do apodans probably eat? Would you expect their intestine to be long or short? Explain.
   c. Describe the processes of fertilization and embryo development in apodans. How are the ways apodans reproduce an adaptation to their environment? Explain.
   d. What other inferences can you make about the habits of apodans? Explain.
2. **Relating concepts** When threatened, a certain harmless salamander stands on tiptoe, touches the tip of its tail to the top of its head to display a bright red underside, and remains perfectly still. Explain this behavior.
3. **Relating cause and effect** Certain deep-sea fishes "explode" if they are rapidly brought up from the depths at which they usually live. Using your knowledge of fish anatomy, explain why the fishes explode. (*Hint:* Pressure increases one atmosphere for every 10 meters of sea water.)
4. **Applying concepts** Explain why even the best-camouflaged prey cannot escape detection by a predator that can sense electricity.
5. **Relating concepts** A scuba diver notices that the surface of the water overhead seems light in color, whereas the water below seems quite dark. How might this observation relate to the fact that most fish have countershading (they are dark-colored on the dorsal side and light-colored on the ventral side)?
6. **Using the writing process** Humans depend mostly on vision. Sharks depend mostly on smell and electric sense. Describe a place you know as a shark might sense it.

---

three-chambered.
5. The excretory systems must be able to switch between a saltwater mode, by conserving water and excreting excess salts, and a freshwater mode, by excreting excess water and conserving salts.

CRITICAL AND CREATIVE THINKING

1. a. Most apodans are terrestrial animals that burrow in soil or in leaf litter. They use their shovellike head to push through the soil. b. Apodans are carnivores that feed on insects, worms, and other small animals. Because they are carnivores, their intestines are relatively short. c. Accept all logical answers. All apodans exhibit internal fertilization; they may be oviparous, ovoviviparous, or viviparous; a few have gilled aquatic larvae. The majority of apodans bear their young alive—an adaptation to life on land. d. Accept all logical answers.
2. The salamander is displaying a warning posture. As the problem states, it is a harmless salamander, and students should be able to infer from this that the salamander is just mimicking the coloring and warning behavior of a poisonous species.
3. Gases in the swim bladder expand as pressure decreases, causing this organ to burst. In addition, dissolved gases in the blood come out of solution under rapid decompression, producing air bubbles that can also cause the fish to explode.
4. Nerve impulses are a form of electricity. A predator with electric sense can detect these impulses whether or not it sees its prey.
5. Countershading is a form of camouflage. Dark objects (such as a fish's back) are hard to see against a dark background (such as water seen from above). Light objects (such as a fish's belly) are hard to see against a light background (such as water seen from below).
6. Students' essays should demonstrate an understanding of the concept of different types of senses. They should also be well-written and creative.

CHAPTER 32 Reptiles and Birds

Section	Laboratory Investigations and Activities
32–1 Reptiles pages 707–719 THEMES: Evolution, Unity and Diversity	**Teacher Edition** DEMONSTRATION: Reptiles, p. 706D **Laboratory Manual** Adaptations in Lizards, p. 409 **Teaching Resources** HANDS-ON ACTIVITY: An Eggs-aggeration, p. 155
32–2 The Evolution of Temperature Control pages 720–723 THEMES: Evolution, Stability	
32–3 Birds pages 723–731 THEMES: Evolution, Scale and Structure	**Student Edition** LABORATORY INVESTIGATION: Observing Terrestrial Adaptations of a Chicken Egg, p. 732 **Teacher Edition** DEMONSTRATION: Birds, p. 706D **Laboratory Manual** Examining Bird Adaptations, p. 415 **Teaching Resources** HANDS-ON ACTIVITY: Do Oil and Water Mix? p. 157 HANDS-ON ACTIVITY: Strictly for the Birds, p. 159
Chapter Review pages 733–735	

Teacher Resources

Books

Bakker, Robert. *The Dinosaur Heresies.* William Morrow, 1986.

Bradshaw, S. D. *Ecophysiology of Desert Reptiles.* Academic Press, 1986.

Horner, Jack, and James Gorman. *Digging Dinosaurs.* Workman, 1988.

Mace, Alice, ed. *The Birds Around Us.* Ortho, 1986.

Mattison, Christopher. *Snakes of the World.* Facts on File, 1986.

Raup, David M. *Extinction: Bad Genes or Bad Luck?* Norton, 1991.

CHAPTER PLANNING GUIDE

Other Activities	Media and Technology
Study Guide Section 32–1, p. 305	**Interactive Videodisc** On Dry Land: The Desert Biome **Videodisc** Aquatic Ecosystems: Freshwater Wetlands and Freshwater **Biology Transparencies** 43: The Internal Anatomy of a Turtle 44: The Anatomy of a Snake **Biology Media Guide** pages 42–43
Teaching Resources ACTIVITY: Investigating How Reptiles Regulate the Temperatures of Their Bodies, p. 3 **Study Guide** Section 32–2, p. 308	**Videodisc** Aquatic Ecosystems: Freshwater Wetlands and Freshwater Aquatic Ecosystems: Estuaries and Marine Super Scents **Video/Videodisc** Seeing Sense **Biology Transparencies** 45: The Internal Anatomy of a Bird **Biology Media Guide** pages 43–45
Teaching Resources VOCABULARY REVIEW: Crossword, p. 1 ACTIVITY: Exploring the Body Temperatures of Birds, p. 5 BIOLOGY CASE STUDY: Weather and Nesting Success in Whooping Cranes, p. 11 **Study Guide** Section 32–3, p. 309	**Computer Test Bank** Chapter Test, p. 337
Teaching Resources Chapter Test A, p. 11 Chapter Test B, p. 15	

Audiovisuals

Remarkable Reptiles. Film or video. Marty Stouffer Productions.
Reptiles, rev. Film or video. Coronet.
Birds, rev. Film or video. Coronet.

Birds of the Arctic, Birds of Shore and Marsh, American Bald Eagle, The Great Blue Heron, Birds in Winter, Birds That Hunt. Film or video. Coronet.

What Is a Bird? Film. Encyclopaedia Britannica.

CHAPTER 32 Reptiles and Birds

CHAPTER OVERVIEW

Reptiles are vertebrate animals that have adaptations such as lungs, dry and scaly skin, and amniotic eggs that enable them to live their entire lives out of water. Reptiles entered a period of great adaptive radiation during the Permian Period, when the Earth's surface and climate changed dramatically, making it difficult for water-dependent life to survive. Some reptile species are oviparous, laying eggs that develop outside the mother's body. Other species are ovoviviparous, bearing live young. Virtually all reptiles practice internal fertilization.

Modern reptiles can be divided into four orders: tuataras, lizards and snakes, crocodilians, and turtles. The tuatara retains many of the features of the ancient reptiles from which it evolved. All lizards have legs, clawed toes, external ears, and movable eyelids. Snakes are lizards that have lost their legs during their evolution. Crocodilians are large reptiles that live in tropical climates. Turtles have some sort of shell that covers and protects their bodies.

The control of body temperature is very important for animals, particularly in habitats where temperatures vary widely with the time of day and season. Reptiles are ectotherms, or animals that obtain heat from outside their bodies. Birds and mammals are endotherms, or animals that generate heat inside their bodies. Ectothermy and endothermy are survival strategies that have advantages and disadvantages in different environments.

Birds are endothermic reptilelike animals with an outer covering of feathers, two legs used for walking or perching, and front limbs modified into wings. Fossils confirm the evolution of birds from ancient reptiles.

Bird feathers are generally of three different types: contour, down, and powder. Birds' wings show many variations that involve flight. Hollow bones and large chest muscles are found in flying birds. Birds practice internal fertilization. They lay eggs that are usually incubated until they hatch.

32-1 REPTILES

Section Focus 32-1

The purpose of this section is to describe the physical and behavioral characteristics of the different types of reptiles. Reptiles are vertebrate animals that have special adaptations that enable them to live their entire lives on land. Reptilian skin is dry, leathery, and often covered with scales to prevent water loss. Reptiles produce amniotic eggs that contain a protective environment in which the embryo can develop. Reptiles also have efficient respiratory and circulatory systems to aid their life on land.

Reptiles began their great period of adaptive radiation during the Permian Period. One type of reptile, the dinosaur, appeared during the Triassic Period, about 195 million years ago, and died out at the end of the Cretaceous Period, about 65 million years ago.

Because of their adaptation to many different terrestrial environments, reptiles exhibit numerous variations in structure and behavior. All body systems of the reptile are adapted to a successful life on land. There are four orders of modern reptiles: tuataras, lizards and snakes, crocodilians, and turtles and tortoises.

Performance Objectives 32-1

1. Identify the distinguishing characteristics of reptiles.
2. Describe the structure of an amniotic egg.
3. Describe internal fertilization.
4. Compare the similarities and differences of some specific reptiles.

Science Terms 32-1

reptile   p. 707
amniotic egg   p. 708
transition fossil   p. 709
internal fertilization   p. 714
carapace   p. 718
plastron   p. 718

32-2 THE EVOLUTION OF TEMPERATURE CONTROL

Section Focus 32-2

The purpose of this section is to describe the mechanism of body-temperature regulation. The control of body temperature is an important mechanism for animals, particularly in habitats with a wide variety of daily and seasonal temperatures. With respect to the generation and control of body temperatures, animals can be divided into two basic groups: ectotherms and endotherms.

Modern reptiles are classified as ectotherms, or animals that gain heat from the environment. Ectotherms have low metabolic rates at rest and tend to lose heat to the environment because they have ineffective body insulation.

Birds and mammals are classified as endotherms, or animals that generate and control heat inside their bodies. Endotherms have high metabolic rates at rest to produce body heat and effective insulation to prevent heat loss.

Neither ectothermy nor endothermy is always superior. For different animals in different environments, each mechanism has its advantages and disadvantages. Although most scientists theorize that the first terrestrial vertebrates were ectotherms, there is much uncertainty about when and how the first endotherms evolved.

CHAPTER PREVIEW

Performance Objectives 32-2

1. Define ectothermic and endothermic.
2. Describe the physical characteristics of ectotherms and endotherms.
3. Describe the advantages and disadvantages of ectothermy and endothermy.

Science Terms 32-2

ectotherm   p. 720
endotherm   p. 720

32-3 BIRDS

Section Focus 32-3

The purpose of this section is to describe the physical and behavioral characteristics of birds. Birds are endothermic reptilelike animals with an outer covering of feathers, hind limbs used for walking or perching, and front limbs modified into wings. There are generally three types of bird feathers: contour, down, and powder. Fossils indicate that birds evolved from ancient reptiles.

Because of their high metabolic rates, birds must eat large amounts of food. They eliminate nitrogenous waste in the form of uric acid. The presence of air sacs causes the bird's respiratory system to be very efficient and its body to be buoyant. Hollow bones and large chest muscles are found in flying birds.

Birds practice internal fertilization and lay eggs that are usually incubated until they hatch. Birds serve many important roles, such as pollinators, seed dispersers, insect eaters, and food sources.

Performance Objectives 32-3

1. Identify the characteristics of birds.
2. Describe the three types of bird feathers.
3. Identify the adaptations that enable birds to fly.
4. List the useful functions served by birds.

Science Terms 32-3

bird   p. 724
contour feather   p. 724
down feather   p. 724
powder feather   p. 724
crop   p. 726
gizzard   p. 726
air sac   p. 727

DISCOVERY LEARNING

TEACHER DEMONSTRATIONS
Modeling

Reptiles

The following demonstration can be used as an introduction to Chapter 32.

Obtain mounted skeletons of a snake, lizard, and turtle. After students have examined each model, ask:

• **To what vertebrate group do all these animals belong?** (Students should identify them as reptiles.)

• **In what ways do these reptile skeletons differ from each other?** (Answers will vary, although such distinguishing traits as the turtle's shell and the absence of legs on snakes should be mentioned.)

• **What modifications help each reptile move and protect itself?** (Accept all logical answers.)

Birds

You may want to perform this experiment after students complete Section 32-3.

Before the demonstration, collect and clean some bones from a roasted chicken. Saw the bones in half. Collect and clean bones from another animal, such as beef or pork bones. Boil the bones to remove excess fat and muscle tissue.

Distribute the bones to students.

• **What are the differences between the two kinds of bones?** (Accept all logical answers.)

• **Which bone do you think came from a bird? Why do you think so?** (Accept all logical answers. Lead students to suggest that the hollow bones are lighter and so should be better for flying.)

• **Which bone do you think came from a large animal? Why do you think so?** (Accept all logical answers. Lead students to suggest that the solid bones are stronger and would support a heavier body.)

CHAPTER **32**

Reptiles and Birds

GUIDED ENQUIRY

Pose the following questions to students and have them record their responses. Point out that they will gain a better understanding of the key concepts if they read the chapter with these basic questions in mind. Upon completion of the chapter, pose the questions again. Ask students to compare their initial responses with those they have developed after reading the chapter.

- How did a change in the Earth's environment lead to the rapid evolution of reptiles?
- How are the characteristics of reptiles related to their success on land?
- How is an animal's mechanism for body-temperature control related to other characteristics of the animal?
- Why is an animal's mechanism for body-temperature control important in determining the type of environment in which the animal can survive?
- What is the evolutionary relationship between reptiles and birds?
- How are a bird's characteristics adapted for flight?
- In what ways are birds useful to humans?

INTRODUCING CHAPTER 32

Using the Visuals

Begin by having students read the chapter-opener text and observe the chapter-opener photograph.
- **What kinds of animals do you see in this drawing?** (Two types of dinosaurs and birdlike creatures.)
- **What can you infer about the climate of the region shown?** (It appears to be tropical.)
- **Does the area appear to be sparsely or heavily populated with living things?** (Heavily populated.)
- **How do the sizes of the animals seem to compare with the surrounding landscape?** (Several of the animals look enormous compared with the surrounding landscape.)
- **How do scientists know what dinosaurs looked like if dinosaurs no longer exist?** (Accept all logical answers.)
- **What are some modern animals that resemble ancient dinosaurs?** (Answers may include lizards, crocodiles, alligators, and turtles.)

Point out that land-dwelling reptiles entered a period of rapid evolution as the Earth's climate became drier. Stress that modern reptiles and birds evolved from dinosaurs and their relatives.

CHAPTER **32**

Reptiles and Birds

The fossil remains of the huge dinosaurs provide overwhelming evidence of their existence.

In the heart of a great forest during the late Carboniferous Period, almost 300 million years ago, giant tree ferns and mosses were everywhere. Amphibians were common.

But the Earth became drier, and as a result, new and different kinds of environments became available. Other animals evolved and filled the new niches. The most successful of these animals were the reptiles, which included the awesome dinosaurs. For more than 200 million years, the dinosaurs were masters of the Earth.

In time, however, the dinosaurs became extinct. They left behind fossil evidence of their passing—evidence that never ceases to amaze those who view the silent testimony of the largest animals ever to walk on Earth. But dinosaurs not only left fossils: They also left several evolutionary lines that were to develop into modern reptiles and birds. In this chapter you will learn more about reptiles and birds: how and why they have been able to survive on Earth long after their giant relatives became extinct.

GUIDE FOR READING

After you read the following sections, you will be able to

32–1 Reptiles
- Define reptile.
- Describe reptile evolution.
- Relate the form and function of reptiles to their success in dry environments.

32–2 The Evolution of Temperature Control
- Define ectothermic and endothermic.
- Compare ectothermic and endothermic strategies.

32–3 Birds
- Describe bird characteristics.
- Relate the structure of birds to flight.
- Explain how birds fit into the natural world.

Journal Activity

YOU AND YOUR WORLD We humans tend to believe that warmbloodedness and viviparousness are superior traits. After all, *we* are warmblooded and viviparous. But would reptiles and birds agree with our notions about what traits are best? Explore your ideas in the form of a cartoon, play, or other creative piece in your journal.

32–1 Reptiles

Guide For Reading
- What are the distinguishing characteristics of reptiles?
- What is the evolutionary history of reptiles?
- How do reptiles perform essential functions?
- How are reptiles classified?

When you think of reptiles, you probably think of snakes slithering through tall grass or lizards scurrying up the trunk of a tree. You may think of a tortoise slowly eating a meal of plant leaves or a crocodile floating noiselessly in a pool of dark water. All these animals are reptiles, and they represent three of the four surviving orders of reptiles. The fourth order is represented by a single species, the tuatara. You will learn more about these orders of reptiles later in this chapter.

What Is a Reptile?

Reptiles are vertebrate animals that have lungs, a scaly skin, and a special type of egg—adaptations that enable them to live their entire life out of water. Today **reptiles** are widely distributed over much of the Earth. The temperate and tropical areas on Earth contain populations of reptiles that are remarkably diverse in appearance and lifestyle. The only places on Earth that lack reptiles are very cold areas. This is due to an important reason that will soon become apparent.

In some ways, reptiles resemble amphibians. However, reptiles are better adapted to life on land. For example, reptilian skin is dry and leathery and is often covered with thick, protective scales. This body covering helps prevent loss of body water in dry environments. But the dry, waterproof skin

Figure 32–1 The chameleon, a modern reptile, moves slowly and deliberately, creeping up to its insect prey. Its eyes are able to move independently of each other, so that one eye can guide its movements and the other can sight the unwitting victim.

TEACHING STRATEGY 32–1

Focus/Motivation
Many students are naturally fearful of reptiles. Snakes in particular have a bad reputation and are the subject of much dislike and many misunderstandings.

- **What have you heard about snakes?** (Accept all answers.)
  As students make their statements, discuss whether each is true or false. All false statements should be refuted.
  The characteristics of a reptile can be introduced by obtaining several live specimens, such as a turtle, small lizard, horned toad, and small nonpoisonous snake. Allow students to observe and gently handle the specimens to become comfortable with them.
- **What characteristics do these different animals have in common?** (Accept all logical answers.)
- **What characteristics make these animals well adapted to life on land?** (Accept all logical answers.)

Content Development
Have students identify the characteristics of reptiles. As students describe each characteristic, explain how the characteristic allows the reptile to be more successful as a land dweller.

COOPERATIVE LEARNING

Using preassigned lab groups or randomly selected teams, have groups produce an illustrated time line that shows geological and biological changes that occurred from the Carboniferous through the Cretaceous periods of Earth's history. The divisions of the time line should be Carboniferous, Permian, Triassic, Jurassic, and Cretaceous periods. Using the information in this chapter, groups should use a minimum of two drawings per period to illustrate the geological and biological changes occurring at that time. This activity could also be completed as a jigsaw assignment. Each group member would be responsible for illustrating one period of the time line. See the front-matter essay on Cooperative Learning for variations on the jigsaw technique.

Journal Activity

YOU AND YOUR WORLD Creativity should be rewarded in this Journal Activity. Later in this chapter, students should realize that body temperature and reproductive strategies are "best" if they confer a survival advantage on an organism. Reptiles and birds show remarkable abilities to survive in various environments.

707

TEACHING SUPPORT

Teaching Resources
- Hands-On Activity: An Eggs-aggeration, p. 155

BACKGROUND INFORMATION
THERAPSID EXTINCTION

Many hypotheses have been suggested to explain the sudden extinction of the mammallike reptiles (therapsids). Until recently, many paleontologists thought that dinosaurs competed ecologically with therapsids and drove them to extinction. There are two problems with this hypothesis: First, the earliest dinosaurs had actually been around for a long time before they took over. Why the delay? Second, 150 million years later, it would be the dinosaurs' turn to become extinct. Mammals would undergo their great adaptive radiation and adapt to many niches previously held by dinosaurs. Why the sudden about-face?

Some biologists suspect worldwide climatic changes. A great many marine animals became extinct in the sea at the same time therapsids were disappearing on land. Another group of paleontologists note that this same period saw major changes in Earth's terrestrial plants. Permian plant life was mostly horsetails, ferns, ginkgoes and some early conifers. In the Triassic Period, these plants were replaced by new groups of cycads and conifers. Perhaps the herbivorous mammallike reptiles were not able to change their diets.

Figure 32–2 Unlike most amphibians, reptiles are able to survive quite well in dry environments, as this snake in a cactus patch shows (right). Their survival is due in part to their dry, scaly skin, which must be shed periodically (left).

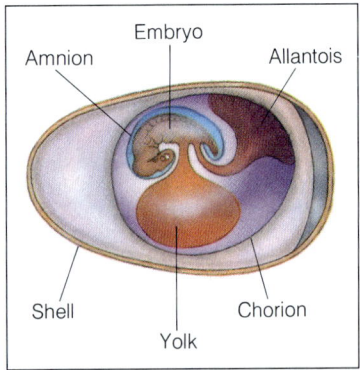

Figure 32–3 The reptile egg shows adaptations to survive the dryness of life on land. This tiny turtle has just hatched and is breaking free of its shell.

can also be a disadvantage to reptiles. The tough, scaly layer of skin does not grow when the rest of the reptile grows, so it must be shed periodically when a reptile increases in size.

Perhaps the most important adaptation of reptiles to life on land is the type of egg they produce. Unlike amphibian eggs, which almost always need to develop in water, reptilian eggs are surrounded by a shell and several membranes that together create a protected environment in which the embryo can develop. See Figure 32–3. In fact, we named these eggs **amniotic** (am-nee-AHT-ihk) **eggs** for one of those membranes. Amniotic eggs are as important to the survival of land animals as the evolution of seeds was to the survival of land plants. In addition to a shell and membranes, amniotic eggs also contain a substantial yolk. The yolk is rich in nutrients that the developing embryo uses until it is ready to hatch. The legacy of an egg adapted to life on land was passed on from early reptiles to their descendants—modern reptiles, birds, and mammals. (Although few mammals lay eggs, they all use the membranes of the amniotic egg, in modified form, in the development of the embryo.)

Another important reason for the success of reptiles on land is the development of a more efficient respiratory system. Remember that most amphibians take in oxygen and give off waste gases primarily through their moist skin. This method of respiration works only as long as the amphibian's skin remains damp. In dry land environments, breathing through moist skin is not an option. The dry skin of reptiles, which prevents water loss, also prevents gases from moving through. To exchange gases with the environment, reptiles have two efficient lungs—or, in the case of certain species of snakes, one lung. All these adaptations contribute to the success of reptiles on land.

32–1 (continued)

Reinforcement/Reteaching
Have students review the characteristics of amphibians and the eggs they produce.

Content Development
Stress that a major advance of reptiles over amphibians is the type of egg they are able to produce.
- **In what way are the eggs of reptiles different from those of amphibians?** (Reptile eggs have a leathery, protective shell.)
- **Why is this covering so important?** (It keeps the eggs from drying out; thus they can be laid on land.)

Point out that the reptile embryo is able to develop completely within the egg before hatching.

Skills Development
Guided Practice
Skill: Making observations
Using the Visuals Have students observe Figure 32–2.
- **What is the reptile in this picture doing?** (Molting or shedding its skin.)

Evolution of Reptiles

Because several fossils show characteristics of both amphibians and reptiles, it is difficult to say exactly when the first true reptiles appeared on Earth. As you have just read, one determining factor that separates living reptiles from amphibians is the type of eggs they produce. But, unfortunately, there are no fossil eggs around for paleontologists to study. Because we cannot tell what kind of eggs these fossil animals laid, they must remain on the amphibian-reptilian borderline—at least for the present time. These animals are often called **transition fossils.** These transition fossils document the slow and steady evolutionary change of amphibianlike ancestors into reptiles over time.

Throughout most of the Carboniferous Period, amphibians greatly outnumbered reptiles. But during the Permian Period, which began about 285 million years ago, the Earth's surface and climate changed dramatically. Mountain ranges such as the Appalachians were formed. The climate became cooler and less humid. Many of the great swamps dried up. These changes made life difficult for a large number of water-dependent amphibians. But such was not the case with the reptiles. It was during this time that they began their great period of adaptive radiation.

One early reptile line developed into a fascinating group of mammallike reptiles that displayed a mix of reptilian and mammalian characteristics. Although these animals were extremely successful at first, they became extinct in just a few million years—nearly overnight on a geological time scale. Toward the end of the Triassic Period, about 195 million years ago, the mammallike reptiles were suddenly replaced in the fossil record by another group of reptiles that had remained in the background for millions of years—the dinosaurs.

Figure 32–4 Although they survived for many millions of years, dinosaurs eventually became extinct. One theory suggests that a meteor struck the Earth, causing the climate to cool. According to this theory, the dinosaurs could not survive this change in temperature.

BACKGROUND INFORMATION
BRONTOSAURUS?

The enormous herbivorous dinosaur known as *Brontosaurus* has been renamed because of an error by the fossil's original discoverers. When the definitive skeleton of this beast was uncovered, its long neck had no head attached to it. Finding a reasonable-looking skull nearby, researchers assumed it belonged to the skeleton they had found and described the animal accordingly.

More recently, however, it has been determined that the head actually belonged to another species. The correct head on the old *Brontosaurus* body has been renamed *Apatosaurus*.

ECOLOGY NOTE
THE USE OF BAIT TO CAPTURE PREY

Students may be familiar with certain species of anglerfishes that use bait to attract prey close enough to be swallowed. Some species of deep-sea fishes have light-producing organs that hang in front of their mouths. These organs act as lures for unsuspecting prey. As the prey approaches the bait, it is suddenly gulped down by the predator!

Certain species of snakes—some vipers and boa constrictors, for example—also use bait to capture prey. Juveniles have brightly colored tails that they dangle in front of small lizards and frogs. These prey animals often mistake the snake's tail for a worm and try to eat it. The snake is able to capture prey that have been lured within striking range. When these snakes grow larger they are able to catch prey without using their tail as a lure. In time the tail fades to a duller color.

During the late Triassic and Jurassic periods, a great adaptive radiation of the dinosaurs, or "terrible lizards," took place. The Triassic Period also saw the appearance of crocodiles and alligators, as well as the first birds.

At the end of the Cretaceous Period, about 65 million years ago, something happened to cause a worldwide mass extinction. Within a few million years, dinosaurs and most other animal and plant groups became extinct. Exactly why this mass extinction occurred is not known for certain. Some biologists think it was caused by the slow process of climatic change that resulted from movements of the continents into their present positions. Other biologists believe that the change in climate that produced the mass extinction occurred more suddenly. Some evidence suggests that a huge meteor struck the Earth, causing an explosion that produced enormous clouds of airborne dust. The explosion may also have caused worldwide forest fires. The smoke from the fires, as well as the clouds of dust, may have blocked the sun's rays, causing Earth's temperature to drop.

Whatever happened at the end of the Cretaceous Period resulted in the death of virtually all the great and terrible lizards. The disappearance of the dinosaurs left open many niches for animals, both on land and in the sea.

Form and Function in Reptiles

Because they have adapted to many different ways of life, reptiles exhibit numerous variations in structure and behavior. Some—for example, turtles, crocodiles, and lizards—move about on four legs. Others move about without legs. Snakes and certain lizards are two examples.

FEEDING There is remarkable diversity in the foods eaten by reptiles and in their modes of feeding. Some reptiles, such as the iguana, are herbivores. Other reptiles are carnivores. Certain carnivorous snakes prey on small animals by grabbing them with their jaws and swallowing them whole. See Figure 32–5. Other snakes live on a diet of birds' eggs. After

Figure 32–5 All snakes are carnivores. Many snakes eat small mammals. Because they are able to stretch their jaws wide, snakes swallow their prey whole. A chameleon obtains its food by flicking out its sticky tongue over a great distance. Any insects within striking range are greedily gobbled up.

32–1 (continued)

Content Development
Point out that the mass extinction of dinosaurs occurred about 65 million years ago. Discuss the different hypotheses that scientists have developed to explain the relatively rapid extinction of the dinosaurs.

Enrichment
Arrange a field trip to a natural history museum that has dinosaur models or skeletons and have students prepare a brief report on their findings. If such a trip is not possible, arrange a guest lecture by a paleontologist from a museum or local university on the uncovering, piecing together, and identifying of dinosaur fossils.

swallowing an egg whole, these snakes crack the egg in their throat by piercing the shell with bony projections of their vertebrae. The snakes swallow the liquid content of the egg, spitting out the cracked shell. Still other snakes, such as the huge king cobra, eat other snakes. Crocodiles and alligators eat fish and land animals, provided they can catch them. If they are successful in snaring a land animal, they pull it under water and drown it, eating what they want by tearing huge chunks off the corpse. Crocodiles and alligators often store the remains of their prey under water by anchoring the rotting body under a tree or rock and returning to it when hunger once again moves them to eat.

Carnivorous reptiles other than snakes feed in similarly unusual ways. Monitor lizards kill their prey with sharp teeth and powerful jaws. Chameleons have sticky tongues as long as their bodies, which flip out to catch insects on the wing. Iguanas tear plant material into shreds with the force of their teeth and jaws. Because herbivorous reptiles do not chew their food, they must swallow large pieces of material. Their long digestive system, however, enables them to digest these large tough, fibrous pieces of food.

Figure 32–6 This anole, a type of lizard, is holding onto the downward hanging flower of a banana. Anoles are quite common in tropical and semitropical areas.

RESPIRATION The lungs of reptiles are better developed than those of amphibians. Because they have muscles around their ribs, many reptiles are able to expand their chest cavity to inhale and collapse the cavity to force air out. Several species of crocodiles also have flaps of skin that can separate the mouth from the nasal passages, thus allowing them to breathe through their nostrils while their mouth remains open.

Although most reptiles have two lungs, some species of snakes have only one. This single lung functions quite well, and fits neatly into their long, thin body. Snakes have another important adaptation that allows them to breathe at the very same time they are swallowing their prey. You probably know from experience that it is impossible for a person to breathe and swallow simultaneously. But snakes have a special tube in the floor of their mouth through which they breathe, so they don't suffocate in the time it takes them to swallow their prey. This tube can be extended out of a snake's mouth while it is dining.

INTERNAL TRANSPORT Reptiles have a well-developed double-loop circulatory system. One of the two loops brings blood to and from the lungs, and the other loop brings blood to and from the rest of the body. Reptile hearts contain two atria and either one or two ventricles. Among reptiles, crocodiles and alligators have the most well-developed heart. Their heart consists of two atria and two ventricles. (This four-chambered heart is found in all birds and mammals.) However, most reptiles have a single ventricle with partial internal walls that help keep oxygenated and deoxygenated blood separate during the pumping cycle.

BACKGROUND INFORMATION
REPTILE'S PUPILS

The shape of a reptile's pupil indicates whether the animal is active at night or during the day. A round pupil indicates that the reptile is active during the day. A slitlike pupil indicates that the reptile is active at night.

32–1 (continued)

Focus/Motivation

Using the Visuals Have students observe Figure 32–7 and read the accompanying caption. Obtain a turtle shell sawed in half so that students can see the fused backbone.

Content Development

Point out that reptiles eliminate liquid wastes in the form of urine produced in the kidneys. Stress that many reptiles are able to conserve water by excreting nitrogen-containing wastes in the form of uric acid.

Content Development

Using the Visuals Have students observe Figures 32–8 and 32–9. Use the following questions to help students distinguish between the ways in which a poisonous snake and a Gila monster poison their prey.

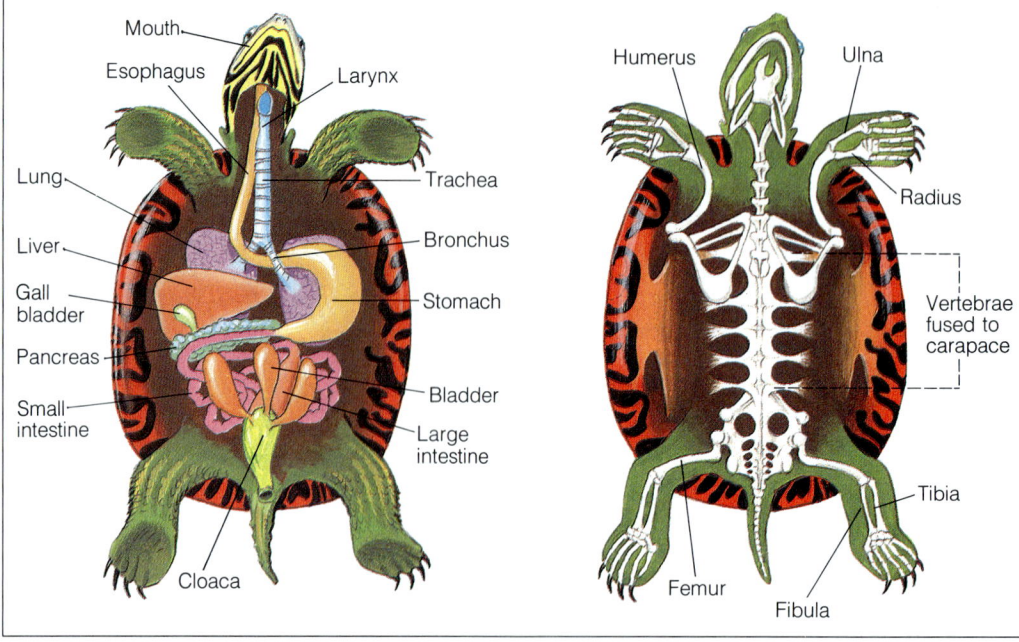

Figure 32–7 This illustration shows the internal organs of a turtle. The turtle's top shell is actually fused to its vertebrae. Other reptiles differ slightly from this body plan.

The circulatory system of reptiles is more well-developed than the circulatory system of amphibians. Because amphibians exchange gases through their moist skin, they do not need to have an efficient circulatory system to deliver oxygen to and remove waste gases from their cells. In reptiles, however, the development of dry skin necessitated the development of a more efficient circulatory system—a system up to the task of moving blood between the small, localized region where gas exchange occurs (the lungs) and the most distant parts of the body.

EXCRETION Reptiles eliminate liquid wastes in the form of urine, which is produced in their kidneys. In some reptiles, urine flows through tubes directly into a cloaca similar to that of amphibians. In other reptiles, a urinary bladder stores urine before it is expelled. In many reptiles, especially desert-dwelling snakes, a large amount of water is removed from the urine in the cloaca and returned to body tissues.

Many reptiles excrete nitrogen-containing wastes in the form of uric acid rather than ammonia. Ammonia is extremely toxic and must be excreted in dilute form to avoid self-poisoning. Excreting dilute ammonia means excreting considerable amounts of water. This is not a problem for aquatic and semi-aquatic animals like fishes and amphibians. But it is a problem for terrestrial animals that must conserve water. Uric acid is less toxic to cells than ammonia is and thus does not have to be diluted to the same extent that ammonia does. In addition, uric acid does not remain dissolved in urine but rather crystallizes out as a solid precipitate. When extra water is removed in the

• **How do snakes release poison into their victims?** (Poisonous snakes have fangs that inject poison.)
• **How does the poisonous Gila monster poison its victim?** (The Gila monster bites its victim and forces poison into the wound from a poison gland in the lower jaw.)

Mention that there are only eight known cases of humans being killed by the bite of a Gila monster. The only other venomous lizard is the beaded lizard of western Mexico.

Content Development

Using the Visuals Have students observe Figure 32–9. Discuss some of a snake's adaptations for locating prey.

• **Some snakes have a small pit on the head in front of the eyes. What is the function of this pit?**

cloaca, urine is reduced to a pasty white material that can be excreted without much water loss. This represents another successful adaptation of reptiles to a land environment.

RESPONSE The reptilian brain shows the same basic pattern as the amphibian brain, although the reptile cerebrum and cerebellum are somewhat larger. Most reptilian sense organs are well-developed, although there are exceptions. Snakes, for example, cannot hear.

Most reptiles that are active during the day have complex eyes that contain several types of photoreceptor cells. Many reptiles can see color quite well. In fact, some turtles can probably see colors better than humans can.

Many snakes have an extremely good sense of smell. Snakes have a pair of nostrils that open near the mouth. They also have a pair of special organs in the roof of the mouth that aid the nose in smelling. Have you ever seen a snake constantly flicking its tongue in and out of its mouth? The tongue picks up molecules from the air and carries them into the mouth and onto this pair of special organs. In this way, the snake gathers information about its environment by "tasting" molecules in the air.

Some reptiles, including most lizards, have simple ears—much like those of amphibians. These ears have an external eardrum, or tympanum, and a single bone that conducts sound to the inner ear. But many other reptiles do not have an eardrum and are completely deaf to sounds carried in the air. Snakes are one example. However, snakes are able to pick up vibrations in the ground through bones in their skull. Tortoises also lack an eardrum, but a thick patch of skin on their head serves the same sound-conducting function.

Some reptiles are able to gather heat information from their environment. Snakes such as rattlesnakes have the extraordinary ability to detect the warmth given off by the body of the small animals they eat. Many vipers prey almost exclusively on small mammals such as mice and rats that have a high body temperature. Some pit vipers have heat-sensitive pits on both sides of their head. Using these pits, vipers obtain a temperature picture of the world around them. Mammals stand out as warm spots in that temperature picture, making them easier for vipers to locate and strike at. This method of locating prey is extremely valuable at night, when low levels of light make it difficult for a reptile to see its prey.

MOVEMENT The reptilian muscle and skeletal systems exhibit many advances over those of amphibians. Compared with amphibians, reptiles with legs have larger, stronger limbs whose movements are well-controlled. For example, the legs of many reptiles are well-adapted to walking, burrowing, swimming or climbing. Snakes, which lost their legs in the course of evolution, move by pressing large ventral scales against the ground. By expanding and contracting the muscles around

Figure 32–8 *Unlike certain snakes, the gila monster does not have fangs to inject its venom. Instead, it bites its prey and lets its venom flow into the open wound.*

Figure 32–9 *The fangs of this rattlesnake are so long they must fold in order for the snake to close its mouth. Note the position of the animal's venom glands.*

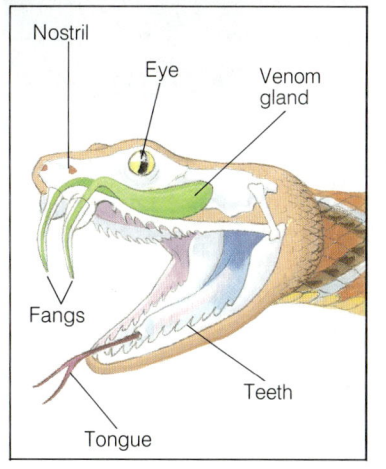

BACKGROUND INFORMATION
CONCERTINA

An accordionlike movement is the form of locomotion used by snakes that hunt prey in narrow burrows. The snake first braces itself in the burrow by making several narrow loops with the front of its body. It then slides its head forward, pushing against the burrow's walls as it moves.

FACTS AND FIGURES

A high school student's heart beats about 85 times per minute. An alligator's heart beats about 40 times per minute, and a turtle's heart beats somewhere around 20 times per minute.

TEACHING SUPPORT

MEDIA AND TECHNOLOGY
Biology Transparencies
- 43: The Internal Anatomy of a Turtle
- 44: The Anatomy of a Snake

Interactive Videodisc
- On Dry Land: The Desert Biome
Use this bar code to introduce students to the ability of reptiles to use camouflage for protection.

Play frames 18546 to 19013

See the Biology Media Guide pages 42–43 for additional bar-code correlations for this section.

(The pit contains a heat-sensitive organ that helps them detect the warmth given off by live prey.)
• **Why do snakes flick their tongues in and out of their mouth?** (The tongue picks up molecules from the air that are given off by nearby prey.)

Explain that the snake transfers the molecules from the air to two cavities called the Jacobson's organ located in the roof of its mouth. The Jacobson's organ is extremely sensitive to odors. Also mention that even though snakes lack outer ears, they do have inner ears that can detect vibrations.

Content Development
Snakes have several adaptations that permit them to swallow prey larger in width than their own body. The lower jaw is not joined directly to the skull. During swallowing, it can be drawn down and forward, allowing the snake to take in large prey. The body can expand considerably during swallowing. All snakes swallow their food whole, and all digestion takes place in the stomach.

713

HISTORICAL NOTE
GECKOS

Geckos, like cats and dogs, were once kept as house pets. At night, they would catch and eat household pests, such as ants, roaches, and spiders.

Figure 32–10 *Although much more at home in trees and shrubs, this chameleon walks gingerly along the ground (top, left). The sidewinder, a type of rattlesnake, is able to move along shifting desert sands quite quickly (top, right). Tiny flaps of skin on its toes enable the gecko to cling to surfaces as smooth as glass (bottom).*

their ribs in waves, snakes dig these ventral scales into the ground and push themselves forward. Because this kind of movement is slow and quiet, it makes snakes masters at stalking prey.

REPRODUCTION As you have already learned, the reptile egg is an important contributing factor to the success of these animals in a land environment. The leathery outer shell protects the delicate tissues inside the egg, and pores in the shell allow oxygen and carbon dioxide to pass through.

Reptiles lay eggs that hatch into animals that resemble small adults. Virtually all reptiles reproduce through **internal fertilization**, which means that a male reptile deposits sperm into the body of the female. In reptiles, as in amphibians, the male and female reproductive systems open into the cloaca. From the outside, it is extremely difficult to tell the sex of a reptile. Almost all male reptiles have a penis that allows them to deliver sperm into the female's cloaca. (Lizards and some snakes have two reproductive structures called hemipenes.) This internal fertilization makes it possible for the female's reproductive system to cover the embryos with protective membranes and a shell after fertilization has occurred.

Once fertilization occurs, reptile species treat their eggs very differently. Many species are oviparous, laying eggs that develop outside the mother's body. Some species, such as sea turtles, come ashore to bury their eggs in sand. After they have covered the eggs, the sea turtles move back into the sea, deserting the eggs and leaving the young turtles without parental care. Other reptiles provide minimal care for their eggs and young. Certain snakes, for example, bask in the sun to warm their bodies and then wrap themselves around their eggs to incubate them. Once the young hatch, however, they are on their own and must catch their own food. Baby cobras are born looking like miniature copies of their parents. They even have fully functioning fangs that inject the venom with which they kill their prey.

32–1 (continued)

Reinforcement/Reteaching
Review the process of amphibian reproduction with students.

Skills Development
Guided Practice
Skill: Making comparisons
Explain to students that a major advance of reptiles over amphibians is their ability to reproduce on land. Internal fertilization and a shelled egg free reptiles from the necessity of returning to water to reproduce.
• **In what important way are the eggs of reptiles different from those of amphibians?** (Reptile eggs have a leathery, protective shell.)
• **Why is this covering so important?** (It keeps the egg from drying out; thus it can be laid on land.)
• **Where are the eggs of most female amphibians fertilized?** (Most are fertilized in water, outside the female's body.)
• **How is fertilization among reptiles different from that of amphibians?** (Lead students to answer that reptiles have internal fertilization.)
• **How is internal fertilization a protective mechanism for reptiles?** (The female's reproductive system is able to cover the developing embryos with protective membranes and a shell after fertilization has occurred.)

Reptiles such as alligators build nests in which they lay their eggs and guard them until they hatch. At that point, alligators provide their young with a certain amount of care. When a female alligator hears her babies squeaking, she bites open the nest, picks up the babies in her mouth, and carries them to the water for protection. Some baby crocodiles and alligators stay with their mother for as long as two years after they hatch.

A number of female snakes and lizards hold their eggs inside their body until the eggs hatch. Thus the young are born alive. During the incubation period, an embryo grows and develops within the egg, using the yolk of the egg as a source of nutrients. Having the eggs carried within the female's body provides several important advantages. The mother snake can protect her young. She can also keep the eggs warm by basking in the sun. Snakes that bear living young are ovoviviparous.

Figure 32–11 *Although reptiles rarely fight, conflicts do occur. Tortoises fight until one is flipped onto its back (left). If neither of the male chameleons retreats from the face-off, a fight will begin (right).*

Tuataras

The tuatara (too-uh-TAH-ruh) is the only surviving member of the order Rhynchocephalia (rihng-koh-suh-FAY-lee-uh). This reptile resembles reptiles that lived during the dinosaur age. The tuatara is of interest to paleontologists because it retains certain features of the ancient reptiles from which it evolved.

Tuataras, which are found only on a few small islands off the coast of New Zealand, lead a leisurely life. Unlike many reptiles, they are active at night, hunting the small animals they eat. In tuataras, the part of the brain called the pineal gland is located on the top of the skull in a place where several skull bones meet. This type of pineal gland is sometimes called the "third eye" because it contains cells that are sensitive to light. However, tuataras do not actually see with the pineal gland. Instead, they use it to detect changes in day length.

Figure 32–13 *The tuatara is a rare lizard that lives in New Zealand. As you can see, it retains many of the features of its prehistoric ancestors.*

Figure 32–12 *Although it appears to be eating a snack, this male crocodile is actually carrying one of its newly hatched offspring to water.*

BACKGROUND INFORMATION
TUATARA

The tuatara, or *Sphenodon punctatum*, looks much like a lizard. The tuatara has dark olive skin and yellow spines on its back. It grows to a length of about 70 centimeters. It lives in a burrow, and it is carnivorous, feeding primarily on insects. The eggs of this reptile require more than a year to develop and hatch.

Content Development
Using the Visuals Have students observe Figure 32–13. Point out that the tuatara is the only remaining living species in the order *Rhynchocephalia* and that it is nearly extinct. Stress that the tuatara is different from other reptiles in that it has a well-developed pineal body, or "third eye," beneath the skin. The pineal body, located in the middle of the head, can detect light and darkness.

Content Development
Point out that some reptile species, such as sea turtles and cobras, are oviparous, laying eggs that develop outside the mother's body. Other species are ovoviviparous, bearing live young. Stress that ovoviviparous reptiles can provide protection for the developing embryos.

Enrichment
Have students do research on the several species of lizards that are parthenogenic. In these all-female reptiles, such as the New Mexico whiptail, eggs can develop directly into adults without being fertilized by males.

Enrichment
Have students do research on the unique parental care displayed by the Nile crocodile.

BACKGROUND INFORMATION
SNAKE VENOM

Snake venoms can be deadly because they cause extensive damage to any living tissue they touch. Many years ago, people thought some snake venoms poisoned nerve tissue specifically, while other venoms caused the death of many different cell types.

New studies show that most venoms contain many toxic compounds that work in a variety of ways. All snake venoms contain powerful protein-digesting enzymes called proteinases that cause massive tissue damage near the point of injection. Several venoms also contain complicated polypeptides that block the transmission of nerve impulses in the prey. Many symptoms of snakebite depend on the location and severity of the bite.

32–1 (continued)

Content Development
Discuss the similarities and differences between lizards and snakes. Stress that although snakes have a bad reputation, only about 200 of the 2500 known species are poisonous. Snakes are actually more helpful than harmful because they kill a large number of rodents.

Enrichment
Have students research the appropriate treatments of snakebite wounds and report their findings to the class.

Enrichment
Have students research the mechanisms of camouflage and mimicry exhibited by certain snake species.

Content Development
Point out that an alligator and a crocodile have some differences in appearance. The snout of an alligator is broader than the snout of a crocodile, which is narrower and more pointed. Also, when a crocodile's mouth is closed, some of its bottom teeth are visible. When an alligator closes its mouth, none of its bottom teeth are visible.

Lizards and Snakes

Modern lizards and snakes belong to the order Squamata (skwah-MAH-tuh). Most lizards have legs, clawed toes, external ears, and movable eyelids. Lizards range in size from tiny geckos a few centimeters long to giant monitor lizards that can be more than 3 meters in length.

Some lizards have evolved into highly specialized forms. See Figure 32–14. For example, African chameleons live exclusively in trees and bushes, eating insects they catch with their incredibly long, coiled tongue. They flick their tongue out of their mouth, and, with unerring aim, they are usually able to catch their insect meal. Gila monsters, large, stocky lizards of the American Southwest and Mexico, have glands in their jaws that produce the venom with which they paralyze small prey. Gila monsters do not inject venom with fangs. Instead, they bite their prey and hold onto it with their teeth while the venom they produce flows into the wound. Another kind of lizard, the iguana, looks ferocious but is almost exclusively herbivorous.

The world's largest living lizards, the monitors, are the only reptiles alive today that provide some idea of what small dinosaurs may have been like. Monitors are quite intelligent and active for reptiles. Many eat birds and mammals. The largest monitor lizards are the Komodo dragons. With a length of as many as 3 meters and a mass of up to 75 kilograms, Komodo dragons can kill and devour animals as large as water buffalo!

Snakes are lizards that have lost both pairs of legs during their evolution into burrowing forms. Millions of years ago, some lizards began to live below ground level. In burrows and cracks, these relatively harmless lizards were safe from predators. Over time, the lizards with smaller legs or with no legs at all were able to burrow most efficiently. These lizards survived,

Figure 32–14 *This male anole is displaying the bright blue patch below its chin (top, left). The Komodo dragon is one of the largest lizards in the world, reaching a length of about 3 meters (top, right). The frilled dragon gets its name from a frill of skin around its neck that it can extend, thus making itself appear larger and more fearsome to other animals (bottom).*

Skills Development
Guided Practice
Skill: Drawing a conclusion
Stress to students that alligators and crocodiles have several traits that are uniquely different from those of other reptiles. List these crocodilian traits on the chalk-

evolving into the legless snakes alive on Earth today. Although being legless may seem to be a disadvantage, snakes are efficient and effective predators in the niches they occupy, In fact, the distribution of snakes on Earth is limited only by temperature.

Snakes vary considerably in size. Some are so small they look like earthworms. Others are so large they're scary. Pythons, for example, can reach lengths of about 10 meters. But perhaps the most fascinating snakes are the venomous snakes —the ones that produce poisonous chemicals they inject into their prey. The ability of certain types of snakes to produce lethal poisons has caused some people to harbor an unjustified fear of all snakes. Actually, more people in the United States die from bee stings than snake bites. And most snakes are happier avoiding people than confronting them!

Crocodilians

Members of the order Crocodilia—such as alligators, crocodiles, caimans, and gavials—split off from the ancient reptiles around the time dinosaurs did, probably at the beginning of the Triassic Period. In the 200 million years since then, alligators and crocodiles have changed little.

Crocodilians are among the largest living reptiles. Some species grow up to 7 meters in length. These animals live only in the tropics and subtropics, where the climate remains warm all year long. Alligators, and their relatives the caimans, live only in fresh water and are found almost exclusively in the Western Hemisphere. Crocodiles, on the other hand, may live in either fresh or salt water and are native to Africa, India, and Southeast Asia.

Figure 32–15 Although they inspire fear in many, most snakes are not aggressive toward people. One can admire the exotic colors and patterns of their scales—and respect their ability to survive in a world where even their shape appears to be a handicap.

FACTS AND FIGURES

The world's largest snakes are the anaconda and the reticulated python. Both are reported to grow up to 10 meters long. Both of these giant snakes kill their prey by constriction. There are occasional reports of their attacking humans, but such incidents are rare. There was, however, a confirmed case of a reticulated python attacking and swallowing a 90-kilogram bear.

One of the world's smallest snakes is the Braminy blind snake. It is only about 15 centimeters long.

TIE-IN SOCIAL STUDIES

Alligators and crocodiles have been hunted for their hides. Have students investigate the uses of the hides of these reptiles. Have students include a description of the penalty imposed on a person caught hunting these creatures today.

TEACHING SUPPORT

Laboratory Manual
- Adaptations in Lizards, p. 409

MEDIA AND TECHNOLOGY
Videodisc
- *Aquatic Ecosystems: Freshwater Wetlands and Freshwater*

Use this bar code to introduce the alligator as a resident of tropical wetlands.

Play frames 27197 to 27544

See the Biology Media Guide pages 42–48 for additional bar-code correlations for this section.

---

board or an overhead projector:
1. The eyes and ears are located on the top of the snout.
2. A fleshy valve at the back of the mouth can tightly cover the passage from the mouth to the lungs.
3. Nostrils on top of the head lead to the air passage behind the throat valve.
4. The nostrils and ears can close.

• **What do these adaptations tell you about an alligator's or crocodile's lifestyle?** (Lead students to conclude that they are adaptations for lying in water with only the ears, eyes, and nostrils above the surface.)

The ability to close the throat, ears, and nostrils allows them to drag prey below the water to drown it.

HISTORICAL NOTE
GALÁPAGOS TORTOISES

The great tortoises of the Galápagos Islands have been so reduced in numbers that they are now in danger of extinction. When the islands were first discovered in the seventeenth century, they were reported to be so numerous that a person could walk long distances atop their shells without touching the ground. In the 1800s, sailors started catching them for their meat and oil. Later, the dogs and cats of settlers further reduced the population when they started eating the eggs and babies of the tortoises.

FACTS AND FIGURES

Turtles and tortoises have a life expectancy greater than most other vertebrates. Most species of turtles live 50 years or more. The record age for any turtle is about 150 years.

Figure 32–16 This painted turtle is one of our most colorful native species. Here you can see the rounded top shell and the flat bottom shell.

Turtles

Turtles and tortoises, members of the order Chelonia (kigh-LOH-nee-uh), also evolved a successful way of life during the Triassic Period and have changed little in the 200 million years since then. The word turtle usually refers to members of this order that live in water; the word tortoise, to those that live on land. All turtles and tortoises have some sort of shell covering their bodies, although in a few species, such as the American soft-shelled turtle, the shell is not very bony. The turtle shells consist of two parts: a dorsal part, or **carapace**, and a ventral part, or **plastron**. The animal's backbone is fused to the inside of the carapace, and its head, legs, and tail stick out through holes where carapace and plastron join.

Tortoises usually have a high, domed carapace and stubby, elephantlike legs. Tortoises pull into their shells to protect their more delicate body parts. In some species, the front end of the plastron is hinged and folds up to further seal out predators.

Turtles are adapted to life in freshwater ponds and lakes or the open sea. The legs and feet of many aquatic turtles have developed into flippers. Certain aquatic species cannot pull back into their shells completely, but they do have powerful jaws that are capable of giving a nasty bite.

Figure 32–17 This tortoise moves quite well, although slowly, on its strong stubby legs (left). The flipper-shaped feet of this sea turtle enable it to swim gracefully in its watery home (right).

32–1 (continued)

Content Development
Stress the fact that turtles and tortoises are distinctly different from other reptiles in several ways. The most obvious difference is the bony shell that encloses the body. In most turtles, the shell is composed of a number of hard plates, but in soft-shelled species, it is composed of a leathery skin. The inner layer of the shell is usually fused with the vertebrae and ribs. Turtles and tortoises are also different in that they lack teeth. They catch and crush their food with horny beaks.

SECTION REVIEW 32–1

1. A scaly skin, lungs, an amniotic egg, and vertebrae.
2. Most of the adaptations of form shown by reptiles enable them to survive in dry land conditions. For example, their skin prevents water loss. Their eggs can be laid on dry land because their shell protects them from drying out. Many reptiles have efficient kidneys that excrete little water with concentrated wastes.
3. Climate conditions during the Permian Period became cooler and much drier. Many of the great swamps dried up. These changes made life for the water-dependent amphibians much more difficult. The reptiles, better adapted for life under dry conditions, entered a period of adaptive radiation.
4. The amniotic egg consists of a shell and membranes that create a protected environment for

Figure 32–18 After they hatch, sea turtles scurry into the water. Male sea turtles will never again walk on land; female sea turtles will return to the shore from which they hatched to lay their own eggs. Although they move effortlessly through water, sea turtles expend a great deal of effort to heave their large bodies along on land.

How Reptiles Fit into the World

The reptiles alive today represent only a few survivors of a group of animals that once ruled the land. They are found in many habitats—from the temperate zone to the tropics and from the tops of rain forest trees to the open ocean.

Reptiles are important predators in many ecosystems. On farms, for example, snakes keep down the large numbers of rats and mice normally attracted to grains. Without snakes, most farms would be overrun by mice and rats. Small lizards eat insects; large lizards eat other small animals. Thus reptiles limit the populations of other animal species.

Sea turtles, once numerous in both the Atlantic and Pacific oceans, are now in danger of extinction for several reasons. Turtle soup and turtle eggs are eaten in many parts of the world. Tortoise shell was once commonly used to manufacture jewelry and as an inlay in furniture. Another reason for the decline in the number of sea turtles is the development of their nesting sites. Sea turtles always return to shore to lay their eggs in sand. They return, however, not to just any patch of beach but to the same spot where they hatched many years before. Seaside development (the construction of houses, hotels, and shopping centers) in many tropical areas threatens the ancient spawning sites of these animals. Today, however, most species of sea turtles are protected by law. This protection may lead to an increase in their dwindling numbers.

32–1 SECTION REVIEW

1. Describe four characteristics of reptiles.
2. How are reptiles adapted for a life on dry land?
3. How did conditions during the Permian Period favor the adaptive radiation of reptiles?
4. Describe the structure of an amniotic egg. Why was the evolution of the amniotic egg critical to the survival of vertebrate life on land?
5. What four orders are found in the reptile class? Give an example of an animal in each order.
6. **Critical Thinking—Making Predictions** What might happen to reptiles if conditions on Earth became permanently warmer and much more damp?

719

the developing embryo. The egg also contains a substantial yolk that provides food for the developing embryo. The amniotic egg was crucial for the survival of reptiles because this egg can be laid on dry land; it does not need to be laid in water.

5. Rhynchocephalia, the tuatura; Squamata, lizards and snakes; Crocodilia, alligators and crocodiles; Chelonia, turtles and tortoises.

6. If Earth became warmer and much more damp, some reptiles might be able to colonize areas that are currently too cold for them. Other reptiles might die out if they could not adapt to damp conditions or if they could not compete successfully with damp-loving competitors such as amphibians. Over time, reptiles might lose some of their adaptations for living in a dry climate, as these adaptations would no longer be favored by natural selection and might even become unfavorable.

TEACHING SUPPORT

Teaching Resources
- Activity: Investigating How Reptiles Regulate the Temperature of Their Bodies, p. 3

BACKGROUND INFORMATION
"THERM" TERMS

There are many terms associated with temperature regulation. Four of these terms—coldblooded, warmblooded, ectothermic, and endothermic—are discussed in this section. Two others that you are likely to encounter are *poikilothermic* and *homeothermic*. Poikilotherms have a variable body temperature that tends to follow the temperature of the surrounding environment. Homeotherms maintain a constant body temperature. Because the term poikilothermic is functionally synonymous with ectotherm, endotherms that because of their tiny size cannot maintain a high body temperature are sometimes said to be heterothermic. For example, at night a hummingbird's body temperature may drop from 40°C to 12°C, the temperature of its surroundings.

A common misconception is that all "lower" animals are ectotherms. Tunas, mako sharks, bumblebees, and sphinx moths are examples of "lower" animals that are true endotherms—they derive heat from their metabolism and maintain a reasonably constant core temperature!

TEACHING STRATEGY 32–2

Focus/Motivation

Obtain some liquid-crystal temperature strips. Have students gently hold the strips against the skin of several reptiles being stored at room temperature and record the temperature readings. Have students place the reptiles in the sun for five minutes and check the temperature again. Repeat this procedure with birds or small mammals and have students compare the results.

Content Development

Stress that the regulation of body temperature is an important animal mechanism, especially in environments with wide daily and seasonal temperature variations. Point out that reptiles are ectotherms, or animals that gain heat from the outside environment. Birds and mammals are endotherms, or animals that generate their own body heat.

Guide For Reading
- What are ectotherms and endotherms?
- What are the advantages and disadvantages of ectothermy? Of endothermy?

32–2 The Evolution of Temperature Control

On a spring morning after a cold night, a turtle lies on a rock basking in the sun. Nearby, a snake crawls out of its burrow beneath a rotting stump. In a tree overhead, a young robin puffs up its feathers into a ball. And as you walk out of the water after an early swim, your skin gets goose bumps, and you shiver. All these activities are examples of the strategies vertebrates use to control their body temperature.

Control of body temperature is important for animals, particularly in habitats where temperature varies widely with time of day and with season. Each animal species has its own preferred "operating range" of temperatures. For example, in order for muscles to move quickly, they must be kept at a certain minimum temperature. Yet if an animal's body gets too hot, muscles tire easily and other body systems are stressed.

In terms of how they generate and control their body heat, animals can be classified into two basic groups: **ectotherms** and **endotherms**. Turtles, snakes, and other modern reptiles are ectotherms, which literally means heat from outside. As a group, these animals have relatively low metabolic rates when they are resting. Thus they do not generate much heat inside their bodies. Because they also lack effective insulation, any heat they do generate is lost to their surroundings. In order to control their body temperature, therefore, these animals must pick up heat from the environment.

Birds and mammals, on the other hand, are endotherms, which literally means heat from inside. Birds and mammals have relatively high metabolic rates that generate a significant amount of heat, even when they are resting. Body fat and either hair or feathers insulate the body, helping to retain that heat so that it is not lost to the environment. Endotherms can move around at night and during cool, cloudy weather more easily than ectotherms can because endotherms do not have to warm up their muscles to operating temperature by basking in the sun.

Ectothermic animals are often incorrectly thought of as coldblooded, and endothermic animals, as warmblooded. These names give the wrong impression of the body temperature of these animals. Coldblooded animals may have a body temperature higher than their surroundings, whereas warmblooded animals may have a temperature lower than their

Figure 32–19 These iguanas position themselves so that they are warmed by the heat of the sun. Then they dive into the cold water to eat algae. They must return to the land again when the water lowers their body temperature. Penguins are able to generate their own body heat. They live in cold Antarctic climates, where their feathers act as insulation.

Reinforcement/Reteaching

Review the concept that amphibians, such as frogs and toads, hibernate during the winter.
- **Are frogs and toads ectotherms or endotherms?** (Ectotherms.)
- **What happens to the body activities of ectotherms when the**

Figure 32–20 *Remaining in the sun for a long period of time would cause the body temperature of a desert tortoise to climb too high. Thus a tortoise spends much of the hot desert days hiding in a shady spot, such as beneath a rock.*

surroundings. It was also long assumed that the body temperature of ectotherms varied much more than that of endotherms. However, a careful study of animal body temperatures shows that these assumptions are incorrect for many animals.

In nature, lizards and snakes warm up when they need to by basking in the sun. When their body reaches the right temperature, they go about their business. The right temperature for reptiles is often higher than that of their surroundings. While they are active, reptile muscles generate heat (just as ours do). If they get too hot during the day, reptiles duck into a cool burrow or under a rock to lose heat. When the sun goes down and the air gets cold, they head for a warm place in which they can conserve their heat overnight. Scientists have implanted radio thermometers inside certain lizards and monitored body temperature as the animals were allowed to be free in their natural habitat. The results showed that these lizards' body temperature was always higher than that of the environment and remained relatively constant as well!

In an absolute sense, neither endothermy nor ectothermy is superior. For different animals in different environments, each strategy has its advantages and disadvantages. For example, ectotherms cannot remain active for long periods. If you watch modern reptiles, you will see that they alternate periods of intense activity with periods of rest. A snake will expend energy in catching a meal, but it will remain relatively inactive while digesting it. And if the meal is large enough, the snake may remain inactive for several weeks. Endotherms, by comparison, remain active for a long time. Think about the amount of energy that is expended by a cheetah as it chases down its prey, or the amount of energy that is expended by a bird during its annual migration.

In climates that remain warm all the time, ectothermy is a way of conserving energy. Ectotherms remain at operating temperature fairly easily by basking in the sun and burrowing in the ground. This is why most of the world's large lizards and amphibians (and many of the largest insects) are today found in the tropics and subtropics. In these warm places, ectothermy is a perfectly adequate way to regulate body temperature. Endotherms, on the other hand, burn lots of calories to generate body heat. Therefore, endotherms need a lot more food than ectotherms of the same size. The amount of food needed to sustain a single cow would be enough to sustain a dozen cow-sized lizards!

Large ectotherms run into trouble, however, in temperate zone habitats where temperature varies a great deal. It takes a

ECOLOGY NOTE
HEAT TOLERANCE IN PENGUINS

Most penguin species are exposed to extremely low temperatures for much of their lives. Their feathers, arranged in interlocking fashion, provide excellent insulation from the cold.

In hot climates, penguins show interesting physiological characteristics that minimize heat stress. Although the insulating characteristic of their feathers does not permit much heat loss, unfeathered parts of the body—such as flippers and feet—act as radiators. Blood vessels in these body parts dilate to allow heat to escape. When temperatures are extremely high, panting allows penguins to evaporate water from their body. This is an efficient way of releasing body heat.

Penguins also exhibit interesting behaviors that help them deal with high temperatures. For example, yellow-eyed penguins in New Zealand, which can be exposed to temperatures as high as 47°C, avoid direct sunlight by nesting in shaded areas. Similarly, penguins that live in areas even closer to the equator frequently nest in burrows or caves in order to escape the heat.

SCIENCE, TECHNOLOGY, AND SOCIETY ISSUE

Dinosaurs: New Light on Old Bones

When discussing this issue, stress to students that there are many differences of opinion in the scientific community about the lives of dinosaurs and other extinct species. American paleontologist Robert T. Bakker is one of the leading paleontologists who has inferred that some dinosaur species may have been agile, fleet footed, and endothermic. Emphasize that although Bakker's findings have not been entirely accepted by the paleontological community, they have provided great food for thought.

- Why are there so many different hypotheses concerning the characteristics of dinosaurs? (Accept all logical answers.)
- What evidence is used to support the hypothesis that some dinosaurs may have been endothermic? (Bones similar to mammalian bones; analysis of predator-prey ratio.)

As an extension of this issue, you may wish to have a group of interested students organize and conduct a formal debate on the different hypotheses of the extinction of dinosaurs. Students who do not participate in the debate can decide which hypothesis is most feasible.

SCIENCE, TECHNOLOGY, AND SOCIETY ISSUE

Dinosaurs: New Light on Old Bones

Until recently, dinosaurs were thought to be lumbering dim-witted beasts scarcely able to support their own weight, let alone move. But more evidence indicates that these animals were probably both more intelligent and more active than was once believed.

New studies of fossil bones show that some dinosaur species stood up on their hind legs and were able to run at reasonable speeds to escape large predators. Scores of fossil footprints indicate that large herbivores such as *Apatosaurus* lived in herds much like some large herbivore species of animals alive today. Other evidence indicates that dinosaurs displayed the same sorts of mating and territorial behaviors seen among large mammals today. Nests containing fossil dinosaur eggs indicate that at least some dinosaur species cared for their young after hatching, not unlike modern-day birds.

A more hotly disputed hypothesis concerns endothermy. Some paleontologists believe that dinosaurs were endothermic. Supporters of this hypothesis point to certain features of dinosaur bones that make them quite similar to mammalian bones. Supporters also rely on an analysis of the ratio of predator to prey animals for evidence. They note that there seems to have been many more prey animals than predators among dinosaurs. Because endotherms require more food than ectotherms do, these researchers argue, a large population of prey animals would be needed to support a small population of endothermic predators. And since this is what the evidence shows, dinosaurs must have been endotherms. If dinosaurs were ectothermic, they argue, a different ratio of predator to prey animals would be expected.

Today most paleontologists believe that dinosaurs had higher body temperatures than modern reptiles. They do not necessarily believe, however, that dinosaurs were endotherms. Instead, these researchers point out that dinosaurs lived during periods when climates were constantly warm. In fact, during the reign of the dinosaurs, much of what is now Europe and North America had a climate similar to that found in the tropics today. For this reason, dinosaurs could grow large and still regulate their body temperature as today's lizards do.

Even as more evidence is being collected, the debate about endothermy in ancient reptiles continues. In the meantime, it is easier to explain the extinction of dinosaurs if it is assumed they really were ectotherms. Large ectotherms, able to survive quite nicely in warm climates, would have been at a disadvantage in the cooler, more variable climates that ended the dinosaur era. Thus they would have disappeared from Earth. Smaller ectotherms, such as modern lizards, would have been better off than their much larger cousins, however.

The Tyrannosaurus rex *was one of the largest carnivores that ever lived. Its fearsome teeth made short work of its hapless prey.*

722

32–2 (continued)

Enrichment

Hibernation is one way in which animals adjust to the environment. Investigate the ways in which snakes and other reptiles prepare for the winter season.

SECTION REVIEW 32–2

1. Endothermy is the ability of an animal to produce heat from within its body. A bird is an endothermic animal.
2. Ectothermy means heat from without. Ectothermic animals are not able to generate much heat from within their body. They are dependent on the heat of the environment to adjust their body temperature.
3. A lizard is an ectothermic animal. It controls its body temperature by basking in the sun to warm up and moving out of the sun to cool down.
4. The winters in North Dakota are too cold for a large lizard. Since large lizards must depend upon their environment as a source of heat, the cold temperatures in North Dakota would prevent a lizard from growing large.

long time for a large animal to warm up in the sun after a cold night. It is also much more difficult for a large animal to find cool shelter during a scorching-hot day. (Can you imagine a burrow big enough for a gigantic dinosaur?) Small ectotherms have much less trouble dealing with hot days and cool nights than large ones do. But even small ectotherms cannot cope with long cold, cloudy winters. They either hibernate or lay resting eggs and die. In certain parts of the United States and Canada, snake dens with many thousands of snakes can be found. These huge numbers of snakes cluster together in the winter in warm dens. Like birds, snakes also migrate. In spring, many snakes leave their winter dens and crawl hundreds of kilometers to return to the area they normally live in during the summer months.

There is little doubt that the first terrestrial vertebrates were ectotherms. But there is some doubt as to when and how endothermy evolved. Some biologists believe that dinosaurs were endotherms, not ectotherms like modern reptiles. Many biologists, however, believe that endothermy evolved much later than the appearance of the dinosaurs, which means that at least some of the dinosaurs were ectotherms. Putting these two hypotheses together means that endothermy evolved more than once—once along the evolutionary line of reptiles that led to birds and once again along the evolutionary line that includes mammallike reptiles and their mammal descendants.

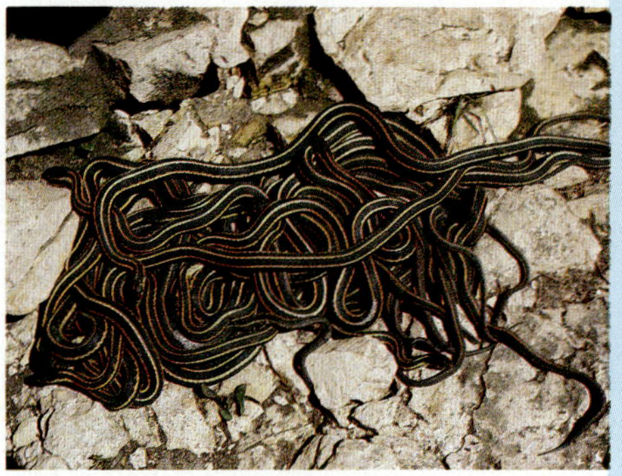

Figure 32–21 *Some species of snakes migrate to special dens during cold winter weather. In these dens, many thousands of snakes keep one another warm. In the spring, the snakes leave the dens and migrate to summer feeding areas—only to return to the dens once again in the fall when air temperatures begin to drop.*

TEACHING SUPPORT

Study Guide
- Section 32–2, p. 308

ESL STRATEGY

Point out that *communicate, migrate,* and *navigate* are terms describing special abilities that birds possess. After making sure that students understand the words, have students write three short paragraphs explaining how birds are able to perform these complex behaviors. Volunteers can read their paragraphs to the class.

32–2 SECTION REVIEW

1. What is endothermy? Give an example of an endothermic animal.
2. What is ectothermy? Give an example of an ectothermic animal.
3. How does a lizard control its body temperature?
4. **Critical Thinking—Making Inferences** Why is it unlikely that you would find a giant lizard living in the wild in North Dakota?

32–3 Birds

Guide For Reading
- What are the distinguishing characteristics of birds?
- How does form and function in birds reflect adaptations for feeding and for flight?
- How do birds fit into the world?

You have probably observed many different kinds of birds during your lifetime. Colorful birds fill woods with song. Flocks of birds fly to backyard feeders. Exotic birds with fanciful feathers inhabit zoological gardens. Yes, there are many birds: about 8700 living species belonging to more than 160 families. And there were even more kinds of birds in the past. Paleontologists estimate that more than 100,000 species of birds have become extinct since the Jurassic Period.

Reinforcement/Reteaching
If students have difficulty with Section Review questions, review the material.

Closure
Before leaving Section 32–2, make certain students understand the concept of body-temperature control.
- **What is the difference between an ectotherm and an endotherm?** (Ectotherms pick up heat from the environment; endotherms generate their own body heat.)
- **What body-temperature control mechanism is displayed by reptiles?** (Ectothermy.)
- **What body-temperature control mechanism is displayed by birds and mammals?** (Endothermy.)
- **In what type of environment would you expect to find the largest reptiles?** (Warm, tropical climates.)

TEACHING STRATEGY 32–3

Focus/Motivation
Obtain a canary or a parakeet for students to observe in class. The bird should be kept in a cage and students should be warned about putting their fingers near the cage. Allow students to observe the characteristics and behaviors of the bird.
- **What are some of the characteristics of birds?** (Accept all logical answers.)
- **What are some behaviors exhibited by birds?** (Accept all logical answers.)
- **What are some of the birds that are indigenous to this area?** (Answers will vary.)
- **How are birds adapted for flight?** (Accept all logical answers.)

BACKGROUND INFORMATION
FEATHER EVOLUTION

We know little about the first stages in the evolution of feathers because we have no fossils that preserve more primitive feathers. There are, however, several hypotheses about feather evolution.

Some biologists think feathers first evolved when birds became endothermic to help them keep warm. Down, after all, is a very good insulator! Later, according to this hypothesis, feathers evolved further to aid in flight. The problem with this idea is that it is hard to imagine how the evolution of a not-quite-useful set of flight feathers would evolve.

Another hypothesis suggests that ancient bird species, such as *Archaeopteryx*, used long feathers on their arms to help them trap insects on the run. According to this line of reasoning, these small dinosaurs captured insects by running after them with outstretched arms. If those arms carried a set of long feathers, *Archaeopteryx* could have used them to corner prey during the chase. Because birds probably evolved from small running carnivorous thecodonts, this is certainly a possible assumption. Long feathers that evolved in this way could then be useful in gliding flights without too many more changes. Full, powered flight could then evolve over time.

BACKGROUND INFORMATION
FILOPLUMES

Another type of feather, called the filoplume, has a long, slender quill with only a few small barbules at the tip. For a long time biologists thought filoplumes had no function. Now we know that these feathers carry many sensory cells that detect vibrations in nearby feathers. It is believed that filoplumes help birds keep their feathers properly adjusted during flight and other activities.

Figure 32-22 Baby owls are covered with a coat of down feathers. The air spaces in these fluffy feathers help insulate the birds from temperature changes. Most of the down feathers will later be shed and a new coat of contour feathers will grow in.

What Is a Bird?

In a group this diverse, it is difficult to find many characteristics that are shared by all members. But we can identify several features **birds** have in common. **Birds are endothermic reptilelike animals with an outer covering of feathers, two legs used for walking or perching, and front limbs modified into wings that usually do not have useful claws.**

The single most important characteristic that separates birds from reptiles is feathers. Feathers help birds fly and also keep them warm. Birds have several different kinds of feathers. Because many feathers are hollow, they are both light and strong.

Contour feathers are large feathers that cover a bird's body and wings. Certain contour feathers, known as flight feathers, are long and stiff. Flight feathers on the wings and tail provide the lifting force and balance needed for flight. From both sides of the long, stiff quill of a flight feather grow side branches called barbs. From the barbs, in turn, grow still smaller structures called barbules. The hooks on a barbule catch on the hooks of nearby barbules, holding the barbs together in flat vanes. Although the barbule hooks can be pulled apart, they can easily be lined up again. You may have seen a bird grooming its feathers by pulling them through its beak. This grooming is called preening. One of the reasons birds preen themselves after each flight is to realign any vanes whose barbules may have been split apart during use. Other contour feathers, known as general body feathers, have fluffy barbules at their base and are not as long and stiff as flight feathers. General body feathers are often brightly colored and help determine the shape of a bird's body.

Down feathers grow underneath and between the contour feathers. Down feathers are short, soft, and fluffy. These feathers trap warm air close to a bird's body, insulating the bird. Baby birds of many species are covered with down feathers for a period of time after they hatch.

Powder feathers are important to birds that live on or in water. As they grow, these feathers release a fine white powder that repels water and keeps it from penetrating the layer of down feathers. Birds also produce a waterproof oil in special glands near their tail. When ducks, geese, or other water birds preen themselves, they rub this oil over their feathers. The oil actually makes water "roll off a duck's back."

Evolution of Birds

Ask many paleontologists what a bird is and they'll reply with a grin, "a hot-blooded dinosaur with feathers." Although that answer may sound odd, there really is reason for it. The first fossil ever found of an early birdlike animal is called *Archaeopteryx* and dates from late in the Jurassic Period. Its skeleton looks much like a small running dinosaur. Unlike

---

32-3 (continued)

Focus/Motivation
The down and other feathers of penguins provide them with nearly perfect insulation. These birds can incubate their eggs during the long polar winter while standing in the open with a wind-chill factor of more than 50 degrees below zero!

Content Development
Point out that feathers grow from little pits in the bird's skin. The soft down feather is the first

modern birds, *Archaeopteryx* had teeth in its beak. It also had toes and claws on its wings. In fact, *Archaeopteryx* would be classified as a dinosaur except for one important feature: It had well-developed feathers covering its entire body. The fossils of the earliest birds are rare, often poorly preserved, and very similar to those of many small dinosaurs. Because of this, there is much controversy over which fossils are those of birds and when birds first appeared on Earth. Although the fine points of bird evolution are hotly debated, one thing is certain—birds evolved from ancient reptiles.

Form and Function in Birds

Many characteristic features of birds—for example, feathers, wings, bones, beaks, and legs—differ dramatically among species adapted to different ways of life. In order to appreciate these differences in features, it is important to study birds that live in different habitats and examine the adaptations they show.

FEEDING Birds have high metabolic rates and burn many calories just to keep warm. For that reason, birds need to eat large amounts of food. They have evolved many specialized organs and behaviors that help them feed in a variety of ways.

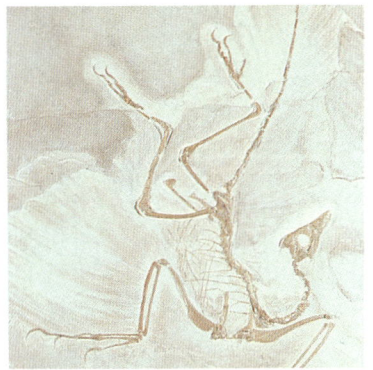

Figure 32–23 Archaeopteryx, which lived about 147 million years ago, is the oldest fossil that has been confirmed to be that of a bird.

Figure 32–24 The huge talons of this soaring eagle help the bird catch fish (bottom, left). Woodpeckers hear insects chewing beneath the bark of a tree. The beak of the woodpecker is strong enough to pierce the bark, revealing the bug beneath it (top, right). The featherless head of this vulture gives the bird a sinister look, but a lack of head feathers serves a useful function: A featherless head is easy to keep clean (bottom, right). Because vultures often put their head into the carcass of dead animals, cleanliness is important.

feather a bird grows and is easily seen on baby birds. It grows close to the skin and acts as insulation.
• **Why is down a good insulator?** (Down "puffs," allowing air to be trapped inside the tiny fibers.)

Point out that the contour feather is the feather that is most visible.
Tell students that large, strong quill feathers are in the wings and tail. They help the bird fly by giving the bird the ability to maneuver.

BACKGROUND INFORMATION
BIRD EVOLUTION

Many early researchers thought that the flying dinosaurs, or pterosaurs, were relatives of birds because they also had wings and hollow bones. But the structure of those wings was very different from that of bird wings. And pterosaurs never showed the slightest trace of feathers.

The next birdlike fossils after *Archaeopteryx* appeared about 10 million years later, in the early Cretaceous Period. Even though 10 million years is a relatively short time, evolutionarily speaking, these fossils were already "real" birds. By the Eocene Period, about 40 million years ago, nearly all modern orders of birds were firmly established. This means that near the end of the Cretaceous Period, birds experienced an extremely rapid adaptive radiation.

TEACHING SUPPORT

Teaching Resources
• Hands-On Activity: Do Oil and Water Mix? p. 157

MEDIA AND TECHNOLOGY
Videodisc
• *Aquatic Ecosystems: Freshwater Wetlands and Freshwater*

Use this bar code to show the underwater feeding techniques of the dipper.

Play frames 8002 to 8660

See the Biology Media Guide pages 43–45 for additional bar-code correlations for this section.

TEACHING SUPPORT

Laboratory Manual
- Examining Bird Adaptations, p. 415

Teaching Resources
- Activity: Exploring Body Temperatures of Birds, p. 5

MEDIA AND TECHNOLOGY

Biology Transparencies
- 45: The Internal Anatomy of a Bird

32–3 (continued)

Content Development
Mention that birds have several structural and physiological adaptations for flight other than their feathers and light bones. First, the respiratory system is designed to provide large amounts of oxygen needed during flight. A series of air sacs extend from the lungs into the body cavity. These sacs increase the total amount of oxygen available for flying by holding oxygen in reserve and also reduce the

Carnivorous birds, such as hawks and eagles, catch prey in razor-sharp talons and slice them to pieces with pointed beaks. Insect-eating birds do everything from picking insects off leaves and branches to catching them on the fly. Some insect eaters, such as woodpeckers, have a complex set of adaptations for drilling into wood and pulling out the insects that live there. Pollen and nectar feeders, such as hummingbirds, have long, probing beaks with which they reach deep into flowers. Their tongues are often equipped with a brushlike structure at the tip for lapping up nectar and fruit juices. Fruit-eating and seed-eating birds may have short, stout beaks or long, sharp ones, depending on the fruits or seeds they commonly eat. Filter feeders such as ducks and flamingoes have broad beaks with strainers built into the upper and side parts of the bill. These birds sift through murky water to filter out plankton a mouthful at a time.

The digestive system of birds, much like that of reptiles, shows specializations for carnivorous and herbivorous diets. Many birds have organs called **crop** and **gizzard**. The crop is an enlarged area of the esophagus, where food can be stored and moistened before it enters the stomach. In some species, the crop stores food that is later regurgitated for feeding to a bird's young. In still other species, the crop actually produces food that is fed to young chicks.

The gizzard is a specialized muscular part of the stomach that often contains small bits of gravel swallowed by a bird. The muscular walls grind the gravel and food together, thus crushing food particles and making them easier to digest. Both a crop and a gizzard are highly developed structures in seed-eating birds.

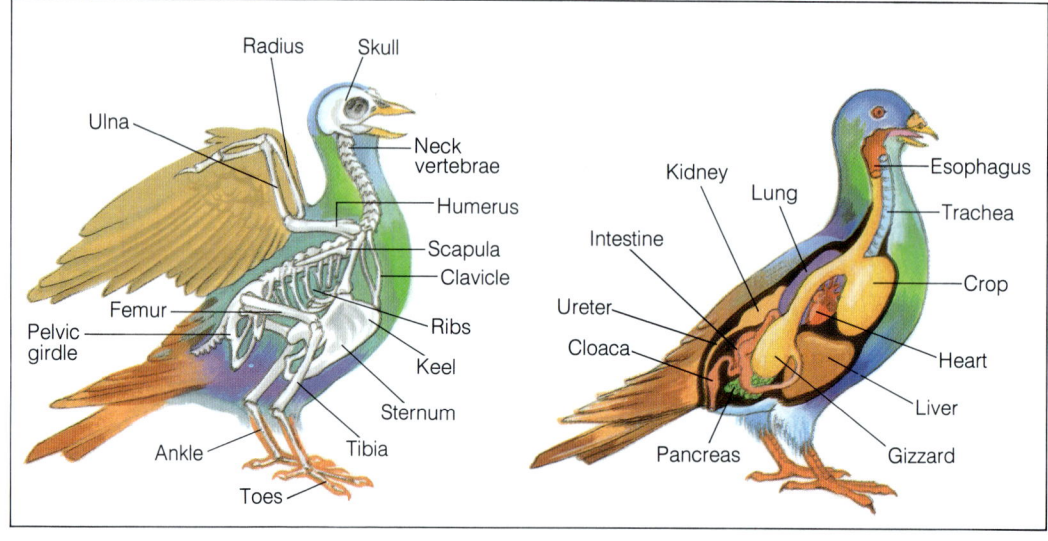

Figure 32–25 *The internal organs of a pigeon are shown here. You might be surprised to learn that there are fewer bones in the long neck of a giraffe than there are in the neck of a bird.*

bird's body weight. Second, birds lack a urinary bladder. Because they do not store urine in the body, their weight is reduced. Another adaptation for reducing weight is the laying of eggs after they are formed, rather than storing them inside the body.

Since flying requires an enormous expenditure of energy, birds have a high rate of metabolism to release the energy needed. As a result, the body temperature of some birds is as high as 43°C. Birds also have very powerful muscles. As much as 50 percent of their body weight is muscle.

Focus/Motivation
Ask students to comment on the often-used phrase, "He or she eats like a bird." Many people erroneously believe that birds are

RESPIRATION The respiratory system of birds is extremely efficient at taking in oxygen and eliminating carbon dioxide. This fact should not surprise you, as the high metabolic rate of birds demands an efficient gas exchange system.

The reason for this efficiency is that bird lungs are connected at both the anterior and posterior to large **air sacs** in the body cavity and bones. When a bird inhales, air travels through passageways that lead into the lungs. Some of this air remains in the lungs, where gas exchange occurs. Most of the air, however, goes through the lungs into posterior air sacs. When a bird exhales, air from the posterior air sacs passes into the lungs for gas exchange. Thus birds are able to remove oxygen from air when they inhale as well as when they exhale. For this reason, gas exchange in birds is more efficient than in other animals. The air sacs serve an additional function. They make a bird's body more buoyant, allowing the bird to fly more easily.

INTERNAL TRANSPORT Birds have a four-chambered heart and two separate circulatory loops. One half of the heart receives oxygen-poor blood from the body and pumps this blood to the lungs. Oxygen-rich blood returns to the other side of the heart to be pumped throughout the rest of the body. This dual-loop system ensures that oxygen collected by the lungs is distributed with maximum efficiency to the tissues that need it. To keep blood moving rapidly, a bird's heart beats quickly—from 150 to more than 1000 beats per minute.

EXCRETION Birds eliminate nitrogenous wastes by filtering them from the blood in the kidneys. Urine, which contains wastes in the form of uric acid, flows to the cloaca through the ureters. Most water is reabsorbed in the cloaca, leaving uric acid crystals in a white pastelike form. These crystals are the familiar "bird droppings."

Bird species that live far from shore or on small islands surrounded by sea water have no source of fresh water. Their diets contain larger amounts of salt than they need. For this reason, many of these species have evolved special salt glands near their eyes. These salt glands work like an extra pair of kidneys, except they specialize in excreting salt.

RESPONSE Despite the derogatory term "bird brain," birds are quite intelligent animals. The bird cerebrum, which controls such behaviors as flying, nest building, care of young, courtship, and mating, is quite large. The cerebellum is also well-developed, as might be expected in an animal in need of precisely coordinated movement. The medulla and spinal cord are much like those of reptiles.

Birds have extraordinarily well-developed eyes. Their excellent eyesight is reflected in a pair of sizable optic lobes in the brain. Birds see color very well—in many cases, better than

Figure 32–26 The huge forward-facing eyes of an owl help this great bird hunt at night. Its eyes are able to spot a tiny mouse foraging among the leaves on the dark floor of a forest.

BACKGROUND INFORMATION
BIRD VISION

Several birds can see in what for us is the ultraviolet part of the spectrum, and many can see well into the infrared region. In addition, the complex construction of their retinas undoubtedly enables many birds to see more subtle shades of colors than we can distinguish. Birds' retinas also enable them to see clearly in three places at once. Human retinas, in contrast, enable us to look closely in only one place at a time.

TEACHING SUPPORT

Teaching Resources
- Biology Case Studies: Weather and Nesting Success in Whooping Cranes, p. 11

MEDIA AND TECHNOLOGY
- *Videodisc*
- *Aquatic Ecosystems: Estuaries and Marine*
- *Videodisc*
- *Super Scents*
- *Video/Videodisc*
- *Seeing Sense*

Use this bar code to introduce students to the keen eyesight of vultures.

Play frames 14625 to 18195

See the Biology Media Guide pages 43–45 for additional bar-code correlations for this section.

32–3 (continued)

Content Development
Point out that feathers help a bird fly. The feathers on the body and wings streamline the bird for flight. Explain that the wings and flight muscles of flying birds are larger and more effective than are those of flightless birds. Point out that the feathers on the wings of flying birds form an airfoil to allow the birds to lift off from a surface.

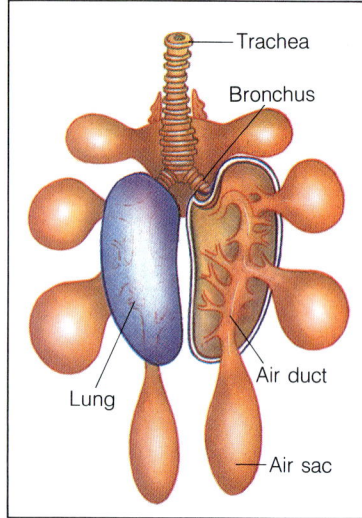

Figure 32–27 Birds have a unique respiratory system. Air sacs allow a bird to exchange oxygen and carbon dioxide when it inhales and when it exhales. The air sacs also make birds more buoyant.

Explain that when a bird flaps its wings, it produces increased pressure on the underside and slightly reduces the pressure on the upperside of the wing. Thus, air spills around the tip of the wing in an upward direction to reduce this difference. The result is an upward lift. When spread, the bird's tail feathers act as a brake.

Point out that airplanes are designed to use an airfoil similar to those of flying birds. Explain that lift and streamlining enable birds and airplanes to fly.

humans. Predatory birds such as hawks and eagles flying high in the air can spot mice on the ground with sight far keener than ours. The senses of taste and smell, however, are not well-developed. The olfactory lobes in a bird's brain are quite small.

Although birds lack external ears, they have ear openings in their head. Many bird species can hear quite well. For example, owls have an extremely acute sense of hearing that they use to find prey at night. Owls can so accurately hear mice crawling through dead leaves that they can swoop down and catch them in total darkness. Many migratory birds can hear the pounding of waves on the shore even when they are many kilometers away.

Some migratory birds use a magnetic sense to navigate. This magnetic sense, located somewhere in the head, operates like a built-in compass, responding to the Earth's magnetic field. Many other migratory birds use a combination of keen eyesight, instinct, and a built-in clock to navigate by the sun and stars.

MOVEMENT The many and diverse species of birds travel through different environments with wings, bodies, legs, and feet adapted for various types of locomotion. Some of the most impressive adaptations in birds involve flight. There are many variations in bird wings, depending on whether the animals soar like eagles, flap their wings steadily as robins do, or hover in place like hummingbirds.

Although the bones in a bird's wing are homologous to the bones in a human's arms, they have changed shape drastically to serve in flight. In flying birds, many large bones are nearly hollow. Hollow bones are not weak, however, because they are strengthened by internal struts similar to those used in the framework of tall buildings and bridges. The air sacs used in respiration extend inside certain bones, making the bones lighter. These air sacs also seem to help "air-condition" a bird's body by getting rid of excess heat generated by the flight muscles. Flying birds have other adaptations that decrease the weight they carry. One example is the shrinking in size of the sex organs during the time the birds are not breeding. As the birds prepare to mate, ovaries and testes grow larger until they reach functioning size.

To power the downward wing stroke necessary for flight, birds have large chest muscles. These muscles attach to a long keel that runs down the front of an enlarged breastbone, or sternum. The sternum, in turn, is firmly attached to the rib cage. In strong flying birds such as pigeons, the chest muscles may account for as much as 30 percent of the animal's mass.

Figure 32–28 This roadrunner comes by its name honestly. It can often be seen running along roads and highways in search of the small snakes and lizards that are a main part of its diet.

Content Development
Explain that the migratory behavior of certain birds is an adaptation for feeding and mating. Migratory birds fly to their winter grounds when food becomes scarce. In spring they fly to other regions to mate. Such behavior

Many birds use their flying ability to migrate, or travel long distances, between summer breeding grounds in the North and winter resting grounds in the South or the tropics. You might be surprised to learn that many of the birds common to North American summers spend their winters in cozy tropical forests side by side with parrots and other exotic species.

A number of birds, however, have lost the ability to fly. Some species, such as ostriches, spend their time walking or running on a powerful pair of hind legs. Their feet usually have three strong toes that make contact with the ground. Their wings are usually much reduced in size and are incapable of lifting them off the ground. These birds can get quite large, as they have no need to minimize their mass. Still other birds have given up flying in favor of swimming. Penguins are a familiar example. Their wings, legs, and feet are so reduced in size that they look quite comical on land. In water, however, their feet and wings are powerful flippers that enable them to "fly" through the water.

REPRODUCTION The reproductive system of birds is similar to that of reptiles. Both male and female reproductive tracts open into the cloaca. In many female birds, only one side of the reproductive tract develops, apparently to minimize body weight. The single functioning ovary, however, is sufficient to provide enough eggs. Male birds have no external reproductive organs. Instead, mating birds press the lips of their cloacas close together to transfer sperm from male to female.

Although bird eggs have hard outside shells, their internal structure and membranes are similar to those of reptiles. Most birds incubate their eggs until the eggs hatch. The time between laying and hatching varies among species from 13 days

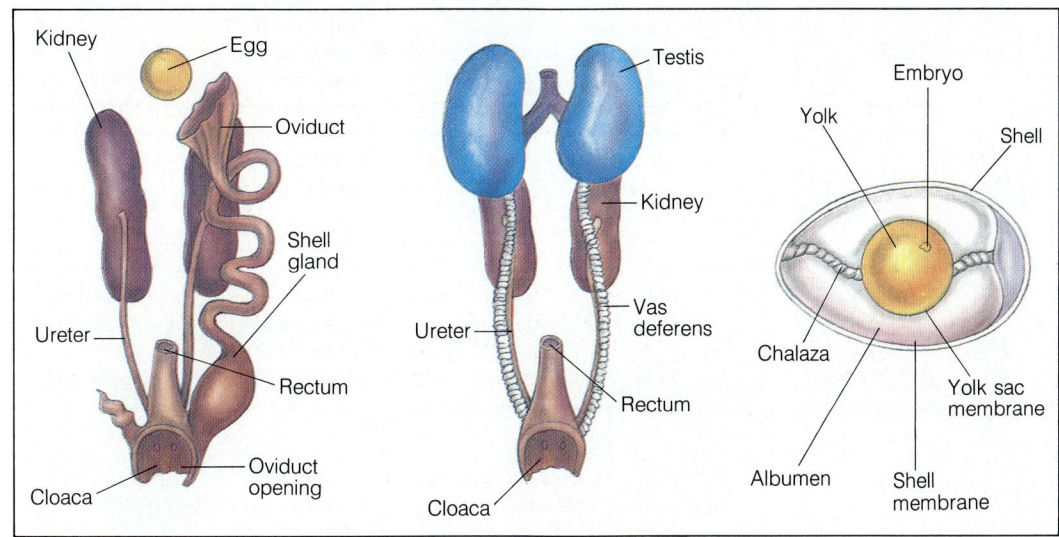

Figure 32-29 In birds, sperm produced by males and eggs produced by females pass through the cloaca. An egg is fertilized before the shell is formed around it.

ECOLOGY NOTE
ARTIFICIAL NESTS

At the beginning of this century, the wood duck was in danger of becoming extinct due to habitat destruction and hunting. A hunting ban in 1918 helped to increase the population of wood ducks slightly. However, the only real improvement in the population of wood ducks occurred when artificial nests were placed in trees. These nesting boxes replaced the scarce natural tree cavities in which these birds normally make their nests.

However, the arrangement of nesting boxes proved important. Female wood ducks that cannot find a suitable nesting place often lay their eggs in the nests of other females of their species. Natural nests are often located in secluded areas where they are relatively inconspicuous. The artificial nests were first arranged so that they were easily found by females. As a result, some nests contained as many as 40 eggs, instead of the normal 10 to 12. A large proportion of these nests were abandoned. It was found that the most effective arrangement was to place the nesting boxes so that one box was not visible from another. When this was done, no single nest received an excessive number of eggs.

TEACHING SUPPORT

Teaching Resources
- Vocabulary Review: Crossword, p. 1
- Hands-On Activity: Strictly for the Birds, p. 159

Study Guide
- Section 32-3, p. 309

32-3 (continued)

Content Development
Explain that all nests are designed to protect the eggs and the young birds as they develop. Point out that the female egg is fertilized by the male sperm inside the body. Explain that after the eggs are laid, the birds incubate the eggs by sitting on them.

Point out that after the eggs hatch, most species of birds are too small to hunt for food and water and need care.

Content Development
Emphasize the varied courtship behaviors of birds. Point out that in most cases these behaviors are instinctive.

- **What is the function of the songs sung by many male birds?** (These songs are used to claim a territory and by some species to attract females.)
- **Why are the feathers of many male birds more attractive than the female's?** (The bright plumage of the male birds attracts the female birds.)

Most students are probably familiar with the peacock's brilliant tail plumage and its strutting behavior. Birds of paradise also have brilliant plumage, and the males of some species carry out their courtship rituals while hanging upside down from trees. Male penguins signal females that they are ready to mate by presenting them with a pebble.

Enrichment
Play tapes of bird songs and discuss how birds learn the songs of their species.

Focus/Motivation
Have students observe microscope or 35 mm slides showing the development of a chick embryo at various stages. As students observe the slides, stress the importance of cell differentiation and the development of specialized tissues and organs within the chick.

to more than 50 days. (The incubation period for chickens is usually about 20 to 22 days.) When a chick is ready to hatch, it uses a small egg tooth on its bill to make a hole in the shell. After much pushing, poking, and prodding by the chick, the eggshell breaks open. After it hatches, the exhausted bird collapses for a while and allows its feathers to dry.

Some birds, such as chickens and ducks, are able to take care of themselves as soon as they hatch. Within hours or a day, they walk on their own and feed themselves. They stay close to their mother for protection, but she rarely feeds them. Other newly hatched birds, such as robins and sparrows, are blind and totally helpless when they hatch. For days or even weeks, they sit in the nest with mouth wide open, screaming for food. Both male and female parents are kept busy providing food for their hungry offspring.

Birds have fascinating courtship and mating behaviors. Some species, such as swans, mate for life. Others pair up only briefly to mate. In such cases, the female wanders off by herself after fertilization to nest and lay her eggs. The male, however, may continue to mate with many other females. In species such as peacocks, the males use brightly colored feathers to attract females and warn off other males during the breeding season. Male canaries and similar species sing to attract females and keep other males out of their territory.

Figure 32-30 Nests are supreme examples of the weavers' art. Birds often make complex nests of straw and other plant materials in which they lay their eggs and raise their young (left and right). Some birds, like the barn swallow, make their nest of mud, which they often attach to a building (center).

Enrichment
Have students do library research on endangered species of birds such as eagles, condors, whooping cranes, ospreys, and certain falcons. Students' reports should include information on how biologists are trying to save these and other species from extinction.

SECTION REVIEW 32-3

1. Birds are endothermic reptilelike animals with an outer covering of feathers. They have

How Birds Fit into the World

Imagine how dull a walk in the country would be without birds in the woods. No chirping. No singing. No graceful forms with brightly colored feathers. Humans the world over have always admired birds for their beauty and their powers of flight. But birds' contributions go far beyond their beauty.

In the course of a long evolutionary relationship with flowering plants, many bird species have come to serve vital functions in fields and forests. Hummingbirds serve as pollinators for a number of temperate and tropical plants. Fruit-eating birds disperse plant seeds when the seeds pass through their digestive tract unharmed.

Birds eat extraordinary numbers of insects. In some parts of the United States, chimney swifts begin their daily hunt for food at twilight. They zoom through the air, flitting back and forth, eating thousands of night-flying mosquitoes. Birds also eat caterpillars. Many bird species nest in the tundra, where huge insect "blooms" occur every spring at about the time baby birds hatch. The insect bloom, coming at hatching time, makes it easy for adult birds to find enough food for their young.

Humans use birds for many purposes. People in cold climates discovered long ago that in addition to being soft and comfortable, down feathers are good insulators. Down feathers are frequently used in making comforters and jackets. Many birds are favorite foods the world over, and raising them is part of the economy in many countries. With its low fat content, bird meat is a healthful source of protein in a balanced human diet. Birds such as chickens and turkeys have been specially bred for their meat. Because domestic strains of chickens and turkeys do not fly, their chest muscles are seldom used, making this part of the bird the juicy and tender "white meat." The leg and thigh muscles of these birds, used constantly for walking and running, are the "dark meat."

Figure 32–31 *This yellow warbler is feeding a group of hungry mouths. Because baby birds do not move from the nest, they use most of the nutrients in their food to grow.*

32–3 SECTION REVIEW

1. What are four characteristics of modern birds?
2. Describe the three different types of bird feathers.
3. In what two ways are a bird's bones adapted for flight?
4. How is a bird's respiratory system adapted to its high metabolic rate?
5. Penguin bones are not hollow like those of other birds. What is a logical explanation for this?
6. **Critical Thinking—Developing a Theory** In what three ways would life on Earth be different if there were no birds?

LABORATORY INVESTIGATION

OBSERVING TERRESTRIAL ADAPTATIONS OF A CHICKEN EGG

Before the Lab
1. Divide the class into groups of 2 to 4 students.
2. Gather all materials at least one day prior to the investigation. You should have enough supplies to meet your class's needs.

Pre-Lab Discussion
Have students read through the complete Procedure. Discuss the investigation by asking questions similar to the following:
- **What is the purpose of this laboratory investigation?** (To determine how birds are adapted to reproduction on land.)
- **What type of fertilization is conducted by birds?** (Internal fertilization.)
- **What is the most obvious function of the chicken egg shell?** (Protection.)
- **What is the incubation period for a chicken egg?** (20 to 22 days.)

Hold up a chicken egg for students to see. Explain to the students that the eggs they will be using in this investigation are unfertilized. Remind students to gently break their eggs to avoid destroying the contents before they have had the opportunity to make their observations.

Skills Development
Students will use the following skills while completing this laboratory investigation.
1. Observing
2. Applying
3. Relating
4. Manipulative
5. Inferring
6. Classifying
7. Safety
8. Comparing

LABORATORY INVESTIGATION

OBSERVING TERRESTRIAL ADAPTATIONS OF A CHICKEN EGG

PROBLEM
How are birds adapted to reproduction on land?

MATERIALS (per group)

400-mL beaker	petri dish
chicken egg (raw)	dissecting needle
hot plate	tongs

PROCEDURE
1. Hold a raw chicken egg in the palm of your hand. Gently close your fingers around the egg. Begin to squeeze the egg slowly, increasing the pressure until you are squeezing quite firmly. Note how much force the egg can withstand.
2. Gently shake the egg back and forth in your hand. Note whether you feel anything moving inside the egg.
3. Place the egg in the 400-mL beaker. Add enough water to barely cover the egg. Heat the water slowly on a hot plate just enough to make the water quite warm but still comfortable to the touch. **CAUTION:** *Do not touch the surface of the hot plate.* Watch the blunt end of the egg to see if tiny bubbles of air escape from the shell. Use the tongs to gently remove the egg from the beaker.
4. Gently crack open the egg over a petri dish. Carefully pour the contents of the egg into the petri dish.
5. Examine the structure and texture of the shell. Look in the blunt end of the shell. Locate the membrane and the air space.
6. Examine the contents of the egg in the petri dish. Refer to Figure 32–29 on page 729 as you work. The germinal disk is a small white spot on the top of the yolk. This is the spot where fertilization and the development of the embryo occur.
7. Find the whitish strands attached to both sides of the yolk. These are the chalazas. When the egg is intact, the chalazas stretch from the yolk to the membrane located just beneath the shell. The chalazas twist when the egg rolls, keeping the germinal disk and the embryo at the top of the egg.
8. The clear fluid in the petri dish is the albumen. In an intact egg, the albumen completely fills the space between the yolk and the membrane beneath the shell.
9. The yolk is the yellow material. Note how the yolk appears to form a slightly flattened sphere. The yolk is surrounded by a membrane that helps it maintain its shape. Puncture the membrane surrounding the yolk with the dissecting needle. Observe what happens.

OBSERVATIONS
1. What happened when you squeezed the intact egg?
2. What did you observe when you shook the egg back and forth?
3. What did you observe when you heated the egg?
4. What are the characteristics of the egg shell?
5. What happened to the yolk when you pierced the membrane?

ANALYSIS AND CONCLUSIONS
1. What characteristics of the egg shell help the egg survive on land?
2. Based on your observations, how does a developing chick get oxygen?
3. Using knowledge of the structures you observed in the egg, explain what happened when you shook the egg back and forth.
4. For what purpose do people use egg yolks? For what purpose do developing chicks use egg yolks?
5. How are birds' eggs adapted to reproduction on land?

732

Safety Tips
Caution students to handle all glassware carefully. Students should be warned not to touch the surface of the hot plate. To avoid cuts, students should use the dissecting needle carefully. Caution students to handle the chicken egg carefully to avoid premature breakage.

Teaching Strategy
1. Circulate through the laboratory to ensure that the groups are following the directions carefully as they work.
2. Discuss how the investigation relates to the chapter concepts by asking questions similar to the following:
- **How are predators able to destroy and eat the eggs of their prey?** (Answers will vary, such as dropping the egg to crack it,

SUMMARIZING THE CONCEPTS

The key concepts in each section of this chapter are listed below to help you review the chapter content. Make sure you understand each concept and its relationship to other concepts and to the theme of this chapter.

32–1 Reptiles

- Reptiles are vertebrate animals that have adaptations such as amniotic eggs, dry skin, and efficient respiratory and circulatory systems that enable them to live their entire lives out of water.
- Reptiles eliminate liquid wastes in the form of urine produced in their kidneys. Many reptiles conserve water by excreting nitrogen-containing wastes in the form of uric acid.
- The tuatara retains many features of the ancient reptiles from which it evolved.
- All lizards have legs, clawed toes, external ears, and movable eyelids. Snakes are lizards that have lost both pairs of legs during their evolution. Crocodilians live in tropic climates and are among the largest living reptiles. Turtles and tortoises are reptiles that have a shell consisting of a carapace and a plastron.

32–2 The Evolution of Temperature Control

- Control of body temperature is important for animals, particularly in habitats where temperature varies widely.
- Modern reptiles are ectotherms; they obtain heat from outside their body. Birds and mammals are endotherms.

32–3 Birds

- Birds are endothermic reptilelike animals with an outer covering of feathers, two legs used for walking or perching, and front limbs modified into wings. Fossils confirm the evolution of birds from ancient reptiles.
- Bird feathers are generally of three different types: contour, down, and powder.
- Because of their high metabolic rates, birds need to eat large amounts of food. The digestive system of many birds contains a crop and a gizzard. Birds eliminate nitrogenous wastes in the form of uric acid.
- Birds have an extremely efficient respiratory system because of the presence of air sacs. These sacs also make a bird's body more buoyant. Birds have a four-chambered heart and two circulatory loops.

REVIEWING KEY TERMS

Vocabulary terms are important to your understanding of biology. The key terms listed below are those you should be especially familiar with. Review these terms and their meanings. Then use each term in a complete sentence. If you are not sure of a term's meaning, return to the appropriate section and review its definition.

32–1 Reptiles
 reptile
 amniotic egg
 transition
  fossil
 internal
  fertilization
 carapace
 plastron

32–2 The Evolution of Temperature Control
 ectotherm
 endotherm

32–3 Birds
 bird
 contour feather
 down feather
 powder feather
 crop
 gizzard
 air sac

733

4. The egg shell is hard and thin.
5. The yolk began to spill out of the membrane.

Analysis and Conclusions

1. The egg shell is hard and strong in order to prevent drying and breakage. It is also thin to permit the passage of air.
2. The developing chick gets oxygen through the egg shell and the membranes.
3. The material inside did not move because the albumin acts as a shock absorber and holds the yolk in place.
4. Both humans and birds use the yolk of a chicken egg for food.
5. The egg shell is hard and strong, making the egg resistant to drying and breakage and able to support its own weight on land. The shell is also thin enough to permit the exchange of gases. The egg contains a large supply of yolk to provide food for a developing embryo.

Going Further: Enrichment

Have students repeat this investigation using the eggs of different types of birds. Have them observe the similarities and differences among the different types of bird eggs and record their observations.

---

pecking at the egg shell, and using the throat to crush the egg shell.)
• **What structures within the chicken egg act as shock absorbers for the developing embryo?** (The chalazas.)
• **How is a chicken egg different from an amphibian egg or a reptile egg?** (An amphibian egg is surrounded by a jellylike material and must remain in water to avoid drying out. A reptile egg has a soft, leathery shell. A chicken egg has a hard outer shell.)

Observations

1. The egg was able to withstand the force without cracking.
2. It was not possible to detect anything moving back and forth inside the egg.
3. Air bubbles escaped through the shell.

733

CHAPTER REVIEW

CONTENT REVIEW

Multiple Choice
1. c 2. a 3. c 4. b
5. d 6. b 7. c 8. c

True or False
1. T
2. F amphibians
3. F Cretaceous Period
4. F Reptiles
5. F Ectothermy
6. T
7. T
8. T

Word Relationships
1. contour feathers
2. crop
3. Reptiles
4. air sacs

CONCEPT MASTERY

1. Reptiles have lungs, lay eggs on land, and have a scaly skin that prevents water loss. These adaptations enable reptiles to live entirely out of water.
2. Scientists believe that the mass extinctions were caused by either slow changes in the worldwide climate or by a quick change in climate that occurred when a meteorite struck the Earth, forcing huge clouds of dust into the air. The clouds of dust blocked sunlight from reaching the Earth's surface.
3. Contour feathers cover the bird's body and give it shape; they also enable the bird to fly. Down feathers are small and fluffy; they trap warm air close to the bird's body. Powder feathers give off a powder that helps make the bird's feathers waterproof.
4. The excretion of nitrogen-containing wastes in the form of uric acid means that little water is used in the excretion of these wastes. By conserving water, reptiles are better able to survive on land.
5. Birds that have no source of fresh water have evolved special salt glands near their eyes. These glands excrete the excess salt that the birds consume in their diet. Thus these birds can rid themselves of salt without using large amounts of water to dilute the salt so it could be excreted in urine.

CHAPTER REVIEW

CONTENT REVIEW

Multiple Choice

Choose the letter of the answer that best completes each statement.

1. Reptiles that lay eggs that develop outside the mother's body are
   a. extinct.   c. oviparous.
   b. ovoviviparous.   d. externally fertilized.
2. A reptile that flips out its sticky tongue to catch insects is the
   a. chameleon.   c. iguana.
   b. monitor lizard.   d. crocodile.
3. A type of grooming in which a bird pulls its feathers through its beak is called
   a. shedding.   c. preening.
   b. contouring.   d. perching.
4. An animal that generates most of its heat inside its body is a (an)
   a. ectotherm.   c. reptile.
   b. endotherm.   d. tuatara.
5. The pineal gland of the tuatara is sensitive to
   a. touch.   c. heat.
   b. vibration.   d. light.
6. The single most important characteristic that separates birds from reptiles is
   a. endothermy.   c. two legs.
   b. feathers.   d. wings.
7. Feathers on the wings and tail of a bird that provide a lifting force and balance are
   a. down feathers.   c. flight feathers.
   b. barbs.   d. powder feathers.
8. The muscular part of a bird's stomach that contains gravel used to crush food particles is the
   a. cloaca.   c. crop.
   b. barbule.   d. gizzard.

True or False

Determine whether each statement is true or false. If it is true, write "true." If it is false, change the underlined word or words to make the statement true.

1. Reptilian eggs are <u>amniotic eggs</u> that are named after one of the membranes that surround the embryo.
2. The changes in the Earth's surface and climate during the Permian Period made life difficult for <u>reptiles</u>.
3. A mass extinction of dinosaurs and many other animals and plants occurred at the end of the <u>Triassic Period</u>.
4. <u>Birds</u> warm up by basking in the sun.
5. <u>Endothermy</u> is a way of conserving energy in warm climates.
6. <u>Down feathers</u> are soft, fluffy feathers that trap warm air close to a bird's body.
7. The <u>air sacs</u> help make a bird more buoyant during flight.
8. Some migratory birds have a <u>magnetic sense</u> that helps them find their way.

Word Relationships

Replace the underlined definition with the correct vocabulary word.

1. Birds have <u>large feathers that cover their bodies and wings.</u>
2. The digestive system of a bird contains a <u>structure where food is moistened before it enters the stomach.</u>
3. <u>Vertebrate animals that have lungs, scaly skin, and amniotic eggs</u> have evolved from amphibians.
4. When a bird inhales, most of the air goes through the lungs into <u>posterior structures</u> that also make a bird's body more buoyant.

CRITICAL AND CREATIVE THINKING

1. Unlike amphibians, reptiles are not able to take in oxygen and give off carbon dioxide through their skin. Reptiles developed efficient respirato-

CONCEPT MASTERY

Use your understanding of the concepts developed in the chapter to answer each of the following in a brief paragraph.

1. What are three adaptations that enable reptiles to live entirely out of water?
2. Discuss two possible explanations for the mass extinction that occurred about 65 million years ago.
3. Identify the three main types of bird feathers.
4. Explain why the excretion of nitrogen-containing wastes in the form of uric acid is another successful adaptation to a land environment shown by reptiles.
5. Describe the special adaptations that have developed in bird species that have no source of fresh water.

CRITICAL AND CREATIVE THINKING

Discuss each of the following in a brief paragraph.

1. **Applying concepts** Reptiles developed efficient respiratory systems after they evolved from water-dependent amphibians. They then developed a more efficient circulatory system. Explain why these systems evolved in this order.
2. **Relating cause and effect** Most of the various species of modern reptiles have changed little since their great period of adaptive radiation about 200 million years ago. Explain why drastic changes have not occurred in reptiles since then.
3. **Making predictions** Suppose you came upon the shore of a tropical island. Predict the method of regulating body temperature that you would expect to find in the animals that live on the island. Explain how this method is an adaptation to this environment.
4. **Drawing conclusions** You are given the description of a certain animal and told that it is endothermic, has two legs, and modified front limbs. You are also told that it has a four-chambered heart, two separate circulatory loops, and a well-developed cerebellum. Describe at least three more characteristics you can add to this description.
5. **Making comparisons** Compare and contrast the characteristics of reptiles with those of amphibians. Explain how each is suited to a particular environment.
6. **Summarizing information** Discuss some of the characteristics of birds that enable them to fly.

7. **Using the writing process** Pretend you are a visitor to Earth from the fictitious planet Chillee. Your ancestors visited planet Earth about 300 million years ago and then again about 55 million years ago. Describe the changes that would have occurred on Earth from the details recorded in the logbook your people maintained on their voyages.

---

ry systems first. They used these systems in a life on land away from water. Amphibians could survive with only a relatively simple circulatory system. With a move away from water and with the development of an efficient respiratory system, reptiles developed a more efficient circulatory system to move oxygen to and from body cells.

2. Modern reptiles have changed little in 200 million years because they are well adapted to live in many different land environments. Since they are widely dispersed and successful, little evolutionary change has occurred.

3. Many of the animals that lived here would be exothermic animals. In a tropical environment, ectothermy is a successful way to live. It is energy efficient. Ectothermic animals can soak up the energy from a warm tropical sun. They have little need to waste food calories in maintaining their body temperature.

4. This description closely fits that of a bird. So feathers, an ability to fly, the ability to lay eggs, and other answers are possible.

5. Reptiles have a dry, scaly skin, lay eggs that are able to survive on dry land, and have a well-developed respiratory system—adaptations needed in a dry environment. Amphibians have a moist skin and a less well-developed circulatory system and must lay eggs in water—adaptations appropriate for a watery environment.

6. Birds have large chest muscles, light yet strong feathers, hollow bones, and air sacs. All help a bird fly.

7. Check students' writings for accuracy and clarity of expression. Answers will vary but should include the following: The climate of the Earth changed greatly over the time that elapsed between the visits. About 300 million years ago, the great adaptive radiation of reptiles occurred. Dinosaurs walked the Earth. About 55 million years ago, the great adaptive radiation of mammals occurred. Most of the reptiles, especially the giant ones, had died out by this time.

CHAPTER 33 *Mammals*

Section	Laboratory Investigations and Activities
33–1 Mammals pages 737–745 THEMES: Scale and Structure, Evolution	**Student Edition** LABORATORY INVESTIGATION: Examining the Typical Mammalian Body Covering, p. 752 **Teacher Edition** DEMONSTRATION: Observation of a Mammal, p. 736D
33–2 Important Orders of Living Mammals pages 746–751 THEMES: Unity and Diversity, Scale and Structure	**Teacher Edition** DEMONSTRATION: Skulls of Mammals, p. 736D **Laboratory Manual** The Most Intelligent Mammal, p. 423
Chapter Review pages 753–755	

Teacher Resources

Books

Anderson, Sydney, ed. *Simon & Schuster's Guide to Mammals.* Simon & Schuster, 1984.

Arseniev, V. A. *Atlas of Marine Mammals.* T.F.H. Publications, 1986.

Burton, John S., and Bruce Pearson. *The Collins Guide to the Rare Mammals.* Stephen Greene Press, 1988.

Corbet, G. B., and J. E. Hill. *A World List of Mammalian Species.* Cornell University Press, 1980.

Dietz, Tim. *Whales and Man: Adventures with the Giants of the Deep.* Yankee Books, 1987.

Evans, Peter G. H. *The Natural History of Whales and Dolphins.* Facts on File, 1987.

Macdonald, David, ed. *Encyclopedia of Mammals.* Facts on File, 1984.

Weyler, Rex. *Song of the Whale.* Anchor Press/Doubleday, 1986.

Wild Animals of North America, rev. ed. National Geographic Society, 1987.

CHAPTER PLANNING GUIDE

Other Activities	Media and Technology
Biotechnology Workbook 　Animals in Research, p. 141 **Study Guide** 　Section 33–1, p. 317	**Interactive Videodisc** 　In the Company of Whales **Video/Videodisc** 　Seeing Sense **Videodisc** 　Sound Sense **Biology Media Guide** 　pages 46–48
Teaching Resources 　VOCABULARY REVIEW: Word Game, p. 1 　ACTIVITY: Classifying Mammals, p. 3 　ACTIVITY: Which Animal Doesn't Belong? p. 5 **Study Guide** 　Section 33–2, p. 320	**Interactive Videodisc** 　In the Company of Whales **Interactive Videodisc/CD ROM** 　Amazonia 　The Virtual BioPark **Videodisc** 　Aquatic Ecosystems: Freshwater Wetlands and Freshwater **Biology Media Guide** 　pages 48–49
Teaching Resources 　Chapter Test A, p. 11 　Chapter Test B, p. 15	**Computer Test Bank** 　Chapter Test, p. 347

Audiovisuals

Mammals, rev. Film or video. Coronet.
The Mammals. 5 filmstrips with cassettes. Encyclopaedia Britannica Educational Corporation.
The Great Whales. Video. National Geographic Society.
Introduction to the Mammals. Film or video. Carolina.
Man-eaters of India. Video. National Geographic Society.
Mammals of the Sea. Video. PBS Video.

CHAPTER 33 Mammals

CHAPTER OVERVIEW

The group of animals known as mammals includes many organisms that vary greatly in size and habits. However, they all have certain characteristics in common. They are all endothermic. All mammals maintain a constant body temperature using some combination of fur, hair, and subcutaneous fat to conserve body heat. Many mammals also have a mechanism for getting rid of excess heat when necessary. With a few exceptions, mammals are viviparous, bearing living young. The young are nourished with milk from the mammary glands of the females. This is the characteristic that gives mammals their name.

The teeth of mammals vary greatly and are used as the basis for putting various species into orders. All mammals are air-breathers and have well-developed four-chambered hearts. Another characteristic of mammals is the presence of the diaphragm, which is used in breathing and also separates the chest cavity from the abdomen.

Mammals evolved over time. By the end of the Cretaceous Period, there were three distinct groups of mammals. They include the monotremes, the marsupials, and the placental mammals.

Mammals have evolved many shapes and special adaptations that allow them to function in a variety of different environments. The feeding habits of mammals are extremely varied. Their teeth, organ systems, and body shapes are adapted to their particular modes of nutrition. Mammals have well-developed respiratory, transport, and excretory systems. They have the most highly developed brains and sense organs of any animals. They depend on their nervous system to provide them with information about their environment and the ability to react appropriately to allow them to survive. All mammals have four limbs, but these limbs have evolved into many varied structures, including strong legs for running, limbs adapted for grasping and climbing, wings, and flippers.

Reproduction varies among the three groups of mammals. The primitive monotremes are egg-layers. There are only a few examples of these oviparous mammals still in existence. The marsupials are viviparous, but the young are nourished with yolk from the egg sac instead of a placenta. Because of this, the young are born at a very early stage of development and migrate into a pouch for further development. The placental mammal has a longer period of development within the body of the female, but even after birth, there is usually a period of care provided for the young.

The characteristics used to classify mammals include tooth structure, the bones of the head, and the method of reproduction. Monotremes are egg-layers, marsupials are the pouched mammals, and there are sixteen orders of living placental mammals. The placental mammals are much more abundant than their marsupial or monotreme relatives. Twelve of the sixteen orders of placental mammals are discussed in this chapter.

Order Insectivora, which contains the most primitive placental mammals, includes shrews, hedgehogs, and tree shrews. Bats belong to order Chiroptera. Although not strictly toothless, members of order Edentata—such as armadillos and anteaters—have greatly reduced teeth. Order Rodentia includes many familiar—and sometimes unpopular—animals such as rats and gophers. Although rabbits and hares greatly resemble rodents, their dentition is different, so they are placed in a separate order, Lagomorpha. Meat-eaters such as dogs, cats, and seals belong to order Carnivora. Order Cetacea contains whales, dolphins, and porpoises. Although the members of order Sirenia—manatees and dugongs—resemble cetaceans, they are actually more closely related to elephants (order Proboscidea). Ungulates with cloven hooves or an even number of toes, such as goats and hippopotami, belong to order Artiodactyla. Ungulates with an odd number of toes, such as horses and rhinoceroses, belong to order Perissodactyla. The order Primates includes humans, monkeys, and a number of other animals with well-developed cerebrums and complicated behaviors.

Mammals evolved from early reptile ancestors and are today found throughout the world. Humans and other mammals live in almost any kind of environment. Herbivorous mammals are major consumers of plant materials in many parts of the world. Mammals are found in the air and the water as well as on land. Mammals have both positive and negative effects on one another and the environment.

33-1 MAMMALS

Section Focus 33-1

The purpose of this section is to introduce students to the characteristics of the animal group known as mammals. The group is large and diverse in appearance and habit. Mammals are found on land, in the air, and in the waters of the Earth. All mammals are endothermic. They generate enough heat to allow themselves to maintain a constant body temperature. They have a number of devices to help them conserve heat and eliminate excess heat. All mammals have some combination of fur, hair, and subcutaneous fat that helps to conserve heat. Many mammals have sweat glands that allow them to eliminate excess heat when necessary.

With the exception of the primitive monotremes, all mammals are viviparous. Viviparous animals develop for a period of time within the body of the mother and then are born alive. Female mammals have mammary glands that produce milk for the nourishment of the young. The name of the group comes from these structures. The teeth of mammals are adapted to their feeding habits. This characteristic is used by scientists to classify mammals into various orders. Mammals are all air-breathers with well-developed breathing muscles, including a diaphragm that separates the chest cavity from the abdomen. All mammals have well-developed four-chambered hearts with two completely separate cir-

CHAPTER PREVIEW

culatory loops that form an efficient system for the exchange of gases with the environment and within the body.

According to the fossil record, the first mammals were probably very small and probably active primarily at night. By the end of the Cretaceous Period, the mammals had split into three groups. The most primitive group is the monotremes. The second group includes the opossums and kangaroos. These pouched animals are known as marsupials. The third group, the placental mammals, includes the mammals that are most familiar. The fossil record is incomplete, so many details are missing. However, we do know that placental mammals experienced a period of adaptive radiation in North America and Europe while the marsupials were experiencing a similar evolutionary explosion in Australia, South America, and Antarctica.

The organ systems and limbs of mammals have evolved to serve many functions in different environments. Just a few examples are discussed in this section. Carnivorous mammals have strong, sharp teeth that are well adapted for eating flesh and possess physical characteristics and behaviors that allow them to hunt and capture prey. Herbivorous mammals have flat teeth adapted for grinding plant material as well as adaptations in their digestive tracts that make it possible for them to exist on a diet of vegetation. Some mammals have unique adaptations for feeding on blood or filtering tiny microorganisms from vast quantities of ocean water.

All mammals have powerful breathing muscles. Some are able to make sounds as the air is exhaled over the vocal cords. The mammalian circulatory system efficiently transports gases and nutrients throughout the body of the animal. The highly developed kidneys of mammals efficiently remove waste from the body as well as control the overall composition of all body fluids. These wonderful organs have allowed mammals to survive in many habitats they could not otherwise have inhabited.

Mammals have the most highly developed brains of any animals. They also depend on well-developed sense organs that give them detailed information about the environment.

Mammals vary greatly in their methods of reproduction. The young of the most primitive mammals, the monotremes, develop in eggs incubated outside the body of the mother. After hatching, however, the young are nourished on milk from the mammary glands. The marsupial mammals are viviparous. The young develop for a time inside the body of the mother, where they are nourished by yolk from an egg sac. Early in the embryonic development, the young migrate to a pouch, where they attach to a nipple and continue to develop. The placental mammals develop within the uterus of the mother, where they are nourished by an organ called a placenta.

Performance Objectives 33-1

1. Discuss the characteristics of mammals.
2. Explain how mammals maintain a constant body temperature.
3. Compare the structural adaptations shown by carnivorous and herbivorous mammals.
4. Discuss some of the features of mammals' internal body systems.

Science Terms 33-1

mammary gland   p. 738
monotreme   p. 739
marsupial   p. 739
placental mammal   p. 739
rumen   p. 740
cecum   p. 741
marsupium   p. 744
placenta   p. 744

33-2 IMPORTANT ORDERS OF LIVING MAMMALS

Section Focus 33-2

The purpose of this section is to introduce the important orders of mammals and explain how mammals fit into the world. Scientists use several characteristics to classify mammals. The structure of teeth, the number and kind of bones in the head, and the method of reproduction are among the characteristics.

The monotremes are the egg-laying mammals. There are only six living species of monotremes today. The most familiar is the duckbill platypus, but the group also includes the spiny anteater, or echidna. The marsupials are the pouched mammals, such as kangaroos and koalas. The only North American marsupial is the opossum. The living placental mammals are placed in sixteen orders. Twelve of them are discussed in this section.

Performance Objectives 33-2

1. Compare reproduction and care of the young among the monotremes, marsupials, and placental mammals.
2. Name and give examples of several important orders of mammals.

Science Terms 33-2

prehensile tail   p. 750

DISCOVERY LEARNING

TEACHER DEMONSTRATIONS
Modeling

Observation of a Mammal
This activity can be used to introduce Chapter 33.

Bring an example of a living mammal to class. Allow students some time to observe the animal. As a class, make a list of the characteristics of the animal. Go back through the list and have the students decide whether or not the characteristics that are listed are those of all mammals or just certain ones. Add any important mammalian characteristics that may have been omitted. Discuss with students some of the characteristics that might have been unique to a smaller group of mammals, such as an order.

Skulls of Mammals
This demonstration can be used to introduce Section 33-2.

Bring several skulls of mammals to class. Have students examine the teeth and the bones of the head and jaw. Explain that these are some of the important characteristics used to classify mammals into different orders.

CHAPTER 33
Mammals

GUIDED ENQUIRY
Pose the following questions to students and have them record their responses. Point out that they will gain a better understanding of the key concepts if they read the chapter with these basic questions in mind. Upon completion of the chapter, pose the questions again. Ask students to compare their initial responses with those they have developed after reading the chapter.

- How are the characteristics of mammals related to their lifestyles?
- How is endothermy related to the success of mammals?
- What evidence is there to indicate that mammals are descended from ancient reptiles?
- How can mammals be classified into groups?
- Why are mammals important?

INTRODUCING CHAPTER 33
Using the Visuals
Tell the students that you are going to make a list of all of the animals they can name in one minute. After you perform this exercise, check to see how many of the named animals are mammals. Usually, a good percentage will be. Discuss with the students that the mammals are the group of animals that humans belong to and with which they have the most contact. Have students look at the chapter-opener illustration and read the caption. Ask a student to read the opening paragraph below the illustration. Ask them to think about the questions at the end of the paragraph and to formulate some tentative answers to the questions. Record the answers for later discussion.

CHAPTER 33
Mammals

These now-extinct mammoths, like many other mammals, evolved layers of fat and thick fur coats, which served them well during ice ages. Today, mammals include some of the largest and most intelligent animals that have ever lived.

Around the end of the Cretaceous Period, Earth's climate changed dramatically. The Rocky Mountains and other large mountain ranges arose, blocking the flow of warm, moist air from the oceans over the continent. Inland seas and swamps on the side of the mountain ranges away from the ocean dried up, winters became colder, and summers became hotter and drier.

These conditions proved lucky for mammals, whose small ancestors had spent millions of years scrambling around in the shadow of the giant dinosaurs. As mammals evolved, they became accomplished endotherms that could survive cooler and more variable climates. So as dinosaurs vanished, the great mammalian radiation began. In this chapter you will see how successful mammals have been and examine some of the adaptations that make them so fascinating.

TEACHING STRATEGY 33–1
Focus/Motivation
A good way to illustrate the point of diversity among mammals is with a collage of as many types of mammals as possible. Using a large piece of paper or poster board, begin a collage with this chapter. You find the first few examples of pictures or drawings of mammals. Encourage the students to bring one picture of a different mammal each day during this study. (You can use them as admission tickets for class or

33–1 Mammals

Guide For Reading

- What are the characteristics of mammals?
- What are the three main groups of mammals?
- How do mammals perform essential life functions?

The group of animals called mammals includes many diverse species that vary greatly in appearance. Mammals range in size from a tiny mouse nibbling its way along a corn cob several times its size to a huge elephant uprooting a gigantic tree with its tusks and trunk. Mammals can be found flying in the air, running along the ground, and swimming in the sea. Although they differ in size and habits, however, all members of the class Mammalia share certain characteristics.

What Is a Mammal?

Mammals are endothermic animals, which means they are able to generate substantial body heat internally. Most species are experts at maintaining a constant body temperature. Mammals use various combinations of fur, hair, and subcutaneous fat to conserve body heat. Subcutaneous fat is fat located under the skin (*sub-* means under; *cutaneous* refers to the skin). (We may not always appreciate having a layer of subcutaneous fat, but it is there for a reason!) Many mammals also have sweat glands that help cool the body. Sweat produced by sweat glands evaporates from the skin, lowering body temperature whenever necessary.

Figure 33–1 One characteristic that unites all mammals is hair. This brown bear and her cubs can sleep through winter's cold insulated by their thick coats and a layer of fat beneath their skin.

TEACHING SUPPORT

Biotechnology Workbook
- Animals in Research, p. 141

MEDIA AND TECHNOLOGY

Interactive Videodisc
- *In the Company of Whales*
Use this bar code to show mammalian characteristics in a representative mammal—the whale.

Play frames 4384 to 5035

See the Biology Media Guide pages 46–48 for additional bar-code correlations for this section.

ECOLOGY NOTE
SAVING RHINOS

Excessive hunting of black rhinos for their horns has meant that there are now only about 2000 remaining worldwide. In an effort to increase the shrinking population, the World Wide Fund for Nature (formerly World Wildlife Fund) began projects to dehorn rhinos in Namibia in 1988, and Zimbabwe in 1992. The aim of these projects was to remove poachers' incentive to kill the rhinos. As well as being relatively safe from hunters, a dehorned rhino retains its ability to ward off other predators, for it is the rhino's size, not its horns, that deters attacks.

Although there have been some incidences in which spiteful poachers deliberately killed dehorned rhinos, dehorning seems to be having the desired effect. Along with translocation of rhinos to breeding sanctuaries and increased guarding of rhino habitats, dehorning should help to prevent the extinction of this spectacular animal.

Figure 33–2 All female mammals nurse their young, feeding them milk they produce in mammary glands. These glands are the source of the class Mammalia's name.

Figure 33–3 Be they lions or whales, all mammals breathe air. Breathing is easy for land mammals, but sea mammals must return to the surface to breathe.

With the exception of several very primitive species that lay eggs, all mammals are viviparous. This means that young mammals develop within the mother for a time and then are born alive. Female mammals have **mammary glands**, which produce milk to nourish the young for some time after they are born. **Mammary glands, which give mammals their name, are probably the most important characteristic that scientists use to include an animal in class Mammalia.**

Mammals have several kinds of teeth. Combined with their jaws, the teeth of mammals bite, chew, and grind food efficiently. The teeth and jaws of various mammalian species take many forms, depending upon the species' feeding habits. Scientists use the teeth of a mammal to classify it in one of the mammalian orders. You will read about these different orders in the next section.

Mammals have well-developed breathing muscles, including a diaphragm that separates the chest cavity from the abdomen. The diaphragm, along with other muscles in the chest, pulls air into the lungs by expanding the chest cavity. Mammals have a four-chambered heart consisting of two atria and two ventricles. Each side of the heart, consisting of an atrium and a ventricle, is part of a completely separate circulatory circuit. One circuit moves blood to and from the lungs, and the other circuit moves blood to and from cells in the rest of the body. The two circuits make up an efficient system for the transfer of gases with the environment and for the delivery and removal of gases and other materials to and from body cells.

33–1 (continued)

Content Development
Discuss the term *viviparous*. Point out that the presence of mammary glands is the most important characteristic of mammals. All animals that actually produce milk from mammary glands are mammals.
- Why are echidnas, which lay eggs, considered to be mammals? (Because they have mammary glands.)
- What major characteristic do scientists use for classifying mammals into orders? (The teeth.)

Enrichment
If you have students who are particularly interested in anatomy, encourage them to obtain a cow, pig, or sheep heart from the meat

Evolution of Mammals

The first mammals were very small and, according to fossil evidence, resembled species of tree shrews alive today. They were probably nocturnal, which means they were active primarily at night. Because they were endotherms, they did not need to obtain heat from the environment in order to remain active—thus making them able to function well after dark. The ectothermic dinosaurs, on the other hand, would probably have been rather sluggish after sunset. (Even today there are very few nocturnal lizards, and those that do exist live only in the tropics.)

By the end of the Cretaceous Period, the mammals had split into three groups. The first group, and the most primitive, is the **monotremes**. Today only three species of monotremes survive. The most familiar species are the duckbill platypus and the Australian spiny anteater.

The second group, the **marsupials**, includes opossums, kangaroos, wombats, and koalas. Each of these species has a pouch in which its young live for a time.

The third group, the **placental mammals**, include the mammals you are most familiar with. Mice, cats, whales, elephants, and humans are just a few examples of placental mammals.

Because the fossil record of the earliest mammals is incomplete, it is hard to say precisely where and when each of these three groups appeared. However, we do know that placental mammals experienced a period of adaptive radiation in North America and Europe. Marsupials experienced a period of adaptive radiation in Australia, South America, and Antarctica (which was a good deal warmer then than it is today).

Form and Function in Mammals

Mammals have limbs and organ systems that have evolved many shapes to serve many functions in different environments. The specialized adaptations are far too numerous to explore here, so we shall mention only a few of the more interesting ones.

FEEDING Carnivorous mammals, such as cats and dogs, have strong, sharp teeth called incisors and canines that are used for biting and ripping flesh from their prey. Some extinct carnivores, such as saber-toothed cats, had enormous canines. Even the molars of carnivorous mammals are sharp, for they are used to slice meat into small pieces to speed digestion. Carnivores use an up-and-down chopping movement of their jaws to chew their food.

The behavioral and physical characteristics of many mammals allow them to capture prey. For example, some carnivores have sharp claws on their feet with which they grab onto prey. Their bodies are built to produce the quick bursts of speed

Figure 33–4 The fossil record shows that the first mammals resembled this tree shrew. Tree shrews are omnivores; they eat both plants and animals.

TIE-IN EARTH SCIENCE

Continental drift, the slow movement of Earth's landmasses over time, has had a profound impact on the evolution of Earth's living things, including mammals. For example, Africa split off from Gondwanaland (which consisted of just about all the landmasses except North America and Eurasia) about 100 million years ago, separating populations of ancestral primates and setting the stage for the evolution of New World and Old World monkeys.

Continental drift also explains past and present distributions of marsupials. These mammals originated in the Americas and were able to spread to Antarctica and Australia because these continents were joined together with South America until roughly 45 million years ago. After the three southern continents drifted apart, marsupials underwent an adaptive radiation in Australia. In South America, marsupials dominated the carnivore and omnivore niches. But when the isthmus of Panama rose from the sea about 3 million years ago and connected South America to North America, the North American placentals out-competed many of the South American marsupials, which soon became extinct. Of all South America's marsupials, only the common opossum succeeded in colonizing North America.

TEACHING SUPPORT

MEDIA AND TECHNOLOGY

Video/Videodisc
- *Seeing Sense*

Use this bar code to introduce carnivore adaptations.

Play frames 9072 to 10323

See the Biology Media Guide pages 46–48 for additional bar-code correlations for this section.

ECOLOGY NOTE
SIBERIAN TIGERS

Although the Siberian tiger, an endangered species, has been protected by law since the Russian Revolution, this magnificent carnivore is far from safe. In fact, tigers are being slaughtered in the very nature reserves set up to protect them!

In some reserves, such as the Sikhote-Alin Biosphere National Preserve, tigers are fitted with radio collars so that their movements can be tracked. It might be thought that a radio collar would also protect the animal. However, even tagged tigers have been killed. The poachers avoid detection by removing the collar that allows the tiger to be traced.

Many poachers kill tigers and sell their parts for badly needed money—a tiger's meat and bones are worth about $3000 (six times the yearly income of the average Russian) in South Korea and other Asian countries that use tiger products for traditional medicines. But gangsters are also involved in poaching, due to the lucrative nature of the animal trade.

It is clear that the Siberian tiger remains in danger of extinction. As a result, some scientists feel that the last resort is to take orphaned cubs captive so that the species can be saved.

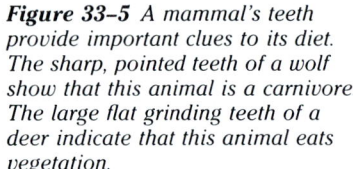
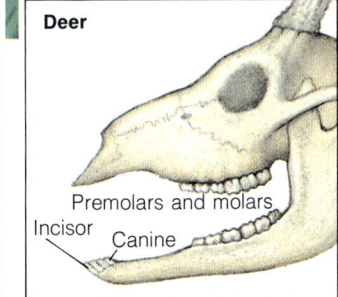

Figure 33–5 A mammal's teeth provide important clues to its diet. The sharp, pointed teeth of a wolf show that this animal is a carnivore. The large flat grinding teeth of a deer indicate that this animal eats vegetation.

they need to chase prey. Fish-eating mammals such as dolphins and killer whales have many sharp teeth to grasp and hold onto their slippery prey. Some whales and dolphins can produce loud bursts of sound that may stun nearby fish, thus making it easier for the fish to be caught.

Herbivorous mammals, from cows to giraffes, eat plants that are tough and require thorough chewing in order to be digested. Herbivorous animals have evolved strong lips and flat-edged incisors that grasp and tear this tough vegetation. They chew by moving their jaws from side to side, using flattened molars to grind the plant food into a pulp.

Despite this efficient chewing, the cellulose that most plant tissues contain is impossible for mammals to digest on their own. The vertebrate digestive system has never evolved the ability to produce enzymes that digest cellulose. To help in the digestion of plant material, many grazing mammals have a chamber in their digestive tract called the **rumen**, in which newly swallowed plant food is stored and processed for a time.

33–1 (continued)

Content Development
Be sure to point out that the jaws of herbivorous animals have not only different tooth structure but different action as well. Use the visuals on this page to help you make your points. The concept of animals eating vegetation and not being able to digest the major component, the cellulose, may seem strange to students. You will want to spend a few minutes discussing the rumen and the cecum.

Enrichment
Students might enjoy picking a mammal that they are particularly interested in learning more about and doing a library research report on the animal. You may want to standardize the information that should be

The rumen contains thriving colonies of symbiotic bacteria that produce enzymes needed to break down cellulose. After a certain amount of time, the mammal regurgitates the plant food from the rumen into its mouth. There the partially digested food is again chewed and mixed with saliva. This is the process being described when mammals such as cows, goats, and deer "chew their cud." The second time the food is swallowed, it moves through the rest of the digestive tract, where digestion is completed and nutrients are absorbed.

Some herbivores, such as rabbits, lack a rumen but have a large dead-end sac, or **cecum** (SEE-kuhm), forming part of their intestines. Many of the same kinds of microorganisms that digest cellulose are found in the cecum. The ancestors of modern humans had a cecum, but over time it has shrunk to the small, sometimes troublesome pouch we call the appendix.

Various other mammals have strange ways of harvesting food. Blood drinkers, such as vampire bats, have razor-sharp incisors that easily slice through the skin of larger mammals. Vampire bats also have a chemical in their saliva that keeps blood from clotting as they feed on it. Filter feeders, such as the giant blue whale, have teeth that do not resemble those we are familiar with. Their teeth are modified into huge stiffened plates called baleen, which act like giant filters. Baleen whales swallow huge mouthfuls of water that contains zooplankton and small fishes. They then force the water out of their mouth through the baleen plates, which strain out small organisms. When all the water is completely expelled, the whale swallows the small organisms that remain in its mouth.

RESPIRATION All mammals, even sea mammals, use lungs powered by two sets of muscles. Chest muscles pull air in and push air out by moving the ribs up and down to increase and decrease the size of the chest cavity. When the large muscle known as the diaphragm contracts, it pulls the bottom of the chest cavity downward, further increasing the cavity size and causing air to rush into the lungs. Many mammals are able to use exhaled air to vibrate their vocal cords and produce a variety of sounds, such as a roar, a bark, or even a song.

INTERNAL TRANSPORT The mammalian circulatory system is a wondrous arrangement of pumps and blood vessels. The main pump, a four-chambered heart, sends deoxygenated blood to the lungs. After it leaves the lungs, the now oxygenated blood returns to the heart and is pumped throughout the rest of the body via blood vessels. The two separate circuits—one to and from the lungs, the other to and from the rest of the body— efficiently transport gases and nutrients to every cell of a mammal's body.

EXCRETION Mammals have the most highly developed kidneys of all vertebrates, an important feature since the kidneys control the composition of all body fluids. Mammalian

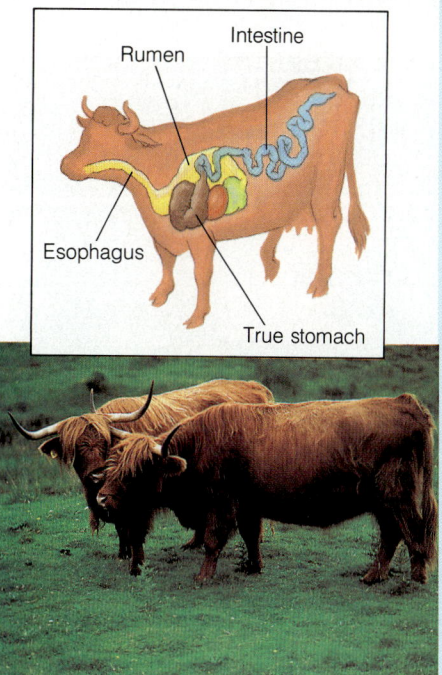

Figure 33–6 *It is difficult for a mammal to digest plant material. A cow has several stomachs through which its food must pass in order for the nutrients it needs to be extracted.*

BACKGROUND INFORMATION
GENES, ENZYMES, AND RUMINANT DIGESTION

Our story of ruminant digestion leaves out one step. After a ruminant has grown a culture of cellulose-digesting bacteria, it must digest those microorganisms to harvest the energy they've extracted. But bacterial cell walls are almost as tricky to digest as cellulose! In fact, only one mammalian enzyme, lysozyme, can do this job. Lysozyme is ordinarily found in tears and other body secretions around openings to the body through which infectious microorganisms might enter. There, lysozyme serves as a first defense against bacteria. But ruminants have extra copies of the lysozyme gene—copies that have mutated in two vital ways. First, the promoter region of these genes turns them on in the stomach. Second, another mutation alters the chemical structure of the protein so that it can survive and function under the acidic conditions in the stomach.

ESL STRATEGY

Have students complete the following analogies.
1. rumen: grazing mammals:: _____ : rabbits
2. monotreme: ovoviviparous:: marsupial: _____
3. carnivore: _____ :: herbivore: plants
4. cat: placental mammal :: koala: _____
5. incisor: front:: molar: _____

---

included in the report. Be sure to include tooth structure and feeding habits.

Content Development
Most students realize that all mammals, even the aquatic varieties, are air-breathers, but it never hurts to remind them. Emphasize that the movement of air in or out is directly related to the increase or decrease in the size of the chest cavity. The production of sounds is usually not something that students automatically associate with the breathing process unless they have had some kind of music training. If you have students in your class who have been taught to breathe properly in order to perform vocally or with wind instruments, you might have them share some information with other students.

TEACHING SUPPORT

MEDIA AND TECHNOLOGY
Videodisc
- *Sound Sense*

Use this bar code to introduce the topic of hearing in mammals.

Play frames 4591 to 7900

See the Biology Media Guide pages 46–48 for additional bar-code correlations for this section.

BACKGROUND INFORMATION
NOCTURNAL MAMMALS

Nocturnal mammals, such as cats, have eyes specially designed to see in the dark. Their pupils open wide to let in as much light as possible. Their retinas do not respond strongly to color but are sensitive to dim light. Behind the retina is a mirror made from crystals of the amino acid guanine. This mirror reflects light that passes through the eye back onto the photoreceptors to give them a second chance to detect it. This mirror produces the green shine in cats' eyes when flashlights or headlights strike them at night.

BACKGROUND INFORMATION
ADAPTING TO WEATHER

Mammals' behavior and physiology help protect them from extremes of weather and major changes in food availability. Some mammals, such as whales, make migrations that cover thousands of kilometers from rich summer feeding grounds in the northern oceans to protected breeding grounds in the tropics. Other mammals, such as squirrels and bears, escape winter's cold by holing up in a convenient cave or burrow and hibernating. During hibernation, the animal's metabolic rate is lowered and its body temperature drops. In this way it can live off stored fat reserves for several months when food is not available.

Figure 33–7 This desert fox survives in areas that have scant supplies of water. Mammals that live in desert regions have very efficient kidneys. Thus their urine contains little water.

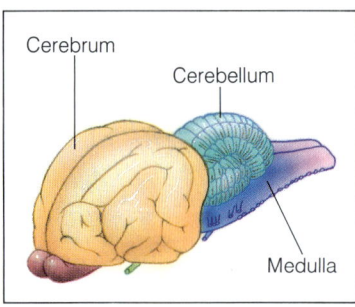

Figure 33–8 Compared with other animals, a mammal has a large brain. Each antelope, alert to danger, is processing a great deal of information about the environment in its brain. If there is real danger, a delay of a fraction of a second could mean that the antelope will not survive.

kidneys extract nitrogenous wastes from the blood in the form of urea. Urea, water, and other wastes form urine. From the kidneys, urine flows to a urinary bladder, where it is stored until it is eliminated. But mammalian kidneys do more than just filter urea from blood. Under the control of several hormones, the kidneys can excrete excess liquid or retain it. They can also retain salts, sugars, and other compounds the body cannot afford to lose. Because they are so efficient at controlling the composition and levels of body fluids, the mammalian kidneys have allowed mammals to live in many habitats they could not otherwise have inhabited (such as deserts).

RESPONSE Mammals have the most highly developed brains of any animals. The brain consists of three parts: cerebrum, cerebellum, and medulla. The large mammalian cerebrum makes such complicated behaviors as thinking, learning, and even understanding a textbook possible. The cerebellum coordinates movements such as the flight of a bat, the dive of a dolphin, and the wondrous somersaults of a human gymnast. The medulla regulates body functions such as breathing and heart rate, which are not under conscious control.

Mammals depend on highly developed senses to provide themselves with information about their external environment. Certain mammal species have well-developed senses of sight, hearing, and smell.

Eyes vary a great deal from one mammal species to another. With the exception of apes, monkeys, and humans, most mammals do not see color well. Biologists believe this is because the first mammals were nocturnal. Monkeys and humans, however, are active during the day and have accurate color vision. Most Old World apes, such as chimpanzees and gorillas, seem to have color vision much like our own.

Although all mammalian ears are built on the same basic plan, they also vary a great deal in their abilities. For example, human ears are not very sensitive when compared with those

33–1 (continued)

Content Development
Most students have little appreciation for the mammalian kidney that is such an important adaptation for survival in a number of environments. To set the stage for a more in-depth study later, begin now to relate hormone function, the action of the kidney, and overall fluid volume. Let this be a brief preview of coming attractions.

Content Development
Most students do realize that the brain and sense organs of mammals are highly developed. They may not know exactly what that means or the extent of variation among species. Since they tend to think that humans are best

of dogs, cats, and bats—all of which can detect sounds of much higher frequencies than humans can. Other mammals, such as elephants, can hear sounds of much lower frequencies.

The senses of smell and taste are also often more highly developed in other mammals than in humans. You probably know that dogs and cats can easily identify people by recognizing their particular body odor. Some mammal species seem to use a combination of smell and hearing to obtain information they cannot get through their sight. For example, antelopes can detect the scent and sounds of a predator from afar.

More than any other animal group, mammals depend on complex behaviors for protection. Many herbivores are able to run long distances to escape a predator. When cornered, or when their young are threatened, some herbivores use horns and hooves to strike their attackers. Others band together in herds or groups and work to repel an attack by predators.

MOVEMENT From the four limbs they inherited from their ancestors, mammals have evolved different structures for movement. Running mammals such as horses and antelope can achieve great speeds on level ground. Climbing mammals such as monkeys have hands and feet with flexible digits that can grasp vines and branches. Some monkeys and other mammals also use a strong flexible tail as an additional hand for grasping branches. Flying mammals such as bats have arms modified to support flaps of skin that form wings. Aquatic mammals such as dolphins have arms modified into flippers, which they use to control their speed and direction in the water.

REPRODUCTION The three groups of mammals differ greatly in their methods of reproduction. These differences illustrate how mammals are linked with their reptilian ancestors

Figure 33–9 This baby monkey swinging from a tree branch shows only some of the movements mammals are capable of. Its long tail is useful in maintaining balance as it moves through the trees.

Figure 33–10 A platypus is a strange animal indeed. A female platypus feeds milk to her young— but only after the young have hatched from eggs laid by the mother.

ECOLOGY NOTE
CHEETAHS

Cheetahs were once present in much of Africa, Central Asia, and the Middle East. Now however, they are present only in central and southern Africa, and they are threatened with extinction.

The threats from habitat destruction and hunting have been greatly reduced by the establishment of nature reserves, an international ban on spotted-cat furs, and the use of donkeys to chase cheetahs away from livestock. However, other problems now place the cheetah at risk of extinction.

Lions and hyenas, which also reside in nature preserves, find cheetah cubs easy targets, even if their mother is nearby. In fact, cheetah cubs have a better chance of surviving outside the preserves, where predators have been wiped out by hunters.

The greatest threat, however, lies within the cheetahs themselves. Cheetah populations are inbred and thus lack genetic diversity. In wild and captive populations, the lack of genetic diversity has resulted in poor sperm quality, high rates of juvenile mortality, and susceptibility to disease.

Much research is now being carried out so that the number of cheetahs worldwide can be increased. Tissues from diseased cheetahs are being examined so that more can be learned about the spread of disease. In the near future, researchers hope to collect and genetically analyze sperm from wild cheetahs. Through artificial insemination and other breeding programs, it may be possible to infuse much-needed genetic diversity into wild and captive populations and thereby increase their fitness.

---

at everything, a brief discussion of eyes, noses, and ears will be enlightening for them. The inability of most mammals to see in color may also be a new idea for some students. Spend a little time discussing the idea of complex behaviors as survival mechanisms; the students will find this interesting.

Focus/Motivation
The adaptations of limbs in mammals is a topic that you could spend hours discussing. The examples given in the text are just a few that could have been presented. Encourage students to think of others with which they are familiar. Emphasize, as always, the relationship between structure and function.

TEACHING SUPPORT

Study Guide
- Section 33–1, p. 317

Figure 33–11 The baby kangaroo, safely hitching a ride in its mother's pouch, views the world around it with amusement. If danger threatens, mother and baby can quickly hop away.

Figure 33–12 This orangutan mother will care for her baby for some time. Rarely setting foot on the ground, the baby clings tightly to its mother's fur.

33–1 (continued)

Content Development
Begin your discussion of mammalian reproduction with a brief review of sexual reproduction. Emphasize that any form of reproduction is successful if it guarantees the survival of the species.

SECTION REVIEW 33–1

1. Mammals have hair, are endothermic, and feed their young milk.
2. Mammals are endothermic and thus can maintain their body temperature regardless of changes in the temperature of the environment. Most mammals give birth to live young and thus can reproduce in many different environments and do not have to incubate their eggs. Mammals have highly efficient kidneys that are able to conserve water, salts, and other compounds.
3. The Earth became colder and drier. These climate changes enabled mammals, which are much more efficient at maintaining their body temperature, to experience a period of adaptive radiation as other organisms died out.
4. Mammals have well-developed brains that are able to respond to a great many stimuli. Most mammals have well-developed eyesight that enables them to see danger from afar. Mammals have well-developed hearing, which enables them to gather information about the environment. Well-developed senses and a well-developed brain contribute to the success of mammals.
5. Monotremes lay eggs. Marsupials give birth to young that are not fully developed. The young develop further in the marsupium, or pouch. Placental mammals give birth to young that are more mature than those of marsupials. All mammals care for their young for a

---

through a series of stages in the evolution of reproductive techniques.

Egg-laying mammals, the monotremes, are the most primitive mammals and reproduce much like reptiles. Monotremes such as the duckbill platypus are oviparous. A female platypus lays eggs that are incubated outside the mother's body. Once the young hatch, however, they nurse on milk provided by the mother. Thus the platypus and its relatives show both reptilian and mammalian reproductive habits.

Marsupials are viviparous and bear their young alive. The fertilized egg grows into an embryo inside the mother's reproductive tract. The embryo is supplied with nourishment by a yolk sac on the egg. But because this yolk is not large enough to nourish the embryo through its entire developmental period, the embryo must leave its mother's womb very early. At such an early stage of development, the embryo is unable to survive alone. Instinctively, it crawls across its mother's fur into a pouch called the **marsupium** (mahr-SOO-pee-uhm). Inside the marsupium, the young quickly locates a nipple and grabs onto it with its jaws. It spends the next several months attached there, growing sufficiently large and independent so that it can leave the pouch.

The early stages of placental embryos are much like those of marsupials. But in placental mammals, the embryo's chorion, amnion, and allantois develop differently. Tissues from these membranes join with tissues from the mother's uterus to form an organ called the **placenta**. Nutrients, oxygen, carbon dioxide, and wastes are exchanged between embryo and mother through the placenta. The placenta allows the embryo to develop for a much longer time inside the mother. During this time the mother is free to move about and feed while still protecting the developing embryo. The time the embryo spends inside the uterus is called the gestation period. The gestation period of mammals ranges from a few weeks in mice and rats to as long as two years in elephants. The gestation period in humans is nine months.

After birth, most placental mammals provide their young with a period of care. The duration of this parental care varies among different species. For example, a newborn fawn is awkward and barely able to stand. Within a few hours, however, it can see well and walk around. Despite this, the mother protects its fawn and feeds it milk for some time. The young of mammals such as monkeys and humans are helpless at birth and for quite some time thereafter. These infants depend on their mother for food and protection, and they often spend several years growing up before they are able to live on their own. During the time infant and mother live together, the infant learns a great deal about its surroundings from its mother. Many biologists believe that this long learning period is one of the most important benefits of the prolonged childhood of many mammals. It may also be an important reason for the evolutionary success of these animals.

SCIENCE, TECHNOLOGY, AND SOCIETY CONNECTION

The Smallest Paleontologists

Many early mammals are known from their fossilized teeth. Because teeth are the hardest and most durable body part, they fossilize better than any other body part, including bones. As you might imagine, collecting teeth and bones from fossil animals not much bigger than mice is no easy job. Farish Jenkins of Harvard University, who has found many important early mammal fossils, must often search long and hard. Over kilometers of badlands in the Midwest, he and his research team choose rocks that seem likely to contain fossils. They pick these rocks apart with needles, cleaning the pieces with toothbrushes in order to expose tiny fossil teeth and bones.

Sometimes paleontologists enlist thousands of unusually tiny helpers for this tedious job. It just so happens that one ant species often dig large nests in places that are rich in early mammal fossils. As these ants tunnel, they uncover fossil teeth and sometimes entire jaws of early mammals. When the ants encounter such items, they treat them as they would anything that blocks their way—they dig them out and carry them to the surface. There, in a mound around the nest entrance, humans can sift through them easily!

Ants are remarkable excavators, moving material to the surface when they enlarge their underground homes. Sometimes they bring small fossils to the surface, making the job of paleontologists easier.

33–1 SECTION REVIEW

1. What are three important characteristics of all mammals?
2. What are two adaptations of mammals that enable them to live in many more different kinds of environments than reptiles do?
3. What changes occurred in the Earth's climate that finally allowed the mammalian adaptive radiation to occur?
4. In what three ways have well-developed senses contributed to the success of mammals?
5. How do the reproductive methods of the three mammalian groups differ? How are they similar?
6. **Critical Thinking—Making Comparisons** The big cats of Africa are known to sleep between kills. In fact, lions sleep up to 20 hours out of every 24! Large herbivores, on the other hand, seem to eat all the time. What characteristics of the foods eaten by these animals might be responsible for these differences in carnivore and herbivore behaviors?

745

SCIENCE, TECHNOLOGY, AND SOCIETY CONNECTION

The Smallest Paleontologists

The teeth of mammals are the hardest and most durable parts of their bodies. Because of this, they fossilize better than other body parts. However, finding and collecting teeth from very small animals can be a tedious job. Farish Jenkins, of Harvard University, and his research team look for tiny teeth with needles and toothbrushes.

Another way that paleontologists collect fossil teeth is by enlisting the help of ants. One ant species digs large nests in areas where early mammalian fossils are found. In the digging process many tiny teeth are removed from the rock and brought to the surface. The mounds around the nest entrances are treasure troves to the paleontologist.

Ask students to think for just a few minutes about how paleontologists learned that the ants removed the teeth. Then think about other ways that animals naturally help people or have been taught to perform tasks. You might get answers like horses pulling wagons, dogs fetching newspapers, goats acting as natural lawn mowers, beetles cleaning skulls for mammalogists.

Just for fun, have students write an essay about human-animal interaction that results in a service being performed by the animal that benefits humans in some way. You will need to set some guidelines about how serious or fanciful you want them to be.

---

period and feed their young milk produced in special glands.
6. Plants are difficult for mammals to digest. It takes a great deal of time and effort for a cow to remove nutrients from the grass it eats. Herbivores must also eat a great deal of plant material to get the nutrients they need. Carnivores are able to digest meat more easily, extracting from comparatively small quantities the proteins and other nutrients they need.

Reinforcement/Reteaching
If you have some students who are having difficulty answering one or more of the Section Review questions, at this time turn back to the appropriate material and review it.

Closure
Turn to the Chapter Preview at the beginning of the chapter. Read each of the Performance Objectives for Section 33–1. Ask students in the class to perform each task. This should be a good review of the section.

Guide For Reading
- How do the methods of reproduction in mammals differ?
- What are the most important orders of placental mammals?
- How do mammals fit into the world?

33–2 Important Orders of Living Mammals

Scientists use several important characteristics to classify mammals. The structure of teeth and the number and kinds of bones in the head are two important features by which mammals are classified. But perhaps the most important characteristic used to classify mammals is the method of reproduction. As you have already learned, mammal species show three different methods of reproduction. Some mammals lay eggs. Some mammals give birth to young that are not well developed at birth and must therefore spend time developing in a special pouch in their mother's body. Some mammals—in fact, most—retain embryos within the mother's body, where the embryos grow and develop, nourished by the mother. At birth, these young are more well developed than the young of mammals that spend time in their mother's pouch.

Monotremes

Monotremes, or egg-laying mammals, are very rare. In fact, only three species of monotremes exist today, living in isolated parts of Australia and New Guinea. You have already learned about the duckbill platypus, the most familiar monotreme. The spiny anteater, or echidna, is another monotreme. It has a jaw shaped like a bird's bill, strong clawed feet with which it can dig quickly, and long sharp spines among its body hairs.

Marsupials

Marsupials are pouched mammals. A number of fascinating marsupials, such as kangaroos and koalas, are found in Australia. Kangaroos and their close relatives, the wallabies, are herbivores that feed primarily on range grasses. Koalas spend most of their time in trees and eat only the leaves of eucalyptus trees.

Opossums are the only marsupials found in North America today. Active mostly at night, opossums spend the day sleeping in protected spots. These animals eat a variety of insects, birds, and small mammals. A newborn opossum is about the size of a bee. Just after birth, it uses its tiny hands and feet to grasp onto its mother's hair and crawl to the safety of her pouch.

In the past, many other marsupials lived in South America. These included many species similar in form to wolves, bears, camels, and moles (none of which are actually marsupials).

Placentals

Living placental mammals are placed in sixteen orders, most of which contain animals familiar to you. In addition to having different reproductive habits than marsupials, placental

Figure 33–13 The echidna is known as the spiny anteater (top). Its spines protect it from enemies. Its long sticky tongue laps up the ants it eats. These baby opossums are hitching a ride on their mother's back, where they are safe (bottom). When their mother locates food, the babies climb off to eat.

mammals also have slightly higher metabolic rates. Placental mammals are much more abundant today than their marsupial cousins. The most important orders of placental mammals are briefly described here.

ORDER INSECTIVORA This order, whose name means insect eaters, includes tree shrews, hedgehogs, shrews, and moles. Shrews, several of which are about the size of a mouse, have extremely high metabolic rates and must eat almost constantly to stay alive. Biologists believe that the first mammals looked and behaved much like certain modern tree shrews. Remember, however, that this does not mean that mammals evolved from tree shrews as we know them today. It simply means that living tree shrews have many characteristics of primitive mammals.

ORDER CHIROPTERA This order contains the many different species of bats, which in fact account for one quarter of all mammal species. Bats are closely related to insectivores, although different types of bats eat many different kinds of foods. Some bats eat only insects, whereas others eat only fruits. Still others, such as vampire bats, feed on the blood of other mammals. Many bats are active only at night. Night-flying bats use echolocation to help them navigate while in flight. By emitting high-pitched sounds that bounce off objects, the bats can calculate distances and determine locations of tiny insect prey. Some fruit-eating bats, like the giant flying foxes, feed during the day. Colonies of these bats sleep together in trees, hanging upside down and wrapping their wings around their body.

Figure 33–14 Moles dig tunnels under ground (left). In the subterranean darkness, these animals eat the insect larvae that live in the soil. Bats are the only true flying mammals (right). They glide along, effortlessly flapping their delicate wings. Like moles, many species of bats also eat insects, which they catch in the air. The bats in this photograph, however, eat frogs.

mals not discussed in the textbook. These orders, which contain only a few obscure species, are order Dermoptera (colugo or "flying lemur"), order Pholidota (pangolin), order Tubulidentata (aardvark), and order Hyracoidea (hyraxes).

You may want to point out that different sources may use slightly different mammal classification systems. For example, elephant shrews and tree shrews, classified as insectivores in this textbook, are considered by some to be separate orders—Macroscelidea and Scandentia, respectively. The carnivore suborder Pinnepedia, which contains seals, sea lions, and the walrus, is sometimes elevated to order status.

FACTS AND FIGURES

Bats give off their ultrasonic (above human hearing) signals in two ways. Some bats send the signals out through their mouths while others send signals out through their nostrils.

BACKGROUND INFORMATION
TREE SHREWS

Tree shrews were once generally classified as primates. After a period of more intensive study, primatologists removed them from that order. Today, they are sometimes placed within the Insectivora with the other shrews and hedgehogs. Some taxonomists place these animals, which most closely resemble the oldest mammalian fossils, in a separate order, Scandentia.

Content Development

Insectivores are diverse and interesting animals. If students ever come in contact with a shrew, they will probably mistake it for a small mouse unless they look at it carefully. Because of their high metabolic rate and their need to eat constantly, these animals appear rather vicious. The many sharp teeth add to the image. Hedgehogs have some interesting physical and behavioral adaptations. Encourage students to investigate this group.

Content Development

Bats always fascinate students. Look carefully at all the information given in the textbook to help dispell the misconception that all bats are vampires. Have a student research echolocation and explain it to the class. Students might also be able to find some interesting information about the seasonal habits of some bat species.

TEACHING SUPPORT

Laboratory Manual
- The Most Intelligent Mammal, p. 423

Teaching Resources
- Activity: Classifying Mammals, p. 3

MEDIA AND TECHNOLOGY
- Interactive Videodisc/CD ROM
- *Amazonia*
- *The Virtual BioPark*
- Videodisc
- *Aquatic Ecosystems: Freshwater Wetlands and Freshwater*
- Interactive Videodisc
- *In the Company of Whales*

Use this bar code to introduce cetaceans.

Play frames 6761 to 7424

See the Biology Media Guide pages 48–49 for additional bar-code correlations for this section.

33-2 (continued)

Content Development

Using the Visuals Order Edentata is a relatively small order containing only about 30 species. The armadillo in Figure 33–15 is just one of the odd-looking members of this group. For protection, an armadillo simply rolls up so that only its bony armored plates are exposed. Armadillos are thought to be the only non-human animals that can catch Hansen disease, or leprosy. Nine-banded armadillos are often used in research of this disease because the four offspring in a litter are genetically identical.

Figure 33–15 Common in certain parts of the United States, armadillos are in the order Edentata (top). The porcupine depends upon its many sharp quills for protection (center). One function of the huge ears of a rabbit is to give off excess body heat to the environment through the many blood vessels in these organs (bottom).

ORDER EDENTATA The name of this order means without teeth, although some of the mammals included here have small teeth. The edentates include such odd animals as sloths, anteaters, and armadillos. Sloths are South American animals that live most of their life hanging upside down in trees. Many sloths move so slowly that at times they scarcely seem alive. Anteaters have long tapered snouts and powerful front legs with sharp claws. They feed by ripping open ant nests and collecting the scurrying insects with a long and sticky tongue.

ORDER RODENTIA This order includes many amusing mammals, as well as many destructive ones. Mice, rats, squirrels, beavers, porcupines, and gophers are all rodents. Rodents have two long front teeth, which they use for chewing wood and other tough plant material. These two front teeth continue to grow during a rodent's life. The constant gnawing on tough plant material wears the teeth down. Most rodents are small and have a short gestation period. Rats and mice are very adaptable animals that eat a wide variety of foods. Long ago, both these rodents moved in with humans and have traveled with us all over the world.

ORDER LAGOMORPHA The familiar rabbits and hares comprise this order. In many ways, lagomorphs resemble rodents. They have sharp front teeth and eat plant material. Their gestation period is short, and the number of young they produce is high. Many of these animals compete with humans for food. Several species, such as jack rabbits and cottontails, are widely distributed across the United States.

ORDER CARNIVORA Carnivores are meat eaters. Many familiar animals—including cats, dogs, wolves, bears, weasels, hyenas, and seals—are in this order. Most are terrestrial, stalking and chasing their prey by running and pouncing, and then killing them with sharp teeth and claws. Carnivores such as seals and walruses had ancestors that at one time lived on land, but these mammals have since returned to the ocean, where they feed on fish, mollusks, and sea birds. Although quite agile in water, aquatic carnivores return to land to breed and bear their young. On land these animals move with difficulty. Their appendages, so useful in the sea, are not very effective at moving their body from place to place on land.

ORDER CETACEA This order contains truly aquatic mammals: whales, dolphins, and porpoises. Although they still breathe air, these mammals have lungs and a circulatory system designed to permit long, deep dives. Their thick layer of subcutaneous fat, called blubber, keeps them warm in even the coldest water. Cetaceans have lost both their external ears and their hind legs. These animals mate and bear their young in the water. On land, they are completely helpless. All cetaceans are carnivores. A few, such as the great blue whale and the humpback whale, are filter feeders and live by eating plankton.

Some prehistoric edentates were enormous. Certain giant ground sloths grew to the size of modern elephants; they were about 6 meters long. *Glyptodon*, an armadillolike edentate, was about 5 meters long and carried a 3-meter shell on its back.

Content Development
Students are probably familiar with the harmful effects of rodents, which include eating foods intended for humans and livestock and carrying disease. However, they may not be aware that rabbits—usually thought of as cute and cuddly—can also be harmful to humans. Describe the environmental damage done by hungry rabbits in Australia, Hawaii, and other places. Remind students that the loss of plants also results in the loss of other organisms. Discuss the economic

Figure 33–16 Carnivores such as the walrus (left) and the hyena (right) are placed in the same order. The walrus moves along the ocean bottom, using its tusks to find the clams and other shellfish it eats. The hyena is an efficient hunter and often kills small antelopes.

ORDER SIRENIA These strange aquatic mammals are related to elephants. They are peaceful, slow-moving herbivores that live in rivers and streams in parts of Africa, South America, and Florida. Some species are also found in tropical oceans and the Caribbean. The manatee, or sea cow, lives in quiet waters in southern Florida, where it is often injured by careless boaters. The propellers of boat engines cut the manatee's back as it swims along just below the surface of the water.

ORDER ARTIODACTYLA This order contains the large grazing animals: cattle, sheep, goats, hippopotami, giraffes, and pigs. For mammals in this order, the original five toes on each foot have been reduced to two. Thus artiodactyls are called even-toed ungulates. The word ungulate means hoofed mammal.

ORDER PERISSODACTYLA Horses, zebras, tapirs, and rhinoceroses—the odd-toed ungulates—make up this order. Many grazing animals with habits similar to those of even-toed mammals are included in this order. Some odd-toed ungulates have hooves formed from the center toe of each foot.

ORDER PROBOSCIDEA These are the mammals with trunks, the great elephants. Some time ago, this order had a reasonably large adaptive radiation that produced about thirty species. Included in those species were the mammoths and mastodons, which today are extinct. Only two species, the Indian elephant and the African elephant, presently survive. Both species are in danger of becoming extinct.

ORDER PRIMATES This order, which includes our own species, is closely related to the ancient insectivores. Of all the animals, primates have the most highly developed cerebrum and the most complicated behaviors. The most primitive living primates, the lemurs, are small tree dwellers. The primates most people call monkeys or apes represent two main branches within this order.

Figure 33–17 Manatees live in tropical water, often floating. Their diet consists of water plants.

MULTICULTURAL STRATEGY

Tell students the Cherokee legend of Awi Usdi, the Little Deer, which stresses the Native Americans' respect for animals:

In early times, people and animals could communicate with each other, and they lived in harmony. People killed animals only as needed for food and clothing. After the development of the bow and arrow, things changed, and some animals were slaughtered almost to extinction. The animals got together to see what they could do to remedy things. Awi Usdi, the Little Deer, had a suggestion. She would whisper in the ears of the hunters that they must first ask permission of the animals they were hunting and then ask for pardon from the spirit of the animal they had to kill for food. If the hunters failed to do so, Little Deer would cast a magic spell on them and cause them to be crippled. The next day, some hunters heeded the advice of Little Deer, and others did not. The ones who did not became crippled. That is why, to this day, Native Americans who follow traditional ways hunt only animals that they need for their food and clothing and give thanks to the animals that they hunt.

FACTS AND FIGURES

The largest mammal of all, and the largest animal ever to have lived on Earth, is the blue whale. The largest blue whale ever caught was 33.58 meters long and had a mass of about 140,000 kilograms. The largest land animal, the African elephant, is tiny by comparison. Its trunk-to-tail length is about 7 meters, and it may have a mass of about 7500 kilograms.

---

impact of rodents and lagomorphs in your area.

Focus/Motivation

The aquatic mammals also have great appeal for students, probably none more so than the killer whale. If students are writing reports, be sure that someone takes this animal. Killer whales frequently travel in groups called pods. These groups work cooperatively to hunt and kill prey far larger than themselves. This cooperative effort contributes to the reputation of the killer whale. Other whales are docile plankton-eating creatures. The idea of these huge animals feeding exclusively on microscopic organisms is amazing.

ECOLOGY NOTE
SEAL THEIR FATE

The following can be used as a springboard for discussion or given as a writing assignment:

At the beginning of the eighteenth century, there were about 9 million harp seals in the North Atlantic Ocean. Overhunting once drove the harp seal to the edge of extinction, but strict hunting laws allowed the population to recover. Today, harp seals have a population of about 4 million, which is growing by about 5 percent each year.

For the first two weeks of life, young harp seals have beautiful white coats. In the past, large numbers of seal pups were killed for their fur, usually by being clubbed to death in the presence of their mothers. Conservation and humane groups still strongly oppose the killing of the seals. They think that killing young seals for their fur is cruel and may have a negative effect on the seal population and the environment. Seal hunters, facing economic difficulty, want permission to kill greater numbers of young seals. They believe that the seal population is no longer in danger and that the young seals can be killed humanely. Should seal hunters be permitted to harvest greater numbers of seal pups? What is your opinion and why?

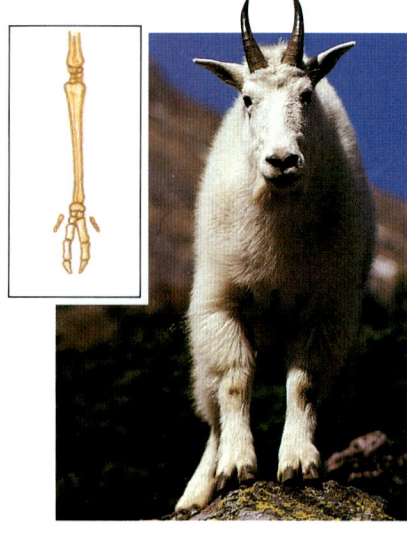

Figure 33–18 The mountain goat (left) has two toes and is included with other two-toed mammals in the order Artiodactyla. The tapir (right) has three toes and is grouped with other animals that have an odd number of toes.

Very early in their evolutionary history, primates (as a group) were split apart by the moving continents. One branch, the New World monkeys, includes the squirrel monkey and the spider monkey. These monkeys live almost entirely in trees. They have long arms for swinging from branch to branch and long **prehensile tails**, which they use for grasping while climbing.

The other branch, the Old World monkeys, include chimpanzees, gorillas, and the ancestors of humans. Many Old World monkeys still spend much of their time in trees, but they all lack prehensile tails.

How Mammals Fit into the World

Mammals evolved from early reptile ancestors during the Mesozoic Era about two hundred million years ago. During this era, and later during the Cenozoic Era, mammals underwent massive adaptive radiations. Today mammals are distributed throughout the world. Humans, one of the important mammalian species, inhabit areas that range from the very cold polar

Figure 33–19 Among the most intelligent animals, the black lemur (left) and the chimpanzee (right) are both primates. Today, chimpanzees and many species of lemurs are endangered.

33–2 (continued)
Content Development
Discuss the two groups of primates mentioned in the text. Point out the physical and behavioral differences. Particularly discuss the prehensile tail and how important it is in the life of these primates. (Wouldn't it be useful to have a third "hand" sometimes?)

Skills Development
Guided Practice
Skill: Organizing mammals
Divide the class into small working groups. Have each group prepare a web illustrating the positive and negative ways in which mammals interact with humans.

SECTION REVIEW 33–2

1. A placental mammal produces young that remain within the female and get food and exchange wastes through a special membrane, the placenta. A pouched mammal does not have a placenta. The young of pouched mammals develop for a time and are born in an immature state. After they are born, the young travel to the mother's pouch, where they continue to develop.
2. Like reptiles, monotremes reproduce by laying eggs.
3. Chiroptera, the bats, are one order—flying foxes or vampire bats are examples. Rodentia, the rodents, are another mam-

regions to the warmest equatorial regions. And in all these regions, other mammal species live alongside humans.

In many parts of the world, herbivorous mammals are major consumers of plant material. For example, huge herds of grazing zebras and wildebeests eat their way across the savannas of Africa. Herds of reindeer and musk oxen move across the tundra, eating small plants and lichens. These peaceful grazers are, in turn, food for carnivorous mammals. Lions, leopards, cheetahs, wild dogs, wolves, and other carnivores hunt and kill plant eaters for food.

In the air, flying bats and gliding squirrels move gracefully in search of food. Bats eat enormous numbers of mosquitoes and other insects. Gliding squirrels feed on nuts and seeds.

Mammals also inhabit the oceans. Whales, the largest animals to have lived on Earth, are probably the most familiar example. Despite their huge size, many species are today in danger of becoming extinct.

Domesticated mammals such as dogs, cows, sheep, and goats have a significant influence on human culture. Many of these animals provide food in the form of meat and dairy products such as milk, butter, and cheese. Others help humans find food. Dogs are used to hunt, and monkeys are used to harvest coconuts and other fruits.

Some mammals have a negative impact on human life. Carnivores prey on domesticated animals. Rodents such as rats and mice damage crops and eat stored food. Some mammals carry diseases that can affect humans. For example, rats harbor fleas that can spread the plague. Dogs, squirrels, and other wild animals can transmit the virus that causes rabies.

Mammals can have a profound effect on the environment. Elephants destroy huge numbers of trees as they feed. Overgrazing by cattle and rabbits can turn an area of prime farmland into a virtual dust bowl. Beavers flood areas and can create ponds when they build their dams. Humans in particular are capable of altering the environment in many ways—both good and bad.

Figure 33–20 People used natural fibers and fur to weave cloth long before synthetic fibers were developed. In this photograph, sheep are being shorn of their fine wool.

Figure 33–21 Animals often make profound changes to the natural environment. This beautiful dam was constructed by beavers, not by engineers. The pond that formed behind the dam provides safety for the beavers and a home for many other kinds of animals.

33–2 SECTION REVIEW

1. What is a placental mammal? How does a placental mammal differ from a pouched mammal?
2. How do monotremes provide evidence for the evolution of mammals from ancient reptiles?
3. Name five different orders of placental mammals and give an example of each.
4. In what four ways do mammals influence human life?
5. **Critical Thinking—Making Predictions** How could you predict the diet of a mammal by looking at its teeth?

---

malian order—mice and rats are examples. Cetacea, the whales, are an order of mammals—dolphins and killer whales are examples. Proboscidea, mammals with trunks, are an order—elephants are an example. Primates, monkeys and apes, are an order—baboons and gorillas are examples. Other answers are possible.
4. Mammals influence human life in many ways. Certain mammals are eaten by humans. Other mammals perform work for humans. Rats and mice compete with humans for food. Some mammals spread disease. Other answers are possible.
5. Sharp cutting teeth probably belong to a mammal that is a carnivore. Large, flat teeth are found in herbivores that must grind up the plants they eat. A mixture of sharp and flat teeth probably means that the mammal is omnivorous, eating both plant and animal foods.

LABORATORY INVESTIGATION

EXAMINING THE TYPICAL MAMMALIAN BODY COVERING

Before the Lab
1. Assemble the materials needed for each group.
2. Purchase some inexpensive combs. (You can ask students to furnish their own, but some will forget!) For hygienic reasons, students should not be allowed to share combs or brushes.
3. Flat toothpicks work better than rounded ones for this lab, but they may be harder to find.
4. Check the dilution information on the methylene blue container. Too strong a solution makes it more difficult to see the cells.
5. Prepare methylene blue stain by diluting 10 milliliters of stock solution with 90 milliliters of water for every 100 milliliters of stain. Stock solutions of methylene blue can be prepared by adding 1.48 grams of methylene blue powder to 100 milliliters of 95% ethyl alcohol and letting it stand for two days. During this time, stir frequently. After two days, filter.

Pre-Lab Discussion
For reference, review the body coverings of all vertebrates, including the mammals.

Review the process involved in correctly preparing a wet-mount slide, reminding students that they are using methylene blue instead of water.

Remind students that methylene blue is a stain, so they should take care not to spill it on their clothing or on countertops where they might get it on their sleeves or skin.

Demonstrate the way students are to scrape cheek cells without injuring themselves. Remind them that they will not be able to see any of the cells on the toothpick.

LABORATORY INVESTIGATION

EXAMINING THE TYPICAL MAMMALIAN BODY COVERING

PROBLEM
What are the characteristics of hair?

MATERIALS (per group)

comb or brush
glass slide
coverslip
electric light or bright sunlight
medicine dropper
methylene blue
microscope
scissors
toothpick
hand lens

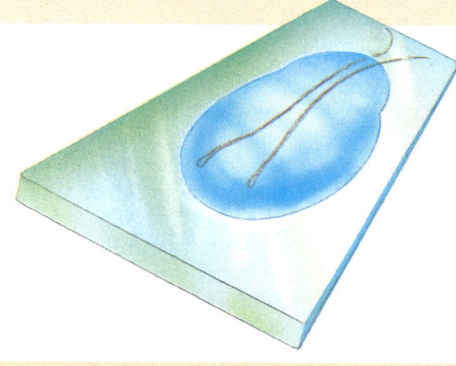

PROCEDURE

1. Using a medicine dropper, put two drops of methylene blue in the center of a clean glass slide.
2. Comb or brush your hair vigorously to remove a few loose hairs.
3. From your comb or brush, select two hairs that each have a root attached. Look for a small bulb-shaped swelling at one end of the hair. This is the root.
4. Using the scissors, trim the other end of the hairs so they will be short enough to fit on the slide. Place the trimmed hairs in the drops of methylene blue on the slide.
5. Gently rub the inside of your cheek with a toothpick.
6. Stir the material from the inside of your cheek in the methylene blue between the hairs on the glass slide. Cover the specimens with a coverslip.
7. Use the low-power objective to locate the hairs and some of the material from the inside of your cheek. Then switch to the high-power objective and focus with the fine adjustment.
8. When the hair appears to be in perfect focus, you are looking at the inside of the hair strand. The inside of the hair is made of keratin, a secretion of epidermal cells in the hair follicles of your scalp. Make a sketch of the hair strand.
9. Turn the fine adjustment toward you (counterclockwise) to focus on the upper surface of the hair. At one point in your focusing, the hair will appear to be covered by overlapping structures that look like shingles. Draw these structures, which are actually cells.
10. Locate some of the epidermal cells scraped from the inside of your cheek. Compare the size and shape of these cells to the overlapping cells on the hair strand. (Remember the layers of overlapping cells on the strand of hair are only partially visible.)
11. Under bright light, use the hand lens to examine a portion of your skin that does not seem to be covered with hair. (Do not examine the palms of your hands or the soles of your feet.)

OBSERVATIONS
1. Describe the appearance of a strand of hair under magnification.
2. How does the appearance of the epidermal cells from your cheek compare to the cells covering the strand of hair?
3. What did you observe when you examined the surface of your skin under bright light?

ANALYSIS AND CONCLUSIONS
1. What covers the surface of your body and the body of most mammals?
2. Based on your observations, what is hair?

Skills Development
The students will use the following skills in the completion of this laboratory investigation.
1. Manipulative
2. Observing
3. Diagramming
4. Communicating
5. Comparing
6. Inferring

Safety Tips
There are no real health hazards involved in this lab. Students should be cautioned to take care using the scissors, toothpicks, glass slides, and coverslips. The methylene blue presents more of a hazard to their clothing than to their persons.

Teaching Strategy
Students should be monitored to ensure that they are carefully fol-

STUDENT STUDY GUIDE

SUMMARIZING THE CONCEPTS

The key concepts in each section of this chapter are listed below to help you review the chapter content. Make sure you understand each concept and its relationship to other concepts and to the theme of this chapter.

33-1 Mammals

- The group of mammals includes a variety of endothermic animals.
- Monotremes—the egg-laying mammals—are oviparous and show both reptilian and mammalian reproduction habits.
- Marsupials are viviparous animals whose young are born in an immature state.
- Mammals have well-developed breathing muscles that include a diaphragm that changes the size of the chest cavity to pull air in and push air out.
- Mammalian circulatory systems consist of a four-chambered heart that moves blood through the rest of the body.
- The digestive system of herbivorous mammals contains either a rumen or cecum that houses symbiotic bacteria that produce enzymes that break down cellulose.
- Mammals have highly developed kidneys that function to extract nitrogenous wastes from the blood.
- The mammalian brain consists of a cerebrum, cerebellum, and medulla. Mammals depend on highly developed senses to provide information about their external environment.

33-2 Important Orders of Living Mammals

- Mammals are classified by structure of teeth, number and kinds of head bones, and methods of reproduction.
- Members of the order insectivora are relatively small mammals with high metabolic rates. The order Chiroptera contains the many species of bats. The order Edentata includes mammals that lack teeth or have very small teeth. Small mammals with long front teeth and short gestation periods belong to the order Rodentia. Rabbits and hares belong to the order Lagomorpha. Mostly terrestrial meat-eaters belong to the order Carnivora. Whales and dolphins, truly aquatic animals, belong in the order Cetacea. Slow-moving herbivorous aquatic mammals belong in the order Sirenia. Even-toed ungulates are in the order Artiodactyla, whereas odd-toed ungulates are in the order Perissodactyla. Mammals with trunks are found in the order Proboscidea. The order Primates contains mammals with the most developed cerebrum, including our own species.

REVIEWING KEY TERMS

Vocabulary terms are important to your understanding of biology. The key terms listed below are those you should be especially familiar with. Review these terms and their meanings. Then use each term in a complete sentence. If you are not sure of a term's meaning, return to the appropriate section and review its definition.

33-1 Mammals
- mammary gland
- monotreme
- marsupial
- placental mammal
- rumen
- cecum
- marsupium
- placenta

33-2 Important Orders of Living Mammals
- prehensile tail

753

---

lowing the directions. Some of them might need some assistance with placing the hairs and the cheek cells on the slides. You might need to help them find and focus on the cells. They might have difficulty making the comparisons between the cells of the hair shaft and the cheek cells. You might want them to work with another group—one focusing on the cheek cells, the other on the hair cells. Emphasize that they are to look for similarities and differences.

Observations
1. The inside appears to be made up of parallel layers of some uniform material. The outside is covered with overlapping layers of thin cells.
2. The cells are very similar. Both types are flat and irregularly shaped.
3. Even areas that appear to have no hair are covered with fine hairs.

Analysis and Conclusions
1. Hair.
2. Hair is a tube of keratin surrounded by dead epidermal cells.

Going Further: Enrichment
Obtain hair from several common animals. Dogs, cats, rabbits, hamsters, gerbils, horses, and cows make good sources. Prepare these hair samples as you did the human hair. Compare the various samples to the human hair and ask the following:
- Do the hairs appear to be about the same thickness when you look at them with your eye?
- What about with the microscope?
- How do the cells that make up the hair shaft compare in size and shape?
- If you can see the root areas, do they appear similar?
- What other similarities or differences do you observe?

(Answers to all of the questions will vary depending on the type of hair observed.)

CHAPTER REVIEW

CONTENT REVIEW

Multiple Choice
1. b 2. a 3. a 4. d
5. a 6. c 7. b 8. d

True or False
1. F New World monkeys
2. T
3. T
4. F filter-feeding cetaceans
5. F fat
6. F monotreme
7. F lower
8. T

Word Relationships
1. mammary glands
2. cecum
3. marsupial

CONCEPT MASTERY

1. Three important characteristics used to classify mammals are method of reproduction, the structure and kinds of teeth, and the number and kinds of bones in the head.
2. They emit high-pitched sounds that bounce off objects and return to the bats' ears. Using this kind of echolocation, bats can calculate the distance to prey animals with great accuracy.
3. In animals with a four-chambered heart, the oxygenated blood and the deoxygenated blood are kept separate from each other. Thus the body circulation can more effectively bring oxygen to cells and carry waste products away from cells.
4. The cerebrum controls thought processes. The cerebellum controls balance and movement. The medulla controls actions of the body that are not under conscious control, such as breathing.
5. Sharp teeth, such as canines and incisors, are used for biting and ripping flesh from prey. Most carnivores also have sharp molars that are used to slice meat into small pieces for easy digestion. Herbivores have flat-edged incisors that are used to grasp and tear plants. They also have flattened molars to grind plant food.
6. Cows have a series of four stomach chambers. They regurgitate their food and rechew it. This breaks up the partially digested food even more. Cows also have populations of bacteria in their digestive system that help break down cellulose.

CHAPTER REVIEW

CONTENT REVIEW

Multiple Choice

Choose the letter of the answer that best completes each statement.

1. Certain whales have teeth modified into huge stiffened plates called
   a. rumen.   c. cecums.
   b. baleen.   d. monotremes.
2. Carnivorous mammals have strong
   a. sharp incisors.   c. flat-edged incisors.
   b. baleens.   d. flattened molars.
3. In humans, the cecum has evolved into the
   a. appendix.   c. kidney.
   b. small intestine. d. liver.
4. The mammalian circulatory system sends deoxygenated blood to the
   a. heart.   c. brain.
   b. internal organs. d. lungs.
5. The duckbill platypus and other egg-laying mammals are
   a. monotremes.   c. placental mammals.
   b. marsupials.   d. extinct.
6. Many biologists believe that mammals alive today do not see color well because early mammals
   a. lacked eyes.
   b. lived in cold regions.
   c. were nocturnal.
   d. were blind.
7. Bacteria in the digestive tracts of grazing animals produce enzymes that
   a. speed digestion.
   b. break down cellulose.
   c. break down fats.
   d. digest food for a second time.
8. In humans, nutrients, oxygen, carbon dioxide, and wastes are exchanged between embryo and mother through the
   a. rumen.   c. marsupium.
   b. uterus.   d. placenta.

True or False

Determine whether each statement is true or false. If it is true, write "true." If it is false, change the underlined word or words to make the statement true.

1. <u>Old World monkeys</u>, such as the squirrel monkey, have long prehensile tails used in climbing.
2. The odd-toed ungulates belong to the order <u>Perissodactyla</u>.
3. Night-flying bats use echolocation to <u>calculate distances from objects</u>.
4. The great blue whale and the humpback whale are <u>carnivores</u>.
5. Aquatic mammals such as whales and dolphins have a thick layer of <u>skin</u> called blubber.
6. The duckbill platypus is an example of a <u>marsupial</u>.
7. Elephants can hear sounds of <u>higher</u> frequencies than humans can.
8. Cottontail rabbits are in the order <u>Lagomorpha</u>.

Word Relationships

Replace the underlined definition with the correct vocabulary word.

1. Female mammals have <u>structures that produce milk for their young</u>.
2. Grazing animals have a <u>chamber in their digestive tract</u> that contains bacteria that break down cellulose.
3. Koalas are an example of a <u>pouched mammal</u>.

CRITICAL AND CREATIVE THINKING

1. The embryos of marsupials are supplied with nourishment by a yolk sac on the egg. But because the yolk sac is small, a marsupial is born at

CONCEPT MASTERY

Use your understanding of the concepts developed in the chapter to answer each of the following in a brief paragraph.

1. Describe three important characteristics used to classify mammals.
2. Explain how night-flying bats find tiny insect prey in total darkness.
3. How does a four-chambered heart add to the efficiency of the circulatory system of mammals?
4. What are the three parts of a mammal's brain and what functions does each control?
5. Describe the different types of teeth that have evolved in carnivorous and herbivorous mammals. How is each adapted to the animal's particular diet?
6. The digestive system of cows does not produce enzymes that break down cellulose. Explain how cows digest plant tissues that contain large amounts of cellulose.

CRITICAL AND CREATIVE THINKING

Discuss each of the following in a brief paragraph.

1. **Making comparisons** Compare and contrast the embryos of marsupials with those of placental mammals.
2. **Relating cause and effect** Explain how the change in world climate at the end of the Cretaceous Period led to the great mammalian radiation.
3. **Making inferences** In what ways have well-developed senses contributed to the success of mammals?
4. **Relating facts** Explain the relationship between a human's ability to speak and the movement of the diaphragm.
5. **Classifying organisms** You are given the following descriptions of three placental mammals. A is a flying mammal that has sharp teeth and a liquid diet. B is a mammal that has sharp front teeth and eats plant material. It has a short gestation period and produces many offspring. C is a carnivorous mammal that has a layer of subcutaneous fat. It lacks external ears and hind legs, and it mates and bears its young in water. Using your knowledge of mammals, classify each organism in its proper order. Give your reasoning for each classification.
6. **Identifying relationships** Early humans had a functioning appendix. Describe how the appendix changed over time. What reasons can you give to explain this change?
7. **Using the writing process** You are a writer for a small wildlife magazine. Your boss tells you that he has decided not to feature mammals any longer because he finds them dull. Write a memo to him in order to change his mind. Include the characteristics that separate mammals from organisms in other classes and describe the diversity that exists within the class.

755

an early stage of development. After it is born, a marsupial continues to develop in its mother's pouch. Tissues from the embryo of a placental mammal join with tissues from its mother to form a placenta. Nutrients are exchanged between mother and embryo across the placenta. Wastes are also exchanged between the embryo and the mother through the placenta.

2. The climate at the end of the Cretaceous Period became much colder and drier. The exothermic reptiles did not fare well with this change in temperature. The endothermic mammals, which remained in the shadow of the reptiles, were better able to adapt to the changing climates. They began a period of adaptive radiation at this time.

3. Well-developed senses enable mammals to be aware of dangers, and well-developed brains enable mammals to react much more quickly to the dangers they perceive.
4. The diaphragm is the sheet of muscle that separates the lung cavity from the stomach cavity. The diaphragm can be used to control the passage of air over the vocal cords, thus causing them to vibrate and produce sound.
5. The first mammal is a vampire bat, order Chiroptera. The second is a mouse or a rat, order Rodentia. The third mammal is a species of whale, order Cetacea.
6. Over time, the appendix has become a vestigial organ, an organ that no longer has a function. Scientists speculate that early humans ate a diet that consisted mostly of plant material. In time, this diet changed. The lack of a need for a functioning appendix to aid in the digestion of plant materials meant that this organ had little survival value.
7. Check all memos for clarity and correctness. Listing some of the wide range of mammals, from bats to whales, and their interesting lifestyles should convince even the most recalcitrant boss that mammals are indeed interesting. Offering the boss a mirror should convince the boss that he or she will find at least one mammal interesting.

CHAPTER 34 Humans

Section	Laboratory Investigations and Activities
34–1 Primates and Human Origins pages 757–759 THEMES: Evolution, Unity and Diversity	**Teacher Edition** DEMONSTRATION: Humans and Geologic Time, p. 756D **Laboratory Manual** The Hands of Primates, p. 439
34–2 Hominid Evolution: Human Ancestors and Relatives pages 759–765 THEMES: Evolution, Unity and Diversity	**Student Edition** LABORATORY INVESTIGATION: Comparing Primates, from Gorillas to Humans, p. 766 **Teacher Edition** DEMONSTRATION: Hands, p. 756D **Laboratory Manual** Comparing Primates, p. 431
Chapter Review pages 767–769	

Teacher Resources
Books

Brace, C. Loring. *The Stages of Human Evolution*, 4th ed. Prentice Hall, 1991.

Ciochon, Russell L., and John G. Fleagle. *The Human Evolution Source Book*. Prentice Hall, 1993.

Day, M. H. *Guide to Fossil Man*, 4th ed. University of Chicago Press, 1986.

Fossey, Dian. *Gorillas in the Mist*. Houghton Mifflin, 1983.

Goodall, Jane. *The Chimpanzees of Gombe: Patterns of Behavior*. Harvard University Press, 1986.

Leakey, Mary. *Disclosing the Past: An Autobiography*. Doubleday, 1984.

Lewin, Roger. *Bones of Contention: Controversies in the Search for Human Origins*. Simon & Schuster, 1988.

———. *In the Age of Mankind: A Smithsonian Book of Human Evolution*. Smithsonian Institute Press, 1988.

Milner, Richard. *The Encyclopedia of Evolution: Humanity's Search for Its Origins*. Facts on File, 1990.

Poirer, Frank E. *Understanding Human Evolution*, 3rd ed. Prentice Hall, 1993.

Premack, David, and Ann James Premack. *The Mind of an Ape*. Norton, 1983.

Richard, A. F. *Primates in Nature*. Freeman, 1985.

Strum, Shirley C. *Almost Human: A Journey into the World of Baboons*. Random House, 1987.

Weiss, M. L., and A. E. Mann. *Human Biology and Behavior*, 4th ed. Little, Brown, 1985.

CHAPTER PLANNING GUIDE

Other Activities	Media and Technology
Teaching Resources ACTIVITY: Observing Primate Structures, p. 3 **Study Guide** Section 34–1, p. 325	
Teaching Resources VOCABULARY REVIEW: Word Scramble, p. 1 ACTIVITY: Interpreting Information About the Evolution of Primates, p. 7 **Study Guide** Section 34–2, p. 327	**Biology Transparencies** 46: Comparison of the Skeletal Systems of a Human and an Ape
Teaching Resources Chapter Test A, p. 13 Chapter Test B, p. 17	**Computer Test Bank** Chapter Test, p. 357

Audiovisuals

The Fossil History of Man. Filmstrip or video. Carolina Biological Supply Company.

The Evolution of Man. 6 filmstrips. Ward.
Mammals III: Primates. Larousse Animal Life Series. Slides. Ward.

Mankind's Relatives—The Primates. Educational Images. Slides. Ward.

CHAPTER 34 Humans

CHAPTER OVERVIEW

The focus of this chapter is a topic that most of us find fascinating—humans. In this chapter students will explore primates and human origins. They will also explore the evolution of human ancestors and relatives.

Although the study of human origins has evoked controversy, there are certain basic facts on which all researchers agree. One of these facts is that humans evolved from common ancestors that we share with other primates.

Students will learn about the characteristics of primates and how early primates split into several evolutionary lines. They will discover that early primates eventually gave rise to an evolutionary line called hominoids, which in turn gave rise to a group of species called hominids. Scientists now recognize hominids as being our closest relatives.

Hominids displayed many characteristics that distinguished them from other hominoids. Students will learn about the evolution of various species of hominids that eventually led to the appearance of our own species.

34-1 PRIMATES AND HUMAN ORIGINS

Section Focus 34-1

The purpose of this section is to discuss the study of human origins in general and to focus on the importance of primates in particular. Students will begin by learning how paleontologists, anthropologists, archaeologists, and biochemists have all worked to bring together pieces of the puzzle of human origins. Despite varying degrees of controversy, certain basic facts about human origins are agreed upon by all researchers.

Students will learn that one thing all scientists agree on is that humans share common ancestors with other primates such as chimpanzees. Students will learn that as a group, primates display certain important adaptations, such as binocular vision, flexible fingers, and a large and complicated cerebrum.

Students will discover that early in their history, primates split into several evolutionary lines. One of these lines was the anthropoids. The anthropoids gave rise to the two major groups of monkeys—New World monkeys and Old World monkeys. The Old World monkeys eventually gave rise to the group called hominoids, from which humans are descended.

Performance Objectives 34-1

1. Discuss some of the branches of science that have contributed to knowledge of human origins.
2. List the basic facts about human origins on which all researchers agree.
3. Describe the distinguishing characteristics of primates.
4. Explain how primates evolved.
5. Compare New World monkeys with Old World monkeys.

Science Terms 34-1

binocular vision   p. 758
anthropoid   p. 758
hominoid   p. 759

34-2 HOMINID EVOLUTION: HUMAN ANCESTORS AND RELATIVES

Section Focus 34-2

The purpose of this section is to describe the characteristics of hominids and to explain how trends in their evolution eventually gave rise to modern humans. Students will learn that hominids showed several characteristics that distinguished them from other hominoids. First, they were omnivores. Second, they were able to walk erect. Third, their opposable thumb enabled them to grasp tools more effectively than other primates. Fourth, they displayed a remarkable increase in brain size.

Students will learn that most of our evidence for hominid evolution comes from fossils found in a small area of Africa. The first hominids were of the genus *Australopithecus*. Later hominid species were similar enough to humans to be placed in the genus *Homo*. These species included *Homo habilis*, *Homo erectus*, and the archaic *Homo sapiens*.

Students will learn that the first hominids truly identical to modern humans appeared roughly 100,000 years ago. These people, called Cro-Magnon, are now classified as modern humans, *Homo sapiens sapiens*.

Performance Objectives 34-2

1. Describe adaptations of hominids.
2. Discuss basic evolutionary trends in hominids.
3. List and describe various hominid species.
4. Explain how the evolution of hominids gave rise to modern humans.

Science Terms 34-2

hominid   p. 759
bipedal locomotion   p 760
opposable thumb   p. 760

Discovery Learning

TEACHER DEMONSTRATIONS
Modeling

Humans and Geologic Time

The following demonstration can be used as an introduction to Section 34–1.

Draw on the chalkboard or on an overhead transparency a time line showing the major divisions of geologic time. Include on the line the four eras: Precambrian, Paleozoic, Mesozoic, and Cenozoic. Divide the last era into the Tertiary Period and the Quaternary Period. Label on the time line the number of years ago that each era and period began and ended.

Point out the Tertiary Period to students. Explain that it was during this period that the first primates and humanlike creatures appeared. Then point out the Quaternary Period. Explain that this is the period in which we are now living; it was during this period that the first modern humans appeared.

Emphasize to students that humans have been on Earth for only a fraction of the Earth's history. The following analogy should help to underscore this point. Display a typical one-month wall calendar and ask students to imagine that this one month represents the age of the Earth so far. Point out that the appearance of humans would take place in the last second of the last day of the month—that is how short our time on Earth is compared to the age of the Earth.

Hands

This demonstration can be performed as students study Section 34-2.

Ask students to place their hands in front of them and spread their fingers apart. Explain that such a movement is called divergent movement. Students can also try performing divergent movements with their toes.

Next ask students to cup their hands, as if they were filled with water. Explain that this type of movement is called convergent movement. Then ask students to wrap their fingers around objects such as rulers or chalkboard erasers. Explain that such movements are called prehensile movements. Finally, have students hold a pen or pencil in their hands as if they were about to write. Explain that this type of movement, in which the thumb is touching one or more of the tips of the fingers, is called opposable movement.

Explain to students that divergent movements, convergent movements, prehensile movements, and opposable movements are all possible because clawed paws gave rise to hands among certain primates. Also explain that as humanlike creatures evolved into modern humans, the hand bones became modified and these four types of movements became more and more refined.

CHAPTER 34
Humans

GUIDED ENQUIRY
Pose the following questions to students and have them record their responses. Point out that they will gain a better understanding of the key concepts if they read the chapter with these basic questions in mind. Upon completion of the chapter, pose the questions again. Ask students to compare their initial responses with those they have developed after reading the chapter.

- How have many kinds of scientists contributed to knowledge about human origins?
- In what ways do all scientists agree about our species' past?
- Why are primates important in the study of human origins?
- How are hominids different from other primates?
- How did the evolution of hominids give rise to modern humans?

INTRODUCING CHAPTER 34

Using the Visuals
Have students observe the chapter-opener paragraph.
- **What do you see in the large picture?** (An astronaut.)
- **What special abilities and characteristics of humans are evident in a picture such as this one?** (Possible answers: high intelligence; the ability to develop sophisticated technology; the ability to invent; curiosity and the desire to explore the unknown; the ability to perform complex physical feats.)

Now have students look at the small inset photograph.
- **What do you see in this smaller picture?** (A stone. Students should read the caption and realize that it is a stone tool.)
- **When do you think this tool was made?** (Answers may vary. Students should recognize that this tool probably was made by early humans.)
- **What do you think the tool was used for?** (Possible answers: hunting; cutting and tearing animal meat; carving.)
- **What does a tool like this tell you about the characteristics of early humans?** (Possible answers: They had the dexterity to make and use tools; they had the intelligence to figure out what type of tool was needed for a particular purpose; they were involved in shaping and changing their environment.)
- **Why have humans changed from the makers of stone tools to astronauts who explore outer space?** (Accept all reasonable answers.)

CHAPTER 34
Humans

From the creation of stone tools to a trip through space, it is the nature of humans to question and wonder.

Humans are unique on this planet. No other organism lives in as many different habitats and does as many different things. No other organism has written and spoken language and keeps records of the past. We as a species have created art, music, philosophy, and science. Where did we come from? What were our ancestors like? How did we come to be as we are today?

There are few chapters of evolutionary history that are more fascinating (and more controversial) than the origin of our own species. More than a century before Darwin, Carolus Linnaeus realized that he should classify *Homo sapiens* in the same group of mammals into which he placed the apes. He did not do so, however, primarily because he feared that such an act would cause a great furor. Darwin himself shied away from the topic at first. But after the success of *On the Origin of Species*, he wrote two books dealing with human evolution. In the years since Darwin's work, our understanding of our species' history has come a long way.

756

GUIDE FOR READING

After you read the following sections, you will be able to

34–1 Primates and Human Origins
- Describe the characteristics of primates.
- Compare New World monkeys with Old World monkeys.

34–2 Hominid Evolution: Human Ancestors and Relatives
- Describe the importance of various hominid adaptations.
- Discuss the evolutionary trends in hominids that led to *Homo sapiens*.

Journal Activity

YOU AND YOUR WORLD

Space—The Final Frontier. It is a long voyage through time from the origins of humans to space travel. Pretend for a moment that you are the astronaut shown in the picture on the facing page. What thoughts would go through your mind as you take the first step from the safety of your spaceship to the chilly emptiness of space? Write these thoughts in your journal. Perhaps the future will give you the opportunity to experience space travel firsthand!

34–1 Primates and Human Origins

Guide For Reading
- What characteristics are shared by all primates?
- In what ways do Old World monkeys differ from New World monkeys?

The study of human origins is an exciting, almost frantic, search for our past. To piece together this complicated story in detail requires the skills of many kinds of scientists. Paleontologists study fossil primates and compare them with living forms. Archaeologists and anthropologists study ancient human tools and cultures, trying to piece together a story of human history. Molecular biologists examine the DNA of different species, looking for similarities and differences that show whether or not the species are closely related. All of these kinds of scientists and the methods they use—as you have read in Chapters 13 and 14—have made important contributions to the study of human evolution.

Research into human origins has always been spiced with competition among scientists, many of whom have different interpretations of the data gathered on our species' past. But all researchers agree on certain basic facts. We know, for example, that humans evolved from common ancestors we share with other living primates such as chimpanzees and apes. Our species almost certainly evolved in Africa and then spread around the world. We know that the first *Homo sapiens* appeared around 500,000 years ago, practically the day before yesterday on an evolutionary time scale. This means that humans did not appear until dinosaurs had been extinct for more than 60 million years (TV shows such as *The Flintstones* to the contrary).

Figure 34–1 One adaptation of primates are eyes that point forward, as you can see in the ring-tailed lemur (left) and the orangutan (right).

COOPERATIVE LEARNING

Divide students into five groups. Assign one of the following hominids to each group: Australopithecus, Homo habilis, Homo erectus, Homo sapiens neanderthalensis, *or* Homo sapiens sapiens. *Based on chapter descriptions, each group should develop a plan for a museum exhibit that showcases their assigned hominid. Money is no object! Their exhibit should illustrate both the physical appearance and the way of life of their assigned hominid. Students might be encouraged to use resources in addition to the textbook to add anthropological and archaeological details to their exhibits.*

Journal Activity

YOU AND YOUR WORLD

Ask volunteers to share their essays with the class. Have students keep their journal assignments in their portfolios.

that time there may not have been enough scientific evidence to support his idea; people may have reacted negatively to the idea of humans being placed in the same category as apes and monkeys.)

Content Development

As you study this chapter, be aware that some students may have strong personal views about the topic of human evolution. It is important to respect every student's point of view, particularly those whose convictions are based on religious beliefs. However, it is equally important that you present clearly the facts that have been backed up by scientific evidence, particularly those facts on which virtually all scientists agree.

TEACHING STRATEGY 34–1

Focus/Motivation

Have students read the chapter-opener text.

• **Why do you think that the study of human origins has been so controversial?** (Possible answers: A great deal of evidence has been gathered and not all scientists have agreed on the interpretation of the evidence; the study of humans evokes strong emotional responses because in studying human origins, people are studying themselves; some people have objected to the study of evolution because they feel that it conflicts with religious beliefs.)

• **Why do you think that Linnaeus felt that classifying humans with the apes would cause a furor?** (Possible answers: At

TEACHING SUPPORT

Laboratory Manual
- The Hands of Primates, p. 439

Teaching Resources
- Activity: Observing Primate Structures, p. 3

Study Guide
- Section 34–1, p. 325

ECOLOGY NOTE
PRIMATE BEHAVIOR

Have students watch the movie *Gorillas in the Mist*, which is based on the adventures of anthropologist Dr. Dian Fossey. Interested students may want to research books, magazines, and journal articles to learn more about primate societies and behavior.

Figure 34–2 Flexible primate fingers permit this gibbon to swing from branch to branch (top). Primates such as this chimpanzee have large cerebrums and complex social behaviors, including play (center). This loris, like other prosimian primates, is nocturnal and has large eyes that are well-adapted to seeing in the dark (bottom).

What Are Primates?

As a group, primates share several important adaptations, many of which are extremely suitable to a life spent mainly in trees. In general, primates have faces that are much flatter than those of other mammals. Primate eyes point forward, and their snout is very much reduced. (Compare your nose to that of your favorite dog.) These features together allow both eyes to inspect the same area at the same time. Information gathered by those eyes is processed by highly developed visual centers in the brain to produce what is known as **binocular vision**, or stereoscopic vision. Binocular vision equips primates with a three-dimensional view of the world (a handy adaptation when trying to judge accurately the locations of tree branches, from which many primates swing).

All primates have flexible fingers (and some have flexible toes) that can curl around objects. This allows many primates to hold objects in either their hands or their feet. It also enables many of them to run along branches and swing from branch to branch with ease. Primates' arms also are well-adapted to swinging and climbing because they can rotate in broad circles around the shoulder.

Finally, primates have a large and complicated cerebrum. For that reason, they display far more complex behaviors than other animals. For example, primate mothers take care of their young for a much longer time than most other mammalian species. Many primate species also have complicated social behaviors that include friendships, protective relationships among relatives, adoption of orphans, and—unfortunately—warfare between rival primate troops.

How Did Primates Evolve?

Very early in their history, primates split into several evolutionary lines. Those that evolved from two of the earliest branches look very little like typical "monkeys" and are called prosimians. Living prosimians, which are odd but interesting animals, include lemurs, lorises, and aye-ayes. With a few exceptions, prosimians are almost entirely nocturnal and have large eyes adapted for seeing in the dark. Members of the more familiar primate group that includes monkeys, apes, and humans are called **anthropoids**, or humanlike primates. This group, in turn, has given rise over time to several major primate branches.

Two anthropoid branches—the two major groups of monkeys and apes—separated around 45 million years ago when the continents on which they lived moved apart and were no longer connected by land bridges. One anthropoid group, known as New World monkeys, evolved into the monkeys found today in Central and South America. (The term New World comes from the days of Columbus when the Americas were called the New World.) These animals are virtually all tree

---

34–1 (continued)

SECTION REVIEW 34–1

1. Faces flatter than most mammals; eyes point forward and snout reduced; highly developed visual centers in brain enabling binocular vision; flexible fingers; arms that can rotate in broad circles around the shoulder; highly developed cerebrum.
2. New World monkeys live in Central and South America, are tree dwellers, and many have prehensile tails. Old World monkeys live in Africa and Asia, are not exclusively tree dwelling, and do not have prehensile tails.
3. When the continents separated, various lines of monkeys were geographically isolated. Geographic isolation and different environmental pressures caused divergent evolution among the monkeys, eventually leading to Old World monkeys and New World monkeys.

Reinforcement/Reteaching

If students are having trouble answering the Section Review questions, go back to the part of the section that contains the material with which they are having difficulty.

dwellers, and many of them have grasping (prehensile) tails that aid in balance while moving through tree branches. The other anthropoid group evolved into the Old World monkeys and the great apes that are found today in areas that extend from Africa all the way across Asia to Indonesia and Japan. The Old World monkeys, which do not have prehensile tails, include baboons and macaques (muh-KAHKS). Some of these animals live in trees, whereas others spend a good deal of time on the ground. The great apes, also called **hominoids**, include gorillas, gibbons, orangutans, chimpanzees, and *Homo sapiens*.

Figure 34–3 *Many New World monkeys spend their entire lives among the branches of trees. This spider monkey uses its prehensile tail like a fifth hand (left). Old World monkeys, such as this macaque, do not have prehensile tails and may spend much of their time on the ground (right).*

34–1 SECTION REVIEW

1. What characteristics are shared by all primates?
2. Compare New World with Old World monkeys.
3. **Critical Thinking—Applying Concepts** How did the separation of the continents contribute to the development of New World and Old World monkeys?

34–2 Hominid Evolution: Human Ancestors and Relatives

Some time between 4 and 9 million years ago, the hominoid line in Africa gave rise to a small group of species that we now recognize as our closest relatives. These species, called **hominids**, were not yet human, but they showed several evolutionary trends that distinguish them from other hominoids.

Guide For Reading
- What is the importance of bipedal locomotion?
- Why is an opposable thumb important in the evolution of hominids?
- What evolutionary trends in hominids led to *Homo sapiens*?

TIE-IN LITERATURE

For thousands of years, poets, philosophers, and theologians have attempted to explain the essential nature of humans. The following quotations reflect a variety of viewpoints.

What a chimera, then, is man! What a novelty, what a monster, what a chaos . . . the glory and the shame of the universe!
—Blaise Pascal

Herein lies the tragedy of the age . . . that men know so little of men.
—W.E.B. Du Bois

Perhaps the only true dignity of man is his capacity to despise himself.
—George Santayana

We are a spectacular, splendid manifestation of life. . . . We have genes for usefulness, and usefulness is about as close to a "common goal" of nature as I can guess at. And finally, and perhaps best of all, we have music.
—Dr. Lewis Thomas*

What is man, that thou art mindful of him? . . . Thou hast given him dominion over the works of thy hands; thou hast put all things under his feet.
—The Book of Psalms

Created half to rise, and half to fall;
Great lord of all things, yet a prey to all;
Sole judge of truth, in endless error hurled:
The glory, jest, and riddle of the world!
—Alexander Pope

It is possible that our race may be an accident, in a meaningless universe, living its brief life uncared-for, on this dark, cooling star; but even so—and all the more—what marvelous creatures we are!
—Clarence Day

What a piece of work is a man!
—William Shakespeare

---

* From "The Youngest and Brightest Thing Around" by Lewis Thomas, originally appeared on July 2, 1978, and is reprinted by permission of the New York Times. Copyright © 1978 by The New York Times.

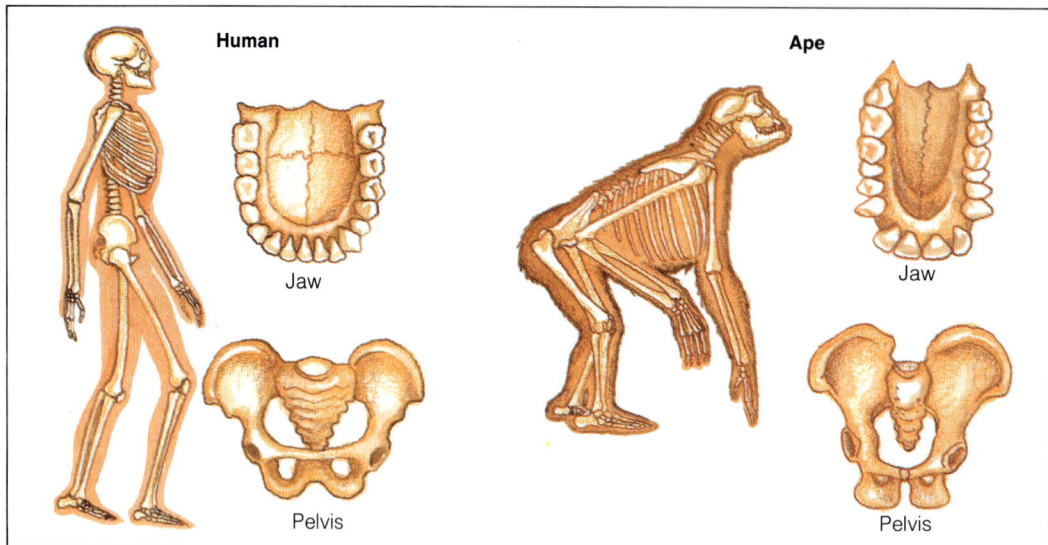

Figure 34–4 Compare the skeletal system in a human and an ape. Note the shape of the jaws, the pelvic structure, and the way the spine enters the skull.

What Are Hominids?

Hominids were omnivores that ate both meat and vegetable foods, as modern humans do. As time progressed, the spinal column, hip bones, and leg bones of these animals changed shape in ways that made it easier for them to walk upright on two legs. The evolution of this **bipedal**, or two-foot, **locomotion** was very important. Because our ancestors could walk erect, their hands were free to use tools more often. At the same time, the thumb of the hominid hand became more and more independent from the other fingers. The evolution of an **opposable thumb** enabled ancient hominids to grasp objects and use them as tools more effectively than other primates.

Hominids also displayed a remarkable increase in brain size. Hominid brains are exceptionally large, even for primates. Chimpanzees, our closest living relatives among the apes, have a brain size of about 280 to 450 cubic centimeters. The brain of *Homo sapiens*, on the other hand, ranges in size from 1200 to 1600 cubic centimeters! Most of the difference in brain size results from the enormously expanded human cerebrum. The cerebrum is the "thinking" area of the brain.

How Did Hominids Evolve?

To follow the story of human evolution, we need fossils of human ancestors. Much of our most recent evidence for hominid evolution comes from a small area in eastern Africa between Tanzania and Ethiopia. There, several researchers have found fossils of several species of hominids dating from about 4 million to about 1.5 million years ago.

***AUSTRALOPITHECUS*: THE FIRST HOMINIDS** The first hominid fossil to be found, a nearly complete skull of a young child, was discovered in South Africa in 1924. This specimen was placed in a new genus called *Australopithecus* (aw-stray-loh-PIHTH-uh-kuhs), or Southern Ape. Because the skull belonged to a child, it could not be used to determine how adults of the species looked. But 12 years later, investigators in Africa found fossils of adult australopithecines. One of these fossils was part of a hip bone, indicating that *Australopithecus* walked upright. Walking erect was an essential step in the evolution of our species from an apelike ancestor.

Since those discoveries, researchers have found many more complete hominid fossils. In 1974, a team led by Donald Johanson and Tim White made a truly exciting find—a nearly complete *Australopithecus* skeleton. From the shape of the pelvic bone, it was clear that this skeleton had been that of a female, and the fossil has since been called Lucy. (They took the name, by the way, from the Beatles' song "Lucy in the Sky with Diamonds," which they had listened to in their camp the night after the discovery.)

In 1977, anthropologist Mary Leakey made an equally exciting discovery: a set of fossil hominid footprints. The mud in which the prints were preserved has been dated at 4 million years old! Those footprints formed a fossilized record of two hominids walking together. From the size of the prints, they were probably a parent and an offspring. This record is clear evidence that the animals that made the footprints walked erect on two legs, as humans do. No stone tools have been found among *Australopithecus* fossils, but they may have used twigs and stones as tools in a way similar to that of chimpanzees today.

In recent years, other hominid fossils have been placed in the genus *Australopithecus*. Most current studies suggest that there were at least four species of *Australopithecus*: *A. boisei*, *A. robustus*, *A. afarensis*, and *A. africanus*. (The letter *A.* represents the genus name.) These species all lived between 4 and 1.5 million years ago, walked upright, and had much smaller brains than present-day humans. Many questions as to how these species were related to one another, as well as to human evolution, still remain to be answered.

HOMO HABILIS For a while, australopithecines were the only known links in the chain of human evolution. Then anthropologist Richard Leakey found another hominid fossil with a smaller face and significantly larger brain than the australopithecines. Leakey felt this species was similar enough to humans to be placed in our own genus, *Homo*. Fossils of this hominid were found along with tools made of stone and bone. As if to emphasize that fact, scientists have called these hominids *Homo habilis* (HAB-ih-lihs), which, appropriately, means handy man.

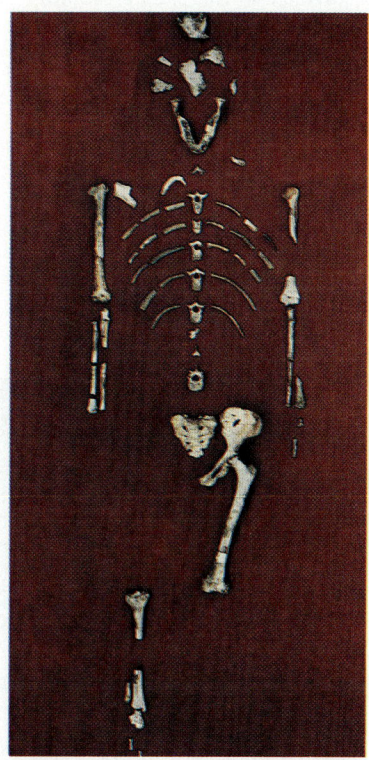

Figure 34–5 This nearly complete *Australopithecus* skeleton was nicknamed Lucy.

FACTS AND FIGURES

Human DNA differs in only about 2% of its nucleotide order from the DNA of chimpanzees.

HISTORICAL NOTE
HOMO HABILIS

Anthropologists have pieced together evidence from fossils and relics to create a picture of what daily life must have been like for *Homo habilis*, the earliest human ancestor. Their findings include the following:

1. Homes: The first humans probably made their homes in trees. A recently discovered fossil skeleton reveals that *H. habilis* had long arms adapted for tree climbing.
2. Food: *Homo habilis* ate plant parts such as berries, roots, and tubers. They also scavenged meat from the kills of other carnivores.
3. Family: The young matured twice as fast as modern humans, needing much less extensive parental care.
4. Social Structure: Males were larger than females, suggesting a social structure in which groups of females were allied with a single male.
5. Speech: Although lacking the language abilities of modern humans, *Homo habilis* had the neurological structures in their brains associated with speech, as well as a larynx adapted to making a broad range of sounds.

---

gers; separating the pages of a book.)

Reinforcement/Reteaching

Make sure that students do not confuse the terms *hominid* and *hominoid*. Point out that hominids are a small group of species that grew out of the hominoid line in Africa.

Content Development

Explain that when Charles Darwin wrote *On the Origin of Species*, no fossils of ancient humans had yet been found. Darwin based his hypothesis on the idea that if other living things had evolved from earlier forms, then humans must have evolved as well.

Also emphasize that neither Darwin nor any other scientist proposed that humans evolved from apes or chimpanzees. Scientists have simply concluded that humans and animals such as apes and monkeys share common primate ancestors.

Figure 34–6 This skull of Homo habilis *was found in Kenya and is about 1.8 million years old.*

Near one of these fossil finds, in a valley in Kenya called Olduvai Gorge, is the oldest human settlement yet discovered. The settlement was found at a level in the rock dated at 1.9 million years ago. The main site is a circular stone structure about 4 meters in diameter. Inside, the floor is littered with animal bones and stone tools. Just what *Homo habilis* used these tools for is not clear. Some scientists think this species ate meat and the tools were used for hunting prey. More recent evidence indicates that *Homo habilis* was basically a vegetarian species that may have followed in the paths of other carnivores, stealing whatever parts of the kill they could find. Although not quite so glorious a past, it is one that seems to be real.

HOMO ERECTUS Evidence suggests that within a few hundred thousand years *Homo habilis* disappeared and was replaced by a larger brained species called *Homo erectus* (ee-REHK-tuhs). By 1 million years ago, this species had spread over most of the Old World, from Africa to Europe to Asia.

With a cranial capacity of more than 800 cubic centimeters, *Homo erectus* was an excellent toolmaker. Carefully chipped and balanced hand axes have been found with *Homo erectus* fossils throughout the world. In caves in China that are at least half a million years old, charred animal bones have been found around fire sites. This shows that *Homo erectus* must have used fire for cooking. From a site in France dated at about 400,000 years old, the remains of primitive huts have been discovered —huts not too different from some still in use in parts of the world today.

Figure 34–7 Homo erectus *replaced* Homo habilis, *spreading throughout Europe, Africa, and Asia by about 1 million years ago. Evidence strongly suggests that* Homo erectus *used and controlled fire.*

PROBLEM SOLVING IN BIOLOGY

ANALYZING *HOMO ERECTUS* BEHAVIOR

Scientists have found fossils of *Homo erectus* in many places on Earth. In one fossil find in Spain, a puzzling event seems to have occurred. On top of a cliff, the remains of ancient brush fires have been unearthed. At the base of the cliff, the bones of an entire herd of elephants have been found. Scattered among the elephant bones were stone tools.

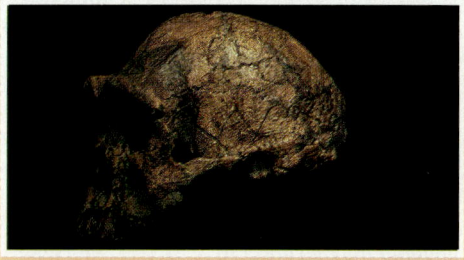

Analysis: Using the information in this chapter, formulate a hypothesis that would explain this event in ancient history. Then explain how your hypothesis can be used to make some logical assumptions about the behavior of *Homo erectus*.

HOMO SAPIENS About 500,000 years ago, the first hominids assigned to our own species (*Homo sapiens*) appeared. These hominids, often called archaic *Homo sapiens*, would not be easily recognizable as modern humans. Little is known about this species.

Around 150,000 years ago, a new hominid walked on Earth. First discovered in the Neander valley in Germany, this species was called Neanderthal man, or *Homo neanderthalensis*, for many years. Now, based on more complete fossil evidence, Neanderthals have been placed in our own species and are called *Homo sapiens neanderthalensis*. Although you have probably seen movies in which Neanderthal man is depicted as primitive looking, hunched over, and covered with hair, this depiction is totally inaccurate. Neanderthal man could probably walk down a busy street today and not be noticed (assuming he was dressed in modern clothes)! These early members of our species were quite successful for a time and became common throughout Europe and the Middle East by 70,000 years ago.

The first hominids truly identical to modern humans appeared in locations scattered throughout the Old World roughly 100,000 years ago. These large-brained people, called Cro-Magnon (kroh-MAG-nuhn), were more slender than the Neanderthals and had a more complex culture. They made a wide

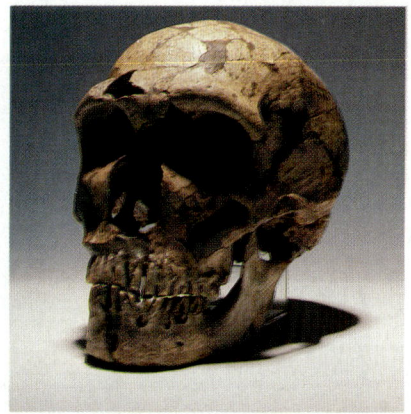

Figure 34–8 This Neanderthal skull is between 35,000 and 50,000 years old.

PROBLEM SOLVING

Analyzing *Homo erectus* Behavior

Have students work in pairs or small groups. First, have students make a list of the pertinent facts of the fossil find. They also may find it helpful to illustrate these facts with a drawing or diagram. Next, have students use the information in the chapter to develop several possible explanations for the facts on their list. Encourage students to neither immediately adopt nor immediately discard any of their ideas. Instead, tell them to list each of their possible explanations and use the list to synthesize a hypothesis that explains all aspects of the fossil find.

Answer

The brush fire and stone tools suggest that *Homo erectus* hominids killed the elephants. The hominids intentionally set the brush fires to surround the elephants, and the fires panicked the elephants into stampeding off the cliff. The stone tools were used not as weapons but to cut open and carve the elephant kill.

This event illustrates many qualities and abilities of *Homo erectus*. They were able to formulate a plan and cooperate with one another to achieve a goal. They could control fire and make stone tools. They also killed their own food, rather than scavenging the kills of other animals, as did *H. habilis*.

---

the large carnivores had scavenged the remains.

However, looking at these bones in a new way—under a scanning electron microscope—reveals evidence against the original hypothesis. The scanning electron microscope shows that on many bones, the tool marks fall on top of the tooth marks—the tooth marks were made first. This suggests that the large carnivores were the predators and the hominids were the scavengers. This new theory presents a very different picture of *Homo habilis* and probably a more accurate one.

Discuss with students how anthropologists developed and changed their thinking about *H. habilis*. Also discuss how scientists use new technology, such as the scanning electron microscope, for a surprising variety of purposes. Students may be interested in reviewing the scanning electron microscope, discussed on pages 36–37 of the textbook.

TEACHING SUPPORT

Teaching Resources
- Vocabulary Review: Word Scramble, p. 1

Study Guide
- Section 34–2, p. 327

HISTORICAL NOTE
NEANDERTHAL

The first fossils of human ancestors were discovered in 1857 when workers were clearing a limestone cave in a part of Germany called the Neander valley. The workers dumped the bones, which looked to them like human remains, into a pile with debris from the quarry. There they were found by a teacher named Dr. Fuhlrott, who passed them along to an anatomy professor. The anatomy professor realized that the skullcap and limb bones he had been given were humanlike but very, very old.

In time, more specimens like these were discovered all over Europe. All of them were called Neanderthal man, named after the Neander valley in which the first bones were discovered. However, scientists did not agree on the age of the bones. Some thought that the bones belonged to primitive ancestors of humans, whereas others believed the bones to be only a few hundred years old. Today, thanks to radioactive dating, scientists know that the Neanderthal bones are between 100,000 and 150,000 years old.

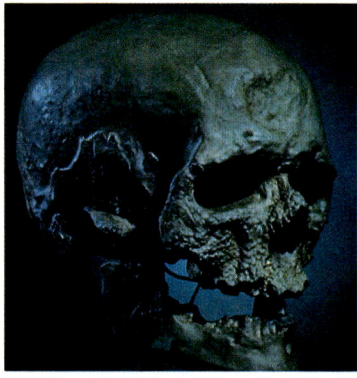

Figure 34–9 Cro-Magnon was the first hominid truly identical to modern humans.

Figure 34–10 Neanderthal buried their dead with tools, animal bones, and even flowers, probably indicating some sort of belief in an afterlife. In several caves around the world, animal skulls and bones were laid out on piles of stone and in nooks in the walls, which may have been altars to primitive gods.

variety of stone and bone tools, including spear points, knives, chisels, and needles. They were also talented artists. Fossils of Cro-Magnon are now classified as modern humans, *Homo sapiens sapiens*.

Most paleontologists interpret the dates of Cro-Magnon fossils found throughout the world as indicating that modern humans originated in Africa and from there spread out over the rest of the world. However, there are those who argue that modern humans evolved from Neanderthals in several regions, including Europe, the Middle East, and Asia. At this time, there is no clear resolution to this complex debate.

However and wherever Cro-Magnons originated, there is ample fossil evidence that they lived side by side with Neanderthals in several locations for some time. Then, around 30,000 years ago, the Neanderthals disappeared. Some scientists believe that Cro-Magnons interbred with Neanderthals, blending their characteristics. Others believe that the more intelligent newcomers slaughtered their older relatives. In either case, only *Homo sapiens sapiens* remained to populate the rest of the world.

34–2 SECTION REVIEW

1. What is the importance of the evolution of bipedal locomotion? Of the opposable thumb?
2. Why is *Homo habilis* aptly named?
3. Which genus was the first hominid to walk erect?
4. **Critical Thinking—Making Inferences** Based on your knowledge of evolution, why is it more likely that the ancestors of *Homo sapiens sapiens* evolved in a single area rather than in many places throughout the world?

34–2 (continued)

SECTION REVIEW 34–2

1. Bipedal locomotion helped free the arms and hands from use in walking and allowed primates to use their hands for other purposes. The opposable thumb permitted grasping of objects and was vital for the construction and use of tools.
2. *Homo habilis* is the first human ancestor known to use tools. As such, the name *Homo habilis*, which means handy man, is quite appropriate.
3. *Australopithecus*.
4. Since humans throughout the world are of one species, it is likely that all humans evolved from a single common ancestor in a particular place. If humans had evolved from different ancestors in different parts of the world, it is likely that there would be more than one human species, which we know there is not.

SCIENCE, TECHNOLOGY, AND SOCIETY ISSUE

Women and the Development of Human Society

Until recently, the study of human evolution has been a male-dominated science. A simple look at fossil names tells you that. We learn, after all, about Neanderthal man and Cro-Magnon man. But where are the Neanderthal and Cro-Magnon women in all this? Equally important to this concept of male domination are older studies of primates and early hominids that constantly place males in positions superior to those of females. By studying the social groups of other living primates, such as baboons and chimpanzees, biologists often try to infer the sorts of behaviors australopithecines might have displayed. And by examining certain isolated "stone age" human tribes, researchers try to imagine what the first human societies were like.

Both of these research techniques are useful, but they must be applied with caution. Male researchers, it seems, have found an overwhelming body of evidence that points toward universal male superiority among primates and early humans. It was male hunting, they say, that led to civilization. Another interpretation places males in control of most primate groups because "man the hunter" is strong, aggressive, and intelligent. "Woman the gatherer" is weaker, less coordinated, and always busy caring for infants.

But the old story is not necessarily the true story. For as more and more female researchers have entered the field of anthropology and other related fields (and, to be fair, as more open-minded males have joined in the pursuit), new information has come to light. It now seems likely that early hominids depended more on vegetable material than on meat. If this is true, the women who gathered vegetable foods would have been in powerful positions! The ability of females to cooperate in gathering and distributing food may have been a vital ingredient in the founding of the earliest civilizations. Women's skills might well have led to the development of agriculture, a cultural innovation that made large permanent human settlements possible. At the very least, women's contributions were just as important as men's ability to organize hunting parties.

It is also true that females of other primate species are not nearly as male dominated as researchers once reported them to be. Females, it appears, often control the length of time certain primate troops spend in particular feeding areas. In some species, a dominant male's position often depends on the support of female allies, even if he "rules the roost." And in several species, newborn infants inherit their rank within the troop not from their father, but from their mother.

These and many other studies remind us that the story of our species involves the evolution of both men and women, for neither could have survived alone.

SCIENCE, TECHNOLOGY, AND SOCIETY ISSUE

Women and the Development of Human Society

After students have read the feature, ask:
- **Would you say that the authors feel that women are inferior to men, superior to men, or equal in importance to men?** (Equal in importance to men.)
- **Do you agree with this point of view? Why or why not?** (Accept all reasonable responses.)
- **Do the authors believe that men and women in early society were essentially the same? Explain your answer.** (No, the authors stress that men and women made different contributions and performed different tasks.)
- **Do you think that the same thing is true of men and women today? Support your answer with evidence.** (Accept all logical responses.)

After students have discussed the questions, you may wish to have them use the feature as a jumping-off point for a class debate or for essays or speeches in which they can express their own points of view on a particular aspect of women and society. Students may choose such topics as women's rights, the feminist movement in America, sexist language, media stereotypes of males and females, and male and female roles in the family. Students should be encouraged to use additional reference sources in preparing their debates, papers, or speeches.

Reinforcement/Reteaching
If students are having trouble answering the Section Review questions, go back to the part of the section that contains the material with which they are having difficulty.

Closure
Have students work in groups of six to eight to simulate round-table discussions on the topic of human origins. Students may use information from outside reference sources as well as the material in this chapter.

LABORATORY INVESTIGATION

COMPARING PRIMATES, FROM GORILLAS TO HUMANS

Before the Lab
At least a day prior to the investigation, gather enough materials for your class, assuming 6 students per group.

Pre-Lab Discussion
Read or have a student read the Problem at the beginning of the investigation. Have students use information from the chapter to list as many changes as they can think of that occurred as humans evolved from early hominids. Some of these changes may be clearly stated in the text; others may be inferred. Some of the changes students might list include changes in face, teeth, and jaw; increase in brain size and complexity; change to bipedal locomotion, which involved changes in pelvic and leg bones; development of an opposable thumb; change from scavengers to hunters; loss of much body hair.

Point out to students that in this investigation they will be exploring the changes that occurred in the jaw and hand. Have students state in the form of a hypothesis how they think the jaw and hand changed as humans evolved from early hominids.

Have students read through the entire Procedure. Discuss with students how they might set up their data tables. Also make sure that students understand how to use a protractor to draw a perpendicular line.

Direct students' attention to the drawings of the *Australopithecus* and gorilla jaws. Point out the dimensions that will be used to calculate the jaw index. If math skills are a problem for more than a few students in your class, you may wish to perform a sample calculation of jaw index.

Demonstrate to students the correct way to measure the length of the thumb and index finger. Write on the chalkboard and discuss the formula for calculating the thumb index.

LABORATORY INVESTIGATION

COMPARING PRIMATES, FROM GORILLAS TO HUMANS

PROBLEM
What changes occurred as humans evolved from earlier hominids?

MATERIALS (per group)

| metric ruler | protractor |
| clean paper | scissors |

PROCEDURE

1. After reading the investigation, prepare a data table to record your observations.
2. Use the scissors to cut a strip of clean paper about 6-cm wide and 9-cm long.
3. Insert the strip of paper lengthwise into your mouth. Place the paper over your tongue so that it covers all your teeth, including your back molars. Bite down hard enough to make an impression of your teeth on the paper. Remove the paper from your mouth.
4. Draw a line on the paper from the center of the impression of the left back molar to the center of the right back molar. Mark the midpoint of this line. Use the protractor to draw a perpendicular line from the midpoint of the line connecting the back molars to the front teeth.
5. Measure the width of the jaw by measuring the length of the line between the back molars. Measure the length of the jaw by measuring the line from the back of the mouth to the front teeth. Record your measurements.

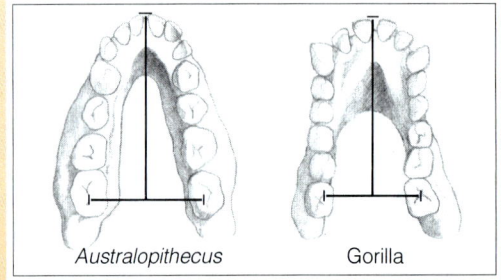
Australopithecus Gorilla

6. Calculate the jaw index by multiplying the width of the jaw by 100 and then dividing this number by the length of the jaw. Record the jaw index.
7. Repeat steps 5 and 6 using the drawings of the gorilla jaw and the *Australopithecus* jaw. Record your results.
8. Find the indentation at the bottom of your palm at the ball of your thumb. Measure the length of your thumb from the indentation to the tip. Measure the length of your index finger from the indentation to its tip. Record your measurements.
9. Calculate the thumb index by multiplying the length of the thumb by 100 and dividing by the length of the index finger. Record the thumb index.
10. Repeat steps 8 and 9 using the drawings of the thumb and index finger.

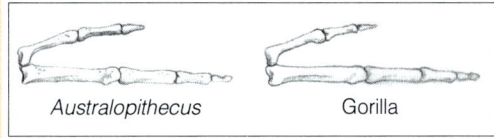
Australopithecus Gorilla

OBSERVATIONS
1. What trend did you observe regarding the relative length of the jaw? The relative length of the thumb and index fingers?

ANALYSIS AND CONCLUSIONS
1. What evidence is there that *Australopithecus* was an organism with characteristics intermediate to gorillas and humans?
2. Based on the thumb index, what adaptive change occurred in the evolution of humans? What was the advantage of this change?
3. Based on your observations, what other change occurred as humans evolved?

Skills Development
Students will use the following skills while completing this laboratory investigation.
1. Measuring
2. Observing
3. Manipulative
4. Predicting
5. Hypothesizing
6. Comparing
7. Relating
8. Applying
9. Calculating
10. Analyzing
11. Inferring

STUDENT STUDY GUIDE

SUMMARIZING THE CONCEPTS

The key concepts in each section of this chapter are listed below to help you review the chapter content. Make sure you understand each concept and its relationship to other concepts and to the theme of this chapter.

34-1 Primates and Human Origins

- Humans (*Homo sapiens*) evolved from common ancestors we share with other living primates such as chimpanzees and apes.
- Primates share certain characteristics: flat faces, reduced snouts, eyes that face forward and allow for binocular vision, flexible fingers and toes, arms that can rotate in broad circles around the shoulders, and a large cerebrum.
- Anthropoid primates include monkeys, apes, and humans.
- Two anthropoid branches separated around 45 million years ago when the continents shifted and were no longer connected by land bridges.
- One anthropoid branch evolved into the New World monkeys, which are primarily tree dwellers and, in general, have tails.
- The other anthropoid branch evolved into the Old World monkeys, which include baboons and the great apes.
- The great apes, also called hominoids, include gibbons, orangutans, chimpanzees, and *Homo sapiens*.

34-2 Hominid Evolution: Human Ancestors and Relatives

- Hominid adaptations include changes in the spinal column, hip bones, and leg bones that allow bipedal (two-foot) locomotion. Hominids also have a much larger brain than other primates.
- The first recognized hominids were the australopithecines, all of which walked erect.
- The first species to be classifed in the genus *Homo* was *Homo habilis*, or handy man.
- *Homo habilis* was replaced by *Homo erectus*, which spread throughout much of the world by around 1 million years ago.
- The first species to resemble modern humans seems to have evolved around 150,000 years ago and is called *Homo sapiens neanderthalensis*.
- The first fossils of modern humans date to about 100,000 years ago. These humans, called Cro-Magnons, are in the genus *Homo sapiens sapiens*, as are modern humans.
- Cro-Magnons probably evolved in Africa and spread throughout most of the world.

REVIEWING KEY TERMS

Vocabulary terms are important to your understanding of biology. The key terms listed below are those you should be especially familiar with. Review these terms and their meanings. Then use each term in a complete sentence. If you are not sure of a term's meaning, return to the appropriate section and review its definition.

34-1 Primates and Human Origins
binocular vision
anthropoid
hominoid

34-2 Hominid Evolution: Human Ancestors and Relatives
hominid
bipedal locomotion
opposable thumb

---

garding their own jaw and thumb indexes. Have the class observe the data and discuss any differences that they see among members of the class. Have students offer possible explanations for these differences.

Observations

1. The jaw became progressively shorter in going from *Australopithecus* to humans. But the thumb and index fingers became progressively longer.

Analysis and Conclusions

1. The jaw index and thumb index are intermediate to those of the gorilla and the human.
2. The thumb became relatively longer, making it more able to reach across the hand and grasp objects.
3. The jaw became shorter, producing a more V-shaped dental arch.

Going Further: Enrichment

Have students conduct outside research to find out how tooth size has changed throughout human evolution. Then have them use this information to design an investigation in which they compare tooth sizes of modern humans to those of several earlier species.

An interesting part of such an investigation would be observing the different size teeth among members of the class. Scientists believe that variation among humans today is part of a trend toward smaller teeth—meaning that human teeth are still changing. People with smaller teeth are on the "cutting edge" of this evolutionary change.

Safety Tips

Make sure that students observe reasonable sanitary precautions when putting paper in their mouths. Stress that the paper must be clean, and that once paper has been placed in a student's mouth, it should remain with that student and not be used or handled by other students.

Teaching Strategy

1. As students are working, circulate throughout the room and help any groups that are having trouble. Check especially that data tables are correctly set up, measurements are being properly taken, and calculations are being carried out correctly.
2. When the groups are finished, gather and display all the data that students obtained re-

CHAPTER REVIEW

CONTENT REVIEW

Multiple Choice
1. b 2. a 3. b 4. b
5. c 6. d 7. c 8. d

True or False
1. T
2. F anthropoid
3. F Old World
4. F omnivores
5. F *Homo habilis*
6. T
7. T
8. T

Word Relationships
1. great apes; lemur is a prosimian
2. hominids; hominoid is the collective term for great apes and hominids
3. *Australopithecus*; *H. erectus*
4. *Homo sapiens*; *Homo habilis*

CONCEPT MASTERY

1. An opposable thumb enabled primates to grasp branches, and binocular vision allowed them to judge distances accurately when leaping from tree to tree.
2. *Homo erectus* had a larger brain than *Homo habilis*, was more humanlike in appearance than *H. habilis*, and used fire and built huts, which *H. habilis* did not do.
3. In biological terms, a species is successful if its members survive and reproduce for a long period of time. Since Neanderthal man existed for only a short period of time, it was not a successful species.
4. Answers will vary. Examples might include molecular biologists who study DNA and other molecular aspects of early peoples; paleontologists who study fossils of early humans; and anthropologists who study ancient dwellings in order to learn more about the cultures of early peoples.
5. From a behavioral point of view, *H. erectus* used fire and built huts and *H. habilis* did not. From a physical point of view, *H. erectus* had a larger cerebrum and was more humanlike in appearance than *H. habilis*.
6. Anthropoids are humanlike primates that gave rise to New World monkeys and Old World monkeys. Hominoids, or great apes, are one anthropoid line and include gorillas, chimpanzees, and humans. Hominids are our closest living relatives and exhibit common human characteristics such as bipedal locomotion and greatly enlarged cerebrums.
7. Two possible causes have been proposed. First, Neanderthals may have been killed off by Cro-Magnons. Second, Neanderthals may have interbred with Cro-Magnons.

CHAPTER REVIEW

CONTENT REVIEW

Multiple Choice

Choose the letter of the answer that best completes each statement.

1. Modern humans are included in the genus
   a. *sapiens*.
   b. *Homo*.
   c. hominid.
   d. hominoid.
2. Old World monkeys include
   a. baboons.
   b. prosimians.
   c. lemurs.
   d. lorises.
3. Great apes do not include
   a. chimpanzees.
   b. macaques.
   c. orangutans.
   d. gibbons.
4. Hominids were
   a. carnivores.
   b. omnivores.
   c. herbivores.
   d. saprophytes.
5. The first hominid known to use tools was
   a. *Australopithecus boisei*.
   b. Cro-Magnon.
   c. *Homo habilis*.
   d. *Homo erectus*.
6. The first species to be considered *Homo sapiens* is believed to have evolved
   a. 100,000 years ago.
   b. 150,000 years ago.
   c. 1.3 to 4 million years ago.
   d. 500,000 years ago.
7. All of these are hominids except
   a. *Homo habilis*.
   b. *Australopithecus afarensis*.
   c. hominoids.
   d. *Homo sapiens*.
8. Which of these is not a characteristic of hominids?
   a. spinal cord
   b. bipedal locomotion
   c. opposable thumb
   d. enlarged cerebrum

True or False

Determine whether each statement is true or false. If it is true, write "true." If it is false, change the underlined word or words to make the statement true.

1. <u>Binocular vision</u> is an adaptation found in most primates.
2. Two <u>hominoid</u> branches split off around 45 million years ago when the continents separated.
3. Gibbons, chimpanzees, and baboons are all <u>New World</u> monkeys.
4. Hominids were primarily <u>carnivores</u>.
5. The first toolmakers were <u>Homo erectus</u>.
6. Cro-Magnon is included in the species <u>Homo sapiens sapiens</u>.
7. Primitive huts have been found in <u>Homo erectus</u> campsites.
8. Modern humans probably originated in <u>Africa</u> and spread out over the rest of the world from there.

Word Relationships

In each of the following sets of terms, three of the terms are related. One term does not belong. Determine the characteristic common to three of the terms and then identify the term that does not belong.

1. gibbon, chimpanzee, lemur, orangutan
2. hominid, *Homo sapiens*, *Homo erectus*, hominoid
3. *H. erectus*, Lucy, *A. boisei*, *A. robustus*
4. Cro-Magnon, Neanderthal, archaic *Homo sapiens*, *Homo habilis*

CONCEPT MASTERY

Use your understanding of the concepts developed in the chapter to answer each of the following in a brief paragraph.

1. Describe how various primate adaptations made them successful tree dwellers.
2. Compare *Homo habilis* and *Homo erectus*.
3. Explain why, in biological terms, Neanderthal was not a successful species.
4. Choose three different fields of biology and explain how each field can contribute to the study of human origins.
5. Why might *Homo erectus* be considered more advanced than *Homo habilis*?
6. Compare anthropoids, hominoids, and hominids.
7. Discuss possible reasons for the disappearance of Neanderthal.

CRITICAL AND CREATIVE THINKING

Discuss each of the following in a brief paragraph.

1. **Making inferences** Suggest some reasons for the fact that Cro-Magnon buried their dead in a ritualistic manner.
2. **Making predictions** Describe how society might have evolved if hominids had not developed an opposable thumb.
3. **Interpreting data** There are many theories regarding the specifics of human evolution and our common ancestors, two of which are shown in the accompanying illustration. Use the illustration to describe the major differences between these two versions of human evolutionary trends.
4. **Distinguishing fact and opinion** People often say that humans evolved from monkeys and apes. Explain why such a statement does not conform with modern evolutionary theory.
5. **Relating concepts** Most scientists agree that language did not develop prior to the appearance of Cro-Magnon. Describe some of the benefits language provides society.
6. **Using the writing process** With the exception of modern humans, choose a hominid and write a short story entitled "A Day in the Life of a Hominid."

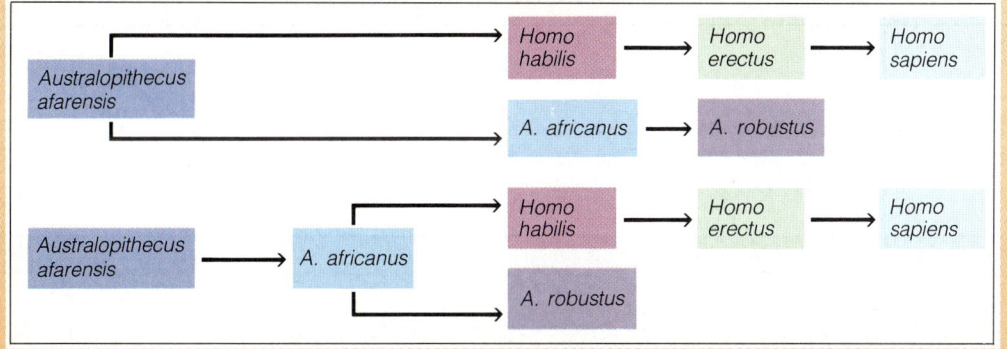

CRITICAL AND CREATIVE THINKING

1. Answers will vary. Students may suggest that Cro-Magnons buried their dead because of some belief in an afterlife.
2. Answers will vary. Students may point out that society would probably be much less advanced technologically than it is now and that tools, if they were developed, would be radically different from tools now in use. Some students may even point out that they would have difficulty turning the pages of a book.
3. In the top version, two main lines of hominids evolved from *Australopithecus afarensis*. One line eventually led to modern humans. In the bottom version, *A. afarensis* evolved into *A. africanus*, which then separated into two evolutionary lines, one of which evolved into modern humans.
4. Evolutionary theory in no way states that humans evolved from monkeys and apes. This is a serious misrepresentation of the theory of evolution. Rather, the theory states that humans, monkeys, and apes evolved from a common ancestor.
5. Answers will vary. Students will point out that language allows communication and perhaps was even vital to the development of organized societies.
6. Short stories should be well-written, imaginative, and reflect the information presented in the chapter.

CHAPTER 35 Animal Behavior

Section	Laboratory Investigations and Activities
35–1 Elements of Behavior pages 771–775 THEMES: Unity and Diversity, Patterns of Change	**Student Edition** LABORATORY INVESTIGATION: Exploring Mealworm Behavior, p. 780 **Teacher Edition** DEMONSTRATION: Examples of Animal Behavior, p. 770D **Laboratory Manual** Animal Behavior, p. 443
35–2 Communication: Signals for Survival pages 775–777 THEMES: Unity and Diversity, Systems and Interactions	**Teacher Edition** DEMONSTRATION: A Courtship Ritual, p. 770D **Teaching Resources** ECOLOGY INVESTIGATION: Soil Burrowing in Gerbils, p. 45
35–3 The Evolution of Behavior pages 778–779 THEMES: Evolution, Systems and Interactions	**Laboratory Manual** Competition or Cooperation? p. 449
Chapter Review pages 781–783	

Teacher Resources

Books

Alcock, J. *Animal Behavior: An Evolutionary Approach*, 3rd rev. ed. Sinauer Associates, 1983.

Anderson, E. W. *Animals as Navigators*. Van Nostrand Reinhold, 1983.

Brown, Vinson. *The Secret Languages of Animals*, rev. ed. Prentice Hall, 1987.

Dawkins, M. S. *Unravelling Animal Behavior*. Wiley, 1986.

Grier, James. *Biology of Animal Behavior*, 2nd ed. Mosby, 1989.

McFarland, David. *Animal Behavior*. Benjamin-Cummings, 1985.

Slater, Peter J. B., ed. *The Encyclopedia of Animal Behavior*. Facts on File, 1987.

CHAPTER PLANNING GUIDE

Other Activities	Media and Technology

Teaching Resources
    ACTIVITY: Identifying Methods of Learning, p. 3
    BIOLOGY CASE STUDY: "Learning" in Ants, p. 13
Study Guide
    Section 35–1, p. 333	

Teaching Resources
    ACTIVITY: Signals for Survival, p. 5
Study Guide
    Section 35–2, p. 336
	Interactive Videodisc
	    In the Company of Whales
	Videodisc
	    Sound Sense
	    Super Scents
	Biology Media Guide
	    pages 50–51

Teaching Resources
    VOCABULARY REVIEW: Cross-a-Clue, p. 1
Study Guide
    Section 35–3, p. 338	

Teaching Resources
    Chapter Test A, p. 13
    Chapter Test B, p. 17
	Computer Test Bank
	    Chapter Test, p. 367

Audiovisuals

A Thousand Million Million Ants. Video. Carolina Biological Supply Company.
Designs for Defense. Film or video. Marty Stouffer Productions.
Animal Behavior: The Mechanism of Imprinting. Film or video. Coronet.

Dwellings. Video. Carolina Biological Supply Company.
Camouflage in Nature: Form, Color, Pattern Matching, 2nd ed., rev. Film or video. Coronet.

Animal Homes. Film or video. Coronet.
Adaptations. Film or video. Coronet.

CHAPTER 35 Animal Behavior

CHAPTER OVERVIEW

Animal behavior is as important to animal survival and species continuation as any physical characteristic. Behavior has also evolved in many different ways, just as have physical characteristics. Also, behavior, like physical characteristics, can have adaptive value.

Some behaviors are automatic from birth and others are more flexible and capable of change through experience. These behaviors are divided into two main categories: instincts and learning. Instincts are behaviors that can be called inborn. They are built into the animal's nervous system and cannot be changed during the animal's lifetime. Some instinctive behaviors are simple and others are very complex.

Learned behaviors, unlike instincts, are shaped by experiences. What animals can learn is largely determined by their genetic makeup. Among some species little learning occurs. In others, especially vertebrates, learning takes place in a variety of ways, such as habituation, classical conditioning, and operant conditioning. Habituation, one of the simplest ways in which animals learn, is a decrease in response to a stimulus that neither rewards nor harms an animal. Classical conditioning occurs when an animal makes a mental connection between a stimulus and some good or bad event. In operant conditioning, sometimes called trial-and-error learning, an animal learns to behave in a certain way to receive a reward or to avoid punishment. Insight learning occurs when an animal applies something it has already learned to a new situation, without a trial-and-error period. Imprinting is behavior that combines instincts and learning.

If animal behavior involves more than one individual, some type of communication is necessary. This involves the passing of information from one animal to another. There are many ways in which animals communicate based on how they perceive the world through their different senses. Animals communicate for a variety of reasons, for example, to attract a mate and to indicate the availability of food and the presence of danger. Animals communicate with members of the same species and with other species. Communication takes many different forms, including visual, sound, chemical, and electrical signals.

Behaviors, like physical characteristics, are controlled by genes or groups of genes. Genetic variation and natural selection increase the frequency of the most beneficial genes. In behavioral terms, if social behavior offers an advantage for an animal, natural selection will favor the evolution of social behavior in a species.

35–1 ELEMENTS OF BEHAVIOR

Section Focus 35–1

The purpose of this section is to familiarize students with the basic types of animal behavior. Animal behavior is just as important to survival as any physical trait that the animal might have. Animal behavior is rooted in the genetic makeup of the individual animal. Behaviors can have adaptive value. Some behaviors must be performed automatically soon after birth if the animal is to survive. Other behaviors must be more flexible and capable of being changed by experience. Behaviors can be grouped into two main categories: instincts and learning.

Instincts are often called inborn behaviors. They are part of the animal's nervous system and cannot be changed. Sometimes a pattern of instinctive behavior, once begun, must continue through a sequence of events, even if it is no longer logical. The example given in the text is the egg-rolling behavior of the graylag goose. Some instinctive behaviors are rather simple and others, such as web-building behaviors in spiders, are fairly complex.

Learned behaviors, unlike instincts, are shaped by experience. In some animals, little learning occurs. Other animals can learn a great deal. Learning is important to animals because it can improve chances of survival. There are several different ways that learning occurs: habituation, classical conditioning, and operant conditioning. Types of learning, and examples of each, are discussed in this section. The section concludes with an explanation of imprinting, which is an example of a behavior combining instinct and some learning.

Performance Objectives 35–1

1. Define instinct and give an example.
2. Discuss the different types of learned behaviors.
3. Describe imprinting.

Science Terms 35–1

instinct   p. 772
learning   p. 773
habituation   p. 773
classical conditioning   p. 774
operant conditioning   p. 774
insight learning   p. 774

35–2 COMMUNICATION: SIGNALS FOR SURVIVAL

Section Focus 35–2

The purpose of this section is to introduce the topic of animal communication. Anytime a behavior involves more than one animal, some type of communication is necessary. Communication is the passing of information from one animal to another. Animals use a variety of techniques to communicate since animals receive and perceive information from the outside world in many ways.

Animals communicate with one another for a variety of reasons, including courtship, to pass information about food location, and to signal danger. Methods of communication are quite varied. Visual, sound, chemical, and electrical signals are all ways in which animals communicate with other members of the same species as well as with members of other species. If communication becomes complex and sophisticated, it can be termed *language*.

Performance Objectives 35–2

1. Discuss what animal communication is and why it is important.

CHAPTER PREVIEW

2. Explain why animals would need to communicate with one another.
3. Describe the various methods of communication that animals might use.

Science Terms 35–2

communication   p. 775
visual signal   p. 776
sound signal   p. 777
chemical signal   p. 777
pheromone   p. 777
electrical signal   p. 777

35–3 THE EVOLUTION OF BEHAVIOR

Section Focus 35–3

The purpose of this section is to explain the relationship of behavior to an animal's genetic makeup, therefore showing how evolution can affect behaviors. Behaviors are largely controlled by DNA, just as physical traits are. Variations in genes and the process of natural selection allow beneficial genes and the behaviors they code for to increase. If social behavior, for example, is beneficial to the organism in some way, it is favored by natural selection because it enhances the animal's chances of surviving and reproducing.

Performance Objectives 35–3

1. Discuss how behaviors are controlled by DNA.
2. Describe how social behavior can increase the chances for survival in a species.

DISCOVERY LEARNING

TEACHER DEMONSTRATIONS
Modeling

Examples of Animal Behavior
This activity can be used to introduce Chapter 35.

Make a list of common animal behaviors with which the students may be familiar. Have them decide if the behaviors are learned or inborn. Ask them to discuss why they decided as they did. If there are any answers that are disputable, leave them at this point and return to them after studying Section 35–1. Some possible behaviors might include housebreaking a pet, a squirrel hiding nuts, a kitten nursing, a bird building a nest, a hummingbird taking nectar from different flowers, a spider building a web, the family dog obeying the command "sit and stay," and any others that you can think of—you might even want to include student behaviors such as raising their hand to speak.

A Courtship Ritual
This activity can be used to introduce or conclude Section 35–2.

Students sometimes think that animal behavior, particularly courtship behavior, is peculiar. Actually, there is a fixed pattern of actions and reactions that must be rigidly followed if the species is to survive. As you describe the elaborate courtship behavior in three-spined stickleback fish, have students try to anticipate the reaction of the opposite partner. Give them some time to think about this. Stress that courtship ritual is a stimulus-response chain. (**Note**: Encourage students to treat this as a serious lesson and not as an opportunity to giggle or act immaturely.)

Male sticklebacks in breeding condition develop bright-red bellies. They select a territory, build a nest, and wait for a suitable female. How might they recognize a suitable female? When a female stickleback is carrying eggs, her belly becomes swollen. The sight of a red-bellied male attracts the female. On sighting such a male, she swims toward him with her head pointed up in the water. What do you think this position allows the male to clearly see? This position exposes her swollen belly. What does the male do? The male, who has found a suitable female, does a peculiar zig-zag dance. What does the female do? What any curious female would do! She follows him! The male points to the nest and the female enters. The male places his snout near the female's body and vibrates it. This causes the female to raise her tail and release eggs into the nest. The male then chases the female out of the nest and fertilizes the eggs. End of romance!

As you can see, any break in the chain would prevent mating. This sort of complicated behavior is found in many animals. It is quite unlikely that two different species would have the same long set of patterns. For this reason, the mating behavior makes it unlikely that males and females of different species will ever mate. This complicated behavior, though different in each case, occurs in hundreds of animal species.

CHAPTER 35
Animal Behavior

GUIDED ENQUIRY

Pose the following questions to students and have them record their responses. Point out that they will gain a better understanding of the key concepts if they read the chapter with these basic questions in mind. Upon completion of the chapter, pose the questions again. Ask students to compare their initial responses with those they have developed after reading the chapter.

- Why do animals do the things they do?
- Are behaviors and physical characteristics related?
- How do animals learn?
- How do animals communicate with one another?
- Why do some animals show social behaviors whereas others do not?

CHAPTER 35
Animal Behavior

Animals communicate through their behavior. It might not be obvious to a human what the male frigate bird wants to communicate, but the cheetah's message is quite clear!

On a tropical island, a group of male frigate birds flap their wings, clack their bills, screech, and puff up their enormous red throat pouches. Females fly overhead, inspecting the group's display.

In a freshwater marsh in Massachusetts, male redwing blackbirds perch on the tips of tall marsh reeds. There they sit for hours, singing loudly and flashing their red wing patches at other birds.

In the ice-cold waters of an Oregon stream, a meter-long salmon fights its way upstream. After living for several years in the vast Pacific Ocean, this salmon—and countless others like it—has returned to the very stream in which it hatched several years before.

Why do birds act the way they do? How do salmon find the right stream to return to? In other words, how can the behaviors of animals—puzzling and often bizarre—be explained? The answers to these questions and a look at the fascinating field of animal behavior lie in the pages that follow.

INTRODUCING CHAPTER 35

Using the Visuals
Ask the students to write down a definition for behavior. Read and discuss a few of their answers. Ask them to think about some of their own behaviors. Are they learned or unlearned? After discussing this point for a few minutes, direct their attention to the chapter-opener photographs. Read the caption. Ask the same question about the behavior of male frigate birds and cheetahs and discuss briefly. Read the chapter-opener information with the students. Consider the questions posed in the paragraph.

GUIDE FOR READING

After you read the following sections, you will be able to

35–1 Elements of Behavior
- Describe the adaptive value of certain behaviors.
- Contrast instincts and learned behavior.

35–2 Communication: Signals for Survival
- Relate communication to animal behavior.
- Describe ways animals communicate with each other.
- Recognize the importance of communication to survival.

35–3 The Evolution of Behavior
- Explain how behaviors are controlled by an animal's DNA.
- Describe how social behavior can increase the survival of a species.

Journal Activity

YOU AND YOUR WORLD
Imagine that you are a behavioral scientist from a galaxy far, far away. In your journal, record your observations of the alien creatures that call themselves humans. Include your interpretation of the behaviors you observe.

35–1 Elements of Behavior

Guide For Reading
- How is animal behavior related to evolution?
- What are instincts, learning, and imprinting?

Animals are able to do some amazing things: Some bats fly in total darkness, navigating by means of high-pitched sounds, much as submarines do using sonar. Sea turtles spend many years swimming vast distances in the open sea, far from the beach on which they hatched. But when they are mature and able to reproduce, the sea turtles return to their birthplace to lay their eggs. Chimpanzees have a complex social system in which they love, tease, and help their friends and relatives—but fight, cheat, steal from, and even kill their enemies.

Behavior and Survival

Bats, turtles, chimpanzees, and other animals behave as they do for the same reasons that birds have feathers, frogs have sticky tongues, and giraffes have long necks. **The behavior of an animal is just as important to its survival and reproduction as any of its physical characteristics. For that reason, animal behaviors have evolved in many different ways, just as animal physical characteristics have.** All animal behaviors have their roots in the genetic makeup of the individual animal. Just as certain characteristics can enhance an animal's ability to survive, so can behaviors. Behaviors, like physical characteristics, can have adaptive value. For example, the fangs and sharp claws of a lion would have little survival value if a lion did not have clever hunting strategies.

Figure 35–1 The songs of many birds are beautiful to our ears. But birds sing for serious reasons. The song of this three-wattled bell bird pierces through thick jungle foliage, warning other birds to stay away from the territory it has claimed for itself.

771

COOPERATIVE LEARNING

Using preassigned lab groups or randomly selected teams, have each group brainstorm for 2 minutes, listing stimuli to which they respond in a given day. Groups should cluster stimuli into categories of their choosing. They might categorize according to the sense to which stimuli appeal or according to the pleasure/pain sensation.

This activity can be used as a springboard for each group to design an experiment that will illustrate how to condition a mammal of their choice to respond to a stimulus. Each group should follow the scientific method to produce an outline description of their experiment.

Journal Activity
YOU AND YOUR WORLD

This Journal Activity offers a chance for student imagination to soar. Students should reread this journal entry after they have completed the chapter. They should then decide if they should change any of the conclusions they based on their "observations."

TEACHING STRATEGY 35–1
Focus/Motivation

Ask the students to list some classroom behaviors that they have learned. Put this list on the board or overhead so everyone can see it. The text explains that animal behaviors are important to survival. For fun, label the classroom behaviors that have been listed. The two categories are Helps Survival in the Classroom or Works Against Survival in the Classroom.

Content Development
Using the Visuals Students may have a difficult time understanding that behavior is controlled or influenced by an animal's genetic makeup. Take a few minutes to discuss this point. Also, be sure that students remember what adaptive value means. The relationship of physical characteristics to behaviors is clearly illustrated by the example of the lion, which should help you to emphasize that point. Finally, direct the students' attention to Figure 35–1 because it explains the answer to one of the questions in the chapter-opener paragraph.

771

BACKGROUND INFORMATION
NATURE VERSUS NURTURE

Between inborn behaviors and those that are totally learned, there is a large area of controversy. Thus, the "nature–nurture" argument was born.

On one side of the debate were most of the classically trained ethologists, whose approach to animal behaviors was based on either field studies or lab experiments mimicking nature. Ethologists believed strongly that many behaviors throughout the animal kingdom were innate, or genetically preprogrammed. The other side was composed largely of American psychologists who tested learning in carefully controlled laboratory environments. Some of these behaviorists believed that genes had nothing to do with behavior. In their view, behavior consisted of simple responses tied together by learning. The battle raged throughout the 1950s and 1960s.

Now the two groups often work together. The general outcome of the struggle has been the realization that there is a gradient of behavior that ranges from totally inborn responses through genetically guided forms of learning, such as imprinting, all the way through to the insight learning shown in the most neurally complex organisms.

TEACHING SUPPORT

Laboratory Manual
- Animal Behavior, p. 443

Teaching Resources
- Biology Case Study: "Learning" in Ants, p. 13

Figure 35–2 This yawning lioness reveals one reason for her hunting success—huge fangs. Although male lions are well-fed, they rarely hunt. Relying on the hunting abilities of the females, the males grow fat on the efforts of others.

There are some behaviors that animals must perform automatically in order to survive. For example, many carnivorous animals must "know" how to hunt soon after they are born. They have no chance to learn basic hunting techniques, for if they do not catch some food, they will starve. And newborn dolphins must know in advance that they have to hold their breath under water, or they will no doubt drown!

Other behaviors must be more flexible and capable of being changed by experience. Flexible behaviors allow animals to adjust to a changing environment. For example, hummingbirds must learn to find food in different kinds of flowers at different times of the year. For if they insisted on feeding on just a single type of flower that blooms only in May, they would starve when those flowers disappear in June.

A variety of automatic and flexible behaviors exist in the animal kingdom. Scientists separate these behaviors into two main categories: instincts and learning.

Instincts

Instincts are behaviors that can be called inborn. Instincts are built into an animal's nervous system and cannot be changed during the animal's lifetime, even by learned experiences. For example, newly hatched birds beg for food instinctively within moments after hatching. Similarly, all newborn mammals instinctively suckle at their mother's breast. Because animals perform these behaviors without any previous experience, the behaviors must be controlled by the "wiring" in the animal's nervous system. And because the nervous system is "assembled" under the instructions contained in an animal's DNA, instinctive behaviors are genetically controlled.

Figure 35–3 Baby birds beg for food automatically when one of their parents approaches, and their pleadings rarely go unanswered (left). Nurturing and protecting its infant, this monkey makes an excellent parent (right).

35–1 (continued)

Content Development
Talk with students about behaviors that are automatic and those that are flexible.
- **What are the two examples of automatic behaviors given in the textbook?** (The hunting behavior of carnivores and baby dolphins holding their breath. Students may also mention suckling behavior, which will be studied in the next section.)
- **Can you think of other examples?** (Answers will vary widely based on students' prior knowledge.)
- **What is meant by a flexible behavior?** (One that changes with experience.)
- **How might flexible behaviors be important to animals?** (They allow them to adjust to a changing environment.)
- **What are the automatic behaviors called?** (Instincts.)
- **What are the flexible behaviors called?** (Learned behavior, or learning.)

Many instinctive behaviors consist of actions that always continue in a certain order once they have begun. An example of this kind of fixed instinctive behavior is shown by the graylag goose. A graylag goose makes its nest on the ground. If a graylag goose sitting on her eggs sees an egg outside her nest, she stares at it closely. Instinctively, she slowly stands up, stretches out her neck, and carefully uses her beak to roll the egg back into the nest. This routine is always performed in the same way. By performing a simple experiment, researchers have proved that this sort of behavior is completely automatic: Just as a goose started to stretch out her neck to reach an egg placed outside her nest, the researchers pulled the egg away. To their amazement, the goose still went through all the motions of retrieving an egg—rolling it closer and closer to the nest and then settling down to incubate it—even though the egg had been taken away and was no longer there! It is interesting to note that further studies show nesting graylag geese will perform the same egg-rolling behavior even if the object outside the nest only barely resembles an egg. Employing the same techniques used to retrieve an errant egg, these geese have tried to roll objects as large as volleyballs to their nest.

Although some instinctive behaviors are relatively simple, others can be very complex. Web-building behavior in spiders is a complex fixed instinctive behavior. Other examples of fixed instinctive behaviors are seen in many of the complicated courtship and nest-building behaviors found in insects, fishes, birds, and mammals.

Learning

Unlike instinctive behaviors, learned behaviors are shaped by experience. Animals would not survive in the world if they were unable to modify their behavior. **Learning** is the way animals change their behavior as a result of experience. In invertebrates such as insects, little learning occurs. The things these animals can learn are strongly determined by their genetic makeup. For example, bees can learn to tell certain colors apart but only if those colors are associated with food such as sugar water. In vertebrate animals, however, learning is much less controlled by genetic makeup. Humans, for example, can learn to differentiate many colors under a wide variety of circumstances.

Learning is valuable to an animal because it may enhance the animal's chances of survival and, thus, its chances of reproducing and passing on its genes to another generation. Today, biologists who study learning recognize several different ways in which animals learn: habituation, classical conditioning, operant conditioning, and insight learning.

Habituation is a decrease in response to a stimulus that neither rewards nor harms an animal. It is one of the simplest ways in which animals learn. An example of habituation is

Figure 35–4 This graylag goose carefully rolls an egg back to her nest. So instinctive is this behavior that she will continue her rolling actions even if the egg is removed from her sight.

BACKGROUND INFORMATION
GRAYLAG GEESE

Further experiments with graylag geese provided more interesting information. Any number of objects that somehow resemble an egg could set off, or release, the egg-rolling behavior. The confused goose would retrieve beer cans, baseballs, and even volleyballs. The researchers thus hypothesized that the goose had a built-in behavioral program, called an innate releasing mechanism, or IRM. The IRM enabled the goose to recognize certain characteristics of an object near the nest and start off the egg-rolling behavior. The characteristics of the object that triggered the IRM were called releasers.

IRMs have since been found in a great number of animals. In many lower animals, no amount of experience can ever alter IRMs. They are literally programmed into the nervous system, just like a simple computer program. *Tritonia*, a simple sea slug, for example, always responds to the odor of a predatory starfish by thrashing around and swimming away. In higher animals, there are some circumstances in which experience or the situation in which an animal finds itself can alter an IRM. A female herring gull who is not breeding, for example, will eat the same egg she would roll into her nest if she were incubating her own eggs.

Content Development
Using the Visuals Discuss the characteristics of instincts. Talk about each of the examples presented in the text and in Figures 35–3 and 35–4.
• **Explain how each behavior is related to survival.** (Answers will vary.)

Begin the discussion of learned behavior by defining what it is. Next, compare learned behavior with instincts, so that students will not confuse the two. Point out the variation that exists among species in terms of their ability to learn. There are great differences in how much and by what processes animals learn. List the ways in which animal learning occurs. Define and discuss the simplest type of learning, habituation.

Figure 35–5 Bright colors often provide a warning of danger. This brightly colored poison arrow frog warns potential predators that a hasty meal may prove fatal.

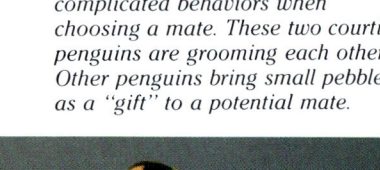

Figure 35–6 Animals exhibit complicated behaviors when choosing a mate. These two courting penguins are grooming each other. Other penguins bring small pebbles as a "gift" to a potential mate.

found in certain young birds. Very young ducks and geese are frightened of any shadow that moves overhead. Within a few days of hatching, however, the young birds find that some shadows moving overhead—for example, the shadows of adult geese and ducks—mean nothing. They soon habituate to these shadows and no longer try to escape from them.

Classical conditioning occurs when an animal makes a mental connection between a stimulus and some kind of good or bad event. One famous example of classical conditioning is the work of the Russian biologist Ivan Pavlov. Pavlov discovered that if he always switched on a light before he fed his dogs, the dogs would begin to salivate whenever he turned on the light. The dogs had learned to associate the light (stimulus) with the arrival of food (good event). In a similar manner, newly hatched ducks and geese learn to associate one kind of overhead shadow with fear. When eagles or hawks fly overhead, adult ducks and geese become terrified and try to find a safe place to hide. The young birds quickly learn to respond to the overhead shadows of eagles and hawks by scurrying to find a safe place to hide. The ducklings and goslings learn which shadows to avoid by observing the actions of the adults.

Operant conditioning is sometimes called trial-and-error learning. In operant conditioning, an animal learns to behave in a certain way in order to receive a reward or to avoid punishment. There are many examples of operant conditioning in nature. Recall that many insects and amphibians carry bad-tasting chemicals or poisons in their bodies. Remember also that most of these bad-tasting animals are brightly colored. If a bird eats a brightly colored butterfly and gets sick soon afterward, it will avoid those butterflies in the future. In this example of operant conditioning, a predator learns not to eat a particular prey in order to avoid an unpleasant experience.

In **insight learning**, an animal applies something it has already learned to a new situation—without a period of trial and error. Insight learning is rare among most animals, including many mammals. Insight learning is common only in primates and is found most often among humans.

A dog on a long leash will often accidentally wrap the leash around a tree or a pole. In such a situation, the dog is usually completely helpless. Running in circles and tugging on the leash, the dog is not able to free itself by retracing its steps in the opposite direction. A chimpanzee in the same predicament, however, may be able to apply insight learning to unwind the leash and get free.

Instinct and Learning Combined

Some behaviors, although primarily instinctive, cannot occur without some learning on the part of the animal. For example, newborn ducks and geese have a built-in urge to follow their mother. But this instinct to follow does not include a

---

TEACHING SUPPORT

Teaching Resources
- Activity: Identifying Methods of Learning, p. 3

Study Guide
- Section 35–1, p. 333

35–1 (continued)

Content Development

Using the Visuals Discuss imprinting as an example of behavior that combines instinct and some learning. Use Figure 35–7 to discuss the imprinting experiments performed by Konrad Lorenz.

SECTION REVIEW 35–1

1. An instinct is a behavior that is inborn or built into an animal's nervous system. It is not a learned behavior and cannot be changed by learning. Instincts necessary for survival are found in newborns. For example, human babies know how to suckle. This is an important instinctive behavior. A newborn infant would probably starve if it had to learn to suckle.

2. *Habituation* is a decrease in response to a stimulus that neither rewards nor harms an animal. An example of habituation is shown in young birds that are afraid of a shadow that moves overhead. By observing older birds they learn to fear only certain shadows. *Classical conditioning* occurs when an animal makes a mental connection between a stimulus and some kind of good or bad event. An example is shown by Pavlov's dog, which learned to salivate at a flash of light. *Operant conditioning* is often called trial-and-error learning. An animal learns to behave in a certain way in order to receive a reward or to avoid punishment. An example is a bird learning to avoid eating butterflies that resemble the "bad-tasting" monarch butterfly. In *insight learning*, an animal applies something it has already learned to a new situation. Insight learning is common in primates and is most often found in humans. For example, a young child might use climbing and object-moving behaviors in a new way to escape from a playpen.

3. In imprinting, an animal combines instinctive behavior with learned behavior. For example, baby ducks often follow the first thing they see after hatching—that is an instinct. But if the first thing they see is not their mother, they will still follow it even if it is an inanimate object.

picture of what their mother looks like. This picture must be provided by experience in a process called **imprinting**. The newborn bird will follow the first large slowly moving object it sees. In nature, birds will almost always imprint on their mother, for it is she who is usually closest at the critical time during which imprinting takes place. In the laboratory, however, birds can be imprinted on humans, on wooden models, or even on a slow-moving watering can! See Figure 35-7. It is an important characteristic of imprinting that once the built-in circuitry of the nervous system is given a picture, the behavior cannot be changed. Once a baby bird decides who or what its mother is, it will follow "her"—even if a more appropriate mother appears on the scene.

Figure 35-7 These geese, swimming with the famous animal behaviorist Konrad Lorenz, have decided that he is their mother. They have even followed him around on land—trailing along behind him in single file.

35-1 SECTION REVIEW

1. What is an instinct? Why do newborn animals show many instinctive behaviors?
2. Describe the three kinds of learning, giving examples.
3. **Critical Thinking—Relating Concepts** How does imprinting combine instinct and learning?

35-2 Communication: Signals for Survival

Any time animal behavior involves more than one individual, some form of **communication** is involved. **Communication is the passing of information from one animal to another. Animals use many varied techniques to communicate with one another.** Some of these methods are quite different from anything humans are familiar with.

Guide For Reading
- What is communication?
- How do animals communicate with one another?
- How does communication affect survival?

775

TEACHING SUPPORT

Teaching Resources
- Ecology Investigation: Soil Burrowing in Gerbils, p. 45
- Activity: Signals for Survival, p. 5

Study Guide
- Section 35–2, p. 336

MEDIA AND TECHNOLOGY

 Videodisc
- *Super Scents*
- *Sound Sense*

 Interactive Videodisc
- *In the Company of Whales*
Use this bar code to introduce the students to the social behavior of whales.

Play frames 13588 to 14988

See the Biology Media Guide pages 50–51 for additional bar-code correlations for this section.

35–2 (continued)

Content Development
Using the Visuals The idea that not all people see the world the way they see it comes as a surprise to most students. The idea that different species perceive the world entirely differently will be an intriguing thought. Use the examples in the text and in Figure 35–8 to make your point. You might even want to talk about people who have some impairment, such as blindness. They have a very different perception of the world around them.

Enrichment
Many students find animal communication fascinating. They might be interested in doing some library research or designing a project that involves tape recording and analyzing animal sounds.

Figure 35–8 *The clear, piercing eyes of an eagle enable this animal to see a fish swimming beneath the surface of a mountain lake (top). This grass snake gathers information about its environment by tasting it. Its tongue carries molecules from the air to special sensory organs in its mouth (bottom).*

Sensing the Natural World

No two animal species sense the world in the same way. Each animal species has a unique way of gathering and transmitting information. Because animals perceive the world in different ways through different senses, each obtains a different image of the world. Understanding the differences between our sensory world and that of animals is important in the study of animal behavior. Unless scientists understand the sorts of events an animal can detect, they will not be able to understand its behavior. Today, many scientists study animals under natural conditions. These scientists are called ethologists, and the study of animal behavior under natural conditions is called ethology. It is important to keep in mind that because animals react to stimuli in the natural world in different ways, they may also communicate using signals that human senses cannot detect.

Why Animals Communicate

Scientists have found that animals communicate with one another for a variety of reasons. For example, correctly choosing a mate is essential to species survival. Thus, courtship behavior is one of the most important types of communication in almost all animal species. Animals have developed a fairly complicated system of signals between males and females to make certain that the choice is correct. Penguins carry tiny rocks to potential mates, offering the pebbles as a choice bit of nesting material. Other birds have elaborate courtship displays that may occur over a considerable period of time as potential mates indicate their desire to pair. Because choosing a mate is so important, many birds, like the mute swan, pair for life.

Food is another reason why animals communicate. Parents and offspring often need to communicate about the location and availability of food. Parents also warn offspring about potential dangers. Adult animals, too, often warn one another of danger. Many animals, such as lions, communicate with one another when they band together in packs to hunt and kill prey. Animals that live in groups have developed complex and efficient ways to communicate with one another.

How Animals Communicate

Animals communicate with other members of their species and with other species. The ways in which they communicate are limited only by the kinds of stimuli their senses can detect.

Animals with good eyesight often use **visual signals** such as movement and color to communicate. For example, wolves raise their ears and arch their back to communicate anger. The males of many bird and fish species use brightly colored body parts as signals to attract females during courtship and to warn potential rivals to keep away.

Content Development
Spend some time talking about why animals would need to communicate. Be sure to relate each example to the survival of the individual and/or species.

Content Development
The first thing that comes to mind when most students think about communication is sound. If you have any students who know sign language, have them give a brief demonstration as a common example of visual communication. Chemical and electrical signals are less common but interesting to discuss.

Content Development
Language among animals is interesting and varied. Have students consider what makes

Some animals use **sound signals** to communicate. Crickets, frogs, and birds are all able to make sounds to attract mates. Dolphins and other marine mammals signal to one another by making special sounds under water. Scientists have recorded many fascinating songs made by whales as they communicate over vast distances in the oceans.

Some animals with a well-developed sense of smell use **chemical signals** to communicate. These animals produce special chemicals called **pheromones** that transmit information. For example, when an ant finds food some distance from the nest, she walks back to the nest dragging her abdomen on the ground. As she does so, she releases a trail pheromone. Other ants quickly follow the scent of the trail pheromone to reach the food.

Some fishes can generate and detect **electrical signals**. These fishes communicate with other members of their school by changing the electrical signal they produce.

Language

Some forms of animal communication, such as the dance language of the bees, are more complicated than any of the signals just described. The dance developed by honeybees can pinpoint a food source with a great degree of accuracy.

Chimpanzees, one of the most intelligent of the great apes, have a reasonably complicated language composed of sounds, gestures, and facial expressions. However, human language is the most complicated form of communication. Some researchers feel they can teach chimpanzees to communicate with humans. Chimpanzees cannot talk, but chimpanzees do seem able to learn a language composed of hand signs. Some scientists believe the chimps are really intelligent and actually learn to communicate by using these signals. Other scientists insist the chimps are just learning which signs they need to use to get food and attention. According to these researchers, the chimps do not really understand the signs. The debate continues as both sides argue about this fascinating topic.

Figure 35–9 Animals often use sound to communicate with each other. This frog has filled special sacs in its throat with air (top). The air is then expelled, causing sounds to be produced when tissues in the frog's throat vibrate. The naked mole rat spends its life in dark underground tunnels (bottom). The society of mole rats is similar in some ways to insect societies. The "queen" mole rat is the only individual that reproduces. She excretes pheromones in her urine that control the behavior of other mole rats in the colony.

35–2 SECTION REVIEW

1. What is communication? Why is it important?
2. What are four ways in which animals communicate with one another? How do these methods of communication depend upon an animal's sensory abilities?
3. **Critical Thinking—Making Inferences** Suppose that you visited a planet populated by people very different in appearance from humans. How could an observation of the sense organs of these people help you to communicate with them?

SECTION REVIEW 35–2

1. Communication is the passing of information from one animal to another. Communication is important when animals choose a mate, search for food, establish territory, and hunt. Animals that live in groups have developed complex and efficient ways to communicate with each other.
2. Animals use visual signals, sound signals, chemical signals, and electrical signals to communicate with each other. Animals with good eyesight often communicate with visual signals. Animals with good hearing often communicate with sound signals. Animals with a well-developed sense of smell often communicate through chemical signals. Animals with special receptors that receive electrical signals often communicate with each other in this way. The way animals communicate with each other is often determined to a large extent by the kinds of sensory receptors they have.
3. A keen observer would notice the kinds of sensory organs these people had. For example, if the people seemed to sniff the air, it might be assumed that they were searching for chemical methods of communication. If they had ears that moved in response to sounds, it might be assumed that these people communicated with sounds.

TEACHING SUPPORT

Laboratory Manual
- Competition or Cooperation? p. 449

Teaching Resources
- Vocabulary Review: Cross-a-Clue, p. 1

Study Guide
- Section 35–3, p. 338

BACKGROUND INFORMATION
ANIMAL BREEDING

Animal breeders often produce strains of pets and livestock that have specific dispositions. In Spain and Mexico, bulls have been bred especially for bullfighting for many years. The most desirable strains are very aggressive but not too bright. You probably already know that different dog breeds have different behavior patterns. Some are high-strung and nervous, and others are quiet and shy. Some make good family pets, and others are easily trained as watchdogs. Many of these personality traits are genetically controlled.

TEACHING STRATEGY 35–3

Focus/Motivation

Using the Visuals Have students look at Figure 35–10 and read the caption. This rather unusual behavior has undoubtedly evolved over many years because of some benefit to the organism.

- **Can you propose a possible explanation?** (Answers will vary, but one will probably involve protection.)

Content Development

Emphasize the relationship between behavior, genes, and natural selection. Discuss the example of hybrid behavior given in the text. Discuss some of the circumstances that would favor social behavior.

SECTION REVIEW 35–3

1. Nesting behavior is under genetic control. The genes of some species program the birds to build a certain kind of nest.
2. Cooperation between social animals may enhance the survival ability of an individual. Social animals may cooperate in hunting techniques. Hunting as a member of a group may be more successful than hunting alone. This kind of cooperation may increase the chances that an individual will survive.
3. In many cases a social animal driven from its group will not survive. For example, an animal such as a wolf can be quite successful as a pack animal, shar-

---

Guide For Reading
- What is the relationship between genes and behavior?
- How does social behavior affect an individual's fitness?

Figure 35–10 This male hornbill has sealed his mate within a hollow tree. Inside the tree, the female lays her eggs and tends her young. The male supplies food for both mother and babies. Later, when the babies have matured, the male frees mother and babies from their tree prison.

35–3 The Evolution of Behavior

As you have learned in Chapters 9 and 10, the physical structures in organisms develop according to a program contained in their DNA. Different characteristics are coded in different genes or groups of genes. Variations in these genes lead to inheritable variations in the characteristics of the animals that carry them. This genetic variation is the raw material of evolutionary change. Natural selection increases the relative frequency of the most beneficial genes in the population.

Although it may be difficult to believe, genes code for behaviors as well as for physical characteristics. Thus a series of base pairs on a DNA molecule can cause a newborn goose to follow a watering can or a bee to perform a waggle dance. You have seen that genetic control of characteristics can be demonstrated by crossing individuals with different physical traits. In the same way, evidence for genetic control of behavior can be demonstrated by crossing closely related animals that show different behaviors.

A good example of hybrid behavior is found in lovebirds. A researcher studied two closely related species of lovebirds that build nests of leaves they rip into shreds and carry back to their roosts. One species of lovebird instinctively carries the leaf strips in its beak. The other species tucks the strips neatly between its tail feathers. In the laboratory, a lovebird that normally carries the strips in its beak was crossed with a lovebird that normally carries the strips in its tail. The nest-building attempts of the resulting hybrid birds were quite comical to watch. Hybrid birds were totally unable to build a nest! First they would try to carry the leaf strips in their beak. Then, failing that and acting quite confused, they would try to tuck the strips into their tail feathers. This attempt also met with failure. The birds spent a great deal of time trying to decide how to carry the strips, and when they finally took action, they wound up dropping the strips. Other examples of genetically controlled behaviors have been studied in laboratories in breeding experiments that involved crickets, bees, and fruit flies.

In some situations the fitness of an individual is not affected by cooperation with other members of its species. In such species, social behavior would not be favored by natural selection.

For some species, however, social behavior offers great survival advantages. The evolutionary fitness of an individual is increased if it forms some type of social group with others of its kind. Natural selection would favor the evolution of social behavior in such species. There are many ways that social behavior can increase an individual's fitness. Some animals, such as wolves, hunt together in groups much more successfully than they can as individuals. Other animals, such as many grazing animals, band together because they are safer from predators when they are part of a group than when alone.

778

SCIENCE, TECHNOLOGY, AND SOCIETY

BREAKTHROUGH

The Benefits of Living with Relatives

Over the last twenty years, researchers studying animal societies have made some interesting discoveries. They have found that in many animal societies, all or most members of each social group are related to one another. And in many social groups of insects, birds, and mammals (including lions, elephants, and monkeys), all members are either parents, brothers, or sisters.

The theory of kin selection offers an explanation of the phenomena. Evolutionary fitness is now defined as the number of copies of an individual's genes that are inherited by the next generation. Kin selection theory states that it makes no difference whether those genes are provided by a particular individual itself or by that individual's relatives.

Think about this for a moment. A bird receives half of its genes from each parent. Therefore it shares many genes with its brothers and sisters. If for some reason a particular bird is unable to mate, that bird can still ensure that extra copies of some of its genes get into the next generation by helping its parents, brothers, or sisters raise more offspring. The bird's fitness is thus improved by becoming what is called a helper at the nest.

Such helpers are, in fact, commonly found among birds such as Florida scrub jays.

Similarly, prides of lions and groups of many primate species are composed of close relatives that help one another. By helping related members of a group to survive, an individual animal, even if it is unable to breed, can contribute to the evolutionary fitness of the group. Thus natural selection can favor the evolution of behaviors that help other individuals.

All the lions in this photograph are closely related. Because they are, each young lion has many similar genes.

35–3 SECTION REVIEW

1. Describe animal behavior that is under genetic control.
2. How does social behavior contribute to the survival of an individual animal?
3. **Critical Thinking—Making Predictions** What do you predict would happen to a social animal that was driven from its group? Explain your answer.

LABORATORY INVESTIGATION

EXPLORING MEALWORM BEHAVIOR

Before the Lab

1. Assemble all the materials for each group. Some of these materials, such as the paintbrushes and lamps, might not be found in sufficient quantity in the biology laboratory, so provision should be made ahead of time.
2. Mealworms can be purchased from scientific supply firms, pet stores, or bait shops. Be sure that you have several extras in case of classroom fatalities.
3. You might want to perform this investigation ahead of time if you are unfamiliar with mealworm behavior.

Pre-Lab Discussion

1. Review the types of behavior common among animals. Tell students to be aware of the different kinds of behavior as they make their observations. They will be able to label mealworm behavior when they finish the lab.
2. Caution students to handle the mealworms with care. They are rather fragile living organisms.
3. Read through the directions with the students and ask them to follow each one step by step.
4. Give students some specific instructions about how to make observations—especially tell them to watch patiently and continuously and then record carefully and accurately.
5. You might want students to observe the mealworms for a few minutes on a flat surface without any experimentation so that they will be familiar with normal mealworm behavior.

LABORATORY INVESTIGATION: EXPLORING MEALWORM BEHAVIOR

PROBLEM

How do mealworms respond to their environment?

MATERIALS (per person)

aluminum foil
150-mL beaker
dry cereal
small paintbrush
compass
cellophane tape
dissecting tray
lamp (or sunlight)
10 mealworms
petri dish
ruler
scissors
sheet of unlined paper

PROCEDURE

1. Use the scissors to cut a sheet of paper to fit into the bottom of the dissecting tray.
2. Find the exact center of the paper. Use a compass to draw a circle with a 2.5-cm radius at the center of the sheet of paper. Use the ruler to find the centers of the two longest edges of the paper. Mark these two points. Connect the points with a straight line.
3. Place the sheet of paper in the bottom of the dissecting tray. Tape the sheet of paper to the tray along all edges of the paper.
4. Place 10 mealworms into a 150-mL beaker.
5. Put a petri dish under one edge of the dissecting tray to raise it about 1 cm. Place the mealworms into the circle in the center of the tray.
6. Observe the mealworms' behavior for 10 minutes. Note whether they crawl uphill or downhill. Observe what they do when they reach the edge of the tray. Record your observations. Remove the mealworms from the tray.
7. Place a piece of aluminum foil over half of the tray. Line up one edge of the foil with the line you drew down the center of the paper. Shine the lamp directly over the tray. Place the mealworms back in the circle. Observe the mealworms' behavior for 10 minutes. Record your observations.
8. Remove the aluminum foil. Place a small handful of dry cereal in one corner of the tray.

780

Brush the mealworms back into the circle. Observe the mealworms' behavior for 10 minutes. Record your observations.

OBSERVATIONS

1. How many mealworms remained within the circle in the tilted dissecting tray? How many climbed uphill? How many climbed downhill?
2. In response to light, how many mealworms remained within the circle? How many crawled under the aluminum foil? How many crawled into the light?
3. When cereal was placed in the dissecting tray, how many mealworms remained within the circle? How many crawled into the cereal? How many did not?
4. What did the mealworms do when they reached the edge of the tray?

ANALYSIS AND CONCLUSIONS

1. Do mealworms prefer to climb uphill or downhill? How might this behavior be adaptive?
2. Do mealworms prefer to be in light or in shade? How might this be adaptive behavior?
3. Do mealworms prefer to be in or out of the cereal? How might this behavior be adaptive?
4. How is the mealworms' behavior at the edge of the tray adaptive?

Skills Development

Students will use the following skills in completing the laboratory investigation.
1. Manipulative
2. Measuring
3. Observing
4. Communicating
5. Comparing
6. Relating
7. Inferring

Safety Tips

This lab is not particularly dangerous for the students, but they should be cautioned to take care when using scissors. They should also be alerted to the potential for burns from the lamp if it overheats.

SUMMARIZING THE CONCEPTS

The key concepts in each section of this chapter are listed below to help you review the chapter content. Make sure you understand each concept and its relationship to other concepts and to the theme of this chapter.

35–1 Elements of Behavior

- The behavior of an animal is just as important to its survival and reproduction as any of its physical characteristics.
- Instincts are behaviors that are built into an animal's nervous system and cannot be changed during the animal's lifetime.
- Learning is the way animals change their behavior as a result of experience. Animals learn in several different ways.
- Some behaviors, although primarily instinctive, cannot occur without some learning on the part of the animal. In the process of imprinting, for example, newborn ducks and geese combine their natural instinct to follow their mother with an image obtained by experience.

35–2 Communication: Signals for Survival

- Communication is the passing of information from one animal to another. Animals use many varied techniques to communicate with each other.
- Animals communicate with each other for many reasons. For example, animals communicate in order to choose a mate, transmit information about the location and availability of food, and to warn others about potential dangers.
- The ways in which animals communicate is only limited by the kinds of stimuli their senses can detect.

35–3 The Evolution of Behavior

- The DNA in genes codes for behaviors as well as for physical characteristics. Like physical characteristics, variations in the genetic material that codes for behavior can be inherited if the behavior contributes to the animal's survival.
- In some species, the fitness of an individual is affected by cooperating with other members of its species. When social behavior offers great survival advantage, natural selection favors the evolution of such behavior.

REVIEWING KEY TERMS

Vocabulary terms are important to your understanding of biology. The key terms listed below are those you should be especially familiar with. Review these terms and their meanings. Then use each term in a complete sentence. If you are not sure of a term's meaning, return to the appropriate section and review its definition.

35–1 Elements of Behavior
instinct
learning
habituation
classical conditioning
operant conditioning
insight learning
imprinting

35–2 Communication: Signals for Survival
communication

visual signal
sound signal
chemical signal
pheromone
electrical signal

---

place or crawl to the darkness. None should have crawled toward the light.

3. Answers will vary, but most mealworms will have crawled to the cereal.

4. They lifted the front portion of their bodies and peered around.

Analysis and Conclusions

1. Uphill, which might increase their chances of escaping from an unsuitable place.

2. Shade, which helps them hide and also helps them to get food.

3. They show a strong preference for burrowing into the cereal since this is how they obtain food.

4. Lifting the front portion of the body probably helps them to spot and crawl around obstacles.

Going Further: Enrichment

Design your own experiments to test the mealworms' reactions to cold, to water, and to sugar. Observe the mealworms for 10 minutes in each experiment and make some conclusions based on your observations.

Teaching Strategy

1. Monitor the students to be sure they are following all the directions.

2. Check students' handling of the mealworms.

3. As you go around the room while students are observing, ask them questions to check their understanding of what they are seeing. Make sure they are accurately recording their observations.

Observations

1. Answers will vary. Students should note that about equal numbers remained in the same place or crawled uphill. Fewer mealworms crawled downhill.

2. Answers will vary, but about equal numbers will remain in

CHAPTER REVIEW

CONTENT REVIEW

Multiple Choice
1. b. 2. a 3. d 4. c
5. d 6. c 7. b 8. a

True or False
1. F   as
2. F   Learned
3. T
4. T
5. F   Habituation
6. F   operant conditioning
7. F   DNA or genes
8. T

Word Relationships
1. terms associated with learning; an instinct is an unlearned, automatic behavior
2. methods of communication; imprints
3. types of conditioning; insight learning

CONCEPT MASTERY

1. Behavior is the way an organism acts. There are two main types of behaviors: instinctive and learned. Behaviors contribute to an organism's survival.
2. Habituation is a decrease in response to a stimulus that neither rewards nor harms an animal. Habituation enables animals to react properly to dangerous stimuli and to ignore harmless ones. The correct reaction to stimuli may help an organism to survive.
3. Behavior can contribute to an animal's survival in many ways. Behaviors can help an animal to escape from danger, find food, or find a mate. Behaviors can help some animals to learn ways to react to certain stimuli. Some animals may even be able to apply learned behaviors to new situations.
4. Social behavior contributes to the survival of each animal in the group. In a group, social animals are more effec-

tive at hunting, for example. Being part of a successful hunting group may contribute to the survival of the individual animal.
5. DNA determines the physical structures present in an organism. Genes also determine

some of the ways an organism will behave. This has been proven by crossing animals that show two dramatically different behaviors. The offspring often show behaviors that are unlike either parent. They often seem confused

about appropriate ways to behave.
6. An instinct is a type of behavior that does not have to be learned. Instincts are built into an animal's nervous system. For example, a human baby knows how to suckle soon af-

CHAPTER REVIEW

CONTENT REVIEW

Multiple Choice

Choose the letter of the answer that best completes each statement.

1. The fact that many carnivorous animals know how to hunt soon after they are born is an example of a (an)
   a. learned behavior.
   b. automatic behavior.
   c. flexible behavior.
   d. conditioned behavior.
2. The ability of an animal to make a mental connection between a stimulus and some kind of good or bad event is
   a. classical conditioning.
   b. operant conditioning.
   c. habituation.
   d. insight learning.
3. Crossing individuals with different behavioral characteristics demonstrates evidence for
   a. imprinting.
   b. conditioned learning.
   c. habituation.
   d. genetic control of behavior.
4. Behaviors that are naturally built into an animal's nervous system are
   a. imprints.   c. instincts.
   b. habits.   d. insights.
5. A newborn duck may decide that a wooden object or a human is its mother through the process of
   a. insight learning.   c. habituation.
   b. communication.   d. imprinting.
6. Scientists who study animals in their natural environments are called
   a. ornithologists.   c. ethologists.
   b. imprinters.   d. entomologists.
7. The use of pheromones to transmit information is an example of a (an)
   a. sound signal.   c. electrical signal.
   b. chemical signal. d. visual signal.
8. An animal that can gather information from electric currents is a
   a. shark.   c. dog.
   b. bat.   d. goose.

True or False

Determine whether each statement is true or false. If it is true, write "true." If it is false, change the underlined word or words to make the statement true.

1. Physical characteristics of an animal are more important than its behavior to its survival and reproduction.
2. Automatic behaviors can be altered by experience and allow animals to adjust to a changing environment.
3. Instinctive behaviors cannot be changed and are said to be genetically controlled.
4. Learning is the way in which animals change their behavior as a result of experience.
5. Classical conditioning is a decrease in response to an unimportant stimulus that neither rewards nor harms an animal.
6. Trial-and-error learning is also known as habituation.
7. In humans, sugar molecules code for behaviors as well as for physical characteristics.
8. Genes code for behaviors as well as for physical characteristics.

Word Relationships

In each of the following sets of terms, three of the terms are related. One term does not belong. Determine the characteristic common to three of the terms and then identify the term that does not belong.

1. flexible behavior, learning, instinct, habituation
2. movement and color, songs of whales, imprints, pheromones
3. insight learning, training a pet, operant conditioning, trial and error

CONCEPT MASTERY

Use your understanding of the concepts developed in the chapter to answer each of the following in a brief paragraph.

1. What is behavior?
2. What is habituation? How does habituation contribute to an animal's survival?
3. How does behavior contribute to the survival of an individual?
4. How does social behavior contribute to the survival of a group of animals?
5. How do an animal's genes influence its behavior?
6. What is an instinct? Give an example of instinctive behavior in an animal.
7. What is learning? Give an example of a learning behavior shown by an animal.
8. What is insight learning? What group of animals commonly shows insight learning?
9. What is communication? What are two reasons why animals communicate with each other?

CRITICAL AND CREATIVE THINKING

Discuss each of the following in a brief paragraph.

1. **Applying concepts** In the past, zoos often exhibited a single social animal, like a wolf, in a cage. People observing the actions of this animal came away with a distorted view of a wolf's behavior. Today, most zoos exhibit social animals in a group. How does this present a more accurate picture of wolf behavior in the wild?
2. **Applying concepts** A pride of lions consists of several males, many females, and their offspring. About every two years new male lions drive the old males from the pride. One of the first acts of the new males is to kill all of the lion cubs. How could you explain the action of the new male lions?
3. **Relating concepts** Birds often build nests near outdoor bird feeders. What kinds of behaviors are shown by this action?
4. **Designing an experiment** Some people train a dog by giving the animal a treat every time it performs a trick successfully. Other people provide a treat occasionally. Design an experiment to determine which method is more effective.
5. **Using the writing process** You have been placed in charge of a program that hopes to wipe out drug abuse in the United States. Prepare a guidebook that details ways in which the behaviors of people who abuse drugs can be changed.

CRITICAL AND CREATIVE THINKING

1. Wolves, being social animals, would be more likely to show natural behaviors in a group. The social interactions between animals would be apparent to an observer. A single wolf in a cage would not exhibit natural behavior. In fact, it would probably be a pretty miserable animal, pacing the floor waiting for its next meal.
2. Like other placental mammals, lionesses are not fertile while nursing their cubs. When the cubs are killed, the lionesses become ready to breed, and the new lions gain the opportunity to mate with the lionesses and thus pass on their genes. Killing the cubs sired by the previous pride males also removes a threat to the new lions' future offspring—if allowed to live, the old cubs would compete with the new cubs for food and other resources.
3. This is a learned behavior and could probably be considered an example of classical conditioning. The birds learn that a supply of food is available and take advantage of this situation by building a nest close by.
4. Accept all answers that seem reasonable. For your information, studies have shown that an occasional treat is a more effective way to train a dog than giving a treat each time a trick is completed.
5. Check students' writing for clarity of expression. Students will suggest many ways to alter behaviors. Some may seem novel. Do not make judgments about which methods seem more appropriate. You may want to encourage students to work together to make this a class project. You may even want to circulate the finished guidebook in your district.

---

ter it is born. It does not have to learn this behavior.
7. Learning is the shaping of behaviors by experience. Humans can learn to tell many different colors apart.
8. In insight learning, an animal applies behavior it has already learned to new situations. Most examples of insight learning are found in primates. Insight learning is common in humans.
9. Communication is the passing of information from one animal to another. Animals communicate to warn of danger, to find a mate, and so on.

CHAPTER 36 *Comparing Vertebrates*

Section	Laboratory Investigations and Activities
36–1 Evolution of the Vertebrates pages 785–789 THEMES: Evolution, Patterns of Change	**Student Edition** LABORATORY INVESTIGATION: Comparing Vertebrates, p. 800 **Teacher Edition** DEMONSTRATION: How Are Vertebrates Related to One Another? p. 770D
36–2 Form and Function in Vertebrates pages 789–799 THEMES: Scale and Structure, Unity and Diversity	**Teacher Edition** DEMONSTRATION: Comparing Live Vertebrates, p. 770D **Laboratory Manual** Vertebrate Skeletons, p. 453
Chapter Review pages 801–803	

Teacher Resources
Books

Jamesone, E. W., Jr. *Patterns of Vertebrate Biology.* Springer-Verlag, 1981.

Jeffries, R. P. *The Ancestry of Vertebrates.* Cambridge University Press, 1987.

Kent, George. *Comparative Anatomy of the Vertebrates*, 7th ed. Mosby–Year Book, 1992.

Orr, R. T. *Vertebrate Biology,* 5th ed. Saunders, 1982.

Radinsky, Leonard B. *The Evolution of Vertebrate Design.* University of Chicago Press, 1987.

Romer, A. S., and T. S. Parsons. *The Vertebrate Body*, 4th ed. Saunders, 1986.

Walker, Warren F. *Functional Anatomy of Vertebrates.* Saunders, 1987.

Young, J. Z. *The Life of Vertebrates*, 3rd ed. Oxford University Press, 1981.

CHAPTER PLANNING GUIDE

Other Activities	Media and Technology

Teaching Resources
  ACTIVITY: Finding the Oddball, p. 3
Study Guide
  Section 36–1, p. 341	

Teaching Resources
  VOCABULARY REVIEW: Word Scramble, p. 1
  ACTIVITY: Form and Function in Vertebrates, p. 5
Study Guide
  Section 36–2, p. 343
	Interactive Videodisc/CD ROM
	  Amazonia
	  The Virtual BioPark
	Biology Transparencies
	  47: The Digestive Systems of Vertebrates
	  48: The Respiratory Systems of Vertebrates
	  49: The Heart Structure of Vertebrates
	  50: The Brain Structure of Vertebrates
	  51: The Reproductive Systems of Vertebrates

Teaching Resources
  Chapter Test A, p. 11
  Chapter Test B, p. 15
  Performance-Based Assessment, Unit 8
	Computer Test Bank
	  Chapter Test, p. 379

Audiovisuals

Animals With Backbones, 2nd ed. Film or video. Coronet.

The Vertebrates. 6 filmstrips with cassettes. Ward.

Chordates: Diversity in Structure. Film. Syracuse University.

An Outline of Vertebrate Evolution. Carolina Biological Supply Company.

CHAPTER 36 Comparing Vertebrates

CHAPTER OVERVIEW

In this chapter major facts and concepts discussed in previous chapters in this unit are brought together and presented in a new way. The purpose of this is to assist students in constructing a cognitive framework for understanding the evolution of vertebrates as a cohesive whole now that their study of the previous chapters has equipped them with a knowledge base upon which this framework can be built.

36–1 EVOLUTION OF THE VERTEBRATES

Section Focus 36–1

The purpose of this section is to introduce students to a phylogenetic tree that represents one hypothesis about the evolution of vertebrates and to discuss major evolutionary trends and innovations in vertebrates. Students will learn how different vertebrate groups are thought to be related to one another. They will be introduced to the concept of convergent evolution and will review the concept of divergent evolution. Students will also review the concept of body temperature control—one of the most significant evolutionary innovations—in vertebrates.

Performance Objectives 36–1

1. Interpret a phylogenetic tree of the vertebrates.
2. Describe when major innovations occurred in vertebrate evolution.
3. Compare convergent and divergent evolution.
4. Discuss body temperature control in vertebrates.

Science Terms 36–1

divergent evolution   p. 787
convergent evolution   p. 787

36–2 FORM AND FUNCTION IN VERTEBRATES

Section Focus 36–2

The purpose of this section is to present information about the essential life functions in vertebrates in a way that helps students understand some of the basic trends in the evolution of vertebrate systems and that encourages students to compare the ways different vertebrates carry out life functions. Students will develop a better appreciation of the diversity of ways in which vertebrates carry out the functions of movement, feeding, respiration, internal transport, excretion, response, and reproduction.

Performance Objectives 36–2

1. Describe amphibian, reptilian, and mammalian stance.
2. Explain how the mouthparts and digestive systems of vertebrates are adapted for different feeding habits.
3. Discuss trends in the evolution of the vertebrate lung, heart, and brain.
4. Relate the single- and double-loop circulatory systems to respiration.
5. Compare excretion in aquatic and terrestrial vertebrates.
6. Discuss different vertebrate reproductive strategies.

Science Terms 36–2

single-loop circulatory system   p. 794
double-loop circulatory system   p. 794

CHAPTER PREVIEW

DISCOVERY LEARNING

TEACHER DEMONSTRATIONS
Modeling

How Are Vertebrates Related to One Another?
The following demonstration can be used as an introduction to Chapter 36.

Obtain photographs and pictures of vertebrates that are similar in form. Possible combinations of organisms include kangaroo, *Tyrannosaurus rex*, gerbil; killer whale, seal, penguin, shark, ichthyosaur, tuna; and snake, eel, ferret. Obtain more than one group of similarly shaped vertebrates.

- **How are these vertebrates related to one another?** (Accept all logical answers.)
- **Which of these organisms are most closely related?** (Accept all logical answers.)
- **Why are these organisms similar in form?** (Accept all logical answers.)

Rearrange the pictures so that animals in the same class are now grouped together.

- **Why are these related organisms quite different in form?** (They are adapted to different conditions.)
- **What information would you need to determine how closely these animals are related to one another?** (Accept all logical answers.)

Comparing Live Vertebrates
You may wish to perform this demonstration when students study Section 36–2.

Obtain an assortment of small, live vertebrates. If possible, try to have a fish, an amphibian, a reptile, a bird, and a mammal—a goldfish, salamander, anole, finch, and mouse, for example. You may be able to arrange for students to bring in appropriate pets for this demonstration. Make sure that all the animals are in secure cages.

Display the animals together on a demonstration table or another appropriate location. Have students gather in this location to observe the animals. If you have a large class, you may wish to divide the class into small groups and have each group in turn briefly observe the animals.

Have students compare the displayed vertebrates by asking questions such as the following:

- **What do all these animals have in common?** (Accept all logical answers. Students should recognize that all the animals are vertebrates.)
- **How are these animals different from one another?** (Accept all logical answers.)
- **How does the location of the fins differ from that of the limbs on land animals?** (The fins are on the sides of the fish's body, whereas the limbs of land animals tend to be located beneath the body.)
- **How do these animals control their body temperature?** (Fishes, amphibians, and reptiles are ectotherms, which obtain heat from the environment and regulate body temperature primarily through behavior. Birds and mammals are endotherms, which generate body heat through metabolic activity.)
- **Would you expect this animal to have a single-loop or a double-loop circulatory system? Why?** (The fish breathes with gills and thus has a single-loop system. All the other animals—with the possible exception of the salamander, which may belong to a lungless species—have a double-loop system.)

CHAPTER 36
Comparing Vertebrates

GUIDED ENQUIRY

Pose the following questions to students and have them record their responses. Point out that they will gain a better understanding of the key concepts if they read the chapter with these basic questions in mind. Upon completion of the chapter, pose the questions again. Ask students to compare their initial responses with those they have developed after reading the chapter.

- How are the vertebrate classes related to one another?
- Why do very distantly related vertebrates sometimes look similar to one another?
- How does a vertebrate's posture affect its movement?
- How are the ways vertebrates carry out their life functions similar? How are they different?
- How is the way a vertebrate performs a specific essential life function adaptive?
- How is the way a vertebrate performs a specific essential life function appropriate for that animal and inappropriate for another animal?

CHAPTER 36
Comparing Vertebrates

In the hot, dry African savanna, a rhinoceros and her calf wait patiently as tiny oxpecker birds peck at parasites attached to their skin. In the cool coniferous forest of North America, a bull moose nibbles on an appetizing plant.

Along the floor of the Atlantic Ocean, a rattail fish slithers through ice-cold water. In Brazil, a brightly colored tree frog peeps its mating song. In the South Pacific, a three-meter-long Komodo dragon dashes from its hiding place to ambush its prey. High above a New England forest, a bald eagle soars majestically. And atop a sheer cliff in Oregon, a mountain goat hops nimbly from rock to rock.

Each of these animals represents a major vertebrate group. Although each is visibly different from the others, they all share many features inherited from common ancestors that lived hundreds of millions of years ago. The ways in which the similarities and differences among vertebrates have evolved provide some of the most fascinating stories in biology . . . stories that you will read in the pages that follow, as you compare the systems that carry out essential life functions in vertebrates.

INTRODUCING CHAPTER 36

Using the Visuals

Have students observe the photographs and read the text on page 784.

• **What kinds of animals are shown in the photographs?** (Rhinoceroses, oxpecker birds, moose.)

• **What do these animals have in common?** (They are vertebrate, terrestrial, and endothermic.)

• **How are these animals different?** (Accept all logical answers. Students should note that the rhinoceroses and the moose are herbivorous mammals and the oxpeckers are carnivorous birds. Some students may point out differences in the ways these animals carry out their life functions—for example, birds are oviparous and rhinoceroses and moose are viviparous.)

• **Which vertebrate classes are not represented in the photographs?** (Hagfishes, lampreys, cartilaginous fishes, bony fishes, amphibians, reptiles.)

• **Which vertebrate classes are described?** (Rattail: bony fishes. Frog: amphibians. Komodo dragon: reptiles. Bald eagle: birds. Mountain goat: mammals.)

36–1 Evolution of the Vertebrates

Guide For Reading

- How do divergent and convergent evolution differ?
- What are some ways in which vertebrates control their body temperatures?

Ever since the first vertebrates appeared more than 500 million years ago, they have been evolving. During this continual process of change, vertebrates have developed many new and unusual features. Some of these features—sharper claws or longer hair, for example—were relatively simple. Others—such as paired front and rear limbs or an amniotic egg—were far more complex. All these features were tested and shaped by natural selection in a constantly changing world.

The Vertebrate Family Tree

The evolutionary relationships between different groups of vertebrates are shown graphically as a phylogenetic tree in Figure 36–2 on page 786. This tree represents just one hypothesis about how vertebrates are related. Although most (but not all) scientists will agree that amphibians evolved from lobe-finned fish ancestors, reptiles evolved from amphibian ancestors, and birds and mammals evolved from reptile ancestors, there is still much debate about the exact details of vertebrate evolution. For example, some experts think that birds are direct descendants of early dinosaurs, whereas others think that birds are descended from relatives of dinosaurs.

Figure 36–1 Taking its first look at the world, the hatching green snake pushes its head through the leathery shell of the amniotic egg in which it developed.

BACKGROUND INFORMATION
THE PHYLOGENETIC TREE

The most recent geological period, the Quaternary, is not shown on this diagram because it comprises a minuscule fraction of time.

The early jawless fish depicted in this diagram is a heterostracan. Heterostracans are the earliest known vertebrates, with fossils dating back to the Upper Cambrian Period.

The early jawed fish is the placoderm *Dunkleosteus*, a specialized form that lived during the Upper Devonian Period. *Dunkleosteus* was chosen to depict jawed fishes because its appearance reinforces the concepts of jaws and armor. In addition, the earliest jawed fishes are known from fragmentary fossils and are difficult to reconstruct.

The unlabeled fish on the branch that leads to hagfishes and lampreys is the anaspid ostracoderm, *Jamoytius*, which lived during the Silurian Period. *Jamoytius* is thought by some experts to be the ancestor of modern jawless fishes.

The early amphibian depicted on the tree is the labyrinthodont *Ichthyostega*, which may well be the first species of amphibian. *Ichthyostega* was approximately 90 centimeters in length.

Crocodiles, birds, and dinosaurs are descended from thecodont reptiles; mammals are descended from therapsid (mammallike) reptiles. Both the thecodont and the therapsid shown on this tree are Lower Triassic forms from South Africa. The thecodont is the pseudosuchian *Euparkeria*; the therapsid is the cynodont carnivore *Cynognathus*.

The dinosaur depicted on the tree belongs to the group called the coelurosaurs. Some experts who hypothesize that birds evolved from dinosaurs think that coelurosaurs are the ancestors of birds.

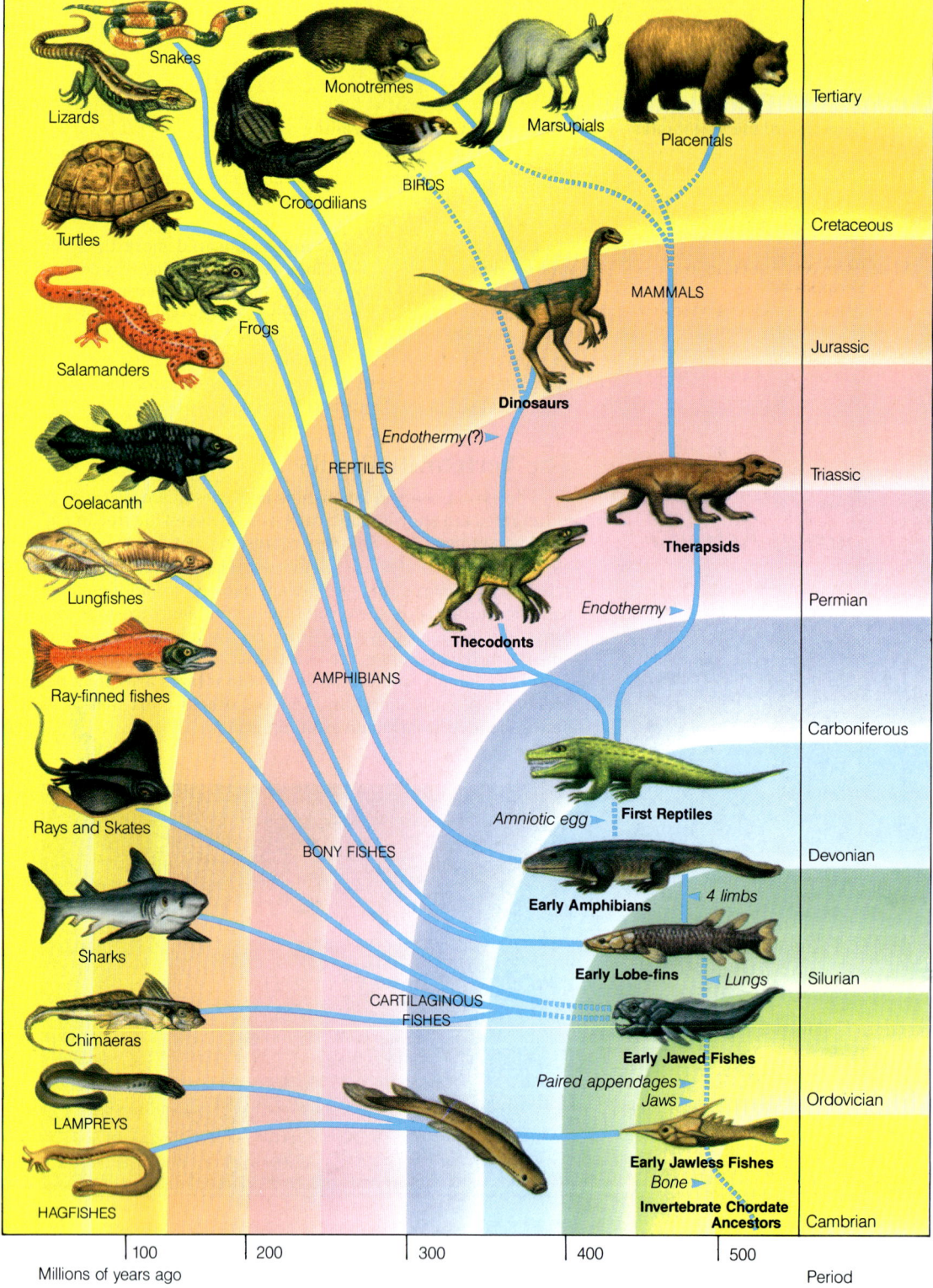

36–1 (continued)

Content Development
Using the Visuals Have students study Figure 36–2 and read the caption.
• **During which period did bone evolve?** (Cambrian.)
• **Describe the evolution of endothermy in vertebrates.** (According to this diagram, endothermy evolved twice, once in the evolutionary line leading to birds and dinosaurs and once in the evolutionary line leading to mammals. It is not known exactly when endothermy first appeared in the evolutionary line leading to birds and dinosaurs.)
• **What modern animals are descended from the first reptiles?** (Lizards, snakes, crocodilians, birds, mammals.)
• **When did birds first appear?**

Trends in Vertebrate Evolution

Two general trends appear repeatedly during the course of vertebrate evolution. The first trend can be stated as follows: **If closely related evolutionary lines are subjected to different forces of natural selection, they tend to become more dissimilar as they evolve.** The second trend is related to the first: **If evolutionary lines encounter extremely similar forces of natural selection, they tend to become more similar to one another as they evolve.**

The animals represented by the vertebrate family tree have branched off into an enormous diversity of habitats and lifestyles. Although the animals produced by the vertebrate adaptive radiation may look incredibly different from the outside, they are remarkably similar inside. For as different vertebrate groups evolved, they utilized the same basic sets of body parts for many uses. The pattern of evolution known as adaptive radiation is also known as **divergent evolution**.

Convergent evolution is the opposite of divergent evolution. When animals from different groups evolve in ways that cause them to resemble one another, we say that they have experienced convergent evolution. One of the many examples of convergent evolution among vertebrates is illustrated in Figure 36-3. As you can see, as certain marsupials and placentals adapted to similar environmental conditions over time, many came to resemble one another a great deal.

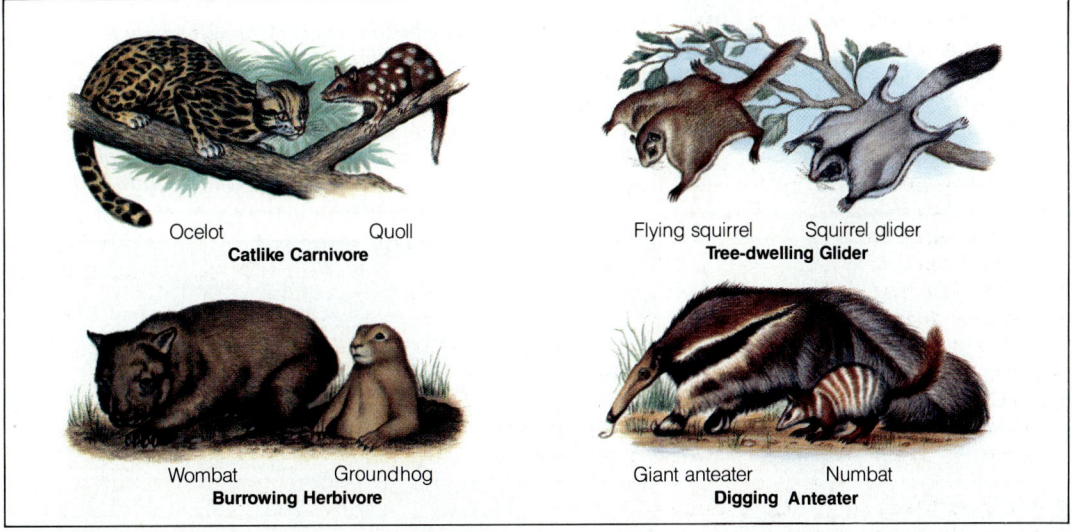

Figure 36–3 *Many examples of convergent evolution can be seen between marsupials from Australia and placentals from elsewhere in the world. The ocelot, flying squirrel, groundhog, and giant anteater are placentals.*

Figure 36–2 *(Opposite page) In addition to illustrating one hypothesis about the evolutionary relationships among vertebrate groups, this phylogenetic tree also shows approximately when important evolutionary innovations occurred. When did lungs appear? Which groups are endothermic?* (1)

ANNOTATION KEY

(1) During the Silurian Period, about 420 million years ago. Dinosaurs, birds, therapsids, modern mammals (monotremes, marsupials, placentals). (Interpreting diagrams)

(Late Jurassic.)

• **When did the amniotic egg evolve?** (Middle of the Carboniferous Period, between Early Amphibians and First Reptiles.)

• **Are all the reptiles that ever existed ectothermic? Explain.** (No. Therapsid reptiles were endothermic, and dinosaurs were probably endothermic.)

Reinforcement/Reteaching

Stress that modern vertebrates are neither better nor worse than one another. Emphasize that modern animal groups are not descended from other modern animal groups.

• **Did mice evolve from fishes?** (Yes, the branch of the phylogenetic tree that leads to placental mammals can be traced back to early lobe-fins.)

• **Did mice evolve from goldfish?** (No.)

• **Do modern fishes and mice share a common ancestor?** (Yes.)

Enrichment

The paleontologist Robert Bakker has proposed that the standard version of the phylogenetic tree be modified to include a branch called Dinoaves, which would include birds and certain dinosaurs.

• **What does this arrangement imply?** (Accept all logical answers. Some implications are that birds are descended from dinosaurs, the dinosaurs' evolutionary line is not really extinct, and dinosaurs may have been similar in habit and metabolism to birds and other endotherms.)

Skills Development

Guided Practice

Skill: Interpreting diagrams

Using the Visuals Direct students' attention to Figure 36–3.

• **How are the marsupials similar to their placental counterparts?** (Accept all logical answers. The marsupials are similar in habits and appearance.)

• **How are they different?** (Accept all logical answers. The marsupials are quite different in terms of reproduction.)

TEACHING SUPPORT

Teaching Resources
- Activity: Finding the Oddball, p. 3

Study Guide
- Section 36–1, p. 341

36–1 (continued)

Skills Development
Guided Practice
Skill: Developing a hypothesis
A student notices that in the morning a pet duck likes to sit in the middle of a sunny lawn with its wings spread out. On hot days, the duck sits in the shade of a tree with its beak open or swims in a pond.
- **How might you explain these behaviors?** (Accept all logical answers. Students will probably recognize that these behaviors have something to do with temperature regulation. Do not be alarmed if a few students suggest that ducks are ectothermic.)
- **How might you test your hypothesis?** (Accept all logical answers. Many experiments will have something to do with taking the duck's temperature and seeing if it can maintain its body temperature.)

Tell students to suppose that they found that a duck's body temperature remains constant even when the outside temperature is somewhat higher or lower.
- **What conclusions can you draw about the duck's behavior from these data?** (It is endothermic but uses behaviors to make itself more comfortable.)

SECTION REVIEW 36–1

1. In divergent evolution, closely related evolutionary lines that are subjected to different forces of natural selection become more dissimilar over time. For example, carnivores such as dogs, seals, bears, and weasels are descended from a common ancestor but evolved to fit different niches. Students' examples will vary.
2. Marsupial wombats resemble placental groundhogs. Aquatic mammals such as whales, manatees, and seals have a streamlined, rather fish-like body shape. Anteaters, numbats, and pangolins, which are adapted for eating ants and termites, have narrow snouts, long tongues, and strong claws for tearing apart insect nests. Students' examples will vary.
3. A turtle is ectothermic and controls its body temperature primarily through behavior. A mouse is endothermic; its body heat is produced through metabolic activity.
4. A low body temperature will cause animals to slow down.

Body Temperature Control

The ability to control body temperature is an enormous asset in the struggle for existence. Why? Recall from Chapter 4 that many chemical reactions, including those important to living things, work better at certain temperatures. And essential life functions can be carried out smoothly and efficiently when an animal's internal body temperature is within its preferred "operating range." When the body temperature is too low, animals slow down or become immobile. (This is why ectotherms such as snakes and frogs are easier to catch in the early morning, before they have had a chance to warm up. It is also the reason that such animals necessarily enter a dormant state during cold winters.) When the body temperature is too high, body systems are stressed and fail to function properly. Thus it is not surprising that many techniques of temperature control have appeared in vertebrates. All these techniques incorporate three important features: a source of heat for the body, a way to conserve that heat, and a method of eliminating excess heat when necessary.

As you may recall from Chapter 32, ectotherms must obtain the heat they need from their environment. Endotherms, on the other hand, generate all the heat they need through metabolic activity.

Ectotherms usually rely on specific behaviors to conserve heat and to avoid overheating. For example, lizards bask in the

Figure 36–4 A horned lizard (right) is an ectotherm that regulates its body temperature in an unusual way. When its body temperature is low, it changes color to a heat-absorbing dark brown. When its body temperature is high, it changes to a heat-reflecting light brown. A snow monkey (left) is an endotherm. Thick gray fur helps conserve heat produced in metabolic processes and allows the monkey to frolic in the snow without lowering its body temperature.

sun to heat their body and hide in the shade or in an underground burrow to cool themselves. A few ectotherms have body structures that enable them to maintain parts of their body at optimal, elevated temperatures. For example, swordfish have special muscles attached to the eye that produce heat and thus keep the brain at a constant warm temperature.

Endotherms, on the other hand, tend to rely on body structures and physiological functions to regulate their internal temperature. Birds and mammals generate body heat internally through metabolic activity. Birds conserve that heat primarily through the use of insulating feathers; mammals use fat and fur as insulation. Mammals can get rid of excess heat either by panting (as dogs do) or by sweating (as humans do). Because it is difficult to get rid of excess heat, many endotherms also employ behaviors that keep their body from overheating when the outside temperature is greater than their body temperature. Some animals rest in the shade, swim in cool water, or retreat to cool underground burrows (or, in the case of many humans, to air-conditioned buildings).

36–1 SECTION REVIEW

1. What is divergent evolution?
2. Give three examples of convergent evolution among mammals.
3. Compare the ways in which body temperature is controlled in a turtle and in a mouse.
4. **Connection—Human Physiology** People suffering from hypothermia, or low body temperature, often have slurred speech and uncoordinated movements. Explain this observation.

36–2 Form and Function in Vertebrates

As you learned in previous chapters, vertebrates perform the essential functions of life with a variety of body structures. In this section you will compare those structures and examine how evolutionary processes have modified certain basic structures over time. As you examine form and function in vertebrates, keep this general evolutionary trend in mind: **As you move through the vertebrate classes from fishes to mammals, organ systems tend to become increasingly complex.** Note that the term increasingly complex does not mean better. The various organs and systems, with their varying degrees of complexity, are simply different ways of performing the functions essential to life.

Guide For Reading
- What are some trends in the evolution of vertebrate body systems?
- How do vertebrates carry out their essential life functions?

a brief explanation of why mammals are unlikely to evolve intelligent forms.

TEACHING STRATEGY 36–2

Focus/Motivation
Obtain some mounted vertebrate skeletons.
- **What kind of vertebrate is this skeleton from?** (Answers will vary, depending on the skeleton. If students have trouble identifying the skeleton, provide some hints by pointing out distinguishing features—beaks, claws, and limbs, for example.)
- **How are these vertebrates similar?** (Accept all logical answers.)
- **How are they different?** (Accept all logical answers.)
- **How does the structure of this vertebrate's skeleton help it survive?** (Accept all logical answers.)
- **What sorts of things can you determine from a vertebrate's skeleton?** (Accept all logical answers.)
- **What sorts of things can you not determine from a vertebrate's skeleton?** (Accept all logical answers.)

Reinforcement/Reteaching
If students cannot answer the Section Review questions, review the material in the section that is giving them trouble.

Closure
Ask students to imagine an Earth on which the dinosaurs had remained the dominant species and on which one line of dinosaurs had developed into an intelligent tool-making and book-writing species. Have students imagine that they are some of those latter-day dinosaurs learning about vertebrate evolution in school. Ask them to write a brief essay about the evolution of vertebrates from their new point of view. This essay should include

Content Development
You may wish to have more students compare the structure of homologous bones in vertebrate skeletons and develop hypotheses about the reasons for the differences in the shapes of the bones.

BACKGROUND INFORMATION
VERTEBRATE STANCE

The three types of posture, or stance, seen in four-legged vertebrates—sprawl, semiupright, and fully upright—are sometimes known as amphibian stance, reptilian stance, and mammalian stance, respectively. Although these terms link the type of stance with the group of animals that best typify it, they should not be interpreted as meaning that only certain animals have a particular kind of stance. Many reptiles, such as modern chameleons and many dinosaurs, have an exclusively mammalian stance. And crocodilians, which usually have a distinctly sprawling stance, are capable of holding their limbs close to their bodies and walking or even galloping with a mammalian posture.

36–2 (continued)
Content Development
• What are some examples of vertebrates in which the limbs are not the most important body structures in movement? (Fishes, amphibian larvae, salamanders, legless amphibians, snakes, legless lizards, whales, dolphins. Note that whales and dolphins do not have blocks of muscles as fishes do and that the tail in cetaceans moves up and down, rather than side to side.)

Content Development
Point out that mammals are not the only animals that have an upright stance—there are a number of reptiles that also stand upright.
    Tell students that the idea that all reptiles were sluggish animals with a sprawling stance was firmly entrenched in the minds of many nineteenth-century biologists—so much so that many early dinosaur fossils were mounted incorrectly to reflect this view.
• How were the limbs positioned in these incorrect reconstructions of dinosaurs? (Sticking out from the sides of the body, holding the body only partially upright.)
• How could people tell that the limbs were positioned incorrectly? (An examination of the structure of the limb girdles and joints indicated that the limbs should be placed directly beneath the body.)
• How were the limbs positioned in dinosaurs? (Directly beneath the body.)
• What is the moral of this story? (Accept all logical answers. Students should recognize that it is important to question commonly held assumptions. Students should also point out that

Movement

As you may recall, all vertebrates except jawless fishes have a vertebral column, or backbone, made of numerous individual bones called vertebrae. These vertebrae are connected to one another by tough ligaments that allow the vertebral column to bend to a certain extent. Two pairs of limbs are attached to this basic supporting structure by sets of bones called limb girdles. Most of the bones in the body can be made to move through the contraction of muscles that are attached to the bones.

In certain vertebrates, such as many fishes and snakes, the main body muscles are arranged into blocks that are positioned on either side of the vertebral column. These muscle blocks contract in waves, one after another, first on one side of the body and then on the other. Because the vertebral column cannot be compressed, these contractions make the body bend rapidly back and forth. This is how many fishes and snakes develop the forward thrust they need to move.

In many (though not all) amphibians, reptiles, birds, and mammals, the muscles and bones of the limbs are the most important body structures involved in movement. There are two interesting evolutionary trends that can be seen as we move from amphibians to mammals. First, the position of the limbs relative to the body shifts toward the center. Second, the movement of the vertebral column when the animal runs changes from a predominantly side-to-side motion to an up-and-down motion. Refer to Figure 36–5 as you read about movement in some representative vertebrates.

Figure 36–5 *The positions of the pectoral and pelvic girdles and the limb bones differ among vertebrates. More primitive vertebrates such as salamanders have limbs that stick out from the sides of their body (top). The limbs of reptiles such as marine iguanas allow the body to be lifted off the ground more (left). In mammals such as lions, the limbs are positioned directly beneath the body (right).*

In many amphibians, such as salamanders, the limbs stick out sideways from the body, making it difficult for them to support much weight on land. This gives amphibians their characteristic sprawl. Although the limb muscles in these animals can create some movement, salamanders bend their bodies from side to side (fishlike) to help themselves move farther with each step. In many reptiles, the limbs point more directly down to the ground, allowing them to support weight more efficiently. The body curves somewhat during walking, but not as much as it does in amphibians. Many mammals can stand erect with their legs straight under them, whether they walk on two legs or on four. When walking, most mammals make minimal side-to-side movements with their body. But when running, many flex their backbones up and down a great deal.

Feeding

The heads of vertebrates show many adaptations for feeding. For example, the long bill of the hummingbird and the narrow snout of the honey possum are both adaptations to feeding

Figure 36–6 As the shape of their jaws and teeth indicate, mammals have adapted to many different feeding habits during the course of evolution. How do the jaws and teeth of a filter feeder compare with those of a gnawing herbivore? How is a walrus specialized for its feeding habits?

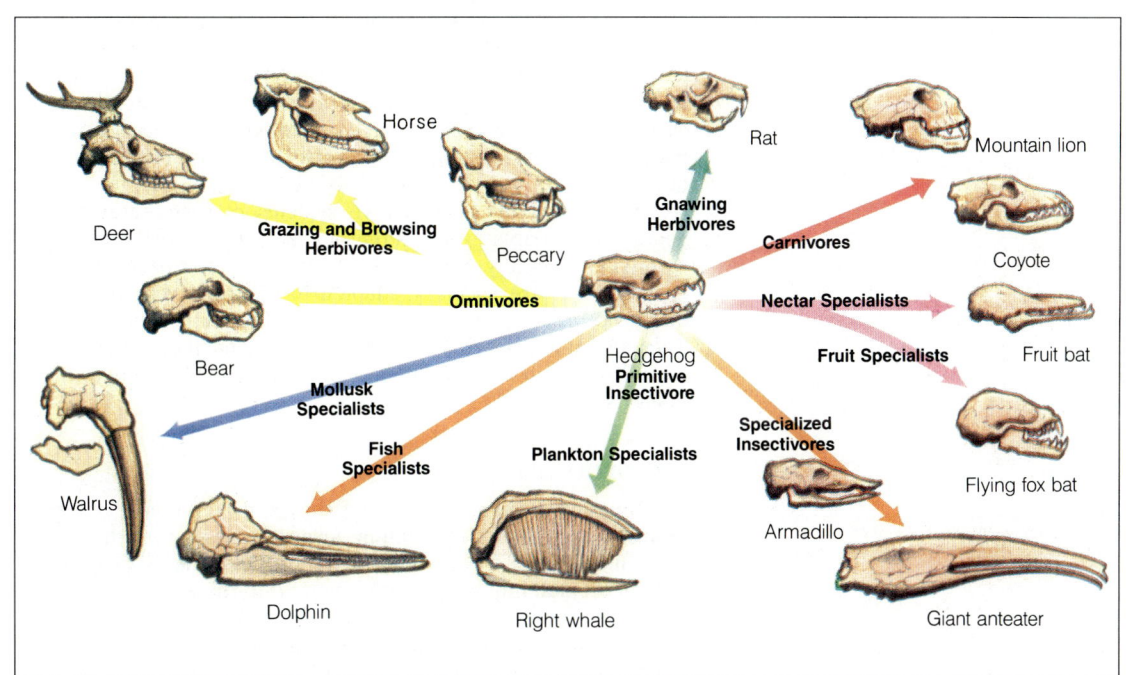

ANNOTATION KEY

❶ Lamprey. (Making observations) Plant material is difficult to digest.
(Applying concepts)

BACKGROUND INFORMATION
DIGESTION IN HERBIVORES

Horses and koalas are among the many herbivores that have hindgut fermentation: Cellulose in the food is broken down in the large intestine and ceca. Hindgut fermenters successfully digest only about 45% of the cellulose they eat.

Although rodents and lagomorphs are hindgut fermenters, they are able to break down about 80% of the cellulose they eat because they pass their food through the digestive tract twice. Food that has been through the digestive tract once is eliminated as a special type of feces known as cecal pellets, which the animal then eats.

Ruminants utilize about 60% of the cellulose they eat. Food travels from the esophagus to the rumen, where cellulose fermentation takes place. Undigested plant matter (cud) can be regurgitated from the rumen and chewed again. When the cud is swallowed, it once again enters the rumen. Digested food passes from the rumen to the omasum, which serves in the further mechanical breakdown of the food. The food then travels to the abomasum (true stomach), where more digestion takes place. Further digestion and absorption occur in the intestine and ceca.

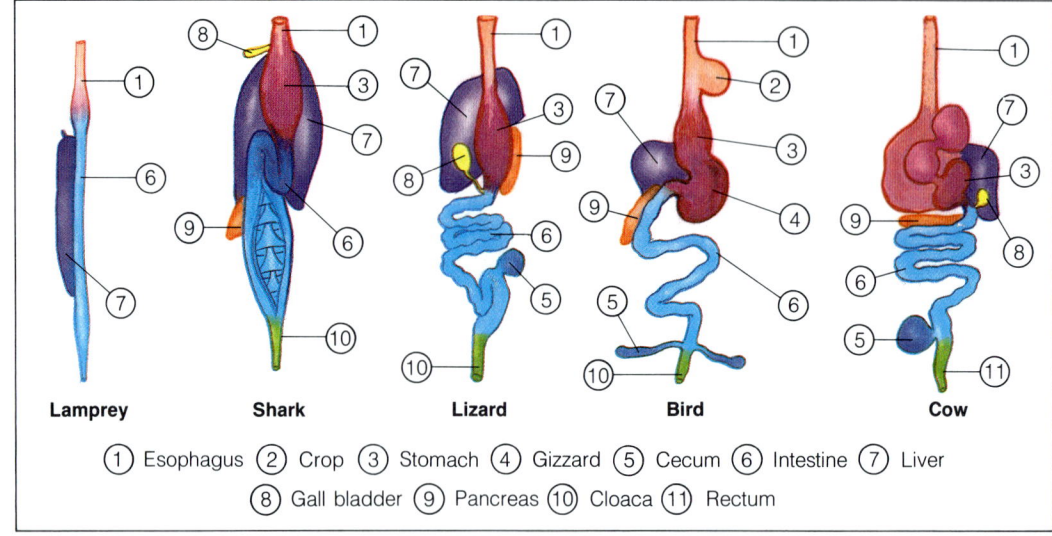

Figure 36–7 The digestive systems of vertebrates vary in their complexity and are adapted for a variety of modes of feeding. Which vertebrate has the most primitive digestive system? Why does a cow require four "stomachs" to digest its food? ❶

on nectar. These animals also have long tongues, the better to sip the sweet liquid from flowers. Filter feeders—such as baleen whales, flamingoes, and manta rays—have sievelike structures that enable them to strain food from the water. And carnivores—such as sharks, eagles, and leopards—have sharp mouthparts that help them tear chunks of meat from the body of their prey.

The organs of the digestive systems of vertebrates are equally well adapted for different feeding habits. For example, carnivores typically have short digestive tracts that produce enzymes that are specially adapted to the rapid digestion of meat. Herbivores, on the other hand, often have long intestines and stomachs that harbor colonies of bacteria or protozoans helpful in the digestion of the tough cellulose fibers in plant tissues.

Respiration

Aquatic vertebrates such as fishes and amphibian larvae typically use gills for respiration. As water passes over the gill filaments, oxygen diffuses into the blood in the capillaries and carbon dioxide diffuses from the blood into the water.

Terrestrial vertebrates—and a few aquatic vertebrates such as porpoises and sea turtles—typically use lungs to breathe. When the animals inhale, oxygen-rich air enters the lungs. The

36–2 (continued)

Content Development

Using the Visuals Have students observe Figure 36–7. Point out that the organs in the diagram are spread out so that they can be seen. In live animals, they are arranged quite compactly. You may want to mention that the structure in the shark intestine is the spiral valve, which increases the surface area for the absorption of food.

• **Why are the liver, gall bladder, and pancreas considered part of the digestive system even though food does not pass through them?** (Accept all logical answers. These organs secrete chemicals that are involved in the digestion of food.)

• **Why are the stomach and gizzard of the bird the same basic color on the diagram?** (Accept all logical answers. The stomach of a bird is divided into two parts: the proventriculus—labeled the stomach on the diagram—and the gizzard.)

• **Why are the cow's four "stomachs" different colors on the diagram?** (Accept all logical answers. Only the fourth stom-

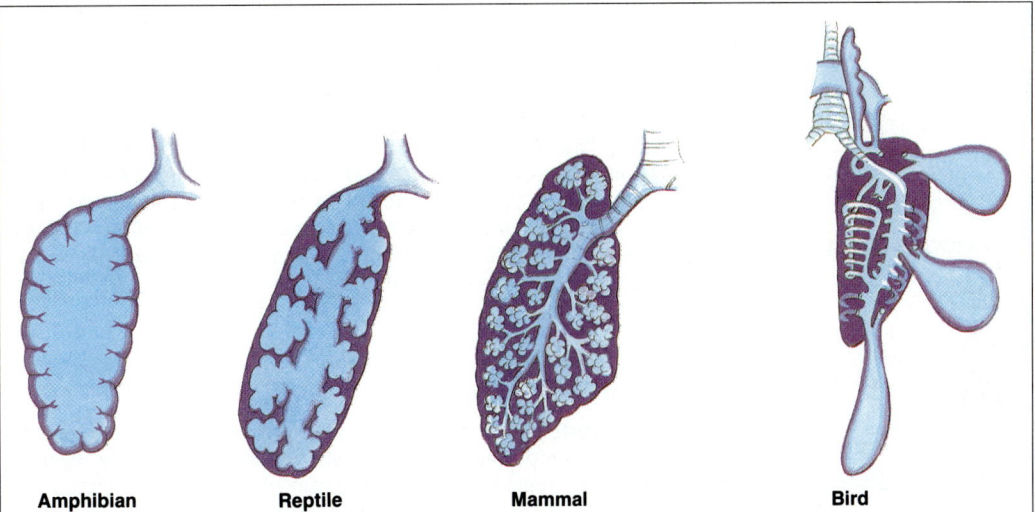

Figure 36-8 As you move from amphibians to mammals, the branching of the bronchi, or air tubes, increases. In birds, the air tubes do not terminate in "dead-end" alveoli. Instead, tiny one-way tubes carry a continuous flow of fresh air past the respiratory tissues.

oxygen in the air diffuses into the blood contained within the capillaries of the lungs; the carbon dioxide diffuses into the space within the lungs. When the animals exhale, carbon dioxide-rich air is expelled. Refer to Figure 36-8 as you read about the evolution of the lungs in terrestrial vertebrates.

The typical amphibian lung is little more than a sac with some ridges that increase its surface area slightly. The simplicity of amphibian lungs is not surprising. Because amphibians are ectothermic and not normally very active, their oxygen requirements are fairly low. In addition, many amphibians can also use their moist skin and the linings of their mouth and pharynx in respiration.

Although reptiles are also ectotherms and thus have fairly low oxygen requirements, their lungs are better developed than those of amphibians. Because lungs are typically the only structures used for gas exchange, it is important that their structure ensures efficiency. In more complex reptiles—such as turtles, crocodilians, and certain lizards—the lungs are divided into many large chambers that are in turn divided into smaller chambers. This greatly increases the surface area available for gas exchange.

In mammals, bronchi branch extensively, and the entire volume of the lungs is filled with many thousands of bubblelike structures called alveoli. These alveoli provide an enormous surface for gas exchange. The structure of the lungs allows

BACKGROUND INFORMATION
RESPIRATION IN BIRDS

A bird's respiratory system is more efficient than that of other air-breathing vertebrates because fresh air flows in one direction over the respiratory surfaces and there is little mixing of fresh and "used" air. In other vertebrates, fresh air travels through the air tubes to the respiratory surfaces and used air returns along the same route. In addition, a certain amount of air remains in the lungs after exhalation, so fresh air is always mixed with used air.

In birds, the trachea branches into two primary bronchi, each of which leads to a lung. Once inside a bird's lung, a bronchus is called a mesobronchus. The mesobronchus divides into five or six major branches, each of which connects with an air sac. The mesobronchus is also connected to a number of paired secondary bronchi that are connected by extremely fine tubes called parabronchi, in which gas exchange occurs. Note that in Figure 36-8, only one set of the many parabronchi is shown so that the connections between the branches of the mesobronchus and the air sacs may be seen more clearly.

When a bird inhales, air travels through the mesobronchus to the posterior air sacs. This displaces fresh air already in the posterior air sacs (little, if any, gas exchange is thought to occur in the air sacs), which flows through a small tube into the network of parabronchi. Air travels from the lungs to the anterior air sacs, where it serves a cooling function. When a bird exhales, air from the posterior air sacs is pumped over the respiratory surfaces and used air from the anterior air sacs is exhaled.

Note that air flows continuously in one direction during both inhalation and exhalation in a bird's lungs; fresh air is constantly flowing over the respiratory surfaces in the walls of the parabronchi. Since every major passage in a bird's lung is open at both ends, air circulates freely.

---

ach, the abomasum, is considered a true stomach. The first three stomachs are analogous to a bird's crop.)

Content Development
- **What are the two features common to all respiratory systems?** (The systems possess structures that maximize the respiratory surface area and have some way of keeping the gas-exchange surfaces moist.)
- **How do vertebrate lungs show these features?** (There are ridges and folds inside the lungs that increase the surface area. The lungs are located inside the body, where it is moist.)
- **What trends are seen in the evolution of the vertebrate lung?** (Reptiles, mammals, and birds have increasingly complex folds and passageways.)

endotherms to obtain the large amounts of oxygen required by their metabolism. However, the structure of the lungs is still somewhat inefficient. Because air must move in and out of the same passageways, there is always some old air trapped in the lungs after each breath.

The lungs of birds are even more efficient than those of humans and other mammals. A system of tubes within the lungs and air sacs attached to the lungs ensures that air is moved through the lung tissues in only one direction. Thus gas exchange surfaces constantly come in contact with fresh air. Unlike other vertebrates, birds never have old air sitting in their lungs.

Internal Transport

Vertebrates that use gills for respiration, such as fishes and larval amphibians, have a **single-loop circulatory system**. Blood travels from the heart to the gills to the body and back to the heart. The heart in a single-loop circulatory system is simple. It consists of two chambers: an atrium that receives blood and a ventricle that pumps blood to the body.

Vertebrates that use lungs for respiration have a **double-loop circulatory system**. The first loop carries blood between the heart and the lungs. Oxygen-poor blood from the heart is pumped to the lungs; oxygen-rich blood from the lungs returns to the heart. The second loop carries blood between the heart and the body. Oxygen-rich blood from the heart is pumped to the body; oxygen-poor blood from the body returns to the heart. During the course of vertebrate evolution, the heart developed chambers and partitions that help separate the blood traveling in the two loops of the circulatory system. See Figure 36–10.

In lungfishes, which rely both on gills and on primitive lungs, we can see the first step toward the development of a separate circulatory loop for the lungs. Blood from the lungs has a direct connection back to the heart, and there are partial partitions in the atrium and ventricle. Despite these partitions, a great deal of mixing of oxygen-rich and oxygen-poor blood takes place in the chambers of the heart.

In frogs and toads there are two atria. This means that oxygen-rich and oxygen-poor blood cannot mix in the atria. However, oxygen-rich and oxygen-poor blood can mix in the ventricle. Thus the concentration of oxygen in the blood traveling to the body is not as high as it could be.

Unlike frogs and toads, most reptiles have a partial partition in their ventricle. This partial partition minimizes the mixing of oxygen-rich and oxygen-poor blood, but it does not completely eliminate it.

Birds, mammals, and crocodilian reptiles have hearts that are completely partitioned into four chambers. This means that the lung loop and the body loop of the circulatory system are

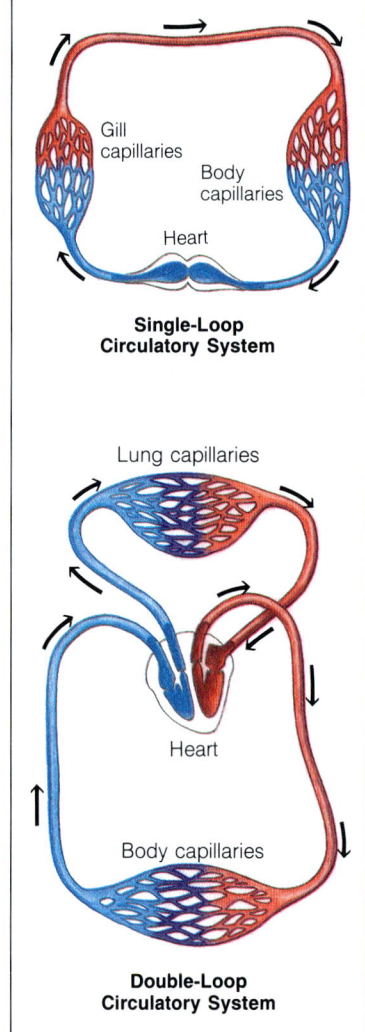

Figure 36–9 Most vertebrates that use gills for respiration have a single-loop circulatory system. Vertebrates that use lungs have a double-loop system.

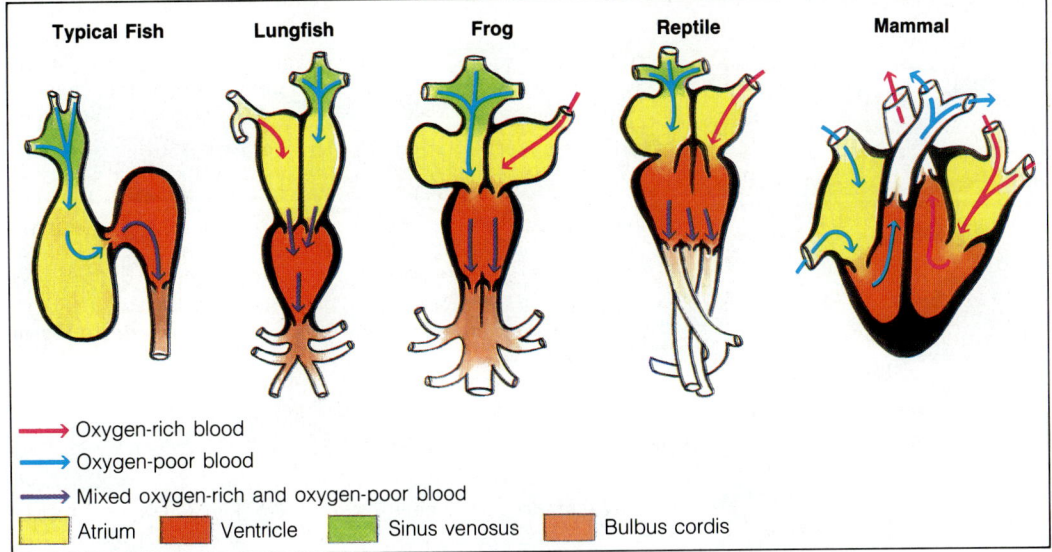

Figure 36–10 *In most fishes, the atrium and ventricle act as a single pump that forces blood around the body. An examination of the hearts of vertebrates that use lungs for respiration reveals various stages in the evolution of a double-pump (four-chambered) heart. Note that oxygen-rich and oxygen-poor blood do not mix in a four-chambered heart.*

completely separated: There is no mixing of oxygen-rich and oxygen-poor blood in the heart. The four-chambered heart is sometimes described as a double pump. One pump moves blood through the lung loop. The other pump moves blood through the body loop.

Excretion

Excretory systems eliminate nitrogenous wastes and regulate the amount of water in the body. In fishes, the gills play an important role in excretion. However, most vertebrates rely on kidneys to carry out the process of excretion. (Recall that a kidney is an excretory organ that is made up of tubules that filter nitrogenous wastes from the blood.)

Nitrogenous wastes are first produced in the form of ammonia, a highly toxic compound that must quickly be either eliminated from the body or changed into a less poisonous form. In aquatic amphibians and most fishes, ammonia diffuses from the gills into the surrounding water, which dilutes the ammonia and carries it away. In mammals, terrestrial amphibians, and cartilaginous fishes, the ammonia is changed into urea before it is excreted. In reptiles and birds, the ammonia is changed into uric acid.

The production of urea or uric acid is an adaptation to marine and terrestrial environments—environments in which the

---

BACKGROUND INFORMATION
ONTOGENY RECAPITULATES PHYLOGENY

This famous phrase is, in many cases, an extreme oversimplification. But with regard to the development of the human heart, individual development does indeed echo evolutionary development. The heart in a three-week-old embryo is a pulsating vessel similar to the specialized blood vessel that serves as a heart in modern cephalochordates and presumably similar to the heart's precursor in the invertebrate ancestors of the vertebrates. Later, the embryo's heart develops into a line of four chambers similar to the sinus venosus, atrium, ventricle, and conus arteriosus in fishes. (In human embryos, as well as in frogs and many other animals, the posterior final chamber is called the bulbus cordis.) As in fishes, the sinus venosus of the embryo serves as the pacemaker. As the embryo continues to develop, the sinus venosus and conus arteriosus become incorporated into the heart and disappear. Eventually, the partitions that separate the right and left atria and the right and left ventricles grow into place.

BACKGROUND INFORMATION
NITROGENOUS WASTES

The breakdown of nitrogen-containing compounds, such as amino acids and nucleic acids, results in the formation of nitrogenous wastes. The deamination of an amino acid, for example, produces a molecule of a substance such as pyruvic acid, which can be further broken down to release energy or used to synthesize biologically important compounds, and an amino ($-NH_2$) group. The amino group reacts with H^+ ions from water to produce ammonia (NH_3) and ammonium ions (NH_4^+). (Under the pH conditions found in the bodies of most animals, most of the ammonia is converted to ammonium ions.)

- **What are some examples of vertebrates that have a double-loop system?** (Accept all logical answers.)
- **How is the double-loop system an adaptation for breathing air?** (The second loop connects to the lungs.)
- **What major trend is seen in the evolution of the vertebrate heart?** (In air-breathing vertebrates, the separation of the heart into right and left halves becomes more complete as you move from amphibians to birds and mammals; the number of chambers in the heart increases from two to four as you move from fishes to birds and mammals.)

ANNOTATION KEY

❶ Crocodile: uric acid and ammonia. Turtle: uric acid and urea. Frog, salamander: urea and ammonia.
(Interpreting diagrams)

ESL STRATEGY

Explain that tables, charts, and other graphical organizers are useful when trying to assimilate information in science and in other school subjects. So that they can see how one such organizer works, have students construct a table that compares the vertebrate classes and selected groups within these classes with regard to the seven essential life functions and gives examples of animals in each of the groups listed in the table. On one axis, the table should have the following heads: Examples, Movement, Feeding, Internal Transport, Respiration, Excretion, Response, Reproduction. On the other axis, the table should list the important groups of vertebrates. For example: Hagfishes, Lampreys, Cartilaginous Fishes, Bony Fishes, Amphibians, Reptiles, Birds, Monotreme Mammals, Marsupial Mammals, Placental Mammals. Instruct students to use their textbook to fill in the spaces in the table. Point out that entries in the table should contain just enough to jog the memory—terms or short phrases, rather than complete sentences.

36–2 (continued)

Content Development
Using the Visuals Direct students' attention to Figure 36–11.
• What are the three main kinds of nitrogenous wastes? (Ammonia, urea, uric acid.)
• What kind of nitrogenous waste was probably produced by the earliest vertebrates? (Ammonia.)
• Which type of waste requires the most water to excrete? (Ammonia.)
• Which type of waste requires the least water to excrete? (Uric acid.)
Tell students that uric acid production can be seen as an adaptation to development inside a land egg.
• Why is the production of uric acid useful to animals that develop in a land egg? (Wastes produced by the embryo can be safely stored as uric acid without tying up the limited water supply in the egg.)
• Why don't the embryos of viviparous vertebrates need to excrete uric acid? (The embryos' wastes can be removed by the mother's body.)
• What are the main organs of excretion in vertebrates? (Gills and kidneys.)

Figure 36–11 The phylogenetic distribution of nitrogenous wastes is shown in this illustration. Which modern-day animals excrete more than one kind of nitrogenous waste? ❶

body tends to lose water through osmosis. Because urea and uric acid are less toxic than ammonia, they can be concentrated. Thus it takes less water to flush these forms of nitrogenous waste from the body. This helps conserve water.

Response

All vertebrates display a high degree of cephalization—their head contains a well-developed brain and bears many sense organs. Refer to Figure 36–12 on page 798 as you read about the evolution of the brain in vertebrates.

Note that the size and complexity of the cerebrum and cerebellum increase as you move through the vertebrate classes from fishes to mammals. The cerebrum can be thought of as the "thinking" region of the brain. It receives, interprets, and determines the response to sensory information. It is also involved in learning, memory, and conscious thought. In fishes, amphibians, and reptiles, the cerebrum is relatively small. In birds and mammals—especially primates—the cerebrum is enormously enlarged and may contain many folds that increase its area. The cerebellum, which coordinates movement and controls balance, is also best-developed in birds and mammals.

SCIENCE, TECHNOLOGY, AND SOCIETY CONNECTION

The Third Eye

Some parts of the brain have taken on a variety of forms and functions in different vertebrate groups. One such part is the pineal gland, a structure that controls the secretion of the hormone melatonin. Under normal conditions in most animals, the pineal gland seems to secrete melatonin at night but not during the day.

In some fishes and reptiles, the pineal gland is located at the top of the brain in a place where several skull bones meet. In some of these animals, such as the tuatara, the pineal gland is sometimes called the "third eye" because it contains cells that are sensitive to light. Despite this name, animals do not see with their pineal gland. Instead, they use it to detect the amount of time they are exposed to either light or darkness. When days are long, the pineal gland secretes less melatonin than when days are short. By reacting to the amount of melatonin secreted by the pineal gland, an animal's body somehow manages to determine changes in season and to prepare itself for such seasonal activities as breeding or hibernation.

Birds and mammals also have a pineal gland, although the gland is buried deep in the brain where it could not possibly be exposed to light. Yet in these animals—including humans—the pineal gland still secretes differing amounts of melatonin in response to changes in day length. Just how the pineal gland receives this information at its position deep in the brain is not certain. But it is clear that changes in melatonin levels from one season to another can have profound effects on human mood. For example, some people become extremely depressed during the long dark months of winter. By exposing such people to extremely bright light, physicians can simulate the effects of spring and cure the depression.

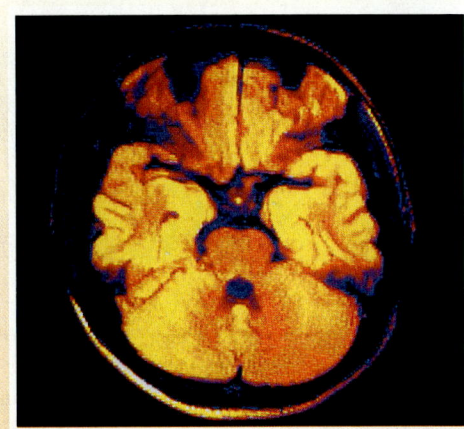

The pineal gland appears as a tiny yellow dot in the center of this NMR (nuclear magnetic resonance) image of a human brain.

Because humans are considered to have the best developed brain of all animals, people often tend to think that humans can see, hear, and generally detect the world around them better than most animals can. But nothing could be further from the truth. Fishes, certain turtles, and many birds can see colors and patterns much better than humans can. Dogs, bats, and dolphins all can hear sounds of much higher pitch than humans can; and elephants can hear very low-pitched sounds. Humans have an extremely poor sense of smell compared to dogs, cats, and, in fact, most other mammals. As well as having senses that are more keenly developed than ours,

Skills Development
Guided Practice
Skill: Applying concepts
Ask students to create a fictional hero or villain that is based on a real vertebrate animal. Have them write a short story describing how their character used its special sensory abilities to solve a mystery, catch a criminal, or commit the perfect crime.

Enrichment
Have interested students do library research to find out more about the specific functions controlled by the various regions of the brain in vertebrates.

SCIENCE, TECHNOLOGY, AND SOCIETY CONNECTION

The Third Eye
After students have read the feature, pose the following questions:

- **Where is the pineal gland located?** (In the brain.)
- **With which essential life function is the pineal gland involved?** (Response.)
- **What substance is produced by the pineal gland?** (Melatonin.)
- **What effect does melatonin have on humans?** (Melatonin levels affect mood. High melatonin levels can cause depression in some people.)
- **Why might exposure to bright light cure melatonin-related depression?** (It might stimulate the pineal gland to turn off melatonin production.)

Tell students that the highest relative number of cases of depression occur during the winter in places such as Scandinavia, the northern part of Russia, northern Canada, and Alaska.

- **Why might this be so?** (During the winter in areas near the poles, the days are very short and the nights are very long; melatonin levels thus are quite high, and people who are susceptible to melatonin-related depression are likely to become depressed.)

Ask students to imagine the following situation: A person from Finland moves to Texas. In January, this person, who is usually grumpy and lazy all winter long is cheerful and active.

- **Why might this person no longer hate winter?** (Accept all logical answers. One reason for the person's improved attitude might be that there is less seasonal change in day length in Texas, which is closer to the equator than is Finland. Thus the person's melatonin levels never reach a depression-causing level.)

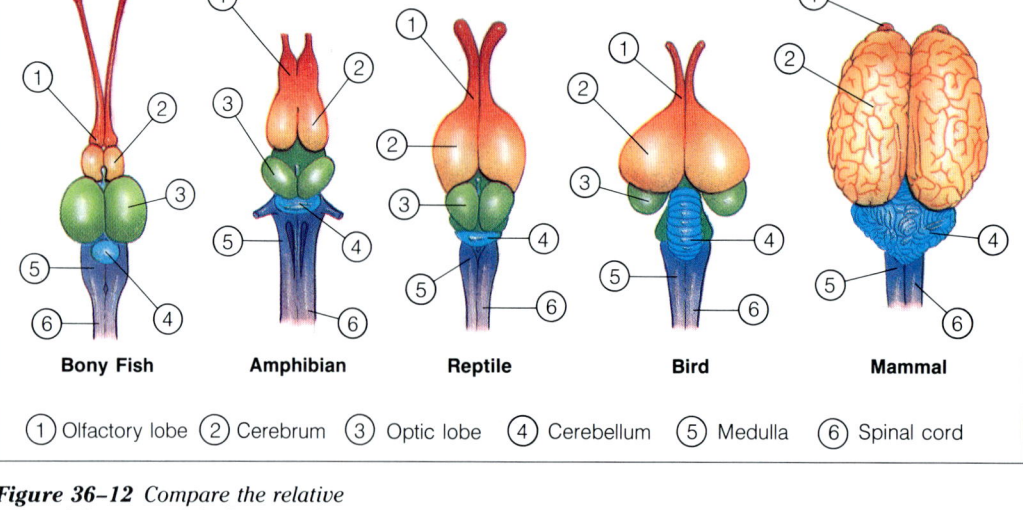

1 Olfactory lobe 2 Cerebrum 3 Optic lobe 4 Cerebellum 5 Medulla 6 Spinal cord

Figure 36–12 *Compare the relative sizes of the parts of the brain in the fish, amphibian, bird, reptile, and mammal.*

animals also possess senses that we lack. For example, rattlesnakes can detect infrared radiation (heat). And certain fishes and the duckbill platypus can detect the presence of extremely weak electric fields!

Reproduction

Almost all vertebrates reproduce sexually. (However, there are a few species of lizards, fishes, and amphibians that develop from unfertilized eggs.) In some vertebrates, such as codfish and frogs, fertilization is external. In others—reptiles, birds, mammals, cartilaginous fishes, and certain amphibians, for example—fertilization occurs inside the body of the female. As you move through the vertebrate classes from fishes to mammals, there is a general trend from external fertilization to internal fertilization, although there are some exceptions.

Figure 36–13 *The reproductive systems in an amphibian, a bird, and a mammal are shown here. What adaptations to requirements of flight are shown by a female bird's reproductive system?*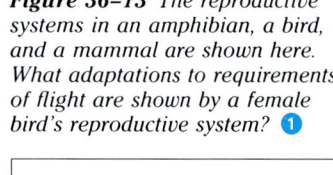

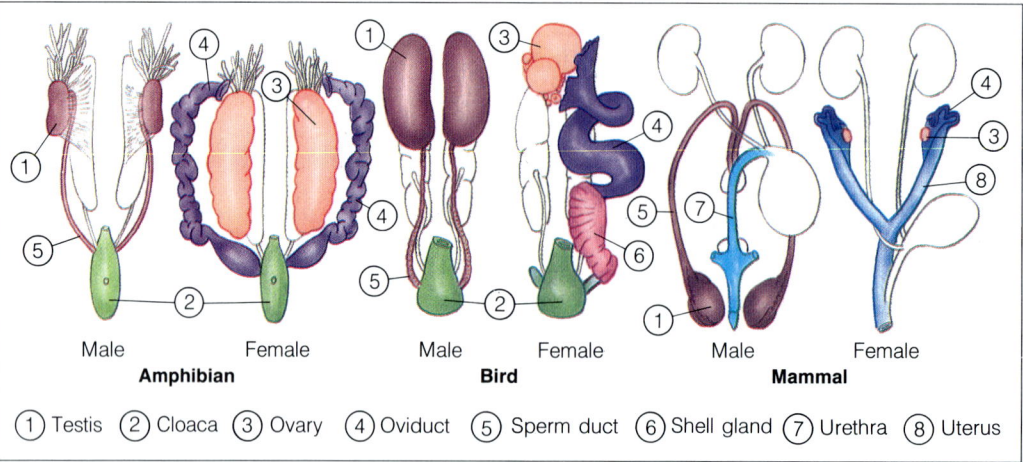

1 Testis 2 Cloaca 3 Ovary 4 Oviduct 5 Sperm duct 6 Shell gland 7 Urethra 8 Uterus

SECTION REVIEW 36–2

1. The vertebrate stance tends to become more upright; for the most part, the legs are positioned far to the sides of the body in amphibians, slightly to the sides in reptiles, and directly beneath the body in mammals. Side to side movement of the spine tends to become less important in movement; in many mammals, the spine flexes with an up-and-down motion when the animals run.

2. In all three vertebrates, the respiratory system contains a large air tube (trachea) that divides into two smaller air tubes (bronchi) that lead to the lungs. A frog's lungs are simple sacs that have some internal pockets that slightly increase the area of the respiratory surface. In a hu-

Vertebrates show three different modes of development: They may be oviparous, ovoviviparous, or viviparous. In oviparous species, the eggs develop outside the mother's body. In ovoviviparous species, the eggs develop inside the mother's body but the embryos receive the nutrients they need from the yolk that surrounds them, not from the mother directly. In viviparous species, the developing embryos obtain nutrients directly from the mother's body.

36–2 SECTION REVIEW

1. Describe two evolutionary trends involving the function of movement in vertebrates.
2. Compare the respiratory systems of a frog, a human, and a bird.
3. Compare the circulatory systems of a typical fish and a crocodile.
4. Explain why aquatic animals can eliminate wastes in the form of ammonia whereas land animals cannot.
5. **Critical Thinking—Applying Concepts** Some biologists say that they can tell more about an animal by looking at certain of its teeth than they can by looking at almost any bone. Why do you think this could be so?

Figure 36–14 Many vertebrates care for their offspring. The fuzzy gray chinstrap penguin chick is begging for food. Its taps on the parent's beak signal the parent to release food stored in its crop.

PROBLEM SOLVING IN BIOLOGY

IDENTIFYING VERTEBRATES

Read the following descriptions of animals and identify the vertebrate class to which each animal belongs. (*Hint:* The vertebrate classes are hagfishes, lampreys, cartilaginous fishes, bony fishes, amphibians, reptiles, birds, and mammals.)

a. Two-chambered heart; single-loop circulatory system; excretes ammonia; herbivore; vertebral column
b. Endothermic; one-way respiratory system; excretes uric acid; single ovary; oviparous
c. Endothermic; limbs located directly beneath the body; excretes urea; highly developed cerebrum
d. Parasite; ectothermic; gills; two-chambered heart; closed circulatory system; no vertebral column
e. Ectothermic; well-developed lungs; excretes uric acid; three-chambered heart; no limbs
f. Four-chambered heart; feet not directly beneath the body when standing; oviparous; ectothermic
g. Three-chambered heart; breathes through the lungs, the lining of the mouth, and the skin; excretes urea; carnivore
h. Single-loop circulatory system; excretes urea; internal fertilization; viviparous

LABORATORY INVESTIGATION  COMPARING VERTEBRATES

PROBLEM
What similarities and differences are there among the classes of vertebrates?

MATERIALS (per group)

assorted vertebrate pictures or specimens

PROCEDURE

1. Your teacher will provide you with pictures and/or specimens of vertebrates. Each vertebrate will have an identification number.
2. Examine one vertebrate at a time. Determine to which class each vertebrate belongs by using the key below.
3. To use the key, start at step 1 and read descriptions A and B. Only one of the two descriptions correctly describes the vertebrate. At the end of the correct description is either the class to which the vertebrate belongs or directions to go to another step.
4. All steps work the same as step 1. Continue following the directions in each step until you find the proper classification of each vertebrate.
5. Record its identification number and classification. Repeat this procedure for all the vertebrates.

OBSERVATIONS

1. What did the vertebrates you examined have in common?
2. How were the classes of vertebrates you examined different from one another?
3. Did any two vertebrates with visible differences belong to the same class?

ANALYSIS AND CONCLUSIONS

1. How many classes of organisms are generally called fish? Why are they usually grouped together? Why are they placed in separate classes?
2. Why does a snake belong to the same class as a lizard even though it looks more like an eel? What problems does this present in devising a classification key?
3. Although a salamander looks more like a lizard than a frog, frogs and salamanders are in the same vertebrate class. What does this tell you about their ancestry?

Classification Key

1.   A. Adult has gills and gill openings . . . . Go to 2
     B. Adult has no gills or gill openings . . . . Go to 5
2.   A. Has no jaw . . . . Go to 3
     B. Has a jaw . . . . Go to 4
3.   A. Wormlike with no eyes . . . . Hagfish
     B. Snakelike with a circular mouth . . . . Lamprey
4.   A. Mouth on ventral surface . . . . Chondrichthyes
     B. Mouth in front . . . . Osteichthyes
5.   A. No covering on skin, no claws on toes . . . . Amphibians
     B. Skin covered with scales, feathers, or fur . . . . Go to 6
6.   A. Scaly skin . . . . Reptiles
     B. Skin not scaly . . . . Go to 7
7.   A. Skin covered with feathers . . . . Birds
     B. Skin covered with hair or fur . . . . Mammals

LABORATORY INVESTIGATION

COMPARING VERTEBRATES

Before the Lab

1. Gather all materials at least one day before the investigation. Science and nature magazines often have good photographs of vertebrates. If you wish to use pictures in books or in magazines that you do not want to cut up, you may want to make "frames" of opaque paper that cover the whole page except for the picture you wish to show. Collect as many stuffed or preserved specimens as you can. Try to have a few specimens representing each class. If you keep classroom pets, you may wish to have students classify them as well.

    Since many vertebrates are familiar to students, students may find it easy to classify them without properly following the key. Try to use some less familiar or less obvious specimens. For example, the frog is an obvious amphibian, but the salamander can be easily mistaken for a lizard if careful observations are not made.

2. Number the specimens in the lab and prepare an answer key for yourself based on your numbering.

Pre-Lab Discussion

Have students read the complete laboratory procedure.

- **What is the purpose of this investigation?** (To classify vertebrate specimens and to examine the similarities and differences among classes of vertebrates.)
- **How do you use a classification key?** (For each step, choose the statement that is true for the organism being classified and follow the directions associated with that statement. Continue making choices until the organism is identified.)

Stress that it is important to try to make observations without letting them be colored by preconceived notions. Remind students that many scientific "facts" of the past have been false.

Skills Development

Students will use the following skills in this investigation.

1. Observing
2. Comparing
3. Relating
4. Inferring
5. Classifying

Safety Tips

Caution students that preserved animals contain chemicals that can stain clothes and that often have an unpleasant smell or are toxic. Warn students of specimens that must be handled with care or that should not be

SUMMARIZING THE CONCEPTS

The key concepts in each section of this chapter are listed below to help you review the chapter content. Make sure you understand each concept and its relationship to other concepts and to the theme of this chapter.

36–1 Evolution of the Vertebrates

- In divergent evolution, related evolutionary lines become more dissimilar as they are subjected to different forces of natural selection. Divergent evolution is also known as adaptive radiation.
- In convergent evolution, evolutionary lines that are subjected to similar forces of natural selection become more similar to one another as they evolve.
- Ectotherms must obtain the heat they need from their environment. They typically rely primarily on behavior to regulate their body temperature.
- Endotherms generate all the heat they need through metabolic activity. They typically rely on physiological mechanisms to regulate their body temperature. They also use a number of behaviors to prevent overheating.

36–2 Form and Function in Vertebrates

- As you move through the vertebrate classes from fishes to mammals, organ systems tend to become increasingly complex.
- In more primitive vertebrates, the limbs stick out from the sides of the body. In more advanced vertebrates, the limbs tend to be positioned directly beneath the body.
- Some vertebrates use gills for respiration; others use lungs. Lungs increase in efficiency as you move from amphibians to reptiles to mammals. Birds have the most advanced respiratory system of all vertebrates.
- Vertebrates that have a single-loop circulatory system also have a two-chambered heart. Double-loop circulatory systems are associated with lungs.
- Fishes have a two-chambered heart. Frogs and toads have a three-chambered heart. Most reptiles have a three-chambered heart that has a partial partition in the ventricle. Birds, mammals, and crocodilians have a four-chambered heart.
- Most fishes and aquatic amphibians excrete nitrogenous wastes in the form of ammonia. Mammals and cartilaginous fishes excrete urea. Birds and reptiles excrete uric acid.
- As you move through the vertebrate classes, the relative size and complexity of the cerebrum and cerebellum increase.
- Vertebrates may be oviparous, ovoviviparous, or viviparous.

REVIEWING KEY TERMS

Vocabulary terms are important to your understanding of biology. The key terms listed below are those you should be especially familiar with. Review these terms and their meanings. Then use each term in a complete sentence. If you are not sure of a term's meaning, return to the appropriate section and review its definition.

36–1 Evolution of the Vertebrates
divergent evolution
convergent evolution

36–2 Form and Function in Vertebrates
single-loop circulatory system
double-loop circulatory system

---

touched. If you have live specimens, do not allow students to disturb them or handle them during the investigation. Remind students that live animals should be treated with care; warn them that animals may scratch or bite if disturbed.

Teaching Strategy
1. Demonstrate how to use the key with an unnumbered specimen.
2. Encourage students to rely on their own observations, rather than on previously learned facts to identify the vertebrates.

Stress that science often makes progress when someone questions previously held assumptions.

Observations
1. Bilateral symmetry, an obvious anterior and posterior end (a dorsal nerve cord, but this can't be observed).
2. Some breathed with gills and some had smooth skin, scaly skin, feathers, or fur.
3. Yes.

Analysis and Conclusions
1. Four. They are lumped together because they all live in water and breathe with gills. They are placed in different classes because of fundamental differences. Hagfishes are jawless scavengers with a number of unique traits, such as an open circulatory system. Lampreys are jawless and are parasites as adults. Members of the class Chondrichthyes have skeletons made of cartilage. Members of the class Osteichthyes have skeletons made of bone.
2. Unlike eels, snakes breathe with lungs, reproduce on land, and have dry, scaly skin like lizards. External characteristics are often used to devise keys, but they are not sufficient for proper classification since superficially similar organisms may be fundamentally different. As a result, external characteristics used in keys must be selected carefully.
3. The way salamanders, frogs, and lizards are classified shows that salamanders and frogs are more closely related than are salamanders and lizards.

Going Further: Enrichment
Have students create classification keys that allow them to place vertebrates into the major orders they have studied, as well as into classes. Have students use and informally evaluate keys devised by their classmates. Encourage students to suggest ways that will help others improve their keys.

CHAPTER REVIEW

CONTENT REVIEW

Multiple Choice
1. b 2. d 3. d 4. b
5. d 6. a 7. b 8. d

True or False
1. F similar
2. F reptiles
3. F Ectotherms
4. F oviparous
5. F uric acid
6. F Birds
7. T
8. F Carnivores

Word Relationships

A.
1. cerebrum
2. single-loop circulatory system

B.
3. divergent
4. double-loop circulation
5. uric acid
6. three-chambered heart

CONCEPT MASTERY

1. The earliest vertebrates were fishes that used gills for respiration and thus had single-loop circulatory systems and two-chambered hearts. With the advent of air-breathing, the atrium and ventricle began to develop partitions. This intermediate stage in the evolution of the vertebrate heart can be seen in frogs, which have a three-chambered heart, and in most modern reptiles, which have a three-chambered heart and a partial septum in the ventricle. The final stage in the evolution of the vertebrate heart can be seen in crocodilian reptiles, birds, and mammals, which have four-chambered hearts and thus have a complete separation of the two circulatory loops.
2. Aquatic vertebrates can excrete ammonia, a toxic compound that needs to be diluted and washed away by large amounts of water. Terrestrial vertebrates excrete urea or uric acid—compounds that are less toxic than ammonia and that require less water to be transported and expelled from the body.
3. The cerebrum and cerebellum become more complex and larger relative to the rest of the brain.
4. A salamander uses its spine more than its weak, stubby legs in movement; it basically walks by throwing its body into S-shaped curves to move its legs around. Because an alligator's legs are positioned more directly beneath its body and are much stronger than those of a salamander, it relies more on its legs in movement. However, its spine still contributes somewhat to movement by moving side to side when it walks. The legs

---

CHAPTER REVIEW

CONTENT REVIEW

Multiple Choice

Choose the letter of the answer that best completes each statement.

1. Flying squirrels and flying lizards have flaps of skin that help them glide from tree to tree. This is an example of
   a. phylogenetic evolution.
   b. convergent evolution.
   c. divergent evolution.
   d. adaptive radiation.
2. How many chambers are in a bird's heart?
   a. one   c. two
   b. three   d. four
3. Adaptive radiation is also known as
   a. endothermy.
   b. ectothermy.
   c. convergent evolution.
   d. divergent evolution.
4. If a vertebrate has lungs, it probably
   a. has single-loop circulation.
   b. has double-loop circulation.
   c. has a four-chambered heart.
   d. excretes ammonia.
5. If a terrestrial animal's limbs stick out from the sides of its body, it probably
   a. is viviparous.
   b. has a two-chambered heart.
   c. is endothermic.
   d. excretes urea.
6. Seals have flippers, bats have wings, and zebras have legs. This is an example of
   a. divergent evolution.
   b. convergent evolution.
   c. phylogenetic evolution.
   d. ectothermy.
7. Animals that generate all the body heat they need through metabolic activity are
   a. oviparous.   c. viviparous.
   b. endothermic.   d. ectothermic.
8. Which term is not associated with life functions in a typical mammal?
   a. large cerebrum   c. endotherm
   b. urea   d. oviparous

True or False

Determine whether each statement is true or false. If it is true, write "true." If it is false, change the underlined word or words to make the statement true.

1. Evolutionary lines subjected to similar forces of natural selection tend to become more <u>dissimilar</u> as they evolve.
2. According to the phylogenetic tree, birds are most closely related to <u>mammals</u>.
3. <u>Endotherms</u> rely on behavior to regulate body temperature.
4. All birds are <u>ovoviviparous</u>.
5. Most reptiles excrete <u>urea</u>.
6. <u>Mammals</u> have the most advanced respiratory systems of all vertebrates.
7. Most reptiles have a heart with <u>three</u> chambers.
8. <u>Herbivores</u> have relatively short intestines and special enzymes for digesting meat.

Word Relationships

A. *Replace the underlined definition with the correct vocabulary word.*

1. The <u>part of the brain involved with learning, memory, making decisions, and interpreting information</u> is well developed in mammals.
2. Fishes have a <u>type of circulatory system in which blood travels from the heart to the gills to the body and then back to the heart.</u>

B. An analogy is a relationship between two pairs of words or phrases generally written in the following manner: a:b::c:d. The symbol : is read "is to," and the symbol :: is read "as." For example, cat:animal::rose:plant is read "cat is to animal as rose is to plant."

In the analogies that follow, a word or phrase is missing. Complete each analogy by providing the missing word or phrase.

3. internal:external::convergent:_____
4. gills:single-loop circulation::lungs:_____
5. shark:urea::chicken:_____
6. crocodile:frog::four-chambered heart:_____

CONCEPT MASTERY

Use your understanding of the concepts developed in the chapter to answer each of the following in a brief paragraph.

1. Describe the evolution of the vertebrate heart.
2. How is a vertebrate's excretory product related to its environment?
3. Discuss the trends in the evolution of the vertebrate brain.
4. Compare movement in a salamander, an alligator, and a leopard.

CRITICAL AND CREATIVE THINKING

Discuss each of the following in a brief paragraph.

1. **Applying concepts** North American hummingbirds and Hawaiian honeycreepers have long, thin bills adapted for drinking nectar from flowers. How would you go about determining whether their similarities were the result of convergent evolution?

2. **Applying concepts** While traveling along a muddy stream bank, you notice the trail of a small animal. This trail consists of a wide squiggly groove located between footprints that are to either side. To which group of vertebrates does the trail-making animal belong? Explain.
3. **Assessing concepts** Suppose some friends tell you that they have discovered a new evolutionary trend in vertebrates: As you move from fishes to mammals, the body coverings of animals go from scales to fur. Explain whether or not you consider this a useful concept. What kind of information would you need to better evaluate this concept?
4. **Using the writing process** Write a humorous story in which an endotherm and an ectotherm debate the advantages and disadvantages of their particular lifestyles.

CAREERS IN BIOLOGY

Nature Photographer, Marine Biologist, Veterinarian

Students interested in a biological career should be instructed to write for further information to the address listed beneath each career description. However, in consideration of the organizations that provide career information, please have only one student write to an organization. Do not instruct the entire class to write to every organization for the same career information. You may want to use the information provided to start a biology career file for the use of all students.

UNIT 8

CAREERS IN BIOLOGY

Nature Photographer

The wide variety of nature photographs that you see in magazines, books, and research articles does more than simply catch your attention. The photographs help you visualize and understand the subject matter. These pictures are the work of nature photographers.

Although no formal education is required to become a nature photographer, a knowledge of photographic techniques and equipment is essential. A strong background in biology or ecology would also be helpful in understanding the subjects being photographed.

To receive information write to the Professional Photographers of America, 1090 Express Way, Des Plaines, IL 60018.

Marine Biologist

Marine biologists study organisms that live in water—from the smallest marine bacteria to the largest whales. On sea expeditions marine biologists collect samples that are later identified and analyzed. Their conclusions provide information about how marine life is affected by environmental conditions. Most marine biologists specialize in either plants or animals, and many select a particular species to study.

Marine biologists are required to have a Bachelor's degree in the biological sciences or chemistry.

For additional information write to the Virginia Institute of Marine Science, Gloucester Point, VA 23062.

Veterinarian

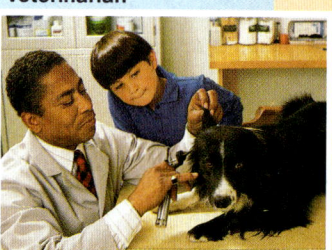

Veterinarians diagnose animal illnesses, treat diseased or injured animals, and give advice on animal care and breeding. Some veterinarians work exclusively with large animals, mostly in rural areas. Other veterinarians work in small animal hospitals, where they specialize in domestic animals.

Veterinarians must complete a four-year course of study after college to receive a Doctor of Veterinary Medicine degree. A state license is also required for private practice.

For information write to the American Veterinary Medical Association, 930 N. Meacham Rd., Schaumburg, IL 60196.

HOW TO WRITE A RÉSUMÉ

A résumé is a short summary of your background and accomplishments. Whether or not you will be granted an interview is dependent upon how you present yourself in your résumé. Your résumé, therefore, must highlight your skills in a clear and impressive manner.

Your résumé must first include your name, address, and telephone number. Your high school and the major courses you took as well as your past employers and the positions you held should be listed next. You can also include any personal accomplishments such as awards or scholarships. Any organizations to which you belong that might be of importance to an employer should be added.

Your résumé, like your cover letter, must be neatly typed without errors. If possible, your résumé should not exceed one page.

FROM THE AUTHORS

As a biologist, I am extremely fond of just about all the plants and animals I've ever met—with the possible exception of mosquitoes and slugs, which I think the world could do very well without. But I admit that I particularly like vertebrates and am unusually fond of fish. Perhaps that's because my love affair with the animal kingdom really began at the age of nine when I won a goldfish in a little glass jar with a lucky toss of a coin at the county fair. When I showed up on my parents' doorstep carrying "Oscar" in a plastic bag, my folks were kind enough—or perhaps foolish enough—to allow us both to stay. You see, we lived in an apartment building that didn't allow any other sorts of pets, so I could never have a dog, cat, or even a bird. But in less than a year, Oscar's bowl had grown into a ten-gallon aquarium, and things soon got out of control.

Later on, when my parents sent me off to college, they thought I was planning to go to medical school. Instead, several globe-trotting expeditions and two advanced degrees in biology later, I finally received my Ph.D. from the Fish Department at Harvard University. Along the way, I have managed to develop a liking for lots of other living things, many of which I have either worked with in my research or kept as pets. And to this day, some of my most enjoyable working days are spent designing exhibits for really big fish tanks—the kind they have in major public aquariums. Care to feed the sharks, anyone?

Joe Levine

FROM THE AUTHORS

Joseph Levine

In this feature author Joseph Levine describes how he "disappointed" his parents by deciding to study marine organisms instead of going to medical school. Levine points out that it all began with Oscar, a fish he won at a county fair. Oscar must have made quite an impression, for Levine currently designs aquariums throughout the United States. Suggest that students may have a pet at home that could provide an impetus for a career in science.

UNIT 9
Human Biology

UNIT OVERVIEW

In Unit 9 students are introduced to the anatomy, physiology, and development of the human body. They begin with nine of the organ systems and related processes: nervous system; skeletal, muscular, and integumentary systems; digestive system and nutrition; respiratory system; circulatory and excretory systems; and endocrine system. They then study reproduction and development. Next they read about human diseases and the immune system. Finally, they learn how drugs, alcohol, and tobacco affect the human body.

UNIT OBJECTIVES

1. Explain how cells join together to form tissues, organs, and organ systems and classify different types of tissue.
2. Describe the features and functions of the body's important organ systems.
3. Relate nutrition to digestion.
4. Describe human reproduction and development.
5. Compare human diseases and agents of disease and relate them to the activities of the immune system.
6. Differentiate among drugs, alcohol, and tobacco and explain the effects of each.

UNIT 9
Human Biology

Pause for a moment and think of all the things your body can do. Perhaps you thought of running a race, reading a book, building a go-cart, tasting a ripe apple, or solving a math problem. The human body is truly remarkable!

Even in its earliest stages, the human body is capable of doing astounding things. Although young babies may not seem able to do much of anything, their bodies are performing in much the same way as those of the most impressive athletes. The babies' senses allow them to perceive the world. Their brains process information and enable them to respond. Their muscles pull against their bones to produce movement. Their organs work together to extract energy from food, obtain oxygen, distribute materials throughout the body, get rid of wastes, and fight off disease. In time, their systems will develop further, and they will acquire new abilities—just as you did as you grew up.

The internal performance of the human body is shared by all humans—in fact, it is taking place in your own body at this very moment. It is required for every single action you perform—from tying your shoes to skating on ice. The interaction of the body's various systems defines human life and makes survival, growth, development, and reproduction possible.

DISCOVERY LEARNING
A MATTER OF TASTE

1. For this activity you will need a friend, a blindfold, and two small, peeled, seedless pieces of each of the following: apple, cucumber, and raw potato. **CAUTION:** *If you or your friend are allergic to one of these foods, substitute a different type of food that has a similar texture to the ones listed.*
2. Blindfold your friend. Have your friend pinch his or her nose closed.
3. Feed your friend a piece of one of the foods. Have him or her try to identify the food.
4. Repeat step 3 with the other two food samples.
5. Now it's your turn to take the taste test.
   - What do you observe? Explain your observations.
   - After you complete this unit, review the results of the taste tests. Can you better explain what you observed?
   - Why do you think you have trouble tasting your food when you have a bad cold?

INTRODUCING UNIT 9

Using the Visuals
Have students examine the photograph on pages 806 and 807.
- **What do you see in this photograph?** (Several human babies.)
- **What are the babies doing?** (Answers will vary. Students may say that the babies are not doing much of anything.)

Tell students that the babies are actually performing many activities—although many are not apparent to the casual observer. Have students read the unit-opening text.
- **What is going on inside the babies' bodies?** (The babies are extracting energy from food, breathing, pumping blood, eliminating waste, fighting diseases, and assimilating information from their senses.)

Point out that babies require no special training to perform any of these activities. Even in the youngest—and oldest—individuals, the human body is a remarkably complex, self regulating, and well-organized organism.

CHAPTERS

37 *Nervous System*

38 *Skeletal, Muscular, and Integumentary Systems*

39 *Nutrition and Digestion*

40 *Respiratory System*

41 *Circulatory and Excretory Systems*

42 *Endocrine System*

43 *Reproduction and Development*

44 *Human Diseases*

45 *Immune System*

46 *Drugs, Alcohol, and Tobacco*

CHAPTER DESCRIPTIONS

37 Nervous System
Chapter 37 describes basic structures and functions of the nervous system.

38 Skeletal, Muscular, and Integumentary Systems
Chapter 38 first deals with the structure and development of bones and joints. Types of muscle tissue, the mechanism of muscle contraction, and the interaction of muscles and bones are then discussed.

39 Nutrition and Digestion
In Chapter 39 the process of digestion, including the various organs involved, is explained in detail and related to nutrition.

40 Respiratory System
Chapter 40 explores the characteristics and functions of the respiratory system.

41 Circulatory and Excretory Systems
Chapter 41 discusses these two closely interrelated systems.

42 Endocrine System
The glands of the endocrine system and the hormones they produce are discussed in Chapter 42.

43 Reproduction and Development
Chapter 43 begins with a comparison of the male and female reproductive systems. The development of the embryo, the process of childbirth, and the responses of the body after childbirth are then discussed.

44 Human Diseases
Chapter 44 describes the various agents of disease.

45 Immune System
Nonspecific defenses of the body, such as the skin, are discussed first in Chapter 45. Specific defenses and the activities of the immune system are then explored.

46 Drugs, Alcohol, and Tobacco
Chapter 46 opens with a comparison of the various drugs and their effects on the body. The effects of alcohol- and tobacco-related diseases are then discussed.

DISCOVERY LEARNING
A MATTER OF TASTE

If students correctly perform this activity, they should not be able to distinguish the different foods from one another. Afterward, have students discuss the activity and their results.

• **When you couldn't smell or see the food, what sense did you have left?** (The sense of taste.)
• **How did each food taste?** (Students should report that each food had the same, slightly sweet taste.)
• **How do you explain this result?** (The sense of taste by itself is a very basic one. A food's smell contributes a great deal to how we taste and enjoy it.)

807

CHAPTER 37 Nervous System

Section	Laboratory Investigations and Activities

37–1 The Nervous System
pages 809–815
THEMES: Scale and Structure, Patterns of Change
	Teacher Edition
	   DEMONSTRATION: Bridging the Gap, p. 808D
	Laboratory Manual
	   Constructing a Model of a Nerve Cell, p. 461

37–2 Divisions of the Nervous System page 816 THEME: Scale and Structure	
37–3 The Central Nervous System pages 817–824 THEME: Scale and Structure	
37–4 The Peripheral Nervous System pages 825–826 THEMES: Scale and Structure, Systems and Interactions	

37–5 The Senses
pages 827–831
THEMES: Scale and Structure, Systems and Interactions
	Student Edition
	   LABORATORY INVESTIGATION: Measuring Reaction Time, p. 832
	Teacher Edition
	   DEMONSTRATION: A Watery Balance, p. 808D
	Laboratory Manual
	   Skin Sensitivity, p. 467
	Teaching Resources
	   HANDS-ON ACTIVITY: A Gentle Touch, p. 161
	   HANDS-ON ACTIVITY: Colored "Sandwiches," p. 165
	   HANDS-ON ACTIVITY: More Than Meets the Eye, p. 169
	   HANDS-ON ACTIVITY: Some Sound Reasons, p. 171
	   TEACHER DEMONSTRATION: Analyzing Afterimages, p. 23
	Product Testing Activities
	   Testing Bubble Gum Testing Yogurts
	   Testing Cereals

Chapter Review pages 833–835	

CHAPTER PLANNING GUIDE

Other Activities	Media and Technology
Study Guide Section 37–1, p. 349	**Video/Videodisc** Human Body: Nervous System and Systems Working Together **Biology Transparencies** 52: The Human Control Systems **Biology Media Guide** pages 52–53
Study Guide Section 37–2, p. 352	
Teaching Resources ACTIVITY: Examining the Human Nervous System, p. 3 **Biotechnology Workbook** Unlocking the Trapped Mind, p. 133 **Study Guide** Section 37–3, p. 354	**Video/Videodisc** Human Body: Nervous System and Systems Working Together **Biology Transparencies** 57: The Structure of the Human Brain **Biology Media Guide** pages 53–54
Teaching Resources ACTIVITY: The Nervous System in Everyday Life, p. 5 **Study Guide** Section 37–4, p. 356	**Video/Videodisc** Human Body: Nervous System and Systems Working Together **Biology Media Guide** page 54
Teaching Resources VOCABULARY REVIEW: Word Game, p. 1 **Study Guide** Section 37–5, p. 358	**Video/Videodisc** Human Body: Nervous System and Systems Working Together **Biology Transparencies** 58: The Structure of the Human Eye 59: The Structure of the Human Ear **Biology Media Guide** pages 54–55
Teaching Resources Chapter Test A, p. 11 Chapter Test B, p. 15	**Computer Test Bank** Chapter Test, p. 389

CHAPTER 37 Nervous System

CHAPTER OVERVIEW

In this chapter students will learn about the human nervous system. The first two sections of the chapter discuss the nervous system in general, and the third and fourth sections describe the central and peripheral nervous systems in particular. In the fifth section students will explore the five senses.

Students will learn that the nervous system controls and coordinates all the essential functions of the human body. The nervous system receives and relays information about activities within the body and monitors and responds to internal and external changes. Students will learn how messages in the form of nerve impulses are carried throughout the body by specialized cells called neurons.

Students will discover that the central nervous system consists of the brain and spinal cord, and the peripheral nervous system consists of all the nerves and associated cells that are not part of the brain and spinal cord. Students will learn that the peripheral nervous system is divided into two main parts, the sensory division and the motor division, and that the motor division is further divided into the somatic nervous system and the autonomic nervous system.

Students will discover that neurons called sensory neurons respond directly to stimulation from the environment. These sensory receptors are contained in the sense organs. Each of the five senses—sight, hearing, smell, taste, and touch—has a specific sense organ associated with it.

37–1 THE NERVOUS SYSTEM

Section Focus 37–1

The purpose of this section is to describe the function of the nervous system and to explain how nerve impulses are transmitted throughout the body. Students will learn that nerve impulses are carried by specialized cells called neurons. They will learn about the structure of a neuron and discover how this structure is related to its function.

Students will discover that a nerve impulse is a flow of electrical charges along the cell membrane of a neuron. The minimum level of a stimulus required to produce a nerve impulse is called the threshold. Students will learn that the points of contact at which impulses are passed from one cell to another are known as synapses. The release of neurotransmitters will be discussed, as will be the role of myelin in speeding up nerve impulses.

Performance Objectives 37–1

1. Describe the function of the nervous system.
2. Describe the changes that occur across the cell membrane of a neuron during the transmission of a nerve impulse.
3. Explain what happens at synapses when nerve impulses are passed from one neuron to another.

Science Terms 37–1

nervous system   p. 809
neuron   p. 810
impulse   p. 810
sensory neuron   p. 810
motor neuron   p. 810
interneuron   p. 810
cell body   p. 810
dendrite   p. 810
axon   p. 810
axon terminal   p. 810
resting potential   p. 811
action potential   p. 812
myelin   p. 813
threshold   p. 814
receptor   p. 814
effector   p. 814
synapse   p. 814
neurotransmitter   p. 814

37–2 DIVISIONS OF THE NERVOUS SYSTEM

Section Focus 37–2

The purpose of this section is to briefly outline the two major divisions of the human nervous system. Students will learn that these two divisions are the central nervous system and the peripheral nervous system.

Students will learn that the central nervous system consists of the brain and spinal cord. They will discover that the purpose of this part of the nervous system is to act as a central control unit. They will also discover that the central nervous system does not come in contact with the environment but that this is the function of the peripheral nervous system. The peripheral nervous system consists of all the nerves and associated cells that are not part of the brain and spinal cord.

Performance Objectives 37–2

1. Identify the two major divisions of the nervous system.
2. Describe the principal functions of the central nervous system.
3. Describe the chief function of the peripheral nervous system.

Science Terms 37–2

central nervous system   p. 816
peripheral nervous system   p. 816

37–3 THE CENTRAL NERVOUS SYSTEM

Section Focus 37–3

The purpose of this section is to describe the structure and function of the central nervous system. The parts of the brain will be discussed in detail, as will various functions of the brain, such as memory.

Students will learn that the brain consists of the cerebrum, cerebral cortex, cerebellum, brainstem, thalamus, and hypothalamus. Students will learn that many of the functions we associate with the brain are performed in the gray matter of the cerebral cortex.

Performance Objectives 37–3

1. List and describe the parts of the brain.
2. Discuss the functions of the brain and spinal cord.
3. Discuss the phenomena of memory and sleep.

CHAPTER PREVIEW

Science Terms 37-3

cerebrum p. 818
cerebral cortex p. 818
cerebral medulla p. 818
cerebellum p. 818
brainstem p. 819
medulla oblongata p. 819
pons p. 819
midbrain p. 819
thalamus p. 820
hypothalamus p. 820
reflex p. 823

37-4 THE PERIPHERAL NERVOUS SYSTEM

Section Focus 37-4

The purpose of this section is to describe the structure and function of the peripheral nervous system. Students will learn that the peripheral nervous system can be divided into two parts: the sensory division and the motor division. Students will also discover that the motor division is further divided into the somatic nervous system and the autonomic nervous system.

Performance Objectives 37-4

1. Describe the motor division and the sensory division of the peripheral nervous system.
2. Describe and compare the somatic nervous system and the autonomic nervous system.
3. Describe and compare the sympathetic nervous system and the parasympathetic nervous system.

Science Terms 37-4

somatic nervous system p. 825
autonomic nervous system p. 825
reflex arc p. 825
sympathetic nervous system p. 826
parasympathetic nervous system p. 826

37-5 THE SENSES

Section Focus 37-5

The purpose of this section is to discuss the five senses and to describe the sense organs that are associated with them.

Performance Objectives 37-5

1. Describe the major parts of the eye and explain how the eye enables us to see.
2. Identify the parts of the ear that make possible hearing and balance.
3. Compare the senses of smell and taste.
4. Discuss the sense of touch and identify the various types of sense receptors in the skin.

Science Terms 37-5

sense organ p. 827

DISCOVERY LEARNING

TEACHER DEMONSTRATIONS
Modeling

Bridging the Gap

The following demonstration can be used as an introduction to Chapter 37.

For the demonstration, you will need two lengths (about 30 centimeters each) of extension cord wire. Fray one end of each piece of wire and spread out the pieces to look like dendrites.

Show students the wires. Touch the cut end of one wire to the frayed dendrite end of the other wire.

• **With the ends together like this, could an electrical impulse travel from one wire to the other?** (Yes.)
• **Why?** (The ends are touching.)

Now move the two wires so they are about a centimeter apart.

• **What would happen to an electrical impulse now if it were intended to travel from one wire to the other?** (It would not be able to go beyond the end of the first wire.)
• **What would have to happen in order for an impulse to travel from the cut end of this one wire to the frayed ends of this other wire?** (Answers may vary. Some students may say that the two ends would have to be made to touch each other again; others may realize that something would have to serve as a link between the ends of the two pieces of wire.)
• **What could serve as a link between the two wires?** (Accept all answers.)

Explain to students that these wires represent the situation that exists between nerve cell endings in the body. Tell students that as they study the first section of this chapter they will find out how the nervous system "bridges the gap" between nerve cells that are not touching.

A Watery Balance

This demonstration can be performed as students learn about the ear and balance in Section 37-5.

For the demonstration you will need a clear glass object such as a bottle or vase.

Fill the glass object about half full with water. You may wish to color the water to make it more visible. Place the glass on a table or desk.

• **Relative to the floor, how would you describe the base of the glass object?** (It is parallel to the floor.)
• **How would you describe the surface of the water?** (It is also parallel to the floor.)

Lift the glass object off the table and hold it in the air, still keeping the base parallel to the floor.

• **How would you describe the base of the object and the surface of the water relative to the floor?** (Both are still parallel.)

Now begin to tilt the glass object so that the base is no longer parallel to the floor.

• **Is the base of the object still parallel to the floor?** (No.) **What about the surface of the water?** (The surface of the water is still parallel to the floor.)

Continue tipping the object as far as you can without spilling the water.

• **What happens to the surface of the water as I tip the container more and more?** (It remains parallel to the floor.)

Point out that although the glass object was tilted, the water maintained a surface parallel to the floor. Explain that this is what the fluid in the tubes of the semicircular canal of the inner ear does. Whenever your head changes position, the fluid presses against tiny hairs on the inner surface of the tubes. Which hairs are pressed on by the fluid depends on the position of the tubes relative to the floor. The hairs produce nerve impulses that notify your brain about which way your head is moving and what your muscles have to do to help you keep your balance.

CHAPTER 37
Nervous System

GUIDED ENQUIRY

Pose the following questions to students and have them record their responses. Point out that they will gain a better understanding of the key concepts if they read the chapter with these basic questions in mind. Upon completion of the chapter, pose the questions again. Ask students to compare their initial responses with those they have developed after reading the chapter.

- How does the nervous system control and coordinate the functions of the human body?
- How is electrical activity vital to the transmission of a nerve impulse?
- What makes a neuron capable of carrying messages throughout the nervous system?
- What relationship exists between the major divisions of the nervous system and the environment?
- How is the central nervous system like a main switching unit?
- How does the peripheral nervous system control voluntary and involuntary activities of the body?
- How does the structure of the eye enable you to see?
- How does the inner ear enable you to maintain balance?

INTRODUCING CHAPTER 37

Using the Visuals
Have students examine the photographs on this page.
- **What is shown in the large photograph?** (A nerve cell placed on top of an integrated circuit from a computer.)
- **Why would someone make such an unusual photograph?** (Answers may vary. The photograph suggests that nerves in some way resemble computer circuitry.)

Have students read the caption and the chapter-opener text.
- **What comes to mind when you hear the words *coordinate, react, respond*?** (Answers will vary. Some possible answers include the coordinated movements of athletes; the activities of an air-traffic controller as he or she instructs planes; the efforts of a political campaign headquarters to coordinate diverse efforts of people working for the candidate.)

Point out that in the human body, even the simple action of swatting a mosquito requires as much—or more—organization and control as any of the situations just described. The reason is that a single action involves thousands of cells. Without some

CHAPTER 37
Nervous System

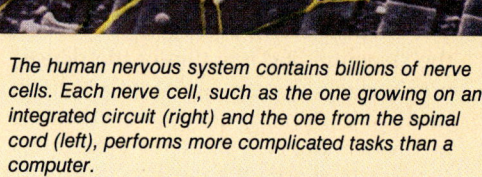

The human nervous system contains billions of nerve cells. Each nerve cell, such as the one growing on an integrated circuit (right) and the one from the spinal cord (left), performs more complicated tasks than a computer.

Because it's such a beautiful summer day, you decide to visit one of your favorite places: the beach, a park, a woods, or a favorite spot in your own yard. Suddenly, while engrossed in reading a book, you feel a tingle on your shoulder. Quickly you place your book down next to you and swat at a mosquito just as it flies away.

What has happened? You and the mosquito have just engaged in combat. Fortunately for the mosquito, it got away. (It usually does.)

But how did you know the mosquito was there? And how were you able to coordinate the actions of thousands of your cells in order to react and respond to the presence of the mosquito? In this chapter you will find the answers to these questions. You will also learn how one of the systems of the human body—the nervous system—responds to, monitors, interprets, regulates, and adjusts to changes in the environment in a matter of milliseconds.

GUIDE FOR READING

After you read the following sections, you will be able to

37–1 The Nervous System
- Describe the function of the nervous system.
- Relate the structure of a neuron to its function.
- Explain the changes that occur across a neuron during the transmission of a nerve impulse.

37–2 Divisions of the Nervous System
- Identify the two major divisions of the nervous system.

37–3 The Central Nervous System
- Describe the parts of the brain.
- Discuss the functions of the brain and the spinal cord.

37–4 The Peripheral Nervous System
- Describe the function and structure of the peripheral nervous system.

37–5 The Senses
- Describe the major parts of the eye and their function.
- Identify the parts of the ear responsible for hearing and balance.
- Compare the senses of smell and taste.
- Identify the various sense receptors in the skin.

Journal Activity

YOU AND YOUR WORLD

Have you had an experience in which the sense of smell recalled the memory of a past event? In your journal, describe the experience.

Figure 37–1 Athletes such as Kathy Johnson have raised their coordinated movement to an art form. Their movements, like the hundreds we make every day, are made possible by the nervous system, with its intricate network of nerves (inset).

37–1 The Nervous System

Guide For Reading

■ What is the function of the nervous system?
■ How is the structure of a neuron related to its function?
■ How is a nerve impulse transmitted?

Most of us have played softball—a game that requires not only athletic skill but also control and communication. The catcher and the pitcher must agree on the pitch before it is thrown. The batter must be alert for the signal to bunt, take a pitch, or swing away. On-base runners must know if they are expected to steal a base. From the dugout, the coaches send in signals about these and other important matters. Communication is vital to a team's success.

Communication is also vital to the survival of living organisms. In order to interact with their environment, multicellular organisms have developed a communication system at the cellular level. Within these organisms, specialized cells allow messages to be carried from one cell to another so that communication among all body parts is smooth and efficient. In humans, these cells make up the **nervous system**. The nervous system controls and coordinates all the essential functions of the human body. **The nervous system receives and relays information about activities within the body and monitors and responds to internal and external changes.**

COOPERATIVE LEARNING

Have groups of students complete one of the following activities:
1. Design a warning label (complete with logo) to be used on records, cassette tapes, and compact disks that warns users that excessive volume can impair hearing.
2. Write a 30-second public service announcement for TV that warns viewers of the danger of excessive noise levels to hearing. This public service announcement is to be aired on the music video stations.

Groups should be reminded to include factual information about hearing in their final product.

Journal Activity

YOU AND YOUR WORLD

Marcel Proust begins *Remembrance of Things Past* with just such a recollection. Ask volunteers to share their experiences with the class. Have students keep their journal entries in their portfolio.

answer: noticing that the weather is getting cold; interpreting that warm clothes should be worn.)

TEACHING STRATEGY 37–1

Focus/Motivation

Describe the nervous system's role as the body's communication network.

• **What messages are you sending to or receiving from your body right now?** (Answers will vary. Students may realize that even answering this question requires communication from brain to vocal cords.)

Emphasize that the nervous system continuously relays an enormous number of messages, most of which the brain processes automatically. You also may emphasize that vision and hearing can be thought of as messages from the eyes and ears, respectively.

kind of central control, each cell would "do its own thing."

• **In the last sentence of the text, the word *monitors* is used. What does it mean to say that the nervous system monitors changes in the environment?** (It keeps a constant watch on the environment and is instantly aware when a change occurs.)

• **What does it mean to say that the nervous system "interprets" changes in the environment?** (It makes a judgment about the meaning and significance of a change.)

• **What might be your nervous system's interpretation of the realization that a mosquito is buzzing around you?** (Possible answer: Get rid of it!)

• **Give another example of the nervous system monitoring and interpreting change.** (Possible

809

TEACHING SUPPORT

Laboratory Manual
- Constructing a Model of a Nerve Cell, p. 461

MEDIA AND TECHNOLOGY

Biology Transparencies
- 52: The Human Control Systems

Video/Videodisc
- *Human Body Series: Nervous System and Systems Working Together*

Use this bar code to introduce neurons and the nervous system.

Play frames 1447 to 1789

See the Biology Media Guide pages 52–55 for additional bar-code correlations for this chapter.

The Neuron

The cells that carry messages throughout the nervous system are called **neurons**. Because the messages take the form of electrical signals, they are known as **impulses**. Neurons can be classified into three types according to the directions in which these impulses move. **Sensory neurons** carry impulses from the sense organs to the brain and spinal cord. **Motor neurons** carry impulses from the brain and spinal cord to muscles or glands. And **interneurons** connect sensory and motor neurons and carry impulses between them.

Although neurons come in all shapes and sizes, they have enough features in common that we can draw a typical neuron. See Figure 37–3. The largest part of the neuron is the **cell body**. The cell body contains the nucleus and much of the cytoplasm. Most of the metabolic activity of the cell, including the generation of ATP and the synthesis of proteins, takes place in the cell body. Spreading out from the cell body are short branched extensions called **dendrites**. Dendrites carry impulses from the environment or from other neurons toward the cell body. The long fiber that carries impulses away from the cell body is called the **axon**. The axon ends in a series of small swellings called **axon terminals**, which are located some distance from the cell body. Neurons may have dozens or even hundreds of dendrites but usually only one axon.

Figure 37–2 One of the body's billions of neurons can be seen in this electron micrograph. Note the ropelike axon at the bottom of the photograph.

Figure 37–3 In a typical neuron, the dendrite and cell body receive the stimulus, which then travels through the axon to the axon terminals.

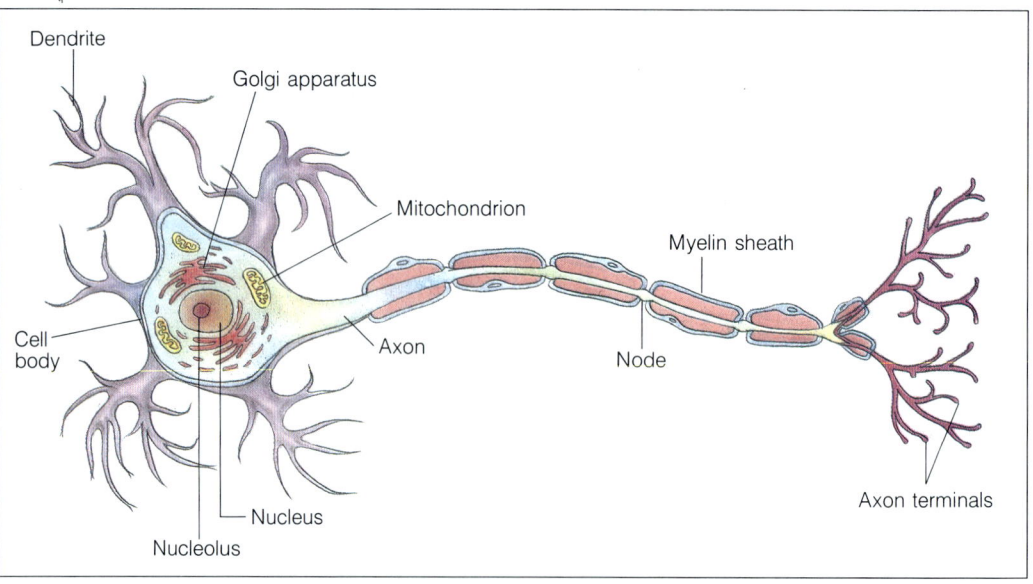

810

37–1 (continued)

Content Development

Using the Visuals Have students look at Figure 37–3.
- **How many cells are shown in this figure?** (Only one! With its extended axon, the typical neuron is an unusually long cell.)
- **What is the function of the dendrites?** (Dendrites are the neuron's receivers. When stimulated, they transmit an impulse to the cell body.)
- **What is the function of the axon?** (Axons transmit impulses down the length of the neuron.)
- **Would you guess that every dendrite stimulation is translated into an impulse down the axon?** (In general, dendrites act in concert to stimulate the axon. By itself, the impulse from a single dendrite receptor would probably dissipate at the cell body. In addition, some types of receptors actually inhibit axon stimulation, instead of promoting it.)

As students will discover later in this section, neurons follow an "all-or-none" principle of trans-

The Nerve Impulse

More than 150 years ago, the Italian scientist Luigi Galvani, a pioneer in the study of electricity, found that nervous tissue (groups of cells that conduct impulses) displays electrical activity. This electrical activity is in the form of a nerve impulse, which is a flow of electrical charges along the cell membrane of a neuron. This flow is due to movement of ions across the membrane.

RESTING POTENTIAL As shown in Figure 37–4, a nerve cell has an electrical potential (also known as a voltage) across its cell membrane because of a difference in the number of positively and negatively charged ions on each side of the cell membrane. This potential is approximately 70 millivolts (mV). One millivolt is equal to 0.001 volt. (By contrast, the potential between the poles of a flashlight battery is 1500 millivolts, or 1.5 volts.) What is the source of this potential? Proteins in the neuron known as sodium-potassium pumps move sodium ions (Na^+) out of the cell and actively pump potassium ions (K^+) into the cell.

As a result of this active transport, the cytoplasm of the neuron contains more K^+ ions and fewer Na^+ ions than the surrounding medium. The cytoplasm also contains many negatively charged protein molecules and ions. However, K^+ ions leak back out across the cell membrane more easily than Na^+ ions leak in, and the negatively charged protein molecules and ions do not leak in or out. The net result of the leakage of positively charged ions out of the cell is a negative charge on the inside of the neuron's cell membrane. This charge difference is known as the **resting potential** of the neuron's cell membrane. The neuron, of course, is not actually resting because it must produce a constant supply of ATP to fuel the sodium-potassium pump.

As a result of its resting potential, the neuron is said to be polarized: that is, negatively charged on the inside of the cell membrane, and positively charged on the outside. See Figure 37–4. A neuron maintains this polarization until it is stimulated.

Figure 37–4 *The resting potential across the neuron cell membrane is established when the protein pump in the cell membrane pumps potassium ions (K^+) in one direction and sodium ions (Na^+) in the other (A). As the protein pump continues to work for a while, a large number of K^+ ions enter the cell and a large number of Na^+ ions leave the cell (B). The cell membrane, however, is more leaky to K^+ ions than it is to Na^+ ions. As a result, more K^+ ions leak out of the cell, and fewer Na^+ ions leak into the cell. This leakage causes an excess of positive charges on the outside of the membrane and an excess of negative charges on the inside (C).*

Key
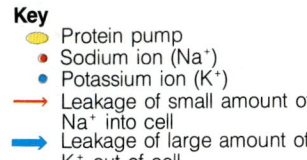
- Protein pump
- Sodium ion (Na^+)
- Potassium ion (K^+)
- Leakage of small amount of Na^+ into cell
- Leakage of large amount of K^+ out of cell

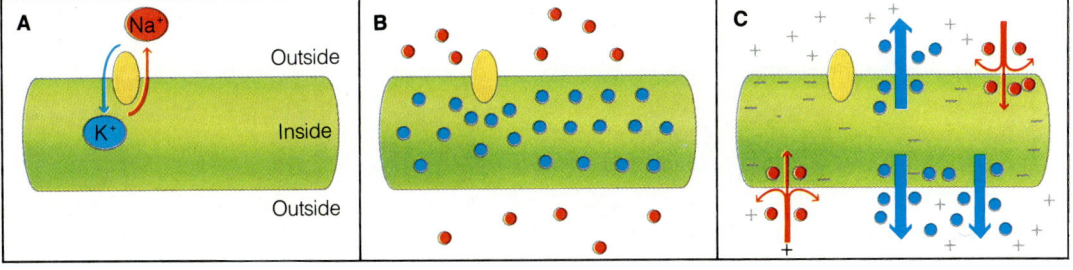

BACKGROUND INFORMATION
LUIGI GALVANI

Luigi Galvani (1737–98) was an Italian physician, surgeon, and researcher in comparative anatomy. During some of his experiments, Galvani observed that a frog's leg contracted when touched by two different metals in a moist environment. His observation of this electrical activity caused him to conclude erroneously that animal tissues generate electricity. Although Galvani's hypothesis was wrong, his observation stimulated research in electrotherapy and electric currents and led to the understanding we have today of the electrical nature of nerve impulses. Many electrical terms, such as *galvanometer*, are derived from Galvani's name.

mission. Only when a neuron's threshold is reached—when its various receptors are stimulated to the proper degree—will a neuron "fire" and send an impulse down its axon.

Skills Development
Guided Practice
Skill: Making diagrams
Have students make diagrams to show the relationship that exists among the three types of neurons: sensory neurons, interneurons, and motor neurons. Their diagrams should show clearly that sensory neurons carry messages from sense organs to the brain and spinal cord; motor neurons carry messages away from the brain and spinal cord to the muscles and glands; and interneurons, which are located in the brain and spinal cord, connect and coordinate messages between the sensory neurons and the motor neurons.

You may wish to have students diagram the functions of the neurons in general or a specific example, such as the process involved in swatting a fly or shivering in response to cold.

BACKGROUND INFORMATION
VERTEBRATE AND INVERTEBRATE NERVES

We humans are used to muscles that respond almost instantly to our commands. We select a movement—such as wiggling our fingers or toes—and our muscles receive the message an undetectable moment later. As discussed in the textbook, this speed of nerve transmission comes from the myelin that is wrapped around nerve axons. Without myelin, no animal as large as a human could efficiently control its distant muscles.

But only vertebrates make myelin, a fact that partially explains the size and body plan of most invertebrates. Most invertebrates have small, compact bodies that never grow beyond a certain size. Moreover, invertebrates such as clams and starfishes have centrally located nerve centers (ganglia), which help them to minimize the distance to their outer muscles.

However, at least two invertebrate species—squids and octopi—grow significantly larger than the invertebrate norm. These animals have adapted a different way of increasing the speed of nerve transmission: They have very wide nerves. The speed with which an impulse travels down an axon is proportional to the diameter of the axon. (In other words, wide nerves carry impulses faster.) With wide axons running the length of their bodies, squids and octopi can control tentacles some 50 centimeters away from their head. The giant axon of a squid may be more than a millimeter in diameter—a significant width for a single cell of any kind.

It is interesting to note that the giant squid is the experimental system from which neurobiologists have learned much of what we know today about nerve impulses. This animal's great size makes it easy to study with electrodes and recording devices.

THE MOVING IMPULSE A nerve impulse is similar to a ripple passing along the surface of a pond. Instead of a splash, the impulse causes a movement of ions across the cell membrane of a neuron. How does this movement occur?

The cell membrane of a neuron contains thousands of tiny molecules, known as gates, that allow either sodium or potassium ions to pass through. Generally, the gates are closed. At the leading edge of an impulse, however, the sodium gates open, allowing positively charged Na+ ions to flow inside the cell membrane. The inside of the membrane temporarily becomes more positive than the outside. The membrane is now said to be depolarized. As the impulse passes, the potassium gates open, allowing positively charged K+ ions to flow out. The membrane is now said to be repolarized, which means that it is once again negatively charged on the inside of the cell membrane and positively charged on the outside.

The depolarization and repolarization of a membrane produce an **action potential**. The nerve impulse can be defined as an action potential traveling along the membrane.

There are several important facts about impulses (action potentials) that you should keep in mind. First, an impulse is not an electric current. Instead, it is a wave of depolarization and repolarization that passes along the neuron. Second, an impulse is much slower than an electric current. Electric currents move almost instantaneously, whereas action potentials usually travel at speeds ranging from 10 centimeters per second to 1 meter per second. Third, unlike an electric current, the strength of an impulse is always the same—there is either an impulse in response to a stimulus or there is not.

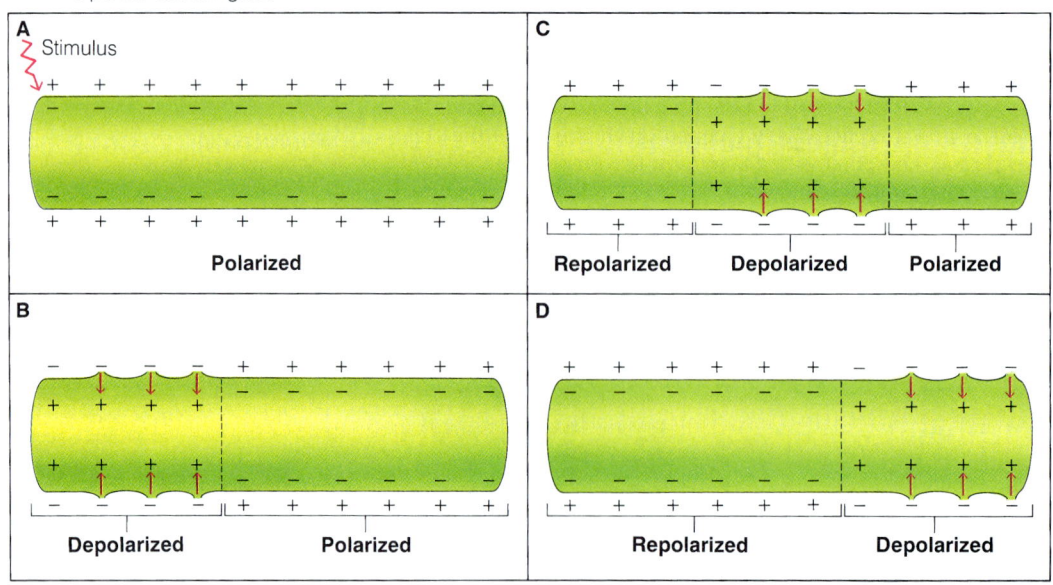

Figure 37–5 When the inside of a neuron's cell membrane is negatively charged with respect to the outside, it is said to be polarized. If a stimulus is applied to the membrane, electrical changes occur across the cell membrane and may result in an impulse (A). At the leading edge of the impulse, a small part of the membrane becomes depolarized. When this happens, sodium gates open, the membrane becomes more permeable to Na+ ions, and an action potential occurs (B). As the action potential passes, potassium gates open, allowing K+ ions to flow out. This outward flow of K+ ions restores the resting potential, and the membrane is said to be repolarized (C). The action potential continues to move along the axon in the direction of the nerve impulse (D).

37–1 (continued)

Reinforcement/Reteaching

Students may find it helpful to picture the structure and function of a neuron in the following way. The dendrites can be thought of as the "input zone" of the neuron—the area where an impulse originates and is passed to the cell body. The length of the axon can be thought of as the "conducting zone" of the neuron because the axon conducts an impulse in much the same way as a wire conducts an electric current. Finally, the nerve impulse will eventually reach the end of the axon, or the axon terminals. The axon terminals can be thought of as the "output zone" of the neuron. From the output zone, the impulse can move to another nerve cell.

PROPAGATION OF THE IMPULSE Until now, we have discussed the nerve impulse as if it occurs in only one place on the membrane. However, an impulse is self-propagating. That is, an impulse at any point on the membrane causes an impulse at the next point along the membrane. We might compare the flow of an impulse to the fall of a row of dominoes. As each domino falls, it causes its neighbor to fall. Unlike dominoes, however, the impulse can restore itself. Imagine dominoes that can set themselves back up and wait to fall again!

Although the nerve impulse is self-propagating, it can move in only one direction. This is because the part of the membrane behind the impulse has a brief period during which its sodium gates will not open. As a result, the impulse cannot go backward.

THE ROLE OF MYELIN As you just read, impulses can move along the membrane of a neuron at rates as fast as 1 meter per second. Although this rate is impressive, it is not practical for large animals. For example, a giraffe might have to wait three or four seconds for impulses to travel from its feet to its brain. Such delays would make large animals hopelessly uncoordinated. But as you probably know, giraffes are graceful and efficient in their movements.

What improves the rate of impulses along an axon? The answer is a substance known as **myelin**. Myelin, which is composed of 80 percent lipid and 20 percent protein, forms an insulated sheath, or wrapping, around the axon.

The most important feature of myelin is that there are small nodes, or gaps, between adjacent sheaths along the axon. As an impulse moves down a myelinated (covered with myelin) axon, the impulse jumps from node to node instead of moving continuously along the membrane. This jumping greatly increases the speed of the impulse. Some large myelinated axons conduct impulses as rapidly as 200 meters per second. This speed is significant when compared with speeds of only a few millimeters per second in small unmyelinated axons.

The formation of myelin around axons can be thought of as a crucial event in evolution. Because of myelin, the propagation of the nerve impulse is faster in vertebrates than in invertebrates.

Figure 37–6 Because many of the axons of vertebrates, such as the giraffe, are wrapped in myelin, nerve impulses travel more rapidly than they do in invertebrates.

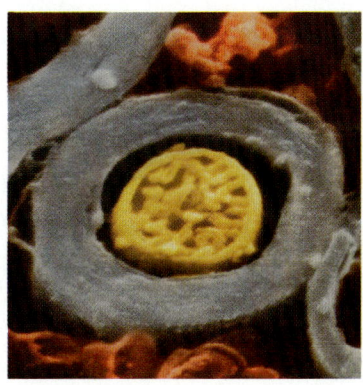

Figure 37–7 Most nerve fibers, such as the human auditory nerve (right), are wrapped in myelin, which forms a thick outer covering. In myelinated fibers, the action potential jumps from node to node (left).

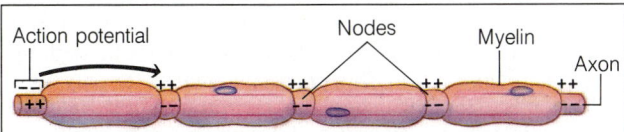

TEACHING SUPPORT

Study Guide
- Section 37-1, p. 349

BACKGROUND INFORMATION
NERVE GAS

Nerve gas is the popular name for a class of very lethal poisons, all of which inactivate a certain important enzyme of nerve transmission. Nerve gas killed many in World War I, and more recently, Iraq threatened to use it against Israel in the Gulf War of 1991.

Normally, neurotransmitters either diffuse away, are recycled, or are inactivated in some manner. The latter is the case for the neurotransmitter acetylcholine, which is inactivated by enzymes called cholinesterases. Once released, cholinesterases inactivate acetylcholine very quickly, which allows for very efficient, precise synaptic transmissions.

What nerve gas does is bind to and inhibit the cholinesterases, which allows acetylcholine to remain in synapses almost indefinitely. The result is continually stimulated nerves, which typically cause uncontrollable convulsions, muscular contractions, and death.

THE THRESHOLD Recall that the strength of an impulse is always the same—either there is an impulse in response to a stimulus or there is not. In other words, a stimulus must be of adequate strength to cause a neuron to conduct an impulse. The minimum level of a stimulus that is required to activate a neuron is called the **threshold**. Any stimulus that is weaker than the threshold will produce no impulse; any stimulus that is stronger than the threshold will produce an impulse. Thus a nerve impulse follows the all-or-none principle.

We can illustrate the all-or-none principle by again using a row of dominoes. Suppose we were to give the first domino in the row a slight push. If the push was really slight, the domino would not move at all. If we were to push a little harder, the domino would teeter back and forth a bit, touching the next domino. A slightly stronger push would cause the domino to fall, hitting the next domino. We have succeeded in reaching the domino's threshold and the row of dominoes would continue to fall.

It is important to mention that the all-or-none principle is not restricted to impulses as they travel along neurons. It also occurs when impulses move from one neuron to another and when information from the environment causes a nerve impulse to occur.

The Synapse

As you may recall, the axon ends with many small swellings called axon terminals. At these terminals the neuron may make contact with the dendrites of another neuron, with a **receptor**, or with an **effector**. Receptors are special sensory neurons in sense organs that receive stimuli from the external environment. Effectors are muscles or glands that bring about a coordinated response. The points of contact at which impulses are passed from one cell to another are known as **synapses**.

The axon terminals at a synapse contain tiny vesicles, or sacs. These tiny vesicles are filled with chemicals known as **neurotransmitters**. A neurotransmitter is a substance that is used by one neuron to signal another.

When an impulse moves down the axon and arrives at the axon terminal, dozens of vesicles fuse with the cell membrane and discharge the neurotransmitter into the small gap between the two cells. See Figure 37-8. The molecules of the neurotransmitter diffuse across the gap and attach themselves to special receptors on the membrane of the neuron receiving the impulse.

When the neurotransmitter becomes attached to the cell membrane of the adjacent nerve cell, it changes the permeability of that membrane. As a result, Na^+ ions diffuse through the membrane into the cell. This process continues for only a few milliseconds, stopping when the neurotransmitter detaches from the membrane. However, if enough neurotransmitter is

814

37-1 (continued)

SECTION REVIEW 37-1

1. The nervous system receives and relays information about activities within the body and monitors and responds to internal and external changes.

2. The largest part of the neuron is the cell body, which contains the nucleus and much of the cytoplasm. Spreading out from the cell body are short, branched extensions called dendrites. The long fiber that extends from the cell body is the axon, which ends in a series of small swellings called axon terminals.

3. A nerve impulse can be defined as an action potential traveling along the nerve's membrane. When it meets the action potential, the nerve's membrane becomes more permeable to Na^+ ions. These ions rush inside the cell, making it positively charged on the inside, or depolarized. The cell repolarizes almost immediately afterward, when the membrane allows K^+ ions to rush out. This region of depolarizing and repolarizing membrane travels as a wave down the length of the nerve.

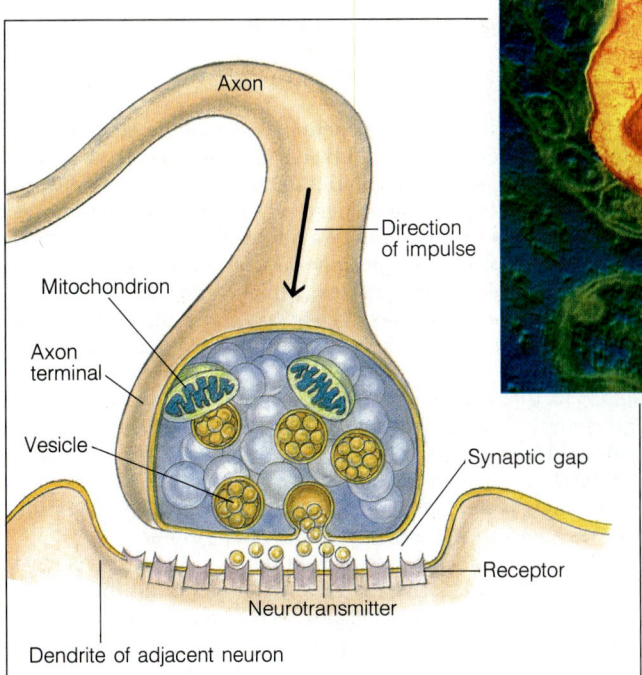

Figure 37-8 When an impulse arrives at the axon terminal, dozens of vesicles fuse with the axon membrane, releasing neurotransmitter molecules into the synaptic gap. These molecules diffuse across the gap and combine with receptors on the membrane of the adjacent neuron. Compare the structures in the diagram with those in the electron micrograph.

released by the axon terminal, so many Na^+ ions diffuse into the neuron that the neuron becomes depolarized. A threshold is reached and an impulse (action potential) begins in the second cell.

After the neurotransmitter detaches from the membrane of the cell, it is rapidly removed or destroyed, thus halting its effect. The molecules of the neurotransmitter may be broken down by specific enzymes, taken up again by the axon terminal and recycled, or they may simply diffuse away.

37–1 SECTION REVIEW

1. What is the function of the nervous system?
2. Describe the structure of a typical neuron.
3. Describe a nerve impulse in terms of an action potential.
4. What is a neurotransmitter? How is its release controlled by a nerve impulse?
5. **Critical Thinking—Relating Concepts** Why is the rapid removal or destruction of neurotransmitters important in controlling the activities of the nervous system?

BACKGROUND INFORMATION
EXCITATORY AND INHIBITORY ACTION AT SYNAPSES

The release of neurotransmitters at a synapse can either trigger an action potential or inhibit an action potential. Which one depends on whether a particular receptor is excitatory or inhibitory. Both types of receptors exist near each other on adjacent regions of a neuron membrane.

At each neuron in the nervous system, excitatory and inhibitory signals compete for control of the membrane. The effects of competing receptors are combined in a process called synaptic integration. Synaptic integration is the moment-by-moment combining of excitatory and inhibitory signals acting on adjacent membrane regions of a neuron. In simple terms, this means that several signals can totally or partially cancel each other out or they can augment each other's effect. The net outcome will depend on the strength, direction, and location of each signal.

In a way, synaptic integration provides a kind of "checks and balances" in the nervous system. What would happen if these checks and balances did not exist? The disease tetanus (also known as lockjaw) provides a clue. The bacterium *Clostridium tetani* produces a toxin that interferes with inhibitory receptors on motor neurons in the brain and spinal cord. The result is an unbalanced excitation of muscle cells, causing excessive contraction. Muscles cannot be released from contraction, and prolonged spastic paralysis often leads to death.

---

4. A neurotransmitter is a substance that is used by one neuron to signal another. When an impulse moves down an axon and arrives at the axon terminal, dozens of vesicles fuse with the cell membrane and discharge the neurotransmitter into the synapse. The neurotransmitter then diffuses across the synapse and binds to receptors on the membrane of the neuron receiving the impulse.
5. If neurotransmitters were not removed, they would continually stimulate receptors.

Reinforcement/Reteaching
If students are having trouble answering the Section Review questions, review the pertinent material.

Closure
Have students work in small groups to create a test that covers the material in this section. When all the tests are finished, have each group trade tests with another group.

37–2 Divisions of the Nervous System

Guide For Reading

■ What are the two major divisions of the nervous system?

Neurons, which are the functional units of the nervous system, do not act alone as individual cells. Instead, they are joined together to form a complicated communication network that gives rise to the human nervous system. **The human nervous system is divided into two major divisions: the central nervous system and the peripheral nervous system.**

The **central nervous system**, which serves as the control center of the body, consists of the brain and the spinal cord. Both the brain and the spinal cord are encased in and protected by bone. The functions of the central nervous system are similar to those of the central processing unit of a computer. The central nervous system relays messages, processes information, and compares and analyzes information. But the central nervous system does not come in contact with the environment. This job is left to the other major division of the nervous system—the **peripheral nervous system**.

The peripheral nervous system lies outside of the central nervous system. This means that it consists of all the nerves (bundles of axons) and associated cells that are not part of the brain and the spinal cord. Included here are all the cranial (pertaining to the brain) and spinal nerves and ganglia (GANG-glee-uh; singular: ganglion). Ganglia are a collection of nerve cell bodies. You will read more about the two major divisions of the nervous system in the following sections.

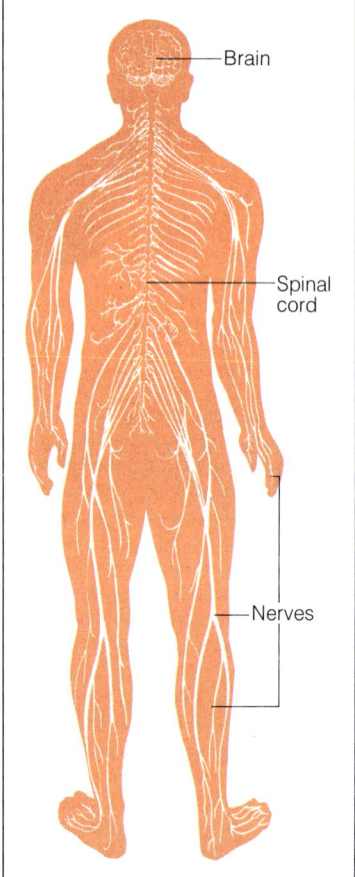

Figure 37–9 The human nervous system is made up of the central nervous system and the peripheral nervous system. The central nervous system contains the brain and the spinal cord. The peripheral nervous system contains all the nerves that carry information to and from the central nervous system.

Figure 37–10 The human nervous system is divided into many subdivisions. What are the two major subdivisions of the nervous system? ❶

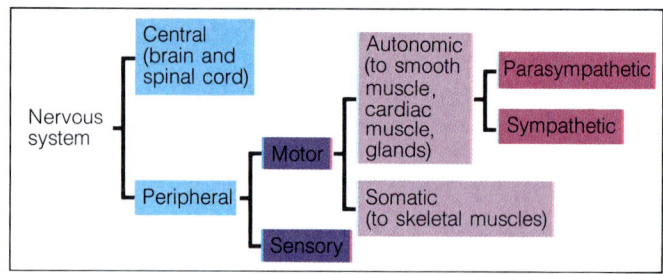

37–2 SECTION REVIEW

1. What are the two major divisions of the nervous system? What structures make up each of these systems?
2. **Critical Thinking—Making Comparisons** How do the two major divisions of the nervous system compare?

37–3 The Central Nervous System

Guide For Reading

- What are the parts of the brain?
- What are the functions of the brain and the spinal cord?

As you have just read, the central nervous system consists of the brain and the spinal cord. **The brain is the main switching unit of the central nervous system; it is the place to which impulses flow and from which impulses originate. The spinal cord provides the link between the brain and the rest of the body.**

The brain is a highly organized organ that contains approximately 35 billion neurons and has a mass of 1.4 kilograms. In addition to being protected by a bony covering called the skull, the brain is also wrapped in three layers of connective tissue known as the meninges (muh-NINH-jeez). Connective tissue, as its name implies, connects one tissue to another. The innermost layer, which covers and is bound to the surface of the brain, is called the pia mater. It is a fibrous layer made up of many blood vessels, which help to carry food and oxygen to the spinal cord. The outermost layer, called the dura mater, is composed of thick connective tissue. The arachnoid (uh-RAK-noid) is the thin, cobweblike layer between the pia mater and the dura mater. Between the pia mater and the arachnoid is a space that is filled with cerebrospinal fluid. The cerebrospinal fluid protects the brain from mechanical injury by acting as a shock absorber.

In order for the brain to perform its functions, it must have a constant supply of food and oxygen. If the oxygen supply to the brain is cut off even for a few minutes, the brain will usually suffer enormous damage. Such damage may result in death.

The spinal cord is continuous with the brain and emerges from the opening at the base of the skull. The spinal cord stretches downward for approximately 42 to 45 centimeters. Like the brain, the spinal cord is protected by bone (vertebral column), by the meninges, and by cerebrospinal fluid.

Figure 37–11 The brain and the spinal cord are wrapped in three layers of connective tissue called the meninges. The innermost layer is called the pia mater; the outermost layer is called the dura mater; the middle layer is called the arachnoid. The photograph shows the intricate network of blood vessels that constantly supply the brain with food and oxygen.

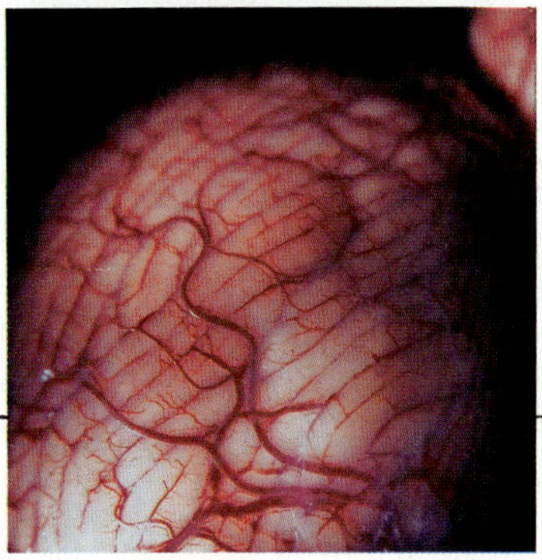

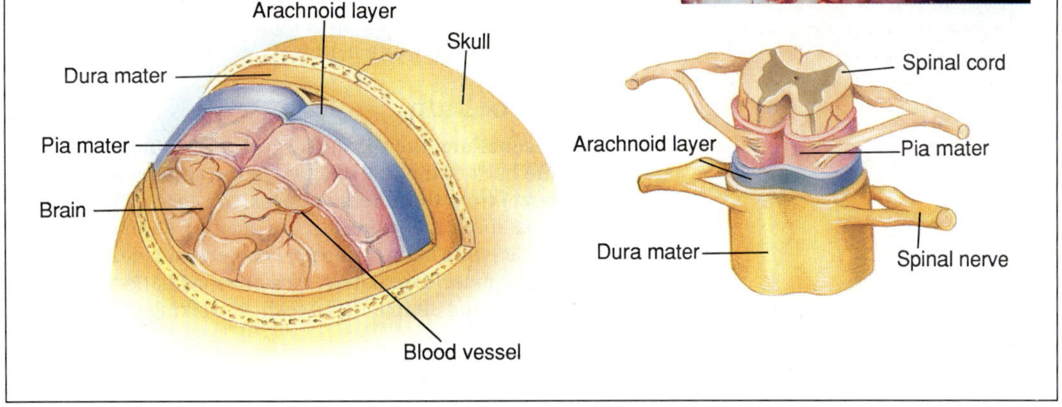

MULTICULTURAL STRATEGY

Some interesting research suggests that the different styles of music preferred by different cultures may be related to cerebral lateralization. For example, the atonality of Asian music and the structure of a European fugue are very different and may relate to a preference for the artistic side of the brain by one culture and the logical/mathematical orientation of another culture. Ask students to consider the merits of this hypothesis.

cars—who come into direct contact with the environment. They report on what is happening in each part of the city and take direct action in various situations.

Content Development

- **If we compare the city police department to the human nervous system, what does central headquarters represent?** (The central nervous system.)
- **What do the officers in the field represent?** (The peripheral nervous system.)
- **What do the various beats of these officers represent?** (Various parts of the body.)

SECTION REVIEW 37–2

1. The central nervous system consists of the brain and the spinal cord. The peripheral nervous system consists of all the nerves and associated cells that are not part of the brain and the spinal cord.
2. The central nervous system relays messages, processes information, and compares and analyzes information. The peripheral nervous system carries messages to and from the central nervous system.

Reinforcement/Reteaching

If students are having trouble answering the Section Review question, go over the parts of the section that are giving them difficulty.

Closure

Have students make up their own analogies to describe the divisions of the nervous system and the relationship that exists between them. Students may present their ideas in writing, in visual form, or orally, or they can work together in groups to dramatize their analogies.

TEACHING STRATEGY 37–3

Focus/Motivation

- **Have you ever made a call to a switchboard?** (Accept all responses.)
- **How would you describe the function of a switchboard?** (To take all incoming calls for a large establishment such as a corporation, then route each call to the proper person or department.)

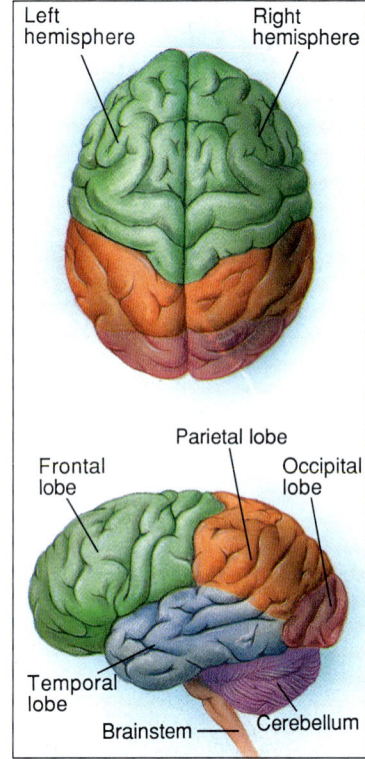

Figure 37–12 *The cerebrum is divided into the left and right hemispheres. Each hemisphere contains four lobes: frontal, parietal, temporal, and occipital.*

The Cerebrum

The largest and most prominent part of the human brain is the **cerebrum**. The cerebrum is responsible for all of the voluntary (conscious) activities of the body. In addition, it is the site of intelligence, learning, and judgment. The cerebrum takes up most of the space in the cavity that houses the brain. The cerebrum is divided into two hemispheres—the left hemisphere and the right hemisphere—by a deep groove. The hemispheres are connected in a region known as the corpus callosum (KOR-puhs kuh-LOH-suhm).

The most obvious feature on the surface of each hemisphere are the numerous folds. These folds and the grooves associated with them greatly increase the surface area of the cerebrum. The increased surface area permits the large number of neurons contained in the cerebrum to fit easily within the confines of the skull. Each hemisphere of the cerebrum is divided into regions called lobes. These lobes are named for the skull bones that cover them.

Remarkably, each half of the cerebrum deals with the *opposite* side of the body. Sensations from the left side of the body go to the right hemisphere of the cerebrum, and those from the right side of the body go to the left hemisphere. Commands to move muscles are generated in the same way—the left hemisphere controls the body's right side and the right hemisphere controls the body's left side.

There may be more than a simple left-right division of labor between the hemispheres. Some studies have suggested that the right hemisphere is associated with creativity and artistic ability, whereas the left hemisphere is associated with analytical and mathematical ability.

The Cerebral Cortex

The cerebrum consists of two surfaces. The outer surface is called the **cerebral cortex** and consists of gray matter. The gray matter is composed of densely packed nerve cell bodies that make it gray in appearance. The cerebral cortex is an extremely important part of the brain. Its functions will be discussed in more detail later in the section. The inner surface is called the **cerebral medulla**. The cerebral medulla consists of white matter, which is made up of bundles of myelinated axons. The myelin gives the white matter its white color.

The Cerebellum

The **cerebellum**, the second largest part of the brain, is located at the back of the skull. Although the commands to move muscles come from the cerebral cortex, the cerebellum coordinates and balances the actions of the muscles so that the body can move gracefully and efficiently. People with a damaged cerebellum suffer muscle weakness, lack of coordination, and difficulty in performing simple tasks such as walking and running.

TEACHING SUPPORT

Teaching Resources
• Activity: Examining the Human Nervous System, p. 3

MEDIA AND TECHNOLOGY
Biology Transparencies
• 57: The Structure of the Human Brain

BACKGROUND INFORMATION
NEURONAL PLASTICITY

For many years biologists believed that all of the nerve connections in the brain were genetically determined. Recently, however, experiments carried out by David Hubel and Torsten Wiesel of Harvard University have shown that this is not the case.

Hubel and Wiesel studied the visual cortex of kittens in which one eye had been kept closed since birth. They discovered that the nerve connections to the closed eye had failed to develop, although nerve connections to the open eye had developed normally. Further experiments proved that the connections between nerve cells in the optic cortex are not genetically determined. Instead, the pattern of connection depends on the kinds of visual stimulation that an animal experiences early in its development.

Thus it was concluded that the abnormal nerve connections in the kittens were the result of environmental rather than genetic influence. There is every reason to believe that environmental influence plays an important role in the development of other regions of the brain as well. Some scientists have named this phenomenon neuronal plasticity. This term emphasizes the idea that the architecture of parts of the nervous system is "plastic" and can be strongly influenced by experience and sensory input.

37–3 (continued)

Content Development
Continue the Motivation discussion by explaining to students that in some ways the brain is like a switchboard. It is the central unit to which impulses from the rest of the nervous system flow. Part of its job is to sort out and relay the information that it receives. Yet the brain is far more complex than any switchboard or central computer processing unit. Even a sophisticated computer can only process information according to the set of instructions that it is given. Yet the human brain is capable of thinking, reasoning, solving problems in unique ways, formulating verbal expression, and making value judgments.

Point out to students that most

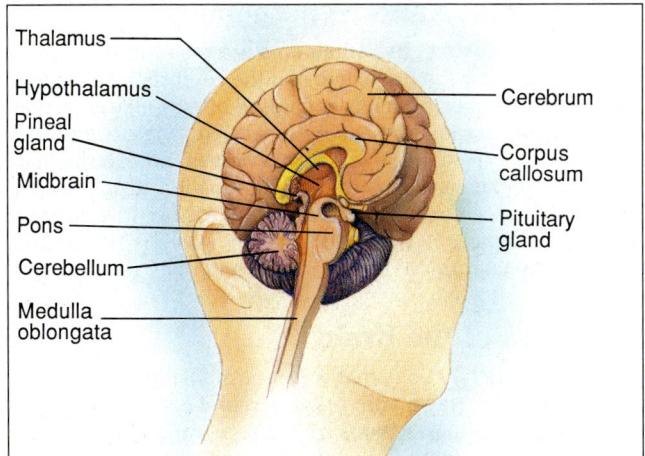

Figure 37–13 The human brain consists of the cerebrum, cerebellum, and brainstem.

A major part of learning how to perform physical activities seems to be related to training the cerebellum to coordinate the proper muscles. When you practice throwing a football, performing a pirouette, or executing a difficult dive, you are helping to develop connections in the cerebellum that will make the activity easier to do the next time. Because the functioning of the cerebellum is involuntary (not under conscious control), learning a completely new physical activity can be very difficult.

The Brainstem

The **brainstem** connects the brain to the spinal cord. The brainstem not only coordinates and integrates all incoming information, it also serves as the place of entry or exit for ten of the twelve cranial nerves.

The lowest part of the brainstem is the **medulla oblongata** (sometimes just called the medulla). It contains white matter that conducts impulses between the spinal cord and brain. The medulla controls involuntary functions that include breathing, blood pressure, heart rate, swallowing, and coughing.

The medulla also contains some of the cells of the reticular activating system. The reticular activating system actually helps to alert, or awaken, the upper parts of the brain, including the cerebral cortex. Such action keeps the brain alert and conscious.

Just above the medulla oblongata, the brainstem enlarges to form the **pons**. Pons means bridge, and this area of the brainstem contains mostly white matter that provides a link between the cerebral cortex and the cerebellum.

Above the pons and continuous with it is the **midbrain**, the smallest division of the brainstem. Areas in the midbrain are involved in hearing and vision.

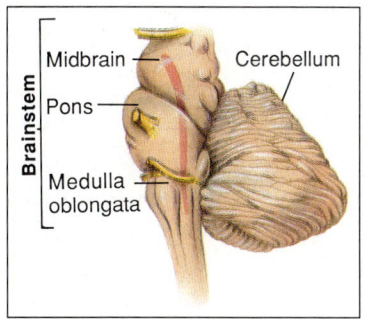

Figure 37–14 The brainstem, which consists of the midbrain, pons, and medulla oblongata, links the brain to the spinal cord and lies next to the cerebellum.

TIE-IN MEDICINE

Dance therapy is a form of psychiatric therapy that is practiced in many mental hospitals and mental health facilities. Dance therapists have found that there is often marked improvement in a patient's mental condition as a result of learning certain movement skills and developing the capacity for expression through movement and dance.

Although this mind–body connection may seem mystical at first glance, it is really quite logical if one looks at the workings of the cerebellum. The cerebellum is the area of the brain that must be activated if a person wishes to learn and perfect skills that require coordination and balance. Yet scientists know that the functioning of the cerebellum is not under conscious control. This means that mastering movements that require the development of connections in the cerebellum must of necessity impinge on the patient's unconscious. Since most mental disturbances involve unconscious factors, dance therapy may be able to reach aspects of a patient's illness that would prove elusive to more verbal forms of therapy.

---

motor impulses of the cerebrum to make a person's movements smooth and coordinated. The cerebellum also controls balance and posture.

- **Suppose that you are practicing a jazz dance routine. Which part of your brain initiates the basic movements of the dance?** (The cerebrum.)
- **Suppose that you are trying to learn a difficult dance move, such as a turn that lands on one leg. Which part of the brain would you have to "develop" in order to master the movement? Why?** (The cerebellum, because performing the movement requires coordination and balance.)

---

of the mental activities that make humans different from other animals take place in the cerebrum. Located in the cerebrum are the nerve impulses that enable a person to think, remember, and speak. The cerebrum also controls most voluntary muscle contractions and identifies the information gathered by the senses. It is the activities of the cerebrum that enable a person to respond appropriately to environmental stimuli.

Further point out that an interesting relationship exists between the functions of the cerebrum and those of the cerebellum. Explain that the cerebrum generates the motor impulses that cause muscles to move. Yet these impulses alone would produce sudden, jerky movements. The cerebellum fine-tunes the

BACKGROUND INFORMATION
REGIONS OF THE CORTEX

Different regions of the cerebral cortex include the motor centers, the primary receiving centers, and the association centers. In the motor centers, instructions for motor responses are coordinated. Parts of the motor cortex connect directly with descending motor tracts, and stimulation of different points on the motor cortex triggers muscles in different body parts to contract. It is significant that a relatively large area of the motor cortex is devoted to muscles that control thumb and tongue movements. This reflects the amount of control that is required for intricate hand movements and speech.

The primary receiving centers receive sensory input, such as that from the eyes and ears. The somatic sensory cortex, which is located just behind the motor cortex, is a primary receiving center for sensory input from the skin and joints.

Separate from the motor centers and the primary receiving centers are the association centers. The association centers are connected to the motor and sensory regions by neural pathways. In the association centers, information from memory is added to primary sensory information to give it fuller meaning.

FACTS AND FIGURES

The extensive folding of the gray matter of the cerebral cortex has led some scientists to suggest that the evolution of the mammalian cerebrum outpaced the development of the skull bones that contain it.

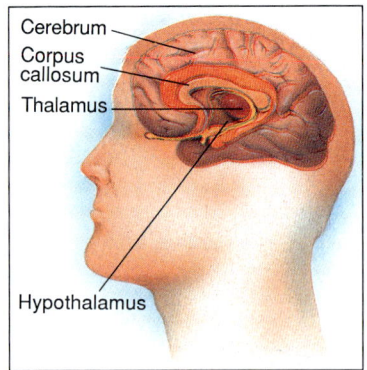

Figure 37–15 The thalamus, located within the cerebrum, is the main relay center between the brainstem and the cerebrum. Lying below the thalamus is the hypothalamus, which coordinates the activities associated with hunger, thirst, fatigue, anger, and body temperature.

The Thalamus and Hypothalamus

The **thalamus** and **hypothalamus** are found in the part of the brain between the brainstem and the cerebrum. See Figure 37–15. The thalamus, which is composed of gray matter, serves as a switching station for sensory input. With the exception of smell, each sense channels its sensory nerves through the thalamus. The thalamus seems to pass information to the proper region of the cerebrum for further processing. Immediately below the thalamus is the hypothalamus, which is the control center for hunger, thirst, fatigue, anger, and body temperature.

Functions of the Brain

Many of the functions that we associate with the brain are performed in the gray matter of the cerebral cortex. Figure 37–16 shows the various regions of the cerebral cortex and the parts of the body that they control. Some regions of the cerebral cortex are associated with sensory input; others, with motor output. Still other regions in the cerebral cortex are responsible for specific skills, such as the complex series of movements necessary for speech and the understanding of speech itself.

For some time, scientists believed that many functions of the body were controlled by specific regions of the cerebral cortex. They had good reasons for their belief. In the 1940s and

Figure 37–16 This cutaway view of the cerebrum shows the motor cortex of one hemisphere and the sensory cortex of the other. The large areas devoted to the face and hands explain why these body parts are so sensitive.

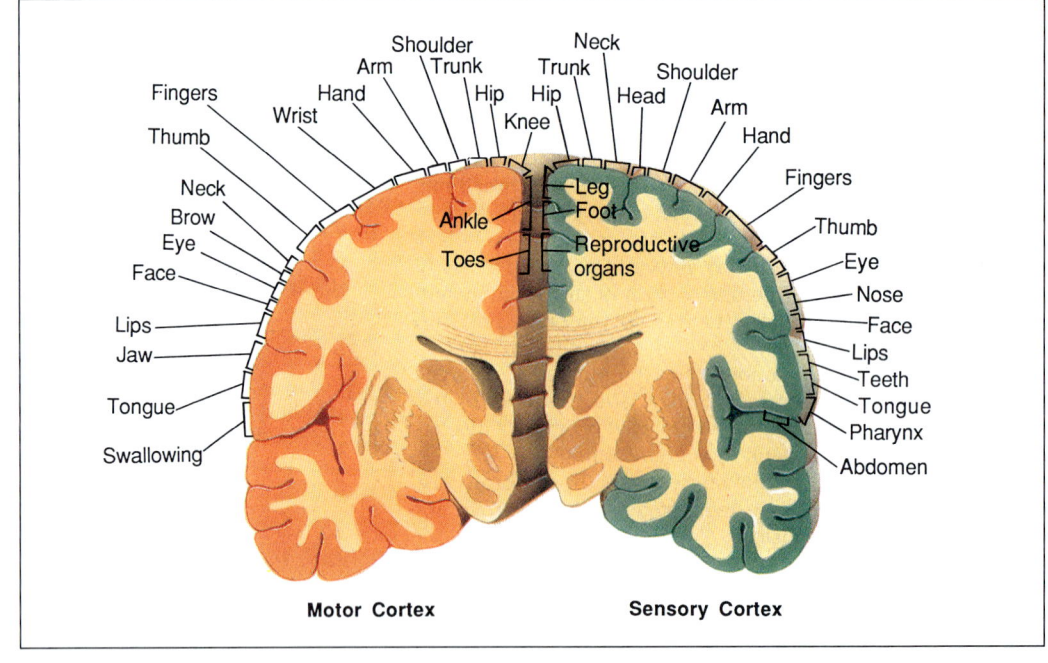

37–3 (continued)

Reinforcement/Reteaching
Using the Visuals Use Figures 37–13 through 37–16 to review with students the three main parts of the brain.
• **Which part of the brain is the largest?** (The cerebrum.)
• **Where is the cerebrum located?** (It takes up most of the skull cavity, particularly the front and top.)
• **What is distinctive about the cerebrum?** (It is divided into two hemispheres; it consists of an outer surface called the cerebral cortex and an inner surface called the cerebral medulla.)
• **What is the next largest part of the brain?** (The cerebellum.)
• **Where is it located?** (Toward the back of the brain below the cerebrum.)
• **What is the area at the base of the brain called?** (The brainstem.)
• **What parts make up the brainstem?** (The medulla oblongata, the pons, and the midbrain.)
• **Which of these three parts is the lowest part of the**

1950s, the Canadian neurosurgeon Wilder Penfield, along with other scientists, began to perform delicate experiments in which the brain of a patient was surgically exposed and different regions of the cerebral cortex were stimulated with weak electric currents. Experiments such as Penfield's are possible because of an interesting reason: Although the brain is packed with neurons, it does not have any pain receptors on its surface. Thus the brain cannot sense pain!

Using local anesthetics (pain killers), Penfield was able to perform surgery on his patients while they were awake. Stimulating one part of the cortex at a time, Penfield asked his patients to describe the sensations they experienced. In some places, stimulations caused muscles to contract. These areas, Penfield concluded, form a motor cortex that controls movement. In other places, stimulations caused the sensations of taste, touch, and sound. These regions of the brain form the sensory cortex. In still other places, stimulations caused his wide-awake patients to have vivid memories of past events, scenes, people, and places. These were the physical locations of memory, he concluded.

In recent years, the actual story has turned out to be quite a bit more complex. You might think from this brief discussion that sensory neurons are connected directly to the appropriate part of the sensory cortex. Unfortunately, this is not the case. There is no direct connection. Instead, sensory neurons synapse in the spinal cord, and neurons located in the spinal cord carry the impulses to the thalamus. The thalamus then relays the impulses to the sensory cortex. This is not the whole story, however. If it were, then in cases where parts of the spinal cord have been destroyed due to injury or surgery, sensation to certain parts of the body would be lost as well. But sometimes sensation is not affected—even though there seem to be no cells to carry the impulses to the cortex.

As you might imagine, the picture of the brain that is emerging is not a simple one. Scientists argue that information and control may be shared between different regions of the cerebral cortex in a complex way, but many more experiments will be required to understand more fully the workings of this vital part of the brain.

BRAIN WAVES Because the brain contains so many neurons, each one capable of maintaining an action potential, it is a source of electrical activity. If voltage-sensitive electrodes are placed on the scalp, a weak electrical signal can be recorded.

When a recording of electrical activity is made at a number of places on the scalp, the result is a record called an electroencephalogram (EEG). Although EEGs show the average activities of thousands of neurons, they cannot provide the specifics about any one cell. The EEGs can, however, give a general idea of the activity of the brain.

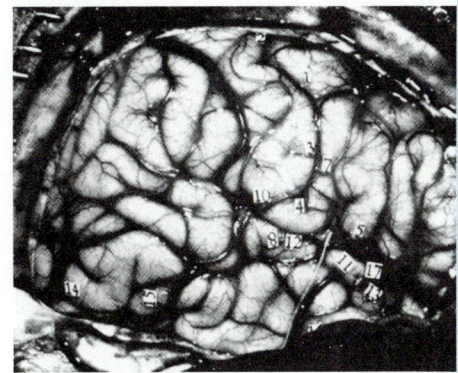

Figure 37-17 During one of his operations, Doctor Penfield used numbered tags to mark the different areas of the cerebral cortex that were being electrically stimulated. When the area marked 13 was stimulated, the person recalled a circus.

TEACHING SUPPORT

Study Guide
- Section 37–3, p. 354

Biotechnology Workbook
- Unlocking the Trapped Mind, p. 133

BACKGROUND INFORMATION
LEVELS OF CONSCIOUSNESS

Levels of consciousness range from the mindless drift of a coma to total alertness. In between are such states as sleeping, dozing, meditating, and daydreaming. All levels of consciousness are governed by the central nervous system. They can also be altered by psychoactive drugs.

Changes in a person's level of consciousness are controlled by the reticular formation in the brainstem. Messages routed from the reticular activating system, a group of neurons within the reticular formation, arouse the brain and maintain wakefulness. Damage to parts of the reticular formation can lead to unconsciousness and coma.

Present in the reticular formation are sleep centers. One sleep center contains neurons that release serotonin, a transmitter substance that inhibits reticular activating system neurons. This means that high serotonin levels are associated with drowsiness and sleep. Another sleep center is located in a part of the reticular formation that lies within the pons. This center has been linked with REM sleep. It releases substances that are believed to counteract serotonin and cause the brain to maintain a more active state.

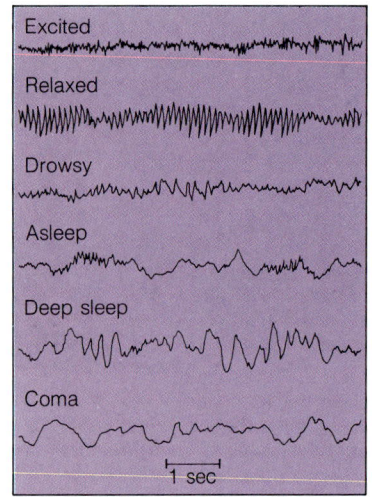

Figure 37–18 *The electrical activity of the brain is recorded as an electroencephalogram, or EEG. Notice the differences in the waves during excitement through coma.*

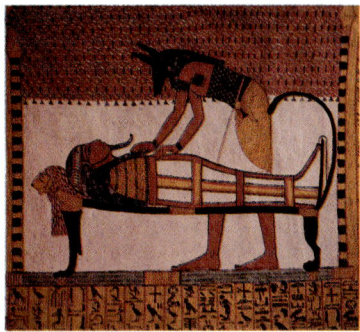

Figure 37–19 *Because the Egyptians thought the brain was an unimportant organ, they discarded it during the embalming process.*

Brain-wave recordings, as EEGs are sometimes called, are shown in Figure 37–18. As you can see, brain waves vary during sleep and consciousness.

SLEEP When the activity of the cerebral cortex falls to the lowest possible level, a person becomes unconscious. Forms of unconsciousness can range from a deep, unresponsive state to a light sleep. Sleep is a state of unconsciousness in which a person can be awakened by normal sensory stimulation, such as a gentle nudge.

As you read earlier in this chapter, the network of neurons in the brainstem, known as the reticular activating system, helps to control consciousness. When sleep begins, the level of activity in the reticular activating system drops off. A special group of neurons in the brainstem seems to activate light sleep, as when we first close our eyes and lose consciousness. During deep sleep, other groups of neurons cause a decline in heart rate, blood pressure, respiratory rate, and use of energy. During rapid eye movement (REM) sleep, active dreaming occurs.

MEMORY In a computer, information is coded in numerical form and stored as bits in a special memory region of the circuit system. We might wonder whether the same thing happens in the brain. Interestingly, the answer is both yes and no.

Scientists now believe that there are at least two different kinds of memory: short-term (primary) and long-term. Each kind is stored in the brain in a different way. Short-term memories contain small amounts of information, such as a person's name or a telephone number. When you memorize a list of spelling words just before a test, you are making use of short-term memory. Short-term memory, as its name implies, is not permanent—you can easily forget some of the details of your last class or a story you read yesterday. There is some evidence that short-term memory is stored as a pattern of nerve impulses in the cerebral cortex. Generally, short-term memories vanish within a few days, except for the interesting ones we make an effort to remember.

Long-term memories are more permanent memories. Some may last for a lifetime. Some may fade with time and require considerable effort to recall. And some seem to be part of a person's consciousness, such as one's name. Is there a special place in the brain where long-term memories are stored? Probably not. Many patients with severe brain injuries suffer no loss of long-term memory, even when parts of the cerebral cortex have been destroyed. Some scientists have proposed that long-term memories are stored in the structure of the brain itself, not in any one place at any one time. This is an unusual proposal, but one that makes the following fact all too obvious: The human brain has, as yet, failed to figure itself out completely.

37–3 (continued)

SECTION REVIEW 37–3

1. The cerebrum is the largest and most prominent part of the brain and is responsible for all of the voluntary activities of the body. The cerebellum, the second largest part of the brain, is located at the back of the skull. The cerebellum coordinates and balances the actions of the muscles. The brainstem connects the brain to the spinal cord and coordinates and integrates all incoming information and serves as the place for the entry and exit of the cranial nerves. The brainstem is divided into the medulla oblongata, the pons, and the midbrain.
2. The brain is the main switching unit of the central nervous system and the place to which impulses flow and from which impulses originate. The spinal cord provides the link between the brain and the rest of the body.
3. The cerebrum is more developed in humans than it is in any

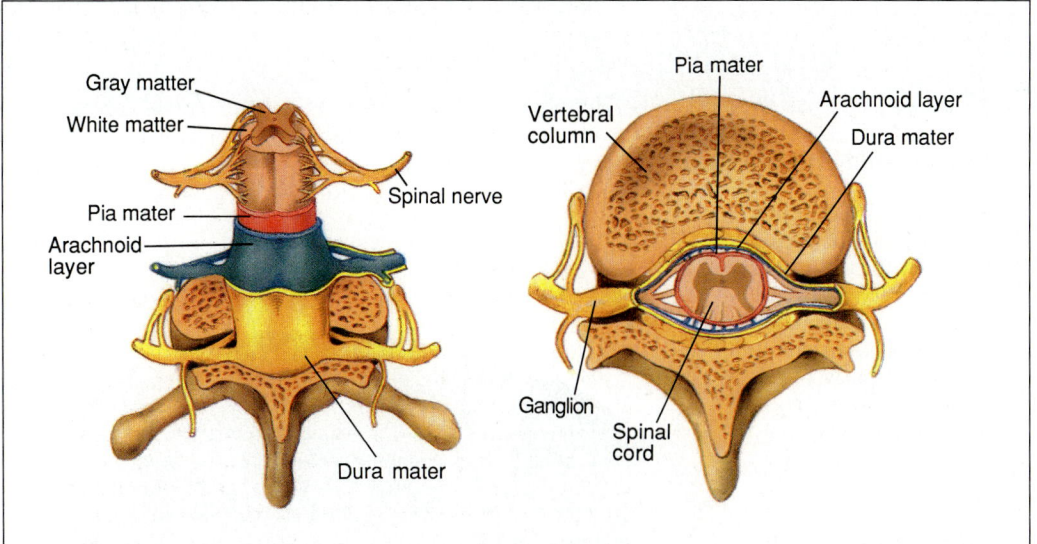

The Spinal Cord

The spinal cord acts as a communications link between the brain and the peripheral nervous system. In addition to carrying impulses to and from the brain, the spinal cord regulates **reflexes**. A reflex is the simplest response to a stimulus. Sneezing and blinking are two examples of reflexes. Thirty-one pairs of spinal nerves originate in the spinal cord and branch out to both sides of the body. These nerves carry messages to and from the spinal cord.

Figure 37-20 shows a cross section of the spinal cord. Notice that it consists of two types of nerve tissue. The central portion of the cord is H-shaped and is made up of gray matter. Gray matter, as you may recall, consists of nerve cell bodies and unmyelinated axons. The outer portion of the cord consists of white matter, which is made up of myelinated axons. Sensory neurons carry impulses from receptors to the spinal cord, and motor neurons carry impulses from the spinal cord to the effectors. Within the spinal cord, motor and sensory neurons are connected by interneurons.

Figure 37-20 The spinal cord, which provides the critical link between the brain and the rest of the body, is protected by the vertebral column and the meninges. The butterfly-shaped gray matter within the spinal cord is composed mainly of interneurons.

37-3 SECTION REVIEW

1. Describe the parts of the brain.
2. What are the functions of the brain and the spinal cord?
3. **Critical Thinking—Applying Concepts** Explain why the cerebrum is more developed in humans than it is in any other vertebrate.

BACKGROUND INFORMATION
EEG PATTERNS

The predominant wave pattern for someone who is relaxed with eyes closed is an alpha rhythm. In an alpha rhythm, EEG waves are recorded at a frequency of about 10 cycles per second. Alpha waves predominate when a person is meditating.

When a person moves from the alpha state to sleep, the wave cycles become larger, slower, and more erratic. This pattern, called slow-wave sleep, shows up in about 80% of the total sleeping time for adults. During slow-wave sleep, sensory input is low and the mind is like an engine that is idling. The person experiencing slow-wave sleep is usually not dreaming.

Slow-wave sleep is punctuated by brief periods of REM, or rapid eye movement sleep. The term *REM* refers to rapid eye movements in which the eyes jerk beneath closed lids. REM sleep is also characterized by irregular breathing, faster heartbeat, and twitching fingers. Most people experience vivid dreams during REM sleep.

When a person moves from a state of sleep or deep relaxation to a state of alertness, EEG patterns show a shift to lower amplitudes and higher frequencies. This transition, which is called EEG arousal, is characterized by increased blood flow and oxygen uptake in the cortex. EEG arousal occurs when conscious effort is made to focus on external stimuli or on one's own thoughts.

---

other vertebrate because it is the site of intelligence, learning, and judgment, which are uniquely human characteristics.

Reinforcement/Reteaching
If some students are having trouble answering the Section Review questions, go back to the material with which they are having difficulty.

Closure
Have students list the names, locations, and functions of each part of the brain. Have students work in pairs or small groups to make drawings or posters that illustrate the section material.

SCIENCE, TECHNOLOGY, AND SOCIETY
BREAKTHROUGH

Nitric Oxide and the Brain

Smog! Acid rain! You hear about these forms of air pollution almost every day. One substance that contributes to both smog and acid rain is nitric oxide (NO), a colorless, odorless, and very poisonous gas. Some nitric oxide is produced naturally. For instance, the crack of a bolt of lightning produces a bit of nitric oxide as the lightning sears through the oxygen and nitrogen of the atmosphere. But nitric oxide is of great concern to environmentalists because it is released in large, air-polluting quantities from the exhaust pipes of cars and other gasoline-powered vehicles. Nitric oxide is a very reactive compound—it is unstable and is toxic to many cells, even in small amounts. In fact, careful studies indicate that nitric oxide generally lasts no more than 6 to 10 seconds before it reacts with oxygen and water to produce more stable nitrogen compounds. (The nitric acid in acid rain is produced when nitric oxide reacts with water.) Because of its instability and toxicity, nitric oxide does not seem to be a very likely compound to play an important role in living organisms.

In the last few years, however, studies of the nervous system have revealed something remarkable. In 1982, Takeo Deguchi of Japan discovered that in order for brain cells to respond to a particular neurotransmitter, they required arginine, a nitrogen-rich amino acid. Some years later, researchers determined that arginine was being used by an enzyme to produce nitric oxide.

What is the nitric oxide doing? The best evidence is that it is acting as a neurotransmitter. In fact, some experiments seem to show that blocking nitric oxide production interferes with the development of long-term memory. Therefore, nitric oxide, a gas that can be toxic in large amounts, may establish the nerve connections in the brain that are responsible for memory and learning. Other studies, in fact, now suggest that nitric oxide may be just the first of a number of gases that are important in carrying messages from one nerve cell to the next.

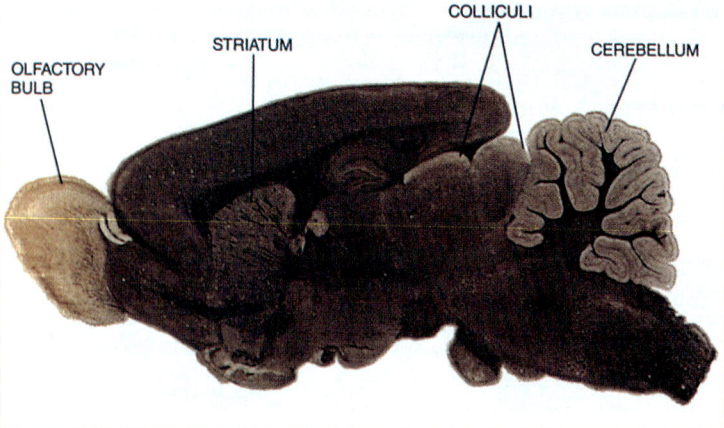

A chemical stain reveals the nitric oxide-producing regions of a rat brain as light-colored areas in this photograph. These areas include regions of the olfactory bulb and colliculi, which are involved with the senses, and the striatum and cerebellum, which are involved with the coordination of body movements.

37–4 The Peripheral Nervous System

The peripheral nervous system, the link between the central nervous system and the rest of the body, consists of the cranial and spinal nerves and ganglia. The peripheral nervous system can be divided into two divisions: the sensory division and the motor division. **The sensory division of the peripheral nervous system transmits impulses from sense organs—such as the ears and taste buds—to the central nervous system. The motor division transmits impulses from the central nervous system to the muscles or glands (effectors).** The motor division is further divided into the **somatic nervous system** and the **autonomic nervous system**.

The Somatic Nervous System

The somatic nervous system regulates activities that are under conscious control, such as the movement of the skeletal muscles. Every time you lift your finger or wiggle your toes, you are using the motor neurons of the somatic nervous system. However, many nerves within this system are parts of reflexes and as such can act automatically.

If you accidentally step on a tack with your bare foot, your leg may recoil before you are aware of the pain. This rapid reflex is possible because receptors in the skin stimulate the sensory neurons to carry the impulse to the spinal cord. Even before the information is relayed to your brain, a group of neurons in the spinal cord automatically activates motor neurons. These motor neurons cause the muscles (effectors) in your leg to contract, pulling your foot away. The receptor, sensory neuron, motor neuron, and effector that are involved in this quick response are together known as a **reflex arc**.

Guide For Reading
■ What is the function and structure of the peripheral nervous system?

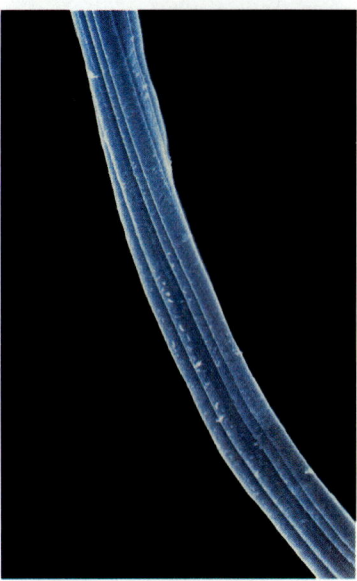

Figure 37–21 Axons of the peripheral nervous system form cables that bring information to and from the brain.

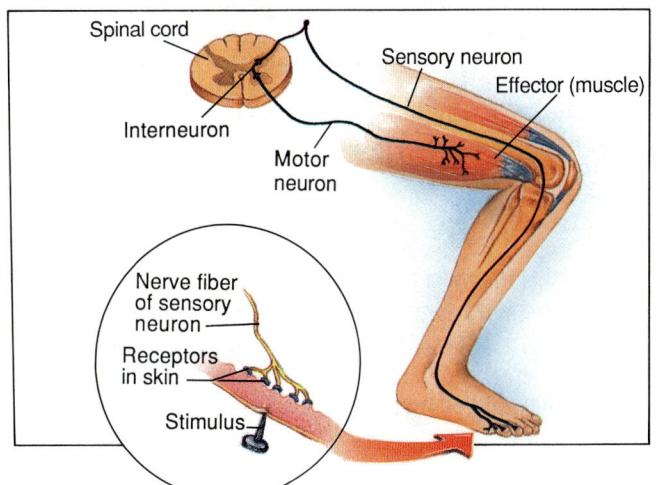

Figure 37–22 In a reflex arc, sensory receptors stimulate a sensory neuron, which relays the signal to an interneuron within the spinal cord. The signal is then sent to a motor neuron, which in turn stimulates an effector.

825

TEACHING SUPPORT

Study Guide

- Section 37–4, p. 356

MEDIA AND TECHNOLOGY

Biology Transparencies

- 58: The Structure of the Human Eye

BACKGROUND INFORMATION

CONTROLLING THE AUTONOMIC NERVOUS SYSTEM

The autonomic nervous system is usually outside a person's conscious control. But many people claim that certain techniques, such as biofeedback, let them control their heart rate, blood pressure, and other autonomic functions. Have interested students research biofeedback and other techniques of controlling autonomic response.

Figure 37–23 *The autonomic nervous system consists of the sympathetic and the parasympathetic nervous systems, which usually have complementary functions. In general, the sympathetic nervous system is involved in fight-or-flight reactions and the parasympathetic nervous system stimulates calmer functions, such as digestion.*

Part of Body	Effect of Sympathetic Nervous System	Effect of Parasympathetic Nervous System
Pupil of eye	Dilated	Constricted
Liver	Glucose released	None
Urinary bladder muscle	Relaxed	Contracted
Muscle of heart	Increased rate and force	Slowed rate
Bronchi of lungs	Dilated	Constricted

The Autonomic Nervous System

The autonomic nervous system regulates activities that are automatic, or involuntary. The nerves of the autonomic nervous system control functions of the body that are not under conscious control, such as the contractions in the heart and the movement of smooth muscles surrounding the blood vessels and the digestive system.

The autonomic nervous system is further subdivided into two parts that have opposite effects on the organs they control. The two parts are known as the **sympathetic nervous system** and the **parasympathetic nervous system**. Most organs controlled by the autonomic nervous system are under the control of both sympathetic and parasympathetic neurons. For example, heart rate is speeded up by the sympathetic nervous system, whereas it is slowed down by the parasympathetic nervous system.

Why is it important to have two systems that have opposite effects on the organs they control? Wouldn't one system make more sense? To answer this question, imagine a situation in which a single system could increase the rate of activity of an organ by releasing a single type of neurotransmitter. This system would be much like a car that had an accelerator pedal but no brake pedal. Getting the car to move would be easy, but stopping it would present a problem! The same would be true of an organ regulated by a single nervous system. The dual-control system puts an on-off switch on every organ and thus ensures precise control of a dozen organs or more.

37–4 SECTION REVIEW

1. What is the role of the peripheral nervous system?
2. What are the parts of the peripheral nervous system?
3. **Critical Thinking—Relating Concepts** Is a reflex arc completely within the peripheral nervous system? Explain your answer.

37–4 (continued)

SECTION REVIEW 37–4

1. The peripheral nervous system carries impulses to and from the central nervous system.
2. The peripheral nervous system consists of the sensory division and the motor division.
3. The reflex arc consists of a receptor, sensory neuron, motor neuron, and effector. The reflex arc is not completely within the peripheral nervous system because the spinal cord is involved in this action. Sensory neurons carry the impulses from receptors to the spinal cord, and motor neurons carry the impulses from the spinal cord to muscles or glands.

Reinforcement/Reteaching

If students are having trouble answering the Section Review questions, go back to the part of the section that contains the material with which they are having difficulty.

Closure

Divide the class into 6 groups. On separate index cards write each of the following terms: Sensory Division, Motor Division, Somatic Nervous System, Autonomic Nervous System, Sympathetic Nervous System, Parasympathetic

37-5 The Senses

There are millions of neurons in the body that do not receive impulses from other neurons. Instead, these neurons, which are called sensory receptors, react directly to stimulation from the environment. Examples of stimulation include light, sound, motion, chemicals, pressure, or changes in temperature. Once these sensory receptors are stimulated, they transform one form of energy from the environment (light, sound) into another form of energy (action potential) that can be transmitted to other neurons. Eventually these action potentials (impulses) reach the central nervous system.

The sensory receptors are contained in the sense organs. **Each of the five senses (sight, hearing, smell, taste, and touch) has a specific sense organ associated with it.** Specialized cells within each **sense organ** enable it to respond to particular stimuli.

Guide For Reading

- What are the major parts of the eye? What is the function of each part?
- What parts of the ear are responsible for hearing and balance?
- How do the senses of smell and taste compare?
- What are the various sense receptors found in the skin?

Vision

The world around us is bathed in light. The sense organs that we use to sense light are the eyes. Each eye is composed of three layers. The outer layer consists of the sclera and the cornea. The middle layer contains the choroid, ciliary body, and iris. The inner layer consists of the retina.

The sclera, or white of the eye, consists of tough white connective tissue. The sclera helps maintain the shape of the eye and also provides a means of attachment for the muscles that move the eye. In the front of the eye, the sclera forms a transparent layer called the cornea. The cornea is the part of the eye through which light enters.

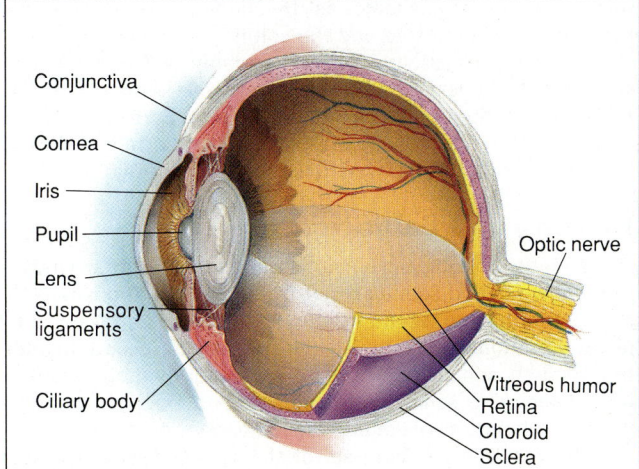
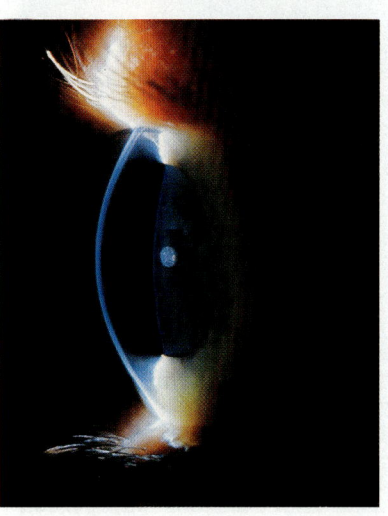

Figure 37–24 The eye is a complicated sense organ composed of three layers of tissue: sclera, choroid, and retina (left). As light enters the eye, it passes through the thin transparent cornea (right).

MULTICULTURAL STRATEGY

Why do some people like pizza and others like chop suey? Why do some people prefer tacos whereas some love mincemeat pies? How is food preference developed? All humans possess the same basic taste organs, but their preferences for foods have an environmental (and hence a cultural) origin. Discuss with students some of their food likes and dislikes. Ask them to consider whether there are some foods that they once hated that they now like.

Nervous System. Have each group choose a card at random. Then ask each group to lead the class in a review discussion about the part of the peripheral nervous system that they have chosen.

TEACHING STRATEGY 37–5

Focus/Motivation

Ask students to imagine that they are at a picnic on a beautiful summer day. Their picnic baskets are full of their favorite foods. Have students write a paragraph or two describing the scene that they are imagining. Their descriptions should include their surroundings, the weather, and the foods they are eating.

When students are finished writing, have them use different colored pencils to underline the things in their description that pertain to sight, hearing, touch, smell, and taste.

TEACHING SUPPORT

Teaching Resources
- Hands-On Activity: Colored "Sandwiches," p. 165
- Hands-On Activity: More Than Meets the Eye, p. 169
- Hands-On Activity: Some Sound Reasons, p. 171

BACKGROUND INFORMATION
NEARSIGHTEDNESS AND FARSIGHTEDNESS

When light strikes the retina, the light image is normally focused on the fovea. Sometimes, however, the lens of the eye focuses light in front of, rather than on, the fovea. When this happens, nearsightedness results. A nearsighted person finds it difficult to focus on distant objects.

Farsightedness results when the lens focuses an image behind the fovea. Farsighted people have trouble focusing on objects that are close up. (This is the reason farsighted people often wear reading glasses.) Because nearsightedness and farsightedness are both caused by defects in the lens of the eye, corrective lenses (glasses or contact lenses) remedy both problems.

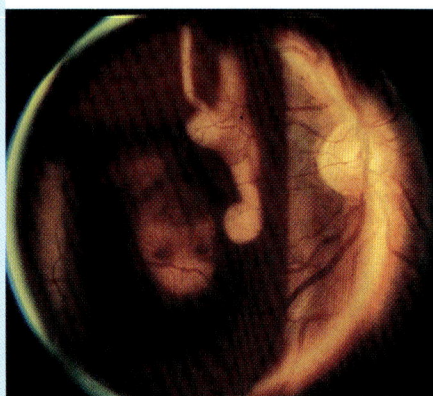

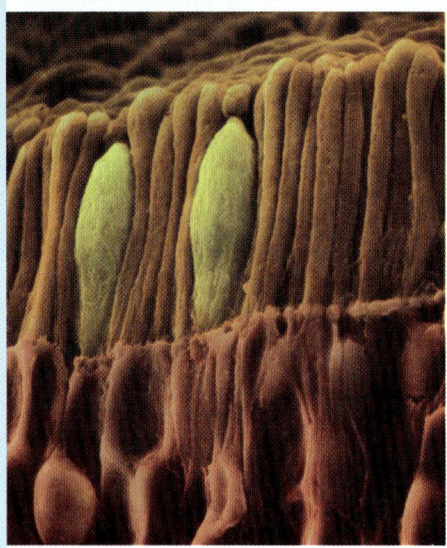

Figure 37–25 The retina, which is the innermost layer of the eye, contains the photoreceptor cells called rods and cones (bottom). Notice the upside-down image that was superimposed on a photograph of the retina taken through an ophthalmoscope (top).

Just inside the cornea is a small chamber (anterior chamber) filled with a fluid known as the aqueous humor. (The word humor means any fluid within the body.) At the back of this chamber, the pigmented choroid, which contains the blood vessels of the eye, becomes the disklike structure called the iris. The iris is the portion of the eye that gives your eye its color. In the middle of the iris is a small opening called the pupil, through which light enters the eye. The pupil appears as the small black disk in the center of the eye. Tiny muscles in the iris regulate the size of the pupil and thus the amount of light that enters. In dim light the pupil opens to increase the amount of light entering the eye. In bright light the pupil closes to decrease the amount of light entering the eye, thus preventing damage to the delicate structures within the eye.

Just behind the iris is the lens. The cells that form the lens contain a special protein called crystalin. Crystalin is almost transparent and thus allows light to pass through. Small muscles attached to the lens cause it to bend slightly. This bending enables the normal eye to focus on close and distant objects. Behind the lens is a large chamber (vitreal chamber) filled with a transparent jellylike fluid called vitreous humor.

Special light-sensitive receptor cells, or photoreceptors, are arranged in a layer in the retina, which is located at the back of the eye. The photoreceptors convert light energy into impulses that are carried to the central nervous system. There are two types of photoreceptors: rods and cones.

Photoreceptors contain a pigment called rhodopsin (also called visual purple) that can respond to most wavelengths of light. Rods are extremely sensitive to all colors of light, but they do not distinguish different colors. Cones are less sensitive than rods, but they do respond differently to light of different colors, producing color vision. In very dim light, when only rods are activated, objects may be clearly seen but their colors may not be distinguishable. As the amount of light increases, the cones are stimulated and the colors become clear.

The impulses assembled by this complicated layer of interconnected cells leave each eye by way of an optic nerve. The optic nerves then carry the impulses to a part of the brain known as the optic lobe. Here the brain interprets the visual images and provides information about the external world.

Hearing and Balance

Sound is nothing more than vibrations in the air around us. Deep, low-pitched sounds result from slow vibrations—100 to 500 vibrations per second. High-pitched sounds are caused by faster vibrations—1000 to 5000 vibrations per second. In addition to pitch, sounds differ from one another by their loudness, or volume. The sense organs that can distinguish both the pitch and loudness of sounds are the ears.

The external ear consists of the visible fleshy part that helps to collect sounds and funnel them into the auditory

---

37–5 (continued)

Content Development
Use the Motivation exercise to illustrate how nearly everything we perceive in the world around us depends on the five senses—sight, hearing, taste, smell, and touch. Emphasize that each of these senses depends on both specialized nerves and unique sensory apparatus. Each sense will be discussed in detail in this section.

Reinforcement/Reteaching
Review with students the structures that make up the outer, middle, and inner layers of the eye.
- **Which layer contains the white of the eye?** (The outer layer.)
- **What is the white of the eye called?** (The sclera.)
- **What other structure is found in the outer layer of the eye?** (The cornea.)
- **What is the opening called through which light enters the eye?** (The pupil.)
- **In what structure and in which**

PROBLEM SOLVING IN BIOLOGY

THE "EYES" HAVE IT!

The retina contains two types of photoreceptors: rods and cones. Rods are responsible for black-and-white vision; cones are responsible for color vision. In humans, there are three types of cones: blue cones, red cones, and green cones. Each type of cone is sensitive to a specific portion of the visible spectrum. The combined stimulation of these cones produces all the colors you see.

As the accompanying graph shows, the sensitivities of blue, red, and green cones overlap to some degree. Because their sensitivities overlap, we are able to distinguish all colors. For example, red light stimulates red cones and leaves the other cones unaffected. Thus we see the color red. Green light stimulates all three cones, but each to a certain degree. Depending upon the combinations, we see many other colors.

To determine the type of cone(s) stimulated and the percentage of stimulation produced by a certain light, you can analyze the graph. For example, if orange light were to strike the retina, about 99 percent of the red cones and about 40 percent of the green cones would be stimulated. No blue cones would be stimulated.

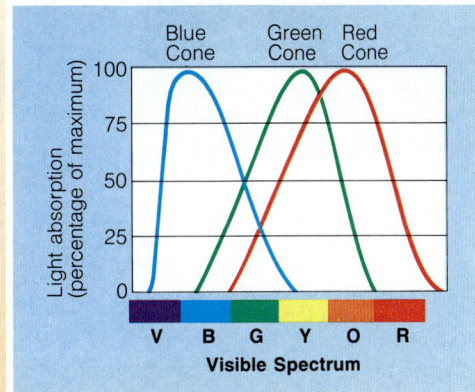

Now try your hand at a few problems.
1. In order to see the color green, what cone(s) must be stimulated?
2. What type of cone(s) and approximately what percentage of the cone(s) is (are) stimulated when blue light strikes the retina?
3. What cone(s) must be stimulated in order to see the color yellow?
4. What type of cone(s) and approximately what percentage of the cone(s) is (are) stimulated when red light strikes the retina?

PROBLEM SOLVING
The "Eyes" Have It!
Rod cells and cone cells are so named because of their shapes. Rods are typically abundant in the periphery of the retina, whereas cones are concentrated near the center of the retina. Cones are densely packed in the fovea, a funnel-shaped depression in the central part of the retina where nerve tissue is thinner. The fovea is about 1 millimeter in diameter.

Answers to Problems
1. Red, green, and blue.
2. Blue cones—about 90% and green cones—about 10%
3. Green, red, and sometimes blue.
4. Red cones—about 50%.

BACKGROUND INFORMATION
RHODOPSIN

Rhodopsin molecules consist of a protein, *opsin*, with an attached side group called *cis-retinal*. When this side group absorbs light energy, it is temporarily converted to a slightly different form called *trans-retinal*. In this altered form, the side group breaks away from the protein. The breakdown of rhodopsin leads to a change in the voltage difference across the membrane of the photoreceptor (rod or cone) in which the protein is embedded. This change signals the presence of light to neighboring neurons, which then relay signals to ganglion cells. The axons of these cells converge to form the optic nerve.

---

canal. The auditory canal contains tiny hairs and wax-producing glands that prevent foreign objects from entering the ear.

The auditory canal extends into the bones of the head but stops at the eardrum, or tympanum. The eardrum is the beginning of the middle ear. Sound vibrations strike the eardrum and are then transmitted through three tiny bones: the malleus (hammer), incus (anvil), and stapes (stirrup).

The stapes vibrates against a thin membrane covering an opening called the oval window. This membrane transmits the vibrations to the cochlea, which begins the inner ear. The cochlea is a snail-shaped fluid-filled cavity. When the fluid vibrates, tiny hair cells lining the cochlea are pushed back and forth, providing stimulation that is turned into nerve impulses. These impulses are carried to the brain by the acoustic nerve.

---

layer of the eye is the pupil found? (In the iris, which is in the middle layer of the eye.)
• What structure is located just behind the iris? (The lens.) What does this structure do? (Focuses on close and distant objects.)
• What structure makes up the inner layer of the eye? (The retina.)
• What is the function of the retina? (To convert light energy into impulses, which are then carried to the brain.)

Have students make simple diagrams to show the passage of light as it enters the eye through the pupil in the iris, passes through the lens, and is focused onto the retina. Also have them show how on the retina, light energy is converted into impulses that are conducted to the brain via the optic nerve.

TEACHING SUPPORT

Laboratory Manual
- Skin Sensitivity, p. 467

Teaching Resources
- Vocabulary Review: Word Game, p. 1
- Hands-On Activity: A Gentle Touch, p. 161

Study Guide
- Section 37–5, p. 358

Product Testing Activities
- Testing Bubble Gum
- Testing Cereals
- Testing Yogurts

MEDIA AND TECHNOLOGY

Biology Transparencies
- 59: The Structure of the Human Ear

37–5 (continued)

Skills Development
Guided Practice
Skill: Comparing

Have students compare the structure and function of the ear with the structure and function of the eye.

- How many layers or parts does the eye have? What are they? (Three: outer layer, middle layer, and inner layer.)
- How many basic sections or parts does the ear have? What are they? (Three: outer ear, middle ear, and inner ear.)
- How does the function of the outer layer of the eye compare with the function of the outer ear? (Both collect stimuli from the environment: The cornea of the outer eye allows light to enter the eye; the visible external part of the ear collects sounds and funnels them into the auditory canal.)
- How is the lens of the eye similar to the eardrum of the ear? (The lens is located in the middle layer of the eye and the eardrum is located in the middle ear. The lens transmits and focuses light in the eye; the eardrum transmits sound waves in the ear.)
- Compare the function of the inner layer of the eye with the function of the inner ear. (The inner layer of the eye, which is the retina, converts light into impulses that travel to the brain to produce vision; the cochlea of the inner ear converts sound waves into impulses that travel to the brain to produce hearing.)

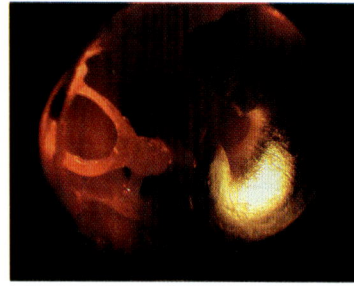

Figure 37–26 Sound waves enter the ear and are converted into impulses that are carried to the brain. The photograph of the middle ear shows the eardrum, which is tinted yellow, and the three small ear bones.

Figure 37–27 The semicircular canals (top) are set at right angles to one another so that they react to up-and-down, side-to-side, and tilting motions. Tiny crystals called otoliths (bottom) also play a part in maintaining balance.

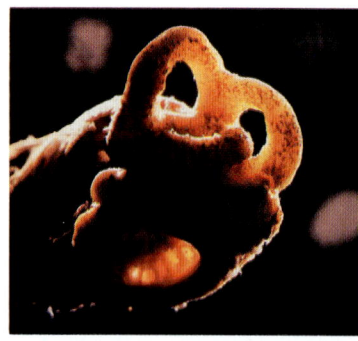

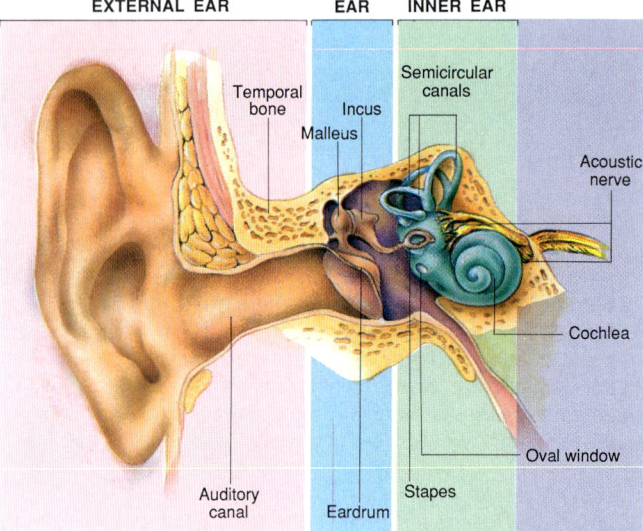

In addition to enabling us to hear, the ears contain structures for detecting stimuli that make us aware of our movements and allow us to maintain our balance. Located within the inner ear just above the cochlea are three tiny canals that lie at right angles to each other. They are called the semicircular canals because each makes half a circle. The semicircular canals and the two tiny sacs located behind them help us sense balance, or equilibrium.

The semicircular canals and the sacs are both filled with fluid and lined with hair cells (ciliated cells). The hair cells of each sac, however, are embedded in a gelatinlike substance that contains tiny grains of calcium carbonate and protein called otoliths (hearing stones). Otoliths roll back and forth in response to gravity, acceleration, and deceleration. Together, the movement of fluid and the otoliths bend the hair on the hair cells. This action, in turn, sends impulses to the brain that enable it to determine body motion and position.

Smell

Because the sense of smell is a chemical sense, the cells that are responsible for smell are called chemoreceptors. These cells are located in the upper part of the nasal (pertaining to the nose) cavity. See Figure 37–28. These chemoreceptors contain cilia that extend into the air passageways of the nose and react to chemicals in the air. Chemicals that come into contact with these chemoreceptors stimulate them, causing impulses to be sent to the brain.

Unfortunately, relatively little is known about the sense of smell. Although tens of thousands of different odors can be distinguished, it is not understood how one odor is distinguished from another. A challenge for biologists who study the sensory systems is to determine the basis of scent discrimination.

SECTION REVIEW 37–5

1. The sclera maintains the shape of the eye and attaches the muscles that move the eye. The cornea is the part through which light enters the eye. The choroid contains the blood vessels. The iris is the colored portion of the eye and contains a small opening in the middle called the pupil through which light enters the eye. Behind the

Taste

Like the sense of smell, the sense of taste is a chemical sense. And the cells that are stimulated by the chemicals are also called chemoreceptors. The sense organs that detect taste are the taste buds. Most taste buds are located between small projections on the tongue. However, taste buds are also found on the roof of the mouth and on the lips and throat (especially in children).

The tastes detected by taste buds are of four main kinds: sweet, salty, sour, and bitter. Each taste bud shows a particular sensitivity to one of these tastes.

Because many of the sensations associated with taste are actually smell sensations, humans depend upon both senses to detect flavors in food. Perhaps you are already aware of this fact. When you have a cold and your smell receptors are covered by mucus, food seems to have little, if any, flavor.

Touch and Related Senses

The sense of touch, unlike the other senses you have just read about, is not found in one particular place. All regions of the skin are sensitive to touch. In this respect, your largest sense organ is your skin.

There are several distinct types of sensory receptors that are present just below the surface of the skin. Two types of sensory receptors respond to heat and cold; two other types respond to touch; one type responds to pain. See Figure 37–29.

Sensory receptors for heat and cold are scattered directly below the surface of the skin. In general, there are three to four warm receptors for every cold receptor. Sensory receptors for touch are much more concentrated in some areas of the body than in others. For example, the most touch-sensitive areas are located on the fingers, toes, and lips. Pain receptors are located throughout the skin. Depending upon what type of sensory neurons are stimulated, the sensation of pain can be experienced as either prickling pain (fast pain) or burning and aching pain (slow pain).

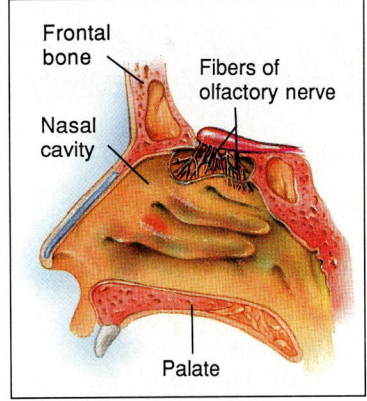

Figure 37–28 A tiny patch of specialized tissue located over the top of each nasal cavity is responsible for the sense of smell.

Figure 37–29 The skin contains many types of sensory receptors that provide information about the external environment.

1. Receptors for touch and pressure
2. Hair
3. Receptor for touch or pressure
4. Receptor for deep pressure
5. Receptors for light touch
6. Dermis
7. Free nerve endings
8. Epidermis

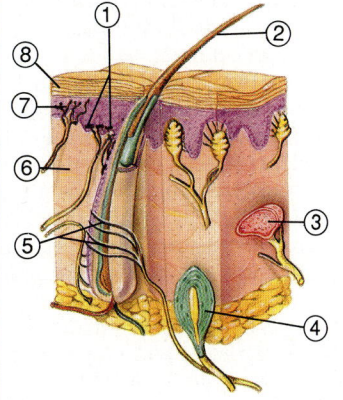

37–5 SECTION REVIEW

1. Identify the parts of the eye and the function of each.
2. What ear parts are responsible for hearing? For balance?
3. Compare the senses of smell and taste.
4. Describe the location and function of sensory receptors in the skin.
5. **Connection—You and Your World** Explain why it is sometimes helpful to inhale slowly and deeply through the nose when trying to identify an odor.

---

iris is the lens, which focuses the light onto the retina. The retina contains the photoreceptors, which convert light into impulses.

2. The parts for hearing are the eardrum, malleus, incus, stapes, and cochlea. The parts for balance are the semicircular canals.

3. Both senses rely on chemoreceptors. The chemoreceptors for smell are located in the upper part of the nasal cavity. The chemoreceptors for taste are the taste buds, most of which are located on the tongue.

4. Sensory receptors for heat and cold are scattered directly below the surface of the skin. Sensory receptors for touch are found throughout the skin but are concentrated on the fingers, toes, and lips. Pain receptors are located throughout the skin.

ESL STRATEGY

Provide students with worksheets illustrating the parts of the eye and ear. Using different-colored pencils, have them color and label the different parts of each sense organ. Ask students to exchange their worksheets and check for accuracy.

5. Inhaling slowly and deeply draws a large amount of air into the nasal passages, whereas not as much air enters during normal breaths. Sniffing, or the rapid, repeated intake of air, is effective for the same reason.

Reinforcement/Reteaching

If students are having trouble answering the Section Review questions, go back to the material with which they are having difficulty.

Closure

List on the chalkboard various stimuli that affect each of the five senses. Some possible examples include looking at a sunset, listening to a symphony, smelling bread baking, tasting Mexican food, touching a hot stove, pricking your finger with a pin. Have students take turns describing in detail how each stimulus produces a sensation.

LABORATORY INVESTIGATION

MEASURING REACTION TIME

Before the Lab
1. At least one day prior to the investigation, make sure that you have enough metersticks for your class.

Pre-Lab Discussion
- **What is reaction time?** (Answers may vary. In this investigation, reaction time refers to how long it takes for a person to respond to a stimulus.)
- **What factors do you think affect a person's reaction time?** (Answers may vary. Possible answers include age; degree of alertness; ability to concentrate; eye-hand coordination; type of stimulus.)

Skills Development
Students will use the following skills while completing this laboratory investigation.
1. Observing
2. Measuring
3. Recording
4. Analyzing
5. Inferring
6. Relating
7. Predicting
8. Comparing

Teaching Strategy
During the investigation, circulate throughout the classroom and help any groups that are having trouble.

Observations
1. In most cases, distances will vary with distractions. Distraction distances should be longer than those without distractions.
2. Answers will vary, but most students should note a longer distance when distracted.
3. Answers will vary.

Analysis and Conclusions
1. It gave a relative measure of the amount of time it took the person to react to a particular

---

LABORATORY INVESTIGATION: MEASURING REACTION TIME

PROBLEM
How can reaction time be measured? What factors affect reaction time?

MATERIALS (per pair of students)

meterstick
biology textbook

PROCEDURE
1. On a separate sheet of paper, construct a data table similar to the one shown here.
2. While you hold the 100-centimeter end of the meterstick, have your partner position his or her thumb and forefinger around the zero end. Make sure that your partner does not touch the meterstick.
3. Instruct your partner to concentrate on catching the meterstick as soon as you drop it. Your partner should move only the thumb and forefinger to catch the meterstick.
4. When you are ready, drop the meterstick between your partner's fingers.
5. Note the measurement on the meterstick at the point at which your partner's fingers caught the meterstick. Record this distance in centimeters in the data table.
6. Repeat steps 1 through 5 four more times.
7. Switch roles with your partner and repeat steps 1 through 6.
8. Now work with another pair of students to determine whether reaction rate is affected by distractions. Have the other pair of students select a chapter from your biology textbook that your class has already studied.
9. Repeat steps 1 through 6, but this time have the distractors ask the student who is trying to catch the meterstick questions from the Chapter Review section at the end of the selected chapter.
10. Switch roles and repeat step 9 until all four students have caught the meterstick.

832

OBSERVATIONS

Trial	Distance Ruler Falls (cm)	
	Without Distractions	With Distractions
1		
2		
3		
4		
5		
Average		

1. Did your distances vary in the trials without distractions? With distractions? Describe the variations in each case, if any.
2. How did your average distance without distractions compare with your average distance with distractions?
3. How did your distances without and with distractions compare with those of your partner? With those of the other pair of students?

ANALYSIS AND CONCLUSIONS
1. Why did measuring the distance the meterstick falls give a relative measure of reaction time?
2. Why was it necessary to have each person perform five trials?
3. What might account for variations in your reaction time without distractions? With distractions?
4. How did the distractions affect your reaction times?
5. Explain why your reaction times might be different from the reaction times of your partner and the other pair of students.

---

controlled situation. It did not give an absolute measure of reaction time in general but only of reaction time to catch the meterstick. As such, the distances are relative since they can be compared to those of other students.

2. To obtain an average reaction time. During a single trial, hidden distractions or other variables could come into play.
3. For both cases, variations are to be expected since many factors come into play during the experiment, and it is not to be expected that the timing with or without distractions will be the same for each trial.
4. For most students, distractions should slow down their reaction time.
5. Each student is an individual with his or her own particular

832

STUDENT STUDY GUIDE

SUMMARIZING THE CONCEPTS

The key concepts in each section of this chapter are listed below to help you review the chapter content. Make sure you understand each concept and its relationship to other concepts and to the theme of this chapter.

37-1 The Nervous System
- The nervous system receives and relays information about activities within the body and monitors and responds to internal and external changes.
- Neurons transmit messages in the form of electrical signals known as impulses. A neuron consists of a cell body, dendrites, an axon, and axon terminals.

37-2 Divisions of the Nervous System
- The human nervous system is divided into two major divisions: the central nervous system and the peripheral nervous system.

37-3 The Central Nervous System
- The brain is the main switching unit of the central nervous system. The spinal cord provides the link between the brain and the rest of the body.

37-4 The Peripheral Nervous System
- The peripheral nervous system is made up of the sensory division and the motor division.

37-5 The Senses
- Each of the five senses (sight, hearing, smell, taste, and touch) has a specific sense organ associated with it.

REVIEWING KEY TERMS

Vocabulary terms are important to your understanding of biology. The key terms listed below are those you should be especially familiar with. Review these terms and their meanings. Then use each term in a complete sentence. If you are not sure of a term's meaning, return to the appropriate section and review its definition.

37-1 The Nervous System
nervous system
neuron
impulse
sensory neuron
motor neuron
interneuron
cell body
dendrite
axon
axon terminal
resting potential
action potential
myelin
threshold
receptor
effector
synapse
neurotransmitter

37-2 Divisions of the Nervous System
central nervous system
peripheral nervous system

37-3 The Central Nervous System
cerebrum
cerebral cortex
cerebral medulla
cerebellum
brainstem
medulla oblongata
pons
midbrain
thalamus
hypothalamus
reflex

37-4 The Peripheral Nervous System
somatic nervous system
autonomic nervous system
reflex arc
sympathetic nervous system
parasympathetic nervous system

37-5 The Senses
sense organ

CHAPTER REVIEW

CONTENT REVIEW

Multiple Choice
1. a 2. b 3. c 4. c
5. b 6. d 7. b 8. a

True or False
1. T
2. F action
3. F threshold
4. F neurotransmitters
5. F peripheral nervous system
6. F cerebrum
7. T
8. T

Word Relationships
1. Dendrite, axon, and cell body are all parts of a neuron; the synapse is the gap between adjacent neurons
2. Ciliary body, iris, and choroid make up middle layer of the eye; sclera makes up the outer layer of the eye
3. Cranial nerves, spinal nerves, and ganglia make up the peripheral nervous system; the cerebrum is part of the brain
4. Malleus, stapes, and incus are the tiny bones in the ear; a ganglion is a collection of nerve cell bodies
5. motor neuron
6. optic nerve
7. vitreous humor
8. brainstem or medulla

CONCEPT MASTERY

1. The cell body contains the nucleus and is the site of most of the metabolic activity of the cell. Dendrites are short, branched extensions that carry impulses to the cell body. Axons are long fibers that carry impulses away from the cell body. Axon terminals are swellings located at the end of axons.
2. Receptors are special sensory neurons in sense organs that receive stimuli from the external environment. Effectors are muscles or glands that bring about a coordinated response.

---

coordination, skills, and reaction time.

Going Further: Enrichment
Challenge students to design an investigation to answer the question: Can reaction time be improved by practice?

3. During the resting potential, positively charged ions leak out of the cell. This causes the neuron to become negatively charged on the inside of the cell membrane and positively charged on the outside. During an action potential, the polarity of the cell membrane is reversed. The inside of the cell membrane is positively charged and the outside is negatively charged.
4. The sodium–potassium pump maintains the electrical potential (difference in electrical charge) across the nerve-cell membrane. It does this by moving sodium ions out of the cell and actively pumping potassium ions into the cell.
5. When a neuron is stimulated, the sodium gates at the leading edge of an impulse open, allowing positively charged sodium ions to flow inside the membrane. The inside of the membrane temporarily becomes more positive than the outside. The membrane is now said to be depolarized. As the impulse passes, the potassium gates open, allowing positively charged potassium ions to flow out. The membrane is now said to be depolarized, or negatively charged on the inside and positively charged on the outside.
6. In order for a neuron to conduct an impulse, a threshold, or minimum level of a stimulus, must be reached to activate the neuron. Any stimulus that is weaker than the threshold will produce no impulse; any stimulus that is stronger than the threshold will produce only one impulse. This is known as the all-or-none principle.
7. The skull protects the brain, and the vertebrae protect the spinal cord. Meninges protect both the brain and the spinal cord. Cerebrospinal fluid protects the nervous tissue against mechanical shock.
8. The cerebrum is divided into the left and right hemispheres.

CHAPTER REVIEW

CONTENT REVIEW

Multiple Choice

Choose the letter of the answer that best completes each statement.

1. Which type of neuron is responsible for transmitting impulses to the central nervous system?
   a. sensory neuron   c. interneuron
   b. motor neuron   d. receptor neuron
2. The central nervous system consists of the
   a. somatic system and the autonomic system.
   b. brain and the spinal cord.
   c. cerebrum, cerebellum, and medulla.
   d. spinal cord and the peripheral nerves.
3. Coordination and balance are controlled principally in the
   a. spinal cord.   c. cerebellum.
   b. medulla.   d. cerebrum.
4. The autonomic nervous system controls
   a. thinking.   c. digestion.
   b. walking.   d. hearing.
5. Which is not usually involved in a simple reflex?
   a. receptor   c. spinal cord
   b. cerebrum   d. effector
6. The outer layer of the eye consists of the
   a. choroid, ciliary body, and iris.
   b. cochlea.
   c. retina.
   d. sclera and the cornea.
7. The malleus, incus, and stapes conduct vibrations from the eardrum to the
   a. outer ear.   c. brain.
   b. inner ear.   d. middle ear.
8. The largest sense organ is the
   a. skin.   c. eyes.
   b. ears.   d. nose.

True or False

Determine whether each statement is true or false. If it is true, write "true." If it is false, change the underlined word or words to make the statement true.

1. <u>Impulses</u> are a flow of electrical charges along the cell membrane of a neuron.
2. A nerve impulse is the <u>resting</u> potential traveling along the membrane.
3. The minimum level of stimulus that is required to activate a neuron is called the <u>conductor</u>.
4. Axon terminals contain tiny vesicles that are filled with molecules of <u>myelin</u>.
5. The central nervous system and the <u>sympathetic nervous system</u> are the two main divisions of the human nervous system.
6. The largest and most prominent part of the brain is the <u>cerebellum</u>.
7. The <u>brainstem</u> connects the spinal cord to the brain.
8. The <u>iris</u> is the colored portion of the eye.

Word Relationships

A. *In each of the following sets of terms, three of the terms are related. One term does not belong. Determine the characteristic common to three of the terms and then identify the term that does not belong.*

1. dendrite, synapse, axon, cell body
2. sclera, ciliary body, iris, choroid
3. cerebrum, cranial nerves, spinal nerves, ganglia
4. malleus, ganglion, stapes, incus

834

The cerebral cortex, or outer surface, has numerous folds and consists of gray matter. Gray matter is made up of densely packed nerve cell bodies. The cerebral cortex performs sensory and motor functions. The inner surface is called the cerebral medulla and consists of white matter. White matter is made up of bundles of myelinated axons.
9. The spinal cord consists of an H-shaped central portion made up of gray matter and an outer portion of white matter.
10. Light entering the eye passes through the cornea, pupil, aqueous humor, lens, and vitreous humor and forms an image on the retina at the back of the eye.
11. The rods and cones are photoreceptors. Rods are very

B. An analogy is a relationship between two pairs of words or phrases generally written in the following manner: a:b::c:d. The symbol : is read "is to," and the symbol :: is read "as." For example, cat:animal::rose:plant is read "cat is to animal as rose is to plant."

In the analogies that follow, a word or phrase is missing. Complete each analogy by providing the missing word or phrase.

5. receptor:sensory neuron::effector:_____
6. ear:acoustic nerve::eye:_____
7. anterior chamber:aqueous humor::vitreal chamber:_____
8. intelligence:cerebrum::breathing:_____

CONCEPT MASTERY

Use your understanding of the concepts developed in the chapter to answer each of the following in a brief paragraph.

1. Describe the structure and function of a neuron.
2. What is the relationship between receptors and effectors?
3. What changes occur in the neuron during the resting potential? During an action potential?
4. What is the function of the sodium-potassium pump?
5. Explain what happens when a nerve is stimulated.
6. How does the all-or-none principle relate to the transmission of a nerve impulse?
7. How is the central nervous system protected?
8. Describe the structure and function of the cerebrum.
9. Describe the structure of the spinal cord.
10. Trace the path of light through the eye.
11. What are the functions of the rods and cones?
12. Trace the path of sound through the ear.

CRITICAL AND CREATIVE THINKING

Discuss each of the following in a brief paragraph.

1. **Making predictions** Suppose a portion of an axon is cut so that it is no longer connected to its nerve cell body. Predict the effect this would have on the transmission of impulses.
2. **Applying concepts** What might happen if the cornea becomes inflamed so that fluid accumulates there?
3. **Relating concepts** Constant exposure to loud noises may cause loss of hearing. What part of the ear may be damaged?
4. **Relating cause and effect** Explain why you feel dizzy after spinning around for a few seconds.
5. **Relating facts** Hydrocephalus is a condition in which there is an abnormal increase of cerebrospinal fluid on the brain, causing the head to enlarge. How might this condition interfere with the proper functioning of the brain?
6. **Relating facts** Can a person who has normally functioning eyes still be blind? Explain your answer.
7. **Using the writing process** Imagine that you have to do without one of your sense organs for one day. Which one would you choose to give up? In an essay, describe how the absence of this sense organ would affect your life.

CRITICAL AND CREATIVE THINKING

1. Impulses would not be able to be transmitted beyond the cut portion of the axon.
2. Fluid accumulation in the cornea would increase its water content, which would decrease its transparency and interfere with normal vision.
3. Constant exposure to loud noises would damage the delicate structures of the inner ear, the cochlea and the membrane covering the oval window.
4. The fluids and otoliths in the semicircular canals move back and forth during the spinning motion, thus stimulating the hair to bend on the hair cells. Because balance is controlled, in part, by the movement of these materials in the semicircular canals, a person will feel dizzy until these materials stop swirling about.
5. The increase in cerebrospinal fluid would cause an increase in pressure on the brain, thereby injuring its delicate parts.
6. Yes. Even though a person's eyes may be functioning normally, the visual center in the brain may not be. Also, the optic nerves could be cut or damaged.
7. Make sure that students accurately describe the functioning of the chosen sense organ.

sensitive to all colors of light, but they do not distinguish different colors. Cones respond differently to light of different colors, thereby producing color vision.

12. Sound waves collected by the outer ear pass down the ear canal to the eardrum, causing it to vibrate. The vibrations are transmitted through the malleus, incus, and stapes of the middle ear to the oval window. This causes hair cells in the cochlea to move back and forth, which stimulate the acoustic nerve.

CHAPTER 38 Skeletal, Muscular, and Integumentary Systems

Section	Laboratory Investigations and Activities
38–1 The Skeletal System pages 837–842 THEMES: Scale and Structure, Unity and Diversity	**Teacher Edition** DEMONSTRATION: The Human Skeleton, p. 836D DEMONSTRATION: Joints, p. 836D **Laboratory Manual** Bone Composition and Structure, p. 473 The Skeletal System, p. 477
38–2 The Muscular System pages 842–846 THEMES: Scale and Structure, Systems and Interactions	**Student Edition** LABORATORY INVESTIGATION: Examining Skeletal Muscle, p. 850 **Laboratory Manual** The Muscular System, p. 483 **Teaching Resources** HANDS-ON ACTIVITY: Under Tension, p. 181
38–3 The Integumentary System pages 847–849 THEME: Scale and Structure	**Teaching Resources** HANDS-ON ACTIVITY: How Fast Do Your Nails Grow? p. 175 HANDS-ON ACTIVITY: Lip Service? p. 177 **Product Testing Activities** Testing Bandages Testing Lip Balms Testing Shampoos
Chapter Review pages 851–853	

Teacher Resources

Books

Anthony, Catherine, and Gary Thibodeau. *Structure and Function of the Body*, 8th ed. Mosby, 1988.

Curry, J. *The Mechanical Adaptations of Bones*. Princeton University Press, 1984.

Huxley, A. F. *Reflections on Muscle*. Princeton University Press, 1980.

McMahon, Thomas A. *Muscles, Reflexes, and Locomotion*. Princeton University Press, 1984.

Shipman, Pat, et al. *The Human Skeleton*. Harvard University Press, 1985.

Steele, D. Gentry, and Claude A. Bramblett. *The Anatomy and Biology of the Human Skeleton*. Texas A&M University Press, 1988.

CHAPTER PLANNING GUIDE

Other Activities	Media and Technology
Teaching Resources 　Activity: Skeletal Bones and Where They Meet, p. 3 **Study Guide** 　Section 38–1, p. 363	**Video/Videodisc** 　Human Body: Skeletal System and Muscular System **Biology Transparencies** 　53: The Human Skeletal System 　60: The Structure of a Human Bone 　61: Five Types of Human Skeletal Joints **Biology Media Guide** 　pages 56–57
Study Guide 　Section 38–2, p. 366	**Video/Videodisc** 　Human Body: Skeletal System and Muscular System **Biology Transparencies** 　54: The Human Muscular System **Biology Media Guide** 　pages 57–58
Teaching Resources 　Vocabulary Review: Crossword, p. 1 　Activity: Describing Skin and Bones, p. 5 **Study Guide** 　Section 38–3, p. 369	**Biology Transparencies** 　62: The Structure of Human Skin
Teaching Resources 　Chapter Test A, p. 11 　Chapter Test B, p. 15	**Computer Test Bank** 　Chapter Test, p. 401

Audiovisuals

Bones and Structure. Film or video. MTI Coronet.
Skeletal System, 2nd ed. Film or video. Coronet.

Muscular System, 2nd ed. Film or video. Coronet.
The Skeletal System. Sound filmstrip. Science Software Systems, Inc.

Your Body: Series 1. 3 films with cassettes. Focus Media.

CHAPTER 38 Skeletal, Muscular, and Integumentary Systems

CHAPTER OVERVIEW

All the movements of the human body—from the fine brush strokes of an artist to the precisely executed maneuvers of an Olympic gymnast—are made possible by the coordinated activity of the muscles and bones of the human body. The skeleton not only protects and supports the internal organs, but allows motion to occur by acting as a series of levers. The force that causes this motion to occur is provided by the contraction of the skeletal muscles and is transmitted to the bones through the tendons.

The bones have an additional vital function: It is within the soft tissue known as red bone marrow that red and white blood cells, as well as the platelets necessary for clotting, are produced.

Besides the skeletal muscles, the two other types of muscles found in the human body are the smooth muscles and the cardiac muscles. Smooth muscles are generally involuntary and are found in the internal organs; cardiac muscle is found only in the heart.

The integumentary system, consisting of skin and its associated structures, is the largest organ in the human body. It provides the internal organs with a sensory link to, as well as a protective barrier from, the external environment.

38-1 THE SKELETAL SYSTEM

Section Focus 38-1

The purpose of this section is to describe the structure and function of the skeletal system and the structure and development of the bone tissue of which the human skeleton is composed.

The human skeletal system consists of the bones and the associated tissues of cartilage, tendons, and ligaments. Besides giving shape to the human body and allowing for its wide and varied range of motion, the skeletal system also provides support and protection for internal organs and is the site for the storage of minerals and the production of blood cells.

Bones are living tissue; as such, they are served by nerves and blood vessels that reach the bone tissue through the periosteum and penetrate into the bones through the Haversian canals. Within the bones is soft tissue, or marrow. Red marrow, found at the ends of the long bones, is responsible for the production of red and white blood cells, as well as the platelets vital for blood clotting.

Bones grow in one of two ways. All bones are initially formed from cartilage or other connective tissue. In the developing fetus and the young child, mineral matter consisting of calcium phosphate gradually replaces this connective tissue. In addition, the long bones in the arms and legs have cartilage "growth plates" near their ends. As new cartilage is produced, older material is ossified, lengthening the bone. This process continues until adulthood is reached.

Joints, the junction of separate bones, allow for the proper positioning and motion of bones. The three types of joints are freely movable, slightly movable, and immovable. At the joints, spaces called bursa act to reduce friction.

Performance Objectives 38-1

1. List the structures and functions of the skeletal system.
2. Describe the structure of bone.
3. Explain the different ways in which bone tissue is formed.
4. List the different kinds of joints and describe the range of motion of each.

Science Terms 38-1

skeletal system　p. 837
periosteum　p. 838
Haversian canals　p. 838
osteocyte　p. 838
bone marrow　p. 838
cartilage　p. 839
ossification　p. 839
joint　p. 840
ligament　p. 841

38-2 THE MUSCULAR SYSTEM

Section Focus 38-2

The purpose of this section is to describe the three types of muscles in the muscular system. Although all muscle tissue is specialized for contraction, each of the three types has a different structure and performs a specific function.

Skeletal muscles, consciously controlled by the central nervous system, are attached to bones and cause them to move. Smooth muscles, generally involuntary, are found in the internal organs and in the walls of blood vessels. Cardiac muscle is found only in the heart and may contract even without stimulation from the central nervous system.

The mechanism of muscle contraction is described in the sliding filament theory. This theory explains muscle contraction as the interaction of thick bands of the protein myosin with thin bands of the protein actin.

Muscle contraction is controlled by nerve impulses that cause the secretion of the neurotransmitter acetylcholine. Acetylcholine triggers the occurrence of other events within the muscle, ultimately leading to contraction.

Muscles are not directly attached to bone. The tendons, a type of connective tissue, form the link between muscle and bone. The contraction of muscles results in the exertion of force by the tendons on the bones, causing them to act as levers.

The controlled movement of our bodies results from the coordinated action of paired muscle groups that have opposite effects on a particular part of the skeleton.

Even when a muscle is not actively contracting, some of its fibers will remain contracted, producing muscle tone. One of the benefits of regular exercise is the increase in muscle tone.

Performance Objectives 38-2

1. Describe the structure and function of each of the three types of muscles.
2. State several specific locations of each of the three muscle types in the human body.
3. Describe the mechanism of muscle contraction.
4. Describe the interaction of muscles, bones, and tendons.

Science Terms 38-2

muscular system   p. 842
skeletal muscle   p. 843
smooth muscle   p. 843
cardiac muscle   p. 844
myosin   p. 844
actin   p. 844
cross-bridge   p. 844
sliding filament theory   p. 845
tendon   p. 846

38-3 THE INTEGUMENTARY SYSTEM

Section Focus 38-3

The purpose of this section is to describe the structure and function of the integumentary system, which serves mainly as a protective barrier between the internal organs and the outside world. It also functions to regulate body temperature and to mediate the reception of sensations from the environment.

The integumentary system consists of skin, hair, nails, and their associated glands. The skin consists of two main layers: the outer epidermis and the inner dermis. The epidermis forms the skin's flexible, waterproof covering. The dermis contains blood vessels, sensory nerve endings, and glands.

Hair and nails both consist of dead cells containing the protein keratin. Hair is produced by cells in the hair follicles, while nails grow from the nail matrix near the tips of fingers and toes.

Performance Objectives 38-3

1. Identify the structures of which the integumentary system is composed.
2. Describe the structure and functions of the skin.
3. Describe the structure of the hair and nails.
4. Describe how hair and nails grow.

Science Terms 38-3

integumentary system   p. 847
epidermis   p. 847
dermis   p. 848
hair follicle   p. 848
nail matrix   p. 848

CHAPTER PREVIEW

DISCOVERY LEARNING

TEACHER DEMONSTRATIONS
Modeling

The Human Skeleton
The following demonstration can be used as an introduction to Chapter 38.

Obtain an articulated skeleton and a separate human skull. (An actual human skeleton, if available, is far more effective than is the plastic kind.) Point out some of the general features of the skeleton:

• **How do different parts of the skeleton protect the internal organs?** (The skull encloses the brain; the ribs enclose the heart, lungs, and upper intestinal tract; the pelvis encloses the reproductive organs and the lower intestines.)

• **Is this skeleton that of a male or a female?** (This will be more obvious in a real skeleton; the female's pelvic outlet is wider to allow for pregnancy and childbirth and the pubic arch is U-shaped, compared to the Y-shape of the male. There are other differences as well.

• **Notice the skull. Are the bones of uniform thickness?** (The skull is thickest in front, to protect the cerebrum; it is extremely thin at the bridge of the nose.)

Joints

You may want to use this demonstration while discussing Section 38–1.

Using the articulated skeleton and skull, ask students to point out the location of the different types of joints. (Answers may include, but are not limited to, the skull sutures, shoulder, knee, hip, elbow, ankle, and wrist.)

Ask students to demonstrate, insofar as the skeleton will allow, the range of motion at each of the joints. (There is no motion along the skull sutures, while only slight motion is allowed along the vertebrae. A hinge joint, such as the knee and elbow, allows motion in one plane only; and the ball-and-socket joints at the hips and shoulders allow the widest range of motion. There are many other motions that you may want students to observe.)

CHAPTER 38
Skeletal, Muscular, and Integumentary Systems

GUIDED ENQUIRY

Pose the following questions to students and have them record their responses. Point out that they will gain a better understanding of the key concepts if they read the chapter with these basic questions in mind. Upon completion of the chapter, pose the questions again. Ask students to compare their initial responses with those they have developed after reading the chapter.

- What are some of the functions of the skeletal system?
- What is the structure of bone tissue?
- How does the human skeleton move?
- What are the different kinds of muscles in the human body?
- How do the skeletal muscles interact with the bones to cause motion?
- What are the structures and functions of the integumentary system?

INTRODUCING CHAPTER 38

Using the Visuals

Ask students to look at the photograph that introduces Chapter 38 and read the accompanying caption. Then ask them to list the words that describe the different motions of the human body.
- **What structures in the human body allow it to accomplish all these different motions?** (Bones and muscles.)
- **How do the bones and muscles interact to allow these motions to occur?** (Muscles are attached to bones via the tendons; the contraction of muscles causes the bones to move as if they were levers.)
- **How do bones interact with each other?** (There are different kinds of joints between the bones that allow different amounts and directions of motion.)
- **What are some examples of joints?** (Accept all correct answers.)
- **Where in the human body besides attached to the bones are muscles found?** (Muscles are found in the intestinal tract, the urinary bladder, the walls of arteries, and the heart, among other places.)
- **What is the function of the skin and its associated structures such as hair, nails, and glands?** (The skin protects the internal organs and prevents

---

CHAPTER 38

Skeletal, Muscular, and Integumentary Systems

The actions involved in playing tennis require the combined use of the bones, joints, and muscles.

Compared to animals that have claws, sharp teeth, and protective scales, humans appear to be somewhat fragile. But you need only observe a group of people playing football, a rider taking a horse over a jump, dancers leaping through the air, a pole vaulter clearing the bar, or a swimmer moving through the water and the image of fragile humans begins to fade. In truth, we are a large, strong, and remarkably agile species.

The basis for this agility is the way in which the human body is organized. This organization makes it possible for humans to perform a variety of physical activities that is truly astonishing. In this chapter you will discover how the body's foundation—its bones, muscles, and skin—provides structural support, mobility, and protection.

GUIDE FOR READING

After you read the following sections, you will be able to

38–1 The Skeletal System
- List the parts and functions of the skeletal system.
- Describe the structure and development of bone.
- Identify three classes of joints and describe the action of each.

38–2 The Muscular System
- Describe the function and components of the muscular system.
- Discuss the mechanism of muscle contraction.
- Explain the relationship among muscles, bones, and joints.

38–3 The Integumentary System
- Identify the structures that make up the integumentary system.
- List the functions of the skin.
- Describe the structure of the skin, hair, and nails.

Journal Activity

YOU AND YOUR WORLD

Are you "into" body building or weight training? Have you occasionally basked in the sun trying to get a tan? Have you ever broken a bone, sprained an ankle, or skinned your knee? In your journal, describe a memorable incident in your life that has something to do with your bones, muscles, or skin.

Figure 38–1 The step-by-step movements of the human skeleton during jumping and walking have a ghostly appearance in this computer image.

38–1 The Skeletal System

Guide For Reading

- What are the parts of the skeletal system and what are their functions?
- What is the structure of a typical bone? How does a bone grow?
- How are joints classified? How do the different types of joints work?

In order to retain their shape and form, living things need some type of support. In single-celled organisms, this support is provided by the cell membranes. In multicellular animals, the support is provided by some form of skeleton. Among multicellular animals, skeletons range from the exoskeletons (outside skeletons) of arthropods to the endoskeletons (inside skeletons) of vertebrates.

Like that of all vertebrates, the skeleton of humans is composed of a special type of connective tissue (tissue that joins other tissues together) called bone. **Bones and their associated tissues—cartilage, tendons, and ligaments—make up the skeletal system.**

The bones that make up the **skeletal system** serve several important functions. They support and shape the body much as the internal wooden frame does a house. Just as the house could not stand without its wooden frame, the human body would collapse without its bony skeleton. Bones protect the delicate internal organs of the body. For example, the bones of the cranium form a protective shell around the brain, and the ribs form a basketlike cage that protects the heart and lungs.

Bones also provide a system of levers (rigid rods that can be moved about a fixed point) on which a group of specialized

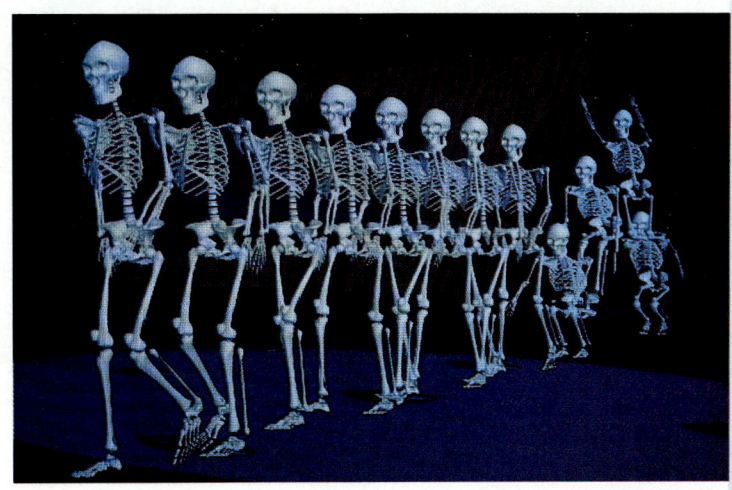

COOPERATIVE LEARNING

Students are aware that a common advertising practice used by producers of suntanning products is to place billboards that appeal to young people near water recreation areas. Using preassigned lab groups or randomly selected teams, have groups design a billboard warning the public of the dangers of unprotected exposure to the sun. Remind groups that their billboards must be factual but appealing to their audience. Billboards can be produced on butcher paper or a chalkboard that has been divided into sections for each group. You may want to make felt-tip pens and colored chalk available for the groups' use.

Journal Activity

YOU AND YOUR WORLD

Encourage students to share their experiences with the class. Students should keep their descriptions in their portfolio.

Content Development

Ask students to recall the exoskeletons of arthropods.
- **What are some of the advantages of having an internal skeleton?** (It's easier to move; growth may occur from within, eliminating the necessity of periodic molting.)
- **State some specific examples of how bones protect the internal organs.** (Examples may include, but are not limited to, the skull surrounding the brain and the ribs surrounding the heart and lungs.)

---

disease-causing organisms from entering the body.)

TEACHING STRATEGY 38–1

Focus/Motivation

Have students observe an actual skeleton or, if one is not available, a chart showing the human skeleton.
- **What is the main purpose of the skeleton?** (To support and protect the internal organs and allow motion to occur.)
- **What other tissues are associated with the bones?** (Cartilage, ligaments, and tendons are all associated with bone.)
- **How do bones act as levers?** (The interaction of the bones of the skeleton with other tissues causes the bones to move in certain allowed directions at the joints.)

837

BACKGROUND INFORMATION
FLOPPY WISHBONES

Bone not only consists of dead mineral material but contains large amounts of two proteins, collagen and glycoprotein. Although the role of glycoprotein is not well understood, collagen is known to give elasticity to the bone (enhancing its shock-absorbing ability) while providing a matrix on which the calcium phosphate crystallizes.

Students may enjoy soaking a chicken bone in vinegar for a few days to make a "rubber bone." The vinegar dissolves the calcium, and the soft material that remains consists mostly of collagen.

TEACHING SUPPORT

Laboratory Manual
- Bone Composition and Structure, p. 473

MEDIA AND TECHNOLOGY
📖 Biology Transparencies
- 53: The Human Skeletal System

📹 Video/Videodisc
- *Human Body: Skeletal System and Muscular System*
Use this bar code to introduce students to the skeletal system.

Play frames 696 to 1328

See the Biology Media Guide pages 56–57 for additional bar-code correlations for this section.

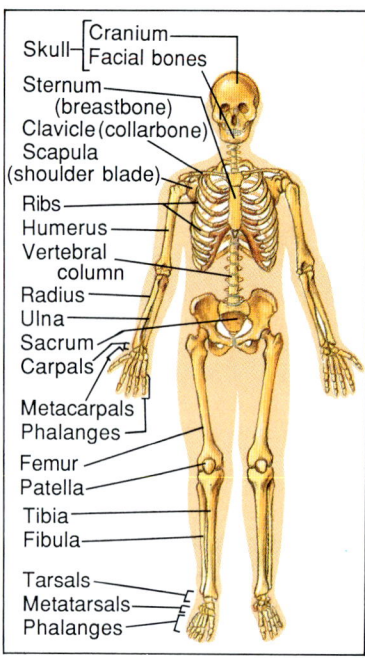

Figure 38–2 The human skeleton consists of 206 bones.

tissues act to produce movement. The movements produced range from the gentle motion of a fingertip to the powerful actions that cause a change in the position of the entire body.

In addition, bones contain enormous reserves of minerals, especially calcium and phosphorus, which are important to many body processes. Finally, bones are the sites of blood-cell formation. Most blood cells are produced within the soft tissue that fills the internal cavities in bones.

Structure of Bones

It is easy to think of bones as nonliving. After all, most of the mass of bone is mineral—largely calcium and phosphorus. But bones are living tissue. Bones are a solid network of living cells and fibers that are supported by deposits of calcium salts.

Each bone is surrounded by a tough membrane called the **periosteum** (per-ee-AHS-tee-uhm). Blood vessels pass through the periosteum, carrying oxygen and nutrients to the bone. Beneath the periosteum is a thick layer of compact bone. Although compact bone is dense and similar in texture to ivory, it is far from being solid. Running through compact bone is a network of tubes called **Haversian** (huh-VER-zhuhn) **canals** that contain blood vessels and nerves.

Inside the layer of compact bone is spongy bone. Spongy bone is not soft and spongy, as its name implies, but actually quite strong. Near the ends of bones where force is applied, spongy bone is organized into structures that resemble the supporting girders in a bridge. This structure of spongy bone helps to add strength to bone without adding mass.

Embedded in compact and spongy bone are cells known as **osteocytes** (AHS-tee-oh-sights) that can either deposit the calcium salts in bone or absorb them again. Osteocytes are responsible for bone growth and changes in the shapes of bones. Within bones are cavities that contain a soft tissue called **bone marrow**. There are two types of bone marrow: yellow and red. Most bone contains yellow marrow, which is made up of blood vessels, nerve cells, and fat cells. Red marrow produces red blood cells as well as special white blood cells (lymphocytes) and other elements of blood (platelets).

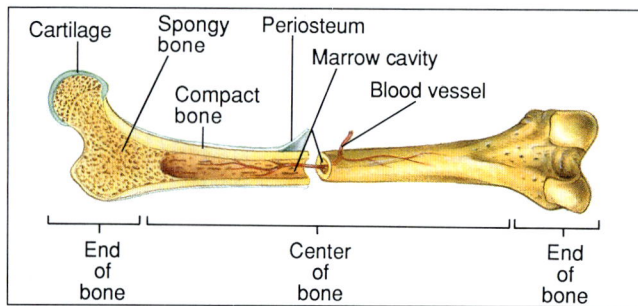

Figure 38–3 The structure of a typical bone, such as the femur, consists of spongy and compact bone surrounded by a tough membrane called the periosteum.

38–1 (continued)

Content Development
• What are some other functions of bone? (Bones serve as depositories of calcium and phosphorus and are the site of blood cell formation.)

Focus/Motivation
Using the Visuals Show students a bone that has been dissected in cross or longitudinal section; if this is not possible, have them refer to Figure 38–3. Emphasize that although bone tissue consists mostly of nonliving minerals, it is a living tissue.

Content Development
Discuss the importance of the periosteum (which will not be visible on the bone specimen) to nourish the living tissue. Also point out the difference in the appearance of compact and spongy bone. In long bones, such as the humerus and femur, most of the spongy bone is at the ends of the bone and has a "grain," or pattern, that enables it to withstand the stresses placed on it by the body's weight and motion.

Development of Bones

Many bones are formed from the cells of a type of connective tissue called **cartilage**. Unlike bone, cartilage does not contain blood vessels. Cartilage cells must rely on the diffusion of nutrients from the tiny blood vessels (capillaries) in surrounding tissues. The cells that make up cartilage are scattered in a network of fibers composed of an elastic protein called collagen. Cartilage is dense and fibrous, can support weight, but is still extremely flexible.

Many bones in the skeleton of a newborn baby are composed almost entirely of cartilage. The cartilage is replaced by bone during **ossification** (ahs-uh-fih-KAY-shuhn), or the process of bone formation. Ossification begins to take place up to seven months before birth as mineral (calcium and phosphorus) deposits are laid down near the center of the bone. Gradually, bone tissue forms as osteocytes secrete mineral deposits that replace the cartilage.

Many long bones, such as those of the arms and legs, have growth (epiphyseal) plates at either end in which the growth of cartilage causes the bones to lengthen. Gradually, this new growth of cartilage is ossified (replaced by bone), and the bones become larger and stronger. This process usually continues until a person reaches the age of 18 or 20. At that time the growth plates disappear, the bones become completely ossified, and the person "stops growing."

In adults, cartilage is found in those parts of the body where flexibility is needed. Such places include the tip of the

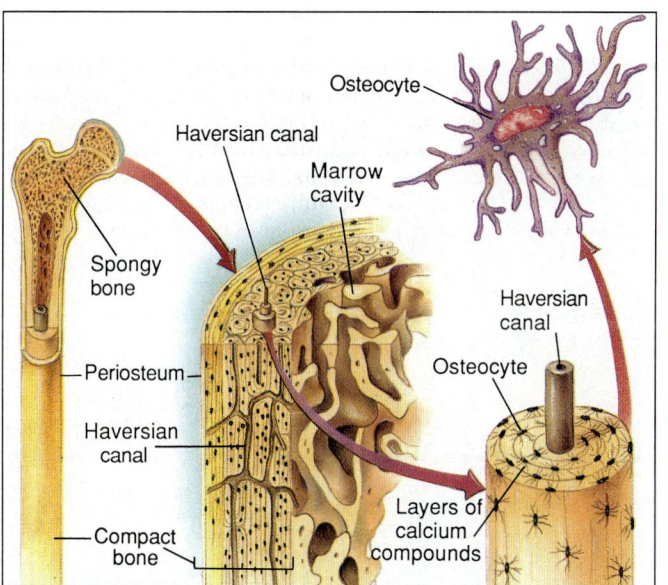

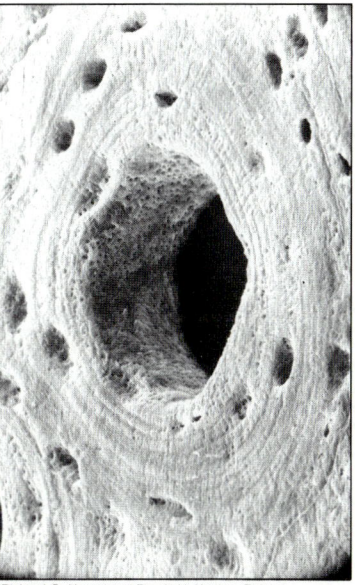

Figure 38-4 The illustration shows the structures of compact and spongy bone. The opening in the center of the electron micrograph of compact bone is a single Haversian canal.

Richard G. Kessel and Randy H. Kardon. *Tissues and Organs: A Text Atlas of Scanning Electron Microscopy.* W. H. Freeman & Co. © 1979.

BACKGROUND INFORMATION
CALCIUM IN SPACE

The deposition of calcium in bone is at least partially dependent on the application of force to the bone as it is developing. Bone also has a "grain," much like the wood of a tree, and the application of force causes the grain to align itself in such a way as to enable the bone to tolerate the constant stresses applied to it. But what happens when there is no force?

Studies of Soviet and American astronauts have shown that the absence of gravity in space causes, over a long period of time, severe loss of calcium from bones. Long-term space flight so weakens the skeleton that space scientists regard this loss as a major problem that future space explorers will face.

"soft spots" on a baby's head). Throughout childhood and adolescence or as long as a person is still growing, the growth plates near the ends of the bones remain cartilaginous.

Reinforcement/Reteaching
Ask students to compare and contrast cartilage and bone as to their microscopic structure, location in the body (both in children and adults), and texture.

Enrichment
You may want students to investigate bone disorders, such as those caused by improper growth or infection. Or students may want to investigate the process by which bone heals after a fracture.

Enrichment
Have some students investigate the growth of bone. As bone grows, it lengthens and changes in shape. Osteocytes remove and redeposit mineral material, "remodeling" the bone as it grows.

Focus/Motivation
If possible, show students a diagram of the changes in the human skeleton from birth to adulthood. Have them observe that the bones not only lengthen but change in shape.

Content Development
Discuss the process of ossification. Point out that although ossification begins before birth, many bones in a newborn are still mostly cartilage (example: the feet) or membranous connective tissue (example: the

BACKGROUND INFORMATION
HEALING OF FRACTURES

Bones occasionally break, and when they do, the main requirement for successful healing is for the broken edges to be in contact with one another. The "fit" does not have to be perfect; that is, the fractured ends do not have to match exactly. As healing occurs, the periosteum produces osteocytes, which form a callus, a sleevelike structure around the fracture. The callus acts like an internal splint, holding the broken ends of the bone together. As the bone heals, it is molded and reshaped and the callus is completely reabsorbed. Years later, it may be difficult for anyone except a medical specialist to tell that the bone was ever broken.

TEACHING SUPPORT

Laboratory Manual
- The Skeletal System, p. 477

Teaching Resources
- Activity: Skeletal Bones and Where They Meet, p. 3

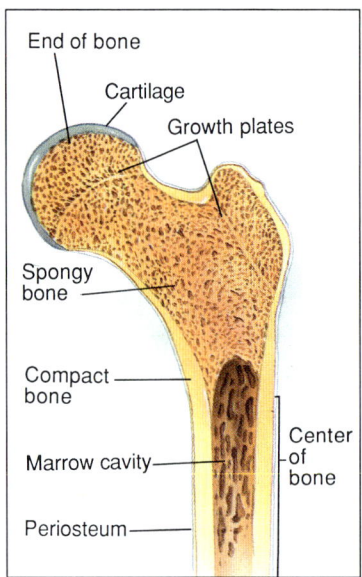

Figure 38–5 A growth plate, which is located between the end and the center of a long bone, is a site at which new bone tissue is produced, causing the bone to grow in length.

Figure 38–6 Of the 28 bones that make up the skull, 8 of them form the cranium, or bony covering surrounding the brain. The photograph shows an interlocking immovable joint between two bones in the cranium.

nose, the external ears, the voice box (larynx) and walls of the windpipe (trachea), and the ends of the bones where joints are formed. Cartilage is also found where the ribs are attached to the breastbone (sternum), thus allowing the rib cage to move during breathing. In all these places, cartilage provides an important combination of strength and flexibility. You may recall from Chapter 31 that animals such as sharks and rays have skeletons composed entirely of cartilage.

Joints: Where Two Bones Meet

Joints, or places where two bones come together, permit the bones to move without damaging each other. In other words, joints are responsible for keeping the bones far enough apart so they do not rub against each other as they move. At the same time, joints hold the bones in place.

Because of the presence of joints, the human body is capable of a wide variety of movements, ranging from extensive movement (at the shoulder joint) to no movement at all (at the joints of the skull). Depending on their type of movement, joints are classified as either immovable, slightly movable, or freely movable.

IMMOVABLE JOINTS These joints, often called fixed joints, allow no movement between bones. This is because the bones at an immovable joint are interlocked and held together by connective tissue, or they are fused. The places where the bones in the skull meet are examples of immovable joints. Skull bones do not need to move because their main function is to protect the brain and the sense organs located in the head.

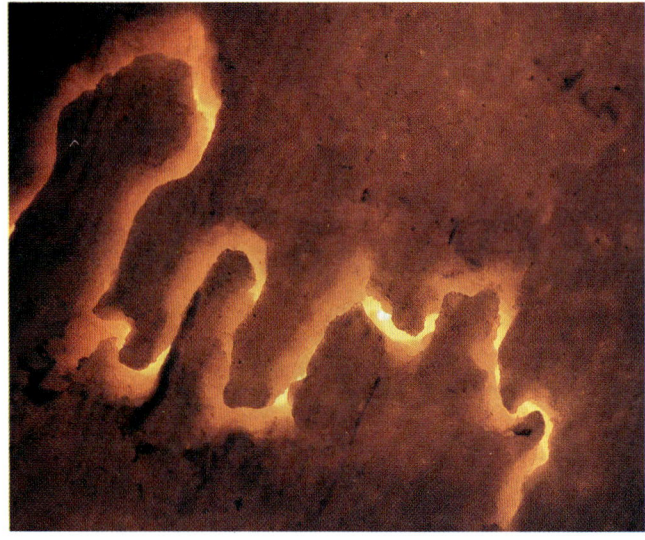

38–1 (continued)

Focus/Motivation
Ask students to observe a model skeleton, paying particular attention to the joints. Have a student demonstrate the various ranges of motion of which the different types of joints are capable (without reference to specific terminology).

Content Development
Most of the body's joints are movable to some degree, and the ends of the bones at most of these joints are covered by cartilage with a thin film of lubricating synovial fluid between them. The bursae—small sacs containing synovial fluid—are other lubricating and shock-absorbing devices between joints where pressure may be exerted, such as the shoulder, elbow, hip, and knee.

As the different types of freely movable joints are discussed, have students locate specific examples of each type on the articulated skeleton and demonstrate the range of motion of which each is capable.

SLIGHTLY MOVABLE JOINTS These joints permit a small amount of movement. Unlike the bones of immovable joints, the bones of slightly movable joints are farther apart from each other. The joints between the two bones of the lower leg (tibia and fibula) and the joints of the vertebrae (bones of the spinal column) are examples of slightly movable joints.

FREELY MOVABLE JOINTS Most of the joints of the body are freely movable joints. In freely movable joints, the ends of the bones are covered with a layer of cartilage that provides a smooth surface at the joint. The joints are also surrounded by a fibrous joint capsule that helps hold the bones together and at the same time allows for movement. The joint capsule consists of two layers.

One of the layers of the joint capsule may thicken to form strips of tough connective tissue called **ligaments**. Ligaments are attached to the membranes that surround bones and hold the bones together. The other layer of the joint capsule produces synovial (sih-NOH-vee-uhl) fluid, which forms a thin lubricating film over the surface of a joint. This lubricating film enables the cartilage found on the ends of the bones to slip past each other more smoothly as the joint moves.

In some freely movable joints, small pockets of synovial fluid called bursae (BER-see; singular: bursa) form. A bursa reduces the friction between the bones of a joint and also acts as a tiny shock absorber. Sometimes when a joint is injured, too much fluid moves into the bursa, causing it to swell and become painful. When this happens at a joint, a condition called bursitis results.

A more serious disorder that affects the joints is arthritis, or inflammation of the joint. There are at least 20 different types of arthritis that affect approximately 10 percent of the world's population. You will learn more about arthritis in Chapter 45.

Freely movable joints are grouped according to the shapes of the surfaces of the adjacent bones. There are six types of freely movable joints. Refer to Figure 38-9 on page 842 as you read about each type of joint. Locate each joint on your body and duplicate its movement as you read about it.

A ball-and-socket joint permits circular movement—the widest range of movement. The shoulder joint and the hip joint are examples of a ball-and-socket joint. A hinge joint permits a back-and-forth motion, much like the opening and closing of a door. Examples of a hinge joint are the knee and the elbow. A pivot joint allows for rotation of one bone around another. Examples of this type of joint are found between the first two neck vertebrae and between the bones of the lower arm (ulna and radius).

A gliding joint permits a sliding motion of one bone over another. Gliding joints are found at the ends of the collarbones (clavicles), between wrist bones (carpals), and between ankle

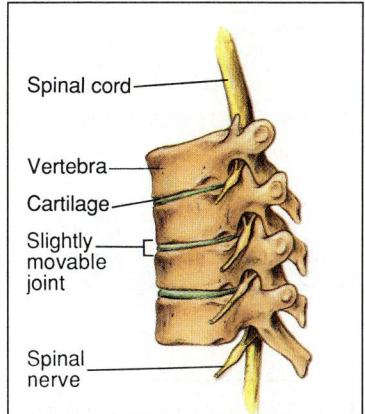

Figure 38-7 Slightly movable joints are found between the vertebrae, or bones that make up the spinal column.

Figure 38-8 A hinge joint, such as the knee, allows for movement in one plane only.

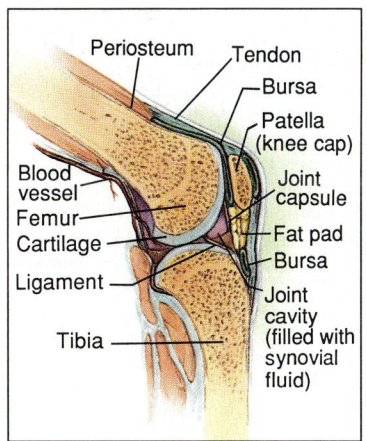

BACKGROUND INFORMATION
RHEUMATOID ARTHRITIS

One of the worst forms of arthritis—rheumatoid arthritis—can occur at any age, even in young children. Although its exact cause is unknown, it seems to involve a malfunction of the immune system, in which the synovial joints are attacked as if they were foreign matter, causing inflammation. In some cases, the symptoms disappear by themselves, and they may or may not return. In other cases, the disease may progress to the point where the articular cartilage that covers the "heads" of the long bones may be destroyed, or the adjacent bones forming the fingers may fuse together. At present, there is no cure for this crippling disease, although some relief from the symptoms may be obtained from anti-inflammatory medications.

ESL STRATEGY

Ask students if they, or someone they know, have ever had a sprain, a bone fracture, or a dislocation. Have them write a short description about how it happened, how it felt, and what medical treatment the person received. Have students exchange papers and check for clarity. Then ask them to use these accident accounts as the basis for a brief oral presentation.

Reinforcement/Reteaching
It may be useful for students to summarize the types of joints in a four-column chart. Label the first column Name of Joint, the second column Type of Joint, the third column Range of Motion, and the fourth, Examples.

Enrichment
Some students may be interested in knowing the terms that describe the motions of the joints. These terms include *flexion*, or bending so the parts are brought closer together; *extension*, or straightening out, the opposite of flexion; *rotation*, or partial turning of a body part about its long axis; *abduction*, the movement away from the middle of the body, and *adduction*, the opposite of abduction.

BACKGROUND INFORMATION
TYPES OF SKELETAL MUSCLE

Red muscle, so called because it contains the red pigment myoglobin, has a rich blood supply and many mitochondria. Myoglobin is also capable of storing oxygen. Red muscle contracts a little more slowly than does white muscle but can work for much longer periods of time. Red muscle makes up the dark meat of chicken legs and its the dominant muscle tissue in long-distance runners.

White muscle has a smaller blood supply, fewer mitochondria, and little myoglobin. White muscle can contract quickly and powerfully, but because it relies on anaerobic fermentation for the production of ATP, it rapidly builds up lactic acid, resulting in fatigue. This kind of muscle is found in chicken breast meat and in the muscles of weight lifters.

TEACHING SUPPORT

Study Guide
- Section 38–1, p. 363

38–1 (continued)

SECTION REVIEW 38–1

1. The parts of the skeletal system include the bones and their associated tissues—cartilage, tendons, and ligaments.
2. The skeletal system supports and shapes the body, protects the internal organs, provides a system of levers on which muscles act to produce movement, provides reserves of minerals, and is the site of blood-cell formation.
3. Bone is surrounded by a tough membrane called the periosteum. Beneath the periosteum is a thick layer of compact bone. Running through compact bone is a network of tubes called Haversian canals. Inside the compact bone is spongy bone. Many bones are formed from cartilage cells that undergo ossification. Ossification begins when mineral deposits are laid down near the center of bone. Gradually, bone tissue forms as osteocytes secrete mineral deposits that replace cartilage.
4. A joint is a place where two bones come together. Immovable, slightly movable, and freely movable are three types of joints.
5. See Figure 38–9 on page 842.
6. Bone is living tissue that contains bone cells, blood vessels, and nerves. Because of the presence of nerves, any injury to the bone, such as a fracture, will cause pain.

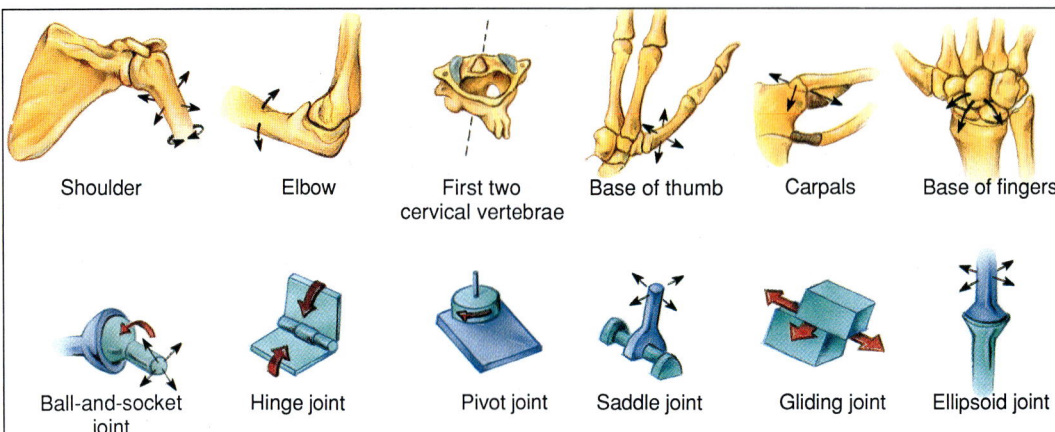

Figure 38–9 *Freely movable joints permit a wide range of motion. Notice the movements of six types of freely movable joints.*

bones (tarsals). A saddle joint permits movements in two planes. This type of joint is found at the base of the thumb. An ellipsoid (ee-LIHP-soid) joint allows for a hinge-type movement in two directions. The joints that connect the fingers with the palms of the hands and the toes with the soles of the feet are examples of ellipsoid joints.

38–1 SECTION REVIEW

1. Identify the parts of the skeletal system.
2. What are the functions of the skeletal system?
3. Describe the structure and development of bone.
4. What is a joint? List the three types of joints.
5. Describe six types of freely movable joints.
6. **Critical Thinking—Making Inferences** Explain why a bone that fractures, or breaks, is painful.

Guide For Reading

- What is the muscular system? What is its function?
- How are the three types of muscle tissue similar? How are they different?
- How do muscles contract?
- What are the relationships among muscles, nerves, bones, and joints?

38–2 The Muscular System

As you have just read, the skeleton and its joints support, protect, and provide flexibility for the body. But the skeleton cannot move by itself. That job is performed by the muscle tissue that makes up the **muscular system**. Muscle tissue consists of groups of cells that are specialized for contraction. Approximately 40 to 50 percent of the mass of the human body is composed of muscle tissue—and that is true whether or not one is an athlete. **The muscular system is composed of muscle tissue that is highly specialized to contract, or shorten, when stimulated.** The word muscle is derived from the Latin word *mus,* meaning mouse. The Greeks and Romans, fascinated by the distribution of muscles in the body, compared the contractions of muscles to the movements of mice beneath the skin.

Muscle tissue is found everywhere within the body—not only beneath the skin but deep within the body, surrounding many internal organs and blood vessels. The size and location of muscle tissue helps determine the shape of our body, the way we move, and even the way we smile!

Types of Muscle Tissue

There are three different types of muscle tissue, or muscles: skeletal, smooth, and cardiac. Each type of muscle has a different structure and plays a different role in the body.

SKELETAL MUSCLES **Skeletal muscles** are generally attached to bones and are at work every time we lift a finger, wink an eye, chew something, or stand up. Skeletal muscles are responsible for voluntary (conscious) movement. When viewed under a microscope, skeletal muscles appear to have striations (bands or stripes). For these reasons, skeletal muscles are also known as voluntary or striated muscles. Most skeletal muscles are consciously controlled by the central nervous system.

Skeletal muscle cells are large, have more than one nucleus, and vary in length from 1 millimeter to as many as 30 to 60 centimeters. Because skeletal muscle cells are long and slender, they are often called muscle fibers rather than muscle cells. The muscle fibers together with the connective tissues, blood vessels, and nerves associated with them form a skeletal muscle such as a leg muscle.

SMOOTH MUSCLES **Smooth muscles** are usually not under voluntary control. Smooth muscle cells are spindle-shaped, have individual nuclei, and are not striated. Smooth muscles are found in many internal organs and in the walls of many blood vessels. Most smooth muscle cells can contract

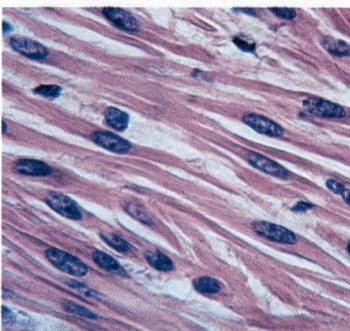
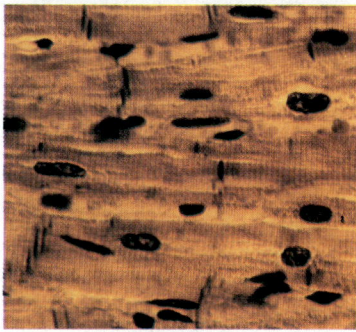

Figure 38–10 Muscle tissue is specialized for contraction. There are three types of muscle tissue: skeletal muscle (top), smooth muscle (center), and cardiac muscle (bottom).

Figure 38–11 This drawing shows the structure of the biceps in the upper arm, a skeletal muscle. Notice that skeletal muscle is made up of bundles of muscle fibers.

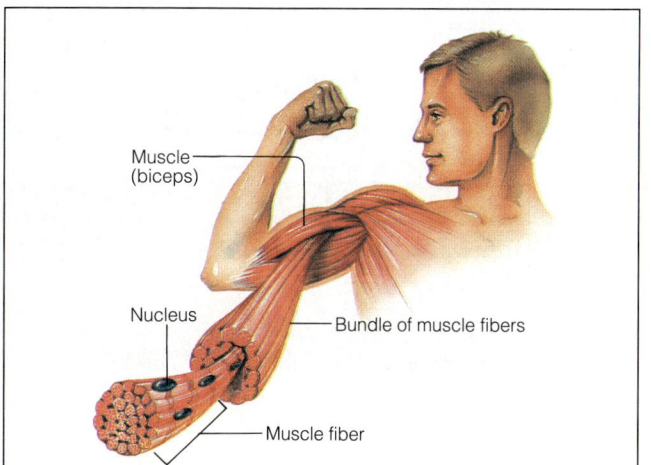

TEACHING SUPPORT

Laboratory Manual
- The Muscular System, p. 483

Teaching Resources
- Hands-On Activity: Under Tension, p. 181

MEDIA AND TECHNOLOGY

Biology Transparencies
- 54: The Human Muscular System

Video/Videodisc
- *Human Body: Skeletal System and Muscular System*

Use this bar code to introduce students to how skeletal muscles attach to and move bones.

Play frames 25215 to 25749

See the Biology Media Guide pages 57–58 for additional bar-code correlations for this section.

FACTS AND FIGURES

Even under ideal conditions, the efficiency of muscles is less than 50%, which means that more than half the total chemical energy used by muscles is lost in the form of heat.

How hot can you get? A 62-kilogram person who plays squash for just 7 minutes produces enough heat to increase the temperature of 91 liters of water by 1°C.

38-2 (continued)

Content Development
As its name implies, cardiac muscle is found only in the heart. Ask students to compare and contrast cardiac muscle with skeletal and smooth muscle. They should realize that cardiac muscle has some of the characteristics of each of the other muscle types.

Reinforcement/Reteaching
• What causes a muscle to contract? (A motor nerve cell stimulates it. The nerve divides into several branches, all of which stimulate separate muscle cells simultaneously. All the muscle cells stimulated by the same nerve are known as a motor unit.)

Content Development
Discuss the sliding filament theory of muscle contraction. The effect of the attachment of the myosin cross-bridges to successive sites on the actin filament may be compared to pulling a resisting rope toward oneself hand over hand. The combined action of thousands of these sliding actin and myosin filaments in one muscle cell and of thousands of muscle cells contracting simul-

---

without nervous stimulation. These contractions in smooth muscles move food through our digestive tract, control the way blood flows through our circulatory system, and decrease the size of the pupils of our eyes in bright light.

CARDIAC MUSCLES The only place in the body where **cardiac muscles** are found is in the heart. Although cardiac muscle cells are striated, they are not under voluntary control. Unlike skeletal muscles, cardiac muscles contract without direct stimulation by the nervous system. The cardiac muscle cells contain one nucleus located near the center of the cell. Adjacent cells form branching fibers that allow nerve impulses to pass from cell to cell.

Mechanism of Muscle Contraction

Muscle contraction is one of the best-understood processes in the human body. We know as much as we do about muscle contraction because the regular structure of striated muscle cells has made them easy to study. When viewed through the electron microscope, the muscle cells are seen to contain both thick and thin filaments. The overlapping patterns of the thick and thin filaments are responsible for the light and dark bands seen in skeletal (striated) muscle. See Figure 38–12.

In the 1950s, the British scientist Hugh Huxley noted that when a muscle cell contracts, the light and dark bands contained in the muscle cell get closer together. We now know that the thick filaments are made up of a protein called **myosin**, and the thin filaments are composed of a protein called **actin**. When myosin filaments and actin filaments come near each other, many knoblike projections in each myosin filament form **cross-bridges** with an actin filament. When the muscle is stimulated to contract, the cross-bridges move, pulling the two filaments

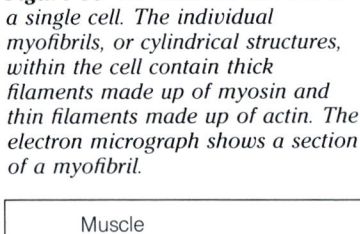

Figure 38–12 Each muscle fiber is a single cell. The individual myofibrils, or cylindrical structures, within the cell contain thick filaments made up of myosin and thin filaments made up of actin. The electron micrograph shows a section of a myofibril.

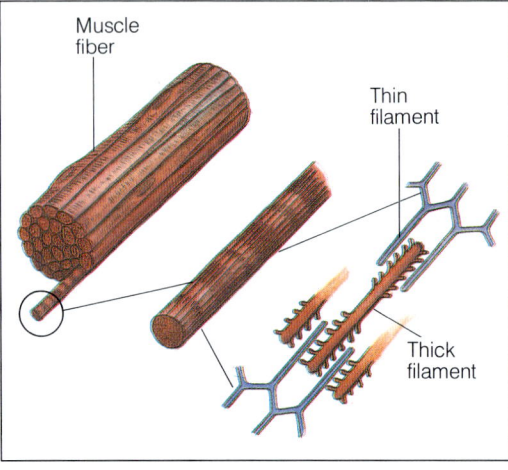
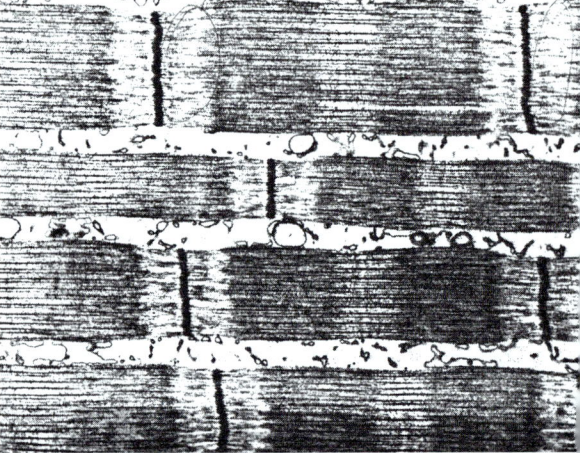

844

past each other. After each cross-bridge has moved as far as it can, it releases the actin filament and returns to its original position. The cross-bridge then attaches to the actin filament at another place and the cycle is repeated. When thousands of actin and myosin filaments interact in this way, the entire muscle cell shortens. This concept is the **sliding filament theory**.

ATP provides the energy necessary to make and break connections between actin and myosin filaments. ATP also powers the movement of the cross-bridges by attaching a molecule of itself to each myosin cross-bridge. Each time a cross-bridge goes through the cycle, the ATP is broken down to ADP and inorganic phosphate.

When a muscle cell contracts in the manner just described, large amounts of ATP are used up as the thick and thin filaments slide past each other. What is the source of all this ATP? Cells have two methods by which they make ATP from carbohydrates (glucose): by the aerobic process of cellular respiration and by the anaerobic process of fermentation. Muscle cells use both these methods to generate the ATP they need.

Control of Muscle Contraction

Muscles are useful only if they contract in a controlled fashion. In Chapter 37 you learned that motor neurons connect the central nervous system to skeletal muscle cells (effectors). Impulses (action potentials) from motor neurons control the contraction of skeletal muscle cells.

Figure 38–14 shows how a motor neuron makes contact with a typical skeletal muscle cell. The point of contact is called the neuromuscular junction. Vesicles, or pockets, in the axon terminals of the motor neuron release molecules of the neurotransmitter acetylcholine (as-ih-tihl-KOH-leen). These molecules diffuse across the synapse, producing an impulse in the cell membrane of the muscle cell. The impulse causes the release of calcium ions (Ca^{2+}) within the cell. The calcium ions affect regulatory proteins that allow actin and myosin filaments to interact and form cross-bridges. From the time a nerve impulse reaches a muscle cell, it is only a matter of a few milliseconds before this series of events takes place and the muscle cell contracts.

A muscle cell will remain in a state of contraction until the production of acetylcholine stops. An enzyme called acetylcholinesterase (as-ih-tihl-koh-leen-EHS-ter-ayz), also produced at the neuromuscular junction, destroys acetylcholine, permits the reabsorption of calcium ions into the muscle cell, and terminates the contraction.

Is there a difference between a strong contraction and a weak contraction? Recall that each muscle contains hundreds of muscle cells. When your brain sends a message for a weak contraction, such as blinking the eyes, only a few muscle cells are stimulated. When you exert maximum effort, trying to lift a heavy weight perhaps, most muscle cells are stimulated.

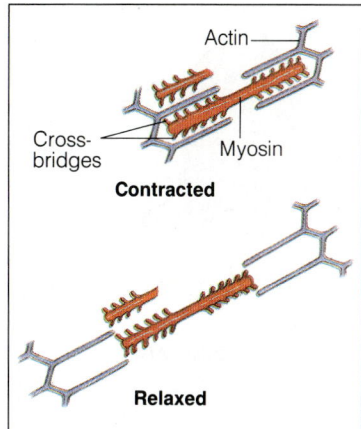

Figure 38–13 When the muscle is stimulated to contract, cross-bridges on the myosin filaments move and pull the actin filaments past each other. These actions cause the muscle cells to shorten. When there is no stimulation, the cross-bridges let go and the muscle relaxes.

Figure 38–14 The faintly striped regions in this electron micrograph are muscle fibers, and the dark branches are motor nerve fibers. The dark, brushlike ovals are formed by the many axon terminals at the neuromuscular junctions.

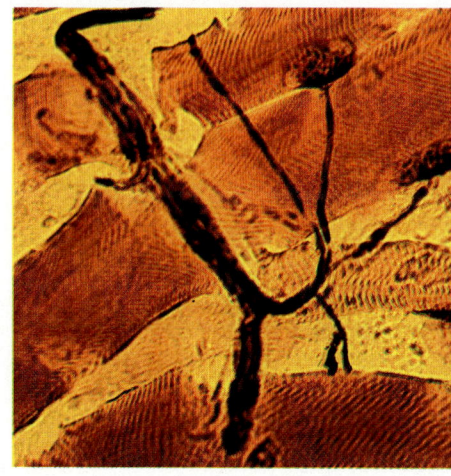

BACKGROUND INFORMATION
OXYGEN DEBT

During strenuous exercise, such as running, the powerful contractions of the leg muscles use every molecule of available ATP within a few seconds. Oxygen cannot be supplied rapidly enough to produce ATP aerobically by the oxidation of glucose, so anaerobic fermentation, producing lactic acid along with the needed ATP, occurs. Too much lactic acid in the blood can dangerously lower blood pH, so the liver converts lactic acid back to carbohydrates, but it needs oxygen to do so. While running, even though breathing heavily, the runner cannot get enough oxygen to handle all the lactic acid. Therefore, an oxygen debt builds up. This oxygen debt is "paid back" at the end of the race when the runner breathes heavily for some time, acquiring the oxygen the liver needs to do its job.

The increase in lactic acid is one of the reasons muscles may begin to hurt during vigorous exercise. Although regular exercise improves the ability of the lungs and circulatory system to get oxygen to the muscles, a well-trained athlete learns to tolerate the inevitable pain.

---

pulse to the muscle cell? (Acetylcholine released from vesicles in the cytoplasm of the axons causes the release of calcium ions in the cell, resulting in muscle contraction. Within a few milliseconds, acetylcholine is destroyed by acetylcholinesterase, and the muscle cell relaxes and is ready to contract again.)

• **What is the difference between a strong and a weak muscle contraction?** (The all-or-none rule applies; a stimulated muscle cell contracts completely or not at all. The strength of a contraction depends, therefore, on the number of cells that contract, not on the degree of contraction.)

---

taneously provides the force necessary to move the skeleton.

Reinforcement/Reteaching
Using the Visuals Have students refer to Figures 38–12 and 38–13 as the sliding filament theory is discussed.

Content Development
Using the Visuals As you discuss the control of muscle contraction, you may wish to ask questions such as the following:
• **How are muscles able to contract in a coordinated way?** (Refer students to Figure 38–14. A motor neuron typically divides into several branches, with each branch going to a separate muscle cell. A nerve impulse traveling down a neuron will cause all the muscle cells its branches contact to contract simultaneously.)
• **What transmits a nerve im-**

TIE-IN PHYSICS

The movement of the skeleton by the muscles results in the accomplishment of work, defined as the movement of a force through distance. In this respect, our skeletomuscular system is similar in action to many of the tools we use, which are simple machines called *levers.* However, there is an important difference. Most of our tools function as first- or second-class levers, enabling us to use less effort force to move a larger resistance force. In contrast, the body's long bones function as third-class levers, in which a large effort force is needed by the muscles to move the resistance (the bones) through a much greater distance. For example, as the biceps muscle forcefully contracts a few centimeters, the hand moves a distance of about a meter. This explains why in most people, the muscles and bones of the thigh are the largest in the body. In walking or running, the large motions of the femur in relation to both the pelvis and the knee joint require powerful forces.

38–2 (continued)

Focus/Motivation
If students are not familiar with levers and their operation, you may want to demonstrate to them the operation of such levers as a crowbar, a can opener, a fork, and a baseball bat. Use the terms *fulcrum, effort,* and *resistance;* compare the motions of the levers to the motions of bones.

Content Development
Most muscles are not directly attached to bone but act through tendons that transmit the pull of the muscle to the bone. Made of extremely strong fibers under extreme strain, tendons will tear away from the bone before they themselves tear.

Few body motions are the result of a single muscle acting alone; usually muscles work in antagonistic pairs or in groups.

SECTION REVIEW 38–2

1. The muscular system allows for movement of the body.
2. Skeletal muscles have striations, and their cells are long and slender with more than one nucleus. Smooth muscles do not have striations, and their cells are spindle-shaped with individual nuclei. Cardiac muscles are striated, and their cells contain one nucleus located near the cell's center.
3. When a muscle is stimulated to contract, the cross-bridges, which connect the myosin filaments to the actin filaments, move, pulling the two filaments past each other. After each cross-bridge has moved as far

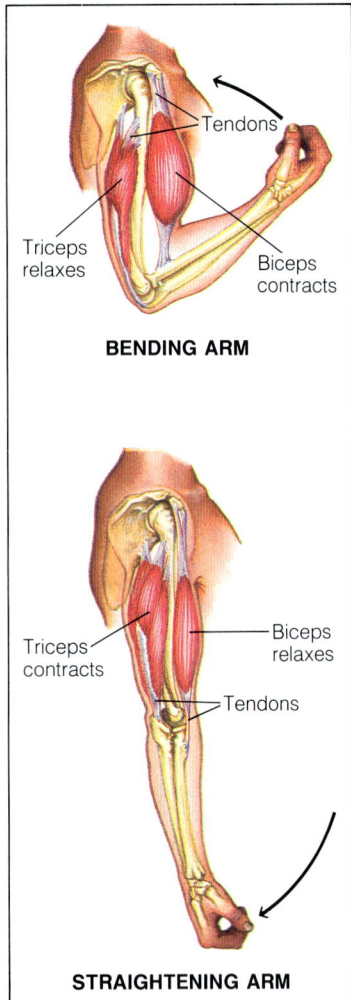

Figure 38–15 Muscles, which are attached to bones by tendons, usually work in pairs. The biceps and the triceps work together to move your elbow joint. Which muscle contracts when you "make a muscle"? Which muscle relaxes? Why are the biceps and triceps called opposing muscles? ❶

How Muscles and Bones Interact

You have just learned that skeletal muscles generate force and produce movement only by contracting, or pulling on body parts. Individual muscles can only pull; they cannot push. Yet you know from experience that your arms and legs can push when you want them to. How is this possible?

Skeletal muscles are joined to bones by tough connective tissues called **tendons.** Tendons are attached in such a way that they pull on the bones and make them work like levers. The movements of the muscles and joints enable the bones to act as levers. The joint functions as a fulcrum (fixed point around which the lever moves), and the muscles provide the force to move the lever. See Figure 38–15. Usually there are several muscles surrounding each joint that pull in different directions.

Most skeletal muscles work in pairs. When one muscle or set of muscles contracts, the other relaxes. The muscles of the upper arm are a good example of this dual action. When the biceps muscle (on the front of the upper arm) contracts, it bends, or flexes, the elbow joint. When the triceps muscle (on the back of the upper arm) contracts, it opens, or extends, the elbow joint. A controlled movement, however, requires contraction by both muscles. To hold a tennis racket or a violin, for example, both the biceps and triceps must contract in balance. This is why the training of athletes and musicians is so difficult. The brain must learn how to work opposing muscle groups to just the right degree to get the joint to move precisely.

A normal characteristic of all skeletal muscles is that they remain in a state of partial contraction. At any given time, some muscles are being stimulated while others are not. This causes a tightened, or firmed, muscle and is known as muscle tone. Muscle tone is responsible for keeping the back and legs straight and the head upright even when you are relaxed. Regular exercise is good for your body because it increases your muscle tone. Muscles that are exercised regularly stay firm and increase in size by adding more materials to the inside of the muscle fibers. Muscles that are not used at all get weak, flabby, and actually decrease in size.

38–2 SECTION REVIEW

1. What is the function of the muscular system?
2. Name and describe the three types of muscles.
3. Describe the mechanism of muscle contraction. How is muscle contraction controlled?
4. How do muscles, bones, and joints produce movement?
5. **Critical Thinking—Relating Concepts** Do the number of bands in a striated muscle cell increase or decrease during contraction? Explain.

38–3 The Integumentary System

"Good fences make good neighbors," wrote the American poet Robert Frost, as he explained the importance of property boundaries. Living things have their own "fences," and none is as important as the skin—the boundary that separates the human body from the outside world.

Skin, the single largest organ in the body, is part of the **integumentary** (ihn-tehg-yoo-MEHN-ter-ee) **system**. The word integument comes from a Latin word that means to cover, reflecting the fact that the skin and its accessory structures form a covering over the entire body. **Skin and its accessory organs—the hair, nails, and a variety of glands—make up the integumentary system.**

The most important function of the integumentary system is protection. It performs this function by serving as a barrier against infection and injury; helping to regulate body temperature; removing waste products from the body; and providing protection against ultraviolet radiation from the sun. Because the main component of the integumentary system, the skin, contains several types of sensory receptors, it serves as the gateway through which sensations such as pressure, heat, cold, and pain are transmitted to the nervous system.

Epidermis

The skin consists of two main layers. The outermost layer is known as the **epidermis**. Most of the cells of the epidermis undergo rapid cell division (mitosis). As new cells are produced, they push older cells to the surface of the skin. Here the older cells become flattened, lose their cellular contents, and begin making keratin, a tough fibrous protein. In humans, keratin forms the basic structure of hair, nails, and calluses. In

Guide For Reading
- What structures make up the integumentary system?
- What are the functions of the skin?
- What is the structure of the skin, hair, and nails?

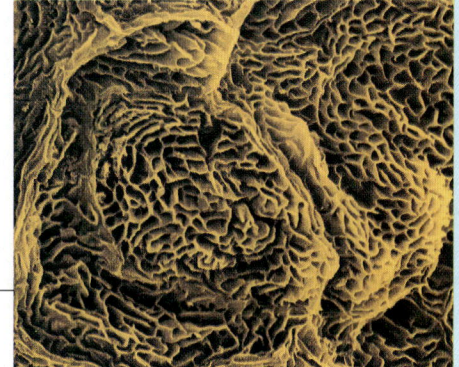

Figure 38–16 The skin is the body's largest organ, covering an area of almost 2 square meters. About 6.5 square centimeters of skin may contain 20 blood vessels, 650 sweat glands, and more than 1000 nerve endings. Grooves and ridges formed by epidermal cells on the fingertips cause fingerprint patterns, as seen in the electron micrograph.

TEACHING SUPPORT

Product Testing Activities
- Testing Bandages
- Testing Lip Balms
- Testing Shampoos

Teaching Resources
- Activity: Describing Skin and Bones, p. 5
- Hands-On Activity: How Fast Do Your Nails Grow? p. 175
- Hands-On Activity: Lip Service? p. 177
- Vocabulary Review: Crossword, p. 1

Study Guide
- Section 38-3, p. 369

38-3 (continued)

Content Development

As they move toward the surface of the skin, epidermal cells flatten, lose their ability to reproduce, and produce granules containing keratin. Eventually they die, forming a protective waterproof covering.

In discussing the dermis, emphasize that most of what we can "feel" as skin consists of dermis. It consists mainly of connective tissue called collagen. As a person ages, the collagen layer thins out, causing the skin to sag and wrinkle. The dermis also contains blood vessels and nerve endings, which are absent in the epidermis.

The sweat glands play an important role in the regulation of body temperature: In strenuous exercise, heat might elevate body temperature to dangerous levels if it were not for the cooling effect of the evaporation of perspiration.

Enrichment

You may want to discuss with students the claims made by skin-care products that "nourish" and firm the skin. In light of what they have learned about the structure and function of the skin, are these claims justified?

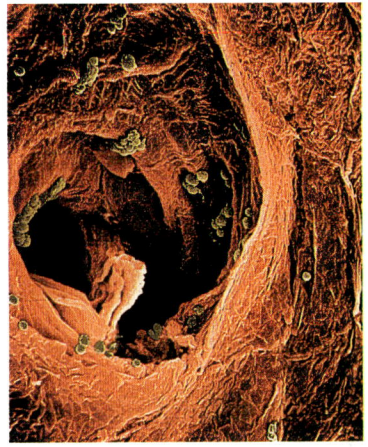

Figure 38-17 The skin of an adult contains approximately 3 million sweat glands, one of which is shown in this photograph. The tiny green circular objects inside this sweat gland are bacteria.

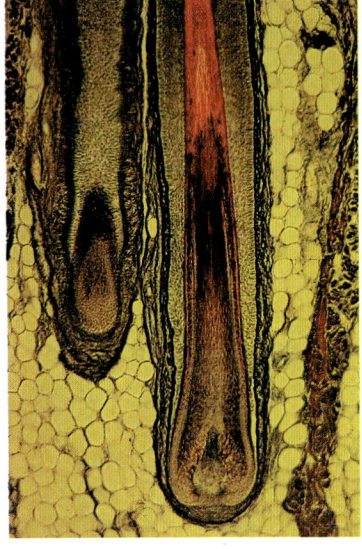

Figure 38-18 In humans, hair is found almost everywhere in skin. Hair is produced in tubelike structures called hair follicles.

848

Content Development

Hair, like the outer layers of epidermis, is nonliving material that consists of the protein keratin and the pigment that gives hair its color. Discuss with students the growth process of hair.

Nails are made of compact and tough keratin. The capillaries under them cause them to appear pink. The "white crescent" at the base of the nail is the actively growing area.

Encourage students to do research on the effects of ultraviolet light on the skin and on the dangers of overexposure.

other animals, keratin is more versatile—forming cow horns, reptile scales, bird feathers, and porcupine quills.

Eventually, the keratin-producing cells (keratinocytes) die and form a tough, flexible waterproof covering on the surface of the skin. This outer layer of dead cells is shed or washed away at a surprising rate—once every 14 to 28 days.

The epidermis also contains melanocytes, or cells that produce melanin, a dark pigment. Although light-skinned and dark-skinned people have roughly the same number of melanocytes, the difference in their skin color is caused by the amount of melanin the melanocytes produce and distribute. There are no blood vessels in the epidermis, which explains why a slight scratch will not cause bleeding.

Dermis

The **dermis** is the innermost layer of the skin. It lies beneath the epidermis and contains blood vessels, nerve endings, glands, sense organs, smooth muscles, and hair follicles. When the body needs to conserve heat on a cold day, the blood vessels in the dermis narrow, helping to limit heat loss. On hot days, the blood vessels widen, warming the skin and increasing heat loss. Beneath the dermis is the hypodermis, a layer of fat and loose connective tissue that insulates the body.

The dermis contains two major types of glands: sweat glands and sebaceous (suh-BAY-shuhs), or oil, glands. These glands pass through the epidermis and release their products at the surface of the skin. Sweat glands produce the watery secretion known as sweat, which contains salts, water, and other compounds. These secretions are stimulated by nerve impulses that cause the production of sweat when the temperature of the body is raised. Sebaceous glands produce an oily secretion known as sebum that spreads out along the surface of the skin and keeps the keratin-rich epidermis flexible and waterproof.

Hair and Nails

Hair is produced by cells at the base of structures called **hair follicles**. Hair follicles are tubelike pockets of epidermal cells that extend into the dermis. Individual hairs are actually large columns of cells that have filled with keratin and then died. Rapid cell growth at the base of the hair follicle causes the hair to grow longer. Hair follicles are in close contact with sebaceous glands. The oily secretions of these glands help maintain the condition of each individual hair.

Nails grow from an area of rapidly dividing cells known as the **nail matrix**. The nail matrix is located near the tips of the fingers and toes. During cell division, the cells of the nail matrix fill with keratin and produce a tough, strong platelike nail that covers and protects the tips of the fingers and toes. Nails grow at an average rate of 0.5 to 1.2 millimeters per day, with fingernails growing more rapidly than toenails.

SECTION REVIEW 38-3

1. The skin, hair, and nails.
2. The skin protects the body by serving as a barrier against

SCIENCE, TECHNOLOGY, AND SOCIETY

BREAKTHROUGH

Finding the Gene for Duchenne Muscular Dystrophy

Muscular dystrophy (MD) is a serious genetic disorder that results in a gradual wasting away of muscle tissue. The most common severe form of the disorder is Duchenne muscular dystrophy. Duchenne MD is a sex-linked disorder that occurs in 1 out of 3500 male babies in the United States. Males who carry the gene for Duchenne MD begin to suffer its first symptoms in early childhood. By the time these males reach early adulthood, the gradual degeneration of muscle tissue causes paralysis and then death.

Using the techniques of molecular biology, a group of researchers, led by Louis Kunkel of Harvard University, announced in 1987 that they had found the location of the Duchenne MD gene on the X chromosome. This gene codes for an unknown protein that was soon to bear the name dystrophin. In healthy people, dystrophin is a tiny fraction, less than 0.002 percent, of the total protein in muscle tissue. But in people with Duchenne MD, dystrophin is completely absent or nonfunctional. Genetic studies now explain this phenomenon: MD sufferers are missing portions of the dystrophin gene.

Before the discovery of the MD gene, the existence of dystrophin was unknown. At present, scientists are trying to determine what role dystrophin plays in normal muscle. Their first experiments show that dystrophin can bind actin, the protein found in thin filaments, and that the absence of dystrophin causes the gradual weakening of muscle tissue that results in MD.

Will it ever be possible to cure MD? Some researchers think so. They are experimenting with replacing the defective dystrophin gene in muscle cells with a healthy one, then injecting the modified cells into muscle tissue in the hope that the modified cells will replace MD-weakened cells. It is too early to say for certain, but such techniques may one day make MD a thing of the past.

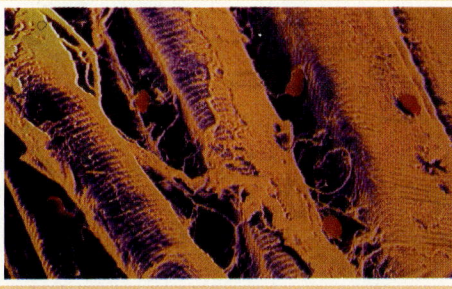
Normal skeletal muscle

38–3 SECTION REVIEW

1. What structures make up the integumentary system?
2. What are the functions of the skin?
3. Describe the epidermis and the dermis.
4. In what way is the growth of hair and nails similar?
5. **Critical Thinking—Relating Cause and Effect** Explain why your nose and ears turn red on cold winter days.

LABORATORY INVESTIGATION

EXAMINING SKELETAL MUSCLE

Before the Lab
Gather and prepare all materials. It is best to use fresh beef, to be sure that the muscle tissue has not begun to degenerate. It would also be useful, if possible, to obtain an entire (small) muscle, so that students can see the shape of the muscle and observe its attached tendons.

Pre-Lab Discussion
- **What kind of muscle is this? How do you know?** (Skeletal muscle—its large size and parallel fibers would indicate this.)
- **What do you expect to see when you observe this muscle under the microscope?** (Striations, multinucleated cells, parallel muscle fibers.)

Skills Development
Students will use the following skills while completing this laboratory investigation.
1. Hypothesizing
2. Manipulative
3. Observing
4. Recording
5. Inferring
6. Safety
7. Comparing
8. Applying
9. Relating
10. Concluding

Safety Tips
Dissecting needles should be used with caution; emphasize that horseplay with the dissecting needles will not be tolerated.

Teaching Strategy
Emphasize that the piece of muscle fiber needs to be extremely thin if the student's slide preparation is to be successful. Demonstrate the method of obtaining a thin muscle fiber for your students before allowing them to begin the lab. It will probably be necessary to tease apart the muscle fiber you obtain from the large piece of muscle (using two dissecting needles) to obtain a thin enough fiber. Do so before attempting to stain the tissue. If the coverslip sits awkwardly on the piece of muscle fiber, it is too thick.

LABORATORY INVESTIGATION

EXAMINING SKELETAL MUSCLE

PROBLEM
How are muscle cells suited to their function?

MATERIALS (per group)

small piece of raw beef	toluidine blue
dissecting tray	clock with second hand
dissecting needle	coverslip
glass slide	microscope
2 medicine droppers	paper towels

PROCEDURE

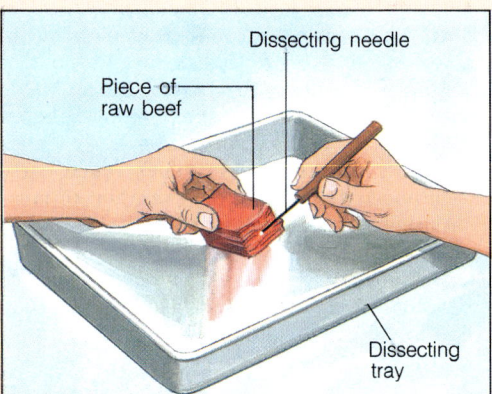

1. Place the piece of beef in the dissecting tray.
2. To remove a muscle fiber from the beef, pass the tip of the dissecting needle over the beef's surface a few times. **CAUTION:** *Be careful when using a dissecting needle.*
3. Transfer the muscle fiber to a glass slide.
4. Using one of the medicine droppers, place two drops of toluidine blue on top of the muscle fiber. Allow the muscle fiber to remain in the toluidine blue for two minutes.
5. After two minutes, use a piece of paper towel to absorb the excess toluidine blue.
6. Using the second medicine dropper, place two drops of water on the muscle fiber. Allow the muscle fiber to remain in the water for two minutes.
7. After two minutes, use a piece of paper towel to absorb the excess water.
8. Place two additional drops of water on the muscle fiber and cover with a coverslip.
9. With the low-power objective of the microscope, locate a transparent, lightly stained section of the muscle fiber.
10. Switch to the high-power objective to focus on a few muscle cells.
11. Note the general shape of the muscle cells. Observe the light and dark patterns, the number of nuclei, and the arrangement of the muscle cells with respect to one another. Make a labeled sketch of the structures that you see.

OBSERVATIONS
1. What is the general shape of a muscle cell?
2. Describe the appearance of a muscle cell.
3. How many nuclei does a muscle cell have?
4. How are the muscle cells arranged with respect to one another?

ANALYSIS AND CONCLUSIONS
1. What microscopic evidence do you have that shows muscle cells contract by the overlapping of actin and myosin filaments?
2. How should the muscle fibers be arranged in order to provide for their movement in the same direction?
3. How are muscle cells suited to their function?

Observations
1. Long and slender.
2. The muscle cell appears striated (has bands or stripes).
3. Many.
4. They are arranged parallel to one another.

Analysis and Conclusions
1. The striated appearance is caused by the areas where the filaments overlap.
2. To provide for movement in the same direction, the fibers should be arranged parallel to one another.

STUDENT STUDY GUIDE

SUMMARIZING THE CONCEPTS

The key concepts in each section of this chapter are listed below to help you review the chapter content. Make sure you understand each concept and its relationship to other concepts and to the theme of this chapter.

38-1 The Skeletal System

- Bones and their associated tissues—cartilage, tendons, and ligaments—make up the skeletal system.
- Bones of the skeletal system support and shape the body, provide a system of levers on which muscle tissues act to produce movement, contain reserves of minerals and are sites of blood-cell formation.
- During ossification, or bone formation, cartilage is replaced by bone.
- Joints, or places where two bones come together, permit bones to move without damaging each other.

38-2 The Muscular System

- The muscular system is composed of muscle tissue that is highly specialized to contract, or shorten, when stimulated.
- There are three different types of muscle tissue: skeletal, smooth, and cardiac.

38-3 The Integumentary System

- Skin and its accessory organs—the hair, nails, and a variety of glands—make up the integumentary system.
- The integumentary system has many functions: It protects the body by serving as a barrier against infection and injury, regulates body temperature, removes waste products, provides protection against ultraviolet radiation, and serves as a gateway through which sensations are sent to the nervous system.
- The outermost layer of the skin is the epidermis. The innermost layer is the dermis.
- Hair is produced by cells at the base of hair follicles. The nail matrix produces a tough, strong platelike nail.

REVIEWING KEY TERMS

Vocabulary terms are important to your understanding of biology. The key terms listed below are those you should be especially familiar with. Review these terms and their meanings. Then use each term in a complete sentence. If you are not sure of a term's meaning, return to the appropriate section and review its definition.

38-1 The Skeletal System
skeletal system
periosteum
Haversian canal
osteocyte
bone marrow
cartilage
ossification
joint
ligament

38-2 The Muscular System
muscular system
skeletal muscle
smooth muscle
cardiac muscle
myosin
actin
cross-bridge
sliding filament theory
tendon

38-3 The Integumentary System
integumentary system
epidermis
dermis
hair follicle
nail matrix

851

3. Muscle cells are suited to their function because of their long, slender shape, which causes movement when they contract or relax.

Going Further: Enrichment
Students should observe commercially prepared slides of smooth and cardiac muscles and compare their microscopic appearance to that of the skeletal muscle fiber observed in this investigation.

CHAPTER REVIEW

CONTENT REVIEW

Multiple Choice
1. b   2. c   3. b   4. d
5. a   6. b   7. d   8. c

True or False
1. F living
2. F periosteum
3. F hinge
4. F ball-and-socket joint
5. F Ligaments
6. T
7. F myosin
8. F keratin

Word Relationships
1. names for the one type of muscle that is under voluntary control; smooth muscle is under involuntary control
2. muscle systems that consist of smooth muscle; arm consists of skeletal muscle
3. freely movable joints; vertebrae are slightly movable joints
4. parts of the integumentary system; muscle makes up the muscular system
5. osteocyte
6. tendon
7. ball-and-socket

CONCEPT MASTERY

1. Bone is an almost solid network of living cells and fibers supported by deposits of calcium salts. Blood vessels supply the bone tissue with oxygen and nutrients, and nerves help the bone tissue experience sensation.
2. Compact bone is dense and lies beneath the periosteum. Running through compact bone is a collection of tubes called Haversian canals. Spongy bone, on the other hand, resembles the supporting girders in a bridge. Both bone types contain osteocytes.
3. In freely movable joints, the ends of the bones are covered with a layer of cartilage that provides a smooth surface at the joint. The joints are also surrounded by a fibrous joint capsule that helps hold the bones together and at the same time allows for movement. The joint capsule consists of two layers. One layer forms the ligaments and the other layer produces synovial fluid, which lubricates the surface of the joint.
4. Skeletal muscles are made up of long, slender multinucleate muscle cells that are known as muscle fibers. These muscles are responsible for voluntary movement and are found attached to bone. Smooth muscles are composed of spindle-shaped individually nucleated unstriated muscle cells. These muscles are responsible for involuntary movement and are found in many internal organs and in the walls of many blood vessels. Cardiac muscle is found only in the heart and is involuntarily controlled.
5. During muscle contraction, cross-bridges pull the actin and myosin filaments over one another, increasing the amount of overlap between the filaments and shortening the fiber. When the muscle relaxes, the cross-bridges disengage and the filaments slide apart.
6. A tendon connects muscle to bone. A ligament connects bone to bone.
7. Skeletal muscles are joined to bones by tendons, which are attached in such a way that they pull on bones and make them work like levers. The joint functions as a fulcrum, and the muscles provide the force to move the lever.
8. When the biceps muscle contracts, it bends, or flexes, the elbow joint. When the triceps muscle contracts, it opens, or extends, the elbow joint.
9. The skin, along with its accessory organs, covers the entire body, forming a boundary that separates the internal body structures from the external world. This boundary serves as a barrier against infection and injury.

CHAPTER REVIEW

CONTENT REVIEW

Multiple Choice

Choose the letter of the answer that best completes each statement.

1. Which is not found in the human skeleton?
   a. calcium
   b. cellulose
   c. phosphorus
   d. living tissue
2. Bone cells are called
   a. melanocytes.
   b. lymphocytes.
   c. osteocytes.
   d. keratinocytes.
3. During ossification, cartilage is replaced by
   a. ligament.
   b. bone.
   c. tendon.
   d. muscle.
4. Bones are connected to muscles by
   a. ligaments.
   b. skin.
   c. cartilage.
   d. tendons.
5. Which serves as a lever for muscles?
   a. bones
   b. ligaments
   c. spinal cord
   d. tendons
6. Cardiac muscle tissue is located in the
   a. brain.
   b. heart.
   c. arms.
   d. digestive tract.
7. The outermost layer of skin is called the
   a. dermis.
   b. melanin.
   c. hypodermis.
   d. epidermis.
8. Melanocytes are located in the
   a. bones.
   b. dermis.
   c. epidermis.
   d. muscles.

True or False

Determine whether each statement is true or false. If it is true, write "true." If it is false, change the underlined word or words to make the statement true.

1. Bone is <u>nonliving</u> tissue.
2. The <u>Haversian canal</u> is a tough membrane that surrounds bone.
3. The elbow is a <u>pivot</u> joint.
4. An example of a <u>hinge</u> joint is the shoulder.
5. <u>Tendons</u> attach bones to bones.
6. Two types of striated muscle are <u>cardiac</u> and skeletal.
7. The thick filaments in a muscle cell are made up of <u>actin</u>.
8. Individual hairs are actually large columns of cells that have filled with <u>collagen</u>.

Word Relationships

A. *In each of the following sets of terms, three of the terms are related. One term does not belong. Determine the characteristic common to three of the terms and then identify the term that does not belong.*

1. striated, voluntary, skeletal, smooth
2. digestive tract, blood vessel, arm, pupil
3. hip, vertebrae, elbow, knee
4. muscle, hair, skin, nail

B. *An analogy is a relationship between two pairs of words or phrases generally written in the following manner: a:b::c:d. The symbol : is read "is to," and the symbol :: is read "as." For example, cat:animal::rose:plant is read "cat is to animal as rose is to plant."*

In the analogies that follow, a word or phrase is missing. Complete each analogy by providing the missing word or phrase.

5. skin:melanocyte::bone:_____
6. bone to bone:ligament::muscle to bone:_____
7. elbow:hinge::hip:_____

CONCEPT MASTERY

Use your understanding of the concepts developed in the chapter to answer each of the following in a brief paragraph.

1. Explain why bone is considered living.
2. Compare compact bone and spongy bone.
3. Describe a freely movable joint.
4. Compare the structure, function, and location of the three types of muscles.
5. Describe the sliding filament theory.
6. Describe a tendon and a ligament.
7. Using the terms fulcrum, lever, and force, explain how the contraction of a muscle results in movement.
8. What happens to the biceps and triceps when the arm is extended? When the arm is flexed?
9. Explain how the integumentary system provides protection for the body.
10. What is the importance of the sweat glands? Of sebaceous glands?
11. Explain why you are able to move the bones of your finger but not the bones of your skull.

CRITICAL AND CREATIVE THINKING

Discuss each of the following in a brief paragraph.

1. **Applying concepts** Suggest a reason why there are more joints in the hands and feet than in most other parts of the body.
2. **Making generalizations** People usually have more calluses, pads of thickened skin, on the soles of their feet or the palms of their hands. Explain the reason for this.
3. **Relating facts** Examine the photograph of coarse and fine hairs. Is hair alive or dead? Explain your answer.
4. **Making comparisons** Why do bones heal faster in children than they do in adults?
5. **Developing a hypothesis** You have a habit of leaning on your elbow while reading. As a result, you have developed a noticeable swelling at your elbow. Explain what has happened.
6. **Identifying relationships** Few nerves and blood vessels enter a tendon or a ligament. Therefore, the supply of nutrients to these tissues is poor. Explain why injured tendons and ligaments take a long time to heal.
7. **Identifying patterns** Osteoporosis is a disease that usually occurs in older women. It involves a loss and weakening of bone tissue. Doctors recommend that women over the age of 45 eat more foods that contain calcium. How is this helpful in preventing osteoporosis?
8. **Using the writing process** Develop an advertising campaign for the dairy industry based on the relationship between milk and healthy bone development. Make it a full media blitz!

CRITICAL AND CREATIVE THINKING

1. The hands and feet are two parts of the body that perform the most complex movements and movements that require fine motor coordination. To accomplish such diverse movements, a wide variety of bones and joints is necessary.
2. The soles of the feet and the palms of the hands usually have more calluses because these areas are subjected to more friction and pressure than any other part of the body. As a result, the layers of skin are thickened in these areas to protect underlying structures.
3. Hair is composed of large columns of cells that have filled with keratin and then died.
4. The bones of young children are still developing and heal faster because they are in the formative stages. In adults, the bones have become calcified and heal less quickly because they are no longer growing.
5. The constant pressure of leaning on your elbow causes an irritation of the joint capsule. Eventually, the bursa in the capsule becomes inflamed and swells.
6. In the absence of a good blood supply, nutrients, oxygen, and cells that are involved in tissue repair enter the tissue of tendons and ligaments very slowly. As a result, the ability of these tissues to undergo repair is poor.
7. Calcium is necessary for bone growth and to keep bones strong. The weakening of bones due to osteoporosis can be somewhat alleviated if the body has a plentiful supply of calcium, which can be used to make the bones harder.
8. Students' advertising campaigns should show knowledge of the relationship between milk and the development of healthy bone.

10. Sweat glands produce the watery secretion called sweat, which helps to keep the internal temperature of the body stable. Sebaceous glands produce an oily secretion called sebum, which keeps the surface of the skin flexible and waterproof.
11. The joints located between the bones of the finger are freely movable joints and permit more movement than the joints found between the bones of the skull. These joints are immovable.

CHAPTER 39 Nutrition and Digestion

Section	Laboratory Investigations and Activities
39–1 Food and Nutrition pages 855–866 THEMES: Unity and Diversity, Energy	**Student Edition** LABORATORY INVESTIGATION: Testing the Nutrient Content of Foods, p. 876 **Teacher Edition** DEMONSTRATION: Determining the Energy Content of Foods, p. 854D **Laboratory Manual** Measuring Food Energy, p. 493 **Teaching Resources** HANDS-ON ACTIVITY: Getting the Iron Out, p. 185 TEACHER DEMONSTRATION: Digestion of Carbohydrates, p. 25 **Product Testing Activities** Testing Bottled Water Testing Cereals Testing Popcorn
39–2 The Process of Digestion pages 867–875 THEMES: Patterns of Change, Systems and Interactions	**Teacher Edition** DEMONSTRATION: The Esophagus, p. 854D **Laboratory Manual** Mechanical and Chemical Digestion, p. 489 **Teaching Resources** HANDS-ON ACTIVITY: Toasting to Good Health, p. 183 HANDS-ON ACTIVITY: Going Crackers, p. 189 **Product Testing Activities** Testing Antacids
Chapter Review pages 877–879	

Teacher Resources

Books

Calloway, D. H., and K. O. Carpenter. *Nutrition and Health*. Saunders, 1981.

Magee, D. F., and A. F. Dalley. *Digestion and the Structure and Function of the Gut*. Karger, 1986.

Marieb, Elaine. *Essentials of Human Anatomy and Physiology,* 2nd ed. Benjamin-Cummings, 1988.

Nilsson, Lennart. *Behold Man*. Little, Brown, 1973.

CHAPTER PLANNING GUIDE

Other Activities	Media and Technology
Teaching Resources ACTIVITY: Calculating Nutrients Available in Fast Foods, p. 3 **Study Guide** Section 39–1, p. 373	**Video/Videodisc** Human Body: Endocrine System and Digestive System **Biology Media Guide** page 59
Teaching Resources VOCABULARY REVIEW: Cross-a-Clue, p. 1 ACTIVITY: The Effects of Food on Digestion, p. 7 **Study Guide** Section 39–2, p. 377	**Video/Videodisc** Human Body: Endocrine System and Digestive System **Biology Transparencies** 55: The Human Digestive System 63: The Structure of a Human Tooth **Biology Media Guide** pages 59–61
Teaching Resources Chapter Test A, p. 13 Chapter Test B, p. 17	**Computer Test Bank** Chapter Test, p. 411

Audiovisuals

Digestion. Film or video. MTI Coronet.
Digestive System, 2nd ed. Film or video. Coronet.
The Chemistry of Digestion. Film or video. Coronet.
Nutrition and Metabolism. Film or video. Coronet.
Digestion: A Tough Dirty Job; It Takes a Lot of Guts. Video. University of Wisconsin—Milwaukee.
Digestive System: Structure and Function. Sound filmstrip. Ward.
Introduction to Nutrition. Sound filmstrip. Ward.
The Digestive System. Film. Encyclopaedia Britannica Ed. Corp.

854B

CHAPTER 39 Nutrition and Digestion

CHAPTER OVERVIEW

The focus of this chapter is on a favorite human activity—eating. In the first section of the chapter, food will be looked at from the nutritional perspective. In the second section, the process of digestion will be explored.

Students will learn that food is as important to the human body as fuel is to a machine. In fact, food is the fuel that gives the human "machine" energy. But providing energy is not the only purpose of food. Food must also provide the substances that are needed for body growth, maintenance, and repair.

The human body must have certain essential nutrients if it is to survive and thrive. These nutrients fall into six categories: water, carbohydrates, fats, proteins, vitamins, and minerals. Some of these nutrients, such as water, must be consumed in fairly large amounts. Others, such as vitamins and minerals, are needed in much smaller amounts.

Once food is ingested, it must be broken down, absorbed, and transported to various parts of the body. The first two of these processes take place in the digestive tract. The human digestive tract consists of six organs: mouth, pharynx, esophagus, stomach, small intestine, and large intestine. Also essential to digestion are the salivary glands, the pancreas, and the liver.

39–1 FOOD AND NUTRITION

Section Focus 39–1

The purpose of this section is to emphasize the importance of food and to discuss the nutritional requirements of the human body. Students will learn that food is needed to supply the human body with energy. Food is also the source of substances that are needed to maintain body functions and to provide the raw materials needed for growth and repair.

The human body requires six types of essential nutrients. These are water, minerals, carbohydrates, proteins, fats, and vitamins. Students will learn which foods supply the various minerals and vitamins that their body requires. They will also learn how the body uses carbohydrates, fats, and proteins to provide energy, to build and rebuild tissue, and to maintain body structure and function.

In the last part of the section, the importance of a balanced diet will be discussed. Students will learn that in order to achieve good nutrition and health, they should select a variety of foods from the basic food groups identified in the food guide pyramid. They will learn that they should also strive to adhere to certain percentages of fats, carbohydrates, and proteins in their diet.

Performance Objectives 39–1

1. Discuss the importance of food for human survival and well-being.
2. Explain how the energy content of food is measured.
3. Identify the essential nutrients and tell how each is important to the body.
4. Discuss the different criteria for a well-balanced diet.

Science Terms 39–1

nutrition   p. 855
calorie   p. 855
mineral   p. 857
essential amino acid   p. 863
vitamin   p. 863

39–2 THE PROCESS OF DIGESTION

Section Focus 39–2

The purpose of this section is to describe the process of digestion in the human body. The section discusses in detail the function of each organ in the digestive tract, as well as the action of several important glands.

During digestion, food is broken down both mechanically and chemically. Mechanical digestion is the breakdown of food from larger into smaller pieces. This process begins in the mouth with the action of the teeth. Chemical digestion takes place when digestive enzymes and other chemicals act on food. During chemical digestion, special enzymes break down food molecules into smaller, simpler molecules that can be used by the body.

After food has been acted on mechanically and chemically, the products of digestion are absorbed into the bloodstream. Absorption takes place in the small intestine. Villi increase the surface of the small intestine, making absorption proceed more rapidly.

In the last part of the section, the role of the large intestine is discussed. The main task of the large intestine, or colon, is to remove water from undigested materials so that they can pass out of the body as solid wastes.

CHAPTER PREVIEW

Performance Objectives 39–2

1. Identify the organs of the digestive system and explain their functions.
2. Identify several important glands associated with the digestive system and explain their functions.
3. Compare mechanical and chemical digestion.
4. Discuss the roles of specific enzymes in the digestive process.
5. Explain how the products of digestion are absorbed into the bloodstream and how waste materials are prepared for elimination.

Science Terms 39–2

digestion   p. 867
gastrointestinal tract   p. 867
bolus   p. 870
pharynx   p. 870
epiglottis   p. 870
esophagus   p. 870
peristalsis   p. 870
sphincter   p. 870
stomach   p. 871
chyme   p. 871
pyloric valve   p. 871
duodenum   p. 872
small intestine   p. 872
pancreas   p. 872
liver   p. 873
gallbladder   p. 873
villus   p. 874
large intestine   p. 875
colon   p. 875

DISCOVERY LEARNING

TEACHER DEMONSTRATIONS
Modeling

Determining the Energy Content of Food

The following demonstration can be used as an introduction to Chapter 39.

For the demonstration you will need a Styrofoam cup or container; a small metal can that will fit easily inside the cup; water; a needle; a cork; matches; a piece of food such as bread, candy, meat, or potato; a balance scale; a graduated cylinder; and a thermometer.

Explain to students that different foods contain different amounts of energy. Energy is measured in a unit called the calorie. A calorie is defined as the amount of energy needed to raise the temperature of 1 gram of water 1 degree Celsius. Also explain that 1000 calories (1 kilocalorie) equals one food calorie (1 Calorie), which is customarily written with a capital C.

Measure or have a student measure the mass of the food item to be burned. Record the value in grams on the chalkboard. Place the can inside the Styrofoam cup and determine how much water is needed to fill the area between the can and the cup so that the water is about 2.5 centimeters from the top of the can. Record the mass of the water on the chalkboard. (Remind students that the volume of the water can be used to determine the mass, since 1 milliliter of water has a mass of 1 gram.) Then measure and record on the chalkboard the temperature of the water.

Once the can is inside the Styrofoam cup and water has been poured into the area between the can and the cup, insert a needle into the food substance to be burned. Insert the other end of the needle into a cork, then place the cork inside the can to serve as a mount for the piece of food. Ignite the food substance and let it burn completely. As soon as it is completely burned, have a student measure the temperature of the water. Record the value on the chalkboard.

Multiply the change in temperature of the water by the mass of the water to get the number of calories released by the burning of the food. Divide this number by 1000 to find the equivalent number of Calories (food calories). Finally, divide the number of Calories by the mass of the food to determine the number of Calories per gram.

Students may wish to compare the results with the established Calorie content of the food given in a Calorie chart.

The Esophagus

This demonstration can be performed as students study Section 39–2.

For the demonstration you will need a length of plastic or rubber tubing, a marble, and some mineral or cooking oil.

Display the length of tubing and explain that it represents the esophagus. Insert the marble into the tubing and pinch the sides of the tube to push the marble through.

- **What movement in the esophagus does this wavelike movement represent?** (Peristalsis.)
- **What does peristalsis accomplish?** (It pushes food through the esophagus into the stomach.)

Next, lubricate the marble by rolling it in oil. Repeat the demonstration and ask:

- **What effect does the oil have on the marble?** (It makes the marble pass through the tube more easily.)

Explain that mucus and saliva lubricate food so that it passes through the digestive tract more easily.

CHAPTER 39

Nutrition and Digestion

Eating, necessary to maintain life, can also be one of life's great pleasures. Eating a meal that contains a variety of nutrients is important for good health.

Most of us would agree that eating can be one of life's great pleasures. After all, what can compare with the delight of drinking an ice-cold glass of orange juice on a hot summer afternoon or savoring a turkey leg on Thanksgiving Day?

But just because eating is an enjoyable activity doesn't mean it is one to be taken lightly. You have probably been told many times that proper eating habits and good nutrition are important to your health. But what is good nutrition? And why aren't candies and snack foods as beneficial to you as a well-balanced meal? In order to answer these and other questions about healthful eating, you need to know what food is and how your body processes food through digestion. These are the subjects on the pages that follow.

39–1 Food and Nutrition

Guide For Reading

- Why is food important?
- What main nutrients does the body need?
- Why is a balanced diet important?

Humans, like most other animals, are heterotrophs. Heterotrophs eat other organisms for food. But why must living things eat food? **Food contains nutrients, or molecules that provide energy and material for growth. Nutrients include water, minerals, carbohydrates, fats, proteins, and vitamins.** Although these nutrients are available in food, we cannot always be certain that the foods we eat will supply us with the nutrients we need. That's where the science of **nutrition** comes in. Nutrition is the study of how our bodies obtain energy, build tissue, and control body functions using materials supplied in the food we eat.

Why Food Is Important

Like an automobile or other machine that does work, our body needs fuel to supply energy. Food is our body's fuel. It supplies us with energy. We need energy not only to do work but to generate the heat that maintains our body temperature. As you may remember from Chapter 6, animals obtain energy primarily by oxidizing food to make ATP. To measure the amount of energy that can be obtained from food, biologists and chemists use the unit known as a **calorie**. A calorie is the amount of energy needed to raise the temperature of 1 gram of water by 1 degree Celsius. Because the energy needs of the body are great, nutritionists usually refer to the energy content of food in terms of the kilocalorie, which is 1000 calories. When books on diet and nutrition refer to food energy, they substitute the unit Calorie for kilocalorie. Notice that the Calorie used in place of the kilocalorie is spelled with a capital C and thus means 1000 of the calories spelled with a small c.

Figure 39–1 Doing work requires energy. These kayakers expend a great deal of energy keeping their canoes on course in the white water of the river.

855

ANNOTATION KEY

① Calcium: milk products, green leafy vegetables; iron: liver, red meats, grains, raisins, nuts. (Interpreting charts)

TEACHING SUPPORT

Teaching Resources
- Hands-On Activity: Getting the Iron Out, p. 185

Product Testing Activities
- Testing Bottled Water

MEDIA AND TECHNOLOGY

▶ Video/Videodisc
- *Human Body Series: Endocrine System and Digestive System*

Use this bar code to introduce the digestive system.

Play frames 28820 to 29295

See the Biology Media Guide pages 59–61 for additional bar-code correlations for this chapter.

BACKGROUND INFORMATION
RECOMMENDED DIETARY ALLOWANCE

The American Academy of Arts and Sciences in Washington, DC, publishes the U.S. Recommended Dietary Allowance (USRDA) for each mineral that is required by the human body. The USRDA value gives, in milligrams, the amount of each mineral that should be ingested daily by the average adult in good health. The USRDAs for the minerals listed in Figure 39–4 are as follows:

MINERAL	RDA (milligrams)
Calcium	800
Chlorine	2000
Magnesium	350
Potassium	2500
Phosphorus	800
Sodium	2500
Iron	10
Fluorine	2
Iodine	0.14

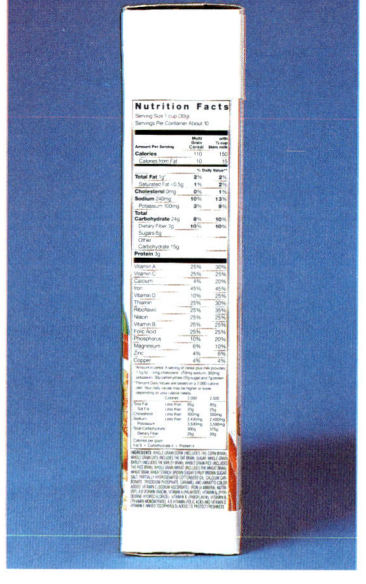

Figure 39–2 Food labels provide valuable information for consumers.

Figure 39–3 Animals that live in dry environments get the water they need from the foods they eat, as this tortoise is doing by eating a piece of cactus.

The basic energy needs of an average-sized adult human are about 1500 Calories per day. But energy needs vary depending upon the kind of work you do, how active you are, and your gender. Men generally have higher energy needs than women.

In addition to energy, food supplies building materials—the substances required by the cells in our body for proper growth and development. Even after we have reached adult size, tissues throughout the body must be repaired and replaced. Proteins and nucleic acids cannot be synthesized unless key compounds are supplied by a complete diet.

In the first part of this chapter you will learn about the kinds of food we need to eat to guarantee a supply of the vital nutrients. In the second part of the chapter you will see how the design of the digestive system ensures the proper delivery of those nutrients to the tissues that require them.

Water

Water is one of the simplest of the essential nutrients and also the most important. Animals will die from a lack of water long before they will starve from a lack of food. Most of the weight of the human body is water. So although a popular saying tells you that "you are what you eat," it would be closer to the truth if it were "you are as long as you drink."

Blood plasma—the liquid part of blood—is more than 90 percent water. Water is the solvent in which food and enzymes are dissolved in the digestive system. Water dissolves the waste materials that are eliminated in urine. On hot days, or when you take part in strenuous exercise, sweat glands remove

39–1 (continued)

Content Development
Use the Motivation discussion to illustrate that food is important to people—important enough that we often think about it and talk about it several times a day.

Yet although the average person may think of food in terms of enjoyment, appetite, or the need to count calories, the most important role of food is to maintain the function and structure of the human body.

Emphasize to students that food must serve the body in two ways. First, it must provide energy in the form of calories. Second, it must provide specific substances that are needed for body growth, maintenance, and repair. Point out that if the body needed only calories for energy,

water from your tissues and use it to moisten the surface of your body. As the water in sweat evaporates, it cools the body.

Each time you take a breath, you lose water from the inner surfaces of the lungs. (That's why a mirror fogs up if you hold it in front of your nose and mouth.) Because water is constantly being lost from the body, a steady supply of this vital liquid is required. If the water lost from the body is not replaced, essential functions will slow down, the blood will begin to thicken, and a serious medical crisis will develop.

Although most of the water lost from the body is replaced by drinking liquids, we do obtain small quantities of water in two other ways. The foods we eat contain water. And water is released in cellular respiration when carbohydrates are broken down to produce energy ($C_6H_{12}O_6 + 6O_2 \rightarrow 6CO_2 + 6H_2O$).

Minerals

Minerals are inorganic substances required by the body. Some of the most important molecules in the body contain minerals. For example, hemoglobin, the protein in red blood cells that carries oxygen, contains four atoms of the mineral iron. DNA, RNA, and ATP all contain the mineral phosphorus. Calcium, another important mineral, is a major component of bones and teeth. It is also important for the normal functioning of nerves and muscles and for the normal clotting of blood. Without magnesium, another mineral, neither nerves nor muscle tissue will function properly.

Although the body does not destroy the minerals it takes in, it does lose many of them in sweat, urine, and other waste

Figure 39-4 Minerals, which make up an important part of our diet, are used by the body in many ways. For example, your bones contain the mineral calcium; your red blood cells, the mineral iron. What is a good source of each of these minerals?

MINERALS		
Mineral	**Source**	**Use**
Calcium	Milk products, green leafy vegetables	Important component of bones and teeth; needed for normal blood clotting and for normal cell functioning
Chlorine	Table salt, many foods	Important for fluid balance
Magnesium	Milk products, meat, many foods	Needed for normal muscle and nerve functioning; metabolism of proteins and carbohydrates
Potassium	Grains, fruits, many foods	Normal muscle and nerve functioning
Phosphorus	Meats, nuts, whole grains, many foods	Component of DNA, RNA, ATP, and many proteins; part of bone tissue
Sodium	Many foods, table salt	Nerve and muscle functioning; water balance in body
Iron	Liver, red meats, grains, raisins, nuts	Important part of hemoglobin molecule
Fluorine	Water (natural and added)	Part of bones and teeth
Iodine	Seafood, iodized table salt	Part of hormones that regulate rate of metabolism

FACTS AND FIGURES

The human body is made up of 60% to 80% water. The average adult loses about 2.5 liters of water every day through respiration, perspiration, and excretion.

ESL STRATEGY

About half of all English words are of Latin origin, and many scientific terms have Latin roots. Point out that *calorie* is taken from a Latin word meaning "to be hot." Explain that when the original meaning of the word *calorie* is known, it is easier to understand its scientific usage—the amount of energy needed to raise the temperature of 1 gram of water by 1°C. Calorie units are measurements of the amount of energy we receive from the nutrients in foods we eat.

it would not really matter if a person ate a chocolate bar or a steak. Yet because the body requires various substances that are found in certain groups of foods, the type of food a person eats is of great importance. Therefore, when counting calories, quality and quantity both count.

Skills Development
Guided Practice
Skills: Interpreting charts, applying information
Using the Visuals Direct students' attention to the chart of minerals in Figure 39-4.
• Which minerals are specifically found in meats? (Iron, phosphorus, magnesium.)
• Which minerals can be obtained by eating grains? (Potassium, iron.)
• Which minerals are involved in nerve function? (Calcium, magnesium, potassium, sodium.)
• Suppose a person is suffering from an abnormal rate of metabolism. What mineral might be lacking in that person's diet? (Iodine.)
• A disease called osteoporosis afflicts many women in the middle to later years of their life. The symptoms of osteoporosis include brittle bones that break easily. The therapy for this disease involves giving the patients dietary supplements of a particular mineral. Which mineral are they given? (Calcium.)
• Why might a person suffering from anemia be told to eat liver? (Liver contains iron, which is needed to maintain the ability of red blood cells to carry oxygen.)

products. How then are these important chemicals replaced? Obviously, nibbling on automobiles to replenish the supply of iron or swallowing chalk dust to obtain calcium is not a reasonable solution. Luckily, most of these elements are found in the living tissues of plants and other animals. Plants absorb minerals from the soil and incorporate them into their tissues. Herbivores take in these minerals when they eat plants. Fruits, whole grains, meats, and vegetables contain iron, phosphorus, calcium, and magnesium. These foods also contain a wide range of other "trace elements" that the body needs in small amounts.

The minerals sodium and chlorine make up the compound commonly known as table salt and are also found in familiar foods such as milk, cheese, meat, and canned vegetables. Because the body loses sodium, potassium, and chlorine in sweat, athletes and people who perform hard physical work on hot days often need to supplement their diet with extra salt. Iodine, another mineral, is a natural component of seafood and of the drinking water in many areas. In the past, iodine deficiency, or goiter, was a serious problem in certain parts of the United States. Today, the addition of small amounts of iodine to table salt has all but eliminated this problem.

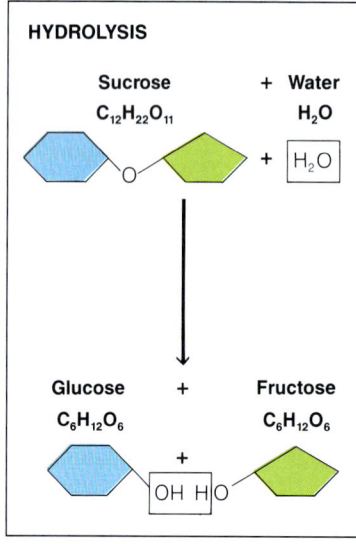

Figure 39–5 Sucrose is another name for table sugar, the sugar used to sweeten coffee, lemonade, pies, and cakes. In order to be used by the body, sucrose is first hydrolyzed to release the simple sugars glucose and fructose.

Carbohydrates

In Chapter 6 you learned that cells obtain energy from glucose in a process known as respiration. We get most of our glucose from carbohydrates: organic compounds composed entirely of carbon, hydrogen, and oxygen. Carbohydrates are the body's fuel.

Glucose is a simple sugar molecule that contains 6 carbon atoms, 12 hydrogen atoms, and 6 oxygen atoms ($C_6H_{12}O_6$). Glucose is a monosaccharide, or single sugar. Single sugars like glucose, fructose, and galactose are found in fruit juices, honey, and vegetables. Because they can be used directly by the body's cellular respiratory machinery, these simple sugars do not have to be digested, or broken down. They are therefore good sources of quick energy.

Many naturally occurring sugars are composed of two simple sugars that are linked together. These double sugars are called disaccharides. Maltose, or malt sugar, is a disaccharide made up of two glucose molecules. Another common disaccharide is sucrose, or table sugar, which is the sugar obtained from sugar cane. The disaccharide lactose is the sugar found in milk. In order for a disaccharide to be used for energy, enzymes must break the bond that holds the two simple sugars together.

Plants and animals store sugars as polysaccharides, long chains of monosaccharides. Polysaccharides are complex carbohydrates that are part of many foods. Two important polysaccharides found in plants are starch and cellulose.

Starches have always been the most important carbohydrate in the diets of humans. The starch in grains used to make bread has nourished civilizations in Europe and the Middle East. Corn provided starch to Native Americans in the United States and in Central and South America. Starch in rice provides needed energy to billions of people in Asia. Potatoes, an important crop plant throughout the world, are another important source of starch in the diets of many people.

Before starches can be used in cellular respiration, however, they must be broken down. This is accomplished by our digestive system, which first breaks the long polysaccharide chain into disaccharides. The disaccharides are then broken apart to yield simple sugars such as glucose.

In plants, starch is usually found in the form of rather large granules that do not dissolve in water. Because enzymes in the digestive system are water soluble and work only on molecules in solution, it is difficult for them to break down insoluble starch granules. Fortunately, when starch granules are treated with hot water they swell and burst, releasing water-soluble starch molecules. This is one reason why many vegetables are more easily digested after they have been cooked.

Cellulose, another important polysaccharide, is part of the cell wall of plants. However, the sugar molecules in cellulose are hooked together in such a way that humans (and many other animals) cannot take them apart. For humans, cellulose is indigestible. Our digestive system cannot extract the energy contained in wood or blades of grass. But even though the body cannot digest cellulose, we need to include a substantial amount of it in our diets. Cellulose provides our digestive system with bulk, or roughage. Roughage stimulates the muscles of the stomach and intestines. This stimulation causes our digestive system to work more efficiently. Foods such as lettuce, celery, whole grain breads, and bran are rich in cellulose.

Figure 39–6 Sugars and starches are an important part of the human diet. The source of these nutrients, however, varies in different parts of the world. These Peruvian natives eat some foods you would recognize and probably many more that you would not.

Figure 39–7 Like this beaver, many animals are herbivores. Their digestive systems have special adaptations that make it possible for them to extract nutrients from foods that the human digestive system cannot easily break down, such as these leaves.

TEACHING SUPPORT

Laboratory Manual
- Measuring Food Energy, p. 493

Product Testing Activities
- Testing Cereals
- Testing Popcorn

HISTORICAL NOTE
EARLY DIETS

Anthropologists and scientists have concluded that the earliest members of the human lineage lived on a diet of fruits, seeds, and other plant material. It is food for thought that these early ancestors knew nothing of the fats, sugars, and food additives that are so much a part of the human diet in many parts of the world today. It is also interesting that modern efforts to eat a more healthful diet—with a greater emphasis on fiber, nonmeat products, and natural foods—have taken us right back to the eating habits of the first humans.

---

diverticular disease. Scientists are currently studying whether fiber can prevent or treat colon cancer, heart disease, obesity, and diabetes.

Medical scientists have not yet determined exactly how much and what types of fiber are needed daily in the average diet. Most experts agree, however, that most Americas should increase their dietary fiber. Some common fiber foods include whole-grain breads; whole-grain breakfast cereals; whole-wheat pasta; vegetables, especially with edible skins, stems, and seeds; dried peas and beans; whole fruits, especially with edible skins and seeds; and nuts.

Point out to students that the term *whole grain* refers to products that contain the entire grain, or all of the grain that is edible. This includes the bran and germ portions as well as the starchy endosperm. Some examples of whole grains are whole wheat, cracked wheat, bulgur, oatmeal, whole cornmeal as used in corn tortillas, popcorn, brown rice, whole rye, and scotch barley.

PROBLEM SOLVING

Reading a Food Label

Answers to Practice Problems

1. 9 Calories of fat; 4 Calories of protein; 4 Calories of carbohydrate.
2. Less than 65 g total fat; less than 2400 mg sodium; 25 g fiber.
3. Whole-grain corn
4. Sugar, brown sugar syrup, and brown sugar. Students may speculate that by dividing sugar into three categories, the cereal makers avoid presenting sugar as the first or second most plentiful ingredient.
5. 4% without milk; 20% with milk.
6. 30 g; 10%
7. Total fat: about 1%; Saturated fat: 1%; Cholesterol: 1%; Sodium: 3%; Potassium: 6%; Total carbohydrate: 2%; Sugar: 6 g (Daily Value cannot be determined from the information given); Protein: 4 g (Daily Value cannot be determined from the information given); Vitamin A: 5%; Calcium: 16%; Vitamin D: 15%; Thiamine: 5%; Riboflavin 10%; Phosphorus 10%; Magnesium 6%; Zinc: 2%.
8. 18 g starch, 6 g sugar
9. 1500 Calories, which is 25% or 40% fewer Calories than what appears on food labels. The food labels' diet may be better for students because they are still growing, have higher metabolic rates than adults, or are more physically active.

PROBLEM SOLVING IN BIOLOGY

READING A FOOD LABEL

What are the ingredients of your favorite snack? Is that canned chili as good for you as its advertising would have you believe? Is that "lite" soup really low in salt and Calories? Which breakfast cereal is the most nutritious?

You can find the answers to these and other important nutrition questions by reading the food labels that are found on almost all packaged foods. At one time, food labels were more difficult to use. Nutrients were measured simply in grams or milligrams—units that are meaningless to many people; serving sizes on nearly identical products often differed, making it difficult to make comparisons; and some important nutrients were not listed on the label at all. But by mid-1994, due to new government regulations, manufacturers standardized serving sizes and changed the food labels on most packaged foods to a new format that is easier to use and understand.

Take a look at the food label on this page. Notice that the major nutrients are listed both in grams (or milligrams) and as a percentage of the total recommended intake for someone eating 2000 calories a day. For example, the 1 gram of fat in the cereal provides about 2% of the suggested daily allowance. The label also provides some general information about nutrition, telling you the Daily Values of certain nutrients and the Calories per gram for protein, carbohydrate, and fat.

To become a health-conscious and well-informed consumer, use the information on food labels to compare similar foods on the basis of their proportion of nutrients to Calories. When you choose a food, it should be high in nutrition and low in Calories.

Now practice your food-label skills on the typical cereal-box label shown here. You may wish to use a calculator or computer to help you answer some of the following questions. Keep in mind that medical considerations may mean that your own diet should contain nutrients in different amounts than those suggested.

Nutrition Facts
Serving Size: 1 cup (30g)
Servings Per Container: About 10

Amount Per Serving	Multi Grain Cereal	with ½ cup Skim milk
Calories	110	150
Calories from Fat	10	15
	% Daily Value**	
Total Fat 1g*	2%	2%
Saturated Fat <0.5g	1%	2%
Cholesterol 0mg	0%	1%
Sodium 240mg	10%	13%
Potassium 100mg	3%	9%
Total Carbohydrate 24g	8%	10%
Dietary Fiber 3g	10%	10%
Sugars 6g		
Other Carbohydrate 15g		
Protein 3g		
Vitamin A	25%	30%
Vitamin C	25%	25%
Calcium	4%	20%
Iron	45%	45%
Vitamin D	10%	25%
Thiamin	25%	30%
Riboflavin	25%	35%
Niacin	25%	25%
Vitamin B_6	25%	25%
Folic Acid	25%	25%
Phosphorus	10%	20%
Magnesium	6%	10%
Zinc	4%	6%
Copper	4%	4%

*Amount in cereal. A serving of cereal plus milk provides 1.5g fat, < 5mg cholesterol, 310mg sodium, 300mg potassium, 30g carbohydrate (12g sugar) and 7g protein.
**Percent Daily Values are based on a 2,000 calorie diet. Your daily values may be higher or lower depending on your calorie needs.

	Calories:	2,000	2,500
Total Fat	Less than	65g	80g
Sat Fat	Less than	20g	25g
Cholesterol	Less than	300mg	300mg
Sodium	Less than	2,400mg	2,400mg
Potassium		3,500mg	3,500mg
Total Carbohydrate		300g	375g
Dietary Fiber		25g	30g

Calories per gram:
Fat 9 • Carbohydrate 4 • Protein 4

INGREDIENTS: WHOLE GRAIN CORN (INCLUDES THE CORN BRAN), WHOLE GRAIN OATS (INCLUDES THE OAT BRAN), SUGAR, WHOLE GRAIN BARLEY (INCLUDES THE BARLEY BRAN), WHOLE GRAIN RICE (INCLUDES THE RICE BRAN), WHOLE GRAIN WHEAT (INCLUDES THE WHEAT BRAN), WHEAT BRAN, WHEAT STARCH, BROWN SUGAR SYRUP, BROWN SUGAR, SALT, PARTIALLY HYDROGENATED COTTONSEED OIL, CALCIUM CARBONATE, TRISODIUM PHOSPHATE, CARAMEL AND ANNATTO COLOR ADDED. VITAMIN C (SODIUM ASCORBATE), IRON (A MINERAL NUTRIENT), A B VITAMIN (NIACIN), VITAMIN A (PALMITATE), VITAMIN B_6 (PYRIDOXINE HYDROCHLORIDE), VITAMIN B_2 (RIBOFLAVIN), VITAMIN B_1 (THIAMIN MONONITRATE), A B VITAMIN (FOLIC ACID) AND VITAMIN D. VITAMIN E (MIXED TOCOPHEROLS) ADDED TO PROTECT FRESHNESS.

39–1 (continued)

Reinforcement/Reteaching

Remind students that carbohydrates come in many forms, including the simple sugars, or monosaccharides; double sugars, or disaccharides; and starches, which are polysaccharides. What we categorize as fiber includes cellulose and other types of polysaccharides that humans cannot digest.

- **What are some examples of simple sugars?** (Glucose, fructose, galactose.)
- **In what foods are these sugars found?** (Honey, fruit juices, vegetables.)
- **What are some examples of double sugars?** (Maltose, sucrose, lactose.)
- **What are some sources of these sugars?** (Table sugar, milk, malt.)
- **What must happen to these sugars before the body can use them for energy?** (They must be split into simple sugars.)
- **What are some sources of starch?** (Potatoes, grains such as wheat and corn.)
- **Can humans digest cellulose**

PROBLEM SOLVING CONTINUED

1. How many Calories are in a gram of fat? Of protein? Of carbohydrate?
2. On a 2000-Calorie diet, what is Daily Value for total fat? For sodium? For fiber?
3. Ingredients are listed in order of greatest quantity to least quantity. Which ingredient is used in the greatest quantity?
4. Which ingredients are used in the third, ninth, and tenth greatest quantities? What is your opinion about this?
5. What percent of the Daily Value of calcium is provided by 1 cup of the cereal? By 1 cup of the cereal with 1/2 cup skim milk?
6. How many grams of carbohydrate are in a 1-cup serving of the cereal with 1/2 cup skim milk? What percentage of the Daily Value of a 2000-Calorie diet does this represent?
7. What percentages of a 2000-Calorie diet's Daily Values are provided by 1/2 cup of skim milk?
8. In the dry cereal, how many grams of carbohydrates were in the form of starches? In the form of sugars?
9. What are the basic energy needs of an average-sized adult human? (*Hint:* If you need help with this question, see page 856.) How does this compare to the standard diets that appear on the food labels? Why are your personal energy needs probably better met by the diets on the food labels?

BACKGROUND INFORMATION
LACTOSE INTOLERANCE

The most common sugar in milk is lactose, and the enzyme that breaks down lactose is called lactase. Because milk is their principal food, all mammalian infants produce lactase in their digestive systems.

However, most mammals do not drink milk after they have been weaned from their mothers. As a result, most mammals stop producing lactase after infancy. The genes that are responsible for producing lactase are "turned off" by an unknown mechanism in these mammals, and no amount of milk drinking will turn on these genes again.

Intriguingly, researchers have discovered the same phenomenon in many human populations. In East Asia and many parts of Africa, human adults have never consumed much milk. So just as other mammals do, they stop producing lactase as they grow older. If they do try to drink milk, their intestinal bacteria ferment the lactose instead of digesting it. This produces acid and carbon dioxide, which causes cramps, vomiting, and diarrhea.

Those who suffer this syndrome when they drink milk are described as lactose intolerant. In the United States, many of these people have tried a product called acidophilus milk. This milk has been infused with a lactase-producing bacteria, *Lactobacillus acidophilus*.

Fats

Because our society places great emphasis on a trim physical appearance, the word fat has come to have a negative connotation. But fats in reasonable quantities are important to good health—just as important as other nutrients. Fats perform several vital functions in the body. Fats protect vital organs and joints and help keep the skin from drying out. Lipids, which are a kind of fat, are important parts of cell membranes. A layer of fat just beneath the skin helps insulate the body against changes in environmental temperature.

THE STRUCTURE OF FATS Fats, like carbohydrates, are organic compounds composed of carbon, oxygen, and hydrogen. But fats contain those elements in different proportions. Many fat molecules have a glycerol backbone to which are attached up to three fatty acid molecules. Depending upon the type of bond between adjacent carbon atoms and on the number of hydrogen atoms they contain, fatty acid molecules may be classified as either saturated or unsaturated. Fatty acids that contain single bonds and thus the maximum number of hydrogen atoms are saturated. Saturated fatty acids are usually solid at room temperature. Typical saturated fats include butter, lard, and other animal fats. Unsaturated fats have fatty acid molecules that contain one or more double bonds and thus have fewer hydrogen atoms. Unsaturated fats are usually liquid at room temperature. Most vegetable oils contain unsaturated fatty acid molecules that have multiple double bonds between adjacent carbon atoms and thus the fewest hydrogen atoms.

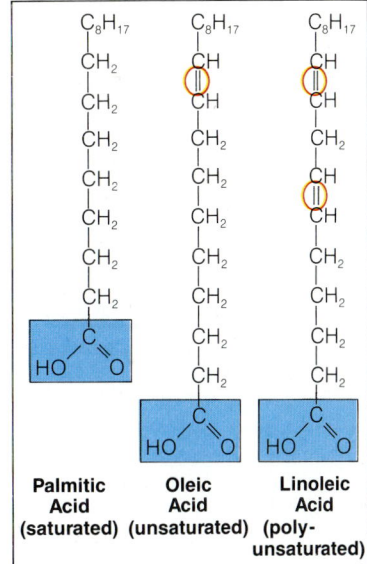

Figure 39-8 Fats are an important nutrient found in many foods. Most people should restrict the amount, as well as the kinds, of fat they eat. Physicians recommend eating limited amounts of unsaturated or polyunsaturated fats.

---

and other types of fiber? (No.)
• **Why is fiber important in the diet?** (Fiber stimulates and helps to move food through the digestive tract.)
• **What are some foods that contain cellulose?** (Lettuce, bran, celery, whole-grain products.)

Focus/Motivation
Show students samples of several common foods such as butter, chocolate, bread, pasta, bacon, lean meat, and carrots.
• **Which of these foods do you think are high in fat?** (Accept all answers.)

A simple hands-on qualitative test that can be used to determine whether a food is high in fat is to rub the food on a piece of brown wrapping paper. Allow the paper to dry. If the food contains a fair amount of fat, a greasy spot will remain on the paper.

Students have probably noticed similar grease spots on boxes containing bakery products or paper plates that held French fries.

BACKGROUND INFORMATION

PROTEINS AND ESSENTIAL AMINO ACIDS

There are eight essential amino acids in the human diet: tryptophan, methionine, valine, threonine, phenylalanine, leucine, isoleucine, and lysine. Histidine is also believed to be essential, but the exact requirement has not yet been established. Arginine is essential for infants and growing children, but not for adults.

A person following a strictly vegetarian diet must combine foods carefully because most plant foods do not contain all essential amino acids in a usable form. For example, corn is deficient in lysine and isoleucine. Beans are deficient in tryptophan and methionine. Yet together, these two foods supply all the essential amino acids. That is why a combination of beans and tortillas is a satisfactory protein meal.

To compare proteins from different food sources, nutritionists use a measure called the Net Protein Utilization, or NPU. NPU values indicate how efficient a single protein source is in meeting minimum daily requirements. NPU values range from 100 (perfect) to 0 (useless). The fact that a food contains all the essential amino acids does not automatically give it an NPU of 100. An NPU of 100 would mean that all essential amino acids are present in ideal proportions. The following chart shows NPU values for some common foods.

PROTEIN SOURCE	NPU
Eggs	97
Milk	82
Fish	80
Cheese	70
Meat	68
Soybean flour	60
Soybeans	60
Corn	50
Kidney beans	40

Figure 39-9 One reason to limit the amount and kinds of fat you eat is dramatically illustrated in this photograph of fat deposits in an artery. Fat buildup can cause the opening in the blood vessel to narrow, restricting the flow of blood.

These are polyunsaturated fatty acid molecules. The fats that contain them, such as many vegetable oils, are called polyunsaturated fats.

According to nutritionists, the diets of most Americans include too high a proportion of fats. A high level of dietary fats and cholesterol (a fatty sterol) can produce fatty deposits in the arteries. Physicians therefore recommend limiting the total amount of fats eaten and replacing saturated fats with unsaturated fats in the diet whenever possible. Unsaturated and polyunsaturated fats are not considered to be as harmful to the body's circulatory system as saturated fats.

FATS AND ENERGY Fats provide twice as many calories per gram as carbohydrates. For that reason, fats are an excellent way to store energy for future use. When a person eats more food than is needed, the body stores the extra energy by producing fat. This fat is deposited in a layer under the skin. When the body needs more energy than its glucose and glycogen supply can provide, it relies on the energy stored in that fat. Many mammals (bears, for example) prepare for long winters when their normal food supply is unavailable by purposely overeating during the late summer and autumn. These animals convert the energy in the extra food to fat. They use the stored fat during their long winter sleep.

Proteins

The organic nutrients we have discussed so far are important because they provide the body with energy. They do not, however, provide the body with the building materials it needs for growth and repair. That is the job of proteins. As you learned in Chapter 4, proteins are the complex macromolecules that make up critical parts of muscles, skin, and internal organs. Proteins also make up the enzymes that keep cell processes going.

39-1 (continued)

Content Development

Explain to students that most of the fats we eat are triglycerides. Triglycerides are formed when three fatty acids combine with a molecule of glycerol. Triglycerides are broken down in the intestines into parts that the body can absorb.

Emphasize that fatty acids are necessary in the human diet, especially linoleic and linolenic acids. These fatty acids, which are sometimes called the essential fatty acids, cannot be synthesized by the body. Usually obtaining these two acids is not a problem, because they are abundant in many plant foods, in fish, and in poultry.

Point out that because of the body's need for certain fatty

ESSENTIAL AMINO ACIDS Amazingly enough, the thousands of different proteins in our body are built from only 20 or so different amino acids. Your body can manufacture 12 of these amino acids; the other 8 must be obtained from your diet. These 8 are called the **essential amino acids**—not because they are any more important in making up proteins than the other amino acids, but because it is essential that they are included in the food you eat.

Amino acids are released in the body when protein-rich food is digested. The proteins in meat and dairy products usually contain all 8 essential amino acids and are called complete proteins. Most plant proteins lack one or more essential amino acids and are called incomplete proteins. If any essential amino acids are missing from the diet, synthesis of important proteins stops completely. For this reason, the diets of vegetarians must include a careful balance of different plant foods so that all of the essential amino acids are obtained.

PROTEINS AND ENERGY The amino acids obtained from the digestion of proteins may be used by the body as a source of energy. However, before most cells can do this, the amino acids the proteins contain must be converted into carbohydrates. This conversion is accomplished in the liver as an amino group (NH_2) is removed from the amino acid in a process called deamination. Deamination produces ammonia, which is poisonous to cells. The ammonia is immediately converted by the liver to urea, a compound that is much less poisonous. Urea then enters the bloodstream and is later removed by the kidneys.

Figure 39–10 Weightlifting is a demanding sport, and its participants must eat a diet rich in amino acids to supply the raw materials needed for the building of muscle protein.

Vitamins

The organic nutrients you have read about so far are compounds needed by the body in large amounts. **Vitamins** are complex molecules that are needed in small amounts and that usually cannot be synthesized by the body. Under certain conditions, however, some vitamins can be manufactured by the body. For example, vitamin D can be made in the skin under direct sunlight.

There is nothing magical about vitamins. If you are tired, vitamins will not make you feel wide awake. By themselves vitamins cannot make you grow bigger or stronger. Taking vitamins cannot substitute for a well-balanced diet. With a few exceptions, taking enormous doses of vitamins will not make a healthy person any healthier. In fact, taking large doses of certain vitamins can actually make you ill.

What then do vitamins do? And if they are so important, why do we need only tiny quantities of them? The answers are not simple, for different vitamins have different functions. Many vitamins are cofactors for enzyme-catalyzed chemical reactions. Cofactors are organic catalysts that assist enzymes in

ANNOTATION KEY

❶ Vitamins A, D, E, and K are fat soluble. The B and C vitamins, niacin, folic acid, and pantothenic acid are water soluble. (Interpreting charts)

BACKGROUND INFORMATION

VITAMINS SYNTHESIZED WITHIN THE BODY

Several of the vitamins that humans need are synthesized within the body by nonpathogenic bacteria that live in the intestines. The most important of these vitamins is vitamin K, which aids in blood-clotting reactions. Vitamin K is produced by the *E. coli* bacterium.

Vitamin-producing bacteria normally function very well in human intestines, but they are susceptible to oral antibiotics. For most short-term therapies, this susceptibility is temporary and not serious. Typically, the bacteria start to produce vitamins again once the antibiotic treatment has ended. But over a months-long course of antibiotics, or when exposed to especially strong antibiotics, the loss of the vitamin-producing bacteria can be more serious and long-lasting. In these cases, physicians usually prescribe vitamin supplements.

39-1 (continued)

Focus/Motivation

• How many of you have ever taken vitamin supplements? (Accept all responses.)
• Why did you start taking vitamins? (Answers will vary. Some may say that their parents told them to take vitamins; others may say that a doctor prescribed them or that they had vitamin therapy such as injections for a specific disease; still others may say that they started taking them on their own.)

VITAMINS

Vitamin	Source	Use
A (carotene)	Yellow and green vegetables, fish-liver oil, liver, butter, egg yolks	Important for growth of skin cells; important for vision
D (calciferol)	Fish oils, liver, made by body when exposed to sunlight, added to milk	Important for the formation of bones and teeth
E (tocopherol)	Green leafy vegetables, grains, liver	Proper red blood cell structure
K	Green leafy vegetables, made by bacteria that live in human intestine	Needed for normal blood clotting
B_1 (thiamine)	Whole grains, liver, kidney, heart	Normal metabolism of carbohydrates
B_2 (riboflavin)	Milk products, eggs, liver, whole grain cereal	Normal growth; part of electron transport chain
Niacin	Yeast, liver, milk, whole grains	Important in energy metabolism
B_6 (pyridoxine)	Whole grains, meats, poultry, fish, seeds	Important for amino acid metabolism
Pantothenic acid	Many foods, yeast, liver, wheat germ, bran	Needed for energy release
Folic acid	Meats, leafy vegetables	Proper formation of red blood cells
B_{12} (cyanocobalamin)	Liver, meats, fish, made by bacteria in human intestine	Proper formation of red blood cells
C (ascorbic acid)	Citrus fruits, tomatoes, green leafy vegetables	Strength of blood vessels; important in the formation of connective tissue; important for healthy gums

Figure 39-11 There are two main kinds of vitamins: water-soluble vitamins and fat-soluble vitamins. Which vitamins are soluble in fat? Which are soluble in water? ❶

catalyzing reactions. If we think of growth as a building process, we can say that vitamins are some of the tools needed to piece together amino acids and the body's other building materials. So although vitamins are essential, they are needed in only very small amounts.

Because they can be stored in the fatty tissues of the body, vitamins A, D, E, and K are known as fat-soluble vitamins. The body can build up small stores of these vitamins for future use. The water-soluble vitamins, which include vitamin C and the B vitamins, cannot be stored. Therefore they should be included in a balanced diet every day.

Like other essential nutrients, most vitamins can be obtained naturally by eating a balanced diet that includes fresh fruits, vegetables, and meats. However, when the body does not receive a sufficient supply of vitamins, it can develop vitamin-deficiency diseases. One such disease is scurvy, a painful disease whose victims suffer bleeding gums, loss of teeth, aching muscles, and even death. Scurvy was once

• Do you feel that taking vitamin supplements has helped you? (Accept all responses.)

Content Development

Emphasize to students that although vitamins are essential in the human diet, popular ideas about dietary vitamin supplements are often incorrect. Usually, a person can obtain all the vitamins he or she needs by eating a balanced diet.

Skills Development

Guided Practice

Skills: Collecting data, expressing an opinion

Divide the class into small groups. Have each group interview several professionals in the fields of health and nutrition to

common among sailors. Many years before the discovery of vitamins, the British naval surgeon James Lind found that a supply of fresh fruit or fruit juice would prevent scurvy. (In fact, it was because of the lime juice they had to drink that British sailors came to be called "limeys.") We now know that citrus fruits and tomatoes prevent scurvy because they supply the body with large quantities of vitamin C.

Health and a Balanced Diet

How can you be sure that you get all the nutrients you need? The answer is relatively simple. You should select a variety of foods from the five food groups:

- Milk, yogurt, and cheese
- Bread, cereal, rice, and pasta
- Meat, poultry, fish, dry beans, eggs, and nuts
- Fruits
- Vegetables

If fats, oils, and sugars are added to the diet, they should be used sparingly. Although the typical American diet already includes foods from all five food groups, more often than not the proportions are wrong. Many people obtain far too many Calories from the meat and milk food groups and too few Calories from the vegetable and fruit food groups. Nutritionists suggest that an ideal diet would derive no more than 30 percent of its Calories from fats, a reduction from the nearly 40 percent that is common today. The key word here is balance: Increasing the amount of fruits, vegetables, and grains in your diet is a simple step that can result in better nutrition and health.

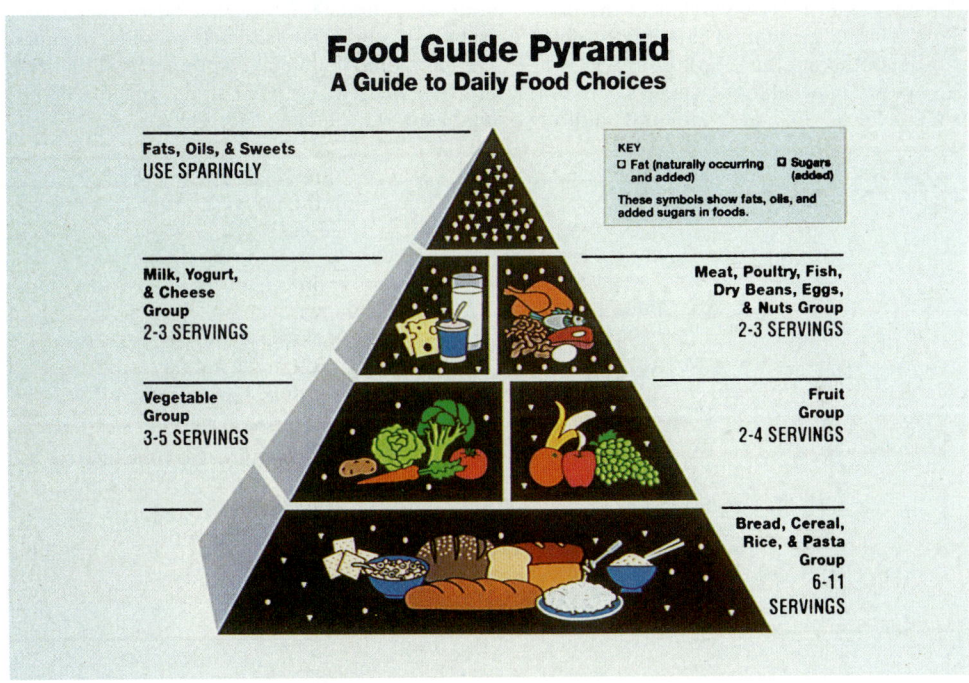

Figure 39–12 In the food guide pyramid, fats, oils, and sugar are placed at the top of the pyramid with the words "use sparingly" near them. The remainder of the pyramid is divided into the five food groups.

HISTORICAL NOTE
BERIBERI

Beriberi, a serious disease that is caused by a deficiency in thiamine, first appeared in Southeast Asia and the East Indies. Victims of the disease became extremely weak and suffered from uncontrollable muscle spasms. Just before 1900, Christiaan Eijkman, a Dutch physician, noticed that beriberi did not occur in populations of prisoners who were fed ordinary brown rice. This was surprising, since beriberi was common among prisoners who were fed the more expensive polished white rice. Eijkman thought about this and realized that the hulls of rice seeds are removed from white rice but not from brown rice. The hulls, he reasoned, must contain some factor that prevents beriberi.

The Polish chemist Casimir Funk studied the chemical basis of Eijkman's "factor." He incorrectly concluded that the factor that prevents beriberi belonged to a group of compounds known as amines. Since this factor was vital to life, he named it vitamin, or "vital amine." Funk's chemistry was wrong, but the name stuck. The term *vitamin* is now part of the scientific vocabulary.

Enrichment
Have students investigate some of the many fad diets that are currently on the market. Have students compare these diets with the requirements for a well-balanced diet, then give their own answers to the following questions: Is the diet healthful? Will the diet work? Is the person on the diet likely to keep the weight off or gain it back as soon as he or she goes off the diet? Encourage students to find and read any magazine or newspaper articles about the diet they selected.

find out what these people think about the effectiveness of dietary vitamin supplements. Possible people to interview include doctors, pharmacists, nutritionists, and owners of health food stores. Have students pool and analyze their data. Encourage students to predict which professionals would be the most and least likely to recommend vitamins, then see if their data prove these predictions true.

Have students present their findings, analyses, and conclusions in a short paper. Encourage students to include their own opinions in their reports and to discuss which arguments about vitamins they find most convincing.

SCIENCE, TECHNOLOGY, AND SOCIETY

BREAKTHROUGH

Fat Rats

The simplest answer to the problem of obesity has been around for years: Lose weight by decreasing the amount of food eaten (without sacrificing essential nutrients and vitamins) and increasing the amount of exercise (to burn up energy stored in the form of fat). Even while following these guidelines, certain people seem to have a tendency toward being obese. Researchers have wondered why these people seem predisposed to gain weight. Although a theory that explains obesity in humans has still not been completely formulated, scientists are studying obesity in other animals with interesting results.

Scientists have developed strains of "genetically" obese laboratory rats—rats that inherit the tendency to become extremely overweight. In all rats, fat cells release a protein called adipsin into the bloodstream. In normal rats, as well as in rats that have been force-fed to gain weight, the levels of adipsin are relatively high. But in genetically obese rats, there is as little as 1 percent of the normal level of this protein. What exactly does adipsin do? Although there is little direct evidence as yet, speculation centers around adipsin's effects on the appetite and/or metabolism—both of which are controlled by the brain. High levels of adipsin in the blood may suppress appetite; low levels may increase appetite.

This genetically obese rat (left) is much larger than a normal one (right).

We cannot be sure that an identical system works in humans, but researchers recently have discovered a crucial link with the rodent system: Human fat cells also produce adipsin. Future research will attempt to link the problem of obesity in humans with the production of adipsin. In time, this research may lead to a treatment for obesity.

39–1 SECTION REVIEW

1. What are the six nutrients and their uses?
2. What are the five food groups and how do they help to produce a balanced diet?
3. **Critical Thinking—Relating Concepts** When proteins are used for energy, the amount of urea released from the body increases. Why?

39-2 The Process of Digestion

Guide For Reading

■ How do mechanical and chemical digestion differ?
■ What organs make up the digestive system?
■ What is the function of each organ of the digestive system?

Digestion is the breakdown of food into simpler molecules that can be absorbed by the body. As you can see from Figure 39-13, the digestive system is actually a long, hollow tube called the **gastrointestinal tract**, or **GI tract**. **The digestive system includes the mouth, pharynx, esophagus, stomach, small intestine, and large intestine. Several major glands, including the salivary glands, the pancreas, and the liver, add their secretions to the digestive system.**

The first task of the digestive system is to break down food into a fine pulp. When food is in the form of a fine pulp, its surface area is increased and more food molecules are exposed to the action of digestive chemicals. The next task of the digestive system is to chemically act on the food, breaking it down into smaller and smaller molecules. For example, starches must be reduced to simple sugars before the cells of the body can oxidize them for energy. Proteins must be broken down into amino acids before they can be used to build new proteins. The

Figure 39-13 The human digestive system is made up of a number of different organs. These organs work together to break down food and absorb the resulting nutrients into the blood for distribution throughout the body.

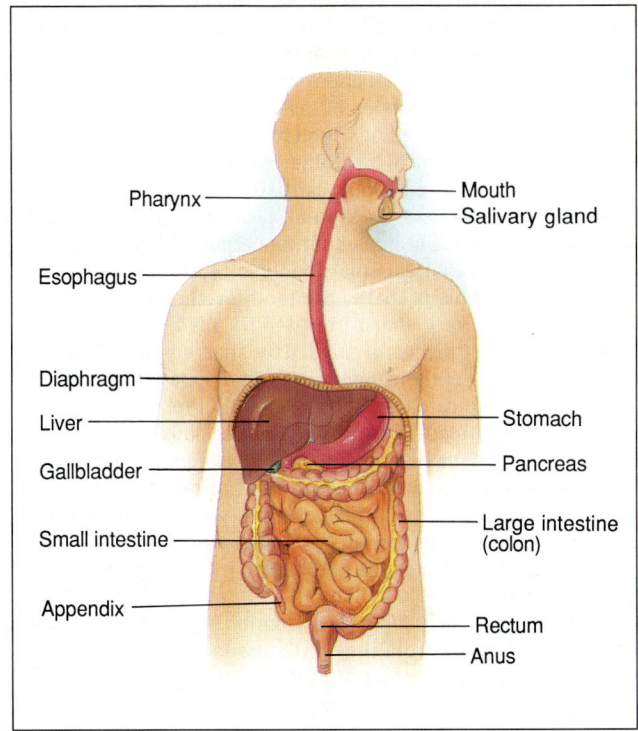

an amino acid. This process produces ammonia, which is poisonous to body cells. The ammonia is converted into urea. Thus, more urea is produced when the body breaks down proteins for energy than when it breaks down carbohydrates.

Reinforcement/Reteaching
If students are having difficulty answering the Section Review questions, go back to the part of the section that contains the material with which they are having difficulty.

Closure
Have students work in small groups. Ask each group to imagine that they are renting a house together and must plan breakfast, lunch, and dinner for the group. Have each group plan one day's meals, taking into account the nutritional requirements for a healthy diet as well as group members' likes and dislikes. Students should consider such factors as including foods from the five basic food groups; meeting group members' calorie requirements; providing appropriate percentages of protein, carbohydrate, and fat; and serving foods that include the essential vitamins and minerals. After each group has planned their meals, have them post their menus on a bulletin board. Ask members of the class to read each menu and comment on whether they think each one is nutritionally sound and why. Groups should be prepared to defend their food choices if necessary.

TEACHING STRATEGY 39-2

Focus/Motivation
Pass out small pieces of plain bread to each student. Ask students to chew the bread longer than usual before swallowing it.
• **Did you notice anything happening to the taste of the bread as you chewed it longer?** (Answers may vary, but students should notice that the bread begins to taste sweet.)
• **Why do you think the bread begins to taste sweet?** (Accept all responses.)

BACKGROUND INFORMATION
DIGESTIVE SYSTEM CONTROL

Most controls over body functions operate in response to an internal stimulus, such as the oxygen concentration of the extracellular fluid. In the digestive system, however, controls operate in response to an external stimulus—the amount and composition of food in the digestive tract.

Receptors in the gut wall respond to such stimuli as solute concentrations and acidity. The receptors are part of a short reflex pathway that bypasses the central nervous system. Thus signals from the receptors travel through nerve plexuses in the gut wall to directly influence wall contractions and secretions. The receptors in the gut wall are also connected to the central nervous system by a long-distance reflex pathway. Either the short reflex pathway, the long-distance pathway, or both can act to control the digestive system.

39–2 (continued)

Content Development
Explain that the bread began to taste sweet because enzymes in the saliva had begun to break down the bread's starch into maltose. Point out that people usually do not experience this sweet taste because they swallow the bread before their saliva has the chance to thoroughly digest it.

Content Development
Show students a model or diagram of a set of human teeth. Point out the location of the four types of teeth: incisors, canines, premolars, and molars. Also encourage students to identify these different types of teeth in their own mouths or in the mouths of partners. Point out that incisors and canines cut and tear food while premolars and molars crush and grind food.
- **How does the shape of each type of tooth indicate its function?** (Canines and incisors are long, sharp, and pointed; molars and premolars are broader and flatter.)
- **Why do humans need to have different types of teeth?** (Because humans eat a variety of different foods.)
- **Which teeth would we probably not need if we were herbivores? Why?** (We would probably not need the incisors and the canines because we would not need to cut and tear meat.)

---

last task of the digestive system is to absorb these small molecules and pass them along to the bloodstream for distribution to the rest of the body.

In our survey of the animal kingdom, we encountered many animals that we were able to classify as either herbivores (plant eaters) or carnivores (meat eaters). These animals have digestive systems that are well adapted to digest the particular food they eat. But humans do not fit neatly into either of these categories. Humans are omnivores who eat both plant and animal food. The human digestive system is adapted to process both vegetable and animal materials.

The Mouth

The mouth is the organ in which the process of digestion begins. The mouth prepares food for entry into the gastrointestinal tract. The remarkably well-adapted parts of the mouth give it a variety of abilities that most of us take for granted.

The lips, cheeks, and tongue work in a carefully coordinated manner to place food between the teeth for chewing. The teeth break up the food as part of the process known as mechanical digestion. Mechanical digestion is the physical breaking up of food by the teeth and other parts of the digestive system. The teeth themselves are composed of the hardest materials found in the body. The inside lining of the cheeks contains special mucous glands. The secretions of these glands help lubricate the food and make it easier to swallow.

As food is passed back and forth in the mouth it comes in contact with a large number of taste buds located on the surface of the tongue. Taste buds send messages to the brain about the nature of the food being eaten. Taste buds can determine whether a food is sweet, sour, salty, or bitter. By passing this information to the nervous system, taste buds help us decide whether or not the contents of our mouth should be swallowed.

All mammals that eat solid food use their teeth to cut and grind the food. This prepares the food for chemical digestion. Carnivores have very sharp front teeth called incisors. These teeth can tear off pieces of meat. The sharp, pointed teeth at the front corners of their mouth are called canines. Carnivores use these teeth to catch, hold onto, and kill their prey. In a carnivore, the rear teeth, or molars, act like knives to help cut meat into pieces small enough to be swallowed. Herbivores, on the other hand, have teeth that are used primarily to grind the tough, fibrous plant food they eat into a fine pulp. Because humans are omnivores, eating both plant and animal materials, it is not surprising that some of their teeth look like the teeth of a carnivore and some look like the teeth of a herbivore.

Individually, each tooth is a wonder of biological design. Although teeth grow from the upper and lower jaw bones, they are constructed from material that is much harder than ordinary bone. During chewing, the jaw muscles close the mouth

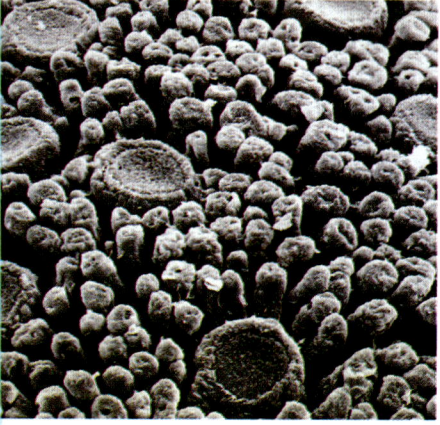

Figure 39–14 Taste buds are scattered over different areas on the surface of the tongue. The taste of food is actually a combination of taste and smell. Thus foods lose much of their taste when the nose is blocked by a cold.

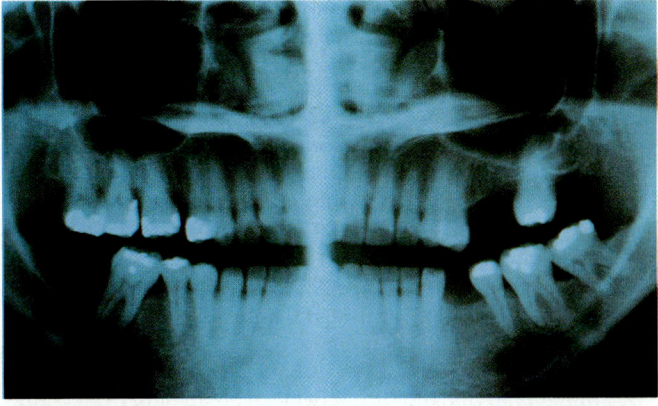

Figure 39–15 *A dentist uses X-rays to see parts of the teeth that cannot be seen with the unaided eye. The brighter areas in the teeth are fillings. You can also see spaces where teeth have been lost. The illustration shows a cross section of a tooth. A cavity occurs when places in the normally hard enamel decay. As an untreated cavity continues to increase in size, it may reach down into the pulp of the tooth, where nerves and blood vessels are located.*

with great force, pounding the teeth against the food and against one another with enough force to crack ordinary bones. To help in resisting this constant wear and tear, the crown, or top, of each tooth is covered by a layer of enamel, an extremely hard substance. Underneath the enamel is another hard material called dentin that makes up the bulk of the tooth. The enamel and dentin have no nerve cells in them.

Each tooth is anchored by long, pointed roots that extend into the jaw bone. A tough, fibrous periodontal membrane holds these roots in the jaw. The hollow center of the tooth, called the pulp cavity, contains the blood vessels and nerve cells that serve the tooth. If they are disturbed by heat, cold, or dental decay, it is these nerve cells that send the messages to the brain that signal "toothache."

While the teeth cut and grind the food into a pulp in preparation for swallowing, the salivary glands secrete the first digestive enzymes of the GI tract. The release of saliva is triggered by the nervous system. Often the mere sight or smell of food is enough to cause saliva to be secreted. Just think about your reaction to a whiff of bread baking in an oven or the sight of food at a holiday picnic!

Saliva has three important functions. One, saliva dissolves some foods and combines with mucus to speed the passage of food through the digestive system. Two, saliva contains enzymes that attack many of the potentially dangerous microorganisms that can enter the mouth. Three, saliva contains the

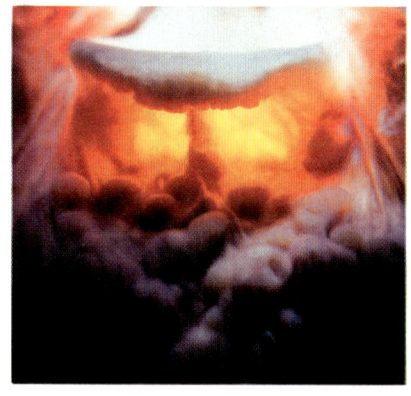

Figure 39–16 The epiglottis is the flap of skin that folds over the trachea to prevent foods from moving down the "wrong pipe" when they are swallowed.

enzyme salivary amylase, which breaks up long starch molecules into maltose. If you chew a soda cracker, which is made up of starches, for several minutes, it will begin to taste sweet. The starches in the cracker are broken down into sugars by the action of salivary amylase.

Once the teeth and salivary glands have completed the initial processing, the food is ready to be swallowed. Gathering the food together in a ball called a **bolus**, the tongue pushes it toward the back of the mouth. The back of the tongue sends the bolus down to the part of the throat called the **pharynx**.

In the pharynx, the GI tract and the respiratory tract cross each other. See Figure 39–16. At this point, food moving down the GI tract could enter the respiratory tract by mistake, causing a person to choke. However, as the tongue moves the food into the pharynx, it presses down on a small flap of cartilage called the **epiglottis**. When the epiglottis is depressed, it closes the entrance to the respiratory tract and guides the food down the GI tract.

Occasionally, if we try to talk and eat at the same time, the epiglottis does not have a chance to close properly. When that happens, we feel that "something went down the wrong pipe," which is an appropriate description. For some of the food we tried to swallow slipped into the passageway leading to the lungs instead of the one leading to the stomach. We usually begin coughing to clear the food out of the respiratory tract.

The Esophagus

The bolus next passes down the **esophagus**, a tube about 25 centimeters long that connects the pharynx with the stomach. The walls of the esophagus are made up of rings of muscles that circle the tube. As the bolus enters the tube, the circular muscles just above it begin to contract in waves that travel down the length of the esophagus. Each wave squeezes the bolus along ahead of it. These rhythmic muscular contractions are called **peristalsis**. See Figure 39–17. The esophagus passes through the diaphragm—the thick sheet of muscle that separates the chest cavity from the abdominal cavity—and empties into the stomach. A thick ring of muscle called a **sphincter** is found where the esophagus joins the stomach. The sphincter acts like a valve, allowing food to pass into the stomach but usually not letting it move back up into the esophagus.

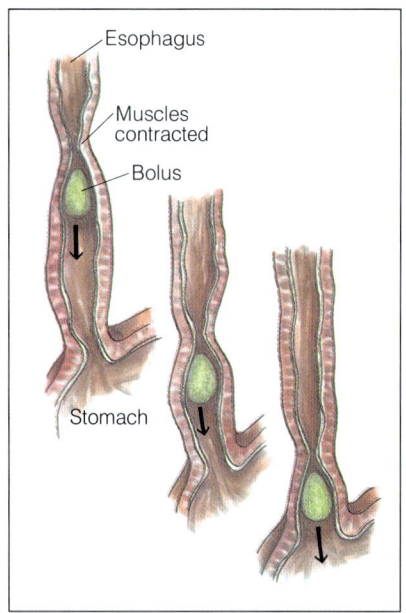

Figure 39–17 Muscles in the walls of the esophagus contract in waves. Each wave pushes a bolus of food in front of it. Eventually the bolus is pushed into the stomach.

The Stomach

The partially digested food is now in the **stomach**, a thick muscular sac located just below the diaphragm in the abdomen. Three sets of glands in the stomach lining produce secretions called gastric fluids. One set of stomach glands produces more mucus, keeping the food well lubricated and protecting the walls of the stomach from being digested by its own secretions. Another set of glands secretes hydrochloric acid, which makes the contents of the stomach very acidic. Yes, despite certain commercial messages to the contrary, the stomach is supposed to be acidic! The acid helps break down food. In fact, if all the acid in the stomach were to be completely neutralized, the stomach would not be able to digest food. The digestive enzyme pepsin, secreted by a third set of stomach glands, works best under the acidic conditions present in the stomach. Together with hydrochloric acid, pepsin in the stomach begins the complex process of protein digestion. Pepsin breaks proteins into smaller polypeptides.

As chemical digestion continues, the muscular walls of the stomach contract powerfully, churning and mixing food with the gastric fluids. After two or three hours of this action, the food has been changed into a pasty mixture called **chyme**. At this point, the **pyloric valve** between the stomach and the small intestine opens and the contents of the stomach are allowed to enter the small intestine, as peristalsis forces the mixture through the open valve.

By the time chyme leaves the stomach, most proteins have been broken down into smaller polypeptides. Sugars, however, have not yet been chemically altered, nor have fats. Some starch molecules have been broken down into disaccharides by the action of amylase, but other starch molecules have not been affected.

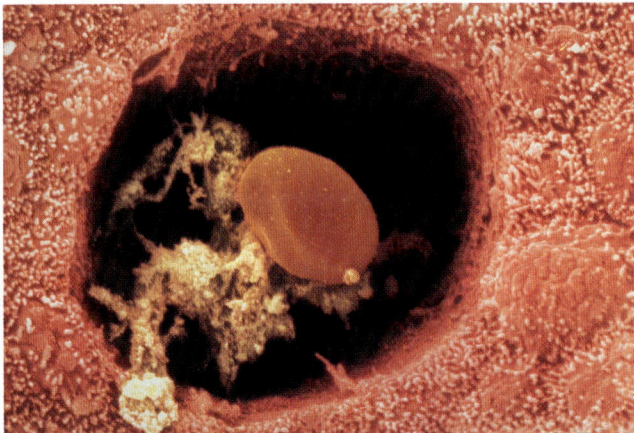
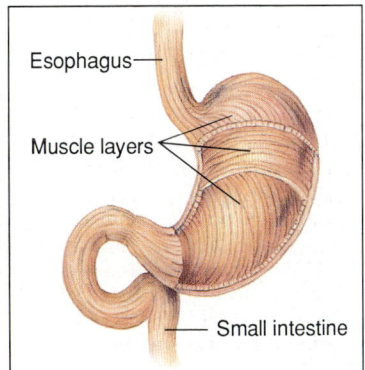

Figure 39–18 *Gastric pits are located in the lining of the stomach wall (left). Cells at the base of the gastric pits secrete enzymes, hydrochloric acid, and other substances that aid in digestion. Several layers of muscle make up the stomach wall (right). When these muscles contract, food and chemicals within the stomach are mixed together.*

BACKGROUND INFORMATION
THE STOMACH

The stomach is made up of several layers of muscles. In the first muscle layer, the fibers run lengthwise. In the second muscle layer, the fibers are circular. In the third muscle layer, the fibers are diagonal.

Food stays in the stomach anywhere from 2 to 6 hours before entering the small intestine. Three factors determine how fast the stomach empties. First, a larger meal tends to cause food to move out of the stomach faster than a smaller meal. This happens because a large amount of food distends the stomach, activating mechanoreceptors in the stomach wall. These receptors trigger reflexes that increase the force of contraction to empty the stomach. Second, foods that are high in fat content or high in acidity tend to make the stomach empty more slowly. This happens because receptors in the duodenum trigger the release of hormones such as CCK and GIP. These hormones "tell" the stomach to slow down because the small intestine cannot handle so much fat and acid at one time. As a result, greasy or fatty foods often feel "heavy" in the stomach for quite some time after they have been eaten. The third factor that affects stomach motility is a person's emotional state. Emotions such as fear and depression can trigger signals from the nervous system that slow down the emptying of the stomach. That is one reason why a person who is depressed or anxious often finds that he or she has little appetite.

Explain that mechanical digestion takes place not only in the mouth but in the stomach as well. Muscle layers of the stomach churn the food. This churning action breaks up the food and mixes it with gastric fluids.

Point out that chemical digestion takes place in the mouth, the stomach, and the small intestine. It is only in the small intestine, however, that all types of foods are acted on chemically. In the mouth, only starches are digested by the enzyme salivary amylase. In the stomach, proteins are acted on chemically and broken down into peptides and amino acids. In the small intestine, single and double sugars, peptides, amino acids, and fats are acted upon by various enzymes. Point out that the small intestine is the only part of the digestive tract where fats are chemically digested. Fats must pass through the mouth and stomach before the body provides any enzymes to break them down.

The Small Intestine

As chyme is pushed through the pyloric valve, it enters the **duodenum** (doo-oh-DEE-nuhm). The duodenum is the first part of the **small intestine**. In the duodenum, chyme is flooded with a variety of enzymes and other digestive fluids that break down additional food molecules. The enzymes and digestive fluids that are active in the small intestine come from three separate sources: the intestine itself, the pancreas, and the liver.

Glands lining the duodenum release peptidases—enzymes that continue the process of protein digestion begun in the stomach. Other enzymes produced by intestinal glands attack complex carbohydrates, breaking them down into their component sugars. The enzyme maltase, for example, breaks the disaccharide maltose into two molecules of glucose. Lactase breaks down lactose (milk sugar) into glucose and galactose. Sucrase breaks down sucrose into fructose and glucose.

The **pancreas** is a long organ located behind the stomach. Most of the tissue in the pancreas consists of glands. When chyme enters the small intestine, glands in the pancreas are stimulated to release their secretions, called pancreatic fluid. Pancreatic fluid enters a duct that empties into the duodenum. Enzymes in pancreatic fluid are responsible for the digestion of carbohydrates, fats, and proteins. Amylases and proteases complete the breakdown of carbohydrates and proteins, respectively. At the same time, lipase begins to break down fats into glycerol and fatty acids.

Pancreatic fluid also contains sodium bicarbonate. This compound neutralizes the hydrochloric acid produced in the stomach, which entered the small intestine along with the chyme. Unlike enzymes that work in the stomach, pancreatic

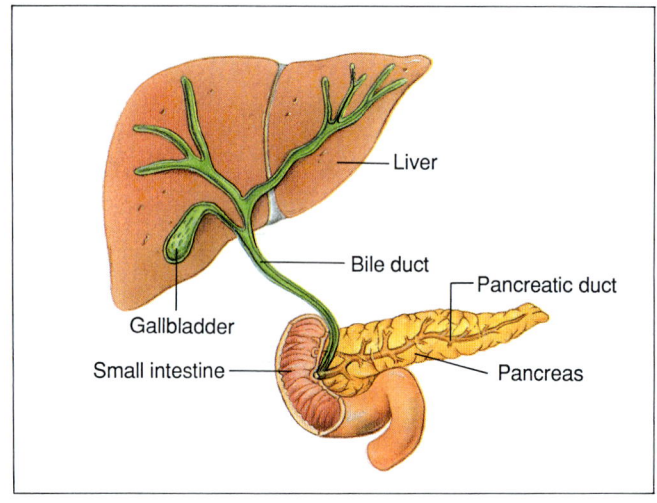

Figure 39–19 Various organs of the body also add chemicals to the digestive system. The liver secretes bile, a chemical that breaks up large fat molecules. The pancreas secretes pancreatic fluid, which helps break down carbohydrates, proteins, and fats.

enzymes work best near neutral pH, usually 7.0 to 8.0. In addition to secreting these enzymes, the pancreas also secretes a chemical that regulates the level of sugar in the blood. You will learn more about this action of the pancreas in Chapter 42.

The **liver** is a large brownish organ that lies above the stomach in the abdomen. The liver may weigh as much as 1500 grams. One of the functions of the liver is to secrete a yellow-brown liquid called bile. Bile is stored in a small sac called the **gallbladder**. The entrance of food into the small intestine stimulates the release of bile from the liver through the bile duct into the duodenum.

Although it is an extremely important digestive fluid, bile does not contain enzymes. Bile is a complicated mixture of cholesterol, colored pigments, and chemicals called bile salts. Bile salts play a major role in aiding the enzyme lipase to properly digest fats. Because fats do not dissolve in water, they tend to stick together in large globules—much like drops of oil in salad dressing. These fat globules are too large for lipase to work on them efficiently. Bile salts emulsify fats, or break them into smaller and smaller droplets. Bile salts act on fats in much the same way as a good dish-washing detergent would—it breaks them up into smaller and smaller particles. When fat particles are emulsified by bile salts, lipase can work on them more efficiently.

SOME DIGESTIVE ENZYMES

	Enzyme	Action
Proteins	Pepsin (produced by glands in stomach)	Breaks peptide bonds at NH_2 group
	Trypsin (pancreas)	Breaks peptide bonds
	Chymotrypsin (pancreas)	Breaks peptide bonds
	Carboxypeptidase (pancreas)	Breaks COOH terminal peptide bonds
Fats	Pancreatic lipase (pancreas)	Triacylglycerols
	Intestinal lipase (small intestine)	Tri-, di-, and monoglycerides
Carbohydrates	Ptyalin (saliva)	Starch
	Amylase (pancreas)	Starch
	Intestinal amylase (small intestine)	Starch
	Maltase (small intestine)	Maltose
	Sucrase (small intestine)	Sucrose
	Lactase (small intestine)	Lactose

Figure 39–20 Many enzymes act in the digestive system to break down foods and make nutrients available to the body. Where in the body does the digestion of carbohydrates begin? ❶

BACKGROUND INFORMATION
GASTRIC FLUIDS

Scattered throughout the stomach mucosa are exocrine cells that release hydrochloric acid, pepsinogens, mucus, and other substances into the stomach cavity. In addition, endocrine cells in one part of the mucosa release hormones, such as gastrin, which travel in the bloodstream to target cells. These target cells secrete substances that also contribute to the gastric fluids.

The hydrochloric acid that is secreted into the stomach has several functions. First, it dissolves bits of food to produce chyme. Second, it kills most of the microbes that enter the body in food. Third, it acts on proteins and on protein-degrading enzymes to aid in protein digestion.

Sometimes part of the stomach mucosa becomes damaged by the digestive action of gastric fluids. When this happens, a peptic ulcer forms. The surface of the stomach breaks down, and hydrogen ions from the hydrochloric acid diffuse into the mucosa and trigger the release of histamine. Histamine may then trigger vasodilation and increased capillary permeability and HCl secretion. The result is tissue damage, which may lead to bleeding into the stomach and abdomen— what people often refer to as a bleeding ulcer.

---

dissolve in water.)
• **Why is bile needed to break fats into smaller particles in the small intestine?** (Fats do not dissolve in water, so they tend to stick together in globules as they travel through the digestive system. They need to be broken down into smaller pieces so that digestion can continue.)

Skills Development
Guided Practice
Skill: Interpreting charts
Using the Visuals Direct students' attention to Figure 39–20.

• **Which enzymes are secreted by the pancreas?** (Amylase, trypsin and chymotrypsin, carboxypeptidase, and lipase.)
• **What kinds of substances do each of these enzymes break down?** (Amylase breaks down starch; trypsin and chymotrypsin break down proteins; carboxypeptidase breaks down simple proteins; and lipase breaks down fats.)
• **Which digestive enzymes are secreted by the small intestine?** (Amylase, lactase, maltase, sucrase, and lipase.)
• **What substances do lactase, maltase, and sucrase break down?** (Complex sugars.)
• **What is the original source of these complex sugars?** (Some were ingested as complex sugars; others were broken down from starches by the action of amylase.)

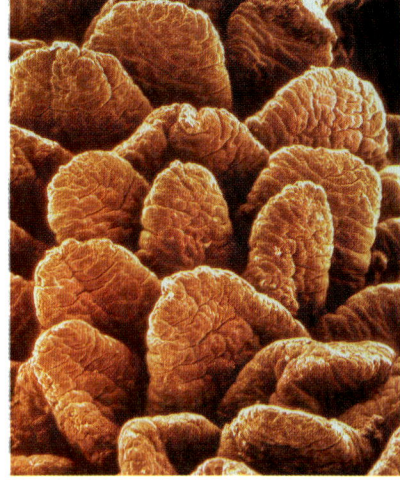

Figure 39–21 *The villi that line the wall of the small intestine greatly increase its surface area, and thus the area over which food molecules are absorbed.*

The small intestine is very long, often as long as 7 meters in an adult. Spending several hours traveling from one end to the other, food is pushed slowly through the small intestine by peristalsis. After leaving the duodenum, the now mostly digested food passes into the other parts of the small intestine, the jejunum and the illium. Here most digestive activities are completed, and many of the end products of digestion are absorbed into the bloodstream.

The inside walls of the small intestine are folded in such a way as to greatly increase their surface area. Under a microscope, the surface of the folds resembles wall-to-wall carpeting because of countless tiny fingerlike projections called **villi** (singular: villus). See Figure 39–21. The cells that make up the surfaces of the villi are folded into thousands of even tinier folds called microvilli.

What does all this folding accomplish? The folding increases the surface area of the small intestine, making the process of absorption proceed much more rapidly than it would if intestinal walls were smooth. The villi in the small intestine are richly supplied with blood vessels to absorb and carry away nutrients. Each tiny villus has its own set of arteries and veins with a dense network of capillaries connecting them. Among the capillaries are a number of very fine lymph vessels.

Figure 39–22 *Microvilli are the tiny projections that increase the surface area of the villi (right). Inside each microvilli are tiny blood vessels that begin the process in which nutrients are circulated to the rest of the body's cells (left).*

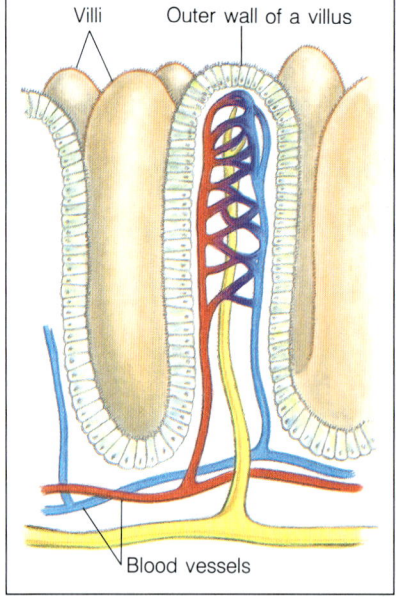
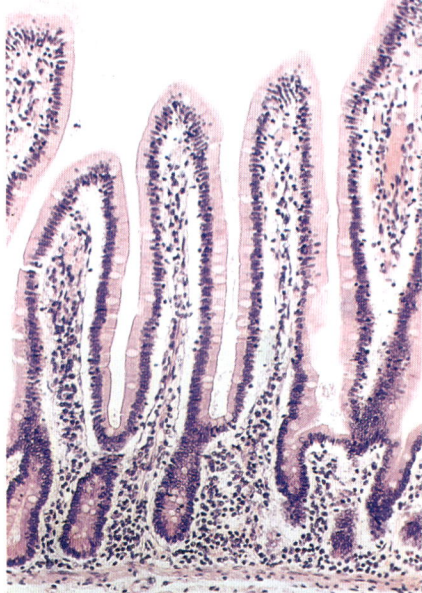

sorption takes place.
- **What role do villi play in the absorption process?** (They increase the surface area of the small intestine so absorption can take place more rapidly.)
- **What happens to nutrients once they are in the bloodstream?** (They are carried to cells in all parts of the body.)

SECTION REVIEW 39–2

1. The mouth is where the teeth begin mechanical digestion and where saliva begins breaking down carbohydrates. The esophagus is a muscular tube that connects the mouth to the stomach, where the food is churned and proteins are partially broken down. The small intestine is where a wide variety of enzymes further digest

(You will learn more about lymph in Chapter 41.) Most of the products of carbohydrate and protein digestion are absorbed into the capillaries in the villi. Molecules of undigested fat and some fatty acids may enter the lymph vessels directly.

By the time it is ready to leave the small intestine, the food is basically nutrient-free. The complex organic molecules have been digested and absorbed, leaving only water, cellulose, and other undigestible substances behind. As this material leaves the small intestine, it passes by a small saclike organ called the appendix. The appendix is a vestigial organ in humans—it has virtually no function. In other mammals, the appendix is large and is used to store cellulose and other materials that the digestive enzymes cannot break down. But the only time we pay attention to the appendix is when it becomes clogged and inflamed, causing appendicitis. The only remedy for appendicitis is to remove the infected organ by surgery—as quickly as possible.

The Large Intestine

When food leaves the small intestine, it passes into the **large intestine**, or **colon**. The main job of the colon is to remove water from the undigested materials passing through it. Normally, the colon removes water efficiently and reduces the undigested materials to solid waste products called feces.

The feces move through the colon by peristalsis, finally collecting at the end of the colon in the rectum. Here valvelike anal sphincter muscles prevent the feces from being released until it is convenient for us to expel them through the anus.

Sometimes the normal function of the large intestine is disturbed and it does not absorb as much water as it should. This leaves a great deal of water in the feces, causing diarrhea.

The colon also contains large amounts of bacteria that aid in the final stages of digestion. Some of the bacteria even manufacture vitamins, notably vitamin K. These vitamins are absorbed across the intestinal wall into the bloodstream.

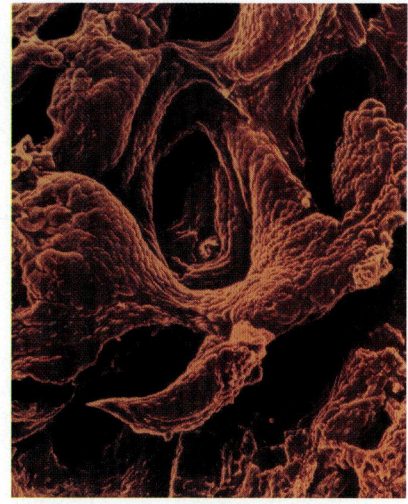

Figure 39–23 *This photograph shows the lining of the large intestine. By the time digested food reaches this organ, all the nutrients have been removed. The large intestine removes water from the remains of the digestive process.*

39–2 SECTION REVIEW

1. How do mechanical and chemical digestion work together to break down foods?
2. List the organs of the digestive system. What is the function of each organ?
3. How does bile aid in the digestion of fats? Where are fats absorbed from the GI tract?
4. **Critical Thinking—Making Inferences** What could you infer about the diet of an animal that has a large appendix?

ESL STRATEGY

Give each student six small cards. Ask them to number and copy one of the following terms on each card:

1. Mouth   4. Liver
2. Esophagus 5. Pancreas
3. Stomach   6. Small intestine

Then have them write the correct phrase from the list below on the back of each card.

- Most digestion takes place here
- Epiglottis and windpipe
- Tube between mouth and stomach
- Taste bud
- Ptyalin
- Bile
- Digestion helper and insulin maker
- Digestion helper and largest organ
- Gastric juice—pepsin, hydrochloric acid, and mucus

Have students exchange cards and check for accuracy. Later, the cards can be used as individual study tools.

---

with a large appendix probably eats mostly plants.

Reinforcement/Reteaching

If students are having trouble answering the Section Review questions, go back to the part of the section that contains the material with which the students are having difficulty.

Closure

Divide the class into six groups. Assign each group one of the following parts of the digestive system: mouth; esophagus and pharynx; stomach; pancreas and liver; small intestine; large intestine. Ask each group to prepare a presentation that reviews important information about the part of the digestive system they have been assigned.

---

nutrients and where the nutrients are absorbed into the blood. The large intestine removes water from the undigested food, which is passed as feces from the rectum.
2. In mechanical digestion, food particles are broken down into smaller and smaller pieces. This action allows enzymes to access the food, and these enzymes chemically digest the food into their component nutrients.
3. Bile emulsifies or breaks fats down into relatively small globules. The digestive enzyme lipase works more efficiently on the smaller globules. Fats are absorbed in the small intestine.
4. Animals such as the koala have a large appendix. The diet of a koala consists exclusively of eucalyptus leaves. Therefore, you can infer that an animal

LABORATORY INVESTIGATION

TESTING THE NUTRIENT CONTENT OF FOODS

Before the Lab

1. At least one day prior to the lab, gather enough materials for your class, assuming six students per group.
2. To prepare Lugol's iodine solution, dissolve 10 grams of potassium iodine in 100 milliliters of distilled water. Then add 5 grams of iodine crystals and stir until the solids are dissolved. Keep refrigerated until needed.
3. To prepare the dextrose solution, dissolve 100 grams of dextrose in 500 milliliters of distilled water.
4. Biuret reagent can be purchased commercially or prepared by dissolving 2.5 grams of copper sulfate crystals in 1 liter of water. Then prepare 10M sodium hydroxide by placing 400 grams of sodium hydroxide in a volumetric flask and bringing the total volume of the solution to 1 liter with water. Complete the preparation by mixing 25 milliliters of the copper sulfate solution with the 1 liter of 10M sodium hydroxide.
5. Prepare albumin solution by mixing the whites of two raw eggs with 250 milliliters of water just prior to the laboratory period. Keep refrigerated when not in use. Discard the unused portion at the end of the day.
6. Prepare the starch solution by adding a small amount of cold water to 1 gram of starch and stirring the mixture into a paste. Mix the paste in 100 milliliters of boiling water and bring it back to a boil. Then cool.

Pre-Lab Discussion

Display the foods that students will be testing in this investigation. For each item, have students raise their hands to indicate if they think the food contains protein, sugar, or starch. Then have students write down their predictions for each item. Ask students to save these predictions and compare them with the results of the investigation.

Skills Development

Students will use the following skills while completing this laboratory investigation.

1. Observing
2. Predicting
3. Comparing
4. Relating
5. Applying
6. Inferring
7. Analyzing

Safety Tips

1. Review with students the rules for heat, glassware, and electrical safety.

LABORATORY INVESTIGATION: TESTING THE NUTRIENT CONTENT OF FOODS

PROBLEM

How can you be certain you are eating a balanced diet?

MATERIALS (per group)

starch suspension	400-mL beaker
Benedict's solution	10-mL graduated
Biuret reagent	cylinder
dextrose solution	glass-marking pencil
Lugol's iodine solution	hot plate
albumin solution	medicine dropper
assorted food items	test tubes
(milk, nuts, hard	test tube holder
candy, and so on)	test tube rack

PROCEDURE

1. Add 200 mL of water to the beaker. Place the beaker on the hot plate and heat to barely simmering. This is a water bath. Check the level of the water bath periodically, adding more water if necessary.
2. Use the clean graduated cylinder to measure 3 mL of tap water. Pour this into a clean test tube that has been labeled CONTROL. Pour about 3 mL of the starch suspension into another clean test tube. Label this test tube STARCH. Be sure to rinse the graduated cylinder well after each use.
3. Use the medicine dropper to add two or three drops of Lugol's iodine solution to these two test tubes. Record any changes you observe. Rinse out the test tubes.
4. Use the clean graduated cylinder to measure 3 mL of tap water. Pour this water into a clean test tube that has been labeled CONTROL. Pour about 3 mL of dextrose solution into another test tube. Label this test tube SUGAR.
5. Using the clean graduated cylinder, measure out and pour 3 mL of Benedict's solution into each test tube.
6. Use the test tube holder to place both test tubes in the water bath. **CAUTION:** *Use care in placing the test tubes in the water.* Record any changes you observe in the contents of the test tubes. Use the test tube holder to remove the test tubes from the water. Place them in the test tube rack to cool.
7. Use the clean graduated cylinder to pour 3 mL of tap water into another clean test tube labeled CONTROL. Use the graduated cylinder to pour 3 mL of albumin solution into a clean test tube labeled PROTEIN.
8. Using a clean graduated cylinder, measure out and pour 3 mL of Biuret reagent into these two test tubes. **CAUTION:** *Biuret reagent contains a strong base. Do not let this reagent come into contact with your skin or clothing.* Observe any changes that occur in the test tubes and record your observations.
9. Test a selection of foods for the presence of starch, sugar, and protein, following the same procedures you used in steps 2 through 9. Crush small pieces of each solid food and dissolve it in 3 mL of water. Record your observations.

OBSERVATIONS

1. What color change is a positive test for starch? For sugar? For protein?
2. Which of the nutrients tested are present in your food samples?

ANALYSIS AND CONCLUSIONS

1. Based on your observations, would eating only nuts or hard candies provide a balanced diet? Explain.
2. Would eating a diet of nuts and hard candy provide more nutrients than either food alone? Explain.
3. What nutrients are present in milk? Is milk a more complete food than nuts or hard candy?
4. How can you be sure you are eating a balanced diet?

876

STUDENT STUDY GUIDE

SUMMARIZING THE CONCEPTS

The key concepts in each section of this chapter are listed below to help you review the chapter content. Make sure you understand each concept and its relationship to other concepts and to the theme of this chapter.

39-1 Food and Nutrition

- Water, one of the simplest essential nutrients, is also the most important.
- Minerals are inorganic substances that are needed by the body to make many important compounds.
- Carbohydrates, sugars, and starches are organic nutrients the body uses for energy.
- In reasonable quantities, fats are an important nutrient for humans.
- Proteins provide the body with building materials needed for growth and repair. Proteins also make up the enzymes that keep cell processes going.
- Vitamins are complex molecules the body needs in small amounts. Vitamins are important cofactors for enzyme-catalyzed chemical reactions.

39-2 The Process of Digestion

- The teeth begin the process of digestion by breaking up food into small pieces. The pieces are mixed with the enzyme salivary amylase, which begins starch digestion.
- The esophagus connects the pharynx to the stomach. Rhythmic contractions of the muscular rings of the esophagus push the bolus of food down toward the stomach.
- The stomach is a muscular sac located in the abdomen. Glands in the wall of the stomach secrete gastric fluids, which include mucus, hydrochloric acid, and the enzyme pepsin. In the stomach, the food becomes a pasty mixture called chyme.
- Digestion continues in the small intestine as nutrients are absorbed through the villi and enter the bloodstream.
- The final stages of digestion occur in the large intestine, or colon, where water is removed from the materials that have passed through the small intestine.

REVIEWING KEY TERMS

Vocabulary terms are important to your understanding of biology. The key terms listed below are those you should be especially familiar with. Review these terms and their meanings. Then use each term in a complete sentence. If you are not sure of a term's meaning, return to the appropriate section and review its definition.

39-1 Food and Nutrition
nutrition
calorie
mineral
essential amino acid
vitamin

39-2 The Process of Digestion
digestion
gastrointestinal tract
bolus
pharynx
epiglottis
esophagus
peristalsis
sphincter
stomach
chyme
pyloric valve
duodenum
small intestine
pancreas
liver
gallbladder
villus
large intestine
colon

Analysis and Conclusions

1. No, even combined they lack essential nutrients.
2. Yes, since the nuts provide protein and the candies provide sugar.
3. Milk contains sugar and proteins. As such, it is a more complete food than nuts or candy, each of which provides only one main nutrient.
4. Eat a variety of foods from different food groups.

Going Further: Enrichment

Have students design an investigation that tests foods for the presence of fats. Students may want to use reference sources to find a suitable test or they may wish to devise their own test.

2. Caution students not to taste the foods they are testing.

Teaching Strategy

Make sure that each group has clearly determined the color change that is obtained for each nutrient. If there is confusion or doubt about any of the changes, you may wish to demonstrate the change in question for the class.

Observations

1. Starch: Iodine changes from amber to blue. Sugar: Benedict's solution changes from blue to yellow (or green, orange, or red). Protein: Biuret reagent turns from blue to purple.
2. Answers will vary depending on foods used. Milk and milk products usually contain sugar and protein, nuts contain protein, and candies contain sugar.

CHAPTER REVIEW

CONTENT REVIEW

Multiple Choice
1. c 2. b 3. d 4. d
5. d 6. c 7. c 8. d

True or False
1. T
2. F fats
3. F mouth
4. T
5. T
6. F glucose or carbohydrates
7. T
8. T

Word Relationships

A.
1. proteins
2. mouth, stomach, or small intestine
3. vitamin A, D, E, or K
4. amylase

B.
5. types of nutrients; roughage is indigestible
6. sources of digestive enzymes or fluid; esophagus is the tube that connects the pharynx with the stomach
7. types of teeth; enamel forms the outermost layer of a tooth
8. stomach functions; sugars are digested by the small intestine and pancreas

CONCEPT MASTERY

1. A balanced diet includes a selection of foods from all five food groups. A balanced diet is important because it provides all of the nutrients the body needs for energy, growth, and repair.
2. When food enters the stomach it is bathed in gastric fluids. Mucus keeps the food well lubricated and also protects the walls of the stomach itself. Acid helps break down foods in the stomach. The digestive enzyme pepsin begins the digestion of proteins. Sugars and fats pass through the stomach without being chemically altered.
3. Roughage stimulates the walls of the digestive tract to contract and relax. Peristalsis is rhythmic contractions of the digestive tract's muscles, which are coordinated to propel food through the tract.
4. The large intestine absorbs much of the water from food and it absorbs many vitamins. It also stores wastes until they can be expelled.
5. Saturated fats are usually solid at room temperature. Saturated fats contain the maximum number of hydrogen atoms. Unsaturated fats are usually liquid at room temperature. Unsaturated fats have multiple double bonds between carbon atoms and thus contain fewer hydrogen atoms. Many vegetable oils are polyunsaturated fats and contain even fewer numbers of hydrogen atoms.
6. A tiny flap of tissue, the epiglottis, covers the entrance to the trachea, or breathing tube. When you swallow food, the epiglottis flips down to cover the breathing tube, preventing food from entering.
7. Water is an important constituent of most body tissues. Water acts as a solvent in which food and enzymes are dissolved in the digestive system. Water also dissolves the body's waste materials. Water, in the form of sweat, cools the body during strenuous exercise.

CHAPTER REVIEW

CONTENT REVIEW

Multiple Choice

Choose the letter of the answer that best completes each statement.

1. The body derives its energy from
   a. fats.   c. carbohydrates.
   b. proteins.   d. water.
2. The breakdown of a disaccharide results in the production of
   a. fats.   c. proteins.
   b. monosaccharides.   d. roughage.
3. Complex molecules our body usually cannot make that are needed in very small amounts are
   a. minerals.   c. proteins.
   b. fats.   d. vitamins.
4. Which one of the following vitamins is made in the large intestine?
   a. A   b. B   c. C   d. K
5. A balanced diet contains a variety of foods from the
   a. complex proteins.   c. essential amino acids.
   b. complex sugars.   d. four food groups.
6. The amount of energy in foods is measured in
   a. ATP.   c. calories.
   b. disaccharides.   d. carbohydrates.
7. Cellulose, an important part of plant cell walls, is made up of molecules of
   a. water.   c. starch.
   b. amino acids.   d. sugar.
8. The part of the digestive system in which digested materials are absorbed is the
   a. mouth.   c. esophagus.
   b. stomach.   d. small intestine.

True or False

Determine whether each statement is true or false. If it is true, write "true." If it is false, change the underlined word or words to make the statement true.

1. The surface area of the small intestine is increased by the presence of <u>villi</u>.
2. Bile helps the body digest <u>proteins</u>.
3. Digestion begins in the <u>stomach</u>.
4. Water is removed from <u>digested</u> food in the <u>large intestine</u>.
5. Humans are <u>omnivores</u>.
6. The body gets most of its energy from <u>maltose</u>.
7. <u>Lipids</u>, which are a kind of fat, are important parts of cell membranes.
8. <u>Saturated fatty acids</u> are usually solid at room temperature.

Word Relationships

A. *An analogy is a relationship between two pairs of words or phrases generally written in the following manner: a:b::c:d. The symbol : is read "is to," and the symbol :: is read "as." For example, cat:animal::rose:plant is read "cat is to animal as rose is to plant."*

  In the analogies that follow, a word or phrase is missing. Complete each analogy by providing the missing word or phrase.

1. sugars:starches::amino acids:_____
2. mechanical digestion:mouth::chemical digestion:_____
3. water soluble:vitamin C::fat soluble:_____
4. pancreas:proteases::mouth:_____

B. *In each of the following sets of terms, three of the terms are related. One term does not belong. Determine the characteristic common to three of the terms and then identify the term that does not belong.*

5. proteins, roughage, fats, starches
6. pancreas, liver, small intestine, esophagus
7. incisors, enamel, canines, molars
8. produces acid, breaks down proteins, digests sugars, produces mucus

CONCEPT MASTERY

Use your understanding of the concepts developed in the chapter to answer each of the following in a brief paragraph.

1. What is a balanced diet? Why is a balanced diet important?
2. Briefly describe what happens to food in the stomach.
3. How do peristalsis and roughage help move food through the digestive tract?
4. What are the functions of the large intestine?
5. Explain the differences between saturated, unsaturated, and polyunsaturated fatty acids.
6. The respiratory and digestive systems meet in the area of the pharynx. How is food prevented from entering the respiratory system?
7. Why is water an important nutrient?

CRITICAL AND CREATIVE THINKING

Discuss each of the following in a brief paragraph.

1. **Applying concepts** Heartburn occurs when stomach acid enters the esophagus. Use your knowledge of the digestive system to explain how this might happen.
2. **Relating facts** Although the body gets most of its energy from glucose, it can also get energy from proteins and fats. Explain how this is so.
3. **Relating cause and effect** In what way is a Calorie, the unit of energy in food, related to an increase in body weight?
4. **Relating concepts** How is the pH of certain body organs important to digestion?
5. **Designing an experiment** Your friend tells you that the digestion of starches begins in the stomach. You suggest that starch digestion begins in the mouth. Design an experiment to show who is correct.
6. **Relating concepts** Some animals that live in the desert rarely drink water. How do they satisfy their need for this nutrient?
7. **Applying concepts** The diets of people in other parts of the world vary considerably from ours. Yet these people are for the most part healthy. How can you explain this?
8. **Using the writing process** A children's television workshop wants to make a movie that explains the process of digestion to young students. You have been asked to write a script that describes the travels of a hamburger and a glass of milk through the human digestive system. Write a brief but concise outline of your script, including information about what happens to each nutrient in each part of the digestive system.

do so. This is a more complex and difficult process and results in the production of various wastes that must be removed from the body.

3. A person uses a certain number of Calories in a day to operate his or her body. People who have strenuous jobs use more Calories; those with a sedentary job use fewer. If you take in more Calories than your body needs for its daily operation, your body will store the extra Calories as fat—you will gain weight. If you use more Calories than you eat, your body will use stored fats for energy and you will lose weight.
4. Enzymes work best at a specific pH. Pepsin, the enzyme that begins the breakdown of proteins in the stomach, works best under acidic conditions. Other enzymes—such as those in the small intestine—work best in a neutral or basic pH.
5. Test an unsweetened cracker for the presence of starch and sugars. You will find starches present but no sugars. Chew the cracker for a few minutes, spit some of it into a test tube and test again. You will see that there are now some sugars present. Digestion of starch begins in the mouth, as some of the starches are broken down into sugars.
6. Desert herbivores get the water they need from plants. Desert carnivores get their moisture from the animals they eat.
7. Although diet may vary in different parts of the word, if people eat a variety of foods from different food groups, they will maintain their health.
8. Check outlines for correctness and clarity.

CRITICAL AND CREATIVE THINKING

1. If the sphincter that separates the stomach from the esophagus does not close completely, acid produced in the stomach can move back into the esophagus. Unlike the tissues in the stomach, the esophagus is not protected by a coating of mucus. Therefore the acid from the stomach actually burns the tissues of the esophagus, causing the pain generally referred to as heartburn. Heartburn has nothing to do with the heart itself. The name refers to the general area in which the pain is felt.
2. The body is able to break down proteins and fats for energy, but it must convert these nutrients into sugars to

CHAPTER 40 Respiratory System

Section	Laboratory Investigations and Activities
40–1 The Importance of Respiration pages 881–882 THEMES: Energy, Scale and Structure	**Teacher Edition** DEMONSTRATION: Air Pressure, p. 880D **Teaching Resources** ACTIVITY: Changes in Pitch, p. 3
40–2 The Human Respiratory System pages 882–889 THEMES: Scale and Structure, Systems and Interactions	**Student Edition** LABORATORY INVESTIGATION: Constructing a Model of the Respiratory System, p. 892 **Teacher Edition** DEMONSTRATION: Adam's Apple, p. 880D **Laboratory Manual** Measuring Lung Capacity, p. 497 **Teaching Resources** HANDS-ON ACTIVITY: Taking a Breather, p. 193
40–3 Control of the Respiratory System pages 890–891 THEME: Systems and Interactions	**Laboratory Manual** The Effect of Exercise on Respiration, p. 503
Chapter Review pages 893–895	

Teacher Resources

Books

Altman, Philip L., and Dorothy S. Dittmer, eds. *Respiration and Circulation*, rev. ed. Pergamon, 1983.

Cherniack, Reuben M., and Louis Cherniack. *Respiration in Health and Disease*, 3rd ed. Saunders, 1983.

Martini, Frederic. *Fundamentals of Anatomy and Physiology*. Prentice Hall, 1992.

Pallot, David J. *Control of Respiration*. Oxford University Press, 1983.

Weibel, Ewald R. *The Pathway for Oxygen*. Harvard University Press, 1984.

Weinstein, Alan. *Asthma*. McGraw-Hill, 1987.

Whipp, B. J., ed. *The Control of Breathing in Man*. University of Pennsylvania Press, 1987.

CHAPTER PLANNING GUIDE

Other Activities	Media and Technology
Study Guide Section 40–1, p. 383	**Video/Videodisc** Human Body: Respiratory System and Circulatory System **Biology Media Guide** page 62
Teaching Resources ACTIVITY: Understanding the Physics of Respiration, p. 5 **Study Guide** Section 40–2, p. 385	**Video/Videodisc** Human Body: Respiratory System and Circulatory System **Biology Transparencies** 56: The Human Respiratory System 64: The Structure of the Human Lungs **Biology Media Guide** pages 62–63
Teaching Resources VOCABULARY REVIEW: Word Game, p. 1 **Study Guide** Section 40–3, p. 389	**Video/Videodisc** Human Body: Respiratory System and Circulatory System **Biology Media Guide** page 63
Teaching Resources Chapter Test A, p. 13 Chapter Test B, p. 17	**Computer Test Bank** Chapter Test, p. 421

Audiovisuals

Respiration. Film or video. MTI Coronet.
The Lungs: An Inside Story. Film or video. Coronet.
Respiratory System: Structure and Function. Sound filmstrip. Ward.
Respiratory System, 2nd ed. Film or video. Coronet.
Respiration in Animals. Film or video. Coronet.

CHAPTER 40 Respiratory System

CHAPTER OVERVIEW

Respiration is a vital function of all living things. The common function of all types of respiration is the transfer of gases between structures. The method by which respiration occurs differs in single-celled and multicellular organisms.

In humans, respiration occurs through the interaction of various structures. Structures that are located in the upper region of the respiratory system filter harmful contaminants from air drawn into the body. Respiratory trees housed in the lungs contain structures that facilitate an exchange of gases with the circulatory system.

Breathing refers to drawing air into and out of the lungs. Movement of the diaphragm creates unequal air pressure between the lungs and the chest cavity.

The medulla oblongata of the brain controls breathing. Carbon dioxide concentrations in the blood trigger particular responses by the nervous system. Although an individual may stop breathing for a brief period, eventually the brain takes over to restart the process.

40-1 THE IMPORTANCE OF RESPIRATION

Section Focus 40-1

The purpose of this section is to discuss the general functions of the respiratory system. Respiration occurs at two different levels. Internal respiration occurs at the level of the cell. External respiration occurs within a group of organs working together to exchange carbon dioxide and oxygen with the environment. Single-celled organisms obtain dissolved oxygen directly from the environment. Therefore, they can survive without a respiratory system.

Performance Objectives 40-1

1. Define respiratory system.
2. Compare internal and external respiration.
3. Explain why single-celled organisms can survive without a respiratory system.

Science Terms 40-1

internal respiration   p. 881
external respiration   p. 881
respiratory system   p. 881

40-2 THE HUMAN RESPIRATORY SYSTEM

Section Focus 40-2

The purpose of this section is to introduce the structure and function of the human respiratory system. The human respiratory system consists of the nose, nasal cavity, pharynx, larynx, trachea, smaller conducting passageways, and lungs. The nose, nasal cavity, pharynx, and trachea all possess structures that filter harmful contaminants from the air. The larynx contains highly elastic folds of tissue that vibrate and produce sound as air rushes past.

The lungs house respiratory trees in which gases are transferred between the alveoli and the bloodstream. Along the bottom of the rib cage lies a large, flat muscle called the diaphragm. The motion of the diaphragm causes air to be drawn into and out of the lungs.

In the lungs, oxygen is removed from inhaled air and enters the alveoli. At the same time, carbon dioxide from the blood enters the alveoli. In an exchange of gases, oxygen enters the blood and carbon dioxide is exhaled as a waste product. Hemoglobin in red blood cells carries the oxygen to all parts of the body.

Performance Objectives 40-2

1. Describe the structures of the respiratory system.
2. Compare inhalation and exhalation.
3. Compare the levels of gases in inhaled air with those in exhaled air.
4. Explain how gases are transferred between the respiratory and circulatory systems.
5. Explain why hemoglobin is important.

Science Terms 40-2

pharynx   p. 882
trachea   p. 882
larynx   p. 883
vocal cord   p. 883
bronchus   p. 885
lung   p. 885
bronchiole   p. 885
alveolus   p. 885
inhalation   p. 886
exhalation   p. 886
diaphragm   p. 886
hemoglobin   p. 888

CHAPTER PREVIEW

40-3 CONTROL OF THE RESPIRATORY SYSTEM

Science Focus 40-3

The purpose of this section is to discuss how the body controls respiration. Breathing is an involuntary action controlled by the medulla oblongata of the brain. The level of carbon dioxide in the blood triggers specific responses in sensory receptors. These responses, in turn, trigger a response by the brain.

The maximum amount of air that can be moved into and out of the respiratory system is called the vital capacity of the lungs. Exercise can increase an individual's vital capacity.

Performance Objectives 40-3

1. Describe how breathing is controlled.
2. Identify the relationship between the levels of carbon dioxide in the blood and breathing.
3. Explain why the vital capacity of an athlete may be greater than that of a nonathlete.

Science Terms 40-3

vital capacity   p. 891

DISCOVERY LEARNING

TEACHER DEMONSTRATIONS
Modeling

Air Pressure

The following demonstration can be used as an introduction to Chapter 40.

Blow up a balloon and hold the mouth closed with your fingers. Ask students to explain why the balloon inflated. (You put gas into it.) Then remove your fingers from the balloon and observe. Ask students what caused the balloon to deflate. (The gases were released from the balloon.) Discuss what caused the release of gas. (Greater air pressure outside the balloon than inside.) Ask students what would happen if you continuously put air into the balloon. (It would expand and then break.) Discuss what would cause the balloon to break. (Greater pressure than the balloon could withstand.) Develop the idea that air moves from a place of higher pressure to a place of lower pressure.

Adam's Apple

You may want to perform this activity after students have completed Section 40-2.

Tell students to place their hand gently over their throat. Then ask students to swallow. Have students describe what they feel. (They should feel a structure moving upward and then down.) Challenge students to identify the structure they felt moving. (Larynx.)

Explain to students that when they swallow, breathing is momentarily stopped to prevent the passage of food into the trachea. Muscles contract and raise the larynx under a fold of tissue called the epiglottis. By closing off the opening to the larynx, the epiglottis prevents food from entering the trachea. The motion the students feel in their throat is actually the raising of the larynx, which is sometimes called the Adam's apple.

CHAPTER 40
Respiratory System

GUIDED ENQUIRY
Pose the following questions to students and have them record their responses. Point out that they will gain a better understanding of the key concepts if they read the chapter with these basic questions in mind. Upon completion of the chapter, pose the questions again. Ask students to compare their initial responses with those they have developed after reading the chapter.

- How does a single-celled organism, such as a diatom, obtain dissolved oxygen?
- How are solid particles prevented from entering the lungs?
- What causes air to enter and exit the body?
- How does oxygen from the air enter the bloodstream?
- How is breathing regulated?

CHAPTER 40
Respiratory System

The main job of the respiratory system is to get oxygen into the body and waste gases out of the body. Oxygen combines with food to produce energy, which is required for all life functions, including such activities as skiing. The respiratory tree (inset) contains the main organs of respiration.

Watch traffic on a crowded city street on a cold winter day and what do you see? Plumes of whitish steam rising from the exhausts of dozens of cars. Where does this exhaust come from? Your immediate thought might be that it comes from the engines. But there is more to it. For every liter of gasoline each car burns, hundreds of liters of air must enter the engine and hundreds of liters of warm, moist exhaust must be expelled. Burning even a small amount of fuel requires an enormous amount of oxygen.

Our bodies do not burn gasoline, but they do undergo a similar process. The constant idling of our metabolic "engines," which run in every cell of the body, requires an enormous amount of oxygen and produces large quantities of waste gases. How is oxygen efficiently supplied to all the cells of the body? And how are wastes effectively removed? Such questions are the subject of this chapter, for it is the respiratory system that performs these vital functions.

880

INTRODUCING CHAPTER 40

Using the Visuals
Begin your teaching of the chapter by having students examine the inset photograph on page 880. The photograph shows a respiratory tree that contains the main organs of respiration.

• **What organs are contained in the structure shown in the photograph?** (Bronchi, bronchioles, alveoli.)

• **What structure contains a respiratory tree?** (Lung.)

Have students read the chapter opener. Make a class list comparing a car engine with the human body. (Air enters and exits both structures; both structures require energy to operate; both structures need oxygen.) Develop the idea that an exchange of gases occurs in both structures.

TEACHING STRATEGY 40–1

Focus/Motivation
Using the Visuals Have students study Figure 40–1 and read the caption. Ask students to compare the organisms shown in the figure. (Diatoms are single-celled

880

GUIDE FOR READING

After you read the following sections, you will be able to

40–1 The Importance of Respiration
- Compare internal and external respiration.
- Define respiratory system and explain its importance to multicellular organisms.

40–2 The Human Respiratory System
- Identify the parts of the human respiratory system and their functions.
- Explain the mechanisms of inhalation and exhalation.
- Describe the process of gas exchange.

40–3 Control of the Respiratory System
- Identify the conditions that regulate breathing.
- Relate exercise and lung capacity.

Journal Activity

YOU AND YOUR WORLD
The effects of tobacco smoke have caused much controversy. In your journal, explain how you feel about this issue.

40–1 The Importance of Respiration

Guide For Reading
- How do internal and external respiration compare?
- What is the respiratory system?
- Why is a respiratory system important to multicellular organisms?

Respiration is a vital function of all living things. We can think of respiration as a process that occurs at two different levels. One level is the level of the cell. Here, in the mitochondria of eukaryotic cells, respiration requires oxygen, releases carbon dioxide, and produces large amounts of ATP. This level of respiration is called **internal respiration,** or cellular respiration. Cellular respiration was discussed in detail in Chapter 6.

The other level at which respiration occurs is the level of the organism. Here, an organism must get oxygen into its cells (and thereby into the mitochondria) and carbon dioxide back out. This level of respiration is called **external respiration** because the exchange of gases takes place with the external environment. External respiration involves the **respiratory system.**

A respiratory system is a group of organs working together to bring about the exchange of oxygen and carbon dioxide with the environment. A single-celled organism living in sea water gets oxygen directly from its surroundings. There usually is plenty of dissolved oxygen in the sea water, and it diffuses easily across the cell membrane into the cell. Carbon dioxide diffuses out of the cell just as easily. Thus single-celled organisms do not need a respiratory system.

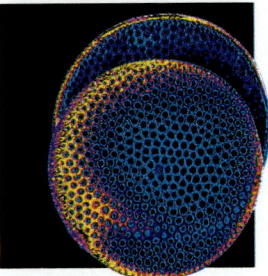

Figure 40–1 Single-celled organisms, such as diatoms, respire through their cell membranes, which are in direct contact with the environment. Most of the cells of the grizzly bear are not in direct contact with the atmosphere and therefore need a respiratory system to bring them oxygen and take away carbon dioxide.

COOPERATIVE LEARNING

Using preassigned lab groups or randomly selected teams, have groups design a survey instrument to determine public opinion in response to the question, Should smoking be banned in all public areas? It would be helpful to show the groups examples of several different types of surveys and questionnaires. The following procedure should be followed in completing this activity:
- Determine the type of survey instrument to be used.
- Identify the target population to survey.
- Develop the survey instrument (suggest a minimum number of questions or items the survey must contain).

As time permits:
- Conduct the survey.
- Organize and analyze the data.
- Communicate the information from the survey in graph or chart form.

Journal Activity

YOU AND YOUR WORLD
Most students will have strong feelings about this issue. Because the relatives of some students and some students themselves may smoke, make sure nonsmoking students treat smokers' feelings with sensitivity. You might want to have students see if their opinions about tobacco smoke change after they have completed this chapter.

---

organisms. A grizzly bear is a multicellular organism.) Develop the idea that although these organisms are quite different, they all require oxygen for life.

Content Development
Guide students' understanding that respiration is a vital function of all living things. All types of respiration involve an exchange of gases.
- **What is internal respiration?** (Respiration that occurs in a cell.)
- **What gases are exchanged in internal respiration?** (Cells take in oxygen and release carbon dioxide.)
- **What is external respiration?** (Respiration that occurs within a group of organs.)
- **What gases are exchanged in external respiration?** (Oxygen is taken into an organism from its environment. Carbon dioxide is released from the organism into the environment.)

881

ANNOTATION KEY

① Nitrogen and oxygen. (Interpreting graphs)
② Oxygen is used during cellular respiration. (Relating facts)

TEACHING SUPPORT

Teaching Resources
- Activity: Changes in Pitch, p. 3

Study Guide
- Section 40–1, p. 383

MEDIA AND TECHNOLOGY

Biology Transparencies
- 56: The Human Respiratory System
- 64: The Structure of the Human Lungs

Video/Videodisc
- *Human Body: Respiratory System and Circulatory System*

Use this bar code to introduce students to the human respiratory system.

Play frames 700 to 1005

See the Biology Media Guide pages 62–63 for additional bar-code correlations for this section.

BACKGROUND INFORMATION
RESPIRATORY SYSTEM OF A BIRD

Because air flows into and out of our lungs through the same passageway, only about 10% of the air in the lungs is replaced with each breath. Birds have a system of air sacs that lightens their bodies and directs air through the lungs in a one-way pattern. This system allows air to flow through the lungs in one direction, ensuring that the lungs have fresh, oxygen-rich air available for gas exchange at all times. The lungs of birds are one of the physiological adaptations that makes them capable of the physical effort required for flying.

But this is not the case with most multicellular organisms. Within these organisms, each cell consumes oxygen and produces carbon dioxide. Without a system to exchange these gases with the atmosphere quickly and efficiently, large multicellular organisms could not survive. A respiratory system ensures the effective exchange of gases—and thus survival—every time an organism takes a breath.

40–1 SECTION REVIEW

1. What is a respiratory system?
2. Why do most multicellular organisms need a respiratory system?
3. What is the distinction between internal respiration and external respiration? How are they related?
4. **Critical Thinking—Making Inferences** Why are more complex respiratory systems found in larger animals?

Guide For Reading
- What are the parts of the respiratory system? What is the function of each part?
- What are the mechanics of inhalation and exhalation?
- What is the process of gas exchange?

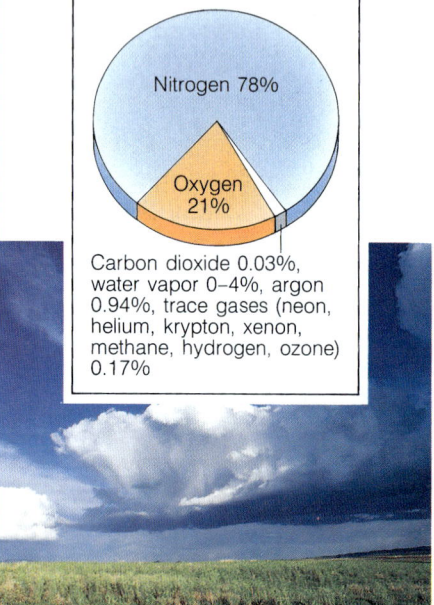

Figure 40–2 Which two gases comprise the largest percentage of the Earth's atmosphere? ①

40–2 The Human Respiratory System

The atmosphere of planet Earth is approximately 78 percent nitrogen and 21 percent oxygen. The remaining 1 percent is made up of carbon dioxide, water vapor, argon, and other trace gases (neon, helium, krypton, hydrogen, and ozone are a few). Humans, like most animals, are air-breathers. Our respiratory system, as well as theirs, has adapted to these concentrations of gases in the atmosphere. Indeed, if the amount of available oxygen falls much below 15 percent, the human respiratory system cannot provide enough oxygen to support cellular respiration.

Respiratory Structures

The human respiratory system consists of the nose, nasal cavity, pharynx, larynx, trachea, smaller conducting passageways, and lungs. If we follow the passage of air from the time it enters the respiratory system to the time it leaves, you will be able to see how each structure of the system functions. Refer to Figure 40–3 often as you read.

Air enters the body through the mouth or nose. Air entering the nose passes into the nasal cavity. The nasal cavity is richly supplied with arteries, veins, and capillaries, which bring nutrients and water to its cells. As air passes back from the nasal cavity, it enters the **pharynx**. The pharynx is located in the back of the mouth and serves as a passageway for both air and food. From the pharynx, the air moves into the **trachea**

40–1 (continued)

Content Development
Compare respiration in single-celled organisms with that in multicellular organisms.

- **What is a respiratory system?** (A group of organs working together to bring about the exchange of oxygen and carbon dioxide with the environment.)
- **How can single-celled organisms exist without a respiratory system?** (Single-celled organisms obtain oxygen directly from the environment.)
- **Identify the importance of a respiratory system to multicellular organisms.** (A respiratory system ensures the exchange of gases vital to the survival of a multicellular organism.)

882

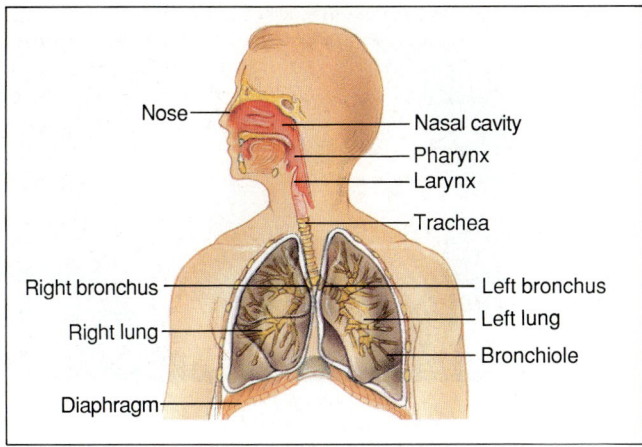

Figure 40–3 Each component of the respiratory system has a specific function to perform. Working together, they ensure the exchange of oxygen and carbon dioxide with the environment. Why is oxygen required by the body? ❷

(windpipe), which leads directly into the lungs. All these passageways provide a direct connection between the outside air and some of the most delicate tissues in the body. Therefore, these passageways must filter out dust, dirt, smoke, bacteria, and a variety of other contaminants found in ordinary air.

The first filtering is done in the nose. The nasal airways are lined with hair and kept moist by mucous secretions. The combination of hair and mucus helps to filter out all sorts of solid particles from the air that passes through the nose. The moisture in the nose helps to humidify the air, increasing the amount of water vapor the air entering the lungs contains. This helps to keep the air that enters the nose from drying out the lungs and other parts of the respiratory system. When air enters the respiratory system through the mouth, much less filtering is done. Therefore, it is generally better to take in air through the nose.

At the top of the trachea is the **larynx**. The larynx is made up of several pieces of cartilage, the largest of which you can feel as your Adam's apple. Because the larynx produces sound, it is sometimes known as the voice box. Inside the larynx are two highly elastic folds of tissue called the **vocal cords**. Air being released from the lungs rushes past the vocal cords. When muscles cause the vocal cords to contract, the air passing between them vibrates and produces sound. The pitch of the sound produced by the vocal cords depends on the length of the vocal cords and their tension (how tightly stretched they are). Male and female children have short vocal cords and so produce high-pitched sounds. Adult males generally develop longer vocal cords than females and thus produce lower pitched sounds.

From the larynx, the air passes downward into the chest cavity through the trachea. The walls of the trachea are made up of C-shaped rings of tough, flexible cartilage. These rings of cartilage protect the trachea, make it flexible, and keep it from

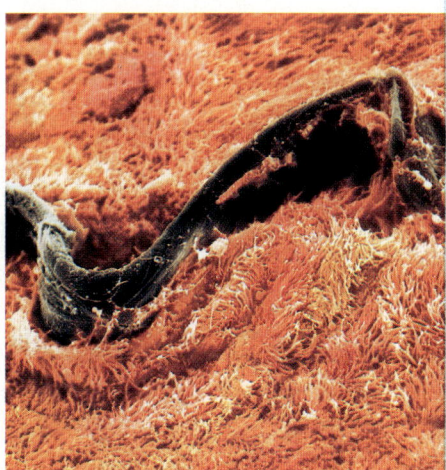

Figure 40–4 Cilia lining the nose, nasal passages, and trachea filter out dust, dirt, smoke, bacteria, and other contaminants found in ordinary air. As the particles get trapped, the nose and trachea often respond to the irritation by producing tiny explosions—sneezes and coughs!

SECTION REVIEW 40-1

1. The respiratory system is a group of organs working together to bring about the exchange of oxygen and carbon dioxide with the environment.
2. Multicellular organisms need a respiratory system so that each cell is able to obtain oxygen and release carbon dioxide quickly and efficiently.
3. Internal respiration is the process by which oxygen and carbon dioxide are exchanged at the level of the cell. External respiration is the process by which the exchange of gases takes place with the external environment at the level of the organism. Both processes involve respiration, or the process that utilizes oxygen to provide energy.

BACKGROUND INFORMATION
SAVING A LIFE

The pharynx is a structure of both the digestive and the respiratory systems. Food passes through the pharynx into the esophagus and then to the stomach. Air passes through the pharynx into the larynx, the trachea, and then to the lungs. Food is prevented from entering the larynx by the motion of a flap of tissue called the epiglottis. Whenever food is swallowed, the epiglottis folds over the larynx, closing off the passageway. Unfortunately, sometimes this mechanism fails and food goes down the "wrong pipe."

Sometimes a large piece of food gets stuck in the air passageway. The effects can be tragic. In the wrong place, the stuck food can shut off the flow of air completely. Victims are suddenly unable to breathe, speak, or call for help. They struggle, become unconscious, and may die from suffocation in a matter of minutes.

First-aid workers use an effective technique to dislodge the food. In the so-called Heimlich maneuver, the victim is grabbed around the waist from behind. Then the person attempting the maneuver forces his or her fists and forearms up against the bottom of the rib cage. This action compresses the lungs and usually exerts enough force to blast the food out of the passageway with a stream of air.

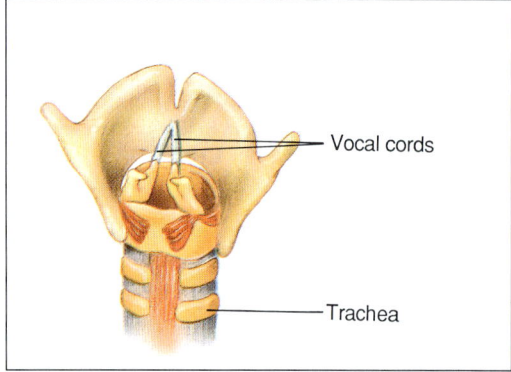
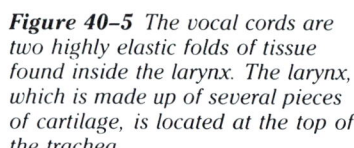
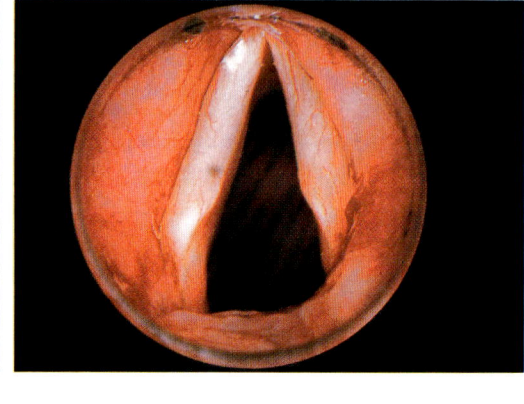

Figure 40–5 The vocal cords are two highly elastic folds of tissue found inside the larynx. The larynx, which is made up of several pieces of cartilage, is located at the top of the trachea.

collapsing or overexpanding. The cells that line the trachea produce mucus. This mucus is swept out of the air passageway by tiny cilia on other cells. In this way, particles trapped in the mucus are carried to the upper part of the trachea and swept down into the digestive system.

This means of protecting the respiratory system works quite well, except in those unfortunate cases in which people sabotage it! When tobacco smokers breathe in tobacco smoke, millions of tiny particles become lodged in the trachea. In time, the cilia of the trachea may actually become paralyzed and cease their cleansing action. When this happens, smokers begin to feel irritation in the chest and develop a persistent smokers' cough. Like the other problems associated with smoking, such coughs are self-induced and only the first sign of more serious problems to follow.

Figure 40–6 It is through a bronchus (left) that air entering the lung is conducted to bronchioles. The feathery appearance of the interior surface of a bronchus is due to the presence of cilia. From bronchioles, air passes into tiny alveoli (right), where gas exchange with the bloodstream takes place.

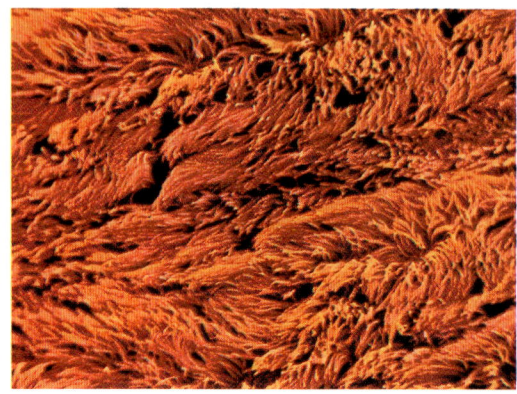
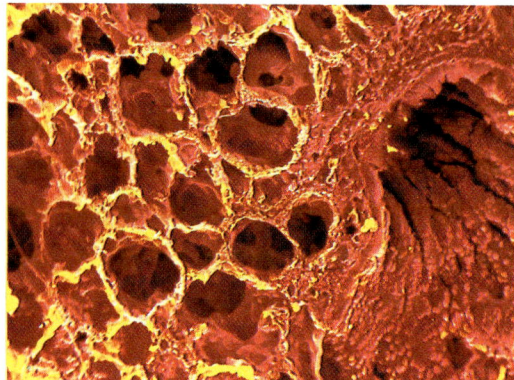

884

40–2 (continued)

Content Development
Guide students' understanding of the danger that cigarette smoking poses to the respiratory system.
- **What effect does cigarette smoke have on the trachea?** (Cigarette smoke contains harmful contaminants that may become lodged in the trachea.)

- **What causes "smoker's cough"?** (Smoker's cough is the result of paralysis of cilia, causing a buildup of irritants in the trachea. By continuous coughing, the body attempts to rid itself of the contaminants.)

Enrichment
Two diseases of the respiratory system are emphysema and lung cancer. Have students research symptoms of each disease as well as the incidence of these diseases in cigarette smokers.

Content Development
Develop students' understanding that the lungs are the main organs of the respiratory system. Each lung houses a respiratory tree in which gas exchange occurs.
- **What respiratory structures are connected by the bronchi?**

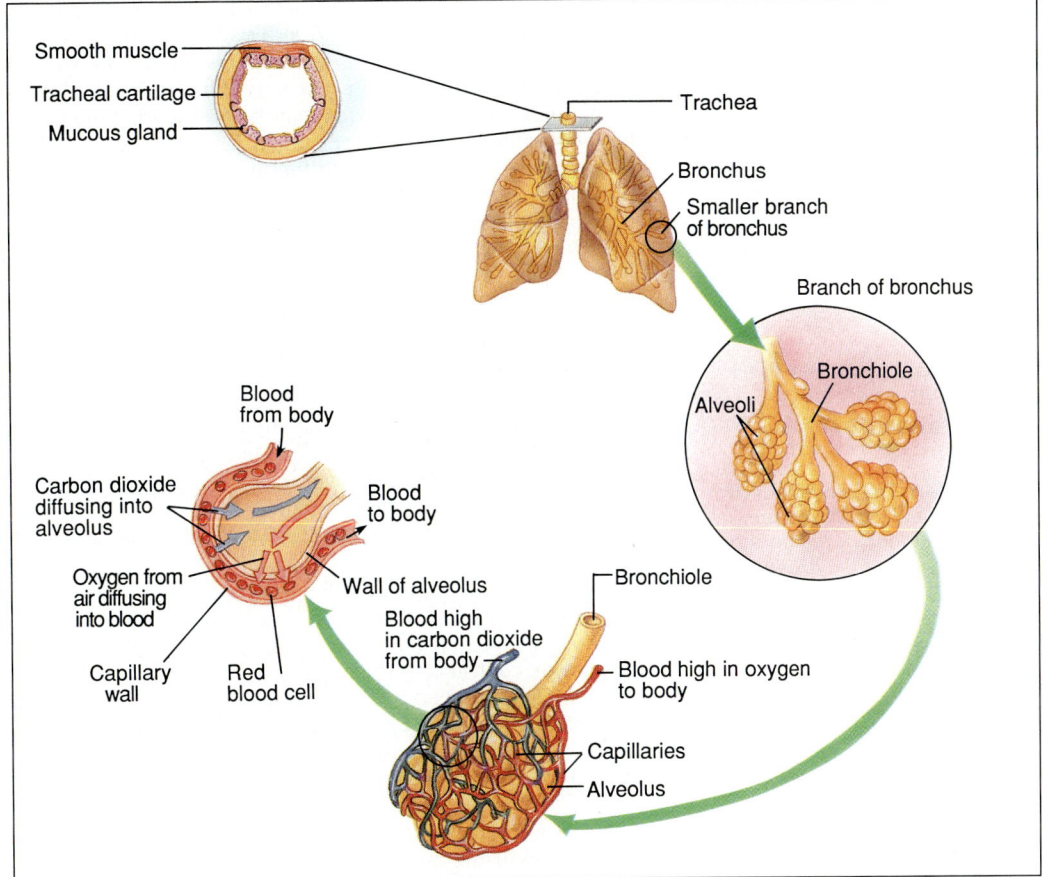

Figure 40-7 *Oxygen and carbon dioxide are exchanged with the blood in the tiny alveoli, or air sacs, of the lungs. Notice the extensive network of capillaries contained in the thin, flexible membrane of an alveolus.*

Within the chest cavity, the trachea divides into two branches, the right and left **bronchi** (singular: bronchus). Each bronchus enters the **lung** on its respective side. See Figure 40-7. The lungs are the main organs of the respiratory system. Like a superhighway splitting into smaller highways and eventually into country roads, the bronchi divide into smaller and smaller passageways. These passageways finally lead into even smaller passageways called **bronchioles**. Both bronchi and bronchioles contain smooth muscle tissue in their walls. This muscle tissue controls the size of the air passages.

The bronchioles continue to subdivide until they finally end in clusters of tiny hollow air sacs called **alveoli** (al-VEE-uh-ligh; singular: alveolus). As you can see from Figure 40-7, groups of alveoli look like bunches of grapes. The alveoli consist of thin, flexible membranes that contain an extensive network of capillaries. The membranes separate a gas

TIE-IN MEDICINE

From 5% to 10% of secondary school students suffer from asthma. During an asthmatic attack, muscles surrounding the air passages that lead to the lungs contract. This contraction reduces the diameter of the bronchioles, which makes it difficult for large columns of air to pass through.

Most asthma sufferers are treated with inhalers that spray drugs in the form of a mist to make breathing easier. A drug commonly found in an inhaler is adrenaline (epinephrine). Adrenaline is nearly identical to a neurotransmitter that causes these muscles to contract. With a few puffs of an inhaler, the muscles generally stop contracting and the attack is under control.

MULTICULTURAL STRATEGY

Garett A. Morgan (1877–1963) is noted for inventing a gas inhaler that became famous following a tunnel explosion in the Cleveland Waterworks. While wearing the gas inhalers, Morgan and three other volunteers were able to descend into the smoky, gas-filled tunnel and rescue several of the workers trapped inside.

Because of the success of this rescue effort, fire companies from all over the nation placed orders for Morgan's invention. However, many of these orders were cancelled when people learned that Morgan was an African American. In order to sell his invention, Morgan was forced to enlist the services of a Caucasian to demonstrate his invention.

(Trachea and lungs.)
- **What structures does air pass through upon entering the lungs?** (Bronchi, bronchioles, alveoli.)
- **What controls the diameter of the bronchi and bronchioles?** (Smooth muscle tissue in the walls of these air passages expand and contract, adjusting the diameter of these structures.)

Content Development
Develop the idea that the structure of the alveoli permits an exchange of gases.

- **Why are alveoli thin membranes?** (Alveoli must be thin for oxygen and carbon dioxide to diffuse through easily.)
- **What structures of the circulatory system surround each alveolus?** (Capillaries.)

from a liquid. The gas is the air that we take in through our respiratory system, and the liquid is blood. Oxygen and carbon dioxide must diffuse across these delicate membranes, so it follows that the membranes must be very thin. However, the fragile nature of the alveoli means that the air entering them must be thoroughly filtered and moistened. Thus the entire respiratory system—from the nose to the lungs—is a device for getting clean, fresh air into the alveoli. Each alveolus acts like a tiny loading station for gas exchange, a process we shall examine shortly.

The Mechanics of Breathing

Every single time you take a breath, or move air in and out of your lungs, two major actions take place. During **inhalation** (also called inspiration), air is pulled into the lungs. During **exhalation** (also called expiration), air is pushed out of the lungs. These two actions deliver oxygen to the alveoli and remove carbon dioxide. The continuous cycles of inhalation and exhalation are known as breathing. Most people breathe 12 to 16 times per minute.

Exactly what happens when you breathe? Inhale deeply. Now exhale. Does it seem as if you are expanding and then contracting your lungs? Perhaps it does, but the lungs cannot expand or contract by themselves. Inhalation and exhalation are actually produced by the movements of the large flat muscle called the **diaphragm** and the intercostal (between the ribs) muscles. The diaphragm is located along the bottom of the rib cage and separates the chest cavity from the abdominal cavity.

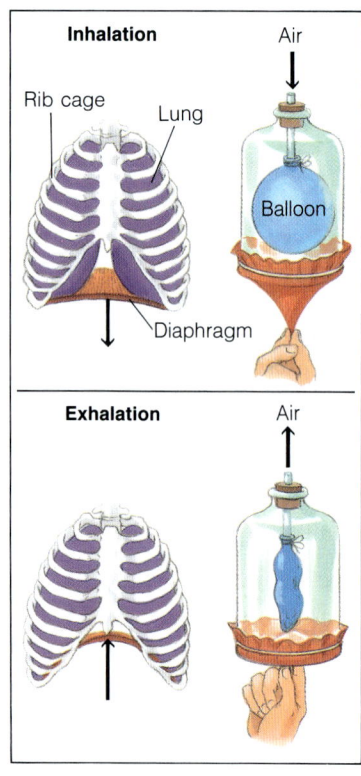

Figure 40–8 The mechanics of inhalation and exhalation are shown in this illustration. What does the flexible bottom of the bell-jar model represent? The balloon?

The mechanics of inhalation and exhalation can be best illustrated by using the apparatus shown in Figure 40–8. A special jar called a bell jar has a flexible bottom and a balloon suspended from a glass tube inserted in a rubber stopper. Before inhalation, the diaphragm is curved upward into the chest cavity. During inhalation, the diaphragm contracts and moves down, causing the volume of the chest cavity to increase. Our bell-jar apparatus shows why this action causes the lungs to fill up with air. When the flexible bottom (diaphragm) moves down, the volume of the jar (chest cavity) increases and the pressure inside it decreases. The air outside the jar is still at atmospheric pressure, however. To equalize the pressure inside and out, the air rushes through the tubing (trachea) into the jar. This inflates the balloon (lungs). The lungs inflate in much the same way. Our lungs have a total capacity of about 6 liters. A normal breath exchanges only one tenth (about 600 milliliters) of this air volume.

When the diaphragm relaxes, it returns to its curved position. This action causes the volume of the chest cavity to decrease. As the volume decreases, the pressure in the chest cavity outside the lungs increases. This increased pressure causes the lungs to decrease in size. The air in the lungs is pushed out, or exhaled.

Although we generally breathe with the diaphragm and intercostal muscles, under extreme conditions we use other muscles to breathe. These muscles can be used to expand and forcibly contract the chest cavity. Both methods work just fine. The diaphragm and intercostal muscles are used most frequently at rest. Breathing by using accessory muscles occurs during vigorous activity for rapid and deep breathing.

The process of exhalation also involves the alveoli. The walls of these tiny air sacs contain many elastic fibers. During inhalation, the fibers are stretched. When inhalation ends and exhalation begins, the elastic fibers pull back to their original size, helping to force air out of the lungs.

The lungs can work properly only if the space around them is sealed. When the diaphragm contracts, the expanded volume in the chest cavity quickly fills as air rushes into the lungs. If there is a hole (even a small one) in the chest cavity, the system will not work. Air will rush into the cavity through the hole, upset the pressure relationship, and possibly cause the collapse of the lungs. This is one of the reasons why puncture wounds in the chest area are extremely serious, even if the lungs themselves are not damaged.

Gas Exchange

As you have read, the atmosphere of the Earth is about 78 percent nitrogen (N_2) and 21 percent oxygen (O_2). What happens to this air when we inhale? The results of a simple experiment provide some immediate answers. Chemical analysis of the gases that are inhaled and exhaled offers these data:

Gas	Inhaled	Exhaled
O_2 (oxygen)	20.71%	14.6%
CO_2 (carbon dioxide)	0.04%	4.0%
H_2O (water)	1.25%	5.9%

As you can see from these figures, three important things happen to the air we inhale: In the lungs, oxygen is removed from the inhaled air and carbon dioxide and water vapor are added to it. Let's take a close look at a single alveolus to understand how all this occurs.

There are nearly 300 million alveoli in a healthy lung, each alveolus similar to the one shown in Figure 40–9. Blood flowing from the heart enters the capillaries surrounding each alveolus. This blood contains a large amount of carbon dioxide and very little oxygen. It spreads around the surface of the alveolus like water streaming around a balloon. The alveolus is filled with fresh oxygen-rich inhaled air. The concentrations of the gases in the blood and in the alveolus are not equal.

Because there is much more carbon dioxide in the blood than in the alveolus, carbon dioxide diffuses out of the blood

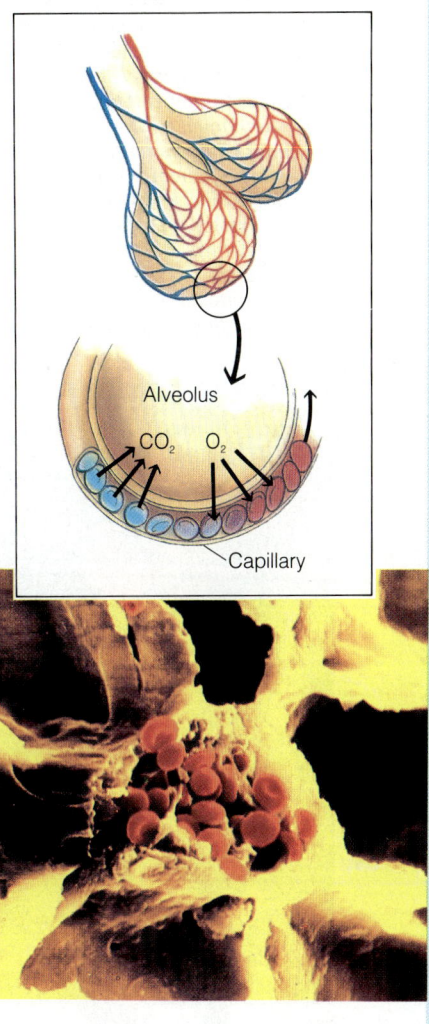

Figure 40–9 The membrane of each alveolus contains a network of capillaries. As blood flows through these capillaries, oxygen diffuses out of the alveolus and into the blood. Carbon dioxide diffuses in the opposite direction—out of the blood and into the alveolus. In the scanning electron micrograph of a section of the lung, you can see a capillary containing red blood cells surrounded by alveoli.

> **BACKGROUND INFORMATION**
> **HEMOGLOBIN**

It's almost impossible to overstate the importance of hemoglobin. Only about 2% of the oxygen needed by the body can be carried directly in solution. What hemoglobin does, in essence, is to expand the oxygen-carrying capacity of the blood. As an example of how important this function is, ask students to imagine what the human body would be like *without* hemoglobin. Without an oxygen carrier, our lungs would have to be much larger and a far greater proportion of our body weight would be taken up by the blood required to deliver oxygen to the tissues. In fact, the body would have to contain about 50 times the blood volume it does now. Since a typical human contains about 6 liters of blood, this would mean an expansion to 300 liters, weighing 300 kilograms—just for blood alone!

> **TEACHING SUPPORT**
>
> **Study Guide**
> • Section 40–2, p. 385

> **40–2 (continued)**
>
> **Enrichment**
>
> Anemia is a condition in which the blood lacks a sufficient quantity of hemoglobin. This condition is not uncommon in adolescence and at other times when the body's need for iron increases. Have students research symptoms of anemia. They may also enjoy researching the way medical treatment of anemia has changed during the past century.

SECTION REVIEW 40–2

1. Air enters through the nose and passes into the nasal cavity, which warms, moistens, and filters the air. The air then enters the pharynx, which serves as a passageway for both air and food. From the pharynx, air moves into the trachea, which leads directly into the lungs. At the top of the trachea is the larynx, the structure that contains the vocal cords. The inside of the trachea is lined with mucus-producing cells and cilia that trap particles and filter the air, respectively. The trachea divides into two bronchi, each of which enters the lungs and divides into smaller branches called bronchioles. The bronchioles divide into smaller and smaller branches that end in clusters of air sacs called alveoli. The exchange of gases occurs in the alveoli.

2. During inhalation, the diaphragm contracts and moves down, causing the volume of the chest cavity to increase and pressure in the chest cavity to decrease. As a result, air rushes

and into the alveolus. Because there is much more oxygen in the alveolus than in the blood, oxygen diffuses across the thin membrane of the sac and into the blood. Thus the blood that leaves the alveolus is oxygen-rich. In fact, it has nearly tripled the total amount of oxygen it originally carried.

Two special molecules help this process of gas exchange work effectively. One of the molecules has an action quite similar to that of soap when it mixes with water. Perhaps you have noticed that when soap is dissolved in water, the water does a much better job of coating surfaces and can even form intricate structures such as soap bubbles. Soaplike macromolecules, consisting of phospholipid and protein, coat the inner surface of each tiny alveolus. This coating keeps the membranes open and makes it easier for oxygen gas to diffuse into the blood.

The other molecule is a special oxygen-carrying molecule that is a natural component of the blood. Without this molecule the blood would be able to carry only about 2 percent of the oxygen needed by the body. This oxygen-carrying molecule is the red-colored protein **hemoglobin**, which is found in red blood cells. A single molecule of hemoglobin has four sites to which oxygen atoms can bind. As the blood flows through the capillaries surrounding an alveolus, oxygen diffusing into the blood is picked up by the hemoglobin in the red blood cells. This allows still more oxygen to diffuse into the blood.

To appreciate just how important hemoglobin is, imagine what the human body would be like without it. Without this oxygen-carrying protein, our lungs would have to be much larger and a far greater proportion of our body weight would have to be taken up by the blood required to deliver oxygen to the tissues. In fact, the body would have to contain about 50 times the blood volume it does now. Since a typical human contains approximately 6 liters of blood, this would mean an increase to 300 liters—with a mass of 300 kilograms!

Figure 40–10 At altitudes much higher than sea level, atmospheric pressure drops. This means that the total amount of gases is much less, and it is more difficult to obtain the required amount of oxygen. Thus mountain climbers must use oxygen masks to help them breathe (right). At higher pressures below the ocean's surface, the story is quite different. Divers must carry air tanks, which release oxygen-containing air at a pressure that matches the pressure of the ocean water on the diver's body (left).

SCIENCE, TECHNOLOGY, AND SOCIETY CONNECTION

An Invisible Poison

When a substance burns completely (that is, in a sufficient amount of oxygen), carbon dioxide (CO_2) is given off into the atmosphere. If burning occurs with only a limited oxygen supply, large amounts of a different gas are produced. This gas is carbon monoxide (CO). Carbon monoxide is produced when gasoline is burned in automobile engines, in charcoal fires, and in some wood-burning stoves.

Carbon monoxide is harmless to most cells. However, the carbon monoxide molecule can bind very tightly to the oxygen-carrying protein hemoglobin. When this happens, oxygen is prevented from attaching to hemoglobin. The carbon monoxide-hemoglobin combination is so stable that only a small amount of the gas is enough to inactivate much of the hemoglobin in the blood, making it unable to carry oxygen to all the parts of the body. As a result, the tissues of the body, deprived of oxygen, begin to suffocate and die. Death can occur after only a few minutes of carbon monoxide poisoning. One of the symptoms of carbon monoxide poisoning is rather surprising. The carbon monoxide-hemoglobin combination is a more brilliant red than normal hemoglobin. Thus, victims of carbon monoxide poisoning often have a flushed, ruddy appearance—what might at first be taken for a healthy look. They should immediately be exposed to fresh air and given as much oxygen as possible in an effort to force oxygen into the blood.

Nearly 500 people die of carbon monoxide poisoning every year in the United States. Auto exhaust systems, stove pipes, and indoor heaters should be carefully checked to make sure they are not leaking dangerous carbon monoxide fumes into living spaces.

The protein hemoglobin is the oxygen- and carbon dioxide-transporting molecule in red blood cells. This computer representation of part of a hemoglobin molecule shows the binding site for oxygen, an iron atom, in red.

40-2 SECTION REVIEW

1. In the order in which air passes through, identify the parts of the respiratory system and their functions.
2. What happens during an inhalation? An exhalation?
3. Explain the process of gas exchange in the lungs. What is the role of hemoglobin in this process?
4. **Connection—Medicine** When a blockage in the upper airways of the respiratory system cannot be removed, an emergency operation called a tracheotomy, which opens the trachea to the outside, is often performed. How does this operation solve the immediate problem of blockage?

in, inflating the lungs. During exhalation, the diaphragm relaxes and moves up, causing the volume of the chest cavity to decrease and pressure in the chest cavity to increase. As a result, air is pushed out, deflating the lungs.

3. Blood flowing from the heart enters the capillaries surrounding each alveolus. This blood contains a large amount of carbon dioxide and little oxygen. The alveolus is filled with fresh, oxygen-rich, inhaled air. Because there is more carbon dioxide in the blood than in the alveolus, carbon dioxide diffuses out of the blood into the alveolus. Because there is more oxygen in the alveolus than in the blood, oxygen diffuses into the blood. Thus, the blood leaving the alveolus is oxygen rich.

SCIENCE, TECHNOLOGY, AND SOCIETY CONNECTION

An Invisible Poison

Instances of carbon monoxide poisoning often occur during cold winter months, usually when a car is left running inside a closed garage. While keeping the garage door shut does keep out cold air, it also keeps automobile exhaust inside the garage. Carbon monoxide gas contained in the exhaust fumes builds up to a lethal level. Occupants of either the car or the garage may die within minutes.

Stress the importance of proper ventilation when working on a running automobile. You may wish to invite a car mechanic to visit the class and discuss proper safety precautions to avoid carbon monoxide poisoning.

Hemoglobin is a special oxygen-carrying molecule that has four sites to which oxygen can bind. As the blood flows through the capillaries surrounding an alveolus, oxygen diffusing into the blood is picked up by the hemoglobin in red blood cells.

4. The tracheotomy permits air to enter directly into the trachea, thereby bypassing the blockage in the upper airways.

Reinforcement/Reteaching

If some students are experiencing difficulty answering the Section Review questions, review the pertinent material in the section.

Closure

Write the terms *bronchi, bronchiole, alveolus,* and *hemoglobin* on the board. Have students identify the role of each of these structures in respiration.

TEACHING SUPPORT

Laboratory Manual
- The Effect of Exercise on Respiration, p. 503

Teaching Resources
- Vocabulary Review: Word Game, p. 1

Study Guide
- Section 40-3, p. 389

MEDIA AND TECHNOLOGY

Video/Videodisc
- *Human Body: Respiratory System and Circulatory System*

Use this bar code to introduce students to how levels of CO_2 in the blood regulate breathing and heart rate.

Play frames 22248 to 23571

See the Biology Media Guide page 63 for additional bar-code correlations for this video/videodisc.

TEACHING STRATEGY 40-3

Focus/Motivation

Using the Visuals Have students observe Figure 40-11 and read the caption. Develop the idea that the amount of pressure increases as the ocean depth increases.
- **Where is the amount of pressure exerted on a body greatest, at sea level or 100 kilometers below sea level? Why?** (Pressure is greater at 100 kilometers below sea level because of the weight of sea water.)
- **What adaptation must creatures of the deep sea possess for survival?** (Creatures of the deep sea must have adaptations that regulate great differences in pressure within and surrounding their body.)

Content Development

Develop the idea that breathing is an involuntary action controlled by the brain.

Guide For Reading
■ What are the conditions that regulate breathing?
■ What is the relationship between exercise and lung capacity?

- **What is another involuntary action of the body?** (One example: the beating of the heart.)
- **How does the nervous system determine the level of gases in the blood?** (Sensory neurons in the carotid arteries of the neck and near the aorta are sensitive to the levels of gases in the blood.)
- **How does the level of carbon dioxide in the blood trigger a response by the nervous system?** (Dissolved carbon dioxide in the blood forms carbonic acid. When this acid breaks down, it releases a hydrogen ion that changes the acidity of blood. The change in acidity triggers a response by the nervous system.)
- **What happens when you hold your breath?** (The level of carbon dioxide gas in the blood increases, causing an increase in

40-3 Control of the Respiratory System

Hold your breath for just a minute or so and then let it go. Did you notice that at first everything seemed just fine, but then you began to develop an urge to breathe. Your chest probably felt tight. Your throat burned. If you held your breath any longer, the pain would have been too much to bear—you just had to breathe!

Breathing is such an important function that your nervous system simply will not let you have complete control of it! **Breathing is an involuntary action under the control of the medulla oblongata in the lower part of the brain.** Sensory neurons (nerve cells) in this region control motor neurons in the spinal cord, and these motor neurons make direct connections with the diaphragm muscle. These neurons do not act alone, however. They are influenced by other nerve cells that detect the stretching action in the lungs. Although breathing can be consciously controlled to a limited extent—such as when you hold your breath or blow up a balloon—it cannot be consciously suppressed. The need to supply oxygen to the cells and remove carbon dioxide is a powerful one indeed.

Obviously, then, the nervous system must have a way of determining whether enough oxygen is getting into the blood. Two special sets of sensory neurons constantly check the levels of gases in the blood. One set is located in the carotid arteries in the neck, which carry blood to the brain. The other set is located near the aorta, the large artery that carries blood from the heart to the rest of the body. **These special sensory receptors are sensitive to the levels of gases in the blood, especially the level of carbon dioxide.** Perhaps this comes as a surprise to you, since you are accustomed to thinking it is the level of oxygen that is most important. Let's see why carbon dioxide is the determining factor.

When carbon dioxide dissolves in the blood, it forms an acid known as carbonic acid.

$$CO_2 + H_2O \rightarrow H_2CO_3$$

Carbonic acid is so unstable that it immediately breaks down into a hydrogen ion (H^+) and a bicarbonate ion (HCO_3^-).

$$H_2CO_3 \rightarrow H^+ + HCO_3^-$$

The hydrogen ions formed by this reaction change the acidity of the blood (its pH). It is to the change in acidity that the special sensory cells respond.

What happens when you hold your breath? The special sensory receptors begin to send messages to the breathing center in the brain. You can override the impulse to breathe at first. But as the messages become stronger due to rising carbon dioxide levels (and increased acidity of the blood), it becomes

890

more difficult to hold your breath. Eventually the medulla oblongata sends such severe sensations to the conscious part of the brain that you are forced to yield—and gasp for air. Phew!

The lungs of an average person have a total air capacity of about 6.0 liters. Only about 0.6 liter is exchanged during normal breathing, however. That is about all the air we need when we are at rest. During vigorous exercise, the situation is considerably different. Deep breathing forces out much more of the total lung capacity. In normal people, as much as 4.5 liters of air can be inhaled and exhaled with effort. The maximum amount of air that can be moved into and out of the respiratory system is known as the **vital capacity** of the lungs. The vital capacity is always about 1 to 1.5 liters less than the total capacity because the lungs cannot be completely deflated without serious damage.

The extra capacity of the lungs allows us to exercise vigorously for long periods of time. Rather than breathing about 12 times a minute, as most of us do at rest, a runner may breathe as often as 50 times a minute.

Figure 40-11 Both the Weddell seal (right) and the gray whale (left) have a sophisticated way of holding their breath known as the diving reflex. The diving reflex enables these mammals to dive to great depths and remain submerged for long periods of time. As the animal dives, blood vessels in most parts of the body constrict, limiting the flow of blood to primarily the heart and the brain—the organs that need oxygen the most.

40-3 SECTION REVIEW

1. What part of the brain controls breathing? How is this control regulated?
2. What gas has the greatest influence on the control of breathing? Explain why this is so.
3. What is vital capacity? Would you expect a trained athlete to have a greater or lesser vital capacity than an average person? Explain your answer.
4. **Critical Thinking—Applying Concepts** Explain the importance of the location of each set of gas-monitoring sensory receptors.

LABORATORY INVESTIGATION

CONSTRUCTING A MODEL OF THE RESPIRATORY SYSTEM

Before the Lab
1. Gather all the materials at least one day before the investigation. Be sure you have enough supplies for each group to make a model of the respiratory system.
2. You may wish to have student volunteers stretch the balloons before the investigation to make sure the balloons are pliable.

Pre-Lab Discussion
Review the structures of the respiratory system. Be sure students understand that air is drawn into the lungs and expelled from the lungs because of unequal air pressure. You may wish to illustrate the effect of unequal pressure by attaching a vacuum pump to a metal can. Use the pump to draw air from the can and observe the results. Discuss with the class why the sides of the can cave inward. (Greater air pressure outside the can than inside the can.)

Skills Development
Students will use the following skills while completing the laboratory investigation.
1. Observing
2. Comparing
3. Relating
4. Applying
5. Inferring

Observations
1. The balloon inflated.
2. The balloon deflated.

Analysis and Conclusions
1. The volume of the bottle increased when the large balloon was moved down and decreased when the large balloon was moved up.
2. When the large balloon was moved down and the volume in the bottle increased, the air pressure inside decreased, causing air to rush in and expand the small balloon. When the large balloon was moved up and volume in the bottle decreased, the air pressure inside decreased, causing air to rush out and deflate the small balloon.
3. The model is similar to the human respiratory system because the small balloon expands and contracts when the large balloon moves as the lungs expand and contract when the diaphragm moves.
4. Humans breathe by moving the diaphragm up and down to increase and decrease the volume of the chest cavity. This action causes changes in the pressure of the chest cavity, causing air to move in and out of the lungs.

LABORATORY INVESTIGATION

CONSTRUCTING A MODEL OF THE RESPIRATORY SYSTEM

PROBLEM
How do humans breathe?

MATERIALS (per group)

#2 one-hole rubber stopper
200-mL polyethylene bottle
round balloon (large)
round balloon (small)
scissors

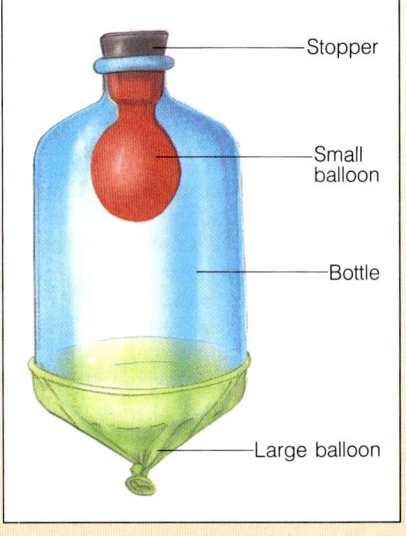

PROCEDURE
1. Place a small polyethylene bottle on its side. Press one point of a scissors through the side of the bottle about 1 cm from the bottom.
2. Using the scissors, cut off the bottom of the bottle by cutting all the way around.
3. Stretch a small balloon and blow it up several times to make it pliable.
4. Pull the lip of the small balloon over the bottom of a #2 one-hole rubber stopper.
5. Insert the balloon through the mouth of the polyethylene bottle. Press the stopper tightly into the bottle so it holds the lip of the balloon in place.
6. Stretch a large balloon and blow it up several times to make it pliable.
7. Using the scissors, cut off about 1 cm from the rounded closed end of the large balloon. Tie the other end closed.
8. Stretch the large balloon far enough over the cut end of the polyethylene bottle so it does not slip off. See the accompanying figure.
9. As you watch the small balloon, pull down on the knot of the large balloon. Then, still watching the small balloon, press up on the large balloon.

OBSERVATIONS
1. What happened to the small balloon when you pulled down on the large balloon?
2. What happened to the small balloon when you pressed up on the large balloon?

ANALYSIS AND CONCLUSIONS
1. What happened to the volume inside the bottle when the large balloon was moved up and down?
2. What caused the small balloon to expand and contract?
3. How is the model you constructed similar to the human respiratory system?
4. How do humans breathe?

STUDENT STUDY GUIDE

SUMMARIZING THE CONCEPTS

The key concepts in each section of this chapter are listed below to help you review the chapter content. Make sure you understand each concept and its relationship to other concepts and to the theme of this chapter.

40–1 The Importance of Respiration
- Internal respiration occurs at the level of cells. External respiration occurs at the level of the organism.
- A respiratory system is a group of organs working together to bring about the exchange of oxygen and carbon dioxide with the environment.

40–2 The Human Respiratory System
- Air enters the respiratory system through the nose and travels through the nasal cavity to the pharynx, the trachea, and the lungs.
- The larynx, located at the top of the trachea, contains the vocal cords.
- The trachea divides into the right bronchus and the left bronchus. Each bronchus enters a lung and further divides into bronchioles, finally ending in groups of alveoli.
- Breathing consists of inhalation and exhalation, which are both produced by movements of the diaphragm, a large flat muscle at the bottom of the rib cage.
- Due to unequal concentrations of gases, carbon dioxide diffuses out of the blood into the air in the alveoli and oxygen diffuses from the air in the alveoli into the blood. Hemoglobin in red blood cells increases the amount of oxygen that can diffuse into the blood.

40–3 Control of the Respiratory System
- Breathing is an involuntary action under the control of the medulla oblongata in the brain.
- The vital capacity of the lungs is the maximum amount of air that can be moved into and out of the respiratory system.

REVIEWING KEY TERMS

Vocabulary terms are important to your understanding of biology. The key terms listed below are those you should be especially familiar with. Review these terms and their meanings. Then use each term in a complete sentence. If you are not sure of a term's meaning, return to the appropriate section and review its definition.

40–1 The Importance of Respiration
- internal respiration
- external respiration
- respiratory system

40–2 The Human Respiratory System
- pharynx
- trachea
- larynx
- vocal cord
- bronchus
- lung
- bronchiole
- alveolus
- inhalation
- exhalation
- diaphragm
- hemoglobin

40–3 Control of the Respiratory System
- vital capacity

CHAPTER REVIEW

CONTENT REVIEW

Multiple Choice
1. b 2. c 3. b 4. a
5. c 6. b 7. d 8. d

True or False
1. T
2. F hemoglobin
3. F diaphragm
4. F medulla oblongata
5. T
6. F external
7. T
8. T

Word Relationships
1. gases that are inhaled and exhaled; helium is not involved in respiration
2. respiratory structures; esophagus is not
3. terms that pertain to alveoli; rings of cartilage compose the trachea
4. activities that are performed on the air that enters the body; decomposing is not performed on the air

CONCEPT MASTERY

1. Air enters through the nose and then passes into the nasal cavity, which warms, moistens, and filters the air. The air then enters the pharynx, which serves as a passageway for both air and food. From the pharynx, air moves into the trachea, which leads directly into the lungs. At the top of the trachea is the larynx, the structure that contains the vocal cords. The inside of the trachea is lined with mucus-producing cells and cilia that trap particles and filter the air, respectively. The trachea divides into two bronchi, each of which enters the lungs and divides into smaller branches that end in clusters of air sacs called alveoli. The exchange of gases occurs in the alveoli.

Going Further: Enrichment
The model can be further extended by comparing the parts of the model with structures of the respiratory system.
- **What respiratory structure is modeled by the small balloon?** (Lung.)
- **What respiratory structure is modeled by the large balloon?** (Diaphragm.)
- **What structure is modeled by the bottle?** (Rib cage.)

The investigation can be further extended by removing the stopper and adding 50 milliliters of water to the bottle. (Be sure the large balloon is securely fastened to the bottom of the bottle.) Repeat Procedure step 9 to observe the effect of the water.

2. The process of diffusion takes place during gas exchange in the alveoli. Because there is a greater concentration of oxygen in the alveoli than in the surrounding capillaries, oxygen diffuses from the alveoli into the capillaries. Because there is a greater concentration of carbon dioxide in the blood vessels surrounding the alveoli than in the alveoli, carbon dioxide diffuses from the blood into the alveoli.

3. Breathing is the process whereby air is taken into the lungs (inhalation) and air is pushed out of the lungs (exhalation). During inhalation, the diaphragm contracts and moves down, causing the volume of the chest cavity to increase. The increased volume in the chest cavity causes a decrease in pressure in the chest cavity. As a result, air rushes in, inflating the lungs. During exhalation, the diaphragm relaxes and moves up, causing the volume of the chest cavity to decrease. The decreased volume of the chest cavity causes an increase in pressure in the chest cavity. As a result, air is pushed out, deflating the lungs.

4. When tobacco smokers breathe tobacco smoke, millions of tiny particles become lodged in the trachea. In time, the trachea's cilia may become paralyzed and cease their cleansing action. When this happens, smokers begin to feel irritation in the chest and develop a persistent smoker's cough.

5. Special sensory receptors located in the carotid artery in the neck and near the aorta are sensitive to the level of carbon dioxide in the blood. When the carbon dioxide dissolves in the blood, it forms carbonic acid, which breaks down into a hydrogen ion and a bicarbonate ion. The hydrogen ions change the acidity of the blood (its pH). This change in acidity is what causes the special sensory receptors to respond. In turn, these receptors send messages to the breathing center in the brain.

6. The structure of the vocal cords enables them to vibrate as air passes through them. The vibration of the vocal cords produces sound.

CHAPTER REVIEW

CONTENT REVIEW

Multiple Choice

Choose the letter of the answer that best completes each statement.

1. Internal respiration is at the level of the
   a. organism.   c. lungs.
   b. cell.   d. pharynx.
2. Sound is produced by vibrations of the
   a. notochords.   c. vocal cords.
   b. bronchi.   d. tracheal rings.
3. The passageway leading directly from the pharynx to the lungs is the
   a. esophagus.   c. bronchus.
   b. trachea.   d. larynx.
4. The order of air movement within the lungs is best described as
   a. bronchi to bronchioles to alveoli.
   b. bronchi to alveoli to bronchioles.
   c. bronchioles to bronchi to alveoli.
   d. trachea to bronchi to alveoli.
5. The substance in blood whose level regulates breathing is
   a. carbon monoxide.   c. carbon dioxide.
   b. oxygen.   d. nitrogen.
6. During inhalation, the contraction of the diaphragm and intercostal muscles causes the chest cavity's
   a. volume to decrease and pressure to increase.
   b. volume to increase and pressure to decrease.
   c. volume to increase and pressure to increase.
   d. volume to decrease and pressure to decrease.
7. Air in the alveoli following an inhalation has the highest concentration of which gas?
   a. nitrogen   c. carbon dioxide
   b. water vapor   d. oxygen
8. The two gases exchanged by the respiratory system are
   a. CO_2 and N_2.   c. O_2 and N_2.
   b. CO and O_2.   d. O_2 and CO_2.

True or False

Determine whether each statement is true or false. If it is true, write "true." If it is false, change the underlined word or words to make the statement true.

1. The part of the throat located behind the mouth is the pharynx.
2. The oxygen-carrying protein found in blood is mucus.
3. The large muscle whose movements control breathing is the trachea.
4. The cerebellum controls breathing.
5. Vital capacity refers to the maximum amount of air that can be inhaled and exhaled.
6. Respiration at the level of the organism is also called internal respiration.
7. The larynx is also known as the voice box.
8. Nitrogen and oxygen account for 99 percent of the Earth's atmosphere.

Word Relationships

In each of the following sets of terms, three of the terms are related. One term does not belong. Determine the characteristic common to three of the terms and then identify the term that does not belong.

1. carbon dioxide, oxygen, helium, water vapor
2. nose, pharynx, trachea, esophagus
3. rings of cartilage, alveoli, gas exchange, thin elastic membranes
4. filtering, humidifying, decomposing, warming

CRITICAL AND CREATIVE THINKING

1. The waste products carbon dioxide and water vapor are produced by cellular respiration. These gases are exhaled. When water vapor (a gas) comes into contact with a cold surface, it condenses.
2. A person who has emphysema would have difficulty breathing because oxygen entering the lungs would not be easily able to pass through the damaged alveoli and en-

CONCEPT MASTERY

Use your understanding of the concepts developed in the chapter to answer each of the following in a brief paragraph.

1. Describe the passage of air through the respiratory system. Indicate the function of each structure of the system.
2. What process occurs during gas exchange in the alveoli? Explain why this process takes place.
3. What is breathing? Explain the mechanism of breathing by describing the relationship between movement of the diaphragm and volume-pressure changes.
4. What damage to the respiratory system is done by tobacco smoke? Why is this dangerous?
5. Explain how the concentration of carbon dioxide in the blood affects the rate of breathing.
6. How is the structure of the vocal cords related to their function?

CRITICAL AND CREATIVE THINKING

Discuss each of the following in a brief paragraph.

1. **Drawing conclusions** You exhale onto a cold surface and notice the condensation of water vapor. Where does this water vapor come from? Why?
2. **Identifying patterns** Emphysema is a disease of the respiratory system that mainly affects the alveoli. In the early stages, the elastic fibers in the walls of the tiny air sacs become damaged. In the later stages, the walls are destroyed. Describe the possible symptoms of a person suffering from this disease. Explain your reasoning.
3. **Relating cause and effect** The condition known as hypoxia is caused by a shortage of oxygen in the body tissues. Explain why the following two changes take place in a person who has hypoxia: (1) The bone marrow produces more red blood cells. (2) The respiratory center of the brain is stimulated.
4. **Relating facts** The air that patients who are breathing on a respirator (tube providing air directly into the trachea) receive must be filtered and humidified externally. Explain why this is so.
5. **Applying concepts** Drugs that artificially lower the pH of the blood can be dangerous. Explain this fact in light of the effects of pH on the breathing center.
6. **Drawing conclusions** Observing people in a crowded room shows that their breathing rates are elevated. Explain why this might be so.
7. **Identifying relationships** Metabolism is the sum of all the chemical activities carried out by a living organism, including cellular respiration. Explain why metabolism can be measured in terms of oxygen consumed.
8. **Relating facts** What physical problems can produce a cough? Why can coughing be considered a defensive action?
9. **Using the writing process** Reliable statistical and experimental evidence indicates that 85 to 90 percent of all lung cancers result directly from cigarette smoking. Write an advertising campaign in which you urge people, especially young people, to avoid smoking. Be sure to include information about smoking's effect on the respiratory system.

---

through the nasal cavity, it must be filtered and humidified externally.

5. Normally, low blood pH is caused by increased carbon dioxide. Low pH triggers the breathing center to increase the breathing rate, and this returns both the blood's carbon dioxide and its pH to normal levels. But if drugs lowered the blood's pH artificially, the breathing center would be stimulated continuously. This results in hyperventilation, a state of excessive, rapid, prolonged breathing.
6. In a crowded room, the amount of carbon dioxide in the air is increased because all the people are exhaling carbon dioxide. The increased carbon dioxide in the air causes the people in the room to inhale more carbon dioxide, which raises the carbon dioxide level in their blood. As a result, the people in the room breathe at a faster rate.
7. Cellular respiration is the process by which the cells take in oxygen, produce energy, and release carbon dioxide. Because metabolism includes cellular respiration, the rate of cellular respiration can be determined by measuring the amount of oxygen consumed.
8. Smoking, colds, pneumonia, and influenza are a few examples of physical problems that can produce a cough. Coughing helps to remove irritants from the breathing passages.
9. Students' campaigns should show an understanding of the respiratory system, as well as the effects of smoking on the respiratory system.

---

ter the bloodstream. People with emphysema would also have much more trouble exhaling than inhaling because the damaged walls of the alveoli are not as elastic as they should be.

3. In hypoxia, the bone marrow produces more red blood cells to carry much-needed oxygen to the body tissues. Red blood cells contain the oxygen-carrying molecule called hemoglobin. The respiratory center of the brain is also stimulated so that the breathing rate is increased, thereby increasing the amount of oxygen entering the body.
4. The nasal cavity is the place in which air is filtered and humidified. Because the air that enters a person who is on a respirator does not pass

CHAPTER 41 Circulatory and Excretory Systems

Section	Laboratory Investigations and Activities
41–1 The Circulatory System pages 897–904 THEMES: Scale and Structure, Systems and Interactions	**Teacher Edition** DEMONSTRATION: One-Way Blood Flow, p. 896D DEMONSTRATION: How to Determine Blood Pressure, p. 896D **Teaching Resources** HANDS-ON ACTIVITY: A Pulsating Question, p. 197 HANDS-ON ACTIVITY: The Squeeze Is On, p. 201 HANDS-ON ACTIVITY: You've Got to Have Heart, p. 205
41–2 Blood pages 905–907 THEMES: Scale and Structure, Systems and Interactions	**Student Edition** LABORATORY INVESTIGATION: Observing Circulation in a Fish Tail, p. 912 **Laboratory Manual** Simulating Blood Typing, p. 507 **Product Testing Activities** Testing Sports Drinks
41–3 The Excretory System pages 908–911 THEMES: Scale and Structure, Systems and Interactions	**Laboratory Manual** Simulating Urinalysis, p. 511
Chapter Review pages 913–915	

Teacher Resources

Books

Gilles, R., ed. *Circulation, Respiration, and Metabolism.* Springer-Verlag, 1985.

Gray, Henry. *Gray's Anatomy.* Running Press, 1987.

Kinne, R. K., ed. *Structure and Function of the Kidney.* S. Karger, 1989.

Little, Robert C. *Physiology of the Heart and Circulation,* 4th ed. Year Book Medical Pubs., 1988.

Marsh, D. J. *Renal Physiology.* Raven Press, 1983.

Tiger, Steven. *Heart Disease.* Julian Messner, 1986.

CHAPTER PLANNING GUIDE

Other Activities	Media and Technology
Teaching Resources ACTIVITY: Analyzing the Human Heart, p. 3 **Biotechnology Workbook** A Change of Heart, p. 119 **Study Guide** Section 41–1, p. 393	**Video/Videodisc** Human Body: Respiratory System and Circulatory System **Biology Transparencies** 65: The Structure of the Human Heart **Biology Media Guide** pages 64–65
Biotechnology Workbook Clot Busters, p. 155 **Study Guide** Section 41–2, p. 396	**Video/Videodisc** Human Body: Respiratory System and Circulatory System **Biology Media Guide** pages 65–66
Teaching Resources VOCABULARY REVIEW: Cross-a-Clue, p. 1 ACTIVITY: The Inner Workings of the Kidneys, p. 5 **Study Guide** Section 41–3, p. 398	**Video/Videodisc** Human Body: Reproductive System and Excretory System **Biology Transparencies** 66: The Human Kidney **Biology Media Guide** pages 66–67
Teaching Resources Chapter Test A, p. 11 Chapter Test B, p. 15	**Computer Test Bank** Chapter Test, p. 431

Audiovisuals

Circulation. Film or video. Coronet.
Excretory System, 2nd ed. Film or video. Coronet.
Circulatory System, 2nd ed. Film or video. Coronet.

Circulatory System: Structure and Function. Sound filmstrip. Ward.

The Heart: An Inside Story. Film or video. Coronet.

CHAPTER 41 Circulatory and Excretory Systems

CHAPTER OVERVIEW

The human circulatory system consists of the heart, the blood vessels, and the blood. The heart is the pump that circulates the blood through blood vessels to all parts of the body. The heart is really a double pump: The right side pumps deoxygenated blood to the lungs, and the left side circulates oxygenated blood to all the organ systems of the body.

Blood vessels form a network throughout the body. Arteries carry blood away from the heart, branching into smaller and smaller blood vessels until they are the size of capillaries. It is at the capillaries that the exchange of nutrients, oxygen, water, salts, carbon dioxide, and wastes occurs. Capillaries mesh into larger and larger blood vessels until they form the large veins that lead back to the heart.

Blood serves many vital homeostatic functions for the human body. It is a type of tissue composed of a liquid portion and a cellular portion. Plasma, red blood cells, white blood cells, and platelets are all found in human blood, and each has its own specific function to perform.

The main organs of the human excretory system are the kidneys. The job of the kidneys is to filter the blood and then to reabsorb any substances the body needs. The kidneys control the composition of the blood by controlling the rate of reabsorption of many of the filtered substances.

Other organs of the excretory system include the urinary bladder, which temporarily stores urine, and the ureters, tubes that lead from each kidney to the urinary bladder. The lungs and the skin also perform excretory functions.

41-1 THE CIRCULATORY SYSTEM

Section Focus 41-1

This section introduces the human circulatory system. The circulatory system is composed of the heart, blood vessels, and blood. The heart is the pump that forces blood through the circulatory system. It functions as a double pump: The right side sends blood into the lungs; the left side circulates blood to the rest of the body. The heart contracts in a wavelike pattern; the atria contract first, then the ventricles. Cardiac muscle initiates its own stimulation for contraction, although the autonomic nervous system does influence the heart rate.

Blood travels throughout the body through a network of blood vessels. The arteries have thick muscular walls that enable them to withstand the high pressure required to carry blood away from the heart. Arteries branch into arterioles, which further branch out into capillaries, the smallest blood vessels in the body. The walls of capillaries are only one cell layer thick, and it is here where the exchange of nutrients and oxygen, carbon dioxide, and waste products occurs between the blood and the tissues. The capillaries regroup to form venules, then form larger vessels called veins that carry blood back to the heart.

Blood travels through the circulatory system because it is under pressure from the contraction of the heart and the muscles surrounding the blood vessels. The body has its own internal mechanisms for regulating blood pressure. However, problems may arise, and blood pressure may become too high or too low.

Fluid that leaks from the blood vessels into the tissue makes up lymph. A system of vessels—the lymphatic system—collects this liquid and, after filtering it through lymph nodes at several points, returns the fluid to the circulatory system.

Performance Objectives 41-1

1. Explain how blood circulates through the chambers of the heart.
2. Describe the pumping action of the heart.
3. Describe the types of blood vessels involved in human circulation and the function of each type.
4. Compare pulmonary circulation with systemic circulation.
5. Explain the regulation of blood pressure and the consequences of low and high blood pressure.
6. Describe the functioning of the lymphatic system.

Science Terms 41-1

heart   p. 898
circulatory system   p. 898
atrium   p. 899
ventricle   p. 899
vena cava   p. 899
artery   p. 900
aorta   p. 900
capillary   p. 901
vein   p. 901
pulmonary circulation   p. 902
systemic circulation   p. 902
blood pressure   p. 903
lymphatic system   p. 904

41-2 BLOOD

Section Focus 41-2

This section introduces the components of human blood. Blood is a tissue that performs many vital homeostatic functions for the body.

The liquid portion of blood is plasma. Its function is to transport salts, sugars, fats, and the plasma proteins. The cellular portion of blood consists of red blood cells, white blood cells, and platelets. Each has its own special function.

Red blood cells are the most numerous. They contain hemoglobin, a protein that transports oxygen from the lungs to the tissues of the body. Hemoglobin gives these cells their red color.

White blood cells protect the body against infection and disease. Some of these cells use phagocytosis to destroy invading bacteria or foreign cells, while other white blood cells produce antibodies.

Platelets are fragments of cells that initiate the clotting process. They clump together at the site of a wound and release clotting factors, setting off a chain reaction to stop the loss of blood.

CHAPTER PREVIEW

Performance Objectives 41-2

1. Explain the homeostatic functions performed by blood tissue.
2. Describe blood plasma.
3. Describe the appearance of red blood cells and explain their function.
4. Describe white blood cells and explain their functions.
5. Discuss how platelets initiate the sequence of blood-clotting reactions.

Science Terms 41-2

blood p. 905
plasma p. 905
red blood cell p. 905
white blood cell p. 905
platelet p. 906

41-3 THE EXCRETORY SYSTEM

Section Focus 41-3

This section describes the excretory system and discusses its most important component, the kidneys.

The functional unit of the kidney is the nephron. At the beginning of the nephron, almost all of the blood is filtered into the Bowman's capsule. Then, as the filtrate courses through the nephron, the kidney reabsorbs most of the blood's components. What remains in the filtrate—urea, some salts, and water—is called urine and eventually is passed through the ureters, bladder, and urethra.

The kidneys also serve to regulate the blood's volume and salinity. With its complex filtering and reabsorbing apparatus, the kidneys can selectively eliminate both water and salts from the blood.

Performance Objectives 41-3

1. Describe the parts of the human excretory system.
2. Describe the nephron's structure.
3. Explain how filtration and reabsorption take place in the nephron.
4. Explain how the kidneys control the composition of the blood.

Science Terms 41-3

excretory system p. 908
kidney p. 908
ureter p. 908
urinary bladder p. 908
nephron p. 908
glomerulus p. 908
Bowman's capsule p. 908
filtration p. 909
reabsorption p. 909
urethra p. 909

DISCOVERY LEARNING

TEACHER DEMONSTRATIONS
Modeling

One-Way Blood Flow

Use this demonstration to introduce the chapter.

William Harvey (1578–1657) published a short book entitled *Anatomical Dissertation Concerning the Motion of the Heart and Blood* in 1628. He clearly described the concept of circulation of the blood and the parts of the circulatory system. It is an amazing work, since Harvey had never even seen a capillary.

You can use one of Harvey's demonstrations to show that blood will flow in only one direction through a blood vessel. Ask the students to examine their forearms and locate one of the large veins found there. Select a student volunteer who has large veins that will be visible to the class. The other students may observe or try the same demonstration on themselves.

Use your fingers to press down on the vein in the region near the wrist. Run a fingertip along the vein toward the elbow. The vein will disappear. Blood will not flow back into it until you release pressure near the wrist.

- **In which direction is the blood flowing in this blood vessel?** (Blood is flowing up the arm.)

Try the opposite effect. Use your fingers to press down on the vein in the region near the elbow. Run a finger tip along the vein toward the wrist. The vein will keep refilling with blood.

- **Why does this blood vessel keep on refilling?** (Because blood is flowing upward from the wrist.)
- **What does this demonstration show about the direction of blood flow?** (Blood flows in only one direction in this blood vessel.)
- **Why was the discovery of blood capillaries so important in confirming Harvey's theory of circulation?** (Arteries carry blood away from the heart; veins carry blood towards the heart. Some type of blood vessel must connect the two. Today, we know that capillaries perform this function.)

How to Determine Blood Pressure

Perform this demonstration after students have completed Section 41-1. Be sure to practice this demonstration before performing it for your class. You also may ask the school nurse or another health care provider to perform this demonstration.

Display a stethoscope and a sphygmomanometer for the students to examine. Explain how these instruments are used to determine blood pressure.

The rubber cuff of the sphygmomanometer is attached around the upper arm and inflated until all blood flow through the artery at the crook of the elbow stops. The reading on the dial or column of mercury on the sphygmomanometer indicates how much pressure is being exerted by the cuff. A stethoscope is placed over the artery just below the cuff, and pressure is released slowly. When the pressure of the cuff falls to equal the pressure in the artery, blood will begin to flow and a pulse sound can be heard. This pressure reading, known as systolic pressure, is the first number recorded.

- **Describe what the heart is doing at systole.** (The heart is contracting.)

As more pressure is released from the cuff, the pulse sounds stop. This pressure reading is diastolic pressure, the second number recorded.

- **Describe what the heart is doing at diastole.** (The heart is relaxing.)

CHAPTER 41
Circulatory and Excretory Systems

GUIDED ENQUIRY
Pose the following questions to students and have them record their responses. Point out that they will gain a better understanding of the key concepts if they read the chapter with these basic questions in mind. Upon completion of the chapter, pose the questions again. Ask students to compare their initial responses with those they have developed after reading the chapter.

- How does the heart pump blood to all parts of the body?
- How do nutrients travel from the digestive system to all the cells of the body?
- How does oxygen travel from the lungs to all the cells of the body?
- How are wastes, such as urine and excess salts and water, removed from the body?

INTRODUCING CHAPTER 41
Using the Visuals
Ask students to cover the caption to the photograph that opens this chapter.
- **Can you identify what is being shown in the photograph?** (Answers may vary.)

If students cannot answer, give them a few hints:

Hint 1. This is a very small part of a very great river.

Hint 2. This river is found inside you.

Hint 3. This river is red in color.

Rivers serve as an essential transportation system for human societies.

- **What transportation system does the human body have that functions in a similar way to river transportation?** (The circulatory system.)
- **What tissue performs the functions of water in the river?** (Blood.)
- **What structures serve a similar function to the riverbank?** (Blood vessels, such as arteries and veins.)
- **What organ keeps pumping blood through the blood vessels?** (The heart.)

CHAPTER 41
Circulatory and Excretory Systems

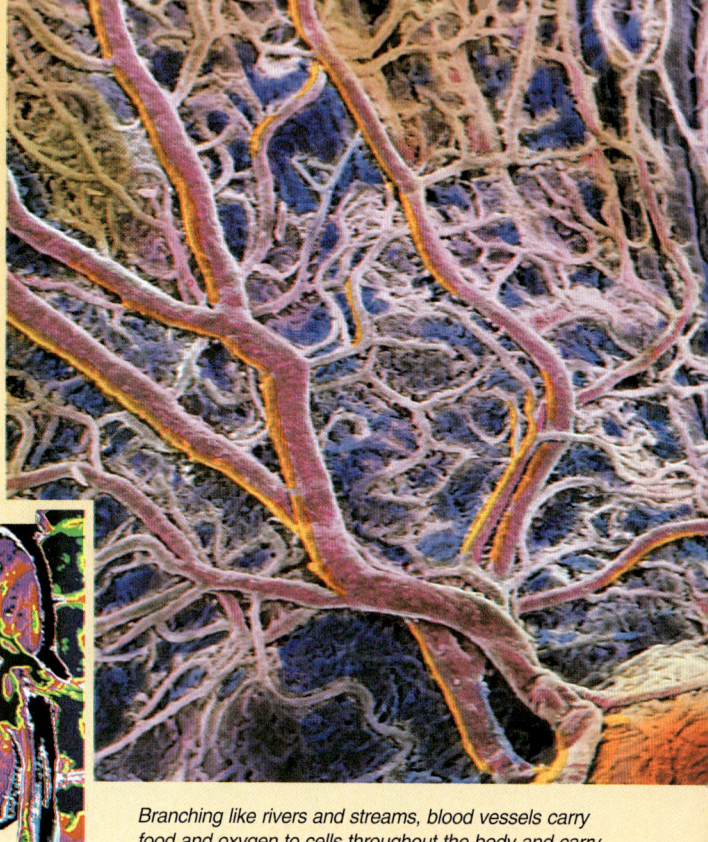

Branching like rivers and streams, blood vessels carry food and oxygen to cells throughout the body and carry wastes away. These wastes are removed from the blood by the kidney (inset).

Large cities are often located near rivers because rivers provide fresh water, food, and, most importantly, a means of transportation for people living there. A complex society cannot develop and thrive without a way to move raw materials, food, and finished products from one place to another. A flowing river provides the means of transportation that holds a society together.

Within the complex society of cells that make up the human body, the needs of living tissues are remarkably similar to those of a city. Each cell needs a steady supply of food and oxygen. That supply is provided by the body's great river—the circulatory system. The basic material of this system is a living tissue called blood.

In its travels, the circulatory system collects poisons and wastes from the cells of the body. These materials are removed by yet another system—the excretory system. You will now discover how these two systems allow billions of cells to live side by side in the most complex society ever constructed—the human body.

896

GUIDE FOR READING

After you read the following sections, you will be able to

41-1 The Circulatory System
- List the parts of the circulatory system.
- Describe the structure and operation of the heart.
- Identify the three types of blood vessels and their functions.
- Compare pulmonary circulation and systemic circulation.

41-2 Blood
- List the functions of blood.
- Describe the components of blood.
- Discuss the role of platelets in blood clotting.

41-3 The Excretory System
- List the parts of the excretory system.
- Describe the structure and function of the kidneys.
- Relate the parts of the nephron to their function.

Journal Activity

YOU AND YOUR WORLD

There are many issues surrounding the transplants of hearts, kidneys, and other organs: Who should get the transplants? Should people be allowed to sell their own organs? In your journal, discuss the issue that most interests you. (You are not limited to the issues mentioned.)

Figure 41–1 The red wavy line that is superimposed on a computer graphic grid of the human heart is a portion of an electrocardiogram (EKG). An electrocardiogram is a record of the electrical activities of the heart.

41–1 The Circulatory System

Guide For Reading

- What are the parts of the circulatory system? What are their functions?
- What structures are found in the heart? How do they function?
- How do blood and lymph circulate through the body?

For thousands of years people have wondered about the nature of blood. The Roman physician Galen believed that blood was formed from food that passed from the stomach into the liver, then flowed through vessels to all parts of the body, returning slowly to the liver through the same vessels.

An accurate understanding of the nature of blood had to wait until 1628, when the English physician William Harvey showed that blood circulated throughout the body in one-way vessels. According to Harvey, blood was pumped out of the heart and into the tissues through one type of vessel and back to the heart through another type of vessel. The blood, in other words, moved in a closed cycle through the body.

Harvey was not sure why blood circulated in this way. Today we know the answer: Blood is the body's internal transportation system. Pumped by the heart, blood travels through a network of vessels, carrying materials such as oxygen, nutrients, and hormones to and waste products from each of the one hundred trillion cells in the human body. **Blood, the heart, and blood vessels make up the circulatory system.**

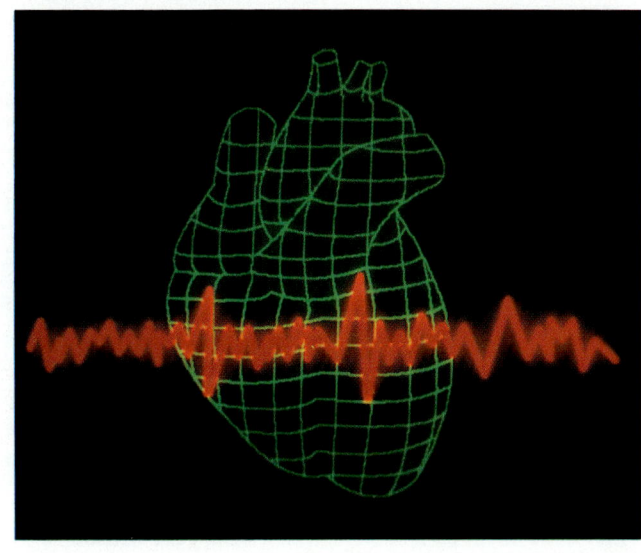

897

COOPERATIVE LEARNING

Using preassigned lab groups or randomly selected teams, have groups prepare and perform a two-minute scene from the soap opera "As the Heart Beats" or "When the Kidneys Excrete." The script may include references to the following:
- Parts of the organ system selected
- Processes of circulation or excretion
- Health problems
- Organ transplants

Encourage groups to be creative in their storyline and to use exaggerated emotions and body movements.

Journal Activity

YOU AND YOUR WORLD

You may want to use the Journal Activity as the starting point for a class discussion. Ask students if they know anyone who has received an organ transplant or who is in need of one. Have students keep their journal assignments in their portfolio.

TEACHING STRATEGY 41–1

Focus/Motivation

Have the class repeat a famous demonstration. The directions for Harvey's demonstration of one-way blood flow can be found on page 896D.

Content Development

Briefly describe the parts of the human circulatory system: heart, blood vessels (arteries, capillaries, veins), and blood. Point out that students will learn about each of these parts as they read the chapter.

Have students read the paragraphs that introduce this section.
- **What does blood do?** (It travels throughout the body, transporting compounds to and from the body's cells.)
- **What does blood deliver to the body's cells?** (Oxygen, nutrients, and hormones.)
- **What does blood take away from the cells?** (The cells' waste products.)

As discussed in Chapter 40, one of the waste products of cells is carbon dioxide. The blood carries this gas to the lungs, where it exchanges it for oxygen. The kidneys remove the other cellular wastes that the blood carries, wastes that include the remains of proteins and other cellular compounds. These wastes are excreted as urine.

897

BACKGROUND INFORMATION
HEART SIZE

The size of a healthy human heart is proportional to the size of the person's body. The heart is roughly as large as the person's fist.

TEACHING SUPPORT

Teaching Resources
- Activity: Analyzing the Human Heart, p. 3
- Hands-On Activity: You've Got to Have Heart, p. 205

MEDIA AND TECHNOLOGY

📺 **Biology Transparencies**
- 65: The Structure of the Human Heart

▶ **Video/Videodisc**
- *Human Body: Respiratory System and Circulatory System*

Use this bar code to introduce the circulatory system.

Play frames 26673 to 27120

See the Biology Media Guide page 64–65 for additional bar-code correlations for this section.

41–1 (continued)
Focus/Motivation
Display a model, transparency, or chart of the human heart. Explain that the heart is made of cardiac muscle tissue sandwiched between two layers of epithelial tissue.
- **How many chambers are found in the human heart?** (Four.)
- **What name is given to the upper chambers?** (Atria.)
- **What name is given to the lower chambers?** (Ventricles.)

Identify each chamber on the model.

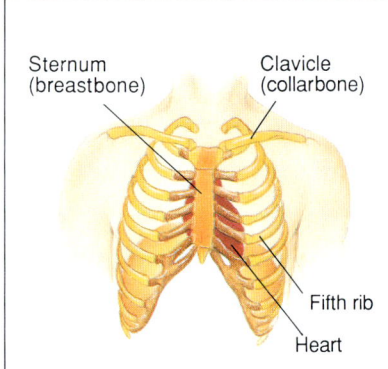

Figure 41–2 The heart is located in the chest cavity and extends almost the length of the sternum (breastbone).

Figure 41–3 The muscular heart is divided into halves by a septum, or wall. On each side of the septum are two chambers; the upper chambers are called atria and the lower chambers are called ventricles.

The Heart

The **heart** is a hollow, muscular organ that contracts at regular intervals, forcing blood through the **circulatory system**. The walls of the heart are made up of three layers of tissue. The outer and inner layers are epithelial tissue, which covers and protects other tissue. The middle layer is cardiac muscle tissue.

As you may recall from Chapter 38, cardiac muscle tissue is not under conscious control of the nervous system. It can contract on its own. As you will soon see, this tendency is important in understanding how the heart works. Cardiac muscle tissue has a rich supply of blood, which ensures that it gets plenty of oxygen. It also has special connections between cells that allow impulses to travel from one cell to another. The cells that make up cardiac muscle tissue are loaded with mitochondria, guaranteeing that each cell has a constant supply of ATP. Whether you are asleep or awake, the heart contracts about once every second every day of your life. The only time the heart rests is between beats.

How the Heart Works

The heart can really be thought of as two pumps sitting side by side. The right side of the heart pumps blood from the body into the lungs, where oxygen-poor (deoxygenated) blood gives up carbon dioxide and picks up oxygen. The left side of the heart pumps oxygen-rich (oxygenated) blood from the lungs to the rest of the body.

The heart is enclosed in a protective sac of tissue called the pericardium (per-ih-KAHR-dee-uhm). Dividing the right side from the left side is a septum, or wall. The septum prevents the mixing of oxygen-poor and oxygen-rich blood.

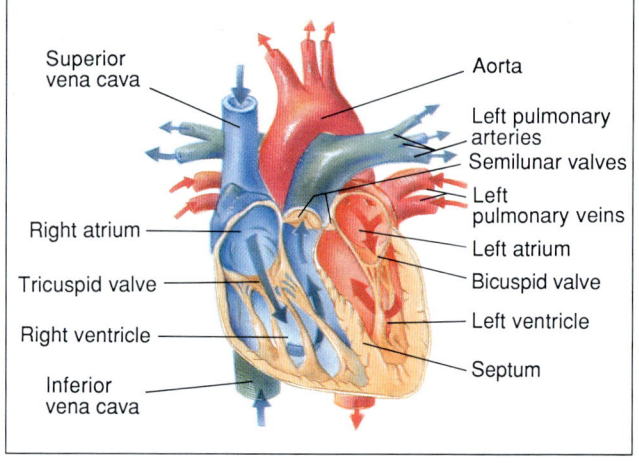

- **Which chambers have the thickest walls?** (Ventricles.)

Content Development
Locate the right side and the left side of the heart for the students.
- **Why do we identify the chambers with this set of right and left directions?** (Right and left do not refer to the observer's right and left but, rather, to the right and left side of the heart's owner.)
- **In the adult heart, is it possible for blood to travel directly from the chambers on the right side to the chambers on the left side?** (No, the heart is divided by a wall, the septum.)

The heart really functions as a double pump; the right side pumps blood to the lungs, and the left side pumps blood out to the other organs of the body.

On each side of the septum are two chambers. The upper chambers, which are called **atria** (AY-tree-uh; singular: atrium), receive blood coming into the heart. The lower chambers, which are called **ventricles**, pump blood out of the heart.

THE RIGHT SIDE OF THE HEART Blood from the body enters the right side of the heart through two large blood vessels called **venae cavae** (veh-nee KAY-vee; singular: vena cava). The superior (upper) vena cava brings blood from the upper part of the body to the heart. The inferior (lower) vena cava brings blood from the lower part of the body to the heart. Both venae cavae empty into the right atrium. When the heart relaxes (between beats), pressure in the circulatory system causes the atrium to fill up with blood. When the heart contracts, blood is squeezed from the right atrium into the right ventricle through an opening surrounded by flaps of tissue. These flaps of tissue form the tricuspid valve, which prevents blood from flowing back into the right atrium.

Each contraction of the heart muscle forces blood out of the right ventricle through blood vessels (pulmonary arteries) into the lungs. At the base of these blood vessels is a valve that prevents the blood from traveling back into the right ventricle. Both this valve and the tricuspid valve ensure that blood flows in only one direction.

THE LEFT SIDE OF THE HEART Oxygen-rich blood leaves the lungs and returns to the heart by way of blood vessels (pulmonary veins). As the blood enters the left atrium, it passes through the bicuspid, or mitral, valve into the left ventricle. Powerful contractions force blood from the left ventricle into a large blood vessel (aorta) that carries it to every part of the body. At the base of this large blood vessel is another valve that prevents blood from flowing back into the left ventricle.

The Heartbeat

Although the heart is a single muscle, it does not contract in a single motion. Instead, the contraction spreads out over the heart like a wave. This wave begins in a small bundle of cells embedded in the right atrium. Because this structure initiates each heartbeat, and thus sets the pace for the heart rate, it is called the pacemaker.

Figure 41–5 on page 900 shows how the impulse spreads from the pacemaker through the cardiac muscle cells in the right and left atria, causing both atria to contract almost simultaneously. When the impulse reaches the walls of the right and left ventricles, they too contract almost simultaneously. As the impulse spreads from one end of a chamber to the other, it stimulates the cardiac muscle cells to contract. This contraction causes the chambers to squeeze the blood, pushing it in the proper direction along its path.

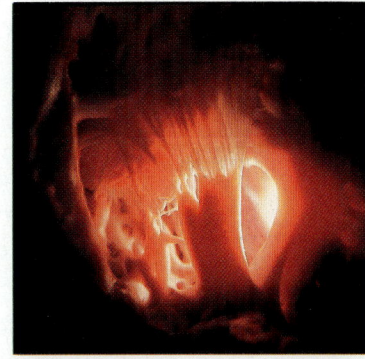
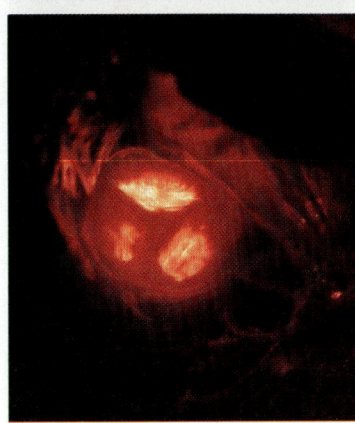

Figure 41–4 Valves control the passage of blood through the heart. The tricuspid valve (top) is located between the right atrium and the right ventricle. The aortic valve (bottom) is found at the base of the aorta.

ECOLOGY NOTE
TRYPANOSOMA CRUZI

Trypanosoma cruzi is a parasitic protozoan that causes Chagas' disease, an often lethal heart disease that is prevalent throughout Latin America.

T. cruzi enters its human host through the disquieting actions of the reduviid bug (*Triatomidae*). The bugs hide in the cracks of rickety houses and bite humans during the night. The bugs defecate while they feed, and they pass *T. cruzi* through their feces. When humans scratch the bug bites, *T. cruzi* has the chance to invade the humans' bloodstream. *T. cruzi* may spread throughout the body, but proves especially lethal when it encysts in heart muscle. In severe cases, the infection causes fever, heart failure, and death.

Identify the venae cavae; the blood these veins carry is returning from circulating among the tissues of the body.
• **Is this blood rich or poor in oxygen? Why?** (Poor in oxygen because it has given up its oxygen to the cells of the body.)

Reinforcement/Reteaching
Locate the region of the pacemaker on the model of the heart. Describe how a wave of contraction spreads over the heart, away from the region of the pacemaker.
• **As this wave spreads over the heart, which chambers of the heart would be the first to contract?** (The atria.)
• **What will happen to the blood that is inside the atria?** (It will be forced through the valves into the ventricles.)
• **What part or parts of the heart will be affected next as this wave of contraction spreads?** (The ventricles will contract.)
• **What will happen to the blood in the ventricles?** (It will be forced out of the heart. The blood on the right side will be sent to the lungs and the blood on the left side will be sent out to the body.)

ESL STRATEGY

Explain that the word *cardiovascular* is taken from a Greek word meaning "heart" and a Latin word that when referring to the heart means "a tube for circulating blood."

After discussing the role cholesterol plays in cardiovascular disease, have students collect as many labels of food products as possible that show cholesterol content. This will help them become aware of the variation in cholesterol levels that exist in everyday foods.

FACTS AND FIGURES

On the average, the heart pumps about 7500 liters of blood every day.

41–1 (continued)

Content Development

Unlike other muscles, the heart does not need stimulation from a motor neuron to contract. It initiates its own contraction. But the rate at which it contracts is controlled by the autonomic nervous system.
• **Which portion of the autonomic nervous system is responsible for speeding up heart rate?** (Sympathetic nervous system.) **What is its counterpart that slows down heart rate?** (Parasympathetic nervous system.)

Content Development

Arteries are the blood vessels that carry blood away from the heart to all parts of the body, including the heart itself. Arteries have thick walls made of smooth muscle and elastic fibers that make them strong but flexible.
• **Why are arteries constructed in this manner?** (When blood is pumped out of the ventricles, it is under great pressure.)

The blood vessel that carries blood away from the left ventricle is the aorta. The aorta is the largest artery and divides into many smaller arteries.
• **Why is the aorta the largest artery?** (It carries blood to the entire body.)

The smallest arteries are called arterioles. The arterioles divide even further into smaller vessels known as capillaries.

Enrichment

A pulse is the alternating expansion and relaxation of an artery. It is due to the elasticity of the arterial walls responding to the regular pumping of blood into the aorta. The pulse can be felt wherever an artery lies near the surface or over a bone.

Content Development

Capillaries branch out to form

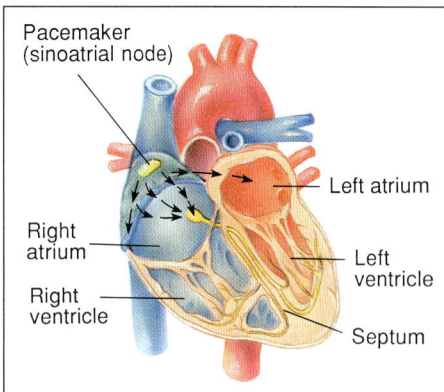
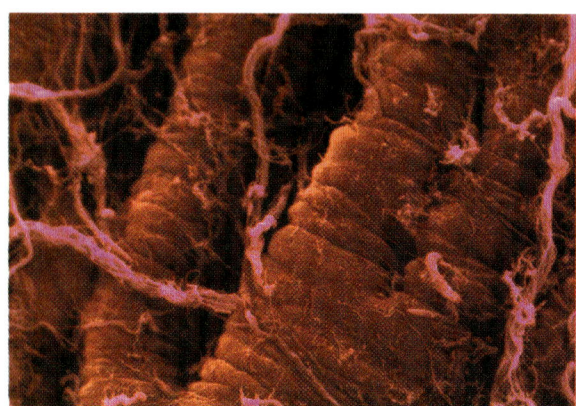

Figure 41–5 *The drawing shows the path an impulse takes as it spreads through the heart. The photograph shows a network of nerves that line a portion of the ventricle.*

You may recall from Chapter 38 that most muscles contract only when stimulated by a motor neuron. This is not the case with cardiac muscle, which initiates the stimulation by itself. However, the autonomic nervous system does influence heart rate. The sympathetic nervous system increases heart rate, and the parasympathetic system decreases it.

In most people at rest, the heart rate is between 60 and 80 beats per minute. During intense exercise, the rate can increase to as many as 200 beats per minute. In this way, the heart can pump oxygen-rich blood through the body more quickly.

Blood Vessels

After oxygen-rich blood leaves the heart, it passes through a network of blood vessels that lead to different parts of the body. The three types of blood vessels that form this network are arteries, capillaries, and veins.

With the exception of capillaries and tiny veins, blood vessels have walls made up of three layers of tissue. The inner layer is epithelial tissue. The middle layer is smooth muscle tissue and elastic fibers. The outer layer is connective tissue.

ARTERIES Arteries carry blood from the heart to all the tissues of the body. In general, the walls of arteries are thicker than those of veins. The smooth muscle cells and elastic fibers that make up these walls help make arteries tough and flexible. These characteristics enable arteries to withstand the high pressure of blood as it is pumped from the heart. Except for the pulmonary arteries, all arteries carry oxygen-rich blood.

The artery that carries oxygen-rich blood from the left ventricle to all parts of the body is the **aorta**. The aorta, with a diameter of 2.5 centimeters, is the largest artery in the body. As the aorta travels away from the heart, it branches into smaller and smaller arteries so that all parts of the body are supplied with blood. The smallest arteries are called arterioles.

CAPILLARIES Arterioles branch into networks of very small blood vessels called **capillaries**. It is in the thin-walled capillaries that the real work of the circulatory system is done. The walls of the capillaries consist of only one layer of cells, making it easy for oxygen and nutrients to diffuse from the blood into the tissues. The forces of diffusion drive carbon dioxide and waste products from the tissues into the capillaries. Capillaries are extremely narrow blood vessels—so narrow, in fact, that blood cells moving through them must pass in single file.

VEINS The flow of blood moves from capillaries into **veins**. Veins form a system that collects blood from every part of the body and carries it back to the heart. The smallest veins are called venules. Like arteries, veins are lined with smooth muscle. The walls of veins, however, are thinner and less elastic than those of arteries. Although the walls are less elastic, they are more flexible and are able to stretch out readily. This is important because it reduces the resistance the flow of blood encounters on its way back to the heart. Large veins contain valves that keep blood from flowing backward. These valves play an important role because blood must frequently flow against the force of gravity.

Blood flowing through veins gets quite a push from the contractions of skeletal muscles, especially those in the arms and the legs. When these muscles contract, they squeeze against veins and help force blood toward the heart. When these muscles are not used for long periods of time, this extra push is lost and blood accumulates in different parts of the body. This is what happens to people who are confined to bed for an extended period of time. In order to prevent this accumulation of blood, health-care workers often try to get these patients to walk— even for just a few minutes a day.

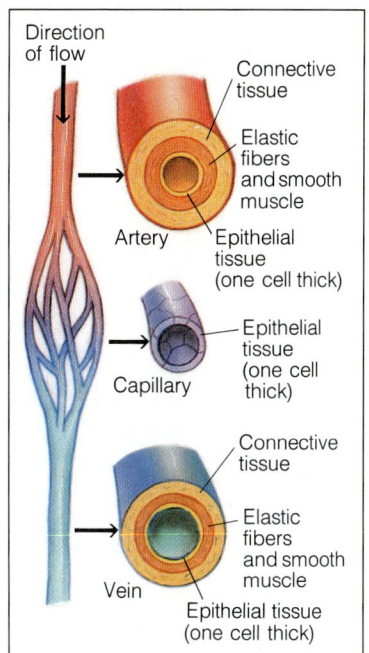

Figure 41–6 Although arteries, capillaries, and veins are all blood vessels, they each have a unique structure that helps them perform their particular function.

Figure 41–7 Note the thick, muscular wall of the artery (left) and the thin, extremely flexible wall of the vein (right).

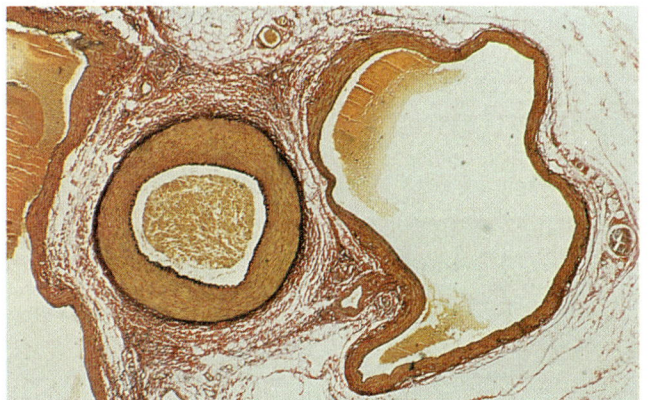

FACTS AND FIGURES

Gaining weight can put additional strain on your heart. It is estimated that each kilogram of fat requires 660 kilometers of capillaries to supply it.

41-1 (continued)

Focus/Motivation

Since the heart functions as a double pump, there are two major pathways of circulation.
- Toward what organs does the right side of the heart pump blood? (Lungs.)
- Toward what organs does the left side of the heart pump blood? (Toward every other organ in the body except the lungs.)

Content Development

Using the Visuals The pathway that blood follows from the heart to the lungs and back to the heart is known as pulmonary circulation.

Use a diagram of the circulatory system (Figure 41-8) to trace the pathway of pulmonary circulation starting at the right ventricle, through the pulmonary artery, through the lungs, through the pulmonary veins, to the left atrium.

Reinforcement/Reteaching

- What is the oxygen content of the blood in the pulmonary arteries? (Deoxygenated.)
- How does this compare with what we have learned previously about the arteries? (It is an exception. Arteries usually carry oxygenated blood.)
- What is the oxygen content of the blood in the pulmonary veins. (Oxygenated.)
- How does this compare with what we have learned previously about veins? (It is also an exception. Veins usually carry deoxygenated blood.)

Content Development

Using the Visuals The pathway that blood follows from the heart, out through the organs of the body, and back to the heart is known as systemic circulation.

Use the diagram of the circulatory system to trace the pathway of systemic circulation starting at the left ventricle, through the aorta, out to the many organ systems of the body, and through the venae cavae into the right atrium.

Each major organ of the body has arteries to supply it with oxygenated blood and veins to transport deoxygenated blood away from it.

Pathways of Circulation

Blood moves through the body in a continuous pathway, of which there are two major parts. **Pulmonary circulation** carries blood between the heart and the lungs. This circulation begins at the right ventricle and ends at the left atrium. **Systemic circulation,** which starts at the left ventricle and ends at the right atrium, carries blood to the rest of the body.

PULMONARY CIRCULATION Oxygen-poor blood is pumped out of the right ventricle of the heart into the lungs through the pulmonary arteries. These are the only arteries in the body that carry deoxygenated blood. In fact, the blood in the pulmonary arteries contains the lowest percentage of oxygen of all other blood vessels. Blood returns to the heart through the pulmonary veins, which are the only veins in the body that carry oxygen-rich blood. The lungs are the only organs directly connected to both chambers of the heart.

SYSTEMIC CIRCULATION Oxygen-rich blood leaving the heart passes through the aorta and into arteries that supply blood to every part of the body. Systemic circulation supplies every major organ with blood, including the heart.

You might think that the heart muscle can get oxygen and nutrients from the enormous amounts of blood that pass through it, but this is not the case. Instead, a pair of coronary arteries leading from the aorta carry blood through the tissues of the heart. These arteries branch into arterioles and then into capillaries, forming a lacy network throughout the heart. See Figure 41-9. The capillaries lead into veins through which blood returns to the right atrium.

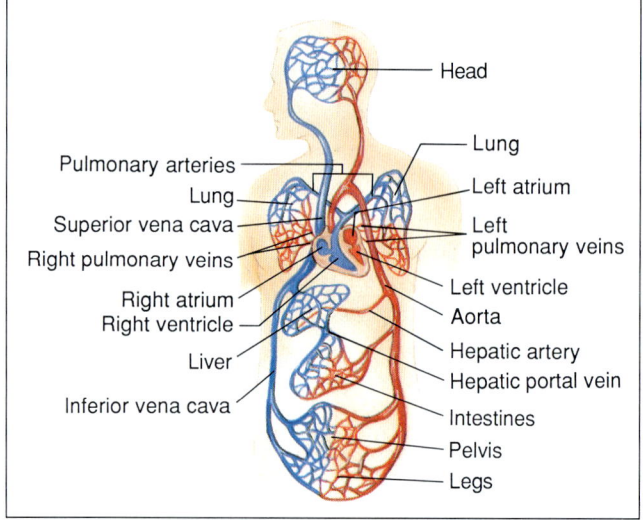

Figure 41-8 Blood moves through the body in a continuous path. The path of oxygen-rich blood is shown in red; the path of oxygen-poor blood is shown in blue.

Content Development

- Since oxygenated blood flows directly through the heart, why does the heart need its own coronary circulatory system? (Heart muscle is very thick. Oxygen and nutrients cannot diffuse to all the layers of heart muscle

In general, blood travels through only one set of capillaries before it returns to the heart. However, there is a special circulation known as the hepatic portal system that is an exception to this rule. Capillaries that line the walls of the digestive system absorb food, minerals, and water. Blood in these capillaries is collected into veins, but these veins do not lead directly to the heart. Instead, they are carried to the liver by the hepatic portal vein.

The hepatic portal vein carries blood directly from the digestive system into capillaries that pass through the liver. Excess sugar and other nutrients are removed in the liver. Veins leading out of the liver collect the blood and return it to the general circulation through the heart. In this way, the hepatic portal system allows the liver to act as a gatekeeper, carefully regulating blood content.

Blood Pressure

Blood moves through the vessels of the circulatory system because it is under pressure. The pressure is produced by the contraction of the heart and by the muscles that surround blood vessels. A measure of the force that blood exerts against a vessel wall is called **blood pressure**.

Blood pressure is regulated by the body in two ways. At several places in the body a special system of sensory neurons attached to blood vessels measures blood pressure. When blood pressure is too low, the autonomic nervous system constricts the smooth muscles in the walls of arteries. This reduces their diameter, and helps to raise blood pressure. When blood pressure is too high, the same system relaxes those smooth muscles, helping to lower pressure. The kidneys are the other means by which blood pressure is regulated. When blood pressure is high, the kidneys remove more water from the blood, lowering the total amount of fluid in the circulatory system. The loss of fluid from the blood lowers the blood pressure, much like the loss of air from a balloon lowers the air pressure inside the balloon.

Problems can result when blood pressure is either too low or too high. Low blood pressure slows down the rate at which blood flows through the body. The parts of the body that are far away from the heart, such as the hands, feet, and the head, do not receive enough blood. If this happens in cold weather, these body parts may become easily chilled and injured.

Hypertension, or excessively high blood pressure, also has serious medical consequences. When blood pressure is high, the heart works much harder to pump blood, causing the heart muscle to weaken. People with high blood pressure are also more likely to develop problems in the arteries outside the heart. Like garden hoses under high pressure, the arteries are more likely to develop leaks.

Although scientists do not know all the causes of hypertension, they are sure of one—obesity. As fat tissue is added to the

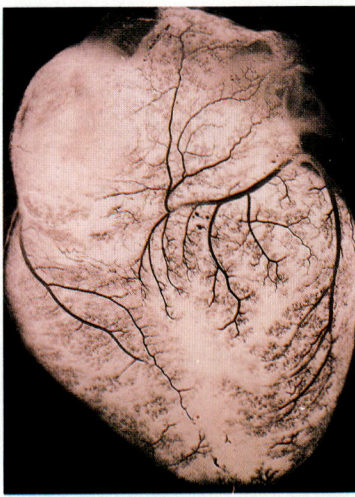

Figure 41–9 The blood vessels covering the surface of the heart bring oxygen and nutrients to the heart muscle and carry wastes away.

Figure 41–10 When arteries become narrowed by deposits of cholesterol, atherosclerosis results.

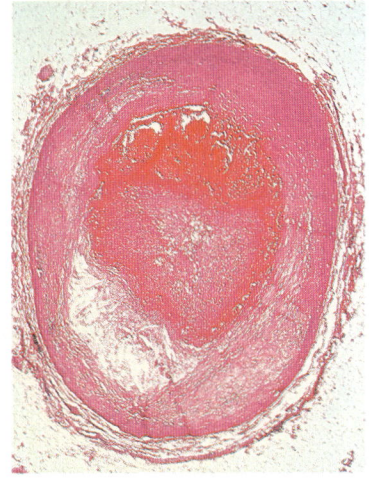

> **TEACHING SUPPORT**
>
> **Study Guide**
> • Section 41–1, p. 393

> **41–1 (continued)**
>
> **Content Development**
> The lymphatic system collects excess fluid, known as lymph, and returns it to the circulatory system. The lymphatic system consists of lymphatic capillaries, which lead into small lymph vessels, which join into larger lymph vessels, which finally empty into the circulatory system at a vein under the left collarbone. Lymph nodes—small bean-shaped filters—are located along major lymph vessels. Larger lymph vessels have valves.
>
> • **What function do valves perform in lymph vessels?** (They prevent the backward flow of lymph.)
>
> • **What other vessels have valves that serve a similar function?** (Veins.)
>
> **SECTION REVIEW 41–1**
>
> 1. The blood, heart, and blood vessels make up the circulatory system.
> 2. The heart is enclosed in a sac called the pericardium. On each side of the septum, which divides the heart into two halves, there are two upper chambers called atria and two lower chambers called ventricles. Blood from the body enters the right atrium through the venae cavae. Blood then moves into the right ventricle through the tricuspid valve. Blood in the right ventricle is sent through the pulmonary artery to the lungs, where the blood picks up oxygen. Oxygen-rich blood leaves the lungs and enters the left atrium by way of the pulmonary vein. As blood enters the left atrium, it passes through the bicuspid, or mitral, valve into the left ventricle. Blood is then forced out of the left ventricle into the aorta

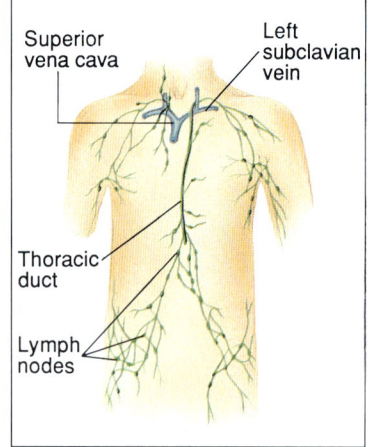

Figure 41–11 The lymphatic system is made up of vessels, capillaries, nodes, and glands that recirculate any fluid that leaves the circulatory system.

body, additional blood vessels are required in order to supply the tissues with oxygen and nutrients. The addition of more blood vessels causes the heart to work harder. As a result, blood pressure increases.

Scientists have developed a number of drugs to lower blood pressure and prevent some of its serious consequences. However, hypertension is easier to prevent than to cure. Exercising regularly, eating sensibly, watching one's weight, and avoiding smoking are the most sensible means of preventing hypertension.

The Lymphatic System

As blood circulates throughout the body, fluid from the blood leaks into the tissues. This action helps to maintain an efficient flow of nutrients and salts from the blood into the tissues. The total leakage of fluid from the circulatory system amounts to approximately 3 liters per day. If this leakage were to continue unchecked, the body would begin to swell with fluid—not a very pleasant prospect!

Fortunately this does not happen. A network of vessels known as the **lymphatic** (lihm-FAT-ihk) **system** collects the fluid that is lost by the blood and returns it to the circulatory system. This fluid, which is known as lymph (LIHMF), collects in lymphatic capillaries and slowly flows into larger and larger lymph vessels. Like veins, lymph vessels contain valves that prevent lymph from flowing backward.

The slowly moving fluid is returned to the circulatory system at an opening in a vein located under the left clavicle, or collarbone, just below the shoulder. Along the length of the lymph vessels are small bean-shaped enlargements called lymph nodes. Lymph nodes act as filters and producers of special white blood cells that prevent harmful material from invading the body cells.

41–1 SECTION REVIEW

1. What are the parts of the circulatory system?
2. Describe the structure of the heart and the circulation of blood through it.
3. How is the heartbeat controlled?
4. Identify the three types of blood vessels. What is the function of each?
5. What is blood pressure?
6. Describe the lymphatic system.
7. **Critical Thinking—Relating Concepts** Why does regular exercise promote good circulation?

> through another valve and is carried to every part of the body except the lungs.
> 3. The heartbeat is controlled by a small bundle of cells, called the pacemaker, that are embedded in the right atrium.
> 4. Arteries carry blood from the heart to all body tissues. Capillaries permit the diffusion of materials between the blood and body tissues. Veins carry blood from all parts of the body back to the heart.
> 5. Blood pressure is a measure of the force that blood exerts against a blood vessel wall.
> 6. The lymphatic system is a network of vessels that collect fluid lost by the blood and return it to the circulatory system.
> 7. Regular exercise strengthens the heart. In addition, muscular contractions in the arms and

41–2 Blood

The function of the circulatory system is to transport material in a fluid medium throughout the body. This fluid medium is called **blood.** Blood is a type of liquid connective tissue that has many functions. **Blood transports nutrients, dissolved gases (oxygen and carbon dioxide), enzymes, hormones, and waste products; blood regulates body temperature, pH, and electrolytes (ions in solution that conduct electric current); blood protects the body from invaders; and blood clots to prevent its own loss after an injury.**

Although blood is referred to as the river of life, the human body contains only 4 to 6 liters of this precious fluid—only 8 percent of the total mass of the body. Approximately 45 percent of the blood consists of living cells. The remaining 55 percent is a fluid called **plasma.**

Blood Plasma

Plasma is a straw-colored fluid that is 90 percent water and 10 percent dissolved fats, salts, sugars, and proteins called plasma proteins. The plasma proteins, which perform a number of vital functions, are divided into three types: albumins, globulins, and fibrinogen. Albumins help regulate osmotic pressure. Globulins include antibodies that help fight off infection. Fibrinogen is responsible for the ability of blood to clot, a process that will be discussed in more detail later in the chapter. Nutrients, hormones, and waste products are also carried in the plasma.

Blood Cells

The cellular portion of the blood includes several types of highly specialized red blood cells, white blood cells, and cell fragments known as platelets.

RED BLOOD CELLS Red blood cells, or erythrocytes (eh-RIHTH-roh-sights), are the most numerous of the blood cells. One microliter of blood contains approximately 5 million red blood cells.

Red blood cells are best described as biconcave disks, narrower in the center than along the edges. These cells are produced from cells in the bone marrow that gradually become filled with hemoglobin, forcing out their nucleus and other organelles. Thus mature red blood cells do not have a nucleus. Hemoglobin is the iron-containing protein that carries oxygen from the lungs to the tissues of the body. Hemoglobin gives the red blood cells their characteristic color.

Red blood cells normally stay in circulation for approximately 120 days before they are destroyed by special white

Guide For Reading

■ What are the functions of blood?
■ What are the components of blood? What are their functions?

Figure 41–12 These packets of frozen plasma have been laid out to thaw so that a necessary clotting factor can be extracted from the plasma. This clotting factor is then used to treat hemophilia.

TEACHING SUPPORT

Laboratory Manual
- Simulating Blood Typing, p. 507

Study Guide
- Section 41–2, p. 396

Product Testing Activities
- Testing Sports Drinks

Biotechnology Workbook
- Clot Busters, p. 155

MEDIA AND TECHNOLOGY

Video/Videodisc
- *Human Body: Respiratory System and Circulatory System*

See the Biology Media Guide pages 65–66 for additional bar-code correlations for this section.

MULTICULTURAL STRATEGY

Have students research the work of Dr. Charles Richard Drew (1904–1950). Dr. Drew developed a procedure for long-term preservation of blood plasma.

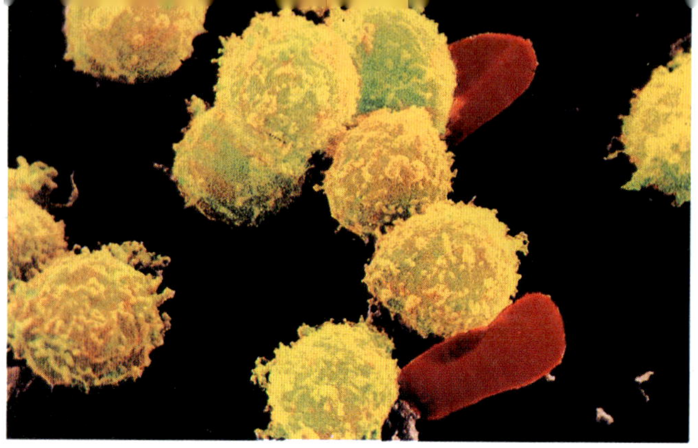

Figure 41–13 *The large yellow objects in this photograph are the smallest types of white blood cells. The two red flattened objects are red blood cells.*

blood cells in the liver and the spleen. At this moment, red blood cells in your body are dying at a rate of about 2 million per second. To replace them, new red blood cells are being formed in the bone marrow at the same rate.

WHITE BLOOD CELLS **White blood cells**, or leukocytes (LOO-koh-sights), are outnumbered by red blood cells almost 500 to 1. White blood cells are produced in the bone marrow, are larger than red blood cells, almost colorless, and do not contain hemoglobin. Unlike red blood cells, white blood cells have a nucleus and can live for many months or years.

The main function of white blood cells is to protect the body against invasion by foreign cells or substances. What enables white blood cells to do this job is, in part, their ability to move out into the surrounding tissue like amebas. Some white blood cells can destroy bacteria and foreign cells by phagocytosis (large particles are taken inside a cell and digested). Others make special proteins called antibodies (globulin-type plasma proteins). Still others release special chemicals that help the body fight off disease and resist infection.

White blood cells respond quickly to infection. Physicians are often able to detect the presence of a serious infection by counting the number of white blood cells in the blood. When an infection such as appendicitis occurs, the number of white blood cells may increase from 10,000 to 30,000 per microliter.

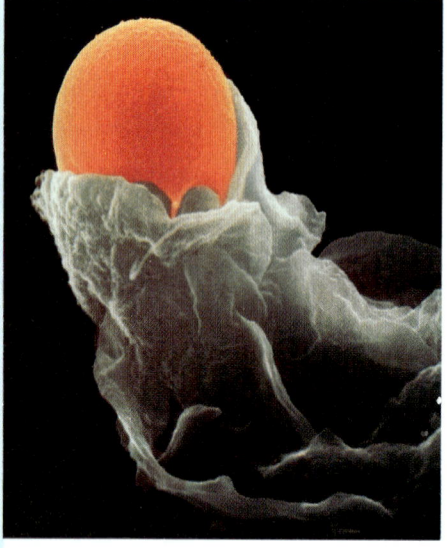

Figure 41–14 *When a red blood cell is about 120 days old, its life draws to a close. Here a special type of white blood cell engulfs and destroys an aging red blood cell.*

PLATELETS AND BLOOD CLOTTING **Platelets** are not cells; rather, they are tiny fragments of other cells. Platelets are formed when small pieces of cytoplasm are pinched off the large cells called megakaryocytes (mehg-uh-KAHR-ee-oh-sights), which are found in the bone marrow. One microliter of blood contains between 250,000 and 400,000 platelets. The life span of a platelet is approximately 5 to 9 days.

Platelets play an important role in preventing the loss of blood by beginning a chain of reactions that result in blood clotting. When you cut or scratch yourself, blood will flow from the wound. Because of the action of platelets, the blood will clot and bleeding will stop after a short time.

41–2 (continued)

Content Development

White blood cells are colorless, are larger than red blood cells, and have a nucleus. They are much less numerous than are red blood cells, but, like red blood cells, they are produced in the bone marrow.

Explain how white blood cells combat disease by removing foreign cells or substances through phagocytosis and the production of antibodies.

Enrichment

Five different types of white blood cells are normally found in the blood. Some leukocytes have dense granules in their cytoplasm (neutrophils, eosinophils, and basophils). Other leukocytes do not have granules (lymphocytes and monocytes.) All white blood cells, except lymphocytes, are phagocytic. In the presence of a foreign antigen, lymphocytes undergo changes to secrete antibodies. There is evidence that although the granular white blood cells are formed in the bone marrow, lymphocytes and monocytes are formed from lymphatic tissue.

Some white blood cells are more common than are others. The normal percentages are neu-

Figure 41–15 This electron micrograph shows the three types of blood cells that make up the cellular portion of the blood. The three large furry objects are white blood cells, the tiny spherical objects in the foreground are platelets, and the beret-shaped object is a red blood cell.

Platelets help the clotting process by clumping together and forming a plug at the site of a wound and then releasing proteins called clotting factors. These proteins start a series of chemical reactions that are extremely complicated. In one reaction, a clotting factor called thromboplastin (thrahm-boh-PLAS-tihn) converts prothrombin, which is found in the plasma of blood, into thrombin. Thrombin is an enzyme that converts the soluble plasma protein fibrinogen into a sticky meshwork of fibrin filaments that stop the bleeding by producing a clot.

The clotting process is extremely complex, and every step of it must go smoothly if a clot is to form. If one of the clotting factors is missing or defective, the clotting process does not work. A serious genetic disorder known as hemophilia results from defects in one of the clotting factor genes. Because they lack one of the clotting factors, hemophilia sufferers may bleed uncontrollably from even small cuts or scrapes. People suffering from hemophilia are given transfusions of clotting factors and platelets in order to treat their disorder.

The clotting of blood is not always a good thing. Sometimes a small clot will form in an unbroken blood vessel, blocking the flow of blood to the cells. If this happens in the brain, brain cells may begin to die, causing a stroke. A stroke may result in the loss of motor functions, such as speech or muscle control, or in death.

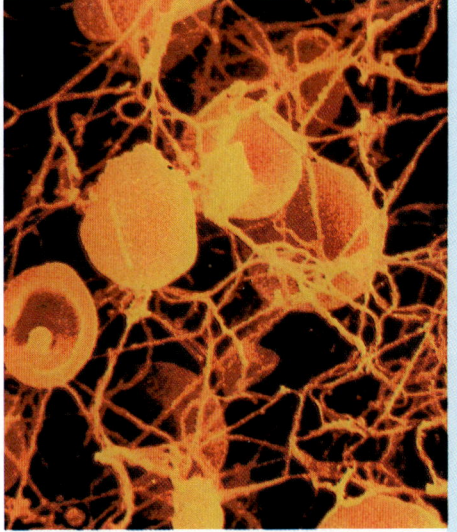

Figure 41–16 In the clotting process, a network of fibrin threads forms over a wound, trapping the blood cells.

41–2 SECTION REVIEW

1. What are the functions of blood?
2. Describe the components of blood.
3. What role do platelets play in blood clotting?
4. **Critical Thinking—Relating Cause and Effect** Infections sometimes block the flow of fluid through the lymphatic system, causing severe swelling in the tissues. Explain why this happens.

BACKGROUND INFORMATION
BLOOD CLOTTING CHAIN REACTION

The actual reactions involved in blood clotting are extremely complicated. When a blood vessel is broken, the damaged tissue releases a protein factor known as *thromboplastin*. Thromboplastin helps to convert *prothrombin* into *thrombin*. Prothrombin is found in the blood plasma. Thrombin, which is formed from prothrombin, is an enzyme that converts still another plasma protein, *fibrinogen*, into *fibrin*.

Why should this pathway be so complicated? To make the system sensitive! Even a few molecules of thromboplastin need to be able to create a clot consisting of thousands of molecules of fibrin. By activating a series of reactions in a "cascade," each step in the pathway involves greater and greater numbers of proteins. This makes the clotting system sensitive enough to stop the bleeding from even the tiniest of wounds—something for which we can all be grateful!

4. The lymphatic system is a network of vessels that collect the fluid lost by the blood. If a blockage occurs in any of the lymphatic vessels, fluid from these vessels moves into the surrounding tissue, causing swelling, or edema.

Reinforcement/Reteaching

If students are having difficulty understanding the many functions of blood, return to the summary of these functions at the beginning of Section 41–2.

Closure

Use prepared slides of blood smears made with Wright's stain for your students to view using their student microscopes. Have them identify and draw as many different kinds of blood cells as they can find.

trophils, 62.0%; eosinophils, 2.3%; basophils, 0.4%; monocytes, 5.3%; lymphocytes, 30.0%

SECTION REVIEW 41–2

1. Blood transports nutrients, dissolved gases, enzymes, hormones, and waste products; regulates body temperature, pH, and electrolytes; protects the body from invaders; and restricts fluid loss.
2. Plasma: liquid portion of blood; red blood cells (erythrocytes): transport oxygen and carbon dioxide; white blood cells (leukocytes): fight infection; platelets: play a role in blood clotting.
3. Platelets first clump together and form a plug at the wound site, and then they release clotting factors.

TEACHING SUPPORT

Laboratory Manual
- Simulating Urinalysis, p. 511

Teaching Resources
- Activity: The Inner Workings of the Kidneys, p. 5

MEDIA AND TECHNOLOGY
📺 Biology Transparencies
- 66: The Human Kidney

📹 Video/Videodisc
- *Human Body: Reproductive System and Excretory System*

Use this bar code to introduce the excretory system.

Play frames 34182 to 36389

See the Biology Media Guide pages 66–67 for additional bar-code correlations for this section.

TEACHING STRATEGY 41–3

Focus/Motivation
Display a model (or a chart) of the human excretory system for the students to examine. Identify the main organs of the excretory system: two kidneys, two ureters, the urinary bladder, and the urethra.

Content Development
- What kinds of metabolic wastes does the body produce? (Carbon dioxide, excess water and salts, urea.)
- By what process do living organisms rid themselves of these wastes? (Excretion.)
- What other organs in the human body are considered to be organs of excretion? (Lungs and skin.)
- What wastes do each of these organs excrete? (The lungs excrete carbon dioxide; the skin excretes water, salts, and a small amount of urea.)

Guide For Reading
■ What are the parts of the excretory system? What are their functions? Why are they important?
■ How are filtration and reabsorption accomplished in the kidneys?

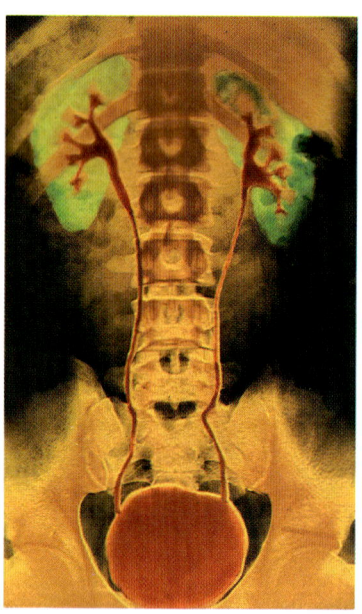

Figure 41–17 The main organs of the excretory system include the two kidneys, which in this X-ray are the green-colored bean-shaped organs located on either side of the spinal column. The tube that leaves each kidney is the ureter. It carries fluid to the urinary bladder, which is the round structure at the bottom of the X-ray.

41–3 The Excretory System

As a normal consequence of being alive, every cell in the body produces metabolic wastes. Examples of metabolic wastes include excess water and salts, carbon dioxide, and urea. Urea is a toxic compound that is produced when amino acids are used for energy. The process by which these metabolic wastes are removed from the body is called excretion.

You have already learned about two organs of excretion—the skin and the lungs. The skin excretes excess water and salts, as well as a small amount of urea. The lungs excrete carbon dioxide. The remaining organs of excretion are the kidneys. **Together, the skin, lungs, and kidneys—along with their associated organs—make up the excretory system.**

The Kidneys

The main organs of the **excretory system** are the **kidneys**. The kidneys are two bean-shaped organs, one located on either side of the spinal column near the lower back. Each kidney is about the size of a tightly clenched fist. Two blood vessels, the renal artery and the renal vein, enter and leave each kidney, respectively. A third vessel, the **ureter** (yoo-REET-er), leaves each kidney, carrying fluid to the **urinary bladder**.

What does the kidney do? As waste-laden blood enters the kidney through the renal artery, urea, excess water, and other waste products are removed and then passed to the ureter. The blood—now clean and filtered—leaves the kidney through the renal vein and returns to circulation.

KIDNEY STRUCTURE If a kidney is cut in half, two distinct regions can be seen. The inner part is called the renal medulla; the outer part is called the renal cortex. The renal cortex contains the **nephrons** (NEHF-rahnz), the basic functional units of the kidneys. Each nephron is a small independent filtering unit. In each kidney there are about 1 million nephrons. Figure 41–19 shows the relationship of a single nephron to the two regions of the kidney.

Each nephron has its own blood supply: an arteriole, a venule, and a network of capillaries connecting them. In addition, each nephron has its own collecting tubule, which leads to the ureter. As blood enters a nephron through an arteriole, impurities are filtered out and emptied into the collecting tubule. Purified blood leaves the nephron through a venule. The actual mechanism of blood purification is rather complex, involving two separate processes—filtration and reabsorption.

FILTRATION As blood enters a nephron through an arteriole, it flows into a small network of 50 separate capillaries known as a **glomerulus** (gloh-MER-yoo-luhs). The glomerulus is encased in the upper end of the nephron by a cup-shaped structure called **Bowman's capsule**.

Focus/Motivation
Obtain a beef or lamb kidney from a butcher. Cut it in half, lengthwise. Use it to show the parts of the kidney: the renal medulla, renal cortex, and renal pelvis. Locate the area containing the nephrons in the renal cortex.

Content Development
Using the Visuals Explain that kidney function comes from the 1 million nephrons found in each kidney. Use Figure 41–19 to identify the parts of the nephron: arteriole, venule, glomerulus, Bowman's capsule, loop of Henle,

908

Because the blood is under pressure and the walls of the capillaries and Bowman's capsule are permeable, much of the fluid from the blood filters into Bowman's capsule. This process is known as **filtration**. The materials that are filtered from the blood are collectively called the filtrate. The filtrate contains water, urea, glucose, salts, amino acids, and some vitamins. Because plasma proteins, cells, and platelets are too large to pass through the membrane, they remain in the blood.

REABSORPTION Almost 180 liters of filtrate pass from the blood into the collecting tubules each day. This volume is equivalent to 90 2-liter bottles of soft drink! Needless to say, not all of the 180 liters is excreted. Most of the material removed from the blood at Bowman's capsule makes its way back into the blood by a process known as **reabsorption**.

A number of materials—including salts, amino acids, fats, and sugars—are removed from the filtrate by active transport and reabsorbed by the capillaries. Because water follows these materials by osmosis, almost 99 percent of the water that enters Bowman's capsule is reabsorbed into the blood. This process is akin to cleaning out your room by tossing everything into the hall, then taking back only the things you want to keep. When you're done, you throw away the unwanted things in the hall—and this is basically what the kidney does, too!

The material that remains after reabsorption, called urine, is emptied into a collecting tube. Urine, which contains urea and excess salts and water, is concentrated in the loop of Henle (HEHN-lee), a section of the nephron that conserves water and minimizes the volume of urine.

As the kidney works, purified blood is returned to circulation, while urine is collected in the urinary bladder. Urine is stored in the bladder until it can be released from the body through a tube called the **urethra**.

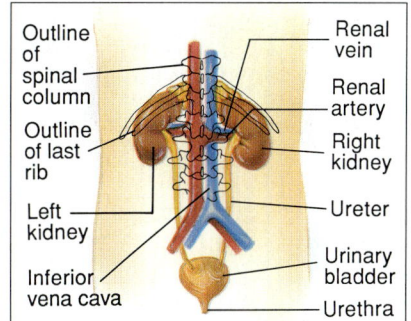

Figure 41–18 This drawing shows the posterior, or back, view of the kidneys and their related structures.

Figure 41–19 Inside each kidney, more than one million nephrons function as filters. A nephron consists of a funnellike Bowman's capsule surrounding a tuft, or ball, of capillaries called a glomerulus. The electron micrograph shows a glomerulus.

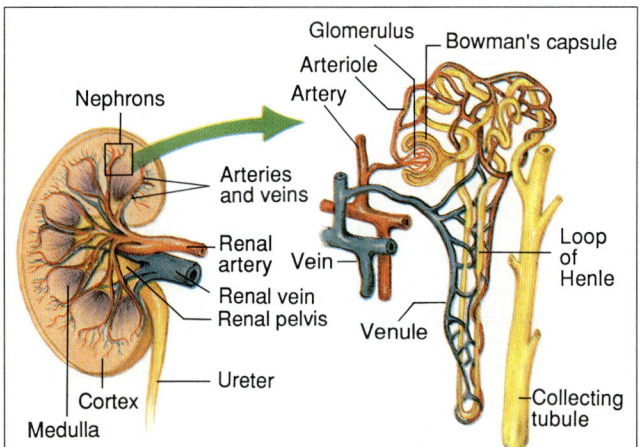

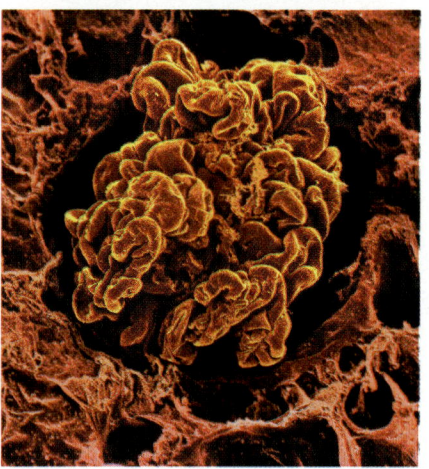

FACTS AND FIGURES

Each nephron is about 3 centimeters long, yet it is so narrow that it cannot be seen by the unaided eye.

FACTS AND FIGURES

Blood passes through the kidneys about 20 times each hour.

FACTS AND FIGURES

Over 99% of the filtrate is usually reabsorbed in the tubules.

collecting tubule.
 Describe filtration of blood in the nephron.
• **What substances move from the blood into the Bowman's capsule of the nephron?** (Water, urea, glucose, salts, amino acids, some vitamins.)

• **Why do plasma proteins, blood cells, and platelets remain in the blood?** (They are too large to pass through the membranes of the capillaries.)

Explain that not all the filtrate is used to form urine. Most of the filtrate is reabsorbed by active transport. Urine becomes more concentrated as active transport reabsorbs more and more of the filtrate.

Describe how purified blood is returned through the renal vein and urine is collected in the urinary bladder.

Reinforcement/Reteaching

Trace the pathway of renal circulation for the students: heart→aorta→renal artery→capillaries of the kidneys→renal vein→interior vena cava→heart.

Trace the pathway of urine collection and excretion: kidneys→ureter→urinary bladder→urethra.

TEACHING SUPPORT

Teaching Resources
- Vocabulary Review: Cross-a-Clue, p. 1

Study Guide
- Section 41-3, p. 398

TIE-IN HUMOR

In one of his comedy routines, Woody Allen claimed that he recently donated a kidney to his brother. "Now I have one kidney," said Woody, "and my brother has three!"

Such jokes notwithstanding, donated kidneys have saved numerous lives, as well as countless hours on dialysis machines. (See the CONNECTION feature on page 911.)

41-3 (continued)

Focus/Motivation

It has been said that the composition of the blood is determined not by what the mouth ingests but by what the kidneys keep. Discuss this idea briefly with the students. Ask them to explain it in their own words.

Content Development

Explain the role of the kidneys in maintaining a homeostatic balance in the body's internal environment.
- How would drinking a large volume of water affect the reabsorption of water by the kidneys? (The rate of water reabsorption would decrease.)
- How would eating very salty food affect the reabsorption of salt by the kidneys? (The rate of salt reabsorption would decrease.)

SECTION REVIEW 41-3

1. The skin, lungs, and kidneys make up the excretory system.
2. The kidney is a bean-shaped organ that has an inner part called the renal medulla and an outer part called the renal cortex. A ureter leaves each kidney, carrying fluid to the urinary bladder.
3. In the process of filtration,

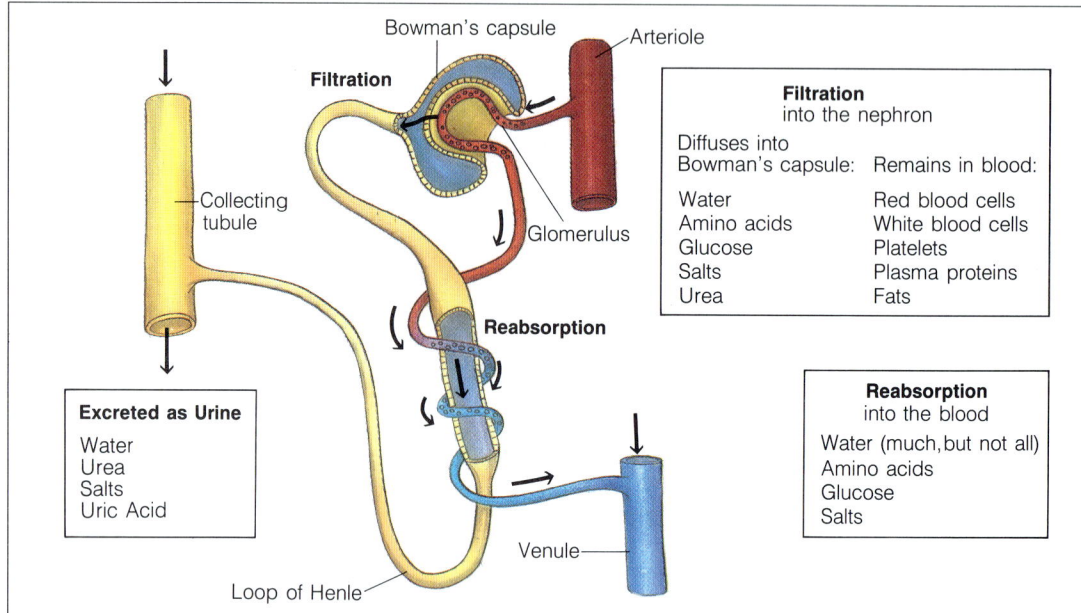

Figure 41-20 *The nephrons remove wastes from the blood through the processes of filtration and reabsorption. Filtration occurs in the glomerulus, and reabsorption occurs near the loop of Henle.*

Control of Kidney Function

The activity of the kidneys is controlled in a number of ways, one of which is the composition of blood itself. Another is the action of hormones that are released in response to the composition of blood. You will learn more about these hormones in Chapter 42. Both of these control mechanisms ensure that the kidneys will keep the composition of blood the same.

You have probably had the experience of occasionally drinking glass after glass of liquid. This liquid is quickly absorbed into the blood through the digestive system. As a result, the concentration of water in the blood increases. If it were not for your kidneys, this increased concentration of water in the blood would force water into cells and tissues by osmosis, causing the body to swell.

As the amount of water in the blood increases, the rate of water reabsorption in the kidneys decreases. Thus less water is returned to the blood, and the excess water is sent to the urinary bladder, to be excreted as urine.

If you eat salty food now and then, your kidneys will respond appropriately to the increased level of salt in your blood. As an increase in salt is detected by your kidneys, they will respond by returning less salt to the blood by reabsorption. The excess salt the kidneys keep is excreted in urine. The composition of the blood remains constant.

The kidneys are the master chemists of the blood supply. If anything goes wrong with the kidneys, serious medical problems soon follow. Fortunately, we have two kidneys and can survive with only one. Medical science is working rapidly to improve the efficiency of artificial kidneys and to develop alternate techniques of treating kidney problems.

such substances as water, urea, glucose, salts, amino acids, and some vitamins pass from the glomerulus into the Bowman's capsule. In the process of reabsorption, useful substances (some salts, amino acids, fats, and sugars) pass out of the col-

SCIENCE, TECHNOLOGY, AND SOCIETY CONNECTION

Replacing a Kidney

A functioning kidney is absolutely essential to life. Fortunately, most of us have built-in insurance against a threat to our survival because of kidney damage. We have two kidneys—one more than is necessary to live a healthy, normal life. Thus you might think of the second kidney as a spare.

Unfortunately, many people develop severe kidney diseases that destroy both of their kidneys. Until recently, these people could not survive for more than a few weeks or months. Today, however, machines that purify blood by a process known as dialysis are reversing the situation and giving hope of a longer life.

Renal dialysis machines mimic some of the functions of the nephron. During dialysis, blood flows through a series of tubes composed of a selectively permeable membrane. Surrounding these tubes is a dialysis fluid that has a composition similar to that of blood, except that the concentration of waste products is low. The dialysis fluid flows in a direction opposite to that of blood. As a result, waste products such as urea diffuse from the blood into the dialysis fluid. Blood is usually taken from an artery, passed through the tubes of the dialysis machine, and then returned through a vein.

Patients experiencing kidney failure may visit a dialysis center several times a week to have their circulation cleansed by the machine. Although these machines work well, patients must use them nearly 20 hours per week, and the treatment often can be painful. Fortunately, a permanent cure for many people is available—a kidney transplant.

When trying to find a kidney donor, doctors must make sure that the donor's tissues are compatible with the patient's, so that the tissues of the donated kidney will not be attacked by the patient's immune system. Only then can the kidney from the donor be transplanted into the body of the patient. If the patient's body does not reject the donated kidney, the kidney will begin to function and both donor and recipient will live normal lives with only one kidney each.

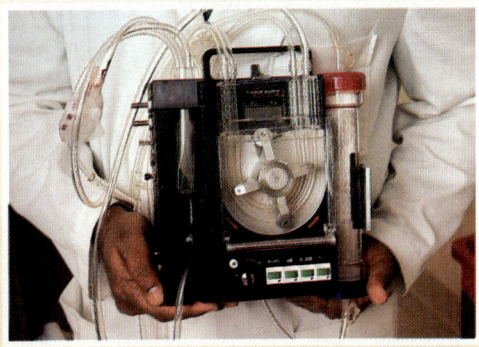

Recently designed dialysis machine

41–3 SECTION REVIEW

1. What are the parts of the excretory system?
2. Describe the structure and function of the kidney.
3. Describe the processes of filtration and reabsorption.
4. **Critical Thinking—Making Inferences** Explain why a person loses weight after changing to a low-salt diet even without reducing caloric intake.

LABORATORY INVESTIGATION

OBSERVING CIRCULATION IN A FISH TAIL

Before the Lab
Gather all materials at least one day prior to having students perform the investigation.

Pre-Lab Discussion
1. Remind students that they are working with living organisms and that the fish must be treated humanely and with great care.
2. Live fish must be kept in a suitably large fish tank.
3. Advise students to return the fish to the tank as soon as possible.

Skills Development
Students will use the following skills while completing this laboratory investigation.
1. Observing
2. Manipulative
3. Safety
4. Comparing
5. Relating
6. Applying
7. Inferring
8. Concluding

Teaching Strategy
1. Circulate around the room to ensure that students do not mistreat the fish.
2. Assist students who are having trouble placing the fish on the microscope as directed.

Observations
1. The blood cells are disk shaped and red.
2. Red blood cells in the blood move in single file through the capillary. Blood flow is in only one direction.
3. Blood pulsates in arterioles (small arteries) and flows smoothly and at a uniform rate in venules (small veins).
4. The blood vessels are tube-like structures that branch and form an intricate network.

Analysis and Conclusions
1. Their size helps the capillaries allow substances to diffuse into and out of the blood.
2. The blood flows unevenly because it is being moved by the beating of the heart. This uneven blood flow is seen as a pulsating movement in the arterioles.
3. Arterioles carry blood toward the tip of the tail, while venules carry blood back toward the heart. Capillaries connect arterioles and venules.

Going Further: Enrichment
Have students design an experiment to test the effects of various substances on the rate of flow of blood in the fish tail. Such substances as caffeine and aspirin could be suggested. Remind students that they are not to run

---

LABORATORY INVESTIGATION

OBSERVING CIRCULATION IN A FISH TAIL

PROBLEM
How does blood flow in the tail of a goldfish?

MATERIALS (per group)

150-mL beaker	glass plate
goldfish in an aquarium	2 glass slides
medicine dropper	fish net
absorbent cotton	microscope
petri dish	

PROCEDURE

1. Fill a 150-mL beaker almost full with water from the aquarium. Keep the water and medicine dropper near you at your laboratory table.
2. Using the medicine dropper, moisten two pieces of cotton with aquarium water.
3. Place one piece of moistened cotton in the petri dish and the other piece on the glass plate. Put the glass plate aside for now.
4. Place one of the glass slides on the side of the petri dish opposite the moistened cotton.
5. Using the fish net, remove a goldfish from the aquarium. **CAUTION:** *Be careful when handling live animals.* Gently place the head of the goldfish on the cotton in the petri dish. Be sure that the tail is on the glass slide.
6. Place the second piece of moistened cotton over the goldfish, leaving the mouth and tail uncovered. Place the other slide on top of the thin part of the tail. **Note:** *Be sure to keep the goldfish moist by adding aquarium water to the cotton as it begins to dry out.*
7. Remove the stage clips from the microscope and put them in a safe place.
8. When the goldfish has calmed down, place the petri dish on the stage of the microscope. Position the petri dish so the goldfish's tail is over the opening in the stage.
9. Examine the goldfish's tail under low power only. Move the petri dish around until you see blood moving in the blood vessels.
10. Locate a blood vessel in which blood cells are passing in single file. This is a capillary. Note the direction of the flow of blood.
11. Trace the capillary in the direction opposite the blood flow to where it joins a slightly larger vessel (arteriole). Then trace the capillary in the direction of blood flow until it joins a slightly larger vessel (venule). Draw and label the different types of blood vessels. Use arrows to show the direction of blood flow.
12. Return the goldfish to the aquarium as soon as you have completed the investigation. Replace the stage clips on the microscope.

OBSERVATIONS
1. Describe the blood cells' shape and color.
2. Describe the movement of blood through a capillary.
3. In which type of blood vessel does the blood seem to pulsate? In which type of blood vessel does the blood seem to flow smoothly and at a uniform rate?
4. Describe the appearance of the blood vessels.

ANALYSIS AND CONCLUSIONS
1. Explain why the capillaries are so small.
2. Why does blood in arterioles flow unevenly?
3. How does blood travel through capillaries, arterioles, and venules?

SUMMARIZING THE CONCEPTS

The key concepts in each section of this chapter are listed below to help you review the chapter content. Make sure you understand each concept and its relationship to other concepts and to the theme of this chapter.

41-1 The Circulatory System

- Blood, the heart, and blood vessels make up the circulatory system.
- The heart, which pumps blood through the circulatory system, has four chambers: two atria and two ventricles.
- Arteries carry blood from the heart, and veins carry blood to the heart. Capillaries connect arteries and veins.
- Pulmonary circulation carries blood between the heart and the lungs. Systemic circulation carries blood between the heart and the rest of the body.
- The lymphatic system collects the fluid that is lost by blood and returns it to the circulatory system.

41-2 Blood

- Blood transports nutrients, dissolved gases, enzymes, hormones, and waste products; regulates body temperature, electrolytes, and pH; protects the body from invaders; and restricts the loss of fluid.
- Blood, which is a type of connective tissue, is made up of a liquid portion called plasma and a cellular portion composed of red blood cells, white blood cells, and platelets.

41-3 The Excretory System

- The skin, lungs, and kidneys—along with their associated organs—make up the excretory system, which removes metabolic wastes from the body.
- The kidneys, which are the main organs of the excretory system, regulate the fluid and salt balance in the body.
- The basic functional unit of the kidney is the nephron, which is composed of the glomerulus, Bowman's capsule, and tubules. Within the nephron, metabolic wastes are removed by the processes of filtration and reabsorption.

REVIEWING KEY TERMS

Vocabulary terms are important to your understanding of biology. The key terms listed below are those you should be especially familiar with. Review these terms and their meanings. Then use each term in a complete sentence. If you are not sure of a term's meaning, return to the appropriate section and review its definition.

41-1 The Circulatory System
heart
circulatory system
atrium
ventricle
vena cava
artery
aorta
capillary
vein
pulmonary circulation
systemic circulation
blood pressure
lymphatic system

41-2 Blood
blood
plasma
red blood cell
white blood cell
platelet

41-3 The Excretory System
excretory system
kidney
ureter
urinary bladder
nephron
glomerulus
Bowman's capsule
filtration
reabsorption
urethra

CHAPTER REVIEW

CONTENT REVIEW

Multiple Choice
1. c   2. c   3. c   4. d
5. b   6. b   7. a   8. c

True or False
1. T
2. F platelets
3. F aorta
4. F ureters
5. F glomerulus
6. F urethra
7. T
8. F veins

Word Relationships
1. vein
2. pulmonary artery
3. blood clotting
4. atrium

CONCEPT MASTERY

1. The smooth muscle cells and elastic fiber that make up arterial walls help make the arteries tough and flexible. These characteristics enable the arteries to withstand the high pressure of blood as it is pumped from the heart. Although the walls of the veins are less elastic than are those of the arteries, they are more flexible and are able to stretch out readily. This flexibility is important because it reduces the resistance the blood flow encounters on its way back to the heart. Capillaries are thin walled so that oxygen and nutrients easily diffuse into the body tissues and carbon dioxide and waste products easily diffuse out.
2. The lymphatic system collects fluid that has leaked out of the blood vessels and returns it to the circulatory system. In addition, lymph nodes act as filters and producers of special white blood cells that prevent harmful material from invading body cells.
3. In pulmonary circulation, blood leaves the right ventricle through the pulmonary

their experiment on a living organism but simply to design the experiment. All designs must contain a control and a clearly designated variable.

artery, is oxygenated in the lungs, and returns to the heart through the pulmonary vein. In systemic circulation, blood is pumped from the left ventricle into the aorta. The blood travels through arteries and arterioles into capillaries, where gases, nutrients, and wastes transfer between blood and body tissue. The blood returns to the heart through the veins.

4. Like other connective tissues, blood is homogeneously organized, has very few metabolic requirements, and serves to connect other body tissues. Blood contains too many cells and other components to be thought of as a simple fluid or liquid.

5. Platelets clump together and form a plug at the wound site. Then clotting factors, one of which is called thromboplastin, convert prothrombin into thrombin. Thrombin converts the plasma protein fibrinogen into fibrin filaments that stop the bleeding by producing a clot.

6. Red blood cells are the most numerous blood cells because they transport the much-needed oxygen throughout the body. White blood cells are less numerous because they are needed only when the body is under attack by foreign cells or substances. Red blood cells have a short life span because they do not have a nucleus that controls cellular activities.

7. Blood pressure is highest in the aorta because the aorta is closest to the heart. Because the venae cavae are farthest away from the heart, blood pressure is lowest in them.

8. The heartbeat is controlled by small bundles of cells, called the pacemaker.

9. Unlike the circulatory system, the lymphatic system contains mostly water and produces some special white blood cells. It is also an open system.

10. Glucose and urea enter each kidney through the renal artery, which branches into an arteriole. The arteriole forms a glomerulus, which is encased by the Bowman's capsule. Here, the glucose and urea are filtered out of the glomerulus into the Bowman's capsule. These substances move into the collecting tubule, which is surrounded by a network of capillaries. As glucose moves through the collecting tubule, it is reabsorbed into the blood. The urea is not reabsorbed and travels into the ureter to the urinary bladder, where it is stored with other substances, collectively called urine, until it is released through the urethra.

11. Ventricles are important because they are the pumping chambers. If they are not replaced in a heart transplant,

CHAPTER REVIEW

CONTENT REVIEW

Multiple Choice

Choose the letter of the answer that best completes each statement.

1. About 90 percent of plasma is composed of
   a. salts. c. water.
   b. proteins. d. tissues.
2. The color of red blood cells comes from a pigment called
   a. thrombin. c. hemoglobin.
   b. globulin. d. albumin.
3. Blood traveling to capillaries in the arm leaves the heart from the
   a. left atrium. c. left ventricle.
   b. right atrium. d. right ventricle.
4. If the blood flow in a vessel is toward the heart, then the vessel is known as a (an)
   a. atrium. c. artery.
   b. ventricle. d. vein.
5. Circulation to and from the lungs is
   a. lymphatic circulation.
   b. pulmonary circulation.
   c. coronary circulation.
   d. systemic circulation.
6. Which are excretory organs?
   a. skin and heart c. liver and pancreas
   b. lungs and kidneys d. lungs and heart
7. The kidney's basic functional unit is the
   a. nephron. c. ureter.
   b. Bowman's capsule. d. glomerulus.
8. Filtration of the blood in the kidney occurs at the
   a. urethra. c. Bowman's capsule.
   b. urinary bladder. d. loop of Henle.

True or False

Determine whether each statement is true or false. If it is true, write "true." If it is false, change the underlined word or words to make the statement true.

1. The liquid portion of the blood is called <u>plasma</u>.
2. The blood cell fragments that play a role in blood clotting are the <u>leukocytes</u>.
3. The largest artery in the body is the <u>pulmonary artery</u>.
4. Urine is carried from the kidneys to the urinary bladder by the <u>urethras</u>.
5. A network of capillaries in a Bowman's capsule is known as <u>nephron</u>.
6. Urine is eliminated from the urinary bladder through the <u>ureter</u>.
7. The fluid that is lost by the blood is called <u>lymph</u>.
8. The blood vessels that contain valves are the <u>capillaries</u>.

Word Relationships

An analogy is a relationship between two pairs of words or phrases generally written in the following manner: a:b::c:d. The symbol : is read "is to," and the symbol :: is read "as." For example, cat:animal::rose:plant is read "cat is to animal as rose is to plant."

In the analogies that follow, a word or phrase is missing. Complete each analogy by providing the missing word or phrase.

1. away from the heart:artery::to the heart: _____
2. left ventricle:aorta::right ventricle: _____
3. hemoglobin:carrying oxygen::fibrinogen: _____
4. pumping chamber:ventricle::receiving chamber: _____

CONCEPT MASTERY

Use your understanding of the concepts developed in the chapter to answer each of the following in a brief paragraph.

1. How is the structure of an artery, a vein, and a capillary adapted to its function?
2. Explain how the lymphatic system protects the body.
3. Trace the path that blood follows in pulmonary circulation and in systemic circulation.
4. Explain why blood is considered a connective tissue.
5. Briefly describe the events of the clotting process.
6. Explain why red blood cells are the most numerous of the blood cells and why they have such a short life span.
7. Why is the blood pressure highest in the aorta and lowest in the venae cavae?
8. How is heartbeat rate controlled?
9. How does the lymphatic system differ from the circulatory system?
10. Trace the path of glucose and urea through the kidney.
11. Explain why it is important to replace the ventricles in a heart transplant.
12. Why does the left ventricle have a thicker wall than the right ventricle?
13. Explain why the tricuspid and bicuspid valves are also known as atrioventricular valves.

CRITICAL AND CREATIVE THINKING

Discuss each of the following in a brief paragraph.

1. **Applying concepts** Some infants are born with a small hole in the septum of the heart. Explain how this situation will affect the infant. How do you think this problem could be corrected?
2. **Relating facts** Atherosclerosis is a disorder in which deposits of cholesterol and fatty materials collect on the inner walls of the arteries. How does this disorder affect the circulatory system?
3. **Making inferences** Explain why it is possible that kidney failure can be fatal even though an individual's heart is still strong.
4. **Relating cause and effect** Would urine contain more or less water on a hot day? Explain your answer.
5. **Making predictions** Imagine that you are a medical student and your professor has asked you what would happen to a patient whose renal arteries have become greatly narrowed. How would you respond?

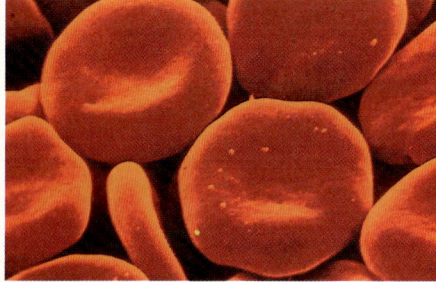

6. **Sequencing events** Trace the course of a single red blood cell from the right ventricle to the right atrium.
7. **Relating concepts** Could a person adrift on a life raft survive by drinking sea water? Explain your answer.
8. **Using the writing process** Pretend that you have hitched a ride on a red blood cell. Write a short description of your adventures through the circulatory system.

CHAPTER 42 Endocrine System

Section	Laboratory Investigations and Activities
42–1 Endocrine Glands pages 917–926 THEMES: Scale and Structure, Systems and Interactions	**Teacher Edition** DEMONSTRATION: Growth Hormone, p. 916D DEMONSTRATION: The Effects of Stress, p. 916D **Laboratory Manual** Observing Human Growth, p. 517
42–2 Control of the Endocrine System pages 927–929 THEMES: Systems and Interactions, Stability	**Student Edition** LABORATORY INVESTIGATION: Simulating the Negative-Feedback Process, p. 930
Chapter Review pages 931–933	

Teacher Resources

Books

Bennett, G., and S. A. Whitehead. *Mammalian Neuroendrocrinology*. Oxford University Press, 1983.

Berczi, I., and K. Kovacs, eds. *Hormones and Immunity*. Kluwer Academic Press, 1987.

Davis, J. *Endorphins: New Waves in Brain Chemistry*. Doubleday, 1984

Hadley, M. E. *Endocrinology*, 2nd ed. Prentice Hall, 1988

Sims, Dorothea. *Diabetes: Reach for Health and Freedom*. Mosby, 1984.

Underwood and Sherman. *Human Growth Hormone: Progress and Challenges*. Dekker, 1987.

Weller, H., and R. L. Wiley. *Basic Human Physiology*, 2nd ed. Prindle, Weber, and Schmidt, 1985.

CHAPTER PLANNING GUIDE

Other Activities	Media and Technology
Teaching Resources ACTIVITY: Analyzing the Endocrine Functions of the Pancreas, p. 3 ACTIVITY: Studying the Composition of the Endocrine System, p. 5 **Biotechnology Workbook** Hormones That Make You Grow, p. 25 **Study Guide** Section 42–1, p. 403	
Teaching Resources VOCABULARY REVIEW: Crossword, p. 1 **Study Guide** Section 42–2, p. 406	
Teaching Resources Chapter Test A, p. 11 Chapter Test B, p. 15	**Computer Test Bank** Chapter Test, p. 441

Audiovisuals

Endocrine System. Film or video. Coronet.
Homeostasis: Maintaining the Stability of Life. Sound-slide set. Science and Mankind.
Diabetes: Restoring the Balance. Film. Pennsylvania State University A-V Services.

Diabetes Understood. Filmstrip with cassette. Carolina Biological Supply Company.

Your Body: Series 3. 4 filmstrips with cassettes. Focus Media.

CHAPTER 42 Endocrine System

CHAPTER OVERVIEW

The nervous system is one of the body's regulatory systems. The endocrine system is the other. It is composed of glands that secrete substances called hormones into the bloodstream. Hormones travel through the bloodstream and bind to receptors on target cells. They affect the behavior of those cells. Hormone action regulates body function but is much slower than the nervous system. The major hormone-producing organs of the endocrine system are the endocrine glands. Glands that transport their secretions through ducts are called exocrine glands.

The individual endocrine glands and their hormones are discussed in this chapter, beginning with the thyroid gland. The major hormone produced by the thyroid gland is called thyroxine. Thyroxine regulates the rate of metabolism in nearly all the cells of the body. The thyroid secretes a second hormone in specialized cells of the gland. This hormone, calcitonin, regulates the level of calcium in the blood. Attached to the back surface of the thyroid gland are the four parathyroid glands. These glands secrete the parathyroid hormone, which also regulates calcium levels in the blood. The adrenal glands, situated on top of the kidneys, are divided into two parts. The adrenal cortex produces more than two dozen hormones essential for normal body functioning. The adrenal medulla secretes the neurohormones called adrenaline and noradrenaline. The action of these hormones results in a general increase in body activity.

The reproductive glands, the gonads, produce sex hormones that affect cells throughout the body. The female hormones, the estrogens and progesterone, are responsible for the proper functioning of the female reproductive system and the development of physical characteristics associated with females. The male hormones, the androgens, perform similar functions in the male. The pancreas is an unusual organ in that it has an exocrine function and also acts as an endocrine gland. The hormones produced by the pancreas include insulin and glucagon. These two hormones regulate glucose levels in the blood.

The pituitary gland is divided into two parts. The posterior pituitary secretes two hormones. Antidiuretic hormone helps in the control of the body's fluid balance, and oxytocin serves a function during childbirth. The anterior pituitary produces seven hormones. These hormones directly regulate many body functions as well as the release of hormones by several other endocrine glands. Disorders of the pituitary can have serious medical implications. A part of the brain known as the hypothalamus is attached to the posterior portion of the pituitary gland. This is the primary point at which the nervous system and the endocrine system interact. The hypothalamus controls the secretions of pituitary hormones. In turn, the hypothalamus is influenced by the hormone levels in the blood.

The endocrine system is regulated by a negative-feedback mechanism which helps to maintain homeostasis. This mechanism can be compared to a thermostatically controlled heating and cooling system. There are many examples of negative feedback that control the levels of many hormones.

Hormones produced by endocrine glands can be placed in two general groups. Polypeptide hormones are large proteins, and steroid hormones are lipids produced from cholesterol. The basic action of these two types of hormones is quite different. Prostaglandins are sometimes referred to as local hormones because although they are produced in many cells and tissues of the body, they function only within the cells in which they are produced.

42-1 ENDOCRINE GLANDS

Section Focus 42-1

The purpose of this section is to discuss the general functioning of the endocrine system and to discuss the function of each of the endocrine glands. The nervous system has already been described as one of the regulatory systems of the body. The endocrine system is the other. The endocrine system is composed of glands that secrete the products, called hormones, directly into the bloodstream. These chemical messengers bind to receptors on cells some distances away. They then affect the behavior of these target cells. Hormone action is generally slower and longer lasting than that of the nervous system. The major hormone-producing organs of the endocrine system are called endocrine glands. The body also has other organs that produce secretions. Those that use ducts to deliver their products to target sites are called exocrine glands.

The thyroid gland is a major endocrine gland located at the base of the neck. The thyroid produces a hormone called thyroxine, an amino acid hormone to which iodine is attached. Thyroxine affects nearly all the cells of the body by regulating their rate of metabolism. Several abnormalities of the thyroid gland are discussed in this section. A second hormone secreted by the thyroid gland is calcitonin. Its major function is to regulate the level of calcium in the blood.

The parathyroid glands are attached to the back surface of the thyroid gland. The parathyroids produce the hormone that regulates the level of calcium in the blood by increasing the reabsorption of calcium in the kidneys and by increasing the uptake of calcium from the digestive system. This hormone works with calcitonin and vitamin D to promote proper nerve and muscle function.

The adrenal glands sit on top of the kidneys. Each gland is composed of two very different types of tissue, divided into two very distinct parts: the adrenal cortex and the adrenal medulla. The large, outer part of the gland is the cortex. It produces more than two dozen hormones called corticosteroids, which are essential for normal body function. Some of the individual hormones and their functions are discussed in this section as well as some of the disorders associated with the adrenal cortex.

CHAPTER PREVIEW

The adrenal medulla is considered a specialized part of the sympathetic nervous system. Therefore, the adrenaline and noradrenaline produced in the medulla are called neurohormones. These hormones act to increase the level of body activity, which can serve as preparation for intense physical response when faced with a "fight-or-flight" situation.

The gonads are the body's reproductive glands. They serve two important functions. They produce the sex cells, the egg and sperm, and they produce sex hormones that affect cells throughout the body. The female sex hormones, the estrogens and progesterone, are necessary for the proper functioning of the reproductive system and are also important in the development of physical characteristics associated with the female body. The testes produce androgens, male sex hormones, which perform a similar function in the male body.

The pancreas is an unusual gland. It has an exocrine function, producing pancreatic fluid to the digestive tract through a duct. The hormone-producing portion of the pancreas consists of cells called the islets of Langerhans. The beta cells of the islets secrete insulin, and the alpha cells secrete glucagon. These two hormones have opposite effects and together regulate the level of glucose in the blood. The functioning of insulin and glucagon is discussed, as is diabetes mellitus.

The pituitary gland is a bean-sized structure at the base of the skull. The gland is divided into two parts: the anterior and the posterior. The posterior portion of the pituitary secretes two hormones. Antidiuretic hormone stimulates the kidneys to reabsorb more water. Oxytocin stimulates the contractions of the uterine muscles during childbirth and causes the release of milk from the breast of a nursing mother. The anterior pituitary secretes seven hormones. Each of these hormones is discussed briefly. Generally, the hormones of the anterior pituitary control many body functions directly and also regulate the release of hormones by several other endocrine glands. Some disorders of the pituitary are discussed in this section. Attached to the posterior pituitary is the hypothalamus. This part of the brain controls the secretions of the pituitary gland. In turn, the hypothalamus is influenced by the levels of hormones in the blood, as well as sensory information entering the central nervous system. The hypothalamus and the pituitary gland are the major areas where the nervous system and the endocrine system interact.

Performance Objectives 42-1

1. Briefly describe the structure and function of the endocrine system.
2. Identify the major endocrine glands and the functions of each.
3. Name and discuss some important disorders of the endocrine system.
4. Discuss the function of the hypothalamus and explain its importance.

Science Terms 42-1

endocrine system   p. 917
hormone   p. 917
target cell   p. 917
endocrine gland   p. 918
exocrine gland   p. 918
thyroid gland   p. 918
parathyroid gland   p. 919
adrenal gland   p. 920
adrenal cortex   p. 920
adrenal medulla   p. 921
gonad   p. 921
ovary   p. 921
testis   p. 921
pancreas   p. 921
islets of Langerhans   p. 922
pituitary gland   p. 923
posterior pituitary   p. 924
anterior pituitary   p. 924
hypothalamus   p. 925

42-2 CONTROL OF THE ENDOCRINE SYSTEM

Section Focus 42-2

The purpose of this section is to explain the negative-feedback process and the actual functioning of hormones. Most of the systems in the body are regulated by a negative-feedback mechanism. The endocrine system is no exception. This mechanism acts to maintain an internal balance, or homeostasis. The mechanism is automatic and self-regulating. There are many such processes that control the levels of hormones.

Performance Objectives 42-2

1. Explain how the negative-feedback mechanism works.
2. Discuss the ways in which polypeptide hormones function.
3. Describe the function of steroid hormones in the body.
4. Describe the action of prostaglandins in the body.

Science Terms 42-2

negative-feedback mechanism   p. 927
prostaglandin   p. 927

DISCOVERY LEARNING

TEACHER DEMONSTRATIONS
Modeling

Growth hormone

This demonstration can be used to introduce Chapter 42.

Ask all of the students in the class to stand. Tell them to look at the various heights of their classmates. This is one example of the influence of a hormone. Briefly talk about the growth hormones. Tell the students that during the study of this chapter they will learn about several other hormones and their action.

The Effects of Stress

This activity can be done in conjunction with the study of the adrenal glands.

Have students think of a stressful situation they have experienced recently. Ask one or two students to tell about the experience in detail. Then ask them to try to remember how they felt—physically and mentally—at the time. Explain that adrenaline is the hormone responsible for that overall reaction.

CHAPTER 42
Endocrine System

GUIDED ENQUIRY

Pose the following questions to students and have them record their responses. Point out that they will gain a better understanding of the key concepts if they read the chapter with these basic questions in mind. Upon completion of the chapter, pose the questions again. Ask students to compare their initial responses with those they have developed after reading the chapter.

- Why is the endocrine system considered a regulatory system?
- How do glands "know" when enough of a hormone has been produced?
- How could hormones be helpful in a crisis situation?
- What happens when too much or too little of a hormone is produced?
- How are the endocrine system and the nervous system related?

CHAPTER 42
Endocrine System

The rush of fear and excitement experienced by these people as they ride the roller coaster stimulates the release of the hormone adrenaline, which is seen as crystals in the inset.

Biological communications systems can be compared to those in human societies. We are all familiar with broadcast communications such as radio and television. The waves carrying the messages travel through the air to special receivers that tune in to the frequency of the messages. In this way, a radio or television broadcast can be sent to thousands or millions of homes at the same time.

The biological communications system that resembles a radio or television communications system is the endocrine system. Like broadcast signals, the messages sent by the endocrine system can affect millions of cells that are some distance away. How does the endocrine system "broadcast" its messages, and what effects do the messages have on the body's cells? Read on to find out the answers to these questions.

INTRODUCING CHAPTER 42

Using the Visuals
Ask the students to define stress. After discussing it for a few minutes, question whether stress can ever be positive. List several causes of stress. Talk briefly about the physical results of stress. Direct students' attention to the chapter-opener photograph and caption. Explain that adrenaline is one example of a chemical messenger that regulates some aspects of body functioning. Read the chapter-opener information. If the telephone network represents the nervous system, then the radio or television represents the endocrine system.

TEACHING STRATEGY 42–1

Focus/Motivation
Have students name all the hormones they can. Write the names in a column on the chalkboard or an overhead transparency. Name all of the endocrine glands with

GUIDE FOR READING

After you read the following sections, you will be able to

42–1 Endocrine Glands
- Describe the function of the endocrine system.
- Identify the major endocrine glands and the function of each.

42–2 Control of the Endocrine System
- Describe the negative-feedback mechanism.
- Compare polypeptide and steroid hormones.
- Discuss the action of prostaglandins.

Journal Activity

YOU AND YOUR WORLD
In your journal, describe how you felt when someone startled you. What other types of experiences have you had that caused similar reactions? Why do you think it is advantageous for your body to react this way?

42–1 Endocrine Glands

Guide For Reading
- What is the function of the endocrine system?
- What are the major endocrine glands?
- What is the function of each major endocrine gland?

As you may recall from Chapter 37, the nervous system regulates many of the body's activities. The nervous system, however, is only one of the body's regulatory systems. The **endocrine system** is the other regulatory system. **The endocrine system is composed of glands that secrete their products, called hormones, into the bloodstream.**

The endocrine system produces **hormones**, or chemical messengers, that travel through the bloodstream and exert their effects some distance from where they are produced. Hormones bind to receptors on the cells and affect the behavior of the cells. Cells that have receptors for a particular hormone are called **target cells**.

Although the endocrine system and the nervous system regulate the activities of the body, they do so in different ways. In general, the body's responses to hormones are slower and longer lasting than the responses to nerve impulses. For example, it may take several minutes, several hours, or even several days for a hormone to have its full effect on its target cells. A nerve impulse, on the other hand, may take only a fraction of a second to reach and affect its target cells.

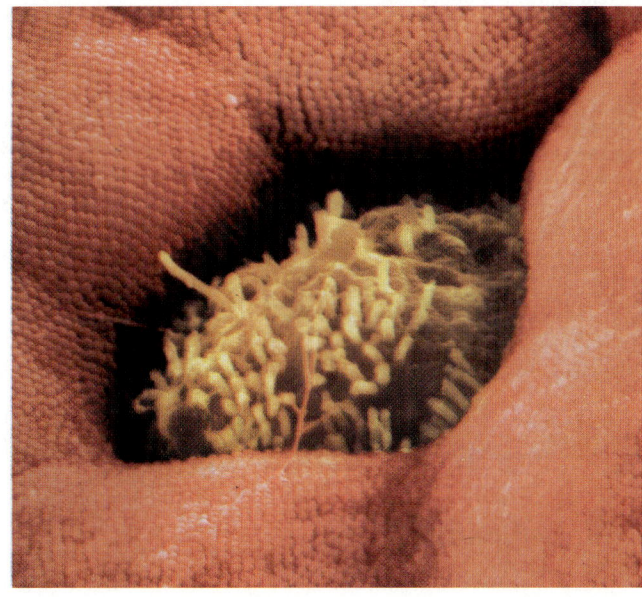

Figure 42–1 The tiny rod-shaped filaments sticking out of this target cell in the stomach contain receptors to which the hormone gastrin binds. Gastrin stimulates the production of gastric juice.

COOPERATIVE LEARNING

Divide the class into five groups. Each group is to imagine that they are the editorial staff of a national daily newspaper. Using the issue of steroid use by athletes, have groups produce the editorial page for that newspaper. When completed, the page should contain the following components:

- Background information (Refer to the STS Issue feature on page 926.)
- Editorial cartoon (pro or con)
- Guest editorial in support of STS Issue position
- Guest editorial expressing a different viewpoint
- Hypothetical quotes from "man-on-the-street" surveys

Authors of guest editorials and respondents of man-on-the-street surveys should be identified. It might be helpful to show students the actual editorial pages of local and national newspapers as models. If time does not permit having each group complete the entire page, you might randomly assign each of the five components to one group to produce a class page.

Journal Activity

YOU AND YOUR WORLD
Most students have been startled at one time or another—if not by another person, then by a scary movie, book, or television show. Students should suggest that the body's reactions to being scared might help to protect a person in a truly dangerous situation. Students should keep their descriptions in their portfolio.

---

which they are familiar. Write them in another column. Draw lines to connect the two lists. Tell students that these and other hormones and glands will be discussed in this chapter.

Content Development
Compare the endocrine and nervous systems. Check students' understanding with questions such as the following:
- **What is meant by the term *regulatory system*?** (It refers to a body system that exerts a direct control over other parts of the body.)
- **What are hormones?** (Chemical messengers produced by endocrine glands.)
- **How do hormones travel through the body?** (Through the bloodstream.)

- **What is a target cell?** (A cell that has receptor sites to which the hormones bind.)
- **Compare the functioning of the nervous and endocrine systems.** (Answers will vary, but in general the nervous system causes more direct, quick but short-lived reactions, while the endocrine system accounts for more widespread, slower but more long-lasting responses.)

917

TEACHING SUPPORT

Teaching Resources
- Activity: Studying the Composition of the Endocrine System, p. 5

BACKGROUND INFORMATION
GOITER

Goiter sometimes becomes so pronounced that it disfigures the neck. The thyroid swells during iodine deficiency in response to very low levels of iodine in the blood. The thyroid produces more and more tissue, enabling it to capture every available molecule of of the element to produce thyroxine. Physicians refer to goiter as a "compensatory hypertrophy," meaning that the gland swells in an effort to compensate for the dietary deficiency. Unfortunately, with a diet completely lacking in iodine, no amount of swelling will produce the necessary thyroxine.

42–1 (continued)

Content Development
Using the Visuals Discuss the similarities and differences between endocrine glands and exocrine glands. Point out Figure 42–3 and talk about the various endocrine glands. Remind students that proximity to target cells is not as necessary for endocrine glands as it is for exocrine glands because the hormones are carried in the bloodstream rather than through ducts.

Content Development
Using the Visuals Use Figure 42–4 to point out the location of the thyroid gland in the body. Students usually begin to feel their own necks, so you might want to direct them to the approximate location. The structure of the thyroxine molecule is interesting because of its direct dietary implications. You might want to bring a box of iodized salt to show the students. That product has been around for so long that we tend to take it for granted. Amazingly, students are sometimes completely unaware of it. (This might be a good time to remind students to be aware of the world around them; to be wise consumers of scientific advances such as iodized salt and fluoridated water.)

You may have to spend a little time talking about metabolic rates to refresh some memories. The prefix *hyper-* should be familiar to students, so they should easily understand too much or too little thyroid secretion. You may, however, have to explain what that really means. Goiter is not as

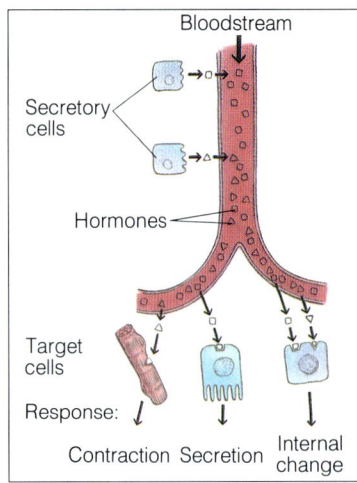

Figure 42–2 *In the endocrine system, hormones diffuse into the bloodstream and are carried throughout the body to target cells.*

The major hormone-producing organs of the endocrine system are the **endocrine glands**. A gland is an organ that produces a secretion, or a substance made inside a cell and released from that cell. Endocrine glands are only one group of glands that produce secretions. The other group is called the **exocrine glands**. Exocrine glands release their secretions through tubelike structures called ducts. For this reason, exocrine glands are also known as duct glands. Two examples of exocrine glands are the sweat glands and the digestive glands. Unlike exocrine glands, endocrine glands do not have ducts. Thus they are called ductless glands. Endocrine glands release their secretions (hormones) directly into the bloodstream.

Figure 42–3 shows the location of the major endocrine glands. As you can see, the endocrine glands are scattered throughout the body and generally do not have direct connections to one another. Like signals from a broadcast station that are beamed throughout the country, the hormones released from the endocrine glands travel throughout the body, reaching almost every cell.

Thyroid Gland

The **thyroid gland** is located at the base of the neck and wraps around the upper part of the trachea just below the larynx. See Figure 42–4. The thyroid gland produces a hormone called thyroxine, which is an amino acid to which four iodine atoms are attached. Because iodine is needed for the production of thyroxine, it must be included in the diet. Fish and other seafoods are good sources of iodine. But if these foods are not your favorites, do not worry. You can get the small amount of iodine your body needs from table salt, provided it is iodized salt, or salt to which small amounts of iodine have been added.

Thyroxine affects nearly all the cells of the body by regulating their metabolic rate. Increased levels of thyroxine cause an increase in the cellular respiration rate, which means the cells produce more energy and become more active. Decreased levels of thyroxine cause a decrease in the cellular respiration rate. Cells produce less energy and become less active.

If the thyroid gland produces too much thyroxine, a condition called hyperthyroidism (high-per-THIGH-roid-ihz-uhm) occurs. Hyperthyroidism results in nervousness, elevated body temperature, increased heart and metabolic rates, increased blood pressure, and weight loss. This condition can be treated by the surgical removal of part of the gland or by the use of certain drugs that slow down the production of thyroxine.

If the thyroid gland does not make enough thyroxine, a condition called hypothyroidism (high-poh-THIGH-roid-ihz-uhm) results. Lower metabolic rates and body temperature, lack of energy, and weight gain are characteristics of this

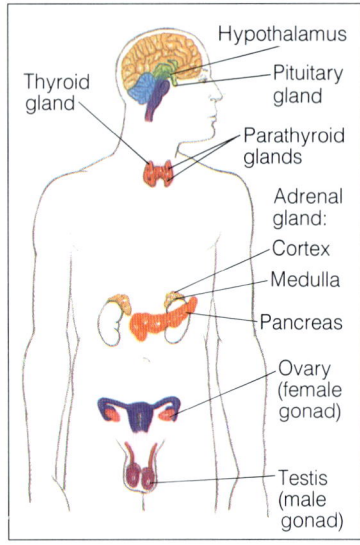

Figure 42–3 *According to this illustration, which hormone-producing gland of the endocrine system is located above each kidney?* ❶

918

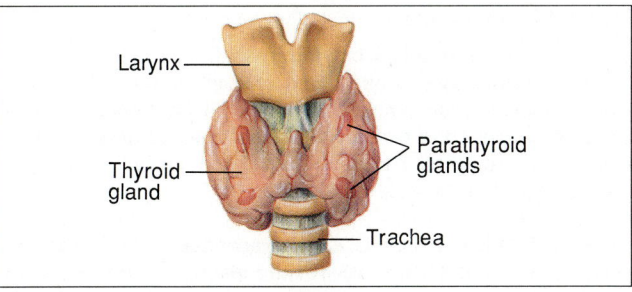

Figure 42–4 *The thyroid gland is a butterfly-shaped organ that wraps around the front of the trachea (windpipe). The electron micrograph shows a section of thyroid tissue in which the hormone thyroxine is formed.*

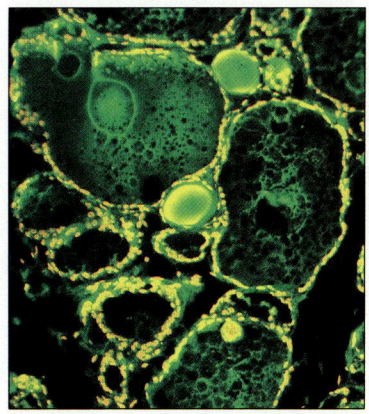

condition. Fortunately, hypothyroidism can be successfully treated by administering thyroxine. In some cases, hypothyroidism is associated with goiter (GOIT-er), or an enlargement of the thyroid gland.

The detection of hypothyroidism in infants is especially important. In infancy, hypothyroidism affects the normal development of the skeletal, muscular, and nervous systems and results in a condition called cretinism (KREET-'n-ihz-uhm). Cretinism is characterized by dwarfism and mental retardation. If detected in time, the most serious aspects of the disorder can be prevented by treatment with thyroxine.

In addition to thyroxine, the thyroid gland secretes a hormone called calcitonin (kal-sih-TOH-nihn). Calcitonin is produced by specialized cells (C cells) within the thyroid gland. The major action of calcitonin is to regulate the level of calcium in the blood.

Parathyroid Glands

The **parathyroid glands** are attached to or embedded in the back surface of the thyroid gland. See Figure 42–4. There are usually four parathyroid glands. These glands secrete a hormone called parathyroid hormone (PTH). PTH regulates the calcium levels in the blood by increasing the reabsorption of calcium in the kidneys (less calcium is excreted) and by increasing the uptake of calcium from the digestive system. Together with calcitonin and vitamin D, parathyroid hormone is important in promoting proper nerve and muscle function as well as maintaining bone structure.

As part of the treatment for hyperthyroidism, the parathyroid glands are sometimes surgically removed with the thyroid gland. This action causes a drop in the level of calcium in the blood, which may result in violent muscular spasms (tetany) and contractions. These symptoms can be relieved by the administration of PTH and injections of calcium.

Figure 42–5 *Embedded within the thyroid gland, the parathyroids produce the parathyroid hormone that regulates the amount of calcium in the blood. What thyroid hormone has an effect opposite to that of the parathyroid hormone?* ❷

Hormone	Action
Thyroid	
Thyroxine, other thyroxinelike hormones	Stimulate and maintain metabolic activities
Calcitonin	Inhibits release of calcium from bone
Parathyroid	
Parathyroid hormone (PTH)	Stimulates release of calcium from bone

FACTS AND FIGURES

The thymus is sometimes considered an endocrine gland because it produces a hormone called thymosin. This hormone helps to promote the development of certain types of white blood cells. Because its effects are restricted to the immune system, it is sometimes considered part of that system rather than part of the endocrine system.

HISTORICAL NOTE
BARBARA BUSH

In 1989, Barbara Bush underwent radiation therapy to destroy her thyroid because she suffered from hyperthyroidism. Since the treatment, she is required to take regular doses of synthetic thyroxine to maintain normal metabolic activity.

ANNOTATION KEY

❶ Adrenal gland. (Interpreting illustrations)
❷ Calcitonin. (Interpreting charts)

Adrenal Glands

The **adrenal glands** are pyramid-shaped structures that sit on top of the kidneys, one gland on each kidney. The word adrenal, in fact, means near the kidneys. Because each adrenal gland is composed of two very different types of tissue, it is divided into two parts. The outer part is the adrenal cortex and the inner part is the adrenal medulla.

THE ADRENAL CORTEX The **adrenal cortex** makes up about 80 percent of the mass of the gland. It produces more than two dozen hormones called corticosteroids (kor-tih-koh-STIHR-oidz), which are essential for normal body function. One of these hormones is aldosterone (al-DAHS-tuh-rohn), which regulates the reabsorption of sodium and the excretion of potassium by the kidneys. Another hormone called cortisol (KORT-uh-sohl) helps to control the rate of metabolism of carbohydrates, fats, and proteins. In addition, cortisol helps people cope with stress.

A decrease in the activity of the adrenal cortex can result in a condition known as Addison disease. Some symptoms of this condition include weight loss, low blood pressure, and general weakness. In some cases, death may occur because of heart failure. Former President John F. Kennedy suffered from Addison disease and received regular doses of adrenal cortical hormones to control the symptoms.

An increase in the activity of the adrenal cortex can result in a condition called Cushing syndrome. Symptoms of this condition include obesity, increased blood sugar levels, high blood pressure, and weakening of bones. Treatment of Cushing syndrome involves decreasing the secretion of the hyperactive hormone, if possible.

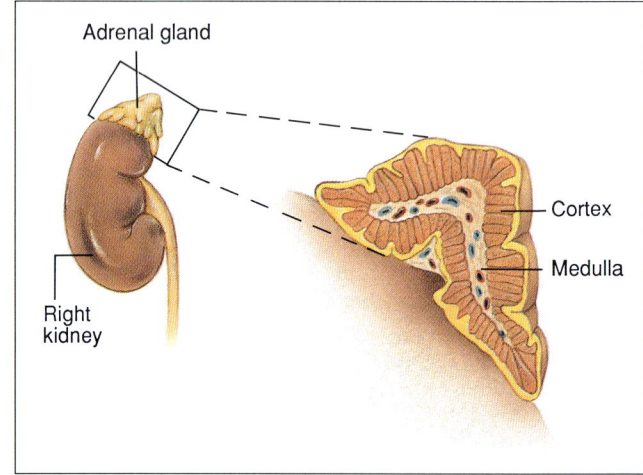

Figure 42–6 The adrenal glands, each of which is located atop a kidney, are divided into an outer part called the adrenal cortex and an inner part called the adrenal medulla.

THE ADRENAL MEDULLA The **adrenal medulla** is a specialized part of the sympathetic nervous system. For this reason, the adrenaline and noradrenaline it secretes are called neurohormones. Adrenaline is sometimes known as epinephrine (ehp-uh-NEHF-rihn). The words adrenaline and epinephrine are derived from Latin and Greek words, respectively, and mean "near the kidneys." Adrenaline, which is more powerful in its action than noradrenaline, makes up about 80 percent of the total secretion of the adrenal medulla.

When the body is confronted with a threatening "fight or flight" situation, nerve impulses from the sympathetic neurons enter and stimulate cells of the adrenal medulla. The result is an increase in the secretion of adrenaline and noradrenaline. Adrenaline increases heart rate, blood pressure, and blood supply to skeletal muscles. It also increases the conversion of glycogen to glucose and stimulates the rate of metabolism. Noradrenaline stimulates the heart muscle. These actions result in a general increase in body activity, which can serve as preparation for intense physical activity. The next time you take an exam, remember that your rapidly beating heart and perspiring hands are the results of your adrenal medulla at work!

Reproductive Glands

The **gonads** are the body's reproductive glands. They serve two important functions. The female gonads, or **ovaries**, produce eggs (ova). The male gonads, or **testes** (TEHS-teez; singular: testis), produce sperm. The gonads also produce sex hormones that affect cells throughout the body.

The ovaries produce the female sex hormones, the estrogens (EHS-truh-jehnz) and progesterone (proh-JEHS-ter-ohn). Estrogens are required for the development of ova and for the formation of the physical characteristics associated with the female. These characteristics include the development of the female reproductive system, widening of the hips, and development of the breasts. Progesterone prepares the uterus for the arrival of a developing embryo.

The testes produce androgens, or the male sex hormones. Androgens are required for normal sperm production and the development of physical characteristics associated with the male. These characteristics include the growth of facial hair, increase in body size, and deepening of the voice.

The male and female sex hormones have important functions in the process of reproduction. For this reason they will be discussed in more detail in Chapter 43.

Pancreas

The **pancreas** is a most unusual gland. Located just behind the stomach, the pancreas seems to be a single gland—but its appearance is deceiving! You may recall from Chapter 39 that

Hormone	Action
Adrenal cortex	
Aldosterone	Affects water and salt balance by the reabsorption of sodium and the excretion of potassium
Cortisol, other corticosteroids	Affect carbohydrate, protein, and fat metabolism
Adrenal medulla	
Adrenaline and noradrenaline	Increase blood glucose level, dilate or constrict specific blood vessels, increase rate and strength of heartbeat
Ovary	
Estrogens	Develop and maintain female sex characteristics, initiate buildup of uterine wall
Progesterone	Promotes continued growth of uterine lining
Testis	
Androgens	Support sperm development, develop and maintain male sex characteristics

Figure 42–7 The hormones secreted by the adrenal glands and reproductive glands (ovary and testis) regulate many of the body's activities.

FACTS AND FIGURES

How many different types of hormones are there? It depends on who is doing the choosing. They can easily be divided into lipid types (steroids and prostaglandins) and peptide types (polypeptides and amine derivatives). Other classification systems eliminate prostaglandins and group the peptide types together because they have the same mode of action.

MULTICULTURAL STRATEGY

In 1901, a Japanese industrial chemist, Jokichi Takamine, became the first person to isolate adrenaline. Takamine was also the first person to separate the starch-hydrolyzing enzyme, diastase. With this accomplishment, the production of enzymes became commercially feasible.

Jokichi Takamine worked in the United States for many years. In 1912, as a tribute to the United States and Japan, Takamine convinced the mayor of Tokyo to send cherry trees as a gift to the United States. Today, these cherry trees bloom along the Potomac River in Washington, DC.

---

explain part of the phenomenon in biological terms.

Enrichment
The diseases and malfunctions of the endocrine system, such as Addison disease and Cushing syndrome, are interesting to research. If you have students who choose to do this research, ask them to share the information.

Content Development
The section dealing with reproductive glands is short, but there is a lot of interesting and important information here. Help students to understand where reproductive hormones are produced and what they do; it will make Chapter 43 much easier. Remind students that sex hormones are steroids; therefore, they have a powerful influence on cell behavior. The reason will be explained later in the chapter. (The use of anabolic steroids has become common among some students, so they will probably notice the word *steroid*. Be sure they also hear the words *powerful*, even *dangerous*.)

Reinforcement/Reteaching
Using the Visuals Be sure to take advantage of the chart in Figure 42–7 to review the material.

BACKGROUND INFORMATION
DIABETES

Diabetes mellitus is often simply called diabetes. This is technically incorrect. The word diabetes refers to any disorder marked by excessive urination and thirst. Diabetes insipidus, for example, is a disorder that results from the body's failure to make ADH. Lacking ADH, the kidneys remove water from the blood, producing as much as 10 liters of urine a day and requiring sufferers to drink uncontrollably to prevent dehydration. Diabetes insipidus, like diabetes mellitus, cannot be cured. Insipidus is treated by ADH injections.

TEACHING SUPPORT

Teaching Resources
- Activity: Analyzing the Endocrine Functions of the Pancreas, p. 3

42–1 (continued)

Content Development
Using the Visuals Students should remember studying the pancreas in Chapter 39. Review the exocrine function performed by this unusual gland. Point out the islets of Langerhans in Figure 42–8. Be sure that students understand that these are not the same cells that produce the pancreatic fluid used in digestion. As you explain the action of insulin (which means island) and glucagon, use Figure 42–9 to help clarify your explanation. After you have discussed it, ask one or more students to explain it in their own words. (Negative feedback will be discussed later in the chapter. If the students understand it now, they will have a head start.)

Check students' understanding with questions such as the following.
- **Where is insulin produced?** (In the beta cells of the pancreas.)
- **What causes the production of insulin?** (High glucose levels in the blood.)
- **What does insulin do?** (It stimulates the ability of its target cells to take up and use glucose.)
- **What are the major target cells for insulin?** (Liver, skeletal muscles, and fat tissues.)
- **What happens to excess glucose in adipose tissue?** (It is converted to fat.)
- **What is the result of insulin production that benefits the body?** (Prevents the level of glucose in the blood from rising after a meal and also ensures that excess glucose is stored for later use.)
- **Where is glucagon produced?** (In the alpha cells of the pancreas.)
- **What causes the production of glucagon?** (A drop in blood glucose levels.)
- **In general, what does glucagon do?** (It stimulates the cells of the liver and skeletal muscles to

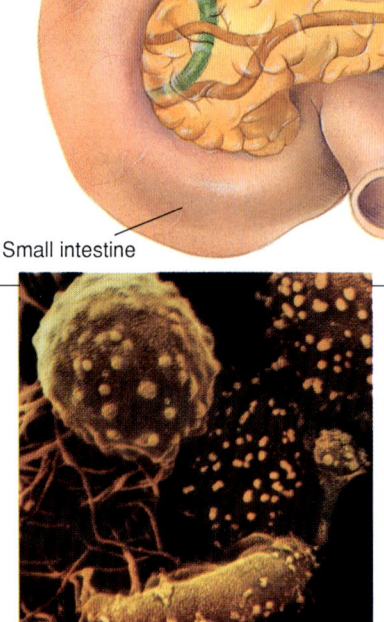

Figure 42–8 Within the pancreas are groups of cells called the islets of Langerhans (inset). An islet of Langerhans can be seen as a large circular object in the electron micrograph. Each islet is composed of beta cells, which secrete insulin, and alpha cells, which secrete glucagon.

the pancreas produces pancreatic fluid, which leaves the pancreas by means of a duct and empties into the digestive system. Once in the digestive system, the pancreatic fluid helps break down food. Because the pancreas has a duct, it is an exocrine gland. However, the pancreas also releases hormones into the blood. Therefore, it is also an endocrine gland.

The hormone-producing portion of the pancreas consists of clusters of cells that resemble islands. For this reason, their discoverer, the German anatomist Paul Langerhans, called them **islets of Langerhans.**

Each islet is composed of beta cells, which secrete insulin, and alpha cells, which secrete glucagon. Insulin and glucagon regulate the metabolism of blood glucose (sugar). Because these two hormones have opposite effects, it is vital that a proper balance between them is maintained by the body.

INSULIN When blood glucose levels rise, such as after eating a meal, beta cells release insulin. Insulin stimulates the ability of its target cells to take up and use glucose. Insulin's major target cells are those of the liver, skeletal muscles, and fat (adipose) tissue. Glucose molecules that are not immediately used as an energy source are stored as glycogen in the liver and skeletal muscles. In fat tissue, excess glucose molecules are converted to fat. Thus insulin prevents the level of glucose in the blood from rising immediately after a meal. It also ensures that excess glucose will be stored by the body for further use.

GLUCAGON Within one or two hours after eating, when the level of blood glucose drops, alpha cells release glucagon from the pancreas. In general, glucagon stimulates the cells of the liver and skeletal muscles to break down glycogen and increase glucose levels in the blood. Glucagon also causes fat cells to break down fats so that they can be used for the production of carbohydrates. These actions make more chemical energy available to the body and help raise the blood glucose level to normal.

DIABETES MELLITUS When there is an undersecretion of insulin, a condition called diabetes mellitus (digh-uh-BEET-eez muh-LIGHT-uhs) occurs. In diabetes mellitus, the amount of glucose in the blood is so high that the kidneys cannot reabsorb all of the glucose. As a result, glucose is excreted in the urine. You may recall from Chapter 41 that glucose is not normally present in urine. Thus its presence is a symptom that is used by doctors to confirm diabetes mellitus.

There are two principal types of diabetes mellitus. Juvenile-onset diabetes, as its name implies, most commonly develops in people before the age of 25. In this type of diabetes, also known as Type I, there is little or no secretion of insulin. The treatment includes a combination of strict diet control and daily injections of insulin. Adult-onset, or Type II, diabetes, most commonly develops in people after the age of 40. Although people with adult-onset diabetes produce normal amounts of insulin, their cells are unable to respond to the hormone properly because the cells lack the necessary number of insulin receptors. Adult-onset diabetes, especially in its early stages, can be controlled by diet.

Pituitary Gland and Hypothalamus

The **pituitary gland** is a bean-sized structure that dangles on a slender stalk of tissue at the base of the skull. The gland is divided into two parts: the anterior pituitary and the posterior pituitary. See Figure 42–10. The pituitary gland secretes nine major hormones that directly regulate many body functions. It also controls the release of hormones by several other endocrine glands.

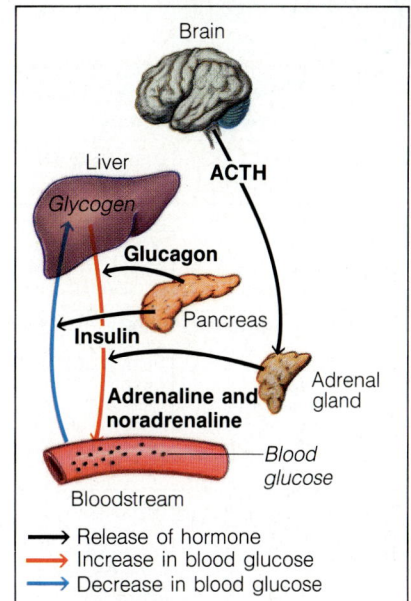

Figure 42–9 This illustration shows the regulation of blood glucose by hormones. When blood glucose levels are low, the pancreas releases glucagon, which stimulates the breakdown of glycogen and the release of glucose from the liver. When blood glucose levels are high, the pancreas releases insulin, which removes glucose from the blood by increasing its uptake by cells and its conversion to glycogen.

Figure 42–10 The pituitary gland, which lies under the hypothalamus, is located at the base of the brain in the center of the skull.

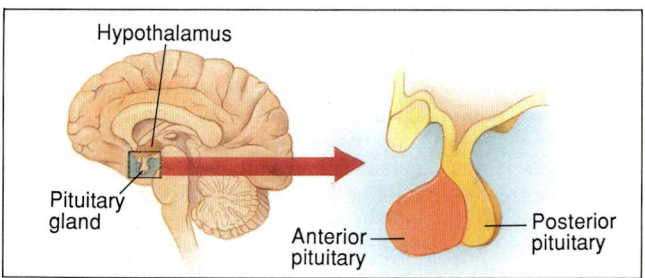

FACTS AND FIGURES

Diabetes mellitus affects about 11 million people in the United States. The National Institute of Diabetes and Digestive and Kidney Diseases says that Type II accounts for 95 percent of the cases. The remaining 5 percent are cases of juvenile-onset diabetes. Diabetes mellitus can lead to blindness, nerve and kidney damage, and amputations—the result of gangrene in the extremities due to circulatory problems. There is much research in order to determine the cause of this disease. Recently some breakthroughs have been reported in *The New England Journal of Medicine*. Scientists may have discovered a molecule responsible for choking off the cells of the pancreas, which make the hormone insulin. Research is continuing.

MULTICULTURAL STRATEGY

In 1969, a British chemist and crystallographer Dorothy Crowfoot Hodgkin determined the three-dimensional structure of insulin. This accomplishment was among the highlights in a distinguished career that spanned more than 50 years.

In 1933, when crystallography was a new and limited field of study, Hodgkin made the first X-ray diffraction photograph of a protein molecule. Later, in 1948, she worked with other scientists to determine the structure of penicillin. In 1964, Dorothy Crowfoot Hodgkin won the Nobel prize for chemistry and the British Order of Merit for her work in determining the crystalline structure of vitamin B_{12}.

---

break down glycogen and increase glucose levels in the blood. Also causes fat cells to break down fats so that they can be used for the production of carbohydrates.)

• **What is the result of glucagon action?** (The level of glucose in the blood is raised back to normal.)

Content Development
Point out the location and size of the pituitary gland. Be sure that students understand that this is another gland that is divided into two parts. There are a total of nine hormones that directly control body functions or control the release of hormones by other endocrine glands. Explain that there are several examples of one hormone controlling the production of another, as with the insulin-glucagon interaction. In the case of the pituitary, however, the hormones produced come from different endocrine glands.

ANNOTATION KEY

❶ Stimulates adrenal cortex. (Interpreting charts)

ESL STRATEGY

The theme of this chapter is the communication or message function of the endocrine system. To reinforce this idea, provide pairs or groups of students with 6 envelopes and 20 to 25 small index cards or small slips of paper that will fit in the envelopes.

On the outside of each envelope, have students print the name of a gland. (Point out the chapter subheadings that are printed in blue type.) On each slip of paper or card, have students print a specific hormone on one side and the message or action it produces on the other side. The prepared envelopes can be used for review and practice.

42–1 (continued)

Content Development

The hormones secreted by the posterior pituitary include ADH and oxytocin. Explain the function of each hormone. (If students are doing any outside reading, you might want to alert them to the fact that ADH is sometimes called vasopressin.) Insufficient amounts of ADH result in a condition known as diabetes insipidus.

Hormone	Action
Posterior Pituitary	
Antidiuretic hormone (ADH)	Controls water excretion
Oxytocin	Stimulates uterine contraction, milk release
Anterior Pituitary	
Follicle-stimulating hormone (FSH)	Stimulates follicle maturation in females and sperm production in males
Lutenizing hormone (LH)	Stimulates ovulation and corpus luteum formation in females and androgen secretion in males
Thyroid-stimulating hormone (TSH)	Stimulates and maintains metabolic activities
Adrenocorticotropic hormone (ACTH)	Stimulates adrenal cortex
Growth hormone (GH), or somatotropin	Stimulates bone growth, inhibits glucose oxidation, promotes fatty acid breakdown
Prolactin	Stimulates milk production
Melanocyte-stimulating hormone (MSH)	Increases synthesis of melanin

Figure 42–11 The pituitary gland is divided into the anterior pituitary and the posterior pituitary. What effect does ACTH have on the body? ❶

Content Development
The hormones of the anterior pituitary have diverse functions, but many of them have the common trait of exerting direct influence on the production of other hormones by other endocrine glands. The two hormones involved in the menstrual cycle will be discussed in greater detail in the next chapter. The growth hormone is one that is usually of interest to students, particularly if you discuss it in conjunction with disorders associated with too much or too little of the hormone. Several of the better known stars of horror films have been victims of acromegaly. There are many good photographs available. If you are showing photographs, try to obtain one that shows the changes that occur in the hands—acromegaly is particularly evident there.

POSTERIOR PITUITARY The **posterior pituitary** secretes antidiuretic (an-tigh-digh-yoo-REHT-ihk) hormone (ADH) and oxytocin (ahks-ih-TOH-sihn). ADH stimulates the kidneys to reabsorb more water from the collecting tubules that comprise the organs. In females, oxytocin stimulates the contractions of smooth muscles in the uterus, which help push a baby out of the mother during childbirth. Oxytocin also causes the release of milk from the breasts of a nursing mother. Oxytocin is found in males, too, but its function is not known.

ANTERIOR PITUITARY The remaining seven hormones are produced by the **anterior pituitary**. Two of these—follicle-stimulating hormone (FSH) and luteinizing (LOOT-ee-ihn-ighz-ihng) hormone (LH)—control the growth, development, and functioning of the ovaries and testes (gonads).

The third hormone secreted by the anterior pituitary is thyroid-stimulating hormone (TSH). TSH stimulates the synthesis and release of thyroxine from the thyroid gland. Adrenocorticotropic (uh-dree-noh-kor-tih-koh-TROHP-ihk) hormone, also known as ACTH, is the fourth hormone secreted by the anterior pituitary. ACTH stimulates the release of hormones from the adrenal cortex.

Growth hormone (GH), the fifth anterior pituitary hormone, is also known as somatotropin (suh-maht-uh-TROHP-uhn). Growth hormone stimulates protein synthesis and growth in cells throughout the body. Although almost all cells respond to levels of GH, skeletal muscle cells and cartilage cells are particularly sensitive. As you might expect, GH is extremely important during the first 15 years of life, when its production ensures normal body growth.

The sixth hormone secreted by the anterior pituitary is prolactin, which plays an important role in the production of milk in pregnant females. Prolactin has no known function in males. As its name indicates, melanocyte-stimulating hormone (MSH), the seventh anterior pituitary hormone, stimulates the melanocytes of the skin, increasing their production of the dark pigment melanin.

PITUITARY GLAND DISORDERS The release of growth hormone (GH) from the anterior pituitary is associated with the normal growing years, from birth to age 15 or 16. If too much GH is produced, the body will grow too fast and a condition called giantism will result. If too little GH is produced, a condition known as pituitary dwarfism occurs. Although people who suffer from pituitary dwarfism are proportionately smaller than the average-sized person, the development of their nervous system is not affected. The symptoms of pituitary dwarfism can be treated in childhood by administering growth hormone. Prior to the advent of genetic engineering, growth hormone was in short supply. Today, however, genetically engineered bacteria are able to produce GH in large quantities.

Normally, the production of growth hormone decreases dramatically in the late teens, when body growth is complete. Sometimes, however, the anterior pituitary fails to stop producing GH and a disorder known as acromegaly (ak-roh-MEHG-uh-lee) results. In this disorder, there is no increase in height because the growth plates of bones have ossified. However, there is an increase in the diameter of fingers, toes, hands, and feet. There is also formation of bony ridges over the eyes and enlargement of the jaw. Treatment for acromegaly involves reducing the levels of growth hormone by surgery, radiation, or hormone therapy.

Other pituitary gland disorders can lead to underactivity of the adrenal cortex, problems with thyroid activity, or failure to produce sex hormones. Each of these disorders has serious medical consequences.

THE HYPOTHALAMUS Attached to the posterior pituitary is a portion of the brain called the **hypothalamus** (high-poh-THAL-uh-muhs). The hypothalamus controls the secretions of the pituitary gland. In turn, the activity of the hypothalamus is influenced by the levels of the hormones in the blood and by sensory information that enters the central nervous system. The hypothalamus and the pituitary gland are the major areas where the nervous system and the endocrine system interact. Indeed, a major portion of the pituitary gland is an extension of the hypothalamus.

In the hypothalamus, special neurons called neurosecretory cells extend their axons into the posterior pituitary. When the neurosecretory cells are stimulated, the vesicles located at the ends of the axon terminals release their contents (ADH and oxytocin). Following their release from the neurosecretory cells in the posterior pituitary, ADH and oxytocin diffuse into capillaries, thus entering the general circulation of blood.

The neurosecretory cells not only allow the hypothalamus to control the posterior pituitary, they also automatically coordinate its activity with the rest of the body. Does the anterior pituitary work in the same way?

For many years, scientists searched for a connection between the hypothalamus and the anterior pituitary. Their efforts were seemingly in vain, for they found no connection. About 30 years ago, however, some scientists remembered that there was a tiny blood vessel that passed through the hypothalamus on its way to the anterior pituitary. They hypothesized that the hypothalamus might control the anterior pituitary by releasing substances through this blood vessel into the anterior pituitary.

Today we know that the hypothalamus produces tiny amounts of special hormones called releasing hormones. The releasing hormones are secreted directly into capillaries and then into veins to yet another network of capillaries. Here the releasing hormones enter the anterior pituitary, where they affect the production of anterior pituitary hormones.

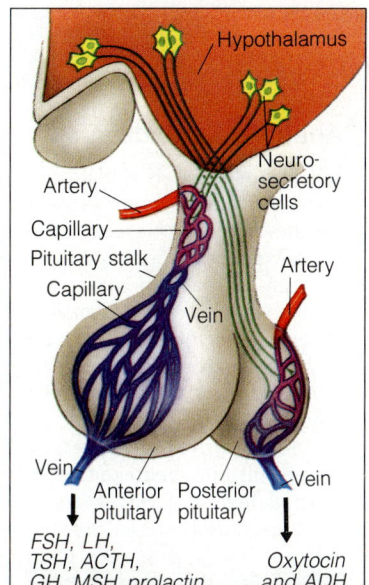

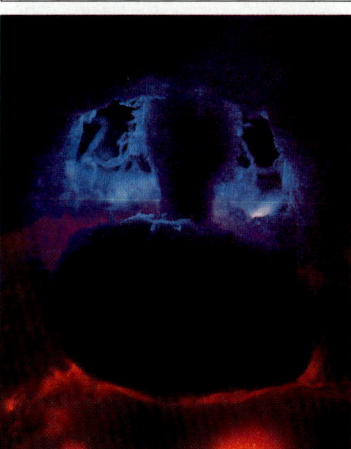

Figure 42–12 The bean-shaped structure in the photograph is the pituitary gland, which is connected to the hypothalamus by a stalk. The hypothalamus coordinates the activities of the endocrine and nervous systems.

SCIENCE, TECHNOLOGY, AND SOCIETY ISSUE

Anabolic Steroids

Testosterone is the male sex hormone that promotes protein synthesis that results in increased muscle mass. It is one of several hormones known as anabolic steroids. Operating on the theory that "if some is good, then more is better," some athletes have been taking synthetic anabolic steroids. In large amounts over long periods of time, anabolic steroids have serious side effects. They can cause damage to many body systems, sometimes resulting in death. Not only do steroids endanger the lives of athletes, they also cripple their chances of winning competitions.

This is a timely topic that probably affects some of your students on a personal level. A class discussion involving the pros and cons of steroid use might be beneficial to your students. It would be enhanced by having students do some individual research to better support their positions. One way to formalize the discussion and make it more interesting would be to have four to six students research the topic and act as a panel of experts on steroid use. After the information is presented, the panel could accept questions from the audience.

42–1 (continued)

SECTION REVIEW 42–1

1. The endocrine system regulates many activities of the body by producing hormones, which travel through the bloodstream, exerting their influence on target cells.
2. Refer to Figures 42–5, 42–7, and 42–11 on pages 919, 921, and 924, respectively, for a list of the major endocrine glands and their functions.
3. Endocrine glands are ductless glands that secrete hormones directly into the bloodstream. Exocrine glands release their secretions through ducts.
4. When stimulated, special neurons called neurosecretory cells in the hypothalamus release their contents (ADH and oxytocin) into the posterior pituitary. The hypothalamus regulates the anterior pituitary by producing releasing hormones, which enter the anterior pituitary and affect production of its hormones.
5. An athlete is usually excited and tense before a meet. This excitement causes an increase in the functioning of the adrenal medulla, thus increasing the amount of adrenaline released into the blood. The adrenaline causes an increase in blood pressure, heart rate, blood supply to skeletal muscles, and

---

SCIENCE, TECHNOLOGY, AND SOCIETY ISSUE

Anabolic Steroids

Testosterone is a male sex hormone that promotes protein synthesis in most tissues of the body. As a result, skeletal muscle can increase in mass. Because this process stimulates growth and synthesis, it is said to be anabolic. Thus testosterone is known as an anabolic steroid.

Some athletes, especially weight lifters, believe that if normal levels of testosterone increase muscle mass, then increased levels will work even better. By ingesting synthetic anabolic steroids, these athletes have been risking their health and even their lives in the hope of gaining an edge over their competitors.

Taken in large quantities over long periods of time, anabolic steroids have grave side effects. In athletes of both sexes, anabolic steroids can stunt growth. They can damage the liver, and they may cause cancer. Their use is also linked to strokes and heart disease. In males, anabolic steroids sometimes stop the production of testosterone, resulting in a decrease in the size of the testes and the inability to produce sperm. In females, anabolic steroids can interfere with the menstrual cycle and may lead to the development of male sex characteristics, including a deep voice, excessive facial hair, and enlarged muscles.

Athletes who use anabolic steroids endanger their lives as well as their chances of winning. The sprinter Ben Johnson was stripped of his Olympic Gold Medal in 1988 when traces of synthetic steroids were found in his system.

42–1 SECTION REVIEW

1. What is the function of the endocrine system?
2. List the major endocrine glands and give the function of each.
3. How does an endocrine gland differ from an exocrine gland?
4. Describe how the hypothalamus regulates the anterior and posterior pituitary.
5. **Connection—Sports** The heartbeat of a swimmer increases significantly both before and during a swim meet. Explain this.

42–2 Control of the Endocrine System

Guide For Reading
- What is the negative-feedback mechanism?
- How do polypeptide and steroid hormones differ?
- What is the function of prostaglandins?

As powerful as they are, the concentrations of hormones must be closely monitored in order to keep the functions of different organs in balance. Even though we may think of the endocrine system as one of the master regulators of the body, it too must be controlled. **Like most systems of the body, the endocrine system is regulated by a negative-feedback mechanism that functions to maintain homeostasis, or the maintenance of a relatively constant internal environment.**

Negative-Feedback Mechanism

A **negative-feedback mechanism** occurs when an increase in a substance inhibits the process leading to the increase. Although this sounds complicated, it is actually rather simple.

Consider a thermostatically controlled heating and cooling system found in many homes. A thermostat controls the system in order to keep the temperature within acceptable limits. Suppose you set the thermostat to 20°C. If the temperature in your home rises above 20°C, the thermostat turns on the air-conditioner. The cooling effect produced by the air-conditioner brings the temperature back down to 20°C. Now the thermostat shuts the air-conditioner off. If the temperature falls below 20°C, the thermostat turns on the heater rather than the air-conditioner. The heater, which has a warming effect, remains on until the temperature again returns to the desired level (20°C). In this way, the thermostat controls the internal temperature of your home. In a manner similar to this, a negative-feedback mechanism automatically controls the levels of hormones in the body. The following example will help you understand exactly how this mechanism works.

Recall that the production of hormones in the anterior pituitary is regulated by the hypothalamus. One of the hormones, thyroid-stimulating hormone (TSH), stimulates the activity of the thyroid gland, causing it to release thyroxine into the bloodstream. An increase in the level of thyroxine in the blood stimulates cells throughout the body to increase their metabolic activity. When the level of thyroxine in the blood drops, the cells' metabolic activity decreases.

The hypothalamus is sensitive to body temperature as well as to the level of thyroxine in the blood. Body temperature decreases as metabolic activity decreases. When the hypothalamus is activated, either by lowered body temperature or lowered thyroxine level, it produces a releasing hormone, or factor, that acts on the anterior pituitary. The anterior pituitary, in turn, releases TSH into the bloodstream.

As the level of TSH in the blood increases, the thyroid gland is stimulated and steps up the rate at which it releases thyroxine into the bloodstream. Increasing levels of thyroxine

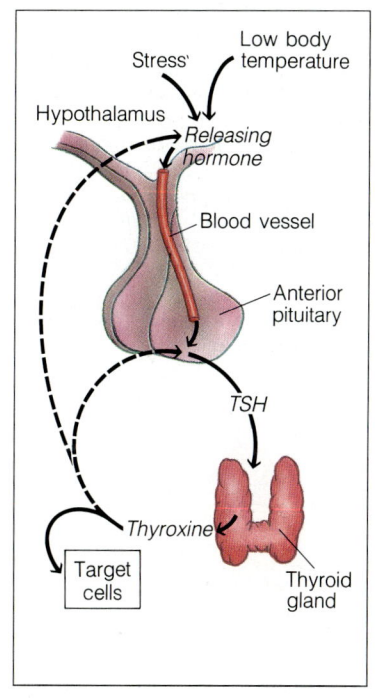

Figure 42–13 The production of many hormones is regulated by the complicated negative-feedback mechanism involving the pituitary gland and the hypothalamus.

927

in the blood speed up metabolism in cells throughout the body. As a result, there is an increase in body temperature. These changes in the body cause the hypothalamus to react negatively by lowering the rate at which it sends releasing hormone to the anterior pituitary.

As you can see, the negative-feedback mechanism is automatic and self-regulating. Similar negative-feedback mechanisms exist to control the levels of other hormones.

Hormone Action

The hormones that are produced by the endocrine system fall into two general groups. Polypeptide hormones, such as glucagon and thyroxine, are large proteins composed of chains of amino acids. Steroid hormones, such as progesterone, are lipids that are produced from cholesterol. How do these two groups of hormones affect the functions of their target cells? There are two basic patterns of hormone action that we understand well enough to explain here.

Polypeptide hormones do not enter their target cells. Instead, they bind to receptors on the cell membrane. See Figure 42–14. The binding of these hormones activates enzymes attached to the inner surface of the cell membrane. Because the polypeptide hormones are the first substances to carry a signal to the cell membrane, they are called first messengers. The activated enzymes then convert ATP into molecules known as cyclic AMP (cAMP), which functions as a second messenger.

From the cell membrane, cAMP diffuses through the target cell's cytoplasm, binding to and activating other enzymes. As more hormones bind to their receptors on the cell membrane, the level of cAMP within the target cell increases. Because different target cells respond in different ways to the change in cAMP levels, cells may be stimulated to speed up or slow down certain cellular activities. For example, the binding of adrenaline to liver cell membranes causes an increase in cAMP. This increase activates the enzymes that break down glycogen to glucose and then release glucose into the bloodstream. As you can

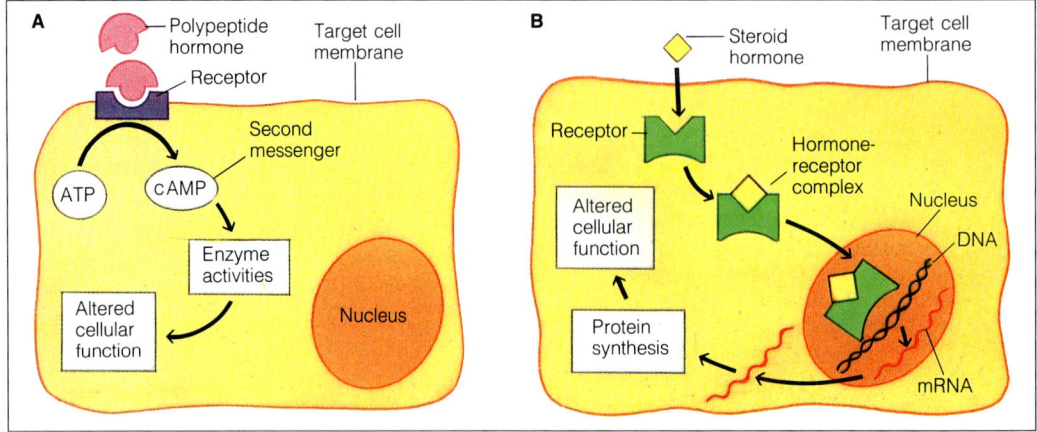

Figure 42–14 The hormones produced by the endocrine system fall into two groups, depending upon their mechanism of hormone action. In A, polypeptide hormones affect the target cell by means of receptor molecules embedded in the cell membrane. In B, steroid hormones pass through the cell membrane and combine with receptor molecules in the cytoplasm, forming a hormone-receptor complex.

cluding the way that aspirin actually works.

SECTION REVIEW 42–2

1. A negative-feedback mechanism automatically controls the levels of hormones in the body.
2. Polypeptide hormones do not enter their target cells. Instead, they bind to receptors on the cell membrane. The binding of these hormones activates enzymes attached to the inner surface of the cell membrane. Steroid hormones easily diffuse through the lipid portion of the cell membrane and enter the cytoplasm of their target cells. Here they become attached to receptor molecules, forming a hormone-receptor complex.
3. Prostaglandins are hormone-like substances found in cells

see, cAMP is one of the most important second messengers. Another one is the calcium ion.

Unlike polypeptide hormones, steroid hormones can easily diffuse through the lipid portion of the cell membrane and enter the cytoplasm of their target cells. Here they become attached to receptor molecules, which are usually proteins that float freely in the cytoplasm. The hormones then bind with the receptor molecules, forming a hormone-receptor complex. This complex drifts through the cell until it makes its way into the cell nucleus. Within the nucleus, the hormone-receptor complex can affect gene expression by attaching tightly to certain gene sequences. This action can cause the activation of genes that previously were not expressed.

Because they do not use second messengers, steroid hormones control gene expression directly. This causes steroid hormones to produce dramatic changes in cell behavior.

Prostaglandins: Local Hormones

Until recently, the endocrine system was thought to have a monopoly on the production of hormones. In the last decade this idea has changed. All sorts of cells and tissues have been shown to produce small amounts of hormonelike substances called **prostaglandins** (prahs-tuh-GLAN-dihnz).

Prostaglandins are so named because they were originally discovered in the secretions of the prostate gland, a structure in the male reproductive system. Prostaglandins, which are fatty acids, function only within the same cells in which they are produced. Thus prostaglandins are called "local hormones."

Among the many effects of prostaglandins is their ability to cause contractions in smooth muscles, such as those in the uterus, bronchioles, and walls of blood vessels. One group of prostaglandins was found to be involved in the sensation of pain. As it turns out, the discovery of these prostaglandins helped provide a solution to another puzzle. Although the pain-killing effect of aspirin was well-known, its action was not understood—at least not until it was discovered that aspirin stops the synthesis of the pain-causing prostaglandins.

42–2 SECTION REVIEW

1. Describe the negative-feedback mechanism.
2. Compare polypeptide and steroid hormones.
3. What are prostaglandins?
4. **Critical Thinking—Relating Cause and Effect** Suppose the secretion of a certain hormone causes an increase in the concentration of substance X in the blood. What is the effect on the rate of hormone secretion if an abnormal condition causes the level of X in the blood to remain very low?

BACKGROUND INFORMATION
PLANT AND ANIMAL HORMONES

Within the animal kingdom there are many examples of hormonal control. These responses, particularly to the outside environment, are usually very slow. There are some cases, however, where they are rapid. Those involving color change are good examples of rapid change. Many animals can change color and are thus camouflaged in their environment. An animal changes color by altering the size of special color-containing cells in the skin called chromatophores. When the change occurs, the overall color and pattern of the animals becomes different. In other animals, these changes are controlled by the nervous system. In some fish, amphibians, and reptiles, the reaction is hormonally controlled. In the vertebrates, the pattern of light reaching the retina of the eye controls the release of the melanocyte-stimulating hormone that acts to change the size of the chromatophores.

Plants also have a number of important hormones. Plant hormones control growth patterns, flowering, leaf structure, and seed production. Scientists are not certain how plant hormones act at the cellular level. However, they do agree that in some respects, hormones may be more important in plants than they are in animals. Because plants do not have nervous systems, the hormones that they produce may be the plant's only mechanism for communication between one part of the organism and another.

---

and tissues.

4. If the concentration of substance X is lower than normal, the secretion of the hormone should increase. Normally the increase in hormone would cause an increase in substance X and would restore homeostasis. As long as the abnormal condition keeps the level of substance X below normal, the rate of hormone secretion will remain high.

Reinforcement/Reteaching
At this time, if students are having trouble answering the Section Review questions, return to the appropriate part of the section for review.

Closure
Use the diagrams in this section of the chapter to assist you in reviewing this material. As you point out each drawing, ask one or more students to explain the diagram. Ask any necessary questions to bring out all of the important points in Section 42–2.

LABORATORY INVESTIGATION

SIMULATING THE NEGATIVE-FEEDBACK PROCESS

Before the Lab
1. This lab requires very little lab preparation and few materials.
2. Assemble the materials needed for each group. Be sure to have extra unlined paper and graph paper.
3. Perform the lab before the students do. Since they have little experience with feedback mechanisms, they may be confused by the simulation. You will need to be very familiar with the lab so that you can help dispel their confusion.

Pre-Lab Discussion
Review the explanation of the feedback process in the textbook. Be sure that students understand the point of the lab. Remind students to follow the instructions carefully, step by step.

Check to be sure that students understand all of the parts of the simulation by questioning them on these points. Go over any points of confusion that you discovered when you performed the lab activity yourself.

Skills Development
Students will use the following skills in completing this laboratory investigation.
1. Manipulative
2. Observing
3. Graphing
4. Inferring
5. Relating
6. Analyzing

Safety Tips
There are no hazards associated with this lab activity, but students should be reminded always to take care when using scissors.

Teaching Strategy
1. Monitor the students to be sure that they are following directions carefully.
2. Question the students to determine their understanding of the feedback process
3. Clear up any points of confusion as students read and interpret the instructions. (The directions are rather long and complex and this might be a problem for some students.)

Observations
1. Increase; decrease.
2. Decrease; increase.

Analysis and Conclusions
1. Thyroxine levels increase because of the increase in TSH.
2. TSH levels decrease because of the increase in thyroxine.
3. Hormone levels in the blood are controlled by a process of negative feedback in which interac-

---

LABORATORY INVESTIGATION

SIMULATING THE NEGATIVE-FEEDBACK PROCESS

PROBLEM
How are hormone levels in the blood controlled?

MATERIALS (per group)

penny	scissors
3 sheets of unlined white paper	2 small boxes
	sheet of graph paper

PROCEDURE

1. On one sheet of paper, construct a data table similar to the one shown here.
2. Draw 22 circles on another sheet of white paper using the penny as a guide. With the scissors, cut out the circles.
3. Place the letters TSH on one side of 11 circles and the letter T on one side of the remaining 11 circles. The letters TSH and T represent thyroid-stimulating hormone and thyroxine, respectively. Each circle represents a level of that particular hormone.
4. Label one box ANTERIOR PITUITARY and the other THYROID. Place all the TSH circles in the ANTERIOR PITUITARY box and all the T circles in the THYROID box. Label the third sheet of white paper BLOODSTREAM. The boxes and the sheet of paper represent the anterior pituitary, the thyroid, and the bloodstream, respectively.
5. To represent the starting level of each hormone in the bloodstream, remove 5 TSH circles and 0 T circles from their boxes. Place the TSH circles in the "bloodstream." The starting number for each hormone has been recorded for you in the data table.
6. To adjust the level of thyroxine in the bloodstream, follow these directions:
   When the level of thyroxine in the blood is low, TSH production is stimulated and the level of TSH in the blood increases. To simulate this action, if the number of T's is below 5, remove 1 TSH from the anterior pituitary and place it in the bloodstream. When T = 5, there is no change in TSH. Record.

   When the level of thyroxine in the blood is high, TSH production is inhibited and the level of TSH in the blood decreases. To simulate this action, if the number of T's is above 5, remove 1 TSH from the bloodstream and place it back in the anterior pituitary. Record.
7. To adjust the level of TSH in the bloodstream, follow these directions:
   When the level of TSH in the blood is low, the production of thyroxine is inhibited and the level of thyroxine in the blood decreases. To simulate this action, if the number of TSH's is below 5, remove 1 T from the bloodstream and put it back in the thyroid. Record.

   When the level of TSH in the blood is high, thyroxine production is stimulated and the level of the thyroxine in the blood increases. To simulate this action, if the number of TSH's is above 5, remove 1 T from the thyroid and place it in the bloodstream. When TSH = 5, there is no change in T. Record.
8. Repeat steps 6 and 7 for a total of 30 trials.
9. On the sheet of graph paper, construct a graph of your results.

OBSERVATIONS

Trial	Hormone Level in Bloodstream (number of circles)	
	TSH	Thyroxine (T)
0 (start)	5	0
⋮		
30		

1. What happens to levels of thyroxine as TSH levels increase? As TSH levels decrease?
2. What happens to levels of TSH as thyroxine levels increase? As thyroxine levels decrease?

ANALYSIS AND CONCLUSIONS
1. What causes thyroxine levels to increase?
2. What causes TSH levels to decrease?
3. How are blood hormone levels controlled?

STUDENT STUDY GUIDE

SUMMARIZING THE CONCEPTS

The key concepts in each section of this chapter are listed below to help you review the chapter content. Make sure you understand each concept and its relationship to other concepts and to the theme of this chapter.

42-1 Endocrine Glands

- The endocrine system is composed of glands that secrete their products, called hormones, into the bloodstream.
- The thyroid gland regulates the cells' rates of metabolism and the blood's calcium level.
- The parathyroid glands control the level of calcium in the blood.
- The adrenal cortex regulates reabsorption of sodium, excretion of potassium, and rate of metabolism of carbohydrates, fats, and proteins. The adrenal medulla increases heart rate, blood pressure, and blood supply to skeletal muscles, converts glycogen to glucose, and stimulates the heart muscle.
- The gonads—ovaries and testes—stimulate the development of the secondary sex characteristics in both females and males.
- The pancreas stimulates the uptake and use of glucose and the conversion of glycogen to glucose.
- The posterior pituitary gland stimulates the kidneys to reabsorb more water and the uterus to contract during childbirth.
- The anterior pituitary gland controls the growth, development, and functioning of the gonads, the functioning of the thyroid gland and adrenal glands, the growth of the body, the production of milk in females, and the production of melanin.
- The hypothalamus, a portion of the brain, controls the secretions of the pituitary gland.

42-2 Control of the Endocrine System

- Like most systems of the body, the endocrine system is regulated by a negative-feedback mechanism that functions to maintain homeostasis, or the maintenance of a relatively constant internal environment.
- Hormonelike substances called prostaglandins are synthesized and secreted by the tissues upon which they act.

REVIEWING KEY TERMS

Vocabulary terms are important to your understanding of biology. The key terms listed below are those you should be especially familiar with. Review these terms and their meanings. Then use each term in a complete sentence. If you are not sure of a term's meaning, return to the appropriate section and review its definition.

42-1 Endocrine Glands
- endocrine system
- hormone
- target cell
- endocrine gland
- exocrine gland
- thyroid gland
- parathyroid gland
- adrenal gland
- adrenal cortex
- adrenal medulla
- gonad
- ovary
- testis
- pancreas
- islet of Langerhans
- pituitary gland
- posterior pituitary
- anterior pituitary
- hypothalamus

42-2 Control of the Endocrine System
- negative-feedback mechanism
- prostaglandin

---

tions among hormones help to keep each in balance.

Going Further: Enrichment
Have students demonstrate their understanding of the feedback process by designing and performing a simulation using insulin and glucagon to regulate blood glucose level. They may use the same basic setup as this lab or do something completely different.

CHAPTER REVIEW

CONTENT REVIEW

Multiple Choice
1. d   2. b   3. d   4. d
5. d   6. c   7. b   8. a

True or False
1. T
2. F pancreas
3. F Type II
4. T
5. F posterior
6. F growth hormone (GH)
7. F pituitary
8. T

Word Relationships
1. hormones that are produced by the reproductive glands; cortisol is a hormone produced by the adrenal cortex
2. pituitary gland disorders; cretinism is a thyroid gland disorder
3. pituitary produces hormones that directly stimulate thyroid and adrenal glands; pancreas is not directly stimulated by pituitary hormones
4. pituitary hormones; PTH is a parathyroid hormone

CONCEPT MASTERY

1. Hypothyroidism is a condition in which the thyroid gland does not make enough thyroxine. Hyperthyroidism is a condition in which the thyroid gland produces too much thyroxine.

CHAPTER REVIEW

CONTENT REVIEW

Multiple Choice

Choose the letter of the answer that best completes each statement.

1. Through the activity of which body system do cells receive hormones?
   a. digestive
   b. respiratory
   c. excretory
   d. circulatory
2. Secretions from ductless glands are called
   a. enzymes.
   b. hormones.
   c. digestive fluids.
   d. excretory fluids.
3. A high concentration of calcium in the blood suggests a disorder of the
   a. thyroid gland.
   b. pancreas.
   c. liver.
   d. parathyroid glands.
4. Which gland produces cortisol?
   a. parathyroid
   b. pancreas
   c. thyroid
   d. adrenal
5. If the amount of blood glucose is low, the islets of Langerhans will
   a. secrete insulin.
   b. become inactive.
   c. secrete GH.
   d. secrete glucagon.
6. Which hormone is not involved in the control of glucose metabolism?
   a. adrenaline
   b. insulin
   c. PTH
   d. glucagon
7. A type of self-regulation such as the relationship between the pituitary gland, TSH, the thyroid gland, and thyroxine is known as
   a. cyclosis.
   b. negative-feedback mechanism.
   c. synapsis.
   d. voluntary control.
8. Hormonelike substances that are secreted by all types of cells and are involved in the sensation of pain are called
   a. prostaglandins.
   b. adrenalines.
   c. thyroxines.
   d. insulins.

True or False

Determine whether each statement is true or false. If it is true, write "true." If it is false, change the underlined word or words to make the statement true.

1. <u>Calcitonin</u> is secreted by the thyroid gland.
2. Insulin and glucagon are both produced in the <u>liver</u>.
3. <u>Type I</u> diabetes most commonly develops in people over the age of 40.
4. The <u>hypothalamus</u> is a part of the brain that controls secretions of the pituitary gland.
5. The <u>anterior</u> pituitary gland produces oxytocin.
6. Another name for somatotropin is <u>prolactin</u>.
7. Giantism is caused by an excess of hormones from the <u>thyroid</u> gland.
8. <u>Steroid</u> hormones enter their target cells.

Word Relationships

In each of the following sets of terms, three of the terms are related. One term does not belong. Determine the characteristic common to three of the terms and then identify the term that does not belong.

1. estrogen, cortisol, androgen, progesterone
2. dwarfism, acromegaly, giantism, cretinism
3. pancreas, thyroid, pituitary, adrenal
4. prolactin, PTH, TSH, ACTH

---

2. Adrenaline prepares the body for emergencies by causing an increase in heart rate, blood pressure, blood sugar, breathing rate and blood supply to skeletal muscles.
3. Prostaglandins function only within the cells in which they are produced. Among their many effects is that of causing contractions in smooth muscles.
4. Refer to Figure 42–11 on page 924 for the functions of the anterior and posterior pituitary hormones.
5. Blood glucose levels must be kept within limits for the normal functioning of the body.
6. Aldosterone affects the water and salt balance through the reabsorption of sodium and the excretion of potassium. Cortisol and other corticosteroids affect carbohydrate, protein, and fat metabolism.
7. Cretinism is a condition caused by the undersecretion of thyroxine in infancy. It is characterized by dwarfism and mental retardation. Acromegaly is a condition that is caused by the oversecretion of growth hormones in the late teens. It is characterized by an increase in the diameter of the fingers, toes, hands, and feet and the formation of bony ridges over the eyes and enlargement of the jaw.
8. Glucagon affects glucose metabolism by raising the blood glucose level. Insulin affects glucose metabolism by lowering the blood glucose level.
9. When the parathyroid glands are injured, the calcium and phosphorus levels in the blood drop. The muscles then undergo violent spasms (tetany).
10. The pituitary gland is often called the master gland because it secretes hormones that control most of the other endocrine glands and a variety of body processes.
11. A negative-feedback mechanism returns a condition to its normal state. Because the purpose of homeostasis is to maintain internal stability, a negative-feedback mechanism works toward the homeostatic condition.
12. Polypeptide hormones do not enter their target cells. Instead, they bind to receptors on the cell membrane, which activates enzymes attached to the inner surface of the membrane. These enzymes then convert ATP into cAMP, which diffuses into the target cell's cytoplasm, binding to and activating other enzymes. As more hormones bind to their receptors, the cAMP levels increase. Because the different target cells respond in different ways to the change in cAMP levels, cells may be

CONCEPT MASTERY

Use your understanding of the concepts developed in the chapter to answer each of the following in a brief paragraph.

1. Distinguish between the conditions of hypothyroidism and hyperthyroidism.
2. How does the secretion of adrenaline prepare the body for emergencies?
3. Describe how prostaglandins work.
4. Describe the functions of the hormones of the anterior and posterior pituitary.
5. Why must the blood glucose level be kept fairly constant?
6. Explain the functions of the hormones secreted by the adrenal cortex.
7. Compare cretinism and acromegaly.
8. What are the functions of glucagon and insulin?
9. How may injury to the parathyroid glands affect the muscles?
10. Why has the pituitary gland often been called the master gland?
11. Explain how negative-feedback mechanisms help maintain homeostasis in the body.
12. How do the polypeptide hormones and steroid hormones affect the functions of their target cells?

CRITICAL AND CREATIVE THINKING

Discuss each of the following in a brief paragraph.

1. **Relating facts** Explain why the thyroid gland enlarges in response to a deficiency of iodine in the diet.
2. **Making predictions** Predict the effect of removing the pancreas from a person's body.
3. **Relating concepts** Suppose a person is not getting a sufficient amount of calcium in his or her diet. How does this affect the parathyroid hormone secretion and parathyroid hormone target cells?
4. **Making diagrams** Diagram a negative-feedback mechanism involving hormones that function to regulate the production of thyroid hormones.
5. **Relating cause and effect** Explain why some hormones may be taken orally whereas others, such as insulin, which is a protein, are injected directly into the body.
6. **Interpreting graphs** The graph shows the level of glucose in the blood of two people during a 5-hour period immediately following ingestion of a typical meal. Which line represents an average person? A person with diabetes? Explain.

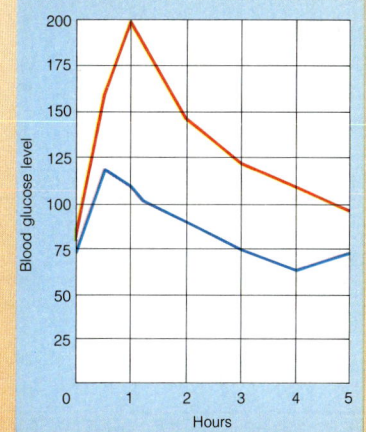

7. **Using the writing process** Recently athletes have been barred from participating in certain sports because they have tested positive for anabolic steroids. Although anabolic steroids have strength-enhancing properties, they can cause irreparable damage to the body. Design a poster in which you inform people of the harmful effects of these substances.

CHAPTER 43 Reproduction and Development

Section	Laboratory Investigations and Activities
43-1 The Reproductive System pages 935-943 THEMES: Scale and Structure, Patterns of Change	**Student Edition** LABORATORY INVESTIGATION: Examining Mammalian Gametes, p. 948 **Teacher Edition** DEMONSTRATION: Sperm and Egg, p. 934D DEMONSTRATION: The Menstrual Cycle, p. 934D **Laboratory Manual** Ovaries and Testes, p. 523 The Human Menstrual Cycle, p. 529
43-2 Human Development pages 943-947 THEME: Patterns of Change	**Laboratory Manual** Examining an Unfertilized Chicken Egg, p. 539
Chapter Review pages 949-951	

Teacher Resources

Books

Browder, L. W. *Developmental Biology,* 2nd ed. Saunders, 1984.

Hafez, E. S., and Kenemans. *An Atlas of Human Reproduction: By Scanning Electron Microscopy.* Kluwer Academic, 1982.

The Incredible Machine. National Geographic Society, 1986.

Kitzinger, Sheila, and Lennart Nilsson. *Being Born.* Putnam, 1986.

Nilsson, Lennart, Axel Ingleman-Sundberg, and Claes Wirsen. *A Child Is Born: The Drama of Life Before Birth.* Dell, 1986.

CHAPTER PLANNING GUIDE

Other Activities	Media and Technology
Teaching Resources ACTIVITY: Studying the Menstrual Cycle, p. 3 **Biotechnology Workbook** Fertility Technology: Hope for Childless Couples and Breeding Supercows, p. 107 **Study Guide** Section 43–1, p. 411	**Video/Videodisc** Human Body: Reproductive System and Excretory System **Biology Transparencies** 67: The Human Male Reproductive System 68: The Human Female Reproductive System **Biology Media Guide** pages 68–70
Teaching Resources VOCABULARY REVIEW: Cross-a-Clue, p. 1 ACTIVITY: Investigating Early Embryonic Development, p. 5 **Study Guide** Section 43–2, p. 415	**Video/Videodisc** Human Body: Reproductive System and Excretory System **Biology Transparencies** 69: The Human Placenta and Related Structures **Biology Media Guide** pages 70–71
Teaching Resources Chapter Test A, p. 13 Chapter Test B, p. 17	**Computer Test Bank** Chapter Test, p. 451

Audiovisuals

Reproductive System, 2nd ed. Film or video. Coronet.
Nova: How Babies Get Made. Film or video. Coronet.
Reproductive Systems: Structure and Function. Sound filmstrip. Ward.
Human Conception and Early Embryology. Sound filmstrip. Ward.
The Fertilization Process. Film. McGraw-Hill.
Hormone Control of Human Reproduction. Film. McGraw-Hill.

CHAPTER 43 Reproduction and Development

CHAPTER OVERVIEW

Plants and animals produce new individuals through the process of reproduction. Reproduction involves special structures that make up the reproductive system. In vertebrates, including humans, the reproductive system produces, stores, nourishes, and releases specialized sex cells known as gametes. The fusion of sperm and egg cells in the process called fertilization results in a zygote that gives rise to all the cells in a human body. About the seventh week of embryonic development, the reproductive organs of the embryo begin to produce sex hormones that cause further development of the reproductive organs. Though these hormones continue to be produced in small amounts for many years, the reproductive system of humans is not functional until the period of growth and sexual maturation known as puberty.

The structures that make up the male reproductive system, the development of secondary sex characteristics, and the production of sperm cells are discussed in detail. The topics that follow include the female reproductive system, the development of the ova, the release of the egg and hormones, and the menstrual cycle.

Fertilization takes place in the Fallopian tube. Implantation occurs about 7 days after fertilization. Eventually three cell layers known as the ectoderm, mesoderm, and endoderm form. These layers, called primary germ layers, give rise to all the organs and tissues of the embryo. During implantation, two important membranes—the amnion and chorion—surround, protect, and nourish the embryo. After about 8 weeks of development, the embryo is known as a fetus. By the end of 3 months, most of the major organs and tissues of the fetus are developed. During the fourth, fifth, and sixth months, the tissues of the fetus continue to become more complex and specialized. About 9 months after fertilization, the fetus is ready for birth. Some of the events surrounding childbirth are still not completely understood. However, it is clear that an interaction between the nervous system, endocrine system, and reproductive system is involved in the entire procedure.

43-1 THE REPRODUCTIVE SYSTEM

Section Focus 43-1

This section introduces the male and female reproductive systems. The reproductive system involves special structures that produce new individuals through the process of reproduction. The reproductive system produces, stores, and releases specialized sex cells known as gametes. The fusion of the male gamete and the female gamete through fertilization produces a fertilized egg, or zygote, that will eventually give rise to all of the cells of the human body. For the first 6 weeks of development, male and female embryos are identical in appearance.

Major changes occur when the embryo begins to produce sex hormones. In the male, the testes produce androgens, which induce certain tissues to become male reproductive organs. In the female, the ovaries produce estrogens, which induce tissues to become female reproductive organs. After birth, the testes and the ovaries continue to produce small amounts of hormones that continue to influence the development of the reproductive organs. However, neither the male nor the female is capable of producing active reproductive cells until the onset of puberty, which occurs between the ages of 9 and 15.

The primary male reproductive organs are the testes. These organs are located in a sac called the scrotum. Because the scrotum is outside the body cavity, the temperature is lower than the normal internal body temperature. Sperm development requires this lower temperature. There are hundreds of tiny tubes called seminiferous tubules that make up the testes. This is where sperm are produced. The sperm travel through the epididymis and the tube known as the vas deferens. Three glands produce the seminal fluid in which the sperm are suspended. The combination of sperm and seminal fluid is known as semen. There are between 100 and 200 million sperm present in 1 milliliter of semen. Eventually the vas deferens merges with the urethra, the tube that leads to the outside of the body through the penis.

The primary reproductive organs in the female are the ovaries. The ovaries usually release one mature ovum per month. There are approximately 400,000 immature ova in the ovaries of the normal female. This is far more than will ever mature in the life of the female. The release of the ovum from the follicle is a process called ovulation. The ovum is swept from the surface of the ovary into the opening of one of the Fallopian tubes. Eventually the egg passes into the uterus. If unfertilized, it passes out of the body through the vagina. The process of ovulation takes place in females from the onset of puberty until menopause.

In women, the interaction of the endocrine and reproductive systems involves a complex series of periodic events called the menstrual cycle. The cycle is controlled by hormones operating on a negative-feedback mechanism. Each of the phases of the cycle involves hormone activity, the development of the ovum, and the preparation of the uterine lining to receive a fertilized egg.

Performance Objectives 43-1

1. Discuss the functions of the reproductive system.
2. Compare sexual development in males and in females.
3. Name and discuss the structures of the male reproductive system.
4. Describe the structures of the female reproductive system.
5. Describe the menstrual cycle.

CHAPTER PREVIEW

Science Terms 43-1

reproductive system p. 935
testis p. 936
androgen p. 936
ovary p. 936
estrogen p. 936
puberty p. 936
gonad p. 936
scrotum p. 937
seminiferous tubule p. 937
testosterone p. 937
secondary sex characteristics p. 937
epididymis p. 937
vas deferens p. 937
seminal fluid p. 937
semen p. 937
urethra p. 938
penis p. 938
ovum p. 938
primary follicle p. 938
ovulation p. 939
Fallopian tube p. 939
uterus p. 939
vagina p. 939
menopause p. 939
menstrual cycle p. 939
corpus luteum p. 940
progesterone p. 941
menstruation p. 942
zygote p. 942

43-2 HUMAN DEVELOPMENT

Section Focus 43-2

The purpose of this section is to discuss human development after fertilization. If an egg is fertilized, the zygote begins a rapid and remarkable process of development. The zygote's first few mitotic cell divisions, or cleavage, take place while it is still in the Fallopian tube. The zygote grows into a hollow structure known as a blastocyst. Six to seven days after fertilization, the blastocyst attaches itself to the wall of the uterus in a process called implantation. Soon after implantation, the blastula undergoes gastrulation. The gastrula is composed of three cell layers called the ectoderm, mesoderm, and endoderm. These layers give rise to all of the organs and tissues of the embryo. Also during this time, two important membranes form that surround and protect the developing embryo. These membranes are the amnion and the chorion. The chorion and the uterine tissue form the placenta. After 8 weeks of development, the embryo is known as a fetus. Development of body systems and structures continues rather rapidly during this time, becoming more specialized and complex. After 9 months of development, the fetus is ready for birth. How this procedure is triggered is not completely understood. It is known that hormones influence the event.

Performance Objectives 43-2

1. Outline the events of development from zygote to birth.
2. Explain the importance of the three germ layers.
3. Describe the process of childbirth.
4. Explain the interaction of the nervous, endocrine, and reproductive systems during embryonic and fetal development, childbirth, and after birth.

Science Terms 43-2

cleavage p. 943
morula p. 943
blastocyst p. 943
implantation p. 943
gastrulation p. 943
ectoderm p. 943
mesoderm p. 943
endoderm p. 943
amnion p. 943
chorion p. 943
placenta p. 944
fetus p. 944
umbilical cord p. 944
amniotic sac p. 944
labor p. 945

DISCOVERY LEARNING

TEACHER DEMONSTRATIONS
Modeling

Sperm and Egg
This activity can be used to introduce Chapter 43.

Bring a large beaker filled with dried peas, small beans, or BB's to the class and set it on a table so that the students can clearly see it. Place one large nut—walnut or pecan—on the table beside the beaker. Ask: If you wanted to crack the shell of the nut with the peas (for example) and could use each one only once (throwing it, blowing it through a straw, and so on), how many do you think it would take to eventually crack the shell?

You can make a loose analogy to the sperm and egg situation. Students often question the need for so many sperm cells in order to fertilize one egg. If you want to illustrate your point, ask a student to toss one of the "sperm" at the "egg." Many will miss entirely, some will roll off the table, others will simply glance off the shell. You might also want to discuss the enzyme produced by the sperm, which facilitates penetration; you can also talk about the rather perilous life of the sperm.

The Menstrual Cycle
This demonstration can be used in conjunction with the study of the menstrual cycle.

Write the stages of the menstrual cycle on the chalkboard or an overhead transparency in haphazard order. Lead the students through a discussion of the cycle, drawing arrows and other notations to indicate relationships. When you have finished, ask the students to organize the information into a web, map, chart, or paragraph explaining the different stages, emphasizing the relationships that exist.

CHAPTER 43
Reproduction and Development

GUIDED ENQUIRY

Pose the following questions to students and have them record their responses. Point out that they will gain a better understanding of the key concepts if they read the chapter with these basic questions in mind. Upon completion of the chapter, pose the questions again. Ask students to compare their initial responses with those they have developed after reading the chapter.

- What is the difference between male and female development?
- How does the endocrine system influence reproduction?
- How does a baby develop from a single cell?
- What happens after fertilization occurs?
- What triggers the events of childbirth?

INTRODUCING CHAPTER 43

Using the Visuals

Have students look at the chapter-opener photograph and read the caption. The chapter-opener paragraph, in contrast to the biological photograph, is rather poetic. You might want to talk briefly about the romance surrounding the concept of new life, which undoubtedly accounts for the continuation of the species. Be sure to consider the questions posed in the paragraph.

CHAPTER 43
Reproduction and Development

As sperm meet egg in the process of reproduction, the continuation of the human species is guaranteed.

The season of spring is synonymous with rebirth. Flowers burst from the soil, the songs of birds are heard after a long winter's absence, and many animals give birth to their young. The symbolism that we apply to this time of year comes, at least in part, from the importance of reproduction to living creatures—ourselves included. For every species of living thing, reproduction forms the next generation. It guarantees the continuation of every form of life. It gives us a personal stake in the world of the future. And it determines many aspects of our lives. How does this extraordinary process take place in humans? What systems interact to produce a new generation of living organisms? And how do single cells become complete human beings millions and millions of times every year? In this chapter you shall learn about the human reproductive system and the way in which new life begins.

TEACHING STRATEGY 43–1

Focus/Motivation

Ask a series of questions similar to the ones that follow to point out the importance of the reproductive system and the process involved.

- What would happen to an organism if the circulatory system was removed or destroyed? (It would die.)
- What would happen to an organism if the reproductive system was removed or destroyed? (It would probably continue to live but it would be unable to reproduce.)
- Why is the reproductive system important if the organism could survive without it? (It is important for the continuation of the species.)
- **Must every member of the**

GUIDE FOR READING

After you read the following sections, you will be able to

43–1 The Reproductive System
- Describe the function and importance of the reproductive system.
- Compare sexual development in males and females.
- Identify the structures of the male and female reproductive systems.

43–2 Human Development
- Trace the development of a fertilized ovum into a fetus.
- Describe the process of childbirth.

Journal Activity

YOU AND YOUR WORLD
Put yourself in another person's shoes for a moment. If you are a girl, pretend you are a boy. If you are a boy, pretend you are a girl. What problems do you think you would have as you mature? How would you deal with them?

43–1 The Reproductive System

Guide For Reading
- What is the function of the reproductive system?
- What changes occur during puberty in females? In males?
- What are the phases in the menstrual cycle?
- How does fertilization occur?

Plants and animals produce new individuals through the process of reproduction. In some organisms, a group of specialized organs known as the **reproductive system** carries out this function. As you have learned in preceding chapters, most of the systems of the body are essential to survival. The loss of the nervous system, the digestive system, or the circulatory system would be fatal to most animals. This is not the case for the reproductive system, because an individual organism can lead a healthy life without reproducing. However, the reproductive system could be thought of as the single most important system to the continuation of a species. Without its reproductive system, no organism could produce the next generation.

In humans, as in other vertebrates, the reproductive system produces, stores, nourishes, and releases specialized sex cells known as gametes. These cells are released in ways that make possible the fusion of sperm and egg, the male and female gametes. All the cells of the human body are derived from the single fertilized egg cell, known as a zygote.

Sexual Development

For the first six weeks after fertilization, human male and female embryos are identical in appearance. Then, during the

Figure 43–1 In generation after generation, similarities and differences between parents and offspring are identifiable. This unity and diversity result from the fact that humans reproduce sexually and genetic material from each parent is contributed to the offspring.

TEACHING SUPPORT

Laboratory Manual
- Ovaries and Testes, p. 523

MEDIA AND TECHNOLOGY

📺 **Biology Transparencies**
- 67: The Human Male Reproductive System

▶ **Video/Videodisc**
- *Human Body: Reproductive System and Excretory System*

Use this bar code to introduce the reproductive system.

Play frames 700 to 1200

See the Biology Media Guide pages 68–71 for additional bar-code correlations for this section.

43–1 (continued)

Content Development
- **What is the primary male reproductive organ?** (Testes.)
- **What hormones do the testes produce?** (Androgens.)
- **What do androgens do?** (Influence the development of the male reproductive system.)
- **What are the primary reproductive organs of the female?** (Ovaries.)
- **What hormones do the ovaries produce?** (Estrogens.)
- **When are the reproductive organs capable of producing active reproductive cells?** (At puberty.)
- **What is puberty?** (Puberty is a period of rapid growth and sexual maturation.)
- **When does puberty occur?** (Usually between the ages of 9 and 15.)
- **What triggers this event?** (It begins with a change in the hypothalamus, which causes the pituitary gland to produce increased levels of two hormones that affect the gonads.)
- **Name the two hormones.** (FSH and LH.)

Figure 43–2 Puberty may occur anytime from age 9 to 15. During this growing period, adolescents display a wide variety of physical characteristics.

seventh week of development, major changes occur. The **testes** (singular: testis), which are the primary reproductive organs of a male embryo, begin to produce steroid hormones known as **androgens**. The tissues of the embryo respond to these hormones by developing into the male reproductive organs.

If the embryo is female, the **ovaries**, or the primary reproductive organs of a female embryo, produce steroid hormones known as **estrogens**. In response to these hormones, the tissues of the embryo develop in a pattern that produces the female reproductive organs. The male and female reproductive organs develop from exactly the same tissues in the embryo.

After birth the testes and the ovaries continue to produce small amounts of androgens and estrogens, respectively. These sex hormones continue to influence the development of the reproductive organs. However, neither testes nor ovaries are capable of producing active reproductive cells (gametes) until the onset of **puberty**. Puberty is a period of rapid growth and sexual maturation during which the reproductive system becomes fully functional. At the completion of puberty, the male and female **gonads**, or reproductive organs, are fully developed. The onset of puberty varies considerably among individuals. It may occur anytime from age 9 to 15. Generally, puberty begins about a year earlier in females than in males.

Puberty begins with a change in the hypothalamus, the part of the brain that regulates the secretions of the pituitary gland. This change causes the pituitary gland to produce increased levels of two hormones that affect the gonads. These hormones, named for their effects on the female, are follicle-stimulating hormone (FSH) and luteinizing hormone (LH).

Figure 43–3 Sperm are produced within the testes in structures known as seminiferous tubules. In the photograph, the rounded cells just inside the rim of the tubule are immature sperm. As they journey to the center of the tubule, they develop into mature sperm whose tangled tails are visible in the center.

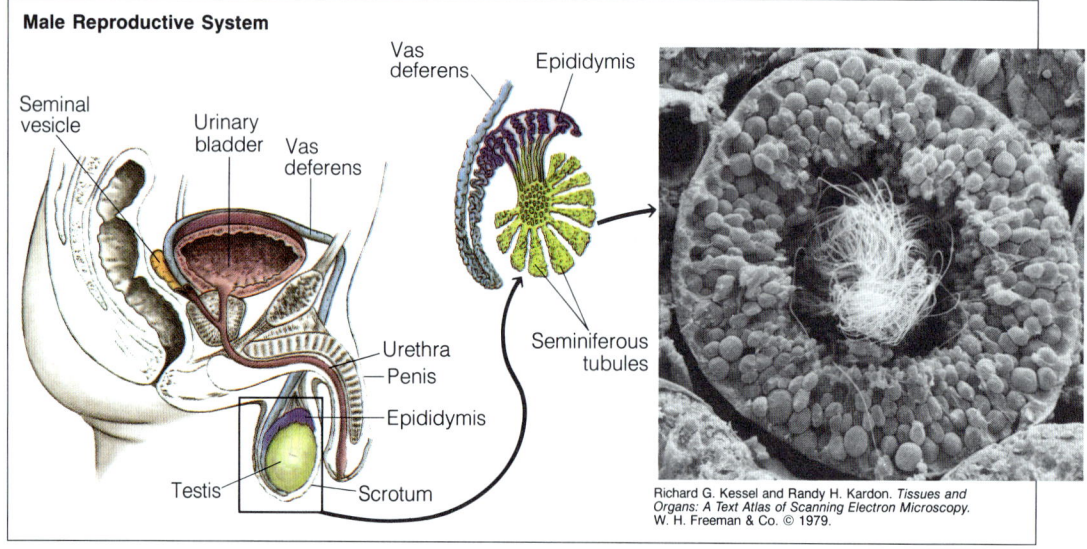

Richard G. Kessel and Randy H. Kardon. *Tissues and Organs: A Text Atlas of Scanning Electron Microscopy.* W. H. Freeman & Co. © 1979.

Content Development
Using the Visuals As you discuss the structures involved in the male reproductive system, use Figures 43–3 and 43–4 to help you in your explanations. Emphasize that all of these structures are uniquely related to function. A

The Male Reproductive System

The primary male reproductive organs, the testes, develop within the abdominal cavity. Just before birth (and sometimes just after) the testes descend through a canal into an external sac called the **scrotum**. The testes remain in the scrotum, outside the body cavity, where the temperature is about 1° to 3°C lower than the internal temperature of the body (37°C). Sperm development in the testes requires the lower temperature.

The testes are clusters of hundreds of tiny tubules called **seminiferous** (sehm-uh-NIHF-er-uhs) **tubules**. The word seminiferous means seed-bearing, making it an appropriate name for these tubules because it is there that sperm are produced. The seminiferous tubules are tightly coiled and twisted together to form a compact organ. See Figure 43–3. As the pituitary gland begins to release FSH and LH, these hormones stimulate the testes to make the principal male sex hormone **testosterone** (tehs-TAHS-ter-ohn).

Cells that can respond to testosterone are found all over the body. Testosterone produces a number of **secondary sex characteristics** that appear in males at puberty. (The primary sex characteristics are those that pertain to the reproductive system itself—the reproductive tract and external features.) A boy's voice becomes deeper, he grows a beard and more body hair, his chest broadens, and he may find it easier to develop large muscles. He will continue to grow for several years after his female classmates have stopped growing.

FSH and testosterone stimulate the development of sperm. When large numbers of sperm have been produced in the testes, the developmental process of puberty is completed. The reproductive system is now functional, meaning that the male can produce and release active sperm.

SPERM DEVELOPMENT Sperm are derived from special cells within the testes that go through the process of meiosis to form the haploid nuclei found in mature sperm. A sperm cell consists of a head, which contains the highly condensed nucleus; a midpiece, which is packed with energy-releasing mitochondria; and a tail, or flagellum, which propels the cell forward. At the tip of the head is a small cap that contains an enzyme vital to the process of fertilization.

Developed sperm travel from the seminiferous tubules into the **epididymis** (ehp-uh-DIHD-ih-mihs), a comma-shaped structure in which they fully mature and are stored. After brief storage in the epididymis, the sperm are forced into a tube known as the **vas deferens** (vas DEHF-uh-rehnz). The vas deferens passes into the abdominal cavity where three glands produce **seminal fluid** in which the sperm are suspended. The combination of sperm and seminal fluid is known as **semen**.

The number of sperm present in even a few drops of semen is astonishing. Between 100 and 200 million sperm are present

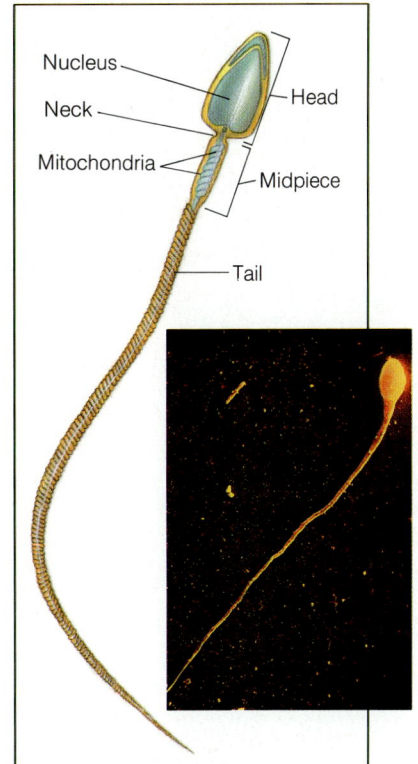

Figure 43–4 A mature sperm cell consists of a head, a midpiece, and a tail. Millions of these male gametes are produced daily by the testes.

BACKGROUND INFORMATION
PUBERTY

Public health scientists have long noted that the onset of puberty is extremely variable. Some girls enter puberty when they are 9 years old, while others are 13 or 14 or even later. Boys are at least as variable. Careful studies of human sexual development show that the age at which puberty starts generally has no significance in later life. You might assure students that if their own experience was earlier or later than the norm, they have nothing to worry about.

TEACHING SUPPORT

Laboratory Manual
- The Human Menstrual Cycle, p. 529

MEDIA AND TECHNOLOGY

Biology Transparencies
- 68: The Human Female Reproductive System

---

in 1 milliliter of semen. That's about 5 million sperm per drop! Eventually, the vas deferens merges with the **urethra**, the tube that leads to the outside of the body through the **penis**.

SPERM RELEASE When the male is sexually excited, the autonomic nervous system prepares the male organs to deliver sperm. Sperm are ejected from the penis by the contractions of smooth muscles lining the vas deferens. This process is called ejaculation. Because ejaculation is regulated by the autonomic nervous system, it is not completely voluntary. Approximately 2 to 3 milliliters of sperm are released in an ejaculation. If the 300 million or so sperm are released in the reproductive tract of a female, the chances of a single sperm fertilizing an ovum, if one is available, are quite good.

The Female Reproductive System

The primary reproductive organs in the female are the ovaries (singular: ovary). The ovaries are located in the abdominal cavity. Whereas the testes may produce several hundred million reproductive cells each day, the ovaries usually produce only one egg, or **ovum**, per month. But in addition to producing ova, the female reproductive system has another important job to perform. Each time an ovum is released, the body of the female must be prepared to nourish a developing embryo.

As in males, puberty in females starts with changes in the hypothalamus that cause the release of FSH and LH from the pituitary gland. These are the same kinds of hormones that are found in males, although their target cells and the effects they produce are quite different.

FSH (follicle-stimulating hormone) stimulates cells within the ovaries to produce the hormones known as estrogens. The estrogens cause the reproductive system to complete its development and also to produce the secondary sex characteristics that appear in females. These characteristics include enlargement of the breasts and reproductive organs, widening of the hips, and growth of hair in the armpits and pubic area.

OVA DEVELOPMENT Each ovary contains about 400,000 **primary follicles**, which are clusters of cells surrounding a single ovum. The function of a follicle is to prepare a single ovum for release into the part of the reproductive system where it can be fertilized.

Ova mature within their follicles. Although a female is born with about 400,000 immature ova (primary follicles) in her ovaries—and does not produce any new ova during her lifetime—fewer than 500 ova will actually be released. Under the influence of FSH, one (or more) ovum completes meiosis and increases in size as nutrients are added to its cytoplasm.

43–1 (continued)

Content Development
As you finish the discussion of the male reproductive system you may find that some of the students in the class are uncomfortable with the topic of sperm release. Concentrating on the numbers of sperm and the primary function of fertilizing the ovum should help take the emphasis away from the physical manifestations of ejaculation.

Content Development
Discuss the various structures and functions of the female reproductive system just as you did with the male system. It is important to discuss the process of ovulation because many students do not understand the

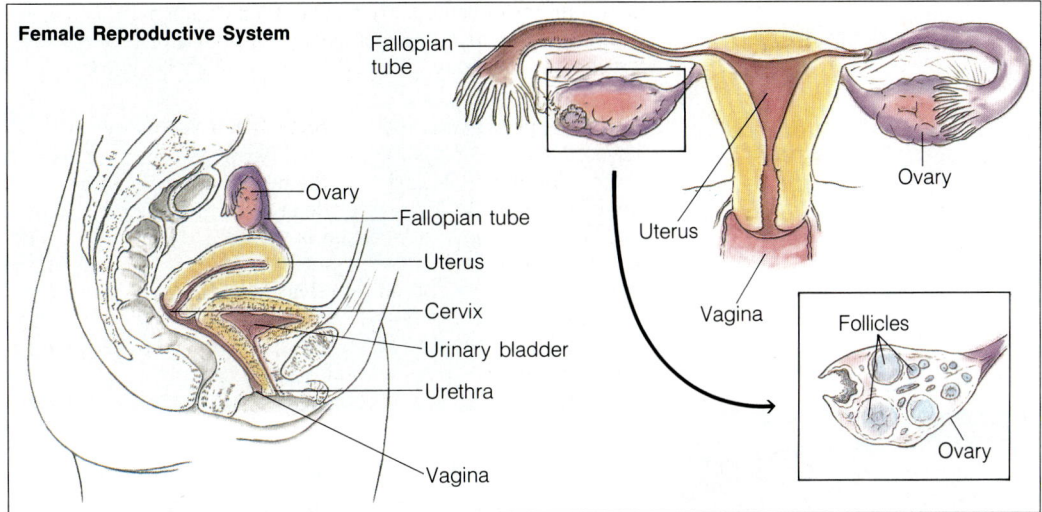

Figure 43–5 Deep within the abdomen, the female reproductive organs perform their dual function. They produce the eggs that could be fertilized by sperm and they prepare a place in which a fertilized egg can grow and develop.

OVULATION When a follicle has completely matured, the ovum is released. This process is called **ovulation**. The follicle literally ruptures, and the ovum is swept from the surface of the ovary into the opening of one of the two **Fallopian tubes**. The ovum moves through the fluid-filled Fallopian tube, pushed along by microscopic cilia attached to the cells that line the walls of the tube. It is during its journey through the Fallopian tube that an egg can be fertilized. After a few days, the ovum passes from the Fallopian tube into a larger cavity known as the **uterus**. The lining of the uterus is specially designed to receive a fertilized ovum, if fertilization has occurred. The uterus opens into a canal known as the **vagina**, which leads to the outside of the body.

Ovulation begins at puberty and usually continues until a female is in her late forties, when **menopause** occurs. After menopause, follicle development no longer occurs and a female is no longer capable of bearing a child.

The Menstrual Cycle

In females, the interaction of the reproductive system and the endocrine system takes the form of a complex series of periodic events called the menstrual cycle. This name is quite appropriate: The cycle takes an average of about 28 days, and the word menstrual comes from the Latin word *mensis*, meaning "month."

The **menstrual cycle**, which is controlled by hormones operating on a negative-feedback mechanism, involves the development and the release of an egg for fertilization and the

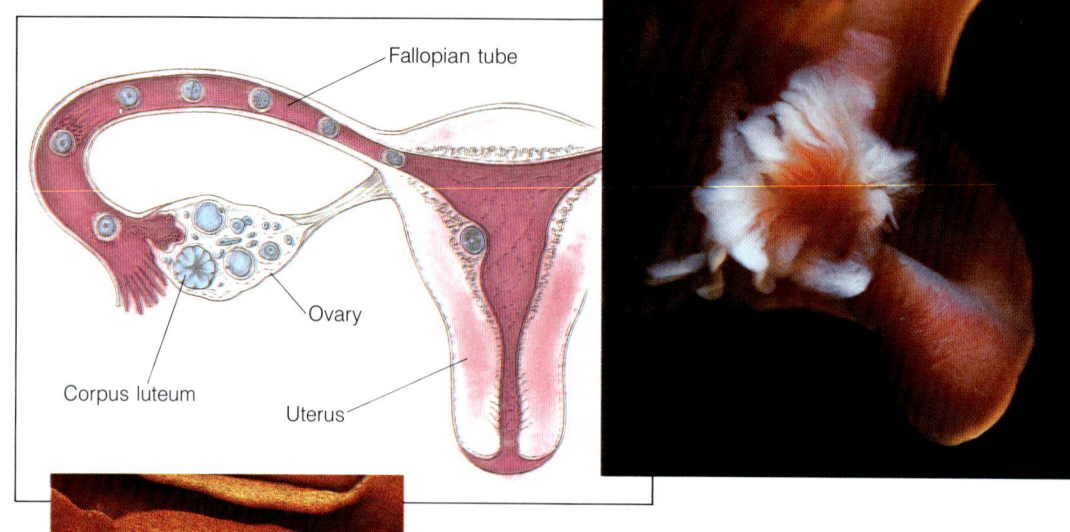

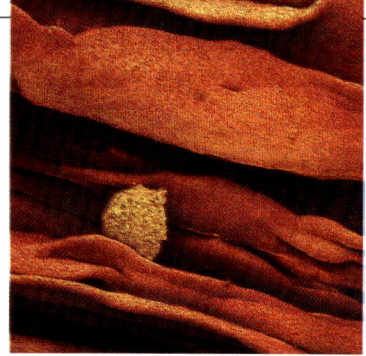

Figure 43–6 During ovulation, an ovum bursts from a ruptured mature follicle (top, left). It is then swept into the funnel-shaped opening of a Fallopian tube by fringelike petals called fimbriae (top, right). The ovum journeys toward the uterus along the folds of the 11-cm-long tube (bottom). It is during this passage that the ovum can be fertilized by a sperm.

preparation of the uterus to receive a fertilized egg. If an egg is not fertilized, it is discharged along with the lining of the uterus. The menstrual cycle has four phases: follicle phase, ovulation, luteal phase, and menstruation.

FOLLICLE PHASE The follicle phase begins when the level of estrogen in the blood is relatively low. The hypothalamus reacts to the low estrogen level by producing a releasing hormone that acts on the pituitary gland. The releasing hormone stimulates the pituitary gland to secrete FSH and LH into the blood. These two pituitary gland hormones travel through the circulatory system to the ovaries, where they cause a follicle to develop to maturity.

As the follicle develops, the cells surrounding the ovum enlarge and begin to produce increased amounts of estrogen. Estrogen levels in the blood rise dramatically as the follicle increases in size and produces more and more of the hormone. Estrogen causes the lining of the uterus to thicken in preparation for receiving a fertilized egg. The development of an ovum in this stage of the cycle takes about 10 days.

OVULATION This phase is the shortest in the cycle, occurring about midway through the cycle (14 days) and lasting about 3 to 4 days. During this phase, something (no one is certain what) causes the hypothalamus to send a large amount of releasing hormone to the pituitary gland. This in turn causes the pituitary gland to produce a sudden rush of FSH and LH. The release of these hormones has a dramatic effect on the follicle: it ruptures, and a mature ovum is released. This is the process known as ovulation.

LUTEAL PHASE The ruptured follicle, now known as the **corpus luteum** (KOR-puhs LOOT-ee-uhm), continues to release estrogen. Corpus luteum means yellow body in Latin, thus describing the color of the follicle after ovulation. Immediately following ovulation, the corpus luteum begins to release a new steroid hormone called **progesterone**. During the first 14 days of the cycle, rising estrogen levels have stimulated cell growth and tissue development in the lining of the uterus. Progesterone adds the finishing touches to that lining. Blood supply is increased, the tissue matures, and the lining is fully prepared to accept a fertilized ovum.

During the first two days of the luteal phase, immediately following ovulation, the chances of an egg being fertilized are greatest. This is usually from 10 to 14 days after the completion of the last menstrual cycle. If an ovum is fertilized by a sperm, the resulting zygote will start to divide. After several divisions, the ball of cells will implant itself in the lining of the uterus. Within a few days of implantation, the uterus and the growing embryo will release hormones that keep the corpus luteum functioning for several weeks. This allows the lining of the uterus to nourish and protect the developing embryo.

MENSTRUATION What happens if fertilization does not occur? Within 2 to 3 days of ovulation, the ovum passes through the uterus without implantation. The corpus luteum, left on the surface of the ovary, begins to disintegrate. As the old follicle breaks down it releases less and less estrogen and progesterone. The result is a decrease in the level of these hormones in the blood.

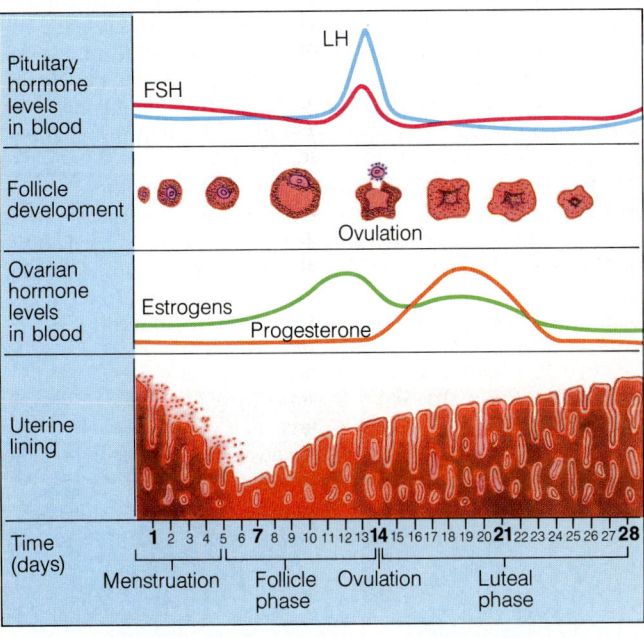

Figure 43–7 The menstrual cycle involves the development and release of an ovum for fertilization and the preparation of the uterus to receive a fertilized ovum. Note the relative concentrations of the various hormones and the corresponding condition of the uterine lining and the follicle. What are the four phases of the menstrual cycle? ❶

BACKGROUND INFORMATION
MENSTRUATION EXPLAINED

Biologists have long wondered about the purpose of menstruation. This event is complexly controlled, limits the time available to reproduce, and costs a woman a good deal of blood and energy. None of these effects appears to offer any benefit or advantage, which begs the question of how and why menstruation evolved.

The recent work of one biologist may provide the answer to this puzzle. Margie Profet of the University of California at Berkeley suggests that menstruation protects a woman from harmful microbes. According to Profet's hypothesis, bacteria and viruses attach themselves to sperm during sexual intercourse, then find a congenial home in a woman's uterus. Through menstruation, women eliminate any pathogens that are infecting their uterine lining.

What is unique about this hypothesis is that it offers a positive explanation for menstruation; previous explanations discounted it as an unnecessary and wasteful side effect. Profet says that the idea for her hypothesis came in a vivid dream: She dreamed of black triangles stuck in deep red tissue.

purpose of the whole cycle. To help students put all of these events into proper order, be sure to talk about the relative time involved for each stage. Stress that these times are not absolutes, and they vary considerably. The times the textbook gives are averages.

Reinforcement/Reteaching

If students have difficulty remembering terms, use flashcards instead of vocabulary lists. Have students put a term on each card, with a definition on the reverse side.

When the level of estrogen falls below a certain point, the lining of the uterus begins to detach from the uterine wall. This tissue along with blood and the unfertilized ovum are sloughed off, or discharged, through the vagina in the last (and externally visible) phase of the cycle, which is called **menstruation**. Menstruation lasts from 3 to 7 days, with an average of 4 days. At the end of menstruation, a new cycle begins.

A few days after menstruation, levels of estrogen in the blood are once again low enough to stimulate the hypothalamus. The hypothalamus produces a releasing hormone that acts on the pituitary gland, which then starts to secrete FSH and LH, and the menstrual cycle begins anew.

Fertilization

In order for an ovum to become fertilized, sperm must be present in the female reproductive tract—more specifically, in a Fallopian tube. Sperm are released during sexual intercourse, when semen is ejaculated through the penis of the male into the vagina of the female. The penis generally enters the vagina to a point just below the cervix, which is the opening that connects the vagina to the uterus. Sperm swim actively through the uterus and up into the Fallopian tubes. Although hundreds of millions of sperm are released during an ejaculation, only a few will reach the ovum in a Fallopian tube. And only a single sperm cell will fertilize the ovum.

The ovum is surrounded by a dense protective layer that contains receptor sites to which sperm bind. This binding causes a vesicle in the sperm head to rupture and release enzymes that break down the protective layer and form a pathway through which the sperm nucleus can reach the ovum. Once the sperm nucleus enters the ovum, the cell membrane of the ovum changes, thereby preventing other sperm from entering the cell.

The fertilized ovum is properly called a **zygote** after the two haploid (N) nuclei (one from the sperm cell and one from the egg cell) fuse to form a single diploid (2N) nucleus. The zygote will go through several rounds of cell division, and the ball of cells formed by those divisions will attach itself to the wall of the uterus and begin to grow into an embryo.

Figure 43–8 *Outnumbering an ovum by nearly 500 million to 1, fewer than 500 sperm survive their journey to the upper part of a Fallopian tube, where they attempt to penetrate the egg (left). Usually only 1 sperm fertilizes an ovum, plunging into its protective layer headfirst (center). As projections on the surface of the ovum pull the sperm inside, the sperm's tail disintegrates. Only the head and midpiece remain within the ovum (right).*

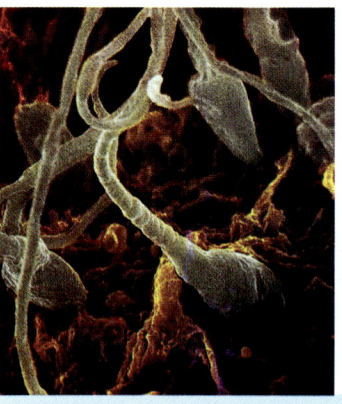
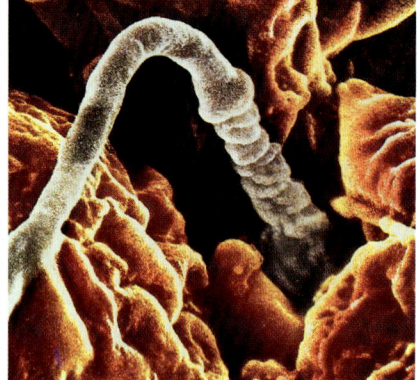
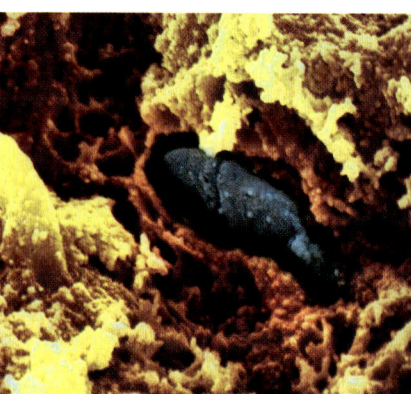

43-1 Section Review

1. What is the function of the reproductive system?
2. At what point in development do male and female embryos begin to differ from each other? What is the nature of this difference? What causes it?
3. Describe the changes that occur in males during puberty. In females. What hormones cause these changes?
4. Describe the phases of the menstrual cycle.
5. How does fertilization occur?
6. **Critical Thinking—Relating Facts** Explain how the shape of a sperm cell helps its function.

43-2 Human Development

When an ovum is fertilized, the remarkable process of human development begins. In the course of this process, a single cell no larger than the period at the end of this sentence will undergo a series of cell divisions that will result in the formation of a new human being.

Early Development

The first few **cleavages**, or mitotic cell divisions of the zygote, take place while the zygote is still in a Fallopian tube. Four days after fertilization, the embryo is a solid ball of about 50 cells that is called a **morula** (MOR-yoo-luh), after the Latin word for raspberry. As the embryo grows, a fluid-filled cavity forms in the center, transforming it into a hollow structure known as a **blastocyst**. About 6 or 7 days after fertilization, the blastocyst attaches itself to the wall of the uterus and begins to grow inward in a process known as **implantation**.

A cluster of cells gradually forms within the cavity of the blastocyst. This cluster sorts itself into two layers, which then produce a third layer, by a process of cell migration known as **gastrulation** (gas-troo-LAY-shuhn). See Figure 43–9. The result of gastrulation is the formation of three cell layers known as the **ectoderm, mesoderm,** and **endoderm**. These three layers are referred to as the primary germ layers because all of the organs and tissues of the embryo will be formed from them. See Figure 43–9.

During implantation, the outer layer of cells of the blastocyst produces two important membranes that surround, protect, and nourish the developing embryo. These membranes are the **amnion** and the **chorion**.

By the end of the third week of development, the nervous system has begun to form and so has a primitive tube that will

Guide For Reading

- What changes occur to a fertilized ovum as it develops into a fetus?
- What happens during childbirth?

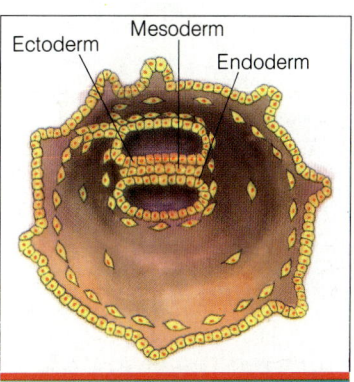

Figure 43–9 During gastrulation, the three primary germ layers, from which all the organs and tissues of the embryo will arise, form.

Primary Germ Layer	Develops into
Ectoderm	Epidermis of skin (hair and nails) Nervous system Tooth enamel Lining of nose and mouth
Mesoderm	Skeleton Muscles Excretory system Circulatory system Gonads
Endoderm	Digestive tract Respiratory system Liver and pancreas

uterine lining. Menstruation: If fertilization does not occur, the uterine lining detaches, and the unfertilized ovum is sloughed off and passes out through the vagina.

5. Fertilization occurs when a sperm passes through the vagina and reaches the egg in the Fallopian tube. The sperm binds to the ovum and releases enzymes that break down the protective layer around the ovum, forming a pathway in which the sperm nucleus can enter the ovum. Once this occurs, no other sperm can enter the ovum.

6. The sperm's long, flagellate tail serves as a powerful propeller. The sperm's compact head and slender shape help it move quickly through the female reproductive tract.

ANNOTATION KEY
① Blastocyst.
(Interpreting diagrams)

HISTORICAL NOTE
CLEAVAGE

Modern biology contains terminology inherited from the past. Cleavage is a good case in point. This term is traditionally used to describe the first few mitoses after fertilization of an ovum. Students sometimes wonder if cleavage is different from mitosis. The answer is no.

The first embryos to be studied in detail were those of the frog. The frog egg is large enough to be seen with a good magnifying glass, and biologists in the eighteenth and nineteenth centuries studied its development in detail. What they saw was a large mass splitting into two, then into four masses, and then into eight. Cleavage, or splitting, was a fair description of what they saw. What they could not know at the time was that this dramatic splitting was just a larger version (because of the massive size of the amphibian egg) of ordinary cell division, mitosis. Traditionally, we continue to apply the word *cleavage* to describe the initial cell division of the much smaller mammalian zygote, even though the word *mitosis* would be just as accurate.

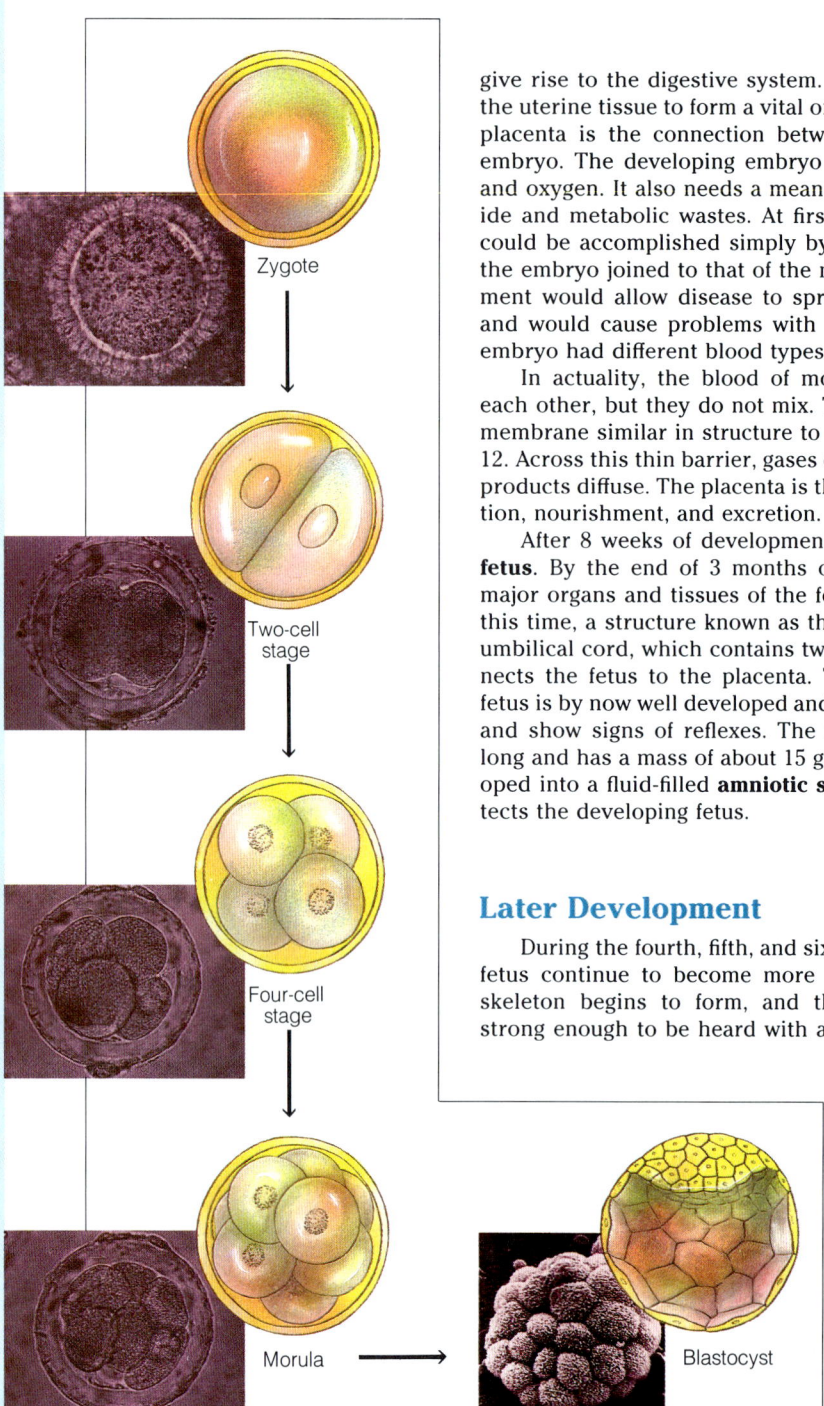

give rise to the digestive system. The chorion has grown into the uterine tissue to form a vital organ called the **placenta**. The placenta is the connection between mother and developing embryo. The developing embryo needs a supply of nutrients and oxygen. It also needs a means of eliminating carbon dioxide and metabolic wastes. At first thought, it seems that this could be accomplished simply by having the blood supply of the embryo joined to that of the mother. But such an arrangement would allow disease to spread from mother to embryo and would cause problems with immunity if the mother and embryo had different blood types.

In actuality, the blood of mother and embryo flow past each other, but they do not mix. They are separated by a thin membrane similar in structure to the one shown in Figure 43–12. Across this thin barrier, gases exchange and food and waste products diffuse. The placenta is the embryo's organ of respiration, nourishment, and excretion.

After 8 weeks of development, the embryo is known as a **fetus**. By the end of 3 months of development, most of the major organs and tissues of the fetus are fully formed. During this time, a structure known as the **umbilical cord** forms. The umbilical cord, which contains two arteries and one vein, connects the fetus to the placenta. The muscular system of the fetus is by now well developed and the fetus may begin to move and show signs of reflexes. The fetus is about 9 centimeters long and has a mass of about 15 grams. The amnion has developed into a fluid-filled **amniotic sac**, which cushions and protects the developing fetus.

Later Development

During the fourth, fifth, and sixth months, the tissues of the fetus continue to become more complex and specialized. A skeleton begins to form, and the fetal heartbeat becomes strong enough to be heard with a stethoscope. A layer of soft

Figure 43–10 As it develops into a fetus, a fertilized ovum, or zygote, goes through a series of mitotic cell divisions known as cleavages. At what stage does the developing embryo implant itself in the uterine wall? ①

43–2 (continued)
Content Development

Discuss the placenta's structure and function. Students may have many misconceptions about the protection and nourishment of the developing human fetus.

Emphasize that the blood of the mother does not intermingle with the blood of the fetus. Instead, the two supplies exist in separate circulatory systems that interface at the placenta. At this interface, oxygen and nutrients diffuse from maternal to fetal blood, and carbon dioxide and other wastes diffuse from fetal blood to maternal. In this simple manner, the placenta assumes the roles of the fetus's lungs, digestive tract, and kidneys.

• **A pregnant woman is said to be "eating for two." Is this description accurate?** (Yes. The fetus gets all of its nutrients directly from the mother. The expressions "breathing for two" and "eliminating waste for two" would also apply.)

Enrichment

If you have access to information, a speaker, or even a demonstration on fetal monitoring,

hair grows over the fetus's skin. As the fetus increases in size, the mother's abdomen swells to accommodate it. The developing fetus will be about 35 centimeters in length and have a mass of approximately 700 grams by the end of 6 months.

In many respects, the fetus is capable of leading a completely independent existence during the final 3 months it spends in the uterus. These months have an important purpose, however. The fetus actually doubles in mass, and the lungs and other organs undergo a series of changes that prepare them for life outside the mother. Premature babies, or those born before 8 months of development have been completed, have severe problems breathing because of incomplete lung development.

Childbirth

About 9 months after fertilization, at the end of a full-term pregnancy, the fetus is ready for birth. Exactly how this process is triggered is not known for certain. However, when the time comes, a hormone known as oxytocin is released from the pituitary gland. As you learned in Chapter 42, oxytocin affects a group of large involuntary muscles that surround the uterus. As these muscles are stimulated, they begin a series of rhythmic contractions known as **labor** that expand the opening of the cervix so that it will be large enough (about 10 centimeters) to allow the baby to pass through it.

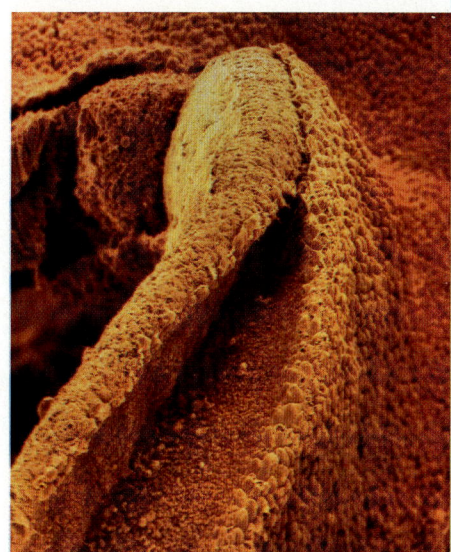

Figure 43–11 Within weeks of fertilization, a blastocyst no larger than a grain of rice begins to form specialized structures and take on shape. A groove that will develop into the nervous system is visible in this 3-week-old embryo.

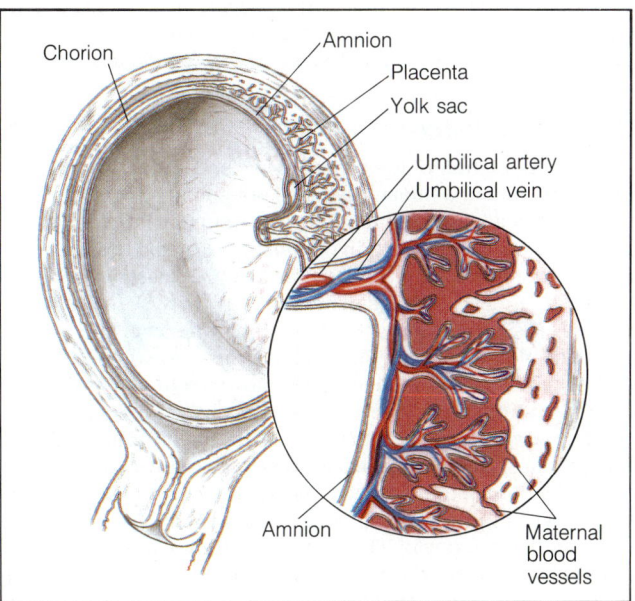

Figure 43–12 The placenta is the lifeline between mother and fetus. Through blood vessels in the umbilical cord, which connects the fetus to the placenta, food and oxygen from the mother and wastes from the fetus are exchanged.

---

follow through with it. This is such an interesting and amazing subject! You and your students will enjoy learning more about it. Prenatal monitoring is becoming fairly common in hospitals, as is monitoring of newborn babies—particularly premature babies. The at-home monitoring of babies with respiratory or neurological problems is also quite interesting and important.

Content Development

Once again, as you talk about childbirth, be sure to emphasize the role played by hormones. The fact that the onset of childbirth is triggered by an unknown stimulus is intriguing to students. Labor is another phenomenon that is not well understood.

TEACHING SUPPORT

Laboratory Manual
- Examining an Unfertilized Chicken Egg, p. 539

Teaching Resources
- Activity: Investigating Early Embryonic Development, p. 5

MEDIA AND TECHNOLOGY
Biology Transparencies
- 69: The Human Placenta and Its Related Structures

MULTICULTURAL STRATEGY

Have students bring in pictures of themselves when they were infants (less than 2 years of age). Make a collage of these pictures without identifying labels. Challenge students to identify as many of their classmates as they can. In what ways do we physically change between infancy and adolescence? What are some of the features that remain constant enough that we can use them for identification?

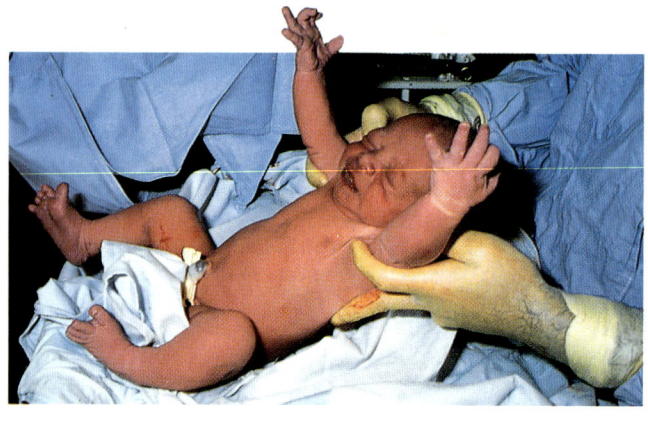

Figure 43–13 *Approximately 9 months after fertilization, the end product of reproduction emerges. Fresh to the world, this newborn is experiencing its first "weighing-in."*

As the contractions continue, they become more powerful and more frequent, occurring once every minute or two. Little by little, in a process that lasts from 2 to 16 hours, the baby is forced toward the vagina as labor continues. The amniotic sac breaks, and the fluid it contains rushes out of the vagina. The baby is finally forced out of the uterus and the vagina, usually head first, still attached to its mother by the umbilical cord.

As the baby meets the outside world, it may begin to cough or cry in order to rid its lungs of the fluid with which they have been filled. Breathing starts almost immediately, and the blood supply to the placenta begins to dry up. The umbilical cord is clamped and cut, leaving a small piece attached to the baby. This piece will soon dry and fall off, leaving a scar known as the navel—or in its more familiar term, the belly button. In a final series of uterine contractions, the placenta itself and the now-empty amniotic sac are expelled from the uterus as the afterbirth. The baby now begins to lead an independent existence.

After Childbirth

The interaction of the reproductive and endocrine systems in the creation of a human life does not end at childbirth, however. Within a few hours, a pituitary hormone known as prolactin stimulates the production of milk in the breast tissues of the mother. If a mother breastfeeds her baby regularly, several remarkable things happen: The milk is always ready when needed, it seldom runs dry, yet it stops whenever the mother decides to end breastfeeding. Exactly how does the mother's body "know" when to release milk and when to stop making it?

Upon stimulation, a series of nerve cells in the breast transmit impulses to the hypothalamus. The hypothalamus causes the pituitary gland to release nearly ten times the normal amount of prolactin. The increased level of prolactin enables milk production to keep up with demand.

Just think about it: The nervous system causes the endocrine system to produce a hormone that is important to the reproductive system. Remarkable!

SECTION REVIEW 43–2

1. About four days after fertilization the embryo consists of a ball of about 50 cells called the morula. As the embryo grows, a fluid-filled cavity forms in the center, transforming the morula into a hollow structure called the blastocyst.
2. Gastrulation is a process of cell migration and differentiation that forms the third of the three primary germ layers. These germ layers are the ectoderm, endoderm, and mesoderm.
3. The placenta grows from the chorion. The placenta is the connection between mother and embryo. Through the placenta, nutrients are passed from mother to embryo and wastes are passed from embryo to mother.
4. There are several reasons. Diseases and poisons could pass from the mother to the embryo if their blood were di-

SCIENCE, TECHNOLOGY, AND SOCIETY CONNECTION

The Intimate Spread of Disease

Unfortunately, the intimate contact that occurs during sexual intercourse provides an excellent opportunity for certain types of disease-causing organisms to be passed from one individual to another. A large number of sexually transmitted diseases (STDs) are spread in this way, and they represent an increasingly serious public health problem, especially to American teenagers.

Syphilis, gonorrhea, and chlamydia, for example, are diseases caused by bacteria that infect the male and female reproductive tracts. More than 1.4 million gonorrhea infections and more than 4 million chlamydia infections were recorded in the United States in 1992. Nearly half of these infections occurred in teenagers. Sexually transmitted diseases can have profound effects on the body. Syphilis can be fatal. Chlamydia and gonorrhea can destroy a woman's ability to have children. And the human papilloma virus, which causes sexually transmitted warts, has been linked to cancer of the cervix.

Many STDs can be treated with antibiotics. However, there is no cure right now for the most dangerous sexually transmitted disease, AIDS. This disease took the lives of more than 40,000 Americans last year.

All of these STDs, and others, are passed from one person to the next during sexual intercourse, and they present a clear danger to your health. How can you protect yourself? The people most vulnerable to STDs are those who have sex with many partners and do nothing to prevent infection during intercourse. Using a condom, which helps to prevent the passage of disease-causing organisms from person to person, provides a measure of protection against these diseases. The surest and safest form of protection, however, is to abstain from sexual intercourse.

Although seemingly carefree, teenage years are often difficult. For many teens, emotional maturity often lags far behind physical maturity.

43–2 SECTION REVIEW

1. Describe the development of a blastocyst.
2. What is gastrulation?
3. What is the placenta? Why is it important?
4. Why is it important to separate the blood supplies of mother and fetus?
5. What is oxytocin? What role does it play in childbirth?
6. **Connection—Health** Explain why it is important for a pregnant woman to have good health practices.

---

rectly connected. Also, if the mother's blood type were different from the embryo's blood type, there could be a reaction in which the embryo's blood would clump, and the embryo would likely die.
5. Oxytocin is a hormone released by the pituitary that causes the muscles surrounding the uterus to begin contracting, which is the onset of labor.
6. It is vital for the mother to have proper nutritional habits, since the embryo obtains all its nutrients from her. In addition, any poisons, drugs, and alcohol that the mother ingests could make their way to the infant through the placenta, possibly causing birth defects or death.

SCIENCE, TECHNOLOGY, AND SOCIETY CONNECTION

The Intimate Spread of Disease

For many teachers, the diseases spread by sexual intimacy are difficult to discuss. Fortunately, there are many sources of information available to students, including health textbooks, libraries, and health care professionals.

You can present this feature as a case study of adaptations—the adaptations certain disease-causing organisms show that enable them to find a new host. Many infectious diseases spread very easily. They move from person to person through food, drinking water, and even in the tiny water droplets spread when someone coughs or sneezes. However, the organisms that cause sexually transmitted diseases (STDs) are exquisitely sensitive to their environment. Vary even slightly the conditions they need to survive and they will die. For the most part, these organisms can spread only during intimate contact, usually sexual intercourse.

Since the development of antibiotics, most STDs are curable. However, one disease, AIDS, is not. Students must learn and practice the ways of avoiding STDs, especially AIDS.

Reinforcement/Reteaching
If students are unable to answer the Section Review questions, turn back to the appropriate material for review.

Closure
Have students prepare a time line to illustrate—from fertilization to birth—the events of early human development. Be sure students use the correct technical terms and discuss events that happen in both the fetus and the mother.

LABORATORY INVESTIGATION

EXAMINING MAMMALIAN GAMETES

Before the Lab
This lab requires very little material and setup. It would be a good idea to look at the slides before beginning the lab so that you will be able to give the students some hints. These are not the easiest slides to view, and students may have difficulty finding what you want them to see.

Pre-Lab Discussion
1. Review the structure and appearance of the sperm and ovary tissue.
2. Demonstrate the way students should use the ruler to measure the diameter of the microscope fields.
3. Give students any hints that will help them find what they are to view. If there are slides from more than one source, point that out to the students and tell them the differences.

Skills Development
Students will use the following skills in the completion of this laboratory investigation.
1. Manipulative
2. Observing
3. Communicating
4. Diagramming
5. Comparing
6. Inferring
7. Analyzing
8. Concluding
9. Synthesizing

Safety Tips
Remind students to be cautious, since the prepared slides are made of glass.

Teaching Strategy
Depending on the level of expertise of the students, this lab may require monitoring and helping students to actually find what they are to draw and describe.

LABORATORY INVESTIGATION: EXAMINING MAMMALIAN GAMETES

PROBLEM
How are mammalian gametes suited to their functions?

MATERIALS (per group)

microscope
transparent plastic metric ruler
prepared slides of a mammalian ovary and of mammalian sperm cells

PROCEDURE

1. Record the magnification of the microscope's low-power and high-power objectives.
2. Place the clear plastic metric ruler on the stage of the microscope. Using the low-power objective, focus on the ruler's millimeter marks. Measure the diameter of the low-power field of view. Record this measurement in decimal form.
3. Calculate the diameter of the high-power field of view according to the procedure described by your teacher. Record this value.
4. Calculate the diameter of the field of view for each objective in micrometers (μm) by multiplying each diameter in millimeters by 1000. Record the results.
5. Examine the slide of the ovary under low power. Notice the many round cavities in the ovary. These are maturing follicles that are swollen with fluid. Note the varying sizes of the ova (eggs) inside the follicles. In the mature follicles, the ovum is toward one side of the follicle and is surrounded by a layer of follicle cells.
6. Locate a mature follicle. Switch to high power and focus with the fine adjustment. Draw a diagram of the mature follicle and ovum.
7. Using the procedure described by your teacher, estimate the size of an ovum. Record the result under your diagram of the ovum.
8. Examine the slide of sperm cells under low power. When you locate some sperm cells, switch to high power and focus with the fine adjustment. Draw a diagram of a sperm cell.
9. Estimate the length of a sperm cell and the width of a sperm head. Record the results under your diagram.

OBSERVATIONS

Power of Objective	Magnification of Objective	Diameter of Field of View	
		Millimeters	Micrometers
Low			
High			

1. Describe the appearance of the ovum. What is the approximate diameter of a mature ovum?
2. How does the size of a mature ovum compare to that of an immature ovum?
3. Describe the appearance of a sperm cell. What is its approximate length? Width?

ANALYSIS AND CONCLUSIONS

1. Why do developing ova come in different sizes? (*Hint:* What is contained inside an ovum?)
2. Why are sperm cells and ova different in size?
3. How does the unique shape of a sperm cell help it perform its function?
4. How are mammalian gametes suited to their functions?

Observations
1. The ovum is large, round and grainy. It is about 120 microns in diameter.
2. The mature ovum is larger.
3. The sperm cell is small with a rounded head and a long tail. The sperm is about 50 microns long, and the head has a diameter of about 5 microns.

Analysis and Conclusions
1. The ova observed are in different stages. As an ovum matures, it fills up with yolk and becomes larger.
2. The egg is much larger because it contains yolk. The sperm's small size helps it move through the vagina toward the egg.
3. The sperm has a rounded head that contains the genetic material and a long flagellum to

STUDENT STUDY GUIDE

SUMMARIZING THE CONCEPTS

The key concepts in each section of this chapter are listed below to help you review the chapter content. Make sure you understand each concept and its relationship to other concepts and to the theme of this chapter.

43-1 The Reproductive System

- The reproductive system produces, stores, nourishes, and releases gametes.
- The testes are the primary male reproductive organs. The primary female reproductive organs are the ovaries.
- The menstrual cycle involves the development and release of an egg for fertilization and the preparation of the uterus to receive a fertilized egg. The menstrual cycle has four phases: follicle phase, ovulation, luteal phase, and menstruation.
- For an ovum to be fertilized, sperm must be present in the Fallopian tube of the female. The fertilized ovum is called a zygote.

43-2 Human Development

- Human development begins when an ovum is fertilized and commences a series of mitotic divisions, or cleavages.
- The placenta supplies the embryo with nutrients and oxygen and eliminates carbon dioxide and metabolic wastes.
- The embryo is known as a fetus after 8 weeks of development. The fetus is ready for birth about 9 months after fertilization.
- Within a few hours of childbirth, a pituitary hormone known as prolactin stimulates the production of milk in the breast tissues of the mother.

REVIEWING KEY TERMS

Vocabulary terms are important to your understanding of biology. The key terms listed below are those you should be especially familiar with. Review these terms and their meanings. Then use each term in a complete sentence. If you are not sure of a term's meaning, return to the appropriate section and review its definition.

43-1 The Reproductive System

reproductive system
testis
androgen
ovary
estrogen
puberty
gonad
scrotum
seminiferous tubule
testosterone
secondary sex characteristic
epididymis
vas deferens
seminal fluid
semen
urethra
penis
ovum
primary follicle
ovulation
Fallopian tube
uterus
vagina
menopause
menstrual cycle
corpus luteum
progesterone
menstruation
zygote

43-2 Human Development

cleavage
morula
blastocyst
implantation
gastrulation
ectoderm
mesoderm
endoderm
amnion
chorion
placenta
fetus
umbilical cord
amniotic sac
labor

CHAPTER REVIEW

CONTENT REVIEW

Multiple Choice
1. c 2. a 3. d 4. b
5. c 6. d 7. a 8. c

True or False
1. F Puberty
2. T
3. T
4. F urethra
5. F testes
6. F prolactin
7. T
8. F gastrulation

Word Relationships
1. primary germ layers; ectotherm is an animal that relies on environmental heat to regulate its body temperature
2. phases of the menstrual cycle; gastrulation is the process in which the blastula infolds to form the gastrula
3. names of hormones; semen is the combination of seminal fluid and sperm
4. structure of the male reproductive system; follicles are small anatomical sacs such as those in which ova develop in the ovary
5. terms relating to an embryo's development; menstruation is the monthly shedding of the uterine lining

---

help it swim.
4. The egg is large and filled with yolk to provide food for a developing embryo. The sperm is small and flagellated, so it is adapted to swimming to and fertilizing the egg.

Going Further: Enrichment
If you have more than one kind of mammalian ovary and sperm slide available, have the students view sperm cells from several types of mammals. Compare them in terms of size and structure. Perform the same activity with the ovary tissue. Have students write a paragraph comparing the cells, noting the similarities and differences.

CONCEPT MASTERY

1. For the first 6 weeks after fertilization, human male and female embryos are identical in appearance. During the seventh week, the gonads start producing the hormones that trigger development of the reproductive system. In male fetuses, the testes produce androgens, which induce certain embryonic tissues to develop into male reproductive organs. In female fetuses the ovaries produce estrogens, and female reproductive organs develop.
2. Puberty is a period of rapid growth and sexual maturation in males and females during which the reproductive system becomes fully functional.
3. Testosterone, produced by the testes, is the principal male sex hormone. It produces a number of secondary sex characteristics that appear in males at puberty. These include the deepening of the voice, growth of facial and body hair, broadening of the chest, and development of large muscles. Testosterone also stimulates the development of sperm, the male gametes.
4. Sperm are produced by special cells in the seminiferous tubules of the testes. They pass into the epididymis for maturation and storage, then enter the vas deferens, where they are combined with seminal fluid from three glands to form semen. The vas deferens merges with the urethra, and sperm are released through the penis during ejaculation. If ejaculation takes place during sexual intercourse, semen is released into the female's vagina just below the cervix. Sperm swim through the uterus and up into the Fallopian tube, where a sperm cell may fertilize an ovum.
5. The menstrual cycle is controlled by hormones (primarily estrogen, FSH, LH, and progesterone) that operate as a negative-feedback mechanism. The rise and fall of the levels of these hormones brings about the complex series of events in the menstrual cycle.
6. In a negative-feedback system, the last step or end product of a series of steps inhibits the first step. If fertilization occurs, for example, estrogen levels remain high. A high level of estrogen prevents the onset of the follicle phase, which begins when the estrogen level is low.
7. Gastrulation is the process of cell migration within the cavity of the blastocyst. Its result is an embryo with three primary germ layers—ectoderm, mesoderm, and endoderm. These layers are important because all the organs and tissues of the embryo will develop from them.

CHAPTER REVIEW

CONTENT REVIEW

Multiple Choice

Choose the letter of the answer that best completes each statement.

1. In male embryos, testes produce steroid hormones known as
   a. oxytocins.   c. androgens.
   b. prolactins.   d. estrogens.
2. A comma-shaped structure in which sperm mature and are stored is the
   a. epididymis.   c. semen.
   b. urethra.   d. vas deferens.
3. The hormone that stimulates cells within the ovaries to produce estrogen is
   a. progesterone.   c. LH.
   b. oxytocin.   d. FSH.
4. The cluster of cells that surrounds a single ovum and prepares it for ovulation is the
   a. ovary.   c. Fallopian tube.
   b. primary follicle.   d. uterus.
5. The development and release of an egg for fertilization and the preparation of the uterus to receive it is
   a. menopause.   c. the menstrual cycle.
   b. puberty.   d. implantation.
6. The chances of an egg being fertilized are greatest during
   a. ovulation.   c. the follicle phase.
   b. menstruation.   d. the luteal phase.
7. Four days after fertilization, the embryo consists of a ball of cells known as the
   a. morula.   c. placenta.
   b. blastocyst.   d. chorion.
8. Mitotic cell divisions of the zygote are
   a. implantations.   c. cleavages.
   b. morulas.   d. blastocysts.

True or False

Determine whether each statement is true or false. If it is true, write "true." If it is false, change the underlined word or words to make the statement true.

1. Menopause is a period of rapid growth and sexual maturation.
2. Male and female reproductive organs develop from the same embryonic tissues.
3. The principal male sex hormone is testosterone.
4. The vas deferens leads to the outside of the male body through the penis.
5. The ovaries consist of seminiferous tubules.
6. After childbirth, the production of milk in the breast tissues of the mother is stimulated by oxytocin.
7. In order for an ovum to become fertilized, sperm must be present in a Fallopian tube.
8. The cluster of cells within the blastocyst sorts itself into layers by implantation.

Word Relationships

In each of the following sets of terms, three of the terms are related. One term does not belong. Determine the characteristic common to three of the terms and then identify the term that does not belong.

1. ectoderm, ectotherm, mesoderm, endoderm
2. ovulation, follicle phase, gastrulation, menstruation
3. semen, FSH, estrogen, LH
4. testes, scrotum, follicles, seminiferous tubules
5. cleavage, blastocyst, implantation, menstruation

CONCEPT MASTERY

Use your understanding of the concepts developed in the chapter to answer each of the following in a brief paragraph.

1. Describe how the male and female reproductive organs develop from the same tissues in the embryo.
2. Describe the period known as puberty.
3. Discuss the importance of and the effects of the hormone testosterone on the male reproductive system.
4. Trace the path a sperm must travel from production to fertilization of an ovum.
5. How does the endocrine system affect the menstrual cycle?
6. Explain why the menstrual cycle is a negative-feedback mechanism.
7. Describe gastrulation and the importance of the primary germ layers.
8. Describe the development and importance of the amnion and chorion.
9. Why is reproduction important?

CRITICAL AND CREATIVE THINKING

Discuss each of the following in a brief paragraph.

1. **Drawing conclusions** The menstrual cycle is suppressed during pregnancy. Explain why this is important to the success of a full-term pregnancy.
2. **Making predictions** If a female does not produce sufficient amounts of FSH and LH, how will her ability to have a baby be affected?
3. **Forming a hypothesis** Two babies born at the same time are called twins. In the photos below, one set of twins is identical (they look exactly alike) and the other set is fraternal (they look as alike or as different as any two children of the same parents). Propose a hypothesis about the formation of each type of twins.
4. **Identifying relationships** Explain the importance of the luteal phase on a female's reproductive system.
5. **Identifying patterns** Although only a single sperm fertilizes an ovum, the presence of many sperm is beneficial to the process of fertilization. Based on what you know about a sperm's effects on an ovum, explain why this is true.
6. **Identifying relationships** Explain why it is important that the membrane of an ovum changes after a sperm has reached its nucleus.
7. **Relating concepts** Describe how each of the following represents an adaptation that helps to ensure successful fertilization: seminal fluid; production and release of millions of sperm; cilia lining the Fallopian tubes; small mass and long tail of a sperm.
8. **Drawing conclusions** Why is it that sperm can survive for only about 24 hours in the female body, whereas they can survive much longer in the male testes?
9. **Using the writing process** Prepare a poster that highlights the dangers of alcohol and drug abuse during pregnancy.

---

8. The amnion and chorion develop from the outer layer of cells of the blastocyst and protect and nourish the developing embryo.
9. An organism can survive without reproducing. However, reproduction is important to the survival of every species on Earth.

CRITICAL AND CREATIVE THINKING

1. At the end of the menstrual cycle—menstruation—the uterus loses its inner lining. An implanted embryo would very probably lose its attachment during this event. Moreover, hormone levels during pregnancy are different from those during a menstrual cycle in which a pregnancy does not occur. These varying hormone levels might destroy the embryo or cause it to form improperly.
2. The woman would probably be unable to become pregnant. Her ova would likely not mature when levels of FSH and LH are too low.
3. Identical twins form when a fertilized ovum splits into two developing embryos during cleavage, forming two embryos with the same genetic material. Fraternal twins develop when two separate ova are fertilized.
4. During the luteal phase, the ruptured follicle continues to release estrogen and begins to release progesterone. The rising levels of these hormones stimulate cell growth and tissue development in the lining of the uterus. Without a prepared uterine lining, a fertilized egg could not be implanted.
5. The release of many sperm better ensures that at least one sperm will make the difficult passage through the uterus into the Fallopian tube.
6. If the membrane did not change, more than one sperm might enter the ovum. This event could produce a polyploid embryo.
7. Seminal fluid: provides a liquid medium in which sperm are suspended; millions of sperm: help to ensure that at least one sperm will reach the ovum; cilia in Fallopian tubes: help sweep sperm toward the ovum as well as help the ovum pass down the Fallopian tube; small mass and long sperm tail: help the sperm to swim to the egg.
8. Within the scrotum, the sperm are kept about 2 degrees cooler than internal body temperature. The temperature within the female reproductive tract is not lower than normal body temperature and sperm can survive only a limited time at the higher temperature.
9. Posters should illustrate the many documented dangers of alcohol and drug abuse during pregnancy.

CHAPTER **44** *Human Diseases*

Section	Laboratory Investigations and Activities
44–1 The Nature of Disease pages 953–955 THEMES: Systems and Interactions, Evolution	**Teacher Edition** DEMONSTRATION: Human Disease and Death: Then and Now, p. 952D **Laboratory Manual** A Model for Disease Transmission, p. 543 **Teaching Resources** ACTIVITY: Investigating Bubonic Plague, p. 3
44–2 Agents of Disease pages 956–959 THEME: Unity and Diversity	**Student Edition** LABORATORY INVESTIGATION: Observing Bacterial Growth, p. 964 **Teacher Edition** DEMONSTRATION: Bacteria and Antibiotics, p. 952D
44–3 Cancer pages 960–963 THEME: Systems and Interactions	**Laboratory Manual** Relating Chronic Diseases and Nutrition, p. 547
Chapter Review pages 965–967	

Teacher Resources

Books

Benenson, Abram S., ed. *Control of Communicable Diseases in Man,* 14th ed. American Public Health, 1985.

Burke, Shirley R. *Human Anatomy for the Health Sciences,* 2nd ed. Wiley, 1985.

Gong, Victor, and Norman Rudnick, eds. *AIDS: Facts and Issues.* Rutgers University Press, 1986.

Jorm, Anthony F. *A Guide to the Understanding of Alzheimer's Disease and Related Disorders.* New York University Press, 1987.

Langone, John. *AIDS: The Facts.* Little, Brown, 1988.

CHAPTER PLANNING GUIDE

Other Activities	Media and Technology

Biotechnology Workbook
  A Protein's Change of Face, p. 35
Study Guide
  Section 44–1, p. 419	

Teaching Resources
  ACTIVITY: Identifying Agents of Disease, p. 5
Study Guide
  Section 44–2, p. 421	

Teaching Resources
  VOCABULARY REVIEW: Word Scramble, p. 1
Study Guide
  Section 44–3, p. 423	

Teaching Resources
  Chapter Test A, p. 11
  Chapter Test B, p. 15
	Computer Test Bank
	  Chapter Test, p. 461

Audiovisuals

The Body Against Disease. 3 filmstrips with cassettes. Ward.
The Skin. Film or video. MTI/Coronet.

Immunology: An Overview. Video. Teaching Films.

CHAPTER 44 Human Diseases

CHAPTER OVERVIEW

Infectious diseases, interruptions of the normal functioning of the body, are caused by various microorganisms called pathogens. Pathogens may include viruses, bacteria, fungi, and protozoans. When a pathogen invades the body of a host organism and causes damage to cells and tissues, an infection is said to have occurred. Pathogens may be spread from one infected person to another by means of the air, contaminated water or food, or direct contact. In some cases, the pathogens may be acquired from infected animals.

Although today we take for granted that microorganisms cause infectious diseases, scientists learned this fact relatively recently. In the nineteenth century, as a result of the work of Louis Pasteur, Robert Koch, and others, the germ theory of infectious disease was developed. At the heart of the theory are Koch's postulates, which may be used to show that a specific microorganism causes a specific disease.

Cancer in human beings is not considered to be an infectious disease because no clear link exists between oncogenes produced by viruses and the occurrence of cancer in humans. However, it is well known that environmental conditions, as well as a hereditary predisposition, may play a part in the acquisition of cancer. Some of these factors include exposure to radiation or carcinogenic chemicals. Cancer is usually treated by several of the following methods: surgery, radiation therapy, and drugs (or chemotherapy). The cure rate is highest in those cases in which the cancer is detected and treatment is obtained early in its development.

44–1 THE NATURE OF DISEASE

Section Focus 44–1

The purpose of this section is to discuss the meaning of the term *disease,* with special emphasis on infectious disease. The methods by which diseases may be spread, as well as the germ theory of infectious disease, are also included.

Disease is defined as a change other than an injury that interferes with the body's normal functioning. Diseases are caused by pathogens, or disease-causing organisms such as viruses, bacteria, fungi, and protozoans that invade the body and injure the host's cells and tissues.

Pathogens may enter the host organism in a variety of ways. For example, they may be transmitted by food, water, or air; they may enter through a wound in the skin; or they may be spread by sexual contact.

In the nineteenth century, the germ theory of infectious disease was developed, stating that infectious diseases are caused by microorganisms. Robert Koch also developed the rules, known as Koch's postulates, for determining the relationship between a disease and a specific causative organism.

Performance Objectives 44–1

1. Define disease.
2. Describe the methods by which diseases are spread.
3. Contrast infectious and noninfectious diseases.
4. Describe the germ theory of infectious diseases.
5. State Koch's postulates.

Science Terms 44–1

disease   p. 953
symptom   p. 953
infectious disease   p. 953
pathogen   p. 953
infection   p. 953
germ theory of infectious disease   p. 954
Koch's postulates   p. 954

44–2 AGENTS OF DISEASE

Section Focus 44–2

The purpose of this section is to describe the various pathogens that cause disease, their method of entry into the human body, and the diseases they cause.

Viruses infect a living cell by attaching to the surface of the host cell and inserting their genetic material (RNA or DNA) into the cell. The genetic material then replicates, becomes enclosed in a new protein coat, and forms many new viruses while destroying the host cell. AIDS, measles, and the common cold are all caused by viral infections.

Most bacteria are nonpathogenic, but the relatively few that are pathogens may either invade the cells of the host directly or may release toxins into the blood of the infected host. The bacteria known as rickettsias can grow only within a living cell and are frequently carried to their host by means of arthropods such as ticks, lice, or fleas. Diseases caused by bacteria include tetanus, typhus, and syphilis.

Fungi normally are nonpathogenic and many reside on the human skin, causing no ill effects. However, these fungi may suddenly begin to grow rapidly, resulting in infections such as ringworm and athlete's foot. Some fungi cause serious infections inside the body.

Infections by protozoans occur mainly in the tropical and subtropical regions of the world that provide the warm, moist environment that these pathogens need. Protozoans cause such diseases as malaria, amebic dysentery, and African sleeping sickness.

Performance Objectives 44–2

1. State the criteria for grouping infectious diseases.
2. List the names of the types of organisms that cause disease.
3. Describe the diseases caused by each type of infectious organism.

Science Terms 44–2

toxin   p. 958

CHAPTER PREVIEW

44-3 CANCER

Section Focus 44-3

The purpose of this section is to describe cancer as a disease, to explain some of the agents that cause cancer, and to describe some of the methods of treatment for cancer.

When cancer occurs, cells multiply uncontrollably and destroy healthy tissue, often forming a malignant tumor. Parts of the tumor may break away and be carried by blood and lymph to other tissue, where they begin to multiply. Eventually, normal tissue function is interfered with, causing a potentially life-threatening condition.

There seem to be not one but several possible causes of cancer, as well as an inherited predisposition toward the disease.

Viruses may play a role in the transmission of cancer by infecting normal cells with oncogenes, or cancer-causing genes, that may, at a later time, cause the abnormal cell growth associated with cancer.

Overexposure to radiation, including ultraviolet, X-ray, and nuclear radiation, may also result in cancer caused by changes in the structure of DNA, or mutations. The mutations produce abnormal cell growth. Overexposure to chemicals called carcinogens may also result in cancer. Carcinogens may be natural substances, such as aflatoxin, a mold that grows on peanuts, or they may be synthetic substances, such as chloroform or benzene.

Cancer is usually treated with some combination of three methods: surgery, radiation, and chemotherapy (use of drugs). The highest success rates for treatment occur when the cancer is detected and treated early.

Performance Objectives 44-3

1. Describe cancer.
2. Describe three likely causes of cancer.
3. List three methods of treating cancer.
4. Explain why cancer is not considered to be an infectious disease.

Science Terms 44-3

cancer   p. 960
tumor   p. 960
oncogene   p. 961
carcinogen   p. 962

DISCOVERY LEARNING

TEACHER DEMONSTRATIONS
Modeling

Human Disease and Death: Then and Now

You may want to perform this demonstration as an introduction to Section 44-1.

Prepare a chart showing the leading causes of death in human beings in the United States in 1900 and at the present time.

- **What were the leading causes of death in 1900?** (Students should see that cardiovascular disease and infectious diseases, such as pneumonia and tuberculosis, as well as accidents, topped the list in 1900.)
- **What are the leading causes of death at the present time?** (Cancer and cardiovascular disease; infectious diseases—aside from AIDS—are far down on the list.)
- **Which diseases have changed places the most since 1900?** (Infectious diseases; if students do not yet know this term, accept all reasonable descriptions.)
- **How can you explain the appearance of cancer at the top of the chart?** (Many types of cancer affect people in old age; people are now living long enough to acquire cancer, whereas they were more likely to die younger of an infectious disease in 1900. Students may also cite environmental pollution and the development of toxic chemicals for the rise of deaths from cancer.)

Bacteria and Antibiotics

You may want to perform this demonstration after students complete Section 44-2.

Obtain prepared petri dishes of nonpathogenic bacteria such as *E. coli, B. subtilis,* or *P. vulgaris.* Cut filter paper into small disks (2 or 3 mm in diameter) and soak these disks in a small amount of antiseptic solution, such as alcohol, iodine, antiseptic soap, or household cleaner. You may also want to obtain commercially prepared disks of antibiotics such as streptomycin, aureomycin, or tetracycline. Using a glass-marking pencil, divide the petri dish into four quadrants by marking the cover. Lifting the cover of the petri dish as little as possible, place a different antiseptic disk in the center of each quadrant using forceps. Close the cover, tape it in place, and mark the quadrant with the name of the antiseptic placed there. Keep the petri dish at room temperature or warmer and observe at least 24 hours later.

- **Why was it important to lift the cover of the petri dish as little as possible?** (To prevent airborne contaminants from falling into the petri dish.)
- **What do you observe where the antiseptic disk was placed?** (There should be a clear area without bacterial growth around at least some of the disks.)
- **What caused the clear area around the disks?** (The antiseptic or antibiotic inhibited the growth of bacteria.)
- **Why are the clear areas different sizes?** (The size of the clear area is an indication of the relative potency of the antiseptic or antibiotic.)

CHAPTER 44
Human Diseases

GUIDED ENQUIRY

Pose the following questions to students and have them record their responses. Point out that they will gain a better understanding of the key concepts if they read the chapter with these basic questions in mind. Upon completion of the chapter, pose the question again. Ask students to compare their initial responses with those they have developed after reading the chapter.

- What is an infectious disease?
- How are infectious diseases spread?
- What are some of the microorganisms that cause infectious diseases?
- What is cancer?
- What causes cancer?
- How is cancer treated?

INTRODUCING CHAPTER 44

Using the Visuals Ask students to look at the illustration that opens Chapter 44 and read the caption.
- **What kind of disease is bubonic plague?** (An infectious disease.)
- **Can you name any other infectious diseases?** (Accept all correct responses.)
- **Do we have epidemics today that are as severe as the Black Death?** (No.)
- **Why not?** (Knowledge of the causes and transmission of diseases, better sanitation and other preventive measures, better medicines available for treatment.)
- **How effective do you think the outfits were that the doctors wore to ward off the plague?** (Accept all reasonable answers. Probably covering the face and hands afforded some protection, but the herbs, perfumes, and spices they carried were probably not effective. Point out that although herbal medicine has been utilized since the beginning of civilization, it is not appropriate or effective in every situation.)

CHAPTER 44
Human Diseases

"Bring out your dead" was often heard throughout Europe as bubonic plague took its toll. Doctors frequently dressed in leather and filled the beaks of their outfits with perfumes and spices to ward off the disease (inset).

> Ring-a-ring o'roses,
> A pocket full of posies,
> A-tishoo! A-tishoo!
> We all fall down.

For all its apparent innocence and playfulness as a child's game, this rhyme originated from something fatally serious—the Black Death, or bubonic plague. This terrible disease ravaged London, England, from 1664 to 1665, killing more than 70,000 people—nearly one fourth of London's population.

In the rhyme, the "ring o' roses" relates to the round rosy rash that is one of the early symptoms of the disease. The phrase "pocket full of posies" stands for the spices or herbs people carried in their pockets to ward off the disease. The last lines, "A-tishoo! A-tishoo! We all fall down," refer to the deadly sneeze that occurred as the victim fell dead.

952

TEACHING STRATEGY 44–1

Focus/Motivation
Ask students to name a disease. Give as many students as possible an opportunity to respond. Write all the names on the chalkboard.

GUIDE FOR READING

After you read the following sections, you will be able to

44–1 The Nature of Disease
- Define disease and describe how disease may spread.
- State the germ theory of infectious disease.
- List Koch's postulates.

44–2 Agents of Disease
- Explain how infectious diseases are grouped.
- List some diseases caused by viruses, bacteria, fungi, and protozoans.

44–3 Cancer
- Define cancer.
- Discuss three main causes of cancer.
- List three methods of cancer treatment.

Journal Activity

YOU AND YOUR WORLD

To control the spread of diseases such as AIDS and drug-resistant tuberculosis, some people have proposed quarantines. Do you think this is a good idea? Why or why not? What practical or ethical problems do you perceive? What are the potential benefits?

44–1 The Nature of Disease

Guide For Reading
- What is disease? What is an infectious disease?
- How are diseases spread?
- What are Koch's postulates? Why are they important?

Any change, other than an injury, that interferes with the normal functioning of the body is a disease. The oldest human documents, some religious and some purely historical, make reference to **disease** and to the burdens that it has placed on human life. Different diseases can be recognized by their **symptoms,** or the changes they produce in the body.

Diseases can be caused by many different kinds of things. **Infectious diseases** are produced by **pathogens**. Pathogens are disease-causing microorganisms—such as viruses, bacteria, fungi, and protozoans. In this chapter we shall discuss several important infectious diseases as well as one noninfectious disease—cancer.

What Is Disease?

When the body is successfully invaded by a pathogen, we say that an **infection** has occurred. The numbers of microorganisms in the world around us are so large that infection is a daily event. But sickness is not a daily event because not all infections produce disease. Infectious disease results only when the growth of a pathogen begins to injure the cells and tissues of an infected person.

The relationship between a pathogen and the organism it infects is essentially that of a parasite and its host. You may recall from Chapter 17 that a parasite is an organism that obtains nutrition from the body of a host organism in a way that harms the host. The parasitic lifestyle of the pathogen enables it to take advantage of the host and to ultimately become dependent upon the host organism for its survival.

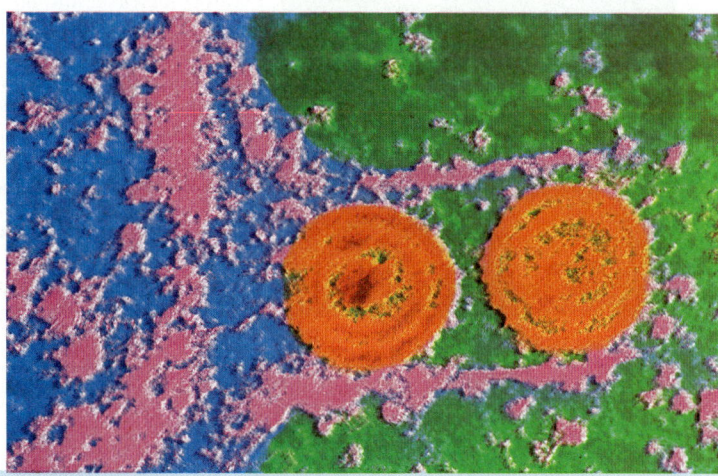

Figure 44–1 The world we live in is full of microorganisms, most of which have little effect on us. Some, however, such as these herpes simplex viruses, can cause disease.

COOPERATIVE LEARNING

Divide the class into five groups. Each group is to assume that they are the editorial staff of a national daily newspaper. Using the issue of banning the sale and use of smokeless tobacco by anyone under age 21, have groups produce the editorial page for their newspaper. When completed, the page should contain the following components:

- Background information (you may want to allow groups to do library research on the topic)
- Editorial cartoon (pro or con)
- Guest editorial supporting a ban on the sale and use of smokeless tobacco by anyone under age 21
- Guest editorial expressing a different viewpoint

It may be helpful to show students actual editorial pages of local and national newspapers as models. If time does not permit having each group complete the entire page, you may randomly assign each of the components to one group to produce a class page.

Journal Activity

YOU AND YOUR WORLD

Students may have strong feelings about quarantining sick people. Have students include their thoughts in their portfolio. You may also want to use this Journal Activity as a basis for a class discussion.

• **Why are these considered diseases?** (They all interfere with the normal functioning of the body but are not caused by injuries.)

• **Are there any differences in the way we acquire these diseases?** (Students should develop the idea that we acquire some of these diseases by "catching" them from others, whereas others are not acquired as a result of being transmitted from one organism to another.)

• **What do we call a disease that is transmitted from one person to another?** (An infectious disease.)

Content Development

Develop the idea that we are surrounded, and constantly invaded by, microorganisms that do not cause us to become sick. Our immune systems frequently enable us to "fight off" invading pathogens. Only when our cells or tissues are changed in sufficient numbers do we begin to feel ill.

Review the criteria for a parasitic relationship and develop the parallel for a pathogen and its host organism.

953

BACKGROUND INFORMATION
KOCH'S POSTULATES

Although Koch's postulates still guide microbiology and epidemiology, there are some important exceptions to these postulates. For example, many pathogens cannot be grown in culture, which makes it impossible to fulfill the second requirement of his postulates. This is true in the cases of syphilis and AIDS, to name just two diseases. Nevertheless, there is usually enough circumstantial evidence regarding each pathogen so that we can be certain of the identity of the microorganism that causes the disease.

TEACHING SUPPORT

Laboratory Manual
- A Model for Disease Transmission, p. 543

Teaching Resources
- Activity: Investigating Bubonic Plague, p. 3

Biotechnology Workbook
- A Protein's Change of Face, p. 35

Study Guide
- Section 44–1, p. 419

44–1 (continued)

Content Development
Discuss the various ways in which disease may be spread from one organism or person to another. Another term for airborne infections caused by coughing or sneezing is *droplet infection* because coughing or sneezing breaks up the mucus carrying the bacteria or virus into small droplets that may travel through the air for long distances as a result of the force of the sneeze. Emphasize that relatively simple sanitary measures, such as wearing shoes, washing cuts, covering the face when coughing or sneezing, and having sanitary water and sewer supplies will help to prevent infecting oneself or others.

Reinforcement/Reteaching
You may want to have your students develop a chart classifying various infectious diseases according to how they are spread.

Enrichment
Assign students to read individual chapters of the now-classic book *Microbe Hunters* by Paul deKruif and report on them to the class. Each chapter chronicles the work of such men as Louis Pasteur, Joseph Lister, Ed-

Figure 44–2 Even when the skin is freshly washed with soap and then rinsed clean, some areas may be home to millions of harmless bacteria (green rod-shaped objects).

How Is Infectious Disease Spread?

Many pathogens are present in the environment and require only the opportunity to enter the body to produce disease. For example, tetanus, often called lockjaw, is produced by the bacterium *Clostridium tetani*, which is present in soil. When the tetanus pathogen enters the body through a cut or puncture in the skin, it grows rapidly in the deepest parts of the wound, causing fever, muscle spasms, and sometimes death.

Some infectious diseases—such as the common cold, measles, mumps, and influenza—are spread from one person to another through coughing and sneezing. Other infectious diseases are spread through contaminated water supplies or through food that has been handled by people infected with a disease. Still other infectious diseases are spread by infected animals such as ticks and mosquitoes. Sexual contact is another way in which infectious diseases are spread.

The Germ Theory of Infectious Disease

For thousands of years people believed that diseases were caused by evil spirits, magic, or miasmas (vapors rising from marshes or decaying plant or animal matter). In fact, the word malaria is taken from the Italian words *mal aria,* meaning bad air. People feared that those who became ill were cursed or had brought bad luck with them.

Fortunately, a new idea developed in the nineteenth century that explained the origins of infectious diseases. Based on the work of the French chemist Louis Pasteur and the German bacteriologist Robert Koch, it was shown that infectious diseases were caused by microorganisms. This idea is now known as the **germ theory of infectious disease.**

Koch was a great pioneer in the study of disease. Not only did he help develop the germ theory of infectious disease, but he also discovered the bacteria that cause tuberculosis, a respiratory disease, and anthrax, a deadly disease that affects farm animals. As he investigated these diseases, Koch realized that the body of an infected person contains dozens of different microorganisms. Which of these microorganisms are responsible for the disease and which are not? Koch's experiments and observations led him to develop a series of rules for proving that a specific type of microorganism causes a specific disease. These rules are called **Koch's postulates:**

- The microorganism should always be found in the body of the host organism and not in a healthy organism.
- The microorganism must be isolated and grown in a pure culture away from the host.

Figure 44–3 This drawing of Robert Koch shows him hard at work in his laboratory investigating rinderpest, a disease that infects cattle.

954

SCIENCE, TECHNOLOGY, AND SOCIETY CONNECTION

The Magic Bullets

When Alexander Fleming discovered penicillin, he opened the door for scores of other scientists to discover weapons against infectious diseases. Penicillin is an antibiotic, or a substance produced by an organism that weakens or kills bacteria. Since Fleming's work in 1928, nearly a hundred different antibiotics have been discovered. Some of them, such as penicillin and tetracycline, are produced by fungi. Others, such as streptomycin, are produced by bacteria.

The action of all antibiotics is based on the differences that exist between bacteria and the cells of the body. For example, penicillin weakens the bacteria's cell walls, causing them to burst under osmotic pressure. Streptomycin and tetracycline, on the other hand, stop protein synthesis on bacterial ribosomes, yet these antibiotics do not affect the cytoplasmic ribosomes of body cells.

Although antibiotics are powerful and effective, they have two serious limitations. First, they act only against bacterial infections. Second, the widespread use of antibiotics has led to the evolution of strains of bacteria that are resistant to antibiotics. This development points out the fact that evolution is a continuing process. It also stresses the need to develop new bacterial-resistant antibiotics and to make wiser use of the ones we already have.

- When the microorganisms grown in pure culture are injected into a new host organism, they produce disease.
- The same microorganisms should be reisolated from the second host and grown in a pure culture, after which the microorganisms should still be the same as the original microorganisms.

Koch's postulates are still in use today. It is now clear, however, that even healthy individuals may harbor small amounts of disease-causing microorganisms without ill effects.

44–1 SECTION REVIEW

1. What is disease? How are diseases spread?
2. What is the germ theory? What are Koch's postulates?
3. **Critical Thinking—Designing an Experiment** Devise an experiment to determine whether a microorganism causes a certain infectious disease.

ANNOTATION KEY

① Spread by droplets or direct contact.
(Interpreting charts)
② *Mycobacterium tuberculosis.*
(Interpreting charts)

ECOLOGY NOTE
IMPORTED PATHOGENS

When foreign organisms are introduced into an ecosystem, the results can be disastrous. Some of the worst problems are caused by organisms that are invisible to the unaided eye—pathogens.

History records many examples of problems caused by introduced pathogens. The fungus that causes chestnut blight damaged both the forests and the timber industry in North America. The protozoan that causes avian malaria has brought native species of birds in Hawaii and Puerto Rico to the brink of extinction. Carried by European explorers, settlers, and missionaries, the viruses that cause smallpox and measles decimated populations of Native Americans and Pacific Islanders.

Guide For Reading
- How are infectious diseases classified?
- What are some examples of diseases caused by bacteria, fungi, and protozoa?

44–2 Agents of Disease

Most microorganisms have little interest in humans. A few microorganisms, however, find the human body an inviting home that is warm, protected, and chock-full of nutrients. These friendly microorganisms settle in certain parts of the body and live in harmony with it.

Unfortunately, some microscopic residents may invade body tissue and multiply within it or travel through the bloodstream to all parts of the body. If left unchecked, these invaders can cause serious, even fatal, illnesses.

Scientists group infectious diseases according to the kind of pathogen that causes them. The most common pathogens are viruses and bacteria. Fungi and protozoans also produce infections.

Viruses

As you may recall from Chapter 17, viruses are noncellular particles that invade living cells. Viruses contain genetic information in the form of RNA or DNA enclosed in a protein coat. Some viruses are also enclosed in a membrane. Nearly all living organisms, including bacteria, plants, insects, and mammals, can be infected by viruses.

Viruses show no lifelike activities unless they infect a living cell. In order to do so, a virus must become attached to the cell's surface and then insert its genetic material into the cell. Once inside, the viral genes may lie dormant for a period of time or they may go right to work producing new viruses and destroying the infected cell. Some serious diseases—such as AIDS (*A*cquired *I*mmune *D*eficiency *S*yndrome), poliomyelitis (polio), smallpox, and measles—are caused by viruses.

Figure 44–4 According to this chart of common viral diseases, by what methods is the virus that causes German measles spread? ①

SOME VIRAL DISEASES		
Disease	**Organism that Causes the Disease**	**Methods of Spreading the Disease**
Chicken pox	One virus	Droplets in air; direct contact with infected person
Common cold	Many viruses	Droplets in air; direct contact with infected person
German measles (rubella)	One virus	Droplet spread; direct contact with infected person
Infectious mononucleosis	Probably one virus	Spread by droplets; may be spread by direct contact
Influenza	Two important types (A,B) of virus and many subtypes	Direct contact with infected person; droplet infection; also may be airborne
Measles (rubeola)	One virus	Droplets in air; direct contact with secretions of infected person
Mumps	One virus	Droplet spread; direct contact with infected person
Pneumonia (viral)	Several viruses	Droplets; oral contact with infected person
Poliomyelitis	Three types of virus	Direct contact with infected person

TEACHING STRATEGY 44–2

Focus/Motivation
Remind students that in the previous section they learned about the development of our knowledge regarding the causes of various diseases, culminating in the germ theory of disease. In this section, specific "germs," or microorganisms, and the diseases they cause will be discussed.

Content Development
Using the Visuals Refer students to Figures 44–4 and 44–6.
• **Against which of the diseases listed have you received inoculations, or "shots"?** (Students should be able to name rubella, measles, mumps, diphtheria, tetanus, and whooping cough. Some may not realize that they had these inoculations as infants. List on the chalkboard the diseases named by the students.)
• **Have you or anyone you know ever had any of these diseases?** (In spite of the requirement that students have these inoculations to attend school, there still may be incidents of occurrence.)

Name the diseases listed in Figures 44–4 and 44–5 that were not mentioned by the students previously.
• **How many of you have had**

956

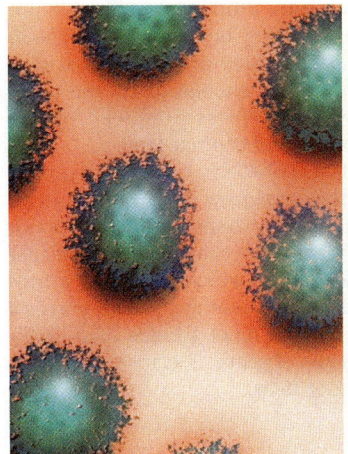
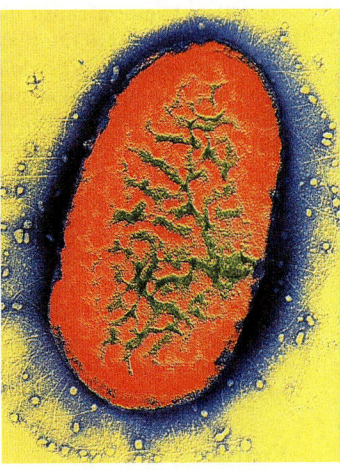
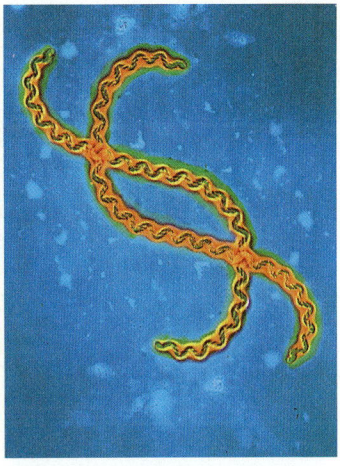

Bacteria

Contrary to popular belief, most bacteria are harmless to humans. The few bacteria that are pathogens produce disease in a variety of ways. Some bacteria infect the tissues of the body directly. For example, the bacterium that causes gonorrhea (*Neisseria gonorrhoeae*) grows in the tissues that line the male and female reproductive tracts. As the bacterium multiplies, it kills the cells in those passageways, causing irritation and bleeding (symptoms of the disease). Gonorrhea may also be transmitted from an infected mother to her child at birth.

Figure 44–5 *Adenoviruses are one of several types of viruses that cause colds (left). The bacterium* Bordetella pertussis *causes whooping cough (center). These spirochete bacteria cause leptospirosis (right).*

Figure 44–6 *According to this chart of bacterial diseases, what organism causes tuberculosis?* ❷

SOME BACTERIAL DISEASES

Disease	Organism That Causes the Disease	Methods of Spreading the Disease
Diphtheria	*Corynebacterium diphtheriae*	Contact with a patient or carrier; contaminated raw milk
Lyme disease	*Borrelia burgdorferi* (in U.S.)	Bite of infected tick
Meningitis	*Listeria monocytogenes*	Direct exposure to the organism
Pneumonia	*Streptococcus pneumoniae*	Droplets in air; direct oral contact with infected person
Rocky mountain spotted fever	*Rickettsia rickettsii*	Bite of infected tick
Scarlet fever	*Streptococcus pyogenes*	Droplets in air; direct contact with infected person or carrier; contaminated milk
Syphilis	*Treponema pallidum*	Sexual contact
Tetanus	*Clostridium tetani*	Dirty wound; usually a puncture wound
Tuberculosis	*Mycobacterium tuberculosis*	Droplets in air; contaminated milk and dairy products
Whooping cough	*Bordetella pertussis*	Droplets in air

(chicken pox, the common cold, pneumonia, . . .)? (Virtually all students will have had at least some of these diseases.)

• Thinking about the diseases for which no inoculation has been developed, why do we get some of them many times, yet we are likely to have others, such as chicken pox, only once? (Diseases such as the common cold and influenza are actually caused by many different viruses. When we catch a cold and recover, we have acquired immunity to only that specific virus.)

• What kinds of medicines were available to help you recover from these illnesses? (Students should be made aware that the antibiotics prescribed for bacterial infections are ineffective against viral infections.)

BACKGROUND INFORMATION
TETANUS

Although tetanus was described by Hippocrates 24 centuries ago, its prevalence has been masked by the fact that it strikes individuals and does not cause the widespread epidemics usually associated with infectious diseases. The organism that causes tetanus is the bacillus *Clostridium tetani*, found mainly in soil. It can enter the body through any break in the skin, from a superficial scratch to a drug addict's needle puncture. However, since it is an anaerobic bacterium, it is most likely to find favorable growth conditions in deeper tissues. It produces one of the most powerful toxins known and affects the nervous system, producing painful muscle contractions, especially those of the neck, jaw, and thoracic muscles, frequently causing death.

ESL STRATEGY

Provide students with a copy of the following exercise. You may also choose to write this exercise on the chalkboard.

A Little Spell of Sickness
Supply the missing letters in the following names of diseases. Then write down what organism causes each disease.

1. t __ t __ n __ s
2. m __ mp __
3. r __ beo __ __
4. __ neumo __ __ __
5. s __ ph __ l __ s
6. wh __ __ ping __ __ __ __
7. __ ip __ t __ eria
8. R __ __ __ y M __ __ __ t __ __ n spot __ __ __ __ ever
9. ty __ __ __ s
10. mal __ __ i __
11. __ ubercul __ __ __ __
12. botu __ __ __ __

Check the dictionary for two meanings for the word *spell*. Why are both meanings appropriate for this activity?

> **BACKGROUND INFORMATION**
> **THE WORLDWIDE NATURE OF DISEASE**
>
> China is home to many new strains of influenza viruses. Some scientists suggest that the pigs and ducks so common on China's farms provide breeding grounds for the viruses that cause influenza. The scientists think that mutations may, in time, enable these animal viruses to infect humans. Today, epidemiologists travel to China hunting not the big-game animals of safaris past but microscopic game that may produce tomorrow's diseases. Actively searching for new strains of influenza viruses may give physicians a head start in the development of vaccines. In a time of rapid worldwide travel, this head start might save lives.

> **TEACHING SUPPORT**
>
> **Teaching Resources**
> - Activity: Identifying Agents of Disease, p. 5
>
> **Study Guide**
> - Section 44–2, p. 421

> **44–2 (continued)**
>
> **Content Development**
> Although rickettsias are usually mentioned only in passing as a result of the small number of diseases for which they are responsible, point out to students that these diseases, including Rocky Mountain spotted fever and typhus, are often fatal, unlike most other bacterial infections. In fact, more people have been killed by rickettsial diseases than by any other single infection except malaria. Since rickettsias cannot live outside cells, they are spread by infected arthropods.
> - **In what parts of the world are pathogenic protozoans most commonly found?** (Tropical regions.)
> - **Explain your answer to the previous question.** (The warm, moist conditions in the tropics are ideal for the survival of protozoan parasites.)
> - **How can we prevent infection by rickettsias, fungi, or protozoans?** (Our best defense against these pathogens is to eliminate the unsanitary conditions in which many of them develop. Where there are arthropod carriers, elimination of the arthropods by pesticides would help to reduce the threat of infection by these organisms.)

Figure 44–7 African sleeping sickness is caused by the protozoan *Trypanosoma*, which is transmitted to humans by the bite of a tsetse fly (top). The discovery of these 35-million-year-old fossils of tsetse flies (bottom) in Colorado is evidence that the flies once existed in an area other than Africa.

Other bacteria cause disease by producing **toxins**, or poisons. In some cases the toxin itself may be enough to cause illness. For example, *Clostridium botulinum* produces a toxin so powerful that just a few drops of it is enough to kill the population of a small city. Botulism, the illness produced by this toxin, attacks the nervous system and is often fatal.

Like other bacteria, rickettsias (rih-KEHT-see-ahz) are prokaryotes. However, like viruses, they can grow only within a living cell. Rickettsias can be transmitted to humans by arthropods (animals with jointed legs). Ticks sometimes carry a disease called Rocky Mountain spotted fever, which produces a high fever, muscle pains, and headaches. These symptoms may last for as long as three weeks. A disease more serious than Rocky Mountain spotted fever is typhus. Typhus is transmitted from person to person by body lice or fleas.

Fungi

Most fungi, which include the molds and mushrooms, do not cause disease. Occasionally, however, the fungi that are normally present on the skin grow so rapidly that they can cause serious infections. The most common fungal infections are those caused by the dermatophytes (skin plants). Dermatophytes include those organisms that produce the rough, irritated patches on the toes known as athlete's foot and the scaly scalp infection known as ringworm. Fungal diseases are not limited to the skin—they can cause serious, even fatal, problems inside the body. Internal fungal diseases include the lung diseases known as San Joaquin Valley fever and histoplasmosis.

Protozoans

More than 30 different species of protozoans cause human disease. The most serious infections produced by these organisms are commonly found in the tropical regions of the world. There, the warm, moist surroundings provide the protozoans with ideal conditions for survival.

Because of the enormous numbers of people infected with malaria, this disease may be considered the most serious health problem affecting the human species. Malaria is caused by the protozoan *Plasmodium*, which lives in the bloodstream. Malaria is spread from one person to another by mosquitoes. The life cycle of the malaria parasite is discussed in Chapter 18.

The protozoan *Entamoeba histolytica* causes a disease known as amebic dysentery. Amebic dysentery affects the intestines and causes abdominal pain and fever. The disease-causing protozoan is found in contaminated water supplies.

African sleeping sickness is another disease caused by a protozoan—in this case, the flagellated protozoan known as *Trypanosoma*. African sleeping sickness is spread by the tsetse fly and is common in tropical regions of Africa. The disease results in an inflammation of the nervous system, including the brain. Once symptoms develop, the disease is usually fatal.

SECTION REVIEW 44–2

1. Infectious diseases are grouped according to the kind

44–2 SECTION REVIEW

1. How are infectious diseases grouped?
2. Give two examples each of diseases caused by viruses, bacteria, fungi, and protozoans.
3. **Critical Thinking—Applying Concepts** Why is it not possible for viruses to grow on agar?

SCIENCE, TECHNOLOGY, AND SOCIETY CONNECTION

Lyme Disease: An Ounce of Prevention

In 1975, Alan Steere and his colleagues at Yale University determined that a puzzling new disease was caused by a spirochete bacterium spread by tiny *Ixodes* ticks. They named the condition Lyme disease (LD) after the Connecticut town in which it was first observed.

Since its discovery, LD has become the most prevalent disease in the United States that is borne by a vector, or organism that transmits a pathogen from one host to another. The 9677 cases of LD reported in 1992 represent more than 90% of all reported vector-borne illnesses in this country.

Within 2 to 40 days of being bitten by an infected tick, a bull's-eye-like rash usually appears in the region of the tick bite. The rash is often accompanied by fatigue, fever, headache, and pains in the joints, back, bones, and muscles. If left untreated, such severe symptoms as arthritis, Bell's palsy (a form of facial paralysis), and inflammations of the heart and eyes may develop.

Fortunately, LD can be treated successfully with antibiotics. But as any person who has had LD can tell you, the best thing to do is avoid catching it in the first place. If you live in or visit any of the areas known to be home to the disease and the ticks that carry it—particularly the Northeast, the upper Midwest, and northern California—you should take certain precautions before venturing into wooded, bushy, or grassy areas. Use repellents according to the manufacturer's directions. Wear light-colored clothing so that you can spot ticks more easily. Wear long pants tucked into your socks and long-sleeved shirts tucked into your pants to minimize skin exposure. Check yourself for ticks and bites several times a day, and make sure you know what the ticks look like and how to remove them safely.

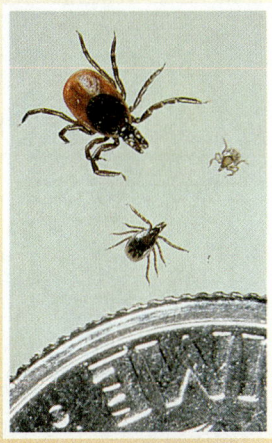

The female Ixodes *tick is larger than the male. Lyme disease is usually transmitted to humans by the even smaller nymph, rather than by the adults.*

Lyme Disease: An Ounce of Prevention

In 1992, almost 10,000 cases of Lyme disease were reported in 45 states. This vector-borne disease occurs most frequently on the northern Atlantic coast, in parts of the upper Midwest, and along most of the Pacific coast. The disease continues to spread as infected ticks increase in number and range each year.

Deer are considered to be the prime agent for spreading infected ticks, although ticks have been found on other animals, especially the white-footed mouse. Attempts to control ticks have been tried. Many of the control attempts have focused on limiting the deer population. This often meets with community resistance, however.

As present, the best offense seems to be a good defense. Be careful when you walk outdoors. Follow the cautions offered in this feature. Meanwhile, work is proceeding on vaccines that underwent limited trials in 1993.

BACKGROUND INFORMATION
CANCER TERMINOLOGY

There are several different types of cancerous tumors, or malignant neoplasms. Those that arise from epithelial tissue are called carcinomas and constitute the majority of all malignant tumors. Sarcomas are malignant neoplasms that originate from connective tissue, muscle, cartilage, fat, or bone. A third type affects the blood cells or the specific tissues that form them and are called leukemias and lymphomas.

FACTS AND FIGURES

Of all the deaths in the United States that are attributable to cancer, more than 60% are caused by only a few forms: leukemia and cancer of the lung, large intestine, breast, pancreas, prostate, and stomach. The remaining almost 40% is spread over more than 100 other forms of cancer. These figures do not necessarily apply to cancer deaths in other countries.

TEACHING SUPPORT
Laboratory Manual
• Relating Chronic Diseases and Nutrition, p. 547

Guide For Reading
■ What is cancer?
■ What causes cancer?
■ How is cancer treated?

44–3 Cancer

Cancer is a life-threatening disease in which cells multiply uncontrollably and destroy healthy tissue. Cancer is a unique disease because the cells that cause it are not foreign cells but rather the body's own cells. This fact has made **cancer** difficult to treat and to understand.

Cancer develops when something goes wrong with the normal controls that exist to regulate cell growth. A single cell or group of cells begins to grow and divide uncontrollably, often resulting in the formation of a **tumor**. A tumor is a mass of tissue. Some tumors are benign (bih-NIGHN), or noncancerous. A benign tumor does not spread to surrounding healthy tissue or to other parts of the body.

Cancerous tumors, on the other hand, are malignant. Malignant tumors invade and eventually destroy surrounding healthy tissue. In some cases, cells from a malignant tumor break away and are carried by blood and lymph to other parts of the body. As the cancer cells spread throughout the body, they absorb the nutrients needed by other cells, block nerve connections, and prevent the organs they invade from functioning properly. Soon the delicate balances that exist in the body are disrupted, and life-threatening illness results. There are many different forms of cancer, most of which take their name from the tissues in which cancer cells originate. Liver cancer, skin cancer, and bone cancer are some examples.

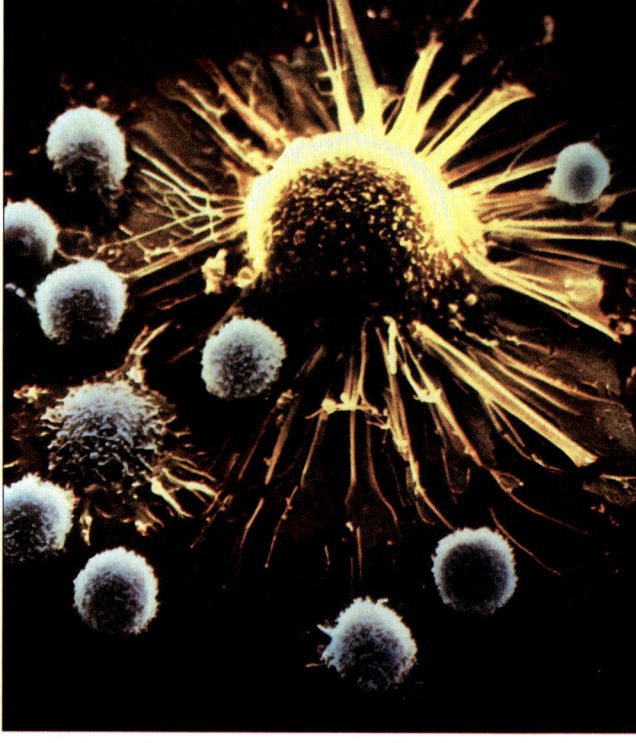

Figure 44–8 The crablike structure in this photograph is a cancer cell. The spherical objects surrounding the cancer cell are special white blood cells that are preparing to attack and destroy it.

TEACHING STRATEGY 44-3
Focus/Motivation
• A hundred years ago most people died of cardiovascular disease or an infectious disease. Now, cardiovascular disease accounts for about 50% of all deaths and cancer for about 20%. How can you explain these changes? (In the past hundred years, improvements in sanitation and the development of vaccines and antibiotics have all but eliminated most infectious diseases as a significant cause of death. People are living longer, and cardiovascular disease and cancer generally are diseases of old age.)

Content Development
•**What is cancer?** (A group of cells that proliferate abnormally and may invade surrounding healthy tissue and/or spread to other tissue through the blood or lymphatic system.)
•**What causes cancer?** (The causes of cancer are complex: In general, cancer seems to be the result of a change, or mutation, in the DNA of the affected cells,

Causes of Cancer

Cancers are caused by defects in the genes that regulate cell growth and division. There are several potential sources of such defects: They may be inherited, they may be caused by viruses, or they may result from mutations in DNA. These mutations in DNA may occur spontaneously, or they may be produced by radiation or chemicals.

VIRUSES Because most cancers seem to arise spontaneously, the causes of cancer have baffled scientists for decades. Many years ago, scientists took the first step toward understanding cancer when they learned that a few types of cancers in animals were caused by viruses. Scientists determined that the viruses contained cancer-causing genes, or **oncogenes.**

Remarkably, viral oncogenes are very similar to normal cellular genes that regulate cell growth, cell division, and metabolic pathways. In fact, you can think of oncogenes as altered or extra copies of normal genes. When oncogenes are brought into normal cells by cancer-causing viruses, the oncogenes upset the normal balances that seem to control cell growth. The virus-infected cells begin to multiply at an abnormally fast rate, and eventually produce tumors.

Despite the association of viruses with cancer, viruses are not the direct causes of most human cancers. Therefore, there is no way of "catching" cancer from a person who has it. The one notable exception is the human papilloma virus, which produces sexually transmitted genital warts and seems to be responsible for many cases of cervical cancer in women.

RADIATION Most forms of radiation—including sunlight, X-rays, and nuclear radiation—can cause cancer. The larger the amount of radiation, the greater the chance of developing cancer. Why does radiation cause cancer?

Radiation produces mutations, or changes in the structure of DNA, in living cells that are exposed to it. If these mutations occur in genes that control cell growth, a normal cell may be transformed into a cancer cell.

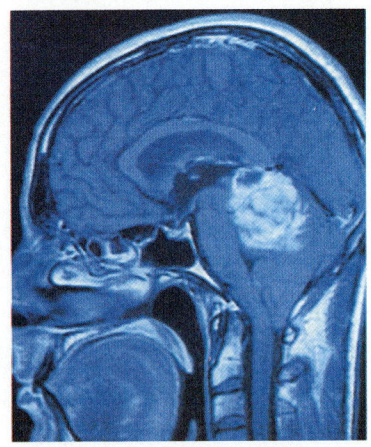

Figure 44–9 In this magnetic-resonance image, a large brain tumor appears as a round, light-colored mass in the cerebellum.

Figure 44–10 Overexposure to the ultraviolet radiation of sunlight on a repeated basis can cause skin cancer.

BACKGROUND INFORMATION
A NOBEL PRIZE

The 1989 Nobel prize in physiology or medicine was awarded to Dr. J. Michael Bishop and Dr. Harold E. Varmus, cancer researchers at the University of California Medical School in San Francisco.

In their research with cancer-causing viruses, or retroviruses, Drs. Bishop and Varmus worked with the Rous sarcoma virus, which causes cancer in chickens. They found that the viruses obtained cancer-causing genes from normal cells that had mutated. The viruses could then reproduce these cellular oncogenes and insert them into the DNA of normal cells, causing them to have the altered or uncontrolled growth pattern associated with malignancies. The initial mutation of the normal cell gene probably came about as a result of exposure to radiation or other environmental carcinogens.

---

causing them to divide at an abnormal rate or to divide at the wrong time. The mutation seems to be triggered by carcinogens in the environment, such as radiation or chemicals.)

• **Can viruses cause cancer?** (Viruses do not cause cancer directly, in the same sense that they cause the common cold. Viruses may carry oncogenes, which alter a normal cell's DNA so it may become a malignant growth. However, the oncogenes are usually suppressed by the cell's defenses until they are "switched on" by some environmental carcinogen.)

• **How does exposure to radiation cause cancer?** (There is a direct relationship between the amount of exposure to nuclear and solar radiation and the occurrence of some types of cancer. Radiation is known to cause mutations in DNA, which may lead to the development of cancerous growths.)

Most cases of skin cancer, for example, are caused by the ultraviolet radiation that is a normal part of sunlight. Over many years, constant exposure of the skin to this type of radiation can produce mutations in the DNA of skin cells. For this reason, it is important to avoid sunburn and excessive exposure to sunlight.

CHEMICALS Chemical **carcinogens**, or cancer-causing compounds, are some of the most important causes of cancer. Like radiation, carcinogens produce cancer by causing mutations in the DNA of normal cells. Some carcinogens, such as aflatoxin, are produced naturally by molds that grow on peanuts. Others, such as chloroform and benzene, are synthetic compounds. Some of these synthetic compounds may pollute the air or drinking water and therefore endanger entire communities. When this occurs, local, state, or federal agencies must step in to stop such practices.

Some of the most powerful chemical carcinogens are found in tobacco smoke. In the United States, cigarette smoking is responsible for nearly half the cancers that occur. Cigarette smoking causes most cases of lung cancer—the most fatal form of cancer.

Cancer Treatment

The most important weapon in the fight against cancer is early detection. If a cancer is detected early, the chances of treating it successfully are as high as 90 percent. **Physicians use three main methods to treat cancer: surgery, radiation**

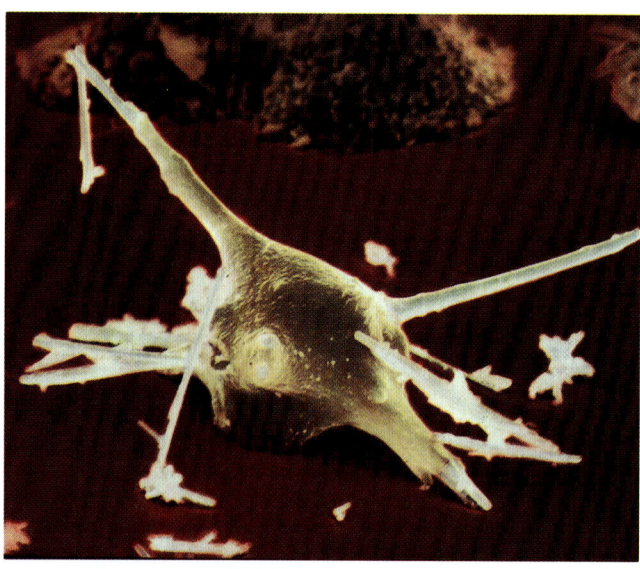

Figure 44–11 Some chemicals, such as asbestos, are carcinogenic, or cancer-causing. The asbestos fibers visible in this photograph are being engulfed by a special white blood cell.

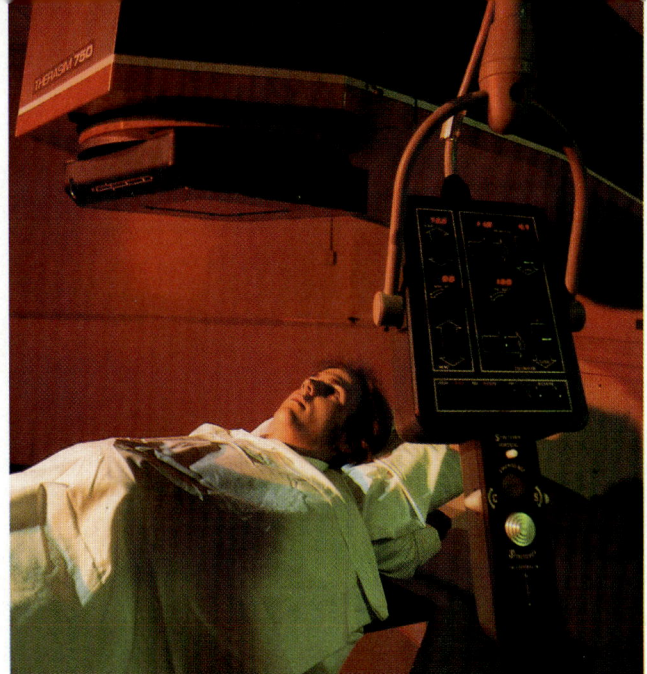

Figure 44–12 Radiation therapy is one method of treating cancer.

therapy, and drug therapy. In many cases, treatment is made up of two or sometimes all three of these methods.

Localized tumors, or those that do not spread, are often removed by surgery. If this can be done before the cancer has spread, the patient may be completely cured. Unfortunately, healthy tissue may also have to be removed to help prevent the disease from spreading.

Radiation therapy involves attacking cancer with radiation. Perhaps this seems confusing, as you just learned that radiation is also a cause of cancer. However, radiation destroys fast-growing cancer cells more quickly than normal cells, making this method a useful form of therapy.

Drug therapy, or chemotherapy, is the use of a combination of chemicals to destroy the fast-growing cancer cells. More than 50 drugs are used against a variety of cancers. Most of the anticancer drugs destroy cancer cells with as little harm to normal cells as possible. However, they do injure normal cells to some degree, producing some side effects such as nausea and high blood pressure.

44–3 SECTION REVIEW

1. What are the three main causes of cancer?
2. What are some methods of treating cancer?
3. **Critical Thinking—Classifying Diseases** Is cancer an infectious disease? Explain your answer.

LABORATORY INVESTIGATION

OBSERVING BACTERIAL GROWTH

Before the Lab
1. The day before the lab, ask students *not* to wash their hands immediately before coming to class.
2. Obtain sufficient laboratory materials for the entire class.

Pre-Lab Discussion
1. Have some prepared nutrient agar plates on which you have already cultured mold, bacteria, or both.
2. Making sure that the plates are taped shut, allow students to observe the plates so they will understand what to expect to see on their own cultures.
3. Discuss the procedure with students.

• **What are the variables that you will investigate in this lab?** (Students will be investigating the effect of temperature and cleanliness on the presence and growth of bacteria.)

• **Why is it necessary to tape shut the agar plates?** (Some of the fungi or bacteria that grow on the plates may be pathogenic, and it might be harmful to expose ourselves to large quantities of these bacteria.)

• **Why is it important to raise the cover only slightly when you touch your thumb to the surface?** (To prevent airborne mold spores or bacteria from falling on the agar.)

You may want to discuss the differences in the morphology of the different types of bacteria: *bacillus*, *spirillum*, and *coccus*.

Skills Development
Students will use the following skills while completing the laboratory investigation.
1. Safety
2. Manipulative
3. Observing
4. Comparing
5. Hypothesizing
6. Applying
7. Relating
8. Recording
9. Inferring

Safety Tips
Be sure to caution students *not* to open the covers of their agar plates once the plates have been inoculated by the thumbprint.

Teaching Strategy
Caution students that results will be better if they work quickly so that the partially raised cover is in this position for only a short time. Also, they do not need to press down hard on the agar; a light touch is sufficient.

Observations
1. The nutrient agar plate that was touched by the unwashed thumb and grown in the incuba-

---

LABORATORY INVESTIGATION

OBSERVING BACTERIAL GROWTH

PROBLEM
What conditions are needed for the growth of bacteria?

MATERIALS (per group)

2 nutrient agar plates
glass-marking pencil
hand soap
paper towel
transparent tape
refrigerator
incubator

PROCEDURE

1. Turn the nutrient agar plates upside down. With a glass-marking pencil, draw a line across the center of the plates dividing each in half. Label one half *WASHED* and the other half *UNWASHED*. Also label one plate *R* for refrigerator and the other *I* for incubator. Write the initials of a group member on each plate.
2. Slightly raise the cover of the agar plate labeled *R* and touch your thumb to the surface on the half labeled *UNWASHED*. **Note:** *Do not completely remove the cover from the plate.* Cover the plate immediately.
3. Repeat step 2 for the plate labeled *I*.
4. Thoroughly wash your hands with soap and water. Rinse and then dry your hands with a clean paper towel.
5. Repeat steps 2 and 3, touching your washed thumb to the surface of the agar on the half of each plate labeled *WASHED*.
6. Tape both plates closed with transparent tape. Place the plate labeled *R* in the refrigerator and the plate labeled *I* in the incubator at 37°C. Be sure that the plates are placed upside down. Allow them to remain undisturbed for 48 hours.
7. After 48 hours, examine both plates. Look for raised patches on the agar where you touched it. The raised patches are colonies of bacteria.
8. Examine the colonies of bacteria. Note the color and general appearance of individual colonies.

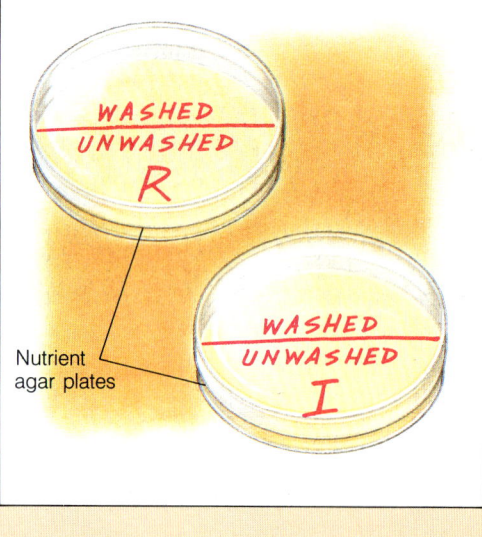

Nutrient agar plates

OBSERVATIONS
1. Which nutrient agar plate has more bacterial growth?
2. Based on color and general appearance, how many different types of bacteria appear to be growing on each agar plate?

ANALYSIS AND CONCLUSIONS
1. Why do you slightly raise the cover of the agar plate rather than remove it completely?
2. What evidence indicates the presence of bacteria on your skin?
3. Are conditions more favorable for the growth of bacteria inside or outside the body? Give evidence to support your answer.
4. In addition to food (which had been added to the agar), what other conditions are necessary for bacterial growth?
5. Describe some ways in which you might stop the growth of bacteria.

STUDENT STUDY GUIDE

SUMMARIZING THE CONCEPTS

The key concepts in each section of this chapter are listed below to help you review the chapter content. Make sure you understand each concept and its relationship to other concepts and to the theme of this chapter.

44–1 The Nature of Disease

- Any change other than an injury that interferes with the normal functioning of the body is a disease. Diseases can be recognized by their symptoms.
- The idea that infectious diseases are caused by microorganisms is known as the germ theory of infectious disease.
- Koch's postulates state: the microorganism should always be found in the body of the host organism and not in a healthy organism; the microorganism must be isolated and grown in a pure culture away from the host; when the microorganisms grown in pure culture are injected into a new host organism, they produce disease; and the same microorganisms should be reisolated from the second host and grown in a pure culture, after which the microorganisms should still be the same as the original microorganisms.

44–2 Agents of Disease

- Infectious diseases are grouped according to the kind of pathogen that causes them. Viruses, bacteria, fungi, and protozoans are some examples of pathogens.
- Some examples of diseases caused by viruses are AIDS, poliomyelitis, smallpox, and measles. Gonorrhea and botulism are examples of bacterial diseases.
- Rocky Mountain spotted fever and typhus are diseases caused by rickettsias.
- Common types of fungal infections include athlete's foot and ringworm.
- Malaria, amebic dysentery, and African sleeping sickness are caused by protozoans.

44–3 Cancer

- Cancer is a life-threatening disease in which cells multiply uncontrollably and destroy healthy tissue.
- Causes of cancer include viral oncogenes, radiation, and chemical carcinogens.
- Physicians use three main methods to treat cancer: surgery, radiation therapy, and drug therapy.

REVIEWING KEY TERMS

Vocabulary terms are important to your understanding of biology. The key terms listed below are those you should be especially familiar with. Review these terms and their meanings. Then use each term in a complete sentence. If you are not sure of a term's meaning, return to the appropriate section and review its definition.

44–1 The Nature of Disease
disease
symptom
infectious disease
pathogen
infection
germ theory of infectious disease
Koch's postulates

44–2 Agents of Disease
toxin

44–3 Cancer
cancer
tumor
oncogene
carcinogen

---

tor has more growth.
2. Answers will vary, but students will likely note a variety of bacteria.

Analysis and Conclusions
1. Slightly raising the cover minimizes the exposure of the agar to air, which contains bacteria.
2. Bacterial growth occurred on the agar where it was touched by the thumb.
3. The warm temperature (37°C) and availability of food and moisture inside the body are extremely favorable conditions for the growth of bacteria—more favorable than those on the skin's surface.
4. Warmth and moisture.
5. Washing with soap, using disinfectants and antiseptics, heating or freezing, and so on.

Going Further: Enrichment
If you have microscopes with oil-immersion lenses available, you may want to obtain cultures of nonpathogenic bacteria and have students stain these cultures with a simple bacterial stain (such as crystal violet) and observe them under high power.

CHAPTER REVIEW

CONTENT REVIEW

Multiple Choice

1. a 2. c 3. d 4. b
5. c 6. a 7. c 8. c

True or False

1. T
2. F bacterium
3. F germ theory of disease
4. F ultraviolet radiation or sunlight
5. F virus
6. T
7. F Carcinogens
8. F malignant

Word Relationships

A.
1. fungus
2. cancer
3. malignant
4. Rocky Mountain spotted fever *or* typhus

B.
5. diseases caused by viruses; gonorrhea is caused by a bacterium
6. diseases caused by microorganisms; cancer is caused by radiation, viruses, and carcinogens
7. methods of treating cancer; viruses are not
8. diseases caused by bacteria; malaria is caused by a protozoan

CONCEPT MASTERY

1. Koch's postulates enabled scientists to determine whether a specific microorganism caused a disease. This information has helped in the development of proper treatment for specific diseases.
2. African sleeping sickness is caused by the flagellated protozoan known as *Trypanosoma*, which is spread by the tsetse fly. The disease results in inflammation of the nervous system. Amebic dysentery is caused by another protozoan, called *Entamoeba histolytica*. Amebic dysentery affects the intestines and causes abdominal pain and fever. The disease-causing protozoan is found in contaminated water supplies.
3. Surgery is the most successful method for the removal of tumors localized in one area. With surgery, only the tumor is removed, with little loss of healthy tissue, and the spread of cancer is usually stopped. A disadvantage of this method is that the tumor may be in an inoperable area. Radiation therapy is advantageous because it can destroy fast-growing cancer cells more quickly than normal cells. However, some normal cells are destroyed. Chemotherapy is used to destroy cancer cells with little injury to normal cells. Nevertheless, the drugs injure normal cells to some

---

CHAPTER REVIEW

CONTENT REVIEW

Multiple Choice

Choose the letter of the answer that best completes each statement.

1. An infectious disease is caused by a (an)
   a. pathogen.   c. infection.
   b. symptom.   d. tumor.
2. What occurs when the body is invaded by a pathogen?
   a. vaccine   c. infection
   b. toxin   d. cancer
3. Which disease is caused by a virus?
   a. botulism   c. malaria
   b. athlete's foot d. measles
4. The poisons produced by some bacteria are called
   a. antibiotics.   c. pathogens.
   b. toxins.   d. oncogenes.
5. Which disease is caused by a rickettsia?
   a. AIDS   c. typhus
   b. cancer   d. mumps
6. A mass of tissue in the body is called a (an)
   a. tumor.   c. infection.
   b. pathogen.   d. toxin.
7. Chemical compounds that are cancer-causing are called
   a. oncogenes.   c. carcinogens.
   b. pathogens.   d. toxins.
8. Which is used in the treatment of cancer?
   a. carcinogens   c. radiation
   b. viruses   d. antibiotics

True or False

Determine whether each statement is true or false. If it is true, write "true." If it is false, change the underlined word or words to make the statement true.

1. Diseases caused by microorganisms are called <u>infectious diseases</u>.
2. Tetanus is caused by a <u>virus</u>.
3. The idea that infectious diseases are caused by microorganisms is known as <u>Koch's postulates</u>.
4. Skin cancer often results from constant overexposure to <u>carcinogens</u>.
5. Smallpox is caused by a <u>protozoan</u>.
6. Ringworm is caused by a <u>fungus</u>.
7. <u>Antibiotics</u> are cancer-causing compounds.
8. Cancerous tumors are also called <u>benign</u> tumors.

Word Relationships

A. *An analogy is a relationship between two pairs of words or phrases generally written in the following manner: a:b::c:d. The symbol : is read "is to," and the symbol :: is read "as." For example, cat:animal::rose:plant is read "cat is to animal as rose is to plant."*

In the analogies that follow, a word or phrase is missing. Complete each analogy by providing the missing word or phrase.

1. tetanus:bacterium::athlete's foot: _____
2. infectious disease:common cold::noninfectious disease: _____
3. noncancerous:benign::cancerous: _____
4. fungus:ringworm::rickettsia: _____

B. *In each of the following sets of terms, three of the terms are related. One term does not belong. Determine the characteristic common to three of the terms and then identify the term that does not belong.*

5. polio, measles, gonorrhea, smallpox
6. cancer, botulism, AIDS, typhus
7. surgery, radiation, drugs, viruses
8. tetanus, malaria, botulism, gonorrhea

CONCEPT MASTERY

Use your understanding of the concepts developed in the chapter to answer each of the following in a brief paragraph.

1. Discuss the importance of Koch's postulates in explaining the origins of infectious disease.
2. Compare the transmission of African sleeping sickness and amebic dysentery.
3. Discuss the advantages and disadvantages of the three methods used to treat cancer.
4. Compare a benign and a malignant tumor.
5. Describe three ways in which infectious diseases are spread.
6. List four types of pathogens that are responsible for the spread of infectious disease. Describe each and give one example of a disease that the specific pathogen may cause.

CRITICAL AND CREATIVE THINKING

Discuss each of the following in a brief paragraph.

1. **Interpreting diagrams** The chart shows the incidence and survival rates of some cancers in the United States. Which type of cancer has the best survival rate? The worst? Why do you think the five-year survival rates increased between 1960 and 1963 and between 1977 and 1983?

Five-Year Survival Rates		
Site of Cancer	1960–63	1977–83
Digestive tract		
Stomach	9.5%	16.0%
Colon and rectum	36.0%	46.0%
Respiratory tract		
Lung and bronchus	6.5%	12.0%
Urinary tract		
Kidney and other urinary structures	37.5%	51.0%
Reproductive system		
Breast	54.0%	68.0%
Ovary	32.0%	38.0%
Testis	63.0%	74.5%
Prostate gland	42.5%	63.0%
Skin (melanoma only)	60.0%	79.0%

2. **Applying concepts** Explain why you should not go to school with influenza.
3. **Developing a hypothesis** A doctor suspects that an apparently new disease is caused by a bacterium. Describe a set of procedures for proving this hypothesis.
4. **Making predictions** Some people think that it will be possible to wipe out all diseases some day. Do you think that this is possible? Explain your answer.
5. **Relating facts** Why is cancer more of a problem today than it was 100 years ago?
6. **Using the writing process** The incidence of sexually transmitted diseases in the United States is on the rise. Prepare an advertising campaign in which you alert people to the serious medical consequences of these diseases.

967

degree and produce undesirable side effects, ranging from nausea to high blood pressure.
4. A benign tumor is a tumor that will not spread; a malignant tumor is a tumor that can spread, or metastasize.
5. Infectious diseases are spread through direct contact with an infected person, contaminated food or water, and infected animals.
6. The four types of pathogens are viruses, bacteria, fungi, and protozoans. Refer to pages 956–959 for an example of a disease caused by each specific pathogen.

CRITICAL AND CREATIVE THINKING

1. Using the more recent data, melanoma of the skin has the best survival rate, whereas cancers of the lungs and bronchus have the worst rates. New and better methods for the treatment of cancer have been discovered and developed, thus increasing the survival rates.
2. Influenza can be transmitted from one person to another. Also, it is important to rest when you have influenza.
3. Students' answers should show a logical development of procedures. They should also follow Koch's postulates.
4. Although answers may vary, students should show some knowledge of disease, its transmission, and treatment.
5. Cancer is more of a problem today than it was 100 years ago because of the increase in carcinogens, especially those found in cigarette smoke. Also, in the past 30 to 40 years, more people have been overexposed to the ultraviolet radiation of sunlight. Furthermore, as people live longer, their chances of developing cancer increase.
6. Students' advertising campaigns should include not only gonorrhea but other sexually transmitted diseases, such as syphilis, AIDS, and herpes simplex type 2.

CHAPTER 45 Immune System

Section	Laboratory Investigations and Activities
45–1 Nonspecific Defenses pages 969–971 THEMES: Unity and Diversity, Systems and Interactions	**Teacher Edition** DEMONSTRATION: People with AIDS, p. 968D **Teaching Resources** HANDS–ON ACTIVITY: It's No Skin Off Your Nose, p. 207 **Product Testing Activities** Testing Bandages
45–2 Specific Defenses pages 972–976 THEMES: Systems and Interactions, Patterns of Change	**Student Edition** LABORATORY INVESTIGATION: Simulating Clonal Selection of Antibodies, p. 982 **Teacher Edition** DEMONSTRATION: B-Cells, Antibodies, and Antigens, p. 968D **Laboratory Manual** Lysis by a Bacteriophage, p. 553 The Effect of Alcohol on the Growth of Microorganisms, p. 559
45–3 Immune Disorders pages 977–981 THEME: Systems and Interactions	
Chapter Review pages 983–985	

Teacher Resources

Books

Davis, Joel. *Defending the Body.* Atheneum, 1988.

Dennis, Frank, ed. *Human Immunity to Viruses.* Academic Press, 1983.

Hood, L. E., I. L. Weissman, and W. B. Wood. *Immunology*, 2nd ed. Benjamin-Cummings, 1982.

Jeljaszewicz, J., and G. Pulverer, eds. *Antimicrobial Agents and Immunity.* Academic Press, 1986.

Kimball, J. W. *Introduction to Immunology.* Macmillan, 1983.

Mizel, Steven B., and Peter Jaret. *The Human Immune System: The New Frontier in Medicine.* Simon & Schuster, 1986.

Nilsson, Lennart, with Jan Lindberg. *The Body Victorious.* Delacorte Press, 1987.

CHAPTER PLANNING GUIDE

Other Activities	Media and Technology
Study Guide Section 45–1, p. 427	
Teaching Resources ACTIVITY: Identifying the Key Cells of the Immune System, p. 3 ACTIVITY: Examining the Development of a Vaccine, p. 5 **Biotechnology Workbook** Measuring with Molecules, p. 87 **Study Guide** Section 45–2, p. 429	
Teaching Resources VOCABULARY REVIEW: Crossword, p. 1 **Study Guide** Section 45–3, p. 431	
Teaching Resources Chapter Test A, p. 11 Chapter Test B, p. 15	**Computer Test Bank** Chapter Test, p. 471

Audiovisuals

Our Immune System. Film or video. National Geographic Society, Educational Services.
The Immune System. 2 filmstrips with cassettes. Educational Dimensions Group.

The Body Against Disease. 3 filmstrips with cassettes. Ward.
The Skin. Film or video. MTI/Coronet.
Immunology: An Overview. Video. Teaching Films, Inc.

The Immune Response and Immunization. Film or video. Benchmark Films, Inc.

968B

CHAPTER 45 Immune System

CHAPTER OVERVIEW

In this chapter students will learn how the immune system protects the human body against disease. The first two sections of the chapter describe the nonspecific and specific defenses that the immune system uses to combat infection. The third section discusses some of the disorders that can result when the immune system does not function properly.

45–1 NONSPECIFIC DEFENSES

Section Focus 45–1

The purpose of this section is to describe the immune system's nonspecific defenses, which are directed at all pathogens that attempt to enter the body. The first line of nonspecific defense includes the skin; mucus and cilia in the nose and throat; gastric juices; and many of the body's secretions, such as saliva, tears, and sweat.

Students will learn that when pathogens manage to enter the body, the second line of nonspecific defense begins. This second line of defense is called the inflammatory response. Bacteria are attacked by phagocytes, which are specialized white blood cells, and a reddish, swollen area develops just beneath the skin. If the infection is serious and pathogens spread throughout the body, fever results. Students will discover that fever serves two purposes: First, it stimulates the action of white blood cells to fight the infection, and second, it produces a temperature range in which many disease-producing microorganisms cannot survive.

Performance Objectives 45–1

1. **Discuss the function of the immune system.**
2. **Describe the body's first line of nonspecific defenses.**
3. **Discuss the role of phagocytes in fighting infection.**
4. **Explain how an inflammatory response defends the body against pathogens.**
5. **Discuss two functions of a fever.**

Science Terms 45–1

immune system   p. 969
nonspecific defense   p. 969
lysozyme   p. 970
inflammatory response   p. 970
phagocyte   p. 970
inflamed   p. 970
lymph node   p. 970
fever   p. 970

45–2 SPECIFIC DEFENSES

Section Focus 45–2

The purpose of this section is to describe the immune system's specific defenses against disease-producing agents. Students will learn that antigens on the surface of viruses, bacteria, and other pathogens trigger the production of antibodies by the immune system.

Students will discover that antibodies are produced by a type of white blood cell called B-lymphocytes. The production of antibodies from the first exposure to an antigen is known as the primary immune response. However, once a person has been exposed to certain diseases, some of the antibody-producing cells remain capable of producing a secondary immune response. This means that the disease never gets a chance to develop again.

It is possible to take advantage of the secondary immune response by injecting a person with a mild or nonfunctioning form of a pathogen in order to activate cells that will produce antibodies. This procedure is called vaccination.

In addition to the production of antibodies, the immune system fights specific pathogens with a type of white blood cell known as killer T-cells. Killer T-cells transfer special proteins into the cell membrane of a pathogen, causing the cell to rupture and die. This immune response is called cell-mediated immunity.

Performance Objectives 45–2

1. **Define antigen and antibody.**
2. **Explain how antibodies are produced by B-lymphocytes.**
3. **Distinguish between active and passive immunity.**
4. **Explain how vaccinations protect a person against specific diseases.**
5. **Discuss the role of killer T-cells in protecting the body against pathogens.**

Science Terms 45–2

specific defense   p. 972
antigen   p. 972
lymphocyte   p. 972
B-lymphocyte   p. 972
antibody   p. 972
antigen-binding site   p. 972
agglutination   p. 972
plasma cell   p. 973
primary immune response   p. 973
T-lymphocyte   p. 973
immune   p. 973
secondary immune response   p. 973
vaccination   p. 973
active immunity   p. 973
passive immunity   p. 975
killer T-cell   p. 975
cell-mediated immunity   p. 975
rejection   p. 975

45–3 IMMUNE DISORDERS

Section Focus 45–3

The purpose of this section is to discuss what can happen when the immune system fails to function properly, or when disease attacks the cells that are essential to the immune system.

Students will learn that allergies are the result of an overreaction of the immune system. Antigens on such substances as plant pollen, dust, mold, and animal fur bind to mast cells, a type of immune cell found especially in the linings of the nasal passages. The activated mast cells release histamines, which produce runny nose, sneezing, and other irritations.

Students will learn that another malfunction of the immune system occurs when the cells of the immune system mistakenly attack the body's own cells. This condition is called autoimmune disease. Students will discover that disorders such as rheumatic fever and multiple sclerosis are caused by autoimmune reactions.

In the last part of this section students will read about the greatest threat to the immune system—a disease called AIDS. Students will learn that the AIDS virus kills and weakens the cells that enable the immune system to function. The section will conclude with a discussion of how AIDS can be prevented.

Performance Objectives 45–3

1. Explain how allergies are caused by an overreaction of the immune system.
2. Describe autoimmune disease and give several examples.
3. Explain how the AIDS virus weakens and destroys the cells of the immune system.
4. Discuss ways in which AIDS can be prevented.

Science Terms 45–3

allergy p. 977
histamine p. 977
asthma p. 977
autoimmune disease p. 978
rheumatic fever p. 978
rheumatoid arthritis p. 978
juvenile-onset diabetes p. 978
multiple sclerosis p. 978
Acquired Immune Deficiency Syndrome p. 979
AIDS p. 979
HIV p. 979

DISCOVERY LEARNING

TEACHER DEMONSTRATIONS
Modeling

People with AIDS
The following demonstration can be used as an introduction to Chapter 45.

Obtain pictures of people of all ages and races. Be sure the pictures include an infant, some children, male and female young adults, male and female middle-aged persons, and some older persons. Display the pictures and ask students,

• **Which of these persons could have AIDS?** (Accept all responses. Depending on the circumstances, any of these persons could have AIDS.)

• **Which persons do you think are most likely to have AIDS?** (Accept all responses. According to statistics, young adults and middle-aged males are most likely to contract AIDS.)

• **How might these other people have contracted AIDS?** (Possible answers: The infant could have been born to a mother with AIDS; the children could have received contaminated blood or been born to a mother with AIDS; any of the adults could be drug users or could have contracted AIDS through sexual contact; they could all have received contaminated blood.)

B-Cells, Antibodies, and Antigens
This demonstration can be performed as students study Section 45–2.

For the demonstration, you will need several pieces of colored poster board, string, scissors, and marking pens.

To prepare for the demonstration, cut five squares of poster board about 30 centimeters on a side. Cut each square in two so that the two resulting pieces fit together like parts of a jigsaw puzzle. (Make sure that each square is cut differently so that the pairing of the pieces will be obvious later on.) Put the puzzle squares aside. Next, make five signs that can be worn around the neck by cutting pieces of poster board about 30 centimeters by 20 centimeters and attaching a loop of string to each piece. On each sign write the name of an infectious bacterial disease, such as Whooping Cough, Strep Throat, Bacterial Pneumonia, Diphtheria, and Tetanus.

Ask for five student volunteers. Give each student a sign to wear and one half of a puzzle square. Tell students to wait at the back of the room. Ask for five more volunteers and give each of them one of the remaining halves of a puzzle square. Ask them to line up across the front of the room.

Tell the class that the students lined up at the front of the room are specialized white blood cells called B-cells and that the pieces of cut poster board are antibodies. Also tell them that the students at the back of the room are pathogens that cause specific diseases and that the pieces of cut poster board they carry are antigens.

Now tell the pathogens to "invade" the classroom. As they do, tell the B-cells to "attack" the pathogen that has the antigen that fits their antibody. Point out to the class that as soon as the B-cell finds the antigen that "fits" its antibody, the pathogen will be destroyed. When a B-cell finds its matching antigen, have students dramatize the demise of the pathogen in whatever (nonviolent!) way they wish.

After all the pathogens have been destroyed, tell students that what they have just witnessed is similar to the immune system's response to specific disease-producing organisms.

CHAPTER 45
Immune System

GUIDED ENQUIRY
Pose the following questions to students and have them record their responses. Point out that they will gain a better understanding of the key concepts if they read the chapter with these basic questions in mind. Upon completion of this chapter, pose the questions again. Ask students to compare their initial responses with those they have developed after reading the chapter.

- Why is the human immune system necessary?
- What is the body's first line of defense against disease?
- How does a fever help to fight infection?
- How does the immune system target specific disease-producing organisms?
- How do different types of white blood cells work together to rid the body of pathogens?
- How does a vaccination produce immunity to a specific disease?
- What happens when the immune system overreacts?
- How does the AIDS virus weaken and impair the function of the immune system?

INTRODUCING CHAPTER 45
Using the Visuals
Begin by having students observe the chapter-opener photograph and read the text.
- **What do you see happening in this picture?** (Children are receiving vaccinations.)
- **Have you ever had a vaccination?** (Accept all responses. Most will probably say yes.)
Point out that most cities and towns require children to have certain vaccinations before they enter school.
- **Why do you think such vaccinations are required?** (In order to prevent a child from bringing a dangerous infectious disease into the school and possibly infecting others.)
- **Have you ever known anyone who has had one of the diseases you have been vaccinated against?** (Answers may vary. Probably no one has had contact with a victim of smallpox or whooping cough. Students may know of some adults who contracted diseases such as polio or measles before vaccines were available.)
- **Why do you think that no one contracts smallpox any longer?** (Because worldwide vaccination programs have essentially eradicated this disease.)

CHAPTER 45
Immune System

The virus that causes polio is only 28 billionths of a meter wide (inset). In 1954, a vaccine was developed that protected people from polio. These children, obviously filled with apprehension, were among the first to be vaccinated.

For parents, springtime had always been a time of mixed feelings: happiness in seeing their children out of doors but terror in knowing that the children could catch a terrible crippling disease called polio.

In 1954, the nation breathed a sigh of relief. Headlines around the country announced the results of an experimental vaccine that had been tested on more than 1 million children. Polio was beaten! Children were free to run and play at last, parents could sleep more easily, and many forgot why they had worried about the coming of springtime. In this chapter you will learn how this disease and many others were conquered. But you will also learn about new diseases that constantly challenge both our social and scientific skills and continue to take their toll in human lives.

968

GUIDE FOR READING

After you read the following sections, you will be able to

45–1 Nonspecific Defenses
- Describe the function of the immune system.
- Identify three nonspecific defenses against infection.

45–2 Specific Defenses
- Define antigen and antibody.
- List several specific defenses of the body.

45–3 Immune Disorders
- Recognize allergies and autoimmune disease as disorders of the immune system.
- Describe the effects of AIDS on the immune system.

Journal Activity

YOU AND YOUR WORLD

Suppose you are a medical researcher who has all the latest medical technology at your disposal. In your journal, explain what you would do with all your knowledge. Do you think that it would be possible to wipe out all disease some day? Why or why not?

45–1 Nonspecific Defenses

Guide For Reading
- Why is the immune system important?
- What are three nonspecific defenses against infection?

Protective mechanisms are found in every living thing. An ameba pulls itself back when it is poked with a pin. Many algae grow thick cell walls to survive in the mud when the lake they live in dries up. A prairie dog dives into its burrow to hide from a hawk circling overhead.

One of our most basic protective devices is our nervous system. Using automatic reflexes and conscious decisions, we protect ourselves from physical injury in many ways. We close our eyelids when an object approaches our face, and we brace for a fall if we lose our balance. Other kinds of protection are built into the human body. Extremely delicate structures, such as the eye, are protected by a casing of bones. Vital organs, such as the lungs and heart, are located deep within the chest cavity. The spinal cord is surrounded by the vertebrae that make up the spinal column.

Skin, muscles, bones, and nerves all protect the body against obvious physical dangers—threats that we can see. But there is an unseen threat that each of us faces every day. It is the threat of infection. As you learned in Chapter 44, an infection occurs when the body is successfully invaded by a pathogen, or disease-causing microorganism. The living world teems with pathogens—viruses, bacteria, rickettsiae, fungi, and protozoans—all capable of producing infectious disease.

The immune system is our primary defense against disease-causing microorganisms. The **immune system** consists of nonspecific and specific defenses against infection. **Nonspecific defenses** are the body's first line against disease.

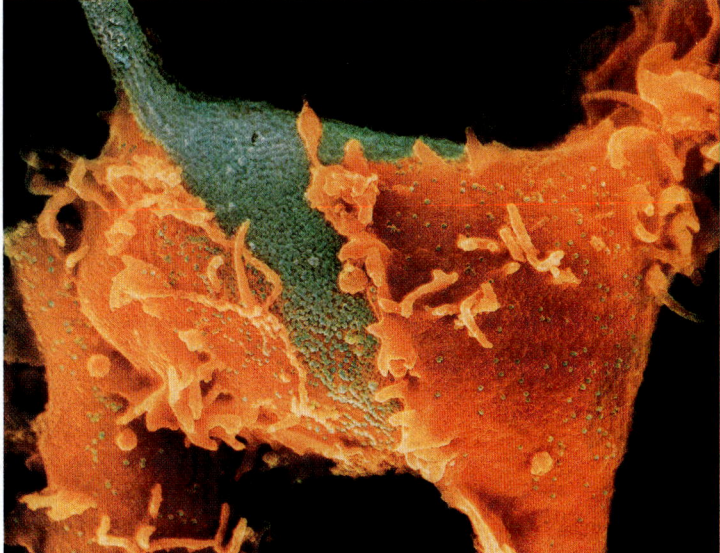

Figure 45–1 *When a virus infects a cell, it takes control of the genetic material of the cell and reproduces. In some cases, thousands of fully formed viruses may burst from the cell as shown here. These new virus particles are then able to infect other cells.*

COOPERATIVE LEARNING

Using preassigned lab groups or randomly selected teams, have groups design a survey to determine the amount of factual knowledge that students in middle or junior high school have about AIDS. The survey should include inquiries about the cause of AIDS, the effect of AIDS on the body, and how AIDS is spread and prevented. Follow the procedure listed below:

1. Remind groups to consider their audience (10- to 13-year-olds) in selecting the type of survey they use.
2. Suggest a minimum number of questions or items their survey must contain.
3. Communicate the information from the survey in graph or chart form.

As an extension of this activity, groups could prepare an informative pamphlet for middle or junior high school students on AIDS. The topics covered in this pamphlet would be determined by the results of each group's survey.

Journal Activity

YOU AND YOUR WORLD

Student answers will vary. With the ability of many disease-producing organisms to mutate, it is doubtful that all disease will be wiped out. You may also want to use this Journal Activity as a basis for a class discussion.

Point out that the last known varioloa virus—the virus that causes smallpox—is kept under very secure conditions at the Center for Disease Control in Atlanta, Georgia. Scientists currently are debating whether or not to continue storing this virus.

• Suppose you could create a vaccine that would prevent one disease that threatens people today. What disease would you choose? (Accept all responses. Many students will probably say AIDS; others may think of diseases such as cancer.)

TEACHING STRATEGY 45–1

Focus/Motivation

• How do you defend yourself against danger? (Accept all reasonable answers. Depending on the type of danger, possible answers include blinking; shielding your face; running; fighting back; hiding; finding a safer place; calling the police; taking certain precautions such as locking the door; eliminating the cause of the danger, such as fixing a gas leak.)

Nonspecific defenses are not directed against a particular pathogen. Rather, nonspecific defenses guard against all infections, regardless of their cause. Specific defenses are attempts by the body to defend itself against particular pathogens. You will read more about specific defenses later in this chapter.

The Skin and Other Barriers

The body's most important nonspecific defense is the skin. As you learned in Chapter 38, skin is a tough, flexible layer that covers most of the body. Very few pathogens can penetrate the layers of dead cells at the skin's outer surface. Oil and sweat glands at the surface of the skin produce an acidic environment that kills many bacteria and other microorganisms. The importance of the skin as a barrier against infection becomes obvious when a small portion of skin is broken or scraped off: Infections almost always follow such cuts and scrapes. These infections are the result of the penetration of the broken skin by microorganisms normally present on the unbroken surface.

Pathogens also enter the body through the mouth and nose, but the body has other nonspecific defenses that protect those openings. For example, mucus and hairs in the nose and throat trap viruses and bacteria. Cilia in the trachea trap bacteria and dust and push them up toward the mouth. Many pathogens that make their way to the stomach are destroyed by stomach acid and digestive enzymes. Finally, many secretions of the body—including mucus, saliva, sweat, and tears—contain **lysozyme**, an enzyme that breaks down the cell walls of many bacteria.

Phagocytes and Inflammation

If large numbers of pathogens do enter the body, however, a second line of defense begins. This second line is called the **inflammatory response**. The bacteria within a wound cause fluid and white blood cells to leak from blood vessels into nearby tissue. Bacteria are attacked by **phagocytes**, which are white blood cells that engulf and destroy bacteria. If the infection remains small and in one place, a reddish swollen area develops just beneath the skin. The area is said to be **inflamed** (literally, "on fire"). Sometimes an infection spreads through the lymphatic system (Chapter 41) to the **lymph nodes**, where it causes swelling and tenderness of the nodes as the battle between pathogen and white blood cells continues.

Serious infections may allow pathogens to spread throughout the body. The immune system now responds in two ways. One, it produces more white blood cells. Two, it releases chemicals that stimulate the action of these white cells by increasing the body's temperature. Thus a **fever** is produced. Physicians know that fever and an increased number of white blood cells are two indications that the body is hard at work fighting infection. Fever also serves another important function: Many

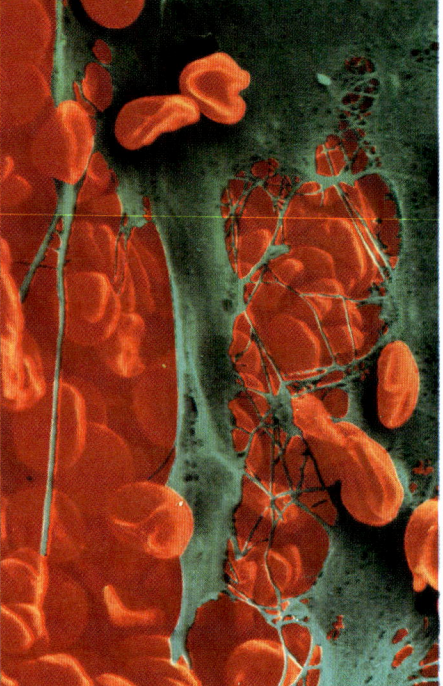

Figure 45–2 Unbroken skin is an effective barrier against most microorganisms. But any break in the skin—even the tiniest—will provide microorganisms with a means of entering the body quite easily. In this photograph, you can see fibrin (a protein) forming a net over a wound. At the same time, white blood cells clean up any debris and attack any invading microorganisms.

SECTION REVIEW 45–1

1. To protect the body from disease-causing microorganisms.
2. Nonspecific defenses are the body's first line against disease. Nonspecific defenses are not directed against a particular pathogen; they guard against all infections regardless of their cause. Specific defenses are attempts by the body to protect itself against specific pathogens.
3. The skin is considered a nonspecific defense because it protects the body from infection by

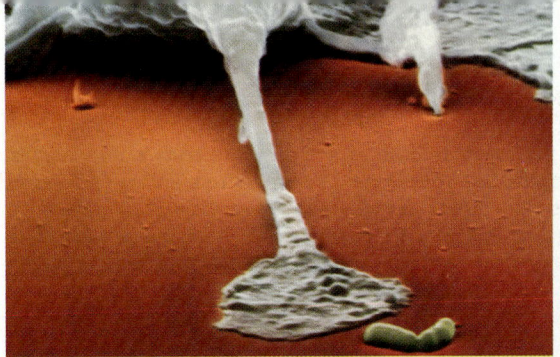
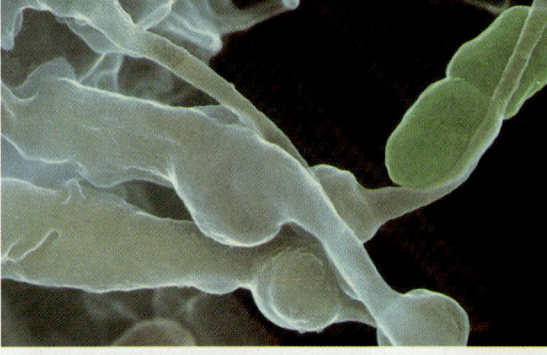
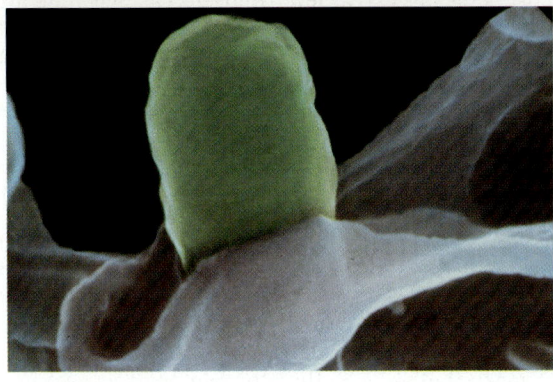
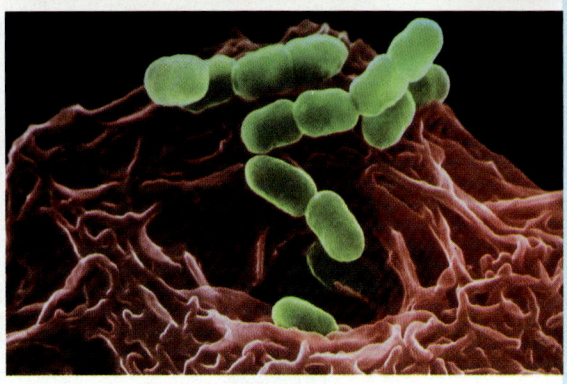

disease-causing microorganisms can survive within only a narrow temperature range. A higher-than-normal temperature often slows down or stops the growth of such microorganisms.

Interferon

In 1957, scientists discovered that virus-infected cells produce a protein that helps other cells resist viral infection. This protein was named **interferon** because it "interferes" with the virus. Interferon inhibits the synthesis of viral proteins in infected cells and helps block viral replication. These effects on a virus slow down the progress of infection and often give the specific defenses of the immune system time to respond.

Figure 45–3 *Macrophages are an extremely important defense against microorganisms. This remarkable sequence of electron micrographs illustrates what happens when a macrophage encounters bacteria. The macrophage first reaches out toward two bacterial cells. Pseudopods from the macrophage then trap the bacteria. Digestive enzymes produced by the macrophage begin to digest one of the bacterial cells. The bacteria will eventually be digested, and their component chemicals will be absorbed by the macrophage.*

45–1 SECTION REVIEW

1. What is the function of the immune system?
2. Compare specific and nonspecific defenses.
3. Why is the skin considered a nonspecific defense?
4. In what two ways does fever help the body fight infection?
5. **Critical Thinking—Applying Concepts** Why should people wash breaks in the skin with soap and water?

971

BACKGROUND INFORMATION
ANTIBODIES

An antibody molecule is a Y-shaped structure that consists of four polypeptide chains joined by disulfide and noncovalent bonds. Two polypeptide chains make up each arm of the Y.

The stem of the Y is essentially the same in every antibody molecule, but the end of each arm has a region that varies. It is this region of the molecule that provides a binding site for a specific antigen. In this area, the two polypeptide chains are folded to form a groovelike cavity that is complementary to the contour and electric charge of the antigen.

These unique antigen-binding sites are created by genetic recombination as B-cells mature in bone marrow. As a result of this genetic recombination, each B-cell becomes precommitted to responding to a particular type of antigen. At any given time, the body has B-cells ready to respond to any of the millions of different antigens that may enter the human body.

Genetic recombination ensures that the production of antibodies is a dynamic, rather than a static, process. Random genetic combinations are constantly producing new antibodies, thus adapting to the body's changing needs. For example, if an entirely new antigen—say, from a mutant of a particular strain of influenza—enters the body, chances are that the millions of possible combinations of genes will have already produced an antibody that can destroy it.

TEACHING STRATEGY 45–2
Focus/Motivation
• Suppose that you are playing a game of tennis. Your opponent hits the ball to the left side of the court. What do you do? (Run to the left side of the court.)
• Suppose that in the next shot your opponent hits the ball close to the net. What do you do? (Run to the net.)
• Suppose that now your opponent lobs the ball to the back of the court. What do you do? (Run to the back of the court.)

Point out that in a game of tennis, each move by one player triggers a specific response by the other player. If the response does not match what is required, the other player loses the point. Explain that specific immune defenses are a lot like a tennis game. Each pathogen that enters the body demands a specific immune response—and without the specific response, the attack on the particular pathogen would be lost.

Guide For Reading
■ What is an antigen? An antibody?
■ How do passive and active immunity differ?
■ What is cell-mediated immunity?

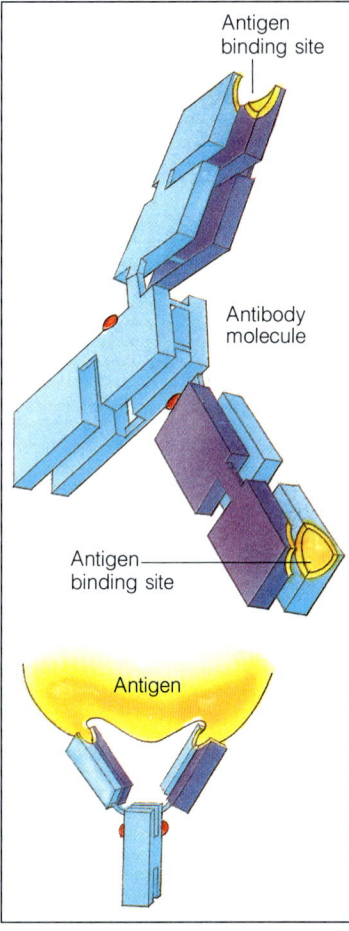

Figure 45–4 *An antibody molecule has two identical antigen-bonding sites. It is at these sites that one or two specific antigens bonds to the antibody.*

45–2 Specific Defenses

If a pathogen is able to get past the body's nonspecific defenses, the immune system reacts with a series of specific defenses that attack the disease-causing agent. A substance that triggers the **specific defenses** of the immune system is known as an **antigen**. Carbohydrates, proteins, and lipids on the surfaces of viruses, bacteria, and other pathogens may serve as antigens that trigger responses by the immune system.

The Immune Response

The key cells of the immune system are **lymphocytes**, a type of white blood cell. **B-lymphocytes**, which mature in the bone marrow, are responsible for producing **antibodies**. Antibodies are special proteins that can bind to the antigens on the surfaces of a pathogen and help destroy it.

ANTIBODIES The antibody molecule is the basic functional unit of the immune response. An antibody molecule is shaped like the letter Y and has two identical **antigen-binding sites** that precisely fit the shape of a particular antigen. See Figure 45–4. These sites allow each antibody to bind to two antigens. Let's suppose that these antigens are proteins found on the surface of the flu virus. Each flu virus particle is covered with scores of such protein antigens. By attaching to the viral antigens, a group of antibody molecules can link the viruses together in a large mass. This process is called **agglutination**.

For a virus, agglutination is bad news. Agglutinated viruses cannot enter cells. In addition, the linked antibody molecules attract phagocytes, which engulf and destroy the entire mass—viruses and antibodies alike. Simply by binding to the correct antigens, antibodies can prevent viruses from infecting cells.

Antibodies can fight bacterial infections as well. Antibodies that bind to bacterial surface antigens can cause agglutination and mark the agglutinated bacteria for destruction by phagocytes and other white blood cells.

ANTIBODY PRODUCTION How does the immune system produce the specific antibodies that bind to antigens on the surfaces of pathogens? The answer involves a little bit of internal genetic engineering. As each B-cell (B-lymphocyte) develops, the genes that code for antibodies rearrange themselves in slightly different ways in each cell. Although each B-cell produces only one type of antibody, the body's population of B-cells can produce hundreds of thousands of different antibodies. This enables the immune system to respond with specific antibodies for almost any antigen.

When a pathogen invades the body, its antigens activate a small fraction of the body's B-cells. These activated B-cells grow and divide rapidly, producing a large population of

Content Development
Emphasize to students that antigens are molecules on the surfaces of pathogens and antibodies are molecules on the surfaces of B-cells. Like pieces of a jigsaw puzzle, each antibody is designed to bind exactly with a particular

specialized B-cells called **plasma cells** that release antibodies into the bloodstream to deal with the infection. The production of antibodies from the first exposure to an antigen is known as the **primary immune response**. Millions of plasma cells may form from just a few dozen B-cells as a result of exposure to an antigen. The activation process is assisted and regulated by lymphocytes that have matured in the thymus gland and are known as **T-lymphocytes** (T-cells).

Immunity

The growth of B-cells and T-cells in response to an infection has an important consequence—one that people have been aware of for more than 2000 years. The Greek physician Hippocrates noted that people who survived certain diseases, such as measles and smallpox, never developed those diseases again. They were permanently **immune**. Today we understand the nature of this immunity. Once the body has been exposed to a disease, a large group of B-cells and T-cells remains capable of producing a **secondary immune response** should the pathogen reappear in the body. A secondary immune response is more powerful than the primary response, producing antibodies so quickly that the disease never gets a chance to develop.

ACTIVE IMMUNITY Smallpox is a serious contagious disease caused by a virus that spreads through the air from one person to another. In 1796, the English country physician Edward Jenner wondered if he could make people immune to smallpox. He knew that there was a mild disease called cowpox that was often contracted by milkmaids. Jenner observed that people who contracted cowpox developed a permanent immunity to smallpox.

Jenner took fluid from one of the sores of a cowpox patient and mixed it into a small cut that he made on the arm of a young farm boy named Jamie Phipps. Jamie developed a mild cowpox infection. Later, to prove that the boy was now immune, Jenner inoculated him with fluid from a smallpox infection. Fortunately for Jamie, the experiment was a success—the boy was indeed immune to smallpox.

The injection of a weakened or mild form of a pathogen to produce immunity is known as **vaccination**. *Vacca* is the Latin word for cow, reflecting the history of Jenner's first vaccination experiment. The immunity produced by a vaccine is known as **active immunity** because the body has the ability to mount an active immune response against the pathogen.

PASTEUR AND THE ANTHRAX VACCINE A different approach was necessary for other diseases. Louis Pasteur, the great French scientist, reasoned that if a weakened or killed disease-causing microorganism was introduced into a person's

THE IMMUNE SYSTEM

Cell	Function
Macrophage	Removes foreign materials and dead and dying cells in the body; also attracts T-cells to foreign organisms
Helper T-cell	Identifies foreign cells in the body; stimulates other cells to fight infection
Killer T-cell	Kills cancerous cells in the body; also kills body cells that have been invaded by pathogenic organisms
Suppressor T-cell	Slows down or stops the activities of B-cells and other T-cells once the danger of infection has passed
B-cell	Produces antibodies; some "remember" the identity of foreign proteins
Antibody	Y-shaped protein molecule that rushes to a site of infection where it neutralizes the enemy or identifies it for attack by other cells or chemicals

Figure 45–5 *Each component of the immune system contributes to the defense of the body. What is the function of killer T-cells?* 1

ANNOTATION KEY

1 Kill cancerous cells, also kill cells in the body that have been invaded by pathogenic organisms. (Interpreting charts)

TEACHING SUPPORT

Teaching Resources

- Activity: Identifying the Key Cells of the Immune System, p. 3

Biotechnology Workbook

- Measuring with Molecules, p. 87

MULTICULTURAL STRATEGY

Dr. Louis Tompkins Wright (1891–1952) was the first African American to be appointed to the staff of New York Hospital and was also the first African American elected to a fellowship in the American College of Surgeons. Noted as one of the most outstanding surgeons and medical researchers in the United States, Dr. Wright carried out the first extensive study of the intradermal method of smallpox vaccination.

antigen.

Explain that when a pathogen manages to break through the body's first line of defense, two things happen. First, some of the pathogenic cells are engulfed and destroyed by phagocytes in a nonspecific response. When this happens, not all of the antigens that were attached to the pathogen are destroyed. Instead, these "remnant" antigens become attached to the surface of the phagocyte, where they act as flags. These antigen flags alert helper T-cells that a dangerous alien has entered the body. Each helper T-cell then activates a B-cell that has the antibody for this particular antigen. The B-cell becomes activated and begins the process of clonal expansion.

In clonal expansion, the B-cell undergoes mitosis and begins to divide and redivide. These divisions give rise to a population of B-cells that produce the same kind of antibody. In time, some of the B-cells differentiate into plasma cells. The plasma cells secrete this same type of antibody into the extracellular fluid. The object of all this activity is, obviously, to saturate the body with the antibody needed to combat the pathogen.

Point out to students that B-cells produce antibodies that remain bound to the surface of the B-cell. Plasma cells, on the other hand, secrete antibodies that separate from the plasma cells. Antibodies secreted by plasma cells do not destroy the pathogen directly; rather, their binding with the antigen alerts macrophages to come and destroy the offending invader.

BACKGROUND INFORMATION
PASSIVE IMMUNITY

One of the most important types of passive immunity is produced immediately after birth. When a newborn baby nurses for the first day or so after its birth, it does not receive milk right away. Instead, the breasts release a watery fluid called colostrum. Colostrum contains a special set of antibodies that are absorbed directly into the baby's bloodstream to help protect it from disease for many weeks.

TEACHING SUPPORT

Laboratory Manual
- Lysis by a Bacteriophage, p. 553

Teaching Resources
- Activity: Examining the Development of a Vaccine, p. 5

FACTS AND FIGURES

A plasma cell secretes about 2000 antibody molecules per second into the extracellular fluid. After devoting nearly all its metabolic energy to this intense rate of antibody production and secretion, the cell dies in less than a week.

---

body, it might be able to produce immunity without causing the disease. Following this reasoning, Pasteur isolated the bacterium responsible for anthrax, a serious disease that affects farm animals and humans. He grew the bacterium in his lab and treated it with heat in order to weaken it. He believed that the treated bacteria from which he would make his vaccine would not cause anthrax. Eager to prove that his approach could prevent disease, Pasteur arranged a public demonstration.

He placed 25 sheep in one pen and 25 in another. All of the animals in the first pen were injected with his vaccine. A week later, Pasteur injected the animals in both pens with anthrax bacteria. The next day every one of the unvaccinated animals was dead. But every vaccinated animal was alive! Pasteur had proved that successful vaccines could be prepared by weakening a pathogen.

RABIES Following his success with anthrax, Pasteur began to work on a vaccine for rabies, a deadly viral disease that can be transmitted by the bite of an infected animal. In 1885, a young boy named Joseph Meister was bitten by a rabid dog. Joseph's parents learned that Pasteur had been working on a vaccine for rabies, and they begged him to try to save their son. At first Pasteur refused. His vaccine was not ready, he argued, and he was afraid that the untested vaccine might be dangerous. The Meisters persisted, and Pasteur finally agreed to try the vaccine on Joseph. Rabies usually takes several weeks to develop, and Pasteur hoped that a series of injections would build up Joseph's immunity before the disease could firmly establish itself in the boy's body. After three desperate weeks, it became apparent that the vaccine had worked. Joseph Meister was the first of many people to be saved from rabies by the Pasteur vaccine.

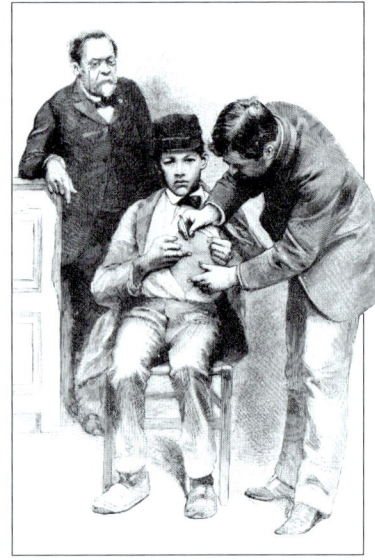

Figure 45–6 In 1885, Louis Pasteur developed a vaccine for rabies. In this photograph, he looks on as an assistant vaccinates a young boy against this serious and often fatal disease.

THE POLIO VACCINE Today, more than 20 serious human diseases can be prevented by vaccination. One of the most important vaccines was developed over 30 years ago by the American physician Jonas Salk. The target of this vaccine was polio, a crippling disease caused by a virus that attacks the nervous system, killing the nerves that carry messages to the muscles of the body. Although it infected people of all ages, polio was most common among children. In 1953, more than 60,000 Americans contracted polio. In some people, the disease destroyed the motor nerves leading to the legs, placing many people in wheelchairs for the rest of their life. In others, polio paralyzed the breathing muscles. Thousands of adults and children were placed in iron lungs—large artificial breathing machines that kept them alive after the ravages of polio.

Salk discovered that three slightly different viruses were responsible for polio. He developed a way to kill these viruses

---

45–2 (continued)

Focus/Motivation

- Suppose that you are shopping at the mall and you discover that your wallet is missing. Later on, you realize how a pickpocket was able to get into your purse or pocket. What are the chances that you will allow yourself to be the victim of a pickpocket the next time you go to the mall? (Answers may vary, but most students will probably say that it would not be likely that they would allow themselves to be pickpocketed again.)

- Why do you think you would not be as likely to be pickpocketed a second time? (Answers will vary. Possible answer: Next time I will be more careful. Now that I know how a pickpocket operates, I won't let my purse or pocket be so accessible.)

Content Development

Use the Focus/Motivation discussion to introduce the idea of secondary immune response. Point out that in the example of the pickpocket, the victim is much better prepared to avoid a second attack after suffering

with the chemical compound formaldehyde. He used the killed viruses, which could not cause infection, as a vaccine. First he tested his experimental vaccine on animals to see if it produced immunity against the viruses. It did. Next he tested the vaccine for safety in humans. His experimental subjects were himself, his wife, and their three sons. The vaccine seemed safe. Then a large group of schoolchildren, potentially at risk for polio, were given the vaccine. The vaccine was nearly 100 percent effective. Finally, in 1955, the vaccine was released for general use. Schoolchildren (including the authors of this textbook) lined up to be vaccinated with the new vaccine. In 1961, Albert Sabin, another American researcher, developed a polio vaccine that could be taken orally, making polio vaccination even easier.

PASSIVE IMMUNITY As you just learned, in active immunity the body makes its own antibodies in response to an antigen. The body can also be protected from disease in another way. If antibodies produced by other animals against a pathogen are injected into the bloodstream, they produce a **passive immunity** against the pathogen as long as they remain in the circulation, usually for several weeks. Before the development of the Salk and Sabin vaccines, polio antibodies were used for temporary protection. Travelers are sometimes given antibodies against tropical diseases before they leave home.

Cell-Mediated Immunity

As you have learned, the function of certain T-cells is to regulate the production of antibodies by B-cells. However, other T-cells can attack antigen-bearing cells directly. The most effective attacking cells in the immune system are **killer T-cells**. These killer cells transfer special proteins into the cell membrane of a pathogen that make the membrane leak fluids from inside the cell. The rapid loss of material from the pathogen cell causes it to rupture and die. This immune response, which is called **cell-mediated immunity**, is particularly important in the case of diseases caused by eukaryotic pathogens, such as fungi and protozoa.

Killer T-cells are also responsible for the rejection of tissue transplants. The cells of your body have a special set of marker proteins on their surfaces that enable the immune system to identify them. If tissue from another individual is transplanted into the body, the immune system recognizes the transplanted tissue as foreign and attacks it. Gradually, the immune system damages and destroys the transplanted organ, a process known as **rejection**. To make rejection less likely, physicians search for an organ donor whose own proteins match the recipient's marker proteins as closely as possible. They may also administer drugs such as cyclosporine, which depress the cell-mediated immune response.

Figure 45–7 Before the Salk polio vaccine was developed, many people with polio spent a great part of their life in this machine. Because polio often paralyzed the diaphragm, this "iron lung" breathed for the person.

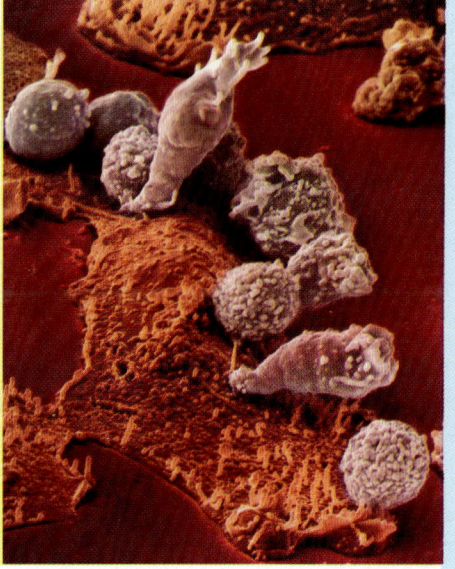

Figure 45–8 Killer T-cells, normally round, become elongated when they are active. The killer T-cells shown here are destroying a cancer cell by breaking down the cancer cell's membrane.

BACKGROUND INFORMATION
FLU VACCINES

The flu is an airborne viral disease that causes periodic epidemics in this country and throughout the world. Most flu infections are mild and last only a few days. However, some forms of the flu can be deadly, especially to the very young and the very old and to persons with chronic disorders or weakened immune systems. In 1968, a flu epidemic infected 50 million people in the United States and killed almost 70,000 people worldwide during a six-week period.

Scientists have prepared many vaccines to help prevent this disease, but the flu virus is an elusive pathogen to pin down. First of all, there is no single virus that causes the flu. Second, mutations in the genetic material of the flu viruses occur frequently—which means that no single flu vaccine remains effective for long.

Once a flu virus is isolated, researchers can prepare a vaccine in six to twelve months. However, the genes that code for the coat proteins of the flu virus mutate at an extraordinarily rapid rate. As a result, varieties of flu viruses with new and different coat proteins appear every couple of years. This means that a vaccine effective against this year's flu strain may prove useless when pitted against next year's flu strain.

---

the first attack. He or she has learned something, and a thief will not have an easy time getting his or her wallet again.

Explain that in a similar way, the body's immune system "learns something" when an attack is mounted against a specific pathogen. In fact, some of the helper T-cells and B-cells that are activated in an antibody response remain circulating in the body after the battle is over and the disease has been cured. These cells are appropriately referred to as "memory cells." Memory cells are so ready to pounce on the offending pathogen that if a subsequent invasion occurs, the immune response is so swift that the disease has no chance to develop again.

• **How does a vaccination take advantage of the body's ability to produce a secondary immune response?** (A vaccination introduces antigens that cause an antibody response, just as if the B-cells and T-cells were attacking the actual disease. Once the "attack" on the injected substance is over, memory cells remain in the body. Should the actual pathogen invade the body, the memory cells would recognize the antigens and swiftly attack the pathogen in a secondary immune response.)

SCIENCE, TECHNOLOGY, AND SOCIETY
BREAKTHROUGH

Activation of Immune Cells Against Cancer

After students have read the feature, pose the following questions for class discussion:

- **Why do you think that the cancer victim's own white blood cells are used for this treatment rather than white blood cells from another source?** (To ensure that the victim's body will not reject the cells.)

- **Why do you think that the cells must be grown and treated outside the patient's body?** (Answers may vary. Possible answers: The cancerous environment of the patient's body may not be conducive to the growth of cells; it may not be possible to target these cells for treatment while they are in the body; perhaps certain aspects of the treatment have to be carried out in a laboratory.)

- **Killer T-cells are usually designed to destroy cells displaying specific antigens. From the feature, what can you infer about the specificity of the killer T-cells once they are injected back into the body?** (The feature states that the cells are injected back into the body "in the hope" that they will attack the tumor; the feature also states that the T-cells may attack the body's own cells. This seems to indicate that there may be some unresolved problems with regards to the specificity of the T-cells.)

SCIENCE, TECHNOLOGY, AND SOCIETY
BREAKTHROUGH

Activation of Immune Cells Against Cancer

One of the primary purposes of the immune system is to protect the body against cancer cells. Scientists believe that most potential cancer cells are destroyed by the immune system before they develop and produce a tumor. Only rarely do cancer cells evade the immune system and produce disease.

Now scientists are experimenting with a radical new treatment designed to boost the immune system and enlist killer T-cells against the most deadly forms of cancer. Led by Steven Rosenberg at the National Institutes of Health Laboratory in Maryland, researchers remove white blood cells from patients stricken with melanoma, a form of skin cancer that is incurable if not treated early. The white blood cells are grown outside the patient's body and treated with interleukin-2, a protein produced by the immune system. Interleukin-2 causes T-lymphocytes to develop into activated killer T-cells. The activated cells are then injected back into the patient in the hope that they will attack the tumor.

The treatment has side effects. The killer T-cells attack many of the body's own cells, producing severe fever and nausea. But tumors in some patients have shrunk by as much as 80 percent, and 10 percent of the patients may have been cured completely by the treatment. Still in its infancy, immune therapy is a promising new treatment for cancer.

The round killer T-cell has done its job. All that is left of the cancer cell are the golden fibers that are part of its cytoskeleton.

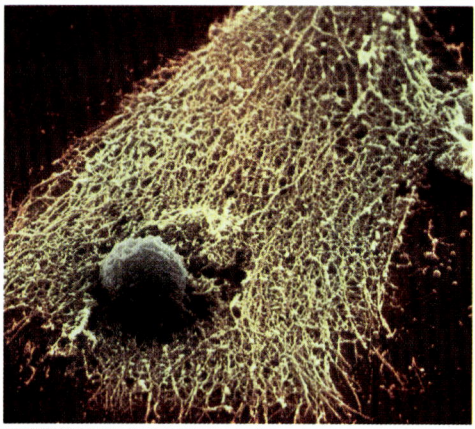

45–2 SECTION REVIEW

1. What is an antigen? Why are antigens important to the immune system?
2. What are the two major types of lymphocytes? What roles do they play in the immune response?
3. Compare active and passive immunity.
4. What is cell-mediated immunity?
5. **Critical Thinking—Making Inferences** Why would a physician use techniques of passive immunity to protect someone from disease?

45–2 (continued)

SECTION REVIEW 45–2

1. An antigen is a substance that triggers the specific defenses of the immune system. Antigens stimulate the body to defend itself. The body recognizes an antigen as a potential threat. The threat is dealt with when the immune system goes into action.

2. The two major types of lymphocytes are T-lymphocytes, or T-cells, and B-lymphocytes, or B-cells. B-cells produce antibodies to specific antigens. T-cells help the body respond to an infection and regulate certain parts of the immune response.

3. Active immunity is produced by the blood in response to a vaccine. The body mounts an active response against a pathogen. In passive immunity, antibodies produced by another organism against a pathogen are introduced into the body. Passive immunity offers protection as long as the introduced antibodies remain within the body.

4. Killer T-cells transfer special proteins into the cell membrane of a pathogen. These special

45–3 Immune Disorders

The impressive power of the immune system to defend the body against a wide range of potential pathogens comes at a price. First, the immune system may overreact to an antigen, producing discomfort or even disease. Second, the cellular nature of the immune response is a potential weak point. What might happen if a disease attacked the lymphocytes that are the heart of the immune system? As we will learn shortly, the consequences are disastrous.

Guide For Reading
- What effects do allergies and autoimmune disease have on the immune system?
- What is AIDS?
- How does AIDS affect the immune system?

Allergies

The most common overreactions of the immune system are known as **allergies**. Allergies result when antigens bind to mast cells, which are a type of immune cell found throughout the body but especially in the linings of the nasal passages. When allergy-causing antigens attach themselves to mast cells, the activated mast cells release chemicals known as **histamines**. Histamines increase the flow of blood and fluids to the surrounding area. Histamines produce the sneezing, runny eyes and nose, and other irritations that make a person with allergies so miserable.

Antigens on plant pollen, dust, molds, and animal fur trigger allergies in as many as 20 percent of the population. One of the most serious allergic reactions is **asthma**, a condition in which smooth muscles contract around the passages leading to the lungs, making breathing difficult.

Scientists do not fully understand the reasons why some individuals become oversensitive to certain antigens. Fortunately, however, asthma and other allergies can usually be treated successfully with antihistamine drugs and other medicines.

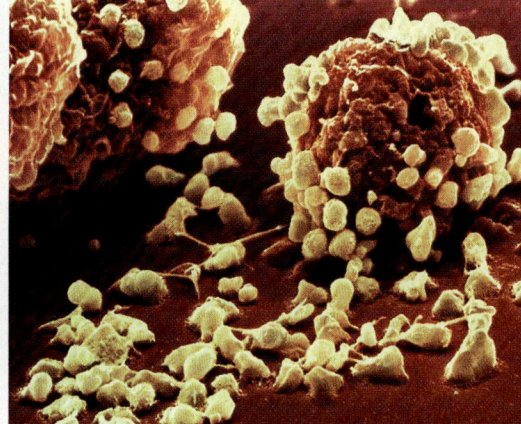

Figure 45–9 These spheres, covered with spikes, are actually microscopic pollen grains produced by ragweed plants (left). Pollen grains often stimulate an allergic response in certain people. Mast cells, a type of immune cell common to the nasal passages, explode when allergy-causing antigens attach themselves (right). The histamine released by the mast cells causes the runny nose, watery eyes, and other unpleasant symptoms associated with allergies.

---

proteins cause a rapid loss of material from the pathogen's cell, causing it to rupture and die. This is cell-mediated immunity. Cyclosporine suppresses the cell-mediated response and is useful in treating patients who have had an organ transplant. By stopping the cell-mediated response, cyclosporine slows down the body's natural rejection process.

5. A physician uses techniques of passive immunity when the patient has been exposed to a disease and the body does not have enough time to develop its own active immunity.

Reinforcement/Reteaching
If some students are having trouble answering the Section Review questions, go back to the part of the section that contains the ma-

BACKGROUND INFORMATION
RECOGNITION OF "SELF" AND "NONSELF"

Cells have many different proteins in their plasma membranes. In human body cells, these proteins include markers that project from the cell's surface and label the cell as "self." Such proteins are called MHC markers (MHC is short for major histocompatibility complex). Cells in the immune system leave cells with MHC markers alone; they know that these are the "good guys."

If cells from another human enter the body—as happens in an organ transplant—the MHC self-markers for this other human are recognized by the immune system of the recipient as foreign. That is why transplanted tissue is so often rejected.

When macrophages engulf foreign substances, they display the invader's antigens along with their own MHC markers. A combination of antigen and self-MHC markers stimulates the immune system to launch a full-scale attack.

Figure 45–10 *The immune system is able to distinguish between the cells that make up an individual and the cells not normally found in the body. During an early stage in their development, these brown mice received an injection of cells from a white mouse. Now as adults, they accept a skin graft from a white mouse. The mice do not recognize the grafted cells as "nonself."*

Autoimmune Disease

The immune system could not defend the body against a host of invading organisms unless it was able to distinguish those organisms from the cells and tissues that belong in the body. In other words, the immune system has the ability to distinguish "self" from "nonself." When this ability breaks down, the immune system may attack the body's own cells, producing an **autoimmune disease.**

Sometimes an infection can produce autoimmune disease by tricking the immune system. This can happen when streptococcus bacteria produce an infection known as strep throat. If the disease is left untreated, the immune system produces antibodies that destroy the bacteria. Because antigens on the surface of the bacteria are so similar to proteins on the surface of some cardiac cells, the immune system, in effect, attacks the heart as well. This results in a condition known as **rheumatic fever.** Antibodies and killer T-cells cause cell death and scarring of the heart lining and the heart valves. Rheumatic fever can be prevented if the streptococcus infection is promptly treated with antibiotics.

Rheumatoid arthritis is a destructive inflammation of the joints. This disease usually first appears between the ages of 30 and 40. The exact cause of rheumatoid arthritis is unknown, but there is clear evidence that the inflammation of the joints is produced by the actions of the immune system.

Juvenile-onset diabetes may be the result of an autoimmune reaction against the insulin-producing cells of the pancreas. **Multiple sclerosis** is a nerve disease that results from an autoimmune destruction of the myelin sheath that surrounds nerve fibers. The first symptoms of multiple sclerosis usually appear between the ages of 20 and 40. There is some evidence that suggests that this disease may result from a viral infection.

45–3 (continued)

Content Development
Explain to students that an allergy is a secondary immune response to what would normally be considered a harmless substance. In other words, it is the immune system's response to a false alarm.

Allergies occur in some individuals because substances such as dust, pollen, animal fur, and certain foods are perceived by the body as antigens. With each new exposure to the antigen, IgE antibodies are produced. IgE antibodies become attached to mast cells which release chemicals such as histamine. Histamine causes mucus secretions and increased capillary permeability. As a result, the allergy sufferer experiences symptoms such as runny nose, sneezing, congestion, and difficulty in breathing.

Skills Development
Guided Practice
Skill: Identifying cause and effect
Point out to students that the immune system often attacks the heart when streptococcus bacteria are present in the body.

AIDS

A dramatic example of what happens when cells of the immune system are weakened by infection is the disease called **Acquired Immune Deficiency Syndrome**, or **AIDS**. The virus that causes AIDS was first discovered in 1984 and was named human immunodeficiency virus, or **HIV.** HIV is a retrovirus, or a virus whose genetic information is contained in RNA. Once HIV enters the body, it attaches to receptors on the surfaces of a type of T-cell known as **helper T-cells**. These cells are so named because they help other lymphocytes respond to the early stages of an infection. Once inside the helper T-cells, HIV replicates and eventually kills the infected cells. Although the body produces antibodies against HIV, the virus grows within cells of the immune system and is thus not affected by the antibodies. Gradually HIV kills off most of the helper T-cell population.

THE EFFECTS OF AIDS The death of the helper T-cells cripples the immune system, making it impossible for it to respond to infection by pathogens that rarely cause disease in healthy people. AIDS patients develop protozoan infections in

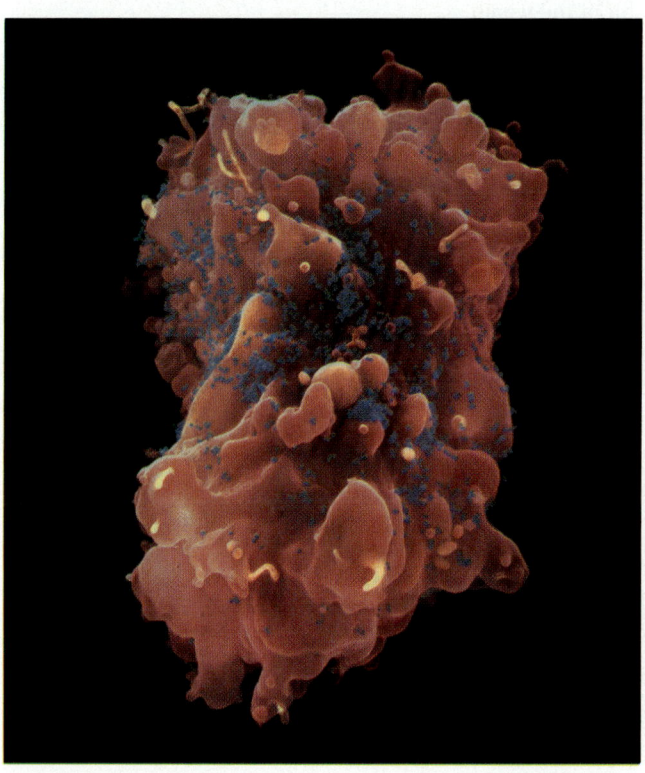

Figure 45–11 The virus that causes AIDS, stained blue in this photograph, has infected a T-cell.

- **Why do streptococcus bacteria cause this action on the part of the immune system?** (Antigens on the strep bacteria trigger the immune system to produce antibodies. These antibodies are designed to destroy the bacteria. However, strep antigens are very similar to proteins on the surface of some cardiac cells. As a result, the immune system may attack the cardiac cells by mistake.)
- **Why does prompt treatment of strep throat prevent heart damage?** (Once the strep bacteria are removed, the immune system stops—or never has a chance to start—producing the antibodies to destroy them. This means that cardiac cells are no longer threatened by these antibodies.)

BACKGROUND INFORMATION
VIOLENT ALLERGIC REACTIONS

In a few individuals, secretions caused by allergic reactions are so extreme that they can be life-threatening. Air passages leading to the lungs undergo massive constriction. Capillaries become so permeable that plasma escapes rapidly, and plummeting blood pressure may lead to circulatory shock. An example of this type of reaction occurs in people who are hypersensitive to bee or wasp venom. Such individuals can die within minutes of a single sting.

It is possible to prevent some violent allergic reactions when the allergy-producing substance has been identified by laboratory tests. The allergy sufferer is given large injections of antigen to stimulate the production of IgG antibodies, or "blocking antibodies." Circulating IgG antibodies bind with the allergy-producing substance and mask its identity before it can be attacked by the immune system.

ESL STRATEGY

Although AIDS is a worldwide problem, it is not known by this acronym around the world. In Spanish, for example, persons are cautioned about the spread of "la SIDA." Ask ESL students to pronounce and spell out on the chalkboard the term used for AIDS in their native language.

Using the library or recent magazine and newspaper articles for additional information, have students develop a short pamphlet to educate people in their native country about AIDS. Students may also draw pictures or produce a poster in this exercise.

BACKGROUND INFORMATION
AIDS

Two types of HIV virus are known: HIV-1 and HIV-2. HIV-2, which may be less virulent than HIV-1, is found mainly in West African countries.

Each HIV particle consists of a protein core that surrounds its RNA and several copies of the enzyme reverse transcriptase. When the virus attaches to a helper T-cell, the protein core becomes wrapped in a lipid envelope derived from the T-cell's plasma membrane.

The virus progresses from the surface of the T-cell to the cell interior. Once the virus is inside the T-cell, reverse transcriptase uses the viral RNA as a template for making DNA. This DNA is then cleverly inserted into a chromosome of the helper T-cell.

Trouble begins when the body is called on to make a specific immune response. The infected T-cell may be activated, and if it is, it will transcribe portions of its own DNA. Of course, the T-cell's DNA now contains the HIV insert. This means that the T-cell inadvertently produces copies of viral RNA, which are then translated into viral proteins. Soon these proteins assemble to form new viruses, which go on to infect and destroy more helper T-cells.

In the United States, AIDS is transmitted most often among male homosexuals and intravenous drug users who share needles. In Africa, AIDS is transmitted primarily among heterosexuals; heterosexual transmission is also on the increase in Latin America. In New York City, AIDS is the leading cause of death among men 25–44 years of age and among women 25–34 years of age.

Scientists are not overly optimistic about the possibility of developing an AIDS vaccine. A major problem is that the AIDS virus mutates rapidly. This means that any vaccine that is developed probably would not work for long and probably would not be effective against the many existing forms of the virus.

the lungs, fungal infections of the mouth, and a rare form of cancer that is normally prevented by the immune system. It is these kinds of repeated, uncontrollable infections that eventually weaken and kill people with AIDS. Although AIDS may take 6 months to 10 years or more to develop from the time of the first HIV infection, no person who has developed the symptoms of AIDS has yet recovered from the disease. At the writing of this book, more than 50 percent of the people infected with HIV have died.

THE SPREAD OF AIDS HIV is present in the blood and body secretions of infected persons. Because it has been found in the semen and vaginal secretions, HIV can be spread during sexual intercourse. It can also be spread from one person to another by contaminated blood. Before 1985, some cases of AIDS were caused by blood that was unknowingly donated by infected individuals and used for transfusions. Since that time, however, all blood has been screened for AIDS. The incidence of AIDS is increasing in sexually active heterosexual teenagers. HIV is also spreading rapidly among intravenous drug users who share the needles used to inject drugs. HIV can pass through the placenta from mother to child, and an increasing number of newborn children have been infected with HIV while still in the womb.

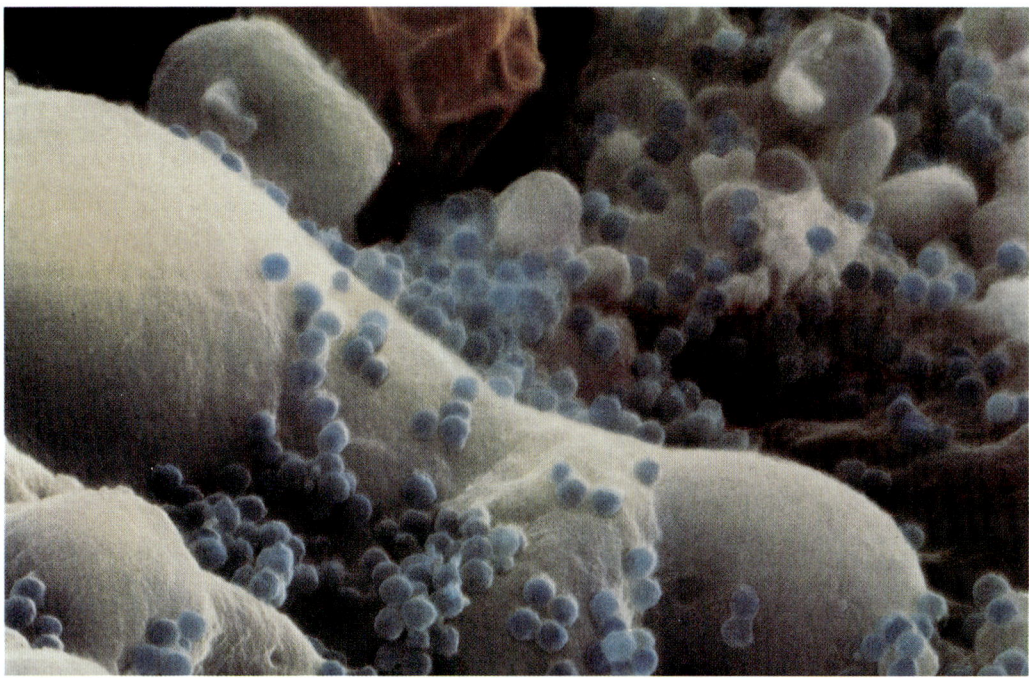

Figure 45–12 By killing T-cells, the AIDS virus increases the body's susceptibility to many infections that would not normally affect a person with a healthy, functioning immune system.

45–3 (continued)

Content Development
Discuss the ways in which the HIV virus weakens and destroys the immune system.
• **What is the role of helper T-cells in fighting infection?** (They recognize the antigens of pathogens and activate the appropriate B-cells to undergo clonal expansion.)
• **What do you think would happen to specific immune responses if helper T-cells were not present?** (There would be a breakdown in the activation of B-cells. This means that needed antibodies would not be produced to combat pathogens and the pathogens would be able to overrun the body.)
• **Why is the HIV virus "safe" inside a helper T-cell?** (Because

AIDS TREATMENT AND PREVENTION The rising number of AIDS cases in the United States has produced a genuine epidemic—by 1992 more than 250,000 cases of AIDS had been identified in the United States. At present there is no cure for AIDS, and experimental vaccines have not yet been proven to prevent HIV infection. Several promising drugs that slow the growth of the virus have been developed, thus enabling AIDS patients to survive longer. There is hope that some of these drugs will prevent the development of AIDS in individuals who have been infected with HIV but who have not yet shown symptoms of the disease. Unfortunately these medicines are extremely expensive. Intensive research in laboratories throughout the world is underway to find a cure for this killer.

Although we have not yet discovered a cure for AIDS, it is a fact that we do know how to prevent it. **YOU CAN PREVENT AIDS BY AVOIDING EXPOSURE TO HIV, THE VIRUS THAT CAUSES THE DISEASE.** AIDS is not spread by casual contact. AIDS is spread by intravenous drug use, and that alone is reason enough to avoid using such drugs. AIDS is also spread by sexual contact, and thus a sexual contact carries with it the risk of contracting an HIV infection. The safest course of conduct, as former Surgeon General of the United States C. Everett Koop pointed out in 1986, is to abstain from sexual intercourse before marriage. According to Dr. Koop, the next safest course is the use of a condom, a sheath that fits around the penis and prevents most sexually transmitted diseases from being passed during intercourse. A condom does not provide 100 percent protection, but it is safer than unprotected sexual intercourse. Every AIDS death is unnecessary. You should be sure that you, your friends, and your classmates understand how to prevent becoming infected with HIV—and how to end this deadly epidemic.

Figure 45–13 Ryan White had hemophilia, a genetic disorder that interferes with the normal ability of the blood to form clots. Dependent upon transfusions, he contracted AIDS from contaminated blood and blood products he needed to treat his disease. Ryan White died on April 8, 1990, but his great courage is an inspiration to many. Today, blood and blood products are tested to ensure that they are free of HIV infection.

45–3 SECTION REVIEW

1. What is an allergy? What are histamines? Name an example of an allergy.
2. What is an autoimmune disease? Name some examples of autoimmune diseases.
3. What is AIDS? What causes it? What happens to the immune system as a result of this disease?
4. What are the three principal ways in which HIV is transmitted from person to person?
5. **Critical Thinking—Relating Cause and Effect** In treating an asthmatic patient, the first thing that many physicians will do is to ask the patient to make a list of times and places they have experienced asthmatic reactions. Why do you suppose they do this?

the virus is inside a cell, the surface of the virus is not exposed. As a result, antibodies cannot bind to the antigens of the HIV virus, and the virus cannot be destroyed.)

SECTION REVIEW 45–3

1. An allergy is an overreaction of the immune system when presented with a harmless allergen. Histamines are chemicals released by mast cells that increase the flow of blood and fluids. People have allergies to plant pollen, dust, and molds. Asthma is a serious type of allergic reaction.
2. An autoimmune disease is a disease that results when the body attacks its own cells. Rheumatoid arthritis, multiple sclerosis, and perhaps juvenile-onset diabetes are autoimmune diseases.
3. AIDS is Acquired Immune Deficiency Syndrome. AIDS is caused by HIV, a retrovirus. In this disease, helper T-cells are invaded by a virus and killed. This results in the immune system being compromised. The body has an impaired ability to defend itself against disease.
4. HIV is transmitted by contaminated blood. Body secretions such as semen may also contain HIV. HIV may also be spread from one person to another by the contaminated needles used by drug abusers. HIV may also be spread from mother to child through the placenta.
5. Physicians may want to check to see what allergens a person has been exposed to that preceded the asthmatic attack. The list will help patients become more aware of the kinds of substances that may trigger an asthma attack.

LABORATORY INVESTIGATION

SIMULATING CLONAL SELECTION OF ANTIBODIES

Before the Lab
1. At least one day before the investigation, gather enough materials for your class, noting that you will need a set of materials for each student.
2. To make the drawing of squares easier, you may wish to use centimeter graph paper in place of the typing paper.

Pre-Lab Discussion
Review with students the material in Section 45–2 on the relationship between antigens and antibodies. Remind students that antibody molecules are equipped with receptor sites that bind with specific antigens. Explain that it is the physical shape of these sites as well as their electrical characteristics that make them "match" a particular antigen. Although the presentation of cloning in this investigation is rather simplified, it is intended to illustrate the way that the shape of the receptor site in a clone is tailored to match the configuration of the antigen molecule.

You may wish at this time to discuss with the class the Clonal Selection Theory, which was first proposed by scientist Macfarlane Burnet in 1955. This theory states that when a lymphocyte is activated by the combining of an antigen with its receptor, the lymphocyte will multiply rapidly and all of its descendants will retain specificity against that antigen only. These descendants constitute a clone of cells that are immunologically identical in their response to the antigen that "selected" them.

Skills Development
Students will use the following skills while completing this laboratory investigation.
1. Manipulative
2. Measuring
3. Comparing
4. Relating
5. Inferring
6. Applying
7. Hypothesizing

LABORATORY INVESTIGATION

SIMULATING CLONAL SELECTION OF ANTIBODIES

PROBLEM
How do plasma cells form antibodies specific for each antigen?

MATERIALS (per person)

protractor	scissors
ruler	typing paper

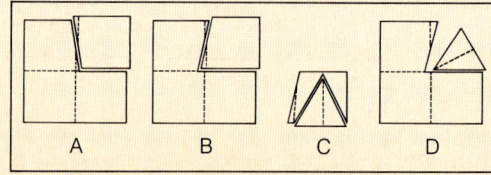

PROCEDURE

1. Obtain a sheet of typing paper. Use a ruler to draw a grid of 5-cm squares. Use the scissors to cut out the squares.
2. Fold two of the squares in half twice to produce four equal quarters. These two folded squares represent plasma cells.
3. Cut one of the folded squares as shown in Figure A. Make the first cut along the horizontal fold. Continue this cut to the vertical fold. Then make a second cut that begins just to the left of the top of the vertical fold and continues to just to the right of the vertical fold at the bottom.
4. Take the other folded square. Cut this square as shown in Figure B. Make the first cut along the horizontal fold. Continue this cut to the vertical fold. Then make a second cut that begins just to the right of the top of the vertical fold and continues to just to the left of the vertical fold at the bottom. The cut corners of the squares represent antigen receptor sites.
5. Take one of the cut-out corners and fold it in half. Cut diagonally from the top point of the fold to each of the opposite corners. This will form a triangle, as shown in Figure C. This triangle represents an antigen.
6. Place the antigen into the receptor site of one of the plasma cells as shown in Figure D. Use a protractor to measure the angle formed by the empty space between the antigen and the receptor site of the plasma cell. Write the angle measurement on the plasma cell. Repeat the procedure with the other plasma cell.
7. The receptor site that forms the smallest angle fits the antigen best. Select the plasma cell that fits best.
8. Take three other squares and fold them in half horizontally. Try to duplicate the receptor site of the plasma cell you selected by cutting partway along the fold. Then cut down toward the fold with scissors. Estimate the correct angle to cut. These three cells represent clones of the selected cell.
9. Insert the antigen into the receptor site of one of the three cells. Measure the angle formed, as you did in step 6. Repeat this procedure with the other two cells.
10. From the original cells or the clones select the plasma cell that fits the antigen best. Try to duplicate this cell using three new pieces of paper (step 8).
11. Repeat steps 9 and 10 until you have used all the cut squares or until the angle formed by the empty space between the antigen and the receptor site equals 0°.

OBSERVATIONS
What happened to the size of the angle with each new generation of cloned cells?

ANALYSIS AND CONCLUSIONS
1. What causes changes to occur in the fit between receptor sites and antigens over time?
2. How do plasma cells form antibodies for each antigen? (*Hint:* Antibodies are formed from receptor sites shed by plasma cell clones.)

Safety Tips
Remind students to be careful when using scissors.

Teaching Strategy
1. Tell students that their grids should consist of 4 rows by 5 rows of 5-centimeter squares. Also tell students that it is important that they measure and cut the squares carefully. Otherwise, they may have trouble making accurate angle measurements later on.
2. If some students are having trouble folding and cutting the

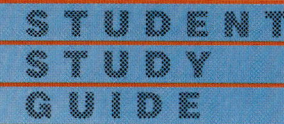

SUMMARIZING THE CONCEPTS

The key concepts in each section of this chapter are listed below to help you review the chapter content. Make sure you understand each concept and its relationship to other concepts and to the theme of this chapter.

45–1 Nonspecific Defenses

- The immune system is the body's primary defense against disease-causing microorganisms. It is composed of specific and nonspecific defense mechanisms. Specific defenses are attempts by the body to defend itself against particular pathogens. Nonspecific defenses guard against all infections.

45–2 Specific Defenses

- If a pathogen is able to get past the body's nonspecific defenses, the immune system reacts with a series of specific defenses that attack the disease-causing agent. A substance that triggers the specific defenses of the immune system is known as an antigen.

45–3 Immune Disorders

- Allergies result when antigens bind to mast cells. The activated mast cells release chemicals known as histamines.
- At present, there is no cure for AIDS, and experimental vaccines have not yet proven to prevent HIV infection. You can prevent AIDS by avoiding exposure to HIV.

REVIEWING KEY TERMS

Vocabulary terms are important to your understanding of biology. The key terms listed below are those you should be especially familiar with. Review these terms and their meanings. Then use each term in a complete sentence. If you are not sure of a term's meaning, return to the appropriate section and review its definition.

45–1 Nonspecific Defenses
immune system
nonspecific defense
lysozyme
inflammatory
  response
phagocyte
inflamed
lymph node
fever
interferon

45–2 Specific Defenses
specific defense
antigen
lymphocyte
B-lymphocyte
antibody
antigen-binding site
agglutination
plasma cell
primary immune
  response
T-lymphocyte
immune
secondary immune
  response
vaccination
active immunity
passive immunity
killer T-cell
cell-mediated
  immunity
rejection

45–3 Immune Disorders
allergy
histamine
asthma
autoimmune disease
rheumatic fever
rheumatoid
  arthritis
juvenile-onset
  diabetes
multiple sclerosis
Acquired Immune
  Deficiency Syndrome
AIDS
HIV
helper T-cell

---

acts as a selective factor for clone formation.

2. The presence of an antigen causes rapid proliferation of those cells whose receptor sites fit it best. By the process of mutation and selection, large numbers of receptor sites that fit the antigen well are formed. These are then shed to serve as antibodies.

Going Further: Enrichment

Have students work in groups. Ask each group to use paper cutouts to simulate the immune system's antibody-mediated attack on a specific pathogen. Students should include in their simulation the role of helper T-cells as well as the activity of B-cells; they should also show how some helper T-cells and B-cells are reserved as memory cells. Students may then simulate a secondary immune response to the pathogen.

CHAPTER REVIEW

CONTENT REVIEW

Multiple Choice
1. c   2. d   3. a   4. c
5. b   6. a   7. d   8. c

True or False
1. F agglutination
2. F Louis Pasteur
3. F Killer T-cells
4. T
5. T
6. T
7. F can
8. T

Word Relationships
1. lysozyme
2. interferon
3. antigens

---

squares, you may wish to use a larger square to demonstrate the technique to the class.

3. Help any student who may be having trouble measuring angles with a protractor.

4. At the end of the investigation, you may wish to have students state their conclusions in writing in the form of a theory or hypothesis. Students may give titles to their theories.

Observations
The size of the angle decreased.

Analysis and Conclusions

1. Mutations cause changes in the receptor site. Those cells with receptor sites that fit the antigen best bind tightly to it and begin to divide, producing a clone of cells with the same receptor site. The degree of fit

CONCEPT MASTERY

1. AIDS is Acquired Immune Deficiency Syndrome. It is a life-threatening disease caused by a retrovirus, HIV. HIV infects certain T-cells in the body that are an important part of the body's immune system. This compromises the immune system, making infection by normally nonpathogenic organisms serious, perhaps even fatal.

2. B-cells produce specific antibodies for a particular disease. They are an important part of the immune response because the antibodies they produce destroy antigens that may cause serious disease. A group of B-cells and T-cells is also responsible for the secondary immune response that occurs if the body is ever exposed to a particular antigen again.

3. It is the proteins and other molecules on the surface of cells that the body's immune system is able to recognize. The proteins that are normally found in the body are recognized as self. Proteins on foreign cells are recognized as nonself. The body's immune system quickly makes antibodies that destroy nonself proteins.

4. A vaccine is a weakened or mild form of a pathogen that is injected or ingested into the body. The weakened or mild form of a pathogen causes the body's immune system to make antibodies; however, the weakened form is not strong enough to cause serious disease. The antibodies the body makes in response to the vaccine are able to defend the body if it becomes infected with a virulent strain of the pathogen.

5. Pasteur made many contributions to human health. Aside from pasteurization—a heat treatment that kills pathogenic organisms in many kinds of food—Pasteur developed vaccines against the virus that causes rabies and the bacteria that cause anthrax. Students might be interested to learn that many foods are pasteurized besides milk. Most of the beer sold in the United States is also pasteurized. Jenner developed the first successful human vaccine. This vaccine worked against smallpox. Salk and Sabin developed vaccines against the polio virus.

6. Active immunity results when the body produces its own antibodies in response to antigens, and it is long-term or permanent. Passive immunity results when antibodies produced by another organism in response to an antigen are injected into the body, and it is temporary.

7. The primary immune response is the body's response to a disease the first time it is

CHAPTER REVIEW

CONTENT REVIEW

Multiple Choice

Choose the letter of the answer that best completes each statement.

1. Special proteins that can bind to antigens on the surfaces of a pathogen and help destroy it are
   a. lymphocytes.   c. antibodies.
   b. plasma cells.   d. phagocytes.
2. A vaccine produces a
   a. primary immune response.
   b. secondary immune response.
   c. passive immunity.
   d. active immunity.
3. Specialized B-cells that release antibodies into the bloodstream to deal with infection are
   a. plasma cells.   c. T-cells.
   b. antigens.   d. histamines.
4. HIV attaches to receptors on
   a. killer T-cells.   c. helper T-cells.
   b. RNA.   d. helper B-cells.
5. The type of immunity that is particularly important against diseases caused by eukaryotic pathogens is
   a. passive immunity.
   b. cell-mediated immunity.
   c. active immunity.
   d. autoimmunity.
6. A disease that results from an autoimmune destruction of the myelin sheath is
   a. multiple sclerosis.   c. polio.
   b. smallpox.   d. rheumatic fever.
7. Allergy-causing antigens cause activated mast cells to release
   a. insulin.   c. antibiotics.
   b. antihistamines. d. histamines.
8. The oral polio vaccine was developed by
   a. Louis Pasteur.   c. Albert Sabin.
   b. Edward Jenner.   d. Jonas Salk.

True or False

Determine whether each statement is true or false. If it is true, write "true." If it is false, change the underlined word or words to make the statement true.

1. The process by which a group of antibody molecules links viruses together in a large mass is rejection.
2. Edward Jenner developed a rabies vaccine.
3. B-cells are responsible for the rejection of tissue transplants.
4. B-lymphocytes, which mature in the bone marrow, are responsible for producing antibodies.
5. The body's most important nonspecific defense is the skin.
6. HIV is present in the blood and body secretions of persons infected with AIDS.
7. Newborn children cannot be infected with HIV while still in the womb.
8. Once exposed to a disease, the body exhibits a secondary immune response if the pathogen reappears.

Word Relationships

Replace the underlined definition with the correct vocabulary word.

1. The secretions of the body contain an enzyme that breaks down the cell wall of many bacteria.
2. Virus-infected cells produce a protein that inhibits the synthesis of viral proteins in infected cells and helps block viral replication.
3. Carbohydrates, proteins, and lipids on the surfaces of pathogens may serve as substances that trigger the specific defenses of the immune system.

CONCEPT MASTERY

Use your understanding of the concepts developed in the chapter to answer each of the following in a brief paragraph.

1. What is AIDS? What causes AIDS? How does AIDS affect the body?
2. What role do B-cells play in the body's immune response?
3. How does the body recognize "self"?
4. What is a vaccine? How does a vaccination protect the body?
5. What contributions to human health care did Pasteur, Jenner, Salk, and Sabin make?
6. Differentiate between active and passive immunity.
7. What is a primary immune response? A secondary immune response?
8. What is the inflammatory response? How does the inflammatory response help the body fight infection?
9. List three nonspecific defenses of the body. Tell how each protects the body.

CRITICAL AND CREATIVE THINKING

Discuss each of the following in a brief paragraph.

1. **Relating cause and effect** The blood of a person with HIV infection often shows decreasing numbers of T-cells each time it is analyzed. How do you explain this observation?
2. **Applying concepts** This photograph shows a child who lives within a sterile environment, an environment free of pathogenic organisms. Offer a probable explanation to explain the way this child must live.
3. **Assessing concepts** The first vaccine was developed against smallpox, a serious and often fatal disease. Today, scientists for the World Health Organization claim that smallpox has been eradicated and the world is free of smallpox infection. What do you think is the basis for this claim? (*Hint:* Smallpox is exclusively a disease of humans.)
4. **Relating concepts** In the past, organ transplants were most effective when organs were transplanted from one close relative to another. Today, organs can be transplanted from one unrelated person to another with success. What is the most probable reason for the dramatic change in transplant success?

5. **Applying concepts** People often view slight fevers with alarm. However, many physicians do not. Why would a slight fever not be viewed with much alarm?
6. **Using the writing process** Pretend you are a kidney that has been transplanted into a person. Write a letter to the body you have been transplanted into explaining why you should not be "rejected."

CRITICAL AND CREATIVE THINKING

1. HIV infects certain T-cells, killing them. In time, the numbers of T-cells decrease to levels too low to offer the body protection from diseases that it could normally fight off.
2. Children who are born without a functioning immune system often must live in a sterile environment. Without the protection of the immune system, a child outside the sterile environment would die from an infection that in a normal child would prove uneventful.
3. The World Health Organization (WHO) claims that since so many people have received smallpox vaccinations, there is no reservoir of human hosts for this disease. In fact, with no cases of smallpox reported, it might be correct to assume that smallpox has been eradicated.
4. The most probable reason is the development of medicines such as cyclosporine, which suppress the immune response and prevent organs from being rejected.
5. A slight fever means that the body is most probably fighting an infection. The slight fever may prove beneficial if it makes the body temperature a little too high for bacteria or other pathogenic organisms to survive.
6. Accept all reasonable letters.

exposed. The secondary immune response occurs if the body is exposed to the same antigens again. The secondary response is much quicker than the primary response.
8. The inflammatory response is a nonspecific defense that involves phagocytes and a rise in body temperature. The phagocytes engulf and destroy bacteria, and the rise in temperature slows or stops the growth of microorganisms.
9. The skin keeps bacteria and other pathogens from entering the body. Special white blood cells called phagocytes are also a nonspecific defense against disease. Phagocytes destroy microorganisms. Interferon is also a nonspecific response of the body in which cells near an infected cell are provided protection due to the release of interferon.

CHAPTER 46 — Drugs, Alcohol, and Tobacco

Section	Laboratory Investigations and Activities
46–1 Drugs pages 987–993 THEMES: Unity and Diversity, Systems and Interactions	**Student Edition** LABORATORY INVESTIGATION: Observing the Effect of a Stimulant on the Heart Rate, p. 998 **Teacher Edition** DEMONSTRATION: A "Puppet" Show, p. 986D
46–2 Alcohol pages 993–995 THEME: Systems and Interactions	**Laboratory Manual** The Effects of Alcohol on Human Reactions, p. 563
46–3 Tobacco pages 995–997 THEME: Systems and Interactions	**Teacher Edition** DEMONSTRATION: Drug Education, p. 986D **Laboratory Manual** The Effects of Tobacco and Alcohol on Seed Germination, p. 567
Chapter Review pages 999–1001	

Teacher Resources

Books

Julien, Robert. *Drugs and the Body.* Freeman, 1988.

Purcey, Russell J., and Jennifer L. DeSantes. *Drugs, Alcohol, Smoking and Nutrition: An Educational Guide.* NEHS Publications, 1988.

Snyder, Solomon H. *Drugs and the Brain.* Freeman, 1987.

Weiss, Roger D., MD, and Steven M. Mirn, MD. *Cocaine.* American Psychiatric Press, 1987.

Winter, Ruth. *The Scientific Case Against Smoking.* Crown, 1980.

Zaridze, B., and R. Petro, eds. *Tobacco: A Major International Health Hazard.* Oxford University Press, 1987.

CHAPTER PLANNING GUIDE

Other Activities	Media and Technology
Teaching Resources ACTIVITY: How Drugs Affect the Body, p. 3 **Study Guide** Section 46–1, p. 435	**Biology Transparencies** 70: The Effects of Crack on the Human Body
Study Guide Section 46–2, p. 438	
Teaching Resources VOCABULARY REVIEW: Word Game, p. 1 ACTIVITY: Understanding the Health Risks of Smoking, p. 5 **Study Guide** Section 46–3, p. 440	
Teaching Resources Chapter Test A, p. 11 Chapter Test B, p. 15 Performance-Based Assessment, Unit 9	**Computer Test Bank** Chapter Test, p. 483

Audiovisuals

Marijuana and Human Physiology. Film or video. AIMS Media.
Tobacco and Human Physiology. Film or video. AIMS Media.
The Body at Risk. Film or video. MTI/Coronet.

Bodywatch: No Butts. Film or video. MTI/Coronet.
Breathing Easy. Film or video. MTI/Coronet.

Bodywatch: A Chemical Called Cocaine. Film or video. MTI/Coronet.
Crack: Cheap and Dangerous. Film or video. MTI/Coronet.

CHAPTER 46 Drugs, Alcohol, and Tobacco

CHAPTER OVERVIEW

A drug is any substance that causes a change in the body in some way. Many helpful drugs are legal and are used to treat disease and infection. Others are powerful and dangerous and illegal to use. Any drug, if used improperly, is potentially harmful. Drugs vary widely in the ways they affect the body. Among the most powerful drugs are those that affect the nervous system and change behavior.

Drug abuse is the use of any drug in a way that would not be medically recommended. Misuse may involve either a legal or an illegal drug. Abuse can cause physical damage or psychological dependence of such force that it can disrupt the family life and career of the abuser. Psychological dependence can have a very powerful effect on an individual's life. The term *addiction,* however, usually applies to physical dependence on a substance. Opiates are examples of drugs that cause addiction. There is an actual cellular basis for this dependence. Scientists now know how and why opiate addiction occurs.

46-1 DRUGS

Section Focus 46-1

The purpose of this section is to introduce various categories of drugs and define drug abuse. A drug is any substance that causes a change in the body. Some drugs are so powerful and dangerous that their possession is illegal. Other drugs, used properly, are helpful and can be used legally under the care of a doctor or even purchased over the counter. All drugs, if used improperly, have the potential to be harmful. Drugs differ widely in the ways that they affect the body. Some of the most powerful are those that affect the nervous system and change behavior.

The most widely abused illegal drug is marijuana. It is commonly called grass or pot. The active ingredient in all forms of marijuana is THC. Smoking or ingesting this chemical can produce a temporary feeling of euphoria and disorientation. Short-term use may not be too damaging to the body, but there is evidence that smoking marijuana is harmful to the lungs. Long-term use can cause more serious physical and psychological problems.

Hallucinogens, such as LSD, affect a user's view of reality. Recently, another powerful hallucinogen, known as PCP, has come into use. High doses of this drug can cause seizures or even heart attacks and can induce violent behavior in users.

Stimulants are drugs that speed up the actions of the nervous system. The most powerful stimulants are a group of drugs called amphetamines. These drugs alter the activities of the nervous system because they are similar to naturally produced neurotransmitters found in the body. Heavy users become so dependent on amphetamines that they are unable to function without them. Depressants, on the other hand, reduce the rate of nervous system function. Among the commonly used and abused depressants are the barbiturates that are often found in sleeping pills. Barbiturate users become psychologically dependent on them. Abusers of these drugs who try to quit often find themselves with some serious medical problems and may need professional help to quit.

Cocaine is an illegal drug that was used during the nineteenth century as a painkiller during surgery. Today, it is smoked, sniffed, or injected directly into the nervous system. Cocaine causes the release of a neurotransmitter in the brain called dopamine. Dopamine, usually associated with well-being, is released in abnormally high concentrations as the result of cocaine use and thus gives the user an intense feeling of pleasure and satisfaction. Psychological dependence on cocaine can be particularly difficult to break. A potent and dangerous form of cocaine that can become addictive after only a few doses is crack.

Some of the most powerful known drugs are opiates. Opium, morphine, and heroin are all examples of this type of drug. They are very effective painkillers. Heroin is the most commonly abused opiate. Because of the intense feeling of well-being induced by this drug, heroin becomes psychologically necessary for many users. Opiates are also physically addicting.

Drug abuse is the misuse of either a legal or an illegal drug. This use of drugs in ways that doctors do not approve has become a serious problem in modern society. Drug abuse causes serious physical damage to the body in some cases. In other cases, psychological dependence can totally disrupt the lives of individuals and their families. The uncontrollable craving for a drug is known as drug addiction. Addiction generally describes physical dependence on a drug. Opiates are examples of drugs that cause a strong physical dependence. Withdrawal from drug use can be a very painful process that frequently requires medical attention. Scientists found that heroin and morphine actually bind to special receptors on the surfaces of nerve cells, which accounts for the physical addiction caused by opiates. In other words, addiction has a cellular basis that the addict simply cannot control.

Performance Objectives 46-1

1. Define the term *drug*.
2. Explain how drugs may be helpful or dangerous.
3. Discuss the ways marijuana use can affect the body.
4. Identify some examples of hallucinogens and explain how they affect the body.
5. Explain why amphetamines are potentially dangerous.
6. Discuss depressants and their effects on the human body.
7. Identify some of the dangers associated with cocaine use.
8. Explain what addiction is by using opiate addiction as an example.

CHAPTER PREVIEW

Science Terms 46-1

drug p. 987
marijuana p. 987
hallucinogen p. 988
LSD p. 988
PCP p. 988
stimulant p. 988
amphetamine p. 988
depressant p. 989
barbiturate p. 989
cocaine p. 989
crack p. 990
opiate p. 990
morphine p. 990
heroin p. 990
drug abuse p. 992
drug addiction p. 992
withdrawal p. 992
endorphin p. 992

46-2 ALCOHOL

Section Focus 46-2

The purpose of this section is to introduce alcohol, the oldest drug known to human culture. The type of alcohol found in alcoholic beverages is ethyl alcohol, which is produced from yeast grown in a sugar-containing liquid in the absence of oxygen. Alcohol is a small molecule that passes through cell membranes easily and is quickly absorbed into the bloodstream. High concentrations are toxic, while low concentrations are used by the body as food. Alcohol affects several body systems, but the most immediate effect is that of a depressant on the nervous system.

One of the dangers of alcohol is its cultural acceptance, which has led to its widespread abuse. Many Americans are killed or injured in alcohol-related incidents every year. More than 50,000 babies are born in the United States every year with alcohol-related birth defects. People who have become addicted to alcohol suffer from a disease called alcoholism. If a person cannot function properly without satisfying the craving for alcohol, that person has an alcohol-abuse problem. Continued heavy use of this powerful drug can result in damage to body systems. Several organizations have been formed to help people and their families deal with alcohol dependency because alcoholics frequently need help to stop drinking.

Performance Objectives 46-2

1. Explain how alcohol affects the body.
2. Discuss some of the social problems associated with alcohol use and abuse.
3. Describe some of the characteristics of alcoholism.
4. Explain why organizations such as Alcoholics Anonymous are important.

Science Terms 46-2

ethyl alcohol p. 993
fetal alcohol syndrome p. 994
alcoholism p. 994

46-3 TOBACCO

Section Focus 46-3

The purpose of this section is to discuss the dangers of tobacco. Tobacco contains many substances that affect the body. Nicotine, carbon monoxide, and tar are some of the substances that have been implicated as the cause of many medical problems associated with smoking tobacco. Tobacco is one of the leading causes of premature death in this country. Lung cancer, other lung disorders, other cancers, and heart disease are some of the diseases associated with tobacco use. There is much evidence to indicate that tobacco smoke also affects nonsmokers. Since smoking clearly reduces the expected life span and is powerfully addictive and difficult to give up, the best advice is never to begin smoking in the first place.

Performance Objectives 46-3

1. Name and discuss some of the chemicals found in tobacco that affect the body.
2. Describe some of the diseases associated with tobacco use.
3. Explain the dangers of passive smoking.

Science Terms 46-3

nicotine p. 995
tar p. 995
bronchitis p. 996
emphysema p. 996

DISCOVERY LEARNING

TEACHER DEMONSTRATIONS
Modeling

A "Puppet" Show
This demonstration can be used to introduce Chapter 46.

Perform or have students or guests perform a short puppet show with puppets on strings. You may use any subject for the show that you desire. After the show, introduce the puppeteers, who are all wearing large placards that say "Drugs." Explain to the students that no matter what activities a drug user is involved in or how seemingly successful the user is, the relationship of puppet to puppeteer exists if drugs and the need for drugs are controlling that person's life. You may wish to give the students some opportunity to discuss this topic now, or you may want to wait until a later time.

Drug Education
This activity may be used at any appropriate point in your discussion of the chapter, depending on the way you wish to handle it.

It is time to call in an expert. Many local police departments have officers who are trained to do drug education presentations. Doctors and drug rehabilitation counselors can often talk quite knowledgeably about the physical or psychological aspects of drug use. Public health officials can bring in the topics of other health concerns that drug users face, such as the threat of infection or AIDS. A representative from Alcoholics Anonymous can be asked to speak about the history of the organization and the services of the family support groups. Any of these speakers can interject interest and information into the drug education unit.

CHAPTER 46

Drugs, Alcohol, and Tobacco

GUIDED ENQUIRY

Pose the following questions to students and have them record their responses. Point out that they will gain a better understanding of the key concepts if they read the chapter with these basic questions in mind. Upon completion of the chapter, pose the questions again. Ask students to compare their initial responses with those they have developed after reading the chapter.

- Why do people use drugs?
- Are there any good drugs?
- How can alcohol be bad if it is so widely used and accepted?
- How do drugs affect the body?

INTRODUCING CHAPTER 46

Using the Visuals

Read or ask a student to read the following poem to the class.

> FROM A HEROIN ADDICT
> Beware, my friend,
> My name is King Heroin. . . .
> I'll make a schoolboy forget
>   about his books
> And I'll cause a beauty to
>   neglect her looks.
> I'll make a good husband cast
>   away his wife
> And send a greedy pusher to
>   prison for life. . . .
>
> Ahh, the police have taken you
>   from under my wing.
> They defy me, I who am King.
> They have taken you from me
>   for a short rest.
> But they can't ride, for I am the
>   best.
> I gave you warning, but you
>   didn't heed.
> So put your foot in my stirrup
>   and ride my steed.
> When you ride me, you'll ride
>   me well.
> On a little white horse of Heroin
> You'll ride me to hell. . . .
> —An anonymous addict

Ask the students for their comments. (Students may point out the poem's sexist imagery.) After you have discussed the poem and listened to students' comments, look at the chapter-opener photograph and read the caption. Next look at the opening paragraph. Students may wish to make additional comments.

CHAPTER 46

Drugs, Alcohol, and Tobacco

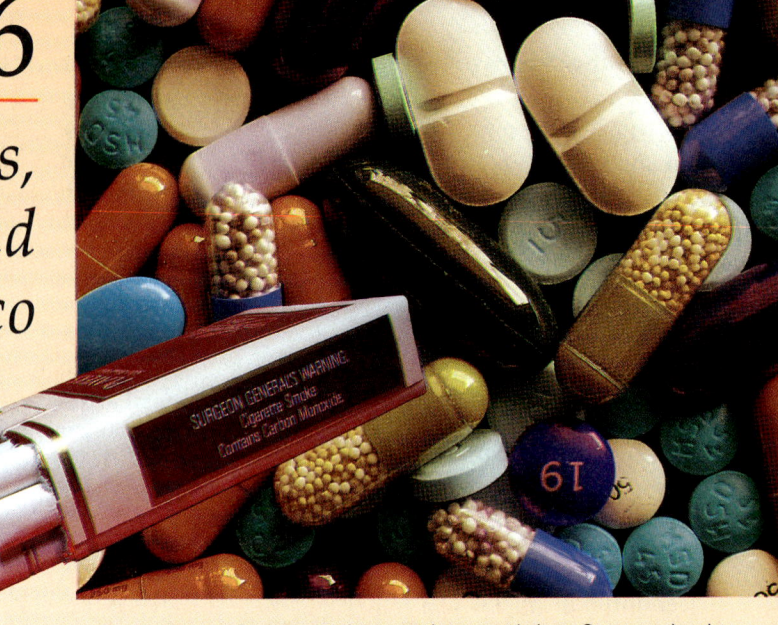

Drugs come in many shapes and sizes. Some are legal and others are not. But any drug, if used improperly, can be dangerous.

It was 1986 and Len Bias was on top of the world. The University of Maryland basketball star had just been drafted by the Boston Celtics. His parents beamed with pride at the success that their son had earned from years of athletic discipline and hard work. Boston sportswriters could barely contain their excitement over what this new star might mean to the team. But one evening, in the space of a few minutes, everything in Len Bias's young life came undone.

Len Bias attended a party with a few friends. Someone took out some cocaine. Bias took a couple of sniffs, moaned, and fell to the floor. A powerful surge of cocaine in his bloodstream had paralyzed his heart muscle. Len Bias was having a heart attack. Before morning, he was dead. Newspaper headlines reported the death of this star athlete. But no headlines were printed for dozens of others who died that same week in exactly the same way. Drug abuse kills!

What are drugs? How do drugs affect the body? And what is drug abuse? Read on to learn the answers.

TEACHING STRATEGY 46–1

Focus/Motivation

Give every student five small pieces of paper (use scratch paper if you have it to avoid waste.) Ask the students to write words representing five things that are

GUIDE FOR READING

After you read the following sections, you will be able to

46–1 Drugs
- Define the term drug and explain why any drug can be harmful.
- Describe the different ways in which drugs can affect the body.

46–2 Alcohol
- Describe the effects of alcohol on the body.
- Compare alcohol use and abuse.

46–3 Tobacco
- Describe the main components of tobacco smoke.
- Discuss the health problems associated with smoking.

Journal Activity

YOU AND YOUR WORLD
Abusing drugs is a habit that sometimes begins when a person is very young. In your journal, write a letter to a youngster warning of the dangers of abusing drugs.

46–1 Drugs

Guide For Reading
- What are some ways in which drugs can affect the body?
- What is drug abuse?

By definition, a **drug** is any substance that causes a change in the body. Many substances fit that definition, including antibiotics that are used to fight infection and aspirin that is used to control pain.

All drugs affect the body in some ways. Some drugs, such as cocaine and heroin, are so powerful and dangerous that their possession is illegal. Other drugs, including penicillin and codeine, are prescription drugs and can be used only under the supervision of a doctor. Still other drugs, including cough and cold medicines, are sold over the counter.

All drugs (legal and illegal) have the potential to do harm if they are used improperly, or abused. In this section we will consider some of the most commonly abused drugs and the ways in which they affect the body.

How Drugs Affect the Body

Drugs differ in the ways in which they affect the body. Some drugs kill bacteria and are useful in treating disease. Other drugs affect a particular system of the body, such as the digestive or circulatory system. Among the most powerful drugs, however, are the ones that affect the nervous system in ways that can change behavior.

MARIJUANA Statistically, the most widely abused illegal drug is **marijuana**. Marijuana comes from a species of hemp plant known as *Cannabis sativa*. Marijuana is commonly called

Figure 46–1 Artist Keith Haring described his attitude toward drug abuse in this anticrack mural.

COOPERATIVE LEARNING

Have groups prepare a pamphlet that explains the dangers of alcohol, tobacco, and drug use during pregnancy. The pamphlet would be available to pregnant women through the local health department. Groups should relate information on substance abuse to the growth and development of the fetus (Chapter 43). Encourage groups to be creative in selecting a design and a writing style for their pamphlet.

Journal Activity

YOU AND YOUR WORLD
Students may want to share their letters with the class. You may also want to use the letters in a class discussion.

dents that when they studied cells, they learned that the body functions through the functioning of the individual cells. Drugs affect cells, sometimes even killing them. Be sure that students understand that many things are considered drugs, including the caffeine in coffee, tea, or soft drinks. Discuss the illegal or improper use of legal drugs, which is one common area of drug abuse.

- **Is it okay to use a drug after the expiration date?** (Answers will vary, but it should be pointed out that doing so could be dangerous for a number of reasons, including the chemical breakdown of the drug.)
- **What about using someone else's prescription to treat your symptoms?** (Answers will vary, but this is an illegal practice.)
- **Have you ever known anyone who uses a drug for a different effect than the recommended use? For example, antihistamines to help one fall asleep or laxatives for weight loss?** (Answers will vary.)

important to them—family, friends, home, car, God, money, and so on. Ask them to wad up the papers and collect them in a waste basket. A person who is dependent on drugs (including alcohol) will actually do what the students have just done symbolically. Ask students to discuss the following questions:
- **How does it feel to think about throwing away your family or college education or God?** (Answers will vary.)
- **More important, how would it be to no longer have the ability to make any free choice?** (Answers will vary.)

Content Development

After you have talked about the definition of the term *drug*, talk about the many ways that drugs can affect the body. Remind stu-

987

TIE-IN LITERATURE

In several of his Sherlock Holmes stories, Sir Arthur Conan Doyle, had his fictional detective experiment with cocaine. Doyle was aware of the drug's value as an anesthetic. It is particularly effective in deadening pain in the nose and throat region. (In fact, it is still used occasionally as a local anesthetic in nose surgery.) More recently, a fictional account of Holmes's alleged addiction to cocaine was published in the novel *The Seven Percent Solution*. It is worth noting that (at least in fiction) even the great Sherlock Holmes needed the help of experts to break the hold of cocaine addiction.

MULTICULTURAL STRATEGY

Perhaps some of the greatest experts at using toxic plants for beneficial purposes are the tribal peoples of Central and South America. The Tarahumara Indians, for example, use toxic plants for a variety of ailments. One such plant is *Ricinus communis,* the source of the laxative castor oil and a number of highly poisonous substances, which is used to treat headaches, boils, and bruises. Another is jimsonweed, which is used as a hallucinogen during rituals and to treat routine ailments such as swellings and headaches.

In most cases, knowledge of the quantity of plant ingested and the way in which it is prepared are essential for the treatment to be safe. The tribal peoples' knowledge is often used by pharmaceutical companies in the development of new drugs for ailments such as heart disease and cancer.

Figure 46–2 Written on this clay tablet are the world's oldest known prescriptions, dating back to about 2000 BC. The prescriptions describe the medicinal uses of certain plants.

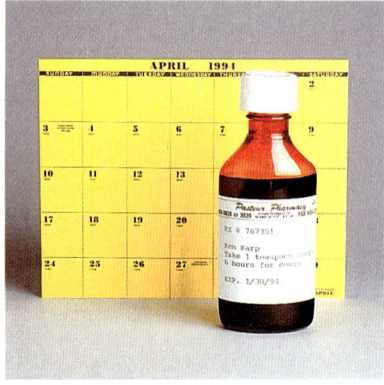

Figure 46–3 Using medication after the prescription has expired is one way that even legal drugs can be abused.

grass or pot. Hashish, or hash, is a potent form of marijuana made from the flowering parts of the plant. The active ingredient in all forms of marijuana is tetrahydrocannabinol (THC). Smoking or ingesting THC can produce a temporary feeling of euphoria and disorientation.

Short-term use of marijuana does not seem to cause immediate physical damage to the body. However, the word *seem* here is deceiving. For there is clear evidence that smoking marijuana is bad for the lungs, although the damaging effects may not be immediate. In fact, smoking marijuana is even more destructive to the lungs than smoking tobacco.

Long-term use of marijuana can result in loss of memory, inability to concentrate, and reduced levels of the hormone testosterone in males. Heavy users develop a psychological dependence (need) on the drug, which can make constructive behavior—work, sports, study, and social activities—almost impossible.

HALLUCINOGENS Some drugs affect a user's view of reality so strongly that they are known as **hallucinogens** (from a Latin word meaning "to dream"). **LSD** (lysergic acid diethylamide) is the most powerful hallucinogen. Acid, as this drug is commonly called, interferes with the normal transmission of nerve impulses in the brain. Although its effects vary from person to person, virtually all people who use LSD regularly have had a "bad trip" in which their hallucinations became frighteningly real. Some LSD users have lost touch with reality after only a single dose of the drug.

In recent years, another hallucinogen has come into use. This powerful drug, called **PCP** (phencyclidine), produces feelings of strength and great power. Also known as angel dust, PCP can result in nightmarish illusions that may last for many days. High doses of PCP produce seizures and even heart attacks. And hospital workers in emergency wards know another side effect of PCP: Users often become extremely violent and are a danger to themselves and others.

STIMULANTS A number of drugs speed up the actions of the nervous system and are therefore known as **stimulants.** The most powerful stimulants are a group of drugs called **amphetamines.** Commonly known as speed or uppers, amphetamines chemically resemble natural neurotransmitters found in the body. You may recall from Chapter 37 that neurotransmitters are compounds that pass nerve impulses from one neuron (nerve cell) to another. When a person takes a dose of amphetamine, the drug floods the body with what the body assumes are natural neurotransmitters. This causes the nervous system to increase its activity, producing a feeling of strength and energy in the user. Fatigue seems to vanish. But there is a dark side to such drugs as well.

The nervous system cannot handle the overstimulation produced by amphetamines. When a dose of the drug wears off,

---

46–1 (continued)

Content Development
As you begin to talk about the effects of drugs—marijuana in particular—on the body, it is appropriate to talk about how different people react to the same drug. The strength of the drug, the duration of its use, the time involved, the presence or absence of food or other drugs, the size and health of the person taking the drug, and tolerance to the drug all influence its overall effect.

Enrichment
Ask interested students to research a particular drug. Ask them to look specifically for information on how the drug biologically affects the body. They may want to access sources available only at a university or

Figure 46–4 Amphetamines are among the most powerful and dangerous drugs known. After it was given a small dose of amphetamine, this orb-weaver spider was unable to spin a normal web.

the user suffers from fatigue and depression. Long-term use causes hallucinations, circulatory problems, and psychological difficulties. Heavy users become so dependent on amphetamines that they are unable to function without them. They have difficulty dealing with other people and fall into a pattern of speeding up and crashing (recovering from the rapid pace of their drug-induced activities).

DEPRESSANTS Drugs that reduce the rate of nervous system activity are called **depressants.** Among the most commonly used (and abused) depressants are the **barbiturates,** a group of compounds often found in sleeping pills. People who abuse downers, as these drugs are called, can quickly become dependent on them. When barbiturates are used with alcohol, the results are often fatal, as the nervous system can become so depressed even breathing stops.

Another danger of barbiturate abuse occurs when a user tries to stop. Unlike virtually all other abused drugs, cutting off the supply of barbiturates to the body can result in serious medical problems that must be treated immediately. Thus a barbiturate abuser needs medical attention when trying to quit.

COCAINE The leaves of the coca plant grown in South America contain a compound known as **cocaine.** In the nineteenth century cocaine was used as a local anaesthetic to deaden pain during surgery. Today, people who abuse cocaine may sniff it, smoke it, or inject it directly into the bloodstream.

Cocaine causes the release of a neurotransmitter in the brain called dopamine. Normally, dopamine release occurs when a basic need, such as hunger or thirst, is satisfied. The release of dopamine in the brain produces a feeling of pleasure (a feeling you have probably experienced after a particularly large Thanksgiving dinner). Cocaine fools the brain into releasing dopamine, producing an intense feeling of pleasure and satisfaction.

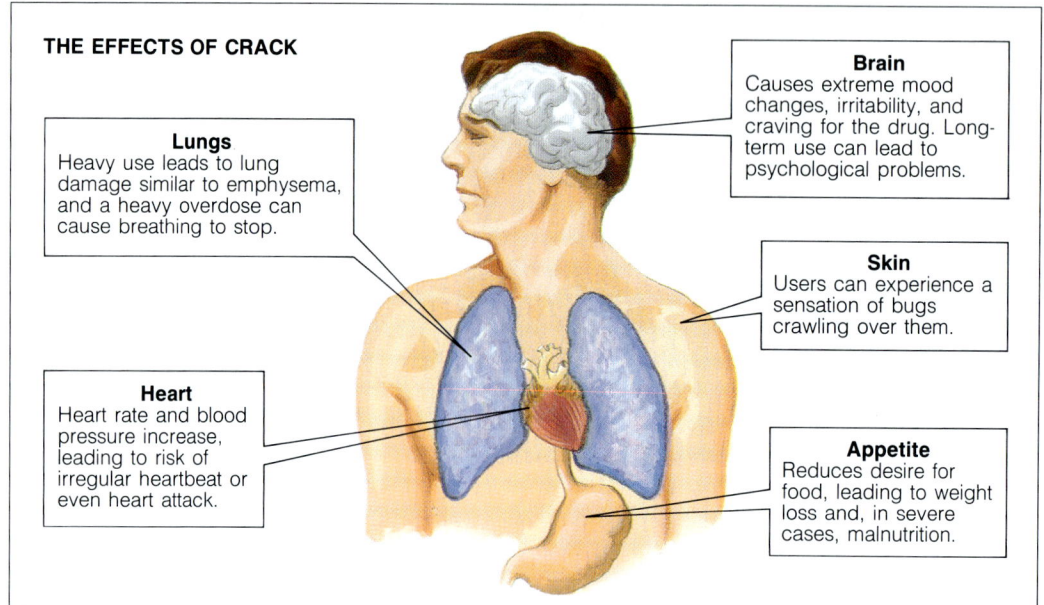

Figure 46-5 *Crack is an extremely powerful form of cocaine. What effect does crack have on the brain?*

The effects of cocaine can be so powerful that a long-term user makes obtaining the drug more important than anything else in life, including food, sleep, and career. So much dopamine is released when the drug is used that not enough is left when it wears off. As a result, users quickly discover that they feel sad and depressed without the drug and seek to use it again and again. Thus the psychological dependence that cocaine produces is particularly difficult to break.

When cocaine reaches the bloodstream, it acts as a powerful stimulant, increasing the heartbeat and blood pressure. The stimulation can be so powerful that the heart is damaged by the drug. In some cases, even a first-time user may experience a heart attack, as happened to basketball star Len Bias.

In the late 1980s cocaine abusers began to use a form of cocaine called **crack,** which can be smoked in a pipe. Crack is a particularly potent and dangerous form of cocaine that can become addictive after only a few doses. Read the Science, Technology, and Society feature in this chapter carefully, for it discusses in more detail the dangers of using crack.

OPIATES Some of the most powerful drugs are the **opiates,** a group of drugs produced from the opium poppy. The most common opiates are opium itself, which is derived directly from the opium poppy, and **morphine** and **heroin,** which are chemically refined forms of opium. All of the opiates, with the exception of heroin, can be used under a doctor's supervision to relieve severe pain. Morphine is particularly effective as a pain killer. However, all of the opiates can also result in death when taken in large doses (overdose).

SCIENCE, TECHNOLOGY, AND SOCIETY ISSUE

Crack: A Cheap Way to Die

For many years, scientists were not certain if cocaine was a physically addictive drug. (Psychological dependence on cocaine is well established.) Cocaine is dangerous and potentially fatal, but true physical addiction to cocaine may take years to develop. However, the same cannot be said for crack, an inexpensive form of cocaine that first appeared in the 1980s. Crack is made by chemically treating cocaine. Crack, which is usually found in the form of small "rocks," can be melted and injected into the bloodstream. However, most users smoke crack. Smoking crack delivers the drug to the nervous system more quickly and in a more powerful form than sniffing even pure cocaine.

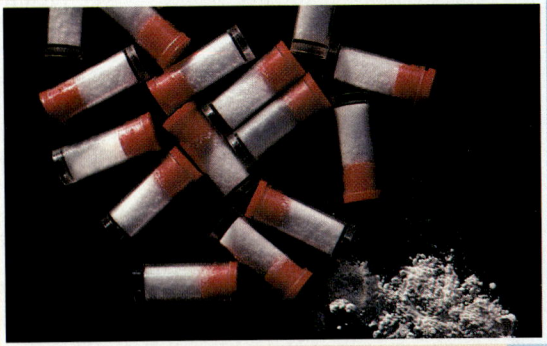

Crack is more powerful than ordinary cocaine because the drug's effect on the nervous system peaks just a few seconds after smoking. The brain is flooded with a concentrated dose of the drug, producing an intense high. The effect of the drug wears off quickly too, leaving the user to seek another dose. After a time, the intense pleasurable effects are no longer felt. Continued use briefly relieves the sense of craving. Crack is among the most addictive drugs known. Crack users can become hooked on the drug in as few as 6 to 10 weeks. (Some studies indicate even less time is required.) And even though an individual dose of crack is inexpensive, the fact that it wears off quickly and is extremely addictive means that continued use is expensive. Where do most crack users get the money to pay for their addiction? The answer is, unfortunately, from violent crime and other illegal acts. Crack also produces disturbed mental behavior in heavy users, which has led to a rash of drug-related murders and assaults.

Scientific work has shown that lab animals supplied with unlimited amounts of cocaine will take the drug again and again until they die from lack of food and water. We humans have the capacity to make better choices than that, and crack is just the latest drug to challenge our ability to say NO.

Knowing what we do about crack, why do people still try the drug—again and again?

Heroin, the most commonly abused opiate, is often injected directly into the bloodstream. Heroin produces a powerful sleepy feeling of well-being that users often crave. As you might expect, heroin can result in strong psychological dependence. Opiates can also cause a strong physical dependence in which the body actually requires the drug in order to function properly. (So can barbiturates, alcohol, and tobacco.) In addition to the many other dangers they face, heroin users run the risk of contracting AIDS from the use of shared needles. (See Chapter 45.)

SCIENCE, TECHNOLOGY, AND SOCIETY ISSUE

Crack: A Cheap Way to Die

For many years scientists questioned whether cocaine was a physically addicting drug. They had long known that it was psychologically addicting. True addiction, if it did occur, might take years to develop. The same thing cannot be said of crack. This inexpensive form of cocaine first appeared in the 1980s. Usually found in the form of small rocks, crack can be melted and injected into the bloodstream, but it is most frequently smoked. This method of delivery gets a concentrated amount of the drug to the nervous system very quickly and results in an intense high. Crack users become hooked on this powerful drug in a very short time.

Even though the individual dose of crack is inexpensive, the results do not last long, so frequent doses are required, which can add up to a big expenditure. Most crack users get the money for their drugs from violent crimes and other illegal acts. The irrational behavior induced by the drug is also associated with murders and assaults.

Lab tests on animals show that animals will take dose after dose of cocaine to the exclusion of food or water until they die. Humans are supposed to be able to make better choices.

Most students, by this point in their education, have had some kind of decision making or refusal training. In small groups or as a class, have students discuss how effective they believe these techniques have been. What additional measures and information do they think would be valuable in helping them, their friends, and especially their younger friends and relatives deal with this very serious problem?

---

some information about heroin users, their general health concerns, and the types of treatment available. You may want to bring in some information about methadone and its successes and problems.

Check students' understanding of the effects of opiates with questions such as the following:
• **Under a doctor's supervision, what is the common use of opiates?** (Relief of pain.)
• **How is heroin most commonly used?** (Injected into the bloodstream.)
• **Is heroin psychologically or physically addictive?** (Both.)
• **Name another health risk that heroin users face.** (Answers will vary, but contracting AIDS is one possibility.)

BACKGROUND INFORMATION
DRUG ABUSE AND LEGAL DRUGS

Drug abuse and addiction need not involve illegal drugs. Some of the most widely abused drugs are prescription tranquilizers, such as Valium and Librium. Others are stimulants. All such drugs have important medical applications and are valuable in treating patients. However, individuals can become dependent on the mood elevation produced by such drugs. They may believe that they need the feeling of calmness and composure that such drugs supply. Dependency on these drugs is dangerous and can produce side effects that cause physical and psychological damage.

46–1 (continued)

Content Development
Ask the students to define the term *abuse*. Abuse can be applied to many things, including food. Ask the students to give some examples of what they think it means to abuse drugs. Somewhere in the discussion, remind them that even legal drugs are frequently abused. You may also want to bring in the topic of steroid use again, since it seems to be a growing problem among young people.

Content Development
Begin the discussion with the following questions:
• **Is addiction the same thing as abuse?** (No.)
• **How are they different?** (Abuse is wrongful use in some way; addiction is an uncontrollable craving for a drug.)
• **What is meant by physical dependence or addiction?** (A biological or chemical need; actually, a demand for a drug.)

Talk about the example given of heroin addiction and the cellular basis for the dependence. Students should be interested in the explanation for the familiar terms "cold turkey" and "kicking the habit."

Figure 46–6 *The round yellow structure in the center of the opium poppy flower is the pod. Notice the sap, from which the opiates are derived, oozing out of the ripe poppy pod.*

Drug Abuse

Each of the drugs we have discussed presents a danger to users. **Drug abuse**—the misuse of either a legal or illegal drug—is a serious problem in modern society. **Drug abuse can be defined as using any drug in a way that most doctors would not approve.** With some drugs, such as cocaine, drug abuse causes serious physical damage to the body. With other drugs, such as marijuana, drug abuse produces psychological dependence that can be strong enough to disrupt family life and schoolwork. Workers under the influence of drugs are unreliable and may commit errors of judgment that place them and their co-workers at risk.

DRUG ADDICTION An uncontrollable craving for a drug is known as a **drug addiction.** As you have read, some drugs cause a strong psychological dependence, or need, in the user, whereas other drugs cause a strong physical dependence. (Many cause both.) In general, the term drug addiction is used to describe a physical dependence on a drug. However, as you now know, even a psychological dependence can have a chemical basis.

Opiates, such as heroin, are examples of drugs that cause a strong physical dependence. All regular users of heroin will eventually become addicted. At that point, their nervous system will become dependent on a steady supply of the drug. Any attempt at **withdrawal,** or stopping the use of the drug, will cause severe pain, nausea, chills, and fever. You may have heard the terms kick-the-habit and cold turkey applied to people who quit using heroin. These terms have a basis in fact. For during heroin withdrawal, a person develops goose bumps that make the skin resemble the skin of a turkey (cold turkey). In addition, the leg muscles of the body may jerk uncontrollably (kicking the habit). The symptoms of withdrawal are so severe that users usually seek another dose of the drug to "cure" them of withdrawal sickness.

Casual users of opiates often believe that they can control their body's need for the drug. But they are nearly always wrong. And now we know the reason why!

In the 1970s, scientists began to look for a cellular basis for opiate addiction. They found that heroin and morphine would bind to special receptors on the surfaces of nerve cells. Why should human nerve cells have receptors for compounds derived from a poppy plant native to Asia?

A group of scientists led by Candace Pert at the National Institutes of Health found the answer: The brain produces its own opiates! These morphinelike chemicals produced by the brain are called **endorphins.** There are several classes of endorphins, and not all endorphin functions are understood. But what is very clear is that endorphins produced by the brain help to overcome pain and produce sensations of pleasure.

SECTION REVIEW 46–1

1. A drug is any substance that causes a change in the body.
2. Drug abuse and addiction are quite different. Drug abuse is the use of any drug in a way most doctors would not approve. Drug addiction is an uncontrollable craving for a particular drug.
3. Stimulants speed up the actions of the nervous system; depressants slow down the actions of the nervous system.
4. Many drugs have a cellular basis for addiction in which the drug either binds to receptors in the nervous system or mimics a natural neurotransmitter. For such drugs, more than willpower or desire may be necessary before the abuser is able to quit.

Now we can understand how opiate addiction occurs. By coincidence, compounds such as morphine and heroin bind to the same receptors as endorphins do, producing a feeling of pleasure and blocking sensations of pain. But the abnormally high levels of opiates reached during drug use upset the normal balance of endorphins and receptors in the brain. Once the body adjusts to the higher levels of opiates, it literally cannot do without them. If the drug is withdrawn, natural endorphins cannot be supplied by the body in large enough amounts to prevent the uncontrollable pain and sickness that are characteristic withdrawal symptoms. Addiction has a cellular basis that the addict simply cannot control!

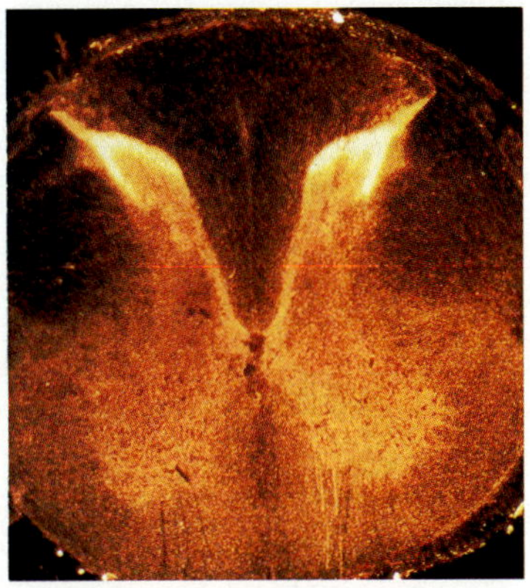

Figure 46–7 The bright spots in this cross section of the spinal cord are some of the receptors in the nervous system to which opiates bind.

46–1 SECTION REVIEW

1. Define the term drug.
2. Distinguish between drug abuse and drug addiction.
3. Compare the actions of stimulants and depressants on the nervous system.
4. **Critical Thinking—Applying Concepts** Explain why withdrawing from an addictive drug is not simply a matter of willpower.

46–2 Alcohol

Guide For Reading

- What type of drug is alcohol?
- What are some effects of alcohol on the body?
- When does alcohol use become alcohol abuse?

Alcohol is a drug—the oldest drug known to human culture. Written records from Egypt and Babylon show that people have made alcoholic beverages for more than 3500 years. Alcohol is produced when yeast grow in a sugar-containing liquid in the absence of oxygen. The yeast ferment the sugar to obtain energy and release alcohol and carbon dioxide as byproducts.

Effects of Alcohol

Alcohol-containing drinks are popular in nearly all cultures. They include fermented drinks, such as beer and wine, and stronger drinks made by distillation, including whiskey, vodka, scotch, and gin. The strength of different drinks depends mainly on the percentage of alcohol. Regardless of the type of drink or its strength, the form of alcohol is always **ethyl alcohol** (C_2H_5OH).

Alcohol is a small molecule that passes through cell membranes easily and is quickly absorbed into the bloodstream. High concentrations of alcohol are toxic. However, very low concentrations of alcohol can be used by the body as a source of food. This is one reason why the effects of alcohol wear off within a few hours after it enters the body.

TEACHING SUPPORT

Study Guide

- Section 46–1, p. 435

Reinforcement/Reteaching

If students are having difficulty answering one or more of the Section Review questions, turn back to the appropriate material and review it.

Closure

List all the drugs named in this section. Ask one or more students to tell you what they know about each drug and its effects on the human body. Also review any terms you want students to remember.

TEACHING STRATEGY 46–2

Focus/Motivation

Bring in several empty containers that once held beer, wine, or other liquor. Point out the government warning label. Discuss why the label has been put on alcoholic beverages.

- **Do you think that this type of warning (there are similar ones on cigarettes) does any good?** (Answers will vary.)
- **Do you think warning labels should be used?** (Answers will vary.)
- **What is your opinion of antidrug and antidrinking advertising on television?** (Answers will vary.)
- **What will work in the war against drugs?** (Answers will vary.)

Content Development

As you talk about alcohol and the history of alcohol use, point out that its social acceptance, in many opinions, is a big part of the problem. Discuss the different types of alcoholic beverages. Sometimes students have the misconception that they cannot really have a drinking problem if they only drink beer. Be sure that they understand that the type of alcohol in all alcoholic beverages is the same. Tell them again, though they may have heard it before, that there is approximately the same amount of ethyl alcohol in a can of beer, a glass of wine, or a standard mixed drink. Talk about the physiological effects of alcohol.

ANNOTATION KEY

① The person becomes unconscious. (Interpreting charts)

TEACHING SUPPORT

Laboratory Manual
- The Effects of Alcohol on Human Reactions, p. 563

Study Guide
- Section 46–2, p. 438

46–2 (continued)

Content Development

Using the Visuals Sometimes students have a difficult time understanding how a depressant can make a person feel "high." Explain that this feeling is a kind of euphoria because nothing is bothering the person and the nervous system is so depressed that it does not "recognize" whatever problems or worries exist. People begin to suffer the side effects of alcohol sometime before they reach the legal level of drunkenness. Depending on the individual and other factors, between the levels 0.01 and 0.04, drinkers may feel dizzy, may say or do things that are uncharacteristic of them, may appear flushed because of the increased rate of blood circulation, and may begin to lose their coordination, making them more prone to accidents. Between 0.05 and 0.09, drinkers cannot think clearly or make wise decisions, their behavior may be uncharacteristic of them, and their loss of coordination is more obvious. Around 0.10, most senses and balance are impaired. About 0.30, vomiting and/or coma may occur. At 0.40 and beyond, drinkers are endangering their lives from alcohol poisoning. Other symptoms are pointed out in Figure 46–8. Remind students again that individual reactions to alcohol vary considerably because of a number of factors.

ALCOHOL'S EFFECTS ON THE BRAIN

BAC	Part of Brain Affected	Behavior
0.05%		Lack of judgment, lack of inhibition
0.1%		Reduced reaction time, difficulty walking and driving
0.2%		Saddened, weeping, abnormal behavior
0.3%		Double vision, inadequate hearing
0.45%		Unconscious
0.65%		Death

Figure 46–8 BAC, or blood alcohol concentration, is a measure of the amount of alcohol in the bloodstream per 100 mL of blood. What happens if the BAC exceeds 0.45 percent? ①

The most immediate effects of alcohol are on the nervous system. **Alcohol is a depressant.** Even small amounts of alcohol slow down the rate at which the nervous system functions. This means that any amount of alcohol slows down reflexes, disrupts coordination, and impairs judgment. Heavy drinking fills the blood with so much alcohol that the nervous system cannot function properly. People who have had three or four drinks in the span of an hour may feel relaxed and confident, but their blood contains as much as 0.10 percent alcohol, making them legally drunk in most states. They usually cannot walk or talk properly, and they are certainly not able to safely control an automobile.

Alcohol is used and accepted by cultures throughout the world. Because of this cultural acceptance, alcohol is the most dangerous and abused drug in the world. More than half of all Americans consume alcoholic beverages. Although many do so quite responsibly, a dangerously large number do not. Alcohol is the drug most commonly abused by teenagers. The abuse of alcohol has a frightening social price. One half of the 50,000 people who die on American highways in a typical year are victims of accidents in which at least one driver has been drinking. One third of all homicides are attributed to the effects of alcohol. At least $25 billion worth of damage is done to the economy in this country alone as a result of alcohol-related accidents that injure workers and damage property.

But the toll of alcohol abuse does not stop there! Women who are pregnant and drink on a regular basis run the risk of **fetal alcohol syndrome,** or damage to the developing baby due to the effects of alcohol. More than 50,000 babies are born in this country every year with alcohol-related birth defects.

Alcohol and Disease

People who have become addicted to alcohol suffer from a disease called **alcoholism.** Some alcoholics may need to have a drink before work or school—every day! They may drink so heavily that they black out and cannot remember what they have done while drinking. Other alcoholics, however, do not necessarily drink to the point where it is obvious that they have an alcohol-abuse problem. **If a person cannot function properly without satisfying the need or craving for alcohol, that person is considered to have an alcohol-abuse problem.**

Repeated bouts of heavy drinking damage the digestive system, through which the alcohol passes on its way into the bloodstream. Alcohol taken in excessive amounts can destroy neurons in the brain. Long-term alcohol use also destroys cells in the liver, where alcohol is broken down. As liver cells die, the liver becomes less able to handle large amounts of alcohol. The formation of scar tissue, known as cirrhosis of the liver, occurs next. The scar tissue blocks the flow of blood through the liver and interferes with its other important functions. Eventually, a heavy drinker may die from chronic liver failure.

Content Development

Many good films and cassettes are available that document the facts of living with alcoholic family members. If students have not had this experience, it is difficult for them to comprehend it. If they have had the experience, they usually prefer not to discuss it. A film or a speaker may be the best way to impress students with the scope of the disease and the problems that it causes. Alcoholics so often refuse to recognize their problem that families become a part of the coverup.

You may be tempted to believe that deaths due to alcoholism are rare. If so, it might surprise you to learn that cirrhosis of the liver is the seventh leading cause of death in the United States! And although attempts have been made in the past to eliminate this drug from our society, alcohol remains with us today. Thus we must each find a way to deal with it.

As with other drugs, dealing with alcohol abuse is not simply a matter of willpower. Alcoholics often need special help and support to quit their drinking habit. Organizations such as Alcoholics Anonymous are available in most communities to help individuals and families deal with the problems created by alcohol abuse. There are even organizations for the relatives of the ten million or so alcoholics in this country. One such organization, Alateen, is for the children of people who have an alcohol problem.

46-2 Section Review

1. What type of drug is alcohol?
2. Distinguish an alcoholic from the millions of Americans who drink but do not have a drinking problem.
3. **Critical Thinking—Making Predictions** Prohibition was a period in our nation's history (1920–1933) during which alcohol was outlawed. Would Prohibition be any more effective today than it was then? Explain.

46-3 Tobacco

Tobacco is a plant native to North America. European explorers learned of the aromatic properties of the tobacco plant and helped to found an industry based on smoking the dried leaves. Today tobacco is used throughout the world. It may be smoked in pipes or in the form of cigarettes and cigars. Tobacco may also be used in the form of chewing tobacco or snuff.

Figure 46-9 More than 25,000 Americans die in car accidents every year in which at least one driver was under the influence of alcohol.

Guide For Reading
- How do the main components of tobacco smoke affect the body?
- What are some health problems caused by smoking tobacco?

Effects of Tobacco

Tobacco contains many substances that affect the body. **Nicotine** in tobacco smoke enters the bloodstream and causes the release of epinephrine, a stimulant that increases the pulse rate and blood pressure. Tobacco smoke also contains carbon monoxide, a poisonous gas that blocks the transport of oxygen by hemoglobin in the blood. (See Chapter 40.) Carbon monoxide decreases the body's supply of oxygen to its tissues, depriving the heart and other organs of the oxygen they need to function. **Tar**, a mixture of complex chemicals in tobacco products, includes a number of compounds that have been shown to cause cancer.

995

Another subject that you may want to discuss because so many students fear it is the idea that the children of alcoholics have a genetic predisposition toward alcoholism. Be sure that students understand that a genetic predisposition does not mean that they will become alcoholics. It simply means that they should be more careful than others. (The children of overweight parents have to be more concerned about their weight than do other people, but they do not all become obese.)

Enrichment
Ask interested students to find out about the Alcoholics Anonymous and family support groups in your community. Ask them to find out when and where these groups meet and how someone could become a participant if he or she wishes to do so. Report this information to the class.

SECTION REVIEW 46-2

1. Alcohol is a depressant.
2. An alcoholic has a craving or need for the drug that impairs judgment and behavior. People who use alcohol but do not have an actual "need" for the drug are not alcoholics.
3. Answers will vary, but most students will likely feel that Prohibition would be no more effective today than it was in the past.

Reinforcement/Reteaching
If students are having difficulty with one or more of the Section Review questions, review the appropriate topics with them.

Closure
Refer to the Performance Objectives for Section 46-2. Ask students to perform each task. Any additional points that you want to review with the students can be formulated as questions and answered by students as well.

TEACHING STRATEGY 46-3

Focus/Motivation
Any of the several simple demonstrations illustrating the residue left by cigarette smoke is effective as a discussion starter. If you wanted to make this a more elaborate demonstration, compare the residue of several brands of cigarettes that make various advertising claims. The bottom line, no matter what they claim, is that all cigarette smoke leaves some residues that end up in a smoker's lungs and respiratory tract.

FACTS AND FIGURES

Alcohol contains many calories but has little food value. The calories are then, in a sense, wasted calories. For this reason, most weight-loss diets severely restrict or eliminate the use of alcohol during the weight-loss period.

Figure 46–10 In addition to these harmful substances, there are over 4000 other compounds given off when a cigarette burns.

When tobacco products are used or smoked regularly, the body develops an addiction to nicotine. For this reason, long-term smokers find it extremely difficult to stop.

Tobacco and Disease

Tobacco is one of the leading causes of premature death in this country. **Lung cancer, the most common form of fatal cancer in the United States, has been directly linked to smoking.** Because lung cancer may take twenty years to develop, smokers may have a false sense of security about their habit. However, even in younger smokers the signs of destruction and illness are unmistakable.

LUNG DISORDERS By age 45, the death rate from lung cancer for smokers is four times that for nonsmokers. By age 65, it is more than ten times greater. For every smoker there is a serious risk of developing lung cancer. But lung cancer is not the only risk! There are other lung disorders associated with smoking. Smoke particles become trapped in the linings of the breathing passageways. At first these linings cleanse themselves. But as smoking continues, the cells of the linings are damaged. In time, smokers may develop a persistent cough as their lungs attempt to clear themselves. Smokers also suffer higher rates of respiratory infections than nonsmokers. They often develop chronic **bronchitis,** an inflammation of the large breathing tubes (bronchi) in the lungs. Long-term smoking can lead to **emphysema,** a stiffening of the normally elastic tissues of the lung.

Figure 46–11 This photograph shows the bronchial walls in the lungs of a nonsmoker (left). Notice the cancer cells invading the bronchial wall of a smoker (right). As you can imagine, it was too late for this person to quit.

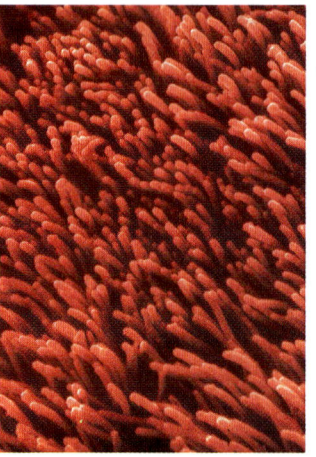 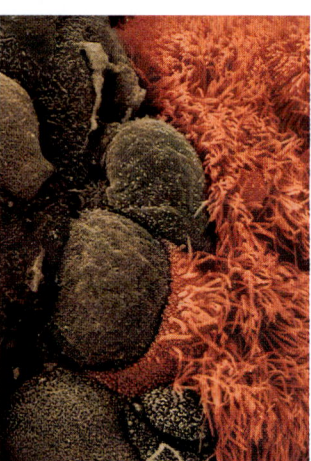

CIRCULATORY DISEASES The steady inhalation of carbon monoxide seems to cause a slow poisoning of the heart. Smoking also constricts, or narrows, blood vessels. Constriction of blood vessels causes blood pressure to rise and makes the heart work harder. It is no real surprise that statistics indicate that smoking doubles the risk of death from heart disease for men between the ages of 45 and 65. Moreover, for every age group and for both sexes, the risk of death from heart disease is greater among smokers than nonsmokers.

OTHER CANCERS Tobacco smokers also suffer from higher than normal levels of mouth and throat cancer, probably due to chemicals in tobacco tar. People who chew tobacco or inhale snuff should not be fooled into thinking these forms of tobacco are safe. Tobacco-chewers, for example, have extremely high rates of oral cancer and often fail to detect these cancers in time to prevent damage to the mouth and face.

SMOKING AND THE NONSMOKER In recent years evidence has clearly shown that the dangers of smoking are not restricted to the smoker. Tobacco smoke in the air is damaging to anyone who inhales it, not just the smoker. For this reason, many states require restaurants to have smoking and nonsmoking sections. And in many parts of the country, smoking in public places has been restricted, if not prohibited.

Passive smoking, or inhaling the smoke of others, is particularly damaging to young children. Studies now indicate that the children of smokers are twice as likely to develop respiratory problems as are children of nonsmokers.

DEALING WITH TOBACCO Only 30 percent of male smokers live to the age of 80. About 55 percent of male nonsmokers do. Clearly, smoking reduces expected life span. Moreover, whatever the age and no matter how long a person has smoked, a person's health can be improved by quitting. But tobacco is a powerful drug with strong addictive qualities that make it very difficult to give up. Thus, considering the cost, the medical dangers, and the chemical power of addiction, the best solution is not to begin smoking. Period!

46-3 SECTION REVIEW

1. Describe the main components of tobacco smoke and their effects on the body.
2. What are some disorders that result from smoking?
3. **Critical Thinking—Relating Cause and Effect** What is the best advice you can give to a smoker who has not shown any ill effects, as yet, from the habit?

SECTION REVIEW 46-3

1. Nicotine: increases pulse rate and blood pressure, poisons many tissue; tar: irritates the lungs and can cause cancer; carbon monoxide: blocks the transport of oxygen by hemoglobin.
2. Lung disorders: lung cancer, persistent cough, bronchitis (inflammation of the breathing passages), emphysema (stiffening of the elastic tissues of the lung). Circulatory diseases: poisoning of heart tissues, constriction of blood vessels, rise in blood pressure. Other cancers: cancer in the mouth, throat, and gums.
3. Stop now, before things get worse.

LABORATORY INVESTIGATION

OBSERVING THE EFFECT OF A STIMULANT ON THE HEART RATE

Before the Lab

1. Gather the materials needed for each group.
2. Check the culture to be sure that there are adequate numbers of *Daphnia*.
3. Prepare the coffee solution in the strength you want to use. It is probably a good idea to perform this lab before class so that you can help students with the pitfalls they are likely to encounter. You may want to vary the concentration of the coffee solution to determine which percentage will give the best results.
4. You may want to provide a dissecting scope to help students get their sample *Daphnia*.

Pre-Lab Discussion

1. Look with the students at the picture or other visual of the *Daphnia* and point out the approximate location of the heart.
2. Familiarize the students with depression slides and how they are to use them.
3. Demonstrate the correct use of the stopwatch.
4. If the microscopes are on unsteady tables, you may want to suggest an alternative to tapping the table (snapping the fingers, quietly clapping the hands, or whatever works for you); tapping the table may move the specimen on the microscope.
5. Go over any points you noticed when you performed the lab at which students are likely to encounter difficulties.
6. Caution students to follow the directions carefully and record data accurately.

Skills Development

Students will use the following skills in the completion of this laboratory investigation.
1. Manipulative
2. Observing
3. Measuring
4. Calculating
5. Applying
6. Relating
7. Inferring

LABORATORY INVESTIGATION

OBSERVING THE EFFECT OF A STIMULANT ON THE HEART RATE

PROBLEM

How does caffeine affect the heart rate of a *Daphnia*?

MATERIALS (per group)

coffee (solution)
coverslip
Daphnia
microscope
depression slide
medicine dropper
paper towel
stopwatch
calculator

PROCEDURE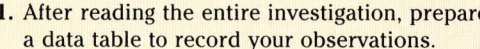

1. After reading the entire investigation, prepare a data table to record your observations.
2. Use the medicine dropper to withdraw a *Daphnia* from the container in which it is stored. Place the *Daphnia* and a few drops of water from the container in the center of a clean depression slide. Cover the *Daphnia* with a coverslip.
3. Examine the *Daphnia* under low power. Because the *Daphnia* is transparent, it is possible to observe its internal organs. Using the accompanying diagram, locate the labeled organs.
4. Set the stopwatch to zero. Locate the beating heart. Gently tap your finger on the table in time with the heartbeat. When you have found the rhythm of the heart, start the stopwatch. Count 20 beats and stop the stopwatch. Record the amount of time it took for the heart to beat 20 times.
5. Repeat step 4 for two more trials. Calculate the average time for 20 beats. Record your results.
6. Calculate the heart rate in beats per minute by dividing 1200 by the average time for 20 beats. Record the result.
7. Use the medicine dropper to put two drops of coffee at one edge of the coverslip. Place a small piece of paper towel at the other edge of the coverslip. As the paper towel soaks up liquid, the *Daphnia* will be exposed to the caffeine in the coffee.
8. Determine the effects of caffeine on the heartbeat by repeating steps 4 through 6.

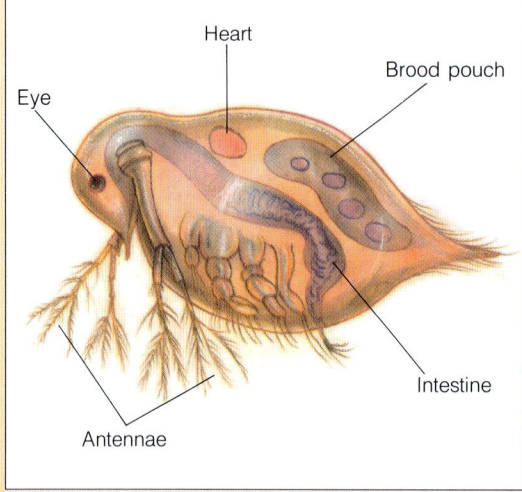

OBSERVATIONS

1. Describe the *Daphnia*'s heartbeat.
2. What was the *Daphnia*'s heart rate before exposure to caffeine? After exposure to caffeine?

ANALYSIS AND CONCLUSIONS

1. Why is the *Daphnia* an ideal organism to use in this type of investigation?
2. What effect did caffeine have on the *Daphnia*'s heart rate?
3. What type of drug is caffeine? How do you know?

Teaching Strategy

Monitor to be sure that students have correctly constructed their data tables. You may have to help them with their laboratory techniques or the operation of the stopwatch. The counting is tricky and may be difficult for some students. (Sometimes altering the lighting makes this lab much easier. If students are having a lot of trouble, you might try that.) Check to be sure that students are following directions and accurately recording the data.

STUDENT STUDY GUIDE

SUMMARIZING THE CONCEPTS

The key concepts in each section of this chapter are listed below to help you review the chapter content. Make sure you understand each concept and its relationship to the theme of this chapter.

46-1 Drugs

- A drug is any substance that causes a change in the body.
- All drugs, when used improperly, have the potential to do harm.
- The most widely abused illegal drug in the United States is marijuana.
- Hallucinogens, such as LSD and PCP, are powerful drugs that change the user's view of reality.
- Stimulants, such as amphetamines, speed up the actions of the nervous system.
- Depressants, such as barbiturates, slow down the actions of the nervous system.
- Cocaine, derived from the coca plant, is a stimulant that produces very strong psychological dependence. Use of cocaine can damage the heart and cause a heart attack. Crack is a form of cocaine that is highly addictive and has many serious effects on the body.
- Opiates, such as heroin and morphine, can be used to reduce pain. Opiates are addictive and lead to strong physical dependence.
- Drug abuse can be defined as using any drug in a way most doctors would not approve.
- People addicted to drugs may suffer withdrawal symptoms when cut off from their drug supply.

46-2 Alcohol

- Alcohol is a depressant that easily passes through cell membranes. Even a small amount of alcohol acts to slow down the actions of the nervous system.
- People who are addicted to alcohol are said to suffer from a disease called alcoholism.

46-3 Tobacco

- Several substances in tobacco—among them nicotine, tar, and carbon monoxide—can lead to serious health problems, including lung cancer, other cancers, respiratory problems, and circulatory problems.

REVIEWING KEY TERMS

Vocabulary terms are important to your understanding of biology. The key terms listed below are those you should be especially familiar with. Review these terms and their meanings. Then use each term in a complete sentence. If you are not sure of a term's meaning, return to the appropriate section and review its definition.

46-1 Drugs			46-2 Alcohol	46-3 Tobacco
drug	amphetamine	heroin	ethyl alcohol	nicotine
marijuana	depressant	drug abuse	fetal alcohol	tar
hallucinogen	barbiturate	drug addiction	syndrome	bronchitis
LSD	cocaine	withdrawal	alcoholism	emphysema
PCP	crack	endorphin		
stimulant	opiate			
	morphine			

Observations

1. The heart appears to be pulsating as it beats.
2. The heart rate will vary, but it will be about 214 before exposure to caffeine and about 240 after exposure to caffeine.

Analysis and Conclusions

1. It is almost transparent, so the heart is easy to observe.
2. It increased the heart rate.
3. Caffeine is a stimulant, since it increases the heart rate, which is a characteristic of stimulants.

Going Further: Enrichment

Now that you have determined the effects of caffeine and the type of drug it is, design an experiment to test the effects of other drugs or types of drugs.

CHAPTER REVIEW

CONTENT REVIEW

Multiple Choice
1. c   2. d   3. b   4. a
5. b   6. a   7. b   8. b

True or False
1. F withdrawal
2. F depressant
3. T
4. F Carbon monoxide
5. T
6. T
7. F endorphins
8. F ethyl alcohol

Word Relationships

A.
1. drugs that speed up the actions of the nervous system; barbiturate is a depressant
2. drugs that reduce the rate of nervous system activity; tobacco contains nicotine, which causes the release of the stimulant epinephrine
3. components of tobacco smoke; ethyl alcohol is the drug found in alcoholic beverages
4. opiates, which are drugs produced from the opium poppy; hallucinogen is a drug that strongly distorts a user's view of reality and may be derived from a variety of sources

B.
5. stimulants
6. dopamine
7. addiction
8. emphysema

CONCEPT MASTERY

1. Alcohol is a depressant. However, when it first acts on the body, it may cause a giddy, happy feeling that can be misinterpreted as a stimulating effect.
2. Alcohol is absorbed through the cell membranes of the digestive system and enters the bloodstream. Its primary effects are on the brain as a depressant, and it slows down the actions of the central nervous system.
3. Psychological dependence occurs when a person needs or desires a drug to achieve a feeling that person believes cannot be achieved without the drug. There is no physical craving for the drug by the body with psychological dependence. Physical dependence, on the other hand, has a physiological cause. In physical dependence, the body needs the drug to function properly.
4. They are incorrect, since tobacco chewing can lead to cancers of the mouth and oral area.
5. Smoking is difficult to quit, even a second time. Furthermore, the damage done by smoking can be alleviated over time by the body as the lungs recover. By starting again, the smoker will merely damage the lungs once again or keep the lungs from healing normally.
6. An alcoholic has a physical or psychological need or craving for the drug. Whether this need is daily or occasional, as long as it is a need that cannot be overcome and affects the person's life, then that person can be considered an alcoholic.
7. Answers will vary but should reflect the content of the chapter. Drug abuse is defined by WHO, the World Health Organization, as the use of any drug in a way most physicians would not approve.

CONTENT REVIEW

Multiple Choice

Choose the letter of the answer that best completes each statement.

1. Over-the-counter drugs include
   a. amphetamines.   c. aspirin.
   b. depressants.   d. codeine.
2. Tetrahydrocannabinol is the active ingredient in
   a. opiates.   c. alcohol.
   b. cocaine.   d. marijuana.
3. Drugs that speed up the actions of the nervous system are
   a. prescription drugs.   c. depressants.
   b. stimulants.   d. opiates.
4. A drug that causes the release of dopamine in the brain is
   a. cocaine.   c. tobacco.
   b. alcohol.   d. marijuana.
5. Cirrhosis is a condition that affects the
   a. heart.   c. lungs.
   b. liver.   d. digestive system.
6. Alcohol is a
   a. depressant.   c. hallucinogen.
   b. stimulant.   d. prescription drug.
7. Which substance is found in tobacco smoke?
   a. PCP   c. LSD
   b. tar   d. THC
8. Morphine binds to the same receptor sites in the brain as
   a. cocaine.   c. THC.
   b. endorphins.   d. LSD.

True or False

Determine whether each statement is true or false. If it is true, write "true." If it is false, change the underlined word or words to make the statement true.

1. When opiate users stop using the drug, they undergo <u>dependence</u>.
2. Alcohol acts as a <u>stimulant</u> on the nervous system.
3. <u>Barbiturates</u> are examples of depressants.
4. <u>Carbon dioxide</u> is a poisonous gas found in cigarette smoke.
5. A <u>drug</u> is any substance that has an effect on the body.
6. The most powerful stimulants are the <u>amphetamines</u>.
7. Morphinelike chemicals produced by the brain are called <u>receptors</u>.
8. The form of alcohol used in beverages is <u>methyl alcohol</u>.

Word Relationships

A. *In each of the following sets of terms, three of the terms are related. One term does not belong. Determine the characteristic common to three of the terms and then identify the term that does not belong.*

1. cocaine, stimulant, barbiturate, amphetamine
2. alcohol, tobacco, barbiturate, depressant
3. ethyl alcohol, tar, nicotine, carbon monoxide
4. hallucinogen, opium, morphine, heroin

B. *Replace the underlined definition with the correct vocabulary word.*

5. Amphetamines are <u>drugs that speed up the actions of the nervous system</u>.
6. Cocaine causes the brain to release <u>a chemical that produces a feeling of pleasure</u>.
7. People who abuse heroin develop <u>an uncontrollable craving for the drug</u>.
8. Long-term smoking of tobacco can lead to <u>stiffening of the elastic tissues in the lungs</u>.

CONCEPT MASTERY

Use your understanding of the concepts developed in the chapter to answer each of the following in a brief paragraph.

1. Many people who drink alcohol state that it makes them more peppy. Explain why that is not the case.
2. Describe the physical effects of alcohol.
3. Compare psychological dependence and physical dependence.
4. Tobacco-chewers often think that their habit is much safer than tobacco smoking. Are they correct? Explain your answer.
5. Suggest some reasons why an ex-smoker should not start smoking again.
6. Not all alcoholics need to drink every day. Explain that statement.
7. Define drug abuse in your own words.
8. It has been said that no one can ever be cured of drug dependence. Explain why.
9. How is cigarette smoke related to respiratory and circulatory problems?

CRITICAL AND CREATIVE THINKING

Discuss each of the following in a brief paragraph.

1. **Relating cause and effect** Based on alcohol's effects on the nervous system, why is drinking and driving an extremely dangerous behavior?

2. **Making comparisons** Compare cocaine and crack.
3. **Expressing an opinion** How might you convince someone not to abuse drugs?
4. **Relating facts** Compare the actions on the nervous system of stimulants and depressants.
5. **Applying concepts** Explain the meaning of this ancient Japanese proverb: "First the man takes a drink, then the drink takes a drink, then the drink takes the man."
6. **Making inferences** In what ways are drug abuse and criminal acts related?
7. **Using the writing process** Choose any commonly abused drug and construct a poster designed to display the dangers of drug abuse.

1001

CRITICAL AND CREATIVE THINKING

1. A person who has drunk even one drink begins to lose some coordination, and his or her reaction time slows down because of the depressant effects of alcohol. Such a person cannot drive properly, and as the person ingests more drinks, it becomes less and less safe for the person to drive.
2. Both are stimulants that cause a release of dopamine by the brain. However, crack is an extremely potent form of cocaine and can lead to physical addiction very quickly. Scientists are still not sure if cocaine can lead to physical addiction, but if it does, it takes much longer than crack.
3. Accept all logical answers.
4. Stimulants speed up the actions of the nervous system; depressants slow down the actions of the nervous system.
5. The proverb implies that even moderate drinking can lead to excess, which can be fatal in some cases.
6. Since drug abuse often leads to psychological or physical dependence (or both), the user must obtain the drug in any way possible. Since drug abuse is also expensive, the user is often led to crime to pay for his or her habit.
7. Students' posters should reflect the content of this chapter. You may want to display imaginative posters in a hallway outside your classroom or even set up a poster competition among your classes.

8. Once a person is dependent on a drug—even if the person quits using that drug—there is always the danger that the person will begin again. And if that person begins again, the dependence will begin anew as well, and very quickly.
9. Chemical compounds in the smoke can damage the respiratory system and possibly lead to lung cancer, bronchitis, and emphysema. Compounds in smoke also increase the heart rate and blood pressure, putting a strain on the heart. Some compounds in smoke are also poisonous and can damage heart tissues.

CAREERS IN BIOLOGY

Physical Therapist Assistant, Dietician, Biomedical Engineer

Students interested in a biological career should be instructed to write for further information to the address listed beneath each career description. However, in consideration of the organizations that provide career information, please have only one student write to an organization. Do not instruct the entire class to write to every organization for the same career information. You may want to use the information provided to start a biology career file for student use.

UNIT 9 — CAREERS IN BIOLOGY

Physical Therapist Assistant

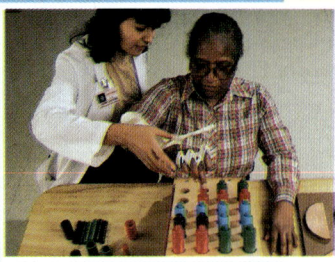

People who have lost partial or total use of a limb or joint due to surgery, injury, or disease can often regain their strength through careful exercise and therapy. Such therapy may be provided by a physical therapist assistant who works under the supervision of a physical therapist. Physical therapist assistants teach patients to exercise properly on different types of equipment. They also observe and record patients' progress.

High school courses in health, biology, and physical education along with some specialized training in physical therapy will help prepare a physical therapist assistant.

For more information write to the American Physical Therapy Association, 1111 N. Fairfax St., Alexandria, VA 22314.

Dietician

Dieticians study the chemical makeup of foods and the effects of foods on the human body. They use this knowledge to plan meals for groups and individuals. Some dieticians work in the management of food equipment and staff, others work directly with patients, and still others perform research on the nutritional needs of the human body.

A Bachelor's degree with a major in foods and nutrition or institutional management is necessary for a career as a dietician.

For more information write to the American Dietetic Association, 430 N. Michigan Ave., Chicago, IL 60611.

Biomedical Engineer

Biomedical engineers use a knowledge of engineering to solve problems in medicine and biology. They are involved in the design of technology used in and on the human body—for example, artificial organs, life support systems, and surgical lasers. Biomedical engineers often participate in research on life systems.

To become a biomedical engineer, a Bachelor's degree in a major field of engineering and an advanced degree in biomedical engineering is needed.

For more information write to the Biomedical Engineering Society, P.O. Box 2399, Culver City, CA 90231.

HOW TO PREPARE FOR A JOB INTERVIEW

A job interview gives you a chance to meet an employer and express your qualifications for a job. The interview is a deciding factor in whether or not you get the job.

You should first find out about the company at which you are having the interview. Information on the history, structure, and financial status of different companies can be found at your local library.

Before the interview, you might want to call to confirm the date and time of the interview. You should dress neatly and properly for any job interview, and you should take along any information that might be requested, such as a résumé. Be prepared to answer questions about yourself, your goals, your background, and why you want to work for that particular company.

FROM THE AUTHORS

FROM THE AUTHORS

Kenneth R. Miller

This letter from Ken Miller is a natural for class discussion. Ken points out the various time periods in which experts in different disciplines would like to have been alive. He also makes the case that for a biologist, today is the most exciting time to study and do research. List various other careers, both in science and in other fields of study, and have students suggest when the best time to be alive would be for each career.

I am always amazed at how many *facts* find their way into a book on biology, a book like this one. Charts, tables, diagrams, lists, and terms seem to occupy every page. It's easy to be overwhelmed. It's also possible to be convinced that biology is a closed science—to think that with so many facts crammed into one book that we must know just about everything. Of all the *mistakes* that you could make as a student, that would be the greatest—to believe that all the great scientists and great discoveries in biology are behind us. The great hidden truth is that we have barely begun to understand life.

Some of my friends who are experts in other subjects have told me that they would like to have lived at a different time. Philosophers sometimes wish that they could have lived in ancient Greece and talked with Plato and Socrates. Physicists wish that they could have worked with Einstein in the early part of the twentieth century when he developed his theories. An artist I know wishes that he could have been in France when the Impressionist movement began in the 1870s.

What time in history would have been the most exciting for a biologist? That's an easy question to answer. Recent history suggests that the greatest achievements for biology are still ahead of us. The last ten years of scientific research have produced one discovery after another that has revolutionized our understanding of living things. Despite all of this, some of the most fundamental questions about how life is produced, adapts to its surroundings, and is passed along to new generations are still unanswered. Right now we are closer to answering those great questions than at any time in human history, and *that* is exciting. As I said, it's easy to pick a favorite time in history. But the best time for a biologist to be alive is right now!

Ken Miller

UNIT 10
Ecological Interactions

UNIT OVERVIEW

In Unit 10 students are introduced to the relationships and interactions of living things with one another and with their environment. They first learn about the different ecosystems and biomes and the movement of energy and nutrients throughout the biosphere. They go on to learn how populations grow and interact and about interactions within and between communities. They study human population growth and its effect on the environment and they learn about the various types of pollution and the effects of each. Finally, they read about the consequences of certain activities that alter the environment and exhaust natural resources, and they consider the important decisions that control the future of the biosphere.

UNIT OBJECTIVES

1. Define ecology and relate it to the biosphere and its ecosystems.
2. Compare the characteristics of the various biomes.
3. Describe the flow of energy and the movement of nutrients through the biosphere.
4. Define population, analyze curves that represent population growth, and discuss factors that control population growth.
5. Describe interactions within and between communities.
6. Discuss natural resources and their conservation and differentiate among the various types of pollution.

UNIT 10
Ecological Interactions

Like all living things, these elk are involved in delicately balanced interactions with one another, with other living things, and with their physical surroundings.

The habitat shown in this photograph has been devastated by a volcanic eruption—in this case, Mount Saint Helens in Washington State. But ecosystems have a way of healing themselves. Borne on the wind and carried on the fur, feathers, and hooves of animals, seeds will take root in soil enriched by volcanic ash. In time, plants will recolonize the area, and animals will return.

Although life may have the potential to re-establish itself after a disaster, there is no guarantee that it will. Because it is not clear how much an ecosystem can take before it is pushed over the edge, it is vitally important that we understand how ecosystems work. Life—including human life—depends on the well-informed management and use of natural resources.

DISCOVERY LEARNING
A WEIGHTY MATTER

Did you ever wonder how much paper goes through your house in a week? Why not find out?

1. Obtain a large cardboard box or a large plastic garbage bag. You will be storing the paper that you collect in this box or bag.
2. Each day, put the unwanted mail, old newspapers and magazines, shopping bags, and other clean waste paper into the box or bag.
3. Every two to three days, weigh the box and its contents. (*Hint:* It might be easier for you to hold the box when you step on a bathroom scale. Have a friend or relative record the reading on the scale. Then subtract your weight to determine the weight of the box and its contents.)
4. At the end of the week, dispose of your paper wastes properly.
   - How much waste paper did you collect in a week?
   - How much paper would you collect in a year?

INTRODUCING UNIT 10

Using the Visuals

Have students look at the photograph on pages 1004 and 1005 and read the paragraphs that introduce the unit.

• **What do you see in this photograph?** (A herd of elk in a barren landscape.)

• **What happened to the elk's environment?** (It was devastated by a volcanic eruption.)

• **Predict how this landscape will look 10 years after this picture was taken.** (Answers may vary. The land will probably be recolonized with grasses, shrubs, and small trees.)

• **What do you think will happen to this region's elk population?** (The landscape as shown cannot support elk or any other large animal life. But if and when

CHAPTERS

47
The Biosphere

48
Populations and Communities

49
People and the Biosphere

CHAPTER DESCRIPTIONS

47 The Biosphere
Chapter 47 begins with a discussion of ecosystems and ecological succession. The various land and aquatic biomes are described in terms of life forms and physical characteristics. The flow of energy through the biosphere, using ecological pyramids, is then explained, and the movement of nutrients through biogeochemical cycles is discussed. Finally, food webs and food chains are described.

48 Populations and Communities
Population growth, including exponential and logistic growth, is described in Chapter 48. Next, density-dependent and density-independent factors, which limit population growth, are discussed. Finally, interactions among organisms, including symbiotic relationships, are explained.

49 People and the Biosphere
Chapter 49 begins with a treatment of human population growth and its effect on the environment. Next, various types of pollution are explained and their effects analyzed with regard to biological magnification. The importance of protecting species from extinction is discussed. Finally, the ramifications of decisions that affect the future of the biosphere, such as conservation of natural resources, are considered.

vegetation returns to the land, elk herds may thrive there again.)

DISCOVERY LEARNING

A WEIGHTY MATTER

To introduce the unit, have students perform this Discovery Activity. Remind students that they are to save only clean waste paper; saving the papers used in food packaging would be unwise and unsanitary.

Have students answer the questions given in the activity, as well as the questions that follow.
• **Assuming that everyone produces roughly the same amount of trash, how much paper trash is produced in one week by your class, school, and town?**
• **What do you think is your weekly production of other solid waste—such as cans, glass bottles, and plastics?**

CHAPTER 47 The Biosphere

Section	Laboratory Investigations and Activities
47–1 Earth: A Living Planet pages 1007–1010 THEMES: Systems and Interactions, Scale and Structure	**Student Edition** LABORATORY INVESTIGATION: Observing Succession in Aged Tap Water, p. 1028 **Teacher Edition** DEMONSTRATION: Observing an Ecosystem, p. 1006D **Teaching Resources** ECOLOGY INVESTIGATION: Ecological Succession in a Rotten Log, p. 55
47–2 Land Biomes pages 1010–1016 THEMES: Unity and Diversity, Stability	**Teacher Edition** DEMONSTRATION: The Earth's Biomes, p. 1006D **Laboratory Manual** Adapting to the Cold, p. 577 **Teaching Resources** HANDS-ON ACTIVITY: Grandeur in the Grass, p. 215 HANDS-ON ACTIVITY: Cutting Down the Rain, p. 217 ECOLOGY INVESTIGATION: Indoor Ecosystem: The Woodland Terrarium, p. 47 ECOLOGY INVESTIGATION: Indoor Ecosystem: The Desert Terrarium, p. 49
47–3 Aquatic Biomes pages 1016–1029 THEMES: Unity and Diversity, Scale and Structure	**Teaching Resources** HANDS-ON ACTIVITY: Hold the Salt! p. 223 ECOLOGY INVESTIGATION: Indoor Ecosystem: The Bog Terrarium, p. 51
47–4 Energy and Nutrients: Building the Web of Life pages 1021–1027 THEMES: Energy, Systems and Interactions	**Laboratory Manual** The Oxygen Cycle, p. 573 **Teaching Resources** HANDS-ON ACTIVITY: A Saucepan Simulation of a Cycle, p. 211 HANDS-ON ACTIVITY: Garbage in the Garden, p. 219 ECOLOGY INVESTIGATION: Comparing the Water-Holding Capacities of Soils, p. 53
Chapter Review pages 1029–1031	

CHAPTER PLANNING GUIDE

Other Activities	Media and Technology
Teaching Resources BIOLOGY CASE STUDY: Temperature Variation, p. 17 BIOLOGY CASE STUDY: Investigating Terrestrial Succession, p. 21 **Study Guide** Section 47–1, p. 445	**Biology Transparencies** 71: The Stages of Ecological Succession on Land
Teaching Resources ACTIVITY: Comparing Biomes of the United States, p. 3 **Study Guide** Section 47–2, p. 448	**Interactive Videodisc** On Dry Land: The Desert Biome **Interactive Videodisc/CD ROM** Amazonia **Video/Videodisc** Biomes: Desert and Tundra Biomes: Temperate Deciduous Forest and Grasslands Biomes: Coniferous Forest and Tropical Rain Forest **Biology Transparencies** 72: The Major Land Biomes **Biology Media Guide** pages 72–73
Teaching Resources ACTIVITY: Finding the Right Biome, p. 5 **Study Guide** Section 47–3, p. 450	**Videodisc** Aquatic Ecosystems: Estuaries and Marine Aquatic Ecosystems: Freshwater Wetlands and Freshwater **Biology Media Guide** pages 73–74
Teaching Resources VOCABULARY REVIEW: Cross-a-Clue, p. 1 **Study Guide** Section 47–4, p. 453	**Interactive Videodisc** In the Company of Whales **Videodisc** Aquatic Ecosystems: Estuaries and Marine **Biology Transparencies** 73: A Food Web **Biology Mcdia Guide** pages 74–75
Teaching Resources Chapter Test A, p. 13 Chapter Test B, p. 17	**Computer Test Bank** Chapter Test, p. 493

CHAPTER 47 The Biosphere

CHAPTER OVERVIEW

With this chapter students begin their study of ecology, the study of interactions of organisms with one another and with their physical surroundings. The chapter is titled "The Biosphere," which is that part of the Earth on which life exists.

Ecologists look at the biosphere in terms of ecosystems, which are a combination of all the biotic and abiotic features in a given area. Many ecosystems change over time in a process of ecological succession, in which an existing community of organisms is replaced by a new community. Eventually, this process leads to a climax community, which is comparatively stable.

Ecologists also look at the biosphere in terms of larger units called biomes. Biomes are environments that have characteristic climax communities. The major land biomes are the tundra, taiga, temperate deciduous forest, grassland, tropical rain forest, and desert. The aquatic biomes include freshwater, marine, and estuary biomes.

Throughout the biosphere, organisms use the same important substances, which are water, nitrogen, carbon, and oxygen. Each of these substances is recycled in the environment.

The chapter concludes with a discussion of food chains, food webs, and energy pyramids. These three types of representations illustrate the feeding and energy relationships within ecosystems.

47–1 EARTH: A LIVING PLANET

Section Focus 47–1

The purpose of this section is to introduce the study of ecology and to define the terms *ecology*, *biosphere*, and *ecosystem*. The section also discusses the process of ecological succession and how it leads to the development of a climax community.

Students will learn that the biosphere, or living globe, is divided into a variety of smaller units called ecosystems. Ecosystems consist of all the living and nonliving factors that surround organisms and affect their way of life. Some examples of ecosystems are pond and forest ecosystems.

Students will learn that environments of organisms are constantly changing in a process called ecological succession. In ecological succession, an existing community is gradually replaced by another community. Students will learn how ecological succession often leads to a relatively stable collection of plants and animals called a climax community.

Performance Objectives 47–1

1. Define and discuss the terms *ecology*, *biosphere*, and *ecosystem*.
2. Distinguish between primary and secondary ecological succession.
3. Explain how ecological succession leads to the development of a climax community.

Science Terms 47–1

ecology   p. 1007
biosphere   p. 1007
ecosystem   p. 1008
community   p. 1008
ecological succession   p. 1009
climax community   p. 1009

47–2 LAND BIOMES

Section Focus 47–2

The purpose of this section is to define the term *biome* and to describe the major land biomes. A biome is an environment that has a characteristic climax community.

The Earth is made up of two main types of biomes: land biomes and aquatic biomes. The major land biomes are the tundra, taiga, temperate deciduous forest, grassland, tropical rain forest, and desert. Students will learn about the climate and physical features of each land biome, as well as the types of plant and animal life that characterize each biome.

Performance Objectives 47–2

1. Define and discuss the term *biome*.
2. Explain how biomes are classified.
3. Describe and compare the characteristics of the major land biomes.

Science Terms 47–2

biome   p. 1010
tundra   p. 1011
taiga   p. 1011
temperate deciduous forest   p. 1012
grassland   p. 1013
tropical rain forest   p. 1014
desert   p. 1015

47–3 AQUATIC BIOMES

Section Focus 47–3

The purpose of this section is to describe the three types of aquatic biomes: freshwater biomes, marine biomes, and estuaries. The section will also discuss some of the abiotic factors that affect the kinds of organisms found in the aquatic biomes.

Rivers, streams, and lakes make up the freshwater biomes of the Earth, and the habitats of the ocean make up the marine biomes. The marine biomes are divided into ecologically distinct zones depending on depth and distance from shore. These zones are the intertidal zone, the neritic zone, the open-sea zone, and the deep-sea zone.

Estuaries exist at the boundary between a freshwater biome and a marine biome. Estuaries are shallow areas that support a variety of life forms.

Performance Objectives 47–3

1. Discuss some abiotic factors that affect aquatic biomes.
2. Describe the three aquatic biomes.
3. Describe and compare the distinct ocean zones that make up marine biomes.
4. Describe estuaries.

CHAPTER PREVIEW

Science Terms 47-3

freshwater biome p. 1016
marine biome p. 1017
estuary p. 1018

47-4 ENERGY AND NUTRIENTS: BUILDING THE WEB OF LIFE

Section Focus 47-4

The purpose of this section is to discuss the recycling of nutrients and the flow of energy in an ecosystem. Essential nutrients are recycled in the environment by the water cycle, the nitrogen cycle, and the carbon and oxygen cycles. The section also discusses how various feeding patterns promote the flow of energy through an ecosystem.

The sun is the ultimate source of energy for all living things and green plants use this energy in the process of photosynthesis. Because photosynthetic plants make their own food, they are called producers. Animals that feed off plants and other animals are called consumers. Energy in an ecosystem flows from the sun to producers and then to consumers.

Each step in the series of organisms eating other organisms is called a trophic level. Ecological pyramids are used to represent the energy relationships among trophic levels. The simplest feeding relationship among animals is a food chain; a more complex feeding relationship that connects organisms is a food web.

Performance Objectives 47-4

1. Explain how energy flows through an ecosystem.
2. Describe the water, nitrogen, carbon, and oxygen cycles.
3. Define the term *trophic* and explain how ecological pyramids are used to represent energy relationships among trophic levels.
4. Define *limiting factor*.
5. Describe a food chain and food web.

Science Terms 47-4

producer p. 1021
consumer p. 1022
decomposer p. 1022
ecological pyramid p. 1022
biogeochemical cycle p. 1023
water cycle p. 1024
nitrogen cycle p. 1024
nitrogen fixation p. 1025
denitrification p. 1025
carbon cycle p. 1025
oxygen cycle p. 1025
limiting factor p. 1025
food chain p. 1026
food web p. 1026

DISCOVERY LEARNING

TEACHER DEMONSTRATIONS
Modeling

Observing an Ecosystem

The following demonstration can be used as an introduction to Chapter 47.

Take the class on a short trip to a park, forest, lake, seashore, or other natural area that exists nearby. Ask students to take paper and pencils with them.

Have students observe the area you have chosen to visit. Ask them to list all the living organisms they can see. Also have them list organisms they believe to be present but cannot see, such as microscopic organisms in water. Then ask students to list all the nonliving factors that make up the area. These would include soil, sunlight, water, rocks, temperature, and elevation.

Next, ask students to choose one of the organisms they have listed. For example, in a park ecosystem, a student might choose a squirrel.

• **How do the nonliving factors you have observed affect the life of the organism you have chosen?** (Answers will vary. For a squirrel, possible answers include: The squirrel needs water to drink; the squirrel stores food and hibernates when the weather becomes cold.)

• **How does the organism you have chosen depend on other organisms in this environment?** (Answers will vary. Possible answers for a squirrel include: The squirrel eats nuts and acorns from trees; the squirrel uses plant parts to build a nest; the squirrel climbs and lives in trees.)

• **How do you think other organisms might depend on this organism?** (Possible answer: Larger carnivores might eat the squirrel.)

Tell students that the environment they have observed—both the living and nonliving factors—is called an ecosystem. Point out that in this chapter they will learn how living things interact with each other and the nonliving environment.

The Earth's Biomes

This demonstration can be performed as an introduction to Section 47-2.

Display a globe. Point out a desert area, such as the southwestern United States.

• **What type of ecosystem would you expect to find in this part of the world?** (Lead students to the correct response. You may find it necessary to give the name of the area you have indicated.)

• **What types of animals would you expect to find in this area?** (Typical southwestern desert animals include jackrabbits, kit foxes, snakes, and lizards. Accept all correct answers.)

• **What types of plants would you expect to find in this area?** (Typical plants include cacti, mesquite, sagebrush, and creosote bushes. Students may have difficulty giving many examples of desert plants.)

Repeat this exercise, selecting areas of the world that typify each of the land biomes. For example, you might wish to select Brazil to discuss the characteristics of the tropical rain forest biome or Kenya to discuss the grasslands biome. To show that major land biomes are found in more than one place, you may wish to select several areas in which the same biome is found.

Point out that the biosphere can be divided into *biomes*, areas that have characteristic climax communities. These climax communities are similar because the areas are similar in climate and other physical factors.

CHAPTER 47
The Biosphere

GUIDED ENQUIRY
Pose the following questions to students and have them record their responses. Point out that they will gain a better understanding of the key concepts if they read the chapter with these basic questions in mind. Upon completion of the chapter, pose the questions again. Ask students to compare their initial responses with those they have developed after reading the chapter.

- How do organisms interact with each other and with the nonliving environment?
- How does a stable collection of plants and animals develop in an area?
- How do the land biomes of the Earth differ from one another?
- How do abiotic factors affect the kinds of organisms found in aquatic biomes?
- How does energy flow through an ecosystem?
- How are nutrients recycled in an ecosystem?
- How are plants and animals connected in feeding relationships?

CHAPTER 47
The Biosphere

The Earth is a planet of striking beauty whose oceans and landmasses are home to a rich variety of life, including this caribou foraging for food in Denali National Park in Alaska.

When the first astronauts traveled beyond Earth and looked back to view their home planet, they were struck by its beauty. Great green continents and sparkling turquoise oceans stand out against the blackness of space. Compared to the other planets, Earth is a special sight.

But it is not appearance alone that makes our planet singular. Earth is different from the other planets because it is teeming with life. This life covers Earth like a huge web whose countless threads spread out and intertwine with one another.

Living things constantly interact with one another and with their surroundings. The invisible processes that occur deep within the cells of all living things form the basis of these interactions and affect the entire Earth. In this chapter you will explore these interactions, which are vital to the life of both the organism and the planet.

INTRODUCING CHAPTER 47
Using the Visuals
Have students study the chapter-opener photograph.
- **What do you see in this picture?** (A small photograph of the Earth from space and a large picture of a caribou searching for food in a national park.)
- **Is this larger picture typical of life on Earth?** (Answers may vary. In some ways it is, because it shows plant and animal life and a relationship that exists between them; in some ways it is not typical, because it shows only one climatic and geographical area—and life in other areas might look quite different.)
- **What part of the Earth does this picture show?** (Alaska, which is a cold area near the polar regions of the Earth.)
- **How might a photograph of life on Earth look different if it were taken in another place on Earth?** (Possible answers: It could show a desert with a cactus plant and a snake; it could show a tropical rain forest with palm trees and jungle animals; it

GUIDE FOR READING

After you read the following sections, you will be able to

47–1 Earth: A Living Planet
- Define ecology, biosphere, and ecosystem.
- Describe the process of ecological succession.

47–2 Land Biomes
- Explain how biomes are classified.
- Describe the characteristics of each land biome.

47–3 Aquatic Biomes
- List some abiotic factors that affect aquatic biomes.
- Describe three aquatic biomes.

47–4 Energy and Nutrients: Building the Web of Life
- Explain how energy flows through an ecosystem.
- Discuss how water, nitrogen, carbon, and oxygen are recycled in the environment.
- Define limiting factor.
- Describe a food chain and a food web.

Journal Activity

YOU AND YOUR WORLD In your journal, discuss the ecological issue of your choice in words and pictures. Explain why you selected this particular issue, why it is important to you, and why you think other people should regard it as important.

47–1 Earth: A Living Planet

Guide For Reading
■ What is ecology?
■ How do ecosystems change over time?

During the last few years, people around the world have finally begun to understand that our remarkable planet is home not only to people but also to other forms of life. And the health of this life is essential to the health of human society.

In order to properly care for our planet, we must understand how the living world operates. To do so, we must study ecology. **Ecology is the study of the interactions of organisms with one another and with their physical surroundings.** The word **ecology** comes from the Greek word *oikos*, which means house. The "house" includes the environment in which organisms live, the interactions of organisms with one another, and the interactions of organisms with the nonliving environment. Scientists who study ecology are called ecologists.

From space, it is easy to see that Earth is a single living system. It is a **biosphere**, or living globe. **The biosphere is that part of the Earth in which life exists.** It includes all the areas of land, air, and water on the planet, as well as all the life that populates these areas. The biosphere extends from about 8 kilometers above the Earth's surface to as far as 8 kilometers below the surface of the ocean. Living organisms are not distributed uniformly throughout the biosphere. For example, few organisms live on top of the ice in polar regions, whereas tropical rain forests swarm with life.

Figure 47–1 A fast-moving stream in Olympic National Park in Washington is an example of how organisms interact with one another and with their physical surroundings. The spray from the stream made it possible for mosses to grow on the surfaces of nearby boulders.

COOPERATIVE LEARNING

Divide the class into nine groups and randomly assign one of the nine biomes to each group. Using the descriptions in the text, each group should produce a drawing of their assigned biome. Remind groups that their illustration should include both the abiotic and biotic features of their biome. You may want to provide groups with a small world map so that they can show the location of their biome as part of their final product. A second illustration should predict what their biome will be like in 100 years. Encourage groups to think about the effects that humans will have on both the abiotic and biotic features of their biome. Group illustrations can be done on poster board, butcher paper, or on a chalkboard divided into sections for each group.

Journal Activity

YOU AND YOUR WORLD Ask volunteers to share their essays with the class. Have students keep their journal assignments in their portfolio.

ronment. They will also learn how various areas on Earth differ according to the kinds of plants and animals that live there.

TEACHING STRATEGY 47–1

Focus/Motivation

Write the word *interdependent* on the chalkboard. Then write or recite these famous lines by poet John Donne:

No man is an island, entire of itself; every man is a piece of the continent, a part of the main.

• **How does the word *interdependent* relate to the lines of this poem?** (The poem is stating that people do not exist in isolation but are part of a greater whole. The word *interdependent* describes a relationship in which people or things rely on one another and need each other. The poem implies that people are interdependent.)

could show a northern forest with deer and tall pine trees; it could show a part of the ocean with fish and aquatic plants.)

Explain to students that life on Earth is extremely diverse and that different types of animals and plants grow in areas that have different climates and other physical factors. However, there are certain principles that describe the growth and interactions of living things everywhere on Earth. For example, the basic feeding relationship between animals and plants—which is shown in this photograph—can be found anywhere on Earth where there is life.

Tell students that in this chapter they will learn about some of the basic principles that describe the way living things interact with each other and their envi-

1007

BACKGROUND INFORMATION
BIOTIC VS. ABIOTIC

Biotic factors may have a powerful effect on the abiotic factors that surround them. For example, the trees in a rain forest hold topsoil with their roots, shade the soil with their leafy canopies, contribute organic matter to the soil in the form of fallen leaves, and return a great deal of water to the atmosphere by both evaporation and transpiration from their leaves. In this example, trees are biotic factors that affect vital abiotic ones, including soil composition, temperature, and humidity.

Point out to students that in any living community, removing biotic elements can dramatically affect an ecosystem's abiotic conditions.

Ecosystems

The biosphere is large, complex, and difficult to study. That's one reason many ecologists work with smaller units called **ecosystems**. An ecosystem consists of a given area's physical features (abiotic factors) and living organisms (biotic factors). For example, a pond ecosystem includes abiotic factors such as water, sunlight, soil type, rocks, temperature, humidity, elevation, and rainfall. It also includes biotic factors such as fishes, frogs, insects, snails, worms, amebas, and waterlilies. The organisms living together in an ecosystem, like the people living together in a human neighborhood, are often referred to as a **community**. A forest community, for example, includes trees, birds, and fungi.

From the way we talk about "a pond" or "a forest," you might get the impression that ecosystems are self-contained and function independently of one another. But that is rarely the case. Why? Because ecosystems are connected by both living and nonliving features. A pond's crayfish may be eaten by a forest's raccoons. Birds and caribou that summer in the barrens of the far north may winter in forests far to the south. A river may emerge in the mountains, meander through forests and grasslands, and end its journey in marshes that border the sea.

Environment

Figure 47–2 *Ecosystems consist of both biotic and abiotic factors. This figure separates the readily observed components of a stream ecosystem into two biotic groups (animals and plants) and an abiotic group (the stream, soil, and air). Of course, ecosystems contain many components that are not easily shown in a drawing, such as microorganisms, humidity, input of solar energy, and so on.*

Biotic Factors   Abiotic Factors

47–1 (continued)

Content Development
Point out that the words written by John Donne over 300 years ago are still true. But today people are aware that not only is no person an island unto himself—no *organism* is an island unto itself. All forms of life on Earth depend on each other and need each other if they are to survive and grow.

• **In what way do you see an awareness of this interdependence in our society today?** (Possible answers: People are more conscious about pollution and other environmental problems; we have recycling and other programs to help conserve resources; many people are devoted to the cause of saving endangered species; in recent years government agencies have been established to protect the environment.)

Point out that much of the study of ecology is motivated by a desire to preserve and protect the biosphere. By understanding how living things interact with

Ecological Succession

You might think that ecosystems—ponds and meadows, for example—remain pretty much the same over time. But if you were able to observe an ecosystem over a long period of time, you might be amazed at how much it changes. Ecosystems change over time because every organism affects environmental conditions around it. For example, burrowing worms change the texture of the soil in which they live. Trees shade the area beneath their branches, making it cooler and darker. Because of such changes, many ecosystems undergo **ecological succession**. In this process, an existing community of organisms is replaced by a different community over periods of time ranging from a few decades to thousands of years.

Sometimes succession occurs in places where no living community existed before. When a new volcanic island arises from the sea, for example, its newly cooled lava is devoid of life. Organisms that colonize such areas are called pioneer species. In a sense, pioneer species are like homesteaders who first built farms and small towns on the frontier. After the pioneers had carved settlements out of the wilderness, different kinds of people with varied skills and living requirements were able to move into the area.

Lichens are typical pioneers on bare rocks. As they grow, lichens produce acids that break down the rock. Over many years, the action of the lichens produces a thin soil of broken-down rock and dead lichens. As rock surfaces break down, mosses appear and further change the soil, enabling other plants to survive and grow. Each species that moves in changes the environment in its own way—sometimes in a way that makes it impossible for earlier settlers to continue living there.

Succession can dramatically transform an ecosystem, as shown in Figure 47–3. Over time, a lake or pond may fill up with silt and organic matter, turning first into a marsh and then into dry land. As physical factors change, living communities also change. Succession can also occur in places where natural disasters (such as the eruption of Mount Saint Helens) or human activities (such as farming) wipe out existing communities. In much of the United States, for example, abandoned farmland is colonized first by grasses, then by other weeds, and later by shrubs and small trees.

Succession often leads to a fairly stable collection of organisms called a **climax community**. Climax communities are often described by the most obvious species they contain. For example, if you tell an ecologist that an area is a "temperate zone beech-maple forest," she will have a good idea of what the area is like. Tell a marine biologist that you visited a "South Pacific coral atoll," and he too will know what you are talking about.

An area's climax community is due partly to chance, for succession can take different paths. An ecosystem may eventually return to the way it was after trees are cut down, land is

Figure 47–3 Ecological succession can occur as a pond fills up with silt and fallen leaves. Over a period of time, the pond turns into a marsh and then into dry land, which soon becomes inhabited by a climax community.

1009

BACKGROUND INFORMATION
OASIS IN THE TUNDRA

Although they are usually snow-covered, tundra biomes are nearly as dry as desert regions. For example, in the Arctic regions north of mainland Canada, less than 15 centimeters of precipitation falls annually. This amount is just about the same as the precipitation in the desert regions of Arizona.

Such dryness, coupled with intense cold, makes the tundra a barren place. Yet there are areas, just like an oasis in the desert, where living things can survive and even thrive. Some of these areas are sheltered valleys that are protected from bitterly cold winds. Others are meadows with an abundant water supply that can support large numbers of animals.

dug up for mining, or any other kind of environmental damage occurs. But it is also possible that succession will take a different path and things will never be the same again.

47-1 SECTION REVIEW

1. Define ecology, biosphere, and ecosystem. How are they related to one another?
2. What are abiotic factors? Biotic factors?
3. Describe the process of ecological succession.
4. What is a climax community?
5. **Critical Thinking—Making Predictions** How would the breakdown of large amounts of organic matter upset the natural balance of a lake ecosystem?

Guide For Reading
- What is a biome?
- What are the distinguishing characteristics of the major land biomes?

47-2 Land Biomes

Areas that are similar in climate and other physical factors develop similar types of climax communities. These areas are called biomes. **A biome is an environment that has a characteristic climax community.** The Earth is made up of two main types of **biomes**: land biomes and aquatic biomes. Aquatic biomes will be discussed in the next section.

Most land biomes are named for their climax community, or the dominant type of plant life. The major land biomes are the tundra, taiga, temperate deciduous forest, grassland, tropical rain forest, and desert.

Figure 47-4 According to this illustration, what type of biome covers most of Europe?

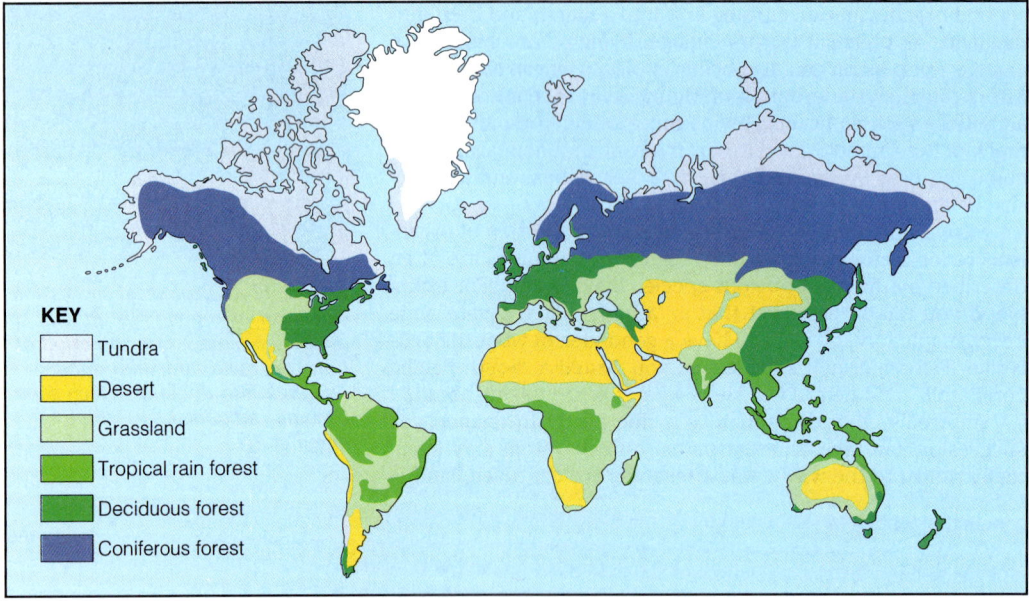

47-1 (continued)

SECTION REVIEW 47-1

1. Ecology is the study of the interactions of organisms with one another and with their physical surroundings. Biosphere is that part of the Earth in which life exists. An ecosystem consists of all the biotic and abiotic factors of a given area. The science of ecology examines the biosphere, which consists of many different ecosystems.
2. Abiotic factors are all the nonliving factors—soil type, rainfall, elevation, temperature, and location on the Earth—that surround organisms and affect them. Biotic factors are all the living organisms.
3. In ecological succession, an existing community of organisms is gradually replaced by another community.
4. A climax community is a relatively stable collection of plants and animals and is usually described by the major forms of plant life.
5. The breakdown of large amounts of organic matter would encourage the growth of algae and other organisms. In turn, the lake would probably be replaced by a marsh and then by dry land.

Reinforcement/Reteaching
If students are having trouble answering the Section Review ques-

Tundra

Northern North America, Asia, and Europe are covered by biomes called **tundra.** The tundra is the northernmost land biome. This nearly treeless biome is covered by mosses, lichens, and grasses. A few stunted trees survive here and there. Many animals migrate into the tundra during the summer to feed on the plants that grow there. Caribou and reindeer are two examples. In turn, wolves, foxes, and hordes of mosquitoes feed on these animals. A great many birds summer here, then fly south in early autumn.

The most characteristic feature of the tundra is permafrost, which is a layer of permanently frozen subsoil. During the summer the ground thaws to a depth of a few centimeters and becomes soggy and wet; in the winter it freezes again. This cycle of thawing and freezing, which rips and crushes plant roots, is what keeps the plants small and stunted.

Taiga

South of the tundra are biomes dominated by great coniferous, or cone-bearing, forests of fir, pine, and spruce. These biomes are called **taiga** (TIGH-guh). The term taiga comes from the Russian word that means primeval forest.

The taiga stretches across much of North America and Asia, with a narrow band reaching into Norway and Sweden (in Europe). The taiga rises to the higher elevations of many mountain ranges in the United States, including the Rocky and Appalachian mountains. The taiga also extends along the coasts of Washington, Oregon, and northern California, where it

Figure 47–5 During the summer months, plants such as grasses and dwarf trees dominate the tundra (center). The caribou (left) and ptarmigan (right) are two types of animals that migrate to this biome in search of food. The polar bear and arctic fox (bottom) live in the tundra throughout the year.

Figure 47–6 Great horned owls and grizzlies are two examples of the animals that inhabit the taiga, which is dominated by coniferous, or cone-bearing, trees.

is home to the giant redwoods—some of the tallest trees in the world. Redwoods can reach heights of more than 60 meters.

Although winters in the taiga are cold, summers are mild enough and long enough to allow many animals and plants to reproduce. The ground thaws during the warmer months, although in some places the thawing does not last for long. Many rivers, ponds, lakes, and bogs provide homes for a variety of living things. Many small birds and mammals live in the taiga and either hibernate or move to warmer regions during the long, cold winters. Typical inhabitants of the taiga include black bears, grizzlies, wolves, moose, elk, and dozens of smaller animals such as voles, wolverines, and grouse.

Temperate Deciduous Forests

Covering the eastern coast of the United States, the southern coast of Canada, most of Europe, and parts of Japan, China, and Australia are biomes that are characterized by changing seasons and leaf fall. These biomes are called **temperate deciduous forests.** A temperate deciduous forest gets its name from its forests of oak, maple, beech, and birch—trees that are deciduous, or shed their leaves in autumn. Although rainfall is sufficient year round, cold winters halt plant growth for several months. Because deciduous trees shed their leaves every autumn, this biome goes through striking seasonal changes. In the spring, many small plants in the forest practically burst out of the ground and grow quickly so that they can flower and bear fruit before they are shaded by the trees.

Although a great number of animals once inhabited these forests, many have been hunted to near extinction. With careful

protection and hunting regulations, deer, moose, gray foxes, and several other species are beginning to reappear. An enormous variety of birds spend their summers in this biome, and chipmunks, raccoons, opossums, and squirrels make it their permanent home.

In the temperate deciduous forest biome, an abundance of organic matter and nutrients are stored in a layer of decaying leaves and twigs called humus (HYOO-muhs). Because humus enriches the soil, these forests make good farmland. And this has encouraged human activities that have greatly altered the biome. For example, forest land in New England was cleared of trees and used for farming. Fortunately, much of the original deciduous forest has since recovered.

Figure 47–7 *Because the trees in a temperate deciduous forest shed their leaves every autumn, this biome goes through striking seasonal changes. The mountain lion and mallard ducks are but two of the many species of animals that make this biome their home.*

Grasslands

Usually found in the interior portions of many continents, **grasslands** are vast areas covered with grasses and small leafy plants. Although this biome may receive significant rainfall (25 to 75 centimeters per year), most of it falls in one season. The grasslands of the world include the plains and prairies of North America, the steppes of Russia, the veld of South Africa, and the pampas of Argentina.

In the midwestern United States, grasslands are characterized by hot summers and cold winters. In some tropical grasslands, however, there is little seasonal change in temperature. Instead, the seasons change from wet to very dry. These tropical grasslands, which are dotted with groves of trees, are called savannas. See Figure 47–8 on page 1014.

ECOLOGY NOTE
THE FLOODS OF 1993

During the summer of 1993, the Mississippi River and its tributaries swelled high over their banks, flooding much of the midwestern United States. In community after community, the floods closed bridges, contaminated water supplies, and damaged or destroyed billions of dollars worth of houses, farms, and property. Before it was over, the United States government declared more than a hundred counties disaster areas, including the entire state of Iowa and large parts of Minnesota, Missouri, and Illinois.

Now, in the wake of this disaster, civil engineers are rethinking their approach to flood management. All along the Mississippi, high flood walls and deep levees saved many communities and a few historical landmarks—such as Mark Twain's Hannibal, Missouri, and the Arch in St. Louis. However, such barriers may have made the flooding worse elsewhere. Perhaps the floods of 1993 would actually have been less destructive *without* the expensive, elaborate structures that had been built to manage just such events.

Have interested students research the floods of 1993 and other floods along the Mississippi. How do engineers, biologists, and other experts think the Mississippi should be managed in the future?

1013

- **Which biome is the second greatest in the United States?** (Temperate deciduous forest.)
- **Which biome is dominant in Canada and the Soviet Union?** (Taiga.)
- **According to type of biome, which part of the United States is most similar to the western European countries of England, Germany, and France?** (Eastern United States.)
- **Where in the world are tropical rain forest biomes found?** (South America, Central American, Africa, Southeast Asia.)
- **Sometimes a small strip of tundra is shown in the middle of a temperate region. Can you think of a possible explanation?** (Tundra conditions often exist at high elevations; the strip of tundra in an otherwise temperate area probably represents a mountain range.)

Focus/Motivation
The temperate deciduous biome is noted for striking fall foliage, particularly in northern regions. Have students collect pictures of fall foliage in an area such as New England. Students may also enjoy obtaining tourist information that describes the best time and locations to view fall foliage.

1013

BACKGROUND INFORMATION
SOIL TYPES

Soil is a mixture of rock, mineral ions, and organic matter. Each land biome tends to have a characteristic soil type.

The soil in desert biomes is dry, brown to reddish-brown, with variable accumulations of clay, calcium carbonate, and soluble salts. A weak humus-mineral mixture exists in a very thin layer of topsoil.

Grassland topsoil tends to be dark, alkaline, and rich in humus. This topsoil layer extends downward for more than a meter. Since topsoil is the most fertile of all soil layers, most of the world's crops are grown on grassland soils. The subsoil consists of clay and calcium compounds.

The top layer of soil in tropical rain forest biomes is acidic, with light-colored humus. The subsoil consists of iron and aluminium compounds mixed with clay.

Soil in the temperate deciduous forest has a surface litter of freshly fallen leaves and organic debris and partially decomposed organic matter. Beneath this litter is topsoil that is a mixture of minerals and humus. The subsoil consists of light grayish-brown silt loam above a layer of dark brown firm clay. Beneath the subsoil is calcareous loam glacial till.

Soil in the taiga has a light-colored acidic surface litter over humus. The soil beneath the humus layer is light-colored and acidic. The subsoil contains iron and aluminum compounds.

Figure 47–8 In Africa, grazing animals, such as zebras and wildebeests, feast on the low-growing plants that make up a savanna, or tropical grassland. Notice the many acacia trees in the background.

In grasslands, as in other biomes, interactions among animals and plants shape the environment. In fact, many grasslands do not undergo ecological succession and thus do not become forests primarily because of the grazing of large animals and periodic fires. On the Serengeti grasslands of Africa, impala, gazelles, wildebeests, and elephants graze.

Wheat, corn, and other grains are heavily farmed in the grasslands of the midwestern United States and in Ukraine. When properly treated, the deep, rich soils of temperate grasslands can support farming for many years. In the 1930s, however, mismanagement of America's great midwestern prairie stripped the land of vegetation and exposed topsoil to the devastating effects of windstorms. Millions of tons of topsoil were blown away, creating what was aptly called the dust bowl.

Tropical Rain Forests

In parts of the world where the temperature stays warm and rain falls year round, **tropical rain forest** biomes are found. Typically, tropical rain forests receive 200 to 400 centimeters of rainfall each year. Temperatures remain constant at about 25°C throughout the year. This biome covers large areas of South America, Southeast Asia, Africa, and Central America.

The tropical rain forests are home to more species of plants and animals than can be found in all the rest of the land biomes combined! Here, many trees grow to a height of 70 meters, and their tops form a dense covering called a canopy. The tall trees also provide surfaces on which many other plants grow. Lianas (lee-AH-nuhz) are large woody vines that use the trees to support their rapid growth.

Animal life in the rain forests is rich and varied. Colorful insects and birds are particularly abundant. Reptiles, small mammals, and amphibians are common inhabitants of this biome. It is not surprising that many of the animals living here are tree dwellers, as the floor of tropical rain forests bristles with danger.

47–2 (continued)

Enrichment

Point out that conditions often vary extensively from one place to another within a given environment. These miniature environments that exist within a larger environment are aptly called microenvironments.

Ask students to find a small area near the school or in their neighborhoods that constitutes a microenvironment. Examples of microenvironments include forest soil and the many small animals such as worms that live in it; a shady side of a rock where mosses are growing; a portion of a garden that is rich in plants, flowers, and insects; a patch of land that is parched and dry because of daily exposure to hot sun; an area that has been altered by human activities such as building or pedestrian traffic. Have students observe their microenvironment and describe its living and nonliving elements; also have them compare it with the larger environment that surrounds it. Then ask students to

Many of the animals and plants that inhabit the rain forests produce chemicals that may be useful in fighting some types of diseases. Unfortunately, the world's tropical rain forests are being destroyed by the rapid growth of the human population. If the destruction continues at its present rate, almost all of the tropical rain forests will disappear by the end of this century! Along with the rain forests will go thousands of plant and animal species found only in this fascinating biome.

Deserts

Deserts are biomes that usually occur in areas where there is less than 25 centimeters of rainfall a year. There are many different kinds of deserts around the world. In deserts such as the Sahara in Africa—the world's largest desert—rain almost never falls, and the wind is hot and dry. Because almost nothing grows in this type of desert, the landscape looks as barren as the surface of the moon. Other deserts are home to many species of lizards, insects, scorpions, snakes, and birds.

In seasonal deserts there is some rainfall during the year. Rapidly growing plants soak up the water as quickly as possible, then grow, flower, fruit, and become dormant until the next rainfall. In the deserts of the southwestern United States and Mexico, rainfall is more even, but it is sparse. Here, sagebrush, cacti, and only a few types of trees survive. Another type of desert is found on mountains and plateaus where the high altitudes cause a decrease in temperature. These deserts are called cold deserts. Cold deserts have a brief rainy season that permits the growth of grasses and shrubs.

If modern science can find a way to bring water to the deserts, they can be made suitable for farming. Desert soil is often very fertile and, of course, receives plenty of sunlight. In parts of the Middle East, archaeologists have uncovered the ruins of waterworks that were used thousands of years ago by desert inhabitants to collect rainwater. In several areas these waterworks have been rebuilt, making farming possible again.

Figure 47–9 Tree ferns, tangled lianas, and towering, slender trees characterize this section of a tropical rain forest in Costa Rica. Most of the animal inhabitants—the sloth (left) and squirrel monkeys (right), for example—live high in the trees, where they stay out of the way of predators that roam the floor of the forest.

1015

HISTORICAL NOTE
THE LOST TAIGA

The taiga once covered great portions of Michigan, but it was completely destroyed by careless and thoughtless logging. There is however, a living tree museum in Michigan. On an estate that once surrounded a logger-baron's house, there are remnants of the forests that once covered the whole region.

BACKGROUND INFORMATION
EPIPHYTIC COMMUNITIES

In the tropical rain forest, a whole community of plants called epiphytes grows on tree branches. The word *epiphyte* means "on top of plants," and that is how epiphytes grow. These plants are not parasites, however; they simply hold onto tree branches with their roots, using the branches as a kind of anchor. The epiphytes then absorb water and animal wastes that fall on the branches.

Epiphytes often give rise to their own little communities. In the water that collects among epiphyte leaves, insects and amphibians lay eggs, grow, and mature. Snakes slither through the branches. Birds perch on the branches, and monkeys swing on the vines. An epiphytic community can be a truly magical place.

Figure 47–10 *The Sonoran desert, which stretches from southern California to western Arizona and south into Mexico, is home to many types of organisms. Some, such as the collared lizard and saguaro cactus, live only in the desert. Others, such as the mountain lion, may live in other biomes as well.*

Guide For Reading
- What are some abiotic factors that affect aquatic biomes?
- What are the distinguishing characteristics of the freshwater, marine, and estuary biomes?

47–2 SECTION REVIEW

1. How are biomes classified?
2. Describe the characteristics of each land biome.
3. **Critical Thinking—Identifying Patterns** What characteristics would you expect tundra animals to have?

47–3 Aquatic Biomes

The aquatic biomes are the water ecosystems. They include the freshwater biome, marine biome, and estuaries. These biomes support more organisms than do the land biomes.

Some of the abiotic factors that affect the kinds of organisms found in the aquatic biomes are light intensity, amounts of oxygen and carbon dioxide dissolved in the water, and the availability of organic and inorganic nutrients. The aquatic biomes do not vary in temperature as much as land biomes do.

Freshwater Biomes

Rivers, streams, and lakes are the lifeblood of our continents and are considered the **freshwater biomes** of the Earth. Not only do they provide much of our drinking water, but they are also an important source of food. Tiny floating plants and animals drift and swim through the water. These organisms are

---

47–2 (continued)

SECTION REVIEW 47–2

1. Biomes are classified according to similarity of climate and types of climax communities.
2. The major land biomes are the tundra, taiga, temperate deciduous forests, grasslands, tropical rain forests, and deserts. Student descriptions should reflect information given in the chapter.
3. Tundra-dwelling animals are covered with heavy fur or have other adaptations that let them withstand extremely cold temperatures.

Reinforcement/Reteaching
If students are having trouble answering the Section Review questions, go back to the part of the section that contains the material with which they are having difficulty.

Closure
Divide the class into small groups and assign each group one of the major land biomes. Ask each group to imagine that they are tour guides whose job is to take people on a worldwide tour of the biome they have been as-

eaten by fishes and amphibians, which also eat the vegetation and insects that fall into the water from overhanging trees. Trout are typical of the fast-swimming fishes that live in mountain streams. Large rivers such as the Amazon in South America and the Nile in Africa are home to many species of insects, fishes, amphibians, reptiles, and mammals.

Unfortunately, people all over the world are using rivers and lakes as dumping grounds for wastes. The results of this carelessness are beginning to catch up with us. We will discuss the problems of water pollution in more detail in Chapter 49.

Marine Biomes

The vast habitats of the ocean, or the **marine biomes**, cover most of the surface of the Earth. Because sunlight penetrates only a short distance before it is absorbed by the water, photosynthesis can take place only in the uppermost region of a marine biome. This region is called the photic (FOHT-ihk) zone. The photic zone may be as shallow as 30 meters in the North Atlantic Ocean or as deep as 200 meters in the South Pacific Ocean. It is in this thin ocean layer that phytoplankton (tiny free-floating photosynthetic organisms) and algae grow.

Oceanographers have divided marine biomes into ecologically distinct zones depending on depth and distance from shore. Each of these zones contains organisms that are adapted to the conditions there.

INTERTIDAL ZONE The intertidal zone is the most difficult zone for organisms to live in. Those that live here must tolerate radical changes in their surroundings: Once or twice a day they are submerged in ocean water; the remainder of the time they are exposed to air and sunlight. Organisms in the intertidal zone have adapted in some way to the pounding and surging of waves. Some organisms, such as clams, burrow into

Figure 47–11 In swift-moving rivers, most organisms live in the shallows, where algae and mosses cling to the surfaces of rocks.

Figure 47–12 The intertidal zone (left), is characterized by organisms such as barnacles and starfish. The neritic zone (right) provides a home to brilliantly colored coral-reef fishes.

TEACHING SUPPORT

MEDIA AND TECHNOLOGY

Videodisc
- *Aquatic Ecosystems: Freshwater Wetlands and Freshwater*
- *Aquatic Ecosystems: Estuaries and Marine*

Use this bar code to introduce marine biomes.

Play frames 25553 to 27277

See the Biology Media Guide pages 73–74 for additional barcode correlations for this section.

BACKGROUND INFORMATION
LAYERS OF PLANT GROWTH

In the temperate deciduous forest there may be up to five layers of plant growth. The tallest trees make up the canopy layer; often this layer consists of only one or two dominant species. Under the canopy is a layer of shorter trees called the understory. Beneath the understory is a shrub layer, which is made up of short branching, woody plants. An herb layer consisting of grasses, ferns, and annual wildflowers grows close to the ground. Finally, there is the ground layer, which consists of such things as mosses, fungi, and leaf litter.

---

signed. Have each group present their guided tour as if the other members of the class are the travelers. Maps and photographs can be used to highlight the presentations.

TEACHING STRATEGY 47–3

Focus/Motivation

Draw on the chalkboard a wavy horizontal line to represent the surface of the ocean. Then draw a diagonal line slanting downward. Mark off on the line depths of 50 meters, 100 meters, 200 meters, 1000 meters, 2000 meters, 6000 meters. Tell students that this simple diagram represents varying depths of ocean water.
• **Do you think that different types of plants and animals live at different depths in the ocean.** (Answers may vary. The correct answer is yes.)
• **What factors do you think determine the kinds of plants and animals that live at different depths?** (Accept all reasonable answers.)

1017

> **BACKGROUND INFORMATION**
> NUTRIENTS IN AQUATIC BIOMES
>
> One problem that occurs in aquatic biomes is that nutrients tend to sink below the photic zone where plants and animals cannot use them. Fortunately, nature has ways of bringing these nutrients back to the surface.
>
> In deep lakes, the spring thaw combines with strong winds to mix lake water thoroughly. In this way, bottom water is brought to the surface. In a time that is called the spring turnover, all nutrients and sunlight are present in abundance. The result is a springtime growth, or bloom, of phytoplankton. Some lakes also have a big turnover in the fall; these lakes may have a fall plankton bloom as well.
>
> In the ocean, the bottom is so far down that no winds are strong enough to "turn over" the ocean like a lake! Luckily, there is another way that bottom water can get up to the surface and bring valuable nutrients with it. The process that does this is called an upwelling. In an upwelling, winds and currents work together near the continental shelf. The winds carry surface water away from land. Water must come from somewhere to replace the water that has been removed. The "somewhere" turns out to be the bottom of the ocean. Bottom water, complete with valuable nutrients, is pulled up into the photic zone to replace the water that has been moved out to sea. Now phytoplankton can thrive and grow at a rapid rate. In turn, the phytoplankton provide the basis for a vigorous and productive food web that includes lobster, cod, hake, bluefish, tuna, and many other important kinds of seafood.
>
> Upwellings occur only in certain places on Earth. One such place is called George's Bank. George's Bank is located in the North Atlantic off the New England coast.

Figure 47–13 The deep-sea zone is home to some of Earth's most bizarre creatures. Although this fish may look monstrous, it could stretch out comfortably on a page of this book. The tube worms belong to an ecological community whose ultimate source of energy is chemicals from volcanic vents—an intriguing exception to the rule that the ultimate source of energy for life on Earth is sunlight.

the sand to keep from being washed out to sea. Others, such as barnacles and seaweed, attach themselves to rocks. Still others, such as snails, sea urchins, and starfish, cling to rocks by their feet or suckers.

NERITIC ZONE The neritic zone is the part of a marine biome that extends from the low-tide line to the edge of the open sea. Large algae (seaweed) are abundant here because this part of the ocean is in the photic zone. For example, off the coast of California grow huge forests of giant kelp (brown algae). In shallow areas of tropical waters, meadows of turtle grass provide food for fishes, invertebrates, and turtles. And along the ocean floor, lobsters and crabs crawl while flounder and rays swim above them.

OPEN-SEA ZONE In the open-sea zone, phytoplankton are responsible for 80 to 90 percent of the Earth's photosynthetic activity. Phytoplankton are in turn eaten by larger animals. Thus the chain of life in the sea begins with these tiny organisms. Swimming rapidly through the open-sea zone are fishes of all shapes and sizes and mammals such as dolphins and whales. The open ocean is also home to sea birds such as albatrosses, which live most of their life at sea.

Because nutrients are scarce in most of the open sea, the growth of phytoplankton is relatively slow. This limits the number of animals that can live there. Closer to the shore, however, nutrients are more abundant, and countless fishes swim there to feed and reproduce. Unfortunately, these rich fishing areas are much more susceptible to pollution than is the open sea.

DEEP-SEA ZONE The deep-sea zone is an area of high pressure, cold temperature, and total darkness. Until recently, biologists thought the deep-sea zone was completely devoid of life. But it is now known that this area is home to some of the Earth's strangest creatures. Gulper eels with mouths that make up almost half of their body and giant squid with glowing spots along their sides inhabit the ocean depths.

Here, too, zooplankton (free-floating microscopic animals) wait for night in order to migrate to the ocean's surface and feed on phytoplankton. Herds of bottom-dwellers, such as sea cucumbers, crawl along the ocean floor. Hardly a day goes by that an interesting life form is not found in the deep-sea zone.

Estuaries

Estuaries (EHS-tyoo-air-eez) are found at the boundary between fresh water and salt water. Salt marshes, mangrove swamps, lagoons, and the mouths of rivers that empty into the ocean are examples of estuaries. These areas contain a mixture of fresh water and salt water.

> **47–3 (continued)**
>
> **Content Development**
> Explain that three main factors affect the various depths at which organisms live in the ocean. These factors are temperature, pressure, and available sunlight.
>
> • **At what depth do you think sunlight is most abundant?** (Near the surface of the ocean.)
> • **How does this factor affect where organisms live.** (Plants that need sunlight for photosynthesis live near the surface; animals that depend on photosynthetic plants for food also tend to live near the surface.)
> • **At what ocean depth would you expect temperatures to be the warmest? Why?** (Near the surface, because sunlight warms the water.)
> • **How do you think pressure**

Figure 47–14 Estuaries are areas where fresh water and salt water meet. This salt marsh in Long Island, New York (top), and the mangrove swamp in Florida (bottom) are examples of estuaries.

Estuaries support a variety of life forms. Because estuaries are usually shallow, sunlight is able to penetrate the water completely. Photosynthesis occurs at all levels, making estuaries a suitable environment for aquatic plants. The abundance of such plants, in turn, supports many types of fishes, shrimps, and crabs. In fact, many fishes and invertebrates spawn, hatch, and nurse their young in estuaries. As the young mature, they head for the open sea, then return to the estuaries to reproduce. Several species of birds use estuaries for nesting, feeding, and resting.

SCIENCE, TECHNOLOGY, AND SOCIETY ISSUE

Seeing the Forest

As shown in this feature, the issues in ecology invariably are entwined with politics, economics, and human psychology. Ecologists are concerned with a wide variety of human practices that threaten biodiversity around the world, and convincing humans to change their behavior has traditionally never been an easy task. For now, you may wish to present students with the following six categories of ecology issues:

- **Habitat loss and fragmentation** As humans develop land for their own use, they destroy natural habitats and the species that live in them.
- **Introduced species** As described in Chapter 48, species that are introduced to new environments may grow in great abundance and displace native species.
- **Overexploitation of plant and animal species** Overharvesting plagues fisheries, forests, and wildlife.
- **Pollution** Discussed in Chapter 49, this category includes the biomagnification of pesticides such as DDT.
- **Global climate changes** This category includes the damaged ozone layer and the greenhouse effect.
- **Hybrid agriculture** Farmers are replacing genetically diverse crop species in favor of single strains.

SCIENCE, TECHNOLOGY, AND SOCIETY ISSUE

Seeing the Forest

Have you ever heard the phrase "You can't see the forest for the trees"? This expression means that it is easy to lose sight of the big picture if you focus too much on the small details.

For many years, the main focus in ecology has been on saving endangered species—individual species that had been pushed to the brink of extinction. This approach has had some success. People are now aware of the plight of endangered species—particularly the "cute" ones such as the giant panda, humpback whale, and Siberian tiger—and are generally willing to support efforts to save these creatures.

This approach, however, has also had some failures. Unexciting or unattractive species are unlikely to win public support and may slip quietly into extinction for lack of funding and concern. In a few cases, certain endangered species have come to symbolize unpopular government regulations and thus have become targets of deliberate destruction.

(Sea otters and northern spotted owls are prime examples.) Most importantly, according to many ecologists, the emphasis on endangered species causes people to forget about what is truly important:

Preserving sizeable chunks of vanishing habitats in which whole ecological communities can thrive.

These ecologists argue that conservation efforts should emphasize habitats rather than species. They point out that saving habitats automatically saves the species within them. Other ecologists think that the current emphasis on species still makes sense. They argue that the preservation of an endangered species' habitat is always a by-product of measures to protect an endangered species in the wild. Should conservation efforts focus on saving endangered species or endangered habitats? What do you think? The future of Earth's priceless and irreplaceable living heritage may depend on this decision.

47–3 SECTION REVIEW

1. What are some abiotic factors that affect aquatic biomes?
2. Describe freshwater, marine, and estuary biomes.
3. Describe the four zones of the marine biome.
4. **Critical Thinking—Making Inferences** In parts of the deep-sea zone far below the depth to which sunlight penetrates, many types of species live. Suggest some special characteristics that would enable species to live in this zone.

47–3 (continued)

Reinforcement/Reteaching
Show students a picture of an area at high tide and a picture of the same area at low tide. Ideal photos of this type can be found of Mont-Saint-Michel in France. Point out that when the tide is high, much more of the beach is covered by water than when the tide is low. It is this portion of the ocean floor—the part that is covered and uncovered as the tides change—that constitutes the intertidal zone.

SECTION REVIEW 47–3

1. These factors include light intensity, amounts of oxygen and carbon dioxide dissolved in the water, and the availability of organic and inorganic nutrients.
2. Student descriptions should reflect information given in the chapter.
3. The four zones of a marine biome are intertidal, neritic, open sea, and deep sea. Student descriptions of each zone should reflect information given in the chapter.

47–4 Energy and Nutrients: Building the Web of Life

Guide For Reading

- How does energy flow through an ecosystem?
- How do water, nitrogen, carbon, and oxygen cycle through the environment?
- What is a limiting factor?
- What are food chains and food webs?

One of the most important factors in any ecosystem is the flow of energy through the ecosystem. Of all the sun's energy that reaches the Earth's surface, only a small amount—approximately 0.1 percent on a worldwide basis—is used by living things. Yet this amount, as small as it is, is responsible for the production of several thousand grams of organic matter per square meter of forest per year.

Approximately one half of the energy plants absorb from the sun is used immediately. The rest is stored in plant tissues in the form of energy-containing compounds (carbohydrates). Animals that eat the plants obtain this energy. But because the animals must use much of this energy to carry on their life activities, they store an even smaller amount. Energy cannot be recycled, or used again. Thus energy in an ecosystem is referred to as a flow rather than a cycle.

Nutrients, on the other hand, are generally recycled through an ecosystem. When an animal dies, its matter does not disappear. Rather, it decomposes and eventually gets used by another organism.

The Flow of Energy

You may recall from Chapter 6 that the sun is the ultimate source of energy for all living things. During photosynthesis, green plants and certain bacteria trap sunlight and use it to assemble carbon dioxide and water into carbohydrates. Because photosynthetic organisms are able to make their own food from inorganic substances, they are called **producers**.

Figure 47–15 Organisms are classified as producers, consumers, or decomposers, depending on how they get their food. Because spruce bud worms feed directly on the jack pine, which is a producer, they are called primary consumers (left). Robins, which feed on the worms, are called secondary consumers (center). Decomposers, such as mushrooms, get their food from the remains of dead organisms (right).

BACKGROUND INFORMATION
THE RULE OF 10

The textbook discusses "the rule of 10," that approximately 10 percent of the energy in one trophic level is available to the animals of the next level. Researchers obtained this rule from early studies of aquatic ecosystems, and it still serves as a good approximation. However, recent studies have shown that energy efficiency varies from one food web to another, as well as between different trophic levels. The approximations of energy efficiency in these studies has ranged from as low as 0.05 percent to as high as 20 percent.

MULTICULTURAL STRATEGY

Many Native American legends deal with the sun. One legend from the Muskogee or Creek people from Oklahoma tells how Grandmother Spider stole the sun. According to the legend, all the animals lived in darkness but knew of a wonderful light called the sun. First the fox tried to capture the sun but burned its mouth, which is why foxes to this day have black mouths. Then the opossum tried to grab the sun but burned its tail, which is why opossums to this day have hairless tails. Finally, Grandmother Spider wove a bag of webbing around the sun and carried it home in the bag.

Animals, on the other hand, are **consumers**. Consumers get their energy either directly or indirectly from producers. Consumers that feed directly on producers are called primary consumers. Primary consumers are also called herbivores (plant-eating animals). Consumers that feed on primary consumers are called secondary consumers. There may be tertiary (third-level) or quaternary (fourth-level) consumers that feed on secondary and tertiary consumers, respectively. Secondary and higher level consumers are usually carnivores (flesh-eating animals). For example, an insect that eats plants is a primary consumer, a frog that eats the insect is a secondary consumer, a snake that eats the frog is a tertiary consumer, and so on. **Energy flows through an ecosystem from the sun to producers and then to consumers.**

When plants and animals in an ecosystem die, their remains do not build up because of the presence of **decomposers**. Decomposers are organisms that obtain their energy from nonliving organic matter. Some examples of decomposers are bacteria and fungi.

Each step in this series of organisms eating other organisms is called a trophic, or feeding, level. The term trophic comes from the Greek word *trophe* which means food. There is no limit to the number of trophic levels in a particular ecosystem. However, at each higher trophic level, less and less of the energy originally captured by the producers is available. This is because the energy obtained from digested food is used to maintain the metabolism of the organism and to power its daily activities. A small amount of the energy taken in by herbivores (primary consumers) is changed into new animal biomass. Biomass is the total mass of all the organisms in a trophic level.

As a rule, approximately 10 percent of the energy at one trophic level can be used by animals at the next trophic level. Thus 10 percent of the energy in plants becomes stored in the tissues of herbivores, and 10 percent of the energy in herbivores becomes stored in the tissues of carnivores. At each successive trophic level, less energy is available to an organism.

Ecological Pyramids

Ecologists use **ecological pyramids** to represent the energy relationships among trophic levels. There are three types of ecological pyramids. A pyramid of energy shows the total amount of incoming energy at each successive level. Notice in Figure 47–15 that energy (in the form of heat) is lost going from one trophic level to another.

The trophic levels of an ecosystem can also be represented by a pyramid of biomass, which shows the total mass of living tissue at each level. See Figure 47–15. This pyramid of biomass shows, for example, that a large amount of grass is needed to feed a single rabbit, and a large number of rabbits is needed to nourish a single hawk.

47–4 (continued)

Content Development

Continue the Focus/Motivation discussion by explaining that photosynthetic plants are the producers in an ecosystem and the animals that eat them or that eat other animals are the consumers.

- **In the food chain you just described, what was the producer?** (The plant or tree that produced the seed.)
- **Who were the consumers?** (The animal that ate the seed and all the animals that followed.)

Point out that energy enters an ecosystem as plants take up the sun's energy for photosynthesis. Energy flows through an ecosystem as one organism eats another organism. Energy finally flows out of an ecosystem as organisms release metabolically generated heat to the surroundings.

It is important for students to understand that although energy flows out of an ecosystem, it is not "lost" in the sense that it disappears. Students should be re-

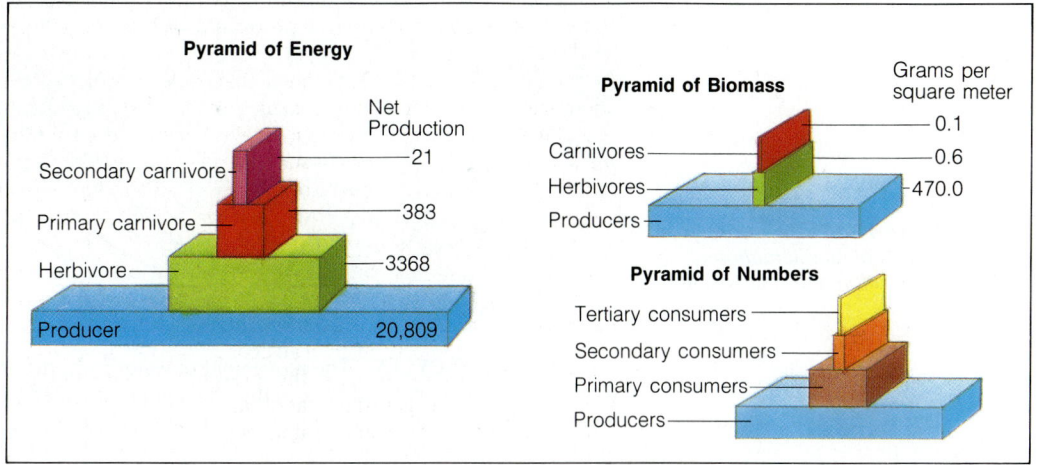

Relationships among trophic levels may also be represented by a pyramid of numbers. A pyramid of numbers illustrates the total number of organisms at each level. In a grassland, for example, a large amount of grass (producers) is needed to support the herbivores (primary consumers). Usually the number of organisms decreases at each successive level. Sometimes, however, this is not the case. In a temperate deciduous forest, one tree (producer) can support a large number of insects (primary consumers).

Like pyramids of biomass, pyramids of numbers show only the amount of organic material present at one time. They do not give the total amount of material produced or the rate at which it is produced, as do pyramids of energy.

Figure 47–16 A pyramid of energy shows that a small amount of energy in an ecosystem is transferred at each trophic level. A pyramid of biomass shows the mass present in each trophic level at any one time. A pyramid of numbers shows the number of organisms in a particular ecosystem.

Biogeochemical Cycles

Although energy moves in a one-way direction through an ecosystem, nutrients are recycled. All organisms require certain essential nutrients in order to grow. Plants need water, carbon dioxide, phosphorus, potassium, and many other elements. Animals require complex compounds (such as proteins and amino acids), several types of vitamins, and many of the same elements plants do.

As members of each trophic level eat members of the level beneath them, they acquire the complex organic molecules and elements they need in addition to energy. Although energy and nutrients move together from one trophic level to the next, they move through the biosphere differently.

Nutrients move through the biosphere in a series of physical and biological processes called **biogeochemical**, or nutrient, **cycles**. They are called cycles because nutrients, unlike energy, may be used over and over again by living systems.

TEACHING SUPPORT

Laboratory Manual
- The Oxygen Cycle, p. 573

Teaching Resources
- Hands-On Activity: Garbage in the Garden, p. 219
- Hands-On Activity: A Saucepan Simulation of a Cycle, p. 211
- Ecology Investigation: Comparing the Water-Holding Capacities of Soils, p. 53

BACKGROUND INFORMATION
PRECIPITATION ON LAND AND OCEAN

The Earth's surface is about 28 percent land and 72 percent water. As one might expect, precipitation falls fairly evenly across the Earth's surface. About 23 percent of it falls on land and 77 percent falls on oceans. However, of the water that evaporates into the atmosphere, 16 percent comes from land and 84 percent comes from oceans. Why are the figures skewed? Because some of the precipitation that falls on land—about 7 percent—runs off into streams and rivers and is carried to the ocean.

47–4 (continued)

Content Development

Point out that in addition to energy, more than 20 different substances must be present in an ecosystem if important life processes are to take place. These elements include phosphorus, sulfur, sodium, potassium, iron, and calcium. But the principal elements that are required for life are hydrogen, oxygen, carbon, and nitrogen. These elements are recycled in the water cycle, the nitrogen cycle, and the carbon and oxygen cycles.

- **What is the meaning of the word *recycle*?** (To use again instead of discarding; to put through a cycle again.)
- **How is this word often used today with regard to environmental protection?** (There is much effort being made to recycle certain materials such as paper, aluminum, and glass, rather than polluting the environment with these substances as waste products.)

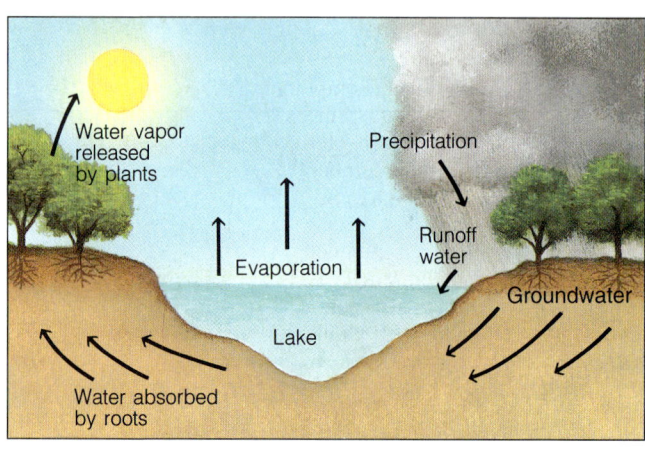

Figure 47–17 The water cycle consists of an alternation of evaporation and condensation.

THE WATER CYCLE The movement of water from the atmosphere to the Earth and back to the atmosphere is called the **water cycle**. The water cycle consists of an alternation of evaporation and condensation. Water molecules enter the air by evaporation from the ocean and other bodies of water. In the air, the water molecules condense (in clouds) and then return to the Earth in the form of precipitation (rain). On land, most of the rainwater runs along the surface of the ground until it enters a river or stream that carries it to a larger body of water. Some water sinks into the ground and is called groundwater. The upper surface of groundwater is known as the water table.

THE NITROGEN CYCLE All organisms require nitrogen to build proteins. Nitrogen is available to organisms in several ways. Free nitrogen gas makes up 78 percent of the atmosphere. Nitrogen is also found in the wastes produced by many organisms and in dead and decaying organisms. The movement of nitrogen through the biosphere is called the **nitrogen cycle**. However, most of this nitrogen cannot be directly used by living things. It must be converted into other forms.

Certain bacteria that live on roots of plants such as legumes (beans, peas, and peanuts) change free nitrogen in the

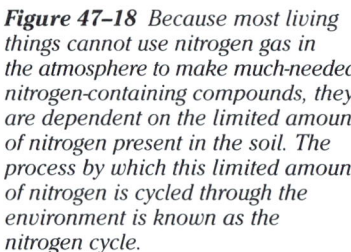

Figure 47–18 Because most living things cannot use nitrogen gas in the atmosphere to make much-needed nitrogen-containing compounds, they are dependent on the limited amount of nitrogen present in the soil. The process by which this limited amount of nitrogen is cycled through the environment is known as the nitrogen cycle.

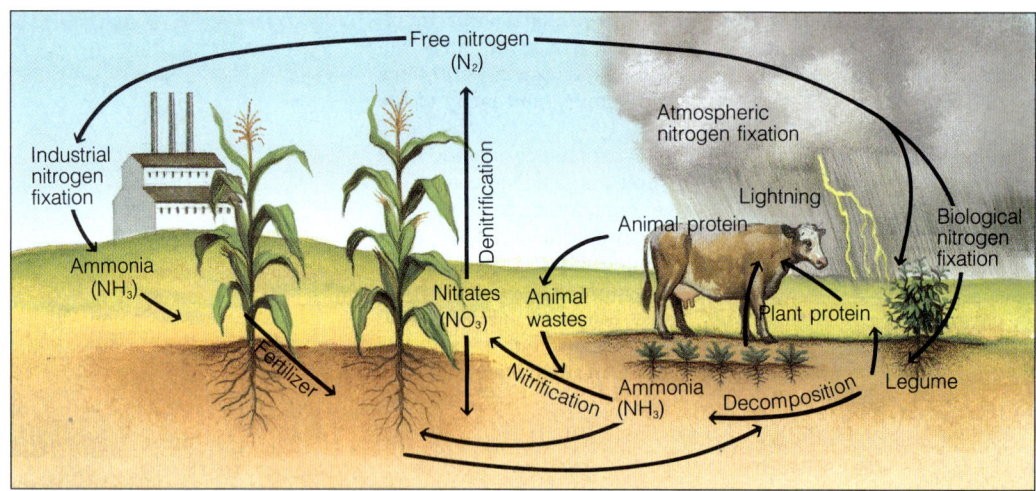

Skills Development
Guided Practice
Skill: Interpreting diagrams
Using the Visuals Direct students' attention to Figure 47–16.
- **According to the diagram, how does water come from the atmosphere to the Earth?** (By precipitation.)
- **What happens to water once it reaches the Earth?** (Some of it runs off the surface to collect in a lake. Some water sinks into the ground to be absorbed by the roots of plants.)
- **How does water leave the**

1024

atmosphere into nitrogen compounds (nitrates and nitrites) that can be used by living things. This process is known as **nitrogen fixation**, and the bacteria are called nitrifying bacteria. Once the nitrogen compounds are available, plants use them to make plant proteins. Animals then eat the plants and use the proteins to make animal proteins. When the plants and animals die, the nitrogen compounds return to the soil.

Eventually other bacteria in the soil break down these nitrogen compounds into free nitrogen in a process called **denitrification** (dee-nigh-trih-fih-KAY-shuhn). These bacteria are called denitrifying bacteria. Through the process of denitrification, free nitrogen is returned to the atmosphere.

THE CARBON AND OXYGEN CYCLES The process by which carbon is moved through the environment is called the **carbon cycle**. During photosynthesis, green plants and algae use carbon dioxide from the atmosphere to form glucose. Consumers and decomposers use glucose in respiration, during which they produce carbon dioxide. Carbon dioxide is then released into the atmosphere, completing the carbon cycle.

The movement of oxygen through the environment is called the **oxygen cycle**. During photosynthesis, water molecules are split, releasing oxygen into the atmosphere. The oxygen is used by most organisms for respiration. During respiration, water is released. The water is absorbed by plants, and the cycle begins again.

Nutrient Limitation

The rate at which producers can capture energy and use it to produce living tissue is controlled by several factors, one of which is the amount of available nutrients. If a nutrient is in short supply—thus limiting an organism's growth—it is called a **limiting factor**.

For example, coastal ocean water often contains sufficient supplies of several nutrients to support much more plant

Figure 47–19 Photosynthesis and respiration are the major biochemical events in the carbon and oxygen cycles.

FACTS AND FIGURES

In order to survive, a 50-kilogram wolf needs to consume approximately 2700 kilograms of moose per year. A moose, in turn, needs to consume about 35,000 kilograms of plants per year.

TEACHING SUPPORT

Teaching Resources
- Vocabulary Review: Cross-a-Clue, p. 1

Study Guide
- Section 47-4, p. 453

MEDIA AND TECHNOLOGY
Interactive Videodisc
- *In the Company of Whales*
Use this bar code to introduce feeding relationships in the ocean.

Search to frame 26068

See the Biology Media Guide pages 74–75 for additional bar-code correlations for this section.

BACKGROUND INFORMATION
BIOGEOCHEMICAL CYCLES

Several general statements may be made about the relationship that exists between biogeochemical cycles and most ecosystems:
1. Mineral elements that serve as nutrients for primary producers are usually in ionized forms.
2. Nutrient reserves in an ecosystem are maintained largely by environmental inputs and by the recycling made possible by the activities of decomposers and detritivores.
3. In any given year, the amount of a nutrient being recycled within an ecosystem is usually greater than the amount of nutrient entering or leaving the ecosystem.
4. Environmental inputs into an ecosystem are by way of precipitation, metabolic fixation such as by nitrogen-fixing bacteria, and the weathering and breakdown of rocks. For land ecosystems, outputs are by way of runoff, evaporation, and soil leaching.

Figure 47-20 *When an algal bloom in a pond begins to cover a large portion of the water's surface, the plants below the surface die from the lack of sunlight. Bacteria then grow at a rapid rate and use up much of the available oxygen. Eventually, the animals in the water suffocate.*

growth than is normally present. The producers in these ecosystems, however, are slowed by the lack of sufficient nitrogen. If nitrogen is added to this system in large amounts, there is a tremendous growth, or bloom, of algae.

Sometimes adding nutrients does not hurt an ecosystem. A little extra fertilizer may even help the system produce more plants and animals for human food. But we must be very careful about tampering with natural ecosystems in this way. If an algal bloom in a lake or river gets too big, it may cover the surface of the water. If that happens, plants below die because they receive no sunlight. Bacteria grow and use up much of the available oxygen. Animals may then suffocate.

Feeding Relationships

Animals and plants in the biosphere are tied together in complicated networks of feeding relationships. The simplest feeding relationship is a **food chain**. In one food chain, a big fish eats little fishes that eat tiny fishes that eat plankton. But nature is almost never that simple.

In nature, plants absorb nutrients and grow. Herbivores eat plants. Carnivores eat herbivores and each other. Scavengers eat dead animals. Bacteria and fungi decompose dead tissue, returning essential elements to the environment. Filter feeders

47-4 (continued)

SECTION REVIEW 47-4

1. Energy flows through an ecosystem from the sun to producers to consumers.
2. A pyramid of energy shows the total amount of incoming energy at each successive trophic level. A pyramid of mass shows the total mass of living tissue at each trophic level. A pyramid of numbers shows the total number of organisms at each trophic level.
3. Water precipitates from the atmosphere, is collected in oceans and other bodies of water, and returns to the atmosphere through evaporation and transpiration. Atmospheric nitrogen is fixed by nitrifying bacteria, and plants use fixed

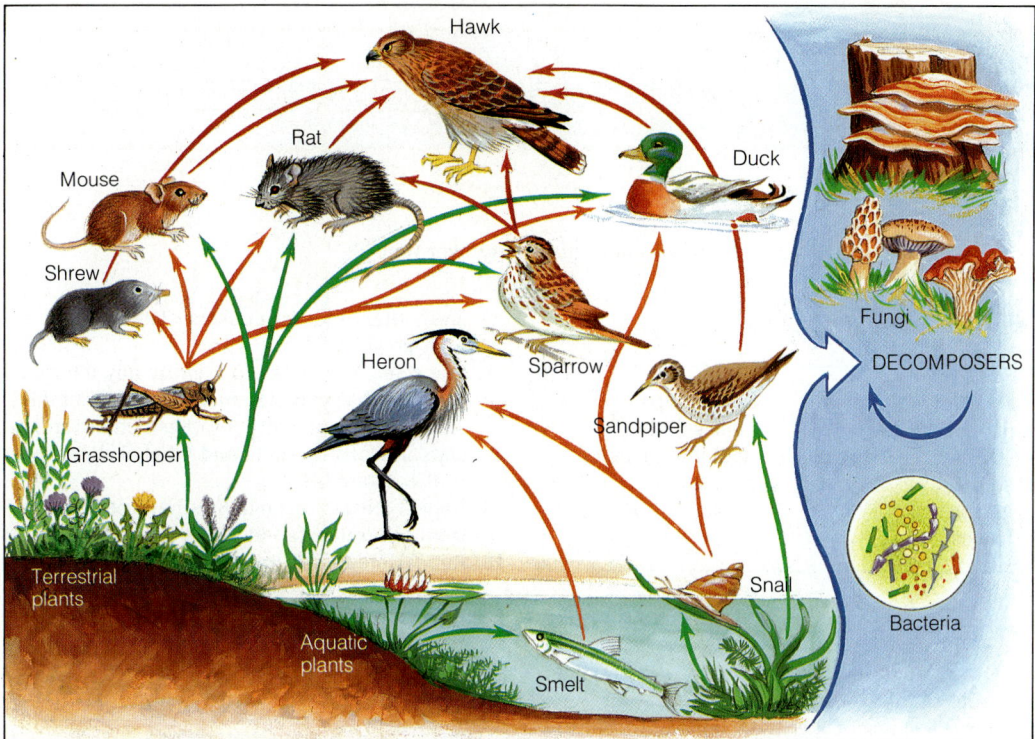

strain floating organisms from the water. Detritus feeders eat bacteria and the wastes of other organisms. Parasites live at the expense of host organisms. All are connected to one another in many complex **food webs.** As Figure 47–21 shows, food webs have many crisscrossing strands.

In the next chapter you will learn that feeding relationships in food webs often contain a built-in system of checks and balances. These checks and balances regulate the number of individuals of each species in the food web. Because of this, many natural ecosystems are remarkably stable.

Figure 47–21 *The living things in an ecosystem are linked by a complicated network of feeding relationships. The arrows in this simplified salt-marsh food web show the transfer of energy and matter as one organism eats another.*

47–4 SECTION REVIEW

1. How does energy flow through an ecosystem?
2. Describe the three types of ecological pyramids.
3. How are water, nitrogen, carbon, and oxygen recycled in the environment?
4. What is a limiting factor? Give an example.
5. Compare a food chain and a food web.
6. **Connections—You and Your World** Describe a food chain in which you are a member.

BACKGROUND INFORMATION
TYPES OF FOOD WEBS

Food webs are divided into two basic types: grazing food webs and detrital food webs. In grazing food webs, herbivores consume living plant tissues and are in turn consumed by a variety of carnivores. In detrital food webs, decomposers and detritivores use organic waste products and the remains of producers and consumers. Bacteria and fungi are examples of decomposers. Earthworms, millipedes, and crabs are examples of detritivores. It is the detrital food webs that enable nutrients to be recycled within an ecosystem.

Reinforcement/Reteaching

If some students are having trouble answering the Section Review questions, go back to the part of the section that contains the material with which they are having difficulty.

Closure

Have students take a closer look at question 6 of the Section Review. On the chalkboard, list several examples of food chains of which humans are a part. In each chain, identify the producers and consumers. You also may discuss humans' roles in the nitrogen, carbon, and oxygen cycles.

nitrogen to make proteins. Denitrifying bacteria return free nitrogen to the atmosphere. Both carbon and oxygen cycle between the atmosphere and organisms through the processes of cellular respiration and photosynthesis.

4. A limiting factor is a nutrient that is in short supply, thus limiting an organism's growth. Water is often an example of a limiting factor.
5. A food chain is a sequence of organisms in which each individual feeds on the next. A food web is the union of interrelated food chains.
6. Student answers will vary but must contain one producer and at least one consumer besides themselves.

LABORATORY INVESTIGATION

OBSERVING SUCCESSION IN AGED TAP WATER

Before the Lab

1. At least one day prior to the investigation gather enough materials for your class, assuming 6 students per group.
2. Be aware that students need two days to age the tap water, then an additional day of waiting after adding the aged water to soil and plant material. Also note that student observations are to take place over a two-week period.

Pre-Lab Discussion

Review with students the material in Section 47–1 on ecological succession. Remind students that the word *succession* means the order in which one group or person displaces another. Point out that in this investigation students will be observing a process of succession in aged tap water.

Have students read the Problem at the top of the page; then have them read through the Procedure.

- **What is the ecosystem that you will be observing in this investigation?** (A microscopic water community.)
- **What types of organisms will you expect to find in this ecosystem?** (Possible answers: bacteria, mold, algae.)
- **What types of organisms do you think will appear first in this community?** (Answers may vary. Students may predict that the simplest organisms will appear first.)
- **How do you predict that the population of the community will change over time?** (Accept all reasonable answers. Students may predict that the community's organisms will increase in size and complexity.)

Have students write several sentences describing the changes they expect to see as they observe the community for two weeks. Then, at the end of the investigation, have them compare their predictions with their observations.

Skills Development

Students will use the following skills while completing this laboratory investigation.

1. Observing
2. Predicting
3. Comparing
4. Relating
5. Applying
6. Inferring
7. Recording
8. Interpreting

Safety Tips

Remind students to be careful when using microscopes and glass slides. Tell students to report any broken slides to the teacher.

LABORATORY INVESTIGATION

OBSERVING SUCCESSION IN AGED TAP WATER

PROBLEM

What changes occur in a microscopic water community over time?

MATERIALS (per group)

sheet of white paper	4 coverslips
600-mL beaker	4 glass slides
1000-mL beaker or jar	medicine droppers
soil	microscope
grasses	reference books
leaves	

PROCEDURE

1. On a separate sheet of paper, draw a data table similar to the one shown.
2. Fill a 600-mL beaker with tap water. Set the beaker aside and allow it to remain undisturbed for 48 hours so that any gases harmful to microscopic organisms can evaporate. After 48 hours, the water is called aged water.
3. Place enough soil in the 1000-mL beaker to cover the bottom. Fill the beaker with a loosely packed mixture of grasses, green leaves, and dried leaves. Pour the aged water over the leaves and the soil.
4. Set the beaker aside in a cool place where it should remain undisturbed for 24 hours.
5. After 24 hours, examine the beaker for signs of life. For example, a strong odor or cloudy water is proof of bacterial growth; fuzzy growths or threads indicate the presence of mold; and a greenish tint is caused by algae.
6. Using a medicine dropper, remove some water from the beaker. Place a drop of the water in the center of a clean glass slide and cover it with a coverslip.
7. Examine the slide under the microscope. Use the low-power objective to locate any microorganisms. If you do not find any microorganisms under low power, focus on some debris, which probably will contain bacteria and other microorganisms. Then switch to high power.
8. Use reference books to identify any microorganisms that you may find. Note the number, size, and complexity of these microorganisms. Record the date and your observations in the data table.
9. Repeat steps 6 through 8 using water samples from several different areas of the beaker.
10. Repeat steps 5 through 9 every day for two weeks. Note any changes in the number or types of organisms in the beaker.

OBSERVATIONS

Date	Observations

1. How do the number and variety of organisms that appear in the beaker change over the two-week period?
2. Describe the size and complexity of the organisms that appear in the beaker over the two-week period.
3. What other changes, if any, occur in the microscopic water community during this time period?

ANALYSIS AND CONCLUSIONS

1. What kind of organisms appeared first in the microscopic water community? Which appeared last? Explain your answers.
2. Explain why the population in the water community changes over time. What is this process of change called?

STUDENT STUDY GUIDE

SUMMARIZING THE CONCEPTS

The key concepts in each section of this chapter are listed below to help you review the chapter content. Make sure you understand each concept and its relationship to other concepts and to the theme of this chapter.

47-1 Earth: A Living Planet

- Ecology is the study of interactions of organisms with one another and with their physical surroundings.
- The biosphere is that part of the Earth in which life exists and is divided into environments that include biotic and abiotic factors.
- Ecological succession is a process by which an existing community is gradually replaced by another community.

47-2 Land Biomes

- A biome is a large area that has a characteristic climax community.
- The major land biomes are the tundra, taiga, temperate deciduous forest, grassland, tropical rain forest, and desert.

47-3 Aquatic Biomes

- The major aquatic biomes include the freshwater biome, marine biome, and estuaries.

47-4 Energy and Nutrients: Building the Web of Life

- Producers make their own food. Consumers feed directly or indirectly on producers. Decomposers break down organisms and organic matter and return them to the environment.
- Water, nitrogen, carbon, and oxygen are recycled within the ecosystem in biogeochemical cycles.
- Networks of feeding relationships called food chains are connected to one another in complex food webs.

REVIEWING KEY TERMS

Vocabulary terms are important to your understanding of biology. The key terms listed below are those you should be especially familiar with. Review these terms and their meanings. Then use each term in a complete sentence. If you are not sure of a term's meaning, return to the appropriate section and review its definition.

47-1 Earth: A Living Planet	47-2 Land Biomes	47-3 Aquatic Biomes	47-4 Energy and Nutrients: Building the Web of Life	
ecology	biome	freshwater biome	producer	water cycle
biosphere	tundra	marine biome	consumer	nitrogen cycle
ecosystem	taiga	estuary	decomposer	nitrogen fixation
community	temperate deciduous forest		ecological pyramid	denitrification
ecological succession	grassland		biogeochemical cycle	carbon cycle
climax community	tropical rain forest			oxygen cycle
	desert			limiting factor
				food chain
				food web

Analysis and Conclusions

1. Bacteria appeared first and complex organisms, such as paramecia and rotifers, appeared last. Simple organisms such as bacteria can survive on organic and inorganic compounds in the water. More complex organisms need to feed on simpler organisms.

2. As soon as an organism moved into the water community, the community changed. This organisms, in turn, served as food for more complex organisms. This process of change is called ecological succession.

Going Further: Enrichment

Have students design an experiment to determine the effects of various abiotic factors on ecological succession. Such factors might include temperature, sunlight, type or amount of soil, or amount of water. Students might perform this investigation by using two separate beakers. Have one beaker be the control community and the other be an experimental community that is kept under different conditions.

Teaching Strategy

1. Check to see that each group has prepared their beaker containing soil, plant matter, and aged tap water correctly before the beakers are set aside for 24 hours.
2. Help any students who may be having trouble preparing slides or using a microscope.
3. You may wish to have students sketch the organisms they observe.

Observations

1. Both the number and variety of organisms increase.
2. The size and complexity of the organisms increase.
3. Answers will vary.

CHAPTER REVIEW

CONTENT REVIEW

Multiple Choice
1. b 2. a 3. b 4. d
5. b 6. a 7. d 8. c

True or False
1. T
2. F biosphere
3. F tropical rain forest
4. F tundra
5. T
6. F denitrification
7. T
8. F decomposers

Word Relationships
1. Inhabitants of a tundra biome; snakes are not found in a tundra biome
2. Names for specific grassland biomes; lianas are large woody vines that grow in a tropical rain forest.
3. Types of land biomes; estuary is an aquatic biome
4. Labels in a food chain; ecosystem is all of the biotic and abiotic factors in an environment

CONCEPT MASTERY

1. Sunlight provides the ultimate source of energy in an ecosystem.
2. Energy is lost between trophic levels because organisms convert only a fraction of their food into tissue mass. Organisms use the bulk of their food for their metabolism and other daily activities.
3. Each trophic level contains a fraction of the energy that is available in the level below it. Consequently, more energy is available at the producer (plant) level than at the consumer (animal) levels.
4. Some limiting factors in the growth of a forest include availability of food, water, and space.
5. Warm temperatures and heavy rainfall allow an abundance of plant life, which in turn results in an abundance of animal life.
6. Because rainfall is scarce.
7. Silt, fallen leaves, and other organic matter will gradually allow more plant life to grow in the pond, which will turn it into first a marsh and then grassland. As shrubs and trees take root, the grassland eventually will become a hardwood forest.
8. Photosynthesis converts the light energy into energy contained in the bonds of the corn plants.
9. Refer to Figure 47–17 and information given in the chapter.
10. Student answers may vary. Unless they were aquatic, such animals would need adaptations to conserve water at high temperatures and perhaps a thick skin to conserve heat at cold temperatures.

CHAPTER REVIEW

CONTENT REVIEW

Multiple Choice

Choose the letter of the answer that best completes each statement.

1. All the biotic and abiotic factors in a pond form a (an)
   a. biosphere.   c. community.
   b. ecosystem.   d. estuary.
2. Which climax community indicates a taiga?
   a. coniferous trees   c. grasses
   b. cacti   d. lichens and mosses
3. Which of the following biomes is the most stable?
   a. desert   c. taiga
   b. marine   d. tropical rain forest
4. A producer-consumer relationship is best illustrated by
   a. foxes eating mice.
   b. leaves growing on trees.
   c. tapeworms living in foxes.
   d. rabbits eating clover.
5. Which are usually the last type of plants to appear in the ecological succession of a forest?
   a. mosses   c. grasses
   b. oaks   d. shrubs
6. Free nitrogen in the atmosphere is converted to nitrogen compounds during
   a. nitrogen fixation.   c. photosynthesis.
   b. denitrification.   d. respiration.
7. Plant-eating animals are known as
   a. producers.   c. decomposers.
   b. carnivores.   d. herbivores.
8. In a food chain, herbivores are known as
   a. secondary consumers.   c. primary consumers.
   b. producers.   d. carnivores.

True or False

Determine whether each statement is true or false. If it is true, write "true." If it is false, change the underlined word or words to make the statement true.

1. The study of the interactions of organisms with one another and with their physical surroundings is called <u>ecology</u>.
2. The <u>ecosystem</u> is that part of the Earth in which life exists.
3. The land biome with the largest variety of plant and animal species is the <u>taiga</u>.
4. Permafrost is characteristic of the <u>tropical rain forest</u>.
5. <u>Estuaries</u> are found at the boundaries of fresh water and salt water.
6. The process by which nitrogen compounds are converted to free nitrogen by bacteria is called <u>nitrogen fixation</u>.
7. A <u>limiting factor</u> is a nutrient, the lack of which can prevent the growth of organisms.
8. Bacteria act as <u>producers</u> in a food web.

Word Relationships

In each of the following sets of terms, three of the terms are related. One term does not belong. Determine the characteristic common to three of the terms and then identify the term that does not belong.

1. mosses, lichens, caribou, snakes
2. prairies, lianas, pampas, velds
3. estuary, tundra, desert, taiga
4. ecosystem, consumer, producer, decomposer

CONCEPT MASTERY

Use your understanding of the concepts developed in the chapter to answer each of the following in a brief paragraph.

1. Why is sunlight needed to maintain an ecosystem?
2. Explain why each trophic level in a food chain contains less energy than the level below it.
3. Why is it more energy efficient for people to eat plants instead of animals?
4. Discuss some of the limiting factors in the growth of a forest.
5. What factors account for the fact that tropical rain forests have the highest biomass of any biome?
6. Why are there no tall trees in the desert?
7. Describe the process of ecological succession in which a pond becomes a forest.
8. Describe what happens to the light energy striking a cornfield.
9. Trace the path of nitrogen through the nitrogen cycle.
10. What are some of the special adaptations that an animal might have in order to survive in areas that have large daily temperature fluctuations?
11. What conditions must be met by an ecosystem for it to be self-sustaining?

CRITICAL AND CREATIVE THINKING

Discuss each of the following in a brief paragraph.

1. **Relating concepts** Why can the tundra be considered a cold desert?
2. **Making predictions** By burning fossil fuels such as coal and oil, carbon dioxide is being added to the atmosphere. In addition, forests are being destroyed at a rapid rate. How do these actions affect the carbon cycle? The oxygen cycle?
3. **Drawing conclusions** Each year, many hectares of wetlands are filled in or paved over for commercial and recreational use. What effect could such development have on surrounding ecosystems?
4. **Making inferences** Water is a vital commodity. What are several ways in which you see water being wasted in your community? Can you offer some suggestions that will help limit the amount of wasted water? Can water consumption be reduced without a change in lifestyle?
5. **Applying concepts** Explain how it is possible to walk through several biomes as you climb up a mountain.

6. **Making diagrams** Draw a food web that includes organisms from a temperate deciduous forest. Identify each organism as a producer or a consumer.
7. **Using the writing process** Pretend that as a reporter for a major newspaper you have been assigned to interview farmers in South America who have cleared areas in tropical rain forests for farming. Prepare a list of questions that you will pose in your interview.

---

11. Ecosystems are self-sustaining if the following conditions are met: There must be a source of energy, which must be changed into a form that can be stored in chemical bonds, such as those in plants. And essential substances, such as water, carbon, oxygen, and nitrogen, must be recycled between the organisms and the environment.

CRITICAL AND CREATIVE THINKING

1. The tundra receives little rainfall and relatively few organisms live there.
2. Plants use carbon dioxide and release oxygen into the environment during photosynthesis. The depletion of plant life added to the burning of fossil fuels would result in a buildup of carbon dioxide in the environment. Oxygen would also become depleted because the reduction of plant life would reduce the amount of oxygen in the atmosphere.
3. These areas, which are all estuaries, are rich in nutrients and diverse in plant and animal life. They are also spawning grounds for many species of fishes. Development of these areas not only disrupts the lives of organisms that live there, it also disrupts any organisms who either directly or indirectly depend on the plants, animals, or nutrients in these areas.
4. Student answers should be logical and should provide some insights into the importance of water conservation.
5. Rainfall and temperature gradations vary considerably from the base of the mountain to the peak. Conditions that produce several different biomes may be present in a mountain ecosystem. For example, the base of the mountain may be desertlike, and the peak may resemble a tundra. The midpoint of the mountain, on the other hand, may contain coniferous forests (taiga).
6. Students' food webs should reflect information given in the chapter.
7. Students' questions should show an understanding of the effects the destruction of tropical rain forests will have on the environment.

CHAPTER 48 Populations and Communities

Section	Laboratory Investigations and Activities
48–1 Population Growth pages 1033–1035 THEMES: Patterns of Change, Stability	**Teacher Edition** DEMONSTRATION: Observing Populations, p. 1032D **Laboratory Manual** Counting a Population, p. 585 **Teaching Resources** HANDS-ON ACTIVITY: How Many Are Too Many? p. 225
48–2 Factors That Control Population Growth pages 1035–1040 THEMES: Systems and Interactions, Patterns of Change	**Student Edition** LABORATORY INVESTIGATION: Examining Patterns of Population Growth in Bacteria, p. 1044 **Teacher Edition** DEMONSTRATION: Food as a Limiting Factor, p. 1032D
48–3 Interactions Within and Between Communities pages 1040–1043 THEMES: Systems and Interactions, Unity and Diversity	**Laboratory Manual** Relationships in an Ecosystem, p. 591 Ecological Succession, p. 595
Chapter Review pages 1045–1047	

Teacher Resources

Books

Allen, T.F.H., and Thomas B. Starr. *Hierarchy: Perspectives for Ecological Complexity.* University of Chicago Press, 1988.

Colinvaux, Paul. *Ecology.* Wiley, 1986.

Diamond, Jared, and Ted J. Case. *Community Ecology.* Harper & Row, 1985.

Ehrlich, Paul R. *The Machinery of Nature.* Simon & Schuster, 1986.

Legge, Allan H., and Sagar V. Krupa, eds. *Air Pollutants and Their Effects on the Terrestrial Ecosystem.* Wiley-Interscience, 1986.

Marchand, Peter J. *Life in the Cold: An Introduction to Winter Ecology.* University Press of New England, 1987.

CHAPTER PLANNING GUIDE

Other Activities	Media and Technology
Study Guide Section 48–1, p. 457	
Teaching Resources ACTIVITY: Predator-Prey Relationships, p. 3 BIOLOGY CASE STUDY: Using Predators to Manage Populations, p. 23 **Study Guide** Section 48–2, p. 459	**Interactive Videodisc** On Dry Land: The Desert Biome **Videodisc** Aquatic Ecosystems: Freshwater Wetlands and Freshwater **Biology Media Guide** page 76
Teaching Resources VOCABULARY REVIEW: Word Game, p. 1 ACTIVITY: Analyzing Ecological Relationships, p. 7 **Study Guide** Section 48–3, p. 461	**Interactive Videodisc** On Dry Land: The Desert Biome **Biology Media Guide** page 76
Teaching Resources Chapter Test A, p. 15 Chapter Test B, p. 19	**Computer Test Bank** Chapter Test, p. 503

Audiovisuals

Partners. Film or video. Coronet Films.
Search for Survival. Video. Coronet Feature Video.
Animal Populations: Nature's Checks and Balances. Film or video. Encyclopaedia Britannica Educational Corp.

Introduction to Ecosystems. Sound filmstrip. Ward.
Population Dynamics. Sound filmstrip. Ward.
Concept of Homeostasis. Sound filmstrip. Ward.

Introduction to Ecological Communities. Sound filmstrip. Ward.
Bacteria and the Ecology of Planet Earth. Filmstrip or video. Ward.
The Living Ocean. Filmstrip or video. National Geographic.

CHAPTER 48 Populations and Communities

CHAPTER OVERVIEW

In this chapter students will continue to study the relationships that exist among organisms. They will do this by learning about two important biological units: populations and communities.

A population is a group of organisms that all belong to the same species and live in a given area. A community consists of all the various populations of organisms living in a given area.

Students will learn how populations of organisms grow and how this growth relates to the carrying capacity of the environment. Different factors control population growth, including predation, competition, and crowding and stress.

Students will learn about the various interactions that occur within and between communities. These interactions include parasitism, commensalism, and mutualism. Students will also learn how ecosystems are related and about the interactions that occur among them.

48–1 POPULATION GROWTH

Section Focus 48–1

The purpose of this section is to describe the growth of populations and to relate population growth to an environment's carrying capacity.

Almost all organisms provided with ideal conditions for growth and reproduction will experience a rapid increase in population. If nothing stops the population from growing, it will experience exponential growth. However, exponential growth does not continue in natural populations for long. Most populations go through a number of growth phases, producing a pattern that is known as logistic growth.

An environment has a certain carrying capacity for a particular species. This means that the environment can support a population of a certain size and no more. Factors such as the amount of food and space, as well as competition among individuals, determine the carrying capacity.

Performance Objectives 48–1

1. Define *population*.
2. Describe and compare exponential growth and logistic growth.
3. Define *steady state*.
4. Define *carrying capacity* and explain its relationship to population growth.

Science Terms 48–1

population   p. 1033
exponential growth curve   p. 1033
logistic growth curve   p. 1034
steady state   p. 1035
carrying capacity   p. 1035

48–2 FACTORS THAT CONTROL POPULATION GROWTH

Section Focus 48–2

The purpose of this section is to discuss the factors that control population growth. These factors enable a population to maintain levels between extinction and overpopulation. The factors that control population can be classified as either density-dependent limiting factors or density-independent limiting factors.

Density-dependent limiting factors usually operate only when a population is large. They include competition, predation, parasitism, and crowding and stress. Density-independent limiting factors affect a population regardless of size. They include such natural occurrences as rainstorms, frost, wind, and temperature changes.

In the last part of this section students will study human population growth. They will learn that the world's human population has been growing exponentially for about the last 500 years but has slowed down in areas such as the United States and Europe.

Performance Objectives 48–2

1. Explain why populations do not grow indefinitely.
2. Define and compare density-dependent limiting factors and density-independent limiting factors.
3. Explain how competition, predation, parasitism, and crowding and stress affect population growth.
4. Explain how natural occurrences affect population growth.
5. Discuss human population growth.

Science Terms 48–2

density-dependent limiting factor   p. 1036
density-independent limiting factor   p. 1039

48–3 INTERACTIONS WITHIN AND BETWEEN COMMUNITIES

Section Focus 48–3

The purpose of this section is to explain how populations within communities interact with one another. This section will also discuss how ecosystems interact with one another.

A community consists of all the populations of organisms living in a given area. An important category of relationships that occurs within a community is symbiosis. Commensalism, mutualism, and parasitism are forms of symbiosis.

Ecosystems do not exist in isolation from one another. Nearly every ecosystem is connected either directly or indirectly with several other ecosystems. This point will be further discussed in the next chapter, when students study the impact of humans on the environment.

CHAPTER PREVIEW

Performance Objectives 48–3

1. Define *community*.
2. Discuss the importance of symbiotic relationships within a community.
3. Define and compare parasitism, commensalism, and mutualism.
4. Discuss how ecosystems interact.

Science Terms 48–3

symbiosis   p. 1040
commensalism   p. 1041
mutualism   p. 1041

DISCOVERY LEARNING

TEACHER DEMONSTRATIONS
Modeling

Observing Populations

The following demonstration can be used as an introduction to Chapter 48.

Take the class on a brief walk around the school grounds or take them to an interesting area near the school. Have students note and write down the various populations of organisms that are present. Depending on whether your school is in an urban, suburban, or rural area, students may see squirrels, horses, rabbits, insects, trees, shrubs, flowers, weeds, grass, frogs, birds—and people! Ask students to think about various factors that might cause each population to increase or decrease in size. For example, if you visit an area that is experiencing a drought, students may realize that grass and plants may decrease in number. In an area where birds and insects exist together, students may realize that the insect population is reduced and controlled by insect-eating birds.

Food as a Limiting Factor

This demonstration may be performed as students study Section 48–2.

Clear an area in the front of the room and place a desk or table in the middle of the area. Ask for five or six volunteers to stand in the area surrounding the desk. Tell these students to imagine that they are very hungry and that their only source of food is whatever they can find on the desk. Now place on the desk a plate of sandwiches or pieces of fruit. Have on the plate more food items than there are students in the group. Ask the rest of the class:

• **What do you think will happen if I tell these people that they are free to take the food?** (Answers may vary, but most students will probably predict that everyone will take only one item first; then some may come back and take seconds.)

• **Do you think that everyone will get enough food?** (Yes.)

Now remove the plate from the desk and replace it with a plate containing the same number of food items as there are people in the group.

• **What do you think will happen now if I tell the people to take the food?** (Possible answer: Probably each person will take one item; possibly an aggressive person might try to take a second item.)

• **Do you think everyone will get enough food?** (Probably, although someone could be left out.)

This time, remove the plate and replace it with a plate containing only two or three food items.

• **What do you think will happen now if I tell these people to take the food?** (Possible answers: The fastest or strongest people will get to the food first; maybe there will be a fight.)

• **Do you think everyone will get enough food?** (No, some people will go hungry, unless the people who take the food share with those who did not get any food—but even then, each person won't get much.)

• **If the food supply does not increase, do you think the same people will get the food every time? Explain.** (Possible answers: Yes, because some people are naturally more clever or stronger than the others; no, because some of those left out may devise ways to get the food or take it from the others.)

• **If the food supply does not increase, what do you think will eventually happen to the people in the group?** (Some may get weak or sick or even die due to lack of food.)

• **How do you think this situation is similar to that which happens among organisms in nature?** (When there is an abundant or adequate amount of food, every organism has enough; but if there is not enough food to go around, competition for food will result. Some organisms will eat and survive, and others will not survive.)

CHAPTER 48
Populations and Communities

GUIDED ENQUIRY
Pose the following questions to students and have them record their responses. Point out that they will gain a better understanding of the key concepts if they read the chapter with these basic questions in mind. Upon completion of the chapter, pose the questions again. Ask students to compare their initial responses with those they have developed after reading the chapter.

- How do groups of organisms change over time?
- What growth patterns are evident in populations?
- How does a particular environment limit the population of a species?
- How do interactions among organisms control populations?
- How do natural occurrences control populations?
- How is the size of the human population changing?
- How are organisms affected by symbiosis?
- How are ecosystems related?

INTRODUCING CHAPTER 48
Using the Visuals
Direct students' attention to the chapter-opener photographs and text.
- **What do you see in these pictures?** (A grizzly bear and her cubs and dandelions.)
- **What type of biome is shown here?** (The bears live in a forest biome, the dandelion in a grassland biome.)
- **What can you tell bout the reproductive rates of the organisms shown in the pictures?** (The bear produces few cubs that need a great deal of care to survive. The dandelion produces many seeds that need good soil and water and no parental care to survive.)
- **What effect do you think bears have on the population of smaller animals in a particular area?** (The bears eat smaller animals and would reduce the population of smaller animals in an area.)
- **What factors do you think might control the number of bears in an area?** (Possible answers: the number of smaller animals for food; the presence of human hunters; changing environmental conditions; disease.)
- **What factors would limit the number of dandelions that live in an area?** (Possible answers: lack of water; insects or other animals that eat plants; changing environmental conditions.)

Point out that planet Earth is inhabited by over 1.6 million

CHAPTER 48
Populations and Communities

Some species, such as grizzly bears, produce just a few offspring at a time. Others, such as dandelions, produce hundreds. But both types of species have the potential to become so numerous that they could overrun the Earth.

Take a look at a dandelion puff, and try to imagine its future. One puff contains so many seeds! What would happen if all the seeds from all of the dandelions eventually produced new dandelions? What would happen if each new dandelion's seeds produced dandelions?

In a suburban lawn or a city park, weeding limits the number of dandelions that survive. But who—or what—does the "weeding" in the wild? Obviously, forces exist in nature that keep the world from being overrun by dandelions—or any other species. Instead of surviving in overwhelming numbers, each species shares the world with the others. How are the numbers of individual species kept in balance? How do different species coexist? These are some of the questions answered through the study of populations and communities, the subjects of this chapter.

GUIDE FOR READING

After you read the following sections, you will be able to

48–1 Population Growth
- Compare exponential growth and logistic growth in a given population.
- Relate population growth to an environment's carrying capacity.

48–2 Factors That Control Population Growth
- Compare density-dependent and density-independent limiting factors.
- Relate competition, predation, parasitism, overcrowding, and natural catastrophes to population growth.

48–3 Interactions Within and Between Communities
- Describe the ways in which populations within a community interact.
- Compare parasitism, commensalism, and mutualism.
- Discuss the interrelationships between ecosystems.

Journal Activity

YOU AND YOUR WORLD
Living communities are found everywhere. In your journal, write a letter to a sixth grader explaining why schoolyards, back yards, parks, and alleys are considered living communities.

Figure 48–1 A population is a group of organisms of the same species that live in a given area and breed with one another. These walruses, packed onto their breeding ground on Alaska's Round Island, are part of one walrus population.

48–1 Population Growth

Guide For Reading
- How do exponential growth and logistic growth differ?
- What is the relationship between population growth and carrying capacity?

In order to study relationships between organisms, ecologists need to know how groups of organisms change over time. How many individuals are born? How many die? How many organisms live in an area at any given time? To answer these questions, ecologists study **populations**. A population is a group of organisms that all belong to the same species and that live in a given area.

Exponential Growth: A Baby Boom

Almost any organism provided with ideal conditions for growth and reproduction will experience a rapid increase in its population. What's more, the larger the population gets, the faster it grows. If nothing stops the population from growing, it will continue to expand faster and faster. The kind of curve this growth pattern produces on a graph is called an **exponential growth curve**. See Figure 48–2 on page 1034.

As you learned in Chapter 13, Charles Darwin realized that this tendency of populations to grow exponentially (doubling and redoubling over time) presented a puzzle to biologists of his time. Among other things, Darwin calculated that if all the offspring of a single elephant couple were to survive and reproduce, in less than 750 years one pair of elephants alone would produce 19 million offspring!

Obviously, exponential growth does not continue in natural populations for long. Most offspring of plants and animals do not survive long enough to reproduce. The question is . . . why?

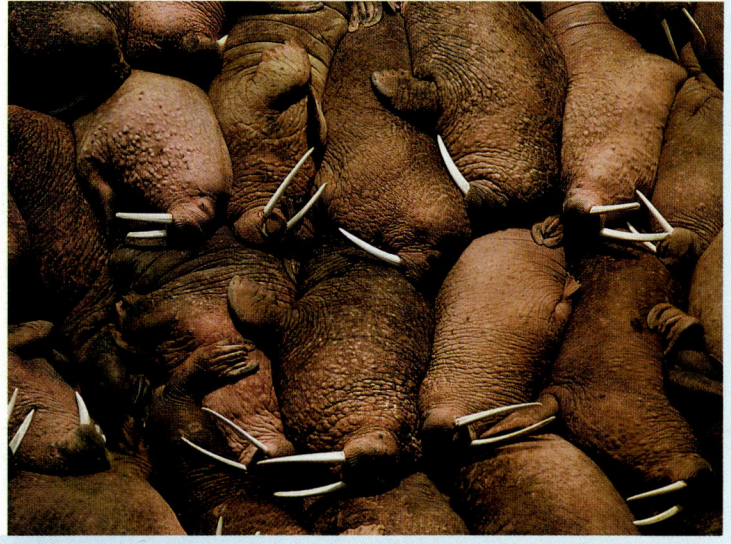

COOPERATIVE LEARNING

Using preassigned lab groups or randomly selected teams, have groups predict what their everyday life would be like if the human population continues to grow exponentially. Groups could present their predictions as:
- A two-minute skit or soap opera
- An illustrated science fiction short story
- A diary entry

In producing their final product, encourage groups to consider the economic, political, and environmental impact that exponential human population growth would have on daily life.

Journal Activity

YOU AND YOUR WORLD
In their letter, students should identify some organisms found in the community they describe. They should include their letter in their portfolio.

known species. Some biologists think that there are as many as 10 million different species of animals and plants. Explain that in a given area, such as a forest in North America, populations of many different species are present. Also explain that these populations live together in what is called a community.

• **What do you think of when you hear the word *community*?** (Possible answers: a group of people who live in the same neighborhood or town; a group of people who share common customs and beliefs; a close-knit group of people such as an extended family.)

Explain that although human communities usually consist of people who are alike in some way, a biological community refers to populations of different species that live in the same area. As in human communities, these species interact in many ways and form many different types of relationships.

TEACHING STRATEGY 48–1

Focus/Motivation
Write on the chalkboard the following number sequence:
    1, 2, 4, 8, 16, 32, 64, 128, 256, . . .

• **What pattern do you see in this sequence?** (Each number is twice the preceding number.)

• **Do you think that the numbers are getting larger slowly or rapidly?** (Slowly at first, then rapidly.)

• **In mathematics, what is this type of sequence called?** (An exponential sequence.)

BACKGROUND INFORMATION
BIOTIC POTENTIAL

The biotic potential of a species is defined as the size a population would reach if all offspring were to survive and produce young. In order for this to happen, conditions would have to be ideal; there would have to be enough food and living space to support the population and there would have to be no factors present that would limit population growth. An illustration of biotic potential is given in the text: two elephants, under ideal conditions, would produce 19 million descendants after 750 years.

In actuality, no population ever reaches its biotic potential. The factors that prevent this ideal growth are called limiting factors, or environmental resistance.

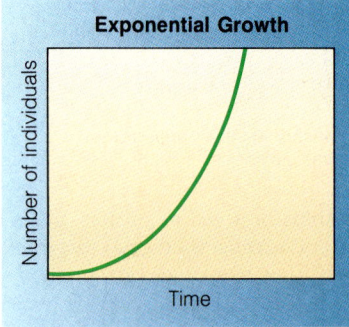

Figure 48–2 Exponential growth is illustrated on this graph. How does the number of organisms change over time during exponential growth? ❶

Logistic Growth: A Step Closer to Reality

The population growth history of a particular species is a bit more complicated than simple exponential growth. **Most populations go through a number of growth phases, which can be represented on a logistic growth curve.** A **logistic growth curve** is shown on the graph in Figure 48–3.

Let's examine an example of logistic growth. Suppose a few animals are introduced into a new environment. At first their numbers will begin to grow slowly. This initial growth is shown by section A in Figure 48–3. Soon, however, the population will begin to grow very rapidly. Here, in section B of the same graph, the population grows exponentially. The population grows quickly because few animals are dying and a great many are being produced.

Exponential growth does not continue for long. Soon the population reaches point C on the graph. Here the speed at which the population grows begins to slow down. Think about this carefully. Notice we did not say that the size of the population drops. The population is still growing, but it is growing at a slower rate. From here on, the population grows more and more slowly, through section D on the curve. How might we explain what is happening?

A population grows when more organisms are produced in a given period of time than die during the same period. In this situation, an ecologist would say that the population's birthrate is greater than its deathrate. Population growth may slow down because either the birthrate decreases or because the deathrate increases or both.

When the birthrate and deathrate are the same, population growth will stop, or reach zero growth. Remember, when we say that population growth is zero, we mean that the number of

Figure 48–3 Most populations undergo logistic growth, which is illustrated in this growth curve. What portion of this graph represents exponential growth? ❷

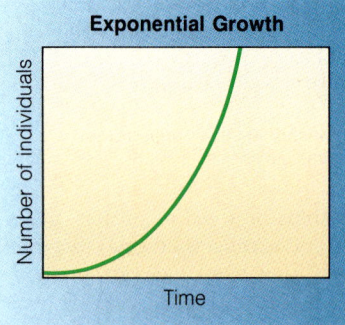

1034

48–1 (continued)

Content Development
Using the Visuals Use the graph in Figure 48–2 to show students how populations grow exponentially. Point out that like the numbers in the mathematical sequence, population increases slowly at first but then suddenly starts to grow very rapidly.

Explain that a population grows or decreases over a given period of time according to four variables: the number of births, the number of deaths, the number of immigrants, and the number of emigrants. Immigrants are members of the same species who join a population from another population, and emigrants are members of a population that leave to go to another population. Often, for the sake of simplicity, biologists discount immigration and emigration. When this is the case, population growth is equal to a population's birthrate minus its deathrate.

SECTION REVIEW 48–1

1. A population is a group of organisms of the same species that live in the same area.
2. At first, growth is exponential, starting slowly and then growing rapidly through phase 2. During phases 3 and 4, the population growth begins to slow down. Finally, when the population reaches the carrying capacity of the ecosystem, the population reaches a steady state.
3. Since the growth of popula-

organisms in the population remains the same. Look again at the logistic growth curve in Figure 48-3. The portion of the curve labeled E is called the **steady state.** During the steady state, the average growth rate is zero. However, it is important to note that the steady state is not really all that steady. The population rises and falls somewhat. In fact, in some populations it rises and falls a great deal. But the rises and falls average out around a certain population size.

If we draw a horizontal line through the middle of the steady state region, as in Figure 48-4, that line will tell us how big the population is in the steady state. Ecologists say this line represents the **carrying capacity** of a particular environment for a particular species.

Once a population reaches the carrying capacity of its environment, certain factors keep the population from growing any further. These factors include a lack of food, overcrowding, and competition among the individuals in the population. If the population does grow larger, either the birthrate will fall or the deathrate will rise. (More individuals will die than will be born and the population will be reduced to the carrying capacity.) If the population falls, either the birthrate will rise or the deathrate will drop. (More individuals will be born than will die and the population will grow once again.)

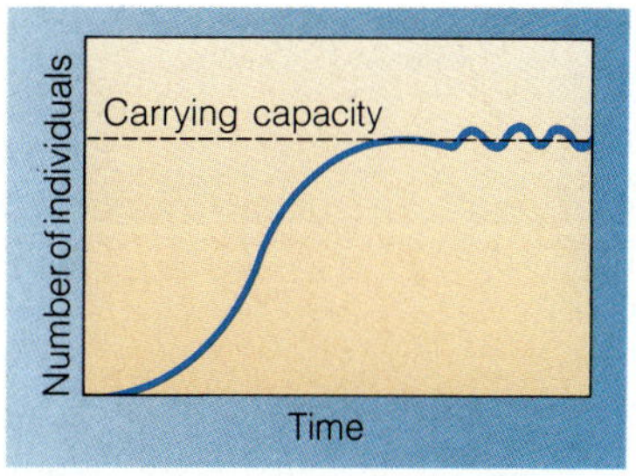

Figure 48-4 *Carrying capacity represents the optimum number of organisms of a particular species that can be supported by a particular environment.*

48–1 SECTION REVIEW

1. What is a population? Give three examples of a population in your area.
2. Describe the five stages of a logistic growth curve.
3. **Critical Thinking—Applying Concepts** Use the concept of carrying capacity to explain the importance of conservation.

48–2 Factors That Control Population Growth

Guide For Reading

- How do density-dependent and density-independent limiting factors differ?
- What effects do competition, predation, parasitism, overcrowding, and natural catastrophes have on population growth?

Why, you might ask, don't populations grow indefinitely? Recall from the previous chapter that the growth of individuals can be controlled by limiting factors. Similarly, both plant and animal populations can be controlled by several factors. Although controls on natural populations do not keep those populations from changing in size, no single species has ever threatened to overpopulate the entire planet. (That is, not until *Homo sapiens* came along.) Let us examine the ways in which natural populations are kept between extinction and overpopulation of their environment.

ANNOTATION KEY

1. Increase over time. (Interpreting graphs)
2. Section B on the graph. (Interpreting graphs)

TEACHING SUPPORT

Laboratory Manual
- Counting a Population, p. 585

Teaching Resources
- Hands-On Activity: How Many Are Too Many? p. 225

Study Guide
- Section 48–1, p. 457

tions in all species is dependent on limiting factors, students should be able to see how conservation (the wise management of resources) will provide for the future growth of the human population.

Reinforcement/Reteaching

If students are having trouble answering the Section Review questions, go back to the material with which they are having difficulty. Review the material; then have students try the questions again.

Closure

Have students compare exponential population growth with logistic population growth. Help students understand that exponential growth would continue indefinitely if a population were exposed to only ideal conditions; however, certain factors in the environment prevent the population from growing this way.

TEACHING STRATEGY 48–2

Focus/Motivation

- **How many students can occupy this classroom the way the room is set up now?** (Answers may vary. Most students will probably say that the number of students that can occupy the classroom is equal to the number of desks and chairs that are in the room.)
- **Suppose we want to increase the number of students in the room. What would we have to do?** (Bring in more desks and chairs.)
- **Do you think there is a limit to the number of students we could have? Why or why not?** (Yes, because beyond a certain point, no more chairs and desks would fit.)
- **Suppose that many more students wanted to be in this room than the room could hold. What do you think would happen?** (Possible answers: Competition for a place in the room might result; some students would be selected to stay and others would have to leave.)

BACKGROUND INFORMATION
POPULATION DENSITY

The number of individuals of a certain species per unit area is called population density. If the population density of a species is higher than the environment's carrying capacity, many individuals may die. There are, however, several advantages to having a high population density that does not exceed the carrying capacity. One such advantage is that in sexually reproducing organisms, there is a greater opportunity for genetic diversity. Such diversity increases the chances of the population adapting to environmental changes; this in turn increases its chances for survival.

When population density becomes too low, the survival of the species is endangered. Scientists estimate that a minimum of 500 individuals is necessary to guarantee long-term survival in nature. A species with a very low population density may become extinct.

TEACHING SUPPORT
Teaching Resources
- Activity: Predator-Prey Relationships, p. 3

MEDIA AND TECHNOLOGY
Videodisc
- Aquatic Ecosystems: *Freshwater Wetlands and Freshwater*
Use this bar code to show two birds competing for the same food source.

Play frames 40633 to 42410

See the Biology Media Guide page 76 for additional bar-code correlations for this section.

48–2 (continued)
Content Development
Use the Focus/Motivation discussion to illustrate the idea that a given environment has only a certain amount of space, food, water, and other life essentials. Because of this, the size of a population will be limited to the number of individuals who can obtain what they need. When there are many more individuals than there are resources, competition results.

- **If an organism is involved in competition for life essentials, what are the possible outcomes for that organism?** (The organism may win the struggle and survive or it may lose the struggle and die; possibly the organism will be able to coexist with its competitor if the differences be-

Density-Dependent Limiting Factors

When factors that control population size operate more strongly on large populations than on small ones, they are called **density-dependent limiting factors.** Density-dependent limiting factors usually operate only when a population is large and crowded. They do not affect small, widely scattered populations much. **Density-dependent limiting factors include competition, predation, parasitism, and crowding.**

COMPETITION When populations become crowded, both plants and animals compete, or struggle, with one another for food, water, space, sunlight, and other essentials of life. It is easy to see why competition among members of the same species is a density-dependent limiting factor. The more individuals there are, the more of them there are to use up the available food, water, space, and other necessities. The fewer individuals around, the less they compete.

Competition between members of different yet similar species is a major force behind evolutionary change. As you learned in Chapter 14, no two organisms can occupy the same niche in the same place at the same time. When two species compete, both find themselves under pressure from natural selection to change in ways that decrease their competition. This idea is important because it ties ecology and evolution together. It is another example of the way in which all the biological sciences are interrelated when you look at them from an evolutionary point of view.

PREDATION Just about every species serves as food for some other species. In most situations, predators and prey coexist over long periods of time. Like tennis partners who have played together for years, predators and prey have become accustomed to each other's strengths and weaknesses.

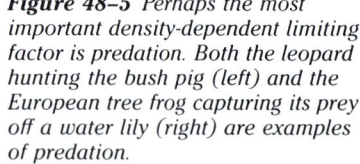

Figure 48–5 Perhaps the most important density-dependent limiting factor is predation. Both the leopard hunting the bush pig (left) and the European tree frog capturing its prey off a water lily (right) are examples of predation.

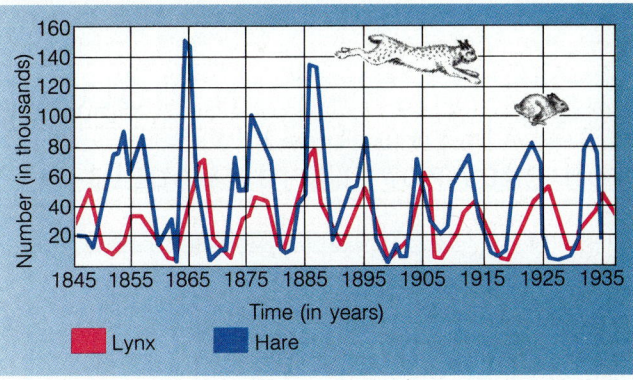

Figure 48-6 *This graph represents the relationship between populations of lynx (predator) and snowshoe hare (prey) over many years. What can you infer from the growth curves of each organism?* ❶

Prey, for example, have evolved defenses against predators. Some plants may produce poisonous chemicals. Some animals may have shells, poisonous skins, or camouflage behaviors and colors that help them hide.

At the same time, predators have evolved counterdefenses. Some herbivores, such as monarch butterfly caterpillars, have evolved the ability to avoid the effects of certain plant poisons. Carnivores have evolved stronger jaws and teeth, powerful digestive enzymes, or extra-keen eyesight.

If we watch populations of predators and prey over time, we almost always find changes in their numbers. Typically, at some point the prey population grows so large that prey are numerous and easy to find. With such a large and available food supply to feast upon, there may soon be almost as many predators as prey. As you probably know by now, this situation cannot last because each predator needs many prey to satisfy its energy needs.

As predators become numerous, they eat more prey than are born. This means that the prey's deathrate becomes higher than its birthrate and the population decreases in size. But as the prey population drops, predators begin to starve, so the predator population drops too. When only a few predators are left, the prey begin reproducing and surviving in large numbers again, and the whole situation repeats itself.

For many years people did not truly understand (as we do now) that predator-prey relationships are important in controlling natural populations. Travelers and farmers took animals from one part of the world to another, releasing them into the wild. There, without a predator to keep their numbers down, the animals became serious pests. One famous example is the introduction of rabbits by Australians to their island continent a number of years ago. The collection of animals native to Australia is unique to that continent. Thus, rabbits had no natural predators. Within a relatively short time, the Australian rabbit population went into exponential growth and remained that way for a long time. Rabbits infested the countryside and devoured much of the natural vegetation. They have been a serious problem in Australia ever since.

ANNOTATION KEY

❶ They are interrelated. As the number of prey increases, the number of predators increases, which then causes the number of prey to decrease. (Relating facts)

BACKGROUND INFORMATION
COMPETITION

When different species in a community compete for resources, the competition is called interspecific competition. When members of the same species compete for resources, the competition is called intraspecific. Intraspecific competition is usually more intense than interspecific competition because the needs of two organisms of the same species will be almost identical.

Competition between like or unlike species can be classified into two different types. One type is called exploitation competition; the other is called interference competition. In exploitation competition, all individuals have equal access to the resources in question but they differ in how fast or how efficiently they can obtain and use them. In this type of competition the fastest or strongest individuals usually succeed. In interference competition, certain individuals limit or prevent others from using certain resources. For example, corals kill neighboring corals by poisoning them, then growing over them; a strangler fig tree surrounds another tree and eventually kills it.

tween them are great enough.)
• **Is there any other alternative for organisms that find themselves in competition with other organisms?** (If the competition is between different yet similar species, the organisms may change in ways that will decrease the competition. In this way, both species may be able to survive.)

Reinforcement/Reteaching
Explain to students that density-dependent factors tend to be biotic and density-independent factors tend to be abiotic. For example, most density-dependent factors have to do with interactions among organisms. These interactions include competition, predation, and reactions to crowding and stress. Most density-independent factors, on the other hand, have to do with interactions between organisms and nature. Some typical density-independent factors are frost, wind, rain, and temperature changes.

PROBLEM SOLVING

Analyzing Predator-Prey Population Models

Point out to students that there are several types of predator-prey relationships. In *stable coexistence*, the two groups maintain steady populations. The populations of both groups may rise and fall in what are referred to as *oscillations* or *cycles*. A prey population may become extinct as a result of predation.

Answers to Questions
1. It will increase; it will appear at a higher level on the graph.
2. The predator population will decline and the prey population will begin to rise; the predator population will gradually increase once the prey population increases.
3. Prey will remain at zero; predator population will decline; eventually the predators will die out unless another food supply is introduced into the environment.

TEACHING SUPPORT

Teaching Resources
- Biology Case Study: Using Predators to Manage Populations, p. 21

48–2 (continued)

Content Development

Point out that parasite-host relationships tend to evolve toward a compromise in which the host is able to tolerate a widespread low-grade infection. This is to the parasite's advantage, because without the host, the parasite has no source of nourishment.

Sometimes, however, environmental conditions may change and become less favorable for the host. If this happens, stress

PROBLEM SOLVING IN BIOLOGY

ANALYZING PREDATOR-PREY POPULATION MODELS

The relationships between predator and prey are often intertwined, particularly in an environment in which each prey has a single predator and vice versa. Examine the accompanying graph, which shows a computer model of the changes in predator and prey populations over time. After analyzing the graph, answer the following questions.

1. A sudden extended cold spell destroys almost the entire predator population at point F on the graph. What will happen to the prey population? How will the next cycle of prey population growth appear on the graph?
2. A bacterial infection kills off most of the prey at point B on the graph. How will this affect the predator and prey growth curves at point C? At point D?
3. A viral infection kills all of the prey at point D on the graph. What effect will this have on the predator and prey growth curves at point E? What will happen in future years to the predator population? How could ecologists ensure the continued survival of the predators in this ecosystem?

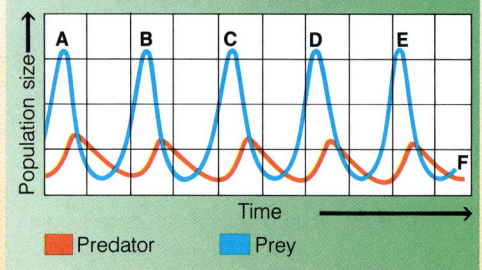

Figure 48–7 The California dodder plant is a parasite that wraps itself around a host plant and uses the host for support and food.

PARASITISM Parasites act like predators in many ways. Parasites live off their hosts, weakening them and causing disease. Like predators, parasites work most effectively if hosts are present in large numbers. Crowding helps parasites travel from one host to another. Stress related to crowding can also reduce a host's resistance to parasites. As a result, parasitism often affects large, concentrated populations more than small, scattered ones. Thus parasitism works as a density-dependent limiting factor on population growth. Note that few parasites kill their hosts—at least not right away. If a parasite kills its host too quickly, the parasite will have no chance to reproduce and spread. It is thus to the parasite's advantage not to be too deadly.

CROWDING AND STRESS Most animals have a built-in behavioral need for a certain amount of space. Both males and females, for example, may need room to hunt for food. They may need a certain amount of space for nesting. Or they may need a territory of a certain size. A number of fishes on coral reefs fit into this latter category. Many of these fishes are

may make some hosts less tolerant of the parasites than they were before; or the sapping of strength by the parasite may make the host less able to adapt to and survive the environmental changes. The result will be a higher deathrate among hosts.

Another factor that can increase the deathrate of hosts is genetic mutations in parasites. If a mutation leads to a strain of parasite that is more virulent, then some hosts will not be able to tolerate the continuing presence of the parasite.

Emphasize to students that population density favors the activity of parasites. The greater the density of the host population, the greater the chance that an uninfected host will come into contact with an infected host. In a crowded situation, parasites

extremely territorial. Each male stakes out a territory and chases away all other males of his species. Young fish do not stand a chance of setting up a territory unless an older male dies or is eaten. In such cases, the number of suitable territories regulates population size in a density-dependent manner.

Certain species fight among themselves if they are overcrowded. Too much fighting can cause high levels of stress. This stress disturbs the finely tuned endocrine system you read about in Chapter 42. Large amounts of adrenaline secreted under conditions of stress upset the body's normal balance. Levels of several other hormones also change due to stress. As a result of these hormonal changes, animals fight more and breed less. Often the immune system is weakened as well. Hormonal changes can so upset a female's behavior that she neglects, kills, or even eats her own offspring. Extreme overcrowding among mice can affect the females' endocrine system so that pregnant females miscarry, or lose the fetus they are carrying. All these factors combine to lower birthrate.

Density-Independent Limiting Factors

Not all populations are controlled by density-dependent limiting factors alone. Many species show what are called boom-and-bust growth curves. Their populations grow exponentially for some time and then suddenly crash. After the crash, the population may build right up again or it may stay low for some time.

Thrips, aphids, and other insects that feed on plant buds and leaves can be washed out by a rainstorm. They may also be harmed by long hot periods of dry weather. Frosts, too, can cause sudden drops in insect populations. For these species, storms, cold weather, dry weather, or other natural occurrences can nearly wipe out the population. Such wipeouts can happen regardless of how large the population is at the time. Because population density does not matter in such cases, these natural occurrences are called **density-independent limiting factors**. As you might expect, the growth of many species is controlled by some combination of density-dependent and density-independent limiting factors.

Human Population Growth

Human populations, like those of all other animals, tend to increase in size with time. If we examine the size of the human population over the course of history, we see that for a long time it grew slowly. Then, about 500 years ago, the world's human population started growing exponentially. See Figure 48–10 on page 1040.

Today, population growth in the United States and parts of Europe has slowed down. But most of the world's people do not live in these countries. Instead, they live in China, India,

Figure 48–8 Floods, such as the ones that devastated the Midwest in 1993, are examples of density-independent limiting factors.

Figure 48–9 The human population grew slowly until between 500 and 1000 years ago, when it began to increase exponentially. The small dip in the graph represents the devastation caused by the Black Death—a plague that killed nearly one third of Europe's population in the Middle Ages.

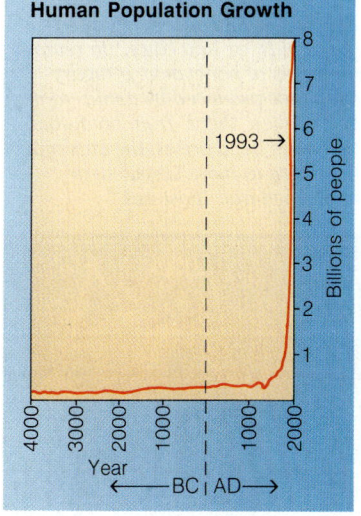

BACKGROUND INFORMATION
HOST-PARASITE RELATIONSHIPS

Host-parasite relationships are continually evolving. For example, hosts develop resistance to parasites such as bacteria or protozoa that cause disease. The parasites then appear in new strains resistant to the old defenses, and the disease reappears.

BACKGROUND INFORMATION
HUMAN POPULATION

Scientists estimate that every minute worldwide, 230 babies are born and 90 people die. This means that the human population is increasing at a rate of about 140 people per minute.

Scientists estimate that the carrying capacity of the Earth is 8 billion to 11 billion people. If this estimate is correct, then the human population density is approaching the upper limit of the carrying capacity of the Earth. At the current annual growth rate of 1.7 percent, the world's human population will reach 9 billion in the year 2020 and 18 billion in the year 2060. In order to support such a large population, new technologies, medical practices, and sources of food will have to be developed, as well as new types of housing, transportation, and communication.

---

often spread through an entire host population. If the host has been weakened by other limiting factors, such as lack of food or an inhospitable climate, then the host is even more susceptible to infection by parasites.

Enrichment
Ask students to develop a demographic profile for the country of their choice. Have them include the population size, birthrate, deathrate, immigration and emigration rates, rate of population change, age and sex distribution, major causes of death, and the life expectancies of males and females.

Skills Development
Guided Practice
Skill: Making a graph
Provide students with the following data on human population density. Then have them make bar graphs to display the information.

Continent	Density (per km²)
Europe	68
Asia	65
Africa	19
N. America	17
S. America	15
Australia	2

BACKGROUND INFORMATION
MUTUALISM

Mutualism can benefit a species in many ways. Three areas of benefit include nutrition, protection, and transport.

One species can benefit another species nutritionally by digesting food for the partner or by providing a supply of nutrients or energy. An example of mutualism in digestion is the relationship that exists between ruminants such as cows and the microbes that live in their stomachs. The microbes digest cellulose from the plant material that the ruminants eat.

A species can protect another species from predators or from environmental conditions. In East Africa, ants live in the thorns of certain acacia trees. The thorns provide the ants with shelter. The stinging, biting ants in turn protect the tree against herbivores.

One species can help another species by transporting materials to a favorable location. An obvious example of this is vector pollination. Butterflies, moths, and bees are among the animals that carry pollen from one flower to another, making it possible for plants to be fertilized.

TEACHING SUPPORT
Study Guide
- Section 48–2, p. 459

48–2 (continued)

SECTION REVIEW 48–2

1. Factors include competition (organisms vying for the same food, space, and so on), predation (organisms feeding on other organisms), and parasitism (one organism living off another organism). Accept other appropriate factors as well.
2. Density-dependent limiting factors (predation and so on) tend to act on large populations, and density-independent limiting factors (storms, floods, and so on) act on populations of any size.
3. It is important to predators that some of the population upon which they feed survive and reproduce. If this did not occur, the prey would eventually decrease, which in turn would decrease the number of predators. Therefore, it is better for the predators that the prey develop adaptations that help them survive, as long as the adaptations do not allow the prey to escape the predators entirely.

---

and parts of Africa and Latin America—places where populations are still growing very rapidly. This population growth poses a serious threat to global ecology, as we shall see in the next chapter.

48–2 SECTION REVIEW

1. Describe three density-dependent limiting factors.
2. Compare density-dependent and density-independent limiting factors.
3. **Critical Thinking—Relating Cause and Effect** How does the evolution of a successful defense mechanism in a species of prey increase the chances of survival for a species of predator that uses the prey for food?

Guide For Reading
- How do populations within a community interact?
- How do parasitism, commensalism, and mutualism differ?
- How do ecosystems interact?

Figure 48–10 The small gray insects are aphids, or "ant cows." In return for meals of honeydew, a sugary substance produced by aphids, ants defend their "herd" from predators. When the need arises, the ants move their herd to more succulent or more sheltered "pastures."

48–3 Interactions Within and Between Communities

After populations, the next larger biological units studied by ecologists are communities. A community consists of all the populations of organisms living in a given area. **Populations in communities interact with one another in many ways.** Plant species, for example, compete for water, nutrients, and sunlight. At the same time, some plants have evolved defenses against herbivores.

Herbivores compete with one another for food and space. There are usually several different herbivore species in any community. Certain of these herbivores may have evolved counterdefenses for the protective mechanisms of one or more plant species.

While this is going on, carnivores are hunting the herbivores. Often there are several carnivores in a community, each of which is best at hunting a particular herbivore.

Symbiosis

Many of the interactions among organisms we have discussed so far involve predation. But there are several other relationships that play an important role in nature. These relationships between organisms are called **symbioses** (*sym-* means together; *-bios* means life; *symbiosis* means living together). Parasitism, which you read about earlier in this chapter, is a symbiosis in which one species benefits and the other is harmed.

There are also many relationships between organisms in which one member benefits and the other is not harmed. This

1040

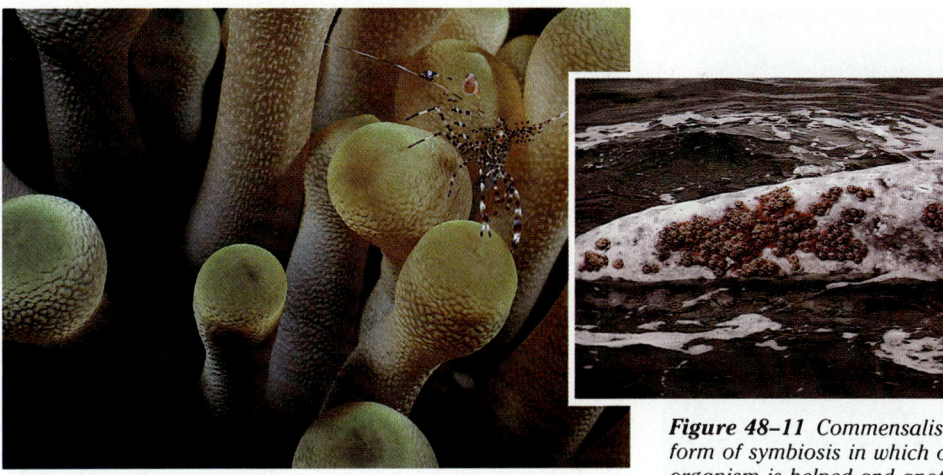

kind of symbiosis is called **commensalism** (kuh-MEHN-suhl-ihz-uhm). A good example of commensalism takes place on a coral reef where shrimp live within the stinging tentacles of sea anemones. The shrimp are not affected by the anemone's poison. As a result, the shrimp are protected from predators that cannot tolerate the anemone's stings. Anemones are not harmed by shrimp living on them, but they are not helped either, which is the definition of commensalism.

In still another kind of symbiosis, two species live together in such a way that both species benefit. This kind of symbiosis is called **mutualism.** Let's return to the coral reef to examine an example of mutualism. Right next to the sea anemone and the shrimp we might find clownfish. Clownfish form a mutualistic symbiosis with sea anemones. Clownfish benefit from living within the stinging tentacles of the sea anemones in the same way shrimp do. However, clownfish also help the anemones by chasing away several species of anemone-eating fish. In this case, both species benefit, which is the definition of mutualism.

Commensal and mutualistic symbioses are everywhere in nature. You may recall that lichens are a mutualistic symbiosis between a blue-green bacteria and a fungus. Many marine animals—such as corals—have symbiotic algae that live inside their tissues. Many land plants can live only with the help of symbiotic fungi on their roots. In fact, practically no organism can live in a world by itself. Each requires other organisms in some way.

Interactions Among Ecosystems

Not only do populations and communities interact, ecosystems also interact with one another in many ways. Consider, for example, a pond in the woods. Certainly that pond contains

Figure 48–11 Commensalism is a form of symbiosis in which one organism is helped and another organism is neither helped nor harmed. The tiny shrimp living within the tentacles of this sea anemone (left) and the barnacles hitching a ride on the back of this gray whale (right) are examples of commensalism.

Figure 48–12 This tiny fish survives on the bits of food it can find by cleaning the teeth of the larger fish. In this case, both organisms benefit, and their symbiotic relationship is known as mutualism.

BACKGROUND INFORMATION
POPULATION INTERRELATIONSHIPS

Because so many organisms are tied together in food webs, population increases and decreases can sometimes have rather complex effects on an ecosystem. For example, a significant decrease in the rabbit population will cause foxes, owls, and hawks to feed more heavily on mice and squirrels. Some of the foxes, owls, and hawks may die of starvation or migrate to another community. If the number of foxes, owls, and hawks significantly declines, then the rabbit, mice, and squirrel populations may grow. This is good for the rabbits, whose numbers were declining—but does the ecosystem need more mice and squirrels? Increased numbers of small animals can damage vegetation and accelerate soil erosion. Thus the effects and countereffects of one group's underpopulation go on and on.

Reinforcement/Reteaching
If students are having trouble answering the Section Review questions, review the material that is causing difficulty.

Closure
Have students work in pairs. Ask each student to write down three review questions based on the material in this section. Then have students exchange questions with their partners and answer them.

TEACHING STRATEGY 48–3
Focus/Motivation
• Suppose that you are good in math and your friend is good in English. If you work on your math and English homework together, who will benefit? (Both will benefit.)
• Suppose that you are good in math and both you and your friend are good in English. Who will benefit if you do your math and English homework together? (Mostly your friend.)
• Will you be harmed by the arrangement? (No, probably not.)
• Suppose that you are good in math and somebody steals your math homework. Who will benefit? (The person who stole the homework, assuming that he or she does not get caught.)
• Who will be harmed? (You, since you lost your homework.)

1041

ANNOTATION KEY

① Accept all logical answers. (Interpreting photographs)

TEACHING SUPPORT

Laboratory Manual
- Relationships in an Ecosystem, p. 591
- Ecological Succession, p. 595

Teaching Resources
- Vocabulary Review: Word Game, p. 1
- Activity: Analyzing Biological Relationships, p. 7

Study Guide
- Section 48–3, p. 461

ESL STRATEGY

Write the following pairs of words on the board or give students a prepared list of these terms. Tell them to write a sentence or two to explain how the terms in each pair differ from each other.

What's the Difference?
1. exponential growth / logistic growth
2. zero population growth / decline in population
3. predator / prey
4. competition / predation
5. maximum / optimum
6. density-dependent limiting factors / density-independent limiting factors
7. population / community
8. symbiosis / parasitism
9. parasitism / mutualism
10. mutualism / commensalism

Figure 48-13 What point was the photographer trying to make when putting together this dramatic shot of the Earth superimposed over a green plant? ①

populations of plants and animals that live only in the pond and not in the woods. But where does the water in the pond come from? Most likely from a stream that flows out of the woods. Where does extra water from the pond go? Probably into another stream or a nearby marsh. What about leaves and insects that fall into the pond from surrounding trees? How about raccoons, birds, and other animals that visit the pond from homes in the woods? What about air and rain that blow in from outside, carrying nutrients and possibly pollution into the pond? And don't forget about migrating birds that travel thousands of kilometers to summer in the pond.

Nearly every ecosystem is connected, either directly or indirectly, with other ecosystems. We can see how some of these connections work if we take an imaginary journey—a journey that begins when a farmer puts fertilizer on a corn field. Let's hitch a ride on a single nitrogen atom in that fertilizer and see where our travels take us.

Rain washes us into a stream. That stream flows through a pond and into a river. The river enters an estuary, where we are taken in by the roots of a plant. Later the plant dies, and its decaying remains are washed out into the coastal waters nearby. As we float along, we are eaten by a shrimp that in turn is eaten by a fish. The fish swims out into the open sea, dies, and sinks to the bottom. On the ocean floor the fish decays, and nutrients (including us) are released. Ocean currents carry us for hundreds of kilometers to a place where upwellings carry us back to the ocean surface. There we are taken in by phytoplankton, which are eaten by zooplankton, which are then eaten by fish. This time the fish is eaten by a bird that picks it up and flies to shore with it.

Are you beginning to get the picture? Our journey could continue indefinitely, taking us from one ecosystem to another all over the globe. Herein lies an important lesson for us all: Winds, rivers, and ocean currents tie Earth's ecosystems together in ways that we are just beginning to understand. That is why our growing human population must be ever more careful about how we dispose of wastes. Disposing of something in one ecosystem may just cause it to show up again somewhere else. We shall discuss this matter in more detail in the next chapter.

48–3 SECTION REVIEW

1. What is a community? How do populations in a community interact?
2. Compare parasitism, commensalism, and mutualism.
3. **Critical Thinking—Relating Concepts** Do ecosystems operate independently of one another? Describe some ways in which ecosystems are interconnected.

1042

48–3 (continued)

Content Development

Use the Focus/Motivation discussion to introduce the idea of symbiotic relationships. Point out that three important types of symbiosis are mutualism, commensalism, and parasitism.

• In the Focus/Motivation discussion, which situation is an example of mutualism? (The first one, in which both people benefit from doing homework together.)
• Which is an example of commensalism? (The second one, in which one person benefits but the other person is not harmed.)
• Which is an example of parasitism? (The last one, in which one person benefits but the other person is harmed.)

SECTION REVIEW 48–3

1. A group of populations living in a given area. Populations within a community interact through competition, predation, and so on.
2. Symbiosis means living to-

SCIENCE, TECHNOLOGY, AND SOCIETY CONNECTION

Gypsy Moths: Nature to the Rescue

In many places around the world, people have unwittingly demonstrated the importance of natural population control mechanisms by introducing into an environment organisms that have no natural predators. Around the turn of the century, a few gypsy moth caterpillars were accidentally introduced into the Northeast. First detected in 1905 in Connecticut, the gypsy moths began chewing their way through the eastern states.

Gypsy moth caterpillars are leaf-eaters and can cause widespread destruction of trees. Without any natural predators, gypsy moths were unaffected by density-dependent limiting factors. The only real check on the gypsy moth population was their breeding cycle: Every 8 to 10 years gypsy moths begin a new breeding cycle, after which they remain to cause problems for about 4 years. Thus gypsy moth populations exhibit a boom-or-bust growth curve characterized by periods in which the gypsy moths present no danger.

The last outbreak of gypsy moths occurred between 1979 and 1983. So in the spring of 1989, scientists prepared themselves for the onset of another breeding cycle and another round of widespread destruction in eastern forests. But something quite unexpected began happening instead. Throughout the Northeast, gypsy moth caterpillars were dying—before they could destroy the trees. Something was killing the gypsy moth caterpillars.

When scientists at the Connecticut Agricultural Experiment Station in New Haven examined the dead caterpillars, they discovered that the animals had been infected by a fungus. It was the fungus that was killing the gypsy moth caterpillars. Researchers now believe that about 80 years ago the fungus was intentionally introduced into the environment by scientists who realized that a similar fungus controlled gypsy moth caterpillars in Japan. Why had it taken so long for the fungus to act?

Scientists now believe that the unusually heavy rains in the Northeast in 1988 and 1989 caused the fungus to suddenly thrive. The thriving fungus, in turn, has destroyed the gypsy moths in record numbers. Due to continued heavy rains, scientists expect the fungus to reduce the gypsy moth population even further in subsequent years.

Will the fungus end the cycle completely and eliminate the gypsy moth population from the northeastern states? Scientists cannot be sure, but they suspect it will not. They know that natural controls tend to keep a population in check rather than destroy it completely. But they do hope to use the fungus to help them develop a biological agent that will control the gypsy moths in future years. However, as Dr. Andreadis at the New Haven experimental station points out, biological controls will not eradicate gypsy moths. "The whole idea behind biological weaponry is to create a balance with nature so that the gypsy moth population will perhaps maintain itself at a very low level."

LABORATORY INVESTIGATION

EXAMINING PATTERNS OF POPULATION GROWTH IN BACTERIA

Before the Lab

1. At least one day prior to the investigation, gather enough materials for your class, assuming six students per group.
2. Note that two days must elapse between steps 2 and 3 and that the investigation must continue for at least four more days.
3. To prepare methylene blue stain, dissolve 0.3 gram of methylene blue in 30 milliliters of 95% alcohol. Add 100 milliliters of distilled water. Set the solution aside for 24 hours before using it.

Pre-Lab Discussion

Discuss with students the various factors that affect the growth of a population. Also review the two types of population growth described in Section 48–1.

Have students read the Problem at the beginning of the investigation. Then have them read through the Procedure.

• **What factors do you think will affect the growth of the bacteria population?** (Accept all reasonable answers.)
• **What growth pattern or patterns do you expect to see over the four- or five-day period?** (Answers may vary. It is reasonable to predict exponential growth at first, since this is the pattern shown by most organisms in a new environment.)

Discuss with students the dilution technique described in step 5. Make sure students keep track of the number of dilutions and enter the dilution factor in their data tables. Explain to students that if the bacteria population grows, they may have to use a higher dilution factor on the second, third, or fourth day than they did on the first day.

Direct students' attention to step 10, at which time they will make a graph.

• **What will be the independent variable in your graph?** (Time.)
• **In what unit is this variable expressed?** (Days.)
• **On which axis on your graph will you plot the independent variable?** (X-axis.)
• **What is the dependent variable?** (Number of bacteria.)
• **On which axis will you plot this variable?** (Y-axis.)

Remind students that when plotting the values for bacteria populations, they must make sure that they have multiplied by the dilution factor.

Skills Development

Students will use the following skills in this investigation.
1. Predicting
2. Observing

LABORATORY INVESTIGATION

EXAMINING PATTERNS OF POPULATION GROWTH IN BACTERIA

PROBLEM

What happens to the number of bacteria in a bean infusion over time?

MATERIALS (per group)

aluminum foil
100-mL beaker
coverslips
glass slides
10-mL graduated cylinder
100-mL graduated cylinder
2 lima beans
2 medicine droppers
methylene blue
microscope
test tube rack
4 test tubes

PROCEDURE

1. Use a 100-mL graduated cylinder to pour 50 mL of water into a 100-mL beaker. Put two lima beans into the water in the beaker. Allow this bean infusion to sit for 48 hours.
2. After 48 hours, use a medicine dropper to transfer a drop of water from the beaker to the center of a clean glass slide. Use the other medicine dropper to add a drop of methylene blue to the drop of water. Cover the sample with a coverslip.
3. Use the low-power objective to examine the slide under a microscope. Switch to high power. Use the fine adjustment to locate some bacteria.
4. Try to count the number of bacteria in the field of view. If there are too many bacteria to count, proceed to step 5. If you are able to count the number of bacteria, do so, then go directly to step 7.
5. In order to reduce the number of bacteria, the bean infusion must be diluted. To do this, proceed as follows. Use a 10-mL graduated cylinder to put 9 mL of water into a test tube. With a medicine dropper, put 1 mL of the bean infusion into the 10-mL graduated cylinder. Pour the bean infusion into the test tube containing 9 mL of water. This procedure, which dilutes the infusion by a factor of 10, can be repeated as often as necessary until the infusion is dilute enough to make the bacteria easy to count. This procedure is known as a serial dilution. Each time you follow the procedure described, you increase the dilution by a factor of 10. If the infusion is not diluted, the dilution factor is 1. Rinse the graduated cylinders when you are finished.
6. Examine each dilution under the microscope. Stop making dilutions when you can count the bacteria in the field of view.
7. After counting the bacteria, record the data, dilution factor, and number of bacteria. To determine the actual number of bacteria present, multiply the number of bacteria observed by the dilution factor.
8. Cover the beaker with aluminum foil and set it aside overnight.
9. Repeat steps 2 through 8 every day for at least four more days.
10. Prepare a graph with the date on the X-axis and the number of bacteria present on the Y-axis.

OBSERVATIONS

1. Describe the appearance of the bacteria.
2. When did the size of the bacterial population change the least?
3. When did the bacterial population grow the fastest?
4. Describe the bacterial population during the course of the observation period.

ANALYSIS AND CONCLUSIONS

1. How can you explain the changes you observed in the bacterial population?
2. What lessons does this investigation have for humans regarding the use of resources?

STUDENT STUDY GUIDE

SUMMARIZING THE CONCEPTS

The key concepts in each section of this chapter are listed below to help you review the chapter content. Make sure you understand each concept and its relationship to other concepts and to the theme of this chapter.

48-1 Population Growth

- A population is a group of organisms that all belong to the same species and that live in a given area.
- Given ideal conditions for growth and reproduction, a population of organisms will grow very rapidly. This rapid growth is shown in an exponential growth curve.
- Most populations go through a series of growth phases, which can be represented on a logistic growth curve.
- On a typical logistic growth curve, a population of organisms grows slowly at first, then grows rapidly during exponential growth, then slows down before reaching a steady state.
- A population with zero growth has reached a steady state. This means the average population over time will not change markedly.
- A population usually achieves a steady state when it reaches the carrying capacity of the environment.

48-2 Factors That Control Population Growth

- Density-dependent limiting factors—which include predation, competition, parasitism, and crowding—operate strongly on large populations with high density.
- Factors that do not depend on population density to control population size are called density-independent limiting factors. Severe storms, dry weather, and other climate conditions that can reduce a species' population are density-independent limiting factors.

48-3 Interactions Within and Between Communities

- A community consists of all the populations of organisms living in an area.
- Although predation is the primary relationship between organisms in a community, symbioses are other important relationships.
- Symbioses include parasitism, commensalism, and mutualism.
- All ecosystems are interconnected.

REVIEWING KEY TERMS

Vocabulary terms are important to your understanding of biology. The key terms listed below are those you should be especially familiar with. Review these terms and their meanings. Then use each term in a complete sentence. If you are not sure of a term's meaning, return to the appropriate section and review its definition.

48-1 Population Growth
population
exponential growth curve
logistic growth curve
steady state
carrying capacity

48-2 Factors That Control Population Growth
density-dependent limiting factor
density-independent limiting factor

48-3 Interactions Within and Between Communities
symbiosis
commensalism
mutualism

---

3. Measuring
4. Recording
5. Graphing
6. Relating
7. Applying
8. Hypothesizing
9. Manipulative
10. Comparing

Safety Tips
Review the rules for glassware safety. Tell students to report cracked or broken glassware immediately.

Teaching Strategy
1. Circulate throughout the room and help groups that are having trouble. Especially be aware of students who may be having trouble counting the bacteria or deciding on how many dilutions to use.

2. At the end of the investigation, you may wish to collect data from all the groups and make a graph of the combined data on the chalkboard.

Observations
1. The bacteria are small and rod-shaped. Some are moving. Some are colonial and others are solitary.
2. The size changes the least between the start of the infusion and the first observation.
3. Answers will vary. Most students should observe the fastest growth between the third and fourth day of observations.
4. The population increases slowly at first, then increases exponentially, and finally decreases.

Analysis and Conclusions
1. The bacteria population increased while the factors needed for growth were present. As these factors decreased, the population growth slowed down and the population decreased. This decrease was due mainly to oxygen depletion and the buildup of bacterial wastes. Students who hypothesize that the population decreased because of a lack of food should be reminded that the food source (lima beans) was not depleted.
2. Overuse of resources can lead to the depletion of resources and to pollution of the environment, either of which can harm the population.

Going Further: Enrichment
Have students design an investigation in which they introduce at least one factor that might affect population growth in bacteria. Such factors may include light, heat, additional beans, the introduction of other organisms, size of container. Have students carry out the investigation using a control setup and an experimental setup.

1045

CHAPTER REVIEW

CONTENT REVIEW

Multiple Choice
1. a   2. d   3. b   4. b
5. d   6. a   7. b   8. c

True or False
1. F population
2. F exponential growth curve
3. T
4. F density-dependent limiting factor
5. T
6. F mutualism
7. F do not kill
8. T

Word Relationships
1. types of symbiosis; predation is the method of existence of predatory animals
2. parts of a logistic growth curve; symbiosis is a close relationship between two species in which one species lives on, in, or near the other and at least one species benefits
3. density-dependent limiting factors; a tornado is a density-independent limiting factor
4. consequences of overpopulation; territoriality is the behavior pattern exhibited by an animal in establishing and maintaining its territory

CONCEPT MASTERY

1. If a natural disaster such as a hurricane strikes, a population may be wiped out even though under normal conditions it is controlled mainly by density-dependent limiting factors.
2. Answers will vary but are likely to include stress due to crowding, increased competition in terms of territoriality, buildup of waste products, and so on.
3. Darwin and others realized that if exponential growth occurred for any length of time, particular species would overpopulate the entire planet.
4. A species' population reaches a steady state when its size reaches the carrying capaci-

CHAPTER REVIEW

CONTENT REVIEW

Multiple Choice

Choose the letter of the answer that best completes each statement.

1. During population growth
   a. birthrate increases.
   b. deathrate increases.
   c. birthrate decreases.
   d. birthrate and deathrate decrease.
2. A population that reaches the carrying capacity of its environment is said to have reached
   a. logistic growth.   c. density dependence.
   b. exponential growth.   d. a steady state.
3. Density-independent limiting factors include
   a. predation.   c. crowding.
   b. hurricanes.   d. parasitism.
4. All of the organisms living in a given area make up a (an)
   a. population.   c. ecosystem.
   b. community.   d. steady state.
5. A relationship in which one organism is helped and another organism is neither helped nor hurt is called
   a. mutualism.   c. symbiosis.
   b. parasitism.   d. commensalism.
6. A form of symbiosis in which both organisms benefit is called
   a. mutualism.   c. commensalism.
   b. parasitism.   d. the carrying capacity.
7. A type of symbiosis in which one organism benefits and the other is harmed is called
   a. mutualism.   c. commensalism.
   b. parasitism.   d. symbiosis.
8. On a logistic growth curve, the portion of the curve in which the population grows rapidly is called
   a. logistic growth.   c. exponential growth.
   b. a steady state.   d. the carrying capacity.

True or False

Determine whether each statement is true or false. If it is true, write "true." If it is false, change the underlined word or words to make the statement true.

1. A <u>community</u> is a group of organisms that belong to the same species and that live in a given area.
2. The rapid growth of a population is best shown by a <u>logistic growth curve</u>.
3. A horizontal line drawn through the middle of the steady state region on a growth curve represents the environment's <u>carrying capacity</u>.
4. Predation is a <u>density-independent limiting factor</u>.
5. <u>Crowding</u> is a density-dependent limiting factor.
6. The sea anemone and the clownfish are an example of <u>commensalism</u>.
7. In general, parasites <u>kill</u> their hosts.
8. When both organisms in a symbiotic relationship benefit, that relationship is called <u>mutualism</u>.

Word Relationships

In each of the following sets of terms, three of the terms are related. One term does not belong. Determine the characteristic common to three of the terms and then identify the term that does not belong.

1. parasitism, mutualism, predation, commensalism
2. exponential growth, steady state, carrying capacity, symbiosis
3. predation, tornado, crowding, competition
4. parasitism, crowding, hormonal imbalance, territoriality

ty of the environment. They are directly related.
5. The waste might destroy plant and animal life, deplete the pond of important nutrients, and so on. Each of these factors would decrease the carrying capacity of the pond.
6. Parasites tend to spread through populations that are crowded, much as an infectious disease is more likely to spread in a crowded area than if the person who is ill is isolated.
7. Parasitism: One organism benefits while the other is harmed. (This is the relationship formed by the fungus and the person with athlete's foot.) Commensalism: One organism benefits and the other is neither helped nor harmed. Mutualism: Both organisms benefit from the relationship.

CONCEPT MASTERY

Use your understanding of the concepts developed in the chapter to answer each of the following in a brief paragraph.

1. Explain how a population normally controlled by density-dependent limiting factors might be affected by a density-independent limiting factor.
2. Given ideal conditions, a population will grow exponentially. What limits does nature place on such exponential growth?
3. Why did biologists such as Darwin question the concept of exponential growth?
4. What is the relationship between steady state and carrying capacity?
5. How might the introduction of a toxic waste in a pond affect the carrying capacity of that pond?
6. Why are parasites considered a density-dependent limiting factor?
7. Describe and compare the three main types of symbiosis. What type of symbiosis is formed by the fungus that causes athlete's foot?

CRITICAL AND CREATIVE THINKING

Discuss each of the following in a brief paragraph.

1. **Applying concepts** Why might a communicable virus that causes a fatal disease be considered a density-dependent limiting factor? What about a virus that is not communicable?
2. **Developing formulas** Based on the information in this chapter, develop a mathematical formula for population growth rate. *Hint:* How are the two axes on a growth curve labeled?
3. **Interpreting graphs** Examine the growth curve shown in the illustration. What does the curve show? How might you interpret this data?
4. **Making comparisons** Describe the growth curve in a small town made up mainly of senior citizens. Compare this growth curve to a small town made up of newly married couples.
5. **Relating concepts** Using the concept of carrying capacity, explain how the growth of both predator and prey are interrelated.
6. **Making inferences** Would a density-independent limiting factor have more of an effect on population size in a large ecosystem or in a small ecosystem?
7. **Using the writing process** The union representing predators cannot reach a labor agreement with the union representing prey. Each side wants a large increase in their steady state population. You are the arbitrator hired to mediate the dispute. What is your learned decision?

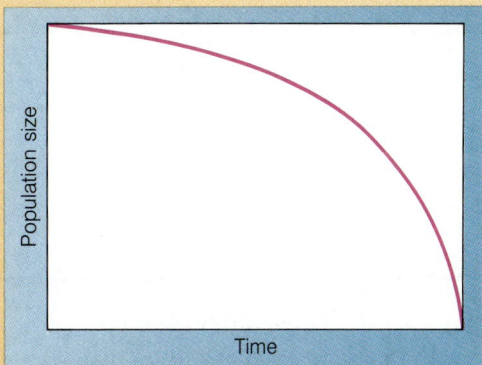

CRITICAL AND CREATIVE THINKING

1. Since the communicable virus is more likely to spread when people are crowded together, it is density-dependent. If it were not communicable, density would not affect its spread and it would be density-independent.
2. Population growth rate equals the change in the number of individuals over time.
3. The curve shows that the population decreases exponentially over time. This could be due to a density-independent factor such as an early frost, or it could be due to a disease or the depletion of prey or available food.
4. If immigration and emigration are not significant, the growth curve in the small town of senior citizens will probably decrease over time, whereas the curve in the town of newly married couples will probably increase over time.
5. As the prey increase, the number of predators increases since the carrying capacity of the environment increases. As the predators increase, they kill off more prey, and the prey population decreases. The carrying capacity of the environment for the predator then decreases as well, and the predator population also decreases. This, in turn, allows the prey population to increase, and the cycle begins again.
6. In most cases it will have a greater effect on a small ecosystem since a small ecosystem will be more susceptible to serious damage from a density-independent limiting factor such as a flood or severe storm.
7. Accept all logical, scientifically accurate answers.

CHAPTER 49 People and the Biosphere

Section	Laboratory Investigations and Activities
49–1 Human Population pages 1049–1052 THEMES: Patterns of Change, Systems and Interactions	**Teacher Edition** DEMONSTRATION: Air Pollution, p. 1048D
49–2 Pollution pages 1052–1063 THEME: Systems and Interactions	**Student Edition** LABORATORY INVESTIGATION: Simulating the Effects of Acid Rain, p. 1072 **Teacher Edition** DEMONSTRATION: Acidity of Coal, p. 1048D **Laboratory Manual** Investigating Air and Water Pollution, p. 601 **Teaching Resources** HANDS-ON ACTIVITY: A Model of Acid Rain, p. 229 HANDS-ON ACTIVITY: Acid Rain Takes Toll on Art, p. 231 ECOLOGY INVESTIGATION: Using a Mini-Ecosystem to Study Pollution, p. 57 **Product Testing Activities** Testing Disposable Cups Testing Toilet Paper Testing Orange Juice
49–3 The Fate of the Earth pages 1063–1068 THEME: Systems and Interactions	**Laboratory Manual** The Effects of Acid Rain on Seed Germination and Plant Growth, p. 607
49–4 The Future of the Biosphere pages 1069–1071 THEME: Systems and Interactions	**Laboratory Manual** Classifying Garbage, p. 613 **Teaching Resources** HANDS-ON ACTIVITY: How Can You Conserve Water? p. 235 HANDS-ON ACTIVITY: Paper Route, p. 239
Chapter Review pages 1073–1075	

CHAPTER PLANNING GUIDE

Other Activities	Media and Technology
Teaching Resources BIOLOGY CASE STUDY: Pesticide Spraying, p. 27 **Study Guide** Section 49–1, p. 465	**Interactive Videodisc** In the Company of Whales **Biology Media Guide** page 77
Teaching Resources ACTIVITY: Waste Disposal and the Environment, p. 3 BIOLOGY CASE STUDY: Effects of Geography and Weather on Air Pollution, p. 31 **Biotechnology Workbook** Scanners On! p. 167 **Study Guide** Section 49–2, p. 467	**Interactive Videodisc/CD ROM** Paul ParkRanger and the Mystery of the Disappearing Ducks **Interactive Videodisc** In the Company of Whales **Biology Media Guide** page 77
Teaching Resources BIOLOGY CASE STUDY: Effects of Crops on the Soil, p. 29 **Study Guide** Section 49–3, p. 470	
Teaching Resources VOCABULARY REVIEW: Crossword, p. 1 ACTIVITY: Investigating Conservation Issues, p. 7 **Study Guide** Section 49–4, p. 473	**Interactive Videodisc** In the Company of Whales **Biology Media Guide** page 77
Teaching Resources Chapter Test A, p. 15 Chapter Test B, p. 19 Performance-Based Assessment, Unit 10	**Computer Test Bank** Chapter Test, p. 515

CHAPTER 49 People and the Biosphere

CHAPTER OVERVIEW

In this final chapter students will focus on the important relationship that exists between people and the biosphere. Although certain human activities damage the Earth, people can work to preserve and care for our planet.

A growing human population poses a threat to the welfare of the Earth. Modern lifestyles often place an inordinate demand on the environment.

Students will learn that many environmental problems are the result of pollution. They will learn how pollution of air, fresh water, and oceans threatens the Earth's survival; they will also learn about the hazards of thermal pollution, sewage contamination, and biological magnification.

Students will come to understand that the survival of humans and human society depends on the survival of other organisms. Loss of forests and endangered species are among the topics that will be discussed.

In the final section of the chapter students will learn how they and others can preserve the Earth by such actions as conservation. Difficult decisions must be made if the Earth is to survive and thrive.

49-1 HUMAN POPULATION

Section Focus 49-1

The purpose of this section is to discuss the impact of a growing human population on planet Earth and explain how modern lifestyles place a heavy demand on the environment. In many parts of the world, the human population is growing rapidly. As a result, more and more land must be used for farming. This often means cutting down valuable forests.

While population growth in industrialized nations has slowed in recent years, the way of life in these countries often has an adverse effect on the environment. Many consumer goods require the use of energy and natural resources—and the production of these goods often causes pollution.

Performance Objectives 49-1

1. Discuss trends in human population growth.
2. Discuss the effects of an increased human population on planet Earth.
3. Relate modern lifestyles to the Earth's environmental problems.

49-2 POLLUTION

Section Focus 49-2

The purpose of this section is to discuss the many ways in which human activities pollute the Earth. Among the topics that are discussed are air pollution, freshwater pollution, ocean pollution, and biological magnification.

Pollution stems largely from the production of wastes and the improper disposal of wastes. Materials released into the environment can be classified as either biodegradable or nonbiodegradable.

The four results of air pollution are smog, acid rain, the greenhouse effect, and damage to the ozone layer. Fresh water becomes polluted by chemical contamination, sewage contamination, and thermal pollution. Much ocean pollution is caused by oil spills and the dumping of wastes.

Sometimes a seemingly minor pollutant enters the environment and causes much greater damage than might have been expected. This happens because the pollutant is passed from one organism to another via food chains and food webs. This phenomenon is called biological magnification.

Performance Objectives 49-2

1. Discuss the impact of pollution on the Earth's environment.
2. Distinguish between biodegradable and nonbiodegradable materials.
3. Define and discuss biological magnification.
4. Discuss the harmful effects of air pollution.
5. Describe ways in which the Earth's waters become polluted.

Science Terms 49-2

biodegradable p. 1053
nonbiodegradable p. 1053
biological magnification p. 1053
smog p. 1055
temperature inversion p. 1055
acid rain p. 1056
greenhouse effect p. 1058
hole in the ozone layer p. 1059
thermal pollution p. 1061

49-3 THE FATE OF THE EARTH

Section Focus 49-3

The purpose of this section is to emphasize the idea that the survival of humans and human society depends on the survival of other organisms. Among the living things that are especially threatened by human activities are forests and endangered species.

Forests are essential to the health of the biosphere. When trees are cut down, many changes occur—and often these changes are harmful to the total environment. Reforestation programs seek to replace the trees that have been cut down by planting new trees.

Some animal and plant species are threatened with extinction due to reduced populations. Such species are called endangered species. The importance of saving endangered species is discussed, as is the importance of protecting all organisms.

Performance Objectives 49-3

1. Discuss why the survival of humans depends on the survival of other organisms in the biosphere.
2. Explain why forests are important to the health of the biosphere.
3. Define endangered species and discuss the importance of saving these species.
4. Relate the destruction of natural habitats of plants to food production.

Science Terms 49-3

reforestation p. 1065
endangered species p. 1065

CHAPTER PREVIEW

49–4 THE FUTURE OF THE BIOSPHERE

Section Focus 49–4

The purpose of this section is to emphasize the need for people to preserve the Earth's environment through such efforts as conservation. Students will learn about the importance of protecting wild plants and animals and conserving natural resources.

Protecting the environment often leads to some difficult decisions. Problems associated with waste disposal are discussed, as are some of the economic factors involved with pollution control.

Performance Objectives 49–4

1. Discuss the importance of protecting the biosphere.
2. Explain the importance of conservation in protecting the environment.
3. Discuss some problems and solutions associated with waste disposal.
4. Explain how economic factors can make the decisions about preserving the environment difficult.

Science Terms 49–4

recycling   p. 1070

DISCOVERY LEARNING

TEACHER DEMONSTRATIONS
Modeling

Air Pollution

The following demonstration can be used as an introduction to Chapter 49. For the demonstration, you will need some kitchen matches, a beaker, and tongs.

Light a match and let it burn until it goes out.

• **What materials, if any, were released into the air as the match burned and as it went out?** (Students will probably observe grayish-white smoke.)

Light another match and hold it 2 or 3 centimeters from a beaker that you or a student is holding with tongs. Allow carbon to accumulate in the beaker.

• **What do you see collecting in the beaker?** (Carbon.)

• **Where is the carbon coming from?** (From the smoke being released as the match is burned.)

Point out that if one tiny match can produce this much carbon, imagine how much more carbon is released into the atmosphere when a whole forest is burned; or how much carbon and other pollutants are released into the atmosphere from the burning of fuels in homes, factories, and motor vehicles.

Acidity of Coal

This demonstration can be performed as students learn about acid rain in Section 49–2. For the demonstration, you will need a 5-gram sample of coal, some water, a graduated cylinder, two beakers, and litmus paper.

Crush the coal sample into a fine powder. Measure about 25 milliliters of water into each of the two beakers. Test the water with litmus paper.

• **What does the litmus paper tell you about the water?** (It is not acidic, for the litmus remains blue.)

Now add the crushed coal to one of the beakers. Point out to students that the other beaker is to be used as a control. Test the water in the beaker containing the coal with litmus.

• **What happens to the litmus paper?** (It is still blue.)

Tell students that you are going to leave the coal in the beaker and test the water in both beakers every day for five days. Have students predict what will happen. (The litmus will gradually turn pink, then red, in the beaker containing the coal. The litmus will remain blue in the beaker containing only water.)

After you have tested the water containing the coal for five days, ask students:

• **What can you conclude about the effect of adding coal to water?** (If it is allowed to remain in water, it makes water acidic.)

CHAPTER 49
People and the Biosphere

GUIDED ENQUIRY

Pose the following questions to students and have them record their responses. Point out that they will gain a better understanding of the key concepts if they read the chapter with these basic questions in mind. Upon completion of the chapter, pose the questions again. Ask students to compare their initial responses with those they have developed after reading the chapter.

- How is human population growth affecting the biosphere?
- What impact do modern lifestyles have on the environment?
- How is the production and disposal of wastes affecting the Earth's environment?
- How is the biosphere damaged by pollutants?
- How does a polluting substance become more harmful by biological magnification?
- How do forests contribute to the health of the biosphere?
- Why is it important to protect endangered species?
- How can people protect the biosphere?

CHAPTER 49
People and the Biosphere

Without adequate conservation measures, the chicken-sized kiwi (inset) would probably have followed its giant cousins, the moas, into extinction.

When humans arrived in New Zealand 1000 years ago, they found various species of flightless birds that they called moas. Unfortunately, the people on the island of Aotearoa discovered that these birds were delicious. A local song tells the rest of the story:

> No moa, no moa,
> In old Ao-tea-roa.
> Can't get 'em.
> They've et 'em;
> They've gone and there ain't no moa!

By the nineteenth century, the moas had become extinct. Humans, like other organisms, have an impact on their surroundings!

When New Zealand was first settled, there were fewer than 500 million people in the world. Today there are more than 6 billion. To prevent environmental disasters far more serious than the extinction of the moas, the people who today inhabit planet Earth must understand how to live with and protect their world.

INTRODUCING CHAPTER 49

Using the Visuals
Direct students' attention to the chapter-opener pictures and text.
- **What do you see in these pictures?** (A moa, a huge bird, and a kiwi, a much smaller bird.)
- **Where did these birds live?** (In New Zealand.)
- **What happened to these birds?** (The large moa is now extinct, the smaller kiwi still survives.)
- **What caused the extinction of the moa?** (People hunted the moa for food.)
- **What effect do you think people have on the population of animals in a particular area?** (Some populations may decline. The animals may be hunted, like the moa, for food. Other animals might be killed if they interfered with the human use of the land. For example, predators might be killed if they ate sheep, cattle, or chickens. Populations of domesticated animals or of pests such as rats and cockroaches may rise. Other answers are possible.)
- **What happens to natural populations in an area when human**

GUIDE FOR READING

After you read the following sections, you will be able to

49–1 Human Population
- Describe the effects of an increased human population on planet Earth.
- Relate the demands of our lifestyle to environmental problems on Earth.

49–2 Pollution
- Distinguish between biodegradable and nonbiodegradable wastes.
- Define biological magnification.
- Describe some of the harmful effects of air pollution on the biosphere.
- List ways in which the Earth's waters are polluted.

49–3 The Fate of the Earth
- Explain the importance of forests to the health of the biosphere.
- Define endangered species.
- List reasons why organisms should be protected.

49–4 The Future of the Biosphere
- Explain why the biosphere must be protected.
- Identify ways in which people can act to protect the environment.
- Explain why protecting the environment is a difficult job.

Journal Activity

YOU AND YOUR WORLD Many people feel that our lifestyle contributes to some of the environmental problems we face. Make a list of lifestyle changes you would be willing to make in order to improve the health of our global environment.

Figure 49–1 The world's human population is growing by leaps and bounds. Parts of the world are already very crowded, as this photograph of New York City clearly shows. As is the case for all organisms, such dense concentrations of individuals place great demands on the environment.

49–1 Human Population

Guide For Reading
- What is the cause of many of the environmental problems facing us today?
- What are some effects of human population growth on the biosphere?
- How does our way of life affect the environment?

In the very distant past, life on Earth was far different from what it is today. Most people lived in small tribes and family groups that hunted animals and gathered plants for food. Because these groups were small and almost always on the move, their effects on the environment were minimal. Natural processes could easily restore depleted food sources and break down wastes once the humans moved on.

When humans formed permanent settlements, however, they began a long history of profoundly affecting the environment. Not only did humans concentrate their effects on a small area for a long period of time, but they made changes in the environment that were not easily undone. For example, large numbers of trees were cut down to provide fuel for fires and lumber for buildings. Plants that were not useful were destroyed to make room for an unnatural number of food-producing plants. Dangerous animals were killed to protect humans and domesticated animals.

Because people did not really understand how the natural environments on Earth functioned, they could not assess the effects of their activities on the environment. For a long time, the human population was unaware that it was harming the world around it.

COOPERATIVE LEARNING

Using preassigned lab groups or randomly selected teams, have groups imagine that they are members of the Consumers' Environmental Awareness Panel, an organization independent of both government and industry. Their task is to design a logo that will identify for the consumer the products that are environmentally safe. This certification system is designed to protect consumers from false claims by product manufacturers. This logo would be used to identify products that are biodegradable, recycled, recyclable, and/or free of hazardous chemicals. Each group should produce a one-page magazine or newspaper ad that introduces the logo and its importance to the public. The ad should explain to the consumer the meaning of any symbols or colors used, as well as why the promotion and use of these products are so important.

Journal Activity

YOU AND YOUR WORLD Students should realize at the conclusion of this exercise that the lifestyle choices they make may have profound effects on the biosphere. Have students reconsider their lists after they have completed this chapter. The lists students generate should become part of their portfolio.

---

populations increase in number? (The natural populations often suffer when human populations increase in number.)

• **What factors could limit the number of humans that could live in an area?** (Possible answers: lack of water; not enough food to support a growing population; pollution of water, air, or land by dangerous substances; not enough room to live.)

TEACHING STRATEGY 49–1

Focus/Motivation
Say to students:
• In the last chapter you learned about carrying capacity. Many biologists feel that in the not-too-distant future, the human population may reach or exceed the carrying capacity of the Earth. What do you think would be the consequences of exceeding the Earth's carrying capacity for humans? (Accept all reasonable answers.)

BACKGROUND INFORMATION
GLOBAL POPULATION GROWTH

According to the World Bank, half of the population in developing countries will be under 15 years of age by year 2000. That means that more than 600 million jobs will be needed in those countries in the next few years. Many cities will spread and join with other cities to form huge, very dense populations of people. For example, Mexico City will double in size from 16 million to 31 million people. Bangladesh, a country the size of Wisconsin, will increase 60 percent in population to reach a total of 157 million people.

BACKGROUND INFORMATION
RESTRICTING POPULATION GROWTH

Some Asian countries, such as the People's Republic of China, have imposed very strict laws to regulate population growth. Couples are told not only how many children they may have but when they may have them. Women who become pregnant after they have had the allowed number of children are forced to have abortions.

TEACHING SUPPORT

Teaching Resources
- Biology Case Study: Pesticide Spraying, p. 27

MEDIA AND TECHNOLOGY
 **Interactive Videodisc**
- *In the Company of Whales*
Use this bar code to introduce the threat to whales posed by the fishing industry.

Play frames 18820 to 21466

See the Biology Media Guide page 77 for additional bar-code correlations for this section.

Figure 49–2 *The diverse human population brings richness to the human family. But a population that is ever-increasing places enormous demands on the resources of planet Earth.*

Over time, the human population has increased dramatically—as has our knowledge of the Earth's fragile environments. Today we are more aware of how our actions affect our world. But at the same time, we are also more capable of permanently damaging not just particular places on Earth but the entire biosphere itself.

Population Growth

Why is there all this environmental gloom and doom? Why have so many environmental disasters happened in our lifetime and not before? Why is quick action necessary? These questions are not easy to answer.

Many environmental problems loom large and seem urgent today because of the rapid increase in human population. The huge numbers of humans on Earth are making enormous demands on the planet. We may, in fact, be reaching the limits of the Earth to support our needs.

Why has the human population grown so quickly? As you may recall from Chapter 48, a population grows when the birthrate is greater than the deathrate. Thus both birthrate and deathrate must be examined in order to understand the growth of the human population.

BIRTHRATE Birthrates vary dramatically from one country to another. They also vary from time to time within a country. In the United States and much of Europe, the birthrate decreased during the 1970s as people had fewer children. In the 1980s, American birthrates rose slightly, although they remain lower than they were in the first half of this century. If current birthrates are maintained, our population will double in less than one hundred years. In some European countries, the growth rate is so low that populations are actually declining.

The situation is quite different in other countries, however. More than two thirds of the world's population live in the tropical countries of Africa, South America, and Asia. In some of these countries, the population continues to grow at an annual rate of about 3 percent. Although this rate of growth appears to be low, it will double the present population in only 23 years!

DEATHRATE Another important factor in determining population growth is the deathrate, which has decreased worldwide. Over the past few hundred years, changes in agriculture have made more and better foods available throughout the world. Improvements in the quality and availability of medical care have wiped out many deadly diseases and contributed to an increase in the life span of the average person. Today physicians can effectively treat diseases that once killed hundreds of thousands of people. Better nutrition and health care have dramatically reduced infant mortality. For example,

---

49–1 (continued)

Content Development
Continue the Focus/Motivation discussion by pointing out that carrying capacity refers to an environment's ability to sustain a particular species. Factors such as available food, water, energy, and space all contribute to carrying capacity. Thus, a human population that exceeds the Earth's carrying capacity would be likely to experience overcrowding and shortages of food, water, and fuel. These problems would no doubt give rise to malnutrition, increased disease, social problems, and, ultimately, the death of many humans. People might, in fact, find themselves in a situation similar to that experienced by earlier civilizations, when a high deathrate,

in the past, a woman in Kenya who gave birth to eight children could probably expect only four to survive beyond infancy. Today, a woman who has eight children will likely see most of them survive and produce families of their own!

EFFECTS OF HUMAN POPULATION GROWTH A high birthrate, a low infant deathrate, and a longer life span all contribute to population growth. But why is human population growth important? And what are its effects on Earth?

As human populations grow, their effects on the environment also grow. Countries with a rapidly growing population are often unable to produce enough food to satisfy the demand. In many parts of the world, all the land suitable for growing crops is already in use. In certain tropical areas, new fields are constantly being carved out of forests and mountains in an attempt to provide new areas for cultivation. Unfortunately, these lands do not remain productive for long. The constant rains characteristic of tropical areas quickly wash nutrients from the soil. After only a few growing seasons, the soil becomes barren.

Today, scientists worry about the destruction of tropical forests. Such devastation seems to be a high price to pay for farmland that is productive for only a few short years. In addition, many of these tropical forests are often cleared by burning. The burning produces large amounts of carbon dioxide, which is added to our atmosphere. As you will learn shortly, increased amounts of carbon dioxide in the air can have serious adverse effects. Many people mistakenly believe that leveling the rain forests produces only local effects. But, as scientists warn, the destruction of tropical forests may produce profound effects on the rest of the world—the world far removed from the rain forest.

Figure 49–3 A hungry population forces many countries to take drastic measures to increase farmable land. In some countries, huge forests are cleared by burning. The effects of forest burning may be felt far away from the burn site.

BACKGROUND INFORMATION
BIOMAGNIFICATION OF MERCURY

Mercury is a byproduct in the production of many electronic components. As a result, it is often introduced into the environment as waste material. Mercury becomes concentrated in food chains, particularly in the ocean. Animals on high trophic levels such as tuna and swordfish can acquire dangerously high concentrations of mercury. If humans eat these fishes, they can become seriously ill. Mercury in humans acts as a powerful poison to the central nervous system, causing paralysis, mental illness, and even death.

An unfortunate episode of mercury poisoning occurred in Minamata Bay in Japan. There, factories producing batteries discharged mercury into the bay. People thought that the mercury would do no harm since it sank to the bottom of the sea. But microorganisms on the ocean bottom changed the mercury into a water-soluble form. Once the mercury became dissolved in the water, it was picked up by the phytoplankton near the top—and became concentrated in the food chain. People in the Minamata Bay area, who eat a lot of fish, inadvertently consumed the mercury. Many became ill and died, and many women who had eaten the contaminated fish gave birth to deformed and retarded children.

TEACHING SUPPORT
Study Guide
- Section 49–1, p. 465

Guide For Reading
- What is the difference between biodegradable and nonbiodegradable materials?
- What is biological magnification?
- What are three kinds of air pollution?
- What are three ways in which waters are polluted?

1052

Effects of Lifestyles

It is not population growth alone that causes environmental problems. After all, the rate of population growth in many industrial countries has actually slowed. **The lifestyle of a population also contributes to the extraordinary environmental demands made on the Earth.**

The population of the United States is a good example. Although our population growth has stabilized at a low level, we use more energy and more natural resources than any other country in the world. We own many consumer goods. Most families in the United States have one or more automobiles. Radios, television sets, and labor-saving devices are staples of most modern homes. Consumers pay a price for these products —and the Earth does too! Natural resources are consumed in the production and use of such goods. And almost all of the factories in which the goods are manufactured produce wastes. In the past, industries could grow rapidly and cheaply because they simply threw their wastes out into the environment. Now we are learning that many of the wastes pollute our air and water, and may be changing the climate of the entire world. It is our responsibility to understand the effects of our actions on the global environment. We must be aware of what we are doing to our planet. And we must learn what we can do to solve the many environmental problems that face us.

49–1 SECTION REVIEW

1. Describe the effects of increased human population on the Earth.
2. What are three factors that have contributed to population overgrowth?
3. How can our way of life contribute to some of the environmental problems on Earth?
4. **Critical Thinking—Relating Concepts** How could you change your lifestyle to lessen your demands on the environment?

49–2 Pollution

You are aware of some of the serious problems that we are presently faced with. We need to make many difficult decisions, some of which will require us to change our way of life. We must learn to produce fewer wastes. We must learn to clean up the wastes we produce. To do these things, we may have to live with fewer of the familiar products we take for granted. As a nation, we may have to live with less economic growth. No one

---

49–1 (continued)

SECTION REVIEW 49–1

1. Increased rate of births, decreased rate of deaths, and longer life span.
2. The increased human population affects the Earth in many ways. It increases the demands for natural resources, requires more food and water to sustain it, and adds more wastes to the environment. Accept other reasonable answers.
3. In order to manufacture the goods people desire, huge amounts of natural resources are used and huge amounts of wastes are produced. Other answers are possible, but students should realize that everything they buy and use comes ultimately from the resources of the Earth and that many of the Earth's resources are finite.
4. By consuming less, by using materials over again, and by becoming wise consumers and buying energy-efficient appliances and cars. Many other answers are possible.

knows how to accomplish this; we only know we must. To help you understand why these difficult decisions are necessary, we shall examine some of our problems from an ecological viewpoint.

Materials released into the environment fall into two broad categories: biodegradable and nonbiodegradable. Materials that can be degraded, or broken down, by microorganisms into the essential nutrients from which they were made are **biodegradable**. Organic wastes such as sewage and scraps of food are biodegradable materials. **Nonbiodegradable** materials cannot be broken down by natural processes or are broken down only very slowly. Examples of nonbiodegradable materials are asbestos, glass, certain plastics and metals, radioactive wastes, and chemicals such as DDT, Dieldrin, and PCB's. Once these materials are released into the environment, they remain there for a long time—sometimes forever.

Biological Magnification

Sometimes pollutants affect the biosphere in ways no one expects. Thus a minor environmental pollutant may develop into a serious threat to life.

Many primary producers (such as plants and algae) pick up nonbiodegradable pollutants from the water. They do not metabolize these compounds, nor do they get rid of them. Instead, they concentrate them and store them in their tissues. When herbivores eat producers, they too concentrate and store these nonbiodegradable compounds in their tissues. In fact, concentrations of these compounds in herbivores may be more than ten times the levels in producers. When carnivores eat herbivores, the compounds are further concentrated. At each step in a food chain, the compounds are concentrated more and more. In other words, the amount of the compounds in each organism in a food chain increases. This phenomenon is known as **biological magnification.** In some animals at the end of a food chain, the concentrations of nonbiodegradable compounds may be 10,000,000 times their original concentration in the environment as a result of biological magnification.

Biological magnification occurs with many pesticides and industrial waste products. The first, and perhaps the most famous case of biological magnification, involved the powerful insecticide DDT. From the outset, DDT was an extremely useful pesticide. It kills many kinds of insects and remains effective for a long period of time. The use of DDT rapidly became widespread. It was used to kill mosquitoes and to help control the diseases malaria and yellow fever in tropical areas. It was used to kill annoying insects around the home. In fact, people were often sprayed from head to foot with DDT to kill body lice. It was also sprayed over huge areas of farmland to control the spread of insect pests.

Figure 49–4 A consumer society produces extremely large quantities of wastes. However, some wastes can prove useful. These automobiles are valuable for the scrap metal they contain. And some household wastes can be recycled for further use.

BACKGROUND INFORMATION
OVERHARVESTING

Many checks and balances exist in nature to prevent predators from overharvesting their prey. As a result, predators rarely cause the extinction of the species they prey upon. Unfortunately, such safeguards do not have much impact on the predatory activities of humans. Because of their ability to use tools and technology, humans often run the risk of diminishing their sources of food by overharvesting.

Humans have contributed to the extinction of certain species, such as the passenger pigeon, heath hen, and great auk, by overhunting. People have also caused severe damage to the numbers of certain fishes—such as Atlantic salmon, cod, haddock, and herring—by overfishing.

The long-term effects of overharvesting will be to lose, rather than gain, valuable food sources. Thus people need to realize that gaining extra food now may result in a significant lack of food later.

mans to preserve the biosphere because all living things are interdependent."

Have students respond to each statement, either orally or in writing. Encourage them to use the knowledge they have gained in this course thus far, as well as their own ideas, in forming their opinions.

After students have finished, ask them to keep their ideas in mind as they continue studying this chapter. Ask them to notice if any of their thoughts change.

Reinforcement/Reteaching
If some students are having trouble answering the Section Review questions, go over the pertinent material.

Closure
Photocopy or write on the chalkboard the following statements:
- "If humans are the most highly evolved animals, then it is right that they should change and shape the Earth as they choose. After all, that is part of natural selection and the survival of the fittest."
- "Humans are the only animals capable of judgment and moral choices. Therefore, they are responsible for the welfare of all living things on Earth."
- "It is in the best interests of hu-

TEACHING STRATEGY 49–2

Focus/Motivation
• **How do you dispose of garbage in your home?** (Most students will probably say that they put it in wastebaskets, garbage cans, or garbage disposals.)

BACKGROUND INFORMATION
AUTOMOBILES AND AIR POLLUTION

Automobiles and other motor vehicles pollute the air in a number of ways. First, like any device that burns fuel, they release carbon dioxide into the atmosphere. Second, they also release the poisonous gas carbon monoxide. This happens because not all of the fuel burned by these engines is oxidized completely into CO_2. A third problem associated with motor vehicles is that their exhausts contain nitrogen oxides. Once in the atmosphere, these oxides may combine with water vapor to form nitric acid. Nitric acid contributes to acid rain, just as sulfuric acid does.

Pollutants from automobiles might be less damaging to the immediate environment if they simply blew away. However, when automobile exhaust is released into the air, it hangs suspended for a time. During this period, people may inhale toxic gases such as carbon monoxide, as well as unburned fuel vapors that contain substances that irritate the eyes and lungs.

Federal laws have been passed to force automobile manufacturers to design vehicles that produce less air pollution. One change that has been helpful is the production of cars that use unleaded gasoline. In the past, lead was added to gasoline to keep it from "knocking." The lead, released into the atmosphere in the exhaust fumes, would remain in the environment and become concentrated in food chains through biomagnification. Since lead is poisonous to humans and other animals, this became a serious problem.

Figure 49–5 If the plants in the pond in which this moose is dining have been contaminated by toxic materials, they may actually poison the moose. The concentrations of toxic chemicals in the plants will increase in the tissues of the moose.

Figure 49–6 The toxic chemical DDT became part of many food chains when it was widely used as a pesticide. These two peregrine falcon eggs show the dramatic effect of DDT poisoning. The red egg is free of contamination; its shell is strong. The shell of the contaminated white egg is much more fragile. This egg will never hatch.

For a while everything seemed fine, and the use of DDT continued to increase. But then people began to notice several disturbing things. Fish in streams started dying. Fish-eating birds such as eagles, pelicans, and ospreys started laying eggs that never hatched. DDT was found in human body fat. And traces of DDT were found in the fat of penguins that lived as far away as Antarctica! What had happened? How had DDT managed to affect so many different organisms in so many diverse locations?

Aerial spraying and runoff from farmlands sprayed with the pesticide had allowed DDT to enter lakes and streams. There, DDT had been picked up by producers and had become part of various food chains. It had concentrated in the tissues of organisms in food chains as a result of biological magnification and had become widespread in food webs in many areas. Fortunately, the adverse effects of DDT were discovered before many people were harmed. Strict federal rules today control the use of insecticides such as DDT. But DDT lasts for a long time. It has entered the biosphere and will remain there for many years to come.

Unfortunately, the case of DDT is not an isolated one. Other compounds have created environmental disasters even worse than that of DDT. Dieldrin is an insecticide that was at one time used to control agricultural pests. Like DDT, Dieldrin builds up in food chains as a result of biological magnification.

PCB's are also dangerous chemical pollutants. PCB's are byproducts of several manufacturing processes. For a long time, PCB's were simply dumped into streams or other bodies of water near the factories that produced them. There they became concentrated in each step in a food chain, reaching dangerous levels in animals at the end of the food chain.

49–2 (continued)

Content Development
Continue the Focus/Motivation discussion by pointing out that people who would never dream of dumping garbage onto their floors at home often think nothing of tossing litter onto city streets or park grounds. In the same way, people often give little thought to the wastes that are dumped into our environment by industry and individual consumers. While the thought of letting garbage pile up endlessly at home seems ridiculous, that is exactly what is happening to the Earth's air and water. Some lakes and streams contain so much garbage that there is hardly room for anything else—including most of the wildlife.

Emphasize that pollution from wastes involves several factors. One factor is that often far too many wastes are produced, and this amount must be reduced; another factor is that wastes are often disposed of improperly; yet another factor is the need to clean up wastes that have al-

Air Pollution

We live in an ocean of air. The air supplies the oxygen our cells need to metabolize food and receives the waste products we give off as a result of our life processes. But the air also contains many other chemicals—chemicals that are not part of the natural composition of the atmosphere. The chemical pollution of the air, though it remains mostly invisible, poses very real problems. For with every breath we take, we are introducing these chemicals into our body.

SMOG If you live in a city, you may be familiar with the dirty-brown haze called **smog**. Smog, which gets its name from a combination of the words smoke and fog, contains different pollutants in different places. But regardless of its makeup, smog is unsightly, unpleasant, and harmful to life.

The causes of smog differ from one location to another. In Los Angeles, most of the smog comes from automobile exhausts. In industrial cities in the Midwest, most smog comes from factory smokestacks.

Weather conditions called **temperature inversions** can make smog a serious health hazard. Normally, cooler air is at higher altitudes than warmer air. The warmer air closer to the Earth's surface contains pollutants. But because it is warm, it is less dense than cool air and rises. As it rises, it slowly cools, and the pollutants it contains are carried away by winds. During an inversion, a layer of cool polluted air is trapped beneath a layer of warm air. Because the cool air is denser than the warm air above it, it cannot rise. As a result, the pollutants are already been introduced into the environment and to repair, if possible, some of the damage that has already been done.

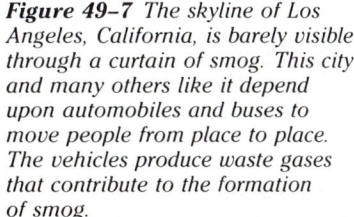

Figure 49-7 The skyline of Los Angeles, California, is barely visible through a curtain of smog. This city and many others like it depend upon automobiles and buses to move people from place to place. The vehicles produce waste gases that contribute to the formation of smog.

HISTORICAL NOTE
KILLER INVERSION

One of the most dramatic examples of air pollution occurred in the 1940s in Donora, Pennsylvania. At that time, the city of Donora had one of the largest steel mills in the world. The people of Donora were proud of the mill, and the economy was thriving—so much so that factories were operating 24 hours a day.

Millions of tons of coal were burned every hour to provide energy for industry. Of course, the sky was usually gray from smoke and soot and the air usually smelled like rotten eggs—but the people reasoned that this was just the price they had to pay for progress.

In October 1948, the "price for progress" became too high. A temperature inversion settled over the city. The air became almost unbreathable, and people could barely see. The sky was so dark at noontime that it seemed like late evening.

The temperature inversion lasted for four days. In that time, 20 people died and thousands required hospitalization. As a result of the Donora disaster, laws were passed in many cities and states to control the release of pollutants from factories and power plants.

Today, thermal inversions periodically produce dangerously high pollution levels in many cities, including New York, Los Angeles, and even Denver—a city that once owed much of its popularity to clean mountain air. When a serious thermal inversion occurs now, health experts warn the elderly and individuals with respiratory problems to stay indoors, preferably in air-conditioned rooms. The rest of the population is advised to avoid strenuous outdoor activities until the air clears.

kept near the ground. In time, this air and the pollutants it contains are warmed enough so that they become lighter and rise.

Temperature inversions can last for hours, days, or even weeks. When a temperature inversion occurs over a large city, pollution levels can rise high enough to threaten human life. Elderly people and those with respiratory problems, such as emphysema or asthma, are especially vulnerable to this kind of air pollution.

ACID RAIN Certain pollutants in the air combine with water vapor to form droplets of acid. When these droplets fall to the Earth in rain, the rain is called **acid rain**. Actually, any form of precipitation—even snow—can contain droplets of acid. Where do the pollutants that form acid rain come from?

In many areas, the coal and oil burned in factories and power plants to generate energy contain large amounts of sulfur. When sulfur is burned, it forms sulfur dioxide (SO_2). The exhausts of gasoline-powered vehicles contain nitric oxide (NO). When sunlight strikes nitric oxide in the air, it causes the nitric oxide to combine with oxygen to form nitrogen dioxide (NO_2). Both SO_2 and NO_2 dissolve in water to form strong acids: SO_2 forms sulfuric acid and NO_2 forms nitric acid.

Often SO_2 and NO_2 are carried on prevailing winds far from where they are produced. Parts of Canada and the Northeast have been and continue to be especially hard hit by acid rain whose origin is pollutants produced by industry in the Midwest. Acid rain also affects parts of Europe and the former Soviet Union.

The effects of acid rain on the environment are numerous and serious to life. Acid rain damages plants directly by

Figure 49–8 *The beauty of this photomicrograph of a drop of acid rain is somewhat misleading. Acid rain is a serious threat to the environment. Its effect on forest trees can be seen in this photograph of evergreens that are now never green. Their needles have been burned off the branches by caustic rains.*

49–2 (continued)

Focus/Motivation

Tie a 4-liter heavy-duty freezer-type plastic bag onto the cold tailpipe of your car and start the engine. Turn the engine off after 10 seconds or so and bring the bag into class. Students may wish to examine the emission particles with a magnifying glass. Point out that this bag full of pollution is from one car that ran for about 10 seconds. Ask students to imagine the number of particles that would be released by hundreds or even thousands of cars during a period of an hour or more.

Point out that one of the problems with motor vehicle exhaust is that it tends to hang in the air for a considerable amount of time after it is emitted. Thus, in

destroying their leaves, which are the food-making organs. And it damages plants indirectly by making the already acidic soil in which they grow even more acidic. This is especially true of many regions in Canada and the Northeast. Plants cannot absorb necessary minerals from such acidic soil, and they eventually die. In addition, acid rain dissolves toxic elements, such as aluminum, found in the soil. Once dissolved, these toxic elements are absorbed by plants. Aluminum in the soil can stunt the growth of plant roots.

Rivers and lakes are also victims of acid rain. In areas plagued by acid rain, lakes have become more and more acidic. Increased acid levels place the entire living community under stress. For example, water with a low pH (low pH means high levels of acid) causes the skeletons of fishes to lose calcium. Fishes become humpbacked, dwarfed, or deformed from the loss of calcium. When the levels of acid become high, fishes and other aquatic organisms die.

The toxic metals dissolved by acid rain—aluminum and mercury, for example—can damage rivers and lakes as well as plant roots. These toxic metals accumulate in the food chain, reaching dangerous levels by biological magnification. Clams and snails are the first organisms to die from aluminum and mercury poisoning. They are followed by aquatic insects, fishes, and amphibians. Soon the entire lake is lifeless. But the destruction does not end there. Animals dependent on these organisms are no longer able to survive without a source of food. Thus an entire region becomes unable to support certain kinds of life. Perhaps this situation sounds a bit exaggerated. Unfortunately, it is not. Entire lakes in Canada, northern New York, and New England have already died from the effects of acid rain. Sadly, many continue to do so.

Mountain forests in areas of Maine and Vermont are dying too. Trees in these areas are surrounded by acid fog. The good health of forest areas is important in protecting the quality of the freshwater supply. When forest areas begin to die, trouble with streams and ground water often follows close behind. Thus acid rain indirectly threatens the water supply in these areas.

THE GREENHOUSE EFFECT You may be surprised to learn that carbon dioxide is considered a pollutant. After all, it is a naturally occurring gas. However, the amount of carbon dioxide in the atmosphere is increasing as a result of human activity in the biosphere. And this increased amount of carbon dioxide may have a major effect on the Earth's climate.

Carbon dioxide is produced when carbon-containing fuels are burned. Carbon-containing fuels include wood and charcoal as well as fossil fuels such as coal, oil, and natural gas. Over the years, the burning of trees and fossil fuels for energy has released vast quantities of carbon dioxide into the atmosphere. In addition, clearing land for farming and building destroys

BACKGROUND INFORMATION
EFFECTS OF ACID RAIN

An effect of acid rain that can damage an entire ecosystem is a change in the pH of the soil. Many microorganisms live in soil, and several critical steps in nutrient cycles are performed by these animals. All of these microorganisms have specific requirements for soil pH. If the soil pH suddenly drops as a result of acid rain, some of the microorganisms may die. The loss of these animals can upset the nutrient cycles and thus endanger the welfare of many organisms in the ecosystem.

TEACHING SUPPORT
Teaching Resources
- Hands-On Activity: A Model of Acid Rain, p. 229
- Hands-On Activity: Acid Rain Takes Toll on Art, p. 231
- Ecology Investigation: Using a Mini-Ecosystem to Study Pollution, p. 57

MULTICULTURAL STRATEGY

A native of Santurce, Puerto Rico, Guadalupe Fortuno is a physicist who received her undergraduate degree from the University of Puerto Rico and her Ph.D. in physics from Harvard University. At the Harvard-Smithsonian Center for Astrophysics, Dr. Fortuno conducted research on the reaction rates of certain radicals with ozone and nitrogen dioxide. These reactions have proven important in the study of the chemical composition of the atmosphere. Encourage students to do a search in the library to locate articles written by Dr. Fortuno about her research.

areas with heavy traffic, the air pollution builds and builds.

Reinforcement/Reteaching
Remind students that a form of "personal" air pollution is smoking. Studies have shown that smoking not only harms the smoker but those around the smoker as well. Encourage students to find out about the effects of smoking on nonsmokers.

FACTS AND FIGURES

Air currents and waterways cause pollutants to travel from one ecosystem to another and eventually from one part of the world to another. Recently, scientists have found that the level of soot in Arctic air and snow is only three or four times lower than the level found in a typical urban environment. Pollution that is in New York City today will reach the Arctic within a week. Similarly, scientists have found DDT in the tissues of penguins in Antarctica, far from any site where this chemical has been used.

BACKGROUND INFORMATION
GOING TO EXTREMES

Students may wonder why anyone would make a fuss about average climate changes that amount to no more than a degree or two. The point to make here is that environmental *extremes*, rather than *averages*, are the problem. Many researchers agree that even slight global warming could cause storms such as hurricanes to increase in strength and could cause shifts in wind patterns that affect local temperatures and rainfall. Depending on the part of the country you teach in, remind students about recent events such as droughts in California, floods in the Mississippi basin, or heat waves in the Northeast, and mention local effects of these climate events on both cities and farms.

49–2 (continued)

Skills Development
Guided Practice
Skill: Relating concepts
Remind students of the demonstration on the acidity of coal found at the beginning of this chapter. (If you have not yet performed this demonstration, you could do so now.)
• **Based on the results of the last several litmus tests, what must be true about coal?** (It must contain acid.)
• **What do you think happens to the acid or acid-producing substance in coal when the coal is burned?** (It is released into the atmosphere.)
Explain that the acid-producing substance in coal is sulfur and that when sulfur is released into the air, it combines with water vapor to form sulfuric acid. This sulfuric acid eventually falls to earth as acid rain.

Content Development
Emphasize the idea that many pollutants are not limited to just air, water, or land but actually affect many or all components of the environment. A perfect example of this is acid rain. The acid that eventually forms acid rain begins as air pollution, as sulfur and nitrogen are released

Figure 49–9 Venus (left), although much like the Earth in many ways, has an atmosphere that consists mostly of carbon dioxide. As a result, Venus is considerably hotter than Earth. Scientists are concerned that the temperature of Earth will increase as levels of carbon dioxide in the atmosphere increase. If this happens, the huge polar icecaps will melt and sea levels will rise, flooding many coastal regions.

plants, which remove carbon dioxide from the air during photosynthesis.

How do increased amounts of carbon dioxide affect the atmosphere? Energy from the sun is absorbed by the Earth and changed into heat. Later, this energy is radiated back from the Earth to the atmosphere. Carbon dioxide and other gases in the atmosphere absorb this heat energy, forming a kind of "heat blanket" around the Earth. This process is called the **greenhouse effect** because the carbon dioxide holds in heat like the glass in a greenhouse. The greenhouse effect makes the Earth a warm, comfortable place to live.

As levels of carbon dioxide in the air increase, however, more heat is absorbed, and the temperature of the Earth increases. The effects of this global warming, which is occurring slowly at present, are unclear. Some scientists believe that if a global warming of even a few degrees occurs, it will cause the polar icecaps to melt. This will release enough water to raise sea levels and flood many coastal areas. A warming of the Earth's climate—by even a few degrees—will also cause major changes in human agriculture to occur. For example, corn grows best within a narrow temperature range. If the Earth's climate becomes warmer, the best areas in the United States for growing corn will shift farther north from the equator. Thus global warming will have a major impact on many farming states.

HOLES IN THE OZONE LAYER A layer of ozone exists far above the Earth in a part of the atmosphere called the stratosphere (16 to 48 kilometers above the Earth's surface). This ozone layer protects the Earth from harmful ultraviolet radiation from the sun. Without the protection of this ozone layer, few living things could survive.

Scientists have recently discovered that the ozone layer is becoming thinner in certain places around the poles—a phenomenon they have called **holes in the ozone layer.** Holes in the ozone layer are not actual holes but rather areas where little ozone is found. The thin spots in the ozone layer indicate that the Earth's protective layer of ozone is being depleted.

The major cause of ozone depletion is a form of chemical air pollution that results primarily from the addition of chlorofluorocarbons to the air. At one time these chemicals were used in most aerosol containers to force the contents from the container. Once the harmful effects of chlorofluorocarbons were known, the United States banned their use as an aerosol propellant. Many other countries did too. However, other sources of these chemicals—whose primary use is now in cooling and refrigeration equipment—continue to pose a threat to the protective ozone layer. If the ozone layer is depleted, more of the sun's harmful ultraviolet radiation will reach the Earth. Scientists predict that if the amount of ultraviolet radiation that reaches the Earth increases, the frequency of certain diseases —such as certain forms of skin cancer—will also increase.

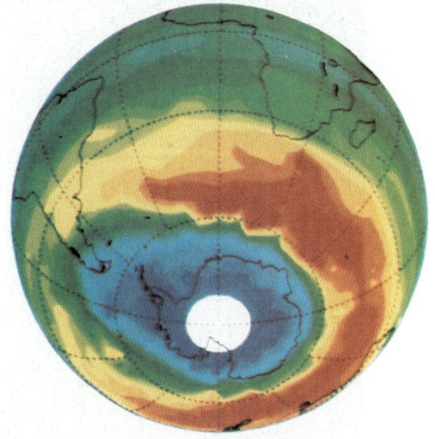

Figure 49–10 In this image, areas of low ozone concentration are blue. Note that the "hole" in the ozone layer extends over Antarctica and part of South America. The white spot indicates an area for which no data were collected.

Water Pollution

In the United States, billions of liters of fresh water are used daily. Water is used for drinking, cooking, bathing, and cleaning. It is also used for industrial processes and irrigation. Although water is a renewable resource, there is a limited amount of fresh water.

Although almost three fourths of the Earth is covered by water, only about 3 percent is fresh water—97 percent is salty ocean water. Of the 3 percent that is fresh water, only a small portion is available for use by living things. The greater portion is locked up in ice, mainly in the polar icecaps and in glaciers. Considering the fact that all living things depend upon water, the pollution of this very small available supply looms as an extremely serious problem facing us today.

CHEMICAL CONTAMINATION Water can be polluted in many ways. **The most common sources of water pollution are chemical wastes, raw sewage, and high temperatures.** Toxic chemicals can pollute water supplies in two ways. The chemicals can enter streams and rivers, which carry them first into lakes and then into oceans. There the toxic chemicals can kill aquatic plants and animals or they can enter the food chain and eventually pose a threat to human health. Even in lakes as large as the Great Lakes, the presence of several pollutants has made eating fishes from these waters dangerous to health. In addition, chemical wastes discarded on land can seep through the

Figure 49–11 In many places in the world, the supply of drinkable water is severely limited. Here in Mali, people fill buckets with water that is pumped from a communal well. These heavy buckets of water must be carried to their home.

into the atmosphere. Once the acid rain falls, however, it runs into lakes and streams, thus becoming water pollution. Some of the acid rain is taken up by soil— land pollution. Soon a whole ecosystem is affected.

Skills Development
Guided Practice
Skills: Conducting an experiment, observing
Students can observe the amount and type of pollutants in the air around them by performing this simple experiment. Have students take several petri dishes and coat the flat surface of each dish with a thin layer of petroleum jelly. Then ask students to place the dishes in various locations where they will be exposed to outside air yet will not be disturbed. Have students leave the dishes for three days.

At the end of three days, have students collect the dishes and use a magnifying glass to count the number of particles. (Placing the dish on graph paper may make the counting easier.) Also have students observe the particles under a microscope and describe what they see. Summarize all students' observations in a class discussion.

BACKGROUND INFORMATION
TRANSFER RATES

A transfer rate is defined as the amount of energy or material that moves from one component of the environment to another within a specified period of time. Until recently, many transfer rates in the global environment have been relatively stable. Human activities, however, are beginning to change that.

For example, without human intervention, the flow of carbon dioxide between the atmosphere and land ecosystems is essentially the same in both directions. This means that, every year, about the same amount of carbon dioxide is taken from the air by plants as is released back into the air by animal and plant respiration. In recent years, human activity has begun to disturb this equilibrium significantly: In a single decade, carbon dioxide levels in the atmosphere increased by about 6%. This increase is attributed primarily to the burning of fossil fuels and the burning of forest lands.

TEACHING SUPPORT
Teaching Resources
- Biology Case Study: Effects of Geography and Weather on Air Pollution, p. 31

MEDIA AND TECHNOLOGY
Interactive Videodisc/ CD ROM
- *Paul ParkRanger and the Mystery of the Disappearing Ducks*

HISTORICAL NOTE
WATER POLLUTION ACTS

The first action taken by the federal government to improve water quality was the Rivers and Harbors Act (also known as the Refuse Act), which was passed in 1899. A major provision of the act was to prohibit the dumping of trash into rivers and harbors. No provision was made, however, to prohibit the discharge of sewage into waterways. The law was not very effective, and was not even used to prosecute violators until the 1970s.

The next action by the federal government to improve water quality was the passage of a series of laws between 1948 and 1970. Because of political pressure, these were weak laws that were difficult to regulate and enforce.

Finally, pressure from concerned voters resulted in the passage by Congress of the Water Pollution Control Act of 1972. This act and its amendments is now called the Clean Water Act. The Clean Water Act empowers the federal government to set minimum water quality standards for rivers and streams; it also allows individual states to pass stricter laws if they so desire.

According to the Clean Water Act, it is illegal to discharge any pollutant into a waterway unless a permit is first obtained from the state. This is important, because it makes possible the control of pollutants from their principal sources. The permit system, which is called the National Pollution Discharge Elimination System (NPDES), sets specific limits of pollutants on a facility-by-facility basis.

The Clean Water Act gives the Environmental Protection Agency (EPA) the power to impose deadlines for compliance with the law and to fine industries and municipalities that do not comply. Fines may go as high as $10,000 per day; repeat offenders may be fined up to $50,000 a day and may receive a prison term.

Figure 49–12 *Although almost three quarters of the Earth is covered by water, the supply of fresh water that is available for human needs is limited. It is extremely important to protect this water from becoming polluted.*

ground and enter the underground water supply, through which they can be carried long distances. Toxic chemicals contaminate wells and the underground water supplies that serve many towns and cities. Because water moves through the ground slowly, toxic chemicals in underground water are difficult to remove.

SEWAGE CONTAMINATION Have you ever wondered what happens to the water after you flush the toilet? Or to the detergents used to wash dishes and clothes? These wastes, like chemical pollutants, are typically added to the Earth's waters. Although these domestic wastes are usually not poisonous to life as are chemical pollutants, they do pose environmental problems.

Sewage consists of large quantities of wastes that contain nitrogen compounds. These compounds are used by bacteria in a process that requires oxygen. If untreated sewage is added to rivers and streams, the number of bacteria increases dramatically. These bacteria use up most of the available oxygen as they break down the nitrogen compounds. Other organisms that live in the water may suffocate because their supply of oxygen is depleted.

In addition to nitrogen compounds, sewage often contains phosphates, or compounds that contain phosphorus. Both nitrogen compounds and phosphates are nutrients that stimulate the growth of plants. In rural areas, where homes are far apart, sewage is usually treated in septic systems. In a septic system, decay bacteria work on the sewage, reducing it to water that is nearly pure. This water, which still contains some dissolved nutrients, seeps out of the septic system and into the ground. This explains why plants near a septic system grow well.

Figure 49–13 *Many cities and towns have sewage-treatment plants that process household waste water. In huge outdoor tanks, water is treated by chemicals and microorganisms and then aerated.*

49–2 (continued)

Focus/Motivation
Ask students if they have ever had a personal experience with water pollution. Some common experiences might include being unable to swim in a particular area due to water pollution or being told not to drink water while traveling in a foreign country. Another common occurrence might be the temporary pollution of a local water supply because of work being done on the water system. This type of pollution is not usually harmful to people, but it can result in water that smells strange or has an odd color. When this happens, municipal governments in the affected area often get hundreds of calls from concerned citizens who wonder if the water is safe to drink. Some-

In cities, however, the large population makes the use of septic systems impossible. City sewage must be treated in sewage-treatment plants. In special ponds and tanks in these plants, some of the organic wastes in sewage are broken down by the actions of bacteria. Once the bacteria have decomposed the organic matter, chemicals that kill harmful microorganisms are added to the sewage and the treated water is released. But the treated water often still contains nutrients that can produce a rapid growth, or bloom, of algae in streams, lakes, coastal bays, and ocean waters. These algal blooms can disrupt the normal balance of organisms in a water community.

Human sewage also contains many potentially harmful microorganisms: bacteria, viruses, and protozoa. A few bacteria in the water usually pose no threat to people. But filter-feeding organisms, such as clams and mussels, ingest the microorganisms and concentrate them in their tissues. When these shellfish are eaten, diseases such as hepatitis, typhoid, and certain forms of dysentery can be spread.

THERMAL POLLUTION Many factories and power plants produce heat as a waste product. In the past, water from nearby rivers, lakes, or the ocean was used to cool such plants. The water was pumped through pipes in the cooling system, where it absorbed heat. The heated water was then pumped back into the environment. In some cases, heated water has no harmful effects on the ecosystem. But often, heated water kills aquatic plants and animals. This kind of water pollution is called **thermal pollution**. Thermal pollution also affects the larvae of many aquatic animals, which are more sensitive than the adults to temperature changes. So even if the adults are unharmed, the larvae may be killed.

Figure 49–14 Water is often used by industries as a coolant. In this nuclear power plant, heated water is returned to the river from which it was pumped, causing a kind of pollution called thermal pollution. Thermal pollution is a danger to some forms of aquatic life.

BACKGROUND INFORMATION
DRILLING MUDS

Another potential problem associated with offshore drilling involves drilling muds. When oil companies drill, they release large amounts of mixtures called drilling muds into the well hole. These muds help lubricate the drill bit as it digs. Drilling muds probably float around for some time and then settle to the bottom. At present, no one knows for sure what effect these substances have on the local marine community.

49–2 (continued)

Enrichment

Ocean pollution along New Jersey beaches made national headlines in 1988. The tourist trade, which is the main source of income for Jersey shore communities in the summer, was all but eliminated. Have interested students find out more about the New Jersey ocean pollution and its causes. Also have students find out what caused the pollution to decrease the following year.

SECTION REVIEW 49–2

1. A biodegradable waste can be broken down by microorganisms into the essential nutrients from which it was made. Sewage and scraps of food are biodegradable wastes. Nonbiodegradable wastes cannot be broken down, or are broken down very slowly. Asbestos, glass, and some plastics are nonbiodegradable wastes.
2. In biological magnification, certain compounds become more concentrated in the tissues of each organism in a food chain. Finally, at the end of a food chain the compounds are many times more concentrated than at the beginning of the food chain. Biological magnification increases the danger of certain pollutants in the following way. A compound may not reach dangerous levels in early stages of a food chain; however, at the end of a food chain, the levels of the compound may reach dangerous levels.
3. Air pollution and water pollution are two kinds of pollution. Because of the introduction of dangerous compounds to the air, people inhale dangerous pollutants along with the oxygen they need to survive. In water pollution, dangerous compounds have been added to the Earth's waters, and these compounds are ingested with each sip of water.
4. The very acidic water that is caused by acid rain changes the pH of streams and lakes. The acidic water kills off certain kinds of algae and bacteria that

OCEAN POLLUTION Water pollution is not limited to the pollution of fresh water. For centuries, people have dumped their wastes into the oceans. Such a practice presented few problems when the number of people was small, the amount of waste was minimal, and the wastes were biodegradable. But today the situation is vastly different. Too many people produce too much waste, a great deal of which is nonbiodegradable. Plastics and other wastes dumped into the oceans may float around for months or even years. Large cities continue to dump so much sewage into the oceans that it cannot be degraded quickly enough. Often, some of this sewage washes up on beaches, offering ample proof that we do indeed foul our nest.

Pollution near the shore is a serious matter. Because so many people live in the great port cities, the sources of pollution are numerous and the problems posed are grave. Some wastes dumped into the oceans—such as containers of disposable medical items—wash back onto shore, threatening the health of beach-goers. Other wastes remain at sea, posing a hazard to ocean life. And along the continental shelves near shore, the oceans' most productive areas are threatened by pollution. Here, many of the world's important fisheries provide food for much of the world's population.

OIL SPILLS Oil pollution in the ocean, too, is serious. Even when oil companies are careful, accidents happen. For example, some oil spills occur when boats run aground. Others occur when oil tankers are damaged during storms. But regardless of the cause, once oil is spilled, it is difficult to remove.

Figure 49–15 The solid wastes produced by consumer societies are often dumped into the oceans. The effects of ocean dumping have become painfully apparent: Washed up by ocean waves, these pollutants despoil once pristine beaches.

Oil slicks are deadly to marine animals that swallow the toxic oil or become coated with it. Oil spilled in the ocean may also enter the estuaries that are the nurseries for the oceans. The larvae and young of many species spend time in estuaries before they move into the open sea. Many plants and animals are killed when oil pollutes estuaries. Toxic chemicals in oil often accumulate in those animals that are not killed immediately. In many cases, these chemicals make the animals sterile, or unable to reproduce. In addition, some of the chemicals that accumulate in animal tissue are potent carcinogens that may cause cancer in the people who eat them. We are still not certain what effects oil pollution has on the plankton that float in the ocean and form the basis of many food chains. Much more research is needed in this area.

Our continental shelves contain large and potentially valuable reserves of crude oil and natural gas. It is clear that precautions must be taken to protect these areas and their aquatic inhabitants if offshore drilling is to take place.

Figure 49–16 *The sea bird in the top photograph was an unlucky victim of a tragic accident. Its body was covered with heavy black oil that spilled when a tanker ran aground. The bird in the bottom photograph was more fortunate. It, too, was covered with oil, but it has been captured and will be washed with special detergents to remove the oil from its feathers. With luck, it will fly again.*

49–2 SECTION REVIEW

1. Compare a biodegradable waste with a nonbiodegradable waste. Give two examples of each kind of waste.
2. What is biological magnification? How does biological magnification increase the dangers posed by certain environmental pollutants?
3. What are two kinds of pollution? How does each pose a danger to human health?
4. **Critical Thinking—Making Inferences** Lakes affected by acid rain often appear clear and blue. Why might this be so?

49–3 The Fate of the Earth

People have always relied upon plants and animals for food, clothing, and shelter. **The survival of humans and human society depends upon the survival of other organisms in the biosphere.** Today, the survival of many of these organisms is threatened.

Forests

You probably do not live in a forest—few people do today. But wherever you live, your life is dependent upon forests in many ways. Trees provide us with many essential products. Wood is used to make everything from pencils to houses. The

Guide For Reading
- What does the survival of humans and human society depend upon?
- In what ways do forests contribute to the health of the biosphere?
- When does a species become endangered?

BACKGROUND INFORMATION
FOREST FIRES

Fire has long been viewed as a terrible threat to forests—and in many cases it is. Forestry scientists have recently realized, however, that in certain cases, fire may be necessary to ensure the health of a forest. Much of what forestry experts are learning about fires has come from studying the effects of the fires that burned large areas of Yellowstone National Park in 1988.

One way fire can actually help a forest is by clearing the way for new growth. Young trees often have trouble growing in the shade of large, older trees. Only when some of these trees are removed can new trees get the sunlight and space they need to grow. While it may seem unfortunate to lose the older trees, one must realize that trees do not live forever. Clearing some older trees to make way for new ones might prove more efficient in the long run.

Nature must have suspected that fire would be a part of the forest environment, for some types of trees, such as the jack pine, have cones that are resistant to fire. In fact, the seeds in these cones will not germinate until they have been through a fire!

paper on which these words are printed is made from wood. Trees are an important source of charcoal used in cooking and in industrial processes.

Trees are also a vital part of many ecosystems. The roots of trees keep the soil loose and at the same time hold it in place, allowing rainwater to penetrate the soil without washing it away. Water that penetrates the ground often becomes part of the underground water table. If you drive through a forested area, you may see signs indicating that you are in a watershed area. This means that the forest is actually helping to gather water for use by people.

When forests are carelessly cut down, many important changes occur. The structure of the soil changes. Microorganisms die. Many small plants and animals can no longer survive. Without trees to hold the soil, heavy rains wash away the fertile topsoil, leaving behind only the less fertile subsoil. Essential nutrients are washed from the soil and carried into lakes, rivers, and streams. The water table drops. Rains bring sudden floods instead of steady streams. The soil cannot store moisture, and so it dries out quickly after each rain. In some cases, new plants cannot grow in what was once fertile soil.

Despite knowledge of these harmful outcomes, many of our country's few remaining old forests are being cut down at an alarming rate. And in tropical countries, rain forests are being destroyed so rapidly that they may disappear completely by the end of this century!

Figure 49–17 *This proud, tall tree is no match for a huge chainsaw. Once cut, it will be used to make many consumer products. A tiny tree will be placed in its stead. Varieties of trees that reach harvesting size in fewer years have been developed.*

49–3 (continued)

Content Development
Explain to students that endangered species are defined as those species of organisms that are considered to be in immediate danger of extinction.

- **What is extinction?** (The dying out of a particular species of plant or animal so that it no longer exists on Earth.)
- **Will extinct species ever reappear?** (No.)

Point out that endangered animals are likely to die out because their populations are incredibly low. The odds are not in favor of enough of these animals mating, reproducing, and having their young survive in order to increase or hold steady their population.

In 1989, the U.S. Fish and Wildlife Service listed 461 foreign species of animals as being endangered, including the African elephant, leopard, giant panda, white rhinoceros, Japanese crane, St. Vincent parrot, and Galápagos tortoise. It also listed 271 domestic species of mammals, birds,

Fortunately, government officials became aware of the problems associated with forest destruction as long ago as 1905. The United States Forestry Service protects and manages our forest resources. Today there is increased awareness on the part of lumber companies to manage forest resources wisely.

Programs that plant new trees when old trees are cut down are called **reforestation** programs. Reforestation programs are vital to the health of the biosphere. But caring for forests while harvesting their wood is not an easy task. When a forest is cut, lumber companies often want to reforest by planting large numbers of a single tree species. However, such plantings cannot adequately replace a natural forest. One obvious reason is that in most natural forests there is a large variety of tree species. A less obvious reason is that the animal and plant communities that develop naturally in forests and help control harmful insects are not easily established in artificially planted forests. Wise planting of various species of trees may help save much former forest land from disaster.

Endangered Species

When an animal or plant species becomes so rare that it is threatened with extinction, it is called an **endangered species**. Species become endangered in several ways. A species that was once quite common can be hunted until few remain. With so few individuals remaining, a harsh winter, a disease epidemic, or any other natural disaster can mean extinction. Some species can become endangered if they are able to survive in only one particular habitat. If that habitat is destroyed, the species may become extinct.

Today, one plant or animal species becomes extinct every hour! It is predicted that by the year 2000, many land animals will be extinct in their natural environment. Just imagine what it will be like for you and your children to live in a world without elephants, giraffes, tigers, or monkeys.

Why Save Endangered Species?

Extinction is forever. Once a species becomes extinct, it will never exist again. Many people feel that we have no right to cause the extinction of other species—as we are presently doing. Many biologists (including the authors of this book) feel this way. But human needs are great, as are the demands on the ecosystem. If someone were to ask you why we should be concerned about protecting other species, could you answer them?

USEFUL PRODUCTS In earlier sections of this textbook we mentioned why other species are important to us. It might

Figure 49–18 Huge herds of bison roamed the Great Plains until hunters brought these awesome animals to the verge of extinction. Today, protected by strict laws, herds of these animals once again roam.

ECOLOGY NOTE
SHY RHINOS

Wherever they are found in the wild, rhinos face extinction. In some places, this is due to hunting. However, this is not the only threat to survival rhinos face. Surprisingly, tourists visiting nature reserves may pose a danger by making shyer rhinos search for cover.

This observation was made by scientists studying the effects of tourism in Nepal's Chitwan National Park. Tourists who visit this park travel by elephant. The elephants, laden with tourists, try to approach the rhinos closely. The more timid rhinos leave the foraging ground in search of dense vegetation for cover. Scientists believe that the shy rhinos are put at a disadvantage because they have less opportunity to feed than those that are not scared off by the tourists. As a result of poor nutrition and stress, the shy rhinos may not produce as many young in the next generation.

49-3 (continued)

Enrichment

One of the more colorful animals in America that is on the list of threatened species is the grizzly bear. The number of grizzly bears in the contiguous United States was over 100,000 in the 1600s and 1700s; today there are less than 1000. Grizzlies have disappeared entirely from California and Colorado. The population of grizzlies is dropping rapidly in Yellowstone National Park. The strongest grizzly population is found in Alaska and western Canada, where there are estimated to be between 18,000 and 38,000 bears.

Because they are listed as a threatened species, grizzlies can be killed only when it is necessary to protect a human life. But since grizzly bears have a reputation for menacing humans in campsites and along wilderness trails, some people are not enthusiastic about protecting grizzlies.

Figure 49-19 Other animals have not been as lucky as the bison. The California condor (left) is close to extinction. Captive breeding programs aim to keep this species alive. The black-footed ferret hunts prairie dogs (center). The population of ferrets is so small that their actual location in the wild is a closely guarded secret. The desert pupfish lives in small desert pools (right). If anything happens to its delicate habitat, it too will be in real danger of becoming extinct.

be useful to the discussion of saving endangered species to summarize a few important points here. Many everyday foods, medicines, and industrial compounds come from wild plant and animal species. Antibiotics, heart drugs, anticancer medications, painkillers, and other important medicines are derived from plants. Yet these plant species represent only a few that have been tested for useful compounds. No one knows the riches that might be hidden in yet unknown plants that grow in tropical rain forests. If these rain forests disappear, the potential contributions of their inhabitants to our society will also vanish.

Many animal species produce compounds that may prove important to human health. Sponges may contain chemical compounds useful in treating viral diseases such as herpes and encephalitis. Chemicals in sea cucumbers are currently being tested to see if they have potential anticancer uses. The blood of the horseshoe crab is presently collected and used in important medical tests.

SECTION REVIEW 49-3

1. Rain forests provide benefits even for organisms that live far from the rain forest. For example, the trees and other plants in a rain forest use up carbon dioxide and give off oxygen. When rain forests are cut, the amount of oxygen in the atmosphere decreases and the amount of carbon dioxide increases. These changes may affect South Dakota, even though the rain forest being destroyed is in South America. Other answers

Food from Plants

Most of the world's population obtains its food from crops grown on farms. But this does not mean that species of wild plants are unimportant in horticulture. They are, in fact, quite important. Here is why.

The crop plants grown in the United States today are the results of generations of selective breeding. When plant breeders develop a better variety, it is produced in enormous numbers and planted all over the country. But planting a single variety can prove dangerous.

Genetically similar plants are susceptible to the same diseases. If only a single variety is planted in huge numbers, entire croplands can be destroyed by a single disease. One famous example of such a disease is the great corn blight that occurred in 1970. This disease was resistant to all existing agricultural defenses. Nothing could save an infected field. That year about 15 percent of the total corn crop was destroyed. In the future, diseases that affect important crop plants could be even more devastating.

How can such situations be prevented? In some cases, people maintain seeds of crop varieties that are no longer commercially grown. The genetic material in these seeds may become important if new crop varieties are wiped out by disease. In other cases, crop breeders are constantly at work developing new disease resistant strains of crop plants. In many cases, they cross crop plants with strains of wild plants that have more genetic variability. (The strains produced by selective breeding have little genetic variability.) So far, wild plants related to crop plants have provided the necessary "new genes" crop breeders want to introduce into already developed strains of crop plants. Scientists all over the world regularly search for new wild species related to food crops. But they face one serious problem: The destruction of natural habitats will surely destroy as yet undiscovered wild plants and thus make their job even harder.

Figure 49–20 The tomato, eaten with relish by many people, was once thought to be poisonous. In the past, the tomato grew in the wild. Today, this plant is commonly grown in many gardens. Who can tell how many other wild plants will be found in remote areas and what uses these plants may serve?

49–3 SECTION REVIEW

1. How does the destruction of a rain forest in South America affect a person living in South Dakota?
2. What is an endangered species? How does a species become endangered?
3. Why should we make an attempt to protect endangered species?
4. **Connection—Agriculture** Planting huge fields of a single crop is often the most economical way to raise a good crop. How does planting a single variety make people who are dependent upon the crop more vulnerable?

SCIENCE, TECHNOLOGY, AND SOCIETY CONNECTION

Ecosystems and Local Economies: Keys to Conservation

Point out that environmental problems do not occur in a vacuum but are often driven by forces at work in human society.

- **Human population and consumption of natural resources are growing.** Human population growth places increasing demands on the environment.
- **World markets are concentrating on fewer species of plants and animals.** Mass production and marketing strategies concentrate on a few species rather than on the broad spectrum of indigenous plants and animals. Farmers abandon the variety of crops they once produced. Biodiversity is lost.
- **Most economic systems undervalue the environment and its resources.** Balance sheets often ignore the value of clean fresh water and fertile topsoil. Instead they focus on the costs of fertilizers, seeds, and equipment used to produce crops or harvest commodities such as lumber. For example, logging companies argue that stopping the cutting of old-growth forests in the Pacific Northwest will cost people jobs. But what will happen to the jobs once the forests are completely cut?
- **We do not know enough about what we are losing.** Many ecosystems have been only partially studied. Many species, especially in the tropics, may be lost before they are even identified.

Ecosystems in both temperate and tropical countries are vital economic resources. It is unrealistic to expect that people will not want to use them in some way. Programs like the one in Zambia help human communities earn a living while enabling natural communities to survive.

SCIENCE, TECHNOLOGY, AND SOCIETY CONNECTION

Ecosystems and Local Economies: Keys to Conservation

These days, nearly everyone is talking about protecting the environment and conserving natural resources. But as world population grows, human needs often conflict with the needs of ecosystems. Living in the United States, it's easy for us to say that people in the tropics should preserve their rain forests.

But when we are faced with similar problems, our response is often quite different. In the Pacific Northwest, for example, a battle rages between loggers and environmentalists. Loggers, who argue that they need to earn money to support themselves and their families, want to cut down much of the region's remaining old-growth forests. Environmentalists want to protect old-growth forests because in addition to being irreplaceable, these forests are home to endangered animals.

Fortunately, both natural ecosystems and human populations can benefit from each other under the right circumstances. One dramatic example of a successful partnership exists in Zambia's Luanga Valley. At one time the killing of elephants for their ivory and black rhinos for their horns posed a serious threat to the survival of these species and, ultimately, to the health of the valley's ecosystem. For years, Zambia's National Parks and Wildlife Service was unable to protect the threatened animals, for they had neither the money nor the staff to adequately police the area. Then, in 1983, the World Wildlife Fund helped to establish an ambitious environmental protection program in the valley. They set up the Wildlife Conservation Revolving Fund, which was supported by money earned from the regulated harvest of big game in the reserve, and from fees charged to safari companies that take hunters into the park. Strict quotas limiting the number of big game animals that could be harvested were established. Safari companies were required to hire a certain number of local people, thus contributing to the local economy. Forty percent of the profits from the fund were given to local village chiefs to use for community projects. The rest of the income earned from the regulated hunting was used for wildlife management.

This carefully planned venture has been a smashing success. Since the fund was set up, the numbers of elephants and rhinos killed each year by illegal hunters has fallen dramatically. In fact, the elephant population has increased so much that the local inhabitants can make money collecting ivory from animals that have died naturally. People now realize that they can benefit personally from this kind of environmental protection program. In fact, the villagers have even established security networks to keep poachers from operating in their territories. Today, what happened in the Luanga Valley serves as an example for environmental planning in other areas.

It is important for all of us to recognize and encourage relationships of mutual respect between human communities and ecological communities. In the long term, this may be the only way conservation efforts can succeed in an ever more crowded world. Who knows, perhaps this kind of approach can be applied in our own Pacific Northwest. A solution that helps loggers while protecting the forests can benefit all involved.

1068

TEACHING STRATEGY 49–4

Focus/Motivation

Show students photographs of national parks.

- **What is the purpose of a national park?** (To preserve certain natural areas and the wildlife found in the areas; to provide a natural setting where people can enjoy recreational activities such as camping and hiking.)
- **Do you think that a national park would remain as it is if it were not set apart as a national park? Explain your answer.** (Possible answer: No, because the land would probably be altered to make room for industry, housing, farming, or other human endeavors. Also, people might tend to destroy wildlife in the area by activities such as hunting.)

49-4 The Future of the Biosphere

Guide For Reading
- Why is cleaning up the environment an important and difficult task?
- How can people protect the environment?

At this point in your reading you might be concerned that we have painted a rather bleak picture of our planet's future. Some people feel that the future is indeed bleak. Other people point with hope to the proven ability of the human community to deal with enormously difficult problems. Almost everyone agrees, however, that because Earth supplies us with all we need for life—food, water, air, and natural resources—we owe our planet the very best care possible.

In the past, people have treated Earth with neglect. Today, however, attitudes are changing. Yet even if we are committed to saving endangered species and protecting the environment, we still face many problems in deciding how to go about it.

Actions for Conservation

Happily, there are people in the world who love wild places, wild plants, and wild animals. There are also many people who understand how important the health of the biosphere is to the health of all species, including humans. These people work together to protect the environment in many different ways.

Towns, counties, states, the federal government, and conservancy groups have all purchased land that is to be set aside for conservation purposes. In many locations this land can simultaneously serve as a watershed and as a home for wildlife. The preservation of land habitats is one of the most important responsibilities we shall assume in the years ahead.

It is more difficult to conserve the resources of the ocean. Ocean currents travel all over the world. Marine animals ignore national boundaries and maritime laws. They are carried along by ocean currents or swim wherever their instincts direct them. Fishing boats in international waters can catch an unlimited number of fish. There is little our government can do to prevent fishing boats from taking too many haddock or too many whales, for example. To conserve ocean species, the countries of the world must join together to protect the oceans and their inhabitants.

Sometimes people work together to protect a single species. For example, sportsfishermen on the east coast of the United States have formed a group called Stripers Unlimited. This group works to protect habitats important to the striped bass, an ocean game fish that is seriously threatened by pollution and by the destruction of estuaries. Striped bass lay eggs in estuaries. PCB's and other chemical pollutants make their way into estuaries and kill the delicate eggs. Members of Stripers Unlimited attend town meetings, support cleanup legislation, and raise money to reclaim estuaries. By these actions, other species that inhabit estuaries are also saved.

Figure 49-21 At one time, the Arabian oryx, hunted for its beautiful horns, was close to extinction. Today, this animal is raised on ranches. Captive-bred animals are returned to their native habitat, where they are protected by strict laws that prohibit hunting. Captive breeding programs offer some hope that endangered animals will survive.

BACKGROUND INFORMATION
NATIONAL PARKS

Congress laid the foundation for the U.S. National Park System in 1872 by establishing Yellowstone National Park. In 1916, the National Park Service was set up to oversee the National Park System. Today, there are 48 national parks. The U.S. National Park System also includes historical sites, lakeshores, seashores, parkways, scenic trails, and rivers.

Reinforcement/Reteaching
Point out to students that conservation is defined as the wise and careful use of resources. Conservation benefits the environment in two ways: First, it makes resources last longer; second, it reduces the pollution that is associated with obtaining and using resources.

Have students list some of the ways resources can be conserved in daily living. Such a list might include making fewer car trips so that less gasoline will be used and less air pollution will result; using less electricity or heat at home; recycling glass and paper products; buying fewer consumer goods that require the operation of factories to produce them; using less water when washing dishes or taking a shower; buying and using energy-efficient appliances.

Content Development
Explain to students that national parks are one of the ways in which the federal government has sought to preserve and protect our nation's natural beauty and wildlife.

Enrichment
Have students work in pairs. Ask each pair to choose a national park that interests them. (A list of national parks can be found in an almanac or encyclopedia.) Then ask students to gather information about the park and present the information in the form of a guided tour. Have students make special mention of any endangered or threatened species that exist in the park.

BACKGROUND INFORMATION
ALTERNATIVE ENERGY SOURCES

One way to reduce pollution is to make use of energy sources that do not involve the burning of fossil fuels. Among these alternative energy sources are solar energy and geothermal energy.

Solar energy involves using sunlight to provide heat or generate electricity. For example, homes designed with solar heating systems have glass-paneled solar collectors that absorb the sun's rays. This energy from the sun is used to heat the house and provide hot water. Photovoltaic cells are cells that produce electricity when they are struck by sunlight. Some students may be familiar with the solar cells that are often used to power pocket calculators.

Geothermal energy is heat from the Earth. Certain places on Earth, called "hot spots," produce geysers and hot springs. At a hot spot, water can be pumped into the ground to be heated. Then this hot water can be used to generate electricity. Most homes in Iceland are heated by geothermal energy.

Since both solar energy and geothermal energy are renewable resources that produce no wastes and no pollution, one might ask why they are not widely used. The main problem is that, so far, the technology does not exist to harvest enough solar or geothermal energy at reasonable prices. Another drawback of these energy sources is that they are limited to certain locations. Solar energy, for example, would not work well in upstate New York where winter days are short and the sun is often hidden by clouds. Geothermal energy is limited to those areas on Earth that have natural hot spots.

Larger organizations such as the Sierra Club, World Wildlife Fund, and Greenpeace work on a national and international level to protect the environment. The actions of these groups recognize the importance of protecting the biosphere. By backing political candidates who support environmental causes and by suing corporations and governments whose actions endanger the environment, these groups demonstrate their commitment to saving planet Earth.

Difficult Decisions

Cleaning up the environment and keeping it clean are not easy jobs. Some cities have realized that they can no longer continue to dump untreated sewage into their rivers and streams and harbors. In some coastal areas, cities have built sewage treatment plants and constructed pipelines to carry sewage far offshore. Other cities are repairing and replacing old-fashioned sewage systems that overflow after heavy rains. These projects are expensive, and all of us must pay for them. This means that the costs of using city water and sewer services will increase. Although no one likes to pay more for such services, increased fees will allow communities to make the environment cleaner. We, and future generations, will benefit.

The problem of solid-waste disposal is probably one that few of us think about. But what happens to our trash after we throw it out? It must all be disposed of, usually by burying it in places called sanitary landfills. However, sanitary landfills, especially those near large cities, are almost filled to overflowing. One solution to the solid-waste problem is **recycling**. In this process, certain kinds of solid wastes—newspapers, bottles, and metal or plastic cans, for example—can be processed and used again. Recycling can make a big difference if enough people participate.

Controlling the dumping of toxic wastes by industry is also important. It is more expensive to clean up toxic wastes after they have been dumped into the environment than it is to dispose of them in a safe way. Once an area has become polluted with toxic chemicals, cleanup can cost millions of dollars.

Up until now, few companies have been willing to pay the costs of cleanup. But public opinion is making more and more companies aware of the need to protect the environment. Consumers should be aware, however, that protecting the environment is often accompanied by increases in the prices of products. So far, most people have not demonstrated that they are willing to pay these extra costs.

But attitudes are changing. A recent poll found that 93 percent of Americans are concerned about pollution of our lakes and rivers by toxic chemicals, and 92 percent are concerned about the disposal of hazardous wastes on land. More than 75 percent of Americans are concerned about air pollution and the problems caused by acid rain.

Figure 49-22 La Guardia airport in New York City indicates how far humans have developed. But its past speaks volumes about how we developed. The land for this airport was once part amusement park and part landfill. Every time a plane takes off or lands, it does so on the wastes produced by other generations of New Yorkers.

49-4 (continued)

SECTION REVIEW 49-4

1. We are part of the biosphere. It is difficult to view humans out of the context of all life on Earth. To protect ourselves and ensure our survival, it is important to protect all life on Earth.
2. People are recycling materials, using machines that are more energy efficient, and cutting down on the amount of wastes produced. Other answers are possible.
3. Many obstacles exist. One important one is education. Educating people on the importance of protecting and cleaning up the environment is essential. People will have to be prepared to pay more for products that are environmentally safe. In the

In overwhelming numbers, Americans are expressing their fears for the future of planet Earth. But they are also expressing their hopes that a healthy biosphere will become a major priority for all the Earth's inhabitants. **You must decide how important the environment is and whether you are willing to pay to keep it healthy.** You must decide, because you are the stewards of the Earth.

Figure 49–23 Planet Earth moves through space carrying its precious cargo of life. The Earth offers us a good home. But we must protect that home so future generations can enjoy their lives here as well.

49–4 SECTION REVIEW

1. Why must the biosphere be protected?
2. Describe three ways in which people are protecting the environment.
3. What are some of the obstacles that stand in the way of cleaning up the environment?
4. **Connection—You and Your World** In what three ways can you protect the environment?

LABORATORY INVESTIGATION

SIMULATING THE EFFECTS OF ACID RAIN

Before the Lab
1. At least one day prior to the investigation, gather enough materials for your class, assuming six students per group.
2. Check that all safety goggles are in good condition.
3. Note that the entire procedure takes place over a three-day period.

Pre-Lab Discussion
Review with students the material on acid rain in Section 49–2. Emphasize the idea that certain elements when combined with water and oxygen produce acids. The acid that is produced when sulfur combines with water vapor is sulfuric acid.

Have students read the Problem at the top of the page, then have them read through the Procedure.

• **Why is sulfur used in this investigation, rather than some other element?** (Fossil fuels contain sulfur.)

• **What is the purpose of burning the sulfur before it is put into the flask?** (Burning the sulfur simulates what happens to sulfur when fuel is burned.)

• **Why is the burning sulfur placed above the water in the flask, rather than in the water?** (Pollutants from burning fuel enter the air first. Then the air pollution affects the water.)

• **What is the purpose of the test tube, beaker, and petri dish labeled WATER?** (To act as a control.)

At this point, stop to ask students what they think will happen to the mustard seeds in the two dishes. Have students write their predictions in the form of a hypothesis.

Skills Development
Students will use the following skills while completing this laboratory investigation.
1. Hypothesizing
2. Predicting
3. Comparing
4. Observing
5. Calculating
6. Applying
7. Relating
8. Measuring
9. Inferring

LABORATORY INVESTIGATION
SIMULATING THE EFFECTS OF ACID RAIN

PROBLEM
How do high-sulfur fuels affect the environment?

MATERIALS (per group)

2 100-mL beakers	pH paper
Bunsen burner	rubber stopper
deflagrating spoon	(#8 ½)
filter paper	safety goggles
1000-mL flask	steel wool
glass-marking pencil	stirring rod
100-mL graduated	sulfur
cylinder	test tube rack
100 mustard seeds	2 test tubes
2 petri dishes	tongs

PROCEDURE

1. Put on safety goggles. Use a graduated cylinder to measure 50 mL of water. Pour the water into a clean 1000-mL flask. Swirl the flask.
2. Use tongs to dip a piece of pH paper in the water. Record the pH of the water.
3. Fill a deflagrating spoon halfway with sulfur. Heat the sulfur with a Bunsen burner until it melts and ignites. **CAUTION:** *Work in a well-ventilated room or under a fume hood. Do not inhale the fumes.*
4. Immediately insert the deflagrating spoon and burning sulfur into the flask. **CAUTION:** *Be sure not to put them into the water.* When the flames fill the flask and begin to escape, insert the rubber stopper into the mouth of the flask so that the deflagrating spoon is held in place.
5. When the sulfur stops burning, remove the deflagrating spoon. Restopper the flask. Swirl the flask gently until the fumes dissolve.
6. Remove the stopper from the flask. Use tongs to dip another piece of pH paper in the water to measure the pH. Record the pH.
7. Label a test tube, a beaker, and a petri dish WATER. Label a second test tube, beaker, and petri dish SULFUR WATER. Put a piece of filter paper in each petri dish.
8. Pour 10 mL of water into the test tube labeled WATER and 30 mL of water into the beaker labeled WATER. Stand the test tube in the test tube rack. Repeat this procedure using the sulfur water in the flask, the test tube, and the beaker labeled SULFUR WATER.
9. Put 50 mustard seeds into each beaker. Set the beakers aside overnight.
10. Place a small piece of steel wool in each test tube. Note any color changes. Check the texture of each piece of steel wool with a stirring rod. Set the test tubes aside overnight. After 24 hours, repeat your observations.
11. Pour the contents of the beakers into the appropriately labeled petri dishes. Set the petri dishes aside overnight.
12. After another 24 hours, count the number of seeds that germinated in each petri dish. Calculate the percentage of seeds that germinated. Record the results.

OBSERVATIONS
1. What was the pH of the water? The pH of the sulfur water?
2. Describe the steel wool in each test tube.
3. How did the percentage of germinating seeds in the petri dishes compare?

ANALYSIS AND CONCLUSIONS
1. What effect does burning high-sulfur fuels have on the acidity of water?
2. Based on your observations, what effect would acid rain have on steel structures?
3. Based on your observations, what effect would acid rain have on living organisms?

Safety Tips
1. Direct students' attention to the four safety symbols at the beginning of the procedure. Have students identify each symbol.
2. Emphasize that the first step of the procedure is to put on safety goggles. DO NOT ALLOW ANY GROUP TO BEGIN WORK IF A MEMBER OF THE GROUP IS NOT WEARING SAFETY GOGGLES.
3. Emphasize that sulfur fumes are poisonous.
4. Point out that steps 3 and 4 require much caution since

STUDENT STUDY GUIDE

SUMMARIZING THE CONCEPTS

The key concepts in each section of this chapter are listed below to help you review the chapter content. Make sure you understand each concept and its relationship to other concepts and to the theme of this chapter.

49–1 Human Population

- The human population has increased dramatically over time. As human populations increase, their effects on the environment also increase. Changes in lifestyles also contribute to environmental problems.

49–2 Pollution

- Some materials that are released into the environment are biodegradable. They can be broken down by microorganisms into the essential elements from which they were made. Other wastes are nonbiodegradable, which means they cannot be broken down by natural processes.
- The ozone layer protects the Earth from the sun's harmful ultraviolet radiation. Scientists have discovered holes in this layer. The holes result mainly from the addition of chlorofluorocarbons to the air.
- Pollution of the Earth's water is another important problem. Industrial chemicals, sewage wastes, and thermal pollution are the three main causes of this problem.

49–3 The Fate of the Earth

- Forest destruction poses a danger to people living far from the forest area. When forests are cut, changes occur in the numbers and kinds of organisms that live in the forest.
- An endangered species is a species whose numbers are so small that the species is threatened with extinction.

49–4 The Future of the Biosphere

- Fortunately, there are people who understand how important the health of the biosphere is to the health of all species. Government agencies have purchased land to be set aside for conservation purposes. To conserve ocean species, the countries of the world must join together to protect the oceans. Sometimes people work together to protect a single species.
- Only an increasing awareness of the problems caused by pollution and the degradation of the environment can ensure that planet Earth will remain a planet full of promise for a better life for all organisms.

REVIEWING KEY TERMS

Vocabulary terms are important to your understanding of biology. The key terms listed below are those you should be especially familiar with. Review these terms and their meanings. Then use each term in a complete sentence. If you are not sure of a term's meaning, return to the appropriate section and review its definition.

49–2 Pollution
biodegradable
nonbiodegradable
biological magnification
smog
temperature inversion
acid rain
greenhouse effect
hole in the ozone layer
thermal pollution

49–3 The Fate of the Earth
reforestation
endangered species

49–4 The Future of the Biosphere
recycling

1073

Observations

1. The pH of water was about 7; the pH of sulfur water should be about 1.
2. The steel wool in water did not change. The steel wool in the sulfur water turned black and crumbled when touched.
3. About 50 to 60 percent of the seeds in regular water should have germinated. None of the seeds in the sulfur water are likely to germinate.

Analysis and Conclusions

1. It increases the acidity of the water.
2. Acid rain could damage steel structures, eventually eroding them away.
3. Acid rain could kill some organisms directly. It could also harm many organisms and prevent the germination of plants.

Going Further: Enrichment

Have students design an experiment to test the effects of other sources of air or water pollution on growing seeds. Two types of pollution they might try to simulate are thermal pollution and carbon dioxide pollution.

CHAPTER REVIEW

CONTENT REVIEW

Multiple Choice
1. b   2. d   3. a   4. b
5. d   6. c   7. c   8. b

True or False
1. F nonbiodegradable
2. T
3. F Thermal pollution
4. T
5. F not leveled
6. T
7. F Acid rain
8. T

---

they involve the use of fire, glassware, and a potentially poisonous substance.

Teaching Strategy

1. Circulate throughout the room and help any group that is having trouble. At the end of the first day's work, check to make sure that each group has properly placed the seeds and steel wool in both experimental and control setups.

2. After all groups have finished, make a chart on the chalkboard listing the percentage of seeds that germinated in each group. Refer to these data as students discuss their observations and conclusions.

Word Relationships

A.
1. biodegradable
2. smog
3. Reforestation programs

B.
4. nonbiodegradable pollutants; food scraps are biodegradable
5. substances involved with acid rain; holes in the Earth's ozone layer may be the result of air pollution
6. forms of water pollution; watershed area collects rainfall and supplies water to an area

CONCEPT MASTERY

1. Sewage acts as a plant fertilizer. When sewage is added to water, it encourages the growth of algae. When algae populations reach very high levels, the level of oxygen in the water decreases. The algae use up the oxygen for respiration. The result is that levels of oxygen can be so diminished that algae and other organisms in the water suffocate.
2. Good medical care has increased the number of children who survive after birth. Good medical care and better nutrition have also increased the life span. These and other factors have increased population growth.
3. Trees release oxygen and use carbon dioxide, thus increasing the amount of oxygen in the atmosphere. Trees also provide a home and food for many animals. When forests are cut, animals may lose their home as well as a source of food. If they are to survive, the animals must move to another area. The roots of trees also hold particles of soil in place. When trees are cut, rains wash particles of soil away, and the land becomes eroded. Other answers are possible.
4. A septic system is a small bacteria-laden tank that treats the sewage produced by a relatively small family unit. Bacteria in a septic system break

CHAPTER REVIEW

CONTENT REVIEW

Multiple Choice

Choose the letter of the answer that best completes each statement.

1. Which of the following is biodegradable?
   a. glass   c. DDT
   b. orange peel   d. mercury
2. If the United States maintains its current rate of population growth, in a hundred years the number of people will
   a. decrease by half.   c. remain the same.
   b. increase by half.   d. double.
3. Wastes that cannot be broken down into essential nutrients by natural processes are
   a. nonbiodegradable.   c. biodegradable.
   b. water soluble.   d. organic compounds.
4. A condition in which smog is trapped close to the surface of the Earth results from a
   a. snowstorm.
   b. temperature inversion.
   c. hurricane.
   d. hole in the ozone layer.
5. Sulfur dioxide and nitric oxide in the air combine with water to form
   a. biological magnification.
   b. smog.
   c. PCB's.
   d. acid rain.
6. The greenhouse effect causes an increase in
   a. carbon dioxide.   c. temperature.
   b. ozone.   d. sulfur.
7. Areas in the ocean where the larvae and young of many species dwell before they move into the open sea are called
   a. sea nurseries.   c. estuaries.
   b. glaciers.   d. tidal pools.
8. An area of a forest where water gathers for use by people is a (an)
   a. rain forest.   c. eroded area.
   b. watershed area.   d. water basin.

True or False

Determine whether each statement is true or false. If it is true, write "true." If it is false, change the underlined word or words to make the statement true.

1. DDT, asbestos, and Dieldrin are examples of <u>biodegradable</u> substances.
2. The burning of fossil fuels and wood can lead to a <u>greenhouse effect</u>.
3. <u>Chemical pollution</u> results when warm water is pumped from factories into the environment.
4. A species that is so rare that its survival is threatened with extinction is an <u>endangered species</u>.
5. The rate of population increase has <u>leveled off</u> in all countries on Earth.
6. <u>Biological magnification</u> occurs with many pesticides and industrial waste products.
7. <u>Smog</u> damages plant leaves, makes water and soil more acidic, and carries salts of aluminum and mercury into rivers and lakes.
8. Chlorofluorocarbons are a major source of the air pollution that depletes the <u>ozone layer</u>.

Word Relationships

A. *Replace the underlined definition with the correct vocabulary word.*

1. Some wastes are <u>materials that can be broken down by microorganisms into the essential nutrients from which they were made</u>.
2. One type of air pollution involves <u>a combination of smoke and fog</u>.
3. <u>Programs that plant new trees when old trees are cut</u> are vitally important.

down sewage materials. A modern sewage treatment plant operates on much the same principle, although on a much larger scale. In a treatment plant, sewage is placed in huge tanks where it is broken down by bacteria. Sewage in a treatment plant may also be aerated, treated with chemicals, and then released.

5. A species may become endangered if it is overhunted. A disease may also kill off many members of a species. Natural disasters such as volcanic eruptions can decimate a population. A species may also become endangered if its habitat is destroyed.
6. Radioactive wastes are extremely dangerous. They give off radiation that may increase cancer rates and may make

B. *In each of the following sets of terms, three of the terms are related. One item does not belong. Determine the characteristic common to three of the terms and then identify the term that does not belong.*

4. food scraps, DDT, asbestos, Dieldrin
5. sulfur, acid rain, ozone, nitric oxide
6. industrial chemicals, sewage, thermal pollution, watershed area.

CONCEPT MASTERY

Use your understanding of the concepts developed in the chapter to answer each of the following in a brief paragraph.

1. Explain how water organisms may suffocate if sewage is added to the water.
2. Discuss some reasons why there has been increased population growth in the past few hundred years.
3. The trees of a forest play an important role in many ecosystems. Discuss the importance of trees and describe the changes that occur when forests are cut.
4. Explain the difference between the treatment of sewage in a septic system and in a sewage treatment plant.
5. How can a species become endangered?
6. What are some of the problems associated with the disposal of radioactive wastes?
7. What is acid rain? How does it affect lakes in upstate New York and in New England?
8. Why is it difficult to replace a forest?

CRITICAL AND CREATIVE THINKING

Discuss each of the following in a brief paragraph.

1. **Identifying relationships** Most people in the United States will never see a tropical rain forest. Why, then, is the destruction of tropical rain forests a global concern?
2. **Relating cause and effect** The owner of a local factory believes that he should be able to dump wastes into a nearby river. His argument is that water in the river will dilute any dangerous chemicals. Present several arguments to convince him that he should find another method to dispose of the wastes.
3. **Making inferences** Oil spills can become large-scale catastrophes that have a major impact on entire ecosystems. Explain how an oil spill can affect an ecosystem.
4. **Using the writing process** Developers

have chosen to build a large factory in a remote area of your state. The land is uninhabited and little is known about the wildlife and plants that exist there. In order to build, the developers plan to clear the area completely. Write a letter to the governor. Explain the type of information that might be lost forever if the developers do not wait at least until the area has been explored and studied.

people sick in other ways. The danger involved in the disposal of radioactive wastes is that the wastes remain potentially injurious for a very long time. If they are disposed of improperly, radioactive wastes may prove a danger to humans and other life forms for many thousands of years.

7. Acid rain is actually any form of precipitation that has a low pH. Acid rain forms when certain pollutants are added to the air as a result of the burning of fossil fuels. When acid rain falls on bodies of water, it lowers the pH of the water. The lowered pH may prove harmful to some forms of aquatic life. If the pH falls very low, all life in the body of water may die.

8. Reforestation is difficult because many different kinds of trees grow in a natural forest and it is desirable to replant the same variety of trees that had grown naturally in an area over time as a result of the forces of natural selection. Students may also point out that it takes a long time for a new forest to grow.

CRITICAL AND CREATIVE THINKING

1. Tropical rain forests perform many functions that affect other parts of the world. The plants in a rain forest give off vast amounts of oxygen and use up vast amounts of carbon dioxide. The rain forests are home to many species. Rain forests produce many decorative woods for use in furniture making. Rain forests are also able to affect climate patterns. The destruction of rain forests would produce many profound effects in areas far removed from the rain forest areas.
2. The wastes may be harmful to plant and animal life in the river. Small amounts of wastes may, through the process of biological magnification, reach dangerous levels. The river may provide drinking water for cities and towns downstream. Other answers are possible.
3. Since oil is lighter than water, oil spills can spread over a vast area. Birds and mammals may pick up oil on feathers and fur. The oil might be ingested by fish and other animals and poison them. Oil spills may kill algae and other plants in the water. The death of plants and sea animals might also affect other animals that normally eat plants and fishes. Thus, even a small oil spill may have far-reaching effects.
4. Check students' letters for clarity of expression and scientific accuracy.

CAREERS IN BIOLOGY

Wastewater-Treatment-Plant Operator, Park Ranger, Biogeographer

Students interested in a biological career should be instructed to write for further information to the address listed beneath each career description. However, in consideration of the organizations that provide career information, please have only one student write to an organization. Do not instruct the entire class to write to every organization for the same career information. You may want to use the information provided to start a biology career file for student use.

UNIT 10 — CAREERS IN BIOLOGY

Wastewater-Treatment-Plant Operator

Waste materials enter our water supply so quickly that if they were not stopped, water would soon become unfit for any use. Wastewater-treatment-plant operators control the equipment and processes that remove wastes from water and restore the water to a condition safe for human consumption.

A high school education is required to work as a wastewater-treatment-plant operator. There are also some two-year programs leading to an associate degree in wastewater technology.

For more information write to the Water Pollution Control Federation, 2626 Pennsylvania Ave., NW, Washington, DC 20037.

Park Ranger

Park rangers enforce the laws and regulations in national, state, and county parks. They are responsible for protecting the park's natural resources and animal and plant life. In addition, rangers inform, guide, and ensure the safety of park visitors.

To become a park ranger, it is necessary to have a Bachelor's degree in park management, forestry, or a related field. Sometimes, however, rangers substitute work experience for part of their education.

To receive information write to the National Park Service, United States Department of the Interior, Washington, DC 21240.

Biogeographer

Biogeographers study the way in which plants and animals are distributed throughout the world. By observing different species and their habitats and then recording and interpreting the data, biogeographers help us understand why organisms exist where they do. Some biogeographers also study the influence of certain human activities on plant and animal life.

Biogeographers must have a Bachelor's degree in geography and a Ph.D. in biogeography. An interest in botany, zoology, and meteorology is also important.

For information write to the Association of American Geographers, Biogeography Specialty Group, 1710 16th St., NW, Washington, DC 20009.

HOW TO FOLLOW UP ON A JOB INTERVIEW

After you have had a job interview, you may have to wait several weeks to hear the results. During that period, however, you should send a letter to the interviewer thanking him or her for the time and consideration. You can also call the person to show your interest in working for that company and to provide any additional information that might be needed.

If you do not get the job that you interviewed for, ask the interviewer for comments or suggestions on how to improve your interviewing skills. Most employers are willing to discuss the interview with you and explain their impressions so that you can be better prepared for your next interview.

FROM THE AUTHORS

As a biologist, I've been lucky enough to travel around the world. I have seen coral reefs in the South Pacific. I have hiked across deserts in Israel. And I have climbed through forests in the mountains of Central America and India. I have worked long hours, studying plants and animals. But I have also sat quietly, drinking in the splendor of a sparkling dawn or a flaming sunset. And I have often taken a few moments to give thanks, in my own way, for the natural beauty that surrounds us.

Yet with all my traveling, one of my favorite places is a wildlife refuge only an hour's drive from my home. I go there when I'm happy. I go there when I'm sad. I go there whenever I have problems to think about or joyful things to celebrate.

This weekend I plan to visit that refuge and wander along its nature trails. I have been there at every season and at every time of day, so I know much of what I'll see. It is autumn, so ripening grasses on the marsh will look like a swirling golden sea in the afternoon sun. Maple trees will be dressing up in red and orange. Flocks of geese will drop in on their long trip south for the winter. Then, in a chorus of honks and squawks, they'll take wing to fly in formation across the sky.

But I also know that there will be something new, something different, to discover. A rare bird, perhaps. A white mushroom, glistening after a rain. Or a flower I've never seen in bloom. That's the way wild places are.

Because I've just finished writing these chapters, I'll be thinking about you—the young people who will inherit the world we older folks have prepared for you. The wildlife refuge I love so much exists thanks to the wisdom of people my parents' age. They realized that such places are important to our society and to our souls. They knew that in order to be sure we will always have wild places, we must protect them from people who are not so wise. I hope that enough people in my generation have the wisdom to protect the health of our planet for you and your children. And I hope that you, too, in your time will be wise enough to care for the living planet that shelters and protects us all.

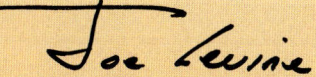

FROM THE AUTHORS

Joseph Levine

In this feature Joe Levine discusses the feelings of both authors regarding the environment. This feature may provoke different opinions among students, depending on whether they read it before or after completing the unit. It is the authors' hope that students will gain an understanding of the fragility of the environment while reading Unit 10 and that they will begin to understand some of the difficult environmental decisions that await them.

For Further Reading

If you have an interest in a specific area of biology or simply want to know more about the topics you are studying, one of the following books may open the door to an exciting learning adventure.

CHAPTER 1 The Nature of Science
Kramer, Stephen P. *How to Think Like a Scientist: Answering Questions by the Scientific Method.* T.Y. Crowell, 1987.
Mannoia, V. James. *What Is Science? An Introduction to the Structure and Methodology of Science.* University Press of America, 1980.
Medawar, P.B. *Advice to a Young Scientist.* Harper, 1984.
Moravcsik, M.J. *How to Grow Science.* Universe Books, 1980.
Morrison, Philip and Phylis. *The Ring of Truth: An Inquiry into How We Know What We Know.* Random, 1987.

CHAPTER 2 Biology as a Science
Campbell, P.N., ed. *Biology in Profile: An Introduction to the Many Branches of Biology.* Pergamon, 1980.
Ford, Brian J. *Single Lens: The Story of the Simple Microscope.* Harper & Row, 1985.
Grave, Eric V. *Discover the Invisible: A Naturalist's Guide to Using the Microscope.* Prentice-Hall, 1984.
Johnson, Gaylord, and Maurice Bleifield. *Hunting with the Microscope,* 3rd ed. Arco, 1980.
Smith, John Maynard. *The Problems of Biology.* Oxford University Press, 1986.

CHAPTER 3 Introduction to Chemistry
Asimov, Isaac. *Asimov on Chemistry.* Doubleday, 1974.
Baker, J.J., and G.E. Allen. *Matter, Energy and Life: An Introduction to Chemical Concepts,* 4th ed. Addison-Wesley, 1981.
Brady, J.E., and G.E. Humiston. *General Chemistry: Principles and Structures,* 2nd ed. Wiley, 1980.
Bruckman, H. J., and A. Cruickshanks. *Understanding Chemistry.* Wiley, 1988.
Puddephatt, R.J., and P.K. Monaghan. *The Periodic Table of Elements.* 2nd ed. Oxford University Press, 1986.

CHAPTER 4 The Chemical Basis of Life
Berger, Jim. *Clear and Simple Chemistry.* Simon and Schuster, 1986.
Berman, W. *Beginning Biochemistry,* rev. ed. Arco, 1980.
Boikess, R.S. *Chemistry for Biologists.* Carolina Biology Reader. Carolina Biological Supply Co., 1987.
Breslow, Ronald. *Enzymes.* Carolina Biology Reader. Carolina Biological Supply Co., 1986.
Carroll, Harvey F. *Preview of General Chemistry.* Wiley, 1988.
Jenson, William B. *The Lewis Acid-Base Concepts: An Overview.* Wiley, 1980.

CHAPTER 5 Cell Structure and Function
Berns, M.W. *Cells.* Holt, Rinehart and Winston, 1977.
De Duve, Christian. *A Guided Tour of the Living Cell.* W.H. Freeman, 1985.
Moner, John. *The Animal Cell.* Carolina Biology Reader. Carolina Biological Supply Co., 1987.
Nachmias, Vivianne T. *Microfilaments.* Carolina Biology Reader. Carolina Biological Supply Co., 1984.
Sheeler, P., and D.E. Bianchi. *Cell Biology: Structure, Biochemistry, and Function.* Wiley, 1983.

CHAPTER 6 Cell Energy: Photosynthesis and Respiration
Amesz, J. *Photosynthesis.* Elsevier, 1987.
Foyer, Christine H. *Photosynthesis.* Wiley, 1984.
Miller, Kenneth. *Energy and Life.* Carolina Biology Reader. Carolina Biological Supply Co., 1988.
Nakatani, Herbert Y. *Photosynthesis.* Carolina Biology Reader. Carolina Biological Supply Co., 1988.

CHAPTER 7 Nucleic Acids and Protein Synthesis
Asimov, Isaac. *How Did We Find Out About DNA?* Walker, 1985.
Dickerson, R., and I. Geis. *The Structure and Action of Proteins,* 2nd ed. W.A. Benjamin, 1985.
Gribbon, John. *In Search of the Double Helix.* Bantam, 1987.
Rosenfield, Israel, Edward Ziff, and Borin Van Loon. *DNA for Beginners.* Writers and Readers, 1983.

CHAPTER 8 Cell Growth and Division
Baserga, Renato. *The Biology of Cell Reproduction.* Harvard University Press, 1985.
Becker, Wayne M. *The World of the Cell.* Benjamin-Cummings, 1986.
John, B., and K.R. Lewis. *Somatic Cell Division,* rev. Carolina Biology Reader. Carolina Biological Supply Co., 1980.
Kornberg, Warren, ed. National Science Foundation Staff. *DNA: The Master Molecule.* Avery Pub., 1982.
Zimmerman, A.M., and A. Forer, eds. *Mitosis/Cytokinesis.* Academic Press, 1981.

CHAPTER 9 Introduction to Genetics
Fincham, John R.S. *Genetics.* Jones and Bartlett, 1983.
Gerbi, Susan A. *From Genes to Proteins.* Carolina Biology Reader. Carolina Biological Supply Co., 1987.
John, Bernard, and Kenneth Lewis. *The Meiotic Mechanism,* rev. Carolina Biology Reader. Carolina Biological Supply Co., 1984.
Moens, Peter B. *Meiosis.* Academic Press, 1987.
Parker, Gary E., et al. *Mitosis and Meiosis,* 2nd ed. Longman Finan, 1979.

CHAPTER 10 Genes and Chromosomes
Arnold, Caroline. *Genetics: From Mendel to Gene Splicing.* Franklin Watts, 1986.
Dillon, Lawrence S. *The Gene: Its Structure, Function, and Evolution.* Plenum Pub., 1987.
Lewin, Benjamin. *Genes.* Wiley, 1987.
McCarty, Maclyn. *The Transforming Principle: Discovering That Genes Are Made of DNA.* Norton, 1986.
Snyder, Hartl F. *Basic Genetics.* Jones and Bartlett, 1987.

CHAPTER 11 Human Heredity
Carlson, Elof A. *Human Genetics.* Heath, 1983.
Cavalli-Sforza, L.L. *The Genetics of Human Races.* Carolina Biology Reader. Carolina Biological Supply Co., 1983.
Fox, L. Raymond, and Paul R. Elliott. *Heredity and You,* 2nd ed. Kendall-Hunt, 1983.
Gardner, Eldon J. *Human Heredity.* Wiley, 1983.
Singer, Sam. *Human Genetics: An Introduction to the Principles of Heredity.* W.H. Freeman, 1986.
Therman, E. *Human Chromosomes,* 2nd ed. Springer-Verlag, 1985.

CHAPTER 12 Genetic Engineering
Lampton, Christopher. *DNA and the Creation of New Life.* Arco, 1985.
Levine, Joseph, and David Suzuki. *The Secret of Life.* WGBH Educational Foundation, 1993.
Nossal, G.J.V. *Reshaping Life. (Key Issues in Genetic Engineering).* Cambridge University Press, 1985.
Perbal, Bernard. *A Practical Guide to Molecular Cloning.* Wiley, 1988.
Sylvester, Edward, and Lynn Klotz. *The Gene Age,* rev. Scribner, 1987.
Watson, James D., and John Tooze. *The DNA Story.* W.H. Freeman, 1983.
Zimmerman, Burke K. *Biofuture: Confronting the Genetic Era.* Plenum Publishers, 1984.

CHAPTER 13 Evolution: Evidence of Change
Barrett, Paul H., et al., eds. *Charles Darwin's Notebooks, 1836–1844: Geology, Transmutation of Species, Metaphysical Enquiries.* Cornell University Press, 1987.
Berry, R.J., and A. Hallam, eds. *The Encyclopedia of Animal Evolution.* Facts on File, 1987.
Case, Gerald Ramon. *A Pictorial Guide to Fossils.* Van Nostrand Reinhold, 1982.
Darwin, Charles. *The Illustrated Origin of Species.* Hill and Wang, 1979.
Lambert, David, and the Diagram Group. *The Field Guide to Prehistoric Life.* Facts on File, 1985.
Steadman, David W., and Steven Zousmer. *Galapagos: Discovery on Darwin's Islands.* Smithsonian Institution Press, 1988.
Taylor, Kenneth N. *What High School Students Should Know About Evolution.* Tyndale, 1983.

CHAPTER 14 Evolution: How Change Occurs
Ayala, F.J. *Origin of Species.* Carolina Biology Reader. Carolina Biological Supply Co., 1983.
Eldredge, Niles, ed. *The Natural History Reader in Evolution.* Columbia University Press, 1987.
Fox, Sidney. *The Emergence of Life: Darwinian Evolution from the Inside.* Basic, 1988.
McMahon, Thomas, and John Bonner. *On Size and Life.* W.H. Freeman, 1983.
Miller, Jonathan. *Darwin for Beginners.* Pantheon, 1982.
Stebbins, G. Ledyard. *Darwin to DNA, Molecules to Humanity.* W.H. Freeman, 1982.

CHAPTER 15 Classification Systems
Hickman, Cleveland J., Jr., et al. *Biology of Animals,* 4th ed. Mosby, 1985.
Jones, Susan, et al. *Classification.* Sabbot-Natural History Books, 1983.
Margulis, Lynn, and Karlene Schwartz. *Five Kingdoms,* 2nd ed. W.H. Freeman, 1987.

CHAPTER 16 The Origin of Life
Asimov, Isaac. *Beginnings: The Story of Origins—of Mankind, Life, the Earth, the Universe.* Walker, 1987.
Bone, Q. *The Origin of Chordates,* 2nd ed. Carolina Biology Reader. Carolina Biological Supply Co., 1979.
Cairns-Smith, A.G. *Seven Clues to the Origin of Life.* Cambridge University Press, 1985.
Cattermole, Peter, and Patrick Moore. *The Story of the Earth.* Cambridge University Press, 1985.
Fisher, David E. *The Origin and Evolution of Our Own Particular Universe.* Macmillan, 1988.
Hagene, Bernard, and Charles Lenay. *The Origin of Life.* Barron, 1987.
Minelli, Giuseppe. *The Evolution of Life.* Facts on File, 1987.
Pellegrino, Charles. *Time Gate: Hurtling Backward Through Time.* TAB Books, 1985.
Woese, Carl R. *The Origin of Life.* Carolina Biology Reader. Carolina Biological Supply Co., 1984.

CHAPTER 17 Viruses and Monerans
Asimov, I. *How Did We Find Out About Germs?* Avon, 1981.
Eron, C. *The Virus That Ate Cannibals.* Macmillan, 1981.

Flint, S. Jane. *Viruses.* Carolina Biology Reader. Carolina Biological Supply Co., 1988.
Lappe, M. *Germs That Won't Die.* Anchor/Doubleday, 1982.
Nourse, Alan E., M.D. *Viruses.* Franklin Watts, 1983.

CHAPTER 18 Protists

Anderson, Dean A. *An Introduction to Microbiology.* Mosby, 1980.
Farmer, J.N. *The Protozoa.* Mosby, 1980.
Gortz, H.D., ed. *Paramecium.* Springer-Verlag, 1988.
Lee, John J., et al., eds. *The Illustrated Guide to the Protozoa.* Allen Press, 1985.
Teasdale, Jim. *Microbes.* Silver Burdett, 1985.

CHAPTER 19 Fungi

Arora, David. *Mushrooms Demystified.* Ten Speed Press, 1986.
Coldrey, Jennifer. *Discovering Fungi.* Bookwright, 1987.
Krieger, Louis C.C. *The Mushroom Handbook.* Dover, 1967.
Miller, Orson, and Hope Miller. *Mushrooms in Color.* Elsevier-Dutton, 1980.
Moore-Landecker, Elizabeth. *Fundamentals of Fungi,* 2nd ed. Prentice-Hall, 1982.
Pearson, Lorentz. *The Mushroom Manual.* Naturegraph, 1987.
Savonius, Moira. *All Color Book of Mushrooms and Fungi.* Octopus Books Limited, 1973.
Smith, Alexander, and Nancy S. Weber. *The Mushroom Hunter's Field Guide.* University of Michigan Press, 1980.

CHAPTER 20 Multicellular Algae

Bold, Harold C., and Michael J. Wynne. *Introduction to the Algae,* 2nd ed. Prentice-Hall, 1985.
Chapman, A.R.O. *Biology of Seaweeds: Levels of Organization.* University Park Press, 1979.
Humm, Harold J., and Susanne Wicks. *Introduction and Guide to the Marine Blue-green algae.* Wiley, 1980.
Kavaler, Lucy. *Green Magic; Algae Rediscovered.* Harper & Row Junior Books, 1983.
Pickett-Heaps, Jeremy. *New Light on the Green Algae.* Carolina Biology Reader. Carolina Biological Supply Co., 1982.
Vinyard, William. *Diatoms of North America.* Mad River, 1979.

CHAPTER 21 Mosses and Ferns

Conrad, Henry S., and Paul L. Redfearn, Jr. *How to Know the Mosses and Liverworts,* 2nd ed. William C. Brown & Co., 1979.
Ferns: A Natural History. The Stephen Green Press, 1981.
Lellinger, David B. *Field Manual of the Ferns and Fern-Allies of the United States and Canada.* Smithsonian, 1985.
Streams, John, ed. *Treasures of Nature: Ferns.* Crossing Press, 1987.
Wexler, Jerome. *From Spore to Spore: Ferns and How they Grow.* Dodd, Mead, 1985.

CHAPTER 22 Plants with Seeds

Bryant, John A. *Seed Physiology.* David R. Murray, ed. Academic Press, 1985.
Burn, Barbara. *North American Trees.* National Audubon Society Collection. Bonanza Books, dist. by Crown Pub., 1984.
———. *North American Wildflowers.* National Audubon Society Collection. Bonanza Books, dist. by Crown Pub., 1984.
Line, Les, Ann Line, and Myron Sutton. *The Audubon Society Book of Trees.* Abrams, 1981.
Meeuse, B.J.D. *Pollination.* Carolina Biology Reader. Carolina Biological Supply Co., 1984.

CHAPTER 23 Roots, Stems, and Leaves

Esau, K. *Plant Anatomy,* 3rd ed. Wiley, 1988.
Grimn-Lacy, Janice, and Peter Kaufman. *Botany Illustrated.* Van Nostrand Reinhold, 1984.
Jensen, W.A., and F.B. Salisbury. *Botany,* 2nd ed. Wadsworth, 1984.
Prance, Ghillean Tulmie. *Leaves.* Crown, 1985.
Raven, Peter, et al. *Biology of Plants,* 4th ed. Worth, 1986.
Stern, Kingsley R. *Introductory Plant Biology,* 4th ed. William C. Brown & Co., 1988.

CHAPTER 24 Plant Growth and Development

Art, Henry. *A Garden of Wildflowers: 101 Species and How to Grow Them.* Storey Communications, Inc., 1986.
Bender, Lionel. *Plants.* Franklin Watts Ltd., ed. Watts, 1988.
Gibbons, Bob. *How Flowers Work: A Guide to Plant Biology.* Blandford Press, 1984.
Hendricks, Sterling B. *Phytochrome and Plant Growth.* Carolina Biology Reader. Carolina Biological Supply Co., 1980.
Wareing, P.F., and I.D.J. Phillips. *Growth and Differentiation in Plants,* 3rd ed. Pergamon, 1981.
Wilkins, Malcolm. *Plantwatching: How Plants Remember, Tell Time, Form Partnerships, and More.* Facts on File, 1988.

CHAPTER 25 Reproduction in Seed Plants

Buckles, Mary Parker. *The Flowers Around Us.* University of Missouri Press, 1985
Hartman, Hudson T., and Dale E. Kester. *Plant Propagation.* Prentice-Hall, Inc., 1983.
Richards, A.J. *Plant Breeding Systems in Seed Plants.* Unwin Hyman, 1986.
Roland, J.C., and F. Roland. *Atlas of Flowering Plant Structure.* Longman, 1981
Stone, Doris. *The Lives of Plants.* Scribner, 1983.

CHAPTER 26 Sponges, Cnidarians, and Unsegmented Worms

Evslin, Bernard. *Thy Hydra (Monsters of Mythology).* Chelsea House, 1989.
Greenberg, Idan. *Field Guide to Marine Invertebrates.* Seahawk Press, 1980.
Simpson, T.L. *The Cell Biology of Sponges.* Springer-Verlag, 1984.
Walls, J.G., ed. *Encyclopedia of Marine Invertebrates.* TFH Publications, 1982.

CHAPTER 27 Mollusks and Annelids

Carstarphen, Dee. *The Conch Book.* Banyan Books, 1981.
Cousteau, Jacques-Yves, and Philippe Diole. *Octopus and Squid: The Soft Intelligence.* Doubleday, 1973.
Lee, Kenneth F. *Earthworms: Their Ecology Relationships with Soils and Land Use.* Academic Press, 1985.
Roberts, Mervin F. *Pearlmakers: The Tidemarsh Guide to Clams, Oysters, Mussels, and Scallops.* Roberts M.F. Enterprises, 1984.
Russell-Hunter, W.D., ed. *Mollusca: Ecology,* Vol 6. Academic Press, 1983.

CHAPTER 28 Arthropods

Carolina Arthropods Manual. Carolina Biological Supply Co., 1982.
Crompton, John. *Ways of the Ant.* Nick Lyons, 1988.
Evans, Howard. *The Pleasures of Entomology: Portraits of Insects and the People Who Study Them.* Smithsonian, 1985.
Headstrom, Richard. *All About Lobsters, Crabs, Shrimps, and Their Relatives.* Dover, 1985.
Nardi, James B. *Close Encounters with Insects and Spiders.* Iowa State University Press, 1988.
Pringle, J.W.S. *Insect Flight,* 2nd ed. Carolina Biology Reader. Carolina Biological Supply Co., 1983.
Saintsing, David. *The World of Butterflies.* Gareth Stevens, 1987.
Taylor, Herb. *The Lobster: Its Life Cycle.* Sterling Publishing Co., Inc., 1984.
Whalley, Paul and Mary. *The Butterfly in the Garden.* Gareth Stevens, 1987.

CHAPTER 29 Echinoderms and Invertebrate Chordates

Lawrence, John. *A Functional Biology of Echinoderms.* Johns Hopkins, 1987.
Lechman, H. Eugene. *Chordate Development,* 2nd ed. Hunter Textbooks, 1983.
Sumich, James L. *An Introduction to the Biology of Marine Life,* 3rd ed. William C. Brown & Co., 1984.

CHAPTER 30 Comparing Invertebrates

Barnes, R.D. *Invertebrate Zoology,* 5th ed. Saunders College Publishing, 1987.
Barth, Robert H., and Broshears, Robert. *The Invertebrate World.* Saunders College Publishing, 1982.
Buchsbaum, Mildred, et al. *Living Invertebrates.* Blackwell Publications, 1987.
Lutz, Paul E. *Invertebrate Zoology.* Benjamin-Cummings, 1985.
Pecknik, Jan A. *Biology of the Invertebrates.* Wadsworth Publications, 1985.

CHAPTER 31 Fishes and Amphibians

Binder, Lionel. *Fish to Reptiles.* Gloucester, 1988.
Dickerson, Mary C. *The Frog Book.* Dover, 1969.
Gibbons, Whit. *Their Blood Runs Cold: Adventures with Reptiles and Amphibians.* University of Alabama Press, 1985.
Mattison, Christopher. *Frogs & Toads of the World.* Facts on File, 1987.
Minelli, Giuseppe. *Amphibians.* Facts on File, 1987.
Nicholls, Richard E. *The Book of Turtles.* Running Press, 1977.
Pyrom, Jay. *Frogs and Toads: A Complete Introduction.* TFH Publications, 1987.
Shoemaker, Hurst, and Herbert S. Zim. *Fishes.* Western Pub., 1987.

CHAPTER 32 Reptiles and Birds

Anderson, Robert. *Snakes.* TFH Publications, 1987.
Burton, Robert. *Bird Behavior.* Knopf, 1985.
Halliday, Dr. Tim, and Dr. Kraig Adler, eds. *Encyclopedia of Reptiles and Amphibians.* Facts on File, 1986.
Mehrtens, John M. *Living Snakes of the World in Color.* Sterling Publishing Co., 1987.
Minelli, Giuseppe. *Reptiles.* Facts on File, 1987.
Sparks, John, and Tony Soper. *Penguins.* Facts on File, 1987.
Terres, John K. *Songbirds in Your Garden.* Harper & Row, 1987.

CHAPTER 33 Mammals

Alden, Peter. *Peterson's First Guide to Mammals.* Houghton Mifflin, 1987.
Bender, Lionel. *Birds and Mammals.* Watts, 1988.
Ferry, G., ed. *The Understanding of Animals.* Basil Blackwell, 1984.
Gallant, Roy A. *The Rise of Mammals.* Franklin Watts, 1986.
Leatherwood, Stephen, and Randall R. Reeves. *The Sierra Club Handbook of Whales and Dolphins.* Sierra Club Books, 1983.
MacDonald, D., ed. *The Encyclopedia of Mammals.* Facts on File, 1984.
Minelli, Giuseppe. *Mammals.* Facts on File, 1988.
Ricciuti, Edward. *Older Than the Dinosaurs: The Origin and Rise of the Mammals.* Harper & Row Junior Books, 1980.

Savage, R.J.G. *Mammal Evolution: An Illustrated Guide.* Facts on File, 1986.
Voelker, William. *The Natural History of Living Mammals.* Plexus Publishing, Inc., 1986.
Walker, Ernest P. *Walker's Mammals of the World* (2 vols.), 4th ed. Johns Hopkins, 1983.

CHAPTER 34 Humans

Asimov, Isaac. *Beginnings.* Walker and Co., 1987.
Day, M.H. *Fossil History of Man,* rev. ed. Carolina Biology Reader. Carolina Biological Supply Co., 1984.
Fossey, Dian. *Gorillas in the Mist.* Houghton Mifflin, 1983.
Goodall, Jane. *In the Shadow of Man.* Houghton Mifflin, 1983.
Johanson, Donald C., and Maitland A. Edey. *Lucy: The Beginnings of Human Kind.* Warner Books, 1982.
Kornberg, W., ed. National Science Foundation Staff. *Human Evolution.* Avery Pub. Group, Inc., 1982.
Leakey, Richard E. *The Making of Mankind.* E.P. Dutton, 1981.

CHAPTER 35 Animal Behavior

Christie, David, et al. *Remarkable Animals.* Sterling Publishing Co. Published by Guinness Superlatives, 1987.
Evans, Peter. *Ourselves and Other Animals.* Pantheon Books, 1987.
Penny, Malcolm. *Animal Homes.* Bookwright, 1987.
———. *Animals and Their Young.* Bookwright, 1987.
———. *Animal Migration.* Bookwright, 1987.
Porter, Keith. *How Animals Behave.* Facts on File, 1987.
Seddon, Tony. *Animal Eyes.* Facts on File, 1988.
Tudge, Colin. *The Environment of Life.* Oxford University Press, 1987.

CHAPTER 36 Comparing Vertebrates

Colbert, Edwin H. *Evolution of the Vertebrates: A History of the Backboned Animals Through Time,* 3rd ed. Wiley, 1980.
Martin, Domm D. *Anatomy of Vertebrates.* William C. Brown, Inc., 1988.
Pough, F. Harvey, et al. *Vertebrate Life,* 3rd ed. Macmillan, 1989.
Rogers, E. *Looking at Vertebrates: A Practical Guide to Vertebrate Adaptations.* Wiley, 1986.
Romer, Alfred S., and Thomas S. Parsons. *The Vertebrate Body,* 6th ed. Saunders College Publishing, 1986.

CHAPTER 37 Nervous System

Bodanis, David. *The Body Book: A Fantastic Voyage to the World Within.* Little, 1984.
Kee, Leong S. *An Introduction to the Human Nervous System.* Ohio University Press, 1987.
Nilson, Lennart. *Behold Man.* Little, Brown, 1973.
Ralston, Diane D., and Henry J. Ralston III. *The Nerve Cell.* Carolina Biology Reader. Carolina Biological Supply Co., 1988.

Restak, Richard, M.D. *The Brain.* Bantam, 1985.
Stafford, Patricia. *Your Two Brains.* Atheneum, 1986.

CHAPTER 38 Skeletal, Muscular, and Integumentary Systems

Buller, A.J., and N.P. Buller. *Contractile Behavior of Mammalian Skeletal Muscle,* 2nd ed. Carolina Biology Reader. Carolina Biological Supply Co., 1980.
Harrington, W.F. *Muscle Contraction.* Carolina Biology Reader. Carolina Biological Supply Co., 1981.
Pritchard, J.J. *Bones,* 2nd ed. Carolina Biology Reader. Carolina Biological Supply Co., 1979.
Steele, D. Gentry, and Claude A. Bramblett. *The Anatomy and Biology of the Human Skeleton.* Texas A&M University Press, 1988.
Tiger, Steven. *Arthritis.* Messner, 1986.

CHAPTER 39 Nutrition and Digestion

Bolt, Robert J., et al. *The Digestive System.* Wiley, 1983.
Calloway, D.H., and K.O. Carpenter. *Nutrition and Health.* Saunders College Publishing, 1981.
Kirschmann, John D. *Nutrition Almanac,* rev. ed. McGraw-Hill, 1985.
Magee, D.F., and A.F. Dalley. *Digestion and the Structure and Function of the Gut.* S. Karger, 1986.
Sproule, Anna. *Bodywatch: Know Your Insides.* Facts on File, 1987.

CHAPTER 40 Respiratory System

Burkart, John, and Loretta Chiarenza. *Human Biology.* Avery Pub. Group, Inc., 1984.
Kittredge, Mary. *The Respiration System.* Chelsea House, 1989.
Nicholls, P. *The Biology of Oxygen.* Carolina Biology Reader. Carolina Biological Supply Co., 1982.
Sebel, Dr. Peter, et al. *The Human Body: Respiration: The Breath of Life.* Torstar Books, 1985.
Whipp, B.J., ed. *The Control of Breathing in Man.* University of Pennsylvania Press, 1987.

CHAPTER 41 Circulatory and Excretory Systems

Asimov, Isaac. *How Did We Find Out About Blood?* Walker and Co., 1986.
Guiness, Alma E., ed. *ABC's of the Human Body.* Random, 1987.
Moffat. *The Control of Water Balance by the Kidney,* 2nd ed. Carolina Biology Reader. Carolina Biological Supply Co., 1978.
Neil, Eric. *The Human Circulation.* Carolina Biology Reader. Carolina Biological Supply Co., 1979.
Nora, James J. *The Whole Heart Book.* Holt, Rinehart, and Winston, 1980.
Ross, Dennis W. *Blood.* Carolina Biology Reader. Carolina Biological Supply Co., 1988.
Tiger, Steven. *Heart Disease.* Messner, 1986.

CHAPTER 42 Endocrine System
Blake, Charles A. *The Pituitary Gland*. Carolina Biology Reader. Carolina Biological Supply Co., 1984.
Bloom, A. *Diabetes Explained*. Kluwer-Boston, Inc., 1982.
Crapo, Lawrence. *Hormones: The Messengers of Life*. W.H. Freeman, 1985.
Elting, Mary. *The Macmillan Book of the Human Body*. Aladdin (Macmillan), 1986.
Villee, Claude A. *Human Hormones*. Carolina Biology Reader. Carolina Biological Supply Co., 1987.

CHAPTER 43 Reproduction and Development
Parker, Gary. *Life Before Birth*. Master Books, 1987.
Smith, Anthony. *The Body*, rev. Viking, 1986.

CHAPTER 44 Human Diseases
Aaseng, Nathan. *The Disease Fighters: The Nobel Prize in Medicine*. Lerner, 1987.
Anderson, Madelyn Klein. *Environmental Diseases*. Watts, 1987.
Eagles, Douglas A. *Nutritional Diseases*. Watts, 1987.
Hughes, Barbara. *Drug-Related Diseases*. Watts, 1987.
Landau, Elaine. *Alzheimer's Disease*. Watts, 1987.
Metos, Thomas H. *Communicable Diseases*. Watts, 1987.

CHAPTER 45 Immune System
Arehart-Treichel, Joan. *Immunity: How Our Bodies Resist Disease*. Holiday House, 1976.
Desowitz, Robert S. *A Thorn in the Starfish: The Immune System and How It Works*. Norton, 1987.
Gershwin, M. Eric, et al. *Nutrition and Immunity*. Academic Press, 1985.
Greenberg, Sylvia. *Immunity and Survival: The Immune System at Work*. Human Sciences Press, 1989.
Nilsson, Lennart, with Jan Lindberg. *The Body Victorious*. Delacorte Press, 1987.
Vidic, Branislav, and Faustino R. Suarez. *Photographic Atlas of the Human Body*. Mosby, 1984.

CHAPTER 46 Drugs, Alcohol, and Tobacco
Alcohol: How it Affects Your Health. Do It Now Foundation, 1988.
Berger, Gilda. *Crack: The New Drug Epidemic*. Watts, 1987.
Chomet, Julian. *Cocaine and Crack*. Watts, 1987.
Cumming, G., and G. Bonsignore, eds. *Smoking and the Lung*. Plenum Publishers, 1985.
Jones, Helen C., and Paul W. Covinger. *The Marijuana Question and Science's Search for an Answer*. Dodd, Mead, 1985.
Stepney, Rob. *Alcohol*. Watts, 1987.

CHAPTER 47 The Biosphere
Bowen, E. *Grasslands and Tundra*. Silver, 1985.
Bramwell, Martyn. *Deserts*. Watts, 1988.
Durrell, Lee. *State of the Ark*. Doubleday, 1986.
Healy, Timothy, and Paul Houle. *Energy and Society*. Boyd and Fraser, 1983.
Perry, Donald. *Life Above the Forest Floor: A Biologist Explores a Strange and Hidden Treetop World*. Simon and Schuster, 1986.
Sagan, Dorion, and Lynn Margulis. *Biospheres: From Earth to Space*. Enslow Pubs., 1989.

CHAPTER 48 Populations and Communities
Begon, Michael, et al. *Ecology: Individuals, Populations and Communities*. Sinauer Associates, 1986.
Clapham, W.B., Jr. *Natural Ecosystems*, 2nd ed. Macmillan, 1983.
Moore, Peter D., ed. *The Encyclopedia of Animal Ecology*. Facts on File, 1987.
Smith, Howard E., Jr. *Small Worlds: Communities of Living Things*. Macmillan, 1987.
Van Lawick, Hugo. *Among Predators and Prey*. Sierra Club Books, 1986.

CHAPTER 49 People and the Biosphere
Adams, Douglas, and Mark Carwardine. *Last Chance to See*. Ballantine Books, 1992.
Albright, Horace M., Russell E. Dickerson, and William Penn Mott, Jr. *National Park Service: The Story Behind the Scenery*. KC Publications, 1987.
Fine, John Christopher. *Oceans in Peril*. Atheneum, 1987.
Lo Pinto, Richard W. *Pollution*. Carolina Biology Reader. Carolina Biological Supply Co., 1987.
Miller, Christina G., and Louise A. Berry. *Acid Rain*. Messner, 1987.
Milne, Louis, and Margery Milne. *A Shovelful of Earth*. Holt, 1987.
Pitt, David C., ed. *The Future of the Environment*. Chapman & Hall, 1988.
Pringle, Laurence. *Restoring Our Earth*. Enslow Pubs., 1987.
Stanley, Steven M. *Extinction*. Scientific American, 1987.

APPENDIX A: The Metric System

The Metric System of measurement is used by scientists throughout the world. It is based on units of ten. Each unit is ten times larger or ten times smaller than the next unit. The most commonly used units of the metric system are given below. After you have finished reading about the metric system, try to put it to use. How tall are you in meters? What is your mass? What is your normal body temperature in degrees Celsius?

COMMONLY USED METRIC UNITS

Length The distance from one point to another

meter(m) A meter is slightly longer than a yard.
　　　　　1 meter = 1000 millimeters (mm)
　　　　　1 meter = 100 centimeters (cm)
　　　　　1000 meters = 1 kilometer (km)

Volume The amount of space an object takes up

liter (L) A liter is slightly more than a quart.
　　　　　1 liter = 1000 milliliters (mL)

Mass The amount of matter in an object

gram (g) A gram has a mass equal to about one paper clip.
　　　　　1000 grams = 1 kilogram (kg)

Temperature The measure of hotness or coldness

degrees　　0°C = freezing point of water
Celsius (°C)　100°C = boiling point of water

METRIC — ENGLISH EQUIVALENTS

2.54 centimeters (cm) = 1 inch (in.)
1 meter (m) = 39.37 inches (in.)
1 kilometer (km) = 0.62 miles (mi)
1 liter (L) = 1.06 quarts (qt)
250 milliliters (mL) = 1 cup (c)
1 kilogram (kg) = 2.2 pounds (lb)
28.3 grams (g) = 1 ounce (oz)
°C = 5/9 x (°F−32)

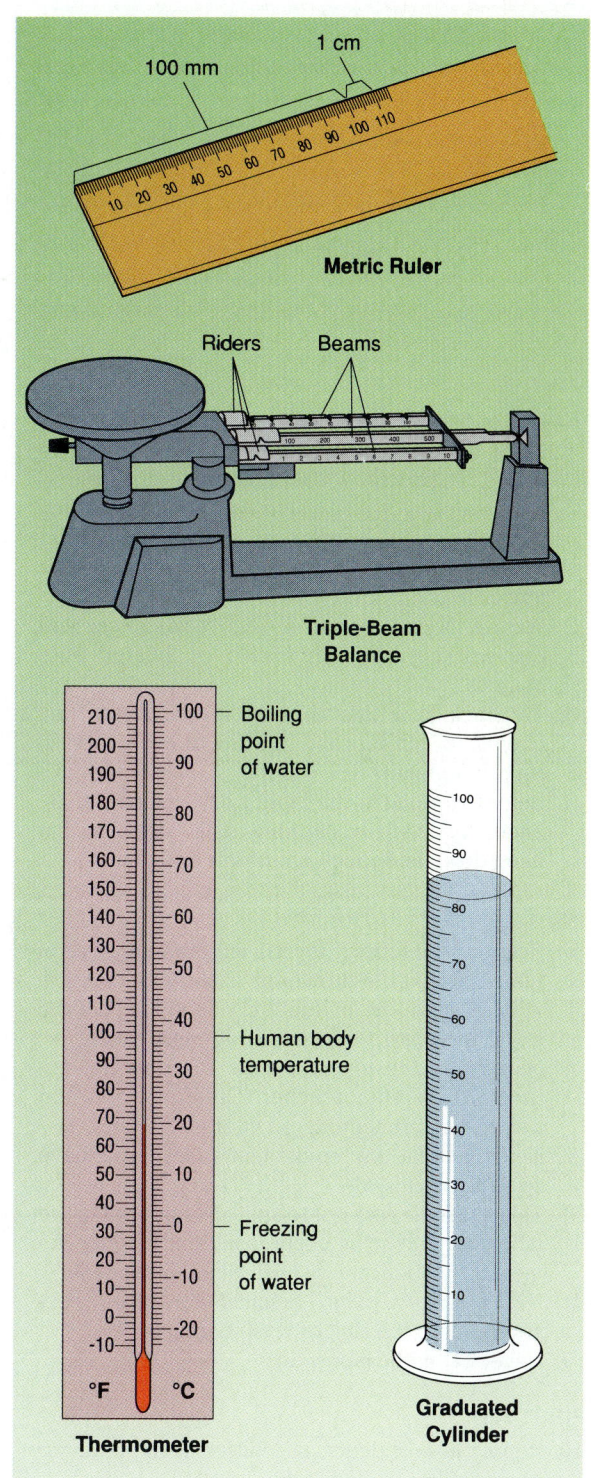

APPENDIX B Science Safety Rules

One of the first things a scientist learns is that working in the laboratory can be an exciting experience. But the laboratory can also be quite dangerous if proper safety rules are not followed at all times. To prepare yourself for a safe year in the laboratory, read over the following safety rules. Then read them a second time. Make sure you understand each rule. If you do not, ask your teacher to explain any rules you don't understand.

DRESS CODE

1. Many materials in the laboratory can cause eye injury. To protect yourself from possible injury, wear safety goggles whenever you are working with chemicals, burners, or any substance that might get into your eyes. Never wear contact lenses in the laboratory.
2. Wear a laboratory apron or coat whenever you are working with chemicals or heated substances.
3. Tie back long hair to keep it away from any chemicals, burners and candles, or other laboratory equipment.
4. Before working in the laboratory, remove or tie back any article of clothing or jewelry that can hang down and touch chemicals and flames.

GENERAL SAFETY RULES

5. Read all directions for an experiment several times. Follow the directions exactly as they are written. If you are in doubt about any part of the experiment, ask your teacher for assistance.
6. Never perform investigations that are not authorized by your teacher. Obtain permission before "experimenting" on your own.
7. Never handle any equipment unless you have specific permission.
8. Take extreme care not to spill any material in the laboratory. If spills occur, ask your teacher immediately about the proper cleanup procedure. Never simply pour chemicals or other substances into the sink or trash container.
9. Never eat in the laboratory.

FIRST AID

10. Report all accidents, no matter how minor, to your teacher immediately.
11. Learn what to do in case of specific accidents, such as getting acid in your eyes or on your skin. (Rinse acids on your body with lots of water.)
12. Be aware of the location of the first-aid kit. Your teacher should administer any required first aid due to injury. Or your teacher may send you to the school nurse or call a physician.
13. Know where and how to report an accident or fire. Find out the location of the fire extinguisher, phone, and fire alarm. Keep a list of important phone numbers such as the fire department and school nurse near the phone. Report any fires to your teacher at once.

HEATING AND FIRE SAFETY

14. Again, never use a heat source such as a candle or burner without wearing safety goggles.
15. Never heat a chemical you are not instructed to heat. A chemical that is harmless when cool can be dangerous when heated.
16. Maintain a clean work area and keep all materials away from flames.
17. Never reach across a flame.
18. Make sure you know how to light a Bunsen burner. (Your teacher will demonstrate the proper procedure for lighting a burner.) If the flame leaps out of a burner toward you, turn the gas off immediately. Do not touch the burner. It may be hot. And never leave a lighted burner unattended!
19. When you are heating a test tube or bottle, point it away from yourself and others. Chemicals can splash or boil out of a heated test tube.
20. Never heat a liquid in a closed container. The expanding gases produced may blow the container apart, causing it to injure yourself or others.
21. Never pick up a container that has been heated without first holding the back of your hand near it. If you can feel the heat on the back of your hand, the container may be too hot to handle. Use a clamp or tongs when handling hot containers.

USING CHEMICALS SAFELY

22. Never mix chemicals for the "fun of it." You might produce a dangerous, possibly explosive substance.
23. Never touch, taste, or smell a chemical that you do not know for a fact is harmless. Many chemicals are poisonous. If you are instructed to note the fumes in an experiment, gently wave your hand over the opening of a container and direct the fumes toward your nose. Do not inhale the fumes directly from the container.

24. Use only those chemicals needed in the investigation. Keep all lids closed when a chemical is not being used. Notify your teacher whenever chemicals are spilled.
25. Dispose of all chemicals as instructed by your teacher. To avoid contamination, never return chemicals to their original containers.
26. Be extra careful when working with acids or bases. Pour such chemicals over the sink, not over your workbench.
27. When diluting an acid, pour the acid into water. Never pour water into the acid.
28. If any acids get on your skin or clothing, rinse them with water. Immediately notify your teacher of any acid spill.

USING GLASSWARE SAFELY

29. Never force glass tubing into a rubber stopper. A turning motion and lubricant will be helpful when inserting glass tubing into rubber stoppers or rubber tubing. Your teacher will demonstrate the proper way to insert glass tubing.
30. Never heat glassware that is not thoroughly dry. Use a wire screen to protect glassware from any flame.
31. Keep in mind that hot glassware will not appear hot. Never pick up glassware without first checking to see if it is hot.
32. If you are instructed to cut glass tubing, fire-polish the ends immediately to remove sharp edges.
33. Never use broken or chipped glassware. If glassware breaks, notify your teacher and dispose of the glassware in the proper trash container.
34. Never eat or drink from laboratory glassware. Thoroughly clean glassware before putting it away.

USING SHARP INSTRUMENTS

35. Handle scalpels or razor blades with extreme care. Never cut material toward you; cut away from you.
36. Notify your teacher immediately if you cut yourself when in the laboratory.

ANIMAL SAFETY

37. No experiments that will cause pain, discomfort, or harm to mammals, birds, reptiles, fishes, and amphibians should be done in the classroom or at home.
38. Animals should be handled only if necessary. If an animal is excited or frightened, pregnant, feeding, or with its young, special handling is required.
39. Your teacher will instruct you as to how to handle each animal species that may be brought into the classroom.
40. Clean your hands thoroughly after handling animals or the cage containing animals.

END-OF-EXPERIMENT RULES

41. When an experiment is completed, clean up your work area and return all equipment to its proper place.
42. Wash your hands before and after every experiment.
43. Turn off all burners before leaving the laboratory. Check that the gas line leading to the burner is off as well.

APPENDIX C: Care and Use of the Microscope

THE COMPOUND MICROSCOPE

One of the most essential tools in the study of biology is the microscope. With the help of different types of microscopes, biologists have developed detailed concepts of cell structure and function. The type of microscope used in most biology classes is the compound microscope. It contains a combination of lenses and can magnify objects normally unseen with the unaided eye.

The eyepiece lens is located in the top portion of the microscope. This lens usually has a magnification of 10 X. A compound microscope usually has two other interchangeable lenses. These lenses, called objective lenses, are at the bottom of the body tube on the revolving nosepiece. By revolving the nosepiece, either of the objectives can be brought into direct line with the body of the tube.

The shorter objective is of low power in its magnification, usually 10 X. The longer one is of high power, usually 40 X or 43 X. The magnification is always marked on the objective. To determine the total magnification of a microscope, multiply the magnifying power of the eyepiece by the magnifying power of the objective being used. For example, the eyepiece magnifying power, 10 X, multiplied by the low-power objective, 10 X, equals 100 X. The total magnification is 100 X.

A microscope also produces clear contrasts to enable the viewer to distinguish between objects that lie very close together. Under a microscope the detail of objects is very sharp. The ability of a microscope to produce contrast and detail is called resolution, or resolving power. Although microscopes can have the same magnifying power, they can differ in resolving power.

Learning the name, function, and location of each of the microscope's parts is necessary for proper use. Use the following procedures when working with the microscope.

1. Remove the microscope from its storage area by placing one hand beneath the base and grasping the arm of the microscope with the other hand.
2. Gently place the microscope on the lab table with the arm facing you. The microscope's base should be resting evenly on the table, approximately 10 centimeters from the table's edge.
3. Raise the body tube by turning the coarse adjustment knob until the objective lens is about 2 centimeters above the opening of the stage.
4. Revolve the nosepiece so that the low-power objective (10 X) is directly in line with the body tube. A click indicates that the lens is in line with the opening of the stage.
5. Look through the eyepiece and switch on the lamp or adjust the mirror so that a circle of light can be seen. This is the field of view. Moving the lever of the diaphragm permits a greater or smaller amount of light to come through the opening of the stage.

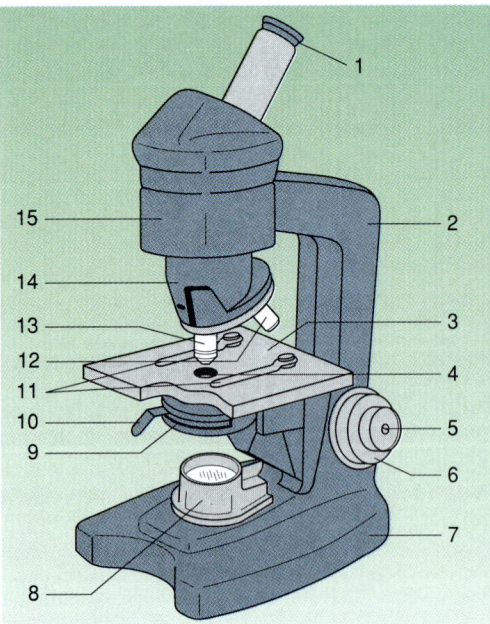

Microscope Parts and Their Function

1. **Eyepiece** Contains a magnifying lens
2. **Arm** Supports the body tube
3. **Stage** Supports the slide being observed
4. **Opening of the stage** Permits light to travel up to the eyepiece
5. **Fine adjustment** Moves the body tube slightly to sharpen the focus
6. **Coarse adjustment** Moves the body tube up and down for focusing
7. **Base** Supports the microscope
8. **Illuminator** Produces light or reflects light up through the body tube
9. **Diaphragm** Regulates the amount of light entering the body tube
10. **Diaphragm lever** Opens and closes the diaphragm
11. **Stage clips** Hold the slide in position
12. **Low-power objective** Provides a magnification of 10 X and is the shorter of the objectives
13. **High-power objective** Provides a magnification of 43 X and is the longer of the objectives
14. **Revolving nosepiece** Contains the low- and high-power objectives and can be rotated to change magnification
15. **Body tube** Maintains a proper distance between the eyepiece and the objective lenses

6. Place a prepared slide on the stage. Place the specimen over the center of the opening of the stage. Fasten the stage clips to hold the slide in position.
7. Look at the microscope from the side. Carefully turn the coarse adjustment knob to lower the body tube until the low-power objective almost touches the slide or until the body tube can no longer be moved. Do not allow the objective to touch the slide.
8. Look through the eyepiece and observe the specimen. If the field of view is out of focus, use the coarse adjustment knob to raise the body tube while looking through the eyepiece. When the specimen comes into view, use the fine adjustment knob to focus the specimen. Be sure to keep both eyes open when viewing a specimen. This helps prevent eyestrain.
9. Adjust the lever of the diaphragm to allow the right amount of light to enter.
10. To view the specimen under high power (43 X), revolve the nosepiece until the high-power objective is in line with the body tube and clicks into place.
11. Look through the eyepiece and use the fine adjustment knob to bring the specimen into focus.
12. After every use remove the slide. Clean the stage of the microscope and the lenses with lens paper. Do not use other types of paper to clean the lenses, as they may scratch the lenses.

PREPARING A WET-MOUNT SLIDE

1. Obtain a clean microscope slide and a coverslip. A coverslip is very thin, permitting the objective lens to be lowered very close to the specimen.
2. Place the specimen in the middle of the microscope slide. The specimen must be thin enough for light to pass through it.
3. Using a medicine dropper, place a drop of water on the specimen.
4. Lower one edge of the coverslip so that it touches the side of the drop of water at a 45° angle. The water will spread evenly along the edge of the coverslip. Using a dissecting needle or probe, slowly lower the coverslip over the specimen and water. Try not to trap any air bubbles under the coverslip. Air bubbles interfere with the view of the specimen. If air bubbles are present, gently tap the surface of the coverslip over the air bubble with a pencil eraser.
5. Remove any excess water at the edge of the coverslip with a paper towel. If the specimen begins to dry out, add a drop of water at the edge of the coverslip.

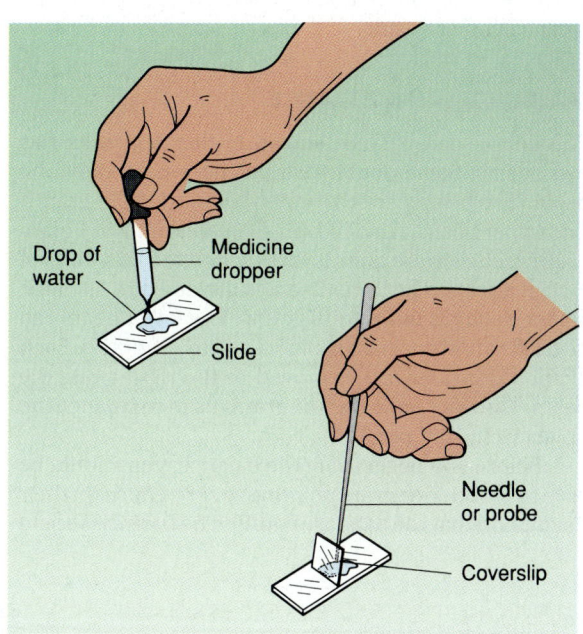

STAINING TECHNIQUES

1. Obtain a clean microscope slide and coverslip.
2. Place the specimen in the middle of the microscope slide.
3. Using a medicine dropper, place a drop of water on the specimen.
4. Place one edge of the coverslip so that it touches the side of the drop of water at a 45° angle. After the water spreads along the edge of the coverslip, use a dissecting needle or probe to lower the coverslip over the specimen.
5. Add a drop of stain at the edge of the coverslip. Using forceps, touch a small piece of lens paper or paper towel to the opposite edge of the coverslip. The paper causes the stain to be drawn under the coverslip and stain the cells. Some common stains are methylene blue, iodine, fuchsin, and Wright's.

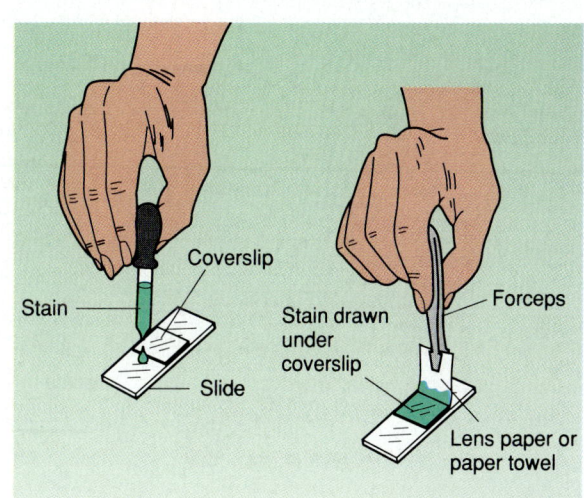

APPENDIX D: The Laboratory Balance

The laboratory balance is an important tool in scientific investigations. You can use the balance to determine the mass of materials that you study or experiment with in the laboratory.

Different kinds of balances are used in the laboratory. One kind of balance is the double-pan balance. Another kind of balance is the triple-beam balance. The balance that you may use in your science class is probably similar to one of the balances illustrated in this Appendix. To use the balance properly, you should learn the name, function, and location of each part of the balance you are using.

THE DOUBLE-PAN BALANCE

The double-pan balance shown in this Appendix has two beams. Some double-pan balances have only one beam. The beams are calibrated, or marked, in grams. The upper beam is divided into 10 major units of 1 gram each. Each of these units is further divided into units of 1/10 of a gram. The lower beam is divided into 20 units, and each unit is equal to 10 grams. The lower beam can be used to find the mass of objects up to 200 grams. Each beam has a rider that is moved to the right along the beam. The rider indicates the grams used to balance the object in the left pan.

Before you begin using the balance, you should be sure that both riders are pointing to zero grams on their beams and that the pans are empty. The balance should be on a flat, level surface. The pointer should be at the zero point. If your pointer does not read zero, slowly turn the adjustment knob until it does.

The following procedure can be used to find the mass of an object with a double-pan balance:

1. Place the object whose mass is to be determined on the left pan.
2. Move the rider on the lower beam to the 10-gram notch.
3. If the pointer moves to the right of the zero point on the scale, the object has a mass less than 10 grams. Return the rider on the lower beam to zero. Slowly move the rider on the upper beam until the pointer is at zero. The reading on the beam is the mass of the object.
4. If the pointer did not move to the right of the zero, move the rider on the lower beam notch by notch until it does. Move the rider back one notch. Then move the rider on the upper beam until the pointer is at zero. The sum of the readings on both beams is the mass of the object.
5. If the two riders are moved completely to the right side of the beams and the pointer remains to the left of the zero point, the object has a mass greater than the total mass that the balance can measure.

The total mass that most double-pan balances can measure is 210 grams. If an object has a mass greater than 210 grams, return the riders to zero.

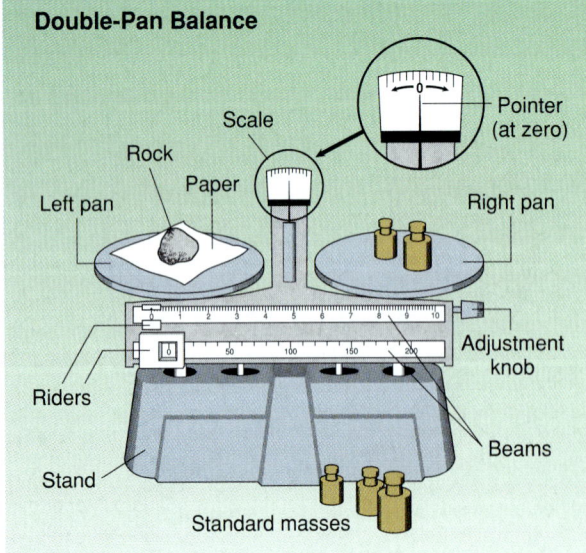

Double-Pan Balance

Scale Graduated instrument along which the pointer moves to show if the balance is balanced

Pointer Marker that indicates on the scale if the balance is balanced

Zero point Center line of the scale to which the pointer moves when the balance is balanced

Adjustment knob Knob used to balance the empty balance

Left pan Platform on which an object whose mass is to be determined is placed

Right pan Platform on which standard masses are placed

Beams Scales calibrated in grams

Riders Movable markers that indicate the number of grams needed to balance an object

Stand Support for the balance

The following procedure can be used to find the mass of an object greater than 210 grams.

1. Place the standard masses on the right pan one at a time, starting with the largest, until the pointer remains to the right of the zero point.
2. Remove one of the large standard masses and replace it with a smaller one. Continue replacing the standard masses with smaller ones until the pointer remains to the left of the zero point. When the pointer remains to the left of the zero point, the mass of the object on the left pan is greater than the total mass of the standard masses on the right pan.
3. Move the rider on the lower beam and then the rider on the upper beam until the pointer stops at the zero point on the scale. The mass of the object is equal to the sum of the readings on the beams plus the mass of the standard masses.

THE TRIPLE-BEAM BALANCE

The triple-beam balance is a single-pan balance with three beams calibrated in grams. The back, or 100-gram, beam is divided into 10 units of 10 grams each. The middle, or 500-gram, beam is divided into 5 units of 100 grams each. The front, or 10-gram, beam is divided into 10 major units of 1 gram each. Each of these units is further divided into units of 1/10 of a gram.

The following procedure can be used to find the mass of an object with a triple-beam balance:

1. Place the object on the pan.
2. Move the rider on the middle beam notch by notch until the horizontal pointer drops below zero. Move the rider back one notch.
3. Move the rider on the front beam notch by notch until the pointer again drops below zero. Move the rider back one notch.
4. Slowly slide the rider along the back beam until the pointer stops at the zero point.
5. The mass of the object is equal to the sum of the readings on the three beams.

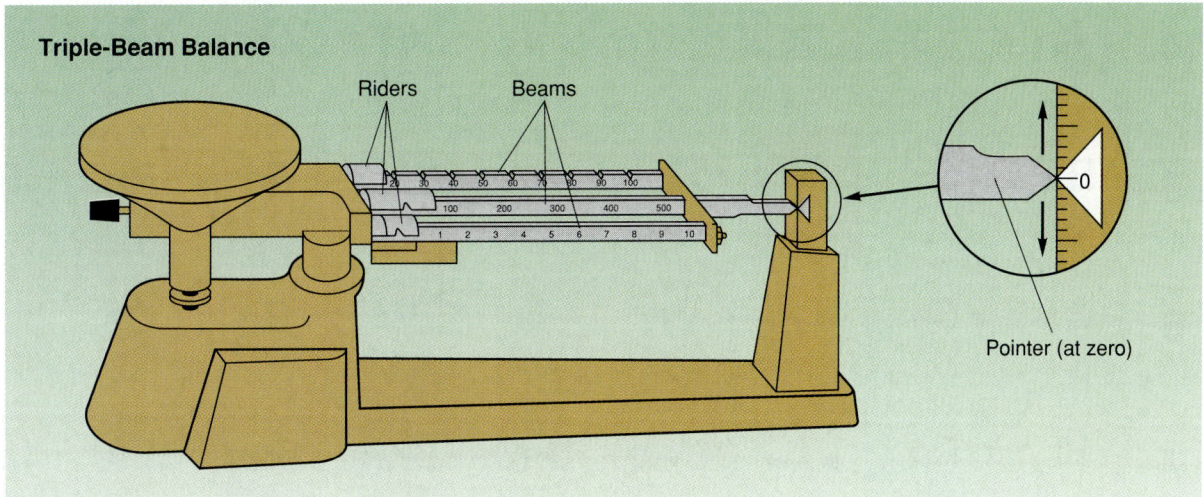

Triple-Beam Balance
Riders   Beams
Pointer (at zero)

APPENDIX E: The Periodic Table of the Elements

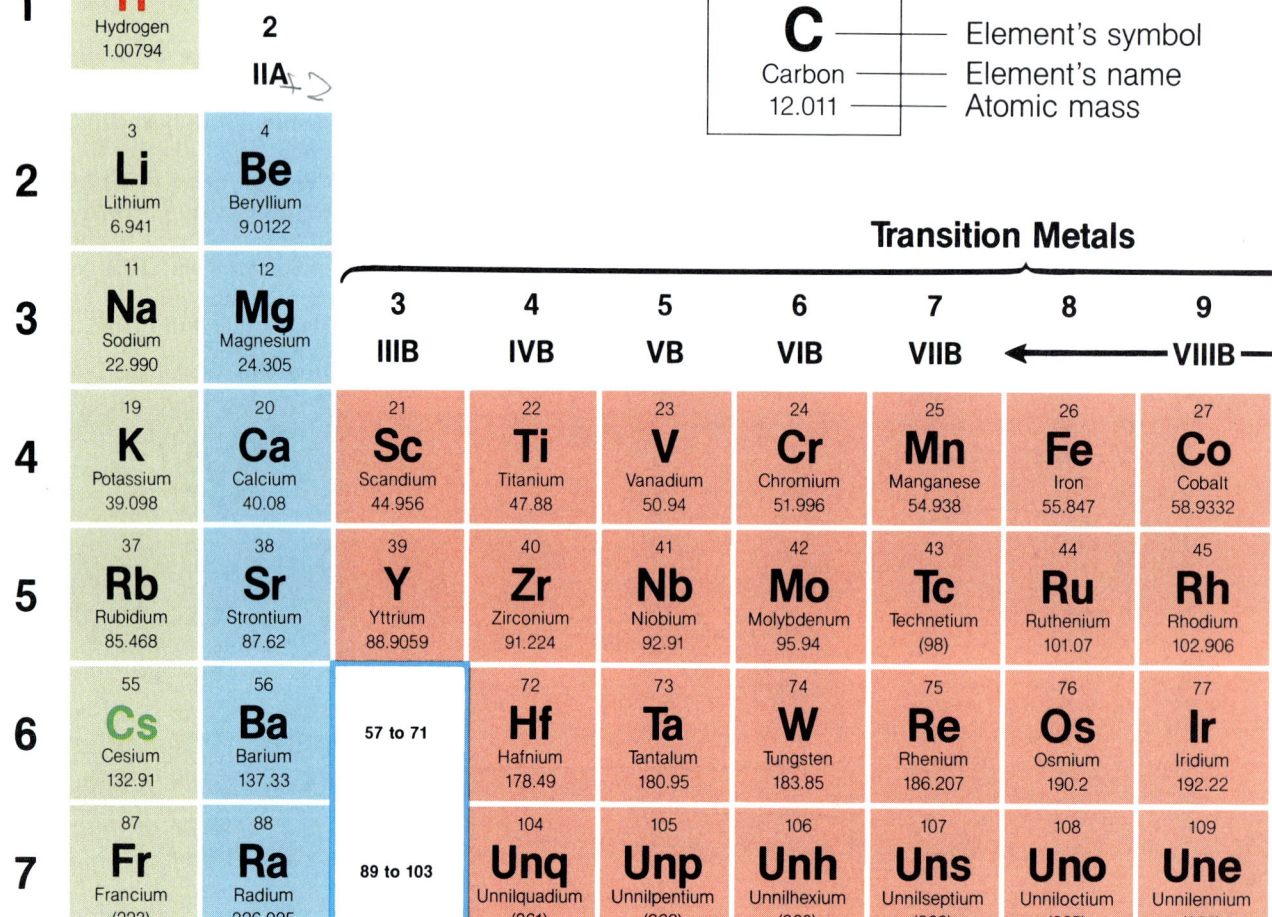

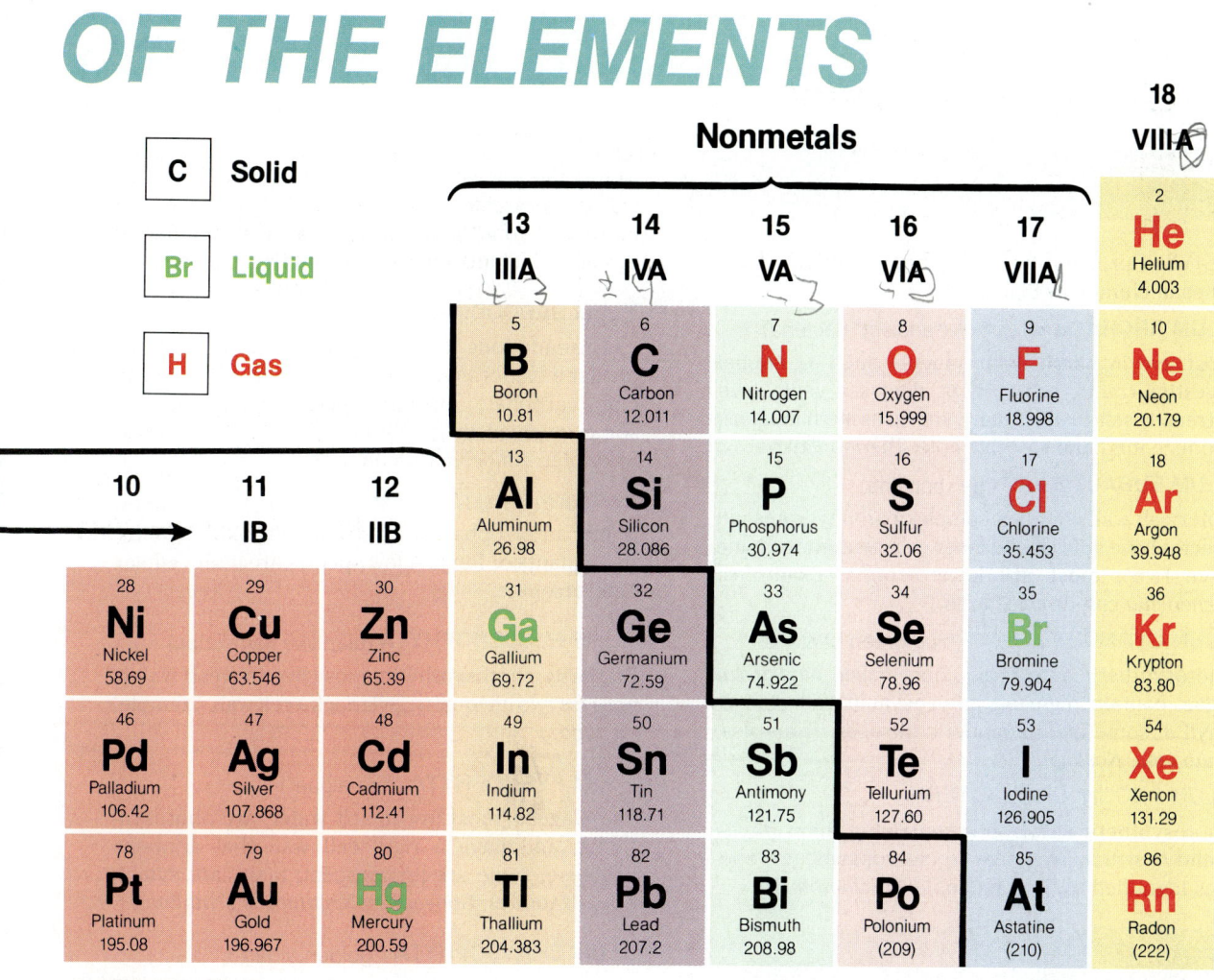

APPENDIX F: The Five-Kingdom Classification System

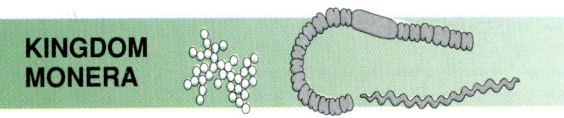

KINGDOM MONERA

Single-celled prokaryotic organisms; sometimes form colonies of clumps or filaments.

PHYLUM ARCHAEBACTERIA ("ancient" bacteria)

Most live in harsh environments such as animal digestive tracts, hot springs, deep-sea vents, and extremely salty water; many, known as methanogens, produce methane gas. Example: *Thermoplasma*.

PHYLUM EUBACTERIA ("true" bacteria)

Outer cell wall contains complex carbohydrates; all species have at least one inner cell membrane (some have two); most are heterotrophs. Examples: *Escherichia coli, Streptococcus*.

PHYLUM CYANOBACTERIA (blue-green bacteria)

Photosynthetic autotrophs, once called blue-green algae; contain pigments phycocyanin and chlorophyll *a*; some fix atmospheric nitrogen. Examples: *Anabaena, Nostoc*.

PHYLUM PROCHLOROBACTERIA

Photosynthetic autotrophs containing chlorophylls *a* and *b*; strikingly similar to chloroplasts; few species identified to date. Example: *Prochloron*.

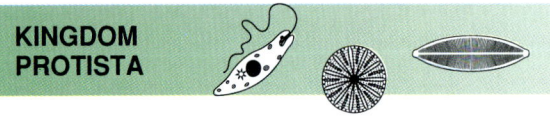

KINGDOM PROTISTA

Eukaryotic; usually unicellular; some multicellular or colonial; heterotrophic or autotrophic organisms.

ANIMALLIKE PROTISTS
Unicellular; heterotrophic; usually motile; also known as protozoa.

PHYLUM CILIOPHORA (ciliates)

All have cilia at some point in development; almost all use cilia to move; characterized by two types of nuclei: macronuclei and micronuclei; most have a sexual process known as conjugation. Examples: *Paramecium, Didinium, Stentor*.

PHYLUM ZOOMASTIGINA (animallike flagellates)

Possess one or more flagella (some have thousands); some are internal symbionts of wood-eating animals. Examples: trypanosomes, *Trichonympha*.

PHYLUM SPOROZOA

Nonmotile parasites; produce small infective cells called spores; life cycles usually complex, involving more than one host species; cause a number of diseases, including malaria. Example: *Plasmodium*.

PHYLUM SARCODINA

Use pseudopods for feeding and movement; some produce elaborate shells that contain silica or calcium carbonate; most free-living; a few parasitic; some involved in formation of sedimentary rock. Examples: *Amoeba*, foraminifers.

PLANTLIKE PROTISTS
Mostly unicellular photosynthetic autotrophs that have characteristics similar to green plants or fungi. A few species are multicellular or heterotrophic.

PHYLUM EUGLENOPHYTA (plantlike flagellates)

Primarily photosynthetic; most live in fresh water; possess two unequal flagella; lack a cell wall. Example: *Euglena*.

PHYLUM PYRROPHYTA ("fire algae")

Two flagella; most live in salt water, are photosynthetic, and have a rigid cell wall that contains cellulose; some are luminescent; many are symbiotic. Examples: *Gonyaulux, Noctilucans scintillans*.

PHYLUM CHRYSOPHYTA ("golden algae")

Photosynthetic; aquatic; mostly unicellular; contain yellow-brown pigments; most are diatoms, which build a two-part cell covering that contains silica. Example: *Thallasiosira*.

PHYLUM ACRASIOMYCOTA (cellular slime molds)

Spores develop into independent free-living amebalike cells that may come together to form a multicellular structure; this structure, which behaves much like a single organism, forms a fruiting body that produces spores. Example: *Dictyostelium*.

PHYLUM MYXOMYCOTA (acellular slime molds)

Spores develop into haploid cells that can switch between flagellated and amebalike forms; these haploid cells fuse to form a zygote that grows into a plasmodium, which ultimately forms spore-producing fruiting bodies. Example: *Physarum*.

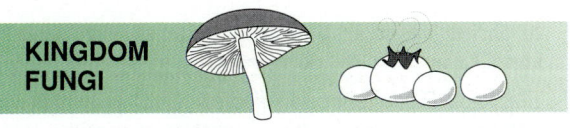

KINGDOM FUNGI

Eukaryotic; unicellular or multicellular; cell walls typically contain chitin; mostly decomposers; some parasites; some commensal or mutualistic symbionts; asexual reproduction by spore formation, budding, or fragmentation; sexual reproduction involving mating types; classified according to type of fruiting body and style of spore formation; heterotrophic.

PHYLUM OOMYCOTA (water molds)

Unicellular or multicellular; mostly aquatic; cell walls contain cellulose or a polysaccharide similar to cellulose; form zoospores asexually and eggs and sperms sexually. Example: *Saprolegnia* (freshwater mold).

PHYLUM ZYGOMYCOTA (conjugation fungi)

Cell walls of chitin; hyphae lack cross walls; sexual reproduction by conjugation produces diploid zygospores; asexual reproduction produces haploid spores; most parasites; some decomposers. Example: *Rhizopus stolonifer* (black bread mold).

PHYLUM ASCOMYCOTA (sac fungi)

Cell walls of chitin; hyphae have perforated cross walls; most multicellular; yeasts unicellular; sexual reproduction produces ascospores; asexual reproduction by spore formation or budding; some cause plant diseases such as chestnut blight and Dutch elm disease. Examples: *Neurospora* (red bread mold), baker's yeast, morels, truffles.

PHYLUM BASIDIOMYCOTA (club fungi)

Cell walls of chitin; hyphae have cross walls; sexual reproduction involves basidiospores, which are borne on club-shaped basidia; asexual reproduction by spore formation. Examples: mushrooms, puffballs, shelf fungi, rusts.

PHYLUM DEUTEROMYCOTA (imperfect fungi)

Cell walls of chitin; sexual reproduction never observed; members resemble ascomycetes, basidiomycetes, or zygomycetes; most thought to be ascomycetes that have lost the ability to form asci. Examples: *Penicillium,* athlete's foot fungus.

KINGDOM PLANTAE

Eukaryotic; overwhelmingly multicellular and nonmotile; photosynthetic autotrophs; possess chlorophylls *a* and *b* and other pigments in organelles called chloroplasts; cell walls contain cellulose; food stored as starch; reproduce sexually; alternate haploid (gametophyte) and diploid (sporophyte) generations; botanists typically use the term division rather than phylum.

PHYLUM CHLOROPHYTA (green algae)

Live in fresh water and salt water; unicellular or multicellular; chlorophylls and accessory pigments similar to those in vascular plants; food stored as starch. Examples: *Ulva* (sea lettuce), *Chlamydomonas, Spirogyra, Acetabularia.*

PHYLUM PHAEOPHYTA (brown algae)

Live almost entirely in salt water; multicellular; contain brown pigment fucoxanthin; food stored as oils and carbohydrates. Examples: *Fucus* (rockweed), kelp, *Sargassum.*

PHYLUM RHODOPHYTA (red algae)

Live almost entirely in salt water; multicellular; contain red pigment phycoerythrin; food stored as carbohydrates. Examples: *Chondrus* (Irish moss), coralline algae.

PHYLUM BRYOPHYTA (bryophytes)

Generally small; multicellular green plants; live on land in moist habitats; lack vascular tissue; lack true roots, leaves, and stems; gametophyte dominant; water required for reproduction. Examples: mosses, liverwort, hornwort.

PHYLUM TRACHEOPHYTA (vascular plants)

Contain xylem and phloem; true roots; sporophyte dominant; in primitive forms, gametophyte independent of sporophyte; in advanced forms, gametophyte dependent on sporophyte.

Subphylum Psilopsida (whisk ferns)   Primitive vascular plants; no differentiation between root and shoot; produce only one kind of spore; motile sperm must swim in water.

Subphylum Lycopsida (lycopods)   Primitive vascular plants; usually small; sporophyte dominant; possess roots, stems, and leaves; water required for reproduction. Examples: club moss, quillwort.

Subphylum Sphenopsida (horsetails)   Primitive vascular plants; stem comprises most of mature plant and contains silica; produce only one kind of spore; motile sperm must swim in water. Only one living genus: *Equisetum.*

Subphylum Pteropsida (ferns)   Vascular plants well-adapted to live in predominantly damp or seasonally wet environments; sporophyte dominant and well-adapted to terrestrial life; gametophyte inconspicuous; reproduction still dependent on water for free-swimming gametes. Examples: cinnamon fern, Boston fern, tree fern, maidenhair fern.

Subphylum Spermopsida (seed plants)   Vascular; use seeds for reproduction: nonmotile; cell walls contain cellulose; almost all photosynthetic; many can live in very wet places but others adapted to live and reproduce in dry environments; do not require water for reproduction.

Gymnosperms Four classes of seed plants—Cycadae, Ginkgoae, Coniferae, and Gnetalae—were once considered to be orders within the class Gymnospermae. Today, botanists realize that members of these groups differ from one another as much as they differ from angiosperms and are therefore not closely related enough to be placed in the same class.

Although the term gymnosperm no longer refers to a taxonomic category, it is still used casually. Gymnosperms are characterized by seeds that develop exposed or "naked" on fertile leaves—there is no ovary wall (fruit) surrounding the seeds. Gymnosperms lack flowers.

Class Cycadae (cycads) Evergreen, slow-growing, tropical and subtropical shrubs; many resemble small palm trees; palmlike or fernlike compound leaves; possess symbiotic cyanobacteria in special roots; sexes are separate—individuals have either male pollen-producing cones or female seed-producing cones.

Class Gingkoae Deciduous trees with fan-shaped leaves; sexes separate; outer skin of ovule develops into a fleshy, fruitlike covering. Only one living species: *Gingko biloba* (gingko).

Class Gnetalae Few species; mostly desert-living; functional xylem cells are alive; sexes separate. Examples: *Welwitschia,* joint fir (*Ephedra*).

Class Coniferae (conifers) Cones predominantly wind-pollinated; most are evergreen; most temperate and subarctic shrubs and trees; many have needlelike leaves; in most species, sexes are not separate. Examples: pine, spruce, cedar, cypress, yew, fir, larch, sequoia.

Angiosperms Unlike the term gymnosperm, the term angiosperm continues to refer to a taxonomic group. Nearly all familiar trees, shrubs, and garden plants are angiosperms.

Class Angiospermae Members of this class are commonly called flowering plants. Seeds develop enclosed within ovaries; fertile leaves modified into flowers; flowers pollinated by wind or by animals, including insects, birds, and bats; occur in many different forms; found in most land and freshwater habitats; a few species found in shallow saltwater and estuarine areas.

Subclass Monocotyledonae (monocots) Embryo with a single cotyledon; leaves with predominantly parallel veination; flower parts in threes or multiples of three; vascular bundles scattered throughout stem. Examples: lily, corn, grasses, iris, palms, tulip.

Subclass Dicotyledonae (dicots) Embryo with two cotyledons; leaves with veination in netlike patterns; flower parts in fours or fives (or multiples thereof); vascular bundles arranged in rings in stem. Examples: rose, maple, oak, daisy, apple.

KINGDOM ANIMALIA

Multicellular; eukaryotic; typical heterotrophs that ingest their food; lack cell walls; approximately 35 phyla; in most phyla cells are organized into tissues that make up organs; most reproduce sexually; motile sperm have flagella; nonmotile egg is much larger than sperm; development involves formation of a hollow ball of cells called a blastula.

Subkingdom Parazoa

Animals that possess neither tissues nor organs; most asymmetrical.

PHYLUM PORIFERA (sponges)

Aquatic; lack true tissues and organs; motile larvae and sessile adults; filter feeders; internal skeleton made up of spongin and/or spicules of calcium carbonate or silica. Examples: Venus' flower basket, bath sponge, tube sponge.

Subkingdom Metazoa

Animals with definite symmetry; definite tissues; most possess organs.

PHYLUM CNIDARIA

Previously known as coelenterates; aquatic; mostly carnivorous; two layers of true tissues; radial symmetry; tentacles bear stinging nematocysts; many alternate between polyp and medusa body forms; gastrovascular cavity.

Class Hydrozoa Polyp form dominant; colonial or solitary; life cycle typically includes a medusa generation that reproduces sexually and a polyp generation that reproduces asexually. Examples: hydra, Portuguese man-of-war.

Class Scyphozoa Medusa form dominant; some species bypass polyp stage. Examples: lion's mane jellyfish, moon jelly, sea wasp.

Class Anthozoa Colonial or solitary polyps; no medusa stage. Examples: reef coral, sea anemone, sea pen, sea fan.

PHYLUM PLATYHELMINTHES (flatworms)

Three layers of tissues (endoderm, mesoderm, ectoderm); bilateral symmetry; some cephalization; acoelomate; free-living or parasitic.

Class Turbellaria Free-living carnivores and scavengers; live in fresh water, in salt water, or on land; move with cilia. Example: planarians.

Class Trematoda (flukes) Parasites; life cycle typically involves more than one host. Examples: *Schistosoma*, liver fluke.

Class Cestoda (tapeworms) Internal parasites; lack digestive tract; body composed of many repeating sections (proglottids).

PHYLUM NEMATODA (roundworms)

Digestive system has two openings—a mouth and an anus; pseudocoelomates. Examples: *Ascaris lumbricoides* (human ascarid), hookworm, *Trichinella*.

PHYLUM MOLLUSCA (mollusks)

Soft-bodied; usually (but not always) posses a hard, calcified shell secreted by a mantle; most adults have bilateral symmetry; muscular foot; divided into seven classes; digestive system with two openings; coelomates.

Class Pelecypoda (bivalves) Two-part hinged shell; wedge-shaped foot; typically sessile as adults; primarily aquatic; some burrow in mud or sand. Examples: clam, oyster, scallop, mussel.

Class Gastropoda (gastropods) Use broad muscular foot in movement; most have spiral, chambered shell; some lack shell; distinct head; some terrestrial, others aquatic; many are cross-fertilizing hermaphrodites. Examples: snail, slug, nudibranch, sea hare, sea butterfly.

Class Cephalopoda (cephalopods) Foot divided into tentacles; live in salt water; closed circulatory system; sexes separate. Examples: octopus, squid, nautilus, cuttlefish.

PHYLUM ANNELIDA (segmented worms)

Body composed of segments separated by internal partitions; digestive system has two openings; coelomate; closed circulatory system.

Class Polychaeta (polychaetes) Live in salt water; pair of bristly, fleshy appendages on each segment; some live in tubes. Examples: sandworm, bloodworm, fanworm, feather-duster worm, plume worm.

Class Oligochaeta (oligochaetes) Lack appendages; few bristles; terrestrial or aquatic. Examples: *Tubifex*, earthworm.

Class Hirudinea (leeches) Lack appendages; carnivores or blood-sucking external parasites; most live in fresh water. Example: medicinal leech (*Hirudo medicinalis*).

PHYLUM ARTHROPODA (arthropods)

Exoskeleton of chitin; jointed appendages; segmented body; many undergo metamorphosis during development; open circulatory system; ventral nerve cord; largest animal phylum.

Subphylum Trilobita (trilobites) Two furrows running from head to tail divide body into three lobes; one pair of unspecialized appendages on each body segment; each appendage divided into two branches—a gill and a walking leg; all extinct.

Subphylum Chelicerata (chelicerates) First pair of appendages specialized as feeding structures called chelicerae; body composed of two parts—cephalothorax and abdomen; lack antennae; most terrestrial. Examples: horseshoe crab, tick, mite, spider, scorpion.

Subphylum Crustacea (crustaceans) Most aquatic; most live in salt water; two pairs of antennae; mouthparts called mandibles; appendages consist of two branches; many have a carapace that covers part or all of the body. Examples: crab, crayfish, pill bug, water flea, barnacle.

Subphylum Uniramia Almost all terrestrial; one pair of antennae; mandibles; unbranched appendages; generally divided into five classes.

Class Chilopoda (centipedes) Long body consisting of many segments; one pair of legs per segment; poison claws for feeding; carnivorous.

Class Diplopoda (millipedes) Long body consisting of many segments; two pairs of legs per segment; mostly herbivorous.

Class Insecta (insects) Body divided into three parts—head, thorax, and abdomen; three pairs of legs and usually two pairs of wings attached to thorax; some undergo complete metamorphosis; approximately 25 orders. Examples: termite, ant, beetle, dragonfly, fly, moth, grasshopper.

PHYLUM ECHINODERMATA (echinoderms)

Live in salt water; larvae have bilateral symmetry; adults typically have five-part radial symmetry; endoskeleton; tube feet; water vascular system used in respiration, excretion, feeding, and locomotion.

Class Crinoidea (crinoids) Filter feeders; feathery arms; mouth and anus on upper surface of body disk; some sessile. Examples: sea lily, feather star.

Class Asteroidea (starfish) Star-shaped; carnivorous; bottom dwellers; mouth on lower surface. Examples: crown-of-thorns starfish, sunstar.

Class Ophiuroidea Small body disk; long armored arms; most have only five arms; lack an anus; most are filter feeders or detritus feeders. Examples: brittle star, basket star.

Class Echinoidea Lack arms; body encased in rigid, boxlike covering; covered with spines; most grazing herbivores or detritus feeders. Examples: sea urchin, sand dollar, sea biscuit.

Class Holothuroidea (sea cucumbers) Cylindrical body with feeding tentacles on one end; lie on their side; mostly detritus or filter feeders; endoskeleton greatly reduced.

PHYLUM CHORDATA (chordates)
Notochord and pharyngeal gill slits during at least part of development; hollow dorsal nerve cord.

Subphylum Urochordata (tunicates) Live in salt water; tough outer covering (tunic); display chordate features during larval stages; many adults sessile, some free-swimming. Examples: sea squirt, sea peach, salp.

Subphylum Cephalochordata (lancelets) Fishlike; live in salt water; filter feeders; no internal skeleton. Example: *Branchiostoma*.

Subphylum Vertebrata Most possess a vertebral column (backbone) that supports and protects dorsal nerve chord; endoskeleton; distinct head with a skull and brain.

Jawless Fishes Characterized by long eellike body and a circular mouth; two-chambered heart; lack scales, paired fins, jaws, and bones; ectothermic; possess a notochord as adults. Once considered a single class, Agnatha, jawless fishes are now divided into two classes: Myxini and Cephalaspidomorphi. Although the term agnatha no longer refers to a true taxonomic group, it is still used informally.

Class Myxini (hagfishes) Mostly scavengers; live in salt water; short tentacles around mouth; rasping tongue; extremely slimy; open circulatory system.

Class Cephalaspidomorphi (lampreys) Larvae filter feeders; adults are parasites whose circular mouth is lined with rasping toothlike structures; many live in both salt water and fresh water during the course of their lives.

Class Chondrichthyes (cartilaginous fishes) Jaw; fins; endoskeleton of cartilage; most live in salt water; typically several gill slits; tough small scales with spines; ectothermic; two-chambered heart; males possess structures for internal fertilization. Examples: shark, ray, skate, chimaera, sawfish.

Class Osteichthyes (bony fishes) Bony endoskeleton; aquatic; ectothermic; well-developed respiratory system, usually involving gills; possess swim bladder; paired fins; divided into two groups—ray-finned fishes (Actinopterygii), which include most living species, and fleshy-finned fishes (Sarcopterygii), which include lungfishes and the coelacanth. Examples: salmon, perch, sturgeon, tuna, goldfish, eel.

Class Amphibia (amphibians) Adapted primarily to life in wet places; ectothermic; most carnivorous; smooth, moist skin; typically lay eggs that develop in water; usually have gilled larvae; most have three-chambered heart; adults either aquatic or terrestrial; terrestrial forms respire using lungs, skin, and/or lining of the mouth.
  ORDER URODELA (salamanders) Possess tail as adults; carnivorous; usually have four legs; usually aquatic as larvae and terrestrial as adults.
  ORDER ANURA (frogs and toads) Adults in almost all species lack tail; aquatic larvae called tadpoles; well-developed hind legs adapted for jumping.
  ORDER APODA (legless amphibians) Wormlike; lack legs; carnivorous; terrestrial burrowers; some undergo direct development; some viviparous.

Class Reptilia (reptiles) As a group adapted to fully terrestrial life, although some live in water; dry, scale-covered skin; ectothermic; most have three-chambered hearts; internal fertilization; amniotic eggs typically laid on land; extinct forms include dinosaurs and flying reptiles.
  ORDER RHYNCHOCEPHALIA (tuatara) "Teeth" formed by serrations of jawbone; found only in New Zealand; carnivorous. One species: *Sphenodon punctatus*.
  ORDER SQUAMATA (lizards and snakes) Most carnivorous; majority terrestrial; lizards typically have legs; snakes lack legs. Examples: iguana, gecko, skink, cobra, python, boa.
  ORDER CROCODILIA (crocodilians) Carnivorous; aquatic or semiaquatic; four-chambered heart. Examples: alligator, crocodile, caiman, gharial.
  ORDER CHELONIA (turtles) Bony shell; ribs and vertebrae fused to upper part of shell; some terrestrial, others semiaquatic or aquatic; all lay eggs on land. Examples: snapping turtle, tortoise, hawksbill turtle, box turtle.

Class Aves (birds) Endothermic; feathered over much of body surface; scales on legs and feet; bones hollow and lightweight in flying species; four-chambered heart; well-developed lungs and air sacs for efficient air exchange; about 27 orders. Examples: owl, eagle, duck, chicken, pigeon, penguin, sparrow, stork.

Class Mammalia (mammals) Endothermic; subcutaneous fat; hair; most viviparous; suckle young with milk produced in mammary glands; four-chambered heart; four legs; use lungs for respiration.

Monotremes (egg-laying mammals)

ORDER MONOTREMATA (monotremes) Exhibit features of both mammals and reptiles; possess a cloaca; lay eggs that hatch externally; produce milk from primitive nipplelike structures. Examples: duck-billed platypus, short-beaked echidna.

Marsupials (pouched mammals)

ORDER MARSUPALIA (marsupials) Young develop in the female's uterus but emerge at very early state of development; development completed in mother's pouch. Examples: opossum, kangaroo, koala.

Placentals Young develop to term in uterus; nourished through placenta; some born helpless, others able to walk within hours of birth; about 16 orders.

ORDER INSECTIVORA (insectivores) Among the most primitive of living placental mammals; feed primarily on small arthropods. Examples: shrew, mole, hedgehog.

ORDER CHIROPTERA (bats) Flying mammals, with forelimbs adapted for flight; most nocturnal; most navigate by echolocation; most species feed on insects, nectar, or fruits; some species feed on blood. Examples: fruit bat, flying fox, vampire bat.

ORDER PRIMATES (primates) Highly developed brain and complex social behavior; excellent binocular vision; quadrupedal or bipedal locomotion; five digits on hands and feet. Examples: lemur, monkey, chimpanzee, human.

ORDER EDENTATA (edentates) Teeth reduced or absent; feed primarily on social insects such as termites and ants. Examples: anteater, armadillo.

ORDER LAGOMORPHA (lagomorphs) Small herbivores with chisel-shaped front teeth; generally adapted to running and jumping. Examples: rabbit, pika, hare.

ORDER RODENTIA (rodents) Mammalian order with largest number of species; mostly herbivorous but some omnivorous; sharp front teeth. Examples: rat, beaver, guinea pig, hamster, gerbil, squirrel.

ORDER CETACEA (cetaceans) Fully adapted to aquatic existence; feed, breed, and give birth in water; forelimbs specialized as flippers; external hindlimbs absent; many species capable of long, deep dives; some use echolocation to navigate; communicate using complex auditory signals. Examples: whale, porpoise, dolphin.

ORDER CARNIVORA (carnivores) Mostly carnivorous; live in salt water or on land; aquatic species must return to land to breed. Examples: dog, seal, cat, bear, raccoon, weasel, skunk, panda.

ORDER PROBOSCIDEA (elephants) Herbivorous; largest land animal; long, flexible trunk.

ORDER SIRENIA (sirenians) Aquatic herbivores; slow-moving; front limbs modified as flippers; hindlimbs absent; little body hair. Examples: manatee, sea cow.

ORDER PERISSODACTYLA (odd-toed ungulates) Hooved herbivores; odd number of hooves; one hoof generally derived from middle digit on each foot; teeth, jaw, and digestive system adapted to plant material. Examples: horse, donkey, rhinoceros, tapir.

ORDER ARTIODACTYLA (even-toed ungulates) Hooved herbivores; hooves derived from two digits on each foot; digestive system adapted to thoroughly process tough plant material. Examples: sheep, cow, hippopotamus, antelope, camel, giraffe, pig.

Glossary

PRONUNCIATION KEY

When difficult names or terms first appear in the text, a pronunciation key follows in parentheses. A syllable in small capital letters receives the most stress.

The key below lists the letters used in the pronunciations. It includes examples of words using each sound and shows how those words would be written.

Symbol	Example	Pronunciation
a	hat	(hat)
ay	pay; late	(pay); (layt)
ah	star; hot	(stahr); (haht)
ai	air; dare	(air); (dair)
aw	law; all	(law); (awl)
eh	met	(meht)
ee	bee; eat	(bee); (eet)
er	learn; sir; fur	(lern); (ser); (fer)
ih	fit	(fiht)
igh	mile; sigh	(mighl); (sigh)
oh	no	(noh)
oi	soil; boy	(soil); (boi)
oo	root; rule	(root); (rool)
or	born; door	(born); (dor)
ow	plow; out	(plow); (owt)

Symbol	Example	Pronunciation
u	put; book	(put); (buk)
uh	fun	(fuhn)
yoo	few; use	(fyoo); (yooz)
ch	chill; reach	(chihl); (reech)
g	go; dig	(goh); (dihg)
j	jet; gently; bridge	(jeht); (JENT-lee); (brihj)
k	kite; cup	(kight); (kuhp)
ks	mix	(mihks)
kw	quick	(kwihk)
ng	bring	(brihng)
s	say; cent	(say); (sehnt)
sh	she; crash	(shee); (krash)
th	three	(three)
y	yet; onion	(yeht); (UHN-yuhn)
z	zip; always	(zihp); (AWL-wayz)
zh	treasure	(TREH-zher)

A

abscission layer: band formed when the cells that join leaf petioles to a stem become weak

absolute dating: method of measuring rates of decay of radioactive materials to determine how long ago an event occurred or an organism lived

accessory pigment: compound other than chlorophyll that absorbs light energy and passes it on to the primary photosynthetic pigment (chlorophyll)

acid: compound that releases hydrogen ions in solution

acid rain: rain that contains droplets of acid formed when certain pollutants in the air combine with water vapor

acoelomate: animal that lacks a body cavity

Acquired Immune Deficiency Syndrome (AIDS): condition in which certain cells of the immune system are killed by infection with HIV

Acrasiomycota (uh-kras-ee-oh-migh-KOH-tuh): phylum containing cellular slime molds

actin: protein that makes up the thin filaments in a muscle cell

action potential: changes in membrane potential that characterize a nerve impulse; essentially the depolarization and repolarization of a neuron

active immunity: type of immunity produced by the body when stimulated by a vaccine or by exposure to a pathogen

active site: region on an enzyme that can bind with a specific substrate or substrates

active transport: energy-requiring process that moves material across a cell membrane against a concentration difference

adaptation: process that enables organisms to become better suited to their environments

adaptive radiation: process, also known as divergent evolution, in which one species gives rise to many species that appear different externally but are similar internally

adenine (AD-uh-neen): nitrogenous base in nucleic acids, belonging to the purines; base pairs with thymine or uracil

adenine triphosphate (ATP): compound that stores energy in cells

adrenal cortex: outer portion of the adrenal gland that produces hormones called corticosteroids

adrenal gland: pyramid-shaped endocrine gland located on top of each kidney that secretes hormones such as adrenaline, noradrenaline, cortisol, and other corticosteroids

adrenal medulla: inner portion of the adrenal gland that secretes adrenaline and noradrenaline and is considered part of the sympathetic nervous system

aerobic: any process that requires oxygen

agglutination: process in which viruses are linked together in a large mass by a group of antibodies

AIDS: Acquired Immune Deficiency Syndrome; condition in which certain cells of the immune system are killed by infection with HIV

air sac: one of several sacs attached to a bird's lungs that allow for the one-way flow of air through the respiratory system and make a bird's body more buoyant

alcoholic fermentation: fermentation that produces alcohol

alcoholism: disease in which a person becomes addicted to alcohol

allele (uh-LEEL): one of a number of different forms of the same gene for a specific trait

allergy: reaction of the immune system that results when antigens bind to mast cells

alternation of generations: the switching back and forth between the production of diploid and haploid cells

alveolus (al-VEE-oh-luhs): air sac at the end of a bronchiole where gas exchange occurs

ameba: flexible active cell without cell walls, flagella, or cilia that moves by means of pseudopods, reproduces by binary fission, and belongs to the phylum Sarcodina

amebocyte (ah-MEE-boh-sight): sponge cell that builds spicules from calcium carbonate or silica

amino acid: substance that has an amino group ($-NH_2$) on one end and a carboxyl group ($-COOH$) on the other; makes up the building blocks for proteins

amniocentesis (am-nee-oh-sehn-TEE-sihs): prenatal diagnostic technique that requires the removal of a small amount of fluid from the sac surrounding the embryo

amnion: membrane that surrounds and protects a developing embryo; in placental mammals such as humans, develops into the amniotic sac

amniotic sac: fluid-filled structure that cushions and protects the developing fetus in placental mammals such as humans

amphetamine: stimulant chemically resembling the body's neurotransmitters that causes the nervous system to increase its activity

amphibian: vertebrate that typically is aquatic in the larval stage and terrestrial as an adult; breathes with lungs as an adult, has a moist skin that contains many glands, and lacks scales and claws

anabolism (uh-NAB-uh-lihz-uhm): process in a living thing that involves synthesizing complex substances from simpler substances

anaerobe: organism that does not require oxygen to survive

anaerobic: any process that does not require oxygen

anal pore: structure in paramecia and certain other protists through which waste materials are emptied into the environment

analogous structures: structures that are similar in appearance and function but have different origins and usually different internal structures

anaphase: third phase of mitosis in which paired chromatids separate

androgen: steroid hormone in males involved in the development of male reproductive structures

angiosperm: flowering plant whose seeds develop within ovaries

Animalia: kingdom that contains multicellular heterotrophic organisms whose cells lack cell walls

annelid: round wormlike animal that has a long segmented body and belongs to the phylum Annelida

annual: plant that grows from seed to maturity, flowers, produces seeds, and dies, all in the course of one growing season

annual ring: tree ring formed by the growth of xylem cells; often indicates the age of a tree

anterior: front end of a bilaterally symmetrical organism

anterior pituitary: part of the pituitary gland that secretes seven hormones: FSH, LH, TSH, GH, prolactin, and MSH

anther: structure in which the male gametophytes are produced in flowering plants

antheridium (an-ther-IHD-ee-uhm): male reproductive structure in some plants, including mosses and liverworts

anthropoid: humanlike primate

antibiotic: drug or natural compound that can attack and destroy certain microorganisms

antibody: special protein that can bind to an antigen on the surface of a pathogen and help destroy it

anticodon: three-nucleotide sequence in transfer RNA that base pairs with a complementary sequence in messenger RNA during protein synthesis

antigen: foreign substance that induces an immune response and interacts with specific antibodies

antigen-binding site: area on an antibody molecule that allows it to bind to an antigen

aorta (ay-OR-tah): large artery in mammals that carries oxygen-rich blood from the left ventricle of the heart to all parts of the body except the lungs

apical meristem: plant structure containing meristematic tissue that divides, allowing stems and roots to grow in length

arachnid: chelicerate arthropod that has four pairs of legs on its cephalothorax and no antennae

archegonium (ahr-kuh-GOH-nee-uhm): female reproductive structure in some plants, including mosses and liverworts

artery: tough, flexible blood vessel that carries blood away from the heart to the tissues of the body

arthropod: animal having a segmented body, an exoskeleton containing chitin, and a series of jointed appendages; belongs to the phylum Arthropoda.

artificial selection: technique in which the intervention of humans allows only selected organisms to produce offspring

ascospore: haploid spore produced within the ascus of ascomycetes

ascus: tiny sac in which fungal spores develop for sexual reproduction in ascomycetes

asexual reproduction: process in which a single organism produces a new organism or organisms identical to itself

asthma: allergic reaction in which smooth muscles contract around the passages leading to the lungs

atom: smallest particle of matter that can exist and still have the properties of a particular kind of matter

atomic number: number of protons in the nucleus of an atom; identifies each element

atrium (AY-tree-uhm): chamber of the heart that receives incoming blood

autoimmune disease: condition in which the immune system attacks the body's own cells

autonomic nervous system: division of the peripheral nervous system that regulates involuntary activities

autosome: chromosome that is not a sex chromosome

autotroph (AW-toh-trohf): organism that is able to use a source of energy to produce its food from inorganic raw materials

auxin: plant hormone that can stimulate elongation of stem cells and inhibit elongation in root cells

axon: long fiber that carries impulses away from the cell body of a neuron

axon terminal: small swelling in a neuron at the end of an axon

B-lymphocyte: white blood cell responsible for producing antibodies

bacillus (buh-SIHL-uhs): rod-shaped bacterium

bacteriophage: virus that invades bacteria and consists of a core of nucleic acid, a capsid, and a tail

bacterium: one-celled prokaryote; chiefly parasitic or saprophytic

barbiturate: depressant often found in sleeping pill medications; can be fatal when combined with alcohol and can result in serious medical problems when supply to the body is discontinued

base: compound that releases hydroxide ions in solution

base pairing: attraction between complementary nitrogenous bases that produces a force that holds the two strands of the DNA double helix together

basidiospore: spore in basidiomycetes that germinates to produce haploid primary mycelia

basidium (buh-SIHD-ee-uhm): specialized reproductive structure in basidiomycetes

biennial: plant that grows roots, stems, and leaves in its first growing season, then flowers and dies in its second season

bilateral symmetry: arrangement of an organism's body parts so that if an imaginary line were drawn down the longitudinal middle of the body, the body's parts would repeat on either side of the line

binary fission: type of asexual reproduction in which an organism divides to produce two identical daughter cells

binocular vision: stereoscopic vision that provides primates with a three-dimensional view of the world

binomial nomenclature (bigh-NOH-mee-uhl NOH-muhn-klay-cher): classification system in which each organism is given a two-part scientific name

biodegradable: description of a substance that can be broken down by microorganisms into the essential nutrients from which it was made

biogeochemical cycle: series of physical and biological processes by which nutrients move through the biosphere

biological magnification: phenomenon in which the concentration of certain compounds in each organism in a food chain increases

biome: environment that has a characteristic climax community

biosphere: part of the Earth in which life exists

bipedal locomotion: upright movement on two legs

bivalve: mollusk that lives within a shell made of two sections that are hinged together

blastocyst: hollow structure in early human embryonic development that results after the morula forms a fluid-filled cavity in the center

blood: fluid medium of transport of the circulatory system

blood pressure: measure of the force that blood exerts against a vessel wall

bloom: enormous growth of algae, protists, and other organisms that results if too much waste is present in a body of water

bolus: ball of partially digested food formed in the mouth after the initial digestive processes of the teeth and salivary glands have acted upon it

bone marrow: soft tissue in the cavities of bones

bony fish: fish whose skeleton is made up of bone; belongs in the class Osteichthyes

Bowman's capsule: cup-shaped structure in the upper end of a nephron that encases the glomerulus and is involved in filtration in the kidneys

brainstem: structure that connects the brain with the spinal cord and is composed of the medulla oblongata, pons, and midbrain; coordinates and integrates all information coming into the brain

bronchiole: one of the finest branches of the bronchial tubes; ends in an air sac

bronchitis: inflammation of the bronchi

bronchus: ringed tube that branches from the trachea and enters each lung

budding: asexual reproduction in sponges

bud scale: thick, waxy structure that wraps around an apical meristem to protect the terminal bud

bulb: modified stem with most of its food stored in layers of short, thick leaves that wrap around the stem

C

calorie: amount of heat energy required to raise the temperature of 1 gram of water 1 degree Celsius

Calvin cycle: name given to the cycle of dark reactions in photosynthesis

calyx: ring of sepals at the base of a flower

capillary: small thin-walled blood vessel that allows materials to diffuse between the blood and the tissues

carapace: in turtles, the dorsal part of the shell; in crustaceans, the part of the exoskeleton that covers the cephalothorax

carbohydrate: organic compound containing carbon, hydrogen, and oxygen in a 1:2:1 ratio; human body's main source of energy

carbon cycle: movement of carbon between organic compounds that make up living tissues and carbon dioxide in the air

carcinogen: cancer-causing compound

cardiac muscle: tissue made up of striated cells not under voluntary control; found only in the heart

carnivore: organism that eats meat

carpel: structure produced from fertile leaves that have rolled up to comprise the centermost circle of flower parts

carrying capacity: size of a population during the steady state portion of a logistic growth curve

cartilage: dense, fibrous connective tissue whose cells are scattered in a network of collagen fibers

cartilaginous fish: fish, such as a shark or one of its relatives, whose skeleton is made up of soft, flexible cartilage instead of bone; belongs to the class Chondrichthyes

Casparian strip: waterproof strip that surrounds each endodermal cell in a root and is involved in the one-way passage of materials into the vascular cylinder in plant roots

catabolism (kuh-TAB-uh-lihz-uhm): breakdown of complex substances into simpler substances

catalyst: substance that speeds up the rate of a chemical reaction without being changed or used up by the reaction

cecum (SEE-kuhm): long sac that forms part of the intestines of some herbivores; contains microorganisms that digest cellulose

cell: basic unit of structure and function in living things

cell body: part of a neuron that contains the nucleus and much of the cytoplasm

cell cycle: period from the beginning of one mitosis to the beginning of the next

cell division: process by which a cell divides into two daughter cells

cell-mediated immunity: immune response in which killer T-cells cause the cells of pathogenic organisms to rupture and die

cell membrane: cell structure that regulates the passage of materials between the cell and its environment; aids in the protection and support of the cell

cell specialization: characteristic of certain cells that makes them uniquely suited to perform a particular function within an organism

cell theory: understanding that all living things are composed of cells and that all cells come from preexisting cells

cell wall: cell structure that surrounds the cell membrane for protection and support in plants, algae, and some bacteria

Celsius (SEHL-see-uhs) (C): metric temperature scale on which water freezes at 0° and boils at 100°

central nervous system: division of the nervous system that consists of the brain and spinal cord

centriole (SEHN-tree-ohl): structure involved in mitosis that contains a microtubule protein called tubulin

centromere (SEHN-troh-meer): structure that holds together each pair of chromatids

cephalization: gathering of sense organs and nerve cells into the head region

cephalopod (SEHF-uh-loh-pahd): marine mollusk whose head is attached to its foot, which is divided into tentacles

cerebellum (sair-uh-BEHL-uhm): part of the brain that coordinates and balances the actions of muscles

cerebral cortex: outer surface of the cerebrum associated with sensory input, motor output, and the complex movements associated with speech

cerebral medulla: inner surface of the cerebrum; composed of bundles of myelinated axons

cerebrum (suh-REE-bruhm): part of the brain responsible for all voluntary activities of the body; largest part of the brain

chelicerate (keh-LIHS-er-ayt): arthropod characterized by a two-part body and mouthparts called chelicerae

chemical bonding: process by which atoms of elements combine to achieve stability

chemical properties: properties that describe a substance's ability to change into a new substance as a result of a chemical reaction

chemical reaction: any process in which a chemical change occurs

chemical signal: stimulus used for communication among animals that involves the sense of smell

chemotrophic autotroph: organism that can obtain energy from inorganic molecules

chemotrophic heterotroph: organism that can obtain energy by taking in organic molecules and then breaking them down

chitin (KIGH-tihn): complex carbohydrate found in arthropod exoskeletons and fungal cell walls

chlorophyll: principal pigment in the cells of photosynthetic autotrophs that captures light energy

chloroplast: organelle that converts sunlight into chemical energy in plants

cholesterol: compound found in animal fats, meats, and dairy products that can build cells but, in excess, can be a risk factor in heart disease

chordate: animal that possesses a notochord, a hollow dorsal nerve cord, and pharyngeal slits at some point in its development

chorion: outermost membrane surrounding a developing reptile, bird, or mammal embryo; forms the placenta in placental mammals

chorionic villus biopsy (kor-ee-AHN-ihk VIHL-uhs BIGH-ahp-see): prenatal diagnostic method in which a sample of embryonic cells is removed from the sac surrounding the embryo

chromatid (KROH-muh-tihd): one of two distinct strands that make up each chromosome

chromatin (KROH-muh-tihn): material in chromosomes that is composed of DNA and protein

chromosomal mutation: change in the number or structure of chromosomes in a cell

chromosome: threadlike structure in a cell that contains the genetic information that is passed on from one generation of cells to the next

chromosome theory of heredity: theory that states that genes are located on chromosomes and that each gene occupies a specific place on a chromosome

Chrysophyta (krihs-uh-FIGHT-uh): phylum containing yellow-green algae, golden-brown algae, and diatoms

chyme: pasty mixture that results after food is partially digested in the stomach

ciliate (SIHL-ee-iht): protist that has many hairlike structures that aid in movement; belongs to the phylum Ciliophora

Ciliophora (sihl-ee-AHF-uh-ruh): phylum containing solitary and colonial ciliates

cilium: short hairlike projection that produces movement in many cells

circulatory system: group of organs including the heart and blood vessels that transports blood to all the body cells

class: group of closely related orders

classical conditioning: type of learning that occurs when an animal makes a mental connection between a stimulus and a positive or negative event

cleavage: mitotic cell division of a zygote or early embryo

climax community: relatively stable collection of plants and animals that results from ecological succession

clone: large population of genetically identical cells derived from one original cell

closed circulatory system: system in which blood always moves inside blood vessels

Cnidaria (nigh-DAIR-ee-ah): phylum that contains soft-bodied animals with stinging tentacles arranged in circles around their mouth

cocaine: addictive drug that causes the brain to release dopamine

coccus (KAHK-uhs): spherical bacterium

codominance: condition in which both alleles of a gene are expressed

codon (KOH-dahn): three-nucleotide sequence on messenger RNA that codes for an amino acid

coelom (SEE-lohm): body cavity that is completely lined with mesoderm

coelomate: animal that has a body cavity that is completely lined with mesoderm

coevolution: process by which two organisms evolve structures and behaviors in response to changes in each other over time

collar cell: one of the cells forming the wall of a sponge's central cavity

colon: digestive organ, also known as the large intestine, that removes water from the undigested materials that pass through it

colony: group of cells that are joined together and show few specialized structures

commensalism: symbiosis in which one member benefits and the other is not harmed

common descent: idea that species have descended from common ancestors

communication: passing of information from one animal to another

community: all the populations of organisms living in a given area

companion cell: special type of phloem cell that is found near a sieve tube element

compound: matter composed of two or more elements chemically bonded

compound light microscope: microscope with more than one lens that uses light to magnify objects

conidiophore (koh-NIHD-ee-uh-for): specialized hypha that produces asexual spores in ascomycetes

conidium: asexual spore produced in an ascomycete

conjugation: process in bacteria and protists that involves an exchange of genetic information

consumer: organism that gets its energy either directly or indirectly from producers

contour feather: one of the large feathers that cover the body and wings of a bird

contractile vacuole: structure in some protists that collects water and discharges it from the cell

control setup: part of an experiment that does not contain the variable

convergent evolution: phenomenon in which adaptive radiations among different organisms produce species that are similar in appearance and behavior; opposite of divergent evolution

cork cambium: meristematic tissue that produces the outer covering of stems

corm: round underground stem that stores food and is surrounded with thin leaves for protection

corolla: combined petals of a flower

corpus luteum (KOR-puhs LOOT-ee-uhm): name given to the follicle after ovulation because of its yellow appearance

cortex: root tissue that transports water and nutrients inward through the root; may store sugar and starches

covalent bond: chemical bond formed by the sharing of electrons

crack: potent and dangerous form of cocaine that can be smoked in a pipe

crop: enlarged area of the esophagus of a bird where food can be stored and moistened before it enters the stomach; or a food-storing organ in an earthworm

cross-bridge: knoblike projection of a myosin filament that connects with an actin filament during muscle contraction

cross-pollination: transfer of pollen from the flower of one plant to the flower of another plant

crossing-over: process by which homologous chromosomes exchange portions of their chromatids during meiosis

crustacean (kruhs-TAY-shuhn): arthropod characterized by a stony exoskeleton, two pairs of antennae, and mandibles

cubic centimeter (cc or cm³): basic metric unit of volume for solids; equal to a milliliter

cuticle: noncellular protective coating on the exterior surface of an organism, such as the waxy covering on leaves and insect exoskeletons that helps prevent water loss through evaporation

cytokinesis (sight-oh-kih-NEE-sihs): process by which a cell's cytoplasm divides to form two distinct cells

cytokinin: hormone manufactured in plant roots that stimulates cell division and causes dormant seeds to sprout

cytoplasm: area between the nucleus and cell membrane of a cell

cytosine (SIGHT-oh-seen): nitrogenous base in nucleic acids belonging to the pyrimidines; base pairs with guanine

cytoskeleton: framework of the cell composed of a variety of filaments and fibers that support cell structure and drive cell movement

dark reactions: reactions of photosynthesis that do not require light but use energy produced and stored during light reactions to make glucose

data: recorded observations and information

decomposer: organism that breaks down and obtains energy from dead organic material

dehydration synthesis: reaction in which small molecules join to form a large molecule, removing water in the process

dendrite: extension from the cell body of a neuron that carries impulses from the environment toward the cell body

denitrification (dee-nigh-trih-fih-KAY-shuhn): process in which bacteria in the soil break down nitrogen compounds into free nitrogen

density-dependent limiting factor: factor that controls population size and operates more strongly on large populations than on small ones

density-independent limiting factor: factor that controls population size regardless of how large the population is at the time

deoxyribonucleic acid (DNA): nucleic acid that stores and transmits genetic information from one generation of an organism to the next by coding for the production of a cell's proteins

depressant: drug that reduces the rate of nervous system activity

dermis: innermost layer of skin beneath the epidermis

desert: biome that receives less than 25 centimeters of rainfall a year

detritus feeder: animal that feeds on tiny bits of decaying plants and animals

deuterostome: organism whose anus formed from its blastopore during early embryonic development

diaphragm: large flat muscle involved in the mechanics of breathing that lies at the bottom of the rib cage

diatom: photosynthetic cell belonging to the phylum Chrysophyta that produces intricate cell walls rich in silicon

dicot: angiosperm whose seed has two cotyledons

diffusion (dih-FYOO-zhuhn): process by which molecules of a substance move from areas of higher concentration of that substance to areas of lower concentration

digestion: breakdown of food into simple molecules that can be absorbed by the body

dinoflagellate: protist belonging to the phylum Pyrrophyta that typically is photosynthetic, moves by means of two flagella, and reproduces by binary fission

diploid: description of a cell that contains a double set of chromosomes, one from each parent

disaccharide: double sugar formed from the combination of two simple sugars

disease: any change, other than an injury, that interferes with the normal functioning of the body

divergent evolution: pattern of evolution, also known as adaptive radiation, in which one species gives rise to many species that appear different externally but are similar internally

division of labor: phenomenon in which groups of specialized cells carry out different tasks in an organism

DNA: nucleic acid that stores and transmits genetic information from one generation of an organism to the next by coding for the production of a cell's proteins

DNA fingerprinting: technique for identifying individuals using repeated sequences in the human genome that produce a pattern of bands that is unique for every individual

dominance: principle that recognizes that some alleles are dominant and some are recessive

dominant: form of a gene that is expressed even if present with a contrasting recessive allele

dormancy: period during which an organism's growth and activity decrease or stop, usually during unfavorable environmental conditions

dorsal: upper side of an organism that has bilateral symmetry

double fertilization: process that occurs in angiosperms, in which one sperm nucleus fuses with the egg nucleus to form the zygote and one sperm nucleus fuses with the two nuclei that flank the egg nucleus to form the endosperm

double-loop circulatory system: system of internal transport in which one loop carries blood between the heart and the lungs and a second loop carries blood between the heart and the body

down feather: short, fluffy feather that traps warm air close to a bird's body

drug: any substance that causes a change in the body

drug abuse: misuse of either a legal or illegal drug

drug addiction: uncontrollable craving for a drug

duodenum (doo-oh-DEE-nuhm): first part of the small intestine

E

echinoderm: spiny-skinned animal that belongs to the phylum Echinodermata and is an invertebrate characterized by five-part radial symmetry, an internal skeleton, a water vascular system, and tube feet

ecological pyramid: diagram used to represent the energy relationships among trophic levels of an ecosystem

ecological succession: process through which an existing community is gradually replaced by another community

ecology: study of the interactions of organisms with one another and with their physical surroundings

ecosystem: division of the biosphere consisting of all the biotic and abiotic factors that surround organisms and affect their way of life

ectoderm: outermost primary germ layer in an animal embryo

ectotherm: animal that does not generate much internal heat and must rely on its external environment for the heat it requires

effector: muscle or gland that brings about a coordinated response to a stimulus

egg: female gamete

egg nucleus: nucleus in the embryo sac of angiosperms that locates itself close to the opening of the ovule and enlarges to become the female gamete

electrical signal: stimulus used for communication among certain fishes involving electrical impulses sent through water

electron: negatively charged subatomic particle located outside the atomic nucleus

electron transport: process in which high-energy electrons are transferred along a series of electron-carrier molecules in a membrane

element: substance consisting entirely of one type of atom

embryo: organism at an early stage of development

embryo sac: female gametophyte within the ovule of a flower

emphysema: stiffening of the normally elastic tissues of the human lung

endangered species: animal or plant that is so rare it is threatened with extinction

endocrine gland: hormone-producing organ of the endocrine system

endocrine system: body system composed of glands that secrete hormones into the bloodstream

endoderm: innermost primary germ layer in an animal embryo

endoplasmic reticulum (ehn-doh-PLAZ-mihk rih-TIHK-yuh-luhm): complex network that transports materials throughout the inside of a cell

endorphin: morphinelike chemical produced by the brain

endoskeleton: skeletal system in which a rigid framework is located inside the body of an animal

endosperm: triploid structure resulting from the fusion of a sperm nucleus with the two nuclei that flank the egg nucleus in the embryo sac of an angiosperm; provides food for the embryo

endospore: type of spore formed when a bacterium produces a thick internal wall that encloses its DNA and a portion of its cytoplasm

Endosymbiont Hypothesis: theory proposed by Lynn Margulis that states that the first eukaryotic cell was formed from a symbiosis among several prokaryotes

endotherm: animal that generates a significant amount of internal heat and has a relatively high metabolic rate

energy level: one of a series of "orbits" in which electrons travel around the nucleus of an atom

enzyme: one of a number of special protein catalysts contained in living organisms

epicotyl: portion of a shoot's stem above the cotyledon(s)

epidermis: in animals, outermost layer of skin; in plants, thin layer of root tissue that takes in water and nutrients

epididymis (ehp-uh-DIHD-ih-mihs): structure in the male reproductive system attached to the seminiferous tubules in which sperm mature and are stored

epiglottis: small flap of cartilage in the back of the mouth that closes the entrance to the respiratory tract when depressed by food

epoch: interval of time in a geologic time scale

equilibrium (ee-kwih-LIHB-ree-uhm): state in which no net change occurs

era: largest interval of time in a geologic time scale

esophagus: tube that connects the pharynx with the stomach

essential amino acid: amino acid that is needed but cannot be manufactured by the body

estrogen: steroid hormone in females involved in the development of the reproductive organs

estuary (EHS-tyoo-air-ee): area between fresh water and salt water

ethyl alcohol: form of alcohol present in alcohol-containing drinks

euglena: cell belonging to the phylum Euglenophyta that contains chlorophyll and has a pouch that contains two flagella at its front end

Euglenophyta: (yoo-glee-nuh-FIGHT-uh): phylum that contains plantlike protists that move by means of flagella and have chloroplasts

eukaryote: (yoo-KAIR-ee-oht): organism made up of cells that have a nucleus

evolution: process by which modern organisms have descended from ancient organisms; any change in the relative frequencies of alleles in the gene pool of a population

excretory system: group of organs, including the skin, lungs, and kidneys, that removes metabolic wastes from the body

exhalation: action in which air is pushed out of the lungs

exocrine gland: gland that releases secretions through tubelike structures called ducts

exon: expressed segment of a gene that is separated from similar segments by unexpressed sequences called introns

exoskeleton: system of supporting structures covering the outside of the body

experimental setup: part of an experiment that contains the variable being tested

exponential growth curve: graph that shows the phase of population growth during which the size of the population doubles regularly within a certain time period

external fertilization: process in which eggs are fertilized outside the body

external respiration: respiration that occurs at the level of the organism in which oxygen is obtained from and carbon dioxide is released to the external environment

extracellular digestion: process in which food is broken down outside the cells

F

facilitated diffusion: diffusion of materials across a cell membrane assisted by carrier molecules

facultative anaerobe: organism that can survive with or without oxygen

Fallopian tube: one of two fluid-filled tubes in human females through which an ovum passes after its release from an ovary; location of fertilization

family: group of closely related genera

fetal alcohol syndrome: damage to a developing baby due to the effects of alcohol in the mother's body

fermentation: process that enables cells to carry out energy production in the absence of oxygen; breakdown of glucose and release of energy in which organic substances are the final electron acceptors

fetus: unborn young of an animal during the later stages of development; in humans, the name given to embryo after eight weeks of development

fever: human body's response to an infection that results in increased body temperature in an effort to kill pathogens with heat

fibrous root: threadlike, branched root developing from a secondary root in some plants

filament: in a plant, the long, thin structure that supports the anther; in algae, a threadlike colony formed by many green algae

filter feeder: aquatic animal that feeds by straining tiny floating plants and animals from the water around it

filtration: process by which fluid from the blood filters into Bowman's capsule in the kidneys

fitness: combination of physical traits and behaviors that help an organism survive and reproduce in its environment

flagellate: organism that has taillike structures that aid in movement

flagellum: long, whiplike projection that aids in movement in some cells

flatworm: simple animal with bilateral symmetry belonging to the phylum Platyhelminthes

flower: reproductive structure in an angiosperm

food chain: series of organisms through which food energy is passed in an ecosystem

food vacuole: membrane-enclosed cavity in protists in which food is digested

food web: complex relationship formed by interconnecting and overlapping food chains

foot: muscular structure in mollusks that usually contains the mouth and other feeding structures

fossil: preserved remains or evidence of an ancient organism

fossil record: collection of fossils that represents the preserved collective history of the Earth's organisms

frameshift mutation: gene mutation involving an addition or deletion that alters every codon from the point of the mutation on

freshwater biome: rivers, streams, and lakes that provide drinking water and food

frond: large leaf of a fern

fruit: protective structure formed from an enlarged, thickened ovary wall that contains angiosperm seeds

Fungi: kingdom that includes heterotrophic organisms that build cell walls that typically do not contain cellulose

fungus: organism made of eukaryotic cells with cell walls that gets its food by absorbing organic substances

G

gallbladder: small sac in which bile is stored

gametangium (gam-uh-TAN-jee-uhm): gamete-forming structure produced when the hyphae of opposing mating types of fungi meet

gamete (GAM-eet): specialized reproductive cell involved in sexual reproduction

gametophyte (gah-MEET-oh-fight): haploid plant that produces gametes

ganglion: small cluster of nerve cells

gastrointestinal tract: long, hollow tube that makes up the digestive system

gastropod (GAS-troh-pahd): mollusk that moves by means of a broad, muscular foot located on its ventral side; usually has a one-piece shell for protection

gastrovascular cavity: digestive cavity in cnidarians with only one opening

gastrulation (gas-troo-LAY-shuhn): process of cell migration during which the primary germ layers are formed in an embryo

gemmule (JEHM-yool): sphere-shaped collection of amebocytes surrounded by spicules, which can grow into a new sponge

gene: segment of DNA that codes for a particular protein

gene mutation: change involving the nucleotides of DNA

gene pool: common group of genes shared by members of a population

generative nucleus: nucleus that results from the mitotic division of the nucleus of the pollen grain in angiosperms; divides to form the two sperm cells involved in double fertilization

genetic code: manner in which cells store the program that they pass from one generation to the next

genetic drift: random change in the frequency of a gene

genetic engineering: technique that directly alters an organism's DNA; altering the structure of a DNA molecule by substituting genes from other DNA molecules

genetics: branch of biology that studies heredity

genome: all the genes possessed by an organism

genotype: genetic makeup of an organism

genus: group of closely related species

geologic time scale: record of the history of life determined by the positions of layers of rock

germ theory of infectious disease: idea that infectious diseases are caused by microorganisms

gibberellin (jihb-er-EHL-ihn): plant hormone that stimulates growth

gill: filamentous respiratory structure in an aquatic animal

gizzard: muscular part of a reptile's or bird's stomach that contains bits of gravel that mix with food in order to make the food easier to digest; organ posterior to the crop in earthworms that is used to grind food

glomerulus (glom-MAIR-yoo-luhs): small network of capillaries surrounded by Bowman's capsule in the nephrons of the kidneys

glucose: sugar with formula $C_6H_{12}O_6$ that is a product of photosynthesis; can be broken down for energy

glycolysis (gligh-KAHL-ih-sihs): production of ATP by the conversion of glucose to pyruvic acid

Golgi apparatus: organelle that modifies, collects, packages, and distributes molecules made at one location of the cell and used at another

gonad: reproductive gland that produces gametes and sex hormones

gradualism: theory that evolutionary change occurs slowly and gradually

grassland: biome consisting of a vast area covered with grasses and small leafy plants

gravitropism: organism's response to gravity

greenhouse effect: condition in which carbon dioxide in the atmosphere absorbs heat radiated from the Earth, forming a kind of heat blanket around the Earth

guanine (GWAH-neen): nitrogenous base in nucleic acids belonging to the purines; base pairs with cytosine

guard cell: specialized epidermal cell that controls the opening and closing of the stomata by responding to changes in water pressure

gullet: indentation on one side of a paramecium that brings food from the outside to the interior of the cell

gymnosperm: plant known as a naked seed plant because its seeds do not develop within ovaries

H

habituation: decrease in response to a stimulus that neither rewards nor harms an animal

hair follicle: tubelike pocket of epidermal cells that extends into the dermis and produces hair

half-life: length of time required for half of the radioactive atoms in a sample to decay

hallucinogen: drug that affects a user's view of reality

haploid: containing a single set of chromosomes

Haversian (huh-VER-zhuhn) **canal**: one of a network of tubes containing blood vessels and nerves that supply blood to bones

heart: hollow muscular organ that contracts at regular intervals, forcing blood through the circulatory system

heartwood: xylem tissue that no longer conducts water but gives strength and support to a stem

helper T-cell: cell of the immune system that helps other lymphocytes respond to the early stages of an infection

hemoglobin: red iron-containing pigment in red blood cells of vertebrates that increases the oxygen-carrying capacity of the blood

herbaceous (her-BAY-shuhs): description of a plant whose stem has little or no woody tissue

herbivore: organism that eats plants

heredity: passing of traits from parents to their young

hermaphrodite: individual that has both male and female reproductive organs

heroin: opiate that is injected directly into the bloodstream, resulting in strong physical and psychological dependence

heterogamy: production of two different kinds of gametes within the same species

heterotroph (HEHT-er-oh-trohf): organism that cannot produce its own food but obtains energy from the food it eats

heterozygous (heht-er-oh-ZIGH-guhs): organism that has two different alleles for the same trait and is said to be hybrid for that particular trait

histamine: chemical released from mast cells when allergy-causing antigens attach themselves to mast cells; responsible for producing allergy symptoms

HIV: human immunodeficiency virus; virus that causes AIDS

holdfast cell: specialized cell that attaches an algal filament to the bottom of a lake or pond

hole in the ozone layer: area where little ozone is found

hollow dorsal nerve cord: cord that runs along the dorsal surface of a chordate and is connected by nerves to internal organs, muscles, and sense organs

homeostasis (hoh-mee-oh-STAY-sihs): process by which organisms keep internal conditions constant despite changes in their external environments

hominid: member of the evolutionary line that produced humans; classified in the family Hominidae; characterized by omnivory, bipedal locomotion, opposable thumbs, and a large brain

hominoid: great ape, such as a gorilla, gibbon, orangutan, chimpanzee, and Homo sapien

homologous: description of chromosomes that occur in pairs; having a corresponding structure

homologous structures: parts of different organisms, often quite dissimilar, that developed from the same ancestral body parts

homozygous (hoh-moh-ZIGH-guhs): organism that has two identical alleles for a particular trait

hormone: chemical substance produced in one part of an organism that affects another part of the organism

hybrid: organism with two different alleles for a particular trait, heterozygous; organism resulting from a cross between dissimilar parents

hybridization: breeding technique that involves a cross between dissimilar individuals

hydrolysis: catabolic reaction that splits apart molecules with the consumption of water

hydrostatic skeleton: skeletal system in which muscles surround and are supported by a water-filled body cavity

hypha (HIGH-fuh): branching filament that makes up a fungus

hypocotyl: portion of a stem below the cotyledon(s) in a newly germinated seed

hypothalamus (high-poh-THAL-uh-muhs): portion of the brain that controls the secretions of the pituitary gland; control center for hunger, thirst, fatigue, anger, and body temperature

hypothesis: possible explanation or conclusion about some event in nature; proposed solution to a scientific problem

I

immune: condition in which a body is able to permanently fight a disease using B-cells and T-cells produced the first time the body was exposed to the disease

immune system: body's primary defense against disease-causing microorganisms

implantation: process in early embryonic development in which the blastocyst attaches itself to the wall of the uterus and begins to grow inward

imprinting: process in which newborn birds add to their natural instinct to follow their mother with an image obtained by experience, usually the first slow-moving object they see

impulse: message carried by the nervous system that takes the form of an electrical signal

inbreeding: method of maintaining desirable characteristics by crossing individuals with similar characteristics who are often closely related

incomplete dominance: inheritance in which an active allele does not entirely compensate for an inactive allele

independent assortment: process by which genes segregate independently

inducer: chemical substance that causes the production of enzymes

infection: condition that results when the body is invaded by a pathogen

infectious disease: disease produced by a pathogen that affects the cells and tissues of the body

inflamed: reddish swollen area of the skin at the site of an infection

inflammatory response: defense mechanism that begins when a number of pathogens enter the body, causing fluid and white blood cells to leak from blood vessels into tissue

inhalation: action in which air is pulled into the lungs

inorganic compounds: compounds that do not contain carbon

insight learning: type of learning in which an animal applies something it has already learned to a new situation

instinct: behavior that is built into an animal's nervous system and cannot be changed

integumentary (ihn-tehg-yoo-MEHN-ter-ee) **system**: protective system formed by the skin and its accessory organs

interferon: protein that helps other cells resist viral infection

internal fertilization: process in which eggs are fertilized inside the female's body

internal respiration: respiration that occurs at the level of the cell

interneuron: cell that connects sensory and motor neurons and carries impulses between them

interphase: period of the cell cycle between cell divisions

intracellular digestion: process in which food is broken down inside the cells

intron: intervening sequence of DNA that does not code for a protein

invertebrate: animal that does not have a backbone

ion: charged particle

ionic bond: chemical bond that involves the transfer of electrons

islet of Langerhans: cluster of cells in the pancreas that produces insulin and glucagon

isogamy (igh-SAHG-ah-mee): condition in which the gametes of a species appear identical

isotope: atom of an element that has a different number of neutrons than other atoms of the same element

J

jawless fish: fish, such as a lamprey or hagfish, that lacks a jaw

joint: place where two bones come together

juvenile-onset diabetes: diabetes that appears in childhood; may be the result of an autoimmune reaction against the insulin-producing cells of the pancreas

K

kidney: organ that filters excess water, urea, and other waste products from the blood and excretes them out of the body

killer T-cell: special type of immune cell that transfers proteins into the cell membrane of a pathogen, causing the pathogen to rupture and die

kilogram (kg): basic metric unit of mass

kingdom: group of closely related phyla

Koch's postulates: series of rules for proving that a specific type of microorganism causes a particular disease

Krebs cycle: continuing series of reactions in cellular respiration that produces carbon dioxide, NADH, and $FADH_2$

L

labor: series of rhythmic contractions that cause the opening of the cervix of the uterus to expand so that it will be large enough to allow the baby to pass through

lactic acid fermentation: anaerobic process of glucose breakdown that produces lactic acid

large intestine: organ, also known as the colon, that removes water from the undigested materials that pass through it

larva: immature stage of an organism that is unlike the adult form in appearance

larynx: structure at the top of the trachea that contains the vocal cords

lateral bud: meristematic area on the side of a stem that gives rise to side branches

learning: way animals change their behavior as a result of experience

leech: annelid worm that typically exists as an external parasite that drinks the blood and body fluids of its host

lichen (LIGH-kuhn): symbiotic partnership between a fungus and a photosynthetic organism

ligament: strip of tough connective tissue in a joint that surrounds bones and holds them together

light reactions: reactions of photosynthesis that require light

limit of resolution: point of magnification in a microscope beyond which images become blurry and lose detail

limiting factor: condition that limits the rate at which energy is produced in an ecosystem by preventing a species from growing to its potential

linkage group: genes that are inherited in a group

linked genes: genes that are inherited together and do not undergo independent assortment

lipid: waxy or oily organic compound that stores energy in its bonds

liter (L): basic metric unit of volume for liquids

liver: large organ that lies above the stomach, secretes bile, and stores excess glucose in the form of glycogen

logistic growth curve: graph of a curve that shows the various growth phases experienced by a given population

LSD: powerful hallucinogen that interferes with the normal transmission of nerve impulses in the brain

lung: organ of respiration specialized for the exchange of gases between the blood and the atmosphere

lymph node: structure in the lymphatic system that acts as a filter and produces special white blood cells

lymphatic (lihm-FAT-ihk) **system**: network of vessels that collects fluid lost by the blood and returns it to the circulatory system

lymphocyte: white blood cell that responds to the presence of antigens

lysogenic (ligh-soh-JEHN-ihk) **infection**: process in which viral DNA is inserted into the DNA of a host cell where it can remain for many generations before becoming active

lysosome (LIGH-suh-sohm): organelle that contains chemicals and enzymes necessary for digesting certain materials in the cell

lysozyme: enzyme that breaks down the cell walls of many bacteria

lytic infection: process in which a host cell is invaded, lysed, and destroyed by a virus

M

macromolecule: large polymer

macronucleus: larger of two types of nuclei in ciliates, which controls the life process of the cell

mammary gland: gland in female mammals that produces milk to nourish the young for some time after they are born

mandible: mouthpart designed for biting and grinding food

mantle: thin, delicate layer of tissue that covers most of a mollusk's body and secretes the shell when one is present

marijuana: drug that comes from a species of hemp plant known as Cannabis sativa

marine biome: ocean habitat

marsupial: nonplacental mammal in which the fetus is born at a very immature stage and completes its development in a pouch on the mother's body

marsupium (mahr-SOO-pee-uhm): pouch in marsupials where the young embryos complete development

mass: measure of the amount of matter in an object

mass extinction: phenomenon in which many species suddenly vanish

mass number: total number of protons and neutrons in the nucleus of an atom

medulla (mih-DUHL-ah): part of the brain that controls internal organ functions and maintains balance

medulla oblongata: part of the brainstem that controls involuntary functions that include breathing, blood pressure, heart rate, swallowing, coughing, and keeping the brain alert

medusa (meh-DOO-sah): motile bell-shaped cnidarian

meiosis: process that produces haploid gametes from diploid cells

menopause: period after which follicle development no longer occurs and a female is no longer capable of bearing a child

menstrual cycle: process that involves the development and release of an egg for fertilization and the preparation of the uterus to receive a fertilized egg

menstruation: last phase of the menstrual cycle during which the lining of the uterus along with blood and the unfertilized ovum are discharged through the vagina

meristematic (mair-ih-steh-MAT-ihk) **tissue**: plant tissue that produces new cells by mitosis

mesoderm: middle primary germ layer in an animal embryo

mesophyll: layer of cells that contains chloroplasts and performs most of a plant's photosynthesis

messenger RNA (mRNA): type of RNA that carries genetic information from the DNA in the nucleus out to the ribosomes in the cytoplasm

metabolism: sum of all chemical reactions in the body; balance of anabolism and catabolism

metamorphosis: series of dramatic changes in body form in the life cycle of some animals

metaphase: second phase of mitosis in which the chromosomes line up across the equator of the cell

meter (m): basic metric unit of length

methanogen: bacterium that produces methane gas

metric system: universal system of measurement scaled on the multiples of ten

microfossil: preserved remains of an ancient microscopic organism

micronucleus: small nucleus in ciliates that undergoes meiosis and mitosis during conjugation and contains more genes than the macronucleus

midbrain: part of the brainstem that is involved in hearing and vision

mineral: inorganic substance required by the body

mitochondrion (might-oh-KAHN-dree-uhn): organelle that changes chemical energy stored in food into compounds that can be used by the cell

mitosis (migh-TOH-sihs): process by which the nucleus of a cell is divided into two nuclei, each with the same number and kinds of chromosomes as the parent cell

mixture: substance composed of two or more elements or compounds that are mixed together but not chemically combined

molecule: collection of two or more atoms covalently bonded

mollusk: soft-bodied invertebrate animal that is characterized by an internal or external shell, a foot, a mantle, and visceral mass; member of the phylum Mollusca

molt: to shed an exterior layer of skin, feathers, or an exoskeleton

Monera: kingdom that includes prokaryotic organisms

monocot: angiosperm whose seeds have one cotyledon

monomer: small compound that can be joined together with other small compounds to form polymers

monosaccharide: simple carbohydrate, also known as single sugar

monotreme: egg-laying mammal that belongs to the order Monotremata

morphine: chemically refined form of opium that is effective as a painkiller in small supervised doses

morula (MOR-yoo-luh): solid ball of cells that makes up an embryo; in humans, this stage occurs four days after fertilization

motor neuron: neuron that carries impulses from the brain and spinal cord to muscles and glands

multicellular: description of an organism consisting of many cells, some of which are typically specialized for particular functions

multiple alleles: three or more alleles of the same gene that code for a single trait

multiple sclerosis: nerve disease that results from autoimmune destruction of the myelin sheath that surrounds nerve fibers

muscular system: system that provides movement to the body and is composed of tissues that contract when stimulated

mutagen: substance or agent that can cause a mutation

mutation: change in the genetic material of a cell

mutualism: symbiosis in which two species live together in such a way that both benefit from the relationship

mycelium (migh-SEE-lee-uhm): thick mass of tangled filaments that make up the body of a fungus

mycorrhiza (migh-koh-RIGH-zuh): symbiotic relationship between a fungus and the roots of a green plant

myelin: substance composed of lipids and protein that forms an insulated sheath around an axon

myosin: protein that makes up the thick filaments of a muscle cell

Myxomycota (mihks-uh-migh-KOH-tuh): phylum containing acellular slime molds

nail matrix: area of rapidly dividing cells located at the tips of the fingers and toes that produces nails

natural selection: process in nature that results in the most fit organisms producing offspring

negative-feedback mechanism: control mechanism whereby an increase in a substance inhibits the process leading to the increase

nematocyst (neh-MAT-oh-sihst): stinging structure on the tentacles of cnidarians that is used to paralyze or kill prey

Nematoda (nee-mah-TOHD-ah): phylum containing roundworms

nephridium (neh-FRIHD-ee-uhm): simple tube-shaped excretory organ used to remove ammonia from the blood and release it from the body

nephron (NEHF-rahn): basic functional unit of the kidneys that filters out impurities from the blood

nervous system: system that receives and relays information about activities within the body and monitors and responds to internal and external changes

neuron: cell that carries impulses throughout the nervous system

neurotransmitter: substance used by one neuron to signal another

neutralization reaction: chemical reaction that occurs when the hydrogen ions of a strong acid react with the hydroxide ions of a strong base to form water and a salt

neutron: subatomic particle that is electrically neutral and is located in the atomic nucleus

niche: (NIHCH): combination of an organism's habitat and its role in that habitat

nicotine: poisonous substance in tobacco that causes the release of epinephrine

nitrogen cycle: movement of nitrogen through the biosphere between organisms and the atmosphere

nitrogen fixation: process by which nitrogen in the atmosphere is converted into a form that can be used by living things

nonbiodegradable: description of material that cannot be broken down by natural processes or is broken down extremely slowly

nondisjunction: failure of homologous chromosomes to separate normally during meiosis

nonspecific defense: defense mechanism of the body that guards against all infections rather than a particular pathogen

notochord: flexible supporting rod that runs along the dorsal surface of the body and is found in all chordates at some point in their development

nuclear envelope: membrane that surrounds the nucleus of a cell

nucleic acid: large, complex organic molecule that stores and transmits genetic information

nucleolus (noo-KLEE-uh-luhs): cell structure that contains RNA and proteins

nucleotide: unit of a nucleic acid that is made up of a 5-carbon sugar, a phosphate group, and a nitrogenous base

nucleus: in atoms, the center, which contains neutrons and protons and accounts for 99.9 percent of the atom's mass; in cells, the organelle that controls the cell's activities and contains DNA

nutrition: study of how bodies obtain energy, build tissue, and control body functions using materials supplied in food

O

obligate aerobe: organism that requires a constant supply of oxygen in order to live

obligate anaerobe: organism that lives only in the absence of oxygen

olfactory bulb: most anterior part of a fish's or amphibian's brain, involved in the sense of smell

oligochaete (AHL-ih-goh-keet): annelid worm belonging to the class that contains common earthworms and related species that live in soil and in water

oncogene: cancer-causing gene

open circulatory system: system in which blood does not always travel inside blood vessels

operant conditioning: trial-and-error learning in which an animal learns to behave in a certain way in order to receive a reward or avoid punishment

operator: region of chromosome near the cluster of genes in an operon to which the repressor binds when the operon is "turned off"

operon: genes and regions of DNA that operate together; consists of a gene cluster and regions involved in the regulation and expression of that cluster

opiate: powerful drug produced from the opium poppy

opposable thumb: thumb that is independent from the rest of the fingers, enabling hominids to grasp objects

optic lobe: in some animals, the part of the brain that processes information from the eyes

order: group of closely related families

organ: group of tissues that work together to perform a specific function

organ system: group of organs that work together to perform a specific function

organelle (or-guh-NEHL): cell structure that performs a specialized function within the cell

organic compounds: primarily those compounds that contain carbon

osculum (AHS-kyoo-luhm): large hole through which water exits the central cavity of a sponge

osmosis (ahs-MOH-sihs): diffusion of water molecules through a selectively permeable membrane from an area of higher water concentration to an area of lower water concentration

ossification (ahs-uh-fih-KAY-shuhn): process in which cartilage is replaced by bone

osteocyte (AHS-tee-oh-sight): cell in compact and spongy bone that is responsible for bone growth and changes in the shape of bones

ovary: in animals, female gonad that produces ova and estrogens; in plants, base of the pistil that contains ovules and developing gametophytes

oviparous (oh-VIHP-ah-ruhs): description of species that lay eggs that develop outside the mother's body

ovoviviparous (oh-voh-vigh-VIHP-ah-ruhs): description of species whose young develop inside the mother's body but are not nourished directly by the mother's body

ovulation: process that involves the release of a mature ovum from the ovary

ovule: specialized reproductive structure in seed plants

ovum: egg produced in an ovary; animal egg cell

oxygen cycle: movement of oxygen through the environment

P

paleontologist: scientist who studies fossils

palisade layer: layer of tall, column-shaped mesophyll cells just beneath the epidermal covering of a leaf

pancreas: organ that is both an exocrine gland and an endocrine gland; secretes digestive fluids and the hormones insulin and glucagon

paramecium: unicellular slipper-shaped ciliate protist

parasite: organism that survives by living and feeding either inside of or attached to outer surfaces of another organism, thus doing harm to the host

parasympathetic nervous system: portion of the autonomic nervous system that controls the internal organs during routine activities

parathyroid gland: gland that produces parathyroid hormone (PTH) and is attached to or embedded in the thyroid gland

parenchyma (puh-REHNG-kuh-muh): tissue composed of thin-walled cells found in roots, stems, and leaves

passive immunity: type of immunity that results when antibodies produced by other animals against a pathogen are injected into the bloodstream

pathogen: disease-causing microorganism

PCP: hallucinogen, also known as angel dust, that produces feelings of strength and power as well as nightmarish illusions and seizures

pellicle: complex living outer layer of certain protists

penis: external male reproductive organ; the organ through which the urethra connects to the outside of the body in humans and certain other animals

peptide bond: covalent bond that joins two amino acids

perennial: plant that grows and reproduces for an indefinite number of years

pericycle: type of cambium that enables roots to grow thicker and branch

period: interval in a geologic time scale that is composed of epochs

periosteum (pair-ee-AHS-tee-uhm): tough membrane that surrounds bone

peripheral nervous system: division of the nervous system that lies outside the central nervous system and contains the cranial and spinal nerves and ganglia

peristalsis: rhythmic muscular contractions that move food through the digestive system

petal: structure located in the second circle of flower parts just inside the sepals

petiole: structure that attaches the leaf blade to the stem

pH scale: measurement system that ranges from 0 to 14 and indicates the relative concentrations of hydrogen ions and hydroxide ions in a substance

phagocyte: white blood cell that engulfs and destroys microorganisms

pharyngeal slit: structure that appears in pairs in the throat region of a chordate's body

pharynx (FAIR-ihnks): muscular tubelike structure located at the back of the mouth that connects the mouth with the rest of the digestive tract

phase: physical property of matter that describes one of a number of different states of the same substance

phenotype: physical characteristics of an organism

pheromone: specific chemical messenger produced by an organism that affects the behavior and/or development of other individuals of the same species

phloem: vascular tissue responsible for the transport of nutrients and the products of photosynthesis throughout the plant

photoperiodicity (foht-oh-peer-ee-uh-DIHS-uh-tee): plant's response to the period of darkness to which it is exposed

photosynthesis: process in which autotrophs make their own food using the energy in light and CO_2 and H_2O

photosynthetic membrane: chlorophyll-containing membrane in chloroplasts that serves as the site of the light reactions

photosystem: cluster of pigment molecules within a photosynthetic membrane

phototrophic autotroph: organism that can trap the energy of sunlight and convert it to organic nutrients

phototrophic heterotroph: organism that is able to use sunlight for energy but also requires organic compounds for nutrition

phototropism: plant's growth responses to light

phylogenetic (figh-loh-juh-NEHT-ihk) **tree**: diagram that shows evolutionary relationships among different groups of organisms

phylum: group of closely related classes

physical property: characteristic of matter that can be observed and measured without permanently changing the identity of the matter

phytoplankton: any small photosynthetic organism found in great numbers near the surface of the ocean

pigment: colored substance that absorbs or reflects light

pistil: female reproductive structure in a flower formed from one or more carpels; consists of the ovary, style, and stigma

pituitary gland: endocrine gland at the base of the skull that secretes hormones that regulate body functions and control other endocrine glands

placenta: organ in placental mammals through which nutrients, oxygen, carbon dioxide, and wastes are exchanged between embryo and mother

placental mammal: mammal in which a placenta forms during development of the embryo

Plantae: kingdom that includes multicellular autotrophic organisms

plasma: liquid portion of the blood that contains water, dissolved fats, salts, sugars, and proteins

plasma cell: specialized B-lymphocyte that releases antibodies into the bloodstream to deal with an infection

plasmid: small circular piece of DNA in some bacterial cells that is often used in genetic engineering

plasmodium: mass of cytoplasm that contains many nuclei, such as the structure produced by acellular slime molds that contains thousands of nuclei enclosed in a single cell membrane

plastid: plant cell organelle involved in the storage of food and pigments

plastron: ventral part of a turtle's shell

platelet: tiny fragment of a cell that plays a part in the process of blood clotting

Platyhelminthes (pla-tee-hehl-MIHN-theez): phylum consisting of acoelomate flatworms such as tapeworms, flukes, and planarians

point mutation: gene mutation that affects a single nucleotide

polar nucleus: one of the two nuclei in the embryo sac of angiosperms that locate themselves in the center of the sac during fertilization

pollen chamber: structure in anther that produces microspore mother cells

pollen cone: male cone in gymnosperms that produces male gametophytes in the form of pollen grains

pollen grain: structure that contains the male gametophyte in seed plants
pollen tube: structure that grows once a pollen grain has landed on the stigma in order to bring the sperm cells to the egg nucleus
pollination: transfer of pollen from the anther of a stamen to the stigma of a pistil
polychaete (PAHL-ee-keet): segmented worm, usually marine, characterized by paired paddlelike appendages on its body segments and a bristly body
polygenic (pahl-uh-JEHN-ihk) **trait**: trait that is controlled by a number of genes
polymer: large compound formed by combinations of monomers
polymerization: process by which large compounds are constructed by joining smaller compounds
polyp (PAH-lihp): sessile flowerlike cnidarian
polyploidy: condition in which an organism has extra sets of chromosomes
polysaccharide: large molecule formed when many monosaccharides link together
pons: part of brainstem that provides a link between the cerebral cortex and the cerebellum
population: collection of individuals of the same species in a given area whose members can breed with one another
Porifera (por-IHF-er-ah): phylum containing the sponges
posterior: back end of a bilaterally symmetrical organism
posterior pituitary: part of the pituitary gland that secretes antidiuretic hormone (ADH) and oxytocin
powder feather: bird feather that releases a fine white powder that repels water and keeps it from penetrating the layer of down feathers
prehensile tail: tail found in New World monkeys that is used for grasping while climbing
primary follicle: cluster of cells that surround an ovum and prepare it for release from the ovary
primary immune response: production of antibodies from the first exposure to an antigen
primary tissue: tissue produced by the apical meristem
probability: likelihood that a particular event will occur
producer: organism that is able to make its own food from inorganic substances
progesterone: steroid hormone released by the corpus luteum
prokaryote (pro-KAHR-ee-oht): single-celled organism whose cells do not have a nucleus
promoter: region of a chromosome next to the operator in an operon to which RNA polymerase binds at the beginning of transcription
prophage: viral DNA attached to a bacterial chromosome
prophase: first phase of mitosis

prostaglandin (prahs-tuh-GLAN-dihn): hormonelike substance that functions only within the same cell in which it is produced
protein: complex polymer of amino acids that builds and repairs cells
prothallium (proh-THAL-ee-uhm): thin heart-shaped structure formed from the gametophyte of a fern
protist: unicellular eukaryotic organism belonging to the kingdom Protista
Protista: kingdom that includes all single-celled eukaryotic organisms
proton: positively charged subatomic particle located in the nucleus
protonema: tangled mass of green filaments in moss that forms during germination
protostome: organism whose mouth was formed from its blastopore during early embryonic development
pseudocoelomate: animal that has a body cavity that is partially lined with mesoderm
pseudopod (SOO-doh-pahd): fingerlike projection of cytoplasm used for movement and feeding
puberty: period of rapid growth and sexual maturation during which the reproductive system becomes fully functional
pulmonary circulation: division of the circulatory system in which blood flows between the heart and the lungs
punctuated equilibria: pattern of long stable periods interrupted by brief periods of change
Punnett square: chart showing the possible combinations of genes in the offspring of a cross
pupa (PYOO-pah): resting stage of metamorphosis in which the tissues of an insect are organized into the adult form
purebred: belonging to a group of organisms that can produce offspring having only one form of a trait in each generation
pyloric valve: valve between the stomach and the small intestine
Pyrrophyta (pigh-roh-FIGHT-uh): phylum containing protists known as dinoflagellates

R

radial symmetry: arrangement of the body parts of an organism in such a way that they repeat around an imaginary line drawn through the center of the organism's body
radicle: region at the base of the hypocotyl of a stem that contains an apical meristem and becomes the primary root
radioactive element: unstable element that decays into a stable element at a steady rate
radula (RAJ-oo-lah): in some mollusks, layer of flexible skin with hundreds of tiny teeth used for feeding

reabsorption: process by which most of the material removed from the blood at Bowman's capsule makes its way back into the blood

receptor: special sensory neuron in a sense organ that receives stimuli from the external environment

recessive: description of a form of a gene (allele) that is only expressed in the homozygous state

recombinant: individual organism with new combinations of genes

recombinant DNA: DNA molecule that forms from the combination of portions of two different DNA molecules

recycling: process by which certain kinds of solid wastes can be processed and used again

red blood cell: blood cell, also known as an erythrocyte, produced in bone marrow and filled with hemoglobin to transport oxygen; in humans, erythrocytes lack a nucleus

reflex: simplest response to a stimulus

reflex arc: receptor, sensory neuron, motor neuron, and effector that are involved in a quick response to a stimulus

reforestation: planting of new trees where old trees have been cut down or burned

rejection: process in which the immune system damages and destroys a transplanted organ

relative dating: technique used to determine the age of fossils by comparing them with other fossils in different layers of rock

relative frequency: number of times an event (allele) occurs compared with the number of times another event (other alleles for the same gene) occurs

replication (rehp-luh-KAY-shuhn): process by which DNA is duplicated before a cell divides

repressor: special protein that binds to the operator and thus turns off an operon

reproductive isolation: separation of populations so that they do not interbreed to produce fertile offspring

reproductive system: body system that produces, stores, nourishes, and releases gametes

respiration: process that involves oxygen and breaks down food molecules to release energy

respiratory system: group of organs working together to bring about the exchange of oxygen and carbon dioxide with the environment

resting potential: difference in charge across a nerve cell membrane resulting from the negative charge on the inside and the positive charge on the outside

restriction enzyme: protein capable of cutting genes at specific DNA sequences

retrovirus: type of virus that contains RNA as its genetic information

rheumatic fever: autoimmune disease that results when antibodies produced to destroy untreated streptococcus bacteria attack cardiac cells as well

rheumatoid arthritis: destructive inflammation of the joints produced by the actions of the immune system

rhizoid: in fungi, small branching hypha growing downward from the stolons that anchors the fungus, releases digestive enzymes, and absorbs digested organic material; in bryophytes, rootlike structure that anchors the plant to the ground

rhizome: thick, fleshy creeping stem that grows either on or just beneath the surface of the ground

ribonucleic acid (RNA): nucleic acid made of a single chain of nucleotides that acts as a messenger between DNA and the ribosome and carries out the process by which proteins are made from amino acids

ribosomal RNA (rRNA): type of RNA that makes up the major part of the ribosomes

ribosome: organelle in which proteins are made

root cap: structure that protects the root as it forces its way through the soil

roundworm: pseudocoelomate animal with a digestive system with two openings that belongs to the phylum Nematoda

rumen: chamber in the digestive tract of some herbivorous animals that contains symbiotic bacteria that produce enzymes needed to break down cellulose

saprophyte: organism that uses the complex molecules of a once-living organism as its source of energy and nutrition

sapwood: area in plants that contains water-conducting xylem tissue

Sarcodina (sahr-kuh-DIGH-nuh): phylum that contains protists that use pseudopods to move and feed

scale: specialized reproductive structure in gymnosperms that forms male or female cones; tough protective structure on the skin of animals such as certain fishes, reptiles, and birds

scanning electron microscope (SEM): microscope that uses a beam of electrons to scan the surface of a specimen and produce a three-dimensional image of the surface by recording the electrons that bounce off

science: process whose goal is to understand the natural world

scientific method: systematic approach to problem solving that involves observation and experimentation

scion: piece of a stem or a lateral bud that is cut from a parent plant and attached to another plant

sclerenchyma (sklih-REHNG-kuh-muh): cell with a tough, thick cell wall that strengthens and supports plant tissues

scrotum: external sac in which the testes are located

secondary immune response: defense mechanism that occurs when a pathogen reappears in the body

secondary phloem: new phloem cells formed on the surface of the vascular cambium that faces the outside of the stem

secondary sex characteristic: sex characteristic that appears at puberty

secondary xylem: new xylem cells formed on the surface of the vascular cambium that faces the center of the stem

sedimentary rock: rock that forms when grains of eroded rock and other materials are carried to the bottom of a body of water and build up under pressure into layers

seed coat: structure that surrounds a plant embryo and protects it and its food supply from drying out

seed dispersal: process of distributing seeds away from the parent plant

segregation: separation; in genetics, the separation of alleles during gamete formation

selective breeding: method of improving a species by choosing animals or plants that have desirable characteristics to produce offspring that have the parents' desirable traits

selectively permeable: description of a biological membrane that allows some substances to pass through but not others

self-pollination: process in which pollen falls from the anther to the stigma of the same flower or between flowers of the same plant

semen: combination of sperm and seminal fluid

seminal fluid: substance in which sperm are suspended that is produced by three glands in the abdominal cavity

seminiferous (sehm-uh-NIHF-er-uhs) **tubule**: one of thousands of tiny tubules that make up the testes

sense organ: structure that contains sensory receptors

sensory neuron: neuron that carries impulses from the sense organs to the brain and spinal cord

sepal: structure in the outermost circle of flower parts that encloses a bud before it opens and protects the flower while it is developing

sex chromosome: chromosome that is different in males and females

sex-influenced: description of a trait that is caused by a gene whose expression differs in males and females

sex-linked: description of a trait that is determined by a gene located on one of the sex chromosomes

sexual reproduction: process in which two cells, normally from different individuals, unite to produce the first cell of a new organism

shell: structure in mollusks made by glands in the mantle that secrete calcium carbonate

sieve tube element: phloem cell that is joined with similar cells end to end to form a continuous tube throughout the plant

single-loop circulatory system: system of internal transport in which blood travels from the ventricle in the heart to the gills to the body to the atrium of the heart

skeletal muscle: tissue, also known as striated muscle, that is attached to bones and is responsible for voluntary movement

skeletal system: system consisting of the bones and their associated tissues that supports, protects, and provides flexibility to the body of humans and many other animals

sliding filament theory: concept in which actin and myosin filaments slide over one another during muscle contraction

slime mold: protist that is amebalike at one stage of its life and at other stages produces moldlike masses that give rise to spores

small intestine: digestive organ in which chyme from the stomach is flooded with enzymes and digestive fluids

smog: air pollutant consisting of a combination of smoke and fog

smooth muscle: tissue made up of spindle-shaped cells that control involuntary activities; found in internal organs and in blood vessel walls

solute: substance that is dissolved in a mixture

solution: homogeneous mixture in which one substance is dissolved in another

solvent: substance in which a solute is dissolved to produce a solution

somatic nervous system: division of the peripheral nervous system that regulates activities that are under conscious control

sorus: large cluster of sporangia on the underside of fern fronds

sound signal: stimulus used for communication among animals that involves the sense of hearing

species: group of organisms that share similar characteristics and can interbreed with one another to produce fertile offspring

specific defense: defense mechanism directed toward a specific disease-causing agent

sperm: male gamete

sperm nucleus: nucleus in angiosperms that results when the generative nucleus within the pollen grain divides

spicule: one of the thin, spiny structures that form the skeleton of a sponge

spindle: meshlike structure of microtubules that appears to guide the movements of chromosomes during mitosis

spirillum (spigh-RIHL-uhm): spiral-shaped bacterium

spongin: protein that makes up the skeleton of some sponges

spongy mesophyll: layer of cells in leaves, arranged in a network with spaces between them, that connect with the stomata

spontaneous generation: hypothesis that life arises from nonlife

sporangium (spoh-RAN-jee-uhm): structure in ferns, some protists, and some fungi that contains spores

sporophyte: diploid plant that produces spores

Sporozoa (spohr-oh-ZOH-uh): phylum containing nonmotile parasitic protists

stamen: male reproductive structure of a flower belonging to the first circle of fertile leaves located just inside the petals

steady state: portion of a logistic growth curve in which the average growth rate is zero

stigma: upper part of a pistil upon which pollen grains are deposited

stimulant: drug that speeds up the actions of the nervous system

stock: plant to which a scion is grafted

stolon: stemlike hypha that grows parallel to a fungus's growth medium

stoma: opening in the leaf epidermis through which water vapor and oxygen pass out of the leaf and carbon dioxide passes into it

stomach: thick muscular sac in which food is partially digested; located just below the diaphragm in humans

style: stalk between the stigma and the ovary in a flower

substrate: reactant affected by an enzyme

survival of the fittest: principle that states that only individuals with characteristics best suited to their environment survive the struggle for existence

suspension: mixture containing nondissolved particles distributed within a solid, liquid, or gas

symbiosis (sihm-bigh-OH-sihs): close relationship between two species in which at least one species benefits from the other

sympathetic nervous system: portion of the autonomic nervous system that controls the internal organs during stressful situations and increased activity

symptom: change in the body as a result of a disease

synapse: point of contact at which an impulse is passed from one cell to another

syngamy (SIHN-gah-mee): the fusing of algal gametes

systemic circulation: division of the circulatory system in which blood flows between the heart and all the tissues of the body other than the lungs

T

T-lymphocyte: lymphocyte that matures in the thymus gland

taiga (TIGH-guh): biome dominated by great coniferous forests

taproot: primary plant root that grows longer and thicker than other roots

tar: mixture of complex chemicals in tobacco products

target cell: cell that has receptors for a particular hormone

target organ: part of an organism affected by a hormone

taxon: group into which organisms are classified

taxonomy: science of naming organisms and assigning them to groups

telophase: final phase of mitosis, during which chromosomes uncoil, a nuclear envelope reforms around the chromatin, and a nucleolus becomes visible in each daughter nucleus

temperate deciduous forest: biome characterized by changing seasons and leaf fall

temperature inversion: condition in which a layer of cool air becomes trapped beneath a layer of warm air, keeping pollutants near the ground

tendon: tough connective tissue that joins skeletal muscle to bone

test cross: cross between an organism of unknown genotype and an organism that is homozygous recessive for a particular trait

testis (TEHS-tihs): male gonad that produces sperm and androgens

testosterone (tehs-TAHS-ter-ohn): principal male sex hormone that stimulates the development of many male sex characteristics

thalamus: structure in the vertebrate brain that serves as a switching station for sensory input

theory: time-tested concept that makes useful and dependable predictions about the natural world

thermal pollution: type of water pollution in which heated water is pumped into the environment, often killing aquatic plants and animals

thigmotropism: response to touch

threshold: minimum level of a stimulus required to activate a neuron

thymine (THIGH-meen): nitrogenous base found in DNA but not in RNA; base pairs with adenine

thyroid gland: gland located at the base of the neck that produces thyroxine and calcitonin

tissue: group of similar cells that perform a particular function

toxin: poison

trachea: tube through which air passes from the pharynx to the lungs; windpipe

tracheid: cell in xylem tissue that conducts water and gives strength to the plant

trait: characteristic that a living thing can pass on to its young

transcription: process by which a molecule of DNA is copied into a complementary strand of RNA

transfer RNA (tRNA): type of RNA that carries amino acids to the ribosomes where the amino acids are joined together to form polypeptides

transformation: process by which genetic material absorbed from the environment is added to or replaces part of a bacterium's DNA

transgenic: description of an organism that contains foreign genes

translation: process in which a message carried by messenger RNA is decoded into a polypeptide chain (protein)

transmission electron microscope (TEM): microscope that uses a beam of electrons to magnify an image onto a fluorescent screen

transpiration: evaporation of water from plant leaves

transpiration pull: force that pulls water from the roots to the leaves as a result of the evaporation of water from the leaves

trichocyst (TRIHK-oh-sihst): flask-shaped structure in the pellicles of some protists used to defend and anchor the organism

trilobite (TRIGH-loh-bight): sea-dwelling organism belonging to the oldest, now extinct, subphylum of arthropods

tropical rain forest: biome in which the temperature stays warm and rain falls year round

tropism (TROH-pihz-uhm): response of an organism to an environmental stimulus

tube foot: suction-cuplike structure connected to the water vascular system of an echinoderm

tube nucleus: nucleus that results from the mitotic division of the nucleus of a pollen grain during the pollination of a flower

tuber: modified underground stem swollen with stored food

tumor: mass of tissue

tundra: northernmost land biome covered by mosses, lichens, and grasses and characterized by permafrost

U

unicellular: description of an organism consisting of only a single cell

uniramian (yoo-nih-RAY-mee-ahn): member of the largest subphylum of arthropods; contains centipedes, millipedes, and insects

unsegmented worm: worm whose body is not divided into special sections by internal partitions; flatworms and roundworms

uracil (YOOR-uh-sihl): nitrogenous base found only in RNA; base pairs with adenine

ureter (yoo-REET-er): tube that carries fluid from the kidney to the urinary bladder or cloaca

urethra: tube through which urine is released from the body

urinary bladder: saclike organ where urine is stored before being excreted

uterus: organ lying between the Fallopian tubes and the vagina in which a fertilized ovum can develop

V

vaccination: injection of a weakened or mild form of a pathogen used to produce immunity

vacuole (VAK-yoo-ohl): organelle that stores materials such as water, salts, proteins, and carbohydrates

vagina: canal that leads from the uterus to the outside of the female body

variable: single factor that is isolated and tested in an experiment

vas deferens (VAS DEHF-uh-rehnz): tube that carries sperm from the epididymis to the urethra

vascular bundle: strand of xylem and phloem cells

vascular cambium: meristematic area that produces vascular tissues and increases the thickness of stems over time

vascular cylinder: central area of a plant root where xylem and phloem tissues are gathered

vascular tissue: specialized tissue that transports water and the products of photosynthesis throughout a plant; xylem and phloem tissue

vector pollination: pollination by the actions of animals

vein: in animals, blood vessel that collects blood from the body and carries it back to the heart; in plants, area in leaves containing one or more bundles of vascular tissue

vena cava (VEE-nah KAY-vah): large blood vessel that brings blood from all parts of the body except the lungs to the heart

ventral: lower side of an organism with bilateral symmetry

ventricle (VEHN-trihk-uhl): muscular chamber that pumps blood out of the heart

vertebral column: backbone, which encloses and protects the nerve cord

vertebrate: animal that has a backbone

vessel element: xylem cell in angiosperms that forms part of a continuous tube through which water can move

vestigial organ: structure that serves no useful purpose or function in an organism

villus: folded projection that increases the surface area of the walls of the small intestine

virus: noncellular particle made up of genetic material and protein that can invade living cells

visceral mass: structure in mollusks that contains the internal organs

visual signal: stimulus used for communication among animals that involves the sense of sight

vital capacity: maximum amount of air that can be moved into and out of the respiratory system

vitamin: complex molecule needed by the body in small amounts that usually cannot be synthesized by the body

viviparous: species that bears living young and in which the unborn young are directly nourished by the mother's body

vocal cord: elastic fold of tissue in the larynx that produces sound when exhaled air is passed by it, causing it to vibrate

W

water cycle: movement of water from the atmosphere to the Earth and back to the atmosphere

water vascular system: internal network of fluid-filled canals involved in feeding, respiration, internal transport, elimination of wastes, and movement in echinoderms

weight: measure of the pull of gravity on a mass

white blood cell: blood cell produced in bone marrow that protects the body against invasion by foreign cells or substances

withdrawal: stopping the use of a drug

X

X chromosome: sex chromosome; in humans, fruit flies, and certain other organisms, females have two X chromosomes and males have only one

xylem: vascular tissue that provides support to a plant and conducts water from the roots to all parts of the plant

Y

Y chromosome: male sex chromosome in humans, fruit flies, and certain other organisms

Z

zone of elongation: area where cell enlargement occurs in plants

zone of maturation: region in plants in which newly lengthened cells differentiate

Zoomastigina (zoh-oh-mas-tuh-GIGH-nuh): phylum consisting of animallike protists that move through the water by means of flagella

zoospore (ZOH-oh-spohr): haploid cell involved in asexual reproduction in algae

zygospore: in fungi, thick-walled zygote formed during sexual reproduction in zygomycetes; in green algae, diploid cells resulting from conjugation

zygote (ZIGH-goht): fertilized egg cell

Index

A

ABO blood groups, 230
Abscission layer, 525, 526
Absolute dating, 276
Accessory pigments, 117, 435
Acetic acid, 125
Acid rain, 1056–1057
Acids, formation of, 65
Acoelomates, 654, 656
Acoustic nerve, 829
Acquired immune deficiency syndrome (AIDS), 979–981
Acrasiomycota, 398–399
Acromegaly, 925
Actin, 844
Action potential, 812
Active immunity, 973–975
Active site, of enzymes, 76
Active transport
   cell membrane, 102–103
   roots, 495–498
Adaptation, evolutionary, 271
Adaptive radiation, 308
Addiction
   alcohol, 995
   drugs, 992–993
   tobacco, 997
Addison disease, 920
Adenine, 141, 146
Adenosine diphosphate (ADP), 116
Adenosine monophosphate (AMP), 116
Adenosine triphosphate (ATP)
   cellular respiration, 127
   photosynthesis, 116–117, 120–121, 124
   structure, 116
Adrenal cortex, 920
Adrenal glands, 920–921
Adrenal medulla, 921
Adrenaline, 921
Adrenocorticotropic hormone (ACTH), 924
Aerobes, obligate, 367
Aerobic metabolism, evolution of, 347–348
Aerobic process, 124
African sleeping sickness, 392, 958–959

Age of Earth, 272–276
   absolute dating, 276
   geological evidence, 273, 274, 293–294
   radioactive dating, 274–276
   relative dating, 274
Agglutination, 230, 972
AIDS, 979–981
Air pollution, 1055–1059
   acid rain, 1056–1057
   greenhouse effect, 1057–1058
   ozone layer and, 1058–1059
   smog, 1055–1056
Air roots, 510
Air sacs, 727, 728
Alcohol abuse
   addiction, 995
   diseases caused by, 994
   effects of, 993–994
Alcoholic fermentation, 131, 419–420
Alcoholics Anonymous, 995
Alcoholism, 994–995
Aldosterone, 920
Algae, 394. *See also* Protists.
   accessory pigments, 435
   adaptation to life under water, 434
   brown algae, 436, 439
   characteristics of, 433–434
   classification of, 434
   forms of chlorophyll in, 434–435
   golden-brown algae, 398
   green algae, 435, 436–438
   red algae, 436, 439
   reproduction, 440–441
   uses, 442–443
   yellow-green, 398
Alleles, 184
Allergies, 977
Alligators, 717
Alternation of generations
   algae, 440
   ferns, 458–459
   mosses, 453–454
   seed plants, 469, 533–535
Alveoli, 885–886, 887
Amebas, 389
   parasitic, 392–393

Amebic dysentery, 392, 958
Amebocyte, 561
Ameboid movement, 389
Amino acids
   essential amino acids, 863
   structure of, 74
Amniocentesis, 240–241
Amnion, 708, 744, 943
Amniotic eggs, 708
Amniotic sac, 944
Amphetamines, 988–989
Amphibians
   circulatory system, 697
   definition, 692–693
   evolution, 694–695
   excretion, 698
   feeding, 695–696
   frogs and toads, 700–701
   nervous and sensory system, 698–699
   parental care, 700
   predators of, 699
   reproduction, 699
   respiration, 696–697
   salamanders, 700
Anabolic steroids, 926
Anabolism, 29
Anaerobes
   facultative anaerobes, 367
   as first cells, 345
   obligate anaerobes, 367
Anaerobic process, 130
Anal pore, 385
Analogous structures, 308
Anaphase, 169
Androgens, 921, 936
Angiosperms, 473–475, 477–481
   carnivorous plants, 510–511
   climbing plants, 510
   coevolution with animals, 476–477, 481
   desert plants, 508–509
   dicots, 474–475
   flowers and reproduction, 534–539
   fruit and flowers, 473
   life cycle, 534–535
   monocots, 474–475
   pollination, 477–479, 537–538

salt-tolerant plants, 510
seed dispersal, 479–481
water plants, 509–510
Animal behavior, 771–778
  genetic coding and, 778
  imprinting, 775
  instincts, 772–773
  learning, 773–774
  survival and, 771–772
  survival and social behavior, 778
Animal communication
  insect communication, 626–627, 777
  methods of, 776–777
  reasons for, 776
Animalia, kingdom, 328
Animals
  basic characteristics, 555
  breeding, 247–248
  cell specialization, 556
  essential life functions, 556–559
  evolution, 559–560
  in genetic engineering, 255–256
Annelids, 594–600
  circulatory system, 596
  definition, 594
  earthworms, 599
  excretion, 596
  feeding, 594–595, 599–600
  importance, 600
  leeches, 599–601
  movement, 597
  nervous and sensory system, 597
  polychaetes, 598–599
  reproduction, 597–598
  respiration, 595–596
Annual rings, trees, 519
Annuals, 517
Anteaters, 748
Anterior, definition, 559
Anterior pituitary, 924
Anther, 535, 537
Antheridia, 410, 453, 458
Anthozoa, 567
Anthropoids, 758–759
Antibiotics, 374, 955
Antibodies, 972–973
Anticodon, 150, 151
Antidiuretic hormone (ADH), 924
Antigens
  antigen-binding sites, 972
  blood groups, 230, 231
Aorta
  in fishes, 684
  in humans, 900

Apes, 749–750, 758–759
Apical meristem, 491
Appendix, 741, 875
Aquatic biomes, 1016–1020
  estuaries, 1018–1019
  freshwater biomes, 1016–1017
  marine biomes, 1017–1018
Aqueous humor, 828
Arachnids, 618–620
Archaebacteria, 362
Archegonia, 453, 458
Arteries, 900
Arthritis, 841
Arthropods
  body structure, 609–610
  centipedes, 622
  circulatory system, 612
  crustaceans, 620–621
  evolution, 608–609
  excretion, 612
  feeding, 610
  growth and development, 615–616
  horseshoe crabs, 617–618
  importance of, 629
  insects, 622–628
  millipedes, 622
  mites and ticks, 619
  movement, 614
  negative aspects, 630–631
  nervous and sensory system, 613–614
  reproduction, 614
  respiration, 611
  scorpions, 619–620
  spiders, 618–619
  subphyla, 607
Artificial selection, 294–295. *See also* Selective breeding.
Artiodactyla, 749
Ascarids, 576–577
Ascomycetes, 411–413
Ascospores, 412, 413
Ascus, 412
Asexual reproduction
  animals, 559, 562, 566, 572, 666
  bacteria, 367–368
  fungi, 408, 409, 411
  plants, 542
  process of, 28, 666
  protists, 385
Astasia, 397
Asters, 168
Asthma, 977
Athlete's foot, 415, 422, 958
Atmosphere of ancient Earth, 343

Atomic number, 48–49
  determination of, 49–50
Atoms, 47–51
  chemical bonding, 52–54
  chemical compounds, 51
  elements, 49–51
  mass, 49
  parts of, 48
  size of, 47
  subatomic particles, 47–48
Atria
  amphibian heart, 697
  fish heart, 684
  human heart, 899
  reptile heart, 711
  vertebrate hearts, 794–795
Auditory canal, 829
AUG (start codon), 149, 151
Australopithecus, 761
Autoimmune disease. *See* Immune disorders.
Autonomic nervous system
  parasympathetic nervous system, 826
  sympathetic nervous system, 826
Autosomes, 209, 210
Autotrophs, 115
  energy sources, 365
Autumn leaves, 117
Auxin, 522–524, 525, 545
Avery, Oswald, 139
Axon, 810, 814
Axon terminals, 810, 814

B

B-lymphocytes (B-cells), 972–973
Bacilli, 364–365
Bacteria, 361–372
  Archaebacteria, 362
  cell wall, 364
  control of, 375
  Cyanobacteria, 362
  decomposition of dead material, 370
  energy sources, 365
  Eubacteria, 361–362
  food poisoning, 366
  genetic engineering, 253–255
  importance, 369–370
  movement, 365
  nitrogen fixation, 371–372, 1024–1025
  Prochlorobacteria, 362
  reproduction, 367–369

respiration, 365–367
saprophytes, 370
shapes of, 363–364
Bacterial diseases
    antibiotics and, 374
    examples of, 374, 957
    mechanisms in, 374, 957–958
Bacterial reproduction
    binary fission, 367–368
    conjugation, 368
    spore formation, 368–369
Bacteriophages, 139, 356
Balance, and ear, 830
Baldcypress trees, 509, 510
Baldness, 238–239
Ball-and-socket joint, 841
Barbiturates, 989
Base pairing, 143
Bases, formation of, 66
Basidiomycetes, 413–414
Basidiospore, 413, 414
Basidium, 413, 414
Bats, 747
Bee "dances," 626–627
Beijerinck, Martinus, 356
Bernal, J., 345
Biennials, 517
Bilateral symmetry, 559, 654, 665
Bile, 873
Binary fission, 348, 387, 389, 396
    bacterial reproduction, 367–368
Binocular vision, 758
Binomial nomenclature, 321
Biochemical taxonomy, 324–325
Biodegradable materials, 1053
Biogeochemical cycles, 1023–1025
    carbon cycle, 1025
    nitrogen cycle, 1024–1025
    oxygen cycle, 1025
    water cycle, 1024
Bioleaching, 371
Biological magnification, 1053–1054
Biologists
    requirements/qualifications for, 33
    role of, 31–32
    types of, 32–33
Biology
    branches of, 32–33
    laboratory techniques used, 37–38
    living things, nature of, 27–31
    microscope, use in, 34–37
Biomes. See Ecosystems; Aquatic biomes; Land biomes.
Biosphere, 1007
Bipedal locomotion, 760

Birds
    circulatory system, 727
    evolution, 724–725
    excretion, 727
    feather types, 724
    feeding, 725–726
    importance, 731
    movement, 728–729
    nervous and sensory system, 727–728
    pollinators of plants, 479
    reproduction, 729–730
    respiration, 727
Birthrate, 1050
Bivalves, 591, 593
Bladderworts, 511
Blastocyst, 943
Blastopore, 655
Blood, 905–907
    plasma, 905
    platelets, 906–907
    red blood cells, 905–906
    white blood cells, 906
Blood groups, 230–231
    ABO groups, 230
    Rh groups, 231
Blood pressure, 903–904
Blood vessels, 900–901
    arteries, 900
    capillaries, 901
    veins, 901
Blooms, of microorganisms, 400, 1026, 1060–1061
Body cavities, 656–657
Bolus, 870
Bone marrow, 838
Bones, 837–842
    bone/muscle interactions, 846
    cartilage, 839–840
    development of, 839–840
    joints, 840–842
    ligaments, 841
    structure of, 838
    tendons, 846
Bony fishes, 680, 689–690
    coelacanth, 690
    lungfishes, 690
    ray-finned fishes, 689–690
Book gill, 611
Book lung, 611
Borel, Jean, 416
Botulism, 367
Bowman's capsule, 908–909
Boysen-Jensen, P., 522
Brain, human, 817–822
    brainstem, 819
    brain waves, 821–822

    cerebellum, 818–819
    cerebral cortex, 818, 820
    cerebrum, 818
    composition, 817
    functions, 820–821
    hypothalamus, 820
    memory and, 822
    sleep and, 822
    thalamus, 820
Brain, invertebrate, 560, 666
    arthropod, 613
    chordate, 645
    flatworm, 571
Brain, vertebrate, 796
    bird, 727–728
    fish, 685
    mammal, 742
    reptile, 713
Brain size, hominids, 760
Brainstem, 819, 822
Branches, effect of auxin on, 523–524
Bread mold, 410–411
Breast milk, 947
Breathing
    mechanics of, 697, 727, 741, 886–887
    respiration and, 129
Breeding, 247–248
    hybridization, 248
    inbreeding, 248
    preservation of genetic diversity and, 250
    selective breeding, 247–248, 250
Brittle stars, 640, 642
Bronchi, 793, 885
Bronchioles, 885
Bronchitis, 996
Brown algae, 436, 439
Brown, Robert, 88, 92
Bryophytes, 451–454
    alternation of generations, 451, 453–454
    characteristics of, 451–452
Budding
    asexual reproduction, 562, 566
    plants, 543–544
Bud scales, 525
Bulbs, 501
Burbank, Luther, 248
Bursitis, 841

C

Cactuses, 509
Caimans, 717
Cairns-Smith, G., 345
Calcitonin, 919

Calcium, role in plant growth, 489
Calorie, 123, 855–856, 860
Calvin, Melvin, 121
Calvin cycle, 121–123
Calyx, 535
Cancer, 960–962
   carcinogens, 962
   immune therapy, 976
   oncogenes, 961
   radiation and, 961–962
   smoking and, 996, 997
   treatment methods, 962–963, 976
   tumors, 960
   uncontrolled cell growth, 162, 164
   as viral disease, 374, 961
Capillaries, definition, 901
Carapace, 620, 718
Carbohydrates
   definition, 70
   dehydration synthesis, 70–71
   disaccharides, 71
   human needs, 858–859
   hydrolysis, 72
   monosaccharides, 70–71
   polysaccharides, 71–72
   starches, 659
   sugars, 858
Carbon, characteristics of, 68–69
Carbon cycle, 1025
Carbon dioxide
   greenhouse effect, 1057–1058
   respiration and, 887, 890
Carbon fixation, 122
Carbon-14 dating, 275, 280
Carbon monoxide, 889
Carcinogens, 962
Cardiac muscle, 844, 898
Carnivora, 748
Carnivores
   amphibians, 696
   birds, 726
   definition, 556–557
   fishes, 682–683
   invertebrate, 587, 594, 599, 610, 618, 619–620, 622, 639
   mammals, 739–740
   reptiles, 710–711, 716
Carnivorous plants, 510–511
Carpels, 536
Carrying capacity, 1035
Cartilage, 839–840
Cartilaginous fishes, 680, 681, 688–689
   rays and skates, 689
   sharks, 688–689

Casparian strip, 497
Catabolism, 30
Catalysts, 75–76
Cecum, 683, 741
Cell biologists, 32
Cell body of neuron, 810
Cell cultures, 38
Cell cycle, 166, 172
Cell division, 161
   cell cycle, 166, 172
   chromosomes and, 164–166
   cytokinesis, 164, 166, 170–171
   interphase, 166–167
   mitosis, 164, 166–170
Cell growth
   controls of, 162
   limits of, 159–161
   rates of, 161
   uncontrolled growth, 162, 164
Cell-mediated immunity, 975
Cell membrane, 90–91, 99–102
   active transport, 102–103
   composition of, 90–91
   diffusion, 99–100
   facilitated diffusion, 102
   function, 90
   osmosis, 100–102
   permeability of, 100
Cells
   cell membrane, 90–91, 99–104
   cell wall, 91–92
   composition of living organisms and, 27–28
   cytoplasm, 93
   cytoplasmic organelles, 94–99, 118
   definition, 87
   nucleus, 92–93
   size variations, 89
Cell specialization, 104
   animals, 556
   light-sensitive cells, 106
Cell theory, 88
Cellular respiration, 124–129, 881
   adenosine triphosphate (ATP) formation, 127
   electron transport, 126–127
   energy sources, 128–129
   Krebs cycle, 125–126
   mitochondria and, 125, 126, 127, 129
Cellulose, 71, 92, 859
Cell wall, 91–92
   of bacteria, 364
   of fungi, 408
Celsius scale, 13
Centipedes, 622

Central nervous system, human, 816, 817–823
   brain, 817–822
   spinal cord, 823
Centrifugation, 38
Centrioles, 98, 168
Centromere, 166, 169
Cephalization, 560, 570, 666
Cephalopods, 592
Cerebellum
   bird, 727
   fish, 685
   human, 818–819
   mammal, 742
   reptile, 713
   vertebrate, 796–797
Cerebral cortex, 818, 820
Cerebral medulla, 818
Cerebrum
   bird, 727
   fish, 685
   human, 818
   mammal, 742
   primate, 758
   reptile, 713
   vertebrate, 796
Cetacea, 748
Chagas disease, 392
Chargaff, Erwin, 143
Chase, Martha, 139–141
Chelicerae, 617
Chelicerates, 607, 617–620
Chemical bonding, 52–54
   covalent bonds, 53–54
   ionic bonds, 53
Chemical change, definition, 46
Chemical compounds
   chemical formula, 51
   formation of, 51
   inorganic compounds, 68
   organic compounds, 68–69
   polymerization, 69
   properties of, 54
   similarities in different organisms, 285
Chemical contamination of environment, 1059–1060
Chemical equation, 56
Chemical properties of matter, 46
Chemical reactions, 55–56
   energy flow and, 56
Chemical signals in animal communication, 777
Chemical symbol, 49
Chemistry
   chemical reactions, 55–56
   matter, 45–54
Chemoreceptors, 686, 830, 831

Chemotherapy, cancer, 963
Chemotrophic autotrophs, 365
Chemotrophic heterotrophs, 365
Childbirth, 945
Chimpanzees, sign language and, 777
Chiroptera, 747
Chitin, 408, 609
Chlorophyll, 115–118, 362,
   algae and, 434–435
   forms of, 435
Chloroplasts, 94–95, 122, 349
Cholesterol, 74, 862
Chondrichthyes, 688
Chondrus crispus, 439
Chordata, 645
Chordates, 645–647
   characteristics of, 645–646
   lancelets, 646
   tunicates, 645–646, 647
   vertebrates, 679
Chorion, 943, 944
Chorionic villus biopsy, 241
Chromatids, 166, 168, 169
Chromatin, 165
Chromosomal mutations, types of, 212–213
Chromosomes, 93, 164–166
   autosomes, 209, 210
   chromosome theory of heredity, 206
   composition of, 162–165
   diploid and haploid cells, 193–194
   genes and, 206
   genetic information, 164–165
   homologous chromosomes, 193, 195, 196
   manipulation of chromosome numbers, 214
   meiosis, 193–196
   mitosis, 168–170
   number in humans, 227–228
   sex chromosomes, 209–211
   structure, 166
Chrysophytes, 398
Chyme, 871
Cilia, 98, 106, 384
Ciliates, 384
Ciliophora, 384–386
Circulation pathways, 902–903
Circulatory diseases, smoking and, 997
Circulatory system, human
   blood, 905–907
   blood pressure, 903–904
   blood vessels, 900–901
   circulation pathways, 902–903
   heart, 898–900
   lymphatic system, 904
   open circulatory system, 662
   single-loop circulatory system, 794
Circulatory system, invertebrate, 557, 588, 596, 612, 662
   closed circulatory system, 588, 662
   open circulatory system, 588, 662
Circulatory system, vertebrate, 684, 697, 711, 727, 741, 794-795
   double-loop circulatory system, 697, 794
   single-loop circulatory system, 697, 794
Cirrhosis of the liver, 994–995
Citric acid, 125
Citric acid cycle, 125
Clams, 589, 591
Classes, in taxonomy, 322
Classical conditioning, 774
Classification, 319–329
   binomial nomenclature, 321
   biological classification, 320–323
   five-kingdom system, 325, 327–328
   Linnaeus's system, 321–322
   reasons for, 319–320, 329
   taxonomy, 321–324
Clay soil, 487
Cleavage, 655, 943
Climax community, 1009, 1010
Climbing plants, 510
Cloning
   DNA, 252–253
   of plants, 545
Closed circulatory system, 588, 662
Club mosses, 456
Cnidaria, 564
Cnidarians, 564–569
   characteristics, 564–566
   corals, 568–569
   hydras, 566–567
   importance, 568–569
   jellyfish, 567
   sea anemones, 567
Cocaine, 989–990
Cocci, 364
Cochlea, 829
Codominance, 216, 230, 232
Codon, 149–150, 151
Coelacanth, 690
Coelom, 657
Coelomates, 654
Coevolution, 476–481, 1037, 1040–1041
Colchicine, 171, 214
Collar cells, 561, 562
Colon, 875
Colony
   algal, 436
   protist, 381
Colorblindness, 237–238
Color vision, 829
Commensalism, 1041
Common descent, 271
Communication, 626–627, 775–777
Communities, 1008, 1040
Companion cells, 493
Competition as population control, 1036
Compound, chemical. *See* Chemical compounds.
Compound light microscope, 34–35
   limit of resolution, 34, 35
   methods for use, 35, 40
Cones, eye, 828, 829
Cones, plant, 471, 472
   pollen cones, 473
   reproductive function, 534
Conidia, 411
Conidiophores, 411
Conifers, 472–473
   leaves, 473
   reproduction, 473
Conjugation
   in bacteria, 368
   in *Paramecium*, 385–386
Conservation, environment, 1069–1071
Consumers, 1022
Contour feathers, 724
Contractile vacuole, 102, 104, 385
Control setup, 8
Convergent evolution, 308, 787
Corals, 568–569
Cork cambium, 491, 500, 518
Cork cells, 500
Corm, 502
Cornea, 827
Corn smut, 422
Corolla, 535
Corpus luteum, 940–941
Cortex, in roots, 494, 497
Corticosteroids, 920
Cortisol, 920

Cotyledons, 474, 539
Covalent bonds, 53–54, 152
Crack, 990, 991
Crick, Francis, 142–143
Crocodiles, 717
Crocodilia, 717
Crop, bird, 724
Crossbridges, 844
Crossing-over of chromosomes, 222
  in meiosis, 195, 196, 208
Cross-pollination, 182, 538
Crowding and stress as population control, 1038–1039
Crustaceans, 607, 620–621
Crystalin, 828
Cubic centimeter, 12
Cushing syndrome, 920
Cuticle, 456, 469
Cuttings, plant, 543
Cuttlefish, 592
Cyanobacteria, 362
Cycads, 472
Cyclic AMP (cAMP), 928
Cyclosporine, 416
Cytokinesis, 164, 166, 170–171
Cytokinins, 524, 545
Cytoplasm, 93
Cytoplasmic organelles
  chloroplasts, 94–95, 118
  cytoskeleton, 97–98
  endoplasmic reticulum, 95–96
  Golgi apparatus, 96
  lysosomes, 97
  mitochondria, 94, 125
  plastids, 97
  ribosomes, 95, 150–152, 146
  vacuoles, 97
Cytosine, 141, 146
Cytoskeleton, 97–98

D

Dark reactions of photosynthesis, 121–123
Darwin, Charles, 33, 392, 522, 600
Darwin's theory, 269–271, 291–297
  adaptation, 271
  artificial selection, 294–295
  common descent, 271, 285–286
  earlier influences in, 292–295
  fitness of organisms, 270, 271, 301, 303
  natural selection, 296–298, 299–300, 305–306, 310
Data, definition, 9

DDT, 1053–1054
Deathrate, 1050
Decomposers, 1022
  bacteria, 370
  fungi, 408, 417
Deep-sea zone of ocean, 1018
Dehydration synthesis
  carbohydrates, 70–71
  peptide bonds, 74
Deletion mutation, 212
Democritus, 47
Dendrites, 810
Denitrification, 1025
Density-dependent limiting factors, 1036–1039
Density-independent limiting factors, 1039
Deoxyribonucleic acid (DNA). *See* DNA.
Depressants, 989
Depression, seasons and, 797
Dermis, skin, 848
Desert plants, 508–509
Deserts, 1015
Detritus feeders, 557, 639, 643
Deuteromycetes, 415, 416
Deuterostomes, 654, 655, 656
Diabetes mellitus, 923
Diaphragm, 886
Diatoms, 398
Dicots
  characteristics, 474–475
  stem growth, 519–520
Diffusion
  cell membrane and, 99–100
  facilitated diffusion, 102
Digestion, 867
Digestive system, human, 867–875
  digestive enzymes, 873–874
  esophagus, 870
  functions, 867–868
  gallbladder, 873
  large intestine, 875
  mouth, 868–870
  pancreas, 872
  pharynx, 867
  small intestine, 872–875
  stomach, 871
Digestive system, invertebrate, 594, 571, 639, 659–661
  extracellular digestion, 659–661
  intracellular digestion, 659–660
Digestive system, vertebrate, 792
  amphibian, 695–696
  bird, 726
  fish, 682–683
  mammal, 740–741

Dimensional analysis, 14–15
Dinoflagellates, 396
  red tide, 400–401
Dinosaurs, 722
Dipeptide, 74
Diploid cells, 193–194, 196, 227–228
Disaccharides, 71, 858
Disease
  bacterial, 957–958
  cancer, 960–962
  fungal, 958
  germ theory of, 954–955
  infectious, 953
  nature of, 953
  protozoan, 958
  rickettsial, 958
  spread of, 954
  viral, 956
Divergent evolution, 308, 787
Division of labor, 556
DNA (deoxyribonucleic acid), 76, 92, 93, 137–139
  genetic code, 137–141
  genetic engineering, 251–259
  replication, 144–145
  structure, 141–144
  as transforming factor, 139
DNA fingerprinting, 256–258
DNA insertion, 252–253
DNA sequencing, 253
Dolphins, 748
Dominance, genetic, 184–185, 215–216
Dopamine, 989
Dormancy
  plants, 500–501, 525–526
  seeds, 541–542
Dorsal, definition, 559
Double fertilization, 538
Double helix, DNA, 143
Double-loop circulatory system, 794
Down feathers, 724
Down syndrome, 239–241
  genetic factors, 239–240
  prenatal diagnosis of, 240–241
Drug, definition, 987
Drug abuse
  addiction, 992–993
  cocaine, 989–990
  crack, 990, 991
  depressants, 989
  hallucinogens, 988
  marijuana, 987–988
  opiates, 990–991

stimulants, 988–989
Duchenne muscular dystrophy, 849
Duckbill platypus, 744, 746
Duct glands, 918
Duodenum, 872
Duplication mutation, 212
Dwarfism, 924

E

Ear, human
  balance and, 830
  hearing and, 829
  parts, 828–829
Eardrum, 829
Earth. *See also* Ecosystems.
  age of Earth, 272–276
  ancient Earth, 342–343
  compared to spaceship, 20–21
Earthworms, 599
Echidna, 746
Echinodermata, 637
Echinoderms, 637–644
  body structure, 638–639
  brittle stars, 642
  circulation, 639–640
  excretion, 640
  feeding, 639
  importance, 644
  movement, 641
  nervous and sensory systems, 640
  predators of, 640–641
  reproduction, 641
  respiration, 639
  sand dollars, 643
  sea cucumbers, 643
  sea lilies, 643–644
  sea urchins, 643
  starfish, 642
  water-vascular system, 638
Ecological pyramids, 1022–1023
Ecological succession, 1009–1010
Ecologists, 33
Ecology, 1007
Ecosystems, 1008
  aquatic biomes, 1016–1020
  biogeochemical cycles, 1023–1025
  ecological pyramids, 1022–1023
  ecological succession, 1009–1010
  energy flow, 1021–1022
  food chain and food webs, 1026–1027
  interactions among, 1041, 1042
  land biomes, 1010–1015
  nutrients in, 1025–1026
Ectoderm, 655, 943
Ectotherms, 720–721, 723, 788–789
Edentata, 748
Effectors, 814
Egg nucleus, 536
Eggs, 193, 196, 441
  amniotic, 708
  bird, 729–730
  fish, 687
  human, 938, 942
  invertebrate, 667–668
  monotreme, 744, 746
  reptile, 714–715, 719
Electrical signals in animal communication, 777
Electric fishes, 686
Electric potential, 811
Electroencephalogram (EEG), 821
Electron, definition, 48
Electron microscopes, 35–36
  scanning electron microscopes (SEM), 36, 37
  transmission electron microscopes (TEM), 36, 37
Electron transport
  cellular respiration, 126–127
  photosynthesis, 119–120
Electrophoresis, 254
Elements
  in chemical compounds, 51
  isotopes, 49–50
  number of elements, 49
  symbols for, 49
Elephantiasis, 578
Elephants, 749
Ellipsoid joint, 842
Embryo
  early development of animal, 655
  human development, 942–943
  of seed, 470
  similarities across species in early development, 283
Embryo sac, 536
Emphysema, 996
Endangered species, 1065–1066, 1068
Endocrine glands, human
  adrenal glands, 920–921
  hypothalamus, 925
  pancreas, 921–923
  parathyroid glands, 919
  pituitary gland, 923
  reproductive glands, 920–921
  thyroid gland, 918–919
Endocrine system, human, 917–929
  negative-feedback mechanisms, 927–928
  polypeptide hormones, 928–929
  prostaglandins, 929
  steroid hormones, 929
Endocytosis, 97, 103, 660
Endoderm, 655, 943
Endodermis, roots, 497–498
  Casparian strip, 497
Endoplasmic reticulum, 95–96
Endorphins, 992–993
Endoskeleton, 558, 641, 659
Endosperm, 538, 539–540
Endospore, 368
Endosymbiont Hypothesis, 382–383
Endotherms, 720, 722, 789
Energy
  chemical reactions and, 56
  use of in living things, 29–30
Energy conversion
  cellular respiration, 124–129
  fermentation, 130–131
  glycolysis, 123–124
  photosynthesis, 113–123
Energy flow in ecosystems, 1021–1022
Energy levels of electrons, 48
Environmental influences
  on human traits, 228–229, 292–293
Enzymes, 75–76
  digestive, 873–874
  function, 76
Epicotyl, 539
Epidermal tissue. *See also* Epidermis.
  leaves, 503
  plants, 491
  roots, 494–496
Epidermis. *See also* Epidermal tissue.
  definition, 847
  human skin, 831, 847–848, 970
  invertebrate, 564
Epididymis, 927
Epiglottis, 870
Epinephrine, 921
Epochs, 276
Equilibrium, 312
Eras, 276
Esophagus, 683, 696, 870

Reference / 49

Essential amino acids, 863
Essential life functions, animal, 556–559
Estrogens, 921, 936
Estuaries, 1018–1019
Ethologists, 32, 33, 776
Ethyl alcohol, 993
Eubacteria, 361–362
Euglena, 394–396, 397
Euglenophyta, 394–396
Eukaryotes, 92, 93, 151
   evolution of, 348, 349
     regulation of gene expression in, 219–220
Evergreens, 472–473
Evolution. *See also* Life, origins of.
   animal, 559–560
   coevolution of plants and animals, 476–477, 481
   convergent evolution, 308, 787
   Darwin's theory, 269–271, 291–297
   divergent evolution, 308, 787
   Earth, age of, 272–276
   fossils, 278–282
   genetic drift, 311–312
   genetics and, 299–302, 311–312
   gradualism, 312
   land plants, 449–451, 455, 456
   punctuated equilibria theory, 313
   rapid evolution, 312–313
   seed plants, 471–475
   similarities of living organisms and, 283–285
   speciation, 304–310
Evolutionary relationships, phylogenetic tree, 653–654, 785
Excretion, 557–558
   forms of nitrogenous waste, 664–665
Excretory system, human
   composition of, 908
   kidneys, 908–912
Excretory system, invertebrate, 562, 612, 640, 664–665
   flame cells, 571
   green glands, 612
   Malpighian tubules, 612
   nephridia, 588, 596
Excretory system, vertebrate, 795–796
   amphibian, 698
   bird, 727
   fish, 684–685
   mammal, 741–742
   reptile, 712–713

Exhalation, mechanics of, 886
Exocrine glands, 918, 922
Exocytosis, 104
Exons, 220
Exoskeleton, 558, 609, 620, 659
Experimental setup, 8
Exponential growth, 1033–1034
Exponential growth curve, 1033
Exponential notation, 163
External fertilization, 667
External respiration, 881
Extinction, endangered species, 1065–1066, 1068
Extracellular digestion, 659–661
Eye, human
   color vision, 829
   parts of, 827
   vision and, 828

F

Facilitated diffusion, 102
Facultative anaerobes, 367
Fallopian tubes, 939, 942
Families, in taxonomy, 322
Fat cells, 866
Fats
   energy and, 862
   human needs, 861–862
   types of, 861–862
Fatty acids, 72–73, 861
Feathers
   contour feathers, 724
   down feathers, 724
   powder feathers, 724
Feces, 875
Feeding, invertebrate, 556–557
   annelid, 594–595, 599–600
   arthropod, 610, 618–624, 629–630
   cnidarian, 565
   echinoderm, 639, 642–643
   flatworm, 571, 574
   invertebrate chordate, 646
   mollusk, 587, 592–593
   roundworm, 576–578
Feeding, modes of, 556–557
Feeding, vertebrate, 791–792
   amphibian, 695–696
   bird, 725–726
   fish, 682–683
   mammal, 739–741
   reptile, 710–711
Female reproductive system, human, 938–942
   fertilization of egg, 942
   menstrual cycle, 939–942

   ova development, 938
   ovulation, 939
   reproductive organs, 938, 939
Fermentation, 130–131
   alcoholic fermentation, 131
   lactic acid fermentation, 130–131
Ferns, 457–461
   alternation of generation, 458
   characteristics, 457
   habitat of, 457, 460
   importance, 460–461
   seed ferns, 471
Fertilization
   external fertilization, 667
   human ovum, 942
   internal fertilization, 667, 714
   seed plants, 538–539
Fetal alcohol syndrome, 994
Fetus, development of human, 943–945
Fever, 970–971
Fibrous roots, 494
Fight or flight, 921
Filamentous green algae, 438
Filaments
   of actin, 844–845
   of algae, 438
   of myosin, 844–845
   of stamen, 535
Filarial worms, 578
Filter feeders, 557, 639, 741, 748
Filtration, kidneys, 908–909
Fishes
   bony fishes, 689–690
   cartilaginous fishes, 680, 681, 688–689
   circulatory system, 684
   classification, 680
   evolution, 680–681
   excretion, 684–685
   feeding and digestion, 682–683
   importance, 691–692
   jawless fishes, 680–681, 687–688
   nervous and sensory system, 685–686
   reproduction, 686–687
   respiration, 684
Fitness of organisms, evolution and, 270, 271, 301, 303
Five kingdoms. *See* Kingdoms.
Fixed joints, 840
Flagella, 98, 365, 386
Flagellates, 386, 394

Flatworms, 570–575
  characteristics, 571–572
  flukes, 573–574
  planarians, 573
  tapeworms, 574–575
Fleming, Alexander, 20, 955
Flight, birds, 728–729
Flowering plants. See Angiosperms.
Flowers and flowering, 473
  control of flowering, 526
  day length and flowering, 527
  parts of flower, 535–536
  reproductive function of flowers, 534–539
Flukes, 573–574
Follicle phase, 940
Follicle-stimulating hormone (FSH), 924, 936, 937, 938
Food, as energy source, 128–129
Food chain, 1026–1027
Food labels, 860
Food and nutrition
  carbohydrates, 858–859
  fats, 861–862
  food guide pyramid, 865
  minerals, 857–858
  proteins, 862–863
  role of food, 855–856
  vitamins, 863–864
  water, 856–857
Food poisoning, 365, 366, 367
Food processing, for bacterial control, 375
Food vacuoles, 384–385
Food webs, 1027
Foot, mollusks, 586
Foraminifers, 389, 390, 391
Forests and reforestation, 1063–1065
Fossils, 273, 278–282
  dating of, 280, 282
  fossil record, 281
  location in rock, 278–279
  microfossils, 343
  problems related to, 279–280
  transition fossils, 709
Fox, Sidney, 344
Frameshift mutations, 213
Franklin, Rosalind, 141, 143
Freely movable joints, 840
Freshwater biomes, 1016–1017
Frisch, Karl von, 626
Frogs, 700–701
  life cycle, 699–700
  structure and function, 695–700
Fronds, 457, 461

Fructose, 70
Fruit, 539
  coevolution with animals, 481
  definition, 473
  role in seed dispersal, 480–481
  unripe, 481
Fruiting bodies, 399, 412, 414
Fucus, 439
  reproduction, 441
Fungal disease
  animal diseases, 423
  corn smut, 422
  human diseases, 422, 958
  potato blight, 421
  wheat rust, 421–422
Fungi
  ascomycetes, 411–413
  basidiomycetes, 413–414
  characteristics, 407–408
  deuteromycetes, 415
  disease-causing, 421–423, 958
  ecological role, 417
  human use, 419–420
  kingdom, 327–328, 408
  oomycetes, 409–410
  reproduction, 408, 409–410, 411–412
  spore dispersal, 417–418
  symbiosis, 418–419
  zygomycetes, 410–411

G

Galactose, 70
Gallbladder, 873
Galvani, Luigi, 811
Gametangium, 408, 412
Gametes, 186, 187, 228, 440
  definition, 186
  formation, 193–196
Gametophyte, definition, 441
Ganglia, 560, 816
Gastrointestinal tract, human, 867
Gastropods, 590–591
Gastrovascular cavity, 564
Gastrulation, 943
Gemmules, 562
Gene expression, regulation of
  in eukaryotes, 219–220
  operon-repressor system, 217–219
  in prokaryotes, 216–219
  RNA transcription, 220
Gene interactions. See also Gene expression, regulation of.
  codominance, 216
  incomplete dominance, 215–216

  jumping genes, 221
  polygenic inheritance, 216
Gene linkage, 206–208
Gene mapping, 208–209
Gene mutations, types, 213
Generative nucleus, 537
Genes. See also Heredity.
  alleles, definition, 184
  dominant and recessive, 184–185
  gene pool, 300–301, 302, 312
  genetic variation, 299–300
  jumping genes, 221
  Mendel's experiments, 182–189
  on sex chromosomes, 210–211
Genetic code
  DNA and, 137–141
  nature of, 148–149
Genetic disorders, 231–233, 239–241
  colorblindness, 237–238
  Down syndrome, 239–241
  hemophilia, 238
  Huntington disease, 231
  muscular dystrophy, 238
  nondisjunction disorders, 235–237, 239
  prenatal diagnosis of, 240–241
  sickle cell anemia, 231–233
Genetic drift, 311–312
Genetic engineering, 250–256
  cloning, 252–253
  DNA fingerprinting, 256–258
  DNA insertion, 252–253
  DNA sequencing, 253
  ethical considerations, 259
  of animals, 255–256
  of bacteria, 253–255
  of humans, 258–259
  of plants, 254–255, 480
  recombinant DNA, 252
  restriction enzymes, 251
Genetics, 181
  breeding strategies, 247–248
  evolutionary theory and, 299–302, 311–312
  mutations, 212–213, 248–249
  probability and, 190–191, 200
  Punnett square, use of, 186, 188, 191, 197
Genome, 256
Genotype, 187, 228. See also Hybrids; Purebreds.
  heterozygous, definition, 187
  homozygous, definition, 187
Genus, in taxonomy, 322

Geological evidence for age of Earth, 273, 274, 293–294
Geologic time scale, 273–274
Germination, seeds, 541
Germ theory of infectious disease, 954–955
Gestation period, mammals, 744
Giantism, 924
Gibberellin, 524
Gills
  invertebrate, 587, 595–596, 611, 639, 645–646, 663
  vertebrate, 685, 792
Ginkgoes, 472
Gizzard, in birds, 724
Gliding joint, 841–842
Global ecologists, 33
Glomerulus, 908
Glucagon, 922–933
Glucose, 70, 115, 858
Glycerol, 73
Glycogen, 71
Glycolysis, 123–124
Golgi apparatus, 96, 97
Gonads, 921, 936
Gonium, 437
Gradualism, 312
Grafting, plants, 543–544
Gram, Christian, 364
Gram-negative bacteria, 364
Gram-positive bacteria, 364
Gram staining, 364
Grasslands, 1013–1014
Gravitropism, 525
Grazing animals, 749
Green algae, 435, 436–438
  colonial, 437–438
  multicellular, 438
  single-celled, 437
  threadlike, 438
Greenhouse effect, 1057–1058
Griffith, Frederick, 138
Growth hormone (GH), 924–925
Guanine, 141, 146
Guard cells, 504
Gullet, 384
Gymnosperms, 471–473
  conifers, 472–473
  cycads, 472
  ginkgoes, 472
  life cycle, 534
  reproduction, 471–472, 473, 534
Gypsy moths, 1043

H

Habituation, 773–774
Hagfishes, 688
Hair, 848
Hair follicles, 848
Half-life, 274–276
Hallucinogens, 988
Haploid cells, 194
Harvey, William, 897
Haversian canals, 838
Hearing, 828, 830
Heart, human, 898–900
  heartbeat, 899–900
  parts of, 898–899
Heart, invertebrate, 662
  annelid, 596
  arthropod, 609, 612
  invertebrate chordate, 646
  mollusk, 588
Heart, vertebrate, 794–795
  amphibian, 697
  bird, 727
  fish, 684
  mammal, 738, 741
  reptile, 711
Heartwood, 519
Heliozoans, 389
Helmont, Jan van, 113–114
Helper T-cells, 979
Hemoglobin, 888
Hemophilia, 238
Herbaceous plants, 518
Herbivores, 556
  mammals, 740
Heredity
  blending inheritance, 181–182
  blood groups, 230–231
  gene linkage, 206–208
  gene mapping, 208–209
  genes and chromosomes, 205–206
  genetic disorders, 231–233, 239–241
  Mendel's contribution, 182–189
  polygenic traits, 234
  recombinants, 208
  sex-linked inheritance, 209–211, 235–238
  sex-influenced inheritance, 238–239
Hermaphrodite, 566, 667
Heroin, 990, 991
Herrick, James, 231
Hershey, Alfred, 139–141
Heterogamy, 441
Heterotrophs, 115, 407
  energy sources, 365
Heterozygous genotype, 187, 191
Hirudinea, 599
Histamines, 977
Histones, 165, 396
HIV, 979–981
Holdfast cell, 438
Holes in the ozone layer, 1058–1059
Hollow dorsal nerve cord, 645
Homeostasis, 30–31
Hominids, 759–764
  Australopithecus, 761
  characteristics, 760
  Cro-Magnon, 763–764
  Homo erectus, 762, 763
  Homo habilis, 761–762
  Homo neanderthalensis, 763
  Homo sapiens, 763–764
  Homo sapiens sapiens, 764
Hominoids, 759
Homo erectus, 762, 763
Homo habilis, 761–762
Homo sapiens, 763–764
Homologous chromosomes, 193, 195, 196
Homologous structures, 284, 286
Homozygous genotype, 187, 191
Hooke, Robert, 88
Hookworms, 577
Hormones. *See also* Endocrine glands; Endocrine system.
  plant hormones, 521–526
  target organ, 522
Hornworts, 451, 452
Horseshoe crabs, 617–618
Horsetails, 456–457
Human development
  postnatal development, 946
  prenatal development, 943–946
Human evolution
  hominids, 759–764
  primates, 757–759
Huntington disease, 231
Hybridization, 248
Hybrids, 184, 248
Hydras, 566–567
Hydrochloric acid, 871
Hydrolysis, carbohydrates, 72
Hydrostatic skeleton, 659
Hydrozoa, 566
Hypertension, 903–904
Hyperthyroidism, 919
Hyphae, 408, 409, 410, 411, 419
Hypocotyl, 539
Hypothalamus, 820, 925, 927
  pituitary control, 925
Hypothesis
  formation of, 8
  testing of, 8
Hypothyroidism, 918–919

I

Immovable joints, 840
Immune disorders, 977–981
   acquired immune deficiency syndrome (AIDS), 979–981
   allergies, 977
   autoimmune disease, 978
Immune system, human, 969–976
   active immunity, 973–975
   antibodies, 972–973
   cell-mediated immunity, 975
   disorders, 977–981
   immune, definition, 973
   inflammatory response, 970–971
   interferon and, 971
   nonspecific defenses, 969–970
   passive immunity, 975
   skin, 970
Immune therapy for cancer, 976
Implantation, 943
Imprinting, 775
Impulses, nerve, 810, 811–815
Inbreeding, 248
Incomplete dominance, 215–216
Independent assortment of genes, 187–189
Inducer, 217, 218, 219
Infection, 953
Infectious disease, 953
Inflammatory response, 970–971
   inflamed, definition, 970
Ingenhousz, Jan, 114
Inhalation, mechanics of, 886
Inorganic compounds, 68
Insecticides, 55
   DDT, 1053–1054
Insectivora, 747
Insect pollination, 477–479
Insects, 622–628
   body structure, 623
   communication, 626–627
   feeding, 623–624
   insect societies, 624–625
   movement, 624
Insight learning, 774
Instincts, 772–773
Insulin, 922, 923
Integumentary system, human, 847–848
   hair, 848
   nails, 848
   skin, 847–848
Interferon, 971
   and viral disease, 373
Interleukin-2, 976
Internal fertilization, 667, 714

Internal respiration. *See* Cellular respiration.
Interneurons, 810
Interphase, 166–167
Intertidal zone, ocean, 1017
Intracellular digestion, 659–660
Introns, 220
Inversion mutation, 213
Invertebrate chordates, 645–646
Invertebrates. *See also* names of specific phyla and classes.
   circulatory system, 662
   definition, 555
   evolution, 654–656
   excretion, 665
   feeding, 659–661
   movement, 659
   nervous and sensory system, 665–666
   parental care, 668
   reproduction, 666–667
   respiration, 662–663
Ion, 53
Ionic bonds, 53
Iris, eye, 828
Islets of Langerhans, 922
Isogamy, 440
Isotopes, 49–50
   radioactive, 50–51, 140
Iwanowski, Dimitri, 356

J

Jacob, François, 217
Jawless fishes, 680–681, 687–688
   hagfishes, 688
   lampreys, 687–688
Jellyfish, 567
Jenner, Edward, 973
Johanson, Donald, 761
Jointed appendages, 610
Joints, 840–842
   freely movable joints, 840
   immovable joints, 840
   slightly movable joints, 840
Jumping genes, 221
Juvenile-onset diabetes, 978

K

Kangaroos, 746
Karyotype, 241
Kettlewell, H.B.D., 298
Kidneys, human, 908–912
   control of function, 910
   filtration, 908–909
   reabsorption, 909

   structure, 908
   transplantation, 911
Kidneys, vertebrate, 795
   amphibian, 698
   bird, 727
   fish, 685
   mammal, 741–742
   reptile, 712
Killer T-cells, 975, 976
Kilocalorie, 123, 855
Kilogram, 13
Kilometer, 12
Kingdoms, 322, 325, 327–328
   Animalia, 328
   Fungi, 327–328
   Monera, 325, 327
   Plantae, 328
   Protista, 325, 327
Kin selection theory, 779
Klinefelter syndrome, 235–236
Koalas, 746
Koch, Robert, 954
Koch's postulates, 954–955
Kölreuter, Josef, 215
Komodo dragons, 716
Krebs cycle, 125–126

L

Labor, and childbirth, 945
Laboratory methods
   cell cultures, 38
   centrifugation, 38
   microdissection, 38
Laboratory safety, 18
Lactic acid fermentation, 130–131
Lactose, 71, 858
Lagomorpha, 748
Lamarck, Jean Baptiste de, 292–293
Lampreys, 687–688
Lancelets, 646
Land biomes, 1010–1015
   deserts, 1015
   grasslands, 1013–1014
   taiga, 1011–1012
   temperate deciduous forests, 1012–1013
   tropical rain forests, 1014–1015
   tundra, 1011
Landsteiner, Karl, 230
Landsteiner blood groups, 230
Large intestine, human, 875
Larva, definition, 559
Larynx, 883
Lateral buds, effect of auxin on, 523–524

Layering, 543
Leakey, Richard and Mary, 761
Learning, 773–774
 classical conditioning, 774
 habituation, 773–774
 insight learning, 774
 operant conditioning, 774
Leaves, 469, 502–505
 epidermal tissue, 503
 mesophyll tissue, 504–505
 stomata, 504
 structure, 502–503
 varieties, 502
 vascular tissue, 504
Leeches, 599–601
Leeuwenhoek, Anton van, 87–88
Length, metric measurement of, 12
Lens, eye, 828
Levels of organization, 106–107
 organ systems, 106
 organs, 107
 tissues, 106–107
Lichens, 418–419
Life, origins of
 aerobic metabolism, 347–348
 eukaryotes, 348, 349
 first cells, 345
 molecules from space, 344
 organic "soup," 344–345
 oxygen, effects of, 347
 photosynthesis, 346
 proto-life, 344
 sexual reproduction, 348
Ligaments, 841
Light reactions, 118–121
Light-sensitive cells, 106, 828, 829
Limiting factors, 1025–1026
Limit of resolution, 34–35
Linkage groups, 208
Linked genes, 206–208
Linnaeus, Carolus, 321
Lipids, 72–74
 formation, 72–73
 phospholipids, 74
 polyunsaturated fats, 73
 saturated fats, 73
 sterols, 74
Liposomes, 74
Liter, definition, 12
Liver, human, 873
Liverworts, 451, 452
Living things
 cells, 27–28
 getting and using energy, 29–30
 growth and development, 28–29
 homeostasis, 30–31
 reproduction, 28
Lizards, 716
Loamy soil, 488
Lobe-finned fishes, 690
Logistic growth, 1034–1035
Logistic growth curve, 1034
Long-term memory, 822
LSD (lysergic acid diethylamide), 988
Lung disorders, 996
Lungfishes, 690
Lungs, human, 885
 breathing, mechanics of, 886–887
 vital capacity, 891
Lungs, invertebrate, 663
 arthropods, 611
Lungs, vertebrate, 792–794
 amphibian, 696–697
 bird, 727
 mammal, 741
 reptile, 711
Luteal phase, 940–941
Luteinizing hormone (LH), 924
Lycophytes, 456
Lyell, Charles, 272, 293–294
Lyme disease, 959
Lymphatic system, 904
Lymph nodes, 970
Lymphocytes, 972–973
Lysogenic infection, 358–359
Lysosomes, 97
Lysozyme, 970
Lytic infections, 358

M

MacLeod, Colin, 139
Macromolecules, 69
Macronucleus, 384
Magnesium, role in plant growth, 489
Malaria, 387–388, 392, 958
Male reproductive system, human, 937–938
 reproductive organs, 937
 sperm development, 937–938
 sperm release, 938
Malpighian tubules, 612, 665
Malthus, Thomas, 295
Malthusian Doctrine, 295
Maltose, 71, 858
Mammalia, 738
Mammalian groups
 Artiodactyla, 749
 Carnivora, 748
 Cetacea, 748
 Chiroptera, 747
 Edentata, 748
 Insectivora, 747
 Lagomorpha, 748
 marsupials, 739, 746
 monotremes, 739, 746
 Perissodactyla, 749
 placentals, 739, 746–747
 Primates, 749–750
 Proboscidea, 749
 Rodentia, 748
 Sirenia, 749
Mammals
 circulatory system, 741
 definition, 737–738
 evolution, 739
 excretion, 741–742
 feeding, 739–741
 movement, 743
 nervous and sensory system, 742–743
 parental care, 744
 reproduction, 743–744
 respiration, 741
Mammary glands, 738
Manatee, 749
Mandibles, 620, 621
Mangrove trees, 509, 510
Mantle, 586
Margulis, Lynn, 349, 382
Marijuana, 987–988
Marine biomes, 1017–1018
Marsupials, 739, 746
Marsupium, 744
Mass
 atomic, 49
 metric system, 12–13
 as property of matter, 45
Mass extinction, 313
Mass number, 49
Matter, 45–54
 atoms, 47–51
 chemical properties, 46
 definition, 45
 interactions, 52–54
 phases of, 46
 physical properties, 45–46
McCarty, Maclyn, 139
McClintock, Barbara, 221
Medulla
 bird, 727
 fish, 685
 human, 819
 mammal, 742

Medulla oblongata, 819
Medusa, 564–567
Meiosis, 193–196, 387
    crossing-over, 195, 196, 208
    compared to mitosis, 196
    nondisjunction, 213, 235
    phases, 194–195
    relationship to genetics, 195–196
Melanocyte-stimulating hormone (MSH), 924
Memory, 822
    long-term, 822
    short-term, 822
Mendel, Gregor, 33, 182–189
Mendel's experiments, 182–189
    dominant and recessive genes, 184–185
    genes, 183–184
    independent assortment of genes, 187–189
    segregation of genes, 185–187
Menopause, 939
Menstrual cycle, 939–942
    follicle phase, 940
    luteal phase, 940–941
    menstruation, 941–942
    ovulation, 940
Menstruation, 941–942
Meristematic tissue, 491
Mesoderm, 656, 943
Mesophyll tissue, 504–505
Messenger RNA (mRNA), 147, 148, 149, 161, 219
Metabolism, 30
Metamorphosis, 559
    in arthropods, 615–616
Metaphase, 168
Meter, 12
Methanogens, 362
Metric system, 11–13
    length, 12
    mass and weight, 12–13
    temperature, 13
    volume, 12
Microbiologists, 32
Microdissection, 38
Microfilaments, 98
Microfossils, 343
Micronucleus, 384
Microscope, 34–37
    compound light microscope, 34–35
    electron microscope, 35–36
    invention, 88
Microtubules, 98, 168, 169, 171
Midbrain, 819

Migratory birds, 728
Miller, Stanley, 343–344
Millipedes, 622
Minerals, human needs, 857–858
Mites, 619
Mitochondria, 94, 125, 349
    cellular respiration, 125, 126, 127, 129
Mitosis, 167–170
    anaphase, 169
    metaphase, 168
    prophase, 167–168
    telophase, 170
Mixtures, 64–65
Molds. *See* Fungi.
Molecular biologists, 32
Molecular formula, 57
Molecules
    drawings, 57
    formation, 54
Mollusca, 585, 587
Mollusks, 585–593
    bivalves, 591
    cephalopods, 592
    circulatory system, 588
    excretion, 588
    feeding, 587
    foot, 586
    gastropods, 590–591
    mantle, 586
    negative aspects, 593
    nervous system, 589
    reproduction, 589
    respiration, 587–588
    uses, 593
    visceral mass, 586
Molting, 615, 616
Monera
    bacteria, 361–372
    kingdom, 325, 327
Monitor lizards, 716
Monkeys, 749–750, 758–759
Monocots, 474–475
    stem growth, 520
Monod, Jacques, 217
Monomers, 69
Monosaccharides, 70–71, 858
Monotremes, 739, 744, 746
Morgan, Thomas Hunt, 206–208, 210–211
Morphine, 990
Morula, 943
Mosses, 452
    alternation of generations, 453–454
    club mosses, 456
    importance, 460–461
Motor neurons, 810

Mouth, human, 868–870
Mucus, 106
Multicellular organisms, 28
    levels of organization, 106–107
Multiple alleles, 230
Multiple sclerosis, 978
Muscular dystrophy, 238, 849
Muscular system, 659, 842–846
    cardiac muscles, 844
    composition, 842
    muscle and bone interaction, 846
    muscle contraction, 844–845
    skeletal muscles, 843
    smooth muscles, 843–844
Mushrooms
    growth, 414
    poisonous, 420
Mutations, 212–213, 248–249
    chromosomal, 212–213
    gene, 213
    mutagenesis, 249
    mutagens, 249
Mutualism, 1041
Mycelium, 408
Mycorrhizae, 419
Myelin, 813
Myosin, 844
Myxomycota, 399

N

NADH, 124
NADP, 120, 121
NADPH, 120, 124
Nail matrix, 848
Nails, 848
Nasal cavity, 882
Natural selection, 296–298, 299–300, 305–306, 310, 778
Nautiluses, 592
Needham, John, 340
Negative-feedback mechanisms, 927–928
Nematocysts, 565
Nematoda, 570, 575
Nephridia, 588, 665
Nephrons, 908
Neritic zone, 1018
Nerve impulses, 811–814
    action potential, 812
    electric potential, 811
    myelin, 813
    propagation, 813
    resting potential, 811
    threshold, 814

Nervous system, human
  central nervous system, 816, 817–823
  function, 809
  nerve impulses, 811–814
  neurons, 810
  neurotransmitters, 814–815
  peripheral nervous system, 816, 825–826
  synapse, 814–815
Nervous system, invertebrate, 558
  arthropod, 613
  annelid, 597
  cnidarian, 565
  echinoderm, 640
  flatworm, 571
  mollusk, 589
  roundworm, 576
Nervous system, vertebrate, 796–798
  amphibian, 698
  bird, 727–728
  fish, 685–686
  mammal, 742–743
  reptile, 713
Neurons, 810
  characteristics, 810
  nerve impulses, 811–814
  types, 810
Neurotransmitters, 814–815
Neutralization reaction, 66
Neutron, 48
Niche, 304–305
Nicotine, 995
Nitrogen, role in plant growth, 489
Nitrogen cycle, 1024–1025
Nitrogen fixation, 371–372, 1025
Nonbiodegradable materials, 1053
Nondisjunction, 213, 235
Nondisjunction disorders, 235–237, 239
Nonspecific defenses, 969–970
Nose, role in respiration, 882, 883
Notochord, 645, 679
Nuclear envelope, 92, 168, 170
Nucleic acids, 76–77, 136–139
  protein synthesis, 148–152
Nucleolus, 92
Nucleosomes, 166
Nucleotides, 76, 141
Nucleus, in atom, 48
Nucleus, in cell, 92–93
  chromosomes, 93
  nuclear envelope, 92, 168, 170
  nucleolus, 92
Nutrient absorption, in roots, 495–496, 497

Nutrients, in ecosystems, 1025–1026
Nutrition. *See* Food and nutrition.

O

Obligate aerobes, 367
Obligate anaerobes, 367
Oceans
  pollution, 1062
  vents, 1020
  zones, 1017–1018
Octopus, 589, 592
Odd-toed ungulates, 749
Oedogonium, 438
Oil spills, 1062
Olduvai Gorge, 762
Olfactory bulbs, 685
Oligochaetes, 599
Olins, Ada, 165
Olins, Don, 165
Oncogenes, 961
Oomycetes, 409–410
Oparin, Alexander, 344
Open circulatory system, 588, 662
Open-sea zone, 1018
Operant conditioning, 774
Operator, 217, 218
Operon, 217–219
Opiates, 990–991
Opossums, 746
Opposable thumb, 760
Optic lobes, in fish, 685
Optic nerve, 828
Order, in taxonomy, 322
Organelles. *See also* Cytoplasmic organelles.
  function, 94
Organic compounds, 68–69
  amino acids, 74
  carbohydrates, 70–72, 858–859
  enzymes, 75–76
  fatty acids, 72, 73
  lipids, 72–74, 861–862
  nucleic acids, 76–77
  proteins, 74–75, 862–863
Organic "soup," 344–345. *See also* Life, origins of.
Organs and organ systems, 106–107
Osculum, 561
Osmosis, 100–102
Osmotic pressure, 101–102
Ossification, 839
Osteichthyes, 689
Osteocytes, 838
Otoliths, 830
Ovaries
  animal, 641

  flower, 536
  human, 921, 936, 938
Oviparous animals, 687, 714
Ovoviviparous animals, 687, 714
Ovulation, 940
Ovule, 196, 536, 539
Ovum, human, 938, 942
Oxygen
  in evolution, 347
  in respiration, 129, 887
  production in photosynthesis, 120
Oxygen cycle, 1025
Oxytocin, 924, 945
Ozone layer, 347
  holes in, 1058–1059

P

Paal, A., 522
Pain, 929
  pain receptors, 831
Paleontologists, 32, 33, 280, 745
Palisade layer, 504
Pancreas, human, 872–873, 921–923
Pancreatic fluid, 872–873, 922
Paper chromatography, 117
Paramecium, 384–386
Parasites, 557
  role in population control, 1038
  unsegmented worms, 571, 572–575, 576–578
  viruses, 359
Parasympathetic nervous system, 826
Parathyroid glands, 919
Parathyroid hormone (PTH), 919
Parenchyma tissue, plant, 492
  stems, 499
Passive immunity, 975
Pasteur, Louis, 341, 374, 954, 973–974
Pathogens, 372, 956, 970
PCB's, 1054
PCP (phencyclidine), 988
Pellicle, 384, 396
Penfield, Wilder, 821
Penicillin, 955
  discovery, 20, 955
Penicillium, 415
Penis, 938, 942
Pepsin, 871
Peptide bond, definition, 74
Perennials, 517–518
Pericardium, 898

Pericycle, 491
Periods of Earth history, 276
Periosteum, 838
Peripheral nervous system, human, 816, 825–826
   autonomic nervous system, 826
   somatic nervous system, 825
Perissodactyla, 749
Peristalsis, 870, 871
Permafrost, 1011
Permeability, 100
Pesticides, 628, 1053
Petals, 535
Petiole, 502
PGAL (Phosphoglyceraldehyde), 122–123, 124
pH scale, 66
Phaeophyta, 439
Phagocytes, 906, 970
Phagocytosis, 103
Pharyngeal slits, 645
Pharynx, animal, 571, 594–595, 645, 646
Pharynx, human
   role in digestion, 870
   role in respiration, 882
Phases of matter, 46
Phenotype, 186–187, 228
   phenotypic variation, 299–300
Pheromones, 626, 777
Phloem, 455, 474, 475, 493
   functions, 469
   growth, 520
   leaves, 504
   phloem cells, 493
   stems, 499–500
   transport of materials, 506–508
Phosphoglyceraldehyde (PGAL), 122–123, 124
Phospholipids, 74
Phosphorus, role in plant growth, 489
Photoperiodicity, 527
Photoreceptors, 828, 829
Photosynthesis, 29–30, 113–123
   Calvin cycle, 121–123
   chlorophyll, 116, 120
   dark reactions, 121–123
   discovery, 114
   electron transport, 119–120
   energy-storing compounds, 116–117, 120
   equation, 115
   evolution, 346
   light reactions, 118–121
   oxygen production, 120
   photosystems, 118
   sunlight, 115–116, 118
Photosynthetic bacteria, 362
Photosynthetic membranes, 118
Photosystems, in green plants, 118
Phototrophic autotrophs, 365, 395
Phototrophic heterotrophs, 365
Phototropism, 525
Phycobilins, 436
Phycocyanin, 362
Phylogenetic tree, 653–654, 785
Phylum, in taxonomy, 322
Physical properties of matter, 45–46
Phytochromes, 527
Phytoflagellates, 394
Phytoplankton, 401
Pigments, 115–117
   accessory pigments, 435
Pineal gland, 715, 797
Pinocytosis, 103
Pistil, 536
Pitcher plants, 511
Pith, stems, 499
Pituitary gland, 923
   anterior pituitary, 924
   disorders, 924–925
   posterior pituitary, 924
Placenta
   humans, 944
   mammals, 744
Placentals, 739, 746–747
Planarians, 573
Plant cells
   cell wall, 91
   chloroplasts, 94–95
   cytokinesis, 171
   photosynthetic membranes, 118
   plastids, 97
   vacuoles, 97
Plant growth
   hormones, 521–526
   roots, 520–521
   stems, 518–520
Plant hormones, 521–526
   auxin, 522–524, 525
   changes in hormone production and effects, 525–526
   cytokinins, 524
   gibberellin, 524
   tropisms, 525
Plantae, kingdom, 328
Plants. *See also* Seed plants.
   annuals, 517
   biennials, 517
   bryophytes, 451–454
   first land plants, 449–451, 456
   in genetic engineering, 254–255, 480
   kingdom Plantae, 328
   perennials, 517–518
   photoperiodicity, 527
   seed plants, 467–481
   tracheophytes, 455–459
Plant tissues, 491–493
Plasma, 905
Plasma cells, 973
Plasmids, 252
Plasmodia, 399
Plasmodium, 387–388
Plastids, 97
Plastron, 718
Platelets, 906–907
Platyhelminthes, 570
Point mutations, 213
Polar nuclei, 536
Polio, 974–975
Pollen, 472
Pollen chambers, 536–537
Pollen cones, 473
Pollen grain, 470, 536–537, 538
Pollen tube, 538
Pollination, 477–479, 537–538
   attracting pollinators, 478
   bird pollination, 479
   cross-pollination, 538
   insect pollination, 477–479
   self-pollination, 537–538
   vector pollination, 478
   wind pollination, 477
Pollution
   air pollution, 1055–1059
   biological magnification, 1053–1054
   water pollution, 1059–1063
Polychaeta, 598
Polygenic inheritance, 216
Polygenic traits, 234
Polymerization, 69, 70
Polymers, 69
Polyp, cnidarian, 564–567
Polypeptides, 74
   action, 928–929
   production, 148–152
Polyploidy, 213
Polysaccharides, 71–72, 858
Polyunsaturated fats, 73, 862
Pons, 819

Population control
  competition, 1036
  crowding and stress, 1038–1039
  density-dependent limiting factors, 1036–1039
  density-independent limiting factors, 1039
  parasitism, 1038
  predation, 1036–1038
Population growth
  birthrate and deathrate, 1050
  exponential growth, 1033–1034
  human population, 1039–1040, 1050–1052
  lifestyle, 1052
  logistic growth, 1034–1035
Population interactions, 1040–1041
Populations, 300
  definition, 1033
Porifera, 560
Porphyra, 439
Porpoises, 748
Portuguese man-of-war, 567
Posterior, definition, 559
Posterior pituitary, 924
Potassium, role in plant growth, 489
Potato blight, 421
Powder feathers, 724
Predation, 1036–1038
Prehensile tails, 750
Prenatal development, human, 943–946
Prenatal testing
  amniocentesis, 240–241
  chorionic villus biopsy, 241
  ethical considerations, 241
Pressure flow hypothesis, 508
Priestley, Joseph, 114
Primary follicles, 938
Primary immune response, 973
Primary tissue, 518
Primates, 749–750, 757–764
  arthropoids, 758–759
  characteristics, 758
  evolution, 758–759
  hominids, 759–764
  hominoids, 759
Probability, 190–191, 200
Proboscidea, 749
Prochlorobacteria, 362
Producers, 1021, 1022, 1023
Progesterone, 921, 941
Prokaryotes, 92, 151, 165, 343
  bacteria, 361–372
  gene expression, 216–219
  kingdom Monera, 327, 361–365
Prolactin, 924
Promoter, 217, 218
Prophage, 358–359
Prophase, 167–168
Prostaglandins, 929
Proteins, 74–75
  amino acids, 74
  energy, 863
  enzymes, 75–76
  essential amino acids, 863
  function, 75
  human needs, 862–863
  structure, 74–75, 77
Protein synthesis, 148–152
  process, 151–152
  ribosomes, 150–152
  translation, 150, 151
Prothallium, 458
Protista, kingdom, 325, 327, 381
Protists
  blooms, 400–401
  chrysophytes, 398
  Ciliophora, 384–386
  classification, 381–382
  diatoms, 398
  dinoflagellates, 396
  Euglena, 394–396, 397
  evolution, 382–383
  importance, 393–394, 401
  microscope observation, 402
  parasites, 387, 392–393, 958
  protozoa, 384
  Sarcodina, 389–390
  slime molds, 398–399
  Sporozoa, 387–388
  symbiosis, 393, 401
  Zoomastigina, 386–387, 392, 393, 394
Proton, 48
Protonema, 454
Protostomes, 654, 655, 656
Protozoan diseases, 958
Pseudocoelomates, 654, 657
Pseudopod, 389, 391
Puberty, 936
Pulmonary circulation, 902
Punctuated equilibria theory, 313
Punnett square, 186, 188, 191, 197–199
Pupa, 616
Purebred, 183, 192. *See also* Genotype.
Pyloric valve, 871
Pyrrophyta, 396
Pyruvic acid, 125, 130, 131

R

Rabbits, 748
Rabies, 974
Radial symmetry, 559, 641
Radiation exposure, cancer and, 961–962
Radiation therapy, 50, 963
Radicle, 539
Radioactive dating
  age of Earth, 274–276
  fossils, 280, 282
Radioactive decay, 274–276
Radioactive element, 50–51, 274
Radioactive isotopes, 50–51, 140
Radiolarians, 389
Radula, 587
Rain forests, destruction of, 326, 1051, 1064
Rapid eye movement (REM) sleep, 822
Ray-finned fishes, 689–690
Reabsorption, 909
Receptors, 814
Recessive genes, 184–185
Recombinants
  genetic, 208
  recombinant DNA, 252, 255
Recycling, 1070
Red algae, 436, 439
Red blood cells, 905–906
Redi, Francesco, 340
Red tide, 400–401
Reed, Walter, 39
Reflex arc, 825
Reflexes, 823
Reforestation, 1063–1065
Regeneration
  definition, 562
  in amphibians, 701
  in flatworms, 572
  in sponges, 562
  in starfish, 641
Rejection, 975
Relative dating, 274
Relative frequency, 300–301
Releasing hormones, 925
Renal dialysis, 911
Replication, DNA, 144–145, 154
Repressor, 218–219
Reproduction, bacteria, 368–369
Reproduction, fungi, 408
  asci, 412
  ascospores, 412
  basidia, 413
  basidiospores, 413

conidia, 411
conidiophores, 411
gametangia, 408
in ascomycetes, 411–413
in basidiomycetes, 413–414
in deuteromycetes, 415
in oomycetes, 409
in zygomycetes, 410–411
sporangia, 408
sporangiophores, 408
zygospore, 410
Reproduction, general
asexual reproduction, 28
gamete formation, 193–196
gametes, 440
hermaphrodite, definition, 667
heterogamy, 441
isogamy, 440
sexual reproduction, 28
Reproduction, invertebrate, 558–559, 666–667
in annelids, 597–598
in arthropods, 614, 625
in cnidarians, 566–567
in echinoderms, 641
in flatworms, 572, 573–574, 575
in mollusks, 589
in roundworms, 576
in sponges, 562
Reproduction, plant
fertilization, angiosperm, 538–539
flower pollination, 477–479, 537–539
in algae, 440–441
in angiosperms (flowering plants), 473, 474, 477–481, 534–540
in bryophytes, 453–454
in conifers, 473
in ferns, 458–459
in gymnosperms, 471–472, 534
in seed plants, 469–470, 533–544
seed dispersal, angiosperm, 479–481
seed formation, angiosperm, 539–540
vegetative (asexual) reproduction, 542–545
Reproduction, protist
in acellular slime molds, 399
in cellular slime molds, 398–399
in chrysophytes, 398
in ciliates, 385–386
in euglenophytes, 396
in pyrrophytes, 396
in sarcodines, 389
in sporozoans, 387–388
in zoomastiginans, 387
Reproduction, vertebrate, 798–799
in amphibians, 699–700
in birds, 729–730
in fishes, 686–687
in mammals, 743–744, 746
in reptiles, 714–715
oviparous animals, definition, 687
ovoviviparous animals, definition, 687
viviparous animals, definition, 687
Reproductive glands, 921
Reproductive isolation, 305
Reproductive system, human
female, 938–942
male, 937–938
sexual development, 925–926
Reptiles
adaptations to land, 707–708
circulatory system, 711–712
crocodilians, 717
evolution, 709–710
excretion, 712–713
feeding and digestion, 710–711
importance, 719
lizards, 716
movement, 713–714
nervous and sensory system, 713
reproduction, 714–715
respiration, 711
snakes, 716–717
tuataras, 715
turtles, 718, 719
Respiration. *See also* Cellular respiration.
bacterial, 365, 367
role of oxygen, 129
types, 127
Respiratory system, human
alveoli, 885–886, 887
bronchi, 885
bronchioles, 885
control, 890
gas exchange, 887–888
larynx, 883
levels, 881
lungs, 885
mechanical aspects, 886
nasal cavity, 882
nose, 882, 883
pharynx, 882
trachea, 882–885
Respiratory systems, invertebrate, 662–663
annelid, 595–596
arthropod, 611
echinoderm, 639
mollusk, 587–588
Repiratory systems, vertebrate, 792–794
amphibian, 696–697
bird, 727
fish, 684
mammal, 738, 741
reptile, 711
Resting potential, 811
Restriction enzymes, 251
Retroviruses, 359
Rh blood groups, 231
Rheumatic fever, 978
Rheumatoid arthritis, 978
Rhizoids, 411, 452
Rhizomes, 457, 459, 501
Rhodophyta, 439
Rhodopsin, 106, 828
Rhynchocephalia, 715
Ribonucleic acid (RNA). *See* RNA.
Ribosomal RNA (rRNA), 150
Ribosomes, 95
protein synthesis, 150–152
Rickettsias, 374
disease, 958
Ringworm, 415, 422, 958
River blindness, 579
RNA (ribonucleic acid), 76, 92, 146–148
codons, 149–150
compared to DNA, 146
messenger RNA, 147, 148, 149, 161, 219
ribosomal RNA, 150
structure, 146
transcription, 146–147, 151
transfer RNA, 150
Rocky Mountain spotted fever, 958
Rodentia, 748
Rodents, 748
Rods, 828
Root cap, 520
Root hairs, 495

Roots, 468, 494–498
   cortex, 494, 497
   effect of auxin on, 523
   endodermis, 497–498
   epidermis, 494–496
   fibrous roots, 494
   growth, 520–521
   nutrient absorption, 495–496, 497
   root pressure, 498, 505
   taproots, 494
   vascular cylinder, 495
   water absorption, 496–498
Rough endoplasmic reticulum, 96
Roundworms, 570, 575–576
Rubisco, 122
Rumen, 740–741
Rusts, 421–422

S

Sabin, Albert, 975
Saddle joint, 842
Salamanders, 700
Saliva, human, 869–870
Salk, Jonas, 974
Salmonella, 365, 366
Salt-tolerant plants, 510
Sand dollars, 643
Sandy soil, 487
Saprophytes, 370, 407
Sapwood, 519
Sarcodina, 389–390
Sargassum, 439
Saturated fats, 73, 861
Scales, plant, 471–472
Scallops, 591
Scanning electron microscopes (SEM), 36
Scanning probe microscopes, 37
Schleiden, Matthias, 88
Schwann, Theodor, 88
Science
   goal, 5–6
   human values, 17–18
   process, 15
   study method, 16–17
Scientific method
   conclusion formation, 9–10
   discovery of yellow fever, 39
   hypothesis formation, 8
   hypothesis testing, 8
   in everyday life, 11
   metric system, 11–13
   observation, 7
   recording and analyzing data, 9
   replication of experiment, 10
Scientific notation, 163
Scientific theory, 10
Scion, 544
Sclera, 827
Sclerenchyma cells, 492
Scorpions, 619–620
Scrotum, 927
Scyphozoa, 567
Sea anemones, 567
Sea cucumbers, 639, 643, 644
Sea lettuce, 438
Sea lilies, 639, 643–644
Sea urchins, 639, 643, 644
Seals, 748
Sebaceous glands, 848
Secondary immune response, 973
Secondary phloem, 519
Secondary sex characteristics, 927
Secondary xylem, 519
Sedimentary rock, 278–279
Seed coat, 470, 539, 540
Seed dispersal, 479–481
   methods, 480–481
Seed ferns, 471
Seed plant reproduction
   adaptations for reproduction, 533–535
   cones, 470, 473, 534
   fertilization, 538–539
   flowers, 470, 473, 534–539
   pollination, 470, 477–479, 537–538
   seed dispersal, 479–481
   seed formation, 539–540
   seeds, 470
   vegetative reproduction, 542–544
Seed plants
   adaptations to land, 468–469
   angiosperms, 473–475, 477–481
   conifers, 472–473
   evolution, 471–475
   gymnosperms, 471–472
   leaves, 469, 502–505
   roots, 468, 494–498
   seed ferns, 471
   specialized tissue, 491–494
   stems, 468–469, 499–502
   transport, 505–508
   vascular tissue, 469, 492–493
Seeds
   development, 541–542
   dormancy, 541–542
   formation, 539–540
   germination, 541
   role in plant reproduction, 470
Segregation of genes, 185–187
Selective breeding, 247–248, 250, 1067
Selectively permeable membrane, 100
Self-pollination, 182, 537–538
Semen, 927
Semicircular canals, 830
Seminal fluid, 927
Seminiferous tubules, 927
Sense organs, human
   ears, 828–830
   eyes, 827–828
   nose, 830
   skin, 831
   taste buds, 831
Senses, human
   hearing, 828–830
   smell, 830
   taste, 831
   touch, 831
   vision, 827–828
Sensory neurons, 810
Sensory receptors, 831
Sepals, 535
Septa, 594
Septum, heart, 898
Sewage
   algae treatment, 443
   bacterial decomposition, 370
   contamination of environment, 1060–1061
Sex chromosomes, 209–211
   genes, 210–211
   X and Y chromosomes, 209–210
Sex-influenced traits, 238–239
Sex-linked inheritance, 235–239
   heredity, 209–211
   sex determination, 235
   sex-linked genetic disorders, 237–238
Sexual development, human, 935–936
Sexual reproduction, 28, 559, 666–667
   evolution, 348
Sharks, 688–689
Sharp, Philip, 219
Shell, mollusk, 586
Shipworms, 593
Short-term memory, 822
Shrews, 747
Sickle cell anemia, 231–233
   cause, 232
   genetics, 232
   incidence, 233
   molecular basis, 232

Sieve tube elements, 493
Single-loop circulatory system, 794
Sinuses, 588
Sirenia, 749
Skates, 689
Skeletal muscles, 843
Skeletal system, animal
   endoskeletons, 659
   exoskeletons, 659
   hydrostatic skeletons, 659
Skeletal system, human, 837–842
   bones, 837–842
   composition, 837
   joints, 840–842
Skin, human, 847–848
   barrier against infections, 970
   dermis, 848
   epidermis, 847–848
   immunity, 970
   role as sense organ, 831
Skin color, inheritance, 234
Sleep, 822
Sliding filament theory, 845
Slightly movable joints, 840
Slime molds, 398–399
   Acrasiomycota, 398–399
   Myxomycota, 399
Sloths, 748
Slugs, 590
Small intestine, 872–875
Smell, 830
Smog, 1055–1056
Smooth endoplasmic reticulum, 95–96
Smooth muscles, 843–844
Snails, 590
Snakes, 716–717
Social insects, 624–627
Soil, 487–489
   essential nutrients, 489
   layers, 488–489
   types, 487–488
Solute, 65
Solutions, 65–66
   acids, 65
   bases, 66
   neutralization reaction, 66
   pH scale, 66
Solvent, 65
Somatic nervous system, 825
Sori, 458
Sound signals, 777
Spallanzani, Lazzaro, 339
Speciation, 304–310
   adaptive radiation, 308, 787
   Darwin's finches, 306–307
   process, 305–306

Species
   definition, 301–302
   niche, 304–305
   in taxonomy, 322–323
Specific defenses, 972–975
Sperm, 196, 228, 235, 942
   algae, 441
   development in humans, 937–938
Sperm nuclei, 538
Spermopsida. *See* Seed plants.
Sphenophytes, 456
Sphincter, 870
Spicules, 561
Spiders, 618–619
Spinal cord, 817, 823
Spindle, 168
Spirilla, 364
Spirogyra, 438
Sponges, 560–564
   characteristics, 561–562
   importance, 563
   reproduction, 562
   structure, 561
Spongin, 561
Spongy mesophyll, 504–505
Spontaneous generation, 339–342, 350
   disproof of hypothesis, 341
   early experiments, 340
Sporangia, 408, 409, 458
Sporangiophores, 408
Spores, 387
   bacterial, 368–369
   dispersal, 417–418
   ferns, 458–459
   fungi, 308–309, 411–413
   mosses, 453–454
Sporophyte, 441
Sporozoa, 387–388
Squamata, 716
Squid, 592, 660
Stamens, 535
Stanley, Wendell, 356
Stapes, 829
Staphylococci, 366
Starches, 859
Starfish, 639, 641, 642
Steady state, 1035
Stems, 468–469, 499–502
   effect of auxin on, 522–523
   functions, 499, 500
   growth, 518–520
   parenchyma tissue, 499

   pith, 499
   stored food, 500–501
   vascular tissue, 499–500
Sterilization, for bacterial control, 375
Steroid hormones, 929
Sterols, 74
Stevens, Nettie, 209
Stigma, 536
Stimulants, 988–989
Stock, 544
Stolon, 411
Stomach, human, 871
Stomata, 504
Stromatolites, 346
Structural formulas, 57
Sturtevant, Alfred, 208
Style, 536
Subsoil, 488
Substrates, 76
Sucrose, 71, 858
Sugars, 858. *See also* Carbohydrates.
Sundews, 511
Survival of the fittest, 297
Suspensions, 66–67
Sutton, Walter, 206
Sweat glands, 848, 918
Swim bladder, 684
Symbiosis, 369
   cnidarians, 568
   fungi, 418–419
   protists, 393, 401
   types, 1040–1041
Symbols, chemical, 49–51
Sympathetic nervous system, 826
Symptom, 953
Synapse, 814–815
Syngamy, 440
Synovial fluid, 841
Systemic circulation, 902–903

T

Tadpoles, 695, 697
Taiga, 1011–1012
Tapeworms, 574–575
Taproots, 494
Target cells, 917
Target organ, 522
Tars, 995
Taste, 831
Taxon, 322
Taxonomy, 321–324
   biochemical taxonomy, 324–325
   categories, 322
   evolution, 324
   modern changes, 323

Teeth, 739–741, 868
Telescope, 87
Telophase, 170
Temperate deciduous forests, 1012–1013
Temperature, 13
Temperature control
  ectotherms, 720–721, 723, 788–789
  endotherms, 720, 722, 789
  hypothalamus, 927–928
  methods, 789
Temperature inversions, 1055
Tendons, 846
Test cross, 191
Testes, 921, 936, 937
Testosterone, 926, 927
Thalamus, 820–821
Theory, definition, 10
Thermal pollution, 1061
Thigmotropism, 525
Threshold, 814
Thymine, 141, 146
Thyroid gland, 918–919
Thyroid-stimulating hormone (TSH), 927
Thyroxine, 918–919, 927
Ticks, 619, 958, 959
Tissues, 106–107
T-lymphocytes (T-cells), 973, 975, 979
Toads, 700–701
Tobacco
  addiction, 997
  diseases, 996–997
  effects, 995–996
Topsoil, 488
Touch, 831
Toxins, 958
Trace elements, role in plant growth, 489
Tracers, 50
Trachea, 882–885
Tracheid, 455, 492
Tracheophytes, 455–459
  club mosses, 456
  ferns, 457–461
  horsetails, 456–457
  vascular system, 455–456
Traits, 183
  acquired traits, 292–293
  environmental factors, 228–229
Transcription
  gene expression, 219–220
  RNA, 146–147, 151
Transfer RNA (tRNA), 150
Transformation, 139

Transgenic organisms, 254
Transition fossils, 709, 725
Translation, 150, 151
Transmission electron microscopes (TEM), 36, 37
Transpiration, 506
Transpiration pull, 506
Transport in plants, 505–508
  phloem, 506–508
  roots, 495–498
  transpiration, 506
  transpiration pull, 506
  xylem, 505–506
Trees
  age estimation, 499, 519
  bark, 500
Trichinosis, 578
Trichocysts, 384
Trilobites, 607, 608
Tropical rain forests, 1014–1015, 1051
Tropisms, 525
Trypanosomes, 392, 958–959
Tsetse fly, 392, 958–959
Tuataras, 715
Tube feet, 638, 639
Tube nucleus, 537
Tubers, 501
Tumors, 960
Tundra, 1011
Tunicates, 645–646, 647
Turner syndrome, 235, 236
Turtles, 718, 719
Typhus, 958

U

Ultraviolet radiation, 347
Ulva, 438
  reproduction, 441
Umbilical cord, 944
Unicellular organisms, 28. *See also* Protists; Bacteria.
  movement, 98
Uniramia, 607, 622
Units of measure
  conversions, 14–15
  metric system, 11–13
Unsaturated fats, 861
Unsegmented worms, 570–579
  flatworms, 570–575
  parasites, 571, 573–575, 576–578
  roundworms, 570, 575–578
Uracil, 146
Ureter, 908
Urethra, 909, 938

Urey, Harold, 343–344
Urinary bladder, 908
Urine, 909
Uterus, 939

V

Vaccination, 973–975
  discovery, 973–974
  polio, 974–975
  rabies, 974
Vacuoles, 97
Vagina, 939, 942
Variable, definition, 8
Vascular bundles, 474
Vascular cambium, 491, 500, 518, 520
Vascular cylinder, 455, 495, 497
Vascular system, tracheophytes, 455–456
Vascular tissue
  leaves, 504
  seed plants, 469, 492–493
  stems, 499–500
  tracheophytes, 455–456
Vas deferens, 927
Vector pollination, 478
Vegetative reproduction, 542–544
  artificial methods, 543–545
Veins
  humans, 901
  leaves, 456, 504
Venae cavae, 899
Ventral, 559
Ventricle
  amphibian, 697
  fish, 684
  human, 899
  reptile, 711
  vertebrates, 794–795
Venus' flytrap, 511
Vertebral column, 679
Vertebrata, 679
Vertebrate evolution
  body temperature control, 788–780
  evolutionary relationships, 785
  trends, 787
Vertebrates, general, 555, 645
  circulatory system, 794–795
  excretion, 795–796
  feeding and digestion, 791–792
  movement, 790–791
  nervous and sensory system, 796–798
  reproduction, 798–799
  respiration, 792–794

Vessel element, 492–493
Vestigial organs, 284–285
Villi, 874
Vines, 510
Viral diseases
　cancer, 374, 961
　interferons, 373
　types, 374
Virchow, Rudolf, 88
Viruses, 355–360
　life cycle, 357–358
　lysogenic infection, 358–359
　lytic infections, 358
　origin, 360
　prophage, 358–359
　retroviruses, 359
　role in cancer, 961
　role in disease, 956
　specificity, 357
　structure, 356–357
Visceral mass, 586
Vision, 827–828
Visual signals, 776
Vital capacity, 891
Vitamins, human needs, 863–864
Viviparous animals, 687, 744, 799
Vocal cords, 883
Volume, 12

W

Walruses, 748
Water, 63–67
　human needs, 856–857
　mixtures, 64–65
　molecule, 63–64
　solutions, 65–66
　suspensions, 66–67
Water absorption. *See* Transport in plants.
Water cycle, 1024
Water mold, 409
Water plants, 509–510
Water pollution, 1059–1063
　chemical contamination, 1059–1060
　ocean pollution, 1062
　oil spills, 1062
　sewage contamination, 1060–1061
　thermal pollution, 1061
Water vascular system, 638
Waterlilies, 509, 510
Watson, John, 142–143
Weight
　metric system, 12–13
　property of matter, 45
Whales, 748, 1068
Wheat rust, 421–422
White blood cells, 906, 970, 972
White, Tim, 761
Wiener, Alexander, 231
Wilkins, Maurice, 142–143
Wind pollution, 477
Withdrawal, 992
Woodcock, Christopher, 165

X

X chromosome, 209–210
X-ray diffraction, 77
Xylem, 455, 474, 475, 492, 498
　functions, 469
　growth, 519
　leaves, 504
　root pressure, 498, 505
　stems, 499–500
　water movement, 505–506
　xylem cells, 492–493

Y

Y chromosome, 210
Yeasts, 411, 412–413, 419–420
Yellow fever, 39

Z

Zone of elongation, 518
Zone of maturation, 518
Zoologists, 32, 33
Zoomastigina, 386–387, 392, 393, 394
Zoospores, 440
Zygomycetes, 410–411
Zygospore, 410, 411
Zygote, 228, 942

Credits

Cover Art: Joseph Cellini

Photo Research: Natalie Goldstein, Yvonne Gerin, Barbara Scott, Martha Conway, Suzi Myers

Design and Layout: Function Thru Form Inc.

Contributing Artists: Joel Ito 817 bottom; 818; 819 top left, bottom left; 820 top left, bottom; 823 top; 825 bottom left; 827 bottom left; 830 top right; 831 top right, bottom right; 838 top left, bottom left; 839 bottom right; 840 top left; 841 top right; 842 top; 846; 847 bottom left; 867; 869 top right; 871 bottom right; 872; 883 top left; 884 top left; 885; 898 top left, bottom right; 900 top left; 902; 904; 909 top right, bottom left; 919 top left; 920; 922 top; 923 bottom left. John Rowe 178; 179; Raymond Smith 246; 264; 265; 268 top; 273; 281 top left; 309; 312; 342; 706; 709; 722; 736; 786; 787; 791; 796; 1024; 1025. All other art by Warren Budd & Associates, Ltd.

Photographs: i: Stephen J. Krasemann/DRK Photo; vi: (left) John Gerlach/Tom Stack & Associates; (right) Wolfgang Kaehler; vii: (top) Eric V. Grave/Phototake; (bottom) Uniphoto; viii: (top) Wayne Lynch/DRK Photo; (bottom) Hans Reinhard/Bruce Coleman, Inc.; ix: (top) Dr. E. R. Degginger; (center) Eric V. Grave/Phototake; (bottom) Manfred Kage/Peter Arnold, Inc.; x: (top) Harvey Lloyd/The Stock Market; (bottom) Kjell B. Sandved; xi: (top) F. Stuart Westmoreland/Tom Stack & Associates; (bottom left) J. H. Robinson/Photo Researchers, Inc.; (bottom right) Denise Tackett/Tom Stack & Associates; xii: (top) Jeffrey L. Rotman; (bottom) Leonard Lee Rue III/Animals Animals/Earth Scenes; xiii: (top) David M. Phillips/Visuals Unlimited; (center) Ronn Maratea/International Stock Photography, Ltd.; (bottom) Prof. P. Motta of Anatomy/University "La Sapienza," Rome/Science Photo Library/Photo Researchers, Inc.; xiv: (top) A. B. Dowsett/Science Photo Library/Photo Researchers, Inc.; (center) Don Carroll/The Image Bank; (bottom) Doug Perrine/DRK Photo; xv: (top) Brian Parker/Tom Stack & Associates; (center) Dean Cornwell/The Bettmann Archive; (bottom) Dwight Kuhn/DRK Photo; xvi: (bottom) Larry Lefever/Grant Heilman Photography; (top) W. H. Hodge/Peter Arnold, Inc.; xvii: Zig Leszczynski/Animals Animals/Earth Scenes; 2: Neil Leifer/Time Magazine; 4: (left) Stephen J. Krasemann/DRK Photo; (right) Art Wolfe/Art Wolfe Incorporated; 5: Michael Coyne/The Image Bank; 6: (top) Suzanne & Nick Geary/Tony Stone Images; (bottom) Wolfgang Kaehler; 7: Giuliano Colliva/The Image Bank; 12: (left) John Gerlach/DRK Photo; (right) CNRI/Science Photo Library/Photo Researchers, Inc.; 13: (left) Willi Dolder/Tony Stone Images; (right) J. F. Preedy/Tony Stone Images; (bottom) Jim Brandenburg/Minden Pictures, Inc.; 15: The Granger Collection; 16: (left) Co Rentmeester/The Image Bank; (top right) Mitch Reardon/Tony Stone Images; (bottom right) Warren & Genny Garst/Tom Stack & Associates; 17: (left) Will McIntyre/Photo Researchers, Inc.; (right) Sepp Seitz/Woodfin Camp & Associates; 18: John Lei/Omni-Photo Communications, Inc.; 20: St. Mary's Hospital Medical School/Photo Researchers, Inc.; 21: (top right) F. S. Mitchell/Tom Stack & Associates; (center) Chris Bjornberg/Photo Researchers, Inc.; (center right) Dick Durrance/Woodfin Camp & Associates; (bottom left) Steve Niedorf/The Image Bank; (bottom right) Jeffrey D. Smith/Woodfin Camp & Associates; 26: (top) Rod Planck/Tony Stone Images; (bottom) Susan McCartney/Photo Researchers, Inc.; 27: (left) Runk/Schoenberger/Grant Heilman Photography; (right) K. H. Switak/Photo Researchers, Inc.; 28: (bottom) David Smart/DRK Photo; (top left) Eric V. Grave/Photo Researchers, Inc.; (top right) Wolfgang Kaehler; 29: (left) Breck P. Kent; 30: (bottom) Merlin B. Tuttle; (top) James H. Carmichael/The Image Bank; 31: Belinda Wright/DRK Photo; 33: (bottom) Fred Ward/Black Star; (top) Frans Lanting/Minden Pictures Inc.; 35: (top) Howard Sochurek/Woodfin Camp & Associates; (bottom) M. I. Walker/Photo Researchers, Inc.; 36: (top) Ted Horowitz/The Stock Market; (center) CNRI/Science Photo Library/Photo Researchers, Inc.; (bottom) CNRI/Science Photo Library/Photo Researchers, Inc.; 37: (bottom) Manfred Kage/Peter Arnold, Inc.; (bottom right) Photo Courtesy of Digital Instruments Corporation, Santa Barbara, CA; (top left) Dr. J.A.L. Cooke; (top right) David Scharf/Peter Arnold, Inc.; 38: (bottom) Hank Morgan/Science Source/Photo Researchers, Inc.; (top) Peter J. Kaplan/Medichrome/The Stock Shop; 39: (bottom) Dwight Kuhn/DRK Photo; (top) Dean Cornwell/The Bettmann Archive; 44: (top) Kjell B. Sandved/Photo Researchers, Inc.; (bottom) D. McCoy/R. Feldman/Rainbow; 45: (left) John Blaustein/Woodfin Camp & Associates; (right) Jeff Foott/Tom Stack & Associates; 46: (top) Paul Silverman/Fundamental Photographs; (left) Richard Megna/Fundamental Photographs; (center) David Sutherland/Tony Stone Images; (right) GJ Images/The Image Bank; 47: (top) Fundamental Photographs; (bottom) Dr. Mitsuo Ohtsuki/Science Photo Library/Photo Researchers, Inc.; 48: Photo Researchers, Inc.; 50: (bottom) Howard Sochurek Inc.; (top) CEA-ORSAY/CNRI/Science Photo Library/Photo Researchers, Inc.; 56: (bottom) Dwight Kuhn Photography; (top) Paul Silverman/Fundamental Photographs 57: Chemical Design Ltd./Science Photo Library/Photo Researchers, Inc.; 62: (bottom) Science Photo Library/Photo Researchers, Inc.; (top) Will McIntyre/Photo Researchers, Inc.; 63: (bottom) Dan Budnik/Woodfin Camp & Associates; (center) Tobias Heldt/The Image Bank; (top) Uniphoto; 64: (left) Dr. E. R. Degginger; (right) Paul Silverman Fundamental Photographs; 65: Hank deLespinasse/The Image Bank; 67: (left) Kip Peticolas/Fundamental Photographs; (right) Dr. Tony Brain/Science Photo Library/Photo Researchers, Inc.; 71: (top) J. Litvay /Visuals Unlimited; (bottom) Larry Ulrich/DRK Photo; 72: Kristen Brochmann/Fundamental Photographs; 73: Wayne Lynch/DRK Photo; 75: (top left and bottom right) Jan Cobb/The Image Bank; (top right) John Gerlach/Tom Stack & Associates; (bottom left) Wolfgang Kaehler; 77: Dave Umberger/Purdue News Photo; 82: (top and center) Robert Frerck/Odyssey Productions; (bottom) A. Boccaccio/The Image Bank; 84: Jeff Lepore/Photo Researchers, Inc.; 86: (left) Jan Hinsch/Science Photo Library/Photo Researchers, Inc.; (right) Dr. Nancy Kedersha/ImmunoGen, Inc.; 87: (top) Eric V. Grave/Phototake; (bottom) Dr. Jean Lorre/SPL/Science Source/Photo Researchers, Inc.; 88: Leonard Lessin/Peter Arnold, Inc.; 92: (top) M.R.J. Salton/Visuals Unlimited; (bottom) S. Elems/Visuals Unlimited; 93: K. G. Murti/Visuals Unlimited; 94: CNRI/Science Photo Library/Photo Researchers, Inc.; 95: Dr. Jeremy Burgess/Science Photo Library/Photo Researchers, Inc.; 96: (top) Dr. Gopal Murti/Science Photo Library/Photo Researchers, Inc.; (bottom left) David M. Phillips/Visuals Unlimited; 97: (bottom) Biophoto Associates/Photo Researchers, Inc.; (top) K. G. Murti/Visuals Unlimited; 98: K. G. Murti/Visuals Unlimited; 99: David M. Phillips/Visuals Unlimited; 101: M. Sheetz/*J. Cell Biology*/University of Connecticut Health Center; 103: M. Abbey/Visuals Unlimited; 105: (top) Jenny Hinshaw/Scripps Research Institute; (bottom) Scripps Research Institute; 112: (bottom) Luis Castaneda/The Image Bank; (top) F. Bergemann/The Image Bank; 113: Wes Thompson Photography/The Stock Market; 114: (left) Pat O'Hara/DRK Photo; (right) Larry Ulrich/DRK Photo; 115: (left) Luis Villota/The Stock Market; (right) Runk/Schoenberger/Grant Heilman Photography; 122: (top) UPI/Bettmann Newsphotos; (bottom) Fig. 14p. 1470 from J. Diesenhoffer and H. Michel, *Science* vol. 245, pp. 1463–1473, Sept. 29, 1989.; 129: Paul J. Sutton/Duomo Photography, Inc.; 130: David Madison/Duomo Photography, Inc.; 131: (top left) Dr. Jeremy Burgees/Science Photo Library/Photo Researchers, Inc.; (top right) Dario Perla/After Image, Inc.; (bottom) Roy Morsch/The Stock Market; 136: (top) Oxford Molecular Biophysics Lab/Science Photo Library/Photo Researchers, Inc.; (bottom) R. Langridge/D. McCoy/USCF Computer Graphics/Rainbow; 137: Petit Format/Photo Researchers, Inc.; 139: Lee Simon/Stammers/Science Photo Library/Photo Researchers, Inc.; 141: Philippe Plailly/Science Photo Library/Photo Researchers, Inc.; 142: Franklin & Gosling B-DNA, NATURE; 143: (bottom) AP/Wide World Photos; (top) Philippe Plailly/Science Photo Library/Photo Researchers, Inc.; 151: Paula Grabowski; 158: (top) © Lennart Nilsson, *The Incredible Machine*, National Geographic Society; (bottom) Ed Reschke/Peter Arnold, Inc.; 159: Miriam Austerman/Animals Animals/Earth Scenes; 161, 162: David M. Phillips/Visuals Unlimited; (top) Brownie Harris/The Stock Market; 164: K. G. Murti/Visuals Unlimited; 165: S. L. McKnight and O. L. Miller, Jr./University of Virginia; 166: Science VU/Visuals Unlimited; 167: Ed Reschke/Peter Arnold, Inc.; 168: (bottom) Dr. Richard Linck/University of Minnesota, Department of Cell Biology and Neuroanatomy; 168: (top left) Carolina Biological Supply; (top right) Ed Reschke/Peter Arnold, Inc.; 169: (top left and bottom left) Carolina Biological Supply; (top right and bottom right) Ed Reschke/Peter Arnold, Inc.; 170: (top left) Carolina Biological Supply; (top right) Ed Reschke/Peter Arnold, Inc.; (bottom) B. A. Palevitz/Tom Stack & Associates; 171: G. I. Bernard/Animals Animals/Earth Scenes; 176: (top) Ted Horowitz/The Stock Market; (center) Gabe Palmer/The Stock Market; (bottom) Steve Dunwell/The Image Bank; 177: Will & Deni McIntyre/Photo Researchers, Inc.; 178: (top) CNRI/Science Photo Library/Photo Researchers, Inc.; (bottom) Dr. Ann Smith/Science Photo Library/Photo Researchers, Inc.; 179: (top) Wayne Lynch/DRK Photo; (bottom left) Uniphoto; (bottom right) Rod Planck/Tom Stack & Associates; 180: (top) M. V. Cooke; (bottom) Gary S. Chapman/The Image Bank; 181: Hans Reinhard/Bruce Coleman, Inc.; 182: Dr. Jeremy Burgess/Science Photo Library/Photo Researchers, Inc.; 189: Dr. E. R. Degginger; 190: Myrleen Ferguson/Photoedit; 192: Walter Chandoha; 193: Runk/Schoenberger/Grant Heilman Photography; 204: (top) Ron Kimball; (bottom) Darwin Dale/Photo Researchers,

Inc.; **205:** David M. Phillips/Visuals Unlimited; **208:** Breck P. Kent; **212:** Runk/Schoenberger/Grant Heilman Photography; **214:** W. H. Hodge/Peter Arnold, Inc.; **215:** Erick Greene/Department of Avian Sciences, University of California; **217:** Hans Reinhard/Bruce Coleman, Inc.; **220:** Dr. Bert O'Malley/Baylor College of Medicine, Houston, Texas; **221:** Nik Kleinberg; **226:** Michel Tcherevkoff; **227:** © Lennart Nilsson, *The Incredible Machine*, National Geographic Society; **228:** (top) John Giannicchi/Science Source/Photo Researchers, Inc.; (center) David M. Phillips/Visuals Unlimited; (bottom) Mary Kate Denny/Photoedit; **229:** Jeanne B. Lawrence, Robert Singer, and Lisa Marselle, University of Massachusetts Medical School, Department of Cell Biology; **230:** Runk/Schoenberger/Grant Heilman Photography; **231:** Dr. E. R. Degginger; **232:** (top) Stanley Flegler/Visuals Unlimited; (bottom) Ed Reschke/Peter Arnold, Inc.; **233:** © Lennart Nilsson, *The Incredible Machine*, National Geographic Society; **234:** Frank Siteman; **236:** Sam Ogden; **238:** Robert Brenner/Photoedit; **240:** (top) Grace Moore/Imagery; (bottom) Mario Ruiz/Picture Group; **247:** R. Van Nostrand/Photo Researchers, Inc.; **248:** (top) Wolfgang Kaehler; (bottom) Hans Reinhard/Bruce Coleman, Inc.; **249:** (top left) Courtesy of Kenneth Miller; (bottom left) Sven-Olaf Lindblad/Photo Researchers, Inc.; (top right) Barry L. Runk/Grant Heilman Photography; (bottom right) Dwight Kuhn Photography; **250:** Vaughn Fleming/Science Photo Library/Photo Researchers, Inc.; **251:** Phil Harrington/Peter Arnold, Inc.; **252:** K. G. Murti/Visuals Unlimited; **254:** Dr. Barbara Stebbins-Boaz; **255:** (top) Hank Morgan/Photo Researchers, Inc.; (bottom left) Alexander R. van der Krol, Peter E. Lenting, Jetty Veenstra, Ingrid M. van der Meer, Ronald E. Koes, Anton G. M. Gerats, Joseph N. M. Mol & Antoine R. Stuitje/Department of Applied Genetics, Vrije Universiteit, Amsterdam; (bottom right) Keith V. Wood/University of California, San Diego, La Jolla; **256:** Jackson Lab/Visuals Unlimited; **258:** (top) Simon Fraser/RVI, Newcastle-Upon-Tyne /Science Photo Library/Photo Researchers, Inc.; (bottom) National Association for Sickle Cell; **264:** (top) Hank Morgan/Rainbow; (center) Robert Frerck/Odyssey Productions; (bottom) Lawrence Migdale/Photo Researchers, Inc.; **265:** Kenneth Miller; **269:** (top) George D. Dodge/Tom Stack & Associates; (center) Brian Parker/Tom Stack & Associates; (bottom) Norman Owen Tomalin/Bruce Coleman, Inc.; **270:** (top) Chip Clark; (bottom) Stephen Dalton/NHPA; **271:** (top left) Dr. E. R. Degginger/Bruce Coleman, Inc.; (top right) D. Cavagnaro/DRK Photo; (bottom) Gary Milburn/Tom Stack & Associates; **272:** Timothy Eagan/Woodfin Camp & Associates; **275:** Kal Muller/Woodfin Camp & Associates; **278:** (left) Dr. E. R. Degginger; (center) Dr. David Schwimmer/Bruce Coleman, Inc.; (right) Jane Burton/Bruce Coleman, Inc.; **279:** Jeff Gnass/The Stock Market; **280:** (top left) Tom McHugh/California Academy of Sciences/Photo Researchers, Inc.; (center left) T. A. Wiewandt/DRK Photo; (right) John Cancalosi/DRK Photo; (bottom left) George Polinar, Jr./University of California at Berkeley; (bottom right) M. D. Maser/Visuals Unlimited; **282:** Tom Bean/DRK Photo; **284:** Frans Lanting/Minden Pictures, Inc; **285:** (top) Don & Pat Valenti/DRK Photo; (bottom) Douglas Kirkland/Woodfin Camp & Associates; **290:** (top) Down House and The Royal College of Surgeons; (bottom) The Granger Collection; **291:** (left) Breck P. Kent; (right) Julie Habel/Woodfin Camp & Associates; **292:** Gerard Lacz/NHPA; **293:** Solarfilma; **294:** (top left) Larry Lefever/Grant Heilman Photography; (center left) Jane Grushow/Grant Heilman Photography; (bottom left) John Colwell/Grant Heilman Photography; (right) Lewis Kemper/DRK Photo; **295:** (top) Dwight Kuhn Photography; (bottom) A. Upitis/The Image Bank; **296:** Dwight Kuhn Photography; **297:** Breck P. Kent; **298:** (top left) Joe McDonald/Tom Stack & Associates; (right) Breck P. Kent; (bottom left) James H. Carmichael/The Image Bank; **299:** Frans Lanting/Minden Pictures, Inc.; **300:** New York Public Library; **301:** M. Philip Kahl/DRK Photo; **302:** Nancy Adams/Tom Stack & Associates; **303:** Hans Wolf/The Image Bank; **304:** Aaron Norman; **305:** (bottom) C. B. & D. W. Frith/Bruce Coleman, Inc.; (top) R. E. Logan/American Museum of Natural History; **307:** (right) Bruce Coleman, Inc.; (left) Jessica Anne Ehlers/Bruce Coleman, Inc.; **310:** (left) Merlin B. Tuttle; (right) G. I. Bernard/NHPA; **311:** (bottom left) Belinda Wright/DRK Photo; (bottom right) Stephen J. Krasemann/DRK Photo; (top) Jeffrey Hutcherson/DRK Photo; **318:** (left) Wolfgang Kaehler; (right) Brian Parker/Tom Stack & Associates; **319:** Kevin Schaefer/Tom Stack & Associates; **320:** (left) Jeanne White/Photo Researchers, Inc.; (center) Larry Lipsky/DRK Photo; (right) Jeffrey L. Rotman; **321:** (top) Stephen J. Krasemann/DRK Photo; (bottom) Marcia Stevens/New York Botanical Garden; **323:** (left) Belinda Wrigh/DRK Photo; (top center) Tom McHugh/Photo Researchers, Inc.; (top right) Stan Osolinski/Tony Stone Images; (bottom center) Lee Green/West Light; (bottom right) Dick Candy/DRK Photo; **324:** (top left) Tom McHugh/Photo Researchers, Inc.; (top right) Michael Holton/Photo Researchers, Inc.; (bottom) Peter Scoones/Colorific; **326:** Asa C. Thoresen/Photo Researchers, Inc.; **327:** (top) Dr. Tony Brain/David Parker/Science Photo Library/Photo Researchers, Inc.; (bottom) Runk/Schoenberger/Grant Heilman Photography; **328:** (top left) Larry Ulrich/DRK Photo; (top right) Wolfgang Kaehler; (bottom left) Breck P. Kent; (bottom right) Steve Powell/Allsport; **334:** (top and bottom) Robert Frerck/Odyssey Productions; (center) Dan Guravich/Photo Researchers, Inc.; **335:** (top) Joe Levine; (bottom) ©1993 Frans Lanting/Minden Pictures, Inc.; **336:** A. B. Dowsett/Science Photo Library/Photo Researchers, Inc.; **338:** NASA; **341:** L. West/Bruce Coleman, Inc.; **343:** (top) NASA; (bottom) Lynn Margulis, Department of Botany/University of Massachusetts; **344:** Roger Ressmeyer/STARLIGHT; **345:** (bottom right) Science VU-WHO/Visuals Unlimited; (top) Science VU/Sidney Fox/Visuals Unlimited; **346:** (bottom) William E. Ferguson; (top) Breck P. Kent; **347:** CNRI/Science Photo Library/Photo Researchers, Inc.; **348:** (top) CNRI/Science Photo Library/Photo Researchers, Inc.; (bottom) Dan McCoy/Rainbow; **349:** Dan McCoy/Rainbow; **354:** (top) R. Feldman/D. McCoy/Rainbow; (bottom) Oxford Molecular Biophysics Lab/Science Photo Library/Photo Researchers, Inc.; **355:** (left) Norm Thomas/Photo Researchers, Inc.; (right) Dennis Kunkel, color by David Wacner/Phototake; **356:** B. Heggeler/Biozentrum, U. of Basel/Science Photo Library/Photo Researchers, Inc.; **358:** Lee D. Simon/Science Source/Photo Researchers, Inc.; **360:** SIU/Photo Researchers, Inc.; **361:** David M. Phillips/Visuals Unlimited; **362:** J. H. Robinson/Science Source/Photo Researchers, Inc.; **363:** Esther R. Angert and Norman R. Pace/Indiana University Art Museum; **364:** (top center, top right, and bottom) David M. Phillips/Visuals Unlimited; (top left) Dr. Tony Brain & David Parker/Science Photo Library/Photo Researchers, Inc.; **365:** Barry Dowsett/Science Photo Library/Photo Researchers, Inc.; **366:** A. B. Dowsett/Science Photo Library/Photo Researchers, Inc.; **367:** George J. Wilder/Visuals Unlimited; **368:** (left) David Scharf/Peter Arnold, Inc.; (right) David M. Phillips/Visuals Unlimited; **369:** (top) Tom Stack & Associates; (bottom) Biophoto Associates/Photo Researchers, Inc.; **370:** Thomas Kitchin/Tom Stack & Associates; **371:** Corale L. Brierley/Visuals Unlimited; **372:** (bottom) C. P. Vance/Visuals Unlimited; (top) Runk/Schoenberger/Grant Heilman Photography; **373:** (top left) Tektoff-RM/CNRI/Science Photo Library/Photo Researchers, Inc.; (top right) CNRI/Science Photo Library/Photo Researchers, Inc.; (bottom) NCI/Science Source/Photo Researchers, Inc.; **374:** Department of Zoology, University of California; **375:** CNRI/Science Photo Library/Photo Researchers, Inc.; **380:** (top) Manfred Kage/Peter Arnold, Inc.; (bottom) Eric V. Grave/Photo Researchers, Inc.; **381:** (left) Manfred Kage/Peter Arnold, Inc. (right) Walker/Photo Researchers, Inc.; **382:** (top) Dr. E. R. Degginger; (bottom) Biophoto Associates/Photo Researchers, Inc.; **384:** (top) John D. Cunningham/Visuals Unlimited; (bottom) Eric V. Grave/Photo Researchers, Inc.; **385:** (top) Eric V. Grave/Photo Researchers, Inc.; (bottom) Michael Abbey/Photo Researchers, Inc.; **386:** (top) Brian Parker/Tom Stack & Associates; (bottom) Eric V. Grave/Photo Researchers, Inc.; **387:** (top left) David M. Phillips/Visuals Unlimited; (top right) Jerome Paulin/Visuals Unlimited; (bottom) CNRI/Science Photo Library/Photo Researchers, Inc.; **388:** © Lennart Nilsson, *The Incredible Machine*, National Geographic Society; **389:** M. Abbey/Visuals Unlimited; **390:** (top and bottom left) Manfred Kage/Peter Arnold, Inc.; (bottom right) Eric V. Grave/Phototake; **392:** Arthur M. Siegelman/Visuals Unlimited; **393:** (left) Eric V. Grave/Photo Researchers, Inc.; (right) Dwight Kuhn Photography; **395:** Biophoto Associates/Photo Researchers, Inc.; **396:** (top) Biophoto Associates/Photo Researchers, Inc.; (bottom) Terry Hazen/Visuals Unlimited; **397:** (top) David M. Phillips/Visuals Unlimited; (bottom) Biophoto Associates/Photo Researchers, Inc.; **398:** (top) Manfred Kage/Peter Arnold, Inc.; (bottom) David M. Phillips/Visuals Unlimited; **399:** (top) CBS/Visuals Unlimited; (bottom left) B. Beatty/Visuals Unlimited; (bottom right) John Shaw/Tom Stack & Associates; **400:** (bottom) David M. Phillips/Visuals Unlimited; (top) Kevin Schaefer/Peter Arnold, Inc.; **401:** (top) Daniel W. Gotshall/Visuals Unlimited; (center) Fred McConnaughey/Photo Researchers, Inc.; (bottom) Roland Birke/Peter Arnold, Inc.; **402:** Manfred Kage/Peter Arnold, Inc.; **406:** Photograph by B. A. Roy, reprinted by permission from *Nature* vol. 362, cover and page 57; **407:** Dr. E. R. Degginger/Bruce Coleman, Inc.; **410:** (top) Ralph C. Eagle/Photo Researchers, Inc.; **411:** (top) W. K. Fletcher/Photo Researchers, Inc.; (bottom) Richard H. Thom/Tom Stack & Associates; **413:** (top left) C. B. & D. W. Frith/Bruce Coleman, Inc.; (bottom left) J. L. Lepore/Photo Researchers, Inc.; (right) Richard Kolar/Animals Animals/Earth Scenes; **414:** Kerry T. Givens/Tom Stack & Associates; **415:** (left) David Scharf/Peter Arnold, Inc.; (right) David M. Phillips/Visuals Unlimited; **416:** R. Wenger/Sandoz Pharmaceutical; **417:** Dick Canby/DRK Photo; **418:** (top left) Noble Proctor/Photo Researchers, Inc.; (bottom left) A. Davies/Bruce Coleman, Inc.; (top right) CBS/Visuals Unlimited; (bottom right) Michael Fogden/Bruce Coleman, Inc.; **419:** (top) Dr. E. R. Degginger; (center) Jack Dermid; (bottom) L. West/Photo Researchers, Inc.; **420:** (top left) Fred Mayer/Woodfin Camp & Associates; (top right) Adam Woolfitt/*Daily Telegraph*

Magazine/Woodfin Camp & Associates; (bottom) Lee Foster/Bruce Coleman, Inc.; **421:** Biophoto Associates/Photo Researchers, Inc.; **423:** Charles Brewer-Carias; **427:** Dr. E. R. Degginger; **428:** (top) Bill Varie/The Image Bank; (bottom) Dan McCoy/Rainbow; (center) Pete Saloutos/The Stock Market; **429:** Oxford Scientific Films/Animals Animals/Earth Scenes; **430:** Robert & Linda Mitchell Photography; **432:** (left) Dr. E. R. Degginger; (right) Frans Lanting/Minden Pictures, Inc.; **433:** (left) Scott Blackman/Tom Stack & Associates; (right) D. P. Wilson/Eric & David Hosking/Photo Researchers, Inc.; **434:** Hal Clason/Tom Stack & Associates; **435:** D. P. Wilson/Eric & David Hosking/Photo Researchers, Inc.; **436:** (top left) G. I. Bernard/Oxford Scientific Films/Animals Animals/Earth Scenes; (top center) L. Sims/Visuals Unlimited; (top right) Doug Wechsler /Animals Animals/Earth Scenes; (bottom) Jeff Foott Productions; **437:** M. I. Walker/Photo Researchers, Inc.; **438:** (top left) John D. Cunningham/Visuals Unlimited; (top center) Michael Abbey/Science Source/Photo Researchers, Inc.; (top right) James Bell/Science Source/Photo Researchers, Inc.; (bottom) Dwight Kuhn Photography; **439:** Jeff Foott Productions; **442:** (top) B. & C. Alexander; (center) Uniphoto; (bottom) R. Calentine/Visuals Unlimited; **443:** Brad Hess/Black Star; **444:** A. M. Siegelman/Visuals Unlimited; **448:** (bottom) L. West/Bruce Coleman, Inc.; (top) D. Cavagnaro/DRK Photo; **449:** Jack Dermid; **450:** (top) Kjell B. Sandved; (bottom) Stephen J. Krasemann/DRK Photo; **452:** Runk/Schoenberger/Grant Heilman Photography; **453:** Dwight Kuhn Photography; **455:** Kjell B. Sandved; **456:** G. R. Roberts; **457:** (top left) E. S. Ross; (top right) Kevin Schaefer/Tom Stack & Associates; (bottom) Kjell B. Sandved; **458:** (top left and bottom right) Kjell B. Sandved; (bottom left) Jeff Foott/DRK Photo; **459:** Stan Elems/Visuals Unlimited; **460:** W. Ormerod/Visuals Unlimited; **461:** (top left) Rod Planck/Tom Stack & Associates; (top right) Derek Fell; (bottom) Pat O'Hara/DRK Photo; **465:** Kjell B. Sandved; **466:** (top) Hans Pfletschinger/Peter Arnold, Inc.; (bottom) Jeff Lepore/Photo Researchers, Inc.; **467:** (top) Wendy Shattil/Bob Rozinski/Tom Stack & Associates; (bottom) Wil Blanche/Uniphoto; **468:** (top) G. R. Roberts; (bottom left) Jack Dermid; (bottom right) Wolfgang Kaehler; **469:** Larry Ulrich/DRK Photo; **470:** Tom Algire; **471:** Dr. E. R. Degginger; **472:** (top left) Derek Fell; (top right) John Trager/Visuals Unlimited; (bottom) G. R. Roberts; **473:** (top) Doug Sokell/Tom Stack & Associates; (bottom) Breck P. Kent/Animals Animals/Earth Scenes; **474:** (top left) D. Cavagnaro/DRK Photo; (top right) Wolfgang Kaehler; (bottom) Kjell B. Sandved; **475:** Dwight Kuhn Photography; **476:** (left) Merlin B. Tuttle/Photo Researchers, Inc.; (right) Paul Skelcher/Rainbow; **477:** Bob & Clara Calhoun/ Bruce Coleman, Inc.; **478:** Thomas Eisner; **479:** (top) John D. Cunningham/Visuals Unlimited; (bottom left) John Gerlach/Tom Stack & Associates; (bottom right) Coco McCoy/Rainbow; **480:** William James Warren/West Light; **481:** (top) Ken W. Davis/Tom Stack & Associates; (bottom) John Serrao/Photo Researchers, Inc.; **485:** Kjell B. Sandved; **486:** (top) Dr. E. R. Degginger; (bottom) Janice Travia/Tony Stone Images; **487:** Peter Katsaros/Photo Researchers, Inc.; **488:** F. C. Earney/Visuals Unlimited; **489:** (top) Holt Studios, Ltd./Animals Animals/Earth Scenes; (bottom) Hans Christian Heap; **490:** Hugh Spencer/Photo Researchers, Inc.; **492:** (left) Dwight Kuhn Photography; (right) E. J. Cable/Tom Stack & Associates; **493:** (top) Biophoto Associates/Photo Researchers, Inc.; (bottom) P. Dayanandan/Photo Researchers, Inc.; **494:** (top) Runk/Schoenberger/ Grant Heilman Photography; (bottom) Dwight Kuhn Photography; **495:** (left) D. Dayanandan/ Photo Researchers, Inc.; (right) J. F. Gennaro/New York Cellular Biology/Photo Researchers, Inc.; **499:** (bottom) Dr. E. R. Degginger; (top) Dwight Kuhn Photography; **500:** Breck P. Kent; **501:** (top left and right) Dwight Kuhn Photography; (bottom left) W. H. Hodge/Peter Arnold, Inc.; (bottom right) Jerome Wexler/Photo Researchers, Inc.; **502:** (top) Jane Burton /Bruce Coleman, Inc.; (bottom) Joe McDonald/Tom Stack & Associates; **503:** John D. Cunningham/Visuals Unlimited; **504:** Dr. Jeremy Burgess/Science Photo Library/Photo Researchers, Inc.; **507:** Jack Dermid; **508:** Alan Pitcairn/Grant Heilman Photography; **509:** (top) Tom Algire; (center) Dr. E. R. Degginger; (bottom) Richard Weiss/Peter Arnold, Inc.; **510:** John D. Cunningham/Visuals Unlimited; **511:** (top left) Breck P. Kent; (top right) Nundsany & Perennou/ Photo Researchers, Inc.; (bottom) Ed Reschke/ Peter Arnold, Inc.; **516:** (top) D. Cavagnaro/DRK Photo; (bottom) Steve Solum/Bruce Coleman, Inc.; **517:** (left) James H. Carmichael, Jr./The Image Bank; (right) James L. Castner; **518:** (left) Derek Fell; (right) Lucy Jones/Visuals Unlimited; **519:** (left) Uniphoto; (right) Dr. William M. Harlow/Photo Researchers, Inc.; **520:** (top) Bill Beatty/Visuals Unlimited; (center and bottom) Wolfgang Kaehler; **522:** Runk/Schoenberger/Grant Heilman Photography; **523:** E. J. Cable/Tom Stack & Associates; **524:** Runk/Schoenberger/Grant Heilman Photography; **525:** Biophoto Associates/Science Source/Photo Researchers, Inc.; **526:** John Sohlden/Visuals Unlimited; **527:** Hans Pfletschinger/Peter Arnold, Inc.; **531:** Tovah Martin; **532:** (left) David M. Phillips/Visuals Unlimited; (right) Tom Bledsoe/DRK Photo; **533:** (left) John Gerlach/Tom Stack & Associates; (right) Breck P. Kent; **535:** Brian Parker/Tom Stack & Associates; **536:** (left) B. Ormerod/Visuals Unlimited; (right) James L. Castner; **538:** (top) Holt Studios/Animals Animals/Earth Scenes; (bottom left) James L. Castner; (bottom right) T. Kitchin/Tom Stack & Associates; **539:** (top) Gabriella Bergamini Mulcahy, University of Massachusetts at Amherst; (center) Runk/Schoenberger/Grant Heilman Photography; (bottom) Dwight Kuhn Photography; **541:** (top) Breck P. Kent; (bottom) C. C. Lockwood/ Animals Animals/Earth Scenes; **543:** (top) Breck P. Kent; (bottom) Runk/Schoenberger/Grant Heilman Photography; **544:** D. Cavagnaro/DRK Photo; **545:** Runk/Schoenberger/Grant Heilman Photography; **549:** William E. Ferguson; **550:** (top) Gary Gladstone/The Image Bank; (center) Benn Mitchell/The Image Bank; (bottom) Sepp Seitz/Woodfin Camp & Associates; **551:** (left) Sydney Thomson/Animals Animals/Earth Scenes; (right) Nicholas Foster/The Image Bank; **552:** Dr. Jeremy Burgess/Science Photo Library/Photo Researchers, Inc.; **554:** (top) C. C. Lockwood/DRK Photo; (bottom) Denise Tackett/ Tom Stack & Associates; **555:** (left) Fran Allan/ Animals Animals/Earth Scenes; (right) Jack Dermid; **556:** (top left) Robert Maier/Animals Animals/Earth Scenes; (top right) Breck P. Kent; (bottom) Polaroid–R. Oldfield/Visuals Unlimited; **557:** (top left) Richard Matthews/Planet Earth Pictures; (top right) Georgette Douwma/Planet Earth Pictures; (bottom) Jeffrey L. Rotman; **558:** (top) David Maitland/Planet Earth Pictures; (center) William E. Ferguson; (bottom left) Robert Arnold/Planet Earth Pictures; (bottom right) Doug Perrine/Innerspace Visions, Miami; **559:** Jeffrey L. Rotman; **560:** Breck P. Kent; **561:** (left and right) Charles Seaborn/Odyssey Productions; (center) Jeffrey L. Rotman; **562:** (top) Oxford Scientific Films/Animals Animals/Earth Scenes; (bottom) Doug Perrine/Innerspace Visions, Miami; **563:** Chris Howes/Planet Earth Pictures; **564:** (top) Breck P. Kent; (bottom left) Charles Seaborn/Odyssey Productions; (bottom right) Jeffrey L. Rotman; **566:** (top) Dwight Kuhn Photography; (bottom) John D. Cunningham/Visuals Unlimited; **567:** Dave B. Fleetham/Tom Stack & Associates; **568:** (top left) Charles Seaborn/Odyssey Productions; (top right) Jeffrey L. Rotman; (bottom) G. I. Bernard/ Oxford Scientific Films/Animals Animals/Earth Scenes; **569:** Denise Tackett/Tom Stack & Associates; **570:** Scott Johnson/Animals Animals/Earth Scenes; **571:** Michael Abbey/Photo Researchers, Inc.; **572:** (top) T. E. Adams/Visuals Unlimited; (bottom left) Charles Seaborn/Odyssey Productions; (bottom right) Jeff Foott Productions; **574:** (top) George J. Wilder/Visuals Unlimited; (bottom) CNRI/Science Photo Library/Photo Researchers, Inc.; **575:** CNRI/Science Photo Library/Photo Researchers, Inc.; **577:** CNRI/ Science Photo Library/Photo Researchers, Inc.; **578:** (top) R. Calentine/Visuals Unlimited; (bottom) Edward Gray/Science Photo Library/Photo Researchers, Inc.; **579:** Eugene Richards/Magnum Photos, Inc.; **584:** (top) Dick Clarke/Planet Earth Pictures; (bottom) Dr. E. R. Degginger/Animals Animals/Earth Scenes; **585:** (left) Ken Lucas/Planet Earth Pictures; (right) Richard LaVal/Animals Animals/Earth Scenes; **587:** G. Alan Solem; **588:** (top left) David Maitland/Planet Earth Pictures; (top right) W. Gregory Brown/Animals Animals/Earth Scenes; (bottom) Mark Mattock/Planet Earth Pictures; **589:** (bottom) Heather Angel/Biofotos; (top left) Leo Collier/Planet Earth Pictures; (top right) Jim Doran/Animals Animals/Earth Scenes; **590:** (top) Breck P. Kent/Animals Animals/Earth Scenes; (bottom left) F. Stuart Westmorland/Tom Stack & Associates; (bottom right) Leo Collier/Planet Earth Pictures; **591:** (top) Heather Angel/Biofotos; (center) John Lythgoe/Planet Earth Pictures; (bottom) Kjell B. Sandved; **592:** (top) P. Herring/Biofotos; (center) Jack Wilburn/Animals Animals/Earth Scenes; (bottom) A. Kerstitch; **594:** Raymond A. Mendez/Animals Animals/Earth Scenes; **595:** (left) Kjell B. Sandved; (right) Larry Lipsky/Tom Stack & Associates; **597:** J. G. James/Planet Earth Pictures; **598:** (left) Kjell B. Sandved; (right) David George/Planet Earth Pictures; **599:** (top) A. Kerstitch; (bottom) Hans Pfletschinger/Peter Arnold, Inc.; **600:** (top) Oxford Scientific Films/Animals Animals/Earth Scenes; (bottom) Oxford Scientific Films/Animals Animals/Earth Scenes; **601:** Geoff Tompkinson/ Aspect Picture Library; **605:** Jeffrey L. Rotman; **606:** (top) J. Carmichael/The Image Bank; (bottom) Stanley Breeden/DRK Photo; **607:** (left) Gary A. Polis/Vanderbilt University; (right) Don & Pat Valenti/DRK Photo; **608:** (top) Richard K. LaVal/Animals Animals/Earth Scenes; (bottom) John Cancalosi/Tom Stack & Associates; **609:** (top left) Jeff Foott/DRK Photo; (top right) Catherine Ellis/Photo Researchers, Inc.; (center) Ken Lucas/Planet Earth Pictures; (bottom) David Maitland/Planet Earth Pictures; **610:** (top) G. I. Bernard/Animals Animals/Earth Scenes; (bottom left) Dwight Kuhn/DRK Photo; (bottom right) James H. Carmichael/The Image Bank; **611:** Joe McDonald/Tom Stack & Associates; **612:** (left) Kjell B. Sandved; (right) Jeffrey L. Rotman; **613:** (top) Kjell B. Sandved; (center) John Lythgoe /Planet Earth Pictures; (bottom left) Raymond A. Mendez/Animals Animals/Earth Scenes; (bottom right) J. H. Robinson/Photo Researchers, Inc.; **614:** (left) L. & D. Klein/Photo Researchers, Inc.; (center) Stephen J. Krasemann/DRK Photo; (right) Dwight Kuhn Photography; **615:** K. G. Preston-Mafham/Premaphotos Wildlife; **616,**

617: Dwight Kuhn Photography; **618:** (bottom left) Dr. C. Andrew Henley; (bottom right) Tom McHugh/Photo Researchers, Inc.; (top) David Maitland/Planet Earth Pictures; **619:** (top) John Gerlach/DRK Photo; (bottom left) Raymond A. Mendez/Animals Animals/Earth Scenes; (bottom right) L. West/Photo Researchers, Inc.; **620:** Donald Specker/Animals Animals/Earth Scenes; **621:** (top) Steinhart Aquarium/Photo Researchers, Inc.; (bottom) Dan Guravich/Photo Researchers, Inc.; **622:** (top) Tom McHugh/Photo Researchers, Inc.; (bottom) Ken Lucas/Planet Earth Pictures; **624:** Stephen Dalton/Photo Researchers, Inc.; **625:** (bottom) Kjell B. Sandved; (top) Richard K. LaVal/Animals Animals/Earth Scenes; **626:** Dwight Kuhn/DRK Photo; **628:** Dwight Kuhn Photography; **629:** (top) Ken W. Davis/Tom Stack & Associates; (bottom) Doug Perrine/DRK Photo; **630:** (top) Catherine Ellis/Photo Researchers, Inc.; (bottom) Stephen Dalton/Photo Researchers, Inc.; **631:** (left) Rod Planck/Tom Stack & Associates; (center) Brian Parker/Tom Stack & Associates; (bottom and top right) Dr. E.R. Degginger; **632:** John Lei/Omni-Photo Communications, Inc.; **636:** (top) Jeffrey L. Rotman; (bottom) Carl Roessler/Animals Animals/Earth Scenes; **637:** (top) A. Kerstitch; (bottom) Robert A. Ross/Dr. E. R. Degginger; **638:** Tom Stack/Tom Stack & Associates; **640:** (left) Herwarth Voightmann/Planet Earth Pictures; (right) Doug Wechsler/Animals Animals/Earth Scenes; **641:** (top) Dave Woodward/Tom Stack & Associates; **642:** (left) Jeffrey L. Rotman; (right) Linda Pitkin/Planet Earth Pictures; **643:** (left) Jeffrey L. Rotman; (top right) Oxford Scientific Films/Animals Animals/Earth Scenes; (bottom right) Kjell B. Sandved; **644:** (top) Bill Wood/Planet Earth Pictures; (center) Charles Seaborn/Odyssey Productions; (bottom) Leo Collier/Planet Earth Pictures; **646:** (bottom) M. Laverack/Planet Earth Pictures; (center) Dick Clarke/Planet Earth Pictures; (top) Carl Roessler/Planet Earth Pictures; **647:** Larry Madin/Planet Earth Pictures; **651:** Jeffrey L. Rotman; **652:** (top) Nancy Sefton/Planet Earth Pictures; (bottom) James H. Carmichael, Jr./The Image Bank; **653:** James L. Castner; **654:** Jeffrey L. Rotman; **656:** (top) Photo Researchers, Inc.; (bottom left) Chuck Nicklin/Sea Library; (bottom right) Oxford Scientific Films/Animals Animals/Earth Scenes; **659:** Chuck Nicklin; **660:** G. I. Bernard/Oxford Scientific Films/Animals Animals/Earth Scenes; **661:** (top) Robert Frerck/Odyssey Productions; (bottom) W. Gregory Brown/Animals Animals /Earth Scenes; **663:** Kjell B. Sandved; **665:** (top) Planet Earth Pictures; (bottom) James M. King/Planet Earth Pictures; **667:** (top left) Dwight Kuhn/DRK Photo; (top right) Chris Prior/Planet Earth Pictures; (bottom) J. H. Robinson/Photo Researchers, Inc.; **668:** Dr. C. Andrew Henley; **670:** Zig Leszczynski/Animals Animals/Earth Scenes; **674:** (top) Ira Block/Woodfin Camp & Associates; (center) Ed Wheeler/The Stock Market; (bottom) Dan McCoy/Rainbow; **675:** (top) Jeff Rotman/Joe Levine; (center) Robert Frerck/Odyssey Productions; (left) Gary Milburn/Tom Stack & Associates; (right) Doug Perrine/DRK Photo; **676:** ©1993 Frans Lanting/Minden Pictures, Inc.; **678:** (top) Herwarth Voigtmann/Planet Earth Pictures; (bottom) David M. Dennis/Tom Stack & Associates; **679:** P. Herring/Biofotos; **680:** (top left) Animals Animals/Earth Scenes; (top right) Jeffrey L. Rotman; (bottom left) Breck P. Kent; **681:** Charles Seaborn/Odyssey Productions; **682:** (top) Peter David/Planet Earth Pictures; (bottom left) Dave B. Fleetham/Tom Stack & Associates; (bottom right) Norbert Wu/Planet Earth Pictures; **684:** (top) Doug Perrine/DRK Photo; (bottom) Walter Deas/Planet Earth Pictures; **686:** Ken Lucas/Planet Earth Pictures; **687:** (top) Jeff Foott/DRK Photo; (center) Breck P. Kent; (bottom) Ken Lucas/Planet Earth Pictures; **688:** (top) Carl Roessler/Planet Earth Pictures; (bottom left) G. I. Bernard/Oxford Scientific Films/Animals Animals/Earth Scenes; (bottom right) Ken Lucas/Planet Earth Pictures; **689:** (top) Charles Seaborn/Odyssey Productions; (center) Larry Lipsky/Tom Stack & Associates; (bottom left) Charles Seaborn/Odyssey Productions; (bottom right) Kjell B. Sandved; **690:** (top) Breck P. Kent; (bottom) Dwight Kuhn/DRK Photo; **691:** Jim Nilsen/Allstock; **692:** Robert Frerck/Odyssey Productions; **693:** (left) Raymond Mendez/Animals Animals/Earth Scenes; (top right) Michael Fogden/DRK Photo; (top center) Juan M. Reujito/Animals Animals/Earth Scenes; (bottom center) Dwight Kuhn Photography; (bottom) Dwight Kuhn/DRK Photo; **697:** Richard Thorn/Tom Stack & Associates; **698:** (top left) Zig Leszczynski/Animals Animals/Earth Scenes; (bottom left) David M. Dennis/Tom Stack & Associates; (right) Stephen Dalton/Animals Animals/Earth Scenes; **699:** (top right) Dwight Kuhn Photography; (center right) Dr. E. R. Degginger/Animals Animals/Earth Scenes; (bottom right) P. J. Palmer/Planet Earth Pictures; (left) Michael Fogden/Animals Animals/Earth Scenes; **700:** Breck P. Kent; **701:** (top) Don and Esther Phillips/Tom Stack & Associates; (bottom) Michael Fogden/DRK Photo; **707:** Breck P. Kent; **708:** (top left) Dwight Kuhn Photography; (top right) Wayne Lynch/DRK Photo; (bottom) David M. Dennis/Tom Stack & Associates; (right) Breck P. Kent; **710:** (left) Dwight Kuhn Photography; **711:** Heather Angel/Biofotos; **713:** Tom McHugh/Photo Researchers, Inc.; **714:** (top left) Kim Taylor/Bruce Coleman, Inc.; (top right) Tom McHugh/Photo Researchers, Inc.; (center) Gary Milburn/Tom Stack & Associates; (bottom) Breck P. Kent; **715:** (bottom left) Paul Kuhn/Tom Stack & Associates; (top right) Art Wolfe Incorporated; (center) Wolfgang Bayer/Bruce Coleman, Inc.; (bottom) Zig Leszczynski/Breck P. Kent; **716:** (top left) Jany Sauvanet/Photo Researchers, Inc.; (top right) Breck P. Kent; (bottom) Belinda Wright/DRK Photo; **717:** (bottom) Michael Fogden/DRK Photo; (top left) Kjell B. Sandved; (top right) Breck P. Kent; **718:** (top) S. Nielsen/DRK Photo; (center) Wayne Lynch/DRK Photo; (bottom left) Stephen J. Krasemann/DRK Photo; (bottom right) Dave B. Fleetham/Tom Stack & Associates; **719:** Frans Lanting/Minden Pictures, Inc.; **720:** (top) Breck P. Kent; (bottom) Kjell B. Sandved; **721:** Jeff Foott Productions; **723:** Jeff Lepore/Photo Researchers, Inc.; **724:** Chuck J. Lamphiear/DRK Photo; **725:** (top) Breck P. Kent/Animals Animals/Earth Scenes; (center) S. Nielsen/DRK Photo; (bottom left) Johnny Johnson/DRK Photo; (bottom right) Tom McHugh/Photo Researchers, Inc.; **727:** Jeff Lepore/Photo Researchers, Inc.; **728:** John Cancalosi/Tom Stack & Associates; **730:** (left) John Shaw/Tom Stack & Associates; (center) John Gerlach/DRK Photo; (right) Belinda Wright/DRK Photo; **731:** Michael Gadomski/Photo Researchers, Inc.; **735:** Johnny Johnson/DRK Photo; **737:** Kennan Ward/DRK Photo; **738:** (top) Stephen J. Krasemann/DRK Photo; (center) Dieter Blum/Peter Arnold, Inc.; (bottom) Francois Gohier/Photo Researchers, Inc.; **739:** Rod Williams/Bruce Coleman, Inc.; **740:** (left) Gary Milburn/Tom Stack & Associates; (right) Thomas Kitchin/Tom Stack & Associates; **741:** Weinberg/Clark/The Image Bank; **742:** (top) Stephen J. Krasemann/DRK Photo; (bottom) Frans Lanting/Minden Pictures, Inc.; **743:** (top) Gregory G. Dimijian/Photo Researchers, Inc.; (bottom) Tom McHugh/Photo Researchers, Inc.; **744:** (top) John Cancalosi/Peter Arnold, Inc.; (bottom) Brian Parker/Tom Stack & Associates; **745:** S. J. Krasemann/DRK Photo; **746:** (top) C. B. & B. W. Frith/Bruce Coleman, Inc.; (bottom) Yeager & Kay/Photo Researchers, Inc.; **747:** (left) Stephen Dalton/Photo Researchers, Inc.; (right) Merlin Tuttle/Photo Researchers, Inc.; **748:** (top) Jeff Foott/Bruce Coleman, Inc.; (center) Brian Parker/Tom Stack & Associates; (bottom) John Shaw/Bruce Coleman, Inc.; **749:** (top left) Carleton Ray/Photo Researchers, Inc.; (top right) Stephen J. Krasemann/DRK Photo; (bottom) Jeff Foott Productions; **750:** (top left) Harry Engels/Photo Researchers, Inc.; (top right) Kevin Schaefer/Tom Stack & Associates; (bottom left) Frans Lanting/Minden Pictures, Inc.; (bottom right) C. Allan Morgan/Peter Arnold, Inc.; **751:** (top) Melinda Berge/DRK Photo; (bottom) Charlie Ott/Photo Researchers, Inc.; **755:** Kevin Schaefer/Peter Arnold, Inc.; **756:** (top) NASA; (bottom) From British Museum/Michael Holford; **757:** (left) Gary Milburn/Tom Stack & Associates; (right) John Cancalosi/Peter Arnold, Inc.; **758:** (top) Tom McHugh/Photo Researchers, Inc.; (center) Dieter & Mary Plage/Bruce Coleman, Inc.; (bottom) Stanley Breeden/DRK Photo; **759:** (left) Jack Swenson/Tom Stack & Associates; (right) Nancy Adams/Tom Stack & Associates; **761:** Cleveland Museum of Natural History; **762, 763:** Margo Crabtree; **764:** American Museum of Natural History; **770:** (left) Peter Lamberti/Tony Stone Images; (right) Sid Bahart/Photo Researchers, Inc.; **771:** Michael Fogden/DRK Photo; **772:** (top) Jonathan Scott/Planet Earth Pictures; (bottom left) Breck P. Kent; (bottom right) Nancy Adams/Tom Stack & Associates; **774:** (top) Breck P. Kent; (bottom) Frans Lanting/Minden Pictures, Inc.; **775:** Nina Leen/Life Picture Service; **776:** (top) Tom & Pat Leeson/DRK Photo; (bottom) Philip Chapman/Planet Earth Pictures; **777:** (top) Michael Fogden/DRK Photo; (bottom) Raymond Mendez/Animals Animals/Earth Scenes; **778:** Bruce Davidson/Animals Animals/Earth Scenes; **779:** P. H. Powers/Sandved & Coleman; **784:** (top) Stephen J. Krasemann/DRK Photo; (bottom) Johnny Johnson/DRK Photo; **785:** Kerry T. Givens/Tom Stack & Associates; **788:** (left) Steve Kaufman/Peter Arnold, Inc.; (right) Breck P. Kent; **790:** (top) Breck P. Kent; (bottom left) Wolfgang Kaehler; (bottom right) Stephen J. Krasemann/DRK Photo; **797:** CNRI/Science Photo Library/Photo Researchers, Inc.; **799:** Wolfgang Kaehler; **803:** (left) J. A.Hancock/Photo Researchers, Inc.; (right) Bob & Clara Calhoun/Bruce Coleman, Inc.; **804:** (top) Stephen J. Krasemann/DRK Photo; (center) Robert Frerck/Odyssey Productions; (bottom) Dan McCoy/Rainbow; **805:** (top left) Ken Lucas/Planet Earth Pictures; (bottom left) Breck P. Kent; (right) Peter Scoones/Planet Earth Pictures; **806:** Melanie Carr/Zephyr Pictures; **808:** (top) Library Science Source/Photo Researchers, Inc.; (bottom) Eric V. Grave/Phototake; **809:** (left) Steven E. Sutton/Duomo Photography, Inc.; (right) Francis Leroy/Photo Researchers, Inc.; **810, 813:** © Lennart Nilsson, *The Incredible Machine*, National Geographic Society; **813:** (top) Steven J. Krasemann/DRK Photo; **815:** CNRI/Science Photo Library/Photo Researchers, Inc.; **817:** © Lennart Nilsson, *The Incredible Machine*, National Geographic Society; **821:** "The Excitable Cortex in Conscious Man"/W. PENFIELD; **822:** Brian Brake/Photo Researchers, Inc.; **824:** Johns Hopkins University; **825:** © Lennart Nilsson, *The Incredible Machine*,

National Geographic Society; **827:** (right) © Lennart Nilsson, *Behold Man*, Little, Brown & Company; **828:** (top) © Lennart Nilsson, *Behold Man*, Little, Brown & Company; (bottom) © Lennart Nilsson, *The Incredible Machine*, National Geographic Society; **830:** (top) © Lennart Nilsson, *Behold Man*, Little, Brown & Company; (center and bottom) © Lennart Nilsson, *The Incredible Machine*, National Geographic Society; **836:** Globus Brothers/The Stock Market; **837:** Dan McCoy/David Seltzer/Rainbow; **839:** (right) Dr. Richard Kessel/University of Iowa; **840:** © Lennart Nilsson, *The Incredible Machine*, National Geographic Society; **843:** (top) Eric Grave/Phototake; (center) Biophoto Associates/Photo Researchers, Inc.; (bottom) John D. Cunningham Visuals Unlimited; **844:** Clara Franzini-Armstrong/Photo Researchers, Inc.; **845:** F. Hossler/Visuals Unlimited; **847:** Veronica Burmeister/Visuals Unlimited; **848:** (top) © Lennart Nilsson, *Behold Man*, Little, Brown & Company; (bottom) John D. Cunningham/Visuals Unlimited; **849:** CNRI/Phototake; **853:** CNRI/Science Photo Library/Photo Researchers, Inc.; **854:** Ken Karp/Omni-Photo Communications, Inc.; **855:** Duomo Photography, Inc.; **856:** (top) Ken Karp Photography; (bottom) Jerry L. Ferrara/Photo Researchers, Inc.; **859:** (top) Wolfgang Kaehler; (bottom) Wayne Lankinen/DRK Photo; **862:** © Lennart Nilsson, *The Incredible Machine*, National Geographic Society; **863:** David Madison/Duomo Photography, Inc.; **865:** United States Department of Agriculture; **866:** Yoav/Phototake; **868:** Omikron/Science Source/Photo Researchers, Inc.; **869:** Howard Sochurek Inc.; **870, 871, 874, 875:** © Lennart Nilsson, *The Incredible Machine*, National Geographic Society; **874:** (right) L. V. Bergman & Associates; **880:** (top) Bob Winsett/Jeff Andrew/Tom Stack & Associates; (bottom) Art Siegel; **881:** (left) Mark Newman/Tom Stack & Associates; (top right) Veronika Burmeister/Visuals Unlimited; (bottom right) David Wagner/Phototake; **882:** Stephen J. Krasemann/DRK Photo; **883:** © Lennart Nilsson, *The Incredible Machine*, National Geographic Society; **884:** (top) Chet Childs/Tony Stone Images; (bottom) CNRI/Science Photo Library/Photo Researchers, Inc.; **887:** CNRI/Science Library/Photo Researchers, Inc.; **888:** (left) Peter Scoones/Planet Earth Pictures; (right) Keith Gunnar/Bruce Coleman, Inc.; **889:** Chemical Design/Science Photo Library/Photo Researchers, Inc.; **891:** (left) Jeff Foott/DRK Photo; (right) M. A. Chappell/Animals Animals/Earth Scenes; **896:** (left) ©1994 by Howard Sochurek/Medical Images, Inc.; (right) Prof. P. Motta of Anatomy/University "La Sapienza," Rome/Science Photo Library/Photo Researchers, Inc.; **897:** David Wagner/Phototake; **899:** (top) © Lennart Nilsson, *Behold Man*, Little, Brown & Company; (bottom) © Lennart Nilsson, *The Incredible Machine*, National Geographic Society; **900:** © Lennart Nilsson, *The Incredible Machine*, National Geographic Society; **901, 903:** Biophoto Associates/Science Source/Photo Researchers, Inc.; **903:** (top) Photo Researchers, Inc.; **905:** Nathan Benn/Woodfin Camp & Associates; **906:** (top) CNRI/Science Photo Library/Photo Researchers, Inc.; (bottom) © Lennart Nilsson, *The Incredible Machine*, National Geographic Society; **907:** (top) Visuals Unlimited; (bottom) Veronika Burmeister/Visuals Unlimited; **908:** CNRI/Science Photo Library/Photo Researchers, Inc.; **909:** © Lennart Nilsson, *The Incredible Machine*, National Geographic Society; **911:** Dan McCoy/Rainbow; **915:** David M. Phillips/Visuals Unlimited; **916:** (top) Co Rentmeester/The Image Bank; (bottom) David Parker/Photo Researchers, Inc.; **917:** © Lennart Nilsson, *The Incredible Machine*, National Geographic Society; **919:** Jan Hinsch/Science Photo Library/Photo Researchers, Inc.; **922:** (top) David Phillips/Science Source/Photo Researchers, Inc.; (bottom) © Lennart Nilsson, *The Incredible Machine*, National Geographic Society; **925:** © Lennart Nilsson, *The Incredible Machine*, National Geographic Society; **926:** David Madison/Duomo Photography, Inc.; **934:** (top) © Lennart Nilsson, *The Incredible Machine*, National Geographic Society; (bottom) Ronn Maratea/International Stock Photography, Ltd.; **935:** Tom Hollyman/Photo Researchers, Inc.; **936:** (top) PH Files; (bottom) Richard G. Kessel & Randy H. Kardon; **937:** Dr. G. Schatten/Science Photo Library/Photo Researchers, Inc.; **940, 942:** © Lennart Nilsson, *The Incredible Machine*, National Geographic Society; **944:** Petit Format/Nestlé/Science Source/Photo Researchers, Inc.; **945:** Bonnier Fakta; **946:** David Leah/Science Photo Library/Photo Researchers, Inc.; **947:** Myrleen Ferguson/Photoedit; **951:** (left) Andree Abecassis/The Stock Market; (right) George Ancona/International Stock Photography, Ltd.; **952:** Mansell Collection; **953:** CNRI/Science Photo Library/Photo Researchers, Inc.; **954:** (top) © Lennart Nilsson, *The Incredible Machine*, National Geographic Society; (bottom) The Granger Collection; **955:** Robert Isear/Photo Researchers, Inc.; **957:** (bottom) CNRI/Science Photo Library/Photo Researchers, Inc.; (left) Omikron/Science Photo Researchers, Inc.; (right) A. B. Dowsett/Science Photo Library/Photo Researchers, Inc.; **958:** (top) Martin Dohrn/Photo Researchers, Inc.; (bottom) Georg Gerster/Comstock; **959:** Russell C. Johnson/University of Minnesota; **960:** © Lennart Nilsson, *The Incredible Machine*, National Geographic Society; **961:** (top) Simon Fraser/Neuroradiology Department/Newcastle General Hospital/Science Photo Library/Photo Researchers, Inc.; (bottom) G. V. Faint/The Image Bank; **962:** © Lennart Nilsson, *The Incredible Machine,* National Geographic Society; **963:** Bill Pierce/Rainbow; **968:** (left) A. B. Dowsett/Science Photo Library/Photo Researchers, Inc.; (right) The Bettmann Archive; **969, 970, 971:** © Lennart Nilsson, *The Incredible Machine*, National Geographic Society; **974:** The Bettmann Archive; **975** (bottom), **976:** © Lennart Nilsson, *The Incredible Machine*, National Geographic Society; **975:** (top) National Library of Medicine; **977:** (left) David M. Phillips/Visuals Unlimited; (right) © Lennart Nilsson, *The Incredible Machine*, National Geographic Society; **978:** Prof. L. Brent/St. Mary's Hospital Medical School; **979, 980:** © Lennart Nilsson, *The Incredible Machine,* National Geographic Society; **981:** Brian Reynolds/Sygma; **985:** NASA/Science Source/Photo Researchers, Inc.; **986:** (top) Gary Cralle/The Image Bank; (bottom) Michael P. Gadomski/Photo Researchers, Inc.; **987:** R. B. Sanchez/The Stock Market; **988:** (top) The University Museum, University of Pennsylvania; (bottom) Ken Karp Photography; **989:** (left) Howard Sochurek Inc.; (right) J. H. Robinson Photo Researchers, Inc.; **991:** Wesley Bocxe/Photo Researchers, Inc.; **992:** (top) Dr. Jeremy Burgess/Science Photo Library/Photo Researchers, Inc.; (bottom) Michael Hardy/Woodfin Camp & Associates; **993:** National Institute of Health; **995:** Bobby Holland/READER'S DIGEST FOUNDATION; **996:** © Lennart Nilsson, *The Incredible Machine*, National Geographic Society; **1001:** Ted Russell/The Image Bank; **1002:** (top) Kay Chernush/The Image Bank; (center) Don Klumpp/The Image Bank; (bottom) Dan McCoy/Rainbow; **1003:** (left) Gary Cralle/The Image Bank; (center) Michael Hardy/Woodfin Camp & Associates; (right) Mark Newman/Tom Stack & Associates **1004:** Pat & Tom Leeson/Photo Researchers, Inc.; **1006:** (bottom) NASA; (top) Johnny Johnson/DRK Photo; **1007:** Wolfgang Kaehler; **1011:** (left, center, and right) Stephen J. Krasemann/DRK Photo; (bottom) Fred Breummer/DRK Photo; **1012:** (left) Charles Krebs/The Stock Market; (center) Michael Giannechini/Photo Researchers, Inc.; (right) Jeff Foott Productions; **1013:** (left) Viviane Moos/The Stock Market; (top right) Jeff Foott Productions; (bottom right) S. Nielsen/DRK Photo; **1014:** Jonathan Scott/Planet Earth Pictures; **1015:** (left and right) Wolfgang Kaehler; (center) Michael Fogden/DRK Photo; **1016:** (center) Roberta Jureit/The Stock Market; (left) Nancy Adams/Tom Stack & Associates; (right) Jeff Foott Productions; **1017:** (top) Uniphoto; (bottom left) Wolfgang Kaehler; (bottom right) Planet Earth Pictures; **1018:** (top) Peter David/Planet Earth Pictures; (bottom) J. Frederick Grassle/Woods Hole Oceanographic Institution; **1020:** (left) John Cancalosi/DRK Photo; (right) Art Wolfe Incorporated; **1021:** (left) Wayne Lynch/DRK Photo; (center) S. Nielsen/DRK Photo; (right) Wolfgang Kaehler; **1026:** Don & Pat Valenti/DRK Photo; **1031:** Pat O'Hara/DRK Photo; **1032:** (left) Robert & Linda Mitchell Photography; (right) Johnny Johnson/DRK Photo; **1033:** Dan Guravich/Photo Researchers, Inc.; **1036:** (left) Warren Garst/Tom Stack & Associates; (right) Kim Taylor/Bruce Coleman, Inc.; **1038:** Stephen J. Krasemann/DRK Photo; **1039:** (bottom) Philippe Giraud/Sygma; (top) Andrew Holbrooke/Gamma-Liaison, Inc.; **1040:** Robert & Linda Mitchell Photography; **1041:** (top left) Doug Perrine/DRK Photo; (top right) Jeff Foott Productions; (bottom) Bill Wood/Bruce Coleman, Inc.; **1042:** Don Carroll/The Image Bank; **1043:** Norman Owen Tomalin/Bruce Coleman, Inc.; **1048:** (left) Hand-colored lithograph drawn by J. G. Keulemans for G. D. Rowley's *Ornithological Miscellany,* 1876–78 (Vol. 1, pl. 3). Published by permission of the Alexander Turnbull Library; (right) The Bettmann Archive; **1049:** John M. Roberts/The Stock Market; **1050:** Richard Hutchings/Photo Researchers, Inc.; **1051:** Frans Lanting/Minden Pictures, Inc.; **1053:** (top) John Bryson/The Image Bank; (bottom) Jack Swenson/Tom Stack & Associates; **1054:** (top) Stephen J. Krasemann/DRK Photo; (bottom) Frans Lanting/Minden Pictures, Inc.; **1055:** (left) Ted Spiegel/Black Star; (right) Melchior DiGiacomo/The Image Bank; **1056:** (right) Breck P. Kent; (left) Roger Cheng; **1058:** (left) NASA; (right) Mark Newman/Earth Images; **1059:** (bottom) Wolfgang Kaehler; (top) NASA; **1060:** (top) Tom Stack/Tom Stack & Associates; (bottom) David Smart/DRK Photo; **1061:** Leif Skoogfors/Woodfin Camp & Associates; **1062:** Ken Sakamoto/Black Star; **1063:** (top) Michael Baytoff/Black Star; (bottom) B. Nation/Sygma; **1064:** Thomas Kitchin/Tom Stack & Associates; **1065, 1066:** Jeff Foott Productions; **1066:** (left) Tom McHugh/Photo Researchers, Inc.; (right) Jeff Foott Productions; **1067:** Larry Lefever/Grant Heilman Photography; **1068:** (right) Frans Lanting/Minden Pictures, Inc.; (top) Steve Jackson/Black Star; **1069:** James Balog/Black Star; **1070:** Joe Azzara/The Image Bank; **1071:** (top left) G. Brimacombre/The Image Bank; (top right) Denis Valentine/The Stock Market; (center) NASA; (bottom left) John Kelly/The Image Bank; (bottom right) Uniphoto; **1075:** Holt Conter/DRK Photo; **1076:** (top) David Smart/DRK Photo; (center) C. C. Lockwood/DRK Photo; (bottom) Robert Frerck/Odyssey Productions; **1077:** NASA